D0821040

BRIDGE ENGINEERING HANDBOOK

BRIDGE ENGINEERING

HANDBOOK

EDITED BY

WAI-FAH CHEN and LIAN DUAN

CRC Press

Boca Raton London New York Washington, D.C.

Acquiring Editor: Nora Konopka
Project Editors: Carol Whitehead, Sylvia Wood
Marketing Managers: Barbara Glunn, Jane Lewis, Arline Massey, Jane Stark
Cover design: Jonathan Pennell
Manufacturing: Carol Slatter

Library of Congress Cataloging-in-Publication Data

Chen, Wai-Fah, Duan, Lian
Bridge engineering handbook / edited by Wai-Fah Chen, Lian Duan.
p. cm.
Includes bibliographical references and index.
ISBN 0-8493-7434-0 (alk. paper)
1. Bridges—Design and construction. I. Chen, Wai-Fah, 1936 - II. Duan, Lian.
TG145 - B85 1999
624'-2 -- d21c 99-3175
CIP

International Standard Book Number 0-8493-7434-0
Library of Congress Card Number 99-3175
Printed in the United States of America 1 2 3 4 5 6 7 8 9 0
Printed on acid-free paper

Foreword

Among all engineering subjects, bridge engineering is probably the most difficult on which to compose a handbook because it encompasses various fields of arts and sciences. It not only requires knowledge and experience in bridge design and construction, but often involves social, economic, and political activities. Hence, I wish to congratulate the editors and authors for having conceived this thick volume and devoted the time and energy to complete it in such short order. Not only is it the first handbook of bridge engineering as far as I know, but it contains a wealth of information not previously available to bridge engineers. It embraces almost all facets of bridge engineering except the rudimentary analyses and actual field construction of bridge structures, members, and foundations. Of course, bridge engineering is such an immense subject that engineers will always have to go beyond a handbook for additional information and guidance.

I may be somewhat biased in commenting on the background of the two editors, who both came from China, a country rich in the pioneering and design of ancient bridges and just beginning to catch up with the modern world in the science and technology of bridge engineering. It is particularly to the editors' credit to have convinced and gathered so many internationally recognized bridge engineers to contribute chapters. At the same time, younger engineers have introduced new design and construction techniques into the treatise.

This Handbook is divided into seven sections, namely:

- Fundamentals
- Superstructure Design
- Substructure Design
- Seismic Design
- Construction and Maintenance
- Special Topics
- Worldwide Practice

There are 67 chapters, beginning with bridge concepts and aesthestics, two areas only recently emphasized by bridge engineers. Some unusual features, such as rehabilitation, retrofit, and maintenance of bridges, are presented in great detail. The section devoted to seismic design includes soil-foundation-structure interaction. Another section describes and compares bridge engineering practices around the world. I am sure that these special areas will be brought up to date as the future of bridge engineering develops.

May I advise each bridge engineer to have a desk copy of this volume with which to survey and examine both the breadth and depth of bridge engineering.

T. Y. Lin
Professor Emeritus, University of California at Berkeley
Chairman, Lin Tung-Yen China, Inc.

Preface

The *Bridge Engineering Handbook* is a unique, comprehensive, and state-of-the-art reference work and resource book covering the major areas of bridge engineering with the theme "bridge to the 21st century." It has been written with practicing bridge and structural engineers in mind. The ideal readers will be M.S.-level structural and bridge engineers with a need for a single reference source to keep abreast of new developments and the state-of-the-practice, as well as to review standard practices.

The areas of bridge engineering include planning, analysis and design, construction, maintenance, and rehabilitation. To provide engineers a well-organized, user-friendly, and easy-to-follow resource, the Handbook is divided into seven sections. *Section I, Fundamentals*, presents conceptual design, aesthetics, planning, design philosophies, bridge loads, structural analysis, and modeling. *Section II, Superstructure Design*, reviews how to design various bridges made of concrete, steel, steel-concrete composites, and timbers; horizontally curved, truss, arch, cable-stayed, suspension, floating, movable, and railroad bridges; and expansion joints, deck systems, and approach slabs. *Section III, Substructure Design*, addresses the various substructure components: bearings, piers and columns, towers, abutments and retaining structures, geotechnical considerations, footings, and foundations. *Section IV, Seismic Design*, provides earthquake geotechnical and damage considerations, seismic analysis and design, seismic isolation and energy dissipation, soil–structure–foundation interactions, and seismic retrofit technology and practice. *Section V, Construction and Maintenance*, includes construction of steel and concrete bridges, substructures of major overwater bridges, construction inspections, maintenance inspection and rating, strengthening, and rehabilitation. *Section VI, Special Topics*, addresses in-depth treatments of some important topics and their recent developments in bridge engineering. *Section VII, Worldwide Practice*, provides the global picture of bridge engineering history and practice from China, Europe, Japan, and Russia to the U.S.

The Handbook stresses professional applications and practical solutions. Emphasis has been placed on ready-to-use materials, and special attention is given to rehabilitation, retrofit, and maintenance. The Handbook contains many formulas and tables that give immediate answers to questions arising from practical works. It describes the basic concepts and assumptions, omitting the derivations of formulas and theories, and covers both traditional and new, innovative practices. An overview of the structure, organization, and contents of the book can be seen by examining the table of contents presented at the beginning, while an in-depth view of a particular subject can be seen by examining the individual table of contents preceding each chapter. References at the end of each chapter can be consulted for more-detailed studies.

The chapters have been written by many internationally known authors from different countries covering bridge engineering practices, research, and development in North America, Europe, and the Pacific Rim. This Handbook may provide a glimpse of a rapidly growing trend in global economy in recent years toward international outsourcing of practice and competition in all dimensions of engineering. In general, the Handbook is aimed toward the needs of practicing engineers, but materials may be reorganized to accommodate undergraduate and graduate level bridge courses. The book may also be used as a survey of the practice of bridge engineering around the world.

The authors acknowledge with thanks the comments, suggestions, and recommendations during the development of the Handbook by Fritz Leonhardt, Professor Emeritus, Stuttgart University, Germany; Shouji Toma, Professor, Horrai-Gakuen University, Japan; Gerard F. Fox, Consulting Engineer; Jackson L. Durkee, Consulting Engineer; Michael J. Abrahams, Senior Vice President, Parsons, Brinckerhoff, Quade & Douglas, Inc.; Ben C. Gerwick, Jr., Professor Emeritus, University of California at Berkeley; Gregory F. Fenves, Professor, University of California at Berkeley; John M. Kulicki, President and Chief Engineer, Modjeski and Masters; James Chai, Senior Materials and Research Engineer, California Department of Transportation; Jinrong Klang, Senior Bridge Engineer, URS Greiner; and David W. Liu, Principal, Imbsen & Associates, Inc.

We wish to thank all the authors for their contributions and also to acknowledge at CRC Press Nora Konopka, Acquiring Editor, and Carol Whitehead and Sylvia Wood, Project Editors.

Wai-Fah Chen
Lian Duan

Editors

Wai-Fah Chen is a George E. Goodwin Distinguished Professor of Civil Engineering and Head of the Department of Structural Engineering, School of Civil Engineering at Purdue University. He received his B.S. in civil engineering from the National Cheng-Kung University, Taiwan, in 1959, M.S. in structural engineering from Lehigh University, Bethlehem, Pennsylvania in 1963, and Ph.D. in solid mechanics from Brown University, Providence, Rhode Island in 1966.

Dr. Chen's research interests cover several areas, including constitutive modeling of engineering materials, soil and concrete plasticity, structural connections, and structural stability. He is the recipient of numerous engineering awards, including the AISC T.R. Higgins Lectureship Award, the ASCE Raymond C. Reese Research Prize, and the ASCE Shortridge Hardesty Award. He was elected to the National Academy of Engineering in 1995, and was awarded an Honorary Membership in the American Society of Civil Engineers in 1997. He was most recently elected to the Academia Sinica in Taiwan.

Dr. Chen is a member of the Executive Committee of the Structural Stability Research Council, the Specification Committee of the American Institute of Steel Construction, and the editorial board of six technical journals. He has worked as a consultant for Exxon's Production and Research Division on offshore structures, for Skidmore, Owings and Merril on tall steel buildings, and for World Bank on the Chinese University Development Projects.

A widely respected author, Dr. Chen's works include *Limit Analysis and Soil Plasticity* (Elsevier, 1975), the two-volume *Theory of Beam-Columns* (McGraw-Hill, 1976–77), *Plasticity in Reinforced Concrete* (McGraw-Hill, 1982), *Plasticity for Structural Engineers* (Springer-Verlag, 1988), and *Stability Design of Steel Frames* (CRC Press, 1991). He is the editor of two book series, one in structural engineering and the other in civil engineering. He has authored or coauthored more than 500 papers in journals and conference proceedings. He is the author or coauthor of 18 books, has edited 12 books, and has contributed chapters to 28 other books. His more recent books are *Plastic Design and Second-Order Analysis of Steel Frames* (Springer-Verlag, 1994), the two-volume *Constitutive Equations for Engineering Materials* (Elsevier, 1994), *Stability Design of Semi-Rigid Frames* (Wiley-Interscience, 1995), and *LRFD Steel Design Using Advanced Analysis* (CRC Press, 1997). He is editor-in-chief of *The Civil Engineering Handbook* (CRC Press, 1995, winner of the Choice Outstanding Academic Book Award for 1996, *Choice Magazine*), and the *Handbook of Structural Engineering* (CRC Press, 1997).

Lian Duan is a Senior Bridge Engineer with the California Department of Transportation, U.S., and Professor of Structural Engineering at Taiyuan University of Technology, China.

He received his B.S. in civil engineering in 1975, M.S. in structural engineering in 1981 from Taiyuan University of Technology, and Ph.D. in structural engineering from Purdue University, West Lafayette, Indiana in 1990. Dr. Duan worked at the Northeastern China Power Design Institute from 1975 to 1978.

Dr. Duan's research interests cover areas including inelastic behavior of reinforced concrete and steel structures, structural stability and seismic bridge analysis and design. He has authored or coauthored more than 60 papers, chapters, and reports, and his research has focused on the development of unified interaction equations for steel beam-columns, flexural stiffness of reinforced concrete members, effective length factors of compression members, and design of bridge structures.

Dr. Duan is also an esteemed practicing engineer. He has designed numerous building and bridge structures. Most recently, he has been involved in the seismic retrofit design of the San Francisco-Oakland Bay Bridge West spans and made significant contributions to the project. He is coeditor of the *Structural Engineering Handbook* CRCnetBase 2000 (CRC Press, 2000).

Contributors

Michael I. Abrahams
Parsons, Brinckerhoff, Quade & Douglas, Inc.
New York, New York

Mohamed Akkari
California Department of Transportation
Sacramento, California

Fadel Alameddine
California Department of Transportation
Sacramento, California

Masoud Alemi
California Department of Transportation
Sacramento, California

S. Altman
California Department of Transportation
Sacramento, California

Rambabu Bavirisetty
California Department of Transportation
Sacramento, California

David P. Billington
Department of Civil Engineering and Operations Research
Princeton University
Princeton, New Jersey

Michael Blank
U.S.Army Corps of Engineers
Philadelphia, Pennsylvania

Simon A. Blank
California Department of Transportation
Walnut Creek, California

Michel Bruneau
Department of Civil Engineering
State University of New York
Buffalo, New York

Chun S. Cai
Florida Department of Transportation
Tallahassee, Florida

James Chai
California Department of Transportation
Sacramento, California

Hong Chen
J. Muller International, Inc.
Sacramento, California

Kang Chen
MG Engineering, Inc.
San Francisco, California

Wai-Fah Chen
School of Civil Engineering
Purdue University
West Lafayette, Indiana

Nan Deng
Bechtel Corporation
San Francisco, California

Robert J. Dexter
Department of Civil Engineering
University of Minnesota
Minneapolis, Minnesota

Ralph J. Dornsife
Washington State Department of Transportation
Olympia, Washington

Lian Duan
California Department of Transportation
Sacramento, California

Mingzhu Duan
Quincy Engineering, Inc.
Sacramento, California

Jackson Durkee
Consulting Structural Engineer
Bethlehem, Pennsylvania

Marc O. Eberhard
Department of Civil and Environmental Engineering
University of Washington
Seattle, Washington

Johnny Feng
J. Muller International, Inc.
Sacramento, California

Gerard F. Fox
HNTB (Ret.)
Garden City, New York

John W. Fisher
Department of Civil Engineering
Lehigh University
Bethlehem, Pennsylvania

Kenneth J. Fridley
Washington State University
Pullman, Washington

John H. Fujimoto
California Department of Transportation.
Sacramento, California

Mahmoud Fustok
California Department of Transportation
Sacramento, California

Ben C. Gerwick, Jr.
Ben C. Gerwick, Inc.
Consulting Engineers
San Francisco, California

Chao Gong
ICF Kaiser Engineers
Oakland, California

Frederick Gottemoeller
Rosales Gottemoeller & Associates, Inc.
Columbia, Maryland

Fuat S. Guzaltan
Parsons, Brickerhoff, Quade & Douglas, Inc.
Princeton, New Jersey

Danjian Han
Department of Civil Engineering
South China University of Technology
Guangzhou, China

Ikuo Harazaki
Honshu-Shikoku Bridge Authority
Tokyo, Japan

Lars Hauge
COWI
Consulting Engineers and Planners
Lyngby, Denmark

Oscar Henriquez
Department of Civil Engineering
California State University
Long Beach, California

Susan E. Hida
California Department of Transportation
Sacramento, California

Dietrich L. Hommel
COWI
Consulting Engineers and Planners
Lyngby, Denmark

Ahmad M. Itani
University of Nevada
Reno, Nevada

Kevin I. Keady
California Department of Transportation
Sacramento, California

Michael D. Keever
California Department of Transportation
Sacramento, California

Sangjin Kim
Kyungpook National University
Taeg, South Korea

F. Wayne Klaiber
Department of Civil Engineering
Iowa State University
Ames, Iowa

Michael Knott
Moffatt & Nichol Engineers
Richmond, Virginia

Steven Kramer
University of Washington
Seattle, Washington

Alexander Krimotat
SC Solutions, Inc.
Santa Clara, California

John M. Kulicki
Modjeski and Masters, Inc.
Harrisburg, Pennsylvania

John Kung
California Department of Transportation
Sacramento, California

Farzin Lackpour
Parsons, Brickerhoff, Quade & Douglas, Inc.
Princeton, New Jersey

Don Lee
California Department of Transportation
Sacramento, California

Fritz Leonhardt
California Department of Transportation
Sacramento, California

Fang Li
California Department of Transportation
Sacramento, California

Guohao Li
Department of Bridge Engineering
Tongji University
Shanghai, People's Republic of China

Xila Liu
Department of Civil Engineering
Tsinghua University
Beijing, China

Luis R. Luberas
U.S.Army Corps of Engineers
Philadelphia, Pennsylvania

M. Myint Lwin
Washington State Department of Transportation
Olympia, Washington

Jyouru Lyang
California Department of Transportation
Sacramento, California

Youzhi Ma
Geomatrix Consultants, Inc.
Oakland, California

Alfred R. Mangus
California Department of Transportation
Sacramento, California

W. N. Marianos, Jr.
Modjeski and Masters, Inc.
Edwardsville, Illinois

Brian Maroney
California Department of Transportation
Sacramento, California

Thomas W. McNeilan
Fugro West, Inc.
Ventura, California

Jack P. Moehle
Department of Civil and Environmental Engineering
University of California at Berkeley
Richmond, California

Serge Montens
Jean Muller International
St.-Quentin-en-Yvelines
France

Jean M. Muller
Jean M. Muller International
St.-Quentin-en-Yvelines
France

Masatsugu Nagai
Department of Civil and Environmental Engineering
Nagaoka University of Technology
Nagaoka, Japan

Andrzej S. Nowak
Department of Civil and Environmental Engineering
University of Michigan
Ann Arbor, Michigan

Atsushi Okukawa
Honshu–Shikoku Bridge Authority
Kobe, Japan

Dan Olsen
COWI
Consulting Engineers and Planners
Lyngby, Denmark

Klaus H. Ostenfeld
COWI
Consulting Engineers and Planners
Lyngby, Denmark

Joseph Penzien
International Civil Engineering Consultants, Inc.
Berkeley, California

Philip C. Perdikaris
Department of Civil Engineering
Case Western Reserve University
Cleveland, Ohio

Joseph M. Plecnik
Department of Civil Engineering
California State University
Long Beach, California

Oleg A. Popov
Joint Stock Company Giprotransmost (Tramos)
Moscow, Russia

Zolan Prucz
Modjeski and Masters, Inc.
New Orleans, Louisiana

Mark L. Reno
California Department of Transportation
Sacramento, California

James Roberts
California Department of Transportation
Sacramento, California

Norman F. Root
California Department of Transportation
Sacramento, California

Yusuf Saleh
California Department of Transportation
Sacramento, California

Thomas E. Sardo
California Department of Transportation
Sacramento, California

Gerard Sauvageot
J. Muller International
San Diego, California

Charles Scawthorn
EQE International
Oakland, California

Charles Seim
T. Y. Lin International
San Francisco, California

Vadim A. Seliverstov
Joint Stock Company Giprotransmost (Tramos)
Moscow, Russia

Li-Hong Sheng
California Department of Transportation
Sacramento, California

Donald F. Sorgenfrei
Modjeski and Masters, Inc.
New Orleans, Louisiana

Jim Springer
California Department of Transportation
Sacramento, California

Shawn Sun
California Department of Transportation
Sacramento, California

Shuichi Suzuki
Honshu-Shikoku Bridge Authority
Tokyo, Japan

Andrew Tan
Everest International Consultants, Inc.
Long Beach, California

Man-Chung Tang
T. Y. Lin International
San Francisco, California

Shouji Toma
Department of Civil Engineering
Hokkai-Gakuen University
Sapporo, Japan

M. S. Troitsky
Department of Civil Engineering
Concordia University
Montreal, Quebec
Canada

Keh-Chyuan Tsai
Department of Civil Engineering
National Taiwan University
Taipei, Taiwan
Republic of China

Wen-Shou Tseng
International Civil Engineering Consultants, Inc.
Berkeley, California

Chia-Ming Uang
Department of Civil Engineering
University of California
La Jolla, California

Shigeki Unjoh
Public Works Research Institute
Tsukuba Science City, Japan

Murugesu Vinayagamoorthy
California Department of Transportation
Sacramento, California

Jinrong Wang
URS Greiner
Roseville, California

Linan Wang
California Department of Transportation
Sacramento, California

Terry J. Wipf
Department of Civil Engineering
Iowa State University
Ames, Iowa

Zaiguang Wu
California Department of Transportation
Sacramento, California

Rucheng Xiao
Department of Bridge Engineering
Tongji University
Shanghai, China

Yan Xiao
Department of Civil Engineering
University of Southern California
Los Angeles, California

Tetsuya Yabuki
Department of Civil Engineering and Architecture
University of Ryukyu
Okinawa, Japan

Quansheng Yan
College of Traffic and Communication
South China University of Technology
Guangzhou, China

Leiming Zhang
Department of Civil Engineering
Tsinghua University
Beijing, China

Rihui Zhang
California Department of Transportation
Sacramento, California

Ke Zhou
California Department of Transportation
Sacramento, California

Contents

SECTION I Fundamentals

SECTION II Superstructure Design

SECTION III Substructure Design

SECTION IV Seismic Design

SECTION VI Special Topics

SECTION VII Worldwide Practice

Section I

Fundamentals

1

Conceptual Bridge Design

M. S. Troitsky
Concordia University

1.1 Introduction

Planning and designing of bridges is part art and part compromise, the most significant aspect of structural engineering. It is the manifestation of the creative capability of designers and demonstrates their imagination, innovation, and exploration [1,2]. The first question designers have to answer is what kind of structural marvel bridge design are they going to create?

The importance of conceptual analysis in bridge-designing problems cannot be emphasized strongly enough. The designer must first visualize and imagine the bridge in order to determine its fundamental function and performance.

Without question, the factors of safety and economy shape the bridge designer's thought in a very significant way. The values of technical and economic analysis are indisputable, but they do not cover the whole design process.

Bridge design is a complex engineering problem. The design process includes consideration of other important factors, such as choice of bridge system, materials, dimensions, foundations, aesthetics, and local landscape and environment. To investigate these issues and arrive at the best solution, the method of preliminary design is the subject of the discussion in this chapter.

0-8493-7434-0/00/$0.00+$.50

1.2 Preliminary Design

1.2.1 Introduction

What is preliminary design? Basically, the design process of bridges consists of two major parts: (1) the preliminary design phase and (2) the final design phase. The first design phase is discussed in this section and the final design phase is discussed in Section 1.3 in more detail.

The preliminary design stage (see Tables 1.1 and 1.2)consists of a comprehensive search of current practical and analytical applications of old and new methods in structural bridge engineering. The final design stage consists of a complete treatment of a new project in all its aspects. This includes any material, steel, or concrete problems. The important argument is that with this approach a significant savings in design effort can be easily achieved, particularly in the final stage.

In order to plan and design a bridge, it is necessary first to visualize it. The fundamental creativity lies in the imagination. This is largely reflected by the designer's creativity and the designer's past experience and knowledge. Also, the designer's concept may be based on knowledge gained from comparisons of different bridge schemes.

Generally, the designer approaches the problem successively, in two steps. In preliminary design, the first and the most important part is the creation of bridge schemes. The second step is to check schemes and sketch them in a drawing. It will then be possible to determine other design needs. An examination process is then carried out for other design requirements (e.g., local conditions, span systems, construction height, profile, etc.). From an economics point of view, choice of span structure, configuration, etc. is very essential. From the cost and aesthetics prospective, the view against the local environment is important. Completing these two steps yields the desired bridge scheme that satisfies the project proposal [3].

In the preliminary design stage it is also required to find a rational scientific analysis scheme for the conceived design. Thus, an essential part of preliminary design is to select and refine various schemes in order to select the most appropriate one. This is not an easy task since there are no existing formulas and solution. It is based mainly on the designer's experience and the requirements dictated by the project.

The final stage requires a detailed study and analysis of structural behavior and stability. Economy and safety are also important aspects in bridge design, but considerable attention must be given to detailed study for the analysis, which involves the final choices of the structural system, dimensions, material, system of spans, location of foundations, wind factor, and many others.

However, the difference in preliminary schemes if all analysis is done accurately should not be substantial. Therefore, it is very important to have, from the first step, the design calculation exact and complete. The designer workload can be dramatically reduced through use of auxiliary coefficients. These coefficients can be used if the chosen scheme needs to be modified.

Design calculation is done on the basis of structural mechanics. Usually the analysis starts with the deck, stringers, and transverse beams which determine the weight of the deck. Final analysis includes a check of the main load-carrying members, determination of various loads and their effects, total weight, and analysis of bearings. Parallel to the analysis, correction of the initial construction scheme is normally carried out.

However, at the preliminary design stage it is only necessary to explain the characteristics of the alternatives. The comparison is normally based upon the weight and cost of the structure. It should also be highlighted that at this stage the weight of the structure cannot be determined with absolute precision. It is normally estimated on the basis of experimental coefficients.

As mentioned earlier, the aim of preliminary design is to compare various design schemes. This can be achieved efficiently by using computers. The designer can create a number of rational schemes and alternatives in a short period of time. A critical comparison between the various schemes should then be made. However, this is not an easy process and it is necessary to go to the next step. Various components of each scheme, such as the deck, the spans, supports, etc. should be compared with each other. It is important at this stage that the designer be able to visualize each component in the scheme, sketch it, and check its rationality, applicability, and economy. Following this, the analysis and drawings can be adjusted and corrected.

Finally, the chosen scheme should undergo a detailed design in order to establish the structure of the bridge. The analysis is applied to each component of the bridge and to the whole structure. Each part should be visualized first by the designer, sketched, analyzed, and checked for feasibility. Then it should be modified if necessary. In each case, the most beneficial alternative should be chosen. It is a very sensitive task because it is not easy to find immediate answers and the required solutions. The problem of making final choices could only be solved on the basis of general considerations and designer's particular point of view, which is undoubtedly based on personal experience and knowledge as well as professional intuition.

The sequence of analysis in detailed final design remains the same as for preliminary design except that it is more complete. The bridge structure at this stage has a physical meaning since each part has been formed and detailed on paper. Finally, the weight is estimated considering the actual volume of the bridge elements and is documented in a special form referred to as "specifications" or a list of weights The specifications generally should be drafted at the end of the project. This sequence leads to the final stage of the project, but the process is still incomplete. The project will reach its final form only at the construction stage. For this reason, it is worth mentioning that the designer should from the beginning give serious consideration to construction problems and provide, in certain cases, complete instructions as well as methods for construction.

1.2.2 General Considerations for the Design of Bridge Schemes

Factually, the structural design scheme of the bridge presents a complex problem for the structural designer despite the presence of modern technology and advanced computer facilities. The scope of such a problem encompasses the determination of general dimensions of the structure, the span system (i.e., number and length of spans), the choice of a rational type of substructure. Also, within this scope, there is a demand to find the most advantageous solution to the problem in order to determine the maximum safety with minimum cost that is compatible with structural engineering principles. Fulfilling these demands will provide the proper solution to the technical and economic parameters, such as structure behavior, cost, safety, convenience, and external view.

Also, during the design of a bridge, crossing the river should take into consideration the cross section under the bridge that provides the required discharge of water. The opening of the bridge is measured from the level of high water as obtained at cross sections between piers, considering the configuration of the river channel, the coefficient of stream compression, and the permissible erosion of the riverbed. By changing the erosion coefficient and the cross-sectional area within the limits permitted by the standards, it is possible to obtain different acceptable dimensions of openings for the same bridge crossing. During the choice of the most expedient alternative, it is necessary also to consider that reducing the bridge opening is connected with increased cost of foundation as a result of the large depth of erosion and the need to apply more-complicated and expensive structures for stream flow. During the design of such structures as viaducts and overpasses, their total length is usually given, which may be determined by the general plan or by the landscape of the location and the relation of the cost of an embankment of great height and the bridge structure.

The design of the bridge usually starts with the development of a series of possible alternatives. By comparing alternatives, considering technical and economic parameters, we try to find the most expedient solution for the local site conditions. At the present time, the development and comparison of alternatives is the only way to find the most expedient solution. Factors influencing the choice of bridge scheme are various and their number is so great that obtaining a direct answer to what bridge scheme is most rational at a given local condition is a challenge. It is necessary to develop a few alternatives based on local conditions (geologic, hydrologic, shipping, construction, etc.) and apply the creative initiative of the designer to the choice of a structural solution. Providing structural schemes of bridge alternatives is a creative act., computers can be used to determine the

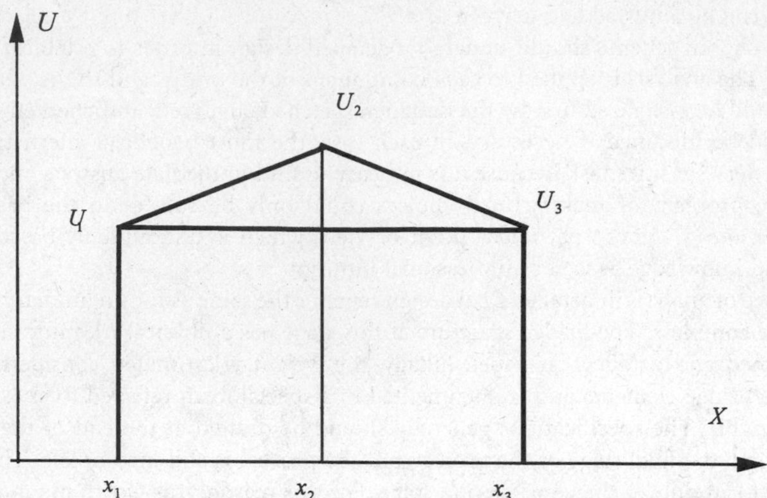

FIGURE 1.1 Quality index of the structure.

most advantageous span length and span system, to find the number of girders on the bridge having a top deck or the number of panels in the truss, and to choose the substructure. However, using computers to make a choice of rational alternatives, considering a comparison of all technical and economic parameters, is impossible. Finding an optimum alternative using different points of view often leads to different conclusions. For example, the alternative may be the most advantageous by cost, but may require great expenditure on metal or require special erection equipment, which cannot be obtained. Some alternatives may not satisfy an architectural requirement, when considering city bridges. When using computers it is still impossible to refute the conventional design method, considering all problems of specific local condition, which are practically impossible to write into a computer program.

1.2.3 Theoretical Basic Method of Preliminary Design

Methods of design cannot be invented on the basis of certain arbitrary principles. They are developed from practice. In a given theoretical study, there are enough proofs that methods of design are changing depending upon the bridge-building practice and its basic problems. Therefore, today's applied method of preliminary design is mainly determined by empirical methods.

To achieve improvement, this method is based on consistency in its exact application and explanation of its logical basis. This advanced method of preliminary design makes it possible to develop a perfect final solution for the project. It is worth mentioning at this point the importance of calculation parameters in the considered design approach.

Using mathematical models, it is possible to express (see Figure 1.1) the quality indexes U of the structure as a function of its parameters; x, y, z, i.e.,

$$U = u\ (x,\ y,\ z,\ \ldots) \tag{1.1}$$

Preliminary design provides means to determine the exact values of parameters and their quality indexes. The problem is similar to finding the limit of a function, as in calculus of variation. This analogy may be used to determine a logical basis for the method of preliminary design. It is clear that the problem of preliminary design cannot be solved in pure mathematics. The quality indexes cannot be expressed by algebraic functions. Note that the majority of parameters from one alternative to another change their size rapidly. Alternatives are shown only for consideration and to show the investigation process in order to prove the correctness of the accepted alternative.

Only in particular cases can a mathematical method be applied to find the limit. For instance, it is known that by this method it is possible to find exact dimensions of span lengths of simple-span trusses or exact heights of steel trusses those with parallel chords because of their behavior of minimal total weight of the structure.

To find the limit of the function *U*, it is possible to find the corresponding values of parameters *x*, *y*, *z* from the following equation:

$$\frac{\partial U}{\partial x} = 0, \quad \frac{\partial U}{\partial y} = 0, \quad \frac{\partial U}{\partial z} = 0 \tag{1.2}$$

These equations provide the tool to investigate the influence of each parameter as it changes the quality indexes of the structure. Leaving all other parameters constant, $\pm\Delta x$ is imposed to study the change of the value *U*. We can then find the value of the parameter for which ΔU changes its sign. This corresponds to the minimum of the function *U*.

Note that the separate parameters are interrelated. If one parameter is changed, it is necessary to modify the others. By exceeding certain limits, the span of the reinforced concrete bridge must change from a beam system to an arched system. Applying this method of preliminary design to bridges, the comparison of Eq. (1.2) leads to composition of alternatives. For each equation in Eq. (1.2), it is necessary to use a minimum of three alternatives. The first equation is formed from certain values of parameters x_1, y_1, z_1, etc. Leaving parameters y_1 and z_1 constant gives a new value of *x*, which is x_2 to compose a second alternative. Comparing this with the first, we establish the change of quality indexes of the bridge. If they have improved, it is necessary to change again the parameter *x* in the same direction, raising it to the new value x_3 to form a third alternative. Then, we compare this with the first two alternatives to determine the change of the quality indexes for the designed bridge. If, for example, they become worse, then their maximum value corresponds to x_2 (see Figure 1.1). If they improve, it is necessary to repeat the investigation for the second equation $\partial U/\partial y = 0$, and so on. All these equations must be solved simultaneously. In preliminary design, this means it is necessary to prepare many alternatives and compare them simultaneously. This process is difficult and tedious. The difficulty is increased because, unlike the purely mathematical method where the function U is given, in preliminary design the type of function is not known and should be determined.

Because of the above-mentioned difficulties, there is enough ground to assume that the first stage of the design process is based on creativity and invention.

How to build a bridge over a certain river? There are number of different answers to this question. The type of bridge can be steel or reinforced concrete, and for each case there are a number of applicable alternatives. If, for the given problem, there are several *known* solutions, there could be just as many or more undeveloped. This shows that building a bridge and creating a design are not easy tasks. Many undetermined problems face the designer. However, engineering science has proved that these difficulties could be solved in a systematic sequence, as illustrated below.

1. Equation (1.2) should be applied to the problem of structural design and solved by a method of successive approximations which follow preliminary design. This method is considered to be technically reliable and has been used in engineering successfully. In order to improve and accelerate this method (of successive approximation), it is of major importance to choose absolute precision. Experience, in a significant way, helps in making such a decision.
2. Preliminary design is generally the first approximation in the creation of a bridge project. It solves the equation for the most important parameters which have great influence on the quality indexes of the structure. Details of the structure may be investigated at a later stage.

Many solutions have been developed in practice for detailed structure. It is worth mentioning that when working with parameters, there are not many basic ones.

3. Some parameters are given and remain constant during design. Others take a limited number of values and this shortens the number of alternatives. Relations among the parameters, their correction, and the importance for the quality indexes of the bridge make it easier to carry out the methods of investigation of alternatives.

If it were possible to solve the problem by pure mathematics, then the solution would be simply to solve the equations. But we should remember that these equations (except those of the first degree) have several roots and arbitrary constants. In application to bridge design, this means, if a few equally valid alternatives are obtained, the investigation should be refined further.

The method of successive approximations should be accepted as the methodological principle because, as the process results in several alternatives for one project, we should consider only the best scientific solution. We may stop at an approximate solution, but only after we have been convinced that, in comparison with other solutions, it is the best scientific approach. This is the best way to generate designer success. The same method is applied for choosing a bridge system, as well as making a final choice for the material of bridge design, and so on.

1.2.4 Choice of Final Alternative for Reinforced-Concrete Bridges

The designer may, for instance, decide that, for a given material (say, reinforced concrete), a third alternative is chosen. Using this reasoning, the following imperfection arises: in the mathematical analogy, it was necessary to solve Eq. (1.2) simultaneously, but here each is solved separately. After determining a certain parameter, it will be kept constant and the choice for the others will follow.

There is an element of sensitivity within this method. The values of parameters are not chosen arbitrarily. The initial values are determined empirically so that their values are as exact as the real ones. The order of invention of individual parameters is also important. First, the parameters that affect the quality indexes of the structure most significantly are investigated. Then, investigation for the less important parameters follows.

Remember, the span structure implies length of span but also requires determination of the type of span structure, its form, its shape, its system, the varied types which could be uniform or unequal span structures, or the number of spans.

Within this method, the chosen alternative determines the system of span structure with minimum weight and maximum economy and safety. Now, a legitimate question may arise. Does it mean that the chosen alternative with varied type of span design (including span length, shape, form, and type) can be considered the *best* choice for the bridge project with maximum economy and safety? Although the answer may sound controversial and theocratically inconsistent, it is not. The answer is factually yes! There is a reason for that. Certain types of bridge projects require structures of various type of spans which represent economically and safely the least choice for bridge design. For refining, continue investigation by the method of successive approximation.

Note, when laying out spans for frame-beam bridges, an equal span system is often used because it provides maximum standardization of elements. However, the application of unequal span construction is also possible and in certain cases more favorable.

In following the above discussion, if an economically feasible span alternative is not satisfactory, say, does not meet the requirements of shipping regulations, or if the bridge length does not permit equal spans, then it is necessary by the method of successive approximation to find more sensitive alternative schemes with a system of spans unequal in shape, form, length, etc.

In conclusion, by changing the span system, the number of spans, or their form or shape or their combination or dimensions, it is possible to obtain a number of alternatives that will satisfy the best given local conditions at minimum cost.

For instance, changing the span system, say, by reducing the number of spans, results into a reexamination of the whole bridge design, and consequently a new bridge scheme should be drafted.

TABLE 1.1 Preliminary Design — Example 1

Design Stages	Beam System
First alternative	System of span structure, deck-type beam; bridge having three spans; construction of span structure from reinforced concrete having four main beams; supports are massive
Second alternative	Same, only two spans
Third alternative	Same, only four spans
Comparison of alternatives	The best alternative is four spans (third alternative); the first and second alternatives are canceled
Subalternatives of third alternative	1. Four-span alternative with two main beams and supports from third alternative two columns
	2. Same, with prestressed concrete
	3. With application of welded reinforcing frame
Comparison of alternatives	The third alternative is chosen; four-span bridges with two main reinforced concrete beams
Fourth alternative	Arch-type, three spans with four separate arches and columns above arches; supports are massive
Fifth alternative	Same, two spans
Sixth alternative	Same, four spans
Comparison of alternatives	Fourth alternative is chosen: three-span bridge with four separate arches
Subalternatives of fourth alternative	1. With two narrow arches and walls above arches
	2. With box-type arches
Comparison of alternatives	Fourth alternative is chosen: three-span bridges with four separate arches

If the weight of the span structure is relieved, the span system should be modified either by decreasing the span length or span shape or form or other aspects of span structure.

It worth mentioning at this point that the significance of choosing the right alternative for the span system should not be underestimated. This is because the choice of material type for the bridge structure (e.g., monolith or prefabricated, conventional or prestressed, or reinforced concrete) is of lesser importance in cost value than the total span system, which consists of length, shape, form, number of spans, etc. Due to the relative simplicity of reinforced concrete shapes of spans and supports, calculating their volume is not a difficult process if their dimensions are given or determined from preliminary calculations.

The greatest advantage of applying theoretical methods is that the process of design is not abstract and is based on scientific analysis and quantifiable information. Therefore, during the process of choosing the best alternatives for solution, there are opportunities for eliminating imperfections for each scheme. Typical for this method is searching for the best solution through detailed investigation for each material, superstructure, bridge system, etc.

The number of alternatives obtained could be large. In the scheme of variation given in Table 1.2 it was decided to compose these alternatives with subalternatives. It takes extensive investigation, which is not always necessary. In some cases, shorter methods may be applied. In the example shown in Table 1.2, it is possible to choose the bridge material first, thus composing one alternative for steel and one for reinforced concrete systems. It is advisable to consider information from experience using an empirical approach for bridge schemes.

Example 1.1: Preliminary Design of Highway Bridge

Given

Clearance, design loads, location, bridge span, type of foundation (wells), and the material is reinforced concrete and steel.

Solution

The detailed design process is shown in Table 1.2. It can be observed that the shortened method is achieved to reduce investigation and the number of alternative projects and to increase the use of

TABLE 1.2 Preliminary Design — Example 2

Design Stages	Beam System
First alternative	Reinforced-concrete beam, three-span bridge of deck system, with four main beams; supports are massive
Second alternative	Steel beams two spans
Comparison of alternatives	The first alternative is chosen: reinforced-concrete bridge
Third alternative	Reinforced-concrete arch, two spans having four separate arches
Comparison of alternatives	After comparison of the first and third alternatives, the first alternative is chosen: beam bridge
Fourth alternative	Two spans, reinforced-concrete beam bridge
Fifth alternative	Four spans, reinforced-concrete beam bridge
Comparison of alternatives	After comparison of the first, fourth, and fifth alternatives, the fifth alternative is chosen: four-span bridge
Subalternatives of fifth alternative	1. With two main beams, monolith 2. Same, with four beams, prestressed concrete prefabricated 3. Same, with welded reinforcing frame
Comparison of alternatives	By comparison of the fifth basic and additional alternatives, fifth subalternative 2 is chosen

existing available data in practice. Thus, new investigations are unnecessary and this results in significant savings in design analysis and endeavor.

1.3 Final Design

1.3.1 Basic Trends in the Design of Bridges

In many aspects, the design of bridges is based on exact analysis and for this reason it is analogous to the solution of mathematical problems, where the results are obtained by examining the problem data and utilizing mathematical methods to arrive at a solution. This approach works well for technical and economic analyses which present very important aspects of bridge design, but it leaves out a significant part of the project.

This is because, first of all, many problems cannot be solved numerically. Second, the analysis may not correspond exactly to the actual situation. Technical analysis is valid for providing information for construction, but not significant for the solution of basic problems: choice of bridge system, choice of material, general dimensions, foundation problems, etc. These problems are solved on the basis of general considerations and the designer's judgment.

For the same problems in technical analysis or basic problems, for a bridge project there could be as many proposals as the number of the participating designers involved in engineering disputes [6]. The final choice of alternative depends to some extent on the attending participants who defend their view and support their arguments technically. It is necessary to analyze the different reasoning and determine which proposal is the most consistent with prevailing and accepted standards in the present circumstances.

The assistance of different methodological trends in bridge design is inevitable considering centuries of steady improvement and progress in bridge engineering. Progress in techniques of bridge construction depends on scientific and technological developments at each historical moment in the creation of a bridge; traditions are preserved and present views are formed.

An investigation of the history of bridges demonstrates that bridge construction has passed through several industrial stages [7]. We can separate these stages into primitive, industrial, architectural, and engineering phases. These can be subdivided still further into simpler forms and characteristics.

The influence of previous centuries on bridge design indicates that, to best understand the present trends, one must study the evolution of bridge engineering. Note that it reflects involvement of materials, the spiritual culture of the society, and the transfer of heritage. Concerning technological advances universities have had a large influence. The future engineers take from their professors the basic knowledge and new trends in design.

1.3.2 Creative Trends

In the 20th century, bridge design has undergone considerable change. With increasing demand for reinforced-concrete bridges, the need for and the creation of a new system was inevitable. The old methods had many limitations and will not be discussed here. They actually presented many obstacles for further developments in bridge engineering. It was necessary to create new specifications for reinforced-concrete structures.

The construction of highway bridges and the application of reinforced concrete presented designers with a basic problem regarding the choice of the bridge system. This created strong demand for preliminary design. This new concept required developing new methods and has put pressure on designers to look at the bridge not as a condensation of essential parts but rather as a monolithic compound unit with interrelated parts.

Because of the growing demand for reinforced-concrete and suspension bridges, the designer had large choice of materials and means to develop new bridge systems and the idea of cable-stayed bridges followed. The new century created strong demand for an analytical approach and necessitated a growing need for preliminary design with more schemes.

The acceptance that for each case there is no *one* solution but, rather, that there are several from which it is possible to choose the one most consistent with prevailing, accepted standards and most effective for the actual project leads to the basic characteristic of the second significant trend in bridge design, which will be called "creative."

Therefore, the design of each bridge is a process of finding a solution to a new problem. If there is no solution available, it must be sought. Considering the role of personal creation, this second trend may provide original new projects. Supporters of such a trend believe that creation of a bridge depends upon personal predisposition, capability, and vision. Design is considered to be a creative process that consists of a combination of structural expressions based on required knowledge and professional intuition.

1.3.3 Practical Trends

Practicability is the main consideration in this trend. The word *practical* goes hand in hand with scientific investigation using modern technology. Designers use both scientific principles and creativity for their designs only in order to solve the actual problem. In this trend, the bridge is considered as part of the highway or railway and its basic purpose is to satisfy the requirements of transportation.

The bridge should satisfy the basic requirements of safety and economic factors. The construction of the bridge should also follow the pattern of successful industrial methods.

Supporters of the creative trend considered highways and railways as areas to apply their creative capabilities and for testing their new inventions. Followers of scientific analysis investigation considered highways as large laboratories for their investigations. Adherents of practical design have borrowed their concepts from both trends, insisting that bridges must be first safe and permitted experimental structures only on secondary highways. Practical designers suggested that the structure should be standardized for industrial preparation because it could lead to faster ways of reconstruction or rehabilitation. Also, practical designers insist on use of construction techniques that require minimum maintenance and do not affect the traffic flow.

1.3.4 Basic Assumptions of Design

Methodological rules compatible with technical and applicable requirements in bridge engineering play a major role in modern progressive methods for designing bridges.

Nowadays, time is an important factor, especially in bridge construction. Progressive methods must satisfy technical swift performance as well as requirements of astute engineering economy. Such majestic structures must function effectively and, in addition, be aesthetically appealing. Bridges play the major role in the transportation system crossing rivers or other obstructions.

At different times, bridges were built for more than one purpose. The following are examples:

1. Roman bridges and those built in the Middle Ages served not only for transportation or for chariots, but also for joyful, exuberant activities for the population. These traditions were continued at later times.
2. Another trend that appeared in the Middle Ages is the construction of bridges for fortresses, castles, and towers as a protective measure against attacks by enemies. An example is the bridge at Avignon, France; also "London Tower Bridge," which was built with towers for aesthetic purposes only.
3. Another trend in the same era was to build chapels on bridges and to collect tolls to maintain them, the same old problem of upkeep (e.g., Italy, Spain, Germany).
4. During the Middle Ages and later, bridges were built to serve as dams for water mills, which were important parts of the economy in those days (e.g., Holland).
5. During the 16th and 17th centuries bridges were built as wide structures for shops and convenience in general. Good examples are London Bridge, England and Ponte Vecchio, Florence, Italy. Construction of these types of bridges was terminated toward the beginning of the 19th century.
6. In Western civilizations, bridges are sometimes built as majestic monuments to commemorate outstanding events or achievements of national importance for an important person or national hero. Examples are the monument to George Washington, the George Washington Bridge, New York City; the monument to Princess Margaret of Great Britain, The Princess Margaret Bridge, Fredericton, New Brunswick, Canada (this bridge was designed by M. S. Troitsky); the monument to the victory at the Battle of Waterloo, The Waterloo Bridge, London, England; the monument to Russian Tzar Alexander the Third, The Alexander IIIrd Bridge, Paris, France (one of the most beautiful cast-iron bridges of imperial style); the monument of the Sarajevo Association, The Gavrilo Princip Bridge, Sarajevo, Yugoslavia; the monument to Napoleon Bonaparte's victory at the battle of Austerlitz, Austerlitz Bridge, Austria.

The 19th century was characterized by industrial growth, and the use of bridges was confined to transportation as a result of the boom in building railways. Later, with Ford promoting "auto-vehicles," the building of bridges for highways became in great demand. This new trend in transport requirements put on pressure to improve safety factors as well. As a result, it is very important in modern bridge engineering to determine the carrying capacity of the bridge or the maximum value of the temporary vertical load that the bridge can bear.

Also, to avoid interruption in traffic flow, the calculations should consider the maximum number of vehicles passing in a given time. For bridges crossing navigable rivers, passing clearance must be considered. Also, similar consideration should be given to underpasses. The carrying capacity of a bridge is defined by the number of lanes, their width, and the accepted lateral clear distances of shoulders and medians required for safety considerations.

To avoid interrupted traffic flow, it is necessary for the width of the bridge to be greater than that required by the calculated carrying capacity. For example, in long bridges, it is necessary to provide an extra parking space for possible emergency cases in order to prevent a traffic jam. As a rule, the width of the roadway on the bridge is equal to the width of the highway. However, there may be deviations from this rule. For instance, although the highway may accommodate three lanes for traffic, the number of lanes on the bridge could be reduced. Also, there are examples of the reversed situation.

The condition of maximum traffic suitability and convenience is not a requirement but is preferred and attention should be paid to this issue during planning the project. Also, this issue could be considered as one of the criteria for the appraisal of the project, provided that the cost is not prohibively excessive.

The most efficient functional bridge structure is considered to be the one that embodies the most requirements of transport, with top safety factors, carrying capacity, that contains extra convenience facilities, that is most effective in labor and material, and that can be completed in a reasonable time. Since Henry Ford's time, extra pressure has been put on the transportation system, primarily on highways and railways, which has directly affected innovation in bridges. Modern-day transport is increasing in number and weight. This means bridges must be designed so that their carrying and passing capacities can accommodate heavier vehicles and larger numbers of vehicles. Designers must be resourceful and have means to overcome difficult situations effectively and to cope with the growing demands of faster and larger moving transport with the greater reserves for future growth, the longer the bridge stands without needing repair or reinforcement.

Note that by increasing the reserves for passing and carrying capacities, the cost of the bridge will increase. Determining the necessary reserve is a problem that needs to be resolved by engineering economy. The Romans did not visualize the fast development of transport and means for transportation, but concentrated their conceptual design on timelessness of the bridge structure and, for this purpose, provided great reserves for passing and carrying capacities.

The property of material is not necessarily the basic factor that defines the service time and safety of the bridge. More often, bridges are reconstructed for other reasons: too small passing and carrying capacity, insufficient clearance under the bridge, straightening of lanes or reduction of the grade.

1.3.5 Basic Requirement of the Bridge under Design

Choosing the right location is crucial for designing and planning a bridge. But above all, safety considerations that govern the technical, functional, economic, efficiencies, expeditiousness, and aesthetic requirements are very important. It is necessary for the bridge and each of its components to be safe, durable, reliable, and stable. This is usually checked by analysis using current specifications. But not all questions of durability, reliability, and stability may be answered by analysis. Therefore, in some cases it is necessary to provide special measures such as testing the performance of the structure and examining its behavior under maximum loading on the construction site.

Specifications and technical requirements should be satisfied because they guarantee the carrying capacity of the structure. From the safety point of view, all bridges designed according to the technical requirements are equal. But practically speaking, different aspects of technical requirements may be satisfied with different margins of safety.

Regarding the various bridge components, it is necessary to know that for engineering structures, the best solution should provide the appropriate material and carrying capacity.

During comparison of projects, the technical requirements should be considered. Because technical requirements may be accomplished using alternatives, consideration should always be given to additional guarantees for safety. Never compromise the safety of the passengers. Essential requirements naturally should have great importance, but they are basically satisfied by accepted clearance. Also, additional consideration must be given to issues other than elementary demands in order to make traffic flow efficiently. Note that the height of the bridge and the elevation of the roadway must be determined at an early stage, because they have influence on the traffic flow. Also, greater or smaller grades of the approaches should be designed earlier in the project. Maximum grades are defined by specifications, but for practical purposes minimum grades are the most convenient. Further, it is important to define the number of joints in the roadway that correlate to the division of the structure in separate sections.

Conditions of minimum wear of the parts of carrying construction under the influence of moving vehicles are also important to consider. Regarding the maintenance of the roadway and the bridge, it is possible to consider this as a general expense and therefore relate it to economic considerations.

Essential requirements indicate that the total cost of the bridge at all conditions should be economically rational. The overall cost of construction and bridge erection is determined in significant part by the quantity of material and the unit price. Yet, the tendency to reduce the quantity of material in order to achieve lower cost does not always lead to minimum overall cost. There are other factors that should be taken into consideration. Take, for example, steel structures: consideration should be given to quantity of steel and on top of that special attention must be given to modern industrial practices in production which in its turn may lead to conveniences in erection resulting from heavy construction with lower cost.

During comparisons of various projects, analysis of their economic criteria may reveal principles of expedience that can be applied to the project under consideration. Construction requirements are connected to economic constraints because, when the amount of material is small, the work is simple and the time required is shorter. Also, the unit price is considered as part of the economic criteria, which implies the cost of preparation and erection. All these factors affect the overall cost.

For conventional bridges to be built from a certain material, construction is carried out by established methods. Therefore, during comparison of alternatives, construction criteria are not so important. In special cases of complicated erection of bridges having large spans, or for urgent work, construction requirements are very important and may influence the choice of the bridge system and material. In these cases, it may be necessary to use a great quantity of materials, thus increasing the cost of construction and ignoring other requirements. For example, during the initial period of application, assembled reinforced-concrete constructions were more expensive than monolithic ones. However, with increased use of these constructions, the application of assembled structures is more rational and economical.

1.3.6 Aesthetic Requirements

Apart from the basic requirements of the bridge design, there are often additional demands. The first is the problem of aesthetics. Beauty should be achieved as a result of good proportions of the whole bridge and its separate parts. In spite of the tendency to build economical structures, we should not forget beauty. The importance of the architecture of the bridge should not be ignored because of economic and technical requirements. In fact, the most famous bridges are remembered by their architectural standards and magnificent structures (examples, Brooklyn Bridge, Verrazano Narrows Bridge, Golden Gate Bridge, Tower Bridge, Alexander IIIrd Bridge, Ponte Vecchio Bridge, Revelstoke Bridge, British Columbia (designed by M.S. Troitsky), Skyway Bridge, Ontario (designed by M.S. Troitsky), etc.).

There are different views regarding aesthetic practices in bridge engineering. Supporters of the rational analytical trend feel that aesthetic demands are not important and not necessary for bridges outside cities. On the other hand, designers of the creative trend consider these aesthetic values to be more important than the economic ones and equivalent to the requirements of strength and longevity.

Because of the conflicting views, this problem requires special consideration. All designers inevitably want their structure to be the most beautiful. This wish is natural and shows love and interest of the work and is necessary in order to make the designed structure head toward perfection.

During the process of design, the engineer is occupied with detailed calculations. The engineer also may be occupied with particularities and may lose sight of the complete structure. By checking the creation from an aesthetic point of view, the engineer gives attention to the wide scope and shape of the structure and has the opportunity to design details and correct if necessary. If the designer is aesthetically unsatisfied with the creation, the designer will improve it and try to find workable solutions. But the designer should always be aware of technical, economic, and safety values of the structure. Note, the architecture of bridges should not contradict either as a whole or in details the purpose of the structure. The designer's ideas should be compatible with the technical concept, surrounding conditions and environment (for example, London Tower Bridge).

It is necessary to be technically literate. Moreover, it is not enough just to design the external view of the bridge. Bridges satisfying demands and requirements of modern engineering requests and properly designed will achieve recognition and will deserve worldwide acknowledgment and credit. If designers are guided by fanciful tastes of their own, regardless of the technical concepts, they will not achieve this goal. A beautiful shape alone cannot be invested and applied to the bridge. The design should consider both the technical concepts and the structural shape.

The critical rules of proportion and the use of purely geometric shapes had, in their time, not so much an aesthetic but a technical basis. Designers based their theories on the principle of initiations and relations that they observed in nature. Historical investigation indicates that many aesthetic rules were preserved from previous centuries when they had a different basis. Even today, a bridge is considered beautiful when it has an even number of supports because it is classic and not easy to achieve with tough natural conditions. According to Palladio [7] it is clear that this rule is accepted because all birds and animals have an even number of extremities which give them better stability. Freeing themselves from prejudice and carrying out independent investigations to find the shape corresponding to the contents should lead designers toward the development of the theory of true aesthetics in bridge engineering. History has shown how the shapes of bridges were changed depending upon the general development of the cultural and economic life of a nation. For this reason, the problem of aesthetics in bridge engineering should be viewed in a historical perspective. A designer should be able to judge the bridge by considering its external view and scheme of construction.

Followers of the historic direction renounced such investigations and by this changed their principles and were more attracted to the design of bridges. However, a joint venture by engineers and architects is not always useful for solving a problem of bridge design. Nowadays, architects specialize in the construction of buildings which is reflected in their aesthetic taste. Although architectural rules and views may be correct for buildings, they may not be applicable to bridges. For example, when designing a building, architects usually use steel construction as a frame for the building which requires certain covering. For the bridge designer, steel construction is a force polygon that clearly demonstrates the transfer of forces. For an attractive external view for the bridge, detailed design and proper accomplishment of the construction are important. The external view may be spoiled by careless work. The technical concepts of structure and the architectural shape should not be separate, but should satisfy the local conditions and cover a wide scope of requirements. By understanding the validity of recognizing special aesthetic criteria a proper alternative can be selected. The final choice of alternative is the solution of some technical problem in correspondence with the basic purpose of the bridge as part of the roadway.

If the bridge is not considered a monument commemorating an outstanding event or an outstanding historic figure or a significant happening in the world, but serves only for traffic for a certain period of time, then it is not necessary to design this bridge as a highly aesthetic creation. We may be satisfied by more modest wishes with regard to its external view. Practice indicates that designers may create, and actually have created, attractive bridges even when they were governed only by the technical and economic requirements during the design process.

A bridge that is properly designed from the technical and economic point of view cannot contradict the basic rules of architecture. The general basis of architecture consists of the idea that masses of material should be distributed expediently. The properties of the material should be used correspondingly, and the whole structure should correspond to its purpose.

Generally, economic considerations of bridge design are the same as those stated above. An economic design is achieved by (1) the expedient distribution of material, choice of the most economical system, cross sections of the members, and considering working conditions; and (2) the use of proper material (members in tension use steel, members in compression use concrete).

Therefore, economic expediency and architectural conception are determined by the same criteria. From this, it is impossible to contrast aesthetic criteria with technical and economic aspects.

For example, it is advisable to reject a beam bridge for an arch in the case when the first by all other properties is better, or to prefer a single-span bridge to the more expedient two spans. Also, it is possible to say that the choice of alternative, considering technical and economic criteria, should not deviate from the proper way to achieve the aesthetic aims.

Finally, the bridge will only be perfect in an aesthetic sense, when its system as a whole and its separate members are chosen not on the basis of personal taste of the designer, but considering technical and economic expedience.

All other proofs that are often applied by the authors of separate projects to defend unsuccessful technical and economic alternatives should be rejected. All these proofs are based on the unstable and changeable bases of personal opinion. Such proofs are only declarations of personal impressions and tend not to prove anything but only to convince people by the use of feeble verbal arguments.

Many definitions are expressed using varied terminology synonymous in meaning, but with drastically different shades in the positive and negative sense. For example, regarding the structure of the bridge, when the deck is at the bottom chord the defender may say that this structure is "expressive," "easily seen," or "stands out with a beautiful shape on the sky." The opponent, however, may object and say that this structure "obstructs view," "hangs on the observer," etc. By the skillful use of such terminology, it is possible to convince the inexperienced that a beautifully presented perspective is not as worthy of praise as a less successful project.

1.3.7 Requirement for Scientific Research

The second additional requirement sometimes asked of bridges under design is called the scientific research or "innovation." This requires that the bridge contain a new achievement due to scientific research or a new invention.

The design of a bridge always contains something new. Even if the project is worked using old examples and applying typical projects, the designer uses new contributions along with the known. Therefore, there is always a certain degree of novelty. A good designer or engineer should not only be familiar with previous designs but should also be updated with modern scientific research and benefit from that by using advanced technical sciences in the design as the project changes.

It is natural for the designer to search for novelty; yet new solutions should be born only from the tendency to reach the best solution by starting from the existing conditions at the project. Therefore, the "novelty" requirement cannot run contrary; they should complement each other.

The history of the evolution of bridge engineering is the progression from simple to more complex, and it was achieved gradually and unevenly. Some periods were distinguished by invention and the appearance of new shapes, systems, and types of bridges; other periods were characterized by mastering and perfecting existing systems and the development of scientific research work. For example, at the end of the 19th century, a great step forward was made in the area of stone bridges. Perhaps the most significant achievement in the modern era was the appearance of the cast-iron arch and iron-suspension systems with different members of trusses and large spans. All these novelties resulted from the impact of growing industry and transportation.

The first 40 or 50 years of the 19th century were spent creating the iron beam bridges, and the second part of the century was devoted to developing expedient systems and improving the construction. Significant periods in later history were devoted to the development of reinforced-concrete bridges. The initial period of trials and creation of the construction was 1880 to 1890, and the period of mastery was 1900 to 1910. However, it is necessary to note that with the general development of science and technology, the role of scientific research is increasingly racing together with novelty, rationalization, and invention. It is obvious that the necessity for novelty results from the general economic conditions and sociocultural requirements.

The attitude toward novelty in bridge engineering has been modified. Adherents of the rational, analytical direction preferred to hold on to some classical models, considering that the search for new shapes should be related only to scientific research work of creative direction, however, tending toward the new and original by ignoring any old pattern. A realistic approach to a new idea should be based on understanding that novelty is not an aim in itself and that the new idea should be a solid ground for improvement.

It is necessary to consider the criterion of novelty because it sometimes appears as an independent factor during appraisal of projects and choice of alternatives. Because novelty is not an aim in itself, it should not be a special criterion, forcing a preference for new construction irrespective of its quality. When by basic conditions the new idea is better and there is no doubt regarding its quality, then it should be adopted and should replace the old. In the opposite case, it should be refused.

Not every novelty leads to progress in bridge engineering. If the novelty is sound, it may be developed to such a degree that it would lead to a new method, but if it is not better than the old method or not yet developed, its development at a later stage may be helped by abstaining from early application. Early application leads to lowering the quality of bridges and may compromise new ideas before they reach full appreciation and are fully evaluated.

The criterion of novelty may be considered independent only in separate cases when economy requires the introduction of a new type of construction. An example is the introduction of prefabricated reinforced-concrete construction. At the present time it is expedient to use prefabricated reinforced-concrete construction, but initially it was more expensive than conventional construction. The criterion of novelty then was contrary to other criteria. It had to be solved for each case, especially when the novelty was not an aim in itself, but was required for economic and commercial demands.

One reason for introducing new construction techniques is related to the necessity of experimental and practical checking of the scientific research work, which is certainly necessary.

Regarding bridges on main highways, however, it is not advisable to subject them to experiment, because their basic designation is to serve transportation. Only separate experimental structures and special controls are permitted. However, in each case, the problems of special scientific research and structural experimentation should be performed at a scientific institution.

1.3.8 Basic Parameters of the Bridge

The quality of the structure is evaluated considering different criteria: technical, functional, economic, construction, and, in addition, the material of the system and the geometric dimensions of the bridge. All these criteria are temporary parameters defining the quality of the structure.

The problem of design generally consists of the way to find the values of these parameters that will correspond to a better quality of the structure. It is necessary to consider first, in detail, basic factors influencing the quality of the structure. All the parameters interact, but their influence on the quality of the structure is different. Their influence on each other is different: one may depend little on another; another may greatly influence the other. For example, basic parameters for material may not influence basic parameters of foundation and so on.

During preliminary design, the determination of basic parameters interacts and has major influence in making decisions about the location of the bridge, the span, the material, the type of foundation, the system of the bridge, the length of separate spans, the type of superstructure, and the type of supports.

The location of the bridge usually does not much depend on other parameters, but does have an impact on them. For small bridges, the location is defined by the intersection of the highway with the river, ravine, etc. For medium to large bridges, it is possible to compare a number of alternatives, such as the basic value of the highway and the cost of approaches and highway installations. The cost of the bridge itself plays a deciding role because its span at all alternatives is usually an unchangeable constant. For this reason, during selection of bridge location it is possible to propose an often-used bridge type without detailed study. However, there are two exceptions to this general rule. First, if the river is not used for shipping and has sandbanks, then at the location of largest curvature the span of the bridge obtained is smaller, but the depth of the water here is greater. Therefore, foundations are complicated and the installation of pile supports may be impossible. On the sandbanks where the span is increased, but the water depth is shallower, it is possible to build a simple viaduct-type bridge supported by the piles. If the bridge is proposed to be built from timber, its location should be chosen over sandbank. Therefore, during choice of crossing, it is necessary to consider both alternative types of bridges.

The second exception is the design of viaducts across mountain ravines. In this case, the change of the crossing has substantial impact on the choice of the span of the bridge and it is reflected in its cost. It is true that the type of the bridge for the first comparison may be left unchanged (e.g., reinforced concrete arch type, etc.), but it may be designed for all alternatives because the cost of the viaduct will have impact on the choice of location of the crossing.

The above exceptions do not occur often and should be considered separately; for this reason the location of the crossing may be chosen before preliminary design and must be made by the investigators with designers' efforts only in order to check the correctness of the choice. The size of the bridge opening is defined by hydraulic and hydroanalytic investigations and is assumed for the design. It some cases, however, during the design process it is possible to change the span. The size of opening, as shown above, depends on the crossing location. It also depends on the type and depth of the foundation. At greater depths, greater washout is permitted, with corresponding diminishing of the opening. At shallow foundations the reverse could occur.

In principle, two opposite solutions may exist:

1. Build bridge supports as safe against washing, squeeze the river by flow-directed dikes, and obtain a minimum opening.
2. Not squeeze the river, cross the whole river during flood, and thus the concern that the supports will wash out will no longer be a problem.

The first solution is used as a rule for rivers on the plain and can be justified economically and technically. Only for a timber bridge is it expedient to cover the whole flood area by the approach viaducts. Here the size of the opening depends upon the bridge material. The second solution may often be expedient for mountain rivers in which the main channel is often changing and threatens to wash out the flood embankment.

Generally, the size of the opening may change a little depending on the type of foundation. If the type of foundation as a whole is determined by the local conditions (e.g., by using caissons or wells), then the size of the opening for all alternatives remains unchanged. Choice of material is the most substantial problem during preliminary design and depends not only on the designer's point of view but also on other conditions that must be considered before preparing the project. Each material has its own area of application and the problem of material choice arises when these areas intersect.

Timber bridges are usually used as temporary structures. Spans greater than 160 ft often present difficulties. For permanent bridges, the choice is usually between reinforced concrete and steel structures. The following are some recommendations concerning the material selection for the bridge:

1. For spans ranging between 65 to 100 ft reinforced-concrete beam-type bridges are mainly used and steel is considered for overpasses and underpasses.
2. For spans ranging between 330 and 500 ft, steel bridges are often preferred.
3. For spans ranging between 650 and 800 ft, it is expedient to use steel bridges.

Therefore, the choice between reinforced concrete and steel bridges is generally for spans ranging between 65 and 330 ft.

The type of foundation for the bridge is determined mainly by the geologic investigation of ground in the riverbanks and in the main channel, and also by the depth and behavior of the water.

Relatively, the type of foundation influences the superstructure, size of separate spans, and type of supports. Foundations built at the present time may be divided into two basic groups:

1. Piler foundation in which timber pilers are used for shallow foundations and reinforced concrete and steel piles are used for deep foundations.
2. Massive, shallow foundation (between others or piles) and deep foundations (caissons and wells). It is obvious that for large spans it is necessary to use a massive foundation.

Shallow pile foundations are possible for viaduct bridges having small spans. Regarding the bridge system, it should be emphasized that pile foundations almost define the beam system and arches. Suspension bridges require a massive foundation and supports, but there might be other alternatives. During design, the following parameters remain constant or are slightly modified for the bridge system:

1. Size of spans (unequal or uniform);
2. Span system;
3. Type of supports.

1.3.9 Bridge System

The bridge system (i.e., beam, arch, suspension) is integrally related to the chosen material. Beam systems are mostly used for small and medium spans. An arch system is mainly used for large spans and a suspension system is used for long spans.

When using reinforced-concrete bridges, the following should be taken into consideration:

1. For spans up to 130 ft, a beam system is recommended.
2. For spans ranging between 130 and 200 ft, either a beam or arch system can be used.
3. For longer spans, an arch system is recommended.

For steel bridges, beam systems are mainly used. The arch system is expedient to use for spans longer than 160 ft. All the above span lengths are approximate and can be used as preliminary guidance in the early stage of the investigation in order to determine the appropriate system to use. The bridge system depends also on other parameters. It is impossible to investigate all other parameters without assuming the material type for the structure and the bridge system in the early stage of the investigation.

1.3.10 Size of Separate System

The size of the separate system greatly influences the cost of bridges. Determination of the span system involves a number of basic problems that need to be solved during the preliminary design.

For beam bridges having steel trusses, a known rule exists. The cost of the main truss with bracing per span should equal the cost of one pier with foundation. For all other cases, the length of span depends upon the type of foundation and pier.

Similarly, the system of the span construction has influence on the system of the span. With arch bridges, the cost of support is generally greater than that of beam type. For this reason (all things being equal), the span of the arch bridge should be greater than a beam bridge. The exceptions are high viaducts having rising high arches which are more economical. The limits of changes to span length are governed by clearances for ships and typical uses of span structures. The clearances for ships regulate the minimum size of the span. Usually the span is greater than the most economical length. For this reason, during crossings of navigable rivers the size of the span at the main channel in most cases is predetermined. It is necessary to change only side and approach spans. When choosing approach spans, it is necessary to consider typical projects because the use of typical construction is more rational and useful.

From this it follows that the length of spans is not arbitrary. They are chosen from defined conditions. The span length is closely connected with the system of span structure. Therefore, it is necessary at the early stage of the project to assume the proper system of span structure, noting that the choice of the system significantly determines the bridge system.

1.3.11 Type of Span Construction

The type of span construction is closely related to the bridge system. After assuming a bridge system, the span structure should be determined. There might be some problems related to the type of structure (e.g., solid or truss type for steel, monolith or prefabricated for reinforced concrete), the number of main girders, the basic dimensions, etc. Detailed study for each case is needed. Many problems common to particular cases can be investigated earlier, during the preparation of typical projects.

The use of typical projects substantially helps the individual design. For example, in the majority of medium-span bridges typical projects may be used. The use of typical projects simplifies fabrication of the structure, reduces the time necessary for design and construction, and makes the structure more economical to execute. However, the immediate use of typical projects should not be considered as a rule. They should be considered as a first solution, which in many cases can be improved. Each project has different circumstances, and typical projects do not provide solutions to all possible design problems. In some projects there might be some local conditions that need to be dealt with and were not addressed in previous projects. This problem is especially recognized in the design and construction of long-span bridges. Examples of already built bridges may provide a rational starting point. Together with this experience in the design and building of bridges, it is possible to establish some useful relations such as the ratio of truss height to span to the number and length of panels, etc.

The design of bridge structures starts with the critical study and the use of existing bridges to prepare the first alternative of the structure and continues during the investigation to separate parameters to prepare the next alternatives.

1.3.12 Type of Support

Supports can be divided into two groups: columns and massive supports. The second group is used in the presence of large floating ice and arch-type span structures. Column-type supports are most expedient with small-beam structures.

1.4 Remarks and Conclusions

A proper design method should meet two basic criteria:

1. First, the design method should be based on scientific engineering research and analysis. From comprehensive research, design derives logical conclusions.
2. Design methods should be achieved by practice and previous experience in the design and construction of bridges. Also, modifications should always be performed to improve the design. This is largely reflected by the designer's creative capability, sense of invention, and innovation.

Therefore, the integrated part of preliminary design is a comprehensive search of scientific, practical findings and analysis.

References

1. Waddell, J. A. L., *Bridge Engineering,* Vol. 1, John Wiley & Sons, New York, 1916, 267–280.
2. Mitropolskii, N. M., *Methodology of Bridges Design,* scientific-technical edition, Avtotransportni Literatury, Moscow, 1958, 215–242 [in Russian].

3. Polivanov, N. I., *Design and Calculation of the Reinforced Concrete and Metal Highway Bridges,* Transport, Moscow, 1970, 5–36 [in Russian].
4. Steinman, D. B. and Watson. S. R., *Bridges and Their Builders,* Dover, New York, 1957, 378–391.
5. DeMare, E., *Your Book of Bridges,* Faber and Faber, London, 1963, 11–28.
6. Vitruvius, *The Ten Books on Architecture,* translated by M. H. Morgan, Dover, New York, 1960, 13–16.
7. Palladio, A., *The Four Books on Architecture,* Stroiizdat, Moscow, 1952 [in Russian].
8. Holgate, A., *The Art in Structural Design,* Clarendon Press, Oxford, 1986, 1–6, 24–30, 187–195.
9. Francis, A. J., *Introducing Structures,* Pergamon Press, New York, 1980, 221–260.

2

Aesthetics — Basics* 1 Introduction

Fritz Leonhardt
Stuttgart University

2.1 Introduction

Aesthetics falls within the scope of philosophy, physiology, and psychology. How then, you may ask, can I as an engineer presume to express an opinion on aesthetics, an opinion which will seem to experts to be that of a layman. Nevertheless, I am going to try.

For over 50 years I have been concerned with, and have read a great deal about, questions concerning the aesthetic design of building projects and judgment of the aesthetic qualities of works in areas of the performing arts. I have been disappointed by all but a few philosophical treatises on aesthetics. I find the mental acrobatics of many philosophers — whether, for example, existence is the existence of existing — difficult to follow. Philosophy is the love of Truth, but truth is elusive and hard to pin down. Books by great building masters are full of observations and considerations from which we can learn in the same way that we study modern natural scientists.

My ideas on aesthetics are based largely on my own observations, the results of years of questioning — why do we find this beautiful or that ugly? — and on innumerable discussions with architects who also were not content with the slogans and "isms" of the times, but tried to think critically and logically.

*Much of the material of this chapter was taken from Leonhardt, F., *Bridges — Aesthetics and Design*, Chapter 2: The basics of aesthetics, OVA, Stuttgart, Germany, 1984, with permission.

0-8493-7434-0/00/$0.00+$.50

The question of aesthetics cannot be understood purely by critical reasoning. It reaches to emotion, where logic and rationality lose their precision. Undaunted, I will personally address these questions, so pertinent to all of us, as rationally as possible. I will confine myself to the aesthetics of building works, of man-made objects, although from time to time a glance at the beauty of nature as created by God may help us reinforce our findings.

I would beg you to pardon the deficiencies that have arisen because of my outside position as a layman. This work is intended to encourage people to study questions of aesthetics using the methods of the natural scientist (observation, experiment, analysis, hypothesis, theory) and to restore the respect and value which it enjoyed in many cultures.

2.2 The Terms

The Greek word *aisthetike* means the science of sensory perception and very early on was attributed to the perception of the beautiful. Here we will define it as follows:

Aesthetics: The science or study of the quality of beauty an object possesses, and communicates to our perceptions through our senses (expression and impression according to Klages [1]).

Aesthetic: In relation to the qualities of beauty or its effects; aesthetic is not immediately beautiful but includes the possibility of nonbeauty or ugliness. Aesthetic is not limited to *forms*, but includes surroundings, light, shadows, and color.

2.3 Do Objects Have Aesthetic Qualities?

Two different opinions were expressed in old philosophical studies of aesthetics:

1. Beauty is not a quality of the objects themselves, but exists only in the imagination of the observer and is dependent on the observer's experience [2]. Smith said in his "Plea for Aesthetics" [3], "Aesthetic value is not an inborn quality of things, but something lent by the mind of the observer, an interpretation by understanding and feeling." But how can we interpret what does not exist? Some philosophers went so far as questioning the existence of objects at all, saying they are only vibrating atoms, and everything we perceive is subjective and only pictured by our sensory organs. This begs the question, then, is it possible to picture the forms and colors of objects on film using a camera? These machines definitely have no human sensory organs.
2. The second school of thought maintains that objects have qualities of beauty. Kant [4] in his *Critique of Pure Reason* said, "Beauty is what is generally and without definition, pleasing." It is not immediately clear what is meant by "without definition," perhaps without explaining and grasping the qualities of beauty consciously. What is "generally pleasing" must mean that the majority of observers "like" it. Paul [5] expressed similar thoughts in his *Vorschule der Aesthetik* and remarked that Kant's constraint "without definition" is unnecessary. Thomas Aquinas (1225–1274) simply said, "A thing is beautiful if it pleases when observed. Beauty consists of completeness, in suitable proportions, and in the luster of colors." At another time, Kant said that objects may arouse pleasure independent of their purpose or usefulness. He discussed "disinterested pleasure," a pleasure free from any interest in objects: "When perceiving beauty, I have no interest in the existence of the object." This emphasizes the subjective aspect of aesthetic perception, but nonetheless bases the origin of beauty in the object.

Is one right? Most would side with Kant and grant that all objects have aesthetic qualities, whether we perceive them or not. Aesthetic value is transmitted by the object as a message or simulation and its power to ourselves depends on how well we are tuned for reception. This example drawn from modern technology should be seen only as an aid to understanding. If a person is receptive to transmissions of beauty, it then depends very largely on how sensitive and developed are the person's senses for aesthetic messages, whether the person has any feeling for quality at all. We will look at this question more closely in Section 2.4.

On the other hand, Schmitz, in his *Neue Phänomenologie* [6], sees in this simple approach "one of the worst original sins in the theory of cognition." ...This *physiologism* limits the information for human perception to messages that reach the sensory organs and the brain in the form of physical signals and are therefore metaphysically raised to consciousness in a strangely transformed shape." We must see the relationships between the object and circumstances, associations, and situations. More important is the situation and observer's background and experience. The observer is "affectively influenced," [6] i.e., the effect depends on the health of the observer's senses, on the observer's mood, on the observer's mental condition; the observer will have different perceptions when sad or happy. The observer's background experience arouses concepts and facts for which the observer is prepared subconsciously or which are suggested by the situation. Such "protensions" [6] influence the effects of the object perceived, and include prejudices which are held by most people and which are often a strong and permanent hindrance to objective cognition and judgment. However, none of this phenomenology denies the existence of the aesthetic qualities of objects.

Aesthetic quality is not limited to any particular fixed value by the characteristics of the object, but varies within a range of values dependent on a variety of characteristics of the observer. Judgment occurs in a process of communication. Bahrdt [7], the sociologist, said, "As a rule aesthetic judgment takes place in a context of social situations in which the observers are currently operating. The observers may be a group, a public audience, or individuals who may be part of a community or public. The situation can arise at work together, during leisure time, or during a secluded break from the rush of daily life. In each of these different situations the observer has a different perspective and interpretation, and thus a different aesthetic experience [impression]."

Aesthetic characteristics are expressed not only by form, color, light, and shadow of the object, but by the immediate surroundings of the object and thus are dependent on object environment. This fact is well known to photographers who can make an object appear much more beautiful by careful choice of light and backdrop. Often a photograph of a work of art radiates a stronger aesthetic message than the object itself (if badly exhibited) in a gallery. With buildings, the effect is very dependent on the weather, position of the sun, and on the foreground and background. It remains undisputed that there is an infinite number and variety of objects (which all normal healthy human beings find beautiful). Nature's beauty is a most powerful source of health for humans, giving credence to the suggestion that we have an inborn aesthetic sense.

The existence of aesthetic qualities in buildings is clearly demonstrated by the fact that there are many buildings, groups of buildings, or civic areas which are so beautifully designed that they have been admired by multitudes of people for centuries, and which today, despite our artless, materialistic attitudes to life, are still visited by thousands and still radiate vital power. We speak of classical beauty. All cultures have such works, and people go to great lengths to preserve and protect them; substantial assistance has come from all over the world to help preserve Venice, whose enchanting beauty is so varied and persuasive.

We can also give negative evidence for the existence of aesthetic qualities in objects in our man-made environment. Think of the ugliness of city slums, or depressing monotonous apartment blocks, or huge blocky concrete structures. These products of the "brutalist" school have provoked waves of protest. This affront to our senses prompted the Swiss architect Rolf Keller to write his widely read book *Bauen als Umweltzerstörung* [8].

All these observations and experiences point to the conclusion that objects have aesthetic qualities. We must now look at the question of how humans receive and process these aesthetic messages.

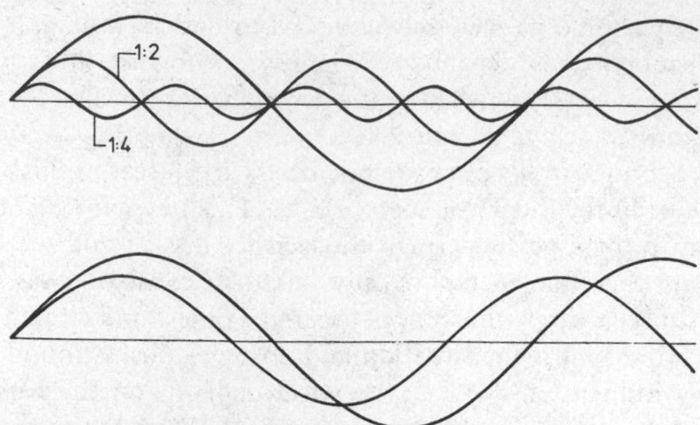

FIGURE 2.1 Wave diagrams for consonant and dissonant tones.

2.4 How Do Humans Perceive Aesthetic Values?

Humans as the receivers of aesthetic messages use all of their senses: they see with their eyes, hear with their ears, feel by touch, and perceive temperature and radiation by sensors distributed in the body, sensors for which there is no one name. Our sensory organs receive different waveforms, wavelengths, and intensities. We read shapes by light rays, whose wavelengths give us information about the colors of objects at the same time. The wavelength of visible light ranges from 400 μm (violet) to 700 μm (red) (1 μm = 1 millionth of 1 mm). Our ears can hear frequencies from about 2 to 20,000 Hz.

The signals received are transmitted to the brain and there the aesthetic reaction occurs — satisfaction, pleasure, enjoyment, disapproval, or disgust. In modern Gestalt psychology, Arnheim [9] explained the processes of the brain as the creation of electrochemical charge fields which are topologically similar to the observed object. If such a field is in equilibrium, the observer feels aesthetic satisfaction, in other cases the observer may feel discomfort or even pain. Much research needs to be done to verify such explanations of brain functions, but they do seem plausible. However, for most of us we do not need to know brain functions exactly.

During the course of evolution, which we assume to have taken many millions of years, the eye and ear have developed into refined sensory organs with varied reactions to different kinds of waveforms. Special tone sequences can stimulate so much pleasure that we like to hear them — they are consonant or in harmony with one another. If, however, the waveforms have no common nodes (Figure 2.1) the result is dissonance or beats, which can be painful to our ear. Dissonances are often used in music to create excitement or tension.

The positive or negative effects are a result not only of the charge fields in the brain, but the anatomy of our ear, a complex structure of drum oscular bones, spiral cochlea, and basilar membrane. Whether we find tones pleasant or uncomfortable would seem to be physiological and thus genetically conditioned. There are naturally individual differences in the sense of hearing, differences which occur in all areas and in all forms of plant and animal life.

There are also pleasant and painful messages for the eye. The effects are partly dependent on the condition of the eye, as, for example, when we emerge from a dark room into light. Color effects of a physiological nature were described in much detail by Goethe in his color theory [10]. In the following, we will discuss the effects of physical colors on the rested, healthy eye, and will not address color effects caused by the refraction or reflection of light.

Some bright chemical colors cause painful reactions, but most colors occurring naturally seem pleasant or beautiful. Again, the cause lies in waves. The monotonous waves of pure spectral colors have a weak effect. The eye reacts more favorably to superimposed waves or to the interaction of two separate colors, especially complementary colors.

We feel that such combinations of complementary colors are harmonious, and speak of "color harmony." Great painters have given us many examples of color harmony, such as the blue and yellow in the coat of Leonardo da Vinci's *Madonna of the Grotto.*

We all know that colors can have different psychological effects: red spurs aggression; green and blue have a calming effect. There are whole books devoted to color psychology and its influence on human moods and attitudes.

We can assume that the eye's aesthetic judgment is also physiologically and genetically controlled, and that harmonic waveforms are perceived as more pleasant than dissonant ones. Our eyes sense not only color but can form images of the three-dimensional, spatial characteristics of objects, which is vital for judging the aesthetic effects of buildings. We react primarily to proportions of objects, to the relationships between width and length and between width and height, or between these dimensions and depth in space. The objects can have unbroken surfaces or be articulated. Illumination gives rise to an interplay of light and shadow, whose proportions are also important.

Here the question of whether there are genetic reasons for perceiving certain proportions as beautiful or whether upbringing, education, or habit play a role cannot be answered as easily as for those of acoustic tone and color. Let us first look at the role proportions play.

2.5 The Cultural Role of Proportions

Proportions exist not only between geometric lengths, but between the frequencies of musical tones and colors. An interplay between harmonic proportions in music, color, and geometric dimensions was discovered very early, and has preoccupied the thinkers of many different cultural eras.

Pythagoras of Samos, a Greek philosopher (571–497 B.C.) noted that proportion between small whole numbers (1:2, 2:3, 3:4, or 4:3, and 3:2) has a pleasing effect for tones and lengths. He demonstrated this with the monochord, a stretched string whose length he divided into equal sections, comparing the tones generated by the portions of the string at either side of an intermediate support or with the open tone [11–13].

In music these harmonic or consonant tone intervals are well known, for example,

String Length	Frequencies	
1:2	2:1	Octave
2:3	3:2	Fifth
3:4	4:3	Fourth
4:5	5:4	Major third

The more the harmonies of two tones agree, the better their consonance; the nodes of the harmonies are congruent with the nodes of the basic tones. Later, different tone scales were developed to appeal to our feelings in a different way depending on the degree of consonance of the intervals; think of major and minor keys with their different emotional effects.

A correspondence between harmonic proportions in music and good geometric proportions in architecture was suggested and studied at an early stage. In Greek temples many proportions corresponding with Pythagoras's musical intervals can be identified. Kayser [14] has recorded these relationships for the Poseidon temple of Paestum.

H. Kayser (1891–1964) dedicated his working life to researching the "harmony of the World." For him, the heart of the Pythagorean approach is the coupling of the tone of the monochord string with the lengths of the string sections, which relates the qualitative (tone perception) to the quantitative (dimension). The monochord may be compared with a guitar. If you pull the string of a guitar, it gives a tone; the height of the tone (quality) depends on the length (dimension = quantity) and the tension of the string. Kayser considered the qualitative factor (tones) as judgment by emotional feeling. It is from this coupling of tone and dimension, of perception and logic, of feeling and knowledge, that the emotional sense for the proportions of buildings originates — the tones of buildings, if you will.

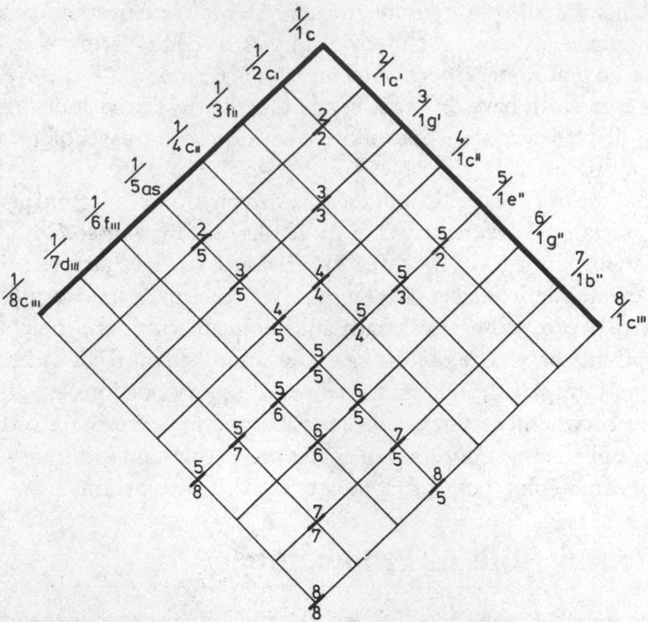

FIGURE 2.2 Giorgio numerical analogy in Λ-shape.

Kayser also had shown that Pythagorean harmonies can be traced back to older cultures such as Egyptian, Babylonian, and Chinese, and that knowledge of harmonic proportions in music and building are about 3000 years old. Kayser's research has been continued by R. Haasse at the Kayser Institute for Harmonic Research at the Vienna College of Music and Performing Arts.

Let us return to our historical survey. In his famous 10 books *De Architectura,* Marcus Vitruvius Pollio (84–14 B.C.) noted the Grecian relationships between music and architecture and based his theories of proportion on them.

Wittkower [12] mentions an interesting text by the monk Francesco Giorgio of Venice. Writing in 1535 on the design of the Church of S. Francesco della Vigna in Venice (shortened extract):

> To build a church with correct, harmonic proportions, I would make the width of the nave nine double paces, which is the square of three, the most perfect and holy number. The length of the nave should be twenty-seven, three times nine, that is an octave and a fifth. … We have held it necessary to follow this order, whose master and author is God himself, the great master builder. … Whoever should dare to break these rules, he would create a deformity, he would blaspheme against the laws of Nature."

So strictly were the laws of harmony, God's harmony, obeyed.

In his book *Harmonia,* Francesco Giorgio represented his mystic number analogies in the form of the Greek letter Λ. Thimus [15] revised this "Lambdoma" for contemporary readers (Figure 2.2).

"Rediscovered" for curing the ills of today's architecture, Andea di Piero da Padova — known to us as *Palladio* [16], was a dedicated disciple of harmonic proportions. He wrote, "The pure proportions of tones are harmonious for the ear, the corresponding harmonies of spatial dimensions are harmonious for the eye. Such harmonies give us feelings of delight, but no-one knows why — except he who studies the causes of things."

Palladio's buildings and designs prove that beautiful structures can be created using these harmonic proportions when they are applied by a sensitive master. Palladio also studied proportions in spatial perspective, where the dimensions are continuously reduced along the line of vision. He confirmed the view already stated by Brunelleschi (1377–1446) that objective laws of harmony also apply to perspective space.

Even before Palladio, Leon Batista Alberti (1404–1472), had written about the proportions of buildings, Pythagoras had said:

> The numbers which thrill our ear with the harmony of tones are entirely the same as those which delight our eye and understanding. … [We] shall thus take all our rules for harmonic relationships from the musicians who know these numbers well, and from those particular things in which Nature shows herself so excellent and perfect.

We can see how completely classical architecture, particularly during the Renaissance, was ruled by harmonic proportions. In the Gothic age master builders kept their canon of numbers secret. Not until a few years ago did the book *Die Geheimnisse der Kathedrale von Chartres* (The Secrets of Chartres Cathedral) by the Frenchman L. Charpentier appear [13], in which he deciphered the proportions of this famous work. It reads like an exciting novel. The proportions correspond with the first Gregorian scale, based on *re* with the main tones of *re-fa-la.* Relationships to the course of the sun and the stars are demonstrated.

Ancient philosophers spent much of their time attempting to prove that God's sun, moon, stars, and planets obeyed these harmonic laws. In his work *Harmonice Mundi* Johannes Kepler (1571–1630) showed that there are a great number of musical harmonies. He discovered his third planetary law by means of harmonic deliberations, the so-called octavoperations. Some spoke of "the music of the spheres" (Boethius, Musica mundana).

Villard de Honnecourt, the 13th-century cathedral builder from Picardy, gave us an interesting illustration of harmonic canon for division based on the upper tone series 1–½–⅓–¼, etc. For Gothic cathedrals he started with a rectangle of 2:1. This Villard diagram (Figure 2.3) [13, 17] was probably used for the design of the Bern cathedral. Whole-number proportions of the fourth and third series can be seen in the articulation of the tower of Ulm Cathedral. A Villard diagram can be drawn for a square, and it then, for example, fits the cross section of the earlier basilica of St. Peter's Cathedral in Rome.

When speaking of proportion, many think of the golden mean, but this does not form a series of whole-number relationships and does not play the important role in architecture which is often ascribed to it. This proportion results from the division of a length $a + b$ where $b < a$ so that

$$\frac{b}{a} = \frac{a}{a+b} \tag{2.1}$$

This is the case if

$$a = \frac{\sqrt{5} + 1}{2} b = 1.618b \tag{2.2}$$

the reciprocal value is $b = 0.618a$, which is close to the value of the minor sixth at $5/8 = 0.625$ or $8/5 = 1.6$. The golden mean is a result of the convergence of the Fibonacci series, which is based on the proportion of $a{:}b$, $b{:}(a + b)$, etc.:

$$\begin{aligned}
a{:}b &= 1{:}\ 2 = 0.500 = \text{octave}\\
b{:}(a+b) &= 2{:}\ 3 = 0.667 = \text{fifth}\\
&\ 3{:}\ 5 = 0.600 = \text{major sixth}\\
&\ 5{:}\ 8 = 0.625 = \text{minor sixth}\\
&\ 8{:}\ 13 = 0.615\\
&\ 13{:}\ 21 = 0.619\\
&\ 21{:}\ 34 = 0.618 = \text{Golden Mean}
\end{aligned}$$

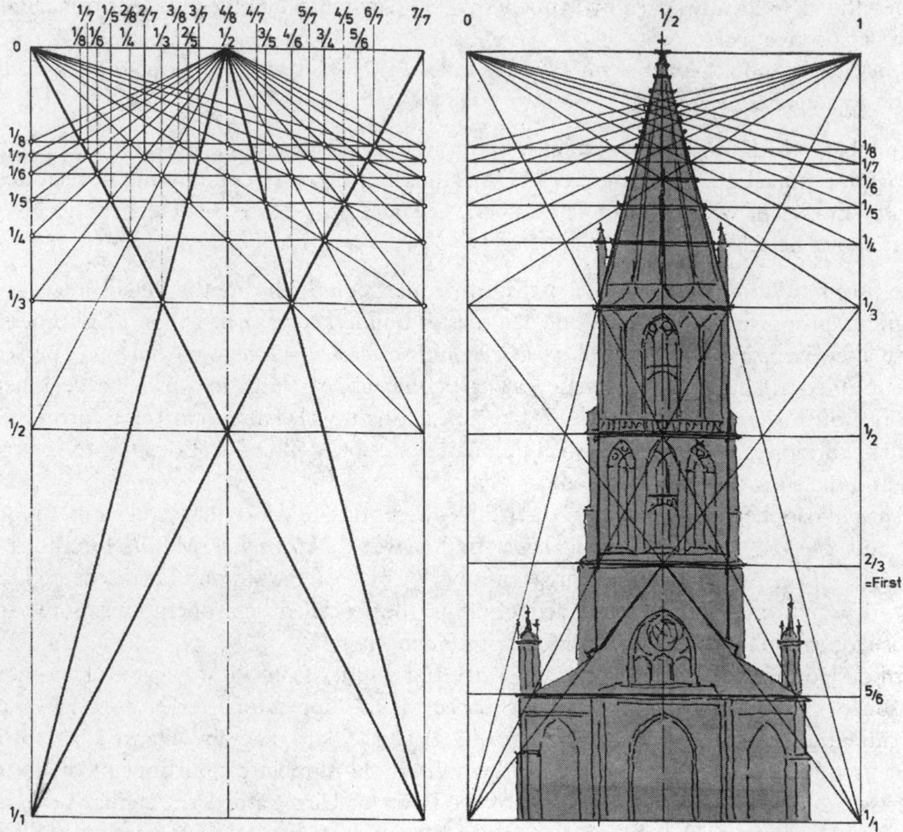

FIGURE 2.3 The Villard diagram for rectangle 2:1.

This numerical value is interesting in that:

$$\frac{1.618}{1.618-1} = \frac{1.618}{0.618} = 2.618$$

and

$$2.168\ (6/5) = 3.1416 = \pi$$

The golden mean thus provided the key to squaring the circle, as can be found in Chartres Cathedral. It can be constructed by dividing the circle into five (Figure 2.4).

The Fibonacci series is also used to construct a logarithmic spiral, which occurs in nature in snail and ammonite shells, and which is considered particularly beautiful for ornaments. Le Corbusier (1887–1965) used the golden mean to construct his "Modulor" based on an assumed body height of 1.829 m but the Modulor is in itself not a guarantee of harmony.

An interesting proportion is $a: b = 1: \sqrt{3} = 1: 1.73$. It is close to the golden mean but for technical applications has the important characteristic that the angles to the diagonals are 30° or 60° (equilateral triangle) and the length of the diagonal is $2a$ or $2b$ (Figure 2.5). A grid with sides in the ratio of $1: \sqrt{3}$ was patented on July 8, 1976 by Johann Klocker of Strasslach. He used this grid to design carpets, which were awarded prizes for their harmonious appearance.

During the last 50 years architects have largely discarded the use of harmonic proportions. The result has been a lack of aesthetic quality in many buildings where the architect did not choose

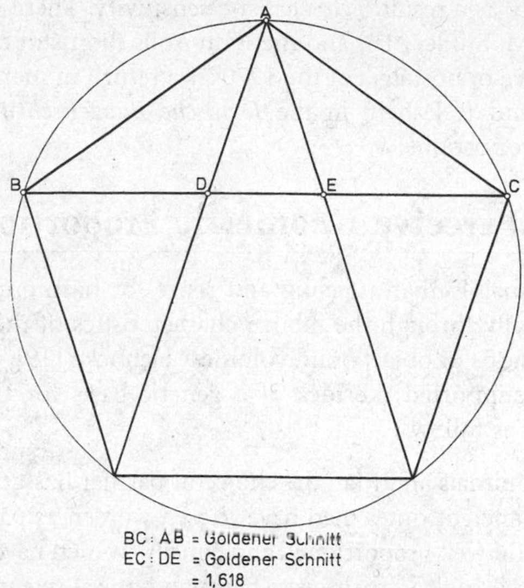

FIGURE 2.4 The golden mean in a pentagon.

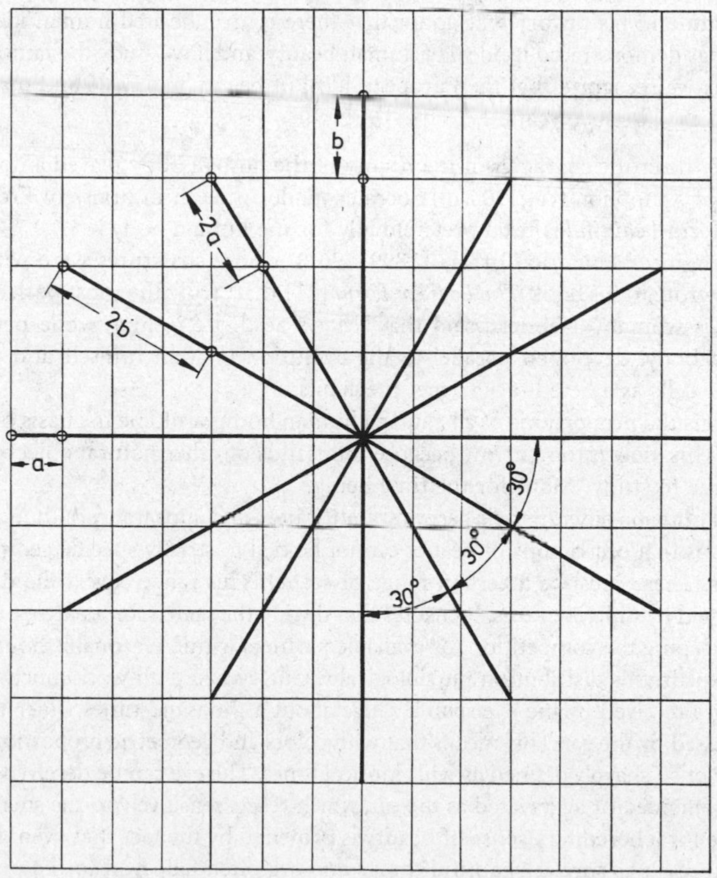

FIGURE 2.5 The Kloecker grid with a:b = 1: $\sqrt{3}$.

good proportions intuitively as a result of his artistic sensitivity. There were exceptions, as always. The Swiss architect Andre M. Studer [18] and the Finn Aulis Blomstdt consciously built "harmonically." One result of the wave of nostalgia of the 1970s is a return in many places to such aesthetics. Kayser in Reference [14] and P. Jesberg in the *Deutsche Bauzeitschrift* DBZ 9/1977 gave a full description of harmonic proportions.

2.6 How Do We Perceive Geometric Proportions?

In music we can assert plausibly that a feeling and sense for harmonic tone series is controlled genetically and physiologically through the inborn characteristics of the ear. What about the proportions of lengths, dimensions of objects, and volumes? Helmcke [19], of the Technical University of Berlin, wholeheartedly supported the idea of a genetic basis for the aesthetic perception of proportions and he argued as follows:

> During the evolution of animals and Man the choice of partner has undoubtedly always played an important role. Since ancient times men have chosen women as partners, who in their eyes were the most beautiful and well proportioned and equally women have chosen men as partners the strongest and most well-built in their eyes. Through natural selection [Darwin] during the evolution of a species this must have led to the evolution of aesthetic perception and feeling and resulted in the development in Man of a genetically coded aesthetic ideal for human partners, passed on from generation to generation. We fall in love more easily with a beautiful partner; love at first sight is directed mostly by an instinctive feeling for beauty, and not by logic. Nobody who knows Man and his history will doubt that there is an inherited human ideal of beauty. Every culture has demonstrated its ideal of human beauty, and if we study the famous sculptures of Greek artists we recognize that the European ideal of beauty in female and male bodies has not changed in the last 3000 years.

For the Greeks the erotic character of the beauty of the human body played a dominant role. At the Symposium of Xenophon (ca. 390 B.C.) Socrates made a speech in praise of Eros. According to Grassi [20], the term *beautiful* is used preferentially for the human body.

The Spanish engineer Eduardo Torroja (1899–1961), whose structures were widely recognized for their beauty, wrote in his book *The Logic of Form* [21] that "truly the most perfect and attractive work of Nature is woman." Helmcke said that "Man's aesthetic feeling, while perceiving certain proportions of a body, developed parallel to the evolution of Man himself and is programmed genetically in our cells as a hereditary trigger mechanism."

According to this the proportions of a beautiful human body would be the basis of our hereditary sense of beauty. This view is too narrow because thousands of other natural objects radiate beauty, but let us continue to study "Man" for the time being.

Fortunately, all humans differ in their hereditary, attributes, and appearance, although generally only slightly. This means that our canons of beauty cannot be tied to strictly specific geometric forms and their proportions. There must be a certain range of scatter. This range covers the differences in the ideals of beauty held by different races. It ensures that during the search for a partner each individual's ideal will differ, keeping the competition for available partners within reasonable bounds.

We can also explain this distribution physiologically. Our eyes have to work much harder than our ears. The messages received by the eye span a range about a thousand times wider than the scale of tones to be processed by the ear. This means that with colors and geometric proportions harmony and disharmony are not so sharply defined as with musical tones. The eye can be deceived more easily and is not as quickly offended or aggravated as the ear, which reacts sensitively to the smallest dissonance.

More evidence for a hereditary sense of beauty is provided by the fact that even during their first year, children express pleasure at beautiful things and are offended, even to the point of weeping, by ugly objects. How children's eyes sparkle when they see a pretty flower.

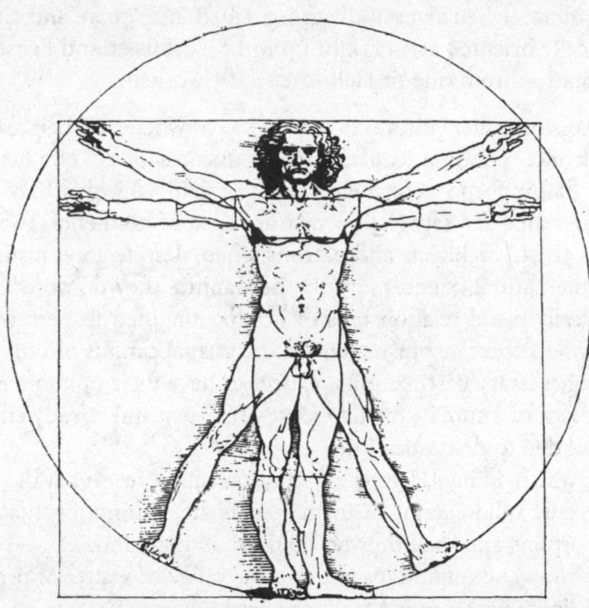

FIGURE 2.6 Image of man in circle and square according to Leonardo da Vinci.

Evidence against the idea that we have a hereditary sense of beauty is suggested by the fact that people argue so much about what is beautiful or ugly, demonstrating a great deal of insecurity in the judgment of aesthetic qualities. We will give this further thought in Section 2.7.

Our ability to differentiate between good and bad using our senses of taste and smell has also developed genetically and with certain variations is the same for most people [22]. With this background of genetic development it is understandable that the proportions of those human bodies considered beautiful have been studied throughout the ages. A Greek sculptor Polyklet of Kikyon (465–420 B.C.) defined the following proportions:

two handbreadths	=	height of the face and height of the breast, distance breast to navel, navel to end of trunk
three handbreadths	=	height of skull, length of foot
four handbreadths	=	distance shoulder to elbow, elbow to fingertips
six handbreadths	=	ear to navel, navel to knee, length of trunk, length of thigh

Plyklet based his "canon for the ideal figure" on these relationships. These studies had the greatest influence on art during the age of humanism, for example, through the *Vier Büecher von menschlicher Proportione* 1528 by Albrecht Dürer (1471–1528).

Vitruvius also dealt with the human body in his books *De Architectura* and used the handbreadth as a unit of measure. Leonardo da Vinci followed Vitruvius's theories when drawing his image of man inscribed in a square (Figure 2.6). Leonardo's friend, the mathematician Luca Pacioli (ca. 1445–1514) began his work *De Divina Proportione,* 1508, with the words:

> Let us first speak of the proportions of Man because all measures and their relationships are derived from the human body and here are to be found all numerical relationships, through which God reveals the innermost secrets of Nature. Once the ancients had studied the correct proportions of the human body they proportioned all their works, particularly the temples, accordingly. (Quoted by Wittkower [12])

The human body with outstretched arms and legs inscribed in a square and circle became a favorite emblem for humanistically oriented artists right up to Le Corbusier and Ernst Neufert. Let us close this section with a quotation from one of Helmcke's [19] works:

> The intellectual prowess of earlier cultures is revealed to us whenever their artists, architects, and patrons succeeded in incorporating, consciously or unconsciously, our hereditary, genetically programmed canon of proportions in their works; in achieving this they come close to our genetically controlled search for satisfaction of our sense of aesthetics. It reveals the spiritual pauperism of today's artists, architects and patrons when, despite good historical examples and despite advances in the natural sciences and the humanities they do not know of these simple biologically, anatomically based relationships or are too ungifted to perceive, understand, and realize them. Those who deprecate our search for the formal canons of our aesthetic feelings as a foolish and thus unnecessary pastime must expect to have their opinion ascribed to arrogant ignorance and to the lack of a sure instinctive sense of beauty, and already ethnologically known as a sign of decadence due to domestication.
>
> The only criticism which, in my [Helmcke's] opinion can be leveled at the thousands of years' old search for universally valid canons of form, lies in the assumption that these canons shall consist of fixed proportions and shall thus be valid for all mankind. …
>
> What is needed is experience of and insight into the range of scatter of proportional relations and insight into the limits within our hereditary aesthetic sense reacts positively, and beyond which it reacts negatively.

2.7 Perception of Beauty in the Subconscious

We are not generally aware of how strongly our world of feelings, our degree of well-being, comfort, disquiet, or rejection is dependent on impressions from our surroundings. Neurologists know that parts of our brain are capable of reacting to external stimuli without reference to the conscious mind and of processing extensive amounts of information. This takes place in the limbic system of the primitive structures of the midbrain and the brain stem. For all those activities of the subconscious which deal with the processing of aesthetic messages, Smith [3] used the phrase "limbic aesthetics" and dedicated a whole chapter of his very readable book to them.

Our subconscious sense of beauty is almost always active, whether we are at home, in the city marketplace, in a church, in a beautiful landscape, or in the desert. Our surroundings affect us through their aesthetic characteristics even if our conscious thoughts are occupied with entirely different matters and impressions.

Smith wrote of the sensory appetite of these primitive parts of the brain for pleasant surroundings, for the magic of the city, and for the beauty of nature. The limbic system reacts to an oversupply of stimuli with rejection or anxiety.

Symbolic values connected with certain parts of our environment also act on the subconscious. The home, the church, school, garden, etc. have always possessed symbolic values created by learning and experience. These are related mostly to basic human situations and cause emotional reactions, without ever reaching the conscious level.

This perception of beauty at the subconscious level plays a particularly strong role in city dwellers. Their basic feeling of well-being is doubtless influenced by the aesthetic qualities of their environment in this way. This has social consequences (see Section 2.10) and underlines our responsibility to care about the beauty of the environment.

2.8 Aesthetic Judgment and Taste

When two observers are not agreed in their judgment of a work of art, the discussion is all too often ended with the old proverb, "De gustibus non disputandum est." We like to use a little Latin to show our classical education, which, as we know, is supposed to include an understanding of art. This "there's no accounting for taste" is an idle avoidance tactic, serving only to show that the speaker has never really made a serious effort to study aesthetics and thus has educational deficiencies in the realm of assessing works of art.

Of course, taste is subject to continual change, which in turn depends on current ideals, fashions, and is dependent on historical and cultural background. The popular taste in any given period of time or even the taste of single individuals is never a reliable measure of aesthetic qualities.

On the other hand, genetic studies have shown that we have a certain basic hereditary sense of beauty. Smith [3] said that this aesthetic perception has developed into one of the highest capabilities of our central nervous system and is a source of deep satisfaction and joy.

The judgment of aesthetic characteristics is largely dependent on feelings which are derived from our sensory perceptions. Beauty, then, despite some theories (Bense, Maser) cannot be rationally measured. When looking at the nature of feelings we must admit the fact that despite all our research and science, we know very little about humanity or about ourselves. We can, however, call upon observations and experiences which are helpful.

We repeatedly experience that the majority of people agree that a certain landscape, great painting, or building is beautiful. When entering a room, for example, in an old church, or while wandering through a street, feelings are aroused which are pleasant, comfortable, even elevating, if we sense a radiation of beauty. If we enter a slum area, we feel revulsion or alarm, as we perceive the disorder and decay. We can be more or less aware of these feelings, depending on how strongly our thoughts are occupied elsewhere. Sensitivities and abilities to sense beauty naturally differ from person to person, as is true of our other talents. This sensitivity is influenced by impressions from our environment, by experience, by relationships with our companions at home, at school, and with our friends. Two people judging the qualities of beauty of an object are likely to give different opinions.

Beautiful surroundings arouse feelings of delight in almost all people, but an ugly, dirty environment causes discomfort. Only the degree of discomfort will differ. In our everyday life such feelings often occur only at a subconscious level and often their cause is only perceived after subsequent reflection.

We can develop a clear capacity for judging aesthetic qualities only when we study the message emanated by an object consciously and ask ourselves whether or not we like a building or a room. Next, we must ask ourselves why. Why do I like this and not that? Only by frequent analysis, evaluation, and consideration of consciously perceived aesthetic values can we develop that capacity of judgment which we commonly call taste — taste about which we must argue, so that we can strengthen and refine it. Taste, then, demands self-education, which can be cultivated by critical discussion with others or by guidance from those more experienced. Good judgment of aesthetic values requires a broad education. It can be compared to an art and requires skill, and like art it takes not only talent, but a lot of work.

We need not be afraid that such analysis will weaken our creative skills; in fact, the opposite is true: the goal of analysis is the discovery of the truth through creative thinking [23]. People have different talents and inclinations since they grow up in different circles with different cultural backgrounds and therefore their tastes will always differ. In any given culture, however, there is a certain polarity on the judgment of beauty. Psychologists call this agreement "normal behavior," "a normal reaction of the majority." This again corresponds with Kant's view that beauty is what is generally thought to be beautiful by the majority of people.

Beauty cannot be strictly proved, however; so we must be tolerant in questions of taste and must give freedom to what is generally felt to be beautiful and what ugly. That there is a generally recognized concept of beauty is proved by the consistent judgment of the classical works of art of all great cultures, visited year after year by thousands of people. Think of the popularity of exhibitions of great historic art today. It is history that has the last word on the judgment of aesthetic values, long after fashions have faded.

Fashions: Artistic creation will never be entirely free from fashion. The drive to create something new is the hallmark of creative beings. If the new becomes popular, it is soon copied, and so fashions are born. They are born of the ambition and vanity of humans and please both. The desire to impress often plays a role. Up to a certain point, fashions are necessary; in certain new directions true art may develop through the fashionable, acquiring stability through a maturing process and enduring beyond the original fashion. Often, such new developments are rejected, because we are strongly influenced by the familiar, by what we are used to seeing, and only later realize the value of the new. Again, history pronounces a balanced judgment.

Confusion is often caused in our sense of judgment by modern artists who deliberately represent ugliness in order to mirror the warped mental state of our industrial society. Some of this work has no real quality, but is nonetheless acclaimed as modern art. The majority dares not question this for fear of rejection, slander, and peer pressure.

Although some works that consciously display ugliness or repulsiveness may well be art, we must seriously question the sanity and honesty of the patrons of primitive smearings, tangles of scrap iron, or old baby baths covered in Elastoplast strips (J. Beuys) when such efforts are exhibited as works of art. Happily, the courage to reject clearly such affronts and to put them in their place is on the increase. We only need to read Claus Borgeest's book, *Das Kunsturteil,* [24], in which he wrote, "the belief in such 'art' is a modern form of self-inflicted immaturity, whose price is the self-deprivation of reason, man's supreme attribute."

In any case, it would be wrong to describe as beautiful works, those haunted by ugliness, even if they have the quality of art. The artist intends to provoke and to encourage deliberation. However, the educational effects of such artistic creations are questionable, because we usually avoid their repeated study. Painters and sculptors, however, should be free to paint and sculpt as hatefully and repulsively as they wish — we do not have to look at their works. It is an entirely different case with buildings; they are not a private affair, but a public one. It follows that the designer has responsibility to the rest of humankind and a duty to produce beautiful buildings so that the designer does not give offence. Rightly, the ancient Greeks forbade public showings of ugliness, because their effects are largely negative.

We seldom find anyone who will hang ugly works of art in his or her home. It is beyond a doubt that in the long term we feel comfortable only in beautiful surroundings and that beauty is a significant requirement for the well-being of our soul; this is much more important for people's happiness than we today care to admit.

2.9 Characteristics of Aesthetic Qualities Lead to Guidelines for Designing

The search for explanations, the analysis of aesthetic values, are bound to lead to useful results, at least for man-made buildings and structures. We will now try to subject matters of feelings, emotions, to the clear light of recognition and understanding.

If we do this, we can certainly find answers to the question, "Why is this beautiful and this ugly?" For recognized masterpieces of architecture generally considered beautiful, there have been answers since ancient times, many of which are given in the quoted literature on proportions. Such buildings reveal certain characteristics of quality and from these we can deduce guidelines for design, such as certain proportions, symmetry, rhythm, repeats, contrasts, and similar factors, The master schools of old had such rules or guidelines, such as those of Vitruvius and Palladio. Today, these rules are surely valid and must be rediscovered for the sake of future architecture. They can prove a valuable aid in the design of building structures and at the very least contribute toward avoiding gross design errors.

Many architects and engineers reject rules, but in their statements about buildings we still find references to harmony, proportion, rhythm, dominance, function, etc. Torroja [21] rejected rules, but he said "the enjoyment and conscious understanding of aesthetic pleasure will without doubt be much greater if, through a knowledge of the rules of harmony, we can enjoy all the refinements and perfections of the building in question." Rules of harmony are based on rules of proportion, and somehow the striving for individual artistic freedom prevents us from recognizing relationships often imposed upon us by ethics.

Let us then attempt to formulate such characteristics, rules, or guidelines as they apply to building structures, particularly bridges.

2.9.1 Fulfillment of Purpose–Function

Buildings or bridge structures are erected for a purpose. The first requirement is that the buildings and bridges be designed to optimally suit this purpose. To meet the specific purpose, a bridge may have different structural types: arches, beams, or suspensions. The structure should reveal itself in a pure, clear form and impart a feeling of stability. We must seek simplicity here. The form of the basic structure must also correspond to the materials used. Brick and wood dictate different forms from those for steel or concrete. We speak of form justified by the material, or of "logic of form" [21]. This reminds us of the architect Sullivan's rule "form follows function" which became an often misunderstood maxim for building design. The function of a building is not only that it stand up. One must fulfill all the various requirements of the people that inhabit the building. These include hygiene, comfort, shelter from weather, beauty, even cosiness. The fulfillment of the functional requirements of buildings includes favorable thermal, climatic, acoustic, and aesthetic qualities. Sullivan undoubtedly intends us to interpret his rule in this sense. For buildings the functional requirements are very complex, but in engineering structures, functions besides load-carrying capacity must be fulfilled, such as adequate protection against weather, limitation of deformation and oscillation, among others, and all these factors affect design. Quality and beauty must be united, and quality takes first priority!

2.9.2 Proportion

An important characteristic necessary to achieve beauty of a building is good, harmonious proportions, in three-dimensional space. Good proportions must exist between the relative sizes of the various parts of a building, between its height, width, and breadth, between masses and voids, closed surfaces and openings, between the light and dark caused by sunlight and shadow. These proportions should convey an impression of balance. Tassios [25] preferred "expressive proportions" which emphasize the desired character of a building (see Section 2.9.8).

For structures it is not sufficient that their design is "statically correct." A ponderous beam can be as structurally correct as a slender beam, but it expresses something totally different. Not only are the proportions of the geometric dimensions of individual parts of the building important, but also those of the masses of the structure. In a bridge, for instance, these relationships may be between the suspended superstructure and the supporting columns, between the depth and the span of the beam, or between the height, length, and width of the openings. Harmony is also achieved by the repetition of the same proportions in the entire structure or in its various parts. This is particularly true in buildings.

Sometimes contrasting proportion can be a suitable element. The detailed discussion is referred to Chapter 4 of my book [26], which shows what good proportions can mean for bridges.

2.9.3 Order

A third important rule is the principle of order in the lines and edges of a building, an order achieved by limiting the directions of these lines and edges to only a few in space.

Too many directions of edges, struts, and the like create disquiet, confuse the observer, and arouse disagreeable emotions. Nature offers us many examples of how order can lead to beauty; just think of the enchanting shapes of snow crystals and of many flowers [27, 28]. Good order must be observed between the proportions occurring in a building; for instance, rectangles of 0.8:1 should not be placed next to slim rectangles of 1:3. Symmetry is a well-tried element of order whenever the functional requirements allow symmetry without constraint.

We can include the repetition of equal elements under the rule of order. Repetition provides rhythm, which creates satisfaction. Too many repetitions, on the other hand, lead to monotony, which we encounter in the modular architecture of many high-rise buildings. Where too many repetitions occur, they should be interrupted by other design elements.

The selection of one girder system throughout the structure provides an element of good order. Interrupting a series of arches with a beam gives rise to aesthetic design problems. Under the principle of order for bridges we may include the desire to avoid unnecessary accessories. The design should be so refined that we can neither remove nor add any element without disturbing the harmony of the whole.

2.9.4 Refining the Form

In many cases, bodies formed by parallel straight lines appear stiff and static, producing uncomfortable optical illusions. Tall bridge piers or towers with parallel sides appear from below to be wider at the top than at the bottom, which would be unnatural. Nor does this uniform thickness conform to our concept of functionality, because the forces decrease with increasing height. For this reason, the Egyptians and Greeks gave the columns of their temples a very slight taper, which in many cases is actually curved. Towers are built tapered or stepped. On high towers and bridge piers, a parabolic taper looks better than a straight taper.

The spans of a viaduct crossing a valley should become smaller on the slopes, and even the depth of the girders or edge fascia can be adjusted to the varying spans. Long beams of which the bottom edge is exactly horizontal look as if they are sagging, and so we give them a slight camber.

We must also check the appearance of the design from all possible vantage points of the future observer. Often the pure elevation on the drawing board is entirely satisfactory, but in skew angle views of unpleasant overlapping are found. We must also consider the effects of light and shadow. A wide cantilever deck slab can throw bridge girders into shadow and make them appear light, whereas similar shadows break the expressive character of an arch. Models are strongly recommended for checking a design from all possible viewpoints.

These refinements of form are based on long experience and must be studied with models from case to case.

2.9.5 Integration into the Environment

As the next rule, we recognize the need to integrate a structure or a building into its environment, landscape, or cityscape, particularly where its dimensional relationships and scale are concerned. In this respect many mistakes have been made during the past decades by placing massive concrete blocks in the heart of old city areas. Many factories and supermarkets also show this lack of sensitive integration. Sometimes long-span bridges with deep, heavy beams spoil lovely valley landscapes or towns with old houses lining the riverbank.

The dimensions of buildings must also be related to the human scale. We feel uneasy and uncomfortable moving between gigantic high-rise buildings. Heavy, brutal forms are often deliberately chosen by architects working with prefabricated concrete elements, but they are simply offensive. It is precisely their lack of scale and proportion that has led to the revolt against the brutality of this kind of architecture.

2.9.6 Surface Texture

When integrating a building with its surroundings, a major role is played by the *choice of materials*, the *texture of the surfaces*, and particularly by *color*. How beautiful and vital a natural stone wall can appear if we choose the right stone. By contrast, how repulsive are many concrete facades; not only do they have a dull gray color from the beginning, but they weather badly, producing an ugly patina and appear dirty after only a few years. Rough surfaces are suitable for piers and abutments; smooth surfaces work well on fascia-beams, girders, and slender columns. As a rule, surfaces should be matte and not glossy.

2.9.7 Color

Color plays a significant role in the overall aesthetic effect. Many researchers have studied the psychological effects of color. Here, too, ancient rules of harmonious color composition apply, but today successful harmonious color schemes are rare. Often, we find the fatal urge for sensation, for startling aggressive effects, which can be satisfied all too easily with the use of dissonant colors, especially with modern synthetic pop — or shocking — colors. We can find, however, many examples of harmonious coloring, generally in town renovation programs. Bavaria has provided several examples where good taste has prevailed.

2.9.8 Character

A building and bridge should have character; it should have a certain deliberate effect on people. The nature of this desired effect depends on the purpose, the situation, the type of society, and on sociological relationships and intentions. Monarchies and dictatorships try to intimidate by creating monumental buildings, which make people feel small and weak. We can hope this belongs to the past. Only large banks and companies still make attempts to impress their customers with monumentalism. Churches should lead inward to peace of mind or convey a sense of release and joy of life as in the Baroque or Rococo. Simple dwellings should radiate safety, shelter, comfort, and warmth. Beautiful houses can stimulate happiness.

Buildings of the last few decades express an air of austere objectivity, monotony, coldness, confinement, and, in cities, confusion, restlessness, and lack of composition; there is too much individuality and egoism. All this dulls people's senses and saddens them.

We seem to have forgotten that people also want to meet with joy in their man-made environment. Modern buildings seem to lack entirely the qualities of cheerfulness, buoyancy, charm, and relaxation. We should once again become familiar with design features that radiate cheerfulness without lapsing into Baroque profusion.

2.9.9 Complexity — Stimulation by Variety

Smith [3] postulated a "second aesthetic order," suggested by findings made by biologists and psychologists [29]. According to this, beauty can be enhanced by the tension between variety and similarity, between complexity and order. Baumgarten expressed this as early as 1750, "Abundance and variety should be combined with clarity. Beauty offers a twofold reward: a feeling of well being both from the perception of newness, originality and variation as well as from coherence, simplicity, and clarity." Leibniz in 1714 demanded for the achievement of perfection as much variety as possible, but with the greatest possible order.

Berlyne [30] considered the sequence of tension and relaxation to be a significant characteristic of aesthetic experience. Venturi [31], a rebel against the "rasteritis" (modular disease) architecture of Mies van der Rohe, said, "A departure from order — but with artistic sensitivity — can create pleasant poetic tension."

A certain amount of excitement caused by a surprising object is experienced as pleasant if neighboring objects within the order ease the release of tension. If variety dominates our orientation, reflex is overtaxed and feelings ranging from distaste to rejection are aroused. Disorder is not beautiful.

This complexity doubtless requires artistic skill to be successful. It can be used well in bridge design if, for instance, in a long, multispan bridge the main span is accented by a variation in the girder form. The interplay of complexity and order is important in architecture, particularly in city planning. Palladio was one of the first to extend the classical understanding of harmony by means of the complexity of architectural elements and ornamentation.

2.9.10 Incorporating Nature

We will always find the highest degree of beauty in nature, in plants, flowers, animals, crystals, and throughout the universe in such a variety of forms and colors that awe and admiration make it extremely difficult to begin an analysis. As we explore deeper into the realm of beauty we also find in nature rules and order, but there are always exceptions. It must also remain possible to incorporate such exceptions in the masterpieces of art made by creative humans [28].

The beauty of nature is a rich source for the needs of the soul, and for humans' psychic well-being. All of us know how nature can heal the effects of sorrow and grief. Walk through beautiful countryside — it often works wonders. As human beings we need a direct relationship with nature, because we are a part of her and for thousands of years have been formed by her.

This understanding of the beneficial effects of natural beauty should lead us to insist that nature again be given more room in our man-made environment. This is already happening in many of our cities, but we must introduce many more green areas and groups of trees. Here we must mention the valuable work of Seifert [32] during the building of the first autobahns in Germany.

2.9.11 Closing Remarks on the Rules

We must not assume that the simple application of these rules will in itself lead to beautiful buildings or bridges. The designer must still possess imagination, intuition, and a sense for both form and beauty. Some are born with these gifts, but they must be practiced and perfected. The act of designing must always begin with individual freedom, which in any case will be restricted by all the functional requirements, by the limits of the site, and not least by building regulations that are usually too strict.

The rules, however, provide us with a better point of departure and help us with the critical appraisal of our design, particularly at the model stage, thus making us aware of design errors.

The artistically gifted may be able to produce masterpieces of beauty intuitively without reference to any rules and without rational procedures. However, the many functional requirements imposed on today's buildings and structures demand that our work must include a significant degree of conscious, rational, and methodical reasoning.

2.10 Aesthetics and Ethics

Aesthetics and ethics are in a sense related; by ethics we mean our moral responsibility to humanity and nature. Ethics also infers humility and modesty, virtues which we find lacking in many designers of the last few decades and which have been replaced by a tendency toward the spectacular, the sensational, and the gigantic in design. Due to exaggerated ambition and vanity and spurred by the desire to impress, unnecessary superlatives of fashions were created, lacking true qualities of beauty. Most of these works lack the characteristics needed to satisfy the requirements of the users of these buildings.

As a responsibility, ethics requires a full consideration of all functional requirements. In our man-made environment we must emphasize the categories of quality and beauty. In his *Acht Todsünden der Menschheit,* Loreanz [33] once said that "the senses of aesthetics and ethics are apparently very closely related, so that the aesthetic quality of the environment must directly affect Man's ethical behavior." He said further, "The beauty of Nature and the beauty of the man-made cultural environment are apparently both necessary to maintain Man's mental and psychic health. Total blindness of the soul for all that is beautiful is a mental disease that is rapidly spreading today and which we must take seriously because it makes us insensitive to the ethically obnoxious."

In one of his last important works, in *To Have or to Be* [34] Erich Fromm also said that the category of "goodness" must be an important prerequisite for the category "beauty," if beauty is to be an enduring value. Fromm goes so far as to say that "the physical survival of mankind is dependent on a radical spiritual change in Man." The demand for aesthetics is only a part of the general demand for changes in the development of "Man." These changes have been called for at least in part and at intervals by humanism, but their full realization in turn demands a new kind of humanism, as well expressed in the appeal by Peccei [35].

2.11 Summary

In order to reach a good capacity of judging aesthetic qualities of buildings or bridge structures, it is necessary to go deep into our human capacities of perception and feelings. The views of many authors who treated aesthetics may help to come to some understanding, which shall help us to design with good aesthetic quality.

References

1. Klages, L., *Grundlagen der Wissenschaft vom Ansdruck*, 9, Auflage, Bonn, Germany, 1970.
2. Hume, H., *On the Standard of Taste*, London, U.K., 1882.
3. Smith, P. F, *Architecktur und Aesthetik*, Stuttgart, Germany, 1981. Original: *Architecture and the Human Dimension*, London, 1979.
4. Kant, I., *Kritik der Urteilskraft, Reklam*, Stuttgart, Germany, 1995.
5. Paul, J., *Vorschule der Aesthetik*, München, Germany, 1974.
6. Schmitz, H., *Neue Phänomenologie*, Bonn, Germany, 1980.
7. Bahrdt, H. P., *Vortrag Hannover*, Stiftung FVS, Hamburg, Germany, 1979.
8. Keller, R., *Bauen als Umweltzerstörung*, Zürich, Switzerland, 1973.
9. Arnheim, R., *Art and Visual Perception*, Berkeley, CA, 1954. German: *Kunst und Sehen*, Berlin, Germany, 1978.
10. Goethe, J. W., *Farbenlehre*, Stuttgart, Germany, 1979.
11. Szabo, I., *Anfänge der griechischen Mathematik*, Berlin, Germany, 1969.
12. Wittlkower, R., *Architectural Principles in the Age of Humanism*, London, 1952. German: *Grundlagen der Architecktur im Zeitalter des Humanismus*, München, Germany, 1969.
13. Charpentier, L., *Die Geheimnisse der Kathedrale von Chartres*, Köln, Germany, 1974.
14. Kayser, H., *Paestum*, Heidelberg, Germany, 1958.
15. Thimus, A. von, *Die Harmonische Symbolik des Altertums*, Köln, Germany, 1968–1976.
16. Puppi, L., *Andrea Palladio*, Stuttgart, Germany, 1977.
17. Strübin, M., Das Villard Diagramm, *Schw. Bauz.*, 1947, 527.
18. Studer, A. M., Architektur, Zahlen und Werte, *Dtsch Bauz.*, 9, Stuttgart, Germany, 1965.
19. Helmcke, J. G., *Ist das Empfinden von aesthetisch schoenen Formen angeboren oder anerzogen*, Heft 3 des SFB 64 der Universität Stuttgart, Germany, 1976, 59; see also *Grenzen menschlicher Anpassung*, IL 14, Universität Stuttgart, Germany, 1975.
20. Grassi, E., *Die Theorie des Schönen in der Antike*, Köln, Germnay, 1980.
21. Torroja, E., *Logik der Form*, München, Germany, 1961.
22. Tellenbach, H., *Geschmack und Atmosphaere*, Salzburg, Austria, 1968.
23. Grimm, C. T., Rationalized esthetics in civil engineering, *J. Struct. Div., ASCE*, 1975.
24. Borgeest, C., *Das Kunsturteil*, Frankfurt, Germany, 1979. Also: *Das sogenannte Schöne*, Frankfurt, Germany, 1977.
25. Tassios, T. P., Relativity and optimization of aesthetic rules for structures, *IABSE Congr. Rep.*, Zürich, Switzerland, 1980.
26. Leonhartdt, F., *Bridges — Aesthetics and Design*, DVA, Stuttgart, Germany, 1984.

27. Heydemann, B., Auswirkungen des angeborenen Schönheitssinnes bei Mensch und Tier, *Nat., Horst Sterns Umweltmag.*, 0, 1980.
28. Kayser, H., *Harmonia Plantarum*, Basel, Switzerland. Also: H. Akroasis, *Die Lehre von der Harmonik der Welt*, Stuttgart, Germany, 1976.
29. Humphrey, N., The illusion of beauty, *Perception Bd.*, 2, 1973.
30. Berlyne, D. E., *Aesthetics and Psycho-Biology*, New York, 1971.
31. Venturi, R., *Complexity and Contradiction in Architecture*, New York, 1966.
32. Seifert, A., *Ein Leben für die Landschaft*, Düsseldorf-Köln, Germany, 1962.
33. Lorenz, K., *Acht Todsünden der zivilisierten Menschheit*, München, Germany, 1973.
34. Fromm, E., *Haben oder Sein*, Stuttgart, Germany, 1976.
35. Peccei, A., *Die Zukunft in unserer Hand*, München, Germany, 1981.

3

Bridge Aesthetics — Structural Art

David P. Billington
Princeton University

Frederick Gottemoeller
Rosales Gottemoeller Associates, Inc.

3.1 Introduction

In recent years it has become apparent that the real problems of bridge design include more than the structural or construction issues relating to the spanning of a gap. The public often expresses concern over the appearance of bridges, having recognized that a bridge's visual impact on its community is lasting and must receive serious consideration.

The public knows that civilization forms around civil works: for water, transportation, and shelter. The quality of public life depends, therefore, on the quality of such civil works as aqueducts, bridges, towers, terminals, and meeting halls: their efficiency of design, their economy of construction, and the visual appearance of their completed forms. At their best, these civil works function reliably, cost the public as little as possible, and, when sensitively designed, become works of art.

Thus, engineers all over the world are being forced to address the issues of aesthetics. Engineers cannot avoid aesthetic issues by taking care of the structural elements and leaving the visual quality to someone else. It is the shapes and sizes of the structural components themselves that dominate the appearance of the bridge, not the details, color, or surfaces. Since they control the shapes and sizes of the structural components, engineers must acknowledge the fact that they are ultimately responsible for the appearance of their structures. Engineers are used to dealing with issues of performance, efficiency, and cost. Now, they must also be prepared to deal with issues of appearance.

0-8493-7434-0/00/$0.00+$.50

FIGURE 3.1 Thomas Telford's Craigellachie Bridge.

3.2 The Engineer's Aesthetic and Structural Art

"Aesthetics" is a mysterious subject to most engineers, not lending itself to the engineer's usual tools of analysis. It is a topic rarely taught in engineering schools. Many contemporary engineers are not aware that a long line of engineers have made aesthetics an explicit element in their work, beginning with the British engineer Thomas Telford. In 1812, Telford defined structural art as the personal expression of structure within the disciplines of efficiency and economy. Efficiency here meant reliable performance with minimum materials, and economy implied the construction with competitive costs and restricted maintenance expenses. Within these bounds, structural artists find the means to choose forms and details that express their own vision, as Telford did in his Craigellachie Bridge (Figure 3.1). The arch is shaped to be an efficient structural form in cast iron, while his diamond pattern of spandrel bars, at a location in the bridge where structural considerations permit many options, is clearly chosen with an eye to its appearance.

Those engineers who were most conscious of the centrality of aesthetics for structure have also been regarded as the best in a purely technical sense. Starting with Thomas Telford (1757–1834), we can identify Gustave Eiffel (1832–1923) and John Roebling (1806–1869) as the undisputed leaders in their fields during the 19th century. They designed the largest and most technically challenging structures, and they were leaders of their professions. Telford was the first president of the first formal engineering society, the Institution of Civil Engineers, and remained president for 14 years until his death. Eiffel directed his own design–construction–fabrication company and created the longest spanning arches and the highest tower; Roebling founded his large scale wire rope manufacturing organization while building the world's longest spanning bridges (Figure 3.2).

In reinforced concrete, Robert Maillart (1872–1940) was the major structural artist of the early 20th century. First in his 1905 Tavanasa Bridge, and later with the 1930 Salginatobel (Figure 3.3) and 1936 Vessy designs, he imagined a new form for three-hinged arches that included his own invention of the hollow box in reinforced concrete. The Swiss engineer Christian Menn (1927–) has demonstrated how a deep understanding of arches, prestressing, and cable-stayed forms can lead to structures worthy of exhibition in art museums. Especially noteworthy are the 1964 Reichenau Arch, the 1974 Felsenau prestressed cantilever, and the 1980 concrete cable-stayed Ganter Bridge. Meanwhile, German engineer

FIGURE 3.2 John Roebling's Brooklyn Bridge.

Jorg Schlaich has developed new ideas for light structures often using cables, characterized by a series of elegant footbridges in and around Stuttgart (Figure 3.4).

The engineers' aesthetic results from the conscious choice of form by engineers who seek the expression of structure. It is neither the unconscious result of the search for economy nor the product of supposedly optimizing calculations. Many of the best structural engineers have recognized the possibility for structural engineering to be an art form parallel to but independent of architecture. These people have, over the past two centuries, defined a new tradition, structural art, which we take here to be the ideal for an engineer's aesthetic.

Although structural art is emphatically modern, it cannot be labeled as just another movement in modern art. For one thing, its forms and its ideals have changed little since they were first expressed by Thomas Telford. It is not accidental that these ideals emerged in societies that were struggling with the consequences not only of industrial revolutions but also of democratic ones. The tradition of structural art is a democratic one.

In our own age the works of structural art provide evidence that the common life flourishes best when the goals of freedom and discipline are held in balance. The disciplines of structural art are efficiency and economy, and its freedom lies in the potential it offers the individual designer for the expression of a personal style motivated by the conscious aesthetic search for engineering elegance. These are the three leading ideals of structural art — efficiency, economy, and elegance.

FIGURE 3.3 Robert Maillert's Salginotobel Bridge.

FIGURE 3.4 One of Jorg Schlaich's footbridges.

3.3 The Three Dimensions of Structure

Its first dimension is a scientific one. Each working structure or machine must perform in accordance with the laws of nature. In this sense, then, technology becomes part of the natural world. Methods of analysis useful to scientists for explaining natural phenomena are often useful to engineers for describing the behavior of their artificial creations. It is this similarity of method that helps to feed the fallacy that engineering is applied science. But scientists seek to discover preexisting form and explain its behavior by inventing formulas, whereas engineers want to invent forms, using preexisting formulas to check their designs. Because the forms studied by scientists are so different from those of engineers, the methods of analysis will differ; yet, because both sets of forms exist in the natural world, both must obey the same natural laws. This scientific dimension is measured by efficiency.

Technological forms live also in the social world. Their forms are shaped by the patterns of politics and economics as well as by the laws of nature. The second dimension of structure is a social one. In the past or in primitive places of the present, completed structures and machines might, in their most elementary forms, be merely the products of a single person; in the civilized modern world, however, these technological forms, although at their best designed by one person, are the products of a society. The public must support them, either through public taxation or through private commerce. Economy measures the social dimension of structure.

Technological objects visually dominate our industrial, urban landscape. They are among the most powerful symbols of the modern age. Structures and machines define our environment. The locomotive of the 19th century has given way to the automobile and airplane of the 20th. Large-scale complexes that include structures and machines become major public issues. Power plants, weapons systems, refineries, river works — all have come to symbolize the promises and problems of industrial civilization.

The Golden Gate, the George Washington, and the Verrazano Bridges carry on the traditions set by the Brooklyn Bridge. The Chicago Hancock and Sears Towers, and the New York Woolworth, Empire State, and World Trade Center Towers all bring the promise of the Eiffel Tower into the utility of city office and apartment buildings. The Astrodome, the Kingdome, and the Superdome carry into the late 20th century the vision of huge permanently covered meeting spaces first dramatized by the 1851 Crystal Palace in London and the 1889 Gallery of Machines in Paris.

Nearly every American knows something about these immense 20th-century structures, and modern cities repeatedly publicize themselves by visual reference to these works. As Montgomery Schuyler, the first American critic of structures, wrote in the 19th century for the opening of the Brooklyn Bridge, "It so happens that the work which is likely to be our most durable monument, and to convey some knowledge of us to the most remote posterity, is a work of bare utility; not a shrine, not a fortress, not a palace but a bridge. This is in itself characteristic of our time."[1].

So it is that the third dimension of technology is symbolic, and it is, of course, this dimension that opens up the possibility for the new engineering to be structural art. Although there can be no measure for a symbolic dimension, we recognize a symbol by its elegance and its expressive power. Thus, the Sunshine Skyway (Figure 3.5) has become a symbol of both Florida's Tampa Bay area and the best of late-20th-century technology.

There are three types of designers who work with forms in space: the engineer, the architect, and the sculptor. In making a form, each designer must consider the three dimensions or criteria we have discussed. The first, or scientific criterion, essentially comes down to making structures with a minimum of materials and yet with enough resistance to loads and environment so that they will last. This efficiency–endurance analysis is arbitrated by the concern for safety. The second, or social criterion, comprises mainly analyses of costs as compared with the usefulness of the forms by society. Such cost–benefit analyses are set in the context of politics. Finally, the third criterion, the symbolic, consists of studies in appearance, along with a consideration of how elegance can be achieved within the constraints set by the scientific and social criteria. This is the aesthetic/ethical basis upon which the individual designer builds his or her work.

For the structural designer the scientific criterion is primary (as is the social criterion for the architect and the symbolic criterion for the sculptor). Yet the structural designer must balance the primary criterion with the other two. It is true that all structural art springs from the central ideal of artificial forms controlling natural forces. Structural forms will, however, never get built if they do not gain some social acceptance. The will of the designer is never enough. Finally, the designer must think aesthetically for structural form to become structural art. All of the leading artists of

FIGURE 3.5 The Sunshine Skyway.

structure thought about the appearance of their designs. These engineers consciously made aesthetic choices to arrive at their final designs. Their writings about aesthetics show that they did not base design only on the scientific and social criteria of efficiency and economy. Within those two constraints, they found the freedom to invent form. It was precisely the austere discipline of minimizing materials and costs that gave them the license to create new images that could be built and endure.

3.4 Structure and Architecture

The modern world tends to classify towers, stadiums, and even bridges as architecture, creating an important, but subtle, fallacy. Even the word is a problem, because *architect* comes from the Greek word meaning chief technician. But, beginning with the Industrial Revolution, structure has become an art form separate from architecture. The visible forms of the Eiffel Tower, Seattle's Kingdome, and the Brooklyn Bridge result directly from technological ideas and from the experience and imagination of individual structural engineers. Sometimes, the engineers have worked with architects just as with mechanical or electrical engineers, but the forms have come from structural engineering ideas.

Structural designers give form to objects that are of relatively large scale and of single use, and these designers see forms as the means of controlling the forces of nature to be resisted. Architectural designers, on the other hand, give form to objects that are of relatively small scale and of complex human use, and these designers see forms as the means of controlling the spaces to be used by people. The prototypical engineering form — the public bridge — requires no architect. The prototypical architectural form — the private house — requires no engineer. Structural engineers and architects learn from each other and sometimes collaborate fruitfully, especially when, as with tall buildings, large scale goes together with complex use. But the two types of designers act predominately in different spheres.

The works of structural art have sprung from the imagination of engineers who have, for the most part, come from a new type of school — the polytechnical school, unheard of prior to the late 18th century. Engineers organized new professional societies, worked with new materials, and stimulated political thinkers to devise new images of future society. Their schools developed curricula that decidedly cut whatever bond had previously existed between those who made architectural forms and those who began to make — out of industrialized metal and later from reinforced concrete — the new engineering forms by which we everywhere recognize the modern world. For these forms the ideas inherited from the masonry world of antiquity no longer applied; new ideas were essential in order to build with the new materials. But as these new ideas broke so radically with conventional taste, they were rejected by the cultural establishment.

This is, of course, a classic problem in the history of art: new forms often offend the academics. In this case, it was beaux arts against structural arts. The skeletal metal of the 19th century offended most architects and cultural leaders. New buildings and city bridges suffered from valiant attempts to cover up or contort their structure into some reflection of stone form. In the 20th century, the use of reinforced concrete led to similar attempts. Although some people were able to see the potential for lightness and energy, most architects tried gamely to make concrete look like stone or, later on, like the emerging abstractions of modern art. There was a deep sense that engineering alone was insufficient.

The conservative, plodding, hip-booted technicians might be, as the architect Le Corbusier said, "healthy and virile, active and useful, balanced and happy in their work, but only the architect, by his arrangement of forms, realizes an order which is pure creation of his spirit . . . it is then that we experience the sense of beauty." The belief that the happy engineer, like the noble savage, gives us useful things but only the architect can make them into art is one that ignores the centrality of aesthetics to the structural artist. In towers, bridges, free-spanning roofs, and many types of industrial buildings, aesthetic considerations provide important criteria for the engineer's design. The best of such engineering works are examples of structural art, made by engineers, and they have appeared with enough frequency to justify the identification of structural art as a mature tradition with a unique character. One of the most recent manifestations is Christian Menn's Sunniberg Bridge (Figure 3.6).

3.5 Application to Everyday Design

Many of today's engineers see themselves as a type of applied scientist, analyzing preexisting structural forms that have been established by others. Seeing oneself as an applied scientist is an unfortunate state of mind for a design engineer. It eliminates the imaginative half of the design process and forfeits the opportunity for the integration of form and structural requirements that can result in structural art. Design must start with the selection of a structural form. It is a decision that can be made well only by the engineer because it must be based on a knowledge of structural forms and how they control forces and movements.

In the case of most everyday bridges the selection of form is based largely on precedents and standards established by the bridge-building agency. For example, the form of a highway overpass may be predetermined by the client agency to be a welded plate girder bridge because that is what the agency prefers or what local steel fabricators are used to or even because the steel industry is a dominant political force in the state. In other cases, the form may be established by an architect or urban designer for reasons outside structural requirements. Thus the form is set without any serious consideration of whether or not that is in fact the best form for that particular site.

Creative form determination consists not of applying free visual imagination alone nor in applying rigorous scientific analysis alone, but of applying both together, at the same time. The art starts with a vision of what might be. The development of that vision is the key. Many engineers call the development of the vision conceptual engineering. It is the most important part of design. It is the stage at which all plausible forms are examined. The examination must include, to a rough level of precision, the whole range of considerations; performance, cost, and appearance. All that follows, including the aesthetic impression the bridge makes, will depend on the quality of the form selected. This stage is often ignored or foreclosed, based on precedents, standards, preconceived ideas or prior experience that may or may not apply.

FIGURE 3.6 Christian Menn's Sunniberg Bridge.

The reasons often given for shortchanging this stage include, "Everybody knows that [steel plate girders, precast concrete girders, cast-in-place concrete] are the most economical structure for this location," or "We always build [steel plate girders, precast concrete girders, cast-in-place concrete] in this state," or "Let's use the same design as we did for [any bridge] last year."

At this point someone will protest that other considerations (costs, the preferences of the local contracting industry, etc.) will indeed differentiate and determine the form. Too often these reasons are based on unexamined assumptions, such as, "The local contracting industry will not adjust to a different form," or "Cost differentials from [a past project] still apply," or "The client will never consider a different idea." Or the belief is based on a misleading analysis of costs which relies too much on assumed unit costs. Or that belief may be simply habit — either the engineer's or the client's — often expressed in the phrase, "We've always done it that way." Accepting these assumptions and beliefs places an unfortunate and unnecessary limitation on the quality of the resulting bridge for, by definition, improvements must come from the realm of ideas not tried before.

As Captain James B. Eads put it in the preliminary report on his great bridge over the Mississippi River at St. Louis:

> Must we admit that because a thing has never been done, it never can be, when our knowledge and judgment assure us that it is entirely practicable?[2].

FIGURE 3.7 MD 18 over U.S. 50.

FIGURE 3.8 Another possibility for MD 18 over U.S. 50.

The engineer's first job is to question all such determinations, assumptions and beliefs. From that questioning will come the open mind that is necessary to develop a vision of what each structure can be at its best.

Unless such questioning is the starting point it is unlikely that the most promising ideas will ever appear. No design will occur. Instead, there will be a premature assumption of the bridge form, and the engineer will move immediately into the analysis of the assumed form. That is why so many engineers mistake analysis for design. Design is more correctly the selection of the form in the first place, which most engineers have not been permitted to do. Design is by far the more important of the two activities.

Engineers also focus on analysis in the belief that the form (shape and dimensions) will be determined by the forces as calculated in the analysis. But, in fact, there are a large number of forms that can be shown by the analysis to work equally well. It is the engineer's option to choose among them, and in so doing to determine the forces by means of the form, not the other way around.

Take the simple example of a two-span continuous girder bridge, using an existing structure, MD 18 over U.S. 50 (Figure 3.7). Here the engineer has a wide range of possibilities such as a girder with parallel flanges, or with various haunches having a wide range of proportions (Figure 3.8). The moments will depend on the stiffness at each point, which in turn will depend on the presence or absence of a haunch and its shape (Figure 3.9). The engineer's choice of shape and dimensions will determine the moments at each point along the girder. The forces will follow the choice of form. Within limits, the engineer can direct the forces.

Let's examine which form the engineer should choose. Allcan support the required load. Depending on the specifics of the local contracting industry, many of them will be essentially equal in cost. All would perform equally well and all are comparable in cost, leaving the engineer a decision that can only be made on aesthetic grounds. Why not pick the one the engineer believes looks best?

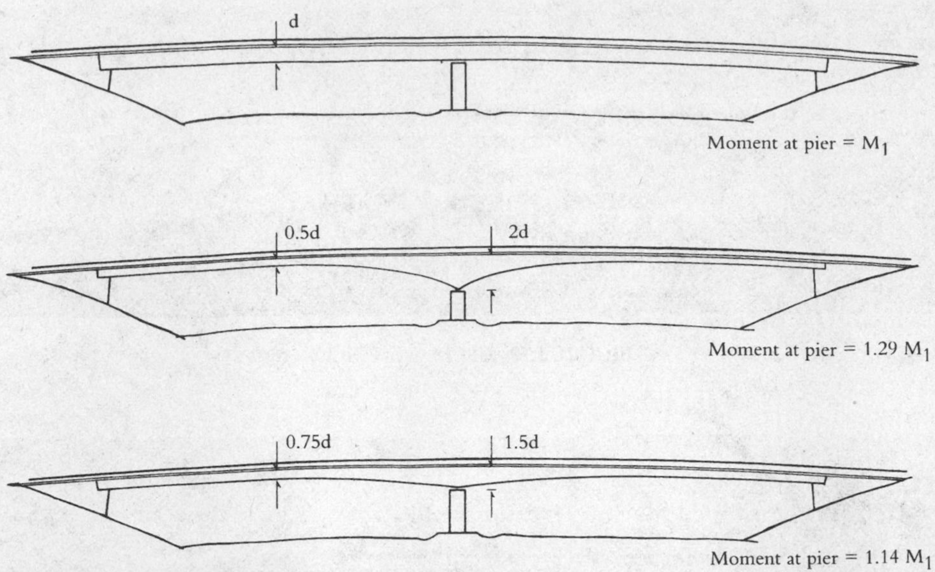

FIGURE 3.9 Forces determined by the engineer's choice of form.

That, in a nutshell, is the process that all of the great engineers have followed. Maillart's work, as one example, shows that the engineer cannot choose form as freely as a sculptor, but the engineer is not restricted to the discovery of preexisting forms as the scientist is. The engineer invents form, and Maillart's career shows that such invention has both a visual and a scientific basis. When either is denied, engineering design ceases. For Maillart, the dimensions were not to be determined by the calculations, and even the results of the calculations could be changed (by adjusting the form) because a designer rather than an analyst is at work. Analysis and calculation are the servants of design. Design, analysis, and must work together. In the words of Spanish engineer Eduardo Torroja,

> The imagination alone cannot reach such [elegant] designs unaided by reason, nor can a process of deduction, advancing by successive cycles of refinement, be so logical and determinate as to lead inevitably to them.[3]

The engineering challenge is not just to find the least costly solution. The engineering challenge is to bring forth elegance from utility: We should not be content with bridges that only move vehicles and people. They should move our spirits as well.

3.6 The Role of Case Studies in Bridge Design

Bridge design, even of highway overpasses, often involves standard problems but always in different situations. Case studies can help in the design of these standard problems by showing models and points of comparison for a large number of bridges without implying that each such bridge be mere imitation.

The primary goals of a case study are to look carefully at all major aspects of the completed bridge, to understand the reasons for each design decision, and to discuss alternatives, all to the end of improving future designs. Such cases help to define more general ideas or principles. Case studies are well recognized by engineers when designing for acceptable performance and low cost; they can be useful when considering appearance as well.

A common organization of these studies will help identify standard problems and make comparisons easier. First comes an overall evaluation of the bridge as a justification for studying it. Is it a good example that can be better? Is it a model of near perfection? Is it a bad example to be avoided?, Second comes a description of the complete bridge, which is divided into parts roughly coinciding with easily identifiable costs and including modifications to each part as suggested improvements. In this major description section there is an order to the parts that implies a priority for the structural engineer: concept and form of the entire structure, superstructure, supports, deck, color, and landscaping.

1. The *concept and form* of the completed bridge goes together with a summary of the bridge performance history (including maintenance) and of its construction cost, usually given per square meter of bridge. Required clearances, foundation conditions, hydraulic requirements, traffic issues, and other general requirements would be covered here.
2. The *superstructure* here includes primarily the main horizontal spanning members such as continuous girders, arches, trusses, etc. In continuous steel girder bridges, the cost is primarily identified with the fabricated steel cost. Modification in design by haunching, changing span lengths, or making girders continuous with columns would be discussed including their influence on cost.
3. The *pier supports* are most frequently columns or frames either in the median or outside the shoulders, or at both places in highway overpasses. These are normally highly visible elements and can have many possible forms. Different designs for the relationship among steel girder, bearings, and columns can make major improvements in appearance without detriment to cost or performance.
4. The *abutment supports* are also highly visible parts of the bridge, which include bearings, cantilever walls, cheek walls, and wing walls.
5. The *deck* includes the concrete slab or orthotropic steel deck, overhangs, railings, parapets, and provisions for drainage, all of which have an influence on performance as well as on the appearance either when seen in profile or from beneath the bridge.
6. The *color* is especially significant for steel structures that are painted, and *texture* can be important for concrete surfaces of piers, abutments, and deck.
7. The *landscaping/guardrail* includes plantings and other features that can have important visual consequences to the design.

The order of these parts is significant because it focuses attention on the engineering design. The performance of a weak structural concept cannot be saved by good deck details. An ugly form cannot be salvaged by color or landscaping. The first four parts are structural, the fifth is in part structural, whereas the last two, while essential for the bridge engineer to consider, involve primarily nonstructural ideas.

Third, the case study can give a critique of the concept and form by comparison with other similar bridges or bridge designs for similar conditions, including those with very different forms, as a stimulus to design imagination.

Fourth and finally, the case should conclude with some discussion of the relationship of this study to a theory of bridge design. Clearly, any such study must be based upon a set of ideas about design which often implicitly bias the writer who should make these ideas explicit. This conclusion should show how the present study illustrates a theory and even at times forces a modification of it. General ideas form only out of specific examples.

3.7 Case Study in Colorado: Buckley Road over I-76

Colorado's Buckley Road over I-76 (Figure 3.10) offers the application of an innovative form to prestressed concrete girders in order to achieve longer than normal spans, with a visually unique result. It is therefore a worthy subject for a case study.

FIGURE 3.10 Buckley Road over I-76.

3.7.1 Description of the Bridge

In *concept* this is a three-span continuous beam bridge with a 47° skew made of precast prestressed girders set onto cast-in-place concrete piers and abutments.

The *superstructure* consists of seven girders spaced approximately 3 m apart and each made up of five precast prestressed concrete segments (Figure 3.11). The main span is approximately 56 m and each side span is approximately 50 m. Segments one and five are 37.8 m long and behave essentially as simply supported beams between the abutments and the cantilever segments two and four. These latter cantilever 12 m into the side spans and 15 m into the main span. Segment three is 25.6 m long and behaves approximately as a simply supported beam within the main span. There are 0.15 m spacings between segments for closure pours. The cantilever segments have a linear haunch of 0.6 m from the girder depth of 1.8 m for the other three segments, which are Colorado BT72 girders.

The two *piers* each consist of a pier cap beam 1.2 m wide by 1.8 m deep and 28 m long supported by three walls each 1.2 m wide, about 7.6 m high, and 6 m long at their tops tapering to 3 m long at the footings. The cap beam extends about 1.5 m beyond the centerline of the exterior girder or about 1.1 m beyond the edge of that girder's bottom flange. The pier next to the railroad has a crash wall built into the three tapered walls.

The *abutments* are shallow concrete beams 0.9 m wide supported on piles and carrying the precast girders. The *deck* is a series of precast pretensioned concrete panels made composite with the precast girders. Bounded by Jersey barriers, the deck is 20.4 m in width and overhangs the exterior girders by 0.96 m or slightly over half the depth of the unhaunched girder segments.

3.7.2 Critique of the Bridge

The *concept* of a fully precast superstructure, a three-span continuous beam, and cast-in-place piers has led to an economical structure and fits well the site conditions of crossing both I-76 and the double-track railroad. Other reasonable concepts include a two-span bridge and a three-span bridge with the cantilever segments two and four cast in place with the piers (as illustrated by the Stewart

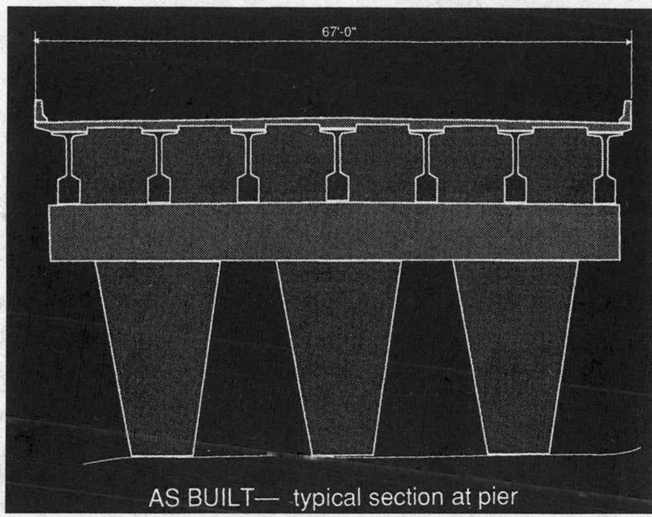

FIGURE 3.11 Typical section of Buckley Road as built.

Park Bridge in Oregon). This critique will confine itself to the present concept, but a comparison of this bridge and the Oregon one will follow. In each case, the ideas of structural art will form the basis for a critique.

The *superstructure* represents an unusual use of a precast bulb T girder whose bulb is extended vertically to create a haunch at the two interior supports. The profile view expresses the increased forces at the interior supports and the construction photos (with temporary walkway) show the lightness achieved by an overhang that is about the same dimension as the girder depth.

The *haunches* would be more effective visually were they deeper and the segments one, three, and five correspondingly shallower. For example, with the Colorado C68 girder, the depth would decrease to 1.7 m and a haunch of 2.6 m would more strikingly express the flow of forces. At the same time, the girder spacing would be reduced to 2.9 m to permit an overhang of 1.4 m.

Another solution would be to retain the Colorado BT72 girders, increase the haunch to 3 m, and reduce the number of girders from seven to six, thus again increasing the overhang. If the six girders were spaced 3.35 m on centers, then the overhang would be 1.7 m or nearly the depth of the BT72 girders.

The *piers* are visually prominent and look heavy. They also have a formal shaping which does not clearly express the structure. Specifically, the horizontal lines of the hammerhead beam separate it from the supporting walls and the 1.8-m depth of that beam is far greater than needed to carry the girder loads over the 2.4-m span between the wide supporting walls below. Since these piers are relatively short compared with the long spans of the girders, their massive appearance is accentuated by the lack of structural expression. It is clear from beneath the bridge that the 6-m-wide walls can easily be made to support all the girders directly without any hammerhead beam (Figure 3.12). The walls will therefore be higher and, if carefully shaped, will form a striking integration with the deck girders. The cast-in-place diaphragm can then be structurally integrated with the walls and the girders to form a cross frame for live loads.

The *abutments* can be improved by eliminating the wall that hides the girder ends and bearings. Along with the lighter-looking girders, this structural expression at the abutments will increase the already striking appearance of the bridge profile.

The *deck* overhang, by being increased, lends lightness to the girders. Otherwise, the system used is good and avoids the staining that can arise when metal slab forms are left in place.

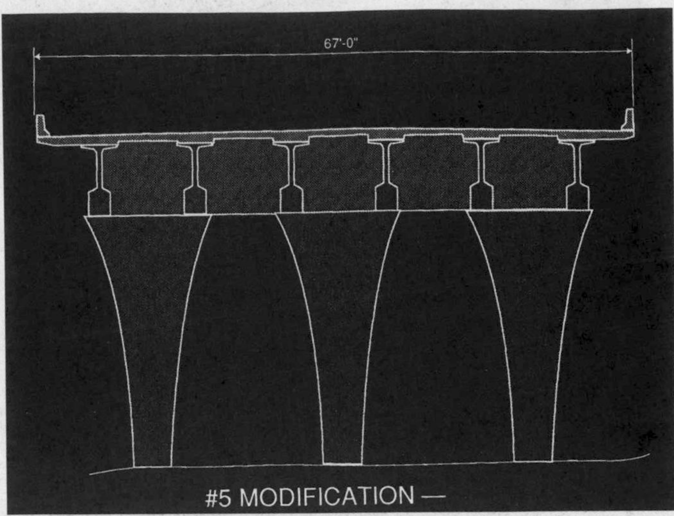

FIGURE 3.12 Possible modification to Buckley Road.

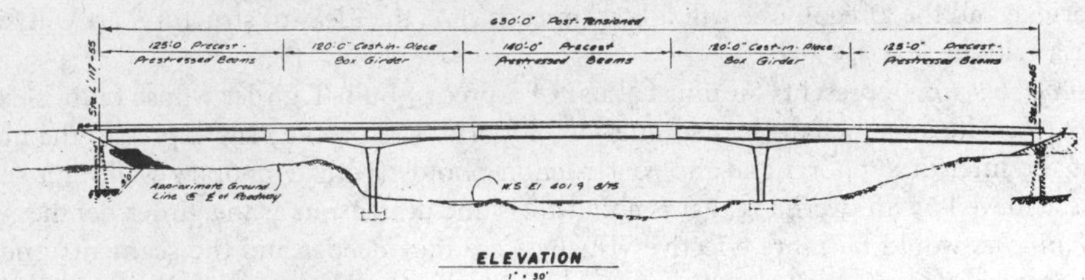

FIGURE 3.13 Elevation of Louis Pierce's Stewart Park Bridge.

3.7.3 The Stewart Park Bridge

The *concept* for this 1978 bridge (Figure 3.13) designed by Louis Pierce is the same as for Buckley Road except for the cantilever segments two and four which are cast-in-place prestressed concrete hollow boxes. Because the spans (56.4, 79.2, 56.4 m) are longer than those for Buckley Road, segments one, three, and five are each made of two separate precast pieces.

The *superstructure* and the *piers* are thus integrated into one form rather than separated into two forms as at Buckley Road. The boxes are haunched from the 2.4 m of the constant section segments to 3.65 m at the two interior supports for a ratio of 1.55. But the boxes are 2.4 m deep along their exterior faces and haunch laterally to 3.65 m over a distance of 2.5 m. Just as at Buckley Road, the *deck* overhang is too short, about 0.8 m for a girder depth of 2.4 m.

The shape of the two piers are walls 7.6 m wide, 2.3 m thick at the top, tapering to 5.8 m wide and 1.4 m thick at the base. The total height is 12.7 m above the footing but only about 7.6 m above the ground line. This shaping of piers, having about the same height as those of Buckley Road, gives an impression of lightness missing from the latter structure (Figure 3.14).

3.7.4 Summary

The Buckley Road bridge represents a good design. A similar concept can be improved in future designs by relatively small changes in the superstructure through stronger haunching and wider deck overhangs and by major changes in the pier form. The use of cast-in-place cantilever sections offers increased possibilities for elegant forms and closer integration of superstructure with piers.

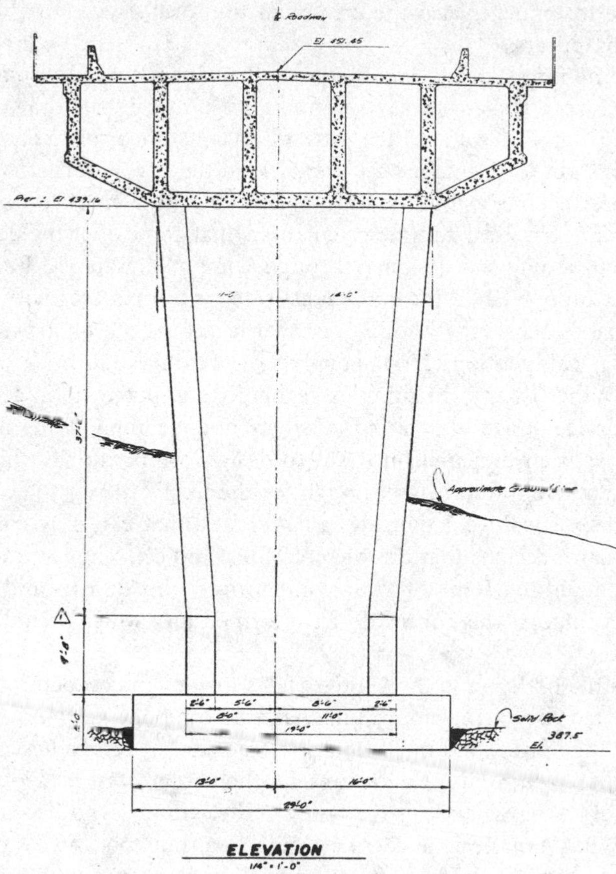

FIGURE 3.14 Typical section of Stewart Park Bridge at pier.

This case study gives an example of how a good bridge can provide an excellent basis for further study and improvement.

3.8 Achieving Structural Art in Modern Society: Computer Analysis and Design Competitions

Most people would agree that the ideals of structural art coincide with those of an urban society: conservation of natural resources, minimization of public expenditures, and the creation of a more visually appealing environment. As the history of structural art shows, some engineers have already turned these ideals into realities. But these are isolated cases. How might they become the rule instead of the exception? We can address this question historically, by identifying the central ideas that have been associated with great structural art. These ideas reflect each of the three dimensions: the scientific, social, and symbolic.

The leading scientific idea might be stated as that of reducing analysis. In structural art, this idea has coexisted with the opposite tendency to overemphasize analysis, which today is typified by the heavy use of the computer for structural calculations. One striking example comes from the design of thin concrete vaults — thin shell roofs. Here, the major advances between 1955 and 1980 — a time of intense analytic developments — were achieved, not by performing complex analyses using computers, but rather by reducing analysis to very simple ideas based on observed physical behavior. Roof vaults characterize this advance and they carry forward the central scientific idea in structural art: the analyst of the form, being also the creator of the form, is free to change shapes so that analytic complexity disappears.

The form controls the forces and the more clearly that designers can visualize those forces the surer they are of their forms. The great early and mid-20th-century structural artists such as Robert Maillart and Pier Luigi Nervi have all written forcefully against the urge to complicate analysis. We see the same arguments put forth by the best designers in the late 20th century. When the form is well chosen, its analysis becomes astoundingly simple. The computer, of course, has become more and more useful as a time saver for routine calculations that come after the design is set. It is also increasingly valuable in aiding the designer through computer graphics. But like any machine, although it can reduce human labor, it cannot substitute for human creativity.

Turning to the social dimension, a leading idea that has come out of structural art is the effectiveness of public design competitions. Design quality arises from the stimulus of competing designs for the same project rather than from complex regulations imposed upon a single designer. The progress of modern bridge design illustrates the benefit and meaning of alternative designs. Many alternative designs have been prepared pursuant to design competitions, which bring the public into the process in a positive way. It is not enough for the public merely to protest the building of ugly, expensive designs. A positive activity is essential, and that can only come about when the public sees the alternative designs that are possible for a project. Thus, governments can ensure better designs by relinquishing some of their control over who designs and on what forms are chosen, and by giving some of this control to an informed jury which includes representatives of the lay public.

Although there is little tradition in the United States for design competitions in bridges, such a tradition is firmly rooted elsewhere, with results that are both politically and aesthetically spectacular. Switzerland has the longest and most intensive tradition of bridge design competitions, and it is no coincidence that, by nearly common consent, the two greatest bridge designers of the first half of the 20th century were Swiss: Robert Maillart (1872–1940), who designed in concrete, and Othmar Amman (1879–1966), designer of the George Washington and Verrazano Bridges, who designed in steel. That Switzerland, one sixth the size of Colorado, and with fewer people than New York City, could achieve such world prominence is due to the centrality of economics and aesthetics for both their engineering teachers and their practicing designers, a centrality which is encouraged by design competitions.

Maillart's concrete arches in Switzerland were often the least expensive proposals in design competitions, and they were later to provide the main focus for the first art museum exhibition ever devoted exclusively to the work of one engineer: the New York City Museum of Modern Art's 1947 exhibition on Maillart's structures. Amman has been similarly honored. His centennial was celebrated by symposia both in Boston and in New York and by an exhibition held in Switzerland. Both Maillart and Amman wrote articulately on the appearance as well as on the economy of bridges. They are prime examples of structural artists.

This Swiss bridge tradition continues today with a large number of striking new bridges in concrete that follow Maillart in principle if not in imitative detail. The most impressive post-World War II works are those of Christian Menn, whose long-span arches and cantilevers extend the new technique of prestressing to its limits, as Maillart's three-hinged and deck-stiffened arches did earlier with reinforced concrete.

Design competitions stimulated these engineers and also educated the general public. To be effective, uch competitions must be accepted by political authorities, judged by engineers and informed lay members whose opinions will be debated in the public press, and controlled by carefully drawn rules.

It is false images of engineering that keep us from insisting on following our normal instinct for open competition. The American politics of public works falsely compares the engineering designer either with a medical doctor or with a building contractor.

Supporters of the first comparison argue that you would never hold a competition to decide who will repair your heart; rather, you would choose professionals on the basis of reputation and then leave them alone to do the skilled work for which they are trained. However, there is a key difference between hearts and bridges. For most people, there is only one heart which will do the job. Picking a "best" heart is not a consideration. On the other hand, for a given bridge site, there are many bridge designs that will solve the problem. The more minds that are put to the problem, the more likely that an outstanding design will emerge. After all, the ultimate goal is to pick the best bridge, not the best bridge designer.

Furthermore, developing the engineer's imagination creates a valuable asset for society. That imagination needs more chances to exercise than there are chances to build, and it is stimulated by competition. However, frustrating it may be to lose a competition, the activity is healthy and maturing, especially when even the losers are compensated financially for their time, as they often are in Switzerland.

For proponents of the second false comparison, design competitions are to be run just as building competitions in which the lowest bid for design cost gets the design contract. In American public structures, design and construction are legally distinct activities. The cost of design is normally 5% or less of the cost of construction. Therefore, a brilliant engineer might spend more preparing a design which, as can often happen, will cost the owners substantially less overall. By the same token, an engineer who cuts the design fee to get the job may have to make a more conservative design which could easily cost the owner more in overall costs. Hence, large amounts of potential savings to the public are lost by a foolish policy of saving a little during the first stage of a project.

In one type of Swiss design competition, a small number of designers are invited to compete, some of their costs are covered, and they get additional prize funds in the order recommended by the jury. The winner usually gets the commission for the detailed design. Only several such competitions a year are needed to stimulate the entire profession and to show the general public the numerous possibilities available as good solutions to any one problem. This method of design award opens up the political process to local people far more than does the cumbersome and largely negative one of protest, legal action, and negation of building that so dominates public action in late-20th-century America.

The state of Maryland is leading the way in the United States. In 1988, Maryland held a design competition for a new structure over the Severn River adjoining the U.S. Naval Academy in Annapolis. The competition was patterned on the Swiss practice. The results of the competition resolved an acrimonious community controversy. The winning structure, by Thomas Jenkins (Figure 3.15), was recognized by the American Institute of Steel Construction as the outstanding medium-span structure constructed in 1995–96. In 1998, Maryland, together with the state of Virginia, the District of Columbia, and the Federal Highway Administration, conducted a competition to select the design of the new Woodrow Wilson Bridge over the Potomac River at Washington, D.C. The winning design (Figure 3.16) was prepared by a team led by the Parsons Transportation Group.

Properly defined design competitions reveal truths about society that are otherwise difficult to define. The resulting designs, therefore, became unique symbols of their time and place. This brings us to the third leading idea that has been associated with great structural art — the idea that its materials and forms possess a particular symbolic significance. Perceptive painters, poets, and writers have recognized in structural art a new type of symbol — first in metal and then in concrete — which fits mysteriously closely both to the engineering possibilities and to the possibilities inherent in democracy. The thinness and openness of the Eiffel Tower, Brooklyn Bridge, and Maillart's arches, as well as the stark contrast between their forms and their surroundings, have a deep affinity to both the political traditions and era in which they arose. They symbolize the artificial rather than the natural, the democratic rather than the autocratic, and the transparent rather than the impenetrable. Their forms reflect directly the inner springs of creativity emerging from contemporary industrial societies.

FIGURE 3.15 Thomas Jenkins's U.S. Naval Academy Bridge over the Severn River.

These forms imply a democratic rather than an autocratic life. When structure and form are one, the result is a lightness, even fragility, which closely parallels the essence of a free and open society. The workings of a democratic government are transparent, conducted in full public view, and although a democracy may be far from perfect, its form and its actual workings (its structure) are inseparable. Furthermore, the public must continually inspect its handiwork: constant maintenance and periodic renewal are essential to its exposed structure. Politicians do not have life tenure; they must be inspected, chastised, and purified from time to time, and replaced when found corrupt or inept. So it is with the works of structural art. They, too, are subject to the weathering and fatigue of open use. They remind us that our institutions belong to us and not to some elite. If we let them deteriorate, as we flagrantly have in our older cities and transportation networks, then that outward sign betokens an inner corruption of the common life in a free democratic society.

3.9 The Engineer's Goal

The ideal bridge is structurally straightforward and elegant. It should provide safe passage and visual delight for drivers, pedestrians, and people living or working nearby. Society holds engineers responsible for the quality of their work, including its appearance. For the same reason engineers would not build a bridge that is unsafe, they should not build one that is ugly. Bridge designers must consider visual quality as fundamental a criterion in their work as performance, cost, and safety.

There are no fast rules or generic formulas conducive to outstanding visual quality in bridge design. Each bridge is unique and should be studied individually, always taking into consideration all the issues, constraints, and opportunities of its particular setting or environment. Nevertheless, by observing other bridges, using case studies and design guidelines, engineers can learn what makes bridges visually outstanding and develop their abilities to make their own bridges attractive. They can achieve outstanding visual quality in bridge design while maintaining structural integrity and meeting their budgets.

FIGURE 3.16 The competition-winning design for the new Woodrow Wilson Bridge over the Potomac River at Washington, D.C

References

1. Schuyler, M., The bridge as a monument, *Harpers Weekly* 27, 326, 1883; reprinted in *American Architecture and Other Writings,* W. H. Jordy and R. Coe, Eds., Atheneum, New York, 1964, 164.
2. Eads, J.B., Report of the Engineer-in-Chief of the Illinois and St. Louis Bridge Company, St. Louis, Missouri Democrat Book and Job Printing House, 1868. Reprinted in *Engineers of Dreams,* H. Petroski, Alfred A. Knopf, New York, 1995, 54.
3. Torroja, E., *The Structures of Eduardo Torroja, an Autobiography of Engineering Accomplishment,* F.W. Dodge, New York, 1958, 7.

Bibliography

Much of the discussion in this chapter is contained in a more complete form in the following four books by the authors:

David P. Billington, *Robert Maillart and the Art of Reinforced Concrete,* MIT Press, Cambridge, MA, 1990.

David P. Billington, *Robert Maillart's Bridges — The Art of Engineering,* Princeton University Press, Princeton, NJ, 1979.

David P. Billington, *The Tower and the Bridge — The New Art of Structural Engineering,* Basic Books, New York, 1983.

Frederick Gottemoeller, *Bridgescape, the Art of Designing Bridges,* John Wiley & Sons, New York, 1998.

The following should be part of the reference library of any engineer interested in bridge aesthetics (not in any particular order):

Fritz Leonhardt, *Brucken,* MIT Press, Cambridge, MA, 1982.

Max Bill, *Robert Maillart, Bridges & Construction,* 1st ed. 1949, Praeger, Westport, CT, 1969.

Steward C. Watson and M. K. Hurd, *Esthetics in Concrete Bridge Design,* American Concrete Institute, Detroit, 1990.

Hans Jochen Oster et al., *Fussgangerbrucken, Jorg Schlaich und Rudolph Bergermann Katalog zur Ausstellung an der ETH Zurich,* Institut für Baustatik und Konstruktion, Zürich, Switzerland, 1994.

Martin P. Burke, Jr. and the General Structures Committee, *Bridge Aesthetics around the World,* Transportation Research Board, Washington, D.C., 1991.

4

Planning of Major Fixed Links

Klaus H. Ostenfeld
COWI, Denmark

Dietrich L. Hommel
COWI, Denmark

Dan Olsen
COWI, Denmark

Lars Hauge
COWI, Denmark

4.1 Introduction

Characteristics of Fixed Links

Within the infrastructure of land transportation, fixed links are defined as permanent structures across large stretches of water allowing for uninterrupted passage of highway and/or railway traffic with adequate safety, efficiency, and comfort.

Traffic services are often provided by ferries before a fixed link is established. Normally, a fixed link offers shorter traveling times and higher traffic capacities than the ferry services. The establishment of a fixed link may therefore have a strong positive impact on the industrial and economic development of the areas to be served by the link. This together with an increased reliability in connection with climatic conditions are the major reasons for considering the implementation of a fixed link.

The waters to be passed by the links are often navigable; the link structures may present obstacles to the vessel traffic and are thus subject to the risk of impact from vessels. If the vessel traffic is important, the link traffic may be better separated from the crossing vessel traffic for general traffic safety. The water flow is often influenced by the link structures and this may affect the environment both near and far from the site. Furthermore, the water stretches and areas to be passed are part of beautiful territories forming important habitats for wildlife fauna and flora. The protection and preservation of the environment will therefore often be a major issue in the political discussions prior to the establishment of the links. These aspects have to be realized and considered in the very beginning of the planning process.

0-8493-7434-0/00/$0.00+$.50

TABLE 4.1 Major Fixed Links Opened Since 1988

Name of Link	Total Length and Types of Structures	Status Early 1998	Traffic Mode
Confederation Bridge, Canada	12.9 km, high-level concrete box girder bridge	Open to traffic 1997	Highway traffic
Vasco da Gama Bridge, Portugal	12.3 km, viaducts and high-level cable-stayed bridge	Open to traffic 1998	Highway traffic
Second Severn Bridge, Great Britain	5.1 km, viaducts and high-level cable-stayed bridge	Open to traffic 1996	Highway traffic
Honshu–Shikoku Connection, Japan			
• Kojima–Sakaide Route	37.3 km, a o high-level suspension bridges	Open to traffic 1988	Highway and railway
• Kobe–Naruto Route	89.6 km, a o high-level suspension bridges	South part open 1998	Highway traffic
• Onomichi–Imabari Route	59.4 km, a o high-level suspension bridges	Under construction	Highway traffic
Lantau Fixed Crossing, Hong Kong	3.4 km High-level suspension bridge	Open to traffic 1997	Highway and railway
• Tsing Ma Bridge	High level cable stayed bridge	Open to traffic 1997	Highway and railway
• Kap Shui Mun Bridge			
Boca Tigris Bridge, China	4.6 km, high-level suspension bridge	Open to traffic 1997	Highway traffic
Great Belt link, Denmark	17.5 km		
• West Bridge	6.6 km, low-level concrete box girder bridge	Open to traffic 1997	Highway and railway
• East Bridge	6.8 km, high-level suspension bridge	Open to traffic 1998	Highway traffic
• East Railway Tunnel	8.0 km, bored tunnel, two tubes	Open to traffic 1997	Railway traffic
Øresund link, Sweden–Denmark	16 km, immersed tunnel, artificial island, high-level cable-stayed bridge, viaducts	Under construction	Highway and railway
Rion–Antirion Bridge, Golf of Corinthe, Greece	2.9 km, high-level cable-stayed bridge, viaducts	Construction started in 1998	Highway traffic
Channel Tunnel, Great Britain–France	50.5 km, bored tunnel, three tubes	Open to traffic 1994	Railway with car and lorry shuttle
Trans-Tokyo Bay Crossing, Japan	15.1 km, bored tunnel, artificial islands, high- and low-level steel box girder bridges	Open to traffic 1997	Highway traffic

Generally, the term *fixed link* is associated with highway and/or railway sections of considerable length and a fixed link may comprise a combination of different civil engineering structures such as tunnels, artificial islands, causeways, and different types of bridges. Selected examples of major fixed links opened or are under construction since 1988 are listed in Table 4.1.

Planning Activities for Major Fixed Links

Major fixed links represent important investments for the society and may have considerable influence on the development potential of the areas they serve.

The political discussions about the decision to design a fixed link may be extended over decades or even centuries. In this period planning activities on a society level are necessary to demonstrate the need for the fixed link and to determine positive and negative effects of the implementation. These early planning considerations are outside the scope of this chapter, but the outcome of the early planning activities may highly influence the tasks in the later planning phases after the final decision is made.

In the early planning phases, basic principles and criteria are dealt with, such as

- Ownership and financing
- Approximate location
- Expected service lifetime
- Necessary traffic capacity
- Considerations for other forms of traffic like vessel traffic and air traffic
- Principles for environmental evaluation
- Risk policy
- International conventions

Section 4.2 will explain the later planning phases by describing major steps in project development with emphasis on the consideration of all relevant aspects. The focus will be on the technical and civil engineering aspects of bridges as fixed links, but most of the methods and principles described can be applied to other types of link structures. In the case of complex fixed link arrangements (comprising more than one type of structure), some of these structures may be alternative solutions. Several combined solutions are therefore studied and for each combination it is normally necessary to perform the planning for the entire link as a whole.

The elements of the project basis for a major fixed link are further detailed in Section 4.3, and examples of major fixed links recently built, under construction, or in the planning stage in the Scandinavian area are described in Section 4.4.

The chapter does not treat aesthetic and environmental issues individually, but assumes that all alternatives are evaluated according to the same principles. Public approval processes are beyond the aim of this chapter; readers are referred to References [1–3].

Fixed links are unique in size and cost, and the political environment differs from project to project. It is thus not possible to provide a recipe for planning major fixed links. The present chapter describes some of the elements, which the authors believe are important in the complex, multidisciplinary planning process of all fixed links.

Many important fixed links still remain to be planned and built. One of the more spectacular ones is the Gibraltar link between Africa and Europe. Figure 4.1 shows an artist's impression of the bridge pylons for the planned Gibraltar link. Examples of other future links are the Messina Strait crossing in Italy, the Mallaca Strait crossing between Malaysia and Indonesia, and the Río de la Plata Bridge connecting Argentina and Uruguay.

4.2 Project Development

4.2.1 Initial Studies

The first step in project development consists of a review of all information relevant to the link and includes an investigation of the most likely and feasible technical solutions for the structures.

The transportation mode, highway and railway traffic, and the amount of traffic is determined based on a traffic estimate. The prognosis of traffic is often associated with considerable uncertainty since fixed links will not only satisfy the existing demands but may also create new demands due to the increased quality of the transport. For railway traffic, it has to be decided whether a railway line will accommodate one or two tracks. Similarly, the highway traffic can either be transported on shuttle trains or the bridge can be accommodated with a carriageway designed to a variety of standards, the main characteristics being the number of lanes.

The decision on the expected traffic demands and the associated traffic solution models is often based on a mix of technical, economic, socioeconomic, and political parameters. The decision may be confirmed at later stages of the planning when more information is available.

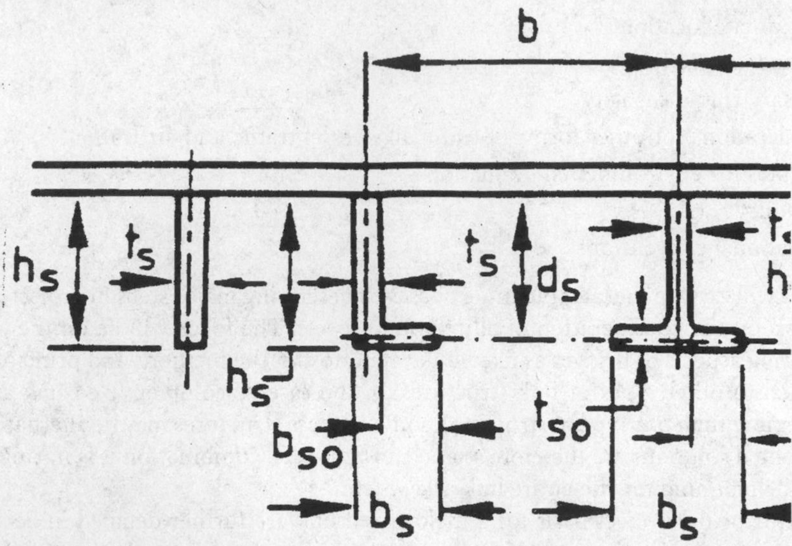

FIGURE 4.1 Artist's impression of 465-m-high pylon on 300 m water for planned Gibraltar link with 3,500 m spans. (Courtesy of Dissing +Weitling, Architects, Denmark.)

A *fixed link concept study* will review alignment possibilities and define an appropriate corridor for further studies. It will consider the onshore interchanges for the anticipated traffic modes and identify potential conflict areas. It describes all feasible arrangements for the structures from coast to coast, and reviews the requirement for special onshore structures. Finally, the study defines the concepts to be investigated in greater depth in subsequent phases.

An *environmental condition study* aims to identify potential effects the structures may have on the environment and to review the legal environmental framework. It also identifies important conflict areas and describes the project study area. It will review the available information on the marine and onshore environment and define the need for additional investigations.

A *technical site condition study* will address the geological, the foundation, the navigation, the climatic, and the hydraulic conditions. It will review the topographic situation and define additional studies or investigations for the following project phase.

A *preliminary design basis study* will review the statutory requirements, codes, and standards and identify the need for relevant safety and durability requirements.

Finally, a *preliminary costing basis study* will define the cost estimation technique and provide first preliminary cost estimates.

Considering the results of these studies a comprehensive investigation program for the next project step — the conceptual study — will be defined.

4.2.2 Conceptual Study

The conceptual study is an iterative process, where all the aspects likely to influence the project should be considered, weighted, and clarified to achieve the most suitable solution for the intended purpose and location. These aspects are cost, construction, structural, navigational, environmental, aesthetic, risk, geological, vessel collision, wind, and earthquake.

After an interim selection of various alternatives, conceptual studies are undertaken for each selected alternative solution. Preliminary site investigations like subsoil investigations in the defined alignment corridor, wind, earthquake, and vessel traffic investigations will be carried out simultaneously with the conceptual studies.

The conceptual study comprises development of a project basis, including

- Defining functional requirements
- Reviewing and defining the navigational aspects
- Establishing risk policy and procedures for risk management
- Specifying design basis including structure-specific requirements
- Developing the costing basis

Each of the selected solutions will be developed in a conceptual design and described through drawings and descriptions. The conceptual design will comply with the project basis and further consider:

- Preliminary site investigations
- Structural aspects
- Architectural aspects
- Environmental aspects
- Mechanical and electrical installations and utilities
- Definitions and constraints for operation and maintenance
- Cost aspects
- Major construction stages

Practically, it is not possible to satisfy all the above requirements, but effort should be made to achieve a balanced solution. The conceptual study phase is concluded by a comparison analysis with predeterminant weighting of parameters, which provides the technical ranking of all alternative solutions.

4.2.3 Project Selection and Procurement Strategy

Project Selection

By using the results from the technical ranking of the solutions, the basis for a project selection has to be established by the owner organization. This requires information from other investigations carried out in parallel with the technical studies that cover:

- Environmental impact assessment including hydraulic studies;
- Traffic demand studies including possible tariff structures;
- Layout, cost, and requirements for connections to the existing network outside the study areas;
- Definition of the project implementation and the tendering procedure.

The information obtained from the above studies may be used as input into a cost–benefit model of the anticipated solutions. These final results usually provide the basis to make a decision that will best consider local and global political viewpoints. Public hearings may be necessary in addition to the investigations. The result of this process is the selection of the solution of choice.

Procurement Strategy

The optimum procurement method should ensure that the work and activities are distributed on and executed by the most qualified party (owner, consultant, contractor) at all phases, to meet the required quality level, at the lowest overall cost. The procurement strategy should clarify tendering procedures with commercial and legal regulations for the region. In the following, three main contracting concepts in the definition of the procurement strategy for a fixed link project are presented.

Contracting Concepts

The three main concepts are as follows:

- *Separate Design and Construction* (*SDC*) is a concept in which the construction contract documents are prepared by the owner, often assisted by an engineering consulting team, and the construction is performed by a contractor.
- *Design-Build* (*DB*) is a concept in which both engineering design and construction responsibilities are assigned to a single entity, most often the contractor.
- *Design-Built-Operate and Transfer* (*BOT*) is a concept in which the financing, design, construction, and operation are assigned to a concessionaire. After an agreed number of years of operation, the link is transferred to the owner. The BOT is not described further because it uses the same design and construction procedure as the DB.
- The *SDC* concept requires that the owner and the consultant participate actively during all phases to influence and control the quality and performance capability of the completed facility. The main differences between the various forms of the SDC are the degree of detailing at the tender stage and whether alternatives will be permitted. Completing the detailed design prior to inviting tenders is good if the strategy is to obtain lump-sum bids in full compliance with the owner's conditions.

Tendering based on a partial design — often 60 to 70% — represents a compromise between initial design costs and definitions of the owner's requirements to serve as a reference for alternative tenders. This procurement strategy has been applied for large construction works from the 1970s. Advantages are that the early start of the construction work can be achieved while completing the design work and that innovative ideas may be developed between the owner, contractor and the consultant, and incorporated in the design. The procedure usually allows contractors to submit alternatives in which case the tender design serves the important purpose of outlining the required quality standards. A disadvantage, however, is the risk for later claims due to the fact that the final design is made after awarding the construction contract. The more aggressive contracting environment and the development of international tender rules have made it desirable to procure on a completely fixed basis.

The *DB* concept assigns a high degree of autonomy to the contractor, and, as a consequence, the owner's's direct influence on the quality and performance of the completed facility is reduced. To ensure that the contractor delivers a project that meets the expectations of the owner, it is necessary to specify these in the tender documents. Aesthetic, functional, maintenance, durability, and other technical standards and requirements should be defined. Also legal, environmental, financial, time, interface, and other more or less transparent constraints to the contractor's freedom of performance should be described in the tender documents in order to ensure comparable solutions and prices. Substantial requirements to the contractor's quality assurance system are essential in combination with close follow-up by the owner.

Tenders for major bridges may be difficult to evaluate if they are based on substandard and marginal designs or on radical and unusual designs. The owner then has the dilemma of either rejecting a low tender or accepting it and paying high additional costs for subsequent upgrading.

Contract Packaging

The total bridge project can be divided into reasonable contract parts:

- Vertical separation (e.g., main bridge, approach bridges, viaducts, and interchanges);
- Horizontal separation (e.g., substructure and superstructure);
- Disciplinary separation (e.g., concrete and steel works).

The application of these general principles depends on the specific situation of each project. Furthermore, the achievement of the intended quality level, together with contract sizes allowing for competitive bidding, should be considered in the final choice. Definition and control of interfaces between the different contractors is an important task for the owner's organization.

4.2.4 Tender Design

The main purpose of a tender design or a bid design is to describe the complexity of the structure and to determine the construction quantities, allowing the contractors to prepare a bid for the construction work. The goal for a tender design is as low cost as possible within the given framework. This is normally identical to the lowest quantities and/or the most suitable method. It is essential that the project basis be updated and completed prior to the commencement of the tender design. This will minimize the risk of contract disputes.

It is vital that a common understanding between consultant and owner is achieved. Assumptions regarding the physical conditions of the site are important, especially subsoil, wind, and earthquake conditions. Awareness that these factors might have a significant impact on the design and thereby on the quantities and complexity is important. The subsoil conditions for the most important structures should always be determined prior to the tender design to minimize the uncertainty.

Determination of the quantities is also necessary. For instance, if splice lengths in the reinforcement are included, if holes or cutouts in the structure are included, what material strengths are assumed. There must be stipulated an estimate of the expected variation of quantities (global or local quantities).

The structures in the tender design shall be constructible. In an SDC contract, the tender design should be based on safe and well-established production and erection procedures. In the case of DB, the tender design is carried out in close cooperation between the, contractor and the consultant. This assures that the design accommodates the contractor's methods and the available equipment.

The tender design is often carried out within a short period of time. It should focus on elements with large cost impact and on elements with large uncertainties in order to arrive as closely as possible at the actual quantities and to describe the complexity of the structure efficiently from a costing point of view. A tender design comprises layout drawings of the main structural elements, detailed drawings of typical details with a high degree of repetition, typical reinforcement arrangement, and material distribution.

Aesthetics are normally treated during the tender design. It is important that extreme event loads such as vessel collision, train derailment, cable rupture, earthquake, and ice impact should be considered in the tender design phase as they often govern the design. Durability, operation, and maintenance aspects should be considered in the tender design. Experience from operation and maintenance of similar bridges allows a proper service life design to be carried out. It is at the early design stages that the construction methods should be chosen which have a significant effect on further operation and maintenance costs.

It is not unusual that a tender design is prepared for more than one solution to obtain the optimal solution. It could, for instance, be two solutions with different materials (concrete and steel) as for the Storebælt East Bridge. It could also be two solutions with traffic arranged differently (one level or two levels) as for the Øresund link. Different structural layouts as cable stayed and suspension bridge could be relevant to investigate under certain conditions. After the designs are prepared to a certain level, a selection can be carried out based on a preliminary pricing, and one or more solutions are brought all the way to tender.

Tender documents to follow the drawings should be prepared. The tender documents comprise bill of quantities, special specifications, and the like.

4.2.5 Tender Evaluation

The objective of tender evaluation is to select the overall most advantageous tender including capitalized owner's risk and cost for operation and maintenance. A basis should be established via a rating system where all tenders become directly comparable. The rating system is predefined by the owner, and should be part of the tender documents.

The tender evaluation activities can be split up in phases:

1. Preparation
2. Compilation and checking of tenders
3. Evaluation of tenders
4. Preparation for contract negotiations
5. Negotiation and award of contract

The *preparation* phase covers activities up to the receipt of tenders. The main activities are as follows:

- Establish the owner's risk for each of the tendered projects, using the owner's cost estimate;
- Define tender opening procedures and tender opening committee;
- Quantify the differences in present value due to function, operation, maintenance, and owner's risk for each of the tendered projects, using the owner's cost estimate.

After receipt of tenders, a summary report, which collects the information supplied in the different tenders into a single summarizing document and presents a recommendation of tenders for detailed review, as a result of *compilation and checking of tenders,* should be prepared. Typical activities are as follows:

- Check completeness of compliance of all tenders, including arithmetical correctness and errors or omissions;
- Identify possible qualifications and reservations;
- Identify parts of tenders where clarification is needed, or more detailed examination required;
- Prepare a preliminary list of questions for clarification by the tenderers;
- Review compliance with requirements for alternative designs;
- Upgrade alternative tender design and pricing to the design basis requirements for tender design.

The *evaluation of tenders* comprises the following:

- Provide initial questionnaires for tender clarification to tenderers, arrange clarification meetings, and request tenderers' written clarification answers;
- Adjust tender prices to a comparable basis taking account of revised quantities due to modified tender design effects of combined tenders, alternatives, options, reservations, and differences in present value;
- Appraise the financial components of the tenders;
- Assess owner's risk;
- Review technical issues of alternatives and their effect on interfaces;
- Review the proposed tender time schedule;
- Evaluate proposed subcontractors, suppliers, consultants, testing institutes, etc.
- Review method statements and similar information;
- Establish list of total project cost.

The assessment of owner's risk concerns exceeding budgets and time limits. An evaluation of the split of financial consequences between contractor and owner should be carried out.

Preparation for contract negotiations should be performed, allowing all aspects for the actual project type to be taken into account. Typical activities are as follows:

- Modify tender design to take current status of the project development into account to establish an accurate contract basis;
- Modify tender design to accommodate alternatives;
- Coordinate with the third parties regarding contractual interfaces;
- Coordinate with interfacing authorities;
- Establish strategies and recommendations for contract negotiations.

The probable extent and nature of the negotiations will become apparent from the tender evaluation. Typical activities during *negotiations and award of contract* are as follows:

- Prepare draft contract documents;
- Clarify technical, financial, and legal matters;
- Finalize contract documents.

4.2.6 Detailed Design

The detailed design is either carried out before (SDC contracts) or after signing of the construction contact (DB contracts). In the case the detailed design is carried out in parallel with the construction work, the completion of the detailed design should be planned and coordinated with the execution. A detailed planning of the design work is required when the parts of the structure, typically the foundation structures, need to be designed and constructed before the completion of the design of the entire structure. Design of temporary works is normally conducted in-house by the contractor, whereas the design of the permanent works is carried out by the consultant.

The purpose of the detailed design is to prepare drawings for construction in accordance with various requirements and specifications. Detailed design drawings define all measures and material qualities for the structure. Shop drawings for steel works are generally prepared by the steel fabricator. Detailed reinforcement arrangements and bar schedules are either prepared by the contractor or the consultant. It is important that the consultant prescribes the tolerance requirements of the design.

The detailed design should consider the serviceability limit state (deflection and comfort), the ultimate limit state (strength and stability), and the extreme event limit state (collapse of the structure). To ensure the adequacy of the design, substantial analyses, including three-dimensional global finite-element analyses, local finite-element analyses, and nonlinear analyses both in geometry and materials, should be carried out. Dynamic calculations, typically response spectrum analyses, are usually performed to determine the response from wind. The dynamic amplifications of traffic loads and cable rupture are determined by a time-history analysis, which is also frequently used for vessel collision and earthquake analyses.

For large cable-supported bridges, wind tunnel testing is conducted as part of the detailed design. Preliminary wind tunnel testing is often carried out in the tender design phase to investigate the aerodynamic stability of the structure. Other tests, such as scour protection and fatigue tests can be carried out to ensure design satisfactions. Detailed subsoil investigations for all foundation locations are carried out prior to, or in parallel with, the detailed design.

The operation and maintenance (O&M) objectives should be implemented in the detailed design in a way which:

- Gives an overall cost-effective operation and maintenance;
- Causes a minimum of traffic restrictions due to O&M works;
- Provides optimal personnel safety;
- Protects the environment;
- Allows for an easy documentation of maintenance needs and results.

In addition, the contractor should provide a forecast schedule for the replacement of major equipment during the lifetime of the bridge.

4.2.7 Follow-Up during Construction

During the construction period the consultant monitors the construction work to verify that it is performed in accordance with the intentions of the design. This design follow-up, or general supervision, is an activity which is carried out in cooperation with (and within the framework of) the owner's supervision organization.

The general supervision activities include review of the contractor's quality assurance manuals, method statements, work procedures, work instructions, and design of temporary structures, as well as proper inspections on the construction site during important construction activities. The quality of workmanship and materials is verified by spot-checking the contractor's quality control documentation.

When the work results in mistakes or nonconformances, the general supervision team evaluates the contractor's proposals for rectification or evaluates whether or not the structural element in question can be used as built, without any modifications. The general supervision team also evaluates proposals for changes to the design submitted by the contractor and issues recommendations on approval of such proposals.

The duties of the general supervision team also include preparation of technical supervision plans, which are manuals used by the supervision organization as a basis for the technical supervision of the construction work. These manuals should be based on inputs from the consultants and experienced engineers to avoid mistakes during the construction work.

The general supervision team monitors the performance of the supervision organization and receives feedback on experience gained by the supervision organization, as in some cases it may be found necessary and advantageous to adjust the design of the project to suit the contractor's actual performance.

The general supervision team provides advice on the necessity for expert assistance, special testing of materials, and special investigations. The general supervision team evaluates the results of such activities and issues recommendations to the owner. Special testing institutes are often involved in the third-party controls which normally are performed as spot checks only. Examples are nondestructive testing of welds, mechanical and chemical analyses of steel materials, and testing of concrete constituents such as cement, aggregates, and admixtures at official laboratories.

The general supervision team assists the supervision organization with the final inspection of the works prior to the contractor's handing over of the works. The general supervision team assists the consultant with the preparation of operation and maintenance manuals and procedures for inspections and maintenance during the operation phase. Some of these instructions are based on detailed manuals prepared by the contractor's suppliers. This can apply to bearings, expansion joints, electrical installations, or special equipment such as dehumidification systems or buffers. Preparation of these manuals by the suppliers is part of their contractual obligations, and the manuals should be prepared in the required language of the country where the project is situated.

4.3 Project Basis

4.3.1 Introduction

The project basis is all the information and requirements that are decisive in the planning and design of a fixed link. The project basis is developed simultaneously with the early design activities, and it is important to have the owner's main requirements defined as early as possible, and to be precise about what types of link solutions are to be included.

4.3.2 Geometric Requirements

Most geometric requirements for the fixed links stem from the operational requirements of traffic and all the important installations. However, geometric requirements may also be necessary to mitigate accidents and to provide the needed space for safety and emergency situations. Geometric considerations should be addressed in the risk analyses.

4.3.3 Structural Requirements

Design Basis

A main purpose of a design basis is to provide a set of requirements to ensure an adequate structural layout, safety, and performance of the load-bearing structures and installations for the intended use.

Structural Design Codes

The structures must resist load effects from self-weight and a variety of external loads and environmental phenomena (climate and degradation effects). To obtain an adequately uniform level of structural safety, the statistical nature of the generating phenomena as well as the structural capacity should be considered. A rational approach is to adapt probabilistic methods, but these are generally inefficient for standard design situations, and consequently it is recommended that a format as used in codes of practice be applied. These codes are calibrated to achieve a uniform level of structural safety for ordinary loading situations, and probabilistic methods can subsequently be used to calibrate the safety factors for loads and/or design situations that are not covered by the codes of practice.

The safety level — expressed as formal probability of failure or exceeding of limit state — is of the order 10^{-6} to 10^{-7}/year for ultimate limit states for important structures in major links.

4.3.4 Environmental Requirements

Fixed links crossing environmentally sensitive water stretches need to be developed with due attention to environmental requirements. Environmental strategies should be directed toward modification of the structural design to reduce any impact and to consider compensation or mitigation for unavoidable impacts. Guidelines for environmental considerations in the structural layout and detailing and in the construction planning are developed by a consultant, and these should typically address the following areas:

- Geometry of structures affecting the hydraulic situation;
- Space occupied by bridge structures, ramps and depot areas;
- Amount and character of excavated soils;
- Amount of external resources (raw materials winning);
- Methodology of earth works (dredging and related spill).

Consequences of the environmental requirements should be considered in the various project phases. Typical examples for possible improvements are selection of spans as large as possible or reasonable, shaping of the underwater part of foundations to reduce their blocking effect, orientation of structures parallel to the prevailing current direction, minimizing and streamlining of protection structures, reduction of embankment length, optimal layout of depot areas close to the shorelines, and reuse of excavated material.

The process should be started at the very early planning stages and continued until the link is completed and the impact on the environment should be monitored and assessed.

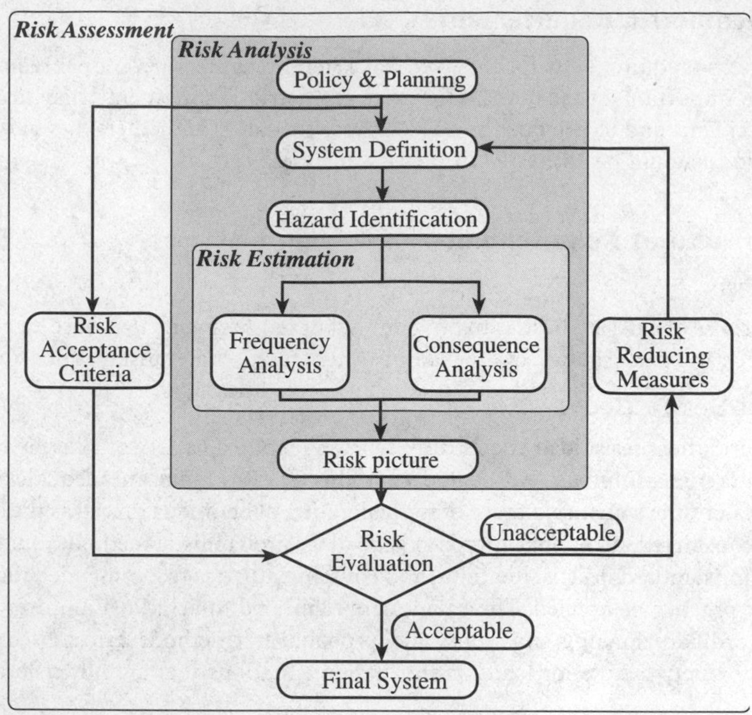

FIGURE 4.2 Risk management components.

4.3.5 Risk Requirements

Types of Risk

Risk studies and risk management have gained a widespread application within the planning, design, and construction of fixed links. Risks are inherent in major transportation links, and therefore it is important for the owner and society that risks are identified and included in the project basis together with the technical and economic aspects. Risks are often studied separately according to the consequences of concern:

- Economic risk (rate of interest, inflation, exceeding of budget, changing traffic patterns);
- Operational risk (accidents, loss of lives, impact to environment, disruption of the traffic, loss of assets, loss of income);
- Construction risks (failure to meet time schedule or quality standards, unexpected ground conditions, accidents).

Economic risks in the project may be important for decisions on whether to initiate the project at all. The construction risk may have important implications on the selection of the structural concept and construction methods.

Risk Management Framework

The main risk management components are shown in Figure 4.2. The risk policy is formulated by the owner in few words: "The safety of the transportation link must be comparable with the safety for the same length of similar traffic on land."

The risk acceptance criteria are an engineering formulation of the risk policy in terms of upper limits of risk. The risk policy also specifies the types of risk selected to be considered, typically user fatalities and financial loss. In some cases, other risks are specifically studied, e.g., risk of traffic disruption and risk of environmental damage, but these risks may conveniently be converted into financial losses.

The risk analysis consists of a systematic hazard identification and an estimation of the two components of the risk, the likelihood and the consequence. Finally, the risk is evaluated against the acceptance criteria. If the risk is found unacceptable, risk-reducing measures are required. It is recommended to develop and maintain an accurate accounting system for the risks and to plan to update the risk assessment in pace with the project development.

In the following, three common risk evaluation methods are discussed: fixed limits, cost efficiency, and ALARP, i.e., as low as reasonably practicable.

Fixed limits is the classical form of acceptance criteria. Fixed limits are also known from legislation and it may easily be determined whether a determined risk is acceptable or not. On the other hand, the determination of limits, which can ensure an optimal risk level, may be difficult.

With a pure *cost efficiency* consideration, an upper limit is not defined, but all cost-efficient risk-reducing measures are introduced. For this cost–benefit consideration it is necessary to establish direct quantification of the consequences in units comparable to costs.

The *ALARP* method applies a cost–benefit consideration in which it, however, is stated that the risk shall be reduced until the cost of the reduction measures is in disproportion with the risk-reducing effect. This will result in a lower risk level than the pure cost efficiency. In ALARP a constraint of the acceptable risk is further introduced as an upper limit beyond which the risk is unconditionally unacceptable.

Often it is claimed that society regards one accident with 100 fatalities as worse than 100 accidents each with one fatality. Such an attitude toward risk aversion can be introduced in the risk policy and the risk acceptance criteria. The aversion against large accidents can also be modeled with aversion factors that are multiplied on the consequences of accidents with many fatalities; the more fatalities, the higher the factor. The sensitivity of the evaluations of risk should be considered by the representation of the uncertainty of the information in the models.

Risk Studies in Different Project Phases

The general result of the risk management is a documentation of the risk level, basis for decisions, and basis for risk communication. The specific aims and purposes for risk management depend on the phase of the project. Here some few examples of the purposes of risk management in conceptual study, tender design, detailed design, and operation, are given.

During initial studies, the risk should be crudely analyzed using more qualitative assessments of the risks. A risk management framework should be defined early in the design process. In the beginning of the project, some investigations should be initiated in order to establish a basis for the more-detailed work in later phases; for example, vessel traffic observations should be performed to provide the basis for the estimation of vessel impact probability. In later phases detailed special studies on single probabilities or consequences may be undertaken.

In the conceptual study the most important activities are to identify all relevant events, focus on events with significant risk contributions and risks with potential impact on geometry (safety, rescue, span width). Extreme event loads are established based on the risk studies. In the tender design phase, the main risks are examined in more detail, in particular risks with potential impact on the project basis. In the detailed design phase, the final documentation of the risk level should be established and modifications to the operational procedures should be made.

4.3.6 Aesthetic Requirements

The final structures and components of a fixed link are a result of a careful aesthetic appraisal and design of all the constituent elements. The purpose of aesthetic requirements is to obtain an optimal technical and sculptural form of individual elements and to obtain an overall aesthetic quality and visual consistency between the elements and the setting. Although difficult, it is recommended to establish guidelines for aesthetic questions.

4.3.7 Navigation Conditions

The shipping routes and the proposed arrangement for a major bridge across navigable waters may be such that both substructure and superstructure could be exposed to vessel collisions. General examples of consequences of vessel collisions are as follows:

- Fatalities and injuries to users of the bridge and to crew and vessel passengers;
- Pollution of the environment, in the case of an accidental release of the hazardous cargo;
- Damage or total loss of bridge;
- Damage or total loss of vessels;
- Economic loss in connection with prolonged traffic disruption of the bridge link.

A bridge design that is able to withstand worst-case vessel impact loads on other piers than the navigation piers is normally not cost-effective. Furthermore, such a deterministic approach does not reduce the risk to the environment and to the vessels. Therefore, a probabilistic approach addressing the main risks in a systematic and comprehensive way is recommended. This approach should include studies of safe navigation conditions, vessel collision risk analysis and vessel collision design criteria, as outlined below.

Navigation risks should be addressed as early as possible in the planning phases. The general approach outlined here is in accordance with the IABSE Green Ship Collision Book [4] and the AASTHO Guide Specification [5]. The approach has been applied in the development of the three major fixed links discussed in Section 4.4.

Safe Navigation Conditions

Good navigation conditions are a prerequisite for the safe passage of the bridge such that vessel collisions with the bridge will not occur under normal conditions, but only as a result of navigation error or technical failure on-board during approach.

The proposed bridge concept should be analyzed in relation to the characteristics of the vessel traffic. The main aspects to be considered are as follows:

- Preliminary design of bridge;
- Definition of navigation routes and navigation patterns;
- Data on weather conditions, currents, and visibility;
- Distribution of vessel movements with respect to type and size;
- Information on rules and practice for navigation, including use of pilots and tugs;
- Records of vessel accidents in the vicinity of the bridge;
- Analysis of local factors influencing the navigation conditions;
- Identification of special hazards from barges, long tows and other special vessels;
- Future navigation channel arrangements;
- Forecast of future vessel traffic and navigation conditions to the relevant study period;
- Identification of largest safe vessel and tow and of preventive measures for ensuring full control with larger passing vessels.

Vessel Collision Analysis

An analysis should be used to support the selection of design criteria for vessel impact. Frequencies of collisions and frequencies of bridge collapse should be estimated for each bridge element exposed to vessel collision. Relevant types of hazards to the bridge should be identified and modeled, hazards from ordinary vessel traffic which is laterally too far out of the ordinary route, hazards from vessels failing to turn properly at a bend near the bridge, and from vessels sailing on more or less random courses.

The frequencies of collapse depend on the design criteria for vessel impact. The overall design principle is that the design vessels are selected such that the estimated bridge collapse frequency fulfills an acceptance criterion.

Vessel Collision Design Criteria

Design criteria for vessel impact should be developed. This includes selection of design vessels for the various bridge elements which can be hit. It also includes estimates of sizes of impact loads and rules for application of the loads. Both bow collisions and sideways collisions should be considered. Design capacities of the exposed girders against impact from a deck house shall be specified.

The vessel impact loads are preferably expressed as load indentation curves applicable for dynamic analysis of bridge response. Rules for application of the loads should be proposed. It is proposed that impact loads will be estimated on the basis of general formulas described in Ref. [4].

4.3.8 Wind Conditions

Bridges exposed to the actions of wind should be designed to be consistent with the type of bridge structure, the overall wind climate at the site, and the reliability of site-specific wind data. Wind effects on traffic could also be an important issue to be considered.

Susceptibility of Bridge Structures to Wind

Winds generally introduce time-variant actions on all bridge structures. The susceptibility of a given bridge to the actions of wind depends on a number of structural properties such as overall stiffness, mass, and shape of deck structure and support conditions.

Cable-supported bridges and long-span beam structures are often relatively light and flexible structures in which case wind actions may yield significant contributions to structural loading as well as influence user comfort. Site-specific wind data are desirable for the design. Engineering codes and standards often provide useful information on mean wind properties, whereas codification of turbulence properties are rare. Guidelines for turbulence properties for generic types of terrain (sea, open farmland, moderately built-up areas) may be found in specialized literature. If the bridge is located in complex hilly/mountainous terrain or in the proximity of large structures (buildings, bridges, dams), it is advisable to carry out field investigations of the wind climate at the bridge site. Important wind effects from isolated obstacles located near the planned bridge may often be investigated by means of wind tunnel model testing.

In general, it is recommended that aerodynamic design studies be included in the designs process. Traditionally, aerodynamic design have relied extensively on wind tunnel testing for screening and evaluation of design alternatives. Today, computational fluid dynamics methods are becoming increasingly popular due to speed and efficiency as compared with experimental methods.

Wind Climate Data

The properties of turbulence in the atmospheric boundary layer change with latitude, season and topography of the site, but must be known with a certain accuracy in order to design a bridge to a desired level of safety. The following wind climatic data should be available for a particular site for design of wind-sensitive bridge structures:

Mean wind:

- Maximum of the 10-min average wind speed corresponding to the design lifetime of the structure;
- Vertical wind speed profile;
- Maximum short-duration wind speed (3-s gust wind speed).

Turbulence properties (along-wind, cross-wind lateral, and cross-wind vertical):

- Intensity;
- Spectral distribution;
- Spatial coherence.

The magnitude of the mean wind governs the steady-state wind load to be carried by the bridge structure and is determinant for the development of aeroelastic instability phenomena. The turbulence properties govern the narrowband random oscillatory buffeting response of the structure, which is similar to the sway of trees and bushes in storm winds.

4.3.9 Earthquake Conditions

Structures should be able to resist regional seismic loads in a robust manner, avoiding loss of human lives and major damages, except for the very rare but large earthquake. The design methods should be consistent with the level of seismicity and the amount of available reliable information. Available codes and standards typically do not cover important lifeline structures such as a fixed link, but they may be used for inspiration for the development of a design basis.

4.3.10 Ice Conditions

The geographic location of a bridge site indicates whether or not ice loads are of concern for that structure. The ice loads may be defined as live loads or extreme event loads (exceptional environmental loads which are not included in live loads).

From recent studies carried out for the Great Belt link, the following main experience was obtained:

- Ice loads have a high dynamic component very likely to lock in the resonance frequencies of the bridge structure;
- Bearing capacity of the soil is dependent on number and type of load cycles, so dynamic soil testing is needed;
- Damping in soil and change of stiffness cause important reductions in the dynamic response;
- If possible, the piers should be given an inclined surface at the water level;
- High ductility of the structure should be achieved.

4.3.11 Costing Basis

The cost estimate is often decisive for the decision on undertaking the construction of the link, for selection of solution models, and for the selection of concepts for tendering. The estimate may also be important for decisions of detailed design items on the bridge.

Cost Uncertainty Estimation

To define the cost uncertainty it may be helpful to divide it into two conceptions which may overlap: (1) the uncertainties of the basis and input in the estimation (mainly on cost and time) and (2) risk of unwanted events. The uncertainty is in principle defined for each single item in the cost estimate. The risks can in principle be taken from a construction risk analysis.

Cost Estimates at Different Project Phases

Cost estimates are made in different phases of the planning of a project. In the first considerations of a project, the aim is to investigate whether the cost of the project is of a realistic magnitude and whether it is worthwhile to continue with conceptual studies. Later, the cost estimates are used to compare solution concepts, and to evaluate designs and design modification until the final cost estimate before the tender is used to evaluate the overall profitability of the project and to compare with the received bids. Different degrees of detailing of the estimates are needed in these stages. In the early phases an "overall unit cost" approach may be the only realistic method for estimating a price, whereas in the later phases it is necessary to have a detailed breakdown of the cost items and the associated risks and uncertainties.

Life Cycle Costs

The life cycle cost is an integration of the entire cost for a bridge from the first planning to the final demolition. The life cycle cost is normally expressed as a present value figure. Hence, the interest rate used is very important as it is a weighting of future expenses against initial expenses.

It shall initially be defined how the lifetime costs are to be considered. For example, disturbance of the traffic resulting in waiting time for the users can be regarded as an operational cost to society whereas it is only a cost for the owner if it influences the users' behavior so that income will be less.

Important contributors to the life cycle cost for bridges are as follows:

- The total construction cost, including costs for the owner's organization;
- Future modifications or expansion of the bridge;
- Risks and major repairs;
- Income from the operation of the bridge;
- Demolition costs.

Comparison Analysis

In the development of a project numerous situations are encountered in which comparisons and rankings must be made as bases for decisions. The decisions may be of different nature, conditions may be developing, and the decision maker may change. The comparisons should be based on a planning and management tool which can rationalize, support, and document the decision making.

A framework for the description of the solutions can be established and maintained. This framework may be modified to suit the purpose of the different situations. It is likely that factual information can be reused in a later phase.

The main components of the comparison can be as follows:

- Establishment of decision alternatives;
- Criteria for evaluation of the alternatives;
- Quantitative assessment of impacts of the various alternatives utilizing an evaluation grid;
- Preference patterns for one or more decision makers with associated importance of criteria;
- Assessment of uncertainties.

The decision maker must define the comparison method using a combination of technical, environmental and financial criteria. Quantitative assessment of all criteria is performed. Decision-making theories from economics and mathematical tools are used.

Establishment of Alternatives and Their Characteristics

All decision alternatives should be identified. After a brief evaluation, the most obviously nonconforming alternatives may be excluded from the study. In complex cases a continued process of detailing of analysis and reduction of number of alternatives may be pursued. The selection of parameters for which it is most appropriate to make more-detailed analyses can be made on the basis of a sensitivity analysis of the parameters with respect to the utility value.

Risks may be regarded as uncertain events with adverse consequences. Of particular interest are the different risk pictures of the alternatives. These risk pictures should be quantified by the use of preferences so that they can be part of the comparison.

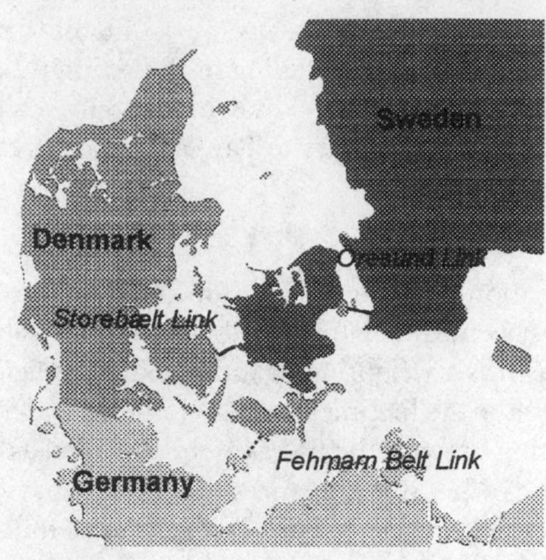

FIGURE 4.3 Denmark and neighboring countries.

Comparisons at Different Project Stages

After the initial identification of all possible solution models, the purpose of the first comparison may be to reduce the number of solution models to be investigated in the later phases. The solution models may here be alternative design concepts. In this first ranking the detail of the analysis should be adequate to determine the least attractive solutions with an appropriate certainty. This will in most cases imply that a relatively crude model can be used at this stage. A partly qualitative assessment of some of the parameters, based on an experienced professional's judgment, can be used.

At a later phase decisions should be made on which models to select for tender design, and later in the tender evaluation, which tenderer to award the contract. In these comparisons the basis and the input should be more well established, as the comparison here should be able to select the single best solution with sufficient certainty.

Weighting the criteria is necessary. Although a strictly rational weighting and conversion of these criteria directly into terms of financial units may not be possible, it is often sufficient if the weighting and selection process are shown to the tenderers before the tender.

An example of the comparison and selection process can be the following, which is performed in stages. Each stage consists of an evaluation and shortlisting of the tenders eliminating the low ranked tenders. At each stage the tenderers not on the shortlist are informed about the weak points and they are given the opportunity of changing their tender within a short deadline. At the last stage the remaining tenderers are requested to state their final offer improving on the technical quality and financial aspects raised by the owner during negotiations. Then the owner can select the financially most advantageous tender.

4.4 Recent Examples of Fixed Links

4.4.1 Introduction

Since the 1980s three major fixed links have been designed or planned in Denmark and neighboring countries, Figure 4.3. A combined tunnel and bridge link for railway and highway traffic has been constructed across the 18-km-wide Storebælt, a 16-km tunnel and bridge link for railway and highway traffic between Denmark and Sweden will be inaugurated in year 2000, and the conceptual study has been completed (1998) for a fixed link across the 19-km-wide Fehmarn Belt between Denmark and Germany.

4.4.2 The Storebælt Link

Over the years, more or less realistic projects for a fixed link across the Storebælt have been presented. At 18-km-wide, the belt is part of the inland sea area and divides Denmark's population and economy into nearly equal halves.

The belt is divided into two channels, east and west, by the small island, Sprogø, which has been as an obvious stepping-stone, an integral part of all plans for fixed link projects. The international vessel traffic between the Baltic Sea and the North Sea navigates the eastern channel, whereas the western channel is a national waterway. To bridge the eastern channel has therefore always been the main challenge of the project.

The first tender design for a combined railway and highway bridge across the eastern channel was prepared in 1977–78. However, only 1½ month short of issuing tender documents and call for bids, the progress of the project was temporarily stopped by the government. This was in August 1978. Several state-of-the-art investigations such as vessel impact, fatigue, and wind loads were carried out for two selected navigation spans: a 780-m main span cable-stayed bridge and a 1416-m main span suspension bridge, both designed for a heavy duty double-track railway and a six lane highway.

The construction of the fixed link was again politically agreed upon on June 12, 1986, and the main principles for the link were set out. It should consist of a low-level bridge for combined railway and highway traffic, the West Bridge, across the western channel; whereas the eastern channel should be crossed by a bored or an immersed tunnel for the railway, the East Tunnel, and a high-level bridge for the highway traffic, the East Bridge.

A company, A/S Storebælt, was established January 23, 1987 and registered as a limited company with the Danish State as sole shareholder. The purpose of the company was to plan, design, implement, and operate the fixed link. The project is financed by government-guaranteed commercial loans to be paid back via user tolls. A/S Storebæltsforbindelsen has published a series of reports on the link structures, see Reference [6–8].

The East Bridge

Project Development

In 1987 conceptual design was carried out for the East Bridge. The main objectives were to develop a global optimization with regard to the following:

- Alignment, profile, and navigation clearance;
- Position of main navigation channel;
- Navigation span solutions, based on robust and proven design and construction technology;
- Constructable and cost-competitive solutions for the approach spans, focusing on repetitive industrialized production methods onshore;
- Master time schedule;
- Master budget.

In 1989–90 pretender studies, tender design, and tender documents were prepared. During the pretender phase, comparative studies of four alternative main bridge concepts were carried out to evaluate thoroughly the technical, financial, and environmental effects of the range of main spans:

Cable-stayed bridge	916 m main span
Cable-stayed bridge	1204 m main span
Suspension bridge	1448 m main span
Suspension bridge	1688 m main span

Navigation risk studies found only the 1688-m main span adequate to cross the existing navigation route without affecting the navigation conditions negatively. This during tender design was reduced to 1624 m, which together with a relocated navigation route, proved to be sufficient and was selected for tender and construction.

The pylons were tendered in both steel and concrete. For the approach span superstructure, 124-m-long concrete spans and 168-m-long steel spans as well as composite steel/concrete concepts were developed. Although an equally competitive economy was found, it was decided to limit the tender designs to concrete and steel spans. The East Bridge was tendered as SDC.

The tender documents were subdivided into four packages to be priced by the contractors: superstructure and substructure inclusive pylons for the suspension bridge (2) and superstructures and substructures for the approach spans. The tender documents were released to prequalified contractors and consortia in June 1990.

In December 1990 the tenders were received. Eight consortia submitted 32 tenders inclusive smaller alternatives and four major alternatives to the basic tender design.

In October 1991, construction contracts were signed with two international consortia; a German, Dutch, and Danish joint venture for the substructures, inclusive of concrete pylons, and an Italian contractor for an alternative superstructure tender where high-strength steel was applied to a more or less unchanged basic cross section, thereby increasing the span length for the approach spans from 168 m to 193 m.

The suspension span is designed with a main cable sag corresponding to ⅑ of span length. The steel bridge girder is suspended from 800-mm-diameter main cables in hangers each 24 m.

The girder is continuous over the full cable-supported length of 2.7 km between the two anchor blocks. The traditional expansion joints at the tower positions are thus avoided. Expansions joints are arranged in four positions only, at the anchor blocks and at the abutments of the approach spans.

The concrete pylons rise 254 m above sea level. They are founded on caissons placed directly on crushed stone beds.

The anchor blocks must resist cable forces of 600,000 tonnes. They are founded on caissons placed on wedge-shaped foundation bases suitable for large horizontal loading. An anchor block caisson covers an area of 6100 m^2.

The caissons for the pylons, the anchor blocks, and the approach spans as well as for the approach span pier shafts have been constructed at a prefabrication site established by the contractor 30 nautical miles from the bridge site. The larger caissons were cast in two dry docks, and the smaller caissons and the pier shafts for the approach spans on a quay area, established for this purpose. A pylon caisson weighed 32,000 tonnes and an anchor block caisson 36,000 tonnes when they were towed from the dry dock by tug boats to their final position in the bridge alignment.

Both the suspension bridge girder and the approach span girders are designed as closed steel boxes and constructed of few basic elements: flat panels with trough stiffeners and transverse bulkhead trusses. The two approach bridges, 2530 and 1538 m, respectively, are continuous from the abutments to the anchor blocks. The suspension bridge girder is 31.0 m wide and 4.0 m deep; the girder for the approach spans is 6.7 m deep. They are fabricated in sections, starting in Italy. In Portugal, on their way by barge to Denmark, a major preassembly yard was established for girder sections to be assembled, before they were finally joined to full-span girders in Denmark. The East Bridge (Figure 4.4), was inaugurated by the Danish Queen on June 14, 1998 and the link was opened to highway traffic.

Project Basis

The project basis was throughout its development reviewed by international panels of experts.

Structural Requirements

Danish codes, standards, rules, and regulations were applied wherever applicable and supplemented with specific additional criteria and requirements, regarding various extreme event loads.

Environmental

The environmental design criteria required that the construction should be executed with no effect on the water flow through the belt. This was achieved by dredging, short ramps, long spans, and hydraulic shaped piers and pylons. The blocking effect to be compensated for was only about 0.5% of the total flow in the belt.

FIGURE 4.4 Storebælt East Bridge.

Risk

Risk acceptance criteria were established early and a series of risk analyses regarding train accidents, fire and explosion, ice loads, and vessel collision were carried out to ensure adequate and consistent safety level for the entire link. The acceptance criteria required that the probability of disruption of a duration of more than 1 month should not exceed a specified level, and that the risk level for fatalities for crossing should be comparable to the risk for a similar length of traffic on land. The analyses were followed up by risk management through the subsequent phases to ensure that the objectives were met.

Navigation

With 18,000 vessel passages each year through the eastern channel, important considerations were given for navigation. Comprehensive vessel simulations and collision analysis studies were performed, leading to an improved knowledge about safe navigation conditions and also to a set of probabilistically based criteria for the required impact resistance of the bridge piers and girders. Vessel impact has been the governing load criterion for all the bridge piers. A vessel traffic service (VTS) system was established mainly for prevention of collision accidents to the low West Bridge.

Wind

The local wind climate at Storebælt was investigated by measurements from a 70-m-high tower on Sprogø. For the East Bridge, aerodynamic investigations were carried out on 16 different highway girder box section configurations in a wind tunnel. The testing determined the critical wind speed for flutter for the selected girder shape to be 74 m/s which was safely above the design critical wind speed of 60 m/s. For the detailed design an aeroelastic full bridge model of 1:200 scale was tested under simulated turbulent wind conditions.

The West Bridge

The 6.6-km West Bridge (Figure 4.5) was tendered in three alternative types of superstructure; a double-deck composite girder, triple independent concrete girders side by side, and a single steel box girder. All three bridge alternatives shared a common gravity-founded sand-filled caisson substructure, topped by pier shafts of varying layout.

FIGURE 4.5 Storebælt West Bridge.

Tender documents were issued to six prequalified consortia in April 1988, and 13 offers on the tender solutions as well as three major alternatives and nine smaller alternatives were received from five groups. Tender evaluation resulted in selecting an alternative design: two haunched concrete box girders with a typical span length of 110.4 m, reduced to 81.75 m at the abutments and the expansion joints. The total length was subdivided into six continuous girders, requiring seven expansion joints.

It was originally intended to tender the West Bridge as an SDC, but as an alternative design was selected, the contract ended up being similar to a DB contract.

Altogether, 324 elements, comprising 62 caissons, 124 pier shafts, and 138 girders, have been cast in five production lines at a reclaimed area close to the bridge site. All the elements were cast, moved by sliding, stored on piled production lines, and later discharged without use of heavy gantry cranes. The maximum weight of an element was 7400 tonnes. The further transportation and installation was carried out by *Svanen,* a large purpose-built catamaran crane vessel.

By this concept, which was originally presented in the tender design, but further developed in the contractor's design, the entire prefabrication system was optimized in regard to resources, quality, and time. The bridge was handed over on January 26, 1994.

The East Tunnel

Two immersed tunnel solutions as well as a bored tunnel were considered for the 9 km wide eastern channel. After tender, the bored tunnel was selected for financial and environmental reasons.

The tunnel consists of two 7.7-m-internal-diameter tubes, each 7412 m long and 25 m apart. At the deepest point, the rails are 75 m below sea level.

Four purpose-built tunnel boring machines of the earth-balance pressure type have bored the tunnels, launched from each end of both tubes. The tunnel tubes are connected at about 250 m intervals by 4.5-m-diameter cross passages which provide safe evacuation of passengers and are the location for all electrical equipment. About 250 m of reinforced concrete cut-and-cover tunnels are built at each end of the bored tubes. The tunnel is lined with precast concrete segmental rings, bolted together with synthetic rubber gaskets. Altogether, 62,000 segments have been produced. A number of protective measures has been taken to ensure a 100-year service life design.

FIGURE 4.6 Storebælt East Tunnel.

On April 7, 1995, the final tunnel lining segment was installed. Thus, the construction of the tunnel tubes was completed, almost 5 years after work commenced. Railway systems were installed and in June 1997 the railway connection (Figure 4.6) was opened to traffic and changes in the traffic pattern between East and West Denmark started.

4.4.3 The Øresund Link

The 16-km fixed link for combined railway and highway traffic between Denmark and Sweden consists of three major projects: a 3.7-km immersed tunnel, a 7.8-km bridge, and an artificial island which connects the tunnel and the bridge.

The tunnel contains a four-lane highway and two railway tracks. The different traffic routes are separated by walls, and a service tunnel will be placed between the highway's two directions. The tunnel will be about 40 m wide and 8 m high. The 20 reinforced concrete tunnel elements, 175 m long and weighing 50,000 tonnes, are being prefabricated at the Danish side, and towed to the alignment.

The owner organization of the Øresund link is Øresundskonsortiet, established as a consortium agreement between the Danish company A/S Øresundsforbindelsen and the Swedish company Svensk-Danska Broförbindelsen on January 27, 1992. The two parties own 50% each of the consortium. The purpose of the consortium is to own, plan, design, finance, construct and operate the fixed link across Øresund.

The project is financed by commercial loans, guaranteed jointly and severally between the Danish and the Swedish governments. The highway part will be paid by user tolls, whereas the railway companies of the two countries will pay fixed installments per year. The revenue also has to cover the construction work expenses for the Danish and Swedish land-based connections.

Prequalified consultants were asked in February 1993 to prepare a conceptual design, as part of a proposal to become the in-house consultant for the owner. Two consultants were selected to prepare tender documents for the tunnel, the artificial island and the bridge, respectively.

The Øresund Bridge

In July 1994, the Øresundkonsortiet prequalified a number of contractors to build the bridge on a design and construct basis.

FIGURE 4.7 Øresund Main Bridge.

The bridge was tendered in three parts; the approach bridge from Sweden, the high-level bridge with a 490-m main span and a vertical clearance of 57 m, and the approach spans toward Denmark. Two solutions for the bridge were suggested: primarily, a two-level concept with the carriageway on the top deck and the two-track railway on the lower deck; secondarily a one-level bridge. Both concepts were based on cable-stayed main bridges.

Five consortia were prequalified to participate in the competition for the high-level bridge, and six consortia for the approach bridges. In June 1995, the bids for the Øresund Bridge were delivered. The two-level concept was selected as the financially most favorable solution. In November 1995, the contract for the entire bridge was awarded to a Swedish–German–Danish consortium.

The 7.8-km bridge includes a 1090-m cable-stayed bridge (Figure 4.7) with a main span of 490 m. The 3013 and 3739 m approach bridges have spans of 140 m. The entire superstructure is a composite structure with steel truss girders between the four-lane highway on the upper concrete deck and the dual-track railway on the lower deck.

Fabrication of the steel trusses and casting of the concrete deck of the approach bridges are carried out in Spain. The complete 140-m-long girder sections, weighing up to 7000 tonnes, are tugged on flat barges to the bridge site and lifted into position on the piers. Steel trusses for the cable-stayed bridge are fabricated in Sweden and transported to the casting yard close to the bridge site, where the concrete decks are cast.

On the cable-stayed bridge the girder will also be erected in 140-m sections on temporary supports before being suspended by the stays. This method is unusual for a cable-stayed bridge, but it is attractive because of the availability of the heavy-lift vessel *Svanen,* and it reduces the construction time and limits vessel traffic disturbance. (*Svanen* was, as mentioned earlier, purpose-built for the Storebælt West Bridge. After its service there *Svanen* crossed the Atlantic to be upgraded and used for the erection works at the Confederation Bridge in Canada. Back again in Europe *Svanen* performs an important job at Øresund). During the construction period two VTS systems have been in operation, the Drogden VTS on the Danish side, and the Flint VTS on the Swedish side. The main tasks for the VTS systems are to provide vessels with necessary information in order to ensure safe navigation and avoid dangerous situations in the vicinity of the working areas. The VTS systems have proved their usefulness on several occasions.

The cable system consists of two vertical cable planes with parallel stays, the so-called harp-shaped cable system. In combination with the flexural rigid truss girder and an efficient pier support in the side spans, a high stiffness is achieved.

The module of the truss remains 20 m both in the approach and in the main spans. This results in stay cable forces of up to 16,000 tonnes which is beyond the range of most suppliers of prefabricated cables. Four prefabricated strands in a square configuration have therefore been adopted for each stay cable.

The concrete pylons are 203.5 m high and founded on limestone. Caissons, prefabricated on the Swedish side of Øresund, are placed in 15 m water depth, and the cast-in-place pylon shafts are progressing. Artificial islands will be established around the pylons and nearby piers to protect against vessel impact. All caissons, piers, and pier shafts are being prefabricated onshore to be assembled offshore. The bridge is scheduled to be opened for traffic in year 2000.

Project Basis

General Requirements

The Eurocode system was selected to constitute the normative basis for the project. Project application documents (PADs) have been prepared as companion documents to each of the Eurocodes. The PADs perform the same function as the national application documents (NADs) developed by the member countries implementing the Eurocodes.

The partial safety and load combination factors are determined by reliability calibration. The target reliability index of $\beta = 4.7$, specified by the owner, corresponds to high safety class as commonly used for important structures in the Nordic countries.

In addition to the Eurocodes and the PADs, general design requirements were specified by the owner to cover special features of a large civil work. This is in line with what is normally done on similar projects. The general design requirements cover the following areas:

- Functional and aesthetic requirements as alignment, gradients, cross sections, and clearance profiles;
- Civil and structural loads, load combinations, and partial safety coefficients; methods of structural analysis and design;
- Soil mechanics requirements to foundation design and construction, including soil strength and deformation parameters;
- Mechanical and electrical requirements to tunnel and bridge installations, including systems for supervision, control and data acquisition (SCADA), power distribution, traffic control, communication.

Risk

LHRisk acceptance criteria were developed such that the individual user risk for crossing the link would be equal to the average risk on a highway and railway on land of similar length and traffic intensity. In addition, the societal risk aspects concerning accidents with larger numbers of fatalities were controlled as well.

The ALARP-principle — as presented in Section 4.3 — was applied to reduce consequences from risks within a cost–benefit approach. Especially the disruption risks were controlled in this way. Risk-reducing measures were studied to reduce the frequency and consequences of hazardous events. The analyses carried out addressed main events due to fire, explosion, toxic releases, vessel collision and grounding, flooding, aircraft crash, and train derailment.

Navigation

Øresund is being used by local vessel traffic and vessels in transit up to a certain limit set by the water depth in the channels Drogden and Flinterännan. The Drogden channel near the Danish coast will be crossed by the immersed tunnel and only requirements regarding accidental vessel impact to tunnel structures have been specified. The Flinterännan near the Swedish coast is being crossed by the bridge, and the navigation route will be improved for safety reasons. Design criteria against vessel impact have been specifically developed on a probabilistic basis, and main piers will be protected by artificial islands.

4.4.4 The Fehmarn Belt Crossing

In 1995, the Danish and German Ministry of Transport invited eight consulting consortia to tender for the preliminary investigations for a fixed link across the 19-km-wide Fehmarn Belt.

Two Danish/German consortia were selected; one to carry out the geological and the subsoil investigations, and the other to investigate technical solution models, the environmental impact, and to carry out the day-to-day coordination of all the investigations.

In the first phase, seven different technical solutions were investigated, and in the second phase five recommended solutions were the basis for a concept study:

- A bored railway tunnel with shuttle services;
- An immersed railway tunnel with shuttle services;
- A combined highway and railway bridge;
- A combined highway and railway bored tunnel;
- A combined highway and railway immersed tunnel.

With a set of more detailed and refined functional requirements, various concepts for each of the five solution models have been studied in more detail than in the first phase. This concept study was finalized in early July 1997 with the submission of an interim report. The conceptual design started in December 1997 and is planned to last 7 months. To provide an adequate basis for a vessel collision study and the associated part of the risk analysis, vessel traffic observations are carried out by the German Navy. In parallel, the environmental investigations are continued, whereas the geological and the subsoil investigations are concluded.

The results of the study will constitute the basis for public discussions and political decisions whether or not to establish a fixed link, and also which solution model should be preferred.

References

1. The Danish Transport Council, Facts about Fehmarn Belt, Report 95-02, February 1995.
2. The Danish Transport Council, Fehmarn Belt. Issues of Accountability, Report 95-03, May 1995.
3. The Interaction between Major Engineering Structures and the Marine Environment, Report from IABSE Colloquium, Nyborg, Denmark, 1991.
4. Ole Damgaard Larsen: Ship Collision with Bridges, IABSE Structural Engineering Documents, 1993.
5. Guide Specification and Commentary for Vessel Collision Design of Highway Bridges, Vol. 1, Final Report, AASHTO, 1991.
6. The Storebælt Publications: East Tunnel, A/S Storebæltsforbindelsen, København, Denmark, 1997.
7. The Storebælt Publications: West Bridge, A/S Storebæltsforbindelsen, København, Denmark, 1998.
8. The Storebælt Publications: East Bridge, A/S Storebæltsforbindelsen, København, Denmark, 1998.

5

Design Philosophies for Highway Bridges

John M. Kulicki

Modjeski and Masters, Inc.

5.1 Introduction

Several bridge design specifications will be referred to repeatedly herein. In order to simplify the references, the "Standard Specifications" means the *AASHTO Standard Specifications for Highway Bridges* [1], and the sixteenth edition will be referenced unless otherwise stated. The "LRFD Specifications" means the *AASHTO LRFD Bridge Design Specifications* [2], and the first edition will be referenced, unless otherwise stated. This latter document was developed in the period 1988 to 1993 when statistically based probability methods were available, and which became the basis of quantifying safety. Because this is a more modern philosophy than either the load factor design method or the allowable stress design method, both of which are available in the Standard Specifications, and neither of which have a mathematical basis for establishing safety, much of the chapter will deal primarily with the LRFD Specifications.

There are many issues that make up a design philosophy — for example, the expected service life of a structure, the degree to which future maintenance should be assumed to preserve the original resistance of the structure or should be assumed to be relatively nonexistent, the ways brittle behavior can be avoided, how much redundancy and ductility are needed, the degree to which analysis is expected to represent accurately the force effects actually experienced by the structure, the extent to which loads are thought to be understood and predictable, the degree to which the designers' intent will be upheld by vigorous material-testing requirements and thorough inspection during construction, the balance between the need for high precision during construction in terms of alignment and positioning compared with allowing for misalignment and compensating for it in the design, and, perhaps most fundamentally, the basis for establishing safety in the design specifications. It is this last issue, the way that specifications seek to establish safety, that is dealt with in this chapter.

0-8493-7434-0/00/$0.00+$.50

5.2 Limit States

All comprehensive design specifications are written to establish an acceptable level of safety. There are many methods of attempting to provide safety and the method inherent in many modern bridge design specifications, including the LRFD Specifications, the Ontario Highway Bridge Design Code [3], and the Canadian Highway Bridge Design Code [4], is probability-based reliability analysis. The method for treating safety issues in modern specifications is the establishment of "limit states" to define groups of events or circumstances that could cause a structure to be unserviceable for its original intent.

The LRFD Specifications are written in a probability-based limit state format requiring examination of some, or all, of the four limit states defined below for each design component of a bridge.

- The *service limit state* deals with restrictions on stress, deformation, and crack width under regular service conditions. These provisions are intended to ensure the bridge performs acceptably during its design life.
- The *fatigue and fracture limit state* deals with restrictions on stress range under regular service conditions reflecting the number of expected stress range excursions. These provisions are intended to limit crack growth under repetitive loads to prevent fracture during the design life of the bridge.
- The *strength limit state* is intended to ensure that strength and stability, both local and global, are provided to resist the statistically significant load combinations that a bridge will experience in its design life. Extensive distress and structural damage may occur under strength limit state conditions, but overall structural integrity is expected to be maintained.
- The *extreme event limit state* is intended to ensure the structural survival of a bridge during a major earthquake, or when collided by a vessel, vehicle, or ice flow, or where the foundation is subject to the scour that would accompany a flood of extreme recurrence, usually considered to be 500 years. These provisions deal with circumstances considered to be unique occurrences whose return period is significantly greater than the design life of the bridge. The joint probability of these events is extremely low, and, therefore, they are specified to be applied separately. Under these extreme conditions, the structure is expected to undergo considerable inelastic deformation by which locked-in force effects due to temperature effects, creep, shrinkage, and settlement will be relieved.

5.3 Philosophy of Safety

5.3.1 Introduction

A review of the philosophy used in a variety of specifications resulted in three possibilities, allowable stress design (ASD), load factor design (LFD), and reliability-based design, a particular application of which is referred to as load and resistance factor design (LRFD). These philosophies are discussed below.

5.3.2 Allowable Stress Design

ASD is based on the premise that one or more factors of safety can be established based primarily on experience and judgment which will assure the safety of a bridge component over its design life; for example, this design philosophy for a member resisting moments is characterized by design criteria such as

$$\Sigma M/S \leq F_y/1.82 \tag{5.1}$$

where
ΣM = sum of applied moments
F_y = specified yield stress
S = elastic section modules

The constant 1.82 is the factor of safety.

The "allowable stress" is assumed to be an indicator of the resistance and is compared with the results of stress analysis of loads discussed below. Allowable stresses are determined by dividing the elastic stress at the onset of some assumed undesirable response, e.g., yielding of steel or aluminum, crushing of concrete, loss of stability, by a safety factor. In some circumstances, the allowable stresses were increased on the basis that more representative measures of resistance, usually based on inelastic methods, indicated that some behaviors are stronger than others. For example, the ratio of fully yielded cross-sectional resistance (no consideration of loss of stability) to elastic resistance based on first yield is about 1.12 to 1.15 for most rolled shapes bent about their major axis. For a rolled shape bent about its minor axis, this ratio is 1.5 for all practical purposes. This increased plastic strength inherent in weak axis bending was recognized by increasing the basic allowable stress for this illustration from 0.55 F_y to 0.60 F_y and retaining the elastic calculation of stress.

The specified loads are the working basis for stress analysis. Individual loads, particularly environmental loads, such as wind forces or earthquake forces, may be selected based on some committee-determined recurrence interval. Design events are specified through the use of load combinations discussed in Section 5.4.1.4. This philosophy treats each load in a given load combination on the structure as equal from the viewpoint of statistical variability. A "commonsense" approach may be taken to recognize that some combinations of loading are less likely to occur than others; e.g., a load combination involving a 160 km/h wind, dead load, full shrinkage, and temperature may be thought to be far less likely than a load combination involving the dead load and the full design live load. For example, in ASD the former load combination is permitted to produce a stress equal to four thirds of the latter. There is no consideration of the probability of both a higher-than-expected load and a lower-than-expected strength occurring at the same time and place. There is little or no direct relationship between the ASD procedure and the actual resistance of many components in bridges, or to the probability of events actually occurring.

These drawbacks notwithstanding, ASD has produced bridges which, for the most part, have served very well. Given that this is the historical basis for bridge design in the United States, it is important to proceed to other, more robust design philosophies of safety with a clear understanding of the type of safety currently inherent in the system.

5.3.3 Load Factor Design

In LFD a preliminary effort was made to recognize that the live load, in particular, was more highly variable than the dead load. This thought is embodied in the concept of using a different multiplier on dead and live load; e.g., a design criteria can be expressed as

$$1.30M_D + 2.17\left(M_{L+I}\right) \le \phi M_u \tag{5.2}$$

where
M_D = moment from dead loads
M_{L+I} = moment from live load and impact
M_u = resistance
ϕ = a strength reduction factor

Resistance is usually based on attainment of either loss of stability of a component or the attainment of inelastic cross-sectional strength. Continuing the rolled beam example cited above, the distinction between weak axis and strong axis bending would not need to be identified because the cross-sectional resistance is the product of yield strength and plastic section modulus in both cases. In some cases, the resistance is reduced by a "strength reduction factor," which is based on the possibility that a component may be undersized, the material may be understrength, or the method of calculation may be more or less accurate than typical. In some cases, these factors have been based on statistical analysis of resistance itself. The joint probability of higher-than-expected loads and less-than-expected resistance occurring at the same time and place is not considered.

In the Standard Specifications, the same loads are used for ASD and LFD. In the case of LFD, the loads are multiplied by factors greater than unity and added to other factored loads to produce load combinations for design purposes. These combinations will be discussed further in Section 5.4.3.1.

The drawback to load factor design as seen from the viewpoint of probabilistic design is that the load factors and resistance factors were not calibrated on a basis that takes into account the statistical variability of design parameters in nature. In fact, the factors for steel girder bridges were established for one correlation at a simple span of 40 ft (12.2 m). At that span, both load factor design and service load design are intended to give the same basic structure. For shorter spans, load factor design is intended to result in slightly more capacity, whereas, for spans over 40 feet, it is intended to result in slightly less capacity with the difference increasing with span length. The development of this one point calibration for steel structures is given by Vincent in 1969 [5].

5.3.4 Probability- and Reliability-Based Design

Probability-based design seeks to take into account directly the statistical mean resistance, the statistical mean loads, the nominal or notional value of resistance, the nominal or notional value of the loads, and the dispersion of resistance and loads as measured by either the standard deviation or the coefficient of variation, i.e., the standard deviation divided by the mean. This process can be used directly to compute probability of failure for a given set of loads, statistical data, and the designer's estimate of the nominal resistance of the component being designed. Thus, it is possible to vary the designer's estimated resistance to achieve a criterion which might be expressed in terms, such as the component (or system) must have a probability of failure of less than 0.0001, or whatever variable is acceptable to society. Design based on probability of failure is used in numerous engineering disciplines, but its application to bridge engineering has been relatively small. The AASHTO "Guide Specification and Commentary for Vessel Collision Design of Highway Bridges" [6] is one of the few codifications of probability of failure in U.S. bridge design.

Alternatively, the probabilistic methods can be used to develop a quantity known as the "reliability index" which is somewhat, but not directly, relatable to the probability of failure. Using a reliability-based code in the purest sense, the designer is asked to calculate the value of the reliability index provided by his or her design and then compare that to a code-specified minimum value. Through a process of calibrating load and resistance factors to reliability indexes in simulated trial designs, it is possible to develop a set of load and resistance factors, so that the design process looks very much like the existing LFD methodology. The concept of the reliability index and a process for reverse-engineering load and resistance factors is discussed in Section 5.3.5.

In the case of the LRFD Specifications, some loads and resistances have been modernized as compared with the Standard Specifications. In many cases, the resistances are very similar. Most of the load and resistance factors have been calculated using a statistically based probability method which considers the joint probability of extreme loads and extreme resistance. In the parlance of the LRFD Specifications, "extreme" encompasses both maximum and minimum events.

5.3.5 The Probabilistic Basis of the LRFD Specifications

5.3.5.1 Introduction to Reliability as a Basis of Design Philosophy

A consideration of probability-based reliability theory can be simplified considerably by initially considering that natural phenomena can be represented mathematically as normal random variables,

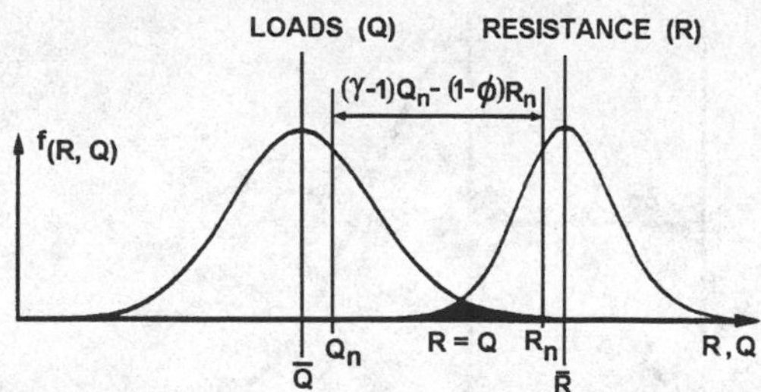

FIGURE 5.1 Separation of loads and resistance. (*Source:* Kulicki, J.M., et al., NH, Course 13061, Federal Highway Administration, Washington, D.C., 1994.)

as indicated by the well-known bell-shaped curve. This assumption leads to closed-form solutions for areas under parts of this curve, as given in many mathematical handbooks and programmed into many hand calculators.

Accepting the notion that both load and resistance are normal random variables, we can plot the bell-shaped curve corresponding to each of them in a combined presentation dealing with distribution as the vertical axis against the value of load, Q, or resistance, R, as shown in Figure 5.1 from Kulicki et al. [7]. The mean value of load, $\overline{Q}$, and the mean value of resistance, $\overline{R}$, are also shown. For both the load and the resistance, a second value somewhat offset from the mean value, which is the "nominal" value, or the number that designers calculate the load or the resistance to be, is also shown. The ratio of the mean value divided by the nominal value is called the "bias." The objective of a design philosophy based on reliability theory, or probability theory, is to separate the distribution of resistance from the distribution of load, such that the area of overlap, i.e., the area where load is greater than resistance, is tolerably small. In the particular case of the LRFD formulation of a probability-based specification, load factors and resistance factors are developed together in a way that forces the relationship between the resistance and load to be such that the area of overlap in Figure 5.1 is less than or equal to the value that a code-writing body accepts. Note in Figure 5.1 that it is the nominal load and the nominal resistance, not the mean values, which are factored.

A conceptual distribution of the difference between resistance and loads, combining the individual curves discussed above, is shown in Figure 5.2. It now becomes convenient to define the mean value of resistance minus load as some number of standard deviations, $\beta\sigma$, from the origin. The variable β is called the "reliability index" and σ is the standard deviation of the quantity $R - Q$. The problem with this presentation is that the variation of the quantity $R - Q$ is not explicitly known. Much is already known about the variation of loads by themselves or resistances by themselves, but the difference between these has not yet been quantified. However, from the probability theory, it is known that if load and resistance are both normal and random variables, then the standard deviation of the difference is

$$\sigma_{(R-Q)} = \sqrt{\sigma_R^2 + \sigma_Q^2} \tag{5.3}$$

Given the standard deviation, and considering Figure 5.2 and the mathematical rule that the mean of the sum or difference of normal random variables is the sum or difference of their individual means, we can now define the reliability index, β, as

$$\beta = \frac{\overline{R} - \overline{Q}}{\sqrt{\sigma_R^2 + \sigma_Q^2}} \tag{5.4}$$

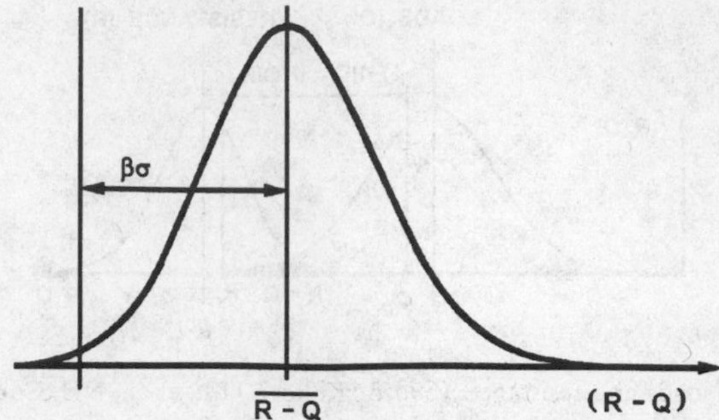

FIGURE 5.2 Definition of reliability index, β. (*Source:* Kulicki, J.M., et al., NH, Course 13061, Federal Highway Administration, Washington, D.C., 1994.)

Comparable closed-form equations can also be established for other distributions of data, e.g., log-normal distribution. A "trial-and-error" process is used for solving for β when the variable in question does not fit one of the already existing closed-form solutions.

The process of calibrating load and resistance factors starts with Eq. (5.4) and the basic design relationship; the factored resistance must be greater than or equal to the sum of the factored loads:

$$\phi R = Q = \Sigma \gamma_i x_i \tag{5.5}$$

Solving for the average value of resistance yields:

$$\overline{R} = \overline{Q} + \beta\sqrt{\sigma_R^2 + \sigma_Q^2} = \lambda R = \frac{1}{\phi}\lambda\Sigma\gamma_i x_i \tag{5.6}$$

By using the definition of bias, indicated by the symbol λ, Eq. (5.6) leads to the second equality in Eq. (5.6). A straightforward solution for the resistance factor, ϕ, is

$$\phi = \frac{\lambda\Sigma\gamma_i x_i}{\overline{Q} + \beta\sqrt{\sigma_R^2 + \sigma_Q^2}} \tag{5.7}$$

Unfortunately, Eq. (5.7) contains three unknowns, i.e., the resistance factor, ϕ, the reliability index, β, and the load factors, γ.

The acceptable value of the reliability index, β, must be chosen by a code-writing body. While not explicitly correct, we can conceive of β as an indicator of the fraction of times that a design criterion will be met or exceeded during the design life, analogous to using standard deviation as an indication of the total amount of population included or not included by a normal distribution curve. Utilizing this analogy, a β of 2.0 corresponds to approximately 97.3% of the values being included under the bell-shaped curve, or 2.7 of 100 values not included. When β is increased to 3.5, for example, now only two values in approximately 10,000 are not included.

It is more technically correct to consider the reliability index to be a comparative indicator. One group of bridges having a reliability index that is greater than a second group of bridges also has more safety. Thus, this can be a way of comparing a new group of bridges designed by some new process to a database of existing bridges designed by either ASD or LFD. This is, perhaps, the most correct and most effective use of the reliability index. It is this use which formed the basis for determining the target, or code specified, reliability index, and the load and resistance factors in the LRFD Specifications, as will be discussed in the next two sections.

The probability-based LRFD for bridge design may be seen as a logical extension of the current LFD procedure. ASD does not recognize that various loads are more variable than others. The introduction of the load factor design methodology brought with it the major philosophical change of recognizing that some loads are more accurately represented than others. The conversion to probability-based LRFD methodology could be thought of as a mechanism to select the load and resistance factors more systematically and rationally than was done with the information available when load factor design was introduced.

5.3.5.2 Calibration of Load and Resistance Factors

Assuming that a code-writing body has established a target value reliability index β, usually denoted β_T, Eq. (5.7) still indicates that both the load and resistance factors must be found. One way to deal with this problem is to select the load factors and then calculate the resistance factors. This process has been used by several code-writing authorities [2–4]. The steps in the process follow:

- Factored loads can be defined as the average value of load, plus some number of standard deviation of the load, as shown as the first part of Eq. (5.6) below.

$$\gamma_i x_i = \bar{x}_i + n\sigma_i = \bar{x}_i + nV_i\bar{x}_i \tag{5.8}$$

Defining the "variance," V_i, as equal to the standard deviation divided by the average value leads to the second half of Eq. (5.8). By utilizing the concept of bias one more time, Eq. (5.6) can now be condensed into Eq. (5.9).

$$\gamma_i = \lambda(1 + nV_i) \tag{5.9}$$

Thus, it can be seen that load factors can be written in terms of the bias and the variance. This gives rise to the philosophical concept that load factors can be defined so that all loads have the same probability of being exceeded during the design life. This is not to say that the load factors are identical, just that the probability of the loads being exceeded is the same.

- By using Eq. (5.7) for a given set of load factors, the value of the resistance factor can be assumed for various types of structural members and for various load components, e.g., shear, moment, etc. on the various structural components. Computer simulations of a representative body of structural members can be done, yielding a large number of values for the reliability index.
- Reliability indexes are compared with the target reliability index. If close clustering results, a suitable combination of load and resistance factors has been obtained.
- If close clustering does not result, a new trial set of load factors can be used and the process repeated until the reliability indexes do cluster around, and acceptably close to, the target reliability index.
- The resulting load and resistance factors taken together will yield reliability indexes close to the target value selected by the code-writing body as acceptable.

The outline above assumes that suitable load factors are assumed. If the process of varying the resistance factors and calculating the reliability indexes does not converge to a suitable narrowly grouped set of reliability indexes, then the load factor assumptions must be revised. In fact, several sets of proposed load factors may have to be investigated to determine their effect on the clustering of reliability indexes.

The process described above is very general. To understand how it is used to develop data for a specific situation, the rest of this section will illustrate the application to calibration of the load and resistance factors for the LRFD Specifications. The basic steps were as follows:

- Develop a database of sample current bridges.
- Extract load effects by percentage of total load.
- Develop a simulation bridge set for calculation purposes.
- Estimate the reliability indexes implicit in current designs.
- Revise loads-per-component to be consistent with the LRFD Specifications.
- Assume load factors.
- Vary resistance factors until suitable reliability indexes result.

Approximately 200 representative bridges were selected from various regions of the United States by requesting sample bridge plans from various states. The selection was based on structural type, material, and geographic location to represent a full range of materials and design practices as they vary around the country. Anticipated future trends should also be considered. In the particular case of the LRFD Specifications, this was done by sending questionnaires to various departments of transportation asking them to identify the types of bridges they are expecting to design in the near future.

For each of the bridges in the database, the load indicated by the contract drawings was subdivided by the following characteristic components:

- The dead load due to the weight of factory-made components;
- The dead load of cast-in-place components;
- The dead load due to asphaltic wearing surfaces where applicable;
- The dead weight due to miscellaneous items;
- The live load due to the HS20 loading;
- The dynamic load allowance or impact prescribed in the 1989 AASHTO Specifications.

Full tabulations for all these loads for the full set of bridges in the database are presented in Nowak [8].

Statistically projected live load and the notional values of live load force effects were calculated. Resistance was calculated in terms of moment and shear capacity for each structure according to the prevailing requirements, in this case the AASHTO Standard Specifications for load factor design.

Based on the relative amounts of the loads identified in the preceding section for each of the combination of span and spacing and type of construction indicated by the database, a simulated set of 175 bridges was developed, comprising the following:

- In all; 25 noncomposite steel girder bridge simulations for bending moments and shear with spans of 9, 18, 27, 36, and 60 m and, for each of those spans, spacings of 1.2, 1.8, 2.4, 3.0, and 3.6 m;
- Representative composite steel girder bridges for bending moments and shear having the same parameters as those identified above;
- Representative reinforced concrete T-beam bridges for bending moments and shear having spans of 9, 18, 27, and 39 m, with spacings of 1.2, 1.8, 2.4, and 3.6 m in each span group;
- Representative prestressed concrete I-beam bridges for moments and shear having the same span and spacing parameters as those used for the steel bridges.

Full tabulations of these bridges and their representative amounts of the various loads are presented in Nowak [8].

The reliability indexes were calculated for each simulated and each actual bridge for both shear and moment. The range of reliability indexes which resulted from this phase of the calibration process is presented in Figure 5.3 from Kulicki et al. [7]. It can be seen that a wide range of values was obtained using the current specifications, but this was anticipated based on previous calibration work done for the Ontario Highway Bridge Design Code (OHBDC) [9].

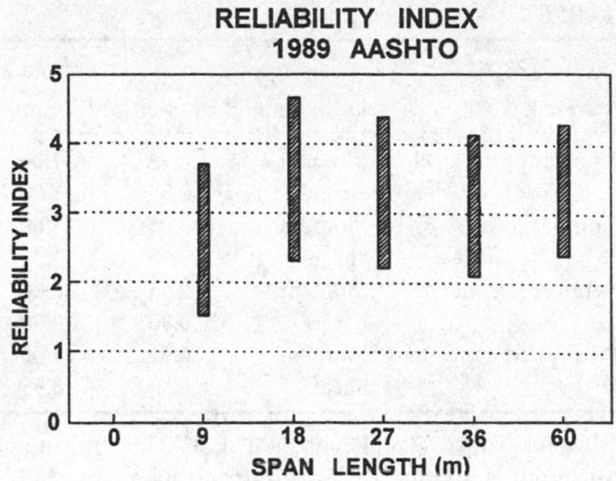

FIGURE 5.3 Reliability indexes inherent in the 1989 AASHTO Standard Specifications. (*Source:* Kulicki, J.M., et al., NII, Course 13061, Federal Highway Administration, Washington, D.C., 1994.)

TABLE 5.1 Parameters of Bridge Load Components

Load Component	Bias Factor	Coefficient of Variation	Load Factor $n = 1.5$	Load Factor $n = 2.0$	Load Factor $n = 2.5$
Dead load, shop built	1.03	0.08	1.15	1.20	1.24
Dead load, field built	1.05	0.10	1.20	1.25	1.30
Dead load, asphalt and utilities	1.00	0.25	1.375	1.50	1.65
Live load (with impact)	1.10–1.20	0.18	1.40–1.50	1.50–1.60	1.60–1.70

Source: Nowak, A.S., Report UMCE 92-25, University of Michigan, Ann Arbor, 1993. With permission.

These calculated reliability indexes, as well as past calibration of other specifications, serve as a basis for selection of the target reliability index, β_T. A target reliability index of 3.5 was selected for the OHBDC and is under consideration for other reliability-based specifications. A consideration of the data shown in Figure 5.3 indicates that a β of 3.5 is representative of past LFD practice. Hence, this value was selected as a target for the calibration of the LRFD Specifications.

5.3.5.3 Load and Resistance Factors

The parameters of bridge load components and various sets of load factors, corresponding to different values of the parameter n in Eq. (5.9) are summarized in Table 5.1 from Nowak [8].

Recommended values of load factors correspond to $n = 2$. For simplicity of the designer, one factor is specified for shop-built and field-built components, $\gamma = 1.25$. For D_3, weight of asphalt and utilities, $\gamma = 1.50$. For live load and impact, the value of load factor corresponding to n = 2 is $\gamma = 1.60$. However, a more conservative value of $\gamma = 1.75$ is utilized in the LRFD Specifications.

The acceptance criterion in the selection of resistance factors is how close the calculated reliability indexes are to the target value of the reliability index, β_T. Various sets of resistance factors, ϕ, are considered. Resistance factors used in the code are rounded off to the nearest 0.05.

Calculations were performed using the load components for each of the 175 simulated bridges using the range of resistance factors shown in Table 5.3. For a given resistance factor, material, span, and girder spacing, the reliability index is computed. Values of β were calculated for live-load factors, $\gamma = 1.75$. For comparison, the results are also shown for live-load factor, $\gamma = 1.60$. The calculations are performed for the resistance factors, ϕ, listed in Table 5.2 from Nowak [8].

Reliability indexes were recalculated for each of the 175 simulated cases and each of the actual bridges from which the simulated bridges were produced. The range of values obtained using the new load and resistance factors is indicated in Figure 5.4.

TABLE 5.2 Considered Resistance Factors

Material	Limit State	Resistance Factors, ϕ	
		Lower	Upper
Noncomposite steel	Moment	0.95	1.00
	Shear	0.95	1.00
Composite steel	Moment	0.95	1.00
	Shear	0.95	1.00
Reinforced concrete	Moment	0.85	0.90
	Shear	0.90	0.90
Prestressed concrete	Moment	0.95	1.00
	Shear	0.90	0.95

Source: Nowak, A.S., Report UMCE 92-25, University of Michigan, Ann Arbor, 1993. With permission.

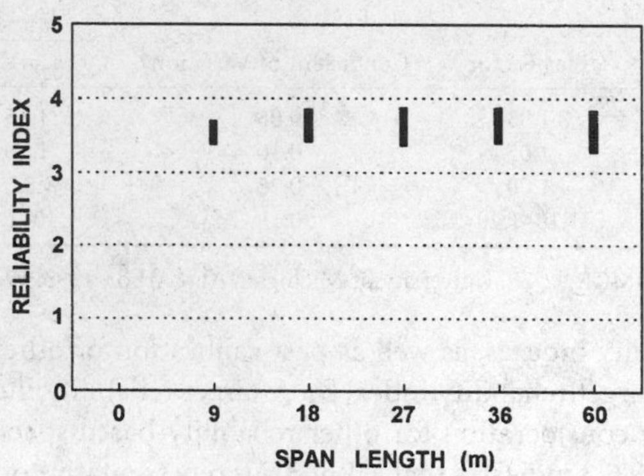

FIGURE 5.4 Reliability indexes inherent in LRFD Specifications. (*Source:* Kulicki, J.M., et al., NH, Course 13061, Federal Highway Administration, Washington, D.C., 1994.)

Figure 5.4 from Kulicki et al. [7] shows that the new calibrated load and resistance factors and new load models and load distribution techniques work together to produce very narrowly clustered reliability indexes. This was the objective of developing the new factors. Correspondence to a reliability index of 3.5 is something which can now be altered by AASHTO. The target reliability index could be raised or lowered as may be advisable in the future and the factors can be recalculated accordingly. This ability to adjust the design parameters in a coordinated manner is one of the strengths of a probabilistically based reliability design.

5.4 Design Objectives

5.4.1 Safety

5.4.1.1 Introduction

Public safety is the primary responsibility of the design engineer. All other aspects of design, including serviceability, maintainability, economics, and aesthetics are secondary to the requirement for safety. This does not mean that other objectives are not important, but safety is paramount.

5.4.1.2 The Equation of Sufficiency

In design specifications the issue of safety is usually codified by an application of the general statement the design resistances must be greater than, or equal to, the design load effects. In ASD, Eq. (5.1) can be generalized as

$$\Sigma Q_i \leq R_E/\text{FS} \tag{5.10}$$

where
Q_i = a load
R_E = elastic resistance
FS = factor of safety

In LFD, Eq. (5.2) can be generalized as

$$\Sigma \gamma_i Q_i \leq \phi R \tag{5.11}$$

where
γ_i = a load factor
Q_i = a load
R = resistance
ϕ = a strength reduction factor.

In LRFD, Eq. (5.2) can be generalized as

$$\Sigma \eta_i y_i Q_i \leq \phi R_n = R_r \tag{5.12}$$

where
η_i = $\eta_D \eta_R \eta_I$: $\eta_i = \eta_D \eta_R \eta_I \geq 0.95$ for loads for which a maximum value of γ_i is appropriate and $\eta_i = 1/(\eta_I \eta_D \eta_R) \leq 1.0$ for loads for which a minimum value of γ_i is appropriate
γ_i = load factor: a statistically based multiplier on force effects
ϕ = resistance factor: a statistically based multiplier applied to nominal resistance
η_i = load modifier
η_D = a factor relating to ductility
η_R = a factor relating to redundancy
η_I = a factor relating to operational importance
Q_i = nominal force effect: a deformation, stress, or stress resultant
R_n = nominal resistance: based on the dimensions as shown on the plans and on permissible stresses, deformations, or specified strength of materials
R_r = factored resistance: ϕR_n

Eq. (5.12) is applied to each designed component and connection as appropriate for each limit state under consideration.

5.4.1.3 Special Requirements of the LRFD Specifications

Comparison of the equation of sufficiency as it was written above for ASD, LFD, and LRFD shows that, as the design philosophy evolved through these three stages, more aspects of the component under design and its relation to its environment and its function to society must be expressly considered. This is not to say that a designer using ASD necessarily considers less than a designer using LFD or LRFD. The specification provisions are the minimum requirements, and prudent designers often consider additional aspects. However, as specifications mature and become more reflective of the real world, additional criteria are often needed to assure adequate safety which may have been provided, albeit nonuniformly, by simpler provisions. Therefore, it is not surprising to find that the LRFD Specifications require explicit consideration of ductility, redundancy, and operational importance in Eq. (5.12), while the Standard Specifications does not.

Ductility, redundancy, and operational importance are significant aspects affecting the margin of safety of bridges. While the first two directly relate to the physical behavior, the last concerns the consequences of the bridge being out of service. The grouping of these aspects is, therefore, arbitrary; however, it constitutes a first effort of codification. In the absence of more precise information, each effect, except that for fatigue and fracture, is estimated as $\pm 5\%$, accumulated geometrically, a clearly subjective approach. With time, improved quantification of ductility, redundancy, and operational importance, and their interaction, may be attained.

Ductility

The response of structural components or connections beyond the elastic limit can be characterized by either brittle or ductile behavior. Brittle behavior is undesirable because it implies the sudden loss of load-carrying capacity immediately when the elastic limit is exceeded. Ductile behavior is characterized by significant inelastic deformations before any loss of load-carrying capacity occurs. Ductile behavior provides warning of structural failure by large inelastic deformations. Under cyclic loading, large reversed cycles of inelastic deformation dissipate energy and have a beneficial effect on structure response.

If, by means of confinement or other measures, a structural component or connection made of brittle materials can sustain inelastic deformations without significant loss of load-carrying capacity, this component can be considered ductile. Such ductile performance should be verified by experimental testing.

Behavior that is ductile in a static context, but that is not ductile during dynamic response, should also be avoided. Examples of this behavior are shear and bond failures in concrete members and loss of composite action in flexural members.

The ductility capacity of structural components or connections may either be established by full- or large-scale experimental testing, or with analytical models that are based on realistic material behavior. The ductility capacity for a structural system may be determined by integrating local deformations over the entire structural system.

Given proper controls on the innate ductility of basic materials, proper proportioning and detailing of a structural system are the key consideration in ensuring the development of significant, visible, inelastic deformations, prior to failure, at the strength and extreme event limit states.

For the fatigue and fracture limit state for fracture-critical members and for the strength limit state for all members:

$\eta_D \geq 1.05$ for nonductile components and connections,

$= 1.00$ for conventional designs and details complying with these specifications

≥ 0.95 for components and connections for which additional ductility-enhancing measures have been specified beyond those required by these specifications

For all other limit states:

$$\eta_D = 1.00$$

Redundancy

Redundancy is usually defined by stating the opposite, e.g., a nonredundant structure is one in which the loss of a component results in collapse or a nonredundant component is one whose loss results in complete or partial collapse. Multiple load path structures should be used, unless there are compelling reasons to the contrary. The LRFD Specifications require additional resistance in order to reduce probability of loss of nonredundant component and to provide additional resistance to accommodate load redistribution.

For the strength limit state:

η_R ≥ 1.05 for nonredundant members
= 1.00 for conventional levels of redundancy
≥ 0.95 for exceptional levels of redundancy

For all other limit states:

$\eta_R = 1.00$

The factors currently specified were based solely on judgment and were included to require more explicit consideration of redundancy. Research is under way by Ghosn and Moses [10] to provide more rational requirements based on reliability indexes thought to be acceptable in damaged bridges which must remain in service for a period of about 2 years. The "reverse engineering" concept is being applied to develop values similar in intent to η_R.

Operational Importance
The concept of operational importance is applied to the strength and extreme event limit states. The owner may declare a bridge, or any structural component or connection, thereof, to be of operational importance. Such classification should be based on social/survival and/or security/defense requirements. If a bridge is deemed of operational importance, η_I is taken as ≥1.05. Otherwise, η_I is taken as 1.0 for typical bridges and may be reduced to 0.95 for relatively less important bridges.

5.4.1.4 Design Load Combinations in ASD, LFD, and LRFD

The following permanent and transient loads and forces are considered in the ASD and LFD using the Standard Specifications, and in LRFD using the LRFD Specifications.

The load factors for various loads, making up a design load combination, are indicated in Table 5.4 and Table 5.5 for LRFD and Table 5.6 for ASD and LFD. In the case of the LRFD Specifications, all of the load combinations are related to the appropriate limit state. Any, or all, of the four limit states may be required in the design of any particular component and those which are the minimum necessary for consideration are indicated in the specifications where appropriate. Thus, a design might involve any load combination in Table 5.4.

In the case of ASD or LFD, there is no direct relationship between the load combinations specified in Table 5.6 and limit states, as the design requirements in the Standard Specifications are not organized in that manner. A design by ASD uses those combinations in Table 5.5 indicated for the allowable stress design method as appropriate for the component under consideration. The load combinations indicated for LFD are not used in conjunction with allowable stress design. The opposite is true for LFD.

The application of the load combinations in Table 5.6 for ASD and LFD has been available to bridge designers for decades and is relatively well understood. Numerous textbooks have dealt with these subjects. For this reason, the remainder of this section will deal primarily with the relatively newer LRFD Specifications.

All relevant subsets of the load combinations in Table 5.4 should be investigated. The factors should be selected to produce the total factored extreme force effect. For each load combination, both positive and negative extremes should be investigated. In load combinations where one force effect decreases the effect of another, the minimum value should be applied to load reducing the force effect. For each load combination, every load that is indicated, including all significant effects due to distortion, should be multiplied by the appropriate load factor.

It can be seen in Table 5.4 that some of the load combinations have a choice of two load factors. The larger of the two values for load factors shown for TU, TG, CR, SH, and SE are to be used when calculating deformations; the smaller value should be used when calculating all other force

TABLE 5.3 Load Designations

Name of Load	LRFD Designation	Standard of Specification Designation
	Permanent Loads	
Downdrag	DD	
Dead load of structural components attachments	DC	D
Dead load of wearing surfaces and utilities	DW	D
Dead load of earth fill	EF	D
Horizontal earth pressure	EH	E
Earth surcharge load	ES	E
Vertical earth pressure	EV	D
	Transient Loads	
Vehicular braking force	BR	LF
Vehicular centrifugal force	CE	CF
Creep	CR	R
Vehicular collision force	CT	—
Vessel collision force	CV	—
Earthquake	EQ	EQ
Friction	FR	—
Ice load	IC	ICE
Vehicular dynamic load allowance	IM	I
Vehicular live load	LL	L
Live–load surcharge	LS	L
Pedestrian live load	PL	L
Settlement	SE	—
Shrinkage	SH	S
Temperature gradient	TG	—
Uniform temperature	TU	T
Water load and stream pressure	WA	SF
Wind on live load	WL	WL
Wind load on structure	WS	W

TABLE 5.4 Load Combinations and Load Factors in LRFD

Limit State Load Combinations	DC DD DW EH EV ES	LL IM CE BR PL LS	WA	WS	WL	FR	TU CR SH	TG	SE	Use One of These at a Time EQ	IC	CT	CV
Strength I	γ_p	1.75	1.00	—	—	1.00	0.50/1.20	γ_{TG}	γ_{SE}	—	—	—	—
Strength II	γ_p	1.35	1.00	—	—	1.00	0.50/1.20	γ_{TG}	γ_{SE}	—	—	—	—
Strength III	γ_p	—	1.00	1.40	—	1.00	0.50/1.20	γ_{TG}	γ_{SE}	—	—	—	—
Strength IV													
EH, EV, ES, DW	γ_p	—	1.00	—	—	1.00	0.50/1.20	—	—	—	—	—	—
DC only	1.5									—	—	—	—
Strength V	γ_p	1.35	1.00	0.40	0.40	1.00	0.50/1.20	γ_{TG}	γ_{SE}	—	—	—	—
Extreme Event I	γ_p	γ_{EQ}	1.00	—	—	1.00	—	—	—	1.00	—	—	—
Extreme Event II	γ_p	0.50	1.00	—	—	1.00	—	—	—	—	1.00	1.00	1.00
Service I	1.00	1.00	1.00	0.30	0.30	1.00	1.00/1.20	γ_{TG}	γ_{SE}	—	—	—	—
Service II	1.00	1.30	1.00	—	—	1.00	1.00/1.20	—	—	—	—	—	—
Service III	1.00	0.80	1.00	—	—	1.00	1.00/1.20	γ_{TG}	γ_{SE}	—	—	—	—
Fatigue LL, IM and CE only	—	0.75	—	—	—	—	—	—	—	—	—	—	—

TABLE 5.5 Load Factors for Permanent Loads, γ_p in LRFD

Type of Load	Load Factor	
	Maximum	Minimum
DC: Component and attachments	1.25	0.90
DD: Downdrag	1.80	0.45
DW: Wearing surfaces and utilities	1.50	0.65
EH: Horizontal earth pressure		
• Active	1.50	0.90
• At rest	1.35	0.90
EV: Vertical earth pressure		
• Overall stability	1.35	N/A
• Retaining structure	1.35	1.00
• Rigid buried structure	1.30	0.90
• Rigid frames	1.35	0.90
• Flexible buried structures other than metal box culverts	1.95	0.90
• Flexible metal box culverts	1.50	0.90
ES: Earth surcharge	1.50	0.75

effects. Where movements are calculated for the sizing of expansion dams, the design of bearing, or similar situations where consideration of unexpectedly large movements is advisable, the larger factor should be used. When considering the effect of these loads on forces that are compatibility generated, the lower factor may be used. This latter use requires structural insight.

Consideration of the variability of loads in nature indicates that loads may be either larger or smaller than the nominal load used in the design specifications. While the concept of variability of permanent loads receives little coverage in ASD, it is codified expressly in LFD. Note that in Table 5.6 the LFD load combinations contain a dead load modifier, indicated as β_E or β_D. These β terms are not to be confused with the reliability index, heretofore referred to as β. The purpose of the modifying factors β_E and β_D is to account for conditions where it is inadvisable to consider either that all of the dead load exists all of the time or that the dead load may be less than the nominal values indicated in the specifications. Thus, for example, the use of the β_D factor 0.75 when checking members for minimum axial load maximum moment means when designing columns and those fixtures which abut the columns, such as footings, it is necessary to evaluate not just the maximum bending moment and the maximum axial load, based on assuming that all the elements of a load combination are thought to obtain their maximum values, but also a load combination in which it is assumed that the dead load is lighter than the nominal load. In the case where the majority of the axial load comes from the dead load and the majority of the bending moment comes from lateral load or live load, this modified combination will tend to produce a maximum eccentricity and hence could control the design of columns and footings.

The specified values of β_E are given below:

β_E 1.00 for vertical and lateral loads on all other structures

β_E 1.3 for lateral earth pressure for retaining walls and rigid frames, excluding rigid culverts; for lateral at-rest earth pressures, $\beta_E = 1.15$

β_E 0.5 for lateral earth pressure when checking positive moments in rigid frames; this complies with Section 3.20

β_E 1.0 for vertical earth pressure

β_D 0.75 when checking member for minimum axial load and maximum moment or maximum eccentricity — for column design

β_D 1.0 when checking member for maximum axial load and minimum moment — for column design

β_D 1.0 for flexural and tension members

β_E 1.0 for rigid culverts

β_E 1.5 for flexible culverts

TABLE 5.6 Table of Coefficients γ and β in ASD and LFD

		Col. No.														
		1	2	3	3A	4	5	6	7	8	9	10	11	12	13	14
			β Factors													
	Group	γ	D	$(L+I)_n$	$(L+I)_p$	CF	E	B	SF	W	WL	LF	R + S + T	EQ	Ice	%
SERVICE LOAD	I	1.0	1	1	0	1	β_E	B	1	0	0	0	0	0	0	100
	IA	1.0	1	2	0	0	0	1	0	0	0	0	0	0	0	150
	IB	1.0	1	0	1	1	β_E	0	1	0	0	0	0	0	0	b
	II	1.0	1	0	0	0	1	1	1	1	0	0	0	0	0	125
	III	1.0	1	1	0	1	β_E	1	1	0.3	1	1	0	0	0	125
	IV	1.0	1	1	0	1	β_E	1	1	0	0	0	1	0	0	125
	V	1.0	1	0	0	0	1	1	1	1	0	0	1	0	0	140
	VI	1.0	1	1	0	1	β_E	1	1	0.3	1	1	1	0	0	140
	VII	1.0	1	0	0	0	1	1	1	0	0	0	0	1	0	133
	VIII	1.0	1	1	0	1	1	1	1	0	0	0	0	0	1	140
	IX	1.0	1	0	0	0	1	1	1	1	0	0	0	0	1	150
	X	1.0	1	1	0	0	β_E	0	0	0	0	0	0	0	0	100
LOAD FACTOR DESIGN	I	1.3	β_D	1.67[a]	0	1.0	β_E	1	1	0	0	0	0	0	0	NOT APPLICABLE
	IA	1.3	β_D	2.20	0	0	0	0	0	0	0	0	0	0	0	
	IB	1.3	β_D	0	1	1.0	β_E	1	1	0	0	0	0	0	0	
	II	1.3	β_D	0	0	0	β_E	1	1	1	0	0	0	0	0	
	III	1.3	β_D	1	0	1	β_E	1	1	0.3	1	1	0	0	0	
	IV	1.3	β_D	1	0	1	β_E	1	1	0	0	0	1	0	0	
	V	1.25	β_D	0	0	0	β_E	1	1	1	0	0	1	0	0	
	VI	1.25	β_D	1	0	1	β_E	1	1	0.3	1	1	1	0	0	
	VII	1.3	β_D	0	0	0	β_E	1	0	0	0	0	0	1	0	
	VIII	1.3	β_D	1	0	1	β_E	1	1	0	0	0	0	0	1	
	IX	1.20	β_D	0	0	0	β_E	1	1	1	0	0	0	0	1	
	X	1.30	1	1.67	0	0	β_E	0	0	0	0	0	0	0	0	

- $(L + I)_n$ = Live load plus impact for AASHTO Highway H or HS loading.
- $(L + I)_p$ = Live load plus impact consistent with the overload criteria of the operation agency.
- % (col. 14) = percentage of basic unit stress.
- No increase in allowable unit stresses shall be permitted for members or connections carrying wind loads only.

[a] 1.25 may be used for design of outside roadway beam when combination of sidewalk live load, and traffic live load plus impact governs the design, but the capacity of the section should not be less than required for highway traffic live load only, using a β factor of 1.67. 1.00 may be used for design of deck slab with combination of loads as described in Article 3.24.2.2.

[b] $$\text{Percentage} = \frac{\text{Maximum Unit Stress (Operating Rating)}}{\text{Allowable Basic Unit Stress}} \times 100.$$

The LRFD Specifications recognize the variability of permanent loads by providing both maximum and minimum load factors for the permanent loads, as indicated in Table 5.5. For permanent force effects, the load factor that produces the more critical combination should be selected from Table 5.4. In the application of permanent loads, force effects for each of the specified six load types should be computed separately. Assuming variation of one type of load by span, length, or component within a bridge is not necessary. For each force effect, both extreme combinations may need to be investigated by applying either the high or the low load factor, as appropriate. The algebraic sums of these products are the total force effects for which the bridge and its components should be designed. This reinforces the traditional method of selecting load combinations to obtain realistic extreme effects.

When the permanent load increases the stability or load-carrying capacity of a component or bridge, the minimum value of the load factor for that permanent load should also be investigated. Uplift, which is treated as a separate load case in past editions of the AASHTO Standard Specifications for Highway Bridges, becomes a Strength I load combination. For example, when the dead-load reaction is positive and live load can cause a negative reaction, the load combination for maximum uplift force would be 0.9DC + 0.65DW + 1.75(LL+IM). If both reactions were negative, the load combination would be 1.25DC + 1.50DW + 1.75(LL+IM).

The load combinations for various limit states shown in Table 5.4 are described below.

Strength I	Basic load combination relating to the normal vehicular use of the bridge without wind.
Strength II	Load combination relating to the use of the bridge by permit vehicles without wind. If a permit vehicle is traveling unescorted, or if control is not provided by the escorts, the other lanes may be assumed to be occupied by the vehicular live load herein specified. For bridges longer than the permit vehicle, addition of the lane load, preceding and following the permit load in its lane, should be considered.
Strength III	Load combination relating to the bridge exposed to maximum wind velocity which prevents the presence of significant live load on the bridge.
Strength IV	Load combination relating to very high ratios of dead load to live load force effect. This calibration process had been carried out for a large number of bridges with spans not exceeding 60 m. Spot checks had also been made on a few bridges up to 180 m spans. For the primary components of large bridges, the ratio of dead and live load force effects is rather high and could result in a set of resistance factors different from those found acceptable for small- and medium-span bridges. It is believed to be more practical to investigate one more load case, rather than requiring the use of two sets of resistance factors with the load factors provided in Strength I, depending on other permanent loads present. This Load Combination IV is expected to govern when the ratio of dead load to live load force effect exceeds about 7.0.
Strength V	Load combination relating to normal vehicular use of the bridge with wind of 90 km/h velocity.
Extreme Event I	Load combination relating to earthquake. The designer-supplied live-load factor signifies a low probability of the presence of maximum vehicular live load at the time when the earthquake occurs. In ASD and LFD the live load is ignored when designing for earthquake.
Extreme Event II	Load combination relating to reduced live load in combination with a major ice event, or a vessel collision, or a vehicular impact.
Service I	Load combination relating to the normal operational use of the bridge with 90 km/h wind. All loads are taken at their nominal values and extreme load conditions are excluded. This combination is also used for checking deflection of certain buried structures and for the investigation of slope stability.
Service II	Load combination whose objective is to prevent yielding of steel structures due to vehicular live load, approximately halfway between that used for Service I and Strength I limit state, for which case the effect of wind is of no significance. This load combination corresponds to the overload provision for steel structures in past editions of the AASHTO Standard Specifications for the Design of Highway Bridges.
Service III	Load combination relating only to prestressed concrete structures with the primary objective of crack control. The addition of this load combination followed a series of trial designs done by 14 states and several industry groups during 1991 and early 1992. Trial designs for prestressed concrete elements indicated significantly more prestressing would be needed to support the loads specified in the proposed specifications. There is no nationwide physical evidence that these vehicles used to develop the notional live loads have caused detrimental cracking in existing prestressed concrete components. The statistical significance

of the 0.80 factor on live load is that the event is expected to occur about once a year for bridges with two design lanes, less often for bridges with more than two design lanes, and about once a day for the bridges with a single design lane.

Fatigue — Fatigue and fracture load combination relating to gravitational vehicular live load and dynamic response, consequently BR and PL need not be considered. The load factor reflects a load level which has been found to be representative of the truck population, with respect to large number of return cycles.

5.4.2 Serviceability

The LRFD Specification treats serviceability from the view points of durability, inspectibility, maintainability, rideability, deformation control, and future widening.

Contract documents should call for high-quality materials and require that those materials that are subject to deterioration from moisture content and/or salt attack be protected. Inspectibility is to be assured through adequate means for permitting inspectors to view all parts of the structure which have structural or maintenance significance. The provisions related to inspectibility are relatively short, but as all departments of transportation have begun to realize, bridge inspection can be very expensive and is a recurring cost due to the need for biennial inspections. Therefore, the cost of providing walkways and other access means and adequate room for people and inspection equipment to be moved about on the structure is usually a good investment.

Maintainability is treated in the specification in a manner similar to durability; there is a list of desirable attributes to be considered.

The subject of live-load deflections and other deformations remains a very difficult issue. On the one hand, there is very little direct correlation between live-load deflection and premature deterioration of bridges. There is much speculation that "excessive" live-load deflection contributes to premature deck deterioration, but, to date (late 1997), no causative relationship has been statistically established.

Rider comfort is often advanced as a basis for deflection control. Studies in human response to motion have shown that it is not the magnitude of the motion, but rather the acceleration that most people perceive, especially in moving vehicles. Many people have experienced the sensation of being on a bridge and feeling a definite movement, especially when traffic is stopped. This movement is often related to the movement of floor systems, which are really quite small in magnitude, but noticeable nonetheless. There being no direct correlation between magnitude (not acceleration) of movement and discomfort has not prevented the design profession from finding comfort in controlling the gross stiffness of bridges through a deflection limit. As a compromise between the need for establishing comfort levels and the lack of compelling evidence that deflection was a cause of structural distress, the deflection criteria, other than those pertaining to relative deflections of ribs of orthotropic decks and components of some wood decks, were written as voluntary provisions to be activated by those states that so chose. Deflection limits, stated as span divided by some number, were established for most cases, and additional provisions of absolute relative displacement between planks and panels of wooden decks and ribs of orthotropic decks were also added. Similarly, optional criteria were established for a span-to-depth ratio for guidance primarily in starting preliminary designs, but also as a mechanism for checking when a given design deviated significantly from past successful practice.

5.4.3 Constructibility

Several new provisions were included in the LRFD Specification related to:

- The need to design bridges so that they can be fabricated and built without undue difficulty and with control over locked-in construction force effects;

- The need to document one feasible method of construction in the contract documents, unless the type of construction is self-evident; and
- A clear indication of the need to provide strengthening and/or temporary bracing or support during erection, but not requiring the complete design thereof.

References

1. American Association of State Highway and Transportation Officials, *Standard Specifications for Highway Bridges,* 16th ed., AASHTO, Washington, D.C., 1996.
2. American Association of State Highway and Transportation Officials, *Load Resistance Factor Design,* AASHTO, Washington, D.C., 1996.
3. Ontario Ministry of Transportation and Communications, *Ontario Highway Bridge Design Code,* OMTC, Toronto, Ontario, Canada, 1994.
4. Canadian Standards Association, *Canadian Highway Bridge Design Code,* Canadian Standards Association, Rexdale, Ontario, Canada, 1998.
5. Vincent, G. S., Load factor design of steel highway bridges, *AISI Bull.,* 15, March, 1969.
6. American Association of State Highway and Transportation Officials, *Guide Specification and Commentary for Vessel Collision Design of Highway Bridges,* Vol. I: Final Report, AASHTO, Washington, D.C., February 1991.
7. Kulicki, J. M., Mertz, D. R., and Wassef, W. G., LRFD Design of Highway Bridges, NHI Course 13061, Federal Highway Administration, Washington, D.C., 1994.
8. Nowak, A. S., Calibration of LRFD Bridge Design Code, Department of Civil and Environmental Engineering Report UMCE 92-25, University of Michigan, Ann Arbor, 1993.
9. Nowak, A. S. and Lind, N. C., Practical bridge code calibration, *ASCE J. Struct. Div.,* 105 (ST12), 2497–2510, 1979.
10. Ghosn, M. and Moses, F., Redundancy in Highway Bridge Superstructures, Draft Report to NCHRP, February 1997.

6

Highway Bridge Loads and Load Distribution

Susan E. Hida
California Department of Transportation

6.1 Introduction

This chapter deals with highway bridge loads and load distribution as specified in the AASHTO Load and Resistance Factor Design (LRFD) Specifications [1]. Stream flow, ice loads, vessel collision loads, loads for barrier design, loads for anchored and mechanically stabilized walls, seismic forces, and loads due to soil–structure interaction will be addressed in subsequent chapters. Load combinations are discussed in Chapter 5.

When proceeding from one component to another in bridge design, the controlling load and the controlling factored load combination will change. For example, permit vehicles, factored and combined for one load group, may control girder design for bending in one location. The standard design vehicular live load, factored and combined for a different load group, may control girder design for shear in another location. Still other loads, such as those due to seismic events, may control column and footing design.

Note that in this chapter, superstructure refers to the deck, beams or truss elements, and any other appurtenances above the bridge soffit. Substructure refers to those components that support loads from the superstructure and transfer load to the ground, such as bent caps, columns, pier walls, footings, piles, pile extensions, and caissons. Longitudinal refers to the axis parallel to the direction of traffic. Transverse refers to the axis perpendicular to the longitudinal axis.

0-8493-7434-0/00/$0.00+$.50

6.2 Permanent Loads

The LRFD Specification refers to the weights of the following as "permanent loads":

- The structure
- Formwork which becomes part of the structure
- Utility ducts or casings and contents
- Signs
- Concrete barriers
- Wearing surface and/or potential deck overlay(s)
- Other elements deemed permanent loads by the design engineer and owner
- Earth pressure, earth surcharge, and downdrag

The permanent load is distributed to the girders by assigning to each all loads from superstructure elements within half the distance to the adjacent girder. This includes the dead load of the girder itself and the soffit, in the case of box girder structures. The dead loads due to concrete barrier, sidewalks and curbs, and sound walls, however, may be equally distributed to all girders.

6.3 Vehicular Live Loads

The design vehicular live load was replaced in 1993 because of heavier truck configurations on the road today, and because a statistically representative, notional load was needed to achieve a "consistent level of safety." The notional load that was found to best represent "exclusion vehicles," i.e., trucks with loading configurations greater than allowed but routinely granted permits by agency bridge rating personnel, was adopted by AASHTO and named "Highway Load '93" or HL93. The mean and standard deviation of truck traffic was determined and used in the calibration of the load factors for HL93. It is notional in that it does not represent any specific vehicle [2].

The distribution of loads per the LRFD Specification is more complex than in the Standard Specifications for Highway Bridge Design [3]. This change is warranted because of the complexity in bridges today, increased knowledge of load paths, and technology available to be more rational in performing design calculations. The end result will be more appropriately designed structures.

6.3.1 Design Vehicular Live Load

The AASHTO "design vehicular live load," HL93, is a combination of a "design truck" or "design tandem" and a "design lane." The design truck is the former Highway Semitrailer 20-ton design truck (HS20-44) adopted by AASHO (now AASHTO) in 1944 and used in the previous Standard Specification. Similarly, the design lane is the HS20 lane loading from the AASHTO Standard Specifications. A shorter, but heavier, design tandem is new to AASHTO and is combined with the design lane if a worse condition is created than with the design truck. Superstructures with very short spans, especially those less than 12 m in length, are often controlled by the tandem combination.

The AASHTO design truck is shown in Figure 6.1. The variable axle spacing between the 145 kN loads is adjusted to create a critical condition for the design of each location in the structure. In the transverse direction, the design truck is 3 m wide and may be placed anywhere in the standard 3.6-m-wide lane. The wheel load, however, may not be positioned any closer than 0.6 m from the lane line, or 0.3 m from the face of curb, barrier, or railing.

The AASHTO design tandem consists of two 110-kN axles spaced at 1.2 m on center. The AASHTO design lane loading is equal to 9.3 N/mm and emulates a caravan of trucks. Similar to the truck loading, the lane load is spread over a 3-m-wide area in the standard 3.6-m lane. The lane loading is not interrupted except when creating an extreme force effect such as in "patch" loading of alternate spans. Only the axles contributing to the extreme being sought are loaded.

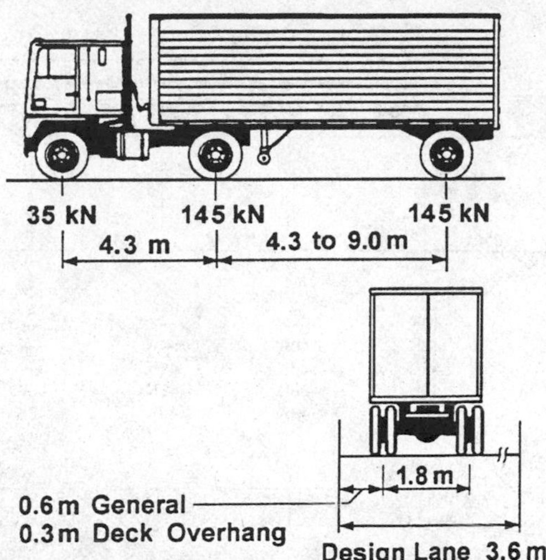

FIGURE 6.1 AASHTO-LRFD design truck. (AASHTO LRFD Bridge Design Specifications 2nd. ed., American Association of State Highway and Transportation Officials. Washington, D.C., 1998. With permission.)

When checking an extreme reaction at an interior pier or negative moment between points of contraflexure in the superstructure, two design trucks with a 4.3-m spacing between the 145-kN axles are to be placed on the bridge with a minimum of 15 m between the rear axle of the first truck and the lead axle of the second truck. Only 90% of the truck and lane load is used. This procedure differs from the Standard Specification which used shear and moment riders.

6.3.2 Permit Vehicles

Most U.S. states have developed their own "Permit Design Vehicle" to account for vehicles routinely granted permission to travel a given route, despite force effects greater than those due the design truck, i.e., the old HS20 loading. California uses anywhere from a 5- to 13-axle design vehicle as shown in Figure 6.2 [4]. Some states use an HS25 design truck, the configuration being identical to the HS20 but axle loads 25% greater.

The permit vehicular live load is combined with other loads in the Strength Limit State II as discussed in Chapter 5. Early editions of the AASHTO Specifications expect the design permit vehicle to be preceded and proceeded by a lane load. Furthermore, adjacent lanes may be loaded with the new HL93 load, unless restricted by escort vehicles.

6.3.3 Fatigue Loads

For fatigue loading, the LRFD Specification uses the design truck alone with a constant axle spacing of 9 m. The load is placed to produce extreme force effects. In lieu of more exact information, the frequency of the fatigue load for a single lane may be determined by multiplying the average daily truck traffic by p, where p is 1.00 in the case of one lane available to trucks, 0.85 in the case of two lanes available to trucks, and 0.80 in the case of three or more lanes available to trucks. If the average daily truck traffic is not known, 20% of the average daily traffic may be used on rural interstate bridges, 15% for other rural and urban interstate bridges, and 10% for bridges in urban areas.

6.3.4 Load Distribution for Superstructure Design

Figure 6.3 summarizes load distribution for design of longitudinal superstructure elements. Load distribution tables and the "lever rule" are approximate methods and intended for most designs.

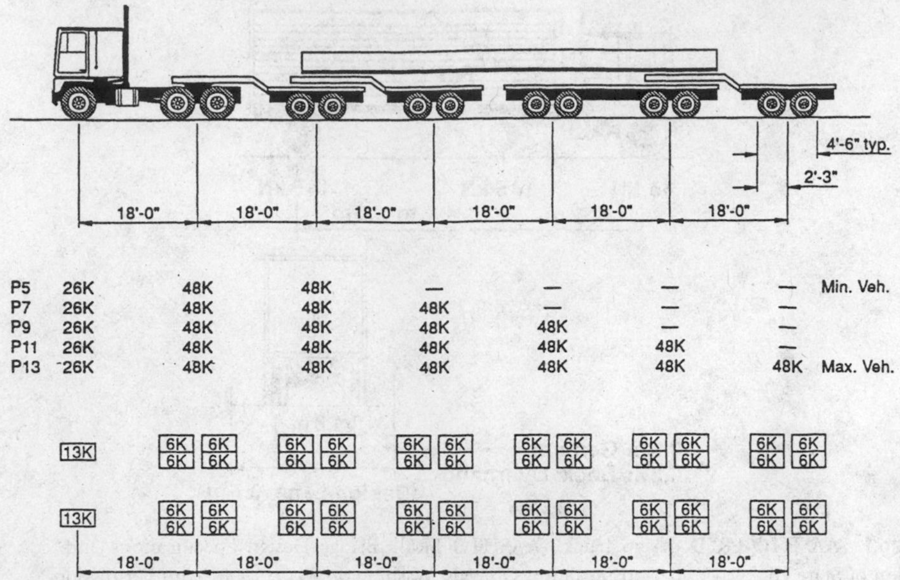

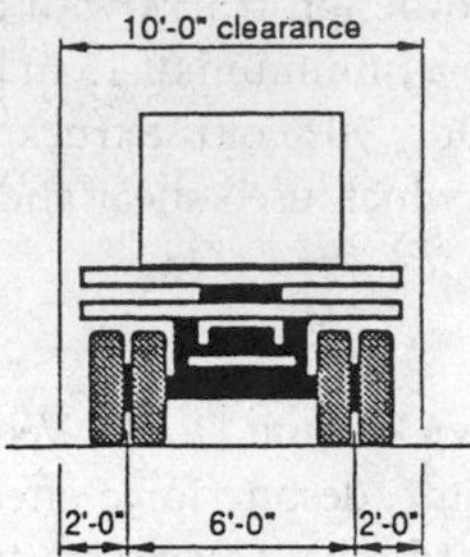

FIGURE 6.2 Caltrans permit truck. (AASHTO LRFD Bridge Design Specifications 2nd. ed., American Association of State Highway and Transportation Officials. Washington, D.C., 1998. With permission.)

The lever rule considers the slab between two girders to be simply supported. The reaction is determined by summing the reactions from the slabs on either side of the beam under consideration. "Refined analysis" refers to a three-dimensional consideration of the loads and is to be used on more complex structures. In other words, classical force and displacement, finite difference, finite element, folded plate, finite strip, grillage analogy, series/harmonic, or yield line methods are required to obtain load effects for superstructure design.

Note that, by definition of the vehicular design live load, no more than one truck can be in one lane simultaneously, except as previously described to generate maximum reactions or negative moments. After forces have been determined from the longitudinal load distribution and the longitudinal members have been designed, the designer may commence load distribution in the transverse direction for deck and substructure design.

6.3.4.1 Decks

Decks may be designed for vehicular live loads using empirical methods or by distributing loads on to "effective strip widths" and analyzing the strips as continuous or simply supported beams.

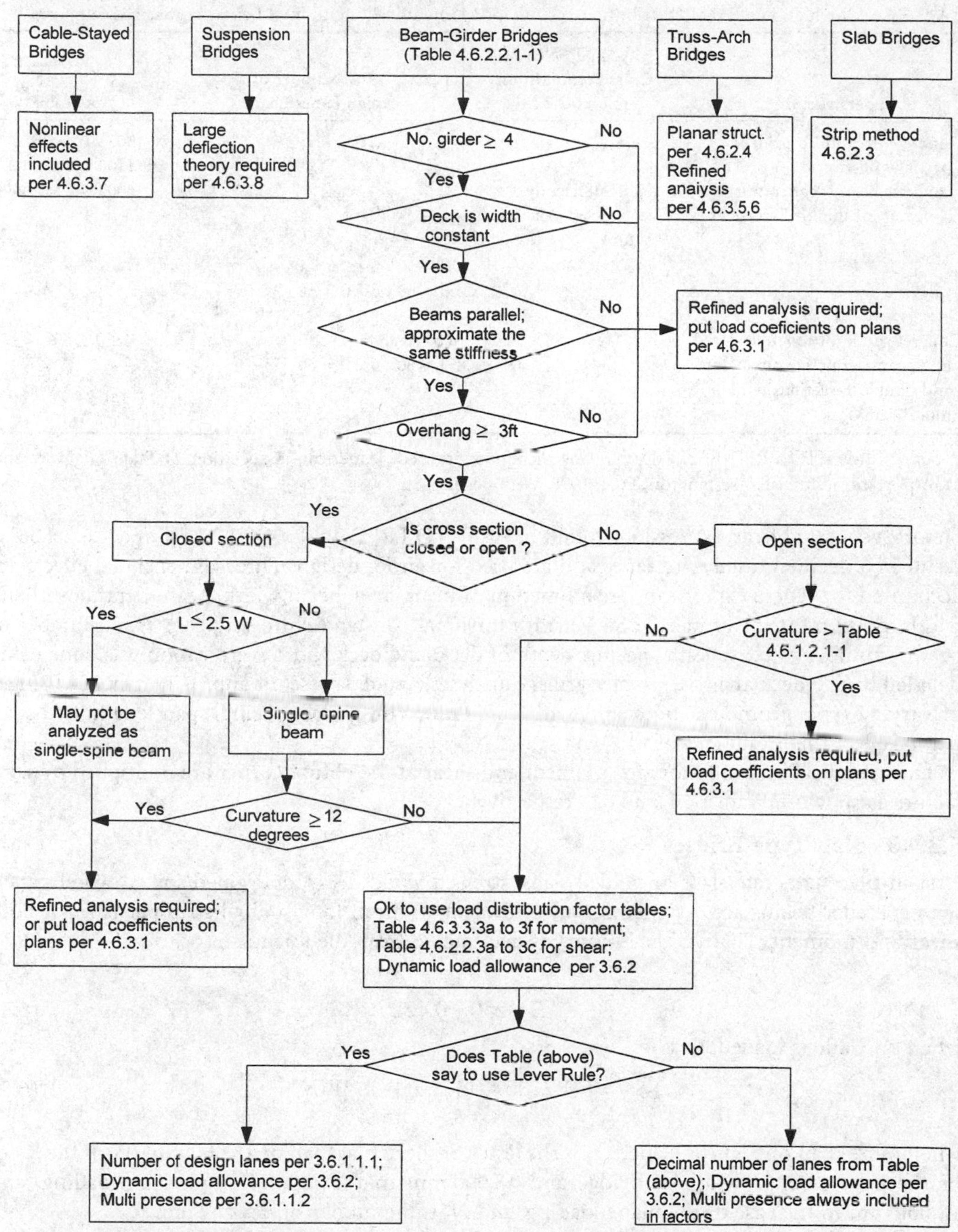

FIGURE 6.3 Live-load distribution for superstructure design.

Empirical methods rely on transfer of forces by arching of the concrete and shifting of the neutral axis. Loading is discussed in Chapter 24, Bridge Decks and Approach Slabs.

6.3.4.2 Beam–Slab Bridges

Approximate methods for load distribution on beam–slab bridges are appropriate for the types of cross sections shown in Table 4.6.2.2.1-1 of the AASHTO LRFD Specification. Load distribution

TABLE 6.1 Reduction of Load Distribution Factors for Moment in Longitudinal Beams on Skewed Supports

Type of Superstructure	Applicable Cross Section from Table 4.6.2.2.1-1	Any Number of Design Lanes Loaded	Range of Applicability
Concrete deck, filled grid, or partially filled grid on steel or concrete beams, concrete T-beams, or double T-sections	a, e, k i, j, if sufficiently connected to act as a unit	$1 - c_1(\tan\theta)^{1.5}$ $c_1 = 0.25\left(\frac{K_g}{Lt_s^3}\right)^{0.25}\left(\frac{S}{L}\right)^{0.25}$ if $\theta < 30°$, then $c_1 = 0$ if $\theta > 60°$, use $\theta = 60°$	$30° \le \theta \le 60°$ $1100 \le S \le 4900$ $6000 \le L \le 73{,}000$ $N_b \ge 4$
Concrete deck on concrete spread box beams, concrete box beams, and double T-sections used in multibeam decks	b, c, f, g	$1.05 - 0.25 \tan\theta \le 1.0$ if $\theta > 60°$, use $\theta = 60°$	$0 \le \theta \le 60°$

Source: AASHTO LRFD Bridge Design Specifications, 2nd. ed., American Association of State Highway and Transportation Officials. Washington, D.C., 1998. With permission.

factors, generated from expressions found in AASHTO LRFD Tables 4.6.2.2.2a–f and 4.6.2.2.3a–c, result in a decimal number of lanes and are used for girder design. Three-dimensional effects are accounted for. These expressions are a function of beam area, beam width, beam depth, overhang width, polar moment of inertia, St. Venant's torsional constant, stiffness, beam span, number of beams, number of cells, beam spacing, depth of deck, and deck width. Verification was done using detailed bridge deck analysis, simpler grillage analyses, and a data set of approximately 200 bridges of varying type, geometry, and span length. Limitations on girder spacing, span length, and span depth reflect the limitations of this data set.

The load distribution factors for moment and shear at the obtuse corner are multiplied by skew factors as shown in Tables 6.1 and 6.2, respectively.

6.3.4.3 Slab-Type Bridges

Cast-in-place concrete slabs or voided slabs, stressed wood decks, and glued/spiked wood panels with spreader beams are designed for an equivalent width of longitudinal strip per lane for both shear and moment. That width, E (mm), is determined from the formula:

$$E = 250 + 0.42\sqrt{L_1 W_1} \tag{6.1}$$

when one lane is loaded, and

$$E = 2100 + 0.12\sqrt{L_1 W_1} \le W/N_L \tag{6.2}$$

when more than one lane is loaded. L_1 is the lesser of the actual span or 18,000 mm, W_1 is the lesser of the edge-to-edge width of bridge and 18,000 mm in the case of single-lane loading, and 18,000 mm in the case of multilane loading, and N_L is the numbr of design lanes.

6.3.5 Load Distribution for Substructure Design

Bridge substructure includes bent caps, columns, pier walls, pile caps, spread footings, caissons, and piles. These components are designed by placing one or more design vehicular live loads on the traveled way as previously described for maximum reaction and negative bending moment, not exceeding the maximum number of vehicular lanes permitted on the bridge. This maximum may be determined by dividing the width of the traveled way by the standard lane width (3.6 m), and "rounding down," i.e., disregarding any fractional lanes. Note that (1) the traveled way need not be

TABLE 6.2 Correction Factors for Load Distribution Factors for Support Shear of the Obtuse Corner

Type of Superstructue	Applicable Cross Section from Table 4.6.2.2.1-1	Correction Factor	Range of Applicability
Concrete deck, filled grid, or partially filled grid on steel or concrete beams, concrete T-beams or double T-sections	a, e, k i, j, if sufficiently connected to act as a unit	— $1.0+2.0\left(\frac{Lt_s^3}{K_g}\right)^{0.3}\tan\theta$	$0° \le \theta \le 60°$ $1100 \le S \le 4900$ $6000 \le L \le 73{,}000$ $N_b \ge 4$
Multicell concrete box beams, box sections	d	$1.0+\left[0.25+\frac{L}{70d}\right]\tan\theta$	$0° \le \theta \le 60°$ $1800 \le S \le 4000$ $6000 \le L \le 73000$ $900 \le d \le 2700$ $N_b \ge 3$
Concrete deck on spread concrete box beams	b, c	$1.0+\frac{\sqrt{Ld}}{6S}\tan\theta$	$0° \le \theta \le 60°$ $1800 \le S \le 3500$ $6000 \le L \le 43{,}000$ $450 \le d \le 1700$ $N_b \ge 3$
Concrete box beams used in multibeam decks	f, g	$1.0+\frac{L\sqrt{\tan\theta}}{90d}$	$0° \le \theta \le 60°$ $6000 \le L \le 37{,}000$ $430 \le d \le 1500$ $900 \le b \le 1500$ $5 \le N_b \le 20$

Source: AASHTO LRFD Bridge Design Specifications, 2nd. ed., American Association of State Highway and Transportation Officials. Washington, D.C., 1998. With permission.

measured from the edge of deck if curbs or traffic barriers will restrict the traveled way for the life of the structure and (2) the fractional number of lanes determined using the previously mentioned load distribution charts for girder design is not used for substructure design.

Figure 6.4 shows selected load configurations for substructure elements. A critical load configuration may result from not using the maximum number of lanes permissible. For example, Figure 6.4a shows a load configuration that may generate the critical loads for bent cap design and Figure 6.4b shows a load configuration that may generate the critical bending moment for column design. Figure 6.4c shows a load configuration that may generate the critical compressive load for design of the piles. Other load configurations will be needed to complete design of a bridge footing. Note that girder locations are often ignored in determination of substructure design moments and shears: loads are assumed to be transferred directly to the structural support, disregarding load transfer through girders in the case of beam–slab bridges. Adjustments are made to account for the likelihood of fully loaded vehicles occurring side-by-side simultaneously. This "multiple presence factor" is discussed in the next section.

In the case of rigid frame structures, bending moments in the longitudinal direction will also be needed to complete column (or pier wall) as well as foundation designs. Load configurations which generate these three cases must be checked:

1. Maximum/minimum axial load with associated transverse and longitudinal moments;
2. Maximum/minimum transverse moment with associated axial load and longitudinal moment;
3. Maximum/minimum longitudinal moment with associated axial load and transverse moment.

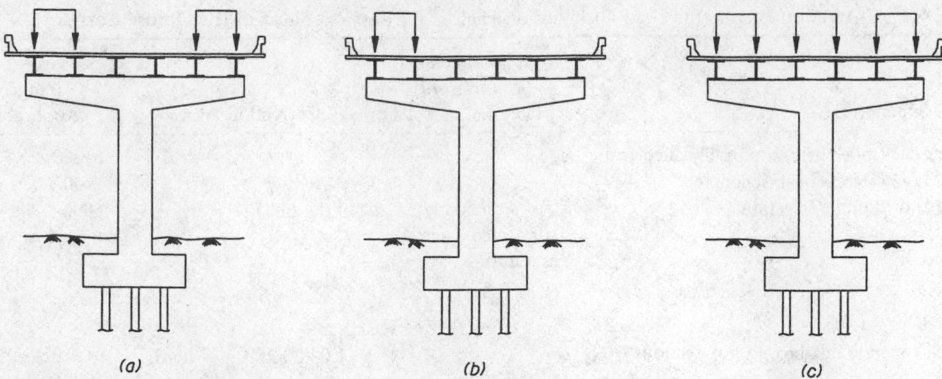

FIGURE 6.4 Various load configurations for substructure design.

TABLE 6.3 Multiple Presence Factors

Number of Loaded Lanes	Multiple Presence Factors *m*
1	1.20
2	1.00
3	0.85
>3	0.65

Source: AASHTO LRFD Bridge Design Specifications, 2nd. ed., American Association of State Highway and Transportation Officials. Washington, D.C., 1998. With permission.

If a permit vehicle is also being designed for, then these three cases must also be checked for the load combination associated with Strength Limit State II (discussed in Chapter 5).

6.3.6 Multiple Presence of Live-Load Lanes

Multiple presence factors modify the vehicular live loads for the probability that vehicular live loads occur together in a fully loaded state. The factors are shown in Table 6.3.

These factors should be applied prior to analysis or design only when using the lever rule or doing three-dimensional modeling or working with substructures. Sidewalks greater than 600 mm can be treated as a fully loaded lane. If a two-dimensional girder line analysis is being done and distribution factors are being used for a beam-and-slab type of bridge, multiple presence factors are not used because the load distribution factors already consider three-dimensional effects. For the fatigue limit state, the multiple presence factors are also not used.

6.3.7 Dynamic Load Allowance

Vehicular live loads are assigned a "dynamic load allowance" load factor of 1.75 at deck joints, 1.15 for all other components in the fatigue and fracture limit state, and 1.33 for all other components and limit states. This factor accounts for hammering when riding surface discontinuities exist, and long undulations when settlement or resonant excitation occurs. If a component such as a footing is completely below grade or a component such as a retaining wall is not subject to vertical reactions from the superstructure, this increase is not taken. Wood bridges or any wood component is factored at a lower level, i.e., 1.375 for deck joints, 1.075 for fatigue, and 1.165 typical, because of the energy-absorbing characteristic of wood. Likewise, buried structures such as culverts are subject to the dynamic load allowance but are a function of depth of cover, D_E (mm):

$$IM = 40(1.0 - 4.1 \times 10^{-4} D_E) \geq 0\% \tag{6.3}$$

6.3.8 Horizontal Loads Due to Vehicular Traffic

Substructure design of vertical elements requires that horizontal effects of vehicular live loads be designed for. Centrifugal forces and braking effects are applied horizontally at a distance 1.80 m above the roadway surface. The centrifugal force is determined by multiplying the design truck or design tandem — alone — by the following factor:

$$C = \frac{4v^2}{3gR} \tag{6.4}$$

Highway design speed, v, is in m/s; gravitational acceleration, g, is 9.807 m/s²; and radius of curvature in traffic lane, R, is in m. Likewise, the braking force is determined by multiplying the design truck or design tandem from all lanes likely to be unidirectional in the future, by 0.25. In this case, the lane load is not used because braking effects would be damped out on a fully loaded lane.

6.4 Pedestrian Loads

Live loads also include pedestrians and bicycles. The LRFD Specification calls for a 3.6×10^{-3} MPa load simultaneous with highway loads on sidewalks wider than 0.6 m. "Pedestrian- or bicycle-only" bridges are to be designed for 4.1×10^{-3} MPa. If the pedestrian- or bicycle-only bridge is required to carry maintenance or emergency vehicles, these vehicles are designed for, omitting the dynamic load allowance. Loads due to these vehicles are infrequent and factoring up for dynamic loads is inappropriate.

TABLE 6.4 Base Wind Pressures, P_B, corresponding to V_B = 160 km/h

Structural Component	Windward Load, MPa	Leeward Load, MPa
Trusses, columns, and arches	0.0024	0.0012
Beams	0.0024	NA
Large flat surfaces	0.0019	NA

Source: AASHTO LRFD Bridge Design Specifications, 2nd. ed., American Association of State Highway and Transportation Officials. Washington, D.C., 1998. With permission.

6.5 Wind Loads

The LRFD Specification provides wind loads as a function of base design wind velocity, V_B equal to 100 mph; and base pressures, P_B, corresponding to wind speed V_B. Values for P_B are listed in Table 6.4. The design wind pressure, P_D, is then calculated as

$$P_D = P_B \left(\frac{V_{DZ}}{V_B} \right)^2 = P_B \frac{V_{DZ}^2}{25{,}600} \tag{6.5}$$

where V_{DZ} is the design wind velocity at design elevation Z in km/h. V_{DZ} is a function of the friction velocity, V_0 (km/h), multiplied by the ratio of the actual wind velocity to the base wind velocity both at 10 m above grade, and the natural logarithm of the ratio of height to a meteorological constant length for given surface conditions:

TABLE 6.5 Values of V_o and Z_o for Various Upstream Surface Conditions

Condition	Open Country	Suburban	City
V_o (km/h)	13.2	15.2	19.4
Z_o (mm)	70	300	800

Source: AASHTO LRFD Bridge Design Specifications, 2nd. ed., American Association of State Highway and Transportation Officials. Washington, D.C., 1998. With permission.

TABLE 6.6 Temperature Ranges, °C

Climate	Steel or Aluminum	Concrete	Wood
Moderate	−18 to 50	−12 to 27	−12 to 24
Cold	−35 to 50	−18 to 27	−18 to 24

Source: AASHTO LRFD Bridge Design Specifications, 2nd. ed., American Association of State Highway and Transportation Officials. Washington, D.C., 1998. With permission.

$$V_{DZ} = 2.5V_0\left(\frac{V_{10}}{V_B}\right)\ln\left(\frac{Z}{Z_0}\right) \tag{6.6}$$

Values for V_0 and Z_0 are shown in Table 6.5.

The resultant design pressure is then applied to the surface area of the superstructure as seen in elevation. Solid-type traffic barriers and sound walls are considered as part of the loading surface. If the product of the resultant design pressure and applicable loading surface depth is less than a lineal load of 4.4 N/mm on the windward chord, or 2.2 N/mm on the leeward chord, minimum loads of 4.4 and 2.2 N/mm, respectively, are designed for.

Wind loads are combined with other loads in Strength Limit States III and V, and Service Limit State I, as defined in Chapter 5. Wind forces due to the additional surface area from trucks is accounted for by applying a 1.46 N/mm load 1800 mm above the bridge deck.

Wind loads for substructure design are of two types: loads applied to the substructure and those applied to the superstructure and transmitted to the substructure. Loads applied to the superstructure are as previously described. A base wind pressure of 1.9×10^{-3} MPa force is applied directly to the substructure, and is resolved into components (perpendicular to the front and end elevations) when the structure is skewed.

In absence of live loads, an upward load of 9.6×10^{-4} MPa is multiplied by the width of the superstructure and applied at the windward quarter point simultaneously with the horizontal wind loads applied perpendicular to the length of the bridge. This uplift load may create a worst condition for substructure design when seismic loads are not of concern.

6.6 Effects Due to Superimposed Deformations

Elements of a structure may change size or position due to settlement, shrinkage, creep, or temperature. Changes in geometry cause additional stresses which are of particular concern at connections. Determining effects from foundation settlement are a matter of structural analysis. Effects due to shrinkage and creep are material dependent and the reader is referred to design chapters elsewhere

TABLE 6.7 Basis of Temperature Gradients

	Concrete		50 mm Asphalt		100 mm Asphalt	
Zone	T_1 (°C)	T_2 (°C)	T_1 (°C)	T_2 (°C)	T_1 (°C)	T_2 (°C)
1	30	7.8	24	7.8	17	5
2	25	6.7	20	6.7	14	5.5
3	23	6	18	6	13	6
4	21	5	16	5	12	6

Source: AASHTO LRFD Bridge Design Specifications, 2nd. ed., American Association of State Highway and Transportation Officials. Washington, D.C., 1998. With permission.

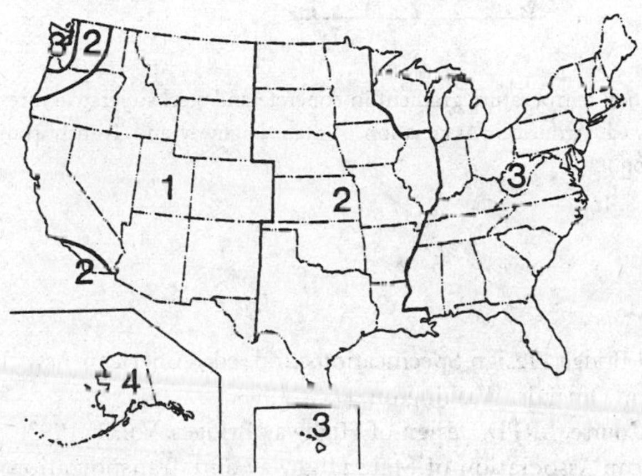

FIGURE 6.5 Solar radiation zones for the United States. (AASHTO LRFD Bridge Design Specifications 2nd. ed., American Association of State Highway and Transportation Officials. Washington, D.C., 1998. With permission.)

in this book. Temperature effects are dependent on the maximum potential temperature differential from the temperature at time of erection. Upper and lower bounds are shown in Table 6.6, where "moderate" and "cold" climates are defined as having fewer or more than 14 days with an average temperature below 0°C, respectively.

By using appropriate coefficients of thermal expansion, effects from temperature changes are calculated using basic structural analysis. More-refined analysis will consider the time lag between the surface and internal structure temperatures. The LRFD Specification identifies four zones in the United States and provides a linear relationship for the temperature gradient in steel and concrete. See Table 6.7 and Figures 6.5 and 6.6.

6.7 Exceptions to Code-Specified Design Loads

The designer is responsible not only for providing plans that accommodate design loads per the referenced Design Specifications, but also for any loads unique to the structure and bridge site. It is also the designer's responsibility to indicate all loading conditions designed for in the contract documents — preferably the construction plans. History seems to indicate that the next generation of bridge engineers will indeed be given the task of "improving" today's new structure. Therefore, the safety of future generations depends on today's designers doing a good job of documentation.

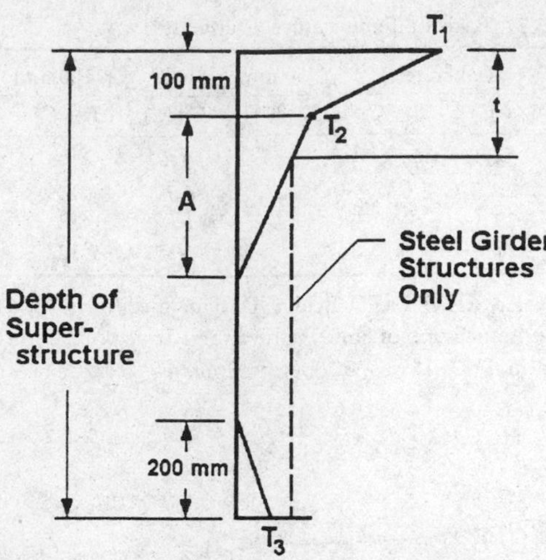

FIGURE 6.6 Positive vertical temperature gradient in concrete and steel superstructures. (AASHTO LRFD Bridge Design Specifications 2nd. ed., American Association of State Highway and Transportation Officials. Washington, D.C., 1998. With permission.)

References

1. AASHTO, LRFD Bridge Design Specifications 2nd. ed., American Association of State Highway and Transportation Officials. Washington, D.C., 1998.
2. FHWA Training Course, LRFD Design of Highway Bridges, Vol. 1, 1993.
3. AASHTO, American Association of State Highway and Transportation Officials, Washington, D.C. Officials, *Standard Specifications for Highway Bridges*, 16th ed., 1996, as amended by the Interim Specifications.
4. Caltrans, Bridge Design Specifications, California Department of Transportation, Sacramento, CA, 1999.

7

Structural Theory

Xila Liu
Tsinghua University, China

Leiming Zhang
Tsinghua University, China

7.1 Introduction

In this chapter, general forms of three sets of equations required in solving a solid mechanics problem and their extensions into structural theory are presented. In particular, a more generally used method, displacement method, is expressed in detail.

7.1.1 Basic Equations: Equilibrium, Compatibility, and Constitutive Law

In general, solving a solid mechanics problem must satisfy equations of equilibrium (static or dynamic), conditions of compatibility between strains and displacements, and stress–strain relations or material constitutive law (see Figure 7.1). The initial and boundary conditions on forces and displacements are naturally included.

From consideration of equilibrium equations, one can relate the stresses inside a body to external excitations, including body and surface forces. There are three equations of equilibrium relating the

0-8493-7434-0/00/$0.00+$.50

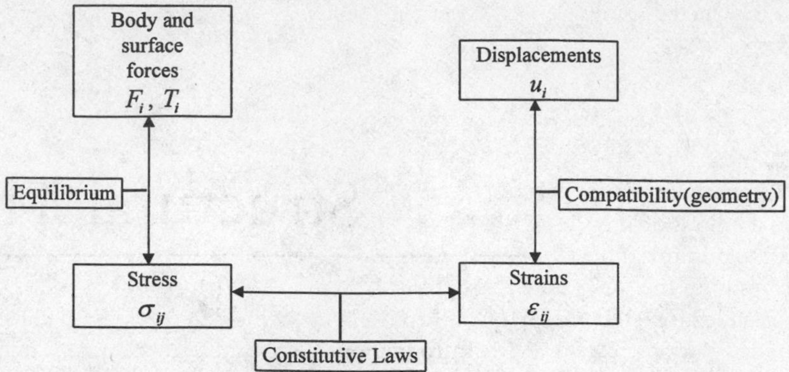

FIGURE 7.1 Relations of variables in solving a solid mechanics problem.

six components of stress tensor σ_{ij} for an infinitesimal material element which will be shown later in Section 7.2.1. In the case of dynamics, the equilibrium equations are replaced by equations of motion, which contain second-order derivatives of displacement with respect to time.

In the same way, taking into account geometric conditions, one can relate strains inside a body to its displacements, by six equations of kinematics expressing the six components of strain (ε_{ij}) in terms of the three components of displacement (u_i). These are known as the strain–displacement relations (see Section 7.3.1).

Both the equations of equilibrium and kinematics are valid regardless of the specific material of which the body is made. The influence of the material is expressed by constitutive laws in six equations. In the simplest case, not considering the effects of temperature, time, loading rates, and loading paths, these can be described by relations between stress and strain only.

Six stress components, six strain components, and three displacement components are connected by three equilibrium equations, six kinematics equations, and six constitutive equations. The 15 unknown quantities can be determined from the system of 15 equations.

It should be pointed out that the principle of superposition is valid only when small deformations and elastic materials are assumed.

7.1.2 Three Levels: Continuous Mechanics, Finite-Element Method, Beam–Column Theory

In solving a solid mechanics problem, the most direct method solves the three sets of equations described in the previous section. Generally, there are three ways to establish the basic unknowns, namely, the displacement components, the stress components, or a combination of both. The corresponding procedures are called the displacement method, the stress method, or the mixed method, respectively. But these direct methods are only practicable in some simple circumstances, such as those detailed in elastic theory of solid mechanics.

Many complex problems cannot be easily solved with conventional procedures. Complexities arise due to factors such as irregular geometry, nonhomogeneities, nonlinearity, and arbitrary loading conditions. An alternative now available is based on a concept of discretization. The finite-element method (FEM) divides a body into many "small" bodies called finite elements. Formulations by the FEM on the laws and principles governing the behavior of the body usually result in a set of simultaneous equations that can be solved by direct or iterative procedures. And loading effects such as deformations and stresses can be evaluated within certain accuracy. Up to now, FEM has been the most widely used structural analysis method.

In dealing with a continuous beam, the size of the three sets of equations is greatly reduced by assuming characteristics of beam members such as plane sections remain plane. For framed structures or structures constructed using beam–columns, structural mechanics gives them a more pithy and practical analysis.

7.1.3 Theoretical Structural Mechanics, Computational Structural Mechanics, and Qualitative Structural Mechanics

Structural mechanics deals with a system of members connected by joints which may be pinned or rigid. Classical methods of structural analysis are based on principles such as the principle of virtual displacement, the minimization of total potential energy, the minimization of total complementary energy, which result in the three sets of governing equations. Unfortunately, conventional methods are generally intended for hand calculations and developers of the FEM took great pains to minimize the amount of calculations required, even at the expense of making the methods somewhat unsystematic. This made the conventional methods unattractive for translation to computer codes.

The digital computer called for a more systematic method of structural analysis, leading to computational structural mechanics. By taking great care to formulate the tools of matrix notation in a mathematically consistent fashion, the analyst achieved a systematic approach convenient for automatic computation: matrix analysis of structures. One of the hallmarks of structural matrix analysis is its systematic nature, which renders digital computers even more important in structural engineering.

Of course, the analyst must maintain a critical, even skeptical, attitude toward computer results. In any event, computer results must satisfy our intuition of what is "reasonable." This qualitative judgment requires that the analyst possess a full understanding of structural behavior, both that being modeled by the program and that which can be expected in the actual structures. Engineers should decide what approximations are reasonable for the particular structure and verify that these approximations are indeed valid, and know how to design the structure so that its behavior is in reasonable agreement with the model adopted to analyze it. This is the main task of a structural analyst.

7.1.4 Matrix Analysis of Structures: Force Method and Displacement Method

Matrix analysis of structures was developed in the early 1950s. Although it was initially used on fuselage analysis, this method was proved to be pertinent to any complex structure. If internal forces are selected as basic unknowns, the analysis method is referred to as force method; in a similar way, the displacement method refers to the case where displacements are selected as primary unknowns. Both methods involve obtaining the joint equilibrium equations in terms of the basic internal forces or joint displacements as primary unknowns and solving the resulting set of equations for these unknowns. Having done this, one can obtain internal forces by backsubstitution, since even in the case of the displacement method the joint displacements determine the basic displacements of each member, which are directly related to internal forces and stresses in the member.

A major feature evident in structural matrix analysis is an emphasis on a systematic approach to the statement of the problem. This systematic characteristic together with matrix notation makes it especially convenient for computer coding. In fact, the displacement method, whose basic unknowns are uniquely defined, is generally more convenient than the force method. Most general-purpose structural analysis programs are displacement based. But there are still cases where it may be more desirable to use the force method.

7.2 Equilibrium Equations

7.2.1 Equilibrium Equation and Virtual Work Equation

For any volume V of a material body having A as surface area, as shown in Figure 7.2, it has the following conditions of equilibrium:

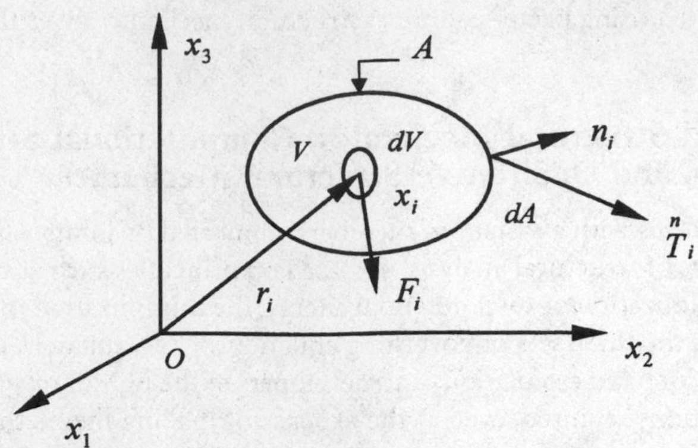

FIGURE 7.2 Derivation of equations of equilibrium.

At surface points

$$T_i = \sigma_{ji} n_j \tag{7.1a}$$

At internal points

$$\sigma_{ji,j} + F_i = 0 \tag{7.1b}$$

$$\sigma_{ji} = \sigma_{ij} \tag{7.1c}$$

where n_i represents the components of unit normal vector **n** of the surface; T_i is the stress vector at the point associated with **n**; $\sigma_{ji,j}$ represents the first derivative of σ_{ij} with respect to x_j; and F_i is the body force intensity. Any set of stresses σ_{ij}, body forces F_i, and external surface forces T_i that satisfies Eqs. (7.1a-c) is a statically admissible set.

Equations (7.1b and c) may be written in (*x,y,z*) notation as

$$\begin{aligned}
&\frac{\partial \sigma_x}{\partial x} + \frac{\partial \tau_{xy}}{\partial y} + \frac{\partial \tau_{xz}}{\partial z} + F_x = 0 \\
&\frac{\partial \tau_{yx}}{\partial x} + \frac{\partial \sigma_y}{\partial y} + \frac{\partial \tau_{yz}}{\partial z} + F_y = 0 \\
&\frac{\partial \tau_{zx}}{\partial x} + \frac{\partial \tau_{zy}}{\partial y} + \frac{\partial \sigma_z}{\partial z} + F_z = 0
\end{aligned} \tag{7.1d}$$

and

$$\tau_{xy} = \tau_{yx}, \quad \text{etc.} \tag{7.1e}$$

where σ_x, σ_y, and σ_z are the normal stress in (*x,y,z*) direction respectively; τ_{xy}, τ_{yx}, and so on, are the corresponding shear stresses in (*x,y,z*) notation; and F_x, F_y, and F_z are the body forces in (*x,y,z*) direction, respectively.

The principle of virtual work has proved a very powerful technique of solving problems and providing proofs for general theorems in solid mechanics. The equation of virtual work uses two

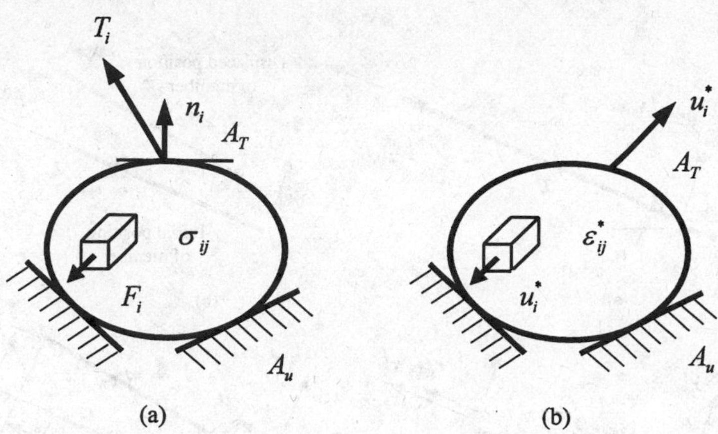

FIGURE 7.3 Two independent sets in the equation of virtual work.

independent sets of *equilibrium* and *compatible* (see Figure 7.3, where A_u and A_T represent displacement and stress boundary, respectively), as follows:

compatible set

$$\int_A T_i u_i^* dA + \int_V F_i u_i^* dV = \int_V \sigma_{ij} \varepsilon_{ij}^* dV \tag{7.2}$$

equilibrium set

or

$$\delta W_{\text{ext}} = \delta W_{\text{int}} \tag{7.3}$$

which states that the *external* virtual work (δW_{ext}) equals the *internal* virtual work (δW_{int}).

Here the integration is over the whole area A , or volume V, of the body. The stress field σ_{ij} , body forces F_i , and external surface forces T_i are a statically admissible set that satisfies Eqs. (7.1a–c). Similarly, the strain field ε_{ij}^* and the displacement u_i^* are a compatible kinematics set that satisfies displacement boundary conditions and Eq. (7.16) (see Section 7.3.1). This means the principle of virtual work applies only to small strain or small deformation.

The important point to keep in mind is that, neither the admissible equilibrium set σ_{ij} , F_i , and T_i (Figure 7.3a) nor the compatible set ε_{ij}^* and u_i^* (Figure 7.3b) need be the actual state, nor need the equilibrium and compatible sets be related to each other in any way. In the other words, these two sets are completely independent of each other.

7.2.2 Equilibrium Equation for Elements

For an infinitesimal material element, equilibrium equations have been summarized in Section 7.2.1, which will transfer into specific expressions in different methods. As in ordinary FEM or the displacement method, it will result in the following element equilibrium equations:

$$\{\overline{F}\}^e = [\overline{k}]^e \{\overline{d}\}^e \tag{7.4}$$

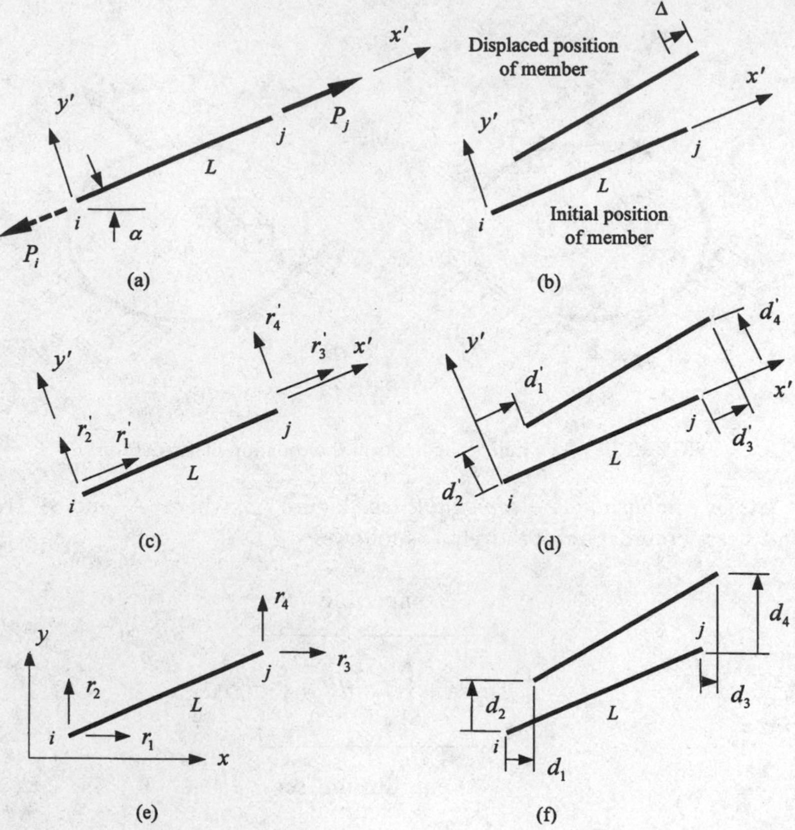

FIGURE 7.4 Plane truss member–end forces and displacements. (*Source*: Meyers, V.J., *Matrix Analysis of Structures*, New York: Harper & Row, 1983. With permission.)

where $\{\bar{F}\}^e$ and $\{\bar{d}\}^e$ are the element nodal force vector and displacement vector, respectively, while $[\bar{k}]^e$ is element stiffness matrix; the overbar here means in local coordinate system.

In the force method of structural analysis, which also adopts the idea of discretization, it is proved possible to identify a basic set of independent forces associated with each member, in that not only are these forces independent of one another, but also all other forces in that member are directly dependent on this set. Thus, this set of forces constitutes the minimum set that is capable of completely defining the stressed state of the member. The relationship between basic and local forces may be obtained by enforcing overall equilibrium on one member, which gives

$$\{\bar{F}\}^e = [L]\{P\}^e \tag{7.5}$$

where $[L]$ = the element force transformation matrix and $\{P\}^e$ = the element primary forces vector. It is important to emphasize that the physical basis of Eq. (7.5) is member overall equilibrium.

Take a conventional plane truss member for exemplification (see Figure 7.4), one has

$$\{\bar{k}\}^e = \begin{bmatrix} EA/l & 0 & -EA/l & 0 \\ 0 & 0 & 0 & 0 \\ -EA/l & 0 & EA/l & 0 \\ 0 & 0 & 0 & 0 \end{bmatrix} \tag{7.6}$$

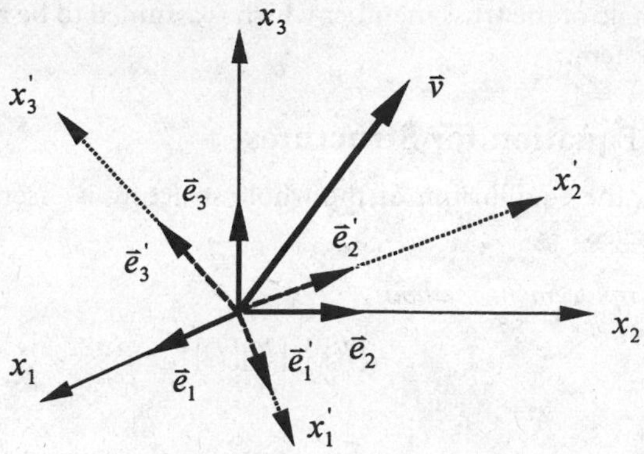

FIGURE 7.5 Coordinate transformation.

and

$$
\begin{aligned}
\{\bar{F}\}^e &= \{r_1' \quad r_2' \quad r_3' \quad r_4'\}^T \\
\{\bar{d}\}^e &= \{d_1' \quad d_2' \quad d_3' \quad d_4'\}^T \\
[L] &= \{-1 \quad 0 \quad 1 \quad 0\}^T \\
\{P\}^e &= \{P\}
\end{aligned}
\tag{7.7}
$$

where EA/l = axial stiffness of the truss member and P = axial force of the truss member.

7.2.3 Coordinate Transformation

The values of the components of vector **V**, designated by v_1, v_2, and v_3 or simply v_i, are associated with the chosen set coordinate axes. Often it is necessary to reorient the reference axes and evaluate new values for the components of **V** in the new coordinate system. Assuming that **V** has components v_i and v_i' in two sets of right-handed Cartesian coordinate systems x_i (old) and x_i' (new) having the same origin (see Figure 7.5), and $\bar{e}_i$, $\bar{e}_i'$ are the unit vectors of x_i and x_i', respectively. Then

$$v_i' = l_{ij} v_j \tag{7.8}$$

where $l_{ji} = \bar{e}_j' \cdot \bar{e}_i = \cos(x_j', x_i)$, that is, the cosines of the angles between x_i' and x_j axes for i and j ranging from 1 to 3; and $[\alpha] = (l_{ij})_{3\times3}$ is called coordinate transformation matrix from the old system to the new system.

It should be noted that the elements of l_{ij} or matrix $[\alpha]$ are not symmetrical, $l_{ij} \neq l_{ji}$. For example, l_{12} is the cosine of angle from x_1' to x_2 and l_{21} is that from x_2' to x_1 (see Figure 7.5). The angle is assumed to be measured from the primed system to the unprimed system.

For a plane truss member (see Figure 7.4), the transformation matrix from local coordinate system to global coordinate system may be expressed as

$$[\alpha] = \begin{bmatrix} \cos\alpha & -\sin\alpha & 0 & 0 \\ \sin\alpha & \cos\alpha & 0 & 0 \\ 0 & 0 & \cos\alpha & -\sin\alpha \\ 0 & 0 & \sin\alpha & \cos\alpha \end{bmatrix} \tag{7.9}$$

where α is the inclined angle of the truss member which is assumed to be measured from the global to the local coordinate system.

7.2.4 Equilibrium Equation for Structures

For discretized structure, the equilibrium of the whole structure is essentially the equilibrium of each joint. After assemblage,

For ordinary FEM or displacement method

$$\{F\} = [K]\{D\} \tag{7.10}$$

For force method

$$\{F\} = [A]\{P\} \tag{7.11}$$

where $\{F\}$ = nodal loading vector; $[K]$ = total stiffness matrix; $\{D\}$ = nodal displacement vector; $[A]$ = total forces transformation matrix; $\{P\}$ = total primary internal forces vector.

It should be noted that the coordinate transformation for each element from local coordinates to the global coordinate system must be done before assembly.

In the force method, Eq. (7.11) will be adopted to solve for internal forces of a statically determinate structure. The number of basic unknown forces is equal to the number of equilibrium equations available to solve for them and the equations are linearly independent. For statically unstable structures, analysis must consider their dynamic behavior. When the number of basic unknown forces exceeds the number of equilibrium equations, the structure is said to be statically indeterminate. In this case, some of the basic unknown forces are not required to maintain structural equilibrium. These are "extra" or "redundant" forces. To obtain a solution for the full set of basic unknown forces, it is necessary to augment the set of independent equilibrium equations with elastic behavior of the structure, namely, the force–displacement relations of the structure. Having solved for the full set of basic forces, we can determine the displacements by backsubstitution.

7.2.5 Influence Lines and Surfaces

In the design and analysis of bridge structures , it is necessary to study the effects intrigued by loads placed in various positions. This can be done conveniently by means of diagrams showing the effect of moving a unit load across the structures. Such diagrams are commonly called influence lines (for framed structures) or influence surfaces (for plates). Observe that whereas a moment or shear diagram shows the variation in moment or shear along the structure due to some particular position of load, an influence line or surface for moment or shear shows the variation of moment or shear at a *particular* section due to a unit load placed anywhere along the structure.

Exact influence lines for statically determinate structures can be obtained analytically by statics alone. From Eq. (7.11), the total primary internal forces vector $\{P\}$ can be expressed as

$$\{P\} = [A]^{-1}\{F\} \tag{7.12}$$

by which given a unit load at one node, the excited internal forces of all members will be obtained, and thus Eq. (7.12) gives the analytical expression of influence lines of all member internal forces for discretized structures subjected to moving nodal loads.

For statically indeterminate structures, influence values can be determined directly from a consideration of the geometry of the deflected load line resulting from imposing a unit deformation corresponding to the function under study, based on the principle of virtual work. This may better be demonstrated by a two-span continuous beam shown in Figure 7.6, where the influence line of internal bending moment M_B at section B is required.

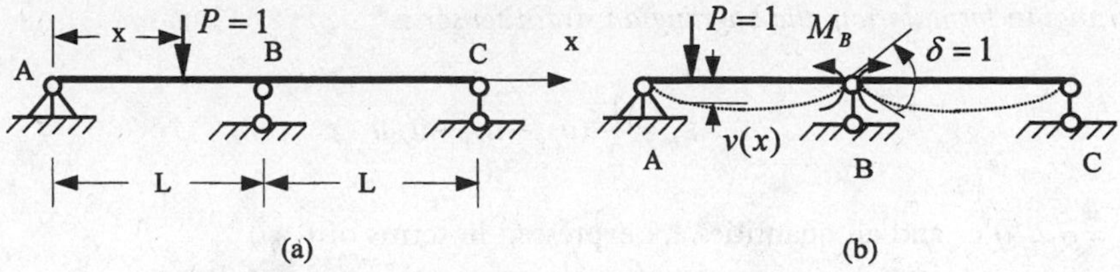

FIGURE 7.6 Influence line of a two-span continuous beam.

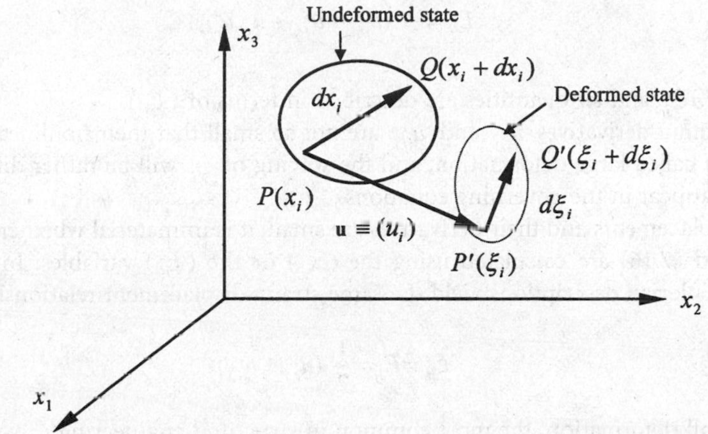

FIGURE 7.7 Deformation of a line element for Lagrangian and Eluerian variables.

Cutting section *B* to expose M_B and give it a unit relative rotation $\delta = 1$ (see Figure 7.6) and employing the principle of virtual work gives

$$M_B \cdot \delta = -P \cdot v(x) \tag{7.13}$$

Therefore,

$$M_B = -v(x) \tag{7.14}$$

which means the influence value of M_B equals to the deflection $v(x)$ of the beam subjected to a unit rotation at joint *B* (represented by dashed line in Figure 7.6b). Solving for $v(x)$ can be carried out easily referring to material mechanics.

7.3 Compatibility Equations

7.3.1 Large Deformation and Large Strain

Strain analysis is concerned with the study of deformation of a continuous body which is unrelated to properties of the body material. In general, there are two methods of describing the deformation of a continuous body, Lagrangian and Eulerian. The Lagrangian method employs the coordinates of each particle in the initial position as the independent variables. The Eulerian method defines the independent variables as the coordinates of each material particle at the time of interest.

Let the coordinates of material particle *P* in a body in the initial position be denoted by x_i (x_1, x_2, x_3) referred to the fixed axes x_i, as shown in Figure 7.7. And the coordinates of the particle after deformation are denoted by ξ_i (ξ_1, ξ_2, ξ_3) with respect to axes x_i. As for the independent variables, Lagrangian formulation uses the coordinates (x_i) while Eulerian formulation employs the coordinates (ξ_i). From motion analysis of line element *PQ* (see Figure 7.7), one has

For Lagrangian formulation, the Lagrangian strain tensor is

$$\varepsilon_{ij} = \frac{1}{2}\,(u_{i,j} + u_{j,i} + u_{r,i}u_{r,j}) \tag{7.15}$$

where $u_{i,j} = \partial u_i / \partial x_j$ and all quantities are expressed in terms of (x_i).

For Eulerian formulation, the Eulerian strain tensor is

$$E_{ij} = \frac{1}{2}\,(u_{i/j} + u_{j/i} + u_{r/i}u_{r/j}) \tag{7.16}$$

where $u_{i/j} = \partial u_i / \partial \xi_j$ and all quantities are described in terms of (ξ_i).

If the displacement derivatives $u_{i,j}$ and $u_{i/j}$ are not so small that their nonlinear terms cannot be neglected, it is called large deformation, and the solving of u_i will be rather difficult since the nonlinear terms appear in the governing equations.

If both the displacements and their derivatives are small, it is immaterial whether the derivatives in Eqs. (7.15) and (7.16) are calculated using the (x_i) or the (ξ_i) variables. In this case both Lagrangian and Eulerian descriptions yield the same strain–displacement relationship:

$$\varepsilon_{ij} = E_{ij} = \frac{1}{2}\,(u_{i,j} + u_{j,i}) \tag{7.17}$$

which means small deformation, the most common in structural engineering.

For given displacements (u_i) in strain analysis, the strain components (ε_{ij}) can be determined from Eq. (7.17). For prescribed strain components (ε_{ij}), some restrictions must be imposed on it in order to have single-valued continuous displacement functions u_i, since there are six equations for three unknown functions. Such restrictions are called compatibility conditions, which for a simply connected region may be written as

$$\varepsilon_{ij,kl} + \varepsilon_{kl,ij} - \varepsilon_{ik,jl} - \varepsilon_{ji,ik} = 0 \tag{7.18a}$$

or, expanding these expressions in the (*x, y, z*) notations, it gives

$$\begin{gathered}
\frac{\partial^2 \varepsilon_x}{\partial y^2} + \frac{\partial^2 \varepsilon_y}{\partial x^2} = 2\frac{\partial^2 \varepsilon_{xy}}{\partial x \partial y} \\
\frac{\partial^2 \varepsilon_y}{\partial z^2} + \frac{\partial^2 \varepsilon_z}{\partial y^2} = 2\frac{\partial^2 \varepsilon_{yz}}{\partial y \partial z} \\
\frac{\partial^2 \varepsilon_z}{\partial x^2} + \frac{\partial^2 \varepsilon_x}{\partial z^2} = 2\frac{\partial^2 \varepsilon_{zx}}{\partial z \partial x} \\
\frac{\partial}{\partial x}\left(-\frac{\partial \varepsilon_{yz}}{\partial x} + \frac{\partial \varepsilon_{zx}}{\partial y} + \frac{\partial \varepsilon_{xy}}{\partial z}\right) = \frac{\partial^2 \varepsilon_x}{\partial y \partial z} \\
\frac{\partial}{\partial y}\left(-\frac{\partial \varepsilon_{zx}}{\partial y} + \frac{\partial \varepsilon_{xy}}{\partial z} + \frac{\partial \varepsilon_{yz}}{\partial x}\right) = \frac{\partial^2 \varepsilon_y}{\partial z \partial x} \\
\frac{\partial}{\partial z}\left(-\frac{\partial \varepsilon_{xy}}{\partial z} + \frac{\partial \varepsilon_{yz}}{\partial x} + \frac{\partial \varepsilon_{zx}}{\partial y}\right) = \frac{\partial^2 \varepsilon_z}{\partial x \partial y}
\end{gathered} \tag{7.18b}$$

Any set of strains ε_{ij} and displacements u_i, that satisfies Eqs. (7.17) and (7.18a) or (7.18b), as well as displacement boundary conditions, is a kinematics admissible set, or a compatible set.

7.3.2 Compatibility Equation for Elements

For ordinary FEM, compatibility requirements are self-satisfied in the formulating procedure. As for equilibrium equations, a basic set of independent displacements can be identified for each member, and the kinematics relationships between member basic displacements and member–end displacements of one member can be given as follows:

$$\{\Delta\}^e = [L]^T\{\bar{d}\}^e \tag{7.19}$$

where $\{\Delta\}^e$ is element primary displacement vector, $[L]$ and $\{\bar{d}\}^e$ have been shown in Section 7.2.2. For plane truss member, $\{\Delta\}^e = \{\Delta\}$, where Δ is the relative displacement of the member (see Figure 7.5). It should also be noted that the physical basis of Eq. (7.19) is the overall compatibility of the element.

7.3.3 Compatibility Equation for Structures

For the whole structure, one has the following equation after assembly process:

$$\{\Delta\} = [A]^T\{D\} \tag{7.20}$$

where $\{\Delta\}$ – total primary displacement vector; $\{D\}$ = total nodal displacement vector; and $[A]^T$ = the transposition of $[A]$ described in Section 7.2.4.

A statically determinate structure is kinematically determinate. Given a set of basic member displacements, there are a sufficient number of compatibility relationships available to allow the structure nodal displacements to be determined. In addition to their application to settlement and fabrication error loading, thermal loads can also be considered for statically determinate structures. External forces on a structure cause member distortions and, hence, nodal displacements, but before such problems can be solved, the relationships between member forces and member distortions must be developed. These will be shown in Section 7.5.1.

7.3.4 Contragredient Law

During the development of the equilibrium and compatibility relationships, it has been noticed that various corresponding force and displacement transformations are the transposition of each other, as shown not only in Eqs. (7.5) and (7.19) of element equilibrium and compatibility relations, but also in Eqs. (7.11) and (7.20) of global equilibrium and compatibility relations, although each pair of these transformations was obtained independently of the other in the development. These special sets of relations are termed the contragredient law which was established on the basis of virtual work concepts. Therefore, after a particular force transformation matrix is obtained, the corresponding displacement transformation matrix would be immediately apparent, and it remains valid to the contrary.

7.4 Constitutive Equations

7.4.1 Elasticity and Plasticity

A material body will produce deformation when subjected to external excitations. If upon the release of applied actions the body recovers its original shape and size, it is called an *elastic* material, or

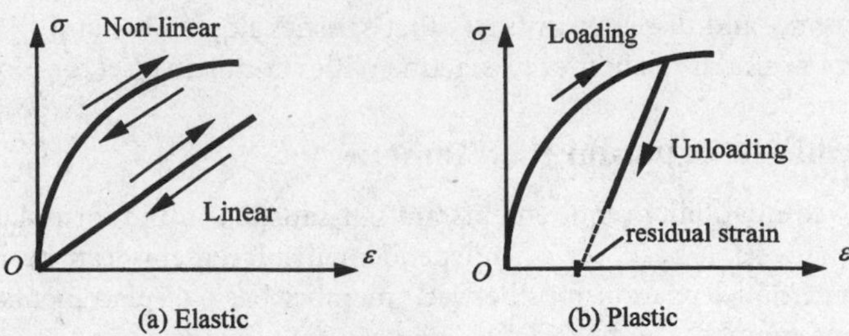

FIGURE 7.8 Sketches of behavior of elastic and plastic materials.

one can say the material has the characteristic of *elasticity*. Otherwise, it is a *plastic* material or a material with *plasticity*. For an elastic body, the current state of stress depends only on the current state of deformation; that is, the constitutive equations for elastic material are given by

$$\sigma_{ij} = F_{ij}(\varepsilon_{kl}) \tag{7.21}$$

where F_{ij} is called the elastic response function. Thus, the elastic material behavior described by Eq. (7.21) is reversible and path independent (see Figure 7.8a), in which case the material is usually termed *Cauchy elastic* material.

Reversibility and path independence are not exhibited by plastic materials (see Figure 7.8b). In general, a plastic material does not return to its original shape; *residual* deformation and stresses remain inside the body even when all external tractions are removed. As a result, it is necessary for plasticity to extend the elastic stress–strain relations into the plastic range where permanent plastic stain is possible. It makes the solution of a solid mechanics problem more complicated.

7.4.2 Linear Elastic and Nonlinear Elastic Behavior

Just as the term *linear* implies, linear elasticity means the elastic response function F_{ij} of Eq. (7.21) is a linear function, whose most general form for a Cauchy elastic material is given by

$$\sigma_{ij} = B_{ij} + C_{ijkl}\varepsilon_{kl} \tag{7.22}$$

where B_{ij} = components of initial stress tensor corresponding to the *initial strain-free* state (i.e., $\varepsilon_{ij} = 0$), and C_{ijkl} = tensor of material elastic constants.

If it is assumed that $B_{ij} = 0$, Eq. (7.22) will be reduced to

$$\sigma_{ij} = C_{ijkl}\varepsilon_{kl} \tag{7.23}$$

which is often referred to as the generalized Hook's law.

For an *isotropic* linear elastic material, the elastic constants in Eq. (7.23) must be the same for all directions and thus C_{ijkl} must be an isotropic fourth-order tensor, which means that there are only two independent material constants. In this case, Eq. (7.23) will reduce to

$$\sigma_{ij} = \lambda\varepsilon_{kk}\delta_{ij} + 2\mu\varepsilon_{ij} \tag{7.24}$$

where λ and μ are the two material constants, usually called *Lame's constants*; δ_{ij} = *Kronecker delta* and ε_{kk} = the summation of the diagonal terms of ε_{ij} according to the *summation convention*, which means that, whenever a subscript occurs twice in the same term, it is understood that the subscript is to be summed from 1 to 3.

If the elastic response function F_{ij} in Eq. (7.21) is not linear, it is called nonlinear elastic, and the material exhibits nonlinear mechanical behavior even when sustaining small deformation. That is, the material elastic "constants" do not remain constant any more, whereas the deformation can still be reversed completely.

7.4.3 Geometric Nonlinearity

Based on the sources from which it arises, nonlinearity can be categorized into material nonlinearity (including nonlinear elasticity and plasticity) and geometric nonlinearity. When the nonlinear terms in the strain–displacement relations cannot be neglected (see Section 7.3.1) or the deflections are large enough to cause significant changes in the structural geometry, it is termed geometric nonlinearity. It is also called large deformation, and the principle of superposition derived from small deformations is no longer valid. It should be noted that for accumulated large displacements with small deformations, it could be linearized by a step-by-step procedure.

According to the different choice of reference frame, there are two types of Lagrangian formulation: the total Lagrangian formulation, which takes the original unstrained configuration as the reference frame, and the updated Lagrangian formulation based on the latest-obtained configuration, which are usually carried out step by step. Whatever formulation one chooses, a geometric stiffness matrix or initial stress matrix will be introduced into the equations of equilibrium to take account of the effects of the initial stresses on the stiffness of the structure. These depend on the magnitude or conditions of loading and deformations, and thus cause the geometric nonlinearity. In beam–column theory, this is well known as the second-order or the P–Δ effect. For detailed discussions, see Chapter 36.

7.5 Displacement Method

7.5.1 Stiffness matrix for elements

In displacement method, displacement components are taken as primary unknowns. From Eqs. (7.5) and (7.19) the equilibrium and compatibility requirements on elements have been acquired. For a statically determinate structure, no subsidiary conditions are needed to obtain internal forces under nodal loading or the displaced position of the structure given the basic distortion such as support settlement or fabrication errors. For a statically indeterminate structure, however, supplementary conditions, namely, the constitutive law of materials constructing the structure, should be incorporated for the solution of internal forces as well as nodal displacements.

From structural mechanics, the basic stiffness relationships for a member between basic internal forces and basic member–end displacements can be expressed as

$$\{P\}^e = [k]^e\{\Delta\}^e \tag{7.25}$$

where $[k]^e$ is the element basic stiffness matrix, which can be termed $[EA/l]$ for a conventional plane truss member (see Figure 7.4).

Substitution of Eqs. (7.19) and (7.25) into Eq. (7.5) yields

$$\begin{aligned}\{\overline{F}\}^e &= [L][k]^e[L]^T\{\overline{d}\}^e \\ &= [\overline{k}]^e\{\overline{d}\}^e\end{aligned} \tag{7.26}$$

where

$$[\overline{k}]^e = [L][k]^e[L]^T \tag{7.27}$$

is called the element stiffness matrix, the same as in Eq. (7.4). It should be kept in mind that the element stiffness matrix $\left[\bar{k}\right]^e$ is symmetric and singular, since given the member–end forces, member–end displacements cannot be determined uniquely because the member may undergo rigid body movement.

7.5.2 Stiffness Matrix for Structures

Our final aim is to obtain equations that define approximately the behavior of the whole body or structure. Once the element stiffness relations of Eq. (7.26) is established for a generic element, the global equations can be constructed by an assembling process based on the law of compatibility and equilibrium, which are generally expressed in matrix notation as

$$\{F\} = [K]\{D\} \tag{7.28}$$

where $[K]$ is the stiffness matrix for the whole structure. It should be noted that the basic idea of assembly involves a minimization of *total* potential energy, and the assembled stiffness matrix $[K]$ is *symmetric* and *banded* or *sparsely populated.*

Eq. (7.28) tells us the capabilities of a structure to withstand applied loading rather than the true behavior of the structure if boundary conditions are not introduced. In other words, without boundary conditions, there can be an infinite number of possible solutions since stiffness matrix $[K]$ is singular; that is, its determinant vanishes. Hence, Eqs. (7.28) should be modified to reflect boundary conditions and the final modified equations are expressed by inserting overbars as

$$\{\bar{F}\} = [\bar{K}]\{\bar{D}\} \tag{7.29}$$

7.5.3 Matrix Inversion

It has been shown that sets of simultaneous algebraic equations are generated in the application of both the displacement method and the force method in structural analysis, which are usually linear. The coefficients of the equations are constant and do not depend on the magnitude or conditions of loading and deformations, since linear Hook's law is generally assumed valid and small strains and deformations are used in the formulation. Solving Eq. (7.29) is, namely, to invert the modified stiffness matrix $\left[\bar{K}\right]$. This requires tremendous computational efforts for large-scale problems. The equations can be solved by using direct, iterative, or other methods. Two steps of elimination and backsubstitution are involved in the direct procedures, among which are Gaussian elimination and a number of its modifications. These are some of the most widely used sets of direct methods because of their better accuracy and small number of arithmetic operations.

7.5.4 Special Consideration

In practice, a variety of special circumstances, ranging from loading to internal member conditions and supporting conditions, should be given due consideration in structural analysis.

Initially strains, which are not directly associated with stresses, result from two causes, thermal loading or fabrication error. If the member with initial strains is unconstrained, there will be a set of initial member–end displacements associated with these initial strains, but nevertheless no initial member–end forces. For a member constrained to act as part of a structure, the general member force–displacement relationships will be modified as follows:

$$\{\bar{F}\}^e = [\bar{k}]^e\left(\{\bar{d}\}^e - \{\bar{d}_0\}^e\right) \tag{7.30a}$$

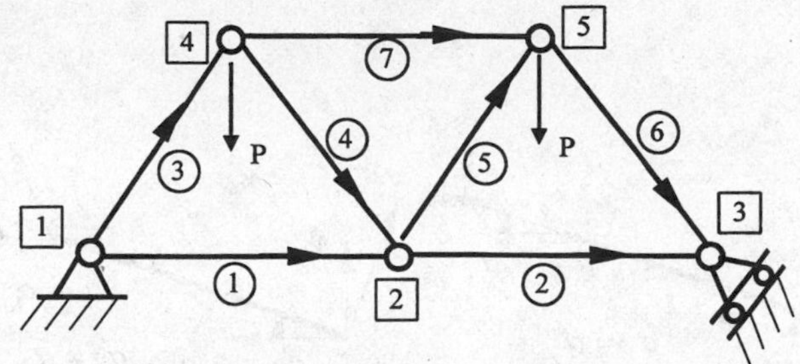

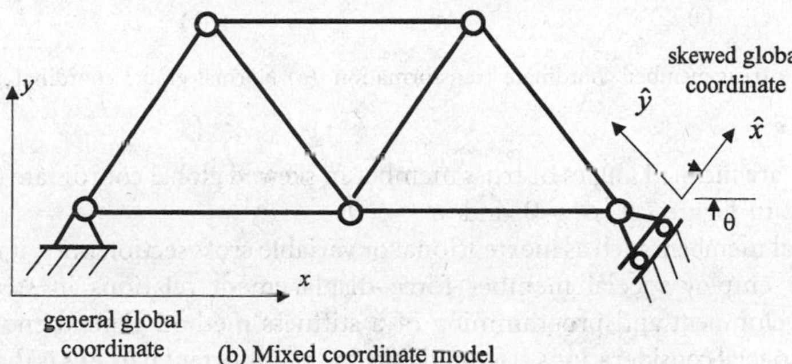

FIGURE 7.9 Plane truss with skewed support.

or

$$\{\overline{F}\}^e = [\overline{k}]^e \{\overline{d}\}^e + \{R_{F0}\}^e \tag{7.30b}$$

where

$$\{R_{F0}\}^e = -[\overline{k}]^e \{\overline{d_0}\}^e \tag{7.31}$$

are fixed-end forces, and $\{\overline{d_0}\}^e$ a vector of initial member–end displacements for the member.

It is interesting to note that a support settlement may be regarded as an initial strain. Moreover, initial strains including thermal loading and fabrication errors, as well as support settlements, can all be treated as external excitations. Hence, the corresponding fixed-end forces as well as the equivalent nodal loading can be obtained which makes the conventional procedure described previously still practicable.

For a skewed support which provides a constraint to the structure in a nonglobal direction, the effect can be given due consideration by adapting a skewed global coordinate (see Figure 7.9) by introducing a skewed coordinate at the skewed support. This can perhaps be better demonstrated by considering a specific example of a plane truss shown in Figure 7.9. For members jointed at a skewed support, the coordinate transformation matrix will takes the form of

$$[\alpha] = \begin{bmatrix} \cos\alpha_i & -\sin\alpha_i & 0 & 0 \\ \sin\alpha_i & \cos\alpha_i & 0 & 0 \\ 0 & 0 & \cos\alpha_j & -\sin\alpha_j \\ 0 & 0 & \sin\alpha_j & \cos\alpha_j \end{bmatrix} \tag{7.32}$$

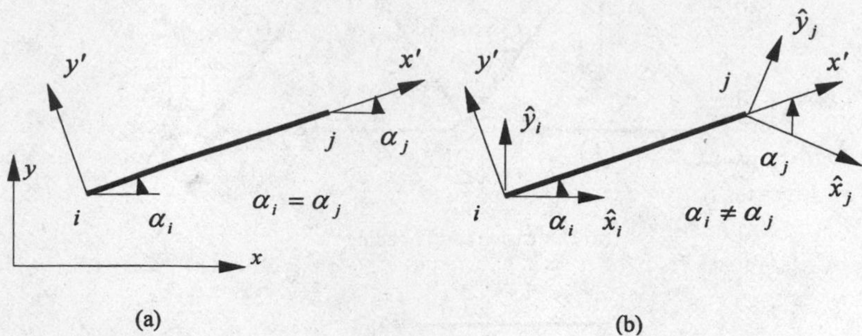

FIGURE 7.10 Plane truss member coordinate transformation. (a) Normal global coordinate; (b) skewed global coordinate.

where α_i and α_j are inclined angles of truss member in skewed global coordinate (see Figure 7.10), say, for member 2 in Figure 7.9, $\alpha_i = 0$ and $\alpha_j = -\theta$.

For other special members such as inextensional or variable cross section ones, it may be necessary or convenient to employ special member force–displacement relations in structural analysis. Although the development and programming of a stiffness method general enough to take into account all these special considerations is formidable, more important perhaps is that the application of the method remains little changed. For more details, readers are referred to Reference. [5].

7.6 Substructuring and Symmetry Consideration

For highly complex or large-scale structures, one is required to solve a very large set of simultaneous equations, which are sometimes restricted by the computation resources available. In that case, special data-handling schemes like static condensation are needed to reduce the number of unknowns by appropriately numbering nodal displacement components and disposition of element force–displacement relations. Static condensation is useful in dynamic analysis of framed structures since the rotatory moment of inertia is usually neglected.

Another scheme physically partitions the structure into a collection of smaller structures called "substructures," which can be processed by parallel computers. In static analysis, the first step of substructuring is to introduce imaginary fixed inner boundaries, and then release all inner boundaries simultaneously, which gives rise to a subsequent analysis of these substructure series in a smaller scale. It is essentially the patitioning of Eq. (7.28) as follows. For the rth substructure, one has

Case (α) : *Introducing inner fixed boundaries*

$$\begin{bmatrix} K_{bb} & K_{bi} \\ K_{ib} & K_{ii} \end{bmatrix}^{(r)} \begin{Bmatrix} 0 \\ D_i^{\alpha} \end{Bmatrix}^{(r)} = \begin{Bmatrix} F_b^{\alpha} \\ F_i \end{Bmatrix}^{(r)} \tag{7.33}$$

Case (β) : *Releasing all inner fixed boundaries*

$$\begin{bmatrix} K_{bb} & K_{bi} \\ K_{ib} & K_{ii} \end{bmatrix}^{(r)} \begin{Bmatrix} D_b \\ D_i^{\beta} \end{Bmatrix}^{(r)} = \begin{Bmatrix} F_b^{\beta} \\ 0 \end{Bmatrix}^{(r)} \tag{7.34}$$

where subscripts b and i denote inner fixed and free nodes, respectively.

Combining Eqs. (7.33) and (7.34) gives the force–displacement relations for enlarged elements — substructures which may be expressed as

$$[K_b]^{(r)}\{D_b\}^{(r)} = \{F_b\}^{(r)} \tag{7.35}$$

which is analogous to Eq. (7.26) and $\{F_b\}^{(r)} = \{F_b^{(r)}\} - [K_{bi}^{(r)}][K_{ii}^{(r)}]^{-1}\{F_i^{(r)}\}$. And thereby the conventional procedure is still valid.

Similarly, in the cases of structural symmetry of geometry and material, proper consideration of loading symmetry and antisymmetry can give rise to a much smaller set of governing equations.

For more details, please refer to the literature on structural analysis.

References

1. Chen, W.F. and Saleeb, A.F., *Constitutive Eqs. for Engineering Materials,* Vols. 1 & 2, Elsevier Science Ltd., New York, 1994.
2. Chen, W.F., *Plasticity in Reinforced Concrete,* McGraw-Hill, New York, 1982.
3. Chen, W.F. and T. Atsuta, *Theory of Beam-Columns,* Vol. 1, *In-Plane Behavior and Design,* McGraw-Hill, New York, 1976.
4. Desai, C.S., *Elementary Finite Element Method,* Prentice-Hall, Englewood Cliffs, NJ, 1979.
5. Meyers, V.J., *Matrix Analysis of Structures,* Harper & Row, New York, 1983.
6. Michalos, J., *Theory of Structural Analysis and Design,* Ronald Press, New York, 1958.
7. Hjelmstad, K.D., *Fundamentals of Structural Mechanics,* Prentice-Hall College Div., Upper Saddle River, NJ, 1996.
8. Fleming, J.F., *Analysis of Structural Systems,* Prentice-Hall College Div., Upper Saddle River, NJ, 1996.
9. Dadeppo, D.A. *Introduction to Structural Mechanics and Analysis,* Prentice-Hall College Div., Upper Saddle River, NJ, 1998.

8

Structural Modeling

Alexander Krimotat
SC Solutions, Inc.

Li-Hong Sheng
California Department of Transportation

8.1 Introduction

Prior to construction of any structural system, an extensive engineering design and analysis process must be undertaken. During this process, many engineering assumptions are routinely used in the application of engineering principles and theories to practice. A subset of these assumptions is used in a multitude of analytical methods available to structural analysts. In the modern engineering office, with the proliferation and increased power of personal computers, increasing numbers of engineers depend on structural analysis computer software to solve their engineering problems. This modernization of the engineering design office, coupled with an increased demand placed on the accuracy and efficiency of structural designs, requires a more-detailed understanding of the basic principles and assumptions associated with the use of modern structural analysis computer programs. The most popular of these programs are GT STRUDL, STAADIII, SAP2000, as well as some more powerful and complex tools such as ADINA, ANSYS, NASTRAN, and ABAQUS.

The objective of the analysis effort is to investigate the most probable responses of a bridge structure due to a range of applied loads. The results of these investigations must then be converted to useful design data, thereby providing designers with the information necessary to evaluate the performance of the bridge structure and to determine the appropriate actions in order to achieve the most efficient design configuration. Additionally, calculation of the structural system capacities is an important aspect in determining the most reliable design alternative. Every effort must be made to ensure that all work performed during any analytical activity enables designers to produce a set of quality construction documents including plans, specifications, and estimates.

The purpose of this chapter is to present basic modeling principles and suggest some guidelines and considerations that should be taken into account during the structural modeling process. Additionally, some examples of numerical characterizations of selected bridge structures and their components are provided. The outline of this chapter follows the basic modeling process. First, the selection of modeling methodology is discussed, followed by a description of the structural geometry, definition of the material and section properties of the components making up the structure, and description of the boundary conditions and loads acting on the structure.

0-8493-7434-0/00/$0.00+$.50

8.2 Theoretical Background

Typically, during the analytical phase of any bridge design, finite-element-based structural analysis programs are used to evaluate the structural integrity of the bridge system. Most structural analysis programs employ sound, well-established finite-element methodologies and algorithms to solve the analytical problem. Others employ such methods as moment distribution, column analogy, virtual work, finite difference, and finite strip, to name a few. It is of utmost importance for the users of these programs to understand the theories, assumptions, and limitations of numerical modeling using the finite-element method, as well as the limitations on the accuracy of the computer systems used to execute these programs. Many textbooks [1, 4, 6] are available to study the theories and application of finite-element methodologies to practical engineering problems. It is strongly recommended that examination of these textbooks be made prior to using finite-element-based computer programs for any project work. For instance, when choosing the types of elements to use from the finite-element library, the user must consider some important factors such as the basic set of assumptions used in the element formulation, the types of behavior that each element type captures, and the limitations on the physical behavior of the system.

Other important issues to consider include numerical solution techniques used in matrix operations, computer numerical precision limitations, and solution methods used in a given analysis. There are many solution algorithms that employ direct or iterative methods, and sparse solver technology for solving the same basic problems; however, selecting these solution methods efficiently requires the user to understand the best conditions in which to apply each method and the basis or assumptions involved with each method. Understanding the solution parameters such as tolerances for iterative methods and how they can affect the accuracy of a solution are also important, especially during the nonlinear analysis process.

Dynamic analysis is increasingly being required by many design codes today, especially in regions of high seismicity. Response spectrum analysis is frequently used and easily performed with today's analysis tools; however, a basic understanding of structural dynamics is crucial for obtaining the proper results efficiently and interpreting analysis responses. Basic linear structural dynamics theory can be found in many textbooks [2,3]. While many analysis tools on the market today can perform very sophisticated analyses in a timely manner, the user too must be more savvy and knowledgeable to control the overall analysis effort and optimize the performance of such tools.

8.3 Modeling

8.3.1 Selection of Modeling Methodology

The technical approach taken by the engineer must be based on a philosophy of providing practical analysis in support of the design effort. Significant importance must be placed on the analysis procedures by the entire design team. All of the analytical modeling, analysis, and interpretation of results must be based on sound engineering judgment and a solid understanding of fundamental engineering principles. Ultimately, the analysis must validate the design.

Many factors contribute to determination of the modeling parameters. These factors should reflect issues such as the complexity of the structure under investigation, types of loads being examined, and, most importantly, the information needed to be obtained from the analysis in the most efficient and "design-friendly" formats. This section presents the basic principles and considerations for structural modeling. It also provides examples of modeling options for the various bridge structure types.

A typical flowchart of the analysis process is presented in Figure 8.1. The technical approach to computer modeling is usually based on a logical progression. The first step in achieving a reliable computer model is to define a proper set of material and soil properties, based on published data and site investigations. Second, critical components are assembled and tested numerically where

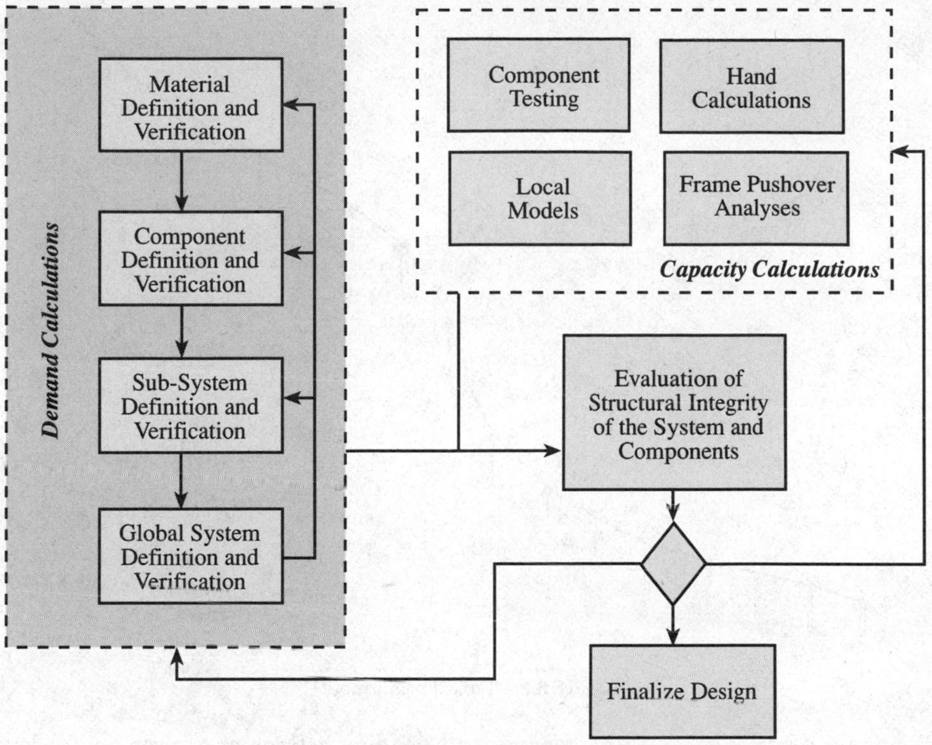

FIGURE 8.1 Typical analysis process.

validation of the performance of these components is considered important to the global model response. Closed-form solutions or available test data are used for these validations.

The next step is the creation and numerical testing of subsystems such as the bridge towers, superstructure elements, or individual frames. Again, as in the previous step, simple procedures are used in parallel to validate computer models. Last, a full bridge model consisting of the bridge subsystems is assembled and exercised. This final global model should include appropriate representation of construction sequence, soil and foundation boundary conditions, structural component behavior, and connection details.

Following the analysis and after careful examination of the analytical results, the data is postprocessed and provided to the designers for the purpose of checking the design and determining suitable design modifications, as necessary. Postprocessing might include computation of deck section resultant forces and moments, determination of extreme values of displacements for columns or towers and deck, and recovery of forces of constraint between structural components. The entire process may be repeated to validate any modifications made, depending on the nature and significance of such modifications.

An important part of the overall analytical procedure is determination of the capacities of the structural members. A combination of engineering calculations, computer analyses, and testing is utilized in order to develop a comprehensive set of component and system capacities. The evaluation of the structural integrity of the bridge structure, its components, and their connections are then conducted by comparing capacities with the demands calculated from the structural analysis.

Depending on the complexity of the structure under investigation and the nature of applied loads, two- or three-dimensional models can be utilized. In most cases, beam elements can be used to model structural elements of the bridge (Figure 8.2), so the component responses are presented in the form of force and moment resultants. These results are normally associated with individual element coordinate systems, thus simplifying the evaluations of these components. Normally, these

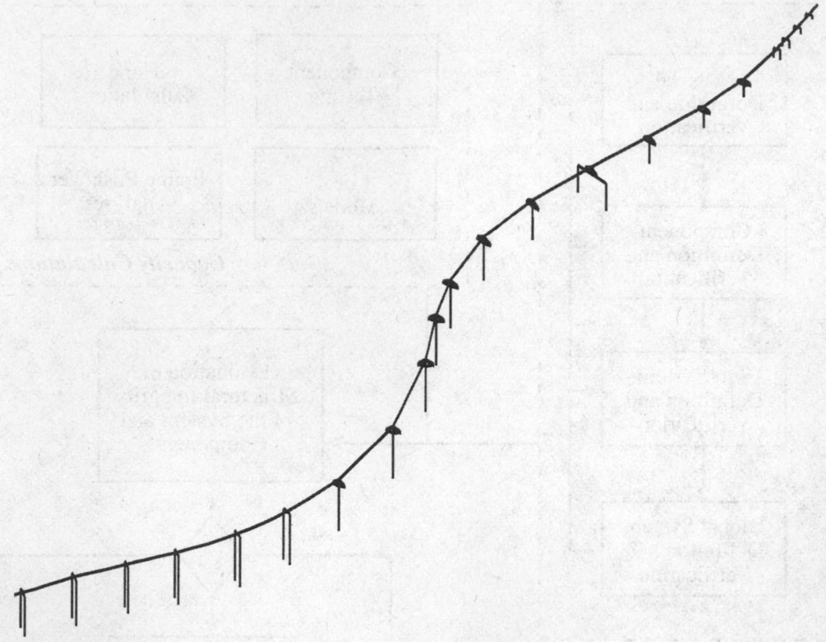

FIGURE 8.2 Typical beam model.

force resultants describe axial, shear, torsion, and bending actions at a given model location. Therefore, it is very important during the initial modeling stages to determine key locations of interest, so the model can be assembled such that important results can be obtained at these locations. While it is convenient to use element coordinate systems for the evaluation of the structural integrity of individual components, nodal results such as displacements and support reactions are usually output in the global coordinate systems. Proper refinement of the components must also be considered since different mesh size can sometimes cause significant variations in results. A balance between mesh refinement and reasonable element aspect ratios must be maintained so that the behavioral characteristics of the computer model is representative of the structure it simulates. Also, mesh refinement considerations must be made in conjunction with the cost to model efficiency. Higher orders of accuracy in modeling often come at a cost of analysis turnaround time and overall model efficiency. The analyst must use engineering judgment to determine if the benefits of mesh refinement justify the costs. For example, for the convenience in design of bridge details such as reinforcement bar cutoff, prestressing cable layouts, and section changes, the bridge superstructure is usually modeled with a high degree of refinement in the dead- and live-load analyses to achieve a well-defined force distribution. The same refinement may not be necessary in a dynamic analysis. Quite often, coarser models (at least four elements per span for the superstructure and three elements per column) are used in the dynamic analyses. These refinements are the minimum guidelines for discrete lumped mass models in dynamic analysis to maintain a reasonable mass distribution during the numerical solution process.

For more complex structures with complicated geometric configurations, such as curved plate girder bridges (Figure 8.3), or bridges with highly skewed supports (Figure 8.4), more-detailed finite-element models should be considered, especially if individual components within the superstructure need to be evaluated, which could not be facilitated with a beam superstructure representation. With the increasing speed of desktop computers, and advances in finite-element modeling tools, these models are becoming increasingly more popular. The main reason for their increased popularity is the improved accuracy, which in turn results in more efficient and cost-effective design.

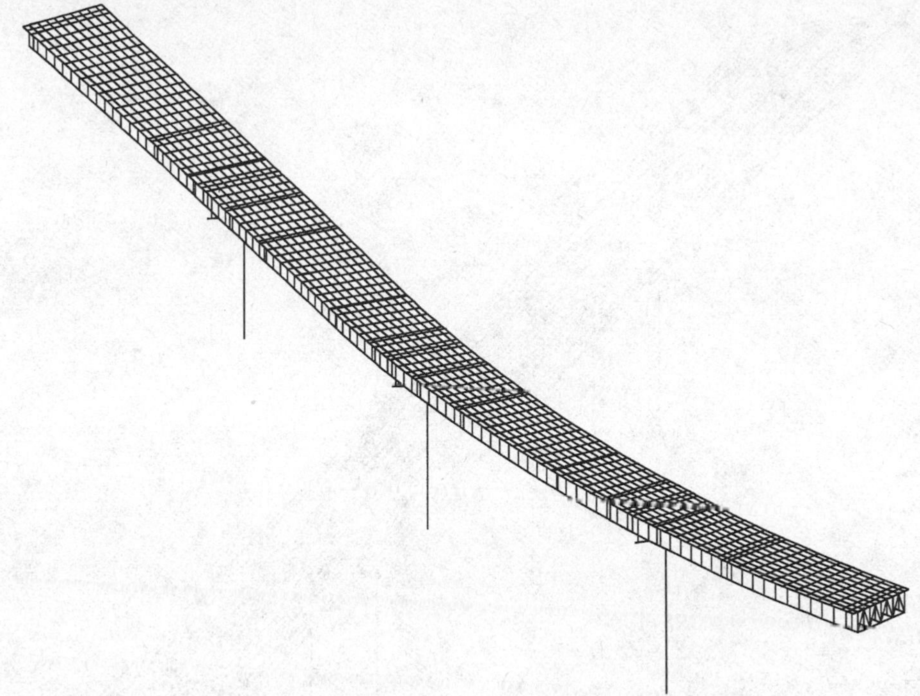

FIGURE 8.3 Steel plate-girder bridge—finite-element model.

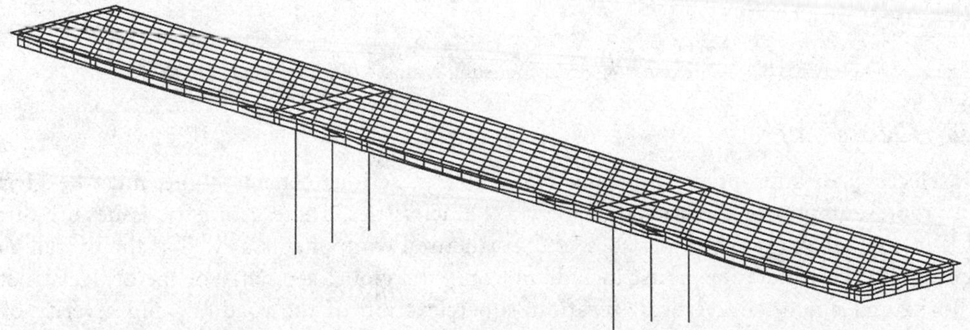

FIGURE 8.4 Concrete box girder with 45° skewed supports finite-element model.

More complex models, however, require a significantly higher degree of engineering experience and expertise in the theory and application of the finite-element method. In the case of a complex model, the engineer must determine the degree of refinement of the model. This determination is usually made based on the types of applied loads as well as the behavioral characteristics of the structure being represented by the finite-element model. It is important to note that the format of the results obtained from detailed models, such as shell and three-dimensional (3D) continuum models) is quite different from the results obtained from beam (or stick) models. Stresses and strains are obtained for each of the bridge components at a much more detailed level; therefore, calculation of a total force applied to the superstructure, for example, becomes a more difficult, tedious task. However, evaluation of local component behavior, such as cross frames, plate girder sections, or bridge deck sections, can be accomplished directly from the analysis results of a detailed finite-element model.

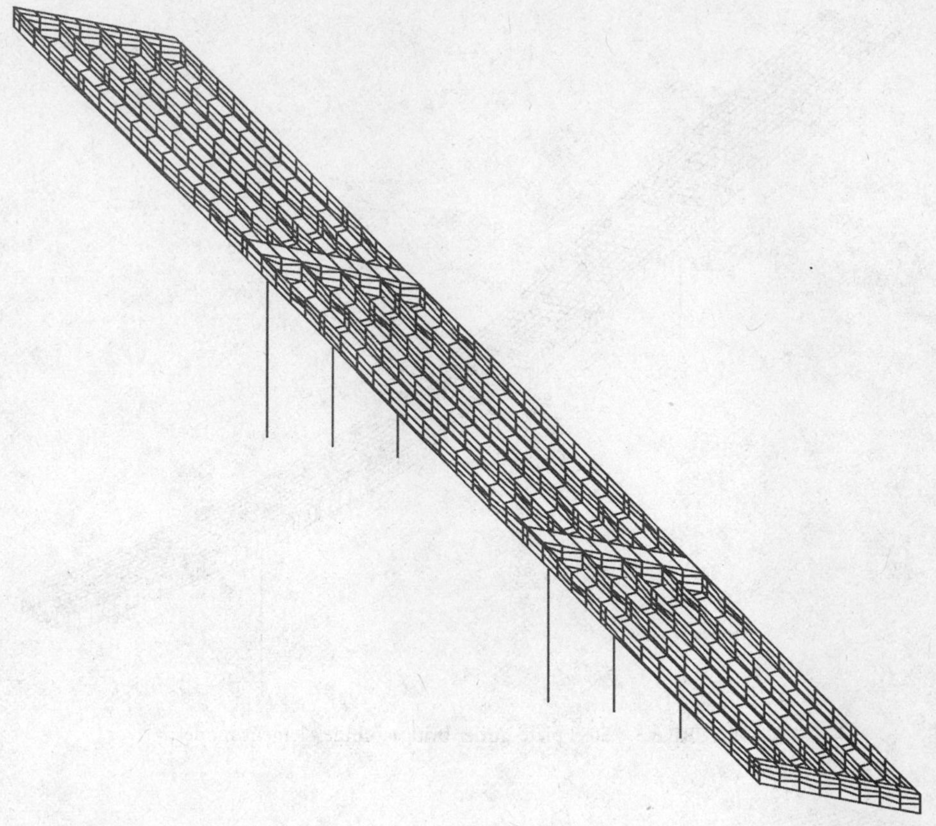

FIGURE 8.5 Concrete box-girder modeling example (deck elements not shown).

8.3.2 Geometry

After selecting an appropriate modeling methodology, serious considerations must be given to proper representation of the bridge geometric characteristics. These geometric issues are directly related to the behavioral characteristics of the structural components as well as the overall global structure. The considerations must include not only the global geometry of the bridge structure, i.e., horizontal alignment, vertical elevation, superelevation of the roadway, and severity of the support skews, but local geometric characterizations of connection details of individual bridge components as well. Such details include representations of connection regions such as column-to-cap beam, column-to-box girder, column-to-pile cap, cap beam to superstructure, cross frames to plate girder, gusset plates to adjacent structural elements, as well as various bearing systems commonly used in bridge engineering practice. Some examples of some modeling details are demonstrated in Figures 8.5 through 8.11.

Specifically, Figure 8.5 demonstrates how a detailed model of a box girder bridge structure can be assembled via use of shell elements (for girder webs and soffit), truss elements (for post-tensioning tendons), 3D solid elements (for internal diaphragms), and beam elements (for columns). Figure 8.6 illustrates some details of the web, deck, and abutment modeling for the same bridge structure. Additionally, spring elements are used to represent abutment support conditions for the vertical as well as back-wall directions. An example of a column and its connection to the superstructure in an explicit finite-element model is presented in the Figure 8.7. Three elements are used to represent the full length of the column. A set of rigid links connects the superstructure to each of the supporting columns (Figures 8.8 and 8.9). This is necessary to properly transmit bending

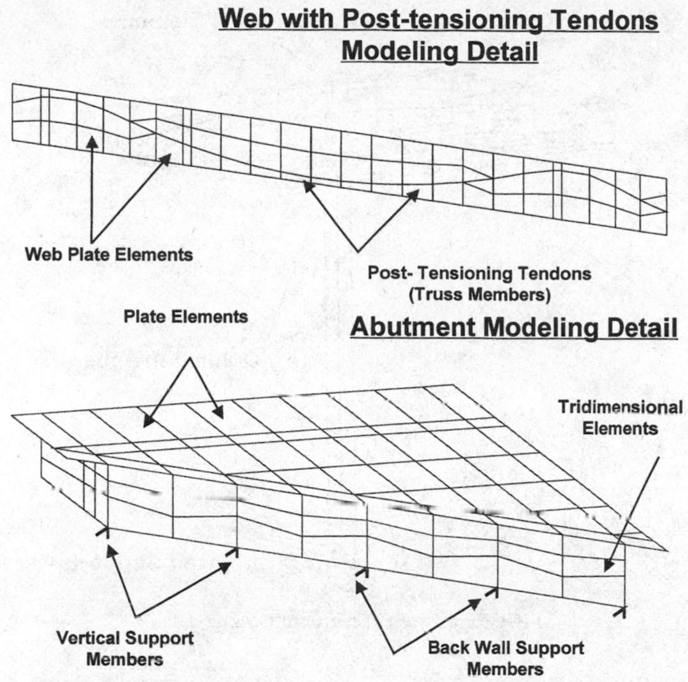

FIGURE 8.6 Selected modeling details.

action of these components, since the beam elements (columns) are characterized by six degrees of freedom per node, while 3D solids (internal diaphragms) carry only three degrees of freedom per node (translations only). In this example post-tensioning tendons are modeled explicitly, via truss elements with the proper drape shape (Figure 8.9). This was done so that accurate post-tensioning load application was achieved and the effects of the skews were examined in detail. However, when beam models are used for the dynamic analysis (Figure 8.2), special attention must be given to the beam column joint modeling. For a box girder superstructure, since cap beams are monolithic to the superstructure, considerations must be given to capture proper dynamic behavior of this detail through modification of the connection properties. It is common to increase the section properties of the cap beam embedded in the superstructure to simulate high stiffness of this connection.

Figure 8.10 illustrates the plate girder modeling approach for a section of superstructure. Plate elements are used to model deck sections and girder webs, while beams are used to characterize flanges, haunches, cross frame members, as well as columns and cap beams (Figure 8.11). Proper offsets are used to locate the centerlines of these components in their proper locations.

8.3.3 Material and Section Properties

One of the most important aspects of capturing proper behavior of the structure is the determination of the material and section properties of its components. Reference [5] is widely used for calculating section properties for a variety of cross-sectional geometry. For 3D solid finite element, the material constitutive law is the only thing to specify whereas for other elements consideration of modification of material properties are needed to match the actual structural behavior. Most structural theories are based on homogeneous material such as steel. While this means structural behavior can be directly calculated using the actual material and section properties, it also indicates that nonhomogeneous material such as reinforced concrete may subject certain limitation. Because of the composite nonlinear performance nature of reinforced concrete, section properties need to be adjusted for the objective of analysis. For elastic analysis, if strength requirement is the objective, section

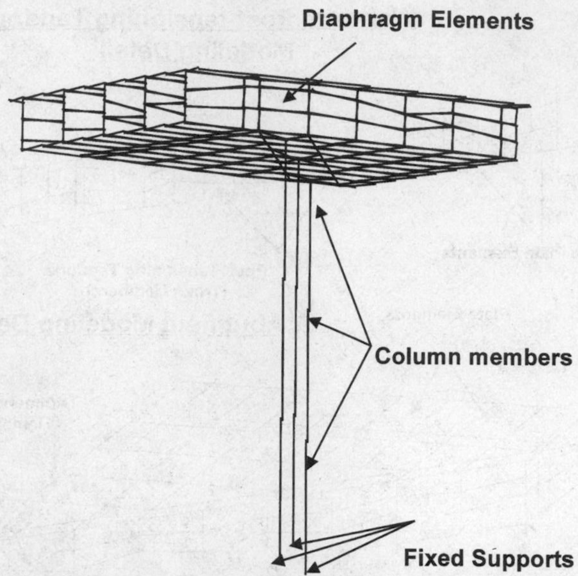

FIGURE 8.7 Bent region modeling detail.

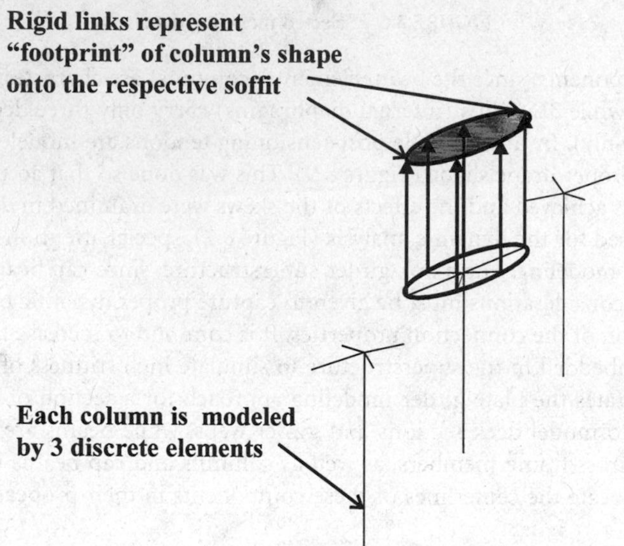

FIGURE 8.8 Column-to-superstructure connection modeling detail.

properties are less important as long as relative stiffness is correct. Section properties become most critical when structure displacement and deformation are objectives. Since concrete cracks beyond certain deformation, section properties need to be modified for this behavior. In general, if ultimate deformation is expected, then effective stiffness should be the consideration in section properties. It is common to use half value of the moment of inertia for reinforced concrete members and full value for prestressed concrete members. To replicate a rigid member behavior such as cap beams, section properties need to be amplified 100 times to eliminate local vibration problems in dynamic analysis.

Nonlinear behaviors are most difficult to handle in both complex and simple finite-element models. When solid elements are used, the constitutive relationships describing material behavior

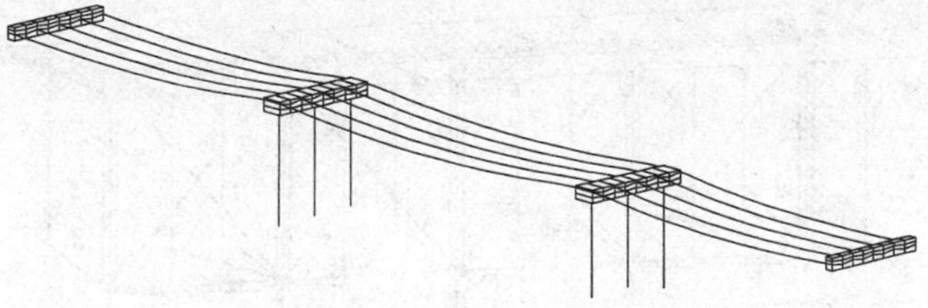

FIGURE 8.9 Post-tensioning tendons, diaphragms, and column-to-diaphragm connection modeling examples.

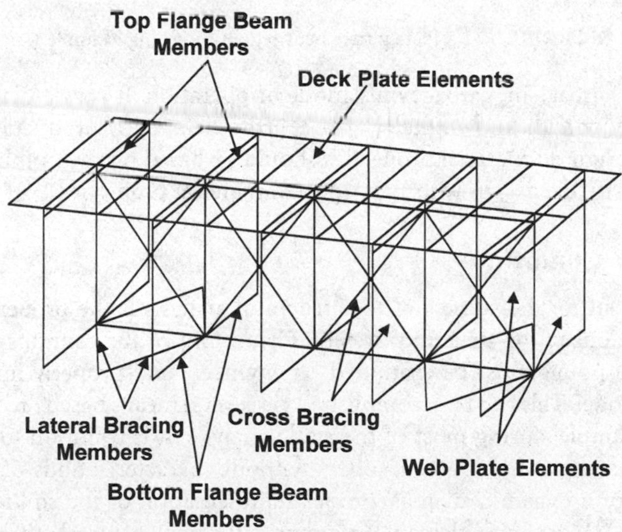

FIGURE 8.10 Plate girder superstructure modeling example.

should be utilized. These properties should be calibrated by the data obtained from the available test experiments. For beam–column-type elements, however, it is essential that the engineer properly estimates performance of the components either by experiments or theoretical detailed analysis. Once member performance is established, a simplified inelastic model can be used to simulate the expected member behavior. Depending on the complexity of the member, bilinear, or multilinear material representations may be used extensively. If member degradation needs to be incorporated in the analysis, then the Takeda model may be used. While a degrading model can correlate theoretical behavior with experimental results very well, elastic–plastic or bilinear models can give the engineer a good estimate of structural behavior without detailed material property parameters.

When a nonlinear analysis is performed, the engineer needs to understand the sensitivity issue raised by such analysis techniques. Without a good understanding of member behavior, it is very

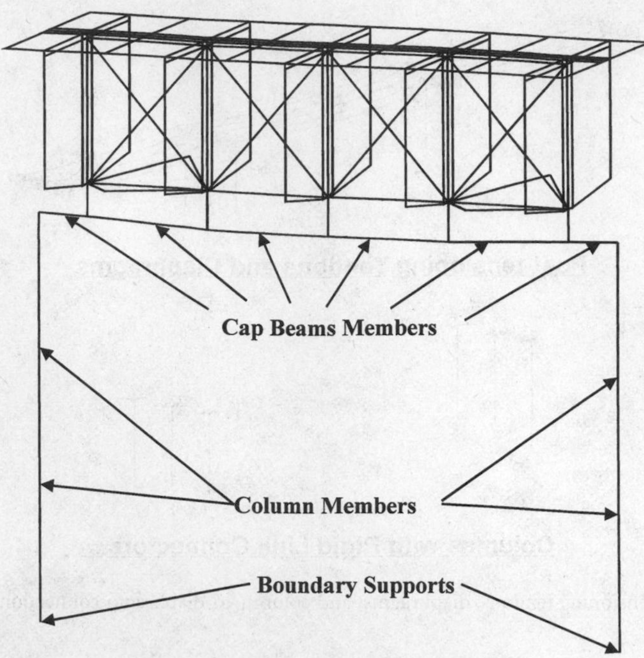

FIGURE 8.11 Plate girder bent region modeling example.

easy to fall into the "garbage in, garbage out" mode of operation. It is essential for the engineer to verify member behavior with known material properties before any production analyses are conducted. For initial design, all material properties should be based on the nominal values. However, it is important to verify the design with the expected material properties.

8.3.4 Boundary Conditions

Another key ingredient for the success of the structural analysis is the proper characterization of the boundary conditions of the structural system. Conditions of the columns or abutments at the support (or ground) points must be examined by engineers and properly implemented into the structural analysis model. This can be accomplished via several means based on different engineering assumptions. For example, during most of the static analysis, it is common to use a simple representation of supports (e.g., fixed, pinned, roller) without characterizations of the soil/foundation stiffness. However, for a dynamic analysis, proper representation of the soil/foundation system is essential (Figure 8.12). Most finite-element programs will accept a $[6 \times 6]$ stiffness matrix input for such system. Other programs require extended $[12 \times 12]$ stiffness matrix input describing the relationship between the ground point and the base of the columns. Prior to using these matrices, it is important that the user investigate the internal workings of the finite-element program, so the proper results are obtained by the analysis.

In some cases it is necessary to model the foundation/soil system with greater detail. Nonlinear modeling of the system can be accomplished via nonlinear spring/damper representation (Figure 8.13) or, in the extreme case, by explicit modeling of subsurface elements and plasticity-based springs representing surrounding soil mass (Figure 8.14). It is important that if this degree of detail is necessary, the structural engineer works very closely with the geotechnical engineers to determine proper properties of the soil springs. **As a general rule it is essential to set up small models to test behavior and check the results via hand calculations.**

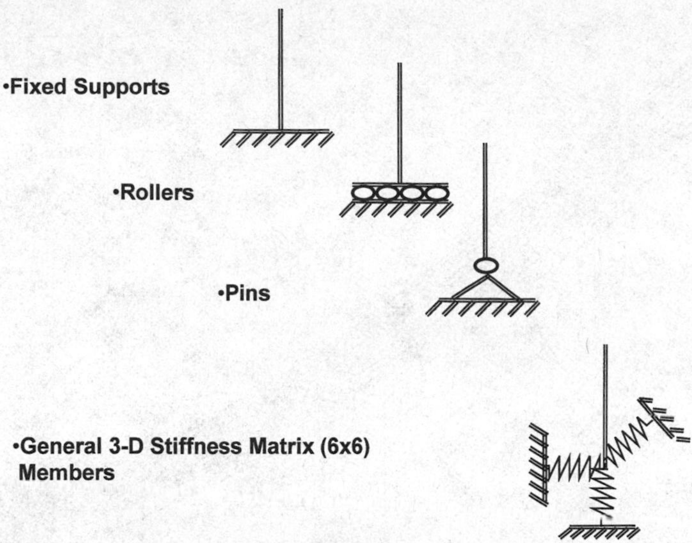

FIGURE 8.12 Examples of foundation modeling.

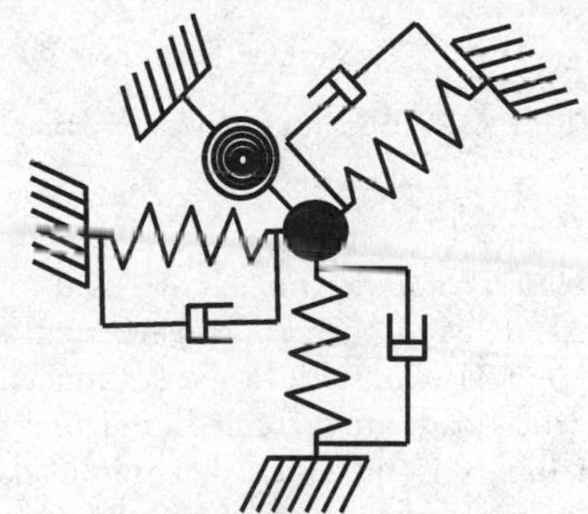

FIGURE 8.13 Nonlinear spring/damper model.

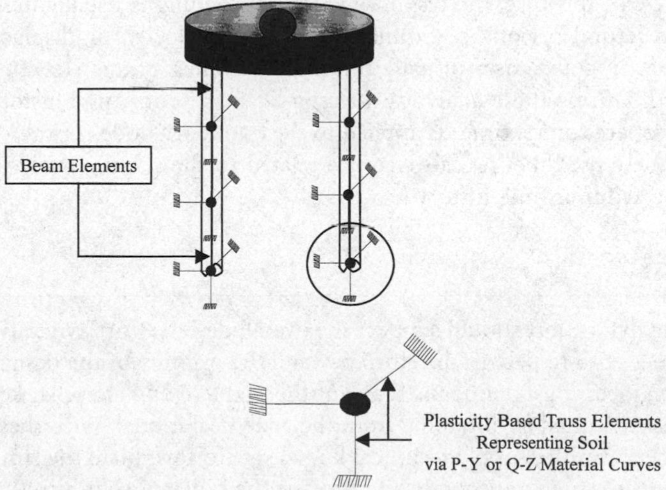

FIGURE 8.14 Soil–structure interaction modeling.

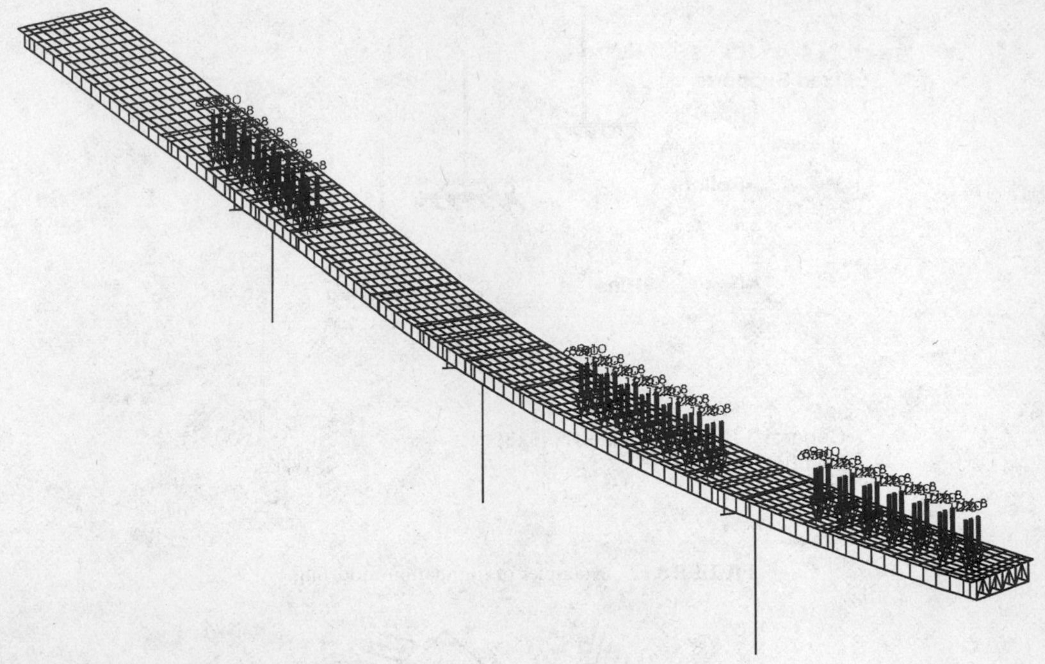

FIGURE 8.15 Truck load application example.

8.3.5 Loads

During engineering design activities, computer models are used to evaluate bridge structures for various service loads, such as traffic, wind, thermal, construction, and other service loads. These service loads can be represented by a series of static load cases applied to the structural model. Some examples of application of the truck loads are presented in Figures 8.15 and 8.16.

In many cases, especially in high seismic zones, dynamic loads control many bridge design parameters. In this case, it is very important to understand the nature of these loads, as well as the theory that governs the behavior of structural systems subjected to these dynamic loads. In high seismic zones, a multimode response spectrum analysis is required to evaluate the dynamic response of bridge structures. In this case, the response spectrum loading is usually described by the relationship of the structural period vs. ground acceleration, velocity, or displacement for a given structural damping. In some cases, usually for more complex bridge structures, a time history analysis is required. During these analytical investigations, a set of time history loads (normally, displacement or acceleration vs. time) is applied to the boundary nodes of the structure. Reference [3] is the most widely used theoretical reference related to the seismic analysis methodology for either response spectrum or time history analysis.

8.4 Summary

In summary, the analysis effort should support the overall design effort by verifying the design and addressing any issues with respect to the efficiency and the viability of the design. Before modeling commences, the engineer must define the scope of the problem and ask what key results and types of data he or she is interested in obtaining from the analytical model. With these basic parameters in mind, the engineer can then apply technical knowledge to formulate the simplest, most elegant model to represent the structure properly and provide the range of solutions that are accurate and fundamentally sound. The engineer must bound the demands on the structure by looking at limiting

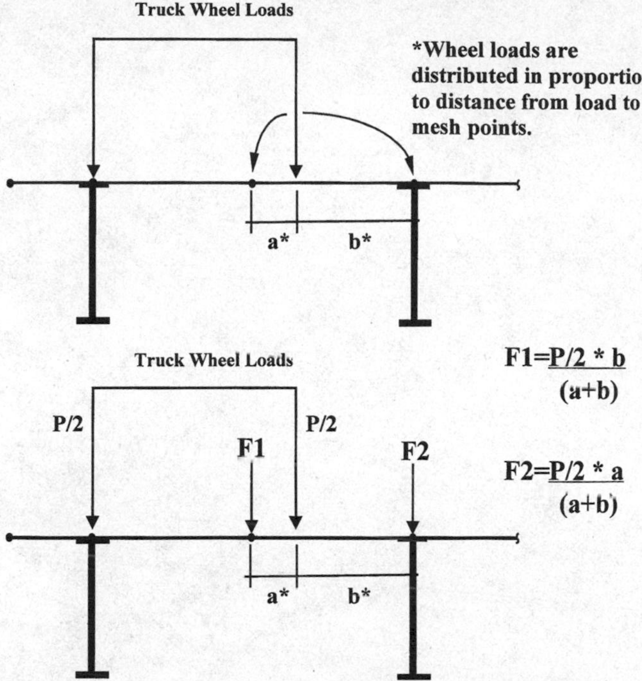

FIGURE 8.16 Equivalent truck load calculation example.

load cases and modifying the structure parameters, such as boundary conditions or material properties. Rigorous testing of components, hand calculations, local modeling, and sound engineering judgment must be used to validate the analytical model at all levels. Through a rigorous analytical methodology and proper use of today's analytical tools, structural engineers can gain a better understanding of the behavior of the structure, evaluate the integrity of the structure, and validate and optimize the structural design.

References

1. Bathe, K.J., *Finite Element Procedures*, 2nd ed., Prentice-Hall, Englewood Cliffs, NJ, 1996.
2. Chopra, A. K., *Dynamics of Structures, Theory and Applications to Earthquake Engineering*, Prentice-Hall, Englewood Cliffs, NJ, 1995.
3. Clough, R. W. and J. Penzien, *Dynamics of Structures*, 2nd ed., McGraw-Hill, New York, 1993.
4. Priestley, M.J.N., F. Seible, and G.M. Calvi, *Seismic Design and Retrofit of Bridges*, John Wiley & Sons, New York, 1996.
5. Young, W.C., *Roark's Formulas for Stress and Strain*, 6th ed., McGraw-Hill, New York, 1989.
6. Zienkiewicz, O.C. and R.L.Taylor, *The Finite Element Method*, Vol. 1, *Basic Formulation and Linear Problems*, 4th ed., McGraw-Hill, Berkshire, England, 1994.

Section II
Superstructure Design

9

Reinforced Concrete Bridges

Jyouru Lyang
California Department of Transportation

Don Lee
California Department of Transportation

John Kung
California Department of Transportation

9.1 Introduction

The raw materials of concrete, consisting of water, fine aggregate, coarse aggregate, and cement, can be found in most areas of the world and can be mixed to form a variety of structural shapes. The great availability and flexibility of concrete material and reinforcing bars have made the reinforced concrete bridge a very competitive alternative. Reinforced concrete bridges may consist of precast concrete elements, which are fabricated at a production plant and then transported for erection at the job site, or cast-in-place concrete, which is formed and cast directly in its setting location. Cast-in-place concrete structures are often constructed monolithically and continuously. They usually provide a relatively low maintenance cost and better earthquake-resistance performance. Cast-in-place concrete structures, however, may not be a good choice when the project is on a fast-track construction schedule or when the available falsework opening clearance is limited. In this chapter, various structural types and design considerations for conventional cast-in-place, reinforced concrete highway bridge are discussed. Two design examples of a simply supported slab bridge and a two-span box girder bridge are also presented. All design specifications referenced in this chapter are based on 1994 AASHTO LRFD (Load and Resistance Factor Design) Bridge Design Specifications [1].

0-8493-7434-0/00/$0.00+$.50

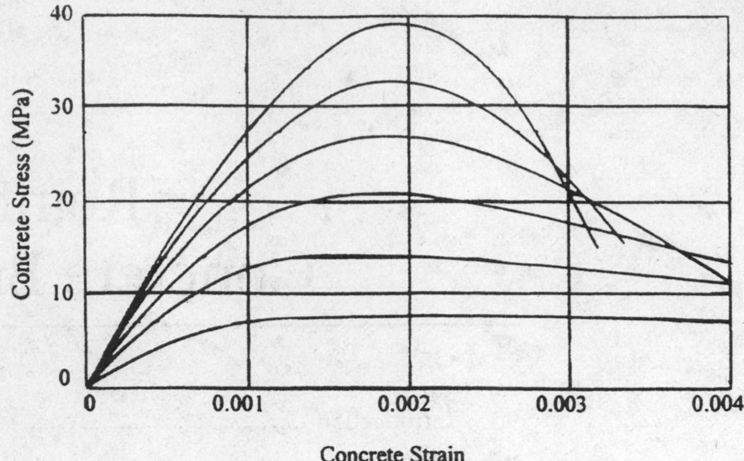

FIGURE 9.1 Typical stress–strain curves for concrete under uniaxial compression loading.

9.2 Materials

9.2.1 Concrete

1 Compressive Strength

The compressive strength of concrete (f_c') at 28 days after placement is usually obtained from a standard 150-mm-diameter by 300-mm-high cylinder loaded longitudinally to failure. Figure 9.1 shows typical stress–strain curves from unconfined concrete cylinders under uniaxial compression loading. The strain at the peak compression stress f_c' is approximately 0.002 and maximum usable strain is about 0.003. The concrete modulus of elasticity, E_c, may be calculated as

$$E_c = 0.043\gamma_c^{1.5}\sqrt{f_c'}\ \text{MPa} \tag{9.1}$$

where γ_c is the density of concrete (kg/m^3) and f_c' is the specified strength of concrete (MPa). For normal-weight concrete ($\gamma_c = 2300$ kg/m^3), E_c may be calculated as $4800\sqrt{f_c'}$ MPa.

The concrete compressive strength or class of concrete should be specified in the contract documents for each bridge component. A typical specification for different classes of concrete and their corresponding specified compressive strengths is shown in Table 9.1. These classes are intended for use as follows:

- Class A concrete is generally used for all elements of structures and specially for concrete exposed to salt water.
- Class B concrete is used in footings, pedestals, massive pier shafts, and gravity walls.
- Class C concrete is used in thin sections under 100 mm in thickness, such as reinforced railings and for filler in steel grid floors.
- Class P concrete is used when strengths exceeding 28 MPa are required.
- Class S concrete is used for concrete deposited under water in cofferdams to seal out water.

Both concrete compressive strengths and water–cement ratios are specified in Table 9.1 for different concrete classes. This is because the water–cement ratio is a dominant factor contributing to both durability and strength, while simply obtaining the required concrete compressive strength to satisfy the design assumptions may not ensure adequate durability.

TABLE 9.1 Concrete Mix Characteristics by Class[1]

Class of Concrete	Minimum Cement Content (kg/m³)	Maximum Water–Cement Ratio (kg/kg)	Air Content Range, %	Coarse Aggregate per AASHTO M43 (square size of openings, mm)	28-day Compressive Strength, f'_c MPa
A	362	0.49	—	25 to 4.75	28
A(AE)	362	0.45	6.0 ± 1.5	25 to 4.75	28
B	307	0.58	—	50 to 4.75	17
B(AE)	307	0.55	5.0 ± 1.5	50 to 4.75	17
C	390	0.49	—	12.5 to 4.75	28
C(AE)	390	0.45	7.0 ± 1.5	12.5 to 4.75	28
P	334	0.49	As specified elsewhere	25 to 4.75 or 19 to 4.75	As specified elsewhere
S	390	0.58	—	25 to 4.75	—
Low-density	334	As specified in the contract documents			

Notes:
1. AASHTO Table C5.4.2.1-1 (From AASHTO LRFD Bridge Design Specifications, ©1994 by the American Association of State Highway and Transportation Officials, Washington, D.C. With permission.)
2. Concrete strengths above 70 MPa need to have laboratory testing verification. Concrete strengths below 16 MPa should not be used.
3. The sum of portland cement and other cementitious materials should not exceed 475 kg/m3.
4. Air-entrained concrete (AE) can improve durability when subjected to freeze–thaw action and to scaling caused by chemicals applied for snow and ice removal.

2. Tensile Strength

The tensile strength of concrete can be measured directly from tension loading. However, fixtures for holding the specimens are difficult to apply uniform axial tension loading and sometimes will even introduce unwanted secondary stresses. The direct tension test method is therefore usually used to determine the cracking strength of concrete caused by effects other than flexure. For most regular concrete, the direct tensile strength may be estimated as 10% of the compressive strength.

The tensile strength of concrete may be obtained indirectly by the split tensile strength method. The splitting tensile stress (f_s) at which a cylinder is placed horizontally in a testing machine and loaded along a diameter until split failure can be calculated as

$$f_s = 2P/(\pi LD) \tag{9.2}$$

where P is the total applied load that splits the cylinder, L is the length of cylinder, and D is the diameter of the cylinder.

The tensile strength of concrete can also be evaluated by means of bending tests conducted on plain concrete beams. The flexural tensile stress, known as the modulus of rupture (f_r) is computed from the flexural formula M/S, where M is the applied failure bending moment and S is the elastic section modulus of the beam. Modulus of rupture (f_r) in MPa can be calculated as

$$f_r = \begin{cases} 0.63\sqrt{f'_c} & \text{for normal-weight concrete} \\ 0.52\sqrt{f'_c} & \text{for sand–low-density concrete} \\ 0.45\sqrt{f'_c} & \text{for all–low-density concrete} \end{cases} \tag{9.3}$$

TABLE 9.2 Steel Deformed Bar Sizes and Weight (ASTM A615M and A706M)

Bar Number	Nominal Dimensions Diameter, mm	Area, mm^2	Unit Weight, kg/m
10	9.5	71	0.560
13	12.7	129	0.994
16	15.9	199	1.552
19	19.1	284	2.235
22	22.2	387	3.042
25	25.4	510	3.973
29	28.7	645	5.060
32	32.3	819	6.404
36	35.8	1006	7.907
43	43.0	1452	11.38
57	57.3	2581	20.24

Both the splitting tensile stress (f_s) and flexural tensile stress (f_r) overestimate the tensile cracking stress determined by a direct tension test. However, concrete in tension is usually ignored in strength calculations of reinforced concrete members because the tensile strength of concrete is low. The modulus of elasticity for concrete in tension may be assumed to be the same as in compression.

3. Creep and Shrinkage

Both creep and shrinkage of concrete are time-dependent deformations and are discussed in Chapter 10.

9.2.2 Steel Reinforcement

Deformed steel bars are commonly employed as reinforcement in most reinforced concrete bridge construction. The surface of a steel bar is rolled with lugs or protrusions called deformations in order to restrict longitudinal movement between the bars and the surrounding concrete. Reinforcing bars, rolled according to ASTM A615/A615M specifications (billet steel) [2], are widely used in construction. ASTM A706/A706M low-alloy steel deformed bars (Grade 420 only) [2] are specified for special applications where extensive welding of reinforcement or controlled ductility for earthquake-resistant, reinforced concrete structures or both are of importance.

1. Bar Shape and Size

Deformed steel bars are approximately numbered based on the amount of millimeters of the nominal diameter of the bar. The nominal dimensions of a deformed bar are equivalent to those of a plain round bar which has the same mass per meter as the deformed bar. Table 9.2 lists a range of deformed bar sizes according to the ASTM specifications.

2. Stress–Strain Curve

The behavior of steel reinforcement is usually characterized by the stress–strain curve under uniaxial tension loading. Typical stress–strain curves for steel Grade 300 and 420 are shown in Figure 9.2. The curves exhibit an initial linear elastic portion with a slope calculated as the modulus of elasticity of steel reinforcement $E_s = 200{,}000$ MPa; a yield plateau in which the strain increases (from ε_y to ε_h) with little or no increase in yield stress (f_y); a strain-hardening range in which stress again increases with strain until the maximum stress (f_u) at a strain (ε_u) is reached; and finally a range in which the stress drops off until fracture occurs at a breaking strain of ε_b.

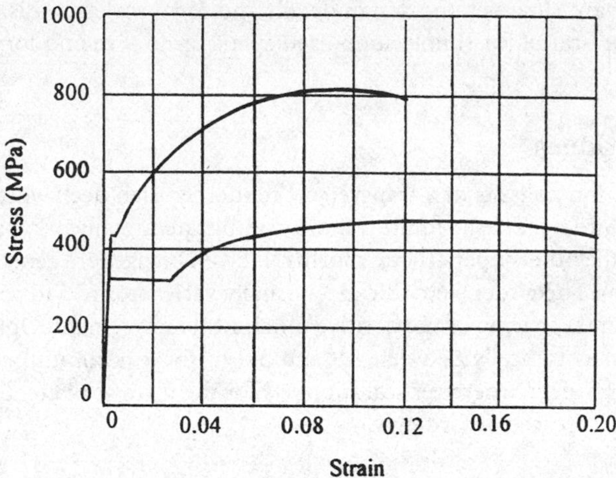

FIGURE 9.2 Typical stress–strain curves for steel reinforcement.

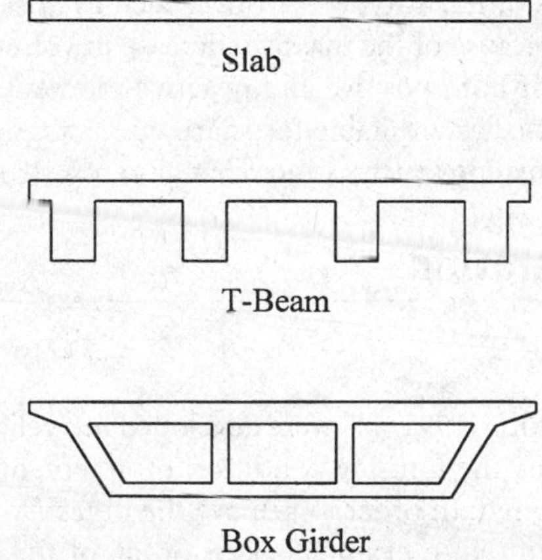

FIGURE 9.3 Typical reinforced concrete sections in bridge superstructures.

9.3 Bridge Types

Reinforced concrete sections, used in the bridge superstructures, usually consist of slabs, T-beams (deck girders), and box girders (Figure 9.3). Safety, cost-effectiveness, and aesthetics are generally the controlling factors in the selection of the proper type of bridges [3]. Occasionally, the selection is complicated by other considerations such as the deflection limit, life-cycle cost, traffic maintenance during construction stages, construction scheduling and worker safety, feasibility of falsework layout, passage of flood debris, seismicity at the site, suitability for future widening, and commitments made to officials and individuals of the community. In some cases, a prestressed concrete or steel bridge may be a better choice.

9.3.1 Slab Bridges

Longitudinally reinforced slab bridges have the simplest superstructure configuration and the neatest appearance. They generally require more reinforcing steel and structural concrete than do girder-type bridges of the same span. However, the design details and formworks are easier and less expensive. It has been found economical for simply supported spans up to 9 m and for continuous spans up to 12 m.

9.3.2 T-Beam Bridges

The T-beam construction consists of a transversely reinforced slab deck which spans across to the longitudinal support girders. These require a more-complicated formwork, particularly for skewed bridges, compared to the other superstructure forms. T-beam bridges are generally more economical for spans of 12 to 18 m. The girder stem thickness usually varies from 35 to 55 cm and is controlled by the required horizontal spacing of the positive moment reinforcement. Optimum lateral spacing of longitudinal girders is typically between 1.8 and 3.0 m for a minimum cost of formwork and structural materials. However, where vertical supports for the formwork are difficult and expensive, girder spacing can be increased accordingly.

9.3.3 Box-Girder Bridges

Box-girder bridges contain top deck, vertical web, and bottom slab and are often used for spans of 15 to 36 m with girders spaced at 1.5 times the structure depth. Beyond this range, it is probably more economical to consider a different type of bridge, such as post-tensioned box girder or steel girder superstructure. This is because of the massive increase in volume and materials. They can be viewed as T-beam structures for both positive and negative moments. The high torsional strength of the box girder makes it particularly suitable for sharp curve alignment, skewed piers and abutments, superelevation, and transitions such as interchange ramp structures.

9.4 Design Considerations

9.4.1 Basic Design Theory

The AASHTO LRFD Specifications (1994) [1] were developed in a reliability-based limit state design format. Limit state is defined as the limiting condition of acceptable performance for which the bridge or component was designed. In order to achieve the objective for a safe design, each bridge member and connection is required to examine some, or all, of the service, fatigue, strength, and extreme event limit states. All applicable limit states shall be considered of equal importance. The basic requirement for bridge design in the LRFD format for each limit state is as follows:

$$\eta \sum \gamma_i Q_i \leq \phi R_n \tag{9.4}$$

where η = load modifier to account for bridge ductility, redundancy, and operational importance, γ_i = load factor for load component i, Q_i = nominal force effect for load component i, ϕ = resistance factor, and R_n = nominal resistance. The margin of safety for a bridge design is provided by ensuring the bridge has sufficient capacity to resist various loading combinations in different limit states.

The load factors, γ, which often have values larger than one, account for the loading uncertainties and their probabilities of occurrence during bridges design life. The resistance factors, ϕ, which are typically less than unity at the strength limit state and equal to unity for all other limit states, account for material variabilities and model uncertainties. Table 9.3 lists the resistance factors in the strength limit state for conventional concrete construction. The load modifiers, η, which are equal to unity for all non-strength-limit states, account for structure ductility, redundancy, and operational importance. They are related to the bridge physical strength and the effects of a bridge being out of service. Detailed load resistance factor design theory and philosophy are discussed in Chapter 5.

TABLE 9.3 Resistance Factors ϕ in the Strength Limit State for Conventional Construction

Strength Limit State	Resistance Factors ϕ
For flexural and tension of reinforced concrete	0.90
For shear and torsion	
Normal weight concrete	0.90
Lightweight concrete	0.70
For axial compression with spirals and ties	0.75
(except for Seismic Zones 3 and 4 at the extreme event limit state)	
For bearing on concrete	0.79
For compression in strut-and-tie models	0.70

Notes:

1. AASHTO 5.5.4.2.1 (From AASHTO LRFD Bridge Design Specifications, ©1994 by the American Association of State Highway and Transportation Officials, Washington, D.C. With permission.)
2. For compression members with flexural, the value of ϕ may be increased linearly to the value for flexural as the factored axial load resistance, ϕP_n, decreases from 0.10 $f_c'A_g$ to 0.

9.4.2 Design Limit States

1. Service Limit States

For concrete structures, service limit states correspond to the restrictions on cracking width and deformations under service conditions. They are intended to ensure that the bridge will behave and perform acceptably during its service life.

a. Control of Cracking

Cracking may occur in the tension zone for reinforced concrete members due to the low tensile strength of concrete. Such cracks may occur perpendicular to the axis of the members under axial tension or flexural bending loading without significant shear force, or inclined to the axis of the members with significant shear force. The cracks can be controlled by distributing steel reinforcements over the maximum tension zone in order to limit the maximum allowable crack widths at the surface of the concrete for given types of environment. The tensile stress in the steel reinforcement (f_s) at the service limit state should not exceed

$$f_{sa} = \frac{Z}{\left(d_c A\right)^{1/3}} \le 0.6 f_y \tag{9.5}$$

where d_c (mm) is the concrete cover measured from extreme tension fiber to the center of the closest bars and should not to be taken greater than 50 mm; A (mm^2) is the concrete area having the same centroid as the principal tensile reinforcement divided by the number of bars; Z (N/mm) should not exceed 30,000 for members in moderate exposure conditions, 23,000 in severe exposure conditions, and 17,500 for buried structures. Several smaller tension bars at moderate spacing can provide more effective crack control by increasing f_{sa} rather than installing a few larger bars of equivalent area.

When flanges of reinforced concrete T-beams and box girders are in tension, the flexural tension reinforcement should be distributed over the lesser of the effective flange width or a width equal to 1/10 of the span in order to avoid the wide spacing of the bars. If the effective flange width exceeds 1/10 of the span length, additional longitudinal reinforcement, with an area not less than 0.4% of the excess slab area, should be provided in the outer portions of the flange.

For flexural members with web depth exceeding 900 mm, longitudinal skin reinforcements should be uniformly distributed along both side faces for a height of $d/2$ nearest the flexural tension reinforcement for controlling cracking in the web. Without such auxiliary steel, the width of the

TABLE 9.4 Traditional Minimum Depths for Constant Depth Superstructures

Bridge Types	Minimum Depth (Including Deck) Simple Spans	Continuous Spans
Slabs	$\frac{1.2(S+3000)}{30}$	$\frac{(S+3000)}{30} \geq 165$ mm
T-beams	$0.070L$	$0.065L$
Box beams	$0.060L$	$0.055L$
Pedestrian structure beams	$0.035L$	$0.033L$

Notes:
1. AASHTO Table 2.5.2.6.3-1 (From AASHTO LRFD Bridge Design Specifications, ©1994 by the American Association of State Highway and Transportation Officials, Washington, D.C. With permission.)
2. S (mm) is the slab span length and L (mm) is the span length.
3. When variable-depth members are used, values may be adjusted to account for change in relative stiffness of positive and negative moment sections.

cracks in the web may greatly exceed the crack widths at the level of the flexural tension reinforcement. The area of skin reinforcement (A_{sk}) in mm²/mm of height on each side face should satisfy

$$A_{sk} \geq 0.001\ (d_e - 760) \leq \frac{A_s}{1200} \tag{9.6}$$

where d_e (mm) is the flexural depth from extreme compression fiber to the centroid of the tensile reinforcement and A_s (mm²) is the area of tensile reinforcement and prestressing steel. The maximum spacing of the skin reinforcement shall not exceed $d/6$ or 300 mm.

b. Control of Deformations

Service-load deformations in bridge elements need to be limited to avoid the structural behavior which differs from the assumed design conditions and to ease the psychological effects on motorists. Service-load deformations may not be a potential source of collapse mechanisms but usually cause some undesirable effects, such as the deterioration of wearing surfaces and local cracking in concrete slab which could impair serviceability and durability. AASHTO LRFD [1] provides two alternative criteria for controlling the deflections:

Limiting Computed Deflections (*AASHTO 2.5.2.6.2*):

Vehicular load, general	Span length/800
Vehicular and/or pedestrian loads	Span length/1000
Vehicular load on cantilever arms	Span length/300
Vehicular and/or pedestrian loads on cantilever arms	Span length/1000

Limiting Span-to-Depth Ratios (*AASHTO 2.5.2.6.3*): For superstructures with constant depth, Table 9.4 shows the typical minimum depth recommendation for a given span length.

Deflections of bridges can be estimated in two steps: (1) instantaneous deflections which occur at the first loading and (2) long-time deflections which occur with time due to the creep and shrinkage of the concrete.

Instantaneous deflections may be computed by using the elastic theory equations. The modulus of elasticity for concrete can be calculated from Eq. (9.1). The moment of inertia of a section can be taken as either the uncracked gross moment of inertia (I_g) for uncracked elements or the effective moment of inertia (I_e) for cracked elements. The effective moment of inertia can be calculated as

$$I_e = \left(\frac{M_{cr}}{M_a}\right)^3 I_g + \left[1 - \left(\frac{M_{cr}}{M_a}\right)^3\right] I_{cr} \leq I_g \tag{9.7}$$

and

$$M_{cr} = f_r \frac{I_g}{y_t} \tag{9.8}$$

where M_{cr} is the moment at first cracking, f_r is the modulus of rupture, y_t is the distance from the neutral axis to the extreme tension fiber, I_{cr} is the moment of inertia of the cracked section transformed to concrete (see Section 9.4.6), and M_a is the maximum moment in a component at the stage for which deformation is computed. For prismatic members, the effective moment of inertia may be calculated at midspan for simple or continuous bridges and at support for cantilevers. For continuous nonprismatic members, the moment of inertia may be calculated as the average of the critical positive and negative moment sections.

Long-time deflections may be calculated as the instantaneous deflection multiplied by the following:

If the instantaneous deflection is based on I_g: 4.0
If the instantaneous deflection is based on I_e: $3.0\text{–}1.2\left(A'_s / A_s\right) \geq 1.6$

where A'_s is area of compression reinforcement and A_s is the area of tension reinforcement.

2. Fatigue Limit States

Fatigue limit states are used to limit stress in steel reinforcements to control concrete crack growth under repetitive truck loading in order to prevent early fracture failure before the design service life of a bridge. Fatigue loading consists of one design truck with a constant spacing of 9000 mm between the 145-kN axles. Fatigue is considered at regions where compressive stress due to permanent loads is less than two times the maximum tensile live-load stress resulting from the fatigue-load combination. Allowable fatigue stress range in straight reinforcement is limited to

$$f_f = 145 - 0.33 f_{min} + 55\left(\frac{r}{h}\right) \tag{9.9}$$

where f_{min} (MPa) is the minimum stress in reinforcement from fatigue loading (positive for tension and negative for compression stress) and r/h is the ratio of the base radius to the height of rolled-on transverse deformations (0.3 may be used if the actual value in not known).

The cracked section properties should be used for fatigue. Gross section properties may be used when the sum of stresses, due to unfactored permanent loads, plus 1.5 times the fatigue load is not to exceed the tensile stress of $0.25\sqrt{f'_c}$.

3. Strength Limit States and Extreme Event Limit States

For reinforced concrete structures, strength and extreme event limit states are used to ensure that strength and stability are provided to resist specified statistically significant load combinations. A detailed discussion for these limit states is covered in Chapter 5.

9.4.3 Flexural Strength

Figure 9.4 shows a doubly reinforced concrete beam when flexural strength is reached and the depth of neutral axis falls outside the compression flange ($c > h_f$). Assume that both tension and compression steel are yielding and the concrete compression stress block is in a rectangular shape. ε_{cu} is the maximum strain at the extreme concrete compression fiber and is about 0.003 for unconfined concrete.

Concrete compression force in the web;

$$C_w = 0.85\ f'_c\ ab_w = 0.85\beta_1\ f'_c\ cb_w \tag{9.10}$$

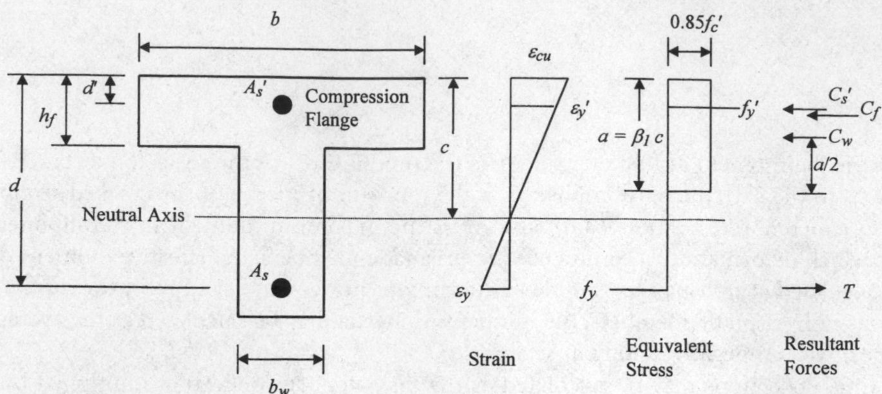

FIGURE 9.4 Reinforced concrete beam when flexural strength is reached.

where

$$a = c\,\beta_1 \tag{9.11}$$

Concrete compression force in the flange:

$$C_f = 0.85\beta_1\ f_c'\ (b - b_w)h_f \tag{9.12}$$

Compression force in the steel:

$$C_s' = A_s' f_y' \tag{9.13}$$

Tension force in the steel:

$$T = A_s f_y \tag{9.14}$$

From the equilibrium of the forces in the beam, we have

$$C_w + C_f + C_s' = T \tag{9.15}$$

The depth of the neutral axis can be solved as

$$c = \frac{A_s f_y - A_s' f_y' - 0.85\beta_1 f_c'(b - b_w)h_f}{0.85\beta_1 f_c' b_w} \geq h_f \tag{9.16}$$

The nominal flexural strength is

$$M_n = A_s f_y\left(d - \frac{a}{2}\right) + A_s' f_y'\left(\frac{a}{2} - d'\right) + 0.85\beta_1 f_c'(b - b_w)h_f\left(\frac{a}{2} - \frac{h_f}{2}\right) \tag{9.17}$$

where A_s is the area of tension steel, A_s' is the area of compression steel, b is the width of the effective flange, b_w is the width of the web, d is the distance between the centroid of tension steel and the most compressed concrete fiber, d' is the distance between the centroid of compression steel and the most compressed concrete fiber, and h_f is the thickness of the effective flange. The concrete stress factor, β_1 can be calculated as

$$\beta_1 = \begin{cases} 0.85 & \text{for } f_c' \leq 28 \text{ MPa} \\ 0.85 - 0.05\left(\dfrac{f_c' - 28}{7}\right) & \text{for } 28 \text{ MPa} \leq f_c' \leq 56 \text{ MPa} \\ 0.65 & \text{for } f_c' \geq 56 \text{ MPa} \end{cases} \tag{9.18}$$

Limits for reinforcement are

- Maximum tensile reinforcement:

$$\frac{c}{d} \leq 0.42 \tag{9.19}$$

When Eq. (9.19) is not satisfied, the reinforced concrete sections become overreinforced and will have sudden brittle compression failure if they are not well confined.

- Minimum tensile reinforcement:

$$\rho_{\min} \geq 0.03\,\frac{f_c'}{f_y'}, \quad \text{where } \rho_{\min} = \text{ratio of tension steel to gross area} \tag{9.20}$$

When Eq. (9.20) is not satisfied, the reinforced concrete sections become underreinforced and will have sudden tension steel fracture failure.

The strain diagram can be used to verify compression steel yielding assumption.

$$f_s' = f_y' \quad \text{if} \quad \varepsilon_s' = \varepsilon_{cu}\left(\frac{c - d'}{c}\right) \geq \frac{f_y'}{E_s} \tag{9.21}$$

If compression steel is not yielding as checked from Eqs. (9.21). The depth of neutral axis, c, and value of nominal flexural strength, M_n, calculated from Eqs. (9.16) and (9.17) are incorrect. The actual forces applied in compression steel reinforcement can be calculated as

$$C_s' = A_s' f_s' = A_s \varepsilon_s' E_s' = A_s \varepsilon_{cu}\left(\frac{d - c}{c}\right) E_s' \tag{9.22}$$

The depth of neutral axis, c, can be solved by substituting Eqs. (9.22) into forces equilibrium Eq. (9.15). The flexural strength, M_n, can then be obtained from Eq. (9.17) with the actual applied compression steel forces. In a typical beam design, the tension steel will always be yielding and the compression steel is close to reaching yielding strength as well.

If the depth of the neutral axis falls within the compression flange ($x \leq h_f$) or for sections without compression flange, then the depth of the neutral axis, c, and the value of nominal flexural strength, M_n, can be calculated by setting b_w equal to b.

9.4.4 Shear Strength

1. Strut-and-Tie Model

The strut-and-tie model should be used for shear and torsion designs of bridge components at locations near discontinuities, such as regions adjacent to abrupt changes in the cross section, openings, and dapped ends. The model should also be used for designing deep footings and pile

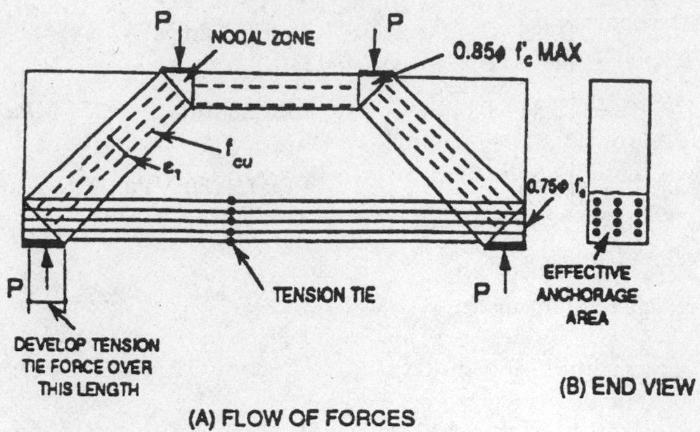

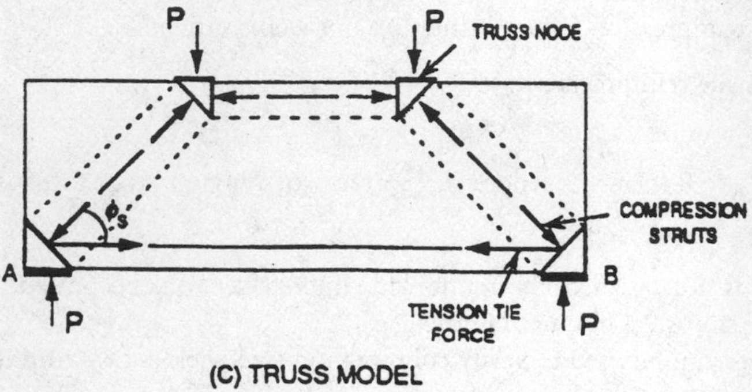

FIGURE 9.5 Strut-and-tie model for a deep beam. (*Source*: AASHTO LRFD Bridge Design Specifications, Figure 5.6.3.2-1, © 1994 by the American Association of State Highway and Transportation Officials, Washington, D.C. With permission.)

caps or in other situations where the distance between the centers of the applied load and the supporting reactions is less than about twice the member thickness. Figure 9.5 shows a strut-and-tie model for a deep beam that is composed of steel tension ties and concrete compressive struts. These are interconnected at nodes to form a truss capable of carrying all applied loads to the supports.

2. Sectional Design Model

The sectional design model can be used for the shear and torsion design for regions of bridge members where plane sections remain plane after loading. It was developed by Collins and Mitchell [4] and is based on the modified compression field theory. The general shear design procedure for reinforced concrete members, containing transverse web reinforcement, is as follows:

- Calculate the effective shear depth d_v:

  Effective shear depth is calculated between the resultants of the tensile and compressive forces due to flexure. This should not be less than the greater of $0.9d_e$ or $0.72h$, where d_e is the effective depth from extreme compression fiber to the centroid of the tensile reinforcement and h is the overall depth of a member.

- Calculate shear stress:

$$v = \frac{V_u}{\phi b_v d_v} \tag{9.23}$$

where b_v is the equivalent web width and V_u is the factored shear demand envelope from the strength limit state.

- Calculate v/f_c', if this ratio is greater than 0.25, then a larger web section needs to be used.
- Assume an angle of inclination of the diagonal compressive stresses, θ, and calculate the strain in the flexural tension reinforcement:

$$\varepsilon_x = \frac{\frac{M_u}{d_v} + 0.5V_u \cot\theta}{E_s A_s} \tag{9.24}$$

where M_u is the factored moment demand. It is conservative to take M_u enveloped from the strength limit state that will occur at that section, rather than a moment coincident with V_u.

- Use the calculated v/f_c' and ε_x to find θ from Figure 9.6 and compare it with the value assumed. Repeat the above procedure until the assumed θ is reasonably close to the value found from Figure 9.6. Then record the value of β, a factor which indicates the ability of diagonally cracked concrete to transmit tension.
- Calculate the required transverse web reinforcement strength, V_s:

$$V_s = \frac{V_u}{\phi} - V_c = \frac{V_u}{\phi} - 0.083\beta\sqrt{f_c'}\, b_v d_v \tag{9.25}$$

where V_c is the nominal concrete shear resistance.

- Calculate the required spacing for the transverse web reinforcement:

$$s \le \frac{A_v f_y d_v \cot\theta}{V_s} \tag{9.26}$$

where A_v is the area of a transverse web reinforcement within distance s.

Check for the minimum transverse web reinforcement requirement:

$$A_v \ge 0.083\sqrt{f_c'}\,\frac{b_v S}{f_y} \quad \text{or} \quad s \le \frac{A_v f_y}{0.083\sqrt{f_c'}\, b_v} \tag{9.27}$$

Check for the maximum spacing requirement for transverse web reinforcements:

$$\text{if } V_u < 0.1\, f_c'\, b_v d_v, \quad \text{then } s \le 0.8 d_v \le 600 \text{ mm} \tag{9.28}$$

$$\text{if } V_u \ge 0.1\, f_c'\, b_v d_v, \quad \text{then } s \le 0.4 d_v \le 300 \text{ mm} \tag{9.29}$$

- Check the adequacy of the longitudinal reinforcements to avoid yielding due to the combined loading of moment, axial load, and shear.

$$A_s f_y \ge \frac{M_u}{d_v \phi} + \left(\frac{V_u}{\phi} - 0.5V_s\right)\cot\theta \tag{9.30}$$

If the above equation is not satisfied, then you need either to add more longitudinal reinforcement or to increase the amount of transverse web reinforcement.

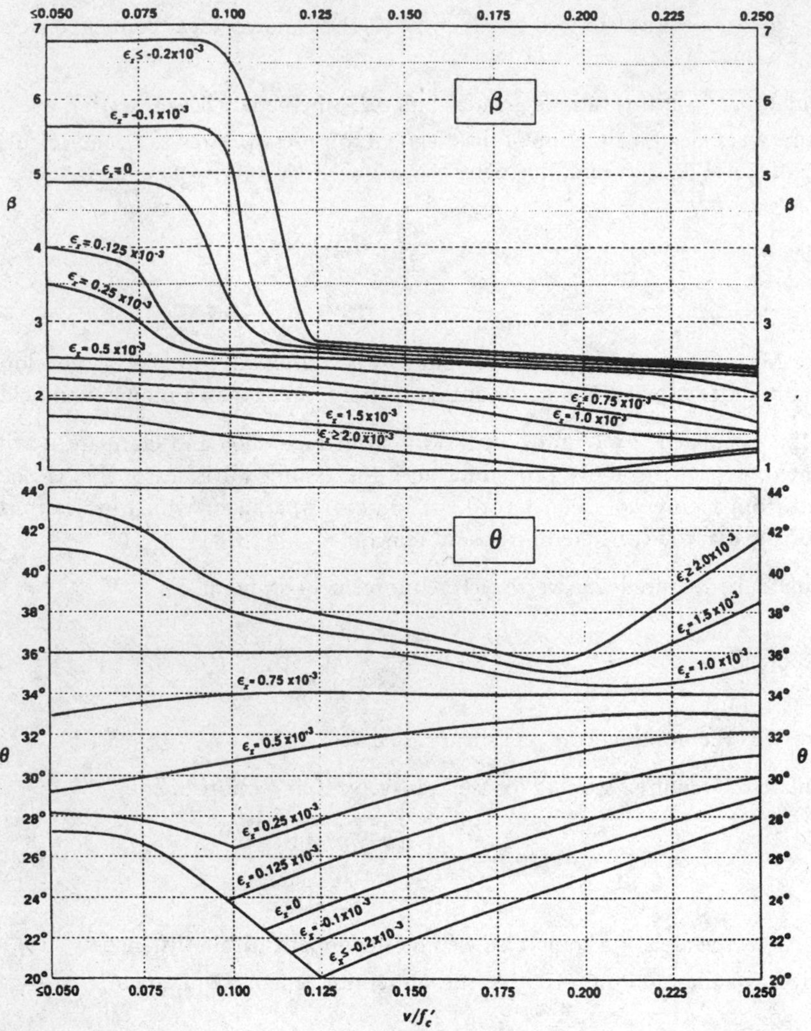

FIGURE 9.6 Values of θ and β for sections with transverse web reinforcement. (*Source*: AASHTO LRFD Bridge Design Specifications, Figure 5.8.3.4.2-1, ©1994 by the American Association of State Highway and Transportation Officials, Washington, D.C. With permission.)

9.4.5 Skewed Concrete Bridges

Shear, in the exterior beam at the obtuse corner of the bridge, needs to be adjusted when the line of support is skewed. The value of the correction factor obtained from AASHTO Table 4.6.2.2.3c-1, needs to be applied to live-load distribution factors for shear. In determining end shear in multibeam bridges, all beams should be treated like the beam at the obtuse corner, including interior beams.

Moment load distribution factors in longitudinal beams on skew supports may be reduced according to AASHTO Table 4.6.2.2.2e-1, when the line supports are skewed and the difference between skew angles of two adjacent lines of supports does not exceed 10°.

9.4.6 Design Information

1. Stress Analysis at Service Limit States [5]

A reinforced concrete beam subject to flexural bending moment is shown in Figure 9.7 and x is the distance between the neutral axis and the extreme compressed concrete fiber. Assume the neutral axis

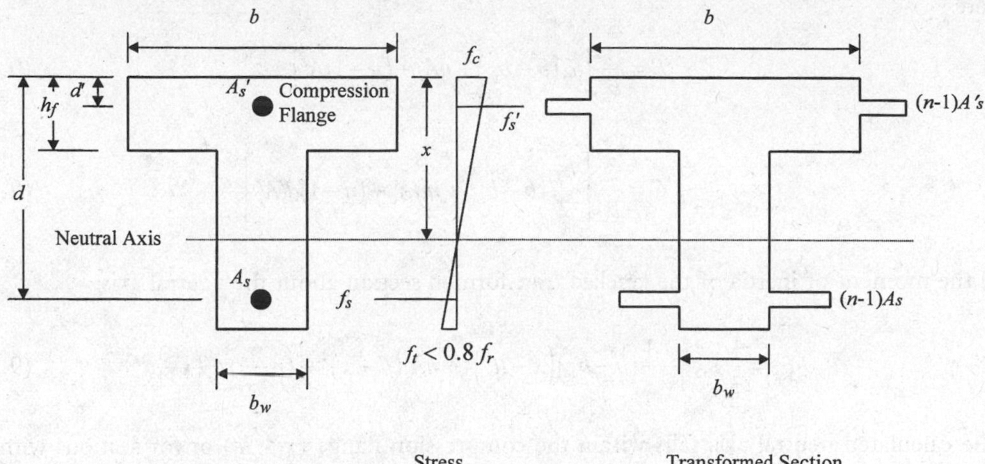

(a) Stress and Transformed Section Before Cracking

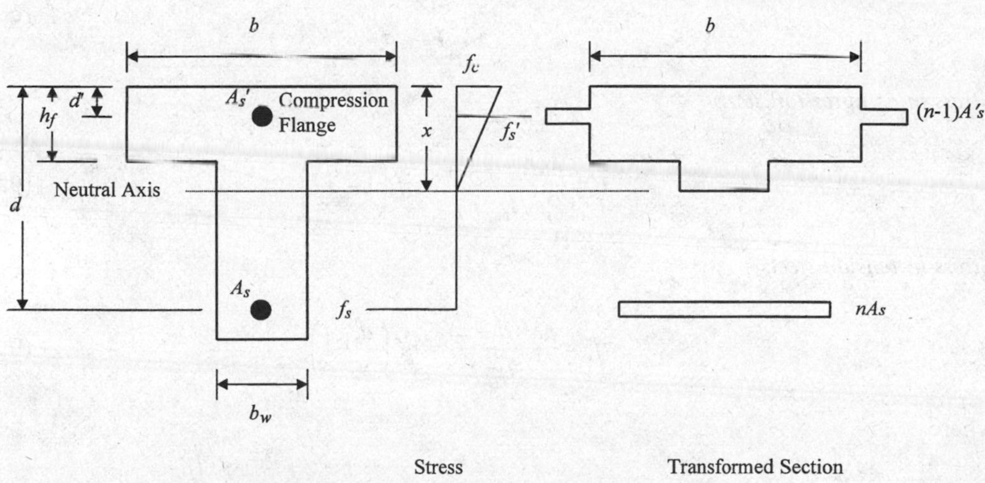

(b) Stress and Transformed Section After Cracking

FIGURE 9.7 Reinforced concrete beam for working stress analysis.

falls within the web ($x > h_f$) and the stress in extreme tension concrete fiber is greater than 80% of the concrete modulus of rupture ($f_t \geq 0.8\, f_r$). The depth of neutral axis, x, can be solved through the following quadratic equation by using the cracked transformed section method (see Figure 9.7).

$$b(x)\left(\frac{x}{2}\right)-(b-b_w)(x-h_f)\left(\frac{x-h_f}{2}\right)+(n-1)A_s'(x-d')=nA_s(d-x) \tag{9.31}$$

$$x=\sqrt{B^2+C}-B \tag{9.32}$$

where

$$B = \frac{1}{b_w}\left[h_f(b - b_w) + nA_s + (n-1)A_s'\right] \tag{9.33}$$

$$C = \frac{2}{b_w}\left[\frac{h_f^2}{2}(b - b_w) + ndA_s + (n-1)d'A_s'\right] \tag{9.34}$$

and the moment of inertia of the cracked transformed section about the neutral axis:

$$I_{\mathrm{cr}} = \frac{1}{3}bx^3 - \frac{1}{3}(b - b_w)(x - h_f)^3 + nA_s(d - x)^2 + (n-1)A_s'(x - d')^2 \tag{9.35}$$

if the calculated neutral axis falls within the compression flange ($x \le h_f$) or for sections without compression flange, the depth of neutral axis, x, and cracked moment of inertia, I_{cr}, can be calculated by setting b_w equal to b.

Stress in extreme compressed concrete fiber:

$$f_c = \frac{Mx}{I_{\mathrm{cr}}} \tag{9.36}$$

Stress in compression steel:

$$f_s' = \frac{nM(x - d')}{I_{\mathrm{cr}}} = nf_c\left(1 - \frac{d'}{x}\right) \tag{9.37}$$

Stress in tension steel:

$$f_s = \frac{nM(d - x)}{I_{cr}} = nf_c\left(\frac{d}{x} - 1\right) \tag{9.38}$$

where

$$n = \frac{E_s}{E_c} \tag{9.39}$$

and M is moment demand enveloped from the service limit state.

2. Effective Flange Width (AASHTO 4.6.2.6)

When reinforced concrete slab and girders are constructed monolithically, the effective flange width (b_{eff}) of a concrete slab, which will interact with girders in composite action, may be calculated as

For interior beams:

$$b_{\mathrm{eff}}^I = \text{the smallest of} \begin{cases} \dfrac{l_{\mathrm{eff}}}{4} \\ 12t_s + b_w \\ \text{the average spacing of adjacent beams} \end{cases} \tag{9.40}$$

TABLE 9.5 Cover for Unprotected Main Reinforcing Steel (mm)

Situation	Cover (mm)
Direct exposure to salt water	100
Cast against earth	75
Coastal	75
Exposure to deicing salt	60
Deck surface subject to tire stud or chain wear	60
Exterior other than above	50
Interior other than above	
• Up to No. 36 Bar	40
• No. 43 and No. 57 Bars	50
Bottom of CIP slab	
• Up to No. 36 Bar	25
• No. 43 and No. 57 Bars	50

Notes:
1. Minimum cover to main bars, including bars protected by epoxy coating, shall be 25 mm.
2. Cover to epoxy-coated steel may be used as interior exposure situation.
3. Cover to ties and stirrups may be 12 mm less than the value specified here, but shall not be less than 25 mm.
4. Modification factors for water:cement ratio, w/c, shall be the following:

   for $w/c \leq 0.40$ modification factor = 0.8
   for $w/c \geq 0.40$ modification factor = 1.2

Source: AASHTO Table C5.12.3 1. (From AASHTO LRFD Bridge Design Specifications, ©1994 by the American Association of State Highway and Transportation Officials, Washington, D.C. With permission.)

For exterior beams:

$$b_{\text{eff}}^{E} = \frac{1}{2} b_{\text{eff}}^{I} + \text{the smallest of} \begin{cases} \dfrac{l_{\text{eff}}}{8} \\ 6t_s + \dfrac{b_w}{2} \\ \text{the width of overhang} \end{cases} \tag{9.41}$$

where the effective span length (l_{eff}) may be calculated as the actual span for simply supported spans. Also, the distance between the points of permanent load inflection for continuous spans of either positive or negative moments (t_s) is the average thickness of the slab, and b_w is the greater of web thickness or one half the width of the top flange of the girder.

3. Concrete Cover (AASHTO 5.12.3)

Concrete cover for unprotected main reinforcing steel should not be less than that specified in Table 9.5 and modified for the water:cement ratio.

9.4.7 Details of Reinforcement

Table 9.6 shows basic tension, compression, and hook development length for Grade 300 and Grade 420 deformed steel reinforcement (AASHTO 5.11.2). Table 9.7 shows the minimum center-to-center spacing between parallel reinforcing bars (AASHTO 5.10.3).

TABLE 9.6 Basic Rebar Development Lengths for Grade 300 and 420 (AASHTO 5.11.2)

Bar	f_c'								
	28 MPa			35 MPa			42 MPa		
Size	Tension	Compression	Hook	Tension	Compression	Hook	Tension	Compression	Hook
Grade 300, f_y = 300 MPa									
13	230	175	240	230	170	215	230	170	200
16	290	220	300	290	210	270	290	210	245
19	345	260	365	345	255	325	345	255	295
22	440	305	420	400	295	375	400	295	345
25	580	350	480	520	335	430	475	335	395
29	735	395	545	655	380	485	600	380	445
32	930	440	610	835	430	550	760	430	500
36	1145	490	680	1020	475	605	935	475	555
43	1420	585	815	1270	570	730	1160	570	665
57	1930	780	1085	1725	765	970	1575	760	885
Grade 420, f_y = 420 MPa									
13	320	245	255	320	235	225	320	235	210
16	405	305	320	405	295	285	405	295	260
19	485	365	380	485	355	340	485	355	310
22	615	425	445	560	410	395	560	410	360
25	810	485	505	725	470	455	665	470	415
29	1025	550	570	920	530	510	840	530	465
32	1300	615	645	1165	600	575	1065	600	525
36	1600	685	710	1430	665	635	1305	665	580
43	1985	820	855	1775	795	765	1620	795	700
57	2700	1095	1140	2415	1060	1020	2205	1060	930

Notes:
1. Numbers are rounded up to nearest 5 mm.
2. Basic hook development length has included reinforcement yield strength modification factor.
3. Minimum tension development length (AASHTO 5.11.2.1). Maximum of (1) basic tension development length times appropriate modification factors (AASHTO 5.11.2.1.2 and 5.11.2.1.3) and (2) 300 mm.
4. Minimum compression development length (AASHTO 5.11.2.2). Maximum of (1) basic compression development length times appropriate modification factors (AASHTO 5.11.2.2.2) and (2) 200 mm.
5. Minimum hook development length (AASHTO 5.11.2.4). Maximum of (1) basic hook development length times appropriate modification factors (AASHTO 5.11.2.4.2), (2) eight bar diameters, and (3) 150 mm.

Except at supports of simple spans and at the free ends of cantilevers, reinforcement (AASHTO 5.11.1.2) should be extended beyond the point at which it is no longer required to resist the flexural demand for a distance of

$$\text{the largest of} \begin{cases} \text{the effective depth of the member} \\ \text{15 times the nominal diameter of a bar} \\ \text{0.05 times the clear span length} \end{cases} \tag{9.42}$$

Continuing reinforcement shall extend not less than the development length beyond the point where bent or terminated tension reinforcement is no longer required for resisting the flexural demand.

TABLE 9.7 Minimum Rebar Spacing for CIP Concrete (mm) (AASHTO 5.10.3)

Bar Size	Minimum Spacing			
		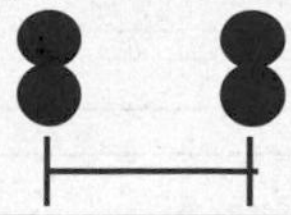		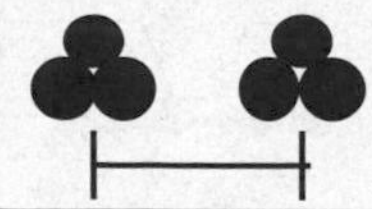
13	51	51	63	63
16	54	56	70	70
19	57	68	76	83
22	60	78	82	96
25	64	90	90	110
29	72	101	101	124
32	81	114	114	140
36	90	127	127	155
43	108	152	152	
57	143	203	203	

Notes:
1. Clear distance between bars should not be less than 1.5 times the maximum size of the course aggregate.
2. Note 1 does not need to be verified when maximum size of the course aggregate grading is less than 25 mm.
3. Bars spaced less than $3d_b$ on center require modification of development length (AASHTO 5.11.2.1.2).

For negative moment reinforcement, in addition to the above requirement for bar cutoff, it must be extended to a length beyond the inflection point for a distance of

$$\text{the largest of} \begin{cases} \text{the effective depth of the member} \\ \text{12 times the nominal diameter of a bar} \\ \text{0.0625 times the clear span length} \end{cases} \tag{9.43}$$

9.5 Design Examples

9.5.1 Solid Slab Bridge Design

Given

A simple span concrete slab bridge with clear span length (S) of 9150 mm is shown in Figure 9.8. The total width (W) is 10,700 mm, and the roadway is 9640 wide (W_R) with 75 mm (d_W) of future wearing surface.

The material properties are as follows: Density of wearing surface $\rho_w = 2250$ kg/m^3; concrete density $\rho_c = 2400$ kg/m^3; concrete strength $f_c' = 28$ MPa, $E_c = 26\ 750$ MPa; reinforcement $f_y = 420$ MPa, $E_s = 200{,}000$ MPa; $n = 8$.

Requirements

Design the slab reinforcement base on AASHTO-LRFD (1994) Strength I and Service I (cracks) Limit States.

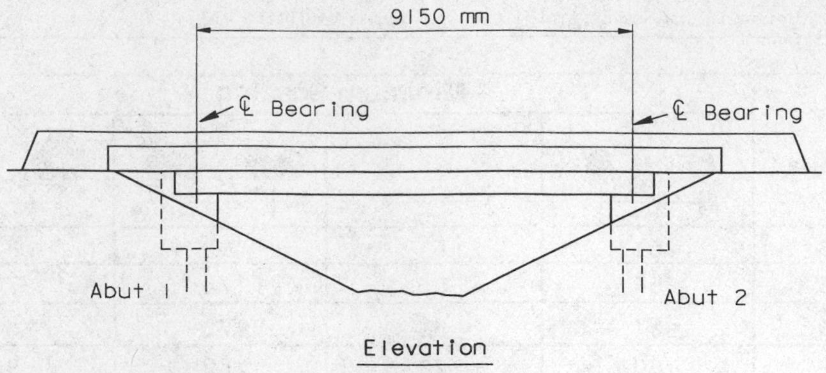

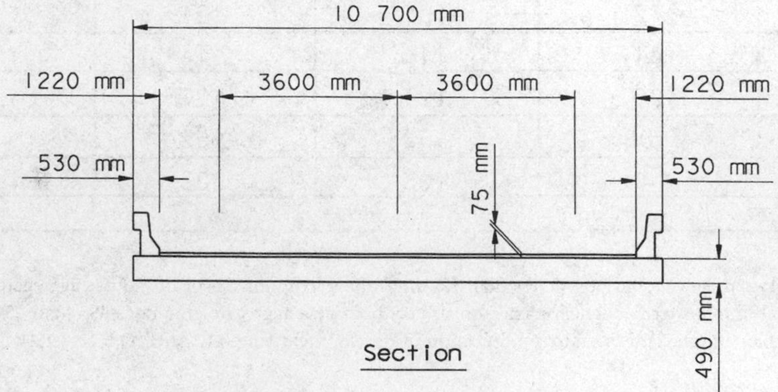

FIGURE 9.8 Solid slab bridge design example.

Solution

1. **Select Deck Thickness (Table 9.4)**

$$h_{min} = 1.2\left(\frac{S+3000}{30}\right) = 1.2\left(\frac{9150+3000}{30}\right) = 486\,\text{mm}$$

$$\underline{\text{Use } h = 490\text{ mm}}$$

2. **Determine Live Load Equivalent Strip Width (AASHTO 4.6.2.3 and 4.6.2.1.4b)**
   a. *Interior strip width*:
      i. Single-lane loaded:

$$E_{\text{interior}} = 250 + 0.42\sqrt{L_1 W_1}$$

L_1 = lesser of actual span length and 18,000 mm

W_1 = lesser of actual width or 9000 mm for single lane loading or 18,000 mm for multilane loading

$$E_{\text{interior}} = 250 + 0.42\sqrt{(9150)(9000)} = 4061\text{ mm}$$

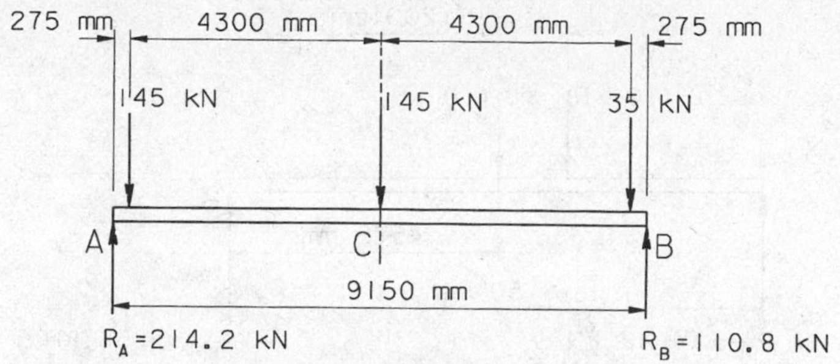

FIGURE 9.9 Position of design truck for maximum moment.

ii. Multilane loaded:

$$N_L = INT\left(\frac{W}{3600}\right) = INT\left(\frac{10,700}{3600}\right) = 2$$

$$\frac{W}{N_L} = \frac{10,700}{2} = 5350 \text{ mm}$$

$$E_{interior} = 2100 + 0.12\sqrt{L_1 W_1} = 2100 + \sqrt{(9150)(10,700)} = 3287 \text{ mm} < 5350 \text{ mm}$$

Use $E_{interior} = 3287$ mm

b. *Edge strip width:*

E_{edge} = the distance between the edge of the deck and the inside face of the barrier + 300 mm + ½ strip width < full strip or 1800 mm

$$E_{edge} = 530 + 300 + \frac{3287}{2} = 2324 \text{ mm} > 1800 \text{ mm}$$

Use $E_{edge} = 1800$ mm

3. Dead Load

Slab: $W_{slab} = (0.49)(2400)(9.81)(10^{-3}) = 11.54$ kN/m^2

Future wearing: $W_{fw} = (0.075)(2250)(9.81)(10^{-3}) = 1.66$ kN/m^2

Assume 0.24 m^3 concrete per linear meter of concrete barrier

Concrete barrier: $W_{barrier} = (0.24)(2400)(9.81)(10^{-3}) = 5.65$ kN/m^2

4. Calculate Live-Load Moments

Moment at midspan will control the design.

a. *Moment due to the design truck* (see Figure 9.9):

$$M_{LL\text{-}Truck} = (214.2)(4.575) - (145)(4.3) = 356.47 \text{ kN·m}$$

b. *Moment due to the design tandem* (see Figure 9.10):

$$M_{LL\text{-}Tandem} = (95.58)(4.575) = 437.28 \text{ kN·m}.$$

Design Tandem Controls

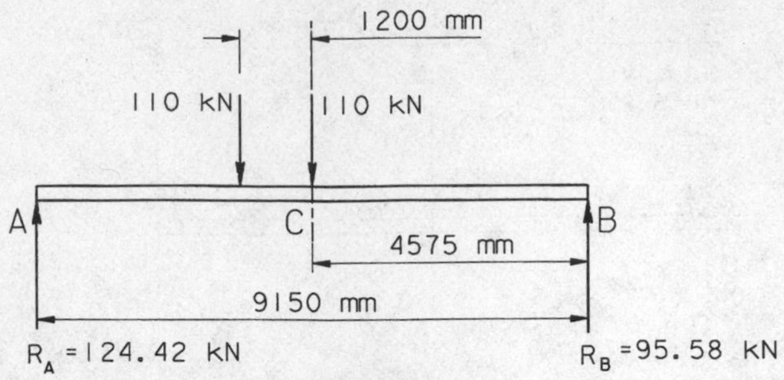

FIGURE 9.10 Position of tandem for maximum moment.

c. *Moment due to lane load:*

$$M_{\text{LL-Lane}} = \frac{(9.3)(9.15)^2}{8} = 97.32 \text{ kN·m}$$

5. Determine Load Factors (AASHTO Table 3.4.1-1) and Load Combinations (AASHTO 1.3.3-5)

a. *Strength I Limit State load factors:*

Weight of superstructure (DC): 1.25
Weight of wearing surface (DW): 1.50
Live Load (LL): 1.75
$\eta_d = 0.95, \eta_R = 1.05, \eta_I = 0.95$
$\eta = (0.95)(1.05)(0.95) = 0.948 \leq 0.95$

Use $\eta = 0.95$

b. *Interior strip moment* (1 m wide) (AASHTO 3.6.2.1 and 3.6.1.2.4):

Dynamic load factor IM = 0.33

Lane load $M_{\text{LL-Lane}} = \left(\frac{97.32}{3.287}\right) = 29.61 \text{ kN·m}$

Live load $M_{\text{LL+IM}} = (1+0.33)\left(\frac{437.28}{3.287}\right) + 29.61 = 206.54 \text{ kN·m}$

Future wearing $M_{\text{DW}} = \frac{W_{fw}L^2}{8} = \frac{(1.66)(9.15)^2}{8} = 17.37 \text{ kN·m}$

Dead load $M_{DC} = \frac{W_{\text{slab}}L^2}{8} = \frac{(11.54)(9.15)^2}{8} = 120.77 \text{ kN·m}$

Factored moment

$$\begin{aligned} M_U &= \eta[1.25(M_{\text{DC}}) + 1.50(M_{\text{DW}}) + 1.75(M_{\text{LL+IM}})] \\ &= (0.95)\,[1.25\,(120.77) + (1.50)\,(17.37) + (1.75)\,(206.54)] \\ &= 511.54 \text{ kN·m} \end{aligned}$$

c. *Edge strip moment* (1 m wide) (AASHTO Table 3.6.1.1.2-1):

End strip is limited to half lane width, use multiple presence factor 1.2 and half design lane load.

Lane load $M_{\text{LL-Lane}} = (1.2)\left(\frac{1}{2}\right)\left(\frac{97.3}{1.8}\right) = 32.44 \text{ kN·m}$

Live load $M_{\text{LL+IM}} = (1 + 0.33)(1.2)\left(\frac{1}{2}\right)\left(\frac{437.28}{1.8}\right) + 32.44 = 226.3 \text{ kN·m}$

Dead load $M_{\text{DC}} = \left(11.54 + \frac{5.65}{1.8}\right)\left(\frac{9.15^2}{8}\right) = 153.63 \text{ kN·m}$

Future wearing $M_{\text{DW}} = (1.66)\left(\frac{1.8 - 0.53}{1.8}\right)\left(\frac{9.15^2}{8}\right) = 12.25 \text{ kN·m}$

Factored moment $M_U = (0.95)[(1.25)(153.63) + (1.50)(12.25) + (1.75)(226.3)] = 579.12 \text{ kN·m}$

6. Reinforcement Design

a. *Interior strip*

Assume No. 25 bars, $d = 490 - 25 - \left(\frac{25}{2}\right) = 452.5$ mm.

The required reinforcements are calculated using Eqs. (9.11), (9.16), and (9.17). Neglect the compression steel and set $b_w = b$ for sections without compression flange.

$$M_u = \phi A_s f_y\left(d - \frac{a}{2}\right) \quad \text{and} \quad a = c\beta_1 = \frac{A_s f_y}{0.85 f'_c b_w}$$

A_s can be solved by substituting a into M_u or

$$R_u = \frac{M_u}{\phi b d^2} = \frac{511.54 \times 10^6}{(0.9)(1000)(452.5)^2} = 2.766 \text{ N/mm}$$

$$m = \frac{f_y}{(0.85) f'_c} = \frac{420}{(0.85)(28)} = 17.647$$

$$\rho = \frac{1}{m}\left[1 - \sqrt{1 - \frac{2mR_u}{f_y}}\right] = \frac{1}{17.647}\left[1 - \sqrt{1 - \frac{2(17.647)(2.776)}{420}}\right] = 0.00705$$

Required reinforced steel $A_s = \rho bd = (0.00705)(1000)(452.5) = 3189 \text{ mm}^2/\text{m}$.

Maximum allowed spacing of No. 25 bar = 510/3189 = 0.160 m.

Try No. 25 bars at 150 mm.

i. Check limits for reinforcement:

$\beta_1 = 0.85$ for $f'_c = 28$ MPa; see Eq. (9.18)

$$c = \frac{A_s f_y}{0.85 \beta_1 f'_c b_w} = \frac{(510)(420)}{0.85(0.85)(28)(150)} = 70.6 \text{ mm}$$

from Eqs. (9.19),

$$\frac{c}{d} = \frac{70.79}{452.5} = 0.156 \leq 0.42 \qquad \text{OK}$$

from Eqs. (9.20),

$$\rho_{\min} = \frac{510}{(150)(452.5)} = 0.007\,51 \geq (0.03)\left(\frac{28}{420}\right) = 0.002 \qquad \text{OK}$$

ii. Check crack control:

Service load moment M_{sa} $= 1.0[1.0(M_{\text{DC}}) + 1.0(M_{\text{DW}}) + 1.0(M_{\text{LL+IM}})]$
$= [120.77 + 17.37 + (176.93 + 29.61)]$
$= 344.68$ kN·m

$$0.8f_r = 0.8(0.63\sqrt{f_c'}) = 0.8(0.63)\sqrt{28} = 2.66 \text{ MPa}$$

$$f_c = \frac{M_{sa}}{S} = \frac{344{,}680}{\frac{1}{6}(490)^2} = 8.61\text{MPa} \geq 0.8f_r;\text{, Section is cracked}$$

Cracked moment of inertia can be calculated by using Eqs. (9.32) to (9.35).

$n = 8$, $b = 150.0$ mm, $A_s = 510$ mm, $d = 452.5$ mm.

$$B = \frac{1}{b}(nA_s) = \frac{1}{150}(8)(510) = 27.2$$

$$C = \frac{2}{b}(ndA_s) = \frac{2}{150}(8)(452.5)(510) = 24616$$

$$x = \sqrt{B^2 + C} - B = \sqrt{(27.2)^2 + (24616)} - (27.2) = 132 \text{ mm}$$

$$I_{cr} = \frac{1}{3}bx^3 + nA_s(d-x)^2 = \frac{1}{3}(150)(132)^3 + (8)(510)(452.5-132)^2 = 534.1\times10^6\text{mm}^4$$

From Eq. (9.38) $f_s = n\dfrac{M_{sa}(d-x)}{I_{cr}} = (8)\dfrac{(344{,}680)(452.5-132)}{534.1\times10^6} = 248$ MPa

Allowable tensile stress in the reinforcement can be calculated from Eq. (9.5) with $Z =$ 23,000 N/mm for moderate exposure and

$$d_c = 25 + \frac{25}{2} = 37.5\text{ mm}$$

$$A = 2d_c \times \text{bar spacing} = (2)(37.5)(150) = 11{,}250 \text{ mm}$$

$$f_{sa} = \frac{Z}{(d_c A)^{1/3}} \leq 0.6f_y$$

$$f_{sa} = \frac{23{,}000}{[(37.5)(11{,}250)]^{1/3}} = \frac{23{,}000}{75} = 307 \text{ MPa} \geq 0.6f_y = 0.6\,(420) = 252 \text{ MPa}$$

$$f_s = 248\text{ MPa} \leq f_{sa} = 252\text{ MPa}, \qquad \text{OK}$$

Use No. 25 Bar @150 mm for interior strip

b. *Edge strip*:

By similar procedure, Edge Strip Use No. 25 bar at 125 mm

7. Determine Distribution Reinforcement (AASHTO 5.14.4.1)
The bottom transverse reinforcement may be calculated as a percentage of the main reinforcement for positive moment:

$$\frac{1750}{\sqrt{L}} \leq 50\,\%, \text{ that is, } \frac{1750}{\sqrt{9150}} = 18.3\% \leq 50\%$$

a. *Interior strip*:
Main reinforcement: No. 25 at150 mm,

$$A_s = \frac{510}{150} = 3.40 \ \text{mm}^2/\text{mm}.$$

Required transverse reinforcement = (0.183)(3.40) = 0.622 mm²/mm

Use No. 16 @ 300 mm transverse bottom bars,

$$A_s = \frac{199}{300} = 0.663 \ \text{mm}^2/\text{mm}$$

b. *End strip*:
Main reinforcement: No. 25 at 125 mm,

$$A_s = \frac{510}{125} = 4.08 \ \text{mm}^2/\text{mm}$$

Required transverse reinforcement = (0.183)(4.08) = 0.746 mm²/mm

Use No. 16 at 250 mm, $A_s = 0.79$ mm²/mm.

For construction consideration, Use No. 16 @250 mm across entire width of the bridge.

8. Determine Shrinkage and Temperature Reinforcement (AASHTO 5.10.8)
Temperature

$$A_s \geq 0.75\frac{A_g}{f_y} = 0.75\frac{(1)(490)}{420} = 0.875 \ \text{mm}^2/\text{mm in each direction}$$

Top layer = 0.875/2 = 0.438 mm²/mm

Use No. 13 @ 300 mm transverse top bars, $A_s = 0.430$ mm²/mm

9. Design Sketch
See Figure 9.11 for design sketch in transverse section.

10. Summary
To complete the design, loading combinations for all limit states need to be checked. Design practice should also give consideration to long-term deflection, cracking in the support area for longer or continuous spans. For large skew bridges, alteration in main rebar placement is essential.

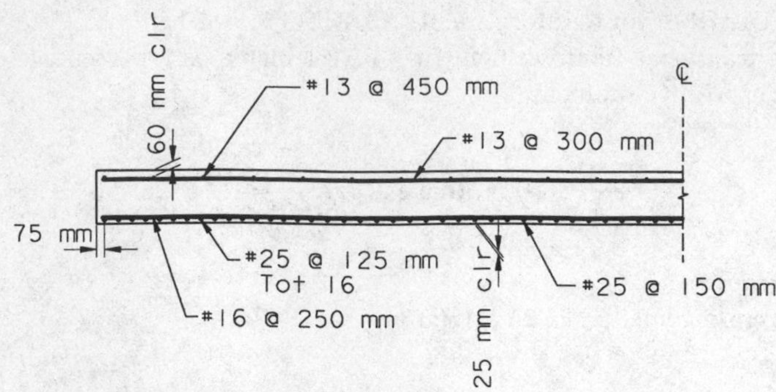

FIGURE 9.11 Slab reinforcement detail.

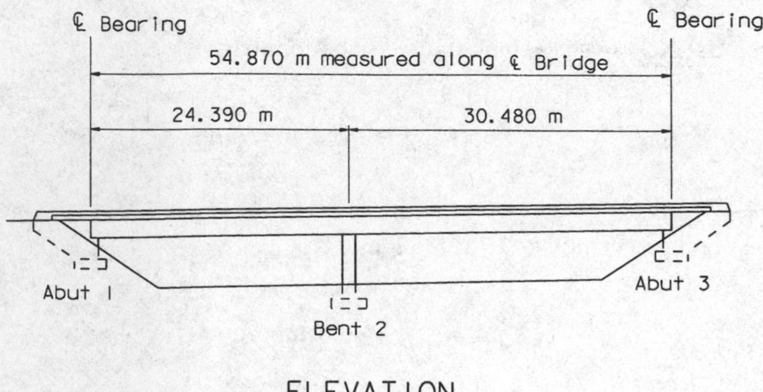

FIGURE 9.12 Two-span reinforced box girder bridge.

9.5.2 Box-Girder Bridge Design

Given

A two-span continuous cast-in-place reinforced concrete box girder bridge, with span length of 24 390 mm (L_1) and 30 480 mm (L_2), is shown in Figure 9.12. The total superstructure width (W) is 10 800 mm, and the roadway width (W_R) is 9730 mm with 75 mm (d_W) thick of future wearing surface.

The material properties are assumed as follows: Density of wearing surface $\rho_w = 2250$ kg/m³; concrete density $\rho_c = 2400$ kg/m³; concrete strength $f_c' = 28$ MPa, $E_c = 26\ 750$ MPa; reinforcement $f_y = 420$ MPa, $E_s = 200\ 000$ MPa.

Requirements

Design flexural and shear reinforcements for an exterior girder based on AASHTO-LRFD (1994) Limit State Strength I, Service I (cracks and deflection), and Fatigue Limit States.

Solution

1. **Determine Typical Section (see Figure 9.13)**
   a. *Section dimensions*:

      Try the following dimensions:

      Overall Structural Thickness, $h =$ <u>1680 mm</u> (Table 9.4)

      Effective length, $s = 2900 - 205 = 2695$ mm

      Design depth (*deck slab*),

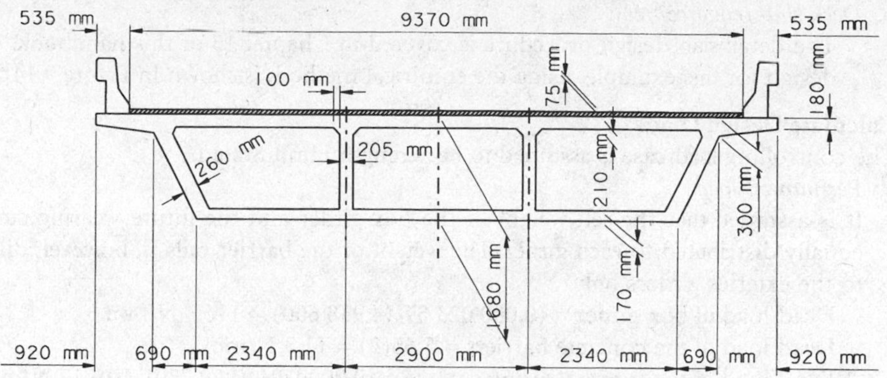

FIGURE 9.13 Typical section.

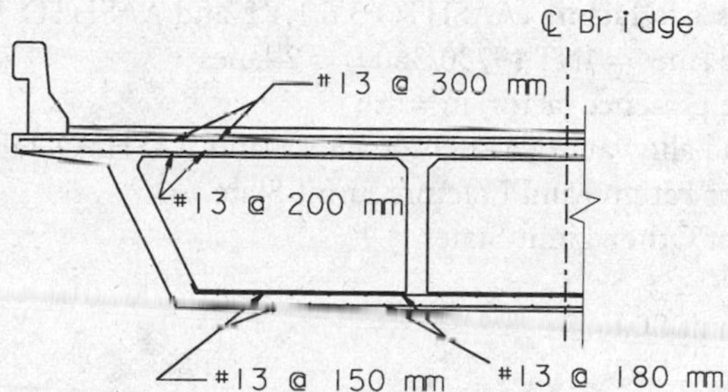

FIGURE 9.14 Slab reinforcement.

$$t_{top} = \underline{210 \text{ mm}}$$

$$> \frac{1}{20}(2900 - 205 - 100 \cdot 2) = 124.8 \text{ mm} \quad \text{(AASHTO 5.14.1.3)}$$

$$\frac{s}{t_{top}} = \frac{2695}{210} = 12.8 < 18 \quad \text{(AASHTO 9.7.2.4)}$$

Bottom flange depth,

$$t_{bot} = \underline{170 \text{ mm}} > 140 \text{ mm} \quad \text{(AASHTO 5.14.1.3)}$$

$$> \frac{1}{16}(2900 - 205 - 100 \cdot 2) = 156 \text{ mm} \quad \text{(AASHTO 5.14.1.3)}$$

Web thickness, $b_w = \underline{205 \text{ mm}} > 200$ mm for ease of construction (AASHTO 5.14.1.3)

b. *Deck slab reinforcement*:

The detail slab design procedure is covered in Chapter 15 of this handbook. The slab design for this example, using the empirical method, is shown in Figure 9.14.

2. **Calculate Design Loads**

The controlling load case is assumed to be Strength Limit State I.

a. *Permanent load*:

It is assumed that the self-weight of the box girder and the future wearing surface are equally distributed to each girder. The weight of the barrier rails is, however, distributed to the exterior girders only.

Dead load of box girder = (0.000 023 57)(4 938 600) = 116.4 N/mm
Dead load of the concrete barriers = 5.65(2) = 11.3 N/mm
Dead load of the future wearing surface = (0.0000221)(729 750) = 16.12 N/mm

b. *Live loads*:

i. Vehicle live loads:
A standard design truck (AASHTO 3.6.1.2.2), a standard design tandem (AASHTO 3.6.1.2.3), and the design lane load (AASHTO 3.6.1.2.4) are used to compute the extreme force effects.

ii. Multiple presence factors (AASHTO 3.6.1.1.2 and AASHTO Table 3.6.1.1.2-1):
No. of traffic lanes = INT (9730/3600) = 2 lanes
The multiple presence factor, m = 1.0

iii. Dynamic load allowance (AASHTO 3.6.2.1 and AASHTO Table 3.6.2.1-1):
IM = 15% for Fatigue and Fracture Limit State
IM = 33% for Other Limit States

c. *Load modifiers*:

For Strength Limit State:

$$\eta_D = 0.95; \quad \eta_R = 0.95; \quad \eta_I = 1.05; \quad \text{and} \quad \eta = \eta_D\eta_R\eta_I = 0.95 \text{ (AASHTO 1.3.2)}$$

For Service Limit State:

$$\eta_D = 1.0; \quad \eta_R = 1.0; \quad \eta_I = 1.0; \quad \text{and} \quad \eta = \eta_D\eta_R\eta_I = 1.0 \text{ (AASHTO 1.3.2)}$$

d. *Load factors*:

$$\gamma_{DC} = 0.9 \sim 1.25; \quad \gamma_{DW} = 0.65 \sim 1.50; \quad \gamma_{LL} = 1.75$$

e. *Distribution factors for live-load moment and shear* (AASHTO 4.6.2.2.1):

i. Moment distribution factor for exterior girders:
For Span 1 and Span 2:

$$W_e = \frac{2900}{2} + 1211 = 2661 \text{ mm} < S = 2900 \text{ mm}$$

$$g_m^E = \frac{W_e}{4300} = \frac{2661}{4300} = 0.619$$

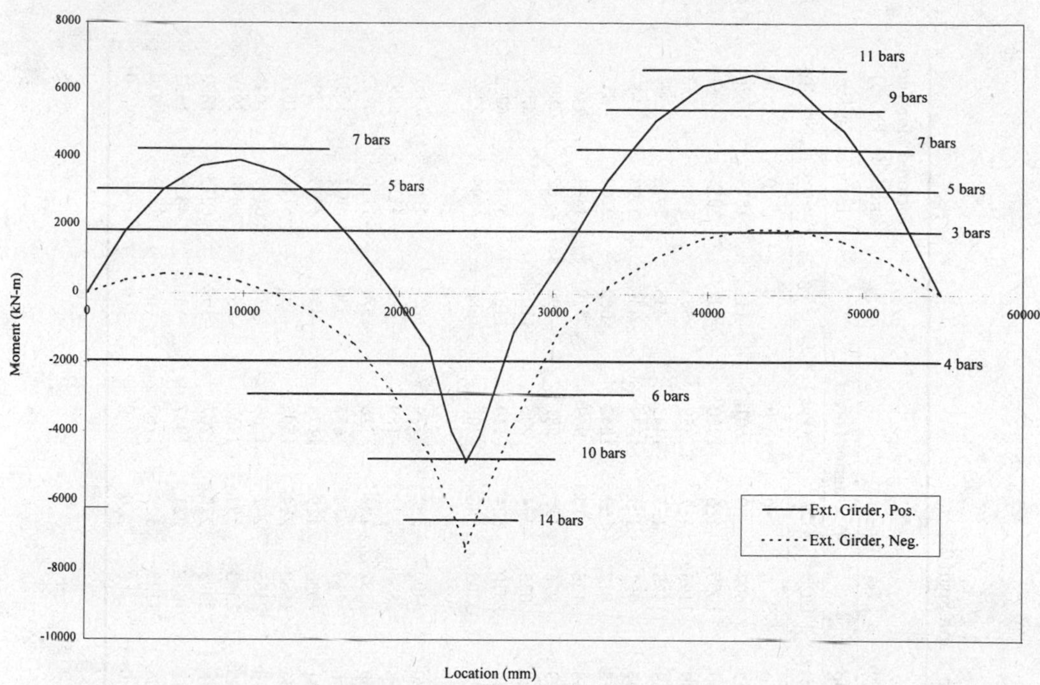

FIGURE 9.15 Design moment envelope and provided moment capacity with reinforcement cut-off.

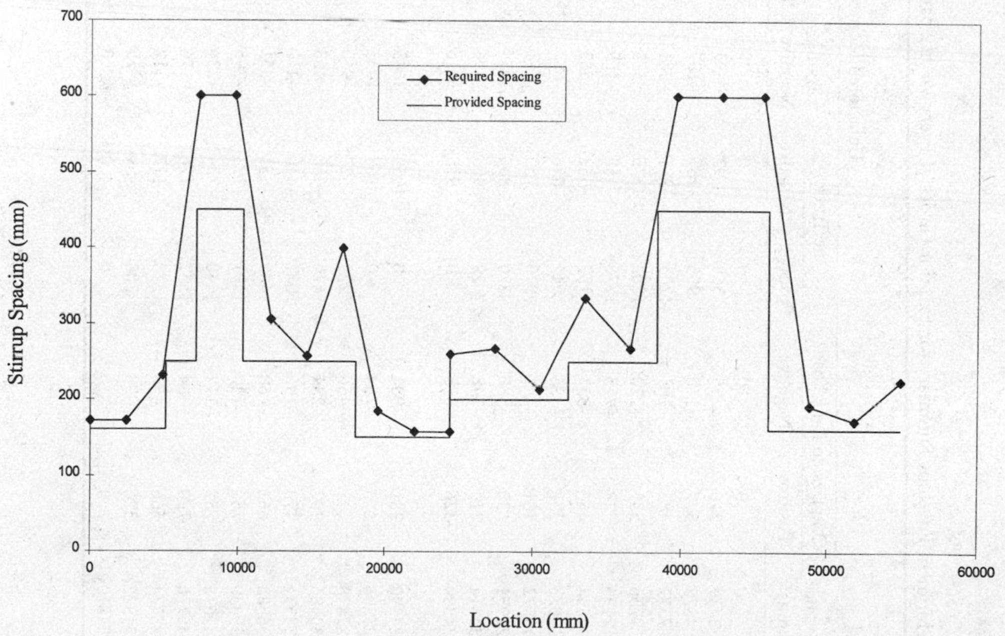

FIGURE 9.16 Shear reinforcement spacing for the exterior girder.

TABLE 9.8 Moment Envelope Summary for Exterior Girder at Every 1/10 of Span Length of Span 1 and Span 2

		Unfactored Moment Envelope (kN-m)												Factored Moment Envelope (kN-m)	
	Distance	One Design Lane Load		One Truck		Train	Live Load Envelope		Exterior Girder					Exterior Girder	
Span	(mm)	Positive	Negative	Positive	Negative	Negative	Positive	Negative	DC	DW	LL (Pos.)	LL (Neg.)		Positive	Negative
0.0 L_1	0	0	0	0	0	0	0	0	0	0	0	0		0	0
0.1 L_1	2439	216	−55	568	−93	−93	971	−179	607	70	601	−111		1819	378
0.2 L_1	4878	377	−110	955	−187	−187	1647	−359	1005	116	1020	−222		3053	561
0.3 L_1	7317	482	−165	1203	−281	−281	2082	−539	1196	137	1289	−333		3758	553
0.4 L_1	9756	533	−220	1303	−375	−375	2266	−719	1178	135	1403	−445		3923	350
0.5 L_1	12195	528	−275	1278	−468	−469	2228	−897	951	109	1379	−556		3577	−43
0.6 L_1	14634	467	−331	1157	−562	−563	2006	−1078	517	59	1242	−668		2762	−632
0.7 L_1	17073	352	−386	912	−656	−657	1565	−1258	−126	−15	969	−779		1494	−1466
0.8 L_1	19512	181	−441	570	−750	−751	939	−1439	−976	−113	581	−890		62	−2800
0.9 L_1	21951	38	−580	220	−844	−1055	331	−1785	−2035	−234	205	−1105		−1545	−4587
0.96 L_1	23414	−216	−748	29	−903	−1396	−177	−2344	−2810	−324	−110	−1451		−3981	−6211
1.0 L_1	24390	−326	−878	0	−938	−1616	−326	−2725	−3303	−380	−202	−1686		−4799	−7267
0.0 L_2	24390	−281	−904	0	−1059	−1643	−281	−2780	−3400	−392	−174	−1721		−4885	−7457
0.03 L_2	25304	−273	−754	0	−868	−1433	−273	−2394	−2835	−327	−169	−1482		−4112	−6295
0.1 L_2	27438	41	−466	224	−527	−960	339	−1569	−1597	−184	210	−971		−1130	−3772
0.2 L_2	30486	202	−234	640	−468	−468	1053	−856	−119	−14	652	−530		974	−1042
0.3 L_2	33534	470	−196	1072	−410	−410	1896	−741	1035	119	1173	−459		3349	195
0.4 L_2	36582	662	−168	1403	−351	−351	2528	−635	1862	214	1565	−393		5118	1071
0.5 L_2	39630	768	−140	1591	−293	−293	2884	−530	2365	272	1785	−328		6163	1645
0.6 L_2	42678	787	−112	1640	−234	−234	2968	−423	2543	292	1837	−262		6490	1919
0.7 L_2	45726	720	−84	1532	−176	−176	2758	−318	2395	275	1707	−197		6073	1890
0.8 L_2	48774	566	−56	1210	−117	−117	2175	−212	1922	221	1347	−131		4835	1562
0.9 L_2	51822	326	−28	708	−59	−59	1268	−106	1123	129	785	−66		2822	930
1.0 L_2	54870	0	0	0	0	0	0	0	0	0	0	0		0	0

TABLE 9.9 Shear Envelope Summary at Every 1/10 of Span 1 and Span 2

		Unfactored Shear Envelope (kN)										Factored Shear Envelope (kN)	
	Distance	One Design Lane Load		One Truck		Live Load Envelope		Exterior Girder				Exterior Girder	
Span	(mm)	Positive	Negative	Positive	Negative	Positive	Negative	DC	DW	LL (Pos.)	LL (Neg.)	Positive	Negative
0.0 L_1	0	100	−23	272	−38	462	−74	292	34	353	−56	981	176
0.1 L_1	2439	77	−23	233	−38	387	−74	206	24	295	−56	770	97
0.2 L_1	4878	55	−23	195	−38	314	−74	120	14	240	−56	562	18
0.3 L_1	7317	32	−23	159	−70	243	−116	36	4	186	−89	358	−115
0.4 L_1	9756	9	−23	124	−105	174	−163	−50	−6	123	−124	175	−274
0.5 L_1	12195	−13	−36	92	−143	109	−226	−136	−16	84	−173	14	−471
0.6 L_1	14634	−23	−59	64	−178	62	−396	−221	−26	48	−226	−126	−675
0.7 L_1	17073	−23	−81	41	−212	32	−363	−306	−35	24	−278	−243	−875
0.8 L_1	19512	−23	−104	23	−243	8	−427	−391	−45	6	−327	−353	−1072
0.9 L_1	21951	−23	−127	8	−271	−12	−487	−477	−55	−9	−373	−660	−1264
1.0 L_1	24390	−23	−149	0	−295	−23	−541	−563	−65	−16	−414	−787	−1449
0.0 L_2	24390	171	9	299	0	569	9	645	74	442	6	1606	882
0.1 L_2	27438	143	9	277	−3	511	5	539	62	383	4	1365	734
0.2 L_2	30486	115	9	250	−15	448	−11	432	50	335	−8	1141	386
0.3 L_2	33534	86	9	220	−38	379	−42	325	38	284	−31	910	249
0.4 L_2	36582	58	9	186	−59	305	−69	218	25	229	−52	675	115
0.5 L_2	39630	30	9	149	−87	228	−107	112	13	171	−80	435	−30
0.6 L_2	42678	9	−8	111	−119	157	−166	5	1	117	−125	202	−202
0.7 L_2	45726	9	−36	75	−154	109	−241	−102	−12	81	−180	41	−437
0.8 L_2	48774	9	−65	48	−192	73	−320	−209	−24	55	−240	−103	−681
0.9 L_2	51822	9	−93	20	−231	36	−400	−316	−36	27	−300	−248	−925
1.0 L_2	54870	9	−121	19	−272	34	−483	−422	−49	26	−362	−348	−1171

ii. Shear distribution factor for exterior girders:

Design Lane	Span 1	Span 2
One design lane loaded	$g_v^E = \dfrac{0.5(1015+2815)}{\left(\dfrac{\sqrt{5}}{2}\right)(2884)} = 0.594$	$g_v^E = \dfrac{0.5(1015+2815)}{\left(\dfrac{\sqrt{5}}{2}\right)(2884)} = 0.594$
Two or more design lanes loaded	$d_e = 1066-535 = 531 < 1500$	$d_e = 1066-535 = 531 < 1500$
	$e = 0.64+\dfrac{531}{3800} = 0.78$	$e = 0.64+\dfrac{531}{3800} = 0.78$
	$g_v^E = 0.78\left(\dfrac{2900}{2200}\right)^{0.9}\left(\dfrac{1680}{24,385}\right)^{0.1}$	$g_v^E = 0.78\left(\dfrac{2900}{2200}\right)^{0.9}\left(\dfrac{1680}{30,480}\right)^{0.1}$
	$= 0.765$	$= 0.749$
Govern	0.765	0.749

f. *Factored moment envelope and shear envelope*:
The moment and shear envelopes for the exterior girder, unfactored and factored based on Strength Limit State I, are listed in Tables 9.8 and 9.9. Figures 9.15 and 9.16 show the envelope diagram for moments and shears based on Strength Limit State I, respectively.

3. **Flexural Design**
   a. *Determine the effective flange width* (Section 9.4.6):
      i. Effective compression flange for positive moments:
      Span 1:
      For interior girder,

$$b^I_{\text{top}} = \text{the smallest of} \begin{cases} \dfrac{1}{4}L_{1,\text{eff}} = \dfrac{1}{4}(0.65)(24,390) = 3963 \text{ mm} \\ 12t_{\text{top}} + b_w = 12(210)+(205) = \underline{2725 \text{ mm}} \;\; \textit{governs} \\ \textit{the average spacing of adjacent beams} = 2900 \text{ mm} \end{cases}$$

For exterior girder,

$$b^E_{top} = \frac{1}{2}b^I_{top} + \text{the smallest of}$$

$$\begin{cases} \dfrac{1}{8}L_{1,eff} = \dfrac{1}{8}(0.65)(24,390) = 1982 \text{ mm} \\ 6t_{\text{top}} + \dfrac{1}{2}b_w = (6)(210) + \dfrac{1}{2}(291) = 1405 \text{ mm} \\ \text{the width of the overhang} = 920 + \dfrac{291}{2} = \underline{1065 \text{ mm}} \;\; \textit{governs} \end{cases}$$

$$= \frac{1}{2}(2724) + 1065$$

$$= 2427 \text{ mm}$$

Span 2: The effective flange widths for Span 2 turns out to be the same as those in Span 1.

ii. Effective compression flange for negative moments:

Span 1:

For interior girder,

$$b^{I}_{\text{bot}} = \text{the smallest of} \begin{cases} \dfrac{1}{4}L_{\text{eff}} = \dfrac{1}{4}[(0.5)(24\ 390) + (0.25)(30\ 480) = 4954 \text{ mm} \\ 12t_{\text{bot}} + b_w = 12(170) + (205) = \underline{2245 \text{ mm}} \ \textit{governs} \\ \textit{the average spacing of adjacent beams} = 2900 \text{ mm} \end{cases}$$

For exterior girder,

$$b^{E}_{\text{bot}} = \frac{1}{2}b^{I}_{\text{bot}} + \text{the smallest of}$$

$$\begin{cases} \dfrac{1}{8}L_{\text{eff}} = \dfrac{1}{8}[(0.5)(24,390) + (0.25)(30,480) = 2477 \text{ mm} \\ 6t_{\text{bot}} + \dfrac{1}{2}b_w = (6)(170) + \dfrac{1}{2}(291) = 1166 \text{ mm} \\ \text{the width of the overhang} = 0 + \dfrac{291}{2} = \underline{146 \text{ mm}} \ \textit{governs} \end{cases}$$

$$= \frac{1}{2}(2245) + 146$$

$$= 1268 \text{ mm}$$

Span 2: The effective flange widths are the same as those in Span 1.

b. *Required flexural reinforcement*:

The required reinforcements are calculated using Eqs. (9.16) and (9.17), neglecting the compression steel

The minimum reinforcement required, based on Eq. (9.20), is

$$\rho_{\min} \geq 0.03\frac{f'_c}{f_y} = 0.03\left(\frac{28}{420}\right) = 0.002$$

$$A_g(\text{Exterior girder}) = 1\ 103\ 530 \text{ mm}^2$$

$$A_{s_{\min}}(\text{Exterior girder}) = (0.002)(1\ 103\ 530) = 2207 \text{ mm}^2$$

Use $A_{s_{\min}} = \underline{2500 \text{ mm}^2}$

The required and provided reinforcements for sections located at 1/10 of each span interval and the face of the bent cap are listed in Table 9.10.

TABLE 9.10 Section Reinforcement Design for Exterior Girder

		Positive Moment					Negative Moment				
Section	Distance from Abut. 1 (mm)	M_u (kN-m)	A_s Required (mm^2)	No. of Reinf. Bars Use #36	A_s (provided) (mm^2)	ϕM_n (provided) (kN-m)	M_u (kN-m)	A_s Required (mm^2)	No. of Reinf. Bars Use #32	A_s (Provided) (mm^2)	ϕM_n (Provided) (kN-m)
0.0 L_1	0	0	0	3	3018	1841	0	0	4	3276	1954
0.1 L_1	2439	1819	2979	3	3018	1841	0	0	4	3276	1954
0.2 L_1	4878	3053	5020	5	5030	3055	0	0	4	3276	1954
0.3 L_1	7317	3758	6194	7	7042	4257	0	0	4	3276	1954
0.4 L_1	9756	3923	6469	7	7042	4257	0	0	4	3276	1954
0.5 L_1	12195	3577	5892	7	7042	4257	43	72	6	4914	2910
0.6 L_1	14634	2762	4537	5	5030	3055	632	1048	6	4914	2910
0.7 L_1	17073	1494	2444	3	3018	1841	1466	2448	6	4914	2910
0.8 L_1	19512	62	101	3	3018	1841	2800	4724	6	4914	2910
0.9 L_1	21951	0	0	3	3018	1841	4587	7848	10	8190	4780
0.96 L_1	23414	0	0	3	3018	1841	6211	10767	14	11466	6544
0.03 L_2	25304	0	0	3	3018	1841	6295	10920	14	11466	6544
0.1 L_2	27438	0	0	3	3018	1841	3772	6412	10	8190	4780
0.2 L_2	30486	974	1590	3	3018	1841	1042	1735	6	4914	2910
0.3 L_2	33534	3349	5512	7	7042	4257	0	0	6	4914	2910
0.4 L_2	36582	5118	8475	9	9054	5449	0	0	4	3276	1954
0.5 L_2	39630	6163	10243	11	11066	6629	0	0	4	3276	1954
0.6 L_2	42678	6490	10799	11	11066	6629	0	0	4	3276	1954
0.7 L_2	45726	6073	10091	11	11066	6629	0	0	4	3276	1954
0.8 L_2	48774	4835	7999	9	9054	5449	0	0	4	3276	1954
0.9 L_2	51822	2822	4637	5	5030	3055	0	0	4	3276	1954
1.0 L_2	54870	0	0	3	3018	1841	0	0	4	3276	1954

c. *Reinforcement layout*:

i. Reinforcement cutoff (Section 9.4.7):

- The extended length at cutoff for positive moment reinforcement, No. 36, is

$$\text{the largest of}\begin{cases}\text{Effective depth of the section} = \underline{1625\text{ mm}} & \textit{governs}\\ 15\ d_b = 537\text{ mm} \\ 0.05\text{ of span length} = 0.05\ (24\ 390) = 1220\text{ mm}\end{cases}$$

From Table 9.6, the stagger lengths for No. 36 and No. 32 bars are

$$l_d\text{ of No. 36 bars } = 1600\text{ mm}$$

$$l_d\text{ of No. 32 bars } = 1300\text{ mm}$$

- The extended length at cutoff for negative moment reinforcement, No. 32, is

$$\text{the largest of}\begin{cases}\text{Effective depth of the section} = \underline{1601\text{ mm}} & \textit{governs}\\ 15\ d_b = 485\text{ mm} \\ 0.05\text{ of span length} = 0.05\ (30\ 480) = 1524\text{ mm}\end{cases}$$

- Negative moment reinforcements, in addition to the above requirement for bar cutoff, have to satisfy Eq. (9.43). The extended length beyond the inflection point has to be the largest of the following:

$$\begin{cases} d = \underline{1601\text{ mm}} \quad \textit{governs for Span 1} \\ 12 d_b = 387.6\text{ mm} \\ 0.0625 \times (\text{clear span length}) = (0.0625)(24\ 390) = 1524\text{ mm} \\ \quad\text{or} \qquad = (0.0625)(30\ 480) = \underline{1905\text{ mm}} \quad \textit{governs for Span 2}\end{cases}$$

ii. Reinforcement distribution (Section 9.4.2):

$$\frac{1}{10}(\text{average adjacent span length}) = \frac{1}{10}(30\ 480 + 24\ 385) = 2743\text{ mm}$$

$$b^E_{\text{top}} = 2427\text{ mm} < 2743\text{ mm}$$

All tensile reinforcements should be distributed within the effective tension flange width.

iii. Side reinforcements in the web, Eq. (9.6)

$$A_{sk} \ge 0.001(d_e - 760) = 0.001(1625 - 760) = 0.865\text{ mm}^2/\text{mm of height}$$

$$A_{sk} \le \frac{A_s}{1200} = \frac{13{,}462}{1200} = 11.21\text{ mm}^2/\text{mm of height}$$

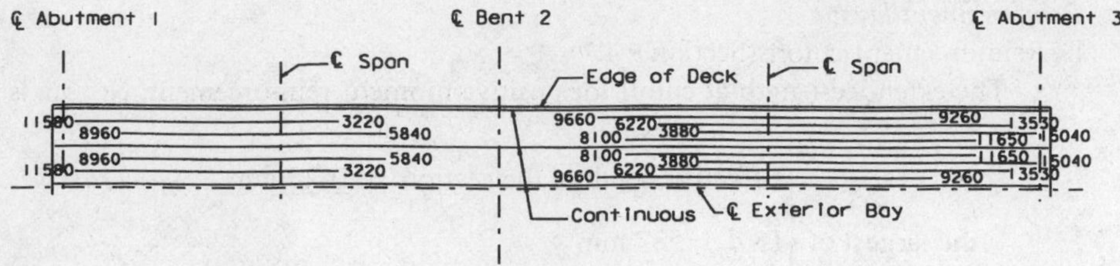

FIGURE 9.17 Bottom slab reinforcement of exterior girder.

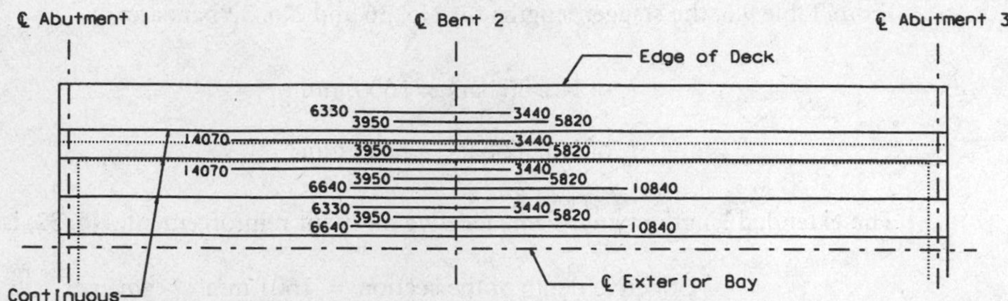

FIGURE 9.18 Top deck reinforcement of exterior girder.

$$A_{sk} = 0.865(250) = 216 \text{ mm}^2$$ Use No. 19 at 250 mm on each side face of the web

The reinforcement layout for bottom slab and top deck of exterior girder are shown in Figure 9.17 and 9.18, respectively. The numbers next to the reinforcing bars indicate the bar length extending beyond either the centerline of support or span.

4. **Shear Design**

From Table 9.9, it is apparent that the maximum shear demand is located at the critical section near Bent 2 in Span 2.

a. *Determine the critical section near Bent 2 in Span 2:*

$$A_s = 11{,}466 \text{ mm}^2,\ b = 1268 \text{ mm}$$

$$a = \frac{A_s f_y}{0.85 f_c' b} = \frac{(11{,}466)(420)}{0.85(28)(1268)} = 160 \text{ mm}$$

$$d_v = \text{ the largest of } \begin{cases} d_e - \dfrac{a}{2} = 1601 - \dfrac{160}{2} = \underline{1521 \text{ mm}} \ \textit{governs} \\ 0.9 d_e = 0.9(1601) = 1441 \text{ mm} \\ 0.72h = 0.72(1680) = 1210 \text{ mm} \end{cases}$$

The critical section is at a distance of d_v from the face of the support, i.e., distance between centerline of Bent 2 and the critical section = 600 + 1521 = 2121 mm = $0.07L_2$.

b. *At the above section, find M_u and V_u, using interpolation from Tables 9.8 and 9.9:*

$$M_u = 3772 + (6295 - 3772)(0.03/0.07) = 4853 \text{ kN·m}$$

$$V_u = 1365 + (1606 - 1365)(0.03/0.1) = 1437 \text{ kN}$$

$$v = \frac{V_u}{\phi_v b_v d_v} = \frac{1437 \cdot (1000)}{(0.9)(291)(1521)} = 3.61 \text{ MPa}$$

$$\frac{v}{f'_c} = \frac{3.61}{28} = 0.129 < 0.25 \quad \text{O.K.}$$

c. *Determine θ and β, and required shear reinforcement spacing:*
Try $\theta = 37.5°$, $\cot\theta = 1.303$, $A_s = 11\ 466$ mm², $E_c = 200$ GPa, from Eq. (9.24).

$$\varepsilon_x = \frac{\frac{4\ 853\ 000}{1521} + 0.5(1437)(1.303)}{200(11\ 466)} = 1.80 \times 10^{-3}$$

From Figure 9.6, we obtain $\theta = 37.5°$, which agrees with the assumption.
Use $\theta = 37.5°$, $\beta = 1.4$, from Eq. (9.25)

$$V_s = \frac{1437}{0.9} - 0.083(1.4)\sqrt{28}(291)(1521) \times 10^{-3} = 1325 \text{ kN}$$

Use No. 16 rebars, $A_v = 199(2) = 398$ mm², from Eq. (9.26)

$$\text{Required spacing, } s \le \frac{(398)(420)(1521)}{1325 \times 10^3}(1.303) = 250 \text{ mm}$$

d. *Determine the maximum spacing required:*
Note that $V_u = 1437$ kN $> 0.1\ f'_c b'_v d_v = 0.1(28)(291)(1521) \times 10^{-3} = 1239$ kN
From Eqs. (9.27) and (9.29):

$$s_{max} = \text{the smallest of} \begin{cases} \dfrac{(398)(420)}{0.083\sqrt{28}(291)} = 1307 \text{ mm} \\ 0.4(1521) = 608 \text{ mm} \\ \underline{300 \text{ mm}} \quad \text{governs} \end{cases}$$

Use $s = 250$ mm < 300 mm OK.

e. *Check the adequacy of the longitudinal reinforcements, using Eq. (9.30):*

$$A_s f_y = (11\ 466)(420) = 4\ 815\ 720 \text{ N}$$

$$\frac{M_u}{d_v \phi_f} + \left(\frac{V_u}{\phi_v} - 0.5V_s\right)\cot\theta = \frac{4853 \times 10^6}{(1521)(0.9)} + \left(\frac{1437 \times 10^3}{0.9} - 0.5(1325 \times 10^3)\right)(1.303)$$

$$= 4\ 762\ 401 \text{ N} < A_s f_y \quad \text{O.K.}$$

Using the above procedure, the shear reinforcements, i.e., stirrups in the web, for each section can be obtained. Figure 9.16 shows the shear reinforcements required and provided in the exterior girder for both spans.

6. **Crack Control Check (Section 9.4.2)**
   For illustration purpose, we select the section located at midspan of Span 1 in this example, i.e. at $0.5\,L_1$
   a. *Check if the section is cracked:*
   Service load moment, $M_{pos} = (1.0)(M_{DC} + M_{DW} + M_{LL+IM})$
   $= (1.0)(951 + 109 + 1379)$
   $= 2439$ kN-m

   Modulus of rupture $f_r = 0.63\sqrt{f'_c} = 0.63\sqrt{28} = 3.33$ MPa, $0.8 f_r = 2.66$ MPa

   $b_{top} = 2427$ mm, $b_{bot} = 1268$ mm, obtain

   $I_g = 4.162 \times 10^{11}$ mm^4 and $\bar{y} = 655$ mm,

   where $\bar{y}$ is the distance from the most compressed concrete fiber to the neutral axis

$$S = \frac{I_g}{(d - \bar{y})} = \frac{4.162 \times 10^{11}}{(1680 - 655)} = 4.06 \times 10^8 \text{ mm}^3$$

$$f_c = \frac{M_{pos}}{S} = \frac{2439 \times 10^6}{4.06 \times 10^8} = 6.01 \text{ MPa} > 0.8 f_r = 2.66 \text{ MPa}$$

   The section is cracked.

   b. *Calculate tensile stress of the reinforcement:*
   Assuming the neutral axis is located in the web, thus applying Eqs. (9.31) through (9.34) with $A_s = 7042$ mm², $A'_s = 0$, and $\beta_1 = 0.85$, solve for x

$$x = 239 \text{ mm} > h_f = b_{top} = 210 \text{ mm} \quad \text{O.K.}$$

   From Eq. (9.35), obtain

$$I_{cr} = \frac{1}{3}(2427)(239)^3 - \frac{1}{3}(2427 - 291)(239 - 210)^3 + 7(7042)(1625 - 239)^2$$

$$= 1.057 \times 10^{11} \text{ mm}^4$$

   and from Eq. (9.38), the tensile stress in the longitudinal reinforcement is

$$f_s = \frac{7(2439 \times 10^6)(1625 - 239)}{1.057 \times 10^{11}} = 224 \text{ MPa}$$

   c. *The allowable stress can be obtained using Eq. (9.5), with $Z = 30\,000$ for moderate exposure and $d_c = 50$ mm*

$$f_{sa} = \frac{Z}{(d_c A)^{1/3}} = \frac{30{,}000}{\left((50)\dfrac{(50 \cdot 2 \cdot 1268)}{7}\right)^{1/3}} = 310 \text{ MPa} > 0.6 f_y = 252 \text{ MPa}$$

   Use $f_{sa} = 252$ MPa $> f_s = 223$ MPa O.K.

   The other sections can be checked following the same procedure described above.

7. **Check Deflection Limit**
   Based on the Service Limit State, we can compute the I_e for sections at 1/10 of the span length interval. For illustration, let the section be at $0.4L_2$

Deflection distribution factor = (no. of design lanes)/(no. of supporting beams) = 2/4 = 0.5
Note that b_{top} = 2424 mm, t_{top} = 210 mm, b_w = 291 mm, h = 1680 mm, d = 1625 mm, b_{bot} = 1268 mm, t_{bot} = 170 mm, and neglecting compression steel

$$A_g = (2427)(210) + (1680 - 210 - 170)(291) + (1268)(170) = 1\ 103\ 530 \text{ mm}^2$$

$$y_t = \left[\frac{(509\ 670)\left(1680 - \dfrac{210}{2}\right) + (378\ 300)\left(170 + \dfrac{1300}{2}\right) + (215\ 560)\left(\dfrac{170}{2}\right)}{1\ 103\ 530}\right] = 1025 \text{ mm}$$

$$I_g = \frac{1}{12}(2427)(210)^3 + (509\ 670)(550)^2 + \frac{1}{12}(291)(1300)^3 + (378\ 300)(205)^2$$

$$+\frac{1}{12}(1268)(170)^3 + (215\ 560)(940)^2$$

$$= 4.16 \times 10^{11} \text{ mm}^4$$

$$M_{cr} = f_r \frac{I_g}{y_t} = (3.33)\frac{4.16 \times 10^{11}}{1025} = 1.35 \times 10^9 \text{ N-mm}$$

Use Eqs. (9.31) throuth (9.35) to solve for x and I_{cr}, with $A_s = 9054$ mm² and $A_s' = 0$, we obtain

$$x = 272 \text{ mm}, \quad I_{cr} = 1.32 \times 10^{11} \text{ mm}^4$$

From Table 9.8:

$$M_a = 1862 + 214 + (0.5)(1565) = 2859 \text{ kN-m}$$

$$\frac{M_{cr}}{M_a} = \frac{1.35 \times 10^9}{2.86 \times 10^9} = 0.47$$

$$I_e = \left(\frac{M_{cr}}{M_a}\right)^3 I_g + \left[1 - \left(\frac{M_{cr}}{M_a}\right)^3\right] I_{cr} = (0.47)^3(4.16 \times 10^{11}) + \left[1 - (0.47)^3\right](1.32 \times 10^{11}) = 1.61 \times 10^{11} \text{ mm}^4$$

The above computation can be repeated to obtain I_e for other sections. It is assumed that the maximum deflection occurs where the maximum flexural moment is. To be conservative, the minimum I_e is used to calculate the deflection.

$$\Delta_{max} = \begin{cases} 19 \text{ mm} & \text{truck load} \\ 13 \text{ mm} & \text{lane + 25\% of truck load} \end{cases} < \frac{L_2}{800} = \frac{30\ 480}{800} = 38 \text{ mm} \quad \text{O.K.}$$

8. **Check Fatigue Limit State**
For illustration purpose, check the bottom reinforcements for the section at $0.7L_1$. For positive moment at this section, A_s = 4024 mm², A_s' = 4095 mm², d = 1625 mm, and d' = 79 mm. Note that the maximum positive moment due to the assigned truck is 757 kN-m, while the largest negative moment 598 kN-m.

$$M_{max} \text{ due to fatigue load} = 0.75(0.619)(757)(1 + 0.15) = 404 \text{ kN-m}$$

Use Eqs. (9.31) through (9.35) and (9.38) to obtain the maximum tensile stress in the main bottom reinforcements as

$$f_{max} = 64 \text{ MPa}$$

The negative moment at this section is

$$M_{min} \text{ due to fatigue load} = 0.75(0.619)(-598)(1 + 0.15) = -319 \text{ kN-m}$$

Using Eqs. (9.31) through (9.35) and (9.38), with $A_s = 4095 \text{ mm}^2$, $A_s' = 4024 \text{ mm}^2$, $d = 1601$ mm, and $d' = 55$ mm, we obtain the maximum compressive stress in the main bottom reinforcements as

$$f_{min} = -7.0 \text{ MPa}$$

Thus, the stress range for fatigue

$$f_{max} - f_{min} = 64 - (-7.0) = 71 \text{ MPa}$$

From Eq. (9.9), allowable stress range

$$f_r = 145 - 0.33(-7.0) + 55(0.3) = 164 \text{ MPa} > 71 \text{ MPa} \qquad \text{OK}$$

Other sections can be checked in the same fashion described above.

9. Summary
The purpose of the above example is mainly to illustrate the design procedure for flexural and shear reinforcement for the girder. It should be noted that, in reality, the controlling load case may not be the Strength Limit State; therefore, all the load cases specified in the AASHTO should be investigated for a complete design. It should also be noted that the interior girder design can be achieved by following the similar procedures described herein.

References

1. AASHTO, *AASHTO LRFD Bridge Design Specifications*, American Association of State Highway and Transportation Officials, Washington, D.C., 1994.
2. ASTM, *Annual Book of ASTM Standards*, American Society for Testing and Materials, Philadelphia, 1996.
3. Caltrans, *Bridge Design Aids Manual*, California Department of Transportation, Sacramento, 1994.
4. Collins, M. P. and Mitchell, D., *Prestressed Concrete Structures*, Prentice-Hall, Englewood Cliffs, NJ, 1991.
5. Caltrans, *Bridge Design Practices*, California Department of Transportation, Sacramento, 1995.
6. ACI Committee 318, Building Code Requirements for Reinforced Concrete (ACI 318-95), American Concrete Institute, 1995.
7. Barker, R. M. and Puckett, J. A., *Design of Highway Bridges*, John Wiley & Sons, New York, 1997.
8. Park, R. and Paulay, T., *Reinforced Concrete Structures*, John Wiley & Sons, New York, 1975.
9. Xanthakos, P. P., *Theory and Design of Bridges*, John Wiley & Sons, New York, 1994.

10

Prestressed Concrete Bridges

Lian Duan
California Department of Transportation

Kang Chen
MG Engineering, Inc.

Andrew Tan
Everest International Consultants, Inc.

10.1 Introduction

Prestressed concrete structures, using high-strength materials to improve serviceability and durability, are an attractive alternative for long-span bridges, and have been used worldwide since the 1950s. This chapter focuses only on conventional prestressed concrete bridges. Segmental concrete bridges will be discussed in Chapter 11. For more detailed discussion on prestressed concrete, references are made to textbooks by Lin and Burns [1], Nawy [2], Collins and Mitchell [3].

10.1.1 Materials

10.1.1.1 Concrete

A 28-day cylinder compressive strength (f_c') of concrete 28 to 56 MPa is used most commonly in the United States. A higher early strength is often needed, however, either for the fast precast method used in the production plant or for the fast removal of formwork in the cast-in-place method. The modulus of elasticity of concrete with density between 1440 and 2500 kg/m^3 may be taken as

$$E_c = 0.043 w_c \sqrt{f_c'} \tag{10.1}$$

where w_c is the density of concrete (kg/m^3). Poisson's ratio ranges from 0.11 to 0.27, but 0.2 is often assumed.

0-8493-7434-0/00/$0.00+$.50

The modulus of rupture of concrete may be taken as [4]

$$f_r = \begin{cases} 0.63\sqrt{f_c'} & \text{for normal weight concrete — flexural} \\ 0.52\sqrt{f_c'} & \text{for sand-lightweight concrete — flexural} \\ 0.44\sqrt{f_c'} & \text{for all-lightweight concrete — flexural} \\ 0.1 f_c' & \text{for direct tension} \end{cases} \tag{10.2}$$

Concrete shrinkage is a time-dependent material behavior and mainly depends on the mixture of concrete, moisture conditions, and the curing method. Total shrinkage strains range from 0.0004 to 0.0008 over the life of concrete and about 80% of this occurs in the first year.

For moist-cured concrete devoid of shrinkage-prone aggregates, the strain due to shrinkage ε_{sh} may be estimated by [4]

$$\varepsilon_{sh} = -k_s k_h \left(\frac{t}{35+t}\right) 0.51\times 10^{-3} \tag{10.3}$$

$$K_s = \left[\frac{\dfrac{t}{26e^{0.0142(V/S)}+t}}{\dfrac{t}{45+t}}\right]\left[\frac{1064-3.7(V/S)}{923}\right] \tag{10.4}$$

where t is drying time (days); k_s is size factor and k_h is humidity factors may be approximated by K_n = (140-H)/70 for $H < 80\%$; K_n = 3(100-H)/70 for $H \geq 80\%$; and V/S is volume to surface area ratio. If the moist-cured concrete is exposed to drying before 5 days of curing, the shrinkage determined by Eq. (10.3) should be increased by 20%.

For stem-cured concrete devoid of shrinkage-prone aggregates:

$$\varepsilon_{sh} = -k_s k_h \left(\frac{t}{55+t}\right) 0.56\times 10^{-3} \tag{10.5}$$

Creep of concrete is a time-dependent inelastic deformation under sustained load and depends primarily on the maturity of the concrete at the time of loading. Total creep strain generally ranges from about 1.5 to 4 times that of the "instantaneous" deformation. The creep coefficient may be estimated as [4]

$$\psi(t,t_1) = 3.5K_c K_f \left(1.58-\frac{H}{120}\right) t_i^{-0.118} \frac{(t-t_i)^{0.6}}{10+(t-t_i)^{0.6}} \tag{10.6}$$

$$K_f = \frac{62}{42+f_c'} \tag{10.7}$$

$$K_s = \left[\frac{\dfrac{t}{26e^{0.0142(V/S)}+t}}{\dfrac{t}{45+t}}\right]\left[\frac{1.8+1.77e^{-0.0213(V/S)}}{2.587}\right] \tag{10.8}$$

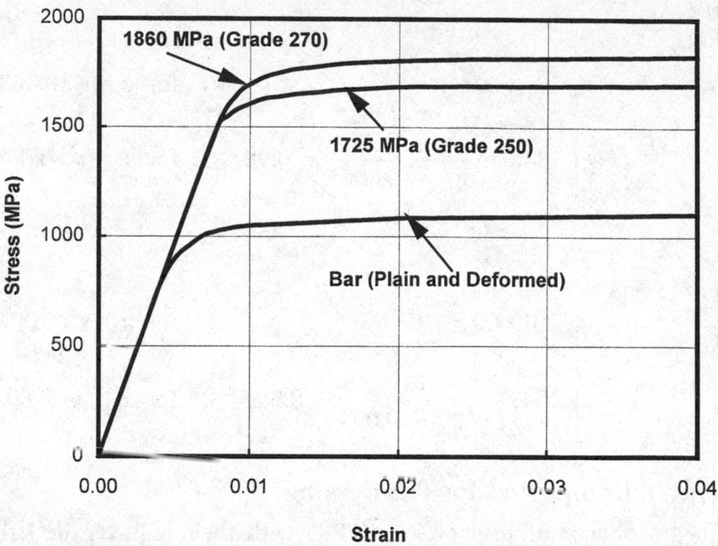

FIGURE 10.1 Typical stress–strain curves for prestressing steel.

where H is relative humidity (%); t is maturity of concrete (days); t_i is age of concrete when load is initially applied (days); K_c is the effect factor of the volume-to-surface ratio; and K_f is the effect factor of concrete strength.

Creep, shrinkage, and modulus of elasticity may also be estimated in accordance with CEB-FIP Mode Code [15].

10.1.1.2 Steel for Prestressing

Uncoated, seven-wire stress-relieved strands (AASHTO M203 or ASTM A416), or low-relaxation seven-wire strands and uncoated high-strength bars (AASHTO M275 or ASTM A722) are commonly used in prestresssed concrete bridges. Prestressing reinforcement, whether wires, strands, or bars, are also called *tendons.* The properties for prestressing steel are shown in Table 10.1.

TABLE 10.1 Properties of Prestressing Strand and Bars

Material	Grade and Type	Diameter (mm)	Tensile Strength f_{pu} (MPa)	Yield Strength f_{py} (MPa)	Modulus of Elasticity E_p (MPa)
Strand	1725 MPa (Grade 250)	6.35–15.24	1725	80% of f_{pu} except 90% of f_{pu}	197,000
	1860 MPa (Grade 270)	10.53–15.24	1860	for low relaxation strand	
Bar	Type 1, Plain	19 to 25	1035	85% of f_{pu}	
	Type 2, Deformed	15 to 36	1035	80% of f_{pu}	207,000

Typical stress–strain curves for prestressing steel are shown in Figure 10.1. These curves can be approximated by the following equations:

For Grade 250 [5]:

$$f_{ps} = \begin{cases} 197{,}000\,\varepsilon_{ps} & \text{for } \varepsilon_{ps} \leq 0.008 \\ 1710 - \dfrac{0.4}{\varepsilon_{ps} - 0.006} < 0.98 f_{pu} & \text{for } \varepsilon_{ps} > 0.008 \end{cases} \tag{10.9}$$

For Grade 270 [5]:

$$f_{ps} = \begin{cases} 197{,}000\,\varepsilon_{ps} & \text{for } \varepsilon_{ps} \le 0.008 \\ 1848 - \dfrac{0.517}{\varepsilon_{ps} - 0.0065} < 0.98 f_{pu} & \text{for } \varepsilon_{ps} > 0.008 \end{cases} \tag{10.10}$$

For Bars:

$$f_{ps} = \begin{cases} 207{,}000\,\varepsilon_{ps} & \text{for } \varepsilon_{ps} \le 0.004 \\ 1020 - \dfrac{0.192}{\varepsilon_{ps} - 0.003} < 0.98 f_{pu} & \text{for } \varepsilon_{ps} > 0.004 \end{cases} \tag{10.11}$$

10.1.1.3 Advanced Composites for Prestressing

Advanced composites–fiber-reinforced plastics (FPR) with their high tensile strength and good corrosion resistance work well in prestressed concrete structures. Application of advanced composites to prestressing have been investigated since the 1950s [6–8]. Extensive research has also been conducted in Germany and Japan [9]. The Ulenbergstrasse bridge, a two-span (21.3 and 25.6 m) solid slab using 59 fiberglass tendons, was built in 1986 in Germany. It was the first prestressed concrete bridge to use advanced composite tendons in the world [10].

FPR cables and rods made of ararmid, glass, and carbon fibers embedded in a synthetic resin have an ultimate tensile strength of 1500 to 2000 MPa, with the modulus of elasticity ranging from 62,055 MPa to 165,480 MPa [9]. The main advantages of FPR are (1) a high specific strength (ratio of strength to mass density) of about 10 to 15 times greater than steel; (2) a low modulus of elasticity making the prestress loss small; (3) good performance in fatigue; tests show [11] that for CFRP, at least three times the higher stress amplitudes and higher mean stresses than steel are achieved without damage to the cable over 2 million cycles.

Although much effort has been given to exploring the use of advanced composites in civil engineering structures (see Chapter 51) and the cost of advanced composites has come down significantly, the design and construction specifications have not yet been developed. Time is still needed for engineers and bridge owners to realize the cost-effectiveness and extended life expectancy gained by using advanced composites in civil engineering structures.

10.1.1.4 Grout

For post-tensioning construction, when the tendons are to be bound, grout is needed to transfer loads and to protect the tendons from corrosion. Grout is made of water, sand, and cements or epoxy resins. AASHTO-LRFD [4] requires that details of the protection method be indicated in the contract documents. Readers are referred to the *Post-Tensioning Manual* [12].

10.1.2 Prestressing Systems

There are two types of prestressing systems: pretensioning and post-tensioning systems. Pretensioning systems are methods in which the strands are tensioned before the concrete is placed. This method is generally used for mass production of pretensioned members. Post-tensioning systems are methods in which the tendons are tensioned after concrete has reached a specified strength. This technique is often used in projects with very large elements (Figure 10.2). The main advantage of post-tensioning is its ability to post-tension both precast and cast-in-place members. Mechanical prestressing–jacking is the most common method used in bridge structures.

FIGURE 10.2 A post–tensioned box–girder bridge under construction.

10.2 Section Types

10.2.1 Void Slabs

Figure 10.3a shows FHWA [13] standard precast prestressed voided slabs. Sectional properties are listed in Table 10.2. Although the cast-in-place prestressed slab is more expensive than a reinforced concrete slab, the precast prestressed slab is economical when many spans are involved. Common spans range from 6 to 15 m. Ratios of structural depth to span are 0.03 for both simple and continuous spans.

10.2.2 I-Girders

Figures 10.3b and c show AASHTO standard I-beams [13]. The section properties are given in Table 10.3. This bridge type competes well with steel girder bridges. The formwork is complicated, particularly for skewed structures. These sections are applicable to spans 9 to 36 m. Structural depth-to-span ratios are 0.055 for simple spans and 0.05 for continuous spans.

10.2.3 Box Girders

Figure 10.3d shows FHWA [13] standard precast box sections. Section properties are given in Table 10.4. These sections are used frequently for simple spans of over 30 m and are particularly suitable for widening bridges to control deflections.

The box-girder shape shown in Figure 10.3e is often used in cast-in-place prestressed concrete bridges. The spacing of the girders can be taken as twice the depth. . This type is used mostly for spans of 30 to 180 m. Structural depth-to-span ratios are 0.045 for simple spans, and 0.04 for continuous spans. The high torsional resistance of the box girder makes it particularly suitable for curved alignment (Figure 10.4) such as those needed on freeway ramps.

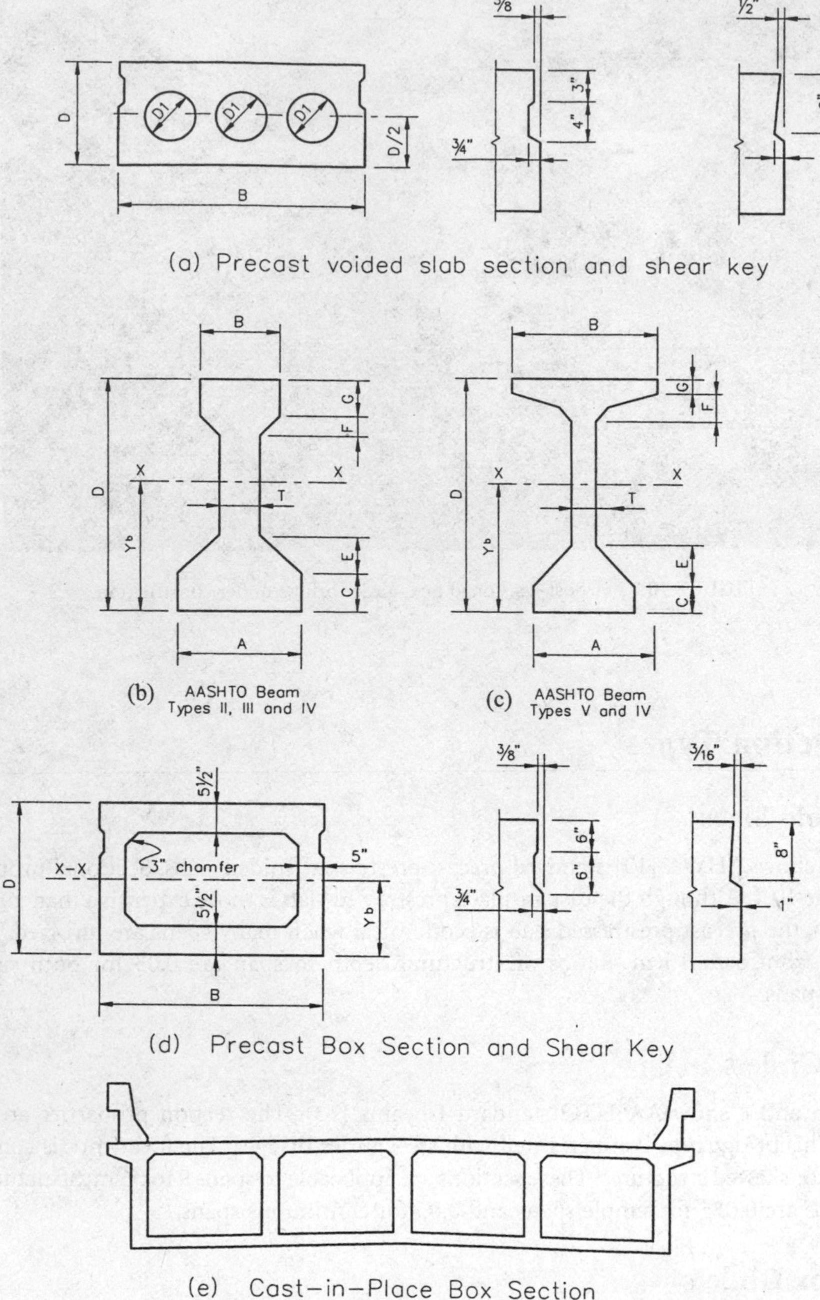

FIGURE 10.3 Typical cross sections of prestressed concrete bridge superstructures.

10.3 Losses of Prestress

Loss of prestress refers to the reduced tensile stress in the tendons. Although this loss does affect the service performance (such as camber, deflections, and cracking), it has no effect on the ultimate strength of a flexural member unless the tendons are unbounded or the final stress is less than $0.5f_{pu}$ [5]. It should be noted, however, that an accurate estimate of prestress loss is more pertinent in some prestressed concrete members than in others. Prestress losses can be divided into two categories:

TABLE 10.2 Precast Prestressed Voided Slabs Section Properties (Fig. 10.3a)

	Section Dimensions				Section Properties		
Span Range, ft (m)	Width *B* in. (mm)	Depth *D* in. (mm)	*D1* in. (mm)	*D2* in. (mm)	*A* in.^{2} (mm^2 10^6)	I_x in.^{4} (mm^4 10^9)	S_x in.^{3} (mm^3 10^6)
25	48	12	0	0	576	6,912	1,152
(7.6)	(1,219)	(305)	(0)	(0)	(0.372)	(2.877)	(18.878)
30~35	48	15	8	8	569	12,897	1,720
(10.1~10.70)	(1,219)	(381)	(203)	(203)	(0.362)	(5.368)	(28.185)
40~45	48	18	10	10	628	21,855	2,428
(12.2~13.7)	(1,219)	(457)	(254)	(254)	(0.405)	(10.097)	(310.788)
50	48	21	12	10	703	34,517	3,287
(15.2)	(1,219)	(533)	(305)	(254)	(0.454)	(1.437)	(53.864)

TABLE 10.3 Precast Prestressed I-Beam Section Properties (Figs. 10.3b and c)

AASHTO	Section Dimensions, in. (mm)							
Beam Type	Depth *D*	Bottom Width *A*	Web Width *T*	Top Width *B*	*C*	*E*	*F*	*G*
II	36 (914)	18 (457)	6 (152)	12 (305)	6 (152)	6 (152)	3 (76)	6 (152)
III	45 (1143)	22 (559)	7 (178)	16 (406)	7 (178)	7.5 (191)	4.5 (114)	7 (178)
IV	54 (1372)	26 (660)	8 (203)	20 (508)	8 (203)	9 (229)	6 (152)	8 (203)
V	65 (1651)	28 (711)	8 (203)	42 (1067)	8 (203)	10 (254)	3 (76)	5 (127)
VI	72 (1829)	28 (711)	8 (203)	42 (1067)	8 (203)	10 (254)	3 (76)	5 (127)

	Section Properties					
	A in.^{2} (mm^2 10^6)	Y_b in. (mm)	I_x in.^{4} (mm^4 10^9)	S_b in.^{3} (mm^4 10^6)	S_t in.^{3} (mm^4 10^6)	Span Ranges, ft (m)
II	369	15.83	50,980	3220	2528	40 ~ 45
	(0.2381)	(402.1)	(21.22)	(52.77)	(41.43)	(12.2 ~ 13.7)
III	560	20.27	125,390	6186	5070	50 ~ 65
	(0.3613)	(514.9)	(52.19)	(101.38)	(83.08)	(15.2 ~ 110.8)
IV	789	24.73	260,730	10543	8908	70 ~ 80
	(0.5090)	(628.1)	(108.52)	(172.77)	(145.98)	(21.4 ~ 24.4)
V	1013	31.96	521,180	16307	16791	90 ~ 100
	(0.6535)	(811.8)	(216.93)	(267.22)	(275.16)	(27.4 ~ 30.5)
VI	1085	36.38	733,340	20158	20588	110 ~ 120
	(0.7000)	(924.1)	(305.24)	(330.33)	(337.38)	(33.5 ~ 36.6)

- Instantaneous losses including losses due to anchorage set (Δf_{pA}), friction between tendons and surrounding materials (Δf_{pF}), and elastic shortening of concrete (Δf_{pES}) during the construction stage;
- Time-dependent losses including losses due to shrinkage (Δf_{pSR}), creep (Δf_{pCR}), and relaxation of the steel (Δf_{pR}) during the service life.

The total prestress loss (Δf_{pT}) is dependent on the prestressing methods.

For pretensioned members:

$$\Delta f_{pT} = \Delta f_{pES} + \Delta f_{pSR} + \Delta f_{pCR} + \Delta f_{pR} \tag{10.12}$$

For post-tensioned members:

$$\Delta f_{pT} = \Delta f_{pA} + \Delta f_{pF} + \Delta f_{pES} + \Delta f_{pSR} + \Delta f_{pCR} + \Delta f_{pR} \tag{10.13}$$

FIGURE 10.4 Prestressed box–girder bridge (I-280/110 Interchange, CA).

TABLE 10.4 Precast Prestressed Box Section Properties (Fig. 10.3d)

	Section Dimensions		Section Properties				
Span ft (m)	Width *B* in. (mm)	Depth *D* in. (mm)	*A* in.^{2} (mm^2 10^6)	Y_b in. (mm)	I_x in^4 (mm^4 10^9)	S_b in.^{3} (mm^3 10^6)	S_t in.^{3} (mm^3 10^6)
50	48	27	693	13.37	65,941	4,932	4,838
(15.2)	(1,219)	(686)	(0.4471)	(3310.6)	(27.447)	(80.821)	(710.281)
60	48	33	753	16.33	110,499	6,767	6,629
(18.3)	(1,219)	(838)	(0.4858)	(414.8)	(45.993)	(110.891)	(108.630)
70	48	39	813	110.29	168,367	8,728	8,524
(21.4)	(1,219)	(991)	(0.5245)	(490.0)	(70.080)	(143.026)	(1310.683)
80	48	42	843	20.78	203,088	9,773	9,571
(24.4)	(1,219)	(1,067)	(0.5439)	(527.8)	(84.532)	(160.151)	(156.841)

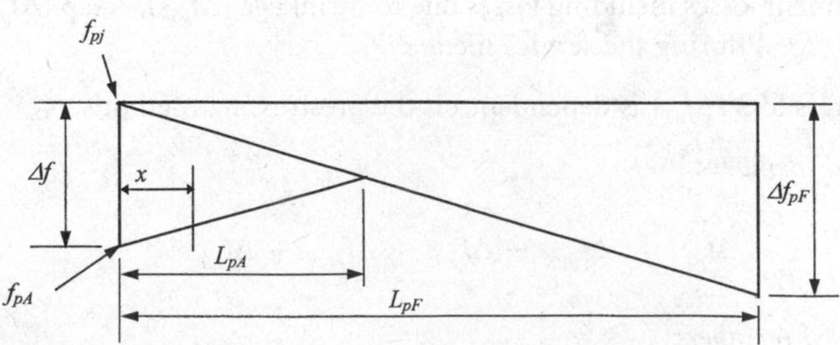

FIGURE 10.5 Anchorage set loss model.

TABLE 10.5 Friction Coefficients for Post-Tensioning Tendons

Type of Tendons and Sheathing	Wobble Coefficient K (1/mm) × (10^{-6})	Curvature Coefficient μ (1/rad)
Tendons in rigid and semirigid galvanized ducts, seven-wire strands	0.66	0.05 ~ 0.15
Pregreased tendons, wires and seven-wire strands	0.98 ~ 6.6	0.05 ~ 0.15
Mastic-coated tendons, wires and seven-wire strands	3.3 ~ 6.6	0.05 ~ 0.15
Rigid steel pipe deviations	66	0.25, lubrication required

Source: AASHTO LRFD Bridge Design Specifications, 1st Ed., American Association of State Highway and Transportation Officials. Washington, D.C. 1994. With permission.

10.3.1 Instantaneous Losses

10.3.1.1 Anchorage Set Loss

As shown in Figure 10.5, assuming that the anchorage set loss changes linearly within the length (L_{pA}), the effect of anchorage set on the cable stress can be estimated by the following formula:

$$\Delta f_{pA} = \Delta f\left(1 - \frac{x}{L_{pA}}\right) \tag{10.14}$$

$$L_{pA} = \sqrt{\frac{E\,(\Delta L)\,L_{pF}}{\Delta f_{pF}}} \tag{10.15}$$

$$\Delta f = \frac{2\,\Delta f_{pF}\,L_{pA}}{L_{pF}} \tag{10.16}$$

where ΔL is the thickness of anchorage set; E is the modulus of elasticity of anchorage set; Δf is the change in stress due to anchor set; L_{pA} is the length influenced by anchor set; L_{pF} is the length to a point where loss is known; and x is the horizontal distance from the jacking end to the point considered.

10.3.1.2 Friction Loss

For a post-tensioned member, friction losses are caused by the tendon profile *curvature effect* and the local deviation in tendon profile *wobble effects.* AASHTO-LRFD [4] specifies the following formula:

$$\Delta f_{pF} = f_{pj}\left(1 - e^{-(Kx + \mu\alpha)}\right) \tag{10.17}$$

where K is the wobble friction coefficient and μ is the curvature friction coefficient (see Table 10.5); x is the length of a prestressing tendon from the jacking end to the point considered; and α is the sum of the absolute values of angle change in the prestressing steel path from the jacking end.

10.3.1.3 Elastic Shortening Loss Δf_{pES}

The loss due to elastic shortening can be calculated using the following formula [4]:

$$\Delta f_{pES} = \begin{cases} \dfrac{E_p}{E_{ci}} f_{cgp} & \text{for pretensioned members} \\[2ex] \dfrac{N-1}{2N}\dfrac{E_p}{E_{ci}} f_{cgp} & \text{for post-tensioned members} \end{cases} \tag{10.18}$$

TABLE 10.6 Lump Sum Estimation of Time-Dependent Prestress Losses

Type of Beam Section	Level	For Wires and Strands with f_{pu} = 1620, 1725, or 1860 MPa	For Bars with f_{pu} = 1000 or 1100 MPa
Rectangular beams and solid slab	Upper bound	200 + 28 PPR	130 + 41 PPR
	Average	180 + 28 PPR	
Box girder	Upper bound	145 + 28 PPR	100
	Average	130 + 28 PPR	
I-girder	Average	$230\left[1.0-0.15\frac{f_c'-41}{41}\right]+41\,\text{PPR}$	130 + 41 PPR
Single–T, double–T hollow core and voided slab	Upper bound	$230\left[1.0-0.15\frac{f_c'-41}{41}\right]+41\,\text{PPR}$	$230\left[1.0-0.15\frac{f_c'-41}{41}\right]+41\,\text{PPR}$
	Average	$230\left[1.0-0.15\frac{f_c'-41}{41}\right]+41\,\text{PPR}$	

Note:
1. PPR is partial prestress ratio = $(A_{ps}f_{py})/(A_{ps}f_{py} + A_sf_y)$.
2. For low-relaxation strands, the above values may be reduced by
   - 28 MPa for box girders
   - 41 MPa for rectangular beams, solid slab and I-girders, and
   - 55 MPa for single–T, double–T, hollow–core and voided slabs.

Source: AASHTO LRFD Bridge Design Specifications, 1st Ed., American Association of State Highway and Transportation Officials. Washington, D.C. 1994. With permission.

where E_{ci} is modulus of elasticity of concrete at transfer (for pretensioned members) or after jacking (for post-tensioned members); N is the number of identical prestressing tendons; and f_{cgp} is sum of the concrete stress at the center of gravity of the prestressing tendons due to the prestressing force at transfer (for pretensioned members) or after jacking (for post-tensioned members) and the self-weight of members at the section with the maximum moment. For post-tensioned structures with bonded tendons, f_{cgp} may be calculated at the center section of the span for simply supported structures, at the section with the maximum moment for continuous structures.

10.3.2 Time-Dependent Losses

10.3.2.1 Lump Sum Estimation

AASHTO-LRFD [4] provides the approximate lump sum estimation (Table 10.6) of time-dependent loses Δf_{pTM} resulting from shrinkage and creep of concrete, and relaxation of the prestressing steel. While the use of lump sum losses is acceptable for "average exposure conditions," for unusual conditions, more-refined estimates are required.

10.3.2.2 Refined Estimation

a. *Shrinkage Loss*: Shrinkage loss can be determined by formulas [4]:

$$\Delta f_{pSR} = \begin{cases} 93 - 0.85H & \text{for pretensioned members} \\ 11 - 1.03H & \text{for post-tensioned members} \end{cases} \tag{10.19}$$

where H is average annual ambient relative humidity (%).

b. *Creep Loss*: Creep loss can be predicted by [4]:

$$\Delta f_{pCR} = 12 f_{cgp} - 7\Delta f_{cdp} \geq 0 \tag{10.20}$$

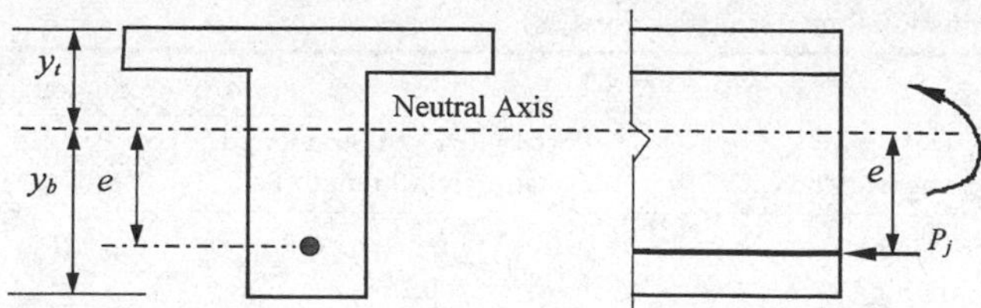

FIGURE 10.6 Prestressed concrete member section at Service Limit State.

where f_{cgp} is concrete stress at center of gravity of prestressing steel at transfer, and Δf_{cdp} is concrete stress change at center of gravity of prestressing steel due to permanent loads, except the load acting at the time the prestressing force is applied.

c. *Relaxation Loss*: The total relaxation loss (Δf_{pR}) includes two parts: relaxation at time of transfer Δf_{pR1} and after transfer Δf_{pR2}. For a pretensioned member initially stressed beyond 0.5 f_{pu}, AASHTO-LRFD [4] specifies

$$\Delta f_{pR1} = \begin{cases} \dfrac{\log 24t}{10}\left[\dfrac{f_{pi}}{f_{py}} - 0.55\right] f_{pi} & \text{for stress-relieved strand} \\ \dfrac{\log 24t}{40}\left[\dfrac{f_{pi}}{f_{py}} - 0.55\right] f_{pi} & \text{for low-relaxation strand} \end{cases} \tag{10.21}$$

For stress-relieved strands

$$\Delta f_{pR2} = \begin{cases} 138 - 0.4\Delta f_{pES} - 0.2(\Delta f_{pSR} + \Delta f_{pCR}) & \text{for pretensioning} \\ 138 - 0.3\Delta f_{pF} - 0.4\Delta f_{pES} - 0.2(\Delta f_{pSR} + \Delta f_{pCR}) & \text{for post-tensioning} \end{cases} \tag{10.22}$$

where t is time estimated in days from testing to transfer. For low-relaxation strands, Δf_{pR2} is 30% of those values obtained from Eq. (10.22).

10.4 Design Considerations

10.4.1 Basic Theory

Compared with reinforced concrete, the main distinguishing characteristics of prestressed concrete are that

- The stresses for concrete and prestressing steel and deformation of structures at each stage, i.e., during prestressing, handling, transportation, erection, and the service life, as well as stress concentrations, need to be investigated on the basis of elastic theory.
- The prestressing force is determined by concrete stress limits under service load.
- Flexure and shear capacities are determined based on the ultimate strength theory.

For the prestressed concrete member section shown in Figure 10.6, the stress at various load stages can be expressed by the following formula:

$$f = \frac{P_j}{A} \pm \frac{P_j e y}{I} \pm \frac{My}{I} \tag{10.23}$$

TABLE 10.7 Stress Limits for Prestressing Tendons

Stress Type	Prestressing Method	Prestressing Tendon Type: Stress Relieved Strand and Plain High-Strength Bars	Low Relaxation Strand	Deformed High-Strength Bars
At jacking, f_{pj}	Pretensioning	$0.72f_{pu}$	$0.78f_{pu}$	—
	Post-tensioning	$0.76f_{pu}$	$0.80f_{pu}$	$0.75f_{pu}$
After transfer, f_{pt}	Pretensioning	$0.70f_{pu}$	$0.74f_{pu}$	—
	Post-tensioning — at anchorages and couplers immediately after anchor set	$0.70f_{pu}$	$0.70f_{pu}$	$0.66f_{pu}$
	Post-tensioning — general	$0.70f_{pu}$	$0.74f_{pu}$	$0.66f_{pu}$
At Service Limit State, f_{pc}	After all losses	$0.80f_{py}$	$0.80f_{py}$	$0.80f_{py}$

Source: AASHTO LRFD Bridge Design Specifications, 1st Ed., American Association of State Highway and Transportation Officials. Washington, D.C. 1994. With permission.

TABLE 10.8 Temporary Concrete Stress Limits at Jacking State before Losses due to Creep and Shrinkage — Fully Prestressed Components

Stress Type	Area and Condition	Stress (MPa)
Compressive	Pretensioned	$0.60\ f'_{ci}$
	Post-tensioned	$0.55\ f'_{ci}$
Tensile	Precompressed tensile zone without bonded reinforcement	N/A
	Area other than the precompressed tensile zones and without bonded auxiliary reinforcement	$0.25\sqrt{f'_{ci}} \leq 1.38$
	Area with bonded reinforcement which is sufficient to resist 120% of the tension force in the cracked concrete computed on the basis of uncracked section	$0.58\sqrt{f'_{ci}}$
	Handling stresses in prestressed piles	$0.415\sqrt{f'_{ci}}$

Note: Tensile stress limits are for nonsegmental bridges only.
Source: AASHTO LRFD Bridge Design Specifications, 1st Ed., American Association of State Highway and Transportation Officials. Washington, D.C. 1994. With permission.

where P_j is the prestress force; A is the cross-sectional area; I is the moment of inertia; e is the distance from the center of gravity to the centroid of the prestressing cable; y is the distance from the centroidal axis; and M is the externally applied moment.

Section properties are dependent on the prestressing method and the load stage. In the analysis, the following guidelines may be useful:

- Before bounding of the tendons, for a post-tensioned member, the net section should be used theoretically, but the gross section properties can be used with a negligible tolerance.
- After bounding of tendons, the transformed section should be used, but gross section properties may be used approximately.
- At the service load stage, transformed section properties should be used.

10.4.2 Stress Limits

The stress limits are the basic requirements for designing a prestressed concrete member. The purpose for stress limits on the prestressing tendons is to mitigate tendon fracture, to avoid inelastic tendon deformation, and to allow for prestress losses. Tables 10.7 lists the AASHTO-LRFD [4] stress limits for prestressing tendons.

TABLET 10.9 Concrete Stress Limits at Service Limit State after All Losses — Fully Prestressed Components

Stress Type	Area and Condition		Stress (MPa)
Compressive	Nonsegmental bridge at service stage		$0.45\,f_c'$
	Nonsegmental bridge during shipping and handling		$0.60\,f_c'$
	Segmental bridge during shipping and handling		$0.45\,f_c'$
Tensile	Precompressed tensile zone assuming uncracked section	With bonded prestressing tendons other than piles	$0.50\sqrt{f_c'}$
		Subjected to severe corrosive conditions	$0.25\sqrt{f_c'}$
		With unbonded prestressing tendon	No tension

Note: Tensile stress limits are for nonsegmental bridges only.
Source: AASHTO LRFD Bridge Design Specifications, 1st Ed., American Association of State Highway and Transportation Officials, Washington, D.C. 1994. With permission.

The purpose for stress limits on the concrete is to ensure no overstressing at jacking and after transfer stages and to avoid cracking (fully prestressed) or to control cracking (partially prestressed) at the service load stage. Tables 10.8 and 10.9 list the AASHTO-LRFD [4] stress limits for concrete.

A prestressed member that does not allow cracking at service loads is called a fully prestressed member, whereas one that does is called a partially prestressed member. Compared with full prestress, partial prestress can minimize camber, especially when the dead load is relatively small, as well as provide savings in prestressing steel, in the work required to tension, and in the size of end anchorages and utilizing cheaper mild steel. On the other hand, engineers must be aware that partial prestress may cause earlier cracks and greater deflection under overloads and higher principal tensile stresses under service loads. Nonprestressed reinforcement is often needed to provide higher flexural strength and to control cracking in a partially prestressed member.

10.4.3 Cable Layout

A cable is a group of prestressing tendons and the center of gravity of all prestressing reinforcement. It is a general design principle that the maximum eccentricity of prestressing tendons should occur at locations of maximum moments. Although straight tendons (Figure 10.7a) and harped multi-straight tendons (Figure 10.7b and c) are common in the precast members, curved tendons are more popular for cast-in-place post-tensioned members. Typical cable layouts for bridge superstructures are shown in Figure 10.7.

To ensure that the tensile stress in extreme concrete fibers under service does not exceed code stress limits [4, 14], cable layout envelopes are delimited. Figure 10.8 shows limiting envelopes for simply supported members. From Eq. (10.23), the stress at extreme fiber can be obtained

$$f = \frac{P_j}{A} \pm \frac{P_j eC}{I} \pm \frac{MC}{I} \tag{10.24}$$

where C is the distance of the top or bottom extreme fibers from the center gravity of the section (y_b or y_t as shown in Figure 10.6).

When no tensile stress is allowed, the limiting eccentricity envelope can be solved from Eq. (10.24) with

$$e_{limit} = \frac{I}{AC} \pm \frac{M}{IP_j} \tag{10.25}$$

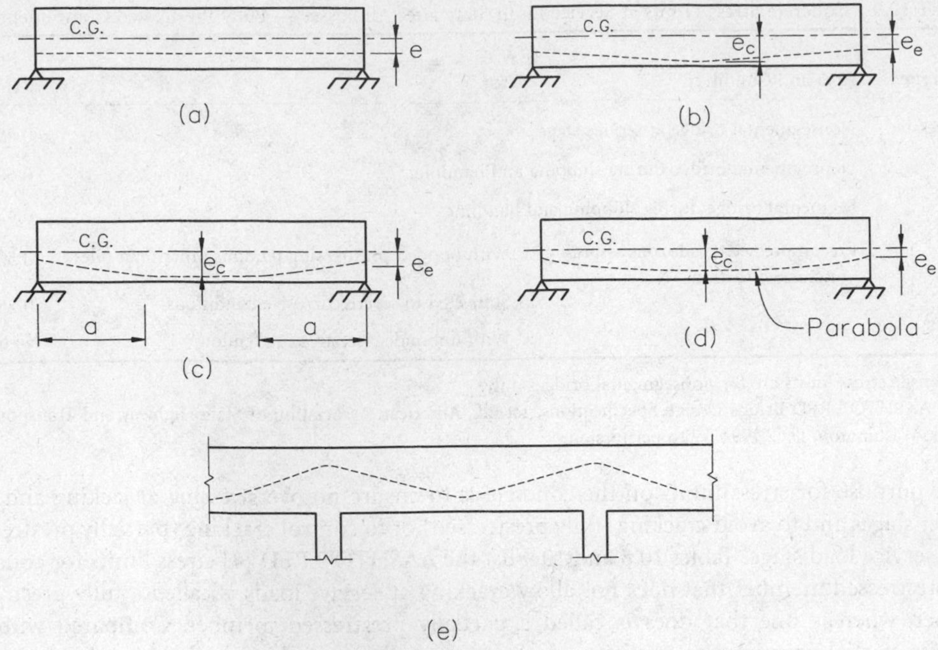

FIGURE 10.7 Cable layout for bridge superstructures.

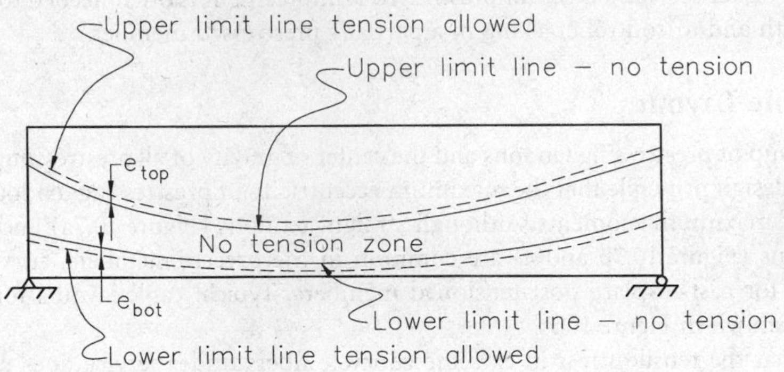

FIGURE 10.8 Cable layout envelopes.

For limited tension stress f_t, additional eccentricities can be obtained:

$$e' = \frac{f_t I}{P_j C} \tag{10.26}$$

10.4.4 Secondary Moments

The primary moment ($M_1 = P_j e$) is defined as the moment in the concrete section caused by the eccentricity of the prestress for a statically determinate member. The secondary moment M_s (Figure 10.9d) is defined as moment induced by prestress and structural continuity in an indeterminate member. Secondary moments can be obtained by various methods. The resulting moment is simply the sum of the primary and secondary moments.

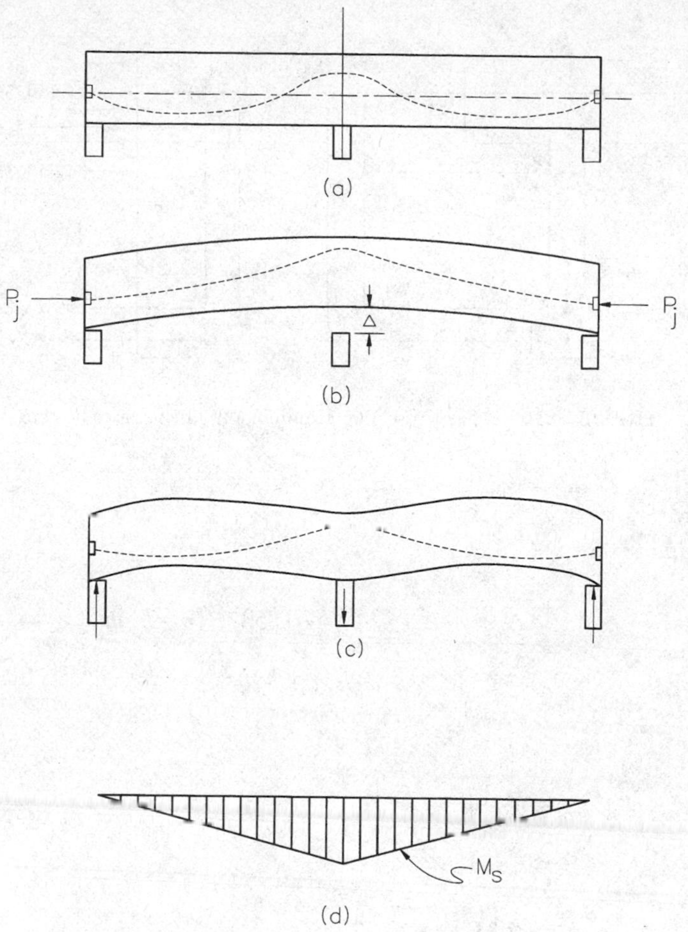

FIGURE 10.9 Secondary moments.

10.4.5 Flexural Strength

Flexural strength is based on the following assumptions [4]:

- For members with bonded tendons, strain is linearly distributed across a section; for members with unbonded tendons, the total change in tendon length is equal to the total change in member length over the distance between two anchorage points.
- The maximum usable strain at extreme compressive fiber is 0.003.
- The tensile strength of concrete is neglected.
- A concrete stress of 0.85 f_c' is uniformly distributed over an equivalent compression zone.
- Nonprestressed reinforcement reaches the yield strength, and the corresponding stresses in the prestressing tendons are compatible based on plane section assumptions.

For a member with a flanged section (Figure 10.10) subjected to uniaxial bending, the equations of equilibrium are used to give a nominal moment resistance of

$$\begin{aligned} M_n &= A_{ps} f_{ps}\left(d_p - \frac{a}{2}\right) + A_s f_y\left(d_s - \frac{a}{2}\right) \\ &\quad - A_s' f_y'\left(d_s' - \frac{a}{2}\right) + 0.85 f_c'(b - b_w)\beta_1 h_f\left(\frac{a}{2} - \frac{h_f}{2}\right) \end{aligned} \tag{10.27}$$

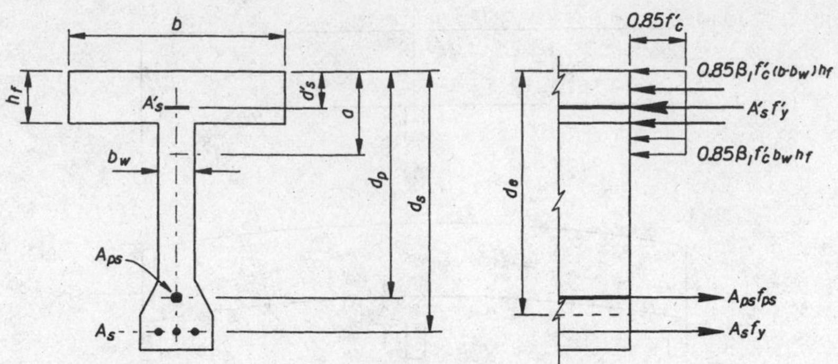

FIGURE 10.10 A flanged section at nominal moment capacity state.

$$a = \beta_1 c \tag{10.28}$$

For bonded tendons:

$$c = \frac{A_{ps} f_{pu} + A_s f_y - A_s' f_y' - 0.85\beta_1 f_c'(b - b_w) h_f}{0.85\beta_1 f_c' b_w + k A_{ps} \dfrac{f_{pu}}{d_p}} \geq h_f \tag{10.29}$$

$$f_{ps} = f_{pu}\left(1 - k\frac{c}{d_p}\right) \tag{10.30}$$

$$k = 2\left(1.04 - \frac{f_{py}}{f_{pu}}\right) \tag{10.31}$$

$$0.85 \geq \beta_1 = 0.85 - \frac{(f_c' - 28)(0.05)}{7} \geq 0.65 \tag{10.32}$$

where A represents area; f is stress; b is the width of the compression face of member; b_w is the web width of a section; h_f is the compression flange depth of the cross section; d_p and d_s are distances from extreme compression fiber to the centroid of prestressing tendons and to centroid of tension reinforcement, respectively; subscripts c and y indicate specified strength for concrete and steel, respectively; subscripts p and s mean prestressing steel and reinforcement steel, respectively; subscripts ps, py, and pu correspond to states of nominal moment capacity, yield, and specified tensile strength of prestressing steel, respectively; superscript ′ represents compression. The above equations also can be used for rectangular section in which $b_w = b$ is taken.

For unbound tendons:

$$c = \frac{A_{ps} f_{pu} + A_s f_y - A_s' f_y' - 0.85\beta_1 f_c'(b - b_w) h_f}{0.85\beta_1 f_c' b_w} \geq h_f \tag{10.33}$$

$$f_{ps} = f_{pe} + \Omega_u E_p \varepsilon_{cu}\left(\frac{d_p}{c} - 1.0\right)\frac{L_1}{L_2} \leq 0.94 f_{py} \tag{10.34}$$

where L_1 is length of loaded span or spans affected by the same tendons; L_2 is total length of tendon between anchorage; Ω_u is the bond reduction coefficient given by

$$\Omega_u = \begin{cases} \dfrac{3}{L/d_p} & \text{for uniform and near third point loading} \\[2ex] \dfrac{1.5}{L/d_p} & \text{for near midspan loading} \end{cases} \tag{10.35}$$

in which L is span length.

Maximum reinforcement limit:

$$\frac{c}{d_e} \le 0.42 \tag{10.36}$$

$$d_e = \frac{A_{ps} f_{ps} d_p + A_s f_y d_s}{A_{ps} f_{ps} + A_s f_y} \tag{10.37}$$

Minimum reinforcement limit:

$$\phi M_n \ge 1.2\ M_{\text{cr}} \tag{10.38}$$

in which ϕ is flexural resistance factor 1.0 for prestressed concrete and 0.9 for reinforced concrete; M_{cr} is the cracking moment strength given by the elastic stress distribution and the modulus of rupture of concrete.

$$M_{\text{cr}} = \frac{I}{y_t}\left(f_r + f_{pe} - f_d\right) \tag{10.39}$$

where f_{pe} is compressive stress in concrete due to effective prestresses; and f_d is stress due to unfactored self-weight; both f_{pe} and f_d are stresses at extreme fiber where tensile stresses are produced by externally applied loads.

10.4.6 Shear Strength

The shear resistance is contributed by the concrete, the transverse reinforcement and vertical component of prestressing force. The modified compression field theory-based shear design strength [3] was adopted by the AASHTO-LRFD [4] and has the formula:

$$V_n = \text{the lesser of} \begin{cases} V_c + V_s + V_p \\ 0.25 f'_c b_v d_v + V_p \end{cases} \tag{10.40}$$

where

$$V_c = 0.083\beta\sqrt{f'_c}\, b_v d_v \tag{10.41}$$

$$V_s = \frac{A_v f_y d_v (\cos\theta + \cot\alpha)\sin\alpha}{s} \tag{10.42}$$

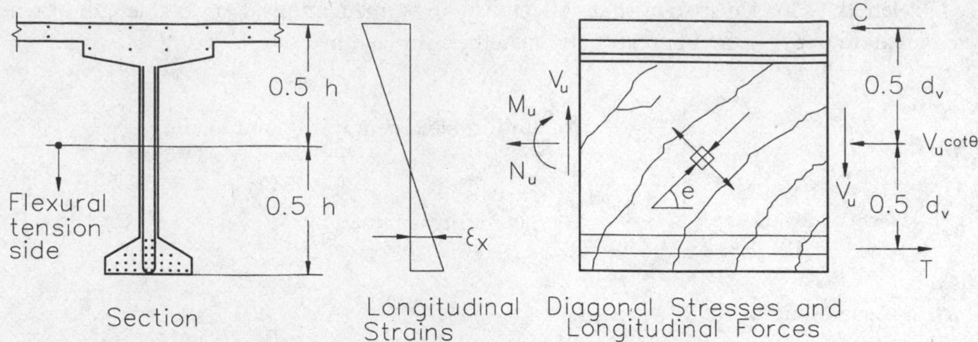

FIGURE 10.11 Illustration of A_c for shear strength calculation. (*Source*: AASHTO LRFD Bridge Design Specifications, 1st Ed., American Association of State Highway and Transportation Officials. Washington, D.C. 1994. With permission.)

TABLE 10.10 Values of θ and β for Sections with Transverse Reinforcement

$\frac{v}{f'_c}$	Angle (degree)	$\varepsilon_x \times 1000$										
		–.02	–0.15	–0.1	0	0.125	0.25	0.50	0.75	1.00	1.50	2.00
≤ 0.05	θ	27.0	27.0	27.0	27.0	27.0	28.5	29.0	33.0	36.0	41.0	43.0
	β	6.78	6.17	5.63	4.88	3.99	3.49	2.51	2.37	2.23	1.95	1.72
0.075	θ	27.0	27.0	27.0	27.0	27.0	27.5	30.0	33.5	36.0	40.0	42.0
	β	6.78	6.17	5.63	4.88	3.65	3.01	2.47	2.33	2.16	1.90	1.65
0.100	θ	23.5	23.5	23.5	23.5	24.0	26.5	30.5	34.0	36.0	38.0	39.0
	β	6.50	5.87	5.31	3.26	2.61	2.54	2.41	2.28	2.09	1.72	1.45
0.127	θ	20.0	21.0	22.0	23.5	26.0	28.0	31.5	34.0	36.0	37.0	38.0
	β	2.71	2.71	2.71	2.60	2.57	2.50	2.37	2.18	2.01	1.60	1.35
0.150	θ	22.0	22.5	23.5	25.0	27.0	29.0	32.0	34.0	36.0	36.5	37.0
	β	2.66	2.61	2.61	2.55	2.50	2.45	2.28	2.06	1.93	1.50	1.24
0.175	θ	23.5	24.0	25.0	26.5	28.0	30.0	32.5	34.0	35.0	35.5	36.0
	β	2.59	2.58	2.54	2.50	2.41	2.39	2.20	1.95	1.74	1.35	1.11
0.200	θ	25.0	25.5	26.5	27.5	29.0	31.0	33.0	34.0	34.5	35.0	36.0
	β	2.55	2.49	2.48	2.45	2.37	2.33	2.10	1.82	1.58	1.21	1.00
0.225	θ	26.5	27.0	27.5	29.0	30.5	32.0	33.0	34.0	34.5	36.5	39.0
	β	2.45	2.38	2.43	2.37	2.33	2.27	1.92	1.67	1.43	1.18	1.14
0.250	θ	28.0	28.5	29.0	30.0	31.0	32.0	33.0	34.0	35.5	38.5	41.5
	β	2.36	2.32	2.36	2.30	2.28	2.01	1.64	1.52	1.40	1.30	1.25

(*Source*: AASHTO LRFD Bridge Design Specifications, 1st Ed., American Association of State Highway and Transportation Officials. Washington, D.C. 1994. With permission.)

where b_v is the effective web width determined by subtracting the diameters of ungrouted ducts or one half the diameters of grouted ducts; d_v is the effective depth between the resultants of the tensile and compressive forces due to flexure, but not to be taken less than the greater of $0.9d_e$ or $0.72h$; A_v is the area of transverse reinforcement within distance s; s is the spacing of stirrups; α is the angle of inclination of transverse reinforcement to longitudinal axis; β is a factor indicating ability of diagonally cracked concrete to transmit tension; θ is the angle of inclination of diagonal compressive stresses (Figure 10.11). The values of β and θ for sections with transverse reinforcement are given in Table 10.10. In using this table, the shear stress v and strain ε_x in the reinforcement on the flexural tension side of the member are determined by

$$v = \frac{V_u - \phi V_p}{\phi b_v d_v} \tag{10.43}$$

$$\varepsilon_x = \frac{\dfrac{M_u}{d_v} + 0.5N_u + 0.5V_u \cot\theta - A_{ps} f_{po}}{E_s A_s + E_p A_{ps}} \le 0.002 \tag{10.44}$$

where M_u and N_u are factored moment and axial force (taken as positive if compressive) associated with V_u and f_{po} is stress in prestressing steel when the stress in the surrounding concrete is zero and can be conservatively taken as the effective stress after losses f_{pe}. When the value of ε_x calculated from the above equation is negative, its absolute value shall be reduced by multiplying by the factor F_ε, taken as

$$F_\varepsilon = \frac{E_s A_s + E_p A_{ps}}{E_c A_c + E_s A_s + E_p A_{ps}} \tag{10.45}$$

where E_s, E_p, and E_c are modulus of elasticity for reinforcement, prestressing steel, and concrete, respectively; A_c is area of concrete on the flexural tension side of the member as shown in Figure 10.11.

Minimum transverse reinforcement:

$$A_{v\min} = 0.083\sqrt{f_c'}\,\frac{b_v s}{f_y} \tag{10.46}$$

Maximum spacing of transverse reinforcement:

$$\text{For } V_u < 0.1 f_c' b_v d_v \qquad s_{\max} = \text{the smaller of } \begin{cases} 0.8 d_v \\ 600 \text{ mm} \end{cases} \tag{10.47}$$

$$\text{For } V_u \ge 0.1 f_c' b_v d_v \qquad s_{\max} = \text{the smaller of } \begin{cases} 0.4 d_v \\ 300 \text{ mm} \end{cases} \tag{10.48}$$

10.4.7 Camber and Deflections

As opposed to load deflection, camber is usually referred to as reversed deflection and is caused by prestressing. A careful evaluation of camber and deflection for a prestressed concrete member is necessary to meet serviceability requirements. The following formulas developed by the moment–area method can be used to estimate midspan immediate camber for simply supported members as shown in Figure 10.7.

For straight tendon (Figure 10.7a):

$$\Delta = \frac{L^2}{8E_c I} M_e \tag{10.49}$$

For one-point harping tendon (Figure 10.7b):

$$\Delta = \frac{L^2}{8E_c I}\left(M_c + \frac{2}{3} M_e\right) \tag{10.50}$$

For two-point harping tendon (Figure 10.7c):

$$\Delta = \frac{L^2}{8E_cI}\left(M_c + M_e - \frac{M_e}{3}\left(\frac{2a}{L}\right)^2\right) \tag{10.51}$$

For parabola tendon (Figure 10.7d):

$$\Delta = \frac{L^2}{8E_cI}\left(M_e + \frac{5}{6}M_c\right) \tag{10.52}$$

where M_e is the primary moment at end, P_je_{end}, and M_c is the primary moment at midspan P_je_c. Uncracked gross section properties are often used in calculating camber. For deflection at service loads, cracked section properties, i.e., moment of inertia I_{cr}, should be used at the post-cracking service load stage. It should be noted that long term effect of creep and shrinkage shall be considered in the final camber calculations. In general, final camber may be assumed 3 times as great as immediate camber.

10.4.8 Anchorage Zones

In a pretensioned member, prestressing tendons transfer the compression load to the surrounding concrete over a length L_t gradually. In a post-tensioned member, prestressing tendons transfer the compression directly to the end of the member through bearing plates and anchors. The anchorage zone, based on the principle of St. Venant, is geometrically defined as the volume of concrete through which the prestressing force at the anchorage device spreads transversely to a more linear stress distribution across the entire cross section at some distance from the anchorage device [4].

For design purposes, the anchorage zone can be divided into general and local zones [4]. The region of tensile stresses is the general zone. The region of high compressive stresses (immediately ahead of the anchorage device) is the local zone. For the design of the general zone, a "strut-and-tie model," a refined elastic stress analysis or approximate methods may be used to determine the stresses, while the resistance to bursting forces is provided by reinforcing spirals, closed hoops, or anchoraged transverse ties. For the design of the local zone, bearing pressure is a major concern. For detailed requirements, see AASHTO-LRFD [4].

10.5 Design Example

Two-Span Continuous Cast-in-Place Box-Girder Bridge

Given

A two-span continuous cast-in-place prestressed concrete box-girder bridge has two equal spans of length 48 m with a single-column bent. The superstructure is 10.4 m wide. The elevation view of the bridge is shown in Figure 10.12a.

Material:

Initial concrete: $f'_{ci} = 24$ MPa, $E_{ci} = 24{,}768$ MPa
Final concrete: $f'_c = 28$ MPa, $E_c = 26{,}752$ MPa
Prestressing steel: $f_{pu} = 1860$ MPa low relaxation strand, $E_p = 197{,}000$ MPa
Mild steel: $f_y = 400$ MPa, $E_s = 200{,}000$ MPa

Prestressing:

Anchorage set thickness = 10 mm
Prestressing stress at jacking $f_{pj} = 0.8\, f_{pu} = 1488$ MPa
The secondary moments due to prestressing at the bent are $M_{DA} = 1.118\ P_j$, $M_{DG} = 1.107\ P_j$

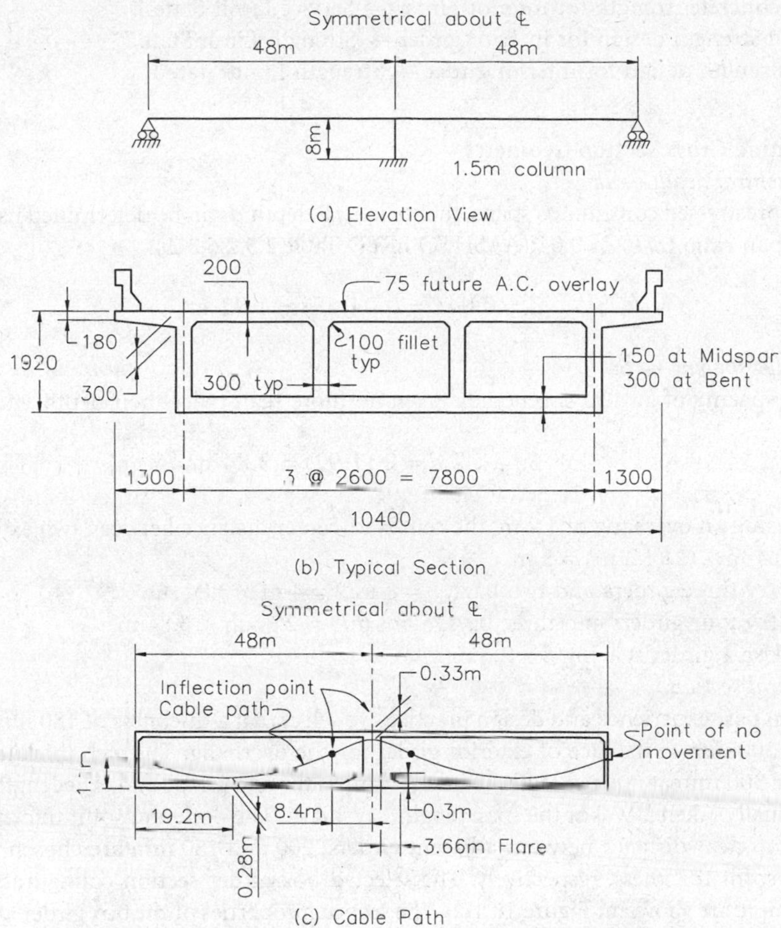

FIGURE 10.12 A two–span continuous prestressed concrete box–girder bridge.

Loads:

Dead Load = self-weight + barrier rail + future wearing 75 mm AC overlay
Live Load = AASHTO HL-93 Live Load + dynamic load allowance

Specification:

AASHTO-LRFD [4] (referred as AASHTO in this example)

Requirements

1. Determine cross section geometry
2. Determine longitudinal section and cable path
3. Calculate loads
4. Calculate live load distribution factors for interior girder
5. Calculate unfactored moment and shear demands for interior girder
6. Determine load factors for Strength Limit State I and Service Limit State I
7. Calculate section properties for interior girder
8. Calculate prestress losses
9. Determine prestressing force P_j for interior girder

10. Check concrete strength for interior girder — Service Limit State I
11. Flexural strength design for interior girder — Strength Limit State I
12. Shear strength design for interior girder — Strength Limit State I

Solution

1. Determine Cross Section Geometry

a. *Structural depth — d:*
For prestressed continuous spans, the structural depth d can be determined using a depth-to-span ratio (d/L) of 0.04 (AASHTO LRFD Table 2.5.2.6.3-1).

$$d = 0.04L = 0.04(48) = 1.92 \text{ m}$$

b. *Girder spacing — S:*
The spacing of girders is generally taken no more than twice their depth.

$$S_{max} < 2\,d = 2\,(1.92) = 3.84 \text{ m}$$

By using an overhang of 1.2 m, the center-to-center distance between two exterior girders is 10.4 m – (2)(1.2 m) = 8 m.

Try three girders and two bays, S = 8 m/2 = 4 m > 3.84 m NG
Try four girders and three bays, S = 8 m/3 = 2.67 m < 3.84 m OK
Use a girder spacing S = 2.6 m

c. *Typical section:*
From past experience and design practice, we select that a thickness of 180 mm at the edge and 300 mm at the face of exterior girder for the overhang. The web thickness is chosen to be 300 mm at normal section and 450 mm at the anchorage end. The length of the flare is usually taken as 1/10 of the span length, say 4.8 m. The deck and soffit thickness depends on the clear distance between adjacent girders; 200 and 150 mm are chosen for the deck and soffit thickness, respectively. The selected box-girder section configurations for this example are shown in Figure 10.12b. The section properties of the box girder are as follows:

Properties	Midspan	Bent (face of support)
A (m^2)	5.301	6.336
I (m^4)	2.844	3.513
y_b (m)	1.102	0.959

2. Determine Longitudinal Section and Cable Path

To lower the center of gravity of the superstructure at the face of the bent cap in the CIP post-tensioned box girder, the thickness of soffit is flared to 300 mm as shown in Figure 10.12c. A cable path is generally controlled by the maximum dead-load moment and the position of the jack at the end section. Maximum eccentricities should occur at points of maximum dead load moments and almost no eccentricity should be present at the jacked end section. For this example, the maximum dead-load moments occur at three locations: at the bent cap, at the locations close to 0.4L for Span 1 and 0.6L for Span 2. A parabolic cable path is chosen as shown in Figure 10.12c.

3. Calculate Loads

a. *Component dead load — DC:*
The component dead load DC includes all structural dead loads with the exception of the future wearing surface and specified utility loads. For design purposes, two parts of the DC are defined as:

*DC*1 — girder self-weight (density 2400 kg/m³) acting at the prestressing stage
*DC*2 — barrier rail weight (11.5 kN/m) acting at service stage after all losses.

b. *Wearing surface load — DW*:
The future wearing surface of 75 mm with a density 2250 kg/m³

$$
\begin{aligned}
DW &= \text{(deck width – barrier width) (thickness of wearing surface) (density)} \\
&= [10.4\text{ m} - 2(0.54\text{ m})](0.075\text{ m})(2250\text{ kg/m}^3)(9.8066\text{ m/s}^2) = 15{,}423\text{ N/m} \\
&= 15.423\text{ kN/m}
\end{aligned}
$$

c. *Live-Load LL and Dynamic Load Allowance — IM*:
The design live load *LL* is the AASHTO HL-93 vehicular live loading. To consider the wheel-load impact from moving vehicles, the dynamic load allowance $IM = 33\%$ [AASHTO LRFD Table 3.6.2.1-1] is applied to the design truck.

4. Calculate Live Load Distribution Factors
AASHTO [1994] recommends that approximate methods be used to distribute live load to individual girders (AASHTO-LRFD 4.6.2.2.2). The dimensions relevant to this prestressed box girder are: depth $d = 1920$ mm, number of cells $N_c = 3$, spacing of girders $S = 2600$ mm, span length $L = 48{,}000$ mm, half of the girder spacing plus the total overhang $W_e = 2600$ mm, and the distance between the center of an exterior girder and the interior edge of a barrier $d_e = 1300 - 535 = 765$ m. This box girder is within the range of applicability of the AASHTO approximate formulas. The live-load distribution factors are calculated as follows:

a. *Live-load distribution factor for bending moments*:
i. Interior girder (AASHTO Table 4.6.2.2.2b-1):
- One design lane loaded:

$$
\begin{aligned}
g_M &= \left(1.75 + \frac{S}{1100}\right)\left(\frac{300}{L}\right)^{0.35}\left(\frac{1}{N_c}\right)^{0.45} \\
&= \left(1.75 + \frac{2600}{1100}\right)\left(\frac{300}{48{,}000}\right)^{0.35}\left(\frac{1}{3}\right)^{0.45} = 0.425 \text{ lanes}
\end{aligned}
$$

- Two or more design lanes loaded:

$$
\begin{aligned}
g_M &= \left(\frac{13}{N_c}\right)^{0.3}\left(\frac{S}{430}\right)\left(\frac{1}{L}\right)^{0.25} \\
&= \left(\frac{13}{3}\right)^{0.3}\left(\frac{2600}{430}\right)\left(\frac{1}{48{,}000}\right)^{0.25} = 0.634 \text{ lanes} \quad \text{(controls)}
\end{aligned}
$$

ii. Exterior girder (AASHTO Table 4.6.2.2.2d-1):

$$
g_M = \frac{W_e}{4300} = \frac{2600}{4300} = 0.605 \text{ lanes}
$$

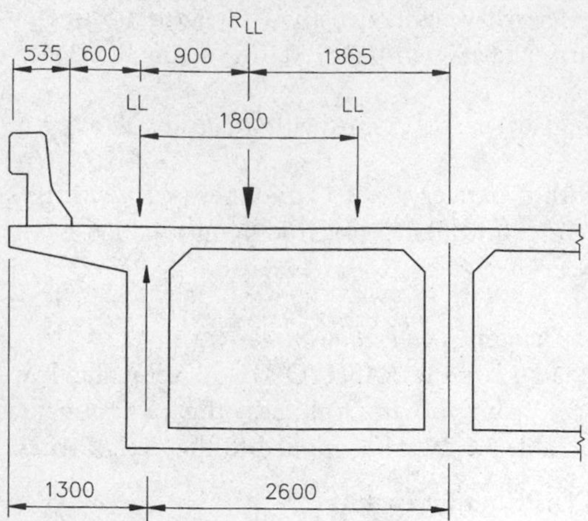

FIGURE 10.13 Live–load distribution for exterior girder — lever rule.

b. *Live-load distribution factor for shear:*
   i. Interior girder (AASHTO Table 4.62.2.3a-1):
      - One design lane loaded:

$$g_V = \left(\frac{S}{2900}\right)^{0.6}\left(\frac{d}{L}\right)^{0.1}$$

$$= \left(\frac{2600}{2900}\right)^{0.6}\left(\frac{1920}{48,000}\right)^{0.1} = 0.679 \text{ lanes}$$

      - Two or more design lanes loaded:

$$g_V = \left(\frac{S}{2200}\right)^{0.9}\left(\frac{d}{L}\right)^{0.1}$$

$$= \left(\frac{2600}{2200}\right)^{0.9}\left(\frac{1920}{48,000}\right)^{0.1} = 0.842 \text{ lanes} \qquad \text{(controls)}$$

   ii. Exterior girder (AASHTO Table 4.62.2.3b-1):
      - One design lane loaded — Lever rule:
        The lever rule assumes that the deck in its transverse direction is simply supported by the girders and uses statics to determine the live-load distribution to the girders. AASHTO-LRFD [4] also requires that when the lever rule is used, the multiple presence factor m should apply. For a one design lane loaded, $m = 1.2$. The lever rule model for the exterior girder is shown in Figure 10.13. From static equilibrium:

$$R = \frac{965 + 900}{2600} = 0.717$$

$$g_v = mR = 1.2(0.717) = 0.861 \qquad \text{(controls)}$$

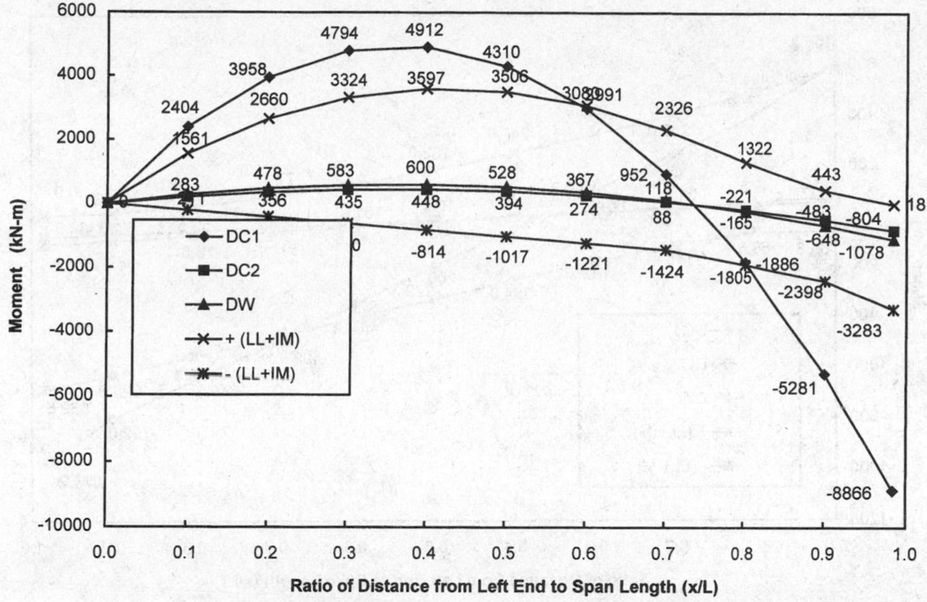

FIGURE 10.14 Moment envelopes for Span 1.

- Two or more design lanes loaded — Modify interior girder factor by *e*:

$$g_V = e g_{V(\text{interior girder})} = \left(0.64 + \frac{d_e}{3800}\right) g_{V(\text{interior girder})}$$

$$= \left(0.64 + \frac{765}{3800}\right)(0.842) = 0.708 \text{ lanes}$$

- The live load distribution factors at the strength limit state:

Strength Limit State I	Interior Girder	Exterior Girder
Bending moment	0.634 lanes	0.605 lanes
Shear	0.842 lanes	0.861 lanes

5. **Calculate Unfactored Moments and Shear Demands for Interior Girder**
   It is practically assumed that all dead loads are carried by the box girder and equally distributed to each girder. The live loads take forces to the girders according to live load distribution factors (AASHTO Article 4.6.2.2.2). Unfactored moment and shear demands for an interior girder are shown in Figures 10.14 and 10.15. Details are listed in Tables 10.11 and 10.12. Only the results for Span 1 are shown in these tables and figures since the bridge is symmetrical about the bent.

6. **Determine Load Factors for Strength Limit State I and Service Limit State I**
   a. *General design equation* (ASHTO Article 1.3.2):

$$\eta \sum \gamma_i Q_i \le \phi R_n \tag{10.53}$$

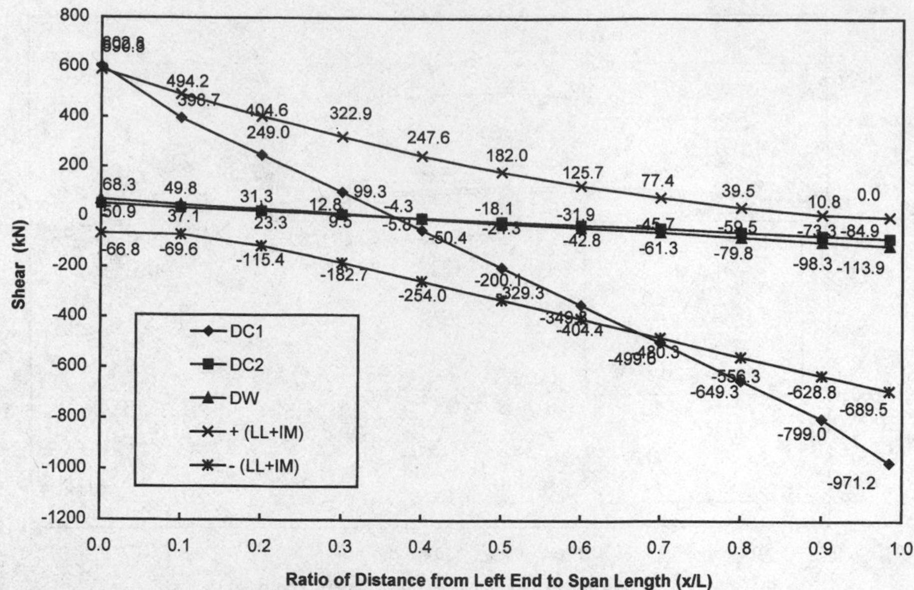

FIGURE 10.15 Shear envelopes for Span 1

TABLE 10.11 Moment and Shear due to Unfactored Dead Load for the Interior Girder

		Unfactored Dead Load					
		*DC*1		*DC*2		DW	
Span	Location (*x/L*)	M_{DC1} (kN-m)	V_{DC1} (kN)	M_{DC2} (kN-m)	V_{DC2} (kN)	M_{DW} (kN-m)	V_{DW} (kN)
	0.0	0	603	0	51	0	68
	0.1	2404	399	211	37	283	50
	0.2	3958	249	356	23	478	31
	0.3	4794	99	435	10	583	13
1	0.4	4912	–50	448	–4	600	–6
	0.5	4310	–200	394	–18	528	–24
	0.6	2991	–350	274	–32	367	–43
	0.7	952	–500	88	–46	118	–61
	0.8	–1805	–649	–165	–59	–221	–80
	0.9	–5281	–799	–483	–73	–648	–98
	Face of column	–8866	–971	–804	–85	–1078	–114

Note:
1. *DC*1 — interior girder self-weight.
2. *DC*2 — barrier self-weight.
3. *DW* — wearing surface load.
4. Moments in Span 2 are symmetrical about the bent.
5. Shears in span are anti-symmetrical about the bent.

where γ_i are load factors and ϕ a resistance factor; Q_i represents force effects; R_n is the nominal resistance; η is a factor related to the ductility, redundancy, and operational importance of that being designed and is defined as:

$$\eta = \eta_D \eta_R \eta_I \geq 0.95 \tag{10.54}$$

TABLE 10.12 Moment and Shear Envelopes and Associated Forces for the Interior Girder due to AASHTO HL-93 Live Load

Span	Location	Positive Moment and Associated Shear		Negative Moment and Associated Shear		Shear and Associated Moment	
	(x/L)	M_{LL+IM} (kN-m)	V_{LL+IM} (kN)	V_{LL+IM} (kN)	M_{LL+IM} (kN-m)	V_{LL+IM} (kN)	M_{LL+IM} (kN-m)
	0.0	0	259	0	–255	497	0
	0.1	1561	312	–203	–42	416	1997
	0.2	2660	249	–407	–42	341	3270
	0.3	3324	47	–610	–42	272	3915
1	0.4	3597	108	–814	–42	–214	3228
	0.5	3506	–25	–1017	–42	–277	3272
	0.6	3080	–81	–1221	–42	–341	2771
	0.7	2326	–258	–1424	–42	–404	1956
	0.8	1322	–166	–1886	–68	–468	689
	0.9	443	–112	–2398	–141	–529	–945
	Face of column	18	–97	–3283	–375	–581	–1850

Note:
1. $LL + IM$— AASHTO HL-93 live load plus dynamic load allowance.
2. Moments in Span 2 are symmetrical about the bent.
3. Shears in Span 2 are antisymmetrical about the bent.
4. Live load distribution factors are considered.

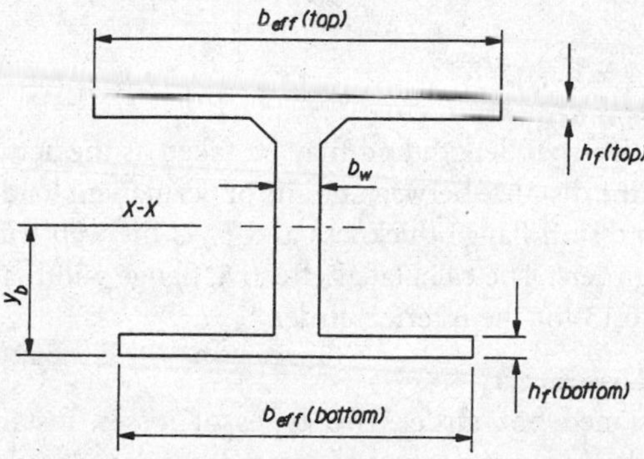

FIGURE 10.16 Effective flange width of interior girder.

For this bridge, the following values are assumed:

Limit States	Ductility η_D	Redundancy η_R	Importance η_I	η
Strength limit state	0.95	0.95	1.05	0.95
Service limit state	1.0	1.0	1.0	1.0

b. *Load factors and load combinations*:
The load factors and combinations are specified as (AASHTO Table 3.4.1-1):
Strength Limit State I: $1.25(DC1 + DC2) + 1.5(DW) + 1.75(LL + IM)$
Service Limit State I: $DC1 + DC2 + DW + (LL + IM)$

7. Calculate Section Properties for Interior Girder

For an interior girder as shown in Figure 10.16, the effective flange width b_{eff} is determined (AASHTO Article 4.6.2.6) by:

TABLE 10.13 Effective Flange Width and Section Properties for Interior Girder

Location	Dimension	Mid span	Bent (face of support)
Top flange	h_f (mm)	200	200
	$L_{eff}/4$ (mm)	9,000	11,813
	$12h_f + b_w$ (mm)	2,700	2,700
	S (mm)	2,600	2,600
	b_{eff} (mm)	**2,600**	**2,600**
Bottom flange	h_f (mm)	150	300
	$L_{eff}/4$ (mm)	9,000	11,813
	$12h_f + b_w$ (mm)	2,100	3,900
	S (mm)	2,600	2,600
	b_{eff} (mm)	**2,100**	**2,600**
Area	A (m²)	1.316	1.736
Moment of inertia	I (m⁴)	0.716	0.968
Center of gravity	y_b (m)	1.085	0.870

Note: L_{eff} = 36.0 m for midspan; L_{eff} = 47.25 m for the bent; b_w = 300 mm.

$$b_{eff} = \text{the lesser of} \begin{cases} \dfrac{L_{eff}}{4} \\ 12h_f + b_w \\ S \end{cases} \tag{10.55}$$

where L_{eff} is the effective span length and may be taken as the actual span length for simply supported spans and the distance between points of permanent load inflection for continuous spans; h_f is the compression flange thickness and b_w is the web width; and S is the average spacing of adjacent girders. The calculated effective flange width and the section properties are shown in Table 10.13 for the interior girder.

8. **Calculate Prestress Losses**

For a CIP post-tensioned box girder, two types of losses, instantaneous losses (friction, anchorage set, and elastic shortening) and time-dependent losses (creep and shrinkage of concrete, and relaxation of prestressing steel), are significant. Since the prestress losses are not symmetrical about the bent for this bridge, the calculation is performed for both spans.

a. *Frictional loss* Δf_{pF}:

$$\Delta f_{pF} = f_{pj}\left(1 - e^{-(Kx + \mu\alpha)}\right) \tag{10.56}$$

where K is the wobble friction coefficient = 6.6×10^{-7}/mm and μ is the coefficient of friction = 0.25 (AASHTO Article 5.9.5.2.2b); x is the length of a prestressing tendon from the jacking end to the point considered; α is the sum of the absolute values of angle change in the prestressing steel path from the jacking end.

For a parabolic cable path (Figure 10.17), the angle change is $\alpha = 2e_p/L_p$, where e_p is the vertical distance between two control points and L_p is the horizontal distance between two control points. The details are given in Table 10.14.

b. *Anchorage set loss* Δf_{pA}:

For an anchor set thickness of ΔL = 10 mm and E = 200,000 MPa, consider the point D where L_{pF} = 48 m and Δf_{pF} = 96.06 MPa:

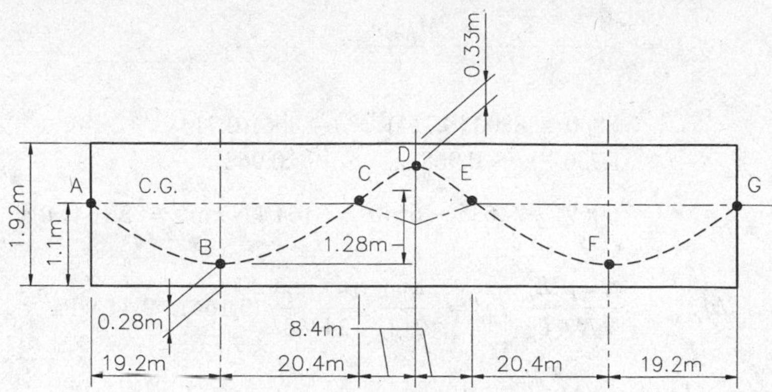

FIGURE 10.17 Parabolic cable path.

TABLE 10.14 Prestress Frictional Loss

Segment	e_p (mm)	L_p (m)	α (rad)	$\Sigma\alpha$ (rad)	ΣL_b (m)	Point	Δf_{pF} (Mpa)
A	0.00	0	0	0	0	A	0.00
AB	820	19.2	0.0854	0.0854	19.2	B	31.44
BC	926	20.4	0.0908	0.1762	39.6	C	64.11
CD	381	8.4	0.0908	0.2669	48.0	D	96.06
DE	381	8.4	0.0908	0.3577	56.4	E	127.28
EF	926	20.4	0.0908	0.4484	76.8	F	157.81
FG	820	19.2	0.0854	0.5339	96.0	G	185.91

$$L_{pA} = \sqrt{\frac{E\,(\Delta L)\,L_{pF}}{\Delta f_{pF}}} = \sqrt{\frac{200{,}000(10)(48{,}000)}{96.06}} = 31\ 613\ \text{mm} = 31.6\ \text{m} \quad < 48\ \text{m} \qquad \text{OK}$$

$$\Delta f = \frac{2\,\Delta f_{pF}\,L_{pA}}{L_{pF}} = \frac{2(96.06)(31.6)}{48} = 126.5\ \text{MPa}$$

$$\Delta f_{pA} = \Delta f\left(1 - \frac{x}{L_{pA}}\right) = 126.5\left(1 - \frac{x}{31.6}\right)$$

c. *Elastic shortening loss* Δf_{pES}.
The loss due to elastic shortening in post-tensioned members is calculated using the following formula (AASHTO Article 5.9.5.2.3b):

$$\Delta f_{pES} = \frac{N-1}{2N}\,\frac{E_p}{E_{ci}}\,f_{cgp} \tag{10.57}$$

To calculate the elastic shortening loss, we assume that the prestressing jack force for an interior girder $P_j = 8800$ kN and the total number of prestressing tendons $N = 4$. f_{cgp} is calculated for face of support section:

$$f_{cgp} = \frac{P_j}{A} + \frac{P_j e^2}{I_x} + \frac{M_{DC1} e}{I_x}$$

$$= \frac{8800}{1.736} + \frac{8800\,(0.714)^2}{0.968} + \frac{(-8866)(0.714)}{0.968}$$

$$= 5069 + 4635 - 6540 = 3164 \text{ kN/m}^2 = 3.164\,\text{MPa}$$

$$\Delta f_{pES} = \frac{N-1}{2N}\frac{E_p}{E_{ci}} f_{cgp} = \frac{4-1}{2(4)}\frac{197{,}000}{24768}(3.164) = 9.44 \text{ MPa}$$

d. *Time-dependent losses* Δf_{pTM}:
AASHTO provides a table to estimate the accumulated effect of time-dependent losses resulting from the creep and shrinkage of concrete and the relaxation of the steel tendons. From AASHTO Table 5.9.5.3-1:

$$\Delta f_{pTM} = 145 \text{ MPa} \quad \text{(upper bound)}$$

e. *Total losses* Δf_{pT}:

$$\Delta f_{pT} = \Delta f_{pF} + \Delta f_{pA} + \Delta f_{pES} + \Delta f_{pTM}$$

Details are given in Table 10.15.

9. Determine Prestressing Force P_j for Interior Girder
Since the live load is not in general equally distributed to girders, the prestressing force P_j required for each girder may be different. To calculate prestress jacking force P_j, the initial prestress force coefficient F_{pCI} and final prestress force coefficient F_{pCF} are defined as:

$$F_{pCI} = 1 - \frac{\Delta f_{pF} + \Delta f_{pA} + \Delta f_{pES}}{f_{pj}} \tag{10.58}$$

$$F_{pCF} = 1 - \frac{\Delta f_{pT}}{f_{pj}} \tag{10.59}$$

The secondary moment coefficients are defined as:

$$M_{sC} = \begin{cases} \dfrac{x}{L}\dfrac{M_{DA}}{P_j} & \text{for Span 1} \\[2ex] \left(1 - \dfrac{x}{L}\right)\dfrac{M_{DG}}{P_j} & \text{for Span 2} \end{cases} \tag{10.60}$$

where x is the distance from the left end for each span. The combined prestressing moment coefficients are defined as:

$$M_{psCI} = F_{pCI}(e) + M_{sC} \tag{10.61}$$

TABLE 10.15 Cable Path and Prestress Losses

Span	Location (x/L)	Prestress Losses (MPa)					Force Coefficient	
		Δf_{pF}	Δf_{pA}	Δf_{pES}	Δf_{pTM}	Δf_{pT}	F_{pCI}	F_{pCF}
	0.0	0.00	126.50			280.94	0.909	0.811
	0.1	7.92	107.28			269.65	0.916	0.819
	0.2	15.80	88.07			258.31	0.924	0.826
	0.3	23.64	68.85			246.94	0.931	0.834
	0.4	31.44	49.64			235.52	0.939	0.842
1	0.5	39.19	30.42	9.44	145.00	224.06	0.947	0.849
	0.6	46.91	11.21			212.56	0.955	0.857
	0.7	54.58	0.00			209.02	0.957	0.860
	0.8	62.21	0.00			216.65	0.952	0.854
	0.9	77.89	0.00			232.33	0.941	0.844
	1.0	96.06	0.00			250.50	0.929	0.832
	0.0	96.06				250.50	0.929	0.832
	0.1	113.99				268.43	0.917	0.820
	0.2	129.10				283.54	0.907	0.809
	0.3	136.33				290.77	0.902	0.805
2	0.4	143.53	0.00	9.44	145.00	297.97	0.897	0.800
	0.5	150.69				305.13	0.892	0.795
	0.6	157.81				312.25	0.888	0.790
	0.7	164.89				319.33	0.883	0.785
	0.8	171.94				326.38	0.878	0.781
	0.9	178.94				333.38	0.873	0.776
	1.0	185.91				340.35	0.869	0.771

Note: $F_{pCI} = 1 - \dfrac{\Delta f_{pF} + \Delta f_{pA} + \Delta f_{pES}}{f_{pj}}$

$F_{pCF} = 1 - \dfrac{\Delta f_{pT}}{f_{pj}}$

$$M_{psCF} = F_{pCF}(e) + M_{sC} \tag{10.62}$$

where e is the distance between the cable and the center of gravity of a cross section; positive values of e indicate that the cable is above the center of gravity, and negative ones indicate the cable is below the center of gravity of the section.

The prestress force coefficients and the combined moment coefficients are calculated and tabled in Table 10.16. According to AASHTO, the prestressing force P_j can be determined using the concrete tensile stress limit in the precompression tensile zone (see Table 10.5):

$$f_{DC1} + f_{DC2} + f_{DW} + f_{LL+IM} + f_{psF} \geq -0.5\sqrt{f_c'} \tag{10.63}$$

in which

$$f_{DC1} = \frac{M_{DC1}C}{I_x} \tag{10.64}$$

$$f_{DC2} = \frac{M_{DC2}C}{I_x} \tag{10.65}$$

TABLE 10.16 Prestress Force and Moment Coefficients

Span	Location (x/L)	Cable Path e (m)	Force Coefficients F_{pCI}	F_{pCF}	Moment Coefficients (m) $F_{pCI}e$	$F_{pCF}e$	M_{sC}	M_{psCI}	M_{psCF}
	0.0	0.015	0.909	0.811	0.014	0.012	0.000	0.014	0.012
	0.1	−0.344	0.916	0.819	−0.315	−0.281	0.034	−0.281	−0.247
	0.2	−0.600	0.924	0.826	−0.554	−0.496	0.068	−0.486	−0.428
	0.3	−0.754	0.931	0.834	−0.702	−0.629	0.102	−0.600	−0.526
	0.4	−0.805	0.939	0.842	−0.756	−0.678	0.136	−0.620	−0.541
1	0.5	−0.754	0.947	0.849	−0.714	−0.640	0.171	−0.543	−0.470
	0.6	−0.600	0.955	0.857	−0.573	−0.514	0.205	−0.368	−0.310
	0.7	−0.344	0.957	0.860	−0.329	−0.295	0.239	−0.090	−0.057
	0.8	0.015	0.952	0.884	0.014	0.013	0.273	0.287	0.286
	0.9	0.377	0.941	0.844	0.355	0.318	0.307	0.662	0.625
	1.0	0.717	0.929	0.832	0.666	0.596	0.341	1.007	0.937
	0.0	0.717	0.929	0.832	0.666	0.596	0.347	1.013	0.943
	0.1	0.377	0.917	0.820	0.346	0.309	0.312	0.658	0.622
	0.2	0.015	0.907	0.809	0.014	0.012	0.278	0.291	0.290
	0.3	−0.344	0.902	0.805	−0.310	−0.277	0.243	−0.067	−0.034
	0.4	−0.600	0.897	0.800	−0.538	−0.480	0.208	−0.330	−0.272
2	0.5	−0.754	0.892	0.795	−0.673	−0.599	0.174	−0.499	−0.426
	0.6	−0.805	0.888	0.790	−0.715	−0.636	0.139	−0.576	−0.497
	0.7	−0.754	0.883	0.785	−0.665	−0.592	0.104	−0.561	−0.488
	0.8	−0.600	0.878	0.781	−0.527	−0.468	0.069	−0.457	−0.399
	0.9	−0.344	0.873	0.776	−0.300	−0.267	0.035	−0.266	−0.232
	1.0	0.015	0.869	0.771	0.013	0.012	0.000	0.013	0.012

Note: e is distance between cable path and central gravity of the interior girder cross section, positive means cable is above the central gravity, and negative indicates cable is below the central gravity.

$$f_{DW} = \frac{M_{DW}C}{I_x} \tag{10.66}$$

$$f_{LL+IM} = \frac{M_{LL+IM}C}{I_x} \tag{10.67}$$

$$f_{psF} = \frac{P_{pe}}{A} + \frac{(P_{pe}e)C}{I_x} + \frac{M_sC}{I_x} = \frac{F_{pCF}P_j}{A} + \frac{M_{psCF}P_jC}{I_x} \tag{10.68}$$

where C (= y_b or y_t) is the distance from the extreme fiber to the center of gravity of the cross section. f_c' is in MPa and P_{pe} is the effective prestressing force after all losses have been incurred. From Eqs. (10.63) and (10.68), we have

$$P_j = \frac{-f_{DC1} - f_{DC2} - f_{DW} - f_{LL+IM} - 0.5\sqrt{f_c'}}{\dfrac{F_{pCF}}{A} + \dfrac{M_{psCF}C}{I_x}} \tag{10.69}$$

Detailed calculations are given in Table 10.17. Most critical points coincide with locations of maximum eccentricity: 0.4L in Span 1, 0.6L in Span 2, and at the bent. For this bridge, the controlling section is through the right face of the bent. Herein, P_j = 8741 kN. Rounding P_j up to 8750 kN gives a required area of prestressing steel of $A_{ps} = P_j/f_{pj}$ = 8750/1488 (1000) = 5880 mm^2.

TABLE 10.17 Determination of Prestressing Jacking Force for an Interior Girder

		Top Fiber					Bottom Fiber				
		Stress (MPa)					Stress (MPa)				
Span	Location (x/L)	f_{DC1}	f_{DC2}	f_{DW}	f_{LL+IM}	Jacking Force, P_j (kN)	f_{DC1}	f_{DC2}	f_{DW}	f_{LL+IM}	Jacking Force P_j(kN)
	0.0	0.000	0.000	0.000	0.000	—	0.000	0.000	0.000	0.000	0
	0.1	2.803	0.246	0.330	1.820	—	−3.642	−0.320	−0.429	−2.365	4405
	0.2	4.616	0.415	0.557	3.103	—	−5.998	−0.540	−0.724	−4.032	6778
	0.3	5.591	0.507	0.680	3.876	—	−7.265	−0.659	−0.884	−5.037	7824
	0.4	5.728	0.522	0.700	4.195	—	−7.442	−0.678	−0.910	−5.450	8101
1	0.5	5.027	0.459	0.616	4.089	—	−6.532	−0.597	−0.800	−5.313	7807
	0.6	3.488	0.319	0.428	3.591	—	−4.532	−0.415	−0.557	−4.667	6714
	0.7	1.110	0.102	0.137	2.712	—	−1.443	−0.133	−0.178	−3.524	3561
	0.8	−2.105	−0.192	−0.258	1.542	2601	2.736	0.250	0.335	−2.004	—
	0.9	−6.159	−0.565	−0.756	0.516	5567	8.003	0.733	0.982	−0.671	—
	1.0	−9.617	−0.872	−1.169	0.020	8406	7.968	0.722	0.969	−0.016	—
	0.0	−9.617	−0.872	−1.169	0.020	8370	7.968	0.722	0.969	−0.016	—
	0.1	−.6.159	−0.564	−0.756	0.516	5661	8.003	0.733	0.982	−0.671	—
	0.2	−2.105	−0.192	−0.258	1.542	2681	2.736	0.250	0.335	−2.004	—
	0.3	1.110	0.102	0.137	2.712	—	−1.443	−0.133	−0.178	−3.524	3974
	0.4	3.488	0.319	0.428	3.591	—	−4.532	−0.415	−0.557	−4.667	7381
2	0.5	5.027	0.459	0.616	4.089	—	−6.532	−0.597	−0.800	−5.313	8483
	0.6	5.728	0.522	0.700	4.195	—	−7.443	−0.678	−0.910	−5.450	**8741**
	0.7	5.591	0.507	0.680	3.876	—	−7.265	−0.659	−0.884	−5.037	8382
	0.8	4.616	0.415	0.557	3.103	—	−5.998	−0.540	−0.724	−4.032	7220
	0.9	2.803	0.246	0.330	1.820	—	−3.642	−0.320	−0.429	−2.365	4666
	1.0	0.000	0.000	0.000	0.000	—	0.000	0.000	0.000	0.000	0

Notes:
1. Positive stress indicates compression and negative stress indicates tension.
2. P_j are obtained by Eq. (10.69).

10. Check Concrete Strength for Interior Girder — Service Limit State I

Two criteria are imposed on the level of concrete stresses when calculating required concrete strength (AASHTO Article 5.9.4.2):

$$\begin{cases} f_{DC1} + f_{psI} \le 0.55 f'_{ci} & \text{at prestressing state} \\ f_{DC1} + f_{DC2} + f_{DW} + f_{LL+IM} + f_{psF} \le 0.45 f'_c & \text{at service state} \end{cases} \tag{10.70}$$

$$f_{psI} = \frac{P_{jI}}{A} + \frac{(P_{jI}e)C}{I_x} + \frac{M_{sI}C}{I_x} = \frac{F_{pCI}P_j}{A} + \frac{M_{psCI}P_jC}{I_x} \tag{10.71}$$

The concrete stresses in the extreme fibers (after instantaneous losses and final losses) are given in Tables 10.18. and 10.19. For the initial concrete strength in the prestressing state, the controlling location is the top fiber at 0.8L section in Span 1. From Eq. (10.70), we have

$$f'_{ci,reg} \ge \frac{f_{DC1} + f_{psI}}{0.55} = \frac{7.15}{0.55} = 13\,\text{MPa}$$

$$\therefore \quad \underline{\text{use } f'_{ci} = 24 \text{ MPa}} \qquad \text{OK}$$

TABLE 10.18 Concrete Stresses after Instantaneous Losses for the Interior Girder

		Top Fiber Stress (MPa)					Bottom Fiber Stress (MPa)				
Span	Location (x/L)	f_{DC1}	F_{pCI^*} P_j/A	M_{psCI^*} $P_j^* Y_t/I$	f_{psI}	Total Initial Stress	f_{DC1}	F_{pCI^*} P_j/A	M_{psCI^*} $P_{j^*} Y_t/I$	f_{psI}	Total Initial Stress
	0.0	0.00	6.04	0.14	6.18	6.18	0.00	6.04	–0.18	5.86	5.86
	0.1	2.80	6.09	–2.87	3.23	6.03	–3.64	6.09	3.72	9.82	6.17
	0.2	4.62	6.14	–4.96	1.18	5.80	–6.00	6.14	6.45	12.59	6.59
	0.3	5.59	6.19	–6.12	0.07	5.66	–7.27	6.19	7.95	14.15	6.88
	0.4	5.73	6.24	–6.32	–0.08	5.65	–7.44	6.24	8.22	14.46	7.02
1	0.5	5.03	6.30	–5.54	0.75	5.78	–6.53	6.30	7.20	13.50	6.97
	0.6	3.49	6.35	–3.76	2.59	6.08	–4.53	6.35	4.88	11.23	6.70
	0.7	1.11	6.36	–0.92	5.44	6.55	–1.44	6.36	1.20	7.56	6.12
	0.8	–2.11	6.33	2.93	9.26	**7.15**	2.74	6.33	–3.81	2.52	5.26
	0.9	–6.16	6.26	6.76	13.02	6.86	8.00	6.26	–8.78	–2.52	5.48
	1.0	–9.62	4.68	9.56	14.24	4.62	7.97	4.68	–7.92	–3.24	4.73
	0.0	–9.62	4.68	9.62	14.30	4.68	7.97	4.68	–7.97	–3.28	4.68
	.1	–6.16	6.10	6.72	12.82	6.66	8.00	6.10	–8.73	–2.63	5.37
	0.2	–2.11	6.03	2.97	9.00	6.90	2.74	6.03	–3.86	2.17	4.90
	0.3	1.11	6.00	–0.69	5.31	6.42	–1.44	6.00	0.89	6.89	5.45
	0.4	3.49	5.97	–3.37	2.60	6.08	–4.53	5.97	4.38	10.34	5.81
2	0.5	5.03	5.93	–5.09	0.84	5.87	–6.53	5.93	6.62	12.55	6.02
	0.6	5.73	5.90	–5.87	0.03	5.75	–7.44	5.90	7.63	13.54	6.09
	0.7	5.59	5.87	–5.73	0.14	5.73	–7.27	5.87	7.44	13.31	6.05
	0.8	4.62	5.84	–4.67	1.71	5.79	–6.00	5.84	6.07	11.90	5.91
	0.9	2.80	5.81	–2.71	3.10	5.90	–3.64	5.81	3.52	9.33	5.69
	1.0	0.00	5.78	0.13	5.91	5.91	0.00	5.78	–0.17	5.60	5.60

Note: Positive stress indicates compression and negative stress indicates tension

For the final concrete strength at the service limit state, the controlling location is in the top fiber at 0.6L section in Span 2. From Eq. (10.70), we have

$$f'_{c,req} \geq \frac{f_{DC1} + f_{DC2} + f_{DW} + f_{LL+IM} + f_{psF}}{0.45} = \frac{11.32}{0.45} = 21.16\ \text{MPa} < 28\,\text{MPa}$$

$$\therefore \quad \underline{\text{choose } f'_c = 28\,\text{MPa}} \qquad \text{OK}$$

11. **Flexural Strength Design for Interior Girder — Strength Limit State I**
AASHTO [4] requires that for the Strength Limit State I

$$M_u \leq \phi M_n$$

$$M_u = \eta \sum \gamma_i M_i = 0.95[1.25(M_{DC1} + M_{DC2}) + 1.5M_{DW} + 1.75M_{LLH}] + M_{ps}$$

where ϕ is the flexural resistance factor 1.0 and M_{ps} is the secondary moment due to prestress. Factored moment demands M_u for the interior girder in Span 1 are calculated in Table 10.20. Although the moment demands are not symmetrical about the bent (due to different secondary prestress moments), the results for Span 2 are similar and the differences will not be considered in this example. The detailed calculations for the flexural resistance ϕM_n are shown in Table 10.21. It is seen that no additional mild steel is required.

TABLE 10.19 Concrete Stresses after Total Losses for the Interior Girder

		Top Fiber Stress (MPa)					Bottom Fiber Stress (MPa)				
Span	Location (x/L)	f_{LOAD}	F_{pCF^*} $P_{j/A}$	M_{psCF^*} $P_{j^*Yt/I}$	f_{psF}	Total Final Stress	f_{LOAD}	F_{pCF^*} $P_{j/A}$	M_{psCF^*} P_{j^*yb}/I	f_{psF}	Total Final Stress
	0.0	0.00	5.39	0.12	5.52	5.52	0.00	5.39	–0.16	5.23	5.23
	0.1	5.20	5.44	–2.52	2.92	8.12	–6.76	5.44	3.28	8.72	1.97
	0.2	8.69	5.49	–4.36	1.13	9.82	–11.29	5.49	5.67	11.16	–0.13
	0.3	10.66	5.55	–5.37	0.17	10.83	–13.85	5.55	6.98	12.52	–1.32
	0.4	11.14	5.60	–5.52	0.07	11.22	–14.48	5.60	7.18	12.77	–1.71
1	0.5	10.19	5.65	–4.79	0.85	11.05	–13.24	5.65	6.23	11.88	–1.37
	0.6	7.83	5.70	–3.16	2.54	10.37	–10.17	5.70	4.11	9.81	–0.36
	0.7	4.06	5.71	–0.58	5.14	9.20	–5.28	5.71	0.75	6.47	1.19
	0.8	–4.75	5.68	2.91	8.60	3.84	6.18	5.68	–3.79	1.89	8.07
	0.9	–10.28	5.61	6.38	11.99	1.72	13.35	5.61	–8.29	–2.68	10.67
	1.0	–15.22	4.19	8.90	13.09	–2.13	12.61	4.19	–7.37	–3.18	9.43
	0.0	–15.22	4.19	8.95	13.14	–2.07	12.61	4.19	–7.42	–3.23	9.38
	0.1	–10.28	5.45	6.34	11.79	1.52	13.35	5.45	–8.24	–2.79	10.56
	0.2	–4.75	5.38	2.96	8.34	3.58	6.18	5.38	–3.84	1.54	7.72
	0.3	4.06	5.35	0.34	5.01	9.07	–5.28	5.35	0.45	5.80	0.52
	0.4	7.83	5.32	–2.77	2.55	10.37	–10.17	5.32	3.60	8.92	–1.25
2	0.5	10.19	5.29	–4.34	0.94	11.13	–13.24	5.29	5.64	10.93	–2.31
	0.6	11.14	5.25	–5.07	0.18	**11.32**	–14.48	5.25	6.59	11.85	**–2.63**
	0.7	10.66	5.22	–4.98	0.24	10.90	–13.85	5.22	6.47	11.69	–2.15
	0.8	8.69	5.19	–4.07	1.12	9.81	11.29	5.19	5.29	10.48	–0.81
	0.9	5.20	5.16	–2.37	2.79	7.99	–6.76	5.16	3.08	8.24	1.48
	1.0	0.00	5.13	0.12	5.25	5.25	0.00	5.13	–0.15	4.97	4.97

Notes:
1. $f_{LOAD} = f_{DC1} + f_{DC2} + f_{DW} + f_{LL+IM}$.
2. Positive stress indicates compression and negative stress indicates tension.

TABLE 10.20 Factored Moments for an Interior Girder

Span	Location (x/L)	M_{DC1} (kN-m) Dead Load 1	M_{DC2} (kN-m) Dead Load 2	M_{DW} (kN-m) Wearing Surface	M_{LL+IM} (kN-m) Positive	M_{LL+IM} (kN-m) Negative	M_{ps} (kN-m) P/S	M_u (kN-m) Positive	M_u (kN-m) Negative
	0.0	0	0	0	0	0	0	**0**	0
	0.1	2404	211	283	1561	–203	298	**6,402**	3,469
	0.2	3958	356	478	2660	–407	597	**10,824**	5,725
	0.3	4794	435	583	3324	–610	895	**13,462**	6,922
	0.4	4912	448	600	3597	–814	1194	**14,393**	7,060
1	0.5	4310	395	528	3506	–1017	1492	**13,660**	6,140
	0.6	2991	274	367	3080	–1221	1790	**11,310**	4,161
	0.7	952	88	118	2326	–1424	2089	**7,358**	1,124
	0.8	–1805	–165	–221	1322	–1886	2387	1,931	**3,403**
	0.9	–5281	–483	–648	443	–2398	2685	–4.348	**9,071**
	1.0	–8866	–804	–1078	18	–3283	2984	–10,005	**–15,492**

Note: $M_u = 0.95[1.25(M_{DC1} + M_{DC2}) + 1.5M_{DW} + 1.75M_{LL+IM}] + M_{ps}$.

12. Shear Strength Design for Interior Girder — Strength Limit State I

AASHTO [4] requires that for the strength limit state I

$$V_u \leq \phi V_n$$

$$V_u = \eta \sum \gamma_i V_i = 0.95[1.25(V_{DC1} + V_{DC2}) + 1.5V_{DW} + 1.75V_{LL+IM}] + V_{ps}$$

TABLE 10.21 Flexural Strength Design for Interior Girder — Strength Limit State I

Span	Location (x/L)	A_{ps} mm²	d_p mm	A_s mm²	d_s mm	b mm	c mm	f_{ps} Mpa	d_e mm	a mm	ϕM_n Mpa	M_u kN-m
	0.0		32.16	0	72.06	104	7.14	253.2	32.16	6.07	5,206	0
	0.1		46.09	0	72.06	104	7.27	258.1	46.09	6.18	7,833	4,009
	0.2		56.04	0	72.06	104	7.33	260.1	56.04	6.23	9,717	6,820
	0.3		61.54	0	72.06	104	7.35	261.0	61.54	6.25	10,759	8,469
	0.4		64.00	0	72.06	104	7.36	261.3	64.00	6.26	11,226	9,012
1	0.5	8.47	62.29	0	72.06	104	7.36	261.1	62.29	6.25	10,903	8,494
	0.6		57.20	0	72.06	104	7.34	260.3	57.20	6.24	9,937	6,942
	0.7		48.71	0	72.06	104	7.29	258.7	48.71	6.20	8,328	4,392
	0.8		38.20	0	71.06	82.5	21.19	228.1	38.20	18.01	–4,965	–1.397
	0.9		53.48	0	71.06	82.5	23.36	237.0	53.48	19.86	–7,822	–5,906
	1.0		62.00	0	71.06	104	8.13	261.0	62.00	6.25	–10,848	–10,716

TABLE 10.22 Factored Shear for an Interior Girder

Span	Location (x/L)	V_{DC1} (kN) Dead Load 1	V_{DC2} (kN) Dead Load 2	V_{DW} (kN) Wearing Surface	V_{LL+IM} (kN) Envelopes	M_{LL+IM} (kN-m) Associated	V_{ps} (kN) P/S	V_u (kN)	M_u (kN-m) Associated
	0.0	602.8	50.9	68.3	497.0	0.0	62.2	1762.0	0
	0.1	398.7	37.1	49.8	416.1	1997.4	62.2	1342.5	7,128
	0.2	249.0	23.3	31.3	340.7	3270.3	62.2	996.5	11,838
	0.3	99.3	9.5	12.8	271.9	3915.3	62.2	661.6	14,446
	0.4	–50.4	–4.3	–5.8	–213.9	3228.4	62.2	–366.6	13,780
1	0.5	–200.1	–18.1	–24.3	–277.3	3271.7	62.2	–692.6	13,270
	0.6	–349.8	–31.9	–42.8	–340.5	2771.1	62.2	–1018.2	10,797
	0.7	–499.6	–45.7	–61.3	–404.4	1955.7	62.2	–1345.0	6,742
	0.8	–649.3	–59.5	–79.8	–468.4	689.4	62.2	–1671.9	879
	0.9	–799.0	–73.3	–98.3	–529.5	–945.3	62.2	–1994.0	–6,655
	1.0	–971.2	–84.9	–113.9	–580.6	–1849.7	62.2	–2319.5	–13,110

Note: $V_v = 0.95[1.25(V_{DC1} + V_{DC2}) + 1.5V_{DW} + 1.75V_{LL+IM}] + V_{ps}$.

TABLE 10.23 Shear Strength Design for Interior Girder Strength Limit State I

Span	Location (x/L)	d_v (mm)	y' (rad)	V_p (kN)	v/f_c'	ε_x (1000)	θ (°)	β	V_c (kN)	S (mm)	ϕV_n (kN)	$\vert V_u\vert$ (kN)
	0.0	1382	0.085	606	0.133	–0.256	21.0	2.68	428	100	1860	1762
	0.1	1382	0.064	459	0.101	–0.382	27.0	5.60	894	300	1513	1342
	0.2	1382	0.043	309	0.078	–6.241	33.0	2.37	378	200	1036	996
	0.3	1503	0.021	156	0.052	–6.299	38.0	2.10	365	300	753	662
	0.4	1555	0.000	0	0.036	–6.357	36.0	2.23	400	600	511	367
1	0.5	1503	0.021	159	0.055	–6.415	36.0	2.23	387	400	710	693
	0.6	1382	0.043	320	0.080	–6.473	30.0	2.48	396	200	1076	1018
	0.7	1382	0.064	482	0.099	–0.401	27.0	5.63	899	300	1538	1345
	0.8	1382	0.085	639	0.120	–0.398	23.5	6.50	1038	300	1813	1672
	0.9	1382	0.091	670	0.152	–6.372	23.5	3.49	557	100	2017	1994
	1.0	1502	0.000	0	0.233	–6.280	36.0	1.00	173	40	2343	2319

where ϕ is shear resistance factor 0.9 and V_{ps} is the secondary shear due to prestress. Factored shear demands V_u for the interior girder are calculated in Table 10.22. To determine the effective web width, assume that the VSL post-tensioning system of 5 to 12 tendon units [VLS, 1994] will be used with a grouted duct diameter of 74 mm. In this example, $b_v = 300 - 74/2 = 263$ mm. Detailed calculations of the shear resistance ϕV_n (using two-leg #15M stirrups $A_v = 400$ mm²) for Span 1 are shown in Table 10.23. The results for Span 2 are similar to Span 1 and the calculations are not repeated for this example.

References

1. Lin, T. Y. and Burns, N. H., *Design of Prestressed Concrete Structure*, 3rd ed., John Wiley & Sons, New York, 1981.
2. Nawy, E. G., *Prestressed Concrete: A Fundamental Approach*, 2nd ed., Prentice-Hall, Englewood Cliffs, NJ, 1996.
3. Collins, M. P. and Mitchell, D., *Prestressed Concrete Structures*, Prentice-Hall, Englewood Cliffs, NJ, 1991.
4. AASHTO, *AASHTO LRFD Bridge Design Specifications*, 1st ed., American Association of State Highway and Transportation Officials, Washington, D.C., 1994.
5. PCI, *PCI Design Handbook – Precast and Prestressed Concrete*, 3rd ed., Prestressed Concrete Institute, Chicago, IL, 1985.
6. Eubunsky, I. A. and Rubinsky, A., A preliminary investigation of the use of fiberglass for prestressed concrete, *Mag. Concrete Res.*, Sept., 71, 1954.
7. Wines, J. C. and Hoff, G. C., Laboratory Investigation of Plastic — Glass Fiber Reinforcement for Reinforced and Prestressed Concrete, *Report 1*, U.S. Army Corps of Engineers, Waterway Experimental Station, Vicksburg, MI, 1966.
8. Wines, J. C., Dietz, R. J., and Hawly, J. L., Laboratory Investigation of Plastic — Glass Fiber Reinforcement for Reinforced and Prestressed Concrete, *Report 2*, U.S. Army Corps of Engineers, Waterway Experimental Station, Vicksburg, MI, 1966.
9. Iyer, S.I. and Anigol, M., Testing and evaluating fiberglass, graphite, and steel prestressing cables for pretensioned beams, in *Advanced Composite Materials in Civil Engineering Structures*, Iyer, S. I. and Sen, R., Eds., ASCE, New York, 1991, 44.
10. Miesseler, H. J. and Wolff, R., Experience with fiber composite materials and monitoring with optical fiber sensors, in *Advanced Composite Materials in Civil Engineering Structures*, Iyer, S. I. and Sen, R., Eds., ASCE, New York, 1991, 167–182.
11. Kim, P. and Meier, U., CFRP cables for large structures, in *Advanced Composite Materials in Civil Engineering Structures*, Iyer, S. I. and Sen, R., Eds., ASCE, New York, 1991, 233–244.
12. PTI, *Post-Tensioning Manual*, 3rd ed., Post-Tensioning Institute, Phoenix, AZ, 1981.
13. FHWA, *Standard Plans for Highway Bridges, Vol. I, Concrete Superstructures*, U.S. Department of Transportation, FHWA, Washington, D.C., 1990.
14. ACI, *Building Code Requirements for Structural Concrete (ACI318-95) and Commentary* (ACI318R-95), American Concrete Institute, Farmington Hills, MI, 1995.
15. CEB-FIP, *Model Code for concrete structures.* (MC-90). Comité Euro-international du Béton (CEB)-Fédération Internationale de la précontrainte (FIP) (1990). Thomas Telford, London, U.K. 1993.

11

Segmental Concrete Bridges

Gerard Sauvageot

J. Muller International

0-8493-7434-0/00/$0.00+$.50

11.1 Introduction

Before the advent of segmental construction, concrete bridges would often be made of several precast girders placed side by side, with joints between girders being parallel to the longitudinal axis of the bridge. With the modern segmental concept, the segments are slices of a structural element between joints which are perpendicular to the longitudinal axis of the structure.

When segmental construction first appeared in the early 1950s, it was either cast in place as used in Germany by Finsterwalder et al., or precast as used in France by Eugène Freyssinet and Jean Muller. The development of modern segmental construction is intertwined with the development of balanced cantilever construction.

By the use of the term *balanced cantilever construction,* we are describing a phased construction of a bridge superstructure. The construction starts from the piers cantilevering out to both sides in such a way that each phase is tied to the previous ones by post-tensioning tendons, incorporated into the permanent structure, so that each phase serves as a construction base for the following one.

The first attempts to use balanced cantilever construction, in its pure form, were made by Baumgart, who in 1929 built the Río Peixe Bridge in Brazil in reinforced concrete, casting the 68-m-long main span in free cantilevering. The method did not really prosper, however, until the post-tensioning technique had been sufficiently developed and generally recognized to allow crack-free concrete cantilever construction.

From 1950, several large bridges were built in Germany with the use of balanced cantilever construction with a hinge at midspan, using cast-in-place segments, such as

- Moselbrücke Koblenz, 1954: Road bridge, 20 m wide, with three spans of 101, 114, 123 m plus short ballasted end spans hidden in large abutments; the cross section is made up of twin boxes of variable depth, connected by the top slab.
- Rheinbrücke Bendorf, 1964: Twin motorway bridges, 1,031 m long, with three main river spans of 71, 208, 71 m, built-in free cantilever construction with variable depth box sections.

In France, the cantilever construction took a different direction, emphasizing the use of precast segments.

Precast segments were used by Eugène Freyssinet for construction of the well-known six bridges over the Marne River in France (1946 to 1950). The longitudinal frames were assembled from precast segments, which were prestressed vertically and connected by dry-packed joints and longitudinal post-tensioning tendons. Precast segments were also used by Jean Muller for the execution of a girder bridge in upstate New York, where longitudinal girders were precast in three segments each, which were assembled by dry-packed joints and longitudinal post-tensioning tendons.

From 1960, Jean Muller systematically applied precast segments to cantilever construction of bridges. It is characteristic for precast segmental construction, in its purest form, that segments are match cast, which means that each segment is cast against the previous one so that the end face of one segment will be an imprint of the neighbor segment, ensuring a perfect fit at the erection. The early milestones were as follows:

- Bridge over the Seine at Choisy-le-Roi in France, 1962: Length 37+55+37 = 130 m; the bridge is continuous at midspan, with glued joints between segments (first precast segmental bridge).
- Viaduc d'Oleron in France, 1964 to 1966: Total length 2862 m, span lengths generally 79 m, with hinges in the quarterpoint of every fourth span; the segments were cast on a long bench (long-line method); erection was by self-launching overhead gantry (first large-scale, industrialized precast bridge construction).

In the same period, precast segmental construction was adopted by other designers for bridge construction with cast-in-place joints. Some outstanding structures deserve mention:

- Ager Brücke in Austria, 1959 to 1962: Precast segments placed on scaffold, cast-in-place joints.
- Río Caroni in Venezuela, 1962 to 1964: Bridge with multiple spans of 96-m each. Precast segments 9.2 m long, were connected by 0.40-m-wide cast-in-place joints to constitute the 480-m-long bridge deck weighing 8400 tons, which was placed by incremental launching with temporary intermediate supports.
- Oosterschelde Bridge in The Netherlands, 1962 to 1965: Precast segmental bridge with a total length of 5 km and span lengths of 95 m; the precast segments are connected by cast-in-place, 0.4-m-wide joints and longitudinal post-tensioning.

Since the 1960s, the construction method has undergone refinements, and it has been developed further to cover many special cases, such as progressive construction of cantilever bridges, span-by-span construction of simply supported or continuous spans, and precast-segmental construction of frames, arches, and cable-stayed bridge decks.

In 1980, precast segmental construction was applied to the Long Key and Seven Mile Bridges in the Florida Keys in the United States. The Long Key Bridge has 100 spans of 36 m each, with continuity in groups of eight spans. The Seven Mile Bridge has 270 spans of 42 m each with continuity in groups of seven spans. The spans were assembled from 5.6-m-long precast segments placed on erection girders and made self-supporting by the stressing of longitudinal post-tensioning tendons. The construction method became what is now known as span-by-span construction.

Comparing cast-in-place segmental construction with precast segmental construction, the following features come to mind:

- Cast-in-place segmental construction is a relatively slow construction method. The work is performed *in situ,* i.e., exposed to weather conditions. The time-dependent deformations of the concrete become very important as a result of early loading of the young concrete. This method requires a relatively low degree of investment (travelers).
- Precast segmental construction is a fast construction method determined by the time required for the erection. The major part of the work is performed in the precasting yard, where it can be protected against inclement weather. Precasting can start simultaneously with the foundation work. The time-dependent deformations of the concrete become less important, as the concrete may have reached a higher age by the time the segments are placed in the structure. This method requires relatively important investments in precasting yard, molds, lifting gear, transportation, and erection equipment. Therefore, this method requires a certain volume of work to become economically viable. Typically, the industrialized execution of the structure leads to higher quality of the finished product.

Since the 1960s, the precast segmental construction method has won widespread recognition and is used extensively throughout the world. Currently, very comprehensive bridge schemes, with more than 20,000 segments in one scheme, are being built as large urban and suburban viaducts for road or rail. It is reasonable to expect that the precast segmental construction method, as introduced by Jean Muller, will contribute extensively to meet the infrastructure needs of humankind well into the next millennium.

11.2 Balanced Cantilever Girder Bridges

11.2.1 Overview

Balanced cantilever segmental construction for concrete box-girder bridges has long been recognized as one of the most efficient methods of building bridges without the need for falsework. This method has great advantages over other forms of construction in urban areas where temporary shoring

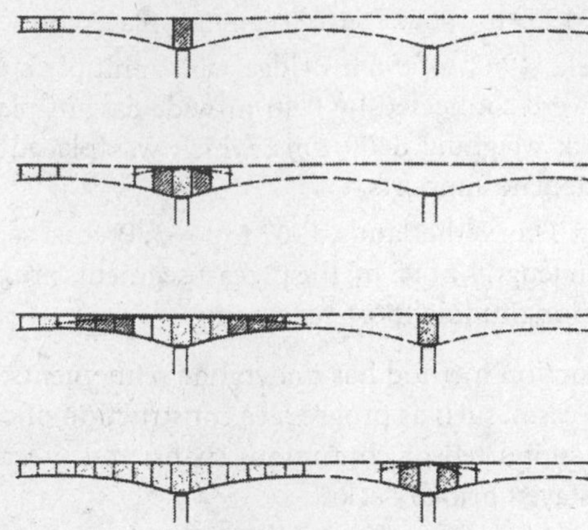

FIGURE 11.1 Balanced cantilever construction.

would disrupt traffic and services below, in deep gorges, and over waterways where falsework would not only be expensive but also a hazard. Construction commences from the permanent piers and proceeds in a "balanced" manner to midspan (see Figure 11.1). A final closure joint connects cantilevers from adjacent piers. The structure is hence self-supporting at all stages. Nominal out-of-balance forces due to loads on the cantilever can be resisted by several methods where any temporary equipment is reusable from pier to pier.

The most common methods are as follows:

- Monolithic connection to the pier if one is present for the final structure;
- Permanent, if present, or temporary double bearings and vertical temporary post-tensioning;
- A simple prop/tie down to the permanent pile cap;
- A prop against an overhead gantry if one is mobilized for placing segments or supporting formwork.

The cantilevers are usually constructed in 3- to 6-m-long segments. These segments may be cast in place or precast in a nearby purpose-built yard, transported to the specific piers by land, water, or on the completed viaduct, and erected into place. Both methods have merit depending on the specific application.

It is usually difficult to justify the capital outlay for the molds, casting yard, and erection equipment required for precast segmental construction in a project with a deck area of less than 5000 m^2. The precasting technique may be viable for smaller projects provided existing casting yard and molds can be mobilized and the segments could be erected by a crane.

11.2.2 Span Arrangement and Typical Cross Sections

Typical internal span-to-depth ratios for constant-depth girders are between 18 and 22. However, box girders shallower than 2 m in depth introduce practical difficulties for stressing operations inside the box and girders shallower than 1.5 m become very difficult to form. This sets a minimum economical span for this type of construction of 25 to 30 m. Constant-depth girders deeper than 2.5 to 3.0 m are unusual and therefore for spans greater than 50 m consideration should be given to varying-depth girders through providing a curved soffit or haunches. For haunch lengths of 20 to 25% of the span from the pier, internal span-to-depth ratios of 18 at the pier and as little as 30 at midspan are normally used.

Single-cell box girders provide the most efficient section for casting – these days multicell boxes are rarely used in this method of construction. Inclined webs improve aesthetics but introduce added difficulties in formwork when used in combination with varying-depth girders. The area of

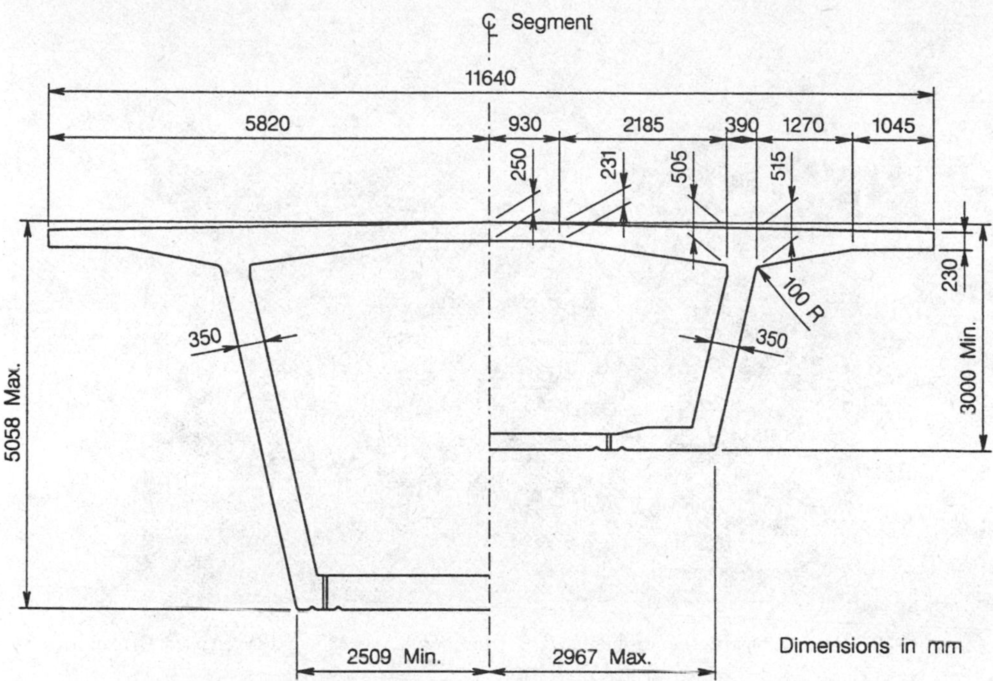

FIGURE 11.2 Typical cross section of a varying-depth girder for a 93-m span.

the bottom slab at the pier is determined by the modulus required to keep bottom fiber compressions below the allowable maximum at this location. In the case of internal tendons local haunches are used at the intersection of the bottom slab and the webs to provide sufficient space for accommodating the required number of tendon ducts at midspan. The distance between the webs at their intersection with the top slab is determined by achieving a reasonable balance between the moments at this node. Web thicknesses are determined largely by shear considerations with a minimum of 250 mm when no tendon ducts internal to the concrete are present and 300 mm in other cases. Figure 11.2 shows the typical dimensions of a varying-depth box girder.

11.2.3 Cast-in-Place Balanced Cantilever Bridges

The cast-in-place technique is preferred for long and irregular span lengths with few repetitions. Bridge structures with one long span and two to four smaller spans usually have a varying-depth girder to carry the longer span, hence making the investment in a mold which accommodates varying-depth segments even more uneconomical. A prime example of application of balanced cantilevering in an urban environment to avoid disruption to existing road services below is the structure of the Bangkok Light Rail Transit System, where it crosses the Rama IV Flyover (see Figure 11.3). The majority of the 26-km viaduct structure is precast, but at this intersection a 60-m span was required to negotiate the existing road at a third level with the flyover in service below. A three-span, 30-, 60-, 30-m structure was utilized with a box-girder depth of 3.5 m at the pier and 2.0 m at midspan and a parabolic curved soffit. The flyover was only disrupted a few nights during concrete placement of the segments directly above as a precaution.

In the above example, the side spans were constructed by balanced cantilevering; however, ideal arrangement of spans normally provides end spans which are greater than half the internal spans. These, therefore, cannot be completed by balanced cantilevering, and various techniques are used to reach the abutments. The most economical and common method is the use of falsework; however,

FIGURE 11.3 Construction of the Bangkok Transit System over Rama IV Flyover, Thailand.

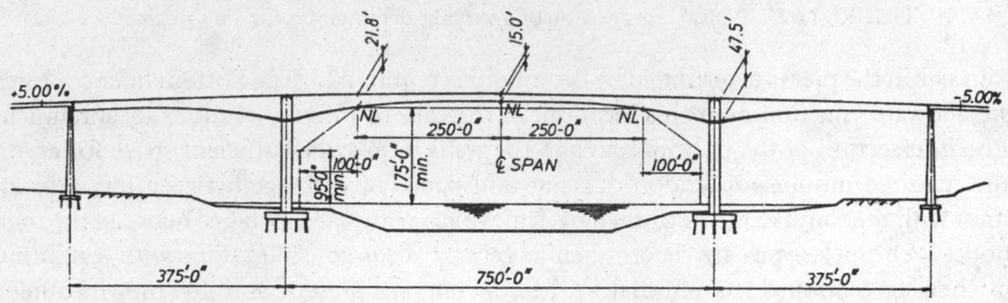

FIGURE 11.4 Houston Ship Channel Bridge, United States.

should the scale of the project justify use of an auxiliary truss to support the formwork during balanced cantilevering, then this could also be used for completing the end spans.

Another example of a cast-in-place balanced cantilever bridge is the Houston Ship Channel Bridge where a three-span, 114-, 229-, 114-m structure was used over the navigation channel (see Figure 11.4). A three-web box girder carrying four lanes of traffic is fixed to the main piers to make the structure a three-span rigid frame. Unusual span-to-depth ratios were dictated by the maximum allowable grade of the approach viaducts and the clearance required for the ship channel. The soffit was given a third-degree parabolic profile to increase the structural depth near the piers in order to compensate for the very limited height of the center portion of the main span. Maximum depth at the pier is 14.6 m, a span-to-depth ratio of 15.3 to enable a minimum depth at midspan of 4.6 m, and a span-to-depth ratio of 49. The box girder is post-tensioned in three dimensions: four 12.7-mm strands at 600-mm centers transversely in the top slab as well as longitudinal and vertical post-tensioning in the webs.

11.2.4 Precast Balanced Cantilever Bridges

Extending segmental construction to balanced cantilevering, and hence eliminating the need for falsework as well as substantial increases in the rate of construction, requires a huge leap in the technology of precasting: match casting. The very first bridge that benefited from match-casting technology was the Choisy-le-Roi Bridge near Paris, designed by Jean Muller and completed in 1964. This method has since grown in popularity and sophistication and is used throughout the world today. The essential feature of match casting is that successive segments are cast against the adjoining segment in the correct relative orientation with each other starting from the first segment away from the pier. The segments are subsequently erected on the pier in the same order, and hence no adjustments are necessary between segments during assembly. The joints are either left dry or made of a very thin layer of epoxy resin, which does not alter the match-cast geometry. Post-tensioning may proceed as early as practicable since there is no need for joints to cure.

The features of this method that provide significant advantages over the cast-in-place method, provided the initial investment in the required equipment is justified by the scale of the project, are immediately obvious and may be listed as follows:

- Casting the superstructure segments may be started at the beginning of the project and at the same time as the construction of the substructure. In fact, this is usually required since the speed of erection is much faster than production output of the casting yard and a stockpile of segments is necessary before erection begins.
- Rate of erection is usually 10 to 15 times the production achieved by the cast-in-place method. The time required for placing reinforcement and tendons and, most importantly, the waiting time for curing of the concrete is eliminated from the critical path.
- Segments are produced in an assembly-line factory environment, providing consistent rates of production and allowing superior quality control. The concrete of the segments is matured, and hence the effects of shrinkage and creep are minimized.

The success of this method relies heavily on accurate geometry control during match casting as the methods available for adjustments during erection offer small and uncertain results. The required levels of accuracy in surveying the segments match-cast against each other are higher than in other areas of civil engineering in order to assure acceptable tolerances at the tip of the cantilevers.

The size and weight of precast segments are limited by the capacity of transportation and placing equipment. For most applications segment weights of 40 to 80 tons are the norm, and segments above 250 tons are seldom economical. An exception to the above is the recent example of the main spans of the Confederation Bridge where complete 192.5-m-long balanced cantilevers weighing 7500 tons were lifted into place using specialized equipment (see Figure 11.5). The 250-m main spans of this fixed link in Atlantic Canada, connecting Cape Tormentine, New Brunswick, and Borden, Prince Edward Island, were constructed by a novel precasting method. The scale of the project was sufficiently large to justify precast segmental construction; however, adverse weather and site conditions provided grounds for constructing the balanced cantilevers, 14 m deep at the piers, in a similar method to cast-in-place construction but in a nearby casting yard. The completed balanced cantilevers were then positioned atop completed pier shafts in a single operation. A light template match-cast against the base of the pier segment allowed fast and accurate alignment control on the spans.

11.2.5 Loads on Substructure

The methods for supporting the nominal out-of-balance forces during balanced cantilevering were described earlier. The following forces should be considered in calculating the possible out-of-balance forces:

- In precast construction, one segment out of balance and the loss of a segment on the balancing cantilever as an ultimate condition;
- In precast construction, presence of a stressing platform (5 to 10 tons) on one cantilever only or the loss of the form traveler in the case of cast-in-place construction;

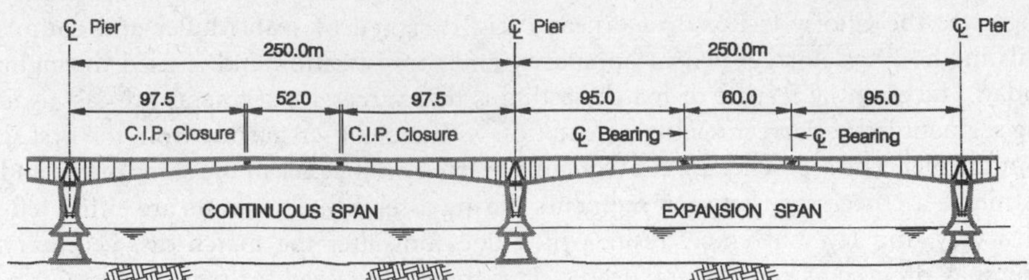

FIGURE 11.5 Main spans of the Confederation Bridge, Canada.

- Live loading on one side of 1.5 kN/m^2;
- Wind loading during construction;
- The possibility of one cantilever having a 2.5% higher dead weight than the other.

The loads on the substructure do not usually govern the design of these elements provided balanced cantilever construction is considered at the onset of the design stage. The out-of-balance forces may provide higher temporary longitudinal moments than for the completed structure; however, in the case of a piled foundation, this usually governs the arrangement and not the number of the piles.

11.2.6 Typical Post-Tensioning Layout

Post-tensioning tendons may be internal or external to the concrete section, but inside the box girder, housed in steel pipes, or both. External post-tensioning greatly simplifies the casting process and the reduced eccentricities available compared with internal tendons are normally compensated by lower frictional losses along the tendons and hence higher forces.

The choice of the size of the tendons must be made in relation to the dimensions of the box-girder elements. A minimum number of tendons would be required for the balanced cantilevering process, and these may be anchored on the face of the segments, on internal blisters, or a combination of both. After continuity of opposing cantilevers is achieved, the required number of midspan tendons may be installed across the closure joint and anchored on internal bottom blisters. Depending on the arrangement and length of the spans, economies may be made by arranging some of the tendons to cross two or more piers, deviating from the top at the piers to the bottom at midspan, thereby reducing the number of anchorages and stressing operations. External post-tensioning is best used for these continuity tendons which would allow longer tendon runs due to the reduced frictional losses. Where the tendons are external to the concrete elements, deviators at piers, quarterspan, and midspan are used to achieve the required profile. An example of a typical internal post-tensioning layout is shown in Figure 11.6.

11.2.7 Articulation and Hinges

The movements of the structure under the effects of cyclic temperature changes, creep, and shrinkage are traditionally accommodated by provision of halving joint-type hinges at the center of various spans. This practice is now discontinued due to the unacceptable creep deformations that occur at these locations. If such hinges are used, these are placed at contraflexure points to minimize the effects of long-term deflections. A development on simple halving joints is a moment-resisting joint, which allows longitudinal movements only. All types of permanent hinges that are more easily exposed to the elements of water and salt from the roadway provide maintenance difficulties and should be eliminated or reduced wherever possible.

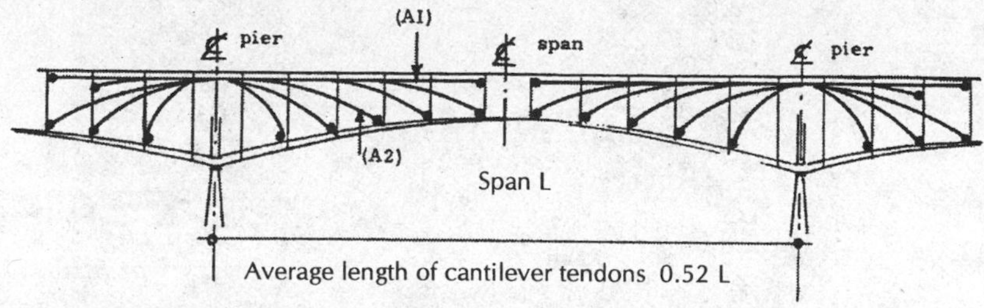

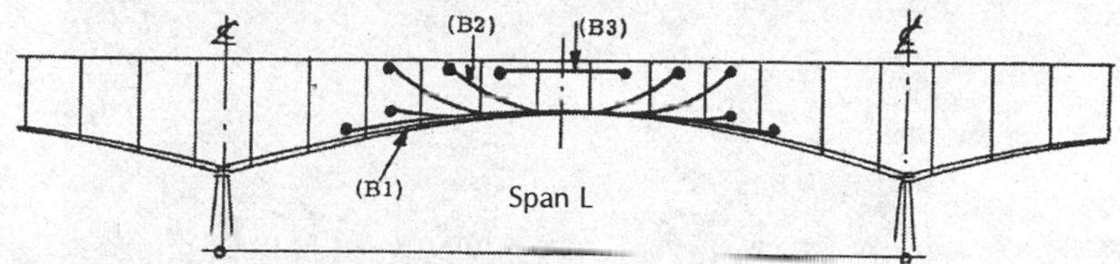

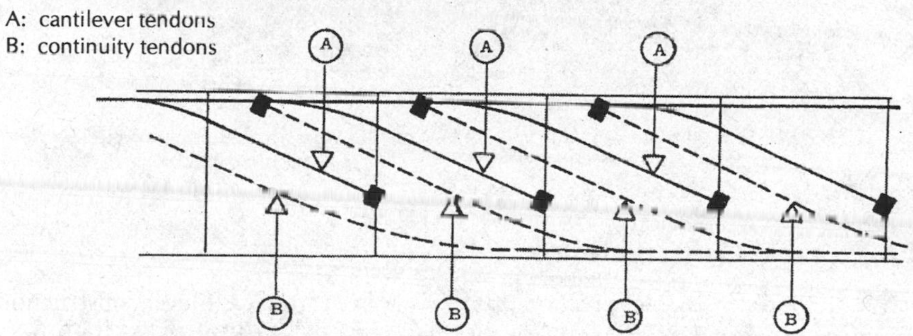

FIGURE 11.6 Typical post-tensioning layout for internal tendons.

If the piers are sufficiently flexible, then a fully continuous bridge may be realized with joints at abutments only. When seismic considerations are not a dominant design feature and a monolithic connection with the pier is not essential, bearings atop of the piers are preferred as they reduce maintenance and replacement cost. In addition, it will allow free longitudinal movements of the deck. A monolithic connection or a hinged bearing at one or more piers would provide a path for transmitting loads to suitable foundation locations.

11.3 Progressive and Span-by-Span Constructed Bridges

11.3.1 Overview

In progressive or span-by-span construction methods, construction starts at one end and proceeds continuously to the other end. Generally, progressive construction is used where access to the ground level is restricted either by physical constraints or by environmental concerns. Deck variable cross sections and span lengths up to 60 m are easily accommodated. In contrast, span-by-span precast segmental construction is used typically where speed of construction is of major concern. Span lengths up to 50 m are most economical as it minimizes the size of the erection equipment.

FIGURE 11.7 Fréburge Viaduct, France—erection with movable stay tower.

11.3.2 Progressive Construction

The progressive method step-by-step erection process is derived from cantilever construction, where segments are placed in a successive cantilever fashion. The method is valid for both precast and cast-in-place segments. Due to the excessively high bending moments the cantilever deck has to resist over the permanent pier during construction, either a temporary bent or a temporary movable tower–stay assembly would have to be used. As shown in Figure 11.7, for precast construction using a temporary tower and stay system, segments are transported over the erected portion of the bridge to the end of the completed portion. Using some type of lifting equipment, e.g., a swivel crane, the segment is placed in position and supported temporarily either by post-tensioning to the previous segment or by stays from a tower.

The advantages of this methods are

- Operations are conducted at deck level.
- Reactions on piers are vertical.
- The method can easily accommodate variable horizontal curves.

The disadvantages are

- The first span is erected on falsework.
- Forces in the superstructure during erection are different from those in the completed structure.
- The piers are temporarily subjected to higher reactions from dead load than in the final structure because of the length of the cantilever erected. However, considering the other loads in the final structure, this case is not generally controlling the pier design.

FIGURE 11.8 Completed Linn Cove Viaduct, United States.

FIGURE 11.9 Linn Cove Viaduct — pier being constructed from the deck level.

The Linn Cove Viaduct (1983) on the Blue Ridge Parkway in North Carolina shown in Figure 11.8, demonstrated the potential progressive placement when one is forced to overcome extreme environmental and physical constraints. Because access at the ground level was limited, the piers were constructed from the deck level, at the tip of an extended cantilever span. Temporary cable stays could not be used due to the extreme horizontal curvature in the bridge. Instead, temporary bent supports were erected between permanent piers. Figure 11.9 shows one temporary support in the background while a permanent precast pier is being erected from the deck level.

FIGURE 11.10 Lifting completed span of the Seven Mile Bridge, Florida, using an *overhead truss.*

11.3.3 Span-by-Span Construction

As with balanced cantilever and progressive placement, span-by-span construction activity is performed primarily at the deck level and typically implemented for long viaducts having numerous, but relatively short spans, e.g., <50 m. It was initially developed as a cast-in-place method of construction, on formwork, with construction joints at joint of contraflexure. The form traveler is supported either on the bridge piers, on the edge of the previously erected span and the next pier or, at times, even at the ground level. With the precast segmental method, segments are placed and adjusted on a steel erection girder spanning from pier to pier, then post-tensioned together in one operation. Although both the cast-in-place and the precast span-by-span construction methods continue to be used, precast segmental has become the method of choice for most applications.

Long Key and Seven Mile Bridges, United States: Two early applications of the precast span-by-span method are the Long Key Bridge (1977) and the Seven Mile Bridge (1978), both located in the Florida Keys. The shorter, 3000-m, 100-span Long Key Bridge is the first application of precast span-by-span construction with dry segment joints and external post-tensioning in the United States.

Essentially the same bridge design concept as Long Key — only much longer — the 10,931-m, 270-span Seven Mile Bridge utilized rectangular precast piers and an overhead truss, as shown in Figure 11.10. The overhead truss allowed easier repositioning from one span to the next one and thus improved overall erection speed.

Bang Na–Bang Pli–Bang Pakong Expressway, Thailand: A number of span-by-span highway and rail mega projects have been either completed recently or currently are under construction in Southeast Asia. Probably, the most innovative of these recent applications is the 54,000-m, 1300-span, Bang Na–Bang Pli–Bang Pakong Expressway. The girder supports segment assembly and span installation activities. This erection process can be regarded as "assembly-line" in that there is no requirement for disassembly and reassembly of the erection girder as it travels from pier to pier. The piers, although designed structurally for the construction process, can also be seen to provide an aesthetically pleasing, somewhat "floating," appearance to the six-lane, 27-m-wide box girder. Figure 11.11 shows one of the erection girders as it lifts a segment. With five erection girders erecting a span every 2 days or 780 m of superstructure per week, construction of the viaduct is expected to last approximately 2 years and be completed in 1999, without interruption of traffic below.

Roize, France: Another innovative example of span-by-span construction is the 112-m, three-span, prestressed composite truss Roize Bridge (1991) in the French Alps, shown in Figure 11.12. The deck is made of prestressed concrete and steel. Each factory-built tetrahedron module and

FIGURE 11.11 Bang Na Expressway, Thailand — launching of girder erection.

FIGURE 11.12 View of the Roize Bridge, France — space truss spans using tetrahedron modules.

precast pretensioned slab is placed on erection beams and adjusted into position. After welding the bottom member joints and casting the closure strips, the modules are post-tensioned together as a completed span. Due to the modular basis, this two-lane bridge represents a new class of super-lightweight, factory-built segments.

Channel Bridge, United States: The first precast, prestressed channel bridge in the United States was built in 1974 in San Diego, California, as a pedestrian crossing at San Diego State University. This concept was reused 18 years later as an experimental study for new bridge standards, initially

FIGURE 11.13 Channel Bridge, France, under construction.

by the French Highway Administration. Figure 11.13 shows the 54-m, two-span Champfeuillet Bridge (1992), under construction along the Rhône Alpine Motorway near Grenoble, France. The most innovative aspect of the concept is the use of the concrete parapets as part of the structure. With the primary longitudinal post-tensioning passing through the barriers, an extremely lightweight, shallow section is possible.

Research and implementation of the Channel Bridge, although continuing in Europe, also has begun recently in the United States. Initiated by the Federal Highway Administration (FHWA) and the Highway Innovative Technology Evaluation Center (HITEC), a branch of the Civil Engineering Research Foundation (CERF), at least two applications of the Channel Bridge concept have been completed recently in the United States for the New York State Department of Transportation (NYSDOT).

The primary benefits of the concept are as follows:

- Lightweight, easily placed segments.
- Fast erection times with small investment in erection equipment.
- Increased vertical clearance beneath the superstructure, because the load-carrying members are above the roadway slab, not below.
- A reduction in the number of bridge overpass piers required, which increases safety levels for traffic lanes below.

Span by span, as used today, utilizes post-tensioning tendons outside the concrete, but inside the box girder for ease of precasting and speed of installation together with dry joints, no epoxy, between segments. The post-tensioning tendons are continuous from pier segment to pier segment.

11.4 Incrementally Launched Bridges

11.4.1 Overview

The incremental launching technique has been used on bridges numbering in the hundreds since its introduction by Professor Fritz Leonhardt in 1961 for the Río Caroni Bridge in Venezuela. It is an effective alternative for the bridge designer to consider when the site meets its particular alignment requirements. The method entails casting the superstructure, or a portion thereof, at a stationary location behind one of the abutments. The completed or partially completed structure is then jacked into place horizontally, i.e., pushed along the bridge alignment. Subsequent segments can then be cast onto the already completed portion and in turn pushed onto the piers. Because all of the casting operations are concentrated at a location easily accessible from the ground, concrete quality of the same level expected from a precasting yard can be achieved. The procedure has the advantage that, like the balanced cantilever technique, it obviates the need for falsework to cast the girder. Moreover, heavy erection equipment, cranes, gantries, and the like, are not necessary, nor is the use of epoxy at segment joints. Usually, the only special equipment required is light steel truss work for a launching nose to reduce the cantilever moments during launching.

11.4.2 Special Requirements

There are two peculiarities associated with the technique, which must be appreciated by the designer. The first is that the alignment must be straight or, if it involves curves, the curvature must be constant. The second is that during launching, every section of the girder will be subjected to both the maximum and minimum moments of the span; and the leading cantilever portion will be subjected to slightly higher moments. This second constraint usually leads to slightly deeper sections, on the order of 1/15 the span, than would otherwise be considered. The girders must also be of constant depth as each section will at sometime be supported on the temporary bearings. Other considerations include the necessity for a large area behind the abutment for the casting operations, the requirement to lift the bridge off of the temporary bearings, and place it on the permanent ones when launching is complete and the need for very careful control of geometry during casting.

Incremental launching is generally considered for long viaducts with many spans of the same length. Spans up to 100 m can be considered; the requirement for constant-depth girders makes longer spans uneconomical. A single long span in the center of a project can be achieved by launching from both abutments and finishing at the long span with two converging cantilevers. The practical length limit for launching in about 1000 m. Bridges of twice this length can be considered by launching from both abutments.

11.4.3 Typical Post-Tensioning Layout

During superstructure launching each section of the girder is subjected to constantly reversing bending moments as it proceeds from temporary support to midspan. Because of the sign change in the applied moments, the efficient use of draped tendons for launching load effects is impossible. The general procedure has therefore been to apply axial prestressing for the launching operation. These tendons are usually straight, being contained in the top and bottom slabs of the girder. The tendons for successive segments must be spliced to these with couplers or stressed in buttresses in an overlapping fashion. This prestressing is subsequently augmented with either draped tendons or short top- and bottom-slab tendons for respective negative and positive moment regions in the completed structure to meet service state requirements. In some instances, permanent draped prestressing has been placed in the configuration required for the final condition, and temporary tendons with an opposing drape are provided to counteract their bending effects during launching. These temporary tendons are then removed when launching is complete.

11.4.4 Techniques for Reducing Launching Moments

As suggested above, the launching moments in the leading spans, especially the first cantilever span, will be greater than those in the following interior spans. If the girder is simply launched to the first pier with no special provision to reduce these moments, they will in fact be on the order of six times the typical negative moment over a pier. The method used most frequently to overcome this problem has been a light structural-steel launching nose attached to the leading cantilever (see Figures 11.14 and 11.15). This nose supports the girder without the weight penalty of the heavier concrete section. In order to be effective, the nose must be both as light and as stiff as possible.

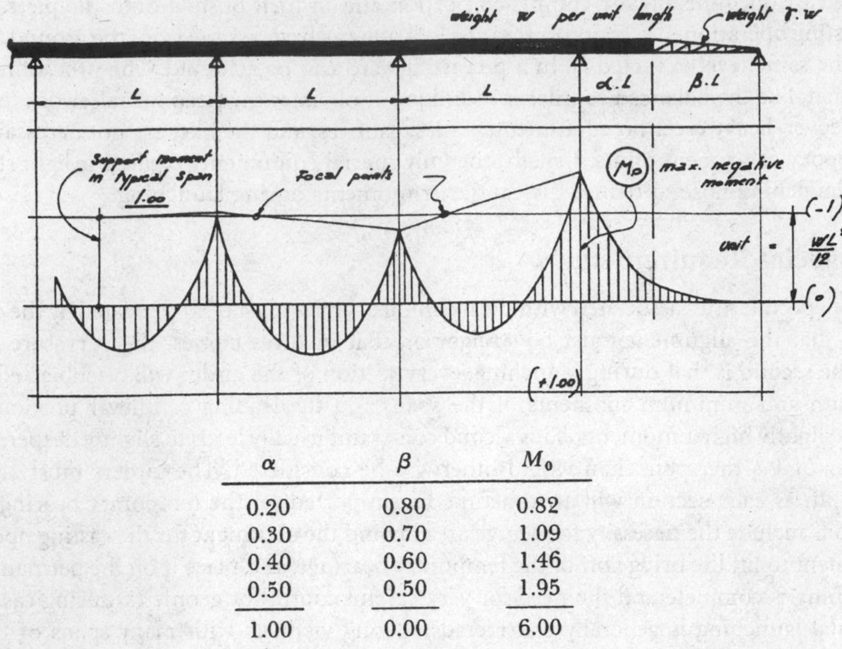

α	β	M_0
0.20	0.80	0.82
0.30	0.70	1.09
0.40	0.60	1.46
0.50	0.50	1.95
1.00	0.00	6.00

FIGURE 11.14 Critical negative moment during launching with nose. $M_1 = [WL^2/12)\ (6\alpha^3 + 6\gamma)\ (1 - \alpha^3)$. Multiplier $= WL^2/12$. For $\gamma = 0.11$.

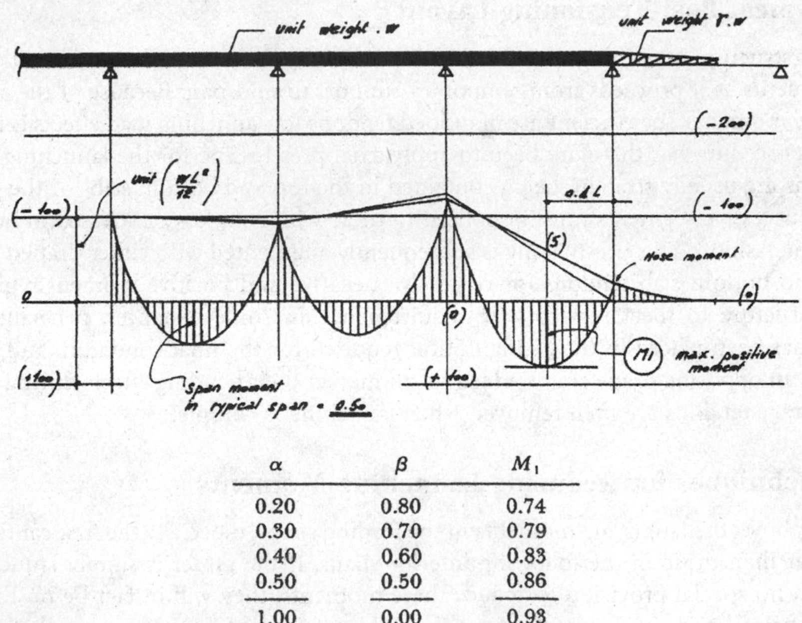

α	β	M_1
0.20	0.80	0.74
0.30	0.70	0.79
0.40	0.60	0.83
0.50	0.50	0.86
1.00	0.00	0.93

FIGURE 11.15 Critical positive moment during launching with nose. $M_1 = [(WL^2/12)\ (0.933 - 2.96\gamma\beta^2)$. Multiplier $= WL^2/12$. For $\gamma = 0.11$.

For longer spans, the steel nose is not as effective, and other methods have been employed to reduce launching moments. Temporary piers are a viable solution when ground conditions are such that the foundation costs are relatively modest and the pier height is not too great. If either of these conditions is not found, the cost can escalate rapidly as a temporary pier will be required in every span.

One last method that has been employed successfully is a temporary pylon attached to the deck at the trailing end of the first span which supports stays connected to the leading end. This device is very efficient in reducing the cantilever moment in the leading span; however, it produces an undesirable positive moment when the pylon is at midspan. For this reason, the stays must be equipped with a jack to adjust the stay force as needed during the various stages of the launching operations.

11.4.5 Casting Bed and Launching Methods

Segment lengths for incrementally launched bridges are generally greater than for other types of segmental bridges. Typical segment lengths range from 15 to 40 m. Usually, a casting area twice the length of the segment is required for actual casting and the ancillary operations that must be conducted there. The casting bed is generally a significant structure itself, as the strict geometry-control requirements of the technique make settlement of the formwork unacceptable.

Launching has been accomplished in the past either by tendons attached to the girder and horizontal jacks bearing on the abutment or by a horizontal jack bearing on the abutment face connected to a vertical jack which slides on a bearing. The upper surface of the vertical jack is fitted with a friction device to bear on the soffit of the box girder. The vertical jack is inflated to provide the normal force required for transferring the launching force by friction.

11.5 Arches, Rigid Frames, and Truss Bridges

11.5.1 Arch Bridges

The first step toward the segmental construction of arches was taken shortly after World War I by Eugéne Freyssinet. He employed hydraulic jacks to lift the completed Villeneuve arch from its falsework by applying an internal thrust at its crown. This departure from the classical method of striking the centering to develop the thrust in the arch opened the door to modern arch construction techniques that do not rely on falsework. It also presented the opportunity to reduce the bending moments in the arch by eliminating the dead load bending associated with axial shortening of the ribs.

11.5.1.1 Arches Erected without Falsework

The development of stay-cable and form-traveler technology has made possible the erection of arches in cantilever fashion without a centering supported from below. One early example of this technique was the suite of viaducts built in Caracas, Venezuela, in 1952 (see Figure 11.16). The first quarter of the arch span was supported by light forms which were in turn supported by stay cables attached to a pilaster at the springing of the arch. The crown portion of the arch was then completed with a light centering supported on the already-completed portion of the arch so that no falsework was required in the valley below.

Several variations on this theme were subsequently developed. The methods employed varied, depending on site conditions, from the use of very high pylons with a single group of stays allowing construction of the arch all the way to the crown to those which used the permanent spandrel columns in conjunction with temporary stay diagonals to form a truss. These methods are summarized in Figure 11.17.

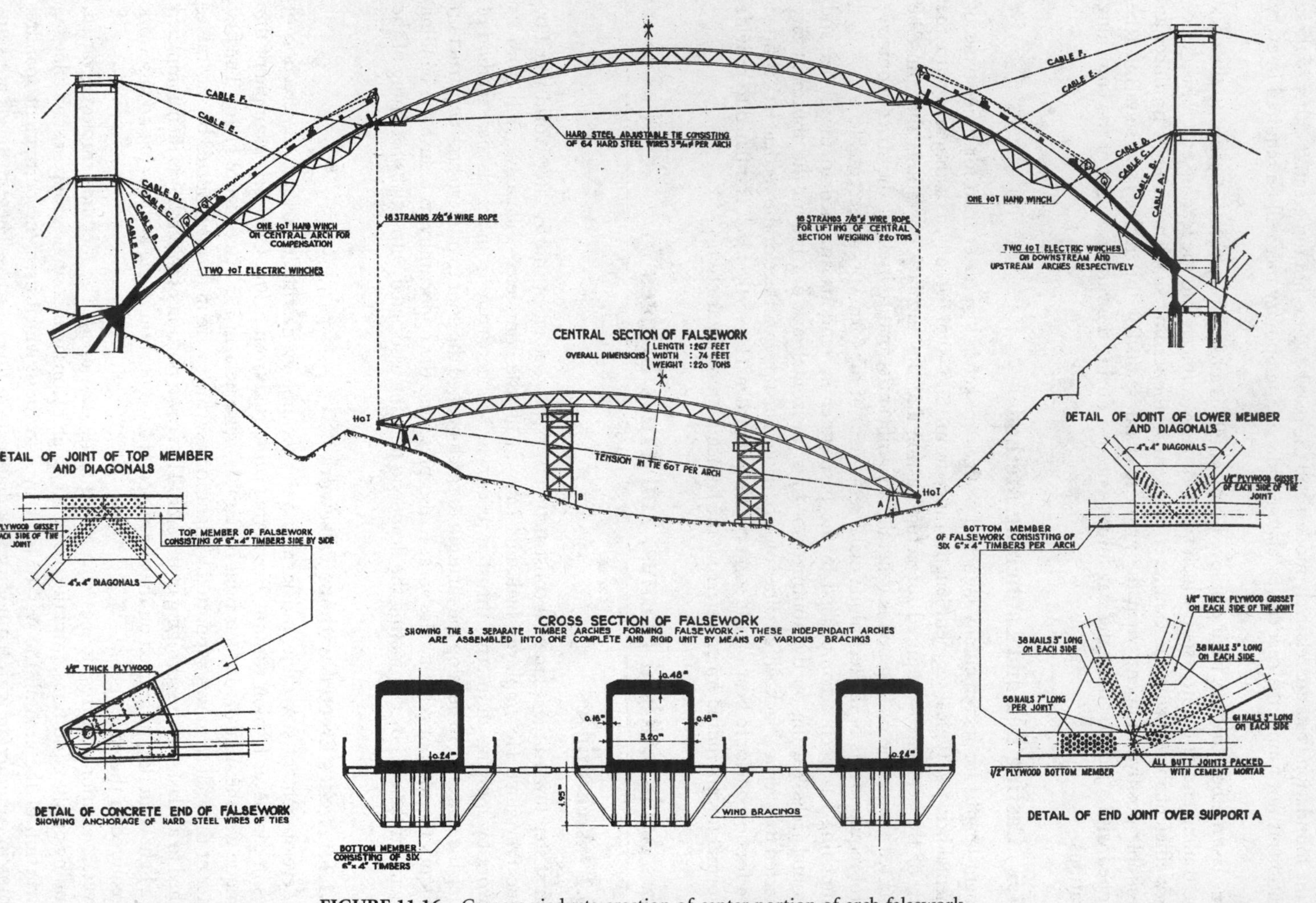

FIGURE 11.16 Caracas viaducts-erection of center portion of arch falsework.

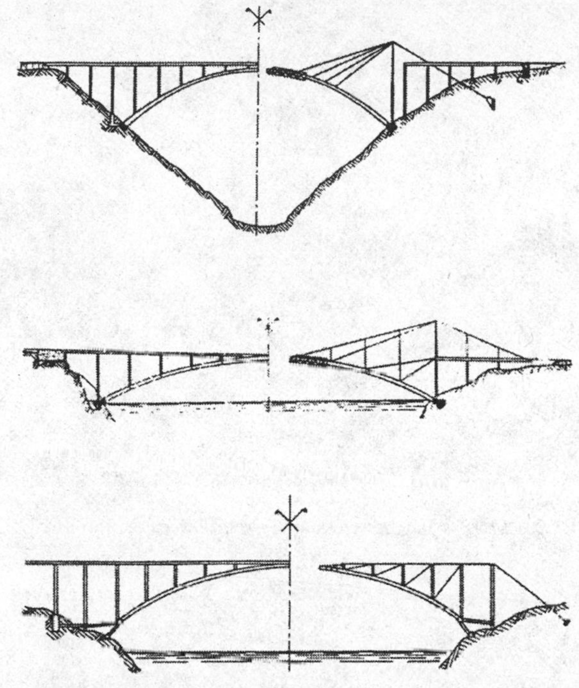

FIGURE 11.17 Various cantilevers—erection techniques for arches.

11.5.1.2 Precast Arches

The first precast segmental arch bridge was built in France in 1948. This bridge at Luzancy over the Marne River is composed of three box-section arches built up from 2.44-m-long precast segments. The finished span length is 55 m. Because of the severe clearance requirements, the arch has a very unusual span-to-rise ratio of 23 to 1. The segments were connected via 20-mm dry-packed joints and prestressing on the approach behind an abutment, with the resulting rib being moved to its final position by an aerial cableway.

Construction of concrete arches without falsework was employed almost exclusively in conjunction with the cast-in-place cantilever technique until the construction of the Natches Trace Bridge in Tennessee in 1993. This precast arch, which originally was designed for erection on a moveable falsework, was the first precast arch to be erected on stays. The unusual design, which omits spandrel columns, results in a slender appearance. There are two arches: one with a span of 177 m and a rise of 44 m, the other with a span of 141 m and a rise of 31 m. The arch segments are 4.9 m wide and vary in depth from 4 m at the springing to 3 m at the crown (see Figure 11.18).

11.5.2 Rigid Frames

Frame bridges can be considered a hybrid of arch and girder forms. They are an appropriate alternative to either of those types for intermediate span lengths. Rigid frame bridges are well suited to segmental construction techniques.

Rigid frame bridges often have some of the same site requirements as arch bridges. They are well suited to valleys and generally will require foundations capable of resisting large horizontal actions. Generally, some form of temporary support will be required until the frame is complete, meaning that construction techniques that eliminate falsework may need slight modification for these structures. One of the most aesthetically convincing applications of the rigid frame is the Bonhomme Bridge in Brittany, France (see Figure 11.19). This slant-leg frame was built using the cast-in-place balanced cantilever technique. Temporary piers were installed below the slant legs to support them

FIGURE 11.18 Natches truss arch—cantilever erection of ribs.

FIGURE 11.19 Bonhomme Bridge in Brittany, France.

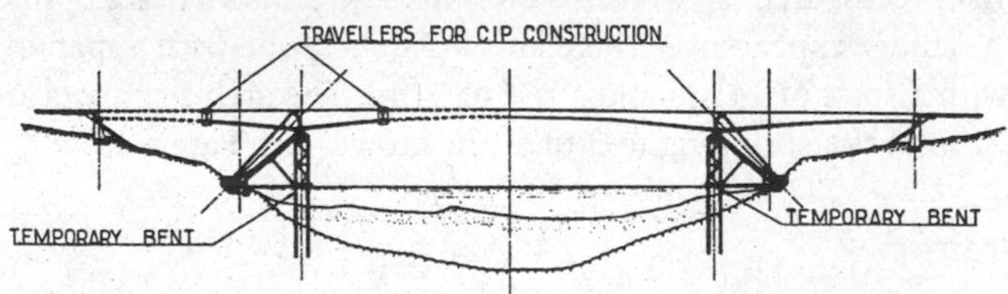

FIGURE 11.20 Temporary support for the Bonhomme Bridge.

before the thrust was developed in the frame (see Figure 11.20). Jacks under the temporary-support piers and at the midspan closure were used to adjust the geometry before closing the span.

11.5.3 Segmental Trusses

Although relatively few examples have been built, segmental trusses are interesting, especially for long spans, in that they offer very efficient use of materials. This economy translates directly into lighter elements and smaller loads to be dealt with during construction, as well as reduced material cost.

FIGURE 11.21 Cantilever erection of the Viaduct des Glacièrs, France.

One of the earliest segmental trusses was the Mangfallbrücke in Austria, which was constructed in 1959. It had a total length of 288 m with a maximum span of 108 m and was constructed by the cast in-place segmental technique in conjunction with temporary piers.

Later examples were developed in precast segmental, of which the Viaduc de Sylans and the Viaduc des Glacieres are the most notable. These sister structures were constructed in balanced cantilever with a self-launching overhead truss (see Figure 11.21). The segments were prestressed in three directions with a combination of external and internal tendons. "X" members for the open webs were precast and subsequently placed in the molds prior to segment casting.

The most recent development in segmental trusses is the composite truss. This concept employs concrete for the top and bottom chords and steel sections for the open webs. In some cases, however, steel is used for the tension chord as well. An excellent example of this type of construction is the bridge over the Roize in France, built with the span-by-span method. This structure was conceived as a truss work of factory-produced steel truss work and precast slabs. These two elements were joined at the site by cast-in-place joints and external tendons (see Figure 11.22). The precast slabs served as the top (compression) chord while a hexagonal steel tube served as the bottom chord. The resulting structure is equally viable as the deck for short-span viaducts and stiffening girder for long-span cable-supported bridges.

11.6 Segmental Cable-Stayed Bridges

11.6.1 Overview

Theories on cable-stayed bridges are presented in another chapter. We shall address here cable-stayed bridges only as they relate to segmental construction. In the majority of segmental cable-stayed bridges, the methods of construction fall in the three following categories, by order of importance:

- Cantilever construction
- In-stage construction
- Push-out construction

FIGURE 11.22 The Roize Bridge, France — erection of steel bottom chord and webs.

The choice of material depends upon many factors and load conditions; it should be remembered that concrete is an excellent material for cable-stayed structures, because of its properties in resisting compression and its mass and damping characteristics in resisting aerodynamic vibrations. For the proposed Ceremonial Bridge in Malaysia, with a main span of 1000 m and a single plane of stays, concrete deck in the pylon area is associated with a composite cross section toward the center of the span. Comparative studies show that the replacement of the composite section with its concrete slab by an orthotropic slab would adversely affect the project because of its lack of mass.

11.6.2 Cantilever Construction

11.6.2.1 Design

It is important to keep the project simple and pay attention to details to achieve economy and efficiency during construction.

The length of segments must be equal and, depending upon the spacing of stays, the segment joints must be such that a stay always falls in the same location within a segment. If the segments are long, the stay should be located toward the free end of the segment. Cross sections must be kept constant as much as possible, the variations being limited to the web and bottom slab thickness. The post-tensioning layout must be repetitive from segment to segment (see Figure 11.23). Erection phases are critical in terms of stability and stresses. Wind effects on the partially built structure must be investigated for static and dynamic effects. A shorter return period is usually used during construction (10 years). Seismic effects must also be investigated in areas prone to earthquakes. To increase stability, temporary cables can be installed at a certain stage of completion.

Stresses in the main elements of the structure often reach a maximum during construction, and the final state of stresses in the finished structure depends greatly on the accuracy of construction. Figure 11.24 shows a typical erection cycle. It is important that all erection phases be reviewed to ensure that the stresses are within allowable limits at each stage.

Stay forces are large and applied on very localized areas of the deck, and their local effects must be analyzed in detail. For instance, the stays apply high, concentrated forces on the section, at the middle, in the case of a single plane of stays, or at the edges with two planes of stays. These forces are not immediately available in the whole cross section, but are spread out at approximately 45°. This shear lag effect is more critical during construction than in service. Construction phases should be checked, assuming a 45° distribution of the horizontal component of the stay force while the vertical component is effectively applied at the stay anchorage (a finite-element computer program will generate the exact cross section stress distribution).

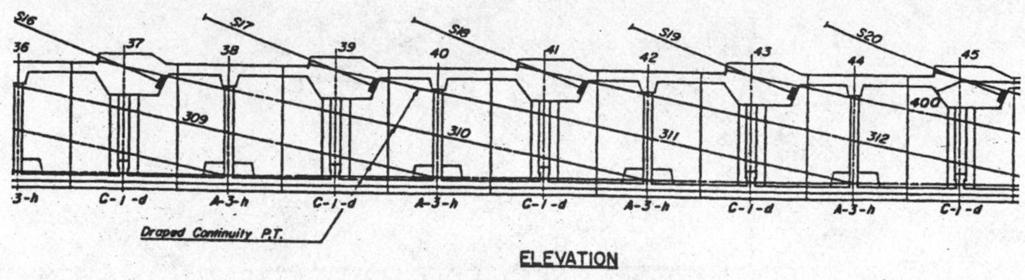

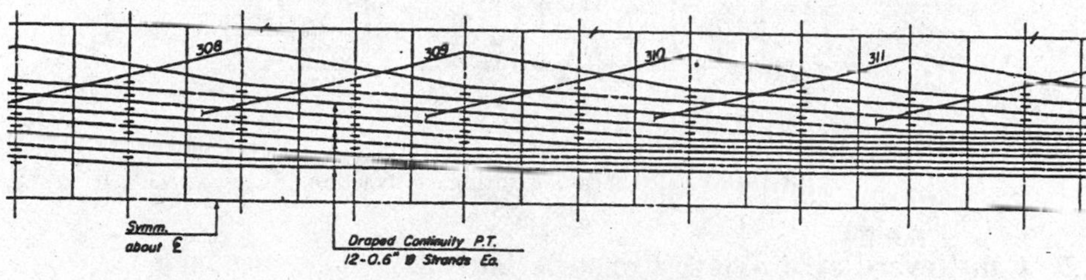

FIGURE 11.23 Sunshine Skyway, Florida — stay cables and post-tensioning layout.

TYPICAL ERECTION PHASES

(1) LIFT SEGMENT

(2) RESTRESS STAY

(3) LIFT SEGMENT

(4) INSTALL AND STRESS STAY

FIGURE 11.24 Typical erection phases.

This analysis usually shows the necessity of adding a temporary post-tensioning system toward the end of the cantilever, in the area outside the stay centerline (see Figure 11.25).

When a stay is anchored in an already constructed deck, such as a backstay anchored in the side span, the horizontal component of the stay force is distributed half in compression in front of the anchor and half in tension in the back of the anchor. This is called the entrainment effect; care must be taken to have enough tension capacity behind the anchor, either rebars or available compression, to prevent cracking or opening of the joint in case of precast construction.

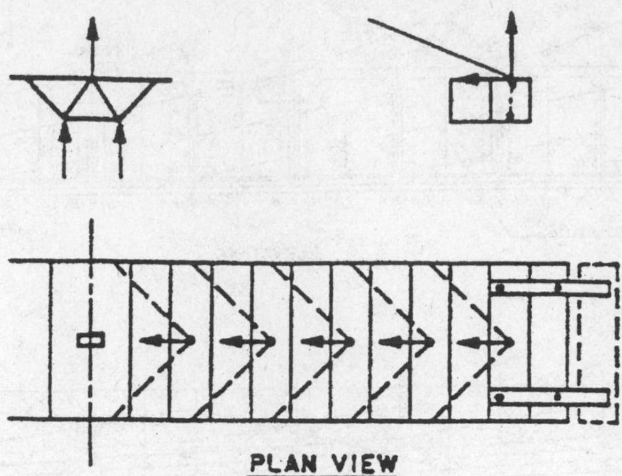

FIGURE 11.25 Shear lag during construction.

11.6.2.2 Cantilever Cast-in-Place Construction

Cast-in-place stayed bridges are built according to the same general principle as a typical box-girder bridge. After the pylon has been built up to the first pylon stay anchorage points and the starting deck segment at the pylon cast, travelers can be installed and cantilever construction started. Temporary stays are sometimes necessary to carry the weight of the traveler plus the newly cast segment before the permanent stay pertaining to that segment can be installed and stressed especially for thin, small inertia decks. A more elegant way is to use the permanent stay, which can be anchored in a precast anchorage block secured to the traveler. The horizontal component of the stay force is carried either by the traveler (see Figure 11.26) or by a precast member, which becomes part of the future segment. The permanent stay can also be anchored in the final deck if the stay anchor structure is staggered ahead of the whole section. This was the case at the Isère Bridge shown in Figures 11.27 and 11.28, with the center spine where the stays are anchored was cast in a first phase and the remainder of the section in a second phase. The phases are as follows:

- Launching of traveler;
- Concreting of the center spine (8 m) and stressing stay to 35% of its final force;
- Launching side forms, connecting bottom slab, and stressing stay to 70% of its final force;
- Concreting top slab and stressing stay to 80% of its final force.

11.6.2.3 Cantilever Precast Construction

Precast segmental bridges become economically feasible for relatively large bridges where the cost associated with setting up a casting yard can be offset by the speed of casting segments and the speed of erection. It is very interesting if the approaches to the main span are also precast segmentally, because then the cost of equipment is written off on an even larger volume.

A great example is the Sunshine Skyway Bridge in Florida, with a main span of 366 m for a total length of 1220 m, where the same cross section is used throughout the high-level bridge (see Figures 11.29 and 11.30). The 120-ton segments were precast in a yard close to the site and delivered by barge. They were lifted into place by beam-and-winch assemblies mounted on the previously completed portion of the deck. The same lifting equipment was used for the high approaches to the main spans. The low-level approaches were made of two parallel box girders.

For the James River Bridge in Virginia, the same twin parallel precast box girders were used from one end of the bridge to the other. For the main span, a single plan of stays was used and the two boxes were connected by a transverse frame at each stay anchor location (see Figures 11.31 and

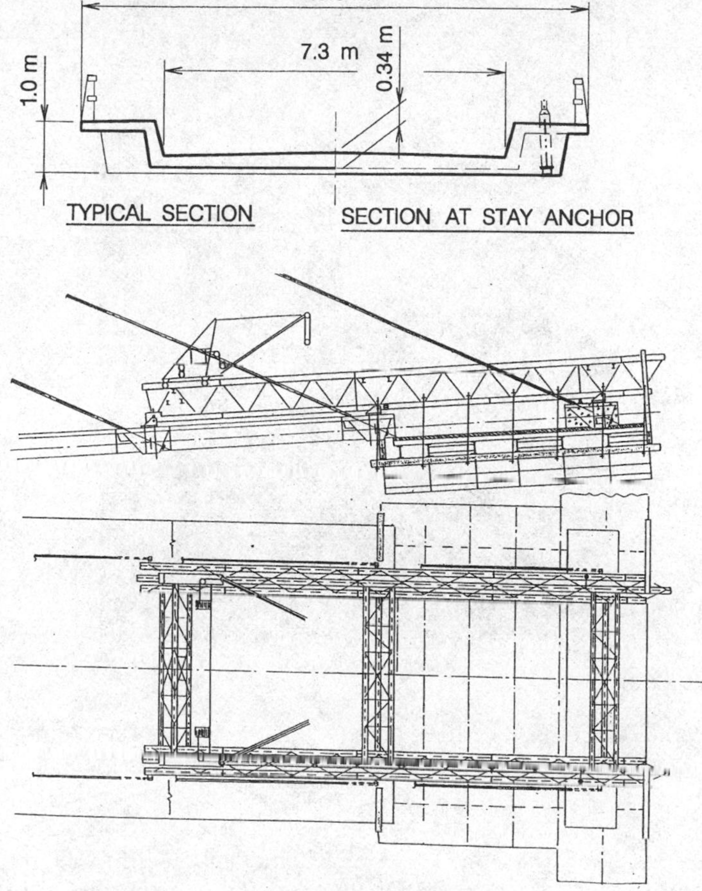

FIGURE 11.26 Santa Rosa Bridge, Bolivia – general view, cross section, and elevation.

11.32). With this scheme, construction can be carried out at deck level, with the segments erected by the span-by-span method. The main span can be built with crane-type lifting equipment mounted on the completed portion of the deck or, if desired, with cranes at ground level or on barges in the river.

11.6.2.4 Structural Steel Segmental Cantilever Construction

Cantilever construction can be applied to steel structures as well, the most recent example being the Normandie Bridge with an 856-m main span and 43.5-m approach spans. Concrete box girders are used for the approaches and part of the main span. The approaches were constructed by incremental launching and the first 116 m out from the pylon by segmental cast-in-place balanced cantilever techniques. Steel segmental construction is used for the remaining 640 m of the main span because of its light weight. The 19.65-m steel segments are barged to the site, lifted in place, secured against the previous segment, and then welded (see Figure 11.33).

11.6.3 In-Stage Construction

With this method, the deck is cast on a fixed soffit, with the side forms moving as the segments are cast. Stays can be installed during the casting, then stressed afterward. The advantage is that the bridge does not go through high-stress-level stages during erection and is practically built in its final stage. This method is only a variation of the cast-in-place scheme.

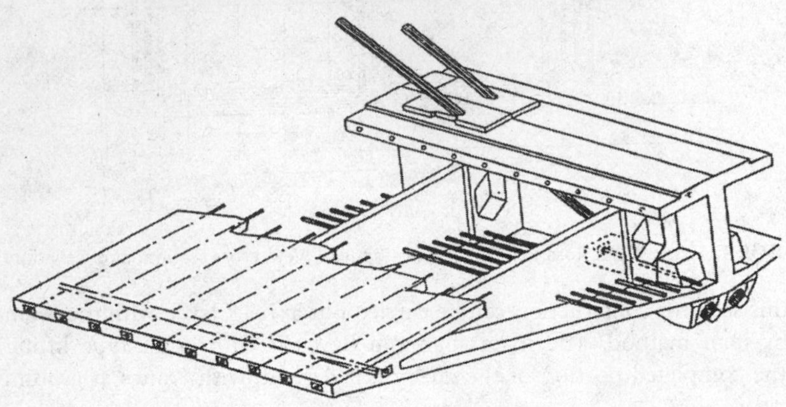

FIGURE 11.27 Isère Bridge, France — general and isometric views.

11.6.4 Push-Out Construction

This method is rarely used and not well adapted to cable-stayed bridges. Its use is restricted to sites where temporary supports can be installed. During pushing, the deck is subjected to large moment variations so steel decks are more suitable.

11.7 Design Considerations

11.7.1 Overview

The intent of this section is to present conditions that the designer should be aware of to produce a satisfactory design. The segmental technique is closely related to the method of construction and the structural system employed. It is usually identified with cantilever construction, but special attention must also be exercised with other methods, such as span-by-span, incremental launching, or progressive placement.

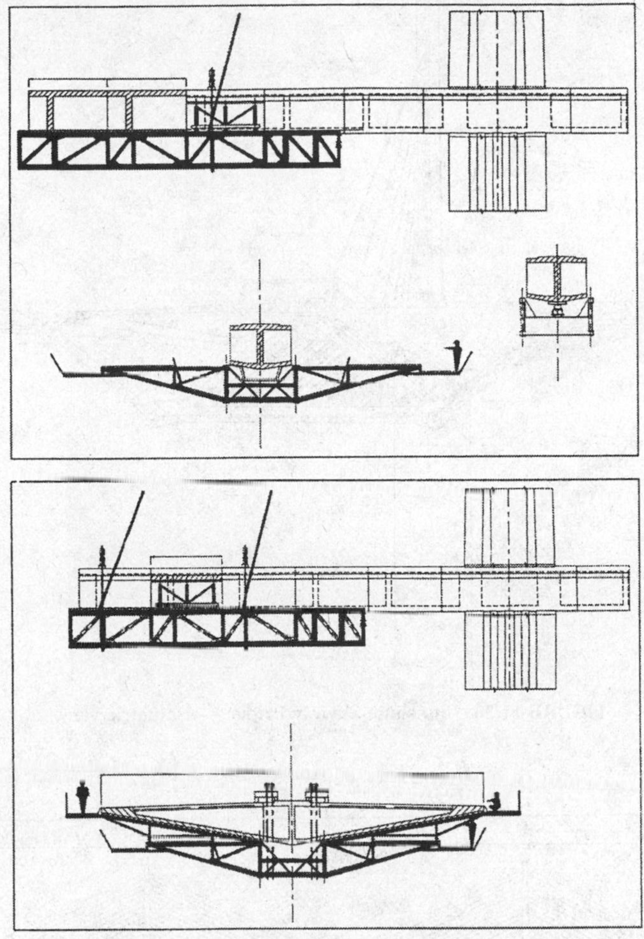

FIGURE 11.28 Isére Bridge — casting sequence.

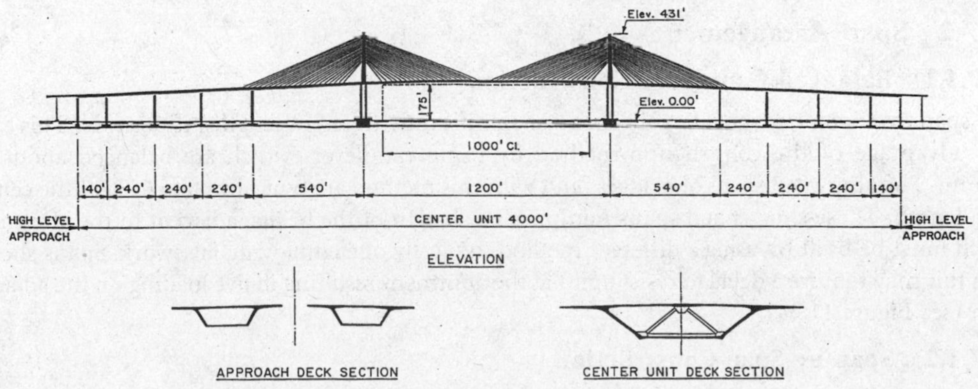

FIGURE 11.29 Sunshine Skyway Bridge, Florida — elevation.

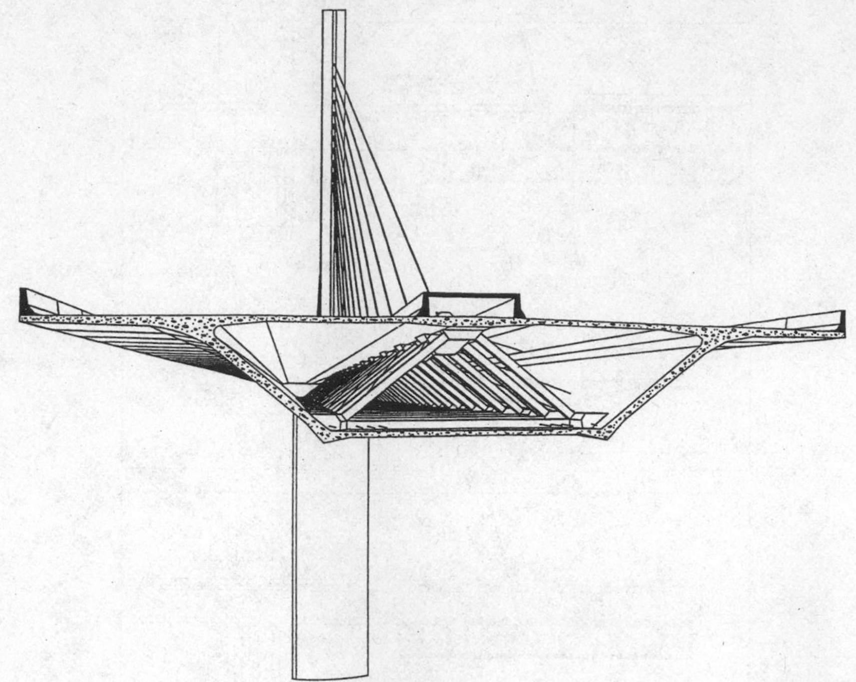

FIGURE 11.30 Sunshine Skyway Bridge — isometric view.

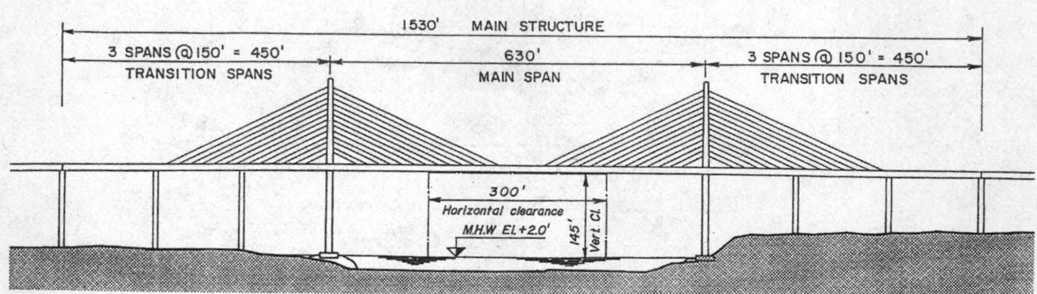

FIGURE 11.31 James River Bridge, Massachusetts – elevation.

11.7.2 Span Arrangement

11.7.1.1 Balanced Cantilever Construction

The span arrangement should avoid spans of significantly different lengths, if possible. This takes best advantage of the construction method by using cantilevers which are balanced about the column. The abutment spans of bridges built with this method are typically 60 to 65% of the central span length. These shorter end spans minimize the length of the bridge adjacent to the abutment, which must be built by using a different method, typically one employing falsework. Spans shorter than this may require a detail to resist uplift at the abutment resulting in live loading on the adjacent span (see Figure 11.34).

11.7.1.2 Span-by-Span Construction

For span-by-span construction the averaging of adjacent span lengths is not required, although it is advantageous to maintain similar span lengths adjacent to one another. The length of the abutment

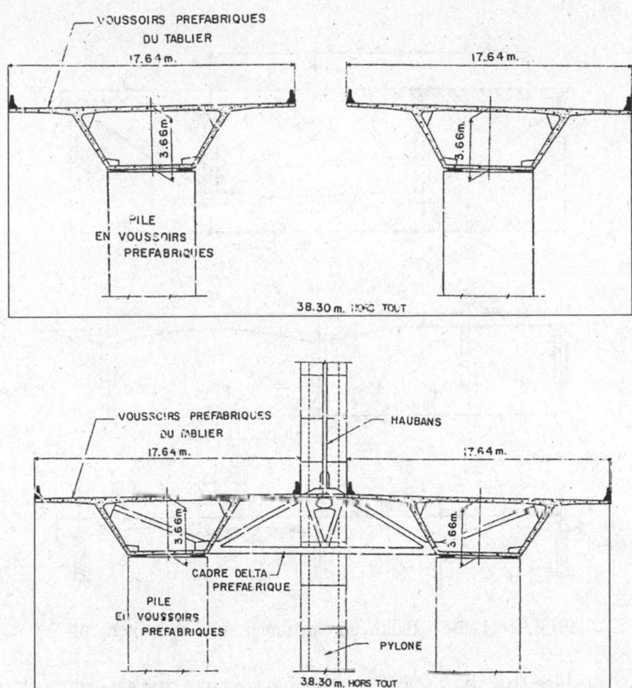

FIGURE 11.32 James River Bridge — cross sections.

FIGURE 11.33 Normandie Bridge, France — lifting of a a steel segment.

or end span is typically kept the same as the interior spans. This is reasonable for this type of construction, since the secondary moments due to post-tensioning in the end spans are less than for the interior spans, and the post-tensioning requirement is therefore similar.

11.7.1.3 Location of Expansion Joints

Concrete bridge decks have been built with a length up to 1220 m between expansion joints and have had acceptable performance. The placement of expansion joints within a longer viaduct may

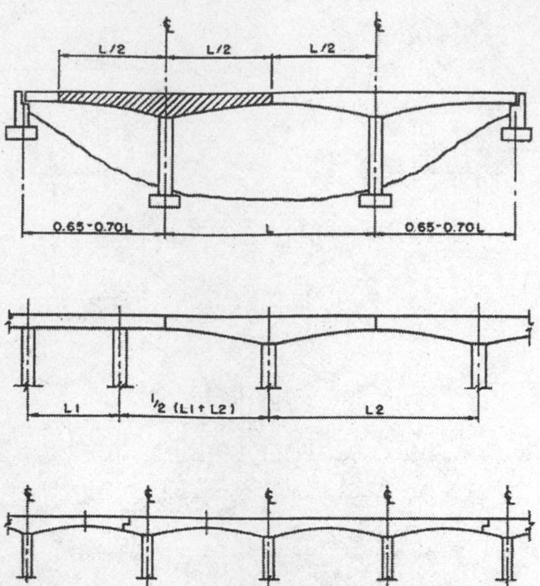

FIGURE 11.34 Balanced cantilever span arrangement.

be necessary to accommodate the change in length of structure due to creep, shrinkage, and thermal changes. The location of the expansion joint within a span will vary, depending on the method of construction.

For balanced cantilever construction, the expansion joints were initially located at the tip of the cantilevers, which is the middle of the span on the completed structure, for ease of construction. Creep effect under dead load plus post-tensioning drives the tip of the cantilever down, resulting in unacceptable angle break at midspan. This disposition is no longer used. An alternative solution is to place the joint at the point of contraflexure of the equivalent continuous span, thus very effectively reducing the angle of break under creep and live load. However, this technique requires expansion segments at midlength of the cantilever, making construction more difficult. The latest technique goes back to the joint at midspan, but with the addition of a stiffening steel beam across the joint, turning the hinged span into a continuous span with expansion capability. Further refinements are introduced such as the capability of controlling the deflection of the span by vertical jacking on the steel beam during the life of the bridge. This technique has been successfully used as it does not interfere with the cantilever erection process (see Figure 11.35).

For spans built with the use of the span-by-span method the expansion joints are typically located at the centerline of a column. The adjacent box-girder spans are both supported by the column with movement allowed between the spans. With this method, the angle break at the expansion joint is minimized, and there is no requirement for temporary moment restraint between the adjacent sections.

11.7.2 Cross-Section Dimensions

11.7.2.1 Overall Box-Girder Dimensions

The overall width of a concrete segmental box-girder bridge is quite adaptable to any requirement. Box-girder spans have been built with widths as low as 3.6 m and as great as 27.50 m, with the configuration of the box girder varying significantly.

The depth of precast segmental box girders is generally somewhat greater than that of similar spans with cast-in-place construction. This increased depth is necessary to offset more stringent requirements for extreme fiber axial stresses and restrictions on the locations of post-tensioning

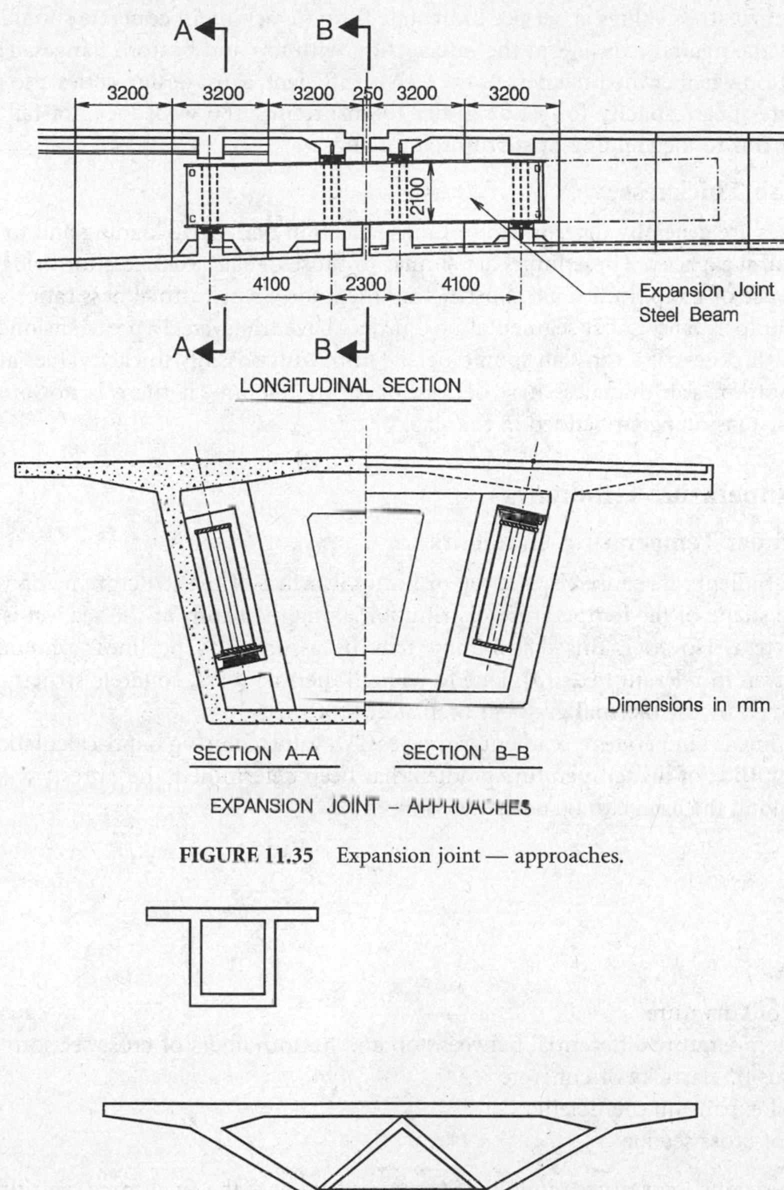

FIGURE 11.35 Expansion joint — approaches.

FIGURE 11.36 Various cross sections.

tendons. Multiple-cell, box-girder bridges will also have more webs to place tendons than a comparable width, single-cell segmental box girder. For span-by-span construction, the span-to-depth ratio should not exceed 25 to 1, and is more comfortable at 20 to 1. For balanced cantilever construction, the span-to-depth ratio at the support should not exceed 18 to 1. However, variable-depth box girders built in balanced cantilever fashion are quite common with straight haunched sections and parabolic extrados. The span-to-depth ratio at midspan of a variable-depth balanced cantilever bridge should not exceed 40 to 1 (see Figure 11.36).

11.7.2.2 Web Thickness

The thickness of the web is generally determined such that the required post-tensioning tendons may be placed without interfering with concrete placement or risking cracking during stressing of tendons. Principal stress values at service limit state for no cracking in concrete should be checked in the webs at the neutral axis and at the intersection with top and bottom flanges. This will give a good indication whether the thickness of the web is sufficient. Most design codes also place a limit on the ultimate shear capacity for a box girder to ensure that the web does not fail in diagonal compression prior to the yielding of stirrup reinforcing.

11.7.2.3 Slab Thickness

Slab thicknesses are generally determined to limit deflection under live loading and to provide the necessary flexural capacity. These limits are similar to those of slab thickness for bridge structures built with the use of more traditional construction methods. Span-to-thickness ratios should be in the range of 30 to 1. Since most segmental box girders have transversely post-tensioned top slabs, the minimum thickness of a top slab should be 200 mm, with possibly thicker values at the tendon anchorages. Bottom slab thickness may be less, down to 180 mm, if there is no longitudinal or transverse post-tensioning embedded in the slab.

11.7.3 Temperature Gradients

11.7.3.1 Linear Temperature Gradients

Temperature gradients are caused by the top or bottom surface of the structure being warmer than the other. The shape of the temperature distribution along the depth of the section is beyond the scope of this text. However, this distribution may be assumed to be linear or nonlinear with magnitudes given in relevant texts [3]. Due to its high thermal mass, concrete structures are more adversely affected by the thermal gradient than steel structures.

Effects of a linear temperature gradient can be easily evaluated using hand-calculation methods. Once the magnitude of the temperature gradient has been determined, the unrestrained curvature at any point along the span can be determined by

$$R = \frac{\Delta T \cdot \alpha \cdot E_c}{h} \tag{11.1}$$

where

R = radius of curvature
ΔT = linear temperature differential between top and bottom fibers of cross section
E_c = Modulus of elasticity of concrete
α = thermal expansion coefficient
h = depth of cross section

Once the unrestrained curvature along the structure is known, the final force distribution can be determined by evaluating the redundant support reactions. It is noted that for a statically determinate structure the linear temperature gradient results in zero effect on the structure.

11.7.3.2 Nonlinear Temperature Gradients

Nonlinear temperature gradients are more difficult to evaluate and are best handled by a well-suited computer program. The general theory is presented here; for a more detailed elaboration see Reference [1]. The nonlinear temperature distribution is determined by field measurements and thermodynamic principles. The general shape may be as shown in Figure 11.37. Assuming that the material has linear stress–strain properties, that plane sections will remain plane (Navier–Bernoulli hypothesis), and that temperature varies only with depth (two-dimensional problem), one can make the following theoretical derivation of the problem: the free thermally induced strain is proportional to the temperature distribution; however, this strain distribution violates the second assumptions above, namely, that plane sections remain plane. In order for the section to remain plane under the

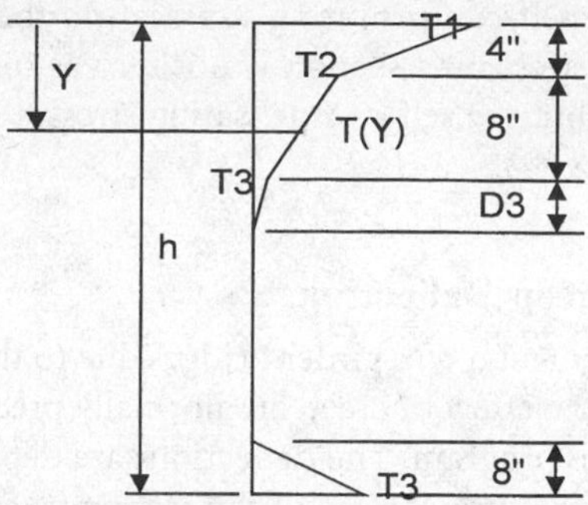

FIGURE 11.37 Nonlinear gradient.

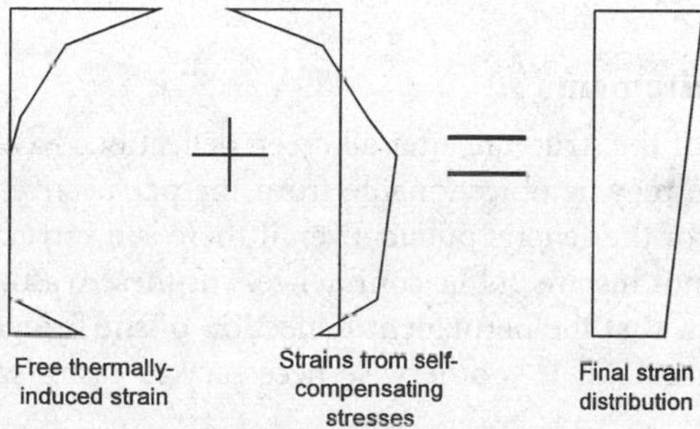

FIGURE 11.38 Self compensating stresses.

effects of the applied temperature gradient, there must be some induced stress on said section. This is termed self-compensating stress. The final strain distribution on the section is, therefore, linear and is the sum of the free thermally induced strain and the strain induced by the self-compensating stresses (see Figure 11.38).

The self-compensating stresses can be derived as

$$\sigma(Y) = E_c \cdot \sigma \cdot T(Y) - \frac{P}{A} - \frac{M \cdot Y}{I} \tag{11.2}$$

where

Y = variable along depth of cross section
$T(Y)$ = temperature at abscissa Y
P = $\int_Y E \cdot a \cdot T(Y) \cdot b(Y) dY$
M = $\int_Y E \cdot a \cdot T(Y) \cdot b(Y) \cdot Y \cdot dY$
$b(Y)$ = width of section at abscissa Y

Similar to the linear gradient, there is now a free unrestrained curvature of the structure along its length. If the structure is continuous, this will result in reactions due to the restraint of the system. The unrestrained curvature at any point along the structure is

$$R = \frac{M}{E_c \cdot I} \tag{11.3}$$

Once the unrestrained curvature along the structure is known, the continuity force distribution can be determined by evaluating the redundant support reactions. The total stress on a section is, therefore, the summation of the self-compensating stresses and the continuity stresses. For a statically determinate structure, the stress on a section is not zero as for a linear temperature gradient; the continuity stresses are zero, but the self-compensating stresses may be significant.

11.7.4 Deflection

11.7.4.1 Dead Load and Creep Deflection

Global vertical deflections of segmental box-girder bridges due to the effects of dead load and post-tensioning as well as the long-term effect of creep are normally predicted during the design process by the use of a computer analysis program. The deflections are dependent, to a large extent, on the method of construction of the structure, the age of the segments when post-tensioned, and the age of the structure when other loads are applied. It can be expected, therefore, that the actual deflections of the structure would be different from that predicted during design due to changed assumptions. The deflections are usually recalculated by the contractor's engineer, based on the actual construction sequence.

11.7.4.2 Camber Requirements

The permanent deflection of the structure after all creep deflections have occurred, normally 10 to 15 years after construction, may be objectionable from the perspective of riding comfort for the users or for the confidence of the general public. Even if there is no structural problem with a span with noticeable sag, it will not inspire public confidence. For these reasons, a camber will normally be cast into the structure so that the permanent deflection of the bridge is nearly zero. It may be preferable to ignore the camber, if it is otherwise necessary to cast a sag in the structure during construction.

11.7.4.3 Global Deflection Due to Live Load

Most design codes have a limit on the allowable global deflection of a bridge span due to the effects of live load. The purpose of this limit is to avoid the noticeable vibration for the user and minimize the effects of moving load impact. When structures are used by pedestrians as well as motorists, the limits are further tightened.

11.7.4.4 Local Deflection Due to Live Load

Similar to the limits of global deflection of bridge spans, there are also limitations on the deflection of the local elements of the box-girder cross section. For example, the AASHTO Specifications limit the deflection of cantilever arms due to service live load plus impact to $1/300$ of the cantilever length, except where there is pedestrian use [1].

11.7.5 Post-Tensioning Layout

11.7.5.1 External Post-Tensioning

While most concrete bridges cast on falsework or precast beam bridges have utilized post-tensioning in ducts which are fully encased in the concrete section, other innovations have been made in precast segmental construction. Especially prevalent in structures constructed using the span-by-span method, post-tensioning has been placed inside the hollow cell of the box girder but not encased in concrete along its length. This is know as external post-tensioning. External post-tensioning is easily inspected at any time during the life of the structure, eliminates the problems associated with internal tendons, and eliminates the need for using expensive epoxy adhesive between precast segments. The problems associated with internal tendons are (1) misalignment of the tendons at segment joints, which causes spalling; (2) lack of sheathing at segment joints; and (3) tendon pull-through on spans with tight curvature (see Figure 11.39). External prestressing has been used on many projects in Europe, the United States, and Asia and has performed well.

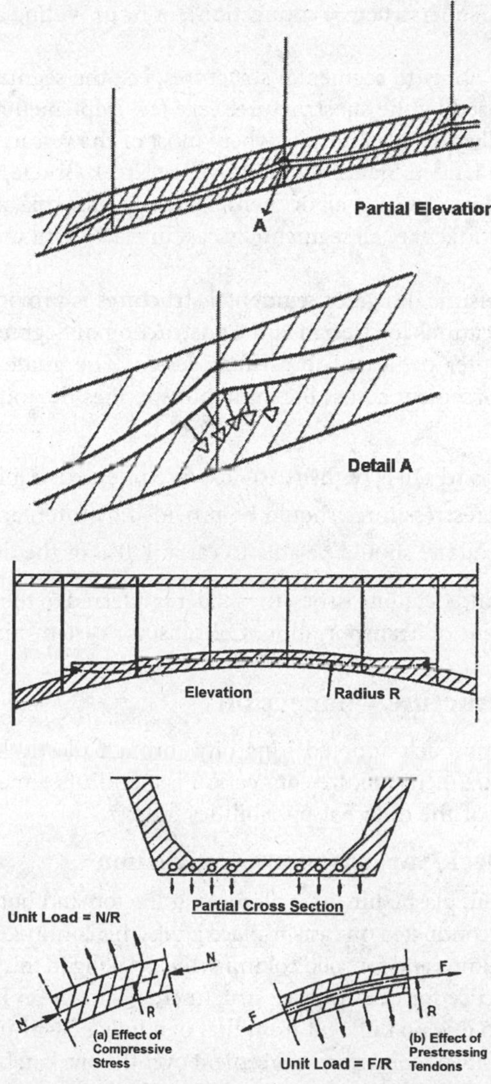

FIGURE 11.39 Problems with internal tendons.

11.7.5.2 Future Post-Tensioning

The provision for the addition of post-tensioning in the future in order to correct unacceptable creep deflections or to strengthen the structure for additional dead load, i.e., future wearing surface, is now required by many codes. Of the positive and negative moment post-tensioning, 10% is reasonable. Provisions should be made for access, anchorage attachment, and deviation of these additional tendons. External, unbonded tendons are used so that ungrouted ducts in the concrete are not left open.

11.8 Seismic Considerations

11.8.1 Design Aspects and Design Codes

Due to typical vibration characteristics of bridges, it is generally accepted that under seismic loads, some portion of the structure will be allowed to yield, to dissipate energy, and to increase the period of vibration of the system. This yielding is usually achieved by either allowing the columns to yield plastically (monolithic deck/superstructure connection), or by providing a yielding or a soft bearing system [6].

The same principles also apply to segmental structures, i.e., the segmental superstructure needs to resist the demands imposed by the substructure. Very few implementations of segmental structures are found in seismically active California, where most of the research on earthquake-resistant bridges is conducted in the United States. The Pine Valley Creek Bridge, Parrots Ferry Bridge, and Norwalk/El Segundo Line Overcrossing, all of them being in California, are examples of segmental structures; however, these bridges are all segmentally cast in place, with mild reinforcement crossing the segment joints.

Some guidance for the seismic design of segmental structures is provided in the latest edition of the AASHTO Guide Specifications for Design and Construction of Segmental Concrete Bridges [2], which now contains a chapter dedicated to seismic design. The guide allows precast-segmental construction without reinforcement across the joint, but specifies the following additional requirements for these structures:

- For Seismic Zones C and D [1], either cast-in-place or epoxied joints are required.
- At least 50% of the prestress force should be provided by internal tendons.
- The internal tendons alone should be able to carry 130% of the dead load.

For other seismic design and detailing issues, the reader is referred to the design literature provided by the California Department of Transportation, Caltrans, for cast-in-place structures [5-8].

11.8.2 Deck/Superstructure Connection

Regardless of the design approach adopted (ductility through plastic hinging of the column or through bearings), the deck/superstructure connection is a critical element in the seismic resistant system. A brief description of the different possibilities follows.

11.8.2.1 Monolithic Deck/Superstructure Connection

For the longitudinal direction, plastic hinging will form at the top and bottom of the columns. Since most of the testing has been conducted on cast-in-place joints, this continues to be the preferred option for these cases. For short columns and for solid columns, the detailing in this area can be readily adapted from standard Caltrans practice for cast-in-place structures, as shown on Figure 11.40. The joint area is then essentially detailed so it is no different from that of a fully cast-in-place bridge. In particular, a Caltrans requirement for positive moment reinforcement over the pier can be detailed with prestressing strand, as shown below. For large spans and tall columns, hollow column sections would be more appropriate. In these cases, care should be taken to confine the main column bars with closely spaced ties, and joint shear reinforcement should be provided according to Reference [3 or 7].

The use of fully precast pier segments in segmental superstructures would probably require special approval of the regulating government agency, since such a solution has not yet been tested for bridges and is not codified. Nevertheless, based upon first principles, and with the help of strut–tie models, it is possible to design systems that would work in practice [6]. The segmental superstructure should be designed to resist at least 130% of the column nominal moment using the strength reduction factors prescribed in Ref. [2].

Of further interest may be a combination of precast and cast-in-place joint as shown in Figure 11.41, which was adapted from Ref. [8]. Here, the precast segment serves as a form for the cast-in-place portion that fills up the remainder of the solid pier cap. Other ideas can also be derived from the building industry where some model testing has been performed. Of particular interest for bridges could be a system that works by leaving dowels in the columns and supplying the precast segment with matching formed holes, which are grouted after the segment is slipped over the reinforcement [9].

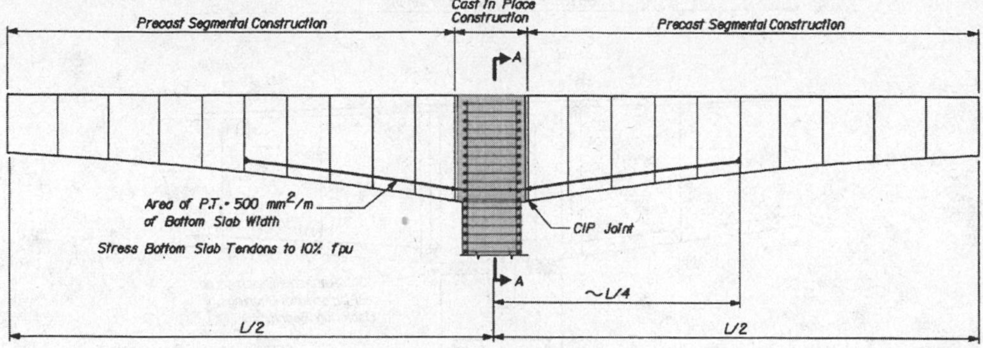

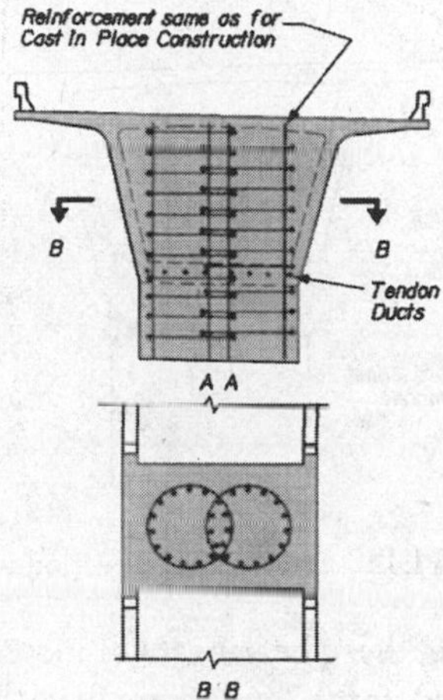

FIGURE 11.40 Deck/pier connection with cast-in-place joint.

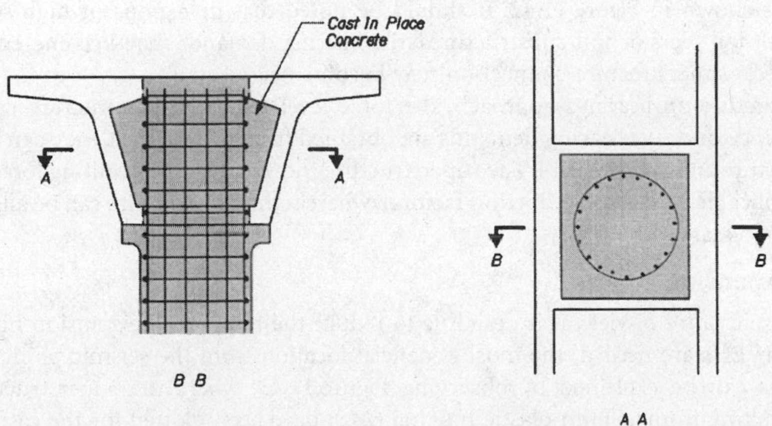

FIGURE 11.41 Combination of precast and cast-in-place joint.

11.8.2.2 Deck/Superstructure Connection via Bearings

Typically, for spans up to 45 m erected with the span-by-span method, the superstructure will be supported on bearings. For action in the longitudinal direction, elastomeric or isolation bearings are preferred to a fixed-end/expansion-end arrangement, since these better distribute the load

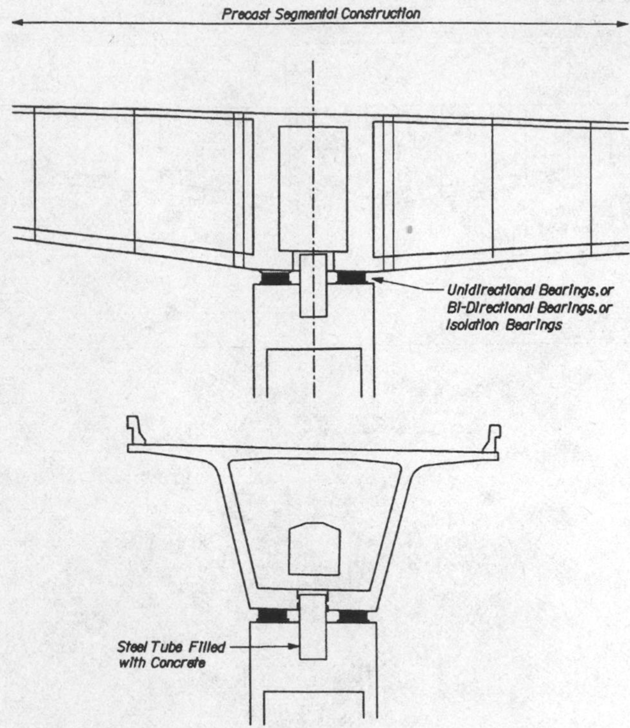

FIGURE 11.42 Deck/pier connection with bearings.

between the bearings. Furthermore, these bearings will increase the period of the structure, which results in an overall lower induced force level (beneficial for higher-frequency structures), and isolation bearings will provide some structural damping as well.

In the transverse direction, the bearings may be able to transfer load between super- and substructure by shear deformation; however, for the cases where this is not possible, shear keys can be provided as is shown in Figure 11.42. It should be noted that in regions of high seismicity, for structures with tall piers or soft substructures, the bearing demands may become excessive and a monolithic deck–superstructure connection may become necessary.

For the structure-on-bearings approach, the force level for the superstructure can be readily determined, since once the bearing demands are obtained from the analysis, they can be applied to the superstructure and substructure. The superstructure should resist the resulting forces at ultimate (using the applicable code force-reduction factors), whereas the substructure can be allowed to yield plastically if necessary.

11.8.2.3 Expansion Hinges

From the seismic point of view, it is desirable to reduce the number of expansion hinges (EH) to a minimum. If EHs are needed, the most beneficial location from the seismic point of view is at midspan. This can be explained by observing Figure 11.43, where the superstructure bending moments, resulting from column plastic hinging (M_p), have been plotted for the case of an EH at midspan and for an EH at quarterspan. For the latter, it can be seen that the moment at the face of the column varies within the range of $\pm 3/4\ M_p$, whereas with the hinge at midspan, the values are only between $\pm 1/2\ M_p$.

The location of expansion hinges within a span, and its characteristics, depends also on the stiffness of the substructure and the type of connection of the superstructure to the piers. Table 11.1 presents general guidelines intended to assist in the selection of location of expansion hinges.

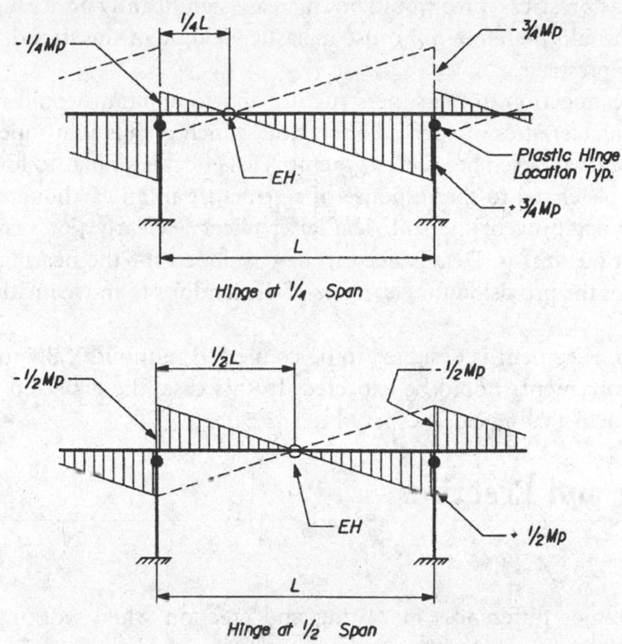

FIGURE 11.43 Longitudinal superstructure seismic moments with hinges at quarterspan and at midspan.

TABLE 11.1 Location of Expansion Hinges in Segmental Bridges

Span Support System	Location of EH: Over Pier	Location of EH: Intermediate Point	Location of EH: Midspan
On bearings	• Standard solution for simple spans • For continuous spans generates moderate superstructure moments at adjacent piers	• Complicated erection for cantilever construction • Generates moderate superstructure moments at adjacent piers • Moderate EH openings requiring restrainers and moderate seat widths, or lock-up devices	• Simplest location for cantilever construction • Will require continuity beam inside cross section • Minimizes superstructure seismic moments • Moderate EH openings requiring adequate gap between the end segments
Monolithic with pier	Not applicable	• Complicated erection for cantilever construction • Generates very large superstructure moments at adjacent piers • If substructure is stiff, expect relatively small EH movements; otherwise expect very large movements, requiring restrainers and large seat widths, or lock-up devices	• Simplest location for cantilever construction • Will require continuity beam inside cross section • Minimizes superstructure seismic moments • If substructure is stiff, expect relatively small EH movements; otherwise expect very large movements, requiring lock-up devices

11.8.2.4 Precast Segmental Piers

Precast segmental piers are usually hollow cross section to save weight. From research in other areas it can be extrapolated that the precast segments of the pier would be joined by means of unbonded prestressing tendons anchored in the footing. The advantage of unbonded over bonded tendons is that for the former, the prestress force would not increase significantly under high column displacement demands, and would therefore not cause inelastic yielding of the strand, which would otherwise lead to a loss of prestress.

The detail of the connection to the superstructure and foundation would require some insight into the dynamic characteristics of such a connection, which entails joint opening and closing—providing that dry joints are used between segments. This effect is similar to footing rocking, which is well known to be beneficial to the response of a structure in an earthquake. This is due to the period shift and the damping of the soil. The latter effect is clearly not available to the precast columns, but the period shift is. Details need to be developed for the bearing areas at the end of the columns, as well as the provision for clearance of the tendons to move relative to the pier during the event.

If the upper column segment is designed to be connected monolithically to the superstructure, yielding of the reinforcement should be expected. In this case, the expected plastic hinge length should be detailed ductile, using closely spaced ties [3,5].

11.9 Casting and Erection

11.9.1 Casting

There are obvious major differences in casting and erection when working with cast-in-place cantilever in travelers or in handling precast segments. There are also common features, which must be kept in mind in the design stages to keep the projects simple and thereby economic and efficient, such as

- Keeping the length of segments equal and segments straight, even in curved bridges;
- Maintaining constant cross section dimensions as much as possible;
- Minimizing the number of diaphragms and stiffeners, and avoiding dowels through formwork.

11.9.1.1 Cast-in-Place Cantilevers

Conventional Travelers

The conventional form traveler supports the weight of the fresh concrete of the new segment by means of longitudinal beams or frames extending out in cantilever from the last segment. These beams are tied down to the previous segment. A counterweight is used when launching the traveler forward. The main beams are subjected to some deflections, which may produce cracks in the joint between the old and new segments. Jacking of the form during casting is sometimes needed to avoid these cracks. The weight of a traveler is about 60% of the weight of the segment. The rate of construction is typically one segment per traveler per week. Precast concrete anchor blocks are used to speed up post-tensioning operations. In cold climates, curing can be accelerated by various heating processes.

Construction Camber Control

The most critical practical problem of cast-in-place construction is deflection control. There are five categories of deflections during and after construction:

- Deflection of traveler frame under the weight of the concrete segment;
- Deflection of the concrete cantilever arm during construction under the weight of segment plus post-tensioning;
- Deflection of cantilever arms after construction and before continuity;
- Short- and long-term deflections of the continuous structure;
- Short- and long-term pier shortenings and foundation settlements.

The sum of the various deflection values for the successive sections of the deck allows the construction of a camber diagram to be added to the theoretical profile of the bridge. A construction camber for setting the elevation of the traveler at each joint must also be developed.

11.9.1.2 Precast Segments

Opposite to the precast girder concept where the bridge is cut longitudinally in the precast segmental methods, the bridge is cut transversally, each slice being a segment. Segments are cast in a casting yard one at a time. Furthermore, the new segment is cast against the previously cast segment so that the faces in contact match perfectly. This is the match-cast principle. When the segments are reassembled at the bridge site, they will take the same relative position with regard to the adjacent segments that they had when they were cast. Accuracy of segment geometry is an absolute priority, and adequate surveying methods must be used to ensure follow-up of the geometry.

Match casting of the segments is a prerequisite for the application of glued joints, achieved by covering the end face of one or both of the meeting segments with epoxy at the erection. The epoxy serves as a lubricant during the assembly of the segments, and it ensures a watertight joint in the finished structure. Full watertightness is needed for corrosion protection of internal tendons (tendons inside the concrete). The tensile strength of the epoxy material is higher than that of the concrete, but, even so, the strength of the epoxy is not considered in the structural behavior of the joint. The required shear capacity is generally provided by shear keys, single or multiple, in combination with longitudinal post-tensioning.

With the introduction of external post-tensioning, where the tendons are installed in PE ducts, outside the concrete but inside the box girder, the joints are relieved of the traditional requirement of watertightness and are left dry. The introduction of external tendons in connection with dry joints greatly enhanced the efficiency of precasting.

11.9.1.3 Casting Methods

There are two methods for casting segments. The first one is the long-line method, where all the segments are cast in their correct position on a casting bed that reproduces the span. The second method, used most of the time, is the short-line method, where all segments are cast in the same place in a stationary form, and against the previously cast segment. After casting and initial curing, the previously cast segment is removed for storage, and the freshly cast segment is moved into place (see Figure 11.44).

11.9.1.4 Geometry Control

A pure translation of each segment between cast and match-cast position results in a straight bridge (Figure 11.45). To obtain a bridge with a vertical curve, the match-cast segment must first be translated and given a rotation α in the vertical plane (Figure 11.46). Practically, the bulkhead is left fixed and the mold bottom under the conjugate unit adjusted. To obtain a horizontal curvature, the conjugate unit is given a rotation β in the horizontal plane (see Figure 11.47). To obtain a variable superelevation, the conjugate unit is rotated around a horizontal axis located in the middle of the top slab (Figure 11.48).

All these adjustments of the conjugate unit can be combined to obtain the desired geometry of the bridge.

11.9.2 Erection

The type of erection equipment depends upon the erection scheme contemplated during the design process; the local conditions, either over water or land; the speed of erection and overall construction schedule. It falls into three categories, independent lifting equipment such as cranes, deck-mounted lifting equipment such as beam and winch or swivel crane, and launching girder equipment.

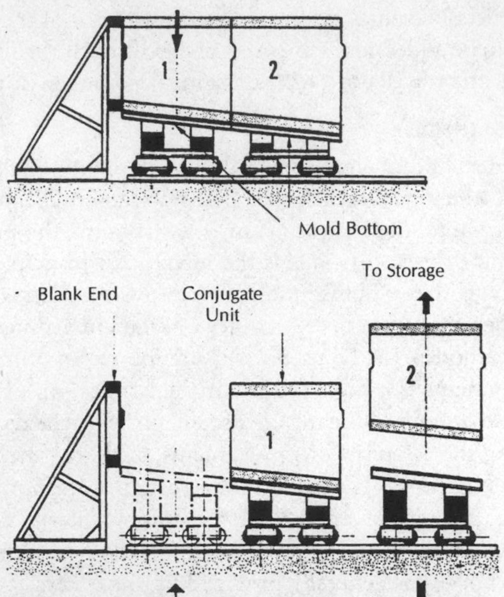

FIGURE 11.44 Typical short-line precasting operation.

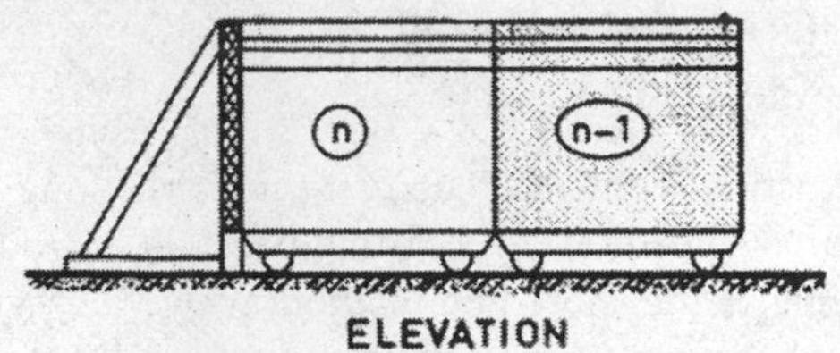

FIGURE 11.45. Straight bridge.

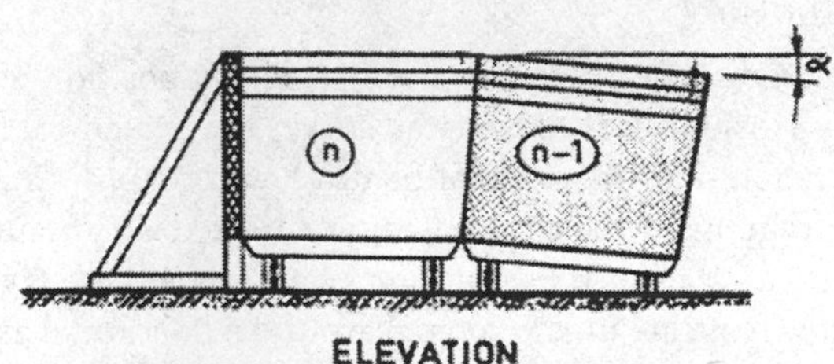

FIGURE 11.46 Bridge with vertical curve.

11.9.2.1 Balanced Cantilever Method

The principle of the method is to erect or cast the pier segment first, then to place typical segments one by one from each side of the pier, or in pairs simultaneously from both sides. Each newly placed precast segment is fixed to the previous one with temporary PT bars, until the cantilever tendons are installed and stressed. The closure joint between cantilever tips is poured in place and continuity tendons installed and stressed.

In order to carry out this erection scheme, segments must be lifted and installed at the proper location. The simplest way is to use a crane, either on land or barge mounted. Many bridges have

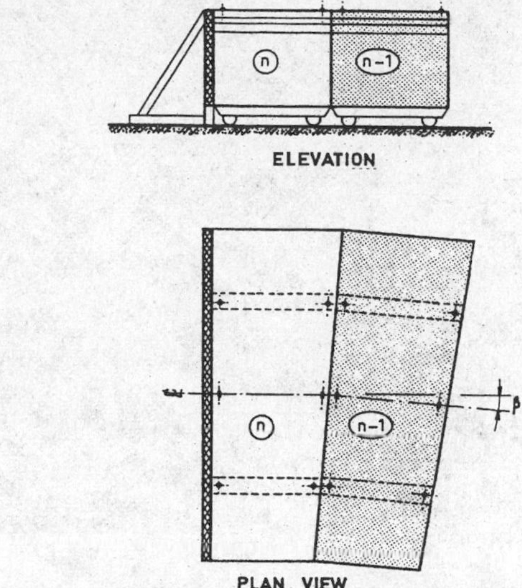

FIGURE 11.47 Bridge with horizontal curve.

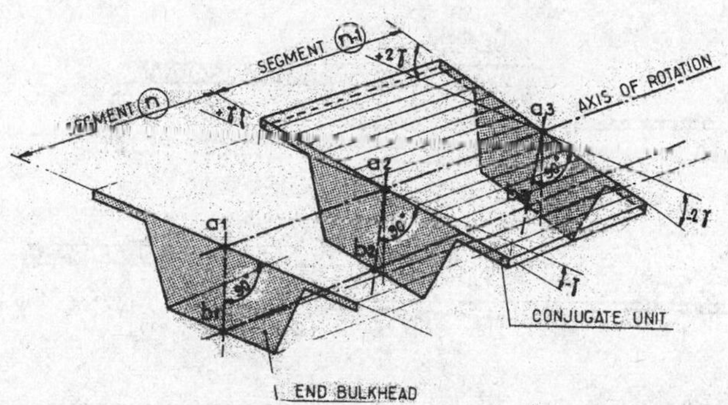

FIGURE 11.48 Bridge with superelevation.

been erected with cranes as they do not require an investment in special lifting equipment. This method is slow. Typically, two to four segments per day are placed. It is used on relatively short bridges. An alternative is to have a winch on the last segment erected. The winch is mounted on a beam fixed to the segment. It picks up segments from below, directly from truck or barge. After placing the segment, the beam and winch system is moved forward to pick up the next segment and so on. Usually, a beam-and-winch system is placed on each cantilever tip. This method is also slow; however, it does not require a heavy crane on the site, which is always very expensive, especially if the segments are heavy.

When bridges are long and the erection schedule short, the best method is the use of launching girders, which then take full advantage of the precast segmental concept for speed of erection.

There are two essential types of self-launching gantries developed for this erection method. The first type is a gantry with a length slightly longer than the typical span (see Figure 11.49). During erection of the cantilever, the center leg rests on the pier while the rear leg rests on the cantilever tip of the previously erected span, which must resist the corresponding reaction. Prior to launching, the back spans must be made continuous. Then, the center leg is moved to the forward cantilever

FIGURE 11.49 French Creek Viaduct, U.S.— single erection truss with portal legs.

tip, which must resist the weight of the gantry plus the weight of the pier segment. This stage controls the design of the gantry, which must be made as light as possible, and of the cantilever.

The second type of gantry has a length that is twice that of the typical span (see Figure 11.50). The reaction from the legs during the erection and launching of the next span is always applied on the piers, so there is no concentrated erection load on the cantilever tip. Each erection cycle consists of the erection of all typical segments of the cantilever and then the placement of the pier segment for the next cantilever, without changing the position of the truss.

The gantries can be categorized by their cross section: single truss, with portal-type legs, and two launching trusses with a gantry across. The twin box girders of the bridge in Hawaii were built with two parallel, but independent trusses (see Figure 11.51), with a typical span of 100.0 m, segment weights of 70 tons; the two bridge structures are 27.5 m apart with different elevations and longitudinal slopes. This system is a refinement of the first type of gantry applied to twin decks with variable geometry.

Normally, the balanced cantilever method is used for spans from 60 to 110 m, with a launching girder. One full, typical cycle of erection is placing segments, installing and stressing post-tensioning tendons, and launching the truss to its next position. It takes about 7 to 10 days, but may vary greatly according to the specifics of a project and the sophistication of the launching girder. With proper equipment and planning, erection of 16 segments per day has been achieved.

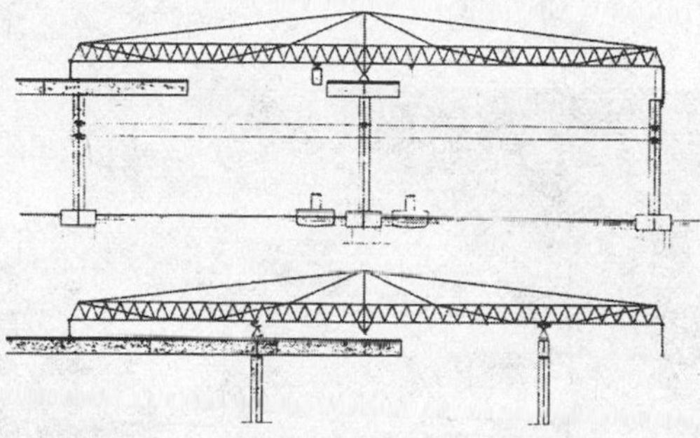

FIGURE 11.50 Río Niteroi, Brazil — two segments being erected simultaneously from one erection truss.

FIGURE 11.51 H-3 Windward Viaduct, Hawaii.

A modification of this method was used to build the 13-km-long Confederation Bridge (typical span 250 m long), linking Prince Edward Island with New Brunswick in Atlantic Canada. The main girder was constructed in a precasting yard as a precast, balanced cantilever. Then the 197-m-long main girder, with a self-weight of 7500 tons, was placed on the pier with a floating crane (see Figure 11.52).

FIGURE 11.52 Confederation Bridge, Canada — floating crane.

FIGURE 11.53 Bang Na–Bang Pli–Bang Pakong Expressway, Thailand — D6 segment erection.

11.9.2.2 Span-by-Span Construction

In the first stage, all precast segments for each span are assembled on an erection girder. The second stage is installing and stressing the tendons, and as a result the span becomes self-supported. In comparison with the balanced cantilever method, those girders have to be designed to carry the load of the entire span. Normally, the duration of one erection cycle is 2 to 3 days per span.

The 56-km-long Bang Na–Bang Pli–Bang Pakong Expressway in Bangkok, carrying six lanes of traffic, totaling 27 m in width, is assembled on erection girders. The girder is placed in the middle of a Y-shaped column. The segments, with self-weight up to 100 tons, are placed on the chassis with a swivel crane and then transported to their final position. Two schemes of erection were developed: a swivel crane mounted to the front of the girder picking up segments from trucks on the highway below and a swivel crane placed on the previously erected span with segments delivered over the deck already built (see Figure 11.53).

Another principle used in erection equipment for the span-by-span method, is the so-called overhang girder. In this case, the girder is above the superstructure, and the precast segments are hung from it.

11.9.2.3 Safety

Due to the inherent character of temporary structures, erection equipment is usually designed to take full advantage of the materials and care must be taken to analyze in-depth all construction stages, anticipating mistakes or shortcuts made on site that always occur, and stay within reasonable safety limits. Overall stability when resting on temporary supports or during launching, reversal of forces, bucking, etc., are the most common problems encountered in these structures. Lifting bars or tie-downs must always be designed with failure or mishandling of one of those elements in mind, and appropriate ultimate resisting paths incorporated in the concept.

11.10 Future of Segmental Bridges

Since their appearance in 1962, precast segmental technologies have been used worldwide in the design and construction of practically all types of bridges. Nevertheless, in the last 5 years, an important further development of these technologies has taken place in Southeast Asia which will have a decisive impact on the way bridges will be built in the next century.

11.10.1 The Challenge

The explosive development of the Southeast Asian economies has been forcing local governments to find new solutions for building infrastructures (build, operate, transfer), which can, when properly managed, become success stories for both the government agencies who organize the projects and for the private groups who develop them.

Privatization of such projects is now accepted by numerous countries as a viable solution for the challenges they face. Large infrastructure projects, worth billions of dollars, are at present being designed, built, financed, and operated by private companies in the region. This trend is expected to extend progressively to the global construction markets. The two key factors of such projects are the amount of toll to be paid by the users and the duration of the concession until the project is transferred to the government agency. For the roadway projects, the tolls vary anywhere from between $1 to $10. The duration of the concessions varies in general from 25 to 35 years.

While basically simple, the scheme presents very complex problems for its implementation. Multidisciplinary skills and a new vision for the design, construction, and operation of the roads and bridges are necessary in order to avoid technical or financial failures. The challenges that private organizations must be able to face can be summarized in just a few words: they need to design and build very competitive projects in the shortest possible time, ensuring the longest possible service life. In light of the experiences gained in the new markets, it is becoming more evident each day that precast segmental technologies often bring the right solutions to these challenges.

11.10.2 Concepts

When Jean Muller invented the precast concrete segmental technology in 1962, his vision was to create an industrialized construction system to build any type of bridge with standard modules, assembled with post-tensioning, without any cast-in-place concrete.

To achieve this objective, he developed the concept of match-cast joints, which allows the transverse slicing of concrete box girders and the assembly of such slices — the segments — in the same order as they were produced, without any need for additional *in situ* concrete to complete the bridge deck [10, 11].

In addition to epoxy-glued joints, the use of dry joints became widespread. In 1978, through the design of the Long Key Bridge in Florida, internal post-tensioning was replaced by external post-tensioning [12]. A number of other concepts invented by Jean Muller allowed further development of the modular construction concept: span-by-span assembly method (Long Key Bridge), progressive

FIGURE 11.54 Bang Na Expressway, Thailand — D6, six-traffic-lane segment at precast yard.

TABLE 11.2 Standard Segments

Segment	Lanes	Widths (m) Min-Max
D2	2	08–12
D3	3	10–15
D4	4	14–20
D6	6	18–30

placing (Linn Cove Bridge), precast segmental construction of the piers, D6 cable-stayed segments (Sunshine Skyway Bridge), delta frames (James River Bridge, C&D Canal Bridge), etc.

These concepts and others developed more recently for the large projects, which are being built in Canada and Thailand, allow the prefabrication of bridge structures with precast modules ranging from 20 tons, in channel-shaped overpasses, to 7500 tons for the main girders of the Confederation Bridge in Canada. In this project, one of the largest bridges ever built, no cast-in-place structural concrete was used in the construction of the main spans [13]. The construction modules are all manufactured in sophisticated and industrialized precasting plants that ensure an unequaled construction rate and quality (see Figure 11.54).

By further standardizing the segments with the number of traffic lanes that they carry, we have developed the modules in Table 11.2. These modules allow the construction of viaducts of any width, ranging from 7 to 8 m up to 30 m. Concurrently, with the effort to standardize the cross sections for precast segmental bridges, there has been significant development of design, shop drawing software, and geometry control systems. This gives us the capability to produce drawings by the thousands for viaducts, interchanges, and merging sections that give, for each segment, the detailed geometry and dimensions of concrete and rebar and the layout of the post-tensioning. Such shop drawings are an essential part of the system and must be integrated into the structural design; no standardization and, hence, no industrialization is possible without them (Figure 11.55).

11.10.3 New Developments

The dynamic business environment of Southeast Asian markets is quite favorable for the introduction of innovative concepts. In recent years, technologies that took over 20 years to develop in Europe and some 10 years to spread throughout the United States were absorbed by countries such

FIGURE 11.55 The Confederation Bridge, Canada — Prince Edward Island precasting yard.

as Thailand, which had limited prior experience in the field of bridge engineering, and already concepts never used before are being developed for new projects, thus giving these countries a leading position in construction innovation. By using the most innovative technologies, the developers involved in the private roadway projects are dramatically changing the very nature of the construction business that, until very recently, was considered one of the most conservative sectors of the industry. The innovative concepts that the industrialization of bridges is introducing cover different areas:

- Reduction of construction time and construction cost;
- Durability of the structures (25 to 50 to 100 years);
- Replaceability of components such as bearings, post-tensioning, stays;
- Earthquake resistance of the structures;
- Staged construction;
- Integrated inspection and surveillance systems;
- Users' comfort and safety.

It is evident that the multiplication of such private projects, where cost, time, and durability are the decisive factors, will open the way to innovation in the bridge business as never before.

11.10.4 Environmental Impact

To prevent private projects from turning into environmental nightmares, private developers need to comply with strict obligations with respect to aesthetics, rights of way, and maintenance of the structures during the duration of the concessions. Government agencies have been developing design, construction, and operation criteria that will progressively become the rules of the BOT projects. As an example, such rules may force structural engineers to conceive structures that can be built in or over crowded areas of cities, with a minimal impact on existing conditions. Or they may impose specific constraints on aesthetics, shapes, or dimensions of structural elements. Further, they may require maintenance costs to be budgeted.

The involvement of the communities in such projects, even if sometimes it may be difficult to manage and may require a profound knowledge of the interests and aspirations of those concerned by the project, is essential for the smooth development of the work. In general, projects that are not well integrated into the context of the local environment or not consistent with the users' expectations run the risk of finishing in disarray or remaining incomplete.

11.10.5 Industrial Production of Structures

The experience acquired in large- or medium-size projects demonstrates that the industrialization of the production of structural elements always brings clear advantages in terms of quality and construction time. What frequently has been less noticed is the advantage that such industrialization can offer as far as the cost of a specific project.

The major change in the contractual conditions of the BOT projects is that the cost of "design + construction time" can now be estimated very precisely. If the completion of a project is delayed by 1 month, for instance, in a project worth U.S.$800,000,000, the cost to the developer is approximately equal to the interest that must be paid on that amount. If the interest is 5%, this represents U.S.$40,000,000/year; thus, every month gained in the duration of the design + construction period represents U.S.$3,300,000, or roughly, U.S.$100,000/day.

In these large projects the industrialization of production and the use of sophisticated systems to transport, erect, and assemble the prefabricated modules is reducing the duration of cycles, which usually may take 6 years when managed by the government agencies, to some 3 years, when managed by private organizations in a fully integrated way.

The introduction of the "assembly-line" approach to bridge building was taken to its limits during the construction of the Northumberland Strait Crossing (Confederation Bridge), Canada, a major bridge project which extends over 13 km of icy strait, with extreme weather conditions. The actual assembly of the components that constitute the 43 spans, 250-m each, took place in only 12 months, whereas to build just a single cast-in-place span of 250 m by traditional means is a difficult venture that takes at least 2 years (see Figure 11.55).

11.10.6 The Assembly of Structures

The production of the structural modules for precast segmental projects represents half of the process. The other half relates to the transport of these modules to the site, to their erection, and to the assembly methods to constitute the structural integrity of the bridge.

Generally, for the transport of current segments weighing from 30 to 100 tons, equipment already available in the market has been used. The transport is commonly by road, using convoys of "low boys," or by water, using barges. In some projects currently being built, 30 to 40 segments weighing between 50 and 60 tons are transported every night from the casting yards situated some 100 km from the large metropolis, to the site in the center of the city. The segments are picked up directly from the trucks by the assembly gantries, between midnight and five o'clock in the morning, to avoid interfering with heavy city traffic during the day.

The erection and assembly of such segments are also performed in a highly industrialized environment. With the span-by-span construction method, spans of 40 m can be assembled in 2 days, with crews working after hours. The cycle is almost independent of the type of segment, from D2 to D6, and therefore, the method is ideal for spans from 30 to 45 m. For cantilever construction, special gantries have been developed to assemble two parallel viaducts mimicking the procedure used in Hawaii, achieving speeds of construction of 3 weeks to complete two double cantilevers of 100 m [14]. This method very competitively covers spans of 80 to 120 m. Progressive placing of segments, using a swivel crane, has also been improved for this type of construction, which allows construction of spans from 45 to 65 m. Finally, for large cable-stayed spans, the use of precast segmental technologies successfully tested in milestone projects, like the Sunshine Skyway and the James River Bridges, is now being developed to cover different cross sections for the segments and to combine space trusses and composite sections [15].

The use of gigantic floating cranes, such as the *Svanen,* to place units as large as 190 m and with weights of 7500 tons, opens new prospects for the construction of bridges over rivers and straits (see Figure 11.52). Bridges that previously were almost impossible to build competitively and within the common constraints of construction schedules can now be conceived, designed, and built in short periods of time, by intensive use of precast technologies.

FIGURE 11.56 View of the completed Second Expressway System Project, Thailand.

Clearly, this evolution is going to accelerate and will become global. Equipment designed to be used anywhere in the world will allow for the reduction of costs charged on a specific project. Furthermore, we can expect improvements in the performance and reliability of equipment specifically conceived to perform heavy lifting and assembly of bridge modules. Bridges will be designed which take into consideration the availability and the characteristics of these machines. Construction methods will then become, more than ever, a decisive factor in the design of structures.

11.10.7 Prospective

Design-and-build projects that were common in the 1960s provided some of the most innovative contributions to bridge engineering. Engineers and contractors working together produced competitive structures that paved the way for the development that has taken place all over the world during the last quarter century (Figure 11.56).

A new wave of innovative bridge concepts is already being generated by the privatization of roadway and bridge projects. This wave, which began in the vibrant business environment of Southeast Asia, will eventually reach the United States and the European markets. This time, engineers and contractors will be seconded by developers, finance specialists, and industrialists to shape the structures that will be built during the next century. The construction industry will also join other key industries in adopting high-technology and innovation as essential ingredients of its renewal [16].

References

1. AASHTO, *Standard Specifications for Highway Bridges*, 16th ed., American Association of State Highway and Transportation Officials, Washington, D.C., 1996.
2. AASHTO, *Guide Specifications for Design and Construction of Segmental Concrete Bridges*, Draft 2nd ed., American Association of State Highway and Transportation Officials, Washington, D.C., August 1997.

3. Imbsen, R. A. et al., *Thermal Effects in Concrete Bridge Superstructures*, National Cooperative Highway Research Program Report 276, Transportation Research Board, Washington, D.C., 1985.
4. Priestley, M. J. N. et al., *Seismic Design and Retrofit of Bridges*, John Wiley & Sons, New York, 1996.
5. California Department of Transportation, *Bridge Design Specifications*, Sacramento.
6. California Department of Transportation, *Bridge Memos to Designers*, Sacramento.
7. California Department of Transportation, *Seismic Design Memo*, Sacramento.
8. Riobóo Martin, J. M., A new dimension in precast prestressed concrete bridges for congested urban areas in high seismic zones, *PCI J.*, 37, (2), 1992.
9. Restrepo, J. I. et al., Design of connections of earthquake resisting precast reinforced concrete perimeter frames, *PCI J.*, 40, (5), 1995.
10. Muller, J., Ten years of experience in precast segmental construction, *J. Precast/Prestressed Concrete Insti.*, 20, (1), 28–61, 1975.
11. Podolny, W., et al., *Construction and Design of Prestressed Concrete Segmental Bridges*, John Wiley & Sons, New York, 1982.
12. Muller, J., *Evolution dans la Construction de Grands Ponts: Montage et Entretien*, IABSE, 11th Congress, Vienna, 1980.
13. Sauvageot, G., Northumberland Strait Crossing, Canada, *4th International Bridge Engineering Conference; Proceedings*, 7, Vol. 1, August, 1995, 238–248.
14. Dodson, B., *Bangkok Second Stage Expressway System Segmental Structures*, in *4th International Bridge Engineering Conference, Proceedings* 7, Vol. 2, August, 1995, 199–204.
15. Sauvageot, G., *Hawaii H-3 Precast Segmental Windward Viaduct*, FIP Congress, Washington, D.C., 1994.
16. Muller, J., *Reflections on cable-stayed bridges*, Rev. Gen. Routes Aérodromes, Paris, October, 1994.

12

Steel-Concrete Composite I-Girder Bridges

Lian Duan
California Department of Transportation

Yusuf Saleh
California Department of Transportation

Steve Altman
California Department of Transportation

12.1 Introduction

An I-section is the simplest and most effective solid section of resisting bending and shear. In this chapter straight, steel–concrete composite I-girder bridges are discussed (Figure 12.1). Materials and components of I-section girders are described. Design considerations for flexural, shear, fatigue, stiffeners, shear connectors, diaphragms and cross frames, and lateral bracing with examples are presented. For a more detailed discussion, reference may be made to recent texts by Xanthakos [1], Baker and Puckett [2], and Taly [3].

12.2 Structural Materails

Four types of structural steels (structural carbon steel, high-strength low-alloy steel, heat-treated low-alloy steel, and high-strength heat-treated alloy steel) are commonly used for bridge structures. Designs are based on minimum properties such as those shown in Table 12.1. ASTM material property standards differ from AASHTO in notch toughness and weldability requirements. Steel meeting the AASHTO-M requirements is prequalified for use in welded bridges.

Concrete with 28-day compressive strength $f_c' = 16$ to 41 MPa is commonly used in concrete slab construction. The transformed area of concrete is used to calculate the composite section properties. The short-term modular ratio n is used for transient loads and long-term modular ratio

0-8493-7434-0/00/$0.00+$.50

FIGURE 12.1 Steel–concrete composite girder bridge (I-880 Replacement, Oakland, California)

TABLE 12.1 Minimum Mechanic Properties of Structural Steel

Material	Structural Steel	High-Strength Low-Alloy Steel		Quenched and Tempered Low-Alloy Steel	High Yield Strength Quenched and TemperedLow-Alloy Steel	
AASHTO designation	M270 Grade 250	M270 Grade 345	M270 Grade 345W	M270 Grade 485W	M270 Grades 690/690W	
ASTM designation	A709M Grade 250	A709M Grade 345	A709M Grade 345W	A709M Grade 485W	M709M Grades 690/690W	
Thickness of plate (mm)	Up to 100 included				Up to 65 included	Over 65–100 included
Shapes	All Groups			Not Applicable		
F_u (MPa)	400	450	485	620	760	690
F_y (MPa)	250	345	485	485	690	620

F_y= minimum specified yield strength or minimum specified yield stress; F_u = minimum tensile strength; E = modulus of elasticity of steel (200,000 MPa).

Source: American Association of State Highway and Transportation Officials, AASHTO LRFD Bridge Design Specifications, Washington, D.C., 1994. With permission.

$3n$ for permanent loads. For normal-weight concrete the short-term ratio of modulus of elasticity of steel to that of concrete are recommended by AASHTO-LRFD [4]:

$$n = \begin{cases} 10 & \text{for } 16 \le f'_c < 20 \text{ MPa} \\ 9 & \text{for } 20 \le f'_c < 25 \text{ MPa} \\ 8 & \text{for } 25 \le f'_c < 32 \text{ MPa} \\ 7 & \text{for } 32 \le f'_c < 41 \text{ MPa} \\ 6 & \text{for } f'_c \le 41 \text{ MPa} \end{cases} \tag{12.1}$$

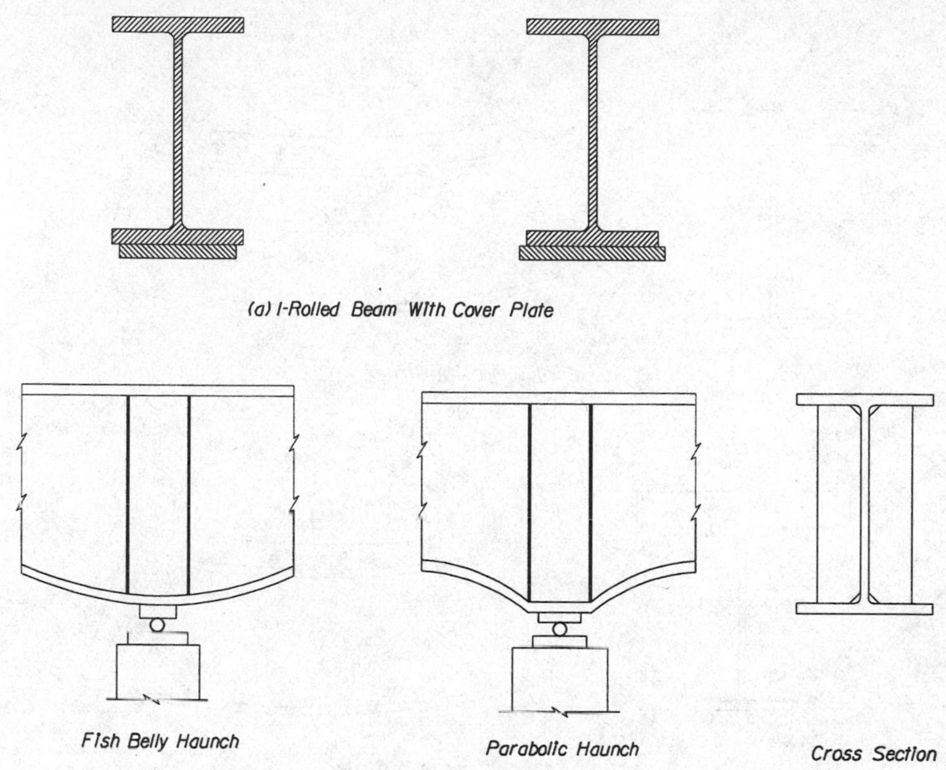

FIGURE 12.2 Typical sections.

12.3 Structural Components

12.3.1 Classification of Sections

I-sectional shapes can be classified in three categories based on different fabrication processes or their structural behavior as discussed below:

1. A steel I-section may be a *rolled* section (*beam*, Figure 12.2a) with or without cover plates, or a *built-up* section (*plate girder*, Figure 12.2b) with or without haunches consisting of top and bottom flange plates welded to a web plate. Rolled steel I-beams are applicable to shorter spans (less than 30 m) and plate girders to longer span bridges (about 30 to 90 m). A plate girder can be considered as a deep beam. The most distinguishing feature of a plate girder is the use of the transverse stiffeners that provide tension-field action increasing the postbuckling shear strength. The plate girder may also require longitudinal stiffeners to develop inelastic flexural buckling strength.
2. I-sections can be classified as *composite* or *noncomposite*. A steel section that acts with the concrete deck to resist flexure is called a composite section (Figure 12.3a). A steel section disconnected from the concrete deck is noncomposite (Figure 12.3b). Since composite sections most effectively use the properties of steel and concrete, they are often the best choice. Steel–concrete composite girder bridges are recommended by AASHTO-LRFD [4] whereas noncomposite members are not and are less frequently used in the United States.

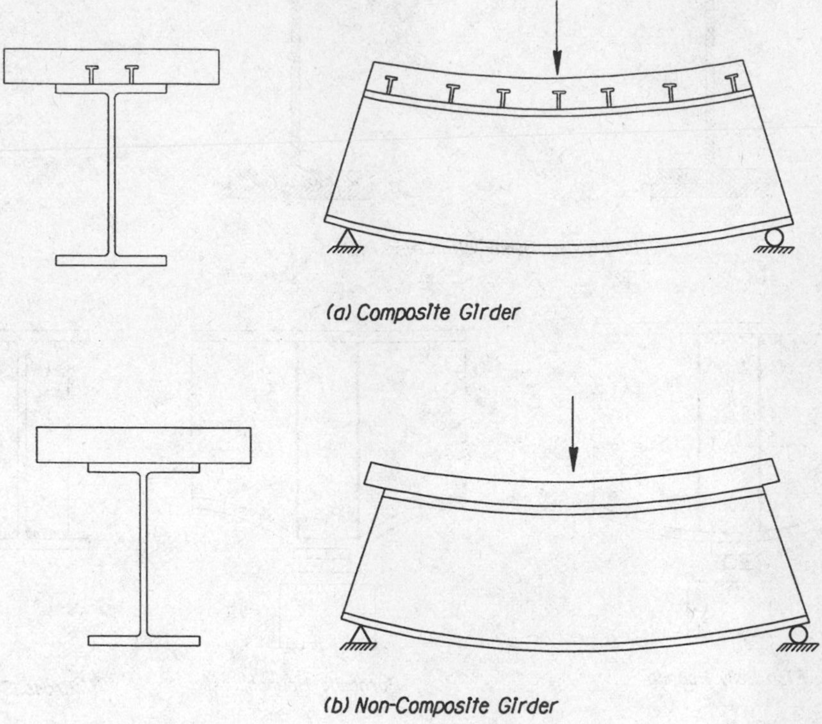

FIGURE 12.3 Composite and noncomposite section.

3. Steel sections can also be classified as *compact, noncompact,* and *slender* element sections [4-6]. A qualified compact section can develop a full plastic stress distribution and possess a inelastic rotation capacity of approximately three times the elastic rotation before the onset of local buckling. Noncompact sections develop the yield stress in extreme compression fiber before buckling locally, but will not resist inelastic local buckling at the strain level required for a fully plastic stress distribution. Slender element sections buckle elastically before the yield stress is achieved.

12.3.2 Selection of Structural Sections

Figure 12.4 shows a typical portion of a composite I-girder bridge consisting of a concrete deck and built-up plate girder I-section with stiffeners and cross frames. The first step in the structural design of an I-girder bridge is to select an I-rolled shape or to size initially the web and flanges of a plate girder. This section presents the basic principles of selecting I-rolled shapes and sizing the dimensions of a plate girder.

The ratio of overall depth (steel section plus concrete slab) to the effective span length is usually about 1:25 and the ratio of depth of steel girder only to the effective span length is about 1:30. I-rolled shapes are standardized and can be selected from a manual such as the AISC-LRFD [7]. It should be noted that the web of a rolled section always meets compactness requirements while the flanges may not. To increase the flexural strength of a rolled section, it is common to add cover plates to the flanges. The I-rolled beams are usually used for simple-span length up to 30 m for highway bridges and 25 m for railway bridges. Plate girder sections provide engineers freedom and flexibility to proportion the flanges and web plates efficiently. Plate girders must have sufficient

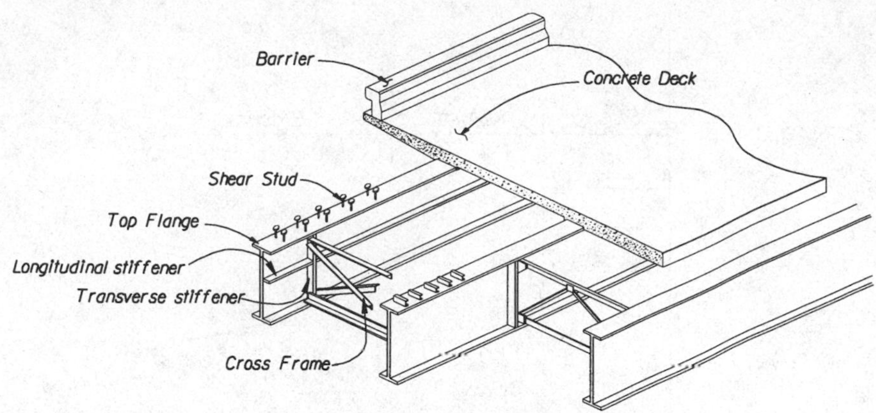

FIGURE 12.4 Typical components of composite I-girder bridge.

flexural and shear strength and stiffness. A practical choice of flange and web plates should not result in any unusual fabrication difficulties. An efficient girder is one that meets these requirements with the minimum weight. An economical one minimizes construction costs and may or may not correspond to the lowest weight alternative [8].

- *Webs*: The web mainly provides shear strength for the girder. The *web height* is commonly taken as 1/18 to 1/20 of the girder span length for highway bridges and slightly less for railway bridges. Since the web contributes little to the bending resistance, its thickness (t) should be as small as local buckling tolerance allows. Transverse stiffeners increase shear resistance by providing tension field action and are usually placed near the supports and large concentrated loads. Longitudinal stiffeners increase flexure resistance of the web by controlling lateral web deflection and preventing the web bending buckling. They are, therefore, attached to the compression side. It is usually recommended that sufficient web thickness be used to eliminate the need for longitudinal stiffeners as they can create difficulty in fabrication. Bearing stiffeners are also required at the bearing supports and concentrated load locations and are designed as compression members.
- *Flanges*: The flanges provide bending strength. The width and thickness are usually determined by choosing the area of the flanges within the limits of the width-to-thickness ratio, b/t, and the requirement as specified in the design specifications to prevent local buckling. Lateral bracing of the compression flanges is usually needed to prevent lateral torsional buckling during various load stages.
- *Hybrid Sections*: The hybrid section consisting of flanges with a higher yield strength than that of the web may be used to save materials; this is becoming more promoted because of the new high-strength steels.
- *Variable Sections*: Variable cross sections may be used to save material where the bending moment is smaller and/or larger near the end of a span (see Figure 12.2b). However, the manpower required for welding and fabrication may be increased. The cost of manpower and material must be balanced to achieve the design objectives. The designer should consult local fabricators to determine common practices in the construction of a plate girder.

Highway bridges in the United States are designed to meet the requirements under various limit states specified by AASHTO-LRFD [4,5] such as strength, fatigue and fracture, service, and extreme events (see Chapter 5). Constructibility must be considered. The following sections summarize basic concepts and AASHTO-LRFD [4,5] requirements for composite I-girder bridges.

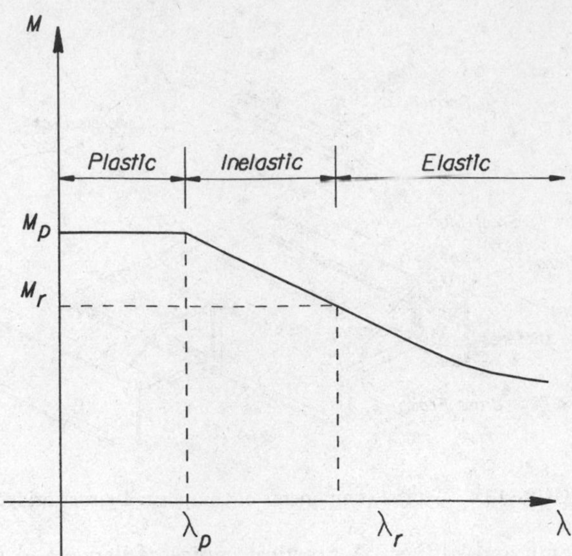

FIGURE 12.5 Three-range design format for steel flexural members.

12.4 Flexural Design

12.4.1 Basic Concept

The flexural resistance of a steel beam/girder is controlled by four failure modes or limit states: yielding, flange local buckling, web local buckling, and lateral-torsional buckling [9]. The moment capacity depends on the yield strength of steel (F_y), the slenderness ratio λ in terms of width-to-thickness ratio (b/t or h/t_w) for local buckling and unbraced length to the radius of gyration about strong axis ratio (L_b/r_y) for lateral-torsional buckling. As a general design concept for steel structural components, a three-range design format (Figure 12.5): plastic yielding, inelastic buckling, and elastic buckling are generally followed. In other words, when slenderness ratio λ is less than λ_p, a section is referred to as compact, plastic moment capacity can be developed; when $\lambda_p < \lambda < \lambda_r$, a section is referred to as noncompact, moment capacity less than M_p but larger than yield moment M_y can be developed; and when $\lambda > \lambda_r$, a section or member is referred to as slender and elastic buckling failure mode will govern. Figure 12.6 shows the dimensions of a typical I-girder. Tables 12.2 and 12.3 list the AASHTO-LRFD [4,5] design formulas for determination of flexural resistance in positive and negative regions.

12.4.2 Yield Moment

The yield moment M_y for a composite section is defined as the moment that causes the first yielding in one of the steel flanges. M_y is the sum of the moments applied separately to the steel section only, the short-term composite section, and the long-term composite section. It is based on elastic section properties and can be expressed as

$$M_y = M_{D1} + M_{D2} + M_{AD} \tag{12.6}$$

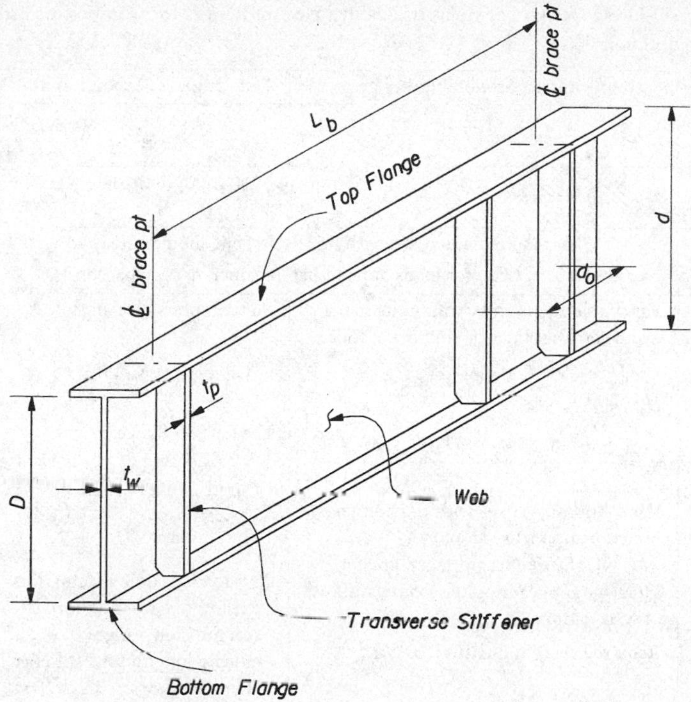

FIGURE 12.6 Typical girder dimensions.

where M_{D1} is moment due to factored permanent loads on steel section; M_{D2} is moment due to factored permanent loads such as wearing surface and barriers on long-term composite section; M_{AD} is additional live-load moment to cause yielding in either steel flange and can be obtained from the following equation:

$$M_{AD} = S_n \left[F_y - \frac{M_{D1}}{S_s} - \frac{M_{D2}}{S_{3n}} \right] \tag{12.7}$$

where S_s, S_n, and S_{3n} are elastic section modulus for steel, short-term composite, and long-term composite sections, respectively.

12.4.3 Plastic Moment

The plastic moment M_p for a composite section is defined as the moment that causes the yielding in the steel section and reinforcement and a uniform stress distribution of 0.85 f'_c in compression concrete slab (Figure 12.7). In positive flexure regions, the contribution of reinforcement in concrete slab is small and can be neglected.

The first step of determining M_p is to find the plastic neutral axis (PNA) by equating total tension yielding forces in steel to compression yield in steel and/or concrete slab. The plastic moment is then obtained by summing the first moment of plastic forces in various components about the PNA. For design convenience, Table 12.4 lists the formulas for $\overline{Y}$ and M_p.

TABLE 12.2 AASHTO-LRFD Design Formulas of Positive Flexure Ranges for Composite Girders (Strength Limit State)

Items	Compact Section Limit, λ_p	Noncompact Section Limit, λ_r	Slender Sections
Web slenderness $2D_{cp}/t_w$	$3.76\sqrt{E/F_{yc}}$	$\alpha_{st}\sqrt{E/f_c}$	N/A
Compression flange slenderness b/t	No requirement at strength limit state		
Compression flange bracing L_b/r_t	No requirement at strength limit state, but should satisfy $1.76\sqrt{E/F_{yc}}$ for loads applied before concrete deck hardens		$>1.76\sqrt{E/F_{yc}}$
Nominal flexural resistance	For simple spans and continuous spans with compact interior support section: For $D_p \le D'$, $M_n = M_p$ If $D' < D_p \le 5D'$ $M_n = \frac{5M_p - 0.85M_y}{4} + \frac{0.85M_y - M_p}{4}\left(\frac{D_p}{D'}\right)$ For continuous spans with noncompact interior support section: $M_n = 1.3R_hM_y$ but not taken greater than the applicable values from the above two equations. **Required section ductility** $D_p / D' \le 5$ $D' = \beta\left(\frac{d+t_s+t_h}{7.5}\right)$ $\beta = \begin{cases} 0.9 \text{ for } F_y = 250 \text{ MPa} \\ 0.7 \text{ for } F_y = 345 \text{ MPa} \end{cases}$	For compression flange: $F_n = R_bR_hF_{yc}$ For tension flange: $F_n = R_bR_hF_{yt}\sqrt{1-3\left(\frac{f_v}{F_{yt}}\right)}$ R_b = load shedding factor, for tension flange = 1.0; for compression flange = 1.0 if either a longitudinal stiffener is provided or $2D_c/t_w \le \lambda_b\sqrt{E/f_c}$ is satisfied; otherwise see Eq. (12.2)	Compression Flange: Eq. (12.4) Tension Flange: $F_n = R_bR_hF_{yt}$

A_{fc} = compression flange area
d = depth of steel section
D_{cp} = depth of the web in compression at the plastic moment
D_p = distance from the top of the slab to the plastic neutral axis
f_c = stress in compression flange due to factored load
f_v = maximum St. Venant torsional shear stress in the flange due to the factored load
F_n = nominal stress at the flange
F_{yc} = specified minimum yield strength of the compression flange
F_{yt} = specified minimum yield strength of the tension flange
M_p = plastic flexural moment

$$R_b = 1 - \left(\frac{a_r}{1200 + 300a_r}\right)\left(\frac{2D_c}{t_w} - \lambda_b\sqrt{\frac{E}{f_c}}\right) \quad (12.2)$$

$$a_r = \frac{2D_ct_w}{A_{fc}} \quad (12.3)$$

M_y = yield flexural moment
R_h = hybrid factor, 1.0 for homogeneous section, see AASHTO-LRFD 6.10.5.4
t_h = thickness of concrete haunch above the steel top flange
t_s = thickness of concrete slab; t_w= web thickness
α_{st} = 6.77 for web without longitudinal stiffeners and 11.63 with longitudinal stiffeners
λ_b = 5.76 for compression flange area ≥ tension flange area, 4.64 for compression area < tension area

TABLE 12.3 AASHTO-LRFD Design Formulas of Negative Flexure Ranges for Composite I Sections (Strength Limit State)

Items	Compact Section Limit, λ_p	Noncompact Section Limit λ_r	Slender Sections
Web slenderness, $2D_{cp}/t_w$	$3.76\sqrt{E/F_{yc}}$	$\alpha_{st}\sqrt{E/F_{yc}}$	N/A
Compression flange slenderness, $b_f/2t_f$	$0.382\sqrt{E/F_{yc}}$	$1.38\sqrt{\dfrac{E}{f_c\sqrt{\dfrac{2D_c}{t_w}}}}$	$>1.38\sqrt{\dfrac{E}{f_c\sqrt{\dfrac{2D_c}{t_w}}}}$
Compression flange unsupported length, L_b	$\left[0.124-0.0759\left(\dfrac{M_l}{M_p}\right)\right]\left[\dfrac{r_y E}{F_{yc}}\right]$	$1.76r_t\sqrt{\dfrac{E}{F_{yc}}}$	$>1.76r_t\sqrt{\dfrac{E}{F_{yc}}}$
Nominal flexural resistance	$M_n = M_p$	$F_n = R_b R_h F_{yf}$	Compression flange Eq. (12.4) Tension flange $F_n = R_b R_h F_{yt}$

b_f = width of compression flange
t_f = thickness of compression flange
M_l = lower moment due to factored loading at end of the unbraced length
r_y = radius of gyration of steel section with respect to the vertical axis (mm)
r_t = radius of gyration of compression flange of steel section plus one third of the web in compression with respect to the vertical axis (mm)

For lateral torsional buckling AASHTO-LRFD 6.10.5.5:

$$F_n = \begin{cases} C_b R_b R_h F_{yc}\left[1.33-0.18\left(\dfrac{L_b}{r_t}\right)\sqrt{\dfrac{F_{yc}}{E}}\right] \le R_b R_h F_{yc} & \text{for } L_p < L_b < L_r \\ C_b R_b R_h\left[\dfrac{9.86E}{(L_b/r_t)^2}\right] \le R_b R_h F_{yc} & \text{for } L_b \ge L_r \end{cases} \tag{12.4}$$

$$C_b = 1.75-1.05\left(\frac{P_1}{P_2}\right)+0.3\left(\frac{P_1}{P_2}\right)^2 \le 2.3 \tag{12.5}$$

P_1 = smaller force in the compression flange at the braced point due to factored loading
P_2 = larger force in the compression flange at the braced point due to factored loading

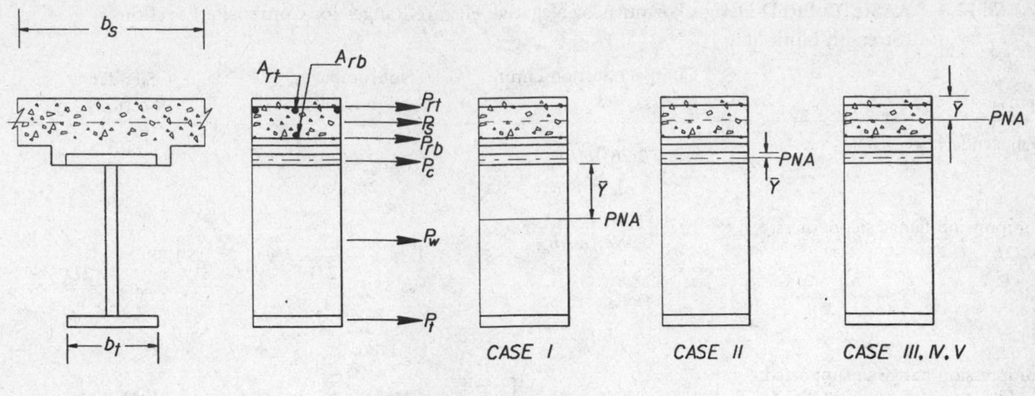

(a) Positive Bending Section

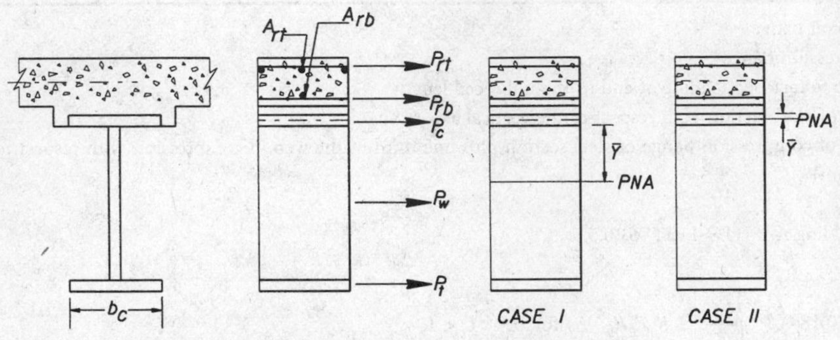

(b) Negative Bending Section

FIGURE 12.7 Plastic moments for composite sections.

Example 12.1: Three-Span Continuous Composite Plate-Girder Bridge

Given

A three-span continuous composite plate-girder bridge has two equal end spans of length 49.0 m and one midspan of 64 m. The superstructure is 13.4 m wide. The elevation, plan, and typical cross section are shown in Figure 12.8.

Structural steel: A709 Grade 345; $F_{yw} = F_{yt} = F_{yc} = F_y = 345$ MPa

Concrete: $f'_c = 280$ MPa ; $E_c = 25{,}000$ MPa; modular ratio $n = 8$

Loads: Dead load = steel plate girder + concrete deck + barrier rail + future wearing 75 mm AC overlay

Live load = AASHTO HL-93 + dynamic load allowance

Deck: Concrete deck with thickness of 275 mm has been designed

Steel section in positive flexure region:

Top flange:	$b_{fc} = 460$ mm	$t_{fc} = 25$ mm
Web:	$D = 2440$ mm	$t_w = 16$ mm
Bottom flange:	$b_{ft} = 460$ mm	$t_{ft} = 45$ mm

Construction: Unshored; unbraced length for compression flange $L_b = 6.1$ m.

TABLE 12.4 Plastic Moment Calculation

Regions	Case	Condition and $\overline{Y}$	$\overline{Y}$ and M_p
	I — PNA in web	$P_t+P_w \ge P_c+P_s+P_{rb}+P_{rt}$ $\overline{Y}=\left(\frac{D}{2}\right)\left[\frac{P_t-P_c-P_s-P_{rt}-P_{rb}}{P_w}+1\right]$	$M_p=\frac{P_w}{2D}\left[\overline{Y}^2+\left(D-\overline{Y}\right)^2\right]$ $+\left[P_s d_s+P_{rt}d_{rt}+P_{rb}d_{rb}+P_c d_c+P_t d_t\right]$
	II — PNA in top flange	$P_t+P_w+P_c \ge P_s+P_{rb}+P_{rt}$ $\overline{Y}=\left(\frac{t_c}{2}\right)\left[\frac{P_w+P_c-P_s-P_{rt}-P_{rb}}{P_c}+1\right]$	$M_p=\frac{P_c}{2t_c}\left[\overline{Y}^2+\left(t_c-\overline{Y}\right)^2\right]$ $+\left[P_s d_s+P_{rt}d_{rt}+P_{rb}d_{rb}+P_w d_w+P_t d_t\right]$
Positive Figure 12.7a	III — PNA in slab, below P_{rb}	$P_r+P_w+P_c \ge \left(\frac{C_{rb}}{t_s}\right)P_s+P_{rb}+P_{rt}$ $\overline{Y}=\left(t_s\right)\left[\frac{P_w+P_c+P_s-P_{rt}-P_{rb}}{P_s}\right]$	$M_p=\left(\frac{\overline{Y}^2P_s}{2t_s}\right)^2$ $+\left[P_{rt}d_{rt}+P_{rb}d_{rb}+P_c d_c+P_w d_w+P_t d_t\right]$
	IV — PNA in slab at P_{rb}	$P_r+P_w+P_c+P_{rb} \ge \left(\frac{C_{rb}}{t_s}\right)P_s+P_{rt}$ $\overline{Y}=C_{rb}$	$M_p=\left(\frac{\overline{Y}^2P_s}{2t_s}\right)^2$ $+\left[P_{rt}d_{rt}+P_c d_c+P_w d_w+P_t d_t\right]$
	V — PNA in slab, above P_{rb}	$P_r+P_w+P_c+P_{rb} \ge \left(\frac{C_{rb}}{t_s}\right)P_s+P_{rt}$ $\overline{Y}=\left(t_s\right)\left[\frac{P_{rb}+P_c+P_w+P_t-P_{rb}}{P_s}\right]$	$M_p=\left(\frac{\overline{Y}^2P_s}{2t_s}\right)^2$ $+\left[P_{rt}d_{rt}+P_{rb}d_{rb}+P_c d_c+P_w d_w+P_t d_t\right]$
Negative Figure 12.7b	I — PNA in web	$P_{cr}+P_w \ge P_c+P_{rb}+P_{rt}$ $\overline{Y}=\left(\frac{D}{2}\right)\left[\frac{P_c-P_{ct}-P_{rt}-P_{trb}}{P_s}+1\right]$	$M_p=\frac{P_w}{2D}\left[\overline{Y}^2+\left(D-\overline{Y}\right)^2\right]$ $+\left[P_{rt}d_{rt}+P_{rb}d_{rb}+P_t d_t+P_c d_c\right]$
	II — PNA in top flange	$P_r+P_w+P_t \ge P_{rb}+P_{rt}$ $\overline{Y}=\left(\frac{t_t}{2}\right)\left[\frac{P_{rb}+P_c-P_w-P_{rb}}{P_t}+1\right]$	$M_p=\frac{P_t}{2t_t}\left[\overline{Y}^2+\left(t_t-\overline{Y}\right)^2\right]$ $+\left[P_{rt}d_{rt}+P_{rb}d_{rb}+P_t d_t+P_c d_c\right]$

P_{rt} $= F_{yrt}A_{rt}$; $P_s = 0.85f'_c b_s t_s$; $P_{rb} = F_{yrb}A_{rb}$

P_c $= F_{yc}b_c t_t$; $P_w = F_{yw}Dt_w$; $P_t = F_{yt}b_t t_t$

A_{rb}, A_{rt} = reinforcement area of bottom and top layer in concrete deck slab

F_{yrb}, F_{yrt} = yield strength of reinforcement of bottom and top layers

b_c, b_t, b_s = width of compression, tension steel flange, and concrete deck slab

t_c, t_t, t_w, t_s = thickness of compression, tension steel flange, web, and concrete deck slab

F_{yt}, F_{yc}, F_{yw} = yield strength of tension flange, compression flange, and web

Source: American Association of State Highway and Transportation Officials, AASHTO LRFD Bridge Design Specifications, Washington, D.C., 1994. With permission.

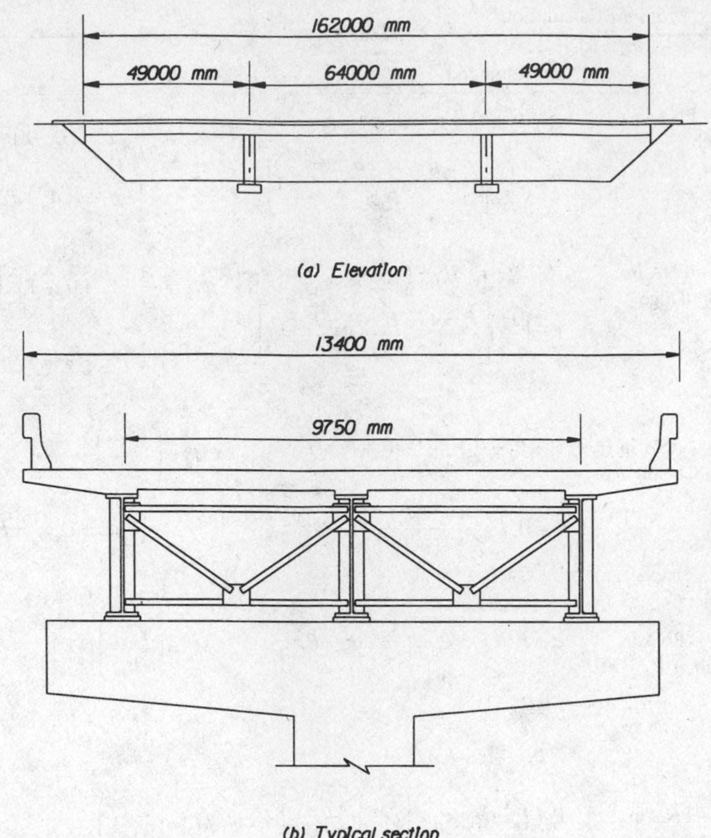

FIGURE 12.8 Three-spans continuous plate-girder bridge.

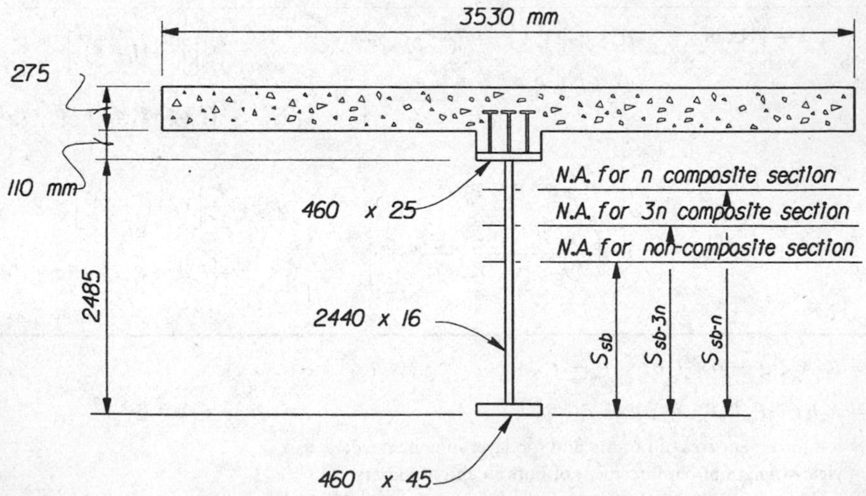

FIGURE 12.9 Cross section for positive flexure region.

Maximum positive moments in Span 1 due to factored loads applied to the steel section, and to the long-term composite section are M_{D1} = 6859 kN-m and M_{D2} = 2224 kN-m, respectively.

Requirement

Determine yield moment M_y, plastic moment M_p, and nominal moment M_n of an interior girder for positive flexure region.

Solutions

1. **Determine Effective Flange Width (AASHTO Article 4.6.2.6)**
For an interior girder, the effective flange width is

$$b_{eff} = \text{the lesser of} \begin{cases} \dfrac{L_{eff}}{4} = \dfrac{35{,}050}{4} = 8763 \text{ mm} \\ 12t_s + \dfrac{b_f}{2} = (12)(275) + \dfrac{460}{2} = 3530 \text{ mm} \quad \text{(controls)} \\ S = 4875 \text{ mm} \end{cases}$$

where L_{eff} is the effective span length and may be taken as the actual span length for simply supported spans and the distance between points of permanent load inflection for continuous spans (35.05 m); b_f is top flange width of steel girder.

2. **Calculate Elastic Composite Section Properties**
For the section in the positive flexure region as shown in Figure 12.9, its elastic section properties for the noncomposite, the short-term composite ($n = 8$), and the long-term composite ($3n = 24$) are calculated in Tables 12.5 to 12.7.

TABLE 12.5 Noncomposite Section Properties for Positive Flexure Region

Component	A (mm²)	y_i (mm)	$A_i y_i$ (mm³)	$y_i - y_{sb}$ (mm)	$A_i(y_i - y_{sb})^2$ (mm⁴)	I_o (in⁴)
Top flange 460 × 25	11,500	2498	28.7 (10)⁶	1395	22.4 (10)⁹	1.2 (10)⁶
Web 2440 × 16	39,040	1265	49.4 (10)⁶	162	3.0 (10)⁸	19.4 (10)⁹
Bottom flange 460 × 45	20,700	22.5	4.7 (10)⁵	−1081	24.2 (10)⁹	3.5 (10)⁶
Σ	71,240	—	78.6 (10)⁶		46.8 (10)⁹	19.4 (10)⁹

$$y_{sb} = \frac{\sum A_i y_i}{\sum A_i} = \frac{78.6(10)^6}{71240} = 1103 \text{ mm} \qquad y_{st} = (45 + 2440 + 25) - 1103 = 1407 \text{ mm}$$

$$I_{girder} = \sum I_o + \sum A_i (y_i - y_{sb})^2$$
$$= 19.4(10)^9 + 46.8(10)^9 = 66.2(10)^9 \text{ mm}^4$$

$$S_{sb} = \frac{I_{girder}}{y_{sb}} = \frac{66.2(10)^9}{1103} = 60.0(10)^6 \text{ mm}^3 \qquad S_{st} = \frac{I_{girder}}{y_{st}} = \frac{66.2(10)^9}{1407} = 47.1(10)^6 \text{ mm}^3$$

3. **Calculate Yield Moment M_y**
The yield moment M_y corresponds to the first yielding of either steel flange. It is obtained by the following formula:

$$M_y = M_{D1} + M_{D2} + M_{AD}$$

TABLE 12.6 Short-Term Composite Section Properties ($n = 8$)

Component	A (mm²)	y_i (mm)	$A_i y_i$ (mm³)	$y_i - y_{sb\text{-}n}$ (mm)	$A_i(y_i - y_{sb-n})^2$ (mm⁴)	I_o (mm⁴)
Steel section	71,240	1103	78.6 (10)⁶	−1027	75.1 (10)⁹	19.4 (10)⁹
Concrete slab 3530/8 × 275	121,344	2733	3.3 (10)⁸	603	44.1 (10)⁹	2.3 (10)⁸
Σ	192,584	—	4.1 (10)⁸	—	119.2 (10)⁹	19.6 (10)⁹

$$y_{sb-n} = \frac{\sum A_i y_i}{\sum A_i} = \frac{4.1(10)^8}{192{,}584} = 2130 \text{ mm} \qquad y_{st-n} = (45+2440+25)-2130 = 380 \text{ mm}$$

$$I_{com-n} = \sum I_o + \sum A_i (y_i - y_{sb-n})^2$$

$$= 19.6(10)^9 + 119.2(10)^9 = 138.8(10)^9 \text{ mm}^4$$

$$S_{sb-n} = \frac{I_{con-n}}{y_{sb-n}} = \frac{138.8(10)^9}{2130} = 65.2(10)^6 \text{ mm}^3 \qquad S_{st-n} = \frac{I_{com-n}}{y_{st-n}} = \frac{138.8(10)^9}{380} = 365.0(10)^6 \text{ mm}^3$$

TABLE 12.7 Long-Term Composite Section Properties ($3n = 24$)

Component	A (mm²)	y_i (mm)	$A_i y_i$ (mm³)	$y_i - y_{sb\text{-}3n}$ (mm)	$A_i(y_i - y_{sb-3n})^2$ (mm⁴)	I_o (mm⁴)
Steel section	71,240	1103	78.6 (10⁶)	–590	24.8 (10⁹)	19.4 (10⁹)
Concrete slab 3530/24 × 275	40,448	2733	1.1 (10⁸)	1040	43.7 (10⁹)	2.3 (10⁸)
Σ	111,688	—	10846.4	—	68.5 (10⁹)	19.6 (10⁹)

$$y_{sb-3n} = \frac{\sum A_i y_i}{\sum A_i} = \frac{88.1(10)^9}{111{,}688} = 1693 \text{ mm} \qquad y_{st-3n} = (45+2440+25)-1693 = 817 \text{ mm}$$

$$I_{com-3n} = \sum I_o + \sum A_i (y_i - y_{sb-3n})^2$$

$$= 19.6(10)^9 + 68.5(10)^9 = 88.1(10)^9 \text{ mm}^4$$

$$S_{sb-3n} = \frac{I_{con-3n}}{y_{sb-3n}} = \frac{88.1(10)^9}{1693} = 52.0(10)^6 \text{ mm}^3 \qquad S_{st-3n} = \frac{I_{com-3n}}{y_{st-3n}} = \frac{88.4(10)^9}{817} = 107.9(10)^6 \text{ mm}^3$$

$$M_{AD} = S_n \left(F_y - \frac{M_{D1}}{S_s} - \frac{M_{D2}}{S_{3n}} \right)$$

$$M_{D1} = 6859 \text{ kN-m}$$

$$M_{D2} = 2224 \text{ kN-m}$$

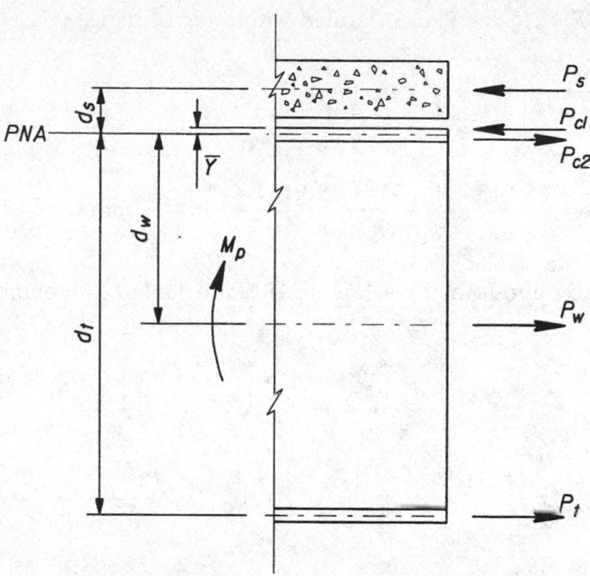

FIGURE 12.10 Plastic moment state.

For the top flange:

$$M_{AD} = (368.4)10^{-3}\left(345(10)^3 - \frac{6859}{47.1(10)^{-3}} - \frac{2224}{108.6(10)^{-3}}\right)$$

$$= 65{,}905 \text{ kN-m}$$

For the bottom flange:

$$M_{AD} = (65.2)10^{-3}\left(345(10)^3 - \frac{6859}{60.0(10)^{-3}} - \frac{2224}{52.1(10)^{-3}}\right)$$

$$= 12{,}257 \text{ kN-m} \quad \text{(controls)}$$

$$\therefore \quad M_y = 6859 + 2224 + 12{,}257 = 21{,}340 \text{ kN-m}$$

4. **Calculate Plastic Moment Capacity M_p**
For clarification, the reinforcement in slab is neglected. We first determine the location of the PNA (see Figure 12.10 and Table 12.4).

$$P_s = 0.85 f'_c b_{\text{eff}} t_s = 0.85(28)(3530)(275) = 23{,}104 \text{ kN}$$

$$P_{c1} = \bar{Y} b_{fc} F_{yc}$$

$$P_{c2} = A_c F_{yc} - P_{c1} = (t_c - \bar{Y}) b_{fc} F_{yc}$$

$$P_c = P_{c1} + P_{c2} = A_{fc} F_{yc} = (460)(25)(345) = 3967 \text{ kN}$$

$$P_w = A_w F_{yw} = (2440)(16)(345) = 13{,}469 \text{ kN}$$

$$P_t = A_{ft} F_{yt} = (460)(45)(345) = 7141 \text{ kN}$$

Since $P_t + P_w + P_c > P_s$, the PNA is located within top of flange (Case II, Table 12.4).

$$\bar{Y} = \frac{t_c}{2}\left(\frac{P_w + P_t - P_s}{P_c} + 1\right)$$

$$= \frac{25}{2}\left(\frac{13,469 + 7141 - 23,104}{3967} + 1\right) = 4.6 \text{ mm} < t_c = 25 \text{ mm}$$

Summing all forces about the PNA (Figure 12.5 and Table 12.4), obtain

$$M_p = \sum M_{\text{PNA}} = P_{c1}\left(\frac{y_{\text{PNA}}}{2}\right) + P_{c2}\left(\frac{t_{fc} - y_{\text{PNA}}}{2}\right) + P_s d_s + P_w d_w + P_t d_t$$

$$= \frac{P_c}{2t_c}\left[\bar{Y}^2 + \left(t_c - \bar{Y}\right)^2\right] + P_s d_s + P_w d_w + P_t d_t$$

$$d_s = \frac{275}{2} + 110 - 25 + 4.0 = 227 \text{ mm}$$

$$d_w = \frac{2440}{2} + 25 - 4.6 = 1240 \text{ mm}$$

$$d_t = \frac{45}{2} + 2440 + 25 - 4.6 = 2483 \text{ mm}$$

$$M_p = \frac{3967}{2(0.025)}\left[0.0046^2 + (0.025 - 0.004)^2\right] + (23,206)(0.227)$$

$$+ (13,469)(1.24) + (7141)(2.483)$$

$$= 39,737 \text{ kN-m}$$

5. **Calculate Nominal Moment**
   a. *Check compactness of steel girder section:*
      - Web slenderness requirement (Table 12.2)

$$\frac{2D_{cp}}{t_w} \leq 3.76\sqrt{\frac{E}{F_{yc}}}$$

      Since the PNA is within the top flange, D_{cp} is equal to zero. The web slenderness requirement is satisfied.
      - It is usually assumed that the top flange is adequately braced by the hardened concrete deck; there is, therefore, no requirements for the compression flange slenderness and bracing for compact composite sections at the strength limit state.

      ∴ The section is a compact composite section.

   b. Check ductility requirement (Table 12.2) $D_p/D' \leq 5$:

The purpose of this requirement is to prevent permanent crashing of the concrete slab when the composite section approaches its plastic moment capacity.

D_p = the depth from the top of the concrete deck to the PNA

$$D_p = 275 + 110 - 25 + 4.6 = 364.6 \text{ mm}$$

$$D' = \beta\left(\frac{d + t_s + t_h}{7.5}\right) = 0.7\left(\frac{2485 + 275 + 110}{7.5}\right) = 267.9 \text{ mm}$$

$$\frac{D_p}{D'} = \frac{364.6}{267.9} = 1.36 < 5 \qquad \text{OK}$$

c. Check moment of inertia ratio limit (AASHTO Article 6.10.1.1):

The flexural members shall meet the following requirement:

$$0.1 \leq \frac{I_{yc}}{I_y} \leq 0.9$$

where I_{yc} and I_y are the moments of inertia of the compression flange and steel girder about the vertical axis in the plane of web, respectively. This limit ensures that the lateral torsional bucking formulas are valid.

$$I_{yc} = \frac{(25)(460)^3}{12} = 2.03(10^8)\ \text{mm}^4$$

$$I_y = 2.03(10^8) + \frac{(2440)(16)^3}{12} + \frac{(45)(460)^3}{12} = 5.69(10^8)\ \text{mm}^4$$

$$0.1 < \frac{I_{yc}}{I_y} = \frac{2.03(10^8)}{5.69(10^8)} = 0.36 < 0.9 \qquad \text{OK}$$

d. *Nominal flexure resistance* M_n (Table 12.2):

Assume that the adjacent interior pier section is noncompact. For continuous spans with the noncompact interior support section, the nominal flexure resistance of a compact composite section is taken as

$$M_n = 1.3\ R_h M_y \leq M_p$$

with flange stress reduction factor $R_h = 1.0$ for this homogenous girder, we obtain

$$M_n = 1.3(1.0)(21{,}340) = 27{,}742\ \text{kN-m} < M_p = 39{,}712\ \text{kN-m}$$

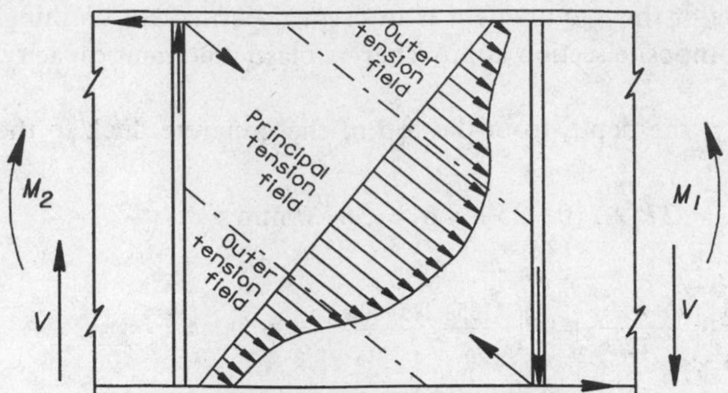

FIGURE 12.11 Tension field action.

12.5 Shear Design

12.5.1 Basic Concept

Similar to the flexural resistance, web shear capacity is also dependent on the slenderness ratio λ in term of width-to-thickness ratio (h/t_w). In calculating shear strength, three failure modes are considered: shear yielding when $\lambda \le \lambda_p$, inelastic shear buckling when $\lambda_p < \lambda < \lambda_r$, and elastic shear buckling when $\lambda > \lambda_r$. For the web without transverse stiffeners, shear resistance is contributed by the beam action of shearing yield or elastic shear buckling. For interior web panels with transverse stiffeners, shear resistance is contributed by both beam action (the first term of the C_s *equation in Table 12.8) and tension field action* (the second term of the C_s equation in Table 12.8). For end web panels, tension field action cannot be developed because of the discontinuous boundary and the lack of an anchor. It is noted that transverse stiffeners provide a significant inelastic shear buckling strength by tension field action as shown in Figure 12.11. Table 12.8 lists the AASHTO-LRFD [4,5] design formulas for shear strength.

Example 12.2: Shear Strength Design — Strength Limit State I

Given

For the I-girder bridge shown in Example 12.1, factored shear V_u = 2026 and 1495 kN are obtained at the left end of Span 1 and 6.1 m from the left end in Span 1, respectively. Design shear strength for the Strength Limit State I for those two locations.

Solutions

1. **Nominal Shear Resistance V_n**
   a. V_n *for an unstiffened web* (Table 12.8 or AASHTO Article 6.10.7.2):
   For D = 2440 mm and t_w = 16 mm, we have

$$\because \quad \frac{D}{t_w} = \frac{2440}{16} = 152.5 > 3.07\sqrt{\frac{E}{F_{yw}}} = 3.07\sqrt{\frac{200{,}000}{345}} = 73.9$$

$$\therefore V_n = \frac{4.55 t_w^3 E}{D} = \frac{4.55\,(16)^3(200{,}000)}{2440} = (1528)10^3 \text{ N} = 1528 \text{ kN}$$

TABLE 12.8 AASHTO-LRFD Design Formulas of Nominal Shear Resistance at Strength Limit State

Unstiffened homogeneous webs	$V_n = \begin{cases} V_p = 0.58F_{yw}Dt_w & \text{for } \frac{D}{t_w} \le 2.46\sqrt{\frac{E}{F_{yw}}} \quad \text{– Shear yielding} \\ 1.48t_w^2\sqrt{EF_{yw}} & \text{for } 2.46\sqrt{\frac{E}{F_{yw}}} < \frac{D}{t_w} \le 3.07\sqrt{\frac{E}{F_{yw}}} \quad \text{– Inelastic buckling} \\ \frac{4.55t_w^3E}{D} & \text{for } \frac{D}{t_w} > 3.07\sqrt{\frac{E}{F_{yw}}} \quad \text{– Elastic buckling} \end{cases}$
Stiffened interior web panels of compact homogeneous sections	$V_n = \begin{cases} C_sV_p & \text{for } M_u \le 0.5\phi_f M_p \\ RC_sV_p \ge CV_p & \text{for } M_u > 0.5\phi_f M_p \end{cases}; \quad C_s = C + \frac{0.87(1-C)}{\sqrt{1+(d_o/D)^2}}$ $R = 0.6+0.4\left(\frac{M_r - M_u}{M_r - 0.75\phi_f M_y}\right) \le 1.0$
Stiffened interior web panels of noncompact homogeneous sections	$V_n = \begin{cases} C_sV_p & \text{for } f_u \le 0.5\phi_f F_y \\ RC_sV_p \ge CV_p & \text{for } f_u \le 0.5\phi_f F_y \end{cases} \quad R = 0.6+0.4\left(\frac{F_r - f_u}{F_r - 0.75\phi_f F_y}\right) \le 1.0$
End panels and hybrid sections	$V_n = CV_p$

d_o = stiffener spacing (mm)
D = web depth
F_r = factored flexural resistance of the compression flange (MPa)
f_u = factored maximum stress in the compression flange under consideration (MPa)
M_r = factored flexural resistance
M_u = factored maximum moment in the panel under consideration
ϕ_f = resistance factor for flexure = 1.0 for the strength limit state
C = ratio of the shear buckling stress to the shear yield strength

$$C = \begin{cases} 1.0 & \text{For } \frac{D}{t_w} \le 1.1\sqrt{\frac{Ek}{F_{yw}}} \\ \frac{1.1}{D/t_w}\sqrt{\frac{Ek}{F_{yw}}} & \text{For } 1.1\sqrt{\frac{Ek}{F_{yw}}} \le \frac{D}{t_w} \le 1.38\sqrt{\frac{Ek}{F_{yw}}} \\ \frac{1.52}{(D/t_w)^2}\sqrt{\frac{Ek}{F_{yw}}} & \text{For } \frac{D}{t_w} > 1.38\sqrt{\frac{Ek}{F_{yw}}} \end{cases} \tag{12.8}$$

$$k = 5 + \frac{5}{(d_o/D)^2}$$

b. *V_n for end-stiffened web panel* (Table 12.8 or AASHTO Article 6.10.7.3.3c):
Try the spacing of transverse stiffeners $d_o = 6100$ mm. In order to facilitate handling of web panel sections, the spacing of transverse stiffeners shall meet (AASHTO Article 6.10.7.3.2) the following requirement:

$$d_o \le D\left[\frac{260}{(D/t_w)}\right]^2$$

$$d_o = 6100 \text{ mm} < D\left[\frac{260}{(D/t_w)}\right]^2 = 2440\left[\frac{260}{2440/16}\right]^2 = 7090 \text{ mm} \qquad \text{OK}$$

Using formulas in Table 12.8, obtain

$$k = 5 + \frac{5}{(d_o/D)^2} = 5 + \frac{5}{(6100/2440)^2} = 5.80$$

$$\because \frac{D}{t_w} = 152.5 > 1.38\sqrt{\frac{Ek}{F_{yw}}} = 1.38\sqrt{\frac{200{,}000(5.8)}{345}} = 80$$

$$\because C = \frac{1.52}{(152.5)^2}\sqrt{\frac{200{,}000(5.80)}{345}} = 0.379$$

$$V_p = 0.58F_{yw}Dt_w = 0.58(345)(2440)(16) = 7812(10)^3 \text{ N} = 7812 \text{ kN}$$

$$V_n = CV_p = 0.379\,(7812) = 2960 \text{ kN}$$

2. **Strength Limit State I**
AASHTO-LRFD [4] requires that for Strength Limit State I

$$V_u \le \phi_v V_n$$

where ϕ_v is the shear resistance factor = 1.0.

a. *Left end of Span* 1:

$$\because V_u = 2026 \text{ kN} > \phi_v V_n \text{ (for unstiffened web)} = 1528 \text{ kN}$$

$$\therefore \text{ Stiffeners are needed to increase shear capacity}$$

$$\phi_v V_n = (1.0)\,2960 = 2960 \text{ kN} > V_u = 2026 \text{ kN} \qquad \text{OK}$$

b. *Location of the first intermediate stiffeners, 6.1 m from the left end in Span* 1:
Since $V_u = 1459$ kN is less than the shear capacity of the unstiffened web $\phi_v V_n = 1528$ kN the intermediate transverse stiffeners may be omitted after the first intermediate stiffeners.

TABLE 12.9 AASHTO-LRFD Design Formulas of Stiffeners

Location	Stiffener	Required Project Width and Area	Required Moment of Inertia
Compression flange	Longitudinal	$b_l \le 0.48\, t_s\sqrt{E / F_{yc}}$	$I_s \ge \begin{cases} 0.125k^3 & \text{for } n=1 \\ 0.07k^3n^4 & \text{for } n=2,3,4 \text{ or } 5 \end{cases}$ k see Table 12.5
	Transverse	Same size as longitudinal stiffener; at least one transverse stiffener on compression flange near the dead load contraflexure point	
Web	Longitudinal	$b_l \le 0.48 t_s\sqrt{E / F_{yc}}$	$I_l \ge Dt_w^3\left[2.4(d_o / D)^2 - 0.13\right]$ $r \ge 0.234 d_o\sqrt{F_{yc} / E}$
	Transverse intermediate	$50 + d / 30 \le b_t \le 0.48 t_s\sqrt{E / F_{ys}}$ $16t_p \ge b_t \ge 0.25 b_f$ $A_s \ge \left[0.15BDt_w \dfrac{(1-C))V_u}{V_r} - 18t_w^2\right]\dfrac{F_{yw}}{F_{ys}}$ $B = 1$ for stiffener pairs, 1.8 for single angle and 2.4 for single plate	$I_t \ge d_o t_w^3 J$ $J = 2.5(D_p / d_o)^2 - 2 \ge 0.5$
	Bearing	$b_t \le 0.48 t_p\sqrt{E / F_{ys}}$ $B_r = \phi_b A_{pn} F_{ys}$	Use effective section (AASHTO-LRFD 6.10.8.2.4) to design axial resistance

b_f = width of compression flange
t_f = thickness of compression flange
f_c = stress in compression flange due to the factored loading
F_{ys} = specified minimum yield strength of the stiffener
ϕ_b = resistance factor of bearing stiffeners = 1.0
A_{pn} = area of the projecting elements of the stiffener outside of the web-to-flange fillet welds, but not beyond the edge of the flange

12.5.2 Stiffeners

For built-up I-sections, the longitudinal stiffeners may be provided to increase bending resistance by preventing local buckling while transverse stiffeners are usually provided to increase shear resistance by the tension field action [10,11]. The following three types of stiffeners are usually used for I-sections:

- *Transverse Intermediate Stiffeners*: These work as anchors for the tension field force so that postbuckling shear resistance can be developed. It should be noted that elastic web shear buckling cannot be prevented by transverse stiffeners. Transverse stiffeners are designed to (1) meet the slenderness requirement of projecting elements to present local buckling, (2) provide stiffness to allow the web to develop its postbuckling capacity, and (3) have strength to resist the vertical components of the diagonal stresses in the web. These requirements are listed in Table 12.9.
- *Bearing Stiffeners*: These work as compression members to support vertical concentrated loads by bearing on the ends of stiffeners (see Figure 12.2). They are transverse stiffeners and connect to the web to provide a vertical boundary for anchoring shear force from tension

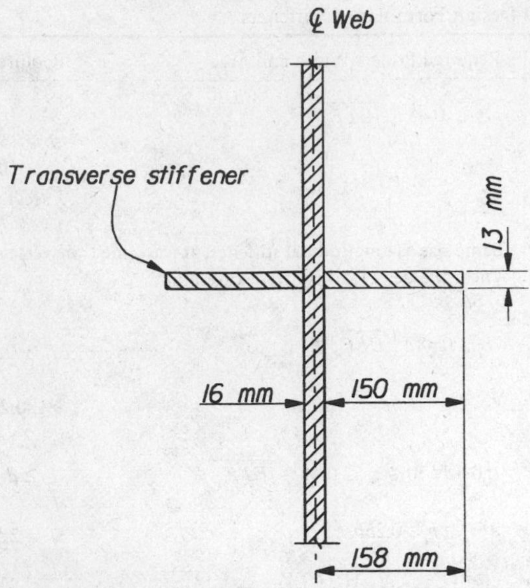

FIGURE 12.12 Cross section of web and transverse stiffener.

field action. They should be placed at all bearing locations and at all locations supporting concentrated loads. For rolled beams, bearing stiffeners may not be needed when factored shear is less than 75% of factored shear capacity. They are designed to satisfy the slenderness, bearing, and axial compression requirements as shown in Table 12.9.

- *Longitudinal Stiffeners*: These work as restraining boundaries for compression elements so that inelastic flexural buckling stress can be developed in a web. It consists of either a plate welded longitudinally to one side of the web, or a bolted angle. It should be located a distance of $2D_c/5$ from the inner surface of the compression flange, where D_c is the depth of web in compression at the maximum moment section to provide optimum design. The slenderness and stiffness need to be considered for sizing the longitudinal stiffeners (Table 12.9).

Example 12.3: Transverse and Bearing Stiffeners Design

Given

For the I-girder bridge shown in Example 12.1, factored shear $V_u = 2026$ and 1495 kN are obtained at the left end of Span 1 and 6.1 m from the left end in Span 1, respectively. Design the first intermediate transverse stiffeners and the bearing stiffeners at the left support of Span 1 using $F_{ys} = 345$ MPa for stiffeners.

Solutions

1. **Intermediate Transverse Stiffener Design**

   Try two 150 × 13 mm transverse stiffener plates as shown in Figure 12.12 welded to both sides of the web.

   a. *Projecting width b_t requirements* (Table12.9 or AASHTO Article 6.10.8.1.2):

   To prevent local buckling of the transverse stiffeners, the width of each projecting stiffener shall satisfy these requirements listed in Table 12.9.

$$b_t = 150 \text{ mm} > \begin{cases} 50 + \dfrac{d}{30} = 50 + \dfrac{2510}{30} = 134 \text{ mm} \\ 0.25 b_f = 0.25(460) = 115 \text{ mm} \end{cases} \quad \text{OK}$$

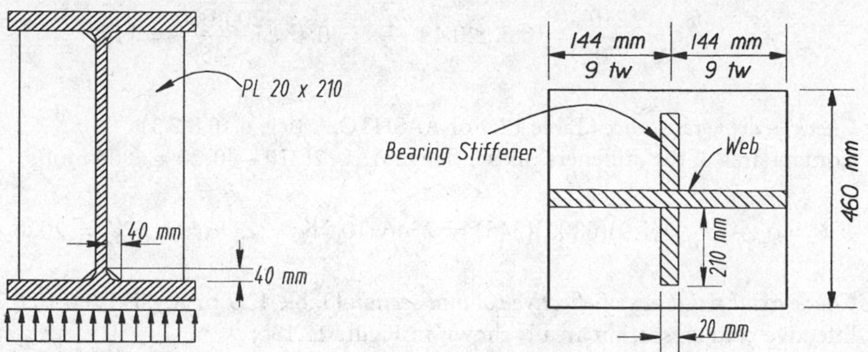

FIGURE 12.13 (a) Bearing stiffeners; (b) effective column area.

$$b_t = 150 \text{ mm} < \begin{cases} 0.48t_p\sqrt{\dfrac{E}{F_{ys}}} = 0.48(13)\sqrt{\dfrac{200{,}000}{345}} = 150 \text{ mm} \\ 16t_p = 16(13) = 208 \text{ mm} \end{cases} \quad \text{OK}$$

b. *Moment of inertia requirement* (Table 12.9 AASHTO Article 6.10.8.1.3):
The purpose of this requirement is to ensure sufficient rigidity of transverse stiffeners to develop adequately a tension field in the web.

$$\because J = 2.5\left(\frac{2440}{6100}\right)^2 - 2.0 = -1.6 < 0.5 \qquad \because \underline{\text{Use } J = 0.5}$$

$$I_t = 2\left(\frac{150^3(13)}{3}\right) = 29.3(10)^6 \text{ mm}^4 > d_o t_w^3 J = (6100)(16)^3(0.5) \quad \text{OK}$$

$$= 12.5(10)^6 \text{ mm}^4$$

c. *Area requirement* (Table 12.9 or AASHTO Article 6.10.8.1.4):
This requirement ensures that transverse stiffeners have sufficient area to resist the vertical component of the tension field, and is only applied to transverse stiffeners required to carry the forces imposed by tension field action. From Example 12.2, we have $C = 0.379$; $F_{yw} = 345$ MPa; $V_u = 1460$ kN; $\phi_v V_n = 1495$ kN; $t_w = 16$ mm; $B = 1.0$ for stiffener pairs. The requirement area is

$$A_{\text{sreqd}} = \left(0.15(1.0)(2440)(16)(1-0.379)\frac{1459}{1495} - 18(16)^2\right)\left(\frac{345}{345}\right) = -1060 \text{ mm}^2$$

The negative value of A_{sreqd} indicates that the web has sufficient area to resist the vertical component of the tension field.

2. **Bearing Stiffener Design**
Try two 20 × 210 mm stiffness plates welded to each side of the web as shown in Figure 12.13a.
a. *Check local buckling requirement* (Table 12.9 or AASHTO Article 6.19.8.2.2):

$$\frac{b_t}{t_p} = \frac{210}{20} = 10.5 \leq 0.48\sqrt{\frac{E}{F_y}} = 0.48\sqrt{\frac{200{,}000}{345}} = 11.6 \qquad \text{OK}$$

b. *Check bearing resistance* (Table 12.9 or AASHTO Article 6.10.8.2.3):
Contact area of the stiffeners on the flange $A_{pn} = 2(210 - 40)20 = 6800\ \text{mm}^2$

$$B_r = \phi_b\, A_{pn} F_{ys} = (1.0)(6800)(345) = 2346\,(10)^3\,\text{N} = 2346\ \text{kN} > V_u = 2026\ \text{kN} \quad \text{OK}$$

c. *Check axial resistance of effective column section* (Table 12.9 or AASHTO 6.10.8.2.4):
Effective column section area is shown in Figure 12.13b:

$$A_s = 2\left[210(20) + 9(16)(16)\right] = 13{,}008\ \text{mm}^2$$

$$I = \frac{(20)(420+16)^3}{12} = 138.14(10)^6\ \text{mm}^4$$

$$r_s = \sqrt{\frac{I}{A_s}} = \sqrt{\frac{138.14(10)^6}{13{,}008}} = 103.1\ \text{mm}$$

$$\lambda = \left(\frac{KL}{r_s\pi}\right)^2 \frac{F_y}{E} = \left(\frac{0.75(2440)}{103.1\pi}\right)^2 \frac{345}{200{,}000} = 0.055$$

$$P_n = 0.66^{\lambda} F_y A_s = 0.66^{0.055}(345)(13{,}008) = 4386\,(10)^3\ \text{N}$$

$$P_r = \phi_c P_n = 0.9(4386) = 3947\ \text{kN} > V_u = 2026\ \text{kN} \qquad \text{OK}$$

Therefore, using two 20 × 210 mm plates are adequate for bearing stiffeners at abutment.

12.5.3 Shear Connectors

To ensure a full composite action, shear connectors must be provided at the interface between the concrete slab and the structural steel to resist interface shear. Shear connectors are usually provided throughout the length of the bridge. If the longitudinal reinforcement in the deck slab is not considered in the composite section, shear connectors are not necessary in negative flexure regions. If the longitudinal reinforcement is included, either additional connectors can be placed in the region of dead load contraflexure points or they can be continued over the negative flexure region at maximum spacing. The two types of shear connectors such as shear studs and channels (see Figure 12.4) are most commonly used in modern bridges. The fatigue and strength limit states must be considered in the shear connector design. The detailed requirements are listed in Table 12.10.

Example 12.4: Shear Connector Design

Given
For the I-girder bridge shown in Example 12.1, design the shear stud connectors for the positive flexure region of Span 1. The shear force ranges V_{sr} are given in Table 12.11 and assume number of cycle $N = 7.844(10)^7$.

TABLE 12.10 AASHTO-LRFD Design Formulas of Shear Connectors

Connector Types	Stud	Channel
Basic requirement	$\frac{h_s}{d_s} \geq 4.0$ $6d_s <$ pitch of connector $p = (nZ_r I)/V_{sr}Q$ < 600 mm Transverse spacing $\geq 4d_s$ Clear distance between flange edge of nearest connector ≥ 25 mm Concrete cover over the top of the connectors ≥ 50 mm and $d_s \geq 50$ mm	Fillet welds along the heels and toe shall not smaller than 5 mm
Special requirement	For noncomposite negative flexure region, additional number of connector: $n_{ac} = (A_r f_{sr})/Z_r$	
Fatigue resistance	$Z_r = \alpha d_s^2 \geq 19 d_s^2$ $\alpha = 238 - 29.5 \log N$	—
Nominal shear resistance	$Q_n = 0.5 A_{sc} \sqrt{f'_c E_c} \leq A_{sc} F_u$	$Q_n = 0.3(t_f + 0.5t_w) L_c \sqrt{f'_c E_c}$
Required shear connectors	$n = \frac{V_h}{\phi_{sc} Q_n}$; $V_h =$ smaller $\begin{cases} 0.85 f'_c b_{eff} t_s \\ \sum A_{si} F_{yi} \end{cases}$ For continuous span between each adjacent zero moment of the centerline of interior support: $V_h = A_r F_{yr}$	

A_{si} = area of component of steel section
b_{eff} = effective flange width
h_s = height of stud
d_s = diameter of stud
n = number of shear connectors in a cross section
E_c = modulus of elasticity of concrete
f_c = stress in compression flange due to the factored loading
f'_c = specified compression strength of concrete
F_{yi} = specified minimum yield strength of the component of steel section
f_{sr} = stress range in longitudinal reinforcement (AASHTO-LRFD 5.5.3.1)
F_u = specified minimum tensile strength of a stud
L_c = length of channel shear connector
Q = first moment of transformed section about the neutral axis of the short-term composite section
I = moment of inertia of short-term composite section
N = number of cycles (AASHTO-LRFD 6.6.1.2.5)
V_{sr} = shear force range at the fatigue limit state
t_s = thickness of concrete slab
t_f = flange thickness of channel shear connector
Z_r = shear fatigue resistance of an individual shear connector

Solutions

1. **Stud Size (Table 12.10 AASHTO Article 6.10.7.4.1a)**
   Stud height should penetrate at least 50 mm into the deck. The clear cover depth of concrete cover over the top of the shear stud should not be less than 50 mm. Try

$$H_s = 180 \text{ mm} > 50 + (110 - 25) = 135 \text{ mm (min)} \qquad \text{OK}$$

$$\text{stud diameter } d_s = 25 \text{ mm} < H_s/4 = 45 \text{ mm} \qquad \text{OK}$$

TABLE 12.11 Shear Connector Design for the Positive Flexure Region in Span 1

Span	Location (x/L)	V_{sr} (kN)	$p_{required}$ (mm)	p_{final} (mm)	$n_{total\text{-}stud}$
	0.0	267.3	253	245	3
	0.1	229.5	295	272	63
	0.2	212.6	318	306	117
	0.3	205.6	329	326	165
1	0.4	203.3	333	326	**210**
	0.4	203.3	333	326	**144**
	0.5	202.3	334	326	99
	0.6	212.6	318	306	51
	0.7	223.7	302	306	3

Notes:

1. $V_{sr} = \left|+(V_{LL+IM})_u\right| + \left|-(V_{LL+IM})_u\right|$.
2. $p_{required} = \dfrac{n_s Z_r I_{com-n}}{V_{sr} Q} = \dfrac{67\,634}{V_{sr}}$.
3. $n_{total\text{-}stud}$ is summation of number of shear studs between the locations of the zero moment and that location.

2. Pitch of Shear Stud, *p*, for Fatigue Limit State

a. *Fatigue resistance Z_r* (Table 12.10 or AASHTO Article 6.10.7.4.2):

$$\alpha = 238 - 29.5\log(7.844\times 10^7) = 5.11$$

$$Z_r = 19d_s^2 = 19(25)^2 = 11{,}875 \text{ N}$$

b. *First moment Q and moment of initial I* (Table 12.6):

$$Q = \left(\frac{b_{\text{eff}}t_s}{8}\right)\left(y_{st-n} + t_h + \frac{t_s}{2}\right)$$

$$= \left(\frac{3530(275)}{8}\right)\left(380+85+\frac{275}{2}\right) = 73.11\ (10^6)\ \text{mm}^3$$

$$I_{com-n} = 138.8(10^9)\ \ \text{mm}^4$$

c. *Required pitch for the fatigue limit state*:
Assume that shear studs are spaced at 150 mm transversely across the top flange of steel section (Figure 12.9) and using $n_s = 3$ for this example and obtain

$$p_{\text{reqd}} = \frac{n_s Z_r I}{V_{sr} Q} = \frac{3(11.875)(138.8)(10)^9}{V_{sr}(73.11)(10)^6} = \frac{67634}{V_{sr}}$$

The detailed calculations for the positive flexure region of Span 1 are shown in Table 12.11.

3. Strength Limit State Check

a. *Nominal horizontal shear force* (AASHTO Article 6.10.7.4.4b):

$$V_h = \text{the lesser of} \begin{cases} 0.85\, f_c' b_{\text{eff}} t_s \\ F_{yw} D t_w + F_{yt} b_{ft} t_{ft} + F_{yc} b_{fc} t_{fc} \end{cases}$$

$$V_{h\text{-}concrete} = 0.5\, f_c' b_{\text{eff}} t_s = 0.85(28)(3530)(275) = 2.31(10)^6 \text{ N}$$

$$V_{h\text{-}steel} = F_{yw} D t_w + F_{yt} b_{ft} t_{ft} + F_{yc} b_{fc} t_{fc}$$

$$= 345[(2440)(16) + (460)(45) + (460)(25)] = 2.458(10)^6 \text{ N}$$

$$\therefore \quad V_h = 23\,100 \text{ kN}$$

b. *Nominal shear resistance* (Table 12.10 or AASHTO Article 6.10.7.4.4c):
Use specified minimum tensile strength $F_u = 420$ MPa for stud shear connectors

$$\because \; 0.5\sqrt{f_c' E_c} = 0.5\sqrt{28(25{,}000)} = 418.3 \text{ MPa} < F_u = 420 \text{ MPa}$$

$$\therefore \; Q_n = 0.5 A_{sc}\sqrt{f_c' E_c} = 418.3\left(\frac{\pi(25)^2}{4}\right) = 205\,332 \text{ N} = 205 \text{ kN}$$

c. *Check resulting number of shear stud connectors* (see Table 12.11):

$$n_{\text{total-stud}} = \begin{Bmatrix} 210 & \text{from left end } 0.4\,L_1 \\ 144 & \text{from } 0.4L_1 \text{ to } 0.7L_1 \end{Bmatrix} > \frac{V_h}{\phi_{sc} Q_n} = \frac{23\,100}{0.85(205)} = 133 \qquad \text{OK}$$

12.6 Other Design Considerations

12.6.1 Fatigue Resistance

The basic fatigue design requirement limits live-load stress range to fatigue resistance for each connection detail. Special attention should be paid to two types of fatigue: (1) load-induced fatigue for a repetitive net tensile stress at a connection details caused by moving truck and (2) distortion-induced fatigue for connecting plate details of cross frame or diaphragms to girder webs. See Chapter 53 for a detailed discussion.

12.6.2 Diaphragms and Cross Frames

Diaphragms and cross frames, as shown in Figure 12.14, are transverse components to transfer lateral loads such as wind or earthquake loads from the bottom of girder to the deck and from the deck to bearings, to provide lateral stability of a girder bridge, and to distribute vertical loads to the longitudinal main girders. Cross frames usually consist of angles or WT sections and act as a truss, while diaphragms use channels or I-sections as a flexural beam connector. End cross frames or diaphragms at piers and abutments are provided to transmit lateral wind loads and/or earthquake load to the bearings, and intermediate one are designed to provide lateral support to girders.

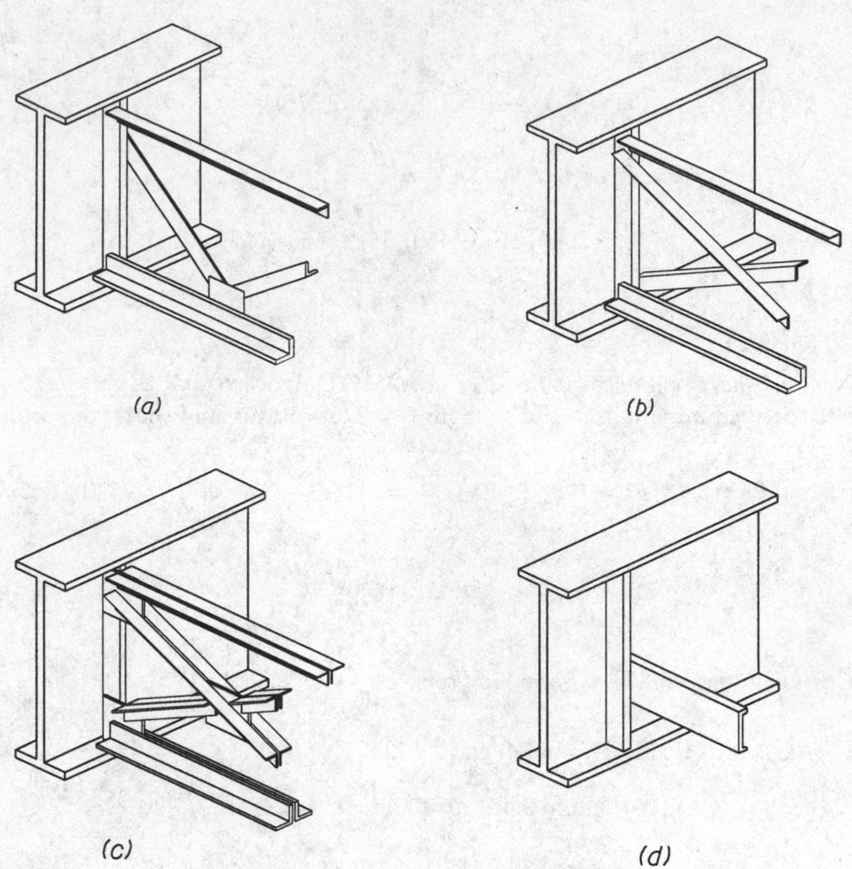

FIGURE 12.14 Cross frames and diaphragms.

The following general guidelines should be followed for diaphragms and cross frames:

- The diaphragm or cross frame shall be as deep as practicable to transfer lateral load and to provide lateral stability. For rolled beam, they shall be at least half of beam depth [AASHTO-LRFD 6.7.4.2].
- Member size is mainly designed to resist lateral wind loads and/or earthquake loads. A rational analysis is preferred to determine actual lateral forces.
- Spacing shall be compatible with the transverse stiffeners.
- Transverse connectors shall be as few as possible to avoid fatigue problems.
- Effective slenderness ratios (KL/r) for compression diagonal shall be less than 140 and for tension member (L/r) less than 240.

Example 12.5: Intermediate Cross-Frame Design

Given

For the I-girder bridge shown in Example 12.1, design the intermediate cross frame as for wind loads using single angles and M270 Grade 250 Steel.

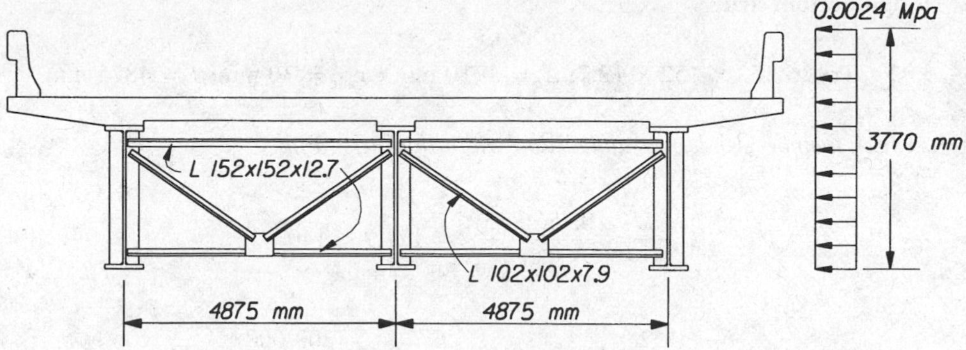

FIGURE 12.15 Wind load distribution.

Solutions:

1. **Calculate Wind Load**
   In this example, we assume that wind load acting on the upper half of girder, deck, and barrier is carried out by the deck slab and wind load on the lower half of girder is carried out by bottom flange. From AASHTO Table 3.8.1.2, wind pressure P_D= 0.0024 MPa, d= depth of structure member = 2,510 mm, and γ = load factor = 1.4 (AASHTO Table 3.4.1-1). The wind load on the structure (Figure 12. 15) is

$$W = 0.0024(3770) = 9.1 \text{ kN/m} > 4.4 \text{ kN/m}$$

Factored wind force acting on bottom flange:

$$W_{bf} = \frac{\gamma p_D d}{2} = \frac{1.4\ (0.0024)(2510)}{2} = 4.21\ \text{kN / m}$$

Wind force acting on top flange (neglecting concrete deck diaphragm):

$$W_{tf} = 1.4(0.0024)\left(3770 - \frac{2510}{2}\right) = 8.45\ \text{kN / m}$$

2. **Calculate forces acting on cross frame**
   For cross frame spacing:

$$L_b = 6.1\ \text{m}$$

Factored force acting on bottom strut:

$$F_{bf} = W_{bf} L_b = 4.21(6.1) = 25.68\ \text{kN}$$

Force acting on diagonals:

$$F_d = \frac{F_{tf}}{\cos\phi} = \frac{8.45(6.1)}{\cos 45^o} = 72.89\ \text{kN}$$

3. Design bottom strut

Try ∠ 152 × 152 × 12.7; A_s = 3710 mm²; $r_{\min}$ = 30 mm; L = 4875 mm

Check member slenderness and section width/thickness ratios:

$$\frac{KL}{r} = \frac{0.75(4875)}{30} = 121.9 < 140 \qquad \text{OK}$$

$$\frac{b}{t} = \frac{152}{12.7} = 11.97 < 0.45\sqrt{\frac{E}{F_y}} = 0.45\sqrt{\frac{200{,}000}{250}} = 12.8 \qquad \text{OK}$$

Check axial load capacity:

$$\lambda = \left(\frac{0.75(4875)}{30.0\pi}\right)^2 \frac{250}{200{,}000} = 1.88 < 2.25 \quad \text{(ASSHTO 6.9.4.1-1)}$$

$$P_n = 0.66^{\lambda} A_s F_y = 0.66^{1.88}(3710)(250) = 424{,}675 \text{ N} = 425 \text{ kN}$$

$$P_r = \phi_c P_n = 0.9(425) = 382.5 \text{ kN} > F_{bf} = 25.68 \text{ kN} \qquad \text{OK}$$

4. Design diagonals

Try ∠ 102 × 102 × 7.9, A_s = 1550 mm²; $r_{\min}$ = 20.1 mm; L = 3450 mm

Check member slenderness and section width/thickness ratios:

$$\frac{KL}{r} = \frac{0.75(3450)}{20.1} = 128.7 < 140 \qquad \text{OK}$$

$$\frac{b}{t} = \frac{102}{7.9} = 12.9 \approx 0.45\sqrt{\frac{E}{F_y}} = 0.45\sqrt{\frac{200{,}000}{250}} = 12.8 \qquad \text{OK}$$

Check axial load capacity:

$$\lambda = \left(\frac{0.75(3450)}{20.1\pi}\right)^2 \frac{250}{200{,}000} = 2.1 < 2.25 \quad \text{(ASSHTO 6.9.4.1-1)}$$

$$P_n = 0.66^{\lambda} A_s F_y = 0.66^{2.1}(1550)(250) = 161{,}925 \text{ N} = 162 \text{ kN}$$

$$P_r = \phi_c P_n = 0.9(162) = 145.8 \text{ kN} > F_d = 72.89 \text{ kN} \qquad \text{OK}$$

5. **Top strut**
   The wind force in the top strut is assumed zero because the diagonal will transfer the wind load directly into the deck slab. To provide lateral stability to the top flange during construction, we select angle $\angle$ 152 × 152 × 12.7 for top struts.

12.6.3 Lateral Bracing

The lateral bracing transfers wind loads to bearings and provides lateral stability to compression flange in a horizontal plan. All construction stages should be investigated for the need of lateral bracing. The lateral bracing should be placed as near the plane of the flange being braced as possible. Design of lateral bracing is similar to the cross frame.

12.6.4 Serviceability and Constructibility

The service limit state design is intended to control the permanent deflections, which would affect riding ability. AASHTO-LRFD [AASHTO-LRFD 6.10.3] requires that for Service II (see Chapter 5) load combination, flange stresses in positive and negative bending should meet the following requirements:

$$f_r = \begin{cases} 0.95\,R_h F_{yf} & \text{for both steel flanges of composite section} \\ 0.80\,R_h F_{yf} & \text{for both flanges of noncomposite section} \end{cases} \tag{12.9}$$

where R_h is a hybrid factor, 1.0 for homogeneous sections (AASHTO-LRFD 6.10.5.4), f_f is elastic flange stress caused by the factored loading, and F_{yf} is yield strength of the flange.

An I-girder bridge constructed in unshored conditions shall be investigated for strength and stability for all construction stages, using the appropriate strength load combination discussed in Chapter 5. All calculations should be based on the noncomposite steel section only.

Splice locations should be determined in compliance with both contructibility and structural integrity. The splices for main members should be designed at the strength limit state for not less than (AASHTO 10.13.1) the larger of the following:

- The average of flexural shear due to the factored loads at the splice point and the corresponding resistance of the member;
- 75% of factored resistance of the member.

References

1. Xanthakos, P. P., *Theory and Design of Bridges*, John Wiley & Sons, New York, 1994.
2. Barker, R. M. and Puckett, J. A., *Design of Highway Bridges*, John Wiley & Sons, New York, 1997.
3. Taly, N., *Design of Modern Highway Bridges*, WCB/McGraw-Hill, Burr Ridge, IL, 1997.
4. AASHTO, *AASHTO LRFD Bridge Design Specifications*, American Association of State Highway and Transportation Officials, Washington, D.C., 1994.
5. AASHTO, *AASHTO LRFD Bridge Design Specifications*, 1996 Interim Revisions, American Association of State Highway and Transportation Officials, Washington, D.C., 1996.
6. AISC., *Load and Resistance Factor Design Specification for Structural Steel Buildings*, 2nd ed., American Institute of Steel Construction, Chicago, IL, 1993.
7. AISC, *Manual of Steel Construction — Load and Resistance Factor Design*, 2nd ed., American Institute of Steel Construction, Chicago, IL, 1994.
8. Blodgett, O. W., *Design of Welded Structures*, James F. Lincoln Arc Welding Foundation, Cleveland, OH, 1966.

9. Galambos, T. V., Ed., *Guide to Stability Design Criteria for Metal Structures*, 5th ed., John Wiley & Sons, New York, 1998.
10. Basler, K., Strength of plates girder in shear, *J. Struct. Div., ASCE*, 87(ST7), 1961, 151.
11. Basler, K., Strength of plates girder under combined bending and shear, *J. Struct. Div., ASCE*, 87(ST7), 1961, 181.

13

Steel–Concrete Composite Box Girder Bridges

Yusuf Saleh
California Department of Transportation

Lian Duan
California Department of Transportation

13.1 Introduction

Box girders are used extensively in the construction of urban highway, horizontally curved, and long-span bridges. Box girders have higher flexural capacity and torsional rigidity, and the closed shape reduces the exposed surface, making them less susceptible to corrosion. Box girders also provide smooth, aesthetically pleasing structures.

There are two types of steel box girders: steel–concrete composite box girders (i.e., steel box composite with concrete deck) and steel box girders with orthotropic decks. Composite box girders are generally used in moderate- to medium-span (30 to 60 m) bridges, and steel box girders with orthotropic decks are often used for longer-span bridges.

This chapter will focus on straight steel–concrete composite box-girder bridges. Steel box girders with orthotropic deck and horizontally curved bridges are presented in Chapters 14 and 15.

13.2 Typical Sections

Composite box-girder bridges usually have single or multiple boxes as shown in Figure 13.1. A single cell box girder (Figure 13.1a) is easy to analyze and relies on torsional stiffness to carry eccentric loads. The required flexural stiffness is independent of the torsional stiffness. A single box girder with multiple cells (Figure 13.1b) is economical for very long spans. Multiple webs reduce the flange

0-8493-7434-0/00/$0.00+$.50

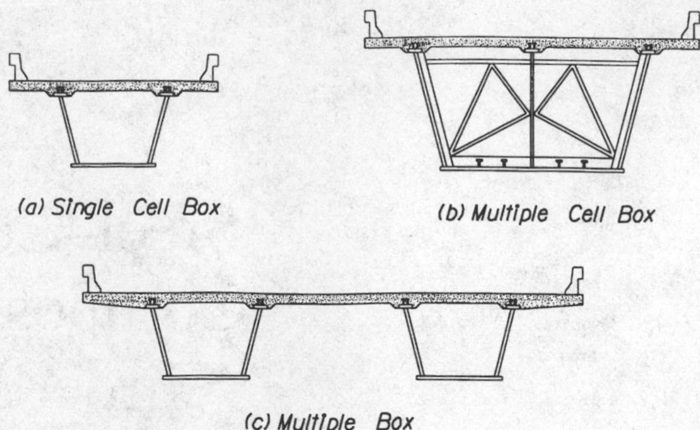

FIGURE 13.1 Typical cross sections of composite box girder.

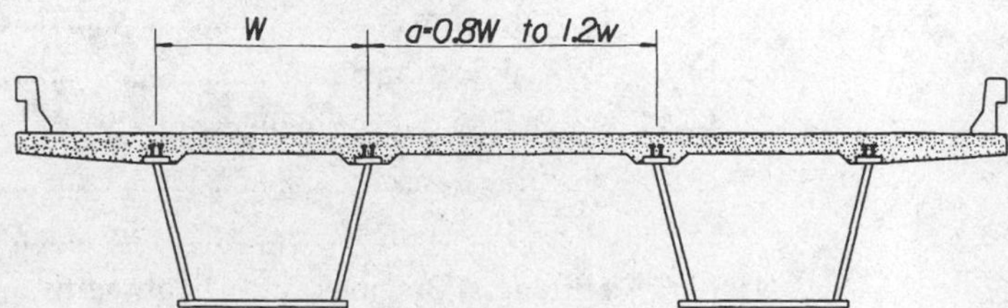

FIGURE 13.2 Flange distance limitation.

shear lag and also share the shear forces. The bottom flange creates more equal deformations and better load distribution between adjacent girders. The boxes in multiple box girders are relatively small and close together, making the flexural and torsional stiffness usually very high. The torsional stiffness of the individual boxes is generally less important than its relative flexural stiffness. For design of a multiple box section (Figure 13.1c), the limitations shown in Figure 13.2. should be satisfied when using the AASHTO-LRFD Specifications [1,2] since the AASHTO formulas were developed from these limitations. The use of fewer and bigger boxes in a given cross section results in greater efficiency in both design and construction [3].

A composite box section usually consists of two webs, a bottom flange, two top flanges and shear connectors welded to the top flange at the interface between concrete deck and the steel section (Figure 13.3). The top flange is commonly assumed to be adequately braced by the hardened concrete deck for the strength limit state, and is checked against local buckling before concrete deck hardening. The flange should be wide enough to provide adequate bearing for the concrete deck and to allow sufficient space for welding of shear connectors to the flange. The bottom flange is designed to resist bending. Since the bottom flange is usually wide, longitudinal stiffeners are often required in the negative bending regions. Web plates are designed primarily to carry shear forces and may be placed perpendicular or inclined to the bottom flange. The inclination of web plates should not exceed 1 to 4. The preliminary determination of top and bottom flange areas can be obtained from the equations (Table 13.1) developed by Heins and Hua [4] and Heins [6].

13.3 General Design Principles

A box-girder highway bridge should be designed to satisfy AASHTO-LRFD specifications to achieve the objectives of constructibility, safety, and serviceability. This section presents briefly basic design principles and guidelines. For more-detailed information, readers are encouraged to refer to several texts [6–14] on the topic.

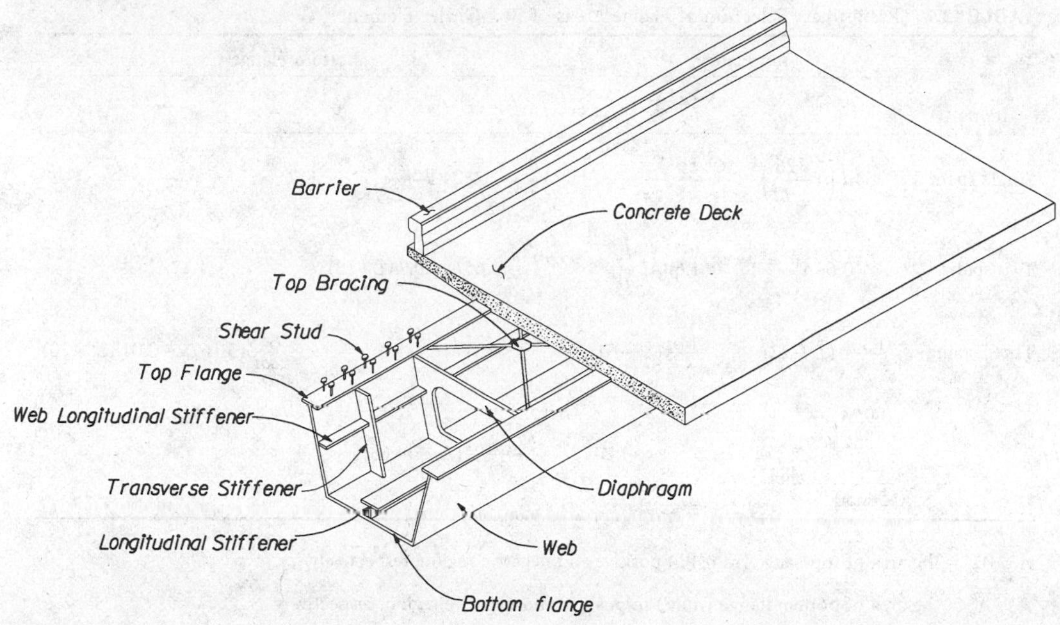

FIGURE 13.3 Typical components of a composite box girder.

In multiple box-girder design, primary consideration should be given to flexure. In single box-girder design, however, both torsion and flexure must be considered. Significant torsion on single box girders may occur during construction and under live loads. Warping stresses due to distortion should be considered for fatigue but may be ignored at the strength limit state. Torsional effects may be neglected when the rigid internal bracings and diaphragms are provided to maintain the box cross section geometry.

13.4 Flexural Resistance

The flexural resistance of a composite box girders depends on the compactness of the cross sectional elements. This is related to compression flange slenderness, lateral bracing, and web slenderness. A "compact" section can reach full plastic flexural capacity. A "noncompact" section can only reach yield at the outer fiber of one flange.

In positive flexure regions, a multiple box section is designed to be compact and a single box section is considered noncompact with the effects of torsion shear stress taken by the bottom flange (Table 13.2). In general, in box girders non-negative flexure regions design formulas of nominal flexure resistance are shown in Table 13.3.

In lieu of a more-refined analysis considering the shear lag phenomena [15] or the nonuniform distribution of bending stresses across wide flanges of a beam section, the concept of effective flange width under a uniform bending stress has been widely used for flanged section design [AASHTO-LRFD 4.6.2.6]. The effective flange width is a function of slab thickness and the effective span length.

13.5 Shear Resistance

For unstiffened webs, the nominal shear resistance V_n is based on shear yield or shear buckling depending on web slenderness. For stiffened interior web panels of homogeneous sections, the postbuckling resistance due to tension-field action [16,17] is considered. For hybrid sections, tension-field action is

TABLE 13.1 Preliminary Selection of Flange Areas of Box-Girder Element

	Top Flange		Bottom Flange	
Items	A_T^+	A_T^-	A_B^+	A_B^-
Single span	$254d\left(1-\frac{26}{L}\right)$	—	$328d\left(1-\frac{28}{L}\right)$	—
Two span	$0.64A_B^+$	$1.60A_B^+\frac{F_y^-}{F_y^+}$	$\frac{645}{k}(1.65L^2-0.74L+13)$	$1.17A_B^+\frac{F_y^-}{F_y^+}$
Three span	$\frac{330n}{k}(L_1-22)$	$\frac{814n}{k}(L_1-31)$	$\frac{423n}{k}(L_1-16)$	$\frac{645}{kn}(3.16L_2-0.018L_2^2-70)$
	$0.95A_T^- - \frac{A_T^{-2}}{58650}-\frac{3484}{k}$		$\frac{211.67n}{k}(L_2-14.63)$	

A_T^+, A_T^- = the area of top flange (mm²) in positive and negative region, respectively

A_B^+, A_B^- = the area of bottom flange (mm²) in positive and negative region, respectively

d = depth of girder (mm)

L, L_1, L_2 = length of the span (m); for simple span $(27 \le L \le 61)$

for two spans $(30 \le L_2 \le 67)$

for three spans $(27 \le L_1 \le 55)$

W_R = roadway width (m)

N_b = number of boxes

n = L_2/L_1

k = $\frac{N_B F_y d}{W_R(344{,}750)}$

F_y = yield strength of the material (MPa)

not permitted and shear yield or elastic shear buckling limits the strength. The detailed AASHTO-LRFD design formulas are shown in Table 12.8 (Chapter 12). For cases of inclined webs, the web depth D shall be measured along the slope and be designed for the projected shear along inclined web.

To ensure composite action, shear connectors should be provided at the interface between the concrete slab and the steel section. For single-span bridges, connectors should be provided throughout the span of the bridge. Although it is not necessary to provide shear connectors in negative flexure regions if the longitudinal reinforcement is not considered in a composite section, it is recommended that additional connectors be placed in the region of dead-load contraflexure points [AASHTO-LRFD 1.10.7.4]. The detailed requirements are listed in Table 12.10.

13.6 Stiffeners, Bracings, and Diaphragms

13.6.1 Stiffeners

Stiffeners consist of longitudinal, transverse, and bearing stiffeners as shown in Figure 13.1. They are used to prevent local buckling of plate elements, and to distribute and transfer concentrated loads. Detailed design formulas are listed in Table 12.9.

TABLE 13.2 AASHTO-LRFD Design Formulas of Nominal Flexural Resistance in Negative Flexure Ranges for Composite Box Girders (Strength Limit State)

Compression flange with longitudinal stiffeners	$$F_n = \begin{cases} R_b R_h F_{yc} & \text{for } \frac{w}{t} \le 0.57\sqrt{\frac{kE}{F_{yc}}} \\ 0.592 R_b R_h F_{yc}\left(1+0.687\sin\frac{c\pi}{2}\right) & \text{for } 0.57\sqrt{\frac{kE}{F_{yc}}} < \frac{w}{t} \le 1.23\sqrt{\frac{kE}{F_{yc}}} \\ 181\,000 R_b R_h k\left(\frac{t}{w}\right)^2 & \text{for } \frac{w}{t} > 1.23\sqrt{\frac{kE}{F_{yc}}} \end{cases}$$ $$c = \frac{1.23 - \frac{w}{t}\sqrt{\frac{F_{yc}}{kE}}}{0.66}$$ $$k = \text{buckling coefficient} = \begin{cases} \frac{8I_s}{wt^3} \le 4.0 & \text{for } n = 1 \\ \frac{14.3I_x}{wt^3 n^4} \le 4.0 & \text{For } n = 2,\ 3,\ 4 \text{ or } 5 \end{cases}$$
Compression flange without longitudinal stiffeners	Use above equations with the substitution of compression flange width between webs, b for w and buckling coefficient k taken as 4
Tension flange	$F_n = R_b R_h F_{yt}$

E = modulus of elasticity of steel
F_n = nominal stress at the flange
F_{yc} = specified minimum yield strength of the compression flange
F_{yt} = specified minimum yield strength of the tension flange
n = number of equally spaced longitudinal compression flange stiffeners
I_s = moment of inertia of a longitudinal stiffener about an axis parallel to the bottom flange and taken at the base of the stiffener
R_b = load shedding factor, R_b = 1.0 — if either a longitudinal stiffener is provided or $2D_c / t_w \le \lambda_b \sqrt{E / f_c}$ is satisfied
R_h = hybrid factor; for homogeneous section, R_h = 1.0, see AASHTO-LRFD (6.10.5.4)
t_h = thickness of concrete haunch above the steel top flange
t = thickness of compression flange
w = larger of width of compression flange between longitudinal stiffeners or the distance from a web to the nearest longitudinal stiffener

13.6.2 Top Lateral Bracings

Steel composite box girders (Figure 13.3) are usually built of three steel sides and a composite concrete deck. Before the hardening of the concrete deck, the top flanges may be subject to lateral torsion buckling. Top lateral bracing shall be designed to resist shear flow and flexure forces in the section prior to curing of concrete deck. The need for top lateral bracing shall be investigated to ensure that deformation of the box is adequately controlled during fabrication, erection, and placement of the concrete deck. The cross-bracing shown in Figure 13.3 is desirable. For 45° bracing, a minimum cross-sectional area (mm²) of bracing of 0.76× (box width, in mm) is required to ensure closed box action [11]. The slenderness ratio (L_b/r) of bracing members should be less than 140.

AASHTO-LRFD [1] requires that for straight box girders with spans less than about 45 m, at least one panel of horizontal bracing should be provided on each side of a lifting point; for spans greater than 45 m, a full-length lateral bracing system may be required.

13.6.3 Internal Diaphragms and Cross Frames

Internal diaphragms or cross frames (Figure 13.1) are usually provided at the end of a span and interior supports within the spans. Internal diaphragms not only provide warping restraint to the box girder, but improve distribution of live loads, depending on their axial stiffness which prevents distortion. Because rigid and widely spaced diaphragms may introduce undesirable large local forces, it is generally good practice to provide a large number of diaphragms with less stiffness than a few very rigid diaphragms. A recent study [18] showed that using only two intermediate diaphragms per span results in 18% redistribution of live-load stresses and additional diaphragms do not significantly improve the live-load redistribution. Inverted K-bracing provides better inspection access than X-bracing. Diaphragms shall be designed to resist wind loads, to brace compression flanges, and to distribute vertical dead and live loads [AASHTO-LRFD 6.7.4].

For straight box girders, the required cross-sectional area of a lateral bracing diagonal member A_b (mm^2) should be less than 0.76× (width of bottom flange, in mm) and the slenderness ratio (L_b/r) of the member should be less than 140.

For horizontally curved boxes per lane and radial piers under HS-20 loading, Eq. (13.1) provides diaphragm spacing L_d, which limits normal distortional stresses to about 10% of the bending stress [19]:

$$L_d = \sqrt{\frac{R}{200L - 7500}} \leq 25 \tag{13.1}$$

where R is bridge radius, ft, and L is simple span length, ft.

To provide the relative distortional resistance per millimeter greater than 40 [13], the required area of cross bracing is as

$$A_b = 750\left[\frac{L_{ds}a}{h}\right]\left[\frac{t^3}{h+a}\right] \tag{13.2}$$

where t is the larger of flange and web thickness; L_{ds} is the diaphragm spacing; h is the box height, and a is the top width of box.

13.7 Other Considerations

13.7.1 Fatigue and Fracture

For steel structures under repeated live loads, fatigue and fracture limit states should be satisfied in accordance with AASHTO 6.6.1. A comprehensive discussion on the issue is presented in Chapter 53.

13.7.2 Torsion

Figure 13.4 shows a single box girder under the combined forces of bending and torsion. For a closed or an open box girder with top lateral bracing, torsional warping stresses are negligible. Research indicates that the parameter ψ determined by Eq. (13.3) provides limits for consideration of different types of torsional stresses.

$$\psi = L\sqrt{GJ / EC_w} \tag{13.3}$$

where G is shear modulus, J is torsional constant, and C_w is warping constant.

For straight box girder (ψ is less than 0.4), pure torsion may be omitted and warping stresses must be considered; when ψ is greater than 10, it is warping stresses that may be omitted and pure

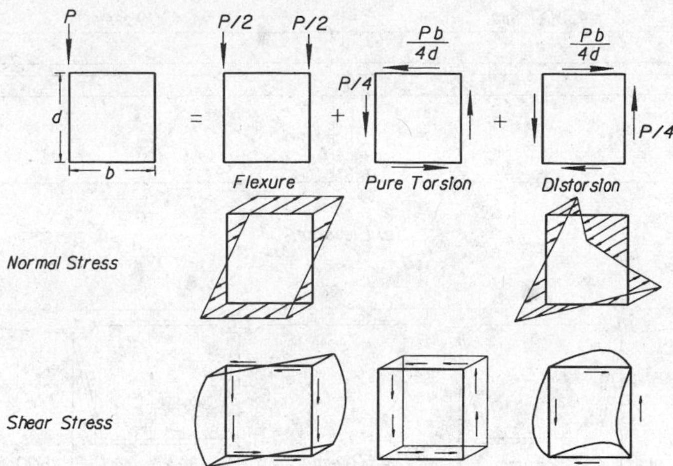

FIGURE 13.4 A box section under eccentric loads.

torsion that must be considered. For a curved box girder, ψ must take the following values if torsional warping is to be neglected:

$$\psi \geq \begin{cases} 10 + 40\theta & \text{for } 0 \leq \theta \leq 0.5 \\ 30 & \text{for } \theta > 0.5 \end{cases} \tag{13.4}$$

where θ is subtended angle (radius) between radial piers.

13.7.3 Constructibility

Box-girder bridges should be checked for strength and stability during various construction stages. It is important to note that the top flange of open-box sections shall be considered braced at locations where internal cross frames or top lateral bracing are attached. Member splices may be needed during construction. At the strength limit state, the splices in main members should be designed for not less than the larger of the following:

- The average of the flexure moment, the shear, or axial force due to the factored loading and corresponding factored resistance of member, and
- 75% of the various factored resistance of the member.

13.7.4 Serviceability

To prevent permanent deflections due to traffic loads, AASHTO-LRFD requires that at positive regions of flange flexure stresses (f_f) at the service limit state shall not exceed $0.95R_hF_{yf}$.

13.8 Design Example

Two-Span Continuous Box-Girder bridge

Given

A two-span continuous composite box-girder bridge that has two equal spans of 45 m. The superstructure is 13.2 m wide. The elevation and a typical cross section are shown in Figure 13.5.

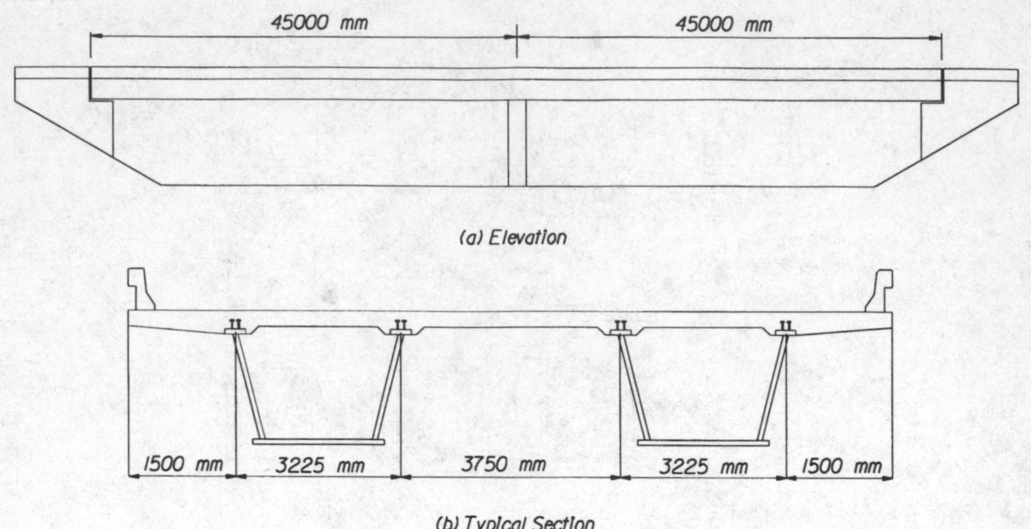

FIGURE 13.5 Two-span continuous box-girder bridge.

Structural steel: AASHTO M270M, Grade 345W (ASTM A709 Grade 345W) uncoated weathering steel with F_y = 345 MPa
Concrete: f'_c = 30.0 MPa; E_c = 22,400 MPa; modular ratio n = 8
Loads: Dead load = self weight + barrier rail + future wearing 75 mm AC overlay
Live load = AASHTO Design Vehicular Load + dynamic load allowance
Single-lane average daily truck traffic ADTT in one direction = 3600
Deck: Concrete slabs deck with thickness of 200 mm
Specification: AASHTO-LRFD [1] and 1996 Interim Revision (referred to as AASHTO)
Requirements: Design a box girder for flexure, shear for Strength Limit State I, and check fatigue requirement for web.

Solution

1. **Calculate Loads**
   a. *Component dead load — DC for a box girder*:
   The component dead-load DC includes all structural dead loads with the exception of the future wearing surface and specified utility loads. For design purposes, assume that all dead load is distributed equally to each girder by the tributary area. The tributary width for the box girder is 6.60 m.
   - *DC*1: acting on noncomposite section
     Concrete slab = (6.6)(0.2)(2400)(9.81) = 31.1 kN/m
     Haunch = 3.5 kN/m
     Girder (steel-box), cross frame, diaphragm, and stiffener = 9.8 kN/m
   - *DC*2: acting on the long term composite section
     Weight of each barrier rail = 5.7 kN/m

   b. *Wearing surface load — DW*:
   A future wearing surface of 75 mm is assumed to be distributed equally to each girder
   - *DW*: acting on the long-term composite section = 10.6 kN/m

2. **Calculate Live-Load Distribution Factors**
   a. *Live-load distribution factors for strength limit state* [AASHTO Table 4.6.2.2.2b-1]:

$$LD_m = 0.05 + 0.85\frac{N_L}{N_b} + \frac{0.425}{N_L} = 0.05 + 0.85\frac{3}{2} + \frac{0.425}{3} = 1.5 \text{ lanes}$$

b. *Live-load distribution factors for fatigue limit state:*

$$LD_m = 0.05 + 0.85\frac{N_L}{N_b} + \frac{0.425}{N_L} = 0.05 + 0.85\frac{1}{2} + \frac{0.425}{1} = 0.9 \text{ lanes}$$

3. **Calculate Unfactored Moments and Shear Demands**
The unfactored moment and shear demand envelopes are shown in Figures 13.8 to 13.11. Moment, shear demands for the Strength Limit State I and Fatigue Limit State are listed in Table 13.3 to 13.5.

TABLE 13.3 Moment Envelopes for Strength Limit State I

Span	Location	M_{DC1} (kN-m);	M_{DC2} (kN-m);	M_{DW} (kN-m);	M_{LL+IM} (kN m)		M_u (kN-m)	
	(x/L)	Dead Load-1	Dead Load-2	Wearing Surface	Positive	Negative	Positive	Negative
	0.0	0	0	0	0	0	0	0
	0.1	3,058	372	681	3338	−442	**10,592**	4,307
	0.2	5,174	629	1152	5708	−883	**18,023**	7,064
	0.3	6,350	772	1414	7174	−1326	**22,400**	8,268
	0.4	6,585	801	1466	7822	−1770	**23,864**	7,917
1	0.5	5,880	715	1309	7685	−2212	**22,473**	6,018
	0.6	4,234	515	943	6849	−2653	**18,369**	2,571
	0.7	1,647	200	367	5308	−3120	**11,540**	−2,472
	0.8	1,882	−229	−419	3170	−3822	2,160	**−9,457**
	0.9	−6,350	−772	−1414	565	−4928	−9,533	**−18,745**
	1.0	11,760	−1430	−2618	−1727	−7640	−22,264	**−32,095**

Notes:
1. Live load distribution factor $LD = 1.467$.
2. Dynamic load allowance $IM = 33\%$.
3. $M_u = 0.95\ [1.25(M_{DC1} + M_{DC2}) + 1.5\ M_{DW} + 1.75\ M_{LL+IM}]$.

TABLE 13.4 Shear Envelopes for Strength Limit State I

Span	Location	V_{DC1} (kN);	V_{DC2} (kN);	V_{DW} (kN);	V_{LL+IM} (kN)		V_u (kN)	
	(x/L)	Dead Load-1	Dead Load-2	Wearing Surface	Positive	Negative	Positive	Negative
	0.0	784	95	87	877	−38	**2626**	1104
	0.1	575	70	64	782	−44	**2158**	784
	0.2	366	44	41	711	−58	**1727**	449
	0.3	157	19	18	601	−91	**1233**	83
	0.4	−53	6	−6	482	−138	724	**−307**
1	0.5	−262	−32	−29	360	−230	208	**−773**
	0.6	−471	−57	−52	292	−354	−216	**−1290**
	0.7	−680	−83	−76	219	−482	−648	**−1815**
	0.8	−889	−108	−99	145	−612	−1083	**−2342**
	0.9	−1098	−133	−122	67	−750	−1524	**−2882**
	1.0	−1307	−159	−145	22	−966	−1910	**−3553**

Notes:
1. Live load distribution factor $LD = 1.467$.
2. Dynamic load allowance $IM = 33\%$.
3. $V_u = 0.95\ [1.25(V_{DC1} + V_{DC2}) + 1.5V_{DW} + 1.75V_{LL+IM}]$.

TABLE 13.5 Moment and Shear Envelopes for Fatigue Limit State

Span	Location (x/L)	M_{LL+IM} (kN-m) Positive	M_{LL+IM} (kN-m) Negative	V_{LL+IM} (kN) Positive	V_{LL+IM} (kN) Negative	$(M_{LL+IM})_u$ (kN-m) Positive	$(M_{LL+IM})_u$ (kN-m) Negative	$(V_{LL+IM})u$ (kN) Positive	$(V_{LL+IM})u$ (kN) Negative
	0.0	0	0	286	–31	0	0	214	–23
	0.1	1102	–137	245	–31	827	–102	184	–23
	0.2	1846	–274	205	–57	1385	–206	154	–38
	0.3	2312	–412	167	–79	1734	–309	125	–59
	0.4	2467	–550	130	–115	1851	–412	98	–86
1	0.5	2405	–687	97	–153	1804	–515	73	–115
	0.6	2182	–824	67	–190	1636	–618	50	–143
	0.7	1716	–962	45	–226	1287	–721	33	–169
	0.8	1062	–1099	25	–257	1796	–824	19	–193
	0.9	414	–1237	9	–286	311	–928	7	–215
	1.0	0	–1373	0	–309	0	–1030	0	–232

Notes:
1. Live load distribution factor $LD = 0.900$.
2. Dynamic load allowance $IM = 15\%$.
3. $(M_{LL+IM})_u = 0.75(M_{LL+IM})_u$ and $(V_{LL+IM})_u = 0.75(V_{LL+IM})_u$.

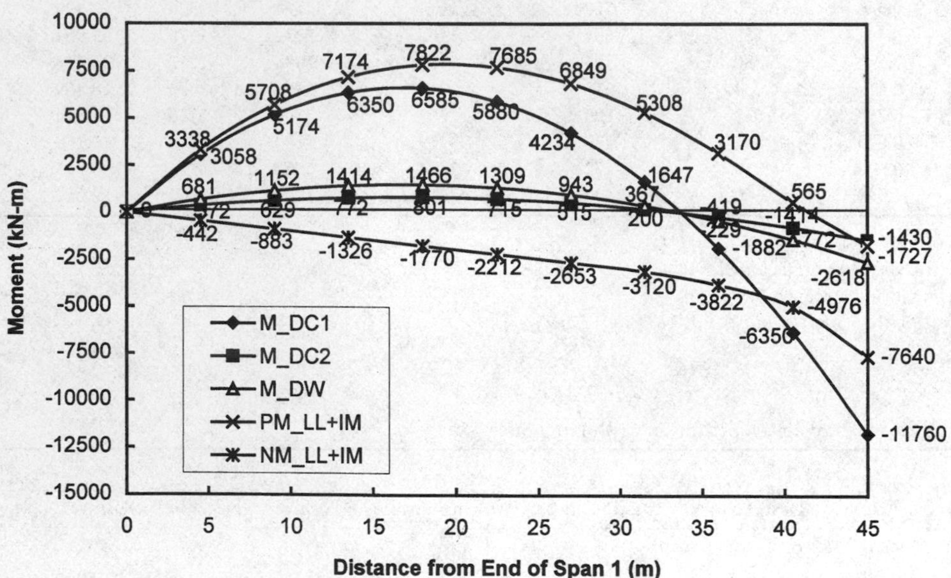

FIGURE 13.6 Unfactored moment envelopes.

4. Determine Load Factors for Strength Limit State I and Fracture Limit State

Load factors and load combinations

The load factors and combinations are specified as [AASHTO Table 3.4.1-1]:

Strength Limit State I: $1.25(DC1 + DC2) + 1.5(DW) + 1.75(LL + IM)$

Fatigue Limit State: $0.75(LL + IM)$

a. *General design equation* [AASHTO Article 1.3.2]:

$$\eta \sum \gamma_i Q_i \le \phi R_n$$

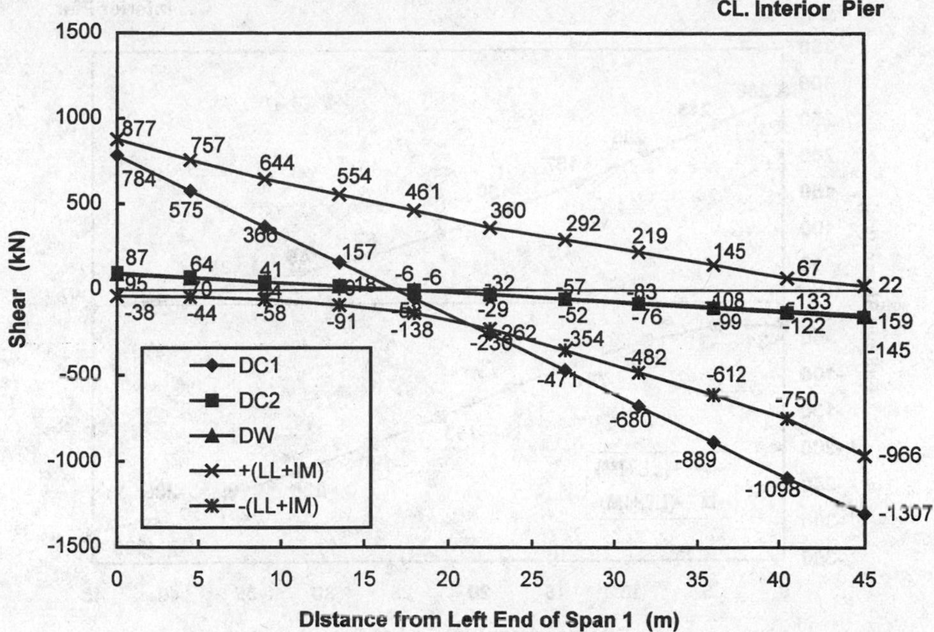

FIGURE 13.7 Unfactored shear envelopes.

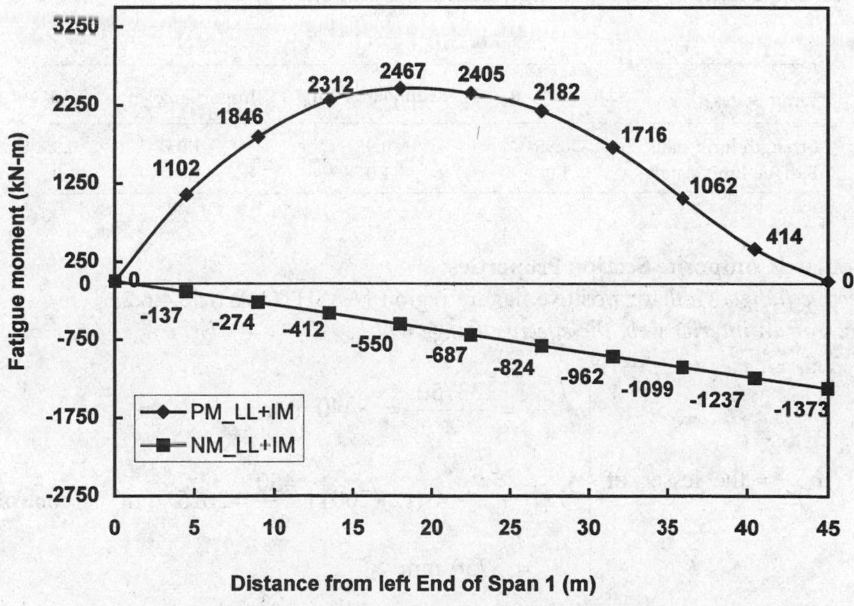

FIGURE 13.8 Unfactored fatigue load moment.

where γ_i is load factor and ϕ resistance factor; Q_i represents force effects or demands; R_n is the nominal resistance; η is a factor related ductility η_D , redundancy η_R , and operational importance η_I of the bridge (see Chapter 5) designed and is defined as:

$$\eta = \eta_D \eta_R \eta_I \geq 0.95$$

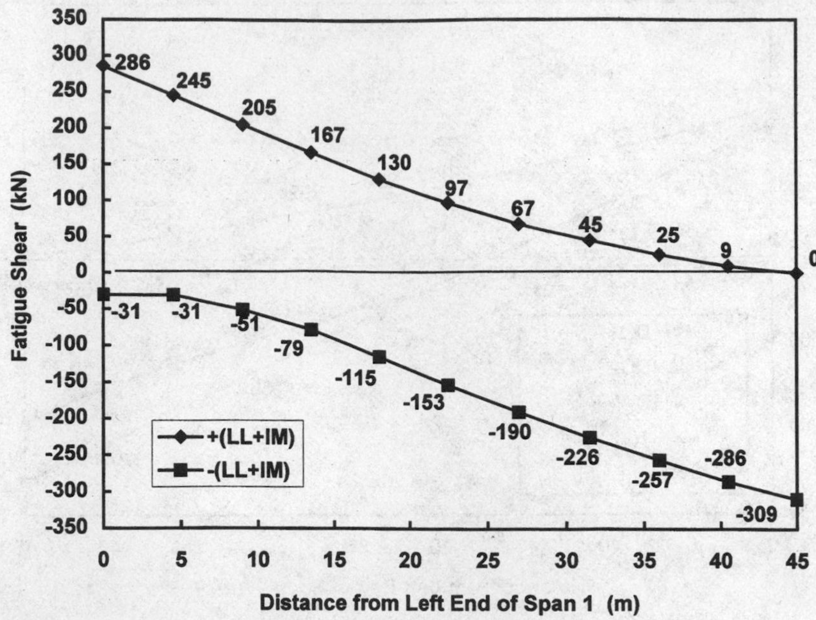

FIGURE 13.9 Unfactored fatigue load shear.

For this example, the following values are assumed:

Limit States	Ductility η_D	Redundancy η_R	Importance η_I	η
Strength limit state	0.95	0.95	1.05	0.95
Fatigue limit state	1.0	1.0	1.0	1.0

5. **Calculate Composite Section Properties:**

Effective flange width for positive flexure region [AASHTO Article 4.6.2.6]

a. For an interior web, the effective flange width:

$$b_{\text{eff}} = \text{the lesser of} \begin{cases} \dfrac{L_{\text{eff}}}{4} = \dfrac{33750}{4} = 8440 \text{ mm} \\ 12t_s + \dfrac{b_f}{2} = (12)(200) + \dfrac{450}{2} = 2625 \text{ mm} \quad \text{(controls)} \\ S = 3750 \text{ mm} \end{cases}$$

b. For an exterior web, the effective flange width:

$$b_{\text{eff}} = \text{the lesser of} \begin{cases} \dfrac{L_{\text{eff}}}{8} = \dfrac{33750}{8} = 4220 \text{ mm} \\ 6t_s + \dfrac{b_f}{4} = (6)(200) + \dfrac{450}{4} = 1310 \text{ mm} \quad \text{(controls)} \\ \textit{The width of the overhang} = 1500 \text{ mm} \end{cases}$$

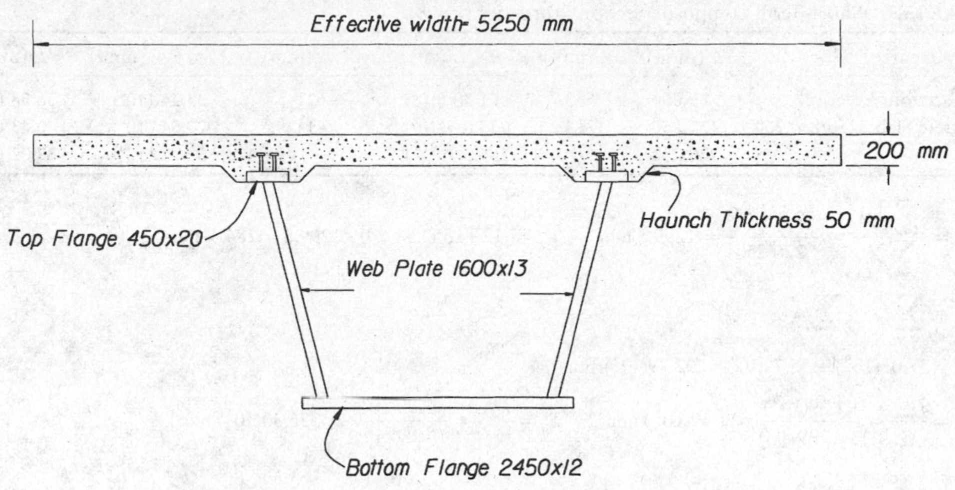

FIGURE 13.10 Typical section for positive flexure region.

$$\textit{Total effective flange width for the box girder} = 1310 + \frac{2625}{2} + 2625 = 5250 \text{ mm}$$

where L_{eff} is the effective span length and may be taken as the actual span length for simply supported spans and the distance between points of permanent load inflection for continuous spans; b_f is top flange width of steel girder.

Elastic composite section properties for positive flexure region:
For a typical section (Figure 13.10) in positive flexure region of Span 1, its elastic section properties for the noncomposite, the short-term composite ($n = 8$), and the long-term composite ($3n = 24$) are calculated in Tables 13.6 to 13.8.

TABLE 13.6 Noncomposite Section Properties for Positive Flexure Region

Component	A (mm²)	y_i (mm)	A_iy_i (mm³)	$y_i - y_{sb}$ (mm)	$A_i(y_i - y_{sb})^2$ (mm⁴)	I_o (mm⁴)
2 top flange 450 × 20	18,000	1574.2	28.34 (10⁶)	885	141 (10⁹)	0.60 (10⁶)
2 web 1600 × 13	41,600	788.1	32.79 (10⁶)	99	0.41 (10⁹)	8.35 (10⁹)
Bottom flange 2450 × 12	29,400	6.0	0.17 (10⁶)	−683	13.70 (10⁹)	0.35 (10⁶)
Σ	89,000	—	61.30 (10⁶)	—	28.23 (10⁹)	8.35 (10⁹)

$$y_{sb} = \frac{\sum A_i y_i}{\sum A_i} = \frac{61.30(10^6)}{89,000} = 688.7 \text{ mm} \qquad y_{st} = (12 + 1552.5 + 20) - 688.7 = 895.5 \text{ mm}$$

$$I_{girder} = \sum I_o + \sum A_i (y_i - y_{sb})^2$$

$$= 8.35(10^9) + 28.23(10^9) = 36.58(10^9) \text{ mm}^4$$

$$S_{sb} = \frac{I_{girder}}{y_{sb}} = \frac{36.58(10^9)}{688.7} = 53.11(10^6)\text{mm}^3 \qquad S_{st} = \frac{I_{girder}}{y_{st}} = \frac{36.58(10^9)}{895.5} = 40.85(10^6)\text{ mm}^3$$

Effective flange width for negative flexure region:
The effective width is computed according to AASHTO 4.6.2.6 (calculations are similar to Step 5a) The total effective of flange width for the negative flexure region is 5450 mm.

TABLE 13.7 Short-Term Composite Section Properties ($n = 8$)

Component	A (mm^2)	y_i (mm)	A_iy_i (mm^3)	$y_i - y_{sb}$ (mm)	$A_i(y_i - y_{sb})^2$ (mm^4)	I_o (mm^4)
Steel section	89,000	688.7	61.30 (10^6)	–611	33.24 (10^9)	36.58 (10^9)
Concrete Slab 5250/8 × 200	131,250	1714.2	225.0 (10^6)	414	22.54 (10^9)	0.43 (10^9)
Σ	220,250	—	386.3 (10^6)	—	55.77 (10^9)	37.02 (10^9)

$$y_{sb} = \frac{\sum A_i y_i}{\sum A_i} = \frac{92.79(10^6)}{220\ 250} = 1299.8 \text{ mm} \qquad y_{st} = (12+1552.5+20) - 1299.8 = 284.4 \text{ mm}$$

$$I_{com} = \sum I_o + \sum A_i (y_i - y_{sb})^2$$

$$= 37.02(10^9) + 55.77(10^9) = 92.79(10^9) \text{ mm}^4$$

$$S_{sb} = \frac{I_{com}}{y_{sb}} = \frac{92.79(10^9)}{1299.8} = 71.39(10^6) \text{ mm}^3 \qquad S_{st} = \frac{I_{com}}{y_{st}} = \frac{92.79(10^9)}{284.4} = 326.30(10^6) \text{ mm}^3$$

TABLE 13.8 Long-Term Composite Section Properties ($3n = 24$)

Component	A (mm^2)	y_i (mm)	A_iy_i (mm^3)	$y_i - y_{sb}$ (mm)	$A_i(y_i - y_{sb})^2$ (mm^4)	I_o (mm^4)
Steel section	89,000	688.4	61.3 (10^6)	–338	10.2 (10^9)	36.58 (10^9)
Concrete slab 5250/24 × 200	43,750	1714.2	75.0 (10^6)	688	20.7 (10^9)	5.40 (10^6)
Σ	132,750	—	136.0 (10^6)	—	30.85 (10^9)	36.59 (10^9)

$$y_{sb} = \frac{\sum A_i y_i}{\sum A_i} = \frac{136.0(10^6)}{132\ 750} = 1026.7 \text{ mm} \qquad y_{st} = (12+1552.5+20) - 1026.7 = 557.5 \text{ mm}$$

$$I_{com} = \sum I_o + \sum A_i (y_i - y_{sb})^2$$

$$= 36.59(10^9) + 136.0(10^9) = 67.43(10^9) \text{ mm}^4$$

$$S_{sb} = \frac{I_{com}}{y_{sb}} = \frac{67.43(10^9)}{1026.7} = 65.68(10^6) \text{ mm}^3 \qquad S_{st} = \frac{I_{com}}{y_{st}} = \frac{67.43(10^9)}{557.5} = 121.0(10^6) \text{ mm}^3$$

Elastic composite section properties for negative flexure region:

AASHTO (6.10.1.2) requires that for any continuous span the total cross-sectional area of longitudinal reinforcement must not be less than 1% of the total cross-sectional area of the slab. The required reinforcement must be placed in two layers uniformly distributed across the slab width and two thirds must be placed in the top layer. The spacing of the individual bar should not exceed 150 mm in each row.

$$A_{s\text{ reg}} = 0.01(200) = 2.00 \text{ mm}^2/\text{mm}$$

$$A_{s\text{ top-layer}} = \frac{2}{3}(0.01)(200) = 1.33 \text{ mm}^2/\text{mm} \ (\#16 \text{ at } 125 \text{ mm} = 1.59 \text{ mm}^2/\text{mm})$$

$$A_{s\text{ bot-layer}} = \frac{1}{3}(0.01)(200) = 0.67 \text{ mm}^2/\text{mm} \ (\text{alternate } \#10 \text{ and } \#13 \text{ at } 125 \text{ mm} = 0.80 \text{ mm}^2/\text{mm})$$

Figure 13.11 shows a typical section for the negative flexure region. The elastic properties for the noncomposite and the long-term composite ($3n = 24$) are calculated and shown in Tables 13.9 and 13.10.

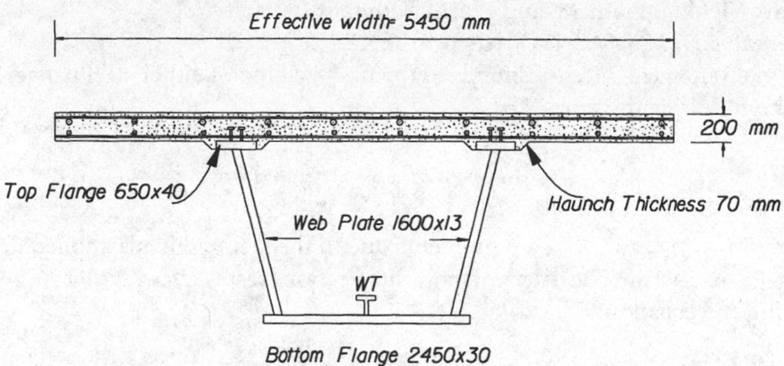

FIGURE 13.11 Typical section for negative flexure region.

TABLE 13.9 Noncomposite Section Properties for Negative Flexure Region

Component	A (mm²)	y_i (mm)	$A_i y_i$ (mm³)	$y_i - y_{sb}$ (mm)	$A_i(y_i - y_{sb})^2$ (mm⁴)	I_o (mm⁴)
2 Top flange 650 × 40	52,000	1602	83.32 (10⁶)	911	43.20 (10⁹)	6.93 (10⁶)
2 Web 1600 × 13	41,600	806	33.53 (10⁶)	115	0.55 (10⁹)	8.35 (10⁹)
Stiffener WT	5,400	224.3	1.21 (10⁶)	–466	1.18 (10⁹)	37.63 (10⁶)
Bottom flange 2450 × 30	73,500	15	1.10 (10⁶)	–676	33.57 (10⁹)	5.51 (10⁶)
Σ	172,500	—	119.2 (10⁶)		78.49 (10⁹)	8.40 (10⁹)

$$y_{sb} = \frac{\sum A_i y_i}{\sum A_i} = \frac{119.2(10^6)}{172500} = 690.8 \text{ mm} \qquad y_{st} = (30+1552.5+40) - 690.8 = 931.4 \text{ mm}$$

$$I_{\text{girder}} = \sum I_o + \sum A_i (y_i - y_{sb})^2$$

$$= 8.40(10^9) + 78.50(10^9) = 86.90(10^9) \text{ mm}^4$$

$$S_{sb} = \frac{I_{\text{girder}}}{y_{sb}} = \frac{86.90(10^9)}{690.8} = 125.8(10^6) \text{ mm}^3 \qquad S_{st} = \frac{I_{\text{girder}}}{y_{st}} = \frac{86.90(10^9)}{931.4} = 93.29(10^6) \text{ mm}^3$$

TABLE 13.10 Composite Section Properties for Negative Flexure Region

Component	A (mm²)	y_i (mm)	$A_i y_i$ (mm³)	$y_i - y_{sb}$ (mm)	$A_i(y_i - y_{sb})^2$ (mm⁴)	I_o (mm⁴)
Steel section	172 500	690.8	119.2 (10⁶)	73.2	0.92 (10⁹)	86.90 (10⁹)
Top reinforcement	8,665	1762.2	15.27 (10⁶)	998.2	8.63 (10⁹)	—
Bottom reinforcement	4,360	1677.2	7.31 (10⁹)	913.2	3.64 (10⁹)	—
Σ	185 525	—	141.7 (10⁶)	—	13.19 (10⁹)	86.90 (10⁹)

$$y_{sb} = \frac{\sum A_i y_i}{\sum A_i} = \frac{141.7(10^6)}{185\,525} = 764 \text{ mm} \qquad y_{st} = (30+1552.5+40) - 764 = 858.2 \text{ mm}$$

$$I_{com} = \sum I_o + \sum A_i (y_i - y_{sb})^2$$

$$= 86.90(10^9) + 13.19(10^9) = 100.09(10^9) \text{ mm}^4$$

$$S_{sb} = \frac{I_{com}}{y_{sb}} = \frac{100.09(10^9)}{764} = 131.00(10^6) \text{ mm}^3 \qquad S_{st} = \frac{I_{com}}{y_{st}} = \frac{100.09(10^9)}{858.2} = 116.63(10^6) \text{mm}^3$$

6. Calculate Yield Moment M_y and Plastic Moment Capacity M_p

a. *Yield moment M_y* [AASHTO Article 6.10.5.1.2]:
The yield moment M_y corresponds to the first yielding of either steel flange. It is obtained by the following formula

$$M_y = M_{D1} + M_{D2} + M_{AD}$$

where M_{D1}, M_{D2}, and M_{AD} are moments due to the factored loads applied to the steel, the long-term, and the short-term composite section, respectively. M_{AD} can be obtained by solving the equation:

$$F_y = \frac{M_{D1}}{S_s} + \frac{M_{D2}}{S_{3n}} + \frac{M_{AD}}{S_n}$$

$$M_{AD} = S_n\left(F_y - \frac{M_{D1}}{S_s} - \frac{M_{D2}}{S_{3n}}\right)$$

where S_s, S_n and S_{3n} are the section modulus for the noncomposite steel, the short-term, and the long-term composite sections, respectively.

$$M_{D1} = (0.95)(1.25)(M_{DC1}) = (0.95)(1.25)(6585) = 7820 \text{ kN-m}$$

$$M_{D2} = (0.95)(1.25M_{DC2} + 1.5M_{DW})$$

$$= (0.95)[1.25(801) + 1.5(1466)] = 3040 \text{ kN-m}$$

For the top flange:

$$M_{AD} = (329.3)10^{-3}\left((345)10^3 - \frac{7.820}{40.85(10)^{-3}} - \frac{3.040}{120(10)^{-3}}\right)$$

$$= 41.912(10)^3 \text{ kN - m}$$

For the bottom flange:

$$M_{AD} = (71.39)10^{-3}\left((345)10^3 - \frac{7.820}{53.11(10)^{-3}} - \frac{3.040}{64.68(10)^{-3}}\right)$$

$$= 10.814(10)^3 \text{ kN-m} \qquad \text{(control)}$$

$$M_y = 7820 + 3040 + 10814 = 21{,}674 \text{ kN-m}$$

b. *Plastic moment M_p* [AASHTO Article 6.1]:
The plastic moment M_p is determined using equilibrium equations. The reinforcement in the concrete slab is neglected in this example.

- Determine the location of the plastic neutral axis (PNA), $\overline{Y}$
From the Equation listed in Table 12.4 and Figure 12.7.

$$P_s = 0.85 f'_c b_{\text{eff}} t_s = 0.85(30)(5250)(200) = 26{,}775 \text{ kN}$$

$$P_c = A_{fc} F_{yc} = 2(450)(20)(345) = 6{,}210 \text{ kN}$$

$$P_w = A_w F_{yw} = 2\ (1600)(13)(345) = 14{,}352 \text{ kN}$$

$$P_t = A_{ft} F_{yt} = 2450(12)(345) = 10{,}143 \text{ kN}$$

$$\because P_t + P_w + P_c = 10{,}143 + 14{,}352 + 6{,}210 = 30\ 705 \text{ kN} > P_s = 26{,}755 \text{ kN}$$

$\therefore$ PNA is located within the top flange of steel girder and the distance from the top of compression flange to the PNA, $\bar{Y}$ is

$$\bar{Y} = \frac{t_{fc}}{2}\left(\frac{P_w + P_t - P_s}{P_c} + 1\right)$$

$$\bar{Y} = \frac{20}{2}\left(\frac{14{,}352 + 10{,}143 - 26{,}775}{6{,}210} + 1\right) = 6.3 \text{ mm}$$

- *Calculate M_p:*
  Summing all forces about the PNA, obtain:

$$M_p = \sum M_{\text{PNA}} = \frac{P_c}{2t_c}\left(\bar{Y}^2 + (t_c - \bar{Y})^2\right) + P_s d_s + P_w d_w + P_t d_t$$

where

$$d_s = \frac{200}{2} + 50 - 20 + 6.3 = 136.3 \text{ mm}$$

$$d_w = \frac{1552.5}{2} + 20 - 6.3 = 789.8 \text{ mm}$$

$$d_t = \frac{12}{2} + 1552.5 + 20 - 6.3 = 1571.9 \text{ mm}$$

$$M_p = \frac{6210}{2(20)}\left(6.3^2 + (20 - 6.3)^2\right) + (26{,}775)(136.3) + (14{,}352)(789.8) + (10{,}143)(1571.9)$$

$$M_p = 30{,}964 \text{ kN-m}$$

7. **Flexural Strength Design — Strength Limit State I:**
   a. *Positive flexure region*:
      - *Compactness of steel box girder*
        The compactness of a multiple steel boxes is controlled only by web slenderness. The purpose of the ductility requirement is to prevent permanent crushing of the concrete slab when the composite section approaches its plastic moment capacity. For this example, by referring to Figures 13.2 and 13.4, obtain:

$\frac{2D_{cp}}{t_w} \leq 3.76\sqrt{\frac{E}{f_c}}$, PNA is within the top flange $D_{cp} = 0$, the web slenderness requirement is satisfied

$D_p = 200 + 50 - 20 + 6.3 = 236.3$ mm

(depth from the top of concrete deck to the PNA)

$$D' = \beta\left[\frac{d + t_s + t_h}{7.5}\right] \quad \beta = 0.7 \text{ for } F_y = 345 \text{ MPa}$$

$$D' = 0.7\left(\frac{1552.5 + 12 + 200 + 50}{7.5}\right) = 169.3 \text{ mm}$$

$$\left(\frac{D_p}{D'}\right) = \left(\frac{236.3}{169.3}\right) = 1.4 \leq 5 \text{ OK}$$

- *Calculate nominal flexure resistance, M_n* (see Table 12.2)

$$1 < \left(\frac{D_p}{D'}\right) = 1.4 < 5$$

$$M_n = \frac{5M_p - 0.85M_y}{4} + \frac{0.85M_y - M_p}{4}\left(\frac{D_p}{D'}\right)$$

$$M_n = \frac{5(30{,}964) - 0.85(21{,}674)}{4} + \frac{0.85(21{,}674) - (30{,}964)}{4}(1.4)$$

$$M_n = 28{,}960 \text{ kN-m} \geq 1.3(1.0)(21674) = 28{,}176 \text{ kN-m}$$

$$\therefore \quad M_n = 28{,}176 \text{ kN-m}$$

From Table 13.3, the maximum factored positive moments in Span 1 occurred at the location of $0.4L_1$.

$$\eta\Sigma\gamma_i M_i \leq \phi_f M_n$$

$$23{,}864 \text{ kN-m} < 1.0\ (28{,}176) \text{ kN-m} \qquad \text{OK}$$

b. *Negative flexure region*:
For multiple and single box sections, the nominal flexure resistance should be designed to meet provision AASHTO 6.11.2.1.3a (see Table 13.2)

i. Stiffener requirement [AASHTO 6.11.2.1-1]:
Use one longitudinal stiffener (Figure 13.11), try WT 10.5 × 28.5.
The projecting width, b_ℓ of the stiffener should satisfy:

$$b_\ell \leq 0.48 t_p \sqrt{\frac{E}{F_{yc}}}$$

where

t_p = the thickness of stiffener (mm)

b_ℓ = the projected width (mm)

$$b_\ell = \frac{267}{2} = 133.5\,\text{mm}$$

$$I_s = 33.4\times10^6 + 4748(190.1)^2 = 20.5\times10^6\,\text{mm}$$

$$133.5 \le 0.48(16.5)\sqrt{\frac{2\times10^5}{345}} = 190\ \text{mm} \qquad \text{OK}$$

ii. Calculate buckling coefficient, k:

For $n - 1$

$$k = \left(\frac{8I_s}{Wt^3}\right)^{1/3} = \left(\frac{8(20.5\times10^6}{1225(24)^3}\right)^{1/3} = 2.13 < 4.0$$

iii. Calculate nominal flange stress (see Table 13.2):

$$0.57\sqrt{\frac{Ek}{F_{yc}}} = 0.57\sqrt{\frac{(2.13)(2)10^5}{345}} = 20.03$$

$$1.23\sqrt{\frac{Ek}{F_{yc}}} = 1.23\sqrt{\frac{(2.13)(2)10^5}{345}} = 43.22$$

$$20.03 < \frac{w}{t} = \frac{1225}{30} = 40.83 < 43.22$$

The nominal flexural resistance of compression flange is controlled by inelastic buckling:

$$F_{nc} = 0.592\ R_b R_h F_{yc}\left(1 + 0.687\sin\frac{c\pi}{2}\right)$$

$$c = \frac{1.23 - \frac{w}{t}\sqrt{\frac{F_{yc}}{kE}}}{0.66} = \frac{1.23 - 40.8\sqrt{\frac{345}{(3.9)2(10^5)}}}{0.66} = 0.56$$

Longitudinal stiffener is provided, $R_b = 1.0$, for homogenous plate girder $R_h = 1.0$:

$$F_{nc} = 0.592.(1.0)(1.0)(345)\left(1 + 0.687\sin\frac{(0.56)\pi}{2}\right) = 313.4\ \text{MPa}$$

For tension flange:

$$F_{nt} = R_b R_h R_{yt} = (1.0)(1.0)(345) = 345 \text{ MPa}$$

iv. Calculate M_{AD} at Interior Support

$$M_{D1} = (0.95)(1.25)(M_{DC1}) = (0.95)(1.25)(11760) = 13{,}965 \text{ kN-m}$$

$$M_{D2} = (0.95)(1.25M_{DC2} + 1.5M_{DW})$$

$$= (0.95)[1.25(1430) + 1.5(2618)] = 5428 \text{ kN-m}$$

$$M_{AD} = S_n\left(F_n - \frac{M_{D1}}{S_s} - \frac{M_{D2}}{S_n}\right)$$

$$M_{AD\text{-comp}} = (0.131)\left(312.4\times 10^3 - \frac{13\ 965}{0.1258} - \frac{5428}{0.131}\right)$$

$$= 20\ 954 \text{ kN-m}$$

$$M_{AD\text{-tension}} = (0.1166)\left(345\times 10^3 - \frac{13\ 965}{9.33(10)^{-2}} - \frac{5428}{0.1166}\right)$$

$$= 17\ 346 \text{ kN-m} \quad (\text{control})$$

- Calculate nominal flexure resistance, M_n:

$$M_n = 13{,}965 + 5{,}428 + 17{,}346 = 36{,}739 \text{ kN-m}$$

From Table 13.3, maximum factored negative moments occurred at the interior support

$$\eta\Sigma\gamma_i M_i \le \phi_f M_n$$

$$\underline{32{,}095 \text{ kN-m} < 1.0\,(36{,}739) \text{ kN-m}} \qquad \text{OK}$$

8. Shear Strength Design — Strength Limit State I

a. *End bearing of Span 1*

- Nominal shear resistance V_n:
  For inclined webs, each web shall be designed for shear, V_{ui} due to factored loads taken as [AASHTO Article 6.11.2.2.1]

$$\therefore\ V_{ui} = \frac{V_u}{\cos\theta} = \frac{2626}{2\cos(14)} = 1353 \text{ kN} \quad (\text{per web})$$

where θ is the angle of the web to the vertical.

$$\because \quad \frac{D}{t_w} = \frac{1600}{13} = 123.1 > 3.07\sqrt{\frac{E}{F_{yw}}} = 3.07\sqrt{\frac{(2.0)10^5}{345}} = 73.9$$

$$\therefore V_n = \frac{4.55t_w^3 E}{D} = \frac{4.55(13)^3(2.0)10^5}{1600} = 1249.5 \text{ kN}$$

$$\because V_{ui} = 1353 \text{ kN} > \phi_v V_n = (1.0)\,(1249.5) \text{ kN}$$

$\therefore$ Stiffeners are required

- V_n for end-stiffened web panel [AASHTO 6.10.7.3.3c]

$$V_n = CV_p$$

$$k = 5 + \frac{5}{(d_o / D)^2}$$

in which d_o is the spacing of transverse stiffeners

$$\text{For } d_o = 2400 \text{ mm} \quad \text{and} \quad k = 5 + \frac{5}{(2400/1600)^2} = 7.22$$

$$\because \quad \frac{D}{t_w} = 123.1 > 1.38\sqrt{\frac{Ek}{F_{yw}}} = 1.38\sqrt{\frac{200,000(7.22)}{345}} = 89.3$$

$$\because \quad C = \frac{152}{(123.1)^2}\sqrt{\frac{200,000(7.22)}{345}} = 0.65$$

$$V_p = 0.58F_{yw}Dt_w = 0.58(345)(1600)(13) = 4162 \text{ kN}$$

$$V_n = CV_p = 0.65(4162) = 2705 \text{ kN} > V_{ui} = 1353 \text{ kN} \qquad \text{OK}$$

b. *Interior support*:
- The maximum shear forces due to factored loads is shown in Table 13.4

$$V_u = \frac{3553}{2} = 1776.5 \text{ kN (per web)} > \phi_v V_n = (1.0)\,(1249.5) \text{ kN}$$

$\therefore$ Stiffeners are required for the web at the interior support.

c. *Intermediate transverse stiffener design*
The intermediate transverse stiffener consists of a plate welded to one of the web. The design of the first intermediate transverse stiffener is discussed in the following.
- Projecting Width b_t Requirements [AASHTO Article 6.10.8.1.2]
To prevent local bucking of the transverse stiffeners, the width of each projecting stiffener shall satisfy these requirements:

$$\begin{Bmatrix} 50 + \dfrac{d}{30} \\ 0.25b_f \end{Bmatrix} \le b_t \le \begin{Bmatrix} 0.48t_p\sqrt{\dfrac{E}{F_{ys}}} \\ 16t_p \end{Bmatrix}$$

where b_f is full width of steel flange and F_{ys} is specified minimum yield strength of stiffener. Try stiffener width, $b_t = 180.0$ mm.

$$b_t = 180 > \begin{cases} 50 + \dfrac{d}{30} = 50 + \dfrac{1600}{30} = 103.3 \text{ mm} \\ 0.25b_f = 0.25(450) = 112.5 \text{ mm} \end{cases} \qquad \text{OK}$$

Try $t_p = 16$ mm

$$b_t = 180 < \begin{cases} 0.48t_p\sqrt{\dfrac{E}{F_{ys}}} = 0.48(16)\sqrt{\dfrac{200\,000}{345}} = 185 \text{ mm} \\ 16t_p = 16(14) = 224 \text{ mm} \end{cases} \qquad \text{OK}$$

Use 180 mm × 16 mm transverse stiffener plates.

- Moment of inertia requirement [AASHTO Article 6.10.8.1.3]
  The purpose of this requirement is to ensure sufficient rigidity of transverse stiffeners to develop tension field in the web adequately.

$$I_t \ge d_o t_w^2 J$$

$$J = 2.5\left(\frac{D_p}{d_o}\right)^2 - 2.0 \ge 0.5$$

where I_t is the moment of inertia for the transverse stiffener taken about the edge in contact with the web for single stiffeners and about the midthickness of the web for stiffener pairs and D_p is the web depth for webs without longitudinal stiffeners.

$$\because J = 2.5\left(\frac{1600}{2400}\right)^2 - 2.0 = -0.89 < 0.5 \quad \because \text{Use } J = 0.5$$

$$I_t = \frac{(180)^3(16)}{3} = 31.1(10)^6 \text{ mm}^4 > d_o t_w^2 J = (2400)(16)^3(0.5) = 4.9(10)^6 \text{ mm}^4 \quad \text{OK}$$

- Area Requirement [AASHTO Article 6.10.8.1.4]:
  This requirement ensures that transverse stiffeners have sufficient area to resist the vertical component of the tension field, and is only applied to transverse stiffeners required to carry the forces imposed by tension-field action.

$$A_s \ge A_{s\min} = \left(0.15BDt_w(1-C)\frac{V_u}{\phi_v V_n} - 18t_w^2\right)\left(\frac{F_{yw}}{F_{ys}}\right)$$

where $B = 1.0$ for stiffener pairs. From the previous calculation:

$C = 0.65$ $\quad F_{yw} = 345$ MPa $\quad F_{ys} = 345$ MPa

$V_u = 1313$ kN (per web) $\quad \phi_f V_n = 1249.5$ kN $\quad t_w = 13$ mm

$$A_s = (180)(16) = 2880 \text{ in.}^2$$

$$> A_{s\min} = \left(0.15(2.4)(1600)(13)(1-0.65)\frac{1313}{1249.5} - 18(13)^2\right)\left(\frac{345}{345}\right)$$

$$= -288 \text{ mm}^2$$

The negative value of $A_{s\min}$ indicates that the web has sufficient area to resist the vertical component of the tension field.

9. Fatigue Design — Fatigue and Fracture Limit State

a. *Fatigue requirements for web in positive flexure region* [AASHTO Article 6.10.4]:

The purpose of these requirements is to control out-of-plane flexing of the web due to flexure and shear under repeated live loadings. The repeated live load is taken as twice the factored fatigue load.

$$D_c = \frac{f_{DC1} + f_{DC2} + f_{DW} + f_{LL+IM}}{\dfrac{f_{DC1}}{y_{st}} + \dfrac{f_{DC2} + f_{DW}}{y_{st-3n}} + \dfrac{f_{LL+IM}}{y_{st-n}}} - t_{fc}$$

$$= \frac{\dfrac{M_{DC1}}{S_{st}} + \dfrac{M_{DC2} + M_{DW}}{S_{st-3n}} + \dfrac{2(M_{LL+IM})_u}{S_{st-n}}}{\dfrac{M_{DC1}}{I_{\text{girder}}} + \dfrac{M_{DC2} + M_{DW}}{I_{\text{com}-3n}} + \dfrac{2(M_{LL+IM})_u}{I_{\text{com}-n}}} - t_{fc}$$

$$D_c = \frac{\dfrac{6585}{40.9(10)^6} + \dfrac{(801+1466)}{121(10)^6} + \dfrac{2(2467)}{326.3(10)^6}}{\dfrac{6585}{36.58(10)^9} + \dfrac{(801+1466)}{67.43(10)^9} + \dfrac{2(2467)}{92.79(10)^9}} - 20$$

$$D_c = 710 \text{ mm}$$

$$\frac{2D_c}{t_w} = \frac{2(710)}{13(\cos 14)} = 113 < 5.76\sqrt{\frac{E}{F_{yc}}} = 5.76\sqrt{\frac{2(10)^5}{345}} = 137.2$$

$$\therefore \; f_{cf} = F_{yw}$$

f_{cf}= maximum compression flexure stress in the flange due to unfactored permanent loads and twice the fatigue loading

$$f_{cf} = \frac{M_{DC1}}{S_{st}} + \frac{M_{DC2} + M_{DW}}{S_{st-3n}} + \frac{2(M_{LL+IM})_4}{S_{st-n}} = 161 + 18.7 + 15.3$$

$$= 195 \text{ Mpa} < F_{yw} = 435 \text{ MPa} \qquad \text{OK}$$

References

1. AASHTO, *AASHTO LRFD Bridge Design Specifications*, American Association of State Highway and Transportation Officials, Washington, D.C., 1994.
2. AASHTO, *AASHTO LRFD Bridge Design Specifications*, 1996 Interim Revisions, American Association of State Highway and Transportation Officials, Washington, D.C., 1996.
3. Price, K. D., Big Steel Boxes, in *National Symposium on Steel Bridge Construction*, Atlanta, 1993, 15-3.
4. Heins, C. P. and Hua, L. J., Proportioning of box girder bridges girder, *J. Struct. Div. ASCE*, 106(ST11), 2345, 1980.
5. Subcommittee on Box Girders of the ASCE-AASHTO Task Committee on Flexural Members, Progress report on steel box girder bridges, *J. Struct. Div. ASCE*, 97(ST4), 1971.
6. Heins, C. P., Box girder bridge design — state of the art, *AISC Eng. J.*, 15(4), 126, 1978.
7. Heins, C. P., Steel box girder bridges — design guides and methods, *AISC Eng. J.*, 20(3), 121, 1983.
8. Wolchuck, R., Proposed specifications for steel box girder bridges, *J. Struct. Div. ASCE*, 117(ST12), 2463, 1980.
9. Wolchuck, R., Design rules for steel box girder bridges, in *Proc. Int. Assoc, Bridge Struct. Eng.*, Zurich, 1981, 41.
10. Wolchuck, R., 1982. Proposed specifications for steel box girder bridges, Discussion, *J. Struct. Div. ASCE*, 108(ST8), 1933, 1982.
11. Seim, C. and Thoman, S., Proposed specifications for steel box girder bridges, discussion, *J. Struct. Div. ASCE*, 118(ST12), 2457, 1981.
12. AISC, *Highway Structures Design Handbook*, Vol. II, AISC Marketing, Inc., 1986.
13. Heins, C. P. and Hall, D. H., *Designer's Guide to Steel Box Girder Bridges*, Bethlehem Steel Corporation, Bethlehem, PA, 1981.
14. Wolchuck, R., Steel-plate deck bridges, in *Structural Engineering Handbook*, 3rd ed., Gaylord, E. H., Jr. and Gaylord, C. N., Eds., McGraw-Hill, New York, 1990, Sect. 113.
15. Kuzmanovic, B. and Graham, H. J., Shear lag in box girder, *J. Struct. Div. ASCE*, 107(ST9), 1701, 1981.
16. Balser, Strength of plate girders under combined bending and shear, *J. Struct. Div. ASCE*, 87(ST7), 181, 1971.
17. Balser, Strength of plate girders in shear, *J. Struct Div. ASCE*, 87(ST7), 151, 1971.
18. Foinquinos, R., Kuzmanovic, B., and Vargas, L. M., Influence of diaphragms on live load distribution in straight multiple steel box girder bridges, in *Building to Last, Proceedings of Structural Congress XV*, Kempner, L. and Brown, C. B., Eds., American Society of Civil Engineers, 1997, Vol. I., 89–93.
19. Olenik, J. C. and Heins, C. P., Diaphragms for curved box beam bridges, *J. Struct. Div.*, ASCE, 101(ST10), 1975.

14

Orthotropic Deck Bridges

Alfred R. Mangus
California Department of Transportation

Shawn Sun
California Department of Transportation

14.1 Introduction

This chapter will discuss the major design issues of orthotropic steel-deck systems. Emphasis will be given to the design of the closed-rib system, which is practicably the only system selected for orthotropic steel deck by the engineers around the world. Examples of short spans to some of the world's long-span bridges utilizing trapezoidal ribs will be presented. The subject of fabrication detailing and fatigue resistant details necessary to prepare a set of contract bridge plans for construction is beyond the scope of this chapter. However, the basic issues of fatigue and detailing are presented. For more detailed discussion, the best references are four comprehensive books on orthotropic steel deck systems by Wolchuk [1], Troitsky [2], and the British Institution of Civil Engineers [3,4].

14.2 Conceptual Decisions

14.2.1 Typical Sections

Modern orthotropic welded steel-deck bridge rib systems were developed by German engineers in the 1950s [1,2]. They created the word *orthotropic* which is from *orthogonal* for *ortho* and *anisotropic* for *tropic.* Therefore, an orthotropic deck has anisotropic structural properties at 90°. Structural steel is used by most engineers although other metals such as aluminum can be used, as well as advanced composite (fiberglass) materials.

0-8493-7434-0/00/$0.00+$.50

The open (torsionally soft) and closed (torsionally stiff) rib-framing system for orthotropic deck bridges developed by the Germans is shown in Figure 14.1. The open-rib and closed-rib systems are the two basic types of ribs that are parallel to the main span of the bridge. These ribs are also used to stiffen other plate components of the bridge. Flat plates, angles, split Ts, or half beams are types of open ribs that are always welded to the deck plate at only one location. A bent or rolled piece of steel plate is welded to the deck plate to form a closed space. The common steel angle can either be used as an open or closed rib depending on how it is welded to the steel deck plate. If the angle is welded at only one leg, then it is an open rib. However, if the angle is rotated to 45° and both legs are welded to the deck plate forming a triangular space or rib, it is a closed rib. Engineers have experimented with a variety of concepts to shape, roll, or bend a flat plate of steel into the optimum closed rib. The trapezoidal rib has been found to be the most practicable by engineers and the worldwide steel industry. Recently, the Japanese built the record span suspension bridge plus the record span cable-stayed bridge with trapezoidal rib construction (see Chapter 65).

The ribs are normally connected by welding to transverse floor beams, which can be a steel hot-rolled shape, small plate girder, box girder, or full-depth diaphragm plate. In Figure 14.1 small welded plate girders are used as the transverse floor beams. The deck plate is welded to the web(s) of the transverse floor beam. When full-depth diaphragms are used, access openings are needed for bridge maintenance purposes. The holes also reduce dead weight and provide a passageway for mechanical or electrical utilities. Since the deck plate is welded to every component, the deck plate is the top flange for the ribs, the transverse floor beam, and the longitudinal plate girders or box girders. All these various choices for the ribs, floor beam, and main girders can be interchanged, resulting in a great variety of orthotropic deck bridge superstructures.

14.2.2 Open Ribs vs. Closed Ribs

A closed rib is torsionally stiff and is essentially a miniature box girder [6]. The closed-rib deck is more effective for lateral distribution of the individual wheel load than the open-rib system. An open rib has essentially no torsional capacity. The open-rib types were initially very popular in the precomputer period because of simpler analysis and details. Once the engineer, fabricator, and contractor became familiar with the flat plate rib system shown in Figure 14.1 and Table 14.1, the switch to closed ribs occurred to reduce the dead weight of the superstructure, plus 50% less rib surface area to protect from corrosion. Engineers discovered these advantages as more orthotropic decks were built. The shortage and expense of steel after the World War II forced the adoption of closed ribs in Europe. The structural detailing of bolted splices for closed ribs requires handholds located in the bottom flange of the trapezoidal ribs to allow workers access to install the nut to the bolt. For a more-detailed discussion on handhold geometry and case histories for solutions to field-bolted splicing, refer to the four comprehensive books [1-4].

Compression stress occurs over support piers when the rib is used as a longitudinal interior stiffener for the bottom flanges of continuous box girders and can be graphically explained [6]. Ribs are usually placed only on the inside face of the box to achieve superior aesthetics and to minimize exterior corrosion surface area that must also be painted or protected. Compression also occurs when the rib is used as a longitudinal interior stiffener for columns, tower struts, and other components. The trapezoidal rib system quite often is field-welded completely around the superstructure cross section to achieve full structural continuity, rather than field bolted.

Table 14.2 [5] shows the greater bending efficiency in load-carrying capacity and stiffness achieved by the trapezoidal (closed) rib. It is readily apparent that a series of miniature box girders placed side by side is much more efficient that a series of miniature T-girders placed side by side. In the tension zones, the shape of the rib can be open or closed depending on the designers' preferences.

A trapezoidal rib can be quickly bent from a piece of steel as shown in Figure 14.2. A brake press is used to bend the shape in a jig in a few minutes. Rollers can also be used to form these trapezoidal ribs.

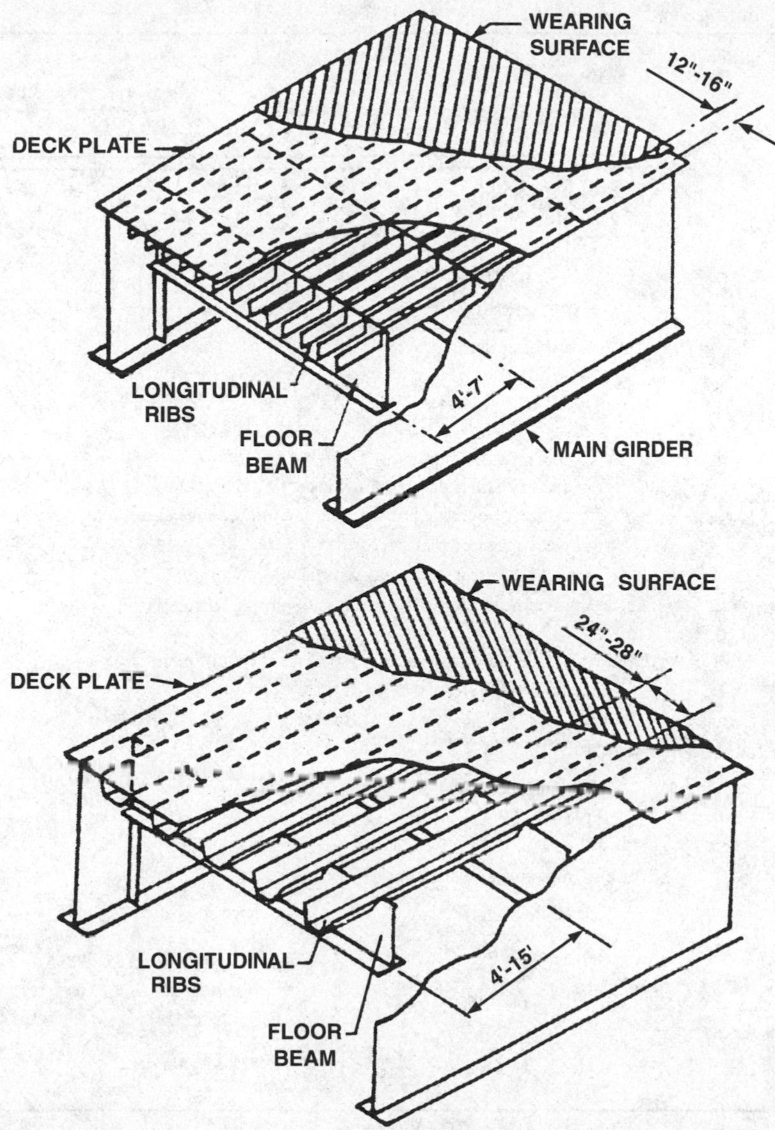

FIGURE 14.1 Typical components of orthotropic deck bridges. (From Troitsky, M. S., *Orthotropic Steel Deck Bridges*, 2nd ed., JFL Arch Welding Foundation, Cleveland, OH, 1987. Courtesy of The James F. Lincoln Arc Welding Foundation.)

A trapezoidal rib can be quickly bent from a piece of steel as shown in Figure 14.2. A brake press is used to bend the shape in a jig in a few minutes. Rollers can also be used to form these trapezoidal ribs. One American steel company developed Table 14.3 to encourage the utilization of orthotropic deck construction. This design aid was developed using main-frame computers in 1970, but due to lack of interest in orthotropic deck by bridge engineers this design aid eventually went out of print; nor was it updated to reflect changes in the AASHTO Bridge Code [5]. Tables 14.4 and 14.5 are excerpts from this booklet intended to assist an engineer quickly to design an orthotropic deck system and comply with minimum deck plate thickness; maximum rib span; and rib-spacing requirements of AASHTO [5]. AASHTO standardization of ribs has yet to occur, but many bridges built in the United States using ribs from Table 14.3 are identified throughout this chapter. The German and Japanese steel companies have developed standard ribs (see Table 14.3)

TABLE 14.1 Limiting Slenderness for Various Types of Ribs

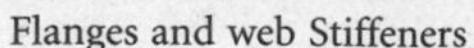

Flanges and web Stiffeners

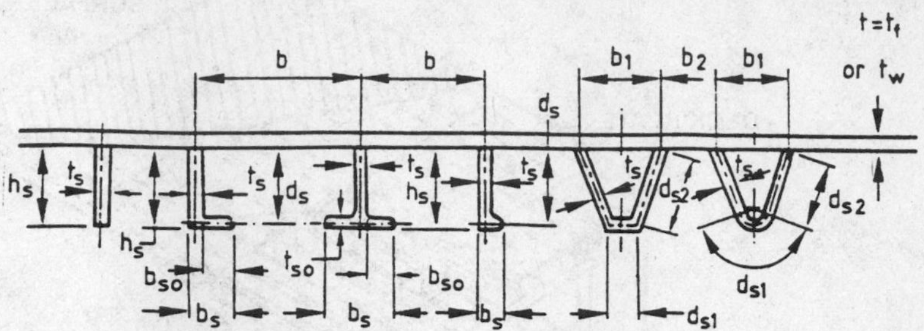

d, h = stiffener depth

b_s = width of angle

t_o, t', t_s = stiffener thickness

t = plate thickness

w, b = spacing of stiffeners

l_s = span of stiffener between supporting members

r_y = radius of gyration of stiffener (without plate) about axis normal to plate

F_y = yield stress of plate, N/mm²

F_{ys} = yield stress of stiffener, N/mm²

F_{max} = maximum factored compression stress, N/mm²

Draft U.S. rules

Effective slenderness coefficient C_s shall meet requirement

$$C_s = \begin{cases} \dfrac{d}{15t_o} + \dfrac{w}{12t} & \text{for flats} \\ \dfrac{d}{1.35t_o + 0.56r_y} + \dfrac{w}{12t} & \text{for Ts or angles} \end{cases} \le \begin{cases} \dfrac{0.4}{\sqrt{F_y / E}} & \text{for } f_{max} \ge 0.5F_y \\ \dfrac{0.65}{\sqrt{F_y / E}} & \text{for } f_{max} \le 0.5F_y \end{cases}$$

For any outstand of a stiffener $\dfrac{b'}{t'} \le \dfrac{0.48}{\sqrt{F_{ys} / E}}$

British Standard 5400

For flats: $\dfrac{h_s}{t_s}\sqrt{\dfrac{F_{ys}}{355}} \le 10$

For angles: $b_s \le h_s$; $\dfrac{b_s}{t_s}\sqrt{\dfrac{F_{ys}}{355}} \le 11$; $\dfrac{h_s}{t_s}\sqrt{\dfrac{F_{ys}}{355}} \le 7$

Source: Galambos, T. V., Ed., *Guide to Stability Design Criteria for Metal Structures*, 4th ed., John Wiley & Sons, New York, 1988. With permission.

TABLE 14.2 AASHTO Effective Width of Deck Plate Acting with Rib

Calculation of	a₀ ; a — a	a₀ + e₀ ; e — a — e
Rib section properties for calculation of deck rigidity and flexural effects due to dead loads	$a_o = a$	$a_o + e_o = a + e$
Rib section properties for calculation of flexural effects due to wheel loads	$a_o = 1.1a$	$a_o + e_o = 1.3(a + e)$

Source: American Association of State Highway and Transportation Officials, *LRFD Bridge Design Specifications*, Washington, D.C., 1994. With permission.

FIGURE 14.2 Press brake forming rib stiffener sections. (Photo by Lawrence Lowe and courtesy of Universal Structural, Inc.)

14.2.3 Economics

Orthotropic deck bridges become an economical alternative when the following issues are important: lower mass, ductility, thinner or shallower sections, rapid bridge installation, and cold-weather construction.

Lower superstructure mass is the primary reason for the use of orthotropic decks in long-span bridges. Table 14.6 shows the mass achieved by abandoning the existing reinforced concrete deck and switching to a replacement orthotropic deck system relationship. The mass was reduced from 18 to 25% for long-span bridges, such as suspension bridges. This is extremely important since dead load causes 60 to 70% of the stresses in the cables and towers [7,8]. The mass is also important for bridge responses during an earthquake. The greater the mass, the greater the seismic forces. The Golden Gate Bridge, San Francisco, California, was retrofitted from a reinforced concrete deck built in 1937 to an orthotropic deck built in 1985 (see Figure 14.3). This retrofit reduced seismic forces in the suspension bridge towers and other bridge components. The engineering statistics of redecking are shown in Table 14.6. The Lions Gate Bridge of Vancouver, Canada was retrofitted in 1975 from a reinforced concrete deck to an orthotropic deck, which increased its seismic durability. Economics or cost of materials can be multiplied against the material saved to calculate money saved by reducing the weight.

A very thin deck structure can be built using this structural system, as shown by the Creitz Road Grade Separation in Figure 14.4 or German Railroad Bridge in Figure 14.5. An orthotropic deck may be the most expensive deck system per square meter in a short-span bridge. So why would the most expensive deck be a standard for the German railroads? The key component in obtaining the thinnest superstructure is the deck thickness. An orthotropic deck is thin because the ribs nest between the floor beams. Concrete decks are poured on top of steel beams. Thin superstructures can be very important for a grade-crossing situation because of the savings to a total project. The two components are bridge costs plus roadway or site costs. High-speed trains require minimal grade changes. Therefore, the money spent on highway or railway approach backfill can far exceed the cost of a small-span bridge. A more expensive superstructure will greatly reduce the backfill work and cost. In urban situations, approach fills may not be possible. The local street may need to be excavated below the railway bridge; therefore a more expensive thin orthotropic deck–floor system may result in the lowest total cost for the entire project.

TABLE 14.3 Properties of Trapezoidal Ribs

American Rib (English Units)

Japanese Rib (Metric)

Depth of Rib *d* (in.)	Width at Top, *a* (in.)	Rib Wall Thickness, t_f (in.)	Weight per Foot, *w* (lb)	Moment of Inertia, I_{xx} (in⁴)	Neutral Axis Location, Y_{xx} (in.)	Sloping Face Length, *h'* (in.)
8.0	11.50	5/16	23.43	46.3	3.09	8.382
		3/8	27.95	54.6	3.12	
		7/16	32.40	62.7	3.14	
9.0	12.12	5/16	25.64	63.8	3.56	9.428
		3/8	30.60	75.5	3.59	
		7/16	35.53	86.8	3.61	
10.0	12.75	5/16	27.88	85.1	4.04	10.477
		3/8	33.29	100.8	4.06	
		7/16	38.66	116.1	4.09	
11.0	13.38	5/16	30.09	110.4	4.52	11.525
		3/8	35.94	131.0	4.54	
		7/16	41.57	151.0	4.57	
12.0	14.00	5/16	32.33	140.2	5.00	12.572
		3/8	38.62	166.4	5.02	
		7/16	44.88	192.1	5.05	
13.0	14.63	5/16	34.53	174.7	5.48	13.621
		3/8	41.31	207.6	5.51	
		7/16	48.01	239.7	5.53	
14.0	15.25	5/16	36.75	214.4	5.97	14.668
		3/8	43.96	254.8	5.99	
		7/16	51.10	294.4	6.02	
Depth of Rib *d* (mm)	**Width at Top, *a* (mm)**	**Rib Wall Thickness, t_f (mm)**	**Weight per Foot, *w* (Kg/m)**	**Moment of Inertia, I_{xx} (cm⁴)**	**Neutral Axis Location, Y_{xx} (mm)**	**Sloping Face Length, *h'* (mm)**
240	320	6	31.6	2460	88.6	246
260	320	6	33.1	3011	99.1	266
242	324	8	42.3	3315	89.9	248
262	324	8	44.3	4055	100.3	268

14.3 Applications

Some of the most notable world bridges were built using an orthotropic steel deck with trapezoidal rib construction. There are about only 50 bridges in North America using orthotropic decks, and eight are built and two more being designed in California. However, there is a vast array of bridge types utilizing the orthotropic deck from very small to some of the longest clear-span bridges of the world. Some orthotropic deck bridges have unique framing systems. Bridges featured and discussed in the following sections were selected to demonstrate the breath of reasons for selecting orthotropic deck superstructure [11-15]. All of these bridges utilize trapezoidal ribs in the deck area or compression zone of the superstructure. These types are: simple span with two plate or box girders, multiple plate girder, single-cell box girder, multicell box girder, wide bridges that have cantilever floor beams supported by struts, a monoarch bridge, a dual-arch bridge, a through-truss

TABLE 14.4 Orthotropic Deck Design Properties — Rigid Floor Beams

H = effective torsional rigidity of orthotropic plate (kip-in.²/in.)
D_y = flexural rigidity of orthotropic plate in y direction (kip-in.²/in.)
H/D_y = rigidity ratio (unitless)
I_r = moment of inertia (in.⁴)
Y_b = centroid (in.)
t_p = deck plate thickness (in.)

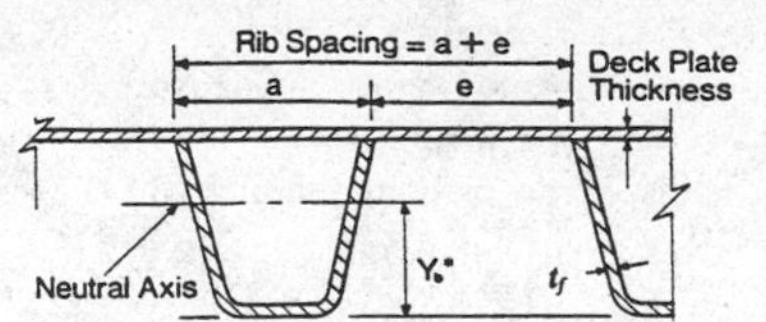

Deck Plate t_p (in.)	$a + e$ (in.)	Rib Wall (in.)	Value	Span (ft)	Rib Depth 8 in.	Rib Depth 9 in.	Rib Depth 10 in.	Span (ft)	Rib Depth 11 in.	Rib Depth 12 in.	Rib Depth 13 in.	Rib Depth 14 in.
9/16	22	5/16	H/D_y	7	0.039	0.034	0.030	10	0.048	0.045	0.042	0.040
			I_r		165	217	278		351	431	520	620
			Y_b		6.45	7.14	7.81		8.54	9.20	9.85	10.49
9/16	26	3/8	H/D_y	11	0.057	0.049	0.043	14	0.056	0.051	0.047	0.044
			I_r		197	259	331		417	512	620	740
			Y_b		6.48	7.18	7.86		8.56	9.23	9.88	10.53
9/16	30	7/16	H/D_y	15	0.066	0.056	0.049	18	0.057	0.051	0.047	0.043
			I_r		226	298	382		480	591	716	855
			Y_b		6.50	7.19	7.88		8.57	9.24	9.89	10.54
5/8	22	5/16	H/D_y	7	0.044	0.038	0.033	10	0.053	0.049	0.046	0.043
			I_r		171	225	288		364	446	539	643
			Y_b		6.59	7.30	7.99		8.73	9.41	10.07	10.72
5/8	30	7/16	H/D_y	15	0.079	0.067	0.058	18	0.067	0.061	0.055	0.051
			I_r		234	309	396		498	612	742	886
			Y_b		6.64	7.35	8.05		8.76	9.44	10.11	10.77
11/16	22	5/16	H/D_y	7	0.048	0.041	0.036	10	0.056	0.052	0.048	0.045
			I_r		177	232	297		375	460	557	664
			Y_b		6.72	7.44	8.14		8.90	9.59	10.27	10.93
11/16	30	7/16	H/D_y	15	0.090	0.078	0.068	18	0.077	0.070	0.064	0.059
			I_r		242	318	408		513	632	765	915
			Y_b		6.76	7.49	8.21		8.93	9.62	10.31	10.98
3/4	22	5/16	H/D_y	7	0.052	0.044	0.038	10	0.059	0.054	0.050	0.047
			I_r		182	239	305		386	474	573	683
			Y_b		6.84	7.57	8.29		9.06	9.76	10.45	11.13
3/4	26	3/8	H/D_y	11	0.084	0.073	0.064	14	0.079	0.072	0.067	0.063
			I_r		216	284	364		458	563	682	815
			Y_b		6.87	7.61	8.34		9.08	9.78	10.48	11.16
3/4	30	7/16	H/D_y	15	101	0.087	0.077	18	0.086	0.078	0.072	0.067
			I_r		248	327	420		528	650	787	941
			Y_b		6.88	7.62	8.35		9.08	9.79	10.49	11.17

(Excerpts from out of print booklet. Permission to reprint granted. Anonymous source as requested.)

bridge, a deck-truss bridge, a monoplane cable-stayed bridge, a dual-plane cable-stayed bridge, a monocable suspension bridge, and a dual-cable suspension bridge.

14.3.1 Plate-Girder Bridges

In the 1960s small orthotropic steel-deck bridges were built in California, Michigan, and for the Poplar Street Bridge as prototypes to examine steel construction systems as well as various wearing surface materials. Each bridge used trapezoidal ribs with a split-beam section as floor beam and two plate girders as the main girders. The California Department of Transportation (Caltrans) built the I-680 over U.S. 580 bridge as their test structure [9,13] in 1968. This bridge has two totally different rib/deck systems including two different wearing surfaces. The two-lane cross section of the bridge and a very similar one were built by the Michigan Department of Transportation. The Creitz Road Bridge is a typical grade crossing built over I-496 near Lansing, Michigan (see

TABLE 14.5 Orthotropic Deck Design Properties — Flexible Floor Beams

Wt. = weight (PSF)
I_r = moment of inertia (in^4)
Y_b = centroid (in.)
t_p = deck plate thickness (in.)

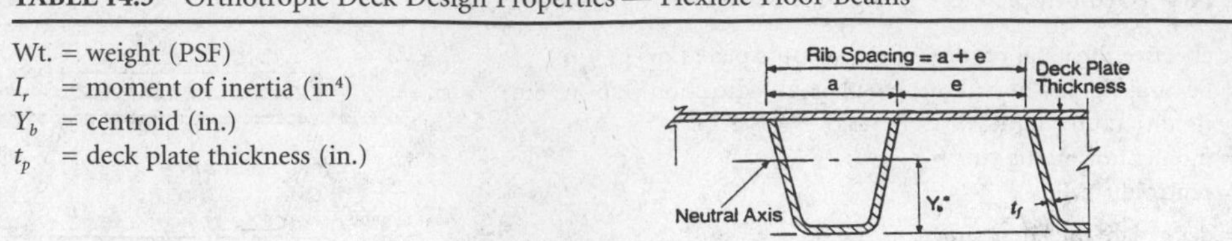

Deck Plate t_p (in.)	$a + e$ (in.)	Rib Wall (in.)	Value	Rib Depth 8 in.	9 in.	10 in.	11 in.	12 in.	13 in.	14 in.
9/16	22	5/16	Wt.	35.7	36.9	38.1	39.4	40.6	41.8	43.0
			I_r	169	222	284	355	437	528	630
			Y_b	6.54	7.24	7.94	8.62	9.29	9.95	10.60
9/16	26	3/8	Wt.	35.8	37.1	38.3	39.5	40.8	42.0	43.2
			I_r	199	262	335	420	517	625	747
			Y_b	6.54	7.24	7.93	8.61	9.28	9.94	10.59
9/16	30	7/16	Wt.	35.9	37.2	38.4	39.7	40.9	42.1	43.4
			I_r	228	301	386	484	595	721	861
			Y_b	6.54	7.24	7.93	8.61	9.28	9.93	10.58
5/8	22	5/16	Wt.	38.3	39.5	0.033	41.9	43.1	44.3	45.6
			I_r	175	230	288	368	452	547	653
			Y_b	6.68	7.40	8.11	8.80	9.49	10.17	10.83
5/8	30	7/16	Wt.	38.5	39.7	41.0	42.2	43.5	44.7	45.9
			I_r	236	311	399	501	616	745	892
			Y_b	6.68	7.40	8.10	8.80	9.48	10.15	10.82
11/16	22	5/16	Wt.	40.8	42.0	43.2	44.5	45.7	46.9	48.1
			I_r	180	237	303	379	466	564	674
			Y_b	6.81	7.54	8.26	8.97	9.67	10.36	11.04
11/16	30	7/16	Wt.	41.0	42.3	43.5	44.8	46.0	47.2	48.5
			I_r	244	321	412	516	636	770	920
			Y_b	6.81	7.54	8.26	8.96	9.66	10.35	11.03
3/4	22	5/16	Wt.	43.4	44.6	45.8	47.0	48.2	49.4	50.7
			I_r	185	243	311	390	479	580	693
			Y_b	6.92	7.67	8.40	9.13	9.84	10.54	11.24
3/4	26	3/8	Wt.	43.5	44.7	46.0	47.2	48.4	49.7	50.9
			I_r	218	287	368	461	567	687	821
			Y_b	6.92	7.66	8.40	9.12	9.83	10.53	11.22
3/4	30	7/16	Wt.	43.6	44.8	46.1	47.3	48.6	49.8	51.0
			I_r	250	330	423	531	653	792	947
			Y_b	6.92	7.67	8.40	9.12	9.83	10.53	11.22

Excerpts from out of print booklet. Permission to reprint granted. Anonymous source as requested.

Figure 14.4). It is a typical two-lane bridge that carries local traffic over an interstate freeway, with two symmetrical spans of 29 m. The bridge uses the 5⁄16-in.-thick, 9-in.-deep, 25.64-plf rib as shown in Table 14.3. The rigid steel bent comprises three welded steel box members aesthetically shaped [14,15]. Caltrans built a weigh station as an orthotropic deck prototype for the Hayward–San Mateo Bridge [17]. All of these short-span orthotropic deck bridges are still in use after 30 years of service, but the wearing surface has been replaced on many of these bridges.

Single-track railroad bridges through steel plate are the most common type for short spans. A two-girder bridge shown in Figure 14.5 is AASHTO fracture critical because, if one girder fractures, the bridge will collapse [18]. Both versions of the current AASHTO codes require the designer to label fracture-critical components, which have more stringent fabrication requirements. Orthotropic bridges can be erected quickly when the entire superstructure is fabricated as a full-width component. Many railroads prefer weathering steel since maintenance painting is not required. The German Federal Railroads have a standard, classic two plate girder with orthotropic deck system

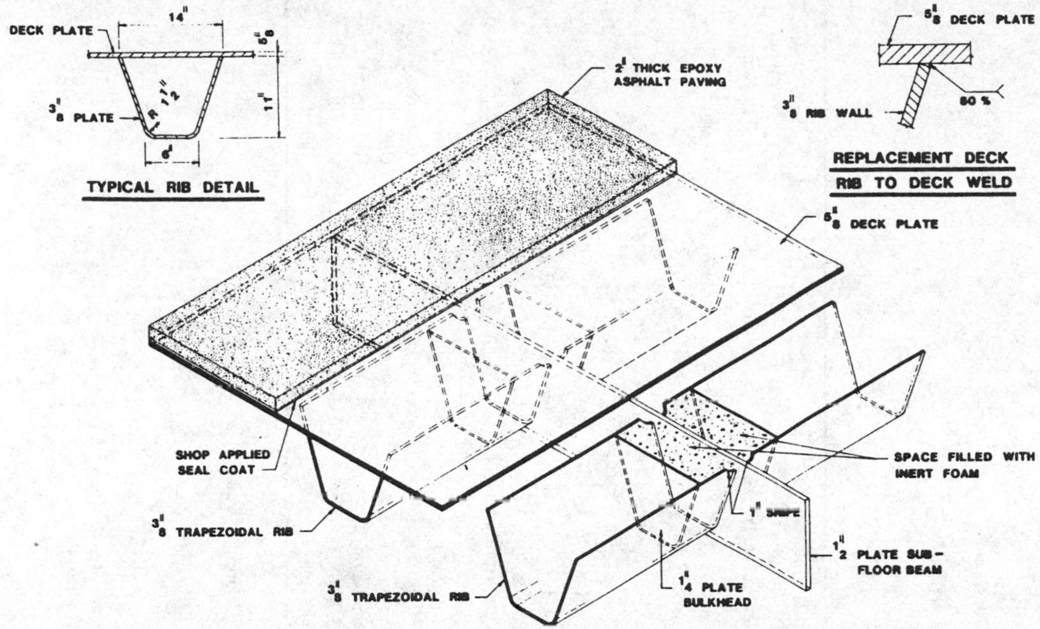

FIGURE 14.3 Golden Gate orthotropic deck details. (From Troitsky, M. S., *Orthotropic Steel Deck Bridges*, 2nd ed., JFL Arc Welding Foundation, Cleveland, OH, 1987. Courtesy of the James F. Lincoln Arc Welding Foundation.)

for their common short-span railroad bridges. Edge plates are used to keep the gravel ballast in place on top of the superstructure (Figure 14.5).

14.3.2 Box-Girder Bridges

Box-girder bridges can be subdivided into three basic categories: the single-cell box, the multicell box, and the box with struts supporting a cantilevered deck. In Figure 14.6, the typical cross section of the Valdez Floating City Dock Transfer Bridge is shown. The entire bridge, only about 3 to 6 m above the waterline, was completed in 1981 using the orthotropic deck with trapezoidal ribs. The two identical bridges were built at each end of the floating dock. Each bridge has only two box girders and has a simple span of 61 m. The transfer bridge provides traffic access to and from the floating dock and serves as the primary mooring tie for dock forces perpendicular to the shoreline. Trapezoidal rubber marine fenders absorb kinetic energy as the floating dock moves with the waves. These bumpers are at each end of the bridge. Box girders are more efficient in transmitting compression forces than plate girders [19]. ASTM A-36 steel was used to meet charpy impact requirements of 15 foot-pounds at –15°F. Automatic flux cored welding was used, and all full-penetration welds were either radiographically or ultrasonically inspected. The bridge uses the 3/8-in.-thick, 12-in.-deep, 38.62-plf rib as shown in Table 14.3. The floor beams are 2 ft deep by 1 ft wide 3/8-in.-thick plate bent in a U shape pattern. The ribs pass through the floor beam.

The typical cross sections of the Yukon River or "E. L. Patton" Bridge are shown in Figure 14.7. The 671-m-long bridge, with spans of 128 m, crosses over the Yukon River and was completed in 1976. The haul road is a gravel road originally built to transport supplies for the pipeline and oil field facilities at Prudhoe Bay, Alaska. The bridge was field-bolted together in cold weather, since the Alaskan winter lasts 6 months. It was important to keep the construction on schedule since the bridge was built to carry the 1.46-m-diameter trans-Alaskan crude oil pipeline. The bridge is the first built in Alaska across the Yukon River [2,20]. It is still the only bridge in Alaska across the

TABLE 14.6 Orthotropic Redecking Statistics Table (weight, deck area, etc.) Metric

Bridge	Lions Gate, Vancouver, BC, Canada	George Washington, New York, NY, USA	Golden Gate, San Francisco, CA, USA	Throngs Neck Viaduct, New York, NY, USA	Ben Franklin, Philadelphia, PA, USA	Champlain, Montreal, PQ, Canada
Bridge type	Girder	Suspension	Suspension + Approaches	Girder	Suspension + Approaches	Trusses
Redecking	Finished 1975	Finished 1978	Finished 1985	Finished 1986	Finished 1987	Finished 1992
Main Spans	13 to 38	186 1,067 186	343 1,280 343	42 to 58	219 534 219	118 215 118
Redecked area (m^2)	8600	40,320	52,680	45,800	55,740	18,620
Rib type	Closed	Open (Ts)	Closed	Closed	Open (bulb section)	Closed
Rib spans (m)	4.12	1.60	7.62	6.10 to 8.50	5.80 to 6.70	6.40 to 9.80
Wearing surface	40 mm epoxy asphalt	40 mm bitum. asphalt	50 mm epoxy asphalt	40 mm bitum. asphalt	32 mm epoxy asphalt + 32 mm bitum. asphalt	50 mm epoxy asphalt
Original concrete deck weight	N/A	517 (kg/m^2)	508 (kg/m^2)	522 (kg/m^2)	601 (kg/m_2)	N/A
Total weight of new deck[a]	300 (kg/m^2)	293 (kg/m^2)	386 (kg/m^2)	406 (kg/m^2)	435 (kg/m^2)	402 (kg/m_2)
Weight savings	N/A	224 (kg/m^2)	122 (kg/m^2)	116 (kg/m^2)	166 (kg/m_2)	N/A
New deck + main members	Yes, integral	No, integral	No, integral	No, integral	Yes, integral	Yes, integral
Cost/m[2b]	U.S. $500	U.S. $460	U.S. $1070	U.S. $770	U.S. $1010	U.S. $402
Redeck	Nighttime	Nighttime	Nighttime	Nighttime	Daytime	Nighttime

[a] Including surfacing, parapets, shear connectors;
[b] Total bid price/deck area. Note that bid prices reflect such variable factors as specific project characteristics, contractors profit margins, etc.
Source: Wolchuk, R., *Structural Engineering International*, IABSE, Zurich, 2(2), 125, 1992. With permission.

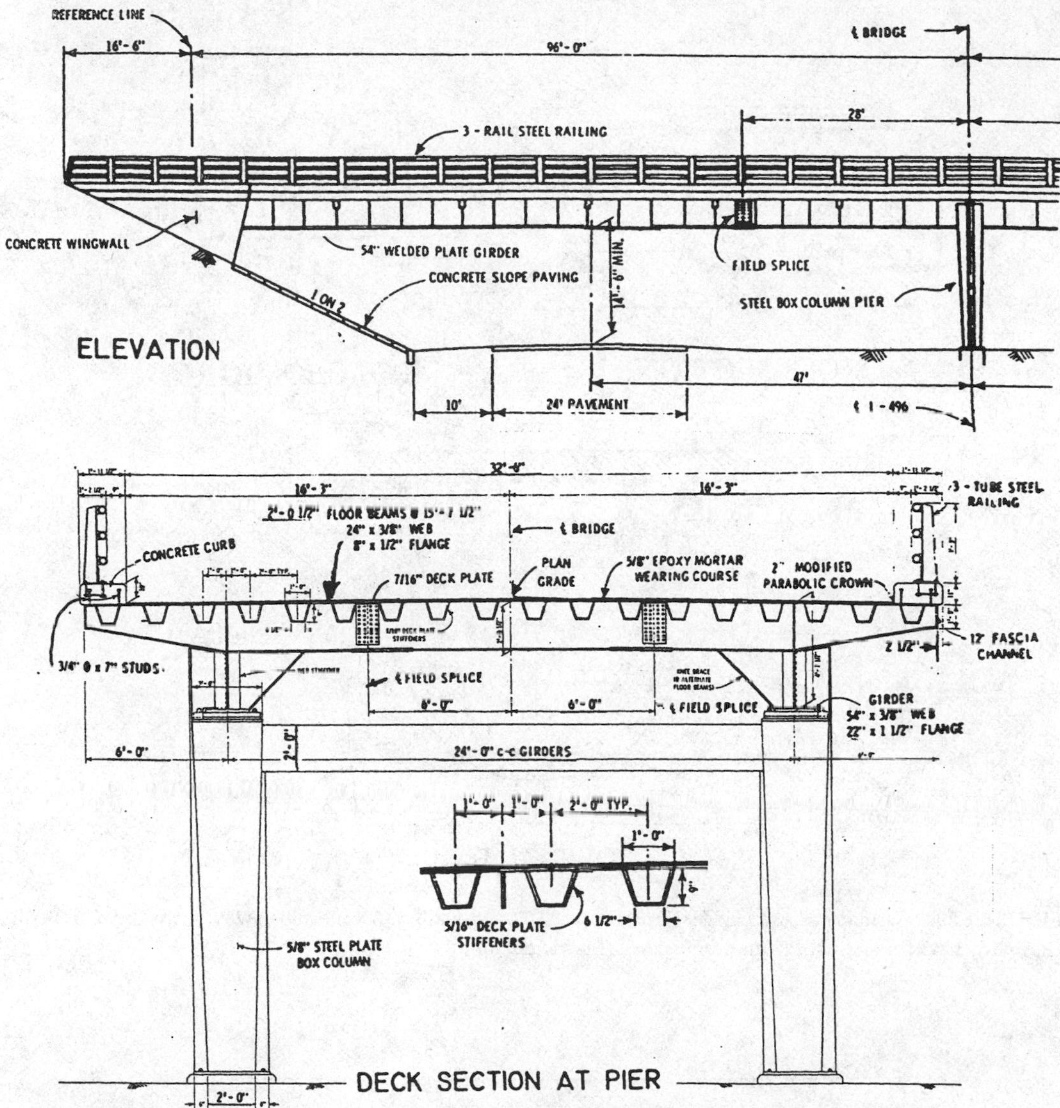

FIGURE 14.4 Typical grade separation — Creitz Road, Lansing, Michigan. (From *Modern Welded Steel Structure,* III, JFL Arc Welding Foundation, Cleveland, OH, 1970, B-10. Courtesy of the James F. Lincoln Arc Welding Foundation.)

Yukon, which has river ice 2 m thick. The superstructure consists of constant-depth twin rectangular box girders which have unique cantilevering brackets that support the trans-Alaskan crude oil pipeline on one side. The future trans-Alaska natural gas line, yet to be built, can be supported on the opposite side of the bridge with these specially designed cantilever support brackets. The bridge, which was fabricated in Japan, uses ribs as shown in Table 14.3. The ⅜-in.-thick, 11-in.-deep, 35.94-plf rib for the deck and the 5⁄16-in.-thick, 8-in.-deep, 23.43-plf rib in other locations are shown in Figure 14.7. Concrete deck construction requires curing temperature of a minimum of 40°F, otherwise, the water freezes during hydration. This was another factor in the selection of the orthotropic deck system, which was erected in temperatures as cold as –60°F [20]. The main bridge components, such as tower columns, tower cross frames, box girders and orthotropic steel deck were stiffened by trapezoidal ribs. The goal of maximizing the number of locations of trapezoidal rib was to reduce

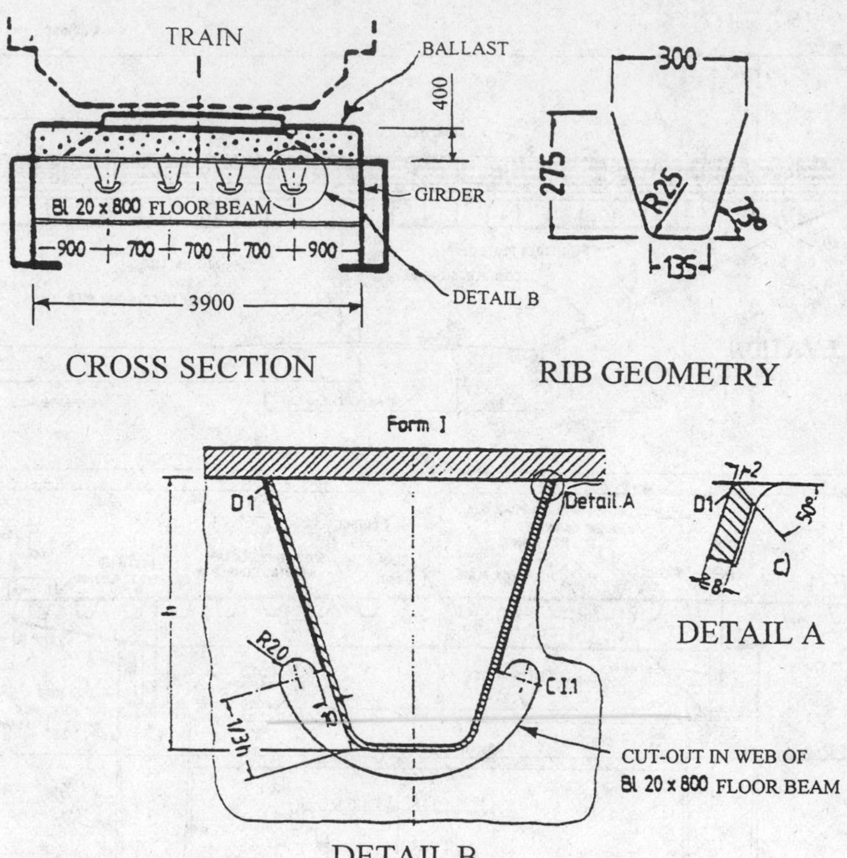

FIGURE 14.5 German railroad bridge. (From Haibach E. and Plasil, I., Der Stahlbau, 269, Ernst & Sohn, Berlin, Germany, 1983 [in German]. With permission.) (Metric units).

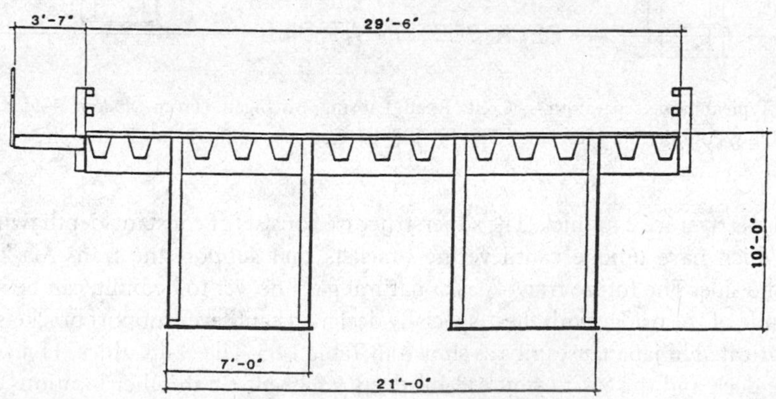

FIGURE 14.6 Valdez Floating City Dock Transfer Bridge. (Courtesy of Berger/ABAM Engineers, Federal Way, WA.) (English units)

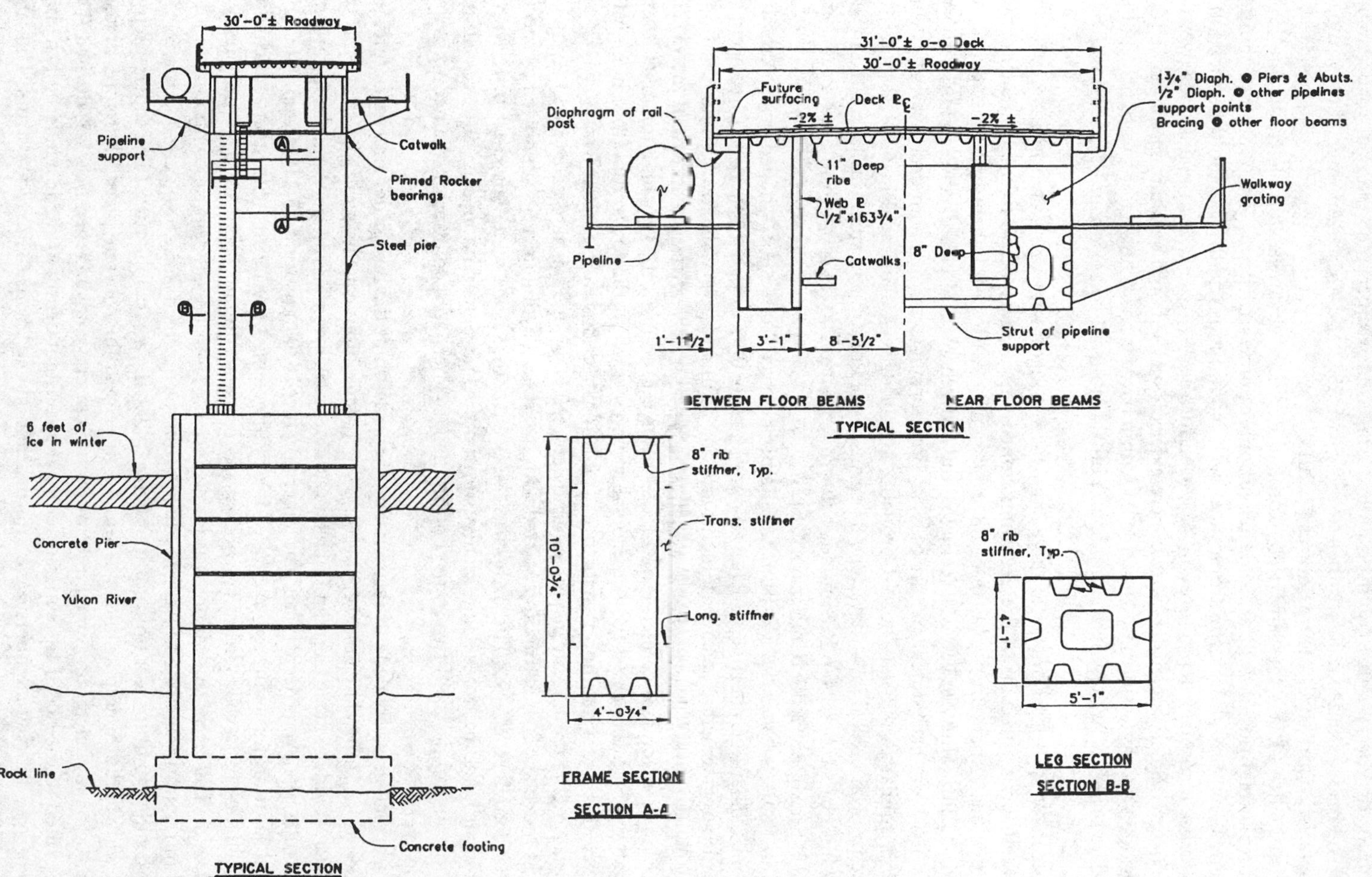

FIGURE 14.7 Yukon River Bridge — orthotropic deck and columns. (Courtesy of Alaska Department of Transportation and Public Facilities.)

the fabricator's setup costs to make a rib. The bridge tower columns and cross frames were built of shop-welded steel and field-bolted splices. The wearing surface of treated timber boards was bolted to the steel bridge because this could be built during cold weather, plus the haul road remains unpaved. The bridge utilized cold-weather steel (ASTM A537 and A514) with high charpy impact test characteristics for subzero temperatures [2, 20].

Shown in Figure 14.8a is the San Diego–Coronado Bridge, which was completed in June 1969. This California toll bridge sweeps around the harbor area of San Diego [3]. The Caltrans engineers selected single-cell box-girder orthotropic steel deck (continuous length of orthotropic portion = 573 m) because a constant-depth box could be used for the 201, -201, 171 m main spans over the shipping channel. Steel plate girders with concrete deck were used on the remaining length of 1690 m. The bridge was erected in these large pieces with a barge crane. The sections were field-bolted together. The bridge is painted on the inside and outside to resist corrosion and carries six lanes of traffic [21–23].

The Queens Way Bridges, identical three span twin bridges, were completed in June 1970 and are near the tourist attraction of the decommissioned Queen Mary ocean liner [25]. Each orthotropic bridge has a main drop-in span of 88 m suspended with steel hanger bars from cantilever side spans of 32 m. Thus, the center to center of the concrete piers or clear span is 152 m, with two side spans of 107 m. Each superstructure cross section is a single-cell box with deck overhangs with components similar to the San Diego–Coronado. The bridge was fabricated in 14 pieces, and the superstructure was erected in 11 days. The 88-m suspended or drop-in span was fabricated in one piece weighing 618 U.S. tons in Richmond, California, floated 700 miles south to Long Beach, and lifted up 15.2 m. by the same barge crane [26].

Shown in Figure 14.8b is the Maritime Off-Ramp a curved "horseshoe"-shaped bridge crossing over I-80 in Oakland, California, which was completed in 1997 as part of the I-880 Replacement Project. This superstructure has a very high radius of 76 m and a very shallow web depth of only 2.13 m for 58 m spans. The bridge has two lanes of traffic that create large centrifugal forces. The box-girder superstructure is divided into three separate cells to resist the very high torsional forces. To reduce the fabrication costs, the trapezoidal ribs were used in the top and bottom box-girder flanges since this was a continuous structure. The bridge sections were erected over busy I-80 on two Saturday nights creating an instant superstructure. The bridge was fabricated in 13 segments weighing as much as 350 U.S. tons and erected with two special hydraulic jacks supported by special multiwheeled trailers [27-29]. The orthotropic superstructure has a wider top deck plate with a 16-mm thickness and narrower flange plate of 19-mm thickness. In addition, each of three cells has four ribs for the top deck and two in the bottom flange. There are two exterior inclined webs and two interior vertical webs.

When a bridge gets very wide in relationship to the depth of the box girder superstructure, the German solutions shown in Figure 14.9 become the most economical. Shown in Figure 14.9a is the Jagst Viaduct, Widdern in the German Interstate system or Heilbronn–Wurzburg Autobaun carrying eight lanes of traffic over a deep valley. The superstructure is 30 m wide and 5.25 m deep. The most economical solution was to brace the ends of the cantilevering floor beams with struts attached to the bottom flange of the box girder. The box system remains constant depth so the strut remains a constant length. This keeps fabrication costs lower since the struts are all identical components [30].

Shown in Figure 14.9b is the Moselle Viaduct, Winningen, which is near Koblenz, Germany. The engineers decided to utilize a bottom soffit or flange following a parabolic curve. The superstructure is 30.5 m wide and 6 m deep at midspan and 8.5 m deep at the concrete piers. Therefore, to keep the struts a constant length, additional interior framing or cross-bracing members was devised. The struts are bolted to the side of the superstructure at a constant depth. At the inside face of the box girder, web cross-bracing members were attached and aligned with the exterior struts. This also produces a more pleasant architectural appearance [30]. The top deck utilizes "martini-glass"-shaped ribs, which consist of two standard shapes welded together. First a V-shaped rib is welded to the deck, second, a split-T is welded to the bottom of a V-shaped rib. It is a hybrid rib because

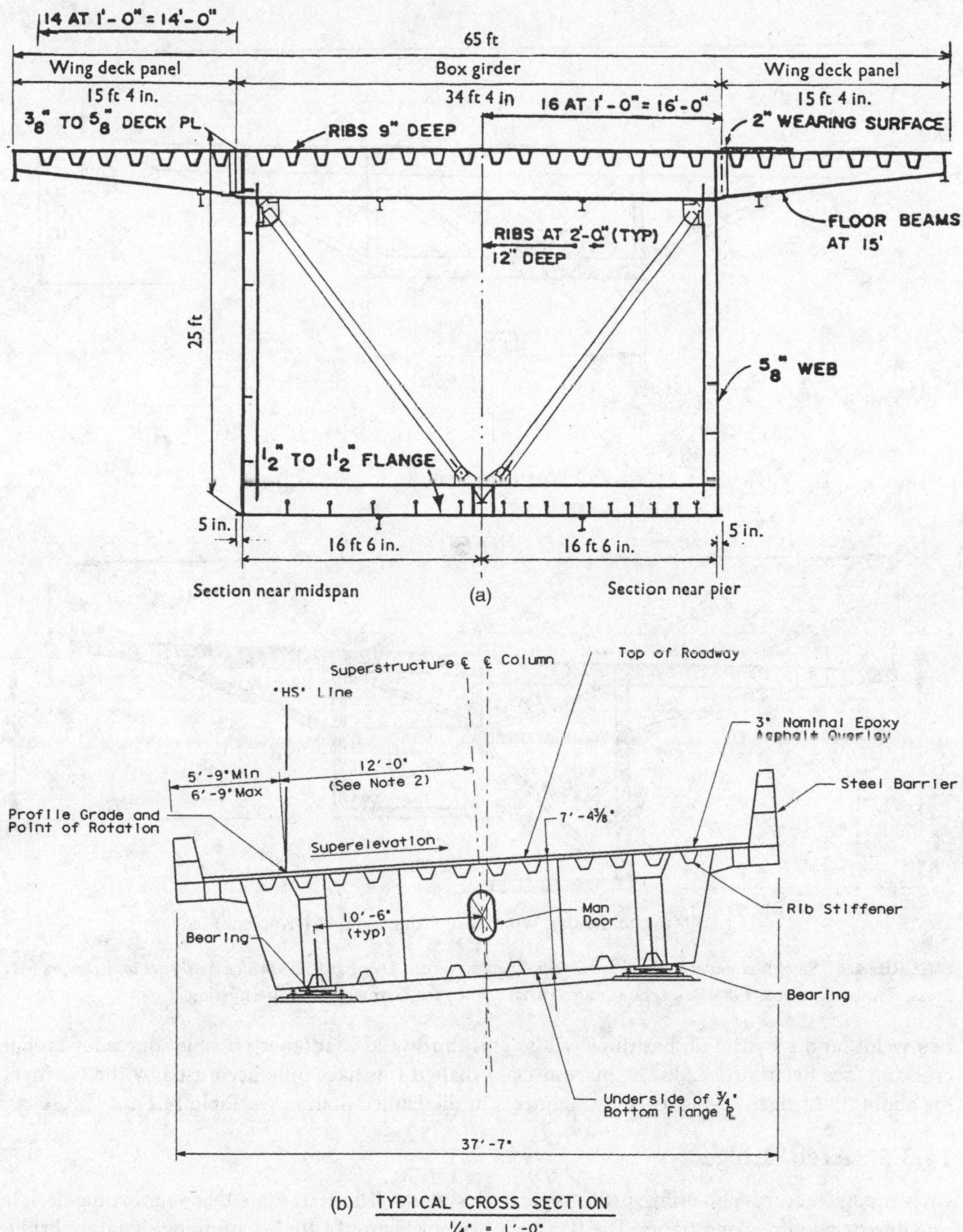

FIGURE 14.8 Steel box-girder bridges (a) San Diego–Coronado Bridge, California. (From Institute of Civil Engineers, Steel box-girder bridges, in *Proceedings of the International Conference*, Thomas Telford Publishing, London, 1973. With permission. (b) Maritime Off-Ramp Bridge, California. (Courtesy of the ICF KAISER Engineers.)

it has characteristics of both the open and closed rib. Some references have categorized it with closed ribs. The top portion provides good torsional stability to the deck, but the lower split-T portion has the buckling and corrosion disadvantages of an open rib. The split-T provides much greater bending strength, but no torsional stiffness to the deck. An open rib has one weld, a closed rib has

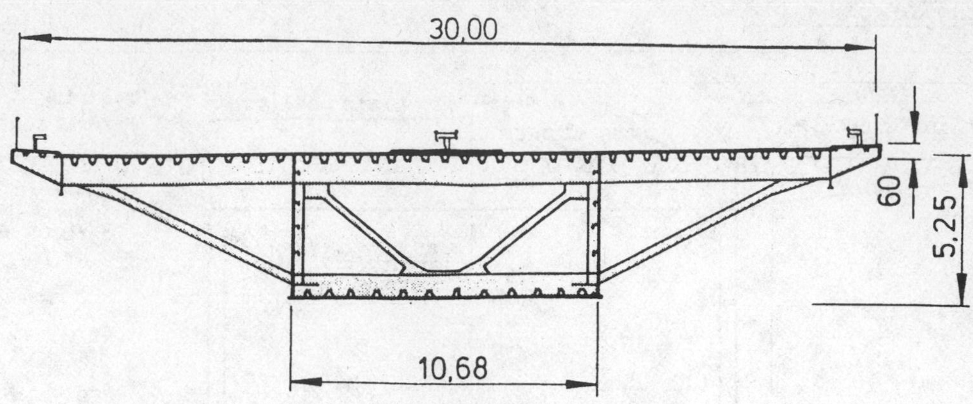

(a) Wurzburg Viaduct

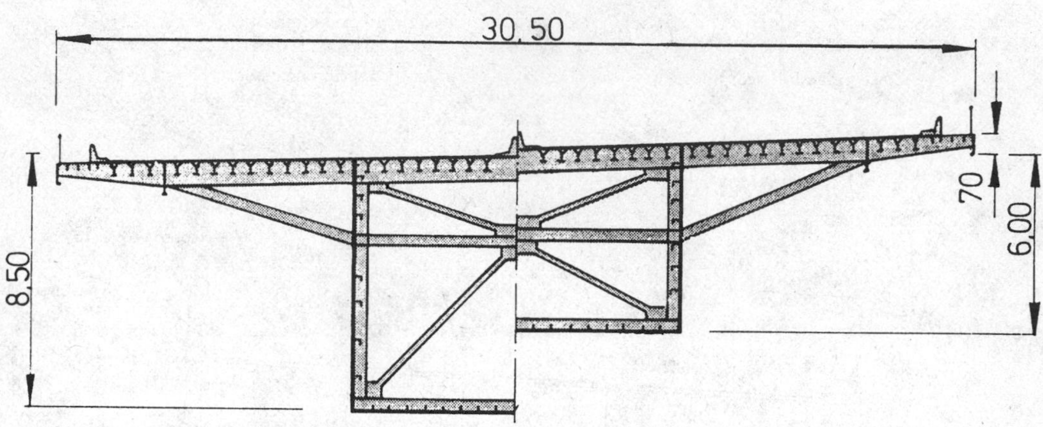

(b) Moseltal Bridge Winninzen (Autobahn A61 Koblenz)

FIGURE 14.9 Steel box girder with strutted deck bridges. (From Leonhardt, F., *Bridge Aesthetics and Design*, MIT Press, Cambridge, MA, 1984. Deutsche Verlags-Anstalt, Stuttgart, Germany. With permission.)

two welds, and a hybrid rib has three welds. This third weld is another possible source for fatigue cracking (see Section 14.4.5). The martini-glass-shaped ribs have only been used by the Germans for about 30 bridges, and ignored by engineers in the United States (See Table 14.1).

14.3.3 Arch Bridges

Arch superstructures also utilize orthotropic deck systems. The arch can either support the deck in one line or two lines of support. The Barqueta Bridge (Figure 14.10a) is a unique signature bridge with high aesthetic appearance built for the Expo 90 Fair held in Seville, Spain [31,32]. This bridge utilizes an aerodynamic system with torsional rigidity because it is supported by a single line of suspender cables from a single arch located at the centerline of the bridge. The center portion of the superstructure cross section is reinforced for the high stress concentrations from the suspender cables. Trapezoidal ribs are also used for the bottom flange or soffit. The bridge was erected on one side of the river and floated in a rotating pattern into a permanent orientation.

The Fremont Bridge of Portland Oregon has an orthotropic deck with trapezoidal ribs for the upper deck with a conventional reinforced concrete slab on steel girders for the lower deck

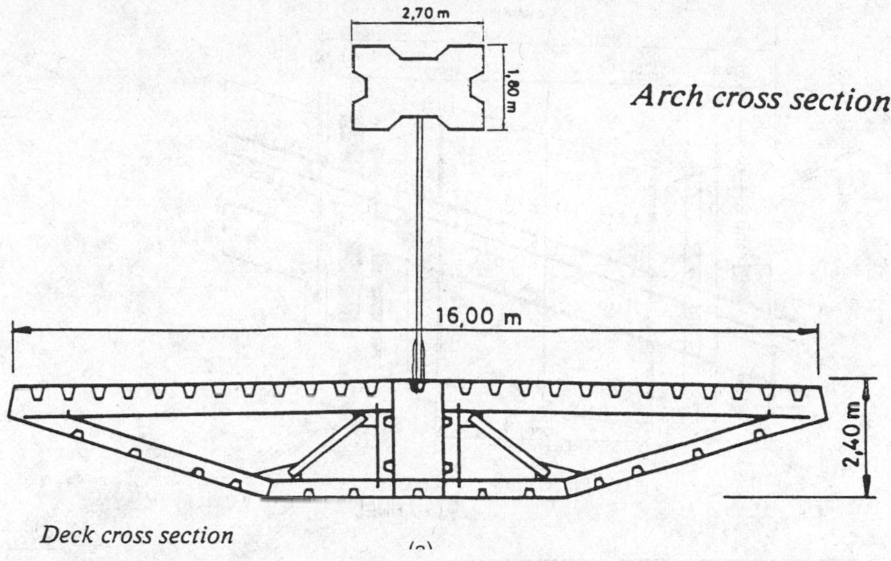

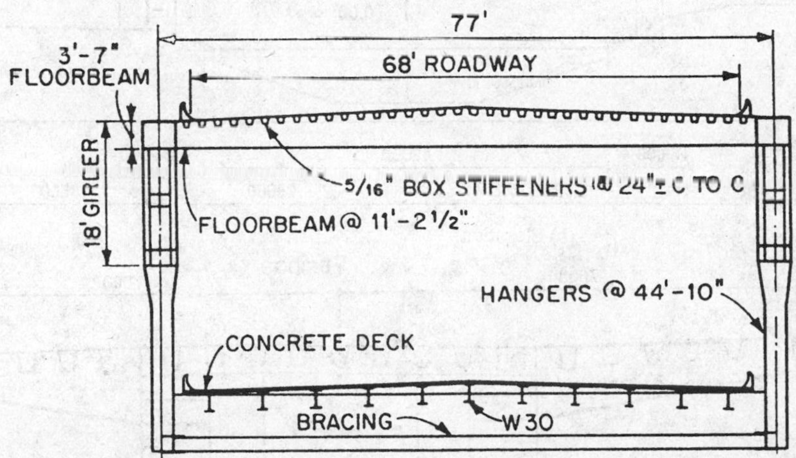

FIGURE 14.10 Arch bridges. (a) Barqueta Bridge, Seville, Spain. (From Arenas, J. J., and Pantaleon, M. J., *Structural Engineering International,* IABSE, 2(4), 251, 1992. With permission.) (b) Fremont Bridge, Portland, Oregon. (From Hedelfine, A. and Merritt, F. S., Ed., *Structural Steel Engineering Handbook,* McGraw Hill, New York, NY., 1972. With permission.)

[Figure 14.10b]. This bridge is a tied arch with a 383-m main span and was the fourth longest arch in the world upon completion in 1973. The deck provides lateral stability to the truss. The truss has two levels of traffic and is painted. The bridge was erected in large pieces with an oceangoing barge crane from the river dividing downtown Portland [33].

14.3.4 Movable Bridges

The orthotropic deck system has the lowest weight for movable bridges, so it is surprising how few examples there are of this excellent system. The swing-bridge design across a 42-m-wide navigation channel near Naestvad, Denmark was selected over 14 proposals from five competing consulting firms (Figure 14.11a). The superstructure is divided into two symmetrical components [34]. The

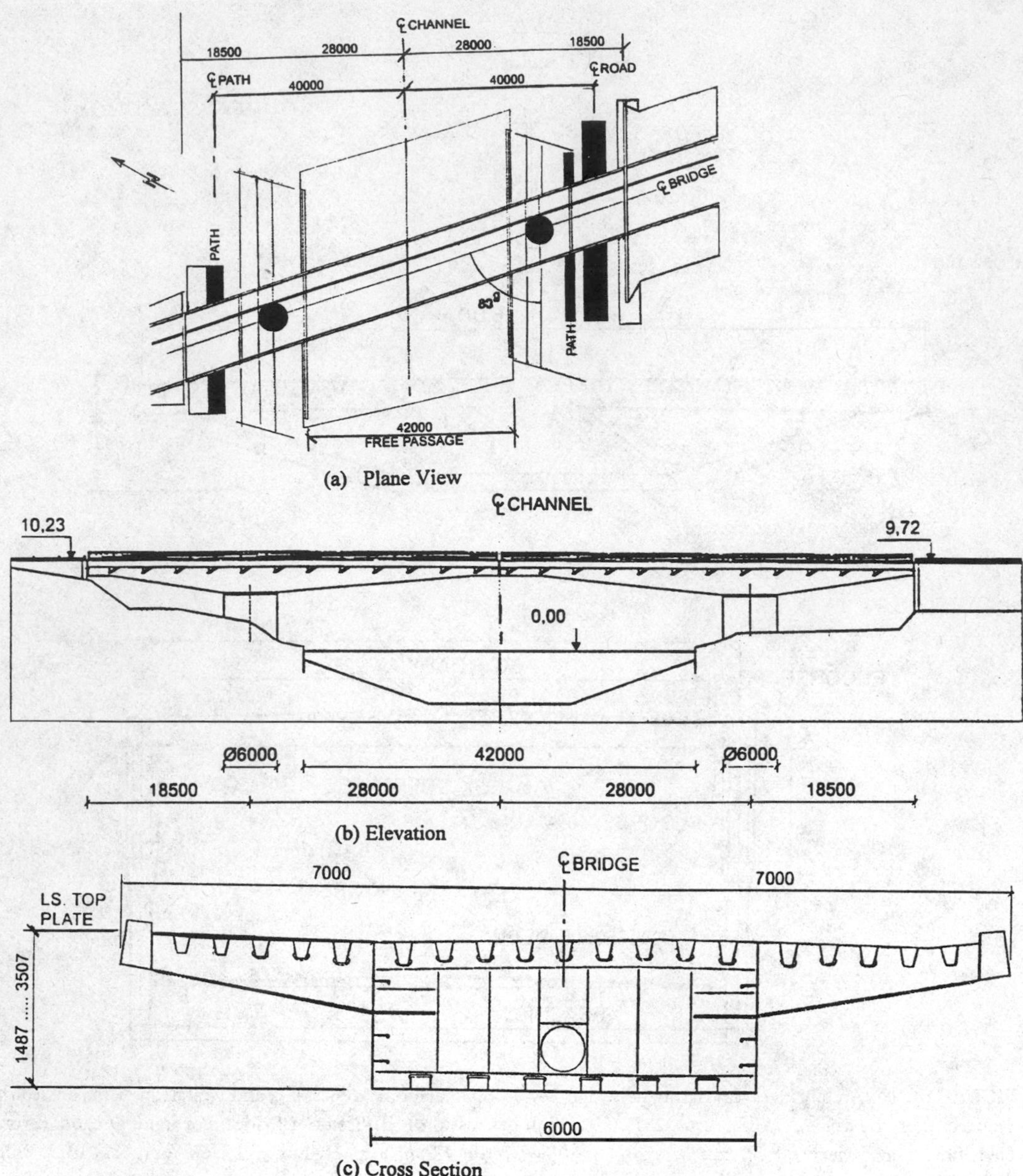

FIGURE 14.11 Swing bridge near Naestved, Denmark. (a) Plane view; (b) elevation; (c) cross section. (Thomsen, K. and Pedersen, K. E, *Structural Engineering International,* IABSE, 8(3), 201, 1998. With permission.)

main component is a 6-m-wide variable-depth box girder with tapered cantilevered floor beams (Figure 14.11b). The top deck plate is stiffened by a cold-rolled trapezoidal rib (Figure 14.11c). The bottom flange plate of the box girder is stiffened by a cold-rolled rectangular-shaped rib. The two exterior web plates of the box girder are stiffened by a bulb-shaped rib. The two sections were fabricated in a shop and barged to the bridge site utilizing the navigation channel. The exterior is painted, and the interior uses dehumidification equipment (see Section 14.4.6) in each of the two sections. The bridge opened to traffic in 1997.

There is an orthotropic swing bridge at the southern mouth of a small slough east of the main channel of the St. Clair River on the Walpole Island Indian Reservation, Ontario, Canada built in

1970. This movable bridge allows small pleasure craft traveling between Lake Huron and Lake St. Clair to pass through. Vehicles can travel to and from the island across this bridge, which was the first movable bridge built with an orthotropic deck in North America. The advantages of this solution are discussed in detail in Reference [35].

The Danziger Vertical Lift Bridge, completed in 1988, is the world's widest vertical-lift bridge and carries seven lanes of traffic on U.S. 90 through downtown New Orleans across the Industrial Canal. The orthotropic deck that is lifted is a 33 m wide × 97 m span supported by three steel 4.26 m deep × 1.82 m wide box beams. The spacing of the box beams is 11.5 m with split-T shaped tapered floor beams at 4.42 m on center. The cantilever on the floor beams is 3.31 m from the face of the box girder. The rectangular boxes are fabricated of ASTM A572 and A588 steel for the main plate and A36 steel for secondary members including all steel median barriers. The ASTM A572 ribs are 5⁄16-in.-thick, 10-in.-deep, 27.88-plf rib as shown in Figure 14.7.

The world's largest double-leaf bascule bridge was opened to traffic in 1969 in Cadiz, Spain [2]. The main girders cantilever 48.3 m, providing a channel between Puerto de Santa Maria and Cadiz of 96.7 m. The orthotropic deck spans 2 m from between split-T-shaped transverse floor beams, which cantilever 2.6 m. At each side of the 12-m deck plate are sidewalks. The two main plate girders are tapered, with maximum depth of 5 m, and are 6 m on center. Sway struts are between floor beam and midspan of plate girders. The signature bridge Erasmus of Rotterdam, the Netherlands utilizes trapezoidal orthotropic deck on both the cable-stayed portion and bascule span. This 33-m-wide by 50-m-long bascule span is skewed at 22°, with an opening of 56 m. The bridge has a very thin, 8-mm wearing surface and was opened in 1996. The Miller Sweeny bridge of Alameda Island California is the only orthotropic bascule bridge in North America.

A unique concept is to have an entire 11.6-m-wide by 33-m-long midsection removed by two cranes to allow ship traffic to pass through once every 2 to 3 years [36]. A conventional concrete box girder bridge supports a drop-in orthotropic box-girder component that has a much smaller mass than concrete. This allows two smaller cranes to move the drop-in unit. The ribs were fabricated from 610-mm-wide plate, a standard plate dimension in the United States. Four of these plates would be cut without waste from warehoused stock plate received directly from the factory. Apparently, there have been no plate optimization studies performed by the steel industry. Also 254-mm-thick urethane foam insulation was sprayed on the bottom face of the steel deck to reduce the tendency of the steel deck to change temperatures more quickly.

A much more dramatic system is planned to move an entire 410-m superstructure in Japan. An all-steel superstructure is planned for the Yumeshima–Maishima Bridge in the "Tech Port Osaka" to be completed in the year 2000. Each end of the all-steel bridge is supported by a 58 × 58 m steel pontoon and will be moved by tugboats. This would be the world's largest movable bridge, with a deck area of 12,000 m^2 [37]. A scale model has been built and the estimated cost of the completed bridge is U.S. $400 million.

A unique civil engineering structure is the curved tidal surge gates of Rotterdam, The Netherlands. The two floating gates, each about the size of the Eiffel Tower, are made of orthotropic deck with trapezoidal ribs. Each gate has 20,500 tons of steel, with 14 mm deck plate. A seawater ballast system is used to adjust this structure to various tidal surge freeboard heights.

14.3.5 Truss Bridges

The German Federal Railways uses its standard orthotropic deck system, shown in Figure 14.5, for one-track bridges using a steel truss superstructure. The lateral bracing for the through truss is provided by the stiffness of the orthotropic deck. The standardization of their steel bridge deck plus floor beams keeps fabrication cost to a minimum [18].

A steel truss is used in both the transverse and longitudinal directions for the double-wall steel pontoon for the port of Iquitos on the Amazon River, in Peru (Figure 14.12a). The orthotropic steel deck with trapezoidal rib is used for the top deck that supports vehicular traffic. The side walls that

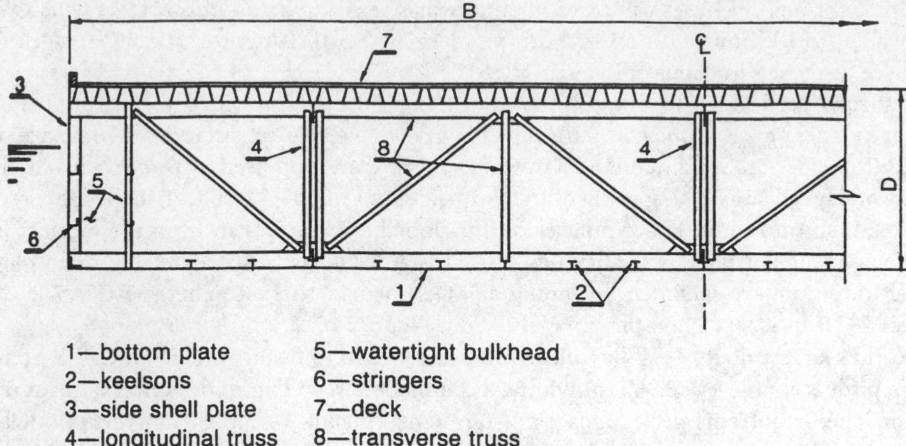

Double-wall steel pontoon (typical cross section).
(a)

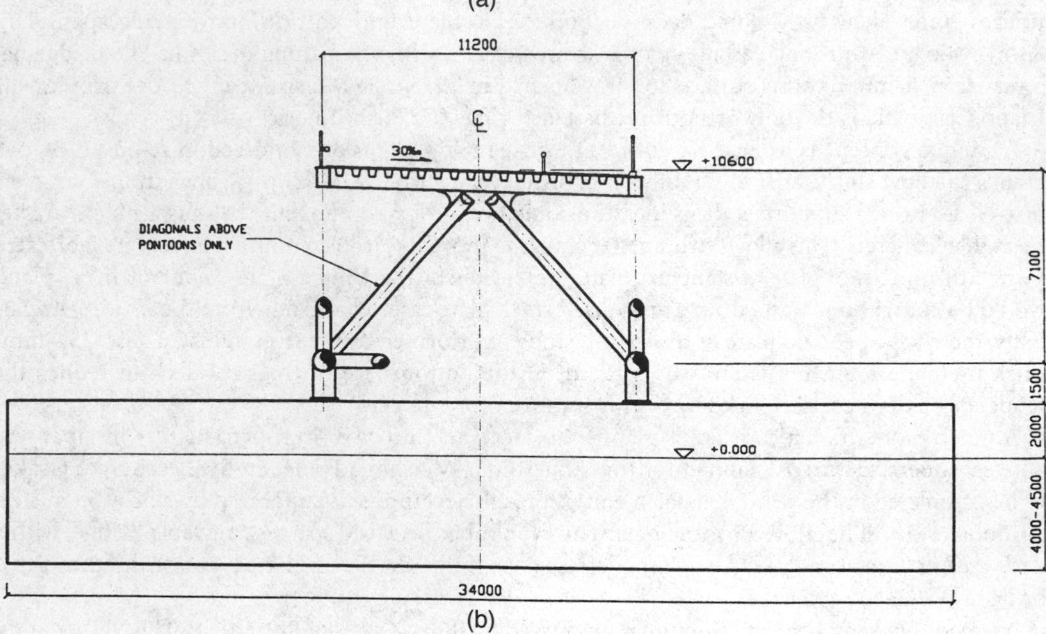

(b)

FIGURE 14.12 Truss bridges. (a) Floating dock on Amazon River. (From Tsinker, G. P., *Floating Ports Design and Construction Practices*, Gulf Publishing Company, Houston, TX, 1986. With permission.) (b) Bergsoysund Floating Bridge. (From Solland, G., Stein, H., and Gustavsen, J. H., *Structural Engineering International*, IABSE, 3(3), 142, 1993. With permission.)

have lower water pressure utilize small angles for the ribs. The bottom plate ribs, or keelsons (nautical terminology), are larger split-Ts because of the higher water pressure. This watertight floating pontoon is completely built of welded steel [38]. This framing system is also known as a "space truss" commonly used for building roofs.

The Bergsoysund Floating Bridge comprises floating concrete pontoons with a painted steel truss superstructure (Figure 14.12b). Floating orthotropic bridges become very economical for Norwegian fjords, which are actually deeper than the adjacent Atlantic Ocean floor. Lateral stability of the entire bridge is provided by an arch shape (in plan view) rather than cables with anchors in the 300-m-deep

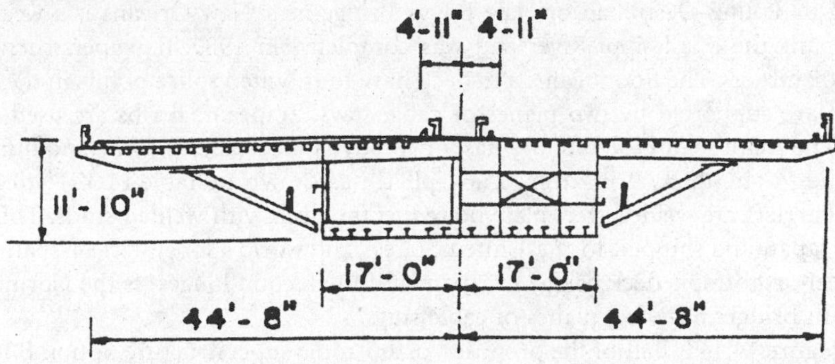

Cross section.

FIGURE 14.13 Papineau–Leblanc cable-stayed bridges. (From Troitsky, M. S., *Orthotropic Steel Deck Bridges,* 2nd ed., JFL Arc Welding Foundation, Cleveland, OH, 1987. Courtesy of the James F. Lincoln Arc Welding Foundation.)

fjord. The lateral stability of the top chords of the trusses is assisted by transverse stiffness of the orthotropic deck. The three-dimensional space truss is built of hollow steel pipe tubular joints, which have the minimum exposed area to resist corrosion. Detailing and design of these joints were based on experience developed for tubular offshore structures built in the North Sea. The closed trapezoidal rib was used, since the bridge is totally exposed to corrosive saltwater spray. Also, it was very important to minimize total bridge weight to reduce the size of the concrete pontoons. To avoid future painting in the ocean water, concrete was selected over orthotropic pontoons similar to Figure 14.12a. This bridge is a state-of-the-art solution utilizing offshore oil platform technology combined with floating bridge design technology [39].

14.3.6 Cable-Stayed Bridges

In cable-stayed bridges, the superstructures can be supported by one or two planes or lines of cables. Additional compression stresses occur in a cable-stayed bridge superstructure where the orthotropic deck is the compression component since the cables are the tension component. The Papineau–Leblanc Bridge, completed in 1969, linking the city of Montreal to Laval Islands, Canada, is a strutted deck box girder supported by a single line of cable stays at the centerline of the superstructure as shown in Figure 14.13. The bridge has spans of 90, 241, and 90 m with a superstructure width of 27.44 m [2,40]. Extra diaphragm plates were located at the cable support locations to transfer the loads from the deck into the cable stay as shown in right side of the split cross section of Figure 14.13. Closed U-shaped ribs were used with the top 7⁄16-in.-thick top deck plate and open ribs for the lower flange plate. The two bridge piers are cone-shaped. The pier face is at a 23° slope to bend the river ice, thus breaking the ice into pieces.

The Bratislava Bridge, completed in 1972 with a main span of 303 m crosses the Danube River, a major transportation river for barges, in Czechoslovakia. The orthotropic superstructure is a double-deck cellular box girder supported by a single line of cable stays at the centerline of the superstructure. The bridge has an anchor span of 75 m, and a superstructure width of 21 m. The feature that makes this bridge a unique signature span is the circular coffeehouse on top of the 85-m-high A-frame tower. Tourists can ride up an elevator in one tower leg to reach the sight-seeing windows. An emergency staircase is located in the other steel tower leg. Another nice feature is the protected pedestrian walkways on each side of the lower orthotropic deck. This feature gives wind and rain protection. The interior of the cross section contains utilities that cross the Danube [40, 41]. The coffeehouse is "saucer shaped," probably inspired by the Seattle Space Needle. The framing consists of steel bowstring trusses for the roof and floor.

The Luling or Luling–Destrehan or Hale Boggs Bridge near New Orleans is a weathering steel bridge that spans the Mississippi River and was completed in 1983. Its superstructure has twin trapezoidal box girders. The floor beams and deck have four bolted splice points in the longitudinal direction and are supported by two planes of cable stays. Trapezoidal ribs are used for the deck system [42]. The main span is 383 m and has an aerodynamic shape to withstand hurricanes. The bridge uses the 5⁄16-in.-thick, 9-in.-deep, 25.64-plf rib as shown in Table 14.3. The center barrier and exterior barriers are welded steel plate bolted to the deck with welded studs. This bridge was fabricated in Japan and shipped to the United States. The world's longest clear-span cable-stayed bridge is a steel orthotropic deck bridge in Japan, and the second longest is the Normandie Bridge in France. Both bridges have two planes of cable stays.

Shown in Figure 14.14 is half of the proposed orthotropic superstructure option B for the cable-stayed replacement bridge for the East Span of the San Francisco–Oakland Bay Bridge. A family of solutions with single and dual towers has reached the 30% design development. The concept illustrated has a divided or separated superstructure connected by steel stiffening trusses to be built of steel tubes. Each half is planned to carry five lanes of traffic and be supported by a single tower. The separated superstructure allows the wind to flow around each side, as well as through the center, reducing wind forces [44].

14.3.7 Suspension Bridges

Suspension bridge superstructures may be supported by one or two planes of cables. Shown in Figure 14.15 is the orthotropic aerodynamic superstructure option for the suspension replacement bridge for the East Span of the San Francisco–Oakland Bay Bridge. A family of solutions with single and dual towers has reached the 30% design development. The concept illustrated has a separated superstructure, and each bridge is planned to carry five lanes of traffic in one direction. Each half is actually an independent bridge, and this superstructure solution is based on the British Severn Bridge completed in 1966. Note how each rib has a different cutout hole to eliminate fatigue cracks depending on the rib shape [44]. The San Francisco–Oakland Bay Bridge East replacement Spans design is currently evolving, and final approved plans have not yet been completed.

The Konohana Bridge in Japan has a main span of 300 m with side spans of 120 m [45]. This is a monocable self-anchoring suspension bridge. Its superstructure is supported only at the centerline of the bridge (Figure 14.16a). This signature span is supported by suspender cables attached to the hanger connection plate. An isometric of the main stiffening plates and diaphragms of the superstructure is shown in Figure 14.16a. Note the concentration of plates needed to distribute cable forces throughout the superstructure. The cantilevering sides of the superstructure mandate an aerodynamic box-girder superstructure. The Japanese rib 242 mm × 324 mm × 8 mm was used (Table 14.3).

Shown in Figure 14.16b is the suspension bridge with two planes of cables and a 762 m main span currently under final design for the Third Carquinez Strait Bridge between San Francisco and Sacramento, California. It is an aerodynamic superstructure with trapezoidal ribs for the deck and bottom soffit [46]. This bridge design has been heavily influenced by the successful bridges built in Europe and Japan.

Shown in Figure 14.17 is a drawing of the world's longest concept suspension bridge, proposed to span the straits of Messina [47, 48]. This clear span between Sicily and Italy has been a historical challenge to engineers. This concept has been given technical approval for the final design by the various Italian authorities and political approval is in progress. European engineers have already performed wind tunnel studies in order to develop contract plans and specifications. The superstructure is supported by two pairs of cables and a 3000-m main span is composed of the middle of welded steel framing system that is 52 m wide. A separated superstructure concept allows wind to flow through the superstructure. The Messina and the option B for the cable-stayed replacement bridge for the East Span of the San Francisco–Oakland Bay Bridge are similar in principle. The

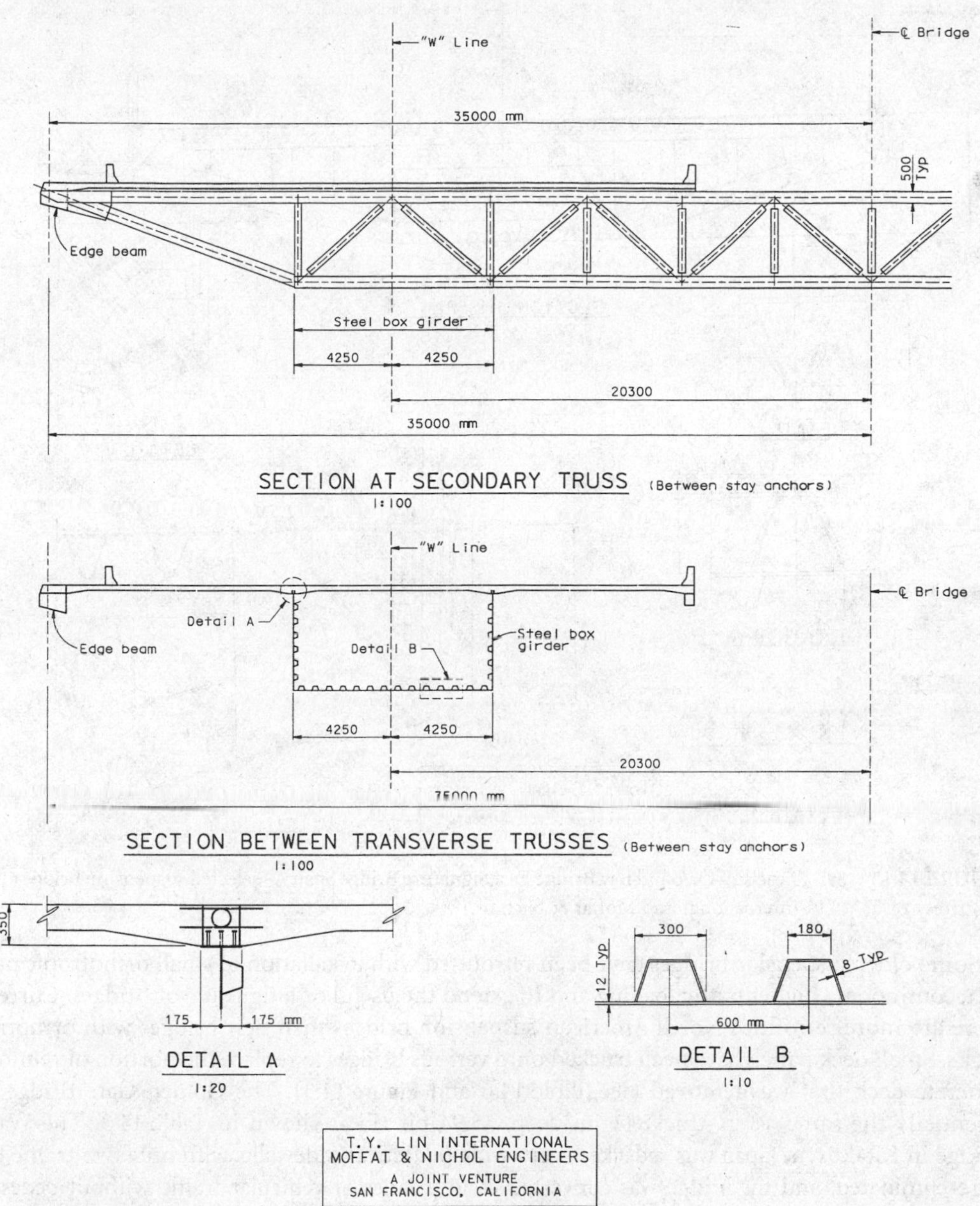

FIGURE 14.14 San Francisco–Oakland Bay Bridge East Signature Bridge Span — proposed cable-stayed bridge option. (Courtesy of T. Y. Lin International and Moffat & Nichols.)

open spaces, while strongly increasing the aeraelastic stability (e.g. flutter stability), reduce wind loading and have a grillage or grating that allow lanes for emergency stop and maintenance vehicle traffic. Three longitudinal independent wing-shaped box girders are linked transversely with very large welded steel box-shaped cross girders. The orthotropic deck is stiffened by trapezoidal ribs for the top deck and open ribs for the bottom soffit. All the barriers have aerodynamic shape and grating to reduce wind forces. New suspension bridges with double-decker superstructures with an upper orthotropic deck for vehicles and a lower orthotropic deck for commuter trains are in use in Asia. Examples are the Tsing Ma Bridge of Hong Kong, China and Yong Jong Grand self-anchoring suspension bridge of Seoul, Korea.

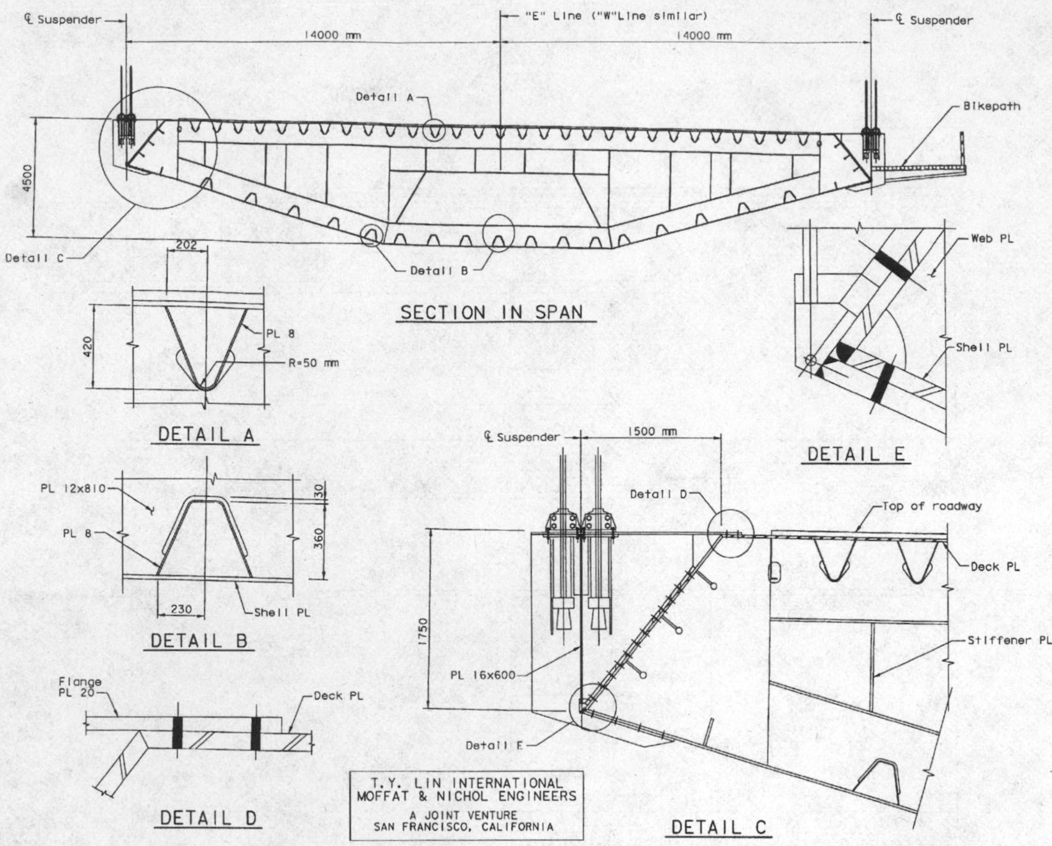

FIGURE 14.15 San Francisco–Oakland Bay Bridge East Signature Bridge Span — selected suspension bridge option. (Courtesy of T. Y. Lin International and Moffat & Nichols.)

Some older suspension bridges have been retrofitted with installation of small orthotropic panels to accommodate higher traffic loading and to extend the useful or fatigue life of bridges. Currently, there are more retrofitted North American suspension bridges than new bridges with orthotropic decks. Small deck panels have been trucked onto various bridges to replace the portion of reinforced concrete deck that was removed (see Table 14.6 and Figure 14.3). The Golden Gate Bridge uses essentially the same 3⁄8-in.-thick, 11-in.-deep, 35.94-plf rib as shown in Table 14.3. The Wakato Bridge in Kitakusyu, Japan was redecked, very wide pedestrian sidewalks with only two traffic lanes were eliminated, and the bridge was converted to four lanes of vehicular traffic without pedestrian access. The historical Williamsburg Suspension Bridge, built in 1903, is the most recent redecking project and, essentially, uses this rib shape again [49]. Extensive testing of a full-size mockup of the designer's concept was performed to verify its durability or fatigue life. An extra internal plate or miniature diaphragm or rib bulkhead aligns with the floor beam web. Welding detail options and cutout holes or scallops were tested and showed that the selected system should have a long fatigue life.

14.4 Design Considerations

14.4.1 General

In contrast to the conventionally designed bridge, where the individual structural elements (stringers, floor beams, and main girders) are assumed to perform separately, an orthotropic steel deck bridge is a complex structural system in which the component members are closely interrelated.

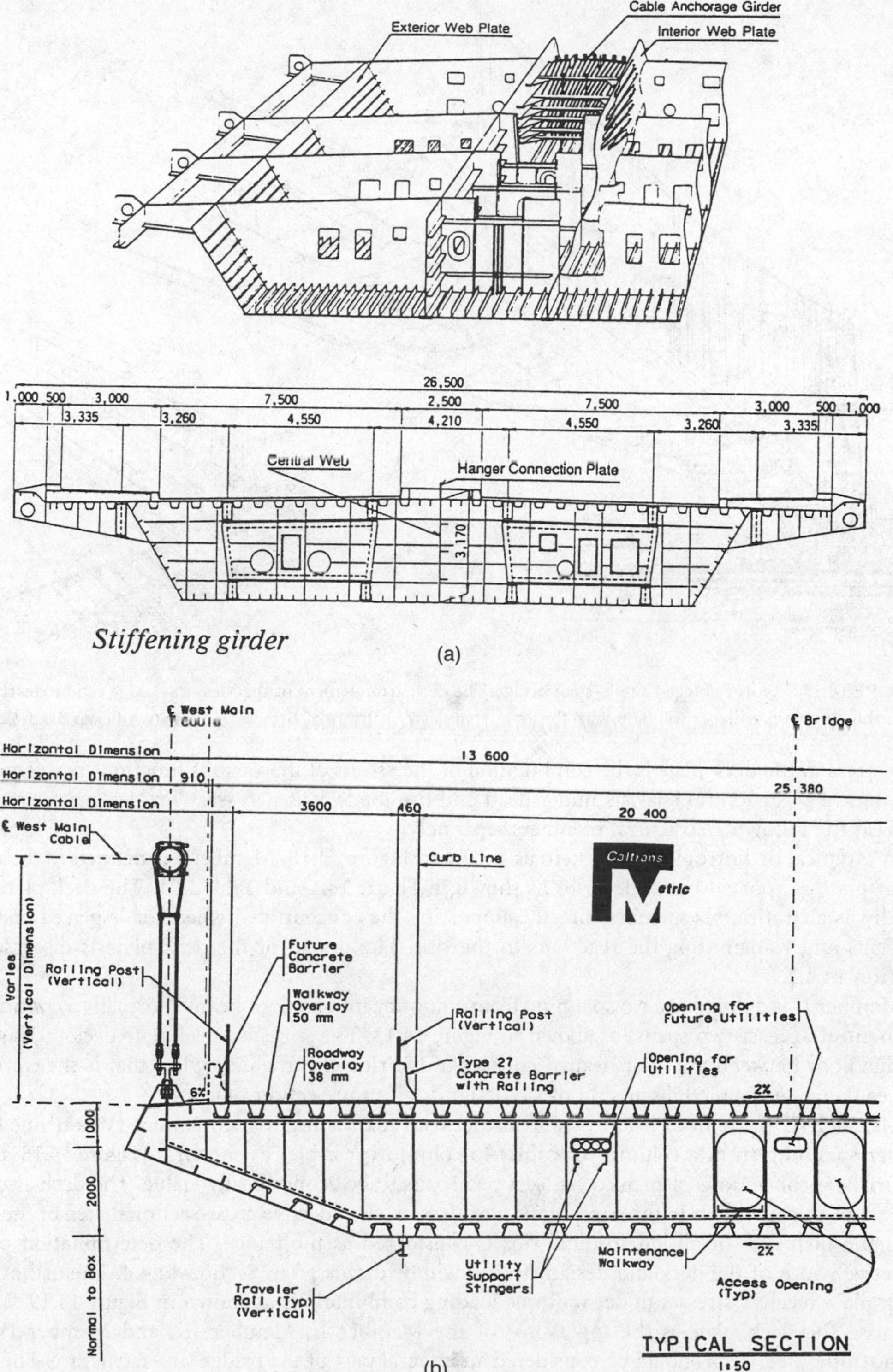

FIGURE 14.16 Suspension bridges (a) Konohana Bridge superstructure. (From Kamei, M., Maruyama, T. and Tanaka H., *Structural Engineering International,* IABSE, 2(1), 4, 1992. With permission.) (b) Third Carquinez Bridge stiffening girder. (Courtesy of De Leuw-OPAC-Steinman.)

FIGURE 14.17 World's biggest single-span bridge. The deck arrangement of the Messina Bridge can be clearly seen from this drawing; grillages will help with the aerodynamics. (Courtesy of Brown, W. at Brown Beech & Associates.)

The stress in the deck plate is the combination of the effects of the various functions performed by the deck. For structural analysis under dead and live loads, it is necessary for design convenience to treat the following structural members separately.

In a typical orthotropic deck system as shown in Figure 14.18, Member I is defined as the deck, a flat plate supported by welded ribs as shown in Figure 14.1 and Table 14.1. The deck plate acts locally as a continuous member directly supporting the concentrated wheel loads placed between the ribs and transmitting the reactions to the ribs. The design of the deck plate is discussed in Section 14.4.2.

Member II is defined as a rib spanning from a floor beam to a floor beam (normally a continuous element of at least two spans) as shown in Figure 14.18. The stiffened steel plate deck (acting as a bridge floor between the floor beams) consists of the ribs plus the deck plate that is the common upper flange. A detailed discussion of rib design is given in Section 14.4.3.

Member III is the floor beam that spans between the main girders. Member IV is defined as a girder spanning from a column (or cable) to column (or cable) as shown in Figure 14.18 and is normally a continuous element of at least two spans to be economically viable. The deck also acts as part of this member. In the computation of stresses, the effective cross-sectional area of the deck plus the inclusion of all longitudinal ribs is considered as the flange. The determination of the effective width of the deck and design stresses will be discussed in Section 14.4.4. The orthotropic deck plate receives stresses under multiple loading combinations as shown in Figure 14.19. This is because the deck plate is the top flange of the Member II, Member III, and Member IV. An orthotropic steel deck should be considered an integral part of the bridge superstructure. The deck plate acts as a common flange of the ribs, the floor beams, and the main longitudinal components of the bridge. Any structural arrangement in which the deck plate is made to act independently from the main components is undesirable.

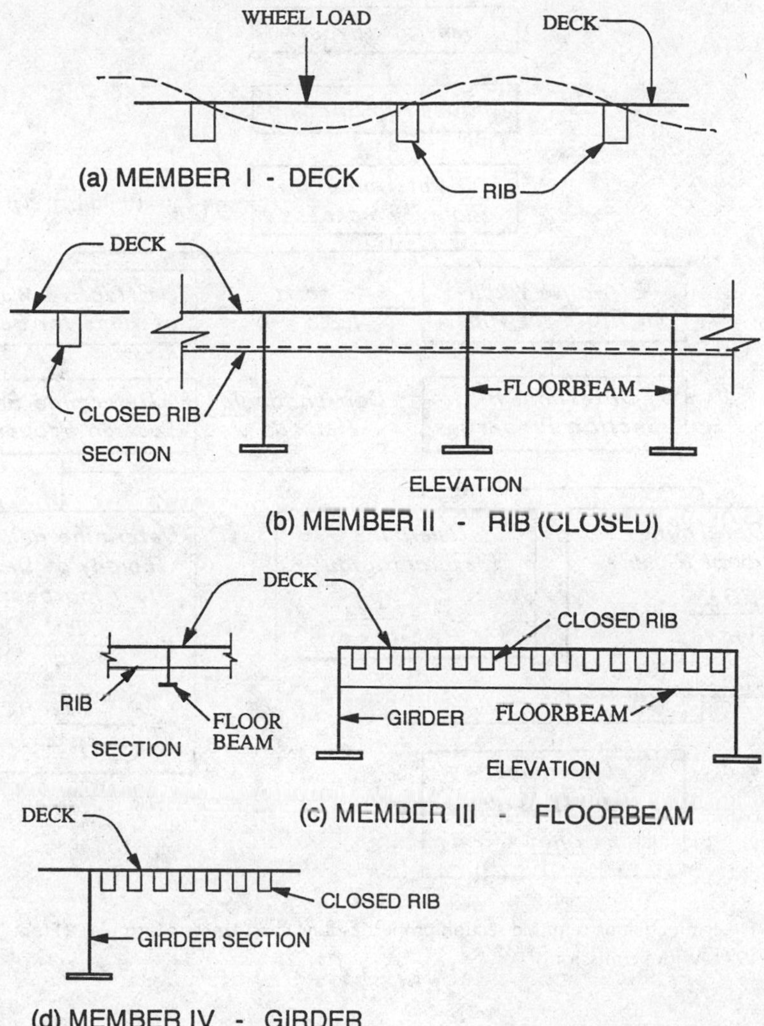

FIGURE 14.18 Four members to be analyzed in an orthotropic deck. (a) Member I — deck; (b) Member II — rib (closed); Member III — floorbeam; (d) Member IV — girder.

When redecking the bridge, if the orthotropic deck is supported by existing floor beams, the connection between the deck and the floor beam should be designed for full composite action, even if the effect of composite action is neglected in the design of floor beams. Where practical, connection suitable to develop composite action between the deck and the main longitudinal components should be provided.

The effects due to global tension and compression should be considered and combined with local effects. When decks are in global tension, the factored resistance of decks subject to global tension, *Pu,* due to the factored loads with simultaneous global shear combined with local flexural must satisfy [5]:

$$\frac{P_u}{P_r} + \frac{M_{lr}}{M_{rr}} \leq 1.33 \tag{14.1}$$

$$P_u = A_{d,\mathrm{eff}}\sqrt{f_g^2 + 3f_{vg}^2} \tag{14.2}$$

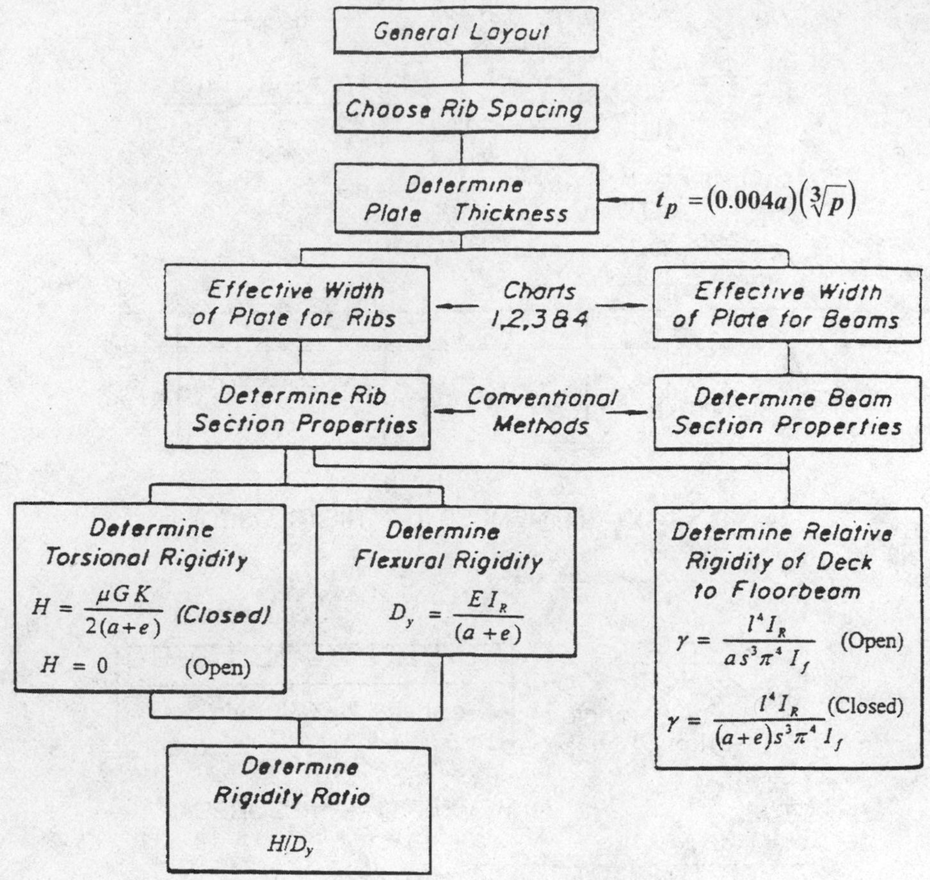

FIGURE 14.19 Determination of required section properties flow chart (metric formula). (From Milek, W. A., Jr., *Eng. J. AISC*, 40, 1974. With permission.)

where

f_g = axial global stress in deck (MPa)
f_{vg} = simultaneous global shear in the deck (MPa)
$A_{d,\,eff}$ = effective cross section area of the deck, including longitudinal ribs (mm^2)
P_r = nominal tensile resistance of the deck with consideration of effective deck width (N)
M_{lr} = local flexural moment of longitudinal rib due to the factored loads (N-mm)
M_{rr} = Flexural resistance of longitudinal rib, governed by yielding in extreme fiber (N-mm)

The effect of simultaneous shear is usually not significant in orthotropic decks of girder or truss bridges, but may be important in decks used as tension ties in arches or in compression for cable-stayed bridges.

When decks are under global compression, longitudinal ribs, including effective width of deck plate, should be designed as individual columns assumed to be simply supported at transverse beams. Buckling formulas for steel decks can be found in the AISC *Design Manual for Orthotropic Steel Plate Deck Bridges* [1].

Diaphragms or cross frames should be provided at each support and should have sufficient stiffness and strength to transmit lateral forces to the bearings and to resist transverse rotation, displacement, and distortion. Intermediate diaphragms or cross frames should be provided at locations consistent with the analysis of the girders and should have sufficient stiffness and strength to resist transverse distortions.

14.4.2 Deck Design

The primary function of the steel deck (Member I) is to directly support the traffic loads and to transmit the reactions to the longitudinal ribs. An important characteristic of the steel deck is its capacity for carrying concentrated loads. When loads approach the ultimate load the deck plate practically acts as a membrane and can carry on the order of 15 to 20 times the ultimate load computed in accordance with the ordinary flexural theory. Thus, the bridge deck plate possesses an ample local overload-carrying capacity.

The minimum thickness of the deck plate may be determined by allowable deflection of the deck plate under a wheel load, which should not exceed 1/300 of the spacing of the deck supports. Based on this criteria, the plate thickness, t_p, may be determined by Kloeppel's formula:

$$t_p \geq (0.004\,a)\left(\sqrt[3]{p}\right) \tag{14.3}$$

where:

a = spacing of the open ribs, or the maximum spacing of the walls of the closed ribs, in mm
p = wheel load unit pressure, under the AASHTO LRFD design tandem wheel load 55 kN, including 33% dynamic load allowance, in kPa. For 50 mm wearing surface, p is 449 kPa.

The distribution of the wheel load is assumed in a 45° footprint from the top of the wearing surface to the top of the deck plate. The AASHTO Specifications, 16th edition [50] tabulates wheel loads and contact area:

Wheel Load (kN)	Width Perpendicular to Traffic (mm)	Length Direction of Traffic (mm)
36	508 + 2t	203 + 2t
54	508 + 2t	203 + 2t
72	610 + 2t	203 + 2t

t = the thickness of the wearing surface in mm.

Using the AASHTO-LRFD [5], the wheel loads and contact area can be tabulated:

Wheel Load (kN)	Width Perpendicular to Traffic (mm)	Length Direction of Traffic (mm)
17.5 (truck)	510 + 2t	53 + 2t
55 (tandem)	510 + 2t	167 + 2t
72.5 (truck)	510 + 2t	220 + 2t

t = the thickness of the wearing surface in mm.

The current AASHTO-LRFD [5] requires that the minimum deck plate thickness, t_p, shall not be less than either 14 mm or 4% of the largest rib spacing. Experience from the durability of previously built bridges shows that this requirement is advisable for both constructibility and long-term bridge life.

For a rib spacing of 300 mm, a 14-mm plate is required per the AASHTO-LRFD, while a 9-mm plate can be derived using Eq. 14.3. For a rib spacing of 380 mm, a 15-mm plate is required per the AASHTO-LRFD, while a 12-mm can be derived using Eq. 14.3.

14.4.3 Rib Design

Table 14.2 gives the effective width of the deck plate acting with a rib. Since most of the ribs used in the current practice are closed type, this section will only discuss closed ribs. The ribs span between and are continuous at floor beams (Figure 14.18). Spacing of ribs depends on the deck plate thickness and usually for closed ribs $(a + e)$ varies between 610 and 760 mm. Rib spans of approximately 4500 mm have been common in North American practice. But, spans up to 8500 mm, required by spacing of existing floor beams, have been used in bridge redeckings. Long rib spans are feasible and may be economical.

The minimum thickness of closed ribs should not be less than 4.75 mm per AASHTO-LRFD. Fatigue tests concluded that local out-of-plane flexural stress in the rib web at the junction with the deck plate should be minimized. It is necessary to limit the stress in the rib web caused by the rotation of the rib–deck plate junction by making the rib webs relatively slender compared with the deck plate. To achieve this, AASHTO-LRFD [5] specifies that the cross-sectional dimensions of an orthotropic steel deck shall satisfy:

$$\frac{t_r a^3}{t_{d,\mathrm{eff}}^3 h'} \leq 400 \tag{14.4}$$

where

t_f = thickness of rib web (mm)

$t_{d,\mathrm{eff}}$ = effective thickness of the deck plate, with consideration of the stiffening effect of the wearing surfacing (mm)

a = largest spacing between the rib webs (mm)

h' = length of the inclined portion of the rib web (mm)

To prevent overall buckling of the deck under compression induced by the bending of the girder, the slenderness, L/r, of a longitudinal rib shall not exceed the value given by the following equation in the AASHTO Standard Specification [50]:

$$\left(\frac{L}{r}\right)_{\max} = 83\sqrt{\frac{1500}{F_y} - \frac{2700F}{F_y^2}} \tag{14.5}$$

where

L = distance between transverse floor beams

r = radius of gyration about the horizontal centroidal axis of the rib including an effective width of the deck plate

F = maximum compressive stress in MPa in the deck plate as a result of the deck acting as top flange of the girder; this stress shall be taken as positive

F_y = yield strength of rib material in MPa

Orthotropic analysis furnishes distribution of loads to ribs and stresses in the member. Despite many simplifying assumptions, orthotropic plate theories that are available and reasonably in accordance with testing results and behaviors of existing structures require long, tedious computations. Computer modeling and analysis may be used to speed up the design. The AISC manual [1] was used when only expensive main-frame computers were available. The flowchart for this process is shown in Figure 14.19. Tables 14.3 to 14.5 were later created to assist those without computers. Today, engineers can write their own software or create the appropriate spreadsheet using a personal computer to expedite the iterative computations. Published tables provide useful databases to check against "bugs" in software. The following method, known as the Pelikan–Esslinger method, has been used in design of orthotropic plate bridges. In this method, the closed-rib decks are analyzed in two stages.

First, the deck with closed ribs is assumed on rigid supports. Only the longitudinal flexural rigidity and the torsional rigidity of the ribs are considered. The transverse flexural rigidity can be negligible. A good approximation of the deflection w may be presented in the form of Huber's differential equation:

$$D_y \frac{\partial^4 w}{\partial y^4} + 2H \frac{\partial^4 w}{\partial x^2 \partial y^2} = p(x, y) \tag{14.6}$$

where

D_y = flexural rigidity of the substitute orthotropic plate in the direction of the rib

H = equivalent torsional rigidity of the substitute orthotropic plate

$p(x,y)$ = load expressed as function of coordinates x and y

In the computation of D_y and H, the contribution of the plate to these parameters must be included. The flexural rigidity in the longitudinal direction usually is calculated as the rigidity of one rib with effective deck width divided by the rib spacing:

$$D_y = \frac{EI_y}{a+e} \tag{14.7}$$

where I_r = moment of inertia, including a rib and effective plate width.

Because of the flexibility of the orthotropic plate in the transverse direction, the full cross section is not completely effective in resisting torsion. Therefore, the formula for computing H includes a reduction factor u.

$$H = \frac{\mu GK}{2(a+e)} \tag{14.8}$$

where

G = shearing modulus of elasticity of steel

K = torsional factor, a function of the cross section

In general, for hollow closed ribs, the torsional factor may be obtained from

$$K = \frac{4A_v^2}{\frac{p_e}{t_r} + \frac{a}{t_p}} \tag{14.9}$$

where

A_r = mean of area enclosed by inner and outer boundaries of ribs

p_r = perimeter of rib, exclusive of top flange

t_r = rib thickness

t_p = plate thickness

The reduction factor u for a trapezoidal rib may be closely approximated by

$$\frac{1}{u} = 1 + \frac{GK}{EI_p} \frac{a^3}{12(a+e)^2} \left(\frac{\pi}{S_e}\right)^2 \left[\left(\frac{e}{a}\right)^3 + \left(\frac{e-b}{a+b} + \frac{b}{a}\right)^2\right] \tag{14.10}$$

where

I_p = moment of inertia = $Et_p^3 / 12(1-u^2)$

s_2 = effective rib span for torsion = $0.81s$

s = rib span

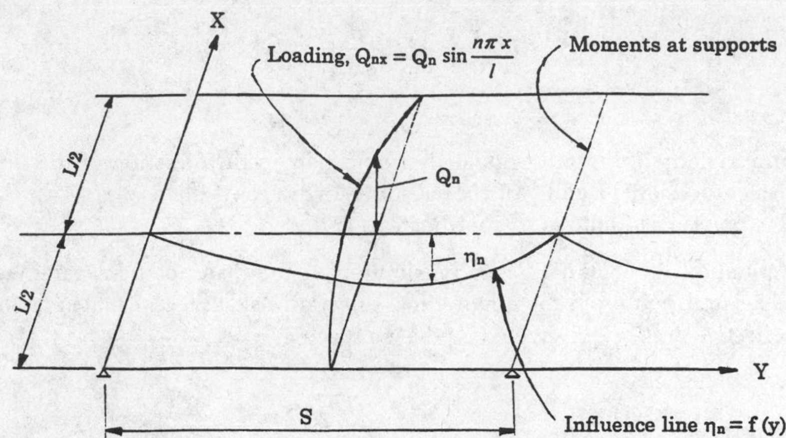

FIGURE 14.20 Computation of bending moments in orthotropic plate.

The values of D_y and H are combined in the relative rigidity coefficient H/D_y, which is a parameter characterizing the load-distributing capacity of the deck in the direction perpendicular to the ribs. For any given rib size, spacing, and deck plate thickness, H does not remain constant but increases with the rib span. Therefore, the parameter H/D_y is also a function of span, and the transverse load distribution of the deck structure improves as the span of the ribs is increased.

The general solution of Eq. (14.4) can only be given as an infinite series:

$$w_n = \left(C_{1n} \sinh a_n y + C_{2n} \cosh a_n y + C_{3n} a_n y + C_{4n}\right) \sin \frac{n\pi x}{l} \tag{14.11}$$

where

n = integer ranging from 1 to ∞ (odd numbers for symmetrical loads)
l = floor beam span
C_{in} = integration constant, determined by boundary conditions

$$a_n = \frac{n\pi}{l} \sqrt{\frac{2H}{D_y}} \tag{14.12}$$

The plate parameter H/D_y can be obtained from Table 14.4 or calculated.

Bending moments in the substitute orthotropic plate due to the given loading can be computed by formulas derived from the above solution. Since the solution can only be given as an infinite series, values of the influence ordinates, bending moments, etc. must be expressed as sums of the component values for each term n of the series of component loads Q_{nx}. The values needed for the design of the ribs are the bending moments in the orthotropic plate over the *floor beams* and at the midspan of the ribs. These moments are obtained by multiplying the values of the component loads Q_{nx} by corresponding influence-ordinate components η_n:

$$M = \sum Q_{nx} \eta_n \tag{14.13}$$

This is shown in Figure 14.20. Formulas and design charts for Q_{nx} and η_n for the various AASHTO loading cases are given in Ref. [1].

The bending moments M_y in the direction of the ribs are obtained in the substitute orthotropic plate system of unit width of the deck. Usually only the maximum moment ordinate $M_{y\max}$ at the center of the loaded rib is computed and the moment acting on one rib is then obtained, conservatively, as

$$M_R = M_{y\max}(a+e) \tag{14.14}$$

In the second stage of the design, the effect of the floor beam flexibility is considered. This effect will result in an increase of bending moment in the middle span and reduction of bending moment at the floor beam. The magnitude of the effect is determined based on the relative rigidity between ribs and floor beam. The effective width of the plate used for the first stage calculations generally can be used for the second stage with small error. Modifications of bending moments and shears in the ribs due the floor beam flexibility may be computed in the formulas and design charts in Ref. [1].

14.4.4 Floor Beam and Girder Design

Dead-load bending moments and shears in a *floor beam* (Member III) are calculated based on its own weight and the weights of tributary area of the deck. Live-load bending moments and shears in a floor beam are computed in two stages. In the first, the floor beams are assumed to act as rigid supports for the continuous ribs. The bending moments and shears are determined using a conventional method.

In the second stage, the effect of the floor beam flexibility is calculated, which tends to distribute the load on a directly loaded floor beam to the adjacent floor beams. The magnitude of this effect is a function of the relative rigidities between ribs and floor beams. The floor beam flexibility will reduce the bending moments and shears calculated in the first stage. The bending moment and shear corrections in the floor beams may be computed by the formulas and charts in Ref. [1].

The floor beam design per AASHTO-LRFD [5] for transverse flexure is

$$\frac{M_{fb}}{M_{rb}} + \frac{M_{ft}}{M_{rt}} \le 1.0 \tag{14.15}$$

where

M_{fb} = applied moment due to the factored loads in transverse beam (N-mm)
M_{rb} = factored moment resistance of transverse beam (N-mm)
M_{ft} = applied transverse moment in the deck plate due to the factored loads as a result of the plate-carrying wheel loads to adjacent longitudinal ribs (N-mm)
M_{rt} = factored moment resistance of deck plate in carrying wheel loads to adjacent ribs (N-mm)

For deck configurations with the spacing of the transverse beams larger than at least three times the spacing of longitudinal rib webs, the second term in Eq. (14.15) may be ignored. It should be noted that the applied moment M_{fb} should be obtained and based on the superstructure configuration.

The methods of analysis and design of the girders (Member IV) are described in other chapters of this book. The effective width of an orthotropic deck acting as the top flange of a girder or a floor beam is a function of the ratio of the span length to the girder or floor beam spacing, the cross-sectional area of the stress-carrying stiffeners, and the type and position of loading. Values of effective widths for the case of uniform loading based on a study by Moffatt and Dowling[58] are given in Figure 14.21 [5]. This figure was originally developed to determine the effective width of deck to be considered active with each web of a box girder, but is also applicable to other types of girders. The cumulative effect of the stresses on a specific orthotropic deck needs to be carefully reviewed by the design engineer and is generalized by Figure 14.22.

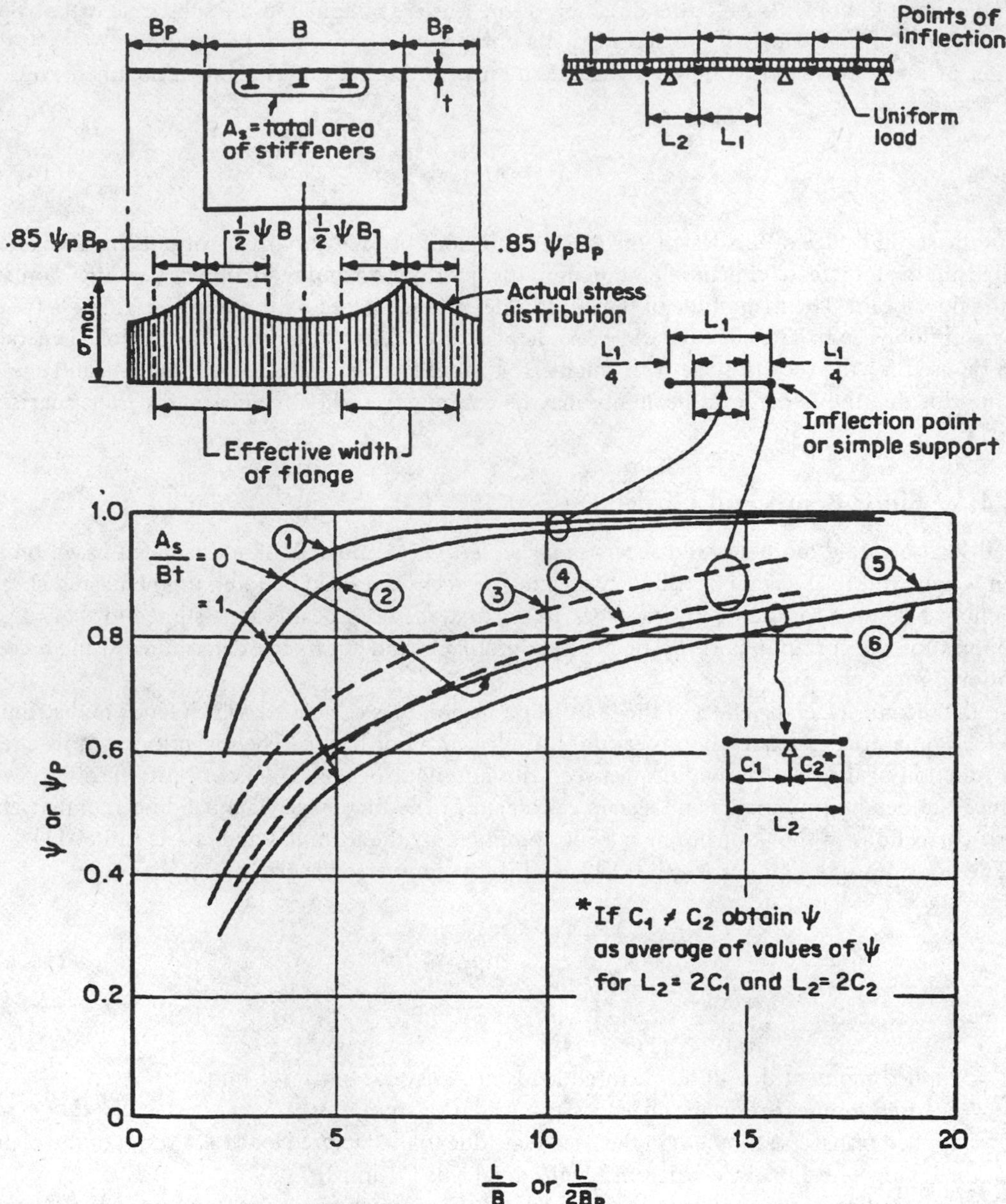

FIGURE 14.21 AASHTO effective deck width. (From American Association of State Highway and Transportation Officials, *LRFD Bridge Design Specifications*, Washington, D.C., 1994.)

14.4.5 Fatigue Considerations

Detailing of an orthotropic bridge is more involved because of all the numerous plates (see Figure 14.16 isometric). In-fill plates and complex geometric detailing make many orthotropic deck bridges one of a kind with unique details. The fatigue strength of the deck plate is very high, but thin deck plates may cause fatigue failures in other components. Fatigue of low-alloy deck plate of usual proportions subject to the AASHTO wheel loads is not considered a critical design factor. Reference [51] shows sketches of several European bridges with steel cracking caused by fatigue stress, and Ref. [52] shows how one bridge with wine-glass ribs was repaired. The original design

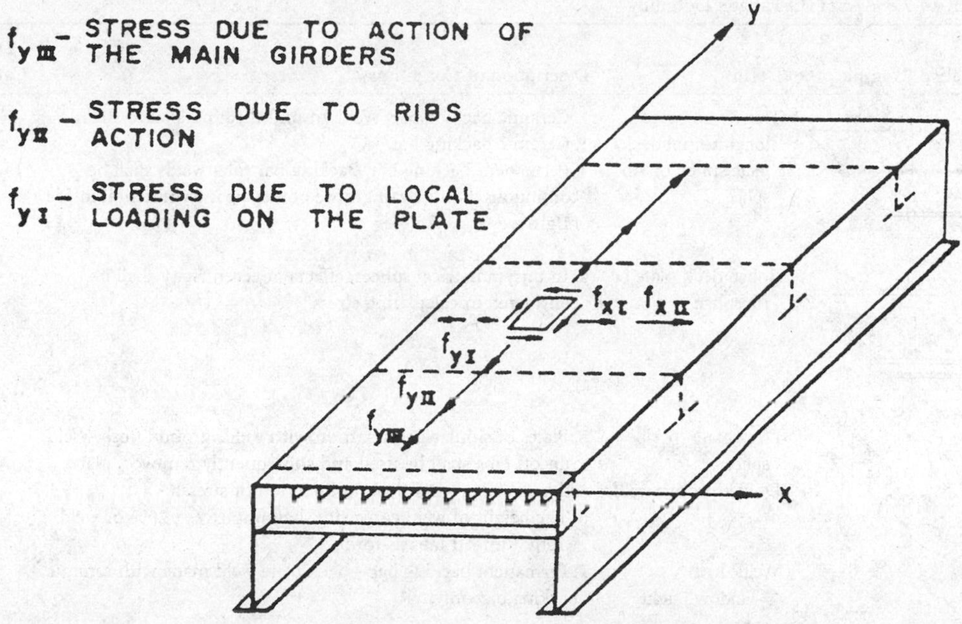

FIGURE 14.22 Resulting stresses in an orthotropic deck. (From Trotsky, M. S., *Orthotropic Steel Deck Bridges*, 2nd ed., JFL Arch Welding Foundation, Cleveland, OH, 1987. Courtesy of the James F. Lincoln Arc Welding Foundation.)

engineers did not fully understand long-term fatigue stress and CAFL (constant-amplitude fatigue life) issues. The AASHTO LRFD commentary [5] states: "Fatigue stress tests indicate that local out-of-plane flexural stress in the rib web at the junction with the deck plate should be minimized. One way to achieve this is to limit the stress in the rib web caused by the rotation of the rib–deck plate junction by making the rib welds relatively slender compared to the deck plate." Fatigue issues add to the complexity of orthotropic box-girder bridge design.

Excessive welding causes shrinkage of the weld metal which may cause additional locked-in stresses. In addition there are more components that are cut to fit between other components. If a component is cut too short this may also cause problems during welding. If additional filler welding is installed to span a larger incorrectly fabricated gap or root opening, then more weld width can increase the probability of shrinkage stresses. Detailing, imperfections in the alignment of a components, and quality control in fabrication have a direct result on the end product. Some designers prefer to let the fabricator select the welding processes (submerged arc welding, gas welding, etc.). Therefore, the welders and management of the fabrication plant are only required to achieve the strength and performance. Both AASHTO specifications require the use of 80% partial penetration welds of the ribs to deck plate (see Figure 14.3).

The AASHTO LRFD [5] maximum allowable stress limit for various details is shown in Table 14.7. The floor beam web cutouts around the rib were developed by the Germans. When ribs are designed to be continuous, it has been found to be best to make the rib go through the floor beam or diaphragm plate. Coping near the bottom of the rib and below greatly reduces the chances of fatigue cracking. There is still research going on about the actual shape of the holes. The cutout pattern per AASHTO is shown in Figure 14.23. However, it should be pointed out that in Figure 14.5 the German Railways has experimented with a different shape. AASHTO does not provide guidance on V-shaped ribs as shown in Figure 14.15. When components align on opposite sides of a plate, AASHTO Detail A must be followed for stress limits. Misaligned plates normally contribute to long-term fatigue failures. Full-scale fatigue testing of the Williamsburg Suspension Bridge has shown that code-acceptable details have a shorter fatigue life than what experts are able to develop with

TABLE 14.7 AASHTO Fatigue Detailing

Illustrative Example	Detail	Description of Condition	Detail Category
	Transverse or longitudinal deck plate splice or rib splice	1. Ceramic backing bar. Weld ground flush parallel to stress.	B
		1. Ceramic backing bar.	C
		3. Permanent backing bar. Backing bar fillet welds shall be continuous if outside of groove or may be intermittent if inside of groove.	D
	Bolted deck plate or rib splice	4. In unsymmetrical splices, effects of eccentricity shall be considered in calculating stress.	B
	Deck plate or rib splice Double groove welds	5. Plates of similar cross-section with welds ground flush. Weld run-off tabs shall be used and subsequently removed, plate edges to be ground flush in direction of stress. 6. The height of weld convexity shall not exceed 20% or weld width. Run-off tabs as for 5.	B
	Welded rib "window" field splice Single groove butt weld	7. Permanent backing bar—deck plate weld made with ceramic backing bar only. Welding gap > rib wall thickness f = axial stress range in bottom of rib	D
	Rib wall at rib/floorbeam intersection Fillet welds between rib and floorbeam web	8. Closed rib with internal diaphragm inside the rib or open rib. Welding gap > rib wall thickness f = axial stress range in rib wall	C
		9. Closed rib, no internal diaphragm inside of rib $f = f_1 = f_2$ f2 = local bending stress range in rib wall due to out-of-plane bending caused by rib–floorbeam interaction, obtained from a rational analysis	C

Source: American Association of State Highway and Transportation Officials, *AASHTO LRFD Bridge Design Specifications,*2nd ed. Washington, D.C., 1998. With permission.

proper funding, as shown in Figure 14.24. This solution was selected as the optimal detail [49]. Termination of a weld is a very common place for a fatigue crack to begin. A runoff tab plate allows the welder to terminate the weld in this piece of steel. The tab plate is cut off and tossed in the recycle bin. The source of a potential flaw is now no longer part of the final structure. The weld tab plate system was introduced to the orthotropic deck detail. This tab plate is cut off with the 1/2-in. radius as shown in Figure 14.24 after welding. This extra step was used on one test panel to compare against test panel without tab plates. These researchers believe that deck plates of 8 to 12 mm are too thin for more than 2 million cycles. For more-detailed discussion of fatigue, see Chapter 53.

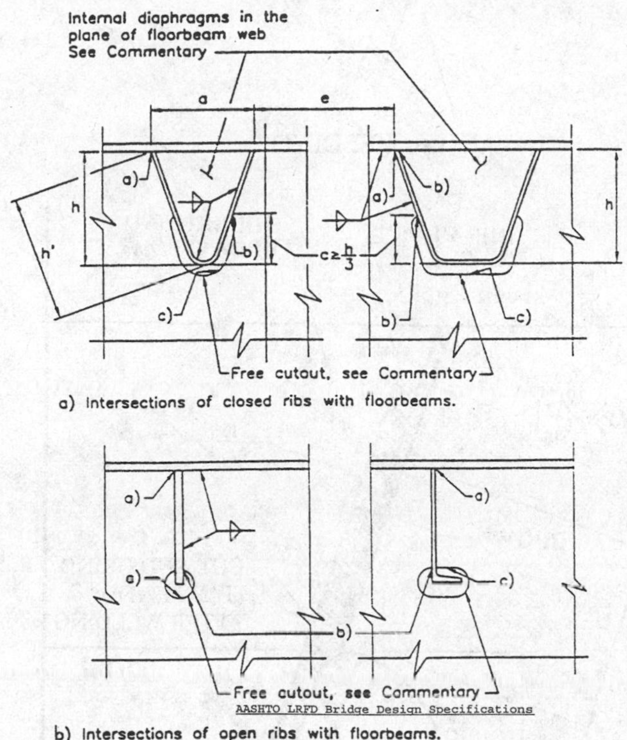

FIGURE 14.23 Detailing requirements for orthotropic deck. (From American Association of State Highway and Transportation Officials, *LRFD Bridge Design Specifications*, 2nd ed., Washington, D.C., 1998.)

14.4.6 Bridge Failures

Unfortunately, from 1969 to 1971 there were four steel box-girder collapses in four different countries, which caused the bridge engineering industry to reevaluate its different design code formulas and methods of erections. The Rhine River Bridge at Koblenz, Germany; the fourth Danube Bridge in Vienna, Austria; the Milford Haven Bridge in Wales, Great Britain; and the Yarra River Bridge in Australia were the four bridges that collapsed, making it a global bridge design issue [53-55]. It is a sobering thought to realize that 35 construction workers died at the Yarra River Bridge collapse [56,57]. This bridge was redesigned with an orthotropic steel deck to reduce its original dead weight. In Great Britain, a Board of Inquiry under the chairmanship of A. W. Merrison produced a list of new design rules and code details for fatigue issues. Extensive testing, research, and symposiums were held in Great Britain [3,4]. The British Institution of Civil Engineers [1979] held a symposium in London to have engineers from around the world share their experiences and ideas. It is very important to remember that codes are imperfect and the long-term fatigue details are still evolving and research continuing [58-64]. When the West Virginia bridge collapsed due to fracture critical failure of a single steel member, 50 individuals died. The FHWA responded by initiating a mandatory bridge maintenance program.

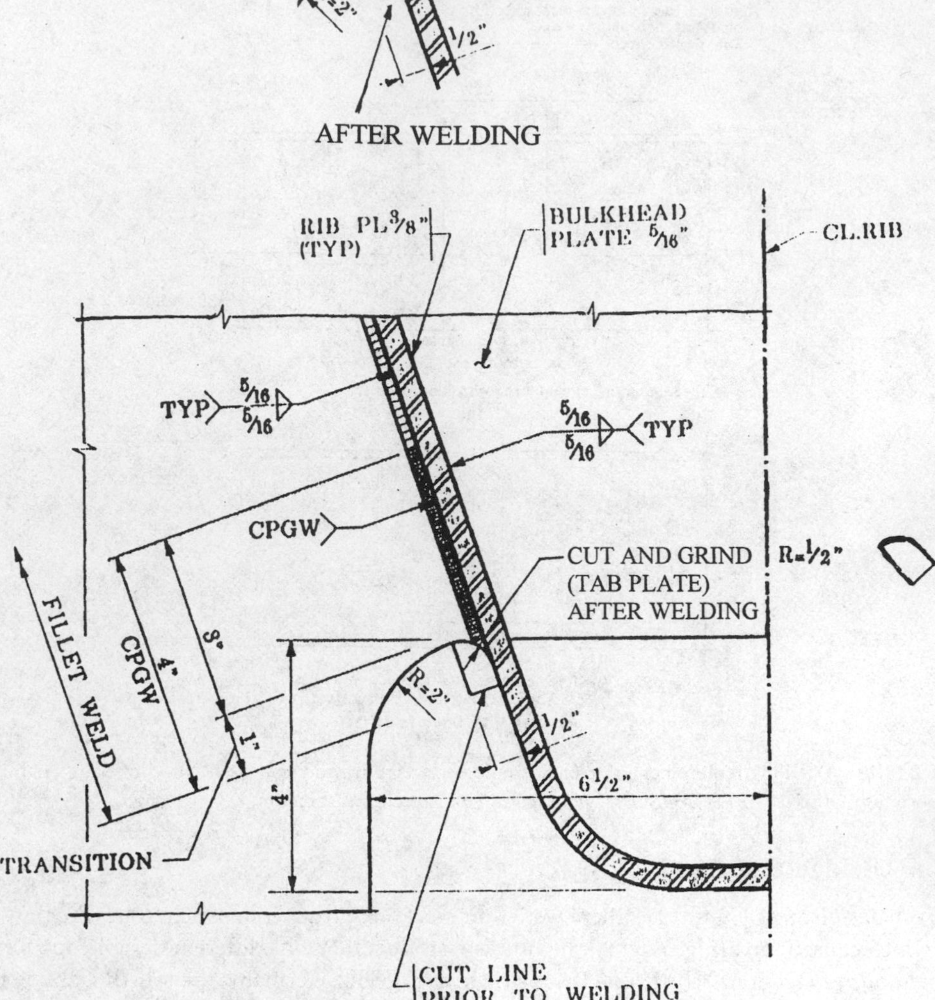

FIGURE 14.24 Fatigue-resistant detail Williamsburg Suspension Bridge. (From Khazen, D., *Civil Eng. ASCE*, June, 1998. With permission.)

14.4.7 Corrosion Protection

The AASHTO LRFD commentary [5] states: " The interior of the closed ribs cannot be inspected and/or repaired. It is essential to hermetically seal them against the ingress of moisture and air." (see Figure 14.3 for a solution). Atmospheric corrosion of steel requires water and a continuous supply of fresh air. Abrasion will speed up the process.

The three different methods that can be used to protect corrosion for new bridges with orthotropic decks are painting, weathering steel, or dehumidification. Painting is the most common method (see Table 14.8 for one Japanese Standard) reference [64]. Weathering steel was invented by the steel industry to eliminate painting. Corrosion can continue if the rusted layer is abraded away exposing bare steel, thus allowing the continuation of corrosion or rusting. Therefore, some designers normally provide an extra thickness of steel with weathering steel in case abrasion occurs. The steel

TABLE 14.8 Japanese Painting Specifications for Steel Bridges

Surface	*N*	Painting in Workshop	Painting in Field	Remarks
A	6	1 Etching primer 2,3 Lead anticorrosive paint[a] 4 Phenolic MIO[b] paint	5,6 Long oil phethalic resin coating 5,6 Chlorinated rubber paint	General location Near sea and over sea
B	3	1 Etching primer 2,3 Epoxy coal-tar paint		
C	5	1 Inorganic zinc-rich primer 2 Organic zinc-rich paint 3 Epoxy MIO paint	4,5 Urethane resin paint 4,5 Chlorinated rubber paint	Painting before surfacing with asphalt Painting after surfacing
D	3	1 Inorganic zinc-rich primer 2,3 Epoxy coal-tar paint	No painting	Inner surface of box girder with steel deck
E	2	1 Inorganic zinc-rich primer 2 Organic zinc-rich paint		

[a] For example, lead suboxide anticorrosive paint (red lead).
[b] MIO = micaceous iron oxide
N = number of painting film
Source: Nakai, H. and Yoo, C.H., *Analysis and Design of Curved Steel Bridges*, MacGraw-Hill, New York, 1968. With permission.

FIGURE 14.25 Dehumidification plant. (Courtesy of Monberg and Thorsen.)

towers for the Luling Bridge are also fabricated of weathering steel; for a color photo of its aesthetic appearance, see Ref. [42]. Another type of steel has been developed for use for contact with salt water or mariner steel.

Inspecting bridges with small access holes is facilitated with an electric-driven inspection cart; the individual rides inside long shallow orthotropic superstructures looking for fatigue cracks or corrosion [65]. The Normandie Bridge in France utilizes a mechanical dehumidification system (see Figure 14.25). The segments were shop-fabricated and then were full butt field-welded at the bridge site. This produced full structural continuity plus completely sealed the bridge superstructure. Exterior access doors for maintenance personnel have gasket seals to eliminate infiltration of air into the superstructure. Dehumidification is popular in European bridges (see Figure 14.11). An air supply system or aspirators are needed for maintenance personnel during bridge inspections. Some engineers are skeptical that dehumidification can actually remove all the moisture. Proponents of dehumidification point out that an air supply system or aspirators are needed for maintenance personnel during future maintenance painting inside the confined space of the box girder. State-of-the-art European technology went into the detailing of this bridge. Rather than paint the interior of the bridge, a mechanical dehumidification process is utilized to prevent corrosion of the interior of the superstructure. A central bridge maintenance walkway gallery was created in a triangular opening. The sections were prefabricated in an assembly-line process, then full butt-welded together at the bridge site.

Hot-dip galvanizing may be utilized for corrosion protection of smaller field-bolted orthotropic deck panels, which are used on deck retrofits and temporary bridges. The fabricated steel component is dipped into a molten zone in a tub. The maximum size of a component is limited by the tub size. A disadvantage is that warping of fabricated components can occur from the hot zinc. Hot-dip galvanizing can be cost-effective for limited-size components, especially near highly corrosive salt water. Drain holes would be needed in any closed ribs to allow molten zinc to drip out. Field welding of hot-dipped galvanizing creates toxic fumes.

14.3.8 Wearing Surface

An orthotropic steel plate deck must be paved with a wearing surface to provide a durable and skid-resistant surface for vehicular traffic. AASHTO [5] states: "The wearing surface should be regarded as an integral part of the total orthotropic deck system and shall be specified to be bonded to the top of the deck plate. For the purpose of designing the wearing surface itself, and its adhesion to the deck plate, the wearing surface shall be assumed to be composite with the deck plate, regardless of whether or not the deck plate is designed on that basis." AREA (American Railroad Engineers Association) specifications should be used for railroad bridges as shown in Figure 14.5. Some materials used in wearing surfaces are proprietary or patented. Aesthetic issues may control for pedestrian bridges or sidewalk areas as shown in Figure 14.16. The surfacing for vehicular traffic performs several functions:

- Exhibits high skid resistance throughout the life of the surfacing;
- Provides a smooth riding surface for the comfort of the drivers using the bridge;
- Helps waterproof the steel deck;
- Resists cracking, delaminating, and displacing for a long service life with low maintenance cost and disruption to traffic for repairs;
- A stiff and well bonded surfacing can also provide some reduction in fatigue stresses in the steel deck, ribs, and welds by dynamic composite action with the deck plate. In addition, surfacing with thickness of about 50 mm provides some distribution of the wheel loads and damping of the steel plate.

The number of wheel load applications during the life of the surfacing can be enormous as each passage of a truck wheel stresses the surfacing and the steel deck. For example, assume a bridge carries 10,000 vehicles a day with 5% being three-axle trucks. If half of these are loaded trucks, the annual full-load applications are about a quarter of a million or 4.5 million for a 20-year service life. A busy bridge carrying a high percentage of trucks can exceed this figure. The wheel load applications for the design of a surfacing should be calculated for the specific bridge site. Vehicle tires also wear and polish the aggregates on the surface of the surfacing causing a reduction in skid resistance. Hard, durable, and polish-resistant aggregates should be selected, preferably with small asperities projecting from the surface. The asperities help to increase the dry skid resistance and also provide a rough surface for improved bonding with the aggregate within the matrix of the surfacing. A long life is required for the surfacing of a heavily traveled bridge as replacement of the surfacing, whether for fatigue cracking, debonding, or for lack of skid resistance, is costly. The surfacing is costly to remove and relay and is also costly to the user as it disrupts and delays traffic.

Treated timber plank wearing surfaces have been used for a few bridges, as shown in Figure 14.7 and described in Section (14.3). Timber planks are not watertight so the steel deck must be painted or other corrosion protection provided. Welded threaded studs or economy head bolts can hold the planks in place. There are three basic classifications of surfacing for orthotropic decks:

1. Thin surfacings from about 4 to 8 mm thick;
2. Surfacings composed of mastics with binders of asphalt or polymer resins usually laid from about 12 to 75 mm thick;
3. Surfacings composed of concertos with binders of asphalt or polymer resins usually laid from about 18 to 60 mm thick.

All three types require a bond coat on the steel deck to hold the surfacing in place against the forces of braking truck wheels and to provide intimate contact with the steel deck for dynamic composite action for each passage of a wheel. The strength of the bond must last for the life of the surfacing.

Thin surfacings are usually laid by flooding the deck with a thermosetting polymer resin and broadcasting a hard aggregate that is locked into the binder to provide skid resistance. Thin surfacings are not appropriate for decks carrying high truck traffic. The repetitive wheel loads will wear away the aggregate exposing the slick resin surface producing a very low skid resistance. The repetitive wheel loads can also wear away the resin and expose the steel deck. However, a thin surfacing can be used as a temporary wearing course for bridge deck replacements. The new orthotropic deck panels can be paved in the shop and installed a panel at a time during replacement of an existing concrete deck. Traffic can run on the temporary surfacing up to 2 years, as was done for the deck of the Golden Gate Bridge. After all the panels are installed, the permanent pavement can be laid as a continuous operation for a smooth riding surface.

Mastics are usually a mixture of aggregates and binder of asphalt, polymer-modified asphalt, or polymer resins. The binder is proportioned in excess of that required to fill the air voids. The strength of mastic surfacing is dependent on the strength of the binder material rather than on the interlock of the aggregates. During placement, the mix is flowed onto the deck and leveled, usually by a vibrating screed, before the binder sets. Hard, durable stones can be broadcast over the surface to improve skid resistance. The high-binder-content polymer resins may cause high shrinkage strains and high bond stress on the deck plate. Gussasphalt (German word for poured asphalt) is a mastic using a low penetration asphalt as a binder heated to a high temperature of about 200°C and applied usually by poring and leveling to a thickness of about 75 mm or more by hand labor. It has been used apparently successfully in Europe and Japan.

Concrete is usually a mixture of polymer resins or polymer-extended asphalt with aggregates with some air voids up to 4% remaining unfilled. Concrete surfacing for steel decks should not be confused with portland cement concrete, which is not suitable for steel-deck surfacing. The strength of concrete surfacings is dependent on both the strength of the interlock of the aggregates and the strength of the binder material. They require compaction usually by a steel roller or vibrating screed. They have the advantage of being mixed, placed, and compacted using conventional paving equipment. Ordinary or modified asphalt concrete surfacings have low first cost, but have not given long, trouble-free service life. If the binder is a thermoset-resin-extended asphalt, such as epoxy asphalt, the cost is increased somewhat but the added strength imparted by the thermoset resin greatly improves the performance and life of the surfacing.

Failure of the wearing surface is common, but the Hayward-San Mateo Bridge still has its original wearing surface after 30 years [24]. Prior to the construction of this bridge a small test panel (truck weight scale) was used to test the durability of the wearing surface under actual truck traffic. The San Diego–Coronado wearing surface was replaced after 25 years. The Caltrans test bridge used two types of wearing surfaces (thin and thick) [13]. The failure of a wearing surface can be caused by the deck plate thickness (stiffness); poor construction practices; installation quality control; bridge deck splice details (bolt heads and splice plates); and/or the temperature range (freeze–thaw action) plus humidity conditions expected at the bridge site. Due to all these factors wearing surfaces can fail very quickly, and few last over 20 years. Flexible orthotropic decks can cause a stiffer wearing surface to pop off. This is one reason AASHTO wants the designer to think of the wearing surface as a composite material that can "structurally" fail, when not deflecting in synchronization with the deck. For case history details on the wearing surfaces of many bridges, the reader can refer to References [1, 2]. The Miller Sweeney bascule bridge's wearing surface failed by creep while the movable span was in the vertical position.

A wearing surface of a sacrificial material is placed on top of the deck since vehicles' wheels cause abrasion. Asphalt and epoxy concrete are the most commonly used materials for wearing surfaces. Timber wearing surfaces have been used as described in Section 14.3. For details on the many possible types of wearing surfaces, the reader may refer to Ref. [1–4]. The proposed test system has been developed for Caltrans [44].

14.3.9 Future Developments

The second generation of orthotropic deck bridges will be better based on lessons learned from the first group of bridges built. Many orthotropic bridges were never built and have remained only a dream. The most interesting is the Ruck-A-Chucky cable-stayed bridge designed by Prof. T. Y. Lin to be built across the flooded American River Canyon after the completion of the Auburn Dam near Sacramento, California [66]. This horseshoe-shaped superstructure was planned to be supported by cable stays anchored into the sloping canyon rock walls. Therefore, no towers would be necessary. Scale models were built for wind tunnel and earthquake shake-table testing. The 1977 orthotropic design featured trapezoidal ribs for a five-cell, 14.61-m-wide superstructure. The bridge plans used a 396-m clear span on a 457-m radius for two lanes of traffic and an equestrian trail.

A promising concept, patent pending to a Redding, California firm, is to have the bridge deck comprise only nested trapezoidal ribs, which are welded together. The three-sided "rectangular" ribs are placed at 90° to the driving surface, so that the sides of the rib become the top and bottom flanges. The result is to achieve a 12-m span that is about 275 mm deep or wide rib is required. This system would compete economically against concrete slab bridges and has been marketed at bridge conventions. Another unique rib system for floating steel bridges or pontoons is the "biserrated-rib" developed by Dr. Arsham Amirikian, consultant to Naval Facilities Engineering Command, Department of the Navy. Portions of the sides of the trapezoidal ribs are removed in a repetitive scallop pattern to reduce the weight of the rib. The structural strength is almost identical to a full rib. The disadvantage would be increased (double) surface area exposed to corrosion.

A panelized orthotropic deck system was developed and built for trapezoidal ribs for a temporary detour bridge in New York [67]. A cross section of this bridge built in 1991 is shown in Figure 14.26a. This system is the engineering evolution of a previous concept for open ribs. A system was developed and engineered for a steel grating company utilizing open ribs for a bridge system [68]. A cross section of the bridge proposed in 1961 is shown in Figure 14.26b. This system would use similar materials stored in the steel grating company warehouse, and allow them to market another product.

The future of orthotropic deck cost reduction lies in the standardization of ribs and details by AASHTO or the steel industry. Such standardization has led to the popularization of precast prestressed concrete girders.

Acknowledgments

The writing of this chapter included the support and suggestions of our co-workers at Caltrans especially Dr. Lian Duan PE, and other individuals plus their employers as identified in the credit lines and references. Mr. Scott Whitaker P.E., Engineer of the Bethlehem Steel Corporation, provided written suggestions for the text, as well as a wealth of articles, many which are included in the references. Special thanks to Mr. Chuck Seim P.E. of T. Y. Lin International for sharing his ideas on wearing surface issues.

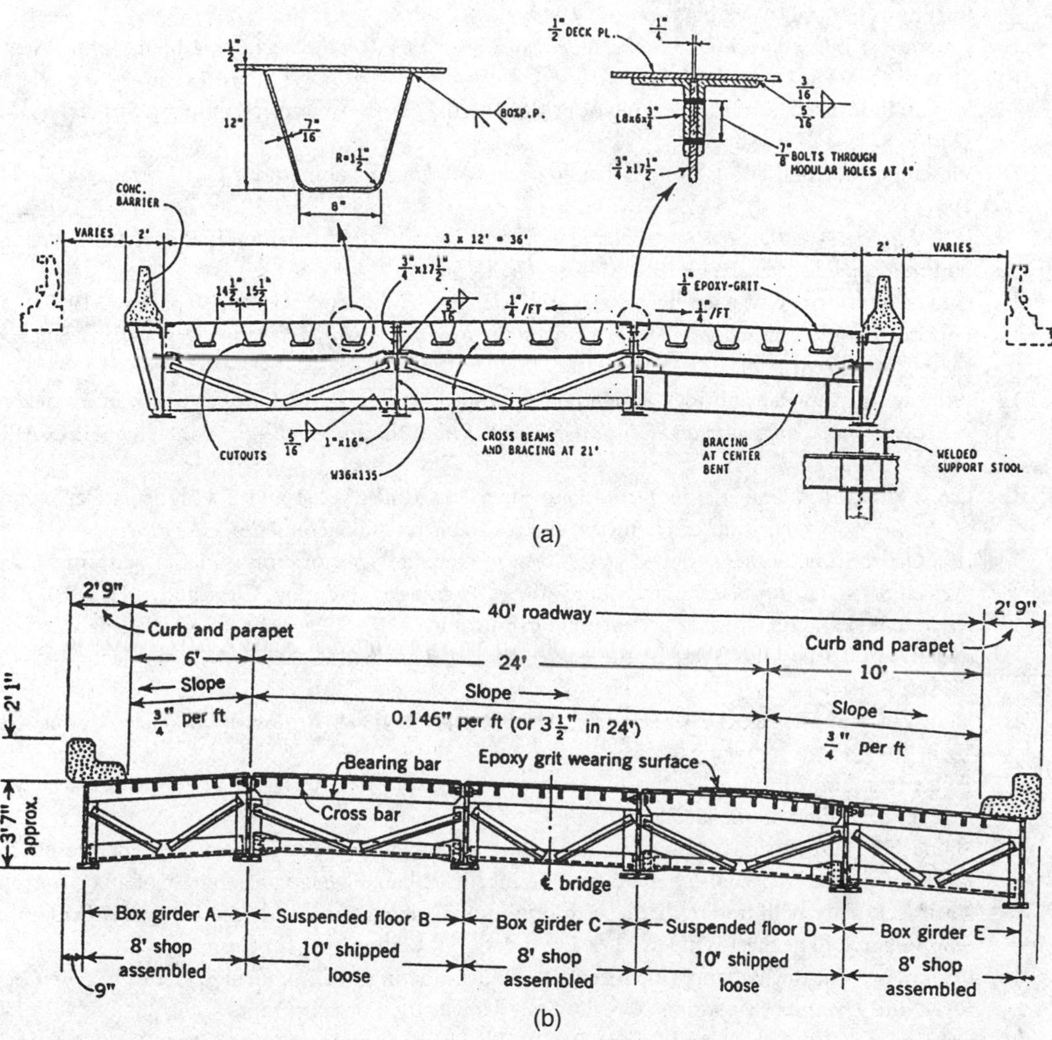

FIGURE 14.26 Prefabricated orthotropic deck panel bridge. (a) A temporary bridge. (From Wolchuk, R., *Welding Innovation Q.*, IX(2), 19, 1992. Courtesy of the James F. Lincoln Arc Welding Foundation.) (b) A short-span bridge. (From Chang, J.C.L., *Civil Eng. ASCE*, Dec. 1961. With permission.)

References

1. Wolchuk, R., *Design Manual for Orthotropic Steel Plate Deck Bridges*, American Institute of Steel Construction, Chicago, 1963.
2. Troitsky, M.S., *Orthotropic Bridges — Theory and Design*, 2nd ed., James F. Lincoln Arc Welding Foundation, Cleveland, OH, 1987.

3. ICE, Steel box girder bridges, in Proceedings of the International Conference, the Institution of Civil Engineers, Thomas Telford Publishing, London, 1972.
4. Cartledge, P., Ed., *Proceedings of the International Conference on Steel Box Girder Bridges*, the Institution of Civil Engineers, Thomas Telford Publishing, London, 1973.
5. AASHTO, *LRFD Bridge Design Specifications*, American Association of State Highway and Transportation Officials, Washington, D.C., 1994.
6. Galambos, T. V. Ed., *Guide to Stability Design Criteria for Metal Structures*, 4th ed., John Wiley & Sons, New York, 1988, Chap. 7.
7. Wolchuk, R., Orthotropic redecking of bridges on the North American continent, *Struct. Eng. Int.*, IABSE, 2(2), 125, 1992.
8. Wolchuk, R., Applications of orthotropic decks in bridge rehabilitation, *Eng. J. AISC*, 24(3), 113, 1987.
9. JFL, *Modern Welded Structure Volume Selection*, Vols. I and II, James F. Lincoln Arc Welding Foundation, Cleveland, OH, 1968, C-1.
10. Guadalajara Bridge, Mexico, *Eng. News Rec.*, McGraw-Hill, New York, Aug 7. page 58, 1969.
11. Popov, O. and Seliverstov, V., Steel bridges on Ankara's Perimeter Motorway, *Struct. Eng. Int.*, IABSE, 8(3), 205, 1998.
12. Ramsay, W., Innovative bridge design concepts steel bridges — the European way, in *Proceedings of the National Symposium on Steel Bridge Constructio*n, November 10–12 Atlanta, AASHTO, AISC, FHWA, 1993.
13. Davis R. E., *Field Testing of an Orthotropic Steel Deck Bridge,*Vol. 1 and 2, California Department of Public Works, Division of Highways, Bridge Department, Sacramento, CA, 1968.
14. JFL, Orthotropic bridge designed as solution to concrete deterioration (Crietz Road), in *Modern Welded Steel Structure*, III, James F. Lincoln Arc Welding Foundation, Cleveland, OH, 1970, B-10.
15. Risch, J. E., Final Report of Experimental Orthotropic Bridge S05 of 23081 A Crietz Road Crossing over I-496 Three Miles West of the City Limits of Lansing, Michigan, DOT project 67 G-157; FHWA Project I 496-7(21), 1971.
16. Heins, C. and Firmage, D. A., *Design of Modern Steel Highway Bridges*, John Wiley & Sons, New York, 1979.
17. Foley, E. R. and Murphy J. P., World's longest orthotropic section feature of San Mateo–Hayward Bridge, *Civil Eng. ASCE*, 38, 54, April 1968.
18. Haibach E. and Plasil, I., September 1983 Untersuchungen zur Betriebsfestigkeit von Stahlleichtfahrbahnen mit Trapezholsteifen im Eisenbahnbruckenbau [The fatigue strength of an orthotropic steel deck plate with trapezoidal closed longitudinal ribs intended for use in railway bridges] in *Der Stahlbau* [The Steel-builder], 269, Ernst & Sohn, Berlin, 1983 [in German].
19. Ozolin, E., Wilson, W., and Hutchison, B., Valdez floating dock mooring system, in *The Ocean Structural Dynamic Symposium*, Oregon State University, Corvallis, Sept., 1982, 381.
20. Carlson, L. A., Platzke, R., and Dreyer, R. C. J., First bridge across the Yukon River, *Civil Eng. ASCE*, 47, August 1976.
21. Hedefine, A., Orthotropic-plate girder bridges, in *Structural Steel Engineering Handbook*, Merritt, F. S., Ed., McGraw Hill, New York, 1972, chap. 11.
22. Merritt, F. S. and Geschwindner, L. F. Analysis of special structures, in *Structural Steel Engineering Handbook*, 2nd ed., Merritt, F. S., Ed., McGraw Hill, New York, 1994, chap. 4.
23. Bouwkamp, J. G., Analysis of the Orthtropic Steel Deck of the San Diego–Coronado Bridge, Report No. 67-20, Structural Engineering Laboratory, University of California, Berekely, 1969.
24. Balala, B., First orthotropic bridge deck paved with epoxy asphalt, (San Mateo-Haywood) *Civil Eng. ASCE*, page 59, April 1968.
25. Curtis, G. N., Design of the Queens Way Bridge, *Modern Welded Structure*, IV, James F. Lincoln Arc Welding Foundation, Cleveland, OH, 1980, A-34.
26. ENR, Prefab steel bridge girders are biggest ever lifted (Queens Way Bridge), *Engineering News Record*, McGraw Hill, New York, August 27, 1970 p. 34.

27. Construction Marketing of Bethlehem Steel Corporation in cooperation with Universal Structural Inc. Steel Bridge Report BG-502 — Cypress Reconstruction — Contract E (Maritime Off- Ramp), Bethlehem Steel Corporation, Bethlehem, PA, 1997.
28. Marquez, T., Huang, C., Beauvoir, C., Benoit, M., and Mangus, A., California's 2356 foot long orthotropic bridge for I-880 Cypress Replacement Project (Maritime Off-Ramp), in *Proceedings of 15th International Bridge Conference*, Pittsburgh, 1998. (IBC 98-22).
29. Marquez, T., Williams, J., Huang, C., Benoit, M., and Mangus, A., Unique steel curved orthotropic bridge for I-880 Cypress Replacement Project (Maritime Off-Ramp), in *Proceedings of International Steel Bridge Symposium*, the National Steel Bridge Alliance, Chicago, IL, 1998.
30. Leonhardt, F., *Bridges Aesthetics and Design*, MIT Press, Cambridge, MA, 1984.
31. Arenas, J. J. and Pantaleon, M. J., Barqueta Bridge Seville, Spain, *Struct. Eng. Int.*, IABSE, 2(4), 251, 1992.
32. Cerver, F. A., Ed., *The Architecture of Bridges*, Barcelona, Spain, 1992, 186.
33. Tokola, A. J. and Wortman, E. J., Erecting the center span of the Fremont Bridge, *Civil Eng. ASCE*, 62, July 1973.
34. Thomsen, K. and Pedersen, K. E, Swing bridge across a navigational channel, Denmark, *Struct. Eng. Int.*, IABSE, 8(3), 201, 1998.
35. Bowen, G. J. and Smith, K. N., Walpole swing span has orthotropic deck, *Heavy Constr. News*, Canada, Feb page 6, 1970.
36. Bender O., Removable section — Sacramento River Bridge at Colusa, in *Arc Welded in Manufacturing and Construction,* II, James F. Lincoln Arc Welding, Cleveland, OH, 1984, page (D-15).
37. Maruyama, T., Watanabe, E., and Tanaka, H., Floating swing bridge with a 280 m span, Osaka, *Struct. Eng. Int.*, IABSE, 8(3), 174, 1998.
38. Tsinker, G. P., *Floating Ports Design and Construction Practices*, Gulf Publishing Company, Houston, TX, 1986.
39. Solland, G., Stein, H., and Gustavsen, J. H., The Bergsoysund Floating Bridge, *Struct. Eng. Int.*, IABSE, 3(3), 142, 1993.
40. Troitsky, M. S., *Cable-Stayed Bridges an Approach to Modern Bridge Design*, 2nd ed., Van Nostrand Reinhold, New York, 1988.
41. Podolny, W., Jr. and Scalzi, J. B., *Construction & Design of Cable-Stayed Bridges*, John Wiley & Sons, New York, 1976.
42. ASCE, Bridge [discusses the Luling Bridge], *Civil Eng. ASCE*, 31, July 1984.
43. ENR, Stayed-girders reaches record [discusses the Luling Bridge], *Engineering News Record*, McGraw-Hill, New York, April 8, 1992, page 33.
44. T. Y. Lin International–Moffat & Nichcols, a joint venture, San Francisco–Oakland Bay Bridge Structure Type Selection Report to Caltrans, San Francisco, CA, May 1998.
45. Kamei, M., Maruyama, T., and Tanaka H., Konohana Bridge Japan, *Struct. Eng. Int.*, 2(1), 4, 1992.
46. DeLeuw Cather–OPAC–Steinman, Third Carquinez Strait Bridge Structure Type Selection Report — Caltrans Contract No. 59A0007, San Francisco, CA, 1997.
47. BD&E, Monster of Messina, in *Bridge Design & Engineering*, Route One Publishing, London, 9, 7, 1997.
48. Gimsing, N., *Cable-Supported Bridges*, John Wiley & Sons, New York, 1997.
49. Kaczinski, M. R., Stokes, F. E., Lugger, P., and Fisher J. W., Williamsburg Bridge Orthotropic Deck Fatigue Test, ATLSS Report No. 97-04, Lehigh University, Bethlehem, PA, 1997.
50. AASHTO, *Standard Specifications for Highway Bridges*, 16th ed., American Association of State Highway and Transportation Officials, Washington, D.C., 1996.
51. Wolchuk, R., Lessons from weld cracks in orthotropic decks on three European bridges, *Welding Innovation Q.*, II(I), 1990.
52. Nather, F., Rehabilitation and strengthening of steel road bridges, *Struct. Eng. Int. IABSE,* 1(2), 24, 1991.

53. *ENR*, Rhine River Bridge Collapse, Koblenz Germany, *Engineering News Record*, McGraw Hill, New York, Nov., 18, 1971, p. 17; Nov., 23, 1972, p. 10; Dec. 20, 1973, p. 26).
54. *ENR*, 4th Danube Bridge Collapse, Austria, *Engineering News Record*, McGraw Hill, New York, Nov. 13 p. 11, and Dec. 4, P. 15, 1969.
55. *ENR*, Milford Haven Bridge Collapse, Great Britain, *Engineering News Record*, McGraw Hill, New York, June 13, p. 11; Dec 4, 69 p. 15, 1970.
56. Kozak, J. and Seim, C., Structural design brings West Gate Bridge failure (Yarra River Melbourne, Australia), *Civil Eng. ASCE*, June 1972, pp 47-50.
57. Wolfram, H. G. and Toakley, A. R., Design modifications to West Gate Bridge, Melbourne, Institution of Engineers, *Civil Engineering Transactions* CE16, Australia, 143, 1974.
58. Moffatt, K. R. and Dowling, D. J., *Parametric Study on the Shear Lag Phenomenon in Steel Box-Girder Bridges*, Engineering Structures Laboratory, Imperial College, London, 1972.
59. Wolchuk, R. and Mayrbourl, R. M. Proposed Design Specification for Steel Box Girder Bridges, RN FHWA-TS 80-205, U.S. Department of Transportation, Federal Highway Administration Washington, D.C., 1980.
60. Milek, W. A., Jr., How to use the AISC Orthotropic Plate Design Manual, *AISC Eng. J.*, 40, April 1964 page 40.
61. Wolchuk, R., Steel-plate-deck bridges and steel box girder bridges, *Structural Engineering Handbook*, 4th ed., Gaylord, E. H., Gaylord C. N., and Stallmeyer, J. E., Eds., McGraw Hill, New York, 1997, chap 19.
62. Xanthakos, P. P., *Theory and Design of Bridges*, John Wiley & Son, New York, 1994.
63. AISC, Orthotropic plate deck bridges, in *Highway Structures Design Handbook*, Vol. 1, American Institute of Steel Construction Marketing, Inc., Chicago, IL, 1992.
64. Nakai, H. and Yoo, C. H., *Analysis and Design of Curved Steel Bridges*, McGraw-Hill, New York, 1988.
65. Thorsen, N. E. and Rouvillain F., *The Design of Steel Parts*, draft of Normandie Bridge France personal correspondence with Mondberg & Thorsen A/S Copenhagen, Denmark, 1997.
66. Lin, T.Y., Kulka, F., Chow P., and Firmage A., Design of Ruck-A-Chucky Bridge, American River, California USA, in *48TH Annual Convention of the Structural Engineers Association of California*, Sacramento, 1979, 133–146.
67. Wolchuk, R., Temporary bridge with orthotropic deck, *Welding Innovation Q.*, IX(2), 19, 1992.
68. Chang, J. C. L., Orthotropic-plate construction for short-span bridges, *Civil Eng. ASCE*, Dec. 1961.
69. Wolchuk, R., Steel Orthotropic Decks — Development in the 1990s, Transportation Research Board 1999 Annual Meeting.
70. Siem, C. and Ferwerda, Fatigue Study of Orthotropic Bridge Deck Welds (for proposed southern crossing of San Francisco Bay), 1972. California Dept. of Public Works Division of Highways and Division of Bay Toll Crossings.

15

Horizontally Curved Bridges

Ahmad M. Itani
University of Nevada at Reno

Mark L. Reno
California Department of Transportation

15.1 Introduction

As a result of complicated geometrics, limited rights of way, and traffic mitigation, horizontally curved bridges are becoming the norm of highway interchanges and urban expressways. This type of superstructure has gained popularity since the early 1960s because it addresses the needs of transportation engineering. Figure 15.1 shows the 20th Street HOV in Denver, Colorado. The structure is composed of curved I-girders that are interconnected to each other by cross frames and are bolted to the bent cap. Cross frames are bolted to the bottom flange while the concrete deck is supported on a permanent metal form deck as shown in Figure 15.2. Figure 15.3 shows the elevation of the bridge and the connection of the plate girders into an integral bent cap. Figure 15.4 shows the U.S. Naval Academy Bridge in Annapolis, Maryland which is a twin steel box-girder bridge that is haunched at the interior support. Figure 15.5 shows Ramp Y at I-95 Davies Blvd. Interchange in Broward County, Florida. The structure is a single steel box girder with an integral bent cap. Figure 15.6 shows a photo of Route 92/101 Interchange in San Mateo, California. The structure is composed of several cast-in-place curved P/S box-girder bridges.

The American Association of Highway and Transportation Officials (AASHTO) governs the structural design of horizontally curved bridges through *Guide Specifications for Horizontally Curved Highway Bridges* [1]. This guide was developed by Consortium of University Research Teams (CURT) in 1976 [2] and was first published by AASHTO in 1980. In its first edition the guide specification included allowable stress design (ASD) provisions that was developed by CURT and load factor design (LFD) provisions that were developed by American Iron and Steel Institute under project 190 [15]. Several changes have been made to the guide specifications since 1981. In 1993 a new version of the guide specifications was released by AASHTO. However, these new specifications did not include the latest extensive research in this area nor the important changes that affected the design of straight I-girder steel bridges.

0-8493-7434-0/00/$0.00+$.50

FIGURE 15.1 Curved I-girder bridge under construction — 20th St. HOV, Denver, Colorado.

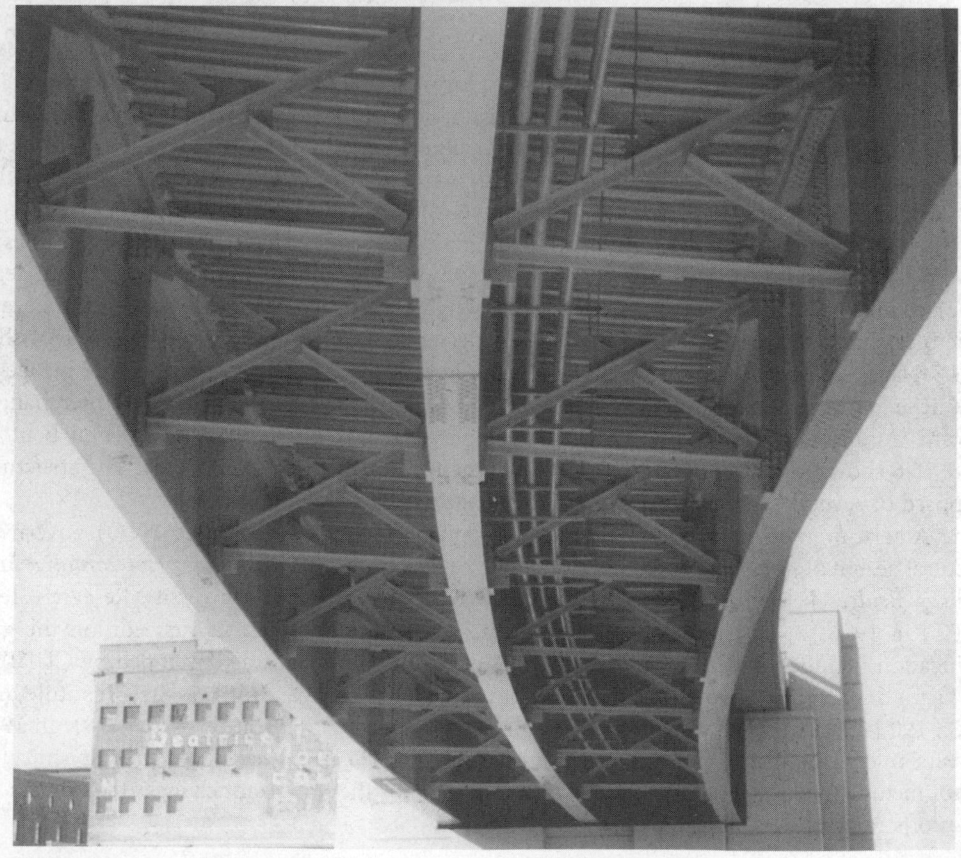

FIGURE 15.2 Bottom view of curved I-girder bridge.

FIGURE 15.3 Curved I-girder bridge with integral bent cap.

The guide specifications for horizontally curved bridges under Project 12-38 of the National Cooperative Highway Research Program (NCHRP) [3] have been modified to reflect the current state-of-the-art knowledge. The findings of this project are fully documented in NCHRP interim reports: "I Girder Curvature Study" and "Curved Girder Design and Construction, Current Practice" [3]. The new "Guide Specifications for Horizontally Curved Steel Girder Highway Bridges" [18] proposed by Hall and Yeo was adopted as AASHTO Guide specifications in May, 1999. In addition to these significant changes, the Federal Highway Administration (FHWA) sponsored extensive theoretical and experimental research programs on curved girder bridges. It is anticipated that these programs will further improve the current curved girder specifications. Currently, the NCHRP 12–50 is developing "LRFD Specifications for Horizontally Curved Steel Girder Bridges" [19].

The guidelines of curved bridges are mainly geared toward structural steel bridges. Limited information can be found in the literature regarding the structural design of curved structural concrete (R/C and P/S) bridges. Curved structural concrete bridges have a box shape, which makes the torsional stiffness very high and thus reduces the effect of curvature on the structural design.

The objective of this chapter is to present guidelines for the design of curved highway bridges. Structural design of steel I-girder, steel, and P/S box-girder bridges is the main thrust of this chapter.

15.2 Structural Analysis for Curved Bridges

The accuracy of structural analysis depends on the analysis method selected. The main purpose of structural analysis is to determine the member actions due to applied loads. In order to achieve reliable structural analysis, the following items should be properly considered:

- Mathematical model and boundary conditions
- Application of loads

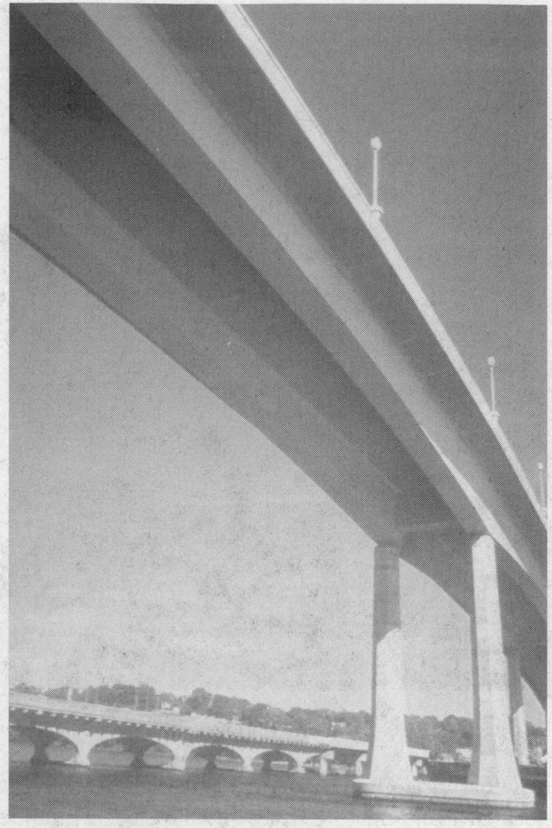

FIGURE 15.4 Twin box-girder bridge — U.S. Naval Academy Bridge, Annapolis, Maryland.

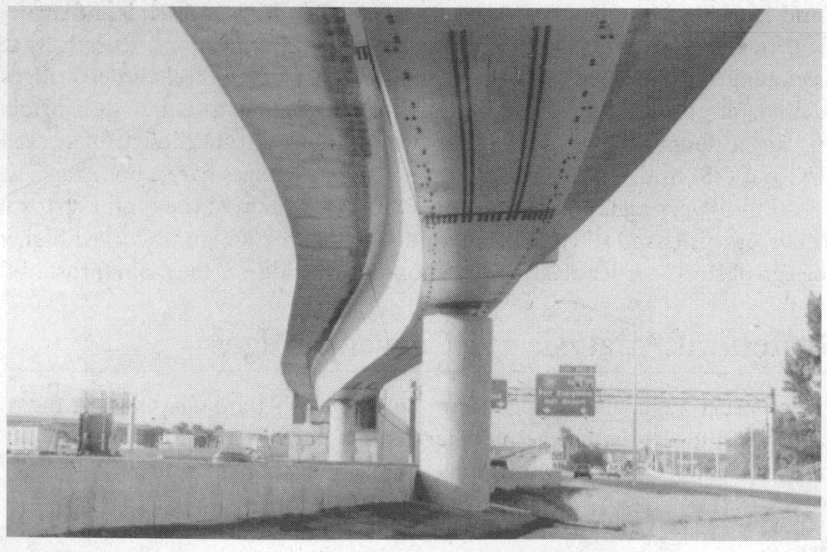

FIGURE 15.5 Single box girder bridge with integral bent cap — Ramp Y, I-95 Davies Blvd., Broward County, Florida.

FIGURE 15.6 Curved concrete box-girder bridges — Route 92/101 Interchange, San Mateo, California.

The mathematical model should reflect the structural stiffness properly. The deck of the superstructure should be modeled in such a way that is represented as a beam in a grid system or as a continuum. The boundary conditions in the mathematical model must be represented properly. Lateral bearing restraint is one of the most important conditions in curved bridges because it affects the design of the superstructure. The deck overhang, which carries a rail, provides a significant torsion resistance. Moreover, the curved bottom flange would participate in resisting vertical load. This participation increases the applied stresses beyond those determined by using simple structural mechanics procedures [3].

Due to geometric complexities, the gravity load will induce torsional shear stresses, warping normal stresses, and flexural stresses to the structural components of horizontally curved bridges. To determine these stresses, special analysis accounting for torsion is required. Various methods were developed for the analysis of horizontally curved bridges, which include simplified and refined analysis methods. The simplified methods such as the *V-Load* method [4] for I-girders and the *M/R* method for box girders are normally used with "regular" curved bridges. However, refined analysis will be required whenever the curved bridges include skews and lateral or rotational restraint. Most refined methods are forms of finite-element analysis. Grillage analysis as well as three-dimensional (3-D) models have been used successfully to analyze curved bridges. The grillage method assumes that the member can be represented in a series of beam elements. Loads are normally applied through a combination of vertical and torsion loads. The 3-D models that represent the actual depth of the superstructure will capture the torsion responses by combining the responses of several bridge elements.

15.2.1 Simplified Method: V-Load

In 1984, AISC Marketing, Inc. published "V-Load Analysis" for curved steel bridges [4]. This report presented an approximate simplified analysis method to determine moments and shears for horizontally

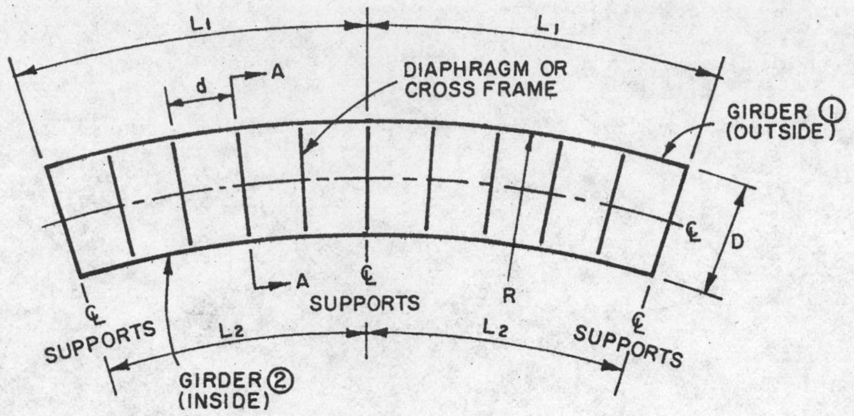

FIGURE 15.7 Plan view of two-span curved bridge.

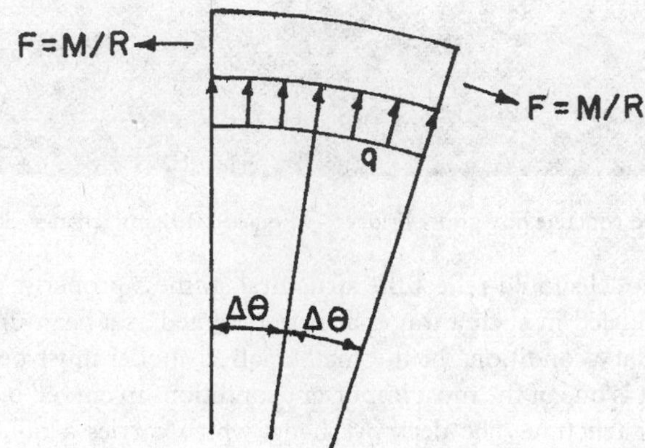

FIGURE 15.8 Plan view of curved bridge top flange.

curved open-framed highway bridges. This method is known as the V-Load method because a large part of the torsion load on the girders is approximated by sets of vertical shears known as "V-Loads." The V-Load method is a two-step process. First, the bridge is straightened out so that the applied vertical load is assumed to induce only flexural stresses. Second, additional fictitious forces are applied to result in final stresses similar to the ones in a curved bridge. The additional fictitious forces are determined so that they result in no net vertical, longitudinal, or transverse forces on the bridge.

Figure 15.7 shows two prismatic girders continuous over one interior support with two equal spans, L_1. Girder 1 has a radius of R and the distance between the girders is D. The cross frames are uniformly spaced at distance equal to d. As shown later, the cross frames in curved bridges are primary members since they are required to resist the radial forces applied on the girder due to bridge curvature.

When the gravity load is applied, the flanges of the plate girder will be subjected to axial forces $F = M/R$, as shown in Figure 15.8. However, due to the curvature of the girder, laterally distributed load q will be applied to flanges of the plate girder in order to achieve equilibrium. By assuming that the flanges resist most of the bending moment, the longitudinal forces in the flanges at any point will be equal to the moment, M, divided by the section height, h. Due to the curvature of the bridge, these forces are not collinear along any given segment of the flange. Thus, radial forces must be developed along the girder in order to maintain equilibrium. The forces cause lateral bending of the girder flanges resulting in warping stresses. The magnitude of the radial forces is equal to M/hR and has the same shape of the bending moment diagram as shown in Figure 15.9.

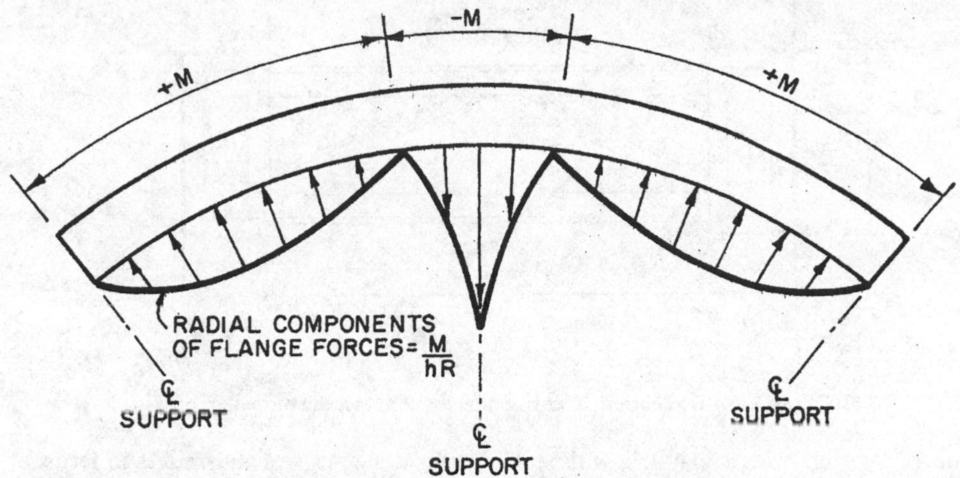

FIGURE 15.9 Lateral forces on curved girder flange.

This distributed load creates equal and opposite reaction forces at every cross frame as shown in Figure 15.10. By assuming the spacing between the cross frames is equal to d, the reaction force at the cross frame is equal to H, which is equal to Md/hR.

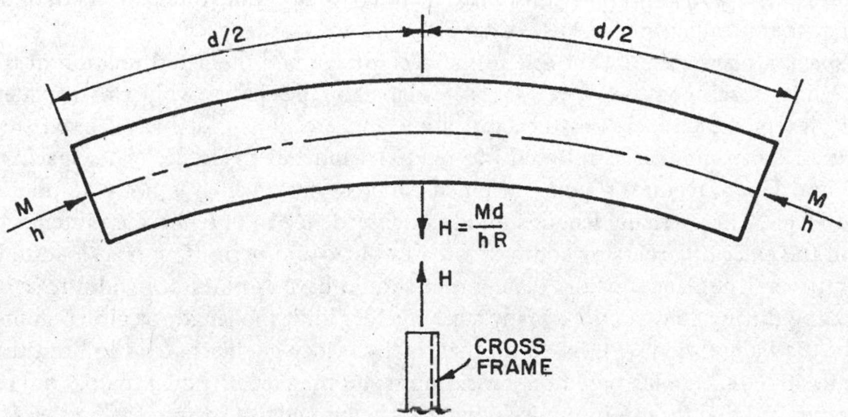

FIGURE 15.10 Reaction at cross frame location.

To maintain equilibrium of the cross frame forces, vertical shear forces must develop at the end of the cross frames as a result of cross frame rigidity and end fixity as shown in Figure 15.11.

15.3 Curved Steel I-Girder Bridges

15.3.1 Geometric Parameters

According to the current AASHTO specifications [13], the effect of curvature may be neglected in determining the primary bending moment in longitudinal members when the central angle of each span in a two or more span bridge is less than 5° for five longitudinal girders. The framing system

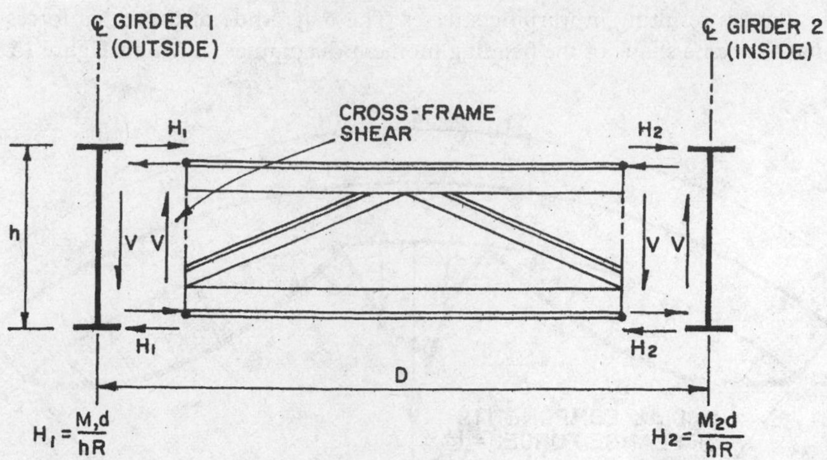

FIGURE 15.11 Equilibrium at cross frame location and the formation of V-loads.

for curved I-girder bridges may follow the preliminary design of straight bridges in terms of span arrangement, girder spacing, girder depth, and cross frame types. The choice of the exterior span length is normally set to give relatively equal positive dead-load moments in the exterior and interior spans. The arrangement results in the largest possible negative moment, which reduces both positive moments and related deflections. Normally, the depth of the superstructure is the same for all spans. Previous successful design showed a depth-to-span ratio equal to 25 for the exterior girder to be adequate. This ratio has been based on vibration and stiffness needed to construct the plate girders. Also, this ratio helps to ensure that the girders do not experience excessive vertical deflections. The uplift of the exterior girder should be prevented as much by extending the span length of the exterior girder rather than dealing with the use of tie-down devices.

Girder spacing plays a significant role in the deck design and the determination of the number of girders. Wider spacing tends to increase the dead load on the girders, while closer spacing requires additional girders, which increases the fabrication and erections costs. For curved steel I-girder bridges, the girder spacing varies between 3.05 m (10 ft) and 4.87 m (16 ft). Wider spacing, common in Europe and Japan, requires a post-tensioned concrete deck, which is not common practice in the United States. The overhang length should not exceed 1.22 m (4 ft) because it tends to increase the load on the exterior girders by adding more dead load and permitting truckload to be applied on the cantilever. The flanges of the plate girder should have a minimum width to avoid out-of-plane buckling during construction. Many steel erectors limit the length of girder shipping pieces to 85 times the flange width [5]. Based on that, many bridge engineers tend to limit the width of the flange to 40.6 mm (16 in) based on a maximum shipping length equal to 36.6 m (120 ft). It is also recommended that the minimum web thickness be limited to 11.1 mm (7/16 in) because of weld distortion problems. The thickness of the web depends on its depth and the spacing of the transverse stiffeners. This represents a trade-off between having extra material or adding more stiffeners. Many bridge engineers use the ratio of D/t=150 to choose the thickness of the web.

The spacing of the cross frame plays an important factor in the amount of force carried out by it and the value of flange lateral bending. Normally, cross-frame spacing is held between 4.57 m (15 ft) and 7.62 m (25 ft).

15.3.2 Design Criteria

The design guidelines, according to the Recommended Specifications for Steel Curved Girder Bridges [3], are established based on the following principles:

- Statics
- Stability

- Strength of materials
- Inelastic behavior

External and internal static equilibrium should be maintained under every expected loading condition. Stability of curved steel girder bridges is a very important issue especially during construction. By their nature, curved girders experience lateral deflection when subjected to gravity loading. Therefore, these girders should be braced at specified intervals to prevent lateral torsional buckling. The compactness ratio of the web and the flanges of curved I-girders are similar to the straight girders. The linear strain distribution is normally assumed in the design of curved girder bridges. The design specification recognizes that compact steel sections can undergo inelastic deformations; however, current U.S. practice does not utilize a compact steel section in the design of curved I-girder bridges.

The design criteria for curved girder bridges can be divided into two main sections.

- Strength
- Serviceability

Limit state design procedures are normally used for the strength design, which includes flexure and shear. Service load design procedures are used for fatigue design and deflection control. The primary members should be designed to be such that their applied stress ranges are below the allowable fatigue stress ranges according to AASHTO fatigue provisions [6]. The deflection check is used to ensure the serviceability of the bridge. According to the recommended specifications for the design of curved steel bridges [3], the superstructure should be first analyzed to determine the first mode of flexural vibration. The frequency of this mode is used to check the allowable deflection of the bridge as indicated in the *Ontario Bridge Code* [7].

15.3.3 Design Example

Following the 1994 Northridge Earthquake in California, the California Department of Transportation (Caltrans) embarked on a task of rebuilding damaged freeways as soon as possible. At the SR 14–I-5 interchange in the San Fernando Valley, several spans of cast-in-place prestressed concrete box girders have collapsed [9]. These were the same ramps that were previously damaged during the 1971 San Fernando Earthquake [8]. Because of the urgency of completion and the restrictions on geometry, steel plate girders were considered a viable replacement alternative. The idea was that the girders could be fabricated while the substructure was being constructed. Once the footings and columns were completed, the finished girders would be delivered to the job site. Therefore, in a period of 5 weeks Caltrans designed two different alternatives for two ramps approximately 396 m (1300 ft) and 457 m (1500 ft) in length. The South Connector Ramp will be discussed in this section. The "As-Built" South Connector was approximately 397 m (1302 ft) in length set on a horizontal curve with a radius of 198 m (650 ft) producing a superelevation of 11%. This ramp was designed utilizing Bridge Software Development International (BSDI) curved girder software package [10] as one frame with expansion joints at the abutments. This computer program is considered one of the most-advanced programs for the analysis and design of curved girder bridges. The program analyses the curved girders based on 3-D finite-element analysis and utilizes the influence surface for live-load analysis. The program has also an interactive postprocessor for performing designs and code checking. The design part of the program follows the 15th edition of AASHTO [13] and the Curved Girder Guide Specifications [1]. The ramp was then checked using DESCUS I [14], another software package, and spot-checked with in-house programs developed by Caltrans. A cross-sectional width of 11.43 m (37.5 ft) was selected for two lanes of traffic (3.66 m, 12 ft), two shoulders (1.52 m, 5 ft), and two concrete barriers (0.533 m, 1.75 ft). This ramp has a 212.7 mm (8⅜ inch) concrete deck, which was composite with four continuous welded plate girders with bolted field splices for erection. The material selected was A709 Grade 50W. The spans ranged from

35.97 m (118 ft) up to 66.44 m (218 ft) in length, which meant the girder depths alone were around 2.2 m (7.25 ft) deep and the composite section was 2.44 m (8 ft) deep. The cross frames were a mixture of inverted K frames and plate diaphragms at the bents. The K frames were inverted so as to place the catwalks between the girders, and the braces were changed to plate sections at the bents to help handle the large seismic forces that are transmitted from the superstructure to the "hammerhead" bent caps both longitudinally and transversely. The bracing was designed for both live-load and seismic-load conditions. Figure 15.12 shows the elevation of intermediate cross frames. The bracing was held to a spacing of less than 6.1 m (20 ft).

The BSDI program works by placing unit loads on a defined geometry pattern of the deck. Then an influence surface is developed so that application of loads for maximum and minimum stresses becomes a simple numerical solution. This program was thoroughly checked utilizing the V-Load method and using an SC-Bridge package that utilizes GT Strudl [11] for the moving load generator. Good correlation was seen by all methods with the exception of the V-Load, which consistently gave more conservative results. As is frequently the case with curved girders, the outside girder ends up being designed heavier than the remaining sections. This difference can be as little as 15%, but as great at 40%, depending on location. It should also be understood that by designing a stiffer girder for the outside, there is the tendency to attract more loads, thereby requiring more material. This is a similar phenomenon to that seen in seismic design. The BSDI system allows the designer to check for construction loads and sequencing. This was absolutely critical on a project like this as the girder sections were often controlled by the sequence of construction load application. Limits on concrete pours were set around limiting stresses on the girders.

Girder plate sizes were optimized both for the design and for the fabrication. A typical span would have five different sections in it. There were two sections at either end over the bents. The top and bottom flanges were very similar at point of maximum negative moment. Then on either side a transition section would be utilized until the inflection point. Finally, a maximum positive section where there is usually a significant difference in the top and bottom flanges was designed. The elevation of the plate girder that shows the different flange dimensions is shown in Figure 15.13. The five different flange dimensions were justified by considering the material costs vs. the welded splice costs. In addition, the "transition" sections were often sized such that the top flange width was the same as the negative moment sections. This way the plates could be welded end to end and then all four girders could be cut on one bed with one operation, saving handling costs. Plate sections were also set based on erection and shipping capabilities.

Steel was a good choice of structure type for this project because of the seismic risk, which exists in this location. Several faults pass in the vicinity of this interchange, and the structure would be subjected to "near-fault" phenomenon. This structure was designed with vertical acceleration. The plate girder with concrete deck superstructure weighs one third as much as the traditional cast-in-place box structure. Some ductile steel details were developed for this project [12]. Since the girders rest on a hammerhead bent cap, the load transfer mechanism is through the bearings and the shear can be as much as the plastic shear of the column. To make this load transfer possible, plate diaphragms were designed at the bent caps. With the plates in place, a concrete diaphragm could be poured that would not only add stiffness, but strength to handle these large seismic forces. The diaphragms were approximately 0.91 m (3 ft) wide by the depth of the girder. The plates were covered with shear studs and reinforcing was placed prior to the concrete. In addition, pipe shear keys were installed in the top of the bent cap on either side of the diaphragm. This structure was redundant in that if the displacements were excessive, the pipes would be engaged.

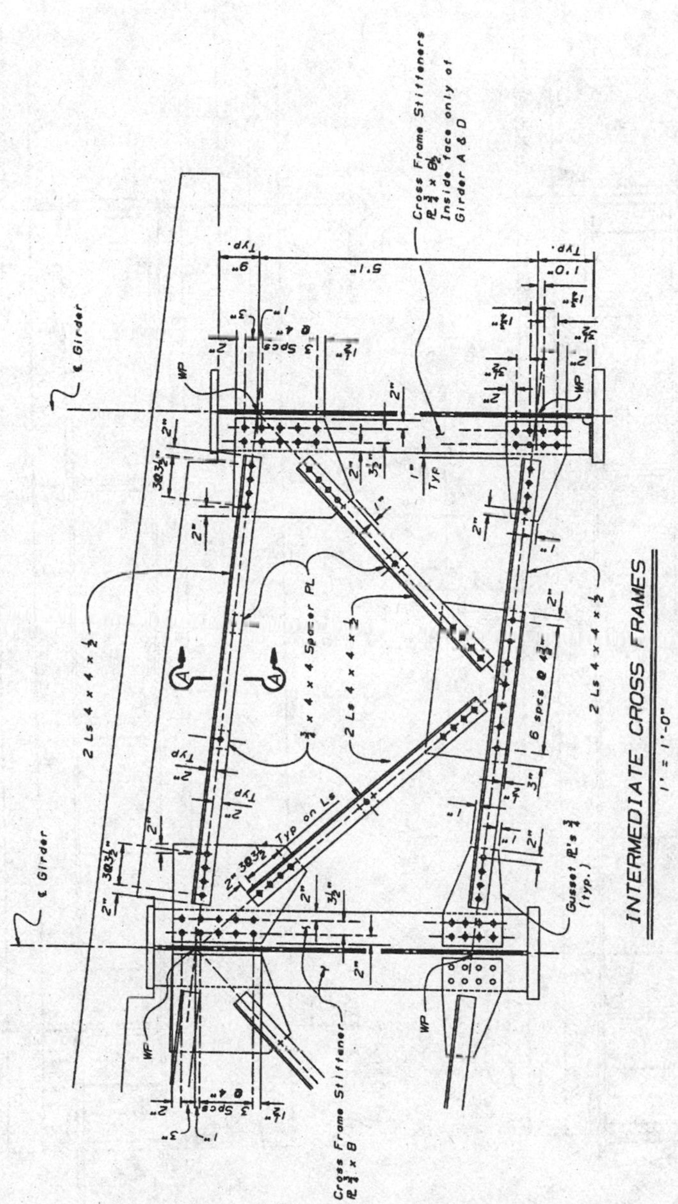

FIGURE 15.12 Elevation of intermediate cross frames.

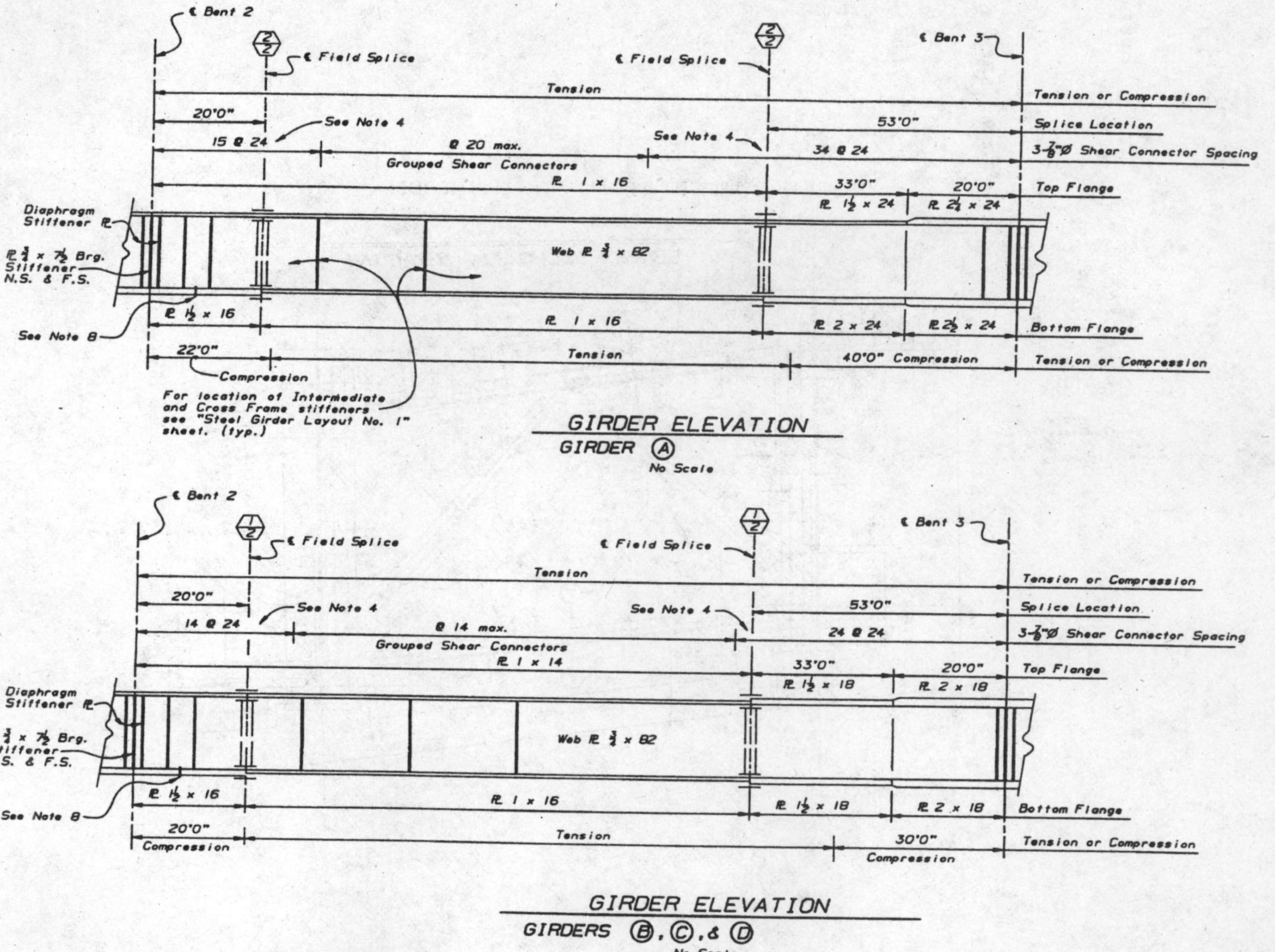

FIGURE 15.13 Elevation of interior and exterior curved plate girder.

15.4 Curved Steel Box-Girder Bridges

The most common type of curved steel box girder bridges are tub girders that consist of independent top flanges and cast-in-place reinforced concrete decks. The design guidelines are covered in the "Recommended Specifications for Steel Curved Girder Bridges"[3]. Normally the tub girder is composed of a bottom plate flange, two web plates, and an independent top flange attached to each web. The top flanges should be braced to become capable of resisting loads until the girder acts in a composite manner. The tub girders require internal bracing because of the distortion of the box due to the bending stresses. Finite-element analysis, which accounts for the distortion, is normally utilized to calculate the stresses and displacement of the box.

The webs of the box girder may be inclined with a ratio of one-to-four, width-to-depth. The AASHTO provisions for straight box girders apply for curved boxes regarding the shear capacity of the web and the ultimate capacity of the tub girders. The maximum bending stresses are determined according to the factored loads with the considerations of composite and noncomposite actions. Bending stresses should be checked at critical sections during erection and deck placement. The bending stresses may be assumed uniform across the width of the box. Prior to curing of concrete, the top flanges of tub girders are to be assumed laterally supported at top flange lateral bracing. The longitudinal warping stresses in the bottom flange are computed based on the stiffness and spacing of internal bracing. It is recommended that the warping stresses should not exceed 15% of the maximum bending stresses.

As mentioned earlier, the *M/R* method is usually used to analyze curved box girder bridges. The basic concept behind this method is the conjugate beam analogy. The method loads a conjugate simple span beam with a distributed loading, which is equal to the moment in the real simple or continuous span induced by the applied load divided by the radius of curvature of the girder. The reactions of the supports are obtained and thus the shear diagram can be constructed representing the internal torque diagram of the curved girder. After the concentrated torque at the ends of the floor beam is known, the end shears are computed from statics. These shears are applied as vertical concentrated loads at each cross frame location to determine the moment of the developed girder. This procedure constitutes a convergence process whereby the *M/R* values are applied until convergence is attained.

15.5 Curved Concrete Box-Girder Bridges

Current curved bridge specifications in the United States do not have any guidelines regarding curved concrete box-girder bridges. It is generally believed that the concrete monolithic box girders have high torsional rigidity, which significantly reduces the effect of curvature. However, during the last 15 years a problem has occurred with small-radius horizontally curved, post-tensioned box-girder bridges. The problem has occurred at two known sites during the construction [16]. The problem can be summarized as, during the prestressing of tendons in a curved box girder, they break away from the web tearing all the reinforcement in the web along the profile of the tendon. Immediate inspection of the failure indicated that the tendons exerted radial horizontal pressure along the wall of the outermost web.

In recognition of this problem, Caltrans has prepared and implemented design guidelines since the early 1980s [17]. Charts and reinforcement details were developed to check girder webs for containment of tendons and adequate stirrup reinforcement to resist flexural bending. Caltrans' Memo-to-Designers 11-31 specifies that designers of curved post-tensioned bridges should consider the lateral prestress force, F, for each girder. This force F is equal to the jacking force, P_j, of each girder divided by the horizontal radius of the girder. If the ratio of $P_j/R > 100$ kN/m per girder or the horizontal radius is equal to 250 m or less, Detail A, as shown in Figure 15.14 should be used. Charts for No. 16 and No. 19 stirrups were developed to be used with the ratio of P_j/R in order to get minimum web thickness and spacing between the No. 16 stirrups, as shown in Figure 15.15.

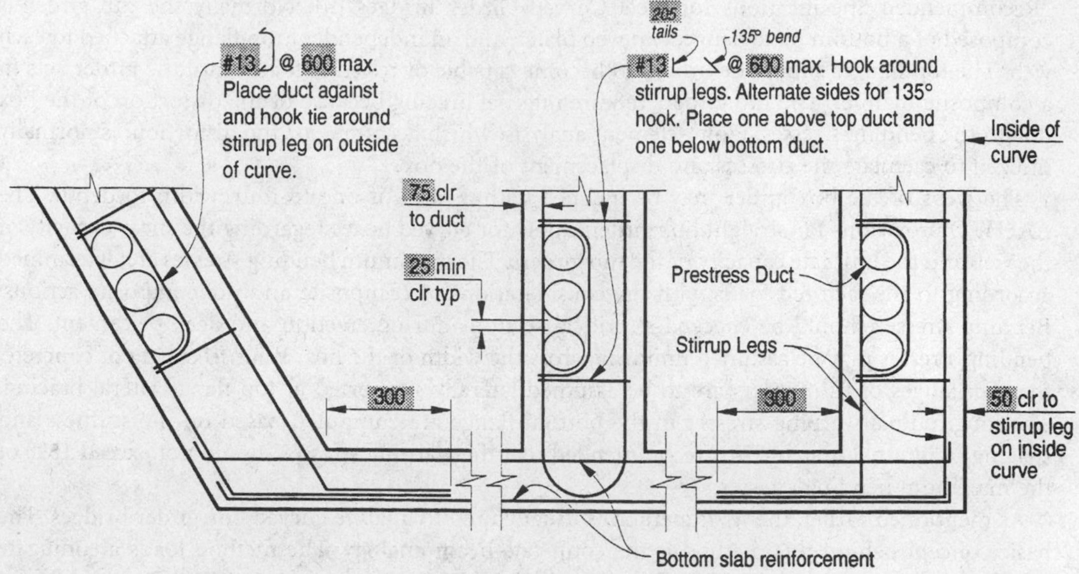

FIGURE 15.14 Caltrans duct detail in curved concrete bridges.

The first step is to enter the chart with the value of F on the vertical axis of the chart and travel horizontally until the height of the web h_c is reached. The chart then indicates the minimum web thickness and the spacing of the No. 16 stirrups.

These charts were developed assuming that the girder web is a beam with a length equal to the clear distance between top and bottom slabs. The lateral force, F, is acting at the center point of the web creating a bending moment in the web. This moment is calculated by the simple beam formula reduced by 20% for continuity between the web and slabs. The value of this bending moment is equal to

$$M_u = 0.8 \frac{P_j}{4R} h_c \tag{15.1}$$

In the commentary of this memo, Caltrans considered the stirrups to be capable of handling the bending and shear stresses for the following reasons:

- M_u is calculated for the maximum conditions of F acting at $h_c/2$. This occurs at only two points in a span due to tendon drape.
- The jacking force, P_j, is used in the calculation of M_u and, at the time P_j is applied, the structure is supported on falsework. When the falsework is removed and vertical shear forces act, the prestressing forces will be reduced by the losses.

In addition, for curve box girders with an inside radius of under 243.8 m (800 ft), intermediate diaphragms are required at a maximum spacing of 24.4 m (80 ft) unless shown otherwise by tests or structural analysis. The code goes further to say that if the inside radius is less than 121.9 m (400 ft), the diaphragm spacing must not exceed 12.2 m (40 ft).

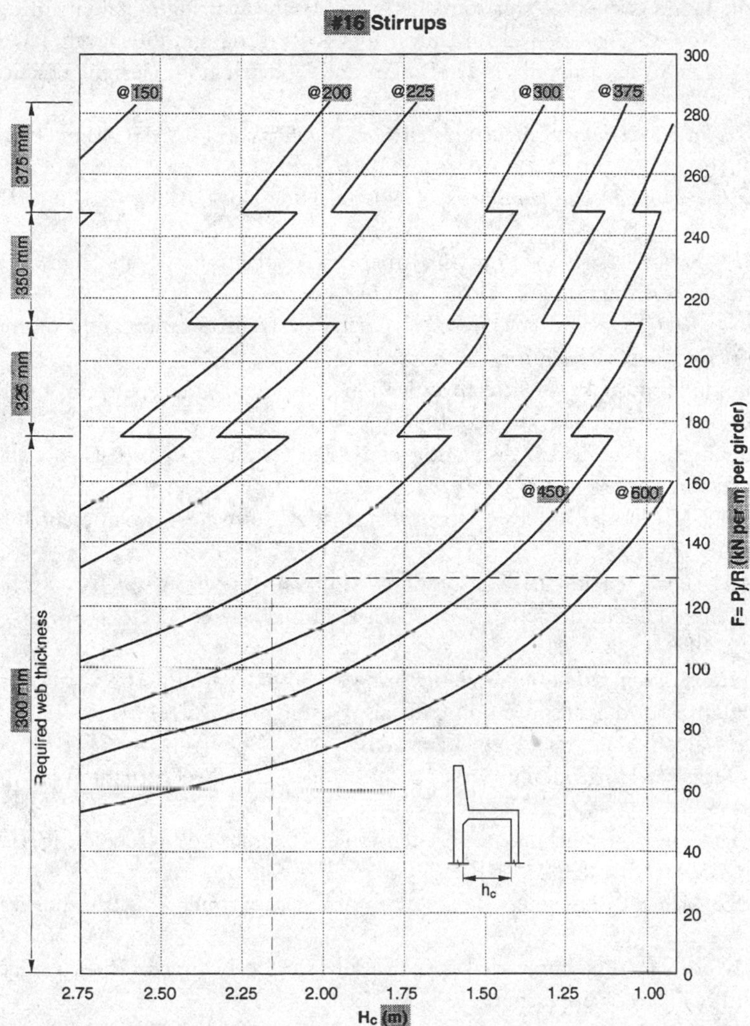

FIGURE 15.15 Caltrans chart design for web thickness and reinforcement.

Acknowledgments

The authors would like to thank Dr. Duan and Prof. Chen for selecting them to participate in this *Bridge Engineering Handbook.* The National Steel Bridge provided the photographs of curved bridges in this document for which the authors are sincerely grateful. The support and the cooperation of Mr. Dan Hall of BSDI are appreciated. Finally, the two authors warmly appreciate the continued support of Caltrans.

References

1. AASHTO, *Guide Specifications for Horizontally Curved Highway Bridges,* American Association of State Highway and Transportation Officials, Washington, D.C., 1993, 1–111.

2. Mozar, J., Cook, J., and Culver, C., Horizontally Curved Highway Bridges-Stability of Curved Plate Girders, Report No. P1, Carnegie-Mellon University, CURT Program, Pittsburgh, PA, Sept. 1971.
3. Hall, D. H. and Yoo, C. H., Curved Girder Design and Construction, Current Practice, NCHRP Project 12-38, 1995, 1–136.
4. V-Load analysis, in *USS Highway Structures Design Handbook*, Vol. 1, AISC Marketing, Inc., Chicago, IL, 1984, chap. 12, 1–56
5. AISC, *Highway Structure Design Handbook*, Newsletter Issue No. 2, AISC Marketing, Chicago, IL, 1991
6. AASHTO, *LRFD Bridge Design Specifications*, American Association of State Highway and Transportation Officials, Washington, D.C, 1994.
7. *Ontario Highway Bridge Design Code*, 3rd ed., Ministry of Transportation and Communications, Highway Engineering Division, Toronto, Ontario, 1991.
8. The San Fernando Earthquake — Field Investigation of Bridge Damage, State of California, Caltrans, Division of Structures, Sacramento, 1991.
9. Northridge Earthquake — Field Investigation of Bridge Damage, State of California, Caltrans, Division of Structures, Sacramento, 1994.
10. Hall, D. H., BSDI 3D System, Internal Document, Bridge Software Development International, Ltd., Coopersburg, PA, 1994.
11. *SC-Bridge, User's Guide*, Version 2.1, SC Solutions, Mountain View, CA, 1994.
12. Itani, A. and Reno, M., Seismic design of modern steel highway connectors, in *ASCE Structures Congress*, Vol. 2, 1995, 1528–1531.
13. AASHTO, *Standard Specifications for Highway Bridges*, 15th ed. with interim's, American Association of State Highway and Transportation Officials, Washington, D.C., 1994.
14. *DESCUS I and II*, Opti-Mate, Inc., Bethlehem, PA.
15. Analysis and Design of Horizontally Curved Steel Bridges, U.S. Steel Structural Report ADUCO 91063, May 1963.
16. Podolny, W., The cause of cracking in post-tensioned concrete box girder bridges and retrofit procedures, *PCI J.*, March 1985.
17. Caltrans, Bridge Memo-to-Designers, Vol. 1, California Department of Transportation, Sacramento, 1996.
18. Hall, D.H. and Yoo, C.H., Recommended Specifications for Steel Curved-Girder Bridges, NCHRP Project 12–38. BSDI, Ltd., Coopersburg, PA, Dec. 1998.
19. NCHRP (12–52), LRFD Specifications for Horizontally Curved Steel Girder Bridges, Transportation Research Board, Washington, D.C.

16

Highway Truss Bridges

John M. Kulicki
Modjeski and Masters, Inc.

16.1 Truss Configurations

16.1.1 Historical

During the 1800s, truss geometries proliferated. *The Historic American Engineering Record* illustrates 32 separate bridge truss geometries in its 1976 print shown in Figure 16.1 [1]. These range from the very short King Post and Queen Post Trusses and Waddell "A" trusses to very complex indeterminate systems, including the Town Lattice and Burr Arch truss. Over a period of years following Squire Whipple's breakthrough treatise on the analysis of trusses as pin-connected assemblies, i.e., two force members, a number of the more complex and less functional truss types gradually disappeared and the well-known Pratt, Howe, Baltimore, Pennsylvania, K truss and Warren configurations came into dominance. By the mid-20th century, the Warren truss with verticals was a dominant form of truss configuration for highway bridges, and the Warren and K trusses were dominant in railroad bridges. *The Historic American Engineering Record* indicates that the Warren truss without verticals may have appeared as early as the mid-1880s, but was soon supplanted by the Warren truss with verticals, as this provided a very convenient way to brace compression chords, reduce stringer lengths, and frame sway frames into the relatively simple geometry of the vertical members.

0-8493-7434-0/00/$0.00+$.50

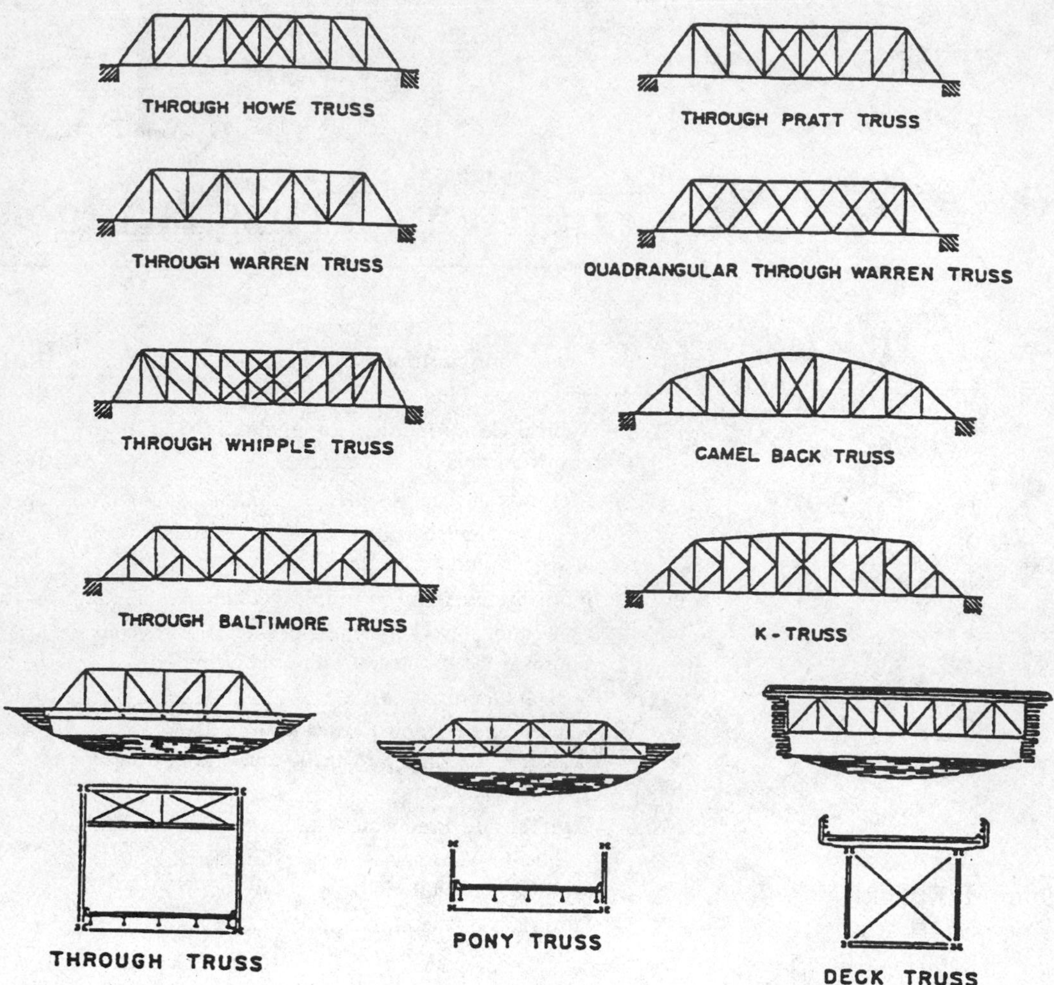

FIGURE 16.1 Historic trusses.

16.1.2 Modern

Few single span trusses are used as highway bridges today, although they are still used for railroad bridges. Modern highway trusses are usually either continuous or cantilever bridges and are typically Warren trusses with or without verticals. Some typical configurations are shown in Figure 16.2.

Throughout the 1980s and 1990s, the Warren truss without verticals has resurfaced as a more aesthetically pleasing truss configuration, especially in the parallel chord configuration, and this has led to a significant simplification in truss detailing, because sway frames are typically omitted in this form of truss, except for portals. The Warren truss without verticals received a great deal of use on the Japanese Railroad System and, more recently, in U.S. highway practice as exemplified in the Cooper River Bridge near Charleston, South Carolina, and the Kanawha River Bridge near Charleston, West Virginia; such a bridge configuration is shown in Figure 16.3 as it was considered one option for the Second Blue Water Bridge (281 m main span) between Port Huron, Michigan, and Point Edward, Ontario, Canada.

The truss bridge behaves much like a closed box structure when it has four planes capable of resisting shear and end portals sufficient to transmit shear back into vertical loads at the bearings. Given the need for a box configuration to resist vertical and lateral loads, it is possible that the configuration could be either rectangular, i.e., four-sided, or triangular, if that geometry is able to

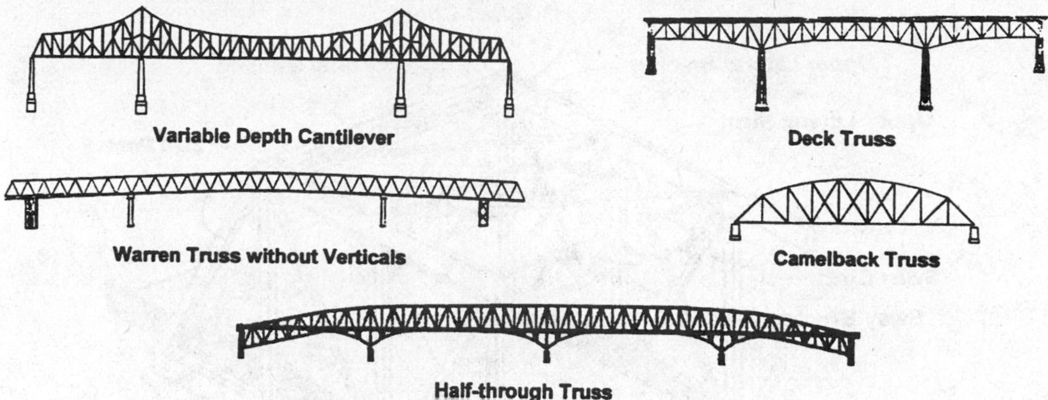

FIGURE 16.2 Typical modern highway truss configuration.

FIGURE 16.3 Second Blue Water Bridge — parallel chord truss study option (on-site photo).

accommodate the roadway clearances. Issues of redundancy should be addressed, either by supplementary load paths, e.g., prestressing, or by sufficiently improved material properties, primarily toughness, to make a triangular configuration acceptable to owners, but it is certainly within the technical realm of reason.

16.2 Typical Components, Nomenclature, and Materials

Components and Nomenclature

The truss bridge is usually characterized by a plethora of bracing and wind-carrying members in addition to those members seen in front elevation. Typical members of a simple single span through-truss are identified in Figure 16.4, taken from Ref. [2].

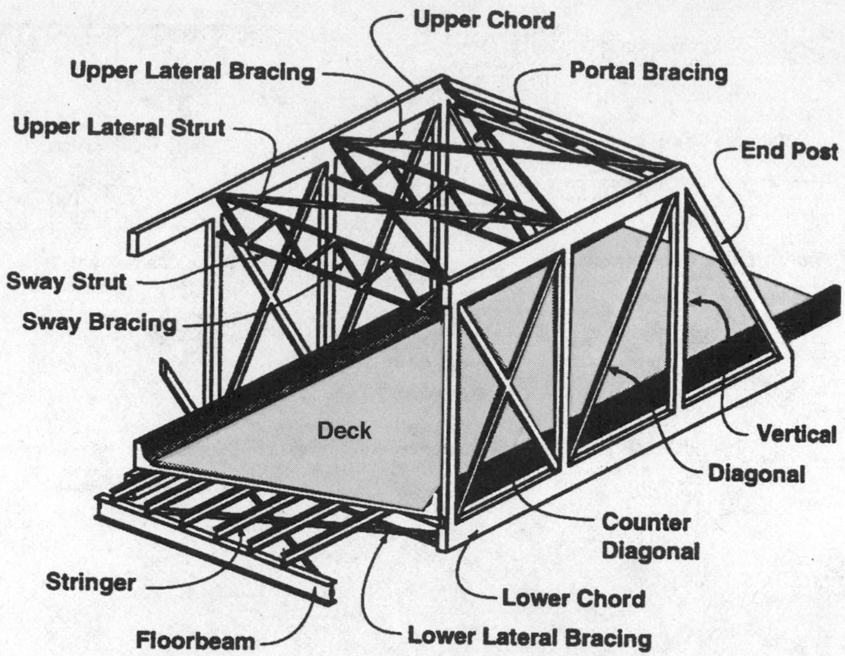

FIGURE 16.4 Typical truss members.

The lateral members in the planes of the top and bottom chords resist wind loads and brace the compression chords. Sway frames are thought to square the truss and increase its torsional rigidity. End portals carry torsional loads resulting from uneven vertical loads and wind loads into the bearings.

It is the visual impact of the various members, especially bracing members, which contribute to aesthetic opposition to many truss designs. However, if unforeseen events cause damage to a main truss member, these bracing members can serve as additional load paths to carry member load around a damaged area.

Truss Members

Some of the cross sections used as modern truss members are shown in Figure 16.5. Truss members have evolved from rods, bars, and eyebars to box and H-shaped members. Generally speaking, the box members are more structurally efficient and resist the tendency for wind-induced vibration better than H-shapes, whereas H-shapes are perceived as being more economical in terms of fabrication for a given tonnage of steel, generally easier to connect to the gusset plates because of open access to bolts, and easier to maintain because all surfaces are accessible for painting. The use of weathering steel offsets these advantages.

Even in the late 1990s, box members are widely used and, in some cases, the apparent efficiency of the H-shape is offset by the need to make the members aerodynamically stable. The choice is clearly project specific, although the H-shaped sections have a relatively clear advantage in the case of tension members because they are easier to connect to gusset plates and easier to paint as indicated above, without the stability design requirements needed for compression members. They are, however, more susceptible to wind-induced vibrations than box shapes. Box shapes have an advantage in the case of compression members because they usually have lower slenderness ratios about the weak axis than a corresponding H-shaped member.

The sealing of box shapes to prevent corrosion on the inside of the members has been approached from many directions. In some cases, box shapes may be fully welded, except at access locations at

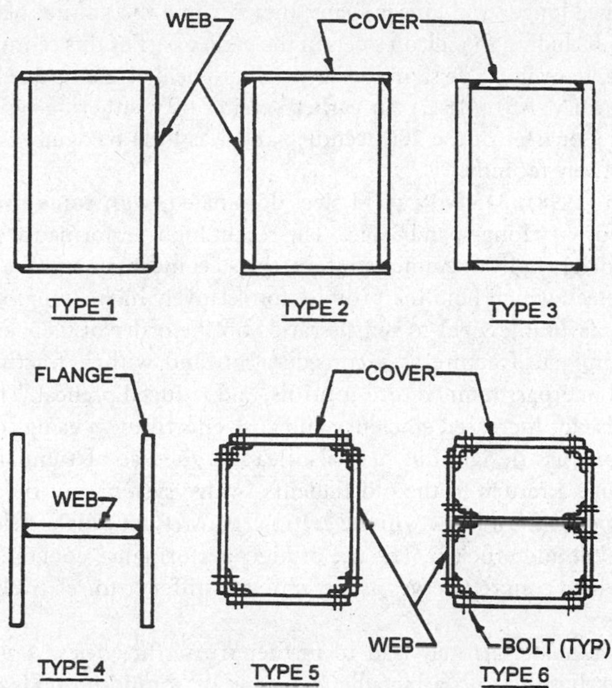

FIGURE 16.5 Cross sections of modern truss members.

the ends used to facilitate connection to gusset plates. Sealing of box members has met with mixed success. In some instances, even box members which have been welded on all four sides and have had welded internal squaring and sealing diaphragms have been observed to collect moisture. The issue is that the member need not be simply watertight to prevent the infiltration of water in the liquid form, it must also be airtight to prevent the natural tendency for the member to "breath" when subjected to temperature fluctuations, which tends to draw air into the member through even the smallest cracks or pinholes in the sealing system. This air invariably contains moisture and can be a recurring source of condensation, leading to a collection of water within the member. In some cases, box members have been equipped with drainage holes, even though nominally sealed, in order to allow this condensate to escape. In some instances, box members have been sealed and pressurized with an inert gas, typically nitrogen, in order to establish that adequate seals have been developed, as well as to eliminate oxygen from the inside of the member, thus discouraging corrosion. Box members have been built with valve stems in order to monitor the internal pressure, as well as to purge and refill the inert gas corrosion protection. Various types of caulking have been used to try to seal bolted joints with mixed success.

Due to the increased interest in redundancy of truss bridges, stitch-bolted members have been used in some cases. Because a bolt does not completely fill a hole, this does leave a path for water ingress, making adequate ventilation and drainage of the member important.

The future of truss member configurations is somewhat dependent on the evolution of new materials as discussed below.

Materials

Early trusses were made of timber. Over a period of time, the combination of timber compression members and wrought iron tension members was evolved and eventually the timber components were replaced by cast iron. After the construction of the Eads Bridge in St. Louis, steel became more widely used and remains today the predominant, and almost exclusive, material for truss construction.

As truss bridges reached longer and longer spans, they became the natural platform for the introduction of new steels, including the silicon steels in the earlier part of this century, weathering steels, copper-bearing steels, low-alloy steels, and even the high-yield-strength quenched and tempered alloy steels, such as ASTM A514/A517. An earlier version of weathering-type steels was used in trusses during the first quarter of the 20th century and was little recognized for their weathering capabilities until relatively recently.

By the current era (1998), 345 MPa yield steel dominated, with some use of higher strength materials, especially for very long-span bridges. The recent high-performance steel (HPS) initiative on the part of the Federal Highway Administration, the steel industry, and the U.S. Navy has led to the development of steels which hold the promise for relatively high strength, e.g., 485 MPa yield and higher, relatively favorable yield-to-tensile ratios on the order of 0.85, extraordinary toughness which could eliminate fracture as a consideration and with it fracture-critical members designation, reduced interpass temperature controls, and reduced preheat. This new material not only holds the promise for increased efficiency and cost-effectiveness using conventional box and I-shaped members in truss design, but may also lead to the use of tubular members and cast joints. This sounds like a return to the old Phoenix bridge system, but the increased efficiency of new materials, as well as the advances in the casting industry, may make field welding of tubular shapes to nodes possible and efficient. The use of high-performance concrete (HPC) truss members may evolve in which concrete serves as a composite stiffener to relatively thin steel plates in compression members.

Advanced composite materials may lead to further truss efficiencies. A pultruded shape as a compression member, butting into a metallic, concrete, or composite node may be a near-term possibility. Composite tension members may require a bonding agent, e.g., glue, before it becomes possible to make a fatigue-resistant joint. The problems associated with mechanical fastener-type connection will probably be solved sooner or later facilitating the use of advanced composites.

16.3 Methods of Analysis

16.3.1 Two-Force Member Methods — Pin-Connected Truss

In the 1840s, a method of analyzing trusses as pin-connected assemblages was developed and is still in wide use today. This method is based on assuming that the truss joints are frictionless pins. This assumption means that as long as loads are applied to the joints and not along the member length, the only bending is caused by self-weight. Thus, the major force in the member is assumed to act along its length. This is often called a "two-force member." The two forces are the axial load at each end of the member.

Throughout the 19th century and even into the early part of the 20th century, it was common to use physical pins in truss joints in order to facilitate the interconnection of components of members, and also to replicate the mathematical assumptions. As a truss deflects under loads, the joints rotate through what are typically very small angles. If the pins truly were frictionless, the truss members would rotate relative to each other and no end moments would be developed on the members. The physical pins never really were friction-free, so some moments developed at the ends in truss members and these were typically regarded as secondary forces. When pin-ended construction gave way to riveted joints and then to bolted or welded joints, the truss joints were detailed so that the working lines of the members intersected either at a common point, so as to reduce eccentricities or to utilize eccentricities to compensate for the bending caused by the dead weight of the members. In either event, it was widely regarded that the pin-connected analysis model was applicable. As will be discussed later, as long as a bridge is properly cambered, it often is an accurate analysis tool.

Two variations of the pin-connected truss model are in common usage; the method of joints and the method of sections. Each of these are illustrated below.

16.3.1.1 Method of Joints

As the name implies, the method of joints is based on analysis of free-body diagrams of each of the truss joints. As long as the truss is determinate, there will be enough joints and equations of equilibrium to find the force in all the members. Consider the simple example shown in Figure 16.6. This six-panel truss supports a load *P* at Joint L3. By taking the summation of the moments about each end of the bridge, it is possible to determine that the left-hand reaction is ⅔ *P* and the right-hand reaction is ⅓ *P*.

Isolating Joint L0, it can be seen that there are two unknowns, the force in Member L0-U1 and the force in Member L0-L1. For this small truss, and as typically illustrated in most textbooks, the truss is assumed to be in a horizontal position, so that it is convenient to take one reference axis through Member L0-L1 and establish an orthagonal axis through L0. These are commonly called the horizontal and vertical axes. The forces parallel to each must be in equilibrium. In this case, that means that the vertical component of the force in Member L0-U1 is equal the reaction RL. By considering the forces in the horizontal direction, the force in Member L0-L1 is equal to the horizontal component of the force in Member L0-U1. Thus, all of the member forces at L0 can be determined.

If we proceed to Joint L1, at which there is no applied load, it is clear that vertical equilibrium of the joint requires that the force in L1-U1 be equal to 0, and that the force in L1-L2 be equal to L0-L1.

Proceeding to Joint U1, it can be seen that, although four members frame into that joint, the force in two of the members are now known, and the force in the other two members can be found from the equation of equilibrium of forces along the two axes.

The analysis continues in this way from joint to joint.

16.3.1.2 Method of Sections

The method of sections proceeds by identifying free-body diagrams which contain only two unknowns, so that equilibrium of the sum of the moment about one joint and the equilibrium of the sum of the shears through a panel are sufficient to determine the two unknown truss forces. Consider Section AA in Figure 16.7 which shows a portion of the same truss shown in Figure 16.6. If we consider the free-body diagram to the left of Section AA, it is clear that the shear in the panel is equal to the reaction R_L and that this can be reacted only by the force L0-U1. Similarly, since the section and hence the free-body diagram is taken just to the left of the Joints L1 and U1, summing moments about Joint U1 or, more accurately, the end of Member L0-U1 an infinitesimally small distance to the left of the section line enables us to compute the force in L0-U1.

If we consider Section BB, it can be seen that the sum of the moments about the lower chord joint enables us to find the force in the top chord, and the shear in this panel enables us to find the force in the diagonal directly.

The analysis then proceeds from section to section along the truss. As a practical matter, a combination of the method of section and the method of joints usually results in the most expeditious calculations.

16.3.1.3 Influence Lines for a Truss

An influence line is a graphical presentation of the force in a truss member as the load moves along the length of the structure.

Influence lines for forces in the members are usually found by applying a unit load at each of the affected chord joints. This information is then shown pictorially, as indicated in Figure 16.8, which shows the influence line for a Chord Force U1-U2 (or U2-U3) and Diagonal Force L2-U3. If the truss is statically determinate, the influence line is a series of straight line segments. Since panel point loading is usually used in a truss, the influence lines for diagonals typically pass through a truss panel, as shown in Figure 16.8. If the truss is statically indeterminate, then the influence lines will be a series of chords to a curve, not a straight line.

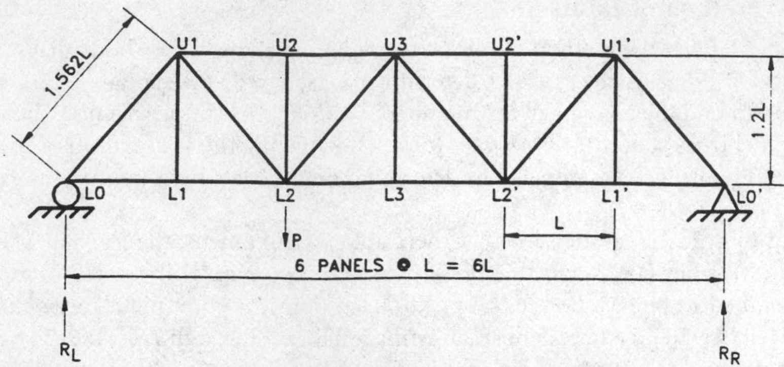

Computer Reactions

Summing Moments About Right End:

$$4PL = 6R_L L$$

$$R_L = \frac{2}{3}P$$

Summing Moments About Left End:

$$2PL = 6R_R L$$

$$R_R = \frac{1}{3}P$$

L0 (LOU1, LOL1, R_L)

From ΣV = 0:

$$\frac{1.2\,LOU1}{1.562} = R_L = \frac{2}{3}P$$

$$LOU1 = 0.868P$$

From ΣH = 0:

$$\frac{LOU1}{1.562} = LOL1$$

$$LOL1 = 0.555P$$

L1 (L1U1, LOL1, L1L2)

From ΣV = 0:

$$L1U1 = O$$

From ΣH = 0:

$$L1L2 = LOL1 = 0.555P$$

U1 (U1U2, LOU1, L1U1, U1L2)

From ΣV = 0:

$$\frac{1.2\,LOU1}{1.562} = \frac{1.2\,U1L2}{1.562}$$

$$U1L2 = LOU1 = 0.868P$$

From ΣH = 0:

$$U1U2 = \frac{LOU1}{1.562} + \frac{U1L2}{1.562} = \frac{2X0.868P}{1.562}$$

$$U1U2 = 1.111P$$

U2 (U1U2, U2U3, L2U2)

From ΣV = 0:

$$L2U2 = 0$$

From ΣH = 0:

$$U2U3 = U1U2 = 1.111P$$

Determined So Far:

$$L2U2 = 0$$

$$U1L2 = 0.858P$$

$$L1L2 = 0.555P$$

L2 (U1L2, L2U2, U3L2, L1L2, L2L3, P)

From ΣV = 0

$$\frac{1.2\,U1L2}{1.562} + \frac{1.2\,U3L2}{1.562} = P$$

$$U3L2 = 0.434P$$

From ΣH = 0

$$L2L3 = L1L2 + \frac{U1L2}{1.562} - \frac{U3L2}{1.562}$$

$$L2L3 = 0.833P$$

FIGURE 16.6 Method of joints.

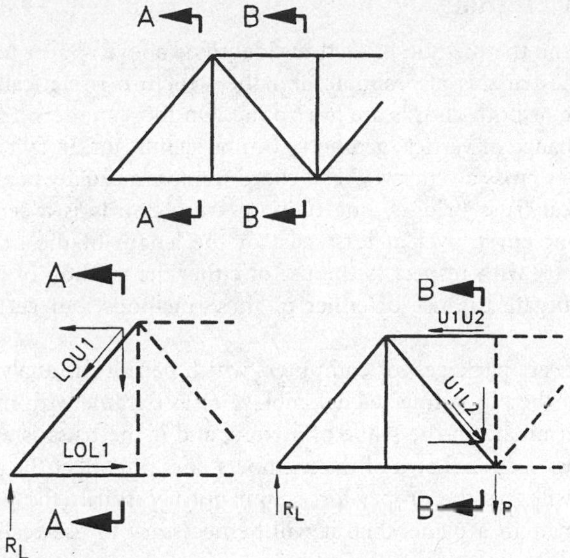

FIGURE 16.7 Method of sections.

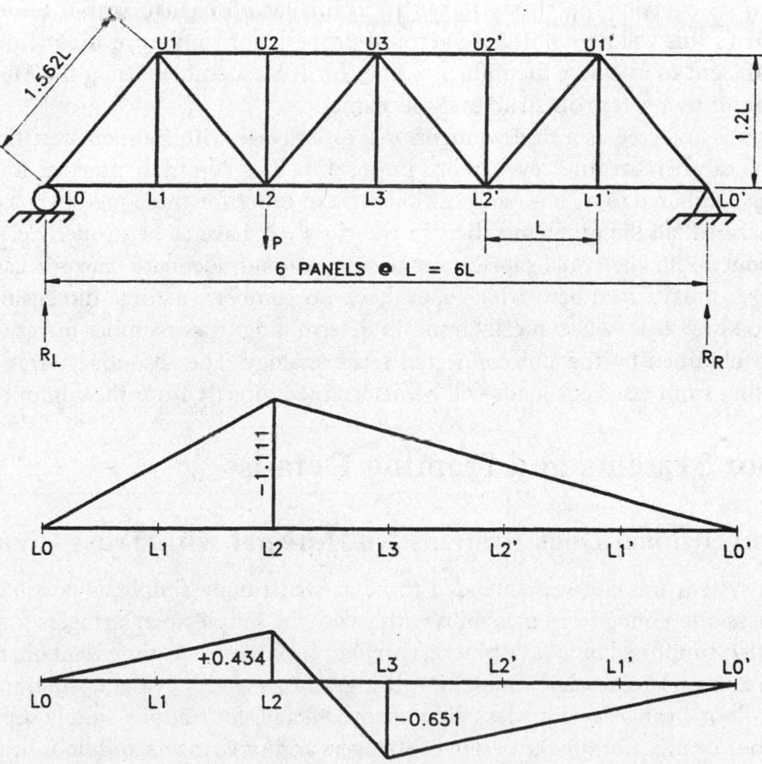

FIGURE 16.8 Influence lines for forces in one chord and one diagonal.

16.3.2 Computer Methods

The method of joints and the method of sections identified above appear to be very simple as long as the geometry of the truss is also simple and the structure is statically determinate. This is particularly true if one or both chords are horizontal. On most modern trusses, the span is sufficiently long that the change in vertical geometry can be significant. In fact, most larger trusses are on vertical curves if they cross a waterway. The chord joints are usually parallel to the deck profile. Thus, in many practical truss bridges, one or both truss chords is a series of chord segments representing a parabolic curve over at least part of the length of the bridge. This significantly complicates the geometry with respect to the use of either the method of joints or the method of sections. It does not negate the use of either of these methods, but certainly makes them less attractive.

There are many software packages for computers which permit the analysis of trusses, as can be done typically as either the pin-connected assemblage or as a frame with moment-resisting joints.

If the bridge is determinate in the plane of a truss, and if the truss is analyzed with two force members, then the cross-sectional area of the members does not affect the analysis. Assuming unit area for all members will give the proper forces, but not necessarily the proper displacements. If the truss is indeterminate in a plane, then it will be necessary to use realistic areas for the truss members and may be important to include the camber of the members in order to get realistic results in some cases. This will be true of the so-called "geometric case" which is usually taken as the state of the bridge under all dead load, at which time it is supposed to have the proper grades and profile. An analysis for a subsequent load, such as unit loads for the assembly of influence lines, or a transient load, does not require inclusion of the member camber. In fact, inclusion of the camber for other than the loads acting in the geometric condition would yield erroneous results for the indeterminate truss.

Where software contains the ability to put in a unit length change within a member and an analysis similar to this will be required to account properly for camber of the members, then it is often found efficient to calculate the influence lines for truss members using the Mueller–Bresslau principle as found in any text on structural mechanics.

When a truss is analyzed as a three-dimensional assemblage with moment-resisting joints, then the method of camber becomes even more important. It is common practice for some of the members to be cambered to a "no-load position" and in order for these members to have no load in them as analyzed, all the other members in the truss will have to be properly cambered in the computer model. With the usual fabrication techniques and adequate care for camber in both primary and secondary member (which may have no camber), a three-dimensional computer analysis of a roadway truss will typically result in determining truss member forces which are very close to those obtained by the pin-connected truss analogy. The secondary stresses from joint rotation resulting from transient loads will be determined directly from the computer analysis.

16.4 Floor Systems and Framing Details

16.4.1 Conventional Deck Systems Not Integral with Truss Chords

Initially, floor system framing was intended to be as structurally simple as possible. In the past, floor beams were often hung from truss pins with yokes, the simple-span stringers framing between floor beams often supported by saddle brackets on floor beam webs. As time went on, the advantages of continuous stringers, particularly in highway bridges, became very evident, and framing involving stringers-over-floor-beams developed, as did improved details for framing simply supported stringers between floor beams. Composite design of stringers and/or stringers and floor beams continued to add strength, stiffness, and robustness to trusses, while simultaneously eliminating many of the

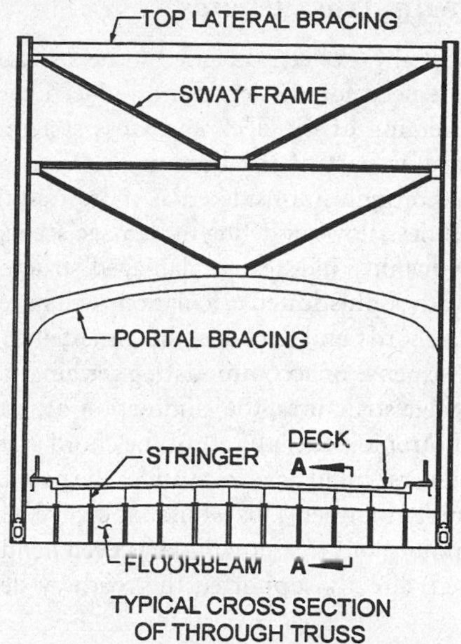

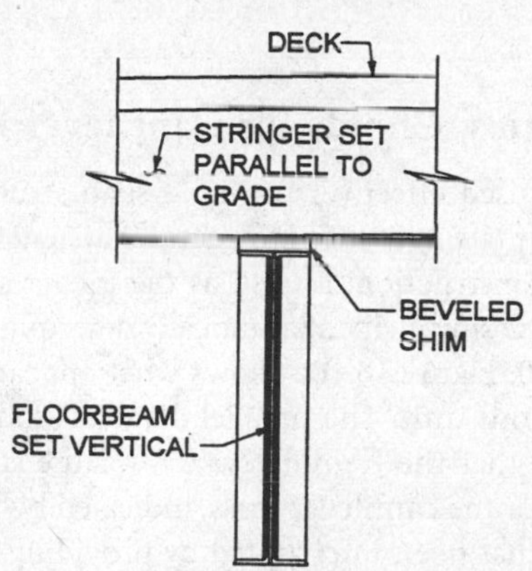

FIGURE 16.9 Typical truss cross-section.

sources of uncontrolled drainage, and hence, corrosion, which have been the perceived source of excessive maintenance in trusses. Currently, floor beams are either vertical or set normal to roadway grade, and stringers are usually normal to crown and parallel to grade. If they are vertical, some sort of beveled fill will be necessary between the floor beam and the stringers. A typical through-truss cross-section is shown in Figure 16.9.

Most modern truss designs continue to use concrete decks, as well as filled grid, or grid and concrete composite systems as efficient durable decks. Relatively little use has been made of orthotropic decks in conjunction with original design (as opposed to rehabilitation) of trusses in the United States but this is certainly a feasible alternative. The use of newer lightweight deck systems, such as the proprietary Aluma-Deck or possibly advanced composite orthotropic deck systems can lead to further reduction in weight and, hence, savings in a competitive environment, as well as holding the potential for significantly reduced maintenance in future trusses.

16.4.2 Decks Integral with Truss Chords

So far, deck systems have almost always been designed to be structurally separate from the main supporting truss systems. As the need for efficiency and reduced cost, as well as increased redundancy, continues, a possible merging of the deck and truss system is a technical possibility. An orthotropic deck has been used as part of the bottom or top chord on some foreign bridges. Redundancy issues should be thoroughly considered as more traditional load paths are reduced. The available computer capabilities allow modeling of damage scenarios and the emerging knowledge on the computation of reliability indexes for damaged structures can provide designs with high levels of confidence, but such sophisticated calculations will have to be justified by cost savings and/or other benefits. Merging chords and deck has the potential to eliminate more joints within the deck system, perhaps at the expense of accommodating certain differential temperature features. Generally, as in all types of bridge structures, the elimination of joints is perceived as a favorable development. If the use of orthotropic decks as part of the chord system, and the lateral system for that matter, were to evolve, designers would have to consider the possibility of using either reinforced or prestressed concrete in a similar manner. This would, of course, tend to lead toward loading the chord at other than the panel points, but this situation has been handled in the past where, in some situations, deck chord members directly supported the roadway deck over their full length, not simply at the panel points.

16.5 Special Details

16.5.1 Hangers and Dummy Chords for Cantilever Bridges

The cantilevered truss has been used effectively on long-span structures since the Firth of Forth Bridge was built in Scotland in the late 1800s. This structural system was developed to provide most of the economy of continuous construction, as well as the longer spans possible with continuity, while simultaneously providing the simplicity of a statically determinate structural system. Consider the system shown in Figure 16.10. Figure 16.10a shows what appears to be a Warren truss configuration for a three-span continuous unit. The parallel diagonal configuration shown in the detail in Figure 16.10a is an indication that the framing system for the standard Warren truss has been interrupted. The statical system for the cantilever truss, indicated by the parallel diagonals, is shown in Figure 16.10b. The continuity has been interrupted by providing two points along the structure where the chords carry no axial force, resulting in a "shear only" connection. This is, by definition, a structural hinge. The unit between the two hinges is commonly referred to as a "suspended span." The remaining portions of the structure are called the "cantilever arms" and the "anchor spans," as indicated in Figure 16.10a. The mechanism for supporting suspended span is shown in concept in Figure 16.10c, which indicates that two chords are missing and hinges have been placed in the strap, or hanger, carrying the load of the suspended span into the anchor arm. The configuration with the link and two hinges allows the portions of the structure to expand and contract relative to each other.

In practice, the unnecessary top and bottom chords are added to the structure to allay public concerns, and are articulated in a manner which prevents them from carrying any axial load. These elements are typically called "false chords, or "dummy chords." A typical top chord joint at the hanger point is shown in Figure 16.11. Figure 16.11a is a plan view of the top chord element, and the corresponding elevation view is shown in Figure 16.11b. The false chord in this case is supported by the anchor arm, utilizing a pin. The false chord is slotted, so that it may move back and forth with expansion and contraction without carrying any load. It simply moves back and forth relative to the pin in the slot provided. The pin carries the vertical weight of the member to the top chord joint. Also shown in Figure 16.11b, and extending further into Figure 16.11c, are details of the hanger assembly and the top pin of the pins in the hanger used to allow it to swing back and forth. Hangers are potentially fracture-critical members, as a failure of this member would almost certainly

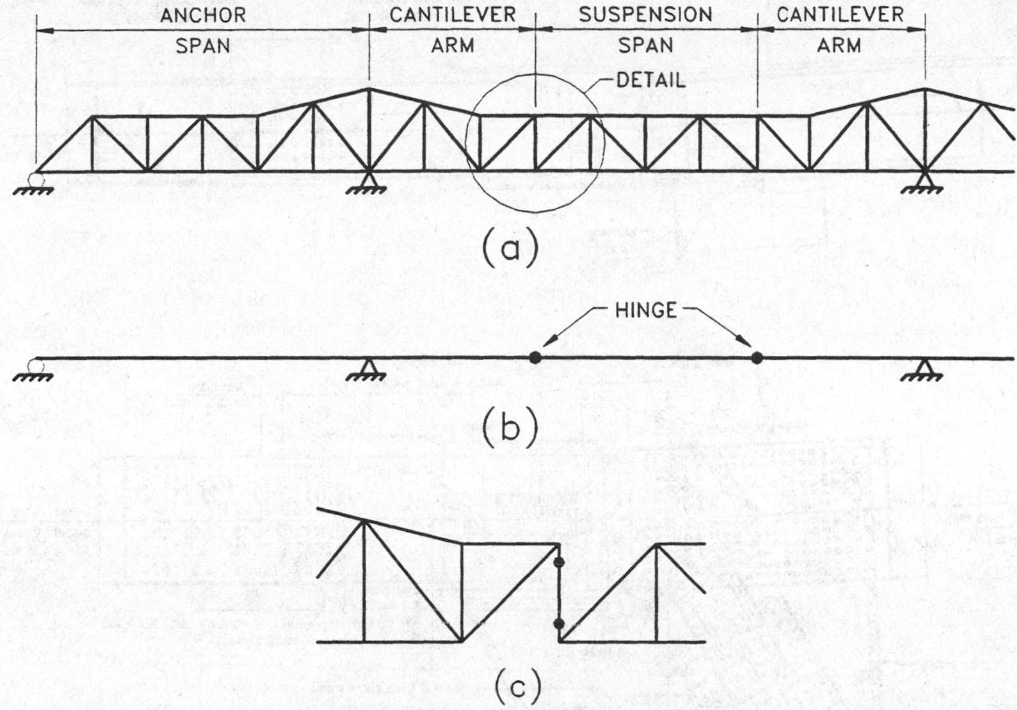

FIGURE 16.10 Cantilever suspension.

result in a collapse of at least the suspended span portion of the structure. These members are usually built of multiple components to add redundancy. The particular assembly shown has multiple plates bolted together to compensate for the hole occupied by the pin. In recent years, many truss bridges have been retrofitted with redundancy-adding assemblies usually consisting of rods or cables parallel to the hanger and attaching to the top and bottom chord. These assemblies are intended to pick up the load if the hanger or pin were to fail. Some of the details for the hanger pin are also shown in Figure 16.11c.

The corresponding portions of the structure at the bottom chord are shown in Figure 16.12. An elevation view of the lower chord joint is shown in Figure 16.12a, and a partial plan view is shown in Figure 16.12b. The concepts are very similar to those utilized in the upper joints, in that there are pins and slots to allow the false chord to move without picking up the axial load and pins and gusset plates to transfer the load from the suspended span into the hanger.

After completion of erection of the anchor span and the cantilever arm, the suspended span may be erected component by component, often referred to as "stick erection," or the entire suspended span may be assembled off site and hoisted into position until it can be brought into bearing at the hanger pins. If stick erection is used, the bridge will sag toward the middle as the cantilevers reach midspan. The bridge will be in the sag position because, once assembled to midspan, the cantilevers will be much shorter than they are at the midspan closure. It will thus be necessary to raise or lower portions of the bridge in order to get the closure members in and to transfer loads to the intended statical system. Also during this time, the false chords have to carry loads to support the cantilevering. With this type of erection, the false chords may be temporarily fixed, and one or both of the chords may have mechanical or hydraulic jacks for transferring load and for repositioning the two cantilevers for closure. Provisions for this type of assembly are often made in the false chord, at least to the point of being certain that the required space is available for jacks of sufficient capacity and that bearing plates to transmit the load are either in place or can be added by the contractor. A

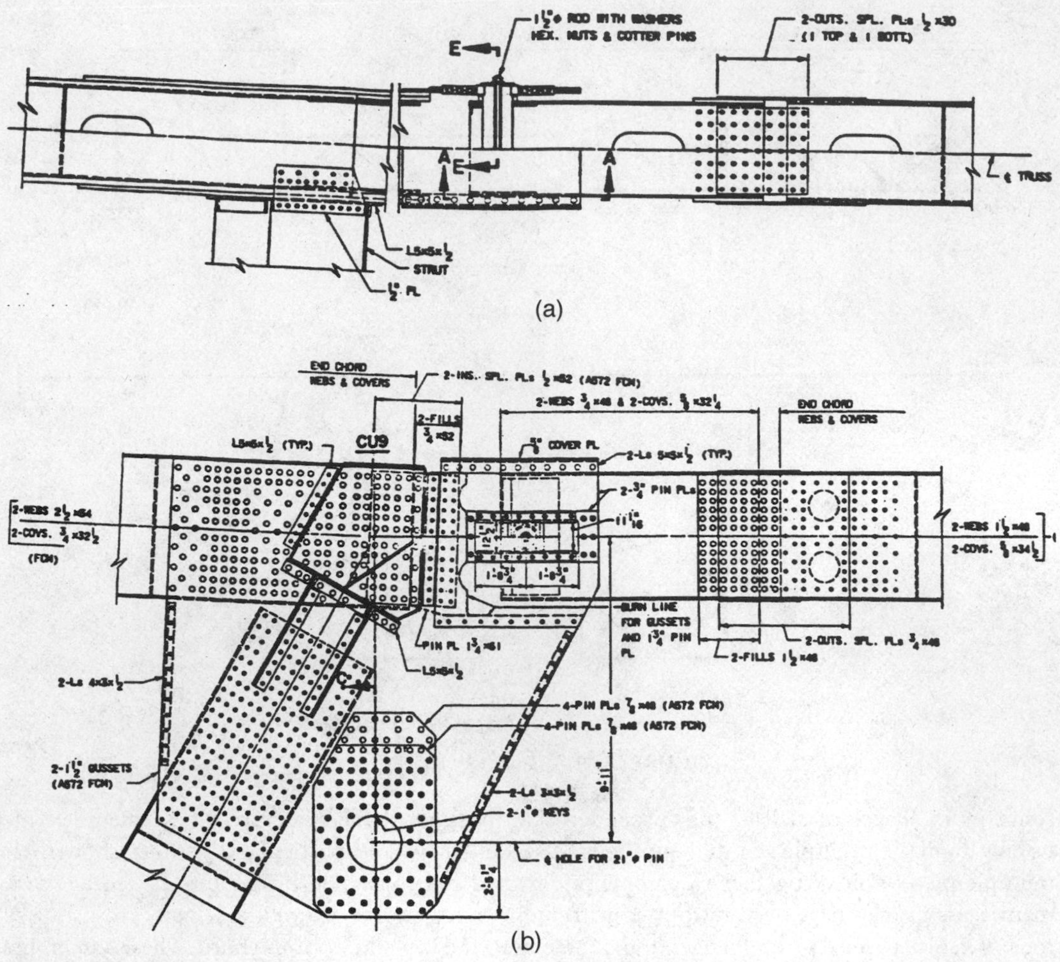

FIGURE 16.11 (a) Top false chord details — plan view; (b) top false chord details — elevation view; (c) hanger details.

typical detail for providing for jack assemblies is shown in Figure 16.13, which indicates how jacks would fit in the bottom false chord shown in Figure 16.12, and bear against the rest of the structure, so as to swing the cantilevered portion of the suspended span upward to facilitate closure.

16.5.2 Bearings

From the viewpoint of bearings, the cantilever form of erection offers several other advantages. The bearings used on the main piers can be fixed to the pier tops, and the chords framing into that point can be pinned into gusset plates. This provides a very simple and relatively maintenance-free connection to carry the major reaction of the bridge. The bearings on the end piers at the ends of the anchor spans are sometimes unique. Depending upon the requirements of the site or to reduce costs, the end spans are sometimes quite short, so that even under dead load, the reaction on the end piers is negative, which is to say an uplift condition exists under the dead load. Under some patterns of live load, this uplift will increase. When this condition exists, hanger assemblies similar to those described in Section 16.5.1 may be used to connect to a bearing fixed to the pier, and connected to an embedded steel grillage or similar device used to engage the weight of the pier to hold down the superstructure. Such a bearing is shown in Figure 16.14. The link accommodates

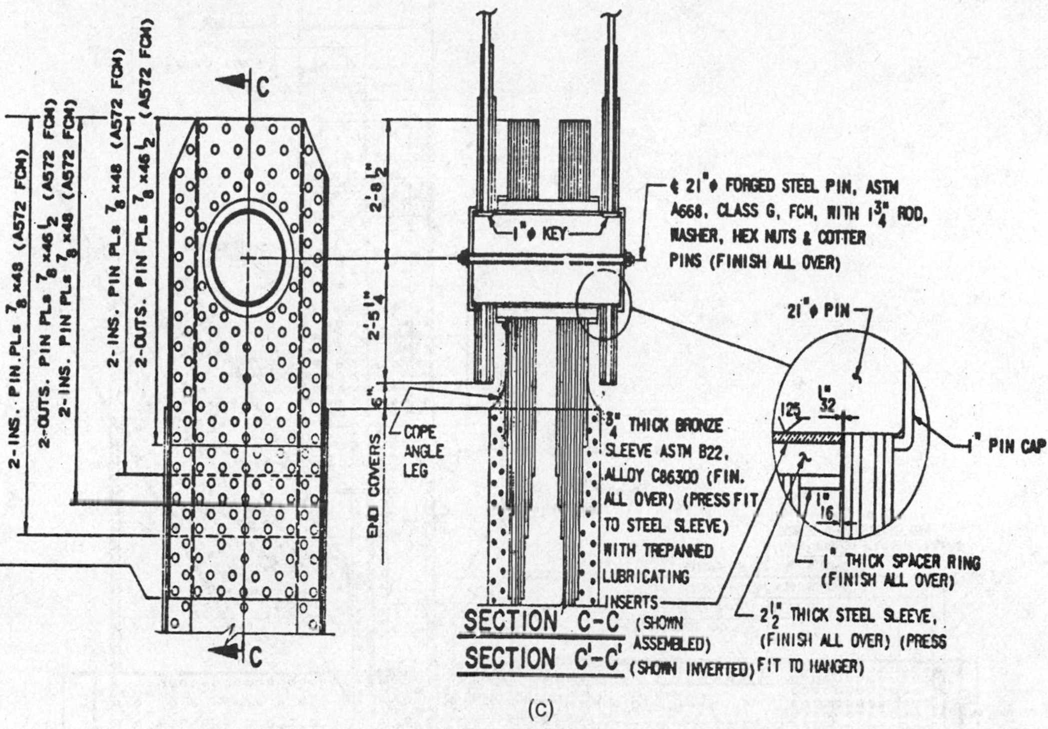

FIGURE 16.11 (continued)

the movement of the superstructure relative to the substructure required by expansion and contraction. The length of the arch swing of the link is designed so that the vertical displacement associated with the swing of the link can be accounted for and accommodated.

Where positive reactions are possible under all loadings, the bearing on the back span pier may be a rocker, roller nest, such as that shown in Figure 16.15, roller and gear assembly, or low-profile modern bearings, such as a pot bearing shown in Figure 16.16 as applied to a girder bridge or the disk bearing shown in Figure 16.17. It will usually be necessary for this bearing to provide for expansion and contraction while minimizing the forces put on the piers and to allow for rotation about the major bending axis of the bridge. Depending upon the designer's preferences, these bearings may or may not also carry the horizontal forces on the structure, such as wind loads, into the piers in the transverse direction. In some instances, the chord bearings serve this function through guide bars or pintles and, in some cases, a separate wind bearing, such as that shown in Figure 16.18, is provided to carry the transverse loads into piers separate from the main chord bearings.

Where structures are continuous, as opposed to cantilevered, it will usually be necessary for three of the four span bearings supporting the typical three-span truss to move. Individual movement will be greater in these bearings than they would be for a comparable-length cantilevered truss, and additional requirements will be placed on the bearing at one of the two main piers which moves, because of the large vertical reaction that will be transmitted at that point. Additionally, the continuous bridge will have two expansion joints, instead of four on the cantilever bridge, which is both an advantage and a disadvantage. These joints will have to be larger for the continuous bridge than for the cantilevered bridge and, therefore, more expensive. On the other hand, the tendency toward minimizing the number of joints in structures in order to reduce damage from deck drainage favors the continuous structure. Generally speaking, the extra points of expansion and contraction, associated deck joints and articulation hardware in the cantilevered bridge have required above-average maintenance.

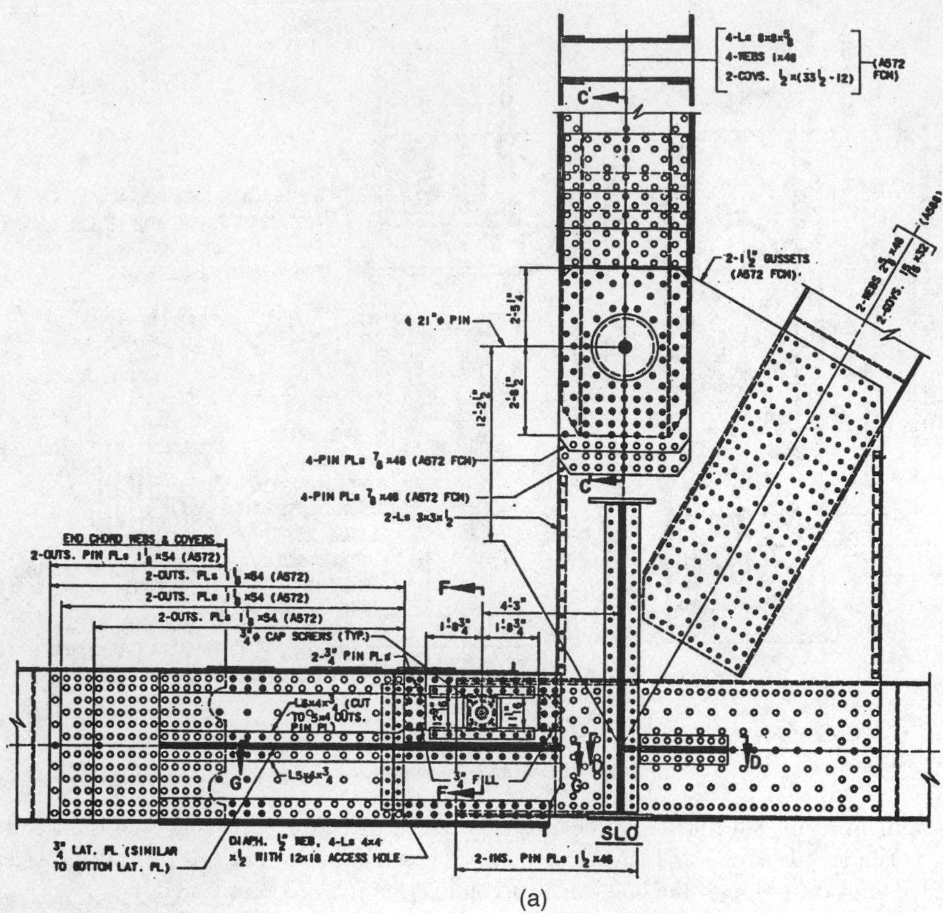

FIGURE 16.12 (a) Bottom false chord details — elevation view; (b) bottom false chord details — plan view.

16.5.3 Wind Tongues and Bearings for Transverse Forces

Wind loads carried by the suspended span in the cantilevered bridge have to be carried to the bearings on the piers. Thus, it is necessary for the wind loads to be carried through the panels framed with the false chords. Typically, in a through-truss, all of the wind loads on the suspended span are reacted by the hangers and by a special-purpose mechanism used to transmit horizontal forces at the lower chord level from the suspended span into the anchor arms. Wind load tributary to the upper lateral truss system in the plane of the top chord joints is carried into the anchor arms as a shear at the lower chord joint and the torque necessary to react to the transfer of loads from the top chord to the lower chord is carried as equal and opposite vertical reactions on the hangers. The horizontal forces at the lower chord level are then transmitted from the suspended span to the cantilever arm by a device called a "wind tongue," shown schematically in Figure 16.19. Because of the offset in chord joints at the suspended span, the horizontal force creates a torque in the plane of the bottom lateral system as the shear is transmitted across the expansion joint. Additionally, because expansion and contraction movements are accommodated at this point, allowance has to be made for some of the lateral members to swing along with that expansion and contraction. Thus, in Figure 16.19, there are four pin assemblies shown in the detail. The two horizontal links thus swing back and forth to accommodate the relative movement occurring at the open joint. The torque caused by the offset shear is reacted by the members framing from the open joint back into

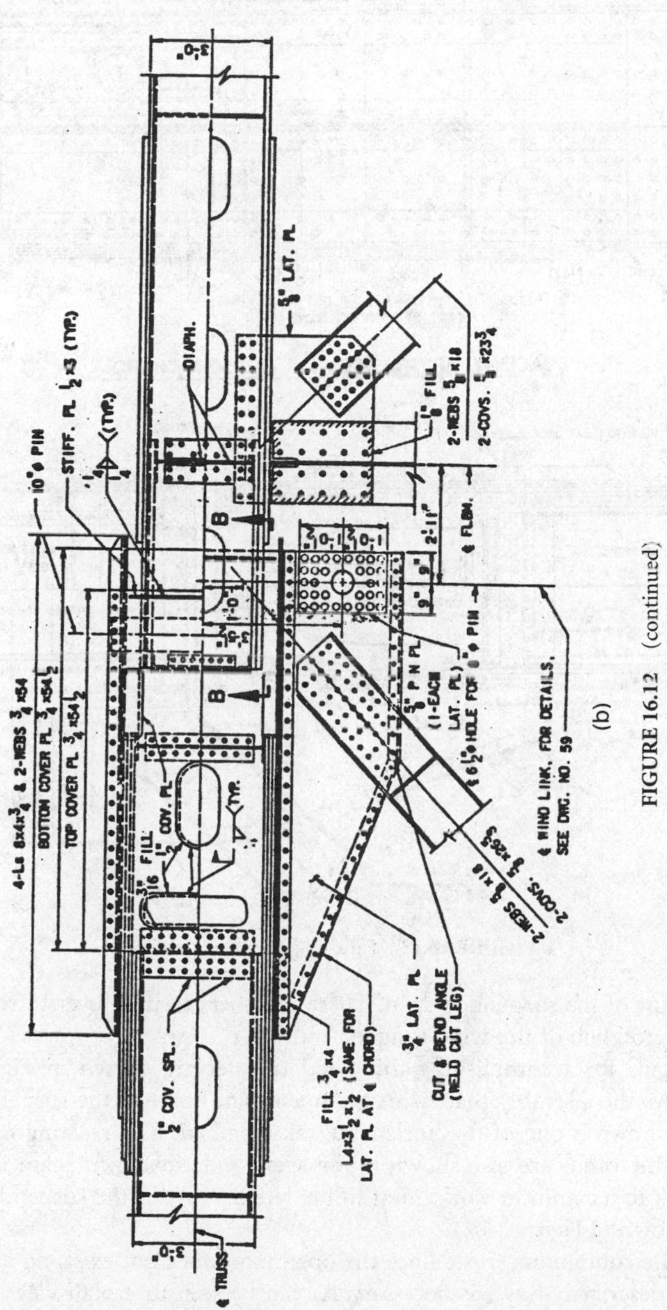

(b)

FIGURE 16.12 (continued)

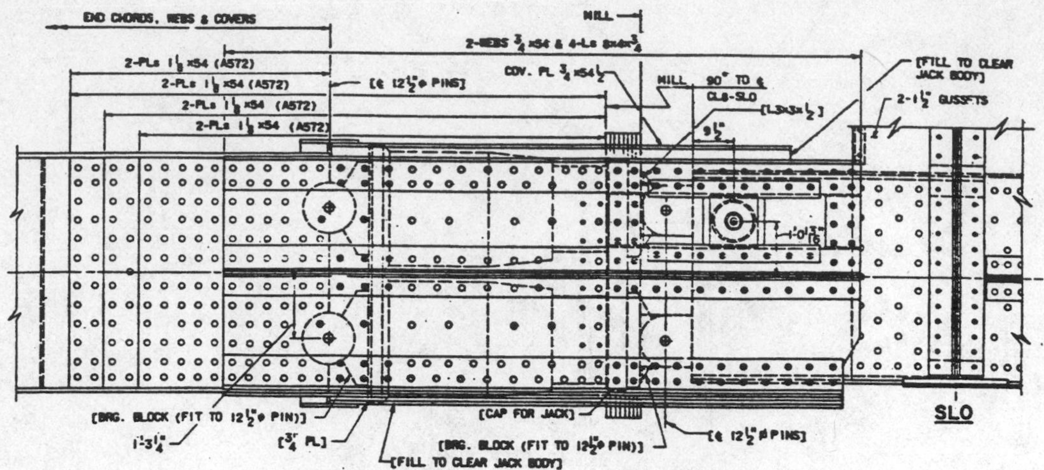

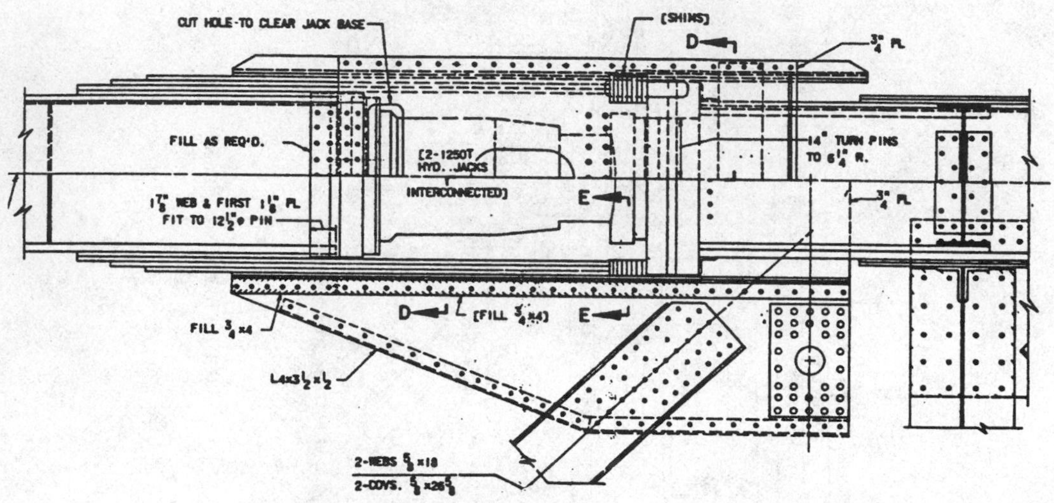

FIGURE 16.13 False chord jacking details.

the next panel point of the suspended span. These members form a lever to react to torque and prevent significant rotation of the wind tongue.

The typical details for accomplishing this wind transfer are shown in Figure 16.20a and b. Figure 16.20a shows the assembly that spans the open joint between the suspended span and the anchor arm. Also shown is one of the horizontal link members. The reacting members that form the lever to react the torque are also shown in this view, and are shown again in Figure 16.20b as they converge back to a common work point in the lateral truss of the suspended span. A typical pin assembly is shown in Figure 16.21.

In the case of the continuous truss, since the open joint does not exist, no assembly similar to the wind tongue, described above, is necessary. As can be seen in a plan view, the colinear force system can be developed to transmit wind and other transverse forces into the piers without creating torque in the bottom lateral system. Despite this, designers will often support the bearing point in order to accommodate accidental eccentricies that might exist. As seen in a vertical plane, there will almost certainly be an eccentricity between the center of transverse forces and the bearing. This will also be typically framed into a triangular system to carry this eccentricity through truss action,

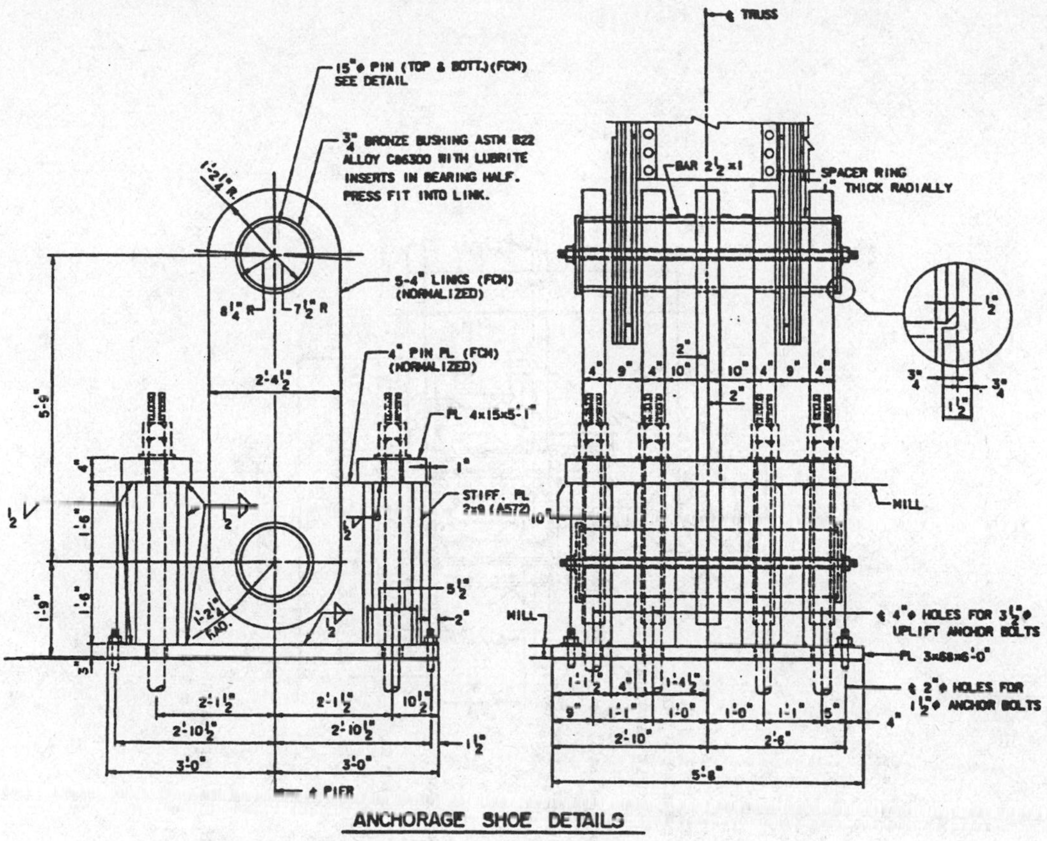

FIGURE 16.14 Link-type tie-down bearing.

rather than bending. These details are usually much simpler than the wind tongue at the suspended span, because it is not necessary to account simultaneously for expansion and contraction in the lateral truss system. This is usually handled by allowing the reaction points to move relative to the bearing while they are stationary relative to the lateral truss.

16.6 Camber and Erection

16.6.1 Camber for Vertical Geometry

It is obvious that all bridges have a theoretical geometric location as determined by the final design drawings. Every member has a theoretical length and location in space. One goal of the designer, fabricator, and erector is to produce a bridge as close to the theoretical position as possible, thereby ensuring actual stresses similar to design stresses.

To accomplish this, main members are usually cambered. Tensioned members that stretch under load are fabricated, such that their unstressed length is shorter than their length under the effect of dead load of the structure. The opposite is true for compression members. The cambered lengths are then accounted for during the erection stress and geometry studies. The state of the bridge when the camber "comes out" is called the geometric position. In this state the loads on the bridge are sufficient to return all of the members to their theoretical length. At any other state, the bridge will be out of shape and additional forces may result from the difference between its shape at any time and the final shape. This is relatively easy to see in the case of a continuous truss because it is clearly statically indeterminant and the shears and moments producing member forces are dependent on

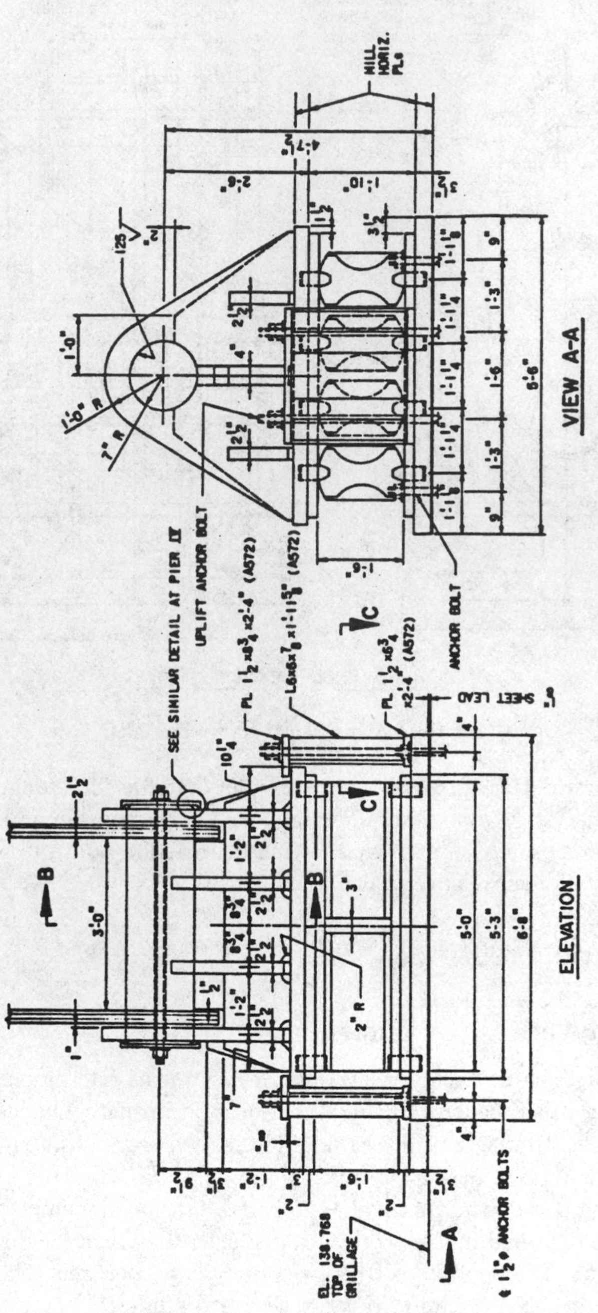

FIGURE 16.15 Roller bearing.

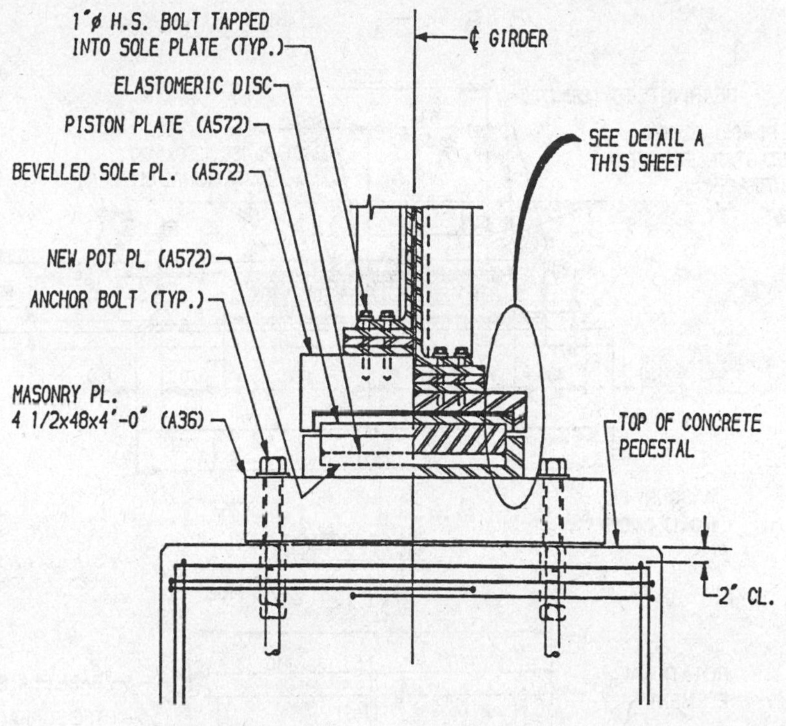

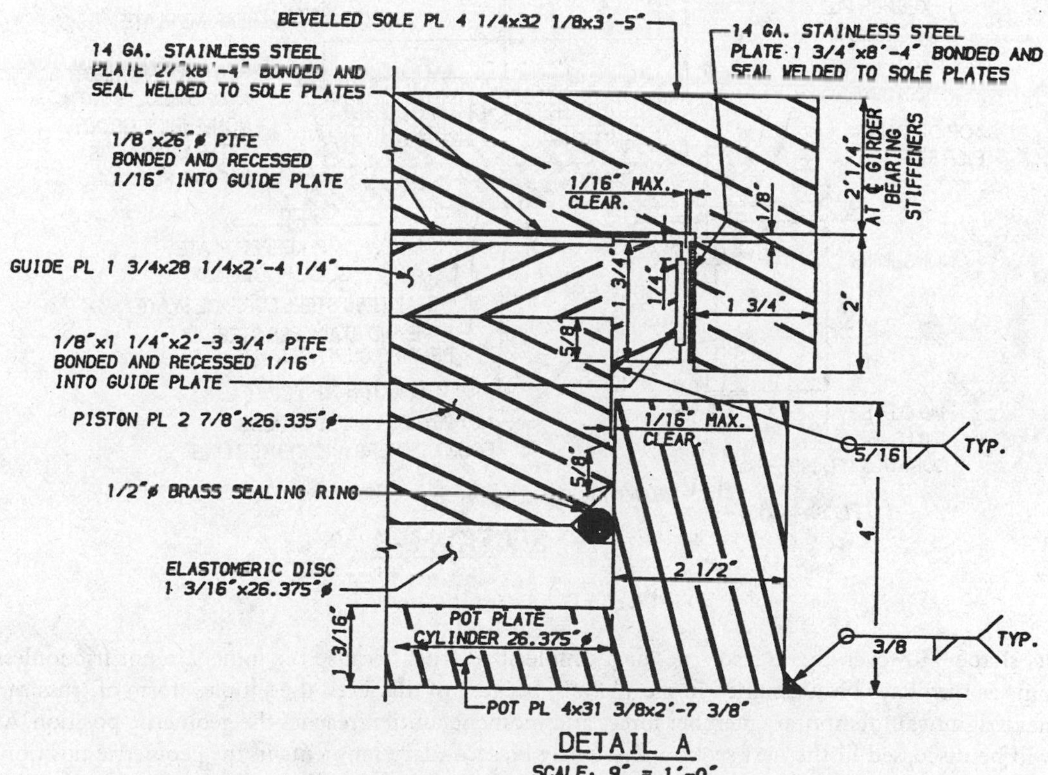

FIGURE 16.16 (a) Typical pot bearing; (b) typical pot bearing details.

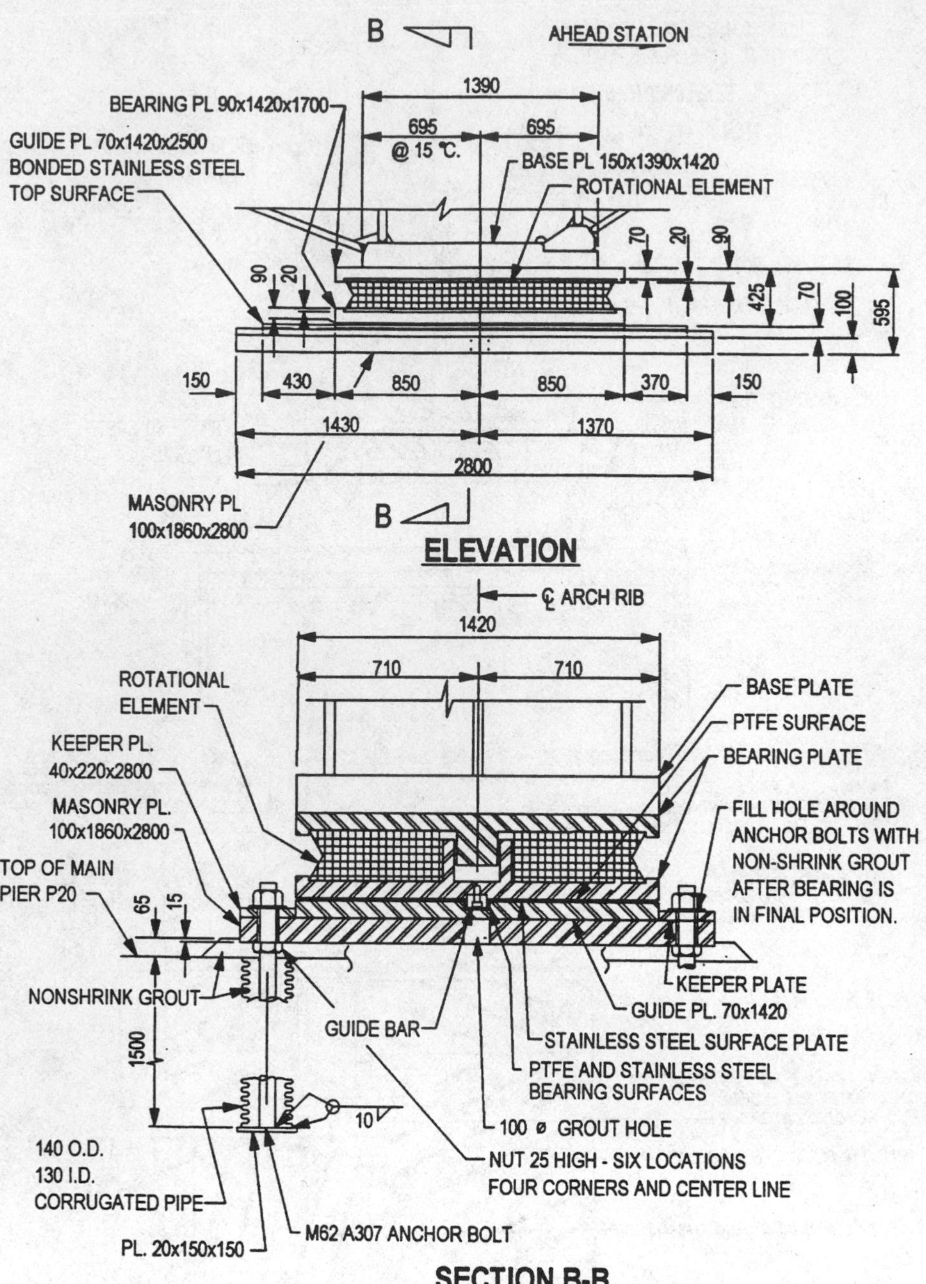

FIGURE 16.17 Typical disc bearing.

its shape. However, this is also true for a simple span truss because the joints are not frictionless pins as may have been assumed in the analysis. Because of this, even the simplest form of truss can have significant temporary member forces and moments until it reaches the geometric position. As will be discussed in the next section, there may be secondary moments in the geometric position, depending on the positioning of connector patterns on member ends in the shop.

Secondary members such as laterals and sway frame members are usually not cambered. They are usually intended to be stress free in the geometric position. Thus, at intermediate stages of erection they may also be subject to temporary forces.

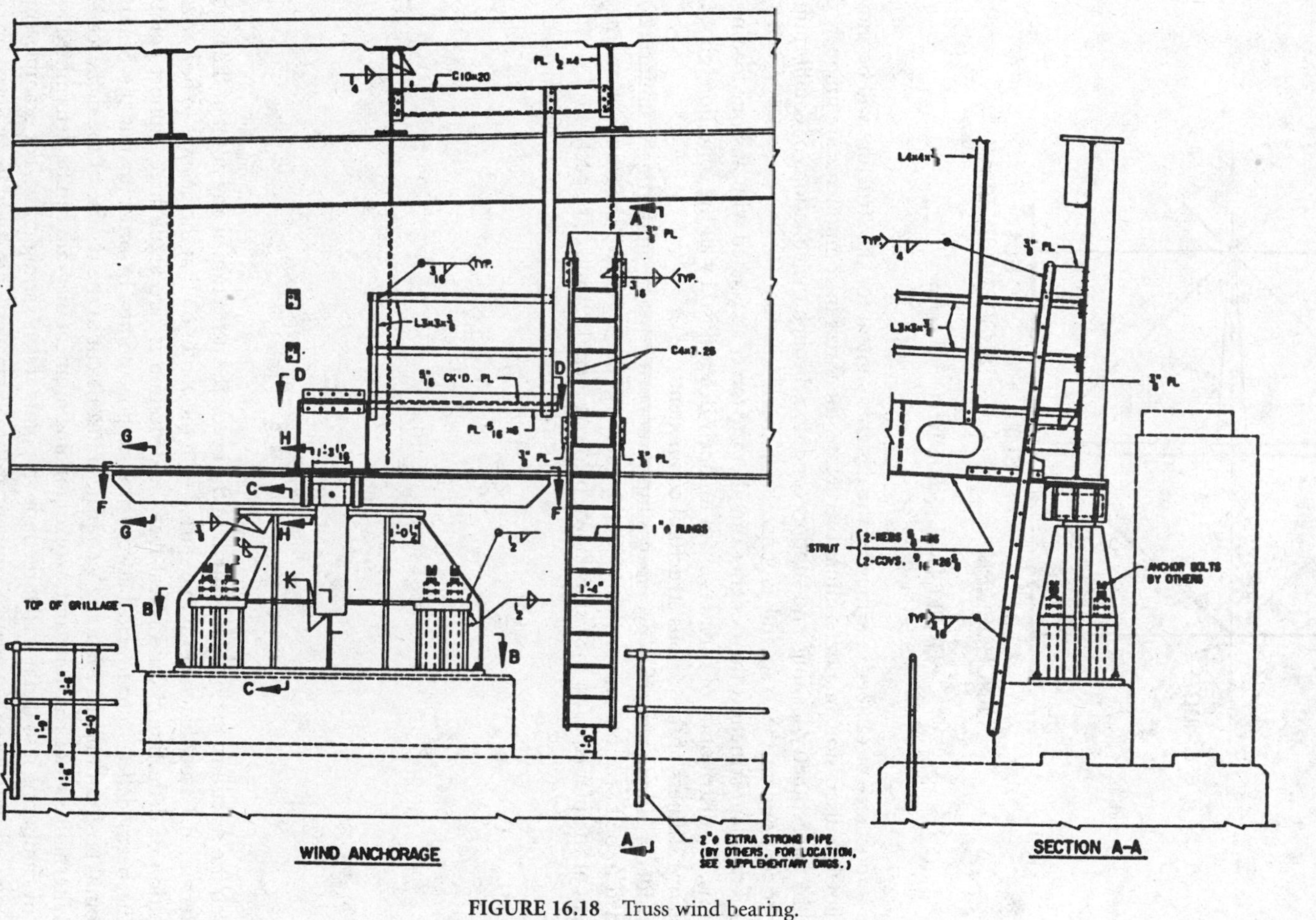

FIGURE 16.18 Truss wind bearing.

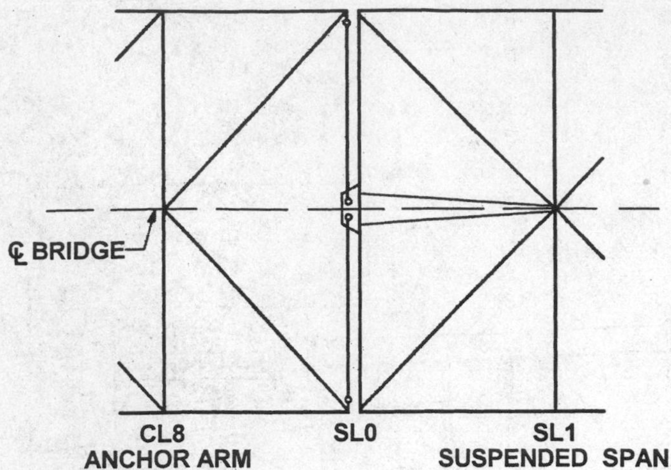

FIGURE 16.19 Schematic of wind tongue.

The importance of camber in achieving the designer's intent for the structure will be shown in the following discussion. In a determinant structure, the forces in components are uniquely determined by the geometry, loading, and support condition through the equations of equilibrium. The designer cannot alter the structural actions.

In a redundant structure, the designer can alter the forces associated with any one loading case. After this, the distribution of forces is again uniquely determined by all the conditions above, plus the relative stiffness of the various structural components.

Consider a simple case. If a two-span simply supported beam supporting a uniform load is intended to be horizontal under that load, the natural order is to have the reactions and negative moment at the pier equal to the value shown below at Piers 1, 2, and 3, respectively.

$$R_1 = \frac{3w\ell}{8} = R_3 \tag{16.1}$$

$$R_2 = \frac{10w\ell}{8} \tag{16.2}$$

$$M_2 = \frac{w\ell^2}{8} \tag{16.3}$$

Given the reactions, the load and length of the beam, the deflection at any point of the beam can be determined. Given that the example started with the condition that the beam is to be horizontal, calculation of deflections would reveal that the deflection is zero at all three support points and that the beam deflects downward between the support points. In order to put the beam on a horizontal position under a load, the web plates must be cut to the reverse of the deflected shape.

An alternative set of reactions and a corresponding moment diagram can be determined for this beam by cambering it so that not all three support points are on a straight line. For example, suppose it was desired to reduce the negative moment at the center pier. This could be achieved by cutting the web plate such that, in the horizontal layout position, the girder would rest such that a straight line between the two external reactions would be below the middle of the beam (with gravity acting downward). Thus, when the beam is assembled in the field, a certain amount of bending will have to take place as a simply supported beam of span equal to $2L$ until such time as the girder touches the middle bearing. After that, the remainder of the loads will be carried as a two-span unit. Unless a loading is sufficient in magnitude and opposite in sense so as to create uplift at the center bearing, all loads will then be handled as a two-span continuous unit. In this way, one might determine that an optimum condition is to have equal dead-load moments in the two spans and over the support. All that is necessary to achieve this is to determine the offset distance off the bearing point in the laydown position. Then cut the beam to that configuration using the cambered shape and erect it.

In truss construction, this adjustment of natural forces is seldom actually done. On occasion, camber to produce a determinant structure at steel closure has been used to facilitate erection. Camber for force control is very common in other types of bridges.

16.6.2 Camber of Joints

When the principal operations on a main member, such as punching, drilling, and cutting are completed, and when the detail pieces connecting to it are fabricated, all the components are brought together to be fitted up, i.e., temporarily assembled with fit-up bolts, clamps, or tack welds. At this time, the member is inspected for dimensional accuracy, squareness, and, in general, conformance with shop detail drawings. Misalignment in holes in mating parts should be detected then and holes reamed, if necessary, for insertion of bolts. When fit-up is completed, the member is bolted or welded with final shop connections.

The foregoing type of shop preassembly or fit-up is an ordinary shop practice, routinely performed on virtually all work. There is another class of fit-up, however, mainly associated with highway and railroad bridges that may be required by project specifications. These may specify that the holes in bolted field connections and splices be reamed while the members are assembled in the shop. Such requirements should be reviewed carefully before they are specified. The steps of subpunching (or subdrilling), shop assembly and reaming for field connections add significant costs. Modern computer-controlled drilling equipment can provide full-size holes located with a high degree of accuracy. AASHTO Specifications, for example, include provisions for reduced shop assembly procedures when computer-controlled drilling operations are used.

16.6.3 Common Erection Methods

The most common construction methods for trusses include cantilever construction, falsework, float-ins, and tiebacks. It is common for more than one method to be used in the construction of any single bridge. The methods selected to erect a bridge may depend on several factors, including the type of bridge, bridge length and height, type and amount of river traffic, water depth, adjacent geographic conditions, cost, and weight, availability, and cost of erection equipment.

Regardless of the method of erection that is used, an erection schedule should be prepared prior to starting the erection of any long-span bridge. The study should include bridge geometry ,member stress, and stability at all stages of erection. Bridges under construction often work completely differently than they do in their finished or final condition, and the character of the stress is changed, as from tension to compression. Stresses induced by erection equipment must also be checked, and, it goes without saying, large bridges under construction must be checked for wind stresses and sometimes wind-induced vibrations. An erection schedule should generally include a fit-up schedule for bolting major joints and a closing procedure to join portions of a bridge coming from opposite directions. Occasionally, permanent bridge members must be strengthened to withstand temporary erection loads. Prior to the erection of any bridge, proper controls for bridge line and elevation must be established and then maintained for the duration of the construction period.

Most long-span bridge construction projects have a formal closing procedure prepared by bridge engineers. The bridge member or assembled section erected to complete a span must fit the longitudinal

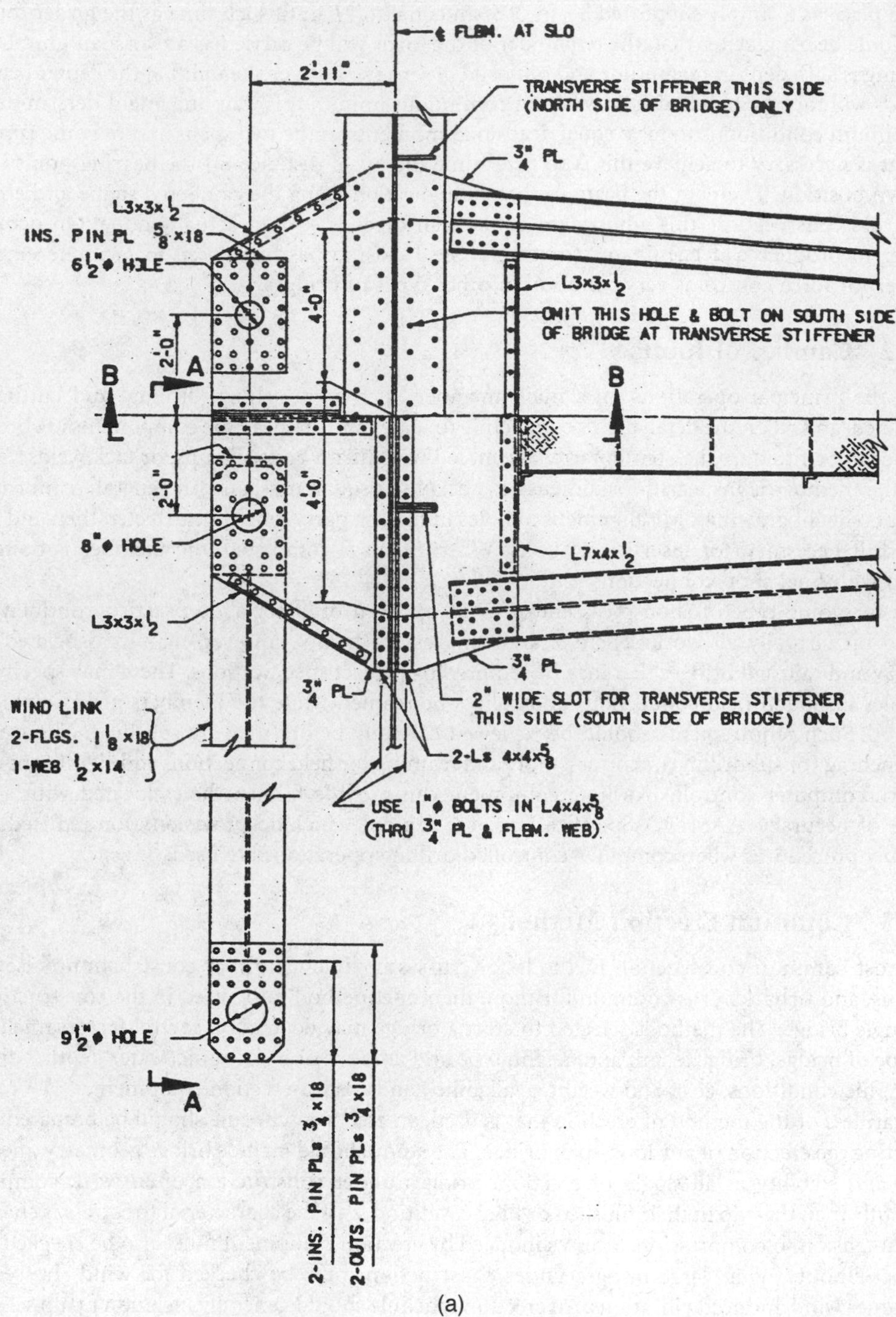

FIGURE 16.20 (a) Details of wind bearing; (b) details of wind tongue.

opening for it, and it must properly align with both adjoining sections of the bridge. Proper alignment of the closing piece is generally obtained by vertical jacking of falsework, lifting the existing bridge with a tieback system, or horizontal jacking of truss chords. At the Greater New Orleans Bridge No. 2, a scissors jack was inserted in place of a dummy top chord member, and it was used to pivot the main span truss for proper alignment.

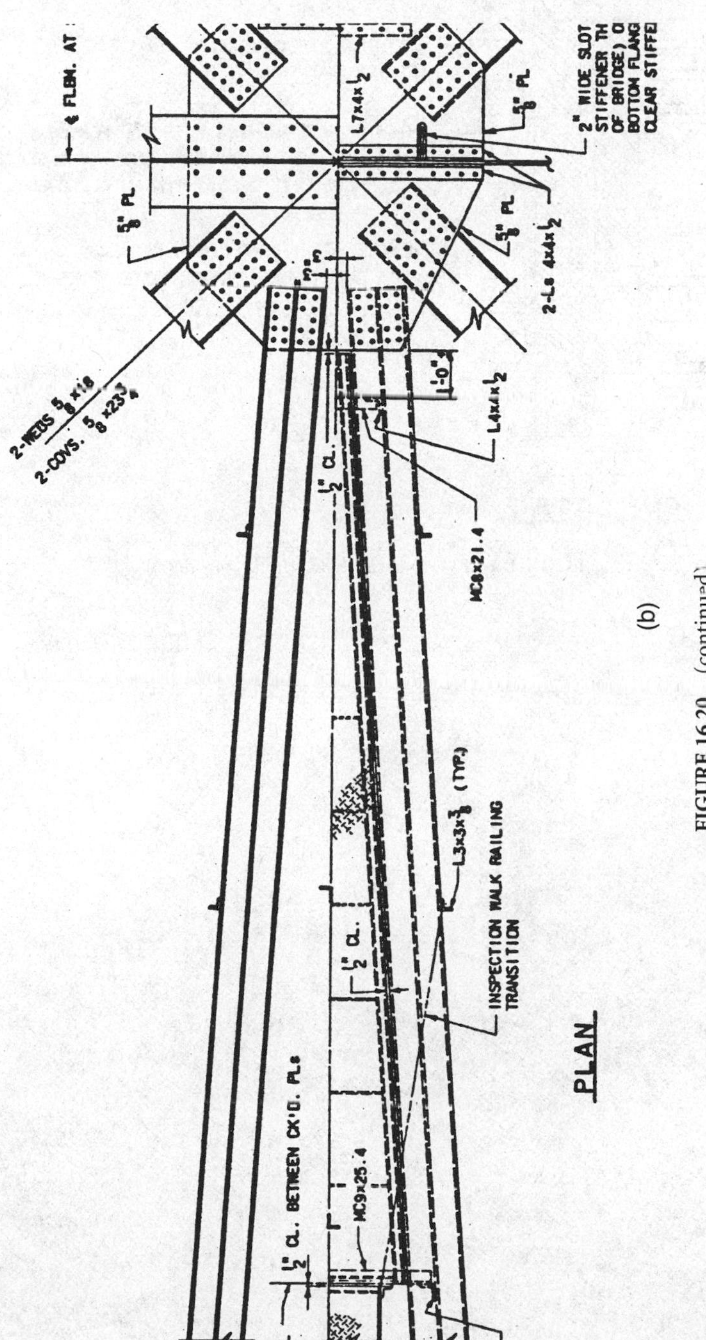

(b)

FIGURE 16.20 (continued)

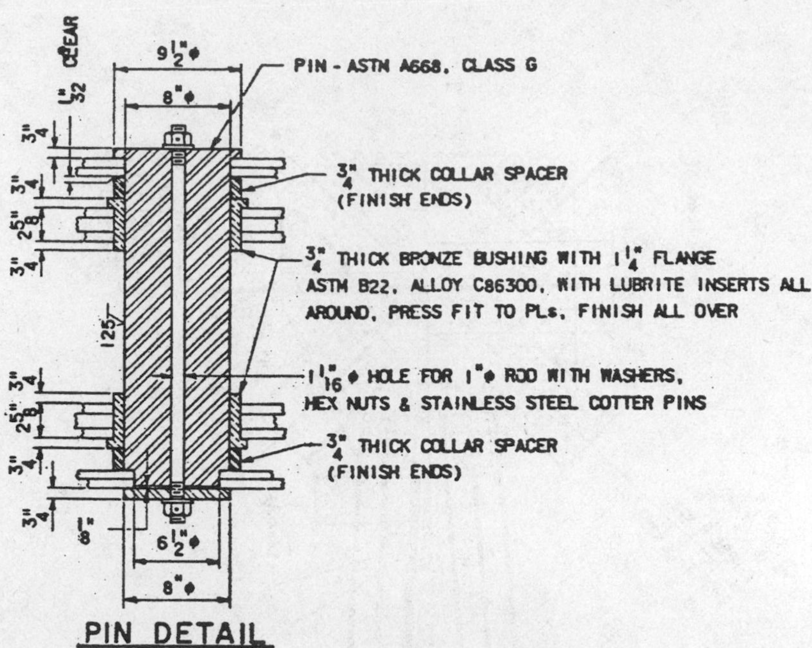

FIGURE 16.21 Typical wind tongue pin details.

FIGURE 16.22 Cantilever erection (on-site photo).

FIGURE 16.23 Early stage.

16.6.3.1 Cantilever Construction

In balanced cantilever construction, a bridge cantilevers in both directions from a single pier, as shown in Figure 16.22. The loads on each side of a pier must be kept reasonably in balance. In this type of construction, the first horizontal member erected in each direction may have to be temporarily supported by a brace, as shown in Figure 16.23.

In other types of cantilever construction, one span of a truss may be complete, and the bridge then cantilevers into an adjacent span, as shown in Figure 16.24. This type of construction is quite common.

16.6.3.2 Falsework

Falsework is commonly used when building medium-span trusses, especially for bridge approaches over land, as shown in Figures 16.25 and 16.26. It is also used in long-span bridge construction, but heavy river traffic, deep water, or poor foundation material may restrict its use. Falsework is sometimes used in the anchor or side span of cantilever trusses.

16.6.3.3 Float-In

Float-in is another commonly used construction method for long-span bridges. In this method, a portion of the bridge is generally assembled on a barge. The barge is then moved to the construction site, and the bridge is then either set into place off the barge or pulled vertically into place from the barge and connected to the part of the bridge already constructed. This method was used on the second Newburgh–Beacon Bridge across the Hudson River as discussed below.

First, a short section of the main span truss was erected on a barge; then sections of the approach spans over the river were assembled on top of the main span truss, floated into position, and erected, as shown in Figures 16.27 and 16.28. The section of bridge floated in is generally higher than its final position, and it is lowered into position with jacks. The two anchor spans were then assembled on a barge and erected the same way, as shown in Figure 16.29.

FIGURE 16.24 Advanced stage of cantilevering (on-site photo).

FIGRE 16.25 Use of falsework — 1 (on-site photo).

The cantilever portion of the bridge was then erected member by member by the cantilever erection method out into the main span to the pin hangers, as shown in Figure 16.30, which also shows the arrival of the suspended span of the main span. The suspended span was then barged into position and hoisted in place as shown in Figure 16.31.

FIGURE 16.26 Use of falsework — 2.

FIGURE 16.27 Float-in — 1.

The forces obtained during various stages of construction may be entirely different from those applicable to the final condition, and, in fact, the entire mechanism for resisting forces within the structure may change. During the erection of a truss, falsework bents may be used to support a portion of the structure. The gravity and lateral loads still act in the sense that they do on the final

FIGURE 16.28 Float-in — 2.

FIGURE 16.29 Float-in — 3.

condition, and the basic internal mechanism of resisting forces, as outlined in Chapter 10 of the Standard Specifications or Section 6 of the AASHTO-LRFD Specifications, remains unchanged, i.e., the primary load path involves axial load in the chords and diagonals, with the vertical components of those forces adding up to equal the applied shear within the panel, and the bending moment accounted for by the horizontal component of those forces. The camber of the truss will not pertain

FIGURE 16.30 Float-in of suspended span.

FIGURE 16.31 Lift of suspended span.

to an intermediate construction case, and, therefore, there is apt to be more joint rotation and, hence, more secondary bending moments within the truss members. Nonetheless, the primary load-carrying mechanism is that of axial forces in members. As the truss is erected, it is entirely possible that members that are in tension in the permanent condition will be under compression during erection, and vice versa. In fact, the state of stress may reverse several times during the erection of the bridge. Clearly, this has to be taken into account, not only in the design of the members, but also in the design of the connections. Compression members are apt to have been designed to transmit part of the forces in bearing. This will not be applicable when the member is in tension during erection. Similarly, a compression member that has tension during erection has to be reviewed for net section provisions and shear lag provisions.

16.7 Summary

Truss bridges have been an effective and efficient force of long-span bridges for over 150 years. As plate girder bridges have been utilized for spans of about 550 ft, box girders for spans of up to 750 ft, segmental concrete box girders for spans of up to about 800 ft, and cable-stayed bridges for spans of about 500 feet to 2000 ft, the use of trusses has declined over the last 25 years. Nonetheless, they remain a cost-effective bridge form, one with which many fabricators and erectors are experienced. Emerging materials and the use of computer analysis to treat the bridge as a three-dimensional structure will keep the truss form viable for the foreseeable future.

References

1. *Historic American Engineering Record,* National Park Service, Washington, D.C., 1976
2. Hartle, R. A., Amrheim, W. J., Willson, K. E., Baughman, D. R., and Tkacs, J. J., Bridge Inspector's Training Manual/90, FHWA-PD-91-015, Federal Highway Administration, Washington, D.C., 1990.

17

Arch Bridges

Gerard F. Fox
HNTB Corporation (Retired)

17.1 Introduction

17.1.1 Definition of Arch

An arch is sometimes defined as a curved structural member spanning an opening and serving as a support for the loads above the opening. This definition omits a description of what type of structural element, a moment and axial force element, makes up the arch. Nomenclature used to describe arch bridges is outlined on Figure 17.1. The true or perfect arch, theoretically, is one in which only a compressive force acts at the centroid of each element of the arch. The shape of the true arch, as shown in Figure 17.2, can be thought of as the inverse of a hanging chain between abutments. It is practically impossible to have a true arch bridge except for one loading condition. The arch bridge is usually subject to multiple loadings (dead load, live load, temperature, etc.) which will produce bending moment stresses in the arch rib that are generally small compared with the axial compressive stress.

17.1.2 Comparison of Arch Bridge with Other Bridge Types

The arch bridge is very competitive with truss bridges in spans up to about 275 m. If the cost is the same or only slightly higher for the arch bridge, then from aesthetic considerations the arch bridge would be selected instead of the truss bridge.

For longer spans, usually over water, the cable-stayed bridge has been able to be more economical than tied arch spans. The arch bridge has a big disadvantage in that the tie girder has to be constructed before the arch ribs can function. The cable-stayed bridge does not have this disadvantage, because deck elements and cables are erected simultaneously during the construction process. The true arch bridge will continue to be built of long spans over deep valleys where appropriate.

0-8493-7434-0/00/$0.00+$.50

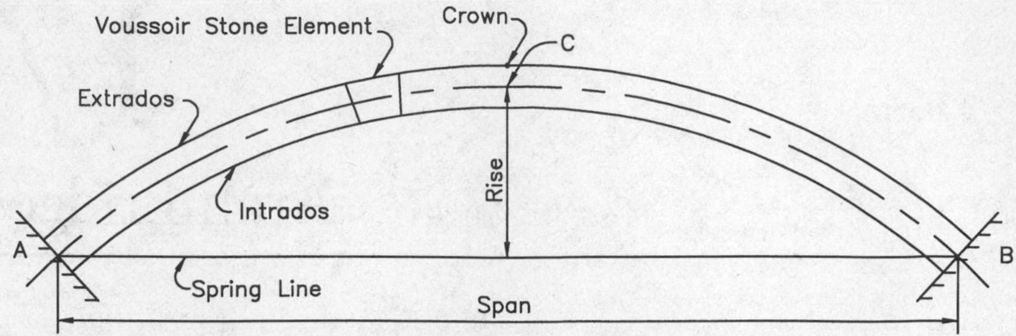

FIGURE 17.1 Arch nomenclature.

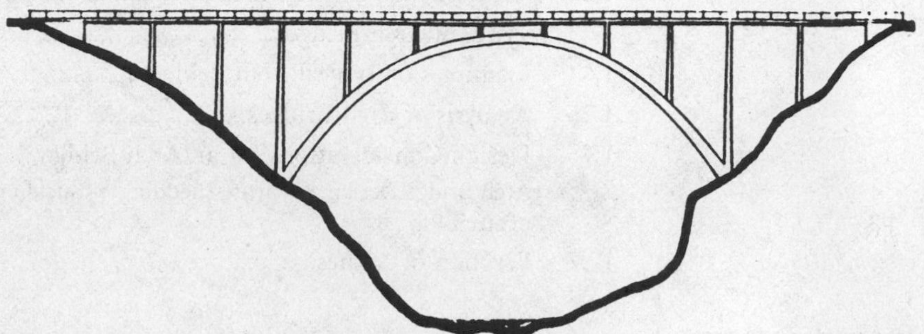

FIGURE 17.2 Concrete true arch.

17.1.3 Aesthetics of Arch Bridges

There is no question that arch bridges are beautiful, functional, and a pleasure for the motorist to drive over. Long-span arch bridges over deep valleys have no competitors as far as aesthetics is concerned. Arch bridges are better looking than truss bridges.

Many of the masonry arch bridges built for the last 2000 years are in the middle of cities whose residents consider these bridges not only necessary for commerce but also for their beautiful appearance. It is regrettable that masonry arches, in general, will not be constructed in today's environment since stonework has become too expensive.

17.2 Short History of Arch Bridges

Even before the beginning of recorded history, humans had a need to cross fast-moving streams and other natural obstacles. Initially, this was accomplished by means of stepping-stones and later by using felled tree trunks that either were supported by stones in the stream or spanned the entire distance between shores. Humans soon discovered that a vine attached to a treetop enabled them to swing across a wide river. This led to the construction of primitive suspension bridges with cables of vines or bamboo strips twisted into ropes. Post and lintel construction utilizing timber and stone slabs, as exemplified by the monument at Stonehenge, soon followed.

The arch came much later as applied to bridge building. The Sumerians, a community that lived in the Tigris-Euphrates Valley, made sunbaked bricks that were used as their main building material. To span an opening, they relied on corbel construction techniques. Around 4000 B.C. they discovered the advantages of the arch shape and its construction and they began to construct arch entranceways and small arch bridges with their sunbaked bricks [1].

Other communities with access to stone soon began to build arches with stone elements. By the time of the Romans most bridges were constructed as stone arches, also known as masonry or Voussoir arches. Empirical rules for dimensioning the shape of the arch and the wedge-shaped stones were developed. The Romans were magnificent builders and many of their masonry bridges are still standing. Probably the most famous is the Pont du Gard at Nimes in France, which is not only a bridge but an aqueduct as well. It stands as a monument to its builders. Excellent descriptions of other great Roman bridges can be found in Reference [1]. Stone arch bridges are very beautiful and much admired. A number have become centerpieces of their cities. There were few failures since stone is able to support very large compressive forces and is resistant to corrosive elements. Also, the arch is stable as long as the thrust line is contained within the cross-sectional area. Masonry arch bridges are very durable and most difficult to destroy [2].

Today, arch bridges are generally constructed of concrete or steel. However, there is still a great deal of research on stone arches directed toward determining their ultimate load capacity, their remaining life, their stability, their maintenance requirements, and also to determine the best methods to retrofit the structures. The reason for this great interest is, of course, that there are thousands of these stone arch bridges all over the world that are still carrying traffic and it would be an enormous cost to replace them all, especially since many of them are national monuments.

In 1779, the first Cast Iron Bridge was constructed at Coalbrookdale, England to span the Severn River. It is a semicircular arch spanning 43 m. By the year 1800, there were very few long-span masonry bridges built because they were not competitive with this new material. The 19th century was really the century of iron/steel bridges, suspension bridges, trusses, large cantilever bridges, viaducts, etc. Gustave Eiffel designed two notable steel arch bridges, a 160 m span at Oporto, Portugal and a 165-m span over the Truyeres River at St. Flour, France. Another notable arch bridge is the Eads steel bridge at St. Louis which has three spans of 155 m. In addition, a very beautiful shallow steel arch, the Pont Alexandre II, was constructed in Paris over the Seine River.

Concrete bridges began to be constructed at the end of the 19th century, and the arch designs of Robert Maillart should be noted since they are original and so beautiful [3].

17.3 Types of Arch Bridges

There are many different types and arrangements of arch bridges. A deck arch is one where the bridge deck which includes the structure that directly supports the traffic loads is located above the crown of the arch. The deck arch is also known as a true or perfect arch. A through-arch is one where the bridge deck is located at the springline of the arch. A half-through arch is where the bridge deck is located at an elevation between a deck arch and a through arch.

A further classification refers to the articulation of the arch. A fixed arch is depicted in Figure 17.1 and implies no rotation possible at the supports, A and B. A fixed arch is indeterminate to the third degree. A three-hinged arch that allow rotation at A, B, and C is statically determinate. A two-hinged arch allows rotation at A and B and is indeterminate to one degree.

A tied arch is shown in Figure 17.3 and is one where the reactive horizontal forces acting on the arch ribs are supplied by a tension tie at deck level of a through or half-through arch. The tension tie is usually a steel plate girder or a steel box girder and, depending on its stiffness, is capable of carrying a portion of the live loads. A weak tie girder, however, requires a deep arch rib and a thin arch rib requires a stiff deep tie girder. Since they are dependent on each other, it is possible to optimize the size of each according to the goal established for aesthetics and/or cost.

While most through or half-through arch bridges are constructed with two planes of vertical arch ribs there have been a few constructed with only one rib with the roadways cantilevered on each side of the rib.

Hangers usually consist of wire ropes or rolled sections. The hangers are usually vertical but truss-like diagonal hangers have also been used as shown in Figure 17.4. Diagonal hangers result in smaller deflections and a reduction in the bending moments in the arch rib and deck.

FIGURE 17.3 Steel tied-arch bridge.

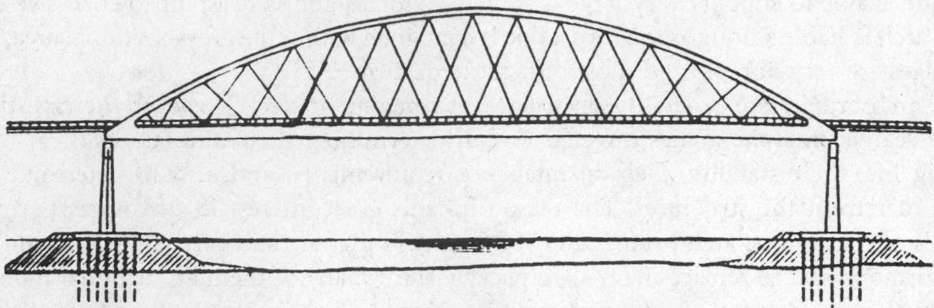

FIGURE 17.4 Arch with diagonal hangers.

There have also been arch bridges constructed with the arch ribs tilted so they can be connected at the crown. This is done for aesthetic reasons but it does add to the lateral stiffness of the arch bridge and could result in reduced bracing requirements.

17.4 Examples of Typical Arch Bridges

The Cowlitz River Bridge in the State of Washington, shown in Figure 17.5, is a typical true concrete arch that consists of a four-rib box section that spans 159 m with a rise of 45 m. Practically all concrete arches are of this type.

Multiple concrete arch spans have also been constructed such as the bridge, shown in Figure 17.6, across the Mississippi River at Minneapolis. The bridge consists of two 168-m shallow arch spans.

The longest reinforced concrete arch span in the world is the Wanxian Yangtze Bridge located in China with a span of 425 m and a rise of 85 m which gives a rise-to-span ratio of 1:5 [4]. It is unusual in that a stiff three-dimensional arch steel truss frame consisting of longitudinal steel tubes filled with concrete as the upper and lower chords was erected to span the 425 m. It served as the steel reinforcing of the arch and supported the cast-in-place concrete that was deposited in stages. This bridge is really a steel–concrete composite structure.

The Chinese have built many long-span concrete arches utilizing steel tubes filled with concrete. They have also constructed one half of an arch rib on each bank of a river, parallel to the river flow, and when completed, rotated both into their final position [5].

The longest steel true arch bridge has a span of 518 m and crosses the New River Gorge at Fayetteville, West Virginia. The arch rib consists of a steel truss. The deck, which is also a steel truss, is supported by transverse braced steel frames that are very slender longitudinally. The arch has rise-to-span ratio of 1:4.6 and was opened to traffic in 1977.

Most tied arch superstructures are of steel construction. A typical tied arch through structure is the Interstate 65 Twin Bridges over the Mobile River in Alabama as shown in Figure 17.7. The bridges are constructed of weathering steel. Also note the good appearance of the Vierendeel bracing between the arch ribs.

The Milwaukee Harbor Bridge is a three-span half-through steel-tied arch as shown in Figure 17.8. The steel tie at deck level is deep and very stiff while the arch rib is very thin. The main span length is 183 m with 82 m flanking spans. Again, note the appearance of the bracing between ribs.

FIGURE 17.5 Cowlitz River concrete arch.

FIGURE 17.6 Two shallow concrete arch spans.

Another three-span half-through steel-tied arch is the Fremont Bridge across the Willamette River in Portland, Oregon, as shown in Figure 17.9. It is a double-deck structure with an orthotropic top deck and a concrete lower deck. The main span of the arch is 383 m between spring points which makes it one of the longest tied-arch spans, if not the longest, in the world.

FIGURE 17.7 Steel twin-tied arches over the Mobile River.

FIGURE 17.8 Milwaukee Harbor half-through arch.

An unusual long-span half-through arch, shown in Figure 17.10, is the Roosevelt Lake Bridge in Arizona. The level of the lake is to be raised, and above the 200-year level of the lake the arch ribs and bracing are constructed of steel and below this level in concrete. To preserve structural continuity at the junction of the steel and concrete, the steel ribs are prestressed into the concrete rib. The arch is not a tied arch and spans 335 m with a rise-to-span ratio of about 1:5. Extensive wind tunnel testing was performed on models of this arch structure during the design phase.

17.5 Analysis of Arch Bridges

The dead load, live load, and temperature loads are covered in Chapter 6 of this Handbook. Wind effects are covered in Chapter 57 and seismic effects in Part IV.

FIGURE 17.9 Fremont Bridge across the Willamette River.

FIGURE 17.10 Roosevelt Lake Bridge, Arizona.

Before the age of structural analysis by computer methods, analysis of arches was not too difficult to accomplish with the help of a slide rule. In general, the analysis was a force method approach. For example, a two-hinged arch is statically indeterminate to the first degree and therefore has one redundant reaction. In the force method, the structure is made determinate, say, by freeing the right horizontal support and letting it move horizontally. The horizontal deflection ($d1$) at the support is then calculated for the applied loads. Next the horizontal deflection ($d2$) at the support is calculated for a horizontal force of 1 N acting at the support. Since the sum of these two deflections must vanish, then the total horizontal reaction (H) at the support for the applied loading must be

$$H = -\frac{(d1)}{(d2)} \tag{17.1}$$

Having the horizontal reaction the moments and axial forces can be calculated for the arch. A good example of early design procedures for a tied two-hinged arch is contained in a paper on the "Design of St. Georges Tied Arch Span" [6]. This arch, completed in 1941, spans the Chesapeake and Delaware Canal and was the first of its type in the United States to have a very stiff tie and shallow rib. The designer of the bridge was Professor Jewell M. Garrelts who was for many years head of the Civil Engineering Department at Columbia University. While this design procedure may be crude compared with modern methods, many fine arches were constructed using such methods.

Modern methods of analysis, of course, utilize a three-dimensional nonlinear finite-element computer program. For information on the finite-element method refer in this Handbook to Chapter 7 on Structural Theory and Chapter 36 on Nonlinear Analysis. Additional information on nonlinear analysis with accompanying computer programs on disk is available in Reference [7]. To use the finite element program it is necessary to have a preliminary design whose properties could then serve as the initial input to the computer program. In a published discussion of the St. Georges paper referred to above, Jacob Karol derived an approximate formula for calculating influence values for the horizontal force in the arch which depends only on the rise-to-span ratio. He also in the same discussion paper gave an approximate formula for the division of the total moment between the tie and the rib depending only on the depths of the rib and the tie girder [8]. These formulas are very useful in obtaining a preliminary design of a tied arch for input to the finite-element program.

17.6 Design Considerations for an Arch Bridge

17.6.1 Arch Bridge Design

Many chapters of this Handbook in Part II, Superstructure Design, have information concerning the design of decks that also apply to the design of decks of arches. By deck is meant the roadway concrete slab or orthotropic steel plate and its structural supports.

The rise-to-span ratio for arches may vary widely because an arch can be very shallow or, at the other extreme, could be a half-circle. Most arches would have rise-to-span ratios within the range of 1:4.5 to 1:6.

After the moments and axial forces become available from the three-dimensional finite-element nonlinear analysis the arch elements, such as the deck, ribs, ties, hangers, and columns can be proportioned. Steel arch ribs are usually made up of plates in the shape of a rectangular box. The ties are usually either welded steel box girders or plate girders.

In the 1970s there were problems in several arch bridges in that cracks appeared in welded tie girders. Repairs were made, some at great cost. However, there were no complete failures of any of the tie girders. Nevertheless, it caused the engineering community to take a new look at the need for redundancy. One proposal for arch bridges is not to weld the plates of the steel tie girders together but rather to use angles to connect them secured by bolts. Another proposal is to prestress the tie girder with post-tensioning cables. Another is to have the deck participate with the tie girder.

17.6.2 Vortex Shedding

Chapter 57 in this Handbook covers Wind Effects on Long Span Bridges. However, it seems appropriate to discuss briefly some problems in arches that are caused by vortex shedding.

Every now and then an arch is identified that is having problems with hanger vibrations especially those with I-section hangers. The vibrations are a result of vortex shedding. The usual retrofit is to

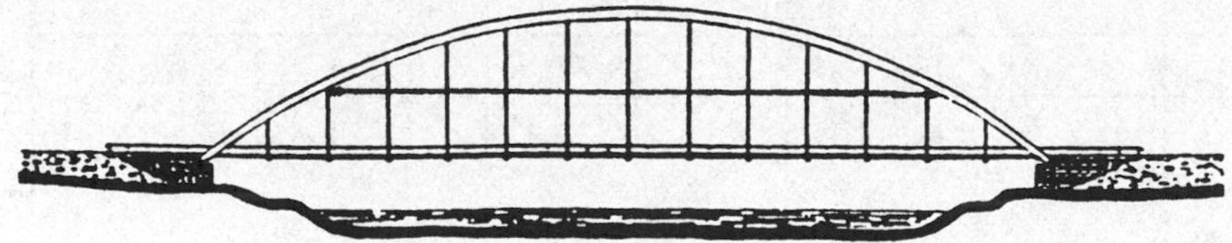

FIGURE 17.11 Horizontal cable connecting hangers.

connect the hangers as shown in Figure 17.11, which effectively reduces the length of the hangers and changes the natural frequency of the hangers. Another method is to add spoiler devices on the hangers [9]. In addition to the hangers, there have also been vortex shedding problems on very long steel columns that carry loads from the arch deck down to the arch rib.

17.6.3 Buckling of Arch Rib

Since the curved rib of the arch bridge is subject to a high axial force, the chance of a failure due to buckling of the rib cannot be ignored and must be accounted for. The subject of stability of arches is very well handled in Reference [10]. Values to use in formulas for critical buckling loads are listed in tables for many different cases of loading and various arch configurations.

As an example, the buckling critical load for a two-hinged parabolic arch rib supporting a uniform vertical deck load distributed on a horizontal projection will be calculated:

Arch span: $L = 120$ m
Arch rise: $S = 24$ m
Rise-to-span ratio = 24/120 = 0.2
Rib moment of inertia: $I = 7.6 \times 10^9$ mm^4
Modulus of elasticity: $E = 20 \times 10^4$ N/mm^2
Horizontal buckling force:

$$H = C_1 \frac{EI}{L^2} \tag{17.2}$$

Uniform load causing buckling:

$$q = C_2 \frac{EI}{L^3} \tag{17.3}$$

$C_1 = 28.8$ and $C_2 = 46.1$ are from Table 17.1 of Reference [10]

$$H = \frac{28.8 \times 20 \times 10^4 \times 7.6 \times 10^9}{\left(120 \times 10^3\right)^2} = 3.04 \text{ MN}$$

$$q = \frac{46.1 \times 3.04 \times 10^3}{28.8 \times 120} = 40.55 \text{ kN/m}$$

The above calculation of critical loads is for buckling in the plane of the arch which assumes very good bracing between ribs to prevent out-of-plane buckling. The bracing types that are generally used include K type bracing shown in Figure 17.12, diamond-shaped bracing shown in Figure 17.13 and Vierendeel type bracing shown in Figure 17.14.

Also the above calculation is for an arch rib without any restraint from the deck. If the deck is taken into account, the buckling critical load will increase.

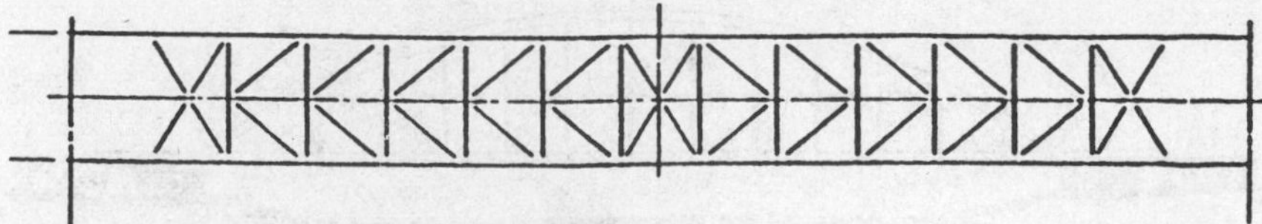

FIGURE 17.12 K-type of bracing.

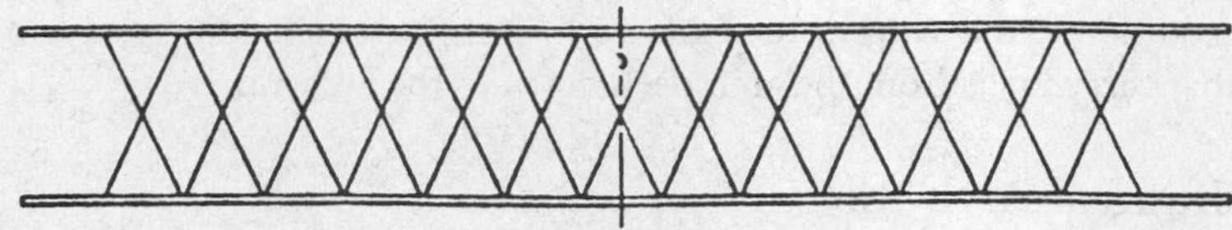

FIGURE 17.13 Diamond type of bracing.

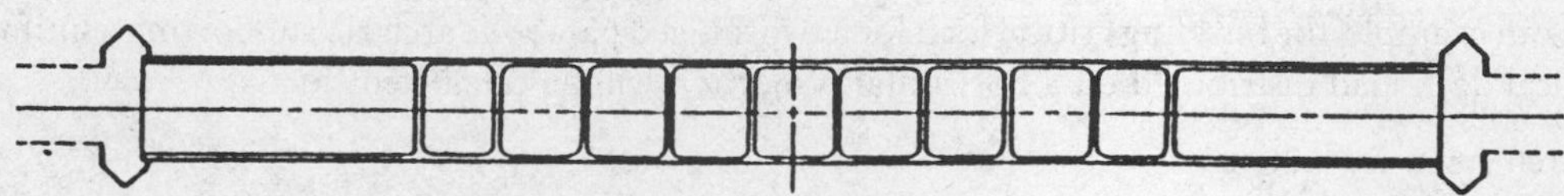

FIGURE 17.14 Vierendeel type of bracing.

17.7 Erection of Arches

Most concrete deck arches have been constructed with the concrete forms and wet concrete being completely supported by falsework. They have also been constructed by means of tieback cables from a tower that is supported by cables anchored into the ground. Each tieback cable would support the forms for a segment length of concrete rib. For multiple arches, the tiebacks would support the forms and wet concrete in balanced cantilever segments off a tower erected on a common pier.

Segmental concrete arches have been erected by cranes that are supported on the ground or on barges. They pick up a concrete segment and connect it to a previously erected segment. Another way of delivering and erecting the concrete segments is by means of a cableway spanning between tower bents.

Steel deck arches have also been erected in segment lengths by means of tieback cables. Another erection method is to have steel rib segments span the distance between temporary erection bents. The New River Gorge steel arch bridge was erected by means of both a cableway and tiebacks. Good examples of steel arch bridge construction are presented in Chapter 45 on Steel Bridge Construction.

For the usual tied-arch span the deck and steel tie can be erected on temporary erection bents. When this operation has been completed, the ribs of the arch, bracing, and hangers can be constructed directly off the deck. An alternative erection scheme that has been used is one in which the deck, steel ties, and ribs are erected simultaneously by means of tieback cables. A more spectacular erection scheme, that is economical when it can be used, involves constructing the tied-arch span on the shore or on piles adjacent and parallel to the shore. When completed the tied arch is floated on barges to the bridge site and then pulled up vertically to its final position in the bridge. For example, Figure 17.15 shows the Fremont Bridge 275 m center-tied arch span being lifted up vertically to connect to the steel cantilevers.

Some arches have had a rib erected on each shoreline in the vertical position as a column. When completed, the ribs are then rotated down to meet at the center between the shorelines.

FIGURE 17.15 Fremont tied arch being lifted into place.

Acknowledgments

My thanks to Lou Silano of Parsons Brinckerhoff for furnishing the two pictures of the Fremont Bridge and thanks to Ray McCabe of HNTB Corp. for the rest of the photos. I appreciate the help I received from Dr. Lian Duan of Caltrans, Professor Wai-Fah Chen of Purdue University, and the people at CRC Press. They are all very patient, for which I am very grateful.

References

1. Steinman, D. B. and Watson, S. R., *Bridges and Their Builders,* G. P. Putnam's Sons, New York, 1941.
2. Heyman, J., *Structural Analysis – A Historical Approach,* Cambridge University Press, Cambridge, 1998.
3. Billington, D. P., *The Tower and the Bridge,* Princeton University Press, Princeton, NJ, 1985.
4. Yan, G. and Yang, Z.-H., Wanxian Yangtze Bridge, China, *Struct. Eng. Int.,* 7, 164, 1997.
5. Zhou, P. and Zhu, Z., Concrete-filled tubular arch bridges in China, *Struct. Eng. Int.,* 7, 161, 1997.
6. Garrelts, J. M., Design of St. Georges tied arch span, *Proc. ASCE,* Dec., 1801, 1941.
7. Levy, R. and Spillers, W. R., *Analysis of Geometrically Nonlinear Structures,* Chapman & Hall, New York, 1995.
8. Karol, J., Discussion of St. Georges tied arch paper, *Proc. ASCE,* April, 593, 1942.
9. Simiu, E. and Scanlan, R. H., *Wind Effects on Structures,* 2nd ed., John Wiley & Sons, New York, 1986.
10. Galambos, T. V., *Stability Design Criteria for Metal Structures,* 5th ed., John Wiley & Sons, New York, 1998.
11. O'Connor, C., *Design of Bridge Superstructures,* John Wiley & Sons, New York, 1971.

Acknowledgments

[illegible]

References

[illegible]

18

Suspension Bridges

Atsushi Okukawa
Shuichi Suzuki
Ikuo Harazaki
Honshu–Shikoku Bridge Authority
Japan

18.1 Introduction

18.1.1 Origins

The origins of the suspension bridge go back a long way in history. Primitive suspension bridges, or simple crossing devices, were the forebears to today's modern suspension bridge structures. Suspension bridges were constructed with iron chain cables over 2000 years ago in China and a similar record has been left in India. The iron suspension bridge, assumed to have originated in the Orient, appeared in Europe in the 16th century and was developed in the 18th century. Although wrought iron chain was used as the main cables in the middle of the 18th century, a rapid expansion of the center span length took place in the latter half of the 19th century triggered by the invention of steel. Today, the suspension bridge is most suitable type for very long-span bridge and actually represents 20 or more of all the longest span bridges in the world.

18.1.2 Evolution of Modern Suspension Bridges

Beginning of the Modern Suspension Bridge

The modern suspension bridge originated in the 18th century when the development of the bridge structure and the production of iron started on a full-scale basis. Jacobs Creek Bridge was constructed by Finley in the United States in 1801, which had a center span of 21.3 m. The bridge's distinguishing feature was the adoption of a truss stiffening girder which gave rigidity to the bridge

0-8493-7434-0/00/$0.00+$.50

to distribute the load through the hanger ropes and thus prevent excessive deformation of the cable. The construction of the Clifton Bridge with a center span of 214 m, the oldest suspension bridge now in service for cars, began in 1831 and was completed in 1864 in the United Kingdom using wrought iron chains.

Progress of the Center Span Length in the First Half of the 20th Century in the United States

The aerial spinning method (AS method) used for constructing parallel wire cables was invented by Roebling during the construction of the Niagara Falls Bridge, which was completed in 1855 with a center span of 246 m. The technology was established in the Brooklyn Bridge, completed in 1883 with a center span of 486 m, where steel wires were first used. The Brooklyn Bridge, which is hailed as the first modern suspension bridge, was constructed across New York's East River through the self-sacrificing efforts of the Roebling family — father, son, and the daughter-in-law — over a period of 14 years.

In 1903, the Manhattan Bridge, with a center span of 448 m, and in 1909 the Williamsburg Bridge, with a center span of 488 m, were constructed on the upper reaches of the river. The first center span longer than 1000 m was the George Washington Bridge across the Hudson River in New York. It was completed in 1931 with a center span of 1067 m. In 1936, the San Francisco–Oakland Bay Bridge, which was twin suspension bridge with a center span of 704 m, and in 1937, the Golden Gate Bridge with a center span of 1280 m were constructed in the San Francisco Bay area.

In 1940, the Tacoma Narrows Bridge, with a center span of 853 m, the third longest in the world at that time, exhibited bending mode oscillations of up to 8.5 m with subsequent torsional mode vibrations. It finally collapsed under a 19 m/s wind just 4 months after its completion. After the accident, wind-resistant design became crucial for suspension bridges. The Tacoma Narrows Bridge, which was originally stiffened with I-girder, was reconstructed in 1950 with the same span length while using a truss-type stiffening girder.

The Mackinac Straits Bridge with a center span of 1158 m was constructed as a large suspension bridge comparable to the Golden Gate Bridge in 1956 and the Verrazano Narrows Bridge with a center span of 1298 m, which updated the world record after an interval of 17 years, was constructed in 1964.

New Trends in Structures in Europe from the End of World War II to the 1960s

Remarkable suspension bridges were being constructed in Europe even though their center span lengths were not outstandingly long.

In the United Kingdom, though the Forth Road Bridge, with a center span of 1006 m, was constructed using a truss stiffening girder, the Severn Bridge, with a center span of 988 m, was simultaneously constructed with a box girder and diagonal hanger ropes in 1966. This unique design revolutionized suspension bridge technology. The Humber Bridge, with a center span of 1410 m, which was the longest in the world before 1997, was constructed using technology similar as the Severn Bridge. In Portugal, the 25 de Abril Bridge was designed to carry railway traffic and future vehicular traffic and was completed in 1966 with a center span of 1013 m.

In 1998, the Great Belt East Bridge with the second longest center span of 1624 m was completed in Denmark using a box girder.

Developments in Asia since the 1970s

In Japan, research for the construction of the Honshu–Shikoku Bridges was begun by the Japan Society of Civil Engineers in 1961. The technology developed for long-span suspension bridges as part of the Honshu–Shikoku Bridge Project contributed first to the construction of the Kanmon Bridge, completed in 1973 with a center span of 712 m, then the Namhae Bridge, completed in 1973 in the Republic of Korea with a center span of 400 m, and finally the Hirado Bridge, completed in 1977 with a center span of 465 m.

The Innoshima Bridge, with a center span of 770 m, was constructed in 1983 as the first suspension bridge of the Honshu–Shikoku Bridge Project, followed by the Ohnaruto Bridge, which was designed to carry future railway traffic in addition to vehicular loads and was completed in 1985 with a center span of 876 m. The center route of the Honshu–Shikoku Bridge Project, opened to traffic in 1988, incorporates superior technology enabling the bridges to carry high-speed trains. This route includes long-span suspension bridges such as the Minami Bisan–Seto Bridge, with a center span of 1100 m, the Kita Bisan–Seto Bridge, with a center span of 990 m, and the Shimotsui–Seto Bridge with a center span of 910 m. The Akashi Kaikyo Bridge, completed in 1998 with the world longest center span of 1991 m, represents the accumulation of bridge construction technology to this day.

In Turkey, the Bosporus Bridge, with a center span of 1074 m, was constructed in 1973 with a bridge type similar to the Severn Bridge, while the Second Bosporus Bridge with a center span of 1090 m, called the Fatih Sultan Mehmet Bridge now, was completed in 1988 using vertical instead of diagonal hanger ropes.

In China, the Tsing Ma Bridge (Hong Kong), a combined railway and roadway bridge with a center span of 1377 m, was completed in 1997. The construction of long-span suspension bridges of 1000 m is currently considered remarkable, the Xi Ling Yangtze River Bridge with a center span of 900 m and the Jing Yin Yangtze River Bridge with a center span of 1385 m are now under construction [1]. Both suspension bridges have a box stiffening girder and concrete main towers. Besides these bridges, additional long-span suspension bridges are planned.

18.1.3 Dimensions of Suspension Bridges in the World

Major dimensions of long-span suspension bridges in the world are shown in Table 18.1.

18.2 Structural System

18.2.1 Structural Components

The basic structural components of a suspension bridge system are shown in Figure 18.1.

1. Stiffening girders/trusses: Longitudinal structures which support and distribute moving vehicle loads, act as chords for the lateral system and secure the aerodynamic stability of the structure.
2. Main cables: A group of parallel-wire bundled cables which support stiffening girders/trusses by hanger ropes and transfer loads to towers.
3. Main towers: Intermediate vertical structures which support main cables and transfer bridge loads to foundations.
4. Anchorages: Massive concrete blocks which anchor main cables and act as end supports of a bridge.

18.2.2 Types of Suspension Bridges

Suspension bridges can be classified by number of spans, continuity of stiffening girders, types of suspenders, and types of cable anchoring.

Number of Spans

Bridges are classified into single-span, two-span, or three-span suspension bridges with two towers, and multispan suspension bridges which have three or more towers (Figure 18.2). Three-span suspension bridges are the most commonly used. In multispan suspension bridges, the horizontal displacement of the tower tops might increase due to the load conditions, and countermeasures to control such displacement may become necessary.

TABLE 18.1 Dimensions of Long-Span Suspension Bridges

No.	Bridge	Country	Year of Completion	Span Lengths (m)	Type	Remarks
1	Akashi Kaikyo	Japan	1998	960+1991+960	3-span, 2-hinged	
2	Great Belt East	Denmark	1998	535+1624+535	Continuous	
3	Humber	U.K.	1981	280+1410+530	3-span, 2-hinged	
4	Jing Yin Yangtze River	China[a]	(1999)	(336.5)+1385+(309.34)	Single-span	
5	Tsing Ma	China[a]	1997	455+1377 (+300)	Continuous	Highway+Railway
6	Verrazano Narrows	U.S.	1964	370.3+1298.5+370.3	3-span, 2-hinged	
7	Golden Gate	U.S.	1937	342.9+1280.2+342.9	3-span, 2-hinged	
8	Höga Kusten	Sweden	1997	310+1210+280	3-span, 2-hinged	
9	Mackinac Straits	U.S.	1957	548.6+1158.2+548.6	3-span, 2-hinged	
10	Minami Bisan–Seto	Japan	1988	274+1100+274	Continuous	Highway+Railway
11	Fatih Sultan Mehmet	Turkey	1988	(210+) 1090 (+210)	Single-span	
12	Bosphorus	Turkey	1973	(231+) 1074 (+255)	Single-span	
13	George Washington	U.S.	1931	185.9+1066.8+198.1	3-span, 2-hinged	
14	3rd Kurushima Kaikyo	Japan	1999	(260+) 1030 (+280)	Single-span	
15	2nd Kurushima Kaikyo	Japan	1999	250+1020 (+245)	2-span, 2-hinged	
16	25 de Abril	Portugal	1966	483.4+1012.9+483.4	Continuous	Highway+Railway
17	Forth Road	U.K.	1964	408.4+1005.8+408.4	3-span, 2-hinged	
18	Kita Bisan–Seto	Japan	1988	274+990+274	Continuous	Highway+Railway
19	Severn	U.K.	1966	304.8+987.6+304.8	3-span, 2-hinged	
20	Shimotsui–Seto	Japan	1988	230+940+230	Single-span with cantilever	Highway+Railway
21	Xi Ling Yangtze River	China[a]	1997	225+900+255	Single-span	
22	Hu Men Zhu Jiang	China[a]	1997	302+888+348.5	Single-span	
23	Ohnaruto	Japan	1985	93+330+876+330	3-span, 2-hinged	Highway+Railway
24	Second Tacoma Narrows	U.S.	1950	335.3+853.4+335.3	3-span, 2-hinged	
25	Askøy	Norway	1992	(173+) 850 (+173)	Single-span	
26	Innoshima	Japan	1983	250+770+250	3-span, 2-hinged	
27	Akinada	Japan	(2000)	255+750+170	3-span, 2-hinged	
28	Hakucho	Japan	1998	330+720+330	3-span, 2-hinged	
29	Angostura	Venezuela	1967	280+712+280	3-span, 2-hinged	
29	Kanmon	Japan	1973	178+712+178	3-span, 2-hinged	
31	San Francisco–Oakland Bay	U.S.	1936	356.9+704.1+353.6 353.6+704.1+353.6	3-span, 2-hinged	

[a] The People's Republic of China.

Continuity of Stiffening Girders

Stiffening girders are typically classified into two-hinge or continuous types (Figure 18.3). Two-hinge stiffening girders are commonly used for highway bridges. For combined highway–railway bridges, the continuous girder is often adopted to ensure train runnability.

Types of Suspenders

Suspenders, or hanger ropes, are either vertical or diagonal (Figure 18.4). Generally, suspenders of most suspension bridges are vertical. Diagonal hangers have been used, such as in the Severn Bridge, to increase the damping of the suspended structures. Occasionally, vertical and diagonal hangers are combined for more stiffness.

Types of Cable Anchoring

These are classified into externally anchored or self-anchored types (Figure 18.5). External anchorage is most common. Self-anchored main cables are fixed to the stiffening girders instead of the anchorage; the axial compression is carried into the girders.

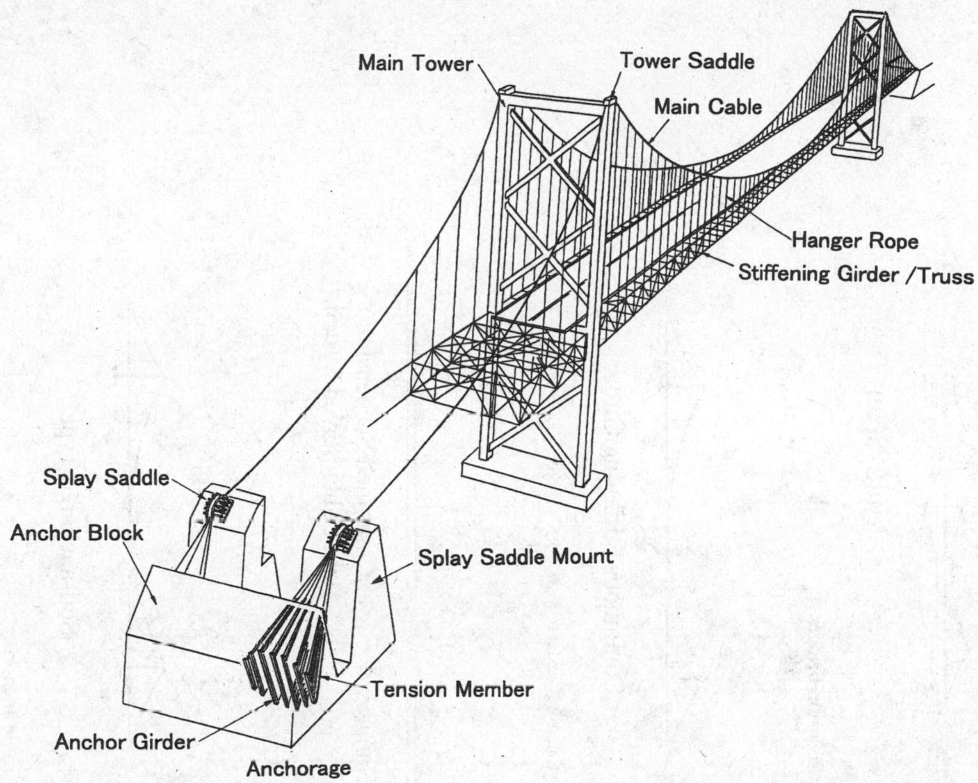

FIGURE 18.1 Suspension bridge components.

18.2.3 Main Towers

Longitudinal Direction

Towers are classified into rigid, flexible, or locking types (Figure 18.6). Flexible towers are commonly used in long-span suspension bridges, rigid towers for multispan suspension bridges to provide enough stiffness to the bridge, and locking towers occasionally for relatively short-span suspension bridges.

Transverse Direction

Towers are classified into portal or diagonally braced types (Table 18.2). Moreover, the tower shafts can either be vertical or inclined. Typically, the center axis of inclined shafts coincides with the centerline of the cable at the top of the tower. Careful examination of the tower configuration is important, in that towers dominate the bridge aesthetics.

18.2.4 Cables

In early suspension bridges, chains, eye-bar chains, or other material was used for the main cables. Wire cables were used for the first time in suspension bridges in the first half of the 19th century, and parallel-wire cables were adopted for the first time in the Niagara Falls Bridge in 1854. Cold-drawn and galvanized steel wires were adopted for the first time in the Brooklyn Bridge in 1883. This type has been used in almost all modern long-span suspension bridges. The types of parallel wire strands and stranded wire ropes that typically comprise cables are shown in Table 18.3. Generally, strands are bundled into a circle to form one cable. Hanger ropes might be steel bars, steel

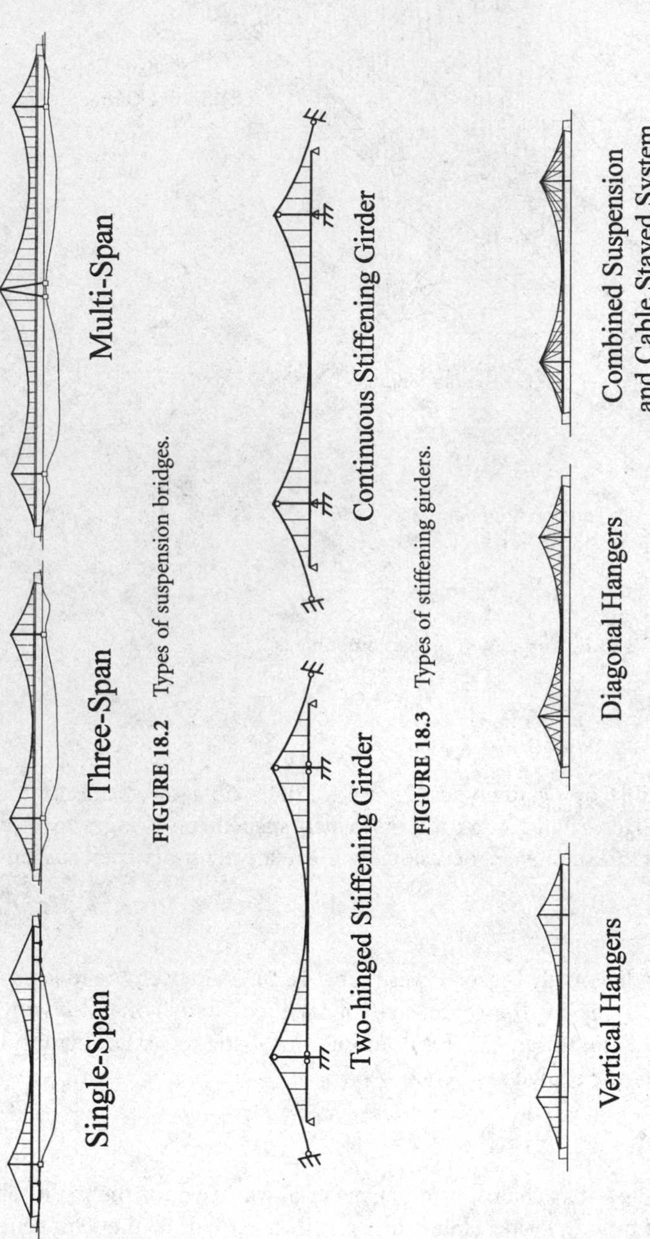

FIGURE 18.2 Types of suspension bridges.

FIGURE 18.3 Types of stiffening girders.

FIGURE 18.4 Types of suspenders.

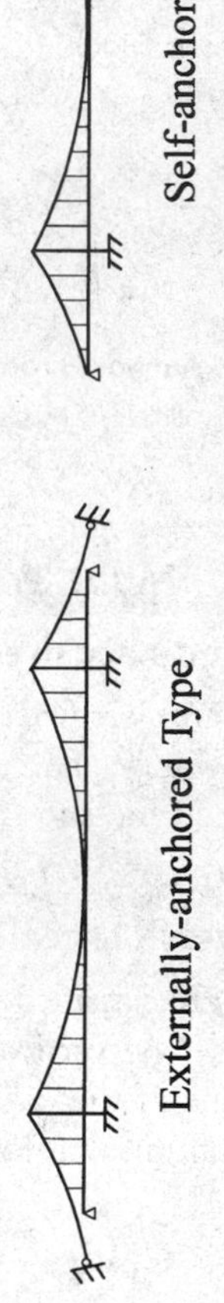

FIGURE 18.5 Types of cable anchoring

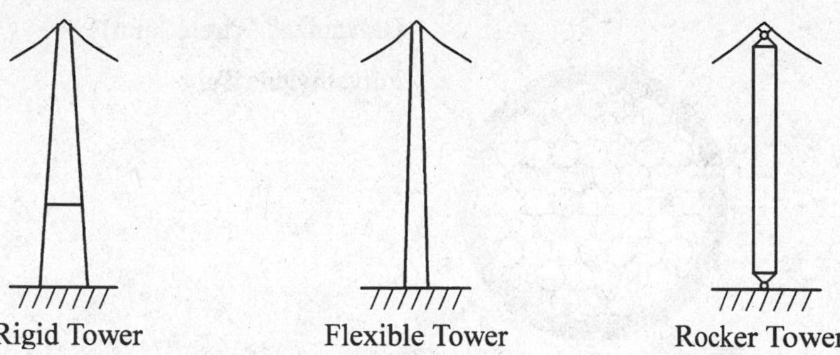

FIGURE 18.6 Main tower structural types.

TABLE 18.2 Types of Main Tower Skeletons

	Truss	Portal	Combined Truss and Portal
Shape			
Bridge	Akashi Kaikyo Forth Road	Great Belt East Humber	Golden Gate Second Tacoma Narrows

TABLE 18.3 Suspension Bridge Cable Types

Name	Shape of section	Structure	Bridge
Parallel Wire Strand		Wires are hexagonally bundled in parallel.	Brooklyn Humber Great Belt East Akashi Kaikyo
Strand Rope		Six strands made of several wires are closed around a core strand.	St.Johns
Spiral Rope		Wires are stranded in several layers mainly in opposite lay directions.	Little Belt Tancarville Wakato
Locked Coil Rope		Deformed wires are used for the outside layers of Spiral Rope.	Kvalsund Emmerich Älbsborg New Köln Rodenkirchen

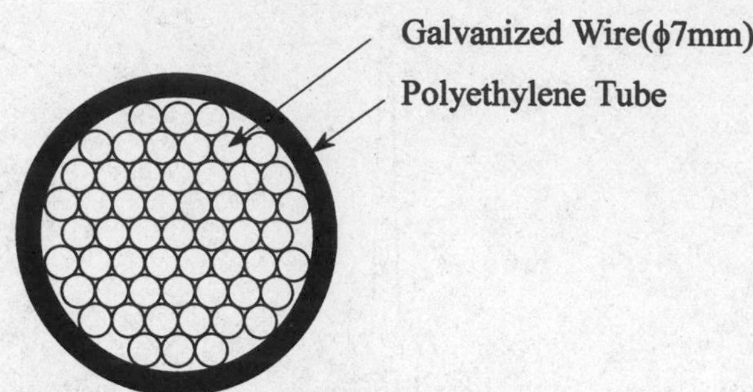

FIGURE 18.7 Parallel wire strands covered with polyethylene tubing.

rods, stranded wire ropes, parallel wire strands, and others. Stranded wire rope is most often used in modern suspension bridges. In the Akashi Kaikyo Bridge and the Kurushima Kaikyo Bridge, parallel wire strands covered with polyethylene tubing were used (Figure 18.7).

18.2.5 Suspended Structures

Stiffening girders may be I-girders, trusses, and box girders (Figure 18.8). In some short-span suspension bridges, the girders do not have enough stiffness themselves and are usually stiffened by storm ropes. In long-span suspension bridges, trusses or box girders are typically adopted. I-girders become disadvantageous due to aerodynamic stability. There are both advantages and disadvantages to trusses and box girders, involving trade-offs in aerodynamic stability, ease of construction, maintenance, and so on (details are in Section 18.3.8).

18.2.6 Anchorages

In general, anchorage structure includes the foundation, anchor block, bent block, cable anchor frames, and protective housing. Anchorages are classified into gravity or tunnel anchorage system as shown in Figure 18.9. Gravity anchorage relies on the mass of the anchorage itself to resist the tension of the main cables. This type is commonplace in many suspension bridges. Tunnel anchorage takes the tension of the main cables directly into the ground. Adequate geotechnical conditions are required.

18.3 Design

18.3.1 General

Naveir [2] was the first to consider a calculation theory of an unstiffened suspension bridge in 1823. Highly rigid girders were adopted for the suspended structure in the latter half of the 19th century because the unstiffened girders which had been used previously bent and shook under not much load. As a result, Rankine in 1858 [3] attempted to analyze suspension bridges with a highly rigid truss, followed by Melan, who helped complete the elastic theory, in which the stiffening truss was regarded as an elastic body. Ritter in 1877 [4], Lévy in 1886 [5], and Melan in 1888 [6] presented the deflection theory as an improved alternative to the elastic theory. Moisseiff realized that the actual behavior of a suspension bridge could not be explained by the elasticity theory in studies of the Brooklyn Bridge in 1901, and confirmed that the deflection theory was able to evaluate the deflection of that bridge more accurately. Moisseiff designed the Manhattan Bridge using the deflection theory in 1909. This theory became a useful design technique with which other long-span suspension bridges were successfully built [7]. Moreover, together with increasing the span length of the suspension bridge, horizontal loads such as wind load and vertical loads came to govern the design of the stiffening

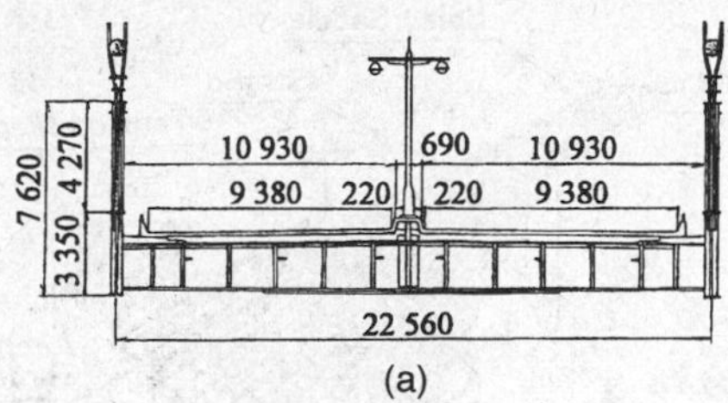

I-girder
(Bronx-Whitestone Bridge)

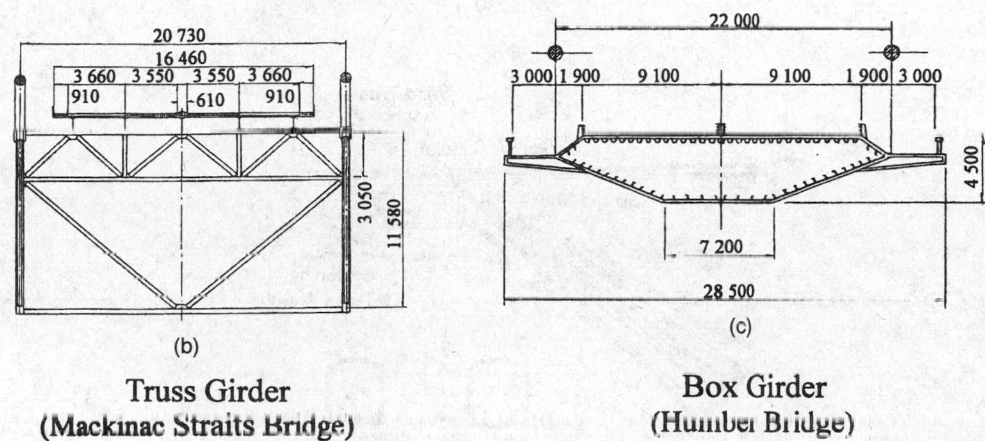

FIGURE 18.8 Types of stiffening girders.

girder. Moisseiff was among the first to establish the out-of-plane analysis method for suspension bridges [8].

Currently, thanks to rapid computer developments and the accumulation of matrix analysis studies on nonlinear problems, the finite deformation theory with a discrete frame model is generally used for the analysis of suspension bridges. Brotton [9,10] was the first to analyze the suspension bridge to be a plane structure in the matrix analysis and applied his findings to the analysis at erection stage for the Severn Bridge with good results. Saafan [11] and Tezcan's [12] thesis, which applied the general matrix deformation theory to the vertical in-plane analysis of a suspension bridge was published almost at the same time in 1966. The Newton–Raphson's method or original iteration calculation method may be used in these nonlinear matrix displacement analyses for a suspension bridge.

18.3.2 Analytical Methods

Classical Theory

Elastic Theory and Deflection Theory

The elastic theory and the deflection theory are in-plane analyses for the global suspension bridge system. In the theories, the entire suspension bridge is assumed a continuous body and the hanger ropes are closely spaced. Both of these analytical methods assume:

- The cable is completely flexible.
- The stiffening girder is horizontal and straight. The geometric moment of inertia is constant.

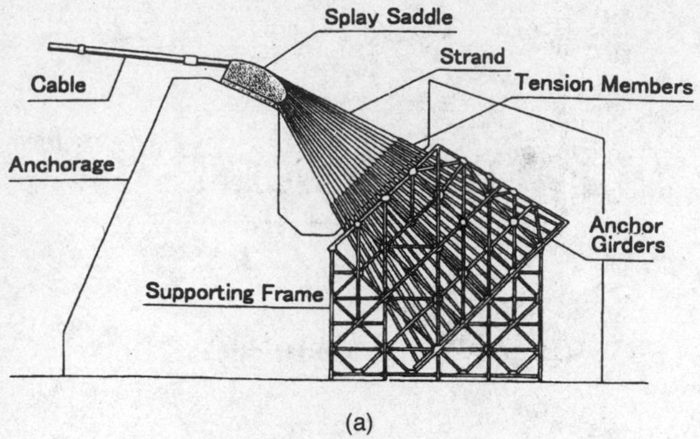

(a)

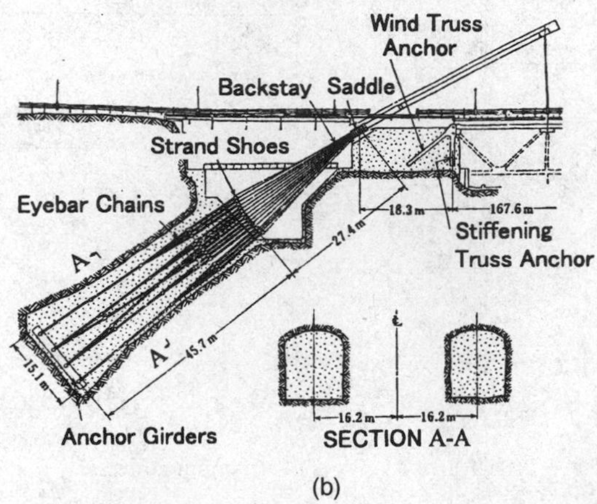

(b)

FIGURE 18.9 Types of anchorages. (a) Gravity, Akashi Kaikyo Bridge; (b) tunnel, George Washington Bridge.

- The dead load of the stiffening girder and the cables is uniform. The coordinates of the cable are parabolic.
- All dead loads are taken into the cables.

The difference between the two theories is whether cable deflection resulting from live load is considered. Figure 18.10 shows forces and deflections due to load in a suspension bridge. The bending moment, $M(x)$, of the stiffening girder after loading the live load is shown as follows:

Elastic Theory:

$$M(x) = M_0(x) - H_p y(x) \tag{18.1}$$

Deflection Theory:

$$M(x) = M_0(x) - H_p y(x) - (H_w + H_p)\eta(x) \tag{18.2}$$

where

$M_0(x)$ = bending moment resulting from the live load applied to a simple beam of the same span length as the stiffening girder
y(x) = longitudinal position of the cable
η(x) = deflection of the cable and the stiffening girder due to live load
H_w, H_p = cable horizontal tension due to dead load and live load, respectively

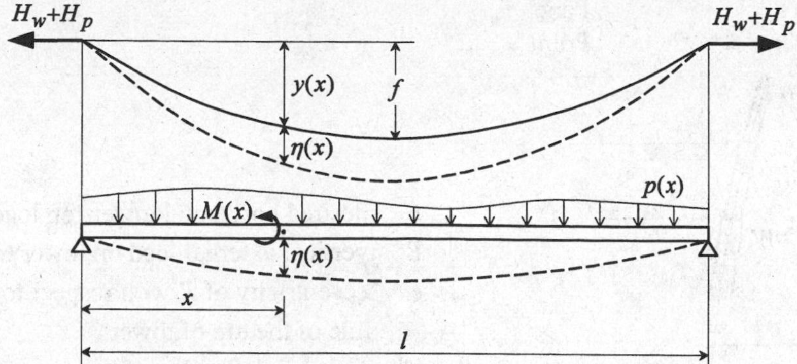

FIGURE 18.10 Deformations and forces of a suspension bridge.

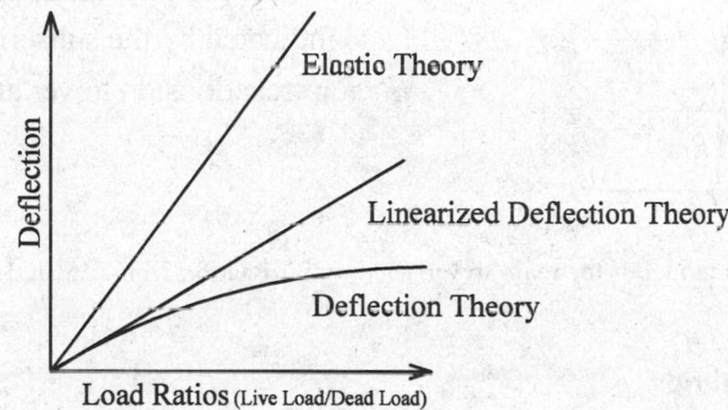

FIGURE 18.11 Deflection–load ratios relations among theories. (*Source:* Bleich, F. et al., *The Mathematical Theory of Vibrations in Suspension Bridges*, Bureau of Public Roads, Washington, D.C., 1950.)

It is understood that the bending moment of the stiffening girder is reduced because the deflection induced due to live load is considered in the last product of Eq. (18.2). Since the deflection theory is a nonlinear analysis, the principle of superposition using influence lines cannot be applied. However, because the intensity of live loads is smaller than that of dead loads for long-span suspension bridges, sufficient accuracy can be obtained even if it is assumed that $H_w + H_p$ is constant under the condition of $H_w \gg H_p$. On that condition, because the analysis becomes linear, the influence line can be used. Figure 18.11 shows the deflection–load ratio relations among the elastic, deflection, and linearized deflection theories [13]. When the ratio of live load to dead load is small, linearized theory is especially effective for analysis. In the deflection theory, the bending rigidity of towers can be neglected because it has no significance for behavior of the entire bridge.

Out-of-Plane Analysis Due to Horizontal Loads

Lateral force caused by wind or earthquake tends to be transmitted from the stiffening girder to the main cables, because the girder has larger lateral deformation than the main cables due to difference of the horizontal loads and their stiffness. Moisseiff [8] first established the out-of-plane analysis method considering this effect.

Out-of-Plane Analysis of the Main Tower

Birdsall [14] proposed a theory on behavior of the main tower in the longitudinal direction. Birdsall's theory utilizes an equilibrium equation for the tower due to vertical and horizontal forces from the cable acting on the tower top. The tower shaft is considered a cantilevered beam with variable cross section, as shown in Figure 18.12. The horizontal load (F) is obtained on the condition that the vertical load (R), acting on the tower top, and the horizontal displacement (Δ) are calculated by using Steinman's generalized deflection theory method [15].

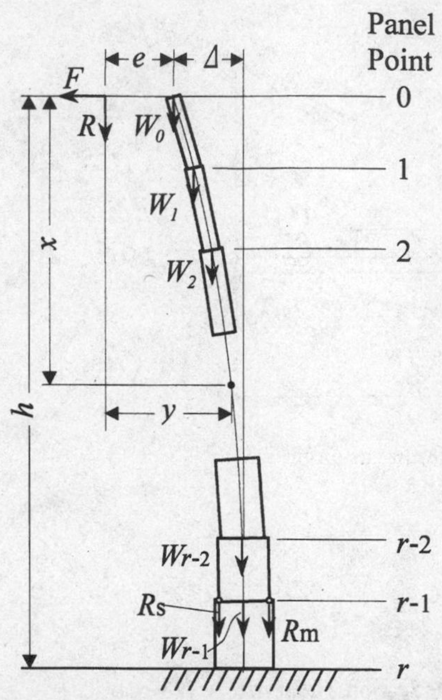

F :desired horizontal tower-top load

R :vertical external load on tower top

e :eccentricity of R with respect to the center line of the top of tower

Δ :required deflection of tower top

$W_0, W_1, \cdots W_{r-1}$:parts of tower weight assumed to be concentrated at the panel points indicated by the subscripts

R_s, R_m :reactions on tower at roadway level

FIGURE 18.12 Analytical model of the main tower. (*Source:* Birdsall, B., *Trans. ASCE*, 1942. With permission.)

Modern Design Method

Finite Deformation Method

With the development of the computer in recent years, finite displacement method on framed structures has come to be used as a more accurate analytical method. This method is used for plane analysis or space frame analysis of the entire suspension bridge structure. The frame analysis according to the finite displacement theory is performed by obtaining the relation between the force and the displacement at the ends of each element of the entire structural system. In this analytical method, the actual behavior of the bridge such as elongation of the hanger ropes, which is disregarded in the deflection theory, can be considered. The suspension bridges with inclined hanger ropes, such as the Severn Bridge, and bridges in the erection stage are also analyzed by the theory. While the relation between force and displacement at the ends of the element is nonlinear in the finite displacement theory, the linearized finite deformation theory is used in the analysis of the eccentric vertical load and the out-of-plane analysis; because the geometric nonlinearity can be considered to be relatively small in those cases.

Elastic Buckling and Vibration Analyses

Elastic buckling analysis is used to determine an effective buckling length that is needed in the design of the compression members, such as the main tower shafts. Vibration analysis is needed to determine the natural frequency and vibrational modes of the entire suspension bridge as part of the design of wind and seismic resistance. Both of these analyses are eigenvalue problems in the linearized finite deformation method for framed structures.

18.3.3 Design Criteria

Design Procedure

A general design procedure for a suspension bridge superstructure is shown in Figure 18.13. Most rational structure for a particular site is selected from the result of preliminary design over various alternatives. Then final detailed design proceeds.

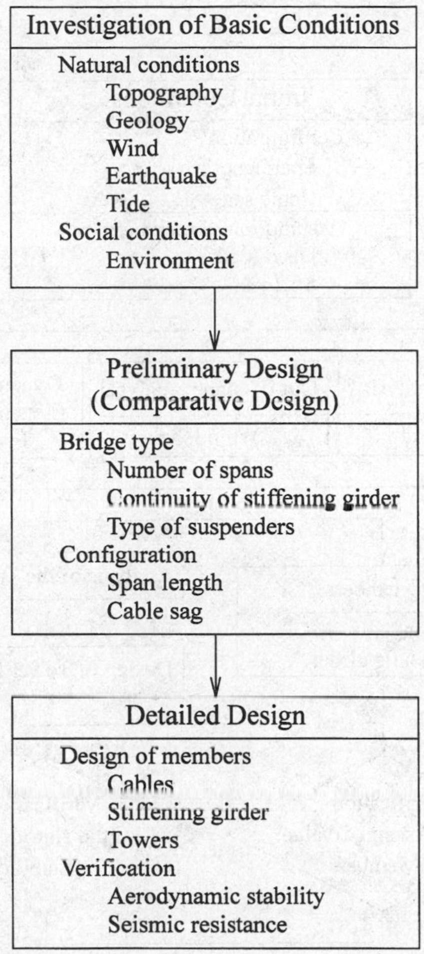

FIGURE 18.13 Design procedure for the superstructure of a suspension bridge.

Design Load

Design loads for a suspension bridge must take into consideration the natural conditions of the construction site, the importance of a bridge, its span length, and its function (vehicular or railway traffic). It is important in the design of suspension bridges to determine the dead load accurately because the dead load typically dominates the forces on the main components of the bridge. Securing structural safety against strong winds and earthquakes is also an important issue for long-span suspension bridges.

1. In the case of wind, consideration of the vibrational and aerodynamic characteristics is extremely important.
2. In the case of earthquake, assumption of earthquake magnitude and evaluation of energy content are crucial for bridges in regions prone to large-scale events.

Other design loads include effects due to errors in fabrication and erection of members, temperature change, and possible movement of the supports.

Analysis Procedure

General procedure used for the design of a modern suspension bridge is as follows (Figure 18.14):

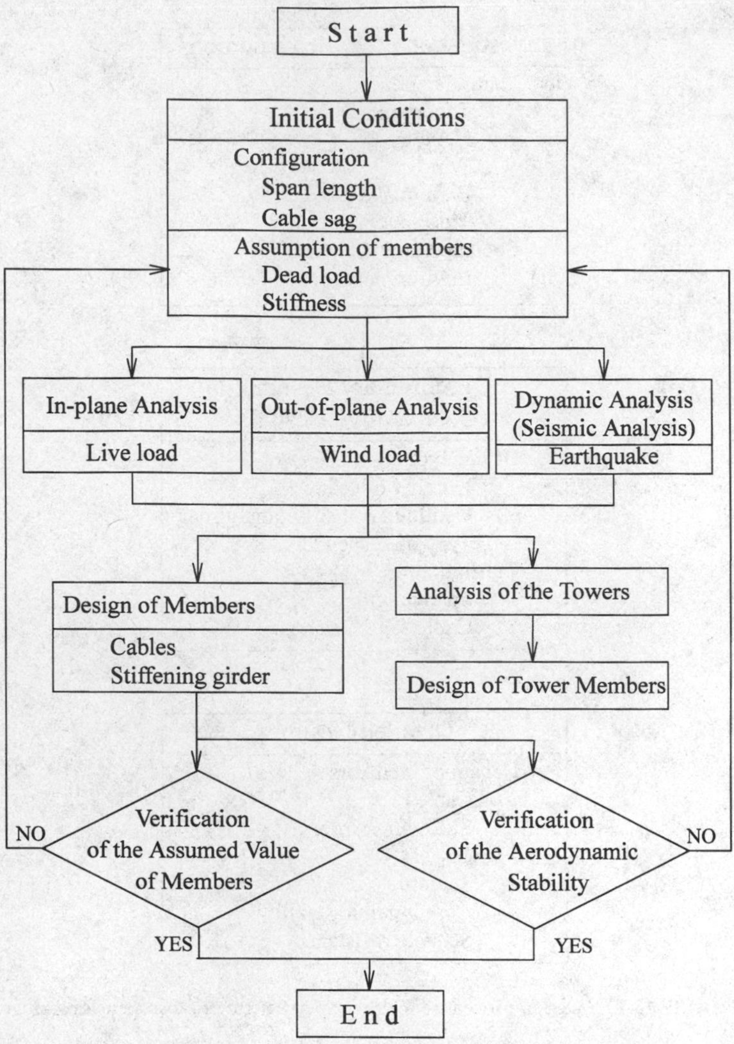

FIGURE 18.14 General procedure for designing a suspension bridge.

1. *Select Initial Configuration*: Span length and cable sag are determined, and dead load and stiffness are assumed.
2. *Analysis of the Structural Model*: In the case of in-plane analysis, the forces on and deformations of members under live load are obtained by using finite deformation theory or linear finite deformation theory with a two-dimensional model. In the case of out-of-plane analysis, wind forces on and deformations of members are calculated by using linear finite deformation theory with a three-dimensional model.
3. *Dynamic Response Analysis*: The responses of earthquakes are calculated by using response spectrum analysis or time-history analysis.
4. *Member Design*: The cables and girders are designed using forces obtained from previous analyses.
5. *Tower Analysis*: The tower is analyzed using loads and deflection, which are determined from the global structure analysis previously described.
6. *Verification of Assumed Values and Aerodynamic Stability*: The initial values assumed for dead load and stiffness are verified to be sufficiently close to those obtained from the detailed analysis. Aerodynamic stability is to be investigated through analyses and/or wind tunnel tests using dimensions obtained from the dynamic analysis.

18.3.4 Wind-Resistant Design

General

In the first half of the 19th century, suspension bridges occasionally collapsed under wind loads because girders tended to have insufficient rigidity. In the latter half of the 19th century, such collapses decreased because the importance of making girders sufficiently stiff was recognized.

In the beginning of the 20th century, stiffening girders with less rigidity reappeared as the deflection theory was applied to long-span suspension bridges. The Tacoma Narrows Bridge collapsed 4 months after its completion in 1940 under a wind velocity of only 19 m/s. The deck of the bridge was stiffened with I-girders formed from built-up plates. The I-girders had low rigidity and aerodynamic stability was very inferior as shown in recent wind-resistant design. After this accident, wind tunnel tests for stiffening girders became routine in the investigation of aerodynamic stability. Truss-type stiffening girders, which give sufficient rigidity and combined partially with open deck grating, have dominated the design of modern suspension bridges in the United States.

A new type of stiffening girder, however, a streamlined box girder with sufficient aerodynamic stability was adopted for the Severn Bridge in the United Kingdom in 1966 [16,17]. In the 1980s, it was confirmed that a box girder, with big fairings (stabilizers) on each side and longitudinal openings on upper and lower decks, had excellent aerodynamic stability. This concept was adopted for the Tsing Ma Bridge, completed in 1997 [18]. The Akashi Kaikyo Bridge has a vertical stabilizer in the center span located along the centerline of the truss-type stiffening girder just below the deck to improve aerodynamic stability [19].

In the 1990s, in Italy, a new girder type has been proposed for the Messina Straits Bridge, which would have a center span of 3300 m [20]. The 60-m-wide girder would be made up of three oval box girders which support the highway and railway traffic. Aerodynamic dampers combined with wind screens would also be installed at both edges of the girder. Stiffening girders in recent suspension bridges are shown in Figure 18.15.

Design Standard

Figure 18.16 shows the wind-resistant design procedure specified in the Honshu–Shikoku Bridge Standard [21]. In the design procedure, wind tunnel testing is required for two purposes: one is to verify the airflow drag, lift, and moment coefficients which strongly influences the static design; and the other is to verify that harmful vibrations would not occur.

Analysis

Gust response analysis is an analytical method to ascertain the forced vibration of the structure by wind gusts. The results are used to calculate structural deformations and stress in addition to those caused by mean wind. Divergence, one type of static instability, is analyzed by using finite displacement analysis to examine the relationship between wind force and deformation. Flutter is the most critical phenomenon in considering the dynamic stability of suspension bridges, because of the possibility of collapse. Flutter analysis usually involves solving the motion equation of the bridge as a complex eigenvalue problem where unsteady aerodynamic forces from wind tunnel tests are applied.

Wind Tunnel Testing

In general, the following wind tunnel tests are conducted to investigate the aerodynamic stability of the stiffening girder.

1. Two-Dimensional Test of Rigid Model with Spring Support: The aerodynamic characteristics of a specific mode can be studied. The scale of the model is generally higher than $^1/_{100}$.
2. Three-Dimensional Global Model Test: Test used to examine the coupling effects of different modes.

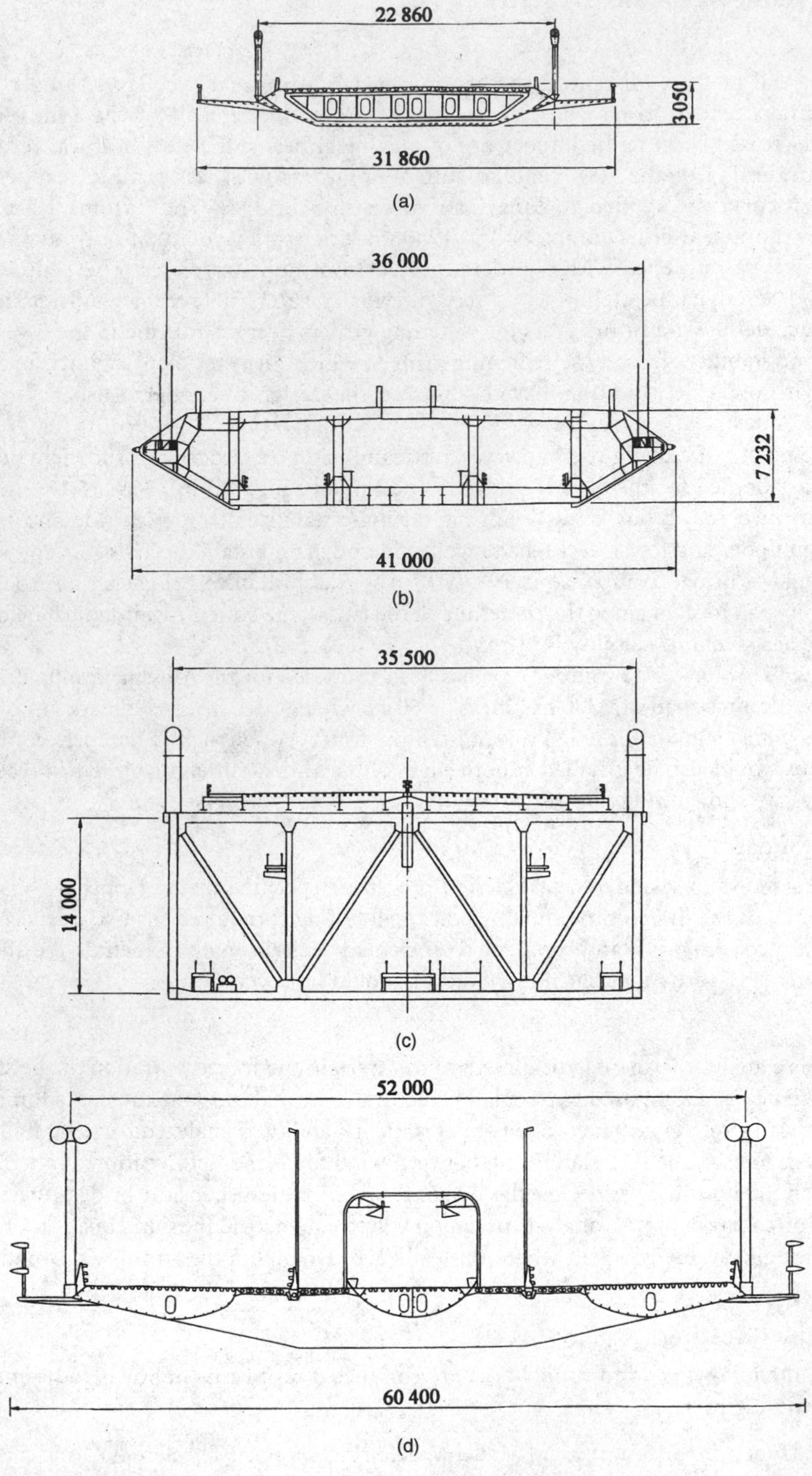

FIGURE 18.15 Cross sections through stiffening girders. (a) Severn Bridge, (b) Tsing Ma Bridge; (c) Akashi Kaikyo Bridge, (d) Messina Straits Bridge.

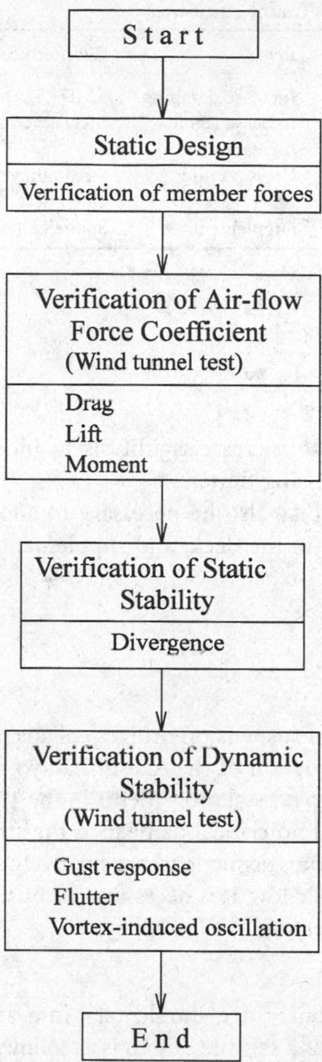

FIGURE 18.16 Procedure for wind–resistant design. (*Source*: Honshu–Shikoku Bridge Authority, Wind–Resistant Design Standard for the Akashi Kaikyo Bridge, HSBA, Japan, 1990. With permission.)

For the Akashi Kaikyo Bridge, a global 1/100 model about 40 m in total length, was tested in a boundary layer wind tunnel laboratory. Together with the verification of the aerodynamic stability of the Akashi Kaikyo Bridge, new findings in flutter analysis and gust response analysis were established from the test results.

Countermeasures against Vibration

Countermeasures against vibration due to wind are classified as shown in Table 18.4.

1. *Increase Structural Damping*: Damping, a countermeasure based on structural mechanics, is effective in decreasing the amplitude of vortex-induced oscillations which are often observed during the construction of the main towers and so on. Tuned mass dampers (TMD) and tuned liquid dampers (TLD) have also been used to counter this phenomenon in recent years. Active mass dampers (AMD), which can suppress vibration amplitudes over a wider frequency band, have also been introduced.

TABLE 18. 4 Vibration Countermeasures

Category	Item	Countermeasures
Structural mechanics	Increase damping	TMD,[a] TLD,[b] AMD[c]
	Increase rigidity	Increase cross-sectional area of girder
	Increase mass	
Aerodynamic mechanics	Cross section	Streamlined box girder
		Open deck
	Supplements	Spoiler, Flap

[a] Tuned mass damper.
[b] Tuned liquid damper.
[c] Active mass damper.

2. *Increase Rigidity*: One way to increase rigidity is to increase the girder height. This is an effective measure for suppressing flutter.
3. *Aerodynamic Mechanics*: It may also be necessary to adopt aerodynamic countermeasures, such as providing openings in the deck, and supplements for stabilization in the stiffening girder.

18.3.5 Seismic Design

General

In recent years, there are no cases of suspension bridges collapsing or even being seriously damaged due to earthquakes. During construction of the Akashi Kaikyo Bridge, the relative location of four foundations changed slightly due to crustal movements in the 1995 Hyogo-ken Nanbu Earthquake. Fortunately, the earthquake caused no critical damage to the structures. Although the shear forces in the superstructure generated by a seismic load are relatively small due to the natural frequency of the superstructure being generally low, it is necessary to consider possible large displacements of the girders and great forces transferring to the supports.

Design Method

The superstructure of a suspension bridge should take into account long-period motion in the seismic design. A typical example of a seismic design is as follows. The superstructure of the Akashi Kaikyo Bridge was designed with consideration given to large ground motions including the long-period contribution. The acceleration response spectrum from the design standard is shown in Figure 18.17 [22]. Time-history analysis was conducted on a three-dimensional global bridge model including substructures and ground springs.

18.3.6 Main Towers

General

Flexible-type towers have predominated among main towers in recent long-span suspension bridges. This type of tower maintains structural equilibrium while accommodating displacement and the downward force from the main cable. Both steel and concrete are feasible material. Major bridges like the Golden Gate Bridge and the Verrazano Narrows Bridge in the United States as well as the Akashi Kaikyo Bridge in Japan consist of steel towers. Examples of concrete towers include the Humber and Great Belt East Bridges in Europe and the Tsing Ma Bridge in China. Because boundary conditions and loading of main towers are straightforward in suspension bridge systems, the main tower can be analyzed as an independent structural system.

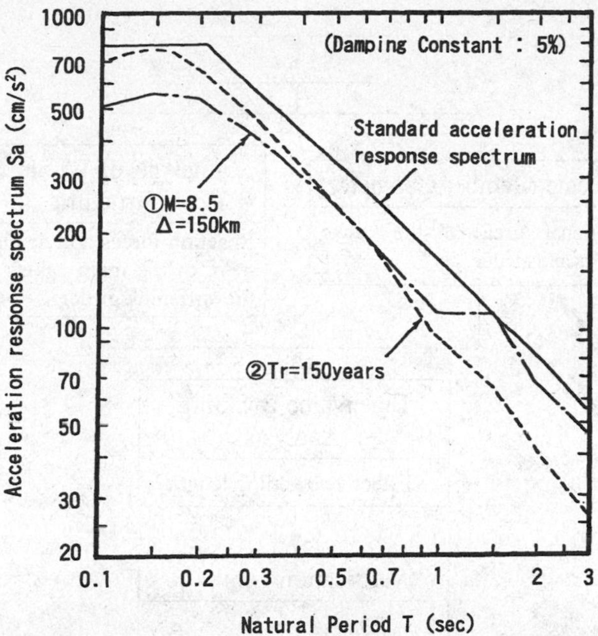

FIGURE 18.17 Design acceleration response spectrum. (*Source:* Honshu–Shikoku Bridge Authority, Seismic Design Standard for the Akashi Kaikyo Bridge, Japan, 1988. With permission.)

Design Method

The design method for steel towers follows. The basic concepts for design of concrete towers are similar. For the transverse direction, main towers are analyzed using small deformation theory. This is permissible because the effect of cable restraint is negligible and the flexural rigidity of the tower is high. For the longitudinal direction, Birdsall's analysis method, discussed in Section 18.3.2, is generally used. However, more rigorous methods, such as finite displacement analysis with a three-dimensional model which allows analysis of both the transverse and longitudinal directions, can be used, as was done in the Akashi Kaikyo Bridge. An example of the design procedure for main towers is shown in Figure 18.18 [23].

Tower Structure

The tower shaft cross section may be T-shaped, rectangular, or cross-shaped, as shown in Figure 18.19. Although the multicell made up of small box sections has been used for some time, cells and single cells have become noticeable in more recent suspension bridges.

The details of the tower base that transmits the axial force, lateral force, and bending moment into the foundation, are either of grillage (bearing transmission type) or embedded types (shearing transmission type), as shown in Figure 18.20. Field connections for the tower shaft are typically bolted joints. Large compressive forces from the cable act along the tower shafts. Tight contact between two metal surfaces acts together with bolted joint to transmit the compressive force across joints with the bearing stresses spread through the walls and the longitudinal stiffeners inside the tower shaft. This method can secure very high accuracy of tower configuration. Another type of connection detail for steel towers using tension bolts was used in the Forth Road Bridge, the Severn Bridge, the Bosporus Bridge, and the first Kurushima Kaikyo Bridge (Figure 18.21).

18.3.7 Cables

General

Parallel wire cable has been used exclusively as the main cable in long-span suspension bridges. Parallel wire has the advantage of high strength and high modulus of elasticity compared with

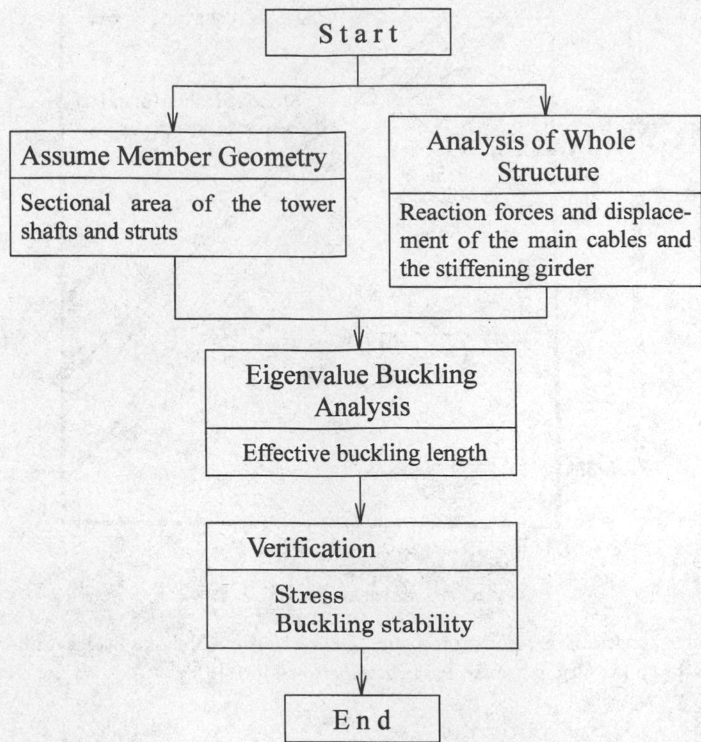

FIGURE 18.18 Design procedure for the main towers. (*Source:* Honshu–Shikoku Bridge Authority, Design Standard of the Main Tower for a Suspension Bridge, HSBA, Japan, 1984. With permission.)

stranded wire rope. The design of the parallel wire cable is discussed next, along with structures supplemental to the main cable.

Design Procedure

Alignment of the main cable must be decided first (Figure 18.22). The sag–span ratios should be determined in order to minimize the construction costs of the bridge. In general, this sag–span ratio is around 1:10. However, the vibration characteristics of the entire suspension bridge change occasionally with changes in the sag–span ratios, so the influence on the aerodynamic stability of the bridge should be also considered. After structural analyses are executed according to the design process shown in Figure 18.14, the sectional area of the main cable is determined based on the maximum cable tension, which usually occurs at the side span face of the tower top.

Design of Cable Section

The tensile strength of cable wire has been about 1570 N/mm^2 (160 kgf/mm^2) in recent years. For a safety factor, 2.5 was used for the Verrazano Narrows Bridge and 2.2 for the Humber Bridge, respectively. In the design of the Akashi Kaikyo Bridge, a safety factor of 2.2 was used using the allowable stress method considering the predominant stress of the dead load. The main cables used a newly developed high-strength steel wire whose tensile strength is 1770 N/mm^2 (180 kgf/mm^2) and the allowable stress was 804 N/mm^2 (82 kgf/mm^2) which led to this discussion. Increase in the strength of cable wire over the years is shown in Figure 18.23. In the design of the Great Belt East Bridge which was done using limit state design methods, a safety factor of 2.0 was applied for the critical limit state [24]. Cable statistics of major suspension bridges are shown in Table 18.5.

Supplemental Components

Figure 18.24 shows the supplemental components of the main cable.

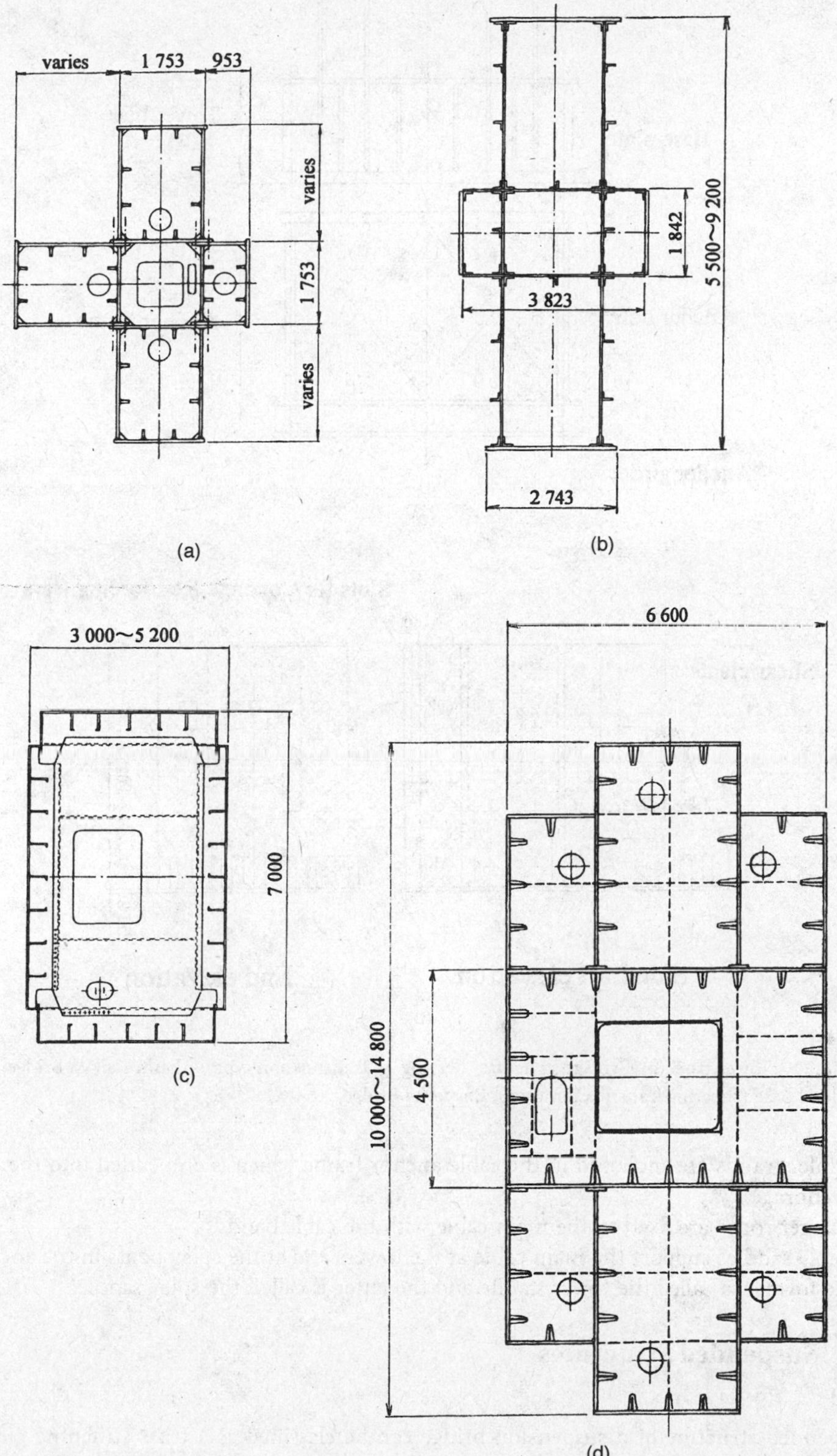

FIGURE 18.19 Tower shaft section. (a) New Port Bridge, (b) 25de Abril Bridge, (c) Bosporus Bridge, (d) Akashi Kaikyo Bridge.

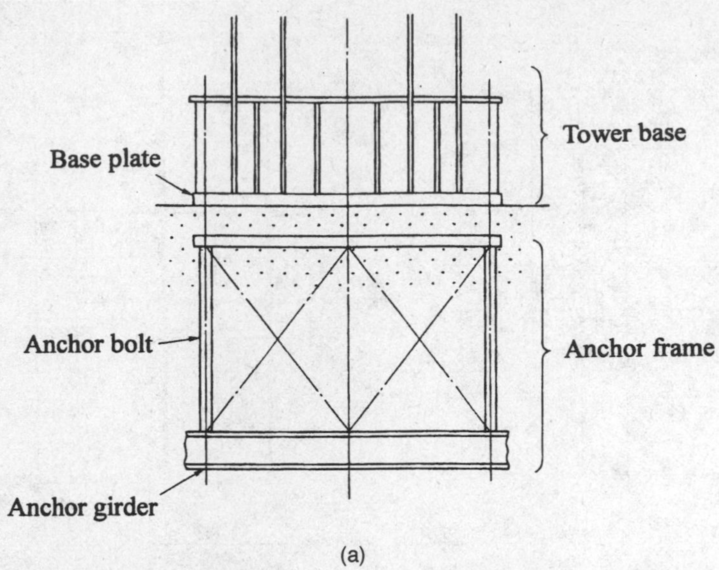

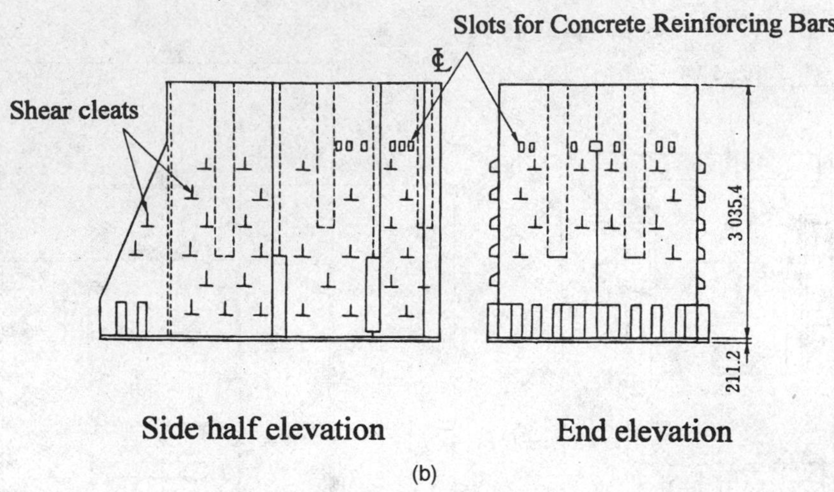

FIGURE 18.20 Tower base. (a) Grillage structure (bearing — transmission type), Akashi Kaikyo Bridge; (b) embedded base (shearing transmission type), Bosporus Bridge.

1. Cable strands are anchored in the cable anchor frame which is embedded into the concrete anchorage.
2. Hanger ropes are fixed to the main cable with the cable bands.
3. Cable saddles support the main cable at the towers and at the splay bents in the anchorages; the former is called the tower saddle and the latter is called the splay saddle.

18.3.8 Suspended Structures

General

The suspended structure of a suspension bridge can be classified as a truss stiffening girder or a box stiffening girder, as described in Section 18.3.4. Basic considerations in selecting girder types are shown in Table 18.6. The length of the bridge and the surrounding natural conditions are also factors.

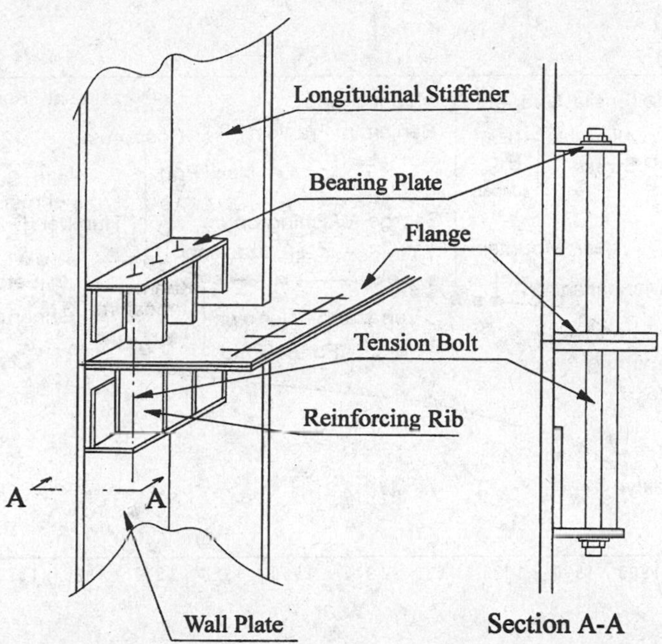

FIGURE 18.21 Connection using tension bolts. (First Kurushima Kaikyo Bridge, Bosporus Bridge.)

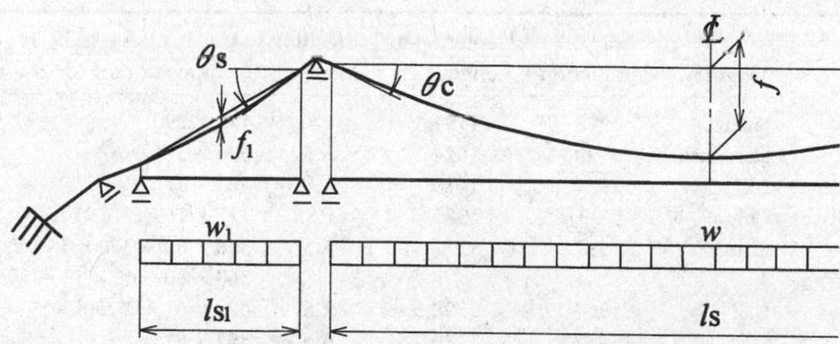

FIGURE 18.22 Configuration of suspension bridge.

Design of the Stiffening Girder

Basic Dimensions

The width of the stiffening girder is determined to accommodate carriageway width and shoulders. The depth of the stiffening girder, which affects its flexural and torsional rigidity, is decided so as to ensure aerodynamic stability. After examining alternative stiffening girder configurations, wind tunnel tests are conducted to verify the aerodynamic stability of the girders.

In judging the aerodynamic stability, in particular the flutter, of the bridge design, a bending–torsional frequency ratio of 2.0 or more is recommended. However, it is not always necessary to satisfy this condition if the aerodynamic characteristics of the stiffening girder are satisfactory.

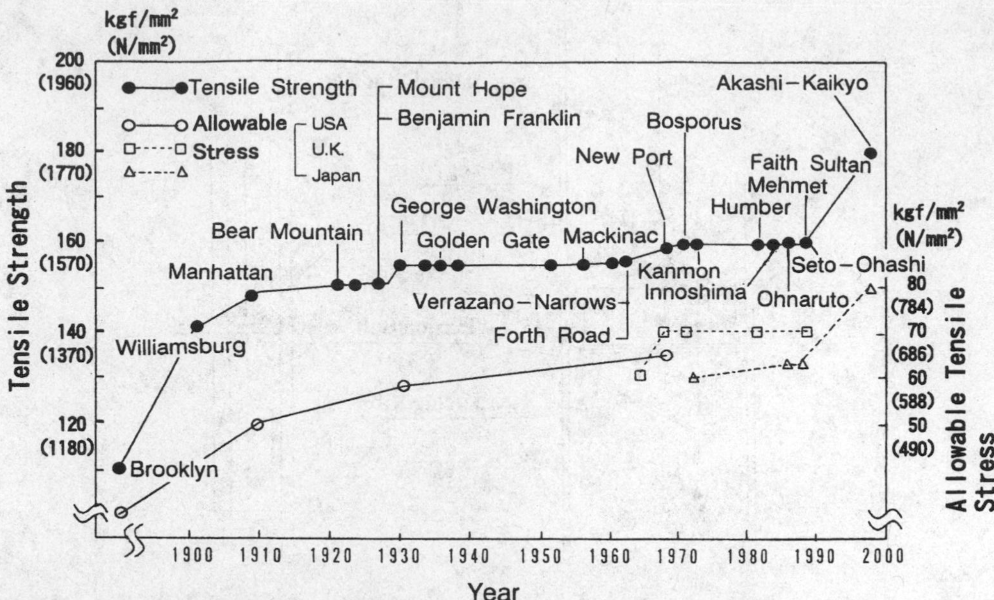

FIGURE 18.23 Increase in strength of cable wire. (*Source:* Honshu–Shikoku Bridge Authority, Akashi Kaikyo Bridge — Engineering Note, Japan, 1992. With permission.)

TABLE 18.5 Main Cable of Long-Span Suspension Bridges

No.	Bridge	Country	Year of Completion	Center Span Length (m)	Erection Method[c]	Composition of Main Cable[d]
1	Akashi Kaikyo	Japan	1998	1991	P.S.	127 × 290
2	Great Belt East	Denmark	1998	1624	A.S.	504 × 37
3	Humber	U.K.	1981	1410	A.S.	404 × 37
4	Jing Yin Yangtze River	China[a]	(1999)[b]	1385	P.S.	127 × 169(c/s), 177(s/s)
5	Tsing Ma	China[a]	1997	1377	A.S.	368 × 80 + 360 × 11 (c/s, Tsing Yi s/s) 368 × 80+360 × 11 + 304 × 6(Ma Wan s/s)
6	Verrazano Narrows	U.S.	1964	1298.5	A.S.	428 × 61 × 2 cables
7	Golden Gate	U.S.	1937	1280.2	A.S.	452 × 61
8	Höga Kusten	Sweden	1997	1210	A.S.	304 × 37(c/s) 304 × 37 + 120 × 4(s/s)
9	Mackinac Straits	U.S.	1957	1158.2	A.S.	340 × 37
10	Minami Bisan-Seto	Japan	1988	1100	P.S.	127 × 271
11	Fatih Sultan Mehmet	Turkey	1988	1090	A.S.	504 × 32(c/s), 36(s/s)
12	Bosphorus	Turkey	1973	1074	A.S.	550 × 19
13	George Washington	U.S.	1931	1066.8	A.S.	434 × 61 × 2 cables
14	3rd Kurushima Kaikyo	Japan	1999[b]	1030	P.S.	127 × 102
15	2nd Kurushima Kaikyo	Japan	1999[b]	1020	P.S.	127 × 102
16	25 de Abril	Portugal	1966	1012.9	A.S.	304 × 37
17	Forth Road	UK	1964	1005.8	A.S.	(304~328) × 37
18	Kita Bisan-Seto	Japan	1988	990	P.S.	127 × 234
19	Severn	UK	1966	987.6	A.S.	438 × 19
20	Shimotsui-Seto	Japan	1988	940	A.S.	552 × 44

[a] The People's Republic of China.
[b] Under construction
[c] P.S.: prefabricated parallel wire strand method A.S.: aerial spinning erection method.
[d] Wire/strand × strand/cable.

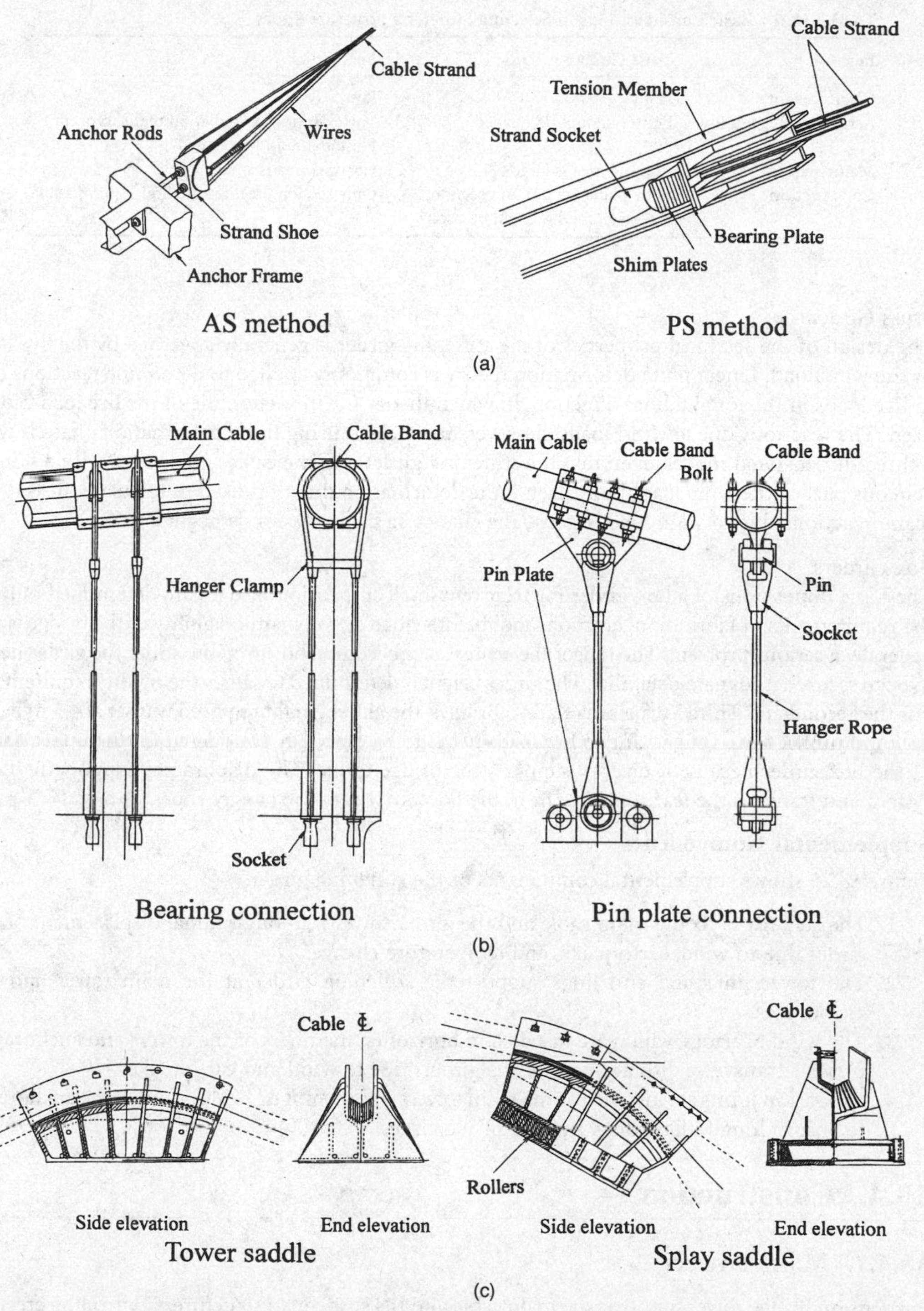

FIGURE 18.24 Supplemental components of the main cable. (a) Strand anchorage of the anchor frame. (*Source:* Japan Society of Civil Engineers, *Suspension Bridge*, Japan, 1996. With permission.) (b) Hanger ropes. (*Source:* Japan Society of Civil Engineers, *Suspension Bridge*, Japan, 1996. With permission.) (c) Cable Saddles. (*Source:* Honshu–Shikoku Bridge Authority, *Design of a Suspension Bridge*, Japan, 1990. With permission.)

TABLE 18.6 Basic Considerations in Selecting Stiffening Structure Types

Item	Truss Girder	Box Girder
Girder height	High	Low
Aerodynamic stability	Flutter should be verified	Vortex-induced oscillation tends to occur Flutter should be verified
Maintenance	Coating area is large	Coating area is small
Construction	Both plane section and section erection methods can be used	Only section erection method is permissible

Truss Girders

The design of the sectional properties of the stiffening girder is generally governed by the live load or the wind load. Linear finite deformation theory is commonly applied to determine reactions due to live loads in the longitudinal direction, in which theory the influence line of the live load can be used. The reactions due to wind loads, however, are decided using finite deformation analysis with a three-dimensional model given that the stiffening girder and the cables are loaded with a homogeneous part of the wind load. Linearized finite deformation theory is used to calculate the out-of-plane reactions due to wind load because the change in cable tension is negligible.

Box Girders

The basic dimensions of a box girder for relatively small suspension bridges are determined only by the requirements of fabrication, erection, and maintenance. Aerodynamic stability of the bridge is not generally a serious problem. The longer the center span becomes, however, the stiffer the girder needs to be to secure aerodynamic stability. The girder height is determined to satisfy the rigidity requirement. For the Second and Third Kurushima Kaikyo Bridges, the girder height required was set at 4.3 m based on wind tunnel tests. Fatigue due to live loads needs to be especially considered for the upper flange of the box girder because it directly supports the bridge traffic. The diaphragms support the floor system and transmit the reaction force from the floor system to the hanger ropes.

Supplemental Components

Figure 18.25 shows supplemental components of the stiffening girder.

1. The stay ropes fix the main cable and the girder to restrict longitudinal displacement of the girder due to wind, earthquake, and temperature changes.
2. The tower links and end links support the stiffening girder at the main tower and the anchorages.
3. The wind bearings, which are installed in horizontal members of the towers and anchorages, prevent transverse displacement of the girders due to wind and earthquakes.
4. Expansion joints are installed at the main towers of two-hinged bridges and at the anchorages to absorb longitudinal displacement of the girder.

18.4 Construction

18.4.1 Main Towers

Suspension bridge tower supports the main cable and the suspended structure. Controlling erection accuracy to ensure that the tower shafts are perpendicular is particularly important. During construction, because the tower is cantilevered and thus easily vibrates due to wind, countermeasures for vibration are necessary. Recent examples taken from constructing steel towers of the Akashi Kaikyo Bridge and concrete towers of the Tsing Ma Bridge are described below.

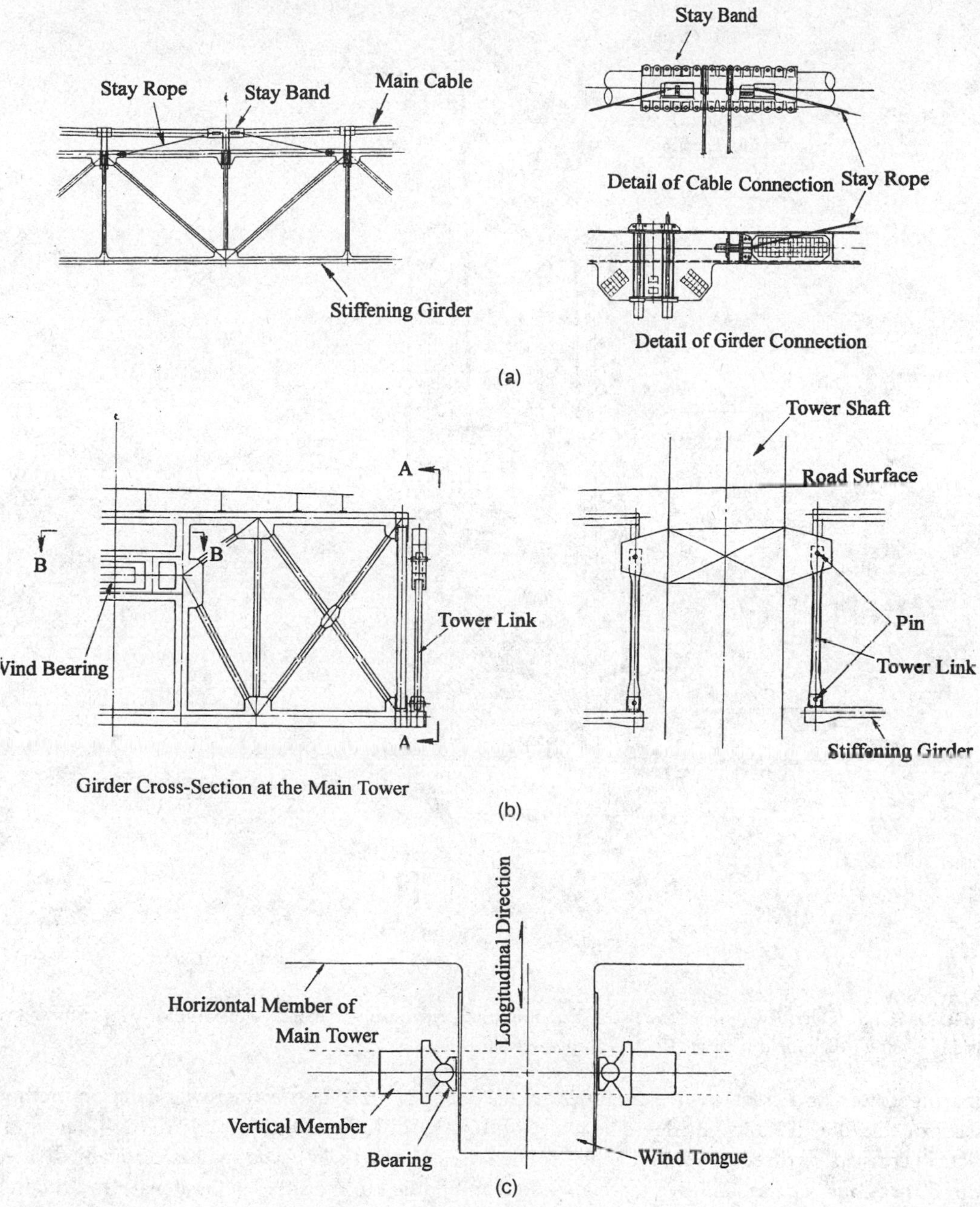

FIGURE 18.25 Supplemental components of the stiffening girder. (a) Center stay; (b) tower link (Section A-A); (c) wind bearing (Section B-B). (*Source:* Honshu–Shikoku Bridge Authority, Design of a Suspension Bridge, Japan, 1990. With permission.)

Steel Towers

Steel towers are typically either composed of cells or have box sections with rib stiffening plates. The first was used in the Forth Road Bridge, the 25 de Abril Bridge, the Kanmon Bridge, and most of the Honshu–Shikoku Bridges. The latter was applied in the Severn Bridge, the Bosporus Bridge, the Fatih Sultan Mehmet Bridge, and the Kurushima Kaikyo Bridges. For the erection of steel towers,

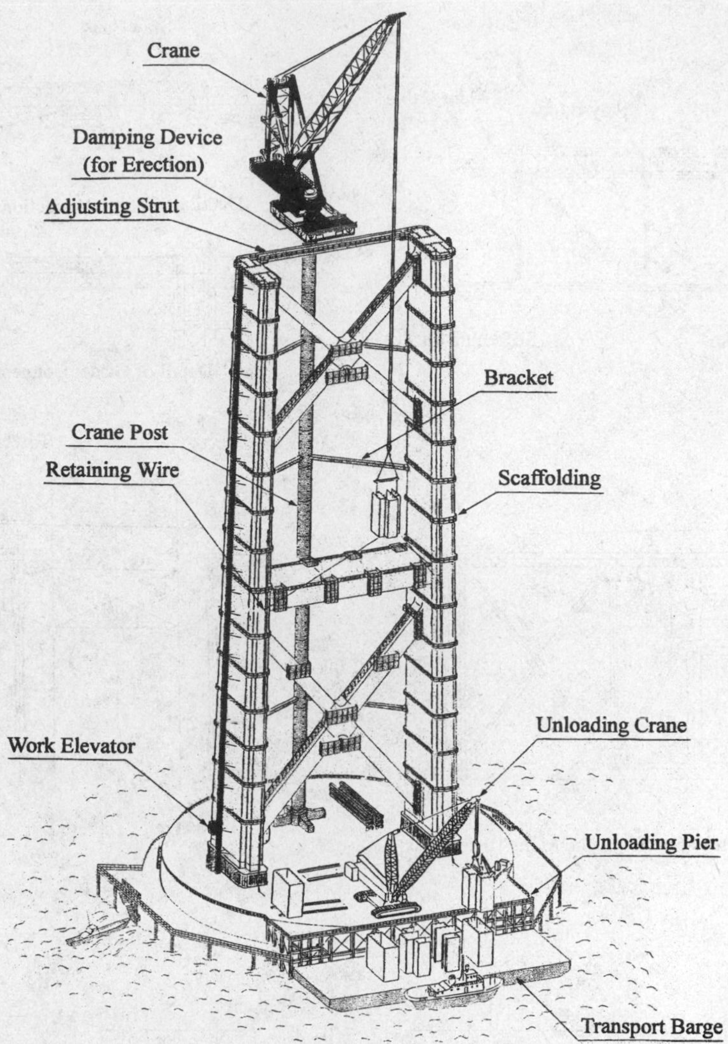

FIGURE 18.26 Overview of main tower construction. (*Source:* Honshu–Shikoku Bridge Authority, Akashi Kaikyo Bridge — Engineering Note, Japan, 1992. With permission.)

floating, tower, and creeper traveler cranes are used. Figure 18.26 shows the tower erection method used for the Akashi Kaikyo Bridge. The tower of the Akashi Kaikyo Bridge is 297 m high. The cross section consists of three cells with clipped corners (see Figure 18.19). The shaft is vertically divided into 30 sections. The sections were prefabricated and barged to the site. The base plate and the first section was erected using a floating crane. The remainder was erected using a tower crane supported on the tower pier. To control harmful wind-induced oscillations, TMD and AMD were installed in the tower shafts and the crane.

Concrete Towers

The tower of the Tsing Ma Bridge is 206 m high, 6.0 m in width transversely, and tapered from 18.0 m at the bottom to 9.0 m at the top longitudinally. The tower shafts are hollow. Each main tower was slip-formed in a continuous around-the-clock operation, using two tower cranes and concrete buckets (Figure 18.27).

FIGURE 18.27 Tower erection for the Tsing Ma Bridge. (Courtesy of Mitsui Engineering & Shipbuilding Co., Ltd.)

18.4.2 Cables

Aerial Spinning Method

The aerial spinning method (AS method) of parallel wire cables was invented by John A. Roebling and used for the first time in the Niagara Falls Bridge which was completed in 1855 with a center span of 246 m (Figure 18.28). He established this technology in the Brooklyn Bridge where steel wire was first used. Most suspension bridges built in the United States since Roebling's development of the AS method have used parallel wire cables. In contrast, in Europe, the stranded rope cable was used until the Forth Road Bridge was built in 1966.

In the conventional AS method, individual wires were spanned in free-hang condition, and the sag of each wire had to be individually adjusted to ensure all were of equal length. In this so-called sag-control method, the quality of the cables and the erection duration are apt to be affected by site working conditions, including wind conditions and the available cable-spinning equipment. It also requires a lot of workers to adjust the sag of the wires.

A new method, called the tension-control method, was developed in Japan (Figure 18.29). The idea is to keep the tension in the wire constant during cable spinning to obtain uniform wire lengths. This method was used on the Hirado, Shimotsui–Seto, Second Bosporus, and Great Belt East Bridges (Figure 18.30). It does require adjustment of the individual strands even in this method.

Prefabricated Parallel Wire Strand Method

Around 1965, a method of prefabricating parallel wire cables was developed to cut the on-site work intensity required for the cable spinning in the AS method. The prefabricated parallel wire strand method (PS method) was first used in the New Port Bridge. That was the first step toward further progress achieved in Japan in enlarging strand sections, developing high-tensile wire, and lengthening the strand.

18.4.3 Suspended Structures

There are various methods of erecting suspended structures. Typically, they have evolved out of the structural type and local natural and social conditions.

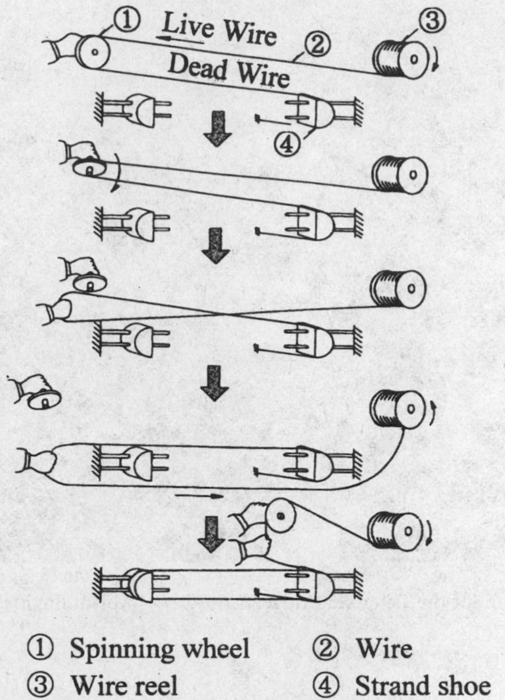

FIGURE 18.28 Operating principle of aerial spinning. (*Source:* Honshu–Shikoku Bridge Authority, Technology of Seto–Ohashi Bridge, Japan, 1989. With permission.)

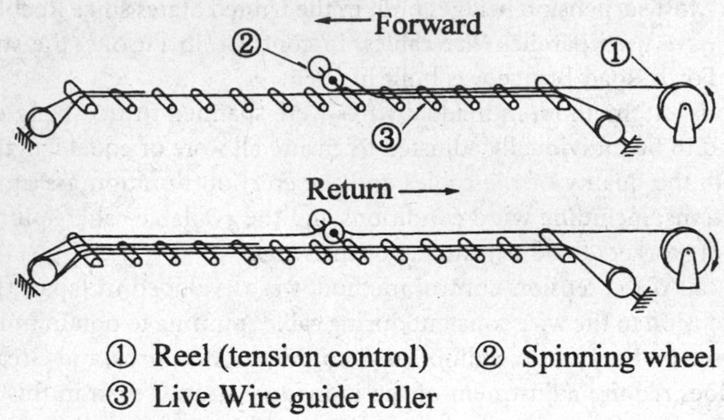

FIGURE 18.29 Operating principle of tension control method.(*Source:* Honshu–Shikoku Bridge Authority, Technology of Set–Ohashi Bridge, Japan, 1989. With permission.)

Girder Block Connection Methods

The connections between stiffening girder section may be classified as one of two methods.

All Hinge Method

In this method the joints are loosely connected until all girder sections are in place in general. This method enables simple and easy analysis of the behavior of the girders during construction. Any temporary reinforcement of members is usually unnecessary. However, it is difficult to obtain enough aerodynamic stability unless structures to resist wind force are given to the joints which were used in the Kurushima Kaikyo Bridges, for example.

FIGURE 18.30 Aerial spinning for the Shimotsui–Seto bridge. (Courtesy of Honshu–Shikoku Bridge Authority.)

Rigid Connection Method

In this method full-splice joints are immediately completed as each girder block is erected into place. This keeps the stiffening girder smooth and rigid, providing good aerodynamic stability and high construction accuracy. However, temporary reinforcement of the girders and hanger ropes to resist transient excessive stresses or controlled operation to avoid overstress are sometimes required.

Girder Erection Methods

Stiffening girders are typically put in place using either the girder-section method or cantilevering from the towers or the anchorages.

Girder-Section Method

The state of the art for the girder-section method with hinged connections is shown in Figure 18.31. At the Kurushima Kaikyo Bridges construction sites, the fast and complex tidal current of up to 5 m/s made it difficult for the deck barges and tugboats to maintain their desired position for a long time. As a result, a self-controlled barge, able to maintain its position using computer monitoring, and a quick joint system, which can shorten the actual erection period, were developed and fully utilized.

Cantilevering Method

A recent example of the cantilevering method of girders on the Akashi Kaikyo Bridge is shown in Figure 18.32. Preassembled panels of the stiffening girder truss were erected by extending the stiffening girders as a cantilever from the towers and anchorages. This avoided disrupting marine traffic, which would have been required for the girder-section method.

18.5 Field Measurement and Coatings

18.5.1 Loading Test

The purpose of loading tests is chiefly to confirm the safety of a bridge for both static and dynamic behavior. Static loading tests were performed on the Wakato, the Kanmon, and the President Mobutu

FIGURE 18.31 Block erection method on the Kurushima Kaikyo Bridge. (Courtesy of Honshu–Shikoku Bridge Authority.)

FIGURE 18.32 Cantilevering method in the Akashi Kaikyo Bridge. (Courtesy of Honshu–Shikoku Bridge Authority.)

Sese–Seko Bridges by loading heavy vehicles on the bridges. Methods to verify dynamic behavior include vibration tests and the measurement of micro-oscillations caused by slight winds. The former test is based on the measured response to a forced vibration. The latter is described in Section 18.5.2. Dynamic characteristics of the bridge, such as structural damping, natural frequency, and mode of vibration, are ascertained using the vibration test. As the real value of structural damping is difficult to estimate theoretically, the assumed value should be verified by an actual measurement. Examples of measured data on structural damping obtained through vibration tests are shown in Table 18.7.

TABLE 18.7 Structural Damping Obtained from Vibration Tests

Bridge	Center Span Length (m)	Logarithmic Decrement[a]
Minami Bisan–Seto	1100	0.020 ~ 0.096
Ohnaruto	876	0.033 ~ 0.112
Kanmon	712	0.016 ~ 0.062
Ohshima	560	0.017 ~ 0.180

[a] Structural damping.

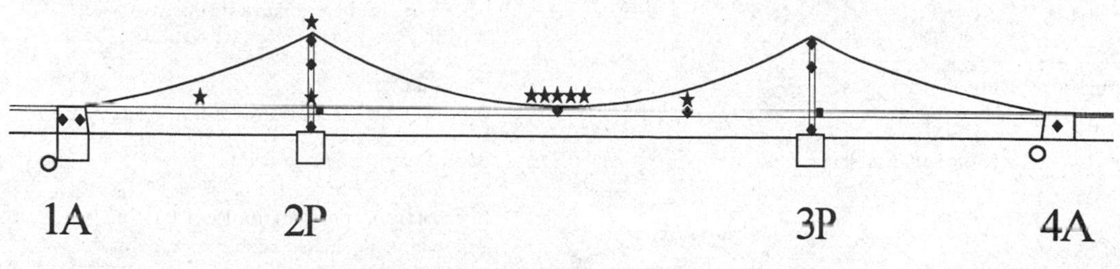

Symbol	Name of instruments
★	wind vane and anemometer
◆	accelerometer, displacement speedmeter
■	deckend displacement gauge
O	seismometer

FIGURE 18.33 Placement of measuring instruments in the Akash Kaikyo Bridge. (*Source:* Abe, K. and Amano, K., Monitoring system of the Akashi Kaikyo Bridge, *Honshi Tech. Rep.*, 86, 29, 1998. With permission.)

18.5.2 Field Observations

Field observations are undertaken to verify such characteristics of bridge behavior as aerodynamic stability and seismic resistance, and to confirm the safety of the bridge. To collect the necessary data for certification, various measuring instruments are installed on the suspension bridge. Examples of measuring instruments used are given in Figure 18.33 [25]. A wind vane and anemometer, which measure local wind conditions, and a seismometer, to monitor seismic activity, gather data on natural conditions. An accelerometer and a displacement speedometer are installed to measure the dynamic response of the structure to wind and earthquake loads. A deck end displacement gauge tracks the response to traffic loads. The accumulated data from these measuring instruments will contribute to the design of yet-longer-span bridges in the future.

18.5.3 Coating Specification

Steel bridges usually get a coating regimen which includes a rust-preventive paint for the base coat, and a long oil-base alkyd resin paint or chlorinated rubber resin paint for the intermediate and top coats. This painting regimen needs to be repeated at several-year intervals. Because long-span suspension bridges are generally constructed in a marine environment, which is severely corrosive, and have enormous painting surfaces, which need to be regularly redone, a heavy-duty coating method with long-term durability is required. The latest coating technology adopted for major suspension bridges is shown in Table 18.8. Previous painting methods relied on oil-base anticorrosive paints or red lead anticorrosive paints for base coats with phthalic resin or aluminum paints as intermediate and top coats. The latest coating specification aimed at long-term durability calls for an inorganic zinc-enriched base

TABLE 18.8 Coating Systems of Major Suspension Bridges

Country	Bridge	Year of Completion	Coating Specification
U.S.	George Washington	1931	Base: oil–based anticorrosive paint Top: phthalic resin paint
	San Francisco–Oakland Bay Golden Gate	1936 1937	Base: red lead anticorrosive paint To: oil–modified phenolic resin aluminum paint
	Mackinac Straits Verrazano Narrows	1957 1965	Base: oil–based anticorrosive paint Top: phthalic resin paint
Canada	Pierre La Porte	1970	Base: basic lead chromate anticorrosive paint Top: alkyd resin paint
Turkey	Bosphorus	1973	Base: zinc spraying Top: phenolic resin micaceous iron oxide paint
	Fatih Sultan Mehmet	1988	Base: organic zinc rich paint Intermediate: epoxy resin paint Intermediate: epoxy resin micaceous iron oxide paint Top: paint chlorinated rubber resin paint
U.K.	Forth Road Severn Humber	1964 1966 1981	Base: zinc spraying Top: phenolic resin micaceous iron oxide paint
Japan	Kanmon	1973	Base: zinc spraying Intermediate: micaceous iron oxide paint Top: chlorinated rubber resin paint
	Innoshima	1983	Base: hi-build inorganic zinc rich paint Intermediate: hi-build epoxy resin paint Top: polyurethane resin paint
	Akashi Kaikyo	1998	Base: hi-build inorganic zinc rich paint Intermediate: hi-build epoxy resin paint Intermediate: epoxy resin paint Top: fluororesin paint

paint, which is highly rust-inhibitive due to the sacrificial anodic reaction of the zinc, with an epoxy resin intermediate coat and a polyurethane resin or fluororesin top coat. Because the superiority of fluororesin paint for long-term durability and in holding a high luster under ultraviolet rays has been confirmed in recent years, it was used for the Akashi Kaikyo Bridge [26].

18.5.4 Main Cable Corrosion Protection

Since the main cables of a suspension bridge are the most important structural members, corrosion protection is extremely important for the long-term maintenance of the bridge. The main cables are composed of galvanized steel wire about 5 mm in diameter with a void of about 20% which is longitudinally and cross-sectionally consecutive. Main cable corrosion is caused not only by water and ion invasion from outside, but also by dew resulting from the alternating dry and humid conditions inside the cable void. The standard corrosion protection system for the main cables ever since it was first worked out for the Brooklyn Bridge has been to use galvanized wire covered with a paste, wrapped with galvanized soft wires and then coated.

New approaches such as wrapping the wires with neoprene rubber or fiberglass acrylic or S-shaped deformed steel wires have also been attempted. A dehumidified air-injection system was developed and used on the Akashi Kaikyo Bridge [27]. This system includes wrapping to improve watertightness and the injection of dehumidified air into the main cables as shown in Figure 18.34. Examples of a corrosion protection system for the main cables in major suspension bridges are shown in Table 18.9.

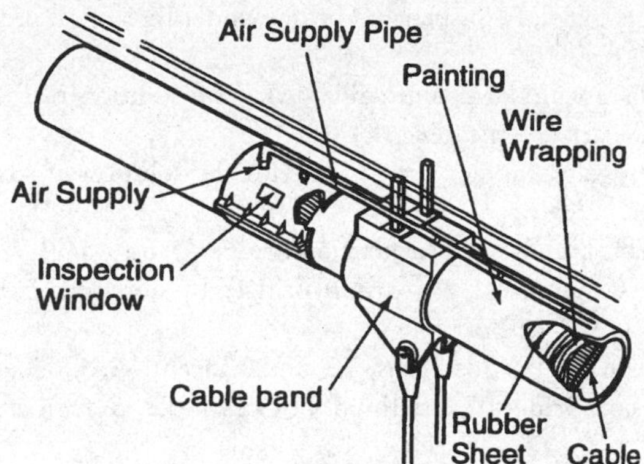

FIGURE 18.34 Dehumidified air–injection system for the main cables of the Akashi Kaikyo Bridge. (Courtesy of Honshu–Shikoku Bridge Authority.)

TABLE 18.9 Corrosion Protection Systems for Main Cable of Major Suspension Bridges

Bridge	Year of Completion	Erection Method	Cable		Wrapping
			Wire	Paste	
Brooklyn	1883	A.S.	Galvanized	Red lead paste	Galvanized wire
Williamsburg	1903	A.S.	—[a]	Red lead paste	Cotton duck + sheet iron coating
Golden Gate	1937	A.S.	Galvanized	Red lead paste	Galvanized wire
Chesapeake Bay II	1973	A.S.	Galvanized	—	Neoprene rubber
Verrazano Narrows	1964	A.S.	Galvanized	Red lead paste	Galvanized wire
Severn	1966	A.S.	Galvanized	Red lead paste	Galvanized wire
New Port	1969	A.S.	Galvanized	—	Glass-reinforced acrylic
Kanmon	1973	P.S.	Galvanized	Polymerized organic lead paste	Galvanized wire
Minami Bisan–Seto	1988	P.S.	Galvanized	Calcium plumbate contained polymerized organic lead paste	Galvanized wire
Hakucho	1998	P.S.	Galvanized	Aluminum triphosphate contained organic lead paste	Galvanized wire (S shape)
Akashi Kaikyo	1998	P.S.	Galvanized	—	Galvanized wire + rubber wrapping

[a] Coated with a raw linseed oil.

References

1. Lu, J., Large Suspension Bridges in China, *Bridge Foundation Eng.*, 7, 20, 1996 [in Japanese].
2. Navier, M., *Papport et Mémoire sur les Ponts Suspendus*, de l' Imprimerie Royale, Paris, 1823.
3. Rankine, W. J. M., *A Manual of Applied Mechanics*, 1858.
4. Ritter, W., Versteifungsfachewerke bei Bogen und Hängebrücken, *Z. bauwesen*, 1877.
5. Lévy, M., Mémoir sur le calcul des ponts suspendus rigides, *Ann. Ponts Chaussées*, 1886.
6. Melan, J., Theorie der eisernen Bogenbrücken und der Hängebrücken, *Handb. Ingenieurwissensch.*, 1888.
7. Moisseiff, L. S., The towers, cables and stiffening trusses of the bridge over the Delaware River between Philadelphia and Camden, *J. Franklin Inst.*, Oct., 1925.

8. Moisseiff, L. S. and Lienhard, F., Suspension bridge under the action of lateral forces, *Trans. ASCE*, 58, 1933.
9. Brotton, D. M., Williamson, N. M., and Millar, M., The solution of suspension bridge problems by digital computers, Part I, *Struct. Eng.*, 41, 1963.
10. Brotton, D. M., A general computer programme for the solution of suspension bridge problems, *Struct. Eng.*, 44, 1966.
11. Saafan, A. S., Theoretical analysis of suspension bridges, *Proc. ASCE*, 92, ST4, 1966.
12. Tezcan, S. S., Stiffness analysis of suspension bridges by iteration, in *Symposium on Suspension Bridges*, Lisbon, 1966.
13. Bleich, F., McCullough, C. B., Rosecrans, R., and Vincent, G. S., The Mathematical Theory of Vibration in Suspension Bridges, Department of Commerce, *Bureau of Public Roads*, Washington, D.C., 1950.
14. Birdsall, B., The suspension bridge tower cantilever problem, *Trans. ASCE*, 1942.
15. Steinman, D. B., A generalized deflection theory for suspension bridges, *Trans. ASCE*, 100, 1935.
16. Walshe, D. E. et al., A Further Aerodynamic Investigation for the Proposed Severn River Suspension Bridge, 1966.
17. Roberts, G., Severn Bridge, Institution of Civil Engineers, London, 1970.
18. Simpson, A. G., Curtis, D. J., and Choi, Y.-L., Aeroelasic aspects of the Lantau fixed crossing, Institution of Civil Engineers, London, 1981.
19. Ohashi, M., Miyata, T., Okauchi, I., Shiraishi, N., and Narita, N., Consideration for Wind Effects on a 1990 m Main Span Suspension Bridge, Pre-report 13th Int. Congress IABSE, 1988, 911.
20. Diana, G., Aeroelastic study of long span suspension bridges, the Messina Crossing, ASCE Structurcs Congress '93, Irvine, CA, 1993.
21. Honshu–Shikoku Bridge Authority, Wind-Resistant Design Standard for the Akashi Kaikyo Bridge, HSBA, Japan, 1990 [in Japanese].
22. Kashima, S., Yasuda, M., Kanazawa, K., and Kawaguchi, K., Earthquake Resistant Design of Akashi Kaikyo Bridge, paper presented as Third Workshop on Performance and Strengthening of Bridge Structures, Tsukuba, 1987.
23. Honshu–Shikoku Bridge Authority, Design Standard of the Main Tower for a Suspension Bridge, HSBA, Japan, 1984 [in Japanese].
24. Petersen, A. and Yamasaki, Y., Great Belt Bridge and design of its cable works, *Bridge Foundation Eng.*,1, 18, 1994 [in Japanese].
25. Abe, K. and Amano, K., Monitoring System of the Akashi Kaikyo Bridge, *Honshi Technical Report*, 86, 29, 1998 [in Japanese].
26. Honshu–Shikoku Bridge Authority, Steel Bridges Coating Standards of Honshu–Shikoku Bridge, HSBA, Japan, 1990 [in Japanese].
27. Ito, M., Saeki, S., and Tatsumi, M., Corrosion protection of bridge cables: from Japanese experiences, in *Proceedings of the International Seminar on New Technologies for Bridge Management*, IABSE, Seoul, 1996.

19

Cable-Stayed Bridges

Man-Chung Tang

T.Y. Lin International

19.1 Introduction

Since the completion of the Stromsund Bridge in Sweden in 1955, the cable-stayed bridge has evolved into the most popular bridge type for long-span bridges. The variety of forms and shapes of cable-stayed bridges intrigues even the most-demanding architects as well as common citizens. Engineers found them technically innovative and challenging. For spans up to about 1000 m, cable-stayed bridges are more economical.

The concept of a cable-stayed bridge is simple. A bridge carries mainly vertical loads acting on the girder, Figure 19.1. The stay cables provide intermediate supports for the girder so that it can span a long distance. The basic structural form of a cable-stayed bridge is a series of overlapping triangles comprising the pylon, or the tower, the cables, and the girder. All these members are under predominantly axial forces, with the cables under tension and both the pylon and the girder under compression. Axially loaded members are generally more efficient than flexural members. This contributes to the economy of a cable-stayed bridge.

At the last count, there are about 600 cable-stayed bridges in the world and the number is increasing rapidly. The span length has also increased significantly [2,7].

Some milestones: the Stromsund Bridge in Sweden, completed in 1955 with a main span of 183 m is usually recognized as the world's first major cable-stayed bridge; the Knie Bridge (320 m) and Neuenkamp Bridge (350 m) in Germany, Figure 19.2, were the longest spans in the early 1970s, until the Annacis Island–Alex Fraser Bridge (465 m) was completed in the mid 1980s. The 602-m-span Yangpu Bridge was a large step forward in 1994 but was surpassed within about half a year by the Normandie Bridge (856 m), Figure 19.3. The Tatara Bridge, with a center span of 890 m, is the world record today. Several spans in the range of 600 m are under construction. Longer spans are being planned.

0-8493-7434-0/00/$0.00+$.50

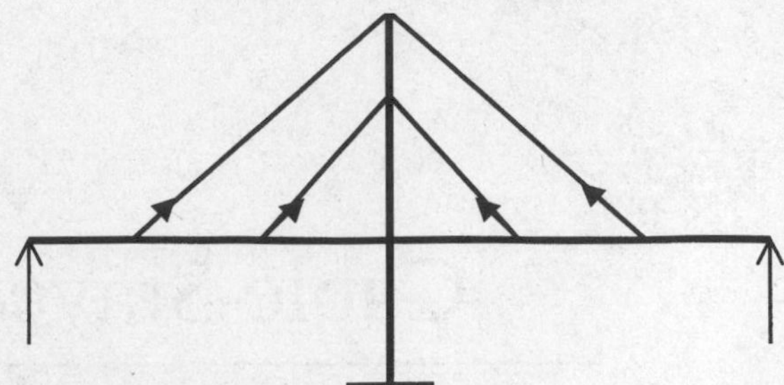

FIGURE 19.1 Concept of a cable-stayed bridge.

FIGURE 19.2 Neuenkamp Bridge.

FIGURE 19.3 Normandie Bridge.

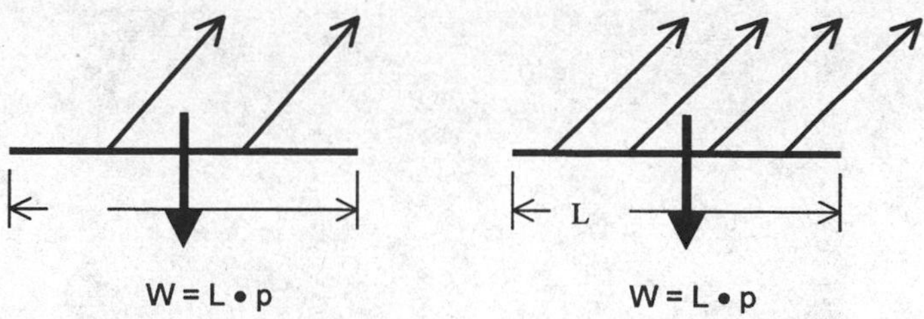

FIGURE 19.4 Cable forces in relation to load on girder.

19.2 Configuration

19.2.1 General Layout

At the early stage, the idea of a cable-stayed bridge was to use cable suspension to replace the piers as intermediate supports for the girder so that it could span a longer distance. Therefore, early cable-stayed bridges placed cables far apart from each other based on the maximum strength of the girder. This resulted in rather stiff girders that had to span the large spacing between cables, in addition to resisting the global forces.

The behavior of a cable-stayed girder can be approximately simulated by an elastically supported girder. The bending moment in the girder under a specific load can be thought of as consisting of a local component and a global component. The local bending moment between the cables is proportional to the square of the spacing. The global bending moment of an elastically supported girder is approximately [5]

$$M = a * p * \sqrt{(I/k)} \tag{19.1}$$

where a is a coefficient depending on the type of load p, I is the moment of inertia of the girder, and k is the elastic support constant derived from the cable stiffness. The global moment decreases as the stiffness of girder, I, decreases.

Considering that the function of the cables is to carry the loads on the bridge girder, which remains the same, the total quantity of cables required for a bridge is practically the same independent of the number of cables, or cable spacing, Figure 19.4. But if the cable spacing is smaller, the local bending moment of the girder between the cables is also smaller. A reduction of the local bending moment allows the girder to be more flexible. A more flexible girder attracts in turn less global moment. Consequently, a very flexible girder can be used with closely spaced cables in many modern cable-stayed bridges. The Talmadge Bridge, Savannah, Figure 19.5, is 1.45 m deep for a 335 m span, The ALRT Skytrain Bridge, Vancouver, Figure 19.6, is 1.1 m deep for a 340 m span and the design of the Portsmouth Bridge had a 84-cm-deep girder for a span of 286 m.

Because the girder is very flexible, questions concerning buckling stability occasionally arose at the beginning. However, as formulated by Tang [4], Eq. (19.2), using the energy method,

$$P(cr) = \left\{ \int EIw''^2 ds + \sum EC * Ac * Lc \right\} \Big/ \left[\int (Ps/Pc) w'^2 ds \right] \tag{19.2}$$

where E is modulus of elasticity, I is moment of inertia, A is area, L is length, w is deflection, and $(\ )'$ is derivative with respect to length s. The buckling load depends more on the stiffness of the cables than on the stiffness of the girder. Theoretically, even if the stiffness of the girder is neglected, a cable-stayed bridge can still be stable in most cases. Experience shows that even for the most flexible girder, the critical load against elastic buckling is well over 400% of the actual loads of the bridge.

FIGURE 19.5 Talmadge Bridge.

FIGURE 19.6 ALRT Skytrain Bridge.

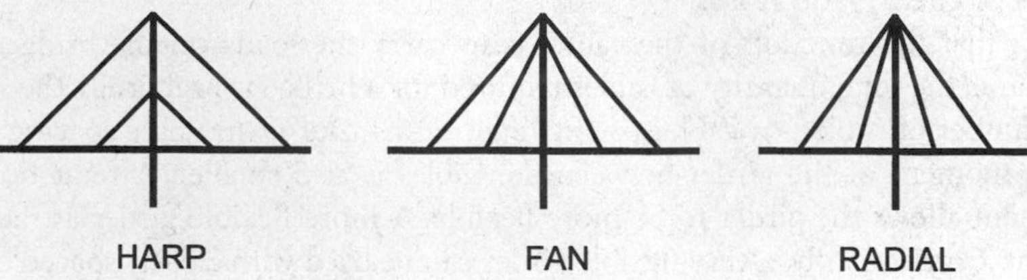

FIGURE 19.7 Cable arrangements.

The recently adopted design requirement that all cables be individually replaceable makes closely spaced cables more desirable. It is usually required that one cable can be detensioned, dismantled, and replaced under reduced traffic loading. The additional bending moment in the girder will not increase excessively if the cable spacing is small.

Availability of ever more powerful computers also helps. The complexity of the analysis increases as the number of cables increases. The computer offers engineers the best tool to deal with this problem.

Harp, radial, fan, Figure 19.7, or other cable configurations have all been used. However, except in very long span structures, cable configuration does not have a major effect on the behavior of the bridge.

A harp-type cable arrangement offers a very clean and delicate appearance because an array of parallel cables will always appear parallel irrespective of the viewing angle. It also allows an earlier

FIGURE 19.8 Nord Bridge.

FIGURE 19.9 Ludwighafen Bridge.

start of girder construction because the cable anchorages in the tower begin at a lower elevation. The Hoechst Bridge and the Dames Point Bridge are examples that fully utilized this advantage.

A fan-type cable arrangement can also be very attractive, especially for a single-plane cable system. Because the cable slopes are steeper, the axial force in the girder, which is an accumulation of all horizontal components of cable forces, is smaller. This feature is advantageous for longer-span bridges where compression in the girder may control the design. The Nord Bridge, Bonn, Figure 19.8, is one of the first of this type.

A radial arrangement of cables with all cables anchored at a common point at the tower is quite efficient. However, a good detail is difficult to achieve. Unless it is well treated, it may look clumsy. The Ludwighafen Bridge, Germany, Figure 19.9, is a successful example. The Yelcho Bridge, Chile, with all cables anchored in a horizontal plane in the tower top, is an excellent solution, both technically and aesthetically.

When the Stromsund Bridge was designed, long-span bridges were the domain of steel construction. Therefore, most early cable-stayed bridges were steel structures. They retained noticeable features from other types of long-span steel bridges.

In the 1960s, Morandi designed and built several relatively long span concrete cable-stayed bridges. His designs usually had few cables in a span with additional strut supports at the towers for the girder. They did not fully utilize the advantages of a cable-stayed system. The concrete cable-stayed bridge in its modern form started with the Hoechst Bridge in Germany, followed by the Brotone Bridge in France and the Dames Point Bridge in the United States, each representing a significant advance in the state of the art.

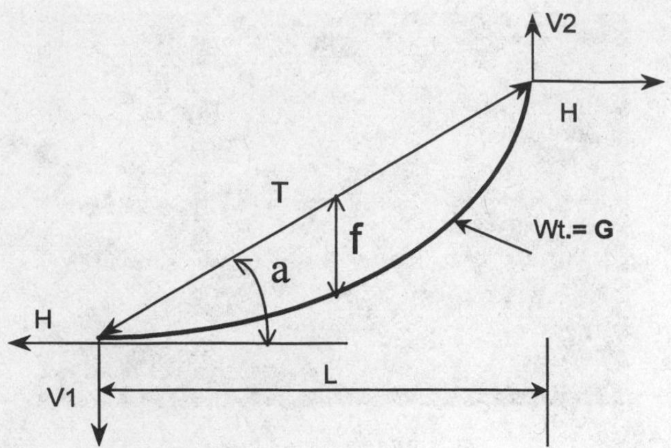

FIGURE 19.10 An inclined cable.

19.2.2 Cables

Cables are the most important elements of a cable-stayed bridge. They carry the load of the girder and transfer it to the tower and the back-stay cable anchorage.

The cables in a cable-stayed bridge are all inclined, Figure 19.10. The actual stiffness of an inclined cable varies with the inclination angle, *a*, the total cable weight, *G*, and the cable tension force, *T* [3]:

$$EA(\text{eff}) = EA/\left\{1 + G^2\ EA\cos^2 a/\left(12\,T^3\right)\right\} \tag{19.3}$$

where *E* and *A* are Young's modulus and the cross-sectional area of the cable. And if the cable tension *T* changes from *T*1 to *T*2, the equivalent cable stiffness will be

$$EA(\text{eff}) = EA/\left\{1 + G^2\ EA\cos^2 a(T1 + T2)/\left(24\,T1^2\,T2^2\right)\right\} \tag{19.4}$$

In most cases, the cables are tensioned to about 40% of their ultimate strength under permanent load condition. Under this kind of tension, the effective cable stiffness approaches the actual values, except for very long cables. However, the tension in the cables may be quite low during some construction stages so that their effectiveness must be properly considered.

A safety factor of 2.2 is usually recommended for cables. This results in an allowable stress of 45% of the guaranteed ultimate tensile strength (GUTS) under dead and live loads [9]. It is prudent to note that the allowable stress of a cable must consider many factors, the most important being the strength of the anchorage assemblage that is the weakest point in a cable with respect to capacity and fatigue behavior.

There have been significant developments in the stay cable system. Early cables were mainly lock-coil strands. At that time, the lock-coil strand was the only cable system available that could meet the more stringent requirements of cable-stayed bridges.

Over the years, many new cable systems have been successfully used. Parallel wire cables with Hi-Am sockets were first employed in 1969 on the Schumacher Bridge in Mannheim, Germany. Since then, the fabrication technique has been improved and this type cable is still one of the best cables commercially available today. A Hi-Am socket has a conical steel shell. The wires are parallel for the entire length of the cable. Each wire is anchored to a plate at the end of the socket by a button head. The space in the socket is then filled with epoxy mixed with zinc and small steel balls.

The Hi-Am parallel wire cables are prefabricated to exact length in the yard and transported to the site in coils. Because the wires are parallel and therefore all of equal length, the cable may sometimes experience difficulty in coiling. This difficulty can be overcome by twisting the cable during the coiling process. To avoid this problem altogether, the cables can be fabricated with a long lay. However, the long lay may cause a very short cable to twist during stressing.

Threadbar tendons were used for some stay cables. The first one was for the Hoechst Bridge over the Main River in Germany. The Penang Bridge and the Dames Point Bridge also have bar cables. They all have a steel pipe with cement grout as corrosion protection. Their performance has been excellent.

The most popular type of cable nowadays uses seven-wire strands. These strands, originally developed for prestressed concrete applications, offer good workability and economy. They can either be shop-fabricated or site-fabricated. In most cases, corrosion protection is provided by a high-density polyethylene pipe filled with cement grout. The technique of installation has progressed to a point where a pair of cables can be erected at the site in 1 day.

In search of better corrosion protection, especially during the construction stage before the cables have been grouted, various alternatives, such as epoxy coating, galvanization, wax and grease have all been proposed and used. Proper coating of strands must completely fill the voids between the wires with corrosion inhibitor. This requires the wires to be loosened before the coating process takes place and then retwisted into the strand configuration.

In addition to epoxy, grease, or galvanization, the strands may be individually sheathed. A sheathed galvanized strand may have wax or grease inside the sheathing. All three types of additional protection appear to be acceptable and should perform well. However, a long-term performance record is not yet available.

The most important element in a cable is the anchorage. In this respect, the Hi-Am socket has an excellent performance record. Strand cables with bonded sockets, similar to the Hi-Am socket, have also performed very well. In a recently introduced unbonded anchorage, all strands are being held in place only by wedges. Tests have confirmed that these anchorages meet the design requirements. But unbonded strand wedges are delicate structural elements and are susceptible to construction deviations. Care must be exercised in the design, fabrication, and installation if such an anchorage is to be used in a cable-stayed bridge. The advantage of an unbonded cable system is that the cable, or individual strands, can be replaced relatively easily.

Cable anchorage tests have shown that, in a bonded anchor, less than half of the cyclic stress is transferred to the wedges. The rest is dissipated through the filling and into the anchorage directly by bond. This is advantageous with respect to fatigue and overloading.

The Post Tensioning Institute's "Recommendations for Stay Cable Design and Testing," [9] was published in 1986. This is the first uniformly recognized criteria for the design of cables. In conjunction with the American Society of Civil Engineers' "Guidelines for the Design of Cable-Stayed Bridges" [10], they give engineers a much-needed base to start their design.

There have been various suggestions for using composite materials such as carbon fiber, etc. as stays and small prototypes have been built. However, actual commercial application still requires further research and development.

19.2.3 Girder

Although the Stromsund Bridge has a concrete deck, most other early cable-stayed bridges have an orthotropic deck. This is because both cable-stayed bridge and orthotropic deck were introduced to the construction industry at about the same time. Their marriage was logical. The fact that almost all long-span bridges were built by steel companies at that time made such a choice more understandable.

A properly designed and fabricated orthotropic deck is a good solution for a cable-stayed bridge. However, with increasing labor costs, the orthotropic deck becomes less commercially attractive except for very long spans.

Many concrete cable-stayed bridges have been completed. In general, there have been two major developments: cast-in-place construction and precast construction.

FIGURE 19.11 Form traveler of the Dames Point Bridge.

FIGURE 19.12 Sunshine Skyway Bridge – precast box and erection.

Cast-in-place construction of cable-stayed bridges is a further development of the free cantilever construction method of box-girder bridges. The typical construction is by means of a form traveler. The box girder is a popular shape for the girder in early structures, such as the Hoechst Bridge, the Barrios de Luna Bridge, and the Waal Bridge. But simpler cross sections have proved to be attractive: the beam and slab arrangement in the Penang Bridge, the Dames Point Bridge, and the Talmadge Bridge, or the solid slab cross section as in the Yelcho Bridge, the Portsmouth Bridge, and the Diepoldsau Bridge are both technically sound and economical. Use of the newly developed cable-supported form traveler, Figure 19.11, makes this type girder much more economical to build [8].

Precast construction can afford a slightly more complicated cross section because precasting is done in the yard. The segments, however, should all be similar to avoid adjustment in the precasting forms. The weight of the segment is limited by the transportation capability of the equipment used. Box is the preferred cross section because it is stiffer and easier to erect. The Brotonne Bridge, the Sunshine Skyway Bridge, Figure 19.12, and the Chesapeake and Delaware Canal Bridge are good examples. However, several flexible girder cable-stayed bridges have been completed successfully, notably the Pasco–Kennewick Bridge, the East Huntington Bridge, and the ALRT Skytrain Bridge.

As concrete technology advances, today's cable-stayed bridges may consider using high-strength lightweight concrete for the girder, especially in high seismic areas.

Although the steel orthotropic deck is too expensive for construction in most countries at this time, the composite deck with a concrete slab on a steel frame can be very competitive. In the Stromsund Bridge, the concrete deck was not made composite with the steel girder. Such a construction is not economical because the axial compressive force in the girder must be taken entirely by

FIGURE 19.13 Baytown–LaPorte Bridge and Yang Pu Bridge.

the steel girder. Making the deck composite with the steel girder by shear studs reduces the steel quantity of the girder significantly. The compressive stress in the concrete deck also improves the performance of the deck slab. The Hootley Bridge was the first major composite bridge designed, but the Annacis Island Bridge was completed first. The Yang Pu Bridge is the longest span today. The Baytown–LaPorte Bridge, Figure 19.13, has the largest deck area.

Precast slab panels are usually used for composite bridges. Requiring the precast panels to be stored for a period of time, say, 90 days, before erection reduces the effect of creep and shrinkage significantly. The precast panels are supported by floor beams and the edge girders during erection. The gaps between the panels are filled with nonshrinking concrete. The detail for these closure joints must be carefully executed to avoid cracking due to shrinkage and other stresses.

Most portions of the girder are under high compression, which is good for concrete members. However, tensile stresses may occur in the middle portion of the center span and at both ends of the end spans. Post-tensioning is usually used in these areas to keep the concrete under compression.

Several hybrid structures, with concrete side spans and steel main span, such as the Flehe Bridge, Germany and the Normandie Bridge, have been completed. There are two main reasons for the hybrid combination: to have heavier, shorter side spans to balance the longer main span or to build the side spans the same way as the connecting approaches. The transition, however, must be carefully detailed to avoid problems.

19.2.4 Tower

The towers are the most visible elements of a cable-stayed bridge. Therefore, aesthetic considerations in tower design is very important. Generally speaking, because of the enormous size of the structure,

FIGURE 19.14 Tower shapes — H (Talmadge), inverted Y (Flehe), diamond (Glebe), double diamond (Baytown), and special configuration (Dan Chiang).

a clean and simple configuration is preferable. The free-standing towers of the Nord Bridge and the Knie Bridge look very elegant. The H towers of the Annacis Bridge, the Talmadge Bridge, and the Nan Pu Bridge are the most logical shape structurally for a two-plane cable-stayed bridge, Figure 19.14. The A shape (as in the East Hungtington Bridge), the inverted Y (as in the Flehe Bridge), and the diamond shape (as in the Baytown–LaPorte and Yang Pu Bridges) are excellent choices for long-span cable-stayed bridges with very flexible decks. Other variations in tower shape are possible as long as they are economically feasible. Under special circumstances, the towers can also serve as tourist attractions such as the one proposed for the Dan Chiang Bridge in Taiwan.

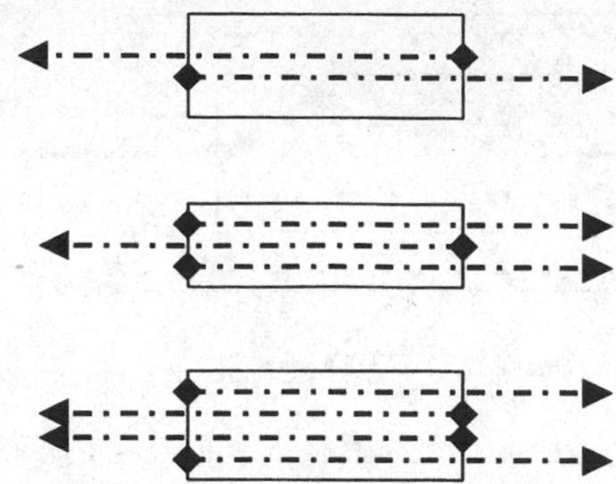

FIGURE 19.15 Crisscrossing cables at tower column.

Although early cable-stayed bridges all have steel towers, most recent constructions have concrete towers. Because the tower is a compression member, concrete is the logical choice except under special conditions such as in high earthquake areas.

Cables are anchored at the upper part of the tower. There are generally three concepts for cable anchorages at the tower: crisscrossing, dead-ended, and saddle.

Early cable-stayed bridges took their anchorage details from suspension bridges that have saddles. Those saddles were of the roller, fixed, or sector types. Roller-type saddles have a roller at the base similar to a bridge bearing. Fixed-type saddles are similar except the base is fixed instead of rolling. Sector-type saddles rotate around a pin. Each of these saddles satisfies a different set boundary condition in the structural system. Cable strands, basically lock-coil strands, were placed in the saddle trough as in a suspension bridge. To assure that the strands do not slide under unequal tension, a cover plate is usually clamped to the saddle trough to increase friction. This transverse pressure increases friction but reduces the strength of the cable. This type of saddle is very expensive and has not been used for recent cable-stayed bridges.

A different type of saddle was used in the Brotonne Bridge and the Sunshine Skyway Bridge. Here the seven-wire strands were bundled into a cable and then pulled through a steel saddle pipe which was fixed to the tower. Grouting the cable fixed the strands to the saddle pipe. However, the very high contact pressure between the strand wires must be carefully considered and dealt with in the design. Because the outer wires of a strand are helically wound around a straight, center king wire, the strands in a curved saddle rest on each other with point contacts. The contact pressure created by the radial force of the curved strands can well exceed the yield strength of the steel wires. This can reduce the fatigue strength of the cable significantly.

Crisscrossing the cables at the tower is a good idea in a technical sense. It is safe, simple, and economical. The difficulty is in the geometry. To avoid creating torsional moment in the tower column, the cables from the main span and the side span should be anchored in the tower in the same plane, Figure 19.15. This, however, is physically impossible if they crisscross each other. One solution is to use double cables so that they can pass each other in a symmetrical pattern as in the case of the Hoechst Bridge. If A-shaped or inverted-Y-shaped towers are used, the two planes of cables can also be arranged in a symmetrical pattern.

If the tower cross section is a box, the cables can be anchored at the front and back wall of the tower, Figure 19.16. Post-tensioning tendons are used to prestress the walls to transfer the anchoring forces from one end wall to the other. The tendons can be loop tendons that wrap around three side walls at a time or simple straight tendons in each side wall independently. The Talmadge, Baytown–LaPorte, and Yang Pu Bridges all employ such an anchoring detail. During the design of the Yang Pu Bridge, a full-scale model was made to confirm the performance of such a detail.

As an alternative, some bridges have the cables anchored to a steel member that connects the cables from both sides of the tower. The steel member may be a beam or a box. It must be connected

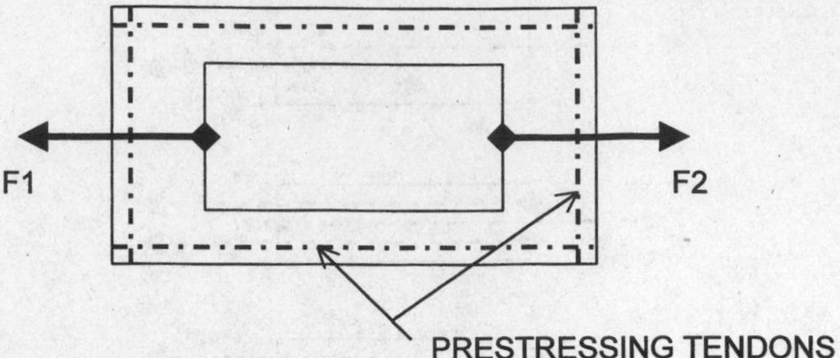

FIGURE 19.16 Tendon layout at the anchorage area.

to the concrete tower by shear studs or other means. This anchorage detail simulates the function of the saddle. However, the cables at the opposite sides are independent cables. The design must therefore consider the loading condition when only one cable exists.

19.3 Design

19.3.1 Permanent Load Condition

A cable-stayed bridge is a highly redundant, or statically indeterminate structure. In the design of such a structure, the treatment of the permanent load condition is very important. This load condition includes all structural dead load and superimposed dead load acting on the structure, all prestressing effects as well as all secondary moments and forces. It is the load condition when all permanent loads act on the structure.

Because the designer has the liberty to assign a desired value to every unknown in a statically indeterminate structure, the bending moments and forces under permanent load condition can be determined solely by the requirements of equilibrium, $\Sigma H = 0$, $\Sigma V = 0$, and $\Sigma M = 0$. The stiffness of the structure has no effect in this calculation. There are an infinite number of possible combinations of permanent load conditions for any cable-stayed bridge. The designer can select the one that is most advantageous for the design when other loads are considered.

Once the permanent load condition is established by the designer, the construction has to reproduce this final condition. Construction stage analysis, which checks the stresses and stability of the structure in every construction stage, starts from this selected final condition backwards. However, if the structure is of concrete or composite, creep and shrinkage effect must be calculated in a forward calculation starting from the beginning of the construction. In such cases, the calculation is a combination of forward and backward operations.

The construction stage analysis also provides the required camber of the structure during construction.

19.3.2 Live Load

Live-load stresses are mostly determined by evaluation of influence lines. However, the stress at a given location in a cable-stayed bridge is usually a combination of several force components. The stress, *f*, of a point at the bottom flange, for example, can be expressed as:

$$f = (1/A) * P + (y/I) * M + c * K \quad (19.5)$$

where A is the cross-sectional area, I is the moment of inertia, y is the distance from the neutral axis, and c is a stress influence coefficient due to the cable force K anchored at the vicinity. P is the axial force and M is the bending moment. The above equation can be rewritten as

$$f = a1 * P + a2 * M + a3 * K \tag{19.6}$$

where the constants $a1$, $a2$, and $a3$ depend on the effective width, location of the point, and other global and local geometric configurations. Under live load, the terms P, M, and K are individual influence lines. Thus, f is a combined influence line obtained by adding up the three terms multiplied by the corresponding constants $a1$, $a2$, and $a3$, respectively.

In lieu of the combined influence lines, some designs substitute P, M, and K with extreme values, i.e., maximum and minimum of each. Such a calculation is usually conservative but fails to present the actual picture of the stress distribution in the structure.

19.3.3 Thermal Loads

Differential temperature between various members of the structure, especially that between the cables and the rest of the bridge, must be considered in the design. Black cables tend to be heated up and cooled down much faster than the towers and the girder, thus creating a significant temperature difference. Light-colored cables, therefore, are usually preferred.

Orientation of the bridge toward the sun is another factor to consider. One face of the towers and some group of cables facing the sun may be warmed up while the other side is in the shadow, causing a temperature gradient across the tower columns and differential temperature among the cable groups.

19.3.4 Dynamic Loads

19.3.4.1 Structural Behavior

The girder of a cable-stayed bridge is usually supported at the towers and the end piers. Depending on the type of bearing or supports used, the dynamic behavior of the structure can be quite different. If very soft supports are used, the girder acts like a pendulum. Its fundamental frequency will be very low. Stiffening up the supports and bearings can increase the frequency significantly.

Seismic and aerodynamics are the two major dynamic loads to be considered in the design of cable-stayed bridges. However, they often have contradictory demands on the structure. For aerodynamic stability a stiffer structure is preferred. But for seismic design, except if the bridge is founded on very soft soil, a more flexible bridge will have less response. Some compromise between these two demands is required.

Because the way these two dynamic loads excite the structure is different, special mechanical devices can be used to assist the structure to adjust to both load conditions. Aerodynamic responses build up slowly. For this type of load the forces in the connections required to minimize the vibration buildup are relatively small. Earthquakes happen suddenly. The response will be especially sudden if the seismic motion also contains large flings. Consequently, a device that connects the girder and the tower, which can break at a certain predetermined force will help in both events. Under aerodynamic actions, it will suppress the onset of the vibrations as the connection makes the structure stiffer. Under seismic load, the connection breaks at the predetermined load and the structure becomes more flexible. This reduces the fundamental frequency of the bridge.

19.3.4.2 Seismic Design

Most cable-stayed bridges are relatively flexible with long fundamental periods in the range of 3.0 s or longer. Their seismic responses are usually not very significant in the longitudinal direction. In the transverse direction, the towers are similar to a high-rise building. Their responses are also manageable.

Experience shows that, except in extremely high seismic areas, earthquake load seldom controls the design. On the other hand, because most cable-stayed bridges are categorized as major structures, they are usually required to be designed for more severe earthquake loads than regular structures.

Various measures have been used to reduce the seismic response of cable-stayed bridges [1]. They range from a simple shock damper with a hydraulic cylinder that freezes at fast motion, to different types of friction dampers. Letting the bridge girder swing for a certain distance like a pendulum is another efficient way of reducing the seismic response. This is especially effective for out-of-phase motions. Because many cable-stayed spans are in the range of the half wavelengths of seismic motions, the out-of-phase motion can create very large reactions in the structure. A partially floating girder is often found to be very advantageous.

19.3.4.3 Aerodynamics

Aerodynamic stability of cable-stayed bridges was a major concern for many bridge engineers in the early years. This was probably because cable-stayed bridges are extremely slender. Lessons learned from aerodynamic problems in suspension bridges lead engineers to worry about cable-stayed bridges.

In reality, cable-stayed bridges, especially concrete cable-stayed bridges, have been found to be surprisingly stable aerodynamically. Although the prediction that cable-stayed bridges could not seriously vibrate under wind due to interference of the different unrelated modes that exist in such a structure was not correct, extremely few cable-stayed bridges were found to be susceptible to wind action after construction. The superior aerodynamic behavior of cable-stayed bridges is one reason for this. But the lessons learned from suspension bridge have educated many engineers so that they are aware of aerodynamic problems and can identify the preferred cross sections against wind actions. The wider deck width of most modern cable-stayed bridges also makes the structure more stable.

Several bridges did require special treatment against aerodynamic action. The Annacis Island Bridge added wind fairing to the main span; the Quincy Bridge added vertical plates to the girder in addition to horizontal fairings. Longs Creek Bridge has a tapered wind nose at each side of the girder.

During the design of the Knie Bridge in the early 1960s, a wind tunnel study was performed to search for a good solution to increase the aerodynamic stability of the bridge in case its responses were found to be unacceptable. Among various alternatives, the tapered nose was the most efficient option. This same idea has been used in many cable-stayed bridges and suspension bridges since then.

Although a cable-stayed bridge is mostly stable in its final condition, it is often vulnerable during construction stages. During the construction of the Knie Bridge, bottom bracing was added to provide the required torsional stiffness in the girder to eliminate flutter possibility. Most high-level cable-stayed bridges in high-wind areas have required wind tie-downs to stabilize the structure against buffeting.

Back in the 1960s, it was thought that a structure would not vibrate in a turbulent flow. However, it was found that by having a certain intensity and frequency content in a turbulent wind, buffeting may occur. In many cases, the buffeting responses of a cable-stayed bridge during construction can be quite severe unless specific measures are taken to stabilize the structure.

The most efficient way to stabilize the structure against buffeting is to increase its fundamental frequency by tie-downs, Figure 19.17. Most tie-downs are simple cables of seven-wire strands anchored to pile foundations, dead weights, or soil anchors such as in the Annacis and Baytown–LaPorte Bridges. Stabilizing the tower by front and back staying cables can also have the same effect as in the ALRT Fraser River and East Huntington Bridges.

Use of tie-downs can also help reduce the unbalanced bending moment in the tower during construction which is inherent in a cantilever method.

The amount of damping can have a decisive effect on the aerodynamic behavior of a cable-stayed bridge, especially on the critical flutter wind speed. Consequently, the assumption of a proper damping ratio is very important in the design. Table 19.1 shows the calculated critical flutter wind speed of the Sidney Lanier Bridge, Georgia, based on sectional model test results.

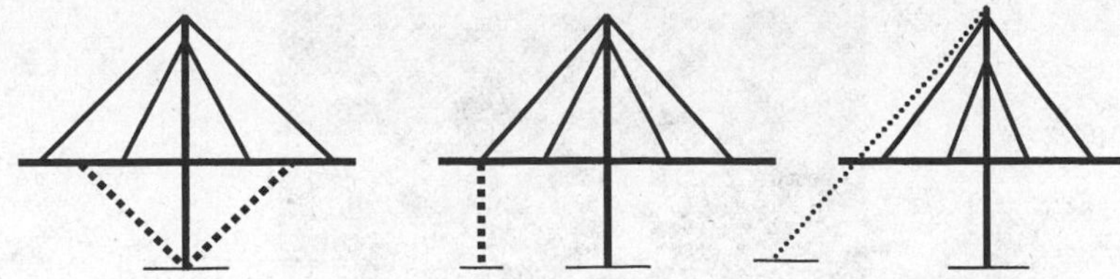

FIGURE 19.17 Temporary stabilization by tie-down cables.

TABLE 19.1 Critical Flutter Wind Speed — Sidney Lanier Bridge

Damping ratio (% of critical)	0.5%	1.0%	2.0%	3.0%	4.0%
Critical flutter wind sped, km/h	160	180	230	340	450

Practically all field measurements of damping ratio of cable-stayed bridges were performed with small amplitudes. Generally, it was found to be between 0.5 and 1.0%. Such measured values correspond to the behavior of low-amplitude votex-shedding responses, which usually happen under relatively low wind speed. However, flutter is a phenomenon represented by large amplitudes; the actual damping coefficient is much higher.

Flutter is considered an extreme natural event that may happen once every 1000 to 2000 years. There will also be no people on the bridge under such high winds. Therefore, very large amplitude oscillations with cracking of concrete and partial yielding of steel are considered acceptable. Under such conditions, the damping increases significantly. A cable-stayed bridge, being a very redundant structure that may allow many plastic hinges to form, offers decisive advantages over regular girder bridges.

A damping ratio of 5% is usually assumed in seismic analysis, which is a similar extreme natural event. A value of 2 to 4% may be conservatively assumed for a concrete cable-stayed bridge. Higher damping can also be achieved by installation of artificial dampers.

The knowledge of cable vibrations has also progressed extensively. The first stay cable vibration problem appeared in the Neuenkamp Bridge. This was a wake vibration of two parallel and horizontally located cables. The problem was new at that time. It was identified, and further vibration was suppressed by connecting the pair of cables together with a damper. This concept was used for several other subsequently built bridges.

Severe cable vibrations were observed in the Brotone Bridge. Dampers were installed and they were successful in suppressing the vibrations. The same concept was used for the Sunshine Skyway.

Rain–wind vibrations were discovered in a few bridges. This phenomenon appears only during light rain in combination with light wind. This problem was found to be caused by the change of shape of rain water mantle on the cable. Increasing the damping of cables can suppress this vibration. Tying cables together by wires, Figure 19.18, and draining the water away from the cable before it accumulates are the more common and effective methods to combat this problem. Adding dimples or spiral-wound ridges on the cable surface have also been found to be effective.

19.4 Superlong Spans

Conceptually, cable-stayed bridges can be used for very long spans. Because a cable-stayed bridge is a closed structural system, similar to a self-anchored suspension bridge, but can be built without temporary supports, it is especially advantageous in areas where the soil condition is not good and anchoring the main suspension cables becomes prohibitively expensive. Span length of over 2000 m is entirely feasible.

Three major details must be properly attended to in the design of superlong-span cable-stayed bridges: the effectiveness of very long cables, the compression in the deck, and the torsional stiffness of the girder for wind stability. When a cable is too long, it becomes ineffective. This can be resolved

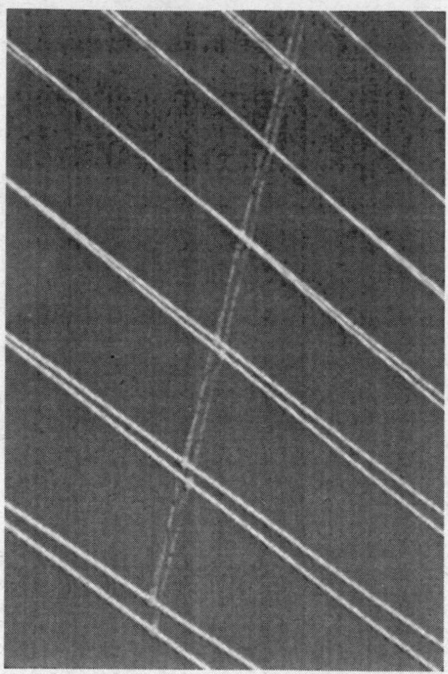

FIGURE 19.18 Tie cables against rain–wind vibrations — Dames Point Bridge.

by providing intermediate supports for the cable to reduce its sag. The compression stress in the deck increases proportionally to the span length. Depending on the type of material used for the deck girder, the maximum span length of a uniform girder is limited by its allowable stresses. This problem can best be solved by using a girder with variable cross section. By increasing the girder section gradually toward the towers, where compression is the highest, the compression stress can be reduced to an acceptable level. The torsional stiffness of the girder can be supplemented by having sufficiently wide towers so that the cables are inclined in the transverse direction.

If the deck is too narrow for the very long span, horizontal staying cables in the transverse direction at the deck level can provide stiffness in that direction.

19.5 Multispan Cable-Stayed Bridges

Most cable-stayed bridges have either three or two cable-stayed spans. The back-staying cables and the anchor pier play an important role in stabilizing the tower. When a bridge has more than three spans, the bending moment in the towers will be very large. One solution is to design the towers to carry the large bending moment [6,11]. This is usually not the most economical solution.

Several methods are available to strengthen the towers of a multispan cable-stayed bridge [6], as shown in Figure 19.19:

1. Tying the tower tops together with horizontal cables;
2. Tying the tower tops to the girder and tower intersection point at the adjacent towers;
3. Adding additional tie-down piers at span centers; and
4. Adding crossing cables at midspans.

19.6 Aesthetic Lighting

Aesthetic lighting is now part of the design of most cable-stayed bridges. Lighting enhances the beauty and visibility of the bridge at night. There are various schemes of lighting a cable-stayed bridge as shown in Figure 19.20.

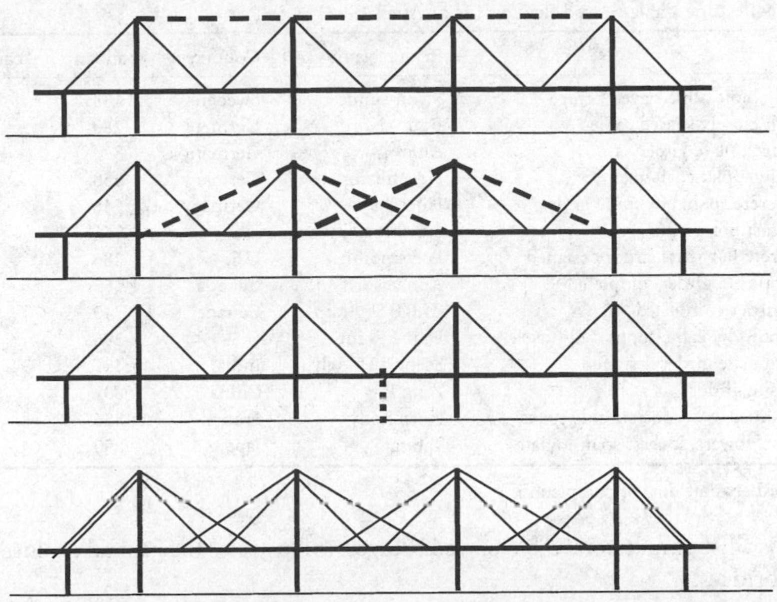

FIGURE 19.19 Multispan cable-stayed bridge system.

FIGURE 19.20 Aesthetic lighting of cable-stayed bridges.

19.7 Summary

Cable-stayed bridges are beautiful structures. Their popular appeal to engineers and nonengineers alike has been universal. In a pure technical sense this bridge type fills the gap of efficient span range between conventional girder bridges and the very long span suspension bridges.

TABLE 19.2 Milestone Cable-Stayed Bridges (as of April 1998)

Features	Bridge Name	Country	Span (m)	Year of Completion
First successfully built cable-stayed bridge	Stromsunde	Sweden	183	1955
First bridge with closely spaced cables	Bonn–Nord[a]	Germany	280	1967
All-steel open-deck plate girder	Knie[a]	Germany	320	1969
Center spine, single plane, all steel	Neuenkamp[a]	Germany	350	1971
First major concrete span, box spine girder	Hoescht	Germany	148	1972
First major precast box girder	Sunshine Skyway	U.S.	366	1986
First solid concrete flat-plate girder designed	Portsmouth	U.S.	286	Not built
First major composite girder completed	Annacis Island[a]	Canada	365	1986
First flat-plate girder completed	ALRT Skytrain	Canada	340	1988
Flexible girder built by cable-supported traveler	Dame Point	U.S.	396	1988
First major composite girder designed	Second Hoogly	India	457	1992
Longest composite girder	Yang Pu[a]	China	602	1993
Hybrid steel main span + concrete side spans	Normandy[a]	France	856	1994
Steel girder, conc. towers, longest span to date	Tatara[a]	Japan	890	1998

[a] World record span at time of completion.

Table 19.2 is a list of milestone cable-stayed bridges. It illustrates the general evolution of modern cable-stayed bridges.

References

1. Abdel-Ghaffar, A.M. Cable-stayed bridges under seismic action, in *Cable-Stayed Bridges*, Elsevier, Amsterdam, 1991.
2. Podolny, W. and Scalzi, J. B. *Construction and Design of Cable-Stayed Bridges*, John Wiley & Sons, New York, 1986.
3. Tang, M. C., Analysis of cable-stayed bridges, *J. Struct. Div. ASCE*, Aug., 1971.
4. Tang, M. C., Buckling of cable-stayed girder bridges, *J. Struct. Div. ASCE*, Sept.,1976.
5. Tang, M. C., Concrete cable-stayed bridges, presented at ACI Convention, Kansas City, Sept.,1983.
6. Tang, M. C. Multispan cable-stayed bridges, in *Proceedings, International Bridge Conference — Bridges into the 21st Century*, Hong Kong, Oct., 1995.
7. Troitsky, M. S., *Cable-Stayed Bridges*, Van Nostrand Reinhold, New York, 1988.
8. Dame Point Bridge reaches for a record, *Eng. News Rec.*, Jan. 7, 1988.
9. Recommendations for Stay Cable Design and Testing, Post-Tensioning Institute, Jan., 1986. Latest rev. 1993.
10. Guidelines for the Design of Cable-Stayed Bridges, American Society of Civil Engineers, 1992.
11. *Festschrift Ulrich Finsterwalder — 50 Jahre für Dywidag*, Dyckerhoff and Widmann, Germany, 1973.

20

Timber Bridges

Kenneth J. Fridley
Washington State University

20.1 Introduction

Wood is one of the earliest building materials, and as such often its use has been based more on tradition than on principles of engineering. However, the structural use of wood and wood-based materials has increased steadily in recent times, including a renewed interest in the use of timber as a bridge material. Supporting this renewed interest has been an evolution of our understanding of wood as a structural material and our ability to analyze and design safe, durable, and functional timber bridge structures.

An accurate and complete understanding of any material is key to its proper use in structural applications, and structural timber and other wood-based materials are no exception to this requirement. This chapter focuses on introducing the fundamental mechanical and physical properties of wood that govern its structural use in bridges. Following this introduction of basic material properties, a presentation of common timber bridge types will be made, along with a discussion of fundamental considerations for the design of timber bridges.

20.1.1 Timber as a Bridge Material

Wood has been widely used for short- and medium-span bridges. Although wood has the reputation of being a material that provides only limited service life, wood can provide long-standing and serviceable bridge structures when properly protected from moisture. For example, many covered bridges from the early 19th century still exist and are in use. Today, rather than protecting wood

Parts of this chapter were previously published by CRC Press in *Handbook of Structural Engineering*, W. F. Chen, Ed., 1997.

0-8493-7434-0/00/$0.00+$.50

by a protective shelter as with the covered bridge of yesteryear, wood preservatives which inhibit moisture and biological attack have been used to extend the life of modern timber bridges.

As with any structural material, the use of wood must be based on a balance between its inherent advantages and disadvantages, as well as consideration of the advantages and disadvantages of other construction materials. Some of the advantages of wood as a bridge material include:

- Strength
- Light weight
- Constructibility
- Energy absorption
- Economics
- Durability, and
- Aesthetics

These advantages must be considered against the three primary disadvantages:

- Decay
- Insect attack, and
- Combustibility

Wood can withstand short-duration overloading with little or no residual effects. Wood bridges require no special equipment for construction and can be constructed in virtually any weather conditions without any negative effects. Wood is competitive with other structural materials in terms of both first costs and life-cycle costs. Wood is a naturally durable material resistant to freeze–thaw effects as well as deicing agents. Furthermore, large-size timbers provide good fire resistance as a result of natural charring. However, if inadequately protected against moisture, wood is susceptible to decay and biological attack. With proper detailing and the use of preservative treatments, the threat of decay and insects can be minimized. Finally, in many natural settings, wood bridges offer an aesthetically pleasing and unobtrusive option.

20.1.2 Past, Present, and Future of Timber Bridges

The first bridges built by humans were probably constructed with wood, and the use of wood in bridges continues today. As recently as a century ago, wood was still the dominant material used in bridge construction. Steel became an economical and popular choice for bridges in the early 1900s. Also during the early part of the 20th century, reinforced concrete became the primary bridge deck material and an economical choice for the bridge superstructure. However, important advances were made in wood fastening systems and preservative treatments, which would allow for future developments for timber bridges. Then, in the mid-20th century, glued-laminated timber (or glulams) was introduced as a viable structural material for bridges. The use of glulams grew to become the primary material for timber bridges and has continued to grow in popularity. Today, there is a renewed interest in all types of timber bridges. Approximately 8% (37,000) of the bridges listed in the National Bridge Inventory in the United States having spans greater than 6.10 m are constructed entirely of wood and 11% (51,000) use wood as one of the primary structural materials [9]. The future use of timber as a bridge material will not be restricted just to new construction. Owing to its high strength-to-weight ratio, timber is an ideal material for bridge rehabilitation of existing timber, steel, and concrete bridges.

20.2 Properties of Wood and Wood Products

It is important to understand the basic structure of wood in order to avoid many of the pitfalls relative to the misuse and/or misapplication of the material. Wood is a natural, cellular, anisotropic,

TABLE 20.1 Moisture Content (%) of Wood in Equilibrium with Temperature and Relative Humidity

Temp. (°C)	Relative Humidity (%) 5	10	15	20	25	30	35	40	45	50	55	60	65	70	75	80	85	90	95	98
0	1.4	2.6	3.7	4.6	5.5	6.3	7.1	7.9	8.7	9.5	10.4	11.3	12.4	13.5	14.9	16.5	18.5	21.0	24.3	26.9
5	1.4	2.6	3.7	4.6	5.5	6.3	7.1	7.9	8.7	9.5	10.4	11.3	12.3	13.5	14.9	16.5	18.5	21.0	24.3	26.9
10	1.4	2.6	3.6	4.6	5.5	6.3	7.1	7.9	8.7	9.5	10.3	11.2	12.3	13.4	14.8	16.4	18.4	20.9	24.3	26.9
15	1.3	2.5	3.6	4.6	5.4	6.2	7.0	7.8	8.6	9.4	10.2	11.1	12.1	13.3	14.6	16.2	18.2	20.7	24.1	26.8
20	1.3	2.5	3.5	4.5	5.4	6.2	6.9	7.7	8.5	9.2	10.1	11.0	12.0	13.1	14.4	16.0	17.9	20.5	23.9	26.6
25	1.3	2.4	3.5	4.4	5.3	6.1	6.8	7.6	8.3	9.1	9.9	10.8	11.7	12.9	14.2	15.7	17.7	20.2	23.6	26.3
30	1.2	2.3	3.4	4.3	5.1	5.9	6.7	7.4	8.1	8.9	9.7	10.5	11.5	12.6	13.9	15.4	17.3	20.8	23.3	26.0
35	1.2	2.3	3.3	4.2	5.0	5.8	6.5	7.2	7.9	8.7	9.5	10.3	11.2	12.3	13.6	15.1	17.0	20.5	22.9	25.6
40	1.1	2.2	3.2	4.1	5.0	5.7	6.4	7.1	7.8	8.6	9.3	10.1	11.1	12.2	13.4	14.9	16.8	20.3	22.7	25.4
45	1.1	2.2	3.2	4.0	4.9	5.6	6.3	7.0	7.7	8.4	9.2	10.0	11.0	12.0	13.2	14.7	16.6	20.1	22.4	25.2
50	1.1	2.1	3.0	3.9	4.7	5.4	6.1	6.8	7.5	8.2	8.9	9.7	10.6	11.7	12.9	14.4	16.2	18.6	22.0	24.7
55	1.0	2.0	2.9	3.7	4.5	5.2	5.9	6.6	7.2	7.9	8.7	9.4	10.3	11.3	12.5	14.0	15.8	18.2	21.5	24.2

Adapted from USDA, 1987 [10].

hygrothermal, and viscoelastic material, and by its natural origins contains a multitude of inclusions and other defects.* The reader is referred to basic texts that present a description of the fundamental structure and physical properties of wood as a material [e.g., Refs. 5, 6, 10].

20.2.1 Physical Properties of Wood

One physical aspect of wood that deserves attention here is the effect of moisture on the physical and mechanical properties and performance of wood. Many problems encountered with wood structures, especially bridges, can be traced to moisture. The amount of moisture present in wood is described by the moisture content (MC), which is defined by the weight of the water contained in the wood as a percentage of the weight of the oven-dry wood. As wood is dried, water is first evaporated from the cell cavities, then, as drying continues, water from the cell walls is drawn out. The point at which *free* water in the cell cavities is completely evaporated, but the cell walls are still saturated, is termed the *fiber saturation point* (FSP). The FSP is quite variable among and within species, but is on the order of 24% to 34%. The FSP is an important quantity since most physical and mechanical properties are dependent on changes in MC below the FSP, and the MC of wood in typical structural applications is below the FSP. Finally, wood releases and absorbs moisture to and from the surrounding environment. When the wood equilibrates with the environment and moisture is not transferring to or from the material, the wood is said to have reached its equilibrium moisture content (EMC). Table 20.1 provides the average EMC as a function of dry-bulb temperature and relative humidity. The *Wood Handbook* [10] provides other tables that are specific for given species or species groups and allow designers better estimates of in-service moisture contents that are required for their design calculations.

Wood shrinks and swells as its MC changes below the FSP; above the FSP, shrinkage and swelling can be neglected. Wood machined to a specified size at an MC higher than that expected in service will therefore shrink to a smaller size in use. Conversely, if the wood is machined at an MC lower than that expected in service, it will swell. Either way, shrinkage and swelling due to changes in MC must be taken into account in design. In general, the shrinkage along the grain is significantly less than that across the grain. For example, as a rule of thumb, a 1% dimensional

*The term *defect* may be misleading. Knots, grain characteristics (e.g., slope of grain, spiral grain, etc.), and other naturally occurring irregularities do reduce the effective strength of the member, but are accounted for in the grading process and in the assignment of design values. On the other hand, splits, checks, dimensional warping, etc. are the result of the drying process and, although they are accounted for in the grading process, they may occur after grading and may be more accurately termed *defects*.

change across the grain can be assumed for each 4% change in MC, whereas a 0.02% dimensional change in the longitudinal direction may be assumed for each 4% change in MC. More-accurate estimates of dimensional changes can be made using published values of shrinkage coefficients for various species, [10].

In addition to simple linear dimensional changes in wood, drying of wood can cause warp of various types. Bow (distortion in the weak direction), crook (distortion in the strong direction), twist (rotational distortion), and cup (cross-sectional distortion similar to bow) are common forms of warp and, when excessive, can adversely affect the structural use of the member. Finally, drying stresses (internal stress resulting from differential shrinkage) can be quite significant and can lead to checking (cracks formed along the growth rings) and splitting (cracks formed across the growth rings).

20.2.2 Mechanical Properties of Wood

The mechanical properties of wood also are functions of the MC. Above the FSP, most properties are invariant with changes in MC, but most properties are highly affected by changes in the MC below the FPS. For example, the modulus of rupture of wood increases by nearly 4% for a 1% decrease in moisture content below the FSP. The following equation is a general expression for relating any mechanical property to MC:

$$P_{\mathrm{MC}} = P_{12}\left(\frac{P_{12}}{P_g}\right)^{(12-\mathrm{MC})/(\mathrm{FSP}-\mathrm{MC})} \tag{20.1}$$

where P_{MC} = property of interest at any MC below the FSP, P_{12} = the property at 12% MC, and P_g = property in the green condition (at FSP).

For structural design purposes, using an equation such as (20.1) would be cumbersome. Therefore, design values are typically provided for a specific maximum MC (e.g., 19%) and adjustments are made for "wet use."

Load history can also have a significant effect on the mechanical performance of wood members. The load that causes failure is a function of the rate and duration of the load applied to the member. That is, a member can resist higher magnitude loads for shorter durations or, stated differently, the longer a load is applied, the less able is a wood member to resist that load. This response is termed *load duration effects* in wood design. Figure 20.1 illustrates this effect by plotting the time-to-failure as a function of the applied stress expressed in terms of the short-term (static) ultimate strength. There are many theoretical models proposed to represent this response, but the line shown in Figure 20.1 was developed at the U.S. Forest Products Laboratory in the early 1950s [11] and is the basis for current design "load duration" adjustment factors.

The design factors derived from the relationship illustrated in Figure 20.1 are appropriate only for stresses and not for stiffness or, more precisely, the modulus of elasticity. Related to load duration effects, the deflection of a wood member under sustained load increases over time. This response, termed *creep effects,* must be considered in design when deformation or deflections are critical from either a safety or serviceability standpoint. The main parameters that significantly affect the creep response of wood are stress level, moisture content, and temperature. In broad terms, a 50% increase in deflection after a year or two is expected in most situations, but can easily be upward of 200% given certain conditions [7]. In fact, if a member is subjected to continuous moisture cycling, a 100 to 150% increase in deflection could occur in a matter of a few weeks. Unfortunately, the creep response of wood, especially considering the effects of moisture cycling, is poorly understood and little guidance is available to the designer.

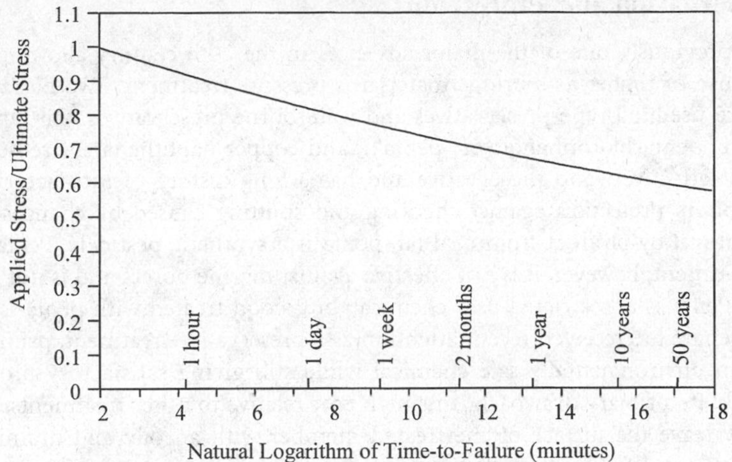

FIGURE 20.1 Load Duration behavior of wood.

Wood, being a fibrous material, is naturally resistant to fatigue effects, particularly when stressed along the grain. However, the fatigue strength of wood is negatively affected by the natural presence of inclusions and other defects. Knots and slope of grain in particular reduce fatigue resistance. Regardless of this, wood performs well in comparison with structural steel and concrete. In fact, the fatigue strength of wood has been shown to be approximately double that of most metals when evaluated at comparable stress levels relative to the ultimate strength of the material [10]. The potential for fatigue-induced failure is considered to be rather low for wood, and thus fatigue is typically not considered in timber bridge design.

20.2.3 Wood and Wood-Based Materials for Bridge Construction

The natural form of timber is the log. In fact, many primitive and "rustic" timber bridges are nothing more than one or more logs tied together. For construction purposes, however, it is simpler to use rectangular elements in bridges and other structures rather than round logs. Solid sawn lumber is cut from logs and was the mainstay of timber bridge construction for years. Solid sawn lumber comes in a variety of sizes including boards (less than 38 mm thick and 38 to 387 mm wide), dimension lumber (38 to 89 mm thick and 38 to 387 mm wide), and timbers (anything greater than 89 by 89 mm). Based on size and species, solid sawn lumber is graded by various means, including visual grading, machine-evaluated lumber (MEL), and machine stress rated (MSR), and engineering design values are assigned.

In the mid-1900s glulam timber began to receive significant use in bridges. Glulams are simply large sections formed by laminating dimension lumber together. Sections as large as 1.5 m deep are feasible with glulams. Today, while solid sawn lumber is still used extensively, the changing resource base and shift to plantation-grown trees has limited the size and quality of the raw material. Therefore, it is becoming increasingly difficult to obtain high-quality, large-dimension timbers for construction. This change in raw material, along with a demand for stronger and more cost-effective material, initiated the development of alternative products that can replace solid lumber such as glulams.

Other engineered products such as wood composite I-joists and structural composite lumber (SCL) also resulted from this evolution. SCL includes such products as laminated veneer lumber (LVL) and parallel strand lumber (PSL). These products have steadily gained popularity and now are receiving widespread use in building construction, and they are beginning to find their way into bridge construction as well. The future may see expanded use of these and other engineered wood composites.

20.2.4 Preservation and Protection

As mentioned previously, one of the major advances in the 20th century allowing for continued and expanded use of timber as a bridge material is pressure treatment. Two basic types of wood preservatives are used: oil-type preservatives and waterborne preservatives. Oil-type preservatives include creosote, pentachlorophenol (or "penta"), and copper naphthenate. Creosote can be considered the first effective wood preservative and has a long history of satisfactory performance. Creosote also offers protection against checking and splitting caused by changes in MC. While creosote is a natural by-product from coal tar, penta is a synthetic pesticide. Penta is an effective preservative treatment; however, it is not effective against marine borers and is not used in marine environments. Penta is a "restricted-use" chemical, but wood treated with penta is not restricted. Copper naphthenate has received recent attention as a preservative treatment, primarily because it is considered an environmentally safe chemical while still giving satisfactory protection against biological attack. Its primary drawback is its high cost relative to other treatments. All these treatments generally leave the surface of the treated member with an oily and unfinishable surface. Furthermore, the member may "bleed" or leach preservative unless appropriate measures are taken.

Most timber bridge applications utilize oil-type preservatives for structural elements such as beams, decks, piles, etc. They offer excellent protection against decay and biological attack, are noncorrosive, and are relatively durable. Oil-type preservatives are not, however, recommended for bridge elements that may have frequent or repeated contact by humans or animals since they can cause skin irritations.

Waterborne preservatives have the advantage of leaving the surface of the treated material clean and, after drying, able to be painted or stained. They also do not cause skin irritations and, therefore, can be used where repeated human and/or animal contact is expected. Waterborne preservatives use formulations of inorganic arsenic compounds in a water solution. They do, however, leave the material with a light green, gray, or brownish color. But again, the surface can be later painted or stained. A wide variety of waterborne preservatives are available, but the most common include chromated copper arsenate (CCA), ammoniacal copper arsenate (ACA), and ammoniacal copper zinc arsenate (ACZA). Leaching of these chemicals is not a problem with these formulations since they each are strongly bound to the wood. CCA is commonly used to treat southern pine, ponderosa pine, and red pine, all of which are relatively accepting of treatment. ACA and ACZA are used with species that are more difficult to treat, such as Douglas fir and larch. One potential drawback to CCA and ACA is a tendency to be corrosive to galvanized hardware. The extent to which this is a problem is a function of the wood species, the specific preservative formulation, and service conditions. However, such corrosion seems not to be an issue for hot-dipped galvanized hardware typical in bridge applications.

Waterborne preservatives are used for timber bridges in applications where repeated or frequent contact with humans or animals is expected. Such examples include handrails and decks for pedestrian bridges. Additionally, waterborne preservatives are often used in marine applications where marine borer hazards are high.

Any time a material is altered due to chemical treatment its microlevel structure may be affected, thus affecting its mechanical properties. Oil-type preservatives do not react with the cellular structure of the wood and, therefore, have little to no effect on the mechanical properties of the material. Waterborne preservatives do react, however, with the cell material, thus they can affect properties. Although this is an area of ongoing research, indications are that the only apparent effect of waterborne preservatives is to increase load duration effects, especially when heavy treatment is used for saltwater applications. Currently, no adjustments are recommended for design values of preservative treated wood vs. untreated materials.

In addition to preservative treatment, fire-retardant chemical treatment is also possible to inhibit combustion of the material. These chemicals react with the cellular structure in wood and can cause significant reductions in the mechanical properties of the material, including strength. Generally, fire retardants are not used in bridge applications. However, if fire-retardant-treated material is used, the designer should consult with the material producer or treater to obtain appropriate design values.

20.3 Types of Timber Bridges

Timber bridges come in a variety of forms, many having evolved from tradition. Most timber bridges designed today, however, are the results of fairly recent developments and advances in the processing and treating of structural wood. The typical timber bridge is a single- or two-span structure. Single-span timber bridges are typically constructed with beams and a transverse deck or a slab-type longitudinal deck. Two-span timber bridges are often beam with transverse decks. These and other common timber bridge types are presented in this section.

20.3.1 Superstructures

As with any bridge, the structural makeup can be divided into three basic components: the superstructure, the deck, and the substructure. Timber bridge superstructures can be further classified into six basic types: beam superstructures, longitudinal deck (or slab) superstructures, trussed superstructures, trestles, suspension bridges, and glulam arches.

Beam Superstructures

The most basic form of a timber beam bridge is a log bridge. It is simply a bridge wherein logs are laid alternately tip-to-butt and bound together. A transverse deck is then laid over the log beams. Obviously, spans of this type of bridge are limited to the size of logs available, but spans of 6 to 18 m are reasonable. The service life of a log bridge is typically 10 to 20 years.

The sawn lumber beam bridge is another simple form. Typically, made of closely spaced 100 to 200-mm wide by 300 to 450-mm-deep beams, sawn lumber beams are usually used for clear spans up to 9 m. With the appropriate use of preservative treatments, sawn lumber bridges have average service lives of approximately 40 years. A new alternative to sawn lumber is structural composite lumber (SCL) bridges. Primarily, laminated veneer lumber (LVL) has been used in replacement of solid sawn lumber in bridges. LVL can be effectively treated and can offer long service as well.

Glulam timber beam bridges are perhaps the most prevalent forms of timber bridges today. A typical glulam bridge configuration is illustrated in Figure 20.2. This popularity is primarily due to the large variety of member sizes offered by glulams. Commonly used for clear spans ranging from 6 to 24 m, glulam beam bridges have been used for clear spans up to 45 m. Transportation restrictions rather than material limitations limit the length of beams, and, therefore, bridges. Since glulam timber can be satisfactorily treated with preservatives, they offer a durable and long-lasting structural element. When designed such that field cutting, drilling, and boring are avoided, glulam bridges can provide a service life of at least 50 years.

Longitudinal Deck Superstructures

Longitudinal deck (or slab) superstructures are typically either glulam or nail-laminated timber placed longitudinally to span between supports. A relatively new concept in longitudinal deck systems is the stress-laminated timber bridge, which is similar to the previous two forms except that continuity in the system is developed through the use of high-strength steel tension rods. In any case, the wide faces of the laminations are oriented vertically rather than horizontally as in a typical glulam beam. Figure 20.3 illustrates two types of glulam longitudinal decks: noninterconnected and interconnected. Since glulam timbers have depths typically less than the width of a bridge, two or more segments must be used. When continuity is needed, shear dowels must be used to provide interconnection between slabs. When continuity is not required, construction is simplified. Figure 20.4 illustrates a typical stress-laminated section.

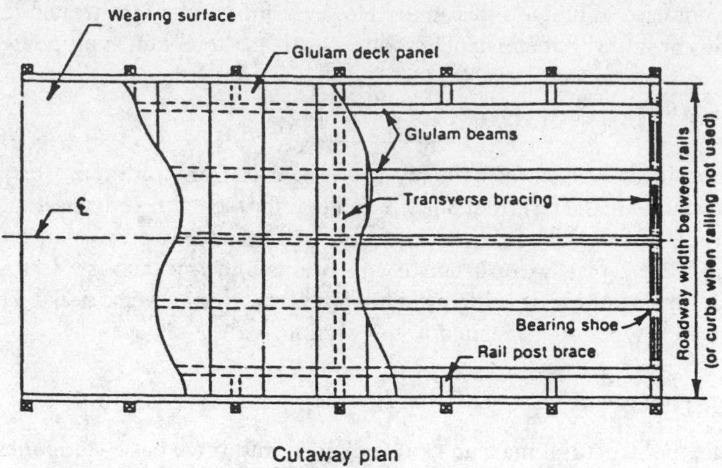

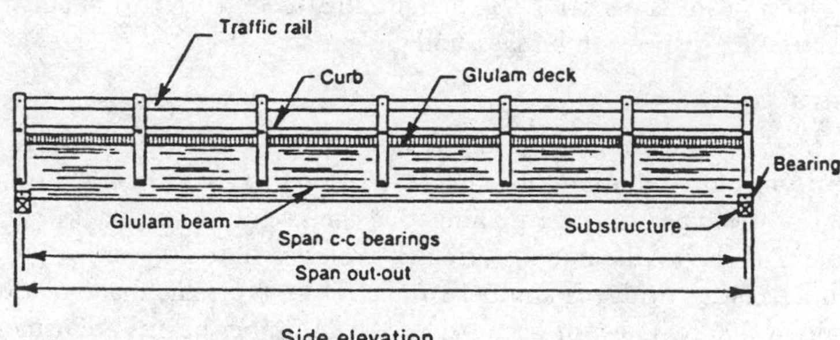

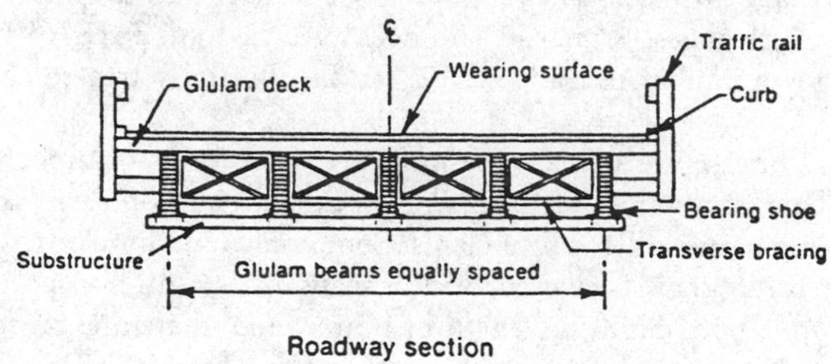

FIGURE 20.2 Glulam beam bridge with transverse deck. (*Source*: Ritter, M.A., EM7700-8, USDA Forest Service, Washington, D.C., 1990.)

Longitudinal deck systems are relatively simple and offer a relatively low profile, making them an excellent choice when vertical clearance is a consideration. Longitudinal decks are economical choices for clear spans up to approximately 10 m. Since the material can be effectively treated, the average service life of a longitudinal timber deck superstructure is at least 50 years. However, proper maintenance is required to assure an adequate level of prestress is maintained in stress-laminated systems.

Trussed Superstructures

Timber trusses were used extensively for bridges in the first half of the 20th century. Many different truss configurations were used including king post, multiple king posts, Pratt, Howe, lattice, long, and bowstring trusses, to name a few. Clear spans of up to 75 m were possible. However, their

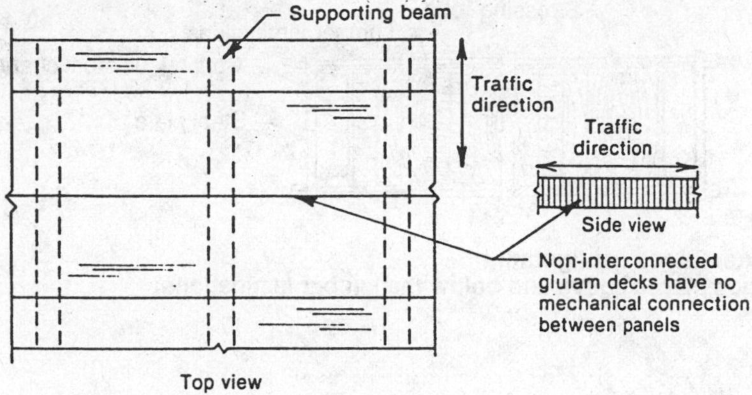

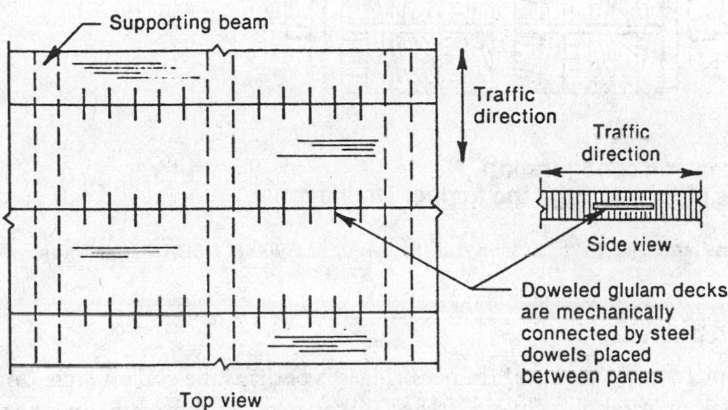

FIGURE 20.3 Glulam longitudinal decks. (*Source*: Ritter, M.A., EM7700-8, USDA Forest Service, Washington, D.C., 1990.)

use has declined due primarily to high fabrication, erection, and maintenance costs. When timber trusses are used today, it is typically driven more by aesthetics than by structural performance or economics.

Trestles

Another form of timber bridge which saw its peak usage in the first half of the 20th century was the trestle. A trestle is a series of short-span timber superstructures supported on a series of closely spaced timber bents. During the railroad expansion during the early to mid 1900s, timber trestles were a popular choice. However, their use has all but ceased because of high fabrication, erection, and maintenance costs.

Suspension Bridges

A timber suspension bridge is simply a timber deck structure supported by steel cables. Timber towers, in turn, support the steel suspension cables. Although there are examples of vehicular timber suspension bridges, the more common use of this form of timber bridge is as a pedestrian bridge. They are typically used for relatively long clear spans, upward of 150 m. Since treated wood can be used throughout, 50-year service lives are expected.

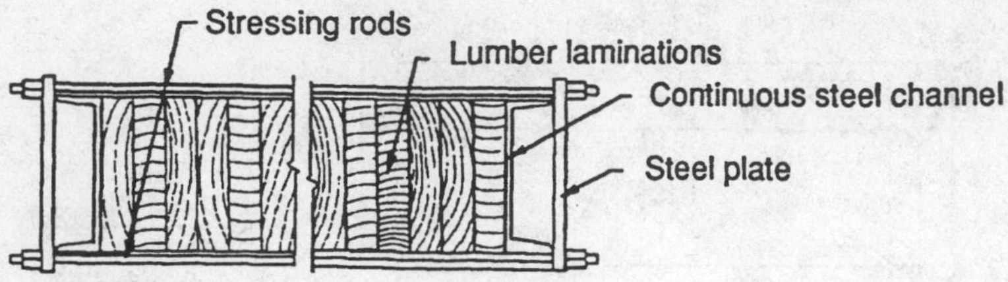

External rod configuration
(rods placed above and below the lumber laminations)

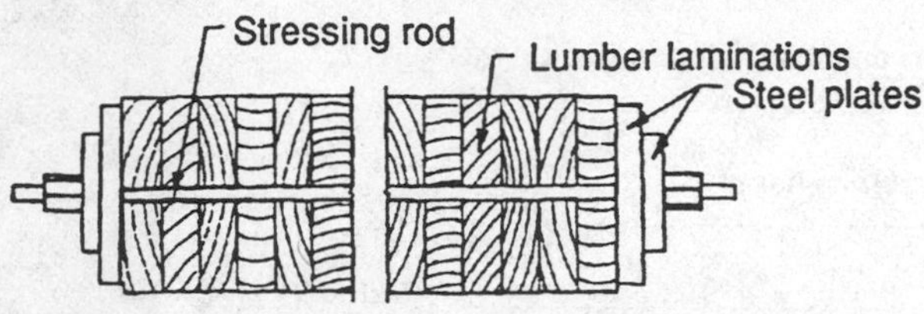

Internal rod configuration
(rods placed through the lumber laminations)

FIGURE 20.4 Stress laminated bridge. (*Source*: Ritter, M.A., EM7700-8, USDA Forest Service, Washington, D.C., 1990.)

Glued Laminated Arches

One of the most picturesque forms of timber bridges is perhaps the glulam arch. Constructed from segmented circular or parabolic glulam arches, either two- or three-hinge arches are used. The glulam arch bridge can have clear spans in excess of 60 m, and since glulam timber can be effectively treated, service lives of at least 50 years are well within reason. Although the relative first and life-cycle costs of arch bridges have become high, they are still a popular choice when aesthetics is an issue.

20.3.2 Timber Decks

The deck serves two primary purposes: (1) it is the part of the bridge structure that forms the roadway, and (2) it distributes the vehicular loads to the supporting elements of the superstructure. Four basic types of timber decks are sawn lumber planks, nailed laminated decks, glulam decks, and composite timber–concrete decks. The selection of a deck type depends mainly on the level of load demand.

Lumber Planks

The lumber plank deck is perhaps the simplest deck type. It is basically sawn lumber, typically 75 to 150 mm thick and 250 to 300 mm wide, placed flatwise and attached to the supporting beams with large spikes. Generally, the planks are laid transverse to the beams and traffic flow, but can be placed longitudinally on cross beams as well. Lumber planks are only used for low-volume bridges. They are also of little use when protection of the supporting members is desired since water freely travels between adjacent planks. Additionally, when a wearing surface such as asphalt is desired, lumber planks are not recommended since deflections between adjacent planks will result in cracking and deterioration of the wearing surface.

Nailed Laminated and Glulam Decks

Nailed laminated and glulam decks are essentially as described previously for longitudinal deck (or slab) superstructures. Nailed laminated systems are typically 38-mm-thick by 89- to 285-mm-deep lumber placed side by side and nailed or spiked together along its length. The entire deck is nailed together to act as a composite section and oriented such that the lumber is laid transverse to the bridge span across the main supporting beams, which are spaced from 0.6 to 1.8 m. Once a quite popular deck system, its use has declined considerably in favor of glulam decks.

A glulam deck is a series of laminated panels, typically 130- to 220-mm thick by 0.9 to 1.5 m wide. The laminations of the glulam panel are oriented with their wide face vertically. Glulam decks can be used with the panels in the transverse or longitudinal direction. They tend to be stronger and stiffer than nailed laminated systems and offer greater protection from moisture to the supporting members. Finally, although doweled glulam panels (see Figure 20.3) cost more to fabricate, they offer the greatest amount of continuity. With this continuity, thinner decks can be used, and improved performance of the wearing surface is achieved due to reduced cracking and deterioration.

Composite Timber–Concrete Decks

The two basic types of composite timber–concrete deck systems are the T-section and the slab (see Figure 20.5). The T-section is simply a timber stem, typically a glulam, with a concrete flange that also serves as the bridge deck. Shear dowels are plates that are driven into the top of the timber stem and develop the needed shear transfer. For a conventional single-span bridge, the concrete is proportioned such that it takes all the compression force while the timber resists the tension. Composite T-sections have seen some use in recent years; however, high fabrication costs have limited their use.

Composite timber–concrete slabs were used considerably during the second quarter of the 20th century, but receive little use today. They are constructed with alternating depths of lumber typically nailed laminated with a concrete slab poured directly on top of the timber slab. With a simple single span, the concrete again carries the compressive flexural stresses while the timber carries the flexural stresses. Shear dowels or plates are driven into the timber slab to provide the required shear transfer between the concrete and the timber.

20.3.3 Substructures

The substructure supports the bridge superstructure. Loads transferred from the superstructures to the substructures are, in turn, transmitted to the supporting soil or rock. Specific types of substructures that can be used are dependent on a number of variables, including bridge loads, soil and site conditions, etc. Although a timber bridge superstructure can be adapted to virtually any type of substructure regardless of material, the following presentation is focused on timber substructures, specifically timber abutments and bents.

Abutments

Abutments serve the dual purpose of supporting the bridge superstructure and the embankment. The simplest form of a timber abutment is a log, sawn lumber, or glulam placed directly on the embankment as a spread footing. However, this form is not satisfactory for any structurally demanding situation. A more common timber abutment is the timber pile abutment. Timber piles are driven to provide the proper level of load-carrying capacity through either end bearing or friction. A backwall and wing walls are commonly added using solid sawn lumber to retain the embankment. A continuous cap beam is connected to the top of the piles on which the bridge superstructure is supported. A timber post abutment can be considered a hybrid between the spread footing and pile abutment. Timber posts are supported by a spread footing, and a backwall and wing walls are added to retain the embankment. Pile abutments are required when soil conditions do not provide adequate support for a spread footing or when uplift is a design concern.

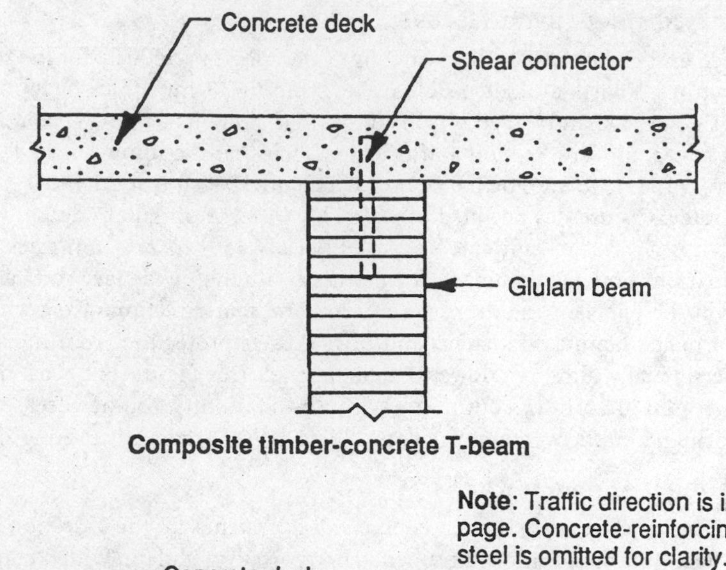

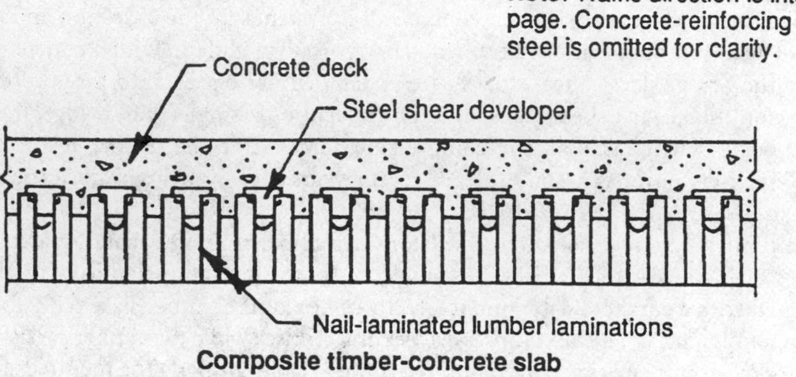

FIGURE 20.5 Composite timber–concrete decks. (*Source*: Ritter, M.A., EM7700-8, USDA Forest Service, Washington, D.C., 1990.)

Bents

Bents are support systems used for multispan bridges between the abutments. Essentially, timber bents are formed from a set of timber piles with lumber cross bracing. However, when the height of the bent exceeds that available for a pile, frame bents are used. Frame bents were quite common in the early days of the railroad, but, due to high cost of fabrication and maintenance, they are not used often for new bridges.

20.4 Basic Design Concepts

In this section, the basic design considerations and concepts for timber bridges are presented. The discussion should be considered an overview of the design process for timber bridges, not a replacement for specifications or standards.

20.4.1 Specifications and Standards

The design of timber bridge systems has evolved over time from what was tradition and essentially a "master-builder" approach. Design manuals and specifications are available for use by engineers involved with or interested in timber bridge design. These include *Timber Bridges: Design, Construction, Inspection, and Maintenance* [8], AASHTO *LRFD Bridge Design Specifications* [1], and AASHTO

Standard Specifications for Highway Bridges [2]. The wood industry, through the American Forest and Paper Association (AF&PA), published design values for solid sawn lumber and glulam timber for both allowable stress design [4] and load and resistance factor design [3] formats. Rather than presenting those aspects of bridge design common to all bridge types, the focus of the following presentation will be on those aspects specific to timber bridge design. Since bridge design is often governed by AASHTO, focus will be on AASHTO specifications. However, AF&PA is the association overseeing the engineering design of wood, much like ACI is for concrete and AISC is for steel, and AF&PA-recommended design procedures will also be presented.

20.4.2 Design Values

Design values for wood are provided in a number of sources, including AF&PA specifications and AASHTO specifications. Although the design values published by these sources are based on the same procedures per ASTM standards, specific values differ due to assumptions made for end-use conditions. The designer must take care to use the appropriate design values with their intended design specification(s). For example, the design should not use AF&PA design values directly in AASHTO design procedures since AF&PA and AASHTO make different end use assumptions.

AF&PA "Reference" Design Values

The AF&PA *Manual for Engineered Wood Construction: Load and Resistance Factor Design* [3] provides nominal design values for visually and mechanically graded lumber, glulam timber, and connections. These values include reference bending strength, F_b; reference tensile strength parallel to the grain, F_t; reference shear strength parallel to the grain, F_v; reference compressive strength parallel and perpendicular to the grain, F_c and $F_{c\perp}$, respectively; reference bearing strength parallel to the grain, F_g; and reference modulus of elasticity, E. These are appropriate for use with the LRFD provisions.

Similarly, the Supplement to the NDS® provides tables of design values for visually graded and machine stress rated lumber, and glulam timber for use in allowable stress design (ASD). The basic quantities are the same as with the LRFD, but are in the form of allowable stresses and are appropriate for use with the ASD provisions of the NDS· Additionally, the NDS provides tabulated allowable design values for many types of mechanical connections.

One main difference between the ASD and LRFD design values, other than the ASD prescribing allowable stresses and the LRFD prescribing nominal strengths, is the treatment of duration of load effects. Allowable stresses (except compression perpendicular to the grain) are tabulated in the NDS and elsewhere for an assumed 10-year load duration in recognition of the duration of load effect discussed previously. The allowable compressive stress perpendicular to the grain is not adjusted since a deformation definition of failure is used for this mode rather than fracture as in all other modes; thus the adjustment has been assumed unnecessary. Similarly, the modulus of elasticity is not adjusted to a 10-year duration since the adjustment is defined for strength, not stiffness. For the LRFD, short-term (i.e., 20 min) nominal strengths are tabulated for all strength values. In the LRFD, design strengths are reduced for longer-duration design loads based on the load combination being considered. Conversely, in the NDS, allowable stresses are increased for shorter load durations and decreased only for permanent (i.e., greater than 10 year) loading.

AASHTO-LRFD "Base" Design Values

AASHTO-LRFD publishes its own design values which are different from those of the AF&PA LRFD. AASHTO publishes base bending strength, F_{b0}; base tensile strength parallel to the grain, F_{t0}; base shear strength parallel to the grain, F_{v0}; base compressive strength parallel and perpendicular to the grain, F_{c0} and $F_{c\perp}$, respectively; and base modulus of elasticity, E_0. While the NDS publishes design values based on an assumed 10-year load duration and the AF&PA LRFD assumes a short-term (20-min) load duration, AASHTO publishes design values based on an assumed 2-month duration.

TABLE 20.2 Factors to Convert NDS-ASD Values to AASTHO-LRFD Values

Material	Property					
	F_b	F_v	F_c	$F_{c\perp}$	F_t	E
Dimension lumber	2.35	3.05	1.90	1.75	2.95	0.90
Beams and stringers, posts and timbers	2.80	3.15	2.40	1.75	2.95	1.00
Glulam	2.20	2.75	1.90	1.35	2.35	0.83

Unfortunately, the AASHTO published design values are not as comprehensive (with respect to species, grades, sizes, as well as specific properties) as thsoe of AF&PA. The AASHTO-LRFD does, however, provide for adjustments from AF&PA-published reference design values so they can be used in AASHTO specifications. For design values not provided in the AASHTO-LRFD, conversion factors are provided from NDS allowable stresses to AASHTO-LRFD base strengths. Table 20.2 provides these adjustments for solid sawn and glulam timbers. The designer is cautioned that these conversion factors are from the NDS allowable stresses, *not* the AF&PA-LRFD strength values.

20.4.3 Adjustment of Design Values

In addition to the providing *reference* or *base* design values, the AF&PA-LRFD, the NDS, and the AASHTO-LRFD specifications provide adjustment factors to determine final *adjusted* design values. Factors to be considered include load duration (termed *time effect* in the LRFD), wet service, temperature, stability, size, volume, repetitive use, curvature, orientation (form), and bearing area. Each of these factors will be discussed further; however, it is important to note that not all factors are applicable to all design values, nor are all factors included in all the design specifications. The designer must take care to apply the appropriate factors properly.

AF&PA Adjustment Factors

LRFD reference strengths and ASD allowable stresses are based on the following specified reference conditions: (1) dry use in which the maximum EMC does not exceed 19% for solid wood and 16% for glued wood products; (2) continuous temperatures up to 32°C, occasional temperatures up to 65°C (or briefly exceeding 93°C for structural-use panels); (3) untreated (except for poles and piles); (4) new material, not reused or recycled material; and (5) single members without load sharing or composite action. To adjust the reference design value for other conditions, adjustment factors are provided which are applied to the published reference design value:

$$R' = R \cdot C_1 \cdot C_2 \cdots C_n \tag{20.2}$$

where R' = adjusted design value (resistance), R = reference design value, and C_1, C_2, ... C_n = applicable adjustment factors. Adjustment factors, for the most part, are common between LRFD and ASD. Many factors are functions of the type, grade, and/or species of material while other factors are common to all species and grades. For solid sawn lumber, glulam timber, piles, and connections, adjustment factors are provided in the AF&PA LRFD manual and the NDS. For both LRFD and ASD, numerous factors need to be considered, including wet service, temperature, preservative treatment, fire-retardant treatment, composite action, load sharing (repetitive use), size, beam stability, column stability, bearing area, form (i.e., shape), time effect (load duration), etc. Many of these factors will be discussed as they pertain to specific designs; however, some of the factors are unique for specific applications and will not be discussed further. The four factors that are applied to all design properties are the wet service factor, C_M; temperature factor, C_t; preservative treatment factor, C_{pt}; and fire-retardant treatment factor, C_{rt}. Individual treaters provide the two treatment factors, but the wet service and temperature factors are provided in the AF&PA LRFD Manual. For example, when considering the design of solid sawn lumber members, the adjustment

TABLE 20.3 AF&PA LRFD Wet Service Adjustment Factors, C_M

	Size Adjusted[a] F_b			Size Adjusted[a] F_c				
Thickness	≤20 MPa	>20 MPa	F_t	≤12.4 MPa	>12.4 MPa	F_v	$F_{c\perp}$	E, E_{05}
≤90 mm	1.00	0.85	1.00	1.00	0.80	0.97	0.67	0.90
>90 mm	1.00	1.00	1.00	0.91	0.91	1.00	0.67	1.00

[a] Reference value adjusted for size only.

TABLE 20.4 AF&PA-LRFD Temperature Adjustment Factors, C_t

	Dry Use		Wet Use	
Sustained Temperature (°C)	E, E_{05}	All Other Prop.	E, E_{05}	All Other Prop.
32 < T ≤ 48	0.9	0.8	0.9	0.7
48 < T ≤ 65	0.9	0.7	0.9	0.5

values given in Table 20.3 for wet service, which is defined as the maximum EMC exceeding 19%, and Table 20.4 for temperature, which is applicable when continuous temperatures exceed 32°C, are applicable to all design values. Often with bridges, since they are essentially exposed structures, the MC will be expected to exceed 19%. Similarly, temperature may be a concern, but not as commonly as MC.

Since, as discussed, LRFD and ASD handle time (duration of load) effects so differently and since duration of load effects are somewhat unique to wood design, it is appropriate to elaborate on it here. Whether using ASD or LRFD, a wood structure is designed to resist all appropriate load combinations – unfactored combinations for ASD and factored combinations for LRFD. The time effects (LRFD) and load duration (ASD) factors are meant to recognize the fact that the failure of wood is governed by a creep–rupture mechanism; that is, a wood member may fail at a load less than its short-term strength if that load is held for an extended period of time. In the LRFD, the time effect factor, λ, is based on the load combination being considered. In ASD, the load duration factor, C_D, is given in terms of the assumed cumulative duration of the design load.

AASHTO-LRFD Adjustment Factors

AASHTO-LRFD base design values are based on the following specified reference conditions: (1) wet use in which the maximum EMC exceeds 19% for solid wood and 16% for glued wood products (this is opposite from the dry use assumed by AF&PA, since typical bridge use implies wet use); (2) continuous temperatures up to 32°C, occasional temperatures up to 65°C; (3) untreated (except for poles and piles); (4) new material, not reused or recycled material; and (5) single members without load sharing or composite action. AASHTO has fewer adjustments available for the designer to consider, primarily but not entirely due to the specific application. To adjust the base design value for other conditions, AASHTO-LRFD provides the following adjustment equation:

$$F = F_0 \cdot C_F \cdot C_M \cdot C_D \tag{20.3}$$

where F = adjusted design value (resistance), F_0 = base design value, C_F = size adjustment factor, C_M = moisture content adjustment factor, C_D = deck adjustment factor, and C_S = stability adjustment factor.

The size factor is applicable only to bending and is essentially the same as that used by AF&PA for solid sawn lumber and the same as the volume effect factor used by AF&PA for glulam timber. For solid sawn lumber and vertically laminated lumber, the size factor is defined as

$$C_F = \left(\frac{300}{d}\right)^{1/9} \leq 1.0 \tag{20.4}$$

TABLE 20.5 AASHTO-LRFD Moisture Content Adjustment Factors, C_M, for Glulam

Property					
F_b	F_v	F_c	F_{cp}	F_t	E
1.25	1.15	1.35	1.90	1.25	1.20

where d = width (mm). The equation implies if lumber less than or equal to 300 mm in width is used, no adjustment is made. If, however, a width greater than 300 mm is used, a reduction in the published base bending design value is required. For horizontally glulam timber, the "size" factor may more appropriately be termed a *volume* factor (per the AF&PA). The size factor for glulam is given as

$$C_F = \left[\left(\frac{300}{d}\right)\left(\frac{130}{b}\right)\left(\frac{6400}{L}\right)\right]^a \leq 1.0 \tag{20.5}$$

where d = width (mm), b = thickness (mm), L = span (mm), and a = 0.05 for southern pine and 0.10 for all other species glulam. As with the previous size adjustment, if the dimensions of the glulam exceed 130 by 300 by 6400, then a reduction in the bending strength is required.

Unlike the size factor, the moisture factor is applicable to all published design values, not just bending strength. The moisture adjustment factor, C_M, is again similar to that provided by AF&PA; however, it is embedded in the published base design values. Unless otherwise noted, C_M should be assumed as unity. The only exception is when glulams are used and the moisture content is expected to be less than 16%. An increase in the design values is then allowed per Table 20.5. A similar increase is not allowed for lumber used at moisture contents less than 19% per AASHTO. This is a conservative approach in comparison with that of AF&PA.

The deck adjustment factor, C_D, is again specific for the bending resistance, F_b, of 50- to 100-mm-wide lumber used in stress-laminated and mechanically (nail or spike) laminated deck systems. For stress-laminated decks, the bending strength can be increased by a factor of C_D = 1.30 for select structural grade lumber, and C_D = 1.5 for No. 1 and No. 2 grade. For mechanically laminated decks, the bending strength of all grades can be increased by a factor of C_D = 1.15.

Since, as discussed, AF&PA LRFD and ASD handle time (duration of load) effects so differently and since duration of load effects are somewhat unique to wood design, it is appropriate to elaborate on it here and understand how time effects are accounted for by AASHTO-LRFD. Implicit in the AASHTO-LRFD Specification, $\lambda = 0.8$ is assumed for vehicle live loads. The published base design values are reduced by a factor of 0.80 to account for time effects. For strength load combination IV, however, a reduction of 75% is required. This load combination is for dead load only. The rationale behind this reduction is found in the AF&PA-LRFD time effects factors. For live-load-governed load combinations, AF&PA requires $\lambda = 0.8$; and for dead load only, $\lambda = 0.6$ is used. The ratio of the dead-load time effect factor to the live-load time effect factor is 0.6/0.8 = 0.75.

20.4.4 Beam Design

The focus of the remaining discussion will be on the design provisions specified in the AASHTO-LRFD for wood members. The design of wood beams follows traditional beam theory. The flexural strength of a beam is generally the primary concern in a beam design, but consideration of other factors such as horizontal shear, bearing, and deflection are also crucial for a successful design.

Moment Capacity

In terms of moment, the AASHTO-LRFD design factored resistance, M_r, is given by

$$M_r = \phi_b M_n = \phi_b F_b S C_s \tag{20.6}$$

where ϕ_b = resistance factor for bending = 0.85, M_n = nominal adjusted moment resistance, F_b = adjusted bending strength, S = section modulus, and C_s = beam stability factor.

The beam stability factor, C_s, is only used when considering strong axis bending since a beam oriented about its weak axis is not susceptible to lateral instability. Additionally, the beam stability factor need not exceed the value of the size effects factor. The beam stability factor is taken as 1.0 for members with continuous lateral bracing; otherwise C_s is calculated from

$$C_s = \frac{1+A}{1.9} - \sqrt{\frac{(1+A)^2}{3.61} - \frac{A}{0.95}} \leq C_F \tag{20.7}$$

where

$$A = \frac{0.438\,EB^2}{L_e dF_b} \quad \text{for visually graded solid sawn lumber} \tag{20.8}$$

and

$$A = \frac{0.609\,Eb^2}{L_e dF_b} \quad \text{for mechanically graded lumber and glulams} \tag{20.9}$$

where E = modulus of elasticity, b = net thickness, d = net width, L_e = effective length, and F_b = adjusted bending strength. The effective length, L_e, accounts for both the lateral motion and torsional phenomena and is given in the AASHTO-LRFD specification for specific unbraced lengths, L_u, defined as the distance between points of lateral and rotations support. For $L_u/d < 7$, the effective unbraced length, $L_e = 2.06L_u$; for $7 \leq L_u/d \leq 14.3$, $L_e = 1.63L_u + 3d$; and for $l_u/d > 14.3$, $L_e = 1.84L_u$.

While the basic adjustment factor for beam stability is quite similar between AASHTO and AF&PA, the consideration of beam stability and size effects combined differs significantly from the approach used by AF&PA. For solid sawn lumber, AF&PA requires both the size factor and the beam stability factor apply. For glulams, AF&PA prescribes the lesser of the volume factor or the stability factor be used. AASHTO compared with AF&PA is potentially nonconservative with respect to lumber elements and conservative with respect to glulam elements.

Shear Capacity

Similar to bending, the basic design equation for the factored shear resistance, V_r, is given by

$$V_r = \phi_v V_n = \phi_v \frac{F_v bd}{1.5} \tag{20.10}$$

where ϕ_v = resistance factor for shear = 0.75, V_n = nominal adjusted shear resistance, F_v = adjusted shear strength, and b and d = thickness and width, respectively. Obviously, the last expression in Eq. (20.10) assumes a rectagular section, the nominal shear resistance could be determined from the relationship

$$V_n = \frac{F_v Ib}{Q} \tag{20.11}$$

where I = monent of inertia and Q = statical moment of an area about the neutral axis.

In timber bridges, notches are often made at the support to allow for vertical clearances and tolerances as illustrated in Figure 20.6; however, stress concentrations resulting from these notches

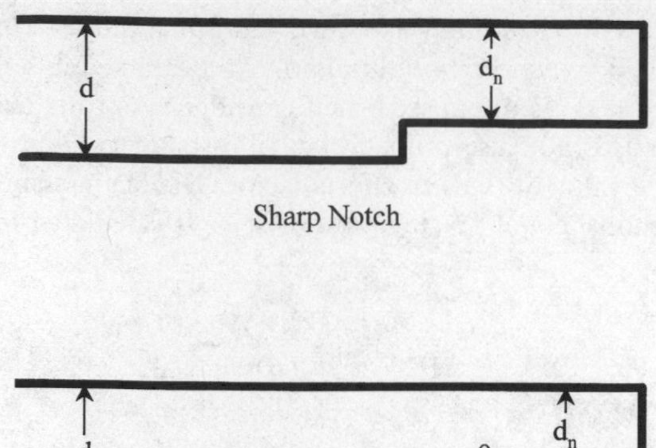

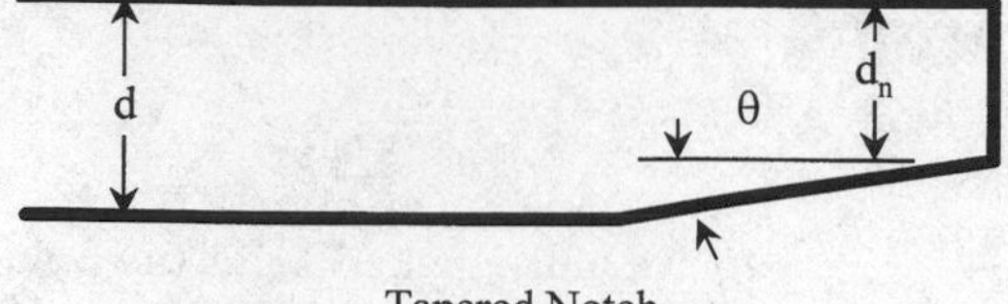

FIGURE 20.6 Notched beam: (a) sharp notch; (b) tapered notch.

significantly affect the shear resistance of the section. AASHTO-LRFD does not address this condition, but AF&PA does provide the designer with some guidance. At sections where the depth is reduced due to the presence of a notch, the shear resistance of the notched section is determined from

$$V' = \left(\frac{2}{3} F_v' b d_n\right)\left(\frac{d_n}{d}\right) \tag{20.12}$$

where d = depth of the unnotched section and d_n = depth of the member after the notch. When the notch is made such that it is actually a gradual tapered cut at an angle θ from the longitudinal axis of the beam, the stress concentrations resulting from the notch are reduced and the above equation becomes

$$V' = \left(\frac{2}{3} F_v' b d_n\right)\left(1 - \frac{(d - d_n)\sin\theta}{d}\right) \tag{20.13}$$

Similar to notches, connections too can produce significant stress concentrations resulting in reduced shear capacity. Where a connection produces at least one half the member shear force on either side of the connection, the shear resistance is determined by

$$V' = \left(\frac{2}{3} F_v' b d_e\right)\left(\frac{d_e}{d}\right) \tag{20.14}$$

where d_e = effective depth of the section at the connection which is defined as the depth of the member less the distance from the unloaded edge (or nearest unloaded edge if both edges are unloaded) to the center of the nearest fastener for dowel-type fasteners (e.g., bolts).

Bearing Capacity

The last aspect of beam design to be covered in this section is bearing at the supports. The governing design equation for factored bearing capacity perpendicular to the grain, $P_{r\perp}$, is

$$P_{r\perp} = \phi_c P_{n\perp} = \phi_c F_{c\perp} A_b C_b \tag{20.15}$$

where ϕ_c = resistance factor for compression = 0.90, P_{np} = nominal adjusted compression resistance perpendicular to the grain, F_{cp} = adjusted compression strength perpendicular to the grain, A_b = bearing area, and C_b = bearing factor.

The bearing area factor, C_b, allows an increase in the compression strength when the bearing length along the grain, l_b, is no more than 150 mm along the length of the member, is at least 75 mm from the end of the member, and is not in a region of high flexural stress. The bearing factor C_b is given by AF&PA as

$$C_b = (l_b + 9.5)/l_b \tag{20.16}$$

where l_b is in mm. This equation is the basis for the adjustment factors presented in the AASHTO-LRFD. For example, if a bearing length of 50 mm is used, the bearing strength can be increased by a factor of (50 + 9.5)/50 = 1.19.

20.4.6 Axially Loaded Members

The design of axially loaded members is quite similar to that of beams. Tension, compression, and combined axial and bending are addressed in AASHTO-LRFD.

Tension Capacity

The governing design equation for factored tension capacity parallel to the grain, P_{rt}, is

$$P_{rt} = \phi_t P_{nt} = \phi_t F_t A_n \tag{20.17}$$

where ϕ_t = resistance factor for tension = 0.80, P_{nt} = nominal adjusted tension resistance parallel to the grain, F_t = adjusted tension strength, and A_n = smallest net area of the component.

Compression Capacity

In terms of compression parallel to the grain, the AASHTO-LRFD design factored resistance, P_{rc} is given by

$$P_{rc} = \phi_c P_{nc} = \phi_c F_c A C_p \tag{20.18}$$

where ϕ_c = resistance factor for bending = 0.85, and P_{nc} = nominal adjusted compression resistance, F_c = adjusted compression strength, A = cross-sectional area, and C_p = column stability factor.

The column stability factor, C_p, accounts for the tendency of a column to buckle. The factor is taken as 1.0 for members with continuous lateral bracing; otherwise, C_p is calculated from one of the following expressions, depending on the material:

For sawn lumber:

$$C_p = \frac{1+B}{1.6} - \sqrt{\frac{(1+B)^2}{2.56} - \frac{B}{0.80}} \tag{20.19}$$

For round timber piles:

$$C_p = \frac{1+B}{1.7} - \sqrt{\frac{(1+B)^2}{2.89} - \frac{B}{0.85}} \tag{20.20}$$

For mechanically graded lumber and glued laminated timber:

$$C_p = \frac{1+B}{1.8} - \sqrt{\frac{(1+B)^2}{3.24} - \frac{B}{0.9}} \tag{20.21}$$

where

$$B = \frac{4.32Ed^2}{L_e^{\,2}F_b} \quad \text{for visually graded solid sawn lumber} \tag{20.22}$$

and

$$B = \frac{60.2Ed^2}{L_e^{\,2}F_b} \quad \text{for mechanically graded lumber and glulams} \tag{20.23}$$

where E = modulus of elasticity, d = net width (about which buckling may occur), L_e = effective length = effective length factor times the unsupported length = KL_u, and F_b = adjusted bending strength.

Combined Tension and Bending

AASHTO uses a linear interaction for tension and bending:

$$\frac{P_u}{P_{rt}} + \frac{M_u}{M_r} \leq 1.0 \tag{20.24}$$

where P_u and M_u = factored tension and moment loads on the member, respectively, and P_{rt} and M_r are the factored resistances as defined previously.

Combined Compression and Bending

AASHTO uses a slightly different interaction for compression and bending than tension and bending:

$$\left(\frac{P_u}{P_{rc}}\right)^2 + \frac{M_u}{M_r} \leq 1.0 \tag{20.25}$$

where P_u and M_u = factored compression and moment loads on the member, respectively, and P_{rc} and M_r are the factored resistances as defined previously. The squared term on the compression term was developed from experimental observations and is also used in the AF&PA LRFD. However, AF&PA includes secondary moments in the determination of M_u, which AASHTO neglects. AF&PA also includes biaxial bending in its interaction equations.

20.4.7 Connections

The final design consideration to be discussed in this section is that of connections. AASHTO-LRFD does not specifically address connections, so the designer is referred to the AF&PA LRFD. Decks must be attached to the supporting beams and beams to abutments such that vertical, longitudinal, and transverse loads are resisted. Additionally, the connections must be easily installed in the field. The typical timber bridge connection is a dowel-type connection directly between two wood components, or with a steel bracket.

The design of fasteners and connections for wood has undergone significant changes in recent years. Typical fastener and connection details for wood include nails, staples, screws, lag screws, dowels, and bolts. Additionally, split rings, shear plates, truss plate connectors, joist hangers, and many other types of connectors are available to the designer. The general LRFD design checking equation for connections is given as follows:

$$Z_u \leq \lambda \phi_z Z' \tag{20.26}$$

where Z_u = connection force due to factored loads, λ = applicable time effect factor, ϕ_z = resistance factor for connections = 0.65, and Z' = connection resistance adjusted by the appropriate adjustment factors.

It should be noted that, for connections, the moisture adjustment is based on both in-service condition and on conditions at the time of fabrication; that is, if a connection is fabricated in the wet condition but is to be used in service under a dry condition, the wet condition should be used for design purposes due to potential drying stresses which may occur. It should be noted that C_M does not account for corrosion of metal components in a connection. Other adjustments specific to connection type (e.g., end grain factor, C_{eg}, group action factor, C_g; geometry factor, C_Δ; penetration depth factor, C_d; toe-nail factor, C_{tn}; etc.) will be discussed with their specific use. It should also be noted that when failure of a connection is controlled by a nonwood element (e.g., fracture of a bolt), then the time-effects factor is taken as unity since time effects are specific to wood and not applicable to nonwood components.

In both LRFD and ASD, tables of reference resistances (LRFD) and allowable loads (ASD) are available which significantly reduce the tedious calculations required for a simple connection design. In this section, the basic design equations and calculation procedures are presented, but design tables are not provided herein.

The design of general dowel-type connections (i.e., nails, spikes, screws, bolts, etc.) for lateral loading are currently based on possible yield modes. Based on these possible yield modes, lateral resistances are determined for the various dowel-type connections. Specific equations are presented in the following sections for nails and spikes, screws, bolts, and lag screws. In general, however, the dowel bearing strength, F_e, is required to determine the lateral resistance of a dowel-type connection. Obviously, this property is a function of the orientation of the applied load to the grain, and values of F_e are available for parallel to the grain, $F_{e\parallel}$, and perpendicular to the grain, $F_{e\perp}$. The dowel bearing strength or other angles to the grain, $F_{e\theta}$, is determined by

$$F_{e\theta} = \frac{F_{e\parallel} F_{e\perp}}{F_{e\parallel} \sin^2 \theta + F_{e\perp} \cos^2 \theta} \tag{20.27}$$

where θ = angle of load with respect to a direction parallel to the grain.

Nails, Spikes, and Screws

Nails, spikes, and screws are perhaps the most commonly used fastener in wood construction. Nails are generally used when loads are light such as in the construction of diaphragms and shear walls; however, they are susceptible to working loose under vibration or withdrawal loads. Common wire nails and spikes are quite similar, except that spikes have larger diameters than nails. Both a 12d (i.e., 12-penny) nail and spike are 88.9 mm in length; however, a 12d nail has a diameter of 3.76 mm while a spike has a diameter of 4.88 mm. Many types of nails have been developed to provide better withdrawal resistance, such as deformed shank and coated nails. Nonetheless, nails and spikes should be designed to carry laterally applied load and not withdrawal. Screws behave in a similar manner to nails and spikes, but also provide some withdrawal resistance.

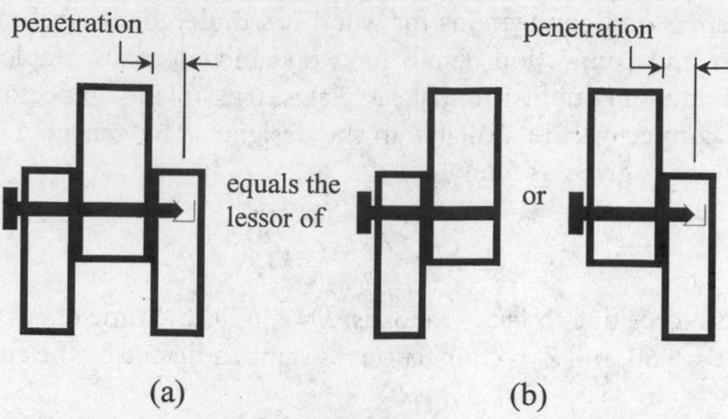

FIGURE 20.7 Double-shear connection: (a) complete connection; (b) left and right shear planes.

Lateral Resistance

The reference lateral resistance of a single nail or spike in single shear is taken as the least value determined by the four governing modes:

$$\mathbf{I_s}: \quad Z = \frac{3.3\, D t_s F_{es}}{K_D} \tag{20.28}$$

$$\mathbf{III_m}: Z = \frac{3.3\, k_1 D p F_{em}}{K_D\left(1 + 2\, R_e\right)} \tag{20.29}$$

$$\mathbf{III_s}: Z = \frac{3.3\, k_2 D t_s F_{em}}{K_D\left(2 + R_e\right)} \tag{20.30}$$

$$\mathbf{IV}: \quad Z = \frac{3.3 D^2}{K_D}\sqrt{\frac{2F_{em}F_{yb}}{3(1+R_e)}} \tag{20.31}$$

where D = shank diameter; t_s = thickness of the side member; F_{es} = dowel bearing strength of the side member; p = shank penetration into member (see Figure 20.7); R_e = ratio of dowel bearing strength of the main member to that of the side member = F_{em}/F_{es}; F_{yb} = bending yield strength of the dowel fastener (i.e., nail or spike in this case); K_D = factor related to the shank diameter as follows: $K_D = 2.2$ for $D \le 4.3$ mm, $K_D = 0.38D + 0.56$ for 4.3 mm $< D \le 6.4$ mm, and $K_D = 3.0$ for $D > 6.4$ mm; and k_1 and k_2 = factors related to material properties and connection geometry as follows:

$$k_1 = -1 + \sqrt{2\left(1 + R_e\right) + \frac{2F_{yb}\left(1 = 2R_e\right)D^2}{3F_{em}p^2}} \tag{20.32}$$

$$k_2 = -1 + \sqrt{\frac{2\left(1 + R_e\right)}{R_e} + \frac{2F_{yb}\left(1 + 2R_e\right)D^2}{3F_{em}t_s^2}} \tag{20.33}$$

Similarly, the reference lateral resistance of a single wood screw in single shear is taken as the least value determined by the three governing modes:

$$\mathbf{I_s}: \quad Z = \frac{3.3\, D t_s F_{es}}{K_D} \tag{20.34}$$

$$\mathbf{III_s}: \quad Z = \frac{3.3\, k_3 D t_s F_{em}}{K_D (2 + R_e)} \tag{20.35}$$

$$\mathbf{IV}: \quad Z = \frac{3.3\, D^2}{K_D} \sqrt{\frac{1.75\, F_{em} F_{yb}}{3(1 + R_e)}} \tag{20.36}$$

where K_D is defined for wood screws as it was for nails and spikes, and $k_3 =$ a factor related to material properties and connection geometry as follows:

$$k_3 = -1 + \sqrt{\frac{2(1 + R_e)}{R_e} + \frac{F_{yb}(2 + R_e) D^2}{2 F_{em} t_s^2}} \tag{20.37}$$

For nail, spike, or wood screw connections with steel side plates, the above equations for yield mode $\mathbf{I_s}$ is not appropriate. Rather, the resistance for that mode should be computed as the bearing resistance of the fastener on the steel side plate. When double shear connections are designed (Figure 20.7a), the reference lateral resistance is taken as twice the resistance of the weaker single shear representation of the left and right shear planes (Figure 20.7b).

For multiple nail, spike, or wood screw connections, the least resistance, as determined from Eqs. (20.28) through (20.31) for nails and spikes or Eqs. (20.34) through (20.36) for wood screws, is simply multiplied by the number of fasteners, n_f, in the connection detail. When multiple fasteners are used, the minimum spacing between fasteners in a row is $10D$ for wood side plates and $7D$ for steel side plates, and the minimum spacing between rows of fasteners is $5D$. Whether a single or a multiple nail, spike, or wood screw connection is used, the minimum distance from the end of a member to the nearest fastener is $15D$ with wood side plates and $10D$ with steel side plates for tension members, and $10D$ with wood side plates and $5D$ with steel side plates for compression members. Additionally, the minimum distance from the edge of a member to the nearest fastener is $5D$ for an unloaded edge and $10D$ for a loaded edge.

The reference lateral resistance must be multiplied by all the appropriate adjustment factors. It is necessary to consider penetration depth, C_d, and end grain, C_{eg}, for nails, spikes, and wood screws. For nails and spikes, the minimum penetration allowed is $6D$, while for wood screws the minimum is $4D$. The penetration depth factor, $C_d = p/12D$, is applied to nails and spikes when the penetration depth is greater than the minimum, but less than $12D$. Nails and spikes with a penetration depth greater than $12D$ assume a $C_d = 1.0$. The penetration depth factor, $C_d = p/7D$, is applied to wood screws when the penetration depth is greater than the minimum, but less than $7D$. Wood screws with a penetration depth greater than $7D$ assume a $C_d = 1.0$. Whenever a nail, spike, or wood screw is driven into the end grain of a member, the end grain factor, $C_{eg} = 0.68$, is applied to the reference resistance. Finally, in addition to C_d and C_{eg}, a toe-nail factor, $C_{tn} = 0.83$, is applied to nails and spikes for "toe-nail" connections. A toe-nail is typically driven at an angle of approximately 30° to the member.

Axial Resistance

For connections loaded axially, tension is of primary concern and is governed by either fastener capacity (e.g., yielding of the nail) or fastener withdrawal. The tensile resistance of the fastener (i.e., nail, spike, or screw) is determined using accepted metal design procedure. The reference withdrawal resistance for nails and spikes with undeformed shanks in the side grain of the member is given by

$$Z_w = 31.6\, D G^{2.5} p n_f \tag{20.38}$$

where Z_w = reference withdrawal resistance in newtons and G = specific gravity of the wood. For nails and spikes with deformed shanks, design values are determined from tests and supplied by fastener manufactures, or Eq. (20.38) can be used conservatively with D = least shank diameter. For wood screws in the side grain,

$$Z_w = 65.3\, DG^2\, pn_f \tag{20.39}$$

A minimum wood screw depth of penetration of at least 25 mm or one half the nominal length of the screw is required for Eq. (20.39) to be applicable. No withdrawal resistance is assumed for nails, spikes, or wood screws used in end grain applications.

The end grain adjustment factor, C_{eg}, and the toe-nail adjustment factor, C_{tn}, as defined for lateral resistance, are applicable to the withdrawal resistances. The penetration factor is not applicable, however, to withdrawal resistances.

Combined Load Resistance

The adequacy of nail, spike, and wood screw connections under combined axial tension and lateral loading is checked using the following interaction equation:

$$\frac{Z_u \cos\alpha}{\lambda\phi_z Z'} + \frac{Z_u \sin\alpha}{\lambda\phi_z Z'_w} \leq 1.0 \tag{20.40}$$

where α = angle between the applied load and the wood surface (i.e., 0° = lateral load and 90° = withdrawal/tension).

Bolts, Lag Screws, and Dowels

Bolts, lag screws, and dowels are commonly used to connect larger-dimension members where larger connection capacities are required. The provisions presented here are valid for bolts, lag screws, and dowels with diameters in the range of 6.3 mm $\leq D \leq$ 25.4 mm.

Lateral Resistance

The reference lateral resistance of a bolt or dowel in single shear is taken as the least value determined by the six governing modes:

$$\mathbf{I_m}: \quad Z = \frac{0.83\, Dt_m F_{em}}{K_\theta} \tag{20.41}$$

$$\mathbf{I_s}: \quad Z = \frac{0.83\, Dt_s F_{es}}{K_\theta} \tag{20.42}$$

$$\mathbf{II}: \quad Z = \frac{0.93 k_1 D F_{es}}{K_\theta} \tag{20.43}$$

$$\mathbf{III_m}: \quad Z = \frac{1.04\, k_2 Dt_m F_{em}}{K_\theta(1+2R_e)} \tag{20.44}$$

$$\mathbf{III_s}: \quad Z = \frac{1.04\, k_3 Dt_s F_{em}}{K_\theta(2+R_e)} \tag{20.45}$$

$$\mathbf{IV}: \quad Z = \frac{1.04\, D^2}{K_\theta} \sqrt{\frac{2F_{em}F_{yb}}{3(1+R_e)}} \tag{20.46}$$

where D = shank diameter; t_m and t_s = thickness of the main and side member, respectively; F_{em} = F_{es} = dowel bearing strength of the main and side member, respectively; R_e = ratio of dowel bearing strength of the main member to that of the side member = F_{em}/F_{es}; F_{yb} = bending yield strength of the dowel fastener (i.e., nail or spike in this case); K_θ = factor related to the angle between the load and the main axis (parallel to the grain) of the member = $1 + 0.25(\theta/90)$; and k_1, k_2, and k_3 = factors related to material properties and connection geometry as follows:

$$k_1 = \frac{\sqrt{R_e + 2R_e^2(1+R_t+R_t^2) + R_t^2R_e^3} - R_e(1+R_t)}{1+R_e} \tag{20.47}$$

$$k_2 = -1 + \sqrt{2(1+R_e) + \frac{2F_{yb}(1+2R_e)D^2}{3F_{em}t_m^2}} \tag{20.48}$$

$$k_3 = -1 + \sqrt{\frac{2(1+R_e)}{R_e} + \frac{2F_{yb}(1+2R_e)D^2}{3F_{em}t_s^2}} \tag{20.49}$$

where R_t = ratio of the thickness of the main member to that of the side member = t_m/t_s.

The reference lateral resistance of a bolt or dowel in double shear is taken as the least value determined by the four governing modes:

$$\mathbf{I_m}:\ Z = \frac{0.83\,Dt_mF_{em}}{K_\theta} \tag{20.50}$$

$$\mathbf{I_s}:\ Z = \frac{1.66\,Dt_sF_{es}}{K_\theta} \tag{20.51}$$

$$\mathbf{III_s}:\ Z = \frac{2.08\,k_3Dt_sF_{em}}{K_\theta(2+R_e)} \tag{20.52}$$

$$\mathbf{IV}:\ Z = \frac{2.08\,D^2}{K_\theta}\sqrt{\frac{2\,F_{em}F_{yb}}{3(1+R_e)}} \tag{20.53}$$

where k_3 is defined by Eq. (20.49)

Similarly, the reference lateral resistance of a single lag screw in single shear is taken as the least value determined by the three governing modes:

$$\mathbf{I_s}:\ Z = \frac{0.83\,Dt_sF_{es}}{K_\theta} \tag{20.54}$$

$$\mathbf{III_s}:\ Z = \frac{1.19\,k_rDt_sF_{em}}{K_\theta(1+R_e)} \tag{20.55}$$

$$\mathbf{IV}:\ Z = \frac{1.11\,D^2}{K_\theta}\sqrt{\frac{1.75\,F_{em}F_{yb}}{3(1+R_e)}} \tag{20.56}$$

where k_4 = a factor related to material properties and connection geometry as follows:

$$k_4 = -1 + \sqrt{\frac{2(1+R_e)}{R_e} + \frac{F_{yb}(2+R_e)D^2}{2\,F_{em}t_s^2}} \tag{20.57}$$

When double shear lag screw connections are designed, the reference lateral resistance is taken as twice the resistance of the weaker single shear representation of the left and right shear planes as was described for nail and wood screw connections.

Wood members are often connected to nonwood members with bolt and lag screw connections (e.g., wood to concrete, masonry, or steel). For connections with concrete or masonry main members, the dowel bear strength, F_{em}, for the concrete or masonry can be assumed the same as the wood side members with an effective thickness of twice the thickness of the wood side member. For connections with steel side plates, the equations for yield modes $\mathbf{I_s}$ and $\mathbf{I_m}$ are not appropriate. Rather, the resistance for that mode should be computed as the bearing resistance of the fastener on the steel side plate.

For multiple bolt, lag screw, and dowel connections, the least resistance is simply multiplied by the number of fasteners, n_f, in the connection detail. When multiple fasteners are used, the minimum spacings, edge distances, and end distances are dependent on the direction of loading. When loading is primarily parallel to the grain, the minimum spacing between fasteners in a row (parallel to the grain) is $4D$, and the minimum spacing between rows (perpendicular to the grain) of fasteners is $1.5D$ but not greater than 127 mm.* The minimum edge distance is dependent on l_m = length of the fastener in the main member for spacing in the main member or total fastener length in the side members for side member spacing relative to the diameter of the fastener. For shorter fasteners ($l_m/D \leq 6$), the minimum edge distance is $1.5D$, while for longer fasteners ($l_m/D > 6$), the minimum edge distance is the greater of $5D$ or one half the spacing between rows (perpendicular to the grain). The minimum end distance is $7D$ for tension members and $4D$ for compression members. When loading is primarily perpendicular to the grain, the minimum spacing within a row (perpendicular to the grain) is typically limited by the attached member but not to exceed 127 mm,* and the minimum spacing between rows (parallel to the grain) is dependent on l_m. For shorter fastener lengths ($l_m/D \leq 2$), the spacing between rows is limited to $2D$; for medium fastener lengths ($2 < l_m/D < 6$), the spacing between rows is limited to $(5l_m + 10D)/8$; and for longer fastener lengths ($l_m/D \geq 6$), the spacing is limited to $5D$; but never should the spacing exceed than 127 mm.* The minimum edge distance is $4D$ for loaded edges and $1.5D$ for unloaded edges. Finally, the minimum end distance for members loaded primarily perpendicular to the grain is $4D$.

The reference lateral resistance must be multiplied by all appropriate adjustment factors. It is necessary to consider group action, C_g, and geometry, C_Δ for bolts, lag screws, and dowels. In addition, penetration depth, C_d, and end grain, C_{eg}, need to be considered for lag screws. The group action factor accounts for load distribution between bolts, lag screw, or dowels when one or more rows of fasteners are used and is defined by

$$C_g = \frac{1}{n_f}\sum_{i=1}^{n_r} a_i \tag{20.58}$$

where n_f = number of fasteners in the connection, n_r = number of rows in the connection, and a_i = effective number of fasteners in row i due to load distribution in a row and is defined by

$$a_i = \left(\frac{1+R_{EA}}{1-m}\right)\left[\frac{m\left(1-m^{2n_i}\right)}{\left(1+R_{EA}m^{n_i}\right)(1+m)-1+m^{2n_i}}\right] \tag{20.59}$$

*The limit of 127 mm can be violated if allowances are made for dimensional changes of the wood.

where

$$m = u - \sqrt{u^2 - 1} \tag{20.60a}$$

$$u = 1 + \gamma \frac{s}{2}\left(\frac{1}{(EA)_m} + \frac{1}{(EA)_s}\right) \tag{20.60b}$$

and where γ = load/slip modulus for a single fastener; s = spacing of fasteners within a row; $(EA)_m$ and $(EA)_s$ = axial stiffness of the main and side member, respectively; R_{EA} = ratio of the smaller of $(EA)_m$ and $(EA)_s$ to the larger of $(EA)_m$ and $(EA)_s$. The load/slip modulus, γ, is either determined from testing or assumed as $\gamma = 0.246D^{1.5}$ kN/mm for bolts, lag screws, or dowels in wood-to-wood connections or $\gamma = 0.369D^{1.5}$ kN/mm for bolts, lag screws, or dowels in wood-to-steel connections.

The geometry factor, C_Δ, is used to adjust for connections in which either end distances and/or spacing within a row does not meet the limitations outlined previously. Defining a = actual minimum end distance, a_{min} = minimum end distance as specified previously, s = actual spacing of fasteners within a row, and s_{min} = minimum spacing as specified previously, the lesser of the following geometry factors are used to reduce the adjusted resistance of the connection:

1. End distance:	for, $a \geq a_{min}$,	$C_\Delta = 1.0$
	for $a_{min}/2 \leq a < a_{min}$,	$C_\Delta = a/a_{min}$
2. Spacing:	for, $s \geq s_{min}$,	$C_\Delta = 1.0$
	for $3D \leq s < s_{min}$,	$C_\Delta = s/s_{min}$

In addition to group action and geometry, the penetration depth factor, C_d, and end grain factor, C_{eg}, are applicable to lag screws (not bolts and dowels). The penetration of a lag screw, including the shank and thread less the threaded tip, is required to be at least $4D$. For penetrations of at least $4D$ but not more than $8D$, the connection resistance is multiplied by $C_d = p/8D$, where p = depth of penetration. For penetrations of at least $8D$, $C_d = 1.0$. The end grain factor, C_{eg}, is applied when a lag screw is driven in the end grain of a member and is given as $C_{eg} = 0.67$.

Axial Resistance

Again, the tensile resistance of the fastener (i.e., bolt, lag screw, or dowel) is determined using accepted metal design procedure. Withdrawal resistance is only appropriate for lag screws since bolts and dowels are "through-member" fasteners. For the purposes of lag screw withdrawal, the penetration depth, p, is assumed as the threaded length of the screw less the tip length, and the minimum penetration depth for withdrawal is the lesser of 25 mm or one half the threaded length. The reference withdrawal resistance of a lag screw connection is then given by

$$Z_w = 92.6\,D^{0.75} G^{1.5} p n_f \tag{20.61}$$

where Z_w = reference withdrawal resistance in newtons and G = specific gravity of the wood.

The end grain adjustment factor, C_{eg}, is applicable to the withdrawal resistance of lag screws and is defined as $C_{eg} = 0.75$.

Combined Load Resistance

The resistance of a bolt, dowel, or lag screw connection to combined axial and lateral load is given by

$$Z'_\alpha = \frac{Z' Z'_w}{Z' \sin^2\alpha + Z'_w \cos^2\alpha} \tag{20.62}$$

where Z'_α = adjusted resistance at an angle and α = angle between the applied load and the wood surface (i.e., 0° = lateral load and 90° = withdrawal/tension).

References

1. American Association of State Highway and Transportation Officials (AASHTO), *AASHTO LRFD Bridge Design Specifications,* AASHTO, Washington, D.C., 1994.
2. American Association of State Highway and Transportation Officials (AASHTO), *Standard Specifications for Highway Bridges,* 16th ed. AASHTO, Washington, D.C., 1996.
3. American Forest and Paper Association (AF&PA), *Manual of Wood Construction: Load and Resistance Factor Design,* AF&PA, Washington, D.C., 1996.
4. American Forest and Paper Association (AF&PA), *National Design Specification for Wood Construction* and *Supplement,* AF&PA, Washington, D.C., 1997.
5. Bodig, J. and Jayne, B., *Mechanics of Wood and Wood Composites,* Van Nostrand Reinhold, New York, 1982.
6. Freas, A.D., Wood properties, in *Wood Engineering and Construction Handbook,* 2nd ed., K.F. Faherty and T.G. Williamson, Eds., McGraw Hill, New York, 1995.
7. Fridley, K.J., Designing for creep in wood structures, *For. Prod. J.,* 42(3):23–28, 1992.
8. Ritter, M.A., Timber Bridges: Design, Construction, Inspection, and Maintenance, EM 7700-8, USDA, Forest Service, Washington, D.C., 1990.
9. Ritter, M.A. and Ebeling, D.W., Miscellaneous wood structures, in *Wood Engineering and Construction Handbook,* 2nd ed., K.F. Faherty and T.G. Williamson, Eds. McGraw-Hill, New York, 1995.
10. U.S. Department of Agriculture (USDA), Wood Handbook: Wood as an Engineering Material, Agriculture Handbook 72, Forest Products Laboratory, USDA, Madison, WI, 1987.
11. Wood, L.W., Relation of Strength of Wood to Duration of Load, USDA Forest Service Report No. 1916, Forest Products Laboratory, USDA, Madison, WI, 1951.

Further Reading

1. American Institute of Timber Construction (AITC), *Glulam Bridge Systems,* AITC, Englewood, CO, 1988.
2. American Institute of Timber Construction (AITC), AITC 117-93, *Design Standard Specifications for Structural Glued Laminated Timber of Softwood Species,* 1993. AITC, Englewood, CO,1993.
3. American Institute of Timber Construction (AITC), *Timber Construction Manual,* Wiley Inter-Science, New York, 1994.
4. Western Wood Products Association (WWPA), *Western Wood Use Book,* 4th ed. WWPA, Portland, OR, 1996.
5. Wipf, T.J., Klaiber, F.W., and Sanders, W.W., Load Distribution Criteria for Glued Laminated Longitudinal Timber Deck Highway Bridges, Transportation Research Record 1053, Transportation Research Board, Washington, D.C.

21

Movable Bridges

Michael J Abrahams
Parsons Brinckerhoff Quade & Douglas, Inc.

21.1 Introduction

Movable bridges have been an integral part of the U.S. transportation system, their development being in concert with that of (1) the development of the railroads and (2) the development of our highway system. While sometimes referred to as draw bridges, movable bridges have proved to be an economical solution to the problem of how to carry a rail line or highway across an active waterway. It is not surprising to learn that movable bridges are found most commonly in states that have low coastal zones such as California, Florida, Louisiana, and New Jersey, or a large number of inland waterways such as Michigan, Illinois, and Wisconsin.

Jurisdiction for movable bridges currently lies with the U.S. Coast Guard. In most instances, marine craft have priority, and the movable span must open to marine traffic upon demand. This precedence is reflected in the terms closed and open, used to describe the position of the movable span(s). A "closed" movable bridge has closed the waterway to marine traffic, while an "open" bridge has opened the waterway to marine traffic. Highway bridges are typically designed to remain in the closed position and only to be opened when required by marine traffic. However, movable railroad bridges can be designed to remain in either the open or closed position, depending on how frequently they are used by train traffic. The difference is important as different wind and seismic load design conditions are used to design for a bridge that is usually open vs. one that is usually closed.

The first specification for the design of movable bridges was published by the American Railway Engineering Association (AREA) in its 1922 *Manual of Railway Engineering* [1]. Until 1938 this specification was used to design both movable highway and railroad bridges, when the American Association of State Highway Officials published its *Standard Specifications for Movable Highway Bridges* [2]. Both specifications are very similar, but have remained separate. Today, movable railroad bridges are designed in accordance with the AREA Manual, Chapter 15, Part 6 [3], and movable

0-8493-7434-0/00/$0.00+$.50

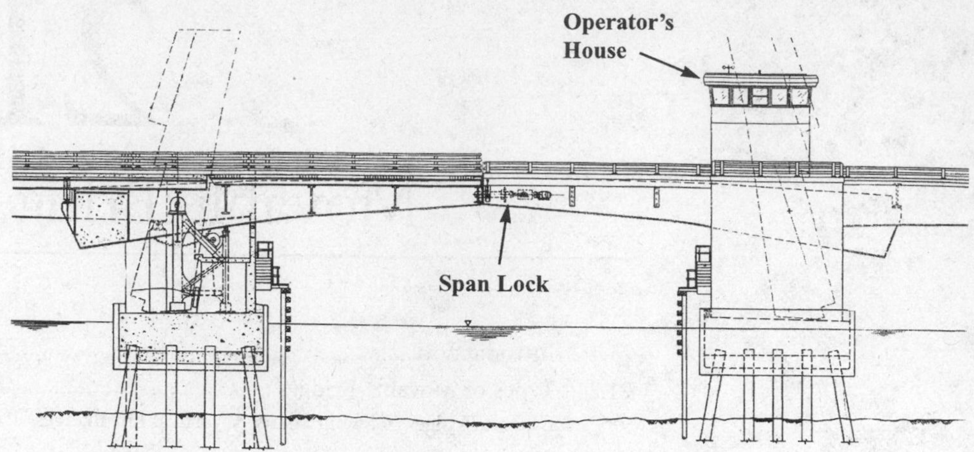

FIGURE 21.1 South Slough (Charleston) Bridge, Coos County, Oregon.

highway bridges are designed in accordance with the American Association of State Highway and Transportation Officials (AASHTO) *Standard Specifications for Movable Highway Bridges* [4]. These specifications primarily cover the mechanical and electrical aspects of a movable bridge; the structural design of the bridge is covered in other parts of the AREA Manual for railroad bridges or the AASHTO *Standard Specifications for Highway Bridges* [5].

21.2 Types of Movable Bridges

The three major categories of movable bridges are swing, bascule, and vertical lift. This list is not exclusive and there are other types, such as jackknife, reticulated, retracting, and floating that are not common and will not be described here. However, the reader should be aware that movable bridges can be crafted to suit specific site needs and are not restricted to the types discussed below.

21.2.1 Bascule Bridges

Bascule bridges are related to medieval drawbridges that protected castles and are familiar illustrations in schoolbooks. The function is the same; the bascule span leaf (or leaves if there are two) rotates from the horizontal (closed) position to the vertical (open) position to allow use of the waterway below. Figure 21.1 illustrates a typical double-leaf deck-girder bascule bridge, the South Slough (Charleston) Bridge, Coos County, Oregon, which spans 126 ft (38.4 m) between trunnions.

This highway bridge includes a number of features. It is a trunnion type as the bascule span rotates about a trunnion. The counterweight, which is at the back end of the leaf and serves to balance the leaf about the trunnion, is placed outside of the pier so that it is exposed. This is advantageous in that it minimizes the width of the pier. Also note that the tail or back end of the leaf reacts against the flanking span to stop the span and to resist uplift when there is traffic (live load) on the span. There is a lock bar mechanism between the two leaves that transfers live-load shear between the leaves as the live load moves from one leaf to the other. The locks (also called center locks to distinguish them from end locks that are provided at the tail end of some bascule bridges) transfer shear only and allow rotation, expansion, and contraction to take place between the leaves. The bridge shown in Figure 21.1 is operated mechanically, with drive machinery in each pier to raise and lower the leaves.

Another feature to note is the operator's house, also referred to as the control house. It is situated so that the operator has a clear view both up and down the roadway and waterway, which is required when the leaves are both raised and lowered. The lower levels of the operator's house typically house

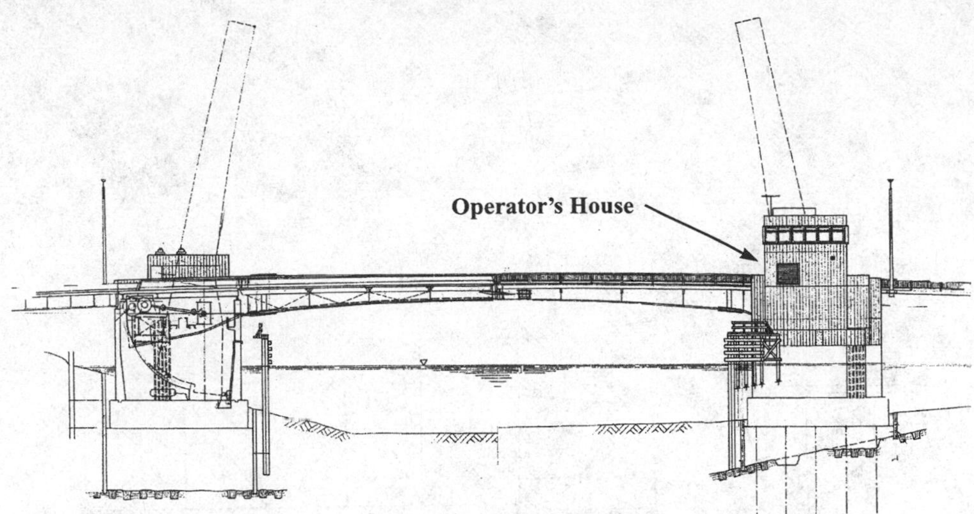

FIGURE 21.2 3rd Street Bridge, Wilmington, Delaware.

the electrical switchgear, emergency generator, bathroom, workshop, and storage space. This bridge has a free-standing fender system that is intended to guide shipping through the channel while protecting the pier from impact. Although not directly related to the bascule bridge, the use of precast footing form and tremie fill shown in the figure can be an excellent solution to constructing the pier as it minimizes the pier depth and avoids excavation at the bottom of the waterway.

Figure 21.2, the 3rd Street Bridge, Wilmington, Delaware, shows a through-girder double-leaf bascule span illustrating other typical bascule span design features. It has a center-to-center trunnion distance of 188 ft (57.3 m). For this bridge the tail or back end of the leaf, including the counterweight, is totally enclosed in the pier and the live-load reaction is located at the front wall of the pier. In addition, this bascule is shown with a mechanical drive. A larger pier is required to protect the enclosed counterweight. The advantage of an enclosed pier is that it allows the counterweight to swing below the waterline within the confines of the bascule pier pit. And, as can be seen, the bascule pier is constructed within a cofferdam. For this bridge there was not enough depth to place a full tremie seal so underwater tie-downs were used to tie the seal to the rock below. Also note the architectural detailing of the cast-in-place concrete substructure, was achieved using form liners. This was done because the bridge is located in a park and needed to be compatible with a parklike setting.

Figure 21.3, the Pelham Bay Bridge, New York, illustrates a single-leaf Scherzer rolling lift bridge or bascule railroad bridge, typical of many movable railroad bridges. The design was developed and patterned by William Scherzer in 1893 and is both simple and widely used. This is a through truss with a span of 81 ft 7 in. (24.9 m). Railroad bascule bridges are always single span, which is required by the AREA Manual, as the heavy live loads associated with heavy rail preclude a joint at midspan. This problem does not occur with light rail (trolley) live loads and combined highway/trolley double-leaf bascule bridges were frequently used in the 1920s. Also, railroad bascule bridges such as the one shown are usually through-truss spans, again due to the heavy rail live loads. This bridge has an overhead counterweight, a typical feature of Scherzer-type bridges. This allows the bridge to be placed relatively close to the water and permits a very simple pier. The track is supported by a steel girder and two simple open piers. As illustrated, the leaf rolls back on a track rather than pivoting about a trunnion. The advantage of this feature is that it is not restricted by the capacity of the trunnion shafts and it minimizes the distance between the front face of the pier and the navigation channel. As the span rotates, it rolls back away from the channel. The drive machinery is located on the moving leaf and typically uses a mechanically or hydraulically driven rack and

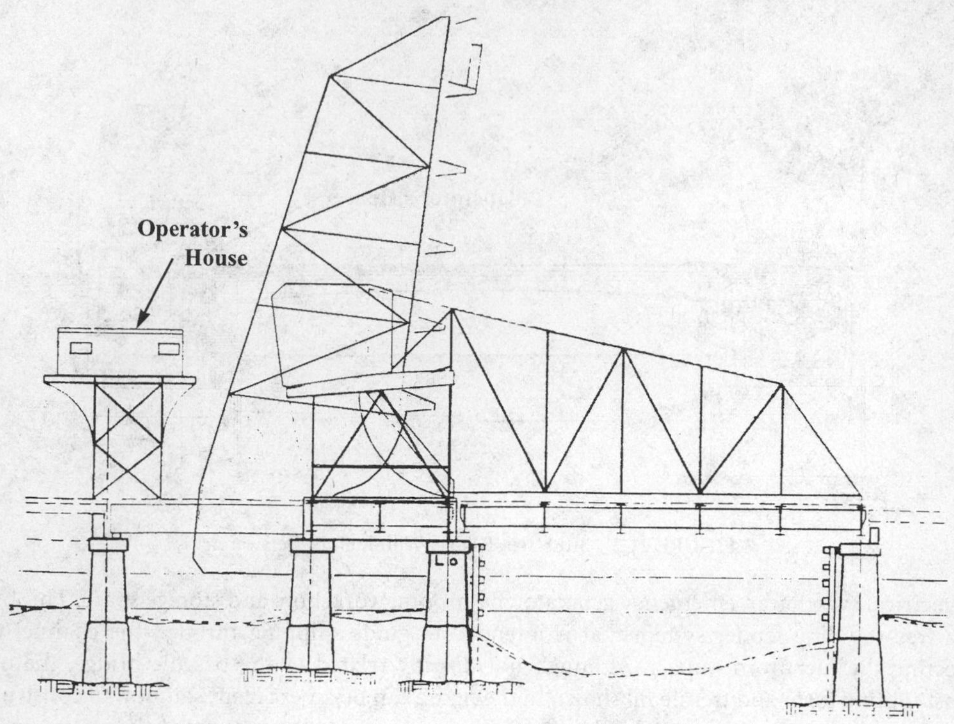

FIGURE 21.3 Pelham Bay Bridge, New York.

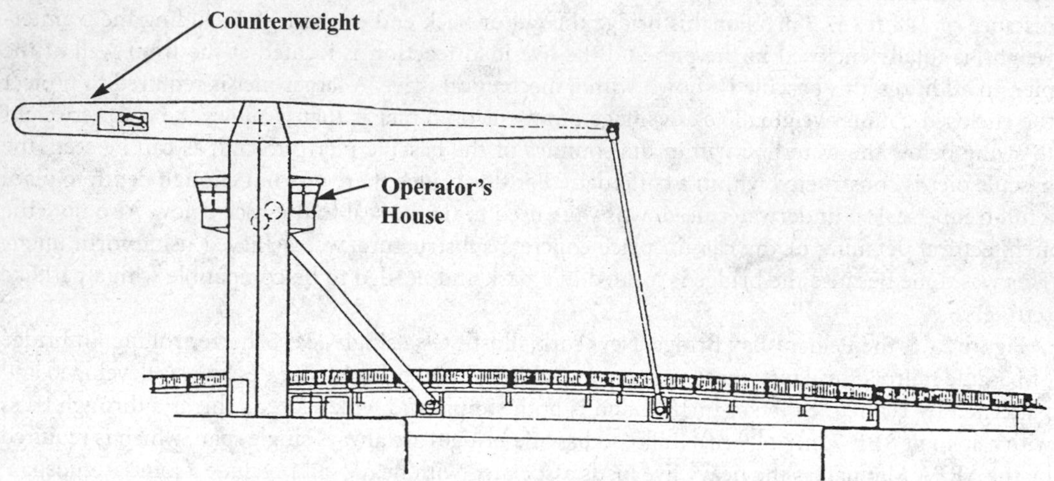

FIGURE 21.4 Manchester Road Bridge at the Canary Wharf, London.

pinion to move the span. The machinery must thus be able to operate as it rotates, and for hydraulic machinery this means the reservoir needs to be detailed accordingly. However, this is not always the case and designs have been developed that actuate the span with external, horizontally mounted hydraulic cylinders. The pier needs to be designed to accommodate the large moving load of the bascule leaf as it rolls back. Conversely, the reaction from the leaf in a trunnion-type span is concentrated in one location, simplifying the design of the pier. More-complicated bascule bridges with overhead counterweight designs have been developed where the counterweight is supported by a scissors-type frame and by trunnions that pivot.

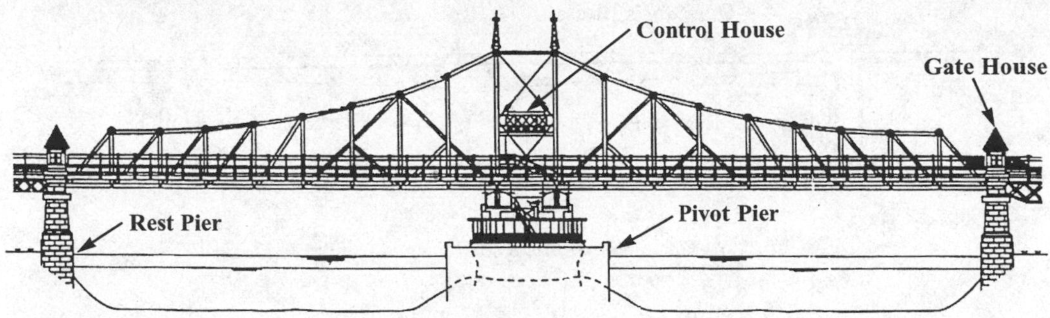

FIGURE 21.5 Macombs Dam Bridge over the Harlem River, New York City.

Figure 21.4, the Manchester Road Bridge at the Canary Wharf, London, illustrates a modern interpretation of an overhead counterweight bascule bridge. It has a span of 109 ft (33.2 m). In addition to being attractive, it is a very practical design with all of the structure above the roadway level, allowing the profile to be set as close to the waterline as desired. The design is not new and is found in many small hand-operated bridges in Holland, perhaps the most famous of which appears in Van Gogh's 1888 painting *The Langlois Bridge.*

21.2.2 Swing Spans

Swing spans were widely used by the railroads. However, they only allowed a limited opening and the center pivot pier was often viewed as a significant impediment to navigation. The pivot pier could also require an elaborate, difficult-to-maintain, and expensive fender system. As a result, swing spans are infrequently used for movable spans. However, they can be a cost-effective solution, particularly for a double-swing span, and should be considered when evaluating options for a new movable bridge.

Figure 21.5 shows a typical through-truss swing span, the Macombs Dam Bridge over the Harlem River in New York City, constructed in 1895. This span is 415 ft (126.5 m) long. The large pivot pier in the middle of the channel illustrates the navigation issue with this design. The piers at either end of the swing span are referred to as the rest piers. By using a through truss, the depth of structure (the distance between the profile grade line and the underside of the structure) is minimized—thus minimizing the height and length of the approaches. The turning mechanism is located at the pivot pier and the entire dead load of the swing span is supported on the pivot pier. As the two arms of the swing span are equal, they are balanced. This bridge is operated with a mechanical drive that utilizes a rack-and-pinion system. There are live-load end lifts at the ends of the swing span that are engaged when the span is closed in order to allow the movable span to act as a two-span continuous bridge under live load. The end lifts, as the name suggests, lift the ends of the swing span, which are free cantilevers when the span operates. The operator's house is typically located on the swing span within the truss but above the roadway, as this location provides good visibility. On older bridges one may also find tenders' houses located at the ends of the swing span. These were for gate tenders who would stop traffic, manually close the traffic gates, and hold horses if necessary. The tenders have been replaced with automatic traffic signals and gates but, on this bridge, their houses remain.

Figure 21.6, the Potato Slough Bridge, San Joaquin County, California, illustrates a good example of a modern highway swing span. This bridge has a 310-ft (94.5-m) swing span that uses simple composite deck, steel girder construction. It is very economical on a square foot basis compared with a bascule or vertical lift bridge, due to its simplicity and lack of a large counterweight. One way of looking at this is that on a bascule or vertical lift bridge a large amount of structure is composed of the counterweight and its supports. These elements do not contribute to effective load-carrying area. The swing span back span, on the other hand, not only acts as a counterweight

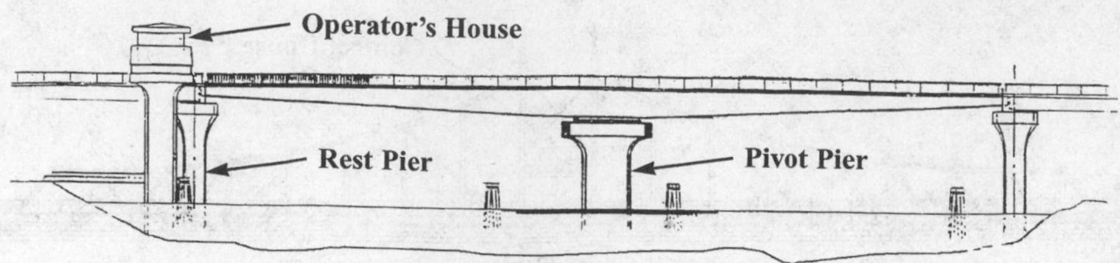

FIGURE 21.6 Potato Slough Bridge, San Joaquin County, California.

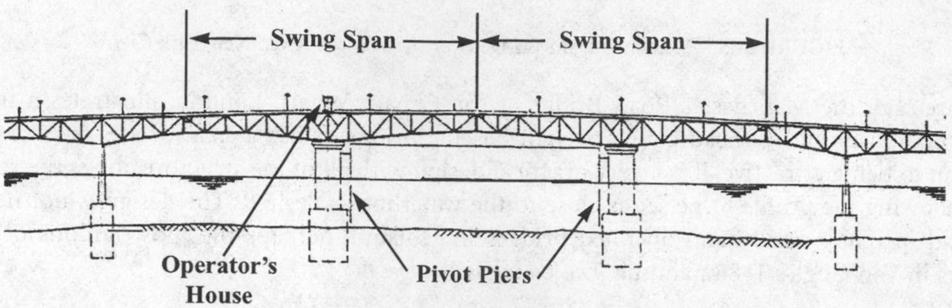

FIGURE 21.7 Coleman Bridge over the York River, Virginia.

but also carries traffic making for a more cost-effective solution. One disadvantage of the deck girder design is that it does not minimize the depth of construction, as does a through-truss or through-girder design. On this bridge the swing span is symmetrical and thus balanced. Nevertheless, some small counterweights may be required to correct any transverse imbalance. The operator's house is located in an adjacent independent structure, again in an area that provides good visibility upstream, downstream, and along the roadway. The pivot pier can accommodate switchgear and a generator. The roadway joints at the ends of the span are on a radius. These could also be detailed as beveled joints, provided that the span only needs to swing in one direction. However, some designers believe it is preferable to design a swing bridge to swing in either direction to allow the bridge to be opened away from oncoming marine traffic and to minimize damage if the structure is struck and needs to swing free.

Figure 21.7 illustrates the double-swing span Coleman Bridge across the York River in Virginia. The two swing spans are each 500 ft (152.4 m) long and provide a 420-ft (128.0-m) wide navigation channel, wide enough to accommodate the range of U.S. Navy vessels that traverse the opening. The bridge is a double-swing deck truss. At this site the river banks are relatively high, so the depth of structure was not a significant issue. Because the bridge is located adjacent to a national park, the low profile of a deck truss was a major advantage. The bridge uses hydraulic motors to drive the span, driving through a rack-and-pinion system similar to that used in large slewing excavators. Unlike the single-swing bridges above, there are lock bars at all three movable span joints. These are driven when the span is in the closed position and function in the same manner as lock bars between the leaves of a double-leaf bascule. There are wedges at each pivot pier to support the live load. As shown, the operator's house is located above one of the swing spans. The control equipment is located inside the operator's house and the generator and switchgear is located on the swing spans below deck. This bridge superstructure was replaced in 1996 and uses a lightweight concrete deck. The piers were constructed in 1952 when the bridge was first built using concrete-filled steel shell caissons that were placed by dredging through open wells.

Figure 21.8, the Tchefuncte River Bridge, Madisonville, Louisiana illustrates a bobtail swing, which is used where only a small channel is required. The structure is a through girder (this minimizes the depth of construction) with a main span of 160 ft (48.8 m). The 80-ft (24.4 m) long

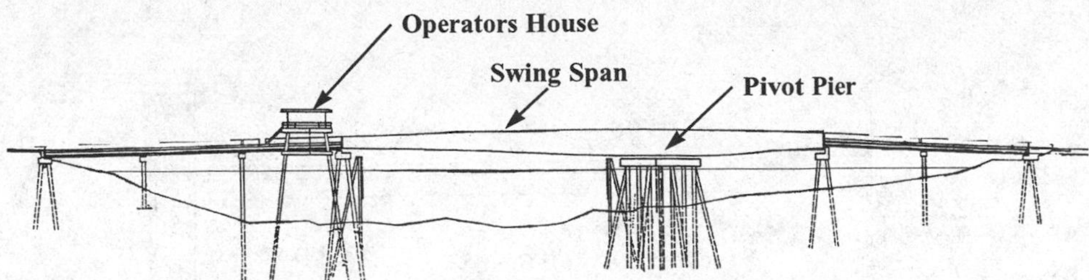

FIGURE 21.8 Tchefuncte River Bridge, Madisonville, Louisiana.

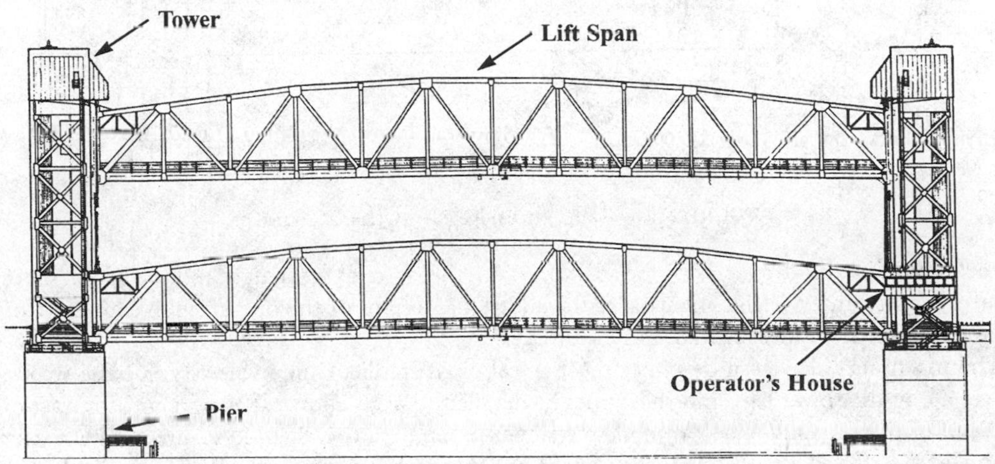

FIGURE 21.9 James River Bridge, Virginia.

bobtail end contains a concrete counterweight that balances the weight of the longer front span. This type of design is particularly well suited where the profile is near the waterline. A relatively simple foundation can support the swing span and no structure is required above deck level. This bridge is operated using hydraulic cylinders on the pivot pier. Girder swing spans tend to be flexible and need a wedge or end lift system that can lift the ends of the span and provide a live-load support when it is in the closed position.

21.2.3 Vertical Lift Bridges

Vertical lift bridges, the last of the three major types of movable bridges, are most suitable for longer spans, particularly for railroad bridges.

Figure 21.9 shows a through-truss highway lift span — the James River Bridge in Virginia — which has a span of 415 ft (126.5 m). The maximum span for this type of design to date is approximately 550 ft (167.7 m) long. The weight of the lift span is balanced by counterweights, one in each tower. Wire ropes that pass over sheaves in the towers are attached to the lift span at one end and the counterweight at the other. A secondary counterweight system is often required to balance the weight of the wire ropes as the span moves up and down and the weight of the wire ropes shifts from one side of the sheaves to the other.

Two types of drive systems are commonly employed, tower drive and span drive. A span drive places the drive machinery in the center of the lift span and, through drive shafts, operates a winch and hauling rope system to raise and lower the span. A tower drive — as the name implies — uses drive machinery in each tower to operate the span. The advantage of the span drive is that it ensures that the two ends lift together, whereas a tower drive requires coordinating the movement

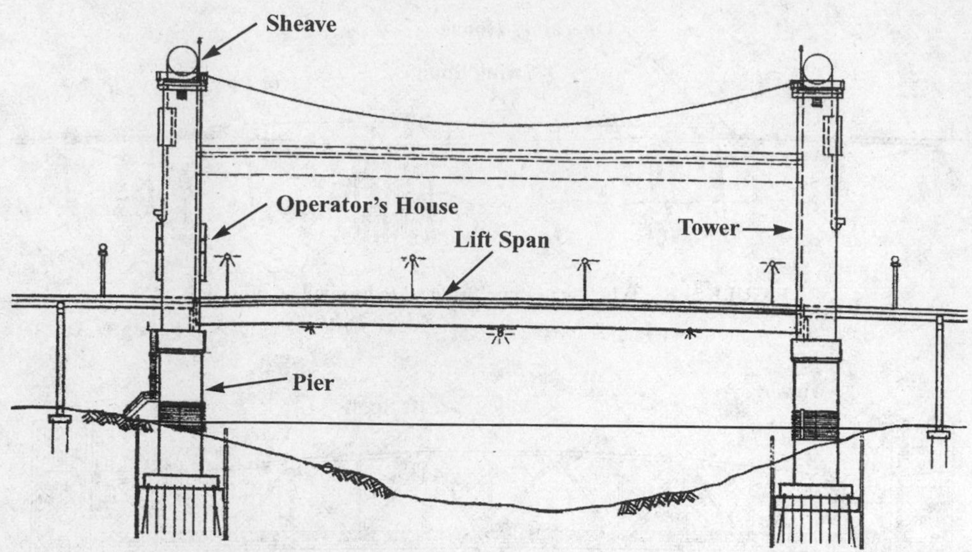

FIGURE 21.10 Danziger Bridge, New Orleans, Louisiana.

at each end. The disadvantages of the span drive are that it tends to be ugly and the lift span, ropes, sheaves, and counterweights must carry the additional weight of the operating machinery. Consequently, tower drives are favored on new bridges.

The machinery drive can be either mechanical or hydraulic. Guide wheels guide the span as it moves along the tower legs, and they must be detailed so as to allow expansion and contraction at one end of the lift span to accommodate changes in temperature. Span locks are used at each end of the lift span to ensure that it does not drift up when in the down (closed) position. If the bridge is normally in the open position, an additional set of span locks needs to be provided. As shown, the operator's house is located on one of the towers. For this bridge, the house partially wraps around the tower to provide good visibility of both the waterway and roadway.

Figure 21.10, the Danziger Bridge, New Orleans, is a vertical lift bridge that uses an orthotropic deck with steel box girders for the lift span and welded steel boxes for the tower. The lift span is 320 ft (97.6 m) long. While the depth of construction is greater than that of an equivalent through truss, the appearance is cleaner, the load to lift should be less and the height of the towers lower than that of an equivalent through truss. The foundations for both of these vertical lift bridges used deep cofferdam construction, which may be advantageous for longer spans because the mass and rigidity of such a foundation should be better able to resist the forces from collision with a large ship.

21.3 Structural Design

21.3.1 Design Criteria

In the closed positions movable bridges are designed for the same design conditions as fixed bridges. However, a movable bridge must also be designed for the following conditions. The load combinations described below are from the AASHTO Specifications [4], and are based on allowable stresses. Similar provisions apply to railroad bridges.

1. *Impact Loads*: Dead load plus 20%. This is applied to structural parts in which the member stress varies with the movement of the span. It is not combined with live-load stresses. For structural parts with stresses caused by machinery or forces applied for moving or stopping the span, 100% impact is used. For end floorbeams, live load plus 100% impact is used.

2. *Wind Loads*:
   a. Movable Span Closed:
      i. Structure to be designed as a fixed span.
   b. Movable Span Open:
      ii. When the movable span is normally left in the closed position, the structure is designed for 30 pounds per square foot (psf) (1.436 kPa) wind load on the structure, combined with dead load, and 20% of dead load to allow for impact, at 1.25 times the allowable unit stresses. For swing bridges, the design is also checked for 30 psf (1.436 kPa) wind load on one arm and 20 psf (0.958 kPa) wind load on the other arm.
      iii. When the movable span is normally left in the open position, the structure is designed for 50 psf (2.394 kPa) wind load on the structure, combined with dead load, at 1.33 times allowable unit stresses. For swing bridges the design is also checked for 50 psf (2.394 kPa) wind load on one arm and 35 psf (1.676 kPa) wind load on the other arm, applied simultaneously).
3. *Ice/Snow Loads*: These are typically not considered in structural design but must be considered in designing the operating machinery.
4. *Bascule Bridges*: The stresses in the main and counterweight trusses or girders are checked for the following load cases:
   a. Case I — Dead load: Bridge open in any position
   b. Case II — Dead load: Bridge closed
   c. Case III — Dead load. Bridge closed with counterweights independently supported
   d. Case IV — Live load plus impact: Bridge closed with live loads thereon
5. *Swing Bridges*: The main trusses or girders are checked for the following load cases:
   a. Case I — Dead load: Bridge open, or closed with end wedges (lifts) not driven
   b. Case II — Dead load: Bridge closed, with its wedges lifted to give positive end reaction, equal to the reaction due to temperature plus 1.5 times the maximum negative reaction of the live load and impact, or the force required to lift the span 1 in. (25 mm) whichever is the greater
   c. Case III — Live load plus impact: Bridge closed, with one arm loaded and considered as a simple span, but with end wedges (lifts) not driven
   d. Case IV — Live load plus impact: Bridge closed and considered as a continuous structure
6. *Vertical Lift Bridges*: The main trusses or girders and towers are checked for the following load cases:
   a. Case I — Dead load: Bridge open
   b. Case II — Dead load: Bridge closed
   c. Case III — Dead load with bridge closed and counterweights independently supported (it should be noted that vertical lift bridges need to include provisions to support the counterweights independently)
   d. Case IV. — Bridge closed with live loads thereon

All of the above applies to the structural design of the moving span and its supports. For design of the operating machinery, there are other load cases contained in the AREA Manual [3] and the AASHTO Specifications [4].

21.3.2 Bridge Balance

Almost all movable bridges are counterweighted so that the machinery that moves the span only needs to overcome inertia, friction, wind, ice, and imbalance. It is prudent to design bridges with a healthy allowance for imbalance as the as-built conditions are never perfect, particularly over time. Recently, at least one bascule bridge and several lift spans have been designed without counterweights, relying instead on the force of the hydraulic machinery to move the span. While this saves the cost of the counterweight and reduces the design dead loads, one needs to compare carefully the reduced construction costs against the present value of the added machinery costs and future annual electric utility demand and service costs (utility rates are based not only on how much energy is consumed but also on how much it costs the utility to be able to supply the energy on demand).

Counterweights are designed to allow for adjustment of the bridge balance, recognizing that during its lifetime, the weight and weight distribution of the bridge can change. The typical reasons for these changes are deck replacement, paint, repairs, or new span locks, among others. Typically, contract drawings show the configuration, estimated concrete volume, and location of the counterweights, but require that the contractor be responsible for balancing the span. This is reasonable as the designer does not know the final weight of the elements to be used, such as the size of the splice plates, the lock bar machinery, concrete unit weight, and other variables. Balance checks can be made during construction or retrofit using detailed calculations accounting for every item that contributes to the weight of the moving span. These calculations need to account for the location of the weight in reference to the horizontal and vertical global axes of the span and, for an asymmetrical span such as a swing span, the transverse axis. For bascule and vertical lift bridges, current practice is to attach strain gauges to the machinery drive shafts and measure the strain in the shafts as the span is actuated through a full cycle, thereby accurately determining the balance. Strain gauge balancing was developed for trunnion bascule type bridges [6,7]. The method has been extended to rolling lift bascule bridges as well as vertical lift bridges.

21.3.3 Counterweights

Figure 21.11 illustrates the typical counterweight configuration for a vertical lift bridge. Both the AREA Manual [3] and AASHTO Specifications [4] require that a pocket be provided in the counterweight for adjustment. The pockets are then partially filled with smaller counterweight blocks, which can be moved by hand to adjust the balance of the bridge. Counterweights are typically made up of a concrete surrounding a steel frame or a reinforced steel box that is filled with normal-weight concrete. Heavyweight concrete can be used to minimize the size of the counterweight. Punchings from bolt holes can be mixed in with concrete to increase its density or concrete can be made using heavyweight aggregate, although this is seldom done due to cost considerations. However, there is at least one vertical lift bridge where cast-iron counterweights were used because the counterweights needed to be as small as possible as they were concealed in the towers. If there is not enough space left for added blocks or if there are no longer any blocks available, counterweight adjustments can always be made by adding steel plates, shapes, or rails.

Figure 21.12 shows the results of a balance check of a rolling lift bridge. In this case, the bridge had been in operation for many years and the owner wanted to replace the timber ties with newer, heavier ties. As shown, the imbalance varied with the position of the span and in the open position the center of gravity was behind the center of rotation. It would be preferable to have all the imbalance on the span side, and to reduce the imbalance. One needs to be careful as an increased imbalance can have a chain reaction and cause an increase in the drive machinery and bridge power requirements.

In general, it is good practice to balance a span so that it is slightly toe heavy for a bascule bridge and slightly span heavy for a vertical lift bridge, the idea being that the span will tend to stay closed under its own weight and will not bounce under live load (although once the span locks are engaged the span cannot rise). The amount of imbalance needs to be included in designing the bridge-operating machinery, so that it can tolerate the imbalance in combination with all the other machinery design loads.

21.3.4 Movable Bridge Decks

An important part of the design of movable bridges is to limit the moving dead load which affects the size of the counterweight, the overall size of the main structural members, and, to a lesser extent,

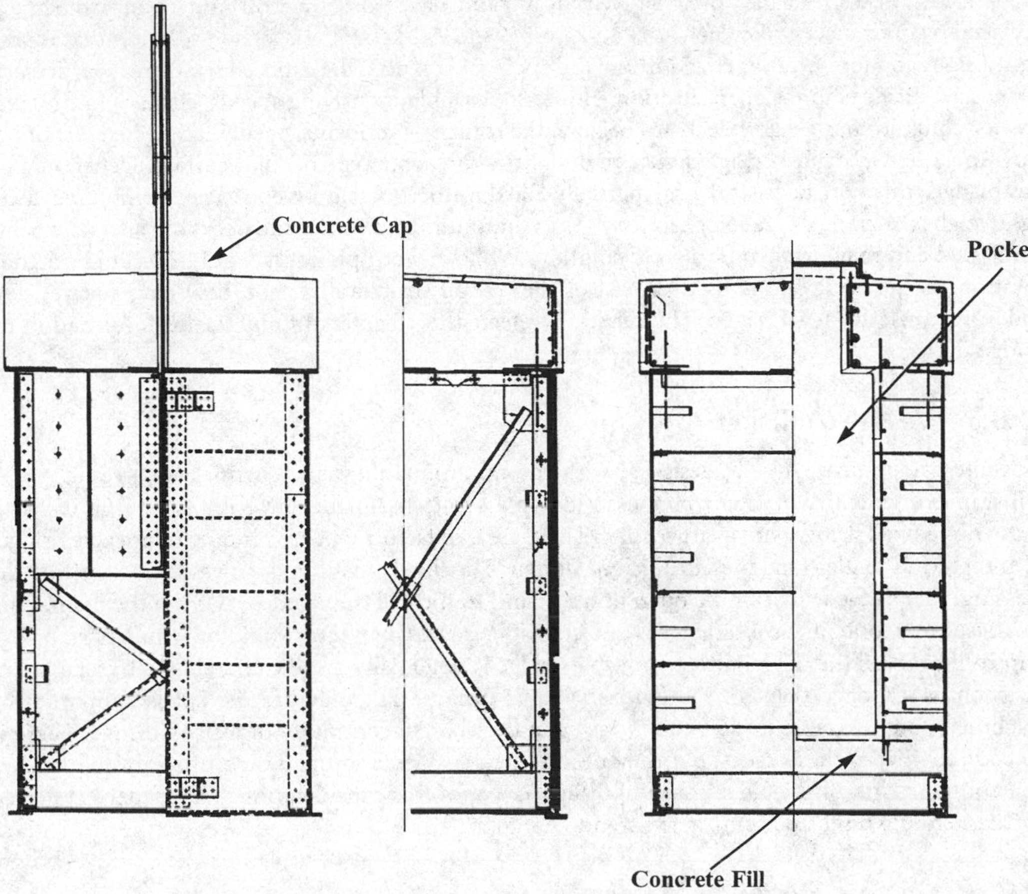

FIGURE 21.11 Typical counterweight configuration for a vertical lift bridge.

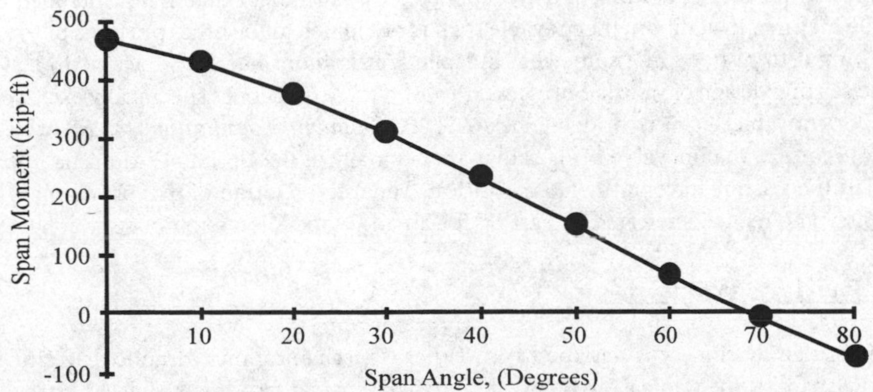

FIGURE 21.12 Results of balance check of a rolling lift bridge.

the machinery depending upon the type of movable bridge. For movable railroad bridges this is typically not a problem, as movable span decks are designed with open decks (timber ties on stringers) and the design live load is such a large part of the overall design load that the type of deck is not an issue. For highway bridges, however, the type of deck needs to be carefully selected to provide a minimum weight while providing an acceptable riding surface. Early movable spans used timber decks, but they are relatively heavy and have poor traction and wear. Timber was replaced by open steel grid, which at 20 to 25 psf (98 to 122 kg/m^2) was a good solution that is both lightweight and long wearing. In addition, the open grid reduced the exposed wind area, particularly for bascule bridges in the open position. However, with higher driving speeds, changes in tires and greater congestion, steel grid deck has become the source of accidents, particularly when wet or icy. Now most new movable bridge decks are designed with some type of solid surface. Depending on the bridge, this can be a steel grid partially filled with concrete or epoxy, an orthotopic deck, lightweight concrete, or the Exodermic system. Aluminum and composite decks are also now being developed and may prove to be a good solution. While orthotropic decks would seem to be a good solution, as the deck can be used as part of the overall structural system, they have not yet seen widespread use in new designs. The reader is referred to Chapters 14 and 24 for information on bridge decks.

21.3.5 Vessel Collision

Movable bridges are typically designed with the minimum allowable channel. As a result, vessel collision is an important aspect as there may be a somewhat higher probability of ship collision than with a fixed bridge with a larger span. There are two factors that are unique to movable bridges with regard to fender (and vessel collision) design. The first is that if a large vessel is transiting the crossing, the bridge will be in the open position and traffic will be halted away from the main span. As a result, the potential consequences of a collision are less than they would be with a fixed bridge ship collision. On the other hand, a movable bridge is potentially more vulnerable to misalignment or extensive damage than a fixed bridge. This is because not only are the spans supported by machinery, but movable spans by their very nature lack the continuity of a fixed bridge. There is no code to govern these issues, but they need to be considered in the design of a movable bridge. The configuration of the piers is an important aspect of this consideration. The reader is referred to Chapter 60, Vessel Collision Analysis and Design.

21.3.6 Seismic Design

The seismic design of movable bridges is also a special issue because they represent a large mass, which may include a large counterweight, supported on machinery that is not intended to behave in a ductile manner. In addition, the movable span is not joined to the other portions of the structure thus allowing it to respond in a somewhat independent fashion. The AREA Manual [3], Chapter 9 covers the seismic design of railroad bridges. However, these guidelines specifically exclude movable bridges. For movable highway bridges, the AASHTO *Standard Specification for Movable Highway Bridges* [4] requires that movable bridges that are normally in the closed position shall be designed for one half the seismic force in the open position. The interpretation of this provision is left up to the designer. The reader is referred to Part IV for an additional discussion of seismic investigation.

21.4 Bridge Machinery

Currently, bridge machinery is designed with either a mechanical or hydraulic drive for the main drive and usually a mechanical drive for the auxiliary machinery items such as span locks and wedges. This is true for all types of movable bridges and the choice of mechanical vs. hydraulic drive is usually based on a combination of owner preference and cost — although other factors may also be considered. Mechanical drives are typically simple configurations based on machinery design principles that were developed long before movable bridges, although now drives use modern enclosed speed reducers and bearings. Overall, these systems have performed very well with sometimes limited maintenance. More recently, hydraulic machinery has been introduced in movable bridge design and it has proved to be an effective solution, as the hydraulics can be closely matched

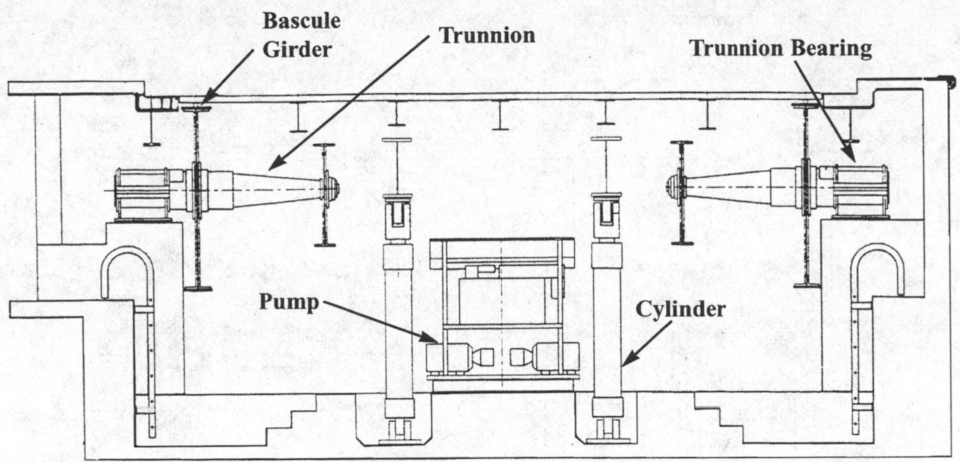

FIGURE 21.13 Section through a bascule pier showing girder trunnions and hydraulic cylinders.

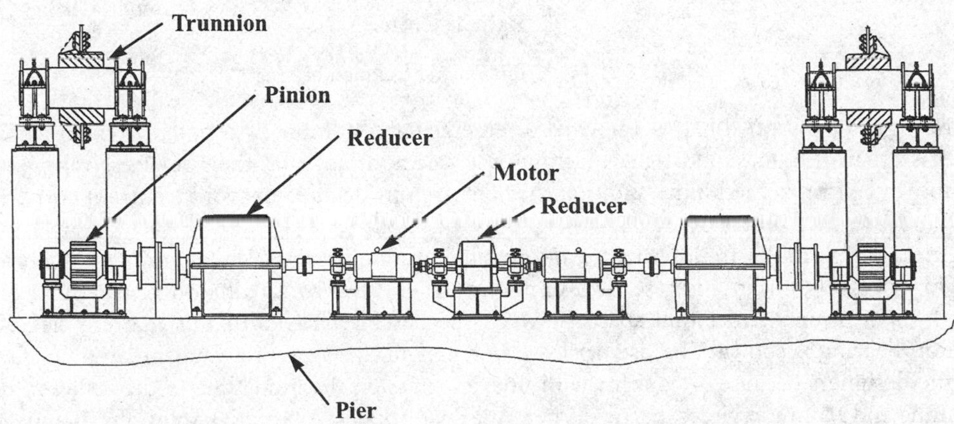

FIGURE 21.14 Section through a bascule pier that utilizes a mechanical drive.

to the power demands, which require good speed control over a wide range of power requirements. Also, there are many firms that furnish hydraulic machinery. However, the systems also require a more-specialized knowledge and maintenance practice than was traditionally the case with mechanical drives.

Figure 21.13 shows a section through a bascule pier illustrating the layout of the bascule girder trunnions (about which the bascule girders rotate) as well as the hydraulic cylinders used to operate the span. Typical design practice is to provide multiple cylinders so that one or more can be removed for maintenance while the span remains in operation. The cylinder end mounts incorporate spherical bearings to accommodate any misalignments. Note that the hydraulic power pack, consisting of a reservoir, motors, pumps, and control valves, is located between the cylinders. Typically redundant motors and pumps are used and the valves can be hand operated if the control system fails. As movable bridges are located in waterways, the use of biodegradable hydraulic fluids is favored in case of a leak or spill.

Figure 21.14 shows a similar section through a bascule pier that utilizes a mechanical drive. What is not shown is the rack attached to the bascule girder. Note the different arrangement here of the trunnions, with bearings on either side of the girders. The central reducer contains a differential, similar to the differential in a vehicle, that serves to equalize the torque in these two drive shafts. As shown, there are two drive motors and typically the span will be designed to operate with only

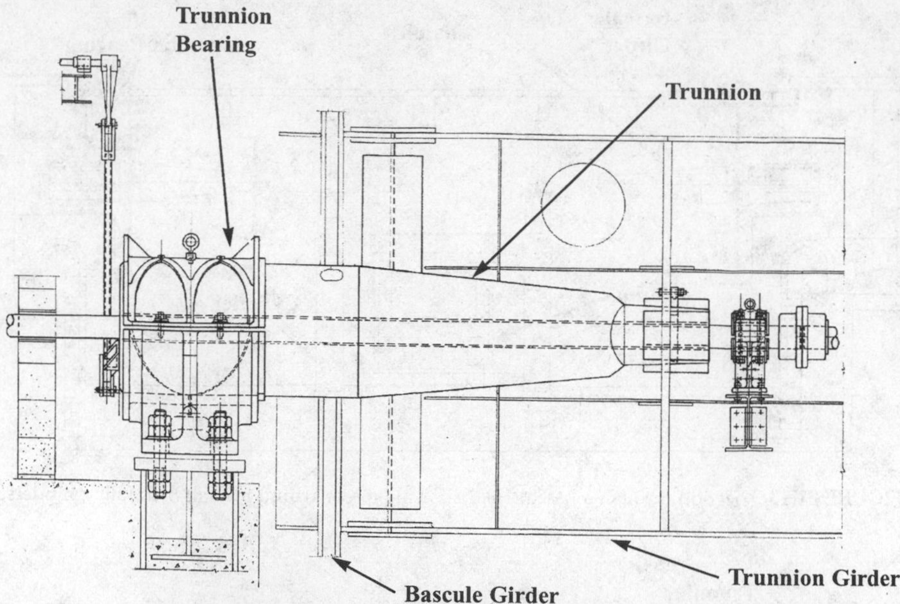

FIGURE 21.15 Trunnion and trunnion bearing.

one motor in operation either as a normal or emergency condition. Also note the extensive use of welded steel frames to support the machinery. It is important that they be stress relieved after assembly but prior to machining and that they be carefully detailed to avoid reentrant corners that could, in time, be a source of cracks.

Figure 21.15 is an illustration of a trunnion and trunnion bearing. The trunnions are fabricated from forged steel and, in this case, are supported on one end by a trunnion bearing and on the other by a trunnion girder that spans between the bascule girders. In this figure a sleeve-type trunnion bearing is shown. The use of sleeve bearings in this type of arrangement is not favored by some designers because of concern with uneven stress on the lining due to deformation of the trunnions and trunnion girder, particularly as the span rotates. Alternative solutions include high-capacity spherical roller bearings and large spherical plain bearings. The crank arrangement shown on the left side of the figure is associated with a position indicator.

Figure 21.16 shows a typical arrangement of the treads for a rolling lift bascule.

Figure 21.17 is a typical drive mechanism for a vertical lift bridge, with a tower drive. The drive is somewhat similar to that used for a bascule bridge except that the pinion drives the rack attached to a sheave rather than a rack attached to a bascule girder. Although a mechanical drive is shown, a similar arrangement could be accomplished with hydraulic motors.

Figure 21.18 is a typical welded sheave used for a vertical lift bridge. As shown, there are 16 rope grooves so this would be associated with a large vertical lift bridge. Typically there are four sheaves for a vertical lift bridge, one at each corner of the lift span. The trunnion bearing is not shown but would be similar to that shown in a bascule bridge trunnion. While sleeve bearings are commonly used, spherical type bearings are also considered to allow for trunnion flexure.

Figure 21.19 shows a span lock typically used between the leaves of a two-leaf bascule bridge. In this case a manufactured unit is illustrated. It incorporates a motor, brake, reducer, and lock bars. Alternative arrangements with a standard reducer are also used, although for this type of an installation the compactness and limited weight favor a one piece unit. It is important that provisions be included for replacement of the wearing surfaces in the lock bar sockets and realignment as they receive considerable wear.

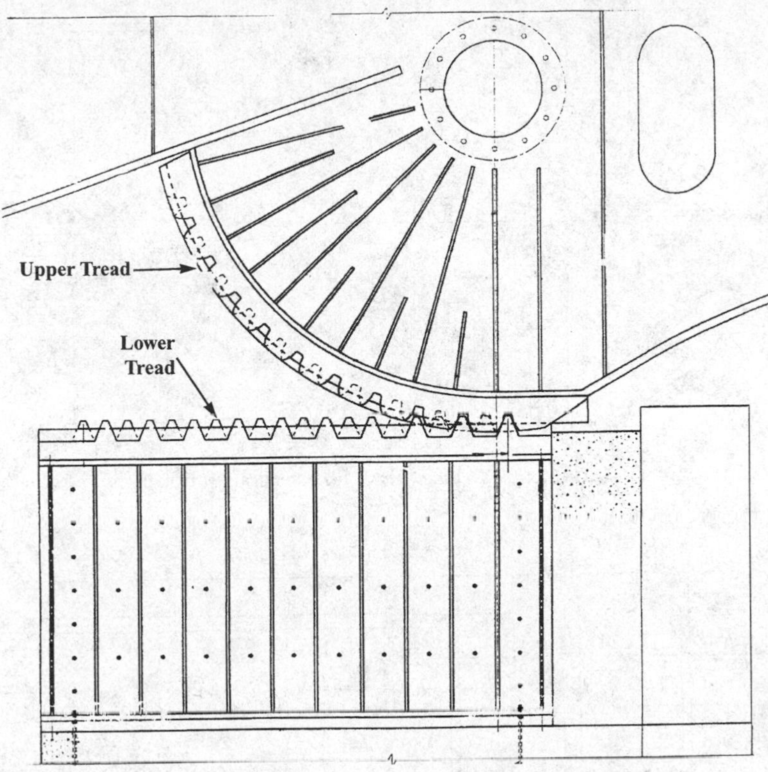

FIGURE 21.16 Treads for a rolling lift bascule.

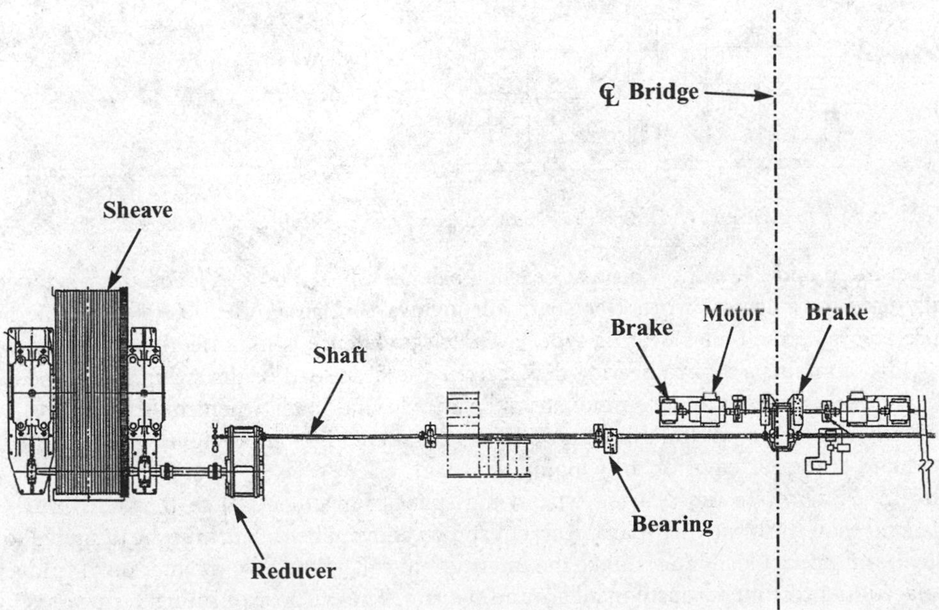

FIGURE 21.17 Drive mechanism for vertical lift bridge.

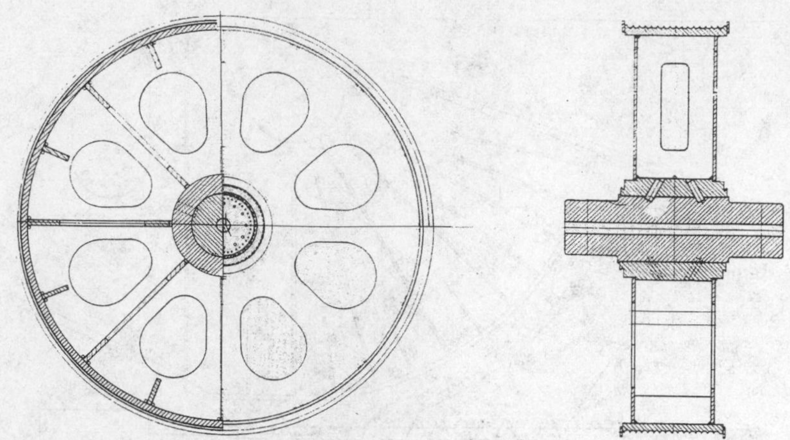

FIGURE 21.18 Welded sheave for a vertical lift bridge.

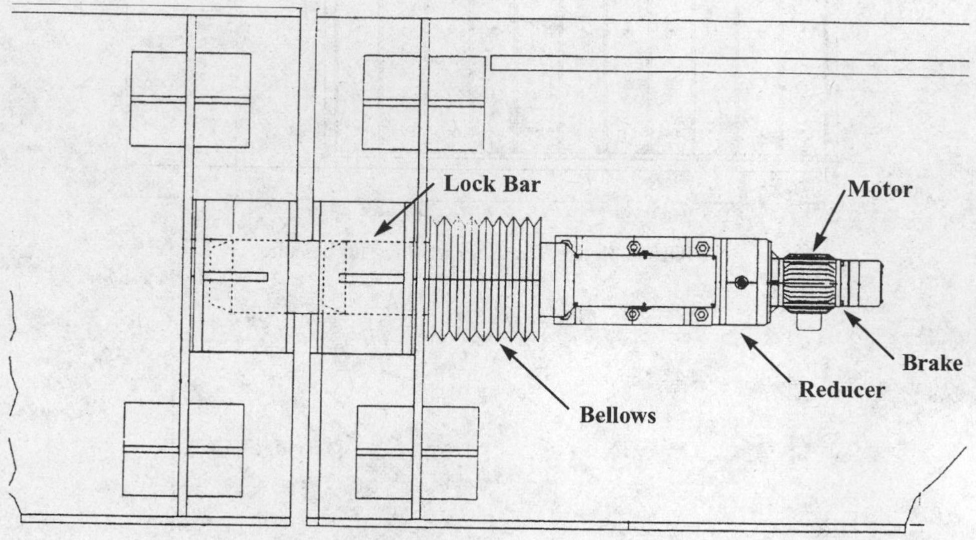

FIGURE 21.19 Span lock between leaves of a two-leaf bascule bridge.

Figures 21.20 and 21.21 show a pivot bearing, balance wheel, and live-load wedge arrangement typically used for a center pivot swing span. For highway bridges AASHTO states, "Swing bridges shall preferably be the center bearing type." No such preference is indicated by AREA. The center pivot, which contains a bronze bearing disk, carries the dead load of the swing span. The balance wheels are only intended to accommodate unbalanced wind loads when the span moves so that they are adjusted to be just touching the roller track. The wedges are designed to carry the bridge live load and are retracted prior to swinging the span.

Figure 21.22 shows a rim-bearing-type swing span arrangement. Note that it is much more complicated than the center pivot arrangement shown above. The rollers must be designed to carry dead, live, and impact loads and, unlike the intermittent rollers used for a center pivot bridge, need to be placed in a continuous fashion all around the rim. The purpose of the center pivot is to keep the rollers centered, and for some bridges to carry a portion of the dead and live load. Figure 21.23 shows an end lift device used for a swing span.

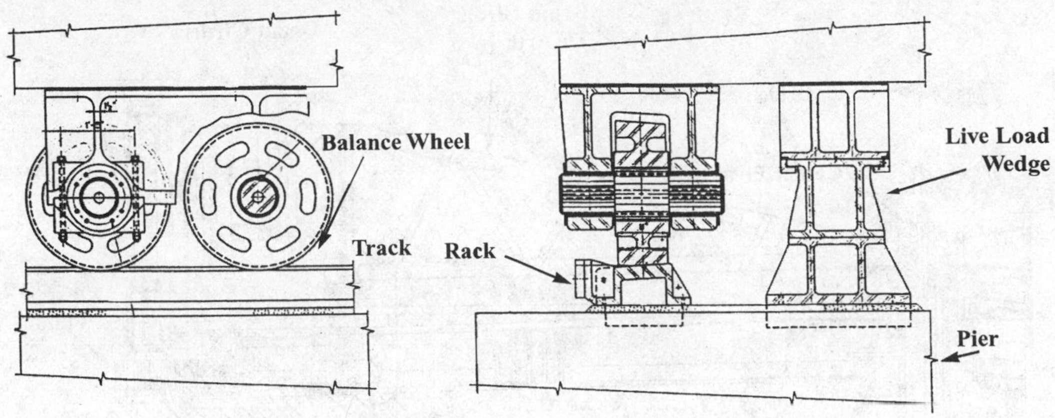

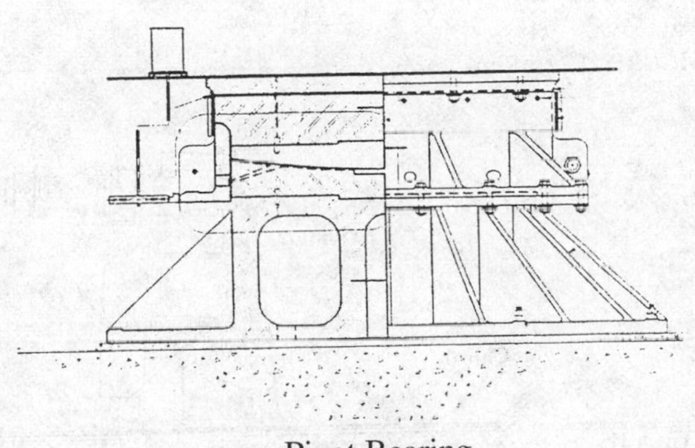

FIGURE 21.20 Balance wheels and pivot bearing for a center pivot swing span.

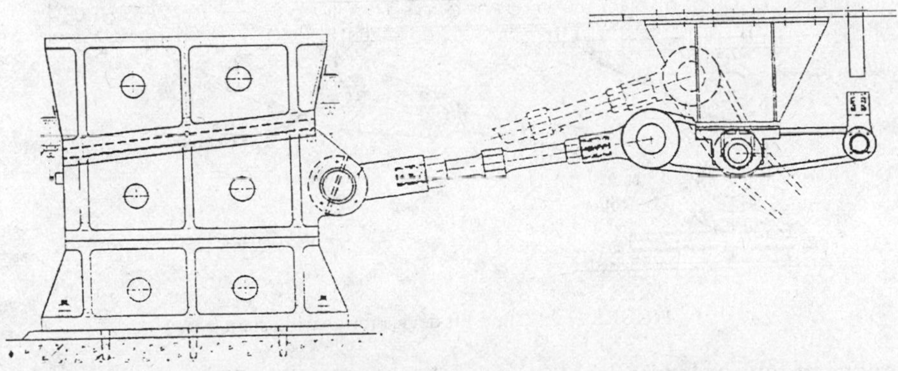

FIGURE 21.21 Live-load wedge arrangement for center pivot swing span.

Figure 21.24 shows a mechanical drive arrangement for a swing span, and similar arrangements can be adapted to both pivot and rim-bearing bridges. A common problem with this arrangement is the pinion attachment to the structural supports as very high forces can be induced in braking the swing span when stopping and these supports tend to be a maintenance problem. Figure 21.25

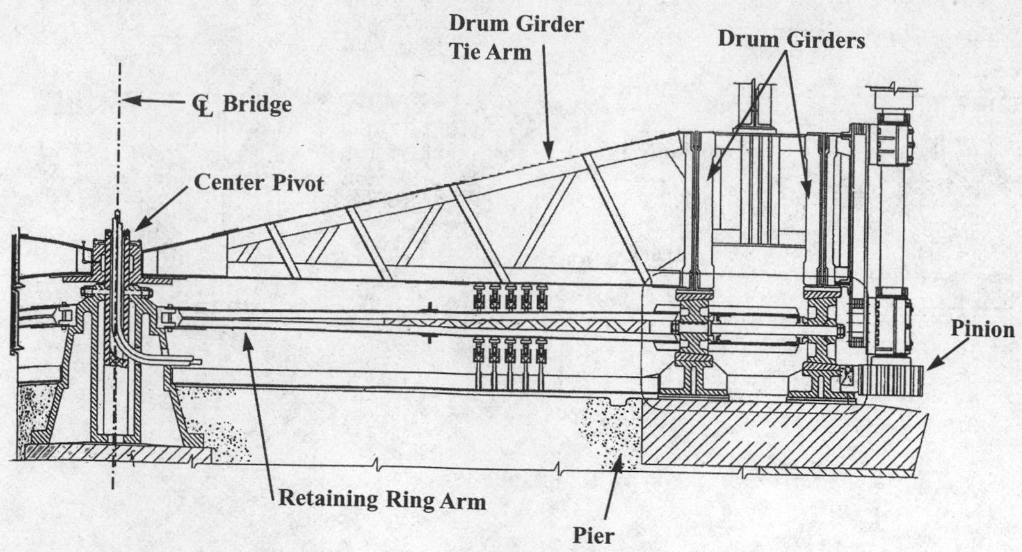

FIGURE 21.22 Rim–bearing swing span arrangement.

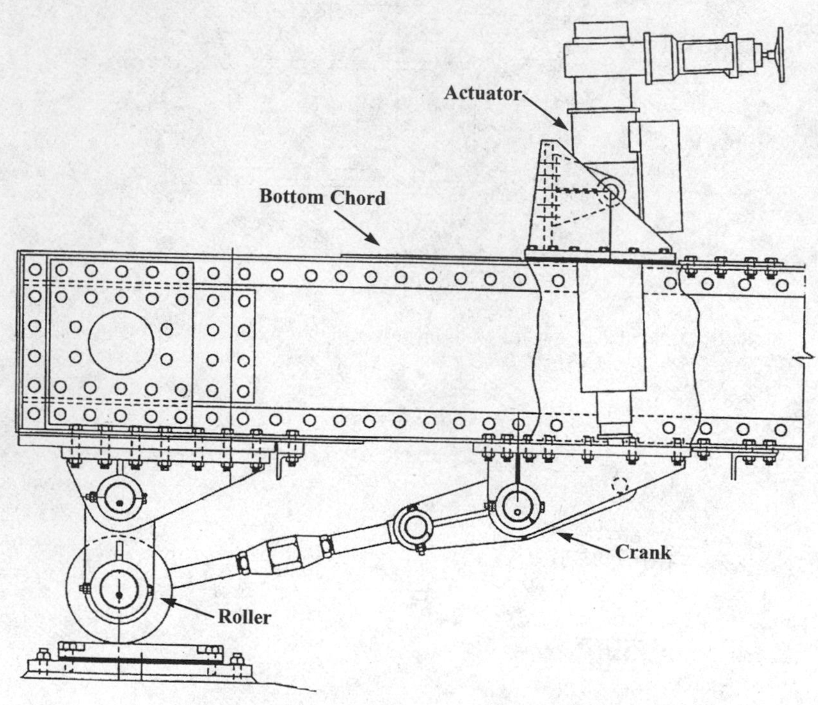

FIGURE 21.23 End lift device for a swing span.

illustrates one of four hydraulic drives from the Coleman Bridge. This drive has an eccentric ring mount so that the pinion/rack backlash can be adjusted.

Figure 21.26 shows a hydraulic drive for a swing span using hydraulic cylinders.

Figure 21.27 shows a typical air buffer. These are provided at the ends of the movable span. With modern control systems, particularly with hydraulics, buffers may not be required to assist in seating. For many years these were custom-fabricated but, if required, one can now utilize off-the-shelf commercial air or hydraulic buffers, as is shown here.

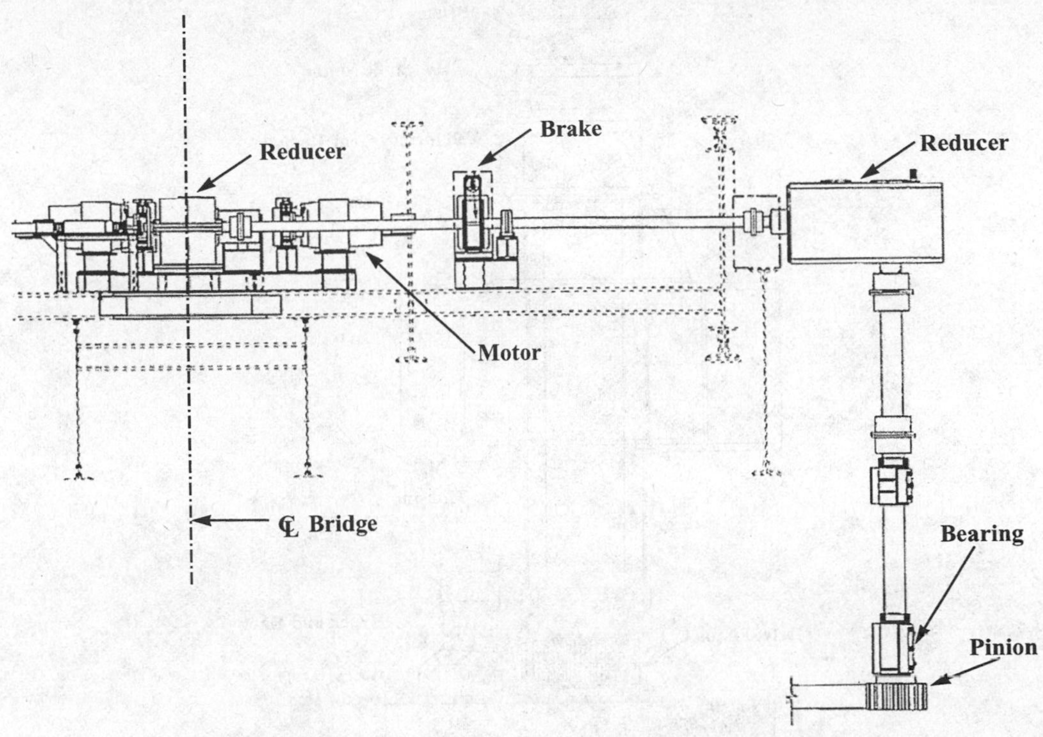

FIGURE 21.24 Mechanical drive arrangement for a swing span.

21.5 Bridge Operation, Power, and Controls

21.5.1 Bridge Operations

Movable bridges are designed to be operated following a set protocol, and this protocol is incorporated into the control system as a series of permissive interlocks. The normal sequence of operation is as follows:

Vessel signals for an opening, usually through a marine radio but it can be through a horn. For a highway bridge the operator sounds a horn, activates the traffic signals, halting traffic, lowers the roadway gates, then lowers the barrier gates. For a rail bridge the operator needs to get a permissive signal from the train dispatcher.

After the barrier gates are lowered, a permissive signal allows the operator to withdraw the locks and/or wedges and lifts and, once that is completed, to open the span. The vessel then can proceed through the opening. To close the bridge, the steps are reversed.

The controls are operated from a control desk and Figure 21.28 shows a typical control desk layout. Note that the control desk includes a position indicator to demonstrate the movable span(s) position as well as an array of push buttons to control the operation. A general objective in designing such a desk is to have the position of the buttons mimic the sequence of operations. Typically, the buttons are lit to indicate their status.

21.5.2 Bridge Power

In the early years, when the streets of most cities had electric trolleys, movable bridges were operated on the 500 VDC trolley power. As the trolleys were removed, rectifiers were installed on the bridges to transform the utility company AC voltage to DC voltage. Many of the historical movable bridges

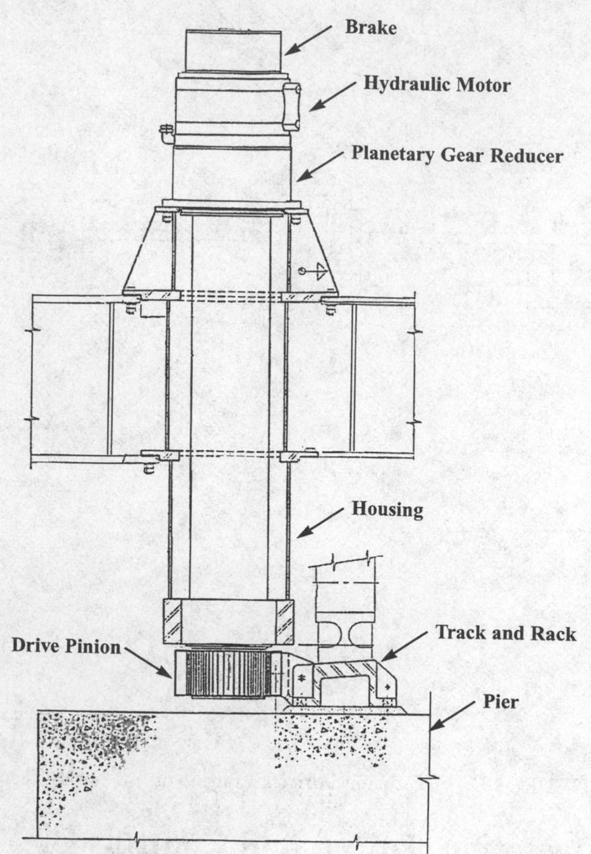

FIGURE 21.25 One of four hydraulic drives from the Coleman Bridge, York River, Virginia.

that are still operating on their original DC motors and drum switch/relay speed controls have these rectifiers. Most of the movable bridges that have been rehabilitated in recent years, but still retain the original DC motors, now have silicon controlled rectifier (SCR) controllers that use AC voltage input and produce a variable DC voltage output directly to the motor.

The most common service voltage for movable bridges is 480 Vac, 3 phase, although in some locations, the service is 240 or 208 Vac, 3 phase. Economics and the utility company policies are the primary determinant factors in what voltage is used. Electrical power, simplified, equals volts times amperes. Thus, for a given horsepower, the motor current at 480 V is one half of that at 240 V. The economics are obvious when one considers the motor frame size, motor controllers, electrical switching equipment, and conductors are all physically smaller for 480 V than for 240 V. However, some utility companies do not normally provide 480 V and they are not willing to maintain a single 480 V service without passing along substantial costs to the bridge owner. If these additional service costs exceed the savings of using 480-V motors and controls, 240-V service becomes more attractive.

The choice of one voltage over the other has no bearing on the cost of power for a movable bridge. Power is power and the rate per kilowatt-hour is the same regardless of voltage. A service cost factor that is sometimes overlooked is the demand charge that utility companies impose on very large intermittent loads. These charges are to offset the utility company cost of reserving power generation and transmission capability to serve the demands of a facility that is normally not online. These charges are based on the peak load, measured at the meter, over a period of, typically, 15 min. The charges are amortized over the year following the last highest reading and added to the billing for the actual amount of electrical power used. In the case of a bridge that has a very high power demand, even if it is opened only once or twice a year, the annual electrical costs are very

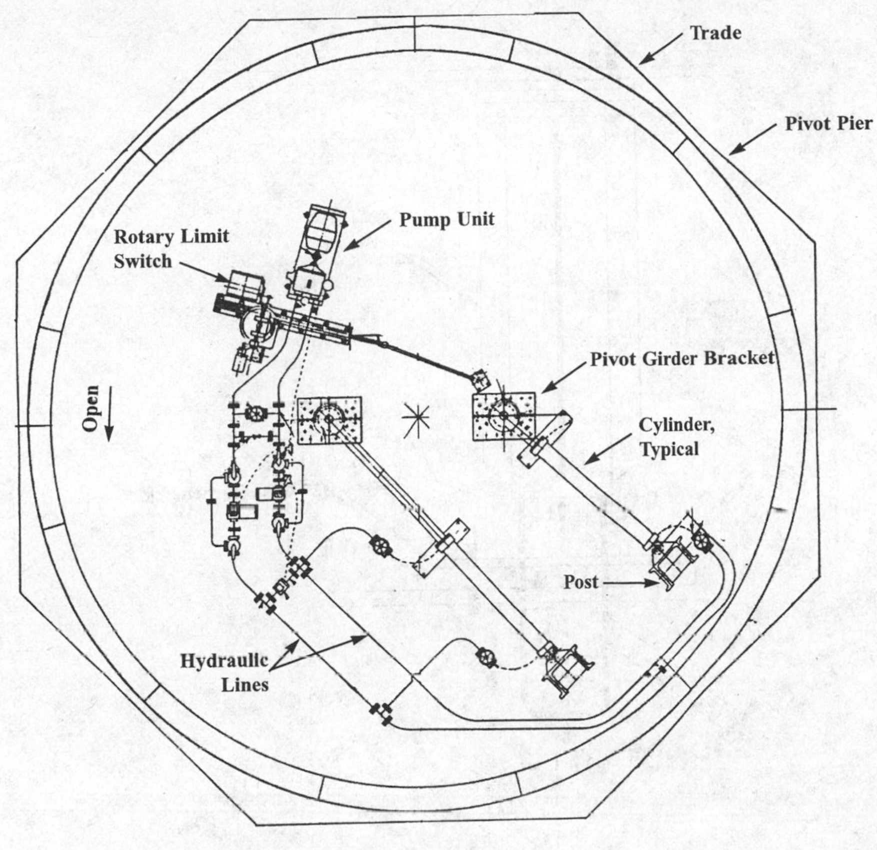

FIGURE 21.26 Hydraulic drive for a swing span using hydraulic cylinders.

high because the owner has to pay for the demand capacity whether it is used or not. Referring to the earlier discussion on counterweights, it is very important that the design of the bridge is such that it is as energy efficient as possible.

Both AREA [3] and AASHTO [4] require a movable bridge to have an emergency means of operation should primary power be lost. Most bridges are designed with a backup engine driven generator and operate the bridge on the normal electrical motor drives. For safety and reliability, diesel engines are preferred by most bridge owners. Hand operation can be provided as backup for auxiliary devices such as locks, gates, and wedges.

However, there are many different types of backup systems, such as the following:

1. Internal combustion engines or air motors on emergency machinery that can be engaged when needed;
2. Smaller emergency electrical motors on emergency machinery to reduce the size of the emergency generator;
3. A receptacle for a portable emergency generator to reduce the capital investment for emergency power for several bridges, as well as other municipality–owned facilities.

21.5.3 Bridge Control

The predominant control system in use in newly constructed or rehabilitated movable bridges is the programmable logic controller (PLC). This is a computer-based system that has been adapted

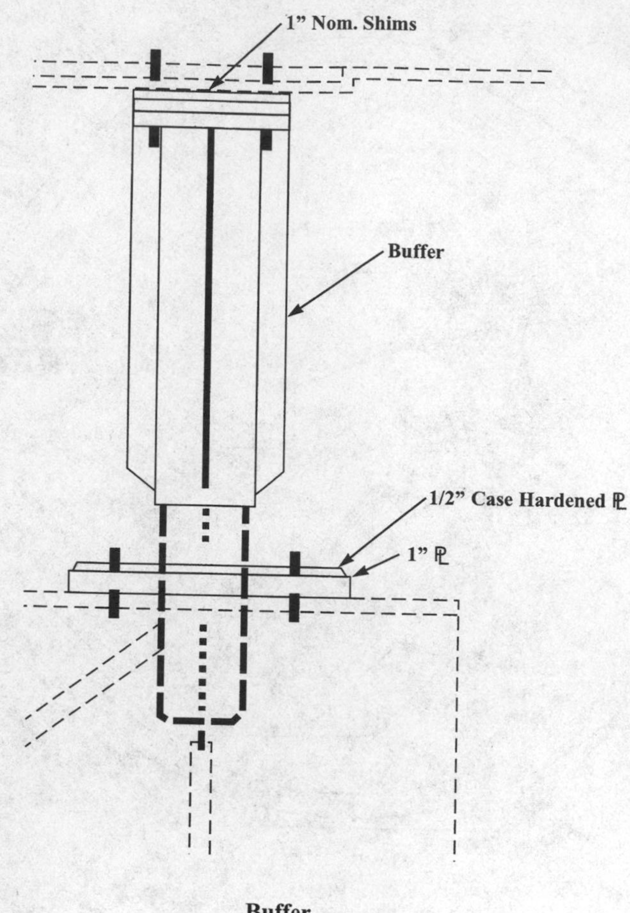

FIGURE 21.27 Typical air buffer.

from other industrial-type applications. The PLC offers the ability to automate the operation of a bridge completely. However, most agencies have used the PLC as a replacement for a relay-based system to reduce the cost of initial construction and to reduce the space required for the control system. Other common applications for the PLC include generation of alarm messages to help reduce time in troubleshooting and maintenance of the systems.

As an example of their widespread use, the New Jersey Department of Transportation has PLCs on all of its bridges and has a proactive training program for its operations and technical staff. However, not all states are using PLCs, as the Florida and Washington State Departments of Transportation are now returning to the relay-based systems because they do not have the technical staff to maintain the PLC.

A more recent development is the use of PLCs for remote operation. For example, the city of Milwaukee, Wisconsin has several bridges that are controlled remotely by means of computer modem links and closed-circuit TV. This reduces the staff to one tender per three bridges. The potential liability of this type of system needs to be carefully evaluated as the bridge operator may not be able to observe adequately all parts of the bridge when operating the span.

Environmental regulations have made the installation permits for submarine cables difficult to obtain. PLC and radio modems have been used in several states to replace the control wiring that would otherwise be in a submarine cable.

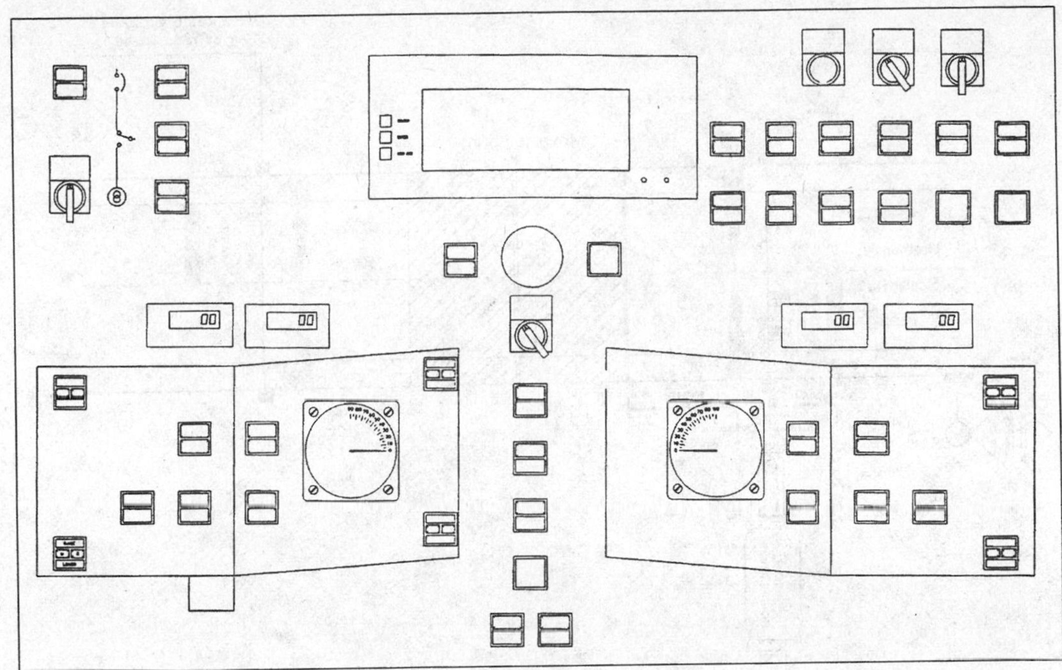

FIGURE 21.28 Layout of a typical control desk.

The selection of a drive system is performance oriented. Reliability and cost are key issues. The most common drives for movable bridges over that past 80 years have been DC and wound rotor AC motors with relays and drum switches. These two technologies remain the most common today although there have been many advances in DC and AC motor controls and the old systems are being rapidly replaced with solid-state drives.

The modern DC drives on movable bridges are digitally controlled, fully regenerative, four-quadrant, SCR motor controllers. In more general terms, this is a solid-state drive that provides infinitely variable speed and torque control in both forward and reverse directions. They have microprocessor programming that provides precise adjustment of operating parameters, and once a system is set up, it rarely needs to be adjusted. Programmable parameters include acceleration, deceleration, preset speeds, response rate, current/torque limit, braking torque, and sequence logic. This type of drive has been proved to provide excellent speed and torque control for bridge-operating conditions.

The wound rotor motor drive technology has also moved into digital control. The new SCR variable voltage controllers are in essence crane control systems. While they are not quite as sophisticated as the DC drives, they have similar speed and torque control capabilities. Most of the movable bridge applications have been retrofitted using the existing motors.

Adjustable frequency controllers (AFC) control speed by varying the frequency of the AC voltage and current to a squirrel cage induction motor. This type of drive has been used on movable bridges with some success but it is not well suited for this type of application. There are two primary reasons. First, this type of drive was designed for the control of pumps and fans, not high-inertia loads. Second, at low speeds, it does not provide sufficient braking torque to maintain control of an overhauling load. This is a significant concern when seating a span with an ice and snow load.

The first flux vector-controled AFC has been in use on a movable bridge for approximately 6 years now. It is a somewhat sophisticated drive system that controls magnetic flux to create slip artificially and thus control torque at any speed including full-rated motor torque at zero speed.

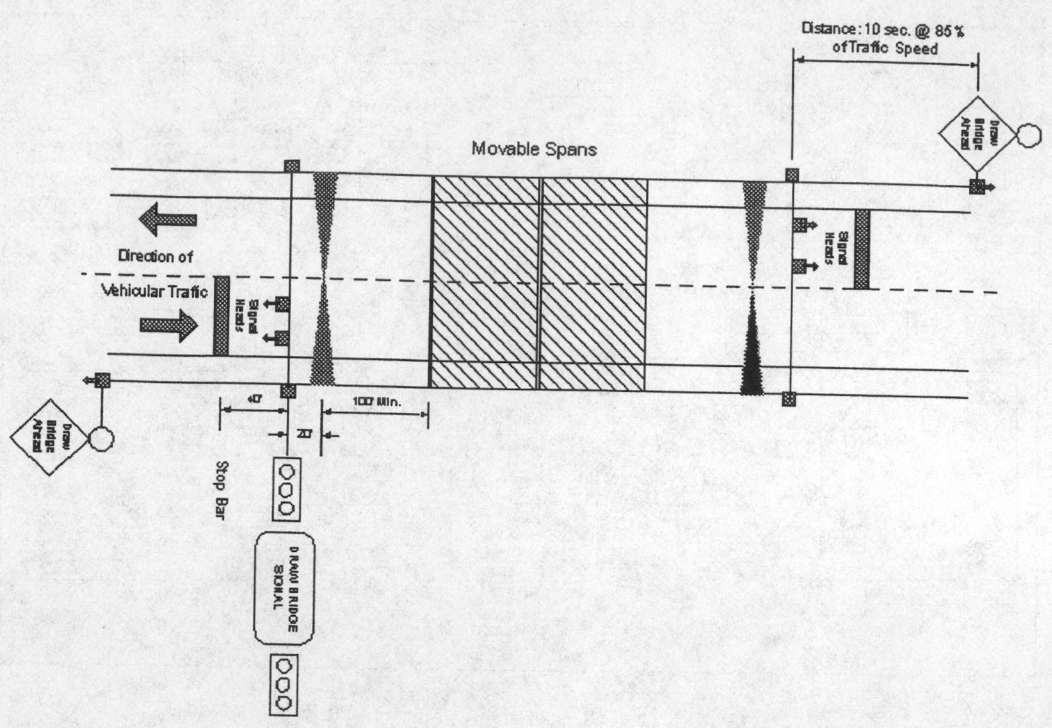

FIGURE 21.29 Typical layout of movable bridge signals and gates.

The drive controller uses input from a digital shaft encoder to locate the motor rotor position and then calculates how much voltage and current to provide to each motor lead. The drive is capable of 100% rated torque at zero speed which gives it excellent motion control at low speeds.

21.6 Traffic Control

Rail traffic control for movable railroad bridges involves interlocking the railroad signal system with the bridge-operating controls. For a movable bridge that is on a rail line that has third rail or catenary power, the interlocking must include the traction power system. In principle, the interlocking needs to be designed so that the railroad signals indicate that the track is closed and the power is deenergized prior to operating the span. However, the particulars of how this is accomplished depends upon the railroad in question and will not be addressed here. For a movable highway bridge, highway traffic control is governed by the AASHTO Movable Highway Bridge Specifications [4], as well as the Manual for Uniform Traffic Control Devices (MUTCD) [8]. Each owner may impose additional requirements but the Manual is typically used in the United States. As a minimum this will include a DRAWBRIDGE AHEAD warning sign, traffic signal, warning (or roadway) gates, and usually resistance (or barrier) gates. One possible arrangement is shown in Figure 21.29 for a two-leaf bascule bridge, note that there are no resistance gates. AASHTO [4] requires that a resistance gate (positive barrier) be placed prior to a movable span opening except where the span itself, such as a bascule leaf, blocks the opening.

For marine traffic, navigation lighting must follow the requirements of the Bridge Permit as approved by the U.S. Coast Guard. The permit typically follows the Coast Guard requirements as found in the U.S. Code of Federal Regulations 33, Part 118, Bridge Lighting and Other Systems [9]. These regulations identify specific types and arrangements for navigation lights depending upon the type of movable bridge.

References

1. American Railway Engineering Association, *Manual of Railway Engineering,* 1922.
2. American Association of State Highway Officials, *Standard Specifications for Movable Highway Bridges,* 1938.
3. *AREA Manual for Railway Engineering,* Chapter 15, Steel Structures, Part 6, Movable Bridges, 1997.
4. AASHTO, *Standard Specifications for Movable Highway Bridges,* 1988.
5. AASHTO, *Standard Specifications for Highway Bridges,* 16th Ed., 1996.
6. Ecale, H., G. Brown, and P. Kocsis, Chicago type bascule balancing: a new technique, Technical Notes, *ASCE J. Struct. Div.,* 103(ST11), 2269–2272, November 1977.
7. Lu, Malvern, Jenkins, Allred and Biwas, Balancing of trunnion type bascule bridges, *ASCE J. Struct. Div.,* V. 108, (ST10), October 1982.
8. Federal Highway Administration, *Manual on Uniform Traffic Control Devices.*
9. U.S. Code of Federal Regulations 33, Part 118, Bridge Lighting and Other Systems.

22

Floating Bridges

M. Myint Lwin

Washington State Department of Transportation

22.1 Introduction

Floating bridges are cost-effective solutions for crossing large bodies of water with unusual depth and very soft bottom where conventional piers are impractical. For a site where the water is 2 to 5 km wide, 30 to 60 m deep and there is a very soft bottom extending another 30 to 60 m, a floating bridge is estimated to cost three to five times less than a long-span fixed bridge, tube, or tunnel.

A modern floating bridge may be constructed of wood, concrete, steel, or a combination of materials, depending on the design requirements. A 124-m-long floating movable wood pontoon railroad bridge was built in 1874 across the Mississippi River in Wisconsin in the United States. It was rebuilt several times before it was abandoned. A 98-m-long wood floating bridge is still in service in Brookfield, Vermont. The present Brookfield Floating Bridge is the seventh replacement structure, and was built by the Vermont Agency of Transportation in 1978. The first 2018-m-long Lake Washington Floating Bridge in Seattle [1,2], Washington, was built of concrete and opened to traffic in 1940 (Figure 22.1). Since then, three more concrete floating bridges have been built [2,3]. These concrete floating bridges form major transportation links in the state and interstate highway systems in Washington State. The Kelowna Floating Bridge on Lake Okanagan in British Columbia, Canada [4], was built of concrete and opened to traffic in 1958. It is 640 m long and carries two lanes of traffic. The 1246-m-long Salhus Floating Bridge and the 845-m-long Bergsoysund Floating Bridge in Norway (Figure 22.2) were constructed of concrete pontoons and steel superstructures [5]. They were opened to traffic in the early 1990s.

0-8493-7434-0/00/$0.00+$.50

FIGURE 22.1 First Lake Washington floating bridge.

Washington State's experience has shown that reinforced and prestressed concrete floating bridges are cost-effective, durable, and low in maintenance as permanent transportation facilities. Concrete is highly corrosion resistant in a marine environment when properly designed, detailed, and constructed. Concrete is a good dampening material for vibration and noise, and is also far less affected by fire and heat than wood, steel, or other construction materials.

22.2 Basic Concept

The concept of a floating bridge takes advantage of the natural law of buoyancy of water to support the dead and live loads. There is no need for conventional piers or foundations. However, an anchoring or structural system is needed to maintain transverse and longitudinal alignments of the bridge.

A floating bridge is basically a beam on an elastic foundation and supports. Vertical loads are resisted by buoyancy. Transverse and longitudinal loads are resisted by a system of mooring lines or structural elements.

The function of a floating bridge is to carry vehicles, trains, bicycles, and pedestrians across an obstacle — a body of water. Inasmuch as a floating bridge crosses an obstacle, it creates an obstacle for marine traffic. Navigational openings must be provided for the passage of pleasure boats, smaller water crafts, and large vessels. These openings may be provided at the ends of the bridge. However, large vessels may impose demands for excessive horizontal and vertical clearances. In such cases, movable spans will have to be provided to allow the passage of large vessels. The Hood Canal Floating Bridge in Washington State has a pair of movable spans capable of providing a total of 183 m of horizontal clearance (Figure 22.3). Opening of the movable spans for marine traffic will cause interruption to vehicular traffic. Each interruption may be as long as 20 to 30 min. If the frequency of openings is excessive, the concept of a floating bridge may not be appropriate for the site. Careful consideration should be given to the long-term competing needs of vehicular traffic and marine traffic before the concept of a highway floating bridge is adopted.

FIGURE 22.2 The Bergsoysund floating bridge.

FIGURE 22.3 Movable spans for large vessels.

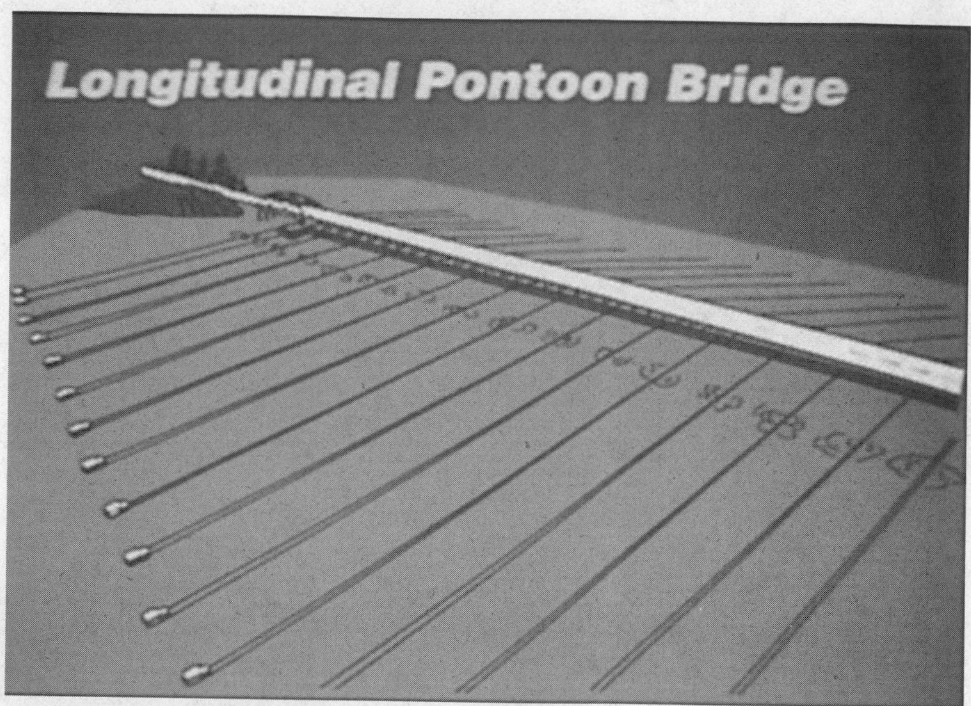

FIGURE 22.4 Continuous pontoon-type structure.

22.3 Types

22.3.1 Floating Structure

Floating bridges have been built since time immemorial. Ancient floating bridges were generally built for military operations [6]. All of these bridges took the form of small vessels placed side by side with wooden planks used as a roadway. Subsequently, designers added openings for the passage of small boats, movable spans for the passage of large ships, variable flotation to adjust for change in elevations, and so on.

Modern floating bridges generally consist of concrete pontoons with or without an elevated superstructure of concrete or steel. The pontoons may be reinforced concrete or prestressed concrete post-tensioned in one or more directions. They can be classified into two types, namely, the continuous pontoon type and the separate pontoon type. Openings for the passage of small boats and movable spans for large vessels can be incorporated into each of the two types of modern floating bridges.

A continuous pontoon floating bridge consists of individual pontoons joined together to form a continuous structure (Figure 22.4). The size of each individual pontoon is based on the design requirements, the construction facilities, and the constraints imposed by the transportation route. The top of the pontoons may be used as a roadway or a superstructure may be built on top of the pontoons. All the present floating bridges in Washington State are of the continuous pontoon floating bridge type.

A separate pontoon floating bridge consists of individual pontoons placed transversely to the structure and spanned by a superstructure of steel or concrete (Figure 22.5). The superstructure must be of sufficient strength and stiffness to maintain the relative position of the separated pontoons. The two floating bridges in Norway are of the separate pontoon floating bridge type.

Both types of floating structures are technically feasible and relatively straightforward to analyze. They can be safely designed to withstand gravity loads, wind and wave forces, and extreme events,

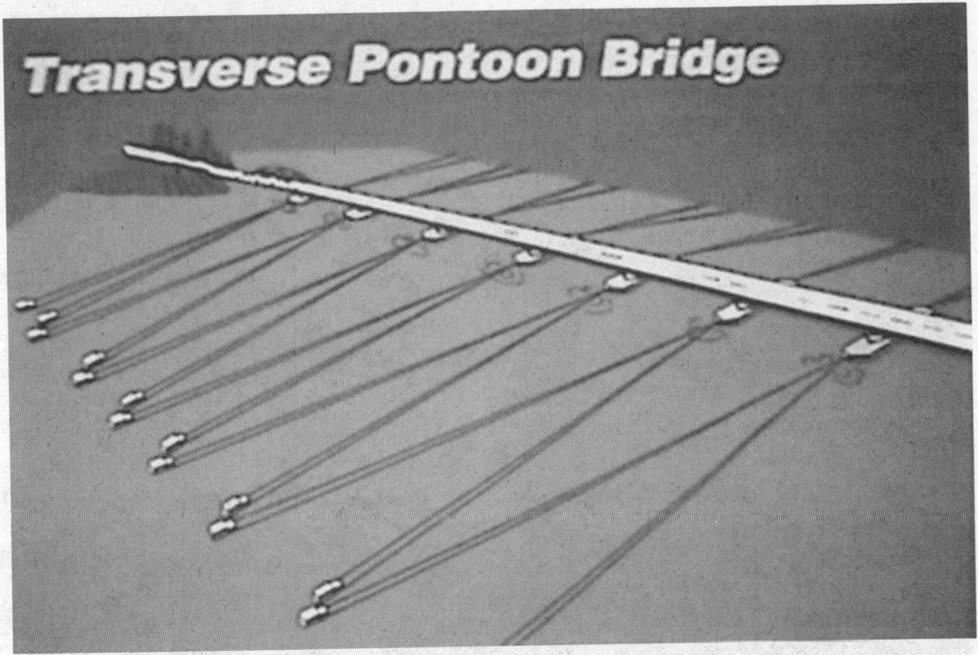

FIGURE 22.5 Separate pontoon-type structure.

such as vessel collisions and major storms. They perform well as highway structures with a high-quality roadway surface for safe driving in most weather conditions. They are uniquely attractive and have low impact on the environment. They are very cost-effective bridge types for water crossing where the water is deep (say, over 30 m) and wide (say, over 900 m), but the currents must not be very swift (say, over 6 knots), the winds not too strong (say, average wind speed over 160 km/h), and the waves not too high (say, significant wave height over 3 m).

22.3.2 Anchoring Systems

A floating structure may be held in place in many ways — by a system of piling, caissons, mooring lines and anchors, fixed guide structures, or other special designs. The most common anchoring system consists of a system of mooring lines and anchors. This system is used in all the existing floating bridges in Washington State. The mooring lines are galvanized structural strands meeting ASTM A586. Different types of anchors may be used, depending on the water depth and soil condition. Four types of anchors are used in anchoring the floating bridges on Lake Washington, Seattle.

Type A anchors (Figure 22.6) are designed for placement in deep water and very soft soil. They are constructed of reinforced concrete fitted with pipes for water jetting. The anchors weigh from 60 to 86 tons each. They are lowered to the bottom of the lake and the water jets are turned on allowing the anchors to sink into the soft lake bottom to embed the anchors fully. Anchor capacity is developed through passive soil pressure.

Type B anchors (Figure 22.7) are pile anchors designed for use in hard bottom and in water depth less than 27 m. A Type B anchor consists of two steel H-piles driven in tandem to a specified depth. The piles are tied together to increase capacity.

Type C anchors (Figure 22.8) are gravity-type anchors, constructed of reinforced concrete in the shape of a box with an open top. They are designed for placement in deep water where the soil is too hard for jetting. The boxes are lowered into position and then filled with gravel to the specified weight.

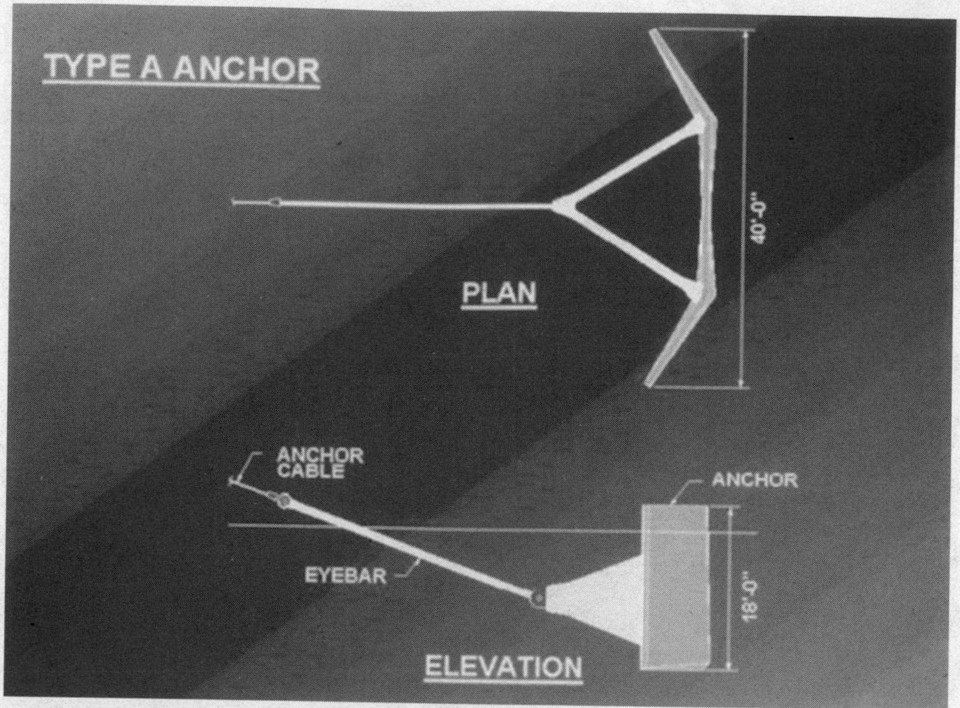

FIGURE 22.6 Type A anchor.

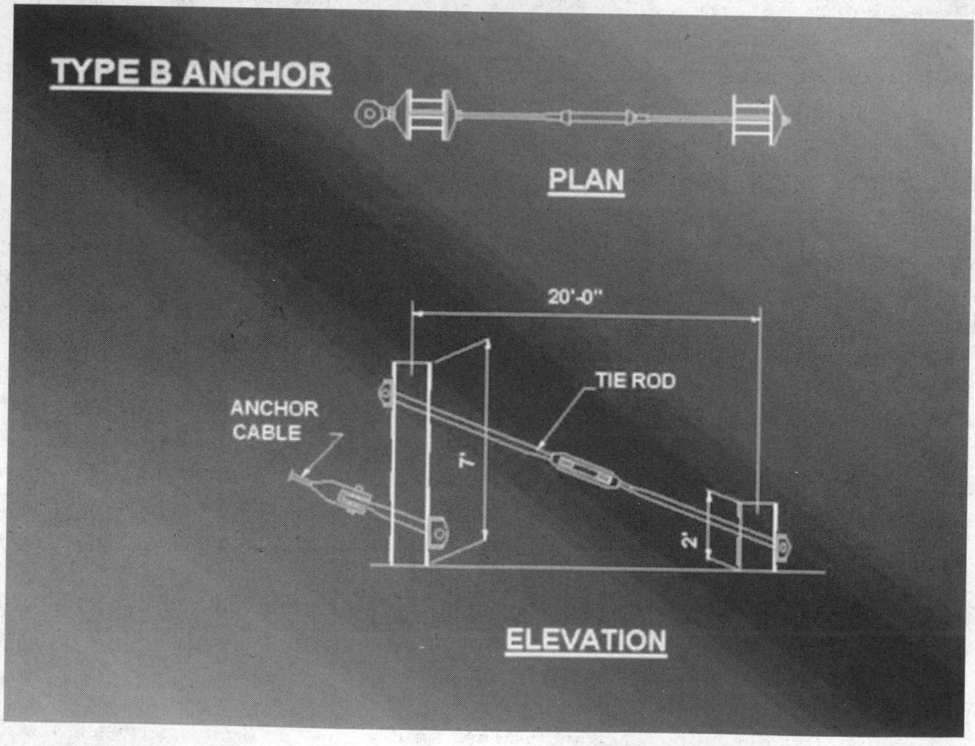

FIGURE 22.7 Type B anchor.

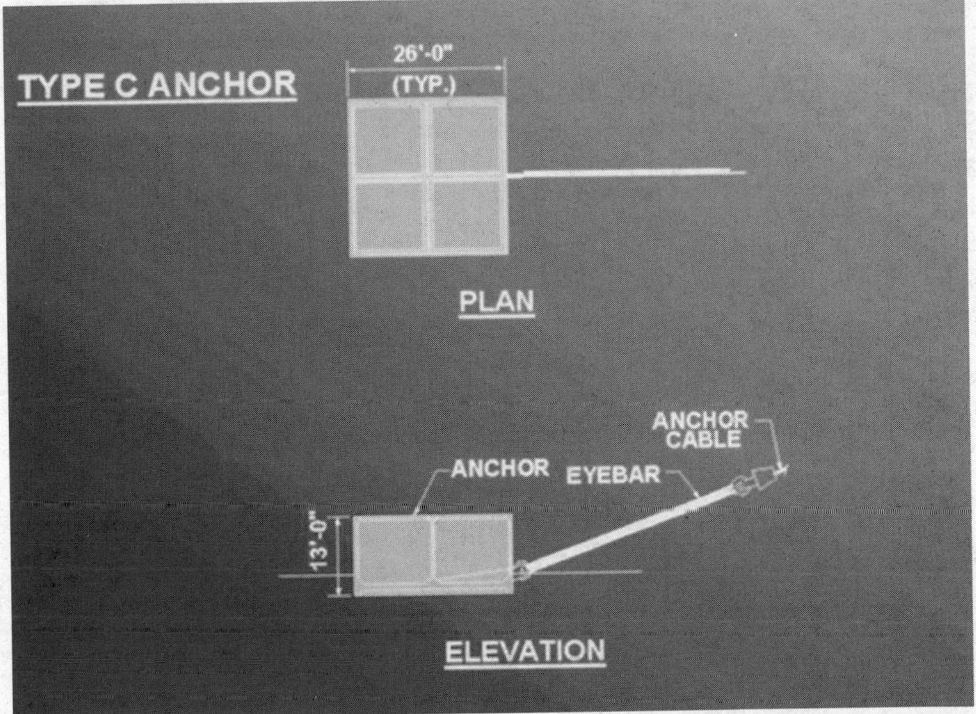

FIGURE 22.8 Type C anchor.

Type D (Figure 22.9) anchors are also gravity-type anchors like the Type C anchors. They consist of solid reinforced concrete slabs, each weighing about 270 tons. They are designed for placement in shallow and deep water where the soil is too hard for water jetting. The first slab is lowered into position and then followed by subsequent slabs. The number of slabs is determined by the anchor capacity required. Type D anchors are the choice over Type B and Type C anchors, because of the simplicity in design, ease in casting, and speed in placement.

22.4 Design Criteria

The design of a floating bridge follows the same good engineering practices as for land-based concrete or steel bridges. The design and construction provisions stipulated in the AASHTO *Standard Specifications for Highway Bridges* [7] or the *LRFD Bridge Design Specifications* [8] are applicable and should be adhered to as much as feasible. However, due to the fact that a floating bridge is floating on fresh or marine waters, the design criteria must address some special conditions inherent in floating structures. The performance of a floating bridge is highly sensitive to environmental conditions and forces, such as winds, waves, currents, and corrosive elements. The objectives of the design criteria are to assure that the floating bridge will

- Have a long service life of 75 to 100 years with low life-cycle cost;
- Meet functional, economical, and practical requirements;
- Perform reliably and be comfortable to ride on under normal service conditions;
- Sustain damage from accidental loads and extreme storms without sinking;
- Safeguard against flooding and progressive failure.

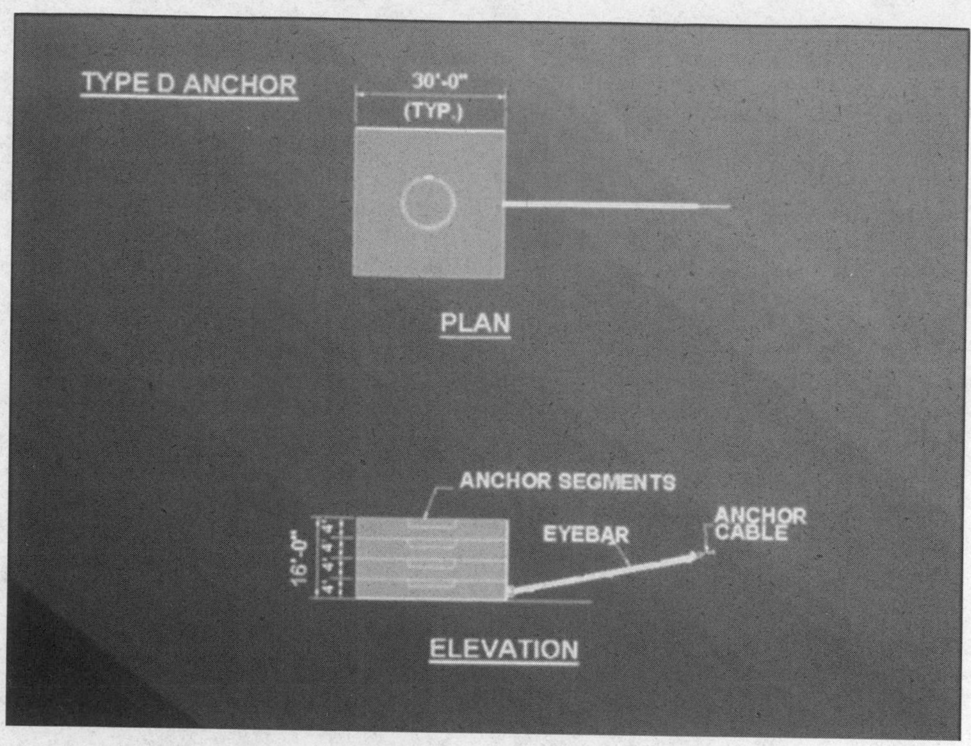

FIGURE 22.9 Type D anchor.

22.4.1 Loads and Load Combinations

The structure should be proportioned in accordance with the loads and load combinations for service load design and load factor design outlined in the AASHTO *Standard Specifications for Highway Bridges* or the AASHTO *LRFD Bridge Design Specifications*, except the floating portion of the structure shall recognize other environmental loads and forces and modify the loads and load combinations accordingly.

Winds and waves are the major environmental loads, while currents, hydrostatic pressures, and temperatures also have effects on the final design. Depending on the site conditions, other loadings, such as tidal variations, marine growth, ice, drift, etc., may need to be considered.

22.4.2 Winds and Waves

Winds and waves exert significant forces on a floating structure (Figure 22.10). Yet these environment loads are the most difficult to predict. Generally, there is a lack of long-term climatological data for a bridge site. A long record of observations of wind data is desirable for developing more accurate design wind speeds and wave characteristics. It is advisable to install instruments on potential bridge sites to collect climatological data as early as possible.

Wind blowing over water generates a sea state that induces horizontal, vertical, and torsional loads on the floating bridge. These loads are a function of wind speed, wind direction, wind duration, fetch length, channel configuration, and depth. Consideration must be given to the normal and extreme storm wind and wave conditions for the site. The normal storm conditions are defined as the storm conditions that have a mean recurrence interval of 1 year, which is the maximum storm that is likely to occur once a year. The extreme storm conditions are defined as the storm conditions that have a mean recurrence interval of 100 years, which is the maximum storm that is likely to occur once in 100 years. These wind and wave forces may be denoted by

FIGURE 22.10 Wind and wave forces (figure computer generated).

WN = Normal Wind on Structure — 1-year Storm
NW = Normal Wave on Structure — 1-year Storm
WS = Extreme Storm Wind on Structure — 100-year Storm
SW = Extreme Storm Wave on Structure — 100-year Storm

The following modifications are recommended for the AASHTO Load Combinations where wind loads are included:

1. Substitute WS + SW for W, and WN + NW for 0.3W.
2. Use one half the temperature loads in combination with WS and SW.
3. Omit L, I, WL, and LF loads when WS and SW are used in the design.

A 20-year wind storm condition is normally used to make operational decisions for closing the bridge to traffic to ensure safety and comfort to the traveling public. This is especially important when there is excessive motion and water spray over the roadway. When there is a movable span in the floating bridge for providing navigation openings, a 20-year wind storm is also used to open the movable span to relieve the pressures on the structure.

Following is an example of a set of wind and wave design data:

Return Interval	Wind Speed (1-min average)	Significant Wave Height	Period
1-year wind storm	76 km/h	0.85 m	3.23 s
20-year wind storm	124 km/h	1.55 m	4.22 s
100-year wind storm	148 km/h	1.95 m	4.65 s

22.4.3 Potential Damage

A floating bridge must have adequate capacity to safely sustain potential damages (DM) resulting from small vessel collision, debris or log impact, flooding, and loss of a mooring cable or component. Considering only one damage condition and location at any one time, the pontoon structure must be designed for at least the following:

1. *Collision*: Apply a 45-kN horizontal collision load as a service load to the pontoon exterior walls. Apply a 130-kN horizontal collision load as a factored load to the pontoon exterior wall. The load may be assumed to be applied to an area no greater than 0.3 × 0.3 m.
2. *Flooding*:
   - Flooding of any two adjacent exterior cells along the length of the structure
   - Flooding of all cells across the width of the pontoon
   - Flooding of all the end cells of an isolated pontoon during towing
   - Flooding of the outboard end cells of a partially assembled structure
3. *Loss of a mooring cable or component.*
4. *Complete separation of the floating bridge by a transverse or diagonal fracture.* This condition should apply to the factored load combinations only.

The above potential DM loadings should be combined with the AASHTO Groups VI, VIII, and IX combinations for Service Load Design and Load Factor Design. If the AASHTO LRFD Bridge Design Specifications are used in the design, the loads from Items 2 to 4 may be considered as extreme events.

Every floating structure is unique and specific requirements must be established accordingly. Maritime damage criteria and practices, such as those for ships and passenger vessels, should be reviewed and applied where applicable in developing damage criteria for a floating structure. However, a floating bridge behaves quite differently than a vessel, in that structural restraint is much more dominant than hydrostatic restraint. The trim, list, and sinkage of the flooded structure are relatively small. With major damage, structural capacity is reached before large deformations occurred or were observed. This is an important fact to note when comparing with stability criteria for ships.

22.4.4 Control Progressive Failure

While water provides buoyancy to keep a floating bridge afloat, water leaking into the interior of a floating bridge can cause progressive failure, eventually sinking the bridge. Time is of the essence when responding to damage or flooding. Maintenance personnel must respond to damage of a floating bridge quickly, especially when water begins to leak into the structure. An electronic cell monitoring system with water sensors to detect water entry and provide early warning to the maintenance personnel should be installed to assure timely emergency response. A bilge piping system should also be installed in the bridge for pumping out water.

It is important to control progressive failure in a floating bridge caused by flooding resulting from structural damage. Damage to the floating bridge could occur from a wind storm, a collision by a boat, severing of mooring lines, or other unforeseen accidents. The interior of the pontoons should be divided into small watertight compartments or cells (Figure 22.11) to confine flooding to only a small portion of the bridge. Access openings in the exterior or interior wall or bulkheads shall be outfitted with watertight doors.

Water sensors may be installed in each watertight compartment for early detection and early warning of water entry. A bilge piping system may be installed in the compartments for pumping out water when necessary. In such cases, pumping ports and quick disconnect couplings should be provided for pumping from a boat or vehicle equipped with pumps.

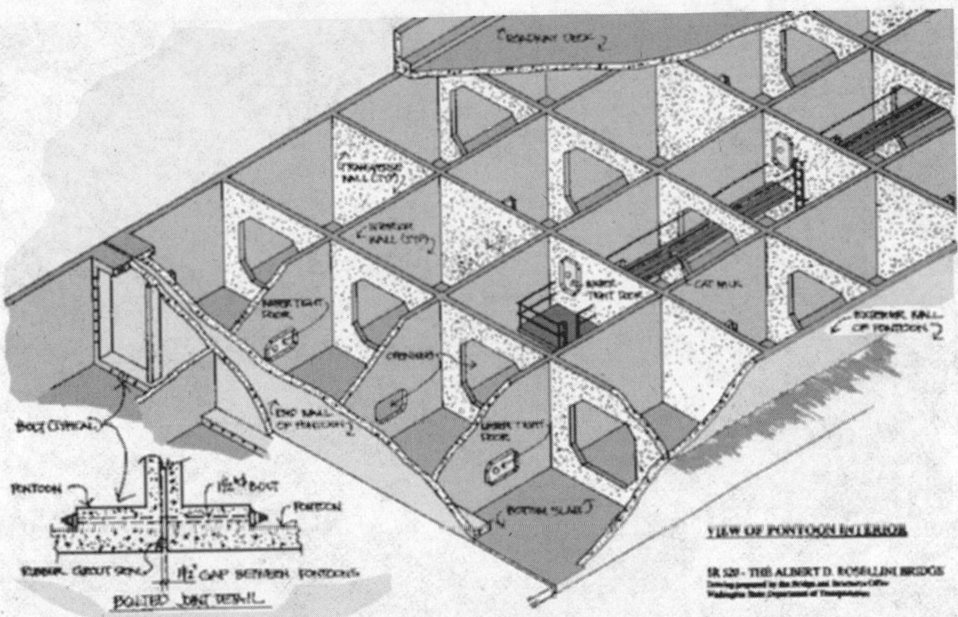

FIGURE 22.11 Small watertight compartments.

22.4.5 Design of Concrete Members

The design of reinforced concrete members should be based on behavior at service load conditions as per AASHTO Standard Specifications, or the service limit state as per AASHTO LRFD Bridge Design Specifications; except sections where reinforcement is to resist sustained hydrostatic forces the allowable stress in the reinforcing steel should not exceed 97 Mpa to limit crack width to a maximum of 0.10mm.

Prestressed members should be designed under the applicable service load and load factor provisions in the AASHTO Standard Specifications or the limit states provisions in the AASHTO LRFD Bridge Design Specifications, except the allowable concrete tensile stress under final conditions in the precompressed tensile zone should be limited to zero.

The ultimate flexural strength of the overall pontoon section should be computed for a maximum crack width of 0.25 mm and should not be less than the loads from the factored load combinations or 1.3 times the cracking moment, M_{cr}.

In a moderate climate, the following temperature differentials between the various portions of a floating bridge may be used:

1. Between the exposed portion and the submerged portion of a pontoon: ±19°C. The exposed portion may be considered as the top slab, and the remaining part of the pontoon as submerged. If the top slab is shaded by an elevated structure, the differential temperature may be reduced to ±14°C
2. Between the top slab and the elevated structure of the pontoon: ±14°C

The effects of creep and shrinkage should be considered while the pontoons are in the dry only. Creep and shrinkage may be taken as zero once the pontoons are launched. The time-dependent effects of creep and shrinkage may be estimated in accordance with the AASHTO Specifications. A final differential shrinkage coefficient of 0.0002 should be considered between the lower portion of the pontoon and the top slab of the pontoon.

High-performance concrete containing fly ash and silica fume is most suitable for floating bridges [9,10]. The concrete is very dense, impermeable to water, highly resistant to abrasion, and relatively

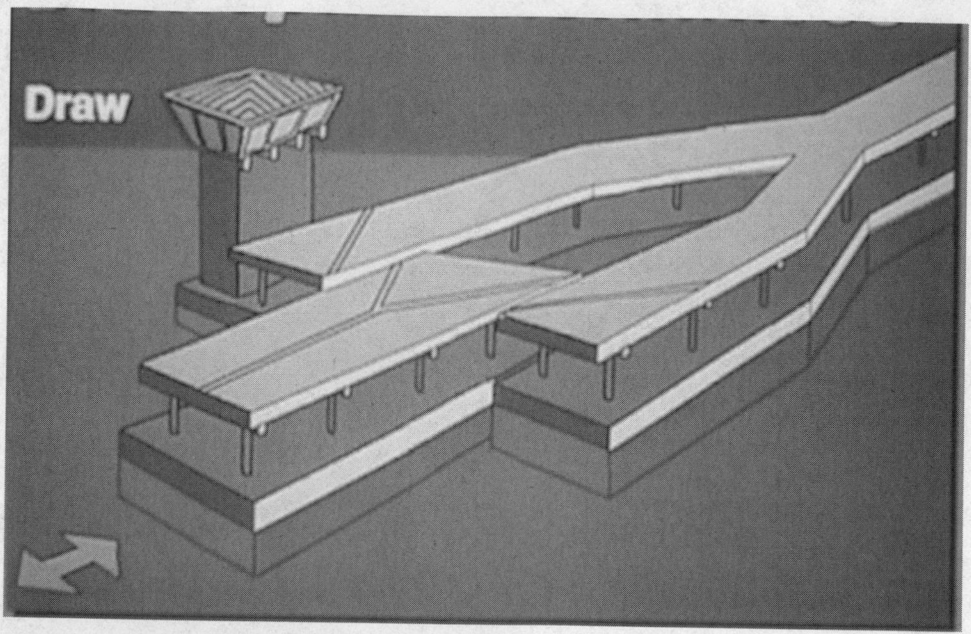

FIGURE 22.12 Draw-type movable span.

crack free. High-performance concrete also has high strength, low creep, and low shrinkage. Concrete mixes may be customized for the project.

Recommended minimum concrete cover of reinforcing steel:

	Fresh Water	Salt Water
Top of roadway slab	65 mm	65 mm
Exterior surfaces of pontoons and barrier	38 mm	50 mm
All other surfaces	25 mm	38 mm

22.4.6 Anchoring System

An anchoring system should be installed in the floating bridge to maintain transverse and horizontal alignment. The anchoring system should be designed to have adequate capacity to resist transverse and longitudinal forces from winds, waves, and current.

Adequate factors of safety or load and resistance factors consistent with the type of anchoring should be included in the design of the components of the system.

22.4.7 Movable Span

A floating bridge creates an obstruction to marine traffic. Movable spans may need to be provided for the passage of large vessels. The width of opening that must be provided depends on the size and type of vessels navigating through the opening. Movable spans of up to a total opening of 190 m have been used.

Two types of movable spans are used in Washington State — the draw-type and the lift/draw type. In the draw type movable span, the draw pontoons retract into a "lagoon" formed by flanking pontoons (Figure 22.12). Vehicles must maneuver around curves at the "bulge" where the "lagoon" is formed. In the lift/draw type of movable span, part of the roadway will be raised for the draw pontoons to retract underneath it (Figure 22.13). As far as traffic safety and flow are concerned, the lift/draw-type movable span is superior over the draw type. Traffic moves efficiently on a straight alignment with no curves to contend with.

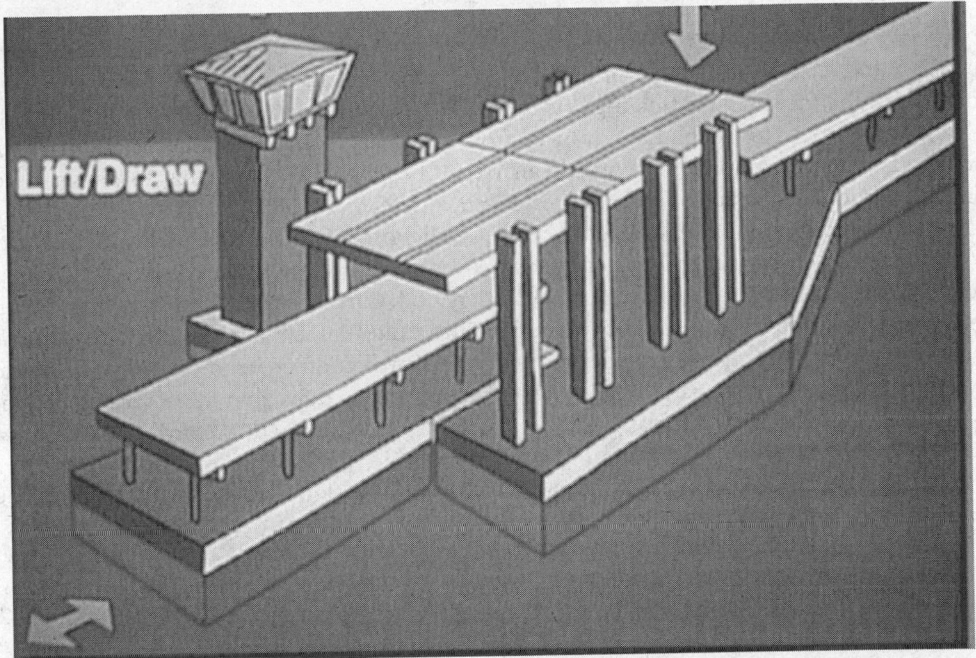

FIGURE 22.13 Lift/Draw-type movable span.

Movable spans may be operated mechanically or hydraulically. The design of the movable spans should be in accordance with the latest AASHTO *Standard Specifications for Movable Highway Bridges* [11].

22.4.8 Deflection and Motion

Floating bridges should be designed so that they are comfortable to ride on during normal storm (1-year storm) conditions and also to avoid undesirable structural effects during extreme storm (100-year storm) conditions. Deflection and motion criteria have been used to meet these objectives. The following deflection and motion limits for normal storm (1-year storm) conditions may be used as guidelines:

Loading Condition	Type of Deflection or Motion	Maximum Deflection	Maximum Motion
Vehicular load	Vertical	$L/800$	
Winds — static	Lateral (drift)	0.3 m	
	Rotation (heel)	0.5°	
Waves — dynamic	Vertical (heave)	±0.3 m	0.5 m/s^2
	Lateral (sway)	±0.3 m	0.5 m/s^2
	Rotation (roll)	±0.5°	0.05 rad/s^2

The objective of the motion limits are to assure that the people will not experience discomfort walking or driving across the bridge during a normal storm. The motion limit for rotation (roll) under the dynamic action of waves should be used with care when the roadway is elevated a significant distance above the water surface. The available literature contains many suggested motion criteria for comfort based on human perceptions [12]. A more-detailed study on motion criteria may be warranted for unusual circumstances.

22.5 Structural Design and Analysis

The design and analysis of floating bridges have gone through several stages of progressive development since the first highway floating bridge was designed and built in the late 1930s across Lake Washington, Seattle. The design has advanced from empirical methods to realistic approach, from the equivalent static approach to dynamic analysis, from computer modeling to physical model testing, and from reinforced concrete to prestressed concrete.

The most difficult part of early designs was the prediction of winds and waves, and the response due to wind–wave–structure interaction. Climatological data were very limited. The wind–wave–structure interaction was not well understood. Current state of the knowledge in atmospheric sciences, computer science, marine engineering, finite-element analysis, structural engineering, and physical model testing provides more accurate prediction of wind and wave climatology, more realistic dynamic analysis of wind–wave–structure interactions, and more reliable designs.

Designing for static loads, such as dead and live loads, is very straightforward using the classical theory on beam on elastic foundation [13]. For example, the maximum shear, moment, and deflection due to a concentrated load, *P*, acting away from the end of a continuous floating structure are given by

$$V_{\max} = \frac{P}{2} \tag{22.1}$$

$$M_{\max} = \frac{P}{4\lambda} \tag{22.2}$$

$$y_{\max} = \frac{P\lambda}{2k} \tag{22.3}$$

where k = modulus of foundation

$$\lambda = 4\sqrt{\frac{k}{4EI}}$$

Designing for the response of the structure to winds and waves is more complex, because of the random nature of these environment loads. To determine the dynamic response of the bridge to wind generated waves realistically, a dynamic analysis is necessary.

22.5.1 Preliminary Design

The design starts with selecting the type, size, and location of the floating bridge (Figure 22.14). Assuming a concrete box section of cellular construction with dimensions as shown in Figure 22.15, the first step is to determine the freeboard required. The height of the freeboard is selected to avoid water spray on the roadway deck from normal storms. The draft can then be determined as necessary to provide the selected freeboard.

The freeboard and draft of the floating structure should be calculated based on the weight of concrete, weight of reinforcing steel, weight of appurtenances, weight of marine growth as appropriate, and vertical component of anchor cable force. The weight of the constructed pontoon is generally heavier than the computed weight, because of form bulging and other construction tolerances. Based on the experience in Washington State, the weight increase varies from 3 to 5% of the theoretical weight. This increase should be included in the draft calculation. Additionally, floating pontoons experience loss in freeboard in the long term. The main reason is due to weight

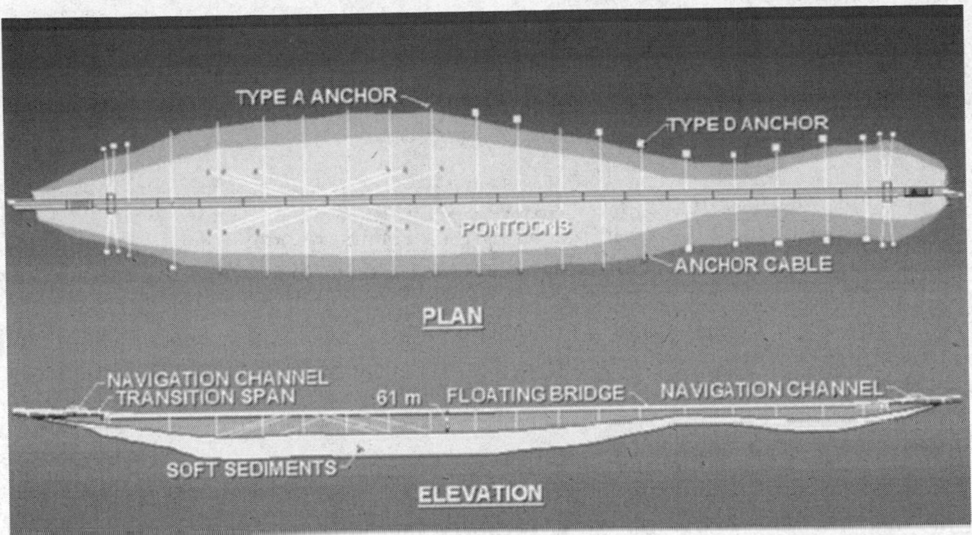

FIGURE 22.14 General layout.

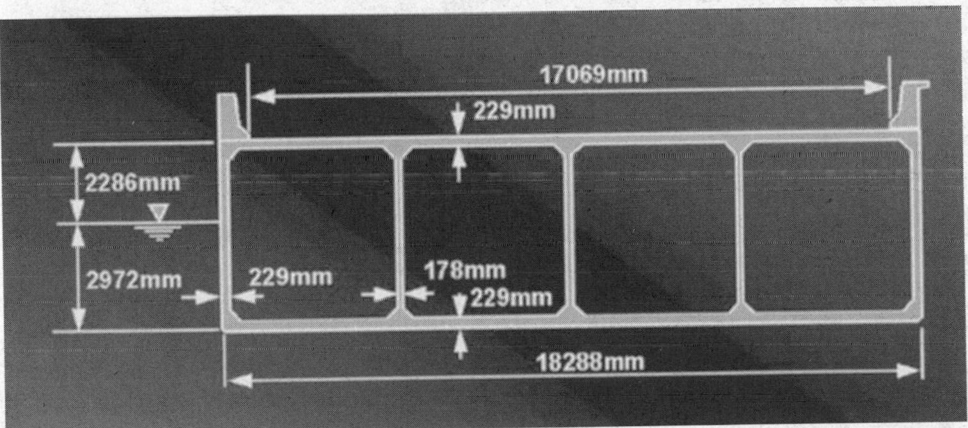

FIGURE 22.15 Typical cross section.

added as a result of modifications in the structural and mechanical elements throughout the service life of the structure. It is prudent to make allowance for this in the design. This can be done by allowing 150 to 230 mm of extra freeboard in the new bridge.

The thicknesses for the walls and slabs should be selected for local and global strength and constructibility. The wall thickness should be the minimum needed to provide adequate concrete cover to the reinforcing steel and adequate space for post-tensioning ducts when used. There must also be adequate room for depositing and consolidating concrete. The objective should be to keep the weight of the structure to a minimum, which is essential for cost-effective design of a floating bridge.

The exterior walls of the pontoons should be designed for wave plus hydrostatic pressures and the collision loads. The bottom slab should be designed for wave plus hydrostatic pressures. The interior walls should be designed for hydrostatic pressures due to flooding of a cell to full height of the wall. The roadway slab should be designed for the live load plus impact in the usual way.

The preliminary design gives the overall cross-sectional dimensions and member thicknesses required to meet local demands and construction requirements. The global responses of the floating bridge will be predicted by dynamic analysis.

22.5.2 Dynamic Analysis

The basic approach to dynamic analysis is to solve the equation of motion:

$$M\ddot{X} + C\ddot{X} + KX = F(t) \tag{22.4}$$

This equation is familiar to structural engineering in solving most structural dynamic problems of land-based structures. However, in predicting the dynamic response of a floating bridge, the effects of water–structure interaction must be accounted for in the analysis. As a floating bridge responds to the incident waves, the motions (heave, sway, and roll) of the bridge produce hydrodynamic effects generally characterized in terms of added mass and damping coefficients. These hydrodynamic coefficients are frequency dependent. The equation of motion for a floating structure takes on the general form:

$$[M + A]\ddot{X} + [C_1 + C_2]\dot{X} + [K + k]X = F(t) \tag{22.5}$$

where

$X, \dot{X}, \ddot{X}$ = generalized displacement, velocity and acceleration at each degree of freedom
M = mass–inertia matrix of the structure
A = added mass matrix (frequency dependent)
C_1 = structural damping coefficient
C_2 = hydrodynamic damping coefficient (frequency dependent)
K = structural stiffness matrix (elastic properties, including effects of mooring lines when used)
k = hydrostatic stiffness (hydrostatic restoring forces)
$F(t)$ = forces acting on the structure

A substantial amount of experimental data has been obtained for the hydrodynamic coefficients for ships and barges [14,15]. Based on these experimental data, numerical methods and computer programs have been developed for computing hydrodynamic coefficients of commonly used cross-sectional shapes, such as the rectangular shape. For structural configurations for which no or limited data exist, physical model testing will be necessary to determine the basic sectional added mass, damping, and wave excitation loads.

Structural damping is an important source of damping in the structure. It significantly affects the responses. A structural damping coefficient of 2 to 5% of critical damping is generally assumed for the analysis. It is recommended that a better assessment of the damping coefficient be made to better represent the materials and structural system used in the final design.

The significant wave height, period, and central heading angle may be predicted using the public domain program NARFET developed by the U.S. Army Coastal Engineering Research Center [16,17]. This program accounts for the effective fetch to a location on the floating structure by a set of radial fetch lines to the point of interest. The Joint North Sea Wave Project (JONSWAP) spectrum is commonly used to represent the frequency distribution of the wave energy predicted by the program NARFET. This spectrum is considered to represent fetch-limited site conditions very well. A spreading function is used to distribute the energy over a range of angles of departure from the major storm heading to the total energy [18]. The spreading function takes the form of an even cosine function, $\cos^{2n}\theta$, where θ is the angle of the incident wave with respect to the central heading angle. Usually $2n$ is 2 or greater. $2n = 2$ is generally used for ocean structures where the structures are relatively small with respect to the open sea. In the case of a floating bridge, the bridge length is very large in comparison to the body of water, resulting in very little energy distributed away from the central heading angle. A larger number of $2n$ will have to be used for a floating bridge. The larger the number of $2n$ the more focused the wave direction near the central heading angle. A $2n$ value of 12 to 16 have been used in analyzing the floating bridges in Washington State. The value of $2n$ should be selected with care to reflect properly the site condition and the wind and wave directions.

The equation of motion may be solved by the time-domain (deterministic) analysis or the frequency-domain (probabilistic) analysis. The time-domain approach involves solving differential equations when the coefficients are constants. The equations become very complex when the coefficients are frequency dependent. This method is tedious and time-consuming. The frequency-domain approach is very efficient in handling constant and frequency-dependent coefficients. The equations are algebraic equations. However, time-dependent coefficients are not admissible and nonlinearities will have to be linearized by approximation. For the dynamic analysis of floating bridges subjected to the random nature of environmental forces and the frequency-dependent hydrodynamic coefficients, frequency-domain analysis involves only simple and fast calculations.

22.5.3 Frequency-Domain Analysis

The frequency-domain dynamic analysis is based on the principles of naval architecture and the strip theory developed for use in predicting the response of ships to sea loads [19,20]. The essence of this approach is the assumption that the flow at one section through the structure does not affect the flow at any other section. Additional assumptions are (1) the motions are relatively small, (2) the fluid is incompressible and invisid, and (3) the flow is irrotational. By using the strip theory, the problem of wave–structure interaction can be solved by applying the equation of motion in the frequency domain [21,22]. By Fourier transform, the equation of motion may be expressed in terms of frequencies, ω, as follows:

$$\left\{-\omega^2[M+A]+i\omega\left[C_1+C_2\right]+[K+k]\right\}=\left\{F(\omega)\right\} \tag{22.6}$$

This equation may be solved as a set of algebraic equations at each frequency and the responses determined. The maximum bending moments, shears, torsion, deflections, and rotations can then be predicted using spectral analysis and probability distribution [23,24]. The basic steps involved in a frequency-domain analysis are

1. Compute the physical properties of the bridge — geometry of the bridge elements, section properties, connections between bridge elements, mass–inertia, linearized spring constants, structural damping, etc.
2. Compute hydrodynamic coefficients — frequency-dependent added mass and frequency-dependent damping.
3. Compute hydrostatic stiffnesses.
4. Calculate wind, wave, and current loads, and other loading terms.
5. Build a finite-element computer model of the bridge [25] as a collection of nodes, beam elements, and spring elements. The nodes form the joints connecting the beam elements and the spring elements, and each node has six degrees of freedom.
6. Solve the equation of motion in the frequency domain to obtain frequency responses, the magnitudes of which are referred to as response amplitude operators (RAOs).
7. Perform spectral analysis, using the RAOs and the input sea spectrum, to obtain the root mean square (RMS) of responses.
8. Perform probability analysis to obtain the maximum values of the responses with the desired probability of being exceeded.
9. Combine the maximum responses with other loadings, such as wind, current, etc., for final design.

22.6 Fabrication and Construction

Concrete pontoons are generally used for building major floating bridges. The fabrication and construction of the concrete pontoons must follow the best current practices in structural and marine engineering in concrete design, fabrication, and construction, with added emphasis on high-quality concrete and watertightness. Quality control should be the responsibility of the fabricators/contractors. Final quality assurance and acceptance should be the responsibility of the owners. In addition to these traditional divisions of responsibilities, the construction of a floating bridge necessitates a strong "partnership" arrangement to work together, contracting agencies and contractors, to provide full cooperation and joint training, share knowledge and expertise, share responsibility, and to help each other succeed in building a quality floating bridge. The contractors should have experience in marine construction and engage the services of naval architects or marine engineers to develop plans for monitoring construction activities and identifying flood risks, and prepare contingency plans for mitigating the risks.

Knowledge is power and safety. The construction personnel, including inspectors from the contracting agencies and the contractors, should be trained on the background of the contract requirements and the actions necessary to implement the requirements fully. Their understanding and commitment are necessary for complete and full compliance with contract requirements that bear on personal and bridge safety.

Construction of floating bridges is well established. Many concrete floating bridges have been built successfully using cast-in-place, precast, or a combination of cast-in-place and precast methods. Construction techniques are well developed and reported in the literature. Owners of floating bridges have construction specifications and other documents and guidelines for the design and construction of such structures.

Floating bridges may be constructed in the dry in graving docks or on slipways built specifically for the purpose. However, construction on a slipway requires more extensive preparation, design, and caution. The geometry and strength of the slipway must be consistent with the demand of the construction and launching requirements. Construction in a graving dock utilizes techniques commonly used in land-based structures. Major floating bridges around the world have been constructed in graving docks (Figure 22.16).

Because of the size of a floating bridge, the bridge is generally built in segments or pontoons compatible with the graving dock dimensions and draft restrictions. The segments or pontoons are floated and towed (Figure 22.17) to an outfitting dock where they are joined and completed in larger sections before towing to the bridge site, where the final assembly is made (Figure 22.18).

It is important to explore the availability of construction facilities and decide on a feasible facility for the project. These actions should be carried out prior to or concurrent with the design of a floating bridge to optimize the design and economy. Some key data that may be collected at this time are

- Length, width, and draft restrictions of the graving dock;
- Draft and width restrictions of the waterways leading to the bridge site;
- Wind, wave, and current conditions during tow to and installation at the job site.

The designers will use the information to design and detail the structural plans and construction specifications accordingly.

22.7 Construction Cost

The construction costs for floating bridges vary significantly from project to project. There are many variables that affect the construction costs. The following construction costs in U.S. dollars for the floating bridges in Washington State are given to provide a general idea of the costs of building

FIGURE 22.16 Construction in a graving dock.

FIGURE 22.17 Towing pontoon.

FIGURE 22.18 Final assembly.

concrete floating bridges in the past. These are original bid costs for the floating portions of the bridges. They have not been adjusted for inflation and do not include the costs for the approaches.

Name of Bridge	Length, m	Width of Pontoon, m	Lanes of Traffic	Cost, U.S. $ million
Original Lacey V. Murrow Bridge[a]	2018	17.9	4	3.25 (1938)
Evergreen Point Bridge[a]	2310	18.3	4	10.97 (1960)
Original Hood Canal Bridge[a]	1972	15.2	2	17.67 (1961)
Homer Hadley Bridge	1771	22.9	5	64.89 (1985)
New Lacey V. Murrow Bridge	2018	18.3	3	73.78 (1991)

[a] These bridges have movable spans which increased construction costs.

22.8 Inspection, Maintenance, and Operation

A floating bridge represents a major investment of resources and a commitment to efficiency and safety to the users of the structure and the waterway. To assure trouble-free and safe performance of the bridge, especially one with a movable span, an inspection, maintenance, and operation manual (Manual) should be prepared for the bridge. The main purpose of the Manual is to provide guidelines and procedures for regular inspection, maintenance, and operation of the bridge to extend the service life of the structure. Another aspect of the Manual is to define clearly the responsibilities of the personnel assigned to inspect, maintain, and operate the bridge. The Manual must address the specific needs and unique structural, mechanical, hydraulic, electrical, and safety features of the bridge.

The development of the Manual should begin at the time the design plans and construction specifications are prepared. This will assure that the necessary inputs are given to the designers to help with preparation of the Manual later. The construction specifications should require the contractors to submit documents, such as catalog cuts, schematics, electrical diagrams, etc., that will be included in the Manual. The Manual should be completed soon after the construction is completed and the bridge is placed into service. The Manual is a dynamic document. Lessons learned and modifications made during the service life of the structure should be incorporated into the Manual on a regular basis.

A training program should be developed for the supervisors and experienced co-workers to impart knowledge to new or inexperienced workers. The training program should be given regularly and aimed at nurturing a positive environment where workers help workers understand and diligently apply and update the guidelines and procedures of the Manual.

22.9 Closing Remarks

Floating bridges are cost-effective alternatives for crossing large lakes with unusual depth and soft bottom, spanning across picturesque fjords, and connecting beautiful islands. For conditions like Lake Washington in Seattle where the lake is over 1610 m wide, 61 m deep, and another 61 m of soft bottom, a floating bridge is estimated to cost three to five times less than a long-span fixed bridge, tube, or tunnel.

The bridge engineering community has sound theoretical knowledge, technical expertise, and practical skills to build floating bridges to enhance the social and economic activities of the people. However, it takes time to plan, study, design, and build floating bridges to form major transportation links in a local or national highway system. There are environmental, social, and economic issues to address. In the State of Washington, the first floating bridge was conceived in 1920, but was not built and opened to traffic until 1940. After over 30 years of planning, studies, and overcoming environmental, social, and economic issues, Norway finally has the country's first two floating bridges opened to traffic in 1992 and 1994. It is never too early to start the planning process and feasibility studies once interest and a potential site for a floating bridge is identified.

Every floating bridge is unique and has its own set of technical, environmental, social, and economic issues to address during preliminary and final engineering:

- *Winds and waves*: Predicting accurately the characteristics of winds and waves has been a difficult part of floating bridge design. Generally there is inadequate data. It is advisable to install instruments in potential bridge sites to collect climatological data as early as possible. Research in the area of wind–wave–structure interactions will assure safe and cost-effective structures.
- *Earthquake*: Floating bridges are not directly affected by ground shakings from earthquakes.
- *Tsunami and seiches*: These may be of particular significant in building floating bridges at sites susceptible to these events. The dynamic response of floating bridges to tsunami and seiches must be studied and addressed in the design where deemed necessary.
- *Corrosion*: Materials and details must be carefully selected to reduce corrosion problems to assure long service life with low maintenance.
- *Progressive failure*: Floating bridges must be designed against progressive failures by dividing the interiors of pontoons into small watertight compartments, by installing instruments for detecting water entry, and by providing means to discharge the water when necessary.
- *Riding comfort and convenience*: Floating bridges must be comfortable to ride on during minor storms. They must have adequate stiffness and stability. They must not be closed to vehicular traffic frequently for storms or marine traffic.
- *Public acceptance*: Public acceptance is a key part of modern civil and structural engineering. The public must be educated regarding the environmental, social, and economic impacts of

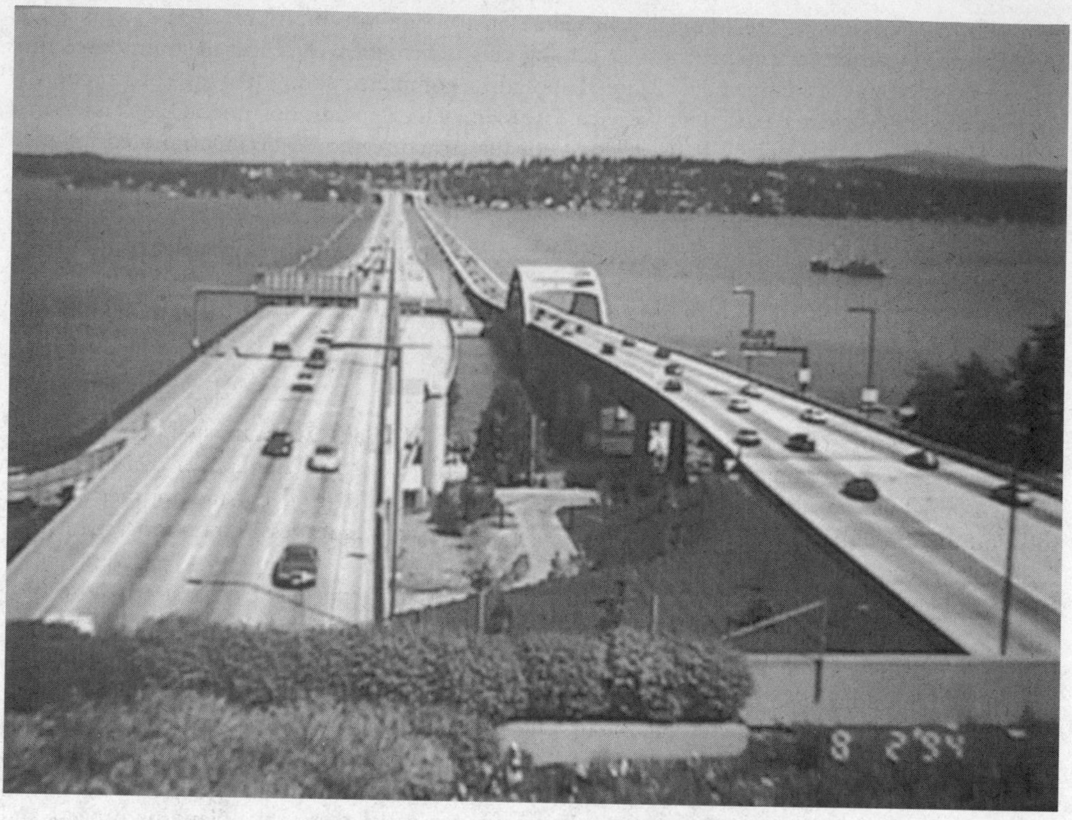

FIGURE 22.19 Floating Bridges on Lake Washington.

proposed projects. Reaching out to the public through community meetings, public hearings, news releases, tours, exhibits, etc. during the early phase of project development is important to gain support and assure success. Many major public projects have been delayed for years and years because of lack of interaction and understanding.

- *Design criteria*: The design criteria must be carefully developed to meet site-specific requirements and focused on design excellence and cost-effectiveness to provide long-term performance and durability. Design excellence and economy come from timely planning, proper selection of materials for durability and strength, and paying attention to design details, constructibility, and maintainability. The design team should include professionals with knowledge and experience in engineering, inspection, fabrication, construction, maintenance, and operation of floating bridges or marine structures.
- *Construction plan*: It is essential to have a good set of construction plans developed jointly by the contracting agency and the contractor to clearly address qualifications, materials control, quality control, quality assurance, acceptance criteria, post-tensioning techniques, repair techniques, launching and towing requirements, weather conditions, flood control and surveillance.

Well-engineered and maintained floating bridges are efficient, safe, durable, and comfortable to ride on. They form important links in major transportation systems in different parts of the world (Figure 22.19).

References

1. Andrew, C. E., The Lake Washington pontoon bridge, *Civil Eng.*, 9(12), 1939.
2. Lwin, M. M., Floating bridges — solution to a difficult terrain, in *Proceedings of the Conference on Transportation Facilities through Difficult Terrain*, Wu, J. T. H. and Barrett, R. K., Eds., A.A. Balkema, Rotterdam, 1993.
3. Nichols, C. C., Construction and performance of hood canal floating bridge, in *Proceedings of Symposium on Concrete Construction in Aqueous Environment*, ACI Publication SP-8, Detroit, MI, 1962.
4. Pegusch, W., The Kelowna floating bridge, in *The Engineering Journey*, the Engineering Institute of Canada, Canada, 1957.
5. Landet, E., Planning and construction of floating bridges in Norway, Proceedings of International Workshop on Floating Structures in Coastal Zone, Port and Harbour Research Institute, Japan, 1994.
6. Gloyd, C. S., Concrete floating bridges, *Concrete Int.*, 10(7), 1988.
7. AASHTO, *Standard Specifications for Highway Bridges*, 16th ed., AASHTO, Washington, D.C., 1996.
8. AASHTO, *LRFD Bridge Design Specifications*, AASHTO, Washington, D.C., 1994.
9. Lwin, M. M., Bruesch, A. W., and Evans, C. F., High performance concrete for a floating bridge, in *Proceedings of the Fourth International Bridge Engineering Conference*, Vol. 1, Federal Highway Administration, Washington, D.C., 1995.
10. Lwin, M. M., Use of high performance concrete in highway bridges in Washington State, in *Proceedings International Symposium on High Performance Concrete*, Prestressed Concrete Institute and Federal Highway Administration, New Orleans, 1997.
11. AASHTO, *Standard Specifications for Movable Highway Bridges*, AASHTO, Washington, D.C., 1988.
12. Bachman, H. and Amman, W., Vibrations in Structure Induced by Men and Machines, Structural Engineering Document No. 3e, International Association for Bridge and Structural Engineering, Zurich, Switzerland, 1987.
13. Hetenyi, M., *Beams on Elastic Foundation*, Ann Arbor, MI, 1979.
14. Frank, W., Oscillation of cylinders in or below the surface of deep fluids, Report No. 2375, Naval Ships Research and Development Center, 1967.
15. Garrison, C. J., Interaction of oblique waves with an infinite cylinder, *Appl. Ocean Res.*, 6(1), 1984.
16. *Shore Protection Manual*, Vol. 1, Coastal Engineering Research Center, Department of the Army, 1984.
17. Program NARFET, Waterways Experiment Station, Corps of Engineers, Coastal Engineering Research Center, Vicksburg, MI.
18. Mitsuyasu, H., Observations of the directional spectrum of ocean waves using a clover leaf buoy, *J. Phys. Oceanogr.*, Vol. 5, 1975.
19. Comstock, J., *Principles of Naval Architecture*, Society of Naval Architects and Marine Engineers, New York, 1975.
20. Salvesen, N., Tuck, E. O., and Faltinsen, O., Ship motions and sea loads, *Trans. Soc. Naval Architects Marine Eng.*, 78, 1970.
21. Engel, D. J. and Nachlinger, R. R., Frequency domain analysis of dynamic response of floating bridge to waves, in *Proceedings of Ocean Structural Dynamics Symposium*, Oregon State University, 1982.
22. Hutchison, B. L., Impulse response techniques for floating bridges and breakwaters subject to short-crested seas, *Marine Technol.*, 21(3), 1984.
23. Marks, W., *The Application of Spectral Analysis and Statistics to Seakeeping*, T&R Bulletin No. 1-24, Society of Naval Architects and Marine Engineers, New York, 1963.
24. Ochi, M. K., On prediction of extreme values, *J. Ship Res.*, 17(1), 1973.
25. Gray, D. L. and Hutchison, B. L., A resolution study for computer modeling of floating bridges, in *Proceedings of Ocean Structural Dynamics Symposium*, Oregon State University, 1986.

23

Railroad Bridges

Donald F. Sorgenfrei
Modjeski and Masters, Inc.

W. N. Marianos, Jr.
Modjeski and Masters, Inc.

23.1 Introduction

23.1.1 Railroad Network

The U.S. railroad network consists predominantly of privately owned freight railroad systems classified according to operating revenue, the government-owned National Railroad Passenger Corporation (Amtrak), and numerous transit systems owned by local agencies and municipalities.

Since the deregulation of the railroad industry brought about by the 1980 Staggers Act, there have been numerous railway system mergers. By 1997 there remained 10 Class I (major) Railroads, 32 Regional Railroads, and 511 Local Railroads operating over approximately 150,000 track miles. The 10 Class I Railroads comprise only 2% of the number of railroads in the United States but account for 73% of the trackage and 91% of freight revenue.

By far the present leading freight commodity is coal, which accounts for 25% of all the carloads. Other leading commodities in descending order by carloads are chemicals and allied products, farm products, motor vehicles and equipment, food and sundry products, and nonmetallic minerals. Freight equipment has drastically changed over the years in container type, size and wheelbase, and carrying capacity. The most predominant freight car is the hopper car used with an open top for coal loading and the covered hopper car used for chemicals and farm products. In more recent years special cars have been developed for the transportation of trailers, box containers, and automobiles. The

It should be noted that much of this material was developed for the American Railway Engineering and Maintenance of Way Association (AREMA) Structures Loading Seminar. This material is used with the permission of AREMA.

0-8493-7434-0/00/$0.00+$.50

average freight car capacity (total number of freight cars in service divided by the aggregate capacity of those cars) has risen approximately 10 tons each decade with the tonnage ironically matching the decades, i.e., 1950s — 50 tons, 1960s — 60 tons, and so on. As the turn of the century approaches, various rail lines are capable of handling 286,000 and 315,000-lb carloads, often in dedicated units.

In 1929 there were 56,936 steam locomotives in service. By the early 1960s they were nearly totally replaced by diesel electric units. The number of diesel electric units has gradually decreased as available locomotive horsepower has increased. The earlier freight trains were commonly mixed freight of generally light railcars, powered by heavy steam locomotives. In more recent years that has given way to heavy railcars, unit trains of common commodity (coal, grain, containers, etc.) with powerful locomotives. Newer locomotives generally have six axles, weigh 420,000 lbs, and can generate up to 8000 Hp.

These changes in freight hauling have resulted in concerns for railroad bridges, many of which were not designed for these modern loadings. The heavy, steam locomotive with steam impact governed in design considerations. Present bridge designs are still based on the steam locomotive wheel configuration with diesel impact, but fatigue cycles from the heavy carloads are of major importance.

The railroad industry records annual route tonnage referred to as "million gross tons" (MGT). An experienced railroader can fairly well predict conditions and maintenance needs for a route based on knowing the MGT for that route. It is common for Class I Railroads to have routes of 30 to 50 MGT with some coal routes in the range of 150 MGT.

Passenger trains are akin to earlier freight trains, with one or more locomotives (electric or diesel) followed by relatively light cars. Likewise, transit cars are relatively light.

23.1.2 Basic Differences between Railroad and Highway Bridges

A number of differences exist between railroad and highway bridges:

1. The ratio of live load to dead load is much higher for a railroad bridge than for a similarly sized highway structure. This can lead to serviceability issues such as fatigue and deflection control governing designs rather than strength.
2. The design impact load on railroad bridges is higher than on highway structures.
3. Simple-span structures are preferred over continuous structures for railroad bridges. Many of the factors that make continuous spans attractive for highway structures are not as advantageous for railroad use. Continuous spans are also more difficult to replace in emergencies than simple spans.
4. Interruptions in service are typically much more critical for railroads than for highway agencies. Therefore, constructibility and maintainability without interruption to traffic are crucial for railroad bridges.
5. Since the bridge supports the track structure, the combination of track and bridge movement cannot exceed the tolerances in track standards. Interaction between the track and bridge should be considered in design and detailing.
6. Seismic performance of highway and railroad bridges can vary significantly. Railroad bridges have performed well during seismic events.
7. Railroad bridge owners typically expect a longer service life from their structures than highway bridge owners expect from theirs.

23.1.3 *Manual for Railway Engineering,* AREMA

The base document for railroad bridge design, construction, and inspection is the American Railway Engineering Maintenance of Way Association (AREMA) *Manual for Railway Engineering* (*Manual*) [1].

Early railroads developed independent specifications governing the design loadings, allowable strains, quality of material, fabrication, and construction of their own bridges. There was a proliferation of specifications written by individual railroads, suppliers, and engineers. One of the earliest general specifications is titled *Specification for Iron Railway Bridges and Viaducts*, by Clarke, Reeves and Company (Phoenix Bridge Company). By 1899 private railroads joined efforts in forming AREMA. Many portions of those original individual railroad specifications were incorporated into the first manual titled *Manual of Recommended Practice for Railway Engineering and Maintenance of Way* published in 1905. In 1911 the Association dropped "Maintenance of Way" from its name and became the American Railway Engineering Association (AREA); however, in 1997 the name reverted back to the original name with the consolidation of several railroad associations.

The *Manual* is not deemed a specification but rather a recommended practice. Certain provisions naturally are standards by necessity for the interchange of rail traffic, such as track gauge, track geometrics, clearances, basic bridge loading, and locations for applying loadings. Individual railroads may, and often do, impose more stringent design requirements or provisions due to differing conditions peculiar to that railroad or region of the country, but basically all railroads subscribe to the provisions of the *Manual*.

Although the *Manual* is a multivolume document, bridge engineering provisions are grouped in the *Structural Volume* and subdivided into applicable chapters by primary bridge material and special topics, as listed:

Chapter 7 Timber Structures
Chapter 8 Concrete Structures and Foundations
Chapter 9 Seismic Design for Railway Structures
Chapter 10 Structures Maintenance & Construction (New)
Chapter 15 Steel Structures
Chapter 19 Bridge Bearings
Chapter 29 Waterproofing

The primary structural chapters each address bridge loading (dead load, live load, impact, wind, seismic, etc.) design, materials, fabrication, construction, maintenance/inspection, and capacity rating. There is uniformity among the chapters in the configuration of the basic live load, which is based on the Cooper E-series steam locomotive. The present live-load configuration is two locomotives with tenders followed by a uniform live load as shown in Fig. 23.1. There is not uniformity in the chapters in the location and magnitude of many other loads due to differences in the types of bridges built with different materials and differences in material behavior. Also it is recognized that each chapter has been developed and maintained by separate committee groups of railroad industry engineers, private consulting engineers, and suppliers. These committees readily draw from railroad industry experiences and research, and from work published by other associations such as AASHTO, AISC, ACI, AWS, APWA, etc.

23.2 Railroad Bridge Philosophy

Railroad routes are well established and the construction of new railroad routes is not common; thus, the majority of railroad bridges built or rehabilitated are on existing routes and on existing right-of-way. Simply stated, the railroad industry first extends the life of existing bridges as long as economically justified. It is not uncommon for a railroad to evaluate an 80- or 90-year-old bridge, estimate its remaining life, and then rehabilitate it sufficiently to extend its life for some economical period of time.

Bridge replacement generally is determined as a result of a lack of load-carrying capacity, restrictive clearance, or deteriorated physical condition. If bridge replacement is necessary, then simplicity, cost, future maintenance, and ease of construction without significant rail traffic disruptions typically govern the design. Types of bridges chosen are most often based on the capability of a railroad to do its own construction work. Low-maintenance structures, such as ballasted deck prestressed concrete box-girder spans with concrete caps and piles, are preferred by some railroads. Others may prefer weathering steel elements.

In a review of the existing railroad industry bridge inventory, the majority of bridges by far are simple-span structures over streams and roadways. Complex bridges are generally associated with crossing major waterways or other significant topographical features. Signature bridges are rarely constructed by railroads. The enormity of train live loads generally preclude the use of double-leaf bascule bridges and suspension and cable-stayed bridges due to bridge deflection and shear load transfer, respectively. Railroads, where possible, avoid designing skewed or curved bridges, which also have inherent deflection problems.

When planning the replacement of smaller bridges, railroads first determine if the bridge can be eliminated using culverts. A hydrographic review of the site will determine if the bridge opening needs to be either increased or can be decreased.

The *Manual* provides complete details for common timber structures and for concrete box-girder spans. Many of the larger railroads develop common standards, which provide complete detailed plans for the construction of bridges. These plans include piling, pile bents, abutments and wing walls, spans (timber, concrete, and steel), and other elements in sufficient detail for construction by in-house forces or by contract. Only site-specific details such as permits, survey data, and soil conditions are needed to augment these plans.

Timber trestles are most often replaced by other materials rather than in kind. However, it is often necessary to renew portions of timber structures to extend the life of a bridge for budgetary reasons. Replacing pile bents with framed bents to eliminate the need to drive piles or the adding of a timber stringer to a chord to increase capacity is common. The replacement of timber trestles is commonly done by driving either concrete or steel piling through the existing trestle, at twice the present timber span length and offset from the existing bents. This is done between train movements. Either precast or cast-in-place caps are installed atop the piling beneath the existing timber deck. During a track outage period, the existing track and timber deck is removed and new spans (concrete box girders or rolled steel beams) are placed. In this type of bridge renewal, key factors are use of prefabricated bridge elements light enough to be lifted by railroad track mounted equipment (piles, caps, and spans), speed of installation of bridge elements between train movements, bridge elements that can be installed in remote site locations without outside support, and overall simplicity in performing the work.

The railroad industry has a large number of 150 to 200 ft span pin-connected steel trusses, many with worn joints, restrictive clearances, and low carrying capacity, for which rehabilitation cannot be economically justified. Depending on site specifics, a common replacement scenario may be to install an intermediate pier or bent and replace the span with two girder spans. Railroad forces have perfected the technique of laterally rolling out old spans and rolling in new prefabricated spans between train movements.

Railroads frequently will relocate existing bridge spans to other sites in lieu of constructing new spans, if economically feasible. This primarily applies to beam spans and plate girder spans up to 100 ft in length.

In general, railroads prefer to construct new bridges online rather than relocating or doglegging to an adjacent alignment. Where site conditions do not allow ready access for direct span replacement, a site bypass, or runaround, called a “shoofly” is constructed which provides a temporary bridge while the permanent bridge is constructed.

The design and construction of larger and complex bridges is done on an individual basis.

23.3 Railroad Bridge Types

Railroad bridges are nearly always simple-span structures. Listed below in groupings by span length are the more common types of bridges and materials used by the railroad industry for those span lengths.

Short spans	to 16 ft	Timber stringers
		Concrete slabs
		Rolled steel beams
	to 32 ft	Conventional and prestressed concrete box girders and beams
		Rolled steel beams
	to 50 ft	Prestressed concrete box girders and beams
		Rolled steel beams, deck and through girders
Medium spans, 80 to 125 ft		Prestressed concrete beams
		Deck and through plate girders
Long spans		Deck and through trusses (simple, cantilever, and arches)

Suspension bridges are not used by freight railroads due to excessive deflection.

23.4 Bridge Deck

23.4.1 General

The engineer experienced in highway bridge design may not think of the typical railroad bridge as having a deck. However, it is essential to have a support system for the rails. Railroad bridges typically are designed as either open deck or ballast deck structures. Some bridges, particularly in transit applications, use direct fixation of the rails to the supporting structure.

23.4.2 Open Deck

Open deck bridges have ties supported directly on load-carrying elements of the structure (such as stringers or girders). The dead loads for open deck structures can be significantly less than for ballast deck structures. Open decks, however, transfer more of the dynamic effects of live load into the bridge than ballast decks. In addition, the bridge ties required are both longer and larger in cross section than the standard track ties. This adds to their expense. Bridge tie availability has declined, and their supply may be a problem, particularly in denser grades of structured timber.

TABLE 23.1 Weight of Rails, Inside Guard Rails, Ties, Guard Timbers, and Fastenings for Typical Open Deck (Walkway not included)

Item	Weight (plf of track)
Rail (136 RE):	
(136 lb/lin. yd × 2 rails/track × 1 lin. yd/3 lin. ft)	91
Inside guard rails:	
(115 lb/lin. yd × 2 rails/track × 1 lin. yd/3 lin. ft)	77
Ties (10 in. × 10 10 ft bridge ties):	
(10 in. × 10 in. × 10 ft × 1 ft^2/144 in.^{2} × 60 lb/ft^3 × 1 tie/14 in. × 12 in./1 ft)	357
Guard Timbers (4 × 8 in.):	
(4 in. × 8 in. × 1 ft × 1 ft^2/144 in.^{3} × 60 lb/1 ft^3 × 2 guard timbers/ft)	27
Tie Plates (7¾ × 14¾ in. for rail with 6 in. base):	
24.32 lb/plate × 1 tie/14 in. × 12 in./ft × 2 plates/tie)	42
Spikes (⅝ × ⅝ in. × 6 in. reinforced throat)	
(0.828 lb/spike × 18 spikes/tie × 1 tie/14 in. × 12 in./1 ft)	13
Miscellaneous Fastenings (hook bolts and lag bolts):	
(Approx. 2.25 lb/hook bolt + 1.25 lb/lag screw × 2 bolts/tie × 1 tie/14 in. × 12 in./ft)	6
Total weight	613

TABLE 23.2 Weight of Typical Ballast Deck

Item	Weight (plf of track)
Rail (136 RE):	
(136 lb/lin. yd. × 2 rails/track × 1 lin. yd/3 lin. ft)	91
Inside Guard Rails:	
(115 lb/lin. yd × 2 rails/track × 1 lin. yd/3 lin. ft)	77
Ties (neglect, since included in ballast weight)	—
Guard Timbers (4 × 8 in.):	
(4 in. × 8 in. × 1 ft × 1 ft^2/144 in.^{2} × 60 lb/1 ft^3 × 2 guard timbers/ft)	27
Tie Plates (7¾ × 14¾ in. for rail with 6 inc. base):	
(24.32 lb/plate × 1 tie/19.5 in. × 12 in./ft × 2 plates/tie)	30
Spikes (⅝ × ⅝ × 6 in. reinforced throat)	
(0.828 lb/spike × 18 spikes/tie × 1 tie/19.5 in. × 12 in./1 ft)	9
Ballast (assume 12 in. additional over time)	
(Approx. 120 lb/ft^3 × 27 in. depth/12 in./1 ft × 16 ft)	4320
Waterproofing:	
(Approx. 150 lb/ft^3 × 0.75 in. depth/12 in./1 ft × 20 ft)	188
Total weight:	4742

23.4.3 Ballast Deck

Ballast deck bridges have the track structure supported on ballast, which is carried by the structural elements of the bridge. Typically, the track structure (rails, tie plates, and ties) is similar to track constructed on grade. Ballast deck structures offer advantages in ride and maintenance requirements. Unlike open decks, the track alignment on ballast deck spans can typically be maintained using standard track maintenance equipment. If all other factors are equal, most railroads currently prefer ballast decks for new structures.

In ballast deck designs, an allowance for at least 6 in. of additional ballast is prudent. Specific requirements for additional ballast capacity may be provided by the railroad. In addition, the required depth of ballast below the tie should be verified with the affected railroad. Typical values for this range from 8 to 12 in. or more. The tie length used will have an effect on the distribution of live-load effects into the structure. Ballast decks are also typically waterproofed. The weight of waterproofing should be included in the dead load. Provisions for selection, design, and installation of waterproofing are included in Chapter 29 of the AREMA *Manual.*

23.4.4 Direct Fixation

Direct fixation structures have rails supported on plates anchored directly to the bridge deck or superstructure. Direct fixation decks are much less common than either open decks or ballast decks and are rare in freight railroad service. While direct fixation decks eliminate the dead load of ties and ballast, and can reduce total structure height, they transfer more dynamic load effects into the bridge. Direct fixation components need to be carefully selected and detailed.

23.4.5 Deck Details

Walkways are frequently provided on railroad bridge decks. They may be on one or both sides of the track. Railroads have their own policies and details for walkway placement and construction.

Railroad bridge decks on curved track should allow for superelevation. With ballast decks, this can be accomplished by adjusting ballast depths. With open decks, it can require the use of beveled ties or building the superelevation into the superstructure.

Continuous welded rail (CWR) is frequently installed on bridges. This can affect the thermal movement characteristics of the structure. Check with the affected railroad for its policy on anchorage of CWR on structures. Long-span structures may require the use of rail expansion joints.

23.5 Design Criteria

23.5.1 Geometric Considerations

Railroad bridges have a variety of geometric requirements. The AREMA *Manual* has clearance diagrams showing the space required for passage of modern rail traffic. It should be noted that lateral clearance requirements are increased for structures carrying curved track. Track spacing on multiple-track structures should be determined by the affected railroad. Safety concerns are leading to increased track-spacing requirements.

If possible, skewed bridges should be avoided. Skewed structures, however, may be required by site conditions. A support must be provided for the ties perpendicular to the track at the end of the structure. This is difficult on open deck structures. An approach slab below the ballast may be used on skewed ballast deck bridges.

23.5.2 Proportioning

Typical depth-to-span length ratios for steel railroad bridges are around 1:12. Guidelines for girder spacing are given in Chapter 15 of the *Manual.*

23.5.3 Bridge Design Loads

23.5.3.1 Dead Load

Dead load consists of the weight of the structure itself, the track it supports, and any attachments it may carry. Dead loads act due to gravity and are permanently applied to the structure. Unit weights for calculation of dead loads are given in AREMA Chapters 7, 8, and 15. The table in Chapter 15 is reproduced below:

Unit Weights for Dead Load Stresses

Type	Pounds per Cubic Foot
Steel	490
Concrete	150
Sand, gravel, and ballast	120
Asphalt-mastic and bituminous macadam	150
Granite	170
Paving bricks	150
Timber	60

Dead load is applied at the location it occurs in the structure, typically as either a concentrated or distributed load.

The *Manual* states that track rails, inside guard rails, and rail fastenings shall be assumed to weigh 200 pounds per linear foot (plf) of track. The 60 pound per cubic foot weight given for timber should be satisfactory for typical ties. Exotic woods may be heavier. Concrete ties are sometimes used, and their heavier weight should be taken into account if their use is anticipated.

In preliminary design of open deck structures, a deck weight of 550 to 650 plf of track can be assumed. This should be checked with the weight of the specific deck system used for final design. Example calculations for track and deck weight for open deck and ballast deck structures are included in this chapter.

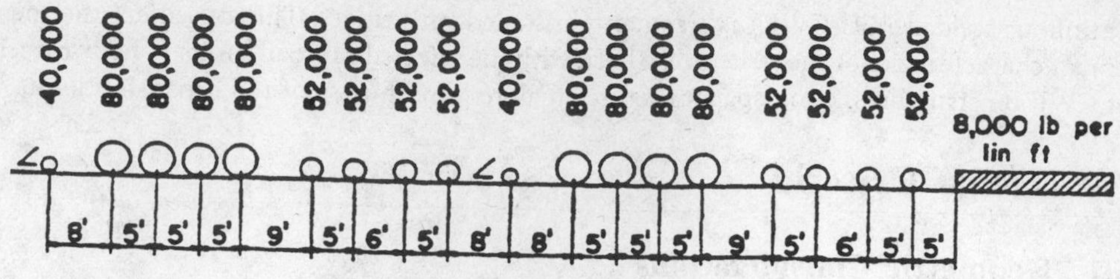

FIGURE 23.1 Cooper E8O live load.

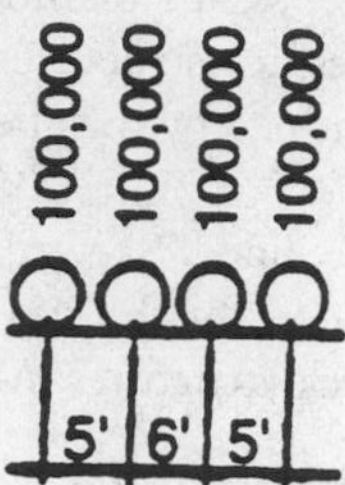

FIGURE 23.2 Alternate live load.

Railroad bridges frequently carry walkways and signal and communication cables and may be used by utilities. Provisions (both in dead load and physical location) may need to be made for these additional items. Some structures may even carry ornamental or decorative items.

23.5.3.2 Live Load

Historically, freight railroads have used the Cooper E load configuration as a live-load model. The Cooper E80 load is currently the most common design live load. The E80 load model is shown in Figure 23.1. The 80 in E80 refers to the 80 kip weight of the locomotive drive axles. An E60 load has the same axle locations, but all loads are factored by 60/80. Some railroads are designing new structures to carry E90 or E100 loads.

The Cooper live-load model does not match the axle loads and spacings of locomotives currently in service. It did not even reflect all locomotives at the turn of the 20th century, when it was introduced by Theodore Cooper, an early railroad bridge engineer. However, it has remained in use throughout the past century. One of the reasons for its longevity is the wide variety of rail rolling stock that has been and is currently in service. The load effects of this equipment on given spans must be compared, as discussed in Section 23.6. The Cooper live-load model gives a universal system with which all other load configurations can be compared. Engineering personnel of each railroad can calculate how the load effects of each piece of equipment compare to the Cooper loading.

The designated steel bridge design live load also includes an "Alternate E80" load, consisting of four 100-kip axles. This is shown in Figure 23.2. This load controls over the regular Cooper load on shorter spans.

A table of maximum load effects over various span lengths is included in Chapter 15, Part 1 of the AREMA *Manual.*

23.5.3.3 Impact

Impact is the dynamic amplification of the live-load effects on the bridge caused by the movement of the train across the span. Formulas for calculation of impact are included in Chapters 8 and 15 of the AREMA manual. The design impact values are based on an assumed train speed of 60 mph. It should be noted that the steel design procedure allows reduction of the calculated impact for ballast deck structures. Different values for impact from steam and diesel locomotives are used. The steam impact values are significantly higher than diesel impact over most span lengths.

Impact is not applied to timber structures, since the capacity of timber under transient loads is significantly higher than its capacity under sustained loads. Allowable stresses for timber design are based on the sustained loads.

23.5.3.4 Centrifugal Force

Centrifugal force is the force a train moving along a curve exerts on a constraining object (track and supporting structure) which acts away from the center of rotation. Formulas or tables for calculation of centrifugal force are included in Chapters 7, 8, and 15 of the AREMA manual. The train speed required for the force calculation should be obtained from the railroad.

Although the centrifugal action is applied as a horizontal force, it can produce overturning moment due to its point of application above the track. Both the horizontal force and resulting moment must be considered in design or evaluation of a structure.

The horizontal force tends to displace the structure laterally:

- For steel structures (deck girders, for example), it loads laterals and cross frames.
- For concrete structures (box girders, for example), the superstructure is typically stiff enough in the transverse direction that the horizontal force is not significant for the superstructure.

For all bridge types, the bearings and substructure must be able to resist the centrifugal horizontal force.

The overturning moment tends to increase the live-load force in members on the outside of the curve and reduce the force on inside members. However, interior members are not designed with less capacity than exterior members. Substructures must be designed to resist the centrifugal overturning moment. This will increase forces toward the outside of the curve in foundation elements. The centrifugal force is applied at the location of the axles along the structure, 6 ft above the top of rail, at a point perpendicular to the center of a line connecting the rail tops. The effect of track superelevation may compensate somewhat for centrifugal force. The plan view location of the curved track on the bridge (since railroad bridge spans are typically straight, laid out along the curve chords) can also be significant. Rather than applying the centrifugal force at each axle location, some railroads simply increase the calculated live-load force by the centrifugal force percentage, factor in the effect of the force location above the top of rail, and use the resulting value for design.

23.5.3.5 Lateral Loads from Equipment

This item includes all lateral loads applied to the structure due to train passage, other than centrifugal force. The magnitude and application point of these loads varies among Chapters 7, 8, and 15. For timber, a load of 20 kips is applied horizontally at the top of rail. For steel, a load of one quarter of the heaviest axle of the specified live load is applied at the base of rail. In both cases, the lateral load is a moving concentrated load that can be applied at any point along the span in either horizontal direction. It should be noted that lateral loads from equipment are not included in design of concrete bridges. However, if concrete girders are supported on steel or timber substructures, lateral loads should be applied to the substructures.

Lateral loads from equipment are applied to lateral bracing members, flanges of longitudinal girders or stringers without a bracing system, and to chords of truss spans. Experience has shown that very high lateral forces can be applied to structures due to lurching of certain types of cars. Wheel hunting is another phenomenon that applies lateral force to the track and structure. Damaged rolling stock can also create large lateral forces.

It should be noted that there is not an extensive research background supporting the lateral forces given in the AREMA *Manual*. However, the lateral loads in the *Manual* have historically worked well when combined with wind loads to produce adequate lateral resistance in structures.

23.5.3.6 Longitudinal Force from Live Load

Longitudinal forces are typically produced from starting or stopping trains (acceleration or deceleration) on the bridge. They can be applied in either longitudinal direction. These forces are transmitted through the rails and distributed into the supporting structure.

Chapters 7, 8, and 15 all take the longitudinal force due to braking to be 15% of the vertical live load, without impact. The chapters differ slightly in their consideration of the acceleration (traction) aspect of the force. Chapter 7 uses 25% of the drive axle loads for traction, while Chapters 8 and 15 use 25% of the axles of the regular Cooper E80 train configuration. In each chapter, the braking and traction forces are compared, and the larger value used in design. Chapters 7, 8, and 15 differ in the point of application of the longitudinal force. Chapter 7 applies it 6 ft above the top of rail. Chapters 8 and 15 apply the braking force at 8 ft above the top of rail and the traction force 3 ft above the top of rail.

All three chapters recognize that some of the longitudinal force is carried through the rails off the structure. (The extent of this transfer depends on factors such as rail continuity, rail anchorage, and the connection of the bridge deck to the span.) Where a large portion of the longitudinal force is carried to the abutments or embankment, Chapter 7 allows neglecting longitudinal force in the design of piles, posts, and bracing of bents. Chapters 8 and 15 allow taking the applied longitudinal force as half of what was initially calculated on short (<200 feet) ballast deck bridges with short spans (<50 feet), if the continuity of members or frictional resistance will direct some of the longitudinal force to the abutments.

Chapters 8 and 15 also state that the longitudinal load is to be applied to one track only, and can be distributed to bridge components based on their relative stiffness and the types of bearings. For multiple-track structures, it may be prudent to include longitudinal force on more than one track, depending on the bridge location and train operation at the site.

Longitudinal force is particularly significant in long structures, such as viaducts, trestles, or major bridges. Large bridges may have internal traction or braking trusses to carry longitudinal forces to the bearings. Viaducts frequently have braced tower bents at intervals to resist longitudinal force.

The American Association of Railroads (AAR) is currently conducting research on the longitudinal forces in bridges induced by the new high-adhesion locomotives now coming into service. In addition, the introduction of new mechanical systems such as the load-empty brake and electronically controlled brakes are affecting the longitudinal forces introduced into the track. Transit equipment can have high acceleration and deceleration rates, which can lead to high longitudinal forces on transit structures.

23.5.3.7 Wind Loading

Wind loading is the force on the structure due to wind action on the bridge and train. Chapters 7, 8, and 15 deal with wind on the structure slightly differently:

1. *Timber*: Use 30 psf as a moving horizontal load acting in any direction.
2. *Concrete*: Use 45 psf as a horizontal load perpendicular to the track centerline.
3. *Steel*: As a moving horizontal load:
    a. Use 30 psf on loaded bridge.
    b. Use 50 psf on unloaded bridge.

The application areas of the wind on structure vary as well:

1. *Timber*: For trestles, the affected area is 1.5 times the vertical projection of the floor system. For trusses, the affected area is the full vertical projection of the spans, plus any portion of the leeward trusses not shielded by the floor system. For trestles and tower substructures, the affected area is the vertical projections of the components (bracing, posts, and piles).
2. *Steel*: Similar to timber, except that for girder spans 1.5 times the vertical projection of the span is used.
3. *Concrete*: Wind load is applied to the vertical projection of the structure. Note that 45 psf = 1.5 (30 psf).

For all materials, the wind on the train is taken as 300 plf, applied 8 ft above the top of rail.

The 30-psf wind force on a loaded structure and 50-psf force on an unloaded structure used in Chapter 15 reflect assumptions on train operations. It was assumed that the maximum wind velocity under which train operations would be attempted would produce a force of 30 psf. Hurricane winds, under which train operations would not be attempted, would produce a wind force of 50 psf.

For stability of spans and towers against overturning due to wind on a loaded bridge, the live load is reduced to 1200 plf, without impact being applied. This value represents an unloaded, stopped train on the bridge.

It should be noted that Chapter 15 has a minimum wind load on loaded bridges of 200 plf on the loaded chord or flange and 150 plf on the unloaded chord or flange.

Virtually every bridge component can be affected by wind. However, wind is typically most significant in design of

1. Lateral bracing and cross frames
2. Lateral bending in flanges
3. Vertical bending in girders and trusses due to overturning
4. Tower piles or columns
5. Foundations

23.5.3.8 Stream Flow, Ice, and Buoyancy

These loads are experienced by a portion of the structure (usually a pier) because of its location in a body of water. These topics are only specifically addressed in Chapter 8, because they apply almost entirely to bridge substructures, which typically consist of concrete.

Buoyancy, stream flow, and ice pressure are to be applied to any portion of the structure that can be exposed to them. This typically includes piers and other elements of the substructure. Buoyancy can be readily calculated for immersed portions of the structure.

While the AREMA *Manual* does not address design forces for stream flow and ice pressure, other design criteria, such as the AASHTO *LRFD Bridge Design Specification* does include procedures for calculating them. The designer can use these sources for guidance until specific forces are included by AREMA.

Spans may be floated off piers due to buoyancy, stream flow, and ice pressure. Loaded ballast cars are sometimes parked on bridges during floods or ice buildup to resist this. Drift or debris accumulation adjacent to bridges can be a significant problem, reducing the flow area through the bridge and effectively increasing the area exposed to force from stream flow.

Two other factors concerning waterways must be considered. The first is vessel collision (or, more correctly) allision with piers. Pier protection is covered in Part 23, Spans over Navigable Streams, of Chapter 8. These requirements should be addressed when designing a bridge across a navigable waterway. The second factor to be considered is scour. Scour is a leading cause of bridge failure. The AASHTO *LRFD Bridge Design Specification* contains scour analysis and protection guidelines. Hydraulic studies to determine required bridge openings should be performed when designing new structures or when hydrologic conditions upstream of a bridge change.

23.5.3.9 Volume Changes

Volume changes in structures can be caused by thermal expansion or contraction or by properties of the structural materials, such as creep or shrinkage. Volume changes in themselves, if unrestrained, have relatively little effect on the forces on the structure. Restrained volume changes, however, can produce significant forces in the structure. The challenge to the designer is to provide a means to relieve volume changes or to provide for the forces developed by restrained changes.

Chapter 7 does not specifically state thermal expansion movement requirements. Due to the nature of the material and type of timber structures in use, it is unlikely that thermal stresses will be significant in timber design. Chapter 15 requires an allowance of 1 in. of length change due to temperature per every 100 ft of span length in steel structures. Chapter 8 provides the following table for design temperature rise and fall values for concrete bridges:

Climate	Temperature Rise	Temperature Fall
Moderate	30°F	40°F
Cold	35°F	45°F

It should be noted that the tabulated values refer to the temperature of the bridge concrete. A specific railroad may have different requirements for thermal movement.

Expansion bearings are the main design feature typically used to accommodate volume changes. Common bearing types include:

1. Sliding steel plates
2. Rocker bearings
3. Roller bearings (cylindrical and segmental)
4. Elastomeric bearing pads

Provision should be made for span length change due to live load. For spans longer than 300 ft, provision must be made for expansion and contraction of the bridge floor system within the trusses.

For concrete structures, provisions need to be made for concrete shrinkage and creep. Specific guidelines are given in Chapter 8, Parts 2 and 17 for these properties. It is important to remember that creep and shrinkage are highly variable phenomena, and allowance should be made for higher-than-expected values. It also should be noted the AREMA *Manual* requires 0.25 in^2/ft minimum of reinforcing steel in exposed concrete surfaces.

Chapter 8 also requires designing for longitudinal force due to friction or shear resistance at expansion bearings. This is in recognition of the fact that most expansion bearings have some internal resistance to movement. This resistance applies force to the structure as the bridge expands and contracts. The AREMA *Manual* contains procedures for calculating the shear force transmitted through bearing pads. Loads transmitted through fixed or expansion bearings should be included in substructure design.

Bearings must also be able to resist wind and other lateral forces applied to the structure. Chapter 19 of the AREMA *Manual for Railway Engineering* covers bridge bearings. It is included in the 1997 *Manual*, and should be applied for bearing design and detailing.

It should be noted that movement of bridge bearings affects the tolerances of the track supported by the bridge. This calls for careful selection of bearings for track with tight tolerances (such as high-speed lines). Maintenance requirements are also important when selecting bearings, since unintended fixity due to freezing of bearings can cause significant structural damage.

23.5.3.10 Seismic Loads

Seismic design for railroads is covered in Chapter 9 of the *Manual*. The philosophical background of Chapter 9 recognizes that railroad bridges have historically performed well in seismic events. This is due to the following factors:

1. The track structure serves as an effective restraint (and damping agent) against bridge movement.
2. Railroad bridges are typically simple in their design and construction.
3. Trains operate in a controlled environment, which makes types of damage permissible for railroad bridges that might not be acceptable for structures in general use by the public.

Item 3 above is related to the post-seismic event operation guidelines given in Chapter 9. These guidelines give limits on train operations following an earthquake. The limits vary according to earthquake magnitude and distance from the epicenter. For example, following an earthquake of magnitude 6.0 or above, all trains within a 100-mile radius of the epicenter must stop until the track and bridges in the area have been inspected and cleared for use. (Note that specific railroad policies may vary.)

Three levels of ground motion are defined in Chapter 9:

- Level 1 — Motion that has a reasonable probability of being exceeded during the life of the bridge.
- Level 2 — Motion that has a low probability of being exceeded during the life of the bridge.
- Level 3 — Motion for a rare, intense earthquake.

Three performance limit states are given for seismic design of railroad bridges. The serviceability limit state requires that the structure remain elastic during Level 1 ground motion. Only moderate damage and no permanent deformations are acceptable. The ultimate limit state requires that the structure suffer only readily detectable and repairable damage during Level 2 ground motion. The survivability limit state requires that the bridge not collapse during Level 3 ground motion. Extensive damage may be allowed. For some structures, the railroad may elect to allow for irreparable damage, and plan to replace the bridges following a Level 3 event.

An in-depth discussion of seismic analysis and design is beyond the scope of this section. Guidelines are given in Chapter 9 of the manual. Base acceleration coefficient maps for various return periods are included in the chapter. It should be noted that no seismic analysis is necessary for locations where a base acceleration of 0.1 *g* or less is expected with a 475-year return period. For most locations in North America, therefore, a seismic analysis would not be needed.

Section 1.4 of Chapter 9 addresses seismic design. Important structures (discussed in its Section 1.3.3) should be designed to resist higher seismic loads than nonimportant structures.

Even if no specific seismic analysis and design is required for a structure, it is good practice to detail structures for seismic resistance if they are in potentially active areas. Specific concerns are addressed in Chapter 9. Provision of adequate bearing areas and designing for ductility are examples of inexpensive seismic detailing.

23.5.4 Load Combinations

A variety of loads can be applied to a structure at the same time. For example, a bridge may experience dead load, live load, impact, centrifugal force, wind, and stream flow simultaneously. The AREMA *Manual* chapters on structure design recognize that it is unlikely that the maximum values of all loads will be applied concurrently to a structure. Load combination methods are given to develop maximum credible design forces on the structure.

Chapter 7, in Section 2.5.5.5, Combined Stresses, states: "For stresses produced by longitudinal force, wind or other lateral forces, or by a combination of these forces with dead and live loads and centrifugal force, the allowable working stresses may be increased 50%, provided the resulting sections are not less than those required for dead and live loads and centrifugal force."

Chapter 15, in Section 1.3.14.3, Allowable Stresses for Combinations of Loads or Wind Loads Only, states:

a. Members subject to stresses resulting from dead load, live load, impact load and centrifugal load shall be designed so that the maximum stresses do not exceed the basic allowable stresses of Section 1.4, of Basic Allowable Stresses, and the stress range does not exceed the allowable fatigue stress range of Article 1.3.13.
b. The basic allowable stresses of Section 1.4, Basic Allowable Stresses, shall be used in the proportioning of members subject to stresses resulting from wind loads only, as specified in Article 1.3.8.

c. Members, except floorbeam hangers, which are subject to stresses resulting from lateral loads, other than centrifugal load, and/or longitudinal loads, may be proportioned for stresses 25% greater than those permitted by paragraph a, but the section of the member shall not be less than that required to meet the provisions of paragraph a or paragraph b alone.
d. Increase in allowable stress permitted by paragraph c shall not be applied to allowable stress in high strength bolts.

Chapter 8, in Part 4 on Pile Foundations, defines primary and secondary loads. Primary loads include dead load, live load, centrifugal force, earth pressure, buoyancy, and negative skin friction. Secondary (or occasional) loads include wind and other lateral forces, ice and stream flow, longitudinal forces, and seismic forces. Section 4.2.2.b allows a 25% increase in allowable loads when designing for a combination of primary and secondary loads, as long as the design satisfies the primary load case at the allowable load.

These three load combination methods are based on service load design. Chapter 8, in Part 2, Reinforced Concrete Design, addresses both service load and load factor design

Chapter 8, Section 2.2.4 gives several limitations on the load combination tables. For example, load factor design is not applicable to foundation design or for checking structural stability. In addition, load factors should be increased or allowable stresses adjusted if the predictability of loads is different than anticipated in the chapter.

For stability of towers, use the 1200 plf vertical live load as described in the Wind Loading section.

As a general rule, the section determined by a load combination should never be smaller than the section required for dead load, live load, impact, and centrifugal force. It is important to use the appropriate load combination method for each material and component in the bridge design. Combination methods from different sections and chapters should not be mixed.

23.5.5 Serviceability Considerations

23.5.5.1 Fatigue

Fatigue resistance is a critical concern in the design of steel structures. It is also a factor, although of less significance, in the design of concrete bridges. A fatigue design procedure, based on allowable stresses, impact values, number of cycles per train passage, fracture criticality of the member, and type of details, is applied to steel bridges. Fatigue can be the controlling design case for many new steel bridges.

23.5.5.2 Deflection

Live load-deflection control is a significant serviceability criterion. Track standards limit the amount of deflection in track under train passage. The deflection of the bridge under the live load accumulates with the deflection of the track structure itself. This total deflection can exceed the allowable limits if the bridge is not sufficiently stiff. The stiffness of the structure can also affect its performance and longevity. Less stiff structures may be more prone to lateral displacement under load and out-of-plane distortions. Specific deflection criteria are given in Chapter 15 for steel bridges. Criteria for concrete structures are given in Chapter 8 using span-to-depth ratios.

Long-term deflections should also be checked for concrete structures under the sustained dead load to determine if any adverse effects may occur due to cracking or creep.

23.5.5.3 Others

Other serviceability criteria apply to concrete structures. Reinforced concrete must be checked for crack control. Allowable stress limits are given for various service conditions for prestressed concrete members.

23.6 Capacity Rating

23.6.1 General

Rating is the process of determining the safe capacity of existing structures. Specific guidelines for bridge rating are given in Chapters 7, 8, and 15 of the AREMA *Manual.* Ratings are typically performed on both as-built and as-inspected bridge conditions. The information for the as-built condition can be taken from the bridge as-built drawings. However, it is important to check the current condition of the structure. This is done by performing an inspection of the bridge and adjusting the as-built rating to include the effects of any deterioration, damage, or modifications to the structure since its construction. Material property testing of bridge components may be very useful in the capacity rating of an older structure.

Structure ratings are normally presented as the Cooper E value live load that thc bridge can safely support. The controlling rating is the lowest E value for the structure (based on a specific force effect on a critical member or section). For example, a structure rating may be given as E74, based on bending moment at the termination of a flange cover plate.

As discussed in the Live Load section, there are a wide variety of axle spacings and loadings for railroad equipment. Each piece of equipment can be rated to determine the maximum force effects it produces for a given span length. The equipment rating is given in terms of the Cooper load that would produce the equivalent force effect on the same span length. Note that this equivalent force effect value will probably be different for shear and moment on each span length.

In addition to capacity ratings, fatigue ratings can be performed on structures to estimate their remaining fatigue life. These are typically only calculated for steel structures. Guidelines for this can be found in the commentary section of Chapter 15.

23.6.2 Normal Rating

The normal rating of the structure is the load level which can be carried by thc bridge for an indefinite time period. This indefinite time period can be defined as its expected service life. The allowable stresses used for normal rating are the same as the allowable stresses used in design. The impact effect calculation, however, is modified from the design equation. Reduction of the impact value to reflect the actual speed of trains crossing the structure (rather than the 60 mph speed assumed in the design impact) is allowed. Formulas for the impact reduction are included in the rating sections of the AREMA *Manual* chapters.

23.6.3 Maximum Rating

The maximum rating of the structure is the maximum load level which can be carried by the bridge at infrequent intervals. This rating is used to check if extraheavy loads can cross the structure. Allowable stresses for maximum rating are increased over the design allowable values.

The impact reduction for speed can be applied as for a normal rating. In addition, "slow orders" or speed restrictions can be placed on the extraheavy load when crossing the bridge. This can allow further reduction of the impact value, thus increasing the maximum rating of the structure. (Note that this maximum rating value would apply only at the specified speed.)

References

AREMA, *Manual for Railway Engineering,* AREMA, Landover, MD, 1997.

American Association of Railroads, *Railroad Facts,* 1997 Edition, Washington, D.C., 1997.

Waddell, J.A.L., *Bridge Engineering,* 1916.

24

Bridge Decks and Approach Slabs

Michael D. Keever
California Department of Transportation

John H. Fujimoto
California Department of Transportation

24.1 Introduction

This chapter discusses bridge decks and structure approach slabs, the structural riding surface that typically is the responsibility of bridge design engineers when developing contract plans and details. Decks and approach slabs are usually not specially designed for each bridge project, but are instead taken from tables and standard plans either developed and provided by the owner or approved for use by the owner. However, as the load and resistance factor design (LRFD) is adopted by various agencies, standards that were developed previously must be reviewed and revised to comply with these new standards.

Not only do decks and approach slabs provide the riding surface for vehicular traffic, but they also serve several structural purposes. The bridge deck distributes the vehicular wheel loads to the girders, which are the primary load-carrying members on a bridge superstructure. The deck is often composite with the main girders and, with reinforcement distributed in the effective regions of the deck, serves to impart flexural strength and torsional rigidity to the bridge. The structure approach slab is a transitional structure between the bridge, which has relatively little settlement, and the roadway approach, which is subject to varying levels of approach settlement, sometimes significant. The approach slab serves as a bridge between the roadway and the primary bridge and is intended to reduce the annoying and sometimes unsafe "bump" that is often felt when approaching and leaving a bridge.

24.2 Bridge Decks

There are several different types of bridge decks, with the most common being cast-in-place reinforced concrete [1]. Other alternative deck types include precast deck panels, prestressed cast-in-place decks, post-tensioned concrete panels, filled and unfilled steel grid, steel orthotropic decks, and timber. These less common types may be used when considering deck rebar corrosion, deck

0-8493-7434-0/00/$0.00+$.50

replacement, traffic, maintenance, bridge weight, aesthetics, and life cycle costs, among other reasons. This chapter will emphasize cast-in-place reinforced concrete decks, including a design example. The alternative deck types will be discussed in less detail later in the chapter.

24.2.1 Cast-in-Place Reinforced Concrete

The extensive use of cast-in-place concrete bridge decks is due to several reasons including cost, acceptable skid resistance, and commonly available materials and contractors to do the work. Despite these advantages, this type of deck is not without disadvantages.

The most serious drawback is the tendency of the deck rebar to corrode. Deicing salts used in regions that must contend with snow and ice have problems associated with corrosion of the rebar in the deck. In these areas, cracks caused by corrosion are aggravated by the results of freeze–thaw action. Damage due to rebar corrosion often results in the cost and inconvenience to the traveling public of replacing the deck. In an effort to minimize this problem, concrete cover can be increased over the rebar, sealants can be placed on the deck, and epoxy-coated or galvanized rebar can be used in the top mat of deck steel. However, these solutions do not prevent the deck from cracking, which initiates the damage.

A common means of reducing deck rebar corrosion is the use of coated rebar. One drawback to this type of rebar is that it is often difficult to protect epoxy-coated and galvanized rebar during construction. Small nicks to the rebar, common when using normal construction methods, may be repaired in the field if they are detected. However, it is difficult to make repairs that are as good as the original coatings.

Despite corrosion problems associated with a cast-in-place reinforced bridge deck, it continues to be the most common type of deck built, and therefore a design example will be included in Section 24.2.1.3.

24.2.1.1 Traditional Design Method

Traditionally, cast-in-place bridge decks are designed assuming that the bridge deck is a continuous beam spanning across the girders, which are assumed to be unyielding supports. Although it is known that the girders do indeed deflect, it greatly simplifies the analysis to assume they do not. By using this method the maximum moments are determined and the deck is designed. This design method, in which the deck is designed as a series of strips transverse to the girders, is referred to as the "approximate strip design method." This method has been refined over time and has now been adapted to LRFD in the 1994 AASHTO-LRFD Specifications [2]. All references to this code will be shown in brackets.

24.2.1.2 Empirical Design Method

More recently, an alternative method of isotropic bridge deck design has been developed for cast-in-place concrete bridge decks in which it is assumed instead that the deck resists the loads using an arching effect between the girders. The 1994 AASHTO-LRFD Specification includes an empirical design method for decks using isotropic reinforcement based on these arch design principles [9.7.2]. Under this method no analysis is required. Instead, four layers of reinforcement are placed with no differentiation between the transverse and longitudinal direction. However, to use this method the conditions outlined under [9.7.2.4] must be satisfied. These conditions include requirements for the effective length between girders, the depth of the slab, the length of the deck overhang, and the specified concrete strength.

The deck overhang itself is not designed using this method. Instead cantilever overhangs are designed using the approximate strip method described above and used in the design example below.

24.2.1.3 Design Example

Given

Consider a cast-in-place conventionally reinforced bridge deck (Figure 24.1). The superstructure has six girders spaced at 2050 mm. The deck width is 11,890 mm wide and the overhang beyond

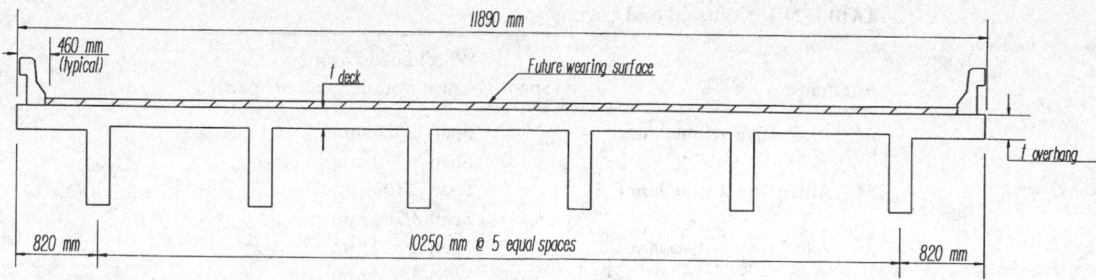

FIGURE 24.1 Typical section.

the exterior girder is 820 mm from the girder centerline. The unit weight of concrete is assumed to be 23.5 kN/m³.

Find

Based on the 1994 AASHTO-LRFD Specifications, use the Approximate Strip Design Method for Decks [4.6.2.1] to determine the deck thickness, design moments, and the detailing requirements necessary to design the bridge deck reinforcement.

Solution:

1. Determine the Deck Thickness [Table 2.5.2.6.3-1] [9.7.1.1]

$$t_{deck} = (S + 3000)/30 = (2050 + 3000)/30 = 168.3 \text{ mm}$$

where S = the girder spacing. The minimum required deck thickness, excluding provisions for grinding, grooving, and sacrificial surface is t_{deck} = 175 mm. ⇐ controls

The deck overhang is often a different thickness than the deck thickness. This may be for aesthetic, structural, or other reasons. For this example, assume the deck overhang is a constant thickness of $t_{overhang}$ = 250 mm.

2. Determine Unfactored Dead Loads

For simplicity the deck will be designed as a 1-m-wide one-way slab. Therefore, all loads will be determined on a per meter width.

Slab:	q_{DS}	$= (23.5 \text{ kN/m}^3)(0.175 \text{ m})(1 \text{ m}) = 4.11 \times 10^{-3} \text{ kN/m}$
Overhang:	q_{DO}	$= (23.5 \text{ kN/m}^3)(0.250 \text{ m})(1 \text{ m}) = 5.88 \times 10^{-3} \text{ kN/m}$
Barrier rail:	P_{DB}	= 5.8 kN per 1 m width
Wearing surface:	q_{DW}	$= (1.70 \text{ kN/m}^2)(1 \text{ m}) = 1.70 \text{ kN/m}$

3. Determine Unfactored Live Loads [4.6.2.1] [3.6.1.3.3] [3.6.1.2.2]

a. *Wheel load:*

Truck axle load = 145 kN/axle.

The axle load of 145 kN is distributed equally such that each wheel load is 72.5 kN. These 72.5-kN wheel loads are moved within each lane, with an edge spacing within the lane of 0.6 m, except at the deck overhang where the edge spacing is 0.3 m as specified in [Figure 3.6.1.2.2-1].

b. *Calculate the number of live load lanes* [3.6.1.1.1]:

Assume for this example the barrier rail width is 460 mm, which implies that the clear distance between the face of rail w = 11,890 mm – 2(460 mm) = 10,970 mm. Therefore, the number of lanes is N = (10,970)/3600 = 3.05, i.e., 3 using just the integer portion of the solution as required.

TABLE 24.1 Wheel Load Layout

Alternative	Wheel Load Layout (Span/Distance from left end of span)
M_{+ve} Alternative 1 (one lane)	Span 2/820 mm
	Span 3/570 mm
M_{+ve} Alternative 2 (one lane)	Span 2/1030 mm
	Span 3/780 mm
M_{+ve} Alternative 3 (two lanes)	Span 2/820 mm
	Span 3/570 mm
	Span 4/820 mm
	Span 5/570 mm
M_{-ve} Alternative 1 (one lane)	Span 2/1150 mm
	Span 3/900 mm

c. *Determine the wheel load distribution* [4.6.2.1.6] [Table 4.6.2.1.3-1] [4.6.2.1.3]:
For cast-in-place concrete decks the strip width, w_{strip}, is
Overhangs:

$$w_{strip} = 1140 + 0.833X$$

where X = the distance from the location of the load to the centerline of the support. If it is assumed that the wheel load is pushed as close to the face of the barrier as permitted by the Code, i.e., 300 mm:

$$X = 820\text{ mm} - 460\text{ mm} - 300\text{ mm} = 60\text{ mm}$$

Therefore, for the overhang $w_{strip} = 1140 + 0.833(60) = 1190$ mm.
Interior slab:

$$\text{Positive moment } (M_{+ve}): w_{strip} = 660 + 0.55S = 660 + 0.55(2050) = 1788\text{ mm}$$

$$\text{Negative moment } (M_{-ve}): w_{strip} = 1220 + 0.25S = 1220 + 0.25(2050) = 1733\text{ mm}$$

where S = the girder spacing of 2050 mm.

d. *Determine the live loads on a 1-m strip*:
The unfactored wheel loads placed on a 1-m strip are

Overhang:	72.5 kN/1.190 m = 60.924 kN/m
Positive moment:	72.5 kN/1.788 m = 40.548 kN/m
Negative moment:	72.5 kN/1.733 m = 41.835 kN/m

4. **Determine the Wheel Load Location to Maximize the Live-Load Moments**
Three alternative wheel load layouts will be considered to determine the maximum positive moment for live load. Three alternatives are investigated to illustrate the method used to determine the maximum moments. Only the controlling location of the wheel loads will be used to determine the maximum factored negative moment. In Table 24.1 , deck spans are numbered from left to right, with the left cantilever being span 1. Distances are measured from the leftmost girder of the span. Wheel loads are placed at the locations listed in the table.

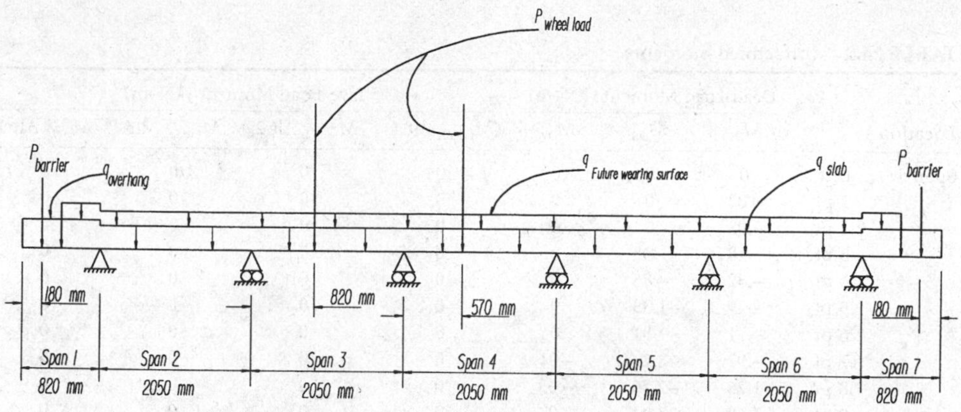

FIGURE 24.2 Loads: M_{+ve} Alternative 1 (one lane).

5. **Calculate Unfactored Moments**
Apply the unfactored loads determined in the steps above to a continuous 1-m-wide beam spanning across the girders (Figure 24.2). Based on these loads, the unfactored moments for the dead load and the live load alternatives are listed in Table 24.2. The bridge is symmetrical about the end of span 3; thus loads are only given for the left half of the bridge. The locations investigated to determine the controlling moments are shown in bold type.
6. **Determine the Load Factors** [1.3.2.1] [3.4.1]

$$Q = \eta \Sigma \gamma_i q_i$$

where
Q = factored load
η = load modifier
γ = load factor
q = unfactored loads

a. *Load modifier*: [1.3.3] [1.3.4] [1.3.5]:

$$\eta = \eta_D \eta_R \eta_i > 0.95$$

where
$\eta_D = 0.95$
$\eta_R = 0.95$
$\eta_i = 0.95$

Therefore, η is the maximum of $\eta = (0.95)(0.95)(0.95) = 0.86$ or $\eta = 0.95 \Leftarrow$ controls.

b. *Load factor*: [Table 3.4.1-1] [Table 3.4.1-2] [3.4.1] [3.3.2]:
γ_{DL}:

Maximum Load Factor	Minimum Load Factor	
$\gamma_{DCmax} = 1.25$	$\gamma_{DCmin} = 0.90$	$\Leftarrow$ slab and barrier rail
$\gamma_{DWmax} = 1.50$	$\gamma_{DWmin} = 0.65$	$\Leftarrow$ future wearing surface

γ_{LL} (Strength-1 Load Combination)

$$\gamma_{LL} = 1.75$$

$$\gamma_{IM} = 1.75$$

TABLE 24.2 Unfactored Moments

Location		Dead-Load Moment (kN–m)			Live-Load Moment (kN–m)			
		M_{DD}	M_{DB}	M_{DW}	$M_{+ve\,LL}$ Alt 1	$M_{+ve\,LL}$ Alt 2	$M_{+ve\,LL}$ Alt3	M_{-veLL} Alt 1
Span 1	Left	0	0	0	0	0	0	0
	.1 pt	–.02	0	0	0	0	0	0
	.2 pt	–.08	0	0	0	0	0	0
	.3 pt	–.18	–.38	0	0	0	0	0
	.4 pt	–.32	–.86	0	0	0	0	0
	.5 pt	–.49	–1.33	0	0	0	0	0
	.6 pt	–.71	–1.81	0	0	0	0	0
	.7 pt	–.97	–2.29	–.01	0	0	0	0
	.8 pt	–1.26	–2.76	–.03	0	0	0	0
	.9 pt	–1.60	–3.24	–.07	0	0	0	0
	Right	–1.98	–3.71	–.11	0	0	0	0
Span 2	Left	–1.98	–3.71	–.11	0	0	0	0
	.1 pt	–1.13	–3.24	.15	3.62	2.64	3.75	2.22
	.2 pt	–.46	–2.77	.34	7.24	5.29	7.51	4.43
	.3 pt	.04	–2.31	.46	10.87	7.93	11.26	6.65
	.4 pt	**.37**	**–1.84**	**.50**	**14.49**	**10.57**	**15.01**	**8.87**
	.5 pt	**.52**	**–1.37**	**.48**	**9.80**	**13.22**	**10.45**	**11.08**
	.6 pt	.50	–.90	.38	5.11	7.75	5.89	9.95
	.7 pt	.31	–.43	.21	.42	2.08	1.33	3.60
	.8 pt	–.05	.04	–.03	–4.28	–3.59	–3.23	–2.76
	.9 pt	–.59	.51	–.34	–8.97	–9.26	–7.79	–9.12
	Right	–1.30	.98	–.72	–13.66	–14.93	–12.35	–15.48
Span 3	Left	**–1.30**	**.98**	**–.72**	**–13.66**	**–14.93**	**–12.35**	**–15.48**
	.1 pt	–.54	.86	–.39	–6.50	–8.61	–5.84	–9.52
	.2 pt	.05	.74	–.12	.67	–2.29	.67	–3.56
	.3 pt	.46	.63	.07	6.00	4.03	5.35	2.40
	.4 pt	.71	.51	.20	4.85	8.72	3.54	8.36
	.5 pt	.78	.39	.25	3.70	6.73	1.74	9.09
	.6 pt	.67	.27	.23	2.55	4.73	–.07	6.47
	.7 pt	.40	.16	.13	1.40	2.74	–1.87	3.86
	.8 pt	–.05	.04	–.03	.25	.74	–3.68	1.24
	.9 pt	–.67	–.08	–.26	–.91	–1.25	–5.48	–1.38
	Right	–1.47	–.20	–.57	–2.06	–3.25	–7.29	–3.99
Span 4	Left	–1.47	–.20	–.57	–2.06	–3.25	–7.29	–3.99

M_{DD}, M_{DB}, M_{DW} represent the moments due to dead loads: deck(including slab and overhang), barrier rail, and future wearing surface moments respectively.

M_{+veLL} and M_{-veLL} represent the positive and negatiave live-load moments, respectively.

c. *Multiple presence factor* [Table 3.6.1.1.2-1]:

$$m_{1lane} = 1.20;\ m_{2lanes} = 1.00;\ m_{3lanes} = 0.85$$

d. *Dynamic load allowance* [3.6.1.2.2] [3.6.2]

$$IM = 0.33$$

7. Calculate the Factored Moments

$$M_u = \eta[\gamma_{DC}(M_{DD})+\gamma_{DC}(M_{DB})+\gamma_{DW}(M_{DW})+(m)(1+IM)(\gamma_{LL})(M_{LL})]$$

TABLE 24.3 AASHTO LRFD Bridge Design Specifications

Type of Deck	Direction of Primary Strip Relative to Traffic	Width of Primary Strip (mm)
Concrete		
• Cast-in-place	Overhang	$1140 + 0.833X$
	Either Parallel or Perpendicular	$+M$: $660 + 0.55S$ $-M$: $1220 + 0.25S$
• Cast-in-place with stay-in-place concrete formwork	Either Parallel or Perpendicular	$+M$: $660 + 0.55S$ $-M$: $1220 + 0.25S$
• Precast, post-tensioned	Either Parallel or Perpendicular	$+M$: $660 + 0.55S$ $-M$: $1220 + 0.25S$
Steel		
• Open grid	Main bars	$0.007P + 4.0S_b$
• Filled or partially filled grid	Main bars	Article 4.6.2.1.8 applies
• Unfilled, composite grids	Main bars	Article 9.8.2.4 applies
Wood		
• Prefabricated glulam		
• Non-interconnected	Parallel	$2.0h + 760$
	Perpendicular	$2.0h + 1020$
• Interconnected	Parallel	$2280 + 0.07L$
	Perpendicular	$4.0h + 760$
• Stress-laminated	Parallel	$0.066S + 2740$
	Perpendicular	$0.84S + 610$
• Spike-laminated	Parallel	$2.0h + 760$
• Continuous decks or interconnected panels	Perpendicular	$2.0h + 1020$
	Parallel	$2.0h + 760$
• Noninterconnected panels	Perpendicular	$2.0h + 1020$
• Planks		Plank width

Note: 1996 Interim Revisions, Table 4.6.2.1.3-1 — Equivalent Strips.

Positive Moment:

a. M_{+ve} *Alternative 1 with one lane of live load (0.4 point of span 2)*:

$$M_u = 0.95[(1.25)(0.37) + (0.90)(-1.84) + (1.50)(0.50) + (1.20)(1.33)(1.75)(14.49)]$$
$$= 38.03 \text{ kNm} \Leftarrow \text{controls positive moment}$$

b. M_{+ve} *Alternative 2 with one lane of live load (0.5 point of span 2)*:

$$M_u = 0.95[(1.25)(0.52) + (0.90)(-1.37) + (1.50)(0.48) + (1.20)(1.33)(1.75)(13.22)]$$
$$= 35.20 \text{ kNm}$$

c. M_{+ve} *Alternative 3 with two lanes of live load (0.4 point of span 2)*:

$$M_u = 0.95[(1.25)(0.37) + (0.90)(-1.84) + (1.50)(0.50) + (1.00)(1.33)(1.75)(15.01)]$$
$$= 32.77 \text{ kNm}$$

Negative Moment:

M_{-ve} *Alternative 1 with one lane of live load (0.0 point of span 3)*:

$$M_u = 0.95[(1.25)(-1.30) + (0.90)(0.98) + (1.50)(-0.72) + (1.20)(1.33)(1.75)(-15.48)]$$
$$= -42.81 \text{ kNm} \Leftarrow \text{controls negative moment}$$

As specified in [4.6.2.1.1] the entire width of the deck shall be designed for these maximum moments.

In reviewing the magnitude of the dead loads in comparison to the live loads it becomes apparent that the total combined dead load plus live-load moment is clearly dominated by the live-load moments. Therefore, performing complex analysis to determine exact dead-load moments is not justified. Using elementary beam formulas or other approximate methods is probably sufficient in most cases.

8. **Determine the Slab Reinforcement Detailing Requirements**
   a. *Determine the top deck reinforcement cover* [Table 5.12.3-1]:
   The top deck requires a minimum cover of 50 mm over the top mat reinforcement, unless environmental conditions at the site require additional cover. This cover does not include additional concrete placed on the deck for sacrificial purposes, grooving, or grinding. The clearance between the bottom mat reinforcement and the bottom of the deck slab is 25 mm, up to a No. 35 bar.
   b. *Determine deck reinforcement spacing requirements* [5.10.3.2]:

$$s_{max} = 1.5(175 \text{ mm}) = 262 \text{ mm} \Leftarrow \text{controls}$$

or

$$s_{max} = 450 \text{ mm}$$

   The minimum spacing of reinforcement is determined by [5.10.3.1] and is dependent on the bar size chosen and aggregate size.
   c. *Determine distribution reinforcement requirements* [9.7.2.3] [9.7.3.2]:
   Reinforcement is needed in the bottom of the slab in the direction of the girders in order to distribute the deck loads to the primary deck slab reinforcement, which is oriented transversely to traffic. The effective span length (S) is dependent on the girder type, which was not specified for this example in order to make the solution general. However, with the girder spacing of 2050 mm used in this example, the maximum value of 67% in the formula $(3840)/(\sqrt{S}) \leq 67\%$ would control. This value is a percentage of the primary slab reinforcement that is to be used for distribution reinforcement in the bottom of the slab and is placed parallel to the main girders.
   d. *Determine the minimum top slab reinforcement parallel to the girders* [5.10.8.2]:
   The top slab reinforcement shall be a minimum as required for shrinkage and temperature of $(0.75)(A_g)/f_y$. The top slab reinforcement may be controlled by the negative moment reinforcement needs of the main girders which would likely be greater than the shrinkage and temperature reinforcement requirements.
9. **Design**
   The entire width of the deck should be designed for the maximum positive and negative moments as specified in [4.6.2.1.1]. The positive and negative reinforcement is designed like a typical concrete beam. Concrete design is covered in many civil engineering texts and it is not the intent of this chapter to cover this topic. See Chapter 9 of this text for a discussion of concrete design methods.

24.2.2 Precast Concrete Bridge Decks

This type of bridge deck (Figure 24.3) has the advantage of not requiring significant curing and setup time prior to being loaded with traffic loads as is required for a cast-in-place deck. Therefore, this type of deck is often used for deck replacement [1]. Work can be done overnight or during off-peak traffic times when traffic can be temporarily detoured around the bridge or when a reduced number of traffic lanes can be provided and the deck is replaced in longitudinal sections while traffic continues in adjacent lanes.

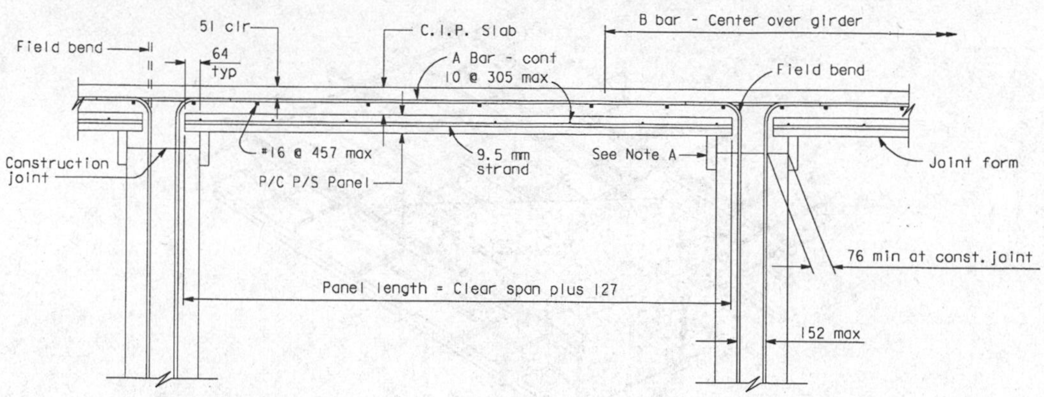

FIGURE 24.3 Precast concrete bridge deck. (From California Department of Transportation, Bridge Standard Detail Sheets, Sacramento, 1993. With permission.)

Precast decks can either serve as the final deck riding surface or a cast-in-place surface can be added on top. A cast-in-place surface uses the precast panels as the deck formwork, which would be placed between the girders. In adding a cast-in-place concrete surface, the problems associated with filling and maintaining the joints between the panels are reduced, and it assists in making the bridge deck composite with the girders. However, this method is at odds with the desire to open the deck to traffic as soon as possible. This led to the development of methods that do not require an additional final surface. If a final concrete surface is not placed on top, the joints between each panel must be successfully filled to avoid leakage and avoid future maintenance problems. This is typically done with expansive grouts or special epoxy crack sealers.

Precast panels may be prestressed, which reduces the depth of the precast panel between the main longitudinal girders or provides for increased spacing between the main girders for a given deck thickness. Perhaps more importantly, prestressing reduces the cracking in the deck. This is especially important for bridges exposed to aggressive environments.

Future widenings of decks using transverse deck prestressing is more difficult than a deck with conventional reinforcement. While a prestressed deck is likely to require less maintenance than a conventionally reinforced deck, repairs that may be required will be more difficult than for a conventionally reinforced deck [1].

24.2.3 Steel Grid Bridge Decks

Steel grids (Figure 24.4) can either be constructed off site as individual panels, constructed at the job site, or can even be assembled on the site of individual components. These grids are then usually welded or mechanically fastened to the supporting components. A significant advantage of open grid decks is their light weight. This can be especially important for existing bridges where the girders would require strengthening if extra dead load were added to the structure, and for movable bridges where dead load is minimized in order to limit loads on the mechanical systems.

However, open grids can result in poor riding quality and a loud whine as traffic crosses them. In addition, rainfall and possible spills fall directly through the deck and cannot be captured or controlled. This can lead to corrosion of components below the bridge and in environmental problems below the bridge because spills can fall directly into waterways. In addition to this, careful detailing is required to avoid fatigue problems associated with steel grid systems [1]. Interlocking grid systems that do not require welding may eliminate stresses that cause these types of failures. Older open steel grid systems have had problems with skid resistance, but newer systems are available that meet today's standards. However, open grids can wear over time, reducing the skid resistance.

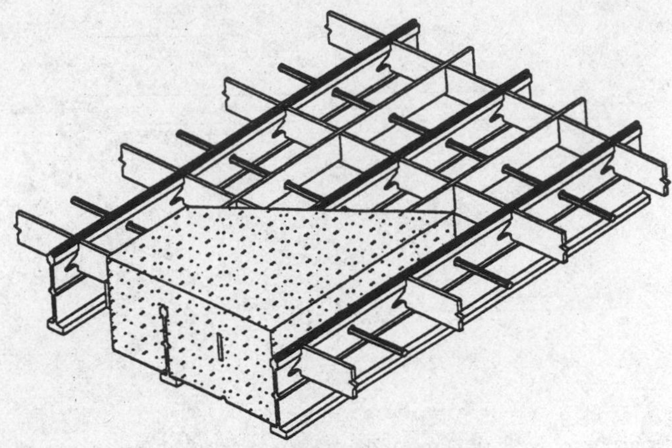

FIGURE 24.4 Steel grid bridge deck. (From American Grid, *Weldless Bridge Deck Systems.* With permission.)

Some of these problems can be solved by filling, or partially filling, the grid with concrete. The grid is then made composite with the concrete and acts as rebar in tension would in a conventionally reinforced-concrete beam. Although this type of deck is heavier than an open grid, it is still lighter than a conventional concrete bridge deck. Typically, an overfill of the grid is specified, which is assumed also to act compositely with the grid. This overfill provides added cover to minimize cracking, reduce corrosion of the steel grid, and improve ridability.

A variation of the filled and unfilled grid type uses an unfilled grid with shear studs. This makes it composite with a reinforced concrete slab which is set on top of the grid in an attempt to combine the advantages of a concrete and a steel grid deck.

24.2.4 Timber Bridge Decks

This was a common deck type prior to the advent of the automobile (Figure 24.5). Because of durability problems this type of deck is rarely used today, except on very low volume bridges, often in rural areas [1]. Timber bridge decks may be constructed using glulam timber panel decks and stress-laminated decks that are post-tensioned together. An asphalt wearing surface may be placed on top of the deck in an attempt to increase the durability of this type of bridge deck.

24.2.5 Steel Orthotropic Bridge Decks

An orthotropic steel deck is a deck plate acting as the flange section and is stiffened with longitudinal ribs and transverse floor beams. A wearing surface is added to act compositely with the deck plate. This subject is covered in greater depth in Chapter 14 of this text.

24.3 Approach Slabs

The structure approach slab provides motorists a smooth transition between the paved highway surface and the riding surface of a bridge. The most common construction consists of reinforced portland cement concrete (PCC). The need for an approach slab is generated by the differential vertical displacement that often occurs between the generally unyielding bridge structure and adjacent fill approaches.

The settlement of the adjacent fill can be gradual, over lengthy periods of time, or sudden, such as when ground motion from an earthquake “liquefies” unconsolidated material in the presence of groundwater. During such a catastrophic event, the role of the approach slab becomes paramount in enabling the passage of emergency vehicles immediately after the earthquake.

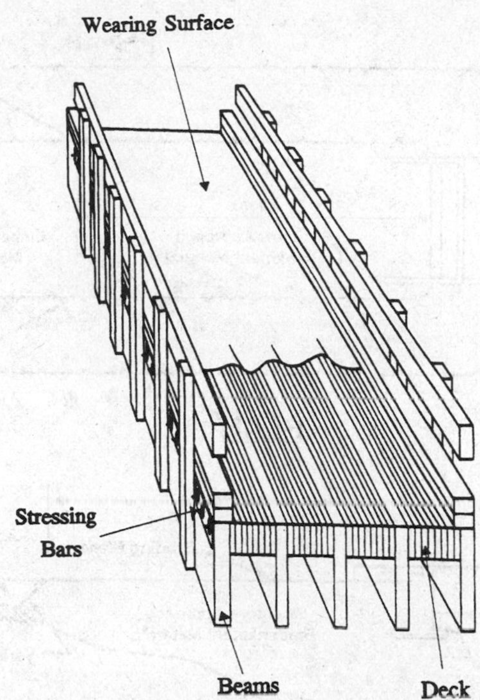

FIGURE 24.5 Timber bridge deck. (From Davalos, J. F., Wolcott, M. P., Dickson, B., and Brokaw J., *Quality Assurance and Inspection Manual for Timber Bridges*, 1992. With permission.)

The design and maintenance of an approach slab is greatly dependent on numerous factors which can affect the amount and rate of settlement that occurs. Careful attention to address these problematic features properly will lead to a serviceable, low-maintenance structure.

24.3.1 Structural Design Considerations

The reinforced-concrete approach slab is designed as a simply supported element, spanning from where it rests on the bridge abutment to the end that meets the roadway. The length of the slab is generally standardized to provide for the majority of applications. Currently, in California, the most commonly used approach slab is one which provides a 9-m transition from roadway to bridge.

The slab itself is usually about 300 mm thick and is reinforced with two-way mats of reinforcing steel both top and bottom. Although the slab is designed to support any passing live load adequately, equal importance is given to preventing the slab from cracking and "breaking up" under the constant cycles of loading by traffic, causing costly maintenance repairs and poor ridability for the motoring public. Reinforcement ratio indexes for the longitudinal bottom reinforcing bars of 0.0110, 0.0022 for the bottom transverse bars, and 0.0031 for the longitudinal top reinforcing bars have historically provided satisfactory performance for the 9-m-long slab. A minimum amount of transverse top reinforcing steel satisfies temperature and shrinkage requirements.

24.3.2 Settlement Problems

Settlement of the adjacent fill abutting a bridge is related to the geologic properties of the fill material and the native soil that underlies it. Considerations should be given to include composition and compaction of the fill material, settlement of the native soil due to the overburden imposed on it, and drainage conditions of the approaches.

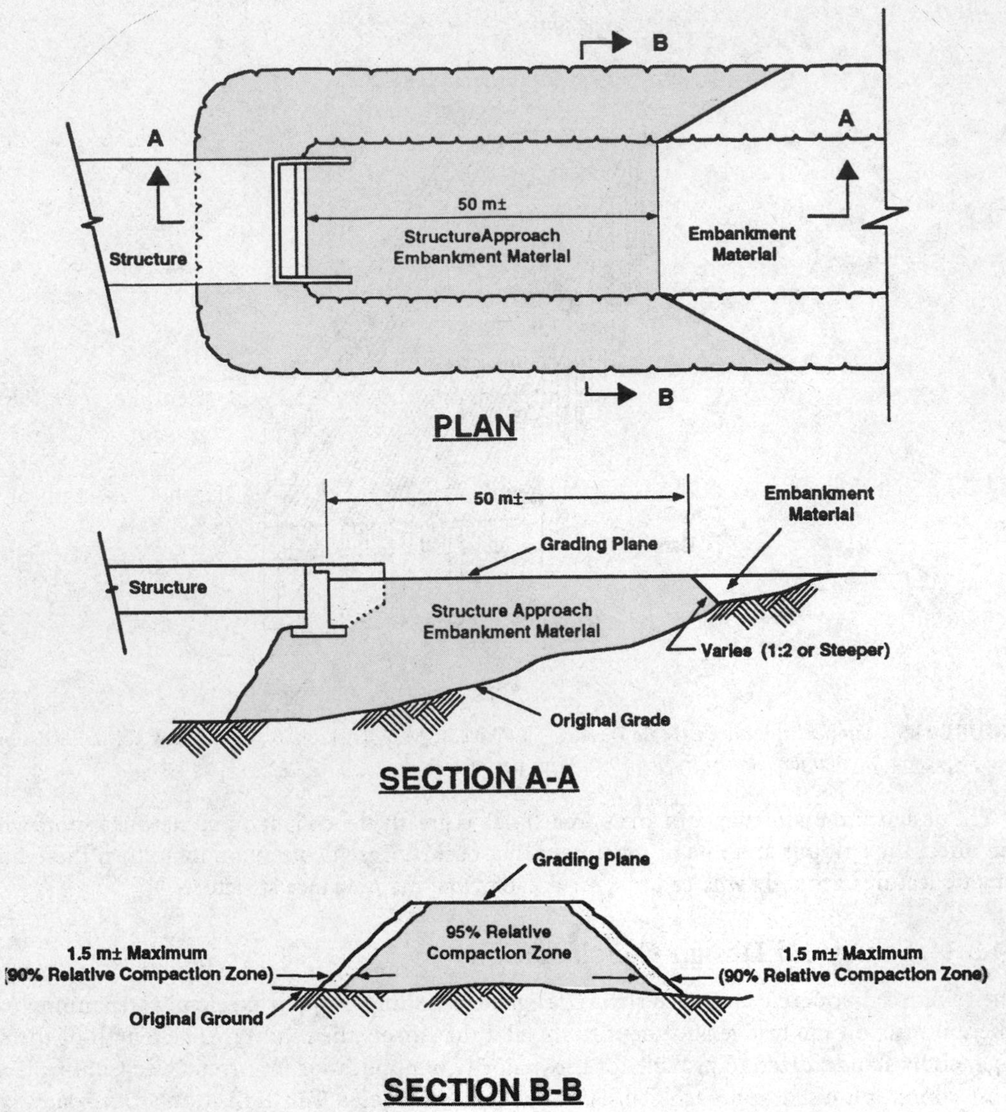

FIGURE 24.6 Limits of structure approach embankment material. (From California Department of Transportation, *Highway Design Manual,* Sacramento, 1995. With permission.)

The selection of fill material that creates the approaches to a structure is critical. Volume changes of the fill material result from rearrangement of soil particles, loss of moisture, gain of moisture, or frost and ice action. The fill material should not consist of soil types that are subject to influence under those conditions, such as cohesionless soils, which tend to settle under the vibration of traffic, or highly plastic soils, which are difficult to work with and highly compressible [3]. Approach fills should be constructed with selected material of slightly cohesive granular soil and extend well beyond the limits of abutments.

Adequate compaction of the fill material will limit the amount of future settlement by removing the potential volume changes of the soil. A minimum relative compaction value of 95% provides a reasonable lower limit that will minimize detrimental magnitudes of settlement of the fill. In the state of California, current practice is to require this type of "structure approach embankment material" 50 m from the back of the abutment (Figure 24.6).

Special attention is required in the areas immediately adjacent to the abutment or wingwalls, where confined spaces may limit the accessibility of large compaction equipment. Typically, small, hand-operated pieces of equipment are resorted to in order to compact the fill in these locations. It is imperative that proper procedures, which should be carefully addressed in the specifications, are adhered to and diligently monitored for quality assurance. This will limit the potential for differential settlement between the fill adjacent to the bridge structure, which is difficult to compact, and the fill away from the structure, which can be compacted by large equipment.

Subsidence of the underlying native soil is another major contribution to the poor performance of a bridge approach. With the additional mass of the overburden in the form of the approach fill, the native soil may not have sufficient strength to support the additional weight and will settle. Several options are possible to alleviate this problem.

The use of a surcharge to preload the approach site and preconsolidate compressible native material can be an effective solution to limit future, postconstruction settlement. The effectiveness of utilizing a surcharge is dependent on the amount of time available for this operation. Naturally, the more time that can elapse before construction of the approaches, the better. Care must be exercised that the amount and rate of applied surcharge does not exceed the shear strength of the native soil. This typically results in the necessity to load the site incrementally [3].

The use of drains in soft and compressible, slow-draining soil can accelerate its consolidation by basically dewatering the site and allowing primary consolidation to occur. This works well in thick, homogeneous layers of clay. Examples of vertical drains are sand drains and wick drains, which allow the migration of water within the soil to the surface where it can be collected and removed.

At sites where relatively shallow thicknesses of unsuitable, compressible soil may exist, a simple solution is to remove this layer of material. The replacement material would then be imported rock or suitable, well-compacted material.

A final alternative may be the use of lightweight fill material to limit the overburden on the native soil. Appropriate materials suitable for this purpose can consist of furnace slag, expanded shale, coal waste refuse, lightweight or cellular concrete, polystyrene foam, and other materials having small unit weights [3].

Proper drainage of the approach site, particularly behind and between the bridge abutment and wingwalls is essential to minimize surface and subsurface erosion and to reduce lateral hydrostatic pressure against these structural elements. An effective drainage system within the approach fill will consist of a medium located immediately adjacent to the abutment and wingwalls that will allow for the migration of water to bottom drains that will move the water away from the area. This can consist of pervious material such as gravel or a geocomposite drain and filter fabric, which is a man-made waffle-like material that channels water vertically downward. The placement of pervious material concurrently with the approach backfill is a difficult and more tedious process than the use of the geocomposite drain. The geocomposite drain system is the preferred method and is applied as shown in Figure 24.7.

24.3.3 Additional Considerations

Some related considerations that appear to have no effect on the performance and serviceability of approach slabs are worth mentioning. These include location of approaches relative to the structure, fill height, and average daily traffic (ADT).

The location of the approach, either leading to or away from the structure, seems to make negligible difference in the performance of the approach slab. One might think that the approach leading away from the structure would experience greater impact from vehicles as they come off the structure, resulting in greater settlement and damage to the approach slab. Studies show, however, essentially no difference between slabs at either end of the structures [5].

The height of approach fill also appears to have little effect on the long-term approach performance. It would seem reasonable to conclude that taller fill heights would exhibit the highest continued

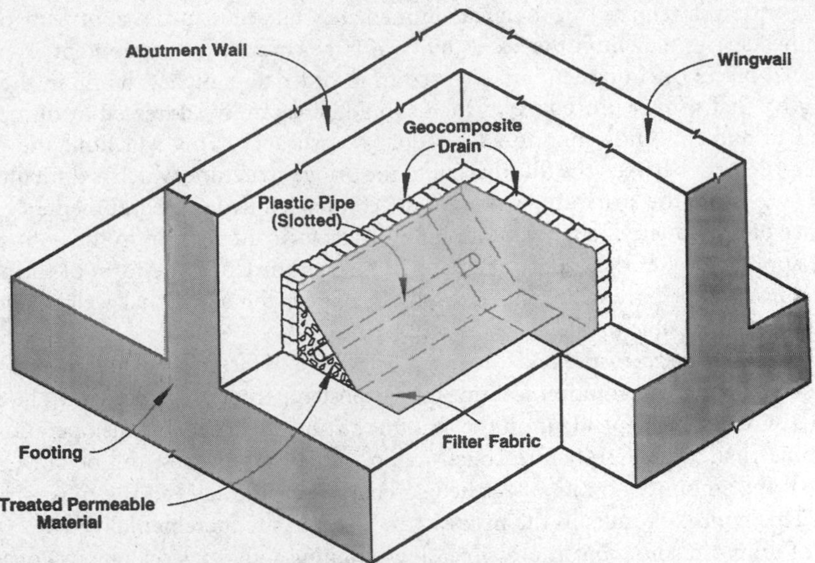

FIGURE 24.7 Abutment drainage details. (From California Department of Transportation, Highway Design Manual, Sacramento, 1995. With permission.)

settlement, but studies again show essentially uniform ground subsidence over all fill heights. A possible explanation is that higher fills are more likely to have been preconditioned in the form of allowing for a settlement period or surcharge of the fill prior to construction of the approach slab [5].

Finally, the volume of average daily traffic similarly exhibits negligible impact to the PCC structure approaches. AC approach data were inconclusive. It appears that the number of cycles of loading has no bearing on the long-term condition of the approach slabs. Maintenance records indicate essentially equal amounts of patching required for varying ADT levels [5].

24.4 Summary

There are several different types of bridge decks available to the design engineer. It is the responsibility of the engineer to determine the most appropriate type of bridge deck for a given site and situation. This can range from a new, short-span bridge with limited traffic to rehabilitation of a long-span bridge over water to be replaced at night because of significant traffic demands.

For most new bridges, cast-in-place concrete bridge decks are chosen as the most appropriate deck type. Typically, these types of decks are designed as a transverse beam supported by the main longitudinal girders. However, just as alternative types of bridge decks are available and are gaining greater acceptance, a new empirical method of designing cast-in-place concrete bridge decks that considers the arching effect between the girders has been introduced. While cast-in-place concrete decks designed as transverse beams have been the standard for decades, bridge deck type and design is continuing to evolve.

The major considerations for approach slabs are settlement of the approach fill, settlement of the underlying native soil, adequate drainage of the fill area behind the abutments and between the wingwalls, and adequate reinforcement; all must be addressed when designing a structure approach slab. Careful selection of suitable fill material and proper consolidation of the fill will limit post-construction settlement. Settlement of the native underlying soil can be eliminated through the practice of applying a surcharge, the use of preconstruction drains in clay soils, the removal of unsuitable compressible soil layers, and, finally, the use of lightweight fills. Proper drainage adjacent to the structure abutments and under the approach slabs will prevent subsurface erosion which would eventually undermine the underlying subbase of the approach slab.

References

1. Bettigole, N. H., and Robison, R., *Bridge Decks*, American Society of Civil Engineers, Danvers, 1997, 3, 40–62.
2. *AASHTO LRFD Bridge Design Specifications*, American Society of State Highway and Transportation Officials, Washington, D.C., 1994.
3. Wolde-Tinsae, A. M., Aggour, M. S., and Chini, S. A., Structural and Soil Provisions for Approaches to Bridges, Report FHWA/MD-89/04, Maryland Department of Transportation, Baltimore, July 1987, 1–48.
4. California Department of Transportation Highway Design Manual, 5th ed., State of California Department of Transportation, Sacramento, 1995, 600-46–600-55.
5. Stewart, C. F., Highway Structure Approaches, Report FHWA/CA/SD-85-05, California Department of Transportation, Sacramento, 1985.
6. Bridge Deck Construction Manual, State of California Department of Transportation, Sacramento, 1991.
7. Barker, R. M. and Puckett, J. A., *Design of Highway Bridges*, John Wiley & Sons, New York, 1997, 504–562.
8. *AASHTO LRFD Bridge Design Specifications*, 1996 Interim Revisions, American Society of State Highway and Transportation Officials, Washington, D.C., 1996.

25

Expansion Joints

Ralph J. Dornsife
Washington State Department of Transportation

25.1 Introduction

Expansion joint systems are integral, yet often overlooked, components designed to accommodate cyclic movements. Properly functioning bridge expansion joint systems accommodate these movements without imposing significant secondary stresses on the superstructure. Sealed expansion joint systems provide barriers preventing runoff water and deicing chemicals from passing through the joint onto bearing and substructure elements below the bridge deck. Water and deicing chemicals have a detrimental impact on overall structural performance by accelerating degradation of bridge deck, bearing, and substructure elements. In extreme cases, this degradation has resulted in premature, catastrophic structural failure. In fulfilling their functions, expansion joints must provide a reasonably smooth ride for motorists.

Perhaps because expansion joints are generally designed and installed last, they are often relegated to peripheral status by designers, builders, and inspectors. As a result of their geometric configuration and the presence of multiple-axle vehicles, expansion joint elements are generally subjected to a significantly larger number of loadings than other structural members. Impact, a consequence of bridge discontinuity inherent at a joint, exacerbates loading. Unfortunately, specific expansion joint systems are often selected based upon their initial cost with minimal consideration for long-term performance, durability, and maintainability. Consequently, a plethora of bridge maintenance problems plague them.

In striving to improve existing and develop new expansion joint systems, manufacturers present engineers with a multitudinous array of options. In selecting a particular system, the designer must carefully assess specific requirements. Magnitude and direction of movement, type of structure, traffic volumes, climatic conditions, skew angles, initial and life cycle costs, and past performance of alternative systems must all be considered. For classification in the ensuing discussion, expansion joint systems will be grouped into three broad categories depending upon the total movement range

0-8493-7434-0/00/$0.00+$.50

accommodated. Small movement range joints encompass all systems capable of accommodating total motion ranges of up to about 45 mm. Medium movement range joints include systems accommodating total motion ranges between about 45 mm and about 130 mm. Large movement range joints accommodate total motion ranges in excess of about 130 mm. These delineated ranges are somewhat arbitrary in that some systems can accommodate movement ranges overlapping these broad categories.

25.2 General Design Criteria

Expansion joints must accommodate movements produced by concrete shrinkage and creep, post-tensioning shortening, thermal variations, dead and live loads, wind and seismic loads, and structure settlements. Concrete shrinkage, post-tensioning shortening, and thermal variations are generally taken into account explicitly in design calculations. Because of uncertainties in predicting, and the increased costs associated with accommodating large displacements, seismic movements are usually not explicitly included in calculations.

Expansion joints should be designed to accommodate all shrinkage occurring after their installation. For unrestrained concrete, ultimate shrinkage strain after installation, β, may be estimated as 0.0002 [1]. More-detailed estimations can be used which include the effect of ambient relative humidity and volume-to-surface ratios [2]. Shrinkage shortening of the bridge deck, Δ_{shrink}, in mm, is calculated as

$$\Delta_{shrink} = (\beta)\cdot(\mu)\cdot(L_{trib})\cdot(1000\text{ mm/m}) \tag{25.1}$$

where

L_{trib} = tributary length of structure subject to shrinkage; m

β = ultimate shrinkage strain after expansion joint installation; estimated as 0.0002 in lieu of more-refined calculations

μ = factor accounting for restraining effect imposed by structural elements installed before slab is cast [1]
= 0.0 for steel girders, 0.5 for precast prestressed concrete girders, 0.8 for concrete box girders and T-beams, 1.0 for flat slabs

Thermal displacements are calculated using the maximum and minimum anticipated bridge deck temperatures. These extreme values are functions of the geographic location of the structure and the bridge type. Thermal movement, in mm, is calculated as

$$\Delta_{temp} = (\alpha)\cdot(L_{trib})\cdot(\delta T)\cdot(1000\text{ mm/m}) \tag{25.2}$$

where

α = coefficient of thermal expansion; 0.000011 m/m/°C for concrete and 0.000012 m/m/°C for steel

L_{trib} = tributary length of structure subject to thermal variation; m

δT = temperature variation; °C

Any other predictable movements following expansion joint installation, such as concrete post-tensioning shortening and creep, should also be included in the design calculations.

25.3 Jointless Bridges

Bridge designers have used superstructure continuity in an effort to avoid some of the maintenance problems associated with expansion joints [3]. This evolution from simple-span construction was facilitated by the development of the moment distribution procedure published by Hardy Cross [4] in 1930.

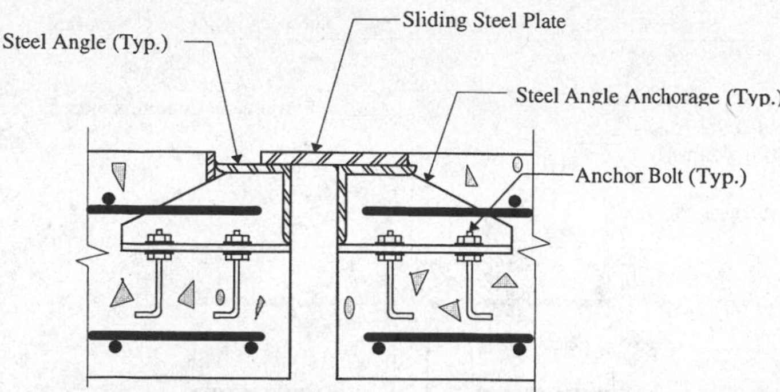

FIGURE 25.1 Sliding plate joint (cross section).

In recent years, some transportation agencies have extended this strategy by developing jointless bridge designs. Jointless bridges are characterized by continuous spans built integrally with their abutments. In many instances, approach slabs are tied to the superstructure slab or to the abutments. The resulting designs are termed *integral* or *semi-integral* depending upon the degree of continuity developed among superstructure, substructure, and approach slab elements. Design methods and details for jointless bridges vary considerably [3,5]. Many transportation agencies have empirically established maximum lengths for jointless bridges [5].

Jointless bridges should not be considered a panacea for addressing expansion jointmaintenance problems. As superstructure movements are restrained in jointless bridges, secondary stresses are induced in superstructure and substructure elements. Stresses may also be induced in approach slabs. If inadequately addressed during design, these stresses can damage structural elements and adjacent asphalt pavements. Damaged structural elements, slabs, and pavements are accompanied by increased probability of moisture infiltration, further exacerbating deterioration. Most jointless bridges have been built relatively recently [5]. Their long-term performance and durability will determine how extensively the jointless bridge concept is applied to future construction.

25.4 Small Movement Range Joints

Many different systems exist for accommodating movement ranges under about 45 mm. These include, but are not limited to, steel sliding plates, elastomeric compression seals, preformed closed cell foam, epoxy-bonded elastomeric glands, asphaltic plug joints, bolt-down elastomeric panels, and poured sealants. In this section, several of these systems will be discussed with an emphasis on design procedures and past performance.

25.4.1 Sliding Plate Joints

Steel sliding plates, shown in Figure 25.1, have been used extensively in the past for expansion joints in both concrete and timber bridge decks. Two overlapping steel plates are attached to the bridge deck, one on each side of the expansion joint opening. They are generally installed so that the top surfaces of the plates are flush with the top of the bridge deck. The plates are generally bolted to timber deck panels or embedded with steel anchorages into a concrete deck. Steel plate widths are sized to accommodate anticipated total movements. Plate thicknesses are determined by structural requirements.

Standard steel sliding plates do not generally provide an effective seal against intrusion of water and deicing chemicals into the joint and onto substructure elements. As a result of plate corrosion and debris collection, the steel sliding plates often bind up, impeding free movement of the superstructure. Repeated impact and weathering tend to loosen or break anchorages to the bridge deck.

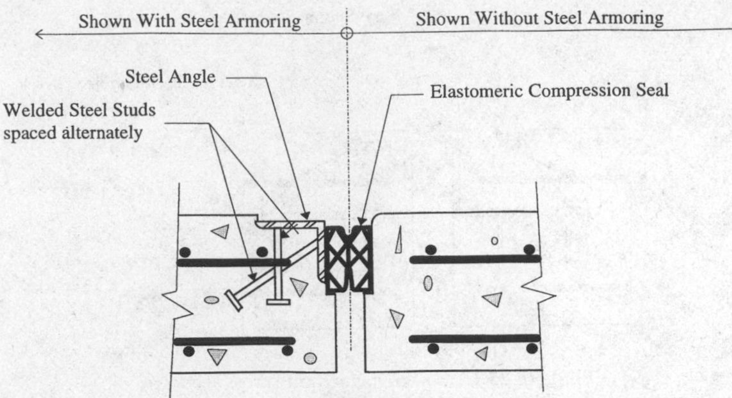

FIGURE 25.2 Compression seal joint (cross section).

Consequently, sliding plate systems are rarely specified for new bridge construction today. Nevertheless, sliding plate systems still exist on many older structures. These systems can be replaced with newer systems providing increased resistance against water and debris infiltration. In situations where the integrity of the deck anchorage has not been compromised, sliding plates can be retrofitted with poured sealants or elastomeric strip seals.

25.4.2 Compression Seal Joints

Compression seals, shown in Figure 25.2, are continuous elastomeric sections, typically with extruded internal web systems, installed within an expansion joint gap to seal the joint effectively against water and debris infiltration. Compression seals are held in place by mobilizing friction against adjacent vertical joint faces. Hence, design philosophy requires that they be sized and installed to be always in a state of compression. Compression seals may be installed against smooth concrete faces or against steel armoring. When installed directly against concrete, polymer concrete nosing material is often used to provide added impact resistance. Combination lubricant/adhesive is typically used to install the seal in its compressed state.

Because compression seals are held in place by friction, their performance is extremely dependent upon the close correlation of constructed joint width and design joint width. If the joint opening is constructed too wide, friction force will be insufficient to prevent the compression seal from slipping out of the joint at wider expansion gap widths. Relaxation of the elastomer and debris accumulation atop the seal contribute to seal slippage. To minimize slippage and maximize compression seal performance, a joint may be formed narrower than the design width, then sawcut immediately prior to compression seal installation. The sawcut width is calculated based upon ambient bridge deck temperature and the degree of slab shrinkage which has already occurred. As an alternative to sawcutting, block outs can be formed on each side of the joint during bridge deck casting. Prior to compression seal installation, concrete is cast into the block outs, often with steel armoring, to form an expansion gap width compatible with ambient conditions.

In design calculations, the maximum and minimum compressed widths of the seal are generally set at 85 and 40% of the uncompressed width [1]. These widths are measured perpendicular to the axis of the joint. It is also generally assumed that the width of the seal at about 20°C is 60% of its uncompressed width. For skewed joints, bridge deck movement must be separated into components perpendicular to and parallel to the joint axis. Shear displacement of the compression seal should be limited to a specified percentage of its uncompressed width, usually set at about 22% [1]. Additionally, the expansion gap width should be set so that the compression seal can be installed over a reasonably wide range of construction temperatures. Manufacturers' catalogues generally specify the minimum expansion gap widths into which specific size compression seals can be installed. The expansion gap width should be specified on the contract drawings as a function of the bridge deck temperature.

Design relationships can be stated as follows:

$$\Delta_{temp\text{-}normal} = \Delta_{temp} \cdot \cos\theta \qquad \text{[thermal movement normal to joint]} \qquad (25.3)$$

$$\Delta_{temp\text{-}parallel} = \Delta_{temp} \cdot \sin\theta \qquad \text{[thermal movement parallel to joint]} \qquad (25.4)$$

$$\Delta_{shrink\text{-}normal} = \Delta_{shrink} \cdot \cos\theta \qquad \text{[shrinkage movement normal to joint]} \qquad (25.5)$$

$$\Delta_{shrink\text{-}parallel} = \Delta_{shrink} \cdot \sin\theta \qquad \text{[shrinkage movement parallel to joint]} \qquad (25.6)$$

$$W_{min} = W_{install} - [(T_{max} - T_{install})/(T_{max} - T_{min})]\ \Delta_{temp\text{-}normal} > 0.40W \qquad (25.7)$$

$$W_{max} = W_{install} + [(T_{install} - T_{min})/(T_{max} - T_{min})]\ \Delta_{temp\text{-}normal} + \Delta_{shrink\text{-}normal} < 0.85W \qquad (25.8)$$

where

θ = skew angle of expansion joint, measured with respect to a line perpendicular to the bridge longitudinal axis; degrees
W = uncompressed width of compression seal; mm
$W_{install}$ = expansion gap width at installation
$T_{install}$ = bridge deck temperature at time of installation; °C
W_{min}, W_{max} = minimum and maximum expansion gap widths; mm
T_{min}, T_{max} = minimum and maximum bridge deck temperatures; °C

Multiplying Eq. (25.7) by –1.0, adding to Eq. (25.8), and rearranging yields:

$$W > (\Delta_{temp\text{-}normal} + \Delta_{shrink\text{-}normal})/0.45 \qquad (25.9)$$

Similarly,

$$W > (\Delta_{temp\text{-}parallel} + \Delta_{shrink\text{-}parallel})/0.22 \qquad (25.10)$$

Now, assuming $W_{install} = 0.6$ W,

$$W_{max} = 0.6W + [(T_{install} - T_{min})/(T_{max} - T_{min})]\ \Delta_{temp\text{-}normal} + \Delta_{shrink\text{-}normal} < 0.85W \qquad (25.11)$$

which, upon rearranging, yields:

$$W > 4\ [(T_{install} - \mathrm{T}_{min})/(\mathrm{T}_{max} - \mathrm{T}_{min}) \cdot (\Delta_{temp\text{-}normal}) + \Delta_{shrink\text{-}normal}] \qquad (25.12)$$

Equations (25.9), (25.10), and (25.12) are used to calculate the required compression seal size. Next, expansion gap widths at various construction temperatures can be evaluated.

25.4.3 Asphaltic Plug Joints

Asphaltic plug joints comprise liquid polymer binder and graded aggregates compacted in preformed block outs as shown in Figure 25.3. The compacted composite material is referred to as polymer modified asphalt (PMA). These joints have been used to accommodate movement ranges up to 50 mm. This expansion joint system was developed in Europe and can be adapted for use with concrete or asphalt bridge deck surfaces. The PMA is installed continuously within a block out centered over the expansion joint opening with the top of the PMA flush with the roadway surface. A steel plate retains the PMA at the bottom of the block out during installation. The polymer

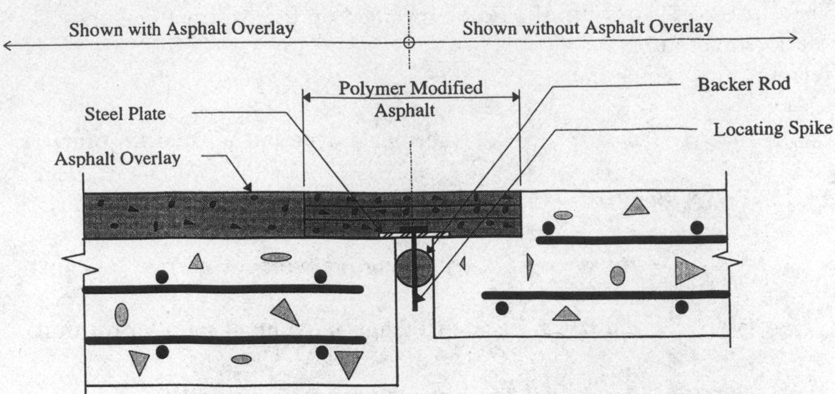

FIGURE 25.3 Asphaltic plug joint (cross section).

binder material is generally installed in heated form. Aggregate gradation, binder properties, and construction quality are critical to asphaltic plug joint performance.

The asphaltic plug joint is designed to provide a smooth, seamless roadway surface. It is relatively easy to repair, is not as susceptible to snowplow damage as other expansion joint systems, and can be cold-milled and/or built up for roadway resurfacing. The performance of asphaltic plug joints in the United States has been somewhat erratic [6]. The material properties of PMA vary with temperature. Asphaltic plug joints have demonstrated a proclivity to soften and creep at warmer temperatures, exhibiting wheel rutting and eventual migration of PMA out of the block outs. In very cold temperatures, the PMA can become brittle and crack at the plug joint-to-pavement interface, making the joint susceptible to water infiltration. Ongoing research is investigating these issues and developing comprehensive design guidelines, material specifications, and installation procedures to improve their performance [6].

As with all expansion joint systems, designers must understand the limitations of asphaltic plug joints. These joints were not designed for, and should not be used to, accommodate differential vertical displacements, as may occur at longitudinal joints. Because of PMA creep susceptibility, asphaltic plug joints should not be used where the roadway is subject to significant traffic acceleration and braking. Examples include freeway off-ramps and roadway sections in the vicinity of traffic signals. Asphaltic plug joints have also performed poorly in highly skewed applications and in applications subjected to large rotations. Maintaining the minimum block-out depth specified by the manufacturer is particularly critical to successful performance. In spite of these limitations, asphaltic plug joints do offer advantages not inherent in other expansion joint systems.

25.4.4 Poured Sealant Joints

Durable low-modulus sealants, poured cold to provide watertight expansion joint seals as shown in Figure 25.4, have been used in new construction and in rehabilitation projects. Properties and application procedures vary between products. Most silicone sealants possess good elastic performance over a wide range of temperatures while demonstrating high levels of resistance to ultraviolet and ozone degradation. Rapid-curing sealants are ideal candidates for rehabilitation in situations where significant traffic disruption from extended traffic lane closure is unacceptable. Other desirable properties include self-leveling and self-bonding capabilities. Installation procedures vary among different products, with some products requiring specialized equipment for mixing individual components. Designers must assess the design and construction requirements, weighing desirable properties against material costs for alternative sealants.

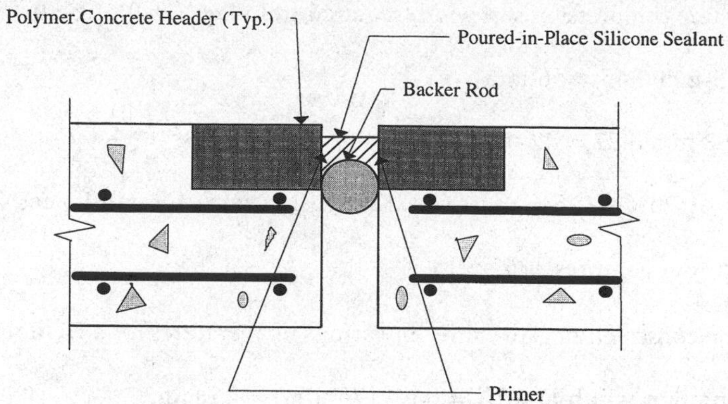

FIGURE 25.4 Poured sealant joint (cross section).

Most sealants can be installed against either concrete or steel. Particularly in rehabilitation projects, it is extremely critical that the concrete or steel substrates be thoroughly cleaned before the sealant is placed. Some manufacturers require application of specific primers onto substrate surfaces prior to sealant placement to enhance bonding. Debonding of sealant from substrate concrete or steel, compromising the integrity of the watertight seal, has previously plagued poured sealant joints. The latest products are relatively new, but have demonstrated good short-term performance and versatility of use in bridge rehabilitation. Their long-term durability will determine the extent of their future application.

Poured sealant joints should be designed based upon manufacturers' recommendations. Maximum and minimum working widths of the poured sealant joint are generally recommended as a percentage of the sealant joint width at installation. A minimum recess is typically required between the top of the roadway surface and the top of the sealant. This recess is critical in preventing tires from contacting and debonding the sealant from its substrate material.

25.4.5 Design Example 1

Given

A reinforced-concrete box-girder bridge has an overall length of 70 m. A compression seal expansion joint at each abutment will accommodate half of the total bridge movement. These expansion joints are skewed 20°. Bridge deck temperatures are expected to range between –15°C and 40°C during the life of the structure.

Find

Compression seal sizes and construction gap widths at 5, 20, and 30°C.

Solution

Step 1: Calculate temperature and shrinkage movement.

$$\text{Temperature: } \Delta_{temp} = (½)(0.000011 \text{ m/m/°C})(55\text{°C})(70 \text{ m})(1000 \text{ mm/m}) = 21 \text{ mm}$$

$$\text{Shrinkage: } \Delta_{shrink} = (½)(0.0002 \text{ m/m})(0.8)(70 \text{ m})(1000 \text{ mm/m}) = \underline{6 \text{ mm}}$$

Total deck movement at the joint: 27 mm

$$\Delta_{temp\text{-}normal} + \Delta_{shrink\text{-}normal} = (27 \text{ mm})(\cos 20°) = 25 \text{ mm}$$

$$\Delta_{temp\text{-}parallel} + \Delta_{shrink\text{-}parallel} = (27 \text{ mm})(\sin 20°) = 9.2 \text{ mm}$$

Step 2: Determine compression seal width required from Eqs. (25.9), (25.10), and (25.12).

$$W > 25\ \text{mm}/0.45 = 56\ \text{mm}$$

$$W > 9.2\ \text{mm}/0.22 = 42\ \text{mm}$$

$$W > 4 \cdot [(20°\text{C} + 15°\text{C})/(40°\text{C} + 15°\text{C}) \cdot (21\ \text{mm}) + 6\ \text{mm}] \cdot \cos 20° = 73\ \text{mm}$$

Use 75 mm compression seal.

Step 3: Evaluate construction gap widths for various temperatures for a 75 mm compression seal.

$$\text{Construction width at } 20°\text{C} = 0.6 \cdot (75\ \text{mm}) = 45\ \text{mm}$$

$$\text{Construction width at } 5°\text{C} = 45\ \text{mm} + [(20°\text{C} - 5°\text{C})/(40°\text{C} + 15°\text{C})] \cdot (21\ \text{mm}) \cdot (\cos 20°) = 50\ \text{mm}$$

$$\text{Construction width at } 30°\text{C} = 45\ \text{mm} - [(30°\text{C} - 20°\text{C})/(40°\text{C} + 15°\text{C})] \cdot (21\ \text{mm}) \cdot (\cos 20°) = 41\ \text{mm}$$

Conclusion
Use a 75-mm compression seal. Construction gap widths for installation temperatures of 5, 20, and 30°C are 50, 45, and 41 mm, respectively.

25.5 Medium Movement Range Joints

Medium movement range expansion joints accommodate movement ranges from about 45 mm to about 130 mm and include sliding plate systems, bolt-down panel joints (elastomeric expansion dams), strip seals, and steel finger joints. Sliding plate systems were previously discussed under small motion range joints.

25.5.1 Bolt-Down Panel Joints

Bolt-down panel joints, also referred to as elastomeric expansion dams, consist of monolithically molded elastomeric panels reinforced with steel plates as shown in Figure 25.5. They are bolted into block outs formed in the concrete bridge deck on each side of an expansion joint gap. Manufacturers fabricate bolt-down panels in varying widths roughly proportional to the total allowable movement range. Expansion is accompanied by uniform stress and strain across the width of the panel joint between anchor bolt rows. Unfortunately, the bolts and nuts connecting bolt-down panels to bridge decks are prone to loosening and breaking under high-speed traffic. The resulting loose panels and hardware in the roadway present hazards to vehicular traffic, particularly motorcycles. Consequently, to mitigate liability, some transportation agencies avoid using bolt-down panel joints.

25.5.2 Strip Seal Joints

An elastomeric strip seal expansion joint system, shown in Figure 25.6, consists of a preformed elastomeric gland mechanically locked into metallic edge rails embedded into concrete on each side of an expansion joint gap. Movement is accommodated by unfolding of the elastomeric gland. Steel studs or reinforcing bars are generally welded to the edge rails to facilitate bonding with the concrete in formed block outs. In some instances the edge rails are bolted in place. Edge rails also furnish armoring for the adjacent bridge deck concrete. Properly installed strip seals have demonstrated relatively good performance. Damaged or worn glands can be replaced with minimal traffic disruptions.

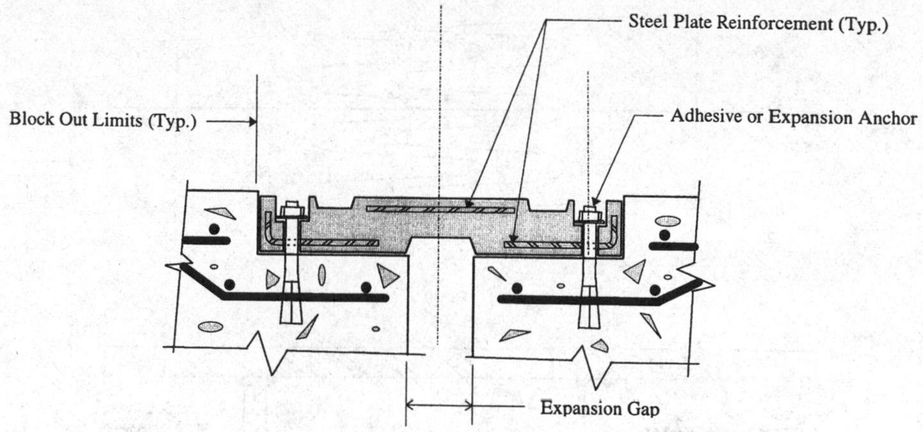

FIGURE 25.5 Bolt-down panel joint (cross section).

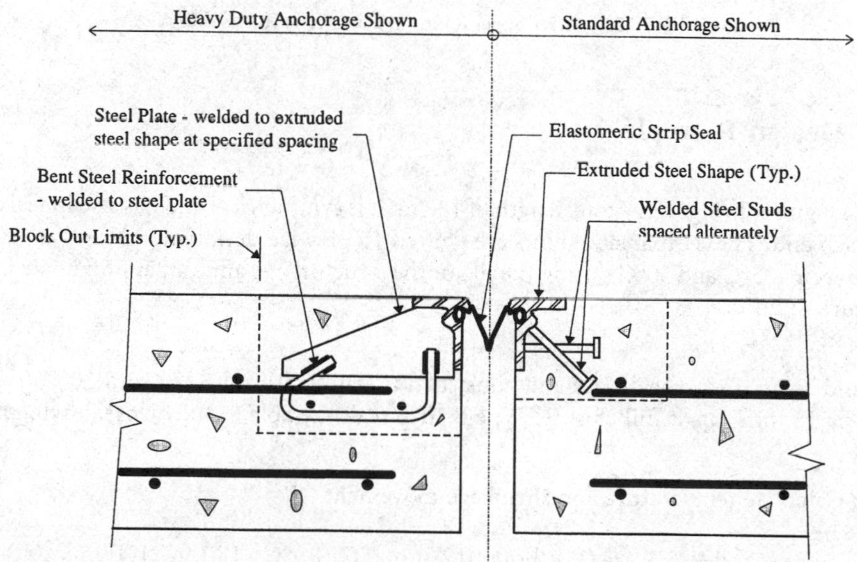

FIGURE 25.6 Elastomeric strip seal joint (cross section).

The elastomeric glands exhibit a proclivity for accumulating debris. In some instances, this debris can resist joint movement and result in premature gland failure.

25.5.3 Steel Finger Joints

Steel finger joints, shown in Figure 25.7, have been used to accommodate medium and large movement ranges. These joints are generally fabricated from steel plate and are installed in cantilever or prop cantilever configurations. The steel fingers must be designed to support traffic loads with sufficient stiffness to preclude excessive vibration. In addition to longitudinal movement, they must also accommodate any rotation or differential vertical deflection across the joint. To minimize the potential for damage from snowplow blade impact, steel fingers may be fabricated with a slight downward taper toward the joint centerline. Generally, steel finger joints do not provide a seal against water intrusion to substructure elements. Elastomeric or metallic troughs can be installed beneath the steel finger joint assembly to catch and redirect water and debris runoff. However, unless regularly maintained, these troughs clog and become ineffective [3].

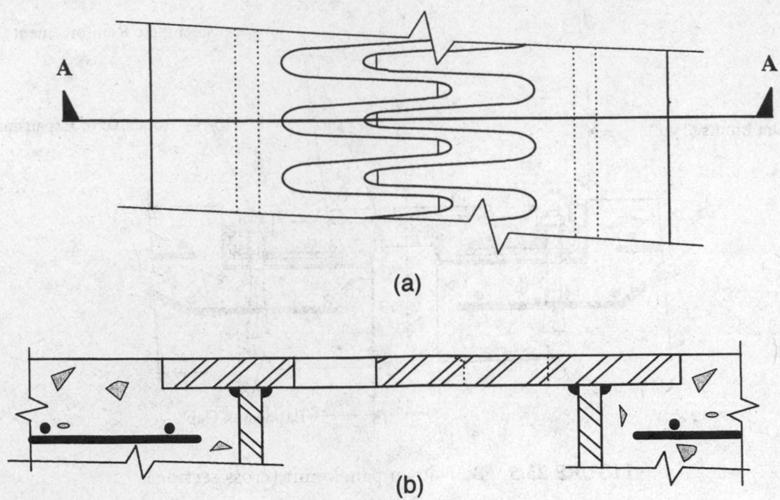

FIGURE 25.7 Steel finger joint. (a) Plan view; (b) section A–A.

25.5.4 Design Example 2

Given

A steel-plate girder bridge has a total length of 180 m. It is symmetrical and has a strip seal expansion joint at each end. These expansion joints are skewed 15°. Bridge deck temperatures are expected to range between –35°C and 50°C during the life of the structure. Assume an approximate installation temperature of 20°C.

Find

Type A and Type B strip seal sizes and construction gap widths at 5, 20, and 30°C. Type A strip seals have a 15 mm gap at full closure. Type B strip seals are able to fully close, leaving no gap.

Solution

Step 1: Calculate temperature and shrinkage movement.

Temperature: $\Delta_{temp} = (½)(0.000012\ m/m/°C)(85°C)(180\ m)(1000\ mm/m) = 92\ mm$

Shrinkage: $\Delta_{shrink} = 0.0$ (no shrinkage, $\mu = 0.0$ for steel bridge) ______

Total deck movement at the joint: 92 mm

$\Delta_{temp\text{-}normal\text{-}closing} = (50°C - 20°C)/(50°C + 35°C)(92\ mm)(\cos 15°) =$ 31 mm

$\Delta_{temp\text{-}normal\text{-}opening} = (20°C + 35°C)/(50°C + 35°C)(92\ mm)(\cos 15°) =$ 58 mm

Step 2: Determine strip seal size required. Assume a minimum construction gap width of 40 mm at 20°C.

Type A: Construction gap width of 40 mm at 20°C will not accommodate 31 mm closing and still allow a 15 mm gap at full closure. Therefore, minimum construction gap width at 20°C must be 31 mm + 15 mm = 46 mm.

Size required = 46 mm + 58 mm = 104 mm → Use 100 mm strip seal

Type B: Construction width of 40 mm at 20°C is adequate.

$$\text{Size required} = 40\text{ mm} + 58\text{ mm} = 98\text{ mm} \rightarrow \underline{\text{Use 100 mm strip seal}}$$

Step 3: Evaluate construction gap widths for various temperatures for a 100 mm strip seal.
Type A: Required construction gap width at 20°C = 15 mm + 31 mm = 46 mm

$$\begin{aligned}\text{Construction gap width at 5°C} &= 46\text{ mm} + (20°C - 5°C)/(20°C + 35°C)(58\text{ mm}) \\ &= 62\text{ mm}\end{aligned}$$

$$\begin{aligned}\text{Construction gap width at 30°C} &= 46\text{ mm} - (30°C - 20°C)/(50°C - 20°C)(31\text{ mm}) \\ &= 36\text{ mm}\end{aligned}$$

Type B: Construction width of 40 mm at 20°C is adequate.

$$\begin{aligned}\text{Construction gap width at 5°F} &= 40\text{ mm} + (20°C - 5°C)/(20°C + 35°C)(58\text{ mm}) \\ &= 56\text{ mm}\end{aligned}$$

$$\begin{aligned}\text{Construction gap width at 30°F} &= 40\text{ mm} - (30°C - 20°C)/(50°C - 20°C)(31\text{ mm}) \\ &= 30\text{ mm}\end{aligned}$$

Conclusion
Use a 100 mm strip seal. Construction gap widths for Type A strip seals at installation temperatures of 5, 20, and 30°C are 62, 46, and 36 mm, respectively. Construction gap widths for Type B strip seals at installation temperatures of 5, 20, and 30°C are 56, 40, and 30 mm, respectively.

25.6 Large Movement Range Joints

Large movement range joints accommodate more than 130 mm of total movement and include bolt-down panel joints (elastomeric expansion dams), steel finger joints, and modular expansion joints. Bolt-down panel and steel finger joints were previously discussed as medium movement range joints.

25.6.1 Modular Bridge Expansion Joints

Modular bridge expansion joints (MBEJ), shown in Figure 25.8, are complex, expensive, structural systems designed to provide watertight wheel load transfer across wide expansion joint openings. These systems were developed in Europe and introduced in the United States in the 1960s [7]. They are generally shipped to the construction site for installation in a completely assembled configuration. MBEJs comprise a series of center beams supported atop support bars. The center beams are oriented parallel to the joint axis while the support bars span parallel to the primary direction of movement. MBEJs can be classified as either single-support bar systems or multiple-support bar systems. In multiple-support bar systems, each center beam is supported by a separate support bar at each support location. Figure 25.8 depicts a multiple-support bar system. In the more complex single-support bar system, one support bar supports all center beams at each support location. This design concept requires that each center beam be free to translate along the longitudinal axis of the support bar as the joint opens and closes. This is accomplished by attaching steel yokes to the underside of the center beams. The support bar passes through the openings in the yokes. Elastomeric springs between the underside of each center beam and the top of the support bar and between the bottom of the support bar and the bottom of the yoke support each center beam and permit it to translate along the longitudinal axis of the support bar.

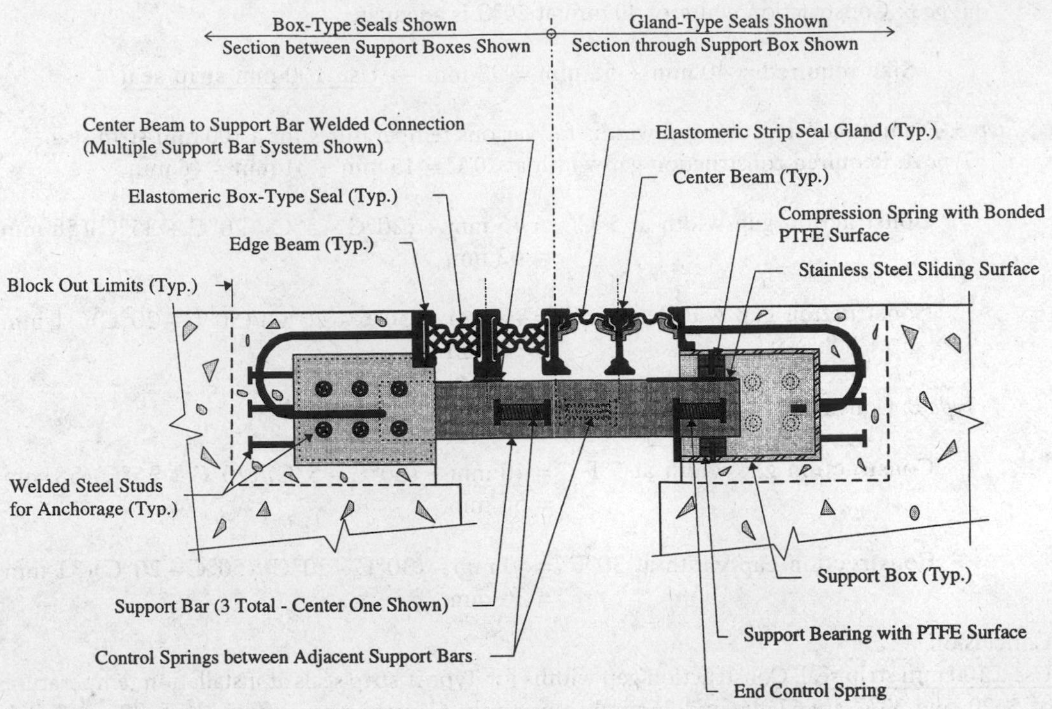

FIGURE 25.8 Modular bridge expansion joint (multiple support bar system), cross section.

The support bars are, in turn, supported on sliding bearings mounted within support boxes. Polytetrafluorethylene (PTFE)-to-stainless-steel interfaces between elastomeric support bearings and support bars facilitate unimpeded translation of the support bars as the expansion gap varies. Control springs between adjacent support bars and between support bars and support boxes of multiple-support bar MBEJs are designed to maintain equal distances between center beams as the expansion gap varies. The support boxes are embedded in bridge deck concrete on each side of the expansion joint. Elastomeric strip seals or elastomeric box-type seals attach to adjacent center beams, providing resistance to water and debris intrusion.

The highly repetitive nature of axle loads predisposes MBEJ components and connections to high fatigue susceptibility, particularly at connections of center beam to support bar. Bolted connections have, generally, performed poorly. Welded connections are preferred, but must be carefully designed, fatigue-tested, fabricated, and inspected to assure satisfactory performance and durability. Field-welded center beam splices are also highly fatigue susceptible, requiring careful detailing, welding, and inspection. A lack of understanding of the dynamic response of these systems, connection detail complexity, and the competitive nature of the marketplace have exacerbated fatigue susceptibility. Fortunately, current research is developing fatigue-resistant structural design specifications in addition to focusing on developing minimum performance standards, performance and acceptance test methods, and installation guidelines for MBEJs [7,8].

Calculated total movements establish MBEJ size. Often, an allowance is made to provide a nominal factor of safety on the calculated movements. Currently available systems permit 75 mm of movement per strip seal element; hence, the total movement rating provided will be a multiple of 75 mm. To minimize impact and wear on bearing elements, the maximum gap between adjacent center beams is limited, typically to about 90 mm [9]. To facilitate installation within concrete block outs, contract drawings should specify the face-to-face distance of edge beams as a function of temperature at the time of installation.

Design relationships can be expressed as:

$$n = \text{MR}/\text{mr} \tag{25.13}$$

$$G_{min} = (n - 1) \cdot (w) + (n) \cdot (g) \tag{25.14}$$

$$G_{max} = G_{min} + \text{MR} \tag{25.15}$$

where

MR = total movement rating of the MBEJ system; mm
mr = movement rating per strip seal element; mm
n = number of seals
$n - 1$ = number of center beams
w = width of each center beam; mm
g = minimum gap per strip seal element at full closure; mm
G_{min} = minimum face-to-face distance of edge beams; mm
G_{max} = maximum face-to-face distance of edge beams; mm

Structural design of MBEJs is generally performed by the manufacturer. Project specifications should require that the manufacturer submit structural calculations, detailed fabrication drawings, and applicable fatigue tests for approval. All elements and connections must be designed and detailed to resist fatigue stresses imposed by repetitive vertical and horizontal wheel loadings. Additionally, MBEJs should be detailed to provide access for inspection and periodic maintenance, including replacement of seals, control springs, and bearing components.

25.6.2 Design Example 3

Given

Two cast-in-place post-tensioned concrete box-girder bridge frames meet at an intermediate pier where they are free to translate longitudinally. Skew angle is 0° and bridge deck ambient temperatures range from –15 to 50°C. A MBEJ will be installed 60 days after post-tensioning operations have been completed. Specified creep is 150% of elastic shortening. Assume that 50% of shrinkage has already occurred at installation time. The following longitudinal movements were calculated for each of the two frames:

	Frame A	Frame B
Shrinkage	30 mm	15 mm
Elastic shortening	36 mm	20 mm
Creep (1.5 × elastic shortening)	54 mm	30 mm
Temperature fall (20 to –15°C)	76 mm	38 mm
Temperature rise (20 to 50°C)	66 mm	33 mm

Find

MBEJ size required to accommodate the total calculated movements and the installation gaps measured face to face of edge beams, "$G_{install}$," at 5, 20, and 30°C.

Solution

Step 1: Determine MBEJ size.

Total opening movement (Frame A) = (0.5)(30 mm) + 54 mm + 76 mm = 145 mm
Total opening movement (Frame B) = (0.5)(15 mm) + 30 mm + 38 mm = 76 mm
Total opening movement (both frames) = 145 mm + 76 mm = 221 mm
Total closing movement (both frames) = 66 mm + 33 mm = 99 mm

Determine size of modular joint, including a 15% allowance:

$$1.15(221 \text{ mm} + 99 \text{ mm}) = 368 \text{ mm} \rightarrow \underline{\text{Use 375 mm movement rating MBEJ}}.$$

Step 2: Evaluate installation gaps measured face to face of edge beams at 5, 20, and 30°C.

MR = 375 mm (MBEJ movement range)
mr = 75 mm (maximum movement rating per strip seal element)
n = 375 mm/75 mm = 5 strip seal elements
$n - 1$ = 4 center beams
w = 65 mm (center beam top flange width)
g = 0 mm
G_{min} = (4)(65 mm) + (4)(0 mm) = 260 mm
G_{max} = 260 mm + 375 mm = 635 mm
G_{20C} = G_{min} + Total closing movement from temperature rise
= 260 mm + 1.15(99 mm) = 374 mm → Use 375 mm.
G_{5C} = 375 mm + [(20°C – 5°C)/(20°C + 15°C)] · (76 mm + 38 mm) = 424 mm
G_{30C} = 375 mm – [(30°C – 20°C)/(50°C – 20°C)] · (66 mm + 33 mm) = 342 mm

Check spacing between center beams at minimum temperature:

$$G_{-15C} = 375 \text{ mm} + 221 \text{ mm} = 596 \text{ mm}$$

$$\text{Maximum spacing} = [596 \text{ mm} - (4) \cdot (65 \text{ mm})]/5 = 67 \text{ mm} < 90 \text{ mm} \quad \text{OK}$$

Check spacing between center beams at 20°C for seal replacement:

$$\text{Spacing} = [375 \text{ mm} - 4(65 \text{ mm})]/5 = 23 \text{ mm} < 40 \text{ mm}$$

Therefore, center beams must be mechanically jacked in order to replace strip seal elements.

Conclusion
Use a MBEJ with a 375 mm movement rating. Installation gaps measured face to face of edge beams at installation temperatures of 5, 20, and 30°C are 424, 375, and 342 mm, respectively.

25.7 Construction and Maintenance

In conjunction with appropriate design procedures, the long-term performance and durability of expansion joint systems require the synergistic application of high-quality fabrication, competent construction practices, assiduous inspection, and routine maintenance. Expansion joint components and connections experience severe loading under harsh environmental conditions. An adequately designed system must be properly manufactured, installed, and maintained to assure adequate performance under these conditions. The importance of quality control must be emphasized. Contract drawings and specifications must explicitly state design, material, fabrication, installation, and quality control requirements. Structural calculations and detailed fabrication drawings should be submitted to the bridge designer for careful review and approval prior to fabrication.

Experience and research will continue to improve expansion joint system technology [10,11]. It is vitally important that design engineers keep abreast of new technological developments. Interdisciplinary and interagency communication facilitates exchange of important information. Maintenance personnel can furnish valuable feedback to designers for implementation in future designs. Designers can provide valuable guidance to maintenance personnel with the goal of increasing service life. Manufacturers furnish designers and maintenance crews with guidelines and limitations for successfully designing and maintaining their products. In turn, designers and maintenance personnel provide feedback to manufacturers on the performance of their products and how they might be improved. Communication among disciplines is key to improving long-term performance and durability.

References

1. Washington State Department of Transportation, Miscellaneous design, in Bridge Design Manual, Washington State Department of Transportation, Olympia, 1997, chap. 8.
2. American Association of State Highway and Transportation Officials, Concrete structures, in *AASHTO LRFD Bridge Design Specifications,* AASHTO Washington, D.C., 1994, sect. 5, 5–14.
3. Burke, M. P., Jr., Bridge Deck Joints, National Cooperative Highway Research Program Synthesis of Highway Practice Report 141, Transportation Research Board, National Research Council, Washington, D.C., 1984.
4. Cross, H., Analysis of continuous frames by distributing fixed-end moments, *Proc. Am. Soc. Civil Eng.,* May, 919, 1930.
5. Steiger, D. J., Field evaluation and study of jointless bridges, in *Third World Congress on Joint Sealing and Bearing Systems for Concrete Structures,* Stoyle, J. E., Ed., American Concrete Institute, Farmingham Hills, MI, 1991, 227.
6. Bramel, B. K., Puckett, J. A., Ksaibati, K., and Dolan, C. W., "Asphalt plug joint usage and perceptions in the United States, draft copy of a paper prepared for the annual meeting of the Transportation Research Board, August 1996.
7. Kaczinski, M. R., Dexter, R. J., and Connor, R. J., Fatigue design and testing of modular bridge expansion joints, in *Fourth World Congress on Joint Sealants and Bearing Systems for Concrete Structures,* Atkinson, B., Ed., American Concrete Institute, Farmingham, Hill, MI, 1996, 97.
8. Dexter, R. J., Connor, R.J., and Kaczinski, M.R., Fatigue Design of Modular Bridge Expansion Joints, National Cooperative Highway Research Program Report 402 , Transportation Research Board, National Research Council, Washington, D.C., April, 1997.
9. Van Lund, J. A., Bridge deck joints in Washington State, in *Third World Congress on Joint Sealing and Bearing Systems for Concrete Structures,* Stoyle, J. E., Ed., American Concrete Institute, Farmingham Hills, MI, 1991, 371.
10. Stoyle, J. E., Ed., *Third World Congress on Joint Sealing and Bearing Systems for Concrete Structures,* American Concrete Institute, Farmingham Hills, MI, 1991.
11. Atkinson, B., Ed., *Fourth World Congress on Joint Sealants and Bearing Systems for Concrete Structures,* American Concrete Institute, Farmingham Hills, MI, 1996.

Section III

Substructure Design

26
Bearings

Johnny Feng
J. Muller International, Inc.

Hong Chen
J. Muller International, Inc.

26.1 Introduction

Bearings are structural devices positioned between the bridge superstructure and the substructure. Their principal functions are as follows:

1. To transmit loads from the superstructure to the substructure, and
2. To accommodate relative movements between the superstructure and the substructure.

The forces applied to a bridge bearing mainly include superstructure self-weight, traffic loads, wind loads, and earthquake loads.

Movements in bearings include translations and rotations. Creep, shrinkage, and temperature effects are the most common causes of the translational movements, which can occur in both transverse and longitudinal directions. Traffic loading, construction tolerances, and uneven settlement of the foundation are the common causes of the rotations.

Usually a bearing is connected to the superstructure through the use of a steel sole plate and rests on the substructure through a steel masonry plate. The sole plate distributes the concentrated bearing reactions to the superstructure. The masonry plate distributes the reactions to the substructure. The connections between the sole plate and the superstructure, for steel girders, are by bolting or welding. For concrete girders, the sole plate is embedded into the concrete with anchor studs. The masonry plate is typically connected to the substructure with anchor bolts.

26.2 Types of Bearings

Bearings may be classified as fixed bearings and expansion bearings. Fixed bearings allow rotations but restrict translational movements. Expansion bearings allow both rotational and translational movements. There are numerous types of bearings available. The following are the principal types of bearings currently in use.

0-8493-7434-0/00/$0.00+$.50

26.2.1 Sliding Bearings

A sliding bearing utilizes one plane metal plate sliding against another to accommodate translations. The sliding bearing surface produces a frictional force that is applied to the superstructure, the substructure, and the bearing itself. To reduce this friction force, PTFE (polytetrafluoroethylene) is often used as a sliding lubricating material. PTFE is sometimes referred to as Teflon, named after a widely used brand of PTFE, or TFE as appeared in AASHTO [1] and other design standards. In its common application, one steel plate coated with PTFE slides against another plate, which is usually of stainless steel.

Sliding bearings can be used alone or more often used as a component in other types of bearings. Pure sliding bearings can only be used when the rotations caused by the deflection at the supports are negligible. They are therefore limited to a span length of 15 m or less by ASHTTO [1].

A guiding system may be added to a sliding bearing to control the direction of the movement. It may also be fixed by passing anchor bolts through the plates.

26.2.2 Rocker and Pin Bearings

A rocker bearing is a type of expansion bearing that comes in a great variety. It typically consists of a pin at top that facilitates rotations, and a curved surface at the bottom that accommodates the translational movements (Figure 26.1a). The pin at the top is composed of upper and lower semi-circularly recessed surfaces with a solid circular pin placed between. Usually, there are caps at both ends of the pin to keep the pin from sliding off the seats and to resist uplift loads if required. The upper plate is connected to the sole plate by either bolting or welding. The lower curved plate sits on the masonry plate. To prevent the rocker from walking, keys are used to keep the rocker in place. A key can be a pintal which is a small trapezoidal steel bar tightly fitted into the masonry plate on one end and loosely inserted into the recessed rocker bottom plate on the other end. Or it can be an anchor bolt passing through a slotted hole in the bottom rocker plate.

A pin bearing is a type of fixed bearings that accommodates rotations through the use of a steel pin. The typical configuration of the bearing is virtually the same as the rocker described above except that the bottom curved rocker plate is now flat and directly anchored to the concrete pier (Figure 26.1b).

Rocker and pin bearings are primarily used in steel bridges. They are only suitable for the applications where the direction of the displacement is well defined since they can only accommodate translations and/or rotations in one direction. They can be designed to support relatively large loads but a high vertical clearance is usually required when the load or displacement is large. The practical limits of the load and displacement are about 1800 kN and ±100 mm, respectively, and rotations of several degrees are achievable [3].

Normally, the moment and lateral forces induced from the movement of these bearings are very small and negligible. However, metal bearings are susceptible to corrosion and deterioration. A corroded joint may induce much larger forces. Regular inspection and maintenance are, therefore, required.

26.2.3 Roller Bearings

Roller bearings are composed of one or more rollers between two parallel steel plates. Single roller bearings can facilitate both rotations and translations in the longitudinal direction, while a group of rollers would only accommodate longitudinal translations. In the latter case, the rotations are provided by combining rollers with a pin bearing (Figure 26.1c).

Roller bearings have been used in both steel and concrete bridges. Single roller bearings are relatively cheap to manufacture, but they only have a very limited vertical load capacity. Multiple roller bearings, on the other hand, may be able to support very large loads, but they are much more expensive.

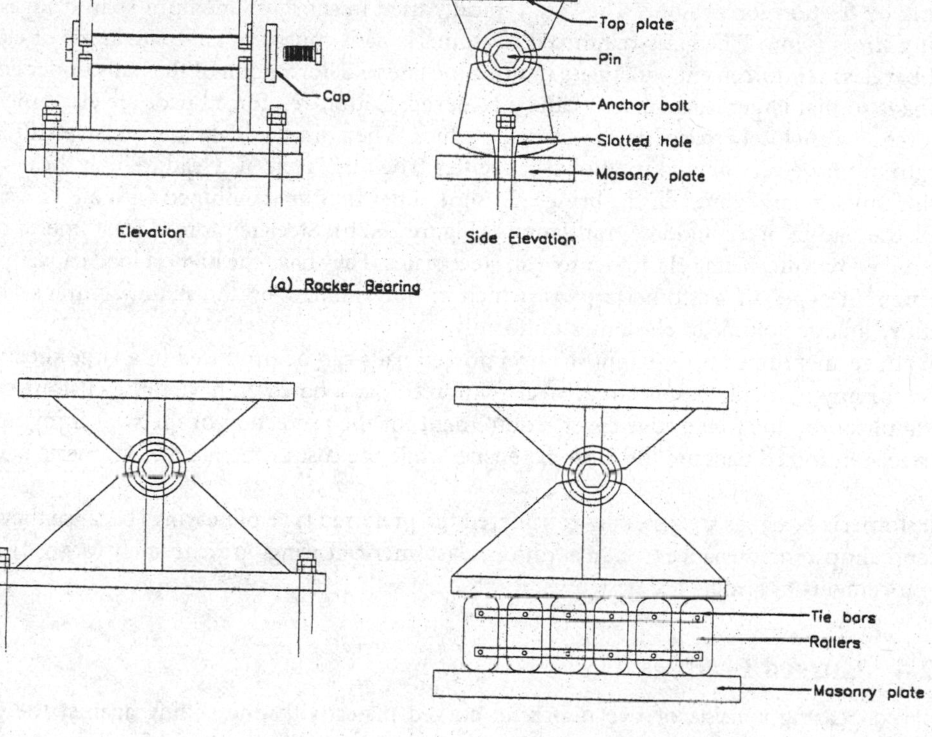

FIGURE 26.1 Typical rocker (a), pin (b), and roller bearings (c).

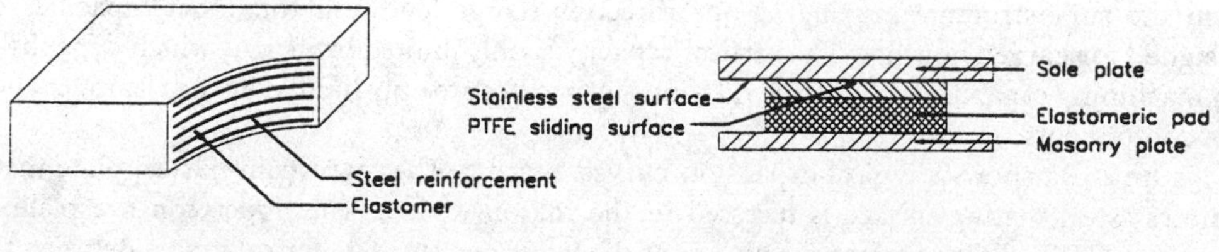

FIGURE 26.2 Elastomeric bearings. (a) Steel-reinforced elastomeric pad; (b) elastomeric pad with PTFE slider.

Like rocker and pin bearings, roller bearings are also susceptible to corrosion and deterioration. Regular inspection and maintenance are essential.

26.2.4 Elastomeric Bearings

An elastomeric bearing is made of elastomer (either natural or synthetic rubber). It accommodates both translational and rotational movements through the deformation of the elastomer.

Elastomer is flexible in shear but very stiff against volumetric change. Under compressive load, the elastomer expands laterally. To sustain large load without excessive deflection, reinforcement is used to restrain lateral bulging of the elastomer. This leads to the development of several types of elastomeric bearing pads — plain, fiberglass-reinforced, cotton duck–reinforced, and steel-reinforced elastomeric pads. Figure 26.2a shows a steel-reinforced elastomeric pad.

Plain elastomeric pads are the weakest and most flexible because they are only restrained from bulging by friction forces alone. They are typically used in short- to medium-span bridges, where bearing stress is low. Fiberglass-reinforced elastomeric pads consist of alternate layers of elastomer and fiberglass reinforcement. Fiberglass inhibits the lateral deformation of the pads under compressive loads so that larger load capacity can be achieved. Cotton-reinforced pads are elastomeric pads reinforced with closely spaced layers of cotton duck. They display high compressive stiffness and strength but have very limited rotational capacities. The thin layers also lead to high shear stiffness, which results in large forces in the bridge. So sometimes they are combined with a PTFE slider on top of the pad to accommodate translations (Figure 26.2b). Steel-reinforced elastomeric pads are constructed by vulcanizing elastomer to thin steel plates. They have the highest load capacity among the different types of elastomeric pads, which is only limited by the manufacturer's ability to vulcanize a large volume of elastomer uniformly.

All above-mentioned pads except steel-reinforced pads can be produced in a large sheet and cut to size for any particular application. Steel-reinforced pads, however, have to be custom-made for each application due to the edge cover requirement for the protection of the steel from corrosion. The steel-reinforced pads are the most expensive while the cost of the plain elastomeric pads is the lowest.

Elastomeric bearings are generally considered the preferred type of bearings because they are low cost and almost maintenance free. In addition, elastomeric bearings are extremely forgiving of loads and movements exceeding the design values.

26.2.4 Curved Bearings

A curved bearing consists of two matching curved plates with one sliding against the other to accommodate rotations. The curved surface can be either cylindrical which allows the rotation about only one axis or spherical which allows the bearing to rotate about any axis.

Lateral movements are restrained in a pure curved bearing and a limited lateral resistance may be developed through a combination of the curved geometry and the gravity loads. To accommodate lateral movements, a PTFE slider must be attached to the bearings. Keeper plates are often used to keep the superstructure moving in one direction. Large load and rotational capacities can be designed for curved bearings. The vertical capacity is only limited by its size, which depends largely on machining capabilities. Similarly, rotational capacities are only limited by the clearances between the components.

Figure 26.3a shows a typical expansion curved bearing. The lower convex steel plate that has a stainless steel mating surface is recessed in the masonry plate. The upper concave plate with a matching PTFE sliding surface sits on top of the lower convex plate for rotations. Between the sole plate and the upper concave plate there is a flat PTFE sliding surface that will accommodate lateral movements.

26.2.5 Pot Bearings

A pot bearing comprises a plain elastomeric disk that is confined in a shallow steel ring, or pot (Figure 26.3b). Vertical loads are transmitted through a steel piston that fits closely to the steel ring (pot wall). Flat sealing rings are used to contain the elastomer inside the pot. The elastomer behaves like a viscous fluid within the pot as the bearing rotates. Because the elastomeric pad is confined, much larger load can be carried this way than through conventional elastomeric pads.

Translational movements are restrained in a pure pot bearing, and the lateral loads are transmitted through the steel piston moving against the pot wall. To accommodate translational movement, a PTFE sliding surface must be used. Keeper plates are often used to keep the superstructure moving in one direction.

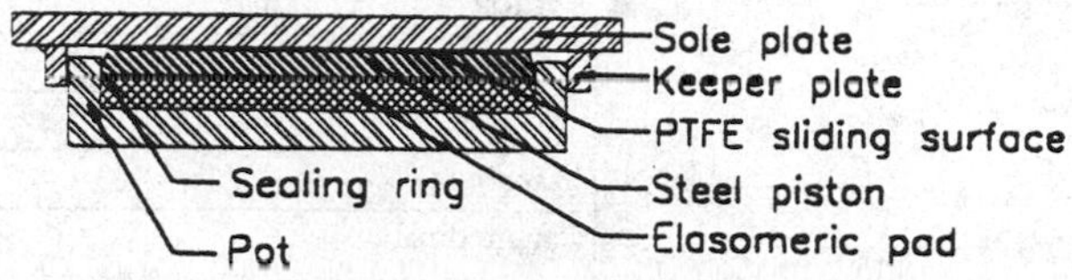

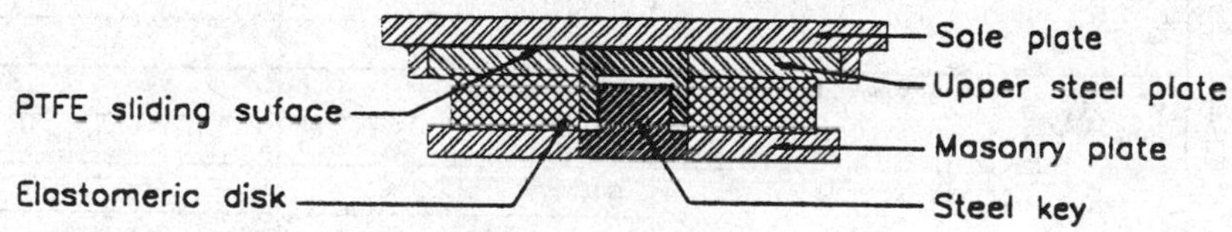

FIGURE 26.3 Typical spherical (a), pot (b), and disk (c) bearings

26.2.6 Disk Bearings

A disk bearing, as illustrated in Figure 26.3c, utilizes a hard elastomeric (polyether urethane) disk to support the vertical loads and a metal key in the center of the bearing to resist horizontal loads. The rotational movements are accommodated through the deformation of the elastomer. To accommodate translational movements, however, a PTFE slider is required. In this kind of bearings, the polyether urethane disk must be hard enough to resist large vertical load without excessive deformation and yet flexible enough to accommodate rotations easily.

26.3 Selection of Bearings

Generally the objective of bearing selection is to choose a bearing system that suits the needs with a minimum overall cost. The following procedures may be used for the selection of the bearings.

26.3.1 Determination of Functional Requirements

First, the vertical and horizontal loads, the rotational and translational movements from all sources including dead and live loads, wind loads, earthquake loads, creep and shrinkage, prestress, thermal and construction tolerances need to be calculated. Table 26.1 may be used to tabulate these requirements.

TABLE 26.1 Typical Bridge Bearing Schedule

Bridge Name of Reference					
Bearing Identification mark					
Number of bearings required					
Seating Material	Upper Surface				
	Lower Surface				
Allowable average contact pressure (PSI)	Upper Surface		Serviceability		
			Strength		
	Lower Surface		Serviceability		
			Strength		
Design Load effects (KIP)	Service limit state		Vertical	max.	
				perm	
				min.	
			Transverse		
			Longitudinal		
	Strength limit state		Vertical		
			Transverse		
			Longitudinal		
Translation	Service limit state	Irreversible	Transverse		
			Longitudinal		
		Reversible	Transverse		
			Longitudinal		
	Strength limit state	Irreversible	Transverse		
			Longitudinal		
		Reversible	Transverse		
			Longitudinal		
Rotation (RAD)	Service limit state	Irreversible	Transverse		
			Longitudinal		
		Reversible	Transverse		
			Longitudinal		
	Strength limit state	Irreversible	Transverse		
			Longitudinal		
		Reversible	Transverse		
			Longitudinal		
Maximum bearing dimensions (IN)	Upper surface		Transverse		
			Longitudinal		
	Lower surface		Transverse		
			Longitudinal		
	Overall height				
Tolerable movement of bearing under transient loads (IN)			Vertical		
			Transverse		
			Longitudinal		
Allowable resistance to translation under service limit state (KIP)			Transverse		
			Longitudinal		
Allowable resistance to rotation under service limit state (K/FT)			Transverse		
			Longitudinal		
Type of attachment to structure and substructure			Transverse		
			Longitudinal		

Source: AASHTO, *LRFD Bridge Design Scecifications*, American Association of State Highway and Transportation Officials, Washington, D.C.

TABLE 26.2 Summery of Bearing Capacities [3,5]

Bearing Type	Load Min. (KN)	Load Max. (KN)	Translation Min. (mm)	Translation Max. (mm)	Rotation Max. (rad)	Costs Initial	Costs Maintenance
Elastomeric pads							
Plain	0	450	0	15	0.01	Low	Low
Cotton duck reinforced	0	1,400	0	5	0.003	Low	Low
Fiberglass reinforced	0	600	0	25	0.015	Low	Low
Steel reinforced	225	3,500	0	100	0.04	Low	Low
Flat PTFE slider	0	>10,000	25	>100	0	Low	Moderate
Disk bearing	1,200	10,000	0	0	0.02	Moderate	Moderate
Pot bearing	1,200	10,000	0	0	0.02	Moderate	High
Pin bearing	1,200	4,500	0	0	>0.04	Moderate	High
Rocker bearing	0	1,800	0	100	>0.04	Moderate	High
Single roller	0	450	25	>100	>0.04	Moderate	High
Curved PTFE bearing	1,200	7,000	0	0	>0.04	High	Moderate
Multiple rollers	500	10,000	100	>100	>0.04	High	High

26.3.2 Evaluation of Bearings

The second step is to determine the suitable bearing types based on the above bridge functional requirements, and other factors including available clearance, environment, maintenance, cost, availability, and client's preferences. Table 26.2 summarizes the load, movement capacities, and relative costs for each bearing type and may be used for the selection of the bearings.

It should be noted that the capacity values in Table 26.2 are approximate. They are the practical limits of the most economical application for each bearing type. The costs are also relative, since the true price can only be determined by the market. At the end of this step, several qualified bearing systems with close cost ratings may be selected [5].

26.3 Preliminary Bearing Design

For the various qualified bearing alternatives, preliminary designs are performed to determine the approximate geometry and material properties in accordance with design specifications. It is likely that one or more of the previously acceptable alternatives will be eliminated in this step because of an undesirable attribute such as excessive height, oversize footprint, resistance at low temperature, sensitivity to installation tolerances, etc. [3].

At the end of this step, one or more bearing types may still be feasible and they will be included in the bid package as the final choices of the bearing types.

26.4 Design of Elastomeric Bearings

26.4.1 Design Procedure

The design procedure is according to AASHTO-LRFD [1] and is as follows:

1. Determine girder temperature movement (Art. 5.4.2.2).
2. Determine girder shortenings due to post-tensioning, concrete shrinkage, etc.
3. Select a bearing thickness based on the bearing total movement requirements (Art. 14.7.5.3.4).
4. Compute the bearing size based on bearing compressive stress (Art. 14.7.5.3.2).
5. Compute instantaneous compressive deflection (Art. 14.7.5.3.3).
6. Combine bearing maximum rotation.
7. Check bearing compression and rotation (Art. 14.7.5.3.5).
8. Check bearing stability (Art. 14.7.5.3.6).
9. Check bearing steel reinforcement (Art. 14.7.5.3.7).

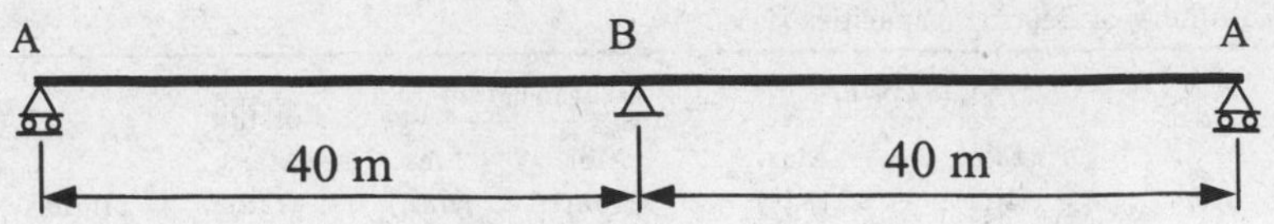

FIGURE 26.4 Bridge layout

26.4.2 Design Example (Figure 26.4)

Given

L	= expandable span length	= 40 m
R_{DL}	= DL reaction/girder	= 690 kN
R_{LL}	= LL reaction (without impact)/girder	= 220 kN
θ_s	= bearing design rotation at service limit state	= 0.025 rad
ΔT	= maximum temperature change	= 21°C
Δ_{PT}	= girder shortening due to post tensioning	= 21 mm
Δ_{SH}	= girder shortening due to concrete shrinkage	= 2 mm
G	= shear modulus of elastomer	= 0.9 ~ 1.38 MPa
γ	= load factor for uniform temperature, etc.	= 1.2
ΔF_{TH}	= constant amplitude fatigue threshold for Category A	= 165 MPa

Using 60 durometer reinforced bearing:

F_y	= yield strength of steel reinforcement	= 350 MPa

Sliding bearing used:

1. **Temperature Movement**
   From Art. 5.4.2.2, for normal density concrete, the thermal coefficient α is

$$\alpha = 10.8 \times 10^{-6}/°\text{C}$$

$$\Delta_{TEMP} = (\alpha)(\Delta T)(L) = (10.8 \times 10^{-6}/°\text{C})(21°\text{C})(40{,}000 \text{ mm}) = 9 \text{ mm}$$

2. **Girder Shortenings**

$$\Delta_{PT} = 21 \text{ mm and } \Delta_{SH} = 2 \text{ mm}$$

3. **Bearing Thickness**
   h_{rt} = total elastomer thickness
   h_{ri} = thickness of ith elastomeric layer
   n = number of interior layers of elastomeric layer
   Δ_S = bearing maximum longitudinal movement = $\gamma \cdot (\Delta_{TEMP} + \Delta_{PT} + \Delta_{SH})$
   $\Delta_S = 1.2 \times (9 \text{ mm} + 21 \text{ mm} + 2 \text{ mm}) = 38.4 \text{ mm}$
   h_{rt} = bearing thickness $\geq 2\Delta_S$ (AASHTO Eq. 14.7.5.3.4-1)
   $h_{rt} = 2 \times (38.4 \text{ mm}) = 76.8$

   Try $h_{rt} = 120$ mm, $h_{ri} = 20$ mm and $n = 5$

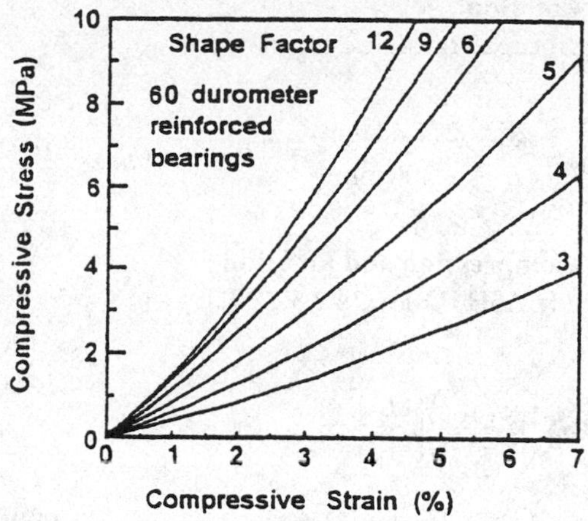

FIGURE 26.5 Stress–strain curves. (From AASHTO, Figure C14.7.5.3.3.1.)

4. **Bearing Size**
   L = length of bearing
   W = width of bearing
   S_i = shape factor of thickness layer of the bearing = $\dfrac{LW}{2h_{ri}(L+W)}$

   For a bearing subject to shear deformation, the compressive stresses should satisfy:

σ_S = average compressive stress due to the total load $\le 1.66GS \le 11$ (AASHTO Eq. 14.7.5.3.2-1)
σ_L = average compressive stress due to the live load $\le 0.66\ GS$ (AASHTO Eq. 14.7.5.3.2-1)

$$\acute{o}_S = \frac{R}{LW} = \frac{1.66GLW}{2h_{ri}(L+W)}$$

Assuming σ_S is critical, solve for L and W by error and trial.

$$L = 300 \text{ mm and } W = 460 \text{ mm}$$

$$S = \frac{LW}{2h_{ri}(L+W)} = \frac{(300\text{ mm})(460\text{ mm})}{2(20\text{ mm})(300\text{ mm} + 460\text{ mm})} = 4.54$$

$$\acute{o}_L - \frac{R_L}{LW} = \frac{(200{,}000\text{ N})}{(300\text{ mm})(460\text{ mm})} = 1.6\text{ MPa}$$

OK

$$\le 0.66\ GS = 0.66(1.0\text{ MPa})(4.54) = 3.0\text{ MPa}$$

5. **Instantaneous Compressive Deflection**
   For σ_S = 6.59 MPa and S = 4.54, one can determine the value of ε_i from Figure 26.5:

$$\varepsilon_i = 0.062$$

$$\ddot{a} = \sum \mathring{a}_i h_{ri} \qquad \text{(AASHTO Eq. 14.7.5.3.3-1)}$$

$$= 6(0.062)(20\text{ mm}) = 7.44\text{ mm}$$

6. **Bearing Maximum Rotation**
   The bearing rotational capacity can be calculated as

$$\acute{a}_{\text{capacity}} = \frac{2\delta}{L} = \frac{2(7.44\,\text{mm})}{300\,\text{mm}} = 0.05\,\text{rad} < \acute{a}_{\text{design}} = 0.025\,\text{rad} \qquad \text{OK}$$

7. **Combined Bearing Compression and Rotation**
   a. *Uplift requirement* (AASHTO Eq. 14.7.5.3.5-1):

$$\acute{o}_{\text{s,uplift}} = 1.0GS\left(\frac{\acute{a}_{\text{design}}}{n}\right)\left(\frac{L}{h_{\text{ri}}}\right)^2$$

$$= 1.0\,(1.2)\,(4.54)\left(\frac{0.025}{5}\right)\left(\frac{300}{20}\right)^2 = 6.13\text{ MPa} < \acute{o}_s = 6.59\text{ MPa} \qquad \text{OK}$$

   b. *Shear deformation requirement* (AASHTO Eq. 14.7.5.3.5-2):

$$\acute{o}_{s,\text{shear}} = 1.875GS\left(1 - 0.20\left(\frac{\acute{a}_{\text{design}}}{n}\right)\left(\frac{L}{h_{\text{ri}}}\right)^2\right)$$

$$= 1.875\,(1.0)\,(4.54)\left(1 - 0.20\left(\frac{0.025}{5}\right)\left(\frac{300}{20}\right)^2\right) = 6.60\text{ MPa} > \acute{o}_s = 6.59\text{ MPa} \qquad \text{OK}$$

8. **Bearing Stability**
   Bearings shall be designed to prevent instability at the service limit state load combinations. The average compressive stress on the bearing is limited to half the predicted buckling stress. For this example, the bridge deck, if free to translate horizontally, the average compressive stress due to dead and live load, σ_s, must satisfy:

$$\sigma_s \le \frac{G}{2A - B} \qquad \text{(AASHTO Eq. 14.7.5.3.6-1)}$$

where

$$A = \frac{1.92\dfrac{h_{\text{rt}}}{L}}{S\sqrt{1 + \dfrac{2.0\,L}{W}}} \qquad \text{(AASHTO Eq. 14.7.5.3.6-3)}$$

$$= \frac{1.92\dfrac{(120\text{ mm})}{(300\text{ mm})}}{(4.54)\sqrt{1 + \dfrac{2.0(300\text{ mm})}{(460\text{ mm})}}} = 0.11$$

$$B = \frac{2.67}{S(S+2.0)\sqrt{1+\frac{L}{4.0W}}}$$

(AASHTO Eq. 14.7.5.3.6-4)

$$= \frac{2.67}{(4.54)\,(4.54+2.0)\sqrt{1+\frac{(300\text{ mm})}{4.0\,(460\text{ mm})}}} = 0.08$$

$$\frac{G}{2A-B} = \frac{(1.0\text{ MPa})}{2\,(0.11)-(0.08)} = 6.87 > \acute{o}_s$$ OK

9. **Bearing Steel Reinforcement**
The bearing steel reinforcement must be designed to sustain the tensile stresses induced by compression of the bearing. The thickness of steel reinforcement, h_s, should satisfy:
a. *At the service limit state:*

$$h_s \geq \frac{3h_{max}\acute{o}_s}{F_y}$$ (AASHTO Eq. 14.7.5.3.7-1)

$$= \frac{3\,(20\text{ mm})\,(6.59\text{ MPa})}{(350\text{ MPa})} = 1.13\text{ mm}$$ (governs)

b. *At the fatigue limit state:*

$$h_s \geq \frac{2h_{max}\acute{o}_L}{\Delta F_y}$$ (AASHTO Eq. 14.7.5.3.7-2)

$$= \frac{2\,(20\text{mm})(1.6\text{MPa})}{(165\text{MPa})} = 0.39\text{mm}$$

where h_{max} = thickness of thickest elastomeric layer in elastomeric bearing = h_{ri}.

Elastomeric Bearings Details
Five interior lays with 20 mm thickness each layer
Two exterior lays with 10 mm thickness each layer
Six steel reinforcements with 1.2 mm each
Total thickness of bearing is 127.2 mm
Bearing size: 300 mm (longitudinal) × 460 mm (transverse)

References

1. AASHTO, *LRFD Bridge Design Specifications,* American Association of State Highway and Transportation Officials, Washington, D.C., 1994.
2. AASHTO, *Standard Specifications for the Design of Highway Bridges,* 16th ed. American Association of State Highway and Transportation Officials, Washington, D.C., 1996.
3. Stanton, J. F., Roeder, C. W., and Campbell, T. I., High Load Multi-Rotational Bridge Bearings, NCHRP Report 10-20A, Transportation Research Board, National Research Council, Washington, D.C., 1993.
4. Caltrans, Memo to Designers, California Department of Transportation, Sacramento, 1994.
5. AISI, Steel bridge bearing selection and design guide, *Highway Structures Design Handbook,* Vol. II, American Iron and Steel Institute, Washington, D.C., 1996, chap. 4.

27

Piers and Columns

Jinrong Wang
URS Greiner

27.1 Introduction

Piers provide vertical supports for spans at intermediate points and perform two main functions: transferring superstructure vertical loads to the foundations and resisting horizontal forces acting on the bridge. Although piers are traditionally designed to resist vertical loads, it is becoming more and more common to design piers to resist high lateral loads caused by seismic events. Even in some low seismic areas, designers are paying more attention to the ductility aspect of the design. Piers are predominantly constructed using reinforced concrete. Steel, to a lesser degree, is also used for piers. Steel tubes filled with concrete (composite) columns have gained more attention recently.

This chapter deals only with piers or columns for conventional bridges, such as grade separations, overcrossings, overheads, underpasses, and simple river crossings. Reinforced concrete columns will be discussed in detail while steel and composite columns will be briefly discussed. Substructures for arch, suspension, segmental, cable-stayed, and movable bridges are excluded from this chapter. Chapter 28 discusses the substructures for some of these special types of bridges.

27.2 Structural Types

27.2.1 General

Pier is usually used as a general term for any type of substructure located between horizontal spans and foundations. However, from time to time, it is also used particularly for a solid wall in order to distinguish it from columns or bents. From a structural point of view, a column is a member that resists the lateral force mainly by flexure action whereas a pier is a member that resists the lateral force mainly by a shear mechanism. A pier that consists of multiple columns is often called a *bent.*

There are several ways of defining pier types. One is by its structural connectivity to the superstructure: monolithic or cantilevered. Another is by its sectional shape: solid or hollow; round, octagonal, hexagonal, or rectangular. It can also be distinguished by its framing configuration: single or multiple column bent; hammerhead or pier wall.

0-8493-7434-0/00/$0.00+$.50

FIGURE 27.1 Typical cross-section shapes of piers for overcrossings or viaducts on land.

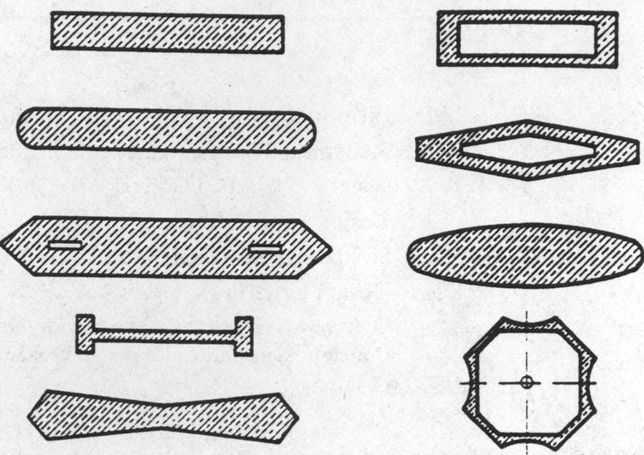

FIGURE 27.2 Typical cross-section shapes of piers for river and waterway crossings.

27.2.2 Selection Criteria

Selection of the type of piers for a bridge should be based on functional, structural, and geometric requirements. Aesthetics is also a very important factor of selection since modern highway bridges are part of a city's landscape. Figure 27.1 shows a collection of typical cross section shapes for overcrossings and viaducts on land and Figure 27.2 shows some typical cross section shapes for piers of river and waterway crossings. Often, pier types are mandated by government agencies or owners. Many state departments of transportation in the United States have their own standard column shapes.

Solid wall piers, as shown in Figures 27.3a and 27.4, are often used at water crossings since they can be constructed to proportions that are both slender and streamlined. These features lend themselves well for providing minimal resistance to flood flows.

Hammerhead piers, as shown in Figure 27.3b, are often found in urban areas where space limitation is a concern. They are used to support steel girder or precast prestressed concrete superstructures. They are aesthetically appealing. They generally occupy less space, thereby providing more room for the traffic underneath. Standards for the use of hammerhead piers are often maintained by individual transportation departments.

A column bent pier consists of a cap beam and supporting columns forming a frame. Column bent piers, as shown in Figure 27.3c and Figure 27.5, can either be used to support a steel girder superstructure or be used as an integral pier where the cast-in-place construction technique is used. The columns can be either circular or rectangular in cross section. They are by far the most popular forms of piers in the modern highway system.

A pile extension pier consists of a drilled shaft as the foundation and the circular column extended from the shaft to form the substructure. An obvious advantage of this type of pier is that it occupies a minimal amount of space. Widening an existing bridge in some instances may require pile extensions because limited space precludes the use of other types of foundations.

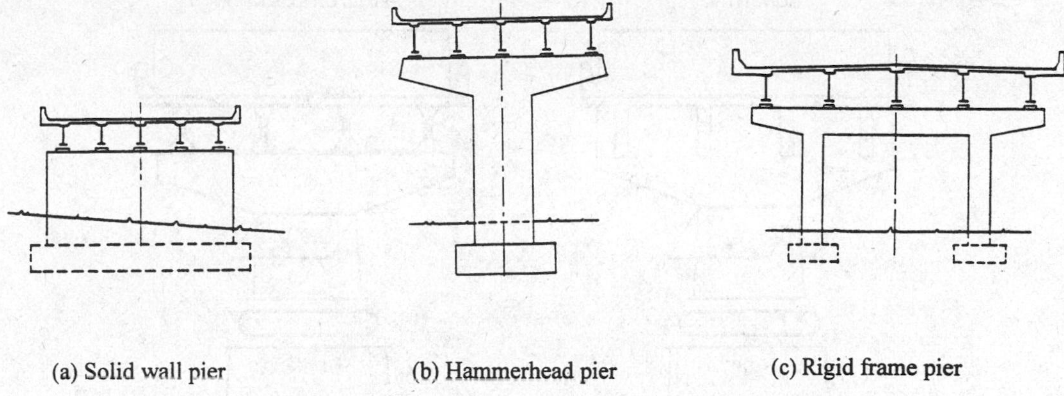

FIGURE 27.3 Typical pier types for steel bridges.

Selections of proper pier type depend upon many factors. First of all, it depends upon the type of superstructure. For example, steel girder superstructures are normally supported by cantilevered piers, whereas the cast-in-place concrete superstructures are normally supported by monolithic bents. Second, it depends upon whether the bridges are over a waterway or not. Pier walls are preferred on river crossings, where debris is a concern and hydraulics dictates it. Multiple pile extension bents are commonly used on slab bridges. Last, the height of piers also dictates the type selection of piers. The taller piers often require hollow cross sections in order to reduce the weight of the substructure. This then reduces the load demands on the costly foundations. Table 27.1 summarizes the general type selection guidelines for different types of bridges.

27.3 Design Loads

Piers are commonly subjected to forces and loads transmitted from the superstructure, and forces acting directly on the substructure. Some of the loads and forces to be resisted by the substructure include:

- Dead loads
- Live loads and impact from the superstructure
- Wind loads on the structure and the live loads
- Centrifugal force from the superstructure
- Longitudinal force from live loads
- Drag forces due to the friction at bearings
- Earth pressure
- Stream flow pressure
- Ice pressure
- Earthquake forces
- Thermal and shrinkage forces
- Ship impact forces
- Force due to prestressing of the superstructure
- Forces due to settlement of foundations

The effect of temperature changes and shrinkage of the superstructure needs to be considered when the superstructure is rigidly connected with the supports. Where expansion bearings are used, forces caused by temperature changes are limited to the frictional resistance of bearings.

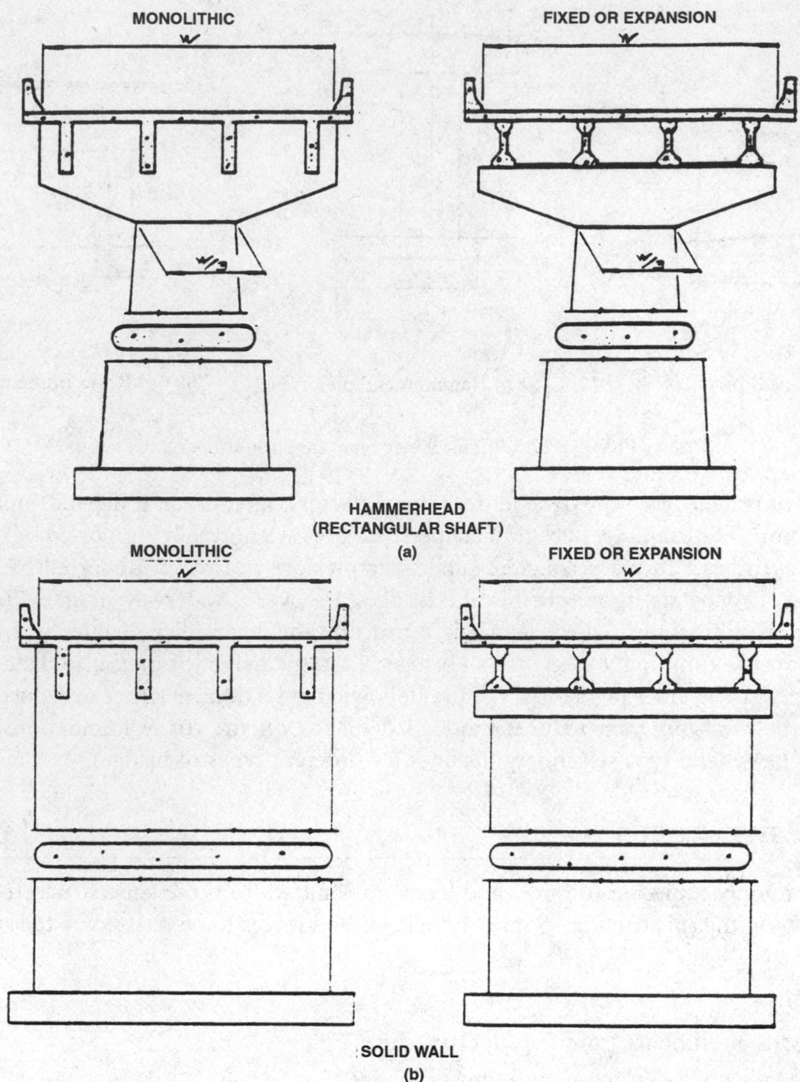

FIGURE 27.4 Typical pier types and configurations for river and waterway crossings.

Readers should refer to Chapters 5 and 6 for more details about various loads and load combinations and Part IV about earthquake loads. In the following, however, two load cases, live loads and thermal forces, will be discussed in detail because they are two of the most common loads on the piers, but are often applied incorrectly.

27.3.1 Live Loads

Bridge live loads are the loads specified or approved by the contracting agencies and owners. They are usually specified in the design codes such as AASHTO LRFD Bridge Design Specifications [1]. There are other special loading conditions peculiar to the type or location of the bridge structure which should be specified in the contracting documents.

Live-load reactions obtained from the design of individual members of the superstructure should not be used directly for substructure design. These reactions are based upon maximum conditions

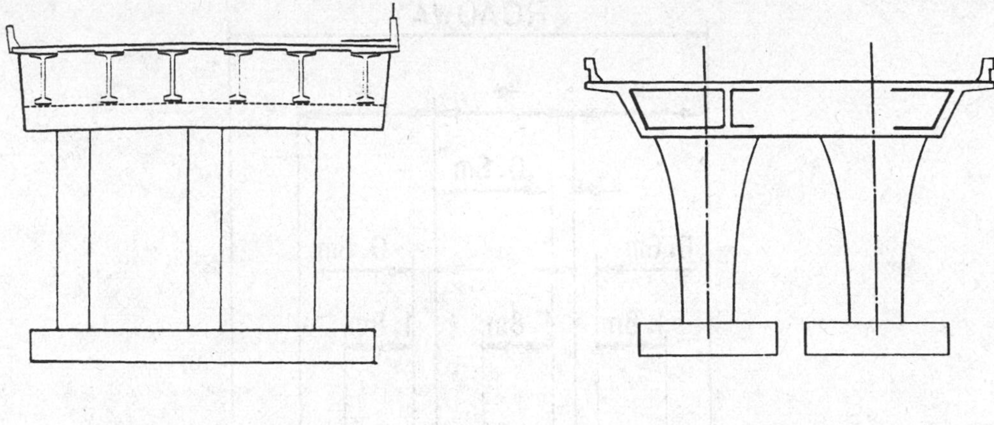

(a) Bent for precast girders (b) Bent for cast-in-place girders

FIGURE 27.5 Typical pier types for concrete bridges.

TABLE 27.1 General Guidelines for Selecting Pier Types

		Applicable Pier Types
		Steel Superstructure
Over water	Tall piers	Pier walls or hammerheads (T-piers) (Figures 27.3a and b); hollow cross sections for most cases; cantilevered; could use combined hammerheads with pier wall base and step tapered shaft
	Short piers	Pier walls or hammerheads (T-piers) (Figures 27.3a and b); solid cross sections; cantilevered
On land	Tall piers	Hammerheads (T-piers) and possibly rigid frames (multiple column bents)(Figures 27.3b and c); hollow cross sections for single shaft and solid cross sections for rigid frames; cantilevered
	Short piers	Hammerheads and rigid frames (Figures 27.3b and c); solid cross sections; cantilevered
		Precast Prestressed Concrete Superstructure
Over water	Tall piers	Pier walls or hammerheads (Figure 27.4); hollow cross sections for most cases; cantilevered; could use combined hammerheads with pier wall base and step-tapered shaft
	Short piers	Pier walls or hammerheads; solid cross sections; cantilevered
On land	Tall piers	Hammerheads and possibly rigid frames (multiple column bents); hollow cross sections for single shafts and solid cross sections for rigid frames; cantilevered
	Short piers	Hammerheads and rigid frames (multiple column bents) (Figure 27.5a); solid cross sections; cantilevered
		Cast-in-Place Concrete Superstructure
Over water	Tall piers	Single shaft pier (Figure 27.4); superstructure will likely cast by traveled forms with balanced cantilevered construction method; hollow cross sections; monolithic; fixed at bottom
	Short piers	Pier walls (Figure 27.4); solid cross sections; monolithic; fixed at bottom
On land	Tall piers	Single or multiple column bents; solid cross sections for most cases, monolithic; fixed at bottom
	Short piers	Single or multiple column bents (Figure 27.5b); solid cross sections; monolithic; pinned at bottom

for one beam and make no allowance for distribution of live loads across the roadway. Use of these maximum loadings would result in a pier design with an unrealistically severe loading condition and uneconomical sections.

For substructure design, a maximum design traffic lane reaction using either the standard truck load or standard lane load should be used. Design traffic lanes are determined according to AASHTO

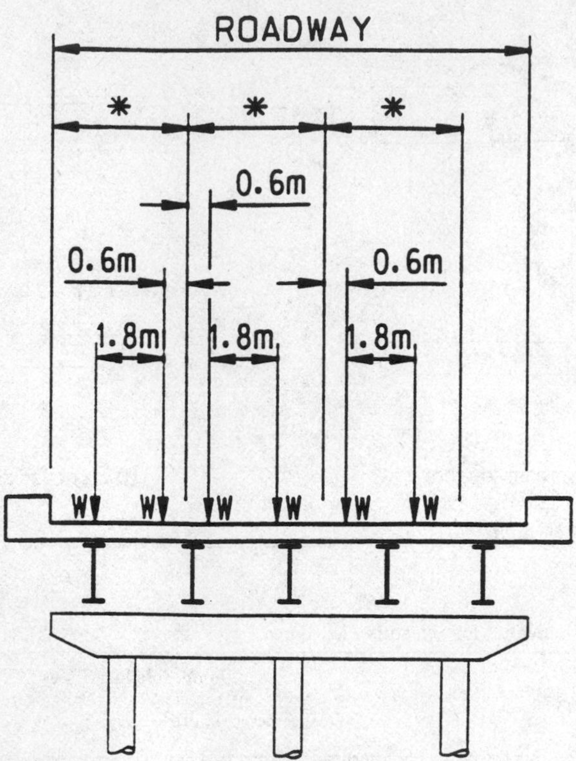

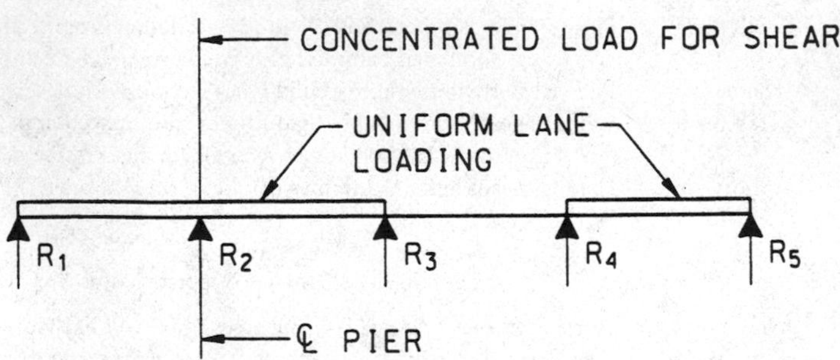

FIGURE 27.6 Wheel load arrangement to produce maximum positive moment.

LRFD [1] Section 3.6. For the calculation of the actual beam reactions on the piers, the maximum lane reaction can be applied within the design traffic lanes as wheel loads, and then distributed to the beams assuming the slab between beams to be simply supported. (Figure 27.6). Wheel loads can be positioned anywhere within the design traffic lane with a minimum distance between lane boundary and wheel load of 0.61 m (2 ft).

The design traffic lanes and the live load within the lanes should be arranged to produce beam reactions that result in maximum loads on the piers. AASHTO LRFD Section 3.6.1.1.2 provides load reduction factors due to multiple loaded lanes.

TABLE 27.2 Dynamic Load Allowance, IM

Component	IM
Deck joints — all limit states	75%
All other components	
• Fatigue and fracture limit state	15%
• All other limit states	33%

Live-load reactions will be increased due to impact effect. AASHTO LRFD [1] refers to this as the *dynamic load allowance, IM.* and is listed here as in Table 27.2.

27.3.2 Thermal Forces

Forces on piers due to thermal movements, shrinkage, and prestressing can become large on short, stiff bents of prestressed concrete bridges with integral bents. Piers should be checked against these forces. Design codes or specifications normally specify the design temperature range. Some codes even specify temperature distribution along the depth of the superstructure member.

The first step in determining the thermal forces on the substructures for a bridge with integral bents is to determine the point of no movement. After this point is determined, the relative displacement of any point along the superstructure to this point is simply equal to the distance to this point times the temperature range and times the coefficient of expansion. With known displacement at the top and known boundary conditions at the top and bottom, the forces on the pier due to the temperature change can be calculated by using the displacement times the stiffness of the pier.

The determination of the point of no movement is best demonstrated by the following example, which is adopted from Memo to Designers issued by California Department of Transportations [2]:

Example 27.1

A 225.55-m (740-foot)-long and 23.77-m (78-foot) wide concrete box-girder superstructure is supported by five two-column bents. The size of the column is 1.52 m (5 ft) in diameter and the heights vary between 10.67 m (35 ft) and 12.80 m (42 ft). Other assumptions are listed in the calculations. The calculation is done through a table. Please refer Figure 27.7 for the calculation for determining the point of no movement.

27.4 Design Criteria

27.4.1 Overview

Like the design of any structural component, the design of a pier or column is performed to fulfill strength and serviceability requirements. A pier should be designed to withstand the overturning, sliding forces applied from superstructure as well as the forces applied to substructures. It also needs to be designed so that during an extreme event it will prevent the collapse of the structure but may sustain some damage.

A pier as a structure component is subjected to combined forces of axial, bending, and shear. For a pier, the bending strength is dependent upon the axial force. In the plastic hinge zone of a pier, the shear strength is also influenced by bending. To complicate the behavior even more, the bending moment will be magnified by the axial force due to the P-Δ effect.

In current design practice, the bridge designers are becoming increasingly aware of the adverse effects of earthquake. Therefore, ductility consideration has become a very important factor for bridge design. Failure due to scouring is also a common cause of failure of bridges. In order to prevent this type of failure, the bridge designers need to work closely with the hydraulic engineers to determine adequate depths for the piers and provide proper protection measures.

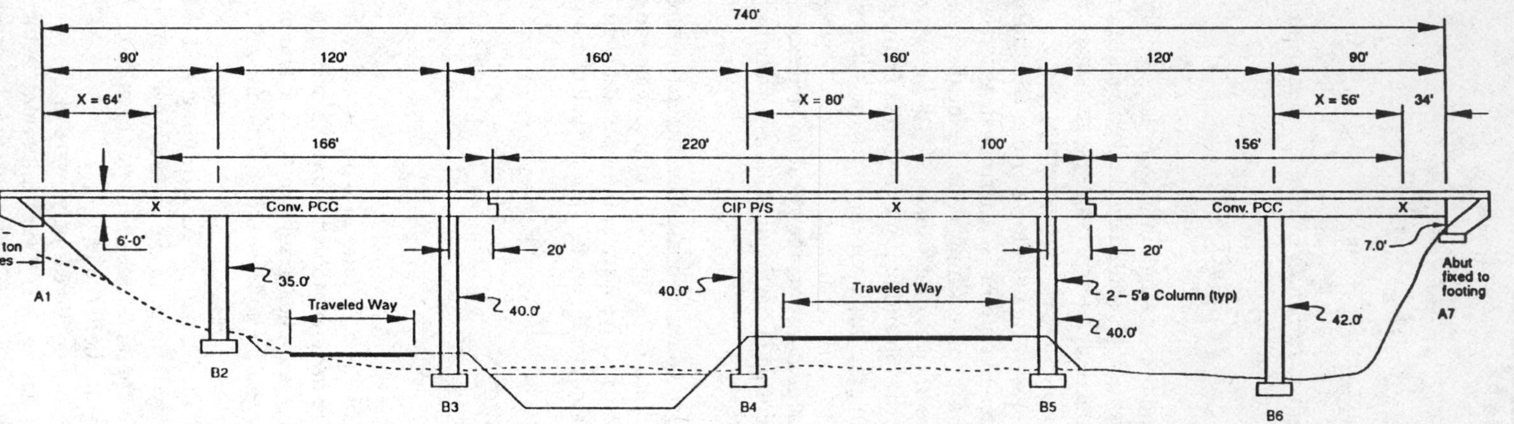

	A1		B2		B3			B4		B5			B6		A7		
I (Ft)[4]	1.38		61.36		61.36			61.36		61.36			61.36		102		
L (Ft)	5.50		35.0		40.0		Sum	40.0		40.0		Sum	42.0		7.0		Sum
P (kips) @ 1" side sway	1200	+	618	+	415	=	2,233	415	+	415	=	830	359 Will slide	+	600	=	959
D (distance from 1st member of frame)	0		90		210			0		160			0		90		
P × D / 100	0	+	556	+	872	=	1,428	0	+	664	=	664	0	+	540	=	540
	$X = \frac{\Sigma(P \times D)/100}{\Sigma P}(100) = \frac{1,428}{2,233}(100) = \underline{64'}$							$\frac{664}{830}(100) = \underline{80'}$					$\frac{540}{959}(100) = \underline{56'}$				

Notes:
- Width of Structure = 78'
- Diameter of Column = 5'-0"
- K/Pile @ 1" deflection = 100 kips
- Point of No Movement = X
- Refer to Properties/Piles Table

Assumptions:
1. Super str. inf. rigid
2. Columns fixed top and bottom
3. Abutment footing will slide @ a force equal to D.W.
4. E (piles) = 4×10^6 psi
   E (columns) = 3×10^6 psi

Fixed/Fixed Condition

$$P\,(\text{Col.}) = 12EI\frac{\Delta}{L^3}$$

$$@\ 1''\ \text{defl.} = \frac{432\,I}{\left(\frac{L}{10}\right)^3}$$

Pinned/Fixed Condition

$$P\,(\text{Col.}) = 3EI\frac{\Delta}{L^3}$$

$$@\ 1''\ \text{defl.} = \frac{108\,I}{\left(\frac{L}{10}\right)^3}$$

D.W. Abut 7 = 600 k (assume linear up to 1" deflection)

$$I\,(\text{abut}) = \frac{78}{12}(2.5)^3 = 102$$

FIGURE 27.7 Calculation of points of no movement.

27.4.2 Slenderness and Second-Order Effect

The design of compression members must be based on forces and moments determined from an analysis of the structure. Small deflection theory is usually adequate for the analysis of beam-type members. For compression members, however, the second-order effect must be considered. According to AASHTO LRFD [1], the second-order effect is defined as follows:

> The presence of compressive axial forces amplify both out-of-straightness of a component and the deformation due to non-tangential loads acting thereon, therefore increasing the eccentricity of the axial force with respect to the centerline of the component. The synergistic effect of this interaction is the apparent softening of the component, i.e., a loss of stiffness.

To assess this effect accurately, a properly formulated large deflection nonlinear analysis can be performed. Discussions on this subject can be found in References [3,4] and Chapter 36. However, it is impractical to expect practicing engineers to perform this type of sophisticated analysis on a regular basis. The moment magnification procedure given in AASHTO LRFD [1] is an approximate process which was selected as a compromise between accuracy and ease of use. Therefore, the AASHTO LRFD moment magnification procedure is outlined in the following.

When the cross section dimensions of a compression member are small in comparison to its length, the member is said to be slender. Whether or not a member can be considered slender is dependent on the magnitude of the slenderness ratio of the member. The slenderness ratio of a compression member is defined as, KL_u/r, where K is the effective length factor for compression members; L_u is the unsupported length of compression member; r is the radius of gyration = $\sqrt{I/A}$; I is the moment of inertia; and A is the cross-sectional area.

When a compression member is braced against side sway, the effective length factor, $K = 1.0$ can be used. However, a lower value of K can be used if further analysis demonstrates that a lower value is applicable. L_u is defined as the clear distance between slabs, girders, or other members which is capable of providing lateral support for the compression member. If haunches are present, then, the unsupported length is taken from the lower extremity of the haunch in the plane considered (AASHTO LRFD 5.7.4.3). For a detailed discussion of the K-factor, please refer to Chapter 52.

For a compression member braced against side sway, the effects of slenderness can be ignored as long as the following condition is met (AASHTO LRFD 5.7.4.3):

$$\frac{KL_u}{r} < 34 - \left(\frac{12M_{1b}}{M_{2b}}\right) \tag{27.1}$$

where

M_{1b} = smaller end moment on compression member — positive if member is bent in single curvature, negative if member is bent in double curvature

M_{2b} = larger end moment on compression member — always positive

For an unbraced compression member, the effects of slenderness can be ignored as long as the following condition is met (AASHTO LRFD 5.7.4.3):

$$\frac{KL_u}{r} < 22 \tag{27.2}$$

If the slenderness ratio exceeds the above-specified limits, the effects can be approximated through the use of the moment magnification method. If the slenderness ratio KL_u/r exceeds 100, however, a more-detailed second-order nonlinear analysis [Chapter 36] will be required. Any detailed analysis should consider the influence of axial loads and variable moment of inertia on member stiffness and forces, and the effects of the duration of the loads.

The factored moments may be increased to reflect effects of deformations as follows:

$$M_c = \delta_b M_{2b} + \delta_s M_{2s} \tag{27.3}$$

where

M_{2b} = moment on compression member due to factored gravity loads that result in no appreciable side sway calculated by conventional first-order elastic frame analysis, always positive

M_{2s} = moment on compression member due to lateral or gravity loads that result in side sway, Δ, greater than $L_u/1500$, calculated by conventional first-order elastic frame analysis, always positive

The moment magnification factors are defined as follows:

$$\delta_b = \frac{C_m}{1 - \dfrac{P_u}{\phi P_c}} \geq 1.0 \tag{27.4}$$

$$\delta_s = \frac{1}{1 - \dfrac{\Sigma P_u}{\phi \Sigma P_c}} \geq 1.0 \tag{27.5}$$

where

P_u = factored axial load

P_c = Euler buckling load, which is determined as follows:

$$P_c = \frac{\pi^2 EI}{(KL_u)^2} \tag{27.6}$$

C_m, a factor which relates the actual moment diagram to an equivalent uniform moment diagram, is typically taken as 1.0. However, in the case where the member is braced against side sway and without transverse loads between supports, it may be taken by the following expression:

$$C_m = 0.60 + 0.40\left(\frac{M_{1b}}{M_{2b}}\right) \tag{27.7}$$

The value resulting from Eq. (27.7), however, is not to be less than 0.40.

To compute the flexural rigidity *EI* for concrete columns, AASHTO offers two possible solutions, with the first being:

$$EI = \frac{\dfrac{E_c I_g}{5} + E_s I_s}{1 + \beta_d} \tag{27.8}$$

and the second, more-conservative solution being:

$$EI = \frac{\dfrac{E_c I_g}{2.5}}{1 + \beta_d} \tag{27.9}$$

where E_c is the elastic modulus of concrete, I_g is the gross moment inertia, E_s is the elastic modules of reinforcement, I_s is the moment inertia of reinforcement about centroidal axis, and β is the ratio of maximum dead-load moment to maximum total-load moment and is always positive. It is an approximation of the effects of creep, so that when larger moments are induced by loads sustained over a long period of time, the creep deformation and associated curvature will also be increased.

27.4.3 Concrete Piers and Columns

27.4.3.1 Combined Axial and Flexural Strength

A critical aspect of the design of bridge piers is the design of compression members. We will use AASHTO LRFD Bridge Design Specifications [1] as the reference source. The following discussion provides an overview of some of the major criteria governing the design of compression members.

Under the Strength Limit State Design, the factored resistance is determined with the product of nominal resistance, P_n, and the resistance factor, ϕ. Two different values of ϕ are used for the nominal resistance P_n. Thus, the factored axial load resistance ϕP_n is obtained using $\phi = 0.75$ for columns with spiral and tie confinement reinforcement. The specifications also allows for the value ϕ to be linearly increased from the value stipulated for compression members to the value specified for flexure which is equal to 0.9 as the design axial load ϕP_n decreases from $0.10 f_c' A_g$ to zero.

Interaction Diagrams

Flexural resistance of a concrete member is dependent upon the axial force acting on the member. Interaction diagrams are usually used as aids for the design of the compression members. Interaction diagrams for columns are usually created assuming a series of strain distributions, and computing the corresponding values of *P* and *M*. Once enough points have been computed, the results are plotted to produce an interaction diagram.

Figure 27.8 shows a series of strain distributions and the resulting points on the interaction diagram. In an actual design, however, a few points on the diagrams can be easily obtained and can define the diagram rather closely.

- Pure Compression:

The factored axial resistance for pure compression, ϕP_n, may be computed by:

For members with spiral reinforcement:

$$P_r = \phi P_n = \phi 0.85 P_o = \phi 0.85 \left[0.85 f_c' \left(A_g - A_{st} \right) + A_{st} f_y \right] \tag{27.10}$$

For members with tie reinforcement:

$$P_r = \phi P_n = \phi 0.80 P_o = \phi 0.80 \left[0.85 f_c' \left(A_g - A_{st} \right) + A_{st} f_y \right] \tag{27.11}$$

For design, pure compression strength is a hypothetical condition since almost always there will be moments present due to various reasons. For this reason, AASHTO LRFD 5.7.4.4 limits the nominal axial load resistance of compression members to 85 and 80% of the axial resistance at zero eccentricity, P_o, for spiral and tied columns, respectively.

- Pure Flexure:

The section in this case is only subjected to bending moment and without any axial force. The factored flexural resistance, M_r, may be computed by

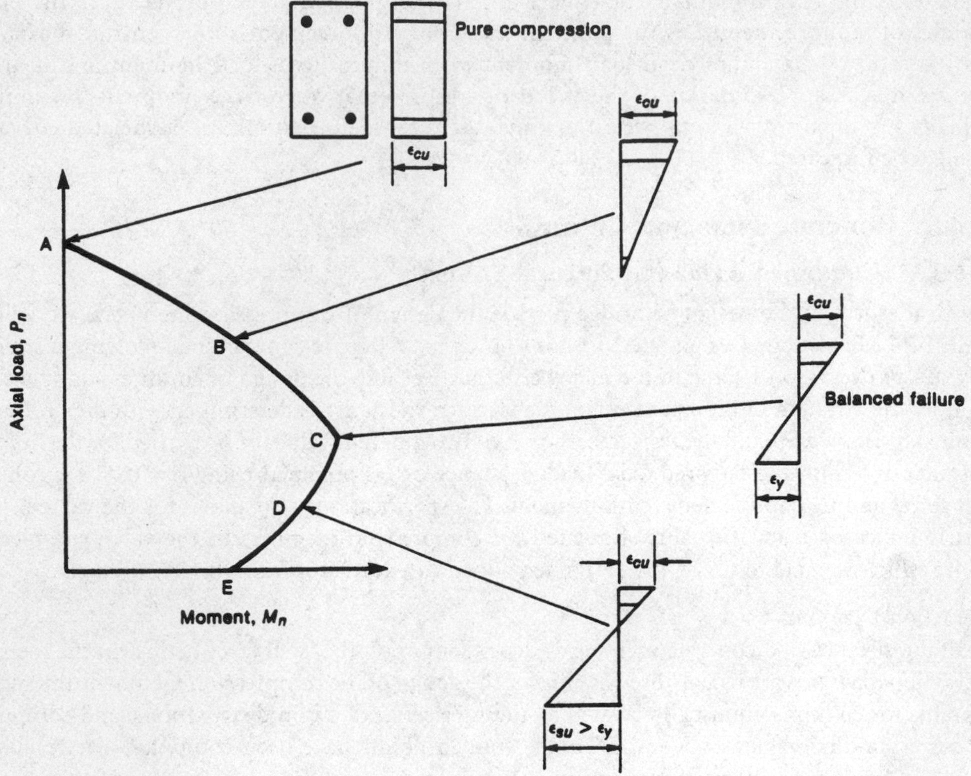

FIGURE 27.8 Strain distributions corresponding to points on interaction diagram.

$$M_r = \phi M_n = \phi\left[A_s f_y d\left(1 - 0.6\rho\frac{f_y}{f_c'}\right)\right]$$
$$= \phi\left[A_s f_y\left(d - \frac{a}{2}\right)\right] \tag{27.12}$$

where

$$a = \frac{A_s f_y}{0.85 f_c' b}$$

- **Balanced Strain Conditions:**

Balanced strain conditions correspond to the strain distribution where the extreme concrete strain reaches 0.003 and the strain in reinforcement reaches yield at the same time. At this condition, the section has the highest moment capacity. For a rectangular section with reinforcement in one face, or located in two faces at approximately the same distance from the axis of bending, the balanced factored axial resistance, P_r, and balanced factored flexural resistance, M_r, may be computed by

$$P_r = \phi P_b = \phi\left[0.85 f_c' b a_b + A_s' f_s' - A_s f_y\right] \tag{27.13}$$

and

$$M_r = \phi M_b = \phi\left[0.85 f_c' b a_b \left(d - d'' - a_b/2\right) + A_s' f_s' \left(d - d' - d''\right) + A_s f_y d''\right] \tag{27.14}$$

where

$$a_b = \left(\frac{600}{600 + f_y}\right)\beta_1 d$$

and

$$f_s' = 600\left[1 - \left(\frac{d'}{d}\right)\left(600 + \frac{f_y}{600}\right)\right] \leq f_y$$

where f_y is in MPa.

Biaxial Bending

AASHTO LRFD 5.7.4.5 stipulates that the design strength of noncircular members subjected to biaxial bending may be computed, in lieu of a general section analysis based on stress and strain compatibility, by one of the following approximate expressions:

$$\frac{1}{P_{rxy}} = \frac{1}{P_{rx}} + \frac{1}{P_{ry}} - \frac{1}{P_o} \tag{27.15}$$

when the factored axial load, $P_u \geq 0.10\phi f_c' A_g$

$$\frac{M_{ux}}{M_{rx}} + \frac{M_{uy}}{M_{ry}} \leq 1 \tag{27.16}$$

when the factored axial load, $P_u < 0.10\phi f_c' A_g$

where

P_{rxy} = factored axial resistance in biaxial flexure

P_{rx}, P_{ry} = factored axial resistance corresponding to M_{rx}, M_{ry}

M_{ux}, M_{uy} = factored applied moment about the *x*-axis, *y*-axis

M_{rx}, M_{ry} = uniaxial factored flexural resistance of a section about the *x*-axis and *y*-axis corresponding to the eccentricity produced by the applied factored axial load and moment, and

P_o = $0.85 f_c'(A_g - A_s) + A_s f_y$

27.4.3.2 Shear Strength

Under the normal load conditions, the shear seldom governs the design of the column for conventional bridges since the lateral loads are usually small compared with the vertical loads. However, in a seismic design, the shear is very important. In recent years, the research effort on shear strength evaluation for columns has been increased remarkably. AASHTO LRFD provides a general shear equation that applies for both beams and columns. The concrete shear capacity component and the angle of inclination of diagonal compressive stresses are functions of the shear stress on the concrete and the strain in the reinforcement on the flexural tension side of the member. It is rather involved and hard to use.

Alternatively, the equations recommended by ATC-32 [5] can be used with acceptable accuracy. The recommendations are listed as follows.

Except for the end regions of ductile columns, the nominal shear strength provided by concrete, V_c, for members subjected to flexure and axial compression should be computed by

$$V_c = 0.165\left(1 + (3.45)\left(10^{-6}\right)\frac{N_u}{A_g}\right)\sqrt{f_c'}A_e \quad \text{(MPa)} \tag{27.17}$$

If the axial force is in tension, the V_c should be computed by

$$V_c = 0.165\left(1 + (1.38)\left(10^{-5}\right)\frac{N_u}{A_g}\right)\sqrt{f_c'}A_e \quad \text{(MPa)} \tag{27.18}$$

(note that N_u is negative for tension),

where

A_g = gross section area of the column (mm^2)
A_e = effective section area, can be taken as $0.8A_g$ (mm^2)
N_u = axial force applied to the column (N)
f_c' = compressive strength of concrete (MPa)

For end regions where the flexural ductility is normally high, the shear capacity should be reduced. ATC-32 [5] offers the following equations to address this interaction.

With the end region of columns extending a distance from the critical section or sections not less than $1.5D$ for circular columns or $1.5h$ for rectangular columns, the nominal shear strength provided by concrete subjected to flexure and axial compression should be computed by

$$V_c = 0.165\left(0.5 + (6.9)\left(10^{-6}\right)\frac{N_u}{A_g}\right)\sqrt{f_c'}A_e \quad \text{(MPa)} \tag{27.19}$$

When axial load is tension, V_c can be calculated as

$$V_c = 0.165\left(1 + (1.38)\left(10^{-5}\right)\frac{N_u}{A_g}\right)\sqrt{f_c'}A_e \quad \text{(MPa)} \qquad (27.18)$$

Again, N_u should be negative in this case.

The nominal shear contribution from reinforcement is given by

$$V_s = \frac{A_v f_{yh} d}{s} \quad \text{(MPa)} \tag{27.20}$$

for tied rectangular sections, and by

$$V_s = \frac{\pi}{2}\frac{A_h f_{yh} D'}{s} \tag{27.21}$$

for spirally reinforced circular sections. In these equations, A_v is the total area of shear reinforcement parallel to the applied shear force, A_h is the area of a single hoop, f_{yh} is the yield stress of horizontal reinforcement, D' is the diameter of a circular hoop, and s is the spacing of horizontal reinforcement.

27.4.3.3 Ductility of Columns

The AASHTO LRFD [1] introduces the term *ductility* and requires that a structural system of bridge be designed to ensure the development of significant and visible inelastic deformations prior to failure.

The term *ductility* defines the ability of a structure and selected structural components to deform beyond elastic limits without excessive strength or stiffness degradation. In mathematical terms, the ductility μ is defined by the ratio of the total imposed displacement Δ at any instant to that at the onset of yield Δ_y. This is a measure of the ability for a structure, or a component of a structure, to absorb energy. The goal of seismic design is to limit the estimated maximum ductility demand to the ductility capacity of the structure during a seismic event.

For concrete columns, the confinement of concrete must be provided to ensure a ductile column. AASHTO LRFD [1] specifies the following minimum ratio of spiral reinforcement to total volume of concrete core, measured out-to-out of spirals:

$$\rho_s = 0.45\left(\frac{A_g}{A_c} - 1\right)\frac{f'_c}{f_{yh}} \tag{27.22}$$

The transverse reinforcement for confinement at the plastic hinges shall be determined as follows:

$$\rho_s = 0.16\frac{f'_c}{f_y}\left(0.5 + \frac{1.25P_u}{A_g f'_c}\right) \tag{27.23}$$

for which

$$\left(0.5 + \frac{1.25P_u}{A_g f'_c}\right) \geq 1.0$$

The total cross-sectional area (A_{sh}) of rectangular hoop (stirrup) reinforcement for a rectangular column shall be either

$$A_{sh} = 0.30ah_c\frac{f'_c}{f_{yh}}\left(\frac{A_g}{A_c} - 1\right) \tag{27.24}$$

or,

$$A_{sh} = 0.12ah_c\frac{f'_c}{f_y}\left(0.5 + \frac{1.25P_u}{A_g f'_c}\right) \tag{27.25}$$

whichever is greater,

where

a = vertical spacing of hoops (stirrups) with a maximum of 100 mm (mm)
A_c = area of column core measured to the outside of the transverse spiral reinforcement (mm^2)
A_g = gross area of column (mm^2)
A_{sh} = total cross-sectional area of hoop (stirrup) reinforcement (mm^2)
f'_c = specified compressive strength of concrete (Pa)
f_{yh} = yield strength of hoop or spiral reinforcement (Pa)
h_c = core dimension of tied column in the direction under consideration (mm)
ρ_s = ratio of volume of spiral reinforcement to total volume of concrete core (out-to-out of spiral)
P_u = factored axial load (MN)

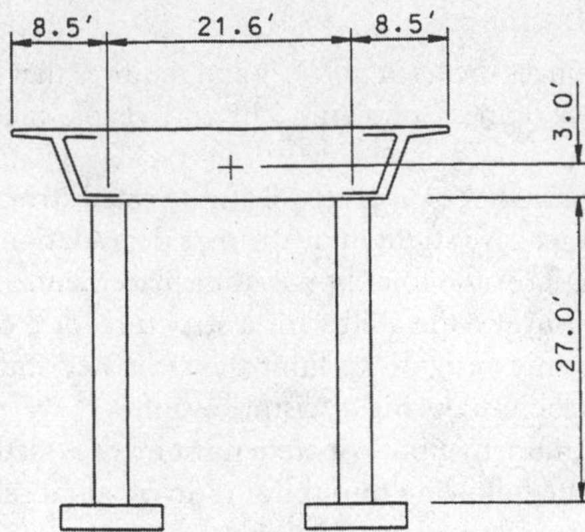

FIGURE 27.9 Example 27.2 — typical section.

TABLE 27.3 Column Group Loads — Service

		Live Load + Impact							
	Dead Load	Case 1 Trans $M_{y\text{-}max}$	Case 2 Long $M_{x\text{-}max}$	Case 3 Axial N-max	Win d	Wind on LL	Long Force	Centrifuga l Force-M_y	Temp.
M_y (k-ft)	220	75	15	32	532	153	208	127	180
M_x (k-ft)	148	67	599	131	192	86	295	2	0
P (k)	1108	173	131	280	44	17	12	23	0

TABLE 27.4 Unreduced Seismic Loads (ARS)

	Case 1 Max. Transverse	Case 2 Max. Longitudinal
M_y — Trans (k-ft)	4855	3286
M_x — Long (k-ft)	3126	3334
P — Axial (k)	–282	–220

Example 27.2 Design of a Two-Column Bent

Design the columns of a two-span overcrossing. The typical section of the structure is shown in Figure 27.9. The concrete box girder is supported by a two-column bent and is subjected to HS20 loading. The columns are pinned at the bottom of the columns. Therefore, only the loads at the top of columns are given here. Table 27.3 lists all the forces due to live load plus impact. Table 27.4 lists the forces due to seismic loads. Note that a load reduction factor of 5.0 will be assumed for the columns.

Material Data

f'_c = 4.0 ksi (27.6 MPa) E_c = 3605 ksi (24855 MPa)

E_s = 29000 ksi (199946 MPa) f_y = 60 ksi (414 MPa)

Try a column size of 4 ft (1.22 m) in diameter. Provide 26-#9 (26-#30) longitudinal reinforcement. The reinforcement ratio is 1.44%.

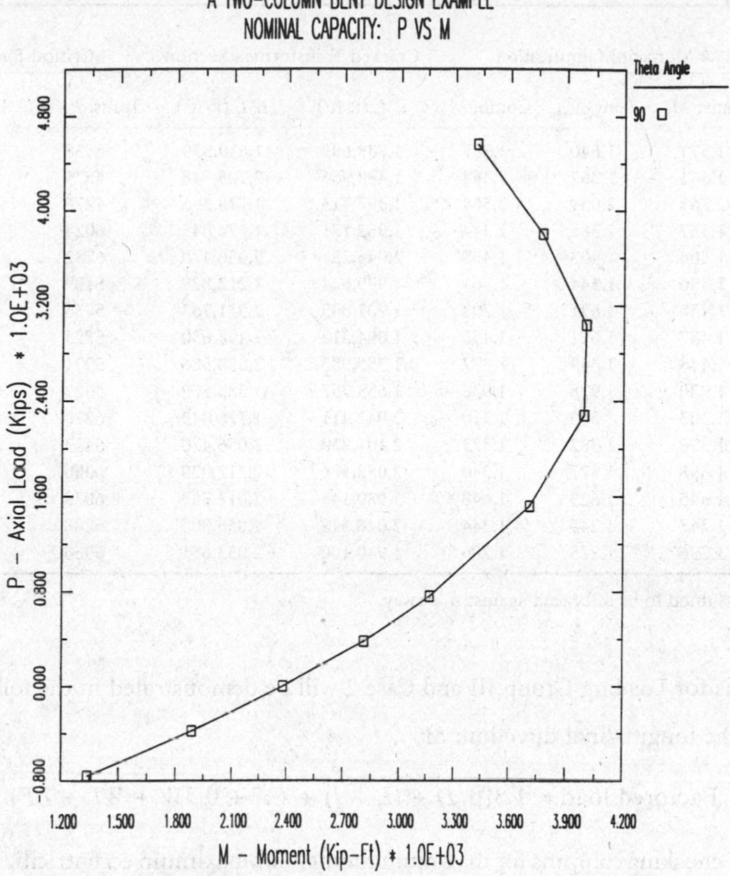

FIGURE 27.10 Example 27.2 — interaction diagram.

Section Properties

A_g = 12.51 ft² (1.16 m²) $\qquad$ A_{st} = 26.0 in² (16774 mm²)

$I_{xc} = I_{yc}$ = 12.46 ft⁴ (0.1075 m⁴) $\qquad$ $I_{xs} = I_{ys}$ = 0.2712 ft⁴ (0.0023 m⁴)

The analysis follows the procedure discussed in Section 27.4.3.1. The moment and axial force interaction diagram is generated and is shown in Figure 27.10.

Following the procedure outlined in Section 27.4.2, the moment magnification factors for each load group can be calculated and the results are shown in Table 27.5.

In which:

$$K_y = K_x = 2.10$$

$$K_y L/R = K_x\ L/R = 2.1 \times 27.0/(1.0) = 57$$

where R = radio of gyration = $r/2$ for a circular section.

$$22 < KL/R < 100 \quad \therefore \text{ \underline{Second-order effect should be considered}.}$$

TABLE 27.5 Moment Magnification and Buckling Calculations

Load	P (k)	Moment Magnification			Cracked Transformed Section		Critical Buckling		Axia l Load
Group	Case	Trans. M_{agy}	Long M_{agx}	Comb. M_{ag}	E^*I_y (k-ft²)	E^*I_x (k-ft²)	Trans. P_{cy} (k)	Long P_{cx} (k)	P(k)
I	1	1.571	1.640	1.587	1,738,699	1,619,399	5338	4972	1455
I	2	1.661	1.367	1.384	1,488,966	2,205,948	4571	6772	1364
I	3	2.765	2.059	2.364	1,392,713	1,728,396	4276	5306	2047
II		1.337	1.385	1.344	1,962,171	1,776,045	6024	5452	1137
III	1	1.406	1.403	1.405	2,046,281	2,056,470	6282	6313	1360
III	2	1.396	**1.344**	1.361	1,999,624	**2,212,829**	6139	**6793**	1305
III	3	1.738	1.671	1.708	1,901,005	2,011,763	5836	6176	1859
IV	1	1.437	1.611	1.455	1,864,312	1,494,630	5723	4588	1306
IV	2	1.448	1.349	1.377	1,755,985	2,098,586	5391	6443	1251
IV	3	1.920	1.978	1.936	1,635,757	1,585,579	5022	4868	1805
V		1.303	1.365	1.310	2,042,411	1,776,045	6270	5452	1094
VI	1	1.370	1.382	1.373	2,101,830	2,056,470	6453	6313	1308
VI	2	1.358	1.327	1.340	2,068,404	2,212,829	6350	6793	1256
VI	3	1.645	1.629	1.640	1,980,146	2,011,763	6079	6176	1788
VII	1	1.243	1.245	1.244	2,048,312	2,036,805	6288	6253	826
VII	2	1.296	1.275	1.286	1,940,100	2,053,651	5956	6305	888

Note: Column assumed to be unbraced against side sway.

The calculations for Loading Group III and Case 2 will be demonstrated in the following:

Bending in the longitudinal direction: M_x

$$\text{Factored load} = 1.3[\beta_D D + (L + I) + CF + 0.3W + WL + LF]$$

$\beta_D = 0.75$ when checking columns for maximum moment or maximum eccentricities and associated axial load. β_d in Eq. (27.8) = max dead-load moment, M_{DL}/max total moment, M_t.

$$M_{DL} = 148 \times 0.75 = 111 \text{ k-ft } (151 \text{ kN·m})$$

$$M_t = 0.75 \times 148 + 599 + 0.3 \times 192 + 86 + 295 + 2 = 1151 \text{ k-ft } (1561 \text{ kN·m})$$

$$\beta_d = 111/1151 = 0.0964$$

$$EI_x = \frac{\frac{E_c I_g}{5} + E_s I_s}{1+\beta_d} = \frac{\frac{3605 \times 144 \times 12.46}{5} + 29{,}000 \times 144 \times 0.2712}{1+0.0964} = 2{,}212{,}829 \text{ k-ft}^2$$

$$P_{cx} = \frac{\pi^2 EI_x}{(KL_u)^2} = \frac{\pi^2 \times 2{,}212{,}829}{(2.1 \times 27)^2} = 6793 \text{ kips } (30{,}229 \text{ kN})$$

$$C_m = 1.0 \text{ for frame braced against side sway}$$

$$\delta_s = \frac{1}{1 - \frac{\Sigma P_u}{\phi \Sigma P_c}} = \frac{1}{1 - \frac{1305}{0.75 \times 6793}} = 1.344$$

The magnified factored moment = 1.344 × 1.3 × 1151 = 2011 k-ft (2728 kN·m)

TABLE 27.6 Comparison of Factored Loads to Factored Capacity of the Column

Group	Case	Applied Factored Forces (k-ft)				Capacity (k-ft)			
		Trans. M_y	Long M_x	Comb. M	Axial P (k)	ϕM_n	ϕ	Ratio M_u/M	Status
I	1	852	475	975	1455	2924	0.75	3.00	OK
I	2	566	1972	2051	1364	2889	0.75	1.41	OK
I	3	1065	981	1448	2047	3029	0.75	2.09	OK
II		1211	546	1328	1137	2780	0.75	2.09	OK
III	1	1622	1125	1974	1360	2886	0.75	1.46	OK
III	2	1402	**2011**	2449	1305	2861	0.75	1.17	OK
III	3	1798	1558	2379	1859	3018	0.75	1.27	OK
IV	1	1022	373	1088	1306	2865	0.75	2.63	OK
IV	2	813	1245	1487	1251	2837	0.75	1.91	OK
IV	3	1136	717	1343	1805	3012	0.75	2.24	OK
V		1429	517	1519	1094	2754	0.75	1.81	OK
VI	1	1829	1065	2116	1308	2864	0.75	1.35	OK
VI	2	1617	1905	2499	1256	2842	0.75	1.14	OK
VI	3	2007	1461	2482	1788	3008	0.75	1.21	OK
VII	1	1481	963	1766	826	2372	0.67	1.34	OK
VII	2	1136	1039	1540	888	2364	0.65	1.54	OK

Notes:
1. Applied factored moments are magnified for slenderness in accordance with AASHTO LRFD.
2. The seismic forces are reduced by the load reduction factor $R = 5.0$.

$L = 27.00$ ft, $f'_c = 4.00$ ksi, $F_y = 60.0$ ksi, $A_{st} = 26.00$ in.^{2}

The analysis results with the comparison of applied moments to capacities are summarized in Table 27.6.

Column lateral reinforcement is calculated for two cases: (1) for applied shear and (2) for confinement. Typically, the confinement requirement governs. Apply Eq. 27.22 or Eq. 27.23 to calculate the confinement reinforcement. For seismic analysis, the unreduced seismic shear forces should be compared with the shear forces due to plastic hinging of columns. The smaller should be used. The plastic hinging analysis procedure is discussed elsewhere in this handbook and will not be repeated here.

The lateral reinforcement for both columns are shown as follows.

For left column:

V_u = 148 kips (659 kN) (shear due to plastic hinging governs)

ϕV_n = 167 kips (743 kN) ∴ No lateral reinforcement is required for shear.

Reinforcement for confinement = $\rho_s = 0.0057$ ∴ Provide #4 at 3 in. (#15 at 76 mm)

For right column:

V_u = 180 kips (801 kN) (shear due to plastic hinging governs)

ϕV_n = 167 kips (734 kN)

ϕV_s = 13 kips (58 kN) (does not govern)

Reinforcement for confinement = $\rho_s = 0.00623$ ∴ Provide #4 at 2.9 in. (#15 at 74 mm)

Summary of design:
4 ft (1.22 m) diameter of column with 26-#9 (26-#30) for main reinforcement and #4 at 2.9 in. (#15 at 74 mm) for spiral confinement.

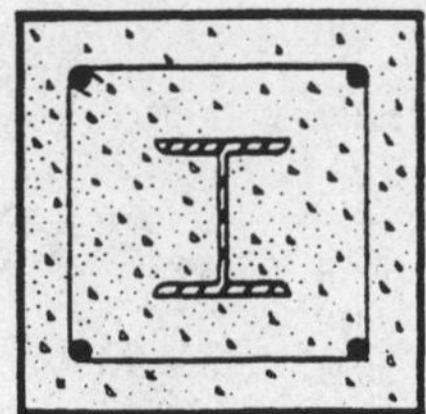

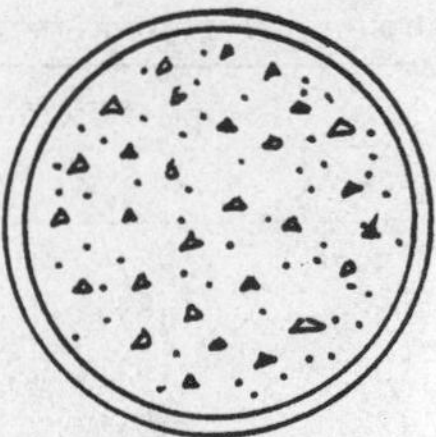

FIGURE 27.11 Typical cross sections of composite columns.

27.4.4 Steel and Composite Columns

Steel columns are not as commonly used as concrete columns. Nevertheless, they are viable solutions for some special occasions, e.g., in space-restricted areas. Steel pipes or tubes filled with concrete known as composite columns (Figure 27.11) offer the most efficient use of the two basic materials. Steel at the perimeter of the cross section provides stiffness and triaxial confinement, and the concrete core resists compression and prohibits local elastic buckling of the steel encasement. The toughness and ductility of composite columns makes them the preferred column type for earthquake-resistant structures in Japan. In China, the composite columns were first used in Beijing subway stations as early as 1963. Over the years, the composite columns have been used extensively in building structures as well as in bridges [6–9].

In this section, the design provisions of AASHTO LRFD [1] for steel and composite columns are summarized.

Compressive Resistance

For prismatic members with at least one plane of symmetry and subjected to either axial compression or combined axial compression and flexure about an axis of symmetry, the factored resistance of components in compression, P_r, is calculated as

$$P_r = \phi_c P_n$$

where
P_n = nominal compressive resistance
ϕ_c = resistance factor for compression = 0.90

The nominal compressive resistance of a steel or composite column should be determined as

$$P_n = \begin{cases} 0.66^{\lambda} F_e A_s & \text{if} \quad \lambda \le 2.25 \\ \dfrac{0.88 F_e A_s}{\lambda} & \text{if} \quad \lambda > 2.25 \end{cases} \tag{27.26}$$

in which

For steel columns:

$$\lambda = \left(\frac{KL}{r_s} \pi \right)^2 \frac{F_y}{E_e} \tag{27.27}$$

For composite column:

$$\lambda = \left(\frac{KL}{r_s}\pi\right)^2 \frac{F_e}{E_e} \tag{27.28}$$

$$F_e = F_y + C_1 F_{yr}\left(\frac{A_r}{A_s}\right) + C_2 f_c\left(\frac{A_c}{A_s}\right) \tag{27.29}$$

$$Ee = E\left[1 + \left(\frac{C_3}{n}\right)\left(\frac{A_c}{A_s}\right)\right] \tag{27.30}$$

where
A_s = cross-sectional area of the steel section (mm^2)
A_c = cross-sectional area of the concrete (mm^2)
A_r = total cross-sectional area of the longitudinal reinforcement (mm^2)
F_y = specified minimum yield strength of steel section (MPa)
F_{yr} = specified minimum yield strength of the longitudinal reinforcement (MPa)
f_c' = specified minimum 28-day compressive strength of the concrete (MPa)
E = modules of elasticity of the steel (MPa)
L = unbraced length of the column (mm)
K = effective length factor
n = modular ratio of the concrete
r_s = radius of gyration of the steel section in the plane of bending, but not less than 0.3 times the width of the composite member in the plane of bending for composite columns, and, for filled tubes,

$$C_1 = 1.0; \quad C_2 = 0.85; \quad C_3 = 0.40$$

In order to use the above equation, the following limiting width/thickness ratios for axial compression of steel members of any shape must be satisfied:

$$\frac{b}{t} \leq k\sqrt{\frac{E}{F_y}} \tag{27.31}$$

where
k = plate buckling coefficient as specified in Table 27.7
b = width of plate as specified in Table 27.7
t = plate thickness (mm)

Wall thickness of steel or composite tubes should satisfy:

For circular tubes:

$$\frac{D}{t} \leq 2.8\sqrt{\frac{E}{F_y}}$$

TABLE 27.7 Limiting Width-to-Thickness Ratios

	k	b
		Plates Supported along One Edge
Flanges and projecting leg or plates	0.56	Half-flange width of I-section Full-flange width of channels Distance between free edge and first line of bolts or welds in plates Full-width of an outstanding leg for pairs of angles on continuous contact
Stems of rolled tees	0.75	Full-depth of tee
Other projecting elements	0.45	Full-width of outstanding leg for single-angle strut or double-angle strut with separator Full projecting width for others
		Plates Supported along Two Edges
Box flanges and cover plates	1.40	Clear distance between webs minus inside corner radius on each side for box flanges Distance between lines of welds or bolts for flange cover plates
Webs and other plates elements	1.49	Clear distance between flanges minus fillet radii for webs of rolled beams Clear distance between edge supports for all others
Perforated cover plates	1.86	Clear distance between edge supports

For rectangular tubes:

$$\frac{b}{t} \leq 1.7\sqrt{\frac{E}{F_y}}$$

where
D = diameter of tube (mm)
b = width of face (mm)
t = thickness of tube (mm)

Flexural Resistance

The factored flexural resistance, M_r, should be determined as

$$M_r = \phi_f M_n \tag{27.32}$$

where
M_n = nominal flexural resistance
ϕ_f = resistance factor for flexure, $\phi_f = 1.0$

The nominal flexural resistance of concrete-filled pipes that satisfy the limitation

$$\frac{D}{t} \leq 2.8\sqrt{\frac{E}{F_y}}$$

may be determined:

$$\text{If } \frac{D}{t} < 2.0\sqrt{\frac{E}{F_y}}, \text{ then } M_n = M_{ps} \tag{27.33}$$

$$\text{If } 2.0\sqrt{\frac{E}{F_y}} < \frac{D}{t} \le 8.8\sqrt{\frac{E}{F_y}}, \text{ then } M_n = M_{yc} \tag{27.34}$$

where
M_{ps} = plastic moment of the steel section
M_{yc} = yield moment of the composite section

Combined Axial Compression and Flexure

The axial compressive load, P_u, and concurrent moments, M_{ux} and M_{uy}, calculated for the factored loadings for both steel and composite columns should satisfy the following relationship:

$$\text{If } \frac{P_u}{P_r} < 0.2, \text{ then } \frac{P_u}{2.0P_r} + \left(\frac{M_{ux}}{M_{rx}} + \frac{M_{uy}}{M_{ry}}\right) \le 1.0 \tag{27.35}$$

$$\text{If } \frac{P_u}{P_r} \ge 0.2, \text{ then } \frac{P_u}{P_r} + \frac{8.0}{9.0}\left(\frac{M_{ux}}{M_{rx}} + \frac{M_{uy}}{M_{ry}}\right) \le 1.0 \tag{27.36}$$

where
P_r = factored compressive resistance
M_{rx}, M_{ry} = factored flexural resistances about x and y axis, respectively
M_{ux}, M_{uy} = factored flexural moments about the x and y axis, respectively

References

1. AASHTO, *LRFD Bridge Design Specifications,* 1st ed., American Association of State Highway and Transportation Officials, Washington, D.C., 1994.
2. Caltrans, Bridge Memo to Designers (7-10), California Department of Transportation, Sacramento, 1994.
3. White, D. W. and Hajjar, J. F., Application of second-order elastic analysis in LRFD: research to practice, *Eng. J.,* 28(4), 133, 1994.
4. Galambos, T. V., Ed., *Guide to Stability Design for Metal Structures,* 4th ed., the Structural Stability Research Council, John Wiley & Sons, New York, 1988.
5. ATC, Improved Seismic Design Criteria for California Bridges: Provisional Recommendations, Applied Technology Council, Report ATC-32, Redwood City, CA, 1996.
6. Cai, S.-H., Chinese standard for concrete-filled tube columns, in *Composite Construction in Steel and Concrete II,* Proc. of an Engineering Foundation Conference, Samuel Easterling, W. and Kim Roddis, W. M., Eds, Potosi, MO, 1992, 143.
7. Cai, S.-H., Ultimate strength of concrete-filled tube columns, in *Composite Construction in Steel and Concrete,* Proc. of an Engineering Foundation Conference, Dale Buckner, C. and Viest, I. M., Eds, Henniker, NH, 1987, 703.
8. Zhong, S.-T., New concept and development of research on concrete-filled steel tube (CFST) members, in *Proc. 2nd Int. Symp. on Civil Infrastructure Systems,* 1996.
9. CECS 28:90, *Specifications for the Design and Construction of Concrete-Filled Steel Tubular Structures,* China Planning Press, Beijing [in Chinese], 1990.

10. AISC, *Load and Research Factor Design Specification for Structural Steel Buildings and Commentary*, 2nd ed., American Institute of Steel Construction, Chicago, IL, 1993.
11. Galambos, T. V. and Chapuis, J., LRFD Criteria for Composite Columns and Beam Columns, Revised Draft, Washington University, Department of Civil Engineering, St. Louis, MO, December 1990.

28
Towers

Charles Seim
T. Y. Lin International

28.1 Introduction

Towers are the most visible structural elements of long-span bridges. They project above the superstructure and are seen from all directions by viewers and by users. Towers give bridges their character and a unifying theme. They project a mnemonic image that people remember as a lasting impression of the bridge itself. As examples of the powerful imagery of towers, contrast the elegant art deco towers of the Golden Gate Bridge (Figure 28.1) with the utilitarian but timeless architecture of the towers of the San Francisco–Oakland Bay Bridge (Figure 28.2). Or contrast the massive, rugged stone towers of the Brooklyn Bridge (Figure 28.3) with the awkward confusing steel towers of the Williamsburg Bridge in New York City (Figure 28.4).

Towers can be defined as vertical steel or concrete structures projecting above the deck, supporting cables and carrying the forces to which the bridge is subjected to the ground. By this definition, towers are used only for suspension bridges or for cable-stayed bridges, or hybrid suspension–cable-stayed structures. The word *pylon* is sometimes used for the towers of cable–stayed bridges. Both *pylon* and *tower* have about the same meaning — a tall and narrow structure supporting itself and the roadway. In this chapter, the word *tower* will be used for both suspension and for cabled-stayed bridges, to avoid any confusion in terms.

Both suspension and cable-stayed bridges are supported by abutments or piers at the point where these structures transition to the approach roadway or the approach structure. Abutments are discussed in Chapter 30. Piers and columns that support the superstructure for other forms of bridge structures such as girders, trusses, or arches, usually do not project above the deck. Piers and columns are discussed in Chapter 27.

The famous bridges noted above were opened in 1937, 1936, 1883, and 1903, respectively, and, if well maintained, could continue to serve for another 100 years. Bridge engineers will not design structures like these today because of changing technologies. These bridges are excellent examples of enduring structures and can serve to remind bridge engineers that well-designed and maintained structures do

0-8493-7434-0/00/$0.00+$.50

FIGURE 28.1 Golden Gate Bridge, San Francisco. (Courtesy of Charles Seim.)

last for 150 years or longer. Robust designs, durable materials, provisions for access for inspection and maintenance, and a well-executed maintenance program will help ensure a long life. The appearance of the bridge, good or bad, is locked in for the life of the facility and towers are the most important visual feature leading to the viewer's impression of an aesthetic structure.

28.2 Functions

The main structural function of the towers of cable-stayed and suspension bridges is carrying the weight of the bridge, traffic loads, and the forces of nature to the foundations. The towers must perform these function in a reliable, serviceable, aesthetic, and economical manner for the life of the bridge, as towers, unlike other bridge components, cannot be replaced. Without reliability, towers may become unsafe and the life of the entire bridge could be shortened. Without serviceability being designed into the structure, which means that it is designed for access and ease of maintenance, the bridge will not provide continuing long service to the user. The public demands that long-span bridges be attractive, aesthetic statements with long lives, so as not to be wasteful of public funds.

28.3 Aesthetics

While the main function of the towers is structural, an important secondary function is visual. The towers reveal the character or motif of the bridge. The bridges used as examples in the introduction are good illustrations of the image of the structure as revealed by the towers. Indeed, perhaps they are famous because of their towers. Most people visualize the character of the Brooklyn Bridge by the gothic, arched, masonry towers alone. The San Francisco–Oakland Bay Bridge and the Golden Gate Bridge give completely different impressions to the viewer as conveyed by the towers. Seim [7] measured the ratios of the visible components of the towers of the latter two bridges and found important, but subtle, diminution of these ratios with height above the tower base. It is the subtle changes in these ratios within the height of the towers that produce the much-admired proportions of these world-renowned bridges. The proportions of the towers for any new long-span bridge

FIGURE 28.2 San Francisco–Oakland Bay Bridge. (Courtesy of Charles Seim.)

should be carefully shaped and designed to give the entire bridge a strong — even robust — graceful, and soaring visual image. The aesthetics of bridges are discussed in greater detail in Chapters 2 and 3 of this volume.

The aesthetics of the array of cables many times are of secondary importance to the aesthetics of the towers. However, the array or form of the cables must be considered in the overall aesthetic and structural evaluation of the bridge. Main cables of suspension bridges always drape in a parabolic curve that most people instinctively enjoy. The large diameter of the cables makes them stand out as an important contribution to the overall visual impression as the supporting element of the roadway.

The cables of cable-stayed bridges are usually of small diameter and do not stand out visually as strongly as do the cables of suspension bridges. However, the array of the stays, such as harp, fan, radiating star, or others, should be considered in context with the tower form. The separated, parallel cables of the harp form, for example, will not be as obtrusive to the towers as will other arrangements. However, the harp cable form may not be appropriate for very long spans or for certain tower shapes. The cables and the towers should be considered together as a visual system.

Billington [2] presents an overview of the importance of the role of aesthetics in the history of the development of modern bridge design. Leonhardt [5] presents many examples of completed bridges showing various tower shapes and cable arrangements for both suspension and cable-stayed bridges.

FIGURE 28.3 Brooklyn Bridge, New York. (Courtesy of Charles Seim.)

28.4 Conceptual Design

Perhaps the most important step in the design of a new bridge is the design concept for the structure that ultimately will be developed into a final design and then constructed. The cost, appearance, and reliability and serviceability of the facility will all be determined, for good or for ill, by the conceptual design of the structure. The cost can be increased, sometimes significantly, by a concept that is very difficult to erect. Once constructed, the structure will always be there for users to admire — or to criticize. The user ultimately pays for the cost of the facility and also usually pays for the cost of maintaining the structure. Gimsing [4] treats the concept design issues of both cable-stayed and suspension bridges very extensively and presents examples to help guide designers.

A proper bridge design that considers the four functions of reliability, serviceability, appearance, and cost together with an erectable scheme that requires low maintenance, is the ideal that the design concept should meet.

A recent trend is to employ an architect as part of the design team. Architects may view a structure in a manner different from engineers, and their roles in the project are not the same. The role of the engineer is to be involved in all four functions and, most importantly, to take responsibility for the structural adequacy of the bridge. The role of the architect generally only involves the function of aesthetics. Their roles overlap in achieving aesthetics, which may also affect the economy of the structure. Since both engineers and architects have as a common objective an elegant and economical bridge, there should be cooperation and respect between them.

Occasional differences do occur when the architect's aesthetic desires conflict with the engineer's structural calculations. Towers, as the most visible component of the bridge, seem to be a target for this type of conflict. Each professional must understand that these differences in viewpoints will occur and must be resolved for a successful and fruitful union between the two disciplines.

While economy is usually important, on occasions, cost is not an objective because the owner or the public desires a "symbolic" structure. The architect's fancy then controls and the engineer can only provide the functions of safety and serviceability.

FIGURE 28.4 Williamsburg Bridge, New York. (Courtesy of Charles Seim.)

28.4.1 Materials

Until the 1970s steel was the predominant material used for the towers of both cable-stayed and suspension bridges. The towers were often rectangular in elevation with a cross-sectional shape of rectangular, cruciform, tee, or a similar shape easily fabricated in steel. Examples of suspension bridge steel tower design are the plain, rectangular steel towers for the two Delaware Memorial Bridges; the first constructed in 1951 and the parallel one in 1968 (Figure 28.5). An example of a cable-stayed bridge that is an exception to the rectangular tower form is the modified A-frame, weathering-steel towers of the Luling Bridge near New Orleans, 1983 (Figure 28.6).

The cross sections of steel towers are usually designed as a series of adjoining cells formed by shop-welding steel plates together in units from 6 to 12 m long. The steel towers for a suspension bridge, and for cable-stayed bridges with stays passing over the top of the tower in saddles, must be designed for the concentrated load from the saddles. The steel cellular towers for a cable-stayed bridge with cables framing in the towers must be designed for the local forces from the numerous anchorages of the cables.

Since the 1970s, reinforced concrete has been used in many forms with rectangular and other compact cross sections. Concrete towers are usually designed as hollow shafts to save weight and to reduce the amount of concrete and reinforcing bars required. As with steel towers, concrete towers must

FIGURE 28.5 Delaware Memorial Bridges. (Courtesy of D. Sailors.)

FIGURE 28.6 Luling Bridge, New Orleans, Louisiana. (Courtesy of Charles Seim.)

be designed for the concentrated load from the saddles at the top, if used, or for the local forces from the numerous anchorages of the cables framing into the tower shafts

Towers designed in steel will be lighter than towers designed in concrete, thus giving a potential for savings in foundation costs. Steel towers will generally be more flexible and more ductile and can be erected in less time than concrete towers. Steel towers will require periodic maintenance painting, although weathering steel can be used for nonmarine environments.

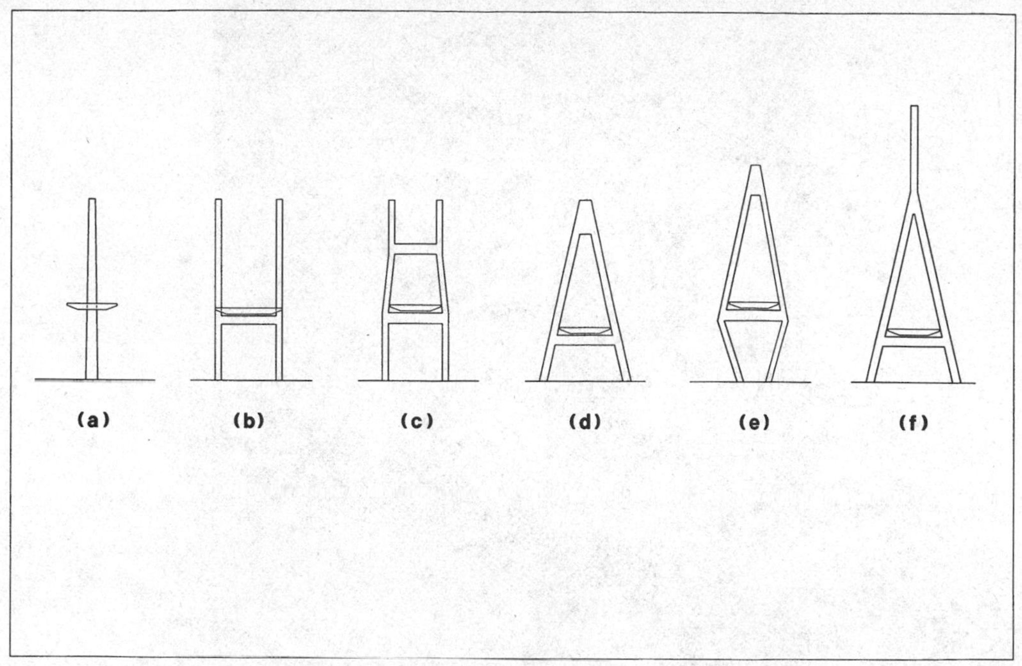

FIGURE 28.7 Generic forms for towers of cable-stayed bridges. (a) Single tower, I; (b) double vertical shafts, H; (c) double cranked shafts; (d) inclined shafts, A; (e) inclined shafts, diamond; (f) inverted Y.

The cost of steel or concrete towers can vary with a number of factors so that market conditions, contractor's experience, equipment availability, and the design details and site-specific influences will most likely determine whether steel or concrete is the most economical material. For pedestrian bridges, timber towers may be economical and aesthetically pleasing.

During the conceptual design phase of the bridge, approximate construction costs of both materials need to be developed and compared. If life-cycle cost is important, then maintenance operations and the frequencies of those operations need to be evaluated and compared, usually by a present-worth evaluation.

28.4.2 Forms and Shapes

Towers of cable-stayed bridges can have a wide variety of shapes and forms. Stay cables can also be arranged in a variety of forms. See Chapter 19. For conceptual design, the height of cable-stayed towers above the deck can be assumed to be about 20% of the main span length. To this value must be added the structural depth of the girder and the clearance to the foundation for determining the approximate total tower height. The final height of the towers will be determined during the final design phase.

The simplest tower form is a single shaft, usually vertical (Figure 28.7a). Occasionally, the single tower is inclined longitudinally. Stay cables can be arranged in a single plane to align with the tower or be splayed outward to connect with longitudinal edge beams. This form is usually employed for bridges with two-way traffic, to avoid splitting a one-way traffic flow. For roadways on curves, the single tower may be offset to the outside of the convex curve of the roadway and inclined transversely to support the curving deck more effectively.

Two vertical shafts straddling the roadway with or without cross struts above the roadway form a simple tower and are used with two planes of cables (Figure 28.7b) The stay cables would incline inward to connect to the girder, introducing a tension component across the deck support system; however, the girders are usually extended outward between the towers to align the cables vertically

FIGURE 28.8 Talmadge Bridge, Georgia. (Courtesy of T. Y. Lin International.)

with the tower shafts. The tower shafts can also be "cranked" or offset above the roadway (Figure 28.7c). This allows the cables to be aligned in a vertical plane and to be attached to the girder, which can pass continuously through the towers as used for the Talmadge Bridge, Georgia (Figure 28.8). A horizontal strut is used between the tower shafts, offset to stabilize the towers.

The two shafts of cable-stayed bridges can be inclined inward toward each other to form a modified A-frame, similar to the Luling Bridge towers (Figure 28.6), or inclined to bring the shafts tops together to form a full A-frame (Figure 28.7d). The two planes of stay cables are inclined outward, producing a more desirable compression component across the deck support system.

The form of the towers of cable-stayed bridge below the roadway is also import for both aesthetics and costs. The shafts of the towers for a modified A-frame can be carried down to the foundations at the same slope as above the roadway, particularly for sites with low clearance. However, at high clearance locations, if the shafts of the towers for a full A-frame or for an inverted Y-frame are carried down to the foundations at the same slope as above the roadway, the foundations may become very wide and costly. The aesthetic proportions also may be affected adversely. Projecting the A-frame shafts downward vertically can give an awkward appearance. Sometimes the lower shafts are inclined inward under the roadway producing a modified diamond (Figure 28.7e), similar to the towers of the Glebe Island Bridge, Sidney, Australia (Figure 28.9). For very high roadways, the inward inclination can form a full diamond or a double diamond as in the Baytown Bridge, Texas (Figure 28.10). For very long spans requiring tall towers, the A-frame can be extended with a single vertical shaft forming an inverted Y shape (Figure 28.7f) as in the Yang Pu Bridge, China

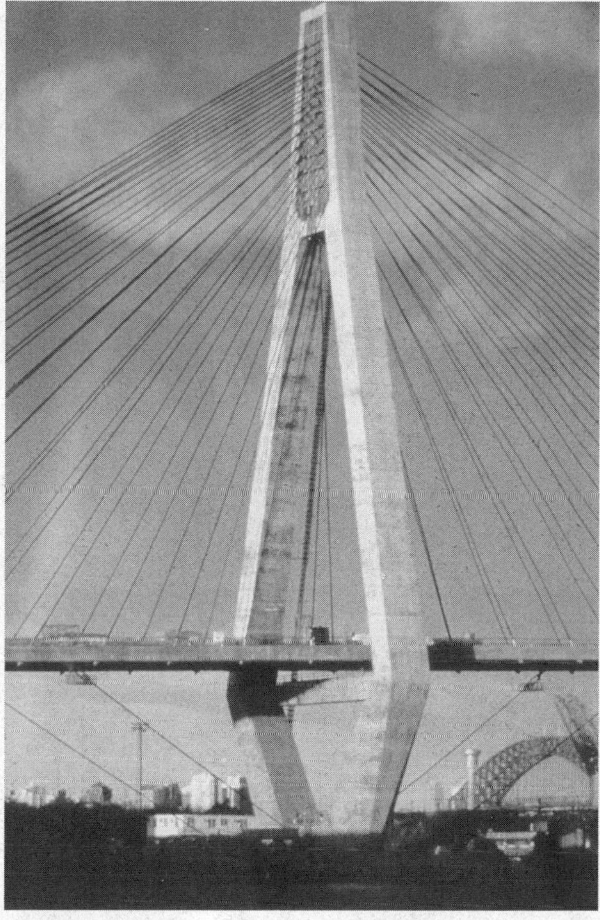

FIGURE 28.9 Glebe Island Bridge, Sidney, Australia. (Courtesy of T. Y. Lin International.)

(Figure 28.11). This form is very effective for very long spans where additional tower height is required and the inclined legs add stiffness and frame action for wind resistance.

The number of shafts or columns within the towers of cable-stayed bridges can vary from one to four. Three-shaft towers generally are not used for cable-stayed bridges except for very wide decks. Four-shaft towers can be used best to support two separate structures instead of a single wide deck. The towers could share a common foundation or each have its own foundation depending on the cost.

Suspension bridges can have from one to four cables depending on structural or architectural needs. Only a few single-cable suspension bridges have been designed with an A or inverted Y form of towers. Usually towers of suspension bridges follow a more traditional design using two vertical shafts and two planes of cables, as illustrated by the steel towers for the Delaware Memorial Bridges (see Figure 28.5). However, concrete towers have recently proved to be economical for some bridges. The very long span (1410 m) Humber Bridge, England, 1983, used uniformly spaced, multi-strut concrete towers (Figure 28.12). The crossing of the Great Belt seaway in Denmark (Figure 28.13), opening in 1999, has concrete towers 254 m high with two struts, one near the midheight and one at the top.

For conceptual designs, the height of suspension bridge towers above the deck depend on the sag-to-span ratio which can vary from about 1:8 to 1:12. A good preliminary value is about 1:10. To this value must be added the structural depth of the deck and the clearance to the foundations to obtain the approximate total tower height. The shafts are usually connected together with several

FIGURE 28.10 Baytown Bridge, Texas. (Courtesy of T. Y. Lin International.)

FIGURE 28.11 Yang Pu Bridge, China. (Courtesy of T. Y. Lin International.)

struts or cross-bracing along the height of the tower, or the shafts are connected at the top with a large single strut. Some form of strut is usually required for suspension bridges as the large cables carry lateral wind and seismic loads to the tops of the tower shafts, which then need to be braced against each other with cross struts to form a tower-frame action.

28.4.3 Erection

During the concept design phase, many different tower forms may be considered, and preliminary designs and cost estimates completed. Each alternative considered should have at least one method

FIGURE 28.12 Humber Bridge, England. (Courtesy of Charles Seim.)

of erection developed during the concept design phase to ensure that the scheme under consideration is feasible to construct. The cost of unusual tower designs can be difficult to estimate and can add significant cost to the project.

28.5 Final Design

The AASHTO Standard Specifications for Highway Bridges [1] apply to bridges 150 m or less in span. For important bridges and for long-span cable-supported bridge projects, special design criteria may have to be developed by the designer. The special design criteria may have to be also developed in cooperation with the owners of the facility to include their operations and maintenance requirements and their bridge-performance expectations after large natural events such as earthquakes. See Chapter 18 for suspension bridge design and Chapter 19 for cable-stayed bridge design. Troitsky [8], Podolny and Salzi [6], and Walther [9] present detailed design theory for cable-stayed bridges.

Design methodology for the towers should follow the same practice as the design methodology for the entire bridge. The towers should be part of a global analysis in which the entire structure is treated as a whole. From the global analyses, the towers can be modeled as a substructure unit with forces and deformations imposed as boundary conditions.

Detailed structural analyses form the basis for the final design of the tower and its components and connections. Both cabled-stayed and suspension bridges are highly indeterminate and require careful analysis in at least a geometric nonlinear program.

28.5.1 Design Loads

The towers are subject to many different loading cases. The towers, as well as the entire structure, must be analyzed, designed, and checked for the controlling loading cases. Chapter 6 presents a detailed discussion of bridge loading.

The weight of the superstructure, including the self-weight of the towers, is obtained in the design process utilizing the unit weights of the materials used in the superstructure and distributed to the tower in accordance with a structural analysis of the completed structure or by the erection equipment during the construction phases.

FIGURE 28.13 Great Belt Bridge, Denmark. (Courtesy of Ben C. Gerwick, Inc.)

Loads from traffic using the bridge such as trains, transit, trucks, or pedestrians are usually prescribed in design codes and specifications or by the owners of the facility. These are loads moving across the bridge and the forces imparted to the towers must be obtained from a structural analysis that considers the moving loading. These are all gravity effects that act downward on the structure, but will induce both vertical and horizontal forces on the towers.

A current trend for spanning wide widths of waterways is to design multispan bridges linked together to form a long, continuous structure. With ordinary tower designs, the multispan cable-stayed girders will deflect excessively under live loads as the towers will not be sufficiently stiffened by the cable stays anchored within the flexible adjacent spans. For multispan suspension bridges with ordinary tower designs, the same excessive live-load deflection can also occur. Towers for multispan cable-supported bridges must be designed to be sufficiently rigid to control live-load deflections.

Towers are also subject to temperature-induced displacements, both from the superstructure and cable framing into the towers, and from the temperature-induced movement of the tower itself. Towers can expand and contract differentially along the tower height from the sun shining on them from morning until sunset. These temperature effects can cause deflection and torsional twisting along the height of the tower.

Wind blowing on the towers as a bluff shape induces forces and displacements in the tower. Forces will be induced into the cables by the pressure of wind on the superstructure, as well as by the wind forces on the cables themselves. These additional forces will be carried to the towers.

For long-span bridges and for locations with known high wind speeds, wind should be treated as a dynamic loading. This usually requires a wind tunnel test on a sectional model of the superstructure in a wind tunnel and, for important bridges, an aeroelastic model in a large wind tunnel. See Chapter 57. Under certain wind flows, the wind can also excite the tower itself, particularly if the tower is designed with light steel components. In the rare instances in which wind-induced excitation of the tower does occur, appropriate changes in the cross section of the tower can be made or a faring can be added to change the dynamic characteristics of the tower.

Forces and deformations of long-span structures from earthquakes are discussed in Chapter 40. The seismic excitation should be treated as dynamic inertia loadings inducing response within the structure by exciting the vibrational modes of the towers. Induced seismic forces and displacement can control the design of towers in locations with high seismic activity. For locations with lower seismic activity, the tower design should be checked at least for code-prescribed seismic loadings. The dynamic analysis of bridges is discussed in Chapter 34.

A full analysis of the structure will reveal all of the forces, displacements, and other design requirements for all loading cases for the final tower design.

28.5.2 Design Considerations

Suspension bridge cables pass over cable saddles that are usually anchored to the top of the tower. A cable produces a large vertical force and smaller, but important, transverse and longitudinal forces from temperature, wind, earthquake, or from the unbalanced cable forces between main and side spans. These forces are transmitted through the cable saddle anchorage at each cable location to the top of the tower. The towers and the permanent saddle anchorages must be designed to resist these cable forces.

The erection of a suspension bridge must be analyzed and the sequence shown on the construction plans. To induce the correct loading into the cables of the side span, the erection sequence usually requires that the saddles be displaced toward the side spans. This is usually accomplished for short spans by displacing the tops of the towers by pulling with heavy cables. For long spans, the saddles can be displaced temporarily on rollers. As the stiffening deck elements are being erected into position and the cable begins to take loads, the towers or saddles are gradually brought into final vertical alignment. After the erection of the stiffening deck elements are completed, the saddles are permanently fastened into position to take the unbalanced cable loads from the center and the side spans.

At the deck level, other forces may be imposed on the tower from the box girder or stiffening truss carrying the roadway. These forces depend on the structural framing of the connection of the deck and tower. Traditional suspension bridge designs usually terminate the stiffening truss or box girder at the towers, which produces transverse, and longitudinal, forces on the tower at this point. Contemporary suspension bridge designs usually provide for passing a box girder continuously through the tower opening which may produce transverse forces but not longitudinal forces. For this arrangement, the longitudinal forces must be carried by the girder to the abutments.

The most critical area of the tower design is the tower-to-foundation connection. Both shear forces and moments are maximum at this point. Anchor bolts are generally used at the base of steel towers. The bolts must be proportioned to transfer the loads from the tower to the bolts. The bolts must be deeply embedded in the concrete footing block to transfer their loads to the footing reinforcement. Providing good drainage for the rainwater running down the tower shafts will increase the life of the steel paint system at the tower base and provide some protection to the anchor bolts.

Concrete towers must be joined to the foundations with full shear and moment connections. Lapped reinforcing bars splices are usually avoided as the lapping tends to congest the connections, the strength of the bars cannot be developed, and lapped splices cannot be used for high seismic areas. Using compact mechanical or welded splices will result in less congestion with easier placement of concrete around the reinforcement and a more robust tower-to-footing connection.

Careful coordination between the foundation designers and tower designers is required to achieve a stable, efficient, and reliable connection.

The cable arrangements for cable-stayed bridges are many and varied. Some arrangements terminate the cables in the tower, whereas other arrangements pass the cable through the tower on cable saddles. Cables terminating in the tower can pass completely through the tower cross section and then anchor on the far side of the tower. This method of anchoring produces compression in the tower cross section at these anchorage points. Cables can also be terminated at anchors within the walls of the tower, producing tension in the tower cross section at the anchorage points. These tension forces require special designs to provide reliable, long-life support for the cables.

Just as for suspension bridges, the erection of cable-stayed bridges must be analyzed and the sequence shown on the construction plans. The girders, as they are erected outward from the towers, are very vulnerable. The critical erection sequence is just before closing the two arms of the girders at the center of the span. High winds can displace the arms and torque the towers, and heavy construction equipment can load the arms without benefit of girder continuity to distribute the loads.

28.6 Construction

Towers constructed of structural steel are usually fabricated in a shop by welding together steel plates and rolled shapes to form cells. Cells must be large enough to allow welders and welding equipment, and if the steel is to be painted, painters and cleaning and painting equipment inside each cell.

The steel tower components are transported to the bridge site and then erected by cranes and bolted together with high-strength bolts. The contractor should use a method of tensioning the high-strength bolts to give constant results and achieve the required tension. Occasionally, field welding is used, but this presents difficulties in holding the component rigidly in position while the weld is completed. Field welding can be difficult to control in poor weather conditions to achieve ductile welds, particularly for vertical and overhead welds. Full-penetration welds require backup bars that must be removed carefully if the weld is subject to fatigue loading.

Towers constructed of reinforced concrete are usually cast in forms that are removed and reused, or jumped to the next level. Concrete placing heights are usually restricted to about 6 to 12 m to limit form pressure from the freshly placed concrete. Reinforcing bar cages are usually preassembled on the ground or on a work barge, and lifted into position by crane. This requires the main load-carrying reinforcing bars to be spliced with each lift. Lapped splices are the easiest to make, but are not allowed in seismic areas.

Slip forming is an alternative method that uses forms that are pulled slowly upward, reinforcing bars positioned and the concrete placed in one continuous operation around the clock until the tower is completed. Slip forming can be economical, particularly for constant-cross-section towers. Some changes in cross section geometry can be accommodated. For shorter spans, precast concrete segments can be stacked together and steel tendons tensioned to form the towers.

Tower designers should consider the method of erection that contractors may use in constructing the towers. Often the design can reduce construction costs by incorporating more easily fabricated and assembled steel components or assembled reinforcing bar cages and tower shapes that are easily formed. Of course, the tower design cannot be compromised just to lower erection costs.

Some engineers and many architects design towers that are not vertical but are angled longitudinally toward or away from the main span. This can be done if such a design can be justified structurally and aesthetically, and the extra cost can be covered within the project budget. The difficulties of the design of longitudinally inclined towers must be carefully considered as well as the more expensive and slower erection, which will create additional costs.

Many towers of cable-stayed bridges have legs sloped toward each to form an A, an inverted Y, a diamond, or similar shapes. These are not as difficult to construct as the longitudinally inclined tower design. The sloping concrete forms can be supported by vertical temporary supports and cross struts that tie the concrete forms together. This arrangement braces the partly cast concrete tower legs against each other for support. Some of the concrete form supports for the double-diamond towers of the Baytown Bridge are visible in Figure 28.9.

As the sloped legs are erected, the inclination may induce bending moments and lateral deflection in the plane of the slope of the legs. Both of these secondary effects must be adjusted by jacking the legs apart by a calculated amount of force or displacement to release the locked-in bending stresses. If the amount of secondary stress is small, then cambering the leg to compensate for the deflection and adding material to lower the induced stress can be used.

The jacking procedure adds cost but is an essential step in the tower erection. Neglecting this important construction detail can "lock-in" stresses and deflections that will lower the factor of safety of the tower and, in an extreme case, could cause a failure.

Tower construction usually requires special equipment to erect steel components or concrete forms to the extreme height of the tower. Suspension bridges and some cable-stayed bridges require cable saddles to be erected on the tower tops. Floating cranes rarely have the capacity to reach to the heights of towers designed for long spans. Tower cranes, connected to the tower as it is erected, can be employed for most tower designs and are a good choice for handling steel forms for the erection of concrete towers. A tower crane used to jump the forms and raise materials can be seen in Figure 28.9. Occasionally, vertical traveling cranes are used to erect steel towers by pulling themselves up the face of the tower following the erection of each new tower component.

The erection sequence for a suspension bridge may require that the towers be pulled by cables from the vertical toward the sides spans or that the cable saddles be placed on rollers and displaced toward the side spans on temporary supports. The tower restraints are gradually released or the rollers pushed toward their final position as the erection of the deck element nears completion. This operation is usually required to induce the design forces into the cables in the side spans. The cable saddles then are permanently anchored to the towers.

Because the tower erection must be done in stages, each stage must be checked for stability and for stresses and deflections. The specifications should require the tower erection to be checked by an engineer, employed by the contractor, for stability and safety at each erection stage. The construction specifications should also require the tower erection stages to be submitted to the design engineer for an evaluation. This evaluation should be thorough enough to determine if the proposed tower erection staging will meet the intent of the original design, or if it needs to be modified to bring the completed tower into compliance.

28.7 Summary

Towers provide the visible means of support of the roadway on which goods and people travel. Being the most visible elements in a bridge, they give the bridge, for good or for ill, its character, its motif, and its identifying aesthetic impression. Towers usually form structural portals through which people pass as they travel from one point to another. Of themselves, towers form an aesthetic structural statement.

Towers are the most critical structural element in the bridge as their function is to carry the forces imposed on the bridge to the ground. Unlike most other bridge components, they cannot be replaced during the life of the bridge. Towers must fulfill their function in a reliable, serviceable, economical, and aesthetic manner for the entire life of the bridge. Towers must also be practicable to erect without extraordinary expense.

Practicable tower shapes for cable-stayed bridges are many and varied. Towers can have one or several legs or shafts arrayed from vertical to inclined and forming A- or inverted Y-shaped frames. Suspension bridge towers are usually vertical, with two shafts connected with one or several struts.

The conceptual design is the most important phase in the design of a long-span bridge. This phase sets, among other items, the span length, type of deck system, and the materials and shape of the towers. It also determines the aesthetic, economics, and constructibility of the bridge. A conceptual erection scheme should be developed during this phase to ensure that the bridge can be economically constructed.

The final design phase sets the specific shape, dimensions, and materials for the bridge. A practical erection method should be developed during this phase and shown on the construction drawings. If an unusual tower design is used, the tower erection should also be shown. The specifications should allow the contractor to employ an alternative method of erection, provided that the method is designed by an engineer and submitted to the design engineer for review. It is essential that the design engineer follow the project into the construction stages. The designer must understand each erection step that is submitted by the contractor in accordance with the specifications, to ensure the construction complies with the design documents. Only by this means are owners assured that the serviceability and reliability that they are paying for are actually achieved in construction.

The successful design of a cable-stayed or a suspension bridge involves many factors and decisions that must be made during the planning, design, and construction phases of the project. Towers play an important role in that successful execution. The final judgment of a successful project is made by the people who use the facility and pay for its construction, maintenance, and long-life service to society.

References

1. AASHTO, *Standard Specifications for Highway Bridges,* American Association of State Highway and Transportation Officials, Washington, D.C., 1994.
2. Billington, D. P., *The Tower and the Bridge, The New Art of Structural Engineering,* Basic Books, New York, 1983.
3. Cerver, F. A., *New Architecture in Bridges,* Muntaner, Spain, 1992.
4. Gimsing, N. J., *Cable-Supported Bridges — Concept and Design,* John Wiley & Sons, New York, 1997.
5. Leonhardt, F., *Bridges, Aesthetics and Design,* MIT Press, Cambridge, MA, 1984.
6. Podolny, W. and Scalzi, J. B., *Construction and Design of Cable Stayed Bridges,* 2nd ed., John Wiley & Sons, New York, 1986.
7. Seim, C., San Francisco Bay's jeweled necklace, *ASCE Civil Eng.,* 66(1), 14A, 1996.
8. Troitsky, M. S., *Cable Stayed Bridges,* Van Nostrand Reinhold, 1988.
9. Walter, R., *Cable Stayed Bridges,* Thomas Telford, U.K., 1988.

29

Abutments and Retaining Structures

Linan Wang
California Transportation Department

Chao Gong
ICF Kaiser Engineers, Inc.

29.1 Introduction

As a component of a bridge, the abutment provides the vertical support to the bridge superstructure at the bridge ends, connects the bridge with the approach roadway, and retains the roadway base materials from the bridge spans. Although there are numerous types of abutments and the abutments for the important bridges may be extremely complicated, the analysis principles and design methods are very similar. In this chapter the topics related to the design of conventional highway bridge abutments are discussed and a design example is illustrated.

Unlike the bridge abutment, the earth-retaining structures are mainly designed for sustaining lateral earth pressures. Those structures have been widely used in highway construction. In this chapter several types of retaining structures are presented and a design example is also given.

29.2 Abutments

29.2.1 Abutment Types

Open-End and Closed-End Abutments

From the view of the relation between the bridge abutment and roadway or water flow that the bridge overcrosses, bridge abutments can be divided into two categories: open-end abutment, and closed-end abutment, as shown in Figure 29.1.

For the open-end abutment, there are slopes between the bridge abutment face and the edge of the roadway or river canal that the bridge overcrosses. Those slopes provide a wide open area for the traffic flows or water flows under the bridge. It imposes much less impact on the environment

0-8493-7434-0/00/$0.00+$.50

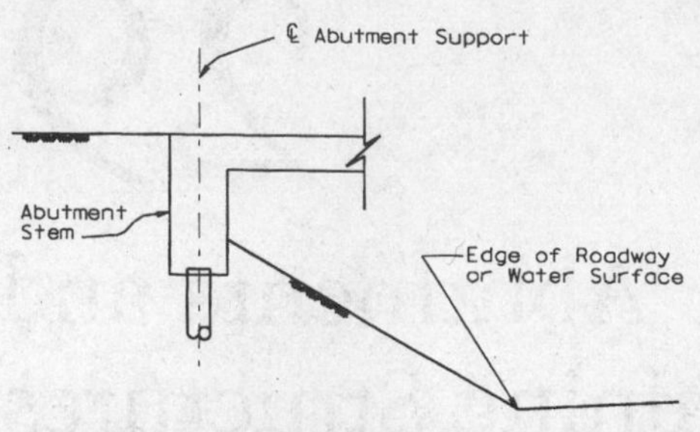

(a) Open End, Monolithic Type Abutment

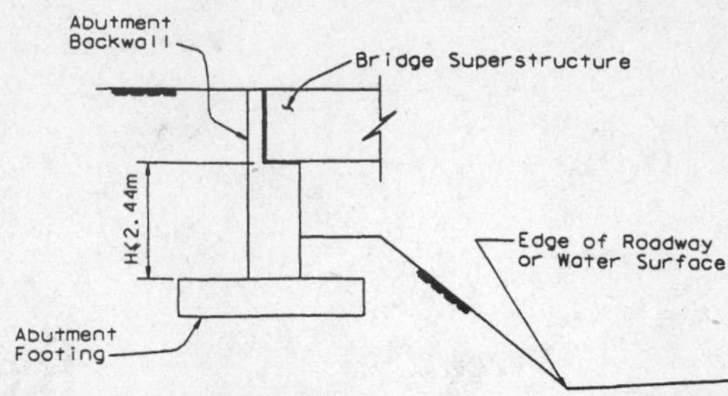

(b) Open End, Short Stem Seat Type Abutment

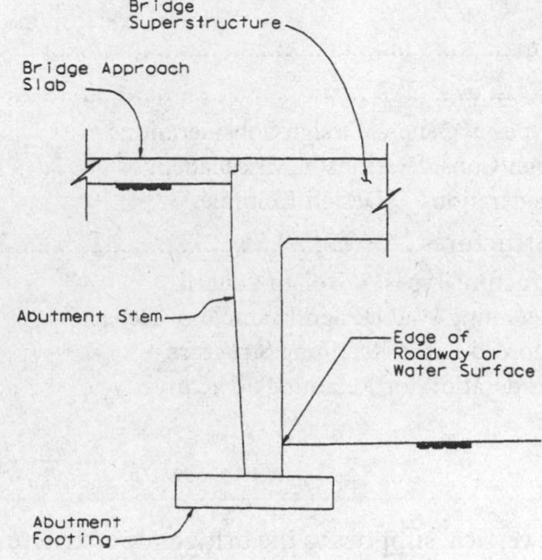

(c) Close End, Monolithic Type Abutment

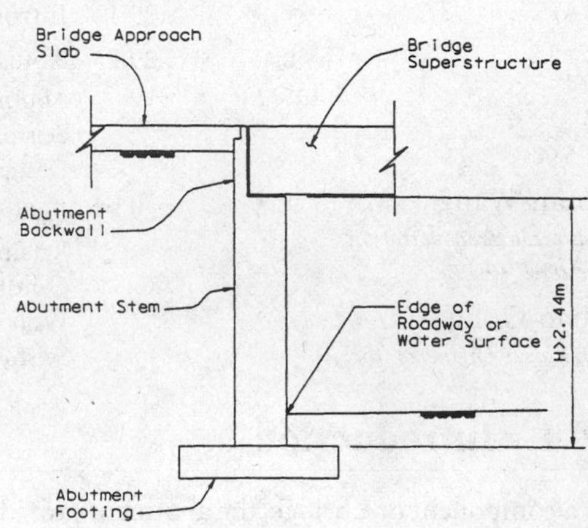

(d) Close End, High Stem Seat Type Abutment

FIGURE 29.1 Typical abutment types.

and the traffic flows under the bridge than a closed-end abutment. Also, future widening of the roadway or water flow canal under the bridge by adjusting the slope ratios is easier. However, the existence of slopes usually requires longer bridge spans and some extra earthwork. This may result in an increase in the bridge construction cost.

The closed-end abutment is usually constructed close to the edge of the roadways or water canals. Because of the vertical clearance requirements and the restrictions of construction right of way, there are no slopes allowed to be constructed between the bridge abutment face and the edge of roadways or water canals, and high abutment walls must be constructed. Since there is no room or only a little room between the abutment and the edge of traffic or water flow, it is very difficult to do the future widening to the roadways and water flow under the bridge. Also, the high abutment walls and larger backfill volume often result in higher abutment construction costs and more settlement of road approaches than for the open-end abutment.

Generally, the open-end abutments are more economical, adaptable, and attractive than the closed-end abutments. However, bridges with closed-end abutments have been widely constructed in urban areas and for rail transportation systems because of the right-of-way restriction and the large scale of the live load for trains, which usually results in shorter bridge spans.

Monolithic and Seat-Type Abutments

Based on the connections between the abutment stem and the bridge superstructure, the abutments also can be grouped in two categories: the monolithic or end diaphragm abutment and the seat-type abutment, as shown in Figure 29.1.

The monolithic abutment is monolithically constructed with the bridge superstructure. There is no relative displacement allowed between the bridge superstructure and abutment. All the superstructure forces at the bridge ends are transferred to the abutment stem and then to the abutment backfill soil and footings. The advantages of this type of abutment are its initial lower construction cost and its immediate engagement of backfill soil that absorbs the energy when the bridge is subjected to transitional movement. However, the passive soil pressure induced by the backfill soil could result in a difficult-to-design abutment stem, and higher maintenance cost might be expected. In the practice this type of abutment is mainly constructed for short bridges.

The seat-type abutment is constructed separately from the bridge superstructure. The bridge superstructure seats on the abutment stem through bearing pads, rock bearings, or other devices. This type of abutment allows the bridge designer to control the superstructure forces that are to be transferred to the abutment stem and backfill soil. By adjusting the devices between the bridge superstructure and abutment the bridge displacement can be controlled. This type of abutment may have a short stem or high stem, as shown in Figure 29.1. For a short-stem abutment, the abutment stiffness usually is much larger than the connection devices between the superstructure and the abutment. Therefore, those devices can be treated as boundary conditions in the bridge analysis. Comparatively, the high stem abutment may be subject to significant displacement under relatively less force. The stiffness of the high stem abutment and the response of the surrounding soil may have to be considered in the bridge analysis. The availability of the displacement of connection devices, the allowance of the superstructure shrinkage, and concrete shortening make this type of abutment widely selected for the long bridge constructions, especially for prestressed concrete bridges and steel bridges. However, bridge design practice shows that the relative weak connection devices between the superstructure and the abutment usually require the adjacent columns to be specially designed. Although the seat-type abutment has relatively higher initial construction cost than the monolithic abutment, its maintenance cost is relatively lower.

Abutment Type Selection

The selection of an abutment type needs to consider all available information and bridge design requirements. Those may include bridge geometry, roadway and riverbank requirements, geotechnical and right-of-way restrictions, aesthetic requirements, economic considerations, etc. Knowledge of the advantages and disadvantages for the different types of abutments will greatly benefit the bridge designer in choosing the right type of abutment for the bridge structure from the beginning stage of the bridge design.

29.2.2 General Design Considerations

Abutment design loads usually include vertical and horizontal loads from the bridge superstructure, vertical and lateral soil pressures, abutment gravity load, and the live-load surcharge on the abutment backfill materials. An abutment should be designed so as to withstand damage from the Earth pressure, the gravity loads of the bridge superstructure and abutment, live load on the superstructure or the approach fill, wind loads, and the transitional loads transferred through the connections between the superstructure and the abutment. Any possible combinations of those forces, which produce the most severe condition of loading, should be investigated in abutment design. Meanwhile, for the integral abutment or monolithic type of abutment the effects of bridge superstructure deformations, including bridge thermal movements, to the bridge approach structures must be

TABLE 29.1 Abutment Design Loads (Service Load Design)

Abutment Design Loads	Case I	II	III	IV	V
Dead load of superstructure	X	X	—	X	X
Dead load of wall and footing	X	X	X	X	X
Dead load of earth on heel of wall including surcharge	X	X	X	X	—
Dead load of earth on toe of wall	X	X	X	X	—
Earth pressure on rear of wall including surcharge	X	X	X	X	—
Live load on superstructure	X	—	—	X	—
Temperature and shrinkage	—	—	—	X	—
Allowable pile capacity of allowable soil pressure in % or basic	100	100	150	125	150

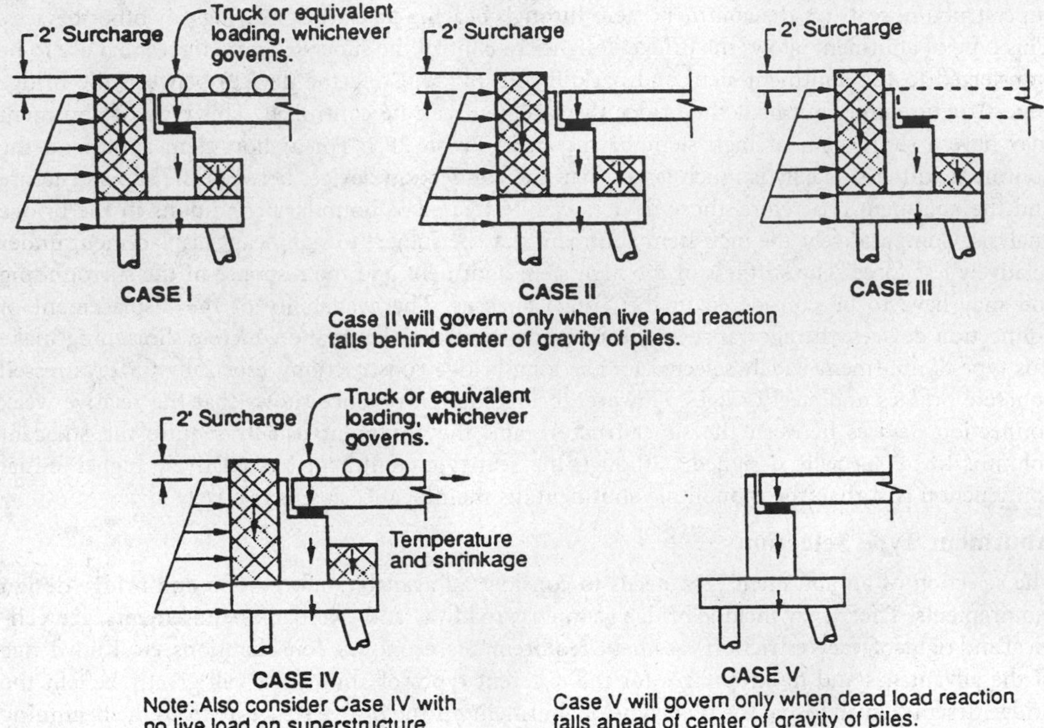

FIGURE 29.2 Configuration of abutment design load and load combinations.

considered in abutment design. Nonseismic design loads at service level and their combinations are shown in Table 29.1 and Figure 29.2. It is easy to obtain the factored abutment design loads and load combinations by multiplying the load factors to the loads at service levels. Under seismic loading, the abutment may be designed at no support loss to the bridge superstructure while the abutment may suffer some damages during a major earthquake.

The current AASHTO Bridge Design Specifications recommend that either the service load design or the load factor design method be used to perform an abutment design. However, due to the uncertainties in evaluating the soil response to static, cycling, dynamic, and seismic loading, the service load design method is usually used for abutment stability checks and the load factor method is used for the design of abutment components.

The load and load combinations listed in Table 29.1 may cause abutment sliding, overturning, and bearing failures. Those stability characteristics of abutment must be checked to satisfy certain restrictions. For the abutment with spread footings under service load, the factor of safety to resist sliding should be greater than 1.5; the factor of safety to resist overturning should be greater than 2.0; the factor of safety against soil bearing failure should be greater than 3.0. For the abutment with pile support, the piles have to be designed to resist the forces that cause abutment sliding, overturning, and bearing failure. The pile design may utilize either the service load design method or the load factor design method.

The abutment deep shear failure also needs to be studied in abutment design. Usually, the potential of this kind of failure is pointed out in the geotechnical report to the bridge designers. Deep pilings or relocating the abutment may be used to avoid this kind of failure.

29.2.3 Seismic Design Considerations

Investigations of past earthquake damage to the bridges reveal that there are commonly two types of abutment earthquake damage — stability damage and component damage.

Abutment stability damage during an earthquake is mainly caused by foundation failure due to excessive ground deformation or the loss of bearing capacities of the foundation soil. Those foundation failures result in the abutment suffering tilting, sliding, settling, and overturning. The foundation soil failure usually occurs because of poor soil conditions, such as soft soil, and the existence of a high water table. In order to avoid these kinds of soil failures during an earthquake, borrowing backfill soil, pile foundations, a high degree of soil compaction, pervious materials, and drainage systems may be considered in the design.

Abutment component damage is generally caused by excessive soil pressure, which is mobilized by the large relative displacement between the abutment and its backfilled soil. Those excessive pressures may cause severe damage to abutment components such as abutment back walls and abutment wingwalls. However, the abutment component damages do not usually cause the bridge superstructure to lose support at the abutment and they are repairable. This may allow the bridge designer to utilize the deformation of abutment backfill soil under seismic forces to dissipate the seismic energy to avoid the bridge losing support at columns under a major earthquake strike.

The behavior of abutment backfill soil deformed under seismic load is very efficient at dissipating the seismic energy, especially for the bridges with total length of less than 300 ft (91.5 m) with no hinge, no skew, or that are only slightly skewed (i.e., <15°). The tests and analysis revealed that if the abutments are capable of mobilizing the backfill soil and are well tied into the backfill soil, a damping ratio in the range of 10 to 15% is justified. This will elongate the bridge period and may reduce the ductility demand on the bridge columns. For short bridges, a damping reduction factor, *D*, may be applied to the forces and displacement obtained from bridge elastic analysis which generally have damped ARS curves at 5% levels. This factor *D* is given in Eq. (29.1).

$$D = \frac{1.5}{40\,C + 1} + 0.5 \tag{29.1}$$

where C = damping ratio.

Based on Eq. (29.1), for 10% damping, a factor $D = 0.8$ may be applied to the elastic force and displacement. For 15% damping, a factor $D = 0.7$ may be applied. Generally, the reduction factor D should be applied to the forces corresponding to the bridge shake mode that shows the abutment being excited.

The responses of abutment backfill soil to the seismic load are very difficult to predict. The study and tests revealed that the soil forces, which are applied to bridge abutment under seismic load, mainly depend on the abutment movement direction and magnitude. In the design practice, the Mononobe–Okabe method usually is used to quantify those loads for the abutment with no restraints on the top. Recently, the "near full scale" abutment tests performed at the University of California at Davis show a nonlinear relationship between the abutment displacement and the backfill soil reactions under certain seismic loading when the abutment moves toward its backfill soil. This relation was plotted as shown in Figure 29.3. It is difficult to simulate this nonlinear relationship between the abutment displacement and the backfill soil reactions while performing bridge dynamic analysis. However, the tests concluded an upper limit for the backfill soil reaction on the abutment. In design practice, a peak soil pressure acting on the abutment may be predicted corresponding to certain abutment displacements. Based on the tests and investigations of past earthquake damages, the California Transportation Department suggests guidelines for bridge analysis considering abutment damping behavior as follows.

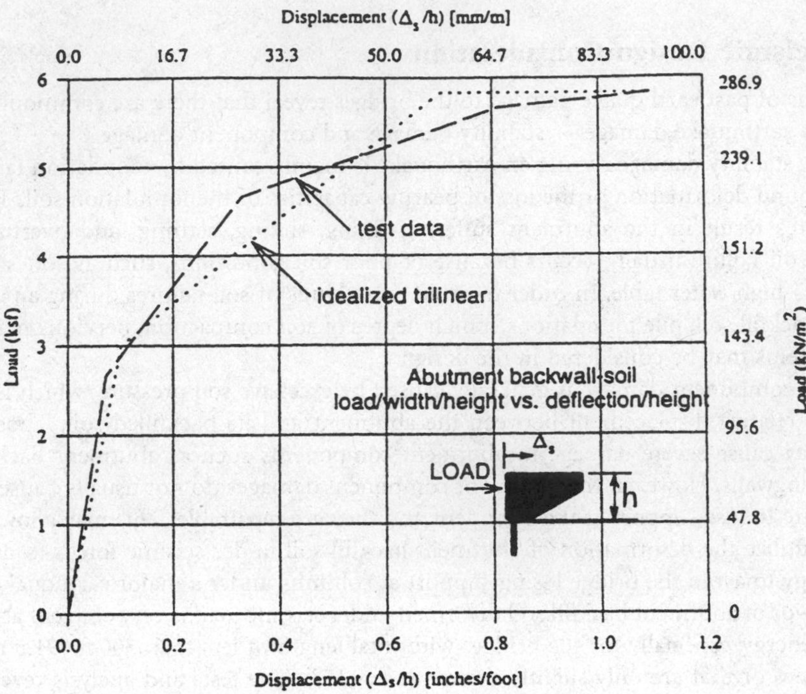

FIGURE 29.3 Proposed characteristics and experimental envelope for abutment backfill load–deformation.

By using the peak abutment force and the effective area of the mobilized soil wedge, the peak soil pressure is compared to a maximum capacity of 7.7 ksf (0.3687 MPa). If the peak soil pressure exceeds the soil capacity, the analysis should be repeated with reduced abutment stiffness. It is important to note that the 7.7 ksf (0.3687 MPa) soil pressure is based on a reliable minimum wall height of 8 ft (2.438 m). If the wall height is less than 8 ft (2.438 m), or if the wall is expected to shear off at a depth below the roadway less than 8 ft (2.438 m), the allowable passive soil pressure must be reduced by multiplying 7.7 ksf (0.3687 MPa) times the ratio of (L/8) [2], where L is the effective height of the abutment wall in feet. Furthermore, the shear capacity of the abutment wall diaphragm (the structural member mobilizing the soil wedge) should be compared with the demand shear forces to ensure the soil mobilizations. Abutment spring displacement is then evaluated against an acceptable level of displacement of 0.2 ft (61 mm). For a monolithic-type abutment this displacement is equal to the bridge superstructure displacement. For seat-type abutments this displacement usually does not equal the bridge superstructure displacement, which may include the gap between the bridge superstructure and abutment backwall. However, a net displacement of about 0.2 ft (61 mm) at the abutment should not be exceeded. Field investigations after the 1971 San Fernando earthquake revealed that the abutment, which moved up to 0.2 ft (61 mm) in the longitudinal direction into the backfill soil, appeared to survive with little need for repair. The abutments in which the backwall breaks off before other abutment damage may also be satisfactory if a reasonable load path can be provided to adjacent bents and no collapse potential is indicated.

For seismic loads in the transverse direction, the same general principles still apply. The 0.2-ft (61-mm) displacement limit also applies in the transverse direction, if the abutment stiffness is expected to be maintained. Usually, wingwalls are tied to the abutment to stiffen the bridge transversely. The lateral resistance of the wingwall depends on the soil mass that may be mobilized by the wingwall. For a wingwall with the soil sloped away from the exterior face, little lateral resistance can be predicted. In order to increase the transverse resistance of the abutment, interior supplemental shear walls may be attached to the abutment or the wingwall thickness may be increased, as shown in Figure 29.4. In some situations larger deflection may be satisfactory if a reasonable load path can be provided to adjacent bents and no collapse potential is indicated. [2]

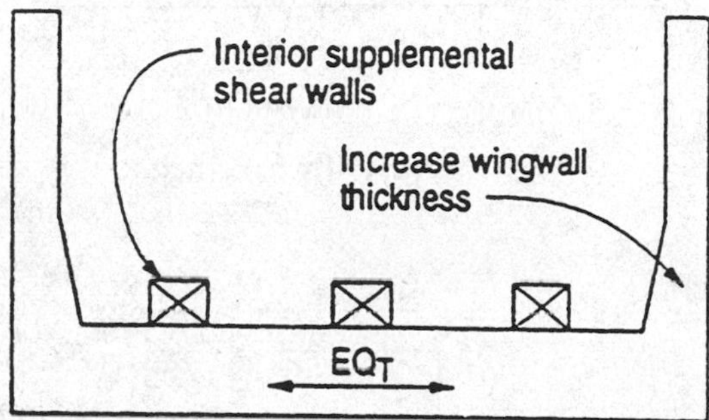

FIGURE 29.4 Abutment transverse enhancement.

Based on the above guidelines, abutment analysis can be carried out more realistically by a trial-and-error method on abutment soil springs. The criterion for abutment seismic resistance design may be set as follows.

Monolithic Abutment or Diaphragm Abutment (Figure 29.5)

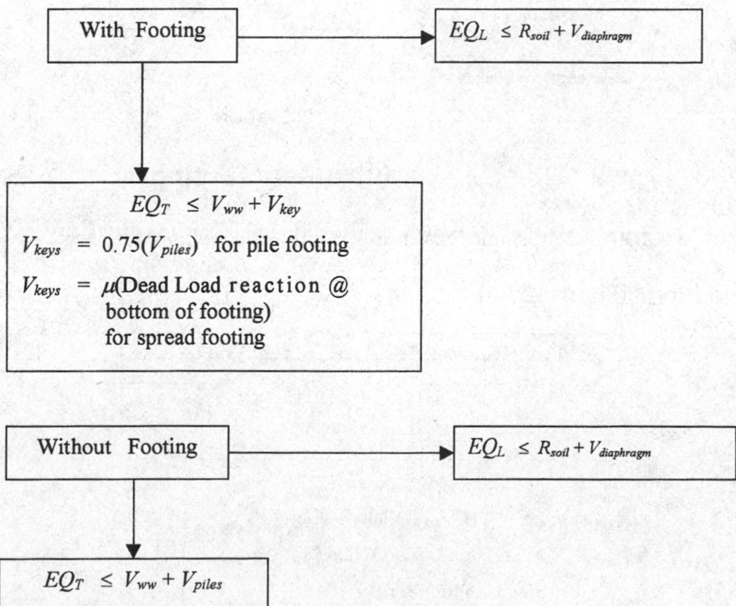

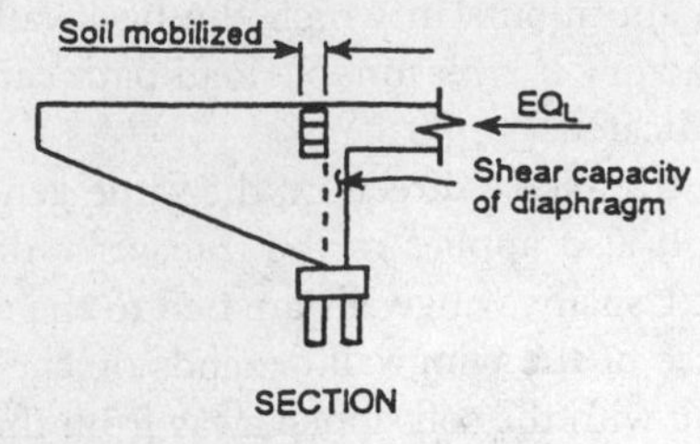

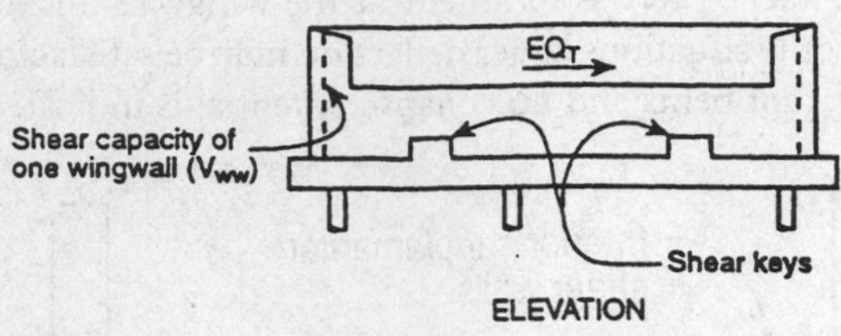

(a)with footing

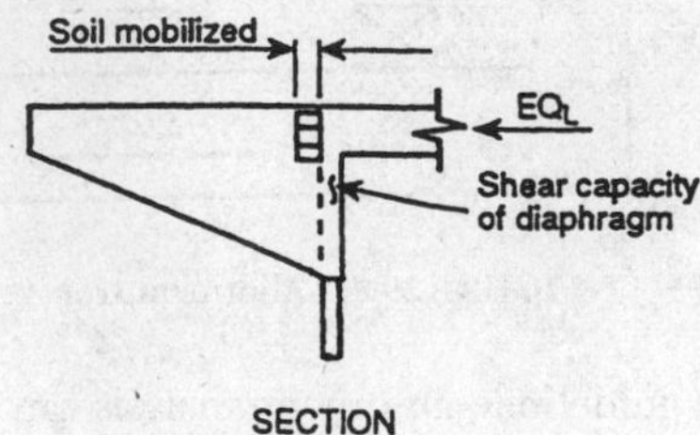

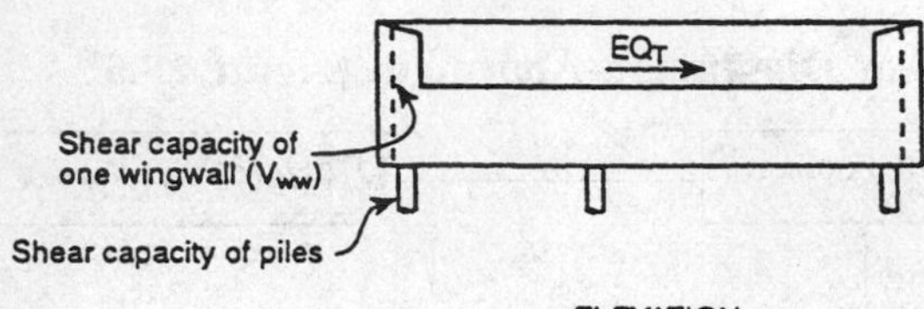

(b)without footing

FIGURE 29.5 Seismic resistance elements for monolithic abutment.

Seat-Type Abutment (Figure 29.6)

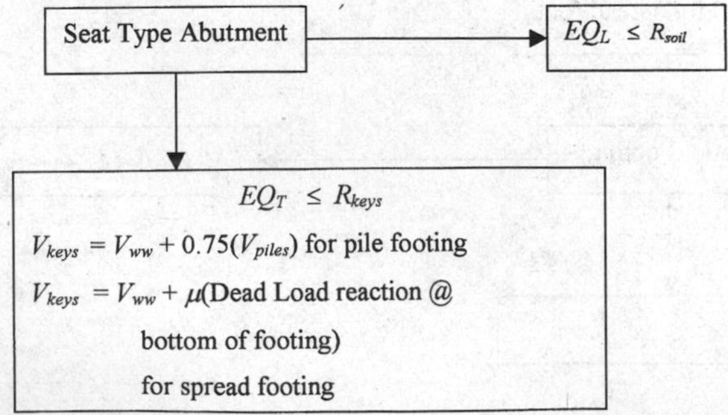

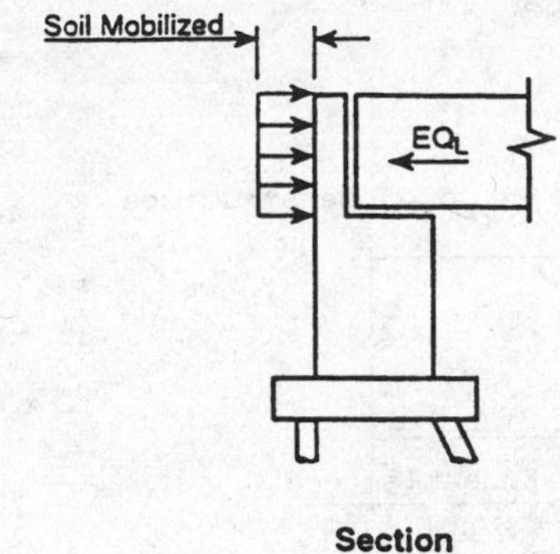

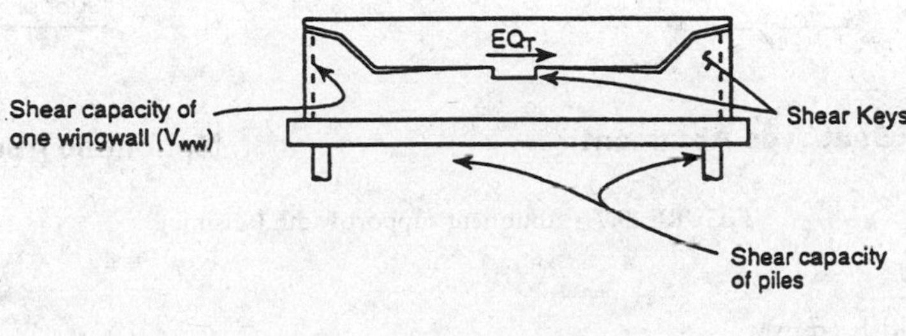

FIGURE 29.6 Seismic resistance elements for seat-type abutment.

where

EQ_L = longitudinal earthquake force from an elastic analysis
EQ_T = transverse earthquake force from an elastic analysis
R_{soil} = resistance of soil mobilized behind abutment
$R_{diaphragm}$ = φ times the nominal shear strength of the diaphragm
R_{ww} = φ times the nominal shear strength of the wingwall
R_{piles} = φ times the nominal shear strength of the piles
R_{keys} = φ times the nominal shear strength of the keys in the direction of consideration
φ = strength factor for seismic loading
μ = coefficient factor between soil and concrete face at abutment bottom

It is noted that the purpose of applying a factor of 0.75 to the design of shear keys is to reduce the possible damage to the abutment piles. For all transverse cases, if the design transverse earthquake force exceeds the sum of the capacities of the wingwalls and piles, the transverse stiffness for the analysis should equal zero ($EQ_T = 0$). Therefore, a released condition which usually results in larger lateral forces at adjacent bents should be studied.

Responding to seismic load, bridges usually accommodate a large displacement. To provide support at abutments for a bridge with large displacement, enough support width at the abutment must be designed. The minimum abutment support width, as shown in Figure 29.7, may be equal to the bridge displacement resulting from a seismic elastic analysis or be calculated as shown in Equation (29-2), whichever is larger:

$$N = (305 + 2.5L + 10H)(1 + 0.002\ S^2) \qquad (29.2)$$

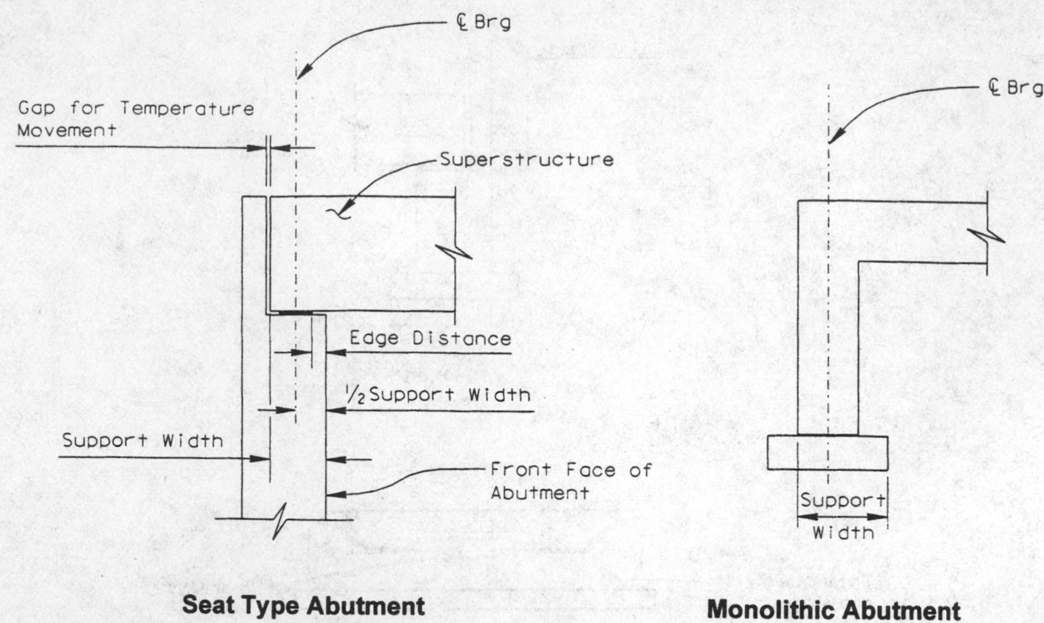

FIGURE 29.7 Abutment support width (seismic).

where

N = support width (mm)
L = length (m) of the bridge deck to the adjacent expansion joint, or to the end of bridge deck; for single-span bridges L equals the length of the bridge deck
S = angle of skew at abutment in degrees
H = average height (m) of columns or piers supporting the bridge deck from the abutment to the adjacent expansion joint, or to the end of the bridge deck; $H = 0$ for simple span bridges

29.2.4 Miscellaneous Design Considerations

Abutment Wingwall

Abutment wingwalls act as a retaining structure to prevent the abutment backfill soil and the roadway soil from sliding transversely. Several types of wingwall for highway bridges are shown in Figure 29.8. A wingwall design similar to the retaining wall design is presented in Section 29.3. However, live-load surcharge needs to be considered in wingwall design. Table 29.2 lists the live-load surcharge for different loading cases. Figure 29.9 shows the design loads for a conventional cantilever wingwall. For seismic design, the criteria in transverse direction discussed in Section 29.2.3 should be followed. Bridge wingwalls may be designed to sustain some damage in a major earthquake, as long as bridge collapse is not predicted.

Abutment Drainage

A drainage system is usually provided for the abutment construction. The drainage system embedded in the abutment backfill soil is designed to reduce the possible buildup of hydrostatic pressure, to control erosion of the roadway embankment, and to reduce the possibility of soil liquefaction during an earthquake. For a concrete-paved abutment slope, a drainage system also needs to be provided under the pavement. The drainage system may include pervious materials, PSP or PVC pipes, weep holes, etc. Figure 29.10 shows a typical drainage system for highway bridge construction.

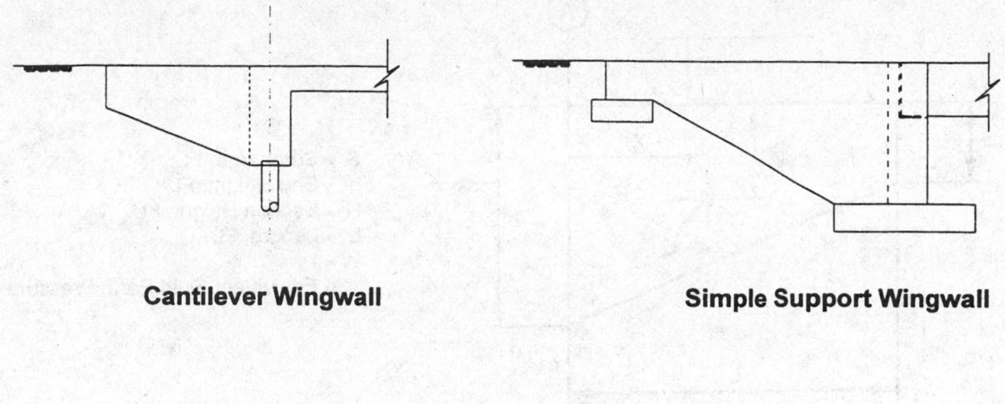

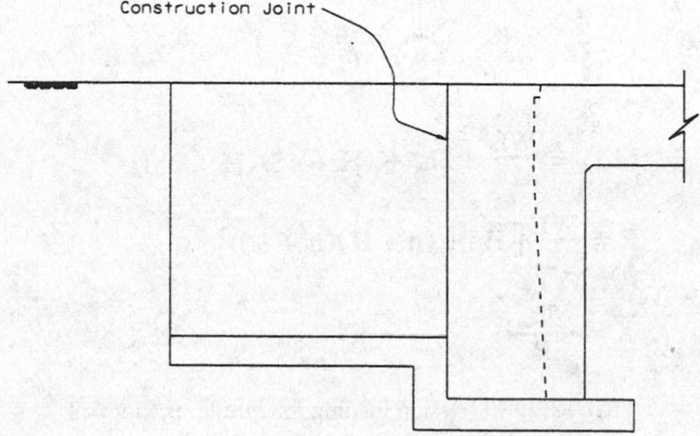

FIGURE 29.8 Typical wingwalls.

TABLE 29.2 Live Load Surcharges for Wingwall Design

Highway truck loading	2 ft 0 in. (610 mm) equivalent soil
Rail loading E-60	7 ft 6 in. (2290 mm) equivalent soil
Rail loading E-70	8 ft 9 in. (2670 mm) equivalent soil
Rail loading E-80	10 ft 0 in. (3050 mm) equivalent soil

Abutment Slope Protection

Flow water scoring may severely damage bridge structures by washing out the bridge abutment support soil. To reduce water scoring damage to the bridge abutment, pile support, rock slope protection, concrete slope paving, and gunite cement slope paving may be used. Figure 29.11 shows the actual design of rock slope protection and concrete slope paving protection for bridge abutments. The stability of the rock and concrete slope protection should be considered in the design. An enlarged block is usually designed at the toe of the protections.

Miscellaneous Details

Some details related to abutment design are given in Figure 29.12. Although they are only for regular bridge construction situations, those details present valuable references for bridge designers.

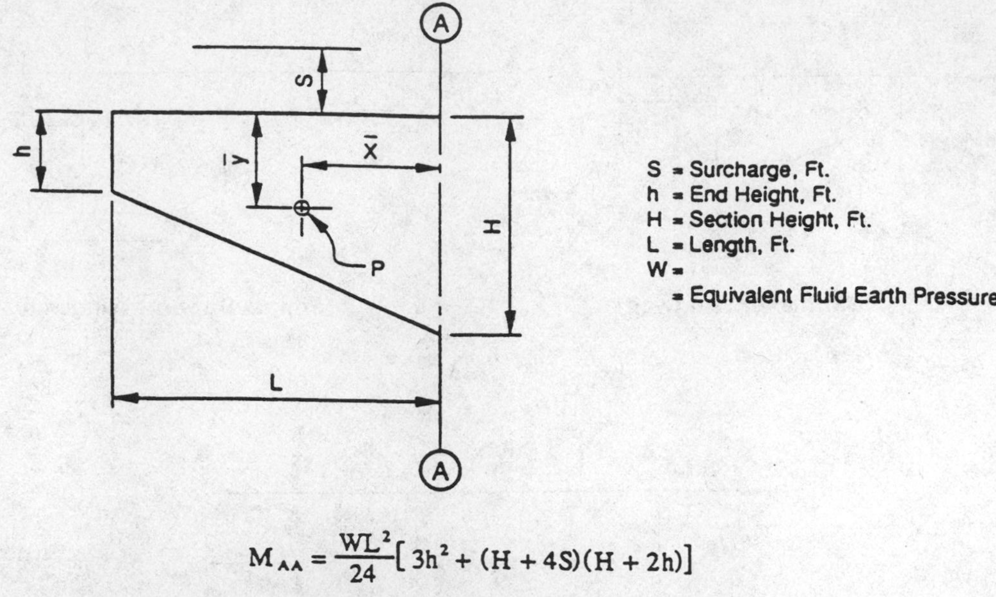

$$M_{AA} = \frac{WL^2}{24}[3h^2 + (H + 4S)(H + 2h)]$$

$$P = \frac{WL}{6}[H^2 + (h + H)(h + 3S)]$$

$$\overline{X} = \frac{M_{AA}}{P}$$

FIGURE 29.9 Design loading for cantilever wingwall.

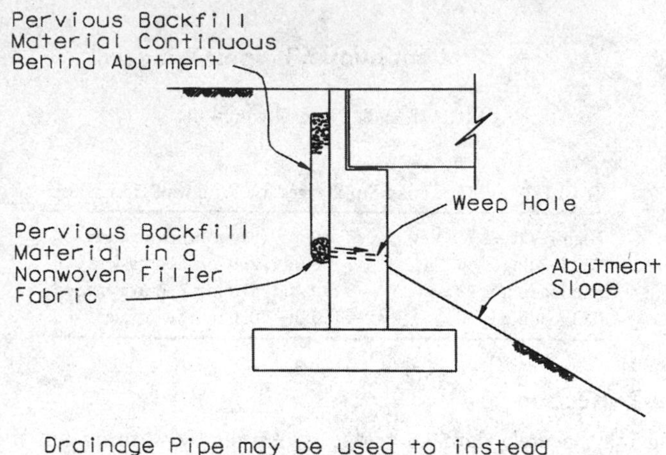

FIGURE 29.10 Typical abutment drainage system.

29.2.5 Design Example

A prestressed concrete box-girder bridge with 5° skew is proposed overcrossing a busy freeway as shown in Figure 29.13. Based on the roadway requirement, geotechnical information, and the details mentioned above, an open-end, seat-type abutment is selected. The abutment in transverse direction is 89 ft (27.13 m) wide. From the bridge analysis, the loads on abutment and bridge displacements are as listed bellow:

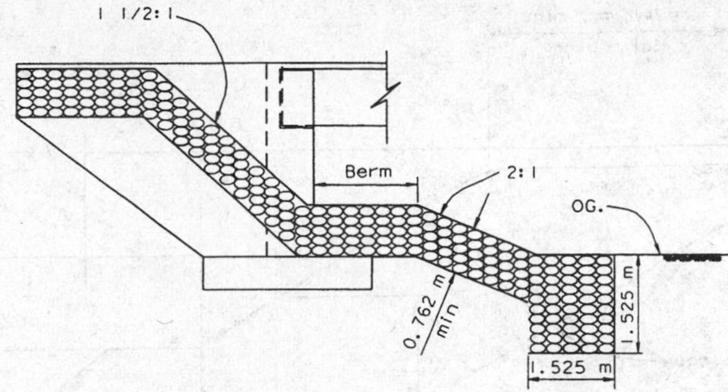

Rock Slope Protection

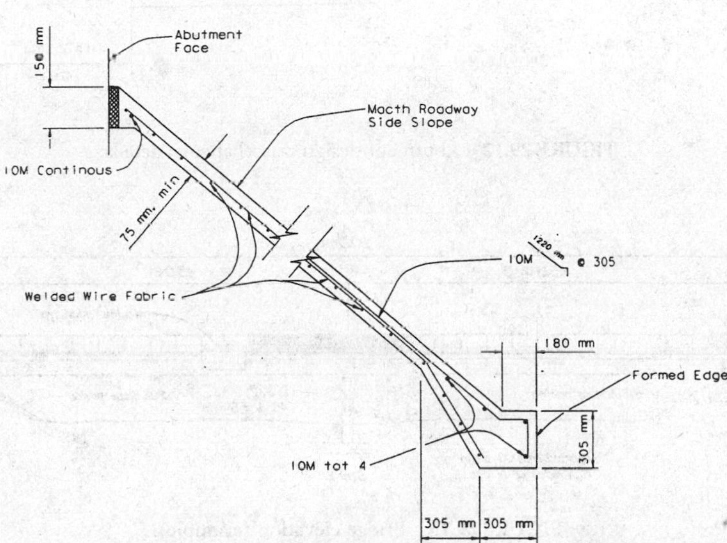

Concrete Slope Protection

FIGURE 29.11 Typical abutment slope protections.

Superstructure dead load	= 1630 kips (7251 kN)
HS20 live load	= 410 kips (1824 kN)
1.15 P-load + 1.0 HS load	= 280 kips (1245 kN)
Longitudinal live load	= 248 kips (1103 kN)
Longitudinal seismic load (bearing pad capacity)	= 326 kips (1450 kN)
Transverse seismic load	= 1241 kips (5520 kN)
Bridge temperature displacement	= 2.0 in. (75 mm)
Bridge seismic displacement	= 6.5 in. (165 mm)

Geotechnical Information

Live-load surcharge	= 2 ft (0.61 m)
Unit weight of backfill soil	= 120 pcf (1922 kg/m^3)

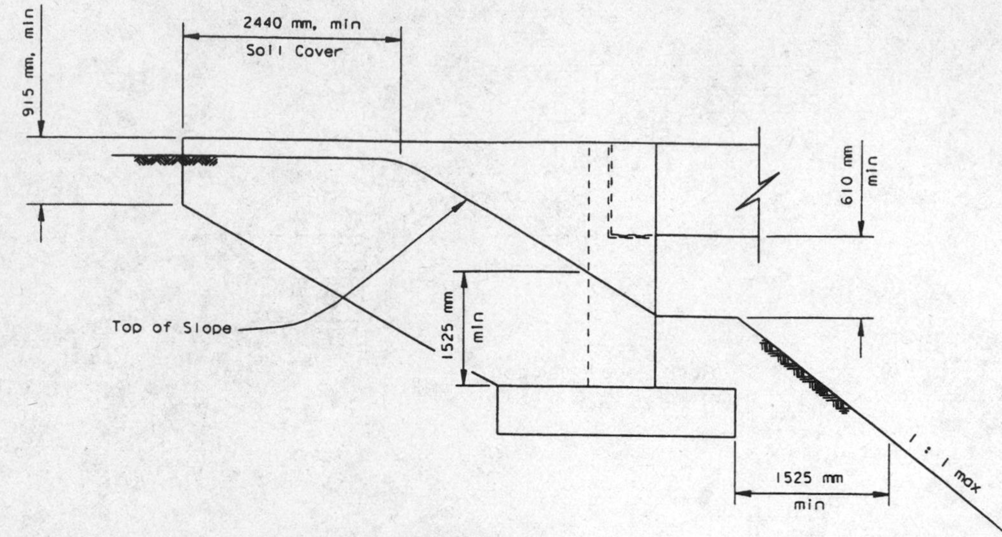

FIGURE 29.12 Abutment design miscellaneous details.

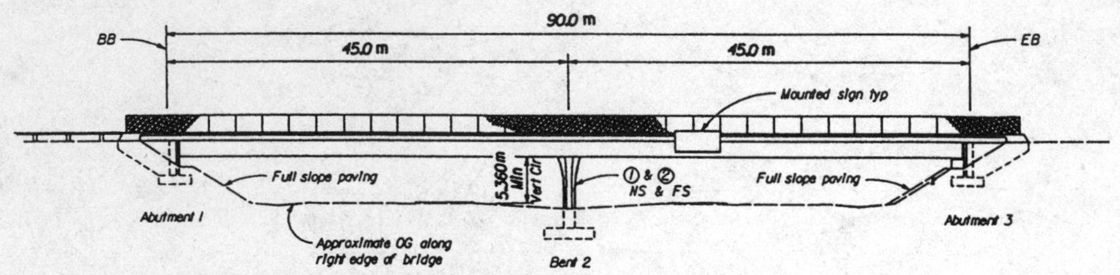

FIGURE 29.13 Bridge elevation (example).

Allowable soil bearing pressure	= 4.0 ksf (0.19 MPa)
Soil lateral pressure coefficient (Ka)	= 0.3
Friction coefficient	= tan 33°
Soil liquefaction potential	= very low
Ground acceleration	= 0.3 g

Design Criteria

Abutment design	Load factor method
Abutment stability	Service load method

Design Assumptions

1. Superstructure vertical loading acting on the center line of abutment footing;
2. The soil passive pressure by the soil at abutment toe is neglected;
3. 1.0 feet (0.305 m) wide of abutment is used in the design;
4. reinforcement yield stress, $f_y = 60000$ psi (414 MPa)
5. concrete strength, $f_c' = 3250$ psi (22.41 MPa)
6. abutment backwall allowed damage in the design earthquake

Solution

1. Abutment Support Width Design
Applying Eq. (29.2) with

$$L = 6.5 \text{ m}$$
$$H = 90.0 \text{ m}$$
$$S = 5°$$

the support width will be $N = 600$ mm. Add 75 mm required temperature movement, the total required support width equals 675 mm. The required minimum support width for seismic case equals the sum of the bridge seismic displacement, the bridge temperature displacement, and the reserved edge displacement (usually 75 mm). In this example, this requirement equals 315 mm, not in control. Based on the 675-mm minimum requirement, the design uses 760 mm, OK. A preliminary abutment configuration is shown in Figure 29.14 based on the given information and calculated support width.

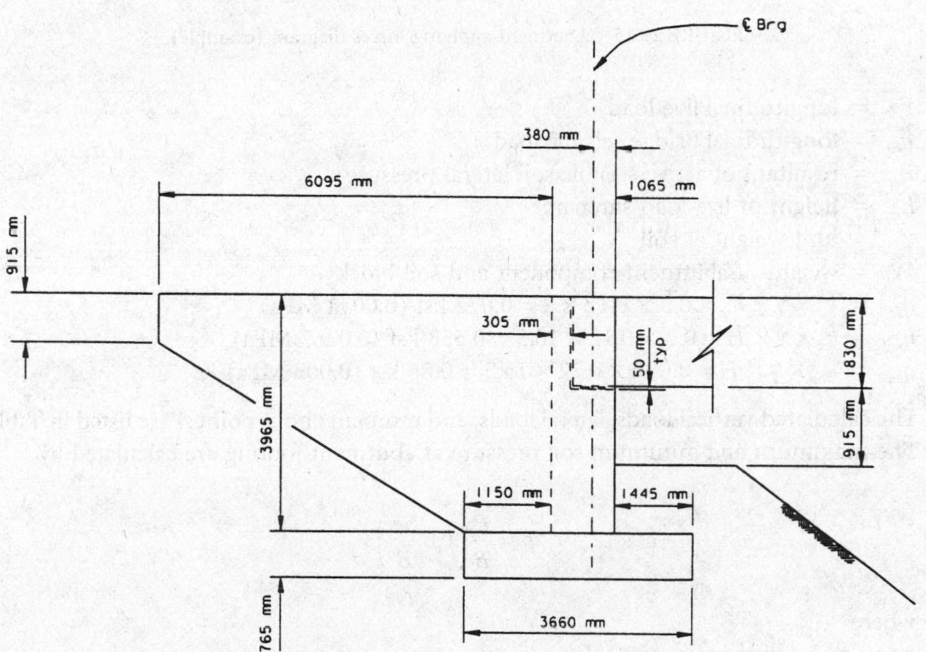

FIGURE 29.14 Abutment configuration (example).

2. Abutment Stability Check
Figure 29.15 shows the abutment force diagram,

where

q_{sc} = soil lateral pressure by live-load surcharge
q_e = soil lateral pressure
q_{eq} = soil lateral pressure by seismic load
P_{DL} = superstructure dead load
P_{HS} = HS20 live load
P_P = permit live load

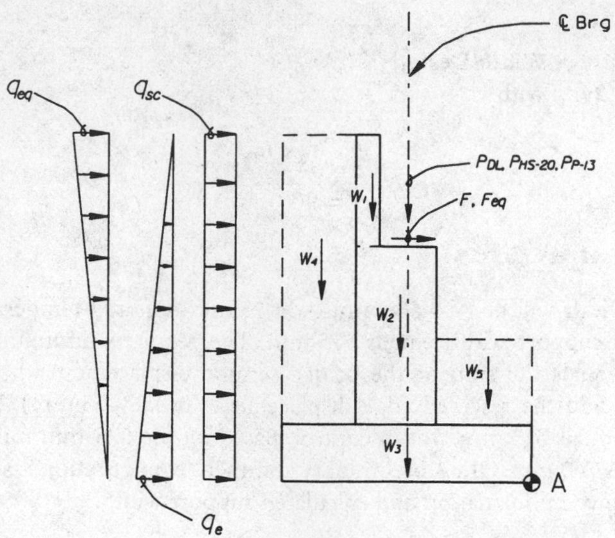

FIGURE 29.15 Abutment applying forces diagram (example).

F = longitudinal live load
F_{eq} = longitudinal bridge seismic load
P_{ac} = resultant of active seismic soil lateral pressure
h_{sc} = height of live-load surcharge
γ = unit weight of soil
W_i = weight of abutment component and soil block
$q_{sc} = k_a \times \gamma \times h_{sc} = 0.3 \times 0.12 \times 2 = 0.072$ ksf (0.0034 MPa)
$q_e = k_a \times \gamma \times H = 0.3 \times 0.12 \times 15.5 = 0.558$ ksf (0.0267 MPa)
$q_{eq} = k_{ae} \times \gamma \times H = 0.032 \times 0.12 \times 15.5 = 0.06$ ksf (0.003 MPa)

The calculated vertical loads, lateral loads, and moment about point A are listed in Table 29.3. The maximum and minimum soil pressure at abutment footing are calculated by

$$p = \frac{P}{B}\left(1 \pm \frac{6e}{B}\right) \tag{29.3}$$

where
p = soil bearing pressure
P = resultant of vertical forces
B = abutment footing width
e = eccentricity of resultant of forces and the center of footing

$$e = 2B - \frac{M}{P} \tag{29.4}$$

M = total moment to point A

Referring to the Table 29.1 and Eqs. (29.3) and (29.4) the maximum and minimum soil pressures under footing corresponding to different load cases are calculated as
Since the soil bearing pressures are less than the allowable soil bearing pressure, the soil bearing stability is OK.

Load Case	p_{max}	p_{min}	$p_{allowable}$ with Allowable % of Overstress	Evaluate
I	3.81	3.10	4.00	OK
II	3.42	2.72	4.00	OK
III	1.84	1.22	6.00	OK
IV	4.86	2.15	5.00	OK
V	2.79	1.93	6.00	OK
Seismic	6.73	0.54	8.00	OK

TABLE 29.3 Vertical Forces, Lateral Forces, and Moment about Point *A* (Example)

Load Description	Vertical Load (kips)	Lateral Load (kips)	Arm to *A* (ft)	Moment to *A* (k-ft)
Backwall W_1	0.94	—	7.75	7.28
Stem W_2	3.54	—	6.00	23.01
Footing W_3	4.50	—	6.00	27.00
Backfill soil	5.85	—	10.13	59.23
	—	4.33	5.17	–22.34
Soil surcharge	—	1.16	7.75	–8.65
Front soil W_4	1.71	—	2.38	4.06
Wingwalls	0.85	—	16.12	13.70
Keys	0.17	—	6.00	1.04
P_{DL}	18.31	—	6.00	110.00
P_{HS}	4.61	—	6.00	27.64
P_P	3.15	—	6.00	18.90
F	—	2.79	9.25	–25.80
F_{eq}	—	3.66	9.25	–33.90
Soil seismic load	—	0.47	9.30	–4.37

Check for the stability resisting the overturning (load case III and IV control):

Load Case	Driving Moment	Resist Moment	Factor of Safety	Evaluate
III	31	133.55	4.3	OK
IV	56.8	262.45	4.62	OK

Checking for the stability resisting the sliding (load case III and IV control)

Load Case	Driving Force	Resist Force	Factor of Safety	Evaluation
III	5.44	11.91	2.18	OK
IV	8.23	20.7	3.26	OK

Since the structure lateral dynamic force is only combined with dead load and static soil lateral pressures, and the factor of safety FS = 1.0 can be used, the seismic case is not in control.

3. **Abutment Backwall and Stem Design**

   Referring to AASHTO guidelines for load combinations, the maximum factored loads for abutment backwall and stem are

Location	*V* (kips)	*M* (k-ft)
Backwall level	1.95	4.67
Bottom of stem	10.36	74.85

Abutment Backwall
Try #5 at 12 in. (305 mm) with 2 in. (50 mm) clearance

$d = 9.7$ in. (245 mm)

$$A_s \times f_y = 0.31 \times 60 \times \frac{12}{16} = 13.95 \text{ kips } (62.05 \text{ kN})$$

$$a = \frac{A_s \cdot f_y}{\phi \cdot f'_c \cdot b_w} = \frac{13.95}{(0.85)(3.25)(12)} = 0.42 \text{ in. } (10.67 \text{ mm})$$

$$M_u = \phi \cdot M_n = \phi \cdot A_s \cdot f_y\left(d - \frac{a}{2}\right) = 0.9 \times 13.95 \times \left(9.7 - \frac{0.42}{2}\right) = 9.33 \text{ k} \cdot \text{ft } (13.46 \text{ kN} \cdot \text{m})$$

$$> 4.67 \text{ k} \cdot \text{ft } (6.33 \text{ kN} \cdot \text{m}) \quad \text{OK}$$

$$V_c = 2\sqrt{f'_c} \cdot b_w \cdot d = 2 \times \sqrt{3250} \times 12 \times 9.7 = 13.27 \text{ kips } (59.03 \text{ kN})$$

$$V_u = \phi \cdot V_c = 0.85 \times 13.27 = 11.28 \text{ kip } (50.17 \text{ kN}) > 1.95 \text{ kips } (8.67 \text{ kN}) \quad \text{OK}$$

No shear reinforcement needed.

Abutment Stem
Abutment stem could be designed based on the applying moment variations along the abutment wall height. Here only the section at the bottom of stem is designed.

Try #6 at 12 in. (305 mm) with 2 in. (50 mm) clearance.

$$A_s \times f_y = 0.44 \times 60 = 26.40 \text{ kips } (117.43 \text{ kN})$$

$d =$ 39.4 in. (1000 mm)

$$a = \frac{A_s \cdot f_y}{\phi \cdot f'_c \cdot b_w} = \frac{26.4}{(0.85)(3.25)(12)} = 0.796 \text{ in } (20.0 \text{ mm})$$

$$M_u = \phi \cdot A_s \cdot f_y\left(d - \frac{a}{2}\right) = 0.9 \times 26.4 \times \left(39.4 - \frac{0.8}{2}\right) = 77.22 \text{ k} \cdot \text{ft } (104.7 \text{kN} \cdot \text{m})$$

$$> 74.85 \text{ k} \cdot \text{ft } (101.5 \text{kN} \cdot \text{m}) \quad \text{OK}$$

$$V_c = 2\sqrt{f'_c}\, b_w\, d = 2 \times \sqrt{3250} \times 12 \times 39.4 = 53.91 \text{ kips } (238 \text{ kN})$$

$$V_u = \phi \cdot V_c = 0.85 \times 53.91 = 45.81 \text{ kips } (202.3 \text{ kN}) > 10.36 \text{ kips } (46.08 \text{ kN}) \quad \text{OK}$$

No shear reinforcement needed.

4. **Abutment Footing Design**
Considering all load combinations and seismic loading cases, the soil bearing pressure diagram under the abutment footing are shown in Figure 29.16.

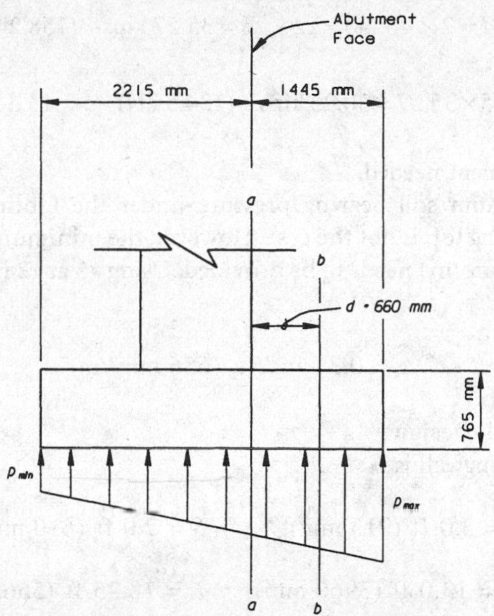

FIGURE 29.16 Bearing pressure under abutment footing (example).

a. *Design forces*:
Section at front face of abutment stem (design for flexural reinforcement):

$$q_{a\text{-}a} = 5.1263 \text{ ksf } (0.2454 \text{ MPa})$$

$$M_{a\text{-}a} = 69.4 \text{ k-ft } (94.1 \text{ kN·m})$$

Section at $d = 30 - 3 - 1 = 26$ in. (660 mm) from the front face of abutment stem (design for shear reinforcement):

$$q_{b\text{-}b} = 5.2341 \text{ ksf } (0.251 \text{ MPa})$$

$$V_{b\text{-}b} = 15.4 \text{ kips } (68.5 \text{ kN})$$

b. *Design flexural reinforcing* (*footing bottom*):
Try #8 at 12, with 3 in. (75 mm) clearance at bottom

$$d = 30 - 3 - 1 = 26 \text{ in. } (660 \text{ mm})$$

$$A_s \times f_y = 0.79 \times 60 = 47.4 \text{ kips} \quad (211 \text{ kN})$$

$$a = \frac{A_s \cdot f_y}{\phi \cdot f_c' \cdot b_w} = \frac{47.4}{(0.85)(3.25)(12)} = 1.43 \text{ in.} \quad (36 \text{ mm})$$

$$M_n = \phi \cdot A_s \cdot f_y \left(d - \frac{a}{2} \right) = 0.9 \times 47.4 \times \left(26 - \frac{1.43}{2} \right) = 89.9 \text{ k·ft} \quad (121.89 \text{ kN·m})$$

$$> 69.4 \text{ k·ft} \quad (94.1 \text{ kN·m}) \qquad \text{OK}$$

$$V_c = 2\sqrt{f'_c} \cdot b_w \cdot d = 2 \times \sqrt{3250} \times 12 \times 26 = 35.57 \text{ kips} \quad (158.24 \text{ kN})$$

$$V_u = \phi \cdot V_c = 0.85 \times 35.57 = 30.23 \text{ kips} \quad (134.5 \text{ kN}) \quad > \quad 15.5 \text{ kips} \quad (68.5 \text{ kN}) \qquad \text{OK}$$

No shear reinforcement needed.

Since the minimum soil bearing pressure under the footing is in compression, the tension at the footing top is not the case. However, the minimum temperature reinforcing, 0.308 in.^{2}/ft (652 mm^2/m) needs to be provided. Using #5 at 12 in. (305 mm) at the footing top yields

$$A_s = 0.31 \text{ in.}^2\text{/ft}, \ (656 \text{ mm}^2\text{/m}) \qquad \text{OK}$$

5. **Abutment Wingwall Design**

The geometry of wingwall is

$$h = 3.0 \text{ ft (915 mm)}; \qquad S = 2.0 \text{ ft (610 mm)};$$

$$H = 13.0 \text{ ft (3960 mm)}; \qquad L = 18.25 \text{ ft (5565 mm)}$$

Referring to the Figure 29.15, the design loads are

$$V_{A-A} = \frac{wL}{6}\left[H^2 + (h+H)(h+3S)\right]$$

$$= \frac{0.36 \times 18.25}{6}\left[13^2 + (3+13)(3+3\times 2)\right] = 34 \text{ kips} \quad (152.39 \text{ kN})$$

$$M_{A-A} = \frac{wL^2}{24}\left[3h^2 + (H+4S)(H+2h)\right]$$

$$= \frac{0.036 \times 18.25^2}{24}\left[3(3)^2 + (13+4\times 2)(12+2\times 3)\right] = 212.8 \text{ k·ft} \quad (3129 \text{ kN·m})$$

Design flexural reinforcing. Try using # 8 at 9 (225 mm).

$$A_s \times f_y = 13 \times (0.79) \times 60 \times \frac{12}{9} = 821.6 \text{ kips} \quad (3682 \text{ kN})$$

$$a = \frac{A_s \cdot f_y}{\phi \cdot f'_c \cdot b_w} = \frac{1280}{(0.85)(3.25)(13)(12)} = 2.97 \text{ in.} \quad (75 \text{ mm})$$

$$d = 12 - 2 - 0.5 = 9.5 \text{ in. (240 mm)}$$

$$M_n = \phi \cdot A_s \cdot f_y\left(d - \frac{a}{2}\right) = 0.9 \times (821.6) \times \left(9.5 - \frac{2.97}{2}\right) = 493.8 \text{ k·ft} \ (7261 \text{ kN}\cdot\text{m})$$

$$> 212.8 \text{ k·ft} \quad (3129 \text{ kN}\cdot\text{m})$$

Checking for shear

$$V_c = 2\sqrt{f_c'} \cdot b_w \cdot d = 2 \times \sqrt{3250} \times 13 \times 12 \times 9.5 = 168 \text{ kips} \quad (757.3 \text{ kN})$$

$$V_u = \varphi \cdot V_c = 0.85 \times 168 = 142 \text{ kips} \quad (636 \text{ kN}) > 34 \text{ kips } (152.3\text{kN}) \qquad \text{OK}$$

No shear reinforcing needed.

Since the wingwall is allowed to be broken off in a major earthquake, the adjacent bridge columns have to be designed to sustain the seismic loading with no wingwall resistance. The abutment section, footing, and wingwall reinforcing details are shown in Figures 29.17a and b.

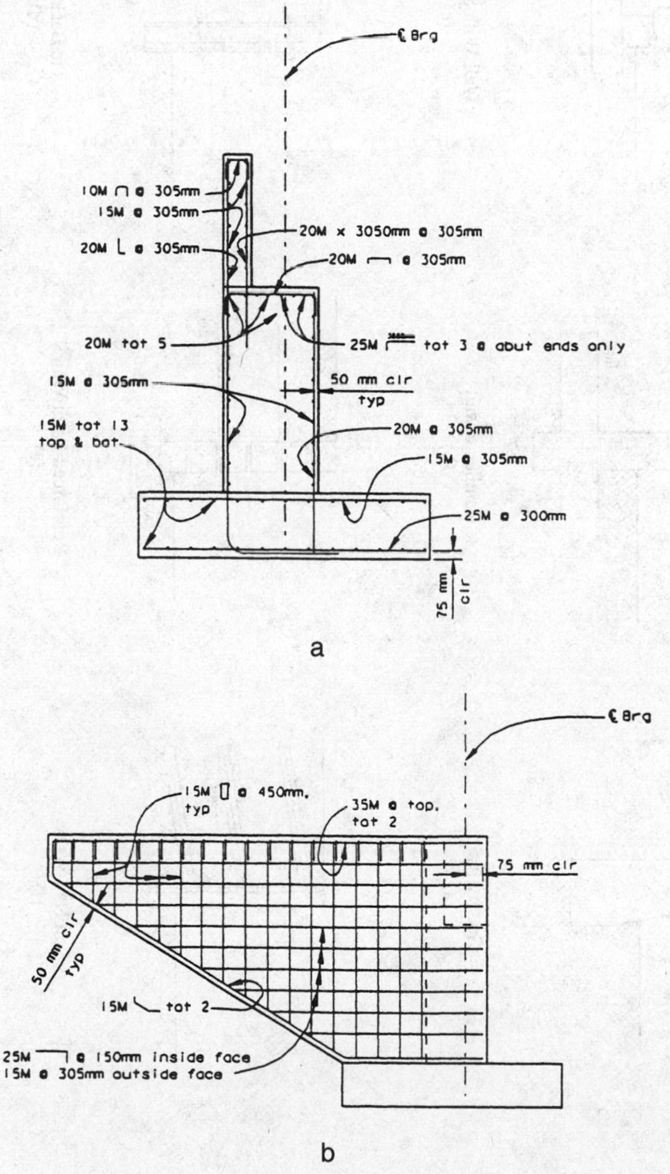

FIGURE 29.17 (a) Abutment typical section design (example). (b) Wingwall reinforcing (example).

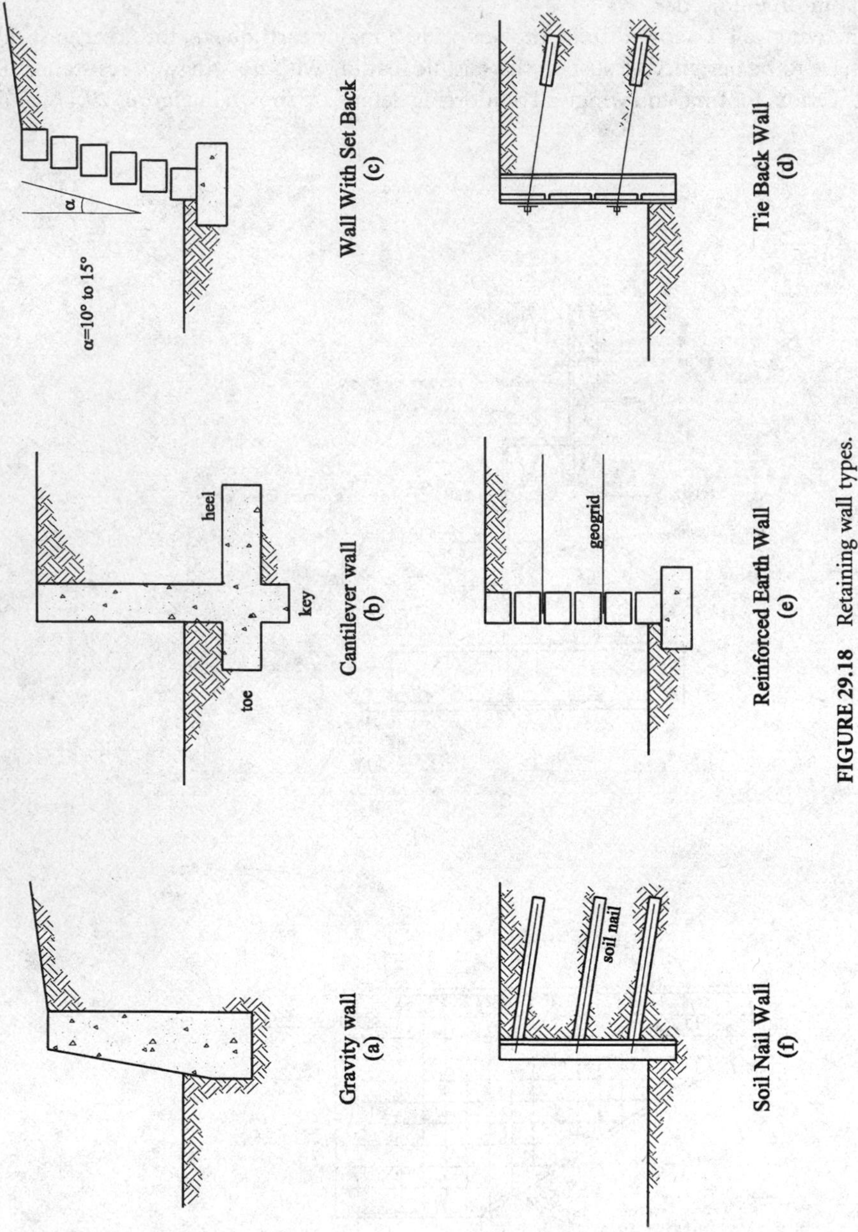

FIGURE 29.18 Retaining wall types.

29.3 Retaining Structures

29.3.1 Retaining Structure Types

The retaining structure, or, more specifically, the earth-retaining structure, is commonly required in a bridge design project. It is common practice that the bridge abutment itself is used as a retaining structure. The cantilever wall, tieback wall, soil nail wall and mechanically stabilized embankment (MSE) wall are the most frequently used retaining structure types. The major design function of a retaining structure is to resist lateral forces.

The cantilever retaining wall is a cantilever structure used to resist the active soil pressure in topography fill locations. Usually, the cantilever earth-retaining structure does not exceed 10 m in height. Some typical cantilever retaining wall sections are shown in Figure 29.18a.

The tieback wall can be used for topography cutting locations. High-strength tie strands are extended into the stable zone and act as anchors for the wall face elements. The tieback wall can be designed to have minimum lateral deflection. Figure 29.18d shows a tieback wall section.

The MSE wall is a kind of "reinforced earth-retaining" structure. By installing multiple layers of high-strength fibers inside of the fill section, the lateral deflection of filled soil will be restricted. There is no height limit for an MSE wall but the lateral deflection at the top of the wall needs to be considered. Figure 29.18e shows an example of an MSE wall.

The soil nail wall looks like a tieback wall but works like an MSE wall. It uses a series of soil nails built inside the soil body that resist the soil body lateral movement in the cut sections. Usually, the soil nails are constructed by pumping cement grout into predrilled holes. The nails bind the soil together and act as a gravity soil wall. A typical soil nail wall model is shown in Figure 29.18f.

29.3.2 Design Criteria

Minimum Requirements

All retaining structures must be safe from vertical settlement. They must have sufficient resistance against overturning and sliding. Retaining structures must also have adequate strength for all structural components.

1. *Bearing capacity*: Similar to any footing design, the bearing capacity factor of safety should be ≥1.0. Table 29.4 is a list of approximate bearing capacity values for some common materials. If a pile footing is used, the soil-bearing capacity between piles is not considered.
2. *Overturning resistance*: The overturning point of a typical retaining structure is located at the edge of the footing toe. The overturning factor of safety should be ≥1.50. If the retaining structure has a pile footing, the fixity of the footing will depend on the piles only.
3. *Sliding resistance*: The factor of safety for sliding should be ≥1.50. The typical retaining wall sliding capacity may include both the passive soil pressure at the toe face of the footing and the friction forces at the bottom of the footing. In most cases, friction factors of 0.3 and 0.4

TABLE 29.4 Bearing Capacity

	Bearing Capacity [N]	
Material	min, kPa	max, kPa
Alluvial soils	24	48
Clay	48	190
Sand, confined	48	190
Gravel	95	190
Cemented sand and gravel	240	480
Rock	240	—

can be used for clay and sand, respectively. If battered piles are used for sliding resistance, the friction force at the bottom of the footing should be neglected.

4. Structural strength: Structural section moment and shear capacities should be designed following common strength factors of safety design procedures.

Figure 29.19 shows typical loads for cantilever retaining structure design.

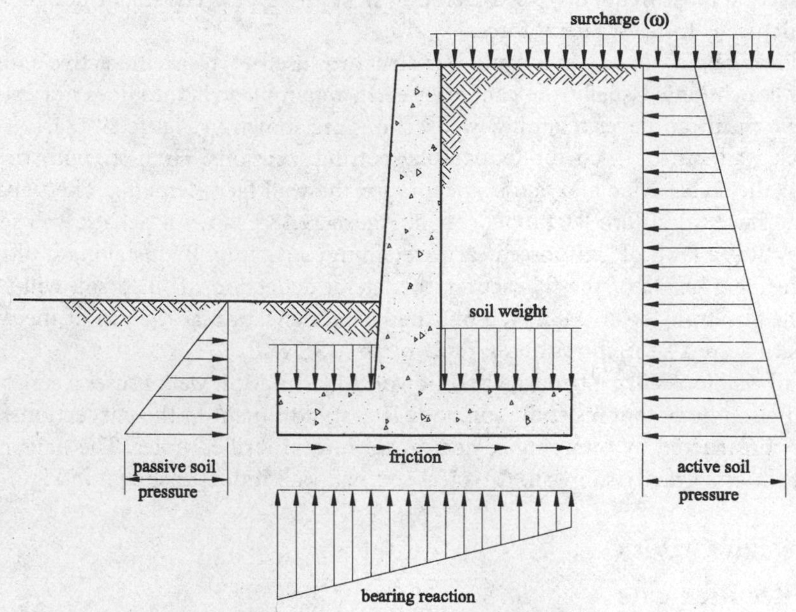

FIGURE 29 19 Typical loads on retaining wall.

Lateral Load

The unit weight of soil is typically in the range of 1.5 to 2.0 ton/m³. For flat backfill cases, if the backfill material is dry, cohesionless sand, the lateral earth pressure (Figure 29.20a) distribution on the wall will be as follows

The active force per unit length of wall (Pa) at bottom of wall can be determined as

$$p_a = k_a \gamma H \tag{29.5}$$

The passive force per unit length of wall (Pa) at bottom of wall can be determined as

$$p_p = k_p \gamma H \tag{29.6}$$

where

H = the height of the wall (from top of the wall to bottom of the footing)
γ = unit weight of the backfill material
k_a = active earth pressure coefficient
k_p = passive earth pressure coefficient

The coefficients k_a and k_p should be determined by a geologist using laboratory test data from a proper soil sample. The general formula is

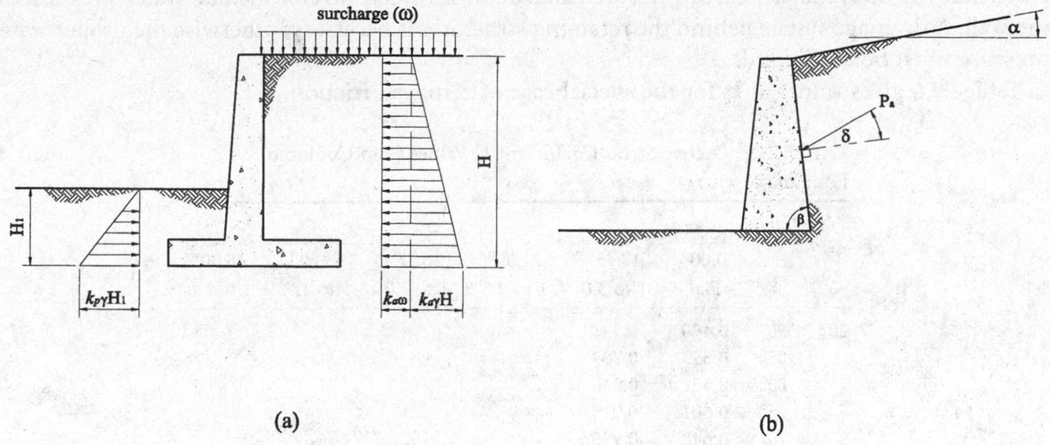

FIGURE 29.20 Lateral Earth pressure.

$$k_a = \frac{1-\sin\phi}{1+\sin\phi} \quad k_p = \frac{1}{k_a} = \frac{1+\sin\phi}{1-\sin\phi} \tag{29.7}$$

where ϕ is the internal friction angle of the soil sample.

Table 29.5 lists friction angles for somc typical soil typcs which can be used if laboratory test data is not available. Generally, force coefficients of $k_a \geq 0.30$ and $k_p \leq 1.50$ should be used for preliminary design.

TABLE 29.5 Internal Friction Angle and Force Coefficients

Material	ϕ(*degrees*)	k_a	k_p
Earth, loam	30–45	0.33–0.17	3.00–5.83
Dry sand	25–35	0.41–0.27	2.46–3.69
Wet sand	30–45	0.33–0.17	3.00–5.83
Compact Earth	15–30	0.59–0.33	1.70–3.00
Gravel	35–40	0.27–0.22	3.69–4.60
Cinders	25–40	0.41–0.22	2.46–4.60
Coke	30–45	0.33–0.17	3.00–5.83
Coal	25–35	0.41–0.27	2.46–3.69

Based on the triangle distribution assumption, the total active lateral force per unit length of wall should bc

$$P_a = \frac{1}{2} k_a \gamma H^2 \tag{29.8}$$

The resultant earth pressure always acts at distance of $H/3$ from the bottom of the wall.

When the top surface of backfill is sloped, the k_a coefficient can be determined by the Coulomb equation: (see Figure 29.20):

$$\mathrm{k_a} = \frac{\sin^2(\phi+\beta)}{\sin^2\beta\,\sin(\beta-\delta)\left[1+\sqrt{\dfrac{\sin(\phi+\delta)\,\sin(\phi-\alpha)}{\sin(\beta-\delta)\,\sin(\alpha+\beta)}}\right]^2} \tag{29.9}$$

Note that the above lateral earth pressure calculation formulas do not include water pressure on the wall. A drainage system behind the retaining structures is necessary; otherwise the proper water pressure must be considered.

Table 29.6 gives values of k_a for the special case of zero wall friction.

TABLE 29.6 Active Stress Coefficient k_a Values from Coulomb Equation ($\delta = 0$)

		α					
		0.00°	18.43°	21.80°	26.57°	33.69°	45.00°
ϕ	β_o	Flat	1 to 3.0	1 to 2.5	1 to 2.0	1 to 1.5	1 to 1.0
20°	90°	0.490	0.731				
	85°	0.523	0.783				
	80°	0.559	0.842				
	75°	0.601	0.913				
	70°	0.648	0.996				
25°	90°	0.406	0.547	0.611			
	85°	0.440	0.597	0.667			
	80°	0.478	0.653	0.730			
	75°	0.521	0.718	0.804			
	70°	0.569	0.795	0.891			
30°	90°	0.333	0.427	0.460	0.536		
	85°	0.368	0.476	0.512	0.597		
	80°	0.407	0.530	0.571	0.666		
	75°	0.449	0.592	0.639	0.746		
	70°	0.498	0.664	0.718	0.841		
35°	90°	0.271	0.335	0.355	0.393	0.530	
	85°	0.306	0.381	0.404	0.448	0.602	
	80°	0.343	0.433	0.459	0.510	0.685	
	75°	0.386	0.492	0.522	0.581	0.781	
	70°	0.434	0.560	0.596	0.665	0.897	
40°	90°	0.217	0.261	0.273	0.296	0.352	
	85°	0.251	0.304	0.319	0.346	0.411	
	80°	0.287	0.353	0.370	0.402	0.479	
	75°	0.329	0.408	0.429	0.467	0.558	
	70°	0.375	0.472	0.498	0.543	0.651	
45°	90°	0.172	0.201	0.209	0.222	0.252	0.500
	85°	0.203	0.240	0.250	0.267	0.304	0.593
	80°	0.238	0.285	0.297	0.318	0.363	0.702
	75°	0.277	0.336	0.351	0.377	0.431	0.832
	70°	0.322	0.396	0.415	0.446	0.513	0.990

Any surface load near the retaining structure will generate additional lateral pressure on the wall. For highway-related design projects, the traffic load can be represented by an equivalent vertical surcharge pressure of 11.00 to 12.00 kPa. For point load and line load cases (Figure 29.21), the following formulas can be used to determine the additional pressure on the retaining wall:

For point load:

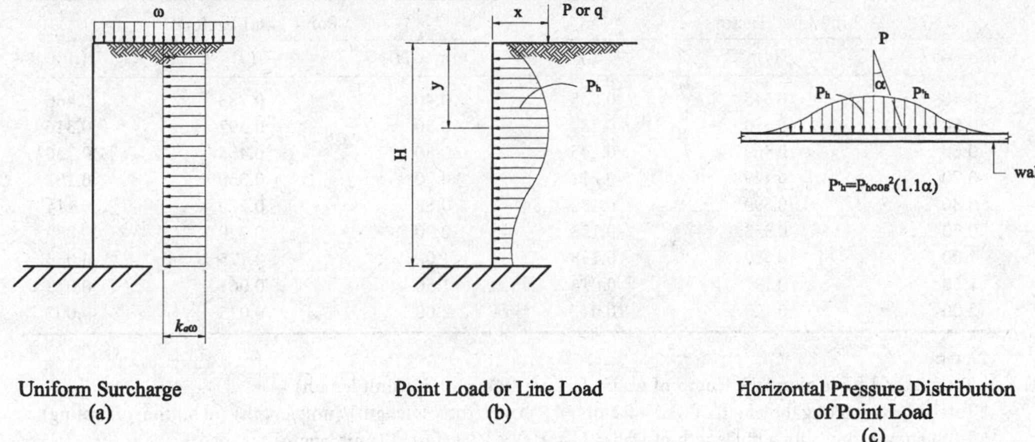

FIGURE 29.21 Additional lateral earth pressure. (a) Uniform surcharge; (b) point or line load; (c) horizontal pressure distribution of point load.

$$p_h = \frac{1.77V}{H^2}\frac{m^2n^2}{\left(m^2+n^2\right)^3} \quad (m \le 0.4) \quad p_h \frac{0.28V}{H^2}\frac{m^2n^2}{\left(0.16+n^3\right)} \quad (m > 0.4) \tag{29.10}$$

For line load:

$$p_h = \frac{\pi}{4}\frac{w}{H}\frac{m^2n}{\left(m^2+n^2\right)^2} \quad (m \le 0.4) \quad p_h = \frac{w}{H}\frac{0.203n}{\left(0.16+n^2\right)^2} \quad (m > 0.4) \tag{29.11}$$

where

$$m = \frac{x}{H}; \quad n = \frac{y}{H}$$

Table 29.7 gives lateral force factors and wall bottom moment factors which are calculated by above formulas.

29.3.3 Cantilever Retaining Wall Design Example

The cantilever wall is the most commonly used retaining structure. It has a good cost-efficiency record for walls less than 10 m in height. Figure 29.22a shows a typical cross section of a cantilever retaining wall and Table 29.8 gives the active lateral force and the active moment about bottom of the cantilever retaining wall.

For most cases, the following values can be used as the initial assumptions in the reinforced concrete retaining wall design process.

- $0.4 \le B/H \le 0.8$
- $1/12 \le t_{bot}/H \le 1/8$
- $L_{toe} \cong B/3$
- $t_{top} \ge 300$ mm
- $t_{foot} \ge t_{bot}$

TABLE 29.7 Line Load and Point Load Lateral Force Factors

Line Load Factors			Point Load Factors		
$m = x/H$	$(f)^a$	$(m)^b$	$m = x/H$	$(f)^c$	$(m)^d$
0.40	0.548	0.335	0.40	0.788	0.466
0.50	0.510	0.287	0.50	0.597	0.316
0.60	0.469	0.245	0.60	0.458	0.220
0.70	0.429	0.211	0.70	0.356	0.157
0.80	0.390	0.182	0.80	0.279	0.114
0.90	0.353	0.158	0.90	0.220	0.085
1.00	0.320	0.138	1.00	0.175	0.064
1.50	0.197	0.076	1.50	0.061	0.019
2.00	0.128	0.047	2.00	0.025	0.007

Notes:

[a] Total lateral force along the length of wall = factor(f) × ω (force)/(unit length).

[b] Total moment along the length of wall = factor(m) × ω × H (force × length)/(unit length) (at bottom of footing).

[c] Total lateral force along the length of wall = factor(f) × V/H (force)/(unit length).

[d] Total moment along the length of wall = factor(m) × V (force × length)/(unit length) (at bottom of footing).

Example

Given

A reinforced concrete retaining wall as shown in Figure 29.22b:

$H_o = 3.0$ m; surcharge $\omega = 11.00$ kPa
Earth internal friction angle $\phi = 30°$
Earth unit weight $\gamma = 1.8$ *ton/m*^{3}
Bearing capacity $[\sigma] = 190$ kPa
Friction coefficient $f = 0.30$

Solution

1. **Select Control Dimensions**
   Try $h = 1.5$ m, therefore, $H = H_o + h = 3.0 + 1.5 = 4.5$ m.
   Use

$$t_{\text{bot}} = 1/10H = 0.45 \text{ m} \Rightarrow 500 \text{ mm};\ t_{top} = t_{bot} = 500 \text{ mm}$$

$$t_{\text{foot}} = 600 \text{ mm}$$

   Use

$$B = 0.6H = 2.70 \text{ m} \Rightarrow 2700 \text{ mm};$$

$$L_{\text{toe}} = 900 \text{ mm; therefore, } L_{\text{heel}} = 2.7 - 0.9 - 0.5 = 1.3 \text{ m} = 1300 \text{ mm}$$

2. **Calculate Lateral Earth Pressure**
   From Table 29.4, $k_a = 0.33$ and $k_p = 3.00$.
   Active Earth pressure:

$$\text{Part 1 (surcharge) } P_1 = k_a \omega H = 0.33(11.0)(4.5) = 16.34 \text{ kN}$$

$$\text{Part 2 } P_2 = 0.5\ k_a \gamma\ H^2 = 0.5(0.33)(17.66)(4.5)^2 = 59.01 \text{ kN}$$

   Maximum possible passive Earth pressure:

$$P_p = 0.5 k_p \gamma h^2 = 0.5(3.00)(17.66)(1.5)^2 = 59.60 \text{ kN}$$

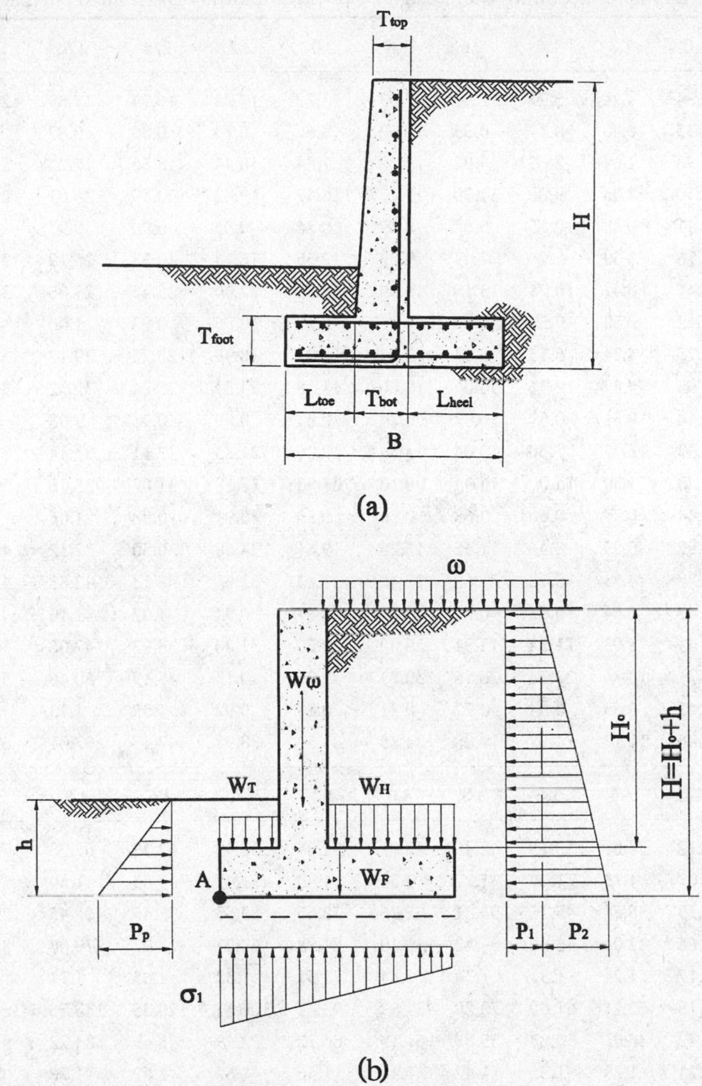

FIGURE 29.22 Design example.

3. Calculate Vertical Loads

Surcharge	W_s (11.00)(1.3)	= 14.30 kN

Use ρ = 2.50 ton/m³ as the unit weight of reinforced concrete

Wall	W_w 0.50 (4.5 – 0.6) (24.53)	= 47.83 kN
Footing	W_f 0.60 (2.70) (24.53)	= 39.74 kN
Soil cover at toe	W_t 17.66 (1.50 – 0.60) (0.90)	= 14.30 kN
Soil cover at heel	W_h 17.66 (4.50 – 0.60) (1.30)	= 89.54 kN
		Total 205.71 kN

TABLE 29.8 Cantilever Retaining Wall Design Data with Uniformly Distributed Surcharge Load

s	h	1.0	1.2	1.4	1.6	1.8	2.0	2.2	2.4	2.6	2.8	3.0
	p	2.94	4.24	5.77	7.53	9.53	11.77	14.24	16.94	19.89	23.06	26.48
0.00	y	0.33	0.40	0.47	0.53	0.60	0.67	0.73	0.80	0.87	0.93	1.00
	m	0.98	1.69	2.69	4.02	5.72	7.84	10.44	13.56	17.24	21.53	26.48
	p	5.30	7.06	9.06	11.30	13.77	16.47	19.42	22.59	26.01	29.65	33.54
0.40	y	0.41	0.48	0.55	0.62	0.69	0.76	0.83	0.90	0.97	1.04	1.11
	m	2.16	3.39	5.00	7.03	9.53	12.55	16.14	20.33	25.19	30.75	37.07
	p	6.47	8.47	10.71	13.18	15.89	18.83	22.00	25.42	29.06	32.95	37.07
0.60	y	0.42	0.50	0.57	0.65	0.72	0.79	0.86	0.93	1.00	1.07	1.14
	m	2.75	4.24	6.15	8.54	11.44	14.91	18.98	23.72	29.17	35.36	42.36
	p	7.65	9.88	12.36	15.06	18.00	21.18	24.59	28.24	32.12	36.24	40.60
0.80	y	0.44	0.51	0.59	0.67	0.74	0.81	0.89	0.96	1.03	1.10	1.17
	m	3.33	5.08	7.30	10.04	13.34	17.26	21.83	27.11	33.14	39.98	47.66
	p	8.83	11.30	14.00	16.94	20.12	23.53	27.18	31.07	35.18	39.54	44.13
1.00	y	0.44	0.53	0.60	0.68	0.76	0.83	0.91	0.98	1.06	1.13	1.20
	m	3.92	5.93	8.46	11.55	15.25	19.61	24.68	30.50	37.12	44.59	52.95
	p	11.77	14.83	18.12	21.65	25.42	29.42	33.65	38.13	42.83	47.77	52.95
1.50	y	0.46	0.54	0.63	0.71	0.79	0.87	0.94	1.02	1.10	1.17	1.25
	m	5.39	8.05	11.34	15.31	20.02	25.50	31.80	38.97	47.06	56.12	66.19
	p	14.71	18.36	22.24	26.36	30.71	35.30	40.13	45.19	50.48	56.01	61.78
2.00	y	0.47	0.55	0.64	0.72	0.81	0.89	0.97	1.05	1.13	1.21	1.29
	m	6.86	10.17	14.22	19.08	24.78	31.38	38.92	47.45	57.01	67.65	79.43

s	h	3.2	3.4	3.6	3.8	4.0	4.2	4.4	4.6	4.8	5.0	5.2
	p	30.12	34.01	38.13	42.48	47.07	51.89	56.95	62.25	67.78	73.55	79.55
0.00	y	1.07	1.13	1.20	1.27	1.33	1.40	1.47	1.53	1.60	1.67	1.73
	m	32.13	38.54	45.75	53.81	62.76	72.65	83.53	95.45	108.45	122.58	137.88
	p	37.66	42.01	46.60	51.42	56.48	61.78	67.31	73.07	79.08	85.31	91.78
0.40	y	1.17	1.24	1.31	1.38	1.44	1.51	1.58	1.65	1.71	1.78	1.85
	m	44.18	52.14	61.00	70.80	81.59	93.41	106.31	120.35	135.56	151.99	169.70
	p	41.42	46.01	50.83	55.89	61.19	66.72	72.49	78.49	84.72	91.20	97.90
0.60	y	1.21	1.28	1.35	1.42	1.49	1.56	1.62	1.69	1.76	1.83	1.90
	m	50.21	58.95	68.63	79.30	91.00	103.79	117.70	132.80	149.11	166.70	185.61
	p	45.19	50.01	55.07	60.37	65.90	71.66	77.66	83.90	90.37	97.08	104.02
0.80	y	1.24	1.31	1.38	1.45	1.52	1.59	1.66	1.73	1.80	1.87	1.94
	m	56.23	65.75	76.25	87.79	100.41	114.17	129.09	145.25	162.67	181.41	201.52
	p	48.95	54.01	59.31	64.84	70.60	76.60	82.84	89.31	96.02	102.96	110.14
1.00	y	1.27	1.34	1.41	1.49	1.56	1.63	1.70	1.77	1.84	1.90	1.97
	m	62.26	72.55	83.88	96.29	109.83	124.54	140.48	157.70	176.23	196.12	217.43
	p	58.37	64.01	69.90	76.02	82.37	88.96	95.79	102.85	110.14	117.67	125.44
1.50	y	1.32	1.40	1.47	1.55	1.62	1.69	1.76	1.84	1.91	1.98	2.05
	m	77.32	89.55	102.94	117.53	133.36	150.49	168.96	188.82	210.12	232.89	257.20
	p	67.78	74.02	80.49	87.19	94.14	101.32	108.73	116.38	124.26	132.38	140.74
2.00	y	1.36	1.44	1.52	1.59	1.67	1.74	1.82	1.89	1.96	2.04	2.11
	m	92.38	106.56	122.00	138.77	156.90	176.44	197.44	219.94	244.00	269.67	296.97

TABLE 29.8 Cantilever Retaining Wall Design Data with Uniformly Distributed Surcharge Load

s	h	5.4	5.6	5.8	6.0	6.2	6.4	6.6	6.8	7.0	7.2	7.4
	p	85.78	92.25	98.96	105.90	113.08	120.50	128.14	136.03	144.15	152.50	161.09
0.00	y	1.80	1.87	1.93	2.00	2.07	2.13	2.20	2.27	2.33	2.40	2.47
	m	154.41	172.21	191.33	211.81	233.70	257.06	281.92	308.33	336.35	366.01	397.36
	p	98.49	105.43	112.61	120.03	127.67	135.56	143.68	152.03	160.62	169.45	178.51
0.40	y	1.92	1.98	2.05	2.12	2.18	2.25	2.32	2.39	2.45	2.52	2.59
	m	188.72	209.11	230.91	254.17	278.94	305.26	333.18	362.74	394.01	427.01	461.80
	p	104.85	112.02	119.44	127.09	134.97	143.09	151.44	160.03	168.86	177.92	187.22
0.60	y	1.96	2.03	2.10	2.17	2.23	2.30	2.37	2.44	2.50	2.57	2.64
	m	205.88	227.56	250.70	275.35	301.55	329.36	358.81	389.95	422.83	457.51	494.02
	p	111.20	118.61	126.26	134.15	142.27	150.62	159.21	168.04	177.10	186.39	195.92
0.80	y	2.01	2.07	2.14	2.21	2.28	2.35	2.41	2.48	2.55	2.62	2.69
	m	223.04	246.01	270.50	296.53	324.17	353.46	384.43	417.16	451.66	488.01	526.24
	p	117.55	125.20	133.09	141.21	149.56	158.15	166.98	176.04	185.33	194.86	204.63
1.00	y	2.04	2.11	2.18	2.25	2.32	2.39	2.46	2.52	2.59	2.66	2.73
	m	240.19	264.46	290.29	317.71	346.79	377.55	410.06	444.36	480.49	518.51	558.46
	p	133.44	141.68	150.15	158.86	167.80	176.98	186.39	196.04	205.93	216.05	226.40
1.50	y	2.12	2.19	2.26	2.33	2.40	2.47	2.54	2.61	2.68	2.75	2.82
	m	283.08	310.59	339.77	370.67	403.33	437.80	474.14	512.38	552.57	594.76	639.00
	p	149.33	158.15	167.21	176.51	186.04	195.81	205.81	216.05	226.52	237.23	248.17
2.00	y	2.18	2.26	2.33	2.40	2.47	2.54	2.62	2.69	2.76	2.83	2.90
	m	325.97	356.72	389.25	423.62	459.87	498.05	538.21	580.39	624.64	671.01	719.55

s	h	7.6	7.8	7.0	8.2	8.4	8.6	8.8	9.0	9.2	9.5	10.0
	p	169.92	178.98	144.15	197.81	207.57	217.58	227.81	238.29	248.99	265.50	294.18
0.00	y	2.53	2.60	2.33	2.73	2.80	2.87	2.93	3.00	3.07	3.17	3.33
	m	430.46	465.35	336.35	540.67	581.21	623.72	668.25	714.86	763.58	840.74	980.60
	p	187.80	197.34	160.62	217.10	227.34	237.82	248.52	259.47	270.65	287.86	317.71
0.40	y	2.65	2.72	2.45	2.85	2.92	2.99	3.06	3.12	3.19	3.29	3.46
	m	498.43	536.94	394.01	619.79	664.23	710.75	759.38	810.17	863.18	946.94	1098.27
	p	196.75	206.51	168.86	226.75	237.23	247.93	258.88	270.06	281.47	299.03	329.48
0.60	y	2.71	2.77	2.50	2.91	2.98	3.04	3.11	3.18	3.24	3.34	3.51
	m	532.41	572.73	422.83	659.36	705.75	754.26	804.94	857.83	912.98	1000.04	1157.11
	p	205.69	215.69	177.10	236.40	247.11	258.05	269.23	280.65	292.30	310.21	341.25
0.80	y	2.75	2.82	2.55	2.96	3.02	3.09	3.16	3.23	3.29	3.39	3.56
	m	566.39	608.53	451.66	698.92	747.26	797.78	850.50	905.49	962.78	1053.14	1215.94
	p	214.63	224.87	185.33	246.05	257.00	268.17	279.59	291.24	303.12	321.39	353.02
1.00	y	2.80	2.87	2.59	3.00	3.07	3.14	3.20	3.27	3.34	3.44	3.61
	m	600.38	644.32	480.49	738.48	788.78	841.29	896.06	953.14	1012.58	1106.24	1274.78
	p	236.99	247.82	205.93	270.17	281.71	293.47	305.48	317.71	330.19	349.34	382.43
1.50	y	2.89	2.96	2.68	3.10	3.17	3.24	3.31	3.38	3.44	3.55	3.72
	m	685.34	733.81	552.57	837.38	892.57	950.08	1009.97	1072.29	1137.07	1238.99	1421.87
	p	259.35	270.76	226.52	294.30	306.42	318.77	331.36	344.19	357.25	377.29	411.85
2.00	y	2.97	3.04	2.76	3.18	3.25	3.32	3.39	3.46	3.53	3.64	3.81
	m	770.30	823.30	624.64	936.28	996.35	1058.87	1123.88	1191.43	1261.57	1371.74	1568.96

Notes:

1. s = equivalent soil thickness for uniformly distributed surcharge load (m).
2. h = wall height (m); the distance from bottom of the footing to top of the wall.
3. Assume soil density = 2.0 ton/m^3.
4. Active earth pressure factor k_a = 0.30.

Hence, the maximum possible friction force at bottom of footing

$$F = f\,N_{tot} = 0.30\ (205.71) = 61.71\ \text{kN}$$

4. **Check Sliding**
Total lateral active force (include surcharge)

$$P_1 + P_2 = 16.34 + 59.01 = 75.35\ \text{kN}$$

Total maximum possible sliding resistant capacity

$$\text{Passive} + \text{friction} = 59.60 + 61.71 = 121.31\ \text{kN}$$

Sliding safety factor = 121.31/75.35 = 1.61 > 1.50 OK

5. **Check Overturning**
Take point *A* as the reference point
Resistant moment (do not include passive force for conservative)

Surcharge	14.30 (1.3/2 + 0.5 + 0.9)	= 29.32 kN·m
Soil cover at heel	89.54 (1.3/2 + 0.5 + 0.9)	= 183.56 kN·m
Wall	47.83 (0.5/2 + 0.9)	= 55.00 kN·m
Soil cover at toe	14.30 (0.9/2)	= 6.44 kN·m
Footing	39.74 (2.7/2)	= 53.65 kN·m
		Total 327.97 kN·m

Overturning moment

$$P_1(H/2) + P_2(H/3) = 16.34\ (4.5)/2 + 59.01\ (4.5)/3 = 125.28\ \text{kN·m}$$

Sliding safety factor = 327.97/125.28 = 2.62 > 1.50 OK

6. **Check Bearing**
Total vertical load

$$N_{tot} = 205.71\ \text{kN}$$

Total moment about center line of footing:

- Clockwise (do not include passive force for conservative)

Surcharge	14.30 (2.70/2 – 1.30/2)	= 10.01 kN·m
Soil cover @ heel	89.54 (2.70/2 – 1.30/2)	= 62.68 kN·m
		72.69 kN·m

- Counterclockwise

Wall	47.83 (2.70/2 – 0.9 – 0.5/2)	= 9.57 kN·m
Soil cover at toe	14.30 (2.70/2 – 0.9/2)	= 12.87 kN·m
Active earth pressure		= 125.28 kN·m
		147.72 kN·m

Total moment at bottom of footing

$$M_{tot} = 147.72\text{-}72.69 = 75.03 \text{ kN·m (counterclockwise)}$$

Maximum bearing stress

$$\sigma = N_{tot}/A \pm M_{tot}/S$$

where

$A = 2.70\ (1.0) = 2.70 \text{ m}^2$

$S = 1.0\ (2.7)^2/6 = 1.22 \text{ m}^3$

Therefore:

$$\sigma_{max} = 205.71/2.70 + 75.03/1.22 = 137.69 \text{ kPa}$$

$$< [\sigma] = 190 \text{ kPa}$$

and

$$\sigma_{min} = 205.71/2.70 - 75.03/1.22 = 14.69 \text{ kPa}$$

$$> 0 \qquad \text{OK}$$

7. **Flexure and Shear Strength**
   Both wall and footing sections need to be designed to have enough flexure and shear capacity.

29.3.4 Tieback Wall

The tieback wall is the proper structure type for cut sections. The tiebacks are prestressed anchor cables that are used to resist the lateral soil pressure. Compared with other types of retaining structures, the tieback wall has the least lateral deflection. Figure 29.23 shows the typical components and the basic lateral soil pressure distribution on a tieback wall.

The vertical spacing of tiebacks should be between 1.5 and 2.0 m to satisfy the required clearance for construction equipment. The slope angle of drilled holes should be 10 to 15° for grouting convenience. To minimize group effects, the spacing between the tiebacks should be greater than three times the tieback hole diameter, or 1.5 m minimum.

The bond strength for tieback design depends on factors such as installation technique, hole diameter, etc. For preliminary estimates, an ultimate bound strength of 90 to 100 kPa may be assumed. Based on construction experience, most tieback hole diameters are between 150 and 300 mm, and the tieback design capacity is in the range of 150 to 250 kN. Therefore, the corresponding lateral spacing of the tieback will be 2.0 to 3.0 m. The final tieback capacity must be proof-tested by stressing the test tieback at the construction site.

A tieback wall is built from the top down in cut sections. The wall details consist of a base layer and face layer. The base layer may be constructed by using vertical soldier piles with timber or concrete lagging between piles acting as a temporary wall. Then, a final cast-in-place reinforced-concrete layer will be constructed as the finishing layer of the wall. Another type of base layer that has been used effectively is cast-in-place "shotcrete" walls.

29.3.5 Reinforced Earth-Retaining Structure

The reinforced earth-retaining structure can be used in fill sections only. There is no practical height limit for this retaining system, but there will be a certain amount of lateral movement. The essential

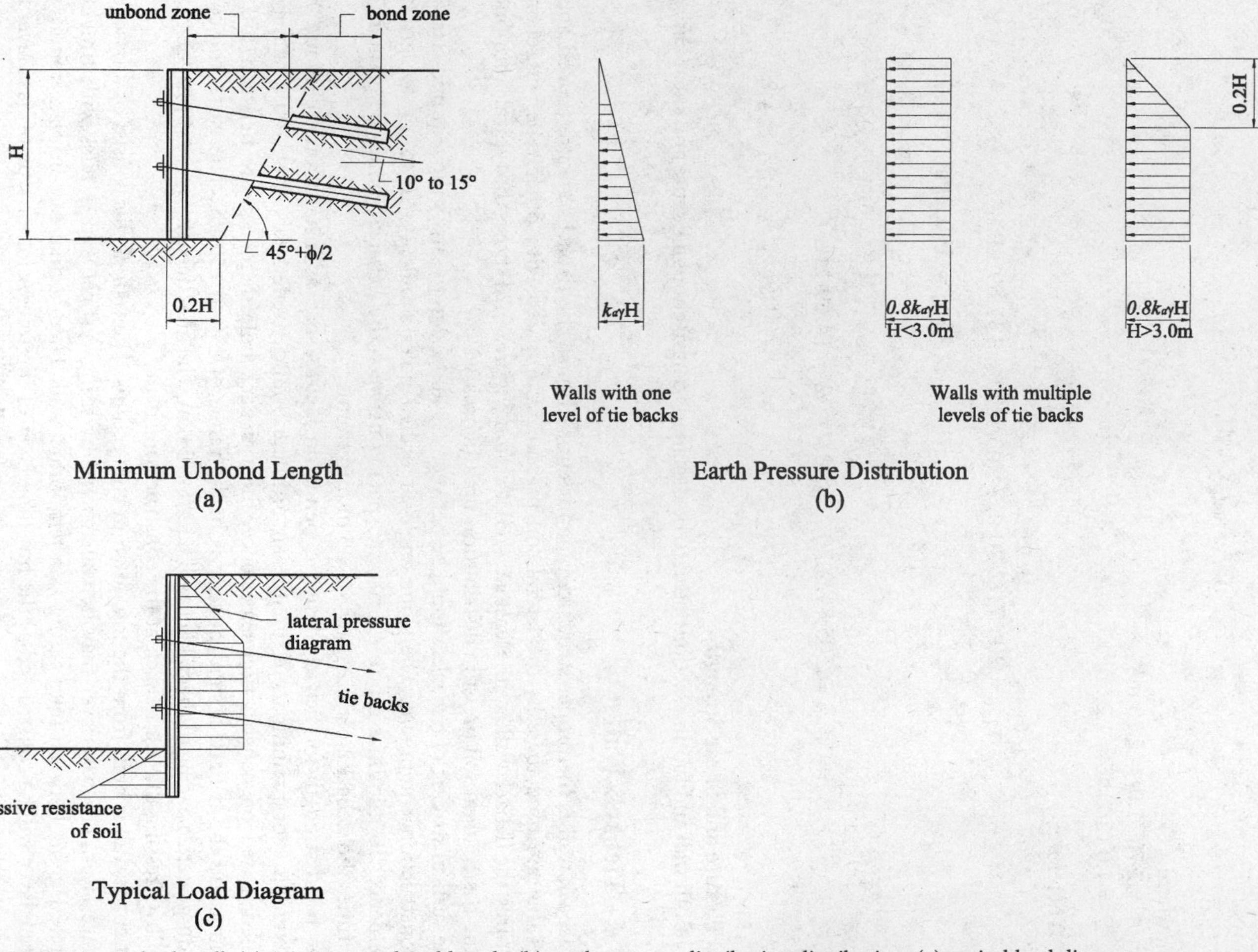

FIGURE 29.23 Tieback wall. (a) Minimum unbond length; (b) earth pressure distribution distribution; (c) typical load diagram.

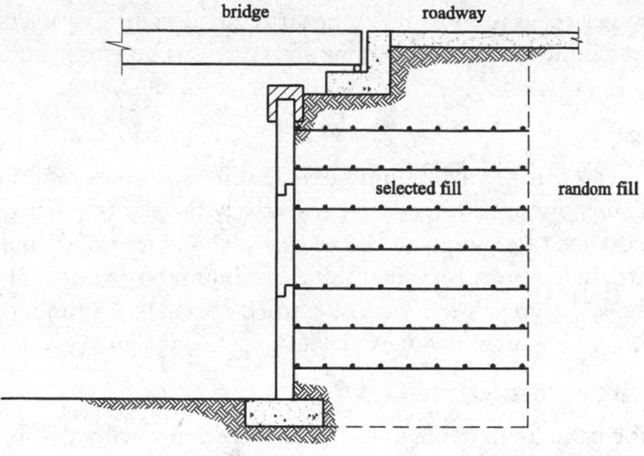

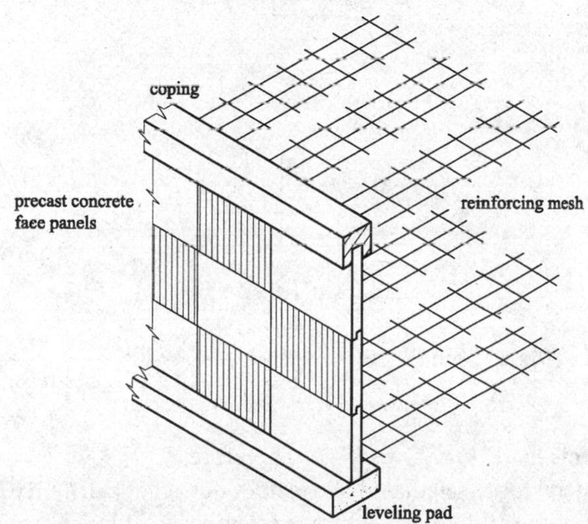

FIGURE 29.24 Mechanical Stabilized Earth (MSE).

concept is the use of multiple-layer strips or fibers to reinforce the fill material in the lateral direction so that the integrated fill material will act as a gravity retaining structure. Figure 29.24 shows the typical details of the MSE retaining structure.

Typically, the width of fill and the length of strips perpendicular to the wall face are on the order of 0.8 of the fill height. The effective life of the material used for the reinforcing must be considered. Metals or nondegradable fabrics are preferred.

Overturning and sliding need to be checked under the assumption that the reinforced soil body acts as a gravity retaining wall. The fiber strength and the friction effects between strip and fill material also need to be checked. Finally, the face panel needs to be designed as a slab which is anchored by the strips and subjected to lateral soil pressure.

29.3.6 Seismic Considerations for Retaining Structures

Seismic effects can be neglected in most retaining structure designs. For oversized retaining structures ($H > 10$ m), the seismic load on a retaining structure can be estimated by using the Mononobe–Okabe solution.

Soil Body ARS Factors

The factors k_v and k_h represent the maximum possible soil body acceleration values under seismic effects in the vertical and horizontal directions, respectively. Similar to other seismic load representations, the acceleration due to gravity will be used as the basic unit of k_v and k_h.

Unless a specific site study report is available, the maximum horizontal ARS value multiplied by 0.50 can be used as the k_h design value. Similarly, k_v will be equal to 0.5 times the maximum vertical ARS value. If the vertical ARS curve is not available, k_v can be assigned a value from $0.1k_h$ to $0.3k_h$.

Earth Pressure with Seismic Effects

Figure 29.25 shows the basic loading diagram for earth pressure with seismic effects. Similar to a static load calculation, the active force per unit length of wall (P_{ac}) can be determined as:

$$P_{ae} = \frac{1}{2} k_{ae} \gamma \left(1 - k_v\right) H^2 \tag{29.12}$$

where

$$\theta' = \tan^{-1}\left[\frac{k_h}{1 - k_v}\right] \tag{29.13}$$

$$k_{ae} = \frac{\sin^2(\phi + \beta - \theta')}{\cos\theta' \sin^2\beta \sin(\beta - \theta' - \delta)\left[1 + \sqrt{\dfrac{\sin(\phi + \delta)\sin(\phi - \theta' - \alpha)}{\sin(\beta - \theta' - \delta)\sin(\alpha + \beta)}}\right]^2} \tag{29.14}$$

Note that with no seismic load, $k_v = k_h = \theta' = 0$. Therefore, $K_{ac} = K_a$.

The resultant total lateral force calculated above does not act at a distance of $H/3$ from the bottom of the wall. The following simplified procedure is often used in design practice:

- Calculate P_{ae} (total active lateral earth pressure per unit length of wall)
- Calculate $P_a = ½\, k_a \gamma H^2$ (static active lateral earth pressure per unit length of wall)

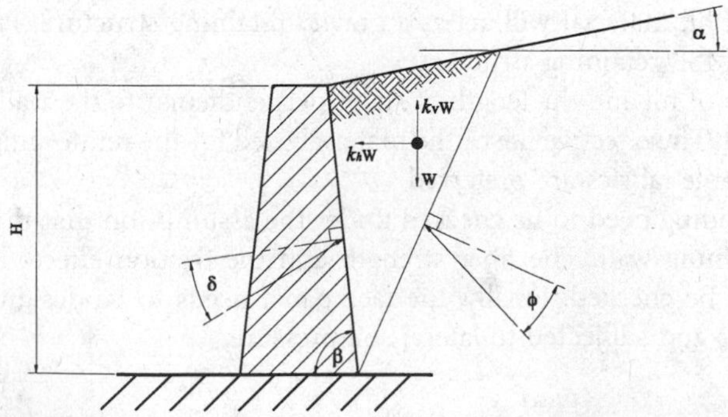

FIGURE 29.25 Load diagram for Earth pressure with seismic effects.

- Calculate $\Delta P = P_{ae} - P_a$
- Assume P_a acts at a distance of $H/3$ from the bottom of the wall
- Assume ΔP acts at a distance of $0.6H$ from the bottom of the wall

The total earth pressure, which includes seismic effects P_{ae}, should always be bigger than the static force P_a. If the calculation results indicate $\Delta P < 0$; use $k_v = 0$.

Using a procedure similar to the active Earth pressure calculation, the passive Earth pressure with seismic effects can be determined as follows:

$$P_{pe} = \frac{1}{2} k_{pe} \gamma (1 - k_v) H^2 \tag{29.15}$$

where

$$\theta' = \tan^{-1}\left[\frac{k_h}{1 - k_v}\right]$$

$$k_{pe} = \frac{\sin^2(\beta + \theta' - \phi)}{\cos\theta' \sin^2\beta \sin(\beta + \theta' + \delta - 90)\left[1 - \sqrt{\dfrac{\sin(\phi + \delta)\sin(\phi - \theta' + \alpha)}{\sin(\beta + \theta' + \delta)\sin(\alpha + \beta)}}\right]^2} \tag{29.16}$$

Note that, with no seismic load, $k_{pc} = k_p$.

References

1. AASHTO, *Standard Specifications for Highway Bridges*, 16th ed., American Association of State Highway and Transportation Officials, Washington, D.C., 1996.
2. *Bridge Memo to Designers Manual*, Department of Transportation, State of California, Sacramento.
3. Brian H. Maroney, Matt Griggs, Eric Vanderbilt, Bruce Kutter, Yuk H. Chai and Karl Romstad, Experimental measurements of bridge abutment behavior, in *Proceeding of Second Annual Seismic Research Workshop*, Division of Structures, Department of Transportation, Sacramento, CA, March 1993.
4. Brian H. Maroney and Yuk H. Chai, Bridge abutment stiffness and strength under earthquake loadings, in *Proceedings of the Second International Workshop of Seismic Design and Retroffitting of Reinforced Concrete Bridges*, Queenstown, New Zealand, August 1994.
5. Rakesh K. Goel, Earthquake behavior of bridge with integral abutment, in *Proceedings of the National Seismic Conference on Bridges and Highways*, Sacramento, CA, July 1997.
6. E. C. Sorensen, Nonlinear soil-structure interaction analysis of a 2-span bridge on soft clay foundation, in *Proceedings of the National Seismic Conference on Bridges and Highways*, Sacramento, CA, July 1997.
7. AEAR, *Manual for Railway Engineering*, 1996.
8. Braja M. Das, *Principles of Foundation Engineering*, PWS-KENT Publishing Company, Boston, MA, 1990.
9. T. William Lambe and Robert V. Whitman, *Soil Mechanics*, John Wiley & Sons, New York, 1969.
10. Gregory P. Tschebotarioff, *Foundations, Retaining and Earth Structures*, 4th ed., McGraw-Hill, New York, 1973.
11. Joseph E. Bowles, *Foundation Analysis and Design*, McGraw-Hill, New York, 1988.
12. Whitney Clark Huntington, *Earth Pressure and Retaining Walls*, John Wiley & Sons, New York.

30

Geotechnical Considerations

Thomas W. McNeilan
Fugro West, Inc.

James Chai
California Department of Transportation

30.1 Introduction

A complete geotechnical study of a site will (1) determine the subsurface stratigraphy and stratigraphic relationships (and their variability), (2) define the physical properties of the earth materials, and (3) evaluate the data generated and formulate solutions to the project-specific and site-specific geotechnical issues. Geotechnical issues that can affect a project can be broadly grouped as follows:

- *Foundation Issues* — Including the determination of the strength, stability, and deformations of the subsurface materials under the loads imposed by the structure foundations, in and beneath slopes and cuts, or surrounding the subsurface elements of the structure.
- *Earth Pressure Issues* — Including the loads and pressures imposed by the earth materials on foundations and against supporting structures, or loads and pressures created by seismic (or other) external forces.

0-8493-7434-0/00/$0.00+$.50

- *Construction and Constructibility Considerations*— Including the extent and characteristics of materials to be excavated, and the conditions that affect deep foundation installation or ground improvement.
- *Groundwater Issues*— Including occurrence, hydrostatic pressures, seepage and flow, and erosion.

Site and subsurface characteristics directly affect the choice of foundation type, capacity of the foundation, foundation construction methods, and bridge cost. Subsurface and foundation conditions also frequently directly or indirectly affect the route alignment, bridge type selection, and/or foundation span lengths. Therefore, an appropriately scoped and executed foundation investigation and site characterization should:

1. Provide the required data for the design of safe, reliable, and economic foundations;
2. Provide data for contractors to use to develop appropriate construction cost estimates;
3. Reduce the potential for a "changed condition" claim during construction.

In addition, the site investigation objectives frequently may be to

1. Provide data for route selection and bridge type evaluation during planning and preliminary phase studies;
2. Provide data for as-built evaluation of foundation capacity, ground improvement, or other similar requirements.

For many projects, it is appropriate to conduct the geotechnical investigation in phases. For the first preliminary (or reconnaissance) phase, either a desktop study using only historical information or a desktop study and a limited field exploration program may be adequate. The results of the first-phase study can then be used to develop a preliminary geologic model of the site, which is used to determine the key foundation design issues and plan the design-phase site investigation.

Bridge projects may require site investigations to be conducted on land, over water, and/or on marginal land at the water's edge. Similarly, site investigations for bridge projects can range from conventional, limited-scope investigations for simple overpasses and grade separations to major state-of-the-practice investigations for large bridges over major bodies of water.

This chapter includes discussions of

- Field exploration techniques;
- Definition of the requirements for and extent of the site investigation program;
- Evaluation of the site investigation results and development/scoping of the laboratory testing program;
- Data presentation and site characterization.

The use of the site characterization results for foundation design is included in subsequent chapters.

30.2 Field Exploration Techniques

For the purpose of the following discussion, we have divided field exploration techniques into the following groupings:

- Borings (including drilling, soil sampling, and rock-coring techniques)
- Downhole geophysical logging
- *In situ* testing— including cone penetration testing (CPT) and vane shear, pressure meter and dilatometer testing)
- Test pits and trenches
- Geophysical survey techniques

FIGURE 30.1 Drilling methods. (a) On land; (b) over water; (c); on marginal land.

30.2.1 Borings and Drilling Methods

Drilled soil (or rock) borings are the most commonly used subsurface exploration technique. The drilled hole provides the opportunity to collect samples of the subsurface through the use of a variety of techniques and samplers. In addition to sample collection, drilling observations during the advancement of the borehole provide an important insight to the subsurface conditions. Drilling methods can be used for land, over water, and marginal land sites (Figure 30.1). It should be noted that the complexity introduced when working over water or on marginal land may require more-sophisticated and more-specialized equipment and techniques, and will significantly increase costs.

30.2.1.1 Wet (Mud) Rotary Borings

Wet rotary drilling is the most commonly used drilling method for the exploration of soil and rock, and also is used extensively for oil exploration and water well installation. It is generally the preferred method for (1) over water borings; (2) where groundwater is shallow; and (3) where the subsurface includes soft, squeezing, or flowing soils.

With this technique, the borehole is advanced by rapid rotation of the drill bit that cuts, chips, and grinds the material at the bottom of the borehole. The cuttings are removed from the borehole by circulating water or drilling fluid down through the drill string to flush the cuttings up through the annular space of the drill hole. The fluids then flow into a settling pit or solids separator. Drilling fluid is typically bentonite (a highly refined clay) and water, or one of a number of synthetic products. The drilling fluids are used to flush the cuttings from the hole, compensate the fluid pressure, and stabilize borehole sidewalls. In broken or fractured rock, coarse gravel and cobbles, or other formations with voids, it may be necessary to case the borehole to prevent loss of circulation. Wet rotary drilling is conducive to downhole geophysical testing, although the borehole must be thoroughly flushed before conducting some types of logging.

30.2.1.2 Air Rotary Borings

The air rotary drilling technology is similar to wet rotary except that the cuttings are removed with the circulation of high-pressure air rather than a fluid. Air rotary drilling techniques are typically used in hard bedrock or other conditions where drill hole stability is not an overriding issue. In very hard bedrock, a percussion hammer is often substituted for the bit. Air rotary drilling is conducive to downhole geophysical testing methods.

30.2.1.3 Bucket-Auger Borings

The rotary bucket is similar to a large- (typically 18- to 24-in.)-diameter posthole digger with a hinged bottom. The hole is advanced by rotating the bucket at the end of a kelly bar while pressing it into the soil. The bucket is removed from the hole to be emptied. Rotary-bucket-auger borings are used in alluvial soils and soft bedrock. This method is not always suitable in cobbly or rocky soils, but penetration of hard layers is sometimes possible with special coring buckets. Bucket-auger borings also may be unsuitable below the water table, although drilling fluids can be used to stabilize the borehole.

The rotary-bucket-auger drilling method allows an opportunity for continuous inspection and logging of the stratigraphic column of materials, by lowering the engineer or geologist on a platform attached to a drill rig winch. It is common in slope stability and fault hazards studies to downhole log 24-in.-diameter, rotary-bucket-auger boreholes advanced with this method.

30.2.1.4 Hollow-Stem-Auger Borings

The hollow-stem-auger drilling technique is frequently used for borings less than 20 to 30 m deep. The proliferation of the hollow-stem-auger technology in recent years occurred as the result of its use for contaminated soils and groundwater studies. The hollow-stem-auger consists of sections of steel pipe with welded helical flanges. The shoe end of the pipe has a hollow bit assembly that is plugged while rotating and advancing the auger. That plug is removed for advancement of the sampling device ahead of the bit.

Hollow-stem-auger borings are used in alluvial soils and soft bedrock. This method is not always suitable where groundwater is shallow or in cobbly and rocky soils. When attempting to sample loose, saturated sands, the sands may flow into the hollow auger and produce misleading data. The hollow-stem-auger drill hole is not conducive to downhole geophysical testing methods.

30.2.1.5 Continuous-Flight-Auger Borings

Continuous-flight-auger borings are similar to the hollow-stem-auger drilling method except that the auger must be removed for sampling. With the auger removed, the borehole is unconfined and hole instability often results. Continuous-flight-auger drill holes are used for shallow exploration above the groundwater level.

30.2.2 Soil-Sampling Methods

There are several widely used methods for recovering samples for visual classification and laboratory testing.

30.2.2.1 Driven Sampling

Driven sampling using standard penetration test (SPT) or other size samplers is the most widely used sampling method. Although this sampling method recovers a disturbed sample, the "blow count" measured with this type of procedure provides a useful index of soil density or strength. The most commonly used blow count is the SPT blow count (also referred to as the N-value). Although the N-value is an approximate and imprecise measurement (its value is affected by many operating factors that are part of the sampling process, as well as the presence of gravel or cementation), various empirical relationships have been developed to relate N-value to engineering and performance properties of the soils.

30.2.2.2 Pushed Samples

A thin-wall tube (or in some cases, other types of samplers) can be pushed into the soil using hydraulic pressure from the drill rig, the weight of the drill rod, or a fixed piston. Pushed sampling generally recovers samples that are less disturbed than those recovered using driven-sampling techniques. Thus, laboratory tests to determine strength and volume change characteristics should preferably be conducted on pushed samples rather than driven samples. Pushed sampling is the preferred sampling method in clay soils. Thin-wall samples recovered using push-sampling techniques can either be extruded in the field or sealed in the tubes.

30.2.2.3 Drilled or Cored Samplers

Drilled-in samplers also have application in some types of subsurface conditions, such as hard soil and soft rock. With these types of samplers (e.g., Denison barrel and pitcher barrel), the sample barrel is either cored into the sediment or rock or is advanced inside the drill rod while the rod is advanced.

30.2.3 Rock Coring

The two rock-coring systems most commonly used for engineering applications are the conventional core barrel and wireline (retrievable) system. At shallow depths above the water table, coring also sometimes can be performed with an air or a mist system.

Conventional core barrels consist of an inner and outer barrel with a bit assembly. To obtain a core at a discrete interval; (1) the borehole is advanced to the top of the desired interval, (2) the drill pipe is removed, (3) the core barrel/bit is placed on the bottom of the pipe, and (4) the assembly is run back to the desired depth. The selected interval is cored and the core barrel is removed to retrieve the core. Conventional systems typically are most effective at shallow depths or in cases where only discrete samples are required.

In contrast, wireline coring systems allow for continuous core retrieval without removal of the drill pipe/bit assembly. The wireline system has a retrievable inner core barrel that can be pulled to the surface on a wireline after each core run.

Variables in the coring process include the core bit type, fluid system, and drilling parameters. There are numerous bit types and compositions that are applicable to specific types of rock; however, commercial diamond or diamond-impregnated bits are usually the preferred bit from a core recovery and quality standpoint. Tungsten carbide core bits can sometimes be used in weak rock or in high-clay-content rocks. A thin bentonite mud is the typical drilling fluid used for coring. Thick mud can clog the small bit ports and is typically avoided. Drilling parameters include the revolutions per minute (RPM) and weight on bit (WOB). Typically, low RPM and WOB are used to start the core run and then both values are increased.

Rock engineering parameters include percent recovery, rock quality designation (RQD), coring rate, and rock strength. Percent recovery is a measure of the core recovery vs. the cored length, whereas RQD is a measure of the intact core pieces longer than 4 in. vs. the cored length. Both values typically increase as the rock mass becomes less weathered/fractured with depth; however, both values are highly dependent on the type of rock, amount of fracturing, etc. Rock strength (which is typically measured using unconfined triaxial compression test per ASTM guidelines) is used to evaluate bearing capacity, excavatability, etc.

30.2.4 *In Situ* Testing

There are a variety of techniques that use instrumented probes or testing devices to measure soil properties and conditions in the ground, the more widely used of which are described below. In contrast to sampling that removes a sample from its *in situ* stress conditions, *in situ* testing is used to measure soil and rock properties in the ground at their existing state of stress. The various *in*

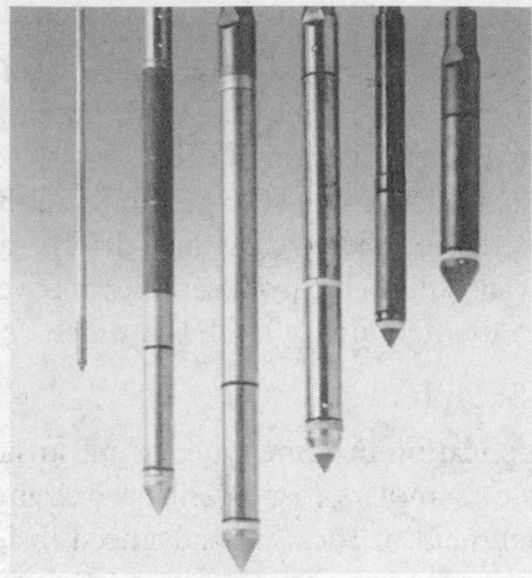

FIGURE 30.2 CPT cones.

situ tests can either be conducted in a borehole or as a continuous sounding from the ground surface. Except as noted, those techniques are not applicable to rock.

30.2.4.1 Cone Penetration Test Soundings

CPT sounding is one of the most versatile and widely used *in situ* test. The standard CPT cone consists of a 1.4-in.-diameter cone with an apex angle of 60°, although other cone sizes are available for special applications (Figure 30.2). The cone tip resistance beneath the 10-cm^2 cone tip and the friction along the 150 cm^2 friction sleeve are measured with strain gauges and recorded electronically at 1- or 2-cm intervals as the cone is advanced into the ground at a rate of about 2 cm/s. In addition to the tip and sleeve resistances, many cones also are instrumented to record pore water pressure or other parameters as the cone is advanced.

Because the CPT soundings provide continuous records of tip and sleeve resistances (and frequently pore pressure) vs. depth (Figure 30.3), they provide a continuous indicator of soil and subsurface conditions that are useful in defining soil stratification. Numerous correlations between the CPT measurements have been developed to define soil type and soil classification. In addition, empirical correlations have been published to relate the cone tip and sleeve friction resistances to engineering behavior, including undrained shear strength of clay soils and relative density and friction of granular soils.

Most land CPTs are performed as continuous soundings using large 20-ton cone trucks (Figure 30.4a), although smaller, more portable track-mounted equipment is also available. CPT soundings are commonly extended down to more than 20 to 50 m. CPT soundings also can be performed over water from a vessel using specialized equipment (Figure 30.4b) deployed by a crane or from a stern A-frame. In addition, downhole systems have been developed to conduct CPTs in boreholes during offshore site investigations. With a downhole system, CPT tests are interspersed with soil sampling to obtain CPT data to more than 100 m in depth.

30.2.4.2 *In Situ* Vane Shear Tests

The undrained shear strength of clay soils can be measured *in situ* using a vane shear test. This test is conducted by measuring the torque required to rotate a vane of known dimensions. The test can be conducted from the ground surface by attaching a vane blade onto a rod or downhole below the bottom of a borehole with a drop-in remote vane (Figure 30.5). The downhole vane is preferable, since the torque required to rotate the active rotating vane is not affected by the torque of the rod. The downhole vane is used both for land borings and over-water borings.

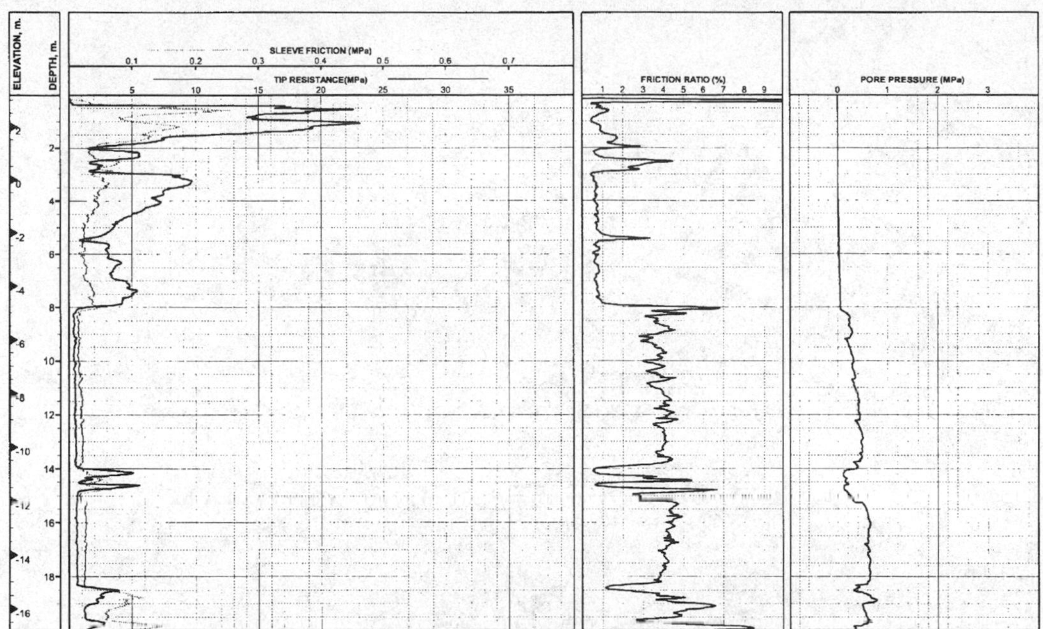

FIGURE 30.3 CPT data provide a continuous record of *in situ* conditions.

30.2.4.3 Pressure Meter and Dilatometer Tests

Pressure meter testing is used to measure the *in situ* maximum and average shear modulus of the soil or rock by inflating the pressure meter against the sidewalls of the borehole. The stresses, however, are measured in a horizontal direction, not in the vertical direction as would occur under most types of foundation loading. A test is performed by lowering the tool to the selected depth and expanding a flexible membrane through the use of hydraulic fluid. As the tool is inflated, the average displacement of the formation is measured with displacement sensors beneath the membrane, which is protected by stainless steel strips. A dilatometer is similar to a pressure meter, except that the dilatometer consists of a flat plate that is pushed into the soil below the bottom of the borehole. A dilatometer is not applicable to hard soils or rock.

30.2.5 Downhole Geophysical Logging

Geophysical logs are run to acquire data about the formation or fluid penetrated by the borehole. Each log provides a continuous record of a measured value at a specific depth in the boring, and is therefore useful for interpolating stratigraphy between sample intervals. Most downhole geophysical logs are presented as curves on grid paper or as electronic files (Figure 30.6). Some of the more prevalent geophysical tools, which are used for geotechnical investigations, are described below.

- *Electrical logs* (*E-logs*) include resistivity, induction, and spontaneous potential (SP) logs. Resistivity and induction logs are used to determine lithology and fluid type. A resistivity log is used when the borehole is filled with a conductive fluid, while an induction log is used when the borehole is filled with a non- or low-conductivity fluid. Resistivity tools typically require an open, uncased, fluid-filled borehole. Clay formations and sands with higher salinity will have low resistivity, while sands with fresh water will have higher resistivity values. Hard rock and dry formations have the highest resistivity values. An SP log is often used in suite with a resistivity or induction log to provide further information relative to formation permeability and lithology.

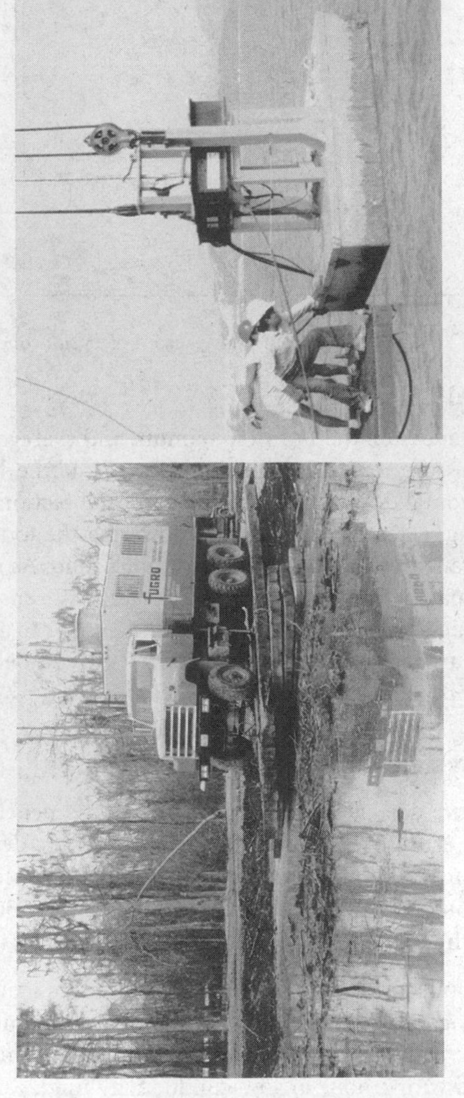

FIGURE 30.4 CPT sounding methods. (a) On land; (b) over water.

FIGURE 30.5 *In situ* vane shear device.

- *Suspension (velocity) logs* are used to measure the average primary, compression wave, and shear wave velocities of a 1-m-high segment of the soil and rock column surrounding the borehole. Those velocities are determined by measuring the elapsed time between arrivals of a wave propagating upward through the soil/rock column. The suspension probe includes both a shear wave source and a compression wave source, and two biaxial receivers that detect the source waves. This technique requires an open, fluid-filled hole.
- *Natural gamma logs* measure the natural radioactive decay occurring in the formation to infer soil or rock lithology. In general, clay soils will exhibit higher gamma counts than granular soils, although decomposed granitic sands are an exception to that generality. Gamma logs can be run in any salinity fluid as well as air, and also can be run in cased boreholes.
- *Caliper logs* are used to measure the diameter of a borehole to provide insight relative to caving and swelling. An accurate determination of borehole diameter also is important for the interpretation of other downhole logs.
- *Acoustic televiewer and digital borehole logs* are conducted in rock to image the rock surface within the borehole (Figure 30.7). These logs use sound in an uncased borehole to create an oriented image of the borehole surface. These logs are useful for determining rock layering, bedding, and fracture identification and orientation.
- *Crosshole, downhole, and uphole shear wave velocity measurements* are used to determine the primary and shear wave velocities either to determine the elastic soil properties of soil and rock or to calibrate seismic survey measurements. With the crosshole technique, the travel time is measured between a source in one borehole and a receiver in a second borehole. This technique can be used to measure directly the velocities of various strata. For downhole and uphole logs, the travel time is measured between the ground surface and a downhole source or receiver. Tests are conducted with the downhole source or receiver at different depths. These measurements should preferably be conducted in cased boreholes.

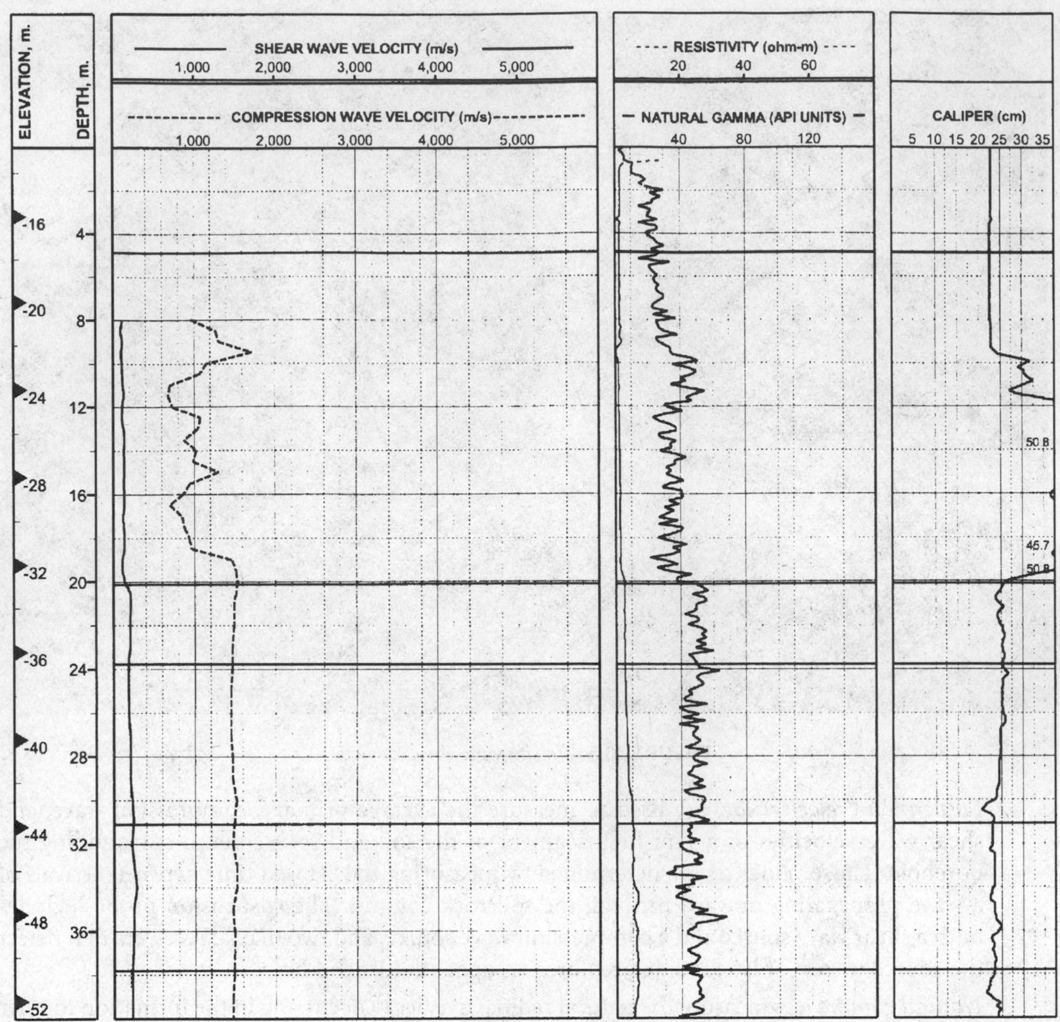

FIGURE 30.6 Example of downhole geophysical log.

30.2.6 Test Pits and Trenches

Where near-surface conditions are variable or problematic, the results of borings and *in situ* testing can be supplemented by backhoe-excavated or hand-excavated test pits or trenches. These techniques are particularly suitable for such purposes as: (1) collecting hand-cut, block samples of sensitive soils; (2) evaluating the variability of heterogeneous soils; (3) evaluating the extent of fill or rubble, (4) determining depth to groundwater, and (5) the investigation of faulting.

30.2.7 Geophysical Survey Techniques

Noninvasive (compared with drilling methods) geophysical survey techniques are available for remote sensing of the subsurface. In contrast to drilling and *in situ* testing methods, the geophysical survey methods explore large areas rapidly and economically. When integrated with boring data, these methods often are useful for extrapolating conditions between borings (Figure 30.8). Techniques are applicable either on land or below water. Some of the land techniques also are applicable for marginal land or in the shallow marine transition zone. Geophysical survey techniques can be used individually or as a group.

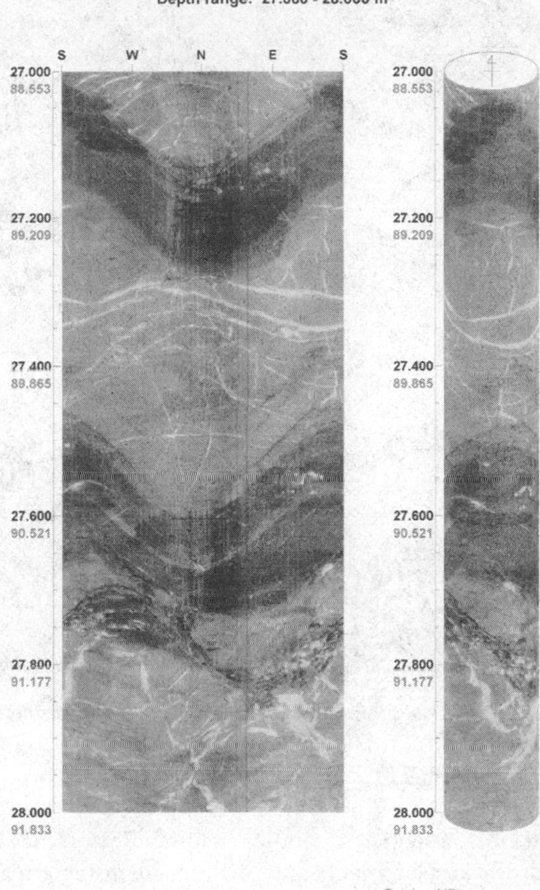

FIGURE 30.7 Example of digital borehole image in rock.

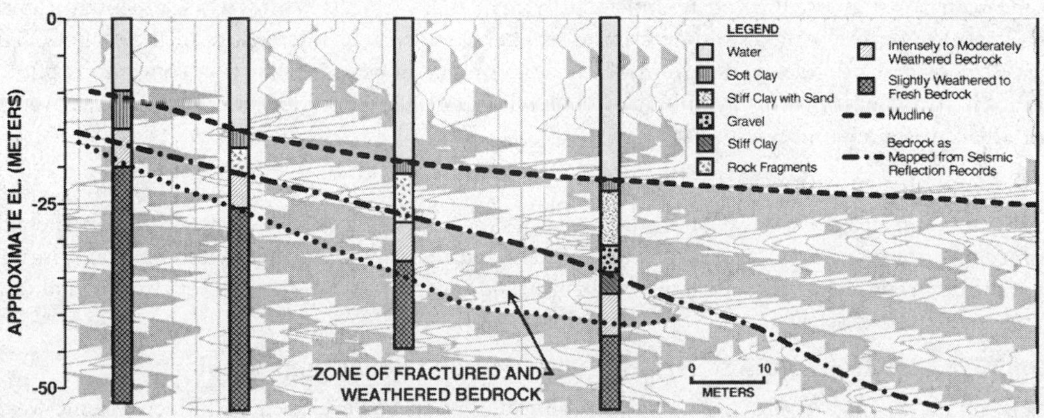

FIGURE 30.8 Example integration of seismic reflection and boring data.

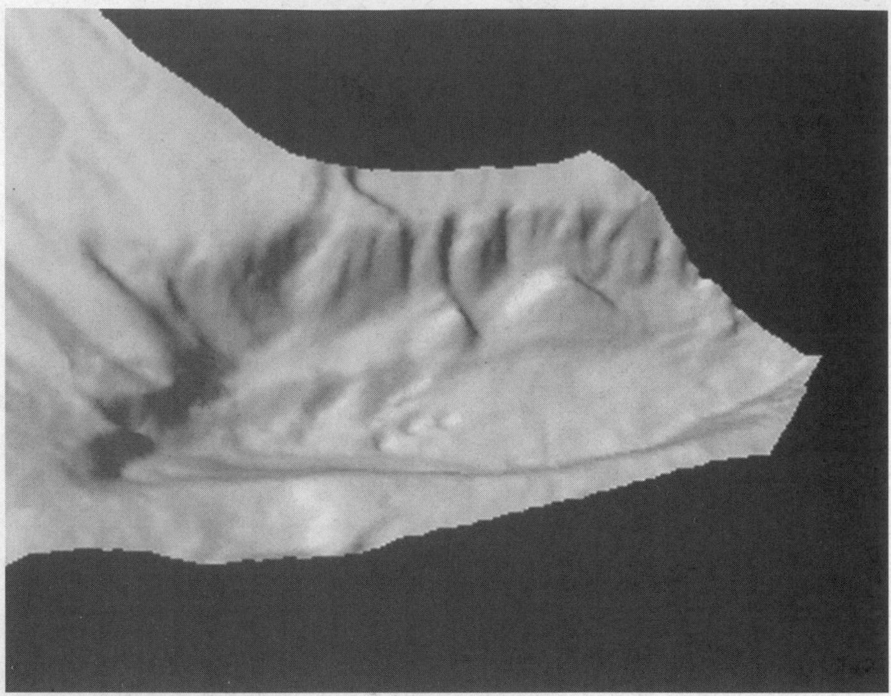

FIGURE 30.9 Multibeam image of river channel bathymetry.

30.2.7.1 Hydrographic Surveys

Hydrographic surveys provide bathymetric contour maps and/or profiles of the seafloor, lake bed, or river bottom. Water depth measurements are usually made using a high-frequency sonic pulse from a depth sounder transducer mounted on a survey vessel. The choice of depth sounder system (single-beam, multifrequency, multibeam, and swath) is dependent upon water depths, survey site conditions, and project accuracy and coverage requirements. The use and application of more-sophisticated multibeam systems (Figure 30.9) has increased dramatically within the last few years.

30.2.7.2 Side-Scan Sonar

Side-scan sonar is used to locate and identify man-made objects (shipwrecks, pipelines, cables, debris, etc.) on the seafloor and determine sediment and rock characteristics of the seafloor. The side-scan sonar provides a sonogram of the seafloor that appears similar to a continuous photographic strip (Figure 30.10). A mosaic of the seafloor can be provided by overlapping the coverage of adjacent survey lines.

30.2.7.3 Magnetometer

A magnetometer measures variations in the earth's magnetic field strength that result from metallic objects (surface or buried), variations in sediment and rock mineral content, and natural (diurnal) variations. Data are used to locate and identify buried objects for cultural, environmental, and archaeological site clearances.

30.2.7.4 High-Resolution Seismic Reflection and Subbottom Profilers

Seismic images of the subsurface beneath the seafloor can be developed by inducing sonic waves into the water column from a transducer, vibrating boomer plate, sparker, or small air or gas gun. Reflections of the sonic energy from the mudline and subsurface soils horizons are recorded to provide an image of the subsurface geologic structure and stratigraphy along the path of the survey

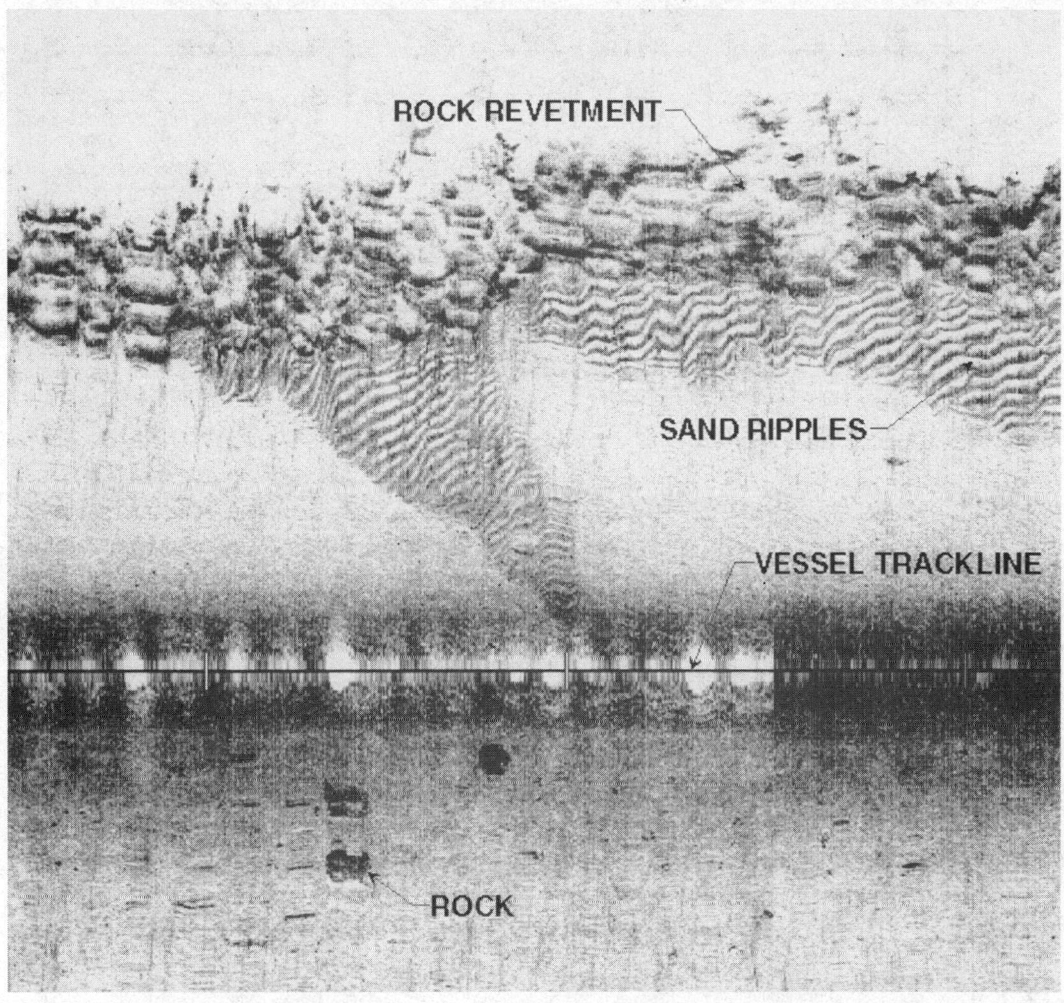

FIGURE 30.10 Side-scan sonar image of river bottom.

vessel. The effective depth of a system and resolution of subsurface horizons depend on a number of variables, including the system energy, output frequency spectrum, the nature of the seafloor, and the subsea sediments and rocks. Seismic reflection data are commonly used to determine the geologic structure (stratigraphy, depth to bedrock, folds, faults, subsea landslides, gas in sediments, seafloor seeps, etc.) and evaluate the horizon continuity between borings (Figure 30.11).

30.2.7.5 Seismic Refraction

Seismic refraction measurements are commonly used on land to estimate depth to bedrock and groundwater and to detect bedrock faulting. Measured velocities are also used for estimates of rippability and excavation characteristics. In the refraction technique, sonic energy is induced into the ground and energy refracted from subsurface soil and rock horizons is identified at a series of receivers laid out on the ground. The time–distance curves from a series of profiles are inverted to determine depths to various subsurface layers and the velocity of the layers. The data interpretation can be compromised where soft layers underlie hard layers and where the horizons are too thin to be detected by refraction arrivals at the surface. The technique also can be used in shallow water (surf zones, lakes, ponds, and river crossings) using bottom (bay) cables.

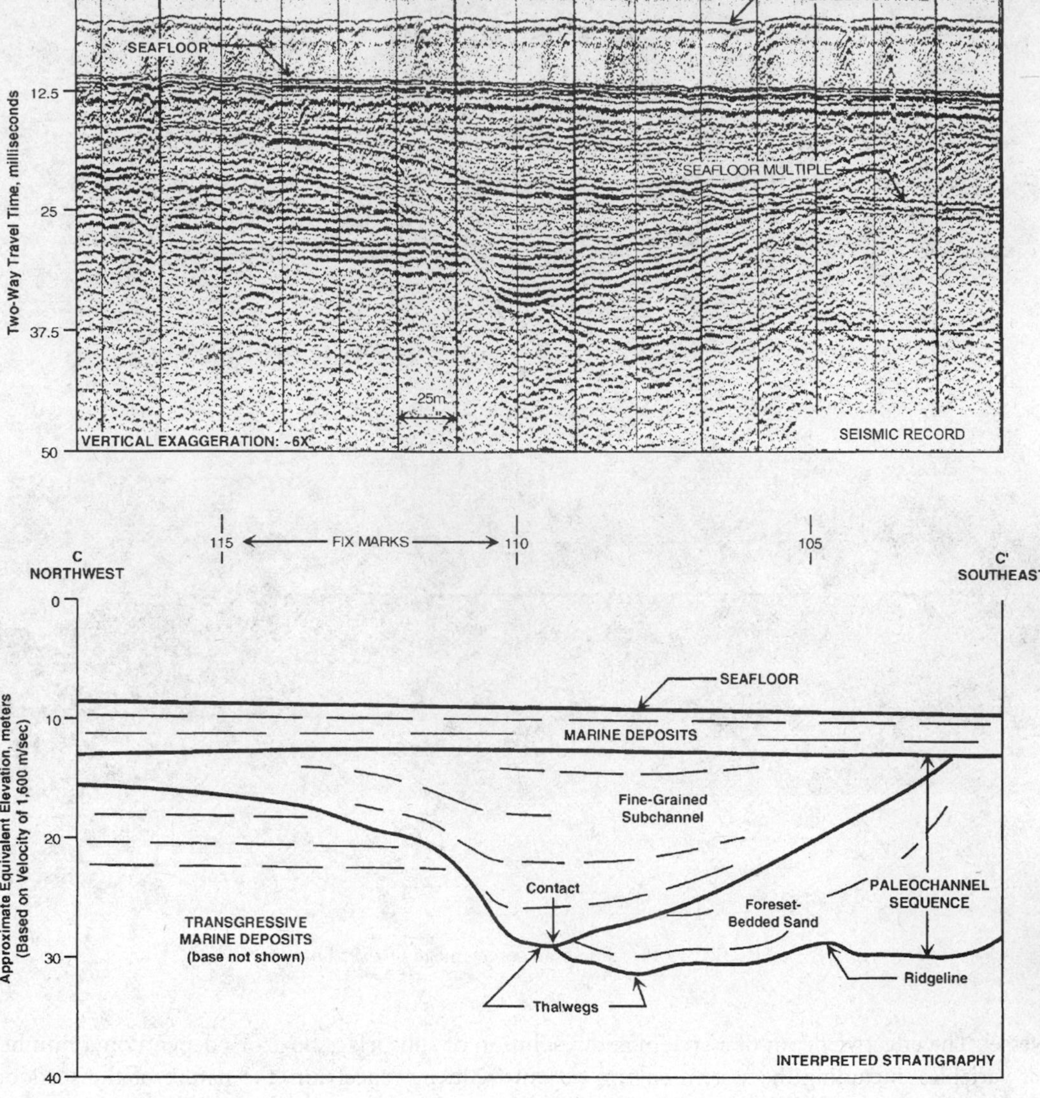

FIGURE 30.11 Interpreted stratigraphic relationships from seismic reflection data.

30.2.7.6 Ground Penetrating Radar Systems

Ground Penetrating Radar (GPR) systems measure the electromagnetic properties of the subsurface to locate buried utilities or rebar, estimate pavement thickness, interpret shallow subsurface stratigraphy, locate voids, and delineate bedrock and landslide surfaces. GPR also can be used in arctic conditions to estimate ice thickness and locate permafrost. Depths of investigation are usually limited to 50 ft or less. Where the surface soils are highly conductive, the effective depth of investigation may be limited to a few feet.

30.2.7.7 Resistivity Surveys

Resistivity surveys induce currents into the ground to locate buried objects and to investigate shallow groundwater. As electrodes are moved in specific patterns of separation, the resistivity is measured

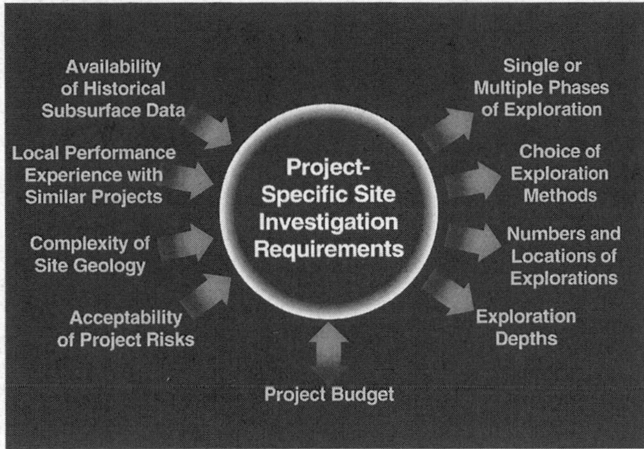

FIGURE 30.12 Key factors to consider when defining site investigation requirements.

and inverted to produce depth sections and contour maps of subsurface resistivity values. This method is used to identify and map subsurface fluids, including groundwater, surface and buried chemical plumes, and to predict corrosion potential.

30.2.8 Groundwater Measurement

Groundwater conditions have a profound effect on foundation design, construction, and performance. Thus, the measurement of groundwater depth (or depth of water when drilling over water) is one of the most fundamentally important elements of the site investigation. In addition to the measurement of the water level, the site investigation should consider and define the potential for artesian or perched groundwater. It is also important to recognize that groundwater levels may change with season, rainfall, or other temporal reasons. All groundwater and water depth measurements should document the time of measurement and, where practical, should determine variations in depth over some period of elapsed time. To determine the long-term changes in water level, it is necessary to install and monitor piezometers or monitoring wells.

30.3 Defining Site Investigation Requirements

Many factors should be considered when defining the requirements (including types, numbers, locations, and depths of explorations) for the site investigation (Figure 30.12). These factors include:

- Importance, uncertainty, or risk associated with bridge design, construction, and performance
- Geologic conditions and their potential variability
- Availability (or unavailability) of historical subsurface data
- Availability (or unavailability) of performance observations from similar nearby projects
- Investigation budget

The following factors should be considered when evaluating the project risk: (1) What are the risks? (2) How likely are the risks to be realized? (3) What are the consequences if the risks occur? Risks include:

- Certainty or uncertainty of subsurface conditions;
- Design risks (e.g., possibility that inadequate subsurface data will compromise design decisions or schedule);
- Construction risks (e.g., potential for changed conditions claims and construction delays);
- Performance risks (e.g., seismic performance).

Two additional requirements that should be considered when planning a subsurface investigation are (1) reliability of the data collected and (2) timeliness of the data generated. Unfortunately, these factors are too often ignored or underappreciated during the site investigation planning process or geotechnical consultant selection process. Because poor-quality or misleading subsurface data can lead to inappropriate selection of foundation locations, foundation types, and/or inadequate or inappropriate foundation capacities, selection of a project geotechnical consultant should be based on qualifications rather than cost. Similarly, the value of the data generated from the subsurface investigation is reduced if adequate data are not available when the design decisions, which are affected by subsurface conditions, are made. All too often, the execution of the subsurface exploration program is delayed, and major decisions relative to the general structure design and foundation locations have been cast in stone prior to the availability of the subsurface exploration results.

Frequently, the execution of the subsurface investigation is an iterative process that should be conducted in phases (i.e., desktop study, reconnaissance site investigation, detailed design-phase investigation). During each phase of site exploration, it is appropriate for data to be reviewed as they are generated so that appropriate modifications can be made as the investigation is ongoing. Appropriate adjustments in the investigation work scope can save significant expense, increase the quality and value of the investigation results, and/or reduce the potential for a remobilization of equipment to fill in missing information.

30.3.1 Choice of Exploration Methods and Consideration of Local Practice

Because many exploration techniques are suitable in some subsurface conditions, but not as suitable or economical in other conditions, the local practice for the methods of exploration vary from region to region. Therefore, the approach to the field exploration program should consider and be tailored to the local practice. Conversely, there are occasions where the requirements for a project may justify using exploration techniques that are not common in the project area. The need to use special techniques will increase with the size of the project and the uniqueness or complexity of the site conditions.

30.3.2 Exploration Depths

The depths to which subsurface exploration should be extended will depend on the structure, its size, and the subsurface conditions at the project location. The subsurface exploration for any project should extend down through unsuitable layers into materials that are competent relative to the design loads to be applied by the bridge foundations. Some of the exploration should be deep enough to verify that unsuitable materials do not exist beneath the bearing strata on which the foundations will be embedded. When the base of the foundation is underlain by layers of compressible material, the exploration should extend down through the compressible strata and into deeper strata whose compressibility will not influence foundation performance.

For lightly loaded structures, it may be adequate to terminate the exploration when rock is encountered, provided that the regional geology indicates that unsuitable strata do not underlie the rock surface. For heavily-loaded foundations or foundations bearing on rock, it is appropriate to verify that the explorations indeed have encountered rock and not a boulder. It is similarly appropriate to extend at least some of the explorations through the weathered rock into sound or fresh rock.

30.3.3 Numbers of Explorations

The basic intent of the site investigation is to determine the subsurface stratigraphy and its variations, and to define the representative soil (or rock) properties of the strata together with their lateral and vertical variations. The locations and spacing of explorations should be adequate to provide a reasonably accurate definition of the subsurface conditions, and should disclose the presence of any important irregularities in the subsurface conditions. Thus, the numbers of explorations will depend on both the project size and the geologic and depositional variability of the site location. When subsurface conditions are complex and variable, a greater number of more closely spaced explorations are warranted. Conversely, when subsurface conditions are relatively uniform, fewer and more widely spaced explorations may be adequate.

30.3.4 The Risk of Inadequate Site Characterization

When developing a site exploration program, it is often tempting to minimize the number of explorations or defer the use of specialized techniques due to their expense. The approach of minimizing the investment in site characterization is fraught with risk. Costs saved by the execution of an inadequate site investigation, whether in terms of the numbers of explorations or the exclusion of applicable site investigation techniques, rarely reduce the project cost. Conversely, the cost saved by an inadequate investigation frequently increases the cost of construction by many times the savings achieved during the site investigation.

30.4 Development of Laboratory Testing Program

30.4.1 Purpose of Testing Program

Laboratory tests are performed on samples for the following purposes:

- Classify soil samples;
- Evaluate basic index soil properties that are useful in evaluating the engineering properties of the soil samples;
- Measure the strength, compressibility, and hydraulic properties of the soils;
- Evaluate the suitability of on-site or borrow soils for use as fill;
- Define dynamic parameters for site response and soil–structure interaction analyses during earthquakes;
- Identify unusual subsurface conditions (e.g., presence of corrosive conditions, carbonate soils, expansive soils, or potentially liquefiable soils).

The extent of laboratory testing is generally defined by the risks associated with the project.

Soil classification, index property, and fill suitability tests generally can be performed on disturbed samples, whereas tests to determine engineering properties of the soils should preferably be performed on relatively undisturbed, intact specimen. The quality of the data obtained from the latter series of tests is significantly dependent on the magnitude of sample disturbance either during sampling or during subsequent processing and transportation.

30.4.2 Types and Uses of Tests

30.4.2.1 Soil Classification and Index Testing

Soil classification and index properties tests are generally performed for even low-risk projects. Engineering parameters often can be estimated from the available *in situ* data and basic index tests using published correlations. Site-specific correlations of these basic values may allow the results of a few relatively expensive advanced tests to be extrapolated. Index tests and their uses include the following:

- Unit weight and water content tests to evaluate the natural unit weight and water content.
- Atterberg (liquid and plastic) limit tests on cohesive soils for classification and correlation studies. Significant insight relative to strength and compressibility properties can be inferred from the natural water content and Atterberg limit test results.
- Sieve and hydrometer tests to define the grain size distribution of coarse- and fine-grained soils, respectively. Grain size data also are used for both classification and correlation studies.

Other index tests include tests for specific gravity, maximum and minimum density, expansion index, and sand equivalent.

30.4.2.2 Shear Strength Tests

Most bridge design projects require characterization of the undrained shear strength of cohesive soils and the drained strength of cohesionless soils. Strength determinations are necessary to evaluate the bearing capacity of foundations and to estimate the loads imposed on earth-retaining structures.

Undrained shear strength of cohesive soils can be estimated (often in the field) with calibrated tools such as a torvane, pocket penetrometer, fall cone, or miniature vane shear device. More definitive strength measurements are obtained in a laboratory by subjecting samples to triaxial compression (TX), direct simple shear (DSS), or torsional shear (TS) tests. Triaxial shear tests (including unconsolidated-undrained, UU, tests and consolidated-undrained, CU, tests) are the most common type of strength test. In this type of test, the sample is subject to stresses that mimic *in situ* states of stress prior to being tested to failure in compression or shear. Large and more high risk projects often warrant the performance of CU or DSS tests where samples are tested along stress paths which model the *in situ* conditions. In contrast, only less-sophisticated UU tests may be warranted for less important projects.

Drained strength parameters of cohesionless soils are generally measured in either relatively simple direct shear (DS) tests or in more-sophisticated consolidated-drained (CD) triaxial tests. In general, few laboratory strength tests are performed on *in situ* specimens of cohesionless soil because of the relative difficulty in obtaining undisturbed specimens.

30.4.2.3 Compaction Tests

Compaction tests are performed to evaluate the moisture–density relationship of potential fill material. Once the relationship has been evaluated and the minimum level of compaction of fill material to be used has been determined, strength tests may be performed on compacted specimens to evaluate design parameters for the project.

30.4.2.4 Subgrade Modulus

R-value and CBR tests are performed to determine subgrade modulus and evaluate the pavement support characteristics of the *in situ* or fill soils.

30.4.2.5 Consolidation Tests

Consolidation tests are commonly performed to (1) evaluate the compressibility of soil samples for the calculation of foundation settlement; (2) investigate the stress history of the soils at the boring locations to calculate settlement as well as to select stress paths to perform most advanced strength tests; (3) evaluate elastic properties from measured bulk modulus values; and (4) evaluate the time rate of settlement. Consolidation test procedures also can be modified to evaluate if foundation soils are susceptible to collapse or expansion, and to measure expansion pressures under various levels of confinement. Consolidation tests include incremental consolidation tests (which are performed at a number of discrete loads) and constant rate of strain (CRS) tests where load levels are constantly increased or decreased. CRS tests can generally be performed relatively quickly and provide a continuous stress–strain curve, but require more-sophisticated equipment.

30.4.2.6 Permeability Tests

In general, constant-head permeability tests are performed on relatively permeable cohesionless soils, while falling-head permeability tests are performed on relatively impermeable cohesive soils. Estimates of the permeability of cohesive soils also can be obtained from consolidation test data.

30.4.2.7 Dynamic Tests

A number of tests are possible to evaluate the behavior of soils under dynamic loads such as wave or earthquake loads. Dynamic tests generally are strength tests with the sample subjected to some sort of cyclic loading. Tests can be performed to evaluate variations of strength, modulus, and damping, with variations in rate and magnitude of cyclic stresses or strains. Small strain parameters for earthquake loading cases can be evaluated from resonant column tests.

For earthquake loading conditions, dynamic test data are often used to evaluate site response and soil–structure interaction. Cyclic testing also can provide insight into the behavior of potentially liquefiable soils, especially those which are not easily evaluated by empirical *in situ* test-based procedures.

30.4.2.8 Corrosion Tests

Corrosion tests are performed to evaluate potential impacts on steel or concrete structures due to chemical attack. Tests to evaluate corrosion potential include resistivity, pH, sulfate content, and chloride content.

30.5 Data Presentation and Site Characterization

30.5.1 Site Characterization Report

The site characterization report should contain a presentation of the site data and an interpretation and analysis of the foundation conditions at the project site. The site characterization report should:

- Present the factual data generated during the site investigation;
- Describe the procedures and equipment used to obtain the factual data;
- Describe the subsurface stratigraphic relationships at the project site;
- Define the soil and rock properties that are relevant to the planning, design, construction, and performance of the project structures;
- Formulate the solutions to the design and construction of the project.

The site data presented in the site characterization report may be developed from the current and/or past field investigations at or near the project site, as-built documents, maintenance records, and construction notes. When historic data are included or summarized, the original sources of the data should be cited.

30.5.2 Factual Data Presentation

The project report should include the accurate and appropriate documentation of the factual data collected and generated during the site investigation and testing program(s). The presentation and organization of the factual data, by necessity, will depend upon the size and complexity of the project and the types and extent of the subsurface data. Regardless of the project size or extent of exploration, all reports should include an accurate plan of exploration that includes appropriate graphical portrayal of surface features and ground surface elevation in the project area.

The boring log (Figure 30.13) is one of the most fundamental components of the data documentation. Although many styles of presentation are used, there are several basic elements that generally should be included on a boring log. Those typical components include:

- Documentation of location and ground surface elevation;
- Documentation of sampling and coring depths, types, and lengths — e.g., sample type, blow count (for driven samples), and sample length for soil samples; core run, recovery, and RQD for rock cores — as well as *in situ* test depths and lengths;
- Depths and elevations of groundwater and/or seepage encountered;
- Graphical representation of soil and rock lithology;
- Description of soil and rock types, characteristics, consistency/density, or hardness;
- Tabular or graphical representation of test data.

In addition to the boring logs, the factual data should include tabulated summaries of test types, depths, and results together with the appropriate graphical output of the tests conducted.

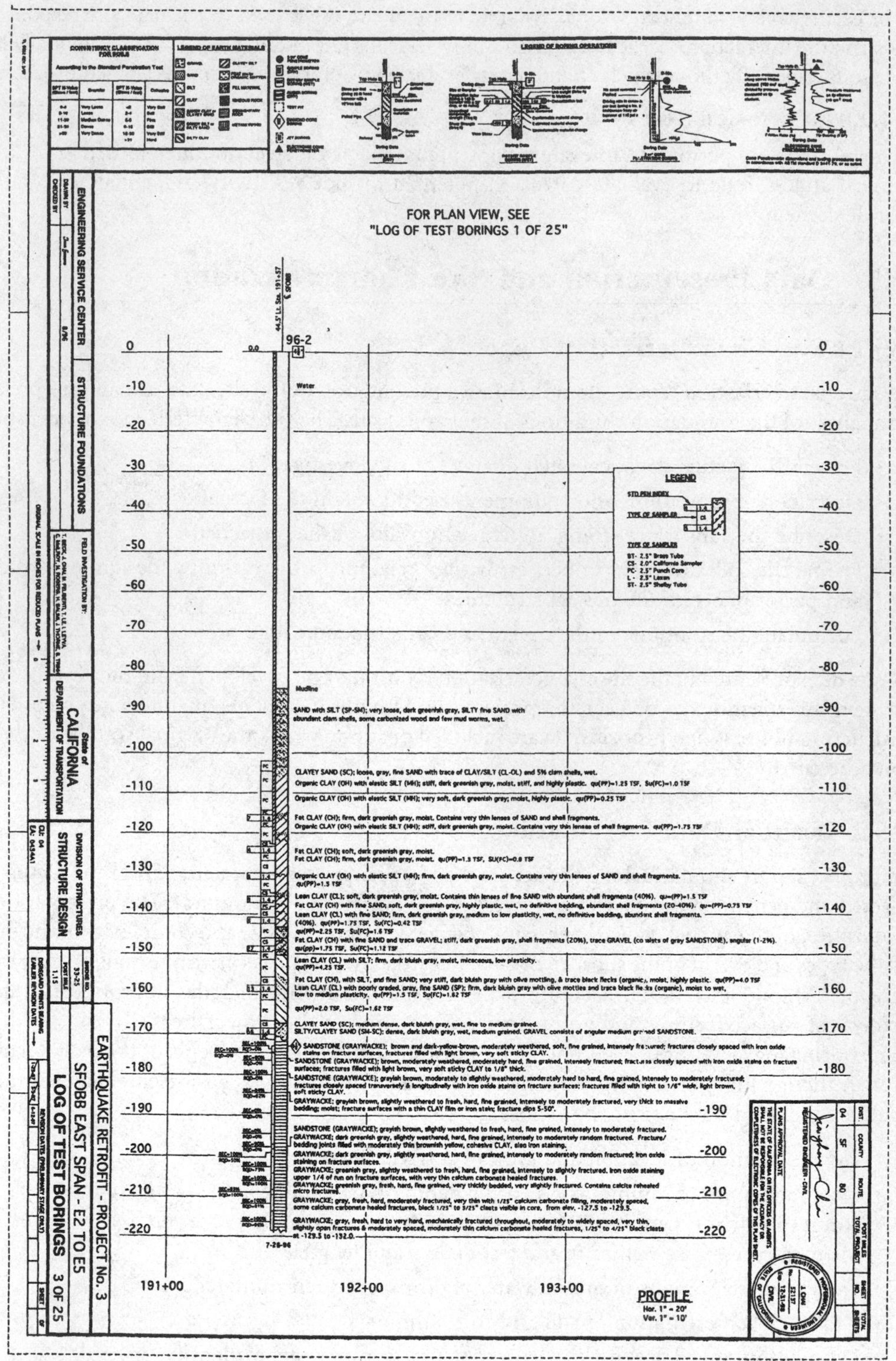

FIGURE 30.13 Typical log of test boring sheet for Caltrans project.

30.5.3 Description of Subsurface Conditions and Stratigraphy

A sound geologic interpretation of the exploration and testing data are required for any project to assess the subsurface conditions. The description of the subsurface conditions should provide users of the report with an understanding of the conditions, their possible variability, and the significance of the conditions relative to the project. The information should be presented in a useful format and terminology appropriate for the users, who usually will include design engineers and contractors who are not earth science professionals.

To achieve those objectives, the site characterization report should include descriptions of

1. Site topography and/or bathymetry,
2. Site geology,
3. Subsurface stratigraphy and stratigraphic relationships,
4. Continuity or lack of continuity of the various subsurface strata,
5. Groundwater depths and conditions, and
6. Assessment of the documented and possible undocumented variability of the subsurface conditions.

Information relative to the subsurface conditions is usually provided in text, cross sections, and maps. Subsurface cross sections, or profiles, are commonly used to illustrate the stratigraphic sequence, subsurface strata and their relationships, geologic structure, and other subsurface features across a site. The cross section can range from simple line drawings to complex illustrations that include boring logs and plotted test data (Figure 30.14).

Maps are commonly used to illustrate and define the subsurface conditions at a site. The maps can include topographic and bathymetric contour maps, maps of the structural contours of a stratigraphic surface, groundwater depth or elevation maps, isopach thickness maps of an individual stratum (or sequence of strata), and interpreted maps of geologic features (e.g., faulting, bedrock outcrops, etc.). The locations of explorations should generally be included on the interpretive maps.

The interpretive report also should describe data relative to the depths and elevations of groundwater and/or seepage encountered in the field. The potential types of groundwater surface(s) and possible seasonal fluctuation of groundwater should be described. The description of the subsurface conditions also should discuss how the groundwater conditions can affect construction.

30.5.4 Definition of Soil Properties

Soil properties generally should be interpreted in terms of stratigraphic units or geologic deposits. The interpretation of representative soil properties for design should consider lateral and vertical variability of the different soil deposits. Representative soil properties should consider the potential for possible *in situ* variations that have not been disclosed by the exploration program and laboratory testing. For large or variable sites, it should be recognized that global averages of a particular soil property may not appropriately represent the representative value at all locations. For that condition, use of average soil properties may lead to unconservative design.

Soil properties and design recommendations are usually presented with a combination of narrative text, graphs, and data presented in tabular and/or bulleted list format. It is often convenient and helpful to reference generalized subsurface profiles and boring logs in those discussions. The narrative descriptions should include such factors as depth range, general consistency or density, plasticity or grain size, occurrence of groundwater, occurrence of layers or seams, degree of weathering, and structure. For each stratigraphic unit, ranges of typical measured field and laboratory data (e.g., strength, index parameters, and blow counts) should be described.

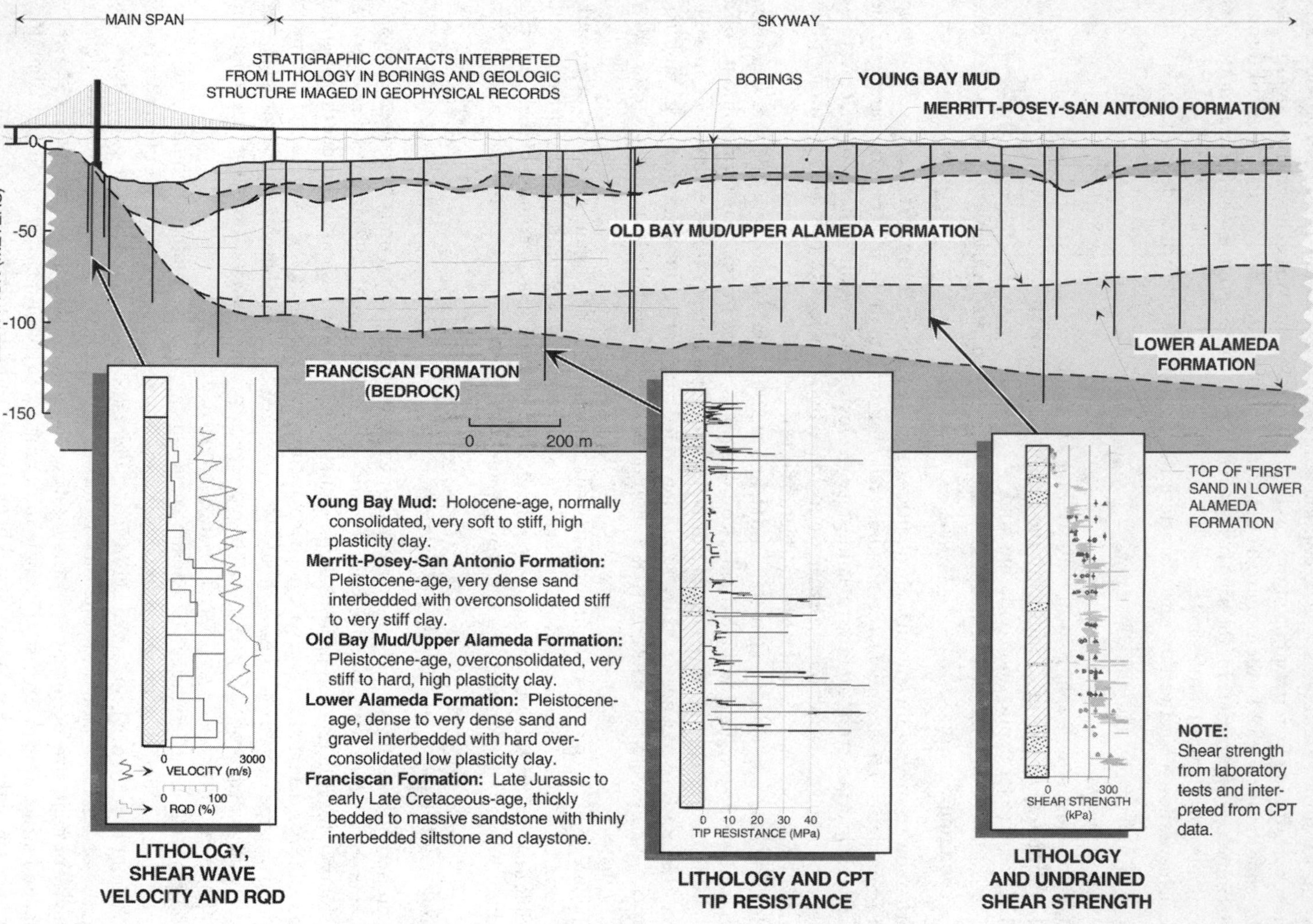

FIGURE 30.14 Subsurface cross section for San Francisco–Oakland Bay Bridge East Span alignment.

30.5.5 Geotechnical Recommendations

The site characterization report should provide solutions to the geotechnical issues and contain geotechnical recommendations that are complete, concise, and definitive. The recommended foundation and geotechnical systems should be cost-effective, performance-proven, and constructible. Where appropriate, alternative foundation types should be discussed and evaluated. When construction problems are anticipated, solutions to these problems should be described.

In addition to the standard consideration of axial and lateral foundation capacity, load–deflection characteristics, settlement, slope stability, and earth pressures, there are a number of subsurface conditions that can affect foundation design and performance:

- Liquefaction susceptibility of loose, granular soils;
- Expansive or collapsible soils;
- Mica-rich and carbonate soils;
- Corrosive soils;
- Permafrost or frozen soils;
- Perched or artesian groundwater.

When any of those conditions are present, they should be described and evaluated.

30.5.6 Application of Computerized Databases

Computerized databases provide the opportunity to compile, organize, integrate, and analyze geotechnical data efficiently. All collected data are thereby stored, in a standard format, in a central accessible location. Use of a computerized database has a number of advantages. Use of automated interactive routines allows the efficient production of boring logs, cross sections, maps, and parameter plots. Large volumes of data from multiple sources can be integrated and queried to evaluate or show trends and variability. New data from subsequent phases of study can be easily and rapidly incorporated into the existing database to update and revise the geologic model of the site.

31

Shallow Foundations

James Chai
California Department of Transportation

31.1 Introduction

A shallow foundation may be defined as one in which the foundation depth (D) is less than or on the order of its least width (B), as illustrated in Figure 31.1. Commonly used types of shallow foundations include spread footings, strap footings, combined footings, and mat or raft footings. Shallow foundations or footings provide their support entirely from their bases, whereas deep foundations derive the capacity from two parts, skin friction and base support, or one of these two. This chapter is primarily designated to the discussion of the bearing capacity and settlement of shallow foundations, although structural considerations for footing design are briefly addressed. Deep foundations for bridges are discussed in Chapter 32.

0-8493-7434-0/00/$0.00+$.50

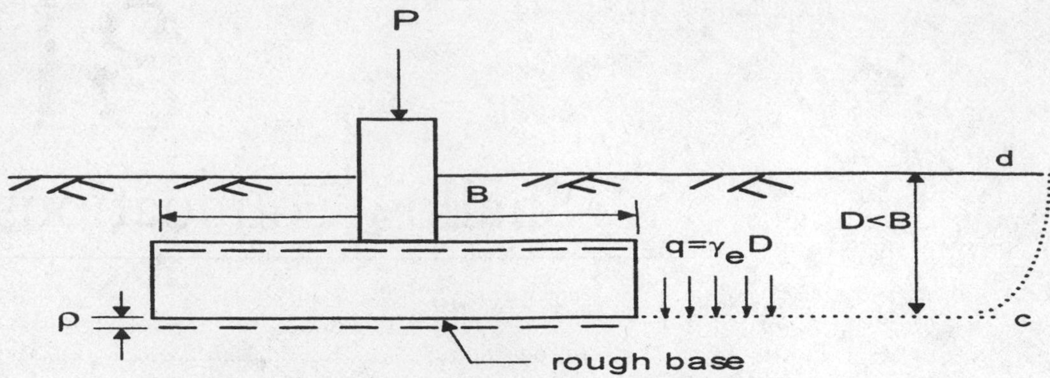

FIGURE 31.1 Definition sketch for shallow footings.

TABLE 31.1 Typical Values of Safety Factors Used in Foundation Design (after Barker et al. [9])

Failure Type	Failure Mode	Safety Factor	Remark
Shearing	Bearing capacity failure	2.0–3.0	The lower values are used when uncertainty in design is small and consequences of failure are minor; higher values are used when uncertainty in design is large and consequences of failure are major
	Overturning	2.0–2.5	
	Overall stability	1.5–2.0	
	Sliding	1.5–2.0	
Seepage	Uplift	1.5–2.0	
	Heave	1.5–2.0	
	Piping	2.0–3.0	

Source: Terzaghi, K. and Peck, R.B., *Soil Mechanics in Engineering Practice*, 2nd ed., John Wiley & Sons, New York, 1967. With permission.

31.2 Design Requirements

In general, any foundation design must meet three essential requirements: (1) providing adequate safety against structural failure of the foundation; (2) offering adequate bearing capacity of soil beneath the foundation with a specified safety against ultimate failure; and (3) achieving acceptable total or differential settlements under working loads. In addition, the overall stability of slopes in the vicinity of a footing must be regarded as part of the foundation design. For any project, it is usually necessary to investigate both the bearing capacity and the settlement of a footing. Whether footing design is controlled by the bearing capacity or the settlement limit rests on a number of factors such as soil condition, type of bridge, footing dimensions, and loads. Figure 31.2 illustrates the load–settlement relationship for a square footing subjected to a vertical load *P*. As indicated in the curve, the settlement *p* increases as load *P* increases. The ultimate load P_u is defined as a peak load (curves 1 and 2) or a load at which a constant rate of settlement (curve 3) is reached as shown in Figure 31.2. On the other hand, the ultimate load is the maximum load a foundation can support without shear failure and within an acceptable settlement. In practice, all foundations should be designed and built to ensure a certain safety against bearing capacity failure or excessive settlement. A safety factor (*SF*) can be defined as a ratio of the ultimate load P_u and allowable load P_u. Typical value of safety factors commonly used in shallow foundation design are given in Table 31.1.

31.3 Failure Modes of Shallow Foundations

Bearing capacity failure usually occurs in one of the three modes described as general shear, local shear, or punching shear failure. In general, which failure mode occurs for a shallow foundation depends on the relative compressibility of the soil, footing embedment, loading conditions, and drainage conditions. General shear failure has a well-defined rupture pattern consisting of three zones, I, II, and III, as shown in Figure 31.3a. Local shear failure generally consists of clearly defined rupture surfaces beneath the footing (zones I and II). However, the failure pattern on the sides of the footing (zone III) is not clearly defined. Punch shear failure has a poorly defined rupture pattern concentrated within zone I; it is usually associated with a large settlement and does not mobilize shear stresses in zones II and III as shown in Figure 31.3b and c. Ismael and Vesic [40] concluded that, with increasing overburden pressure (in cases of deep foundations), the failure mode changes from general shear to local or punch shear, regardless of soil compressibility. The further examination of load tests on footings by Vesic [68,69] and De Beer [29] suggested that the ultimate load occurs at the breakpoint of the load–settlement curve, as shown in Figure 31.2. Analyzing the modes of failure indicates that (1) it is possible to formulate a general bearing capacity equation for a loaded footing failing in the general shear mode, (2) it is very difficult to generalize the other two failure modes for shallow foundations because of their poorly defined rupture surfaces, and (3) it is of significance to know the magnitude of settlements of footings required to mobilize ultimate loads. In the following sections, theoretical and empirical methods for evaluating both bearing capacity and settlement for shallow foundations will be discussed.

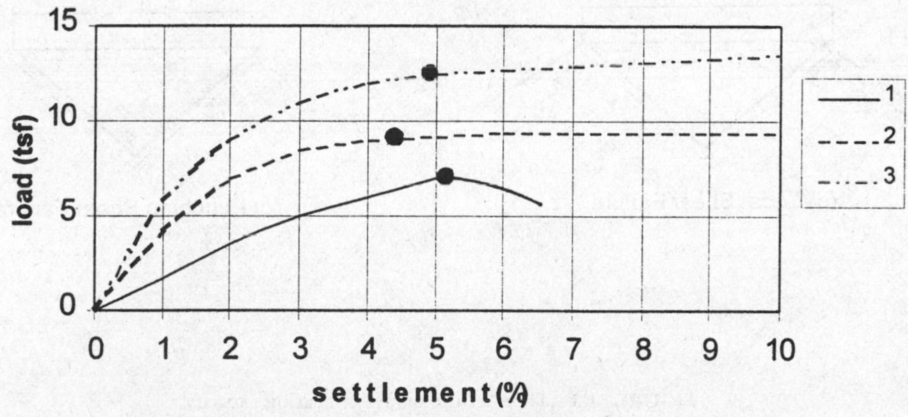

FIGURE 31.2 Load-settlement relationships of shallow footings.

31.4 Bearing Capacity for Shallow Foundations

31.4.1 Bearing Capacity Equation

The computation of ultimate bearing capacity for shallow foundations on soil can be considered as a solution to the problem of elastic–plastic equilibrium. However, what hinders us from finding closed analytical solutions rests on the difficulty in the selection of a mathematical model of soil constitutive relationships. Bearing capacity theory is still limited to solutions established for the rigid-plastic solid of the classic theory of plasticity [40,69]. Consequently, only approximate methods are currently available for the posed problem. One of them is the well-known Terzaghi's bearing capacity equation [19,63], which can be expressed as

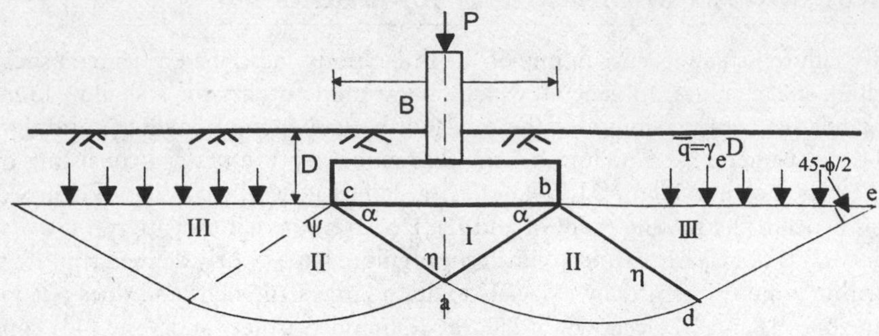

(a) General Shear Failure

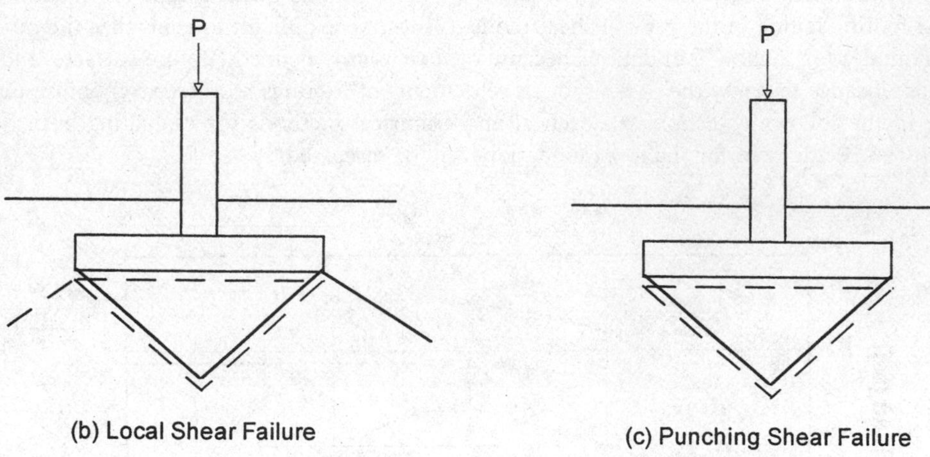

(b) Local Shear Failure

(c) Punching Shear Failure

FIGURE 31.3 Three failure modes of bearing capacity.

$$q_{\text{ult}} = cN_c s_c + \overline{q}N_q + 0.5\gamma BN_\gamma s_\gamma \tag{31.1}$$

where q_{ult} is ultimate bearing capacity, c is soil cohesion, $\overline{q}$ is effective overburden pressure at base of footing ($= \gamma_1 D$), γ is effective unit weight of soil or rock, and B is minimum plan dimension of footing. N_c, N_q, and N_γ are bearing capacity factors defined as functions of friction angle of soil and their values are listed in Table 31.2. s_c and s_r are shape factors as shown in Table 31.3.

These three N factors are used to represent the influence of the cohesion (N_c), unit weight (N_γ), and overburden pressure (N_q) of the soil on bearing capacity. As shown in Figures 31.1 and 31.3(a), the assumptions used for Eq. (31.1) include

1. The footing base is rough and the soil beneath the base is incompressible, which implies that the wedge *abc* (zone I) is no longer an active Rankine zone but is in an elastic state. Consequently, zone I must move together with the footing base.
2. Zone II is an immediate zone lying on a log spiral arc *ad*.

TABLE 31.2 Bearing Capacity Factors for the Terzaghi Equation

ϕ (°)	N_c	N_q	N_γ	$K_{p\gamma}$
0	5.7[a]	1.0	0	10.8
5	7.3	1.6	0.5	12.2
10	9.6	2.7	1.2	14.7
15	12.9	4.4	2.5	18.6
20	17.7	7.4	5.0	25.0
25	25.1	12.7	9.7	35.0
30	37.2	22.5	19.7	52.0
34	52.6	36.5	36.0	—
35	57.8	41.4	42.4	82.0
40	95.7	81.3	100.4	141.0
45	172.3	173.3	297.5	298.0
48	258.3	287.9	780.1	—
50	347.5	415.1	1153.2	800.0

[a] $N_c = 1.5\pi + 1$ (Terzaghi [63], p. 127); values of N_γ for ϕ of 0, 34, and 48° are original Terzaghi values and used to backcompute $K_{p\gamma}$.

After Bowles, J.E., *Foundation Analysis and Design*, 5th ed., McGraw-Hill, New York, 1996. With permission.

TABLE 31.3 Shape Factors for the Terzaghi Equation

	Strip	Round	Square
s_c	1.0	1.3	1.3
s_γ	1.0	0.6	0.8

After Terzaghi [63].

3. Zone III is a passive Rankine zone in a plastic state bounded by a straight line *ed*.
4. The shear resistance along *bd* is neglected because the equation was intended for footings where $D < B$.

It is evident that Eq. (31.1) is only valid for the case of general shear failure because no soil compression is allowed before the failure occurs.

Meyerhof [45,48], Hansen [35], and Vesic [68,69] further extended Terzaghi's bearing capacity equation to account for footing shape (s_i), footing embedment depth (d_1), load inclination or eccentricity (i_i), sloping ground (g_i), and tilted base (b_i). Chen [26] reevaluated *N* factors in Terzaghi's equation using the limit analysis method. These efforts resulted in significant extensions of Terzaghi's bearing capacity equation. The general form of the bearing capacity equation [35,68,69] can be expressed as

$$q_{\text{ult}} = cN_c s_c d_c i_c g_c b_c + \bar{q} N_q s_q d_q b_q + 0.5\gamma B N_\gamma s_\gamma d_\gamma i_\gamma g_\gamma b_\gamma \tag{31.2}$$

when $\phi = 0$,

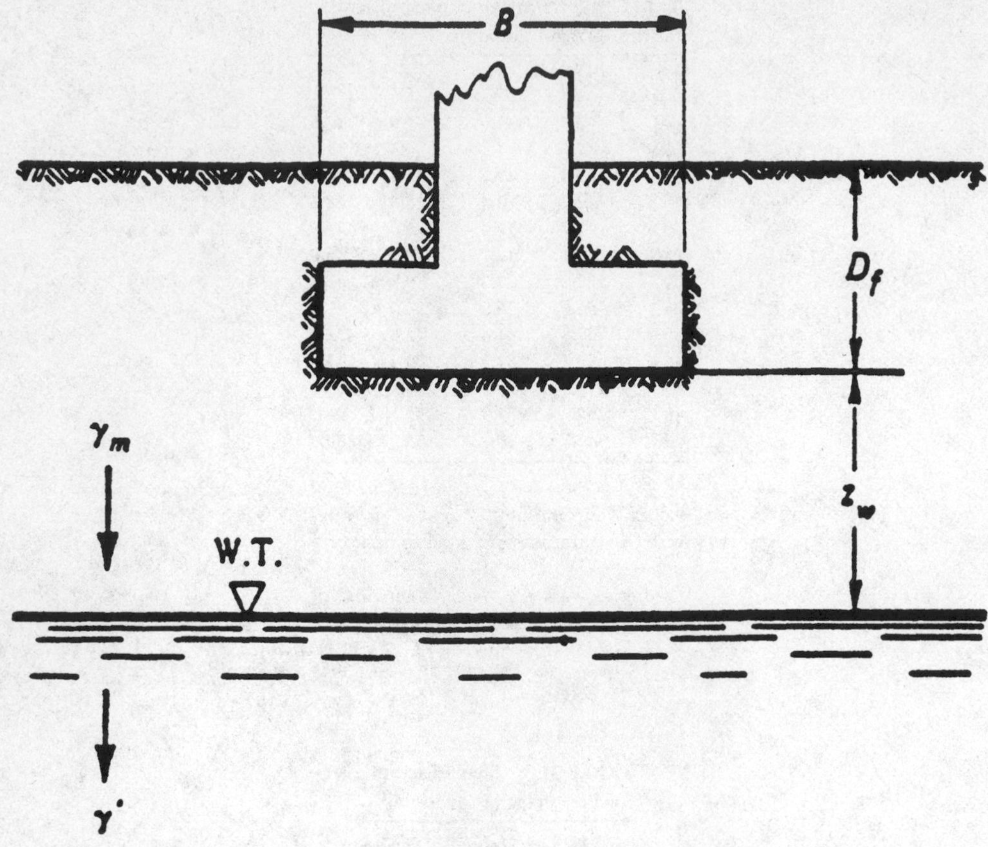

FIGURE 31.4 Influence of groundwater table on bearing capacity. (After AASHTO, 1997.)

$$q_{\text{ult}} = 5.14 s_u \left(1 + s'_c + d'_c - i'_c - b'_c - g'_c\right) + \bar{q} \tag{31.3}$$

where s_u is undrained shear strength of cohesionless. Values of bearing capacity factors N_c, N_q, and N_γ can be found in Table 31.4. Values of other factors are shown in Table 31.5. As shown in Table 31.4, N_c and N_q are the same as proposed by Meyerhof [48], Hansen [35], Vesic [68], or Chen [26]. Nevertheless, there is a wide range of values for N_γ as suggested by different authors. Meyerhof [48] and Hansen [35] use the plain-strain value of ϕ, which may be up to 10% higher than those from the conventional triaxial test. Vesic [69] argued that a shear failure in soil under the footing is a process of progressive rupture at variable stress levels and an average mean normal stress should be used for bearing capacity computations. Another reason causing the N_γ value to be unsettled is how to evaluate the impact of the soil compressibility on bearing capacity computations. The value of N_γ still remains controversial because rigorous theoretical solutions are not available. In addition, comparisons of predicted solutions against model footing test results are inconclusive.

Soil Density

Bearing capacity equations are established based on the failure mode of general shearing. In order to use the bearing capacity equation to consider the other two modes of failure, Terzaghi [63] proposed a method to reduce strength characteristics c and ϕ as follows:

$$c^* = 0.67c \quad \text{(for soft to firm clay)} \tag{31.4}$$

TABLE 31.4 Bearing Capacity Factors for Eqs. (31.2) and (31.3)

ϕ	N_c	N_q	$N_{\gamma(M)}$	$N_{\gamma(H)}$	$N_{\gamma(V)}$	$N_{\gamma(C)}$	N_q/N_c	tan ϕ
0	5.14	1.00	0.00	0.00	0.00	0.00	0.19	0.00
1	5.38	1.09	0.00	0.00	0.07	0.07	0.20	0.02
2	5.63	1.20	0.01	0.01	0.15	0.16	0.21	0.03
3	5.90	1.31	0.02	0.02	0.24	0.25	0.22	0.05
4	6.18	1.43	0.04	0.05	0.34	0.35	0.23	0.07
5	6.49	1.57	0.07	0.07	0.45	0.47	0.24	0.09
6	6.81	1.72	0.11	0.11	0.57	0.60	0.25	0.11
7	7.16	1.88	0.15	0.16	0.71	0.74	0.26	0.12
8	7.53	2.06	0.21	0.22	0.86	0.91	0.27	0.14
9	7.92	2.25	0.28	0.30	1.03	1.10	0.28	0.16
10	8.34	2.47	0.37	0.39	1.22	1.31	0.30	0.18
11	8.80	2.71	0.47	0.50	1.44	1.56	0.31	0.19
12	9.28	2.97	0.60	0.63	1.69	1.84	0.32	0.21
13	9.81	3.26	0.74	0.78	1.97	2.16	0.33	0.23
14	10.37	3.59	0.92	0.97	2.29	2.52	0.35	0.25
15	10.98	3.94	1.13	1.18	2.65	2.94	0.36	0.27
16	11.63	4.34	1.37	1.43	3.06	3.42	0.37	0.29
17	12.34	4.77	1.66	1.73	3.53	3.98	0.39	0.31
18	13.10	5.26	2.00	2.08	4.07	4.61	0.40	0.32
19	13.93	5.80	2.40	2.48	4.68	5.35	0.42	0.34
20	14.83	6.40	2.87	2.95	5.39	6.20	0.43	0.36
21	15.81	7.07	3.42	3.50	6.20	7.18	0.45	0.38
22	16.88	7.82	4.07	4.13	7.13	8.32	0.46	0.40
23	18.05	8.66	4.82	4.88	8.20	9.61	0.48	0.42
24	19.32	9.60	5.72	5.75	9.44	11.17	0.50	0.45
25	20.72	10.66	6.77	6.76	10.88	12.96	0.51	0.47
26	22.25	11.85	8.00	7.94	12.54	15.05	0.53	0.49
27	23.94	13.20	9.46	9.32	14.47	17.49	0.55	0.51
28	25.80	14.72	11.19	10.94	16.72	20.35	0.57	0.53
29	27.86	16.44	13.24	12.84	19.34	23.71	0.59	0.55
30	30.14	18.40	15.67	15.07	22.40	27.66	0.61	0.58
31	32.67	20.63	18.56	17.69	25.99	32.33	0.63	0.60
32	35.49	23.18	22.02	20.79	30.21	37.85	0.65	0.62
33	38.64	26.09	26.17	24.44	35.19	44.40	0.68	0.65
34	42.16	29.44	31.15	28.77	41.06	52.18	0.70	0.67
35	46.12	33.30	37.15	33.92	48.03	61.47	0.72	0.70
36	50.59	37.75	44.43	40.05	56.31	72.59	0.75	0.73
37	55.63	42.92	53.27	47.38	66.19	85.95	0.77	0.75
38	61.35	48.93	64.07	56.17	78.02	102.05	0.80	0.78
39	67.87	55.96	77.33	66.75	92.25	121.53	0.82	0.81
40	75.31	64.19	93.69	79.54	109.41	145.19	0.85	0.84
41	83.86	73.90	113.98	95.05	130.21	174.06	0.88	0.87
42	93.71	85.37	139.32	113.95	155.54	209.43	0.91	0.90
43	105.11	99.01	171.14	137.10	186.53	253.00	0.94	0.93
44	118.37	115.31	211.41	165.58	224.63	306.92	0.97	0.97
45	133.87	134.97	262.74	200.81	271.74	374.02	1.01	1.00
46	152.10	158.50	328.73	244.64	330.33	458.02	1.04	1.04
47	173.64	187.20	414.32	299.52	403.65	563.81	1.08	1.07
48	199.26	222.30	526.44	368.66	495.99	697.93	1.12	1.11
49	229.92	265.49	674.91	456.40	613.13	869.17	1.15	1.15
50	266.88	319.05	873.84	568.56	762.85	1089.46	1.20	1.19

Note: N_c and N_q are same for all four methods; subscripts identify author for N_γ: M = Meyerhof [48]; H = Hansen [35]; V = Vesic [69]; C = Chen [26].

TABLE 31.5 Shape, Depth, Inclination, Ground, and Base Factors for Eq. (31.3)

Shape Factors	Depth Factors
$s_c = 1.0 + \frac{N_q}{N_c}\ \frac{B}{L}$ $s_c = 1.0$ (for strip footing)	$d_c = 1.0 + 0.4k \begin{cases} k = \frac{D_f}{B} & \text{for } \frac{D_f}{B} \le 1 \\ k = \tan^{-1}\left(\frac{D_f}{B}\right) & \text{for } \frac{D_f}{B} > 1 \end{cases}$
$s_q = 1.0 + \frac{B}{L} \tan\phi$ (for all ϕ)	$d_q = 1 + 2\tan\phi\,(1 - \sin\phi)^2 k$ (k defined above)
$s_\gamma = 1.0 - 0.4\frac{B}{L} \ge 0.6$	$d_\gamma = 1.00$ (for all ϕ)

Inclination Factors	Ground Factors (base on slope)
$i'_c = 1 - \frac{mHi}{A_f c_a N_c}$ $(\phi = 0)$	$g'_c = \frac{\beta}{5.14}$ β in radius $(\phi = 0)$
$i_c = i_q - \frac{1 - i_q}{N_q - 1}$ $(\phi > 0)$	$g_c = i_q - \frac{1 - i_q}{5.14 \tan\phi}$ $(\phi > 0)$
$i_q = \left[1.0 - \frac{H_i}{V + A_f c_a \cot\phi}\right]^m$	$g_q = g_\gamma = (1.0 - \tan\beta)^2$ (for all ϕ)
	Base Factors (tilted base)
$i_\gamma = \left[1.0 - \frac{H_i}{V + A_f c_a \cot\phi}\right]^{m+1}$	$b'_c = g'_c$ $(\phi = 0)$
$m = m_B = \frac{2 + B/L}{1 + B/L}$ or	$b_c = 1 - \frac{2\beta}{5.14 \tan\phi}$ $(\phi > 0)$
$m = m_L = \frac{2 + L/B}{1 + L/B}$	$b_q = b_\gamma = (1.0 - \eta \tan\phi)^2$ (for all ϕ)

Notes:

1. When $\gamma = 0$ (and β 'ne 0) use $N_\gamma = 2\sin(\pm\beta)$ in N_γ term
2. Compute $m = m_B$ when $H_i = H_B$ (H parallel to B) and $m = m_L$ when $H_i = H_L$ (H parallel to L); for both H_B and H_L use

$$m = \sqrt{m_B^2 + m_L^2}$$

3. $0 \le i_q, i_\gamma \le 1$
4. $\beta + \eta \le 90°; \beta \le \phi$

where

A_f = effective footing dimension as shown in Figure 31.6
D_f = depth from ground surface to base of footing
V = vertical load on footing
H_i = horizontal component of load on footing with $H_{max} \le V \tan\delta + c_a A_f$
c_a = adhesion to base $(0.6c \le c_a \le 1.0c)$
δ = friction angle between base and soil $(0.5\phi \le \delta \le \phi)$
β = slope of ground away from base with (+) downward
η = tilt angle of base from horizontal with (+) upward

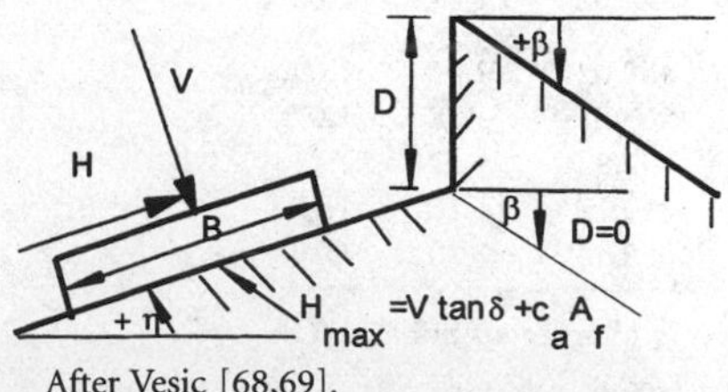

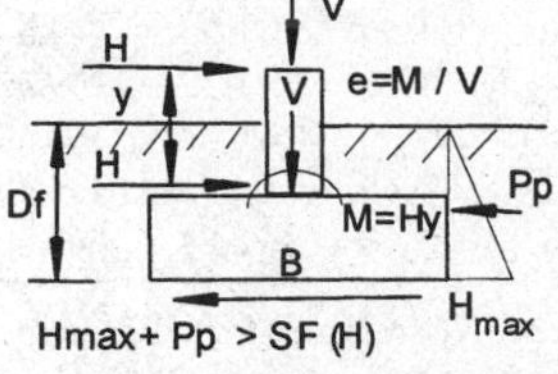

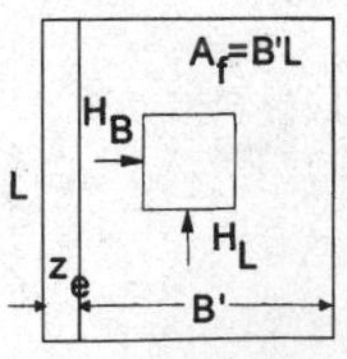

After Vesic [68,69].

$$\phi^* = \tan^{-1}(0.67 \tan\phi) \quad (\text{for loose sands with } \phi < 28°) \tag{31.5}$$

Vesic [69] suggested that a flat reduction of ϕ might be too conservative in the case of local and punching shear failure. He proposed the following equation for a reduction factor varying with relative density D_r:

$$\phi^* = \tan^{-1}\left(\left(0.67 + D_r - 0.75D_r^2\right)\tan\phi\right) \quad (\text{for } 0 < D_r < 0.67) \tag{31.6}$$

Groundwater Table

Ultimate bearing capacity should always be estimated by assuming the highest anticipated groundwater table. The effective unit weight γ_e shall be used in the qN_q and $0.5\gamma B$ terms. As illustrated in Figure 31.5, the weighted average unit weight for the $0.5\gamma B$ term can be determined as follows:

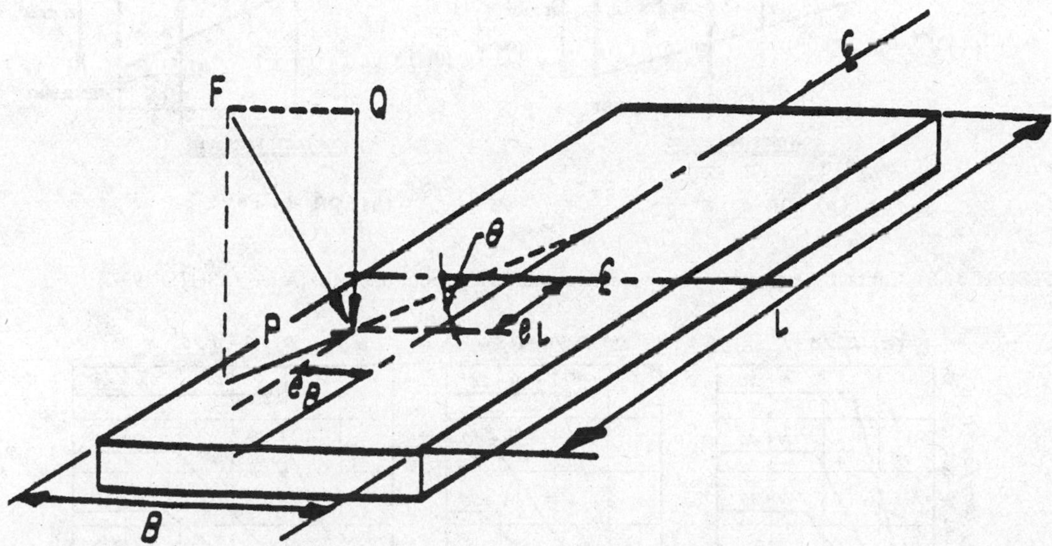

FIGURE 31.5 Definition sketch for loading and dimensions for footings subjected to eccentric or inclined loads. (After AASHTO, 1997.)

$$\gamma = \begin{cases} \gamma_{\text{avg}} & \text{for } d_w \geq B \\ \gamma' + (d_w/B)\left(\gamma_{\text{avg}} - \gamma'\right) & \text{for } 0 < d_w < B \\ \gamma' & \text{for } d \leq 0 \end{cases} \tag{31.7}$$

Eccentric Load

For footings with eccentricity, effective footing dimensions can be determined as follows:

$$A_f = B'L' \tag{31.8}$$

where $L = L - 2e_L$ and $B = B - 2e_B$. Refer to Figure 31.5 for loading definitions and footing dimensions. For example, the actual distribution of contact pressure for a rigid footing with eccentric loading in the L direction (Figure 31.6) can be obtained as follows:

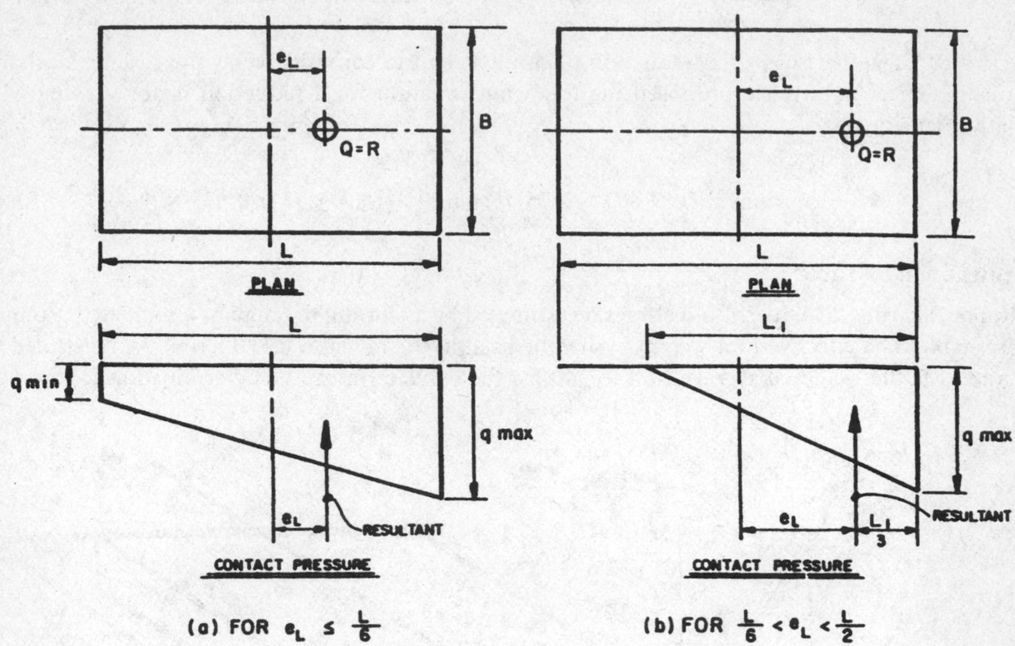

FIGURE 31.6 Contact pressure for footing loaded eccentrically about one axis. (After AASHTO 1997.)

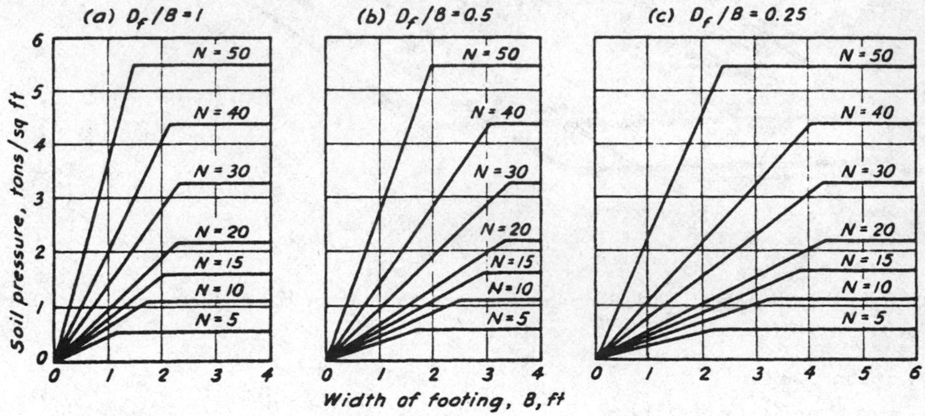

FIGURE 31.7 Design chart for proportioning shallow footings on sand. (a) Rectangular base; (b) round base. (After Peck et al. [53])

$$q_{\substack{\max \\ \min}} = P\left[1 \pm 6e_L/L\right]/BL \quad \left(\text{for } e_L < L/6\right) \tag{31.9}$$

$$q_{\substack{\max \\ \min}} = \begin{cases} 2\,P/\left[3B\left(L/2 - e_L\right)\right] \\ 0 \end{cases} \quad \left(\text{for } L/6 < e_L < L/2\right) \tag{31.10}$$

Contact pressure for footings with eccentric loading in the *B* direction may be determined using above equations by replacing terms *L* with *B* and terms *B* with *L*. For an eccentricity in both directions, reference is available in AASHTO [2,3].

31.4.2 Bearing Capacity on Sand from Standard Penetration Tests (SPT)

Terzaghi and Peck [64,65] proposed a method using SPT blow counts to estimate ultimate bearing capacity for footings on sand. Modified by Peck et al. [53], this method is presented in the form of the chart shown in Figure 31.7. For a given combination of footing width and SPT blow counts, the chart can be used to determine the ultimate bearing pressure associated with 25.4 mm (1.0 in.) settlement. The design chart applies to shallow footings ($D_f \leq B$) sitting on sand with water table at great depth. Similarly, Meyerhof [46] published the following formula for estimating ultimate bearing capacity using SPT blow counts:

$$q_{\mathrm{ult}} = N'_{\mathrm{avg}} \frac{B}{10}\left(C_{w1} + C_{w2}\frac{D_f}{B}\right)R_I \tag{31.11}$$

where R_I is a load inclination factor shown in Table 31.6 ($R_I = 1.0$ for vertical loads). C_{w1} and C_{w2} are correction factors whose values depend on the position of the water table:

TABLE 31.6 Load Inclination Factor (R_I)

For Square Footings — Load Inclination Factor (R_I)

H/V	$D_f/B=0$	$D_f/B=1$	$D_f/B=3$
0.10	0.75	0.80	0.85
0.15	0.65	0.75	0.80
0.20	0.55	0.65	0.70
0.25	0.50	0.55	0.65
0.30	0.40	0.50	0.55
0.35	0.35	0.45	0.50
0.40	0.30	0.35	0.45
0.45	0.25	0.30	0.40
0.50	0.20	0.25	0.30
0.55	0.15	0.20	0.25
0.60	0.10	0.15	0.20

For Rectangular Footings — Load Inclination Factor (R_I)

H/H	$D_f/B=0$	$D_f/B=1$	$D_f/B=5$	$D_f/B=0$	$D_f/B=1$	$D_f/B=5$
0.10	0.70	0.75	0.80	0.80	0.85	0.90
0.15	0.60	0.65	0.70	0.70	0.80	0.85
0.20	0.50	0.60	0.65	0.65	0.70	0.75
0.25	0.40	0.50	0.55	0.55	0.65	0.70
0.30	0.35	0.40	0.50	0.50	0.60	0.65
0.35	0.30	0.35	0.40	0.40	0.55	0.60
0.40	0.25	0.30	0.35	0.35	0.50	0.55
0.45	0.20	0.25	0.30	0.30	0.45	0.50
0.50	0.15	0.20	0.25	0.25	0.35	0.45
0.55	0.10	0.15	0.20	0.20	0.30	0.40
0.60	0.05	0.10	0.15	0.15	0.25	0.35

After Barker et al. [9].

$$\begin{cases} C_{w1} = C_{w2} = 0.5 & \text{for } D_w = 0 \\ C_{w1} = C_{w2} = 1.0 & \text{for } D_w \geq D_f = 1.5B \\ C_{w1} = 0.5 \text{ and } C_{w2} = 1.0 & \text{for } D_w = D_f \end{cases} \tag{31.12}$$

N'_{avg} is an average value of the SPT blow counts, which is determined within the range of depths from footing base to 1.5B below the footing. In very fine or silty saturated sand, the measured SPT blow count (N) is corrected for submergence effect as follows:

$$N' = 15 + 0.5(N - 15) \quad \text{for } N > 15 \tag{31.13}$$

31.4.3 Bearing Capacity from Cone Penetration Tests (CPT)

Meyerhof [46] proposed a relationship between ultimate bearing capacity and cone penetration resistance in sands:

$$q_{\text{ult}} = q_c \frac{B}{40}\left(C_{w1} + C_{w2}\frac{D_f}{B}\right)R_I \tag{31.14}$$

where q_c is the average value of cone penetration resistance measured at depths from footing base to 1.5B below the footing base. C_{w1}, C_{w2}, and R_1 are the same as those as defined in Eq. (31.11).

Schmertmann [57] recommended correlated values of ultimate bearing capacity to cone penetration resistance in clays as shown in Table 31.7.

TABLE 31.7 Correlation between Ultimate Bearing Capacity (q^{ult}) and Cone Penetration Resistance (q_c)

q_c (kg/cm² or ton/ft²)	qult (ton/ft²)	
	Strip Footings	Square Footings
10	5	9
20	8	12
30	11	16
40	13	19
50	15	22

After Schmertmann [57] and Awkati, 1970.

31.4.4 Bearing Capacity from Pressure-Meter Tests (PMT)

Menard [44], Baguelin et al. [8], and Briaud [15,17] proposed using the limit pressure measured in PMT to estimate ultimate bearing capacity:

$$q_{\text{ult}} = r_0 + \kappa(p_1 - p_0) \tag{31.15}$$

where r_0 is the initial total vertical pressure at the foundation level, κ is the dimensionless bearing capacity coefficient from Figure 31.8, p_1 is limit pressure measured in PMT at depths from 1.5B above to 1.5B below foundation level, and p_0 is total horizontal pressure at the depth where the PMT is performed.

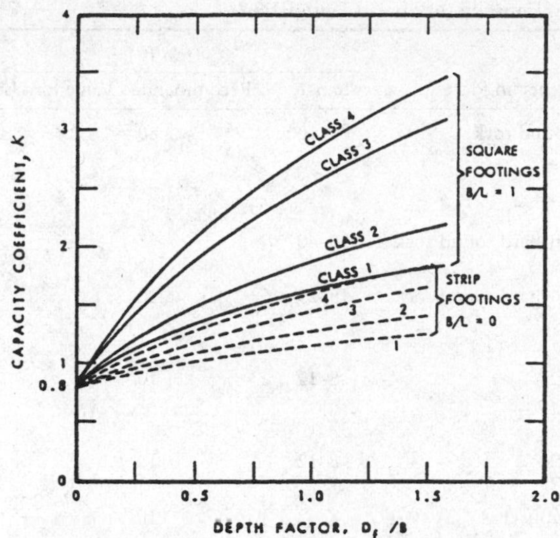

Soil Type	Consistency or Density	$(P_l\text{-}P_0)$ (tsf)	Class
clay	soft to very firm	<12	1
	stiff	8 – 40	2
sand and gravel	loose	4-8	2
	medium to dense	10-20	3
	very dense	30-60	4
silt	loose to medium	<7	1
	dense	12-30	2
rock	very low strength	10-30	2
	low strength	30-60	3
	medium to high strength	60-100$^+$	4

FIGURE 31.8 Values of empirical capacity coefficient, κ. (After Canadian Geotechnical Society [24].)

31.4.5 Bearing Capacity According to Building Codes

Recommendations for bearing capacity of shallow foundations are available in most building codes. Presumptive value of allowable bearing capacity for spread footings are intended for preliminary design when site-specific investigation is not justified. Presumptive bearing capacities usually do not reflect the size, shape, and depth of footing, local water table, or potential settlement. Therefore, footing design using such a procedure could be either overly conservative in some cases or unsafe in others [9]. Recommended practice is to use presumptive bearing capacity as shown in Table 31.8 for preliminary footing design and to finalize the design using reliable methods in the preceding discussion.

31.4.6 Predicted Bearing Capacity vs. Load Test Results

Obviously, the most reliable method of obtaining the ultimate bearing capacity is to conduct a full-scale footing load test at the project site. Details of the test procedure have been standardized as ASTM D1194 [5]. The load test is not usually performed since it is very costly and not practical for routine design. However, using load test results to compare with predicted bearing capacity is a vital tool to verify the accuracy and reliability of various prediction procedures. A comparison between the predicted bearing capacity and results of eight load tests conducted by Milovic [49] is summarized in Table 31.9.

Recently, load testing of five large-scale square footings (1 to 3 m) on sand was conducted on the Texas A&M University National Geotechnical Experimental Site [94]. One of the main objects of the test is to evaluate the various procedures used for estimating bearing capacities and settlements of shallow foundations. An international prediction event was organized by ASCE Geotechnical Engineering Division, which received a total of 31 predictions (16 from academics and 15 from consultants) from Israel, Australia, Japan, Canada, the United States, Hong Kong, Brazil, France, and Italy. Comparisons of predicted and measured values of bearing capacity using various procedures were summarized in Tables 31.10 through 31.12. From those comparisons, it can be argued that the most accurate settlement prediction methods are the Schmertmann-DMT (1986) and the Peck and Bazarra (1967) although they are on the unconservative side. The most conservative

TABLE 31.8 Presumptive Values of Allowable Bearing Capacity for Spread Foundations

Type of Bearing Material	Consistency in Place	q_{all} (ton/ft^2) Range	Recommended Value for Use
Massive crystalline igneous and metamorphic rock: granite, diorite, basalt, gneiss, thoroughly cemented conglomerate (sound condition allows minor cracks)	Hard sound rock	60–100	80
Foliated metamorphic rock: slate, schist (sound condition allows minor cracks)	Medium-hard sound rock	30–40	35
Sedimentary rock: hard cemented shales, siltstone, sandstone, limestone without cavities	Medium-hard sound rock	15–25	20
Weathered or broken bedrock of any kind except highly argillaceous rock (shale); RQD less than 25	Soft rock	8–12	10
Compaction shale or other highly argillaceous rock in sound condition	Soft rock	8–12	10
Well-graded mixture of fine and coarse-grained soil: glacial till, hardpan, boulder clay (GW-GC, GC, SC)	Very compact	8–12	10
Gravel, gravel–sand mixtures, boulder gravel mixtures (SW, SP)	Very compact	6–10	7
	Medium to compact	4–7	5
	Loose	2–5	3
Coarse to medium sand, sand with little gravel (SW, SP)	Very compact	4–6	4
	Medium to compact	2–4	3
	Loose	1–3	1.5
Fine to medium sand, silty or clayey medium to coarse sand (SW, SM, SC)	Very compact	3–5	3
	Medium to compact	2–4	2.5
	Loose	1–2	1.5
Homogeneous inorganic clay, sandy or silty clay (CL, CH)	Very stiff to hard	3–6	4
	Medium to stiff	1–3	2
	Soft	0.5–1	0.5
Inorganic silt, sandy or clayey silt, varved silt-clay-fine sand	Very stiff to hard	2–4	3
	Medium to stiff	1–3	1.5
	Soft	0.5–1	0.5

Notes:

1. Variations of allowable bearing pressure for size, depth, and arrangement of footings are given in Table 2 of NAFVAC [52].
2. Compacted fill, placed with control of moisture, density, and lift thickness, has allowable bearing pressure of equivalent natural soil.
3. Allowable bearing pressure on compressible fine-grained soils is generally limited by considerations of overall settlement of structure.
4. Allowable bearing pressure on organic soils or uncompacted fills is determined by investigation of individual case.
5. If tabulated recommended value for rock exceeds unconfined compressive strength of intact specimen, allowable pressure equals unconfined compressive strength.

After NAVFAC [52].

methods are Briaud [15] and Burland and Burbidge [20]. The most accurate bearing capacity prediction method was the $0.2q_c$ (CPT) method [16].

TABLE 31.9 Comparison of Computed Theoretical Bearing Capacities and Milovic and Muh's Experimental Values

	Test							
Bearing Capacity Method	1	2	3	4	5	6	7	8
	D = 0.0 m	0.5	0.5	0.5	0.4	0.5	0.0	0.3
	B = 0.5 m	0.5	0.5	1.0	0.71	0.71	0.71	0.71
	L = 2.0 m	2.0	2.0	1.0	0.71	0.71	0.71	0.71
	γ = 15.69 kN/m³	16.38	17.06	17.06	17.65	17.65	17.06	17.06
	ϕ = 37°(38.5°)	35.5 (36.25)	38.5 (40.75)	38.5	22	25	20	20
	c = 6.37 kPa	3.92	7.8	7.8	12.75	14.7	9.8	9.8
Milovic (tests)					q_{ult} (kg/cm²) 4.1	5.5	2.2	2.6
Muhs (tests)	q_{ult} (kg/cm²) 10.8	12.2	24.2	33.0				
Terzaghi	9.4*	9.2	22.9	19.7	4.3*	6.5*	2.5	2.9*
Meyerhof	8.2*	10.3	26.4	28.4	4.8	7.6	2.3	3.0
Hansen	7.2	9.8	23.7*	23.4	5.0	8.0	2.2*	3.1
Vesic	8.1	10.4*	25.1	24.7	5.1	8.2	2.3	3.2
Balla	14.0	15.3	35.8	33.0*	6.0	9.2	2.6	3.8

[a]After Milovic (1965) but all methods recomputed by author and Vesic added.

Notes:
1. ϕ = triaxial value ϕ_{tr}; (plane strain value) = 1.5 ϕ_{tr} - 17.
2. * = best: Terzaghi = 4; Hansen = 2; Vesic = 1; and Balla = 1.

Source: Bowles, J.E., *Foundation Analysis and Design*, 5th ed., McGraw-Hill, New York, 1996. With permission.

TABLE 31.10 Comparison of Measured vs. Predicted Load Using Settlement Prediction Method

	Predicted Load (MN) @ s = 25 mm				
Prediction Methods	1.0 m Footing	1.5 m Footing	2.5 m Footing	3.0 m(n) Footing	3.0 m(s) Footing
Briaud [15]	0.904	1.314	2.413	2.817	2.817
Burland and Burbidge [20]	0.699	1.044	1.850	2.367	2.367
De Beer (1965)	1.140	0.803	0.617	0.597	0.597
Menard and Rousseau (1962)	0.247	0.394	0.644	1.017	1.017
Meyerhof CPT (1965)	0.288	0.446	0.738	0.918	0.918
Meyerhof — SPT (1965)	0.195	0.416	1.000	1.413	1.413
Peck and Bazarra (1967)	1.042	1.899	4.144	5.679	5.679
Peck, Hansen & Thornburn [53]	0.319	0.718	1.981	2.952	2.952
Schmertmann CPT (1970)	0.455	0.734	1.475	1.953	1.953
Schmertmann DMT (1970)	1.300	2.165	4.114	5.256	5.256
Schultze and Sherif (1973)	1.465	2.615	4.750	5.850	5.850
Terzaghi and Peck [65]	0.287	0.529	1.244	1.476	1.476
Measured Load @ s = 25mm	0.850	1.500	3.600	4.500	4.500

Source: FHWA, Publication No. FHWA-RD-97-068, 1997.

31.5 Stress Distribution Due to Footing Pressures

Elastic theory is often used to estimate the distribution of stress and settlement as well. Although soils are generally treated as elastic–plastic materials, the use of elastic theory for solving the problems is mainly due to the reasonable match between the boundary conditions for most footings and those of elastic solutions [37]. Another reason is the lack of availability of acceptable alternatives. Observation and experience have shown that this practice provides satisfactory solutions [14,37,54,59].

TABLE 31.11 Comparison of Measured vs. Predicted Load Using Bearing Capacity Prediction Method

	Predicted Bearing Capacity (MN)				
Prediction Methods	1.1 m Footing	1.5 m Footing	2.6 m Footing	3.0m(n) Footing	3.0m(s) Footing
Briaud — CPT [16]	1.394	1.287	1.389	1.513	1.513
Briaud — PMT [15]	0.872	0.779	0.781	0.783	0.783
Hansen [35]	0.772	0.814	0.769	0.730	0.730
Meyerhof [45,48]	0.832	0.991	1.058	1.034	1.034
Terzaghi [63]	0.619	0.740	0.829	0.826	0.826
Vesic [68,69]	0.825	0.896	0.885	0.855	0.855
Measured Load @ s = 150 mm					

Source: FHWA, Publication No. FHWA-RD-97-068, 1997.

TABLE 31.12 Best Prediction Method Determination

		Mean Predicted Load/ Mean Measured Load
	Settlement Prediction Method	
1	Briaud [15]	0.66
2	Burland & Burbidge [20]	0.62
3	De Beer [29]	0.24
4	Menard and Rousseau (1962)	0.21
5	Meyerhof CPT (1965)	0.21
6	Meyerhof SPT (1965)	0.28
7	Peck and Bazarra (1967)	1.19
8	Peck, et al. [53]	0.57
9	Schmertmann — CPT [56]	0.42
10	Schmertmann — DMT [56]	1.16
11	Shultze and Sherif (1973)	1.31
12	Terzaghi and Peck [65]	0.32
	Bearing Capacity Prediction Method	
1	Briaud — CPT [16]	1.08
2	Briaud — PMT [15]	0.61
3	Hansen [35]	0.58
4	Meyerhof [45,48]	0.76
5	Terzaghi [63]	0.59
6	Vesic [68,69]	0.66

Source: FHWA, Publication No. FHWA-RD-97-068, 1997.

31.5.1 Semi-infinite, Elastic Foundations

Bossinesq equations based on elastic theory are the most commonly used methods for obtaining subsurface stresses produced by surface loads on semi-infinite, elastic, isotropic, homogenous, weightless foundations. Formulas and plots of Bossinesq equations for common design problems are available in NAVFAC [52]. Figure 31.9 shows the isobars of pressure bulbs for square and continuous footings. For other geometry, refer to Poulos and Davis [55].

31.5.2 Layered Systems

Westergaard [70], Burmister [21-23], Sowers and Vesic [62], Poulos and Davis [55], and Perloff [54] discussed the solutions to stress distributions for layered soil strata. The reality of interlayer shear is very complicated due to *in situ* nonlinearity and material inhomogeneity [37,54]. Either zero (frictionless) or with perfect fixity is assumed for the interlayer shear to obtain possible

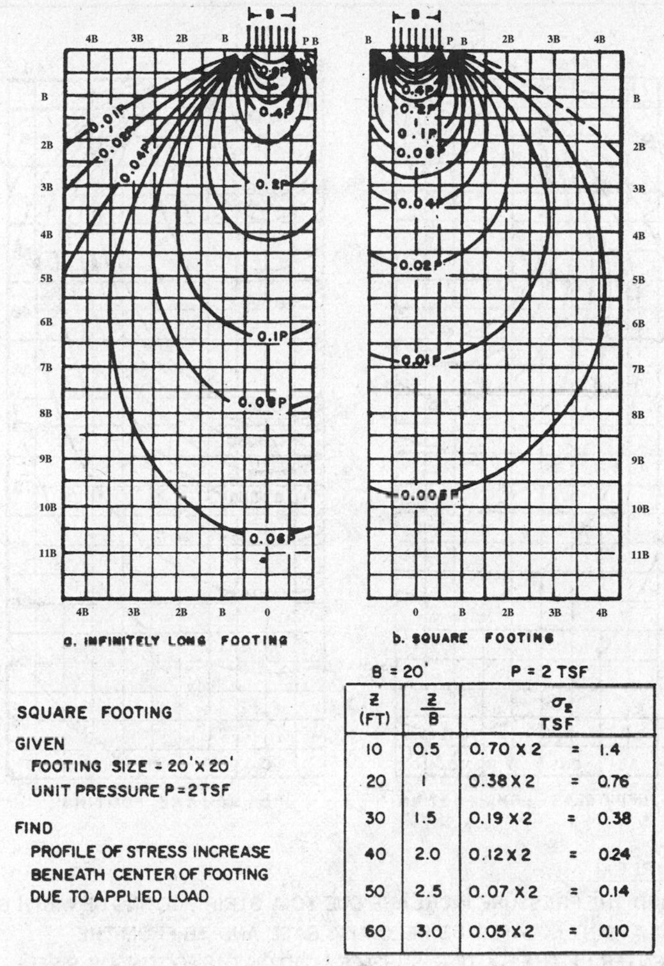

z (FT)	z/B	σ_z TSF
10	0.5	0.70 X 2 = 1.4
20	1	0.38 X 2 = 0.76
30	1.5	0.19 X 2 = 0.38
40	2.0	0.12 X 2 = 0.24
50	2.5	0.07 X 2 = 0.14
60	3.0	0.05 X 2 = 0.10

FIGURE 31.9 Pressure bulbs based on the Bossinesq equation for square and long footings. (After NAVFAC 7.01, 1986].)

solutions. The Westergaard method assumed that the soil being loaded is constrained by closed spaced horizontal layers that prevent horizontal displacement [52]. Figures 31.10 through 31.12 by the Westergaard method can be used for calculating vertical stresses in soils consisting of alternative layers of soft (loose) and stiff (dense) materials.

31.5.3 Simplified Method (2:1 Method)

Assuming a loaded area increasing systemically with depth, a commonly used approach for computing the stress distribution beneath a square or rectangle footing is to use the 2:1 slope method as shown in Figure 31.13. Sometimes a 60° distribution angle (1.73–to–1 slope) may be assumed. The pressure increase Δq at a depth z beneath the loaded area due to base load P is

$$\Delta q = \begin{cases} P/(B+z)(L+z) & \text{(for a rectangle footing)} \\ P/(B+z)^2 & \text{(for a square footing)} \end{cases} \tag{31.16}$$

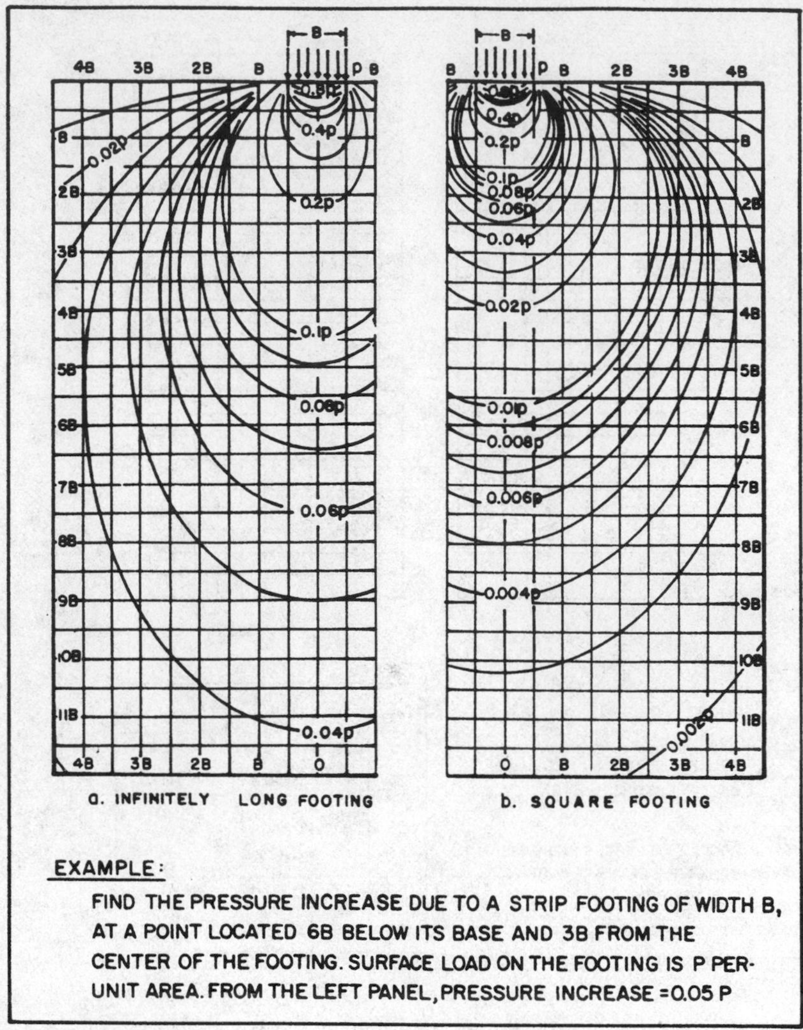

FIGURE 31.10 Vertical stress contours for square and strip footings [Westerqaard Case]. (After NAVFAC 7.01, 1986.)

where symbols are referred to Figure 31.14. The solutions by this method compare very well with those of more theoretical equations from depth z from B to about $4B$ but should not be used for depth z from 0 to B [14]. A comparison between the approximate distribution of stress calculated by a theoretical method and the 2:1 method is illustrated in Figure 31.15.

31.6 Settlement of Shallow Foundations

The load applied on a footing changes the stress state of the soil below the footing. This stress change may produce a time-dependent accumulation of elastic compression, distortion, or consolidation of the soil beneath the footing. This is often termed *foundation settlement*. True elastic deformation consists of a very small portion of the settlement while the major components of the settlement are due to a change of void ratio, particle rearrangement, or crushing. Therefore, very little of the settlement will be recovered even if the applied load is removed. The irrecoverable

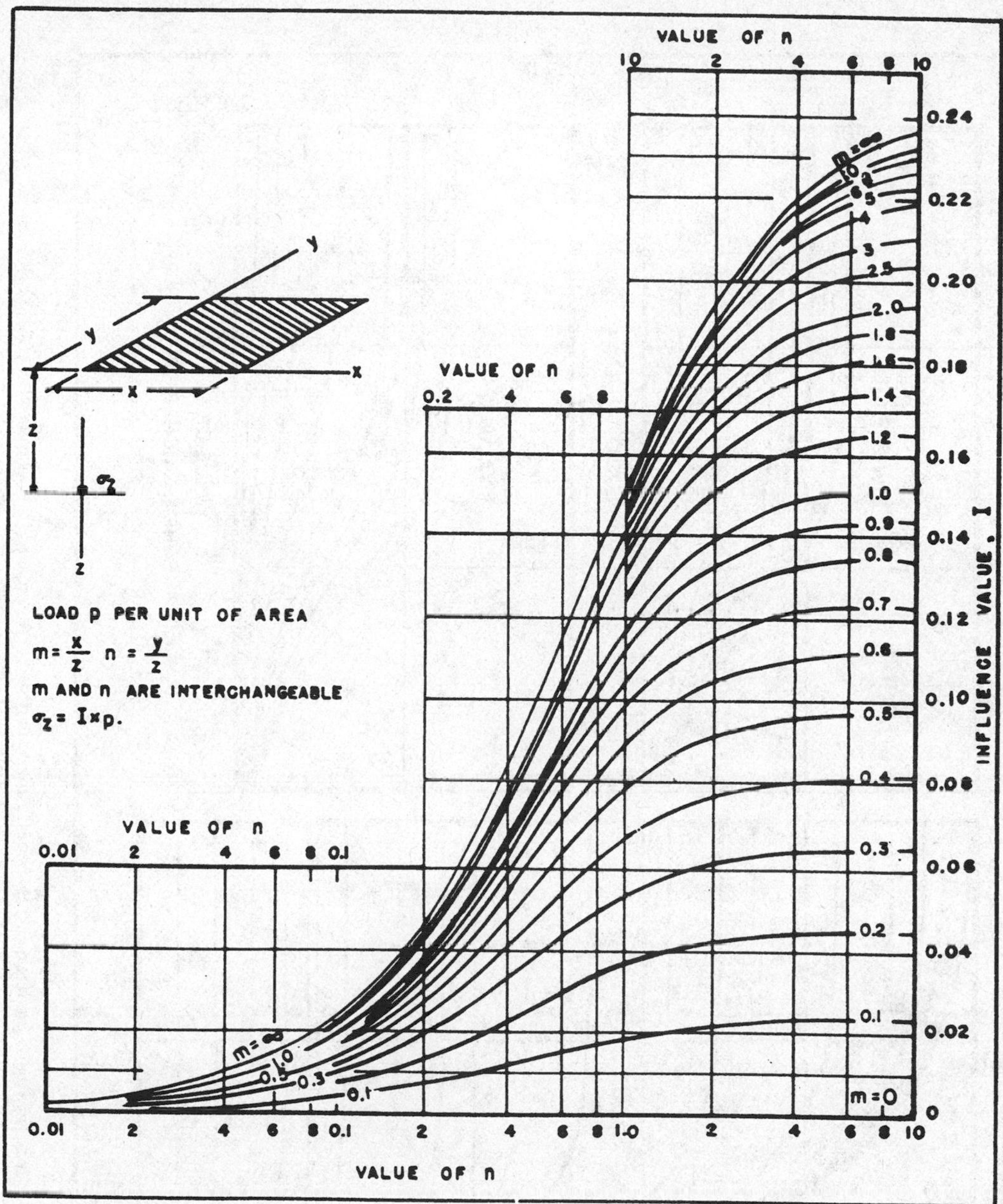

FIGURE 31.11 Influence value for vertical stress beneath a corner of a uniformly loaded rectangular area (Westergaard Case). (After NAVFAC [52].)

deformation of soil reflects its inherent elastic–plastic stress–strain relationship. The reliability of settlement estimated is influenced principally by soil properties, layering, stress history, and the actual stress profile under the applied load [14,66]. The total settlement may be expressed as

$$s = s_i + s_c + s_s \tag{31.17}$$

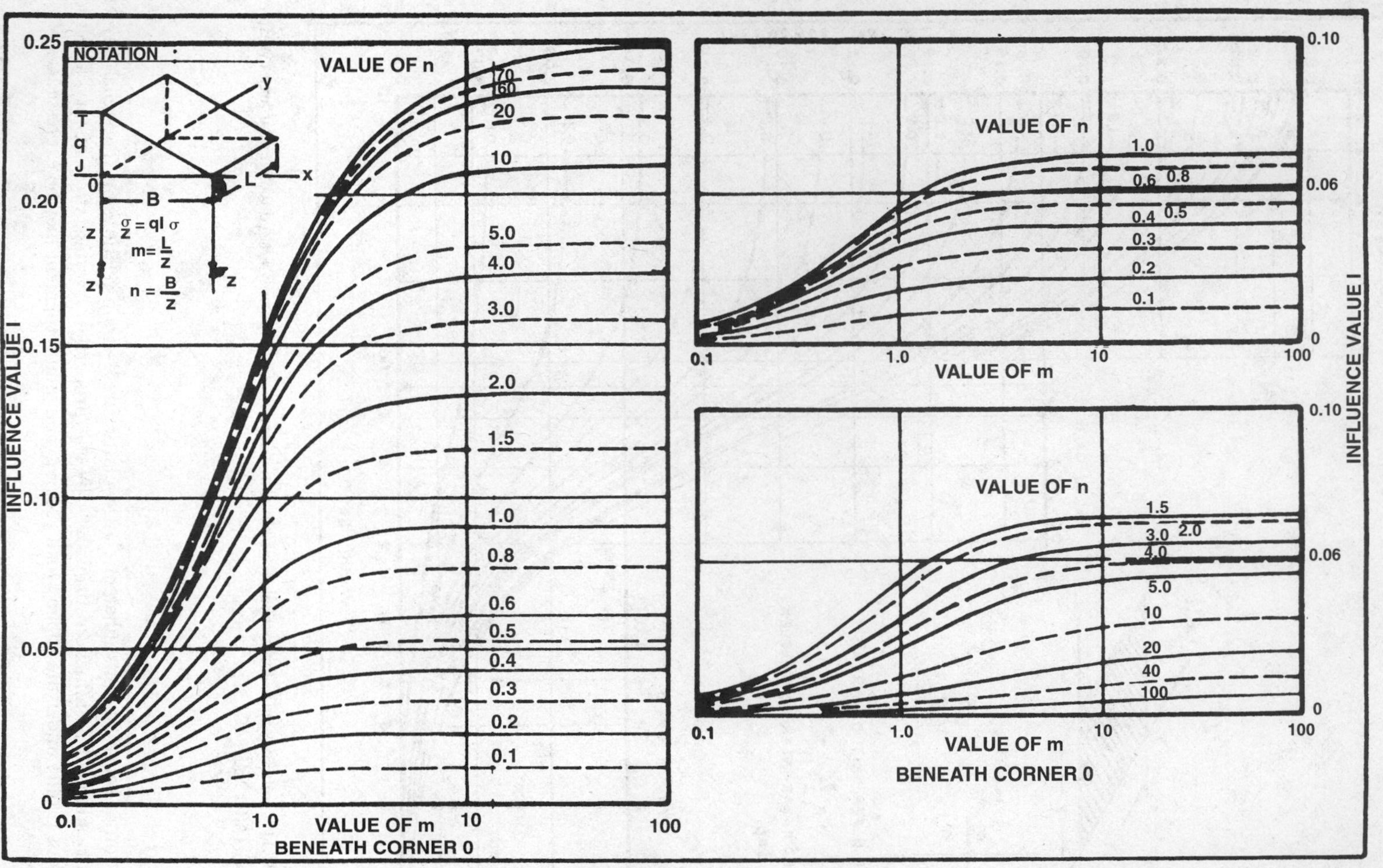

FIGURE 31.12 Influence value for vertical stress beneath triangular load (Westerqaard Case). (After NAVFAC [52].)

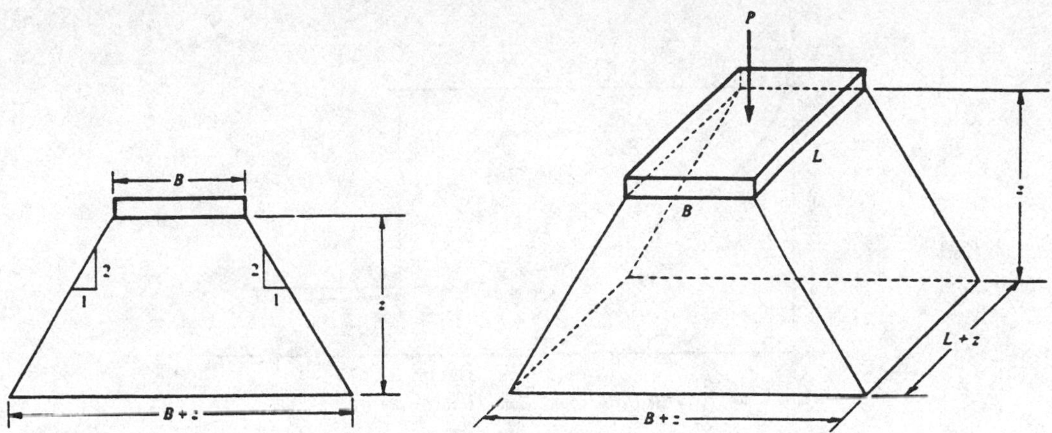

FIGURE 31.14 Approximate distribution of vertical stress due to surface load. (After Perloff [54].)

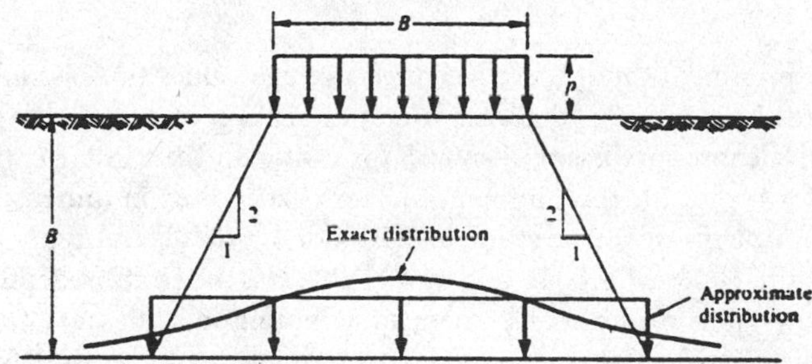

FIGURE 31.15 Relationship between vertical stress below a square uniformly loaded area as determined by approximate and exact methods. (After Perloff [54].)

where s is the total settlement, s_i is the immediate or distortion settlement, s_c is the primary consolidation settlement, and s_s is the secondary settlement. The time-settlement history of a shallow foundation is illustrated in Figure 31.15. Generally speaking, immediate settlement is not elastic. However, it is often referred to as elastic settlement because the elastic theory is usually used for computation. The immediate settlement component controls in cohesionless soils and unsaturated cohesive soils, while consolidation compression dictates in cohesive soils with a degree of saturation above 80% [3].

31.6.1 Immediate Settlement by Elastic Methods

Based on elastic theory, Steinbrenner [61] suggested that immediate settlements of footings on sands and clay could be estimated in terms of Young's modulus E of soils. A modified procedure developed by Bowles [14] may be used for computing settlements of footings with flexible bases on the half-space. The settlement equation can be expressed as follows

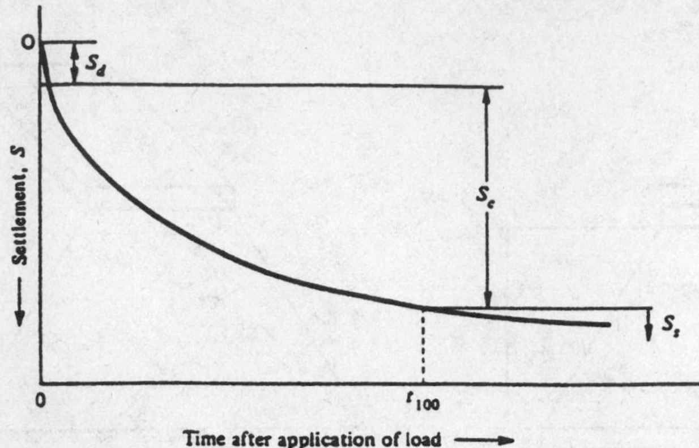

FIGURE 31.15 Schematic time–settlement history of typical point on a foundation. (After Perloff [54].)

$$s_i = q_0 B' \left(1-\mu^2\right) m I_s I_F / E_s \tag{31.18}$$

$$I_s = n\,(I_1 + (1-2\mu) I_2/(1-\mu) \tag{31.19}$$

where q_0 is contact pressure, μ and E_s are weighted average values of Poisson's ratio and Young's modulus for compressive strata, B is the least-lateral dimension of contribution base area (convert round bases to equivalent square bases; $B = 0.5B$ for center and $B = B$ for corner I_i; $L' = 0.5L$ for center and $L' = L$ for corner I_i), I_i are influence factors depending on dimension of footings, base embedment depth, thickness of soil stratum, and Poisson's ratio (I_1 and I_2 are given in Table 31.13 and I_F is given in Figure 31.16; $M = L'/B'$ and $N = H/B'$), H is the stratum depth causing settlement (see discussion below), m is number of corners contributing to settlement (m = 4 at the footing center; m = 2 at a side; and m = 1 at a corner), and n equals 1.0 for flexible footings and 0.93 for rigid footings.

This equation applies to soil strata consisting of either cohesionless soils of any water content or unsaturated cohesive soils, which may be either organic or inorganic. Highly organic soils (both E_s and μ are subject to significant changes by high organic content) will be dictated by secondary or creep compression rather than immediate settlement; therefore, the applicability of the above equation is limited.

Suggestions were made by Bowles [14] to use the equations appropriately as follows: 1. Make the best estimate of base contact pressure q_0; 2. Identify the settlement point to be calculated and divide the base (as used in the Newmark stress method) so the point is at the corner or common corner of one or up to four contributing areas; 3. Determine the stratum depth causing settlement which does not approach to infinite rather at either the depth $z = 5B$ or depth to where a hard stratum is encountered (where F_s in the hard layer is about $10E_s$ of the adjacent upper layer); and 4. Calculate the weighted average E_s as follows:

$$E_{s,\,\text{avg}} = \sum_n^1 H_i E_{si} \Big/ \sum_n^1 H_i \tag{31.20}$$

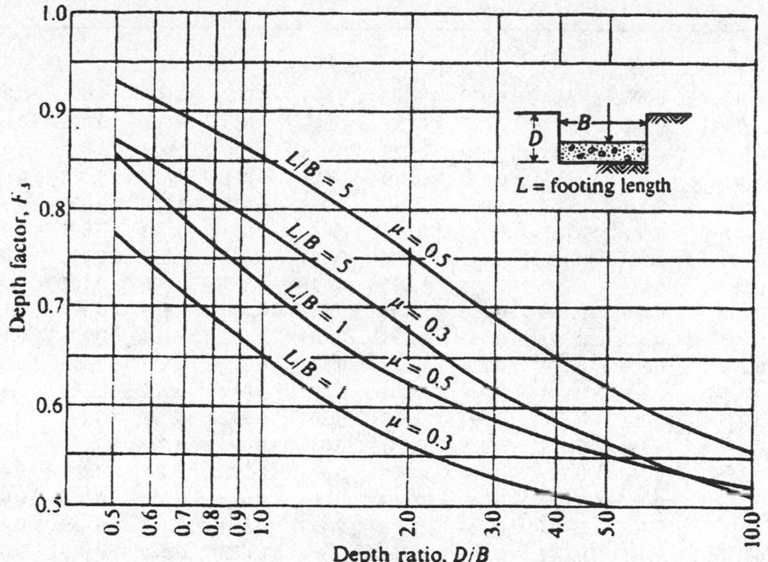

FIGURE 31.16 Influence factor I_F for footing at a depth D (use actual footing width and depth dimension for this D/B ratio). (After Bowles [14].)

31.6.2 Settlement of Shallow Foundations on Sand

SPT Method

D'Appolonio et al. [28] developed the following equation to estimate settlements of footings on sand using SPT data:

$$s = \mu_0 \mu_1 \, pB/M \tag{31.21}$$

where μ_0 and μ_1 are settlement influence factors dependent on footing geometry, depth of embedment, and depth to the relative incompressible layer (Figure 31.17), p is average applied pressure under service load and M is modulus of compressibility. The correlation between M and average SPT blow count is given in Figure 31.18.

Barker et al. [9] discussed the commonly used procedure for estimating settlement of footing on sand using SPT blow count developed by Terzaghi and Peck [64,65] and Bazaraa [10].

CPT Method

Schmertmann [56,57] developed a procedure for estimating footing settlements on sand using CPT data. This CPT method uses cone penetration resistance, q_c, as a measure of the *in situ* stiffness (compressibility) soils. Schmertmann's method is expressed as follows

$$s = C_1 C_2 \Delta p \Sigma \left(I_Z / E_s\right)_i \Delta z_i \tag{31.22}$$

$$C_1 = 1 - 0.5\left(\frac{\sigma'_{v0}}{\Delta p}\right) \geq 0.5 \tag{31.23}$$

$$C_2 = 1 + 0.2 \log\left(t_{yr}/0.1\right) \tag{31.24}$$

TABLE 31.13 Values of I_2 and I_2 to Compute Influence Factors as Used in Eq. (31.21)

N	$M = 1.0$	1.1	1.2	1.3	1.4	1.5	1.6	1.7	1.8	1.9	2.0
0.2	$I_1 = 0.009$	0.008	0.008	0.008	0.008	0.008	0.007	0.007	0.007	0.007	0.007
	$I_2 = 0.041$	0.042	0.042	0.042	0.042	0.042	0.043	0.043	0.043	0.043	0.043
0.4	0.033	0.032	0.031	0.030	0.029	0.028	0.028	0.027	0.027	0.027	0.027
	0.066	0.068	0.069	0.070	0.070	0.071	0.071	0.072	0.072	0.073	0.073
0.6	0.066	0.064	0.063	0.061	0.060	0.059	0.058	0.057	0.056	0.056	0.055
	0.079	0.081	0.083	0.085	0.087	0.088	0.089	0.090	0.091	0.091	0.092
0.8	0.104	0.102	0.100	0.098	0.096	0.095	0.093	0.092	0.091	0.090	0.089
	0.083	0.087	0.090	0.093	0.095	0.097	0.098	0.100	0.101	0.102	0.103
1.0	0.142	0.140	0.138	0.136	0.134	0.132	0.130	0.129	0.127	0.126	0.125
	0.083	0.088	0.091	0.095	0.098	0.100	0.102	0.104	0.106	0.108	0.109
1.5	0.224	0.224	0.224	0.223	0.222	0.220	0.219	0.217	0.216	0.214	0.213
	0.075	0.080	0.084	0.089	0.093	0.096	0.099	0.102	0.105	0.108	0.110
2.0	0.285	0.288	0.290	0.292	0.292	0.292	0.292	0.292	0.291	0.290	0.289
	0.064	0.069	0.074	0.078	0.083	0.086	0.090	0.094	0.097	0.100	0.102
3.0	0.363	0.372	0.379	0.384	0.389	0.393	0.396	0.398	0.400	0.401	0.402
	0.048	0.052	0.056	0.060	0.064	0.068	0.071	0.075	0.078	0.081	0.084
4.0	0.408	0.421	0.431	0.440	0.448	0.455	0.460	0.465	0.469	0.473	0.476
	0.037	0.041	0.044	0.048	0.051	0.054	0.057	0.060	0.063	0.066	0.069
5.0	0.437	0.452	0.465	0.477	0.487	0.496	0.503	0.510	0.516	0.522	0.526
	0.031	0.034	0.036	0.039	0.042	0.045	0.048	0.050	0.053	0.055	0.058
6.0	0.457	0.474	0.489	0.502	0.514	0.524	0.534	0.542	0.550	0.557	0.563
	0.026	0.028	0.031	0.033	0.036	0.038	0.040	0.043	0.045	0.047	0.050
7.0	0.471	0.490	0.506	0.520	0.533	0.545	0.556	0.566	0.575	0.583	0.590
	0.022	0.024	0.027	0.029	0.031	0.033	0.035	0.037	0.039	0.041	0.043
8.0	0.482	0.502	0.519	0.534	0.549	0.561	0.573	0.584	0.594	0.602	0.611
	0.020	0.022	0.023	0.025	0.027	0.029	0.031	0.033	0.035	0.036	0.038
9.0	0.491	0.511	0.529	0.545	0.560	0.574	0.587	0.598	0.609	0.618	0.627
	0.017	0.019	0.021	0.023	0.024	0.026	0.028	0.029	0.031	0.033	0.034
10.0	0.498	0.519	0.537	0.554	0.570	0.584	0.597	0.610	0.621	0.631	0.641
	0.016	0.017	0.019	0.020	0.022	0.023	0.025	0.027	0.028	0.030	0.031
20.0	0.529	0.553	0.575	0.595	0.614	0.631	0.647	0.662	0.677	0.690	0.702
	0.008	0.009	0.010	0.010	0.011	0.012	0.013	0.013	0.014	0.015	0.016
500	0.560	0.587	0.612	0.635	0.656	0.677	0.696	0.714	0.731	0.748	0.763
	0.000	0.000	0.000	0.000	0.000	0.000	0.001	0.001	0.001	0.001	0.001
0.2	$I_1 = 0.007$	0.006	0.006	0.006	0.006	0.006	0.006	0.006	0.006	0.006	0.006
	$I_2 = 0.043$	0.044	0.044	0.044	0.044	0.044	0.044	0.044	0.044	0.044	0.044
0.4	0.026	0.024	0.024	0.024	0.024	0.024	0.024	0.024	0.024	0.024	0.024
	0.074	0.075	0.075	0.075	0.076	0.076	0.076	0.076	0.076	0.076	0.076
0.6	0.053	0.051	0.050	0.050	0.050	0.049	0.049	0.049	0.049	0.049	0.049
	0.094	0.097	0.097	0.098	0.098	0.098	0.098	0.098	0.098	0.098	0.098
0.8	0.086	0.082	0.081	0.080	0.080	0.080	0.079	0.079	0.079	0.079	0.079
	0.107	0.111	0.112	0.113	0.113	0.113	0.113	0.114	0.114	0.014	0.014
1.0	0.121	0.115	0.113	0.112	0.112	0.112	0.111	0.111	0.110	0.110	0.110
	0.114	0.120	0.122	0.123	0.123	0.124	0.124	0.124	0.125	0.125	0.125
1.5	0.207	0.197	0.194	0.192	0.191	0.190	0.190	0.189	0.188	0.188	0.188
	0.118	0.130	0.134	0.136	0.137	0.138	0.138	0.139	0.140	0.140	0.140
2.0	0.284	0.271	0.267	0.264	0.262	0.261	0.260	0.259	0.257	0.256	0.256
	0.114	0.131	0.136	0.139	0.141	0.143	0.144	0.145	0.147	0.147	0.148
3.0	0.402	0.392	0.386	0.382	0.378	0.376	0.374	0.373	0.368	0.367	0.367
	0.097	0.122	0.131	0.137	0.141	0.144	0.145	0.147	0.152	0.153	0.154
4.0	0.484	0.484	0.479	0.474	0.470	0.466	0.464	0.462	0.453	0.451	0.451
	0.082	0.110	0.121	0.129	0.135	0.139	0.142	0.145	0.154	0.155	0.156
5.0	0.553	0.554	0.552	0.548	0.543	0.540	0.536	0.534	0.522	0.519	0.519
	0.070	0.098	0.111	0.120	0.128	0.133	0.137	0.140	0.154	0.156	0.157
6.0	0.585	0.609	0.610	0.608	0.604	0.601	0.598	0.595	0.579	0.576	0.575
	0.060	0.087	0.101	0.111	0.120	0.126	0.131	0.135	0.153	0.157	0.157

TABLE 31.13 (continued) Values of I_2 and I_2 to Compute Influence Factors as Used in Eq. (31.21)

N	M = 1.0	1.1	1.2	1.3	1.4	1.5	1.6	1.7	1.8	1.9	2.0
7.0	0.618	0.653	0.658	0.658	0.656	0.653	0.650	0.647	0.628	0.624	0.623
	0.053	0.078	0.092	0.103	0.112	0.119	0.125	0.129	0.152	0.157	0.158
8.0	0.643	0.688	0.697	0.700	0.700	0.698	0.695	0.692	0.672	0.666	0.665
	0.047	0.071	0.084	0.095	0.104	0.112	0.118	0.124	0.151	0.156	0.158
9.0	0.663	0.716	0.730	0.736	0.737	0.736	0.735	0.732	0.710	0.704	0.702
	0.042	0.064	0.077	0.088	0.097	0.105	0.112	0.118	0.149	0.156	0.158
10.0	0.679	0.740	0.758	0.766	0.770	0.770	0.770	0.768	0.745	0.738	0.735
	0.038	0.059	0.071	0.082	0.091	0.099	0.106	0.122	0.147	0.156	0.158
20.0	0.756	0.856	0.896	0.925	0.945	0.959	0.969	0.977	0.982	0.965	0.957
	0.020	0.031	0.039	0.046	0.053	0.059	0.065	0.071	0.124	0.148	0.156
500.0	0.832	0.977	1.046	1.102	1.150	1.191	1.227	1.259	2.532	1.721	1.879
	0.001	0.001	0.002	0.002	0.002	0.002	0.003	0.003	0.008	0.016	0.031

Source: Bowles, J.E., *Foundation Analysis and Design*, 5th ed., McGraw-Hill, New York, 1996. With permission.

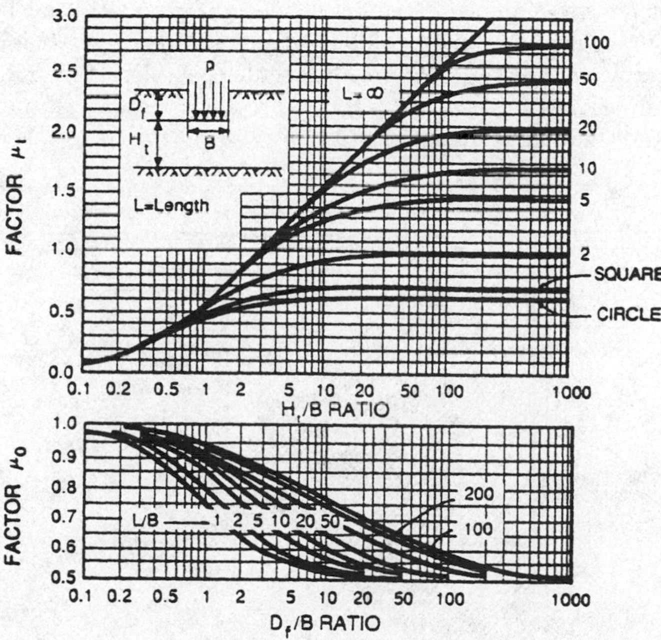

FIGURE 31.17 Settlement influence factors μ_0 and μ_1 for the D'Appolonia et al. procedure. (After D'Appolonia et al [28].)

$$E_s = \begin{cases} 2.5q_c & \text{for square footings (axisymmetric conditions)} \\ 3.5q_c & \text{for continuous footings with } L/B \geq 10 \text{ (plan strain conditions)} \\ [2.5+(L/B-1)/9]q_c & \text{for footings with } 1 \geq L/B \geq 10 \end{cases} \quad (31.25)$$

where $\Delta p = \sigma'_{vf} - \sigma'_{v0}$ is net load pressure at foundation level, σ'_{v0} is initial effective *in situ* overburden stress at the bottom of footings, σ'_{vf} is final effective *in situ* overburden stress at the

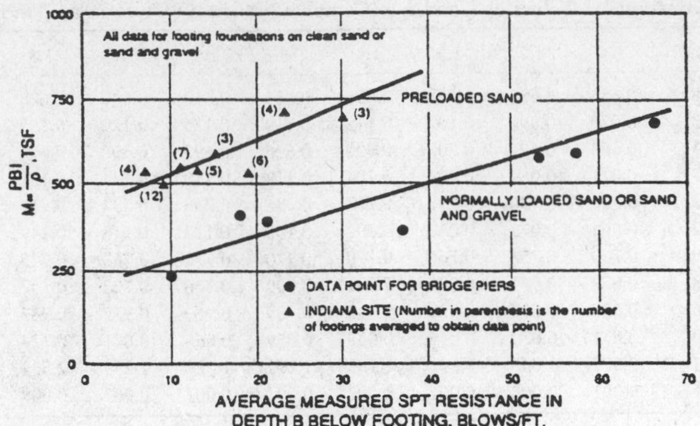

FIGURE 31.18 Correlation between modulus of compressibility and average value SPT blow count. (After D'Appolonia et al [28].)

bottom of footings, I_z is strain influence factor as defined in Figure 31.19 and Table 31.14, E_s is the appropriate Young's modulus at the middle of the *i*th layer of thickness Δz_1, C_1 is pressure correction factor, C_2 is time rate factor (equal to 1 for immediate settlement calculation or if the lateral pressure is less than the creep pressure determined from pressure-meter tests), q_c is cone penetration resistance, in pressure units, and Δz is layer thickness.

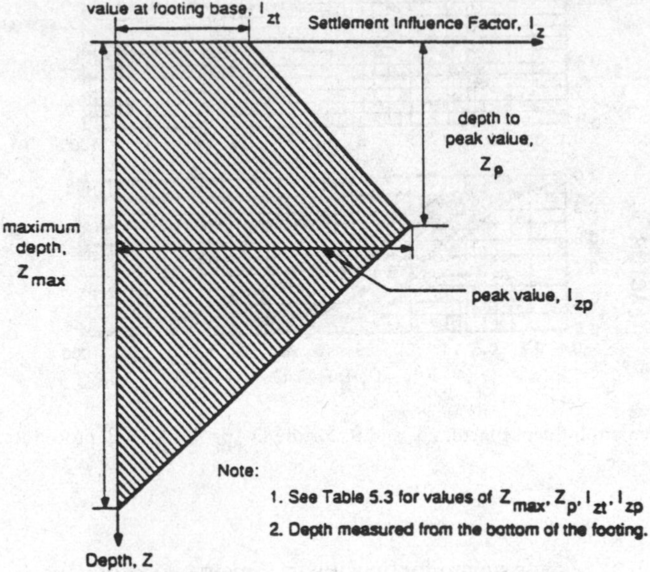

FIGURE 31.19 Variation of Schmertmann's improved settlement influence factors with depth. (After Schmertmann et al [58].)

Recent studies by Tan and Duncan [62] have compared measured settlements with settlements predicted using various procedures for footings on sand. These studies conclude that methods predicting settlements close to the average of measured settlement are likely to underestimate

TABLE 31.14 Coefficients to Define the Dimensions of Schmertmann's Improved Settlement Influence Factor Diagram in Figure 31.20

				Peak Value of Stress Influence Factor I_{zp}			
L/B	Max. Depth of Influence z_{max}/B	Depth to Peak Value z_p/B	Value of I_z at Top I_{zt}	$\frac{\Delta p}{\sigma'_{vp}}=1$	$\frac{\Delta p}{\sigma'_{vp}}=2$	$\frac{\Delta p}{\sigma'_{vp}}=4$	$\frac{\Delta p}{\sigma'_{vp}}=10$
1	2.00	0.50	0.10	0.60	0.64	0.70	0.82
2	2.20	0.55	0.11	0.60	0.64	0.70	0.82
4	2.65	0.65	0.13	0.60	0.64	0.70	0.82
8	3.55	0.90	0.18	0.60	0.64	0.70	0.82
≥10	4.00	1.00	0.20	0.60	0.64	0.70	0.82

Note: σ'_{vp} is the initial vertical pressure at depth of peak influence.

After Schmertmann et al. [57].

settlements half the time and to overestimate them half the time. The conservative methods (notably Terzaghi and Peck's) tend to overestimate settlements more than half the time and to underestimate them less often. On the other hand, there is a trade-off between accuracy and reliability. A relatively accurate method such as the D'Appolonia et al. method calculates settlements that are about equal to the average value of actual settlements, but it underestimates settlements half the time (a reliability of 50%). To ensure that the calculated settlements equal or exceed the measured settlements about 90% of the time (a reliability of 90%), an adjustment factor of two should be applied to the settlements predicted by the D'Appolonia et al. method. Table 31.15 shows values of the adjustment factor for 50 and 90% reliability in settlement predicted using Terzaghi and Peck, D'Appolonia et al., and Schmertmann methods.

TABLE 31.15 Value of Adjustment Factor for 50 and 90% Reliability in Displacement Estimates

		Adjustment Factor	
Method	Soil Type	For 50% Reliability	For 90% Reliability
Terzaghi and Peck [65]	Sand	0.45	1.05
Schmertmann	Sand	0.60	1.25
D'Appolonia et al. [28]	Sand	1.00	2.00

TABLE 31.16 Some Empirical Equations for C_c and C_α

Compression Index	Source	Comment
$C_c = 0.009(LL - 10)$	Terzaghi and Peck [65]	$S_t \leq 5$, $LL < 100$
$C_c = 0.2343e_0$	Nagaraj and Murthy [51]	
$C_c = 0.5G_s(PI/100)$	Worth and Wood [71]	Modified cam clay model
$C_c = 0PI/74$	EPRI (1990)	
$C_c = 0.37(e_0 + 0.003w_L + 0.0004w_N - 0.34)$	Azzouz et al. [7]	Statistical analysis
Recompression Index	**Source**	
$C_r = 0.0463w_LG_s$	Nagaraj and Murthy [50]	

31.6.3 Settlement of Shallow Foundations on Clay

Immediate Settlement

Immediate settlement of shallow foundations on clay can be estimated using the approach described in Section 31.6.1.

Consolidation Settlement

Consolidation settlement is time dependent and may be estimated using one-dimensional consolidation theory [43,53,66]. The consolidation settlement can be calculated as follows

$$S_c \begin{cases} \dfrac{H_c}{1+e_0}\left[C_r \log\left(\dfrac{\sigma'_p}{\sigma''_{vo}}\right) + C_c \log\left(\dfrac{\sigma'_{vf}}{\sigma'_p}\right)\right] & \left(\text{for OC soils, i.e., } \sigma'_p > \sigma'_{v0}\right) \\ \dfrac{H_c}{1+e_0} C_c \log\left(\dfrac{\sigma'_{vf}}{\sigma'_p}\right) & \left(\text{for NC soils, i.e., } \sigma'_p = \sigma'_{v0}\right) \end{cases} \tag{31.26}$$

where H_c is height of compressible layer, e_0 is void ratio at initial vertical effective stress, C_γ is recompression index (see Table 31.16), C_c is compression index (see Table 31.16), σ'_p is maximum past vertical effective stress, σ'_{v0} is initial vertical effective stress, σ'_{vf} is final vertical effective stress. Highly compressible cohesive soils are rarely chosen to place footings for bridges where tolerable amount of settlement is relatively small. Preloading or surcharging to produce more rapid consolidation has been extensively used for foundations on compressible soils [54]. Alternative foundation systems would be appropriate if large consolidation settlement is expected to occur.

TABLE 31.17 Secondary Compression Index

C_α/C_c	Material
0.02 ± 0.01	Granular soils including rockfill
0.03 ± 0.01	Shale and mudstone
0.04 ± 0.01	Inorganic clays and silts
0.05 ± 0.01	Organic clays and silts
0.06 ± 0.01	Peat and muskeg

Source: Terzaghi, I. et al., *Soil Mechanics in Engineering Practice*, 3rd ed., John Wiley & Sons, New York, 1996. With permission.

Secondary Settlement

Settlements of footings on cohesive soils continuing beyond primary consolidation are called secondary settlement. Secondary settlement develops at a slow and continually decreasing rate and may be estimated as follows:

$$s_s = C_\alpha H_t \log\frac{t_{sc}}{t_p} \tag{31.27}$$

where C_α is coefficient of secondary settlement (Table 31.17), H_t is total thickness of layers undergoing secondary settlement, t_{sc} is time for which secondary settlement is calculated (in years), and t_p is time for primary settlement (>1 year).

31.6.4 Tolerable Settlement

Tolerable movement criteria for foundation settlement should be established consistent with the function and type of structure, anticipated service life, and consequences of unacceptable movements on structure performance as outlined by AASHTO [3]. The criteria adopted by AASHTO considering the angular distortion (δ/l) between adjacent footings is as follows:

$$\frac{\delta}{l} \le \begin{cases} 0.008 & \text{for simple-span bridge} \\ 0.004 & \text{for continuous-span bridge} \end{cases} \tag{31.28}$$

where δ is differential settlement of adjacent footings and l is center–center spacing between adjacent footings. These (δ/l) limits are not applicable to rigid frame structures, which shall be designed for anticipated differential settlement using special analysis.

31.7 Shallow Foundations on Rock

Wyllie [72] outlines the following examinations which are necessary for designing shallow foundations on rock:

1. The bearing capacity of the rock to ensure that there will be no crushing or creep of material within the loaded zone;
2. Settlement of the foundation which will result from elastic strain of the rock, and possibly inelastic compression of weak seams within the volume of rock compressed by the applied load;
3. Sliding and shear failure of blocks of rock formed by intersecting fractures within the foundation.

This condition usually occurs where the foundation is located on a steep slope and the orientation of the fractures is such that the blocks can slide out of the free face.

31.7.1 Bearing Capacity According to Building Codes

It is common to use allowable bearing capacity for various rock types listed in building codes for footing design. As provided in Table 31.18, the bearing capacities have been developed based on rock strength from case histories and include a substantial factor of safety to minimize settlement.

31.7.2 Bearing Capacity of Fractured Rock

Various empirical procedures for estimating allowable bearing capacity of foundations on fractured rock are available in the literature. Peck et al. [53] suggested an empirical procedure for estimating allowable bearing pressures of foundations on jointed rock based on the RQD index. The predicted bearing capacities by this method shall be used with the assumption that the foundation settlement does not exceed 12.7 mm (0.5 in.) [53]. Carter and Kulhawy [25] proposed an empirical approach for estimating ultimate bearing capacity of fractured rock. Their method is based on the unconfined compressive strength of the intact rock core sample and rock mass quality.

Wyllie [72] detailed an analytical procedure for computing bearing capacity of fractured rock mass using Hoek–Brown strength criterion. Details of rational methods for the topic can also be found in Kulhawy and Goodman [42] and Goodman [32].

TABLE 31.18 Presumptive Bearing Pressures (tsf) for Foundations on Rock after Putnam, 1981

Code	Year[1]	Bedrock[2]	Sound Foliated Rock	Sound Sedimentary Rock	Soft Rock[3]	Soft Shale	Broken Shale
Baltimore	1962	100	35		10		
BOCA	1970	100	40	25	10	4	(4)
Boston	1970	100	50	10	10		1.5
Chicago	1970	100	100				(4)
Cleveland	1951/1969			25			
Dallas	1968	$0.2q_u$	$2q_u$	0.2qu	$0.2q_u$	$0.2q_u$	$0.2q_u$
Detroit	1956	100	100	9600	12	12	
Indiana	1967	$0.2q_u$	$2q_u$	0.2qu	$0.2q_u$	$0.2q_u$	$0.2q_u$
Kansas	1961/1969	$0.2q_u$	$2q_u$	0.2qu	$0.2q_u$	$0.2q_u$	$0.2q_u$
Los Angeles	1970	10	4	3	1	1	1
New York City	1970	60	60	60	8		
New York State		100	40	15			
Ohio	1970	100	40	15	10	4	
Philadelphia	1969	50	15	10–15	8		
Pittsburgh	1959/1969	25	25	25	8	8	
Richmond	1968	100	40	25	10	4	1.5
St. Louis	1960/1970	100	40	25	10	1.5	1.5
San Francisco	1969	3–5	3–5	3–5			
UBC	1970	$0.2q_u$	$2q_u$	$0.2q_u$	$0.2q_u$	$0.2q_u$	$0.2q_u$
NBC Canada	1970			100			
New South Wales, Australia	1974		33	13	4.5		

Notes:
1. Year of code or original year and date of revision.
2. Massive crystalline bedrock.
3. Soft and broken rock, not including shale.
4. Allowable bearing pressure to be determined by appropriate city official.
5. q_u = unconfined compressive strength.

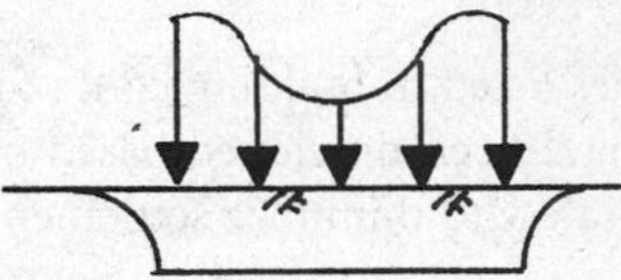

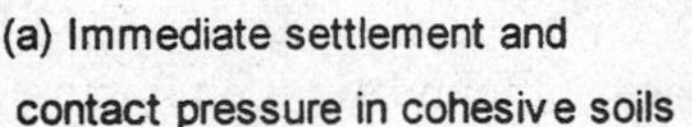

(a) Immediate settlement and
contact pressure in cohesive soils

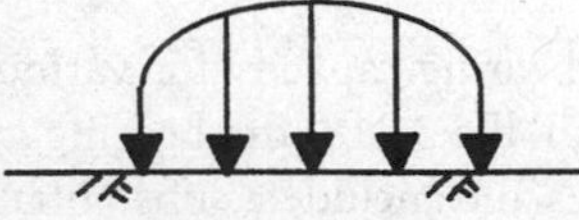

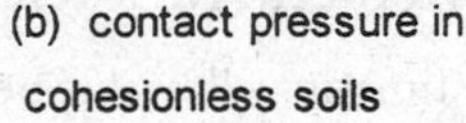

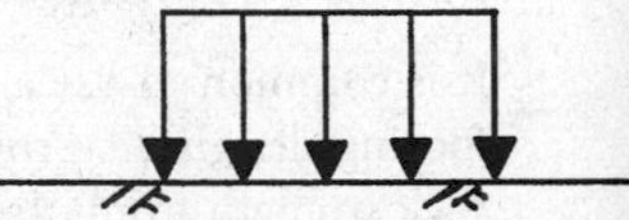

FIGURE 31.20 Contact pressure distribution for a rigid footing. (a) On cohesionless soils; (b) on cohesive soils; (c) usual assumed linear distribution.

31.7.3 Settlements of Foundations on Rock

Wyllie [72] summarizes settlements of foundations on rock as following three different types: 1. Elastic settlements result from a combination of strain of the intact rock, slight closure and movement of fractures and compression of any minor clay seams (less than a few millimeters). Elastic theory can be used to calculate this type of settlement. Detailed information can be found in Wyllie [72], Kulhawy, and AASHTO [3]. 2. Settlements result from the movement of blocks of rock due to shearing of fracture surfaces. This occurs when foundations are sitting at the top of a steep slope and unstable blocks of rocks are formed in the face. The stability of foundations on rock is influenced by the geologic characterization of rock blocks. The information required on structural geology consists of the orientation, length and spacing of fractures, and their surface and infilling materials. Procedures have been developed for identifying and analyzing the stability of sliding blocks [72], stability of wedge blocks [36], stability of toppling blocks [33], or three-dimensional stability of rock blocks [34]. 3. Time-dependent settlement occurs when foundations found on rock mass that consists of substantial seams of clay or other compressible materials. This type of settlement can be estimated using the procedures described in Section 31.6.3. Also time-dependent settlement can occur if foundations found on ductile rocks, such as salt where strains develop continuously at any stress level, or on brittle rocks when the applied stress exceeds the yield stress.

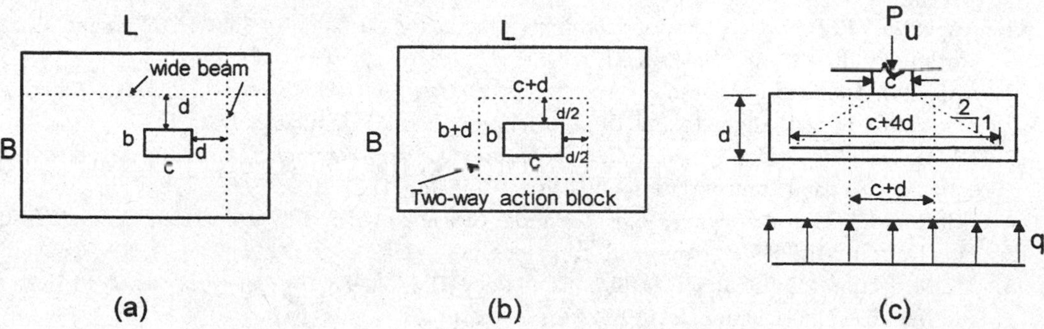

FIGURE 31.21 (a) Section for wide-beam shear; (b) section for diagonal-tension shear; (c) method of computing area for allowable column bearing stress.

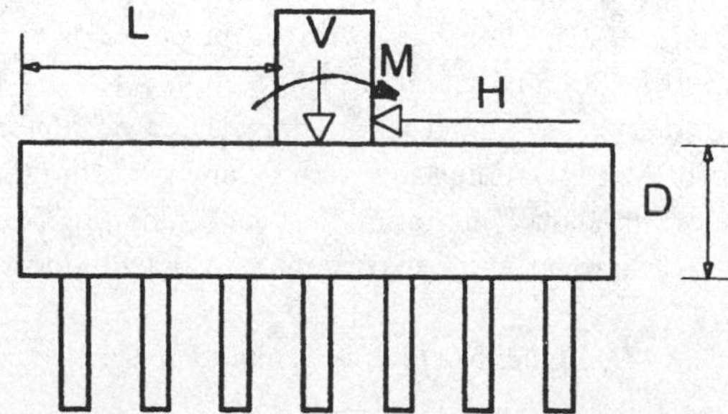

FIGURE 31.22 Illustration of the length-to-thickness ratio of cantilever of a footing or pile cap.

31.8 Structural Design of Spread Footings

The plan dimensions (*B* and *L*) of a spread footing are controlled by the allowable soil pressure beneath the footing. The pressure distribution beneath footings is influenced by the interaction of the footing rigidity with the soil type, stress–state, and time response to stress as shown in Figure 31.20 (a) (b). However, it is common practice to use the linear pressure distribution beneath rigid footings as shown in Figure 31.20 (c). The depth (*D*) for spread footings is usually controlled by shear stresses. Two-way action shear always controls the depth for centrally loaded square footings. However, wide-beam shear may control the depth for rectangular footings when the *L*/*B* ratio is greater than about 1.2 and may control for other *L*/*B* ratios when there is overturning or eccentric loading (Figure 31.21a). In addition, footing depth should be designed to satisfy diagonal (punching) shear requirement (Figure 31.21b). Recent studies by Duan and McBride [30] indicate that when the length-to-thickness ratio of cantilever (*L*/*D* as defined in Figure 31.22) of a footing (or pile-cap) is greater than 2.2, a nonlinear distribution of reaction should be used for footing or pile-cap design. The specifications and procedures for footing design can be found in AASHTO [2], ACI [4], or Bowles [12, 13].

Acknowledgment

I would like to take this opportunity to thank Bruce Kutter, who reviewed the early version of the chapter and provided many thoughtful suggestions. Advice and support from Prof. Kutter are greatly appreciated.

References

1. AASHTO, *LRFD Bridge Design Specifications*, American Association of State Highway and Transportation Officials, Washington, D.C., 1994.
2. AASHTO, *Standard Specifications for Highway Bridges (Interim Revisions)*, 16th ed., American Association of State Highway and Transportation Officials, Washington, D.C., 1997.
3. AASHTO, *LRFD Bridge System Design Specification (Interim Revisions)*, American Association of State Highway and Transportation Officials, Washington, D.C., 1997.
4. ACI, *Building Code Requirements for Reinforced Concrete* (ACI 318-89), American Concrete Institute, Detroit, MI, 1989, 353 pp. (with commentary).
5. ASTM, Section 4 Construction, 04.08 Soil and Rock (I): D420–D4914, American Society for Testing and Materials, Philadelphia, PA, 1997.
6. ATC-32, *Improved Seismic Design Criteria for California Bridges: Provisional Recommendations*, Applied Technology Council, Redwood City, CA, 1996.
7. Azzouz, A.S., Krizek, R.J., and Corotis, R.B., Regression of analysis of soil compressibility, *JSSMFE Soils and Foundations*, 16(2), 19–29, 1976.
8. Baguelin, F., Jezequel, J.F., and Shields, D.H., *The Pressuremeter and Foundation Engineering*, Transportation Technical Publications, Clausthal, 1978, 617 pp.
9. Barker, R.M., Duncan, J.M., Rojiani, K.B., Ooi, P.S.K., Tan, C.K., and Kim, S.G., Manuals for the Design of Bridge Foundations, National Cooperative Highway Research Program Report 343, Transportation Research Board, National Research Council, Washington, D.C., 1991.
10. Bazaraa, A.R.S.S., Use of Standard Penetration Test for Estimating Settlements of Shallow Foundations on Sands, Ph.D. dissertation, Department of Civil Engineering, University of Illinois, Urbana, 1967, 380 pp.
11. Bowles, J.E., *Analytical and Computer Methods in Foundation Engineering*, McGraw-Hill, New York, 1974.
12. Bowles, J.E., Spread footings, Chapter 15, in *Foundation Engineering Handbook*, Winterkorn, H.F. and Fang, H.Y., Eds., Van Nostrand Reinhold, New York, 1975.
13. Bowles, J.E., *Foundation Analysis and Design*, 5th ed., McGraw-Hill, New York, 1996.
14. Briaud, J.L., *The Pressuremeter*, A.A. Balkema Publishers, Brookfield, VT, 1992.
15. Briaud, J.L., Spread footing design and performance, FHWA Workshop at the Tenth Annual International Bridge Conference and Exhibition, 1993.
16. Briaud, J.L., Pressuremeter and foundation design, in *Proceedings of the Conference on Use of in situ tests in Geotechnical Engineering*, ASCE Geotechnical Publication No. 6, 74–116, 1986.
17. Briaud, J.L. and Gibben, R., Predicted and measured behavior of five spread footings on sand, Geotechnical Special Publication No. 41, ASCE Specialty Conference: Settlement 1994, ASCE, New York, 1994.
18. Buisman, A.S.K., *Grondmechanica*, Waltman, Delft, 190, 1940.
19. Burland, J.B. and Burbidge, M.C., Settlement of foundations on sand and gravel, *Proc. Inst. Civil Eng.*, Tokyo, 2, 517, 1984.
20. Burmister, D.M., The theory of stresses and displacements in layered systems and application to the design of airport runways, *Proc. Highway Res. Board*, 23, 126–148, 1943.

21. Burmister, D.M., Evaluation of pavement systems of WASHO road test layered system methods, Highway Research Board Bull. No. 177, 1958.
22. Burmister, D.M., Applications of dimensional analyses in the evaluation of asphalt pavement performances, paper presented at Fifth Paving Conference, Albuquerque, NM, 1967.
23. Canadian Geotechnical Society, *Canadian Foundation Engineering Manual*, 2nd ed., 1985, 456.
24. Carter. J.P. and Kulhawy, F.H., Analysis and Design of Drilled Shaft Foundations Socketed into Rock, Report No. EL-5918, Empire State Electric Engineering Research Corporation and Electric Power Research Institute, 1988.
25. Chen, W.F., *Limit Analysis and Soil Plasticity*, Elsevier, Amsterdam, 1975.
26. Chen, W.F. and Mccarron, W.O., Bearing capacity of shallow foundations, Chap. 4, in *Foundation Engineering Handbook*, 2nd ed., Fang, H.Y., Ed., Chapman & Hall, 1990.
27. D'Appolonia, D.J., D'Appolonia, E., and Brisette, R.F., Settlement of spread footings on sand (closure), *ASCE J. Soil Mech. Foundation Div.*, 96(SM2), 754–761, 1970.
28. De Beer, E.E., Bearing capacity and settlement of shallow foundations on sand, *Proc. Symposium on Bearing Capacity and Settlement of Foundations*, Duke University, Durham, NC, 315–355, 1965.
29. De Beer, E.E., Proefondervindelijke bijdrage tot de studie van het gransdraagvermogen van zand onder funderingen p staal, Bepaling von der vormfactor sb, *Ann. Trav. Publics Belg.*, 1967.
30. Duan, L. and McBride, S.B., The effects of cap stiffness on pile reactions, *Concrete International*, American Concrete Institue, 1995.
31. FHWA, Large-Scale Load Tests and Data Base of Spread Footings on Sand, Publication No. FHWA-RD-97-068, 1997.
32. Goodman, R.E. and Bray, J.W., Toppling of rock slopes, in *Proceedings of the Specialty Conference on Rock Engineering for Foundations and Slopes*, Vol. 2, ASCE, Boulder, CO, 1976, 201–234.
33. Goodman, R.E. and Shi, G., *Block Theory and Its Application to Rock Engineering*, Prentice-Hall, Englewood Cliffs, NJ, 1985.
34. Hansen, B.J., A Revised and Extended Formula for Bearing Capacity, Bull. No. 28, Danish Geotechnical Institute, Copenhagen, 1970, 5–11.
35. Hoek, E. and Bray, J., *Rock Slope Engineering*, 2nd ed., IMM, London, 1981.
36. Holtz, R.D., Stress distribution and settlement of shallow foundations, Chap. 5, in *Foundation Engineering Handbook*, 2nd ed., Fang, H.Y., Ed., Chapman & Hall, 1990.
37. Ismael, N.F. and Vesic, A.S., Compressibility and bearing capacity, *ASCE J. Geotech. Foundation Eng. Div.* 107(GT12), 1677–1691, 1981.
38. Kulhawy, F.H. and Mayne, P.W., Manual on Estimating Soil Properties for Foundation Design, Electric Power Research Institute, EPRI EL-6800, Project 1493-6, Final Report, August, 1990.
39. Kulhawy, F.H. and Goodman, R.E., Foundation in rock, Chap. 55, in *Ground Engineering Reference Manual*, F.G. Bell, Ed., Butterworths, 1987.
40. Lambe, T.W. and Whitman, R.V., *Soil Mechanics*, John Wiley & Sons, New York, 1969.
41. Menard, L., Regle pour le calcul de la force portante et du tassement des fondations en fonction des resultats pressionmetriques, in *Proceedings of the Sixth International Conference on Soil Mechanics and Foundation Engineering*, Vol. 2, Montreal, 1965, 295–299.
42. Meyerhof, G.G., The ultimate bearing capacity of foundations, *Geotechnique*, 2(4), 301–331, 1951.
43. Meyerhof, G.G., Penetration tests and bearing capacity of cohesionless soils, *ASCE J. Soil Mech. Foundation Div.*, 82(SM1), 1–19, 1956.
44. Meyerhof, G.G., Some recent research on the bearing capacity of foundations, *Can. Geotech. J.*, 1(1), 16–36, 1963.
45. Meyerhof, G.G., Shallow foundations, *ASCE J. Soil Mech. and Foundations Div.*, 91, No. SM2, 21–31, 1965.
46. Milovic, D.M., Comparison between the calculated and experimental values of the ultimate bearing capacity, in *Proceedings of the Sixth International Conference on Soil Mechanics and Foundation Engineering*, Vol. 2, Montreal, 142–144, 1965.

47. Nagaraj, T.S. and Srinivasa Murthy, B.R., Prediction of preconsolidation pressure and recompression index of soils, *ASTMA Geotech. Testing J.*, 8(4), 199–202, 1985.
48. Nagaraj. T.S. and Srinivasa Murthy, B.R., A critical reappraisal of compression index, *Geotechnique*, 36(1), 27–32, 1986.
49. NAVFAC, Design Manual 7.02, *Foundations & Earth Structures*, Naval Facilities Engineering Command, Department of the Navy, Washington, D.C., 1986.
50. NAVFAC, Design Manual 7.01, Soil Mechanics, Naval Facilities Engineering Command, Department of the Navy, Washington, D.C., 1986.
51. Peck, R.B., Hanson, W.E., and Thornburn, T.H., *Foundation Engineering*, 2nd ed., John Wiley & Sons, New York, 1974.
52. Perloff, W.H., Pressure distribution and settlement, Chap. 4, in *Foundation Engineering Handbook*, 2nd ed., Fang, H.Y., Ed., Chapman & Hall, 1975.
53. Poulos, H.G. and Davis, E.H., *Elastic Solutions for Soil and Rock Mechanics*, John Wiley & Sons, New York, 1974.
54. Schmertmann, J.H., Static cone to compute static settlement over sand, *ASCE J. Soil Mech. Foundation Div.*, 96(SM3), 1011–1043, 1970.
55. Schmertmann, J.H., Guidelines for cone penetration test performance, and design, Federal Highway Administration, Report FHWA-TS-78-209, 1978.
56. Schmertmann, J.H., Dilatometer to Computer Foundation Settlement, *Proc.* In Situ *'86, Specialty Conference on the Use of* In Situ *Tests and Geotechnical Engineering*, ASCE, New York, 303–321, 1986.
57. Schmertmann, J.H., Hartman, J.P., and Brown, P.R., Improved strain influence factor diagrams, *ASCE J. Geotech. Eng. Div.*, 104(GT8), 1131–1135, 1978.
58. Schultze, E. and Sherif, G. Prediction of settlements from evaluated settlement observations on sand, *Proc. 8th Int. Conference on Soil Mechanics and Foundation Engineering*, Moscow, 225–230, 1973.
59. Scott, R.F., *Foundation Analysis*, Prentice-Hall, Englewood Cliffs, NJ, 1981.
60. Sowers, G.F., and Vesic, A.B., Vertical stresses in subgrades beneath statically loaded flexible pavements, Highway Research Board Bulletin, No. 342, 1962.
61. Steinbrenner, W., *Tafeln zur Setzungberechnung*, Die Strasse, 1943, 121–124.
62. Tan, C.K. and Duncan, J.M., Settlement of footings on sand—accuracy and reliability, in *Proceedings of Geotechnical Congress*, Boulder, CO, 1991.
63. Terzaghi, J., *Theoretical Soil Mechanics*, John Wiley & Sons, New York, 1943.
64. Terzaghi, K. and Peck, R.B., *Soil Mechanics in Engineering Practice*, John Wiley & Sons, New York, 1948.
65. Terzaghi, K. and Peck, R.B., *Soil Mechanics in Engineering Practice*, 2nd ed., John Wiley & Sons, New York, 1967.
66. Terzaghi, K., Peck, R.B., and Mesri, G., *Soil Mechanics in Engineering Practice*, 3rd ed., John Wiley & Sons, New York, 1996.
67. Vesic, A.S., Bearing capacity of deep foundations in sand, National Academy of Sciences, National Research Council, Highway Research Record, 39, 112–153, 1963.
68. Vesic, A.S., Analysis of ultimate loads of shallow foundations, *ASCE J. Soil Mech. Foundation Eng. Div.*, 99(SM1), 45–73, 1973.
69. Vesic, A.S., Bearing capacity of shallow foundations, Chap. 3, in *Foundation Engineering Handbook*, Winterkorn, H.F. and Fang, H.Y., Ed., Van Nostrand Reinhold, New York, 1975.
70. Westergaard, H.M., A problem of elasticity suggested by a problem in soil mechanics: soft material reinforced by numerous strong horizontal sheets, in *Contributions to the Mechanics of Solids*, Stephen Timoshenko Sixtieth Anniversary Volume, Macmillan, New York, 1938.
71. Wroth, C.P. and Wood, D.M., The correlation of index properties with some basic engineering properties of soils, *Can. Geotech. J.*, 15(2), 137–145, 1978.
72. Wyllie, D.C., *Foundations on Rock*, E & FN SPON, 1992.

32

Deep Foundations

Youzhi Ma
Geomatrix Consultants, Inc.

Nan Deng
Bechtel Corporation

32.1 Introduction

A bridge foundation is part of the bridge substructure connecting the bridge to the ground. A foundation consists of man-made structural elements that are constructed either on top of or within existing geologic materials. The function of a foundation is to provide support for the bridge and to transfer loads or energy between the bridge structure and the ground.

A deep foundation is a type of foundation where the embedment is larger than its maximum plane dimension. The foundation is designed to be supported on deeper geologic materials because either the soil or rock near the ground surface is not competent enough to take the design loads or it is more economical to do so.

0-8493-7434-0/00/$0.00+$.50

The merit of a deep foundation over a shallow foundation is manifold. By involving deeper geologic materials, a deep foundation occupies a relatively smaller area of the ground surface. Deep foundations can usually take larger loads than shallow foundations that occupy the same area of the ground surface. Deep foundations can reach deeper competent layers of bearing soil or rock, whereas shallow foundations cannot. Deep foundations can also take large uplift and lateral loads, whereas shallow foundations usually cannot.

The purpose of this chapter is to give a brief but comprehensive review to the design procedure of deep foundations for structural engineers and other bridge design engineers. Considerations of selection of foundation types and various design issues are first discussed. Typical procedures to calculate the axial and lateral capacities of an individual pile are then presented. Typical procedures to analyze pile groups are also discussed. A brief discussion regarding seismic design is also presented for its uniqueness and importance in the foundation design.

32.2 Classification and Selection

32.2.1 Typical Foundations

Typical foundations are shown on Figure 32.1 and are listed as follows:

A *pile* usually represents a slender structural element that is driven into the ground. However, a pile is often used as a generic term to represent all types of deep foundations, including a (driven) pile, (drilled) shaft, caisson, or an anchor. A *pile group* is used to represent various grouped deep foundations.

A *shaft* is a type of foundation that is constructed with cast-in-place concrete after a hole is first drilled or excavated. A *rock socket* is a shaft foundation installed in rock. A shaft foundation also is called a *drilled pier* foundation.

A *caisson* is a type of large foundation that is constructed by lowering preconstructed foundation elements through excavation of soil or rock at the bottom of the foundation. The bottom of the caisson is usually sealed with concrete after the construction is completed.

An *anchor* is a type of foundation designed to take tensile loading. An anchor is a slender, small-diameter element consisting of a reinforcement bar that is fixed in a drilled hole by grout concrete. Multistrain high-strength cables are often used as reinforcement for large-capacity anchors. An *anchor for suspension bridge* is, however, a foundation that sustains the pulling loads located at the ends of a bridge; the foundation can be a deadman, a massive tunnel, or a composite foundation system including normal anchors, piles, and drilled shafts.

A *spread footing* is a type of foundation that the embedment is usually less than its smallest width. Normal spread footing foundation is discussed in detail in Chapter 31.

32.2.2 Typical Bridge Foundations

Bridge foundations can be individual, grouped, or combination foundations. Individual bridge foundations usually include individual footings, large-diameter drilled shafts, caissons, rock sockets, and deadman foundations. Grouped foundations include groups of caissons, driven piles, drilled shafts, and rock sockets. Combination foundations include caisson with driven piles, caisson with drilled shafts, large-diameter pipe piles with rock socket, spread footings with anchors, deadman with piles and anchors, etc.

For small bridges, small-scale foundations such as individual footings or drilled shaft foundations, or a small group of driven piles may be sufficient. For larger bridges, large-diameter shaft foundations, grouped foundations, caissons, or combination foundations may be required. Caissons, large-diameter steel pipe pile foundations, or other types of foundations constructed by using the cofferdam method may be necessary for foundations constructed over water.

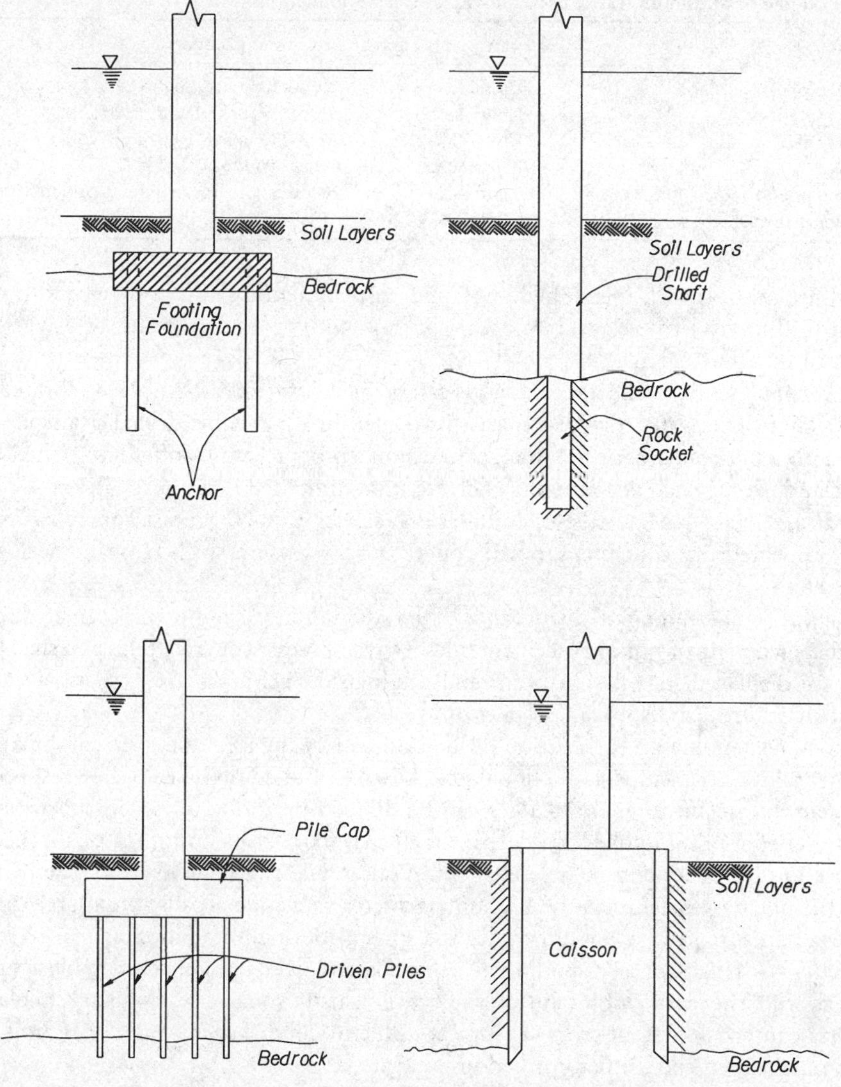

FIGURE 32.1 Typical foundations.

Bridge foundations are often constructed in difficult ground conditions such as landslide areas, liquefiable soil, collapsible soil, soft and highly compressible soil, swelling soil, coral deposits, and underground caves. Special foundation types and designs may be needed under these circumstances.

32.2.3 Classification

Deep foundations are of many different types and are classified according to different aspects of a foundation as listed below:

Geologic conditions — Geologic materials surrounding the foundations can be soil and rock. Soil can be fine grained or coarse grained; from soft to stiff and hard for fine-grained soil, or from loose to dense and very dense for coarse-grained soil. Rock can be sedimentary, igneous, or metamorphic; and from very soft to medium strong and hard. Soil and rock mass may possess predefined weaknesses and

TABLE 32.1 Range of Maximum Capacity of Individual Deep Foundations

Type of Foundation	Size of Cross Section	Maximum Compressive Working Capacity
Driven concrete piles	Up to 45 cm	100 to 250 tons (900 to 2200 kN)
Driven steel pipe piles	Up to 45 cm	50 to 250 tons (450 to 2200 kN)
Driven steel H-piles	Up to 45 cm	50 to 250 tons (450 to 2200 kN)
Drilled shafts	Up to 60 cm	Up to 400 tons (3500 kN)
Large steel pipe piles, concrete-filled; large-diameter drilled shafts; rock rocket	0.6 to 3 m	300 to 5,000 tons or more (2700 to 45000 kN)

discontinuities, such as rock joints, beddings, sliding planes, and faults. Water conditions can be different, including over river, lake, bay, ocean, or land with groundwater. Ice or wave action may be of concern in some regions.

Installation methods — Installation methods can be piles (driven, cast-in-place, vibrated, torqued, and jacked); shafts (excavated, drilled and cast-in-drilled-hole); anchor (drilled); caissons (Chicago, shored, benoto, open, pneumatic, floating, closed-box, Potomac, etc.); cofferdams (sheet pile, sand or gravel island, slurry wall, deep mixing wall, etc.); or combined.

Structural materials — Materials for foundations can be timber, precast concrete, cast-in-place concrete, compacted dry concrete, grouted concrete, post-tension steel, H-beam steel, steel pipe, composite, etc.

Ground effect — Depending on disturbance to the surrounding ground, piles can be displacement piles, low displacement, or nondisplacement piles. Driven precast concrete piles and steel pipes with end plugs are displacement piles; H-beam and unplugged steel pipes are low-displacement piles; and drilled shafts are nondisplacement piles.

Function — Depending on the portion of load carried by the side, toe, or a combination of the side and toe, piles are classified as frictional, end bearing, and combination piles, respectively.

Embedment and relative rigidity — Piles can be divided into long piles and short piles. A long pile, simply called a pile, is embedded deep enough that fixity at its bottom is established, and the pile is treated as a slender and flexible element. A short pile is a relatively rigid element that the bottom of the pile moves significantly. A caisson is often a short pile because of its large cross section and stiffness. An extreme case for short piles is a spread-footing foundation.

Cross section — The cross section of a pile can be square, rectangular, circular, hexagonal, octagonal, H-section; either hollow or solid. A pile cap is usually square, rectangular, circular, or bell-shaped. Piles can have different cross sections at different depths, such as uniform, uniform taper, step-taper, or enlarged end (either grouted or excavated).

Size — Depending on the diameter of a pile, piles are classified as pin piles and anchors (100 to 300 mm), normal-size piles and shafts (250 to 600 mm), large-diameter piles and shafts (600 to 3000 mm), caissons (600 mm and up to 3000 mm or larger), and cofferdams or other shoring construction method (very large).

Loading — Loads applied to foundations are compression, tension, moment, and lateral loads. Depending on time characteristics, loads are further classified as static, cyclic, and transient loads. The magnitude and type of loading also are major factors in determining the size and type of a foundation (Table 32.1).

Isolation — Piles can be isolated at a certain depth to avoid loading utility lines or other construction, or to avoid being loaded by them.

Inclination — Piles can be vertical or inclined. Inclined piles are often called battered or raked piles.

Multiple Piles — Foundation can be an individual pile, or a pile group. Within a pile group, piles can be of uniform or different sizes and types. The connection between the piles and the pile cap can be fixed, pinned, or restrained.

32.2.4 Advantages/Disadvantages of Different Types of Foundations

Different types of foundations have their unique features and are more applicable to certain conditions than others. The advantages and disadvantages for different types of foundations are listed as follows.

Driven Precast Concrete Pile Foundations

Driven concrete pile foundations are applicable under most ground conditions. Concrete piles are usually inexpensive compared with other types of deep foundations. The procedure of pile installation is straightforward; piles can be produced in mass production either on site or in a manufacture factory, and the cost for materials is usually much less than steel piles. Proxy coating can be applied to reduce negative skin friction along the pile. Pile driving can densify loose sand and reduce liquefaction potential within a range of up to three diameters surrounding the pile.

However, driven concrete piles are not suitable if boulders exist below the ground surface where piles may break easily and pile penetration may be terminated prematurely. Piles in dense sand, dense gravel, or bedrock usually have limited penetration; consequently, the uplift capacity of this type of piles is very small.

Pile driving produces noise pollution and causes disturbance to the adjacent structures. Driving of concrete piles also requires large overhead space. Piles may break during driving and impose a safety hazard. Piles that break underground cannot take their design loads, and will cause damage to the structures if the broken pile is not detected and replaced. Piles could often be driven out of their designed alignment and inclination and, as a result, additional piles may be needed. A special hardened steel shoe is often required to prevent pile tips from being smashed when encountering hard rock. End-bearing capacity of a pile is not reliable if the end of a pile is smashed.

Driven piles may not be a good option when subsurface conditions are unclear or vary considerably over the site. Splicing and cutting of piles are necessary when the estimated length is different from the manufactured length. Splicing is usually difficult and time-consuming for concrete piles. Cutting of a pile would change the pattern of reinforcement along the pile, especially where extra reinforcement is needed at the top of a pile for lateral capacity. A pilot program is usually needed to determine the length and capacity prior to mass production and installation of production piles.

The maximum pile length is usually up to 36 to 38 m because of restrictions during transportation on highways. Although longer piles can be produced on site, slender and long piles may buckle easily during handling and driving. Precast concrete piles with diameters greater than 45 cm are rarely used.

Driven Steel Piles

Driven steel piles, such as steel pipe and H-beam piles, are extensively used as bridge foundations, especially in seismic retrofit projects. Having the advantage and disadvantage of driven piles as discussed above, driven steel piles have their uniqueness.

Steel piles are usually more expensive than concrete piles. They are more ductile and flexible and can be spliced more conveniently. The required overhead is much smaller compared with driven concrete piles. Pipe piles with an open end can penetrate through layers of dense sand. If necessary, the soil inside the pipe can be taken out before further driving; small boulders may also be crushed and taken out. H-piles with a pointed tip can usually penetrate onto soft bedrock and establish enough end-bearing capacity.

Large-Diameter Driven, Vibrated, or Torqued Steel Pipe Piles

Large-diameter pipe piles are widely used as foundations for large bridges. The advantage of this type of foundation is manifold. Large-diameter pipe piles can be built over water from a barge, a trestle, or a temporary island. They can be used in almost all ground conditions and penetrate to a great depth to reach bedrock. Length of the pile can be adjusted by welding. Large-diameter pipe piles can also be used as casings to support soil above bedrock from caving in; rock sockets or rock anchors can then be constructed below the tip of the pipe. Concrete or reinforced concrete can be placed inside the pipe after it is cleaned. Another advantage is that no workers are required to work below water or the ground surface. Construction is usually safer and faster than other types of foundations, such as caissons or cofferdam construction.

Large-diameter pipe piles can be installed by methods of driving, vibrating, or torque. Driven piles usually have higher capacity than piles installed through vibration or torque. However, driven piles are hard to control in terms of location and inclination of the piles. Moreover, once a pile is out of location or installed with unwanted inclination, no corrective measures can be applied. Piles installed with vibration or torque, on the other hand, can be controlled more easily. If a pile is out of position or inclination, the pile can even be lifted up and reinstalled.

Drilled Shaft Foundations

Drilled shaft foundations are the most versatile types of foundations. The length and size of the foundations can be tailored easily. Disturbance to the nearby structures is small compared with other types of deep foundations. Drilled shafts can be constructed very close to existing structures and can be constructed under low overhead conditions. Therefore, drilled shafts are often used in many seismic retrofit projects. However, drilled shafts may be difficult to install under certain ground conditions such as soft soil, loose sand, sand under water, and soils with boulders. Drilled shafts will generate a large volume of soil cuttings and fluid and can be a mess. Disposal of the cuttings is usually a concern for sites with contaminated soils.

Drilled shaft foundations are usually comparable with or more expensive than driven piles. For large bridge foundations, their cost is at the same level of caisson foundations and spread footing foundations combined with cofferdam construction. Drilled shaft foundations can be constructed very rapidly under normal conditions compared with caisson and cofferdam construction.

Anchors

Anchors are special foundation elements that are designed to take uplift loads. Anchors can be added if an existing foundation lacks uplift capacity, and competent layers of soil or rock are shallow and easy to reach. Anchors, however, cannot take lateral loads and may be sheared off if combined lateral capacity of a foundation is not enough.

Anchors are in many cases pretensioned in order to limit the deformation to activate the anchor. The anchor system is therefore very stiff. Structural failure resulting from anchor rupture often occurs very quickly and catastrophically. Pretension may also be lost over time because of creep in some types of rock and soil. Anchors should be tested carefully for their design capacity and creep performance.

Caissons

Caissons are large structures that are mainly used for construction of large bridge foundations. Caisson foundations can take large compressive and lateral loads. They are used primarily for over-water construction and sometimes used in soft or loose soil conditions, with a purpose to sink or excavate down to a depth where bedrock or firm soil can be reached. During construction, large boulders can be removed.

Caisson construction requires special techniques and experience. Caisson foundations are usually very costly, and comparable to the cost of cofferdam construction. Therefore, caissons are usually not the first option unless other types of foundation are not favored.

Cofferdam and Shoring

Cofferdams or other types of shoring systems are a method of foundation construction to retain water and soil. A dry bottom deep into water or ground can be created as a working platform. Foundations of essentially any of the types discussed above can be built from the platform on top of firm soil or rock at a great depth, which otherwise can only be reached by deep foundations.

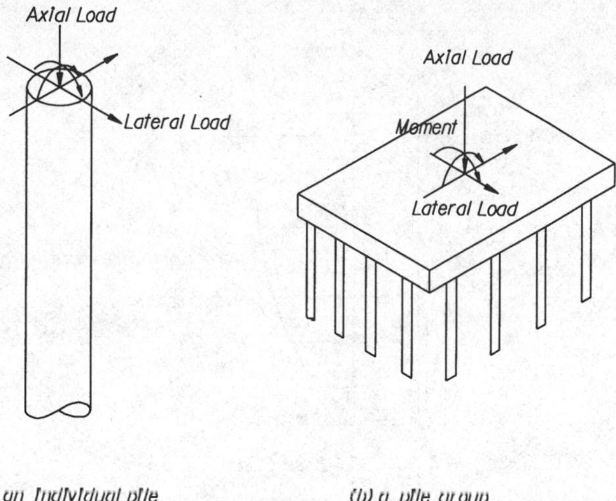

FIGURE 32.2 Acting loads on top of a pile or a pile group. (a) Individual pile; (b) pile group.

A spread footing type of foundation can be built from the platform. Pile foundations also can be constructed from the platform, and the pile length can be reduced substantially. Without cofferdam or shoring, a foundation may not be possible if constructed from the water or ground surface, or it may be too costly.

Cofferdam construction is often very expensive and should only be chosen if it is favorable compared with other foundation options in terms of cost and construction conditions.

32.2.5 Characteristics of Different Types of Foundations

In this section, the mechanisms of resistance of an individual foundation and a pile group are discussed. The function of different types of foundations is also addressed.

Complex loadings on top of a foundation from the bridge structures above can be simplified into forces and moments in the longitudinal, transverse, and vertical directions, respectively (Figure 32.2). Longitudinal and transverse loads are also called horizontal loads; longitudinal and transverse moments are called overturning moments, moment about the vertical axis is called torsional moment. The resistance provided by an individual foundation is categorized in the following (also see Figure 32.3).

End-bearing: Vertical compressive resistance at the base of a foundation; distributed end-bearing pressures can provide resistance to overturning moments;
Base shear: Horizontal resistance of friction and cohesion at the base of a foundation;
Side resistance: Shear resistance from friction and cohesion along the side of a foundation;
Earth pressure: Mainly horizontal resistance from lateral Earth pressures perpendicular to the side of the foundation;
Self-weight: Effective weight of the foundation.

Both base shear and lateral earth pressures provide lateral resistance of a foundation, and the contribution of lateral earth pressures decreases as the embedment of a pile increases. For long piles, lateral earth pressures are the main source of lateral resistance. For short piles, base shear and end-bearing pressures can also contribute part of the lateral resistance. Table 32.2 lists various types of resistance of an individual pile.

For a pile group, through the action of the pile cap, the coupled axial compressive and uplift resistance of individual piles provides the majority of the resistance to the overturning moment loading. Horizontal (or lateral) resistance can at the same time provide torsional moment resistance.

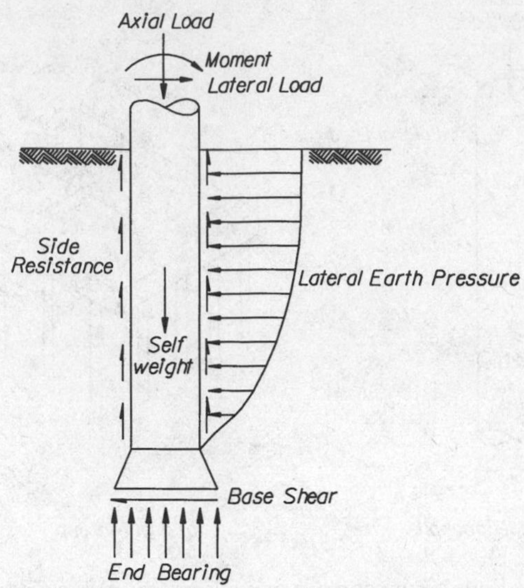

FIGURE 32.3 Resistances of an individual foundation.

TABLE 32.2 Resistance of an Individual Foundation

Type of Foundation	Type of Resistance				
	Vertical Compressive Load (Axial)	Vertical Uplift Load (Axial)	Horizontal Load (Lateral)	Overturning Moment (Lateral)	Torsional Moment (Torsional)
Spread footing (also see Chapter 31)	End bearing	—	Base shear, lateral earth pressure	End bearing, lateral earth pressure	Base shear, lateral earth pressure
Individual short pile foundation	End bearing; side friction	Side friction	Lateral earth pressure, base shear	Lateral earth pressure, end bearing	Side friction, lateral earth pressure, base shear
Individual end-bearing long pile foundation	End bearing	—	Lateral earth pressure	Lateral earth pressure	—
Individual frictional long pile foundation	Side friction	Side friction	Lateral earth pressure	Lateral earth pressure	Side friction
Individual long pile foundation	End bearing; side friction	Side friction	Lateral earth pressure	Lateral earth pressure	Side friction
Anchor	—	Side friction	—	—	—

TABLE 32.3 Additional Functions of Pile Group Foundations

Type of Foundation	Type of Resistance	
	Overturning moment (Lateral)	Torsional moment (Torsional)
Grouped spread footings	Vertical compressive resistance	Horizontal resistance
Grouped piles, foundations	Vertical compressive and uplift resistance	Horizontal resistance
Grouped anchors	Vertical uplift resistance	—

A pile group is more efficient in resisting overturning and torsional moment than an individual foundation. Table 32.3 summarizes functions of a pile group in addition to those of individual piles.

32.2.6 Selection of Foundations

The two predominant factors in determining the type of foundations are bridge types and ground conditions.

The bridge type, including dimensions, type of bridge, and construction materials, dictates the design magnitude of loads and the allowable displacements and other performance criteria for the foundations, and therefore determines the dimensions and type of its foundations. For example, a suspension bridge requires large lateral capacity for its end anchorage which can be a huge deadman, a high capacity soil or rock anchor system, a group of driven piles, or a group of large-diameter drilled shafts. Tower foundations of an over-water bridge require large compressive, uplift, lateral, and overturning moment capacities. The likely foundations are deep, large-size footings using cofferdam construction, caissons, groups of large-diameter drilled shafts, or groups of a large number of steel piles.

Surface and subsurface geologic and geotechnical conditions are another main factor in determining the type of bridge foundations. Subsurface conditions, especially the depths to the load-bearing soil layer or bedrock, are the most crucial factor. Seismicity over the region usually dictates the design level of seismic loads, which is often the critical and dominant loading condition. A bridge that crosses a deep valley or river certainly determines the minimum span required. Over-water bridges have limited options to chose in terms of the type of foundations.

The final choice of the type of foundation usually depends on cost after considering some other factors, such as construction conditions, space and overhead conditions, local practice, environmental conditions, schedule constraints, etc. In the process of selection, several types of foundations would be evaluated as candidates once the type of bridge and the preliminary ground conditions are known. Certain types of foundations are excluded in the early stage of study. For example, from the geotechnical point of view, shallow foundations are not an acceptable option if a thick layer of soft clay or liquefiable sand is near the ground surface. Deep foundations are used in cases where shallow foundations would be excessively large and costly. From a constructibility point of view, driven pile foundations are not suitable if boulders exist at depths above the intended firm bearing soil/rock layer.

For small bridges such as roadway overpasses, for example, foundations with driven concrete or steel piles, drilled shafts, or shallow spread footing foundations may be the suitable choices. For large over-water bridge foundations, single or grouped large-diameter pipe piles, large-diameter rock sockets, large-diameter drilled shafts, caissons, or foundations constructed with cofferdams are the most likely choice. Caissons or cofferdam construction with a large number of driven pile groups were widely used in the past. Large-diameter pipe piles or drilled shafts, in combination with rock sockets, have been preferred for bridge foundations recently.

Deformation compatibility of the foundations and bridge structure is an important consideration. Different types of foundation may behave differently; therefore, the same type of foundations should be used for one section of bridge structure. Diameters of the piles and inclined piles are two important factors to considere in terms of deformation compatibility and are discussed in the following.

Small-diameter piles are more "brittle" in the sense that the ultimate settlement and lateral deflection are relatively small compared with large-diameter piles. For example, 20 small piles can have the same ultimate load capacity as two large-diameter piles. However, the small piles reach the ultimate state at a lateral deflection of 50 mm, whereas the large piles do at 150 mm. The smaller piles would have failed before the larger piles are activated to a substantial degree. In other words, larger piles will be more flexible and ductile than smaller piles before reaching the ultimate state. Since ductility usually provides more seismic safety, larger-diameter piles are preferred from the point of view of seismic design.

Inclined or battered piles should not be used together with vertical piles unless the inclined piles alone have enough lateral capacity. Inclined piles provide partial lateral resistance from their axial capacity, and, since the stiffness in the axial direction of a pile is much larger than in the perpendicular directions, inclined piles tend to attract most of the lateral seismic loading. Inclined piles will fail or reach their ultimate axial capacity before the vertical piles are activated to take substantial lateral loads.

32.3 Design Considerations

32.3.1 Design Concept

The current practice of foundation design mainly employs two types of design concepts, i.e., the permissible stress approach and the limit state approach.

By using the permissible stress approach, both the demanded stresses from loading and the ultimate stress capacity of the foundation are evaluated. The foundation is considered to be safe as long as the demanded stresses are less than the ultimate stress capacity of the foundation. A factor of safety of 2 to 3 is usually applied to the ultimate capacity to obtain various allowable levels of loading in order to limit the displacements of a foundation. A separate displacement analysis is usually performed to determine the allowable displacements for a foundation, and for the bridge structures. Design based on the permissible concept is still the most popular practice in foundation design.

Starting to be adopted in the design of large critical bridges, the limit state approach requires that the foundation and its supported bridge should not fail to meet performance requirements when exceeding various limit states. Collapse of the bridge is the ultimate limit state, and design is aimed at applying various factors to loading and resistance to ensure that this state is highly improbable. A design needs to ensure the structural integrity of the critical foundations before reaching the ultimate limit state, such that the bridge can be repaired a relatively short time after a major loading incident without reconstruction of the time-consuming foundations.

32.3.2 Design Procedures

Under normal conditions, the design procedures of a bridge foundation should involve the following steps:

1. Evaluate the site and subsurface geologic and geotechnical conditions, perform borings or other field exploratory programs, and conduct field and laboratory tests to obtain design parameters for subsurface materials;
2. Review the foundation requirements including design loads and allowable displacements, regulatory provisions, space, or other constraints;
3. Evaluate the anticipated construction conditions and procedures;
4. Select appropriate foundation type(s);
5. Determine the allowable and ultimate axial and lateral foundation design capacity, load vs. deflection relationship, and load vs. settlement relationship;
6. Design various elements of the foundation structure; and
7. Specify requirements for construction inspection and/or load test procedures, and incorporate the requirements into construction specifications.

32.3.3 Design Capacities

Capacity in Long-Term and Short-Term Conditions

Depending on the loading types, foundations are designed for two different stress conditions. Capacity in total stress is used where loading is relatively quick and corresponds to an undrained condition. Capacity in effective stress is adopted where loading is slow and corresponds to a drained condition. For many types of granular soil, such as clean gravel and sand, drained capacity is very close to undrained capacity under most loading conditions. Pile capacity under seismic loading is usually taken 30% higher than capacity under static loading.

Axial, Lateral, and Moment Capacity

Deep foundations can provide lateral resistance to overturning moment and lateral loads and axial resistance to axial loads. Part or most of the moment capacity of a pile group are provided by the axial capacity of individual piles through pile cap action. The moment capacity depends on the axial capacity of the individual piles, the geometry arrangement of the piles, the rigidity of the pile cap, and the rigidity of the connection between the piles and the pile cap. Design and analysis is often concentrated on the axial and lateral capacity of individual piles. Axial capacity of an individual pile will be addressed in detail in Section 32.4 and lateral capacity in Section 32.5. Pile groups will be addressed in Section 32.6.

Structural Capacity

Deep foundations may fail because of structural failure of the foundation elements. These elements should be designed to take moment, shear, column action or buckling, corrosion, fatigue, etc. under various design loading and environmental conditions.

Determination of Capacities

In the previous sections, the general procedure and concept for the design of deep foundations are discussed. Detailed design includes the determination of axial and lateral capacity of individual foundations, and capacity of pile groups. Many methods are available to estimate these capacities, and they can be categorized into three types of methodology as listed in the following:

- Theoretical analysis utilizing soil or rock strength;
- Empirical methods including empirical analysis utilizing standard field tests, code requirements, and local experience; and
- Load tests, including full-scale load tests, and dynamic driving and restriking resistance analysis.

The choice of methods depends on the availability of data, economy, and other constraints. Usually, several methods are used; the capacity of the foundation is then obtained through a comprehensive evaluation and judgment.

In applying the above methods, the designers need to keep in mind that the capacity of a foundation is the sum of capacities of all elements. Deformation should be compatible in the foundation elements, in the surrounding soil, and in the soil–foundation interface. Settlement or other movements of a foundation should be restricted within an acceptable range and usually is a controlling factor for large foundations.

32.3.4 Summary of Design Methods

Table 32.4 presents a partial list of design methods available in the literature.

32.3.5 Other Design Issues

Proper foundation design should consider many factors regarding the environmental conditions, type of loading conditions, soil and rock conditions, construction, and engineering analyses, including:

- Various loading and loading combinations, including the impact loads of ships or vehicles
- Earthquake shaking
- Liquefaction

TABLE 32.4 Summary of Design Methods for Deep Foundations

Type	Design For	Soil Condition	Method and Author
Driven pile	End bearing	Clay	N_c method [67]
			N_c method [23]
			CPT methods [37,59,63]
			CPT [8,10]
		Sand	N_q method with critical depth concept [38]
			N_q method [3]
			N_q method [23]
			N_q by others [26,71,76]
			Limiting N_q values [1,13]
			Value of ϕ [27,30,39]
			SPT [37,38]
			CPT methods [37,59,63]
			CPT [8,10]
		Rock	[10]
	Side resistance	Clay	α-method [72,73]
			α-method [1]
			β-method [23]
			λ-method [28,80]
			CPT methods [37,59,63]
			CPT [8,10]
			SPT [14]
		Sand	α-method [72,73]
			β-method [7]
			β-method [23]
			CPT method [37,59,63]
			CPT [8,10]
			SPT [37,38]
	Side and end	All	Load test: ASTM D 1143, static axial compressive test
			Load test: ASTM D 3689, static axial tensile test
			Sanders' pile driving formula (1850) [50]
			Danish pile driving formula [68]
			Engineering News formula (Wellingotn, 1988)
			Dynamic formula — WEAP Analysis
			Strike and restrike dynamic analysis
			Interlayer influence [38]
			No critical depth [20,31]
	Load-settlement	Sand	[77]
			[41,81]
		All	Theory of elasticity, Mindlin's solutions [50]
			Finite-element method [15]
			Load test: ASTM D 1143, static axial compressive test
			Load test: ASTM D 3689, static axial tensile test
Drilled shaft	End bearing	Clay	N_c method [66]
			Large base [45,57]
			CPT [8,10]
		Sand	[74]
			[38]
			[55]
			[52]
			[37,38]
			[8,10]
		Rock	[10]
		Rock	Pressure meter [10]

TABLE 32.4 (continued) Summary of Design Methods for Deep Foundations

Type	Design For	Soil Condition	Method and Author
	Side resistance	Clay	α-method [52]
			α-method [67]
			α-method [83]
			CPT [8,10]
		Sand	[74]
			[38]
			[55]
			β-method [44,52]
			SPT [52]
			CPT [8,10]
		Rock	Coulombic [34]
			Coulombic [75]
			SPT [12]
			[24]
			[58]
			[11,32]
			[25]
	Side and end	Rock	[46]
			[84]
			[60]
			[48]
			[61,62]
			FHWA [57]
		All	Load test [47]
	Load-settlement	Sand	[57]
		Clay	[57]
			[85]
		All	Load test [47]
All	Lateral resistance	Clay	Broms' method [5]
		Sand	Broms' method [6]
		All	*p–y* method [56]
		Clay	*p–y* response [35]
		Clay (w/water)	*p–y* response [53]
		Clay (w/o water)	*p–y* response [82]
		Sand	*p–y* response [53]
		All	*p–y* response [1]
			p–y response for inclined piles [2,29]
			p–y response in layered soil
			p–y response [42]
		Rock	*p–y* response [86]
	Load-settlement	All	Theory of elasticity method [50]
			Finite-difference method [64]
			General finite-element method (FEM)
			FEM dynamic
	End bearing		Pressure meter method [36,78]
	Lateral resistance		Pressure meter method [36]
			Load test: ASTM D 3966
Group	Theory		Elasticity approach [50]
			Elasticity approach [21]
			Two-dimensional group [51]
			Three-dimensional group [52]
	Lateral g-factor		[10]
			[16]

- Rupture of active fault and shear zone
- Landslide or ground instability
- Difficult ground conditions such as underlying weak and compressible soils
- Debris flow
- Scour and erosion
- Chemical corrosion of foundation materials
- Weathering and strength reduction of foundation materials
- Freezing
- Water conditions including flooding, water table change, dewatering
- Environmental change due to construction of the bridge
- Site contamination condition of hazardous materials
- Effects of human or animal activities
- Influence upon and by nearby structures
- Governmental and community regulatory requirements
- Local practice

32.3.6 Uncertainty of Foundation Design

Foundation design is as much an art as a science. Although most foundation structures are man-made, the surrounding geomaterials are created, deposited, and altered in nature over the geologic times. The composition and engineering properties of engineering materials such as steel and concrete are well controlled within a variation of uncertainty of between 5 to 30%. However, the uncertainty of engineering properties for natural geomaterials can be up to several times, even within relatively uniform layers and formations. The introduction of faults and other discontinuities make generalization of material properties very hard, if not impossible.

Detailed geologic and geotechnical information is usually difficult and expensive to obtain. Foundation engineers constantly face the challenge of making engineering judgments based on limited and insufficient data of ground conditions and engineering properties of geomaterials.

It was reported that under almost identical conditions, variation of pile capacities of up to 50% could be expected within a pile cap footprint under normal circumstances. For example, piles within a nine-pile group had different restruck capacities of 110, 89, 87, 96, 86, 102, 103, 74, and 117 kips (1 kip = 4.45 kN) respectively [19].

Conservatism in foundation design, however, is not necessarily always the solution. Under seismic loading, heavier and stiffer foundations may tend to attract more seismic energy and produce larger loads; therefore, massive foundations may not guarantee a safe bridge performance.

It could be advantageous that piles, steel pipes, caisson segments, or reinforcement steel bars are tailored to exact lengths. However, variation of depth and length of foundations should always be expected. Indicator programs, such as indicator piles and pilot exploratory borings, are usually a good investment.

32.4 Axial Capacity and Settlement — Individual Foundation

32.4.1 General

The axial resistance of a deep foundation includes the tip resistance (Q_{end}), side or shaft resistance (Q_{side}), and the effective weight of the foundation (W_{pile}). Tip resistance, also called end bearing, is the compressive resistance of soil near or under the tip. Side resistance consists of friction, cohesion, and keyed bearing along the shaft of the foundation. Weight of the foundation is usually ignored under compression because it is nearly the same as the weight of the soil displaced, but is usually accounted for under uplift loading condition.

At any loading instance, the resistance of an individual deep foundation (or pile) can be expressed as follows:

$$Q = Q_{end} + \Sigma Q_{side} \pm W_{pile} \tag{32.1}$$

The contribution of each component in the above equation depends on the stress–strain behavior and stiffness of the pile and the surrounding soil and rock. The maximum capacity of a pile can be expressed as

$$Q^c{}_{max} \leq Q^c{}_{end\_max} + \Sigma Q^c{}_{side\_max} - W_{pile} \quad \text{(in compression)} \tag{32.2}$$

$$Q^t{}_{max} \leq Q^t{}_{end\_max} + \Sigma Q^t{}_{side\_max} + W_{pile} \quad \text{(in uplift)} \tag{32.3}$$

and is less than the sum of all the maximum values of resistance. The ultimate capacity of a pile undergoing a large settlement or upward movement can be expressed as

$$Q^c{}_{ult} = Q^c{}_{end\_ult} + \Sigma Q^c{}_{side\_ult} - W_{pile} \leq Q^c_{max} \tag{32.4}$$

$$Q^t{}_{ult} = Q^t{}_{end\_ult} + \Sigma Q^t{}_{side\_ult} + W_{pile} \leq Q^t_{max} \tag{32.5}$$

Side- and end-bearing resistances are related to displacement of a pile. Maximum end bearing capacity can be mobilized only after a substantial downward movement of the pile, whereas side resistance reaches its maximum capacity at a relatively smaller downward movement. Therefore, the components of the maximum capacities (Q_{max}) indicated in Eqs. (32.2) and (32.3) may not be realized at the same time at the tip and along the shaft. For a drilled shaft, the end bearing is usually ignored if the bottom of the borehole is not cleared and inspected during construction. Voids or compressible materials may exist at the bottom after concrete is poured; as a result, end bearing will be activated only after a substantial displacement.

Axial displacements along a pile are larger near the top than toward the tip. Side resistance depends on the amount of displacement and is usually not uniform along the pile. If a pile is very long, maximum side resistance may not occur at the same time along the entire length of the pile. Certain types of geomaterials, such as most rocks and some stiff clay and dense sand, exhibit strain softening behavior for their side resistance, where the side resistance first increases to reach its maximum, then drops to a much smaller residual value with further displacement. Consequently, only a fixed length of the pile segment may maintain high resistance values and this segment migrates downward to behave in a pattern of a progressive failure. Therefore, the capacity of a pile or drilled shaft may not increase infinitely with its length.

For design using the permissible stress approach, allowable capacity of a pile is the design capacity under service or routine loading. The allowable capacity (Q_{all}) is obtained by dividing ultimate capacity (Q_{ult}) by a factor of safety (FS) to limit the level of settlement of the pile and to account for uncertainties involving material, installation, loads calculation, and other aspects. In many cases, the ultimate capacity (Q_{ult}) is assumed to be the maximum capacity (Q_{max}). The factor of safety is usually between 2 to 3 for deep foundations depending on the reliability of the ultimate capacity estimated. With a field full-scale loading test program, the factor of safety is usually 2.

TABLE 32.5 Typical Values of Bearing Capacity Factor N_q

φ^a (degrees)	26	28	30	31	32	33	34	35	36	37	38	39	40
N_q (driven pile displacement)	10	15	21	24	29	35	42	50	62	77	86	120	145
$N_q{}^b$ (drilled piers)	5	8	10	12	14	17	21	25	30	38	43	60	72

[a] Limit φ to 28° if jetting is used.

[b] 1. In case a bailer of grab bucket is used below the groundwater table, calculate end bearing based on φ not exceeding 28°.
2. For piers greater than 24-in. diameter, settlement rather than bearing capacity usually controls the design. For estimating settlement, take 50% of the settlement for an equivalent footing resting on the surface of comparable granular soils (Chapter 5, DM-7.01).

Source: NAVFAC [42].

32.4.2 End Bearing

End bearing is part of the axial compressive resistance provided at the bottom of a pile by the underlying soil or rock. The resistance depends on the type and strength of the soil or rock and on the stress conditions near the tip. Piles deriving their capacity mostly from end bearing are called end bearing piles. End bearing in rock and certain types of soil such as dense sand and gravel is usually large enough to support the designed loads. However, these types of soil or rock cannot be easily penetrated through driving. No or limited uplift resistance is provided from the pile tips; therefore, end-bearing piles have low resistance against uplift loading.

The end bearing of a pile can be expressed as:

$$Q_{\text{end\_max}} = \begin{cases} cN_c A_{\text{pile}} & \text{for clay} \\ \sigma'_v N_q A_{\text{pile}} & \text{for sand} \\ \dfrac{U_c}{2} N_k A_{\text{pile}} & \text{for rock} \end{cases} \tag{32.6}$$

where

$Q_{\text{end\_max}}$ = the maximum end bearing of a pile
A_{pile} = the area of the pile tip or base
N_c, N_q, N_k = the bearing capacity factors for clay, sand, and rock
c = the cohesion of clay
σ'_v = the effective overburden pressure
U_c = the unconfined compressive strength of rock and $\frac{U_c}{2} = S_u$, the equivalent shear strength of rock

Clay

The bearing capacity factor N_c for clay can be expressed as

$$N_c = 6.0\left(1 + 0.2\frac{L}{D}\right) \le 9 \tag{32.7}$$

where L is the embedment depth of the pile tip and D is the diameter of the pile.

Sand

The bearing capacity factor N_q generally depends on the friction angle ϕ of the sand and can be estimated by using Table 32.5 or the Meyerhof equation below.

$$N_q = e^{\pi \tan\varphi} \tan^2\left(45 + \frac{\varphi}{2}\right) \tag{32.8}$$

The capacity of end bearing in sand reaches a maximum cutoff after a certain critical embedment depth. This critical depth is related to ϕ and D and for design purposes is listed as follows:

$$L_c = 7D, \quad \phi = 30^o \text{ for loose sand}$$

$$L_c = 10D, \quad \phi = 34^o \text{ for medium dense sand}$$

$$L_c = 14D, \quad \phi = 38^o \text{ for dense sand}$$

$$L_c = 22D, \quad \phi = 45^o \text{ for very dense sand}$$

The validity of the concept of critical depth has been challenged by some people; however, the practice to limit the maximum ultimate end bearing capacity in sand will result in conservative design and is often recommended.

Rock

The bearing capacity factor N_k depends on the quality of the rock mass, intact rock properties, fracture or joint properties, embedment, and other factors. Because of the complex nature of the rock mass and the usually high value for design bearing capacity, care should be taken to estimate N_k. For hard fresh massive rock without open or filled fractures, N_k can be taken as high as 6. N_k decreases with increasing presence and dominance of fractures or joints and can be as low as 1. Rock should be treated as soil when rock is highly fractured and weathered or in-fill weak materials control the behavior of the rock mass. Bearing capacity on rock also depends on the stability of the rock mass. Rock slope stability analysis should be performed where the foundation is based on a slope. A higher factor of safety, 3 to as high as 10 to 20, is usually applied in estimating allowable bearing capacity for rocks using the N_k approach.

The soil or rock parameters used in design should be taken from averaged properties of soil or rock below the pile tip within the influence zone. The influence zone is usually taken as deep as three to five diameters of the pile. Separate analyses should be conducted where weak layers exist below the tip and excessive settlement or punch failure might occur.

Empirical Methods

Empirical methods are based on information of the type of soil/rock and field tests or index properties. The standard penetration test (SPT) for sand and cone penetration test (CPT) for soil are often used.

Meyerhof [38] recommended a simple formula for piles driven into sand. The ultimate tip bearing pressure is expressed as

$$q_{\text{end\_max}} \leq 4N_{\text{SPT}} \quad \text{in tsf (1 tsf = 8.9 kN)} \tag{32.9}$$

where N_{SPT} is the blow count of SPT just below the tip of the driven pile and $q_{\text{end\_max}} = Q_{\text{end\_max}} / A_{\text{pile}}$. Although the formula is developed for piles in sand, it also is used for piles in weathered rock for preliminary estimate of pile capacity.

Schmertmann [63] recommended a method to estimate pile capacity by using the CPT test:

$$q_{\text{end\_max}} = q_b = \frac{q_{c1} + q_{c2}}{2} \tag{32.10}$$

where

q_{c1} = averaged cone tip resistance over a depth of 0.7 to 4 diameters of the pile below tip of the pile
q_{c2} = the averaged cone tip resistance over a depth of 8 diameters of the pile above the tip of the pile

Chapter 31 presents recommended allowable bearing pressures for various soil and rock types for spread footing foundations and can be used as a conservative estimate of end-bearing capacity for end-bearing piles.

TABLE 32.6 Typical Values of α and f_s

Range of Shear Strength, S_u ksf	Formula to Estimate α	Range of α	Range of f_s ksf[a]	Description
0 to 0.600	$\alpha = 1.0$	1	0–0.6	Soft clay
0.600 to 3	$\alpha = 0.375\left(1+\frac{1}{S_u}\right)$,	1–0.5	0.6–1.5	Medium stiff clay to very stiff clay
3 to 11	$\alpha = 0.375\left(1+\frac{1}{S_u}\right)$,	0.5–0.41	1.5–4.5	Hard clay to very soft rock
11 to 576 (76 psi to 4000 psi)	$\alpha = \frac{5}{\sqrt{2S_u}}$, S_u in psi,	0.41–0.056	4.5–32 (31–220 psi)	Soft rock to hard rock

Note: 1 ksf = 1000 psf; 1 psi = 144 psf; 1 psf = 0.048 kPa; 1 psi = 6.9 kPa
[a] For concrete driven piles and for drilled piers without buildup of mud cakes along the shaft. (Verify if fs ≥ 3 ksf.)

32.4.3 Side Resistance

Side resistance usually consists of friction and cohesion between the pile and the surrounding soil or rock along the shaft of a pile. Piles that derive their resistance mainly from side resistance are termed *frictional piles*. Most piles in clayey soil are frictional piles, which can take substantial uplift loads.

The maximum side resistance of a pile $Q_{\text{side\_max}}$ can be expressed as

$$Q_{\text{side\_max}} = \sum f_s A_{\text{side}} \tag{32.11}$$

$$f_s = K_s \sigma'_v \tan\delta + c_a \tag{32.12}$$

$$c_a = \alpha S_u \tag{32.13}$$

where

$\sum$ = the sum for all layers of soil and rock along the pile
A_{side} = the shaft side area
f_s = the maximum frictional resistance on the side of the shaft
K_s = the lateral earth pressure factor along the shaft
σ'_v = the effective vertical stress along the side of the shaft
δ = the friction angle between the pile and the surrounding soil; for clayey soil under quick loading, δ is very small and usually omitted
c_a = the adhesion between pile and surrounding soil and rock
α = a strength factor, and
S_u = the cohesion of the soil or rock

TABLE 32.7 Typical Values Cohesion and Adhesion f_s

Pile Type	Consistency of Soil	Cohesion, S_u psf	Adhesion, f_s psf
Timber and concrete	Very soft	0–250	0–250
	Soft	250–500	250–480
	Medium stiff	500–1000	480–750
	Stiff	1000–2000	750–950
	Very stiff	2000–4000	950–1300
Steel	Very soft	0–250	0–250
	Soft	250–500	250–460
	Medium stiff	500–1000	480–700
	Stiff	1000–2000	700–720
	Very stiff	2000–4000	720–750

1 psf = 0.048 kPa.
Source: NAVFAC [42].

TABLE 32.8 Typical Values of Bond Stress of Rock Anchors for Selected Rock

Rock Type (Sound, Nondecayed)	Ultimate Bond Stresses between Rock and Anchor Plus (δ_{skin}), psi
Granite and basalt	250–450
Limestone (competent)	300–400
Dolomitic limestone	200–300
Soft limestone	150–220
Slates and hard shales	120–200
Soft shales	30–120
Sandstone	120–150
Chalk (variable properties)	30–150
Marl (stiff, friable, fissured)	25–36

Note: It is not generally recommended that design bond stresses exceed 200 psi even in the most competent rocks. 1 psi = 6.9 kPa.
Source: NAVFAC [42].

TABLE 32.9 Typical Values of earth Pressure Coefficient K_s

	Earth Pressure Coefficients K_s		
Pile Type	K_s [a] (compression)	K_s [a] (tension)	K_s [b]
Driven single H-pile	0.5–1.0	0.3–0.5	—
Driven single displacement pile	1.0–1.5	0.6–1.0	0.7–3.0
Driven single displacement tapered pile	1.5–2.0	1.0–1.3	—
Driven jetted pile	0.4–0.9	0.3–0.6	—
Drilled pile (less than 24-in. diameter)	0.7	0.4	—
Insert pile	—	—	0.7 (compression) 0.5 (tension)
Driven with predrilled hole	—	—	0.4–0.7
Drilled pier	—	—	0.1–0.4

[a] From NAVFAC [42].
[b] From Le Tirant (1979), K_s increases with OCR or D_R.

TABLE 32.10 Typical Value of Pile-Soil Friction Angles δ

Pile Type	δ, °	Alternate for δ
Concrete[a]	—	δ = ¾φ
Concrete (rough, cast-in-place)[b]	33	δ = 0.85φ
Concrete (smooth)[b]	30	δ = 0.70φ
Steel[a]	20	—
Steel (corrugated)	33	δ = φ
Steel (smooth)[c]	—	δ = φ – 5°
Timber[a]	—	δ = ¾φ

[a] NAVFAC [42].
[b] Woodward et al. [85]
[c] API [1] and de Ruiter and Beringen [13]

Typical values of α, f_s, K_s, δ are shown in Tables 32.6 through 32.10. For design purposes, side resistance f_s in sand is limited to a cutoff value at the critical depth, which is equal to about 10*B* for loose sand and 20*B* for dense sand.

Meyerhof [38] recommended a simple formula for driven piles in sand. The ultimate side adhesion is expressed as

$$f_s \leq \frac{N_{\text{SPT}}}{50} \quad \text{in tsf (1 tsf = 8.9 kN)} \tag{32.14}$$

where N_{SPT} is the averaged blow count of SPT along the pile.

Meyerhof [38] also recommended a formula to calculate the ultimate side adhesion based on CPT results as shown in the following.

For full displacement piles:

$$f_s = \frac{q_c}{200} \leq 1.0 \text{ in tsf} \tag{32.15}$$

or

$$f_s = 2 f_c \leq 1.0 \tag{32.16}$$

For nondisplacement piles:

$$f_s = \frac{q_c}{400} \leq 0.5 \text{ in tsf} \tag{32.17}$$

or

$$f_s = f_c \leq 0.5 \tag{32.18}$$

in which

q_c, f_c = the cone tip and side resistance measured from CPT; averaged values should be used along the pile

Downdrag

For piles in soft soil, another deformation-related issue should be noted. When the soil surrounding the pile settles relative to a pile, the side friction, also called the negative skin friction, should be considered when there exists underlying compressible clayey soil layers and liquefiable loose sand layers. Downdrag can also happen when ground settles because of poor construction of caissons in sand. On the other hand, updrag should also be considered in cases where heave occurs around the piles for uplift loading condition, especially during installation of piles and in expansive soils.

32.4.4 Settlement of Individual Pile, *t–z*, *Q–z* Curves

Besides bearing capacity, the allowable settlement is another controlling factor in determining the allowable capacity of a pile foundation, especially if layers of highly compressible soil are close to or below the tip of a pile.

Settlement of a small pile (diameter less than 350 mm) is usually kept within an acceptable range (usually less than 10 mm) when a factor of safety of 2 to 3 is applied to the ultimate capacity to obtain the allowable capacity. However, in the design of large-diameter piles or caissons, a separate settlement analysis should always be performed.

The total settlement at the top of a pile consists of immediate settlement and long-term settlement. The immediate settlement occurs during or shortly after the loads are applied, which includes elastic compression of the pile and deformation of the soil surrounding the pile under undrained loading conditions. The long-term settlement takes place during the period after the loads are applied, which includes creep deformation and consolidation deformation of the soil under drained loading conditions.

Consolidation settlement is usually significant in soft to medium stiff clayey soils. Creep settlement occurs most significantly in overconsolidated (OC) clays under large sustained loads, and can be estimated by using the method developed by Booker and Poulos (1976). In principle, however, long-term settlement can be included in the calculation of ultimate settlement if the design parameters of soil used in the calculation reflect the long-term behavior.

Presented in the following sections are three methods that are often used:

- Method of solving ultimate settlement by using special solutions from the theory of elasticity [50,85]. Settlement is estimated based on equivalent elasticity in which all deformation of soil is assumed to be linear elastic.
- Empirical method [79].
- Method using localized springs, or the so called *t–z* and *Q–z* method [52a].

Method from Elasticity Solutions

The total elastic settlement S can be separated into three components:

$$S = S_b + S_s + S_{sh} \tag{32.19}$$

where S_b is part of the settlement at the tip or bottom of a pile caused by compression of soil layers below the pile under a point load at the pile tip, and is expressed as

$$S_b = \frac{p_b D_b I_{bb}}{E_s} \tag{32.20}$$

S_s is part of the settlement at the tip of a pile caused by compression of soil layers below the pile under the loading of the distributed side friction along the shaft of the pile, and can be expressed as

$$S_s = \sum_i \frac{(f_{si} l_i \Delta z_i) I_{bs}}{E_s} \tag{32.21}$$

and S_{sh} is the shortening of the pile itself, and can be expressed as

$$S_{sh} = \sum_i \frac{(f_{si} l_i \Delta z_i) + p_b A_b (\Delta z_i)}{E_c (A_i)} \tag{32.22}$$

where

p_b = averaged loading pressure at pile tip
A_b = cross section area of a pile at pile tip; $A_b p_b$ is the total load at the tip
D_b = diameter of pile at the pile tip
i = subscript for *i*th segment of the pile
l = perimeter of a segment of the pile
Δz = axial length of a segment of the pile; $L = \sum_i \Delta z_i$ is the total length of the pile.
f_s = unit friction along side of shaft; $f_{si} l_i \Delta z_i$ is the side frictional force for segment i of the pile
E_s = Young's modulus of uniform and isotropic soil
E_c = Young's modulus of the pile
I_{bb} = base settlement influence factor, from load at the pile tip (Figure 32.4)
I_{bs} = base settlement influence factor, from load along the pile shaft (Figure 32.4)

Because of the assumptions of linear elasticity, uniformity, and isotropy for soil, this method is usually used for preliminary estimate purposes.

Method by Vesic [79]

The settlement S at the top of a pile can be broken down into three components, i.e.,

$$S = S_b + S_s + S_{sh} \tag{32.23}$$

Settlement due to shortening of a pile is

$$S_{sh} = (Q_p + \alpha_s Q_s) \frac{L}{AE_c} \tag{32.24}$$

where

Q_p = point load transmitted to the pile tip in the working stress range
Q_s = shaft friction load transmitted by the pile in the working stress range (in force units)
α_s = 0.5 for parabolic or uniform distribution of shaft friction, 0.67 for triangular distribution of shaft friction starting from zero friction at pile head to a maximum value at pile tip, 0.33 for triangular distribution of shaft friction starting from a maximum at pile head to zero at the pile tip
L = pile length
A = pile cross-sectional area
E_c = modulus of elasticity of the pile

Settlement of the pile tip caused by load transmitted at the pile tip is

$$S_b = \frac{C_p Q_p}{D q_o} \tag{32.25}$$

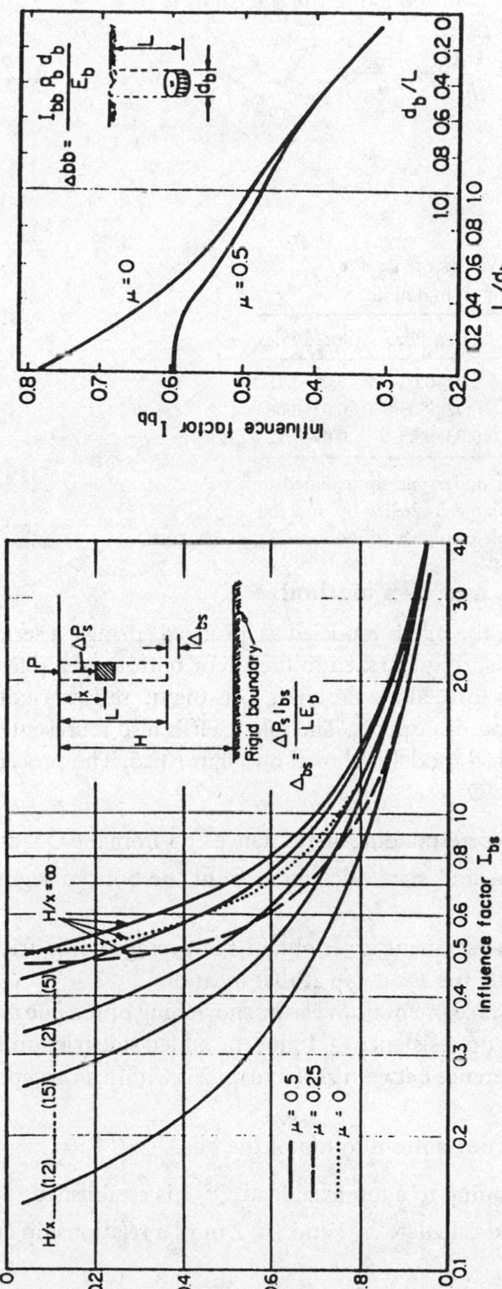

FIGURE 32.4 Influence factors I_{bb} and I_{bs}. [From Woodward, Gardner and Greer (1972)[85], used with permission of McGraw-Hill Book Company]

where

C_p = empirical coefficient depending on soil type and method of construction, see Table 32.11.
D = pile diameter
q_o = ultimate end bearing capacity

and settlement of the pile tip caused by load transmitted along the pile shaft is

$$S_s = \frac{C_s Q_s}{h q_o} \tag{32.26}$$

where

$C_s = (0.93 + 0.16 D / B) C_p$
h = embedded length

TABLE 32.11 Typical Values of C_p for Estimating Settlement of a Single Pile

Soil Type	Driven Piles	Bored Piles
Sand (dense to loose)	0.02–0.04	0.09–0.18
Clay (stiff to soft)	0.02–0.03	0.03–0.06
Silt (dense to loose)	0.03–0.05	0.09–0.12

Note: Bearing stratum under pile tip assumed to extend at least 10 pile diameters below tip and soil below tip is of comparable or higher stiffness.

Method Using Localized Springs: The *t–z* and *Q–z* method

In this method, the reaction of soil surrounding the pile is modeled as localized springs: a series of springs along the shaft (the *t–z* curves) and the spring attached to the tip or bottom of a pile (the *Q–z* curve). *t* is the load transfer or unit friction force along the shaft, *Q* is the tip resistance of the pile, and *z* is the settlement of soil at the location of a spring. The pile itself is also represented as a series of springs for each segment. A mechanical model is shown on Figure 32.5. The procedure to obtain the settlement of a pile is as follows:

- Assume a pile tip movement *zb_1*; obtain a corresponding tip resistance Q_1 from the *Q–z* curve.
- Divide the pile into number of segments, and start calculation from the bottom segment. Iterations:
  1. Assume an averaged movement of the segment *zs_1*; obtain the averaged side friction along the bottom segment *ts_1* by using the *t–z* curve at that location.
  2. Calculate the movement at middle of the segment from elastic shortening of the pile under axial loading *zs_2*. The axial load is the tip resistance Q_1 plus the added side friction *ts_1*.
  3. Iteration should continue until the difference between *zs_1* and *zs_2* is within an acceptable tolerance.

  Iteration continues for all the segments from bottom to top of the pile.
- A settlement at top of pile *zt_1* corresponding to a top axial load *Qt_1* is established.
- Select another pile tip movement *zb_2* and calculate *zt_2* and *Qt_2* until a relationship curve of load vs. pile top settlement is found.

The *t–z* and *Q–z* curves are established from test data by many authors. Figure 32.6 shows the *t–z* and *Q–z* curves for cohesive soil and cohesionless soil by Reese and O'Neil [57].

Although the method of *t–z* and *Q–z* curves employs localized springs, the calculated settlements are usually within a reasonable range since the curves are backfitted directly from the test results. Factors of nonlinear behavior of soil, complicated stress conditions around the pile, and partial

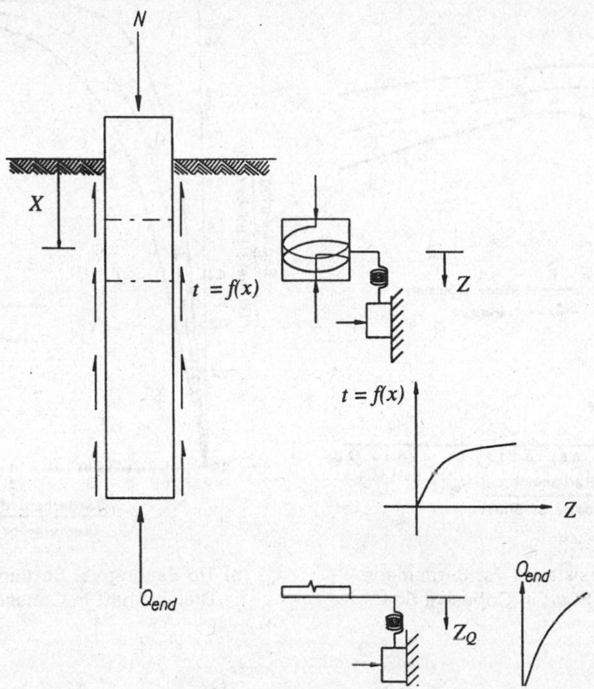

FIGURE 32.5 Analytical model for pile under axial loading with t–z and Q–z curves.

corrections to the Winkler's assumption are embedded in this methodology. Besides, settlement of a pile can be estimated for complicated conditions such as varying pile geometry, different pile materials, and different soil layers.

32.5 Lateral Capacity and Deflection — Individual Foundation

32.5.1 General

Lateral capacity of a foundation is the capacity to resist lateral deflection caused by horizontal forces and overturning moments acted on the top of the foundation. For an individual foundation, lateral resistance comes from three sources: lateral earth pressures, base shear, and nonuniformly distributed end-bearing pressures. Lateral earth pressure is the primary lateral resistance for long piles. Base shear and distributed end-bearing pressures are discussed in Chapter 31.

32.5.2 Broms' Method

Broms [5] developed a method to estimate the ultimate lateral capacity of a pile. The pile is assumed to be short and rigid. Only rigid translation and rotation movements are considered and only ultimate lateral capacity of a pile is calculated. The method assumes distributions of ultimate lateral pressures for cohesive and cohesionless soils; the lateral capacity of piles with different top fixity conditions are calculated based on the assumed lateral pressure as illustrated on Figures 32.7 and 32.8. Restricted by the assumptions, the Broms' method is usually used only for preliminary estimates of the ultimate lateral capacity of piles.

Ultimate Lateral Pressure

The ultimate lateral pressure $q_{h,u}$ along a pile is calculated as follows:

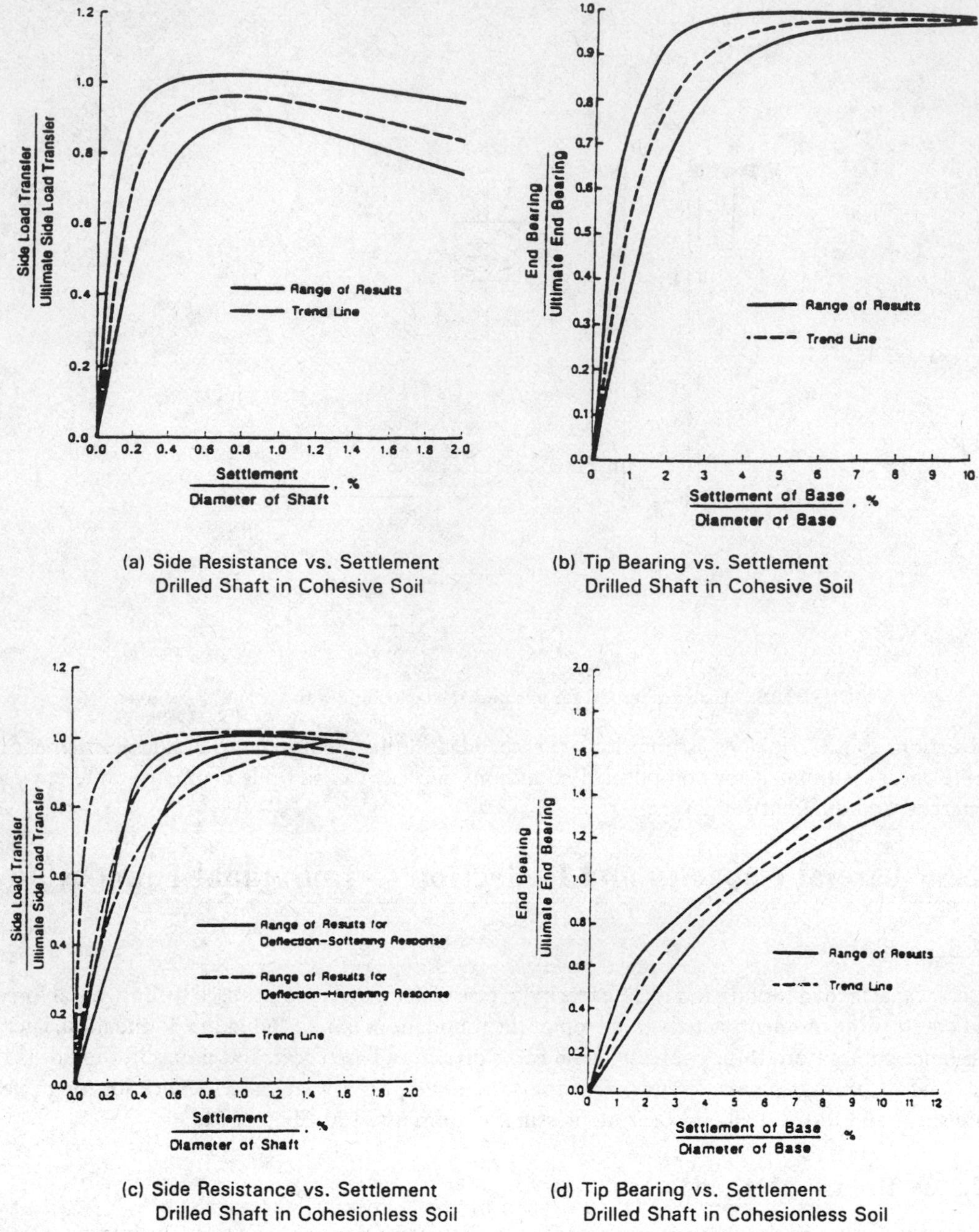

FIGURE 32.6 Load transfer for side resistance (t–z) and tip bearing (Q–z). (a) Side resistance vs. settlement, drilled shaft in cohesive soil; (b) tip bearing vs. settlement, drilled shaft in cohesive soil; (c) side resistance vs. settlement, drilled shaft in cohesionless soil; (d) tip bearing vs. settlement, drilled shaft in cohesionless soil. (From AASHTO LRFD Bridge Design Specifications, First Edition, coyyright 1996 by the American Association of State Highway and Transportation officials, Washington, D.C. Used by permission.)

$$q_{h,u} = \begin{cases} 9c_u & \text{for cohesive soil} \\ 3K_p p_0' & \text{for cohesionless soil} \end{cases} \tag{32.27}$$

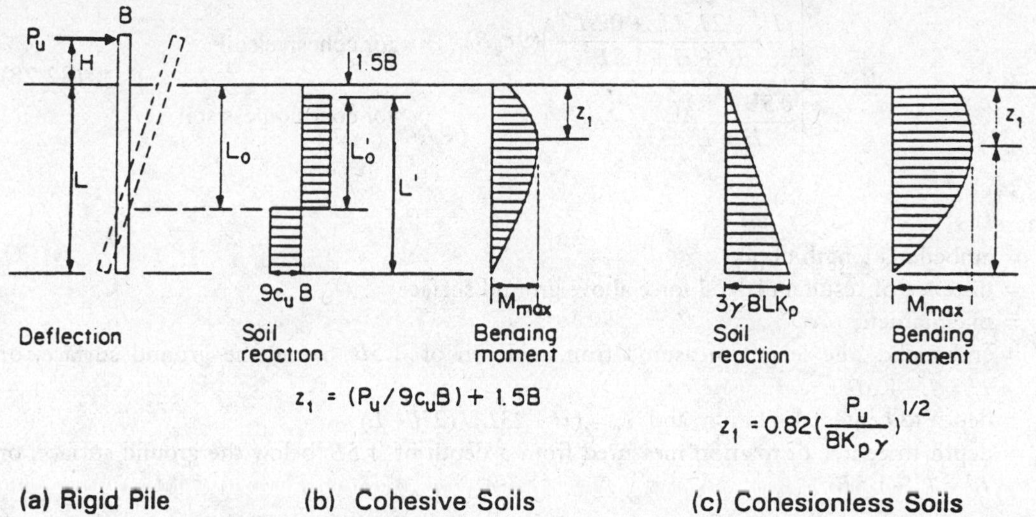

FIGURE 32.7 Free-head, short rigid piles — ultimate load conditions. (a) Rigid pile; (b) cohesive soils; (c) cohesionless soils. [After Broms (1964)[5,6]]

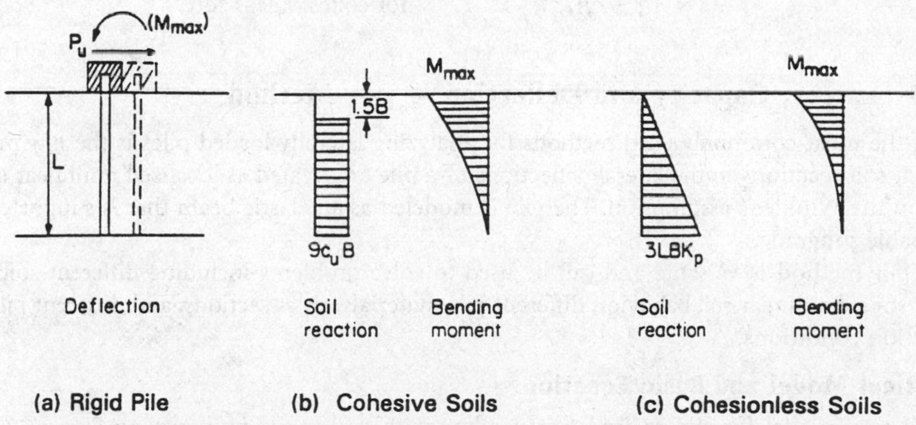

FIGURE 32.8 Fixed-head, short rigid piles — ultimate load conditions. (a) Rigid pile; (b) cohesive soils; (c) cohesionless soils. [After Broms (1964)[5,6]]

where

c_u = shear strength of the soil

K_p = coefficient of passive earth pressure, $K_p = \tan^2(45^o + \varphi/2)$ and φ is the friction angle of cohesionless soils (or sand and gravel)

p_0' = effective overburden pressure, $p_0' = \gamma_z'$ at a depth of z from the ground surface, where γ' is the effective unit weight of the soil

Ultimate Lateral Capacity for the Free-Head Condition

The ultimate lateral capacity P_u of a pile under the free-head condition is calculated by using the following formula:

$$P_u = \begin{cases} \left(\dfrac{L_0'^2 - 2L'L_0' + 0.5L'^2}{L' + H + 1.5B}\right)(9c_u B) & \text{for cohesive soil} \\ \dfrac{0.5BL^3K_p\gamma'}{H+L} & \text{for cohesionless soil} \end{cases} \tag{32.28}$$

where

L = embedded length of pile

H = distance of resultant lateral force above ground surface

B = pile diameter

L' = embedded pile length measured from a depth of $1.5B$ below the ground surface, or $L' = L - 1.5B$

L_0 = depth to center of rotation, and $L_0 = (H + 23L)/(2H + L)$

L_0' = depth to center of rotation measured from a depth of $1.5B$ below the ground surface, or $L_0' = L_0 - 1.5B$

Ultimate Lateral Capacity for the Fixed-Head Condition

The ultimate lateral capacity P_u of a pile under the fixed-head condition is calculated by using the following formula:

$$P_u = \begin{cases} 9c_u B(L - 1.5B) & \text{for cohesive soil} \\ 1.5\gamma' BL^2 K_p & \text{for cohesionless soil} \end{cases} \tag{32.29}$$

32.5.3 Lateral Capacity and Deflection — *p–y* Method

One of the most commonly used methods for analyzing laterally loaded piles is the *p–y* method, in which soil reactions to the lateral deflections of a pile are treated as localized nonlinear springs based on the Winkler's assumption. The pile is modeled as an elastic beam that is supported on a deformable subgrade.

The *p–y* method is versatile and can be used to solve problems including different soil types, layered soils, nonlinear soil behavior; different pile materials, cross sections; and different pile head connection conditions.

Analytical Model and Basic Equation

An analytical model for pile under lateral loading with *p–y* curves is shown on Figure 32.9. The basic equation for the beam-on-a-deformable-subgrade problem can be expressed as

$$EI\frac{d^4y}{dx^4} - P_x\frac{d^2y}{dx^2} + p + q = 0 \tag{32.30}$$

where

y = lateral deflection at point x along the pile

EI = bending stiffness or flexural rigidity of the pile

P_x = axial force in beam column

p = soil reaction per unit length, and $p = -E_s y$; where E_s is the secant modulus of soil reaction.

q = lateral distributed loads

The following relationships are also used in developing boundary conditions:

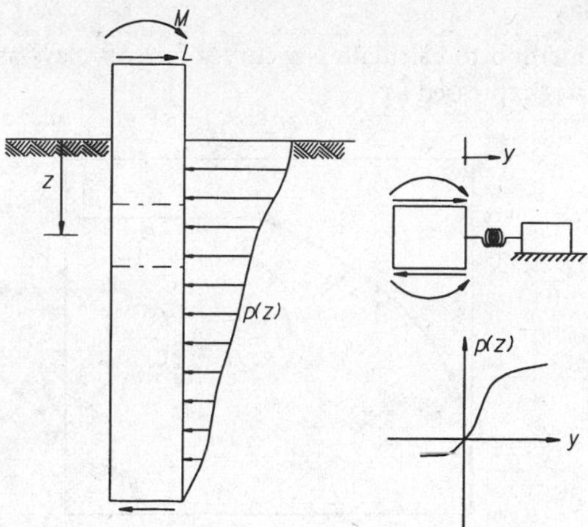

FIGURE 32.9 Analytical model for pile under lateral loading with p–y curves.

$$M = -EI\frac{d^4y}{dx^4} \tag{32.31}$$

$$Q = -\frac{dM}{dx} + P_x\frac{dy}{dx} \tag{32.32}$$

$$\theta = \frac{dy}{dx} \tag{32.33}$$

where M is the bending moment, Q is the shear force in the beam column, θ is the rotation of the pile.

The *p–y* method is a valuable tool in analyzing laterally loaded piles. Reasonable results are usually obtained. A computer program is usually required because of the complexity and iteration needed to solve the above equations using the finite-difference method or other methods. It should be noted that Winkler's assumption ignores the global effect of a continuum. Normally, if soil behaves like a continuum, the deflection at one point will affect the deflections at other points under loading. There is no explicit expression in the *p–y* method since localized springs are assumed. Although *p–y* curves are developed directly from results of load tests and the influence of global interaction is included implicitly, there are cases where unexpected outcomes resulted. For example, excessively large shear forces will be predicted for large piles in rock by using the *p–y* method approach, where the effects of the continuum and the shear stiffness of the surrounding rock are ignored. The accuracy of the *p–y* method depends on the number of tests and the variety of tested parameters, such as geometry and stiffness of pile, layers of soil, strength and stiffness of soil, and loading conditions. One should be careful to extrapolate *p–y* curves to conditions where tests were not yet performed in similar situations.

Generation of *p–y* Curves

A *p–y* curve, or the lateral soil resistance *p* expressed as a function of lateral soil movement *y*, is based on backcalculations from test results of laterally loaded piles. The empirical formulations of *p–y* curves are different for different types of soil. *p–y* curves also depend on the diameter of the pile, the strength and stiffness of the soil, the confining overburden pressures, and the loading conditions. The effects of layered soil, battered piles, piles on a slope, and closely spaced piles are also usually considered. Formulation for soft clay, sand, and rock is provided in the following.

p–y Curves for Soft Clay

Matlock [35] proposed a method to calculate *p–y* curves for soft clays as shown on Figure 32.10. The lateral soil resistance p is expressed as

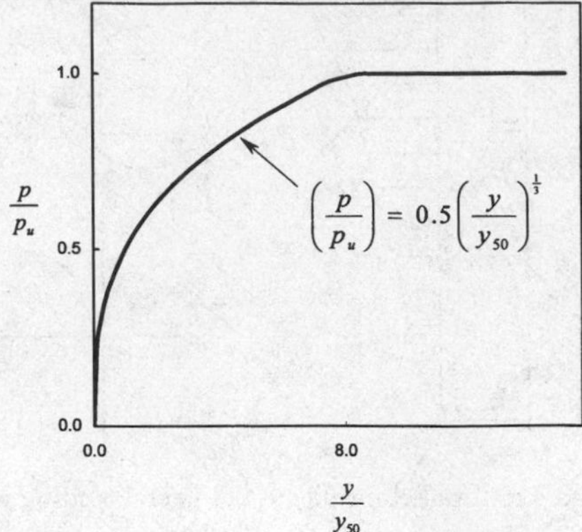

FIGURE 32.10 Characteristic shape of *p–y* curve for soft clay. [After Matlock, (1970)[35]]

$$p = \begin{cases} 0.5\left(\dfrac{y}{y_{50}}\right)^{1/3} p_u & y < y_p = 8y_{50} \\ p_u & y \geq y_p \end{cases} \tag{32.34}$$

in which

p_u = ultimate lateral soil resistance corresponding to ultimate shear stress of soil
y_{50} = lateral movement of soil corresponding to 50% of ultimate lateral soil resistance
y = lateral movement of soil

The ultimate lateral soil resistance p_u is calculated as

$$\mathrm{p_u} = \begin{cases} \left(3 + \dfrac{\gamma' x}{c} + J\dfrac{x}{B}\right)cB & x < x_r = (6B)/\left(\dfrac{\gamma' B}{c} + J\right) \\ 9cB & x \geq x_r \end{cases} \tag{32.35}$$

where γ' is the effective unit weight, x is the depth from ground surface, c is the undrained shear strength of the clay, and J is a constant frequently taken as 0.5.

The lateral movement of soil corresponding to 50% of ultimate lateral soil resistance y_{50} is calculated as

$$y_{50} = 2.5\varepsilon_{50}B \tag{32.36}$$

where ε_{50} is the strain of soil corresponding to half of the maximum deviator stress. Table 32.12 shows the representative values of ε_{50}.

p–y Curves for Sands

Reese et al. [53] proposed a method for developing *p–y* curves for sandy materials. As shown on Figure 32.11, a typical *p–y* curve usually consists of the following four segments:

Segment	Curve type	Range of y	Range of p	p–y curve
1	Linear	0 to y_k	0 to p_k	$p=(kx)y$
2	Parabolic	y_k to y_m	p_k to p_m	$p=p_m\left(\frac{y}{y_m}\right)^n$
3	Linear	y_m to y_u	p_m to p_u	$p=p_m+\frac{p_u-p_m}{y_u-y_m}(y-y_m)$
4	Linear	$\geq y_u$	p_u	$p=p_u$

TABLE 32.12 Representative Values of ε_{50}

Consistency of Clay	Undrained Shear Strength, psf	ε_{50}
Soft	0–400	0.020
Medium stiff	400–1000	0.010
Stiff	1000–2000	0.007
Very stiff	2000–4000	0.005
Hard	4000–8000	0.004

1 psf = 0.048 kPa.

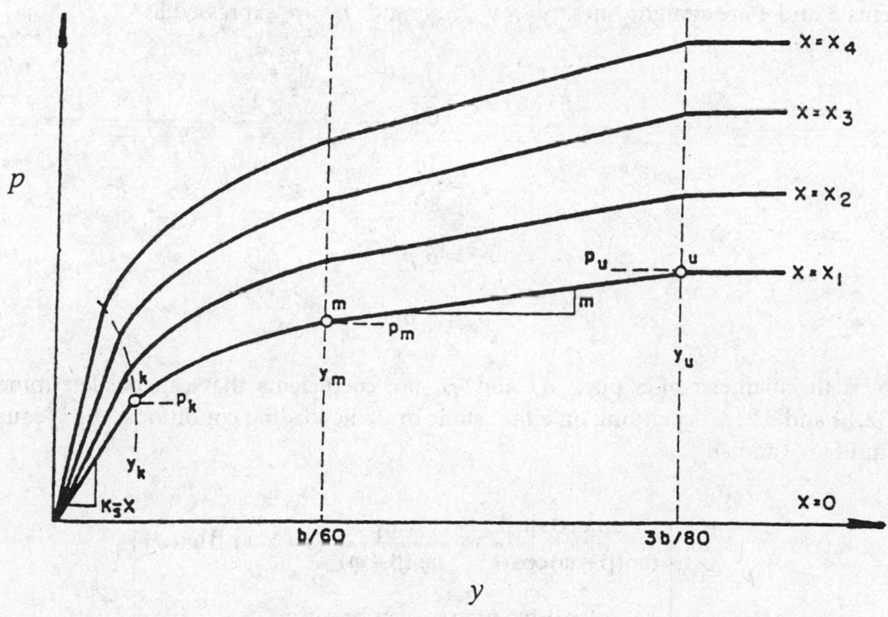

FIGURE 32.11 Characteristic shape of p–y curves for sand. [After Reese, et al. (1974)[53]]

where y_m, y_u, p_m, and p_u can be determined directly from soil parameters. The parabolic form of Segment 2, and the intersection with Segment 1 (y_k and p_k) can be determined based on y_m, y_u, p_m, and p_u as shown below.

Segment 1 starts with a straight line with an initial slope of kx, where x is the depth from the ground surface to the point where the p–y curve is calculated. k is a parameter to be determined based on relative density and is different whether above or below water table. Representative values of k are shown in Table 32.13.

TABLE 32.13 Friction Angle and Consistency

Relative to Water Table	Friction Angle and Consistency 29°–30° (Loose)	30°–36° (Medium Dense)	36°–40° (Dense)
Above	20 pci	60 pci	125 pci
Below	25 pci	90 pci	225 pci

1 pci = 272 kPa/m.

Segment 2 is parabolic and starts from end of Segment 1 at

$$y_k = \left[\frac{p_m / y_m}{(kx)^n}\right]^{1/(n-1)}$$

and $p_k = (kx)y_k$, the power of the parabolic

$$n = \frac{y_m}{p_m}\left(\frac{p_u - p_m}{y_u - y_m}\right)$$

Segments 3 and 4 are straight lines. y_m, y_u, p_m, and p_u are expressed as

$$y_m = \frac{b}{60} \tag{32.37}$$

$$y_u = \frac{3b}{80} \tag{32.38}$$

$$p_m = B_s p_s \tag{32.39}$$

$$p_u = A_s p_s \tag{32.40}$$

where b is the diameter of a pile; A_s and B_s are coefficients that can be determined from Figures 32.12 and 32.13, depending on either static or cyclic loading conditions; p_s is equal to the minimum of p_{st} and p_{sd}, as

$$p_{st} = \gamma x \left[\begin{array}{l} \dfrac{K_o x \tan\varphi \sin\beta}{\tan(\beta-\varphi)\cos\alpha} + \dfrac{\tan\beta}{\tan(\beta-\varphi)}(b + x\tan\beta\tan\alpha) \\ + K_o x \tan\beta(\tan\phi\tan\varphi - \tan\alpha) - K_a b \end{array}\right] \tag{32.41}$$

$$p_{sd} = K_a bx\gamma[\tan^8\beta - 1] + K_o b\gamma x \tan\phi \tan^4\beta \tag{32.42}$$

$$p = \min(p_{st}, p_{sd}) \tag{32.43}$$

in which φ is the friction angle of soil; α is taken as $\varphi/2$; β is equal to $45^o + \varphi/2$; K_o is the coefficient of the earth pressure at rest and is usually assumed to be 0.4; and K_a is the coefficient of the active earth pressure and equals to $\tan^2(45^o - \varphi/2)$.

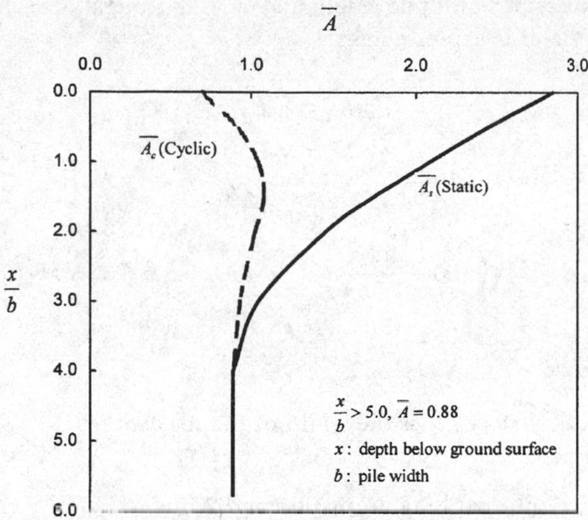

FIGURE 32.12 Variation of A_s with depth for sand. [After Reese, et al. (1974)[53]]

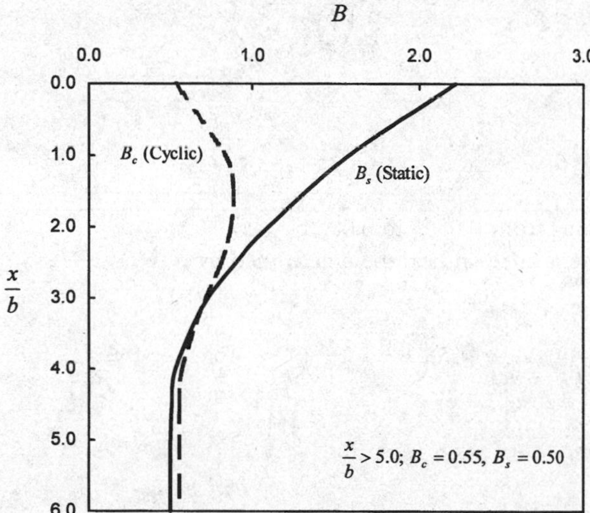

FIGURE 32.13 Variation of B_s with depth for sand. [After Reese, et al. (1974)[53]]

32.5.4 Lateral Spring: *p–y* Curves for Rock

Reese[86] proposed a procedure to calculate *p–y* curves for rock using basic rock and rock mass properties such as compressive strength of intact rock q_{ur}, rock quality designation (RQD), and initial modulus of rock E_{ir}. A description of the procedure is presented in the following.

A *p–y* curve consists of three segments:

$$\text{Segment 1: } p = K_{ir} y \qquad \text{for} \quad y \leq y_a$$

$$\text{Segment 2: } p = \frac{p_{ur}}{2}\left(\frac{y}{y_{rm}}\right)^{0.25} \qquad \text{for} \quad y_a < y < 16 y_{rm} \tag{32.44}$$

$$\text{Segment 3: } p = p_{ur} \qquad \text{for} \quad y \geq 16 y_{rm}$$

where p is the lateral force per unit pile length and y is the lateral deflection.

K_{ir} is the initial slope and is expressed as

$$K_{ir} = k_{ir}E_{ir} \tag{32.45}$$

k_{ir} is a dimensionless constant and is determined by

$$k_{ir} = \begin{cases} \left(100 + \dfrac{400x_r}{3b}\right) & \text{for} \quad 0 \le x_r \le 3b \\ 500 & \text{for} \quad x_r > 3b \end{cases} \tag{32.46}$$

x_r = depth below bedrock surface, b is the width of the rock socket
E_{ir} = initial modulus of rock.

y_a is the lateral deflection separating Segment 1 and 2, and

$$y_a = \left(\frac{p_{ur}}{2y_{rm}^{0.25}K_{ir}}\right)^{1.333} \tag{32.47}$$

where

$$y_{rm} = k_{rm}b \tag{32.48}$$

k_{rm} is a constant, ranging from 0.0005 to 0.00005.

p_{ur} is the ultimate resistance and can be determined by

$$P_{ur} = \begin{cases} a_r q_{ur} b\left(1 + 1.4\dfrac{x_r}{b}\right) & \text{for} \quad 0 \le x_r \le 3b \\ 5.2a_r q_{ur} b & \text{for} \quad x_r > 3b \end{cases} \tag{32.49}$$

where

q_{ur} = compressive strength of rock and α_r is a strength reduction factor determined by

$$\alpha_r = 1 - \frac{\text{RQD}}{150} \qquad 0 \le \text{RQD} \le 100 \tag{32.50}$$

RQD = rock quality designation for rock.

32.6 Grouped Foundations

32.6.1 General

Although a pile group is composed of a number of individual piles, the behavior of a pile group is not equivalent to the sum of all the piles as if they were separate individual piles. The behavior of a pile group is more complex than an individual pile because of the effect of the combination of piles, interactions between the piles in the group, and the effect of the pile cap. For example, stresses in soil from the loading of an individual pile will be insignificant at a certain depth below the pile tip. However, the stresses superimposed from all neighboring piles may increase the level of stress at that depth and result in considerable settlements or a bearing capacity failure, especially if there exists an underlying weak soil layer. The interaction and influence between piles usually diminish for piles spaced at approximately 7 to 8 diameters.

The axial and lateral capacity and the corresponding settlement and lateral deflection of a pile group will be discussed in the following sections.

32.6.2 Axial Capacity of Pile Group

The axial capacity of a pile group is the combination of piles in the group, with consideration of interaction between the piles. One way to account for the interaction is to use the group efficiency factor η_a, which is expressed as:

$$\eta_a = \frac{P_{\text{Group}}}{\sum_i P_{\text{Single\_Pile},i}} \tag{32.51}$$

where P_{Group} is the axial capacity of a pile group. $\sum_i P_{\text{Single\_Pile},i}$ is the sum of the axial capacity of all the individual piles. Individual piles are discussed in detail in Section 32.4. The group efficiency for axial capacity depends on many factors, such as the installation method, ground conditions, and the function of piles, which are presented in Table 32.14.

TABLE 32.14 Group Efficiency Factor for Axial Capacity

Pile Installation Method	Function	Ground Conditions	Expected Group Efficiency	Design Group Efficiency (with minimum spacing equal to 2.5 pile diameter)
Driven Pile	End bearing	Sand	1.0	1.0
	Side friction	Loose to medium dense sand	>1.0, up to 2.0	1.0, or increase with load test
	Side friction	Dense sand	May be ≧ 1.0	1.0
Drilled shaft	All	Sand	<1.0	0.67–1.0
Driven pile and drilled shaft	Side friction	Soft to medium stiff clay	<1.0	0.67–1.0
	End bearing	Soft to medium stiff clay	<1.0	0.67–1.0
	Side friction	Stiff clay	1.0	1.0
	End bearing	Stiff clay	1.0	1.0
	Side friction	Clay	<1.0	Also use "Group Block"
	End bearing	Clay, or underlying clay layers	<1.0	Also use "Group Block"

At close spacings, driven piles in loose to medium dense sand may densify the sand and consequently increase the lateral stresses and frictions along the piles. However, driven piles in dense sand may cause dilation of the sand and consequently cause heave and damage to other piles. The influence of spacing to the end bearing for sand is usually limited and the group efficiency factor η_a is taken as 1.0, under normal conditions.

For drilled piers in loose to medium dense sand, no densification of sand is made. The group efficiency factor η_a is usually less than 1.0 because of the influence of other close piles.

For driven piles in stiff to very stiff clay, the piles in a pile group tend to form a "group block" that behaves like a giant, short pile. The size of the group block is the extent of soil enclosed by the piles, including the perimeter piles as shown on Figure 32.14. The group efficiency factor η_a is usually equal to 1.0. For piles in soft to medium stiff clay, the group efficiency factor η_a is usually less than 1.0 because the shear stress levels are increased by loading from adjacent piles.

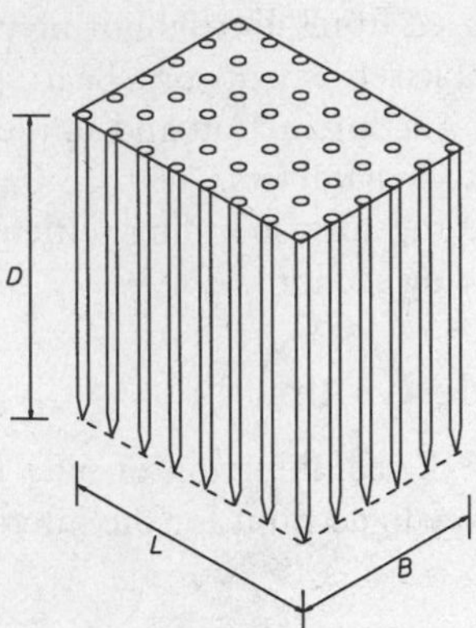

FIGURE 32.14 Block failure model for pile group in clay.

The group block method is also often used to check the bearing capacity of a pile group. The group block is treated as a large deep spread footing foundation and the assumed bottom level of the footing is different depending on whether the pile is end bearing or frictional. For end-bearing piles, the capacity of the group block is examined by assuming the bottom of the footing is at the tip of the piles. For frictional piles, the capacity of the group pile is checked by assuming that the bottom of the footing is located at ⅓ of the total embedded length above the tip. The bearing capacity of the underlying weaker layers is then estimated by using methods discussed in Chapter 31. The smaller capacity, by using the group efficiency approach, the group block approach, and the group block approach with underlying weaker layers, is selected as the capacity of the pile group.

32.6.3 Settlement of a Pile Group

The superimposed stresses from neighboring piles will raise the stress level below the tip of a pile substantially, whereas the stress level is much smaller for an individual pile. The raised stress level has two effects on the settlement of a pile group. The magnitude of the settlement will be larger for a pile group and the influence zone of a pile group will be much greater. The settlement of a pile group will be much larger in the presence of underlying highly compressible layers that would not be stressed under the loading of an individual pile.

The group block method is often used to estimate the settlement of a group. The pile group is simplified to an equivalent massive spread footing foundation except that the bottom of the footing is much deeper. The plane dimensions of the equivalent footing are outlined by the perimeter piles of the pile group. The method to calculate settlement of spread footings is discussed in Chapter 31. The assumed bottom level of the footing block is different depending on either end bearing or frictional piles. For end-bearing piles, the bottom of the footing is at the tip of the piles. For frictional piles, the bottom of the footing is located at ⅓ of total embedded length above the tip. In many cases, settlement requirement also is an important factor in the design of a pile group.

Vesic [79] introduced a method to calculate settlement of a pile group in sand which is expressed as

$$S_g = S_s\sqrt{\frac{B_g}{B_s}} \tag{32.52}$$

where

S_g = the settlement of a pile group
S_s = the settlement of an individual pile

B_g = the smallest dimension of the group block
B_s = the diameter of an individual pile

32.6.4 Lateral Capacity and Deflection of a Pile Group

The behavior of a pile group under lateral loading is not well defined. As discussed in the sections above, the lateral moment capacity is greater than the sum of all the piles in a group because piles would form couples resulting from their axial resistance through the action of the pile cap. However, the capacity of a pile group to resist lateral loads is usually smaller than the sum of separate, individual piles because of the interaction between piles.

The approach used by the University of Texas at Austin (Reese, O'Neil, and co-workers) provides a comprehensive and practical method to analyze a pile group under lateral loading. The finite-difference method is used to model the structural behavior of the foundation elements. Piles are connected through a rigid pile cap. Deformations of all the piles, in axial and lateral directions, and force and moment equilibrium are established. The reactions of soil are represented by a series of localized nonlinear axial and lateral springs. The theory and procedures to calculate axial and lateral capacity of individual piles are discussed in detail in Sections 32.4 and 32.5. A computer program is usually required to analyze a pile group because of the complexity and iteration procedure involving nonlinear soil springs.

The interaction of piles is represented by the lateral group efficiency factors, which is multiplied to the *p–y* curves for individual piles to reduce the lateral soil resistance and stiffness. Dunnavant and O'Neil [16] proposed a procedure to calculate the lateral group factors. For a particular pile *i*, the group factor is the product of influence factors from all neighboring piles *j*, as

$$\beta_i = \beta_0 \prod_{\substack{j=1 \\ j\neq i}}^{n} \beta_{ij} \tag{32.53}$$

where β_i is the group factor for pile *i*, β_0 is a total reduction factor and equals 0.85, β_{ij} is the influence factor from a neighboring pile *j*, and *n* is the total number of piles. Depending on the location of the piles *i* and *j* in relation to the direction of loading, β_{ij} is calculated as follows:

i is leading, or directly ahead of *j* (θ = 0°) $\beta_l = \beta_{ij} = 0.69 + 0.5\log_{10}\left(\frac{S_{ij}}{B}\right) \leq 1$ (32.54)

i is trailing, or directly behead of *j* (θ = 180°) $\beta_t = \beta_{ij} = 0.48 + 0.6\log_{10}\left(\frac{S_{ij}}{B}\right) \leq 1$ (32.55)

i and *j* are abreast, or side-by-side (θ = 90°) $\beta_s = \beta_{ij} = 0.78 + 0.36\log_{10}\left(\frac{S_{ij}}{B}\right) \leq 1$ (32.56)

where S_{ij} is the center-to-center distance between *i* and *j*, *B* is the diameter of the piles *i* and *j*, and θ is the angle between the loading direction and the connection vector from *i* to *j*. When the piles *i* and *j* are at other angles to the direction of loading, β_{ij} is computed by interpolation, as

$$0° < \theta < 90° \qquad \beta_{\theta 1} = \beta_{ij} = \beta_l + (\beta_s - \beta_l)\frac{\theta}{90} \tag{32.57}$$

$$90° < \theta < 180° \quad \beta_{\theta 2} = \beta_{ij} = \beta_t + (\beta_s - \beta_t)\frac{\theta - 90}{90} \tag{32.58}$$

In cases that the diameters of the piles *i* and *j* are different, we propose to use the diameter of pile *j*. To avoid an abrupt change of β_0 from 0.85 to 1.0, we propose to use:

$$\beta_o = \begin{cases} 0.85 & \text{for} \quad \dfrac{S_{ij}}{B_j} \le 3 \\ 0.85 + 0.0375\left(\dfrac{S_{ij}}{B_j} - 3\right) & \text{for} \quad 3 < \dfrac{S_{ij}}{B_j} < 7 \\ 1.0 & \text{for} \quad \dfrac{S_{ij}}{B_j} \ge 7 \end{cases}$$

32.7 Seismic Design

Seismic design of deep bridge foundations is a broad issue. Design procedures and emphases vary with different types of foundations. Since pile groups, including driven piles and drilled cast-in-place shafts, are the most popular types of deep bridge foundations, following discussion will concentrate on the design issues for pile group foundations only.

In most circumstances, seismic design of pile groups is performed to satisfy one or more of the following objectives:

- Determine the capacity and deflection of the foundation under the action of the seismic lateral load;
- Provide the foundation stiffness parameters for dynamic analysis of the overall bridge structures; and
- Ensure integrity of the pile group against liquefaction and slope instability induced ground movement.

32.7.1 Seismic Lateral Capacity Design of Pile Groups

In current practice, seismic lateral capacity design of pile groups is often taken as the same as conventional lateral capacity design (see Section 32.5). The seismic lateral force and the seismic moment from the upper structure are first evaluated for each pile group foundation based on the tributary mass of the bridge structure above the foundation level, the location of the center of gravity, and the intensity of the ground surface acceleration. The seismic force and moment are then applied on the pile cap as if they were static forces, and the deflections of the piles and the maximum stresses in each pile are calculated and checked against the allowable design values. Since seismic forces are of transient nature, the factor of safety required for resistance of seismic load can be less than those required for static load. For example, in the Caltrans specification, it is stipulated that the design seismic capacity can be 33% higher than the static capacity [9].

It should be noted that in essence the above procedure is pseudostatic, only the seismic forces from the upper structure are considered, and the effect of seismic ground motion on the behavior of pile group is ignored. The response of a pile group during an earthquake is different from its response to a static lateral loading. As seismic waves pass through the soil layers and cause the soil layers to move laterally, the piles are forced to move along with the surrounding media. Except for the case of very short piles, the pile cap and the pile tip at any moment may move in different directions. This movement induces additional bending moments and stresses in the piles.

Depending on the intensity of the seismic ground motion and the characteristics of the soil strata, this effect can be more critical to the structural integrity of the pile than the lateral load from the upper structure.

Field measurements (e.g., Tazoh et al. [70]), post-earthquake investigation (e.g., Seismic Advisory Committee, [65]), and laboratory model tests (e.g., Nomura et al., [43]) all confirm that seismic ground movements dictate the maximum responses of the piles. The more critical situation is when the soil profile consists of soft layer(s) sandwiched by stiff layers, and the modulus contrast among the layers is large. In this case, local seismic moments and stresses in the pile section close to the soft layer/hard layer interface may very well be much higher than the moments and stresses caused by the lateral seismic loads from the upper structure. If the site investigation reveals that the underground soil profile is of this type and the bridge is of critical importance, it is desirable that a comprehensive dynamic analysis be performed using one of more sophisticated computer programs capable of modeling the dynamic interaction between the soil and the pile system, e.g., SASSI [33]. Results of such dynamic analysis can provide a better understanding of the seismic responses of a pile group.

32.7.2 Determination of Pile Group Spring Constants

An important aspect in bridge seismic design is to determine, through dynamic analysis, the magnitude and distribution of seismic forces and moments in the bridge structure. To accomplish this goal, the characteristics of the bridge foundation must be considered appropriately in an analytical model.

At the current design practice, the force–displacement relationships of a pile foundation are commonly simplified in an analytical model as a stiffness matrix, or a set of translational and rotational springs. The characteristics of the springs depend on the stiffness at pile head for individual piles and the geometric configuration of piles in the group. For a pile group consisting of vertical piles, the spring constants can be determined by the following steps:

- The vertical and lateral stiffnesses at the pile head of a single pile, K_{vv} and K_{hh}, are first evaluated based on the pile geometry and the soil profile. These values are determined by calculating the displacement at the pile head corresponding to a unit force. For many bridge foundations, a rigid pile cap can be assumed. Design charts are available for uniform soil profiles (e.g., NAVFAC [42]). For most practical soil profiles, however, it is convenient to use computer programs, such as APILE [18] and LPILE [17], to determine the single pile stiffness values. It should be noted that the force–deformation behavior of a pile is highly nonlinear. In evaluating the stiffness values, it is desirable to use the secant modulus in the calculated pile-head force–displacement relationship compatible to the level of pile-head displacement to be developed in the foundation. This is often an iterative process.

  In calculating the lateral stiffness values, it is common practice to introduce a group factor η, $\eta \leq 1.0$, to account for the effect of the other piles in the same group. The group factor depends on the relative spacing *S/D* in the pile group, where *S* is the spacing between two piles and *D* is the diameter of the individual pile. There are studies reported in the literature about the dynamic group factors for pile groups of different configurations. However, in the current design practice, static group factors are used in calculation of the spring constants. Two different approaches exist in determining the group factor: one is based on reduction of the subgrade reaction moduli; the other is based on the measurement of plastic deformation of the pile group. Since the foundation deformations in the analysis cases involving the spring constants are mostly in the small-strain range, the group factors based on subgrade reaction reduction should be used (e.g., NAVFAC [42]).
- The spring constants of the pile group can be calculated using the following formulas:

$$K_{G,x} = \sum_{i=1}^{N} K_{hh,i} \tag{32.59}$$

$$K_{G,y} = \sum_{i=1}^{N} K_{hh,i} \tag{32.60}$$

$$K_{G,z} = \sum_{i=1}^{N} K_{vv,i} \tag{32.61}$$

$$K_{G,yy} = \sum_{i=1}^{N} K_{vv,i} \cdot x_i^2 \tag{32.62}$$

$$K_{G,xx} = \sum_{i=1}^{N} K_{vv,i} \cdot y_i^2 \tag{32.63}$$

where $K_{G,x}$, $K_{G,y}$, $K_{G,z}$ are the group translational spring constants, $K_{G,yy}$, $K_{G,xx}$ are the group rotational spring constants with respect to the center of the pile cap. All springs are calculated at the center of the pile cap; $K_{vv,i}$ and $K_{hh,i}$ are the lateral and vertical stiffness values at pile head of the *i*th pile; x_i, y_i are the coordinates of the *i*th pile in the group; and N is the total number of piles in the group.

In the above formulas, the bending stiffness of a single pile at the pile top and the off-diagonal stiffness terms are ignored. For most bridge pile foundations, these ignored items have only minor significance. Reasonable results can be obtained using the above simplified formulas.

It should be emphasized that the behavior of the soil–pile system is greatly simplified in the concept of "spring constant." The responses of a soil–pile structure system are complicated and highly nonlinear, frequency dependent, and are affected by the inertia/stiffness distribution of the structure above ground. Therefore, for critical structures, it is advisable that analytical models including the entire soil–pile structure system should be used in the design analysis.

32.7.3 Design of Pile Foundations against Soil Liquefaction

Liquefaction of loose soil layers during an earthquake poses a serious hazard to pile group foundations. Field observations and experimental studies (e.g., Nomura et al. [43], Miyamoto et al. [40], Tazoh and Gazetas [69], Boulanger et al. [4]) indicate that soil liquefaction during an earthquake has significant impacts on the behavior of pile groups and superstructures. The impacts are largely affected by the intensity of liquefaction-inducing earthquakes and the relative locations of the liquefiable loose soil layers. If a loose layer is close to the ground surface and the earthquake intensity is moderate, the major effect of liquefaction of the loose layer is to increase the fundamental period of the foundation–structure system, causing significant lateral deflection of the pile group and superstructure. For high-intensity earthquakes, and especially if the loose soil layer is sandwiched in hard soil layers, liquefaction of the loose layer often causes cracking and breakage of the piles and complete loss of capacity of the foundation, thus the collapse of the superstructure.

There are several approaches proposed in the literature for calculation of the dynamic responses of a pile or a pile group in a liquefied soil deposit. In current engineering practice, however, more emphasis is on taking proper countermeasures to mitigate the adverse effect of the liquefaction hazard. These mitigation methods include

- Densify the loose, liquefiable soil layer. A stone column is often satisfactory if the loose layer is mostly sand. Other approaches, such as jet grouting, deep soil mixing with cementing agents, and *in situ* vibratory densification, can all be used. If the liquefiable soil layer is close to the ground surface, a complete excavation and replacement with compacted engineering fill is sometimes also feasible.
- Isolate the pile group from the surrounding soil layers. This is often accomplished by installing some types of isolation structures, such as sheet piles, diaphragm walls, soil-mixing piles, etc., around the foundation to form an enclosure. In essence, this approach creates a huge block surrounding the piles with increased lateral stiffness and resistance to shear deformation while limiting the lateral movement of the soil close to the piles.
- Increase the number and dimension of the piles in a foundation and therefore increase the lateral resistance to withstand the forces induced by liquefied soil layers. An example is 10 ft (3.3 m) diameter cast-in-steel shell piles used in bridge seismic retrofit projects in the San Francisco Bay Area following the 1989 Loma Prieta earthquake.

References

1. API, *API Recommended Practice for Planning, Designing and Constructing Fixed Offshore Platforms*, 15th ed., API RP2A, American Petroleum Institute, 115 pp, 1984.
2. Awoshika, K. and L. C. Reese, Analysis of Foundation with Widely-Spaced Batter Piles, Research Report 117-3F, Center for Highway Research, The University of Texas at Austin, February, 1971.
3. Berezantzev, V. G., V. S. Khristoforov, and V. N. Golubkov, Load bearing capacity and deformation of piled foundations, *Proc. 5th Int. Conf. Soil Mech.*, Paris, 2, 11–15, 1961.
4. Boulanger, R. W., D. W. Wilson, B. L. Kutter, and A. Abghari, Soil–pile-structure interaction in liquefiable sand, *Transp. Res. Rec.*, 1569, April, 1997.
5. Broms, B. B., Lateral resistance of piles in cohesive soils, *Proc. ASCE, J. Soil Mech. Found. Eng. Div.*, 90(SM2), 27–64, 1964.
6. Broms, B. B., Lateral resistance of piles in cohesionless soils, *Proc. ASCE J. Soil Mech. Found. Eng. Div.*, 90(SM3), 123–156, 1964.
7. Burland, J. B., Shaft friction of piles in clay — a simple fundamental approach, *Ground Eng.*, 6(3), 30–42, 1973.
8. Bustamente, M. and L. Gianeselli, Pile bearing capacity prediction by means of static penetrometer CPT, *Proc. of Second European Symposium on Penetration Testing* (ESOPT II), Vol. 2, A. A. Balkema, Amsterdam, 493–500, 1982.
9. Caltrans, Bridge Design Specifications, California Department of Transportation, Sacramento, 1990.
10. CGS, *Canadian Foundation Engineering Manual*, 3rd ed., Canadian Geotechnical Society, BiTech Publishers, Vancouver, 512 pp, 1992.
11. Carter, J. P. and F. H. Kulhawy, Analysis and Design of Drilled Shaft Foundations Socketed into Rock, EPRI Report EI-5918, Electric Power Research Institute, Palo Alto, CA, 1988.
12. Crapps, D. K., Design, construction and inspection of drilled shafts in limerock and limestone, paper presented at the Annual Meeting of the Florida Section of ASCE, 1986.
13. De Ruiter, J. and F. L., Beringen, Pile foundations for large North Sea structures, *Marine Geotechnol.*, 3(2), 1978.
14. Dennis, N. D., Development of Correlations to Improve the Prediction of Axial Pile Capacity, Ph.D. dissertation, University of Texas at Austin, 1982.
15. Desai, C. S. and J. T. Christian, *Numerical Methods in Geotechnical Engineering*, McGraw-Hill Book, New York, 1977.

16. Dunnavant, T. W. and M. W. O'Neil, Evaluation of design-oriented methods for analysis of vertical pile groups subjected to lateral load, *Numerical Methods in Offshore Piling*, Institut Francais du Petrole, Labortoire Central des Ponts et Chausses, 303–316, 1986.
17. Ensoft Inc., Lpile Plus for Windows. Version 3.0. A Computer Program for Analysis of Laterally Loaded Piles, Austin, TX, 1997.
18. Ensoft Inc., Apile Plus. Version 3.0, Austin, TX, 1998.
19. Fellenius, B. H., in an ASCE meeting in Boston as quoted by R. E. Olson in 1991, Capacity of Individual Piles in Clay, internal report, 1986.
20. Fellenius, B. H., The critical depth — how it came into being and why it does not exist, *Proc. of the Inst. Civil Eng., Geotech. Eng.*, 108(1), 1994.
21. Focht, J. A. and K. J. Koch, Rational analysis of the lateral performance of offshore pile groups, in *Proceedings, Fifth Offshore Technology Conference*, Vol. 2, Houston, 701–708, 1973.
22. Geordiadis, M., Development of *p–y* curves for layered soils, in *Proceedings, Geotechnical Practice in Offshore Engineering*, ASCE, April, 1983, 536–545.
23. Goudreault, P. A. and B. H. Fellenius, A Program for the Design of Piles and Piles Groups Considering Capacity, Settlement, and Dragload Due to Negative Skin Friction, 1994.
24. Gupton, C. and T. Logan, Design Guidelines for Drilled Shafts in Weak Rocks in South Florida, Preprint, Annual Meeting of South Florida Branch of ASCE, 1984.
25. Horvath, R. G. and T. C. Kenney, Shaft resistance of rock-socketed drilled piers, in *Symposium on Deep Foundations*, ASCE National Convention, Atlanta, GA, 1979, 182–214.
26. Janbu, N., Static bearing capacity of friction piles, in *Proc. 6th European Conference on Soil Mech. & Found. Eng.*, Vol. 1.2, 1976, 479–488.
27. Kishida, H., Ultimate bearing capacity of piles driven into loose sand, *Soil Foundation*, 7(3), 20–29, 1967.
28. Kraft L. M., J. A. Focht, and S. F. Amerasinghe, Friction capacity of piles driven into clay, *J. Geot. Eng. Div. ASCE*, 107(GT 11), 1521–1541, 1981.
29. Kubo, K., Experimental Study of Behavior of Laterally Loaded Piles, Report, Transportation Technology Research Institute, Vol. 12, No. 2, 1962.
30. Kulhawy, F. H., Transmission Line Structures Foundations for Uplift-Compression Loading, Report No. EL-2870, Report to the Electrical Power Research Institute, Geotechnical Group, Cornell University, Ithaca, NY, 1983.
31. Kulhawy, F. H., Limiting tip and side resistance: fact or fallacy? in *Proc. of the American Society of Civil Engineers, ASCE, Symposium on Analysis and Design of Pile Foundations*, R. J. Meyer, Ed., San Francisco, 1984, 80–89.
32. Kulhawy, F. H. and K. K. Phoon, Drilled shaft side resistance in clay soil or rock, in *Geotechnical Special Publication No. 38, Design and Performance of Deep Foundations: Piles and Piers in Soil to Soft Rock*, Ed. P. P. Nelson, T. D. Smith, and E. C. Clukey, Eds., ASCE, 172–183, 1993.
33. Lysmer, J., M. Tabatabaie-Raissi, F. Tajirian, S. Vahdani, and F. Ostadan, SASSI — A System for Analysis of Soil-Structure Interaction, Report No. UCB/GT/81-02, Department of Civil Engineering, University of California, Berkeley, April, 1981.
34. McVay, M. C., F. C. Townsend, and R. C. Williams, Design of socketed drilled shafts in limestone, *J. Geotech. Eng.*, 118-GT10, 1626–1637, 1992.
35. Matlock, H., Correlations for design of laterally-loaded piles in soft clay, Paper No. OTC 1204, *Proc. 2nd Annual Offshore Tech. Conf.*, Vol. 1, Houston, TX, 1970, 577–594.
36. Menard, L. F., Interpretation and application of pressuremeter test results, *Sols-Soils*, Paris, 26, 1–23, 1975.
37. Meyerhof, G. G., Penetration tests and bearing capacity of cohesionless soils, *J. Soil Mech. Found. Div. ASCE*, 82(SM1), 1–19, 1956.
38. Meyerhof, G. G., Bearing capacity and settlement of pile foundations, *J. Geotech. Eng. Div. ASCE*, 102(GT3), 195–228, 1976.

39. Mitchell, J. K. and T. A. Lunne, Cone resistance as a measure of sand strength, *Proc. ASCE J. Geotech. Eng. Div.*, 104(GT7), 995–1012, 1978.
40. Miyamoto, Y., Y. Sako, K. Miura, R. F. Scott, and B. Hushmand, Dynamic behavior of pile group in liquefied sand deposit, *Proceedings, 10th World Conference on Earthquake Engineering*, 1992, 1749–1754.
41. Mosher, R. L., Load Transfer Criteria for Numerical Analysis of Axially Loaded Piles in Sand, U.S. Army Engineering Waterways Experimental Station, Automatic Data Processing Center, Vicksburg, MI, January, 1984.
42. NAVFAC, Design Manual DM7.02: Foundations and Earth Structures, Department of the Navy, Naval Facilities Engineering Command, Alexandra, VA, September, 1986.
43. Nomura, S., K. Tokimatsu, and Y. Shamoto, Behavior of soil-pile-structure system during liquefaction, in *Proceedings, 8th Japanese Conference on Earthquake Engineering*, Tokyo, December 12–14, Vol. 2, 1990, 1185–1190.
44. O'Neil M. W. and L. C. Reese, Load transfer in a slender drilled pier in sand, ASCE, ASCE Spring Convention and Exposition, Pittsburgh, PA, Preprint 3141, April, 1978, 30 pp.
45. O'Neil, M. W. and S. A. Sheikh, Geotechnical behavior of underrems in pleistocene clay, in *Drilled Piers and Caissons, II*, C. N. Baker, Jr., Ed., ASCE, May, 57–75, 1985.
46. O'Neil, M. W., F. C. Townsend, K. M. Hassan, A. Buller, and P. S. Chan, Load Transfer for Drilled Shafts in Intermediate Geo'materials, FHWA-RD-95-172, November, 184 pp, 1996.
47. Osterberg, J. O., New load cell testing device, in *Proc. 14th Annual Conf.*, Vol. 1, Deep Foundations Institute, 1989, 17–28.
48. Pells, P. J. N. and R. M. Turner, Elastic solutions for the design and analysis of rock-socketed piles, *Can. Geotech. J.*, 16(3), 481–487, 1979.
49. Pells, P. J. N. and R. M. Turner, End bearing on rock with particular reference to sandstone, in Structural Foundations on Rock, *Proc. Intn. Conf. on Structural Found. on Rock*, Vol. 1, Sydney, May 7–9, 1980, 181–190.
50. Poulos, H. G. and E. H. Davis, *Pile Foundation Analysis and Design*, John Wiley & Sons, New York, 1980.
51. Reese, L. C. and H. Matlock, Behavior of a two-dimensional pile group under inclined and eccentric loading, in *Proc. Offshore Exploration Conf.*, Long Beach, CA, February, 1966.
52. Reese, L. C. and M. W. O'Neil, The analysis of three-dimensional pile foundations subjected to inclined and eccentric loads, *Proc. ASCE Conf.*, September, 1967, 245–276.
53. Reese, L. C., W. R. Cox, and F. D. Koop, Analysis of laterally loaded piles in sand, paper OTC 2080, *Proc. Fifth Offshore Tech. Conf.*, Houston, TX, 1974.
54. Reese, L. C., W. R. Cox, and F. D. Koop, Field testing and analysis of laterally loaded piles in stiff clay, paper OTC 2313, in *Proc. Seventh Offshore Tech. Conf.*, Houston, TX, 1975.
55. Reese, L. C. and S. J. Wright, Drilled Shafts: Design and Construction, Guideline Manual, Vol. 1; Construction Procedures and Design for Axial Load, U.S. Department of Transportation, Federal Highway Administration, July, 1977.
56. Reese, L. C., Behavior of Piles and Pile Groups under Lateral Load, a report submitted to the Federal Highway Administration, Washington, D.C., July, 1983, 404 pp.
57. Reese, L. C. and M. W. O'Neil, Drilled Shafts: Construction Procedures and Design Methods, U.S. Department of Transportation, Federal Highway Administration, McLean, VA, 1988.
58. Reynolds, R. T. and T. J. Kaderabek, Miami Limestone Foundation Design and Construction, Preprint No. 80-546, South Florida Convention, ASCE, 1980.
59. Robertson, P. K., R. G. Campanella, et al., Axial Capacity of Driven Piles in Deltaic Soils Using CPT, Penetration Testing 1988, ISOPT-1, De Ruite, Ed., 1988.
60. Rosenberg, P. and N. L. Journeaux, Friction and end bearing tests on bedrock for high capacity socket design, *Can. Geotech. J.*, 13(3), 324–333, 1976.
61. Rowe, R. K. and H. H. Armitage, Theoretical solutions for axial deformation of drilled shafts in rock, *Can. Geotech. J.*, Vol. 24(1), 114–125, 1987.

62. Rowe, R. K. and H. H. Armitage, A design method for drilled piers in soft rock, *Can. Geotech. J.*, 24(1), 126–142, 1987.
63. Schmertmann, J. H., Guidelines for Cone Penetration Test: Performance and Design, FHWA-TS-78-209, Federal Highway Administration, Office of Research and Development, Washington, D.C., 1978.
64. Seed, H. B. and L. C. Reese, The action of soft clay along friction piles, *Trans. Am. Soc. Civil Eng.*, Paper No. 2882, 122, 731–754, 1957.
65. Seismic Advisory Committee on Bridge Damage, Investigation Report on Highway Bridge Damage Caused by the Hyogo-ken Nanbu Earthquake, Japan Ministry of Construction, 1995.
66. Skempton, A. W., The bearing capacity of clay, *Proc. Building Research Congress*, Vol. 1, 1951, 180–189.
67. Skempton, A. W., Cast-*in situ* bored piles in London clay, *Geotechnique*, 9, 153–173, 1959.
68. Sörensen, T. and B. Hansen, Pile driving formulae — an investigation based on dimensional considerations and a statistical analysis, *Proc. 4th Int. Conf. Soil Mech.*, London, 2, 61–65, 1957.
69. Tazoh, T. and G. Gazetas, Pile foundations subjected to large ground deformations: lessons from kobe and research needs, *Proceedings, 11th World Conference on Earthquake Engineering*, Paper No. 2081, 1996.
70. Tazoh, T., K. Shimizu, and T. Wakahara, Seismic Observations and Analysis of Grouped Piles, Dynamic Response of Pile Foundations — Experiment, Analysis and Observation, ASCE Geotechnical Special Publication No. 11, 1987.
71. Terzaghi, K., *Theoretical Soil Mechanics*, John Wiley & Sons, New York, 510 pp, 1943.
72. Tomlinson, M. J., The adhesion of piles in clay soils, *Proc., Fourth Int. Conf. Soil Mech. Found. Eng.*, 2, 66–71, 1957.
73. Tomlinson, M. J., Some effects of pile driving on skin friction, in *Behavior of Piles*, Institution of Civil Engineers, London, 107–114, and response to discussion, 149–152, 1971.
74. Touma, F. T. and L. C. Reese, Load Tests of Instrumented Drilled Shafts Constructed by the Slurry Displacement Method. Research report conducted under Interagency contract 108 for the Texas Highway Department, Center for Highway Research, the University of Texas at Austin, January, 1972, 79 pp.
75. Townsend, F. C., Comparison of deep foundation load test method, in *FHWA 25th Annual Southeastern Transportation Geotechnical Engineering Conference*, Natchez, MS, October 4–8, 1993.
76. Vesic, A. S., Ultimate loads and settlements of deep foundations in sand, in *Proc. Symp. on Bearing Capacity and Settlement of Foundations*, Duke University, Durham, NC, 1967.
77. Vesic, A. S., Load transfer in pile-soil system, in *Design and Installation of Pile Foundations and Cellar Structures*, Fang and Dismuke, Eds., Envo, Lehigh, PA, 47–74, 1970.
78. Vesic, A. S., Expansion of cavities in infinite soil mass, *Proc. ASCE J. Soil Mech. Found. Eng. Div.*, 98(SM3), 1972.
79. Vesic, A. S., Design of Pile Foundations, National Cooperative Highway Research Program Synthesis 42, Transportation Research Board, 1977.
80. Vijayvergiya, V. N. and J. A. Focht, A new way to predict the capacity of piles in clay, *Offshore Technology Conference*, Vol. 2, Houston, TX, 1972, 965–874.
81. Vijayvergiya, V. N., Load-movement characteristics of piles, *Proc. of Ports '77 Conf.*, Long Beach, CA, 1977.
82. Welsh, R. C. and Reese, L. C., Laterally Loaded Behavior of Drilled Shafts, Research Report No. 3-5-65-89, conducted for Texas Highway Department and U.S. Department of Transportation, Federal Highway Administration, Bureau of Public Roads, by Center for Highway Research, the University of Texas at Austin, May, 1972.
83. Weltman, A. J. and P. R. Healy, Piling in boulder clay and other glacial tills, *Construction Industry Research and Information Association*, Report PG5, 1978.

84. Williams, A. F., I. W. Johnson, and I. B. Donald, The Design of Socketed Piles in Weak Rock, *Structural Foundations on Rock, Proc. Int. Conf. on Structural Found. on Rock,* Vol. 1, Sydney, May 7–9, 1980, 327–347.
85. Woodward, R. J., W. S. Gardner, and D. M. Greer, *Drilled Pier Foundations,* McGraw-Hill, New York, 1972.
86. Reese, L. C., Analysis of Laterally Loaded Piles in Weak Rock, *J. of Geotech & Geoenvironmental Engr.*, Vol. 123, No., 11, 1010–1017, 1997.

Section IV

Seismic Design

33
Geotechnical Earthquake Considerations

Steven Kramer
University of Washington

Charles Scawthorn
EQE International

33.1 Introduction

Earthquakes are naturally occurring broad-banded vibratory ground motions, that are due to a number of causes including tectonic ground motions, volcanism, landslides, rockbursts, and man-made explosions, the most important of which are caused by the fracture and sliding of rock along tectonic **faults** within the Earth's crust. For most earthquakes, shaking and ground failure are the dominant and most widespread agents of damage. Shaking near the actual earthquake rupture lasts only during the time when the fault ruptures, a process which takes seconds or at most a few minutes. The seismic waves generated by the rupture propagate long after the movement on the fault has stopped, however, spanning the globe in about 20 min. Typically, earthquake ground

0-8493-7434-0/00/$0.00+$.50

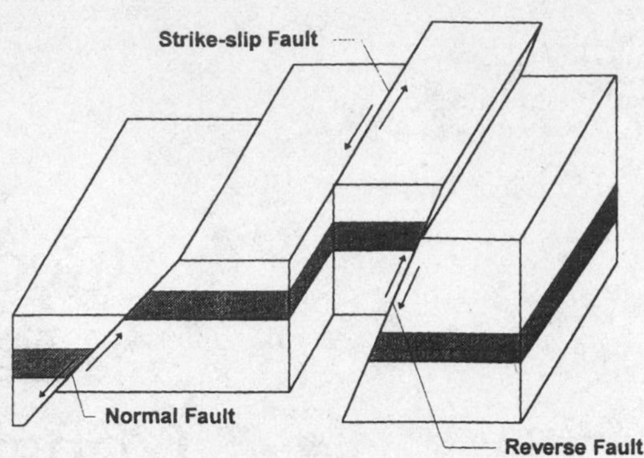

FIGURE 33.1 Fault types.

motions are powerful enough to cause damage only in the near field (i.e., within a few tens of kilometers from the causative fault) — in a few instances, long-period motions have caused significant damage at great distances, to selected lightly damped structures, such as in the 1985 Mexico City earthquake, where numerous collapses of mid- and high-rise buildings were due to a magnitude 8.1 earthquake occurring at a distance of approximately 400 km from Mexico City.

33.2 Seismology

Plate Tectonics: In a global sense, tectonic earthquakes result from motion between a number of large plates comprising the Earth's crust or lithosphere (about 15 in total). These plates are driven by the convective motion of the material in the Earth's mantle, which in turn is driven by heat generated at the Earth's core. Relative plate motion at the fault interface is constrained by friction and/or **asperities** (areas of interlocking due to protrusions in the fault surfaces). However, strain energy accumulates in the plates, eventually overcomes any resistance, and causes slip between the two sides of the fault. This sudden slip, termed **elastic rebound** by Reid [49] based on his studies of regional deformation following the 1906 San Francisco earthquake, releases large amounts of energy, which constitute the earthquake. The location of initial radiation of seismic waves (i.e., the first location of dynamic rupture) is termed the **hypocenter**, while the projection on the surface of the Earth directly above the hypocenter is termed the **epicenter**. Other terminology includes **near-field** (within one source dimension of the epicenter, where source dimension refers to the length of faulting), **far-field** (beyond near-field) and **meizoseismal** (the area of strong shaking and damage). Energy is radiated over a broad spectrum of frequencies through the Earth, in **body waves** and **surface waves** [4]. Body waves are of two types: P waves (transmitting energy via push–pull motion) and slower S waves (transmitting energy via shear action at right angles to the direction of motion). Surface waves are also of two types: horizontally oscillating **Love waves** (analogous to S body waves) and vertically oscillating **Rayleigh waves**.

Faults are typically classified according to their sense of motion, Figure 33.1. Basic terms include **transform** or **strike slip** (relative fault motion occurs in the horizontal plane, parallel to the strike of the fault), **dip-slip** (motion at right angles to the strike, up- or down-slip), **normal** (dip-slip motion, two sides in tension, move away from each other), **reverse** (dip-slip, two sides in compression, move toward each other), and **thrust** (low-angle reverse faulting).

Generally, earthquakes will be concentrated in the vicinity of faults; faults that are moving more rapidly than others will tend to have higher rates of seismicity, and larger faults are more likely than others to produce a large event. Many faults are identified on regional geologic maps, and useful information on fault location and displacement history is available from local and national geologic surveys in areas of high seismicity. An important development has been the growing recognition of **blind thrust faults**, which emerged as a result of the several earthquakes in the 1980s, none of which was accompanied by surface faulting [61].

33.3 Measurement of Earthquakes

Magnitude

An individual earthquake is a unique release of strain energy — quantification of this energy has formed the basis for measuring the earthquake event. C.F. Richter [51] was the first to define earthquake **magnitude**, as

$$M_L = \log A - \log A_o \tag{33.1}$$

where M_L is **local magnitude** (which Richter only defined for Southern California), A is the maximum trace amplitude in microns recorded on a standard Wood–Anderson short-period torsion seismometer at a site 100 km from the epicenter, and log A_o is a standard value as a function of distance, for instruments located at distances other than 100 km and less than 600 km. A number of other magnitudes have since been defined, the most important of which are **surface wave** magnitude M_S, body wave magnitude m_b, and **moment magnitude** M_W. Magnitude can be related to the total energy in the expanding wave front generated by an earthquake, and thus to the total energy release — an empirical relation by Richter is

$$\log_{10} E_s = 11.8 + 1.5 M_s \tag{33.2}$$

where E_S is the total energy in ergs. Due to the observation that deep-focus earthquakes commonly do not register measurable surface waves with periods near 20 s, a body wave magnitude m_b was defined [25], which can be related to M_S [16]:

$$m_b = 2.5 + 0.63 M_S \tag{33.3}$$

Body wave magnitudes are more commonly used in eastern North America, due to the deeper earthquakes there. More recently, **seismic moment** has been employed to define a moment magnitude M_W [26] (also denoted as bold-face **M**) which is finding increased and widespread use:

$$\text{Log } M_o = 1.5\, M_W + 16.0 \tag{33.4}$$

where seismic moment M_o (dyne-cm) is defined as [33]

$$M_o = \mu A \bar{u} \tag{33.5}$$

where μ is the material shear modulus, A is the area of fault plane rupture, and $\bar{u}$ is the mean relative displacement between the two sides of the fault (the averaged fault slip). Comparatively, M_W and M_S are numerically almost identical up to magnitude 7.5. Figure 33.2 indicates the relationship between moment magnitude and various magnitude scales.

From the foregoing discussion, it can be seen that magnitude and energy are related to fault rupture length and slip. Slemmons [60] and Bonilla et al. [5] have determined statistical relations between these parameters, for worldwide and regional data sets, aggregated and segregated by type of faulting (normal, reverse, strike-slip). Bonilla et al.'s worldwide results for all types of faults are

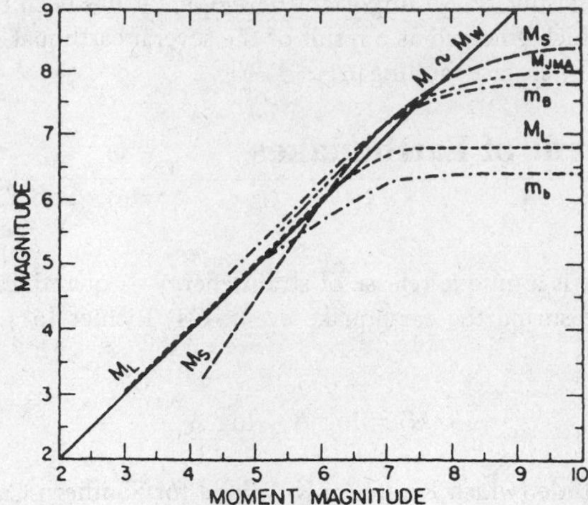

FIGURE 33.2 Relationship between moment magnitude and various magnitude scales. (*Source*: Campbell, K. W., *Earthquake Spectra*, 1(4), 759–804, 1985. With permission.)

$$M_s = 6.04 + 0.708 \log_{10} L \qquad s = 0.306 \tag{33.6}$$

$$\log_{10} L = -2.77 + 0.619 M_s \qquad s = 0.286 \tag{33.7}$$

$$M_s = 6.95 + 0.723 \log_{10} d \qquad s = 0.323 \tag{33.8}$$

$$\log_{10} d = -3.58 + 0.550\, M_s \qquad s = 0.282 \tag{33.9}$$

which indicates, for example, that, for $M_S = 7$, the average fault rupture length is about 36 km (and the average displacement is about 1.86 m). Conversely, a fault of 100 km length is capable of about an $M_S = 7.5$* event (see also Wells and Coppersmith [66] for alternative relations).

Intensity

In general, seismic intensity is a metric of the effect, or the strength, of an earthquake hazard at a specific location. While the term can be generically applied to engineering measures such as peak ground acceleration, it is usually reserved for qualitative measures of location-specific earthquake effects, based on observed human behavior and structural damage. Numerous intensity scales were developed in preinstrumental times — the most common in use today are the Modified Mercalli (MMI) [68] (Table 33.1), the Rossi–Forel (R-F), the Medvedev-Sponheur-Karnik (MSK-64, 1981), and the Japan Meteorological Agency (JMA) scales.

Time History

Sensitive strong motion seismometers have been available since the 1930s, and they record actual ground motions specific to their location, Figure 33.3. Typically, the ground motion records, termed **seismographs** or **time histories**, have recorded acceleration (these records are termed **accelerograms**), for

*Note that $L = g(M_S)$ should not be inverted to solve for $M_S = f(L)$, as a regression for $y = f(x)$ is different from a regression for $x = g(y)$.

TABLE 33.1 Modified Mercalli Intensity Scale of 1931

I	Not felt except by a very few under especially favorable circumstances
II	Felt only by a few persons at rest, especially on upper floors of buildings. Delicately suspended objects may swing.
III	Felt quite noticeably indoors, especially on upper floors of buildings, but many people do not recognize it as an earthquake; standing automobiles may rock slightly; vibration like passing truck; duration estimated
IV	During the day felt indoors by many, outdoors by few; at night some awakened; dishes, windows, and doors disturbed; walls make creaking sound; sensation like heavy truck striking building; standing automobiles rock noticeably
V	Felt by nearly everyone; many awakened; some dishes, windows, etc., broken; a few instances of cracked plaster; unstable objects overturned; disturbance of trees, poles, and other tall objects sometimes noticed; pendulum clocks may stop
VI	Felt by all; many frightened and run outdoors; some heavy furniture moved; a few instances of fallen plaster or damaged chimneys; damage slight
VII	Everybody runs outdoors; damage negligible in buildings of good design and construction, slight to moderate in well-built ordinary structures; considerable in poorly built or badly designed structures; some chimneys broken; noticed by persons driving automobiles
VIII	Damage slight in specially designed structures, considerable in ordinary substantial buildings, with partial collapse, great in poorly built structures; panel walls thrown out of frame structures; fall of chimneys, factory stacks, columns, monuments, walls; heavy furniture overturned; sand and mud ejected in small amounts; changes in well water; persons driving automobiles disturbed
IX	Damage considerable in specially designed structures; well-designed frame structures thrown out of plumb; great in substantial buildings, with partial collapse; buildings shifted off foundations; ground cracked conspicuously; underground pipes broken
X	Some well-built wooden structures destroyed; most masonry and frame structures destroyed with foundations; ground badly cracked; rails bent; landslides considerable from river banks and steep slopes; shifted sand and mud; water splashed over banks
XI	Few, if any (masonry) structures remain standing; bridges destroyed; broad fissures in ground; underground pipelines completely out of service; earth slumps and land slips in soft ground; rails bent greatly
XII	Damage total; waves seen on ground surfaces; lines of sight and level distorted; objects thrown upward into the air

After Wood and Neumann [68].

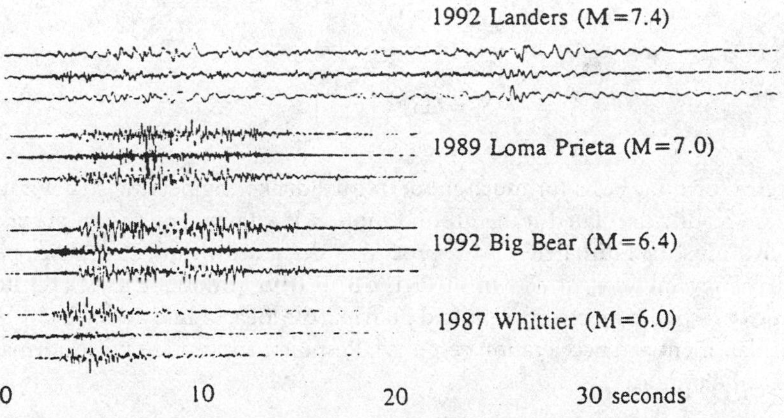

FIGURE 33.3 Typical earthquake accelerograms. (Courtesy of Darragh et al., 1994.)

many years in analog form on photographic film and, more recently, digitally. Analog records required considerable effort for correction due to instrumental drift, before they could be used.

Time histories theoretically contain complete information about the motion at the instrumental location, recording three *traces* or orthogonal records (two horizontal and one vertical). Time histories (i.e., the earthquake motion at the site) can differ dramatically in duration, frequency, content, and amplitude. The maximum amplitude of recorded acceleration is termed the **peak ground acceleration**, PGA (also termed the ZPA, or **zero period acceleration**); peak ground velocity

(PGV) and peak ground displacement (PGD) are the maximum respective amplitudes of velocity and displacement. Acceleration is normally recorded, with velocity and displacement being determined by integration; however, velocity and displacement meters are deployed to a lesser extent. Acceleration can be expressed in units of cm/s^2 (termed *gals*), but is often also expressed in terms of the fraction or percent of the acceleration of gravity (980.66 gals, termed 1 *g*). Velocity is expressed in cm/s (termed *kine*). Recent earthquakes — 1994 Northridge, M_W 6.7 and 1995 Hanshin (Kobe) M_W 6.9 — have recorded PGAs of about 0.8 *g* and PGVs of about 100 kine, while almost 2 g was recorded in the 1992 Cape Mendocino earthquake.

Elastic Response Spectra

If a single-degree-of-freedom (SDOF) mass is subjected to a time history of ground (i.e., base) motion similar to that shown in Figure 33.3, the mass or elastic **structural response** can be readily calculated as a function of time, generating a **structural response time history**, as shown in Figure 33.4 for several oscillators with differing natural periods. The response time history can be calculated by direct integration of Eq. (33.1) in the **time domain**, or by solution of the **Duhamel** integral. However, this is time-consuming, and the elastic response is more typically calculated in the **frequency domain** [12].

For design purposes, it is often sufficient to know only the maximum amplitude of the response time history. If the natural period of the SDOF is varied across a spectrum of engineering interest (typically, for natural periods from 0.03 to 3 or more seconds, or frequencies of 0.3 to 30+ Hz), then the plot of these maximum amplitudes is termed a **response spectrum**. Figure 33.4 illustrates this process, resulting in S_d, the *displacement response spectrum*, while Figure 33.5 shows (a) the S_d, displacement response spectrum, (b) S_v, the *velocity response spectrum* (also denoted PSV, the pseudo-spectral velocity, "pseudo" to emphasize that this spectrum is not exactly the same as the relative velocity response spectrum), and (c) S_a, the *acceleration response spectrum*. Note that

$$S_v = \frac{2\pi}{T} S_d = \varpi S_d \tag{33.10}$$

and

$$S_a = \frac{2\pi}{T} S_v = \varpi S_v = \left(\frac{2\pi}{T}\right)^2 S_d = \varpi^2 S_d \tag{33.11}$$

Response spectra form the basis for much modern earthquake engineering structural analysis and design. They are readily calculated *if* the ground motion is known. For design purposes, however, response spectra must be estimated — this process is discussed below. Response spectra may be plotted in any of several ways, as shown in Figure 33.5 with arithmetic axes, and in Figure 33.6, where the velocity response spectrum is plotted on tripartite logarithmic axes, which equally enables reading of displacement and acceleration response. Response spectra are most normally presented for 5% of critical **damping**.

Inelastic Response Spectra

While the foregoing discussion has been for elastic response spectra, most structures are not expected, or even designed, to remain elastic under strong ground motions. Rather, structures are expected to enter the *inelastic* region — the extent to which they behave inelastically can be defined by the **ductility factor**, μ:

$$\mu = \frac{u_m}{u_y} \tag{33.12}$$

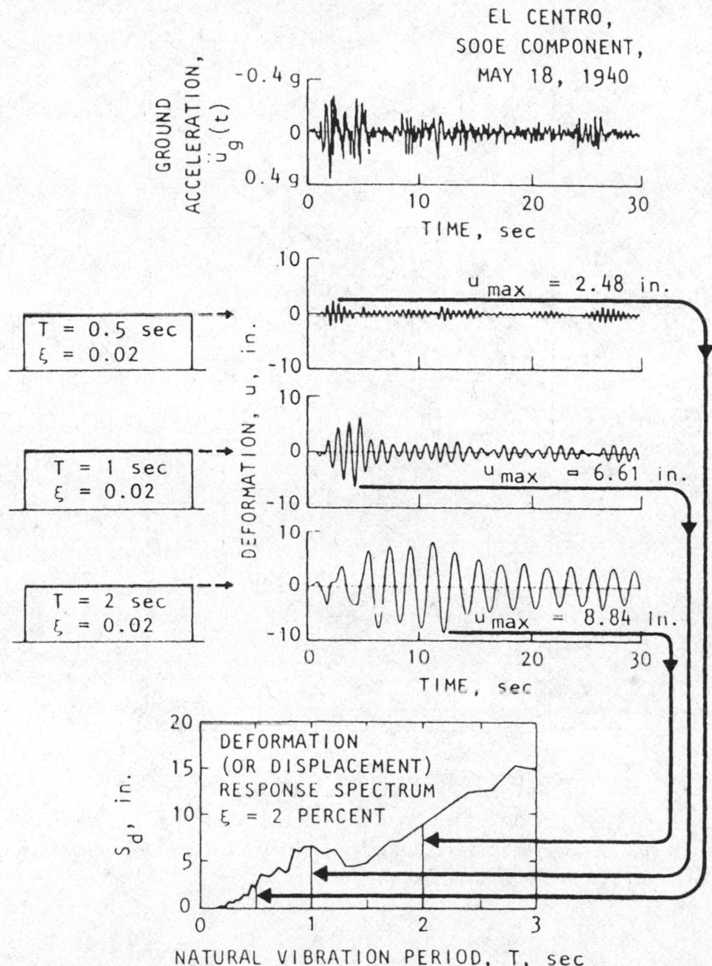

FIGURE 33.4 Computation of deformation (or displacement) response spectrum. (*Source:* Chopra, A. K., *Dynamics of Structures, A Primer,* Earthquake Engineering Research Institute, Oakland, CA, 1981. With permission.)

where u_m is the actual displacement of the mass under actual ground motions, and u_y is the displacement at yield (i.e., that displacement which defines the extreme of elastic behavior). Inelastic response spectra can be calculated in the time domain by direct integration, analogous to elastic response spectra but with the structural stiffness as a nonlinear function of displacement, $k = k(u)$. If elastoplastic behavior is assumed, then elastic response spectra can be readily modified to reflect inelastic behavior, on the basis that (1) at low frequencies (<0.3 Hz) displacements are the same, (2) at high frequencies (>33 Hz), accelerations are equal, and (3) at intermediate frequencies, the absorbed energy is preserved. Actual construction of inelastic response spectra on this basis is shown in Figure 33.9, where $DVAA_o$ is the elastic spectrum, which is reduced to D' and V' by the ratio of $1/\mu$ for frequencies less than 2 Hz, and by the ratio of $1/(2\mu - 1)^{1/2}$ between 2 and 8 Hz. Above 33 Hz, there is no reduction. The result is the inelastic acceleration spectrum ($D'V'A'A_o$), while $A''A_o'$ is the inelastic displacement spectrum. A specific example, for ZPA = 0.16 *g*, damping = 5% of critical and $\mu = 3$ is shown in Figure 33.10.

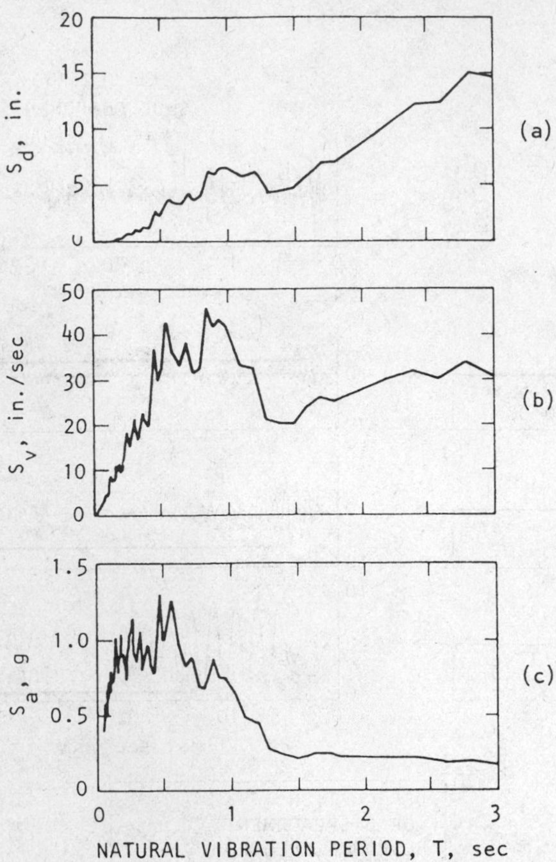

FIGURE 33.5 Response spectra. (*Source*: Chopra, A. K., *Dynamics of Structures, A Primer*, Earthquake Engineering Research Institute, Oakland, CA, 1981. With permission.)

33.4 Strong Motion Attenuation and Duration

The rate at which earthquake ground motion decreases with distance, termed **attenuation**, is a function of the regional geology and inherent characteristics of the earthquake and its source. Campbell [10] offers an excellent review of North American relations up to 1985. Initial relationships were for PGA, but regression of the amplitudes of response spectra at various periods is now common, including consideration of fault type and effects of soil. A currently favored relationship is

Campbell and Bozorgnia [11] (PGA — Worldwide Data)

$$\begin{aligned}\ln(\text{PGA}) = & -3.512 + 0.904M - 1.328\ln\sqrt{\left\{R_s^2 + \left[0.149\exp(0.647M)\right]^2\right\}} \\ & + \left[1.125 - 0.112\ln\left(R_s\right) - 0.0957M\right]F \\ & + \left[0.440 - 0.171\ln\left(R_s\right)\right]S_{sr} + \left[0.405 - 0.222\ln\left(R_s\right)\right]S_{hr} + \varepsilon\end{aligned} \qquad (33.13$$

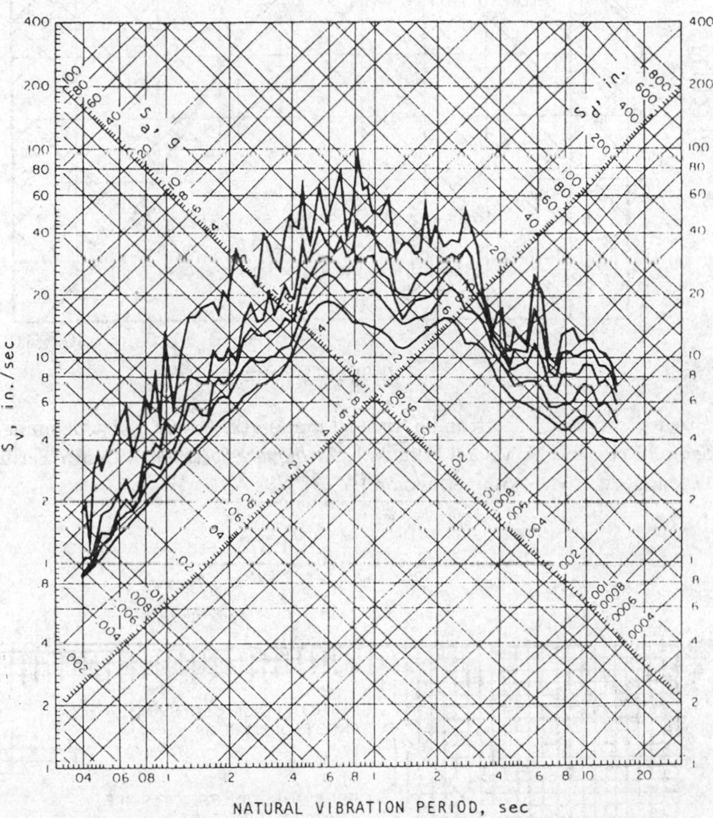

FIGURE 33.6 Response spectra, tripartite plot (El Centro S 0° E component). (*Source*: Chopra, A. K., *Dynamics of Structures, A Primer*, Earthquake Engineering Research Institute, Oakland, CA, 1981. With permission.)

where

PGA = the geometric mean of the two horizontal components of peak ground acceleration (g)

M = moment magnitude (M_w)

R_s = the closest distance to seismogenic rupture on the fault (km)

F = 0 for strike-slip and normal faulting earthquakes, and 1 for reverse, reverse-oblique, and thrust faulting earthquakes

S_{sr} = 1 for soft-rock sites

S_{hr} = 1 for hard-rock sites

$S_{sr} = S_{hr} =$ 0 for alluvium sites

ε = is a random error term with zero mean and standard deviation equal to $\sigma_{\ln}$(PGA), the standard error of estimate of ln(PGA)

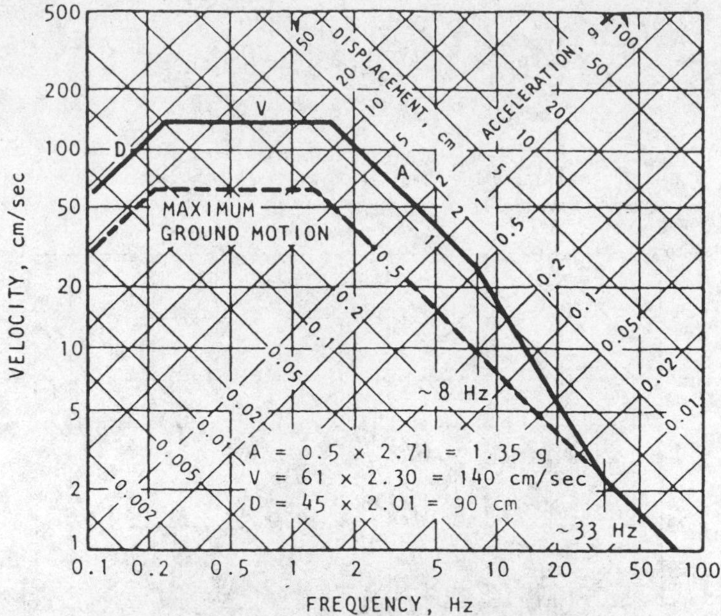

FIGURE 33.7 Idealized elastic design spectrum, horizontal motion (ZPA = 0.5 *g*, 5% damping, one sigma cumulative probability. (*Source*: Newmark, N. M. and Hall, W. J., *Earthquake Spectra and Design,* Earthquake Engineering Research Institute, Oakland, CA, 1982. With permission.)

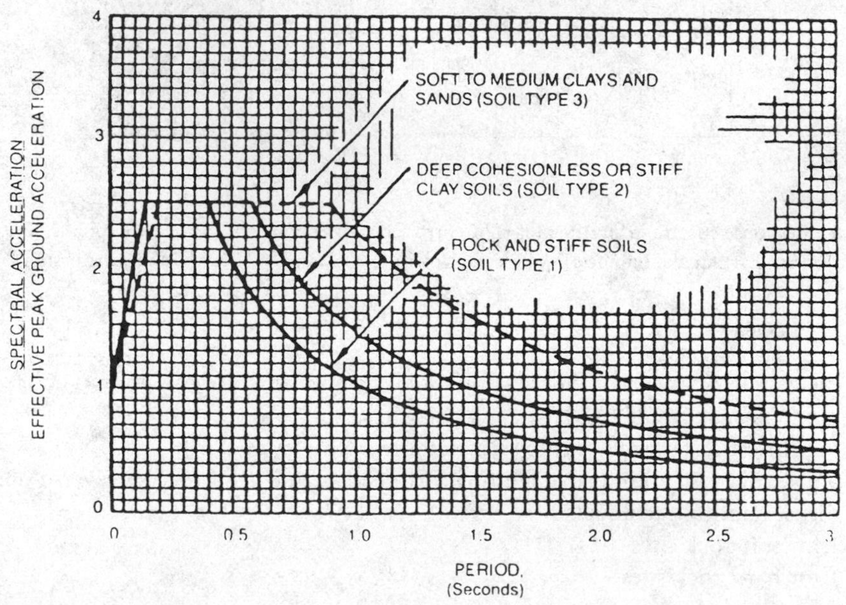

FIGURE 33.8 Normalized response spectra shapes. (*Source*: Uniform Building Code, Structural Engineering Design Provisions, Vol. 2, Intl. Conf. Building Officials, Whittier, 1994. With permission.)

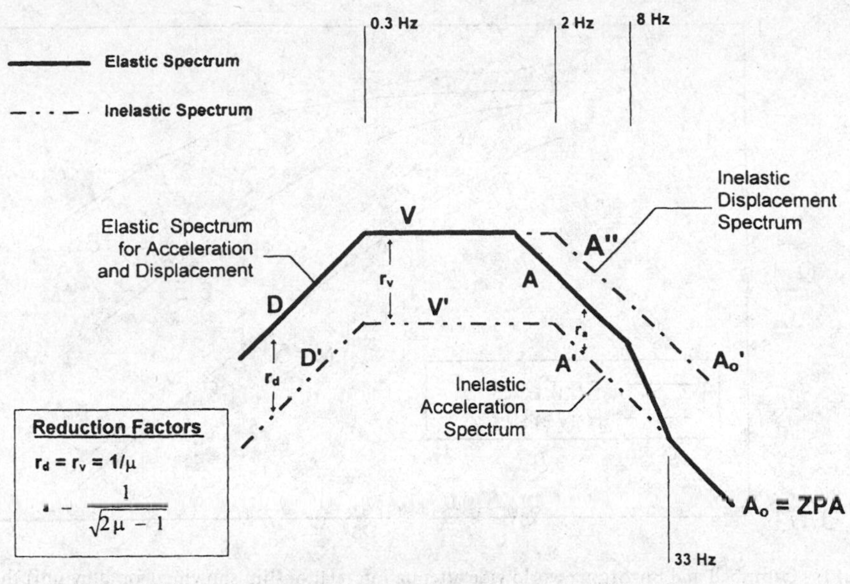

FIGURE 33.9 Inelastic response spectra for earthquakes. (*Source*: Newmark, N. M. and Hall, W. J., *Earthquake Spectra and Design*, Earthquake Engineering Research Institute, Oakland, CA, 1982.)

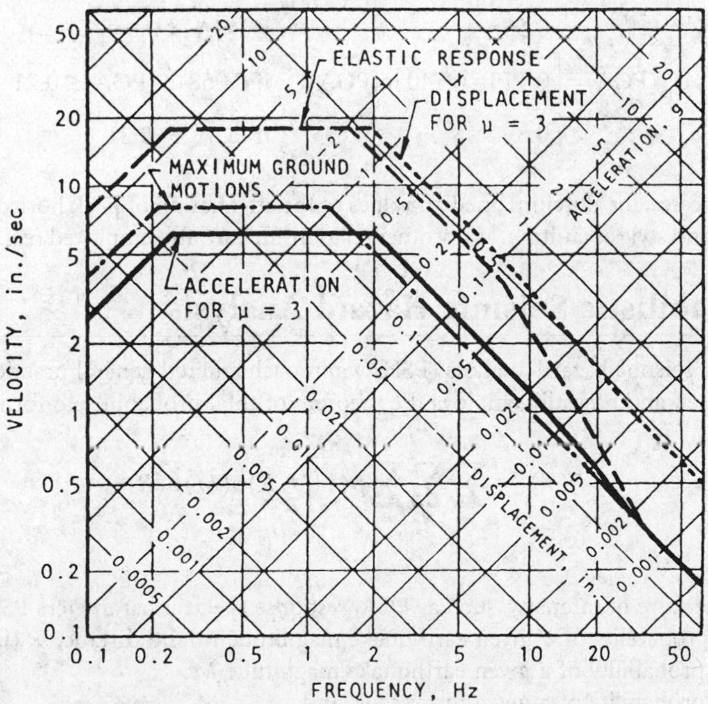

FIGURE 33.10 Example inelastic response spectra. (*Source*: Newmark, N. M. and Hall, W. J., *Earthquake Spectra and Design*, Earthquake Engineering Research Institute, Oakland, CA, 1982.)

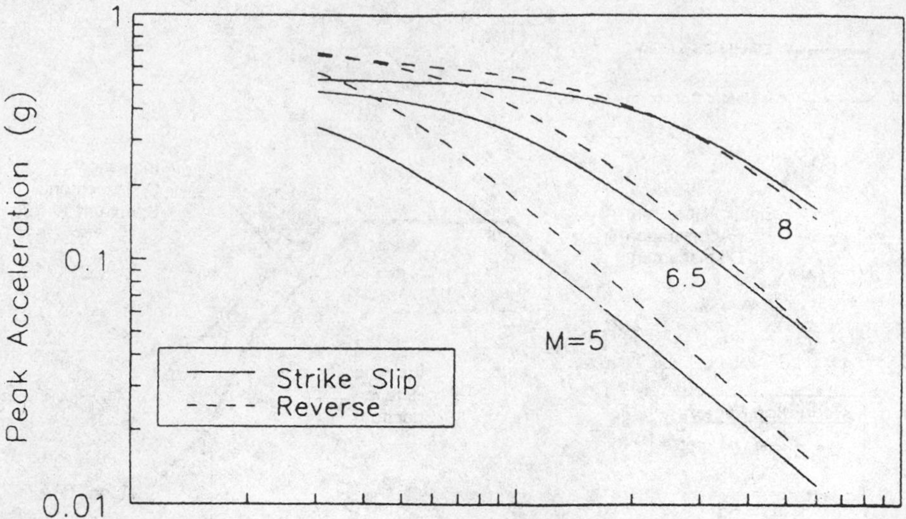

FIGURE 33.11 Campbell and Bozorgnia worldwide attenuation relationship showing (for alluvium) the scaling of peak horizontal acceleration with magnitude and style of faulting. (*Source*: Campbell, K. W. and Bozorgnia, Y., in *Proc. Fifth U.S. National Conference on Earthquake Engineering*, Earthquake Engineering Research Institute, Oakland, CA, 1994. With permission.)

Regarding the uncertainty, ε was estimated as

$$\sigma_{\ln}(\text{PGA}) = \begin{vmatrix} 0.55 & \text{if PGA } < 0.068 \\ 0.173 - 0.140\ \ln(\text{PGA}) & \text{if } 0.068 \le \text{ PGA } \le 0.21 \\ 0.39 & \text{if PGA } > 0.21 \end{vmatrix}$$

Figure 33.11 indicates, for alluvium, median values of the attenuation of peak horizontal acceleration with magnitude and style of faulting. Many other relationships are also employed (e.g., Boore et al.[6]).

33.5 Probabilistic Seismic Hazard Analysis

The probabilistic seismic hazard analysis (PSHA) approach entered general practice with Cornell's [13] seminal paper, and basically employs the theorem of total probability to formulate:

$$P(Y) = \sum_{F}\sum_{M}\sum_{R} p(Y|M, R)p(M)p(R) \tag{33.14}$$

where

Y = a measure of intensity, such as PGA, response spectral parameters PSV, etc.
$p(Y| M,R)$= the probability of Y given earthquake magnitude M and distance R (i.e., attenuation)
$p(M)$ = the probability of a given earthquake magnitude M
$p(R)$ = the probability of a given distance R, and
F = seismic sources, whether discrete such as faults, or distributed

This process is illustrated in Figure 33.12, where various seismic sources (faults modeled as line sources and dipping planes, and various distributed or area sources, including a background source to account for miscellaneous seismicity) are identified, and their seismicity characterized on the basis of historic seismicity and/or geologic data. The effects at a specific site are quantified on the

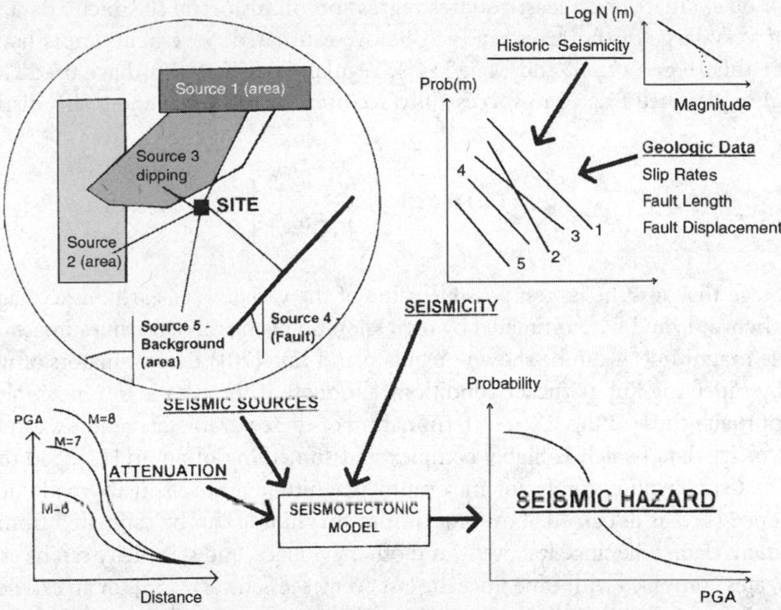

FIGURE 33.12 Elements of seismic hazard analysis — seismotectonic model is composed of seismic sources, whose seismicity is characterized on the basis of historic seismicity and geologic data, and whose effects are quantified at the site via strong motion attenuation models.

basis of strong ground motion modeling, also termed attenuation. These elements collectively are the **seismotectonic model** — their integration results in the **seismic hazard.**

There is an extensive literature on this subject [42,50] so that only key points will be discussed here. Summation is indicated, as integration requires closed-form solutions, which are usually precluded by the empirical form of the attenuation relations. The $p(Y\hat{O}M,R)$ term represents the full probabilistic distribution of the attenuation relation — summation must occur over the full distribution, due to the significant uncertainty in attenuation. The $p(M)$ term is referred to as the **magnitude–frequency relation**, which was first characterized by Gutenberg and Richter [24] as

$$\log \mathbf{N}(m) = a_N - b_N m \tag{33.15}$$

where $\mathbf{N}(m)$ = the number of earthquake events equal to or greater than magnitude m occurring on a seismic source per unit time, and a_N and b_N are regional constants (10^{a_N} = the total number of earthquakes with magnitude >0, and b_N is the rate of seismicity; b_N is typically 1 ± 0.3). The Gutenberg Richter relation can be normalized to

$$\mathbf{F}(m) = 1. - \exp\,[-\,\mathbf{B}_M\,(m - \mathbf{M}_{o})] \tag{33.16}$$

where $F(m)$ is the cumulative distribution function (CDF) of magnitude, $\mathbf{B}_M$ is a regional constant and $\mathbf{M}_o$ is a small enough magnitude such that lesser events can be ignored. Combining this with a Poisson distribution to model large earthquake occurrence [20] leads to the CDF of earthquake magnitude per unit time

$$F(m) = \exp\,[-\exp\,\{-\,a_M\,(m - \mu_M)\}] \tag{33.17}$$

which has the form of a Gumbel [23] extreme value type I (largest values) distribution (denoted $EX_{I,L}$), which is an unbounded distribution (i.e., the variate can assume any value). The parameters

a_M and μ_M can be evaluated by a least-squares regression on historical seismicity data, although the probability of very large earthquakes tends to be overestimated. Several attempts have been made to account for this (e.g., Cornell and Merz [14]). Yegulalp and Kuo [70] have used Gumbel's Type III (largest value, denoted $EX_{III,L}$) to successfully account for this deficiency. This distribution

$$F(m) = \exp\left[-\left(\frac{w-m}{w-u}\right)^k\right] \tag{33.18}$$

has the advantage that w is the largest possible value of the variate (i.e., earthquake magnitude), thus permitting (when w, u, and k are estimated by regression on historical data) an estimate of the source's largest possible magnitude. It can be shown (Yegulalp and Kuo [70]) that estimators of w, u, and k can be obtained by satisfying Kuhn–Tucker conditions although, if the data is too incomplete, the $EX_{III,L}$ parameters approach those of the $EX_{I,L}$. Determination of these parameters requires careful analysis of historical seismicity data (which is highly complex and something of an art [17], and the merging of the resulting statistics with estimates of maximum magnitude and seismicity made on the basis of geologic evidence (i.e., as discussed above, maximum magnitude can be estimated from fault length, fault displacement data, time since last event, and other evidence, and seismicity can be estimated from fault slippage rates combined with time since the last event, see Schwartz [55] for an excellent discussion of these aspects). In a full probabilistic seismic hazard analysis, many of these aspects are treated fully or partially probabilistically, including the attenuation, magnitude–frequency relation, upper- and lower-bound magnitudes for each source zone, geographic bounds of source zones, fault rupture length, and many other aspects. The full treatment requires complex specialized computer codes, which incorporate uncertainty via use of multiple alternative source zonations, attenuation relations, and other parameters [3,19] often using a logic tree format. A number of codes have been developed using the public domain FRISK (Fault RISK) code first developed by McGuire [37].

33.6 Site Response

When seismic waves reach a site, the ground motions they produce are affected by the geometry and properties of the geologic materials at that site. At most bridge sites, rock will be covered by some thickness of soil which can markedly influence the nature of the motions transmitted to the bridge structure as well as the loading on the bridge foundation. The influence of local site conditions on ground response has been observed in many past earthquakes, but specific provisions for site effects were not incorporated in codes until 1976.

The manner in which a site responds during an earthquake depends on the near-surface stiffness gradient and on how the incoming waves are reflected and refracted by the near-surface materials. The interaction between seismic waves and near-surface materials can be complex, particularly when surface topography and/or subsurface stratigraphy is complex. Quantification of site response has generally been accomplished by analytical or empirical methods.

Basic Concepts

The simplest possible case of site response would consist of a uniform layer of viscoelastic soil of density, ρ, shear modulus, G, viscosity, η, and thickness, H, resting on rigid bedrock and subjected to vertically propagating shear waves (Figure 33.13a). The response of the layer would be governed by the wave equation

$$\rho\frac{\partial^2 u}{\partial t^2} = G\frac{\partial^2 u}{\partial z^2} + \eta\frac{\partial^3 u}{\partial z^2 \partial t} \tag{33.19}$$

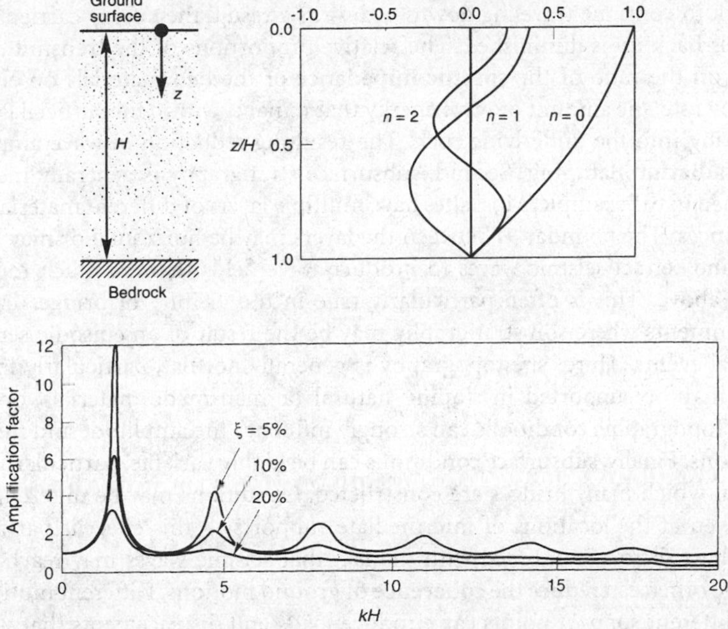

FIGURE 33.13 Illustration of (top) mode shapes and (bottom) amplification function for uniform elastic layer underlain by rigid boundary. (*Source*: Kramer, S.L., *Geotechnical Earthquake Engineering*, Prentice-Hall, Upper Saddle River, NJ, 1996.)

which has a solution that can be expressed in the form of upward and downward traveling waves. At certain frequencies, these waves interfere constructively to produce increased amplitudes; at other frequencies, the upward and downward traveling waves tend to cancel each other and produce lower amplitudes. Such a system can easily be shown to have an infinite number of natural frequencies and mode shapes (Figure 33.13 (top)) given by

$$\omega_n = \frac{v_s}{H}\left(\frac{\pi}{2} + n\pi\right) \quad \text{and} \quad \phi_n = \cos\left[\frac{z}{H}\left(\frac{\pi}{2} + n\pi\right)\right] \tag{33.20}$$

Note that the fundamental, or characteristic site period, is given by $T_s = 2\pi/\omega_o = 4H/v_s$. The ratio of ground surface to bedrock amplitude can be expressed in the form of an **amplification function** as

$$A(\omega) = \frac{1}{\sqrt{\cos^2\left(\omega H / v_s + \left[\xi(\omega H / v_s)\right]^2\right)}} \tag{33.21}$$

Figure 33.13(b) shows the amplification function which illustrates the frequency-dependent nature of site amplification. The amplification factor reaches its highest value when the period of the input motion is equal to the characteristic site period. More realistic site conditions produce more-complicated amplification functions, but all amplification functions are frequency-dependent. In a sense, the surficial soil layers act as a filter that amplifies certain frequencies and deamplifies others. The overall effect on site response depends on how these frequencies match up with the dominant frequencies in the input motion.

The example illustrated above is mathematically convenient, but unrealistically simple for application to actual sites. First, the assumption of rigid bedrock implies that all downward-traveling waves are perfectly reflected back up into the overlying layer. While generally quite stiff, bedrock is not perfectly rigid and therefore a portion of the energy in a downward-traveling wave is transmitted into the bedrock to continue traveling downward — as a result, the energy carried by the reflected wave that travels back up is diminished. The relative proportions of the transmitted and reflected waves depends on the ratio of the **specific impedance** of the two materials on either side of the boundary. At any rate, the amount of wave energy that remains within the surficial layer is decreased by waves radiating into the underlying rock. The resulting reduction in wave amplitudes is often referred to as **radiation damping**. Second, subsurface stratigraphy is generally more complicated than that assumed in the example. Most sites have multiple layers of different materials with different specific impedances. The boundaries between the layers may be horizontal or may be inclined, but all will reflect and refract seismic waves to produce wave fields that are much more complicated than described above. This is often particularly true in the vicinity of bridges located in fluvial geologic environments where soil stratigraphy may be the result of an episodic series of erosional and depositional events. Third, site topography is generally not flat, particularly in the vicinity of bridges which may be supported in sloping natural or man-made materials, or on man-made embankments. Topographic conditions can strongly influence the amplitude and frequency content of ground motions. Finally, subsurface conditions can be highly variable, particularly in the geologic environments in which many bridges are constructed. Conditions may be different at each end of a bridge, and even at the locations of intermediate supports — this effect is particularly true for long bridges. These factors, combined with the fact that seismic waves may reach one end of the bridge before the other, can reduce the **coherence** of ground motions. Different motions transmitted to a bridge at different support points can produce loads and displacements that would not occur in the case of perfectly coherent motions.

Evidence for Local Site Effects

Theoretical evidence for the existence of local site effects has been supplemented by instrumental and observational evidence in numerous earthquakes. Nearly 200 years ago [35], variations in damage patterns were correlated to variations in subsurface conditions; such observations have been repeated on a regular basis since that time. With the advent of modern seismographs and strong motion instruments, quantitative evidence for local site effects is now available. In the Loma Prieta earthquake, for example, strong motion instruments at Yerba Buena Island and Treasure Island were at virtually identical distances and azimuths from the hypocenter. However, the Yerba Buena Island instrument was located on a rock outcrop and the Treasure Island instrument on about 14 m of loose hydraulically placed sandy fill underlain by nearly 17 m of soft San Francisco Bay Mud. The measured motions, which differed significantly (Figure 33.14), illustrate the effects of local site effects. At a small but increasing number of locations, strong motion instruments have been placed in a boring directly below a surface instrument (Figure 33.15a). Because such vertical arrays can measure motions at the surface and at bedrock level, they allow direct computation of measured amplification functions. Such an empirical amplification function is shown in Figure 33.15b. The general similarity of the measured amplification function, particularly the strong frequency dependence, to even the simple theoretical amplification (Figure 33.13) is notable.

Methods of Analysis

Development of suitable design ground motions, and estimation of appropriate foundations loading, generally requires prediction of anticipated site response. This is usually accomplished using empirical or analytical methods. For small bridges, or for projects in which detailed subsurface information is not available, the empirical approach is more common. For larger and more important structures, a subsurface exploration program is generally undertaken to provide information for site-specific analytical prediction of site response.

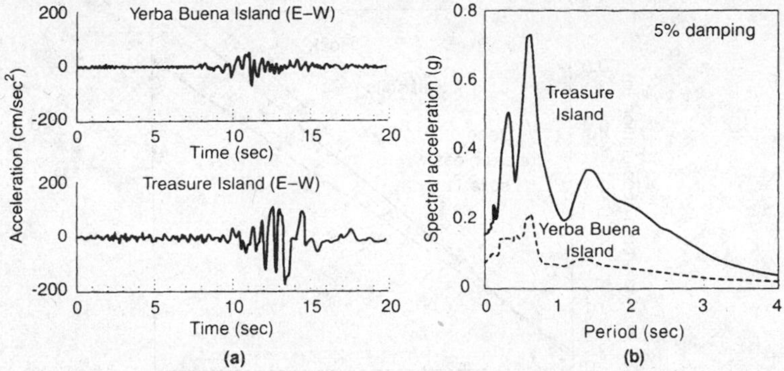

FIGURE 33.14 Ground surface motions at Yerba Buena Island and Treasure Island in the Loma Prieta earthquake. sources Kramer, S.L., *Geotechnical Earthquake Engineering*, Prentice-Hall, Upper Saddle River, NJ, 1996.)

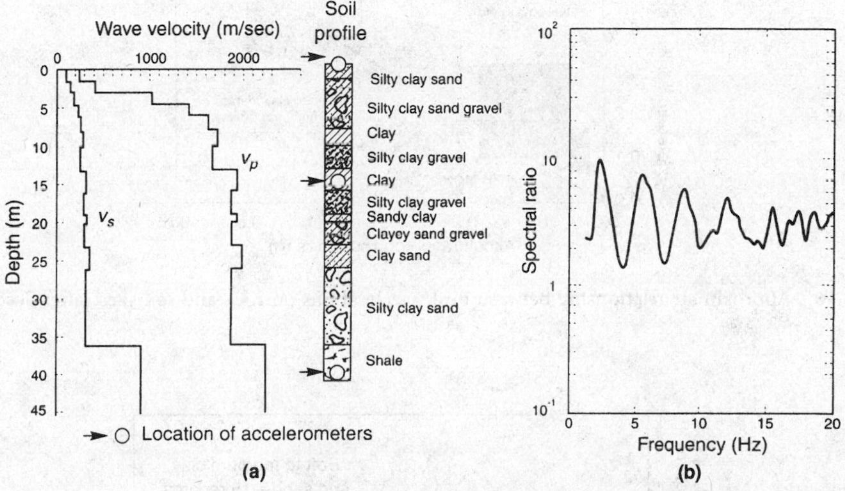

FIGURE 33.15 (a) Subsurface profile at location of Richmond Field Station downhole array, and (b) measured surface/bedrock amplification function in Briones Hills ($M_L = 4.3$) earthquake. sources Kramer, S.L., *Geotechnical Earthquake Engineering*, Prentice-Hall, Upper Saddle River, NJ, 1996.)

Empirical Methods

In the absence of site-specific information, local site effects can be estimated on the basis of empirical correlation to measured site response from past earthquakes. The database of strong ground motion records has increased tremendously over the past 30 years. Division of records within this database according to general site conditions has allowed the development of empirical correlations for different site conditions.

The earliest empirical approach involved estimation of the effects of local soil conditions on peak ground surface acceleration and spectral shape. Seed et al. [59] divided the subsurface conditions at the sites of 104 strong motion records into four categories — rock, stiff soils (<61 m), deep cohesionless soils (>76 m), and soft to medium clay and sand. Comparing average peak ground surface accelerations measured at the soil sites with those anticipated at equivalent rock sites allowed development of curves such as those shown in Figure 33.16. These curves show that soft profiles amplify peak acceleration over a wide range of rock accelerations, that even stiff soil profiles amplify peak acceleration when peak accelerations are relatively low, and that peak accelerations are deamplified at very high

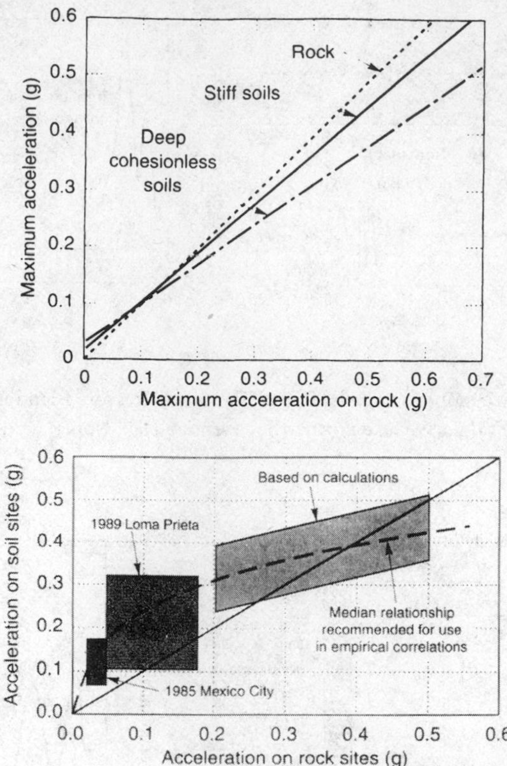

FIGURE 33.16 Approximate relationship between beak accelerations on rock and soil sites (after Seed et al. [59]; Idriss, 1990).

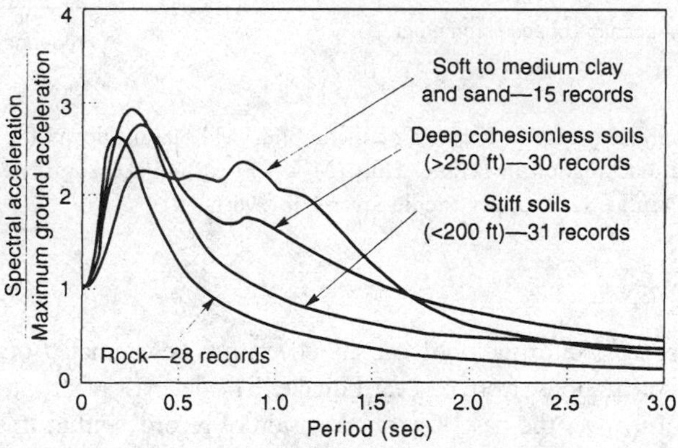

FIGURE 33.17 Average normalized response spectra (5% damping) for different local site conditions (after Seed et al. [59]).

input acceleration levels. Computation of average response spectra, when normalized by peak acceleration (Figure 33.17), showed the significant effect of local soil conditions on spectral shape, a finding that has strongly influenced the development of seismic codes and standards.

A more recent empirical approach has been to include local site conditions directly in attenuation relationships. By developing a site parameter to characterize the soil conditions at the locations of strong motion instruments and incorporating that parameter into the basic form of an attenuation

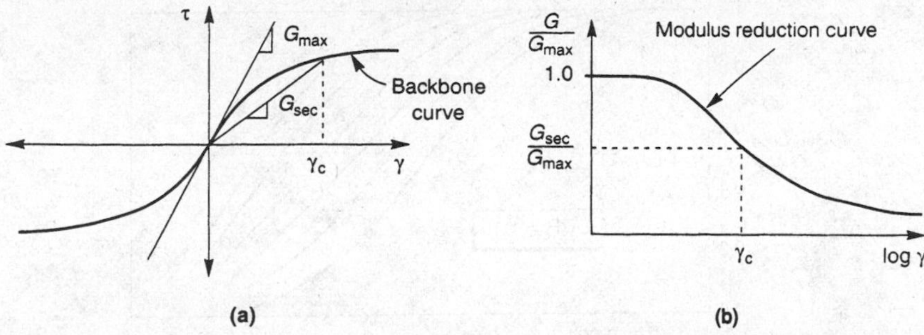

FIGURE 33.18 Relationship between backbone curve and modulus reduction curve.

relationship, regression analyses can produce attenuation relationships that include the effects of local site conditions. In such relationships, site conditions are typically grouped into different site classes on the basis of such characteristics as surficial soil/rock conditions [see the factors S_{sr} and S_{hr} in Eq. (33.13)] or average shear wave velocity within the upper 30 m of the ground surface (e.g., Boore et al. [6]). Such relationships can be used for empirical prediction of peak acceleration and response spectra, and incorporated into probabilistic seismic hazard analyses to produce uniform risk spectra for the desired class of subsurface conditions.

The reasonableness of empirically based methods for estimation of site response effects depends on the extent to which site conditions match the site conditions in the databases from which the empirical relationships were derived. It is important to recognize the empirical nature of such methods and the significant uncertainty inherent in the results they produce.

Analytical Methods

When sufficient information to characterize the geometry and dynamic properties of subsurface soil layers is available, local site effects may be computed by site-specific ground response analyses. Such analyses may be conducted in one, two, or three dimensions; one-dimensional analyses are most common, but the topography of many bridge sites may require two-dimensional analyses.

Unlike most structural materials, soils are highly nonlinear, even at very low strain levels. This nonlinearity causes soil stiffness to decrease and material damping to increase with increasing strain amplitude. The variation of stiffness with strain can be represented in two ways — by nonlinear **backbone** (stress–strain) **curves** or by **modulus reduction curves**, both of which are related as illustrated in Figure 33.18. The modulus reduction curve shows how the secant shear modulus of the soil decreases with increasing strain amplitude. To account for the effects of nonlinear soil behavior, ground response analyses are generally performed using one of two basic approaches: the **equivalent linear approach** or the **nonlinear approach**.

In the equivalent linear approach, a linear analysis is performed using shear moduli and damping ratios that are based on an initial estimate of strain amplitude. The strain level computed using these properties is then compared with the estimated strain amplitude and the properties adjusted until the computed strain levels are very close to those corresponding to the soil properties. Using this iterative approach, the effects of nonlinearity are approximated in a linear analysis by the use of *strain-compatible* soil properties. Modulus reduction and damping behavior has been shown to be influenced by soil plasticity, with highly plastic soils exhibiting higher linearity and lower damping than low-plasticity soils (Figure 33.19). The equivalent linear approach has been incorporated into such computer programs as SHAKE [53] and ProShake [18] for one-dimensional analyses, FLUSH [34] for two-dimensional analyses, and TLUSH [29] for three-dimensional analyses.

In the nonlinear approach, the equations of motion are assumed to be linear over each of a series of small time increments. This allows the response at the end of a time increment to be computed from the conditions at the beginning of the time increment and the loading applied during the time

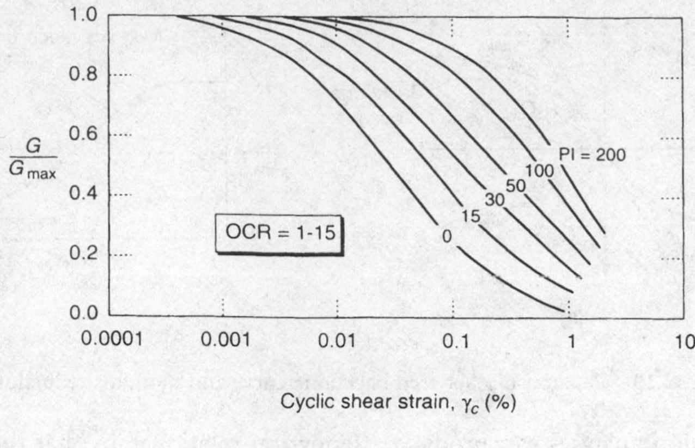

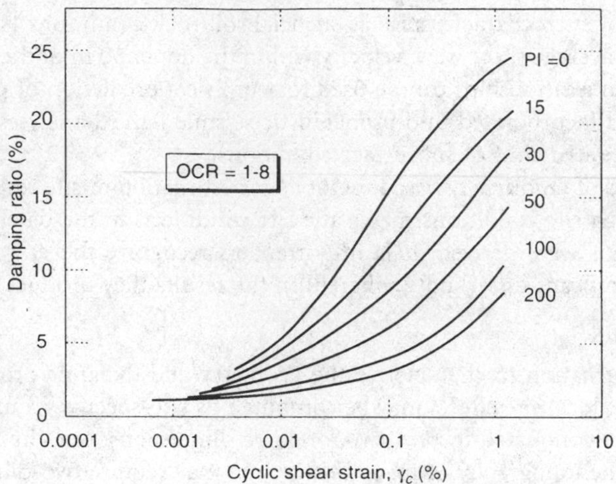

FIGURE 33.19 Equivalent linear soil behavior: (a) modulus reduction curves and (b) damping curves. (*Source:* Vucetic and Dobry, 1991.)

increment. At the end of the time increment, the properties are updated for the next time increment. In this way, the stiffness of each element of soil can be changed depending on the current and past stress conditions and hysteretic damping can be modeled directly. For seismic analysis, the nonlinear approach requires a constitutive (stress–strain) model that is capable of representing soil behavior under dynamic loading conditions. Such models can be complicated and can require calibration of a large number of soil parameters by extensive laboratory testing. With a properly calibrated constitutive model, however, nonlinear analyses can provide reasonable predictions of site response and have two significant advantages over equivalent linear analyses. First, nonlinear analyses are able to predict permanent deformations such as those associated with ground failure (Section 33.7). Second, nonlinear analyses are able to account for the generation, redistribution, and eventual dissipation of porewater pressures which makes them particularly useful for sites that may be subject to liquefaction and/or lateral spreading. The nonlinear approach has been incorporated into such computer programs as DESRA [31], TESS [48], and SUMDES for one-dimensional analysis, and TARA [21] for two-dimensional analyses. General-purpose programs such as FLAC can also be used for nonlinear two-dimensional analyses. In practice, however, the use of nonlinear analyses has lagged behind the use of equivalent linear analyses, principally because of the difficulty in characterizing nonlinear constitutive model parameters.

TABLE 33.2 Site Coefficient

Soil Type	Description	*S*
I	Rock of any characteristic, either shalelike or crystalline in nature (such material may be characterized by a shear wave velocity greater than 760 m/s, or by other appropriate means of classification; or Stiff soil conditions where the soil depth is less than 60 m and the soil types overlying rock are stable deposits of sands, gravels, or stiff clays.	1.0
II	Stiff clay or deep cohesionless conditions where the soil depth exceeds 60 m and the soil types overlying rock are stable deposits of sands, gravels, or stiff clays	1.2
III	Soft to medium-stiff clays and sands, characterized by 9 m or more of soft to medium-stiff clays with or without intervening layers of sand or other cohesionless soils	1.5
IV	Soft clays or silts greater than 12 m in depth; these materials may be characterized by a shear wave velocity less than 150 m/s and might include loose natural deposits or synthetic nonengineered fill	2.0

Site Effects for Different Soil Conditions

As indicated previously, soil deposits act as filters, amplifying response at some frequencies and deamplifying it at others. The greatest degree of amplification occurs at frequencies corresponding to the characteristic site period, $T_s = 4H/v_s$. Because the characteristic site period is proportional to shear wave velocity and inversely proportional to thickness, it is clear that the response of a given soil deposit will be influenced by the stiffness and thickness of the deposit. Thin and/or stiff soil deposits will amplify the short-period (high-frequency) components, and thick and/or soft soil deposits will amplify the long-period (low-frequency) components of an input motion. As a result, generalizations about site effects for different soil conditions are generally based on the average stiffness and thickness of the soil profile.

These observations of site response are reflected in bridge design codes. For example, the 1997 Interim Revision of the 1996 Standard Specifications for Highway Bridges (AASHTO, 1997) require the use of an elastic seismic response coefficient for an SDOF structure of natural period, T, taken as

$$C_s = \frac{1.2AS}{T^{2/3}} \tag{33.22}$$

where A is an acceleration coefficient that depends on the location of the bridge and S is a dimensionless site coefficient obtained from Table 33.2. In accordance with the behavior illustrated in Figure 33.17, the site coefficient prescribes increased design requirements at long periods for bridges underlain by thick deposits of soft soil (Figure 33.20).

33.7 Earthquake-Induced Settlement

Settlement is an important consideration in the design of bridge foundations. In most cases, settlement results from *consolidation*, a process that takes place relatively slowly as porewater is squeezed from the soil as it seeks equilibrium under a new set of stresses. Consolidation settlements are most significant in fine-grained soils such as silts and clays. However, the tendency of coarse-grained soils (sands and gravels) to densify due to vibration is well known; in fact, it is frequently relied upon for efficient compaction of sandy soils. Densification due to the cyclic stresses imposed by earthquake shaking can produce significant settlements during earthquakes. Whether caused by consolidation or earthquakes, bridge designers are concerned with **total settlement** and, because settlements rarely occur uniformly, also with **differential settlement**. Differential settlement can induce very large loads in bridge structures.

While bridge foundations may settle due to shearing failure in the vicinity of abutments (Chapter 30), shallow foundations (Chapter 31), and deep foundations (Chapter 32), this section

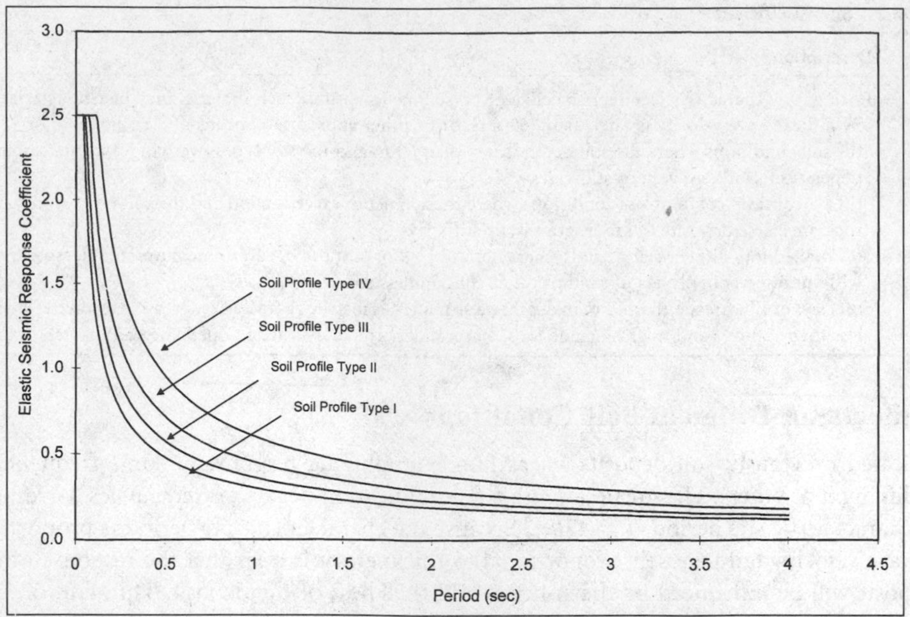

FIGURE 33.20 Variation of elastic seismic response coefficient with period for $A = 0.25$.

deals with settlement due to earthquake-induced soil densification. Densification of soils beneath shallow bridge foundations can cause settlement of the foundation. Densification of soils adjacent to deep foundations can cause downdrag loading on the foundations (and bending loading if the foundations are battered). Densification of soils beneath approach fills can lead to differential settlements at the ends of the bridge that can be so abrupt as to render the bridge useless.

Accurate prediction of earthquake-induced settlements is difficult. Errors of 25 to 50% are common in estimates of consolidation settlement, so even less accuracy should be expected in the more-complicated case of earthquake-induced settlement. Nevertheless, procedures have been developed that account for the major factors known to influence earthquake-induced settlement and that have been shown to produce reasonable agreement with many cases of observed field performance. Such procedures are generally divided into cases of dry sands and saturated sands.

Settlement of Dry Sands

Dry sandy soils are often found above the water table in the vicinity of bridges. The amount of densification experienced by dry sands depends on the density of the sand, the amplitude of cyclic shear strain induced in the sand, and on the number of cycles of shear strain applied during the earthquake. Settlements can be estimated using cyclic strain amplitudes from site response analyses with corrections for the effects of multidirectional shaking [47,58] or by simplified procedures [63]. Because of the high air permeability of sands, settlement of dry sands occurs almost instantaneously.

In the simplified procedure, the effective cyclic strain amplitude is estimated as

$$\gamma_{cyc} = 0.65 \frac{a_{max}}{g} \frac{\sigma_v r_d}{G} \tag{33.23}$$

Because the shear modulus, G, is a function of γ_{cyc}, several iterations may be required to calculate a value of γ_{cyc} that is consistent with the shear modulus. When the low strain stiffness, G_{max} ($= \rho v_s^2$), is known, the effective cyclic strain amplitude can be estimated using Figures 33.21 and 33.22.

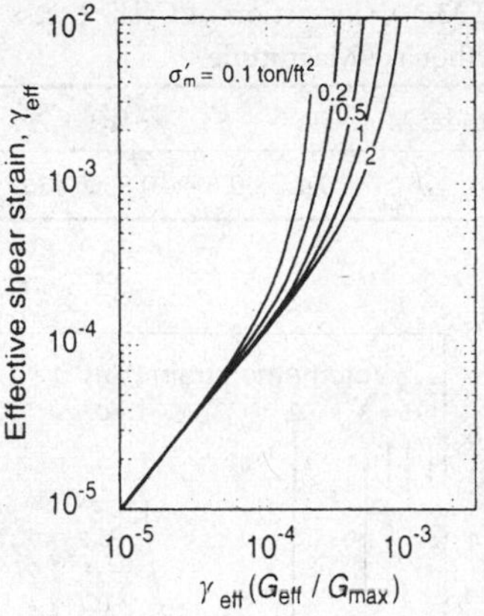

Plot for determination

FIGURE 33.21 Plot for determination of effective cyclic shear strain in sand deposits. (Tokimatsu and Seed [63]).

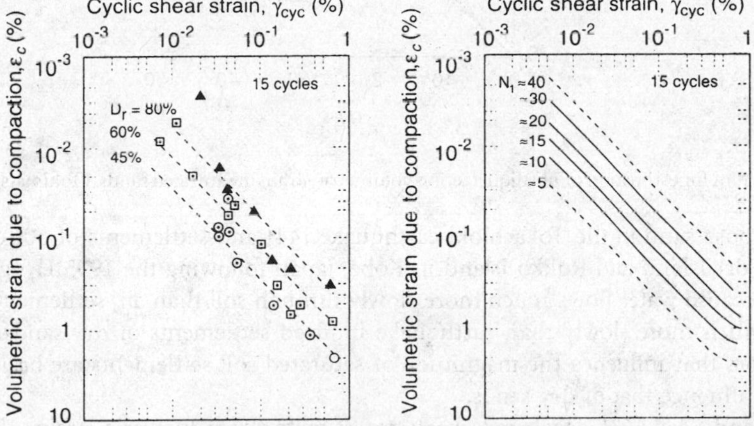

FIGURE 33.22 Relationship between volumetric strain and cyclic shear strain in dry sands as function of (a) relative density and (b) SPT resistance. (Tokimatsu and Seed [63]).

Figure 33.22 then allows the effective cyclic strain amplitude, along with the relative density or SPT resistance of the sand, to be used to estimate the volumetric strain due to densification. These volumetric strains are based on durations associated with a M = 7.5 earthquake, corrections for other magnitudes can be made with the aid of Table 33.3. The effects of multidirectional shaking are generally accounted for by doubling the computed volumetric strain. Because the stiffness, density, and cyclic shear strain amplitude generally vary with depth, a given soil deposit is usually divided into sublayers with the volumetric strain for each sublayer computed independently. The resulting settlement of each sublayer can then be computed as the product of the volumetric strain and thickness. The total settlement is obtained by summing the settlements of the individual sublayers.

Settlement of Saturated Sands

The dissipation of high excess porewater pressures generated in saturated sands (*reconsolidation*) can lead to settlement following earthquakes. Settlements of 50 to 70 cm occurred in a 5-m-thick

TABLE 33.3 Correction of Cyclic Stress Ratio for Earthquake Magnitude

Magnitude, M	5¼	6	6¾	7½	8½
$\varepsilon_{v,M}/\varepsilon_{v,M=7.5}$	0.4	0.6	0.85	1.0	1.25

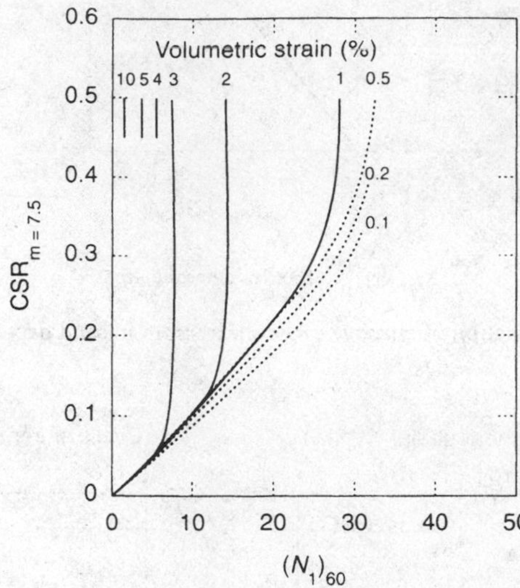

FIGURE 33.23 Plot for estimation of postliquefaction volumetric strain in saturated sands. (Tokimatsu and Seed [63]).

layer of very loose sand in the Tokachioki earthquake [44] and settlements of 50 to 100 cm were observed on Port Island and Rokko Island in Kobe, Japan following the 1995 Hyogo-ken Nambu earthquake. Because water flows much more slowly through soil than air, settlements of saturated sands occur much more slowly than earthquake-induced settlements of dry sands. Nevertheless, the main factors that influence the magnitude of saturated soil settlements are basically the same as those that influence that of dry sands.

Tokimatsu and Seed [63] developed charts to estimate the volumetric strains that develop in saturated soils. In this approach, the volumetric strain resulting from reconsolidation can be estimated from the corrected standard penetration resistance, $(N_1)_{60}$, and the cyclic stress ratio (Figure 33.23). The value of $(N_1)_{60}$ is obtained by correcting the measured standard penetration resistance, N_m, to a standard overburden pressure of 95.8 kPa (1 ton/ft^2) and to an energy of 60% of the theoretical free-fall energy of an SPT hammer using the equation:

$$(N_1)_{60} = N_m C_N \frac{E_m}{0.60\, E_{ff}} \tag{33.24}$$

where C_N is an overburden correction factor that can be estimated as $C_N = (\sigma'_{vo})^{-0.5}$, E_m is the measured hammer energy and E_{ff} is the theoretical free-fall energy. In Figure 33.23, the cyclic stress ratio, defined as $CSR_{M=7.5} = \tau_{cyc}/\sigma'_{vo}$, corresponds to a magnitude 7.5 earthquake. For other magnitudes, the corresponding value of the cyclic stress ratio can be obtained using Table 33.4. As in the case of dry sands, the soil layer is typically divided into sublayers with the total settlement taken as the sum of the products of the thickness and volumetric strain of all sublayers. In some cases, earthquake-induced porewater pressures may be insufficient to cause liquefaction but still may produce post-earthquake settlement. The volumetric strain produced by reconsolidation in such cases may be estimated from Figure 33.24.

TABLE 33.4 Correction of Cyclic Stress Ratio for Earthquake Magnitude

Magnitude, M	5¼	6	6¾	7½	8½
$CSR_M/CSR_{M=7.5}$	1.50	1.32	1.13	1.00	0.89

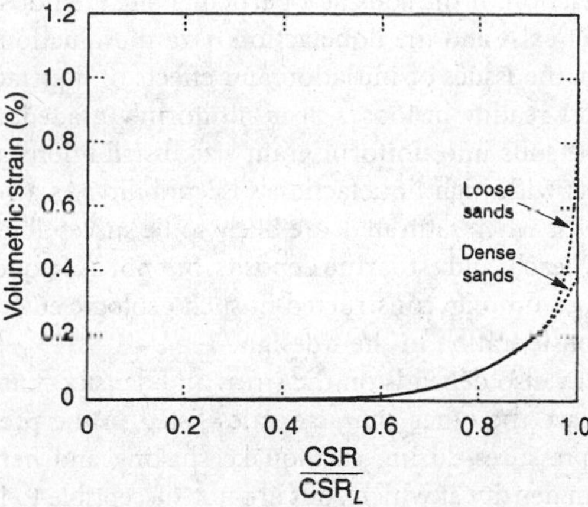

FIGURE 33.24 Plot for estimation of volumetric strain in saturated sands that do not liquefy. (Tokimatsu and Seed [63]).

33.8 Ground Failure

Strong earthquake shaking can produce a dynamic response of soils that is so energetic that the stress waves exceed the strength of the soil. In such cases, ground failure characterized by permanent soil deformations may occur. Ground failure may be caused by weakening of the soil or by temporary exceedance of the strength of the soil by transient inertial stresses. The former case results in phenomena such as liquefaction and lateral spreading; the latter in inertial failures of slopes and retaining wall backfills.

Liquefaction

The term *liquefaction* has been widely used to describe a range of phenomena in which the strength and stiffness of a soil deposit are reduced due to the generation of porewater pressure. It occurs most commonly in loose, saturated sands, although it has also been observed in gravels and non-plastic silts. The effects of liquefaction can range from massive landslides with displacements measured in tens of meters to relatively small slumps or spreads with small displacements. Many bridges, particularly those that cross bodies of water, are located in areas with geologic and hydrologic conditions that tend to produce liquefaction.

The mechanisms that produce liquefaction-related phenomena can be divided into two categories. The first, **flow liquefaction,** can occur when the shear stresses required for static equilibrium of a soil mass is greater than the shear strength of the soil in its liquefied state. While not common, flow liquefaction can produce tremendous instabilities known as **flow failures**. In such cases, the earthquake serves to trigger liquefaction, but the large deformations that result are actually driven by the preexisting static stresses. The second phenomenon, **cyclic mobility,** occurs when the initial static stresses are less than the strength of the liquefied soil. The effects of cyclic mobility lead to deformations that develop incrementally during the period of earthquake shaking, and are commonly called **lateral spreading**.

Lateral spreading can occur on very gentle slopes, in the vicinity of free surfaces such as riverbanks, and beneath and adjacent to embankments. Lateral spreading occurs much more frequently than flow failure, and can cause significant distress to bridges and their foundations.

Liquefaction Susceptibility

The first step in an evaluation of liquefaction hazards is the determination of whether or not the soil is susceptible to liquefaction. If the soils at a particular site are not susceptible to liquefaction, liquefaction hazards do not exist and the liquefaction hazard evaluation can be terminated. If the soil is susceptible, however, the issues of initiation and effects of liquefaction must be considered.

Liquefaction occurs most readily in loose, clean, uniformly graded, saturated soils. Therefore, geologic processes that sort soils into uniform grain size distributions and deposit them in loose states produce soil deposits with high liquefaction susceptibility. As a result, fluvial deposits, and colluvial and aeolian deposits when saturated, are likely to be susceptible to liquefaction. Liquefaction also occurs in alluvial, beach, and estuarine deposits, but not as frequently as in those previously listed. Because bridges are commonly constructed in such geologic environments, liquefaction is a frequent and important consideration in their design.

Liquefaction susceptibility also depends on the stress and density characteristics of the soil. Very dense soils, even if they have the other characteristics listed in the previous paragraph, will not generate high porewater pressures during earthquake shaking and hence are not susceptible to liquefaction. The minimum density at which soils are not susceptible to liquefaction increases with increasing effective confining pressure. This characteristic indicates that, for a soil deposit of constant density, the deeper soils are more susceptible to liquefaction than the shallower soils. For the general range of soil conditions encountered in the field, cohesionless soils with $(N_1)_{60}$ values greater than 30 or normalized cone penetration test (CPT) tip resistances (q_{c1N}, see next section) greater than about 175 are generally not susceptible to liquefaction.

Initiation of Liquefaction

The fact that a soil deposit is susceptible to liquefaction does not mean that liquefaction will occur in a given earthquake. Liquefaction must be triggered by some disturbance, such as earthquake shaking with sufficient strength to exceed the liquefaction resistance of the soil. Even a liquefaction-susceptible soil will have some liquefaction resistance. Evaluating the potential for the occurrence of liquefaction (liquefaction potential) involves comparison of the loading imposed by the anticipated earthquake with the liquefaction resistance of the soil. Liquefaction potential is most commonly evaluated using the cyclic stress approach in which both earthquake loading and liquefaction resistance are expressed in terms of cyclic stresses, thereby allowing direct and consistent comparison.

Characterization of Earthquake Loading

The level of porewater pressure generated by an earthquake is related to the amplitude and duration of earthquake-induced shear stresses. Such shear stresses can be predicted in a site response analysis using either the equivalent linear method or nonlinear methods. Alternatively, they can be estimated using a simplified approach that does not require site response analyses.

Early methods of liquefaction evaluation were based on the results of cyclic triaxial tests performed with harmonic (constant-amplitude) loading, and it remains customary to characterize loading in terms of an equivalent shear stress amplitude,

$$\tau_{cyc} = 0.65\tau_{max} \tag{33.25}$$

When sufficient information is available to perform site response analyses, it is advisable to compute τ_{max} in a site response analysis and use Eq. (33.6) to compute τ_{cyc}. When such information is not available, τ_{cyc} at a particular depth can be estimated as

$$\tau_{cyc} = 0.65 \frac{a_{max}}{g} \sigma_v r_d \tag{33.26}$$

where a_{max} is the peak ground surface acceleration, g is the acceleration of gravity, σ_v is the total vertical stress at the depth of interest, and r_d is the value of a site response reduction factor which can be estimated from

$$r_d = \frac{1.0 - 0.4113z^{0.5} + 0.04052z + 0.001753z^{1.5}}{1.0 - 0.4177z^{0.5} + 0.05729z - 0.006205z^{1.5} + 0.001210z^2} \tag{33.27}$$

where z is the depth of interest in meters. For evaluation of liquefaction potential, it is common to normalize τ_{cyc} by the initial (pre-earthquake) vertical effective stress, thereby producing the **cyclic stress ratio** (CSR)

$$\text{CSR} = \frac{\tau_{cyc}}{\sigma'_{vo}} \tag{33.28}$$

Characterization of Liquefaction Resistance

While early liquefaction potential evaluations relied on laboratory tests to measure liquefaction resistance, increasing recognition of the deleterious effects of sampling disturbance on laboratory test results has led to the use of field tests for measurement of liquefaction resistance. Although the use of new soil freezing and sampling techniques offers considerable promise for acquisition of undisturbed samples, liquefaction resistance is currently evaluated using *in situ* tests such as the standard penetration test (SPT) and the CPT *and* observations of liquefaction behavior in past earthquakes.

Case histories in which liquefaction was and was not observed can be analyzed to obtain empirical estimates of liquefaction resistance. By characterizing each of a series of case histories in terms of a loading parameter, $\mathcal{L}$, and a resistance parameter, $\mathcal{R}$, all combinations of $\mathcal{L}$ and $\mathcal{R}$ can be plotted with symbols that indicate whether liquefaction was observed or was not observed (Figure 33.25).

In this approach, the cyclic stress ratio induced in the soil for each case history is used as the loading parameter and an *in situ* test measurement is used as the resistance parameter. Two *in situ* tests are commonly used — the SPT which produces the resistance parameter $(N_1)_{60}$, and the CPT which produces the resistance parameter, q_{c1N}. Because the value of the cyclic stress ratio given by the curve represents the minimum cyclic stress ratio required to produce liquefaction, it is commonly referred to as the **cyclic resistance ratio**, CRR.

Because liquefaction involves the cumulative buildup of porewater pressure, the ultimate porewater pressure level is a function of the duration of ground shaking. In the development of procedures for evaluation of liquefaction potential, duration was implicitly correlated to earthquake magnitude. As a result, the procedures have been keyed to magnitude 7.5 earthquakes with corrections developed that can be applied for other magnitudes. The procedures have also been keyed to clean sands (<5% fines), again with corrections developed for application to silty sands.

Recent review of SPT-based procedures for characterization of CRR resulted in recommendation of the curve shown in Figure 33.26. This CRR curve is for clean sand and magnitude 7.5 earthquakes. For a silty sand with fines content, FC, an equivalent clean sand SPT resistance can be computed from

$$(N_1)_{60\text{-}cs} = \alpha + \beta\,(N_1)_{60} \tag{33.29}$$

where

$\alpha = 0$	and	$\beta = 1.0$	for FC < 5%
$\alpha = \exp[1.76 - 190/\text{FC}^2]$	and	$\beta = 0.99 + FC^{1.5}/1000$	for 5% < FC < 35%
$\alpha = 5.0$	and	$\beta = 1.2$	for FC > 35%

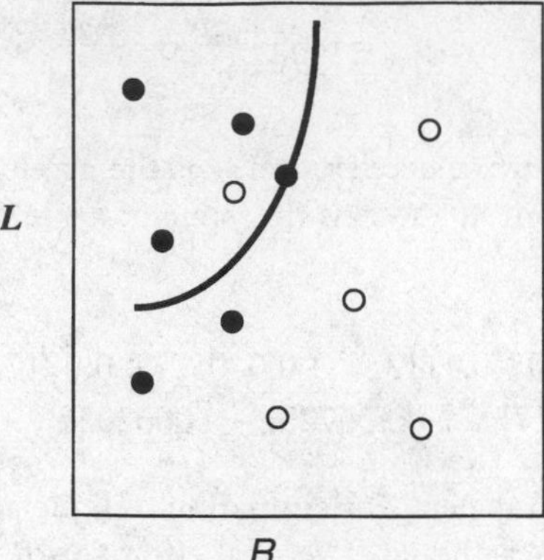

FIGURE 33.25 Discrimination between case histories in which liquefaction was observed (solid circles) and was not observed (open circles). Curve represents conservative estimate of resistance, $\mathcal{R}$, for given level of loading, $\mathcal{L}$.

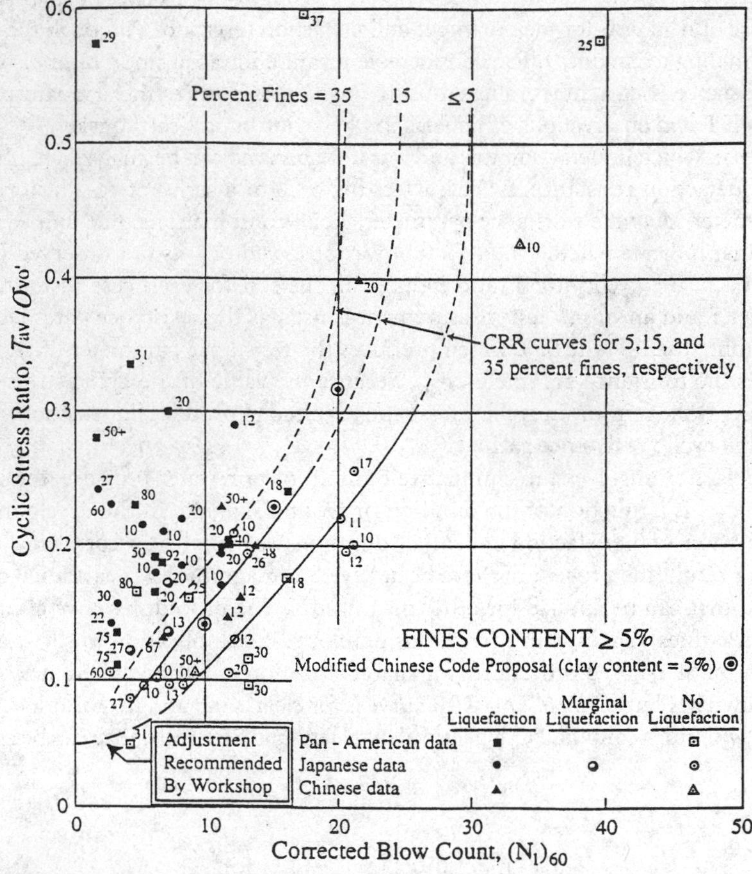

FIGURE 33.26 Relationship between cyclic stress ratios causing liquefaction and $(N_1)_{60}$ values for clean sand (after Youd and Idriss, 1998).

TABLE 33.5 Magnitude Scaling Factor

Magnitude, *M*	MSF
5.5	2.20–2.80
6.0	1.76–2.10
6.5	1.44–1.60
7.0	1.19–1.25
7.5	1.00
8.0	0.84
8.5	0.72

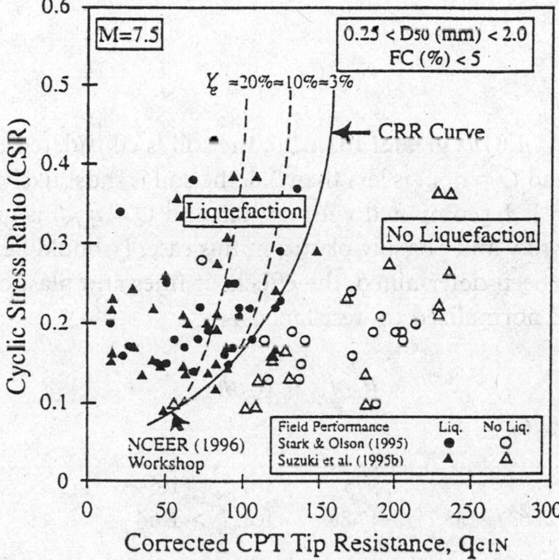

FIGURE 33.27 Relationship between cyclic stress ratios causing liquefaction and $(q_c)_1$ values for clean sand (after Youd and Idriss, 1998).

Correction for magnitudes other than 7.5 is accomplished by correcting the CRR according to

$$CRR_M = CRR_{7.5} \times MSF \tag{33.30}$$

where MSF is a magnitude scaling factor obtained from Table 33.5.

The CPT offers two distinct advantages over the SPT for evaluation of liquefaction resistance. First, the CPT provides a nearly continuous profile of penetration resistance, a characteristic that allows it to identify thin layers that can easily be missed in an SPT-based investigation. Second, the CPT shows greater consistency and repeatability than the SPT. However, the CPT is a more recent development and there is less professional experience with it than with the SPT, particularly in the United States. As more data correlating CPT resistance to liquefaction resistance become available, the CPT is likely to be come the primary *in situ* test for evaluation of liquefaction potential. At present, however, a general consensus on the most appropriate technique for CPT-based evaluation of liquefaction potential has not emerged. One of the most well-developed procedures for CPT-based evaluation of liquefaction potential was described by Robertson and Wride. In this procedure, the measured CPT resistance, q_c, is normalized to a dimensionless resistance

$$q_{c1N} = C_Q \frac{q_c}{p_a} \tag{33.31}$$

where $C_Q = (p_a/\sigma'_{vo})^n$, p_a is atmospheric pressure, and n is an exponent that ranges from 0.5 (clean sand) to 1.0 (clay). A maximum C_Q value of 2.0 is generally applied to CPT data at shallow depths. Soil type can be inferred from CPT tip resistance, q_c, and sleeve resistance, f_s, with the aid of a soil behavior type index

$$I_c = \sqrt{(3.47 - \log Q)^2 + (1.22 + \log F)^2} \tag{33.32}$$

where

$$Q = C_Q \frac{q_c - \sigma_{vo}}{p_a}$$

$$F = \frac{f_s}{q_c - \sigma_{vo}} \times 100\%$$

If I_c (computed with $n = 1.0$) is greater than 2.6, the soil is considered too clayey to liquefy. If I_c (computed with $n = 0.5$ and $Q = q_{c1N}$) is less than 2.6, the soil is most likely granular and nonplastic and capable of liquefying. If I_c (computed with $n = 0.5$ and $Q = q_{c1N}$) is greater than 2.6, however, the soil is likely to be very silty and possibly plastic; in this case, I_c should be recalculated with $n = 0.7$ and $Q = q_{c1N}$. Once I_c has been determined, the effects of fines and plasticity can be considered by computing the clean sand normalized tip resistance

$$q_{c1N\text{-cs}} = K_c \, q_{c1N} \tag{33.33}$$

where

$K_c = 1.0$ for $I_c < 1.64$

$K_c = -0.403 I_c^4 + 5.581 I_c^3 - 21.63 I_c^2 + 33.75 I_c - 17.88$ for $I_c > 1.64$

With the clean sand normalized tip resistance, $CRR_{7.5}$ can be determined using Figure 33.27. For other magnitudes, the appropriate value of CRR can be obtained using the same magnitude scaling factor used for the SPT-based procedure, (Eq. 30). Other procedures for CPT-based evaluation of liquefaction potential include those of Seed and De Alba (57), Mitchell and Tseng [39], and Olson [46].

Liquefaction resistance has also been correlated to other *in situ* test measurements such as shear wave velocity [62,64], dilatometer index, and Becker penetration tests. In addition, probabilistic approaches that yield a probability of liquefaction have also been developed [32].

Lateral Spreading

Lateral spreading has often caused damage to bridges and bridge foundations in earthquakes. Lateral spreading generally involves the lateral movement of soil at and below the ground surface, often in the form of relatively intact surficial blocks riding on a mass of softened and weakened soil. The lateral soil movement can impose large lateral loads on abutments and wingwalls, and can induce large bending moments in pile foundations. The damage produced by lateral spreading is closely related to the magnitude of the lateral soil displacements.

Because cyclic mobility, the fundamental phenomenon that produces lateral spreading, is so complex, analytical procedures for prediction of lateral spreading displacements have not yet reached the point at which they can be used for design. As a result, currently accepted procedures for prediction of lateral spreading displacements are empirically based.

Bartlett and Youd [1] used multiple regression on a large database of lateral spreading case histories to develop empirical expressions for lateral spreading ground surface displacements. Two expressions were developed — a ground slope expression for sites with gentle, uniformly sloping

TABLE 33.6 Range of Verified Values for Eq. 33.32

Input Parameter	Range of Values
Magnitude	$6.0 < M_w < 8.0$
Free-face ratio	$1.0\% < W < 20\%$
Thickness of loose layer	$0.3 \text{ m} < T_{15} < 12 \text{ m}$
Fines content	$0\% < F_{15} < 50\%$
Mean grain size	$0.1 \text{ mm} < (D_{50})_{15} < 1.0 \text{ mm}$
Ground slope	$0.1\% < S < 6\%$
Depth to bottom of section	Depth to bottom of liquefied zone <15 m

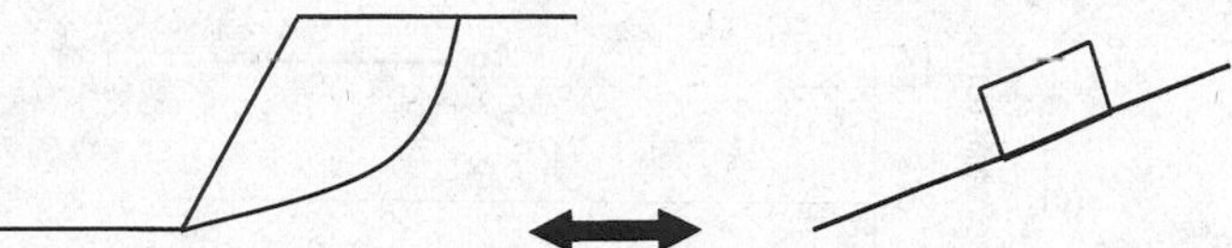

FIGURE 33.28 Illustration of sliding block analogy for evaluation of permanent slope displacements.

surfaces, and a free-face expression for sites near steep banks. For the former, displacements can be estimated from

$$\log D_H = -16.3658 + 1.1782 M_w - 0.9275 \log R - 0.0133 R + 0.4293 \log S \\ + 0.3483 \log T_{15} + 4.5720 \log(100 - F_{15}) - 0.9224 (D_{50})_{15} \tag{33.34a}$$

where D_H is the estimated lateral ground displacement in meters, M_w is the moment magnitude, R is the horizontal distance from the seismic energy source in km, S is the ground slope in percent, T_{15} is the cumulative thickness of saturated granular layers with $(N_1)_{60} < 15$ in meters, F_{15} is the average fines content for the granular layers comprising T_{15} in percent, and $(D_{50})_{15}$ is the average mean grain size for the granular layers comprising T_{15} in millimeters. For free-face sites, displacements can be estimated from

$$\log D_H = -16.3658 + 1.1782 M_w - 0.9275 \log R - 0.0133 R + 0.6572 \log W \\ + 0.3483 \log T_{15} + 4.5720 \log(100 - F_{15}) - 0.9224 (D_{50})_{15} \tag{33.34b}$$

where W is the ratio of the height of the bank to the horizontal distance between the toe of the bank and the point of interest. With these equations, 90% of the predicted displacements were within a factor of two of those observed in the corresponding case histories. The range of parameters for which the predicted results have been verified by case histories is presented in Table 33.6.

Global Instability

Ground failure may also occur due to the temporary exceedance of the shear strength of the soil by earthquake-induced shear stresses. These failures may take the form of large, deep-seated soil failures that can encompass an entire bridge abutment or foundation as illustrated in Figure 33.28. The potential for such failures, often referred to as global instabilities, must be evaluated during design.

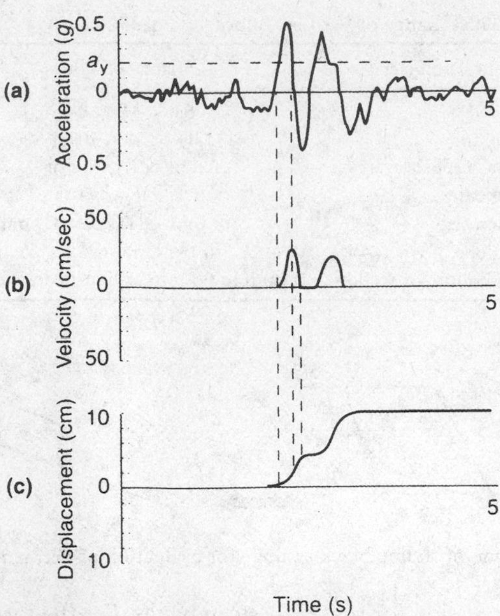

FIGURE 33.29 Illustration of computation of permanent slope displacements using sliding block method.

Historically, inertial failures were evaluated using **pseudo-static methods** in which the transient, dynamic effects of earthquake shaking were represented by constant, pseudo-static accelerations. The resulting destabilizing pseudo-static forces were included in a limit equilibrium analysis to compute a pseudo-static factor of safety. A pseudo-static factor of safety greater than one was considered indicative of stability. However, difficulty in selection of the pseudo-static acceleration, interpretation of the significance of computed factors of safety less than 1, and increasing recognition that serviceability is closely related to permanent deformations led to the development of alternative approaches.

The most common current procedure uses pseudo-static principles to establish the point at which permanent displacements would begin, but then uses a simple slope analogy to estimate the magnitude of the resulting permanent displacements. This procedure is commonly known as the sliding block procedure [43]. By using the common assumptions of rigid, perfectly plastic behavior embedded in limit equilibrium analyses, a potentially unstable slope is considered to be analogous to a block resting on an inclined plane (Figure 33.28) in the sliding block procedure. In both cases, base accelerations above a certain level will result in permanent relative displacements of the potentially unstable mass.

In the sliding block procedure, a pseudo-static analysis is performed to determine the horizontal pseudo-static acceleration that produces a factor of safety of 1.0. This pseudo-static acceleration, referred to as the **yield acceleration**, represents the level of acceleration above which permanent slope displacements are expected to occur. When the input acceleration exceeds the yield acceleration, the shear stress between the sliding block and the plane exceeds the available shear resistance and the block is unable to accelerate as quickly as the underlying plane. As a result, there is a relative acceleration between the block and the plane that lasts until the shear stress drops below the strength long enough to decelerate the block to zero relative acceleration. Integration of the relative acceleration over time yields a relative velocity, and integration of the relative velocity produces the relative displacement between the block and the plane. By this process, illustrated in Figure 33.29, the sliding block procedure allows estimation of the permanent displacement of a slope.

For embankments subjected to ground motions perpendicular to their axes, Makdisi and Seed [36] developed a simplified procedure for estimation of earthquake-induced displacements based

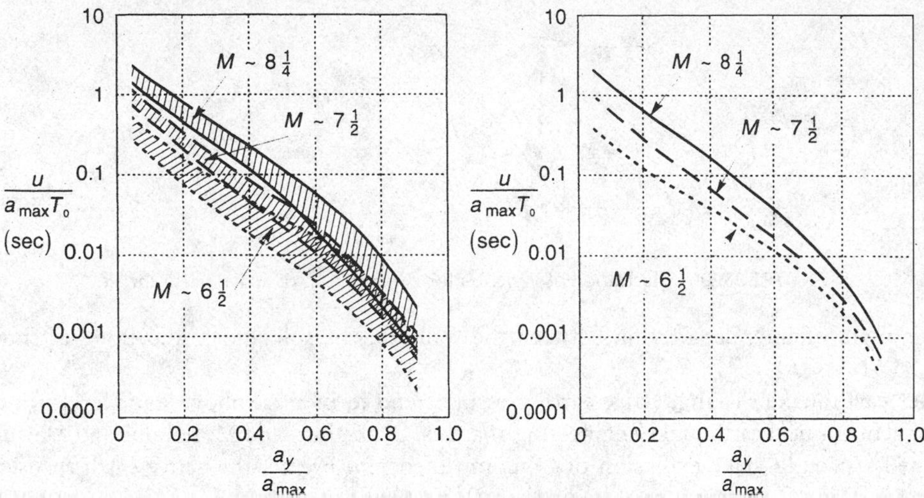

FIGURE 33.30 Variation of normalized permanent slope displacement with yield acceleration for earthquakes of different magnitudes: (a) summary for several different earthquakes and embankments and (b) average values (after Makdisi and Seed [36]).

on sliding block analyses of dams and embankments subjected to several recorded and synthetic input motions. By knowing the yield acceleration of the slope in addition to the peak acceleration and fundamental period of the embankment, Figure 33.30 can be used to estimate permanent slope displacements.

Retaining Structures

Earth-retaining structures are commonly constructed as parts of bridge construction projects and, in the form of abutment walls and wingwalls, as parts of bridge structures themselves. However, there are many different types of retaining structures, several of which have been developed in recent years. Historically, rigid retaining structures have been most commonly used; their static design is based on classical earth pressure theories. However, newer types of retaining structures, such as flexible anchored walls, soil nailed walls, and reinforced walls, have required the development of new approaches, even for static conditions. Under seismic conditions, classical earth pressure theories can be extended in a logical way to account for the effects of earthquake shaking, but seismic design procedures for the newer types of retaining structures remain under development.

Free-standing rigid retaining structures typically maintain equilibrium through the development of active and passive earth pressures that develop as the wall translates and rotates under the action of the imposed stresses. By assuming that static stresses develop through mobilization of the shear strength of the backfill soil on a planar potential failure surface, Coulomb earth pressure theory predicts a static active thrust of

$$P_A = \frac{1}{2} K_A \gamma H^2 \tag{33.35}$$

where

$$K_A = \frac{\cos^2(\phi - \theta)}{\cos^2\theta \cos(\delta + \theta)\left[\dfrac{\sin(\delta + \phi)\sin(\phi - \beta)}{\cos(\delta + \theta)\cos(\beta - \theta)}\right]^2}$$

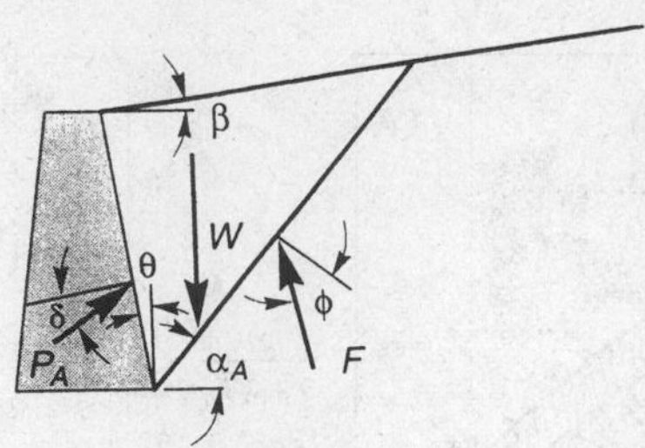

FIGURE 33.31 Illustration of variables for computation of active Earth thrust.

δ is the angle of interface friction between the wall and the soil, and β and θ are as shown in Figure 33.31.

Under earthquake shaking, active earth pressures tend to increase above static levels. In one of the first geotechnical earthquake engineering analyses, Okabe [45] and Mononobe and Matsuo [40] developed a pseudo-static extension of Coulomb theory to predict the active earth thrust under seismic conditions. Assuming pseudo-static accelerations of $a_h = k_h g$ and $a_v = k_v g$ in the horizontal and vertical directions, respectively, the Mononobe–Okabe total thrust is given by

$$P_{AE} = \frac{1}{2} K_{AE} \gamma H^2 (1 - k_v) \tag{33.36}$$

where

$$K_{AE} = \frac{\cos^2(\phi - \theta - \psi)}{\cos\psi \cos^2\theta \cos(\delta + \theta + \psi)\left[1 + \sqrt{\dfrac{\sin(\delta + \phi)\sin(\phi - \beta - \psi)}{\cos(\delta + \theta + \psi)\cos(\beta - \theta)}}\right]^2}$$

where $\phi - \beta \geq \psi$ and $\Psi = \tan^{-1}[k_h/(1 - k_v)]$. Although the assumptions used in the Mononobe-Okabe analysis imply that the total active thrust should act at a height of $H/3$ above the base of the wall, experimental results indicate that it acts at a higher point. The total active thrust of Eq. (33.36) can be divided into a static component, P_A, given by Eq. (33.35), and a dynamic component,

$$\Delta P_{AE} = P_{AE} - P_A \tag{33.37}$$

which acts at a height of approximately $0.6H$ above the base of the wall. On this basis, the total active thrust can be taken to act at a height

$$h = \frac{P_A H/3 + \Delta P_{AE}(0.6H)}{P_{AE}} \tag{33.38}$$

above the base of the wall.

When retaining walls are braced against lateral movement at top and bottom, as can occur with abutment walls, the shear strength of the soil will not be fully mobilized under static or seismic conditions. As a result, the limiting conditions of minimum active or maximum passive conditions cannot be developed. In such cases, it is common to estimate lateral Earth pressures using the elastic solution of Wood [69] for a linear elastic material of height, H, trapped between rigid walls separated by a horizontal distance, L. For motions at less than half the fundamental frequency of the unrestrained

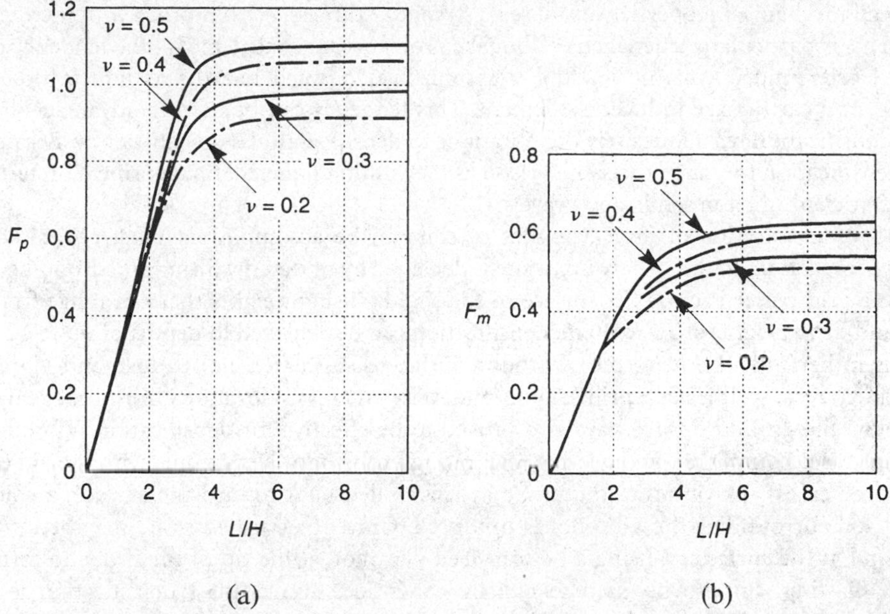

FIGURE 33.32 Charts for determination of (a) dimensionless thrust factor and (b) dimensionless moment factor for various geometries and Poisson ratios (after Wood [69]).

backfill ($f_o = v_s/4H$) the dynamic thrust and dynamic overturning moment (about the base of the wall) can be expressed as

$$\Delta P_{eq} = \gamma H^2 \frac{a_h}{g} F_p \tag{33.39}$$

$$\Delta M_{eq} = \gamma H^3 \frac{a_h}{g} F_m \tag{33.40}$$

where a_h is the amplitude of the harmonic base acceleration and F_p and F_m are dimensionless factors that can be obtained from Figure 33.32. It should be noted that Eqs. 33.39 and 33.40 refer to dynamic thrusts and moments; static thrusts and moments must be added to obtain total thrusts and moments.

33.9 Soil Improvement

When existing subsurface conditions introduce significant seismic hazards that adversely affect safety or impact construction costs, improved performance may be achieved through a program of soil improvement. A variety of techniques are available for soil improvement and may be divided into four main categories: densification, drainage, reinforcement, and grouting/mixing. Each soil improvement technique has advantages and disadvantages that influence the cost and effectiveness under different circumstances. Soil improvement techniques for both seismic and nonseismic areas are described in detail in such references as Welsh [67], Van Impe [65], Hausmann [27], Broms [8], Bell [2], and Mosely [41].

Densification Techniques

Virtually all mechanical properties of soil (e.g., strength, stiffness, etc.) improve with increasing soil density. This is particularly true when earthquake problems are considered — the tendency of loose soils to densify under dynamic loading is responsible for such hazards as liquefaction, lateral spreading, and earthquake-induced settlement. This tendency can be used to advantage, however, as most densification techniques rely on vibrations to densify granular soil efficiently. Because fines inhibit densification for much the same reason as they inhibit liquefaction, densification techniques are most efficient in clean sands and gravels.

Vibratory densification of large volumes of soil can be accomplished most economically by dynamic compaction. In this procedure, a site is densified by repeatedly lifting and dropping a heavy weight in a grid pattern across the surface of the site. By using weights that can range from 53 to 267 kN and drop heights of 10 to 30 m, densification can be achieved to depths of up to 12 m. The process is rather intrusive in terms of ground surface disturbance, noise, dust, and vibration of surrounding areas, so it is used primarily in undeveloped areas. Vibrations from probes that penetrate below the ground surface have also proved to be effective for densification. Vibroflotation, for example, is accomplished by lowering a vibrating probe into the ground (with the aid of water jets, in some cases). By vibrating the probe as it is pulled back toward the surface, a column of densified soil surrounding the vibroflot is produced. Gravel or crushed stone may be introduced into the soil at the surface or, using a bottom-feed vibroflot, at the tip of the probe to form stone columns. Blasting can also be used to densify cohesionless soils. Blast densification is usually accomplished by detonating multiple explosive charges spaced vertically at distances of 3 to 6 m in borings spaced horizontally at distances of 5 to 15 m. The charges at different elevations are often detonated at small time delays to enhance the amplitude, and therefore the densification capacity, of the blast waves. Two or three rounds of blasting, with later rounds detonated at locations between those of the earlier rounds, are often used to achieve the desired degree of densification. Finally, densification may be achieved using static means using compaction grouting. Compaction grouting involved the injection of very low slump (usually less than 25 mm) cementitious grout into the soil under high pressure. The grout forms an intact bulb or column that densifies the surrounding soil by displacement. Compaction grouting may be performed at a series of points in a grid or along a line. Grout points are typically spaced at distances of about 1 to 4 m, and have extended to depths of 30 m.

Drainage Techniques

Excessive soil and foundation movements can often be eliminated by lowering the groundwater table, and construction techniques for dewatering are well developed. The buildup of high porewater pressures in liquefiable soils can also be suppressed using drainage techniques, although drainage alone is rarely relied upon for mitigation of liquefaction hazards. Stone columns provide means for rapid drainage by horizontal flow, but also improve the soil by densification (during installation) and reinforcement.

Reinforcement Techniques

The strength and stiffness of some soil deposits can be improved by installing discrete inclusions that reinforce the soil. Stone columns are columns of dense angular gravel or crushed stone (stone columns) that reinforce the soil in which they are installed. Stone columns also improve the soil due to their drainage capabilities and the densification and lateral stress increase that generally occurs during their installation. Granular soils can also be improved by the installation of compaction piles, usually prestressed concrete or timber, driven in a grid pattern and left in place. Compaction piles can often increase relative densities to 75 to 80% within a distance of 7 to 12 pile diameters. Drilled inclusions such as drilled shafts or drilled piers have been used to stabilize many slopes, although the difficulty in drilling through loose granular soils limits their usefulness for slopes with liquefiable soils. Soil nails, tiebacks, micropiles, and root piles have also been used.

Grouting/Mixing Techniques

The characteristics of many soils can be improved by the addition of cementitious materials. Introduced by injection or mixing, these materials both strengthen the contacts between soil grains and fill the space between the grains. Grouting involves injection of cementitious materials into the voids of the soil or into fractures in the soil; in both cases, the particle structure of the majority of the soil remains intact. In mixing, the cementitious materials are mechanically or hydraulically mixed into the soil, completely destroying the initial particle structure.

Permeation grouting involves the injection of low-viscosity grouts into the voids of the soil without disturbing the particle structure. Both particulate grouts (aqueous suspensions of cement, fly ash, bentonite, microfine cement, etc.) and chemical grouts (silica and lignin gels, or phenolic and acrylic resins) may be used. The more viscous particulate grouts are generally used in coarser-grained soils with large voids such as gravels and coarse sands; chemical grouts can be used in fine sands. The presence of fines can significantly reduce the effectiveness of permeation grouting. Grout pipes are usually arranged in a grid pattern at spacings of 1.2 to 2.4 m and can produce grouted soil strengths of 350 to 2100 kPa. Intrusion grouting involves the injection of more viscous (and hence stronger) cementitious grouts under pressure to cause controlled fracturing of the ground. The first fractures generally follow weak bedding planes or minor principal stress planes; after allowing the initially placed grout to cure, repeated grouting fractures the soil along additional planes, eventually producing a three-dimensional network of intersecting grout lenses.

Using a mechanical system consisting of hollow stem augers and rotating paddles, soil mixing produces an amorphous mixture of soil and cementitious material. The soil-mixing process produces columns of soil–cement that can be arranged in a grid pattern or in a linear series of overlapping columns to produce subsurface walls and/or cellular structures. Soil mixing, which can be used in virtually all inorganic soils, has produced strengths of 1400 kPa and improvement to depths of 60 m. In jet grouting, cement grout is injected horizontally under high pressure through ports in the sides of a hollow rod lowered into a previously drilled borehole. Jet grouting begins at the bottom of the borehole and proceeds to the top. Rotation of the injection nozzle as the process occurs allows the jet to cut through and hydraulically mix columns of soil up to 2.4 m in diameter. Air or air and water may also be injected to aid in the mixing process. Jet grouting can be performed in any type of inorganic soil to depths limited only by the range of the drilling equipment.

Defining Terms

Selected terms used in this section are compiled below:

Amplification function: A function that describes the ratio of ground surface motion to bedrock motion as a function of frequency.

Attenuation: The rate at which earthquake ground motion decreases with distance.

Backbone curve: The nonlinear stress–strain curve of a monotonically loaded soil.

Blind thrust faults: Faults at depth occurring under anticlinal folds — since they have only subtle surface expression, their seismogenic potential can only be evaluated by indirect means [22]. Blind thrust faults are particularly worrisome because they are hidden, are associated with folded topography in general, including areas of lower and infrequent seismicity, and therefore result in a situation where the potential for an earthquake exists in any area of anticlinal geology, even if there are few or no earthquakes in the historic record. Recent major earthquakes of this type have included the 1980 M_w 7.3 El Asnam (Algeria), 1988 M_w 6.8 Spitak (Armenia), and 1994 M_w 6.7 Northridge (California) events.

Body waves: Vibrational waves transmitted through the body of the Earth, and are of two types: p waves (transmitting energy via dilatational or push–pull motion), and slower s waves (transmitting energy via shear action at right angles to the direction of motion).

Characteristic earthquake: A relatively narrow range of magnitudes at or near the maximum that can be produced by the geometry, mechanical properties, and state of stress of a fault. [56].

Coherence: The similarity of ground motions at different locations. The coherence of ground motions at closely spaced locations is higher than at greater spacings. At a given spacing, the coherence of low-frequency (long wavelength) components is greater than that of high-frequency (short-wavelength) components.

Cyclic mobility: A phenomenon involving accumulation of porewater pressure during cyclic loading in soils for which the residual shear strength is greater than the shear stress required to maintain static equilibrium.

Cyclic resistance ratio (CRR): The ratio of equivalent shear stress amplitude required to trigger liquefaction to the initial vertical effective stress acting on the soil.

Cyclic stress ratio (CSR): The ratio of equivalent shear stress amplitude of an earthquake ground motion to the initial vertical effective stress acting on the soil.

Damping: The force or energy lost in the process of material deformation (damping coefficient c = force per velocity).

Damping curve: A plot of equivalent viscous damping ratio as a function of shear strain amplitude.

Differential settlement: The relative amplitudes of settlement at different locations. Differential settlement may be particularly damaging to bridges and other structures.

Dip: The angle between a plane, such as a fault, and the Earth's surface.

Dip-slip: Motion at right angles to the strike, up- or down-slip.

Ductility factor: The ratio of the total displacement (elastic plus inelastic) to the elastic(i.e., yield) displacement.

Epicenter: The projection on the surface of the Earth directly above the hypocenter.

Equivalent linear analysis: An analysis in which the stress–strain behavior of the soil is characterized by a secant shear modulus and damping ratio that, through a process of iteration, are compatible with the level of shear strain induced in the soil.

Far-field: Beyond near-field, also termed teleseismic.

Fault: A zone of the Earth's crust within which the two sides have moved — faults may be hundreds of miles long, from 1 to over 100 miles deep, and may not be readily apparent on the ground surface.

Flow failure: A soil failure resulting from flow liquefaction. Flow failures can involve very large deformations.

Flow liquefaction: A phenomenon that can occur when liquefaction is triggered in a soil with a residual shear strength lower than the shear stress required to maintain static equilibrium.

Hypocenter: The location of initial radiation of seismic waves (i.e., the first location of dynamic rupture).

Intensity: A metric of the effect, or the strength, of an earthquake hazard at a specific location, commonly measured on qualitative scales such as MMI, MSK, and JMA.

Lateral spreading: A phenomenon resulting from cyclic mobility in soils with some nonzero initial shear stress. Lateral spreading is characterized by the incremental development of permanent lateral soil deformations.

Magnitude: A unique measure of an individual earthquake's release of strain energy, measured on a variety of scales, of which the moment magnitude M_w (derived from seismic moment) is preferred.

Magnitude-frequency relation: The probability of occurrence of a selected magnitude — the commonest is $\log_{10} n(m) = a - bm$ [25]

Meizoseismal: The area of strong shaking and damage.

Modulus reduction curve: The ratio of secant shear modulus at a particular shear strain to maximum shear modulus (corresponding to very low strains) plotted as a function of shear strain amplitude.

Near-field: Within one source dimension of the epicenter, where source dimension refers to the length or width of faulting, whichever is less.

Nonlinear approach: An analysis in which the nonlinear, inelastic stress–strain behavior of the soil is explicitly modeled.

Normal fault: A fault that exhibits dip-slip motion, where the two sides are in tension and move away from each other.

Peak ground acceleration (PGA): The maximum amplitude of recorded acceleration (also termed the ZPA, or zero-period acceleration)

Pseudo-static approach: A method of analysis in which the complex, transient effects of earthquake shaking are represented by constant accelerations. The inertial forces produced by these accelerations are considered, along with the static forces, in limit equilibrium stability analyses.

Radiation damping: A reduction in wave amplitude due to geometric spreading of traveling waves, or radiation into adjacent or underlying materials.

Response spectrum: A plot of maximum amplitudes (acceleration, velocity or displacement) of an sdof oscillator, as the natural period of the SDOF is varied across a spectrum of engineering interest (typically, for natural periods from 0.03 to 3 or more seconds or frequencies of 0.3 to 30+ Hz).

Reverse fault: A fault that exhibits dip-slip motion, where the two sides are in compression and move away toward each other.

Seismic hazards: The phenomena and/or expectation of an earthquake-related agent of damage, such as fault rupture, vibratory ground motion (i.e., shaking), inundation (e.g., tsunami, seiche, dam failure), various kinds of permanent ground failure (e.g., liquefaction), fire or hazardous materials release.

Seismic moment: The moment generated by the forces generated on an earthquake fault during slip.

Seismotectonic model: A mathematical model representing the seismicity, attenuation, and related environment.

Specific impedance: Product of density and wave propagation velocity.

Spectrum amplification factor: The ratio of a response spectral parameter to the ground motion parameter (where parameter indicates acceleration, velocity, or displacement).

Strike: The intersection of a fault and the surface of the Earth, usually measured from the north (e.g., the fault strike is N 60° W).

Strike slip fault: See Transform or Strike slip fault.

Subduction: The plunging of a tectonic plate (e.g., the Pacific) beneath another (e.g., the North American) down into the mantle, due to convergent motion.

Surface waves: Vibrational waves transmitted within the surficial layer of the Earth, of two types: horizontally oscillating Love waves (analogous to S body waves) and vertically oscillating Rayleigh waves.

Thrust fault: Low-angle reverse faulting (blind thrust faults are faults at depth occurring under anticlinal folds — they have only subtle surface expression).

Total settlement: The total amplitude of settlement at a particular location.

Transform or strike slip fault: A fault where relative fault motion occurs in the horizontal plane, parallel to the strike of the fault.

Uniform hazard spectra: Response spectra with the attribute that the probability of exceedance is independent of frequency.

Yield acceleration: The horizontal acceleration that produces a pseudo-static factor of safety of 1. Accelerations greater than the yield acceleration are expected to produce permanent deformations.

References

1. Bartlett, S.F. and Youd, T.L., Empirical analysis of horizontal ground displacement generated by liquefaction-induced lateral spread, *Technical Report NCEER-92-0021*, National Center for Earthquake Engineering Research, Buffalo, NY, 1992.
2. Bell, F.G., *Engineering Treatment of Soils*, E & FN Spon, London, 1993, 302 pp.

3. Bernreuter, D.L. et al., Seismic Hazard Characterization of 69 Nuclear Power Plant Sites East of the Rocky Mountains, U.S. Nuclear Regulatory Commission, NUREG/CR-5250, 1989.
4. Bolt, B.A., *Earthquakes*, W.H. Freeman, New York, 1993.
5. Bonilla, M.G. et al., Statistical relations among earthquake magnitude, surface rupture length, and surface fault displacement, *Bull. Seismol. Soc. Am.*, 74(6), 2379–2411, 1984.
6. Boore, D.M., Joyner, W.B., and Fumal, T.E., Estimation of Response Spectra and Peak Acceleration from Western North American Earthquakes: An Interim Report, U.S.G.S Open-File Report 93-509, Menlo Park, CA, 1993.
7. Bozorgnia, Y. and Campbell, K.W., Spectral characteristics of vertical ground motion in the Northridge and other earthquakes, *Proc. 4th U.S. Conf. On Lifeline Earthquake Eng.*, American Society of Civil Engineers, New York, 660–667, 1995.
8. Broms, B., Deep compaction of granular soil, in H.-Y. Fang, Ed., *Foundation Engineering Handbook*, 2nd ed., Van Nostrand Reinhold, New York, 1991, 814–832.
9. BSSC, NEHRP Recommended Provisions for Seismic Regulations for New Buildings, 1994.
10. Campbell, K.W., Strong ground motion attenuation relations: a ten-year perspective, *Earthquake Spectra*, 1(4), 759–804, 1985.
11. Campbell, K.W. and Bozorgnia, Y., Near-source attenuation of peak horizontal acceleration from worldwide accelerograms recorded from 1957 to 1993, *Proc. Fifth U.S. National Conference on Earthquake Engineering*, Earthquake Engineering Research Institute, Oakland, CA, 1994.
12. Clough, R.W. and Penzien, J., *Dynamics of Structures*, McGraw-Hill, New York, 1975.
13. Cornell, C.A., Engineering seismic risk analysis, *Bull. Seismol Soc. Am.*, 58(5), 1583–1606, 1968.
14. Cornell, C.A. and Merz, H.A., Seismic risk analysis based on a quadratic magnitude frequency law, *Bull. Seis. Soc. Am.*, 63(6), 1992–2006, 1973.
15. Crouse, C. B., Ground-motion attenuation equations for earthquakes on the cascadia subduction zone, *Earthquake Spectra*, 7(2), 201–236, 1991.
16. Darragh, R.B., Huang, M.J., and Shakal, A.F., Earthquake engineering aspects of strong motion data from recent California earthquakes, *Proc. Fifth U.S. National Conf. Earthquake Engineering*, V. III, Earthquake Engineering Research Institute, Oakland, CA, 99–108, 1994.
17. Donovan, N.C. and Bornstein, A.E., Uncertainties in seismic risk procedures, *J. Geotech. Div.*, 104(GT7), 869–887, 1978.
18. EduPro Civil Systems, Inc., *ProShake User's Manual*, EduPro Civil Systems, Inc., Redmond, WA, 1998, 52 pp.
19. Electric Power Research Institute, Seismic Hazard Methodology for the Central and Eastern United States, EPRI NP-4726, Menlo Park, 1986.
20. Esteva, L., Seismicity, in *Seismic Risk and Engineering Decisions*, Lomnitz, C. and Rosenblueth, E., Eds., Elsevier, New York, 1976.
21. Finn, W.D.L., Yogendrakumar, M., Yoshida, M., and Yoshida, N., TARA-3: A program to compute the response of 2-D embankments and soil-structure interaction systems to seismic loadings, Department of Civil Engineering, University of Brirish Columbia, Vancouver, Canada, 1986.
22. Greenwood, R.B., Characterizing blind thrust fault sources — an overview, in Woods, M.C. and Seiple. W.R., Eds., The Northridge California Earthquake of 17 January 1994, California Department Conservation, Division of Mines and Geology, Special Publ. 116, 1995, 279–287.
23. Gumbel, E.J., *Statistics of Extremes*, Columbia University Press, New York, 1958.
24. Gutenberg, B. and Richter, C.F., *Seismicity of the Earth and Associated Phenomena*, Princeton University Press, Princeton, NJ, 1954.
25. Gutenberg, B. and Richter, C. F., Magnitude and energy of earthquakes, *Ann. Geof.*, 9(1), p 1–15, 1956.
26. Hanks, T.C. and Kanamori, H., A moment magnitude scale, *J. Geophys. Res.*, 84, 2348–2350, 1979.
27. Hausmann, M.R., *Engineering Principles of Ground Modification*, McGraw-Hill, New York, 1990, 632 pp.

28. Housner, G., Historical view of earthquake engineering, in *Proc. Post-Conf. Volume, Eighth World Conf. on Earthquake Engineering,* Earthquake Engineering Research Institute, Oakland, CA, 1984, 25–39, as quoted by S. Otani [1995].
29. Kagawa, T., Mejia, L., Seed, H.B., and Lysmer, J., TLUSH — a computer program for three-dimensional dynamic analysis of Earth dams, Report No. UCB/EERC-81/14, University of California, Berkeley, 1981.
30. Lawson, A. C. and Reid, H.F., *The California Earthquake of April 18, 1906. Report of the State Earthquake Investigation Commission,* California. State Earthquake Investigation Commission, Carnegie Institution of Washington, Washington, D.C., 1908–1910.
31. Lee, M.K.W and Finn, W.D.L., DESRA-2, Dynamic effective stress response analysis of soil deposits with energy transmitting boundary including assessment of liquefaction potential, Soil Mechanics Series No. 38, University of British Columbia, Vancouver, 1978.
32. Liao, S.S.C., Veneziano, D., and Whitman, R.V., Regression models for evaluating liquefaction probability, *J. Geotech. Eng. ASCE,* 114(4), 389–411, 1988.
33. Lomnitz, C., *Global Tectonics and Earthquake Risk,* Elsevier, New York, 1974.
34. Lysmer, J., Udaka, T., Tsai, C.F., and Seed, H.B., FLUSH — A computer program for approximate 3-D analysis of soil-structure interaction problems, Report No. EERC 75-30, Earthquake Engineering Research Center, University of California, Berkeley, 1975, 83 pp.
35. MacMurdo, J., Papers relating to the earthquake which occurred in India in 1819, *Philos. Mag.,* 63, 105–177, 1824.
36. Makdisi, F.I. and Seed, H.B., Simplified procedure for estimating dam and embankment earthquake-induced deformations, *J. Geotech. Eng. Div. ASCE,* 104(GT7), 849–867, 1978.
37. McGuire, R. K., FRISK: Computer Program for Seismic Risk Analysis Using Faults as Earthquake Sources. U.S. Geological Survey, Reports, U.S. Geological Survey Open file 78-1007, 1978, 71 pp.
38. Meeting on Updating of MSK-64, Report on the ad hoc Panel Meeting of Experts on Updating of the MSK-64 Seismic Intensity Scale, Jene, 10–14 March 1980, *Gerlands Beitr. Geophys.,* Leipzeig 90(3), 261–268, 1981.
39. Mitchell, J.K. and Tseng, D.-J., Assessment of liquefaction potential by cone penetration resistance, *Proceedings, H. Bolton Seed Memorial Symposium,* Berkeley, CA, Vol. 2, J.M. Duncan, Ed., 1990, 335–350.
40. Mononobe, N. and Matsuo, H., On the determination of Earth pressures during earthquakes, *Proceedings, World Engineering Congress,* 9, 1929.
41. Mosely, M.P., ed., *Ground Improvement,* Blackie Academic & Professional, London, 1993, 218 pp.
42. National Academy Press, *Probabilistic Seismic Hazard Analysis,* National Academy of Sciences, Washington, D.C., 1988.
43. Newmark, N., Effects of earthquakes on dams and embankments, *Geotechnique,* 15(2), 139–160, 1965.
44. Ohsaki, Y., Effects of sand compaction on liquefaction during Tokachioki earthquake, *Soils Found.,* 10(2), 112–128, 1970.
45. Okabe, S., General theory of Earth pressures, *J. Jpn. Soc. Civil Eng.,* 12(1), 1926.
46. Olson, R.S., Cyclic liquefaction based on the cone penetrometer test, Proceedings of the NCEER Workshop on Evaluation of Liquefaction Resistance of Soils, Technical Report NCEER-97-0022, T.L. Youd and I.M. Idriss, Eds., National Center for Earthquake Engineering Research, Buffalo, NY, 1997, 280 pp.
47. Pyke, R., Seed, H.B., and Chan, C.K., Settlement of sands under multi-directional loading, *J. Geotech. Eng. Div. ASCE,* 101(GT4), 379–398, 1975.
48. Pyke, R.L., TESS1 User's Guide, TAGA Engineering Software Services, Berkeley, CA, 1985.
49. Reid, H.F., The Mechanics of the Earthquake, The California Earthquake of April 18, 1906, Report of the State Investigation Committee, vol. 2, Carnegie Institution of Washington, Washington, D.C., 1910.

50. Reiter, L., *Earthquake Hazard Analysis, Issues and Insights*, Columbia University Press, New York, 1990.
51. Richter, C. F., An instrumental earthquake scale, *Bull. Seismol. Soc. Am.*, 25 1–32, 1935.
52. Richter, C.F., *Elementary Seismology*, W.H. Freeman, San Francisco, 1958.
53. Schnabel, P.B., Lysmer, J., and Seed, H.B., SHAKE: A computer program for earthquake response analysis of horizontally layered sites, Report No. EERC 72-12, Earthquake Engineering Research Center, University of California, Berkeley, 1972.
54. Scholz, C.H., *The Mechanics of Earthquakes and Faulting*, Cambridge University Press, New York, 1990.
55. Schwartz, D.P., Geologic characterization of seismic sources: moving into the 1990s, in *Earthquake Engineering and Soil Dynamics II — Recent Advances in Ground-Motion Evaluation*, J.L. v. Thun, Ed., Geotechnical Spec. Publ. No. 20., American Society of Civil Engineers, New York, 1988.
56. Schwartz, D.P. and Coppersmith, K.J., Fault behavior and characteristic earthquakes: examples from the Wasatch and San Andreas faults, *J. Geophys. Res.*, 89, 5681–5698, 1984.
57. Seed, H.B. and De Alba, P., Use of SPT and CPT tests for evaluating the liquefaction resistance of soils, *Proceedings, Insitu '86*, ASCE, 1986.
58. Seed, H.B. and Silver, M.L., Settlement of dry sands during earthquakes, *J. Soil Mech. Found. Div.*, 98(SM4), 381–397, 1972.
59. Seed, H.B., Ugas, C., and Lysmer, J., Site-dependent spectra for earthquake-resistant design, *Bull. Seismol. Soc. Am.*, 66, 221–243, 1976.
60. Slemmons, D.B., State-of-the-art for assessing earthquake hazards in the United States, Report 6: Faults and earthquake magnitude, U.S. Army Corps of Engineers, Waterways Experiment Station, Misc. Paper s-73-1, 1977, 129 pp.
61. Stein, R.S. and Yeats, R.S., Hidden earthquakes, *Sci. Am.*, June, 1989.
62. Stokoe, K.H. II, Roesset, J.M., Bierschwale, J.G., and Aouad, M., Liquefaction potential of sands from shear wave velocity, *Proceedings, 9th World Conference on Earthquake Engineering*, Tokyo, Japan, Vol. 3, 1988, 213–218.
63. Tokimatsu, K. and Seed, H.B., Evaluation of settlements in sand due to earthquake shaking, *J. Geotechn. Eng. ASCE*, 113(8), 861–878, 1987.
64. Tokimatsu, K., Kuwayama, S., and Tamura, S., Liquefaction potential evaluation based on Rayleigh wave investigation and its comparison with field behavior, *Proceedings, 2nd International Conference on Recent Advances in Geotechnical Earthquake Engineering and Soil Dynamics*, St. Louis, MO, Vol. 1, 1991, 357–364.
65. Van Impe, W.F., *Soil Improvement Techniques and Their Evolution*, A.A. Balkema, Rotterdam, 1989, 125 pp.
66. Wells, D.L. and Coppersmith, K.J., Empirical relationships among magnitude, rupture length, rupture width, rupture area and surface displacement, *Bull. Seisinol. Soc. Am.*, 84(4), 974–1002, 1994.
67. Welsh, J.P., Soil Improvement: A Ten Year Update, Geotechnical Special Publication No. 12, ASCE, New York, 1987, 331 pp.
68. Wood, H.O. and Neumann, Fr., Modified Mercalli intensity scale of 1931, *Bull. Seismol. Soc. Am.*, 21, 277–283, 1931.
69. Wood, J., Earthquake-Induced Soil Pressures on Structures, Report No. EERL 73-05, California Institute of Technology, Pasadena, 1973, 311 pp.
70. Yegulalp, T.M. and Kuo, J.T., Statistical prediction of the occurrence of maximum magnitude earthquakes, *Bull. Seis. Soc. Am.*, 64(2), 393–414, 1974.
71. Youngs, R R. and Coppersmith, K J., Attenuation relationships for evaluation of seismic hazards from large subduction zone earthquakes, *Proceedings of Conference XLVIII: 3rd Annual Workshop on Earthquake Hazards in the Puget Sound, Portland Area*, March 28–30, 1989, Portland, OR; Hays-Walter-W., Ed., U.S. Geological Survey, Reston, VA, 1989, 42–49.

72. Youngs, R.R. and Coppersmith, K.J., Implication of fault slip rates and earthquake recurrence models to probabilistic seismic hazard estimates, *Bull. Seismol. Soc. Am.*, 75, 939–964, 1987.

34

Earthquake Damage to Bridges

Jack P. Moehle
University of California, Berkeley

Marc O. Eberhard
University of Washington

34.1 Introduction

Earthquake damage to a bridge can have severe consequences. Clearly, the collapse of a bridge places people on or below the bridge at risk, and it must be replaced after the earthquake unless alternative transportation paths are identified. The consequences of less severe damage are less obvious and dramatic, but they are nonetheless important. A bridge closure, even if it is temporary, can have tremendous consequences, because bridges often provide vital links in a transportation system. In the immediate aftermath of an earthquake, closure of a bridge can impair emergency response operations. Later, the economic impact of a bridge closure increases with the length of time the bridge is closed, the economic importance of the traffic using the route, the traffic delay caused by following alternate routes, and the replacement cost for the bridge.

The purpose of this chapter is to identify and classify types of damage to bridges that earthquakes commonly induce and, where possible, to identify the causes of the damage. This task is not straightforward. Damage usually results from a complex and interacting set of contributing variables. The details of damage often are obscured by the damage itself, so that some speculation is required in reconstructing the event. In many cases, the cause of damage can be understood only after detailed analysis, and, even then, the actual causes and effects may be elusive.

Even when the cause of a particular collapse is well understood, it is difficult to generalize about the causes of bridge damage. In past earthquakes, the nature and extent of damage that each bridge

0-8493-7434-0/00/$0.00+$.50

suffered have varied with the characteristics of the ground motion at the particular site and the construction details of the particular bridge. No two earthquakes or bridge sites are identical. Design and construction practices vary extensively throughout the world and even within the United States. These practices have evolved with time, and, in particular, seismic design practice improved significantly in the western United States during the 1970s as a result of experience gained from the 1971 San Fernando earthquake.

Despite these uncertainties and variations, one can learn from past earthquake damage, because many types of damage occur repeatedly. By being aware of typical vulnerabilities that bridges have experienced, it is possible to gain insight into structural behavior and to identify potential weaknesses in existing and new bridges. Historically, observed damage has provided the impetus for many improvements in earthquake engineering codes and practice.

An effort is made to distinguish damage according to two classes, as follows:

Primary damage — Damage caused by earthquake ground shaking or deformation that was the primary cause of damage to the bridge, and that may have triggered other damage or collapse.

Secondary damage — Damage caused by earthquake ground shaking or deformation that was the result of structural failures elsewhere in the bridge, and was caused by redistribution of internal actions for which the structure was not designed.

The emphasis in this chapter is on primary damage. It must be accepted, however, that in many cases the distinction between primary and secondary damage is obscure because the bridge geometry is complex or, in the case of collapse, because it is difficult to reconstruct the failure sequence.

The following sections are organized according to which element in the overall set of contributing factors appears to be the primary cause of the bridge damage. The first three sections address general issues related to the site conditions, construction era, and current condition of the bridge. The next section focuses on the effects of structural configuration, including curved layout, skew, and redundancy. Unseating of superstructures at expansion joints is discussed in the subsequent section. Then, the chapter describes typical types of damage to the superstructure, followed by discussion of damage related to bearings and restrainers supporting or interconnecting segments of the superstructure. The final section describes damage associated with the substructure, including the foundation.

34.2 Effects of Site Conditions

Performance of a bridge structure during an earthquake is likely to be influenced by proximity of the bridge to the fault and site conditions. Both of these factors affect the intensity of ground shaking and ground deformations, as well as the variability of those effects along the length of the bridge.

The influence of site conditions on bridge response became widely recognized following the 1989 Loma Prieta earthquake. Figure 34.1 plots the locations of minor and major bridge damage from the Loma Prieta earthquake [16]. With some exceptions, the most significant damage occurred around the perimeter or within San Francisco Bay where relatively deep and soft soil deposits amplified the bedrock ground motion. In the same earthquake, the locations of collapse of the Cypress Street Viaduct nearly coincided with zones of natural and artificial fill where ground shaking was likely to have been the strongest (Figure 34.2) [10]. A major conclusion to be drawn from this and other earthquakes is the significant impact that local site conditions have on amplifying strong ground motion, and the subsequent increased vulnerability of bridges on soft soil sites. This observation is important because many bridges and elevated roadways traverse bodies of water where soft soil deposits are common.

During the 1995 Hyogo-Ken Nanbu (Kobe) earthquake, significant damage and collapse likewise occurred in elevated roadways and bridges founded adjacent to or within Osaka Bay [2]. Several types of site conditions contributed to the failures. First, many of the bridges were founded on sand–gravel terraces (alluvial deposits) overlying gravel–sand–mud deposits at depths of less than 33 ft (10 m), a condition which is believed to have led to site amplification of the bedrock motions.

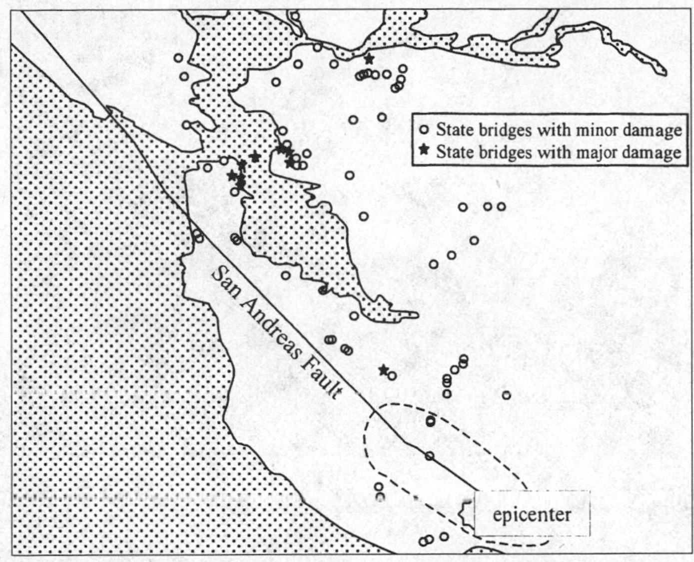

FIGURE 34.1 Incidence of minor and major damage in the 1989 Loma Prieta earthquake [modified from Zelinski, 16].

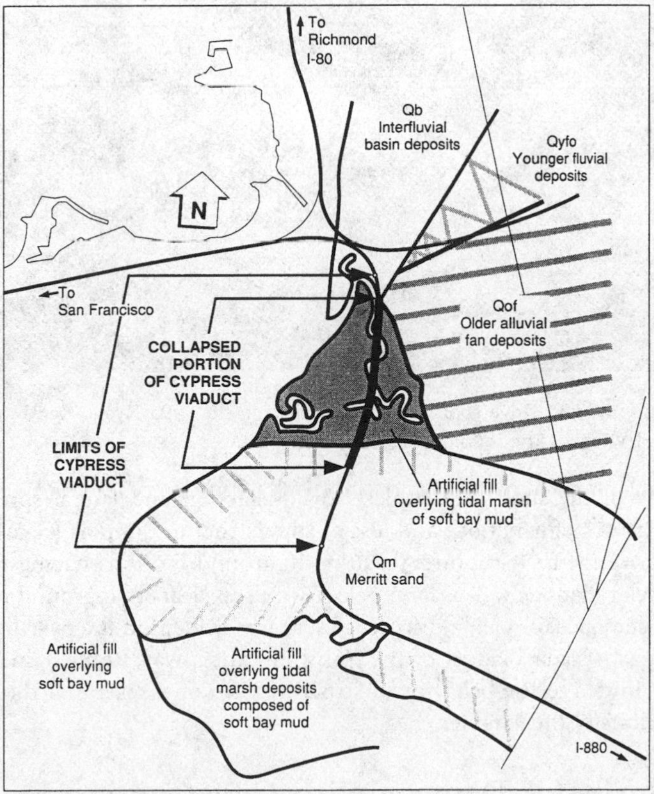

FIGURE 34.2 Geologic map of Cypress Street Viaduct site. (*Source:* Housner, G., Report to the Governor, Office of Planning and Research, State of California, 1990.)

FIGURE 34.3 Nishinomiya-ko Bridge approach span collapse in the 1995 Hyogo-Ken Nanbu earthquake [Kobe Collection, EERC Library, University of California, Berkeley].

Furthermore, many of the sites were subject to liquefaction and lateral spreading, resulting in permanent substructure deformations and loss of superstructure support (Figure 34.3). Finally, the site was directly above the fault rupture, resulting in ground motions having high horizontal and vertical ground accelerations as well as large velocity pulses. Near-fault ground motions can impose large deformation demands on yielding structures, as was evident in the overturning collapse of all 17 bents of the Higashi-Nada Viaduct of the Hanshin Expressway, Route 3, in Kobe (Figure 34.4). Other factors contributed to the behavior of structures in Kobe; several of these will be discussed in subsequent portions of this chapter.

34.3 Correlation of Damage with Construction Era

Bridge seismic design practices have changed over the years, largely reflecting lessons learned from performance in past earthquakes. Several examples in the literature demonstrate that the construction

FIGURE 34.4 Higashi-Nada Viaduct collapse in the 1995 Hyogo-Ken Nanbu earthquake. (*Source:* EERI, The Hyogo-Ken Nanbu Earthquake, January 17, 1995, Preliminary Reconnaissance Report, Feb. 1995.)

era of a bridge is a good indicator of likely performance, with higher damage levels expected in older construction than in newer construction.

An excellent example of the effect of construction era is provided by observing the relative performances of bridges on Routes 3 and 5 of the Hanshin Expressway in Kobe. Route 3 was constructed from 1965 through 1970, while Route 5 was completed in the early to mid-1990s [2]. The two routes are parallel to one another, with Route 3 being farther inland and Route 5 being built largely on reclaimed land. Despite the potentially worse soil conditions for Route 5, it performed far better than Route 3, losing only a single span owing apparently to permanent ground deformation and span unseating (Figure 34.3). In contrast, Route 3 has been estimated to have sustained moderate-to-large-scale damage in 637 piers, with damage in over 1300 spans, and approximately 50 spans requiring replacement (see, for example, Figure 34.4).

The superior performance of newer construction in the Hyogo-Ken Nanbu earthquake and other earthquakes [2,8,10] has led to the use of benchmark years as a crude but effective method for rapidly assessing the likely performance of bridge construction. This method has been an effective tool for bridge assessment in California. The reason for its success there is the rapid change in bridge construction practice following the 1971 San Fernando earthquake [8]. Before that time, California design and construction practice was based on significantly lower design forces and less stringent detailing requirements compared with current requirements. In the period following that earthquake, the California Department of Transportation (Caltrans) developed new design approaches requiring increased strength and improved detailing for ductile response.

The 1994 Northridge earthquake provides an insightful study on the use of benchmarking. Over 2500 bridges existed in the metropolitan Los Angeles freeway system at that time. Table 34.1 summarizes cases of major damage and collapse [8]. All these cases correspond to bridges designed before or around the time of the major change in the Caltrans specifications. It is interesting to note that some bridges constructed as late as 1976 appear in this table. This reflects the fact that the new design provisions did not take full effect until a few years after the earthquake and that these did not govern construction of some bridges that were at an advanced design stage at that time. Some caution is therefore required in establishing and interpreting the concept of benchmark years.

34.4 Effects of Changes in Condition

Changes in the condition of a bridge can greatly affect its seismic performance. In many regions of North America, extensive deterioration of bridge superstructures, bearings, and substructures has

TABLE 34.1 Summary of Bridges with Major Damage — Northridge Earthquake

Bridge Name	Route	Construction Year	Prominent Damage
		Collapse	
La Cienega-Venice Undercrossing	I-10	1964	Column failures
Gavin Canyon Undercrossing	I-5	1967	Unseating at skewed expansion hinges
Route 14/5 Separation and Overhead	I-5/SR14	1971/1974	Column failure
North Connector	I-5/SR14	1975	Column failure
Mission-Gothic Undercrossing	SR118	1976	Column failures
		Major Damage	
Fairfax-Washington Undercrossing	I-10	1964	Column failures
South Connector Overcrossing	I-5/SR14	1971/1972	Pounding at expansion hinges
Route 14/5 Separation and Overhead	I-5/SR14	1971/1974	Pounding at expansion hinges
Bull Creek Canyon Channel Bridge	SR118	1976	Column failures

accumulated. It is evident that the current conditions will lead to reduced seismic performance in future earthquakes, although hard evidence is lacking because of a paucity of earthquakes in these regions in modern times.

Construction modifications, either during the original construction or during the service life, can also have a major effect on bridge performance. Several graphic examples were provided by the Northridge earthquake [8]. Figure 34.5 shows a bridge column that was unintentionally restrained by a reinforced concrete channel wall. The wall shortened the effective length of the column, increased the column shear force, and shifted nonlinear response from a zone of heavy confinement upward to a zone of light transverse reinforcement, where the ductility capacity was inadequate. Failures of this type illustrate the importance of careful inspection during construction and during the service life of a bridge.

34.5 Effects of Structural Configuration

Ideally, earthquake-resistant construction should be designed to have a regular configuration so that the behavior is simple to conceptualize and analyze, and so that inelastic energy dissipation is promoted in a large number of readily identified yielding components. This ideal often is not achievable in bridge construction because of irregularities imposed by site conditions and traffic flow requirements. In theory, any member or joint can be configured to resist the induced force and deformation demands. However, in practice, bridges with certain configurations are more vulnerable to earthquakes than others.

Experience indicates that a bridge is most likely to be vulnerable if (1) excessive deformation demands occur in a few brittle elements, (2) the structural configuration is complex, or (3) a bridge lacks redundancy. The bridge designer needs to recognize the potential consequences of these irregularities and to design accordingly either to reduce the irregularity or to toughen the structure to compensate for it.

A common form of irregularity arises when a bridge traverses a basin requiring columns of nonuniform length. Although the response of the superstructure may be relatively uniform, the deformation demands on the individual substructure piers are highly irregular; the largest strains are imposed on the shortest columns. In some cases, the deformation demands on the short columns can induce their failure before longer, more flexible adjacent columns can fully participate. The Route 14/5 Separation and Overhead structure provides an example of these phenomena. The structure comprised a box-girder monolithic with single-column bents that varied in height depending on the road and grade elevations (Figure 34.6a). Apparently, the short column at Bent 2 failed in shear because of large deformation demands in that column, resulting in the collapse of the adjacent spans (Figure 34.6b).

FIGURE 34.5 Bull Creek Canyon Channel Bridge damage in the 1994 Northridge earthquake.

The effects identified above can be exacerbated in long-span bridges. In addition to changes in subgrade and structural irregularities that may be required to resolve complex foundation and transportation requirements, long bridges can be affected by spatial and temporal variations in the ground motions. Expressed in simple terms, different piers are subjected to different ground motions at any one time, because seismic waves take time to travel from one bridge pier to another. This effect can result in one pier being pulled in one direction while the other is being pushed in the opposite direction. This complex behavior is not accounted for directly in conventional bridge design. An example where this behavior may have resulted in increased damage and collapse is the eastern portion of the San Francisco–Oakland Bay Bridge (Figure 34.7a). This bridge includes a variety of different superstructure and substructure configurations, traverses variable subsoils, and is long enough for spatial and temporal variations in ground motions to induce large relative displacements between adjacent bridge segments. The bridge lost two spans, one upper and one lower, at a location where the superstructure was required to accommodate differential movements of adjacent bridge segments (Figure 34.7b).

34.6 Unseating at Expansion Joints

Expansion joints introduce a structural irregularity that can have catastrophic consequences. Such joints are commonly provided in bridges to alleviate stresses associated with volume changes that occur as a bridge ages and as the temperature changes. These joints can occur within a span (in-span hinges), or they can occur at the supports, as is the case for simply supported bridges.

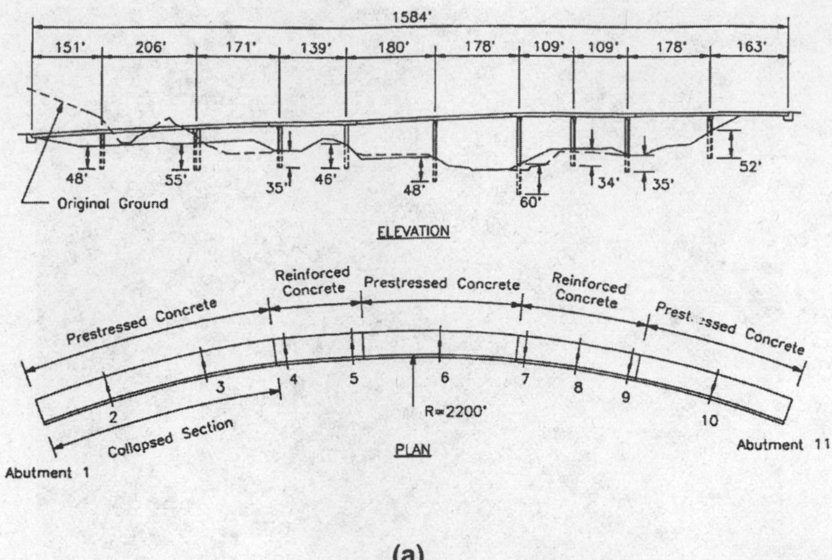

(a)

(b)

FIGURE 34.6 Geometry and collapse of the Route 14/5 Separation and Overhead in the 1994 Northridge earthquake. (a) Configuration [8]; (b) photograph of collapse.

Earthquake ground shaking, or transient or permanent ground deformations resulting from the earthquake, can induce superstructure movements that cause the supported span to unseat. Unseating is especially a problem with the shorter seats that were common in older construction (e.g., References [2,6–8,12]).

Bridges with Short Seats and Simple Spans

In much of the United States and in many other areas of the world, bridges often comprise a series of simple spans supported on bents. These spans are prone to being toppled from their supporting substructures either due to shaking or differential support movement associated with ground

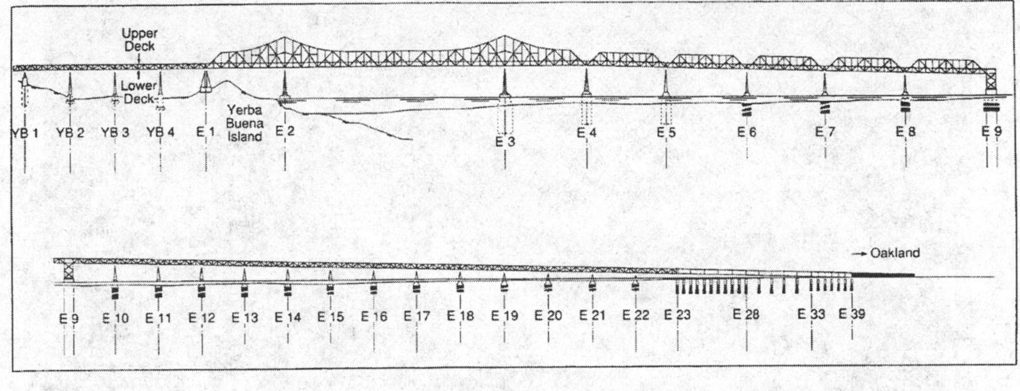

(a)

(b)

FIGURE 34.7 San Francisco–Oakland Bay Bridge, east crossing; geometry and collapse in the 1989 Loma Prieta earthquake. (a) Configuration [10]; (b) photograph of collapse.

deformation. Unseating of simple spans was observed in California in earlier earthquakes, leading in recent decades to development of bridge construction practices based on monolithic box-girder-substructure construction. Problems of unseating still occur with older bridge construction and with new bridges in regions where simple spans are still common. For example, during the 1991 Costa Rica earthquake, widespread liquefaction led to abutment and internal bent rotations, resulting in the collapse of no fewer than four bridges with simple supports [7]. The collapse of the Showa Bridge in the 1964 Niigata earthquake demonstrates one result of the unseating of simple spans (Figure 34.8).

FIGURE 34.8 Showa Bridge collapse in 1964 Niigata earthquake.

Skewed Bridges

Skewed bridges are defined as those having supports that are not perpendicular to the alignment of the bridge. Collisions between a skewed bridge and its abutments (or adjacent frames) can cause a bridge to rotate about a vertical axis. Because the abutments resist compression but not tension, the sense of this rotation is the same (for a given bridge configuration) regardless of whether the

FIGURE 34.9 Rio Bananito Bridge collapse in the 1991 Costa Rica earthquake. (*Source*: EERI, *Earthquake Spectra*, Special Suppl. to Vol. 7, 1991.)

bridge collides with one abutment or the other. If the rotations are large and the seat lengths small, a bridge can come unseated at the acute corners of the decks.

Several examples of skewed bridge damage and collapse can be found in the literature [7,8,12]. A typical example is the Rio Bananito Bridge, in which the bridge and central slab pier were skewed at 30°, which lost both spans off the central pier in the direction of the skew during the 1991 Costa Rica earthquake (Figure 34.9) [7]. Another example of skewed bridge failure is the Gavin Canyon Undercrossing, which failed during the 1994 Northridge earthquake [8]. Both skewed hinges became unseated during the earthquake, resulting in collapse of the unseated spans (Figure 34.10).

Curved Bridges

Curved bridges can have asymmetrical response similar to that of skewed bridges. For loading in one direction, an in-span hinge tends to close, while for loading in the other direction, the hinge opens. An example in which the curved alignment may have contributed to bridge collapse is the curved ramps of the I-5/SR14 interchange, which sustained collapses in both the 1971 San Fernando earthquake [13] and the 1994 Northridge earthquake (see Figure 34.6) [8]. Other factors that may have contributed to the failures include inadequate hinge seats and column deformability.

Hinge Restrainers

Hinge restrainers appear to have been effective in preventing unseating in both the Loma Prieta [10] and Northridge earthquakes [8]. In some other cases, hinge restrainers were not fully effective in preventing unseating. For example, the hinge restrainers in the Gavin Canyon Undercrossing, which were aligned parallel the bridge alignment, did not prevent unseating (see Figure 34.10).

FIGURE 34.10 Gavin Canyon Undercrossing collapse in the 1994 Northridge earthquake.

34.7 Damage to Superstructures

Superstructures are designed to support service gravity loads elastically, and, for seismic applications, they are usually designed to be a strong link in the earthquake-resisting system. As a result, superstructures tend to be sufficiently strong to remain essentially elastic during earthquakes. In general, superstructure damage is unlikely to be the primary cause of collapse of a span.

Instead, damage typically is focused in bearings and substructures. The superstructure may rest on elastomeric pads, pin supports, or rocker bearings, or may be monolithic with the substructure. As bearings and substructures are damaged and in some cases collapse, a wide range of damage and failure of superstructures may result, but these failures are often secondary; that is, they result from failures elsewhere in the bridge. There are, however, some cases of primary superstructure damage as well. Some examples are highlighted below.

With the exception of bridge superstructures that come unseated and collapse, the most common form of damage to superstructures is due to pounding of adjacent segments at the expansion hinges. This type of damage occurs in bridges of all construction materials. Figure 34.11a shows pounding damage at an in-span expansion joint of the Santa Clara River Bridge during the 1994 Northridge earthquake, and Figure 34.11b shows pounding damage at an abutment of the same structure.

Following the 1971 San Fernando earthquake, Caltrans initiated the first phase of its retrofit program, which involved installation of hinge and joint restrainers to prevent deck joints from separating. Both cable restrainers and pipe restrainers (the former intended only to restrain longitudinal movement and the latter intended also to restrain transverse motions) were installed in bridge superstructures. The restrainers extended through end diaphragms that had not been designed originally for the forces associated with restraint. Some punching shear damage to end diaphragms retrofitted with cable restrainers was observed in the I-580/I-980/SR24 connectors following the 1989 Loma Prieta earthquake [15].

FIGURE 34.11 Santa Clara River Bridge pounding damage in 1994 Northridge earthquake. (a) Barrier rail pounding damage; (b) abutment pounding damage.

FIGURE 34.12 Buckling of braces near pier 209 of the Hanshin Expressway in the 1995 Hyogo-Ken Nanbu earthquake.

Steel superstructures commonly comprise lighter framing elements, especially for transverse bracing. These have been found to be susceptible to damage due to transverse loading, especially following failure of bearings [1,8]. Several cases of steel superstructure damage occurred in the Hyogo-Ken Nanbu earthquake. Figure 34.12 shows buckling of cross braces beneath the roadway of a typical steel girder bridge span of the Hanshin Expressway. Figure 34.13 shows girder damage in the same expressway due to excessive lateral movement at the support. Figure 34.14 shows buckled cross-members between the upper chords of the Rokko Island Bridge. That single-span, 710-ft (217-m) tied-arch span bridge slipped from its expansion bearings, allowing the bridge to move laterally about 10 ft (3 m). The movement was sufficient for one end of one arch to drop off the cap beam, twisting the superstructure and apparently resulting in the buckling of the top chord bracing members [2,3].

A spectacular example of steel superstructure failure and collapse is that of the eastern portion of the San Francisco–Oakland Bay Bridge during the 1989 Loma Prieta earthquake (see Figure 34.7) [10]. In this bridge, a 50-ft (15-m) span over tower E9 was a transition point between 506-ft (154-m) truss spans to the west and 290-ft (88-m) truss spans to the east, serving to transmit longitudinal forces among the adjacent spans and the massive steel tower at E9. Failure of a bolted connection between the 290-ft (88-m) span truss and the tower resulted in sliding of the span and unseating of the transition span over tower E9. This collapse resulted in closure for 1 month of this critical link between San Francisco and the East Bay.

34.8 Damage to Bearings

In some regions of the world, the prevalent bridge construction consists of steel superstructures supported on bearings, which, in turn, rest on a substructure. In the United States, this form of

FIGURE 34.13 Girder damage at Bent 351 of the Hanshin Expressway apparently due to transverse movement during the 1995 Hyogo-Ken Nanbu earthquake.

construction is common in new bridges east of the Sierra Nevada Mountains as well as throughout the country in older existing bridges. In such bridges, the bearings commonly consist of steel components designed to provide restraint in one or more directions and, in some cases, to permit movement in one or more directions. Failure of these bearings in an earthquake can cause redistribution of internal forces, which may overload either the superstructure or substructure, or both. Collapse is also possible when bearing support is lost.

The predominant type of bridge construction in Japan involves steel superstructures supported on bearings, which, in turn, are supported on concrete substructures. The Hyogo-Ken Nanbu earthquake provides several examples of bearing failures in these types of bridges [2,3]. One example is provided by the Hamate Bypass, which was a double-deck elevated roadway comprising steel box girders on either fixed or expansion steel bearings. Bearing failure at several locations led to large superstructure rotations that can be seen in Figure 34.15. Another example is provided by the Nishinomiya-ko Bridge, a 830-ft (252-m) span-arch bridge supported on two fixed bearings at one end and two expansion bearings at the other end. The fixed-end bearings, which apparently were designed to have a capacity of approximately 70% of the bridge weight [2], failed, apparently leading to unseating of the adjacent approach span (see Figure 34.3). The failed bearing is shown in Figure 34.16.

FIGURE 34.14 Buckling of cross-members in the upper chord of the Rokko Island Bridge in the 1995 Hyogo-Ken Nanbu earthquake.

FIGURE 34.15 Hamate Bypass superstructure rotations as a result of bearing failures in the 1995 Hyogo-Ken Nanbu earthquake.

FIGURE 34.16 Nishinomiya-ko Bridge bearing failure in the 1995 Hyogo-Ken Nanbu earthquake.

34.9 Damage to Substructures

Columns

Unlike building design, current practice in bridge design is to proportion members of a frame (bent) such that its lateral-load capacity is limited by the flexural strength of its columns. For this strategy to be successful, the connecting elements (e.g., footings, joints, cross-beams) need to be strong enough to force yielding into the columns, and the columns need to be sufficiently ductile (or tough) to sustain the imposed deformations. Even in older bridges, where the "weak column" design approach may not have been adopted explicitly, columns tend to be weaker than the beam–diaphragm–slab assembly to which they connect. Consequently, columns can be subjected to large inelastic demands during strong earthquakes. Failure of a column can result in loss of vertical load-carrying capacity; column failure is often the primary cause of bridge collapse.

Most damage to columns can be attributed to inadequate detailing, which limits the ability of the column to deform inelastically. In concrete columns, the detailing inadequacies can produce flexural, shear, splice, or anchorage failures, or as is often the case, a failure that combines several mechanisms. In steel columns, local buckling has been observed to lead progressively to collapse.

Ideally, a concrete column should be designed such that the lateral load strength is controlled by flexure. However, even if most of the inelastic action is flexural, a column may not be sufficiently tough to sustain the imposed flexural deformations without failure. Such failures are particularly common in older bridges. In the United States, the transverse reinforcement of reinforced concrete columns designed before 1971 commonly consists of #4 hoops (ϕ = 13 mm) or ties at 12-in. (305-mm) spacing. Moreover, the ends of the transverse reinforcement rarely are anchored into the

FIGURE 34.17 San Fernando Road Overhead damage in the 1971 San Fernando earthquake.

core of these columns. This amount and type of reinforcement provides negligible confinement to the concrete, particularly in large columns. Figure 34.17 shows bridge columns that had insufficient flexural ductility to withstand the 1971 San Fernando earthquake. Figure 34.18 shows similar damage in a circular cross section column in the 1995 Hyogo-Ken Nanbu earthquake.

Other detailing practices (in addition to providing little confinement) may lead to flexural failure in reinforced concrete columns. A common practice in Japan has been to terminate some of the longitudinal reinforcement within the column height. The resulting development length of the terminated reinforcement can be inadequate, and may lead to splitting failure along the terminated bars or to flexural and shear distress near the cutoff point. Figure 34.19 illustrates failure of a column with bars terminated near the column midheight. In the case of the Hanshin Expressway Route 3, which collapsed during the 1995 Hyogo-Ken Nanbu earthquake (see Figure 34.4), the curtailment of one third of the main reinforcement was accompanied by the use of gas-pressure butt welding of the continuing longitudinal reinforcement. In tests following the earthquake, approximately half of the undamaged, butt-welded bars failed at the welds [14].

Shear failures of concrete bridge columns have occurred in many earthquakes (e.g., [5,8]). Such failures can occur at relatively low structural displacements, at which point the longitudinal reinforcement may not yet have yielded. Alternatively, because shear strength degrades with inelastic loading cycles, shear failures can occur after flexural yielding. Examples of shear failure can be found in several of the references provided at the end of this chapter. Figure 34.20 illustrates shear failure of a column having relatively light transverse reinforcement typical of bridges constructed in the western United States prior to the mid-1970s. The failure features a steeply inclined diagonal crack and dilation of the core into discrete blocks of concrete. Under the action of several deformation

FIGURE 34.18 Hanshin Expressway, Pier 46, damage in the 1995 Hyogo-Ken Nanbu earthquake.

cycles combined with vertical loads, a column can degrade to nearly complete loss of load-carrying capacity, as suggested by the heavily damaged column in Figure 34.21. Provision of closely spaced transverse reinforcement as required in some modern codes is required to prevent this type of failure.

Shear failures in reinforced concrete columns can be induced by interactions with "nonstructural" elements. These elements can decrease the distance between locations of flexural yielding, and therefore increase the shear demand for a column. Figure 34.5, discussed previously, shows a case

FIGURE 34.19 Failure of column with longitudinal reinforcement cutoffs near midheight in the 1995 Hyogo-Ken Nanbu earthquake.

in which a channel wall restrained the column at the base and forced the location of yielding to occur higher in the column than was anticipated in design [8]. Figure 34.22 shows a case in which an architectural flare strengthened the upper portion of the column, forcing yielding to occur lower than was intended [8]. In both cases, an element that was not considered in designing the column forced failure to occur in a lightly confined portion of the column that was incapable of resisting the force and deformation demands.

Figure 34.23 illustrates the failure of a stout, two-column bent on a spur just to the north of the Hanshin Expressway in the 1995 Hyogo-Ken Nanbu earthquake. The failure involves shattering of the columns, bent cap, and joints, and shatters the notion that strength alone is an adequate provision for bridge seismic design.

Lap splices of longitudinal reinforcement in older reinforced-concrete bridges may be vulnerable because, typically, the splices are short (on the order of 20 to 30 bar diameters), poorly confined, and are located in regions of high flexural demand. In particular, for construction convenience, splices are often located directly above a footing. With these details, the splices may be unable to develop the flexural capacity of the column, and they may be more vulnerable to shear failure. Despite these vulnerabilities, there is little field evidence of lap splice failures at the bases of bridge columns. However, failures associated with welded splices and terminated longitudinal reinforcement were identified in the 1995 Hyogo-Ken Nanbu earthquake (Figures 34.4 and 34.19), as discussed previously.

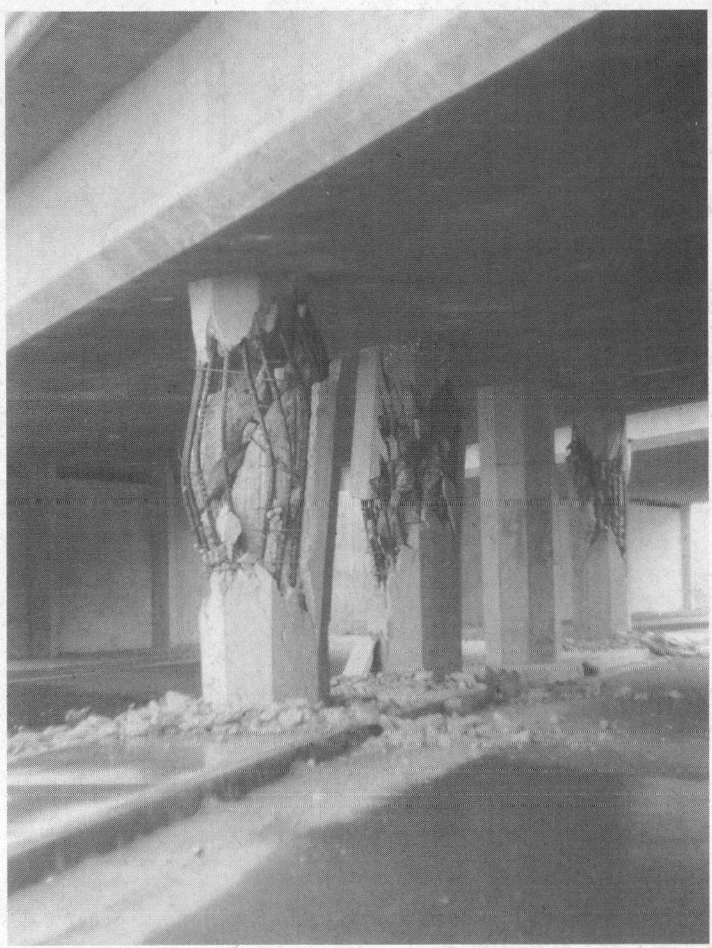

FIGURE 34.20 Failure of columns of the Route 5/210 interchange during the 1971 San Fernando earthquake.

Concrete columns also can fail if the anchorage of the longitudinal reinforcement is inadequate. Such failures can occur both at the top of a column at the connection with the bent cap and at the bottom of a column at the connection with the foundation. Figure 34.24 shows a column that failed at its base during the 1971 San Fernando earthquake (12). The column had been supported on a single 6-ft (1.8-m)-diameter, cast-in-drilled-hole pile. Other columns having hooked longitudinal reinforcement anchored in footings also failed with similar results in that earthquake. The consequences of foundation anchorage failures perhaps are larger in single-column bents than multicolumn bents, because the lateral-force resistance of single-column bents depends on the column developing its flexural strength at the base.

The record of steel column failures is sparse, because few bridges with steel columns have been subjected to strong earthquakes elsewhere than Japan. In the 1995 Hyogo-Ken Nanbu earthquake, failures apparently were associated with local buckling and subsequent splitting at welds or tearing of steel near the buckle. In columns with circular cross sections, the local buckling sometimes occurred at locations where section thicknesses changed. Figure 34.25 illustrates the formation of a local buckle in a circular cross section column accompanied by visible plastic deformation. In rectangular columns, local buckling of web and flange plates was insufficiently restrained by small web stiffeners [14]. Figure 34.26 illustrates the collapse of a rectangular column. A nearby column sustained local buckling at the base and tearing of the vertical welded seam between the two steel plates forming a corner of the column, suggesting the nature of failure that resulted in the collapse shown in Figure 34.26.

FIGURE 34.21 Failure of columns of Interstate 10, La Cienega-Venice Undercrossing in the 1994 Northridge earthquake. (Masonry walls of storage units are supporting the collapsed frame.)

Beams

Beams traditionally have received much less attention than columns in seismic design and evaluation. In many bridges, the transverse beams are stronger than the columns because of gravity load requirements and composite action with the superstructure. Also, in many bridges, the consequences of beam failures are less severe than the consequences of column failures. In bridges with outriggers, however, the beams can be critical components of the bent and can be subjected to loadings that may result in failure. An example illustrating possible damage to an outrigger beam is in Figure 34.27. This outrigger beam was monolithically framed with the superstructure and supporting column such that, under longitudinal load, significant torsion was required to be resisted by the outrigger portion of the beam. In some modern designs, torsion is reduced by providing nominal "pinned" connections between the beams and columns.

Joints

As with beams, joints traditionally have received little attention in seismic design, and they similarly may be exposed to critically damaging actions when the joints lie outside of the superstructure. Although joint failures occurred in previous earthquakes (e.g., Jennings [13]), significant attention was not paid to joints until several spectacular failures were observed following the 1989 Loma Prieta earthquake [6,10]. Figure 34.28 shows joint damage to the Embarcadero Viaduct in San Francisco during the 1989 Loma Prieta earthquake. The occurrence of damage at the relatively large epicentral distance of approximately 60 miles (100 km) is attributed in part to site amplification and focusing of seismic waves as well as the vulnerability of the framing.

The collapse of the Cypress Street Viaduct during the Loma Prieta earthquake had more severe consequences (Figure 34.29). Failure of a concrete pedestal located just above the first-level joint

FIGURE 34.22 Failure of flared column in the Route 118, Mission-Gothic Undercrossing, in the 1994 Northridge earthquake.

led to the collapse of the upper deck on the lower deck, at a cost of 42 lives. Such pedestals are not common, but this collapse demonstrates that each earthquake has the potential to reveal a mode of failure that has not yet been considered routinely.

The Loma Prieta earthquake also identified an apparent weakness of a modern design. For example, damage occurred to the outrigger knee joints of the Route 980/880 connector, which had been constructed just a few years before the earthquake. This damage identified the need for special details in bridge construction, which has been the subject of important studies identified elsewhere in this book.

Abutments

The types of failures that can occur at abutments vary from one bridge to the next. The foundation type varies greatly (e.g., spread footing, pile-supported footing, drilled shafts), and the properties of the soil can be important, particularly if the soil liquefies during an earthquake. The situation is further complicated by the interaction of the backwalls, wingwalls, footings and piles with the surrounding soil. A common practice has been to treat abutments or abutment components as sacrificial elements, acting as fuses to relieve large seismic forces arriving at the stiff abutment. The occurrence of widespread and extensive damage in the 1994 Northridge earthquake [8] suggests that an alternative approach might be economical.

In most seat-type abutments, longitudinal motion is unrestrained, because there is a joint at the interface of the superstructure and abutment backwall. This configuration is attractive, because it

FIGURE 34.23 Failure of a two-column bent in the 1995 Hyogo-Ken Nanbu earthquake.

reduces the superstructure forces induced by temperature and shrinkage-induced displacements. The most important vulnerability of such abutments is span unseating, which can occur when there are large relative displacements between the superstructure and abutment seat. Abutment unseating failures are often attributable to displacement or rotation of the abutment, usually the result of liquefaction or lateral spreading [7].

Shear keys can be damaged also. Shear keys are components that restrain relative displacements (usually in the transverse direction) between the superstructure and the abutments. External shear keys are located outside of the superstructure cross section, while internal shear keys are located within the superstructure cross section. Since these elements are stocky, it is nearly impossible to make them ductile, and they will fail if their strength is exceeded.

Shear key failures were widespread during the 1994 Northridge earthquake [8]. Figure 34.30 shows a typical failure in which the external shear keys failed. Figure 34.31 shows a failed internal shear key. It appears that these failures can occur with small transverse displacement and little energy dissipation. Damage to internal shear keys usually is accompanied by damage to the interlocking backwall. In seat-type abutments, damage has also occurred in seat abutments due to pounding of backwalls by the superstructure. This type of damage is similar to that shown in Figure 34.11.

In monolithic abutments, the superstructure is cast monolithically with the abutments. This configuration is attractive, because it reduces the likelihood of span unseating. However, the abutment can be damaged as the superstructure displaces in the longitudinal direction away from the abutment. Also, depending on the geometry and details of the abutment, the wingwall may serve as an external shear key. In such cases, the wingwall can fail in the same manner as an external shear key.

FIGURE 34.24 Failure at the base of a column supported on a single cast-in-place pile in the 1971 San Fernando earthquake [Steinbrugge Collection, EERC Library, University of California, Berkeley].

Foundations

Reports of foundation failures during earthquakes are relatively rare, with the notable exception of situations in which liquefaction occurred. It is not clear whether failures are indeed that rare or whether many foundation failures are undetected because they remain underground. There are many reasons why older foundations might be vulnerable. Piles might have little confinement reinforcement, yet be subjected to large deformation demands. Older spread and pile-supported footings rarely have top flexural reinforcement or any shear reinforcement.

The 1995 Hyogo-Ken Nanbu earthquake resulted in extensive damage to superstructures and substructures above the ground, as reported elsewhere in this chapter. The occurrence of that damage provided impetus to conduct extensive investigations of the conditions of foundation components [11]. Along the older inland Route 3, an investigation of 109 foundations identified only cases of "small" flexural cracks in piles. Along the newer coastal Route 5, more extensive liquefaction occurred, resulting in lateral spreading in several cases. An investigation of 153 foundations for this route found cases of flexural cracks in piles where large residual displacements occurred, but the investigators found no spalling or reinforcement buckling. The absence of extensive damage was attributed to the spread of deformations along a significant length of the piles.

Foundation damage associated with liquefaction-induced lateral spreading has probably been the single greatest cause of extreme distress and collapse of bridges [12]. The problem is especially critical for bridges with simple spans (see Figure 34.8). The 1991 Costa Rica earthquake provides many examples of foundation damage [7]. For example, Figure 34.32 shows an abutment that

FIGURE 34.25 Local buckling of a circular cross-section column of the Hanshin Expressway in the 1995 Hyogo-Ken Nanbu earthquake.

rotated due to liquefaction and lateral spreading. Figure 34.33 shows a situation in which soil movements have led to extensive damage to the batter piles. Use of batter piles should be considered carefully in design in light of the extensive damage observed in these piles in this and other earthquakes [6].

Approaches

Even if the bridge structure remains intact, a bridge may be placed out of service if the roadway leading to it settles significantly. For example, during the 1971 San Fernando earthquake [12] and the 1985 Chile earthquake [4], settlement of the backfill abutments led to abrupt differential settlements in many locations. Such settlements can be large enough to pose a hazard to the traveling public. Approach or settlement slabs can be effective means of spanning across backfills, as shown in Figure 34.34.

34.10 Summary

This chapter has reviewed various types of damage that can occur in bridges during earthquakes. Damage to a bridge can have severe consequences for a local economy, because bridges provide vital links in the transportation system of a region. In general, the likelihood of damage increases

(a)

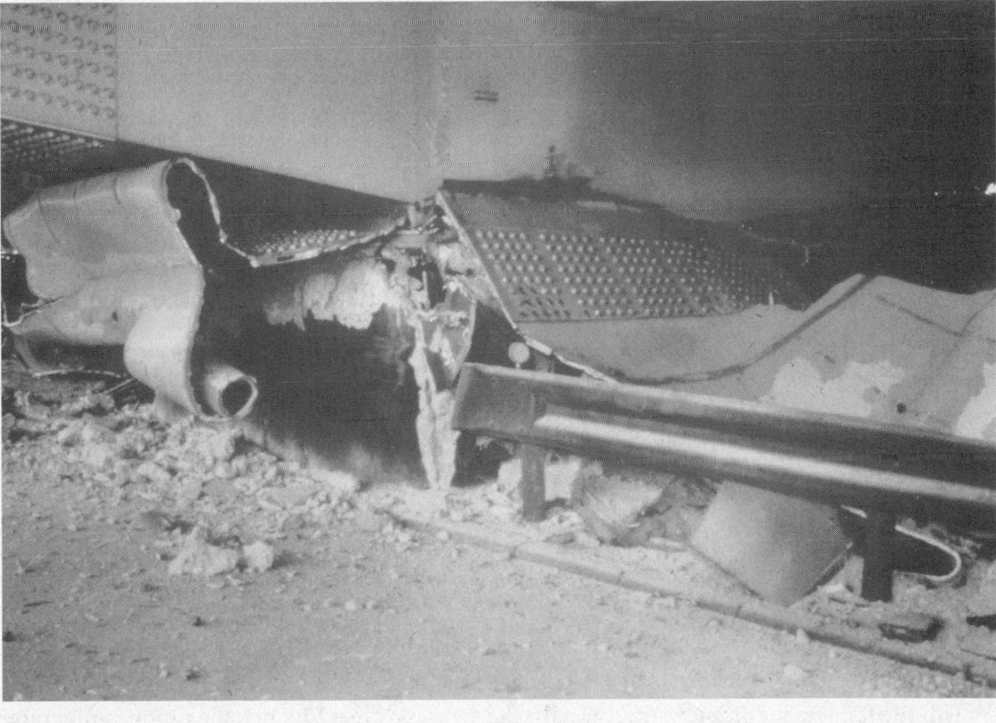

(b)

FIGURE 34.26 Collapse of a rectangular cross section steel column in the 1995 Hyogo-Ken Nanbu earthquake. (a) Collapsed bent and superstructure; (b) close-up of collapsed column.

FIGURE 34.27 Outrigger damage in the 1989 Loma Prieta earthquake.

if the ground motion is particularly intense, the soils are soft, the bridge was constructed before modern codes were implemented, or the bridge configuration is irregular. Even a well-designed bridge can suffer damage if nonstructural modifications and structural deterioration have increased the vulnerability of the bridge.

Depending on the ground motion, site conditions, overall configuration, and specific details of the bridge, the damage induced in a particular bridge can take many forms. Despite these complexities, the record is clear. Damage within the superstructure is rarely the primary cause of collapse. Though exceptions abound, most of the severe damage to bridges has taken one of the following forms:

- Unseating of superstructure at in-span hinges or simple supports attributable to inadequate seat lengths or restraint. The presence of a skewed or curved configuration further exacerbates the vulnerability. For simply supported bridges, these failures are most likely when ground failure induces relative motion between the spans and their supports.
- Column failure attributable to inadequate ductility (toughness). In reinforced-concrete columns, the inadequate ductility usually stems from inadequate confinement reinforcement. In steel columns, the inadequate ductility usually stems from local buckling, which progresses to collapse.
- Damage to shear keys at abutments. Because of their geometry, it is nearly impossible to make these stiff elements ductile.
- Unique failures in complex structures. In the Cypress Street Viaduct, the unique vulnerability was the inadequately reinforced pedestal above the first level. In outrigger column bents, the vulnerability may be in the cross-beam or the beam–column joint.

FIGURE 34.28 Embarcadero Viaduct damage during the 1989 Loma Prieta earthquake.

Acknowledgments

This work was made possible through the reconnaissance work of many individuals identified in the cited references, as well as many not directly cited. The writers acknowledge the significant effort and risk made by those experts, as well as the funding agencies, in particular the U.S. National Science Foundation. Many of the photographs were provided generously by the Earthquake Engineering Research Institute, as cited in the figure captions. Where no citation is given in the caption, the photograph was from the extensive collection of the Earthquake Engineering Research Center Library, Pacific Earthquake Engineering Research Center, University of California, Berkeley. That collection is made possible through funding from the U.S. National Science Foundation and the Federal Emergency Management Agency, and from generous donations from earthquake experts. Photographs in this chapter were made possible by donations from K. Steinbrugge, W. Godden, M. Nakashima, and M. Yashinski.

FIGURE 34.29 Cypress Street Viaduct collapse in the 1989 Loma Prieta earthquake.

FIGURE 34.30 Damage to external shear key in an abutment in the 1994 Northridge earthquake. (*Source*: EERI, *Earthquake Spectra*, Special Suppl. to Vol. II, 1995.)

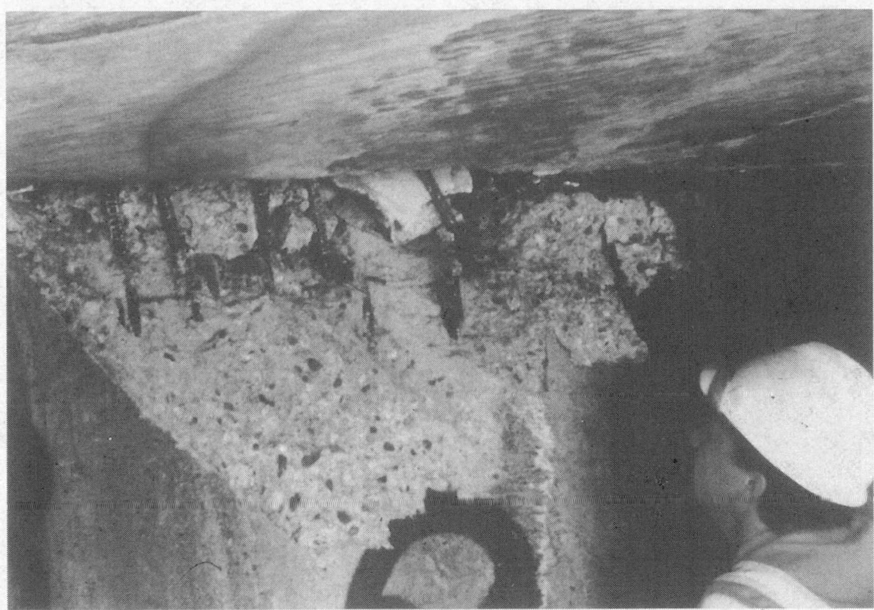

FIGURE 34.31 Damage to internal shear key in an abutment in the 1994 Northridge earthquake. (*Source:* EERI, *Earthquake Spectra*, Special Suppl. to Vol. II, 1995.)

FIGURE 34.32 Rotation of abutment due to liquefaction and lateral spreading during the 1991 Costa Rica earthquake. (*Source:* EERI, *Earthquake Spectra*, Special Suppl. to Vol. 7, 1991.)

FIGURE 34.33 Abutment piles damaged during the 1991 Costa Rica earthquake. (*Source:* EERI, *Earthquake Spectra*, Special Suppl. to Vol. 7, 1991.)

FIGURE 34.34 Settlement slab spanning across slumped abutment fill material at the Rio Quebrada Calderon bridge in the 1991 Costa Rica earthquake. (*Source:* EERI, *Earthquake Spectra*, Special Suppl. to Vol. 7, 1991.)

References

1. Astaneh-Asl, A. et al., Seismic Performance of Steel Bridges during the 1994 Northridge Earthquake: A Preliminary Report, Report No. UCB/CE-STEEL-94/01, Department of Civil Engineering, University of California, Berkeley, April 1994, 296 pp.
2. Chung, R. et al., The January 17, 1995 Hyogoken-Nanbu (Kobe) Earthquake, *NIST Special Publication 901*, National Institute of Standards and Technology, July 1996, 544 pp.
3. EERC, Earthquake Engineering Research Center, Seismological and Engineering Aspects of the 1995 Hyogoken-Nanbu Earthquake, Report No. UCB/EERC-95/10, Nov. 1995, 250 pp.
4. EERI, Earthquake Engineering Research Institute, The Chile earthquake of March 3, 1985, *Earthquake Spectra*, Special Supplement to 2(2), Feb. 1986, 513 pp.
5. EERI, Earthquake Engineering Research Institute, The Whittier Narrows earthquake of October 1, 1987, *Earthquake Spectra*, 4(2), May 1988, 409 pp.
6. EERI, Earthquake Engineering Research Institute, Loma Prieta earthquake reconnaissance report, *Earthquake Spectra*, Special Suppl. to Vol. 6, May 1990, 448 pp.
7. EERI, Earthquake Engineering Research Institute, Costa Rica earthquake reconnaissance report, *Earthquake Spectra*, Special Suppl. to Vol. 7, Oct. 1991, 127 pp.
8. EERI, Earthquake Engineering Research Institute, Northridge earthquake reconnaissance report, *Earthquake Spectra*, Special Suppl. to Vol. 11, April 1995, 523 pp.
9. EERI, Earthquake Engineering Research Institute, The Hyogo-Ken Nanbu Earthquake, January 17, 1995, Preliminary Reconnaissance Report, Feb. 1995, 116 pp.
10. Housner, G., Competing against time, Report to Governor George Deukmejian from the Governor's Board of Inquiry on the 1989 Loma Prieta Earthquake, Office of Planning and Research, State of California, May 1990, 264 pp.
11. Ishizaki, H. et al., Inspection and restoration of damaged foundations due to the Great Hanshin Earthquake 1995, *Proceedings, Third U.S.–Japan Workshop on Seismic Retrofit of Bridges*, Tsukuba, Japan, 1996, 327-341.
12. Iwasaki, T., Penzien, J., and Clough, R. Literature Survey-Seismic Effects on Highway Bridges, Earthquake Engineering Research Report No. 72-11, University of California, Berkeley, November 1972, 397 pp.
13. Jennings, P. C., Ed., Engineering Features of the San Fernando Earthquake of February 9, 1971, Report ERL 71-02, California Institute of Technology, June 1971.
14. Kawashima, K. and Unjoh, S., The damage of highway bridges in the 1995 Hyogo-Ken Nanbu earthquake and its impact on Japanese seismic design, *J. Earthquake Eng.*, 1(3), 505–541, 1997.
15. Saiidi, M., Maragakis, E., and Feng, S., Field performance and design issues for bridge hinge restrainers, *Proceedings, Fifth U.S. National Conference on Earthquake Engineering*, Earthquake Engineering Research Institute, Oakland, CA, Vol. I, 1994, 439–448.
16. Zelinski, R., Post Earthquake Investigation Team Report for the Loma Prieta Earthquake, California Department of Transportation, Division of Structures, Sacramento, 1994.

35
Dynamic Analysis

Rambabu Bavirisetty
California Department of Transportation

Murugesu Vinayagamoorthy
California Department of Transportation

Lian Duan
California Department of Transportation

35.1 Introduction

The primary purpose of this chapter is to present dynamic methods for analyzing bridge structures when subjected to earthquake loads. Basic concepts and assumptions used in typical dynamic analysis are presented first. Various approaches to bridge dynamics are then discussed. A few examples are presented to illustrate their practical applications.

35.1.1 Static vs. Dynamic Analysis

The main objectives of a structural analysis are to evaluate structural behavior under various loads and to provide the information necessary for design, such as forces, moments, and deformations. Structural analysis can be classified as *static* or *dynamic*: while *statics* deals with time-independent loading, *dynamics* considers any load where the magnitude, direction, and position vary with time. Typical dynamic loads for a bridge structure include vehicular motions and wave actions such as winds, stream flow, and earthquakes.

0-8493-7434-0/00/$0.00+$.50

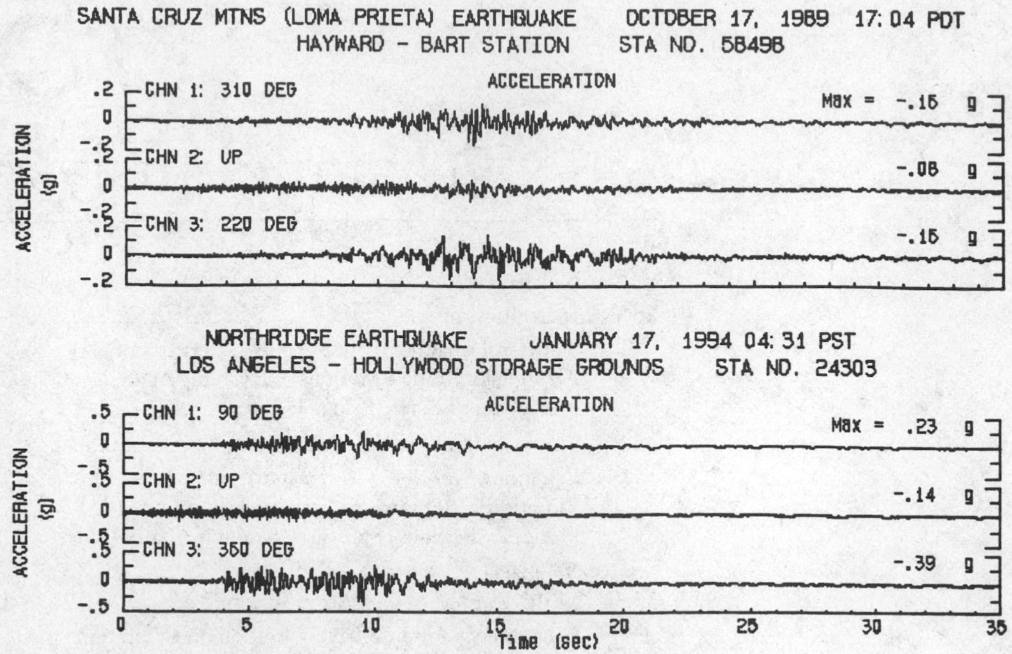

FIGURE 35.1 Ground motions recorded during recent earthquakes.

35.1.2 Characteristics of Earthquake Ground Motions

An earthquake is a natural ground movement caused by various phenomena including global tectonic processes, volcanism, landslides, rock-bursts, and explosions. The global tectonic processes are continually producing mountain ranges and ocean trenches at the Earth's surface and causing earthquakes. This section briefly discusses the earthquake input for seismic bridge analysis. Detailed discussions of ground motions are presented in Chapter 33.

Ground motion is represented by the time history or seismograph in terms of acceleration, velocity, and displacement for a specific location during an earthquake. Time history plots contain complete information about the earthquake motions in the three orthogonal directions (two horizontal and one vertical) at the strong-motion instrument location. Acceleration is usually recorded by strong-motion accelerograph and the velocities and displacements are determined by numerical integration. The accelerations recorded at locations that are approximately the same distance away from the epicenter may differ significantly in duration, frequency content, and amplitude due to different local soil conditions. Figure 35.1 shows several time histories of recent earthquakes.

From a structural engineering view, the most important characteristics of an earthquake are the peak ground acceleration (PGA), duration, and frequency content. The PGA is the maximum acceleration and represents the intensity of a ground motion. Although the ground velocity may be a more significant measure of intensity than the acceleration, it is not often measured directly, but determined using supplementary calculations [1]. The duration is the length of time between the first and the last peak exceeding a specified strong motion level. The longer the duration of a strong motion, the more energy is imparted to a structure. Since the elastic strain energy absorbed by a structure is very limited, a longer strong earthquake has a greater possibility to enforce a structure into the inelastic range. The frequency content can be represented by the number of zero crossings per second in the accelerogram. It is well understood that when the frequency of a regular disturbing force is the same as the natural vibration frequency of a structure (resonance), the oscillation of structure can be greatly magnified and effects of damping become minimal. Although earthquake motions are never as regular as a sinusoidal waveform, there is usually a period that dominates the response.

Since it is impossible to measure detailed ground motions for all structure sites, the rock motions or ground motions are estimated at a fault and then propagated to the Earth surface using a computer program considering the local soil conditions. Two guidelines [2, 3] recently developed by the California Department of Transportation provide the methods to develop seismic ground motions for bridges.

35.1.3 Dynamic Analysis Methods for Seismic Bridge Design

Depending on the seismic zone, geometry, and importance of the bridge, the following analysis methods may be used for seismic bridge design:

- The single-mode method (single-mode spectral and uniform load analysis) [4,5] assumes that seismic load can be considered as an equivalent static horizontal force applied to an individual frame in either the longitudinal or transverse direction. The equivalent static force is based on the natural period of a single degree of freedom (SDOF) and code-specified response spectra. Engineers should recognize that the single-mode method (sometimes referred to as equivalent static analysis) is best suited for structures with well-balanced spans with equally distributed stiffness.
- Multimode spectral analysis assumes that member forces, moments, and displacements due to seismic load can be estimated by combining the responses of individual modes using the methods such as complete quadratic combination (CQC) method and the square root of the sum of the squares (SRSS) method. The CQC method is adequate for most bridge systems [6], and the SRSS method is best suited for combining responses of well-separated modes.
- The multiple support response spectrum (MSRS) method provides response spectra and the peak displacements at individual support degrees of freedom by accurately accounting for the spatial variability of ground motions including the effects of incoherence, wave passage, and spatially varying site response. This method can be used for multiply supported long structures [7].
- The time history method is a numerical step-by-step integration of equations of motion. It is usually required for critical/important or geometrically complex bridges. Inelastic analysis provides a more realistic measure of structural behavior when compared with an elastic analysis.

Selection of the analysis method for a specific bridge structure should not be purely based on performing structural analysis, but be based on the effective design decisions [8]. Detailed discussions of the above methods are presented in the following sections.

35.2 Single-Degree-of-Freedom System

The familiar spring–mass system represents the simplest dynamic model and is shown in Figure 35.2a. When the *idealized, undamped* structures are excited by either moving the support or by displacing the mass in one direction, the mass oscillates about the equilibrium state forever without coming to rest. But, real structures do come to rest after a period of time due to a phenomenon called *damping*. To incorporate the effect of the damping, a massless viscous damper is always included in the dynamic model, as shown in Figure 35.2b.

In a dynamic analysis, the number of displacements required to define the displaced positions of all the masses relative to their original positions is called the number of degrees of freedom (DOF). When a structural system can be idealized with a single mass concentrated at one location and moved only in one direction, this dynamic system is called an SDOF system. Some structures,

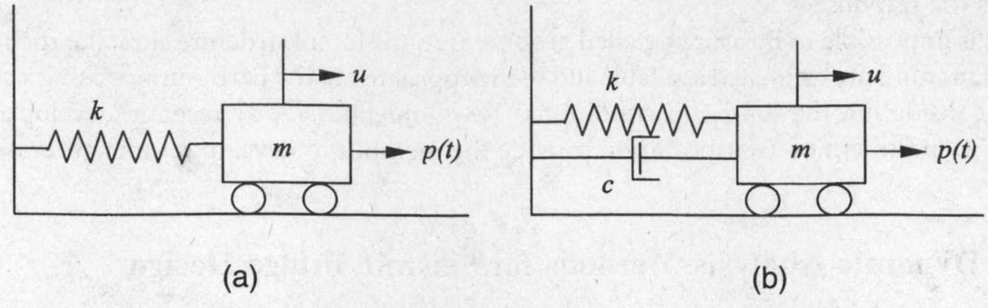

FIGURE 35.2 Idealized dynamic model. (a) Undamped SDOF system; (b) damped SDOF system.

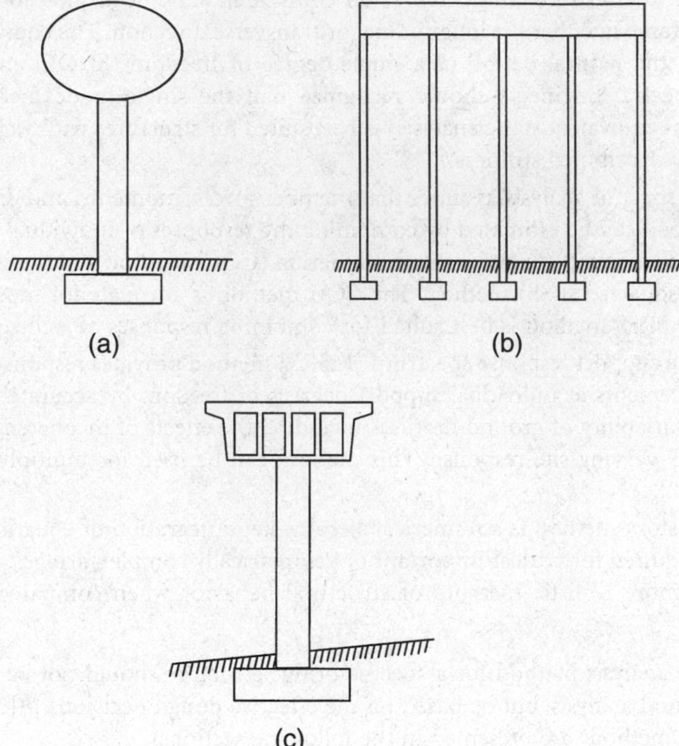

FIGURE 35.3 Examples of SDOF structures. (a) Water tank supported by single column; (b) one-story frame building; (c) two-span bridge supported by single column.

such as a water tank supported by a single-column, one-story frame structure and a two-span bridge supported by a single column, could be idealized as SDOF models (Figure 35.3).

In the SDOF system shown in Figure 35.3c, the mass of the bridge superstructure is the mass of the dynamic system. The stiffness of the dynamic system is the stiffness of the column against side sway and the viscous damper of the system is the internal energy absorption of the bridge structure.

35.2.1 Equation of Motion

The response of a structure depends on its mass, stiffness, damping, and applied load or displacement. The structure could be excited by applying an external force $p(t)$ on its mass or by a ground motion $u(t)$ at its supports. In this chapter, since the seismic loading is induced by exciting the support, we focus mainly on the equations of motion of an SDOF system subjected to ground excitation.

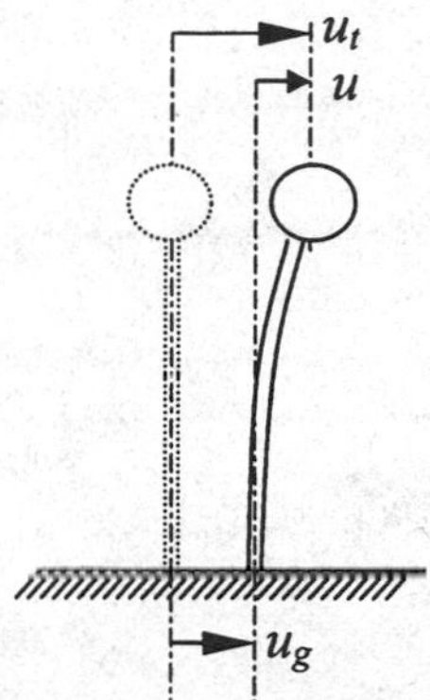

FIGURE 35.4 Earthquake–induced motion of an SDOF system.

The displacement of the ground motion u_g, the total displacement of the single mass u_t, and the relative displacement between the mass and ground u (Figure 35.4) are related by

$$u_t = u + u_g \tag{35.1}$$

By applying Newton's law and D'Alembert's principle of dynamic equilibrium, it can be shown that

$$f_I + f_D + f_S = 0 \tag{35.2}$$

where f_I is the inertial force of the single mass and is related to the acceleration of the mass by $f_I = m\ddot{u}_t$; f_D is the damping force on the mass and related to the velocity across the viscous damper by $f_D = c\dot{u}$; f_S is the elastic force exerted on the mass and related to the relative displacement between the mass and the ground by $f_S = ku$, where k is the spring constant; c is the damping ratio; and m is the mass of the dynamic system.

Substituting these expressions for f_I, f_D, and f_S into Eq. (35.2) gives

$$m\ddot{u}_t + c\dot{u} + ku = 0 \tag{35.3}$$

The equation of motion for an SDOF system subjected to a ground motion can then be obtained by substituting the Eq. (35.1) into Eq. (35.3), and is given by

$$m\ddot{u} + c\dot{u} + ku = -m\ddot{u}_g \tag{35.4}$$

35.2.2 Characteristics of Free Vibration

To determine the characteristics of the oscillations such as the time to complete one cycle of oscillation (T_n) and number of oscillation cycles per second (ω_n), we first look at the *free* vibration of a dynamic system. Free vibration is typically initiated by disturbing the structure from its equilibrium state by an external force or displacement. Once the system is disturbed, the system vibrates without any external input. Thus, the equation of motion for free vibration can be obtained by setting $\ddot{u}_g$ to zero in Eq. (35.4) and is given by

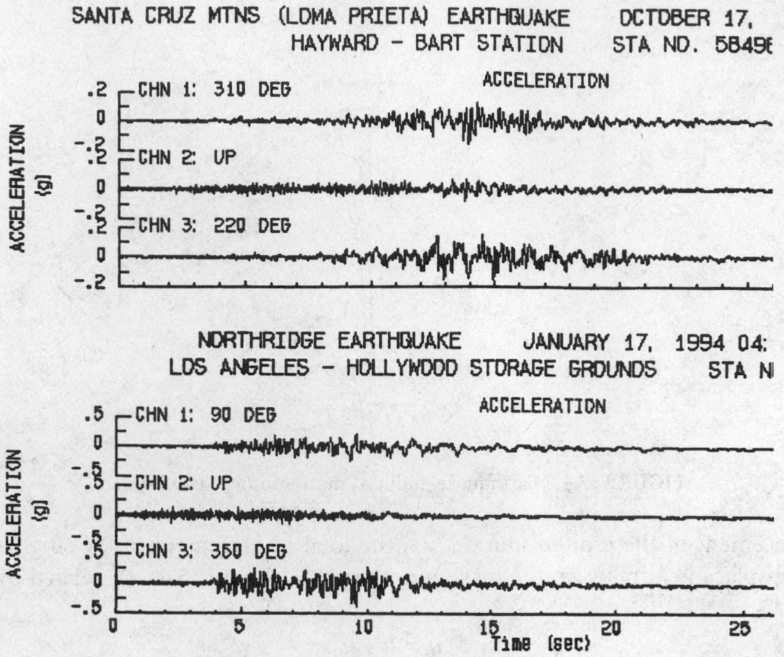

FIGURE 35.5 Typical response of an SDOF system. (a) Undamped; (b) damped.

$$m\ddot{u} + c\dot{u} + ku = 0 \tag{35.5}$$

Dividing the Equation (35.5) by its mass *m* will result in

$$\ddot{u} + \left(\frac{c}{m}\right)\dot{u} + \left(\frac{k}{m}\right)u = 0 \tag{35.6}$$

$$\ddot{u} + 2\xi\omega_n + \omega_n^2 u = 0 \tag{35.7}$$

where $\omega_n = \sqrt{k/m}$ the natural circular frequency of vibration or the undamped frequency; $\xi = c/c_{cr}$ the damping ratio; $c_{cr} = 2m\omega_n = 2\sqrt{km} = 2k/\omega_n$ the critical damping coefficient.

Figure 35.5a shows the response of a typical idealized, *undamped* SDOF system. The time required for the SDOF system to complete one cycle of vibration is called the natural period of vibration (T_n) of the system and is given by

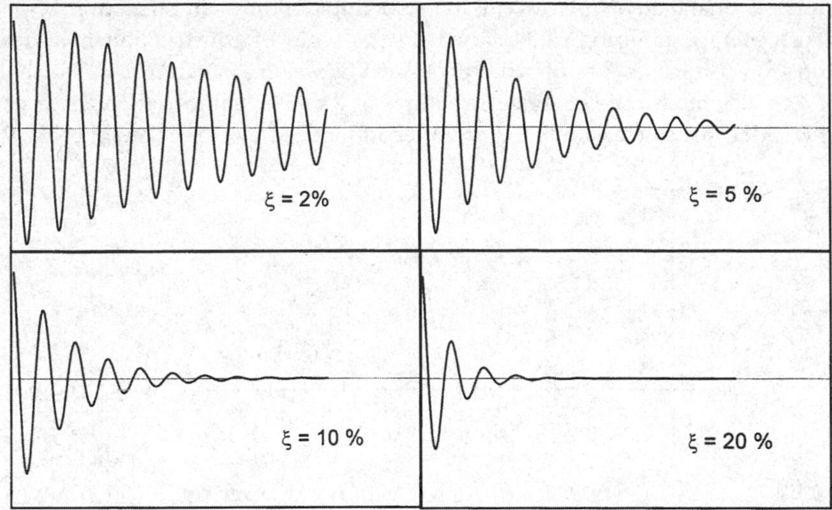

FIGURE 35.6 Response of an SDOF system for various damping ratios.

$$T_n = \frac{2\pi}{\omega_n} = 2\pi\sqrt{\frac{m}{k}} \tag{35.8}$$

Furthermore, the natural cyclic frequency of vibration f_n is given by

$$f_n = \frac{\omega_n}{2\pi} = \frac{1}{2\pi}\sqrt{\frac{k}{m}} \tag{35.9}$$

Figure 35.5b shows the response of a typical *damped* SDOF structure. The circular frequency of the vibration or damped vibration frequency of the SDOF structure, ω_d, is given by $\omega_d = \omega_n\sqrt{1-\xi^2}$.

The damped period of vibration (T_d) of the system is given by

$$T_d = \frac{2\pi}{\omega_d} = \frac{2\pi}{\sqrt{1-\xi^2}}\sqrt{\frac{m}{k}} \tag{35.10}$$

When $\xi = 1$ or $c = c_{cr}$ the structure returns to its equilibrium position without oscillating and is referred to as a critically damped structure. When $\xi > 1$ or $c > c_{cr}$, the structure is *overdamped* and comes to rest without oscillating, but at a slower rate. When $\xi < 1$ or $c < c_{cr}$, the structure is *underdamped* and oscillates about its equilibrium state with progressively decreasing amplitude. Figure 35.6 shows the response of SDOF structures with different damping ratios.

For structures such as buildings, bridges, dams, and offshore structures, the damping ratio is less than 0.15 and thus can be categorized as *underdamped* structures. The basic dynamic properties estimated using damped or undamped assumptions are approximately the same. For example, when $\xi = 0.10$, $\omega_d = 0.995\omega_n$, and $T_d = 1.01T_n$.

Damping dissipates the energy out of a structure in opening and closing of microcracks in concrete, stressing of nonstructural elements, and friction at the connection of steel members. Thus, the damping coefficient accounts for all energy-dissipating mechanisms of the structure and can only be estimated by experimental methods. Two seemingly identical structures may have slightly different material properties and may dissipate energy at different rates. Since damping does not play an important quantitative role except for resonant responses in structural responses, it is common to use average damping ratios based on the types of construction materials. Relative damping ratios for common types of structures, such as welded metal of 2 to 4%, bolted metal structures of 4 to 7%, prestressed concrete structures of 2 to 5%, reinforced-concrete structures of 4 to 7% and wooden structures of 5 to 10%, are recommended by Chmielewski et al. [9].

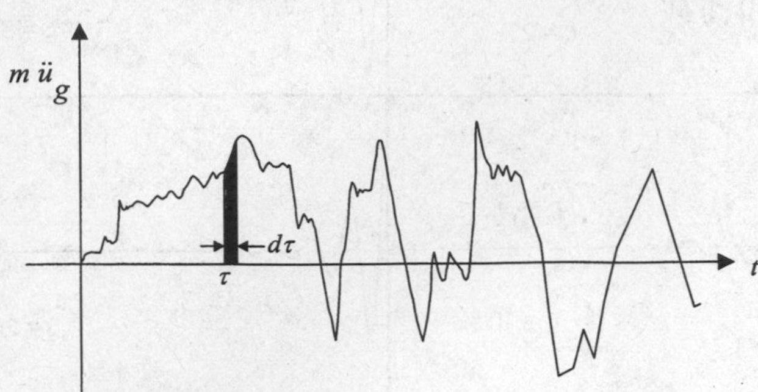

FIGURE 35.7 Induced earthquake force vs. time on an SDOF system.

35.2.3 Response to Earthquake Ground Motion

A typical excitation of an earth movement is shown in Figure 35.7. The basic equation of motion of an SDOF system is expressed in Eq. (35.4). Since the excitation force $m\ddot{u}_g$ cannot be described by simple mathematical expression, closed-form solutions for Eq. (35.4) are not available. Thus, the entire ground excitation needs to be treated as a superposition of short-duration impulses to evaluate the response of the structure to the ground excitation. An impulse is defined as the product of the force times duration. For example, the impulse of the force at time τ during the time interval $d\tau$ equals $-m\ddot{u}_g(\tau)d\tau$ and is represented by the shaded area in Figure 35.7. The total response of the structure for the earthquake motion can then be obtained by integrating all responses of the increment impulses. This approach is sometimes referred to as "time history analysis." Various solution techniques are available in the technical literature on structural dynamics [1,10].

In seismic structural design, designers are interested in the maximum or extreme values of the response of a structure as discussed in the following sections. Once the dynamic characteristics (T_n and ω_n) of the structure are determined, the maximum displacement, moment, and shear on the SDOF system can easily be estimated using basic principles of mechanics.

35.2.4 Response Spectra

The response spectrum is a relationship of the peak values of a response quantity (acceleration, velocity, or displacement) with a structural dynamic characteristic (natural period or frequency). Its core concept in earthquake engineering provides a much more convenient and meaningful measure of earthquake effects than any other quantity. It represents the peak response of all possible SDOF systems to a particular ground motion.

Elastic Response Spectrum

This, the response spectrum of an elastic structural system, can be obtained by the following steps [10]:

1. Define the ground acceleration time history (typically at a 0.02-second interval).
2. Select the natural period T_n and damping ratio ξ of an elastic SDOF system.
3. Compute the deformation response $u(t)$ using any numerical method.
4. Determine u_o, the peak value of $u(t)$.
5. Calculate the spectral ordinates by $D=u_o$, $V=2\pi D/T_n$, and $A=(2\pi/T_n)^2 D$.
6. Repeat Steps 2 and 5 for a range of T_n and ξ values for all possible cases.
7. Construct results graphically to produce three separate spectra as shown in Figure 35.8 or a combined tripartite plot as shown in Figure 35.9.

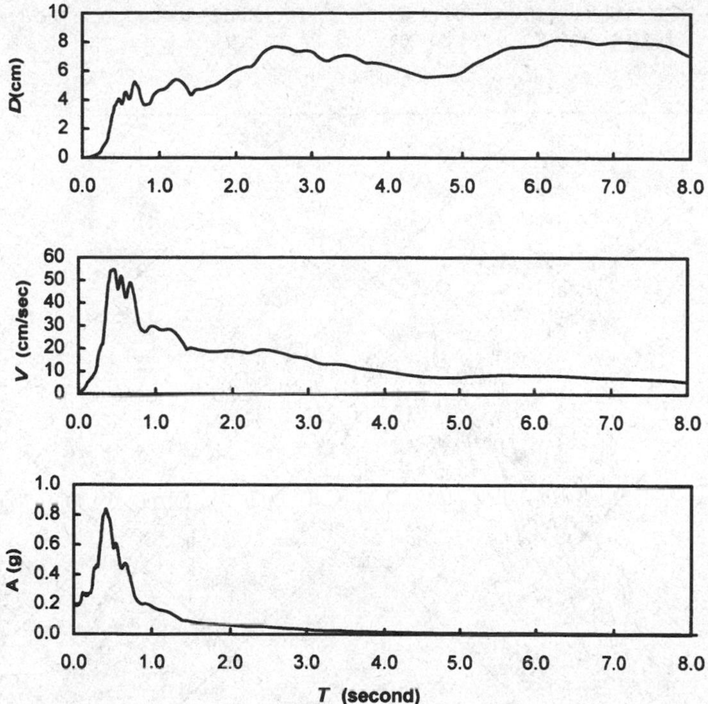

FIGURE 35.8 Example of response spectra (5% critical damping) for Loma Prieta 1989 motion.

It is noted that although three spectra (displacement, velocity, and acceleration) for a specific ground motion contain the same information, each provides a physically meaningful quantity. The displacement spectrum presents the peak displacement. The velocity spectrum is related directly to the peak strain energy stored in the system. The acceleration spectrum is related directly to the peak value of the equivalent static force and base shear.

A response spectrum (Figure 35.9) can be divided into three ranges of periods [10]:

- Acceleration-sensitive region (very short period region): A structure with a very short period is extremely stiff and expected to deform very little. Its mass moves rigidly with the ground and its peak acceleration approximately equals the ground acceleration.
- Velocity-sensitive region (intermediate-period region): A structure with an intermediate period responds greatly to the ground velocity than other ground motion parameters.
- Displacement-sensitive region (very long period region): A structure with a very long period is extremely flexible and expected to remain stationary while the ground moves. Its peak deformation is closer to the ground displacement. The structural response is most directly related to ground displacement.

Elastic Design Spectrum

Since seismic bridge design is intended to resist future earthquakes, use of a response spectrum obtained from a particular past earthquake motion is inappropriate. In addition, jagged spectrum values over small ranges would require an unreasonable accuracy in the determination of the structure period [11]. It is also impossible to predict a jagged response spectrum in all its details for a ground motion that may occur in the future. To overcome these shortcomings, the elastic design spectrum, a smoothened idealized response spectrum, is usually developed to represent the envelopes of ground motions recorded at the site during past earthquakes. The development of an elastic design spectrum is based on statistical analysis of the response spectra for the ensemble of ground motions. Figure 35.10 shows a set of elastic design spectra in Caltrans Bridge Design Specifications [12]. Figure 35.11 shows project-specific acceleration response spectra for the California Sonoma Creek Bridge.

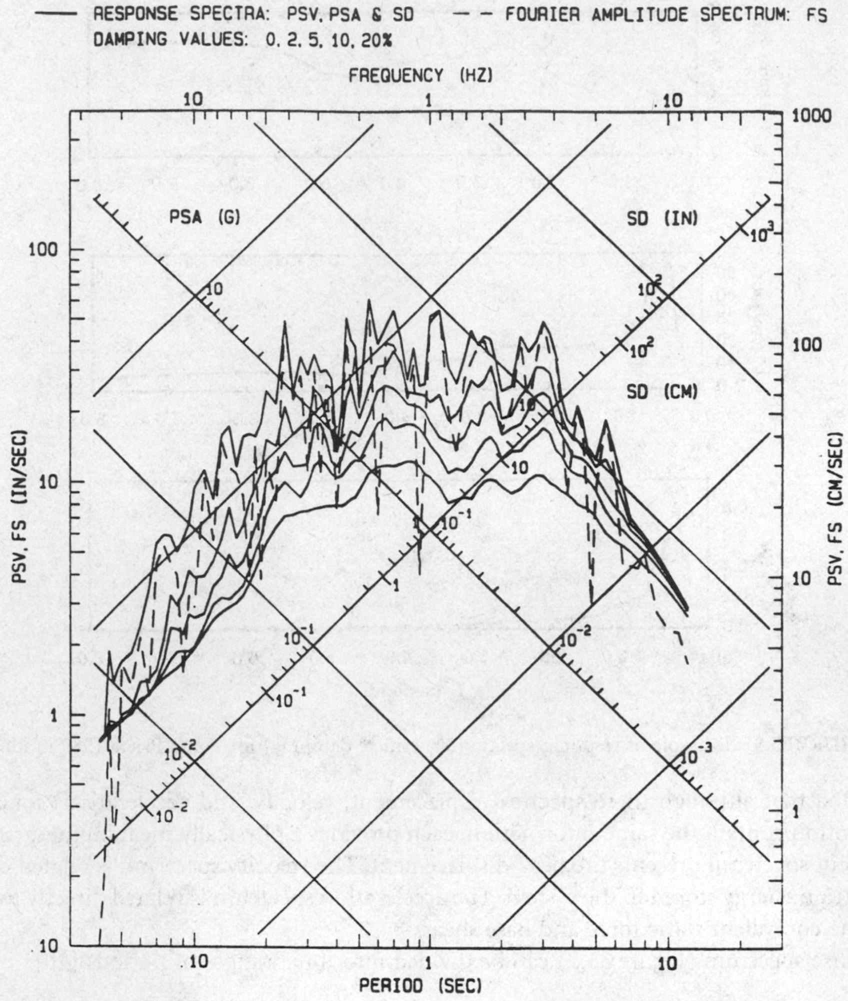

FIGURE 35.9 Tripartite plot–response spectra (1994 Northridge Earthquake, Arleta–Rordhoff Ave. Fire Station).

Engineers should recognize the conceptual differences between a response spectrum and a design spectrum [10]. A response spectrum is only the peak response of all possible SDOF systems due to a particular ground motion, whereas a design spectrum is a specified level of seismic design forces or deformations and is the envelope of two different elastic design spectra. The elastic design spectrum provides a basis for determining the design force and deformation for elastic SDOF systems.

Inelastic Response Spectrum

A bridge structure may experience inelastic behavior during a major earthquake. The typical elastic and elastic–plastic responses of an idealized SDOF to severe earthquake motions are shown in Figure 35.12. The input seismic energy received by a bridge structure is dissipated by both viscous damping and yielding (localized inelastic deformation converting into heat and other irrecoverable forms of energy). Both viscous damping and yielding reduce the response of inelastic structures compared with elastic structures. Viscous damping represents the internal friction loss of a structure when deformed and is approximately a constant because it depends mainly on structural materials. Yielding, on the other hand, varies depending on structural materials, structural configurations, and loading patterns and histories. Damping has negligible effects on the response of structures for

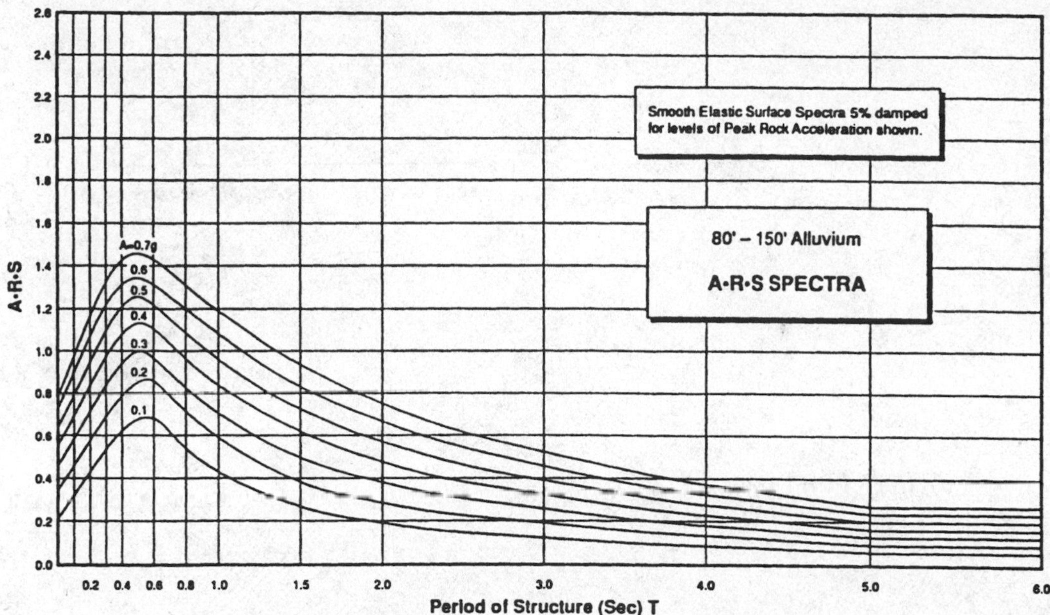

FIGURE 35.10 Typical Caltrans elastic design response spectra.

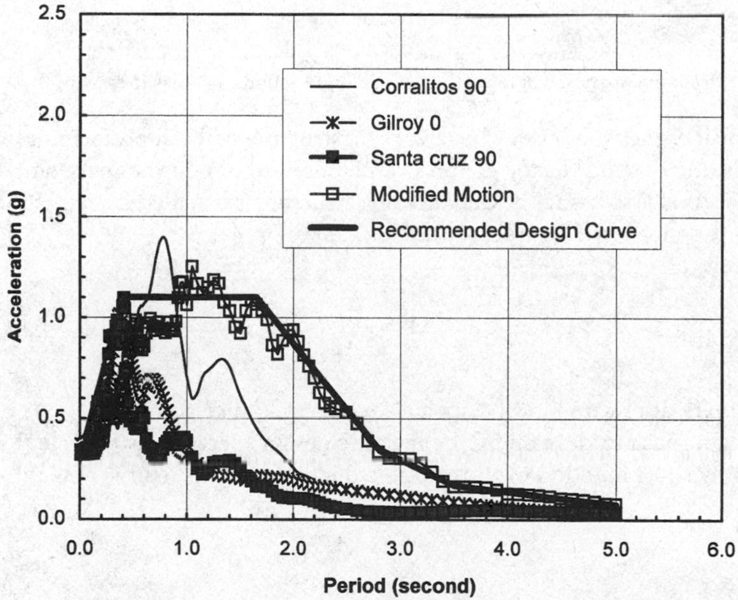

FIGURE 35.11 Acceleration response spectra for Sonoma Creek Bridge.

the long-period and short-period systems and is most effective in reducing response of structures for intermediate-period systems.

In seismic bridge design, a main objective is to ensure that a structure is capable of deforming in a ductile manner when subjected to a larger earthquake loading. It is desirable to consider the inelastic response of a bridge system to a major earthquake. Although a nonlinear inelastic dynamic analysis is not difficult in concept, it requires careful structural modeling and intensive computing

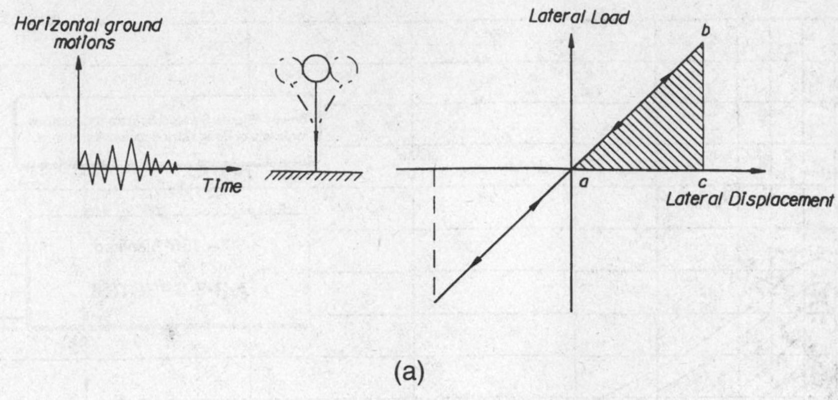

(a)

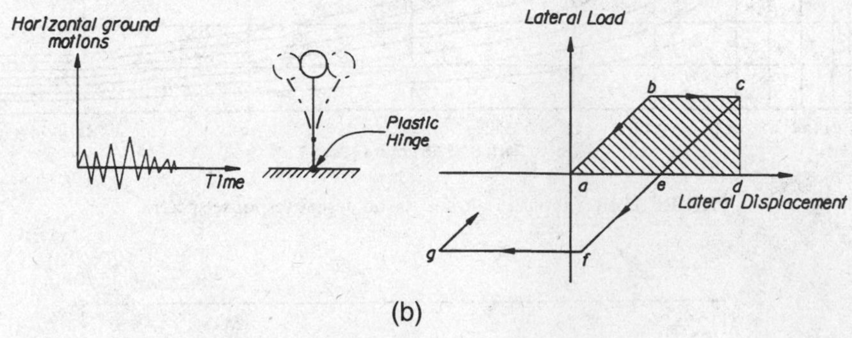

(b)

FIGURE 35.12 Response of an SDOF to earthquake ground motions. (a) Elastic system; (b) inelastic system.

effort [8]. To consider inelastic seismic behavior of a structure without performing a true nonlinear inelastic analysis, the ductility-factor method can be used to obtain the inelastic response spectra from the elastic response spectra. The ductility of a structure is usually referred as the displacement ductility factor μ defined by (Figure 35.13):

$$\mu = \frac{\Delta_u}{\Delta_y} \tag{35.11}$$

where Δ_u is ultimate displacement capacity and Δ_y is yield displacement.

The simplest approach to developing the inelastic design spectrum is to scale the elastic design spectrum down by some function of the available ductility of a structural system:

$$ARS_{\text{inelastic}} = \frac{ARS_{\text{elastic}}}{f(\mu)} \tag{35.12}$$

$$f(\mu) = \begin{cases} 1 & \text{for } T_n \le 0.03 \text{ sec.} \\ 2\mu - 1 & \text{for } 0.03 \text{ sec.} < T_n \le 0.5 \text{ sec.} \\ \mu & \text{for } T_n \ge 0.5 \text{ sec.} \end{cases} \tag{35.13}$$

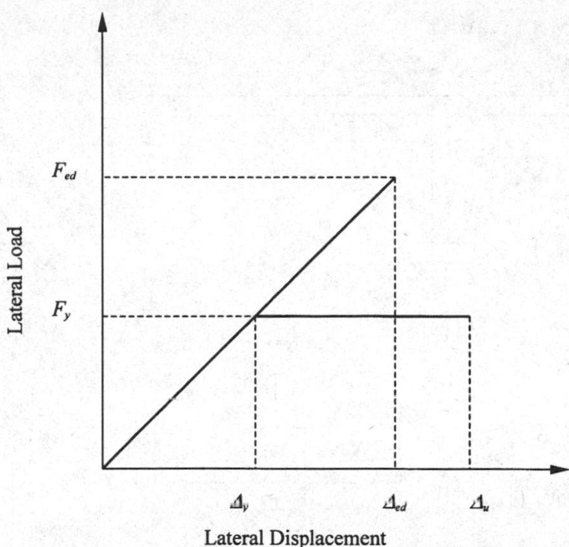

FIGURE 35.13 Lateral load–displacement relations.

For very short period ($T_n \leq 0.03$ sec) in the acceleration-sensitive region, the elastic displacement demand Δ_{ed} is less than displacement capacity Δ_u (see Figure 35.13). The reduction factor $f(\mu) = 1$ implies that the structure should be designed and remained at elastic to avoid excessive inelastic deformation. For intermediate period (0.03 sec $< T_n \leq 0.5$ sec) in the velocity-sensitive region, elastic displacement demand Δ_{ed} may be greater or less than displacement capacity Δ_u and the reduction factor is based on the equal-energy concept. For the very long period ($T_n > 0.5$ sec) in the displacement-sensitive region, the reduction factor is based on the equal displacement concept.

35.2.5 Example of an SDOF system

Given

An SDOF bridge structure is shown in Figure 35.14. To simplify the problem, the bridge is assumed to move only in the longitudinal direction. The total resistance against the longitudinal motion comes in the form of friction at bearings and this could be considered a damper. Assume the following properties for the structure: damping ratio $\xi = 0.05$, area of superstructure $A = 3.57$ m^2, moment of column $I_c = 0.1036$ m^4, E_c of column = 20,700 MPa, material density $\rho = 2400$ kg/m^3, length of column $L_c = 9.14$ m, and length of the superstructure $L_s = 36.6$ m. The acceleration response curve of the structure is given in the Figure 35.11. Determine (1) natural period of the structure, (2) damped period of the structure, (3) maximum displacement of the superstructure, and (4) maximum moment in the column.

Solution

$$\text{Stiffness:}\ \ k = \frac{12E_cI_c}{L_c^3} = \frac{12(20700\times10^6)(0.1036)}{9.14^3} = 33690301\ \text{N/m}$$

$$\text{Mass:}\ \ m = AL_s\rho = (3.57)(36.6)(2400) = 313{,}588.8\ \text{kg}$$

$$\text{Natural circular frequency:}\ \ \omega_n = \sqrt{\frac{k}{m}} = \sqrt{\frac{33{,}690{,}301}{313{,}588.8}} = 10.36\ \ \text{rad/s}$$

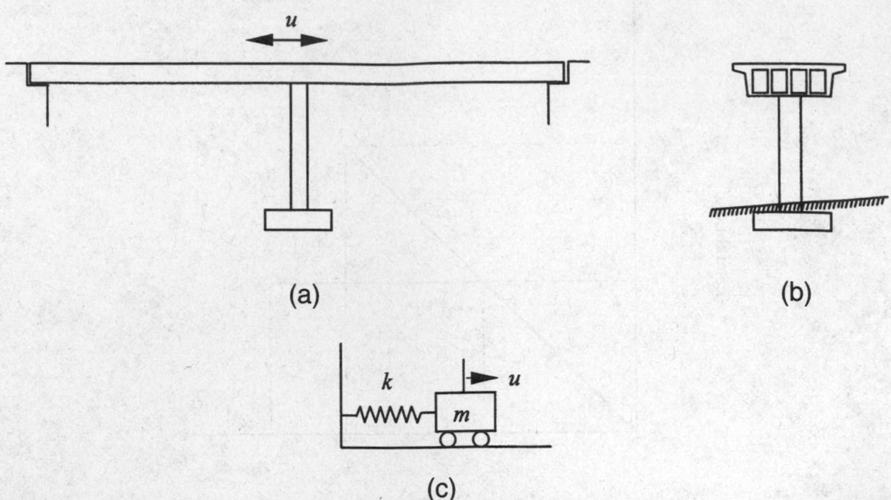

FIGURE 35.14 SDOF bridge example. (a) Two-span bridge schematic diagram; (b) single column bent; (c) idealized equivalent model for longitudinal response.

$$\text{Natural cyclic frequency: } f_n = \frac{\omega_n}{2\pi} = \frac{10.36}{2\pi} = 1.65 \text{ cycles/s}$$

$$\text{Natural period of the structure: } T_n = \frac{1}{f_n} = \frac{1}{1.65} = 0.606 \text{ s}$$

The *damped* circular frequency is given by

$$\omega_d = \omega_n\sqrt{1-\xi^2} = 10.36\sqrt{1-0.05^2} = 10.33 \text{ rad/s}$$

The *damped* period of the structure is given by

$$T_d = \frac{2\pi}{\omega_d} = \frac{2\pi}{10.33} = 0.608 \text{ s}$$

From the ARS curve, for a period of 0.606 s, the maximum acceleration of the structure will be 0.9 g = 1.13 × 9.82 = 11.10 m/s. Then,

$$\text{The force acting on the mass} = m \times 11.10 = 313588.8 \times 11.10 = 3.48 \text{MN}$$

$$\text{The maximum displacement} = \frac{FL^3{}_c}{12EI_c} = \frac{3.48 \times 9.14^3}{12 \times 20700 \times 0.1036} = 0.103 \text{m}$$

$$\text{The maximum moment in the column} = \frac{FL_c}{2} = \frac{3.48 \times 9.14}{2} = 15.90 \text{ MN-m}$$

35.3 Multidegree-of-Freedom System

The SDOF approach may not be applicable for complex structures such as multilevel frame structure and bridges with several supports. To predict the response of a complex structure, the structure is discretized with several members of lumped masses. As the number of lumped masses increases, the number of displacements required to define the displaced positions of all masses increases. The response of a multidegree of freedom (MDOF) system is discussed in this section.

35.3.1 Equation of Motion

The equation of motion of an MDOF system is similar to the SDOF system, but the stiffness **k**, mass **m**, and damping **c** are matrices. The equation of motion to an MDOF system under ground motion can be written as

$$[\mathbf{M}]\{\ddot{u}\}+[\mathbf{C}]\{\dot{u}\}+[\mathbf{K}]\{u\}=-[\mathbf{M}]\{\mathbf{B}\}\ddot{u}_g \tag{35.14}$$

The stiffness matrix $[\mathbf{K}]$ can be obtained from standard static displacement-based analysis models and may have off-diagonal terms. The mass matrix $[\mathbf{M}]$ due to the negligible effect of mass coupling can best be expressed in the form of tributary lumped masses to the corresponding displacement degree of freedoms, resulting in a diagonal or uncoupled mass matrix. The damping matrix $[\mathbf{C}]$ accounts for all the energy-dissipating mechanisms in the structure and may have off-diagonal terms. The vector $\{\mathbf{B}\}$ is a displacement transformation vector that has values 0 and 1 to define degrees of freedoms to which the earthquake loads are applied.

35.3.2 Free Vibration and Vibration Modes

To understand the response of MDOF systems better, we look at the *undamped, free* vibration of an *N* degrees of freedom (*N*-DOF) system first.

Undamped Free Vibration

By setting $[\mathbf{C}]$ and $\ddot{u}_g$ to zero in the Eq. (35.14), the equation of motion of undamped, free vibration of an *N*-DOF system can be shown as:

$$[\mathbf{M}]\{\ddot{u}\}+[\mathbf{K}]\{u\}=0 \tag{35.15}$$

where $[\mathbf{M}]$ and $[\mathbf{K}]$ are $n \times n$ square matrices.

Equation (35.15) could then be rearranged to

$$\left[[\mathbf{K}]-\omega_n^2[\mathbf{M}]\right]\{\phi_n\}=0 \tag{35.16}$$

where $\{\phi_n\}$ is the deflected shape matrix. Solution to this equation can be obtained by setting

$$\left|[\mathbf{K}]-\omega_n^2[\mathbf{M}]\right|=0 \tag{35.17}$$

The roots or eigenvalues of Eq. (35.17) will be the *N* natural frequencies of the dynamic system. Once the natural frequencies (ω_n) are estimated, Eq. (35.16) can be solved for the corresponding *N* independent, deflected shape matrices (or eigenvectors), $\{\phi_n\}$. In other words, a vibrating system with N-DOFs will have N natural frequencies (usually arranged in sequence from smallest to largest), corresponding N natural periods T_n, and N natural mode shapes $\{\phi_n\}$. These eigenvectors are sometimes referred to as natural modes of vibration or natural mode shapes of vibration. It is important to recognize that the eigenvectors or mode shapes represent only the deflected shape corresponding to the natural frequency, not the actual deflection magnitude.

The N eigenvectors can be assembled in a single $n \times n$ square matrix $[\mathbf{\Phi}]$, modal matrix, where each column represents the coefficients associated with the natural mode. One of the important aspects of these mode shapes is that they are orthogonal to each other. Stated mathematically,

$$\text{If} \quad \omega_n \neq \omega_r, \qquad \{\phi_n\}^T[K]\{\phi_r\}=0 \quad \text{and} \quad \{\phi_n\}^T[M]\{\phi_r\}=0 \tag{35.18}$$

$$\left[K^*\right] = [\Phi]^T[K][\Phi] \tag{35.19}$$

$$\left[M^*\right] = [\Phi]^T[M][\Phi] \tag{35.20}$$

where $[\mathbf{K}]$ and $[\mathbf{M}]$ have off-diagonal elements, whereas $\left[\mathbf{K}^*\right]$ and $\left[\mathbf{M}^*\right]$ are diagonal matrices.

Damped Free Vibration

When damping of the MDOF system is included, the free vibration response of the damped system will be given by

$$[\mathbf{M}]\{\ddot{u}\} + [\mathbf{C}]\{\dot{u}\} + [\mathbf{K}]\{u\} = 0 \tag{35.21}$$

The displacements are first expressed in terms of natural mode shapes, and later they are multiplied by the transformed natural mode matrix to obtain the following expression:

$$\left[\mathbf{M}^*\right]\{\ddot{Y}\} + \left[\mathbf{C}^*\right]\{\dot{Y}\} + \left[\mathbf{K}^*\right]\{Y\} = 0 \tag{35.22}$$

where, $\left[\mathbf{M}^*\right]$ and $\left[\mathbf{K}^*\right]$ are diagonal matrices given by Eqs. (35.19) and (35.20) and

$$\left[\mathbf{C}^*\right] = [\mathbf{\Phi}]^T[\mathbf{C}][\mathbf{\Phi}] \tag{35.23}$$

While $\left[\mathbf{M}^*\right]$ and $\left[\mathbf{K}^*\right]$ are diagonal matrices, $\left[\mathbf{C}^*\right]$ may have off diagonal terms. When $\left[\mathbf{C}^*\right]$ has off diagonal terms, the damping matrix is referred to as a *nonclassical* or *nonproportional* damping matrix. When $\left[\mathbf{C}^*\right]$ is diagonal, it is referred to as a *classical* or *proportional* damping matrix. Classical damping is an appropriate idealization when similar damping mechanisms are distributed throughout the structure. Nonclassical damping idealization is appropriate for the analysis when the damping mechanisms differ considerably within a structural system.

Since most bridge structures have predominantly one type of construction material, bridge structures could be idealized as a classical damping structural system. Thus, the damping matrix of Eq. (35.22) will be a diagonal matrix for most bridge structures. And, the equation of nth mode shape or generalized nth modal equation is given by

$$\ddot{Y}_n + 2\xi_n\,\omega_n\,\dot{Y}_n + \omega^2\,Y_n = 0 \tag{35.24}$$

Equation (35.24) is similar to the Eq. (35.7) of an SDOF system. Also, the vibration properties of each mode can be determined by solving the Eq. (35.24).

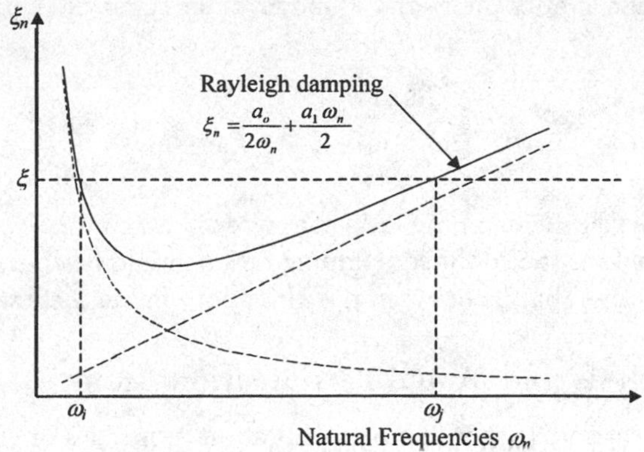

FIGURE 35.15 Rayleigh damping variation with natural frequency.

Rayleigh Damping

The damping of a structure is related to the amount of energy dissipated during its motion. It could be assumed that a portion of the energy is lost due to the deformations, and thus damping could be idealized as proportional to the stiffness of the structure. Another mechanism of energy dissipation could be attributed to the mass of the structure, and thus damping idealized as proportional to the mass of the structure. In Rayleigh damping, it is assumed that the damping is proportional to the mass and stiffness of the structure.

$$[\mathbf{C}] = a_o[\mathbf{M}] + a_1[\mathbf{K}] \tag{35.25}$$

The generalized damping of the nth mode is then given by

$$C_n = a_o M_n + a_1 K_n \tag{35.26}$$

$$C_n = a_o M_n + a_1 \omega_n^{\ 2} M_n \tag{35.27}$$

$$\xi_n = \frac{C_n}{2M_n\omega_n} \tag{35.28}$$

$$\xi_n = \frac{a_o}{2}\frac{1}{\omega_n} + \frac{a_1}{2}\omega_n \tag{35.29}$$

Figure 35.15 shows the Rayleigh damping variation with natural frequency. The coefficients a_o and a_1 can be determined from specified damping ratios at two independent dominant modes (say, i^{th} and j^{th} modes). Expressing Eq. (35.29) for these two modes will lead to the following equations:

$$\xi_i = \frac{a_o}{2}\frac{1}{\omega_i} + \frac{a_1}{2}\omega_i \tag{35.30}$$

$$\xi_j = \frac{a_o}{2}\frac{1}{\omega_j} + \frac{a_1}{2}\omega_j \tag{35.31}$$

When the damping ratio at both the ith and j^{th} modes is the same and equals ξ, it can be shown that

$$a_o = \xi \frac{2\omega_i \omega_j}{\omega_i + \omega_j} \qquad a_1 = \xi \frac{2}{\omega_i + \omega_j} \tag{35.32}$$

It is important to note that the damping ratio at a mode between the ith and j^{th} mode is less than ξ. And, in practical problems the specified damping ratios should be chosen to ensure reasonable values in all the mode shapes that lie between the *i*th and *j*th mode shapes.

35.3.3 Modal Analysis and Modal Participation Factor

In previous sections, we have discussed the basic vibration properties of an MDOF system. Now, we will look at the response of an MDOF system to earthquake ground motion. The basic equation of motion of the MDOF for an earthquake ground motion given by Eq. (35.14) is repeated here:

$$[\mathbf{M}]\{\ddot{u}\} + [\mathbf{C}]\{\dot{u}\} + [\mathbf{K}]\{u\} = -[\mathbf{M}]\,\{B\}\ddot{u}_g$$

The displacement is first expressed in terms of natural mode shapes, and later it is multiplied by the transformed natural mode matrix to obtain the following expression:

$$[\mathbf{M}^*]\{\ddot{Y}\} + [\mathbf{C}^*]\{\dot{Y}\} + [\mathbf{K}^*]\{Y\} = -[\mathbf{\Phi}]^T[\mathbf{M}]\{B\}\,\ddot{u}_g \tag{35.33}$$

And, the equation of the nth mode shape is given by

$$M_n^*\ddot{Y}_n + 2\xi_n\,\omega_n\,M_n^*\dot{Y}_n + \omega^2 M_n^*\,Y_n = L_n\ddot{u}_g \tag{35.34}$$

where

$$M_n^* = \{\phi_n\}^T[\mathbf{M}]\{\phi_n\} \tag{35.35}$$

$$L_n = -\{\phi_n\}^T[\mathbf{M}]\,[\mathbf{B}] \tag{35.36}$$

The L_n is referred to as the *modal participation factor* of the *n*th mode.

By dividing the Eq. (35.34) by M_n^*, the generalized modal equation of the *n*th mode becomes

$$\ddot{Y}_n + 2\xi_n\,\omega_n\,\dot{Y}_n + \omega^2\,Y_n = \left(\frac{L_n}{M_n^*}\right)\ddot{u}_g \tag{35.37}$$

Equation (35.34) is similar to the equation motion of an SDOF system, and thus Y_n can be determined by using methods similar to those described for SDOF systems. Once Y_n is established, the displacement due to the nth mode will be given by $u_n(t)=\phi_n Y_n(t)$. The total displacement due to combination of all mode shapes can then be determined by summing up all displacements for each mode and is given by

$$u(t) = \sum \phi_n Y_n(t) \tag{35.38}$$

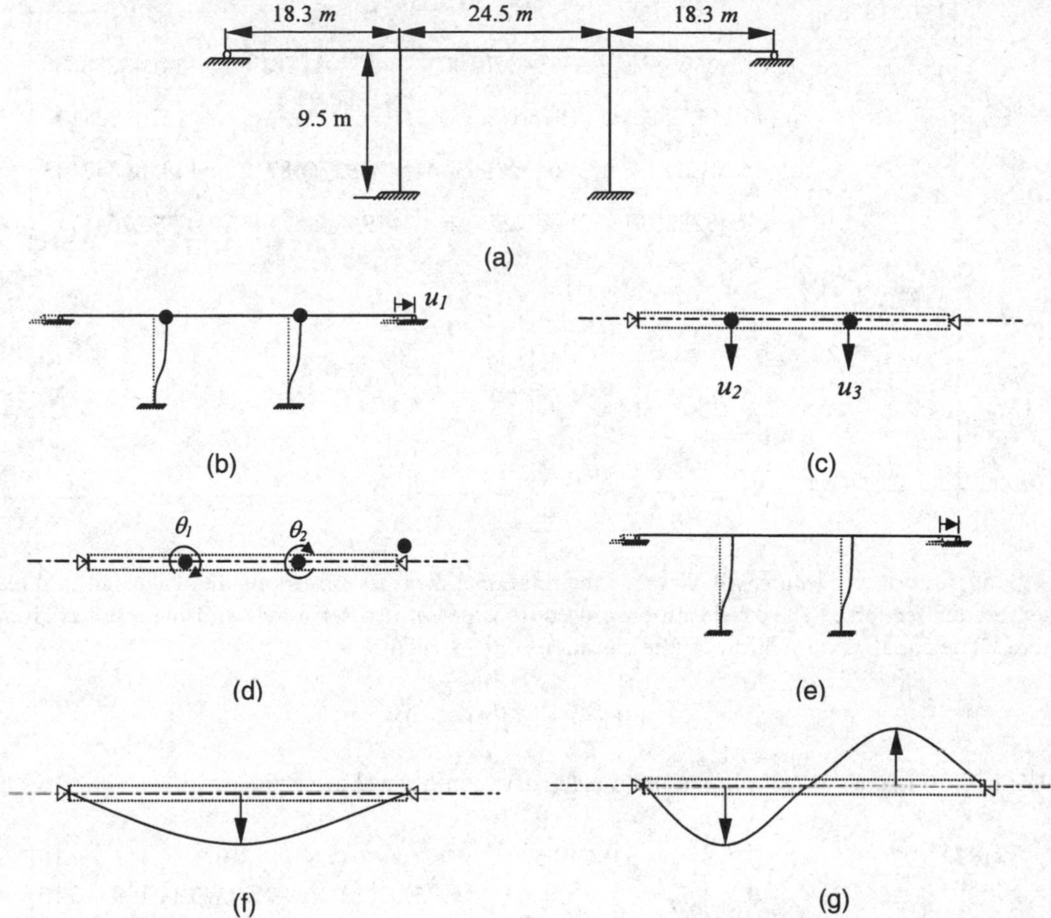

FIGURE 35.16 Three-span continuous framed bridge structure of MDOF example. (a) Schematic diagram; (b) longitudinal degree of freedom; (c) transverse degree of freedom; (d) rotational degree of freedom; (e) mode shape 1; (f) mode shape 2; (g) mode shape 3.

This approach is sometimes referred to as the classical mode superposition method. Similar to the estimation of the total displacement, the element forces can also be estimated by adding the element forces for each mode shape.

35.3.4 Example of an MDOF System

Given

The bridge shown in Figure 35.16 is a three-span continuous frame structure. Details of the bridge are as follows: span lengths are 18.3, 24.5, and 18.3 m.; column length is 9.5 m; area of superstructure is 5.58 m^2; moment of inertia of superstructure is 70.77 m^4; moment of inertia of column is 0.218 m^4; modulus of elasticity of concrete is 20,700 MPa. Determine the vibration modes and frequencies of the bridge.

Solution

As shown in Figures 35.16b, c, and d, five degrees of freedom are available for this structure. Stiffness and mass matrices are estimated separately and the results are given here.

$$[\mathbf{K}]=\begin{bmatrix} 126318588 & 0 & 0 & 0 & 0 \\ 0 & 1975642681 & -1194370500 & -1520122814 & -14643288630 \\ 0 & -1194370500 & 1975642681 & 14643288630 & 1520122814 \\ 0 & -1520122814 & 14643288630 & 479327648712 & 119586857143 \\ 0 & -14643288630 & 1520122814 & 119586857143 & 479327648712 \end{bmatrix}$$

$$[\mathbf{M}]=\begin{bmatrix} 81872 & 0 & 0 & 0 & 0 \\ 0 & 286827 & 0 & 0 & 0 \\ 0 & 0 & 286827 & 0 & 0 \\ 0 & 0 & 0 & 0 & 0 \\ 0 & 0 & 0 & 0 & 0 \end{bmatrix}$$

Condensation procedure will eliminate the rotational degrees of freedom and will result in three degrees of freedom. (The condensation procedure is performed separately and the result is given here.) The equation of motion of free vibration of the structure is

$$[M]\{\ddot{u}\}+[K]\{u\}=\{0\}$$

Substituting condensed stiffness and mass matrices into the above equation gives

$$\begin{bmatrix} 81872 & 0 & 0 \\ 0 & 286827 & 0 \\ 0 & 0 & 286827 \end{bmatrix}\begin{Bmatrix} \ddot{u}_1 \\ \ddot{u}_2 \\ \ddot{u}_3 \end{Bmatrix}+\begin{bmatrix} 126318588 & 0 & 0 \\ 0 & 1975642681 & -1194370500 \\ 0 & -1194370500 & 1975642681 \end{bmatrix}\begin{Bmatrix} u_1 \\ u_2 \\ u_3 \end{Bmatrix}=\begin{Bmatrix} 0 \\ 0 \\ 0 \end{Bmatrix}$$

The above equation can be rearranged in the following form:

$$\frac{1}{\omega^2}[M]^{-1}[K]\{\phi\}=\{\phi\}$$

Substitution of appropriate values in the above expression gives the following

$$\frac{1}{\omega_n^2}\begin{bmatrix} \frac{1}{818172} & 0 & 0 \\ 0 & \frac{1}{286827} & 0 \\ 0 & 0 & \frac{1}{286827} \end{bmatrix}\begin{bmatrix} 126318588 & 0 & 0 \\ 0 & 1518171572 & -1215625977 \\ 0 & -1215625977 & 1518171572 \end{bmatrix}\begin{Bmatrix} \phi_{1n} \\ \phi_{2n} \\ \phi_{3n} \end{Bmatrix}=\begin{Bmatrix} \phi_{1n} \\ \phi_{2n} \\ \phi_{3n} \end{Bmatrix}$$

$$\frac{1}{\omega_n^2}\begin{bmatrix} 154.39 & 0 & 0 \\ 0 & 5292.9 & -4238.2 \\ 0 & -4238.2 & 5292.9 \end{bmatrix}\begin{Bmatrix} \phi_{1n} \\ \phi_{2n} \\ \phi_{3n} \end{Bmatrix}=\begin{Bmatrix} \phi_{1n} \\ \phi_{2n} \\ \phi_{3n} \end{Bmatrix}$$

By assuming different vibration modes, natural frequencies of the structure can be estimated.

Substitution of vibration mode $\{1\ 0\ 0\}^T$ will result in the first natural frequency.

$$\frac{1}{\omega_n^2}\begin{bmatrix}154.39 & 0 & 0\\ 0 & 5292.9 & -4238.2\\ 0 & -4238.2 & 5292.9\end{bmatrix}\begin{Bmatrix}1\\0\\0\end{Bmatrix}=\frac{1}{\omega_n^2}\begin{bmatrix}154.39\\0\\0\end{bmatrix}=\begin{Bmatrix}1\\0\\0\end{Bmatrix}$$

Thus, $\omega_n^2 = 154.39$ and $\omega_n = 12.43\ \text{rad / s}$

By substituting the vibration modes of $\{0\ 1\ 1\}^T$ and $\{0\ 1\ -1\}^T$ in the above expression, the other two natural frequencies are estimated as 32.48 and 97.63 rad/s.

35.3.5 Multiple-Support Excitation

So far we have assumed that all supports of a structural system undergo the same ground motion. This assumption is valid for structures with foundation supports close to each other. However, for long-span bridge structures, supports may be widely spaced. As described in Section 35.1.2, earth motion at a location depends on the localized soil layer and the distance from the epicenter. Thus, bridge structures with supports that lie far from each other may experience different earth excitation. For example, Figure 35.17c, d, and e shows the predicted earthquake motions at Pier W3 and Pier W6 of the San Francisco–Oakland Bay Bridge (SFOBB) in California. The distance between Pier W3 and Pier W6 of the SFOBB is approximately 1411 m. These excitations are predicted by the California Department of Transportation by considering the soil and rock properties in the vicinity of the SFOBB and expected Earth movements at the San Andreas and Hayward faults. Note that the Earth motion at Pier W3 and Pier W6 are very different. Furthermore, Figures 35.17c, d, and e indicates that the Earth motion not only varies with the location, but also varies with direction. Thus, to evaluate the response of long, multiply supported, and complicated bridge structures, use of the actual earthquake excitation at each support is recommended.

The equation of motion of a multisupport excitation would be similar to Eq. (35.14), but the only difference is now that $\{B\}\ddot{u}_g$ is replaced by an displacement array $\{\ddot{u}_g\}$. And, the equation of motion for the multisupport system becomes

$$[\mathbf{M}]\{\ddot{u}\}+[\mathbf{C}]\{\dot{u}\}+[\mathbf{K}]\{u\}=-[\mathbf{M}]\{\ddot{u}_g\} \tag{35.39}$$

where $\{\ddot{u}_g\}$ has the acceleration at each support locations and has zero value at nonsupport locations. By using the uncoupling procedure described in the previous sections, the modal equation of the nth mode can be written as

$$\ddot{Y}_n + 2\xi_n\omega_n\dot{Y}_n + \omega^2 Y_n = -\sum_{l=1}^{N_g}\frac{L_n}{M_n^*}\ddot{u}_g \tag{35.40}$$

where N_g is the total number of externally excited supports.

The deformation response of the nth mode can then be determined as described in previous sections. Once the displacement responses of the structure for all the mode shapes are estimated, the total dynamic response can be obtained by combining the displacements.

35.3.6 Time History Analysis

When the structure enters the nonlinear range, or has nonclassical damping properties, modal analysis cannot be used. A numerical integration method, sometimes referred to as time history analysis, is required to get more accurate responses of the structure.

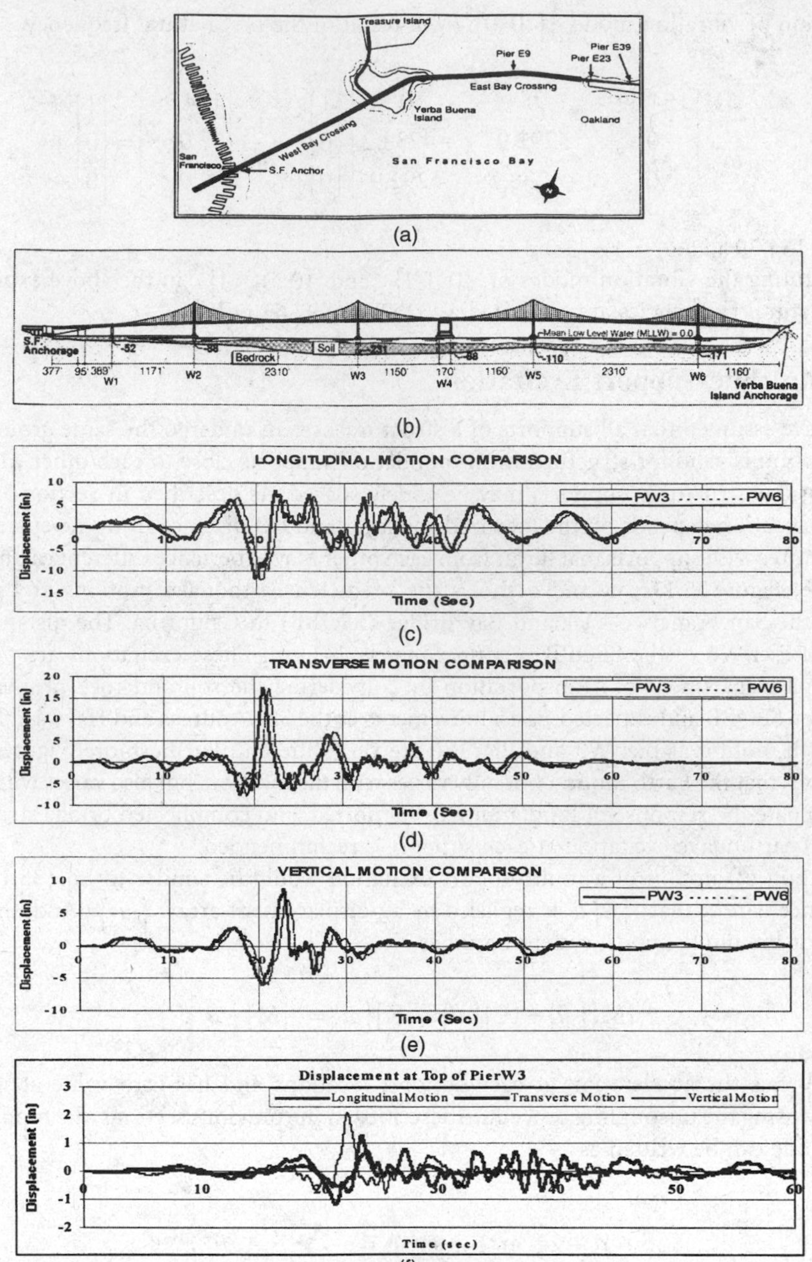

FIGURE 35.17 San Francisco–Oakland Bay Bridge. (a) Vicinity map; (b) general plan elevation; (c) longitudinal motion at rock level; (d) transverse motion at rock level; (e) vertical motion at rock level; (f) displacement response at top of Pier W3.

In a time history analysis, the timescale is divided into a series of smaller steps, d_τ. Let us say the response at i^{th} time interval has already determined and is denoted by $u_i, \dot{u}_i, \ddot{u}_i$. Then, the response of the system at i^{th} time interval will satisfy the equation of motion (Eq. 35.39).

$$[\mathbf{M}]\{\ddot{u}_i\}+[\mathbf{C}]\{\dot{u}_i\}+[\mathbf{K}]\{u_i\}=-[\mathbf{M}]\{\ddot{u}_{gi}\} \tag{35.41}$$

The time-stepping method enables us to step ahead and determine the responses $u_{i+1}, \dot{u}_{i+1}, \ddot{u}_{i+1}$ at the $i+1^{\text{th}}$ time interval by satisfying Eq. (35.39). Thus, the equation of motion at $i+1^{\text{th}}$ time interval will be

$$[\mathbf{M}]\{\ddot{u}_{i+1}\}+[\mathbf{C}]\{\dot{u}_{i+1}\}+[\mathbf{K}]\{u_{i+1}\}=-[\mathbf{M}]\{\ddot{u}_{gi+1}\} \tag{35.42}$$

Equation (35.42) needs to be solved prior to proceeding to the next time step. By stepping through all the time steps, the actual response of the structure can be determined at all time instants.

Example of Time History Analysis

The Pier W3 of the SFOBB was modeled using the ADINA [13] program and nonlinear analysis was performed using the displacement time histories. The displacement time histories in three directions are applied at the bottom of the Pier W3 and the response of the Pier W3 was studied to estimate the demand on Pier W3. One of the results, the displacement response at top of Pier W3, is shown in Figure 35.17f.

35.4 Response Spectrum Analysis

Response spectrum analysis is an approximate method of dynamic analysis that gives the maximum response (acceleration, velocity, or displacement) of an SDOF system with the same damping ratio, but with different natural frequencies, respond to a specified seismic excitation. Structural models with n degrees of freedom can be transformed to n single-degree systems and response spectra principles can be applied to systems with many degrees of freedom. For most ordinary bridges, a complete time history is not required. Because the design is generally based on the maximum earthquake response, response spectrum analysis is probably the most common method used in design offices to determine the maximum structural response due to transient loading. In this section, we will discuss basic procedures of response spectrum analysis for bridge structures.

35.4.1 Single-Mode Spectral Analysis

Single-mode spectral analysis is based on the assumption that earthquake design forces for structures respond predominantly in the first mode of vibration. This method is most suitable to regular linear elastic bridges to compute the forces and deformations, but is not applicable to irregular bridges (unbalanced spans, unequal stiffness in the columns, etc.) because higher modes of vibration affect the distribution of the forces and resulting displacements significantly. This method can be applied to both continuous and noncontinuous bridge superstructures in either the longitudinal or transverse direction. Foundation flexibility at the abutments can be included in the analysis.

Single-mode analysis is based on Rayleigh's energy method — an approximate method which assumes a vibration shape for a structure. The natural period of the structure is then calculated by equating the maximum potential and kinetic energies associated with the assumed shape. The inertial forces $p_e(x)$ are calculated using the natural period, and the design forces and displacements are then computed using static analysis. The detailed procedure can be described in the following steps:

1. Apply uniform loading p_o over the length of the structure and compute the corresponding static displacements $u_s(x)$. The structure deflection under earthquake loading, $u_s(x,t)$ is then approximated by the shape function, $u_s(x)$, multiplied by the generalized amplitude function, $u(t)$, which satisfies the geometric boundary conditions of the structural system. This dynamic deflection is shown as

$$u(x,t) = u_s(x)\ u(t) \tag{35.43}$$

2. Calculate the generalized parameters $\alpha, \beta,$ and γ using the following equations:

$$\alpha = \int u_s(x)\, dx \tag{35.44}$$

$$\beta = \int w(x)\, u_s(x)\; dx \tag{35.45}$$

$$\gamma = \int w(x) \left[u_s(x)\right]^2 dx \tag{35.46}$$

   where $w(x)$ is the weight of the dead load of the bridge superstructure and tributary substructure.
3. Calculate the period T_n

$$T_n = 2\pi \sqrt{\frac{\gamma}{p_o g \alpha}} \tag{35.47}$$

   where g is acceleration of gravity (mm/s²).
4. Calculate the static loading $p_e(x)$ which approximates the inertial effects associated with the displacement $u_s(x)$ using the ARS curve or the following equation [4]:

$$p_e(x) = \frac{\beta C_{sm}}{\gamma} w(x)\; u_s(x) \tag{35.48}$$

$$C_{sm} = \frac{1.2AS}{T_m^{\;2/3}} \tag{35.49}$$

   where C_{sm} is the dimensionless elastic seismic response coefficient; A is the acceleration coefficient from the acceleration coefficient map; S is the dimensionless soil coefficient based on the soil profile type; T_n is the period of the structure as determined above; $p_e(x)$ is the intensity of the equivalent static seismic loading applied to represent the primary mode of vibration (N/mm).
5. Apply the calculated loading $p_e(x)$ to the structure as shown in the Figure 35.18 and compute the structure deflections and member forces.

This method is an iterative procedure, and the previous calculations are used as input parameters for the new iteration leading to a new period and deflected shape. The process is continued until the assumed shape matches the fundamental mode shape.

35.4.2 Uniform-Load Method

The uniform-load method is essentially an equivalent static method that uses the uniform lateral load to compute the effect of seismic loads. For simple bridge structures with relatively straight alignment, small skew, balanced stiffness, relatively light substructure, and with no hinges, the uniform-load method may be applied to analyze the structure for seismic loads. This method is not suitable for bridges with stiff substructures such as pier walls. This method assumes continuity of the structure and distributes earthquake force to all elements of the bridge and is based on the fundamental mode of vibration in either a longitudinal or transverse direction [5]. The period of

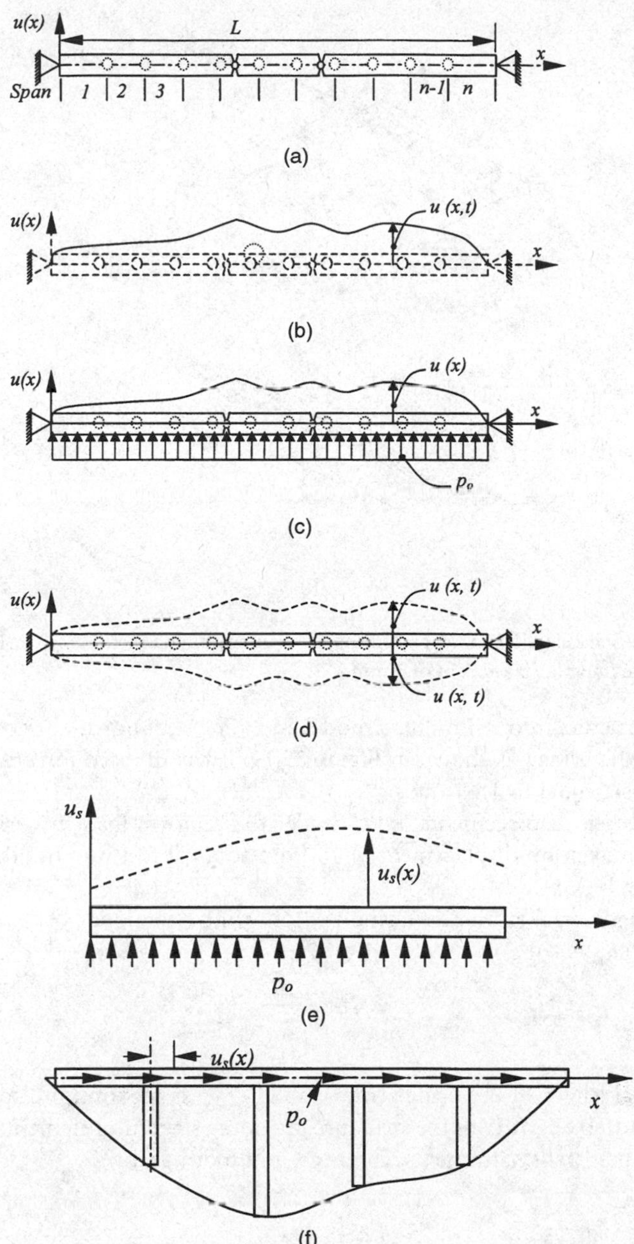

FIGURE 35.18 Single-mode spectral analysis method. (a) Plan view of a bridge subjected to transverse earthquake motion. (b) Displacement function describing the transverse position of the bridge deck. (c) Deflected shape due to uniform static loading. (d) Transverse free vibration of the bridge in assumed mode shape. (e) Transverse loading (f) longitudinal loading.

vibration is taken as that of an equivalent single mass–spring oscillator. The maximum displacement that occurs under the arbitrary uniform load is used to calculate the stiffness of the equivalent spring. The seismic elastic response coefficient C_{sm} or the ARS curve is then used to calculate the equivalent uniform seismic load, using which the displacements and forces are calculated. The following steps outline the uniform load method:

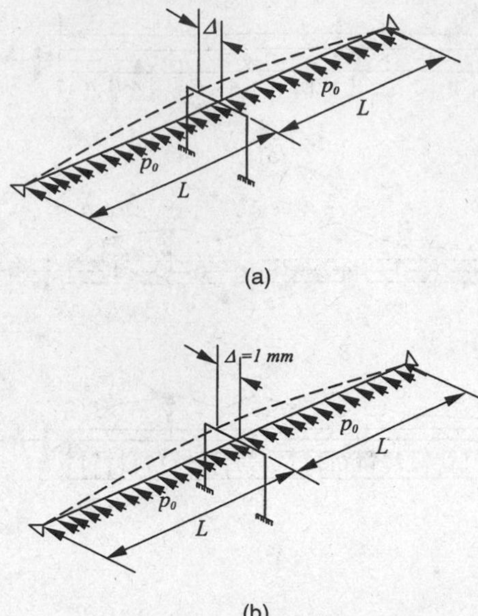

FIGURE 35.19 Structure idealization and deflected shape for uniform load method. (a) Structure idealization; (b) deflected shape with maximum displacement of 1 mm.

1. Idealize the structure into a simplified model and apply a uniform horizontal load (p_o) over the length of the bridge as shown in Figure 35.19. It has units of force/unit length and may be arbitrarily set equal to 1 N/mm.
2. Calculate the static displacements $u_s(x)$ under the uniform load p_o using static analysis.
3. Calculate the maximum displacement $u_{s,\max}$ and adjust it to 1 mm by adjusting the uniform load p_o.
4. Calculate bridge lateral stiffness K using the following equation:

$$K = \frac{p_o L}{u_{s,\max}} \tag{35.50}$$

   where L is total length of the bridge (mm); and $u_{s,\max}$ is maximum displacement (mm).
5. Calculate the total weight W of the structure including structural elements and other relevant loads such as pier walls, abutments, columns, and footings, by

$$W = \int w(x)dx \tag{35.51}$$

   where $w(x)$ is the nominal, unfactored dead load of the bridge superstructure and tributary substructure.
6. Calculate the period of the structure T_n using the following equation:

$$T_n = \frac{2\pi}{31.623}\sqrt{\frac{W}{gK}} \tag{35.52}$$

   where g is acceleration of gravity (m/s²).

7. Calculate the equivalent static earthquake force p_e using the ARS curve or using the following equation:

$$p_e = \frac{C_{sm} W}{L} \tag{35.53}$$

8. Calculate the structure deflections and member forces by applying p_e to the structure.

35.4.4 Multimode Spectral Analysis

The multimode spectral analysis method is more sophisticated than single-mode spectral analysis and is very effective in analyzing the response of more complex linear elastic structures to an earthquake excitation. This method is appropriate for structures with irregular geometry, mass, or stiffness. These irregularities induce coupling in three orthogonal directions within each mode of vibration. Also, for these bridges, several modes of vibration contribute to the complete response of the structure. A multimode spectral analysis is usually done by modeling the bridge structure consisting of three-dimensional frame elements with structural mass lumped at various locations to represent the vibration modes of the components. Usually, five elements per span are sufficient to represent the first three modes of vibration. A general rule of thumb is, to capture the ith mode of vibration, the span should have at least $(2i-1)$ elements. For long-span structures many more elements should be used to capture all the contributing modes of vibration. To obtain a reasonable response, the number of modes should be equal to at least three times the number of spans. This analysis is usually performed with a dynamic analysis computer program such as ADINA [13], GTSTRUDL [14], SAP2000 [15], ANSYS [16], and NASTRAN [17]. For bridges with outrigger bents, C-bents, and single column bents, rotational moment of inertia of the superstructure should be included. Discontinuities at the hinges and abutments should be included in the model. The columns and piers should have intermediate nodes at quarter points in addition to the nodes at the ends of the columns.

By using the programs mentioned above, frequencies, mode shapes, member forces, and joint displacements can be computed. The following steps summarize the equations used in the multimode spectral analysis [5].

1. Calculate the dimensionless mode shapes $\{\phi_i\}$ and corresponding frequencies ω_i by

$$\left[[\mathbf{K}] - \omega^2 [\mathbf{M}] \right] \{u\} = 0 \tag{35.54}$$

where

$$u_i = \sum_{j=1}^{n} \phi_j y_j = \Phi y_i \tag{35.55}$$

y_j = modal amplitude of *j*th mode; ϕ_j = shape factor of jth mode; Φ = mode-shape matrix. The periods for i^{th} mode can then be calculated by

$$T_i = \frac{2\pi}{\omega_i} \quad (i = 1, 2, \ldots, n) \tag{35.56}$$

2. Determine the maximum absolute mode amplitude for the entire time history is given by

$$Y_i(t)_{\max} = \frac{T_i^2 S_a(\xi_i, T_i)}{4\pi^2} \frac{\{\phi_i\}^T [\mathbf{M}]\{B\}\ddot{u}_g}{\{\phi_i\}^T \quad [\mathbf{M}]\{\phi_i\}} \tag{35.57}$$

where $S_a(\xi_i, T_i) = gC_{sm}$ is the acceleration response spectral value; C_{sm} is the elastic seismic response coefficient for mode $m = 1.2AS/T_n^{2/3}$; A is the acceleration coefficient from the acceleration coefficient map; S is the dimensionless soil coefficient based on the soil profile type; T_n is the period of the nth mode of vibration.

3. Calculate the value of any response quantity $Z(t)$ (shear, moment, displacement) using the following equation:

$$Z(t) = \sum_{i=1}^{n} A_i Y_i(t) \tag{35.58}$$

where coefficients A_i are functions of mode shape matrix (Φ) and force displacement relationships.

4. Compute the maximum value of $Z(t)$ during an earthquake using the mode combination methods described in the next section.

Modal Combination Rules

The mode combination method is a very useful tool for analyzing bridges with a large number of degrees of freedom. In a linear structural system, maximum response can be estimated by mode combination after calculating natural frequencies and mode shapes of the structure using free vibration analysis. The maximum response cannot be computed by adding the maximum response of each mode because different modes attain their maximum values at different times. The absolute sum of the individual modal contributions provides an upper bound which is generally very conservative and not recommended for design. There are several different empirical or statistical methods available to estimate the maximum response of a structure by combining the contributions of different modes of vibrations in a spectral analysis. Two commonly used methods are the square root of sum of squares (SRSS) and the complete quadratic combination (CQC).

For an undamped structure, the results computed using the CQC method are identical to those using the SRSS method. For structures with closely spaced dominant mode shapes, the CQC method is precise whereas SRSS estimates inaccurate results. Closely spaced modes are those within 10% of each other in terms of natural frequency. The SRSS method is suitable for estimating the total maximum response for structures with well-spaced modes. Theoretically, all mode shapes must be included to calculate the response, but fewer mode shapes can be used when the corresponding mass participation is over 85% of the total structure mass. In general, the factors considered to determine the number of modes required for the mode combination are dependent on the structural characteristics of the bridge, the spatial distribution, and the frequency content of the earthquake loading. The following list [14] summarizes several commonly used mode combination methods to compute the maximum total response. The variable Z represents the maximum value of some response quantity (displacement, shear, etc.), Z_i is the peak value of that quantity in the ith mode, and N is the total number of contributing modes.

1. *Absolute Sum*: The absolute sum is sum of the modal contributions:

$$Z = \sum_{i=1}^{N} |Z_i| \tag{35.59}$$

2. *SRSS or Root Mean Square (RMS) Method*: This method computes the maximum by taking the square root of sum of squares of the modal contributions:

$$Z = \left[\sum_{i=1}^{N} Z_i^2 \right]^{1/2} \tag{35.60}$$

3. *Peak Root Mean Square (PRMS)*: Absolute value of the largest modal contribution is added to the root mean square of the remaining modal contributions:

$$Z_j = |\max\ Z_i| \tag{35.61}$$

$$Z = \left[\sum_{i=1}^{N} Z_i^2 \right]^{1/2} + Z_j \quad \text{with} \quad i \neq j \tag{35.62}$$

4. *CQC*: Cross correlations between all modes are considered:

$$Z = \left[\sum_{i=1}^{N} \sum_{j=1}^{N} Z_i\ \rho_{ij}\ Z_j \right]^{1/2} \tag{35.63}$$

$$\rho_{ij} = \frac{8\sqrt{\xi_i \xi_j}\left(\xi_i + r\xi_j\right) r^{3/2}}{\left(1-r^2\right)^2 + 4\xi_i\xi_j r\left(1+r^2\right) + 4\left(\xi_i^2 + \xi_j^2\right) r^2} \tag{35.64}$$

where

$$r = \frac{\omega_j}{\omega_i} \tag{35.65}$$

5. *Nuclear Regulatory Commission Grouping Method*: This method is similar to RMS method with additional accounting for groups of modes whose frequencies are within 10%.

$$Z = \left[\sum_{i=1}^{N} Z_i^2 + \sum_{g=1}^{G} \sum_{n=s}^{e} \sum_{m=s}^{e} \left| Z_n^g \times Z_m^g \right| \right]^{1/2} \qquad n \neq m \tag{35.66}$$

where G is number of groups; s is mode shape number where the gth group starts; e is mode shape number where the gth group ends; and Z_i^g is the ith modal contribution in the gth group.

6. *Nuclear Regulatory Commission Ten Percent Method*: This method is similar to the RMS method with additional accounting for all modes whose frequencies are within 10%.

$$Z = \left[\sum_{i=1}^{N} Z_i^2 + 2 \sum \left| Z_n Z_m \right| \right]^{1/2} \tag{35.67}$$

The additional terms must satisfy

$$\frac{\omega_n - \omega_m}{\omega_m} \leq 0.1 \qquad \text{for} \qquad 0.1 \leq m \leq n \leq N \tag{35.68}$$

7. *Nuclear Regulatory Commission Double Sum Method*: This method is similar to the CQC method.

$$Z = \left[\sum_{i=1}^{N} \sum_{j=1}^{N} \left| Z_i Z_j \right| \quad \varepsilon_{ij} \right]^{1/2} \tag{35.69}$$

$$\varepsilon_{ij} = \left[1 + \left\{ \frac{\left(\omega_i' - \omega_j'\right)}{\left(\xi_i'\omega_i + \xi_j'\omega_j\right)} \right\} \right]^{-1} \tag{35.70}$$

$$\omega_i' = \omega\left[1 - \xi_i^2\right]^{1/2} \tag{35.71}$$

$$\xi_i' = \xi_i + \frac{2}{t_d \omega_i} \tag{35.72}$$

where t_d is the duration of support motion.

Combination Effects

Effects of ground motions in two orthogonal horizontal directions should be combined while designing bridges with simple geometric configurations. For bridges with long spans, outrigger bents, and with cantilever spans, or where effects due to vertical input are significant, vertical input should be included in the design along with two orthogonal horizontal inputs. When bridge structures are analyzed independently along each direction using response spectra analysis, then responses are combined either using methods, such as the SRSS combination rule as mentioned in the previous section, or using the alternative method described below. For structures designed using equivalent static analysis or modal analysis, seismic effects should be determined using the following alternative method for the following load cases:

1. *Seismic load case* 1: 100% Transverse + 30% Longitudinal + 30% Vertical
2. *Seismic load case* 2: 30% Transverse + 100% Longitudinal + 30% Vertical
3. *Seismic load case* 3: 30% Transverse + 30% Longitudinal + 100% Vertical

For structures designed using time-history analysis, the structure response is calculated using the input motions applied in orthogonal directions simultaneously. Where this is not feasible, the above alternative procedure can be used to combine the independent responses.

35.4.4 Multiple-Support Response Spectrum Method

Records from recent earthquakes indicate that seismic ground motions can significantly vary at different support locations for multiply supported long structures. When different ground motions are applied at various support points of a bridge structure, the total response can be calculated by superposition of responses due to independent support input. This analysis involves combination of dynamic response from single-input and pseudo-static response resulting from the motion of the supports relative to each other. The combination effects of dynamic and pseudo-static forces due to multiple support excitation on a bridge depend on the structural configuration of the bridge and the ground motion characteristics. Recently, Kiureghian et al. [7] presented a comprehensive study on the multiple-support response spectrum (MSRS) method based on fundamental principles of stationary random vibration theory for seismic analysis of multiply supported structures which accounts for the effects of variability between the support motions. Using the MSRS combination rule, the response of a linear structural system subjected to multiple support excitation can be computed directly in terms of conventional response spectra at the support degrees of freedom and a coherency function describing the spatial variability of the ground motion. This method accounts for the three important effects of ground motion spatial variability, namely, the incoherence effects, the wave passage effect, and the site response effect. These three components of ground motion spatial variability can strongly influence the response of multiply supported bridges and may amplify or deamplify the response by one order of magnitude. Two important limitations of this method are nonlinearities in the bridge structural components and/or connections and the effects of soil–structure interaction. This method is an efficient, accurate, and versatile solution and requires less computational time than a true time history analysis. Following are the steps that describe the MSRS analysis procedure.

1. *Determine the necessity of variable support motion analysis*: Three factors that influence the response of the structure under multiple support excitation are the distance between the supports of the structure, the rate of variability of the local soil conditions, and the stiffness of the structure. The first factor, the distance between the supports, influences the incoherence and wave passage effects. The second factor, the rate of variability of the local conditions, influences the site response. The third factor, the stiffness of the superstructure, plays an important role in determining the necessity of variable-support motion analysis. Stiff structures such as box-girder bridges may generate large internal forces under variable support motion, whereas flexible structures such as suspension bridges easily conform to the variable support motion.
2. *Determine the frequency response function for each support location.* Programs such as SHAKE [18] can be used to develop these functions using borehole data and time-domain site response analysis. Response spectra plots, peak ground displacements in three orthogonal directions for each support location, and a coherency function for each pair of degrees of freedom are required to perform the MSRS analysis. The comprehensive report by Kiureghian [7] provides all the formulas required to account for the effect of nonlinearity in the soil behavior and the site frequency involving the depth of the bedrock.
3. *Calculate the Structural Properties*: such as effective modal frequencies, damping ratios, influence coefficients and effective modal participation factors (ω_i, ξ_i, a_k, and b_{ki}) are to be computed externally and provided as input.
4. *Determine the response spectra plots, peak ground displacements in three directions, and a coherency function for each pair of support degrees of freedom required to perform MSRS analysis*: Three components of the coherency function are incoherence, wave passage effect, and site response effect. Analysis by an array of recordings is used to determine the incoherence component. The models for this empirical method are widely available [19]. Parameters such as shear wave velocity, the direction of propagation of seismic waves, and the angle of incidence are used to calculate the wave passage effect. The frequency response function determined in the previous steps is used to calculate the site response component.

35.5 Inelastic Dynamic Analysis

35.5.1 Equations of Motion

Inelastic dynamic analysis is usually performed for the safety evaluation of important bridges to determine the inelastic response of bridges when subjected to design earthquake ground motions.

Inelastic dynamic analysis provides a realistic measure of response because the inelastic model accounts for the redistribution of internal actions due to the nonlinear force displacement behavior of the components [20–25]. Inelastic dynamic analysis considers nonlinear damping, stiffness, load deformation behavior of members including soil, and mass properties. A step-by-step integration procedure is the most powerful method used for nonlinear dynamic analysis. One important assumption of this procedure is that acceleration varies linearly while the properties of the system such as damping and stiffness remain constant during the time interval. By using this procedure, a nonlinear system is approximated as a series of linear systems and the response is calculated for a series of small equal intervals of time Δt and equilibrium is established at the beginning and end of each interval.

The accuracy of this procedure depends on the length of the time increment Δt. This time increment should be small enough to consider the rate of change of loading $p(t)$, nonlinear damping and stiffness properties, and the natural period of the vibration. An SDOF system and its characteristics are shown in the Figure 35.20. The characteristics include spring and damping forces, forces acting on mass of the system, and arbitrary applied loading. The force equilibrium can be shown as

$$f_i(t)+f_d(t)+f_s(t)=p(t) \tag{35.73}$$

and the incremental equations of motion for time t can be shown as

$$m\,\Delta\ddot{u}(t)+c(t)\,\Delta\dot{u}(t)+k(t)\,\Delta u(t)=\Delta p(t) \tag{35.74}$$

Current damping $f_d(t)$, elastic forces $f_s(t)$ are then computed using the initial velocity $\dot{u}(t)$, displacement values $u(t)$, nonlinear properties of the system, damping $c(t)$, and stiffness $k(t)$ for that interval. New structural properties are calculated at the beginning of each time increment based on the current deformed state. The complete response is then calculated by using the displacement and velocity values computed at the end of each time step as the initial conditions for the next time interval and repeating until the desired time.

35.5.2 Modeling Considerations

A bridge structural model should have sufficient degrees of freedom and proper selection of linear/nonlinear elements such that a realistic response can be obtained. Nonlinear analysis is usually preceded by a linear analysis as a part of a complete analysis procedure to capture the physical and mechanical interactions of seismic input and structure response. Output from the linear response solution is then used to predict which nonlinearities will affect the response significantly and to model them appropriately. In other words, engineers can justify the effect of each nonlinear element introduced at the appropriate locations and establish the confidence in the nonlinear analysis. While discretizing the model, engineers should be aware of the trade-offs between the accuracy, computational time, and use of the information such as the regions of significant geometric and material nonlinearities. Nonlinear elements should have material behavior to simulate the hysteresis relations under reverse cyclic loading observed in the experiments.

The general issues in modeling of bridge structures include geometry, stiffness, mass distribution, and boundary conditions. In general, abutments, superstructure, bent caps, columns and pier walls, expansion joints, and foundation springs are the elements included in the structural model. The mass distribution in a structural model depends on the number of elements used to represent the bridge components. The model must be able to simulate the vibration modes of all components contributing to the seismic response of the structure.

Superstructure: Superstructure and bent caps are usually modeled using linear elastic three-dimensional beam elements. Detailed models may require nonlinear beam elements.

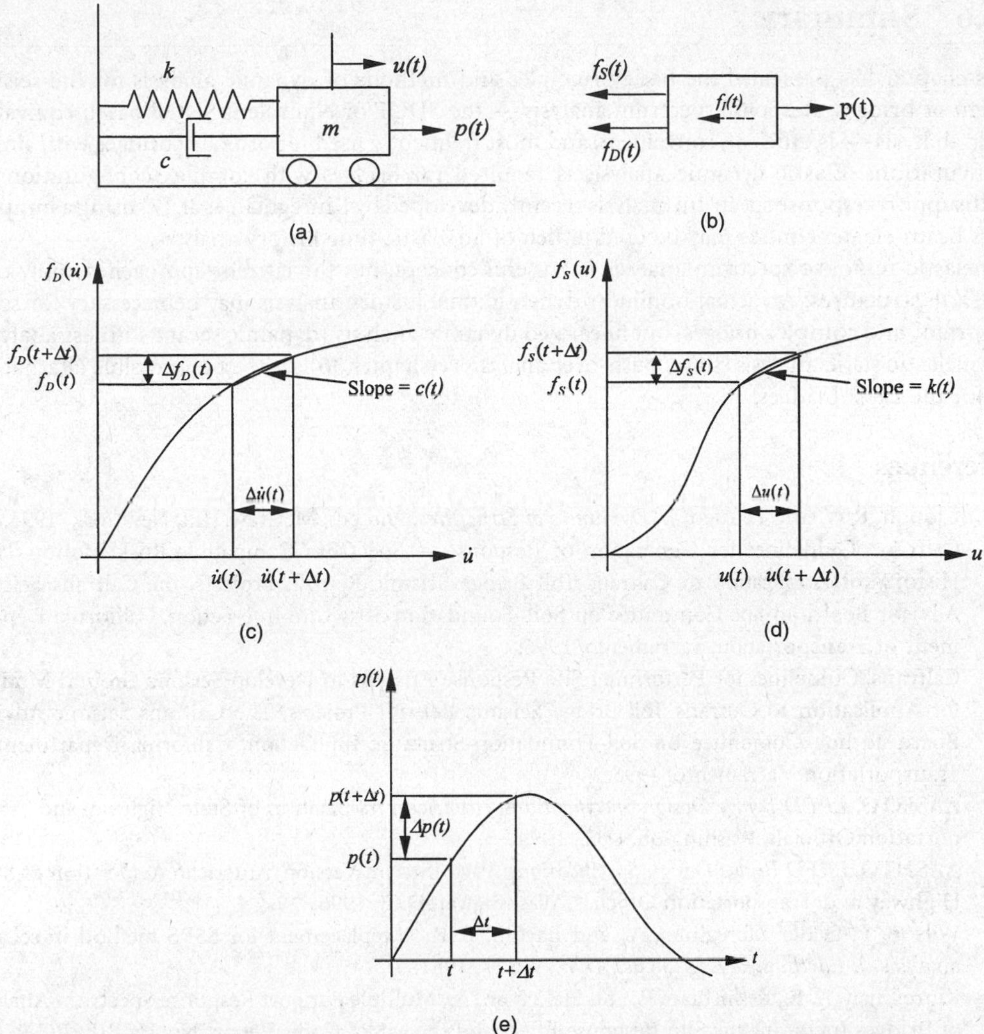

FIGURE 35.20 Definition of a nonlinear dynamic system. (a) Basis SDOF structure; (b) force equilibrium; (c) nonlinear damping; (d) nonlinear stiffness; (e) applied load.

Columns and pier walls: Columns and pier walls are usually modeled using nonlinear beam elements having response properties with a yield surface described by the axial load and biaxial bending. Some characteristics of the column behavior include initial stiffness degradation due to concrete cracking, flexural yielding at the fixed end of the column, strain hardening, pinching at the point of load reversal. Shear actions can be modeled using either linear or nonlinear load deformation relationships for columns. For both columns and pier walls, torsion can be modeled with linear elastic properties. For out-of-plane loading, flexural response of a pier wall is similar to that of columns, whereas for in-plane loading the nonlinear behavior is usually shear action.

Expansion joints: Expansion joints can be modeled using gap elements that simulate the nonlinear behavior of the joint. The variables include initial gap, shear capacity of the joint, and nonlinear load deformation characteristics of the gap.

Foundations and abutments: Foundations are typically modeled using nonlinear spring elements to represent the translational and rotational stiffness of the foundations to represent the expected behavior during a design earthquake. Abutments are modeled using nonlinear spring and gap elements to represent the soil action, stiffness of the pile groups, and gaps at the seat.

35.6 Summary

This chapter has presented the basic principles and methods of dynamic analysis for the seismic design of bridges. Response spectrum analysis — the SDOF or equivalent SDOF-based equivalent static analysis — is efficient, convenient, and most frequently used for ordinary bridges with simple configurations. Elastic dynamic analysis is required for bridges with complex configurations. A multisupport response spectrum analysis recently developed by Kiureghian et al. [7] using a lumped-mass beam element mode may be used in lieu of an elastic time history analysis.

Inelastic response spectrum analysis is a useful concept, but the current approaches apply only to SDOF structures. An actual nonlinear dynamic time history analysis may be necessary for some important and complex bridges, but linearized dynamic analysis (dynamic secant stiffness analysis) and inelastic static analysis (static push-over analysis) (Chapter 36) are the best possible alternatives [8] for the most bridges.

References

1. Clough, R.W. and Penzien, J., *Dynamics of Structures,* 2nd ed., McGraw-Hill, New York, 1993.
2. Caltrans, Guidelines for Generation of Response — Spectrum-Compatible Rock Motion Time History for Application to Caltrans Toll Bridge Seismic Retrofit Projects, the Caltrans Seismic Advisor Board ad hoc Committee on Soil–Foundation-Structure Interaction, California Department of Transportation, Sacramento, 1996.
3. Caltrans, Guidelines for Performing Site Response Analysis to Develop Seismic Ground Motions for Application to Caltrans Toll Bridge Seismic Retrofit Projects, The Caltrans Seismic Advisor Board ad hoc Committee on Soil–Foundation-Structure Interaction, California Department of Transportation, Sacramento, 1996.
4. AASHTO, *LRFD Bridge Design Specifications,* American Association of State Highway and Transportation Officials, Washington, D.C., 1994.
5. AASHTO, *LRFD Bridge Design Specifications,* 1996 Interim Version, American Association of State Highway and Transportation Officials, Washington, D.C., 1996.
6. Wilson, E. L, der Kiureghian, A., and Bayom, E. P., A replacement for SSRS method in seismic analysis, *J. Earthquake Eng. Struct. Dyn.,* 9, 187, 1981.
7. Kiureghian, A. E., Keshishian, P., and Hakobian, A., Multiple Support Response Spectrum Analysis of Bridges Including the Site-Response Effect and the MSRS Code, Report No. UCB/EERC-97/02, University of California, Berkeley, 1997.
8. Powell, G. H., Concepts and Principles for the Application of Nonlinear Structural Analysis in Bridge Design, Report No.UCB/SEMM-97/08, Department of Civil Engineering, University of California, Berkeley, 1997.
9. Chmielewski, T., Kratzig, W. B., Link, M., Meskouris, K., and Wunderlich, W., Phenomena and evaluation of dynamic structural responses, in *Dynamics of Civil Engineering Structures,* W. B. Kratzig and H.-J. Niemann, Eds., A.A. Balkema, Rotterdams, 1996.
10. Chopra, A.K., *Dynamics of Structures,* Prentice-Hall, Englewood Cliffs, NJ, 1995.
11. Lindeburg, M., *Seismic Design of Building Structures: A Professional's Introduction to Earthquake Forces and Design Details,* Professional Publications, Belmont, CA, 1998.
12. Caltrans, Bridge Design Specifications, California Department of Transportation, Sacramento, CA, 1991.
13. ADINA, *User's Guide,* Adina R&D, Inc., Watertown, MA, 1995.
14. GTSTRUDL, *User's Manual,* Georgia Institute of Technology, Atlanta, 1996.
15. SAP2000, *User's Manual,* Computers and Structures Inc., Berkeley, CA, 1998.
16. ANSYS, *User's Manual,* Vols. 1 and 2, Version 4.4, Swanson Analysis Systems, Inc., Houston, TX, 1989.
17. NASTRAN, *User's Manual,* MacNeil Schwindler Corporation, Los Angeles, CA.

18. Idriss, I. M., Sun J. I., and Schnabel, P. B., User's manual for SHAKE91: a computer program for conducting equivalent linear seismic response analyses of horizontally layered soil deposits, *Report of Center for Geotechnical Modeling*, Department of Civil and Environmental Engineering, University of California at Davis, 1991.
19. Abrahamson, N. A., Schneider, J. F., and Stepp, J. C., Empirical spatial coherency functions for application to soil-structure interaction analysis, *Earthquake Spectra*, 7, 1991.
20. Imbsen & Associates, *Seismic Design of Highway Bridges*, Sacramento, CA, 1992.
21. Priestly, M. J. N., Seible, F., and Calvi, G. M., *Seismic Design and Retrofit of Bridges*, John Wiley & Sons, New York, 1996.
22. Bathe, K.-J., *Finite Element Procedures in Engineering Analysis*, 2nd ed., Prentice-Hall, Englewood Cliffs, NJ, 1996.
23. ATC 32, Improved Seismic Design Criteria for California Bridges: Provisional Recommendations, Applied Technology Council, 1996.
24. Buchholdt, H. A., *Structural Dynamics for Engineers*, Thomas Telford, London, 1997.
25. Paz, M., *Structural Dynamics — Theory and Computation*, 3rd ed., Van Nostrand Reinhold, New York, 1991.

36

Nonlinear Analysis of Bridge Structures

Mohammed Akkari
California Department of Transportation

Lian Duan
California Department of Transportation

36.1 Introduction

In recent years, nonlinear bridge analysis has gained a greater momentum because of the need to assess inelastic structural behavior under seismic loads. Common seismic design philosophies for ordinary bridges allow some degree of damage without collapse. To control and evaluate damage, a postelastic nonlinear analysis is required. A nonlinear analysis is complex and involves many simplifying assumptions. Engineers must be familiar with those complexities and assumptions to design bridges that are safe and economical.

Many factors contribute to the nonlinear behavior of a bridge. These include factors such as material inelasticity, geometric or second-order effects, nonlinear soil–foundation–structure interaction, gap opening and closing at hinges and abutment locations, time-dependent effects due to concrete creep and shrinkage, etc. The subject of nonlinear analysis is extremely broad and cannot be covered in detail in this single chapter. Only material and geometric nonlinearities as well as

0-8493-7434-0/00/$0.00+$.50

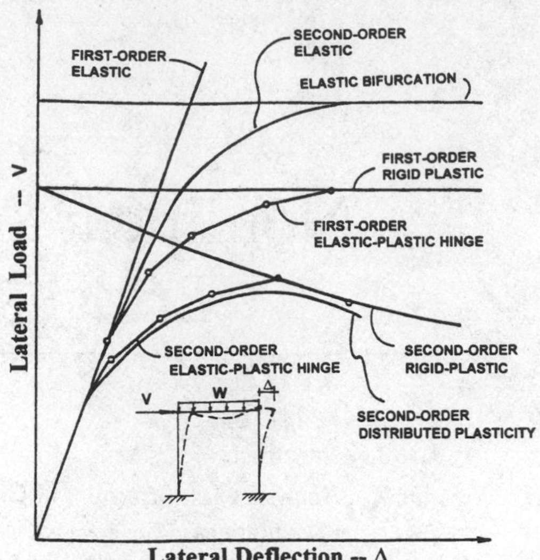

FIGURE 36.1 Lateral load–displacement curves of a frame.

some of the basic formulations of nonlinear static analysis with their practical applications to seismic bridge design will be presented here. The reader is referred to the many excellent papers, reports, and books [1-8] that cover this type of analysis in more detail.

In this chapter, some general guidelines for nonlinear static analysis are presented. These are followed by discussion of the formulations of geometric and material nonlinearities for section and frame analysis. Two examples are given to illustrate the applications of static nonlinear push-over analysis in bridge seismic design.

36.2 Analysis Classification and General Guidelines

Engineers use structural analysis as a fundamental tool to make design decisions. It is important that engineers have access to several different analysis tools and understand their development assumptions and limitations. Such an understanding is essential to select the proper analysis tool to achieve the design objectives.

Figure 36.1 shows lateral load vs. displacement curves of a frame using several structural analysis methods. Table 36.1 summarizes basic assumptions of those methods. It can be seen from Figure 36.1 that the first-order elastic analysis gives a straight line and no failure load. A first-order inelastic analysis predicts the maximum plastic load-carrying capacity on the basis of the undeformed geometry. A second-order elastic analysis follows an elastic buckling process. A second-order inelastic analysis traces load–deflection curves more accurately.

36.2.1 Classifications

Structural analysis methods can be classified on the basis of different formulations of equilibrium, the constitutive and compatibility equations as discussed below.

Classification Based on Equilibrium and Compatibility Formulations

First-order analysis: An analysis in which equilibrium is formulated with respect to the undeformed (or original) geometry of the structure. It is based on small strain and small displacement theory.

TABLE 36.1 Structural Analysis Methods

Methods		Constitutive Relationship	Equilibrium Formulation	Geometric Compatibility
		Features		
First-order	Elastic Rigid–plastic Elastic–plastic hinge Distributed plasticity	Elastic Rigid plastic Elastic perfectly plastic Inelastic	Original undeformed geometry	Small strain and small displacement
Second-order	Elastic Rigid–plastic Elastic–plastic hinge Distributed plasticity	Elastic Rigid plastic Elastic perfectly plastic Inelastic	Deformed structural geometry (P-Δ and P-δ)	Small strain and moderate rotation (displacement may be large)
True large displacement	Elastic Inelastic	Elastic Inelastic	Deformed structural geometry	Large strain and large deformation

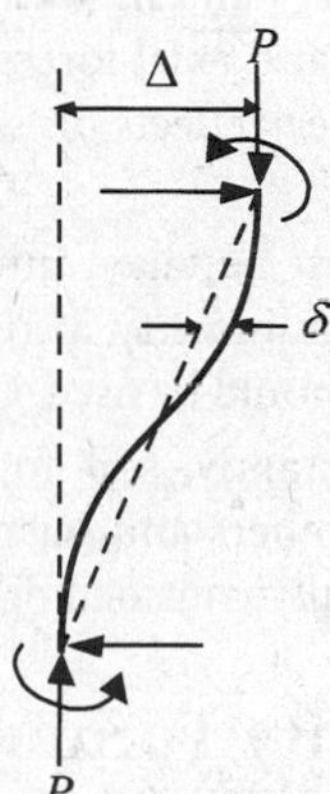

FIGURE 36.2 Second–order effects.

Second-order analysis: An analysis in which equilibrium is formulated with respect to the deformed geometry of the structure. A second-order analysis usually accounts for the P-Δ effect (influence of axial force acting through displacement associated with member chord rotation) and the P-δ effect (influence of axial force acting through displacement associated with member flexural curvature) (see Figure 36.2). It is based on small strain and small member deformation, but moderate rotations and large displacement theory.

True large deformation analysis: An analysis for which large strain and large deformations are taken into account.

Classification Based on Constitutive Formulation

Elastic analysis: An analysis in which elastic constitutive equations are formulated.

Inelastic analysis: An analysis in which inelastic constitutive equations are formulated.

Rigid–plastic analysis: An analysis in which elastic rigid–plastic constitutive equations are formulated.

Elastic–plastic hinge analysis: An analysis in which material inelasticity is taken into account by using concentrated "zero-length" plastic hinges.

Distributed plasticity analysis: An analysis in which the spread of plasticity through the cross sections and along the length of the members are modeled explicitly.

Classification Based on Mathematical Formulation

Linear analysis: An analysis in which equilibrium, compatibility, and constitutive equations are linear.

Nonlinear analysis: An analysis in which some or all of the equilibrium, compatibility, and constitutive equations are nonlinear.

36.2.4 General Guidelines

The following guidelines may be useful in analysis type selection:

- A first-order analysis may be adequate for short- to medium-span bridges. A second-order analysis should always be encouraged for long-span, tall, and slender bridges. A true large displacement analysis is generally unnecessary for bridge structures.
- An elastic analysis is sufficient for strength-based design. Inelastic analyses should be used for displacement-based design.
- The bowing effect (effect of flexural bending on member's axial deformation), the Wagner effect (effect of bending moments and axial forces acting through displacements associated with the member twisting), and shear effects on solid-webbed members can be ignored for most of bridge structures.
- For steel nonlinearity, yielding must be taken into account. Strain hardening and fracture may be considered. For concrete nonlinearity, a complete strain–stress relationship (in compression up to the ultimate strain) should be used. Concrete tension strength can be neglected.
- Other nonlinearities, most importantly, soil–foundation–structural interaction, seismic response modification devices (dampers and seismic isolations), connection flexibility, gap close and opening should be carefully considered.

36.3 Geometric Nonlinearity Formulation

Geometric nonlinearities can be considered in the formulation of member stiffness matrices. The general force–displacement relationship for the prismatic member as shown in Figure 36.3 can be expressed as follows:

$$\{F\} = [K]\{D\} \tag{36.1}$$

where $\{F\}$ and $\{D\}$ are force and displacement vectors and $[K]$ is stiffness matrix.

For a two-dimensional member as shown in Figure 36.3a

$$\{F\} = \{P_{1a}, F_{2a}, M_{3a}, P_{1b}, F_{2b}, M_{3b}\}^T \tag{36.2}$$

$$\{D\} = \{u_{1a}, u_{2a}, \theta_{3a}, u_{1b}, u_{2b}, \theta_{3b}\}^T \tag{36.3}$$

For a three-dimensional member as shown in Figure 36.3b

$$\{F\} = \{P_{1a}, F_{2a}, F_{3a}, M_{1a}, M_{2a}, M_{3a}, P_{1b}, F_{2b}, F_{3b}, M_{1b}, M_{2b}, M_{3b}\}^T \tag{36.4}$$

$$\{D\} = \{u_{1a}, u_{2a}, u_{3a}, \theta_{1a}, \theta_{2a}, \theta_{3a}, u_{1b}, u_{2b}, u_{3b}, \theta_{1b}, \theta_{2b}, \theta_{3b}\}^T \tag{36.5}$$

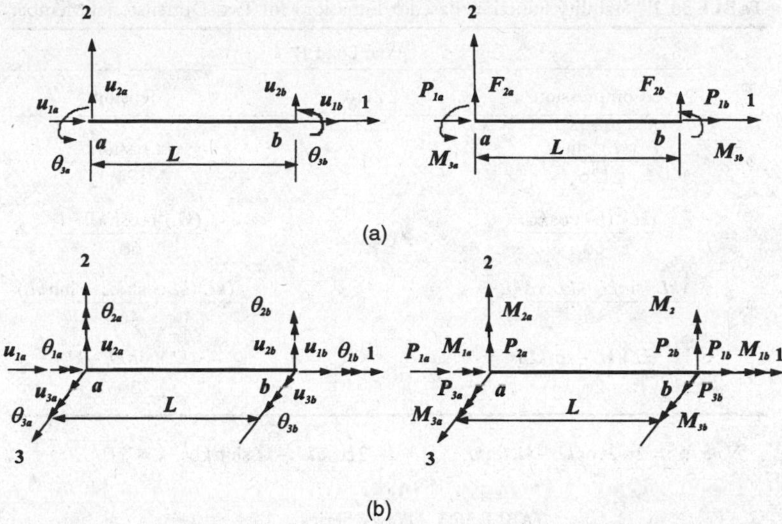

FIGURE 36.3 Degrees of freedom and nodal forces for a framed member. (a) Two-dimensional and (b) three-dimensional members.

Two sets of formulations of stability function-based and finite-element-based stiffness matrices are presented in the following scction.

36.3.1 Two-Dimensional Members

For a two-dimensional prismatic member as shown in Figure 36.3a, the stability function-based stiffness matrix [9] is as follows:

$$[K] = \begin{bmatrix} \frac{AE}{L} & 0 & 0 & -\frac{AE}{L} & 0 & 0 \\ & \frac{12EI}{L^3}\phi_1 & \frac{-6EI}{L^2}\phi_2 & 0 & \frac{-12EI}{L^3}\phi & \frac{-6EI}{L^2}\phi_2 \\ & & 4\phi_3 & 0 & \frac{6EI}{L^2}\phi_2 & 2\phi_4 \\ & & & \frac{AE}{L} & 0 & 0 \\ & & & & \frac{12EI}{L^3}\phi & \frac{6EI}{L^2}\phi_2 \\ & & & & & 4\phi_3 \end{bmatrix} \tag{36.6}$$

where A is cross section area; E is the material modulus of elasticity; L is the member length; ϕ_1, ϕ_2, ϕ_3, and ϕ_4 can be expressed by stability equations and are listed in Table 36.2. Alternatively, ϕ_i functions can also be expressed in the power series derived from the analytical solutions [10] as listed in Table 36.3.

Assuming polynomial displacement functions, the finite-element-based stiffness matrix [11,12] has the following form:

$$[K] = [K_e] + [K_g] \tag{36.7}$$

where $[K_e]$ is the first-order conventional linear elastic stiffness matrix and $[K_g]$ is the geometric stiffness matrix which considers the effects of axial load on the bending stiffness of a member.

TABLE 36.2 Stability Function-Based ϕ_i Equations for Two-Dimensional Member

ϕ	Axial Load P		
	Compression	Zero	Tension
ϕ_1	$\frac{(kL)^3 \sin kL}{12\phi_c}$	1	$\frac{(kL)^3 \sinh kL}{12\phi_t}$
ϕ_2	$\frac{(kL)^2(1-\cos kL)}{6\phi_c}$	1	$\frac{(kL)^2(\cosh kL-1)}{6\phi_t}$
ϕ_3	$\frac{(kL)(\sin kL - kL\cos kL)}{4\phi_c}$	1	$\frac{(kL)(kL\cosh kL - \sinh kL)}{4\phi_t}$
ϕ_4	$\frac{(kL)(kL - \sin kL)}{2\phi_c}$	1	$\frac{(kL)(\sin kL - kL)}{2\phi_t}$

Note: $\phi_c = 2-2\cos kL - kL\sin kL$; $\phi_t = 2-2\cosh kL - kL\sinh kL$; $k = \sqrt{P/EI}$.

TABLE 36.3 Power Series Expression of ϕ_i Equations

ϕ_1	$\dfrac{1+\sum_{n=1}^{\infty}\frac{1}{(2n+1)!}\left[\mp(kL)^2\right]^n}{12\phi}$
ϕ_2	$\dfrac{\frac{1}{2}+\sum_{n=1}^{\infty}\frac{1}{(2n+2)!}\left[\mp(kL)^2\right]^n}{6\phi}$
ϕ_3	$\dfrac{\frac{1}{3}+\sum_{n=1}^{\infty}\frac{2(n+1)}{(2n+3)!}\left[\mp(kL)^2\right]^n}{4\phi}$
ϕ_4	$\dfrac{\frac{1}{6}+\sum_{n=1}^{\infty}\frac{1}{(2n+3)!}\left[\mp(kL)^2\right]^n}{2\phi}$
ϕ	$\frac{1}{12}+\sum_{n=1}^{\infty}\frac{2(n+1)}{(2n+4)!}\left[\mp(kL)^2\right]^n$

Note: minus sign = compression; plus sign = tension.

$$[K] = \begin{bmatrix} \frac{AE}{L} & 0 & 0 & -\frac{AE}{L} & 0 & 0 \\ & \frac{12EI}{L^3} & \frac{-6EI}{L^2} & 0 & \frac{-12EI}{L^3} & \frac{-6EI}{L^2} \\ & & 4 & 0 & \frac{6EI}{L^2} & 2 \\ & & & \frac{AE}{L} & 0 & 0 \\ & \text{sym} & & & \frac{12EI}{L^3} & \frac{6EI}{L^2} \\ & & & & & 4 \end{bmatrix} \tag{36.8}$$

$$
\left[\boldsymbol{K}_g\right] = \mp\frac{P}{L}\begin{bmatrix}
0 & 0 & 0 & 0 & 0 & 0 \\
 & \frac{6}{5} & \frac{-L}{10} & 0 & \frac{-6}{5} & \frac{-L}{10} \\
 & & \frac{2L^2}{15} & 0 & \frac{L}{10} & -\frac{L^2}{30} \\
 & & & 0 & 0 & 0 \\
 & \text{sym.} & & & \frac{6}{5} & \frac{L}{10} \\
 & & & & & \frac{2L^2}{15}
\end{bmatrix} \tag{36.9}
$$

It is noted [13] that Eqs. (36.8) and (36.9) exactly coincide with the stability function-based stiffness matrix when taken only the first two terms of the Taylor series expansion in Eq. (36.6).

36.3.2 Three-Dimensional Members

For a three-dimensional frame member as shown in Figure 36.3b, the stability function-based stiffness matrix has the following form [14]:

$$
[\boldsymbol{K}] = \begin{bmatrix}
\phi_{s1} & 0 & 0 & 0 & 0 & 0 & -\phi_{s1} & 0 & 0 & 0 & 0 & 0 \\
 & \phi_{s7} & 0 & 0 & 0 & \phi_{s6} & 0 & -\phi_{s7} & 0 & 0 & 0 & \phi_{s6} \\
 & & \phi_{s9} & 0 & -\phi_{s8} & 0 & 0 & 0 & \phi_{s9} & 0 & \phi_{s8} & 0 \\
 & & & \frac{GJ}{L} & 0 & 0 & 0 & 0 & 0 & -\frac{GJ}{L} & 0 & 0 \\
 & & & & \phi_{s4} & 0 & 0 & 0 & \phi_{s8} & 0 & \phi_{s5} & 0 \\
 & & & & & \phi_{s2} & 0 & -\phi_{s6} & 0 & 0 & 0 & \phi_{s3} \\
 & & & & & & \phi_{s1} & 0 & 0 & 0 & 0 & 0 \\
 & & & & & & & \phi_{s7} & 0 & 0 & 0 & -\phi_{s6} \\
 & & \text{Sym.} & & & & & & \phi_{s9} & 0 & \phi_{s8} & 0 \\
 & & & & & & & & & \frac{GJ}{L} & 0 & 0 \\
 & & & & & & & & & & \phi_{s4} & 0 \\
 & & & & & & & & & & & \phi_{s2}
\end{bmatrix} \tag{36.10}
$$

where G is shear modulus of elasticity; J is torsional constant; ϕ_{s1} to ϕ_{s9} are expressed by stability equations and listed in Table 36.4.

Finite-element-based stiffness matrix has the form [15]:

$$
[\boldsymbol{K}_e] = \begin{bmatrix}
\phi_{e1} & 0 & 0 & 0 & 0 & 0 & -\phi_{e1} & 0 & 0 & 0 & 0 & 0 \\
 & \phi_{e7} & 0 & 0 & 0 & \phi_{e6} & 0 & -\phi_{e7} & 0 & 0 & 0 & \phi_{e6} \\
 & & \phi_{e9} & 0 & -\phi_{e8} & 0 & 0 & 0 & -\phi_{e9} & 0 & -\phi_{e8} & 0 \\
 & & & \frac{GJ}{L} & 0 & 0 & 0 & 0 & 0 & -\frac{GJ}{L} & 0 & 0 \\
 & & & & \phi_{e4} & 0 & 0 & 0 & -\phi_{e8} & 0 & \phi_{e5} & 0 \\
 & & & & & \phi_{e2} & 0 & -\phi_{e6} & 0 & 0 & 0 & \phi_{e3} \\
 & & & & & & \phi_{e1} & 0 & 0 & 0 & 0 & 0 \\
 & & & & & & & \phi_{e7} & 0 & 0 & 0 & -\phi_{e6} \\
 & & \text{Sym.} & & & & & & \phi_{e9} & 0 & \phi_{e8} & 0 \\
 & & & & & & & & & \frac{GJ}{L} & 0 & 0 \\
 & & & & & & & & & & \phi_{e4} & 0 \\
 & & & & & & & & & & & \phi_{e2}
\end{bmatrix} \tag{36.11}
$$

TABLE 36.4 Stability Function-Based ϕ_{si} for Three-Dimensional Member

ϕ_{si}		Stability Functions S_i: Compression	Stability Functions S_i: Tension
$\phi_{s1} = S_1 \dfrac{EA}{L}$	S_1	$\dfrac{1}{1-\dfrac{EA}{4P^3L^2}\left[H_y + H_z\right]}$	$\dfrac{1}{1-\dfrac{EA}{4P^3L^2}\left[H'_y + H'_z\right]}$
$\phi_{s2} = S_2 \dfrac{(4+\phi_y)EI_Z}{(1+\phi_y)L}$	S_2	$\dfrac{(\alpha L)(\sin\alpha L - \alpha L\cos\alpha L)}{4\phi_\alpha}$	$\dfrac{(\alpha L)(\alpha L\cosh\alpha L - \sinh\alpha L)}{4\phi_\alpha}$
$\phi_{s3} = S_2 \dfrac{(2-\phi_y)EI_Z}{(1+\phi_y)L}$	S_3	$\dfrac{(\alpha L)(\alpha L - \sin\alpha L)}{2\phi_\alpha}$	$\dfrac{(\alpha L)(\sinh\alpha L - \alpha L)}{2\phi_\alpha}$
$\phi_{s4} = S_4 \dfrac{(4+\phi_z)EI_y}{(1+\phi_z)L}$	S_4	$\dfrac{(\beta L)(\sin\beta L - \beta L\cos\beta L)}{4\phi_\beta}$	$\dfrac{(\beta L)(\beta L\cosh\beta L - \sinh\beta L)}{4\phi_\beta}$
$\phi_{s5} = S_2 \dfrac{(2-\phi_z)EI_y}{(1+\phi_z)L}$	S_5	$\dfrac{(\beta L)(\beta L - \sin\beta L)}{2\phi_\beta}$	$\dfrac{(\beta L)(\sinh\beta L - \beta L)}{2\phi_\beta}$
$\phi_{s6} = S_6 \dfrac{6EI_Z}{(1+\phi_y)L^2}$	S_6	$\dfrac{(\alpha L)^2(1-\cos\alpha L)}{6\phi_\alpha}$	$\dfrac{(\alpha L)^2(\cosh\alpha L - 1)}{6\phi_\alpha}$
$\phi_{s7} = S_7 \dfrac{12EI_Z}{(1+\phi_y)L^3}$	S_7	$\dfrac{(\alpha L)^3\sin\alpha L}{12\phi_\alpha}$	$\dfrac{(\alpha L)^3\sinh\alpha L}{12\phi_\alpha}$
$\phi_{s8} = S_8 \dfrac{6EI_y}{(1+\phi_z)L^2}$	S_8	$\dfrac{(\beta L)^2(1-\cos\beta L)}{6\phi_\beta}$	$\dfrac{(\beta L)^2(\cosh\beta L - 1)}{6\phi_\beta}$
$\phi_{s9} = S_9 \dfrac{12EI_y}{(1+\phi_z)L^3}$	S_9	$\dfrac{(\beta L)^3\sin\beta L}{12\phi_\beta}$	$\dfrac{(\beta L)^3\sinh\beta L}{12\phi_\beta}$
$\alpha = \sqrt{P/EI_Z}$	ϕ_α	$2-2\cos\alpha L - \alpha L\sin\alpha L$	$2-2\cosh\alpha L + \alpha L\sinh\alpha L$
$\beta = \sqrt{P/EI_y}$	ϕ_β	$2-2\cos\beta L - \beta L\sin\beta L$	$2-2\cosh\beta L + \beta L\sinh\beta L$

$$H_y = \beta L(M_{ya}^2 + M_{yb}^2)(\cot\beta L + \beta L\,\mathrm{cosec}^2\beta L) - 2(M_{ya} + M_{yb})^2 + 2\beta L M_{ya} M_{yb}(\mathrm{cosec}\beta L)(1 + \beta L\cot\beta L)$$

$$H_z = \alpha L(M_{za}^2 + M_{zb}^2)(\cot\alpha L + \alpha L\,\mathrm{cosec}^2\alpha L) - 2(M_{za} + M_{zb})^2 + 2\alpha L M_{za} M_{zb}(\mathrm{cosec}\alpha L)(1 + \alpha L\cot\alpha L)$$

$$H'_y = \beta L(M_{ya}^2 + M_{yb}^2)(\coth\beta L + \beta L\,\mathrm{cosech}^2\beta L) - 2(M_{ya} + M_{yb})^2 + 2\beta L M_{ya} M_{yb}(\mathrm{cosech}\beta L)(1 + \beta L\coth\beta L)$$

$$H'_z = \alpha L(M_{za}^2 + M_{zb}^2)(\coth\alpha L + \alpha L\,\mathrm{cosech}^2\alpha L) - 2(M_{za} + M_{zb})^2 + 2\alpha L M_{za} M_{zbb}(\mathrm{cosech}\alpha L)(1 + \alpha L\coth\alpha L)$$

$$[K_g] = \begin{bmatrix}
\phi_{g1} & \phi_{g10} & -\phi_{g11} & 0 & 0 & 0 & 0 & -\phi_{g10} & \phi_{g11} & 0 & 0 & 0 \\
 & \phi_{g7} & 0 & \phi_{g12} & \phi_{g13} & \phi_{g6} & -\phi_{g10} & -\phi_{g7} & 0 & \phi_{g14} & -\phi_{g13} & \phi_{g6} \\
 & & \phi_{g9} & \phi_{g15} & -\phi_{g6} & \phi_{g13} & \phi_{g11} & 0 & -\phi_{g9} & \phi_{g16} & -\phi_{g6} & -\phi_{g13} \\
 & & & \phi_{g17} & \phi_{g18} & \phi_{g19} & & -\phi_{g12} & -\phi_{g15} & -\phi_{g17} & -\phi_{g20} & \phi_{g21} \\
 & & & & \phi_{g4} & 0 & 0 & -\phi_{g13} & \phi_{g6} & -\phi_{g20} & -\phi_{g5} & \phi_{g13} \\
 & & & & & \phi_{g2} & 0 & -\phi_{g6} & -\phi_{g13} & \phi_{g21} & -\phi_{g13} & -\phi_{g3} \\
 & & & & & & \phi_{g1} & \phi_{g10} & -\phi_{g11} & 0 & 0 & 0 \\
 & & & & & & & \phi_{g7} & 0 & -\phi_{g14} & \phi_{g13} & -\phi_{g6} \\
 & & \text{Sym.} & & & & & & \phi_{g9} & -\phi_{g16} & \phi_{g6} & \phi_{g13} \\
 & & & & & & & & & \phi_{g17} & \phi_{g18} & \phi_{g19} \\
 & & & & & & & & & & \phi_{g4} & 0 \\
 & & & & & & & & & & & \phi_{g2}
\end{bmatrix} \quad (36.12)$$

where ϕ_{ei} and ϕ_{gi} are given in Table 36.5.

TABLE 36.5 Elements of Finite-Element-Based Stiffness Matrix

Linear Elastic Matrix	Geometric Nonlinear Matrix
$\phi_{e1}=\frac{AE}{L}$; $\phi_{e2}=\frac{4EI_z}{L}$	$\phi_{g1}=0$; $\phi_{g2}=\phi_{g4}=\frac{2F_{xb}L}{15}$; $\phi_{g3}=\phi_{g5}=\frac{F_{xb}L}{30}$
$\phi_{e3}=\frac{2EI_z}{L}$ $\phi_{e4}=\frac{4EI_y}{L}$	$\phi_{g7}=\phi_{g9}=\frac{6F_{xb}}{5L}$; $\phi_{g6}=\phi_{g8}=\frac{F_{xb}}{10}$; $\phi_{g10}=\frac{M_{za}+M_{zb}}{L^2}$
$\phi_{e5}=\frac{2EI_y}{L}$; $\phi_{e6}=\frac{6EI_z}{L^2}$	$\phi_{g11}=\frac{M_{ya}+M_{yb}}{L^2}$; $\phi_{g12}=\frac{M_{ya}}{L}$; $\phi_{g13}=\frac{M_{xb}}{L}$
$\phi_{e7}=\frac{12EI_z}{L^3}$; $\phi_{e8}=\frac{6EI_y}{L^2}$	$\phi_{g14}=\frac{M_{yb}}{L}$; $\phi_{g15}=\frac{M_{za}}{L}$; $\phi_{g16}=\frac{M_{zb}}{L}$
$\phi_{e9}=\frac{12EI_y}{L^3}$	$\phi_{g17}=\frac{F_{xb}I_p}{AL}$; $\phi_{g18}=\frac{M_{zb}}{6}-\frac{M_{za}}{3}$; $\phi_{g19}=\frac{M_{ya}}{3}-\frac{M_{yb}}{6}$
	$\phi_{g20}=\frac{M_{za}+M_{zyb}}{6}$; $\phi_{g21}=\frac{M_{ya}+M_{yb}}{6}$

I_z and I_y are moments of inertia about z–z and y–y axis, respectively; I_p is the polar moment of inertia.

Stiffness matrices considering warping degree of freedom and finite rotations for a thin-walled member were derived by Yang and McGuire [16,17].

In conclusion, both sets of the stiffness matrices have been used successfully when considering geometric nonlinearities (P-Δ and P-δ effects). The stability function-based formulation gives an accurate solution using fewer degrees of freedom when compared with the finite-element method. Its power series expansion (Table 36.3) can be implemented easily without truncation to avoid numerical difficulty.

The finite-element-based formulation produces an approximate solution. It has a simpler form and may require dividing the member into a large number of elements in order to keep the (P/L) term a small quantity to obtain accurate results.

36.4 Material Nonlinearity Formulations

36.4.1 Structural Concrete

Concrete material nonlinearity is incorporated into analysis using a nonlinear stress–strain relationship. Figure 36.4 shows idealized stress–strain curves for unconfined and confined concrete in uniaxial compression. Tests have shown that the confinement provided by closely spaced transverse reinforcement can substantially increase the ultimate concrete compressive stress and strain. The confining steel prevents premature buckling of the longitudinal compression reinforcement and increases the concrete ductility. Extensive research has been made to develop concrete stress–strain relationships [18-25].

36.4.1.1 Compression Stress–Strain Relationship

Unconfined Concrete

A general stress–strain relationship proposed by Hognestad [18] is widely used for plain concrete or reinforced concrete with a small amount of transverse reinforcement. The relation has the following simple form:

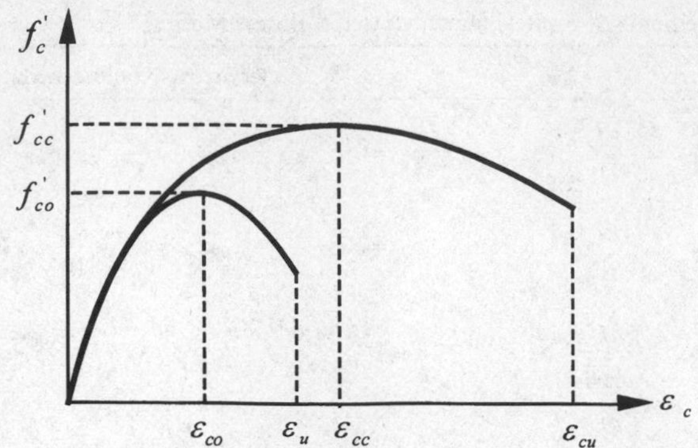

FIGURE 36.4 Idealized stress-strain curves for concrete in uniaxial compression.

$$f_c = \begin{cases} f'_{co}\left[\dfrac{2\varepsilon_c}{\varepsilon_{co}} - \left(\dfrac{\varepsilon_c}{\varepsilon_{co}}\right)^2\right] & \varepsilon_c \le \varepsilon_{co} \\ f'_{co}\left[1-\beta\left(\dfrac{\varepsilon_c - \varepsilon_o}{\varepsilon_u - \varepsilon_{co}}\right)\right] & \varepsilon_{co} < \varepsilon_c \le \varepsilon_u \end{cases} \tag{36.13}$$

$$\varepsilon_{co} = \frac{2f'_{co}}{E_c} \tag{36.14}$$

where f_c and ε_c are the concrete stress and strain; f'_{co} is the peak stress for unconfined concrete usually taken as the cylindrical compression strength f'_c; ε_{co} is strain at peak stress for unconfined concrete usually taken as 0.002; ε_u is the ultimate compression strain for unconfined concrete taken as 0.003; E_c is the modulus of elasticity of concrete; β is a reduction factor for the descending branch usually taken as 0.15. Note that the format of Eq. (36.13) can be also used for confined concrete if the concrete-confined peak stress f_{cc} and strain E_{cu} are known or assumed and substituted for f'_{co} and ε_u, respectively.

Confined Concrete — Mander's Model

Analytical models describing the stress–strain relationship for confined concrete depend on the confining transverse reinforcement type (such as hoops, spiral, or ties) and shape (such as circular, square, or rectangular). Some of those analytical models are more general than others in their applicability to various confinement types and shapes. A general stress–strain model (Figure 36.5) for confined concrete applicable (in theory) to a wide range of cross sections and confinements was proposed by Mander et al. [23,24] and has the following form:

$$f_c = \frac{f'_{cc}(\varepsilon_c / \varepsilon_{cc})r}{r-1+(\varepsilon_c / \varepsilon_{cc})^r} \tag{36.15}$$

$$\varepsilon_{cc} = \varepsilon_{co}\left[1+5\left(\frac{f'_{cc}}{f'_{co}}-1\right)\right] \tag{36.16}$$

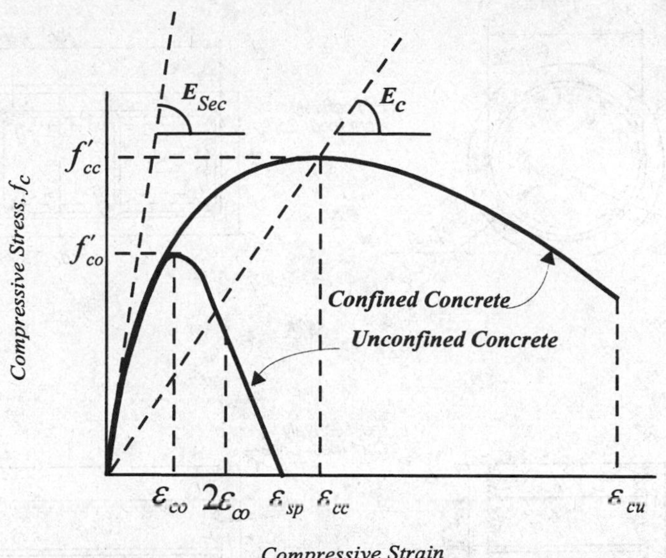

FIGURE 36.5 Stress–strain curves — mander model.

$$r = \frac{E_c}{E_c - E_{\text{sec}}} \tag{36.17}$$

$$E_{\text{sec}} = \frac{f'_{cc}}{\varepsilon_{cc}} \tag{36.18}$$

where f'_{cc} and ε_{cc} are peak compressive stress and corresponding strain for confined concrete. f'_{cc} and ε_{cu} which depend on the confinement type and shape, are calculated as follows:

Confined Peak Stress

1. **For concrete circular section confined by circular hoops or spiral** (Figure 36.6a):

$$f'_{cc} = f'_{co}\left(2.254\sqrt{1 + \frac{7.94 f'_l}{f'_{co}}} - \frac{2 f'_l}{f'_{co}} - 1.254\right) \tag{36.19}$$

$$f'_l = \frac{1}{2} K_e \rho_s f_{yh} \tag{36.20}$$

$$K_e = \begin{cases} \left(1 - s'/2d_s\right)^2 / \left(1 - \rho_{cc}\right) & \text{for circular hoops} \\ \left(1 - s'/2d_s\right) / \left(1 - \rho_{cc}\right) & \text{for circular spirals} \end{cases} \tag{36.21}$$

$$\rho_s = \frac{4A_{sp}}{d_s s} \tag{36.22}$$

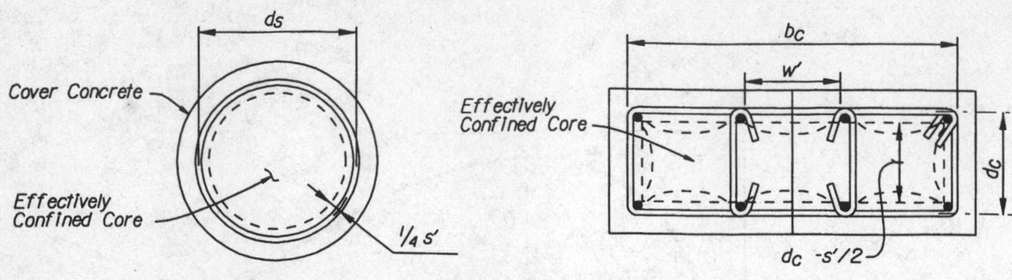

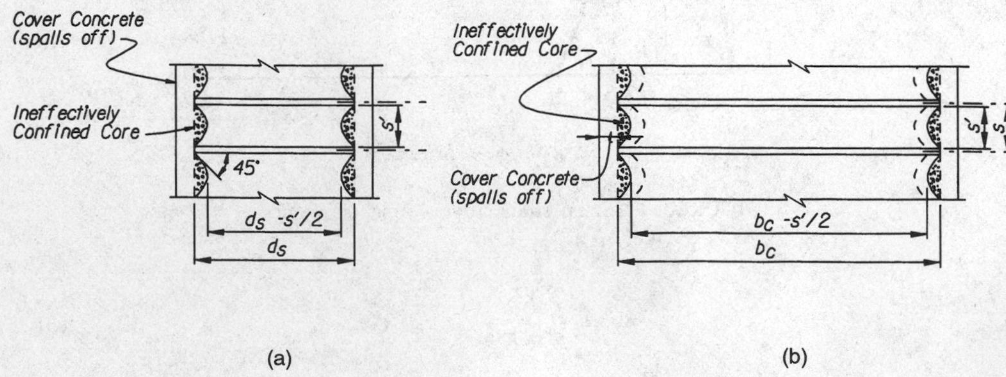

(a) (b)

FIGURE 36.6 Confined core for hoop reinforcement. (a) Circular hoop and (b) rectangular hoop reinforcement.

where f_l' is the effective lateral confining pressure; K_e is confinement effectiveness coefficient, f_{yh} is the yield stress of the transverse reinforcement, s' is the clear vertical spacing between hoops or spiral; s is the center-to-center spacing of the spiral or circular hoops; d_s is the centerline diameter of the spiral or hoops circle; ρ_{cc} is the ratio of the longitudinal reinforcement area to section core area; ρ_s is the ratio of the transverse confining steel volume to the confined concrete core volume; and A_{sp} is the bar area of transverse reinforcement.

2. **For rectangular concrete section confined by rectangular hoops** (Figure 36.6b)

The rectangular hoops may produce two unequal effective confining pressures f_{lx}' and f_{ly}' in the principal x and y direction defined as follows:

$$f_{lx}' = K_e\, \rho_x f_{yh} \tag{36.23}$$

$$f_{ly}' = K_e\, \rho_y f_{yh} \tag{36.24}$$

$$K_e = \frac{\left[1 - \sum_{i=1}^{n} \frac{(w_i')^2}{6 b_c d_c}\right]\left(1 - \frac{s'}{2b_c}\right)\left(1 - \frac{s'}{2d_c}\right)}{\left(1 - \rho_{cc}\right)} \tag{36.25}$$

$$\rho_x = \frac{A_{sx}}{s\, d_c} \tag{36.26}$$

$$\rho_y = \frac{A_{sy}}{s\, b_c} \tag{36.27}$$

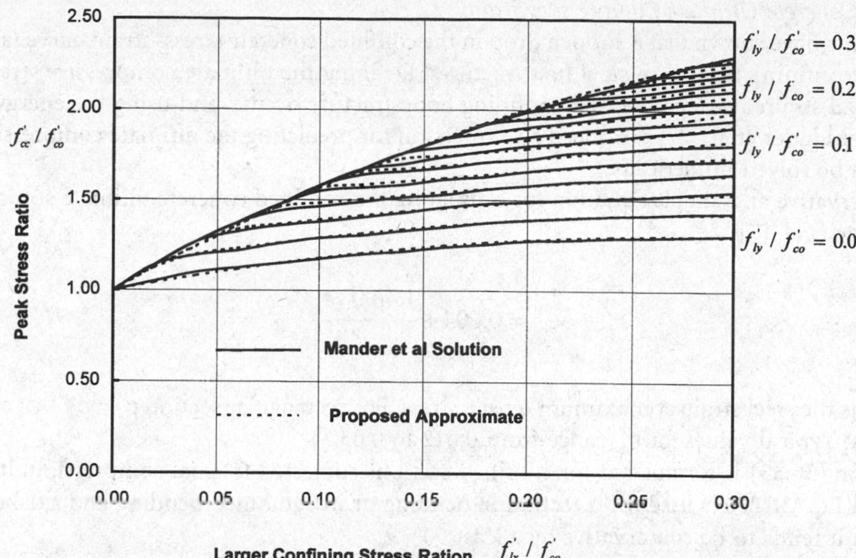

FIGURE 36.7 Peak stress of confined concrete.

where f_{yh} is the yield strength of transverse reinforcement; w_i' is the *i*th clear distance between adjacent longitudinal bars; b_c and d_c are core dimensions to centerlines of hoop in x and y direction (where $b \geq d$), respectively; A_{sx} and A_{sy} are the total area of transverse bars in x and y direction, respectively.

Once f_{lx}' and f_{ly}' are determined, the confined concrete strength f_{cc}' can be found using the chart shown in Figure 36.7 with f_{lx}' being greater or equal to f_{ly}'. The chart depicts the general solution of the "five-parameter" multiaxial failure surface described by William and Warnke [26].

As an alternative to the chart, the authors derived the following equations for estimating f_{cc}':

$$f_{cc} = \begin{cases} Af_{lx}'^2 + Bf_{lx}' + C & f_{ly}' < f_{lx}' \text{ and } f_{ly}' \leq 0.15 \\ \dfrac{f_{lx}' - f_{lx}'}{0.3 - f_{ly}'} D + C & f_{ly}' < f_{lx}' \text{ and } f_{ly}' > 0.15 \\ C & f_{ly}' = f_{lx}' \end{cases} \quad (36.28)$$

$$A = 196.5 f_{lx}'^2 + 29.1 f_{lx}' - 4 \quad (36.29)$$

$$B = -69.5 f_{lx}'^2 - 8.9 f_{lx}' + 2.2 \quad (36.30)$$

$$C = -6.83 f_{lx}'^2 + 6.38 f_{lx}' + 1 \quad (36.31)$$

$$D = -1.5 f_{lx}'^2 - 0.55 f_{lx}' + 0.3 \quad (36.32)$$

Note that by setting $f_l' = 0.0$ in Eqs. (36.19), Eqs. (36.16) and (36.15) will produce to Mander's expression for unconfined concrete. In this case and for concrete strain $\varepsilon_c > 2\,\varepsilon_{co}$, a straight line which reaches zero stress at the spalling strain ε_{sp} is assumed.

Confined Concrete Ultimate Compressive Strain

Experiments have shown that a sudden drop in the confined concrete stress–strain curve takes place when the confining transverse steel first fractures. Defining the ultimate compressive strain as the longitudinal strain at which the first confining hoop fracture occurs, and using the energy balance approach, Mander et al. [27] produced an expression for predicting the ultimate compressive strain which can be solved numerically.

A conservative and simple equation for estimating the confined concrete ultimate strain is given by Priestley et al. [7]:

$$\varepsilon_{cu} = 0.004 + \frac{1.4\rho_s f_{yh}\varepsilon_{su}}{f'_{cc}} \tag{36.33}$$

where ε_{su} is the steel strain at maximum tensile stress. For rectangular section $\rho_s = \rho_x + \rho_y$ as defined previously. Typical values for ε_{cu} range from 0.012 to 0.05.

Equation (36.33) is formulated for confined sections subjected to axial compression. It is noted that when Eq. (36.33) is used for a section in bending or in combined bending and axial compression, then it tends to be conservative by a least 50%.

Chai et al. [28] used an energy balance approach to derive the following expression for calculating the concrete ultimate confined strain as

$$\varepsilon_{cu} = \varepsilon_{sp} + \begin{cases} \rho_s \varepsilon_{su} \dfrac{\gamma_2 f_{yh}}{\gamma_1 f'_{cc}} & \text{confined by reinconcement} \\ \rho_{sj} \varepsilon_{suj} \dfrac{\gamma_2 f_{yj}}{\gamma_1 f'_{cc}} & \text{confined by circular steel jackets} \end{cases} \tag{36.34}$$

where ε_{sp} is the spalling strain of the unconfined concrete (usually = 0.003 to 0.005), γ_1 is an integration coefficient of the area between the confined and unconfined stress–strain curves; and γ_2 is an integration coefficient of the area under the transverse steel stress–strain curve. The confining ratio for steel jackets $\rho_{sj} = 4t_j/(D_j - 2t_j)$; D_i and t_j are outside diameter and thickness of the jacket, respectively; f_{yj} is yield stress of the steel jacket. For high- and mild-strength steels and concrete compressive strengths of 4 to 6 ksi (27.58 to 41.37 MPa), Chai et al. [28] proposed the following expressions

$$\frac{\gamma_2}{\gamma_1} = \begin{cases} \dfrac{2000\rho_s}{\left(1+(1428\rho_s)^4\right)^{0.25}} & \text{for Grade 40 Steel} \\ \dfrac{2000\rho_s}{\left(1+(1480\rho_s)^{0.25}\right)^{0.4}} & \text{for Grade 60 Steel} \end{cases} \tag{36.35}$$

Confined Concrete — Hoshikuma's Model

In additional to Mander's model, Table 36.6 lists a stress–strain relationship for confined concrete proposed by Hoshikuma et al. [25]. The Hoshikuma model was based on the results of a series of experimental tests covering circular, square, and wall-type cross sections with various transverse reinforcement arrangement in bridge piers design practice in Japan.

36.4.1.2 Tension Stress-Strain Relationship

Two idealized stress–strain curves for concrete in tension is shown in Figure 36.8. For plain concrete, the curve is linear up to cracking stress f_r. For reinforced concrete, there is a descending branch

TABLE 36.6 Hoshikuma et al. [25] Stress–Strain Relationship of Confined Concrete

$$f_c = \begin{cases} E_c\varepsilon_c\left[1-\dfrac{1}{n}\left(\dfrac{\varepsilon_c}{\varepsilon_{cc}}\right)^{n-1}\right] & \varepsilon_c \le \varepsilon_{cc} \\ f'_{cc} - E_{des}(\varepsilon_c - \varepsilon_{cc}) & \varepsilon_{cc} < \varepsilon_c \le \varepsilon_{cu} \end{cases}$$

$$n = \frac{E_c\varepsilon_{cc}}{E_c\varepsilon_{cc} - f'_{cc}}; \quad \varepsilon_{cu} = \varepsilon_{cc} + \frac{f'_{cc}}{2E_{des}}; \quad E_{des} = 11.2\frac{f'^2_{co}}{\rho_s f_{yh}}$$

$$\frac{f'_{cc}}{f'_{co}} = \begin{cases} 1.0 + 3.8\dfrac{\rho_s f_{yh}}{f'_{co}} & \text{for circular section} \\ 1.0 + 0.76\dfrac{\rho_s f_{yh}}{f'_{co}} & \text{for square section} \end{cases}$$

$$\varepsilon_{cc} = \begin{cases} 0.002 + 0.033\dfrac{\rho_s f_{sh}}{f'_{co}} & \text{for circular section} \\ 0.002 + 0.013\dfrac{\rho_s f_{sh}}{f'_{co}} & \text{for square section} \end{cases}$$

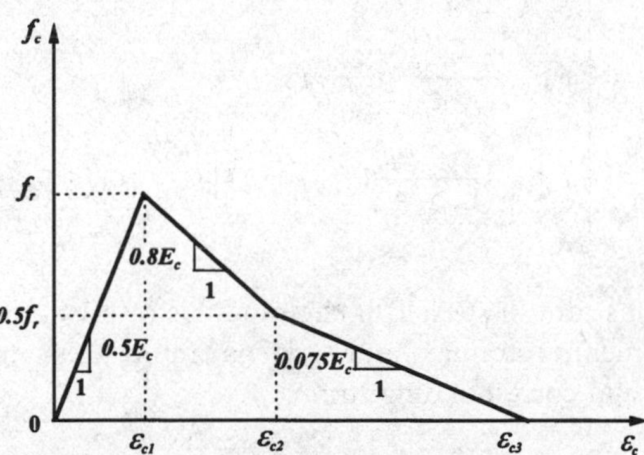

FIGURE 36.8 Idealized stress–strain curve of concrete in uniaxial tension.

because of bond characteristics of reinforcement. A trilinear expression proposed by Vebe et al. [29] is as follows:

$$f_c = \begin{cases} 0.5E_c\varepsilon_c & \varepsilon_c \le \varepsilon_{c1} = 2f_r/E_c \\ f_r[1 - 0.8E_c(\varepsilon_c - 2f_r/E_c)] & \varepsilon_{c1} < \varepsilon_c \le \varepsilon_{c2} = 2.625f_r/E_c \\ f_r[0.5 - 0.075E_c(\varepsilon_c - 2.625f_r/4E_c)] & \varepsilon_c < \varepsilon_{c3} = 9.292f_r/E_c \end{cases} \tag{36.36}$$

where f_r is modulus of rupture of concrete.

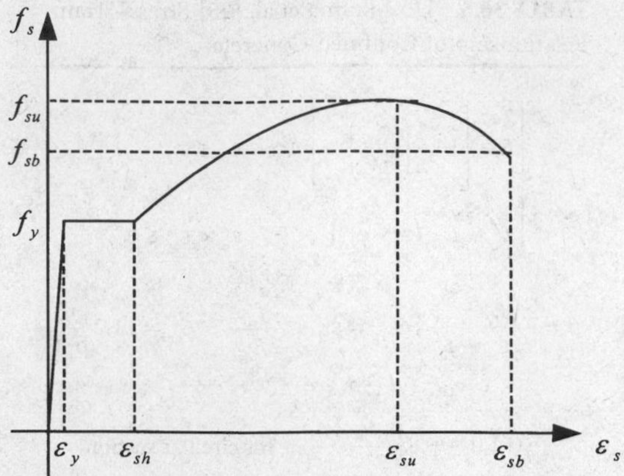

FIGURE 36.9 Idealized stress–strain curve of structural steel and reinforcement.

36.4.2 Structural and Reinforcement Steel

For structural steel and nonprestressed steel reinforcement, its stress–strain relationship can be idealized as four parts: elastic, plastic, strain hardening, and softening, as shown in Figure 36.9. The relationship if commonly expressed as follows:

$$f_s = \begin{cases} E_s\varepsilon_s & 0 \le \varepsilon_s \le \varepsilon_y \\ f_y & \varepsilon_{sy} < \varepsilon_s \le \varepsilon_{sh} \\ f_y + \dfrac{\varepsilon_s - \varepsilon_{sh}}{\varepsilon_{su} - \varepsilon_{sh}}(f_{su} - f_y) & \varepsilon_{sh} < \varepsilon_s \le \varepsilon_{su} \\ f_u\left[1 - \dfrac{\varepsilon_s - \varepsilon_{su}}{\varepsilon_{sb} - \varepsilon_{su}}(f_{su} - f_{sb})\right] & \varepsilon_{cu} < \varepsilon_s \le \varepsilon_{sb} \end{cases} \quad (36.37)$$

where f_s and ε_s is stress of strain in steel; E_s is the modulus of elasticity of steel; f_y and ε_y is yield stress and strain; ε_{sh} is hardening strain; f_{su} and ε_{su} is maximum stress and corresponding strain; f_{sb} and ε_{sb} are rupture stress and corresponding strain.

$$\varepsilon_{sh} = \begin{cases} 14\,\varepsilon_y & \text{for Grade 40} \\ 5\,\varepsilon_y & \text{for Grade 60} \end{cases} \quad (36.38)$$

$$\varepsilon_{su} = \begin{cases} 0.14 + \varepsilon_{sh} & \text{for Grade 40} \\ 0.12 & \text{for Grade 60} \end{cases} \quad (36.39)$$

For the reinforcing steel, the following nonlinear form can also be used for the strain-hardening portion [28]:

$$f_s = f_y\left[\frac{m(\varepsilon_s - \varepsilon_{sh}) + 2}{60(\varepsilon_s - \varepsilon_{sh}) + 2} + \frac{(\varepsilon_s - \varepsilon_{sh})(60 - m)}{2(30r + 1)^2}\right] \quad \text{for } \varepsilon_{sh} < \varepsilon_s \le \varepsilon_{su} \quad (36.40)$$

$$m = \frac{(f_{su}/f_y)(30r+1)^2 - 60r - 1}{15r^2} \tag{36.41}$$

$$r = \varepsilon_{su} - \varepsilon_{sh} \tag{36.42}$$

$$f_{su} = 1.5 f_y \tag{36.43}$$

For both strain-hardening and -softening portions, Holzer et al. [30] proposed the following expression

$$f_s = f_y\left[1 + \frac{\varepsilon_s - \varepsilon_{sh}}{\varepsilon_{su} - \varepsilon_{sh}}\left(\frac{f_{su}}{f_y} - 1\right)\exp\left(1 - \frac{\varepsilon_s - \varepsilon_{sh}}{\varepsilon_{su} - \varepsilon_{sh}}\right)\right] \quad \text{for } \varepsilon_{sh} < \varepsilon_s \le \varepsilon_{sb} \tag{36.44}$$

For prestressing steel, its stress–strain behavior is different from the nonprestressed steel. There is no obvious yield flow plateau in its response. The stress-stress expressions presented in Chapter 10 can be used in an analysis.

36.5 Nonlinear Section Analysis

36.5.1 Basic Assumptions and Formulations

The main purpose of section analysis is to study the moment–thrust–curvature behavior. In a nonlinear section analysis, the following assumptions are usually made:

- Plane sections before bending remain plane after bending;
- Shear and torsional deformation is negligible;
- Stress-strain relationships for concrete and steel are given;
- For reinforced concrete, a prefect bond between concrete and steel rebar exists.

The mathematical formulas used in the section analysis are (Figure 36.10):

Compatibility equations

$$\phi_x = \varepsilon / y \tag{36.45}$$

$$\phi_y = \varepsilon / x \tag{36.46}$$

Equilibrium equations

$$P = \int_A \sigma \, dA = \sum_{i=1}^{n} \sigma_i A_i \tag{36.47}$$

$$M_x = \int_A \sigma y \, dA = \sum_{i=1}^{n} \sigma_i y_i A_i \tag{36.48}$$

$$M_y = \int_A \sigma x \, dA = \sum_{i=1}^{n} \sigma_i x_i A_i \tag{36.49}$$

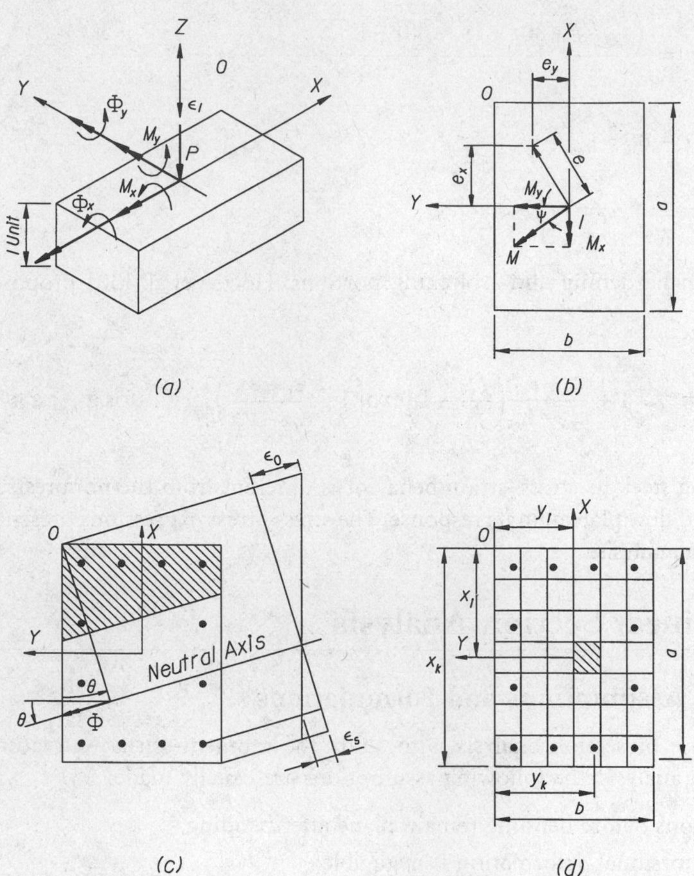

FIGURE 36.10 Moment–curvature–strain of cross section.

36.5.2 Modeling and Solution Procedures

For a reinforced-concrete member, the cross section is divided into a proper number of concrete and steel filaments representing the concrete and reinforcing steel as shown in Figure 36.10d. Each concrete and steel l filament is assigned its corresponding stress–strain relationships. Confined and unconfined stress–strain relationships are used for the core concrete and for the cover concrete, respectively.

For a structural steel member, the section is divided into steel filaments and a typical steel stress–strain relationship is used for tension and compact compression elements, and an equivalent stress–strain relationship with reduced yield stress and strain can be used for a noncompact compression element.

The analysis process starts by selecting a strain for the extreme concrete (or steel) fiber. By using this selected strain and assuming a section neutral axis (NA) location, a linear strain profile is constructed and the corresponding section stresses and forces are computed. Section force equilibrium is then checked for the given axial load. By changing the location of the NA, the process is repeated until equilibrium is satisfied. Once equilibrium is satisfied, for the assumed strain and the given axial load, the corresponding section moment and curvature are computed by Eqs. (36.48) and (36.49).

A moment–curvature (M–Φ) diagram for a given axial load is constructed by incrementing the extreme fiber strain and finding the corresponding moment and the associated curvature. An interaction diagram (M–P) relating axial load and the ultimate moment is constructed by incrementing the axial load and finding the corresponding ultimate moment using the above procedure.

For a reinforced-concrete section, the yield moment is usually defined as the section moment at onset of yielding of the tension reinforcing steel. The ultimate moment is defined as the moment at peak moment capacity. The ultimate curvature is usually defined as the curvature when the extreme concrete fiber strain reaches ultimate strain or when the reinforcing rebar reaches its ultimate (rupture) strain (whichever takes place first). Figure 36.11a shows typical M–P–Φ curves for a reinforced-concrete section.

For a simple steel section, such as rectangular, circular-solid, and thin-walled circular section, a closed-form of M–P–Φ can be obtained using the elastic-perfectly plastic stress–strain relations [4, 31]. For all other commonly used steel section, numerical iteration techniques are used to obtain M–P–Φ curves. Figure 36.11b shows typical M–P–Φ curves for a wide-flange section.

36.5.3 Yield Surface Equations

The yield or failure surface concept has been conveniently used in inelastic analysis to describe the full plastification of steel and concrete sections under the action of axial force combined with biaxial bending. This section will present several yield surface expressions for steel and concrete sections suitable for use in a nonlinear analysis.

36.5.3.1 Yield Surface Equations for Concrete Sections

The general interaction failure surface for a reinforced-concrete section with biaxial bending, as shown in Figure 36.12a can be approximated by a nondimensional interaction equation [32]:

$$\left(\frac{M_x}{M_{xo}}\right)^m + \left(\frac{M_y}{M_{yo}}\right)^n = 1.0 \tag{36.50}$$

where M_x and M_y are bending moments about x–x and y–y principal axes, respectively; M_{xo} and M_{yo} are the uniaxial bending capacity about the x–x and y–y axes under axial load P; the exponents m and n depend on the reinforced-concrete section properties and axial force. They can be determined by a numerical analysis or experiments. In general, the values of m and n usually range from 1.1 to 1.4 for low and moderate axial compression.

36.5.3.2 Yield Surface Equation for Doubly Symmetrical Steel Sections

The general shape of yield surface for a doubly symmetrical steel section as shown in Figure 36.12b can be described approximately by the following general equation [33]

$$\left(\frac{M_x}{M_{pcx}}\right)^{\alpha_x} + \left(\frac{M_y}{M_{pcy}}\right)^{\alpha_y} = 1.0 \tag{36.51}$$

where M_{pcx} and M_{pcy} are the moment capacities about respective axes, reduced for the presence of axial load; they can be obtained by the following formulas:

$$M_{pcx} = M_{px}\left[1 - \left(\frac{P}{P_y}\right)^{\beta_x}\right] \tag{36.52}$$

$$M_{pcy} = M_{py}\left[1 - \left(\frac{P}{P_y}\right)^{\beta_y}\right] \tag{36.53}$$

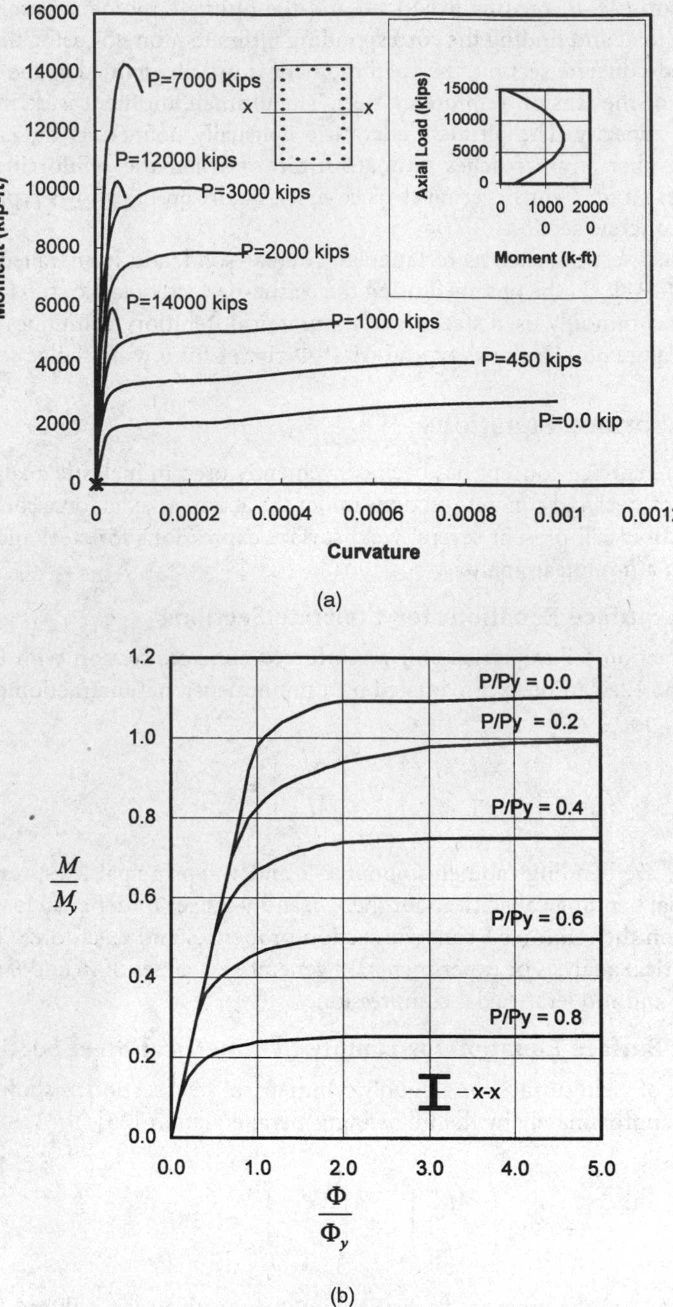

FIGURE 36.11 Moment–thrust–curvature curve. (a) Reinforced concrete section (b) steel I-section.

where P is the axial load; M_{px} and M_{py} are the plastic moments about x–x and y–y principal axes, respectively; α_x, α_y, β_x, and β_y are parameters that depend on cross-sectional shapes and area distribution and are listed in Table 36.7.

Equation (36.51) represents a smooth and convex surface in the three-dimensional stress-resultant space. It is easy to implement in a computer-based structural analysis.

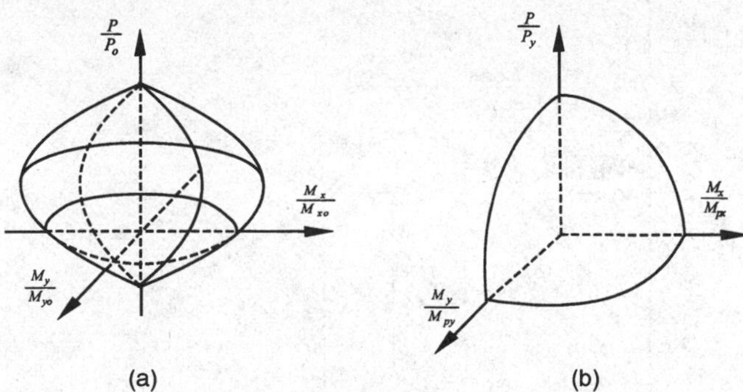

FIGURE 36.12 General yield surfaces. (a) Reinforced concrete section; (b) steel section.

TABLE 36.7 Parameters for Doubly Symmetrical Steel Sections

Section Types	α_x	α_y	β_x	β_y
Solid rectangular	$1.7 + 1.3\ (P/P_y)$	$1.7 + 1.3\ (P/P_y)$	2.0	2.0
Solid circular	2.0	2.0	2.1	2.1
I-shape	2.0	$1.2 + 2\ (P/P_y)$	1.3	$2 + 1.2\ (A_w/A_f)$
Thin-walled box	$1.7 + 1.5\ (P/P_y)$	$1.7 + 1.5\ (P/P_y)$	$2 - 0.5\,\overline{B} \geq 1.3$	$2 - 0.5\,\overline{B} \geq 1.3$
Thin-walled circular	2.0	2.0	1.75	1.75

Where $\overline{B}$ is the ratio of width to depth of the box section with respect to the bending axis.

Orbison [15] developed the following equation for a wide-flange section by trial and error and curve fitting:

$$\begin{aligned} 1.15\left(\frac{P}{P_y}\right)^2 + \left(\frac{M_x}{M_{px}}\right)^2 + \left(\frac{M_y}{M_{py}}\right)^4 + 3.67\left(\frac{P}{P_y}\right)\left(\frac{M_x}{M_{px}}\right)^2 \\ + 3.0\left(\frac{P}{P_y}\right)^2\left(\frac{M_y}{M_{py}}\right)^2 + 4.65\left(\frac{M_x}{M_{px}}\right)^4\left(\frac{M_y}{M_{py}}\right)^2 = 1.0 \end{aligned} \tag{36.54}$$

36.6 Nonlinear Frame Analysis

Both the first-order and second-order inelastic frame analyses can be categorized into three types of analysis: (1) elastic–plastic hinge, (2) refined plastic hinge, and (3) distributed plasticity. This section will discuss the basic assumptions and applications of those analyses.

36.6.1 Elastic–Plastic Hinge Analysis

In an elastic-plastic hinge (lumped plasticity) analysis, material inelasticity is taken into account using concentrated "zero-length" plastic hinges. The traditional plastic hinge is defined as a zero-length point along the structure member which can maintain plastic moment capacity and rotate freely. When the section reaches its plastic capacity (for example, the yield surface as shown in Figures 36.12 or 36.13), a plastic hinge is formed and the element stiffness is adjusted [34, 35] to reflect the hinge formation. For regions in a framed member away from the plastic hinge, elastic behavior is assumed.

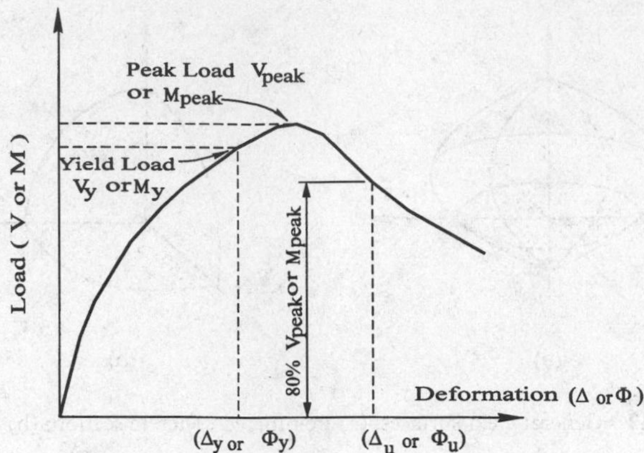

FIGURE 36.13 Load-deformation curves.

For a framed member subjected to end forces only, the elastic–plastic hinge method usually requires only one element per member making the method computationally efficient. It does not, however, accurately represent the distributed plasticity and associated P-δ effects. This analysis predicts an upper-bound solution (see Figure 36.1).

36.6.2 Refined Plastic Hinge Analysis

In the refined plastic hinge analysis [36], a two-surface yield model considers the reduction of plastic moment capacity at the plastic hinge (due to the presence of axial force) and an effective tangent modulus accounts for the stiffness degradation (due to distributed plasticity along a frame member). This analysis is similar to the elastic–plastic hinge analysis in efficiency and simplicity and, to some extent, also accounts for distributed plasticity. The approach has been developed for advanced design of steel frames, but detailed considerations for concrete structures still need to be developed.

36.6.3 Distributed Plasticity Analysis

Distributed plasticity analysis models the spread of inelasticity through the cross sections and along the length of the members. This is also referred to as plastic zone analysis, spread-of-plasticity analysis, and elastoplastic analysis by various researchers. In this analysis, a member needs to be subdivided into several elements along its length to model the inelastic behavior more accurately. There are two main approaches which have been successfully used to model plastification of members in a second-order distributed plasticity analysis:

1. Cross-sectional behavior is described as an input for the analysis by means of moment–thrust–curvature (M–P–Φ) and moment–trust–axial strain (M–P–ε) relations which may be obtained separately from section analysis as discussed in Section 36.5 or approximated by closed-form expressions [31].
2. Cross sections are subdivided into elemental areas and the state of stresses and strains are traced explicitly using the proper stress–strain relations for all elements during the analysis.

In summary, the elastic–plastic hinge analysis is the simplest one, but provides an upper-bound solution. Distributed plasticity analysis is considered the most accurate and is generally computationally intensive for larger and complex structures. Refined plastic hinge analysis seems to be an alternative that can reasonably achieve both computational efficiency and accuracy.

36.7 Practical Applications

In this section, the concept and procedures of displacement-based design and the bases of the static push-over analysis are discussed briefly. Two real bridges are analyzed as examples to illustrate practical application of the nonlinear static push-over analysis approach for bridge seismic design. Additional examples and detailed discussions of nonlinear bridge analysis can be found in the literature [7, 37].

36.7.1 Displacement-Based Seismic Design

36.7.1.1 Basic Concept

In recent years, displacement-based design has been used in the bridge seismic design practice as a viable alternative approach to strength-based design. Using displacements rather than forces as a measurement of earthquake damage allows a structure to fulfill the required function (damage-control limit state) under specified earthquake loads.

In a common design procedure, one starts by proportioning the structure for strength and stiffness, performs the appropriate analysis, and then checks the displacement ductility demand against available capacity. This procedure has been widely used in bridge seismic design in California since 1994. Alternatively, one could start with the selection of a target displacement, perform the analysis, and then determine strength and stiffness to achieve the design level displacement. Strength and stiffness do not enter this process as variables; they are the end results [38, 39].

In displacement-based design, the designer needs to define a criterion clearly for acceptable structural deformation (damage) based on postearthquake performance requirements and the available deformation capacity. Such criteria are based on many factors including structural type and importance.

36.7.1.2 Available Ultimate Deformation Capacity

Because structural survival without collapse is commonly adopted as a seismic design criterion for ordinary bridges, inelastic structural response and some degradation in strength can be expected under seismic loads. Figure 36.13 shows a typical load–deformation curve. A gradual degrading response as shown in Figure 36.13 can be due to factors such as P-Δ effects and/or plastic hinge formulation. The available ultimate deformation capacity should be based on how great a reduction (degradation) in structure load-carrying capacity response can be tolerated [21].

In general, the available ultimate deformation capacity can be referred to as the deformation that a structure can undergo without losing significant load-carrying capacity [40]. It is, therefore, reasonable to define available ultimate deformation as that deformation when the load-carrying capacity has been reduced by an acceptable amount after the peak load, say, 20%, as shown in Figure 36.13. This acceptable reduction amount may vary depending on required performance criteria of the particular case.

The available deformation capacity based on the design criteria requirements needs not correspond to the ultimate member or system deformation capacity. For a particular member cross section, the ultimate deformation in terms of the curvature depends on the shape, material properties, and loading conditions of the section (i.e., axial load, biaxial bending) and corresponds to the condition when the section extreme fiber reaches its ultimate strain (ε_{cu} for concrete and ε_{sp} for steel). The available ultimate curvature capacity ϕ_u can be chosen as the curvature that corresponds to the condition when section moment capacity response reduces by, say, 20%, from the peak moment.

For a framed structure system, the ultimate deformation in terms of the lateral displacement depends on structural configurations, section behavior, and loading conditions and corresponds to a failure state of the frame system when a collapse mechanism forms. The available lateral displacement capacity Δ_u

can be chosen as the displacement that corresponds to the condition when lateral load-carrying capacity reduces by some amount, say, 20%, from its peak load. In current seismic design practice in California, the available frame lateral displacement capacity commonly corresponds to the first plastic hinge reaching its ultimate rotational capacity.

36.7.1.3 Analysis Procedures

Seismic analysis procedures used in displacement-based design can be divided into three groups:

Group I: Seismic displacement and force demands are estimated from an elastic dynamic time history or a response spectrum analysis with effective section properties. For concrete structures, cracked section properties are usually used to determine displacement demands, and gross section properties are used to determine force demands. Strength capacity is evaluated from nonlinear section analysis or other code-specified methods, and displacement capacity is obtained from a static nonlinear push-over analysis.

Group II: Seismic displacement demand is obtained from a specified response spectrum and initial effective stiffness or a substitute structural model [38] considering both the effective stiffness and the effective damping. Effective stiffness and displacement capacity are estimated from a nonlinear static push-over analysis.

Group III: A nonlinear inelastic dynamic time history analysis is performed. Bridge assessment is based on displacement (damage) comparisons between analysis results and the given acceptance criteria. This group of analyses is complex and time-consuming and used only for important structures.

36.7.2 Static Push-Over Analysis

In lieu of a nonlinear time history dynamic analysis, bridge engineers in recent years have used static push-over analyses as an effective and simple alternative when assessing the performance of existing or new bridge structures under seismic loads. Given the proper conditions, this approximate alternative can be as reliable as the more accurate and complex ones. The primary goal of such an analysis is to determine the displacement or ductility capacity which is then compared with displacement or ductility demand obtained for most cases from linear dynamic analysis with effective section properties. However, under certain conditions, the analysis can also be used in the assessment of the displacement demand, as will be illustrated in the examples to follow.

In this analysis, a stand-alone portion from a bridge structure (such as bent-frame with single or multicolumns) is isolated and statically analyzed taking into account whatever nonlinear behavior deemed necessary (most importantly and commonly, material and geometric nonlinear behavior). The analysis can utilize any of the modeling methods discussed in Section 36.6, but plastic hinges or distributed plasticity models are commonly used. The analytical frame model is first subjected to the applied tributary gravity load and then is pushed laterally in several load (or displacement) increments until a collapse mechanism or a given failure criterion is reached. Figure 36.14 shows a flowchart outlining a procedure using static push-over analysis in seismic design and retrofit evaluation.

When applying static push-over analysis in seismic design, it is assumed that such analysis can predict with reasonable accuracy the dynamic lateral load–displacement behavior envelope, and that an elastic acceleration response spectrum can provide the best means for establishing required structural performance.

36.7.3 Example 36.1 — Reinforced Concrete Multicolumn Bent Frame with P-Δ Effects

Problem Statement

The as-built details of a reinforced concrete bridge bent frame consisting of a bent cap beam and two circular columns supported on pile foundations are shown in Figure 36.15. An as-built unconfined

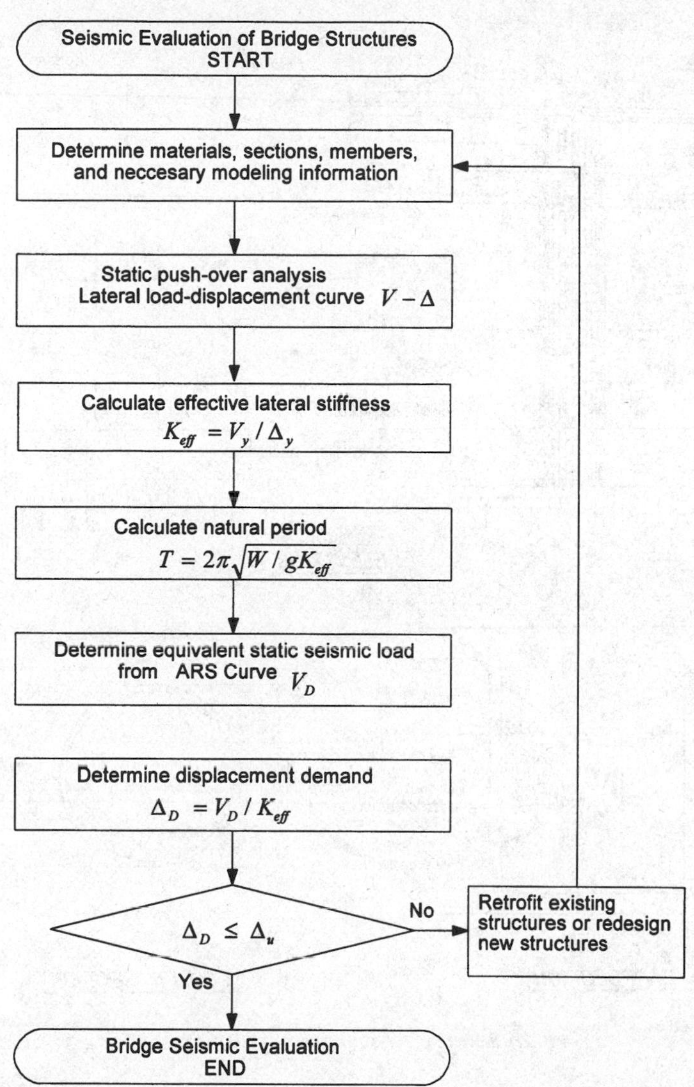

FIGURE 36.14 An alternative procedure for bridge seismic evaluation.

concrete strength of 5 ksi (34.5 MPa) and steel strength of 40 ksi (275.8 MPa) are assumed. Due to lack of adequate column transverse reinforcement, the columns are retrofitted with 0.5-in. (12.7-mm)-thick steel jacket. The bottom of the column is assumed to be fixed, however, since the footing lacks top mat and shear reinforcement, the bottom with a pinned connection is also to be considered. The frame is supported on a stiff pile–foundation and the soil–foundation–structure interaction is to be ignored.

Use static nonlinear push-over analysis to study the extent of the *P*-Δ effect on the lateral response of the bent frame when the columns are assumed fixed at the base in one case and pinned in another case. Assume the columns are retrofitted with steel jacket in both cases and determine if the footing retrofit is also required. Use 0.7 *g* ground acceleration and the ARS spectrum with 5% damping shown in Figure 36.16.

Analysis Procedure

The idealized bent frame, consisting of the cap beam and the two retrofitted column members, is discretized into a finite number of beam elements connected at joints, as shown in Figure 36.17.

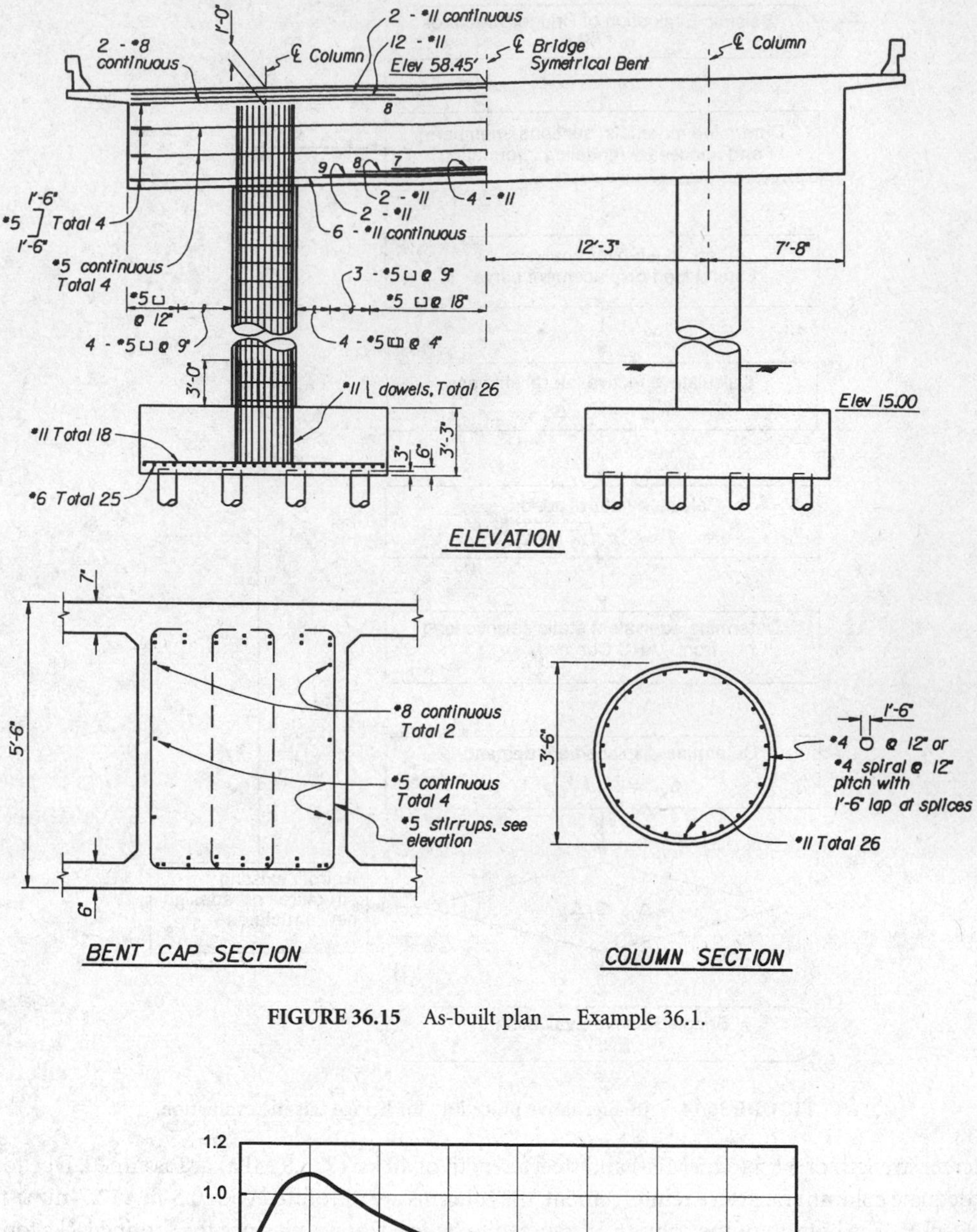

FIGURE 36.15 As-built plan — Example 36.1.

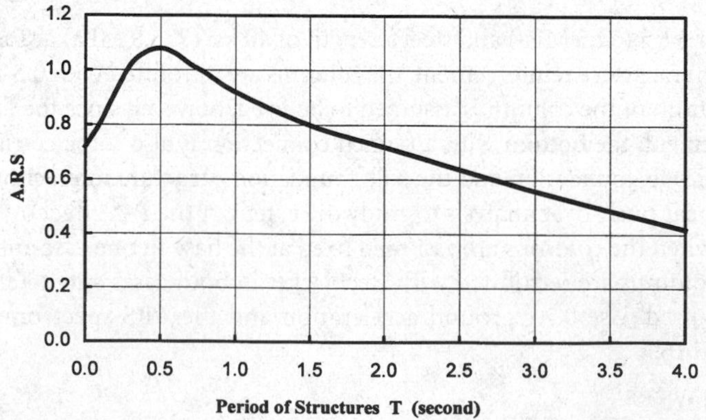

FIGURE 36.16 Specific ars curve — Example 36.1.

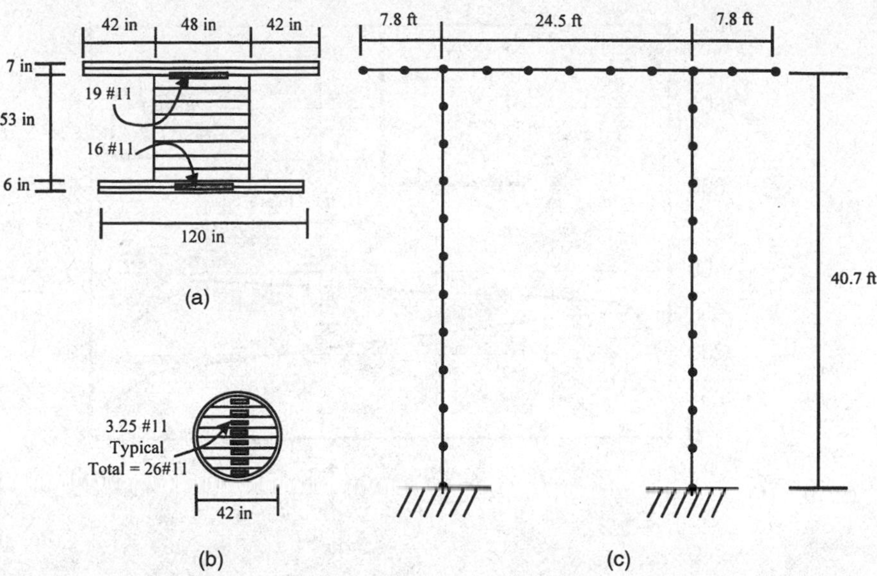

FIGURE 36.17 Analytical model — Example 36.1. (a) Local layered cab beam section: 12 concrete and 2 steel layers; (b) layered column section: 8 concrete and 8 steel layers; (c) discretized frame model.

The idealized column and cap beam cross sections are divided into several concrete layers and reinforcing steel layers as shown. Two different concrete material properties are used for the column and cap beam cross sections. The column concrete properties incorporated the increase in concrete ultimate stress and strain due to the confinement provided by the steel jacket. In this study the column confined ultimate concrete compressive stress and strain of 7.5 ksi (51.7 MPa) and 0.085 are used respectively. The total tributary superstructure dead load of 1160 kips (5160 kN) is applied uniformly along the length of the cap beam. The frame is pushed laterally in several load increments until failure is reached.

For this study, failure is defined as the limit state when one of the following conditions first take place:

1. A concrete layer strain reaches the ultimate compressive strain at any member section;
2. A steel layer strain reaches the rupture strain at any member section;
3. A 20% reduction from peak lateral load of the lateral load response curve (this condition is particularly useful when considering *P*-Δ).

The lateral displacement corresponding to this limit state at the top of the column defines the frame failure (available) displacement capacity.

A nonlinear analysis computer program NTFrame [41, 42] is used for the push-over analysis. The program is based on distributed plasticity model and the P-Δ effect is incorporated in the model second-order member stiffness formulation.

Discussion of the Results

The resulting frame lateral load vs. displacement responses are shown in Figures 36.18 for the cases when the bottom of the column is fixed and pinned. Both cases will be discussed next, followed by concluding remarks.

Column Fixed at Bottom Case

In this case the column base is modeled with a fixed connection. The lateral response with and without the *P*-Δ effect is shown in Figure 36.18a. The sharp drop in the response curve is due to several extreme concrete layers reaching their ultimate compressive strain at the top of the column.

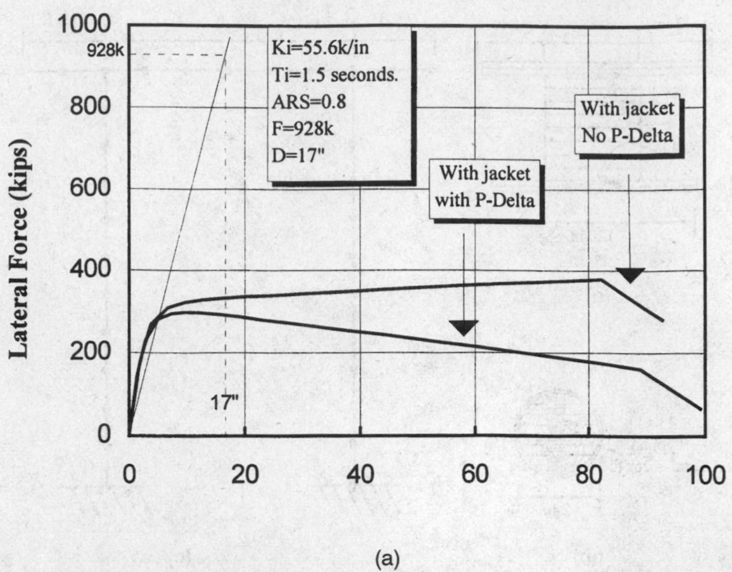

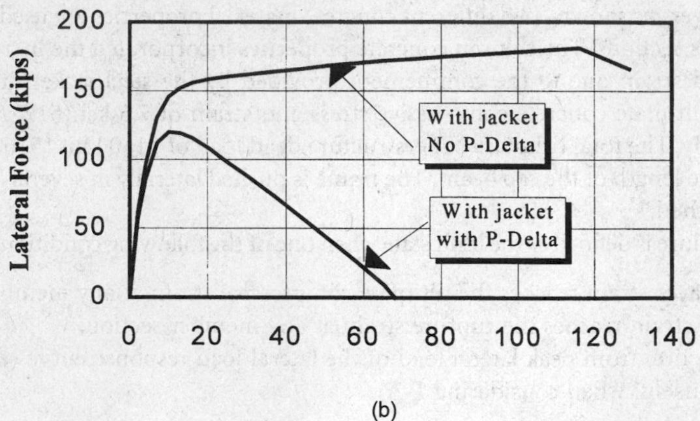

FIGURE 36.18 Lateral load vs. displacement responses — Example 36.1. Lateral response (a) fixed column (b) with penned column.

The effect of P-Δ at failure can be seen to be considerable but not as severe as shown in Figure 36.18b with the pinned connection. Comparing Figures 36.19a and b, one can observe that fixing the bottom of the column resulted in stiffer structural response.

Using the curve shown in Figure 36.18a, the displacement demand for the fixed column case with P-Δ effect is calculated as follows:

Step 1: Calculate the Initial Effective Stiffness K_{eff}

The computer results showed that the first column extreme longitudinal rebar reached yield at lateral force of 928 kips (4128 kN) at a corresponding lateral yield displacement of 17 in. (431.8 mm), therefore

$$K_{eff} = 928/17 = 55 \text{ kips/in. } (9.63 \text{ kN/mm})$$

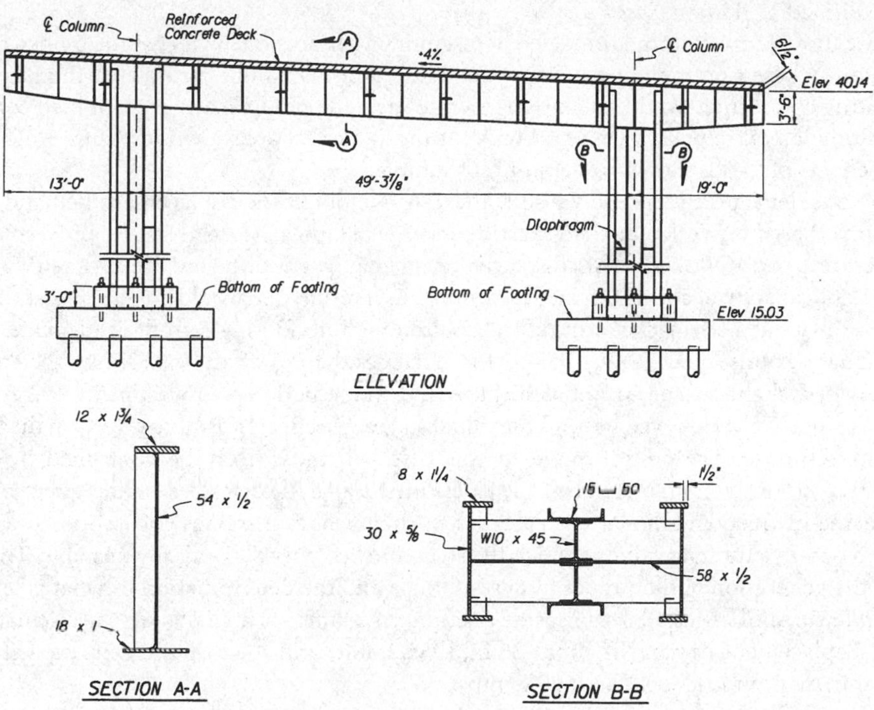

FIGURE 36.19 As-built plane — Example 36.2.

Step 2: Calculate an Approximate Fundamental Period T_f

$$T_f = 0.32\sqrt{\frac{W}{K_{\text{eff}}}} = 0.32\sqrt{\frac{1160}{55}} = 1.5 \text{ s}$$

Step 3: Determine the Damped Elastic Acceleration Response Spectrum (ARS) at the Site in *g*'s

By using the given site spectrum shown in Figure 36.16 and the above calculated period, the corresponding ARS for 5% damping is 0.8.

Step 4: Calculate the Displacement Demand D_d

$$D_d = \frac{ARS(W)}{K_{eff}} = \frac{0.8(1160)}{55} = 16.9 \text{ in. (429.3 m)}$$

(in this case the yield and demand displacements are found to be practically equal).

In much of the seismic design practice in California, the effect of *P*-Δ is usually ignored if the *P*-Δ moment is less than 20% of the design maximum moment capacity. Adopting this practice and assuming the reduction in the moment is directly proportional to the reduction in the lateral force, one may conclude that at displacement demand of 16.9 in. (429.3 mm), the reduction in strength (lateral force) is less then 20%, and as a result the effects of *P*-Δ are negligible.

The displacement demand of 16.9 in. (429.3 mm) is less than the failure state displacement capacity of about 40 in. (1016 mm) (based on a 20% lateral load reduction from the peak). Note that for the fixed bottom case with *P*-Δ, the displacement when the extreme concrete layer at the top of the column reached its ultimate compressive strain is about 90 in. (2286 mm).

Column Pinned at Bottom Case

In this case the column bottom is modeled with a pinned connection. Note that the pinned condition assumption is based on the belief that in the event of a maximum credible earthquake the column/footing connection would quickly degenerate (degrade) and behave like a pinned connection. The resulting lateral responses with and without the *P*-Δ effect are shown in Figure 36.18b. In this case the effects of *P*-Δ is shown to be quite substantial.

When considering the response without the *P*-Δ one obtains a displacement demand of 38 in. (965.2 mm) (based on a calculated initial stiffness of 18.5 kips/in. (3.24 kN/mm) and a corresponding structure period of 2.5 s). This displacement demand is well below the ultimate (at failure) displacement capacity of about 115 in. (2921 mm). As a result, one would conclude that the retrofit measure of placing a steel jacket around the column with no footing retrofit is adequate.

The actual response, however, is the one that includes the *P*-Δ effect. In this case the effect of *P*-Δ resulted in a slight change in initial stiffness and frame period — 15.8 kips/in. (2.77 kN/mm) and 2.7 s, respectively. However, beyond the initial stages, the effects are quite severe on the load–displacement response. The failure mode in this case will most likely be controlled by dynamic instability of the frame. MacRae et al. [43] performed analytical studies of the effect of *P*-Δ on single-degree-of-freedom bilinear oscillators (i.e., single-column frame) and proposed some procedures to obtain a limiting value at which the structure becomes dynamically unstable. The process requires the generation of the proper hysteresis loops and the determination of what is termed the *effective bilinear stiffness factor.* Setting aside the frame dynamic instability issue, the calculated initial stiffness displacement demand is about 38 in. (965.2 mm) and the displacement capacity at 20% reduction from peak load is 24 in. (609.6 mm).

Referring to the curves with *P*-Δ in Figure 36.18, it is of interest to mention, as pointed out by Mahin and Boroschek [44], that continued pushing of the frame will eventually lead to a stage when the frame structure becomes statically unstable. At that stage the forces induced by the *P*-Δ effect overcome the mechanical resistance of the structure. Note that the point when the curve with *P*-Δ effect intersects the displacement-axis in (as shown Figure 36.19b) will determine the lateral displacement at which the structure becomes statically unstable. Dynamic instability limits can be 20 to 70% less than the static instability depending on the ground motion and structural characteristic [44]. Note that dynamic instability is assumed not to be a controlling factor in the previous case with fixed column.

In conclusion, if the as-built column–footing connection can support the expected column moment obtained from the fixed condition case (which is unlikely), then retrofitting the column with steel jacket without footing retrofit is adequate. Otherwise, the footing should also be retrofitted to reduce (limit) the effect of *P*-Δ.

It should be pointed out that in this example the analysis is terminated at the completion of the first plastic hinge (conservative), whereas in other types of push-over analysis such as event-to-event analysis, the engineer may chose to push the frame farther until it forms a collapse mechanism. Also, unlike the substitute structure procedure described by Priestly et al. [7] in which both the effective system stiffness and damping ratio are adjusted (iterated) several times before final displacement demand is calculated, here only the initial effective stiffness and a constant specified structure damping are used.

As a final remark, the *P*-Δ effect in bridge analysis is normally assumed small and is usually ignored. This assumption is justified in most cases under normal loading conditions. However, as this example illustrated, under seismic loading, the *P*-Δ effect should be incorporated in the analysis, when large lateral displacements are expected before the structure reaches its assumed failure state. In the design of a new bridge, the lateral displacement and the effect of *P*-Δ can be controlled. When assessing an existing bridge for possible seismic retrofit, accurate prediction of the lateral displacement with *P*-Δ effects can be an essential factor in determining the retrofit measures required.

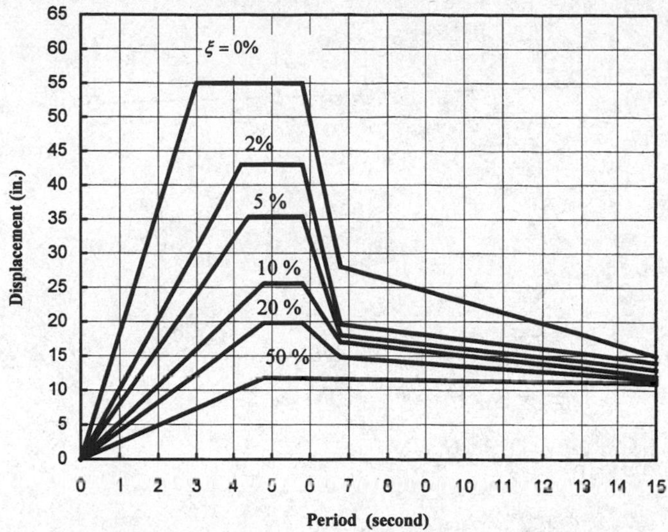

FIGURE 36.20 Displacement Response spectra — Example 36.2.

36.7.4 Example 36.2 — Steel Multicolumn Bent Frame Seismic Evaluation

Problem Statement

The as-built details of a steel bridge bent frame consisting of a bent cap plate girder and two builtup columns supported on a stiff pile–foundation, as shown in Figure 36.19. Steel is Grade 36. Site-specific displacement response spectra are given in Figure 36.20. For simplicity and illustration purposes, fixed bases of columns are assumed and the soil–foundation–structure interaction is ignored.

Evaluate lateral displacement capacity by using static nonlinear push-over analysis. Estimate seismic lateral displacement demands by using the substitute structure approach considering both the effective stiffness and the effective damping. The effective damping ξ can be calculated by Takeda's formula [45]:

$$\xi = 0.05 + \frac{\left(1 - \frac{0.95}{\sqrt{\mu_{\Delta d}}} - 0.05\sqrt{\mu_{\Delta d}}\right)}{\pi} \tag{36.55}$$

$$\mu_{\Delta d} = \frac{\Delta_{ud}}{\Delta_y} \tag{36.56}$$

where $\mu_{\Delta d}$ is displacement ductility demand; Δ_{ud} and Δ_y are displacement demand and yield displacement, respectively.

Analysis Modeling

The bent frame members are divided into several beam elements as shown in Figure 36.21. The properties of beam elements are defined by two sets of relationships for moment–curvature, axial force–strain, and torsion–twist for the cap beam and columns, respectively. The available ultimate curvature is assumed as 20 times yield curvature. The total tributary superstructure dead load of 800 kips (3558 kN) is applied at longitudinal girder locations. A lateral displacement is applied incrementally at the top of the bent column until a collapse mechanism of the bent frame is formed.

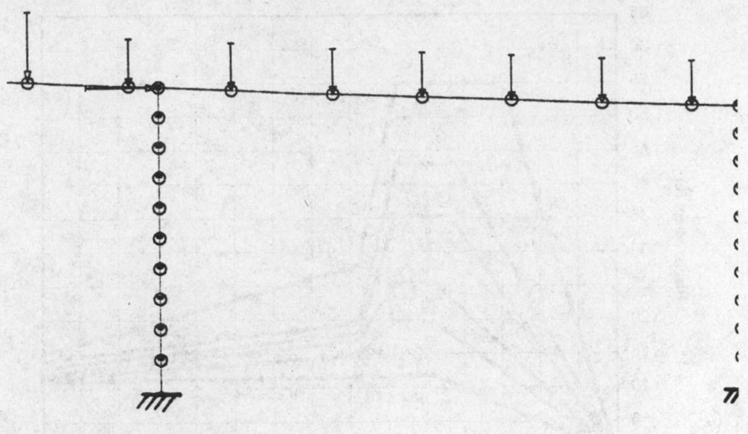

FIGURE 36.21 Analytical model — Example 36.2.

Displacement Capacity Evaluation

The displacement capacity evaluation is performed by push-over analysis using the ADINA [46] analysis program. Large displacements are considered in the analysis. The resulting lateral load vs. displacement response at the top of columns is shown in Figures 36.22. The sudden drops in the response curve are due to the several beam elements reaching their available ultimate curvatures. The yield displacement $\Delta_y = 1.25$ in. (31.8 mm) and the available ultimate displacement capacity (corresponding to a 20% reduction from the peak lateral load) $\Delta_u = 2.61$ in. (66.3 mm) are obtained.

Displacement Demand Estimation

A substitute structure approach with the effective stiffness and effective damping will be used to evaluate displacement demand.

1. Try $\Delta_{ud} = 3$ in. (76.2 mm); from Figure 36.20, Eqs. (36.55) and (36.56), we obtain

$$K_{\text{eff}} = \frac{600}{3} = 200 \text{ kips/in. } (35.04 \text{ kN/mm})$$

$$T_{\text{eff}} = 0.32\sqrt{\frac{W(\text{kips})}{k_{\text{eff}}(\text{kips/in.})}} = 0.32\sqrt{\frac{800}{200}} = 0.64 \text{ (s)}$$

$$\mu_{\Delta d} = \frac{\Delta_{ud}}{\Delta_y} = \frac{3}{1.25} = 2.4$$

$$\xi = 0.05 + \frac{\left(1 - \frac{0.95}{\sqrt{2.4}} - 0.05\sqrt{2.4}\right)}{\pi} = 0.15$$

From Figure 36.20, find $\Delta_d = 2.5$ in. $< \Delta_{ud} = 3$ in. (76.2 mm).

2. Try $\Delta_{ud} = 2.5$ in. (63.5 mm); from Figures 36.20 and 36.22 Eqs. (36.55), and (36.56), we obtain

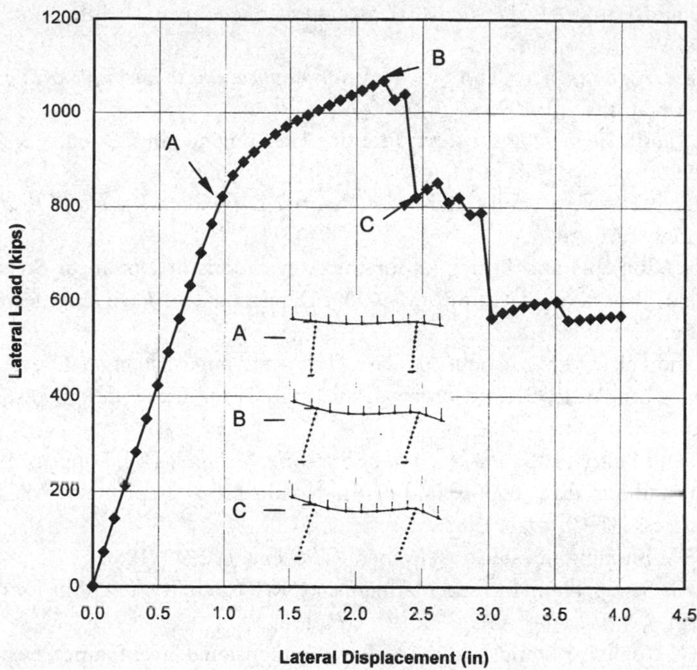

FIGURE 36.22 Lateral load vs. displacement — Example 36.2.

$$K_{\text{eff}} = \frac{830}{2.5} = 332 \text{ kips/in. } (58.14 \text{ kN/mm})$$

$$T_{\text{eff}} = 0.32\sqrt{\frac{W(\text{kips})}{K_{eff}(\text{kips/in.})}} = 0.32\sqrt{\frac{800}{332}} = 0.50$$

$$\mu_{\Delta d} = \frac{\Delta_{ud}}{\Delta_y} = \frac{2.5}{1.25} = 2$$

$$\xi = 0.05 + \frac{\left(1 - \frac{0.95}{\sqrt{2}} - 0.05\sqrt{2}\right)}{\pi} = 0.13$$

From Figure 36.20, find $\Delta_d = 2.45$ in. (62.2 mm) close to $\Delta_{ud} = 2.5$ in. (63.5 mm) OK
Displacement demand $\Delta_d = 2.45$ in. (62.2 mm).

Discussion

It can be seen that the displacement demand Δ_d of 2.45 in. (62.2 mm) is less than the available ultimate displacement capacity of $\Delta_u = 2.61$ in. (66.3 mm). It should be pointed out that in the actual seismic evaluation of this frame, the flexibility of the steel column to the footing bolted connection should be considered.

References

1. Chen, W. F., *Plasticity in Reinforced Concrete*, McGraw-Hill, New York, NY, 1982.
2. Clough, R. W. and Penzien, J., *Dynamics of Structures*, 2nd ed., McGraw-Hill, New York, 1993.
3. Fung, Y. C., *First Course in Continuum Mechanics*, 3rd ed., Prentice-Hall Engineering, Science & Math, Englewood Cliffs, NJ, 1994.

4. Chen, W. F. and Han, D. J., *Plasticity for Structural Engineers*, Gau Lih Book Co., Ltd., Taipei, Taiwan, 1995.
5. Chopra, A. K., *Dynamics of Structures: Theory and Applications to Earthquake Engineering*, Prentice-Hall, Englewood Cliffs, NJ, 1995.
6. Bathe, K. J., *Finite Element Procedures*, Prentice-Hall Engineering, Science & Math, Englewood Cliffs, NJ, 1996.
7. Priestley, M. J. N., Seible, F., and Calvi, G. M., *Seismic Design and Retrofit of Bridges*, John Wiley & Sons, New York, 1996.
8. Powell, G. H., Concepts and Principles for the Applications of Nonlinear Structural Analysis in Bridge Design, Report No. UCB/SEMM-97/08, Department of Civil Engineering, University of California, Berkeley, 1997.
9. Chen, W. F. and Lui, E. M., *Structural Stability: Theory and Implementation*, Elsevier, New York, 1987.
10. Goto, Y. and Chen, W. F., Second-order elastic analysis for frame design, *J. Struct. Eng. ASCE*, 113(7), 1501, 1987.
11. Allen, H. G. and Bulson, P. S., *Background of Buckling*, McGraw-Hill, London, 1980.
12. White, D. W. and McGuire, W., Method of Analysis in LRFD, Reprints of ASCE Structure Engineering Congress '85, Chicago, 1985.
13. Schilling, C. G. Buckling of one story frames, *AISC Eng. J.*, 2, 49, 1983.
14. Ekhande, S. G., Selvappalam, M., and Madugula, M. K. S., Stability functions for three-dimensional beam-column, *J. Struct. Eng. ASCE*, 115(2), 467, 1989.
15. Orbison, J. G., Nonlinear Static Analysis of Three-dimensional Steel Frames, Department of Structural Engineering, Cornell University, Ithaca, NY, 1982.
16. Yang, Y. B. and McGuire, W., Stiffness matrix for geometric nonlinear analysis, *J. Struct. Eng. ASCE*, 112(4), 853, 1986.
17. Yang, Y. B. and McGuire, W., Joint rotation and geometric nonlinear analysis, *J. Struct. Eng. ASCE*, 112(4), 879, 1986.
18. Hognestad, E., A Study of Combined Bending and Axial Load in Reinforced Concrete Members, University of Illinois Engineering Experimental Station, Bulletin Series No. 399, Nov., Urbana, IL, 1951.
19. Kent, D. C. and Park, R., Flexural members with confined concrete, *J. Struct. Div.* ASCE, 97(ST7), 1969, 1971.
20. Popovics, S. A., Review of stress-strain relationship for concrete, *J. ACI*, 67(3), 234, 1970.
21. Park, R. and Paulay, T., *Reinforced Concrete Structures*, John Wiley & Sons, New York, 1975.
22. Wang, W. C. and Duan, L., The stress-strain relationship for concrete, *J. Taiyuan Inst. Technol.*, 1, 125, 1981.
23. Mander, J. B., Priestley, M. J. N., and Park, R., Theoretical stress-strain model for confined concrete, *J. Struct. Eng. ASCE*, 114(8), 1804, 1988.
24. Mander, J. B., Priestley, M. J. N., and Park, R., Observed stress-strain behavior of confined concrete, *J. Struct. Eng. ASCE*, 114(8), 1827, 1988.
25. Hoshikuma, J., et al., Stress-strain model for confined reinforced concrete in bridge piers, *J. Struct. Eng. ASCE*, 123(5), 624, 1997.
26. William, K. J. and Warnke, E. P., Constitutive model for triaxial behavior of concrete, *Proc. IABSE*, 19, 1, 1975.
27. Mander, J. B., Priestley, M. J. N., and Park, R., Seismic Design of Bridge Piers, Research Report No. 84-2, University of Canterbury, New Zealand, 1984.
28. Chai, Y. H., Priestley, M. J. N., and Seible, F., Flexural Retrofit of Circular Reinforced Bridge Columns by Steel Jacketing, Report No. SSRP-91/05, University of California, San Diego, 1990.
29. Vebe, A. et al., Moment-curvature relations of reinforced concrete slab, *J. Struct. Div. ASCE*, 103(ST3), 515, 1977.

30. Holzer, S. M. et al., SINDER, A Computer Code for General Analysis of Two-Dimensional Reinforced Concrete Structures, AFWL-TR-74-228 Vol. 1, Air Force Weapons Laboratory, Kirtland AFB, NM, 1975.
31. Chen, W. F. and Atsuta, T., *Theory of Beam-Columns*, Vol. 1 and 2, McGraw-Hill, New York, 1977.
32. Bresler, B., Design criteria for reinforced concrete columns under axial load and biaxial bending, *J. ACI*, 32(5), 481, 1960
33. Duan, L. and Chen, W. F., A yield surface equation for doubly symmetrical section, *Struct. Eng.*, 12(2), 114, 1990.
34. King, W. S., White, D. W., and Chen, W. F., Second-order inelastic analysis methods for steel-frame design, *J. Struct. Eng. ASCE*, 118(2), 408, 1992.
35. Levy, R., Joseph, F., and Spillers, W. R., Member stiffness with offset hinges, *J. Struct. Eng. ASCE*, 123(4), 527, 1997.
36. Chen, W. F. and S. Toma, *Advanced Analysis of Steel Frames*, CRC Press, Boca Raton, FL, 1994.
37. Aschheim, M., Moehle, J. P., and Mahin, S. A., Design and Evaluation of Reinforced Concrete Bridges for Seismic Resistance, Report, UCB/EERC-97/04, University of California, Berkeley, 1997.
38. Priestley, N., Myths and Fallacies in Earthquake Engineering — Conflicts between Design and Reality, in *Proceedings of Tom Paulay Symposium — Recent Development in Lateral Force Transfer in Buildings*, University of California, San Diego, 1993.
39. Kowalsky, M. J., Priestley, M. J. N., and MacRae, G. A., Displacement-Based Design, Report No. SSRP-94/16, University of California, San Diego, 1994.
40. Duan, L. and Cooper, T. R., Displacement ductility capacity of reinforced concrete columns, *ACI Concrete Int.*, 17(11). 61, 1995.
41. Akkari, M. M., Nonlinear push-over analysis of reinforced and prestressed concrete frames, *Structure Notes*, State of California, Department of Transportation, Sacramento, July 1993.
42. Akkari, M. M., Nonlinear push-over analysis with p-delta effects, *Structure Notes*, State of California, Department of Transportation, Sacramento, November 1993.
43. MacRae, G. A., Priestly, M. J. N., and Tao, J., P-delta design in seismic regions, Structure System Research Project Report No. SSRP-93/05, University of California, San Diego, 1993.
44. Mahin, S. and Boroschek, R., Influence of geometric nonlinearities on the seismic response and design of bridge structures, *Background Report*, California Department of Transportation, Division of Structures, Sacramento, 1991.
45. Takeda, T., Sozen, M. A., and Nielsen, N. N., Reinforced concrete response to simulated earthquakes, *J. Struct. Div. ASCE*, 96(ST12), 2557, 1970.
46. ADINA, *ADINA-IN for ADINA User's Manual*, ADINA R & D, Inc., Watertown, MA, 1994.

37

Seismic Design Philosophies and Performance-Based Design Criteria

Lian Duan
California Department of Transportation

Fang Li
California Department of Transportation

37.1 Introduction

Seismic design criteria for highway bridges have been improving and advancing based on research findings and lessons learned from past earthquakes. In the United States, prior to the 1971 San Fernando earthquake, the seismic design of highway bridges was partially based on lateral force requirements for buildings. Lateral loads were considered as levels of 2 to 6% of dead loads. In 1973, the California Department of Transportation (Caltrans) developed new seismic design criteria related to site, seismic response of the soils at the site, and the dynamic characteristics of bridges. The American Association of State Highway and Transportation Officials (AASHTO) modified the Caltrans 1973 Provisions slightly, and adopted Interim Specifications. The Applied Technology Council (ATC) developed guidelines ATC-6 [1] for seismic design of bridges in 1981. AASHTO adopted ATC-6 [1] as the Guide Specifications in 1983 and later incorporated it into the Standard Specifications for Highway Bridges in 1991.

0-8493-7434-0/00/$0.00+$.50

Since the 1989 Loma Prieta earthquake in California [2], extensive research [3-15] has been conducted on seismic design and retrofit of bridges in the United States, especially in California. The performance-based project-specific design criteria [16,17] were developed for important bridges. Recently, ATC published improved seismic design criteria recommendations for California bridges [18] in 1996, and for U.S. bridges and highway structures [19] in 1997, respectively. Caltrans published the new seismic Design Methodology in 1999. [20] The new Caltrans Seismic Design Criteria [43] is under development. Great advances in earthquake engineering have been made during this last decade of the 20th century.

This chapter first presents the bridge seismic design philosophy and the current practice in the United States. It is followed by an introduction to the newly developed performance-based criteria [17] as a reference guide.

37.2 Design Philosophies

37.2.1 No-Collapse-Based Design

For seismic design of ordinary bridges, the basic philosophy is to prevent collapse during severe earthquakes [21-26]. To prevent collapse, two alternative approaches are commonly used in design. The first is a conventional force-based approach where the adjustment factor Z for ductility and risk assessment [26], or the response modification factor R [23], is applied to elastic member forces obtained from a response spectra analysis or an equivalent static analysis. The second approach is a more recent displacement-based approach [20] where displacements are a major consideration in design. For more-detailed information, reference can be made to a comprehensive discussion in *Seismic Design and Retrofit of Bridges* by Priestley, Seible, and Calvi [15].

37.2.2 Performance-Based Design

Following the 1989 Loma Prieta earthquake, bridge engineers [2] have faced three essential challenges:

- Ensure that earthquake risks posed by new construction are acceptable.
- Identify and correct unacceptable seismic safety conditions in existing structures.
- Develop and implement a rapid, effective, and economic response mechanism for recovering structural integrity after damaging earthquakes.

In the California, although the Caltrans Bridge Design Specifications [26] have not been formally revised since 1989, project-specific criteria and design memoranda have been developed and implemented for the design of new bridges and the retrofitting of existing bridges. These revised or supplementary criteria included guidelines for development of site-specific ground motion estimates, capacity design to preclude brittle failure modes, rational procedures for joint shear design, and definition of limit states for various performance objectives [14]. As shown in Figure 37.1, the performance requirements for a specific project must be established first. Loads, materials, analysis methods, and detailed acceptance criteria are then developed to achieve the expected performance.

37.3 No-Collapse-Based Design Approaches

37.3.1 AASHTO-LRFD Specifications

Currently, AASHTO has issued two design specifications for highway bridges: the second edition of AASHTO-LRFD [23] and the 16th edition of the Standard Specifications [24]. This section mainly discusses the design provisions of the AASHTO-LRFD Specifications.

The principles used for the development of AASHTO-LRFD [23] seismic design specifications are as follows:

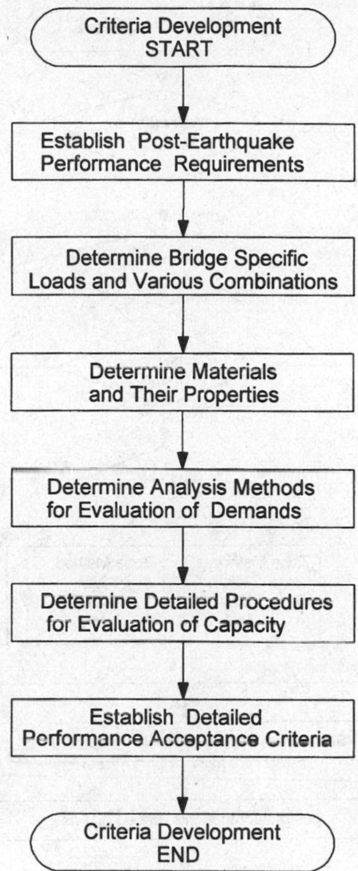

FIGURE 37.1 Development of performance-based seismic design criteria.

- Small to moderate earthquakes should be resisted within the elastic range of the structural components without significant damage.
- Realistic seismic ground motion intensities and forces should be used in the design procedures.
- Exposure to shaking from a large earthquake should not cause collapse of all or part of bridges where possible; damage that does occur should be readily detectable and accessible for inspection and repair.

Seismic force effects on each component are obtained from the elastic seismic response coefficient C_{sm} and divided by the elastic response modification factor R. Specific detailing requirements are provided to maintain structural integrity and to ensure ductile behavior. The AASHTO-LRFD seismic design procedure is shown in Figure 37.2.

Seismic Loads

Seismic loads are specified as the horizontal force effects and are obtained by production of C_{sm} and the equivalent weight of the superstructures. The seismic response coefficient is given as:

$$C_{sm} = \begin{cases} \dfrac{1.25AS}{T_m^{2/3}} \le 2.5A & \\ A(0.8+4T_m) & \text{for Soil III, IV, and nonfundamental } T_m < 0.3s \\ 3AST_m^{0.75} & \text{for Soil III, IV and } T_m > 0.4s \end{cases} \tag{37.1}$$

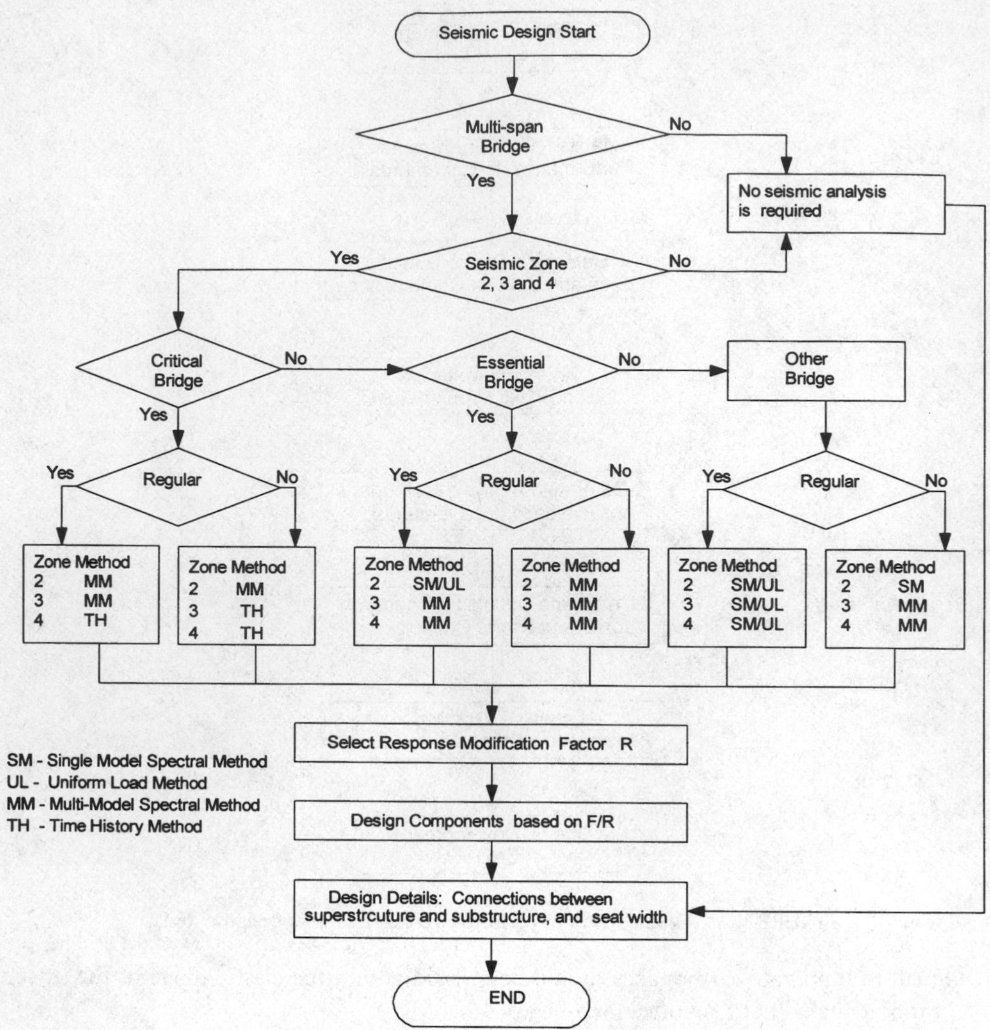

FIGURE 37.2 AASHO-LRFD seismic design procedure.

where A is the acceleration coefficient obtained from a contour map (Figure 37.3) which represents the 10% probability of an earthquake of this size being exceeded within a design life of 50 years; S is the site coefficient and is dependent on the soil profile types as shown in Table 37.1; T_m is the structural period of the mth mode in second.

Analysis Methods

Four seismic analysis methods specified in AASHTO-LRFD [23] are the uniform-load method, the single-mode spectral method, the multimode spectral method, and the time history method. Depending on the importance, site, and regularity of a bridge structure, the minimum complexity analysis methods required are shown in Figure 37.2. For single-span bridges and bridges located seismic Zone 1, no seismic analysis is required.

The importance of bridges is classified as critical, essential, and other in Table 37.2 [23], which also shows the definitions of a regular bridge. All other bridges not satisfying the requirements of Table 37.2 are considered irregular.

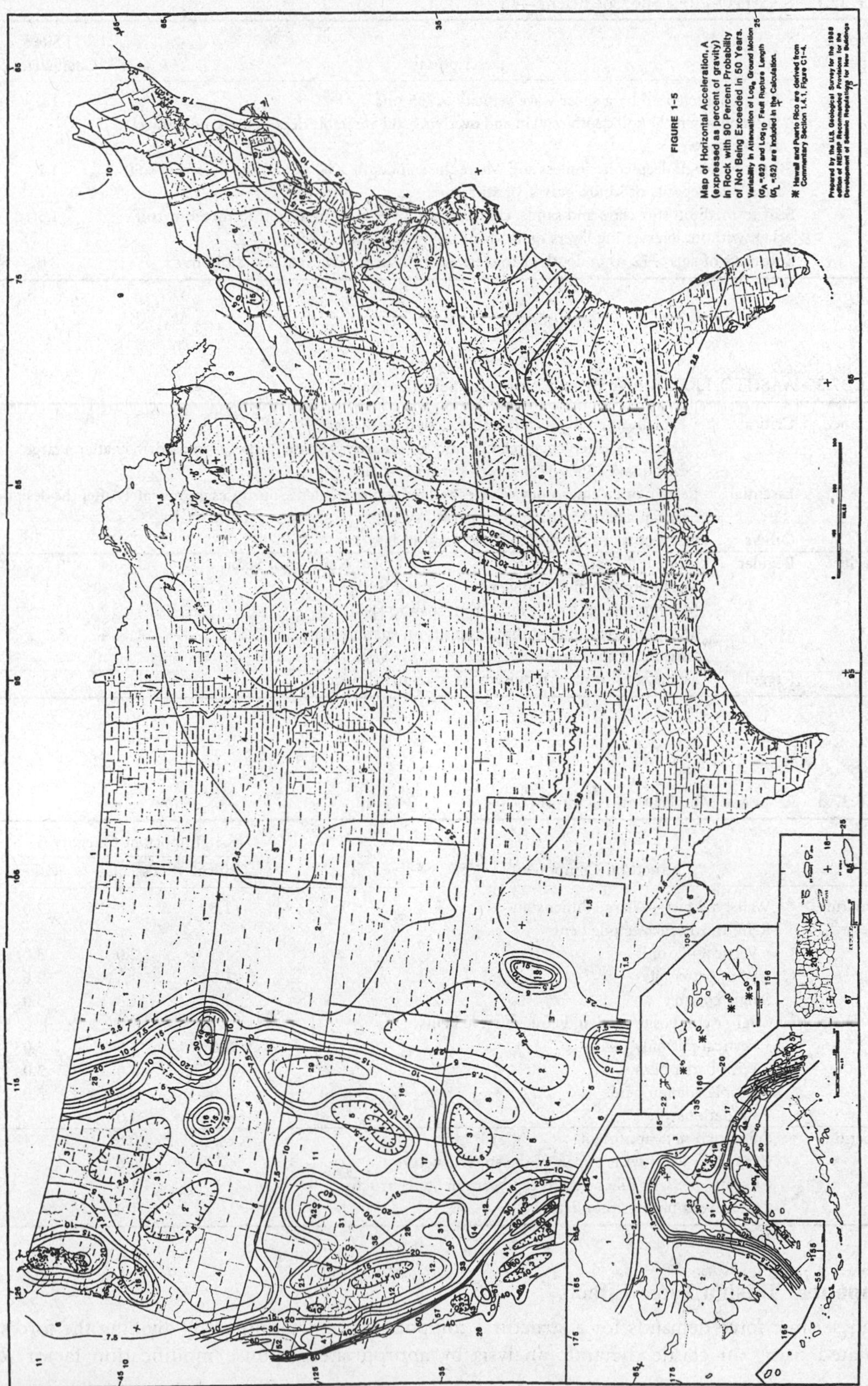

FIGURE 37.3 AASHTO-LRFD seismic contour map.

TABLE 37.1 AASHTO-LRFD Site Coefficient — *S*

Soil Profile Type	Descriptions	Site Coefficient, *S*
I	• Rock characterized by a shear wave velocity > 765 m/s • Stiff soil where the soil depth < 60 m and overlying soil are stable deposits of sands, gravel, or stiff clays	1.0
II	Stiff cohesive or deep cohesionless soil where the soil depth > 60 m and the overlying soil are stable deposits of sands, gravel, or stiff clays	1.2
III	Stiff to medium-stiff clays and sands, characterized by 9 m or more soft to medium-stiff clays without intervening layers of sands or other cohesionless soils	1.5
IV	Soft clays of silts > 12 m in depth characterized by a shear wave velocity < 153 m/s	2.0

TABLE 37.2 AASHTO-LRFD Bridge Classifications for Seismic Analysis

Importance	Critical	• Remain open to all traffic after design earthquake • Usable by emergency vehicles and for security/defense purposes immediately after a large earthquake (2500-year return period event)						
	Essential	Remain open emergency vehicles and for security/defense purposes immediately after the design earthquake (475-year return period event)						
	Others	Not required as critical and essential bridges						
Regularity	Regular	Structural Features	Number of Span	2	3	4	5	6
		Maximum subtended angles for a curved bridge				90°		
		Maximum span length ratio from span to span		3	2	2	1.5	1.5
		Maximum bent/pier stiffness ratio from span to span excluding abutments		—	4	4	3	2
	Irregular	Multispan not meet requirement of regular bridges						

TABLE 37.3 Response Modification Factor, *R*

	Structural Component	Important Category: Critical	Important Category: Essential	Important Category: Others
Substructure	Wall-type pier — Large dimension	1.5	1.5	2.0
	Reinforced concrete pile bent			
	• Vertical pile only	1.5	2.0	3.0
	• With batter piles	1.5	1.5	2.0
	Single column	1.5	2.0	3.0
	Steel or composite steel and concrete pile bents			
	• Vertical pile only	1.5	3.5	5.0
	• With batter piles	1.5	2.0	3.0
	Multiple column bents	1.5	3.5	5.0
	Foundations		1.0	
Connection	Substructure to abutment		0.8	
	Expansion joints with a span of the superstructure		0.8	
	Column, piers, or pile bents to cap beam or superstructure		1.0	
	Columns or piers to foundations		1.0	

Component Design Force Effects

Design seismic force demands for a structural component are determined by dividing the forces calculated using an elastic dynamic analysis by appropriate response modification factor *R*

(Table 37.3) to account for inelastic behavior. As an alternative to the use of R factor for connection, the maximum force developed from the inelastic hinging of structures may be used for designing monolithic connections.

To account for uncertainty of earthquake motions, the elastic forces obtained from analysis in each of two perpendicular principal axes shall be combined using 30% rule, i.e., 100% of the absolute response in one principal direction plus 30% of the absolute response in the other.

The design force demands for a component should be obtained by combining the reduced seismic forces with the other force effects caused by the permanent and live loads, etc. Design resistance (strength) are discussed in Chapter 38 for concrete structures and Chapter 39 for steel structures.

37.3.2 Caltrans Bridge Design Specifications

The current Caltrans Bridge Design Specifications [26] adopts a single-level force-based design approach based on the no-collapse design philosophy and includes:

- Seismic force levels defined as elastic acceleration response spectrum (ARS);
- Multimodal response spectrum analysis considering abutment stiffness effects;
- Ductility and risk Z factors used for component design to account for inelastic effects;
- Properly designed details.

Seismic Loads

A set of elastic design spectra ARS curves are recommended to consider peak rock accelerations (A), normalized 5% damped rock spectra (R), and soil amplification factor (S). Figure 37.4 shows typical ARS curves.

Analysis Methods

For ordinary bridges with well-balanced span and bent/column stiffness, an equivalent static analysis with the ARS times the weight of the structure applied at the center of gravity of total structures can be used. This method is used mostly for hinge restrainer design. For ordinary bridges with significantly irregular geometry configurations, a dynamic multimodal response spectrum analysis is recommended. The following are major considerations in seismic design practice:

- A beam-element model with three or more lumped masses in each span is usually used [25-27].
- A larger cap stiffness is often used to simulate a stiff deck.
- Gross section properties of columns are commonly used to determine force demands, and cracked concrete section properties of columns are used for displacement demands.
- Soil–spring elements are used to simulate the soil–foundation–structure–interaction. Adjustments are often made to meet force–displacement compatibility, particularly for abutments. The maximum capacity of the soil behind abutments with heights larger than 8 ft (2.44 m) is 7.7 ksf (369 kPa) and lateral pile capacity of 49 kips (218 kN) per pile.
- Compression and tension models are used to simulate the behavior of expansion joints.

Component Design Force Effects

Seismic design force demands are determined using elastic forces from the elastic response analysis divided by the appropriate component- and period-based (stiffness) adjustment factor Z, as shown in Figure 38.4a to consider ductility and risk. In order to account for directional uncertainty of earthquake motions, elastic forces obtained from analysis of two perpendicular seismic loadings are combined as the 30% rule, the same as the AASHTO-LRFD [23].

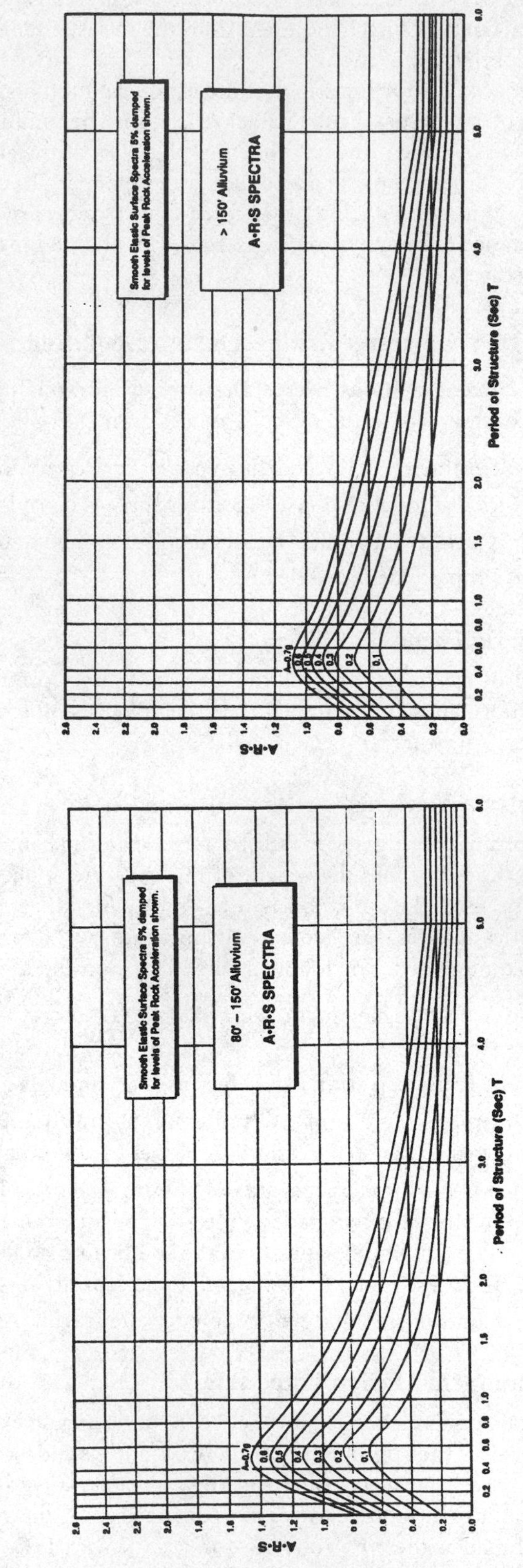

FIGURE 37.4 Caltrans ARS curves.

TABLE 37.4 Caltrans Seismic Performance Criteria

Ground motions at the site	Minimum (ordinary bridge) performance level	Important bridge performance level
Functional evaluation	Immediate service; repairable damage	Immediate service level; minimum damage
Safety evaluation	Limited service level; significant damage	Immediate service level; repairable damage

Definitions:
Important Bridge (one of more of following items present):
- Bridge required to provide secondary life safety
- Time for restoration of functionality after closure creates a major economic impact
- Bridge formally designed as critical by a local emergency plan

(*Ordinary Bridge*: Any bridge not classified as an important bridge.)
Functional Evaluation Ground Motion (*FEGM*): Probabilistic assessed ground motions that have a 40% probability of occurring during the useful lifetime of the bridge. The determination of this event shall be reviewed by a Caltrans-approved consensus group. A separate functionality evaluation is required for important bridges. All other bridges are only required to meet the specified design requirement to assure minimum functionality performance level compliance.

Safety Evaluation Ground Motion (*SEGM*): Up to two methods of defining ground motion may be used:
- Deterministically assessed ground motions from the maximum earthquake as defined by the Division of Mines and Geology Open-File Report 92-1 [1992].
- Probabilistically assessed ground motions with a long return period (approximately 1000–2000 years).

For important bridges both methods should be given consideration; however, the probabilistic evaluation should be reviewed by a Caltrans-approved consensus group. For all other bridges, the motions should be based only on the deterministic evaluation. In the future, the role of the two methods for other bridges should be reviewed by a Caltrans-approved consensus group.

Immediate Service Level: Full access to normal traffic available almost immediately (following the earthquake).
Repairable Damage: Damage that can be repaired with a minimum risk of losing functionality.
Limited Service Level: Limited access (reduced lanes, light emergency traffic) possible with in days. Full service restoration within months.
Significant Damage: A minimum risk of collapse, but damage that would require closure for repairs.

Note: Above performance criteria and definitions have been modified slightly in the proposed provisions for California Bridges (ACT-32, 1996) and the U.S. Bridges (ATC-18, 1997) and Caltrans (1999) MTD 20-1 (920).

37.4 Performance-Based Design Approaches

37.4.1 Caltrans Practice

Since 1989, the design criteria specified in Caltrans BDS [26] and several internal design manuals [20,25,27] have been updated continuously to reflect recent research findings and development in the field of seismic bridge design. Caltrans has been shifting toward a displacement-based design approach emphasizing capacity design. In 1994 Caltrans established the seismic performance criteria listed in Table 37.4. A bridge is categorized as an "important" or "ordinary" bridge. Project-specific two-level seismic design procedures for important bridges, such as the R-14/I-5 Interchange replacement [16], the San Francisco–Oakland Bay Bridge (SFOBB) [17], and the Benicia-Martinez Bridge [28], are required and have been developed. These performance-based seismic design criteria include site-specific ARS curves, ground motions, and specific design procedures to reflect the desired performance of these structures. For ordinary bridges, only one-level safety-evaluation design is required. The following section briefly discusses the newly developed seismic design methodology for ordinary bridges.

37.4.2 New Caltrans Seismic Design Methodology (MTD 20-1, 1999)

To improve Caltrans seismic design practice and consolidate new research findings, ATC-32 recommendations [18] and the state-of-the-art knowledge gained from the recent extensive seismic bridge design, Caltrans engineers have been developing the Seismic Design Methodology [20] and the Seismic Design Criteria (SDC) [43] for ordinary bridges.

Ordinary Bridge Category

An ordinary bridge can be classified as a "standard" or "nonstandard" bridge. An nonstandard bridge may feature irregular geometry and framing (multilevel, variable width, bifurcating, or highly horizontally curved superstructures, different structure types, outriggers, unbalanced mass and/or stiffness, high skew) and unusual geologic conditions (soft soil, moderate to high liquefaction potential, and proximity to an earthquake fault). A standard bridge does not contain nonstandard features. The performance criteria and the service and damage levels are shown in Table 37.4.

Basic Seismic Design Concept

The objective of seismic design is to ensure that all structural components have sufficient strength and/or ductility to prevent collapse — a limit state where additional deformation will potentially render a bridge incapable of resisting its self-weight during a maximum credible earthquake (MCE). Collapse is usually characterized by structural material failure and/or instability in one or more components.

Ductility is defined as the ratio of ultimate deformation to the deformation at first yield and is the predominant measure of structural ability to dissipate energy. Caltrans takes advantage of ductility and postelastic strength and does not design ordinary bridges to remain elastic during design earthquakes because of economic constraints and the uncertainties in predicting future seismic demands. Seismic deformation demands should not exceed structural deformation capacity or energy-dissipating capacity. Ductile behavior can be provided by inelastic actions either through selected structural members and/or through protective systems — seismic isolations and energy dissipation devices. Inelastic actions should be limited to the predetermined regions that can be easily inspected and repaired following an earthquake. Because the inelastic response of a concrete superstructure is difficult to inspect and repair and the superstructure damage may cause the bridge to be in an unserviceable condition, inelastic behavior on most bridges should preferably be located in columns, pier walls, backwalls, and wingwalls (see Figure 38.1).

To provide an adequate margin of strength between ductile and nonductile failure modes, capacity design is achieved by providing overstrength against seismic load in superstructure and foundations. Components not explicitly designed for ductile performance should be designed to remain essentially elastic; i.e., response in concrete components should be limited to minor cracking or limited to force demands not exceeding the strength capacity determined by current Caltrans SDC, and response in steel components should be limited to force demands not exceeding the strength capacity determined by current Caltrans SDC.

Displacement-Based Design Approach

The objective of this approach is to ensure that the structural system and its individual components have enough capacity to withstand the deformation imposed by the design earthquake. Using displacements rather than forces as a measurement of earthquake damage allows a structure to fulfill the required functions.

In a displacement-based analysis, proportioning of the structure is first made based on strength and stiffness requirements. The appropriate analysis is run and the resulting displacements are compared with the available capacity which is dependent on the structural configuration and rotational capacity of plastic hinges and can be evaluated by inelastic static push-over analysis (see Chapter 36). This procedure has been used widely in seismic bridge design in California since 1994. Alternatively, a target displacement could be specified, the analysis performed, and then design strength and stiffness determined as end products for a structure [29,30]. In displacement-based design, the designer needs to define criteria clearly for acceptable structural deformation based on postearthquake performance requirements and the available deformation capacity. Such criteria are based on many factors, including structural type and importance.

Seismic Demands on Structural Components

For ordinary bridges, safety-evaluation ground motion shall be based on deterministic assessment corresponding to the MCE, the largest earthquake which is capable of occurring based on current geologic information. The ARS curves (Figure 37.5) developed by ATC-32 are adopted as standard horizontal ARS curves in conjunction with the peak rock acceleration from the Caltrans Seismic Hazard Map 1996 to determine the horizontal earthquake forces. Vertical acceleration should be considered for bridges with nonstandard structural components, unusual site conditions, and/or close proximity to earthquake faults and can be approximated by an equivalent static vertical force applied to the superstructure.

For structures within 15 km of an active fault, the spectral ordinates of the appropriate standard ARS curve should be increased by 20%. For long-period structures ($T \geq 1.5$ s) on deep soil sites (depth of alluvium ≥ 75 m) the spectral ordinates of the appropriate standard ARS curve should be increased by 20% and the increase applies to the portion of the curves with periods greater than 1.5 s.

Displacement demands should be estimated from a linear elastic response spectra analysis of bridges with effective component stiffness. The effective stiffness of ductile components should represent the actual secant stiffness of the component near yield. The effective stiffness should include the effects of concrete cracking, reinforcement, and axial load for concrete components; residual stresses, out-of-straightness, and axial load for steel components; the restraints of the surrounding soil for pile shafts. Attempts should be made to design bridges with dynamic characteristics (mass and stiffness) so that the fundamental period falls within the region between 0.7 and 3 s where the equal displacement principle applies. It is also important that displacement demands also include the combined effects of multidirectional components of horizontal acceleration (for example, 30% rules).

For short-period bridges, linear elastic analysis underestimates displacement demands. The inability to predict displacements of a linear analysis accurately can be overcome by designing the bridge to perform elastically, multiplying the elastic displacement by an amplification factor, or using seismic isolation and energy dissipation devices to limit seismic response. For long-period ($T > 3$ s) bridges, a linear elastic analysis generally overestimates displacements and linear elastic displacement response spectra analysis should be used.

Force demands for essentially elastic components adjacent to ductile components should be determined by the joint–force equilibrium considering plastic hinging capacity of the ductile component multiplied by an overstrength factor. The overstrength factor should account for the variations in material properties between adjacent components and the possibility that the actual strength of the ductile components exceeds its estimated plastic capacity. Force demands calculated from a linear elastic analysis should not be used.

Seismic Capacity of Structural Components

Strength and deformation capacity of a ductile flexural element should be evaluated by moment–curvature analysis (see Chapters 36 and 38). Strength capacity of all components should be based on the most probable or expected material properties, and anticipated damages. The impact of the second-order P-Δ and P-δ effects on the capacity of all members subjected to combined bending and compression should be considered. Components may require re-design if the P-Δ and P-δ effects are significant.

Displacement capacity of a bridge system should be evaluated by a static push-over analysis (see Chapter 36). The rotational capacity of all plastic hinges should be limited to a safe performance level. The plastic hinge regions should be designed and detailed to perform with minimal strength degradation under cyclic loading.

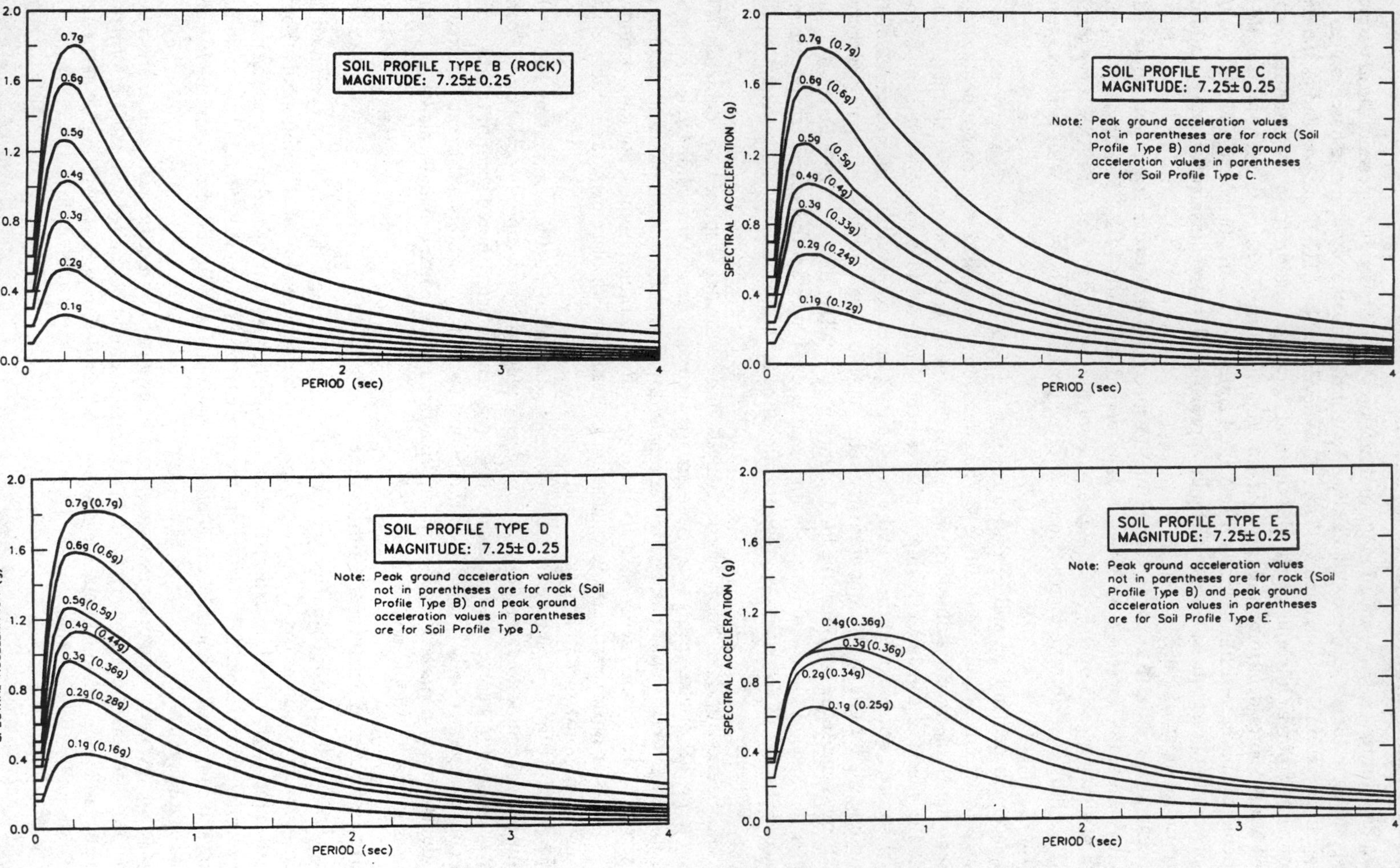

FIGURE 37.5 ATC-32 recommended ARS curves.

Seismic Design Practice

- Bridge type, component selection, member dimensions, and aesthetics should be investigated to reduce the seismic demands to the greatest extent possible. Aesthetics should not be the primary reason for producing undesirable frame and component geometry.
- Simplistic analysis models should be used for initial assessment of structural behavior. The results of more-sophisticated models should be checked for consistency with the results obtained from the simplistic models. The rotational and translational stiffness of abutments and foundations modeled in the seismic analysis must be compatible with their structural and geotechnical capacity. The energy dissipation capacity of the abutments should be considered for bridges whose response is dominated by the abutments.
- The estimated displacement demands under design earthquake should not exceed the global displacement capacity of the structure and the local displacement capacity of any of its individual components.
- Adjacent frames should be proportioned to minimize the differences in the fundamental periods and skew angles, and to avoid drastic changes in stiffness. All bridge frames must meet the strength and ductility requirements in a stand-alone condition. Each frame should provide a well-defined load path with predetermined plastic hinge locations and utilize redundancy whenever possible.
- For concrete bridges, structural components should be proportioned to direct inelastic damage into the columns, pier walls, and abutments. The superstructure should have sufficient overstrength to remain essentially elastic if the columns/piers reach their most probable plastic moment capacity. The superstructure-to-substructure connection for nonintegral caps may be designed to fuse prior to generating inelastic response in the superstructure. The girders, bent caps, and columns should be proportioned to minimize joint stresses. Moment-resisting connections should have sufficient joint shear capacity to transfer the maximum plastic moments and shears without joint distress.
- For steel bridges, structural components should be generally designed to ensure that inelastic deformation only occur in the specially detailed ductile substructure elements. Inelastic behavior in the form of controlled damage may be permitted in some of the superstructure components, such as the cross frames, end diaphragms, shear keys, and bearings. The inertial forces generated by the deck must be transferred to the substructure through girders, trusses, cross frames, lateral bracings, end diaphragms, shear keys, and bearings. As an alternative, specially designed ductile end-diaphragms may be used as structural mechanism fuses to prevent damage in other parts of structures.
- Initial sizing of columns should be based on slenderness ratios, bent cap depth, compressive stress ratio, and service loads. Columns should demonstrate dependable post-yield-displacement capacity without an appreciable loss of strength. Thrust–moment–curvature (P–M–Φ) relationships should be used to optimize the performance of a column under service and seismic loads. Concrete columns should be well proportioned, moderately reinforced, and easily constructed. Abrupt changes in the cross section and the capacity of columns should be avoided. Columns must have sufficient rotation capacity to achieve the target displacement ductility requirements.
- Steel multicolumn bents or towers should be designed as ductile moments-resisting frames (MRF) or ductile braced frames such as concentrically braced frames (CBF) and eccentrically braced frames (EBF). For components expected to behave inelastically, elastic buckling (local compression and shear, global flexural, and lateral torsion) and fracture failure modes should be avoided. All connections and joints should preferably be designed to remain essentially elastic. For MRFs, the primary inelastic deformation should preferably be columns. For CBFs, diagonal members should be designed to yield when members are in tension and to buckle inelastically when they are in compression. For EBFs, a short beam segment designated as a *link* should be well designed and detailed.

TABLE 37.5 ATC-32 Minimum Required Analysis

Bridge Type		Functional Evaluation	Safety Evaluation
Ordinary Bridge	Type I	None required	Equivalent static analysis or elastic dynamic analysis
	Type II	None required	Elastic dynamic analysis
Important Bridge	Type I	Equivalent static analysis or elastic dynamic analysis	
	Type II	Elastic dynamic analysis	Elastic dynamic analysis or inelastic static analysis or inelastic dynamic analysis

- Force demands on the foundation should be based on the most probable plastic capacity of the columns/piers with an appropriate amount of overstrength. Foundation elements should be designed to remain essentially elastic. Pile shaft foundations may experience limited inelastic deformation when they are designed and detailed in a ductile manner.
- The ability of an abutment to resist bridge seismic forces should be based on its structural capacity and the soil resistance that can be reliably mobilized. Skewed abutments are highly vulnerable to damage. Skew angles at abutments should be reduced, even at the expense of increasing the bridge length.
- Necessary restrainers and sufficient seat width should be provided between adjacent frames at all intermediate expansion joints, and at the seat-type abutments to eliminate the possibility of unseating during a seismic event.

37.4.3 ATC Recommendations

ATC-32 Recommendations to Caltrans

The Caltrans seismic performance criteria shown in Table 37.4 provide the basis for development of the ATC-32 recommendations [18]. The major changes recommended for the Caltrans BDS are as follows:

- The importance of relative (rather than absolute) displacement in the seismic performance of bridges is emphasized.
- Bridges are classified as either "important or ordinary." Structural configurations are divided into Type I, simple (similar to regular bridges), and Type II, complex (similar to irregular bridges). For important bridges, two-level design (safety evaluation and function evaluation) approaches are recommended. For ordinary bridges, a single-level design (safety evaluation) is recommended. Minimum analyses required are shown in Table 37.5.
- The proposed family of site-dependent design spectra (which vary from the current Caltrans curves) are based on four of six standard sites defined in a ground motion workshop [31].
- Vertical earthquake design loads may be taken as two thirds of the horizontal load spectra for typical sites not adjacent to active faults.
- A force-based design approach is retained, but some of the inherent shortcomings have been overcome by using new response modification factors and modeling techniques which more accurately estimate displacements. Two new sets of response modification factors Z (Figure 38.4b) are recommended to represent the response of limited and full ductile structural components. Two major factors are considered in the development of the new Z factors: the relationship between elastic and inelastic response is modeled as a function of the natural period of the structure and the predominate period of the ground motion; the distribution of elastic and inelastic deformation within a structural component is a function of its component geometry and framing configuration.

- *P*-Δ effects should be included using inelastic dynamic analysis unless the following relation is satisfied:

$$\frac{V_o}{W} \geq 4\frac{\delta_u}{H} \tag{37.2}$$

where V_o is base shear strength of the frame obtained from plastic analysis; W is the dead load; δ_u is maximum design displacement; and H is the height of the frame. The inequality in Eq. (37.2) is recommended to keep bridge columns from being significantly affected by *P*-Δ moments.

- A adjustment factor, R_d, is recommended to adjust the displacement results from an elastic dynamic analysis to reflect the more realistic inelastic displacements that occur during an earthquake.

$$R_d = \left(1 - \frac{1}{Z}\right)\frac{T}{T^*} + \frac{1}{Z} \geq 1 \tag{37.3}$$

where T is the natural period of the structure, T^* is the predominant period of ground motion, and Z is force-reduction coefficient defined in Figure 38.4b.

- Modification was made to the design of ductile elements, the design of nonductile elements using capacity design approach, and the detailing of reinforced concrete for seismic resistance based on recent research findings.
- Steel seismic design guidelines and detailing requirements are very similar to building code requirements.
- Foundation design guidelines include provisions for site investigation, determination of site stability, modeling and design of abutments and wing-walls, pile and spread footing foundations, drilled shafts, and Earth-retaining structures.

ATC-18 Recommendation to FHWA

The ATC recently reviewed current seismic design codes and specifications for highway structures worldwide and provided recommendations for future codes for bridge structures in the United States [19]. The recommendations have implemented significant changes to current specifications, most importantly the two-level design approach, but a single-level design approach is included. The major recommendations are summarized in Tables 37.6 and 37.7.

37.5 Sample Performance-Based Criteria

This section introduces performance-based criteria as a reference guide. A complete set of criteria will include consideration of postearthquake performance criteria, determination of seismic loads and load combinations, material properties, analysis methods, detailed qualitative acceptance criteria. The materials presented in this section are based on successful past experience, various codes and specifications, and state-of-the-art knowledge. Much of this section is based on the Seismic Retrofit Design Criteria developed for the SFOBB west span [17]. It should be emphasized that the sample criteria provided here should serve as a guide and are not meant to encompass all situations.

The postearthquake performance criteria depending on the importance of bridges specified in Table 37.4 are used. Two levels of earthquake loads, FEGM and SEGM, defined in Table 37.4 are required. The extreme event load combination specified by AASHTO-LRFD [23] should be considered (see Chapter 5).

TABLE 37.6 ATC-18 Recommendations for Future Bridge Seismic Code Development (Two-Level Design Approach)

Level		Lower Level Functional Evaluation	Upper Level Safety Evaluation
Performance Criteria	Ordinary bridges	Service level — immediate Damage level — repairable	Service level — limited Damage level — significant
	Important bridges	Service level — immediate Damage level — minimum	Service level — immediate Damage level — repairable
Design load		Functional evaluation ground motion	Safety evaluation ground motion
Design approach		• Continue current AASHTO seismic performance category • Adopt the two-level design approach at least for important bridges in higher seismic zones • Use elastic design principles for the lower-level design requirement • Use nonlinear analysis — deformation-based procedures for the upper-level design	
Analysis		Current elastic analysis procedures (equivalent static and multimodel)	Nonlinear static analysis
Design force	Ductile component	Remain undamaged	Have adequate ductility to meet the performance criteria
	Nonductile component	Remain undamaged	For sacrificial element — ultimate strength should be close to but larger than that required for the lower-level event For nonsacrificial element — based on elastic demands or capacity design procedure
	Foundation	Capacity design procedure — to ensure there is no damage	
Design displacement		Use the upper-level event Remain current seat width requirements Consider overall draft limits to avoid P-Δ effects on long-period structures	
Concrete and steel design		Use the capacity design procedure for all critical members	
Foundation design		• Complete geotechical analysis for both level events • Prevent structural capacity of the foundations at the lower level event • Allow damage in the upper-level event as long as it does not lead to catastrophic failure	

Functional Evaluation Ground Motion (*FEGM*): Probabilistic assessed ground motions that have a 72 ~ 250 year return period (i.e., 30 to 50% probability of exceedance during the useful life a bridge).

Safety Evaluation Ground Motion (*SEGM*): Probabilistic assessed ground motions that have a 950 or 2475 year return period (10% probability of exceedance for a design life of 100 ~ 250 years).

Immediate Service Level: Full access to normal traffic is available almost immediately (i.e., within hours) following the earthquake (It may be necessary to allow 24 h or so for inspection of the bridge).

Limited Service Level: Limited access (reduced lanes, light emergency traffic) is possible within 3 days of the earthquake. Full service restoration within months.

Minimum Damage: Minor inelastic deformation such as narrow flexural cracking in concrete and no apparent deformations.

Repairable Damage: Damage such as concrete cracking, minor spalling of cover concrete, and steel yield that can be repaired without requiring closure and replacing structural members. Permanent offsets are small.

Significant Damage: Damage such as concrete cracking, major spalling of concrete, steel yield that can be repaired only with closure, and partial or complete replacement. Permanent offset may occur without collapse.

37.5.1 Determination of Demands

Analysis Methods

For ordinary bridges, seismic force and deformation demands may be obtained by equivalent static analysis or elastic dynamic response spectrum analysis. For important bridges, the following guidelines may apply:

1. Static linear analysis should be used to determine member forces due to self-weight, wind, water currents, temperature, and live load.
2. Dynamic response spectrum analysis [32] should be used for local and regional stand-alone models and the simplified global model to determine mode shapes, periods, and initial estimates of seismic force and displacement demands. The analysis may be used on global models prior to a time history analysis to verify global behavior, eliminate modeling errors,

TABLE 37.7 ATC-18 Recommendations for Future Bridge Seismic Code Development (One-Level Approach)

Design philosophy		For lower-level earthquake, there should be only minimum damage For a significant earthquake, collapse should be prevented but significant damage may occur; damage should occur at visible locations The following addition to Item 2 is required if different response modification (R and Z) factors are used for important or ordinary bridges Item 2 as it stands would apply to ordinary bridges For important bridges, only repairable would be expected during a significant earthquake
Design load		Single-level — safety evaluation ground motion — 950 or 2475 year return period for the eastern and western portions of the U.S.
Design approach		• Continue current AASHTO seismic performance category • Use nonlinear analysis deformation-based procedures with strength and stiffness requirements being derived from appropriate nonlinear response spectra
Analysis		Nonlinear static analysis should be part of any analysis requirement At a minimum, nonlinear static analysis is required for important bridges Current elastic analysis and design procedure may be sufficient for small ordinary bridges Incorporate both current R-factor elastic procedure and nonlinear static analysis
Design Force	Ductile component	R-factor elastic design procedure or nonlinear static analysis
	Nonductile component	For sacrificial element, should be designed using a guideline that somewhat correspond to the design level of an unspecified lower-level event, for example, one half or one third of the force required for the upper-level event For nonsacrificial element, should be designed for elastic demands or capacity design procedure
	Foundation	Capacity design procedure — to ensure there is no damage
Design displacement		Maintain current seat width requirements Consider overall draft limits to avoid P-Δ effects on long-period structures
Concrete and steel design		Use the capacity design procedure for all critical members
Foundation design		• Complete geotechical analysis for the upper-level event • For nonessential bridges, a lower level (50% of the design acceleration) might be appropriate

and identify initial regions or members where inelastic behavior needs further refinement and inelastic nonlinear elements. In the analysis:

- Site-specific ARS curves should be used with 5% damping.
- Modal response should be combined using the complete quadratic combination (CQC) method and the resulting orthogonal responses should be combined using either the square root of the sum of the squares (SRSS) method or the "30%" rule as defined by AASHTO-LRFD [1994].

3. Dynamic Time History Analysis: Site-specific multisupport dynamic time histories should be used in a dynamic time history analysis [33].
   - Linear elastic dynamic time history analysis is defined as a dynamic time history analysis with consideration of geometric linearity (small displacement), linear boundary conditions, and elastic members. It should only be used to check regional and global models.
   - Nonlinear elastic dynamic time history analysis is defined as a dynamic time history analysis with consideration of geometric nonlinearity, linear boundary conditions, and elastic members. It should be used to determine areas of inelastic behavior prior to incorporating inelasticity into regional and global models.
   - Nonlinear inelastic dynamic time history analysis, level I, is defined as a dynamic time history analysis with consideration of geometric nonlinearity, nonlinear boundary conditions, inelastic elements (for example, seismic isolators and dampers), and elastic members. It should be used for final determination of force and displacement demands for existing structures in combination with static gravity, wind, thermal, water current, and live loads as specified in AASHTO-LRFD [23].

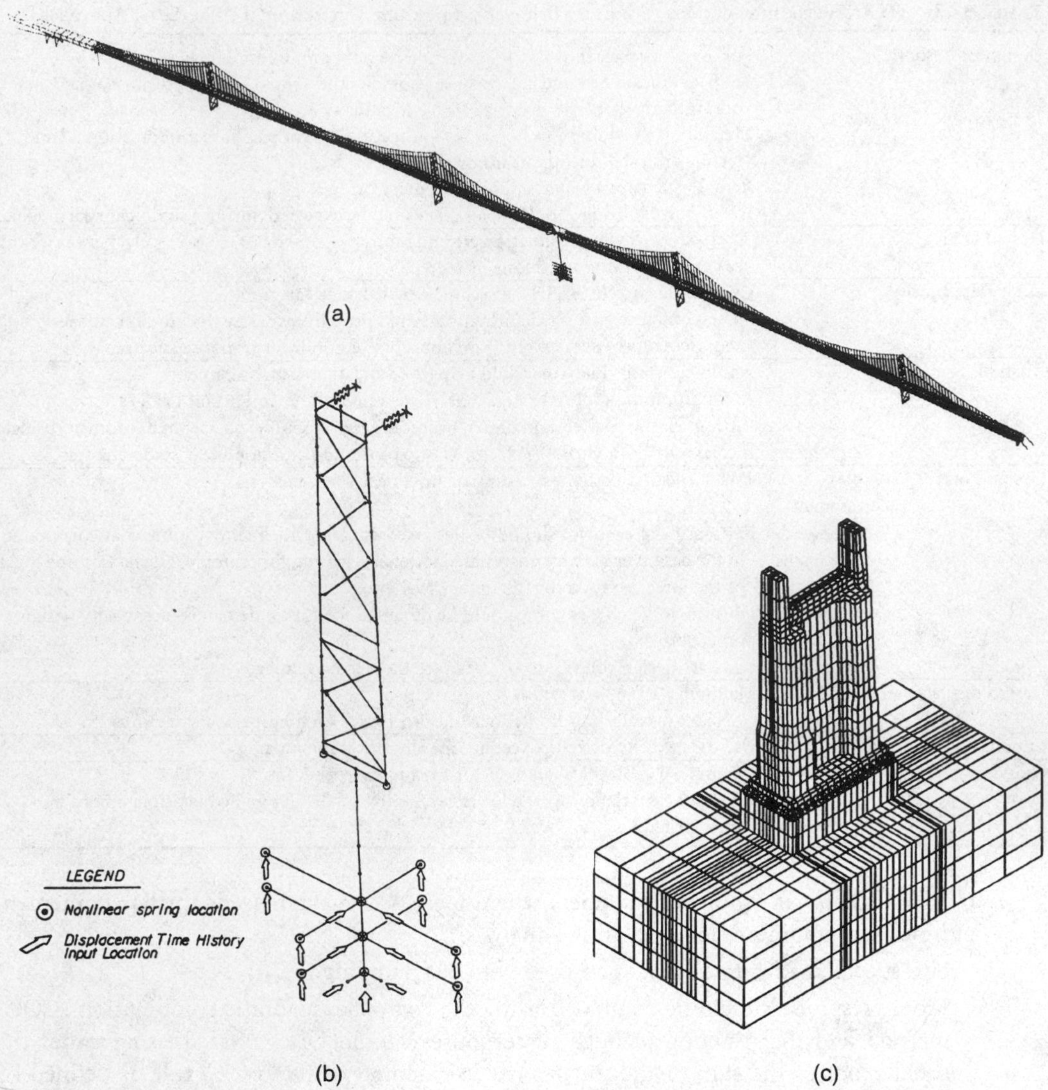

FIGURE 37.6 (a) Global, (b) Regional models for towers, and (c) local model for PW-1 for San Francisco–Oakland Bay Bridge west spans.

- Nonlinear inelastic dynamic time history analysis, level II, is defined as a dynamic time history analysis with consideration of geometric nonlinearity, nonlinear boundary conditions, inelastic elements (for example, dampers), and inelastic members. It should be used for the final evaluation of response of the structures.

Modeling Considerations

1. *Global, Regional, and Local Models*
   The global models consider overall behavior and may include simplifications of complex structural elements (Figure 37.6a). Regional models concentrate on regional behavior (Figure 37.6b). Local models (Figure 37.6c) emphasize the localized behavior, especially complex inelastic and nonlinear behavior. In regional and global models where more than one foundation location is included in the model, multisupport time history analysis should be used.

2. *Boundary Conditions*
   Appropriate boundary conditions should be included in regional models to represent the interaction between the region and the adjacent structure. The adjacent portion is not explicitly modeled but may be simplified using a combination of springs, dashpots, and lumped masses. Appropriate nonlinear elements such as gap elements, nonlinear springs, seismic response modification devices (SRMDs), or specialized nonlinear finite elements should be included where the behavior and response of the structure is sensitive to such elements.
3. Soil–Foundation–Structure Interaction
   This interaction may be considered using nonlinear or hysteretic springs in global and regional models. Foundation springs to represent the properties of the soil at the base of the structure should be included in both regional and global models (see Chapter 42).
4. Damping
   When nonlinear material properties are incorporated in the model, Rayleigh damping should be reduced (perhaps 20%) from the elastic properties.
5. Seismic Response Modification Devices
   The SRMDs should be modeled explicitly with hysteretic characteristics determined by experimental data. See Chapter 41 for a detailed discussion of this behavior.

37.5.2 Determination of Capacities

Limit States and Resistance Factors

The *limit state* is defined as that condition of a structure at which it ceases to satisfy the provisions for which it was designed. Two kinds of limit state corresponding to SEGM and FEGM specified in Table 37.4 apply for seismic design and retrofit. To account for unavoidable inaccuracies in the theory, variation in the material properties, workmanship, and dimensions, nominal strength of structural components should be modified by a resistance factor ϕ specified by AASHTO-LRFD [23] or project-specific criteria to obtain the design capacity or strength (resistance).

Nominal Strength of Structural Components

The strength capacity of structural members should be determined in accordance with specified code formula [23,26, Chapters 38 and 39], or verified with experimental and analytical computer models, or project-specific criteria [19].

Structural Deformation Capacity

Structural deformation capacity should be determined by nonlinear inelastic analysis and based on acceptable damage levels as shown in Table 37.4. The quantitative definition of the damage corresponding to different performance requirements has not been specified by the current Caltrans BDS [26], AASHTO-LRFD [23], and ATC recommendations [18,19] because of the lack of consensus. As a starting point, Table 37.8 provides a quantitative strain and ductility limit corresponding to the three damage levels.

The displacement capacity should be evaluated considering both material and geometric nonlinearities. Proper boundary conditions for various structures should be carefully considered. A static push-over analysis (see Chapter 36) may be suitable for most bridges. A nonlinear inelastic dynamic time history analysis, Level II, may be required for important bridges. The available displacement capacity is defined as the displacement corresponding to the most critical of (1) 20% load reduction from the peak load or (2) the strain limit specified in Table 37.8.

Seismic Response Modification Devices

SRMDs include energy dissipation and seismic isolation devices. Energy dissipation devices increase the effective damping of the structure, thereby reducing reaction forces and deflections. Isolation devices change the fundamental mode of vibration so that the response of the structure is lowered; however, the reduced force may be accompanied by an increased displacement.

TABLE 37.8 Damage Levels, Strain, and Ductility

	Strain		Ductility	
Damage level	Concrete	Steel	Curvature μ_ϕ	Displacement μ_Δ
Significant	ε_{cu}	ε_{sh}	8 ~ 10	4 ~ 6
Repairable	Larger $\begin{cases} 0.005 \\ \dfrac{2\varepsilon_{cu}}{3} \end{cases}$	Larger $\begin{cases} 0.08 \\ \dfrac{2\varepsilon_{y}}{3} \end{cases}$	4 ~ 6	2 ~ 4
Minimum	Larger $\begin{cases} 0.004 \\ \varepsilon_{cu} \end{cases}$	Larger $\begin{cases} 0.03 \\ 15\varepsilon_{y} \end{cases}$	2 ~ 4	1 ~ 2

ε_{cu} = ultimate concrete compression strain depending of confinement (see Chapter 36)
ε_y = yield strain of steel
ε_{sh} = hardening strain of steel
μ_ϕ = curvature ductility (ϕ_u/ϕ_y)
μ_Δ = displacement ductility (Δ_u/Δ_y) (see Chapter 36)

The properties of SRMDs should be determined by the specified testing program. References are made to AASHTO [34], Caltrans [35], and Japan Ministry of Construction (JMC) [36]. Consideration of following items should be made in the test specifications:

- Scales — at least two full-scale test specimens are required;
- Loading (including lateral and vertical) history and rate;
- Durability — design life;
- Deterioration — expected levels of strength and stiffness.

37.5.3 Performance Acceptance Criteria

To achieve the performance objectives in Table 37.4, various structural components should satisfy the acceptable demand/capacity ratios (DC_{accept}) specified in this section. The form of the equation is:

$$\frac{\text{Demand}}{\text{Capacity}} \le DC_{\text{accept}} \tag{37.4}$$

where *demand,* in terms of factored moments, shears, and axial forces, and displacement and rotation deformations, should be determined by a nonlinear inelastic dynamic time history analysis, level I, for important bridges, and dynamic response spectrum analysis for ordinary bridges defined in Section 37.5.1, and *capacity,* in terms of factored strength and deformation capacities, should be obtained according to Section 37.5.2.

Structural Component Classifications

Structural components are classified into two categories: *critical* or *other.* It is the aim that other components may be permitted to function as *fuses* so that the critical components of the bridge system can be protected during the functionality evaluation earthquake (FEE) and the safety evaluation earthquake (SEE). As an example, Table 37.9 shows structural component classifications and their definition for a suspension bridge.

TABLE 37.9 Structural Component Classification

Component Classification	Definition	Example (SFOBB West Spans)
Critical	Components on a critical path that carry bridge gravity load directly The loss of capacity of these components would have serious consequences on the structural integrity of the bridge	Suspension cables Continuous trusses Floor beams and stringers Tower legs Central anchorage A-Frame Piers W-1 and W2 Bents A and B Caisson foundations Anchorage housings Cable bents
Other	All components other than Critical	All other components

Note: Structural components include members and connections.

Steel Structures

1. *General Design Procedure*
   Seismic design of steel members should be in accordance with the procedure shown in Figure 37.7. Seismic retrofit design of steel members should be in accordance with the procedure shown in Figure 37.8.
2. *Connections*
   Connections should be evaluated over the length of the seismic event. For connecting members with force *D*/*C* ratios larger than 1.0, 25% greater than the nominal capacity of the connecting members should be used in connection design.
3. *General Limiting Slenderness Parameters and Width–Thickness Ratios*
   For all steel members (regardless of their force *D*/*C* ratios), the slenderness parameter for axial load dominant members (λ_c) and for flexural dominant members (λ_b) should not exceed the limiting values ($0.9\lambda_{cr}$ or $0.9\lambda_{br}$ for *critical*, λ_{cr} or λ_{br} for *Others*) shown in Table 37.10.
4. *Acceptable Force* D/C *Ratios and Limiting Values*
   Acceptable force *D*/*C* ratios, DC_{accept} and associated limiting slenderness parameters and width–thickness ratios for various members are specified in Table 37.10. For all members with *D*/*C* ratios larger than 1.0, slenderness parameters and width–thickness ratios should not exceed the limiting values specified in Table 37.10. For existing steel members with *D*/*C* ratios less than 1.0, width–thickness ratios may exceed λ_r specified in Table 37.11 and AISC-LRFD [37].

The following symbols are used in Table 37.10. M_u is the factored moment demand; P_u is the vactored axial force demand; M_n is the nominal moment strength of a member; P_n is the nominal axial strength of a member; λ is the width–thickness (b/t or h/t_w) ratio of a compressive element; $\lambda_c = (KL/r\pi)\sqrt{F_y/E}$, the slenderness parameter of axial load dominant members; $\lambda_b = L/r_y$, the slenderness parameter of flexural moment dominant members; $\lambda_{cp} = 0.5$, the limiting column slenderness parameter for 90% of the axial yield load based on AISC-LRFD [37] column curve; λ_{bp} is the limiting beam slenderness parameter for plastic moment for seismic design; $\lambda_{cr} = 1.5$, the limiting column slenderness parameter for elastic buckling based on AISC-LRFD [37] column curve; λ_{br} is the limiting beam slenderness parameter for elastic lateral torsional buckling;

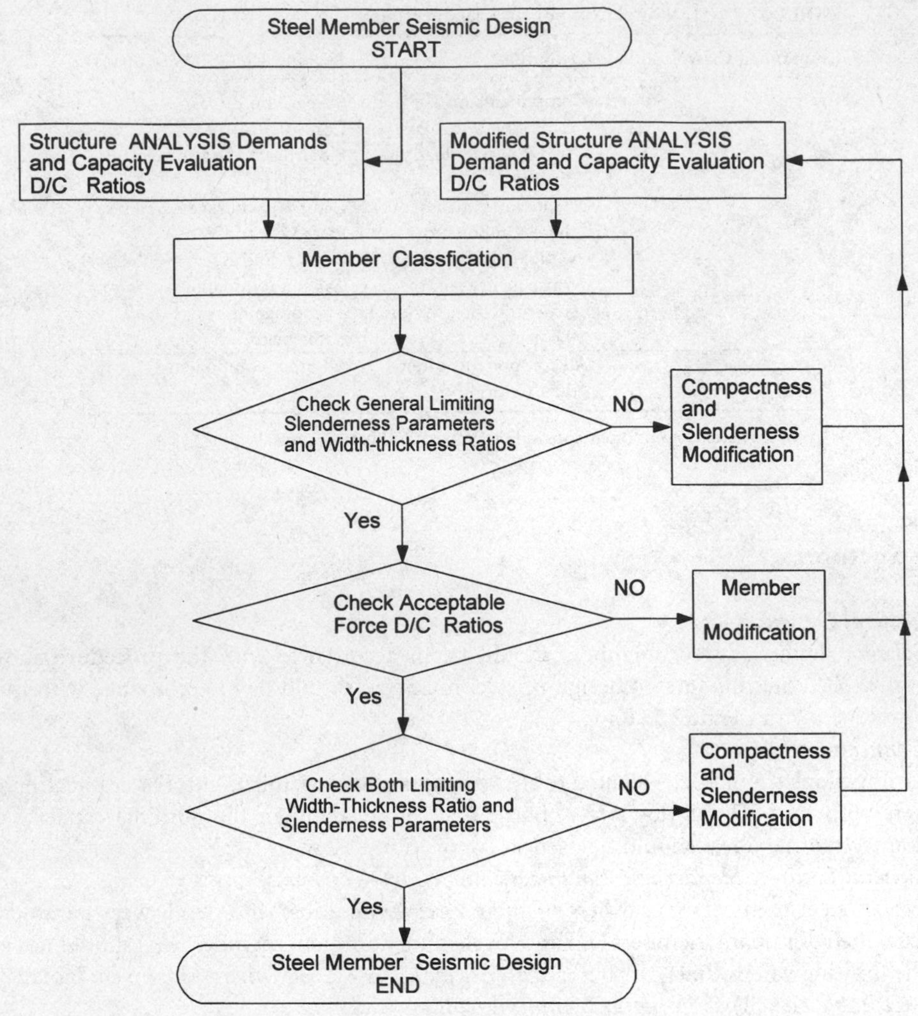

FIGURE 37.7 Steel member seismic design procedure.

$$\lambda_{br} = \begin{cases} \dfrac{57,000\sqrt{JA}}{M_r} & \text{for solid rectangular bars and box sections} \\ \dfrac{X_1}{F_L}\sqrt{1 + \sqrt{1 + X_2 F_L^2}} & \text{for doubly symmetric I-shaped members and channels} \end{cases}$$

$$M_r = \begin{cases} F_L\, S_x & \text{for I - shaped member} \\ F_{yf}\, S_x & \text{for solid rectangular and box section} \end{cases}$$

$$X_1 = \frac{\pi}{S_x}\sqrt{\frac{EGJA}{2}} \qquad X_s = \frac{4\,C_w}{I_y}\left(\frac{S_x}{G\,J}\right) \qquad F_L = \text{smaller}\begin{cases} F_{yw} \\ F_{yf} - F_r \end{cases}$$

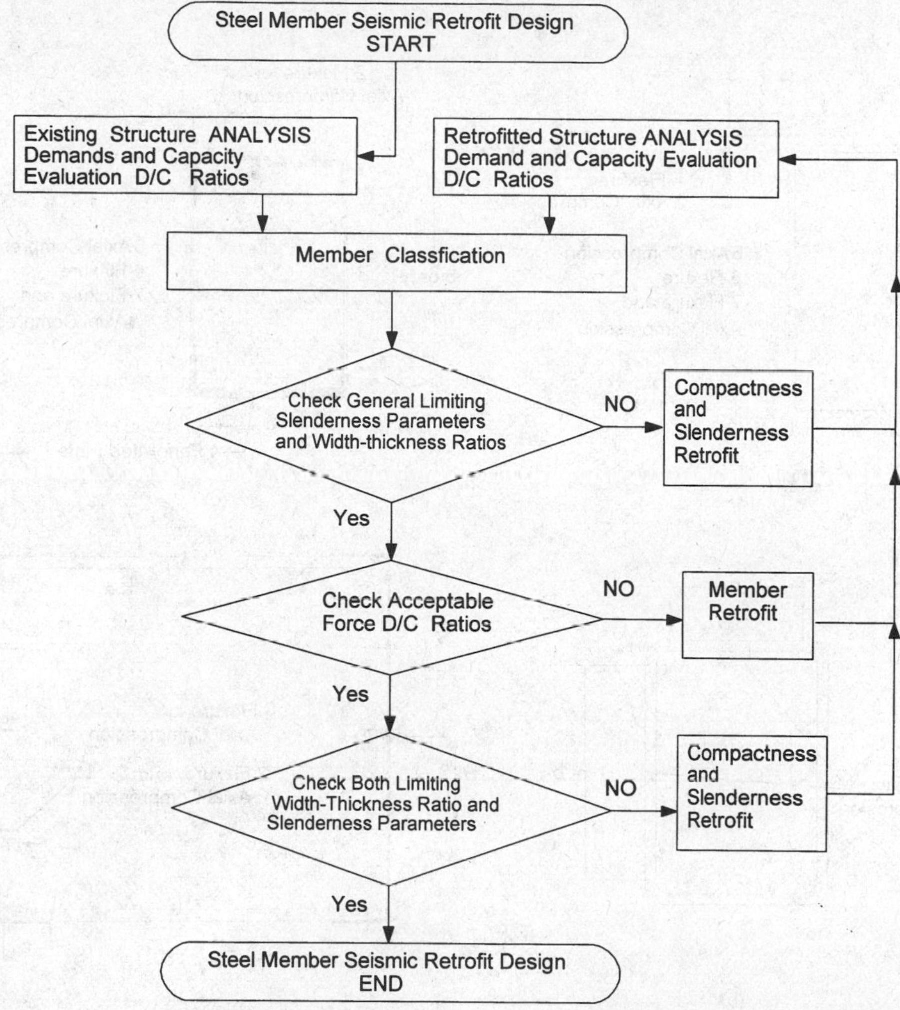

FIGURE 37.8 Steel member seismic retrofit design procedure.

where A is the cross-sectional area, in.^{2}; L is the unsupported length of a member; J is the torsional constant, in.^{4}; r is the radius of gyration, in.; r_y is the radius of gyration about minor axis, in.; F_y is the yield stress of steel; F_{yw} is the yield stress of web, ksi; F_{yf} is the yield stress of flange, ksi; E is the modulus of elasticity of steel (29,000 ksi); G is the shear modulus of elasticity of steel (11,200 ksi); S_x is the section modulus about major axis, in.^{3}; I_y is the moment of inertia about minor axis, in.^{4} and C_w is the warping constant, in.^{6} For doubly symmetric and singly symmetric I-shaped members with compression flange equal to or larger than the tension flange, including hybrid members (strong axis bending):

$$\lambda_{bp} = \begin{cases} \dfrac{\left[3600 + 2200\, M_1/M_2\right]}{F_y} & \text{for } \textit{other} \text{ members} \\[2ex] \dfrac{300}{\sqrt{F_{yf}}} & \text{for } \textit{critical} \text{ members} \end{cases} \tag{37.5}$$

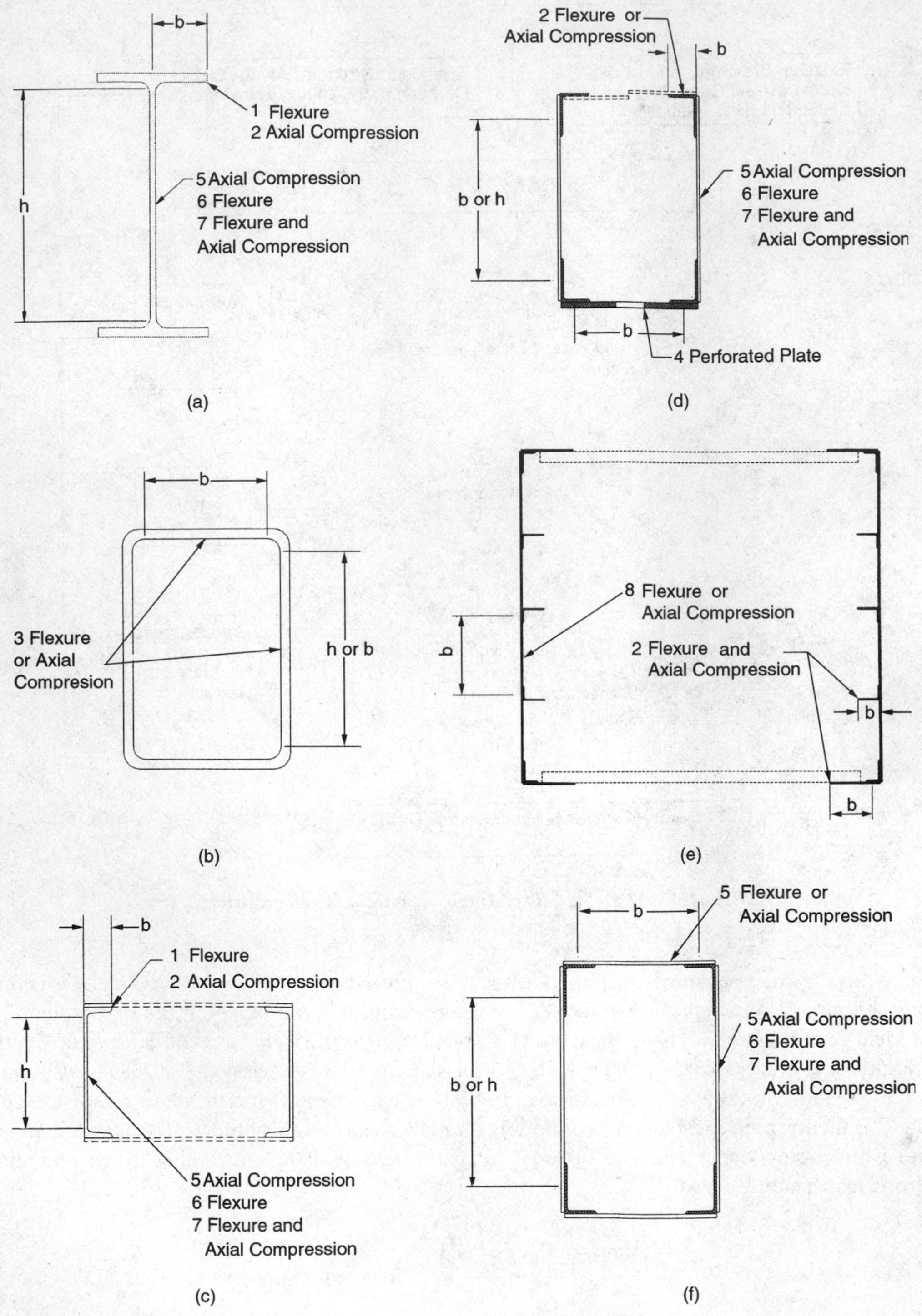

FIGURE 37.9 Typical cross sections for steel members: (a) rolled I section; (b) hollow structured tube; (c) built-up channels; (d) built-up box section; (e) longitudinally stiffened built-up box section; (f) built-up box section.

TABLE 37.10 Acceptable Force Demand/Capacity Ratios and Limiting Slenderness Parameters and Width/Thickness Ratios

Member Classification		Limiting Ratios: Slenderness Parameter (λ_c and λ_b)	Limiting Ratios: Width/Thickness λ (b/t or h/t_w)	Acceptable Force D/C Ratio DC_{accept}
Critical	Axial load dominant	$0.9\lambda_{cr}$	λ_r	$DC_r = 1.0$
	$P_u/P_n \geq M_u/M_n$	λ_{cpr}	λ_{pr}	$1.0 \sim 1.2$
		λ_{cp}	λ_p	$DC_p = 1.2$
	Flexural moment dominant	$0.9\lambda_{br}$	λ_r	$DC_r = 1.0$
	$M_u/M_n > P_u/P_n$	λ_{bpr}	λ_{pr}	$1.2 \sim 1.5$
		λ_{bp}	λ_p	$DC_p = 1.5$
Other	Axial load dominant	λ_{cr}	λ_r	$DC_r = 1.0$
	$P_u/P_n \geq M_u/M_n$	λ_{cpr}	λ_{pr}	$1.0 \sim 2.0$
		λ_{cp}	$\lambda_{p\text{-Seismic}}$	$DC_p = 2$
	Flexural moment dominant	λ_{hr}	λ_r	$DC_r = 1.0$
	$M_u/M_n > P_u/P_n$	λ_{bpr}	λ_{pr}	$1.0 \sim 2.5$
		λ_{bp}	$\lambda_{p\text{-Seismic}}$	$DC_p = 2.5$

in which M_1 is larger moment at end of unbraced length of beam; M_2 is smaller moment at end of unbraced length of beam; (M_1/M_2) is positive when moments cause reverse curvature and negative for single curvature.

For solid rectangular bars and symmetric box beam (strong axis bending):

$$\lambda_{bp} = \begin{cases} \dfrac{5000 + 3000\left(M_1/M_2\right)}{F_y} \geq \dfrac{3000}{F_y} & \text{for } \textit{other} \text{ members} \\[2ex] \dfrac{3750}{M_p}\sqrt{JA} & \text{for } \textit{critical} \text{ members} \end{cases} \tag{37.6}$$

in which M_p is plastic moment $(Z_x F_y)$; Z_x is plastic section modulus about major axis; and λ_r, λ_p, $\lambda_{p\text{–Seismic}}$ are limiting width thickness ratios specified by Table 37.11.

$$\lambda_{pr} = \begin{cases} \left[\lambda_p + \left(\lambda_r - \lambda_p\right)\left(\dfrac{DC_p - DC_{\text{accept}}}{DC_p - DC_r}\right)\right] & \text{for } \textit{critical} \text{ members} \\[2ex] \left[\lambda_{p\text{-Seismic}} + \left(\lambda_r - \lambda_{p\text{-Seismic}}\right)\left(\dfrac{DC_p - DC_{\text{accept}}}{DC_p - DC_r}\right)\right] & \text{for } \textit{other} \text{ members} \end{cases} \tag{37.7}$$

For axial load dominant members $(P_u/P_n \geq M_u/M_n)$

$$\lambda_{cpr} = \begin{cases} \lambda_{cp} + \left(0.9\lambda_{cr} - \lambda_{cp}\right)\left(\dfrac{DC_p - DC_{\text{accept}}}{DC_p - DC_r}\right) & \text{for } \textit{critical} \text{ members} \\[2ex] \lambda_{cp} + \left(\lambda_{cr} - \lambda_{cp}\right)\left(\dfrac{DC_p - DC_{\text{accept}}}{DC_p - DC_r}\right) & \text{for } \textit{other} \text{ members} \end{cases} \tag{37.8}$$

For flexural moment dominant members $(M_u/M_n > P_u/P_n)$

TABLE 37.11 Limiting Width-Thickness Ratios

No	Description of Elements	Examples	Width-Thickness Ratios	λ_r	λ_p	$\lambda_{p\text{-Seismic}}$
			Unstiffened Elements			
1	Flanges of I-shaped rolled beams and channels in flexure	Figure 37.19a Figure 37.19c	b/t	$\dfrac{141}{\sqrt{F_y - 10}}$	$\dfrac{65}{\sqrt{F_y}}$	$\dfrac{52}{\sqrt{F_y}}$
2	Outstanding legs of pairs of angles in continuous contact; flanges of channels in axial compression; angles and plates projecting from beams or compression members	Figure 37.19d Figure 37.19e Figure 37.19f	b/t	$\dfrac{95}{\sqrt{F_y}}$	$\dfrac{65}{\sqrt{F_y}}$	$\dfrac{52}{\sqrt{F_y}}$
			Stiffened Elements			
3	Flanges of square and rectangular box and hollow structural section of uniform thickness subject to bending or compression; flange cover plates and diaphragm plates between lines of fasteners or welds.	Figure 37.19b	b/t	$\dfrac{238}{\sqrt{F_y}}$	$\dfrac{190}{\sqrt{F_y}}$	$110/F_y$ (tubes) $150/F_y$ (others)
4	Unsupported width of cover plates perforated with a succession of access holes	Figure 37.19d	b/t	$\dfrac{317}{\sqrt{F_y}}$	$\dfrac{253}{\sqrt{F_y}}$	$\dfrac{152}{\sqrt{F_y}}$

5	All other uniformly compressed stiffened elements, i.e., supported along two edges.	Figures 37.19a, c,d,f	b/t h/t_w	$\frac{\mathbf{253}}{\sqrt{F_y}}$	$\frac{190}{\sqrt{F_y}}$	$110/\sqrt{F_y}$ (w/lacing) $150/\sqrt{F_y}$ (others)
6	Webs in flexural compression	Figures 37.19a,c,d,f	h/t_w	$\frac{\mathbf{970}}{\sqrt{F_y}}$	$\frac{\mathbf{640}}{\sqrt{F_y}}$	$\frac{\mathbf{520}}{\sqrt{F_y}}$
7	Webs in combined flexural and axial compression	Figures 37.19a,c,d,f	h/t_w	$\frac{\mathbf{970}}{\sqrt{F_y}} \times \left(1 - \frac{\mathbf{0.74}P}{\phi_b P_y}\right)$	For $P_u \le \mathbf{0.125}\,\phi_b P_y$ $\frac{\mathbf{640}}{\sqrt{F_y}}\left(1 - \frac{\mathbf{2.75}P}{\phi_b P_y}\right)$ For $P_u > \mathbf{0.125}\,\phi_b P_y$ $\frac{\mathbf{191}}{\sqrt{F_y}}\left(\mathbf{2.33} - \frac{P}{\phi_b P_y}\right)$ $\ge \frac{\mathbf{253}}{\sqrt{F_y}}$	For $P_u \le \mathbf{0.125}\,\phi_b P_y$ $\frac{\mathbf{520}}{\sqrt{F_y}}\left(1 - \frac{\mathbf{1.54}P}{\phi_b P_y}\right)$ For $P_u > \mathbf{0.125}\,\phi_b P_y$ $\frac{\mathbf{191}}{\sqrt{F_y}}\left(\mathbf{2.33} - \frac{P}{\phi_b P_y}\right)$ $\ge \frac{\mathbf{253}}{\sqrt{F_y}}$
8	Longitudinally stiffened plates in compression	Figure 37.19e	b/t	$\frac{113\sqrt{k}}{\sqrt{F_y}}$	$\frac{95\sqrt{k}}{\sqrt{F_y}}$	$\frac{75\sqrt{k}}{\sqrt{F_y}}$

Notes:

1. Width–thickness ratios shown in **bold** are from AISC-LRFD [1993] and AISC-Seismic Provisions [1997].
2. k = buckling coefficient specified by Article 6.11.2.1.3a of AASHTO-LRFD [AASHTO, 1994]
   for $n = 1$, $k = (8I_s/bt^3)^{1/3} \le 4.0$; for $n = 2, 3, 4$, and 5, $k = (14.3I_s/bt^3n^4)^{1/3} \le 4.0$
   n = number of equally spaced longitudinal compression flange stiffeners
   I_s = moment of inertia of a longitudinal stiffener about an axis parallel to the bottom flange and taken at the base of the stiffener

$$\lambda_{bpr}\begin{cases}\lambda_{bp}+\left(0.9\lambda_{br}-\lambda_{bp}\right)\left(\dfrac{DC_p-DC_{\text{accept}}}{DC_p-DC_r}\right) & \text{for } \textit{critical} \text{ members}\\[2ex] \lambda_{bp}+\left(\lambda_{br}-\lambda_{bp}\right)\left(\dfrac{DC_p-DC_{\text{accept}}}{DC_p-DC_r}\right) & \text{for } \textit{other} \text{ members}\end{cases} \tag{37.9}$$

Concrete Structures

1. *General*
   For all concrete compression members (regardless of D/C ratios), the slenderness parameter (KL/r) should not exceed 60.

   For *critical* components, force $DC_{\text{accept}} = 1.2$ and deformation $DC_{\text{accept}} = 0.4$.
   For *other* components, force $DC_{\text{accept}} = 2.0$ and deformation $DC_{\text{accept}} = 0.67$.

2. *Beam–Column (Bent Cap) Joints*
   For concrete box-girder bridges, the beam–column (bent cap) joints should be evaluated and designed in accordance with the following guidelines [38,39]:
   a. Effective Superstructure Width: The effective width of superstructure (box girder) on either side of a column to resist longitudinal seismic moment at bent (support) should not be taken as larger than the superstructure depth.
      - The immediately adjacent girder on either side of a column within the effective superstructure width is considered effective.
      - Additional girders may be considered effective if refined bent–cap torsional analysis indicates that the additional girders can be mobilized.
   b. Minimum Bent–Cap Width: Minimum cap width outside column should not be less than $D/4$ (D is column diameter or width in that direction) or 2 ft (0.61 m).
   c. Acceptable Joint Shear Stress:
      - For existing unconfined joints, acceptable principal tensile stress should be taken as $3.5\sqrt{f'_c}$ psi $\left(0.29\sqrt{f'_c}\text{ MPa}\right)$. If the principal tensile stress demand exceeds this limiting value, the joint shear reinforcement specified in Item d should be provided.
      - For new joints, acceptable principal tensile stress should be taken as $12\sqrt{f'_c}$ psi ($1.0\sqrt{f'_c}$ MPa).
      - For existing and new joints, acceptable principal compressive stress shall be taken as f'_c.
   d. Joint Shear Reinforcement
      - Typical flexure and shear reinforcement (see Figures 37.10 and 37.11) in bent caps should be supplemented in the vicinity of columns to resist joint shear. All joint shear reinforcement should be well distributed and provided within $D/2$ from the face of column.
      - Vertical reinforcement including cap stirrups and added bars should be 20% of the column reinforcement anchored into the joint. Added bars shall be hooked around main longitudinal cap bars. Transverse reinforcement in the join region should consist of hoops with a minimum reinforcement ratio of 0.4(column steel area)/(embedment length of column bar into the bent cap)^2.
      - Horizontal reinforcement should be stitched across the cap in two or more intermediate layers. The reinforcement should be shaped as hairpins, spaced vertically at not more than 18 in. (457 mm). The hairpins should be 10% of column reinforcement. Spacing should be denser outside the column than that used within the column.
      - Horizontal side face reinforcement should be 10% of the main cap reinforcement including top and bottom steel.

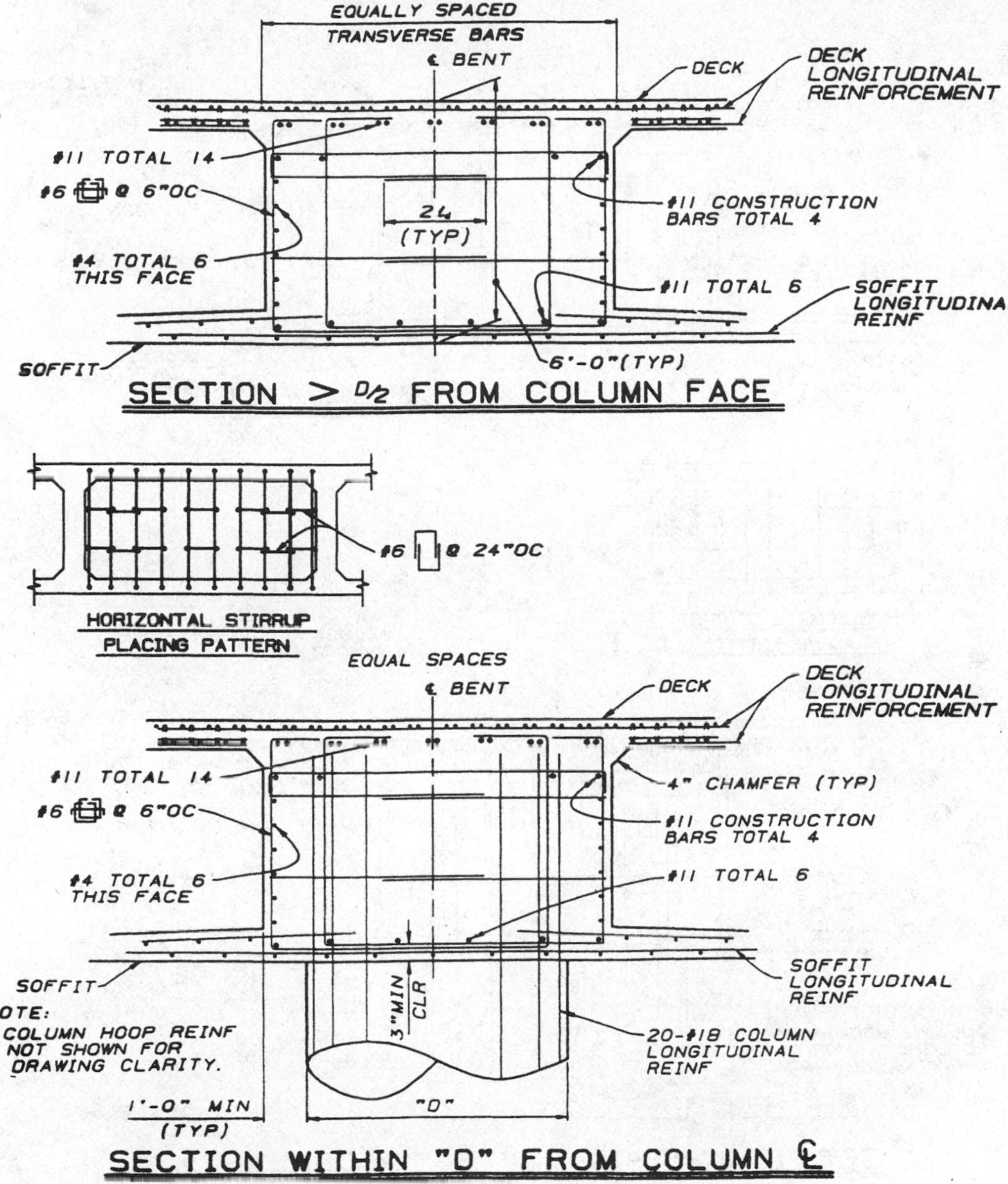

FIGURE 37.10 Example cap joint shear reinforcement — skews 0° to 20°.

- For bent caps skewed greater than 20°, the vertical J-bars hooked around longitudinal deck and bent cap steel should be 8% of column steel (see Figure 37.11). The J-bars should be alternatively 24 in. (600 mm) and 30 in. (750 mm) long and placed within a width of column dimension an either side of the column centerline.
- All vertical column bars should be extended as high as practically possible without interfering with the main cap bars.

Seismic Response Modification Devices

Analysis methods specified in Section 37.5.3 apply for determining seismic design forces and displacements on SRMDs. Properties or capacities of SRMDs should be determined by specified tests.

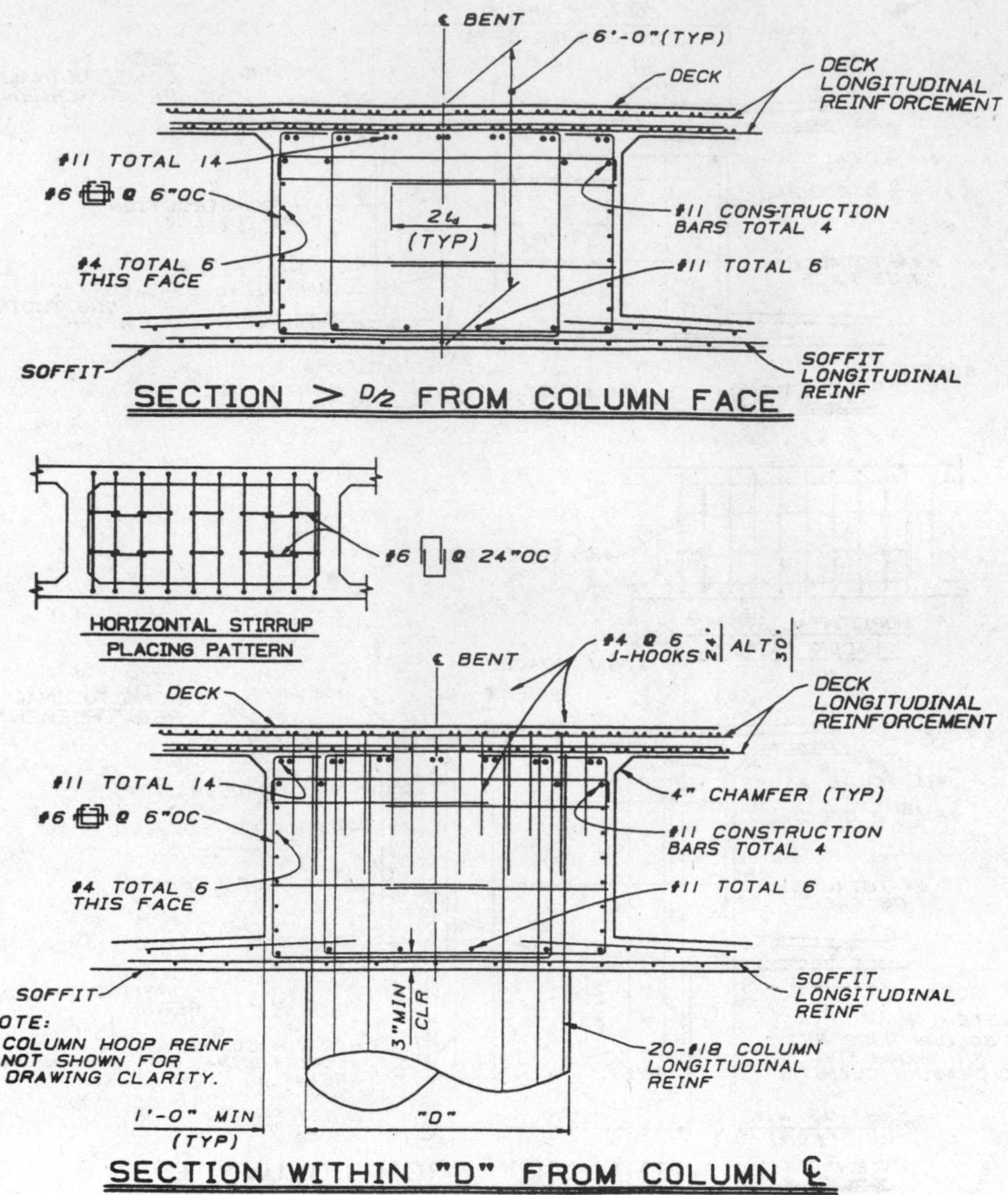

FIGURE 37.11 Example cap joint shear reinforcement — skews > 20°.

SRMDs should be able to perform their intended function and maintain their design parameters for the design life (for example, 40 years) and for an ambient temperature range (for example, from 30 to 125°F). The devices should be accessible for inspection, maintenance, and replacement. In general, SRMDs should satisfy at least the following requirements:

- Strength and stability must be maintained under increasingly large displacement. Stiffness degradation under repeated cyclic load is unacceptable.
- Energy must be dissipated within acceptable design displacement limits, for example, a limit on the maximum total displacement of the device to prevent failure, or the device can be given a displacement capacity 50% greater than the design displacement.

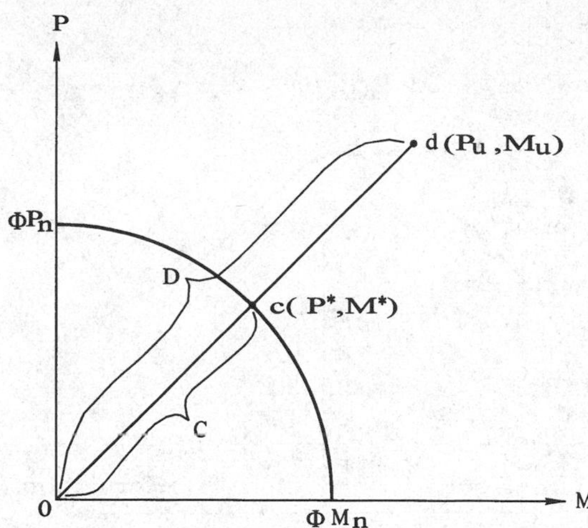

FIGURE 37.12 Definition of force D/C ratios for combined loadings.

- Heat builtup must be withstood and dissipated during "reasonable" seismic displacement time history.
- The device must survive subjected to the number of cycles of displacement expected under wind excitation during the life of the device and continue to function at maximum wind force and displacement levels for at least a given duration.

37.5.4 Acceptable Force *D/C* Ratios and Limiting Values for Structural Members

It is impossible to design bridges to withstand seismic forces elastically and the nonlinear inelastic response is expected. Performance-based criteria accept certain seismic damage in *other* components so the *critical* components will remain essentially elastic and functional after the SEE and FEE. This section presents the concept of acceptable force *D*/*C* ratios, limiting member slenderness parameters, and limiting width–thickness ratios, as well as expected ductility.

Definition of Force Demand/Capacity (D/C) Ratios

For members subjected to a single load, force demand is defined as a factored single force, such as factored moment, shear, or axial force. This may be obtained by a nonlinear dynamic time history analysis, level I, as specified in Section 37.5.1 and capacity is prescribed in Section 37.5.2.

For members subjected to combined loads, the force *D*/*C* ratio is based on the interaction. For example, for a member subjected to combined axial load and bending moment (Figure 37.12), the force demand *D* is defined as the distance from the origin point O(0, 0) to the factored force point $d(P_u, M_u)$, and capacity *C* is defined as the distance from the origin point O(0, 0) to the point $c(P^*, M^*)$ on the specified interaction surface or curve.

Ductility and Load–Deformation Curves

Ductility is usually defined as a nondimensional factor, i.e., the ratio of ultimate deformation to yield deformation [40,41]. It is normally expressed by two forms: (1) curvature ductility ($\mu_\phi = \phi_u/\phi_y$) and (2) displacement ductility ($\mu_\Delta = \Delta_u/\Delta_y$). Representing section flexural behavior, *curvature ductility* is dependent on the section shape and material properties and is based on the moment–curvature diagram. Indicating structural system or member behavior, *displacement ductility* is related to both the structural configuration and section behavior and is based on the load–displacement curve.

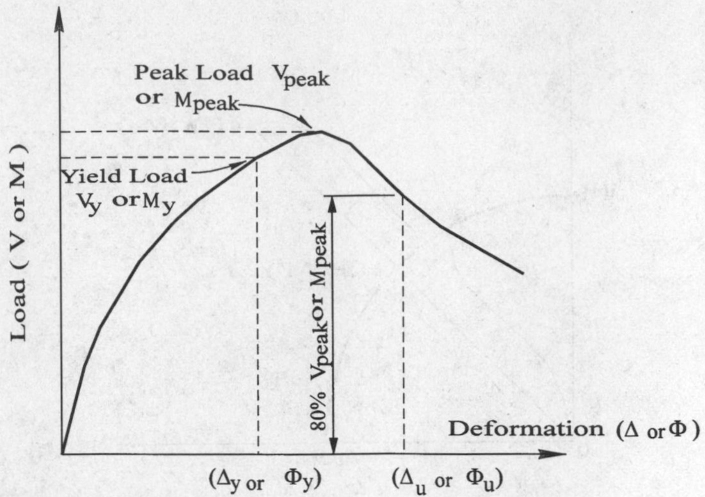

FIGURE 37.13 Load–deformation cCurves.

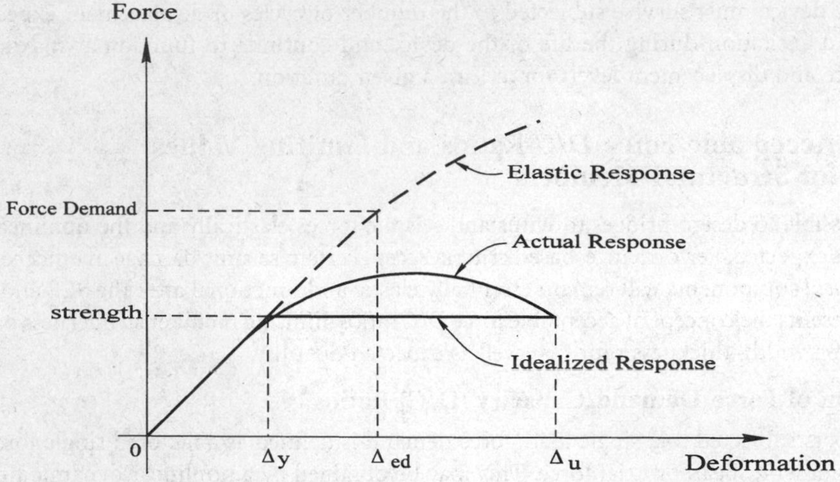

FIGURE 37.14 Response of a single-degree of freedom system.

A typical load–deformation curve, including both ascending and descending branches, is shown in Figure 37.13. The yield deformation (Δ_y or ϕ_y) corresponds to a loading state beyond which the structure responds inelastically. The ultimate deformation (Δ_u or ϕ_u) refers to the loading state at which a structural system or member can sustain without losing significant load-carrying capacity. Depending on performance requirements, it is proposed that the ultimate deformation (curvature or displacement) be defined as the most critical of (1) that deformation corresponding to a load dropping a maximum of 20% from the peak load or (2) that specified strain limit shown in Table 37.8.

Force D/C Ratios and Ductility

The following discussion will give engineers a direct measure of the seismic damage incurred by structural components during an earthquake. Figure 37.14 shows a typical load–response curve for a single-degree-of-freedom system. Displacement ductility is defined as

TABLE 37.12 Force D/C Ratio and Damage Index

Force D/C Ratio	Damage Index D_Δ	Expected System Displacement Ductility μ_Δ
1.0	No damage	No requirement
1.2	0.4	3.0
1.5	0.5	3.0
2.0	0.67	3.0
2.5	0.83	3.0

$$\mu_\Delta = \frac{\Delta_u}{\Delta_y} \tag{37.10}$$

A new term *damage index* is hereby defined as the ratio of elastic displacement demand to ultimate displacement capacity:

$$D_\Delta = \frac{\Delta_{ed}}{\Delta_u} \tag{37.11}$$

When the damage index $D_\Delta < 1/\mu_\Delta$ ($\Delta_{ed} < \Delta_y$), no damage occurs and the structure responds elastically; when $1/\mu_\Delta < D_\Delta < 1.0$, some damage occurs and the structure responds inelastically; when $D_\Delta > 1.0$, the structure collapses completely.

Based on the "equal displacement principle," the following relationship is obtained:

$$\frac{\text{Force Demand}}{\text{Force Capacity}} = \frac{\Delta_{ed}}{\Delta_y} = \mu_\Delta D_\Delta \tag{37.12}$$

It is seen from Eq. (37.12) that the force *D*/*C* ratio is related to both the structural characters in term of ductility $\mu_\Delta$and the degree of damage in terms of damage index D_Δ. Table 37.12 shows this relationship.

General Limiting Values

To ensure that important bridges have ductile load paths, general limiting slenderness parameters and width–thickness ratios are specified in Section 37.5.3.

For steel members, λ_{cr} is the limiting parameter for column elastic buckling and is taken as 1.5 from AISC-LRFD [37]; λ_{br} corresponds to beam elastic torsional buckling and is calculated by AISC-LRFD. For a *critical* member, a more strict requirement, 90% of those elastic buckling limits is proposed. Regardless of the force demand-to-capacity ratios, no members may exceed these limits. For existing steel members with *D*/*C* ratios less than 1, this limit may be relaxed. For concrete members, the general limiting parameter $KL/r = 60$ is proposed.

Acceptable Force *D*/*C* Ratios DC_{accept} for Steel Members

The acceptable force demand/capacity ratios (DC_{accept}) depends on both the structural characteristics in terms of ductility and the degree of damage acceptable to the engineer in terms of damage index D_Δ.

To ensure a member has enough inelastic deformation capacity during an earthquake and to achieve acceptable *D*/*C* ratios and energy dissipation, it is necessary, for steel member, to limit both the slenderness parameter and the width–thickness ratio within the ranges specified below.

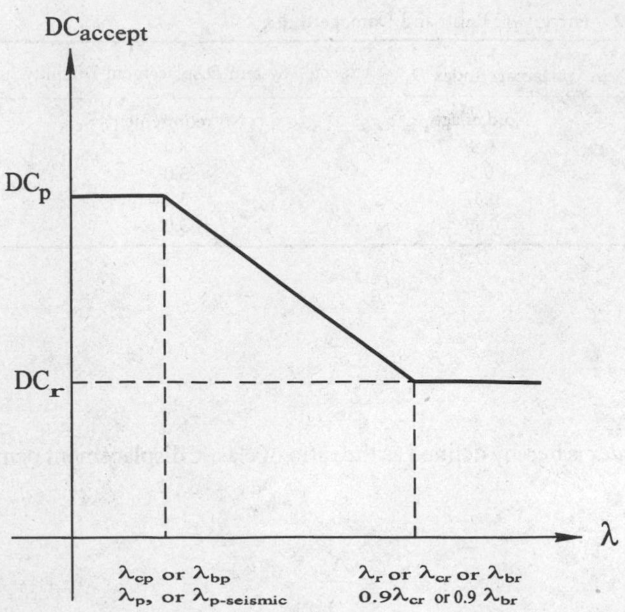

FIGURE 37.15 Acceptable D/C ratios and limiting slenderness parameters and width–thickness ratios.

Upper Bound Acceptable *D/C* Ratio DC_p

1. For *other* members, the large acceptable force *D/C* ratios ($DC_p = 2$ to 2.5) are listed in Table 37.10. The damage index is between 0.67 ~ 0.83 and more damage occurs in *other* members and great ductility is be expected. To achieve this,
   - The limiting width–thickness ratio was taken as $\lambda_{p\text{-Seismic}}$ from AISC-Seismic Provisions [42], which provides flexural ductility of 8 to 10.
   - The limiting slenderness parameters were taken as λ_{bp} for flexure-dominant members from AISC-LRFD [37], which provides flexural ductility of 8 to 10.
2. For *critical* members, small acceptable force *D/C* ratios ($DC_p = 1.2$ to 1.5) are proposed in Table 37.10, as the design purpose is to keep *critical* members essentially elastic and allow little damage (damage index values ranging from 0.4 to 0.5). Thus little member ductility is expected. To achieve this,
   - The limiting width–thickness ratio was taken as λ_p from AISC-LRFD [36], which provides a flexural ductility of at least 4.
   - The limiting slenderness parameters were taken as λ_{bp} for flexure dominant members from AISC [37], providing flexure ductility of at least 4.

3 For axial load dominant members the limiting slenderness parameter is taken as $\lambda_{cp} = 0.5$, corresponding to 90% of the axial yield load by the AISC-LRFD [37] column curve. This limit provides the potential for axial load dominant members to develop inelastic deformation.

Lower Bound Acceptable *D/C* Ratio DC_r

The lower-bound acceptable force D/C ratio $DC_{rc} = 1$ is proposed in Table 37.10. For $DC_{\text{accept}} = 1$, it is not necessary to enforce more strict limiting values for members and sections. Therefore, the limiting slenderness parameters for elastic global buckling specified in Table 27.10 and the limiting width–thickness ratios specified in Table 37.11 for elastic local buckling are proposed.

Acceptable *D/C* Ratios between Upper and Lower Bounds $DC_r < DC_{accept} < DC_p$

When acceptable force *D/C* ratios are between the upper and the lower bounds, $DC_r < DC_{accept} < DC_p$, a linear interpolation (Eqs. 37.7 to 37.9) as shown in Figure 37.15 is proposed to determine the limiting slenderness parameters and width–thickness ratios.

37.6 Summary

Seismic bridge design philosophies and current practice in the United States have been discussed. "No-collapse" based design is usually applied to ordinary bridges, and performance-based design is used for important bridges. Sample performance-based seismic design criteria are presented to bridge engineers as a reference guide. This chapter attempted to address only some of the many issues incumbent upon designers of bridges for adequate performance under seismic load. Engineers are always encouraged to incorporate to the best of their ability the most recent research findings and the most recent experimental evidence learned from past performance under real earthquakes.

References

1. ATC, Seismic Design Guidelines for Highway Bridges, Report No. ATC-6, Applied Technology Council, Redwood City, CA, 1981.
2. Housner, G. W. Competing against Time, Report to Governor George Deuknejian from the Governor's Broad of Inquiry on the 1989 Loma Prieta Earthquake, Sacramento, 1990.
3. Caltrans, The First Annual Seismic Research Workshop, Division of Structures, California Department of Transportation, Sacramento, 1991.
4. Caltrans, The Second Annual Seismic Research Workshop, Division of Structures, California Department of Transportation, Sacramento, 1993.
5. Caltrans, The Third Annual Seismic Research Workshop, Division of Structures, California Department of Transportation, Sacramento, 1994.
6. Caltrans, The Fourth Caltrans Seismic Research Workshop, Engineering Service Center, California Department of Transportation, Sacramento, 1996.
7. Caltrans, The Fifth Caltrans Seismic Research Workshop, Engineering Service Center, California Department of Transportation, Sacramento, 1998.
8. FHWA and Caltrans, *Proceedings of First National Seismic Conference on Bridges and Highways*, San Diego, 1995.
9. FHWA and Caltrans, *Proceedings of Second National Seismic Conference on Bridges and Highways*, Sacramento, 1997.
10. Kawashima, K. and Unjoh, S., The damage of highway bridges in the 1995 Hyogo-Ken Naubu earthquake and its impact on Japanese seismic design, *J. Earthquake Eng.*, 1(2), 1997, 505.
11. Park, R., Ed., Seismic design and retrofitting of reinforced concrete bridges, in *Proceedings of the Second International Workshop*, held in Queenstown, New Zealand, August, 1994.
12. Astaneh-Asl, A. and Roberts, J. Eds., *Seismic Design, Evaluation and Retrofit of Steel Bridges, Proceedings of the First U.S. Seminar*, San Francisco, 1993.
13. Astaneh-Asl, A. and Roberts, J., Ed., *Seismic Design, Evaluation and Retrofit of Steel Bridges, Proceedings of the Second U.S. Seminar*, San Francisco, 1997.
14. Housner, G.W., *The Continuing Challenge — The Northridge Earthquake of January 17, 1994*, Report to Director, California Department of Transportation, Sacramento, 1994.
15. Priestley, M. J. N., Seible, F. and Calvi, G. M., *Seismic Design and Retrofit of Bridges*, John Wiley & Sons, New York, 1996.
16. Caltrans, Design Criteria for SR-14/I-5 Replacement, California Department of Transportation, Sacramento, 1994.

17. Caltrans, San Francisco–Oakland Bay Bridge West Spans Seismic Retrofit Design Criteria, Prepared by Reno, M. and Duan, L., Edited by Duan, L., California Department of Transportation, Sacramento, 1997.
18. ATC, Improved Seismic Design Criteria for California Bridges: Provisional Recommendations, Report No. ATC-32, Applied Technology Council, Redwood City, CA, 1996.
19. Rojahn, C., et al., Seismic Design Criteria for Bridges and Other Highway Structures, Report NCEER-97-0002, National Center for Earthquake Engineering Research, State University of New York at Buffalo, Buffalo, 1997. Also refer as ATC-18, Applied Technology Council, Redwood City, CA, 1997.
20. Caltrans, Bridge Memo to Designers (20-1) — Seismic Design Methodology, California Department of Transportation, Sacramento, January 1999.
21. FHWA, Seismic Design and Retrofit Manual for Highway Bridges, Report No. FHWA-IP-87-6, Federal Highway Administration, Washington, D.C., 1987.
22. FHWA. Seismic Retrofitting Manual for Highway Bridges, Publ. No. FHWA-RD-94-052, Federal Highway Administration, Washington, D.C., 1995.
23. AASHTO, *LRFD Bridge Design Specifications*, 2nd. ed., American Association of State Highway and Transportation Officials, Washington, D.C., 1994 and 1996.
24. AASHTO, *Standard Specifications for Highway Bridges*, 16th ed., American Association of State Highway and Transportation Officials, Washington, D.C., 1996.
25. Caltrans, Bridge Memo to Designers (20-4), California Department of Transportation, Sacramento, 1995.
26. Caltrans, Bridge Design Specifications, California Department of Transportation, Sacramento, 1990.
27. Caltrans, Bridge Design Aids, California Department of Transportation, Sacramento, 1995.
28. IAI, Benicia-Martinez Bridge Seismic Retrofit — Main Truss Spans Final Retrofit Strategy Report, Imbsen and Association, Inc., Sacramento, 1995.
29. Priestley, N., Myths and fallacies in earthquake engineering — conflicts between design and reality, in *Proceedings of Tom Paulay Symposium — Recent Development in Lateral Force Transfer in Buildings*, University of California, San Diego, 1993.
30. Kowalsky, M. J., Priestley, M. J. N., and MacRae, G. A., Displacement-Based Design, Report No. SSRP-94/16, University of California, San Diego, 1994.
31. Martin, G. R. and Dobry, R., Earthquake Site Response and Seismic Code Provisions, NCEER Bulletin, Vol. 8, No. 4, National Center for Earthquake Engineering Research, Buffalo, NY, 1994.
32. Gupta, A. K. *Response Spectrum Methods in Seismic Analysis and Design of Structures*, CRC Press, Boca Raton, FL, 1992.
33. Clough, R. W. and Penzien, J., *Dynamics of Structures*, 2nd ed., McGraw-Hill, New York, 1993.
34. AASHTO, *Guide Specifications for Seismic Isolation Design*, American Association of State Highway and Transportation Officials, Washington, D.C., 1997.
35. Caltrans, Full Scale Isolation Bearing Testing Document (Draft), Prepared by Mellon, D., California Department of Transportation, Sacramento, 1997.
36. Japan Ministry of Construction (JMC), Manual of Menshin Design of Highway Bridges, (English version: EERC, Report 94/10, University of California, Berkeley), 1994.
37. AISC, *Load and Resistance Factor Design Specification for Structural Steel Buildings*, 2nd ed., American Institute of Steel Construction, Chicago, IL, 1993.
38. Zelinski, R., Seismic Design Memo, Various Topics, Preliminary Guidelines, California Department of Transportation, Sacramento, 1994.
39. Caltrans, Seismic Design Criteria for Retrofit of the West Approach to the San Francisco–Oakland Bay Bridge. Prepared by M. Keever, California Department of Transportation, Sacramento, 1996.
40. Park, R. and Paulay, T., *Reinforced Concrete Structures*, John Wiley & Sons, New York, 1975.
41. Duan, L. and Cooper, T. R., Displacement ductility capacity of reinforced concrete columns, *ACI Concrete Int.*, 17(11), 61–65, 1995.
42. AISC, *Seismic Provisions for Structural Steel Buildings*, American Institute of Steel Construction, Chicago, IL, 1991.
43. Caltrans, Seismic Design Criteria, Vers. 1.0, California Department of Transportation, Sacramento, 1999.

38

Seismic Design of Reinforced Concrete Bridges

Yan Xiao
University of Southern California

38.1 Introduction

This chapter provides an overview of the concepts and methods used in modern seismic design of reinforced concrete bridges. Most of the design concepts and equations described in this chapter are based on new research findings developed in the United States. Some background related to current design standards is also provided.

38.1.1 Two-Level Performance-Based Design

Most modern design codes for the seismic design of bridges essentially follow a two-level performance-based design philosophy, although it is not so clearly stated in many cases. The recent document ATC-32

0-8493-7434-0/00/$0.00+$.50

[2] may be the first seismic design guideline based on the two-level performance design. The two level performance criteria adopted in ATC-32 were originally developed by the California Department of Transportation [5].

The first level of design concerns control of the performance of a bridge in earthquake events that have relatively small magnitude but may occur several times during the life of the bridge. The second level of design consideration is to control the performance of a bridge under severe earthquakes that have only a small probability of occurring during the useful life of the bridge. In the recent ATC-32, the first level is defined for functional evaluation, whereas the second level is for safety evaluation of the bridges. In other words, for relatively frequent smaller earthquakes, the bridge should be ensured to maintain its function, whereas the bridge should be designed safe enough to survive the possible severe events.

Performance is defined in terms of the serviceability and the physical damage of the bridge. The following are the recommended service and damage criteria by ATC-32.

1. Service Levels:
   - *Immediate service*: Full access to normal traffic is available almost immediately following the earthquake.
   - *Limited service*: Limited access (e.g., reduced lanes, light emergency traffic) is possible within days of the earthquake. Full service is restorable within months.
2. Damage levels:
   - *Minimal damage*: Essentially elastic performance.
   - *Repairable damage*: Damage that can be repaired with a minimum risk of losing functionality.
   - *Significant damage*: A minimum risk of collapse, but damage that would require closure to repair.

The required performance levels for different levels of design considerations should be set by the owners and the designers based on the importance rank of the bridge. The fundamental task for seismic design of a bridge structure is to ensure a bridge's capability of functioning at the anticipated service levels without exceeding the allowable damage levels. Such a task is realized by providing proper strength and deformation capacities to the structure and its components.

It should also be pointed out that the recent research trend has been directed to the development of more-generalized performance-based design [3,6,8,13].

38.1.2 Elastic vs. Ductile Design

Bridges can certainly be designed to rely primarily on their strength to resist earthquakes, in other words, to perform elastically, in particular for smaller earthquake events where the main concern is to maintain function. However, elastic design for reinforced concrete bridges is uneconomical, sometimes even impossible, when considering safety during large earthquakes. Moreover, due to the uncertain nature of earthquakes, a bridge may be subject to seismic loading that well exceeds its elastic limit or strength and results in significant damage. Modern design philosophy is to allow a structure to perform inelastically to dissipate the energy and maintain appropriate strength during severe earthquake attack. Such an approach can be called ductile design, and the inelastic deformation capacity while maintaining the acceptable strength is called ductility.

The inelastic deformation of a bridge is preferably restricted to well-chosen locations (the plastic hinges) in columns, pier walls, soil behind abutment walls, and wingwalls. Inelastic action of superstructure elements is unexpected and undesirable because that damage to superstructure is difficult and costly to repair and unserviceable.

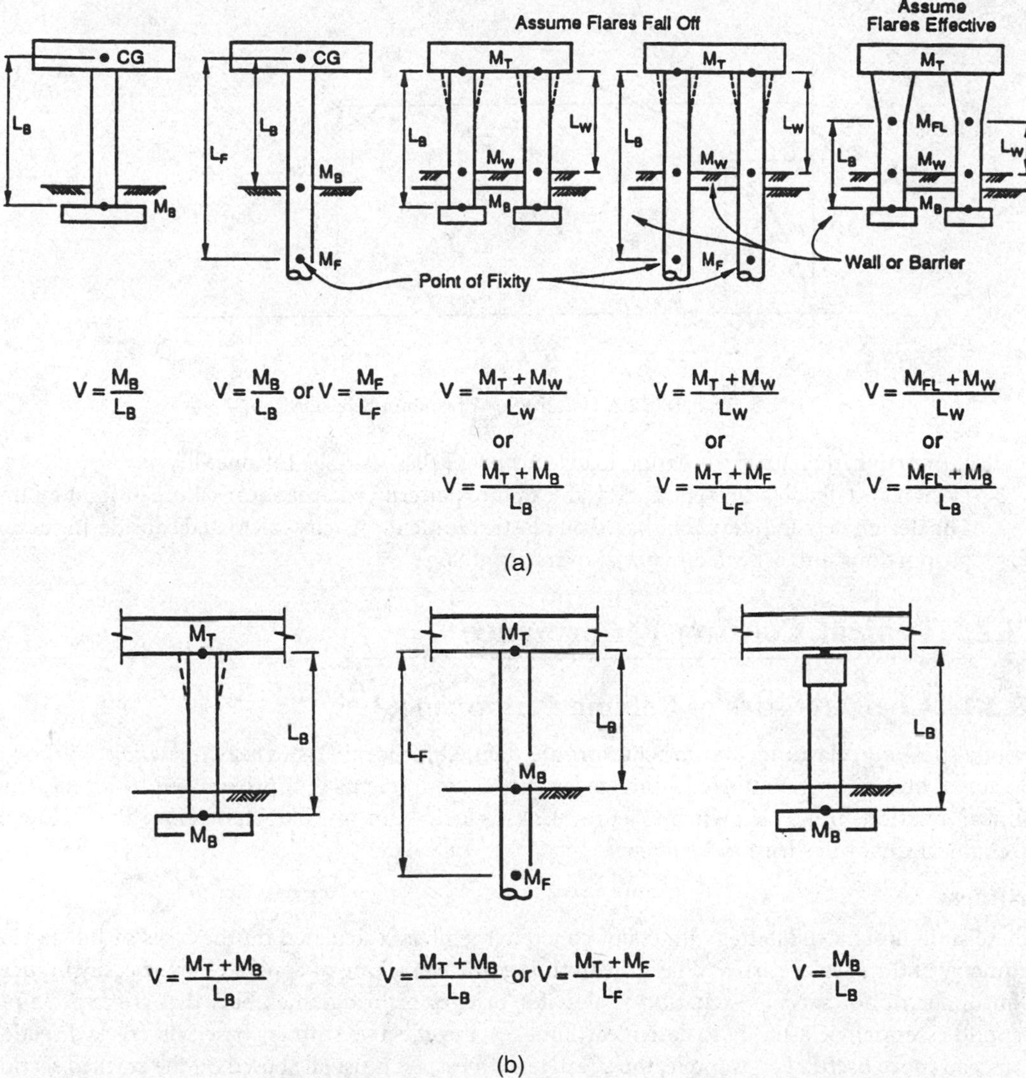

FIGURE 38.1 Potential plastic hinge locations for typical bridge bents: (a) transverse response; (b) longitudinal response. (*Source*: Caltrans, *Bridge Design Specification*, California Department of Transportation, Sacramento, June, 1990.)

38.1.3 Capacity Design Approach

The so-called capacity design has become a widely accepted approach in modern structural design. The main objective of the capacity design approach is to ensure the safety of the bridge during large earthquake attack. For ordinary bridges, it is typically assumed that the performance for lower-level earthquakes is automatically satisfied.

The procedure of capacity design involves the following steps to control the locations of inelastic action in a structure:

1. Choose the desirable mechanisms that can dissipate the most energy and identify plastic hinge locations. For bridge structures, the plastic hinges are commonly considered in columns. Figure 38.1 shows potential plastic hinge locations for typical bridge bents.

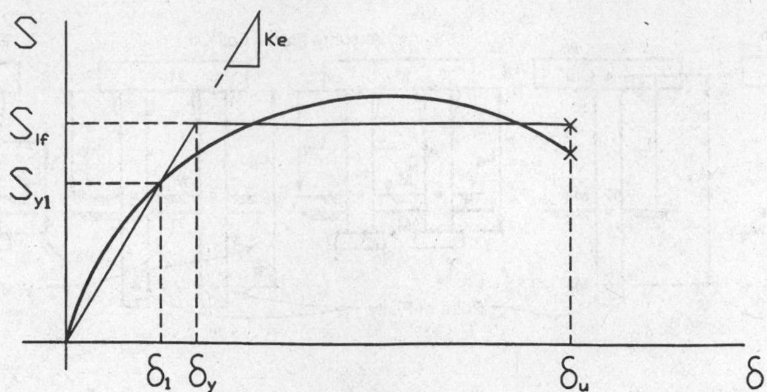

FIGURE 38.2 Idealization of column behavior.

2. Proportion structures for design loads and detail plastic hinge for ductility.
3. Design and detail to prevent undesirable failure patterns, such as shear failure or joint failure. The design demand should be based on plastic moment capacity calculated considering actual proportions and expected material overstrengths.

38.2 Typical Column Performance

38.2.1 Characteristics of Column Performance

Strictly speaking, elastic or plastic behaviors are defined for ideal elastoplastic materials. In design, the actual behavior of reinforced concrete structural components is approximated by an idealized bilinear relationship, as shown in Figure 38.2. In such bilinear characterization, the following mechanical quantities have to be defined.

Stiffness

For seismic design, the initial stiffness of concrete members calculated on the basis of full section geometry and material elasticity has little meaning, since cracking of concrete can be easily induced even under minor seismic excitation. Unless for bridges or bridge members that are expected to respond essentially elastically to design earthquakes, the effective stiffness based on cracked section is instead more useful. For example, the effective stiffness, K_e, is usually based on the cracked section corresponding to the first yield of longitudinal reinforcement,

$$K_e = S_{y1}/\delta_1 \tag{38.1}$$

where, S_{y1} and δ_1 are the force and the deformation of the member corresponding to the first yield of longitudinal reinforcement, respectively.

Strength

Ideal strength S_i represents the most feasible approximation of the "yield" strength of a member predicted using measured material properties. However, for design, such "yield" strength is conservatively assessed using *nominal strength* S_n predicted based on nominal material properties. The *ultimate* or *overstrength* represents the maximum feasible capacities of a member or a section and is predicted by taking account of all possible factors that may contribute to strength exceeding S_i or S_n. The factors include realistic values of steel yield strength, strength enhancement due to strain hardening, concrete strength increase due to confinement, strain rate, as well as actual aging, etc.

Deformation

In modern seismic design, deformation has the same importance as strength since deformation is directly related to physical damage of a structure or a structural member. Significant deformation limits are onset of cracking, onset of yielding of extreme tension reinforcement, cover concrete spalling, concrete compression crushing, or rupture of reinforcement. For structures that are expected to perform inelastically in severe earthquake, cracking is unimportant for safety design; however, it can be used as a limit for elastic performance. The first yield of tension reinforcement marks a significant change in stiffness and can be used to define the elastic stiffness for simple bilinear approximation of structural behavior, as expressed in Eq. (38.1). If the stiffness is defined by Eq. (38.1), then the yield deformation for the approximate elastoplastic or bilinear behavior can be defined as

$$\delta_y = S_{if}/S_{y1}\delta_1 \tag{38.2}$$

where, S_{y1} and δ_1 are the force and the deformation of the member corresponding to the first yield of longitudinal reinforcement, respectively; S_{if} is the idealized flexural strength for the elastoplastic behavior.

Meanwhile, the ductility factor, μ, is defined as the index of inelastic deformation beyond the yield deformation, given by

$$\mu = \delta/\delta_y \tag{38.3}$$

where δ is the deformation under consideration and δ_y is the yield deformation.

The limit of the bilinear behavior is set by an ultimate ductility factor or deformation, corresponding to certain physical events, that are typically corresponded by a significant degradation of load-carrying capacity. For unconfined member sections, the onset of cover concrete spalling is typically considered the failure. Rupture of either transverse reinforcement or longitudinal reinforcement and the crushing of confined concrete typically initiate a total failure of the member.

38.2.2 Experimentally Observed Performance

Figure 38.3a shows the lateral force–displacement hysteretic relationship obtained from cyclic testing of a well-confined column [10,11]. The envelope of the hysteresis loops can be either conservatively approximated with an elastoplastic bilinear behavior with V_{if} as the yield strength and the stiffness defined corresponding to the first yield of longitudinal steel. The envelope of the hysteresis loops can also be well simulated using a bilinear behavior with the second linear portion account for the overstrength due to strain hardening. Final failure of this column was caused by the rupture of longitudinal reinforcement at the critical sections near the column ends.

The ductile behavior shown in Figure 38.3a can be achieved by following the capacity design approach with ensuring that a flexural deformation mode to dominate the behavior and other nonductile deformation mode be prevented. As a contrary example to ductile behavior, Figure 38.3b shows a typical poor behavior that is undesirable for seismic design, where the column failed in a brittle manner due to the sudden loss of its shear strength before developing yielding, V_{if}. Bond failure of reinforcement lap splices can also result in rapid degradation of load-carrying capacity of a column.

An intermediate case between the above two extreme behaviors is shown in Figure 38.3c, where the behavior is somewhat premature for full ductility due to the fact that the tested column failed in shear upon cyclic loading after developing its yield strength but at a smaller ductility level than that shown in Figure 38.3a. Such premature behavior is also not desirable.

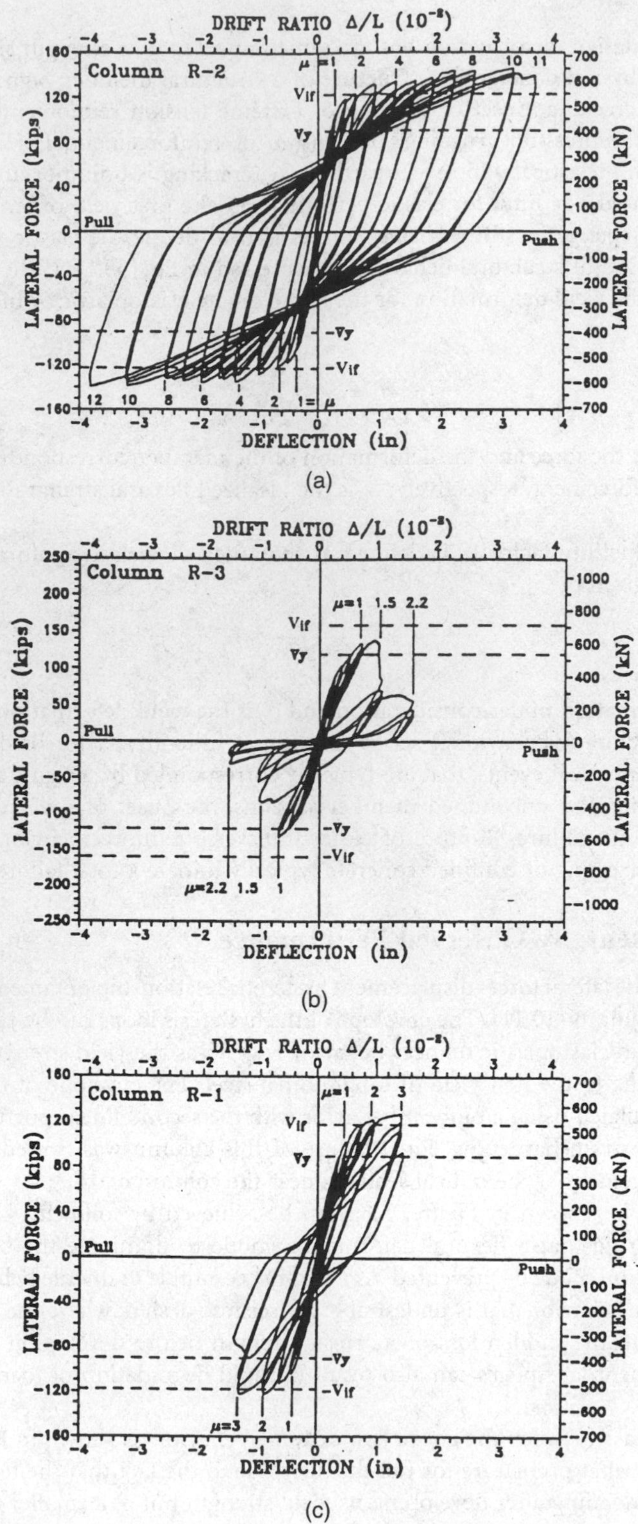

FIGURE 38.3 Typical experimental behaviors for (a) well-confined column; (b) column failed in brittle shear; (c) column with limited ductility. (*Source*: Priestley, M. J. N. et al., *ACI Struct. J.*, 91C52, 537–551, 1994. With permission.)

38.3 Flexural Design of Columns

38.3.1 Earthquake Load

For ordinary, regular bridges, the simple force design based on equivalent static analysis can be used to determine the moment demands on columns. Seismic load is assumed as an equivalent static horizontal force applied to individual bridge or frames, i.e.,

$$F_{eq} = ma_g \tag{38.4}$$

where m is the mass; a_g is the design peak acceleration depended on the period of the structure. In the Caltrans BDS [4] and the ATC-32 [2], the peak ground acceleration a_g is calculated as 5% damped elastic acceleration response spectrum at the site, expressed as ARS, which is the ratio of peak ground acceleration and the gravity acceleration g. Thus the equivalent elastic force is

$$F_{eq} = mg(\text{ARS}) = W(\text{ARS}) \tag{38.5}$$

where W is the dead load of bridge or frame.

Recognizing the reduction of earthquake force effects on inelastically responding structures, the elastic load is typically reduced by a period-dependent factor. Using the Caltrans BDS expression, the design force is found:

$$F_d = W(\text{ARS})/Z \tag{38.6}$$

This is the seismic demand for calculating the required moment capacity, whereas the capability of inelastic response (ductility) is ensured by following a capacity design approach and proper detailing of plastic hinges. Figure 38.4a and b shows the Z factor required by current Caltrans BDS and modified Z factor by ATC-32, respectively. The design seismic forces are applied to the structure with other loads to compute the member forces. A similar approach is recommended by the AASHTO-LRFD specifications.

The equivalent static analysis method is best suited for structures with well-balanced spans and supporting elements of approximately equal stiffness. For these structures, response is primarily in a single mode and the lateral force distribution is simply defined. For unbalanced systems, or systems in which vertical accelerations may be significant, more-advanced methods of analysis such as elastic or inelastic dynamic analysis should be used.

38.3.2 Fundamental Design Equation

The fundamental design equation is based on the following:

$$\phi R_n \geq R_u \tag{38.7}$$

where R_u is the strength demand; R_n is the nominal strength; and ϕ is the strength reduction factor.

38.3.3 Design Flexural Strength

Flexural strength of a member or a section depends on the section shape and dimension, amount and configuration of longitudinal reinforcement, strengths of steel and concrete, axial load magnitude, lateral confinement, etc. In most North American codes, the design flexural strength is conservatively calculated based on nominal moment capacity M_n following the ACI code recommendations [1]. The ACI approach is based on the following assumptions:

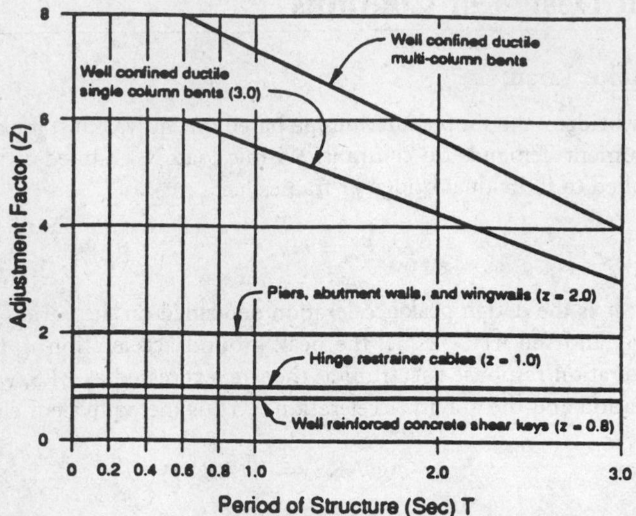

(a)

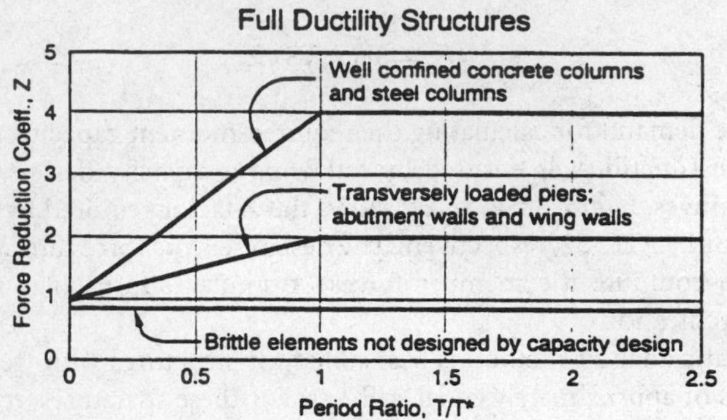

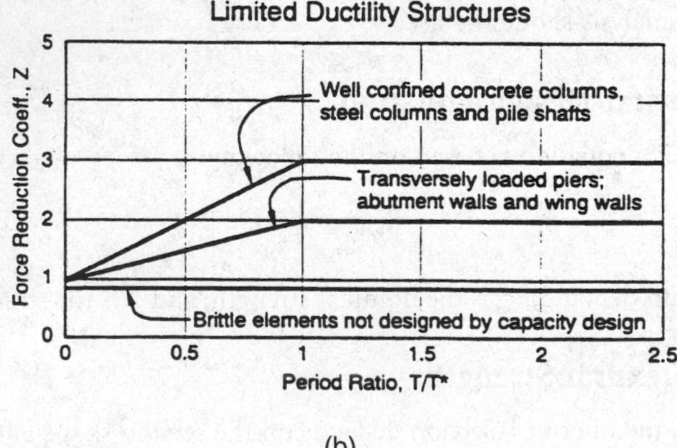

(b)

FIGURE 38.4 Force reduction coefficient Z (a) Caltrans BDS 1990; (b) ATC-32.

1. A plane section remains plane even after deformation. This implies that strains in longitudinal reinforcement and concrete are directly proportional to the distance from the neutral axis.
2. The section reaches the capacity when compression strain of the extreme concrete fiber reaches its maximum usable strain that is assumed to be 0.003.
3. The stress in reinforcement is calculated as the following function of the steel strain

$$f_s = -f_y \quad \text{for} \quad \varepsilon < -\varepsilon_y \tag{38.8a}$$

$$f_s = E_s \varepsilon \quad \text{for} \quad -\varepsilon_y \leq \varepsilon \leq \varepsilon_y \tag{38.8b}$$

$$f_s - f_y \quad \text{for} \quad \varepsilon > \varepsilon_y \tag{38.8c}$$

   where ε_y and f_y are the yield strain and specified strength of steel, respectively; E_s is the elastic modulus of steel.
4. Tensile stress in concrete is ignored.
5. Concrete compressive stress and strain relationship can be assumed to be rectangular, trapezoidal, parabolic, or any other shape that results in prediction of strength in substantial agreement with test results. This is satisfied by an equivalent rectangular concrete stress block with an average stress of 0.85 f_c', and a depth of $\beta_1 c$, where c is the distance from the extreme compression fiber to the neutral axis, and

$$0.65 \leq \beta_1 = 0.85 - 0.05 \frac{f_c' - 28}{7} \leq 0.85 \quad [\, f_c' \text{ in } MPa] \tag{38.9}$$

In calculating the moment capacity, the equilibrium conditions in axial direction and bending must be used. By using the equilibrium condition that the applied axial load is balanced by the resultant axial forces of concrete and reinforcement, the depth of the concrete compression zone can be calculated. Then the moment capacity can be calculated by integrating the moment contributions of concrete and steel.

The nominal moment capacity, M_n, reduced by a strength reduction factor ϕ (typically 0.9 for flexural) is compared with the required strength, M_u, to determine the feasibility of longitudinal reinforcement, section dimension, and adequacy of material strength.

Overstrength

The calculation of the nominal strength, M_n, is based on specified minimum material strength. The actual values of steel yield strength and concrete strength may be substantially higher than the specified strengths. These and other factors such as strain hardening in longitudinal reinforcement and lateral confinement result in the actual strength of a member perhaps being considerably higher than the nominal strength. Such overstrength must be considered in calculating ultimate seismic demands for shear and joint designs.

38.3.4 Moment–Curvature Analysis

Flexural design of columns can also be carried out more realistically based on moment–curvature analysis, where the effects of lateral confinement on the concrete compression stress–strain relationship and the strain hardening of longitudinal reinforcement are considered. The typical assumptions used in the moment–curvature analysis are as follows:

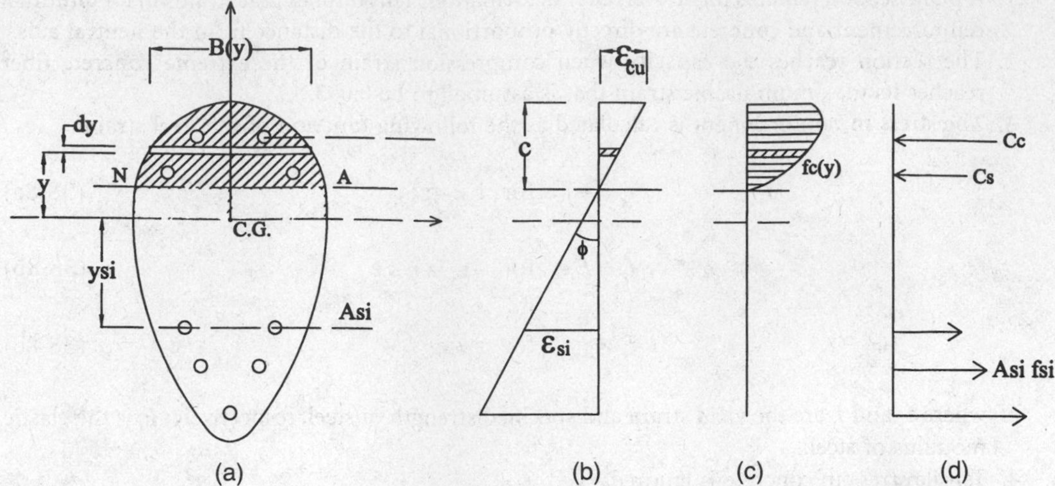

FIGURE 38.5 Moment curvature analysis: (a) generalized section; (b) strain distribution; (c) concrete stress distribution; (d) rebar forces.

1. A plane section remains plane even after deformation. This implies that strains in longitudinal reinforcement and concrete are directly proportional to the distance from the neutral axis.
2. The stress–strain relationship of reinforcement is known and can be expressed as a general function, $f_s = F_s(\varepsilon_s)$.
3. The stress–strain relationship of concrete is known and can be expressed as a general function, $f_c = F_c(\varepsilon_c)$. The tensile stress of concrete is typically ignored for seismic analysis but can be considered if the uncracked section response needs to be analyzed. The compression stress–strain relationship of concrete should be able to consider the effects of confined concrete (for example, Mander et al. [7]).
4. The resulting axial force and moment of concrete and reinforcement are in equilibrium with the applied external axial load and moment.

The procedure for moment–curvature analysis is demonstrated using a general section shown in Figure 38.5a. The distributions of strains and stresses in the cracked section corresponding to an arbitrary curvature, ϕ, are shown in Figure 38.5b, c, and d, respectively.

Corresponding to the arbitrary curvature, ϕ, the strains of concrete and steel at an arbitrary position with a distance of y to the centroid of the section can be calculated as

$$\varepsilon = \phi(y - y_c + c) \tag{38.10}$$

where y_c is the distance of the centroid to the extreme compression fiber and c is the depth of compression zone. Then the corresponding stresses can be determined using the known stress–strain relationships for concrete and steel.

Based on the equilibrium conditions, the following two equations can be established,

$$P = \sum A_{si} f_{si} + \int_A B(y) f_c(y) dy \tag{38.11}$$

$$M = \sum A_{si} f_{si} y_{si} + \int_A B(y) f_c(y) y dy \tag{38.12}$$

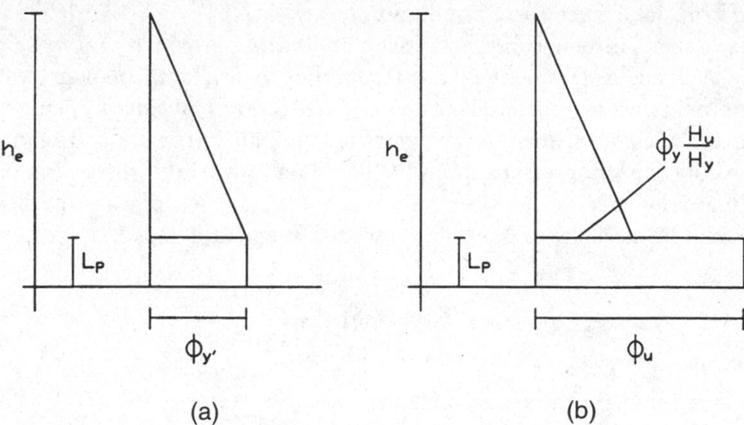

FIGURE 38.6 Idealized curvature distributions: (a) corresponding to first yield of reinforcement; (b) at ultimate flexural failure.

Using the axial equilibrium condition, the depth of the compression zone, c, corresponding to curvature, ϕ, can be determined, and then the corresponding moment, M, can be calculated. The actual computation of moment–curvature relationships is typically done by computer programs (for example, "SC-Push" [15]).

38.3.5 Transverse Reinforcement Design

In most codes, the ductility of the columns is ensured by proper detailing of transverse confinement steel. Transverse reinforcement can also be determined by a trial-and-error procedure to satisfy the required displacement member ductility levels. The lateral displacement of a member can be calculated by an integration of curvature and rotational angle along the member. This typically requires the assumption of curvature distributions along the member. Figure 38.6 shows the idealized curvature distributions at the first yield of longitudinal reinforcement and the ultimate condition for a column.

Yield Conditions: The horizontal force at the yield condition of the bilinear approximation is taken as the ideal capacity, H_{if}. Assuming linear elastic behavior up to conditions at first yield of the longitudinal reinforcement, the displacement of the column top at first yield due to flexure alone is

$$\Delta'_{yf} = \int_0^{h_e} \frac{\phi'_y}{h_e} (h_e - y)^2 dy \quad [f_y \text{ in } MPa] \tag{38.13}$$

where, ϕ'_y is the curvature at first yield and h_e is the effective height of the column. Allowing for strain penetration of the longitudinal reinforcement into the footing, the effective column height can be taken as

$$h_e = h + 0.022 f_y d_{bl} \tag{38.14}$$

where h is the column height measured from one end of a column to the point of inflexion, d_{bl} is the longitudinal reinforcement nominal diameter, and f_y is the nominal yield strength of rebar. The flexural component of yield displacement corresponding to the yield force, H_{if}, for the bilinear approximation can be found by extrapolating the first yield displacement to the ideal flexural strength, giving

$$\Delta_{yf} = \Delta'_{yf} H_{if} / H_y \tag{38.15}$$

where H_y is the horizontal force corresponding to first yield.

Ultimate Conditions: Ultimate conditions for the bilinear approximation can be taken to be the horizontal force, H_u, and displacement, Δ_{uf}, corresponding to development of an ultimate curvature, ϕ_u, based on moment-curvature analysis of the critical sections considering reinforcement strain-hardening and concrete confinement as appropriate. Ultimate curvature corresponds to the more critical ultimate concrete compression strain [7] based on an energy-balance approach, and maximum longitudinal reinforcement tensile strain, taken as $0.7\varepsilon_{su}$, where ε_{su} is the steel strain corresponding to maximum steel stress. The "ultimate" displacement is thus

$$\Delta_{uf} = \theta_p(h - 0.5L_p) + \Delta_y \frac{H_u}{H_{if}} \tag{38.16}$$

where the plastic rotation, θ_p, can be estimated as

$$\theta_p = \left(\phi_u - \phi'_y \frac{H_u}{H_y}\right) L_p \tag{38.17}$$

In Eqs. (38.16) and (38.17), L_p is the equivalent plastic hinge length appropriate to a bilinear approximation of response, given by

$$L_p = 0.08h + 0.022 f_y d_{bl} \quad [f_y \text{ in } MPa] \tag{38.18}$$

Based on the displacement check, the required transverse reinforcement within the plastic hinge region of a column to satisfy the required displacement ductility demand can be determined using a trial-and-error procedure.

Similarly, transverse reinforcement in the potential plastic hinge region of a column can be determined based on curvature ductility requirements. The ATC-32 document [2] suggests the following minimum requirement for the volumetric ratio, ρ_s, of spiral or circular hoop reinforcement within plastic hinge regions of columns based on a curvature ductility factor of 13.0 or larger.

$$\rho_s = 0.16 \frac{f'_{ce}}{f_{ye}} \left[0.5 + \frac{1.25P_e}{f'_{ce} A_g}\right] + 0.13(\rho_l - 0.01) \tag{38.19}$$

where f'_{ce} is the expected concrete compression strength taken as 1.3 f'_c ; f_{ye} is the expected steel yield strength taken as $1.1f_y$; P_e is axial load; A_g is cross-sectional area of column; and ρ_l is longitudinal reinforcement ratio. For transverse reinforcement outside the potential plastic hinge region, the volumetric ratio can be reduced to 50% of the amount given by Eq. (38.19). The length of the plastic hinge region should be the greater of (1) the section dimension in the direction considered or (2) the length of the column portion over which the moment exceeds 80% of the moment at the critical section. However, for columns with axial load ratio $P_e / f'_{ce} A_g > 0.3$, this length should be increased by 50%. Requirements for cross ties and hoops in rectilinear sections can also be found in ATC-32 [2].

38.4 Shear Design of Columns

38.4.1 Fundamental Design Equation

As discussed previously, shear failure of columns is the most dangerous failure pattern that typically can result in the collapse of a bridge. Thus, design to prevent shear failure is of particular importance. The general design equation for shear strength can be described as

$$\phi V_n > V_u \tag{38.20}$$

where V_u is the ultimate shear demands; V_n is the nominal shear resistant; and ϕ is the strength reduction factor for shear strength.

The calculation of the ultimate shear demand, V_u, has to be based on the equilibrium of the internal forces corresponding to the maximum flexural capacity, which is calculated taking into consideration all factors for overstrength. Figure 38.1 also includes equations for determining V_u for typical bridge bents.

38.4.2 Current Code Shear Strength Equation

Following the ACI code approach [1], the shear strength of axially loaded members is empirically expressed as

$$V_n = V_c + V_s \tag{38.21}$$

In the two-term additive equations, V_c is the shear strength contribution by concrete shear resisting mechanism and V_s is the shear strength contribution by the truss mechanism provided by shear reinforcement. Concrete shear contribution, V_c, is calculated as

$$V_c = 0.166\sqrt{f_c'}\left(1 + \frac{P}{13.8A_g}\right)A_e \tag{38.22}$$

where for columns the effective shear area A_e can be taken as 80% of the cross-sectional area, A_g; P is the applied axial compression force. Note that $\sqrt{f_c'} \leq 0.69$ MPa.

The contribution of truss mechanism is taken as

$$V_s = \frac{A_v f_y d}{s} \tag{38.23}$$

where A_v is the total transverse steel area within spacing s; f_y is yield strength of transverse steel; and d is the effective depth of the section.

Comparisons with existing test data indicate that actual shear strengths of columns often exceeds the design shear strength based on the ACI approach, in many cases by more than 100%.

38.4.3 Refined Shear Strength Equations

A refined shear strength equation that agrees significantly well with tests was proposed by Priestley et al. [12] in the following three-term additive expression:

$$V_n = V_c + V_s + V_a \tag{38.24}$$

where V_c and V_s are shear strength contributions by concrete shear resisting mechanism and the truss mechanism, respectively; the additional term, V_a, represents the shear resistance by the arch mechanism, provided mainly by axial compression.

$$V_c = k\sqrt{f_c'}A_e \tag{38.25}$$

and k depends on the displacement ductility factor μ_Δ, which reduces from 0.29 in MPa units (3.5 in psi) for $\mu_\Delta \le 2.0$ to 0.1 in MPa units (1.2 in psi) for $\mu_\Delta \ge 4.0$; A_e is taken as $0.8A_g$. The shear strength contribution by truss mechanism for circular columns is given by,

$$V_s = \frac{\pi}{2}\frac{A_{sp} f_y (d-c)}{s}\cot\theta \qquad (38.26)$$

where A_{sp} is cross-sectional area of spiral or hoop reinforcement; d is the effective depth of the section; c is the depth of compression zone at the critical section; s is spacing of spiral or hoop; and θ is the angle of truss mechanism, taken as 30°. In general, the truss mechanism angle θ should be considered a variable for different column conditions. Note that $(d-x)\cot\theta$ in Eq. (38.26) essentially represents the length of the critical shear crack that intersects with the critical section at the position of neutral axis, as shown in Figure 38.7a [17]. Equation (38.26) is approximate for circular columns with circular hoops or spirals as shear reinforcement. As shown in Figure 38.7b, the shear contribution of circular hoops intersected by a shear crack can be calculated by integrating the components of their hoop tension forces, $A_{sp}f_s$, in the direction of the applied shear, given by [18],

$$V_s = 2\int_{c-d/2}^{d/2} \frac{2A_{sp} f_s}{sd}\cot\theta\sqrt{(d/2)^2 - x^2}\,dx \qquad (38.27)$$

By further assuming that all the hoops intersected by the shear crack develop yield strength, f_y, the integration results in

$$V_s = \frac{A_{sp} f_y d}{s}\cot\theta\left[\left(1-\frac{2c}{d}\right)\sqrt{\frac{c}{d}\left(1-\frac{c}{d}\right)} + \frac{1}{2}\left(1-\frac{2c}{d}\right)^2 \arcsin\left(1-\frac{2c}{d}\right) + \frac{\pi}{4}\right] \qquad (38.28)$$

The shear strength enhancement by axial load is considered to result from an inclined compression strut, given by

$$V_a = P\tan\alpha = \frac{D-x}{2D(M/VD)}P \qquad (38.29)$$

where D is section depth or diameter; x is the compression zone depth which can be determined from flexural analysis; and (M/VD) is the shear aspect ratio.

38.5 Moment-Resisting Connection Between Column and Beam

Connections are key elements that maintain the integrity of overall structures; thus, they should be designed carefully to ensure the full transfer of seismic forces and moments. Because of their importance, complexity, and difficulty of repair if damaged, connections are typically provided with a higher degree of safety and conservativeness than column or beam members. Current Caltrans BDS and AASHTO-LRFD do not provide specific design requirements for joints, except requiring the lateral reinforcement for columns to be extended into column/footing or column/cap beam joints. A new design approach recently developed by Priestley and adopted in the ATC-32 design guidelines is summarized below.

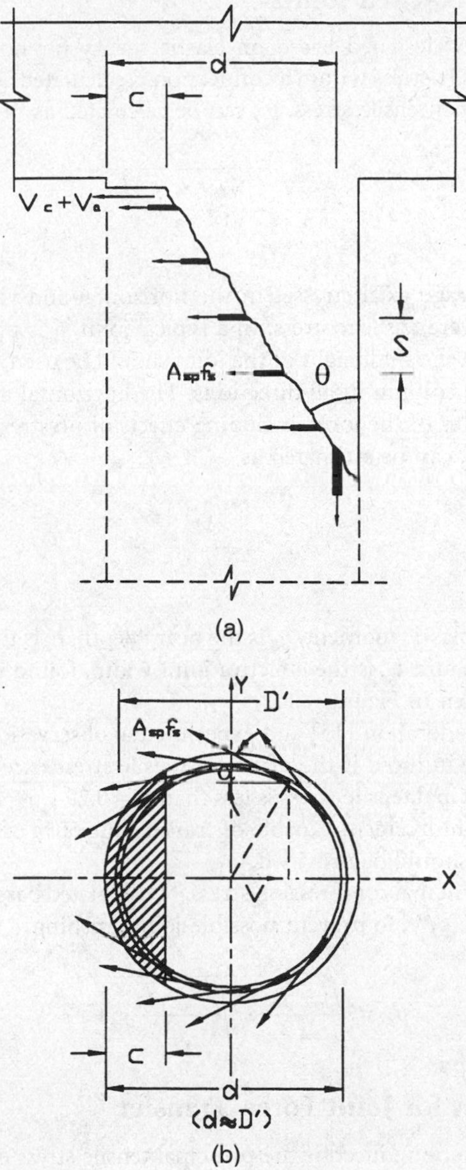

FIGURE 38.7 Shear resisting mechanism of circular hoops or spirals: (a) critical shear section; (b) stresses in hoops or spirals intersected with shear crack. (*Source*: Xiao et al. 1998. With permission.)

38.5.1 Design Forces

In moment-resisting frame structures, the force transfer typically results in sudden changes (magnitude and direction) of moments at connections. Because of the relatively small dimensions of joints, such sudden moment changes cause significant shear forces. Thus, joint shear design is the major concern of the column and beam connection, as well as that the longitudinal reinforcements of beams and columns are to be properly anchored or continued through the joint to transmit the moment. For seismic design, joint shear forces can be calculated based on the equilibrium condition using forces generated by the maximum plastic moment acting on the boundary of connections. In the following section, calculations for joint shear forces in the most common connections in bridge structures are discussed [13].

38.5.2 Design of Uncracked Joints

Joints can be conservatively designed based on elastic theory for not permitting cracks. In this approach, the principal tensile stress within a connection is calculated and compared with allowable tensile strength. The principal tensile stress, p_t, can be calculated as

$$p_t = \frac{f_h + f_v}{2} - \sqrt{\left(\frac{f_h - f_v}{2}\right)^2 + v_{hv}^2} \tag{38.30}$$

where f_h and f_v are the average axial stresses in the horizontal and vertical directions within the connection and v_{hv} is the average shear stress. In a typical joint, f_v is provided by the column axial force P_e. An average stress at the midheight of the joint should be used, assuming a 45° spread away from the boundaries of the column in all directions. The horizontal axial stress f_h is based on the mean axial force at the center of the joint, including effects of prestress, if present.

The joint shear stress, v_{hv}, can be estimated as

$$v_{hv} = \frac{M_p}{h_b h_c b_{je}} \tag{38.31}$$

where M_p is the maximum plastic moment, h_b is the beam depth, h_c is the column lateral dimension in the direction considered, and b_{je} is the effective joint width, found using a 45° spread from the column boundaries, as shown in Figure 38.8.

Based on theoretical consideration [13] and experimental observation , it is assumed that onset of diagonal cracking can be induced if the principal tensile stress exceed $0.29\sqrt{f_c'}$ MPa ($3.5\sqrt{f_c'}$ in psi units). When the principal tensile stress is less than $p_t = 0.29\sqrt{f_c'}$ MPa, the minimum amount of horizontal joint shear reinforcement capable of transferring 50% of the cracking stress resolved to the horizontal direction should be provided.

On the other hand, the principal compression stress, p_c, calculated based on the following equation should not exceed $p_c = 0.25\sqrt{f_c'}$, to prevent possible joint crushing.

$$p_c = \frac{f_h + f_v}{2} + \sqrt{\left(\frac{f_h - f_v}{2}\right)^2 + v_{hv}^2} \tag{38.32}$$

38.5.3 Reinforcement for Joint Force Transfer

Diagonal cracks are likely to be induced if the principal tensile stress exceed $0.29\sqrt{f_c'}$ MPa; thus, additional vertical and horizontal joint reinforcement is needed to ensure the force transfer within the joint. Unlike the joints in building structures where more rigorous dimension restraints exist, the joints in bridges can be reinforced with vertical and horizontal bars outside the zone where the column and beam longitudinal bars anchored to reduce the joint congestion. Figure 38.9 shows the force-resisting mechanism in a typical knee joint.

Vertical Reinforcement

On each side of the column or pier wall, the beam that is subjected to moments and shear will have vertical stirrups, with a total area $A_{jv} = 0.16A_{st}$, where A_{st} is the total area of column reinforcement anchored in the joint. The vertical stirrups shall be located within a distance $0.5D$ or $0.5h$ from the column or pier wall face, and distributed over a width not exceeding $2D$. As shown in Figure 38.9, reinforcement A_{jv} is required to provide the tie force T_s resisting the vertical component of strut $D2$. It is also clear from Figure 38.9, all the longitudinal bars contributing to the beam flexural strength should be clamped by the vertical stirrups.

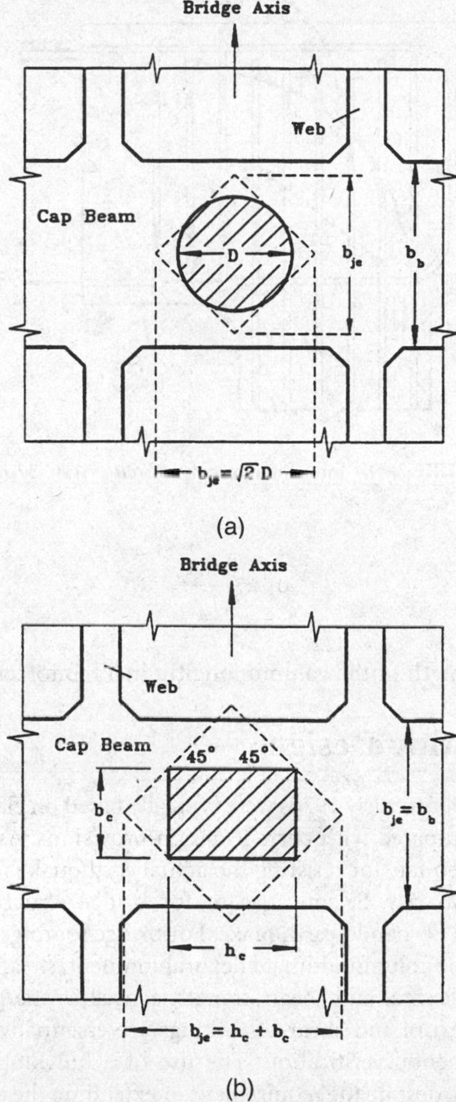

FIGURE 38.8 Effective joint width for joint shear stress calculations: (a) circular column; (b) rectangular column (Priestley et al. 1996; ATC-32, 1996).

Horizontal Reinforcement

Additional cap-beam bottom reinforcement is required to provide the horizontal resistance of the strut D2, shown in Figure 38.9. The suggested details are shown in Figure 38.9. This reinforcement may be omitted in prestressed or partially prestressed cap beams if the prestressed design force is increased by the amount needed to provide an equivalent increase in cap beam moment capacity to that provided by this reinforcement.

Hoop or Spiral Reinforcement

The hoop or spiral reinforcement is required to provide adequate confinement of the joint, and to resist the net outward thrust of struts *D*1 and *D*2 shown in Figure 38.9. The suggested volumetric ratio of column joint hoop or spiral reinforcement to be carried into the joint is

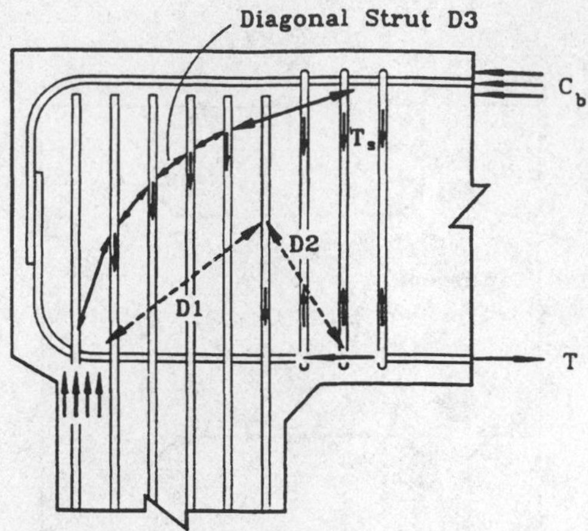

FIGURE 38.9 Joint force transfer. (*Source*: ATC-32 [2]).

$$\rho_s \geq \frac{0.4A_{st}}{l_{ac}^2} \tag{38.33}$$

where l_{ac} is the anchorage length of the column longitudinal reinforcement in the joint.

38.6 Column Footing Design

Bridge footing designs in 1950s to early 1970s were typically based on elastic analysis under relatively low lateral seismic input compared with current design provisions. As a consequence, footings in many older bridges are inadequate for resisting the actual earthquake force input corresponding to column plastic moment capacity. Seismic design for bridge structures has been significantly improved since the 1971 San Fernando earthquake. For bridge footings, a capacity design approach has been adopted by using the column ultimate flexural moment, shear, and axial force as the input to determine the required flexural and shear strength as well as the pile capacity of the footing. However, the designs for flexure and shear of footings are essentially based on a one-way beam model, which lacks experimental verification. The use of a full footing width for the design is nonconservative. In addition, despite the requirement of extending the column transverse reinforcement into the footing, there is a lack of rational consideration of column/footing joint shear in current design [16]. Based on large-scale model tests, Xiao et al. have recommended the following improved design for bridge column footings [16,18].

38.6.1 Seismic Demand

The footings are considered as under the action of column forces, due to the superimposed loads, resisted by an upward pressure exerted by the foundation materials and distributed over the area of the footing. When piles are used under footings, the upward reaction of the foundation is considered as a series of concentrated loads applied at the pile centers. For seismic design, the maximum probable moment, M_p, calculated based on actual strength with consideration of strain hardening of steel and enhancement due to confinement, with the associated axial and lateral loads are applied at the column base as the seismic inputs to the footing. Note that per current Caltrans BDS the maximum probable moment can be taken as 1.3 times the nominal moment capacity of the column, M_n, if the axial load of the column is below its balanced load, P_b. The numbers of piles and the internal forces in the footing are then determined from the seismic input. The internal moment and shear force of the footing can be determined based on the equilibrium conditions of the applied forces at the column base corresponding to the maximum moment capacity and the pile reaction forces.

38.6.2 Flexural Design

For flexural reinforcement design, the footing critical section is taken at the face of the column, pier wall, or at the edge of hinge. In case of columns that are not square or rectangular, the critical sections are taken at the side of the concentric square of equivalent area. The flexural reinforcements near top or the bottom of the footing to resist positive and negative critical moments should be calculated and placed based on the following effective footing width [16].

$$B_{feff} = B_c + 2d_f \quad \text{for rectangular columns} \tag{38.34a}$$

$$B_{feff} = D_c + 2d_f \quad \text{for circular columns} \tag{38.34b}$$

where B_{feff} is the footing effective width; B_c is the rectangular column width; d_f is the effective footing depth; and D_c is the circular column diameter. The minimum reinforcement must satisfy minimum flexural reinforcement requirements. The top reinforcement must also satisfy the requirement for shrinkage and temperature.

38.6.3 Shear Design

For shear, reinforcement of the footing should be designed against the critical shear force at the column face. As with flexure, the effective width of the footing should be used. The minimum shear reinforcement for column footings is vertical No. 5 (nominal diameter = ⅝ in. or 15.9 mm) at 12 in. (305 mm) spacing in each direction in a band between d_f, of the footing from the column surface and 6 in. (152 mm) maximum from the column reinforcement. Shear reinforcement must be hooked around the top and bottom flexure reinforcement in the footing. Inverted J stirrups with a 180° hook at the top and a 90° hook at the bottom are commonly used in California.

38.6.4 Joint Shear Cracking Check

If the footing is sufficiently large, then an uncracked joint may be designed by keeping the principal tensile stress in the joint region below the allowable cracking strength. Average principal tensile stress in the joint region can be calculated from the equivalent joint shear stress, v_{jv}, and the average vertical stress, f_a, by use of Mohr's circle for stress as

$$f_t = -f_a/2 + \sqrt{(f_a/2)^2 + v_{jv}^{\;2}} \tag{38.35}$$

It is suggested that f_a be based on the average effective axial compressive stress at middepth of the footing:

$$f_a = \frac{W_t}{A_{\text{eff}}} \tag{38.36}$$

where the effective area, A_{eff}, over which the total axial load W_t at the column base is distributed, is found from a 45-degree spread of the zone of influence as

$$A_{\text{eff}} = (B_c + d_f)(D_c + d_f) \quad \text{for rectangular column} \tag{38.37a}$$

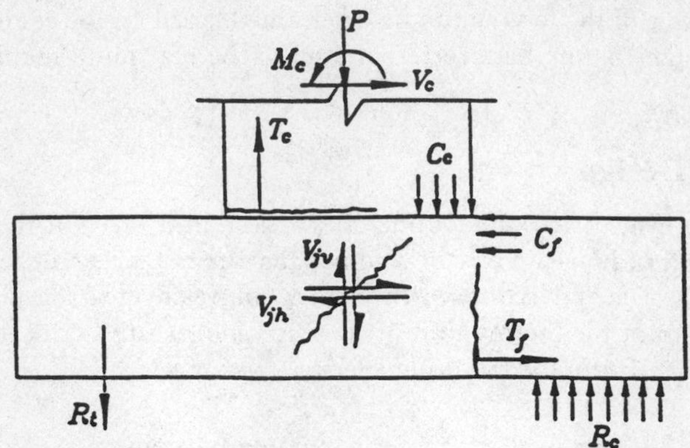

FIGURE 38.10 Shear Force in Column/Footing Joint. (*Source*: Xiao, Y. et al., *ACI Struct. J.*, 93(1), 79–94, 1996. With permission.)

$$A_{\text{eff}} = \pi(D_c + d_f)^2 / 4 \qquad \text{for circular column} \tag{38.37b}$$

where D_c is the overall section depth of rectangular column or the diameter of circular column; B_c is the section width of rectangular column; and d_f is the effective depth of the footing. As illustrated in Figure 38.10, the vertical joint shear V_{jv} can be assessed by subtracting the footing shear force due to the hold-down force, R_t, in the tensile piles and the footing self-weight, W_{fl}, outside the column tension stress resultant, from the total tensile force in the critical section of the column:

$$V_{jv} = T_c - (R_t + W_{fl}) \tag{38.38}$$

Considering an effective joint width, b_{jeff}, the average joint shear stress, v_{jv}, can be calculated as follows:

$$v_{jv} = \frac{V_{jv}}{b_{\text{jeff}} d_f} \tag{38.39}$$

where the effective joint width, b_{jeff}, can be assumed as the values given by Eq. (38.39), which is obtained based on the St. Venant 45-degree spread of influence between the tension and compression resultants in column critical section.

$$b_{\text{jeff}} = B_c + D_c \qquad \text{for rectangular column} \tag{38.40a}$$

$$b_{\text{jeff}} = \sqrt{2} D_c \qquad \text{for circular column} \tag{38.40b}$$

Joint shear distress is expected when the principal tensile stress given by Eq. (38.35) induced in the footing exceeds the direct tension strength of the concrete, which may conservatively be taken as $0.29\sqrt{f_c'}$ MPa (or $3.5\sqrt{f_c'}$ psi) [9], where f_c' is the concrete cylinder compressive strength. The minimum joint shear reinforcement should be provided even when the principal tensile stress is less than the tensile strength. This should be satisfied by simply extending the column transverse confinement into the footing.

38.6.5 Design of Joint Shear Reinforcement

When a footing cannot be prevented from joint shear cracking, additional vertical stirrups should be added around the column. For a typical column/footing designed to current standards, the assumed strut-and-tie model is shown in Figure 38.11a. The force inputs to the footing corresponding to the ultimate moment of the column critical section are the resultant tensile force, T_c, resultant compressive force, C_c, and the shear force, V_c, as shown in Figure 38.11a. The resultant tensile force, T_c, is resisted by two struts, *C*1 and *C*2, inside the column/footing joint region and a strut *C*3, outside the joint. Strut *C*1, is balanced by a horizontal tie, *T*1, provided by the transverse reinforcement of the column inside the footing and the compression zone of the critical section. Struts *C*2, and *C*3, are balanced horizontally at the intersection with the resultant tensile force, T_c. The internal strut *C*2, is supported at the compression zone of the column critical section. The external strut *C*3, transfers the forces to the ties, *T*2, *T*3, provided by the stirrups outside the joint and the top reinforcement, respectively. The forces are further transferred to the tensile piles through struts and ties, *C*4, *C*5, *C*6, and *T*4, *T*5, *T*6, *T*7. In the compression side of the footing, the resultant compressive force, C_c, and the shear force, V_c, are resisted mainly by compression struts, *C*9, *C*10, which are supported on the compression piles. It should be pointed out that the numbers and shapes of the struts and ties may vary for different column/footings.

As shown in Figure 38.11a, the column resultant tensile force, T_c, in the column/footing joint is essentially resisted by a redundant system. The joint shear design is to ensure that the tensile force, T_c, is resisted sufficiently by the internal strut and ties. The resisting system reaches its capacity, R_{ju}, when the ties, *T*1, *T*2, develop yielding. Although tie, *T*3, may also yield, it is not likely to dominate the capacity of the resisting system, since it may be assisted by a membrane mechanism near the footing face.

Assuming the inclination angles of *C*1, *C*2, and *C*3, are 45°, then the resistance can be expressed as

$$R_{ju} = (C1 + C2 + C3)\sin 45° = T1 + 2T2 \tag{38.41}$$

and at steel yielding,

$$R_{ju} = \frac{\pi}{2} n A_{sp} f_y + 2A_{\text{jeff}} \rho_{vs} f_y \tag{38.42}$$

where n is the number of layers of the column transverse reinforcement inside the footing; A_{sp} is the cross-sectional area of a hoop or spiral bar; A_{jeff} is the effective area in which the vertical stirrups are effective to resist the resultant tensile force, T_c; ρ_{vs} is the area ratio of the footing vertical stirrups.

The effective area, A_{jeff}, can be defined based on a three-dimensional crack with 45° slope around the column longitudinal bars in tension. The projection of the crack to the footing surface is shown by the shaded area in Figure 38.11b. The depth of the crack or the distance of the boundary of the shaded area in Figure 38.11b to the nearest longitudinal bar yielded in tension is assumed to be equal to the anchorage depth, d_{af}. The depth of the crack reduces from d_{af} to zero linearly if the strain of the rebar reduces from the yield strain, ε_y, to zero. Thus, A_{jeff} can be calculated as follows,

$$A_{\text{jeff}} = d_{af}(d_{af} + r_c)\arccos\left[\left(1 + \frac{\varepsilon_y}{\varepsilon_c}\right)\frac{x_n}{r_c} - 1\right] + d_{af} r_c \arccos\left(\frac{x_n}{r_c} - 1\right) \tag{38.43}$$

where d_{af} is the depth of the column longitudinal reinforcement inside footing; r_c is the radius of the centroidal circle of the longitudinal reinforcement and can be simply taken as the radius of the column section if the cover concrete is ignored; ε_y is the yield strain of the longitudinal bars; ε_c is

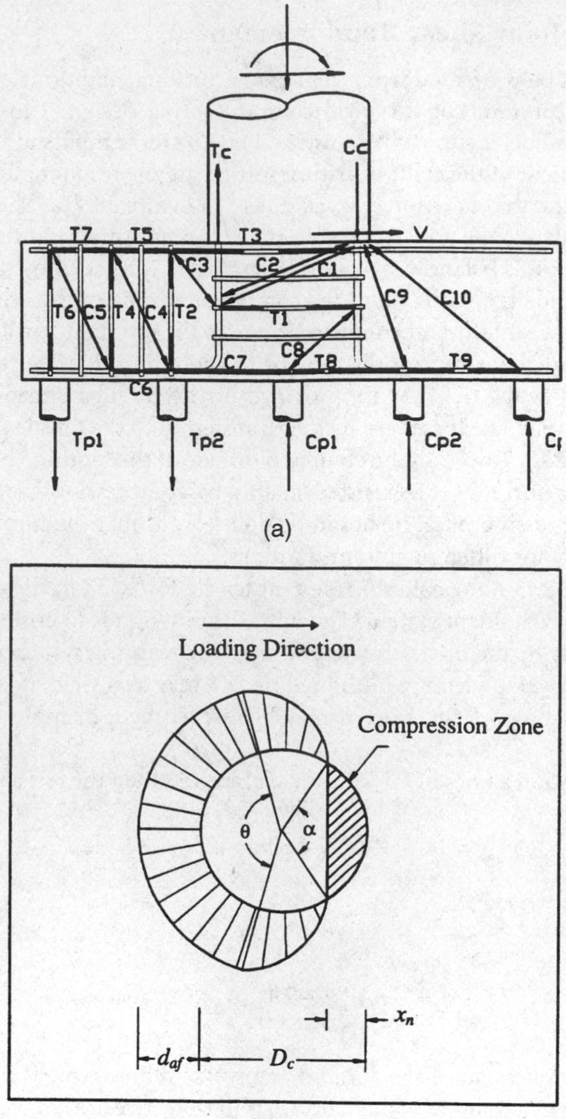

FIGURE 38.11 Column footing joint shear design: (a) force resisting mechanisms in footing; (b) effective distribution of external stirrups for joint shear resistance. (*Source*: Xiao, Y. et al. 1998. With permission.)

strain of the extreme compressive reinforcement or simply taken as the extreme concrete ultimate strain if the cover is ignored; x_n is the distance from the extreme compressive reinforcement to the neutral axis or taken as the compression zone depth, ignoring the cover.

References

1. ACI Committee 318, Building Code Requirements for Reinforced Concrete and Commentary (ACI 318-95/ACI 318R-95), American Concrete Institute, Farmington Hills, MI, 1995.
2. ATC 32, (1996), *Improved Seismic Design Criteria for California Bridges: Provisional Recommendations,* Applied Technology Council, Redwood City, CA, 1996.
3. Bertero, V. V., Overview of seismic risk reduction in urban areas: role, importance, and reliability of current U.S. seismic codes, and performance-based seismic engineering, in *Proceedings of the PRC-USA Bilateral Workshop on Seismic Codes,* Guangzhou, China, December 3–7, 1996, 10–48.

4. Caltrans, Bridge Design Specifications, California Department of Transportation, Sacramento, June, 1990.
5. Caltrans, Caltrans Response To Governor's Board of Inquiry Recommendations and Executive Order of June 2, 1990, 1994: Status Report, Roberts, J. E., California Department of Transportation, Sacramento, January, 26, 1994.
6. Jirsa, O. J. Do we have the knowledge to develop performance-based codes?" *Proceedings of the PRC-USA Bilateral Workshop on Seismic Codes,* Guangzhou, China, December 3–7, 1996, 111–118.
7. Mander, J. B., Priestley, M. J. N., and Park, R., Theoretical Stress-Strain Model for Confined Concrete, *ASCE J. Struct. Eng.*, 114(8), 1827–1849, 1988.
8. Moehle, J. P., Attempts to Introduce Modern Performance Concepts into Old Seismic Codes, in *Proceedings of the PRC-USA Bilateral Workshop on Seismic Codes,* Guangzhou, China, December 3–7, 1996, 217–230.
9. Priestley, M. J. N. and Seible, F., Seismic Assessment and Retrofit of Bridges, M. J. N. Priestley and F. Seible, Eds., University of California at San Diego, Structural Systems Research Project, Report No. SSRP-91/03, July, 1991, 418.
10. Priestley, M. J. N., Seible, F., Xiao, Y., and Verma, R., Steel jacket retrofit of short RC bridge columns for enhanced shear strength — Part 1. Theoretical considerations and test design, *ACI Struct. J.*, American Concrete Institute, 91(4), 394–405, 1994.
11. Priestley, M. J. N., Seible, F., Xiao, Y., and Verma, R., Steel jacket retrofit of short RC bridge columns for enhanced shear strength — Part 2. Experimental results, *ACI Struct. J.*, American Concrete Institute, 91(5), 537–551, 1994.
12. Priestley, M. J. N., Verma, R., and Xiao, Y., Seismic shear strength of reinforced concrete columns, *ASCE J. Struct. Eng.,* American Society of Civil Engineering, 120(8), 2310–2329, 1994.
13. Priestley, M. J. N., Seible, F., and Calvi, M., *Seismic Design and Retrofit of Bridges,* Wiley Interscience, New York, 1996, 686 pp.
14. Priestley, M. J. N., Ranzo, G., Benzoni, G., and Kowalsky, M. J., Yield Displacement of Circular Bridge Columns, in *Proceedings of the Fourth Caltrans Seismic Research Workshop,* July 9–11, 1996.
15. SC-Solution, *SC-Push 3D: Manual and Program Description,* SC-Solutions, San Jose, CA, 1995.
16. Xiao, Y., Priestley, M. J. N., and Seible, F., Seismic Assessment and Retrofit of Bridge Column Footings, *ACI Struct. J.,* 93(1), pp. 79–94, 1996.
17. Xiao, Y. and Martirossyan, A., Seismic performance of high-strength concrete columns, *ASCE J. Struct. Eng.*, 124(3), 241–251, 1998.
18. Xiao, Y., Priestley, M. J. N., and Seible, F., *Seismic Performance of Bridge Footings Designed to Current Standards,* ACI Special Publications on Earthquake Resistant Bridges, in press, 1998.
19. Xiao, Y., H. Wu, and G. R. Martin, (1998) Prefabricated composite jacketing of circular columns for enhanced shear resistance, *ASCE J. Struct. Eng.* 255–264, March, 1999.

39

Seismic Design of Steel Bridges

Chia-Ming Uang
University of California, San Diego

Keh-Chyuan Tsai
National Taiwan University

Michel Bruneau
State University of New York, Buffalo

39.1 Introduction

In the aftermath of the 1995 Hyogo-ken Nanbu earthquake and the extensive damage it imparted to steel bridges in the Kobe area, it is now generally recognized that steel bridges can be seismically vulnerable, particularly when they are supported on nonductile substructures of reinforced concrete, masonry, or even steel. In the last case, unfortunately, code requirements and guidelines on seismic design of ductile bridge steel substructures are few [12,21], and none have yet been implemented in the United States. This chapter focuses on a presentation of concepts and detailing requirements that can help ensure a desirable ductile behavior for steel substructures. Other bridge vulnerabilities common to all types of bridges, such as bearing failure, span collapses due to insufficient seat width or absence of seismic restrainers, soil liquefactions, etc., are not addressed in this chapter.

39.1.1 Seismic Performance Criteria

The American Association of State Highway and Transportation Officials (AASHTO) published both the *Standard Specifications for Highway Bridges* [2] and the *LRFD Bridge Design Specifications* [1], the latter being a load and resistance factor design version of the former, and being the preferred edition when referenced in this chapter. Although notable differences exist between the seismic

0-8493-7434-0/00/$0.00+$.50

design requirements of these documents, both state that the same fundamental principles have been used for the development of their specifications, namely:

1. Small to moderate earthquakes should be resisted within the elastic range of the structural components without significant damage.
2. Realistic seismic ground motion intensities and forces are used in the design procedures.
3. Exposure to shaking from large earthquakes should not cause collapse of all or part of the bridge. Where possible, damage that does occur should be readily detectable and accessible for inspection and repair.

Conceptually, the above performance criteria call for two levels of design earthquake ground motion to be considered. For a low-level earthquake, there should be only minimal damage. For a significant earthquake, which is defined by AASHTO as having a 10% probability of exceedance in 50 years (i.e., a 475-year return period), collapse should be prevented but significant damage may occur. Currently, the AASHTO adopts a simplified approach by specifying only the second-level design earthquake; that is, the seismic performance in the lower-level events can only be implied from the design requirements of the upper-level event. Within the content of performance-based engineering, such a one-level design procedure has been challenged [11,12].

The AASHTO also defines bridge importance categories, whereby essential bridges and critical bridges are, respectively, defined as those that must, at a minimum, remain open to emergency vehicles (and for security/defense purposes), and be open to all traffic, after the 475-year return-period earthquake. In the latter case, the AASHTO suggests that critical bridges should also remain open to emergency traffic after the 2500-year return-period event. Various clauses in the specifications contribute to ensure that these performance criteria are implicitly met, although these may require the engineer to exercise considerable judgment. The special requirements imposed on essential and critical bridges are beyond the scope of this chapter.

39.1.2 The *R* Factor Design Procedure

AASHTO seismic specification uses a response modification factor, R, to compute the design seismic forces in different parts of the bridge structure. The origin of the R factor design procedure can be traced back to the ATC 3-06 document [9] for building design. Since requirements in seismic provisions for member design are directly related to the R factor, it is worthwhile to examine the physical meaning of the R factor.

Consider a structural response envelope shown in Figure 39.1. If the structure is designed to respond elastically during a major earthquake, the required elastic force, Q_e, would be high. For economic reasons, modern seismic design codes usually take advantage of the inherent energy dissipation capacity of the structure by specifying a design seismic force level, Q_s, which can be significantly lower than Q_e:

$$Q_s = \frac{Q_e}{R} \tag{39.1}$$

The energy dissipation (or ductility) capacity is achieved by specifying stringent detailing requirements for structural components that are expected to yield during a major earthquake. The design seismic force level Q_s is the first significant yield level of the structure, which corresponds to the level beyond which the structural response starts to deviate significantly from the elastic response. Idealizing the actual response envelope by a linearly elastic–perfectly plastic response shown in Figure 39.1, it can be shown that the R factor is composed of two contributing factors [64]:

$$R = R_\mu \Omega? \tag{39.2}$$

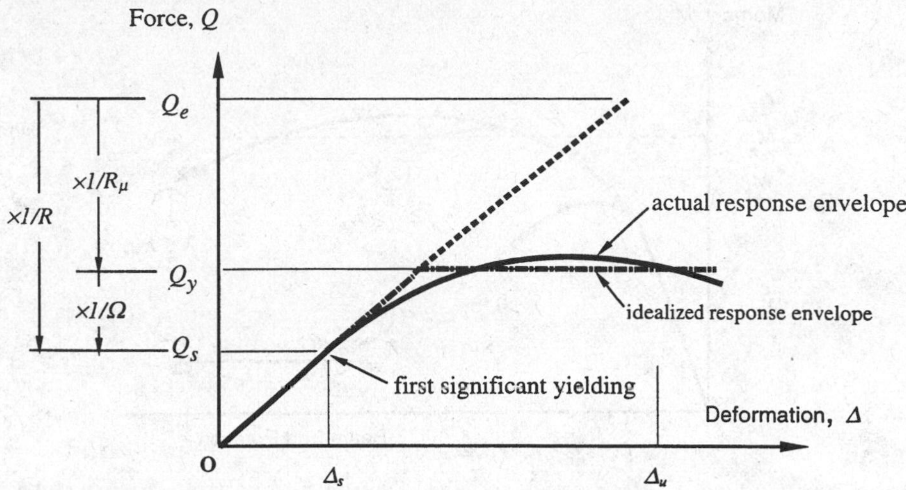

FIGURE 39.1 Concept of response modification factor, *R*.

TABLE 39.1 Response Modification Factor, *R*

Substructure	*R*	Connections	*R*
Single columns	3	Superstructure to abutment	0.8
Steel or composite steel and concrete pile bents		Columns, piers, or pile bents to cap beam or superstructure	1.0
a. Vertical piles only	5		
b. One or more batter piles	3		1.0
Multiple column bent	5	Columns or piers to foundations	

Source: AASHTO, *Standard Specifications for Seismic Design of Highway Bridges*, AASHTO, Washington, D.C., 1992.

The ductility reduction factor, R_μ, accounts for the reduction of the seismic force level from Q_e to Q_y. Such a force reduction is possible because ductility, which is measured by the ductility factor $\mu \Delta_u / \Delta_y$), is built into the structural system. For single-degree-of-freedom systems, relationships between μ and R_μ have been proposed (e.g., Newmark and Hall [43]).

The structural overstrength factor, Ω, in Eq. (39.2) accounts for the reserve strength between the seismic resistance levels Q_y and Q_s. This reserve strength is contributed mainly by the redundancy of the structure. That is, once the first plastic hinge is formed at the force level Q_s, the redundancy of the structure would allow more plastic hinges to form in other designated locations before the ultimate strength, Q_y, is reached. Table 39.1 shows the values of *R* assigned to different substructure and connection types. The AASHTO assumes that cyclic inelastic action would only occur in the substructure; therefore, no *R* value is assigned to the superstructure and its components. The table shows that the *R* value ranges from 3 to 5 for steel substructures. A multiple column bent with well detailed columns has the highest value (= 5) of *R* due to its ductility capacity and redundancy. The ductility capacity of single columns is similar to that of columns in multiple column bent; however, there is no redundancy and, therefore, a low *R* value of 3 is assigned to single columns.

Although modern seismic codes for building and bridge designs both use the *R* factor design procedure, there is one major difference. For building design [42], the *R* factor is applied at the system level. That is, components designated to yield during a major earthquake share the same *R* value, and other components are proportioned by the capacity design procedure to ensure that these components remain in the elastic range. For bridge design, however, the *R* factor is applied at the component level. Therefore, different *R* values are used in different parts of the same structure.

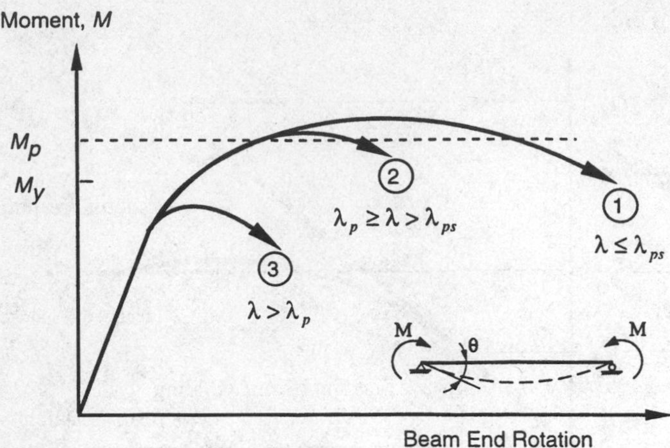

FIGURE 39.2 Effect of beam slenderness ratio on strength and deformation capacity. (Adapted from Yura et al., 1978.)

39.1.3 Need for Ductility

Using an R factor larger than 1 implies that the ductility demand must be met by designing the structural component with stringent requirements. The ductility capacity of a steel member is generally governed by instability. Considering a flexural member, for example, instability can be caused by one or more of the following three limit states: flange local buckling, web local buckling, and lateral-torsional buckling. In all cases, ductility capacity is a function of a slenderness ratio, λ. For local buckling, λ is the width–thickness ratio; for lateral-torsional buckling, λ is computed as L_b/r_y, where L_b is the unbraced length and r_y is the radius of gyration of the section about the buckling axis. Figure 39.2 shows the effect of λ on strength and deformation capacity of a wide-flanged beam. Curve 3 represents the response of a beam with a noncompact or slender section; both its strength and deformation capacity are inadequate for seismic design. Curve 2 corresponds to a beam with "compact" section; its slenderness ratio, λ, is less than the maximum ratio λp for which a section can reach its plastic moment, M_p, and sustain moderate plastic rotations. For seismic design, a response represented by Curve 1 is needed, and a "plastic" section with λ less than λps is required to deliver the needed ductility.

Table 39.2 shows the limiting width–thickness ratios λp and λps for compact and plastic sections, respectively. A flexural member with λ not exceeding λp can provide a rotational ductility factor of at least 4 [74], and a flexural member with λ less than λps is expected to deliver a rotation ductility factor of 8 to 10 under monotonic loading [5]. Limiting slenderness ratios for lateral-torsional buckling are presented in Section 39.2.

39.1.4 Structural Steel Materials

AASHTO M270 (equivalent to ASTM A709) includes grades with a minimum yield strength ranging from 36 to 100 ksi (see Table 39.3). These steels meet the AASHTO Standards for the mandatory notch toughness and weldability requirements and hence are prequalified for use in welded bridges.

For ductile substructure elements, steels must be capable of dissipating hysteretic energy during earthquakes, even at low temperatures if such service conditions are expected. Typically, steels that have $F_y < 0.8F_u$ and can develop a longitudinal elongation of 0.2 mm/mm in a 50-mm gauge length prior to failure at the expected service temperature are satisfactory.

TABLE 39.2 Limiting Width-Thickness Ratios

Description of Element	Width-Thickness Ratio	λ_p	λ_{ps}
Flanges of I-shaped rolled beams, hybrid or welded beams, and channels in flexure	b/t	$65/(\sqrt{F_y})$	$52/(\sqrt{F_y})$
Webs in combined flexural and axial compression	h/tw	for $P_u/\phi bPy \leq 0.125$: $\frac{640}{\sqrt{F_y}}\left(1-\frac{2.75P_u}{\phi_b P_y}\right)$ for $P_u/\phi_b P_y > 0.125$: $\frac{191}{\sqrt{F_y}}\left(2.33-\frac{P_u}{\phi_b P_y}\right) \geq \frac{253}{\sqrt{F_y}}$	for $Pu/\phi bPy > 0.125$: $\frac{520}{\sqrt{F_y}}\left(1-\frac{1.54P_u}{\phi_b P_y}\right)$ for $P_u/\phi_b P_y > 0.125$: $\frac{191}{\sqrt{F_y}}\left(2.33-\frac{P_u}{\phi_b P_y}\right) \geq \frac{253}{\sqrt{F_y}}$
Round HSS in axial compression or flexure	D/t	$\frac{2070}{F_y}$	$\frac{1300}{F_y}$
Rectangular HSS in axial compression or flexure	b/t	$\frac{190}{\sqrt{F_y}}$	$\frac{110}{\sqrt{F_y}}$

Note: F_y in ksi, $\phi_b = 0.9$.

Source: AISC, *Seismic Provisions for Structural Steel Buildings*, AISC, Chicago, IL, 1997.

TABLE 39.3 Minimum Mechanical Properties of Structural Steel

AASHTO Designation	M270 Grade 36	M270 Grade 50	M270 Grade 50W	M270 Grade 70W	M270 Grades 100/100W	
Equivalent ASTM designation	A709 Grade 36	A709 Grade 50	A709 Grade 50W	A709 Grade 70W	A709 Grade 100/100W	
Minimum yield stress (ksi)	36	50	50	70	100	90
Minimum tensile stress (ksi)	58	65	70	90	110	100

Source: AASHTO, *Standard Specification for Highway Bridges*, AASHTO, Washington, D.C., 1996.

39.1.5 Capacity Design and Expected Yield Strength

For design purposes, the designer is usually required to use the minimum specified yield and tensile strengths to size structural components. This approach is generally conservative for gravity load design. However, this is not adequate for seismic design because the AASHTO design procedure sometimes limits the maximum force acting in a component to the value obtained from the adjacent yielding element, per a capacity design philosophy. For example, steel columns in a multiple-column bent can be designed for an R value of 5, with plastic hinges developing at the column ends. Based on the weak column–strong beam design concept (to be presented in Section 39.2), the cap beam and its connection to columns need to be designed elastically (i.e., $R = 1$, see Table 39.1). Alternatively, for bridges classified as seismic performance categories (SPC) C and D, the AASHTO recommends that, for economic reasons, the connections and cap beam be designed for the maximum forces capable of being developed by plastic hinging of the column or column bent; these forces will often be significantly less than those obtained using an R factor of 1. For that purpose, recognizing the possible overstrength from higher yield strength and strain hardening, the AASHTO [1] requires that the column plastic moment be calculated using 1.25 times the nominal yield strength.

Unfortunately, the widespread brittle fracture of welded moment connections in steel buildings observed after the 1994 Northridge earthquake revealed that the capacity design procedure mentioned

TABLE 39.4 Expected Steel Material Strengths (SSPC 1994)

Steel Grade	A36	A572 Grade 50
No. of Sample	36,570	13,536
Yield Strength (COV)	49.2 ksi (0.10)	57.6 ksi (0.09)
Tensile Strength (COV)	68.5 ksi (0.07)	75.6 ksi (0.08)

COV: coefficient of variance.

Source: SSPC, *Statistical Analysis of Tensile Data for Wide Flange Structural Shapes*, Structural Shapes Producers Council, Washington, D.C., 1994.

above is flawed. Investigations that were conducted after the 1994 Northridge earthquake indicate that, among other factors, material overstrength (i.e., the actual yield strength of steel is significantly higher than the nominal yield strength) is one of the major contributing factors for the observed fractures [52].

Statistical data on material strength of AASHTO M270 steels is not available, but since the mechanical characteristics of M270 Grades 36 and 50 steels are similar to those of ASTM A36 and A572 Grade 50 steels, respectively, it is worthwhile to examine the expected yield strength of the latter. Results from a recent survey [59] of certified mill test reports provided by six major steel mills for 12 consecutive months around 1992 are briefly summarized in Table 39.4. Average yield strengths are shown to greatly exceed the specified values. As a result, relevant seismic provisions for building design have been revised. The AISC Seismic Provisions [6] use the following formula to compute the expected yield strength, *Fye*, of a member that is expected to yield during a major earthquake:

$$F_{ye} = R_y F_y \tag{39.3}$$

where *Fy* is the specified minimum yield strength of the steel. For rolled shapes and bars, R_y should be taken as 1.5 for A36 steel and 1.1 for A572 Grade 50 steel. When capacity design is used to calculate the maximum force to be resisted by members connected to yielding members, it is suggested that the above procedure also be used for bridge design.

39.1.6 Member Cyclic Response

A typical cyclic stress–strain relationship of structural steel material is shown in Figure 39.3. When instability are excluded, the figure shows that steel is very ductile and is well suited for seismic applications. Once the steel is yielded in one loading direction, the Bauschinger effect causes the steel to yield earlier in the reverse direction, and the clearly defined yield plateau disappears in subsequent cycles. Where instability needs to be considered, the Bauschinger effect may affect the cyclic strength of a steel member.

Consider an axially loaded steel member first. Figure 39.4 shows the typical cyclic response of an axially loaded tubular brace. The initial buckling capacity can be predicted reliably using the tangent modulus concept [47]. The buckling capacity in subsequent cycles, however, is reduced due to two factors: (1) the Bauschinger effect, which reduces the tangent modulus, and (2) the increased out-of-straigthness as a result of buckling in previous cycles. Such a reduction in cyclic buckling strength needs to be considered in design (see Section 39.3).

For flexural members, repeated cyclic loading will also trigger buckling even though the width–thickness ratios are less than the λps limits specified in Table 39.2. Figure 39.5 compares the cyclic response of two flexural members with different flange *b/t* ratios [62]. The strength of the beam having a larger flange width–thickness ratio degrades faster under cyclic loading as local buckling develops. This justifies the need for more stringent slenderness requirements in seismic design than those permitted for plastic design.

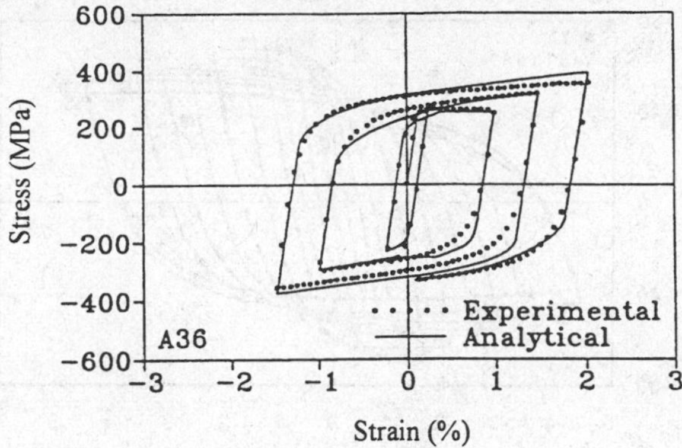

FIGURE 39.3 Typical cyclic stress–strain relationship of structural steel.

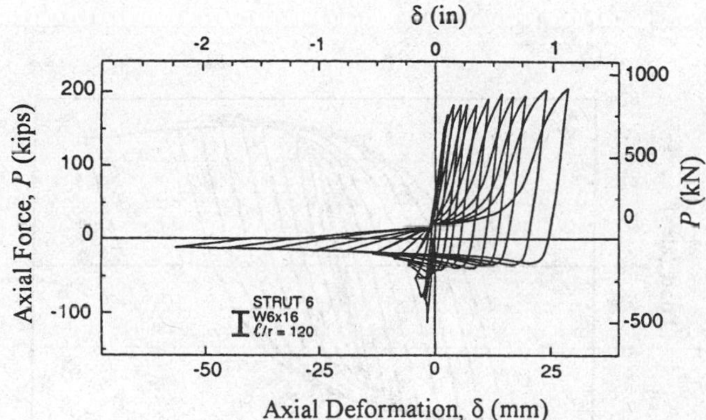

FIGURE 39.4 Cyclic response of an axially loaded member. (Source: Popov, E. P. and Black, W., *J. Struct. Div. ASCE*, 90(ST2), 223-256, 1981. With permission.)

39.2 Ductile Moment-Resisting Frame (MRF) Design

39.2.1 Introduction

The prevailing philosophy in the seismic resistant design of ductile frames in buildings is to force plastic hinging to occur in beams rather than in columns in order to better distribute hysteretic energy throughout all stories and to avoid soft-story-type failure mechanisms. However, for steel bridges such a constraint is not realistic, nor is it generally desirable. Steel bridges frequently have deep beams which are not typically compact sections, and which are much stiffer flexurally than their supporting steel columns. Moreover, bridge structures in North America are generally "single-story" (single-tier) structures, and all the hysteretic energy dissipation is concentrated in this single story. The AASHTO [3] and CHBDC [21] seismic provisions are, therefore, written assuming that columns will be the ductile substructure elements in moment frames and bents. Only the CHBDC, to date, recognizes the need for ductile detailing of steel substructures to ensure that the performance objectives are met when an R value of 5 is used in design [21]. It is understood that extra care would be needed to ensure the satisfactory ductile response of multilevel steel frame bents since these are implicitly not addressed by these specifications. Note that other recent design recommendations [12] suggest that the designer can choose to have the primary energy dissipation mechanism occur in either the beam–column panel zone or the column, but this approach has not been implemented in codes.

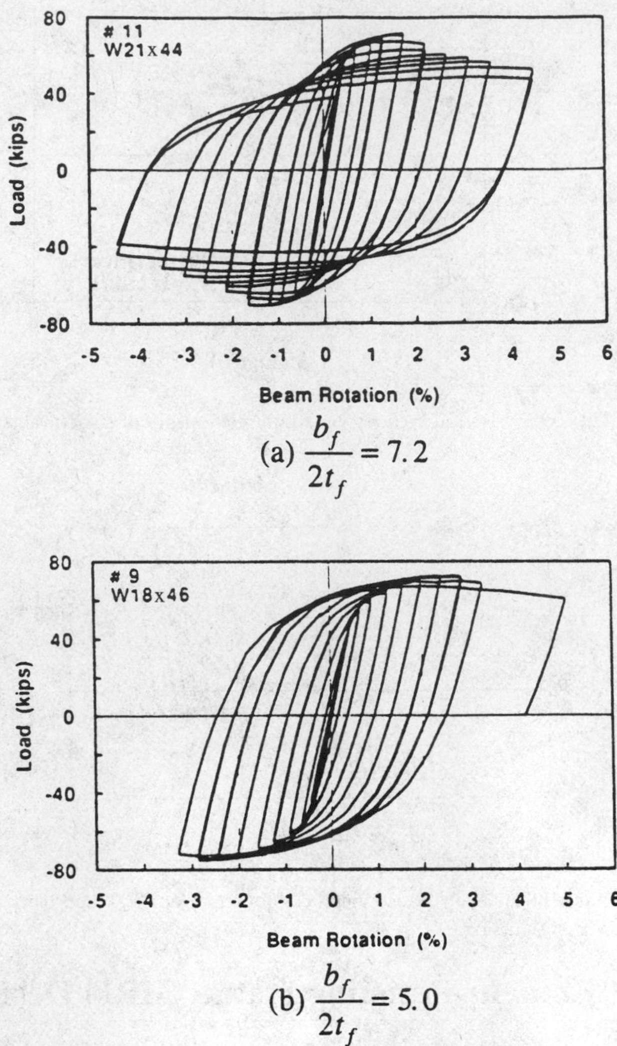

FIGURE 39.5 Effect of beam flange width–thickness ratio on strength degradation. (a) $b_f/2t_f = 7.2$; (b) $b_f/2t_f = 5.0$.

Some detailing requirements are been developed for elements where inelastic deformations are expected to occur during an earthquake. Nevertheless, lessons learned from the recent Northridge and Hyogo-ken Nanbu earthquakes have indicated that steel properties, welding electrodes, and connection details, among other factors, all have significant effects on the ductility capacity of welded steel beam–column moment connections [52]. In the case where the bridge column is continuous and the beam is welded to the column flange, the problem is believed to be less severe as the beam is stronger and the plastic hinge will form in the column [21]. However, if the bridge girder is continuously framed over the column in a single-story frame bent, special care would be needed for the welded column-to-beam connections.

Continuous research and professional developments on many aspects of the welded moment connection problems are well in progress and have already led to many conclusions that have been implemented on an interim basis for building constructions [52,54]. Many of these findings should be applicable to bridge column-to-beam connections where large inelastic demands are likely to

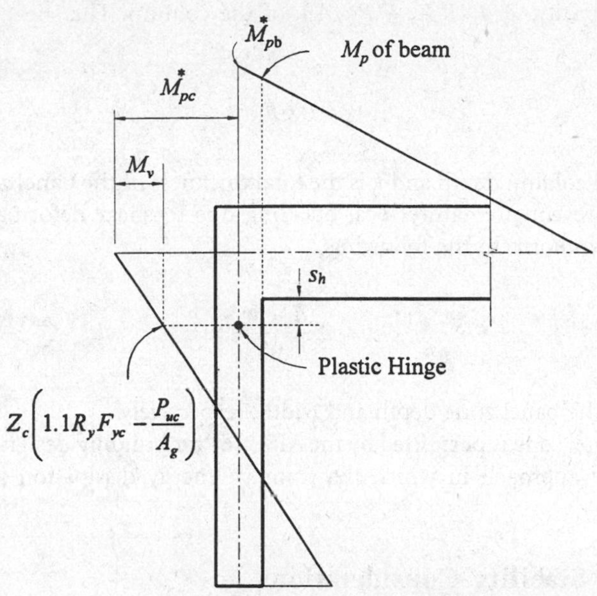

FIGURE 39.6 Location of plastic hinge.

develop in a major earthquake. The following sections provide guidelines for the seismic design of steel moment-resisting beam–column bents.

39.2.2 Design Strengths

Columns, beams, and panel zones are first designed to resist the forces resulting from the prescribed load combinations; then capacity design is exercised to ensure that inelastic deformations only occur in the specially detailed ductile substructure elements. To ensure a weak-column and strong-girder design, the beam-to-column strength ratio must satisfy the following requirement:

$$\frac{\Sigma M^*_{pb}}{\Sigma M^*_{pc}} \geq 1.0 \tag{39.4}$$

where $\Sigma\ M^*_{pb}$ is the sum of the beam moments at the intersection of the beam and column centerline. It can be determined by summing the projections of the nominal flexural strengths, M_p ($= Z_hF_y$, where Z_h is the plastic section modulus of the beam), of the beams framing into the connection to the column centerline. The term $\Sigma\ M^*_{pc}$ is the sum of the expected column flexural strengths, reduced to account for the presence of axial force, above and below the connection to the beam centerlines. The term $\Sigma\ M^*_{pc}$ can be approximated as $\Sigma\ [Z_c(1.1R_yF_{yc}-P_{uc}/A_g)+M_v]$, where A_g is the gross area of the column, P_{uc} is the required column compressive strength, Z_c is the plastic section modulus of the column, F_{yc} is the minimum specified yield strength of the column. The term M_v is to account for the additional moment due to shear amplification from the actual location of the column plastic hinge to the beam centerline (Figure 39.6). The location of the plastic hinge is at a distance s_h from the edge of the reinforced connection. The value of s_h ranges from one quarter to one third of the column depth as suggested by SAC [54].

To achieve the desired energy dissipation mechanism, it is rational to incorporate the expected yield strength into recent design recommendations [12,21]. Furthermore, it is recommended that the beam–column connection and the panel zone be designed for 125% of the expected plastic bending moment capacity, $Z_c(1.1R_yF_{yc} - P_{uc}/A_g)$, of the column. The shear strength of the panel zone, V_n, is given by

$$V_n = 0.6F_y d_c t_p \tag{39.5}$$

where d_c is the overall column depth and t_p is the total thickness of the panel zone including doubler plates. In order to prevent premature local buckling due to shear deformations, the panel zone thickness, t_p, should conform to the following:

$$t_p \geq \frac{d_z + w_z}{90} \tag{39.6}$$

where d_z and w_z are the panel zone depth and width, respectively.

Although weak panel zone is permitted by the AISC [6] for building design, the authors, however, prefer a conservative approach in which the primary energy dissipation mechanism is column hinging.

39.2.3 Member Stability Considerations

The width–thickness ratios of the stiffened and unstiffened elements of the column section must not be greater than the λ_{ps} limits given in Table 39.2 in order to ensure ductile response for the plastic hinge formation. Canadian practice [21] requires that the factored axial compression force due to the seismic load and gravity loads be less than $0.30A_gF_y$ (or twice that value in lower seismic zones). In addition, the plastic hinge locations, near the top and base of each column, also need to be laterally supported. To avoid lateral-torsional buckling, the unbraced length should not exceed $2500r_y/F_y$ [6].

39.2.4 Column-to-Beam Connections

Widespread brittle fractures of welded moment connections in building moment frames that were observed following the 1994 Northridge earthquake have raised great concerns. Many experimental and analytical studies conducted after the Northridge earthquake have revealed that the problem is not a simple one, and no single factor can be made fully responsible for the connection failures. Several design advisories and interim guidelines have already been published to assist engineers in addressing this problem [52,54]. Possible causes for the connection failures are presented below.

1. As noted in Section 39.1.5, the mean yield strength of A36 steel in the United States is substantially higher than the nominal yield value. This increase in yield strength combined with the cyclic strain hardening effect can result in a beam moment significantly higher than its nominal strength. Considering the large variations in material strength, it is questionable whether the bolted web-welded flange pre-Northridge connection details can reliably sustain the beam flexural demand imposed by a severe earthquake.
2. Recent investigations conducted on the properties of weld metal have indicated that the E70T-4 weld metal which was typically used in many of the damaged buildings possesses low notch toughness [60]. Experimental testing of welded steel moment connections that were conducted after the Northridge earthquake clearly demonstrated that notch-tough electrodes are needed for seismic applications. Note that the bridge specifications effectively prohibit the use of E70T-4 electrode.
3. In a large number of connections, steel backing below the beam bottom flange groove weld has not been removed. Many of the defects found in such connections were slag inclusions of a size that should have been rejected per AWS D1.1 if they could have been detected during

the construction. The inclusions were particularly large in the middle of the flange width where the weld had to be interrupted due to the presence of the beam web. Ultrasonic testing for welds behind the steel backing and particularly near the beam web region is also not very reliable. Slag inclusions are equivalent to initial cracks, which are prone to crack initiation at a low stress level. For this reason, the current steel building welding code [13] requires that steel backing of groove welds in cyclically loaded joints be removed. Note that the bridge welding code [14] has required the removal of steel backing on welds subjected to transverse tensile stresses.

4. Steel that is prevented from expanding or contracting under stress can fail in a brittle manner. For the most common type of groove welded flange connections used prior to the Northridge earthquake, particularly when they were executed on large structural shapes, the welds were highly restrained along the length and in the transverse directions. This precludes the welded joint from yielding, and thus promotes brittle fractures [16].
5. Rolled structural shapes or plates are not isotropic. Steel is most ductile in the direction of rolling and least ductile in the direction orthogonal to the surface of the plate elements (i.e., through-thickness direction). Thicker steel shapes and plates are also susceptible to lamellar tearing [4].

After the Northridge earthquake, many alternatives have been proposed for building construction and several have been tested and found effective to sustain cyclic plastic rotational demand in excess of 0.03 rad. The general concept of these alternatives is to move the plastic hinge region into the beam and away from the connection. This can be achieved by either strengthening the beam near the connection or reducing the strength of the yielding member near the connection. The objective of both schemes is to reduce the stresses in the flange welds in order to allow the yielding member to develop large plastic rotations. The minimum strength requirement for the connection can be computed by considering the expected maximum bending moment at the plastic hinge using statics similar to that outlined in Section 39.2.2. Capacity-enhancement schemes which have been widely advocated include cover plate connections [26] and bottom haunch connections. The demand-reduction scheme can be achieved by shaving the beam flanges [22,27,46,74]. Note that this research and development was conducted on deep beam sections without the presence of an axial load. Their application to bridge columns should proceed with caution.

39.3 Ductile Braced Frame Design

Seismic codesfor bridge design generally require that the primary energy dissipation mechanism be in the substructure. Braced frame systems, having considerable strength and stiffness, can be used for this purpose [67]. Depending on the geometry, a braced frame can be classified as either a concentrically braced frame (CBF) or an eccentrically braced frame (EBF). CBFs can be found in the cross-frames and lateral-bracing systems of many existing steel girder bridges. In a CBF system, the working lines of members essentially meet at a common point (Figure 39.7). Bracing members are prone to buckle inelastically under the cyclic compressive overloads. The consequence of cyclic buckling of brace members in the superstructure is not entirely known at this time, but some work has shown the importance of preserving the integrity of end-diaphragms [72]. Some seismic design recommendations [12] suggest that cross-frames and lateral bracing, which are part of the seismic force-resisting system in common slab-on-steel girder bridges, be designed to remain elastically under the imposed load effects. This issue is revisited in Section 39.5.

In a manner consistent with the earthquake-resistant design philosophy presented elsewhere in this chapter, modern CBFs are expected to undergo large inelastic deformation during a severe earthquake. Properly proportioned and detailed brace members can sustain these inelastic deformations and dissipate hysteretic energy in a stable manner through successive cycles of compression buckling and tension yielding. The preferred strategy is, therefore, to ensure that plastic deformation

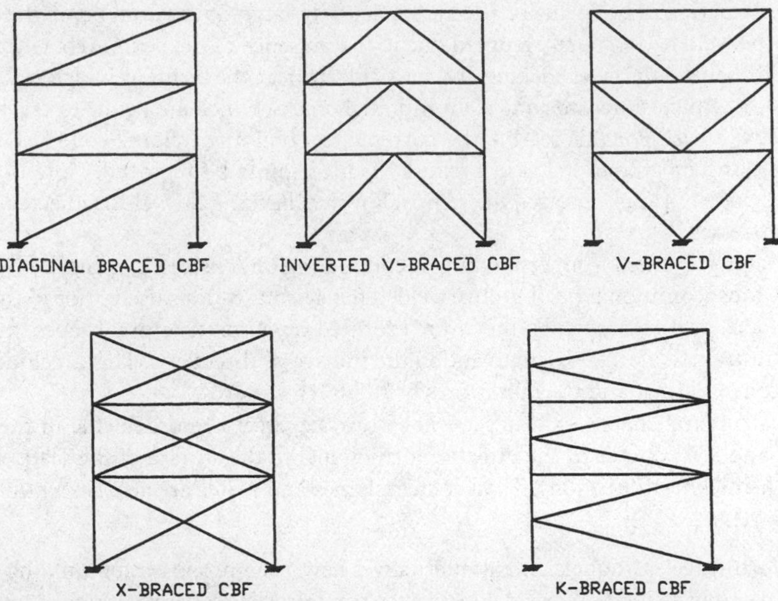

FIGURE 39.7 Typical concentric bracing configurations.

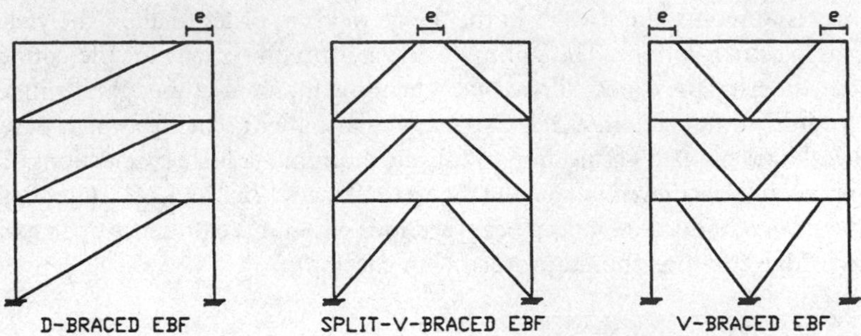

FIGURE 39.8 Typical eccentric bracing configurations.

only occur in the braces, allowing the columns and beams to remain essentially elastic, thus maintaining the gravity load-carrying capacity during a major earthquake. According to the AISC Seismic Provisions [6], a CBF can be designed as either a special CBF (SCBF) or an ordinary CBF (OCBF). A large value of R is assigned to the SCBF system, but more stringent ductility detailing requirements need to be satisfied.

An EBF is a system of columns, beams, and braces in which at least one end of each bracing member connects to a beam at a short distance from its beam-to-column connection or from its adjacent beam-to-brace connection (Figure 39.8). The short segment of the beam between the brace connection and the column or between brace connections is called the link. Links in a properly designed EBF system will yield primarily in shear in a ductile manner. With minor modifications, the design provisions prescribed in the AISC Seismic Provisions for EBF, SCBF, and OCBF can be implemented for the seismic design of bridge substructures.

Current AASHTO seismic design provisions [3] do not prescribe the design seismic forces for the braced frame systems. For OCBFs, a response modification factor, R, of 2.0 is judged appropriate. For EBFs and SCBFs, an R value of 4 appears to be conservative and justifiable by examining the ductility reduction factor values prescribed in the building seismic design recommendations [57].

For CBFs, the emphasis in this chapter is placed on SCBFs, which are designed for better inelastic performance and energy dissipation capacity.

39.3.1 Concentrically Braced Frames

Tests have shown that, after buckling, an axially loaded member rapidly loses compressive strength under repeated inelastic load reversals and does not return to its original straight position (see Figure 39.4). CBFs exhibit the best seismic performance when both yielding in tension and inelastic buckling in compression of their diagonal members contribute significantly to the total hysteretic energy dissipation. The energy absorption capability of a brace in compression depends on its slenderness ratio (KL/r) and its resistance to local buckling. Since they are subjected to more stringent detailing requirements, SCBFs are expected to withstand significant inelastic deformations during a major earthquake. OCBFs are designed to higher levels of design seismic forces to minimize the extent of inelastic deformations. However, if an earthquake greater than that considered for design occurs, structures with SCBF could be greatly advantaged over the OCBF, in spite of the higher design force level considered in the latter case.

Bracing Members

Postbuckling strength and energy dissipation capacity of bracing members with a large slenderness ratio will degrade rapidly after buckling occurs [47]. Therefore, many seismic codes require the slenderness ratio (KL/r) for the bracing member be limited to $720/\sqrt{F_y}$, where F_y is in ksi. Recently, the AISC Seismic Provisions (1997) [6] have relaxed this limit to $1000/\sqrt{F_y}$ for bracing members in SCBFs. This change is somewhat controversial. The authors prefer to follow the more stringent past practice for SCBFs. The design strength of a bracing member in axial compression should be taken as $0.8\,\phi_c P_n$, where ϕ_c is taken as 0.85 and P_n is the nominal axial strength of the brace. The reduction factor of 0.8 has been prescribed for CBF systems in the previous seismic building provisions [6] to account for the degradation of compressive strength in the postbuckling region. The 1997 AISC Seismic Provisions have removed this reduction factor for SCBFs. But the authors still prefer to apply this strength reduction factor for the design of both SCBFs and OCBFs. Whenever the application of this reduction factor will lead to a less conservative design, however, such as to determine the maximum compressive force a bracing member imposes on adjacent structural elements, this reduction factor should not be used.

The plastic hinge that forms at midspan of a buckled brace may lead to severe local buckling. Large cyclic plastic strains that develop in the plastic hinge are likely to initiate fracture due to low-cycle fatigue. Therefore, the width–thickness ratio of stiffened or unstiffened elements of the brace section for SCBFs must be limited to the values specified in Table 39.2. The brace sections for OCBFs can be either compact or noncompact, but not slender. For brace members of angle, unstiffened rectangular, or hollow sections, the width–thickness ratios cannot exceed λ_{ps}.

To provide redundancy and to balance the tensile and compressive strengths in a CBF system, it is recommended that at least 30% but not more than 70% of the total seismic force be resisted by tension braces. This requirement can be waived if the bracing members are substantially oversized to provide essentially elastic seismic response.

Bracing Connections

The required strength of brace connections (including beam-to-column connections if part of the bracing system) should be able to resist the lesser of:

1. The expected axial tension strength ($= R_y F_y A_g$) of the brace.
2. The maximum force that can be transferred to the brace by the system.

In addition, the tensile strength of bracing members and their connections, based on the limit states of tensile rupture on the effective net section and block shear rupture, should be at least equal to the required strength of the brace as determined above.

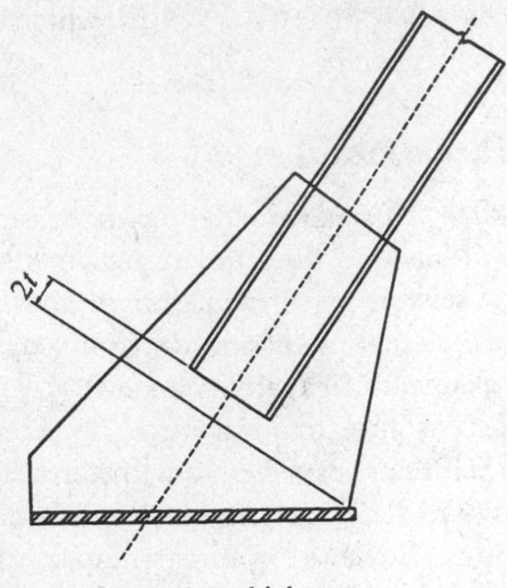

FIGURE 39.9 Plastic hinge and free length of gusset plate.

End connections of the brace can be designed as either rigid or pin connection. For either of the end connection types, test results showed that the hysteresis responses are similar for a given *KL/r* [47]. When the brace is pin-connected and the brace is designed to buckle out of plane, it is suggested that the brace be terminated on the gusset a minimum of two times the gusset thickness from a line about which the gusset plate can bend unrestrained by the column or beam joints [6]. This condition is illustrated in Figure 39.9. The gusset plate should also be designed to carry the design compressive strength of the brace member without local buckling.

The effect of end fixity should be considered in determining the critical buckling axis if rigid end conditions are used for in-plane buckling and pinned connections are used for out-of-plane buckling. When analysis indicates that the brace will buckle in the plane of the braced frame, the design flexural strength of the connection should be equal to or greater than the expected flexural strength ($= 1.1R_yM_p$) of the brace. An exception to this requirement is permitted when the brace connections (1) meet the requirement of tensile rupture strength described above, (2) can accommodate the inelastic rotations associated with brace postbuckling deformations, and (3) have a design strength at least equal to the nominal compressive strength ($= A_gF_y$) of the brace.

Special Requirements for Brace Configuration

Because braces meet at the midspan of beams in V-type and inverted-V-type braced frames, the vertical force resulting from the unequal compression and tension strengths of the braces can have a considerable impact on cyclic behavior. Therefore, when this type of brace configuration is considered for SCBFs, the AISC Seismic Provisions require that:

1. A beam that is intersected by braces be continuous between columns.
2. A beam that is intersected by braces be designed to support the effects of all the prescribed tributary gravity loads assuming that the bracing is not present.
3. A beam that is intersected by braces be designed to resist the prescribed force effects incorporating an unbalanced vertical seismic force. This unbalanced seismic load must be substituted for the seismic force effect in the load combinations, and is the maximum unbalanced vertical force applied to the beam by the braces. It should be calculated using a minimum of P_y for the brace in tension and a maximum of $0.3\ \phi_c\ P_n$ for the brace in compression. This requirement ensures that the beam will not fail due to the large unbalanced force after brace buckling.
4. The top and bottom flanges of the beam at the point of intersection of braces must be adequately braced; the lateral bracing should be designed for 2% of the nominal beam flange strength ($= F_yb_ft_{bf}$).

For OCBFs, the AISC Seismic Provisions waive the third requirement. But the brace members need to be designed for 1.5 times the required strength computed from the prescribed load combinations.

Columns

Based on the capacity design principle, columns in a CBF must be designed to remain elastic when all braces have reached their maximum tension or compression capacity considering an overstrength factor of $1.1R_y$. The AISC Seismic Provisions also require that columns satisfy the λps requirements (see Table 39.2). The strength of column splices must be designed to resist the imposed load effects. Partial penetration groove welds in the column splice have been experimentally observed to fail in a brittle manner [17]. Therefore, the AISC Seismic Provisions require that such splices in SCBFs be designed for at least 200% of the required strength, and be constructed with a minimum strength of 50% of the expected column strength, R_yF_yA, where A is the cross-sectional area of the smaller column connected. The column splice should be designed to develop both the nominal shear strength and 50% of the nominal flexural strength of the smaller section connected. Splices should be located in the middle one-third of the clear height of the column.

39.3.2 Eccentrically Braced Frames

Research results have shown that a well-designed EBF system possesses high stiffness in the elastic range and excellent ductility capacity in the inelastic range [25]. The high elastic stiffness is provided by the braces and the high ductility capacity is achieved by transmitting one brace force to another brace or to a column through shear and bending in a short beam segment designated as a "link." Figure 39.8 shows some typical arrangements of EBFs. In the figure, the link lengths are identified by the letter *e*. When properly detailed, these links provide a reliable source of energy dissipation. Following the capacity design concept, buckling of braces and beams outside of the link can be prevented by designing these members to remain elastic while resisting forces associated with the fully yielded and strain-hardened links. The AISC Seismic Provisions (1997) [6] for the EBF design are intended to achieve this objective.

Links

Figure 39.10 shows the free-body diagram of a link. If a link is short, the entire link yields primarily in shear. For a long link, flexural (or moment) hinge would form at both ends of the link before the "shear" hinge can be developed. A short link is desired for an efficient EBF design. In order to ensure stable yielding, links should be plastic sections satisfying the width–thickness ratios $\lambda\ ps$ given in Table 39.2. Doubler plates welded to the link web should not be used as they do not perform as intended when subjected to large inelastic deformations. Openings should also be avoided as they adversely affect the yielding of the link web. The required shear strength, V_u, resulting from the prescribed load effects should not exceed the design shear strength of the link, $\phi\ V_n$, where ϕ = 0.9. The nominal shear strength of the link is

$$V_n = \min\ \{V_p,\ 2M_p/e\} \tag{39.7}$$

$$V_p = 0.60F_yA_w \tag{39.8}$$

where $A_w = (d - 2t_f)t_w$.

A large axial force in the link will reduce the energy dissipation capacity. Therefore, its effect shall be considered by reducing the design shear strength and the link length. If the required link axial strength, *Pu*, resulting from the prescribed seismic effect exceeds $0.15P_y$, where $P_y = A_gF_y$, the following additional requirements should be met:

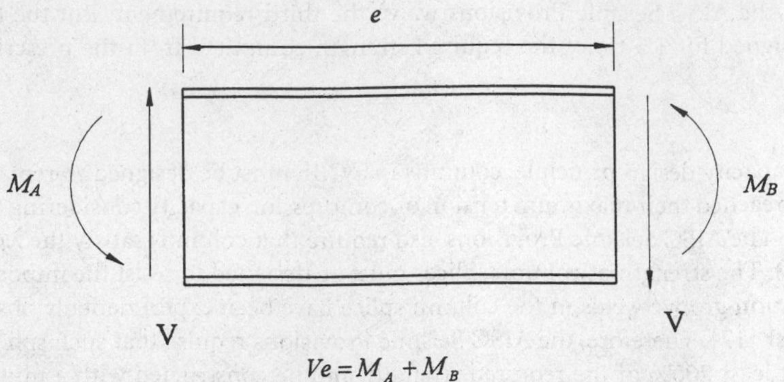

FIGURE 39.10 Static equilibrium of link.

1. The link design shear strength, $\phi\ V_n$, should be the lesser of $\phi\ Vpa$ or $2\ \phi\ M_{Pa}/e$, where V_{pa} and M_{Pa} are the reduced shear and flexural strengths, respectively:

$$V_{pa} = V_p \sqrt{1-(P_u / P_y)^2} \tag{39.9}$$

$$M_{Pa} = 1.18 M_p [1-P_u/P_y)] \tag{39.10}$$

2. The length of the link should not exceed:

$$[1.15 - 0.5\ \rho' (A_w/A_g)] 1.6 M_p/V_p \quad \text{for } \rho' (A_w/A_g) \geq 0.3 \tag{39.11}$$

$$1.6 M_p/V_p \quad \text{for } \rho' (A_w/A_g) < 0.3 \tag{39.12}$$

where $\rho' = P_u/V_u$.

The link rotation angle, γ, is the inelastic angle between the link and the beam outside of the link. The link rotation angle can be conservatively determined assuming that the braced bay will deform in a rigid–plastic mechanism. The plastic mechanism for one EBF configuration is illustrated in Figure 39.11. The plastic rotation is determined using a frame drift angle, θ_p, computed from the maximum frame displacement. Conservatively ignoring the elastic frame displacement, the plastic frame drift angle is $\theta_p = \delta / h$, where δ is the maximum displacement and h is the frame height.

Links yielding in shear possess a greater rotational capacity than links yielding in bending. For a link with a length of $1.6M_p/V_p$ or shorter (i.e., shear links), the link rotational demand should not exceed 0.08 rad. For a link with a length of $2.6M_p/V_p$ or longer (i.e., flexural links), the link rotational angle should not exceed 0.02 rad. A straight-line interpolation can be used to determine the link rotation capacity for the intermediate link length.

Link Stiffeners

In order to provide ductile behavior under severe cyclic loading, close attention to the detailing of link web stiffeners is required. At the brace end of the link, full-depth web stiffeners should be provided on both sides of the link web. These stiffeners should have a combined width not less than $(b_f - 2t_w)$, and a thickness not less than $0.75t_w$ nor ⅜ in. (10 mm), whichever is larger, where b_f and t_w are the link flange width and web thickness, respectively. In order to delay the link web or flange buckling, intermediate link web stiffeners should be provided as follows.

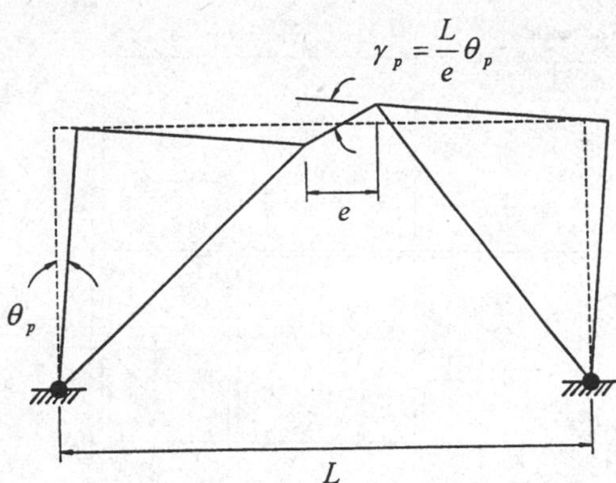

FIGURE 39.11 Energy dissipation mechanism of an eccentric braced frame.

1. In shear links, the spacing of intermediate web stiffeners depends on the magnitude of the link rotational demand. For links of lengths $1.6M_p/V_p$ or less, the intermediate web stiffener spacing should not exceed $(30t_w - d/5)$ for a link rotation angle of 0.08 rad, or $(52tw - d/5)$ for link rotation angles of 0.02 rad or less. Linear interpolation should be used for values between 0.08 and 0.02 rad.
2. Flexural links having lengths greater than $2.6M_p/V_p$ but less than $5M_p/V_p$ should have intermediate stiffeners at a distance from each link end equal to 1.5 times the beam flange width. Links between shear and flexural limits should have intermediate stiffeners meeting the requirements of both shear and flexural links. If link lengths are greater than $5M_p/V_p$, no intermediate stiffeners are required.
3. Intermediate stiffeners shall be full depth in order to react effectively against shear buckling. For links less than 25 in. deep, the stiffeners can be on one side only. The thickness of one-sided stiffeners should not be less than tw or ⅜ in., whichever is larger, and the width should not be less than $(b_f/2) - t_w$.
4. Fillet welds connecting a link stiffener to the link web should have a design strength adequate to resist a force of $A_{st}F_y$, where A_{st} is the area of the stiffener. The design strength of fillet welds connecting the stiffener to the flange should be adequate to resist a force of $A_{st}F_y/4$.

Link-to-Column Connections

Unless a very short shear link is used, large flexural demand in conjunction with high shear can develop at the link-to-column connections [25,63]. In light of the moment connection fractures observed after the Northridge earthquake, concerns have been raised on the seismic performance of link-to-column connections during a major earthquake. As a result, the AISC Seismic Provisions (1997) [6] require that the link-to-column design be based upon cyclic test results. Tests should follow specific loading procedures and results demonstrate an inelastic rotation capacity which is 20% greater than that computed in design. To avoid link-to-column connections, it is recommended that configuring the link between two braces be considered for EBF systems.

Lateral Support of Link

In order to assure stable behavior of the EBF system, it is essential to provide lateral support at both the top and bottom link flanges at the ends of the link. Each lateral support should have a design strength of 6% of the expected link flange strength ($= R_yF_yb_ft_f$).

Diagonal Brace and Beam outside of Link

Following the capacity design concept, diagonal braces and beam segments outside of the link should be designed to resist the maximum forces that can be generated by the link. Considering

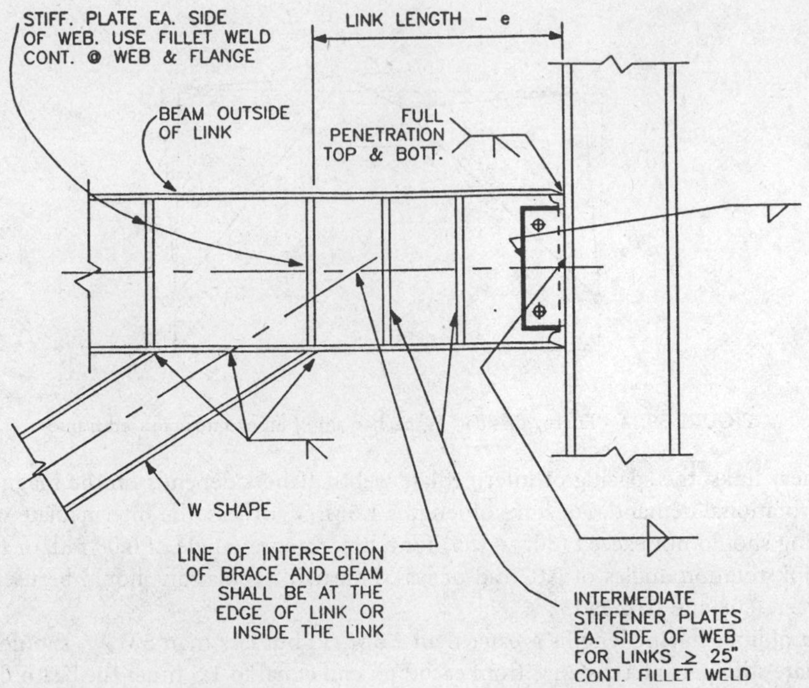

FIGURE 39.12 Diagonal brace fully connected to link. (Source: *AISC, Seismic Provisions for Structural Steel Buildings*; AISC, Chicago, IL, 1992. With permission.)

the strain-hardening effects, the required strength of the diagonal brace should be greater than the axial force and moment generated by 1.25 times the expected nominal shear strength of the link, $R_y V_n$.

The required strength of the beam outside of the link should be greater than the forces generated by 1.1 times the expected nominal shear strength of the link. To determine the beam design strength, it is permitted to multiply the beam design strength by the factor R_y. The link shear force will generate axial force in the diagonal brace. For most EBF configurations, the horizontal component of the brace force also generates a substantial axial force in the beam segment outside of the link. Since the brace and the beam outside of the link are designed to remain essentially elastic, the ratio of beam or brace axial force to link shear force is controlled primarily by the geometry of the EBF. This ratio is not much affected by the inelastic activity within the link; therefore, the ratio obtained from an elastic analysis can be used to scale up the beam and brace axial forces to a level corresponding to the link shear force specified above.

The link end moment is balanced by the brace and the beam outside of the link. If the brace connection at the link is designed as a pin, the beam by itself should be adequate to resist the entire link end moment. If the brace is considered to resist a portion of the link end moment, then the brace connection at the link should be designed as fully restrained. If used, lateral bracing of the beam should be provided at the beam top and bottom flanges. Each lateral bracing should have a required strength of 2% of the beam flange nominal strength, $F_y b_f t_f$. The required strength of the diagonal brace-to-beam connection at the link end of the brace should be at least the expected nominal strength of the brace. At the connection between the diagonal brace and the beam at the link end, the intersection of the brace and the beam centerlines should be at the end of the link or in the link (Figures 39.12 and 39.13). If the intersection of the brace and beam centerlines is located outside of the link, it will increase the bending moment generated in the beam and brace. The width–thickness ratio of the brace should satisfy λ_p specified in Table 39.2.

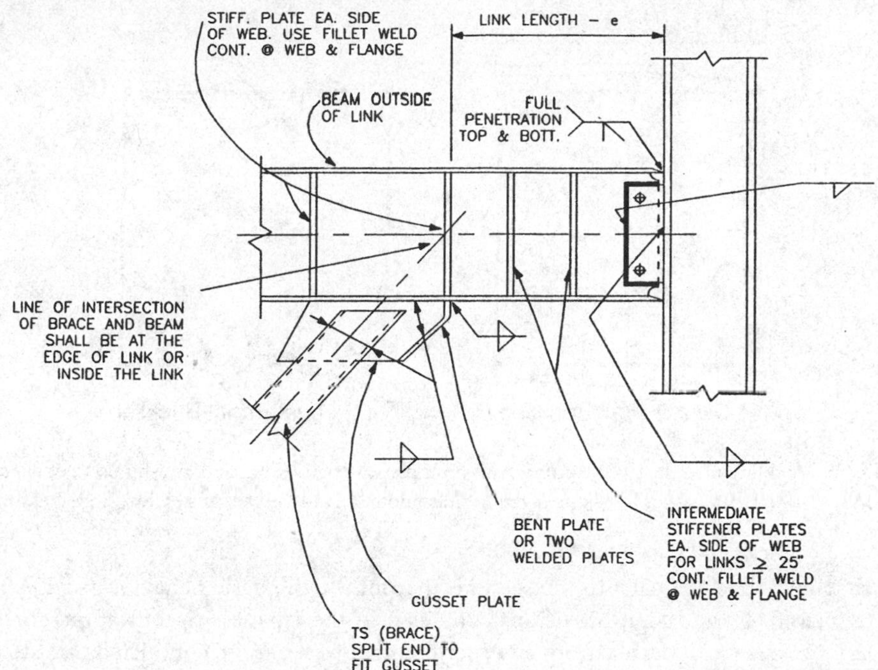

FIGURE 39.13 Diagonal brace pin-connected to link. (Source: AISC, *Seismic Provisions for Structural Steel Buildings*, AISC, Chicago, IL, 1992. With permission.)

Beam-to-Column Connections

Beam-to-column connections away from the links can be designed as simple shear connections. However, the connection must have a strength adequate to resist a rotation about the longitudinal axis of the beam resulting from two equal and opposite forces of at least 2% of the beam flange nominal strength, computed as $F_y b_f t_f$, and acting laterally on the beam flanges.

Required Column Strength

The required column strength should be determined from the prescribed load combinations, except that the moments and the axial loads introduced into the column at the connection of a link or brace should not be less than those generated by the expected nominal strength of the link, $RyVn$, multiplied by 1.1 to account for strain hardening. In addition to resisting the prescribed load effects, the design strength and the details of column splices must follow the recommendations given for the SCBFs.

39.4 Stiffened Steel Box Pier Design

39.4.1 Introduction

When space limitations dictate the use of a smaller-size bridge piers, steel box or circular sections gain an advantage over the reinforced concrete alternative. For circular or unstiffened box sections, the ductile detailing provisions of the AISC Seismic Provisions (1997) [6] or CHBDC [21] shall apply, including the diameter-to-thickness or width-to-thickness limits. For a box column of large dimensions, however, it is also possible to stiffen the wall plates by adding longitudinal and transverse stiffeners inside the section.

Design provisions for a stiffened box column are not covered in either the AASHTO or AISC design specifications. But the design and construction of this type of bridge pier has been common

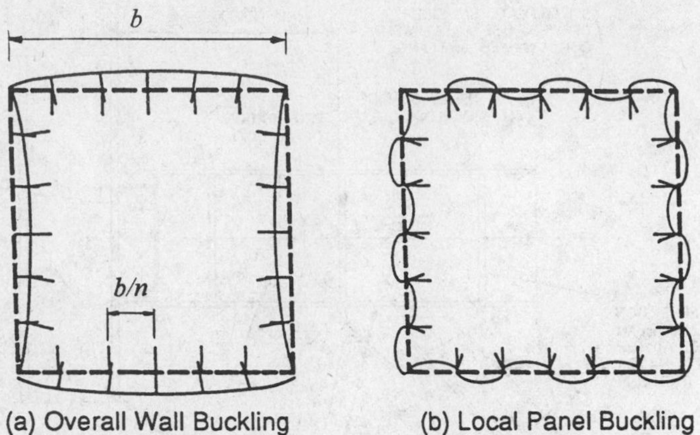

FIGURE 39.14 Buckling modes of box column with multiple stiffeners. (Source: Kawashima, K. et al., in *Stability and Ductility on Steel Structures under Cyclic Loading*, Fulsomoto, Y. and G. Lee, Eds., CRC, Boca Raton, FL, 1992. With permission.)

in Japan for more than 30 years. In the sections that follow, the basic behavior of stiffened plates is briefly reviewed. Next, design provisions contained in the Japanese Specifications for Highway Bridges [31] are presented. Results from an experimental investigation, conducted prior to the 1995 Hyogo-ken Nanbu earthquake in Japan, on cyclic performance of stiffened box piers are then used to evaluate the deformation capacity. Finally, lessons learned from the observed performance of this type of piers from the Hyogo-ken Nanbu earthquake are presented.

39.4.2 Stability of Rectangular Stiffened Box Piers

Three types of buckling modes can occur in a stiffened box pier. First, the plate segments between the longitudinal stiffeners may buckle, the stiffeners acting as nodal points (Figure 39.14b). In this type of "panel buckling," buckled waves appear on the surface of the piers, but the stiffeners do not appreciably move perpendicularly to the plate. Second, the entire stiffened box wall can globally buckle (Figure 39.14a). In this type of "wall buckling," the plate and stiffeners move together perpendicularly to the original plate plane. Third, the stiffeners themselves may buckle first, triggering in turn other buckling modes.

In Japan, a design criterion was developed following an extensive program of testing of stiffened steel plates in the 1960s and 1970s [70]; the results of this testing effort are shown in Figure 39.15, along with a best-fit curve. The slenderness parameter that defines the abscissa in the figure deserves some explanation. Realizing that the critical buckling stress of plate panels between longitudinal stiffeners can be obtained by the well-known result from the theory of elastic plate buckling:

$$F_{cr} = \frac{k_o \pi^2 E}{12(1-\nu^2)(b/nt)^2} \tag{39.13}$$

a normalized panel slenderness factor can be defined as

$$R_P = \sqrt{\frac{F_y}{F_{cr}}} = \left(\frac{b}{nt}\right)\sqrt{\frac{12\ (1-\nu^2)\ F_y}{k_o\ \pi^2 E}} \tag{39.14}$$

where b and t are the stiffened plate width and thickness, respectively, n is the number of panel spaces in the plate (i.e., one more than the number of internal longitudinal stiffeners across the

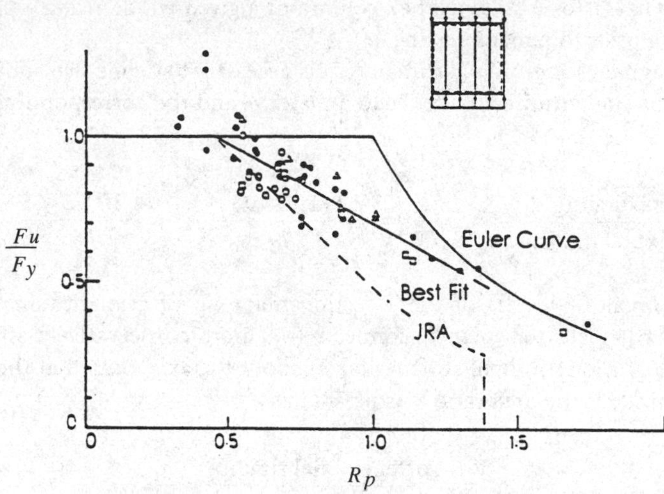

FIGURE 39.15 Relationship between buckling stress and R_p of stiffened plate.

plate), E is Young's modulus, ν is Poisson's ratio (0.3 for steel), and k_o (= 4 in. this case) is a factor taking into account the boundary conditions. The Japanese design requirement for stiffened plates in compression was based on a simplified and conservative curve obtained from the experimental data (see Figure 39.15):

$$\frac{F_u}{F_y} = 1.0 \qquad \text{for } R_p \le 0.5$$

$$\frac{F_u}{F_y} = 1.5 - R_P \qquad \text{for } 0.5 < R_P \le 1.0 \tag{39.15}$$

$$\frac{F_u}{F_y} = \frac{0.5}{R_P^2} \qquad \text{for } R_P > 1.0$$

where Fu is the buckling strength.

Note that values of F_u/F_y less than 0.25 are not permitted. This expression is then converted into the allowable stress format of the Japanese bridge code, using a safety factor of 1.7. However, as allowable stresses are magnified by a factor of 1.7 for load combinations which include earthquake effects, the above ultimate strength expressions are effectively used.

The point $R_P = 0.5$ defines the theoretical boundary between the region where the yield stress can be reached prior to local buckling ($R_P < 0.5$), and vice versa ($R_P \ge 0.5$). For a given steel grade, Eq. (39.14) for $R_P = 0.5$ corresponds to a limiting b/nt ratio:

$$\left(\frac{b}{nt}\right)_o = \frac{162}{\sqrt{F_y}} \tag{39.16}$$

and for a given plate width, b, the "critical thickness," t_o, is

$$t_o = \frac{b\sqrt{F_y}}{162n} \tag{39.17}$$

where F_y is in ksi. That is, for a stiffened box column of a given width, using a plate thicker than t_o will ensure yielding prior to panel buckling.

To be able to design the longitudinal stiffeners, it is necessary to define two additional parameters: the stiffness ratio of a longitudinal stiffener to a plate, γ_l, and the corresponding area ratio, δ_l. As the name implies:

$$\gamma_l = \frac{\text{stiffener flexural rigidity}}{\text{plate flexural rigidity}} = \frac{EI_l}{bD} = \frac{12(1-\nu^2)\, I_l}{bt^3} = \frac{10.92\, I_l}{b\, t^3} \approx \frac{11\, I_l}{b\, t^3} \tag{39.18}$$

where I_l is the moment of inertia of the T-section made up of a longitudinal stiffener and the effective width of the plate to which it connects (or, more conservatively and expediently, the moment of inertia of a longitudinal stiffener taken about the axis located at the inside face of the stiffened plate). Similarly, the area ratio is expressed as

$$\delta_l = \frac{\text{stiffener axial rigidity}}{\text{plate axial rigidity}} = \frac{A_l}{bt} \tag{39.19}$$

where A_l is the area of a longitudinal stiffener.

Since the purpose of adding stiffeners to a box section is partly to eliminate the severity of wall buckling, there exists an "optimum rigidity," γ_l^*, of the stiffeners beyond which panel buckling between the stiffeners will develop before wall buckling. In principle, according to elastic buckling theory for ideal plates (i.e., plates without geometric imperfections and residual stresses), further increases in rigidity beyond that optimum would not further enhance the buckling capacity of the box pier. Although more complex definitions of this parameter exist in the literature [35], the above description is generally sufficient for the box piers of interest here. This optimum rigidity is:

$$\gamma_l^* = 4\alpha^2 n(1 + n\delta_l) - \frac{(\alpha^2+1)^2}{n} \qquad \text{for } \alpha \leq \alpha_o \tag{39.20}$$

and

$$\gamma_l^* = \frac{1}{n}\left\{\left[2n^2(1+n\delta_l)-1\right]^2 - 1\right\} \qquad \text{for } \alpha > \alpha_o \tag{39.21}$$

where α is the aspect ratio, a/b, a being the spacing between the transverse stiffeners (or diaphragms), and the critical aspect ratio α_o is defined as

$$\alpha_o = \sqrt[4]{1 + n\,\gamma_l} \tag{39.22}$$

These expressions can be obtained by recognizing that, for plates of thickness less than t_o, it is logical to design the longitudinal stiffeners such that wall buckling does not occur prior to panel buckling and, consequently, as a minimum, be able to reach the same ultimate stress as the latter. Defining a normalized slenderness factor, R_H, for the stiffened plate:

$$R_H = \sqrt{\frac{F_y}{F_{cr}}} = \left(\frac{b}{t}\right)\sqrt{\frac{12(1-\nu^2)F_y}{k_s\,\pi^2 E}} \tag{39.23}$$

Based on elastic plate buckling theory, k_s for a stiffened plate is equal to [15]:

$$k_s = \frac{(1+\alpha^2)^2 + n\gamma_l}{\alpha^2(1+n\delta_l)} \qquad \text{for } \alpha \le \alpha_o$$
$$k_s = \frac{2\left(1+\sqrt{1+n\gamma_l}\right)}{(1+n\delta_l)} \qquad \text{for } \alpha > \alpha_o \tag{39.24}$$

Letting $R_H = R_P$ (i.e, both wall buckling and panel buckling can develop the same ultimate stress), the expressions for γ_l^* in Eqs. (39.20) and (39.21) can be derived. Thus, when the stiffened plate thickness, t, is less t_o, the JRA Specifications specify that either Eq. (39.20) or (39.21) be used to determine the required stiffness of the longitudinal stiffeners.

When a plate thicker than t_o is chosen, however, larger stiffeners are unnecessary since yielding will occur prior to buckling. This means that the critical buckling stress for wall buckling does not need to exceed the yield stress, which is reached by the panel buckling when $t = t_o$. The panel slenderness ratio for $t = t_o$ is

$$R_{P(t=t_o)} = \frac{b}{nt_o}\sqrt{\frac{12(1-\nu^2)F_y}{k_o\pi^2 E}} \tag{39.25}$$

Equating R_P to R_H in Eq. (39.23), the required γ_l^* can be obtained as follows:

$$\gamma_l^* = 4\alpha^2 n\left(\frac{t_o}{t}\right)^2(1+n\delta_l) - \frac{(\alpha^2+1)^2}{n} \qquad \text{for } \alpha \le \alpha_o \tag{39.26}$$

and

$$\gamma_l^* = \frac{1}{n}\left\{\left[2n^2\left(\frac{t_o}{t}\right)^2(1+n\delta_l)-1\right]^2 - 1\right\} \qquad \text{for } \alpha > \alpha_o \tag{39.27}$$

It is noteworthy that the above requirements do not ensure ductile behavior of steel piers. To achieve higher ductility for seismic application in moderate to high seismic regions, it is prudent to limit t to t_o. In addition to the above requirements, conventional slenderness limits are imposed to prevent local buckling of the stiffeners prior to that of the main member. For example, when a flat bar is used, the limiting width–thickness ratio (λ_r) for the stiffeners is $95/\sqrt{F_y}$.

The JRA requirements for the design of stiffened box columns are summarized as follows.

1. At least two stiffeners of the steel grade no less than that of the plate are required. Stiffeners are to be equally spaced so that the stiffened plate is divided into n equal intervals. To consider the beneficial effect of the stress gradient, b/nt in Eq. (39.14) can be replaced by $b/nt\varphi$, where φ is computed as

$$\varphi = \frac{\sigma_1 - \sigma_2}{\sigma_1} \tag{39.28}$$

In the above equation, σ_1 and σ_2 are the stresses at both edges of the plate; compressive stress is defined as positive, and $\sigma_1 > \sigma_2$. The value of φ is equal to 1 for uniform compression and 2 for equal and opposite stresses at both edges of the plate. Where the plastic hinge is expected to form, it is conservative to assume a φ value of 1.

2. Each longitudinal stiffener needs to have sufficient area and stiffness to prevent wall buckling. The minimum required area, in the form of an area ratio in Eq. (39.19), is

$$\left(\delta_l\right)_{\min} = \frac{1}{10n} \tag{39.29}$$

The minimum required moment of inertia, expressed in the form of stiffness ratio in Eq. (39.18), is determined as follows. When the following two requirements are satisfied, use either Eq. (39.26) for $t \geq t_o$ or Eq. (39.20) for $t < t_o$:

$$\alpha \leq \alpha_o \tag{39.30}$$

$$I_t \geq \frac{bt^3}{11}\left(\frac{1+n\gamma_l^*}{4\alpha^3}\right) \tag{39.31}$$

where *It* is the moment of inertia of the transverse stiffener, taken at the base of the stiffener. Otherwise, use either Eq. (39.27) for $t \geq t_o$ or Eq. (39.21) for $t < t_o$.

39.4.3 Japanese Research Prior to the 1995 Hyogo-ken Nanbu Earthquake

While large steel box bridge piers have been used in the construction of Japanese expressways for at least 30 years, research on their seismic resistance only started in the early 1980s. The first inelastic cyclic tests of thin-walled box piers were conducted by Usami and Fukumoto [65] as well as Fukumoto and Kusama [29]. Other tests were conducted by the Public Works Research Institute of the Ministry of Construction (e.g., Kawashima et al. [32]; MacRae and Kawashima [36]) and research groups at various universities (e.g., Watanabe et al. [69], Usami et al. [66], Nishimura et al. [45]).

The Public Work Research Institute tests considered 22 stiffened box piers of configuration representative of those used in some major Japanese expressways. The parameters considered in the investigation included the yield strength, weld size, loading type and sequence, stiffener type (flat bar vs. structural tree), and partial-height concrete infill. An axial load ranging from 7.8 to 11% of the axial yield load was applied to the cantilever specimens for cyclic testing. Typical hysteresis responses of one steel pier and one with concrete infill at the lower one-third of the pier height are shown in Figure 39.16. For bare steel specimens, test results showed that stiffened plates were able to yield and strain-harden. The average ratio between the maximum lateral strength and the predicted yield strength was about 1.4; the corresponding ratio between the maximum strength and plastic strength was about 1.2. The displacement ductility ranged between 3 and 5. Based on Eq. (39.2), the observed levels of ductility and structural overstrength imply that the response modification factor, *R*, for this type of pier can be conservatively taken as 3.5 ($\approx 1.2 \times 3$). Specimens with a γ_l / γ_l^* ratio less than 2.0 behaved in a wall-buckling mode with severe strength degradation. Otherwise, specimens exhibited local panel buckling.

Four of the 22 specimens were filled with concrete over the bottom one-third of their height. Prior to the Hyogo-ken Nanbu earthquake, it was not uncommon in Japan to fill bridge piers with concrete to reduce the damage which may occur as a result of a vehicle collision with the pier; generally the effect of the concrete infill was neglected in design calculations. It was thought prior to the testing that concrete infill would increase the deformation capacity because inward buckling of stiffened plates was inhibited.

Figure 39.17 compares the response envelopes of two identical specimens, except that one is with and the other one without concrete infill. For the concrete-filled specimen, little plate buckling was observed, and the lateral strength was about 30% higher than the bare steel specimen. Other than

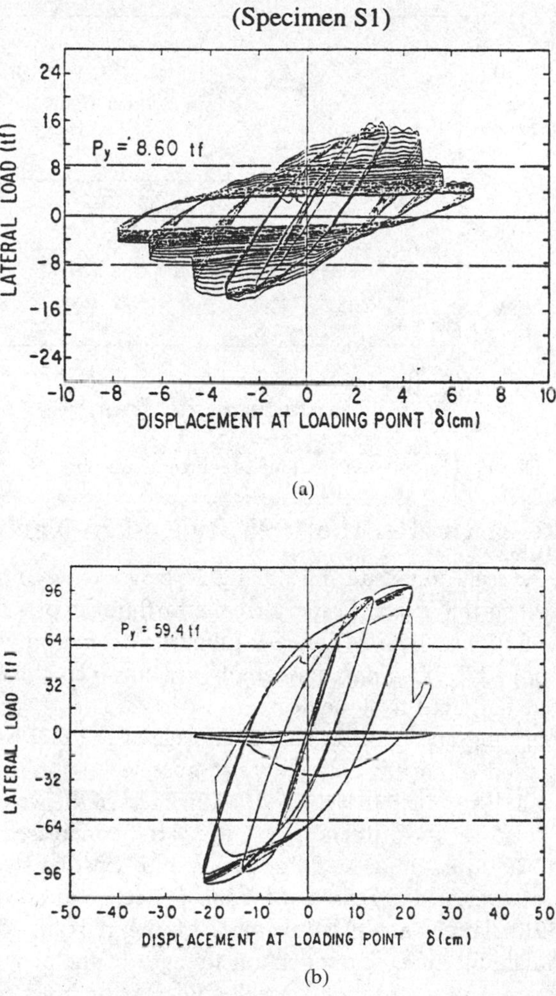

(a)

(b)

FIGURE 39.16 Hysteresis responses of two stiffened box piers. (Source: Kawashima, K. et al., in *Stability and Ductility on Steel Structures under Cyclic Loading*, Fulsomoto, Y. and G. Lee, Eds., CRC, Boca Raton, FL, 1992. With permission.)

exhibiting ductile buckling mode, all the concrete filled specimens suffered brittle fracture in or around the weld at the pier base. See Figure 39.16b for a typical cyclic response. It appears that the composite effect, which not only increased the overall flexural strength but also caused a shift of the neutral axis, produced an overload to the welded joint. As a result, the ductility capacity was reduced by up to 23%. It appears from the test results that the welded connection at the pier base needs to be designed for the overstrength including the composite effect.

A large portion of the research effort in years shortly prior to the Hyogo-ken Nanbu earthquake investigated the effectiveness of many different strategies to improve the seismic performance, ductility, and energy dissipation of those steel piers [34,66,68]. Among the factors observed to have a beneficiary influence were the use of (1) a γ_l / γ_l^* ratio above 3 as a minimum, or preferably 5; (2) longitudinal stiffeners having a higher grade of steel than the box plates; (3) minimal amount of stiffeners; (4) concrete filling of steel piers; and (5) box columns having round corners, built from bend plates, and having weld seams away from the corners, thus avoiding the typically problem-prone sharp welded corners [69].

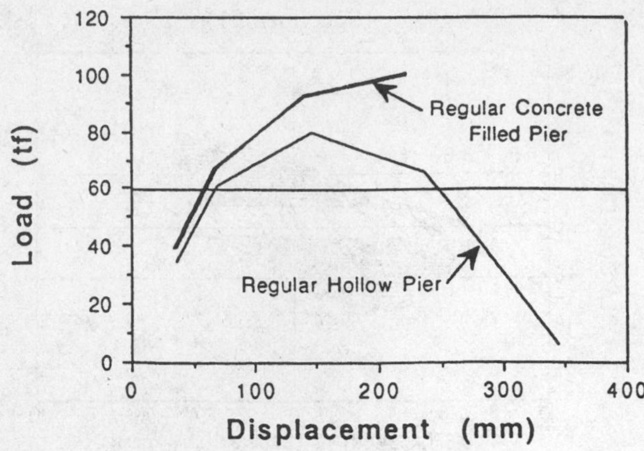

FIGURE 39.17 Effect of infill on cyclic response envelopes. (Source: Kawashima, K. et al., 1992.)

39.4.4 Japanese Research after the 1995 Hyogo-ken Nanbu Earthquake

Steel bridge piers were severely tested during the 1995 Hyogo-ken Nanbu earthquake in Japan. Recorded ground motions in the area indicated that the earthquake of a magnitude 7.2 produced a peak ground velocity of about 90 cm/s (the peak ground acceleration was about a 0.8 *g*). Of all the steel bridge piers, about 1% experienced severe damage or even collapse. But about a half of the steel bridge piers were damaged to some degree.

In addition to wall buckling and panel buckling, damage at the pier base, in the form of weld fracture or plastic elongation of anchor bolts, was also observed. Based on the observed buckling patterns, Watanabe et al. [69] suggested that the width–thickness of the wall plate needs to be further reduced, and the required stiffness of stiffeners, γ_l^*, needs to be increased by three times.

After the Hyogo-ken Nanbu earthquake, filling concrete to existing steel piers, either damaged or nondamaged ones, was suggested to be one of the most effective means to strength stiffened box piers [23,41]. Figure 39.18 shows a typical example of the retrofit. As was demonstrated in Figure 39.17, concrete infill will increase the flexural strength of the pier, imposing a higher force demand to the foundation and connections (welds and anchor bolts) at the base of the pier. Therefore, the composite effect needs to be considered in retrofit design. The capacities of the foundation and base connections also need to be checked to ensure that the weakest part of the retrofitted structure is not in these regions. Since concrete infill would force the stiffened wall plates to buckle outward, the welded joint between wall plates is likely to experience higher stresses (see Figure 39.19).

Because concrete infill for seismic retrofit may introduce several undesirable effects, alternative solutions have been sought that would enhance the deformation capacity while keeping the strength increase to a minimum. Based on the observed buckling of stiffened plates that accounted for the majority of damage to rectangular piers, Nishikawa et al. [44] postulated that local buckling is not always the ultimate limit state. They observed that bridge piers experienced limited damages when the corners of the box section remained straight, but piers were badly deformed when corner welds that fractured could not maintain the corners straight. To demonstrate their concept, three retrofit schemes shown in Figure 39.20a were verified by cyclic testing. Specimen No. 3 was retrofitted by adding stiffeners. Box corners of Specimen No. 4 were strengthened by welding angles and corner plates, while the stiffening angles of Specimen No. 5 were bolted to stiffened wall plates. Response envelopes in Figure 39.20b indicate that Specimen No. 5 had the least increase in lateral strength above the nonretrofitted Specimen No. 2, yet the deformation capacity was comparable to other retrofitted specimens.

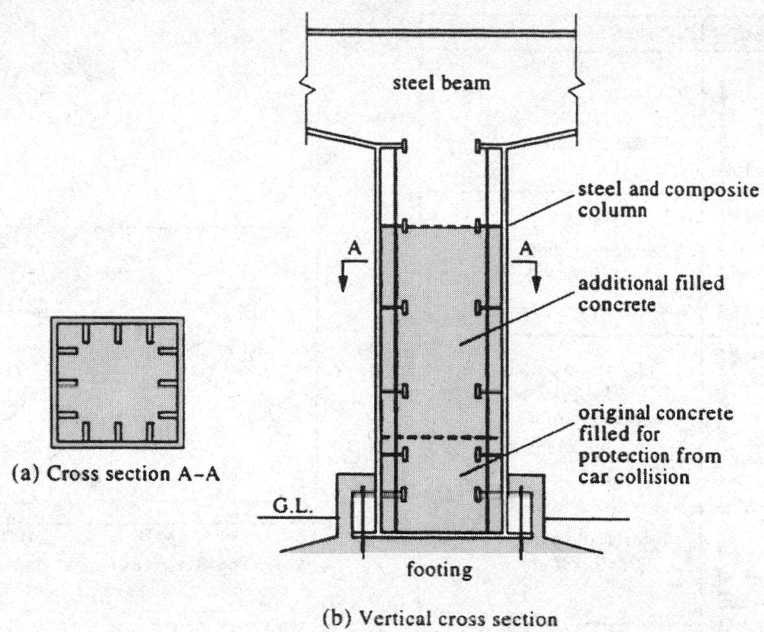

FIGURE 39.18 Retrofitted stiffened box pier with concrete infill. (Source: Fukumoto, Y. et al., in *Bridge Management*, Vol. 3, Thomas Telford, 1996. With permission.)

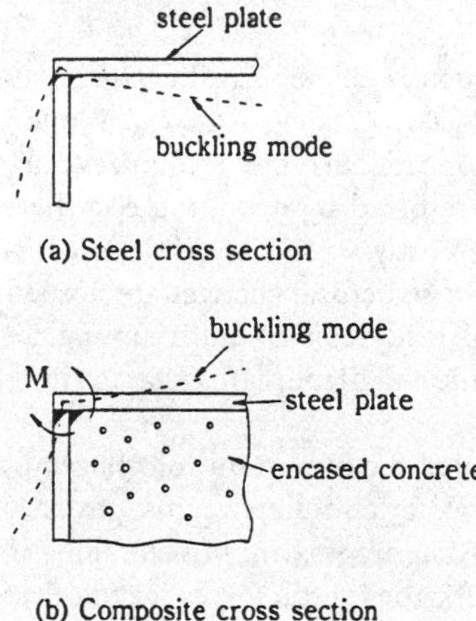

FIGURE 39.19 Effect of concrete infill on welded joint of stiffened box pier. (Source: Kitada, *Eng. Struct.*, 20(4–6), 347-354, 1998. With permission.)

39.5 Alternative Schemes

As described above, damage to substructure components such as abutments, piers, bearings, and others have proved to be of great consequence, often leading to span collapses [8,19,50]. Hence, when existing bridges are targeted for seismic rehabilitation, much attention needs to be paid to these substructure elements. Typically, the current retrofitting practice is to either strengthen or replace the existing nonductile members (e.g., ATC [10], Buckle et al. [20], Shirolé and Malik [58]),

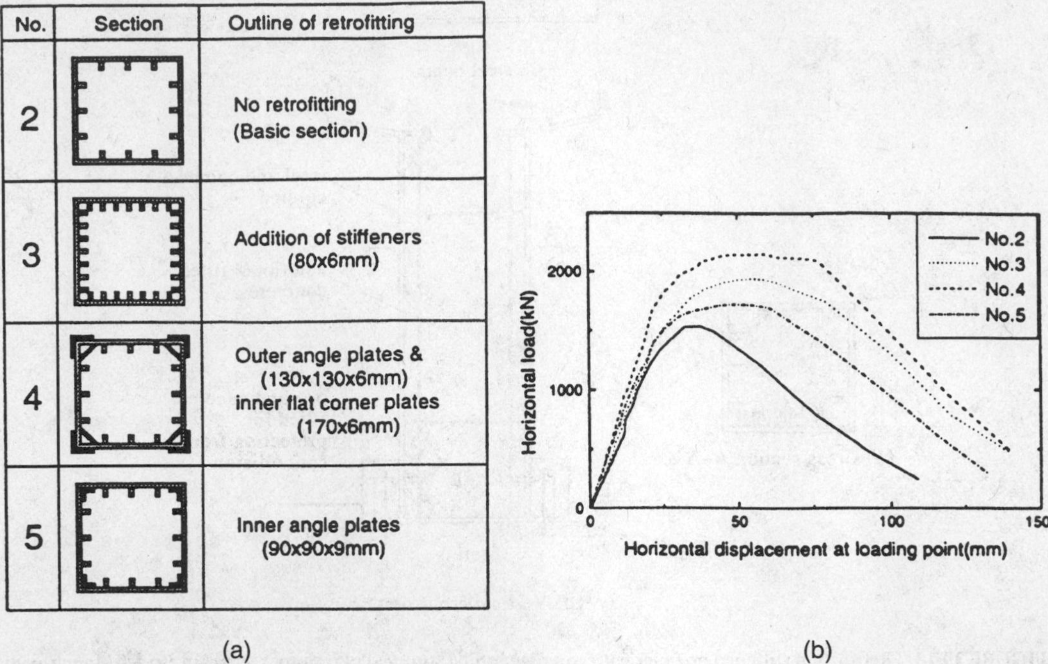

FIGURE 39.20 Seismic retrofit without concrete infill. (a) Retrofit schemes; (b) response envelopes. (Source: Nishikawa, K. et al, *Eng. Struct.*, 2062-6), 540-551, 1998. With permission.)

enhance the ductility capacity (e.g., Degenkolb [24], Priestley et al. [49]), or reduce the force demands on the vulnerable substructure elements using base isolation techniques or other structural modifications (e.g., Mayes et al. [38], Astaneh-Asl [7]). While all these approaches are proven effective, only the base isolation concept currently recognizes that seismic deficiency attributable to substructure weaknesses may be resolved by operating elsewhere than on the substructure itself. Moreover, all approaches can be costly, even base isolation in those instances when significant abutments modifications and other structural changes are needed to permit large displacements at the isolation bearings and lateral load redistribution among piers [39]. Thus, a seismic retrofit strategy that relies instead on ductile end diaphragms inserted in the steel superstructure, if effective, could provide an alternative.

Lateral load analyses have revealed the important role played by the end diaphragms in slab-on-girder steel bridges [72]. In absence of end diaphragms, girders severely distort at their supports, whether or not stiff intermediate diaphragms are present along the span. Because end diaphragms are key links along the load path for the inertia forces seismically induced at deck level, it might be possible, in some cases, to prevent damage from developing in the nonductile substructure (i.e., piers, foundation, and bearings) by replacing the steel diaphragms over abutments and piers with specially designed ductile diaphragms calibrated to yield before the strength of the substructure is reached. This objective is schematically illustrated in Figure 39.21 for slab-on-girder bridges and in Figure 39.22 for deck-truss bridges. In the latter case, however, ductile diaphragms must be inserted in the last lower lateral panels before the supports, in addition to the end diaphragms; in deck-truss bridges, seismically induced inertia forces in the transverse direction at deck level act with a sizable eccentricity with respect to the truss reaction supports, and the entire superstructure (top and lower lateral bracings, end and interior cross-frame bracings, and other lateral-load-resisting components) is mobilized to transfer these forces from deck to supports.

While conceptually simple, the implementation of ductile diaphragms in existing bridges requires consideration of many strength, stiffness, and drift constraints germane to the type of steel bridge

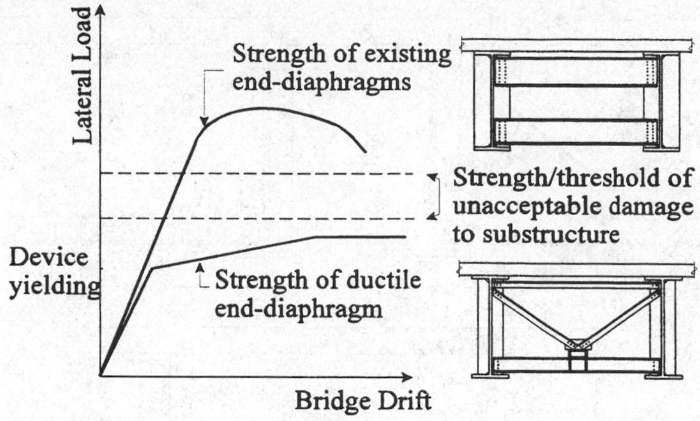

FIGURE 39.21 Schematic illustration of the ductile end-diaphragm concept.

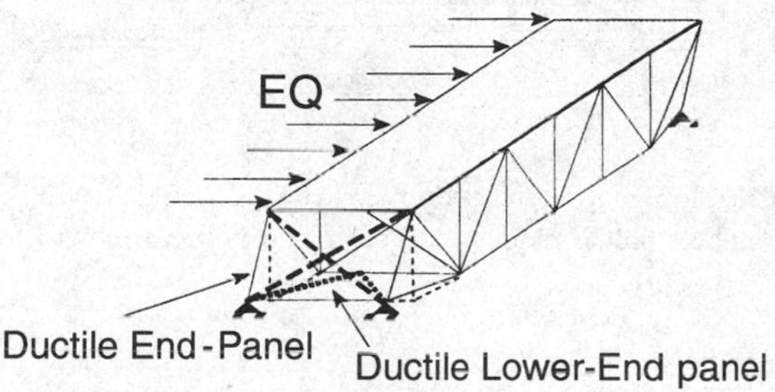

FIGURE 39.22 Ductile diaphragm retrofit concept in a deck-truss.

investigated. For example, for slab-on-girder bridges, because girders with large bearing stiffeners at the supports can contribute non-negligibly to the lateral strength of the bridges, stiff ductile diaphragms are preferred. Tests [73] confirmed that stiff welded ductile diaphragms are indeed more effective than bolted alternatives. As for deck-trusses, both upper and lower limits are imposed on the ductile diaphragm stiffnesses to satisfy maximum drifts and ductility requirements, and a systematic solution strategy is often necessary to achieve an acceptable retrofit [55,56].

Many types of systems capable of stable passive seismic energy dissipation could serve as ductile diaphragms. Among those, EBF presented in Section 39.3.2, shear panel systems (SPS) [28,40], and steel triangular-plate added damping and stiffness devices (TADAS) [61] have received particular attention in building applications. Still, to the authors' knowledge, none of these applications has been considered to date for bridge structures. This may be partly attributable to the absence of seismic design provisions in North American bridge codes. Examples of how these systems would be implemented in the end diaphragms of a typical 40-m span slab-on-girder bridge are shown in Figure 39.23. Similar implementations in deck-trusses are shown in Figure 39.24.

While the ductile diaphragm concept is promising and appears satisfactory for spans supported on stiff substructures based on the results available at the time of this writing, and could in fact be equally effective in new structures, more research is needed before common implementation is possible. In particular, large-scale experimental verification of the concept and expected behavior is desirable; parametric studies to investigate the range of substructure stiffnesses for which this retrofit strategy can be effective are also needed. It should also be noted that this concept only provides enhanced seismic resistance and substructure protection for the component of seismic excitation transverse to the bridge, and must be coupled with other devices that constrain longitudinal seismic displacements, such as simple bearings strengthening [37], rubber bumpers, and the

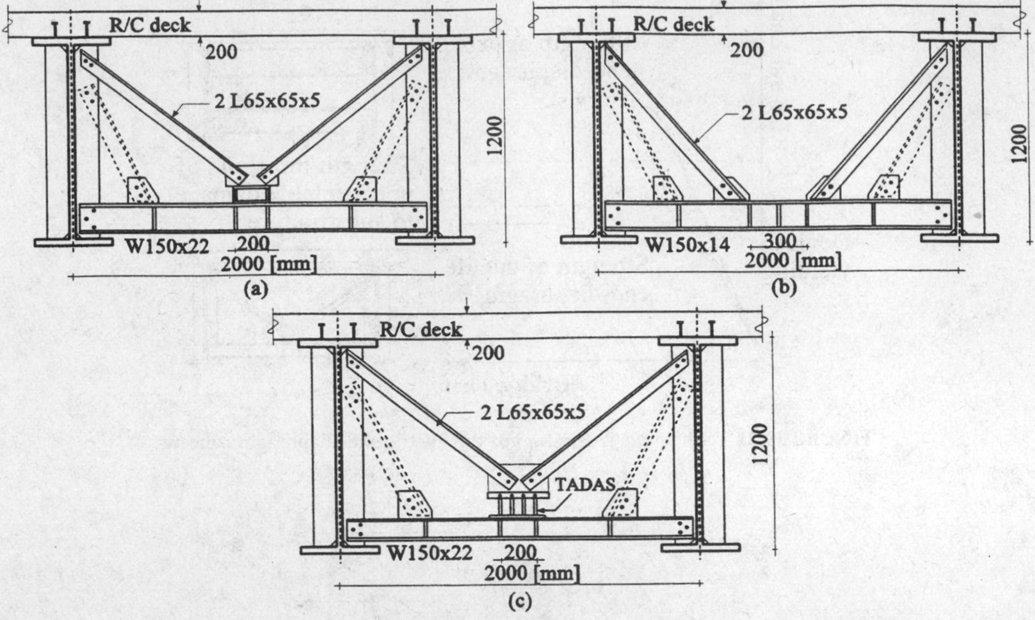

FIGURE 39.23 Ductile end diaphragm in a typical 40-m-span bridge (a) SPS; (b) EBF; (c) TADAS. (Other unbraced girders not shown; dotted members only if required for jacking purposes for nonseismic reasons).

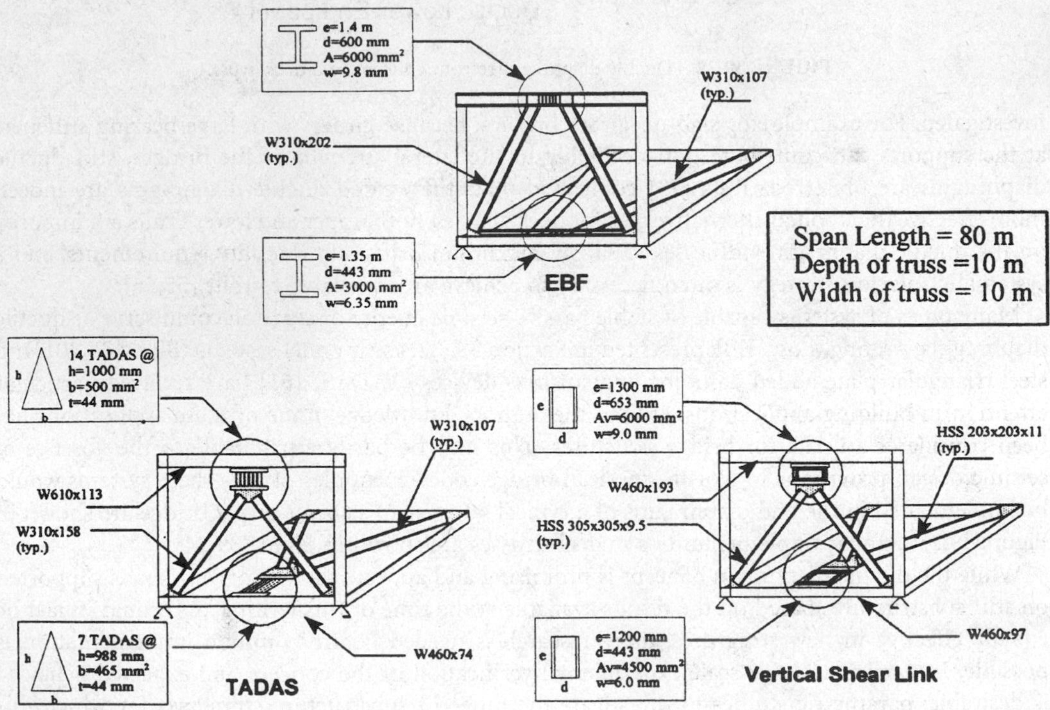

FIGURE 39.24 Examples of ductile retrofit systems at span end of deck-trusses.

like. Transportation agencies experienced in seismic bridge retrofit have indicated that deficiencies in the longitudinal direction of these bridges are typically easier to address than those in the lateral direction.

References

1. AASHTO, *LRFD Bridge Design Specifications*, American Association of State Highways and Transportaion Officials, Washington, D.C., 1994.
2. AASHTO, *Standard Specifications for Highway Bridges*, AASHTO, Washington, D.C., 1996.
3. AASHTO, *Standard Specifications for Seismic Design of Highway Bridges*, Washington, D.C., 1992.
4. AISC, Commentary on highly restrained welded connections, *Eng. J. AISC*, 10(3), 61–73, 1973.
5. AISC, *Load and Resistance Factor Design Specification for Structural Steel Buildings*, AISC, Chicago, IL, 1993.
6. AISC, *Seismic Provisions for Structural Steel Buildings*, AISC, Chicago, IL, 1992 and 1997.
7. Astaneh-Asl, A., Seismic retrofit concepts for the East Bay Crossing of the San Francisco–Oakland Bay Bridge," in *Proc. 1st U.S. Seminar on Seismic Evaluation and Retrofit of Steel Bridges*, San Francisco, CA, 1993.
8. Astaneh-Asl, A., Bolt, B., Mcmullin, K. M., Donikian, R. R., Modjtahedi, D., and Cho, S. W., Seismic Performance of Steel Bridges during the 1994 Northridge Earthquake, Report No. CE-STEEL 94/01, Berkeley, CA, 1994.
9. ATC, Tentative Provisions for the Development of Seismic Design Provisions for Buildings, Report No. ATC 3-06, Applied Technology Council, Palo Alto, CA, 1978.
10. ATC, Seismic Retrofitting Guidelines for Highway Bridges, *Report No. ATC-6-2*, Applied Technology Council, Palo Alto, CA, 1983.
11. ATC, Seismic Design Criteria for Highway Structures, *Report No. ATC-18*, Applied Technology Council, Redwood, CA,1996.
12. ATC, Improved Seismic Design Criteria for California Bridges: Provisional Recommendations, *Report No. ATC-32*, Applied Technology Council, Redwood, CA,1996.
13. AWS, *Structural Welding Code — Steel*, ANSI/AWS D1.1-98, AWS, Miami, FL, 1998.
14. AWS, *Bridge Welding Code*, ANSI/AWS D1.5-96, AWS, Miami, FL, 1996.
15. Ballio, G. and Mazzolani, F. M., *Theory and Design of Steel Structures*. Chapman and Hall, New York, 632 pp, 1983.
16. Blodgett, O. W. and Miller, D. K., Special welding issues for seismically resistant structures, in *Steel Design Handbook*, A. R. Tamboli, Ed., McGraw-Hill, New York, 1997.
17. Bruneau, M. and Mahin, S. A., Ultimate behavior of heavy steel section welded splices and design implications, *J. Struct. Eng. ASCE*, 116(8), 2214–2235, 1990.
18. Bruneau, M., Uang, C.-M., and Whittaker, A., *Ductile Design of Steel Structures*, McGraw-Hill, New York, 1997.
19. Bruneau, M., Wilson, J. W., and Tremblay, R., Performance of steel bridges during the 1995 Hyogo-ken-Nanbu (Kobe, Japan) earthquake, *Can. J. Civ. Eng.*, 23(3), 678–713, 1996.
20. Buckle, I. G., Mayes, R. L., and Button, M. R., Seismic Design and Retrofit Manual for Highway Bridges, Report No. FHWA-IP-87-6, U.S. Department of Transportation, Federal Highway Administration, 1986.
21. CHBDC, Canadian Highway Bridge Design Code, Seismic Provisions, Seismic Committee of the CHBDC, Rexdale, Ontario, Canada, 1998.
22. Chen, S. J., Yeh, C. H., and Chu, J. M., Ductile steel beam-column connections for seismic resistance, *J. Struct. Eng., ASCE*, 122(11), 1292–1299; 1996.
23. Committee on Roadway Bridges by the Hyogoken-Nanbu Earthquake, *Specifications on Retrofitting of Damaged Roadway Bridges by the Hyogoken-Nanbu Earthquake*, 1995. [in Japanese].
24. Degenkolb, O. H., Retrofitting bridges to increase seismic resistance, *J. Tech. Councils ASCE*, 104(TC1), 13–20, 1978.

25. Popov, E. P., Engelhardt, M. D., and Ricles, J. M., Eccentrically braced frames: U.S. practice, *Eng. J. AISC,* 26(2), 66–80, 1989.
26. Engelhardt, M. D. and Sabol, T., Reinforcing of steel moment connections with cover plates: benefits and limitations, *Eng. Struct.*, 20(4–6), 510–520, 1998.
27. Engelhardt, M. D., Winneburger, T., Zekany, A. J., and Potyraj, T. J., "The dogbone connection: Part II, *Modern Steel Construction, AISC,* 36(8), pp. 46–55, 1996.
28. Fehling, E., Pauli, W. and Bouwkamp, J. G., Use of vertical shear-links in eccentrically braced frames, *Proc. 10th World Conf. on Earthquake Eng.*, Madrid, Vol. 9, 1992, 4475–4479.
29. Fukumoto, Y. and Kusama, H., Cyclic bending tests of thin-walled box beams, *Proc. JSCE Struct. Eng./Earthquake Eng.*, 2(1), 117s–127s, 1985.
30. Fukumoto, Y., Watanabe, E., Kitada, T., Suzuki, I., Horie, Y., and Sakoda, H., Reconstruction and repair of steel highway bridges damaged by the Great Hanshin earthquake, in *Bridge Management,* Vol. 3, Thomas Telford, 1996, 8–16.
31. JRA, Specifications of Highway Bridges, Japan Road Association, Tokyo, Japan, 1996.
32. Kawashima, K., MacRae, G., Hasegawa, K., Ikeuchi, T., and Kazuya, O., Ductility of steel bridge piers from dynamic loading tests, in *Stability and Ductility of Steel Structures under Cyclic Loading,* Fukomoto, Y. and G. Lee, Eds., CRC Press, Boca Raton, FL, 1992.
33. Kitada, T., Ultimate strength and ductility of state-of-the-art concrete-filled steel bridge piers in Japan, *Eng. Struct.*, 20(4–6), 347–354, 1998.
34. Kitada, T., Nanjo, A., and Okashiro, S., Limit states and design methods considering ductility of steel piers for bridges under seismic load, in *Proc., 5th East Asia-Pacific Conference on Structural Engineering and Construction,* Queensland, Australia, 1995.
35. Kristek, V. and Skaloud, M., *Advanced Analysis and Design of Plated Structures, Developments in Civil Engineering,* Vol. 32, Elsevier, New York, 1991, 333 pp.
36. MacRae, G. and Kawashima, K., Estimation of the deformation capacity of steel bridge piers, in *Stability and Ductility of Steel Structures under Cyclic Loading,* Fukomoto, Y. and G. Lee, Eds., CRC Press, Boca Raton, FL, 1992.
37. Mander, J. B., Kim, D.-K., Chen, S. S., and Premus, G. J., Response of Steel Bridge Bearings to Reversed Cyclic Loading, Report No. NCEER-96-0014, State University of New York, Buffalo, 1996.
38. Mayes, R. L., Buckle, I. G., Kelly, T. E., and Jones, L. R., AASHTO seismic isolation design requirements for highway bridges, *J. Struct. Eng. ASCE,* 118(1), 284–304, 1992.
39. Mayes, R. L., Jones, D. M., Knight, R. P., Choudhury, D., and Crooks, R. S., Seismically isolated bridges come of age, *Proc., 4th Intl. Conf. on Short and Medium Span Bridges,* Halifax, Nova Scotia, 1994, 1095–1106.
40. Nakashima, M., Strain-hardening behavior of shear panels made of low-yield steel. I: Test, *J. Struct. Eng. ASCE,* 121(12), 1742–1749, 1995.
41. Nanjo, A., Horie, Y., Okashiro, S., and Imoto, I., Experimental study on the ductility of steel bridge piers, *Proc. 5th International Colloquium on Stability and Ductility of Steel Structures — SDSS '97,* Vol. 1, Nagoya, Japan, 1997, 229–236.
42. NEHRP, Recommended Provisions for the Development of Seismic Regulations for New Buildings, Federal Emergency Management Agency, Washington, D.C., 1998.
43. Newmark, N. M. and Hall, W. J., *Earthquake Spectra and Design,* EERI, 1982.
44. Nishikawa, K., Yamamoto, S., Natori, T., Terao, K., Yasunami, H., and Terada, M., Retrofitting for seismic upgrading of steel bridge columns, *Eng. Struct.*, 20(4–6), 540–551, 1998.
45. Nishimura, N., Hwang, W. S., and Fukumoto, Y., Experimental investigation on hysteretic behavior of thin-walled box beam-to-column connections, in *Stability and Ductility of Steel Structures under Cyclic Loading,* Fukomoto, Y. and G. Lee, Eds., CRC, Boca Raton, FL, 163–174, 1992 .
46. Plumier, A., The dogbone: back to the future, *Eng. J., AISC,* 34(2), 61–67, 1997.
47. Popov, E. P. and Black, W., Steel struts under severe cyclic loading, *J. Struct. Div. ASCE,* 90(ST2), 223–256, 1981.

48. Popov, E. P. and Tsai, K.-C., Performance of large seismic steel moment connections under cyclic loads, *Eng. J. AISC*, 26(2), 51–60, 1989.
49. Priestley, M. J. N., Seible, F., and Chai, Y. H., Seismic retrofit of bridge columns using steel jackets, *Proc.* 10th World Conf. on Earthquake Eng., Vol. 9, Madrid, 5285–5290, 1992.
50. Roberts, J. E., Sharing California's seismic lessons, *Modern Steel Constr., AISC*, 32(7), 32–37, 1992.
51. SAC, Interim Guidelines Advisory No. 1, Supplement to FEMA 267, Report No. FEMA 267A/SAC-96-03, SAC Joint Venture, Sacramento, CA, 1997.
52. SAC, Interim Guidelines: Evaluation, Repair, Modification, and Design of Welded Steel Moment Frame Structures, Report FEMA 267/SAC-95-02, SAC Joint Venture, Sacramento, CA, 1995.
53. SAC, Technical Report: Experimental Investigations of Beam-Column Sub-assemblages, Parts 1 and 2, Report No. SAC-96-01, SAC Joint Venture, Sacramento, CA, 1996.
54. SAC, Interim Guidelines Advisory No. 1, Supplement to FEMA 267, Report No. FEMA 267A/SAC-96-03, SAC Joint Venture, Sacramento, CA, 1997.
55. Sarraf, M. and Bruneau, M., Ductile seismic retrofit of steel deck-truss bridges. II: design applications, *J. Struct. Eng. ASCE*, 124 (11), 1263–1271,1998.
56. Sarraf, M. and Bruneau, M., Ductile seismic retrofit of steel deck-truss bridges. II: strategy and modeling, *J. Struct. Eng., ASCE*, 124 (11), 1253–1262, 1998 .
57. SEAOC, *Recommended Lateral Force Requirements and Commentary*, Seismology Committee, Structural Engineers Association of California, Sacramento, 1996.
58. Shirolé, A. M. and Malik, A. H., Seismic retrofitting of bridges in New York State," *Proc. Symposium on Practical Solutions for Bridge Strengthening and Rehabilitation*, Iowa State University, Ames, 123–131, 1993.
59. SSPC, *Statistical Analysis of Tensile Data for Wide Flange Structural Shapes*, Structural Shapes Producers Council, Washington, D.C., 1994.
60. Tide, R. H. R., Stability of weld metal subjected to cyclic static and seismic loading, *Eng. Struct.*, 20(4–6), 562–569, 1998.
61. Tsai, K. C., Chen, H. W., Hong, C. P., and Su, Y. F., Design of steel triangular plate energy absorbers for seismic-resistant construction," *Earthquake Spectra*, 9(3), 505–528, 1993.
62. Tsai, K. C. and Popov, E. P., Performance of large seismic steel moment connections under cyclic loads, *Eng. J. AISC*, 26(2), 51–60, 1989.
63. Tsai, K. C., Yang, Y. F., and Lin, J. L., Seismic eccentrically braced frames, *Int. J. Struct. Design Tall Buildings*, 2(1), 53–74, 1993.
64. Uang, C.-M., "Establishing R (or R_w) and C_d factors for building seismic provisions, *J. Struct. Eng., ASCE*, 117(1), 19–28, 1991.
65. Usami, T. and Fukumoto, Y., Local and overall buckling tests of compression members and an analysis based on the effective width concept, *Proc. JSCE*, 326, 41–50, 1982. [in Japanese].
66. Usami, T., Mizutani, S., Aoki, T., and Itoh, Y., Steel and concrete-filled steel compression members under cyclic loading, in *Stability and Ductility of Steel Structures under Cyclic Loading*, Fukomoto, Y. and G. Lee, Eds., CRC, Boca Raton, 123–138, 1992.
67. Vincent, J., Seismic retrofit of the Richmond–San Rafael Bridge, *Proc. 2nd U.S. Seminar on Seismic Design, Evaluation and Retrofit of Steel Bridges*, San Francisco, 215–232, 1996.
68. Watanabe, E., Sugiura, K., Maikawa, Y., Tomita, M., and Nishibayashi, M., Pseudo-dynamic test on steel bridge piers and seismic damage assessment, *Proc. 5th East Asia-Pacific Conference on Structural Engineering and Construction*, Queensland, Australia, 1995.
69. Watanabe, E., Sugiura, K., Mori, T., and Suzuki, I., Modeling of hysteretic behavior of thin-walled box members, in *Stability and Ductility of Steel Structures under Cyclic Loading*, Fukomoto, Y. and G. Lee, Eds., CRC Press, Boca Raton, FL, 225–236, 1992.
70. Watanabe, E., Usami, T., and Kasegawa, A., Strength and design of steel stiffened plates — a literature review of Japanese contributions, in *Inelastic Instability of Steel Structures and Structural Elements*, Y. Fujita and T. V. Galambos, Ed., U.S.–Japan Seminar, Tokyo, 1981.

71. Yura, J. A., Galambos, T. V., and Ravindra, M. K., The bending resistance of steel beams, *J. Struct. Div. ASCE*, 104(ST9), 1355–1370, 1978.
72. Zahrai, S. M. and Bruneau, M., Impact of diaphragms on seismic response of straight slab-on-girder steel bridges, *J. Struct. Eng. ASCE*, 124(8), 938–947, 1998.
73. Zahrai, S. M. and Bruneau, M, Seismic Retrofit of Steel Slab-on-Girder Bridges Using Ductile End-Diaphragms, Report No. OCEERC 98-20, Ottawa Carleton Earthquake Engineering Research Center, University of Ottawa, Ottawa, Ontario, Canada, 1998.
74. Zekioglu, A., Mozaffarian, H., Chang, K. L., Uang, C.-M., and Noel, S., Designing after Northridge, *Modern Steel Constr. AISC*, 37(3), 36–42, 1997.

40

Seismic Retrofit Practice

James Roberts
California Department of Transportation

Brian Maroney
California Department of Transportation

40.1 Introduction

Until the 1989 Loma Prieta earthquake, most of the United States had not been concerned with seismic design for bridges, although some 37 states have some level of seismic hazard and there are hundreds of bridges in these other states that have been designed to seismic criteria that are not adequate for the seismic forces and displacements that we know today. Recent earthquakes, such as the 1971 San Fernando, California; the 1976 Tangshan, China [3]; the 1989 Loma Prieta, California; the 1994 Northridge, California; and the 1995 Hyogo-ken Nanbu (Kobe), Japan, have repeatedly demonstrated the seismic vulnerability of existing bridges and the urgent need for seismic retrofit.

The California Department of Transportation (Caltrans) owns and maintains more than12,000 bridges (spans over 6 m) and some 6000 other highway structures such as culverts (spans under 6 m), pumping plants, tunnels, tubes, highway patrol inspection facilities, maintenance stations, toll plazas, and other transportation-related structures. There are about an equal number on the City and County systems. Immediately after the February 9, 1971 San Fernando earthquake, Caltrans began a comprehensive upgrading of their *Bridge Seismic Design Specifications,* construction details, and a statewide bridge seismic retrofit program to reinforce the older non-ductile bridges systematically.

The success of the bridge seismic design and retrofit program and the success of future seismic design for California bridges is based, to a large degree, on the accelerated and "problem-focused" seismic research program that has provided the bridge design community with the assurance that the new specifications and design details perform reliably and meet the performance criteria. Caltrans staff engineers, consulting firms, independent peer-review teams, and university researchers have cooperated in this program of bridge seismic design and retrofit strengthening to meet the challenge presented in the June 1990 Board of Inquiry report [4].

0-8493-7434-0/00/$0.00+$.50

This chapter discusses the bridge seismic retrofit philosophy and procedures practiced by the California bridge engineers. Issues addressed in this chapter can be of great benefit to those states and countries that are faced with seismic threats of lesser magnitude, yet have little financial support for seismic retrofitting, and much less for research and seismic detail development.

40.2 Identification and Prioritization

As part of any seismic retrofit program, the first phase should be to identify a list of specific bridges in need of retrofitting. That list of bridges also needs to be prioritized respecting which bridges pose the greatest risk to the community and therefore should be first to enter into a design phase in which a detailed analysis is completed and retrofit construction plans are completed for bidding.

In order to identify and prioritize a group of bridge projects, a type of coarse analysis must be completed. This analysis is carried out to expedite the process of achieving safety at the sites of the greatest risk. This analysis should not be confused with a detailed bridge system analysis conducted as part of the design phase. The process essentially identifies the projects that need to be addressed first. It should be recognized that it is not realistic to evaluate bridge systems to a refined degree in massive numbers simultaneously; however, it is quite possible to identify those bridges that possess the characteristics that have made bridges vulnerable, or at least more vulnerable, during past earthquakes. This coarse analysis is likely to be a collective review of databases of (1) bridge structural parameters that offer insight into the capacity of the systems to withstand earthquake loading and (2) bridge site parameters that offer insight to the potential for a site to experience threatening seismic motions. In case of many parameters to be evaluated, relative measures are possible. For example, if mass is recognized to be a characteristic that leads to poor behavior, then bridge systems can be compared quantitatively to their effective masses.

As the identification and prioritization process is well suited for high-speed computers, the process is vulnerable itself to being refined beyond its effective capacities. It is also vulnerable to errors of obvious omission because of the temptation to finalize the effort without appropriate review of the computer-generated results (i.e., never let a computer make a decision an engineer should make). The results should be reviewed carefully to check if they make engineering sense and are repeatable. In the Caltrans procedure, three separate experienced engineers reviewed each set of bridge plans and there had to be a consensus to retrofit or not. Common sense and experience are essential in this screening process.

40.2.1 Hazard

The seismic threat to a bridge structure is the potential for motions that are large enough to cause failure to occur at the bridge site. These measures of seismic threat eventually develop into the source of the demand side of the fundamental design equation. Such threats are characterized in numerous ways and presented in a variety of formats. One recognized method is to assume a deterministic approach and to recognize a single upper-bound measure of potential event magnitude for all nearby faults, assume motion characterizations for the fault sources, account for motion decay with distance from each fault, and characterize the motions at a site using a selected parameter such as spectral rock acceleration at 1 H. Alternatively, a probabilistic approach can be adopted that in a systematic manner incorporates the probabilities of numerous fault rupture scenarios and the attenuation of the motions generating the scenarios to the site. These motions then can be characterized in a variety of ways, including the additional information of a measure of the probability of occurrence. It is not economical to conduct a probabilistic ground motion study for each bridge. Size, longevity, and unusual foundations will generally determine the need.

Influences of the local geology at various sites are commonly accounted for employing various techniques. The motions can be teamed with the site response, which is often incorporated into the demand side, then called hazard. A hazard map is usually available in the bridge design specifications [1,2].

40.2.2 Structural Vulnerability

The vulnerability of a bridge system is a measure of the potential failure mechanisms of the system. To some degree, all bridge structures are ultimately vulnerable. However, judgment and reason can be applied to identify the practical vulnerabilities. Since the judgment is ideally based upon experience in observing field performances that are typically few in number, observing laboratory tests and considering/analyzing mechanisms, the judgment applied is very important and must be of high quality. Of these foundations upon which to base judgments, field observations are the most influential. The other two are more commonly used to develop or enhance understanding of the potential failure mechanisms.

Much has been learned about bridge performance in previous earthquakes. Bridge site, construction details, and structural configuration have major effects on bridge performance during an earthquake. Local site conditions amplify strong ground motions and subsequently increase the vulnerability of bridges on soft soil sites. The single-column-supported bridges were deemed more vulnerable because of lack of redundancy, based on experience in the 1971 San Fernando earthquake. Structural irregularity (such as expansion joints and C-bents) can cause stress concentration and have catastrophic consequences. Brittle elements with inadequate details always limit their ability to deform inelastically. A comprehensive discussion of earthquake damages to bridges and causes of the damage is presented in Chapter 34.

A designer's ability to recognize potential bridge system vulnerabilities is absolutely essential. A designer must have a conceptual understanding of the behavior of the system in order to identify an appropriate set of assumptions to evaluate or analyze the design elements.

40.2.3 Risk Analysis

A conventional risk analysis produces a probability of failure or survival. This probability is derived from a relationship between the load and resistance sides of a design equation. Not only is an approximate value for the absolute risk determined, but relative risks can be obtained by comparing determined risks of a number of structures. Such analyses generally require vast collections of data to define statistical distributions for all or at least the most important elements of some form of analysis, design, and/or decision equations. The acquisition of this information can be costly if obtainable at all. Basically, this procedure is to execute an analysis, evaluate both sides of the relevant design equation, and define and evaluate a failure or survival function. All of the calculations are carried out taking into account the statistical distribution of every equation component designated as a variable throughout the entire procedure.

To avoid such a large, time-consuming investment in resources and to obtain results that could be applied quickly to the retrofit program, an alternative, level-one risk analysis can be used. The difference between a conventional and level-one risk analysis is that in a level-one analysis judgments take the place of massive data supported statistical distributions.

The level-one risk analysis procedure can be summarized in the following steps:

1. Identify major faults with high event probabilities (priority-one faults)

Faults believed to be the sources of future significant seismic events should be identified by a team of seismologists and engineers. Selection criteria include location, geologic age, time of last displacement (late quarternary and younger), and length of fault (10 km min.). Each fault recognized in this step is evaluated for style, length, dip, and area of faulting in order to estimate potential earthquake magnitude. Faults are then placed in one of three categories: minor (ignored for the purposes of this project), priority two (mapped and evaluated but unused for this project), or priority one (mapped, evaluated, and recognized as immediately threatening). In California, this step was carried out by consulting the California Division of Mines and Geology and the recent U.S. Geological Survey studies.

2. Develop average attenuation relationships at faults identified in Step 1

3. Define the minimum ground acceleration capable of causing severe damage to bridge structures

The critical (i.e., damage-causing) level of ground acceleration is determined by performing nonlinear analyses on a typical highly susceptible structure (single-column connector ramp) under varying maximum ground acceleration loads. The lowest maximum ground acceleration that requires the columns providing a ductility ratio of 1.3 may be defined as the critical level of ground acceleration. The critical ground acceleration determined in the Caltrans study was 0.5 *g*.

4. Identify all the bridges within high-risk zones defined by the attenuation model of Step 2 and the critical acceleration boundary of Step 3

The shortest distance from every bridge to every priority-one fault is calculated. Each distance is compared to the distance from each respective level of magnitude fault to the critical ground acceleration decremented acceleration boundary. If the distance from the fault to the bridge is less than the distance from the fault to the critical acceleration boundary, the bridge shall be determined to lie in the high-risk zone and is added to the screening list for prioritization.

5. Prioritize the threatened bridges by summing weighted bridge structural and transportation characteristic scores

This step constitutes the process used to prioritize the bridges within the high-risk zones to establish the order of bridges to be investigated for retrofitting. It is in this step that a risk value is assigned to each bridge. A specifically selected subset of bridge structural and transportation characteristics of seismically threatened bridges should be prepared in a database. Those characteristics were ground acceleration; route type — major or minor; average daily traffic (ADT); column design single or multiple column bents; confinement details of column (relates to age); length of bridge; skew of bridge, and availability of detour.

Normalized preweight characteristic scores from 0.0 to 1.0 are assigned based on the information stored in the database for each bridge. Scores close to 1.0 represent high-risk structural characteristics or high cost of loss transportation characteristics. The preweight scores are multiplied by prioritization weights. Postweight scores are summed to produce the assigned prioritization risk value.

Determined risk values are not to be considered exact. Due to the approximations inherent in the judgments adopted, the risks are no more accurate than the judgments themselves. The exact risk is not important. Prioritization list qualification is determined by fault proximity and empirical attenuation data, not so much by judgment. Therefore, a relatively high level of confidence is associated with the completeness of the list of threatened bridges. Relative risk is important because it establishes the order of bridges to be investigated in detail for possible need of retrofit by designers.

A number of assumptions are made in the process of developing the prioritized list of seismically threatened bridges. These assumptions are based on what is believed to be the best engineering judgment available. It seems reasonable to pursue verification of these assumptions some time in the future. Two steps seem obvious: (1) monitoring the results of the design departments retrofit analyses and (2) executing a higher-level risk analysis.

Important features of this first step are the ease and cost with which it could be carried out and the database that could be developed highlighting bridge characteristics that are associated with structures in need of retrofit. This database will be utilized to confirm the assumptions made in the retrofit program. The same database will serve as part of the statistical support of a future conventional risk analysis as suggested in the second step. The additional accuracy inherent in a higher-order risk analysis will serve to verify previous assumptions, provide very good approximations of actual structural risk, and develop or evaluate postulated scenarios for emergency responses. It is reasonable to analyze only selected structures at this level. A manual screening process may be used that includes review of "as-built" plans by at least three engineers to identify bridges with common details that appeared to need upgrading.

After evaluating the results of the 1989 Loma Prieta earthquake, Caltrans modified the risk analysis algorithm by adjusting the weights of the original characteristics and adding to the list. The additional characteristics are soil type; hinges, type and number; exposure (combination of length and ADT); height; abutment type; and type of facility crossed.

Even though additional characteristics were added and weights were adjusted, the postweight scores were still summed to arrive at the prioritization risk factor. The initial vulnerability priority lists for state and locally owned bridges were produced by this technique and retrofit projects were designed and built.

In 1992, advances were made in the Caltrans procedures to prioritize bridges for seismic retrofit and a new, more accurate algorithm was developed. The most significant improvement to the prioritization procedure is the employment of the multiattribute decision theory. This prioritization scheme incorporates the information previously developed and utilizes the important extension to a multiplicative formulation.

This multiattribute decision procedure assigns a priority rating to each bridge enabling Caltrans to decide more accurately which structures are more vulnerable to seismic activity in their current state. The prioritization rating is based on a two-level approach that separates out seismic hazard from impact and structural vulnerability characteristics. Each of these three criteria (hazard, impact, and structural vulnerability) depends on a set of attributes that have direct impact on the performance and potential losses of a bridge. Each of the criteria and attributes should be assigned a weight to show their relative importance. Consistent with previous work, a global utility function is developed for each attribute.

This new procedure provides a systematic framework for treating preferences and values in the prioritization decision process. The hierarchical nature of this procedure has the distinct advantage of being able to consider seismicity prior to assessing impact and structural vulnerability. If seismic hazard is low or nonexistent, then the values of impact and structural vulnerability are not important and the overall postweight score will be low because the latter two are added but the sum of those two are multiplied by the hazard rating. This newly developed prioritization procedure is defensible and theoretically sound. It has been approved by Caltrans Seismic Advisory Board.

Other research efforts [5–7] in conjunction with the prioritization procedure involve a sensitivity study that was performed on bridge prioritization algorithms from several states. Each procedure was reviewed in order to investigate whether or not California was neglecting any important principles. In all, 100 California bridges were selected as a sample population and each bridge was independently evaluated by each of the algorithms. The 100 bridges were selected to represent California bridges with respect to the variables of the various algorithms. California, Missouri, Nevada, Washington, and Illinois have thus far participated in the sensitivity study.

The final significant improvement to the prioritization procedure is the formal introduction of varying levels of seismicity. A preliminary seismic activity map for the state of California has been developed in order to incorporate seismic activity into the new prioritization procedure. In late 1992 the remaining bridges on the first vulnerability priority list were reevaluated using the new algorithm and a significant number of bridges changed places on the priority list but there were no obvious trends. Figure 40.1 and Table 40.1 show the new algorithm and the weighting percentages for the various factors.

40.3 Performance Criteria

Performance criteria are the design goals that the designer is striving to achieve. How do you want the structure to perform in an earthquake? How much damage can you accept? What are the reasonable

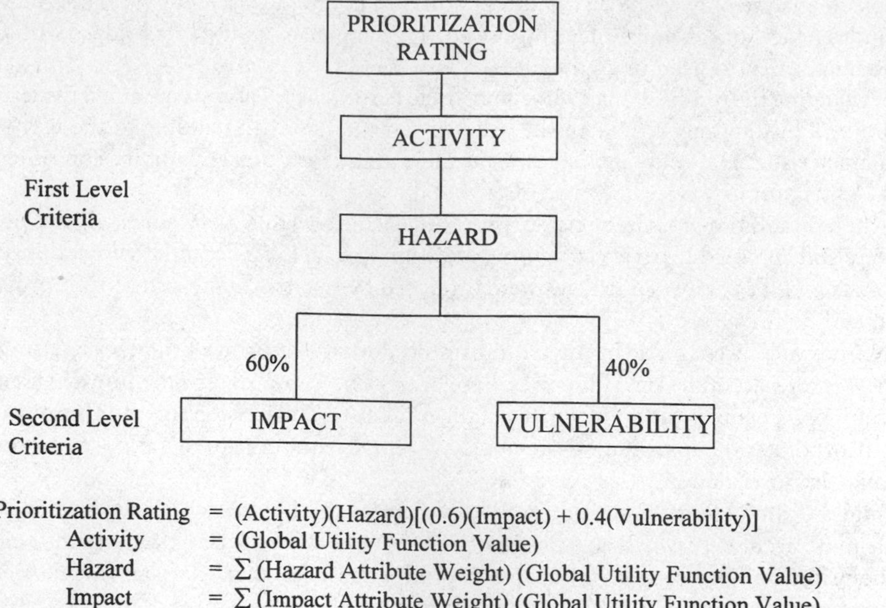

FIGURE 40.1 Risk analysis — multiattribute decision procedure.

TABLE 40.1 Multi-attribute Weights

	Attributes	Weights (%)
Hazard	Soil conditions	33
	Peak rock acceleration	28
	Seismic duration	29
Impact	ADT on structure	28
	ADT under/over structure	12
	Detour length	14
	Leased air space (residential, office)	15
	Leased air space (parking, storage)	7
	RTE type on bridge	7
	Critical utility	10
	Facility crosses	7
Vulnerability	Year designed (constructed)	25
	Hinges (drop-type failure)	16.5
	Outriggers, shared column	22
	Bent redundancy	16.5
	Skew	12
	Abutment type	8

alternate routes? How do you define various levels of damage? How long do you expect for repair of various levels of damage? The form of the performance criteria can take many forms usually depending on the perspective and background of the organization presenting it. The two most common forms are functional and structural [Caltrans 1993]. The functional is the most appropriate form for the performance criteria because it refers to the justification of the existence of the structure. For example, the functional performance criteria of a bridge structure would include measures of post-earthquake capacity for traffic to flow across the bridge. Performance of the structure itself is more appropriately addressed with the structural design criteria. This would be codes, design memorandums, etc. An example of

Ground Motion at Site	Minimum Performance Level	Important Bridge Performance Level
Functional Evaluation	Immediate Service Level Repairable Damage	Immediate Service Level Minimal Damage
Safety Evaluation	Limited Service Level Significant Damage	Immediate Service Level Repairable Damage

DEFINITIONS

Immediate Service Level: Full access to normal traffic available almost immediately.
Limited Service Level: Limited access, (reduced lanes, light emergency traffic) possible within days. Full service restorable within months.

Minimal Damage: Essentially elastic performance.
Repairable Damage: Damage that can be repaired with a minimum risk of losing functionality.
Significant Damage: A minimum risk of collapse, but damage that would require closure for repair.

Important Bridge (one or more of the following items present):

- Bridge required to provide secondary life safety. (example: access to an emergency facility).
- Time for restoration of functionality after closure creates a major economic impact.
- Bridge formally designated as critical by a local emergency plan.

Safety Evaluation Ground Motion (Up to two methods of defining ground motions may be used):

- *Deterministically assessed ground motions from the maximum earthquake as defined by the Division of Mines and Geology Open-File Report 92-1 (1992).*
- *Probabilistically assessed ground motions with a long return period (approx. 1000-2000 years).*

For Important bridges both methods shall be given consideration, however the probabilistic evaluation shall be reviewed by a CALTRANS approved consensus group. For all other bridges the motions shall be based only on the deterministic evaluation. In the future, the role of the two methods for other bridges shall be reviewed by a CALTRANS approved consensus group.

Functional Evaluation Ground Motion:
Probabilistically assessed ground motions which have a 40% probability of occurring during the useful life of the bridge. The determination of this event shall be reviewed by a CALTRANS approved consensus group. **A separate Functional Evaluation is required only for Important Bridges.** All other bridges are only required to meet specified design requirements to assure Minimum Functional Performance Level compliance.

FIGURE 40.2 Seismic performance criteria for the design and evaluation of bridges.

what would be addressed in design criteria would be acceptable levels of strains in different structural elements and materials. These levels of strains would be defined to confidently avoid a defined state of failure, a deformation state associated with loss of capacity to accommodate functional performance criteria, or accommodation of relatively easy-repair.

Performance criteria must have a clear set of achievable goals, must recognize they are not independent of cost, and should be consistent with community planning. Figure 40.2 shows the seismic performance criteria for the design and evaluation of bridges in the California State Highway System [Caltrans 1993].

Once seismic performance criteria are adopted, the important issue then is to guarantee that the design criteria and construction details will provide a structure that meets that adopted performance criteria. In California a major seismic research program has been financed to physically test large-scale and full-sized models of bridge components to provide reasonable assurance to the engineering community that those details will perform as expected in a major seismic event. The current phase of that testing program involves real-time dynamic shaking on large shake tables. In addition, the Caltrans bridge seismic design specifications have been thoroughly reviewed in the ATC-32 project to ensure that they are the most up-to-date with state-of-the-art technology. On important bridges, project-based design criteria have been produced to provide guidance to the various design team members on what must be done to members to ensure the expected performance.

40.4 Retrofit Design

40.4.1 Conceptual Design

Design is the most-impacting part of the entire project. The conceptual design lays out the entire engineering challenge and sets the course for the analysis and the final detailed design. The conceptual design is sometimes referred to as type-selection or, in the case of seismic retrofit, the strategy. A seismic retrofit strategy is essentially the project engineer's plan that lays out the structural behavior to lead to the specified performance. The most important influential earthquake engineering is completed in this early phase of design. It is within this phase that "smart" engineering can be achieved (i.e., work smarter not harder). That is, type-selections or strategies can be chosen such that unreliable or unnecessary analyses or construction methods are not forced or required to be employed. When this stage of the project is completed well, a plan is implemented such that difficulties are wisely avoided when possible throughout not only the analysis, design, specification development, and construction phases, but also the remaining life of the bridge from a maintenance perspective. With such understanding, an informed decision can be made about which structural system and mechanisms should be selected and advanced in the project.

Highway multiple connector ramps on an interchange typically are supported by at least one column in the median of a busy functioning freeway. Retrofit strategies that avoid column retrofitting of the median columns have safety advantages over alternatives. Typically, columns outside the freeway traveled way can be strengthened and toughened to avoid median work and the problems of traffic handing.

On most two- and three-span shorter bridges the majority of seismic forces can be transferred into the abutments and embankments and thus reduce or entirely eliminate the amount of column retrofitting necessary. Large-diameter CIDH piles drilled adjacent to the wingwalls at abutments have been effective in resisting both longitudinal and transverse forces.

For most multiple-column bents the footing retrofits can be reduced substantially by allowing the columns to hinge at the bottom. This reduces the moments transferred into the foundations and lowers total costs. Sufficient testing on footing/pile caps and abutments has been conducted. It is found that a considerable amount of passive lateral resistance is available. Utilizing this knowledge can reduce the lateral force requirement of the structural foundations.

Continuity is extremely important and is the easiest and cheapest insurance to obtain. Well-designed monolithic structures also have the added advantage of low maintenance. Joints and bearings are some of the major maintenance problems on bridges today. If structures are not continuous and monolithic, they must be tied together at deck joints, supports, and abutments. This will prevent them from pulling apart and collapsing during an earthquake.

Ductility in the substructure elements is the second key design consideration. It is important that when you design for ductility you must be willing to accept some damage during an earthquake. The secret to good seismic design is to balance acceptable damage levels with the economics of preventing or limiting the damage. Properly designed ductile structures will perform well during an earthquake as long as the design has accounted for the displacements and controlled or provided for them at abutments and hinges. For a large majority of bridges, displacement criteria control over strength criteria in the design for seismic resistance.

FIGURE 40.3 Hinge joint restrainer.

40.4.2 Retrofit Strategies

Designers of bridge seismic retrofit projects acquire knowledge of the bridge system, develop an understanding of the system response to potential earthquake ground motions, and identify and design modifications to the existing system that will change the expected response to one that satisfies the project performance criteria. This is accomplished by modifying any or all of the system stiffness, energy absorption, or mass characteristics. These characteristics or behavior can commonly be grouped into all structural system types, such as trusses, frames, single-column bent, shear walls, CIDH systems. This section briefly discusses various seismic retrofit strategies used in California. Chapter 43 presents more detailed information.

Hinge Joint Restrainers

Spans dropped off from too narrow support seats and separation of expansion joints were two major causes of bridge collapse during the 1971 San Fernando earthquake. The initial phase of the Caltrans Bridge Seismic Retrofit Program involved installation of hinge and joint restrainers to prevent deck joints from separating (Figure 40.3). Included in this phase was the installation of devices to fasten the superstructure elements to the substructure in order to prevent those superstructure elements from falling off their supports (Figure 40.4). This phase was essentially completed in 1989 after approximately 1260 bridges on the California State Highway System had been retrofitted at a cost of over $55 million.

Figure 40.5 shows the installation of an external hinge extender detail that is designed to prevent the supported section of the superstructure from dropping off its support. Note the very narrow hinge details at the top of this picture, which is common on the 1960s era bridges throughout California.

The Loma Prieta earthquake of October 17, 1989 again proved the reliability of hinge and joint restrainers, but the tragic loss of life at the Cypress Street Viaduct on I-880 in Oakland emphasized the necessity to accelerate the column retrofit phase of the bridge seismic retrofit program immediately with a higher funding level for both research and implementation [8].

Confinement Jackets

The largest number of large-scale tests have been conducted to confirm the calculated ductile performance of older, nonductile bridge columns that have been strengthened by application of

FIGURE 40.4 Hold down devices for vertical acceleration.

FIGURE 40.5 External hinge extenders on Santa Monica freeway structures.

structural concrete, steel plate, prestressed strand, and fiberglass-composite jackets to provide the confinement necessary to ensure ductile performance. Since the spring of 1987 the researchers at University of California, San Diego have completed over 80 sets of tests on bridge column models [9–14]. Figure 40.6 shows reinforcement confinement for a column retrofit. Figure 40.7 shows a completed column concrete jacket retrofit. Figure 40.8 is a completed steel jacket retrofit.

Approximately 2200 of California's 12,000 bridges are located in the Los Angeles area, so it is significant to examine the damage and performance of bridges in the Northridge earthquake of January 17, 1994. About 1200 of these bridges were in an area that experienced ground accelerations greater than 0.25 *g* and several hundred were in the area that experienced ground accelerations of

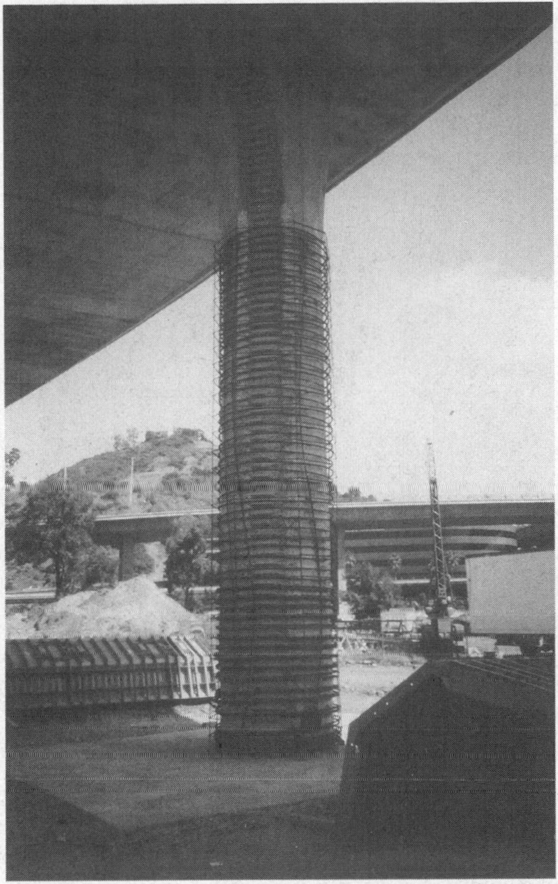

FIGURE 40.6 Reinforcement confinement retrofit.

0.50 *g*. There were 132 bridges in this area with post-San Fernando retrofit details completed and 63 with post-Loma Prieta retrofit details completed (Figure 40.9). All of these retrofitted bridges performed extremely well and most of the other bridges performed well during the earthquake; bridges constructed to the current Caltrans seismic specifications survived the earthquake with very little damage. Seven older bridges, designed for a smaller earthquake force or without the ductility of the current Caltrans design, sustained severe damage during the earthquake. Another 230 bridges suffered some damage ranging from serious problems of column and hinge damage to cracks, bearing damage, and approach settlements, but these bridges were not closed to traffic during repairs.

Link Beams

Link beams may be added to multicolumn bents to provide stiffener frame and reduce the unsupported column length. By using this development combined with other techniques, it may be possible to retrofit older, nonductile concrete columns without extensive replacement. Figure 40.10 shows link beams and installation of columns casings at the points of maximum bending and locations of anticipated plastic hinges on Santa Monica freeway structures. Half-scale models of these columns were constructed and tested under simulated seismic loading conditions to proof-test this conceptual retrofit design.

Ductile Concrete Column Details

Most concrete bridge columns designed since 1971 contain a slight increase in the main column vertical reinforcing steel and a major increase in confinement and shear reinforcing steel over the

FIGURE 40.7 Concrete jacket column retrofit.

pre-1971 designs. All new columns, regardless of geometric shape, are reinforced with one or a series of spiral-wound interlocking circular cages. The typical transverse reinforcement detail now consists of #6 (¾ in. diameter) hoops or continuous spiral at approximately 3-in. pitch over the full column height (Figure 40.11). This provides approximately eight times the confinement and shear reinforcing steel in columns than what was used in the pre-1971 nonductile designs. All main column reinforcing is continuous into the footings and superstructure. Splices are mostly welded or mechanical, both in the main and transverse reinforcing. Splices are not permitted in the plastic hinge zones. Transverse reinforcing steel is designed to produce a ductile column by confining the plastic hinge areas at the top and bottom of columns. The use of grade 60, A 706 reinforcing steel in bridges has recently been specified on all new projects.

Concrete Beam–Column–Bent Cap Details

Major advances have been made in the area of beam–column joint confinement, based on the results of research at both University of California, Berkeley, and San Diego. The performance and design criteria and structural details developed for the I-480 Terminal Separation Interchange and the I-880 replacement structures reflect the results of this research and were reported by Cooper [15]. Research is continuing at both institutions to refine the design details further to ensure ductile performance of these joints.

The concept using an integral edge beam can be used on retrofitting curved alignments, such as the Central Viaduct (U.S. 101) in downtown San Francisco and the Alemany Interchange on U.S. 101 in south San Francisco. The proofing-testing program was reported by Mahin [1991]. The concept using an independent edge beam can be used on retrofitting straight alignments. Figure 40.12 shows a graphic schematic of the proposed retrofit technique and Figure 40.13 shows the field installation of the joint reinforcement steel. Figure 40.14 shows the completed structure after retrofitting for seismic spectra that reach more than 2.0 *g* at the deck level.

FIGURE 40.8 Steel jacket column retrofit.

For outrigger bent cap under combined bending, shear, and torsion, an improved detail of column transverse reinforcement is typically continued up through the joint regions and the joints are further confined for shear and torsion resistance. The details for these joints usually require 1 to 3% confinement reinforcing steel. Thewalt and Stojadinovic [16] of University of California, Berkeley reported on this research. Figure 40.15 shows the complex joint-reinforcing steel needed to confine these joints for combined shear, bending, and torsion stresses. Design of these large joints requires use of the strut-and-tie technology to account properly for the load paths through the joint.

Steel Bridge Retrofit

Despite the fact that structural steel is ductile, members that have been designed by the pre-1972 seismic specifications must be evaluated for the seismic forces expected at the site based on earthquake magnitudes as we know them today. Typically, structural steel superstructures that had been tied to their substructures with joint and hinge restrainer systems performed well. However, we have identified many elevated viaducts and some smaller structures supported on structural steel columns that were designed prior to 1972 and that will require major retrofit strengthening for them to resist modern earthquake forces over a long period of shaking. One weak link is the older rocker bearings that will probably roll over during an earthquake. These can be replaced with modern neoprene, Teflon, pot, and base isolation bearings to ensure better performance in an earthquake. Structural steel columns can be strengthened easily to increase their toughness and ability to withstand a long period of dynamic input.

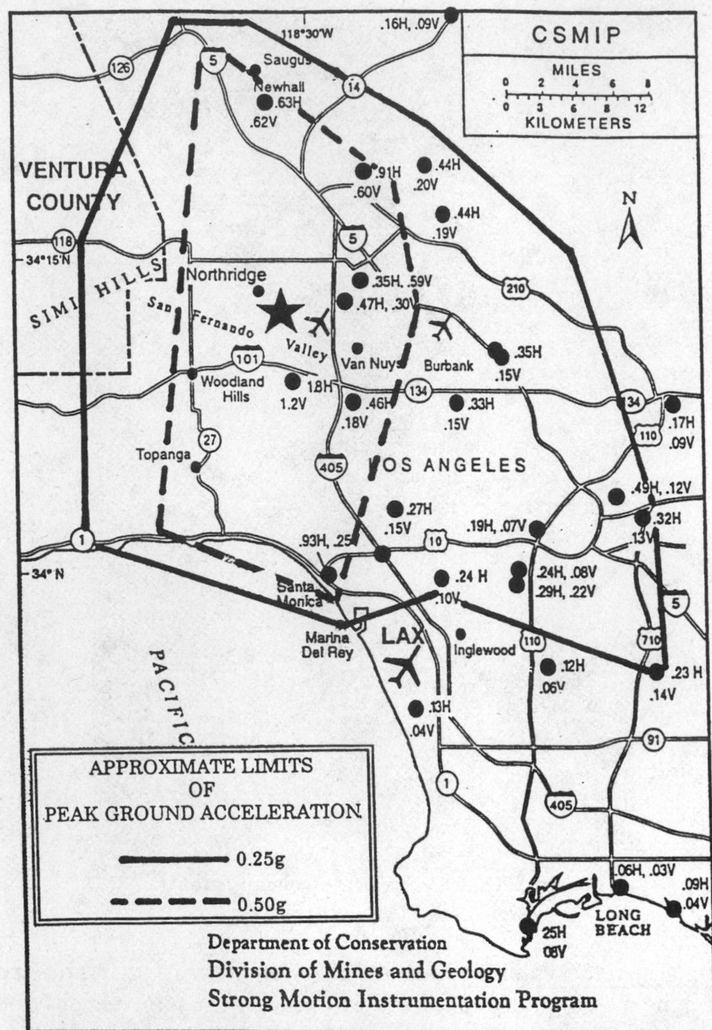

FIGURE 40.9 Peak ground acceleration zones — Northridge earthquake.

FIGURE 40.10 Steel jackets and bond beam — Santa Monica Freeway.

FIGURE 40.11 Column reinforcing steel cage.

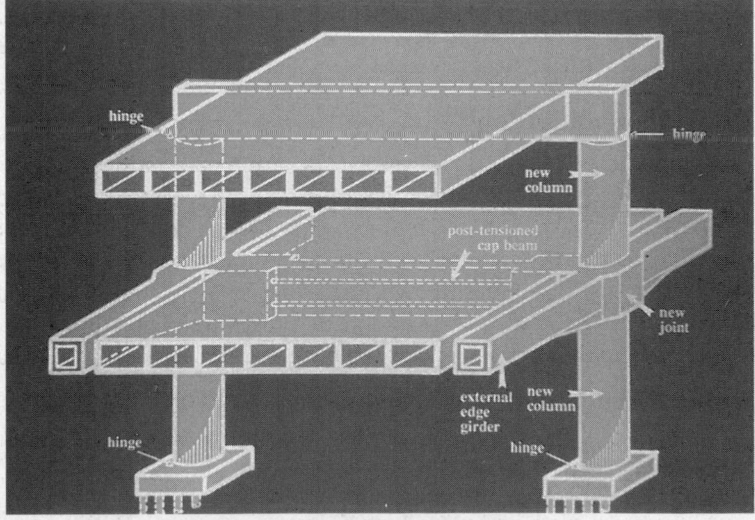

FIGURE 40.12 Graphic of edge beam retrofit scheme.

FIGURE 40.13 Field installation of joint reinforcing steel.

FIGURE 40.14 Completed retrofitted structures.

Footing and Pile Cap Modifications

Bridge column footing details established in 1980 consist of top and bottom mats of reinforcement tied together vertically by closely spaced hooked stirrups (Chapter 43). The column longitudinal rebars rest on the bottom mat, are hooked into the footing with hooks splayed outward, and are confined by spiral or hoop reinforcement between the mats. For pile foundations, the piles are reinforced and securely connected to the pile caps to resist the seismic tensile loads (Figure 40.16). The justifications for these details were widely debated, and strut-and-tie procedure seems to substantiate the need (Chapter 38). However, a proof-test of a footing with typical details performed adequately.

Seismic Isolation and Energy Dissipation Systems

Seismic isolation and supplemental energy dissipation devices have been successfully used in many bridge seismic design and retrofit projects. A detailed discussion is presented in Chapter 41. Extreme caution should be exercised when considering isolation devices. As discussed earlier, good, well-detailed, monolithic moment-resisting frames provide adequate seismic resistance without the

FIGURE 40.15 Reinforcing steel pattern in complex outrigger joint.

FIGURE 40.16 Typical footing and pile cap modification.

inherent maintenance problems and higher initial costs. These devices, however, are excellent for replacing older, rocker bearings.

40.4.3 Analysis

Analysis is the simulation of the structure project engineer's strategy of the bridge response to the seismic motions. A good seismic design is robust and as relatively insensitive to fluctuations in ground motions as possible. Quantitative analysis is the appropriate verification of the capacity of the system and its individual subsystems being greater than the recognized demand.

The more complicated the seismic strategy, the more complicated will be the analysis. If the behavior of the system is to be nearly elastic with minor damage developed, then the analysis is

FIGURE 40.17 Completed seismic retrofit of I-5/710 interchange in Los Angeles.

likely to be simply linear-elastic analysis. However, if the behavior is likely to be complex, changing in time, with significant damage developed and loss of life or important facility loss, then the analysis is likely to be similarly complex.

As a rule of thumb, the complexity of the analysis shadows the complexity of the strategy and the importance of the bridges. However, it should be noted that a very important bridge that is being designed to behave essentially elastically will not require complex analysis. It should always be recognized that analysis serves design and is part of design. Analysis cannot be a separated form. Too many engineers confuse analysis with design. Good design combines the analysis with judgment, common sense, and use of tested details.

40.4.4 Aesthetics

The design approach to bridge architecture, whether it is a proposed new structure or seismical retrofitting of an existing structure, poses a great challenge to the design team. Successful bridge designs are created by the productive and imaginative creations of the bridge architect and bridge engineer working together. The partnership of these talents, although not recognized in many professional societies, is an essential union that has produced structures of notable fame, within immediate identity worldwide.

Why is this partnership considered so essential? There are a number of reasons. Initially, the bridge architect will research the existing structures in the geographic area with respect to the surrounding community's existing visual qualities of the structural elements and recommended materials, forms, and texture that will harmonize with rather than contrast with the built environment. Figure 40.17 shows the architecture success of seismic retrofit of I-5/710 Interchange in SLos Angeles.

The residents of most communities that possess noted historical structures are extremely proud and possessive of their inheritance. So it is incumbent upon the bridge architect to demonstrate the sensitivity that is necessary when working on modifying historically significant bridges. This process often requires presentations at community gatherings or even workshops, where the bridge architect will use a variety of presentation techniques to show how carefully the designer has seismically retrofitted that specific structure and yet preserved the original historic design. This task is by no means easy, because of the emotional attachment a community may have toward its historic fabric.

In addition to the above-noted considerations for aesthetics, the architect and engineer must also take into consideration public safety, maintenance, and constructability issues when they consider seismic retrofit ideas. In addition to aesthetics, any modifications to an existing structure must carefully take into account the other three areas that are paramount in bridge design.

Within the governmental transportation agencies and private consulting firms lies a great deal of talent in both architecture and engineering. One key to utilizing this talent is to involve the bridge architect as early as possible so the engineer can be made aware of the important community and historical issues.

40.5 Construction

Construction is a phase of any retrofit project that is often not respected to an appropriate degree by designers. This is always somewhat of a surprise as construction regularly represents 80 to 90% of the cost of a project. In the authors opinion, a good design is driven by reliable construction methods and techniques. In order to deliver a design package that will minimize construction problems, the design project engineer strives to interact with construction engineers regularly and particularly on issues involving time, limited space, heavy lifts, and unusual specifications.

As mentioned in the section covering design, legal right-of-way access and utilities are very important issues that can stop, delay, or cause tremendous problems in construction. One of the first orders of work in the construction phase is to locate and appropriately protect or relocate utilities. This usually requires a legal agreement, which requires time. A considerable cost is not uncommon. The process required varies as a function of the utility and the owner, but they always take time and money. Access right-of-way is usually available due to existing right-of-way for maintenance. If foundation extensions or additional columns are required, then additional land may need to be acquired or even greater temporary access may be necessary. This issue should be recognized in the design phase, but regularly develops into construction challenges that require significant problem solving by the construction staff. These problems can delay a project many months or even require redesign.

Safety to the traveling public and the construction personnel is always the first priority on a construction site. But most of a structure resident engineer's (SRE) time is invested in assuring the contractors' understanding and adherence to the contract documents. In order to do this well the SRE must first understand well the contract documents, including the plans, construction standard and project special specifications. Then, the SRE must understand well the plan the contractor has to construct the project in such a way as to satisfy the requirements of the contract. It is in understanding the construction plan and observing the implementation of that plan that the SRE ensures that the construction project results in a quality product that will deliver acceptable performance for the life of the structure.

As most transportation structures are in urban areas, traffic handling and safety are important elements of any retrofit project. A transportation management plan (TMP) is a necessary item to develop and maintain. Traffic safety engineers including local highway patrol or police representatives are typically involved in developing such a plan. The TMP clearly defines how and when traffic will be routed to allow the contractor working space and time to complete the required work.

Shop plans are an item that are typically addressed early in the construction phase. Shop plans are structural plans developed by the contractor for structural elements and construction procedures that are appropriately delegated to the contractor by the owner in order to allow for as competitive bids as possible. Examples of typical shop plans include prestress anchorages and steel plate strengthening details and erection procedures.

Foundation modifications have been a major component in the bridge seismic retrofit program the California Department of Transportation has undertaken since the 1989 Loma Prieta earthquake. Considerable problems have been experienced in the reconstruction of many bridge foundations. Most of the construction claim dollars leveled against the state have been associated with foundation-related issues. These problems have included as-built plans not matching actual field conditions, materials, or dimension; a lack of adequate space to complete necessary work (e.g., insufficient overhead clearance to allow for driving or placing piles); damage to existing structural components (e.g., cutting reinforcing steel while coring); splicing of reinforcing steel with couplers or welds; paint specifications and time; and unexpected changes in geologic conditions. Although these items at first appear to have little in common, each of them is founded in uncertainty. That is, the construction problem is based on a lack of information. Recognizing this, the best way to avoid such problems is as follows:

- To invest in collecting factual and specific data that can be made available to the designer and the contractor such as actual field dimensions;
- To consider as carefully as possible likely contractor space requirements given what activities the contractor will be required to conduct;
- To know and understand well the important properties of materials and structural elements that are to be placed into the structure by the contractor; and
- To conduct appropriately thorough foundation investigation which may include field testing of potential foundation systems.

The most common structural modifications to bridge structures in California have been the placement of steel shells around portions of reinforced concrete columns in order to provide or increase confinement to the concrete within the column and increase the shear strength of the column within the dimensions of the steel shell. As part of a construction project, important items to verify in a steel shell column jacket installation are the steel material properties, the placement of the steel shell, the weld material and process, the grouting of the void between the oversized steel shell and the column, and the grinding and painting of the steel shell.

Existing reinforcing steel layouts are designed for a purpose and should not be modified. In some cases they can be modified for convenience in construction. It is important that field engineers be knowledgeable in order to reject modifications to reinforcing steel layouts that could render the existing structural section inadequate.

40.6 Costs

Estimating costs for bridge seismic retrofit projects is an essential element of any retrofit program. For a program to initiate, legislation must typically be passed. As part of the legislation package, funding sources are identified, and budgets are set. The budgets are usually established from estimates. It is ironic that, typically, the word *estimate* is usually dropped in this process. Regardless of any newly assigned title of the estimate, it remains what it is — an estimate. This typical set of circumstances creates an environment in which it is essential that great care be exercised before estimates are forwarded.

The above being stated, methods have been developed to forecast retrofit costs. The most common technique is to calculate and document into a database project costs per unit deck area. When such data are nearly interpolated to similar projects with consistent parameters, this technique can realize success. This technique is better suited to program estimates rather than a specific project estimate.

TABLE 40.2 Approximate Costs of Various Pay Items of Bridge Seismic Retrofit (California, 1998)

Pay times	Approximate Cost	Notes
Access opening (deck)	\$350 to \$1500 per sq. ft.	
Access opening (soffit)	\$400 to \$750 per sq. ft.	
Restrainer cables	\$3.5 to \$6.6 per number	
Restrainer rods	\$2.5 to \$4.5 per number	
Seat extenders	\$1.5 to \$3.3 per number	
Steel shells for columns	\$1.5 to \$2.25 per lb.	
Concrete removal		
Steel removal		
Soil removal	\$40 to \$150 per cy	
Core concrete (6 in.)	\$65 to \$100 per ft.	
Concrete (bridge footing)	\$175 to \$420 per cy	
Concrete (bridge)	\$400 to \$800 per cy	
Minor concrete	\$350 to \$900 per cy	
Structural steel	\$2.50 to \$5 per lb.	
Prestressing steel	\$0.80 to \$1.15 per lb.	
Bar reinforcing steel	\$0.50 to \$1.00 per number	
Precast concrete pile (45T)	\$610 to \$1515 per linear ft.	
CISS piles (24 in.)	\$788 to \$4764 per linear ft.	
Pile shaft (48 in.)	\$170 to \$330 per linear ft.	
Structural backfill	\$38 to \$100 per cy	
Traffic lane closure (day)		
Traffic lane closure (night)		

When applied to a specific project, additional contingencies are appropriate. When an estimate for a specific project is desired, it is appropriate to evaluate the specific project parameters.

Many of the components or pay items of a seismic retrofit project when broken down to pay items are similar to new construction or widening project pay items. As a first estimate, this can be used to approximate the cost of the work crudely. Table 40.2 lists the approximate cost for various pay items in California in 1998. There certainly are exceptions to these general conditions, such as steel shells, very long coring and drilling, and pile installation in low clearance conditions.

40.7 Summary

The two most significant earthquakes in recent history that produced the best information for bridge designers were the 1989 Loma Prieta and the 1994 Northridge events. Although experts consider these to be only moderate earthquakes, it is important to note the good performance of the many bridges that had been designed for the improved seismic criteria or retrofitted with the early-era seismic retrofit details. This reasonable performance of properly designed newer and retrofitted older bridges in a moderate earthquake is significant for the rest of the United States and other countries because that knowledge can assist engineers in designing new bridges and in designing an appropriate seismic retrofit program for their older structures. Although there is a necessary concern for the "Big One" in California, especially for the performance of important structures, it must be noted that many structures that vehicle traffic can bypass need not be designed or retrofitted to the highest standards. It is also important to note that there will be many moderate earthquakes that will not produce the damage associated with a maximum event. These are the earthquake levels that should be addressed first in a multiphased retrofit strengthening program, given the limited resources that are available.

Cost–benefit analysis of retrofit details is essential to measure and ensure the effectiveness of a program. It has been the California experience that a great deal of insurance against collapse can be achieved for a reasonable cost, typically 10% of replacement cost for normal highway bridges. It is also obvious that designing for the performance criteria that provides full service immediately after a major earthquake may not be economically feasible. The expected condition of the bridge approach roadways after a major seismic event must be evaluated before large investments are made in seismic retrofitting of the bridges to the full-service criteria. There is little value to the infrastructure in investing large sums to retrofit a bridge if the approaches are not functioning after a seismic event. Roadways in the soft muds around most harbors and rivers are potentially liquefiable and will require repair before the bridges can be used.

Emerging practices on bridge seismic retrofit in the state of California was briefly presented. The excellent performance of bridges utilizing Caltrans newer design criteria and ductile details gives bridge designers an indication that these structures can withstand a larger earthquake without collapse. Damage should be expected, but it can be repaired in many cases while traffic continues to use the bridges.

References

1. AASHTO, *LRFD Bridge Design Specifications*, 2nd ed., American Association of State Highway and Transportation Officials, Washington, D.C., 1998.
2. Caltrans, Seismic Hazard Map in California, California Department of Transportation, Sacramento, CA, 1996.
3. Xie, L. L. and Housner, G. W., *The Greater Tangshan Earthquake*, Vol. I and IV, California Institute of Technology, Pasadena, CA, 1996.
4. Housner, G. W. (Chairman), Thiel, C. C. (Editor), Competing against Time, Report to Governor George Deukmejian from the Governor's Board of Inquiry on the 1989 Loma Prieta Earthquake, Publications Section, Department of General Services, State of California, Sacramento, June, 1990.
5. Maroney, B., Gates, J., and Caltrans. Seismic risk identification and prioritization in the Caltrans seismic retrofit program, in *Proceedings, 59th Annual Convention,* Structural Engineers Association of California, Sacramento, September, 1990.
6. Gilbert, A., development in seismic prioritization of California bridges, in *Proceedings: Ninth Annual US/Japan Workshop on Earthquake and Wind Design of Bridges,* Tsukuba Science City, Japan, May 1993.
7. Sheng, L. H. and Gilbert, A., California Department of Transportation seismic retrofit program-the prioritization and screening process," *Lifeline Earthquake Engineering: Proceedings of the Third U.S. Conference,* Report: Technical Council on Lifeline Earthquake Engineering, Monograph 4, American Society of Civil Engineers, New York, August, 1991.
8. Mellon, S. et al., Post earthquake investigation team report of bridge damage in the Loma Prieta earthquake, in *Proceedings: ASCE Structures Congress XI,* ASCE, Irvine, CA, April 1993.
9. Priestley, M. J. N., Seible, F., and Chai, Y. H., Flexural Retrofit of Circular Reinforced Bridge Columns by Steel Jacketing, Colret—A Computer Program for Strength and Ductility Calculation, Report No. SSRP-91/05 to the Caltrans Division of Structures, University of California at San Diego, October, 1991.
10. Priestley, M. J. N., Seible, F., and Chai, Y. H., Flexural Retrofit of Circular Reinforced Bridge Columns by Steel Jacketing, Experimental Studies, Report No. SSRP-91/05 to the Caltrans Division of Structures, University of California at San Diego, October, 1991.
11. Priestley, M. J. N. and Seible, F., Assessment and testing of column lap splices for the Santa Monica Viaduct retrofit, in *Proceedings: ASCE Structures Congress,* XI, ASCE, Irvine, California, April 1993.
12. Seible, F., Priestley, M. J. N., Latham, C. T., and Terayama, T., Full Scale Test on the Flexural Integrity of Cap/Column Connections with Number 18 Column Bars, Report No. TR-93/01 to Caltrans, University of California at San Diego, January 1993.

13. Seible, F., Priestley, M. J. N., Hamada, N., Xiao, Y., and MacRae, G. A., Rocking and Capacity Test of Model Bridge Pier, Report No. SSRP-92/06 to Caltrans, University of California at San Diego, August, 1992.
14. Seible, F., Priestley, M. J. N., Hamada, N., and Xiao, Y., Test of a Retrofitted Rectangular Column Footing Designed to Current Caltrans Retrofit Standards, Report No. SSRP-92/10 to Caltrans, University of California at San Diego, November 1992.
15. Cooper, T. R., Terminal Separation Design Criteria: A Case Study of Current Bridge Seismic Design and Application of Recent Seismic Design Research, Seismic Design and Retrofit of Bridges, *Seminar Proceedings,* Earthquake Engineering Research Center, University of California at Berkeley and California Department of Transportation, Division of Structures, Sacramento, 1992.
16. Thewalt, C. R. and Stojadinovic, B. I., Behavior and Retrofit of Outrigger Beams, Seismic Design and Retrofit of Bridges, *Seminar Proceedings,* Earthquake Engineering Research Center, University of California at Berkeley and California Department of Transportation, Division of Structures, Sacramento, 1992.

41

Seismic Isolation and Supplemental Energy Dissipation

Rihui Zhang
California State Department of Transportation

41.1 Introduction

Strong earthquakes impart substantial amounts of energy into structures and may cause the structures to deform excessively or even collapse. In order for structures to survive, they must have the capability to dissipate this input energy through either their inherent damping mechanism or inelastic deformation. This issue of energy dissipation becomes even more acute for bridge structures because most bridges, especially long-span bridges, possess very low inherent damping, usually less than 5% of critical. When these structures are subjected to strong earthquake motions, excessive deformations can occur by relying on only inherent damping and inelastic deformation. For bridges designed mainly for gravity and service loads, excessive deformation leads to severe damage or even collapse. In the instances of major bridge crossings, as was the case of the San Francisco–Oakland

0-8493-7434-0/00/$0.00+$.50

Bay Bridge during the 1989 Loma Prieta earthquake, even noncollapsing structural damage may cause very costly disruption to traffic on major transportation arteries and is simply unacceptable.

Existing bridge seismic design standards and specifications are based on the philosophy of accepting minor or even major damage but no structural collapse. Lessons learned from recent earthquake damage to bridge structures have resulted in the revision of these design standards and a change of design philosophy. For example, the latest bridge design criteria for California [1] recommend the use of a two-level performance criterion which requires that a bridge be designed for both safety evaluation and functional evaluation design earthquakes. A safety evaluation earthquake event is defined as an event having a very low probability of occurring during the design life of the bridge. For this design earthquake, a bridge is expected to suffer limited significant damage, or immediately repairable damage. A functional evaluation earthquake event is defined as an event having a reasonable probability of occurring once or more during the design life of the bridge. Damages suffered under this event should be immediately repairable or immediate minimum for important bridges. These new criteria have been used in retrofit designs of major toll bridges in the San Francisco Bay area and in designs of some new bridges. These design criteria have placed heavier emphasis on controlling the behavior of bridge structural response to earthquake ground motions.

For many years, efforts have been made by the structural engineering community to search for innovative ways to control how earthquake input energy is absorbed by a structure and hence controlling its response to earthquake ground motions. These efforts have resulted in the development of seismic isolation techniques, various supplemental energy dissipation devices, and active structural control techniques. Some applications of these innovative structural control techniques have proved to be cost-effective. In some cases, they may be the only ways to achieve a satisfactory solution. Furthermore, with the adoption of new performance-based design criteria, there will soon come a time when these innovative structural control technologies will be the choice of more structural engineers because they offer economical alternatives to traditional earthquake protection measures.

Topics of structural response control by passive and active measures have been covered by several authors for general structural applications [2–4]. This chapter is devoted to the developments and applications of these innovative technologies to bridge structures. Following a presentation of the basic concepts, modeling, and analysis methods, brief descriptions of major types of isolation and energy dissipation devices are given. Performance and testing requirements will be discussed followed by a review of code developments and design procedures. A design example will also be given for illustrative purposes.

41.2 Basic Concepts, Modeling, and Analysis

The process of a structure responding to earthquake ground motions is actually a process involving resonance buildup to some extent. The severity of resonance is closely related to the amount of energy and its frequency content in the earthquake loading. Therefore, controlling the response of a structure can be accomplished by either finding ways to prevent resonance from building up or providing a supplemental energy dissipation mechanism, or both. Ideally, if a structure can be separated from the most-damaging energy content of the earthquake input, then the structure is safe. This is the idea behind seismic isolation. An isolator placed between the bridge superstructure and its supporting substructure, in the place of a traditional bearing device, substantially lengthens the fundamental period of the bridge structure such that the bridge does not respond to the most-damaging energy content of the earthquake input. Most of the deformation occurs across the isolator instead of in the substructure members, resulting in lower seismic demand for substructure members. If it is impossible to separate the structure from the most-damaging energy content, then the idea of using supplemental damping devices to dissipate earthquake input energy and to reduce structural damage becomes very attractive.

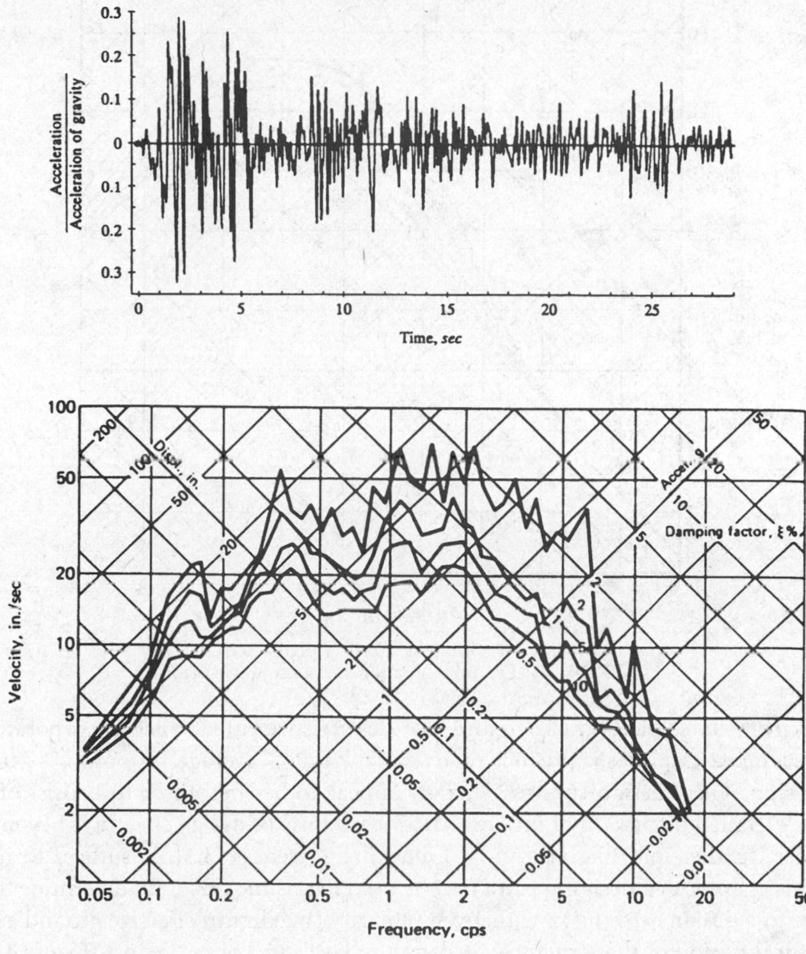

FIGURE 41.1 Acceleration time history and response spectra from El Centro earthquake, May 1940.

In what follows, theoretical basis and modeling and analysis methods will be presented mainly based on the concept of earthquake response spectrum analysis.

41.2.1 Earthquake Response Spectrum Analysis

Earthquake response spectrum analysis is perhaps the most widely used method in structural earthquake engineering design. In its original definition, an earthquake response spectrum is a plot of the maximum response (maximum displacement, velocity, acceleration) to a specific earthquake ground motion for all possible single-degree-of-freedom (SDOF) systems. One of such response spectra is shown in Figure 41.1 for the 1940 El Centro earthquake. A response spectrum not only reveals how systems with different fundamental vibration periods respond to an earthquake ground motion, when plotted for different damping values, site soil conditions and other factors, it also shows how these factors are affecting the response of a structure. From an energy point of view, response spectrum can also be interpreted as a spectrum the energy frequency contents of an earthquake.

Since earthquakes are essentially random phenomena, one response spectrum for a particular earthquake may not be enough to represent the earthquake ground motions a structure may

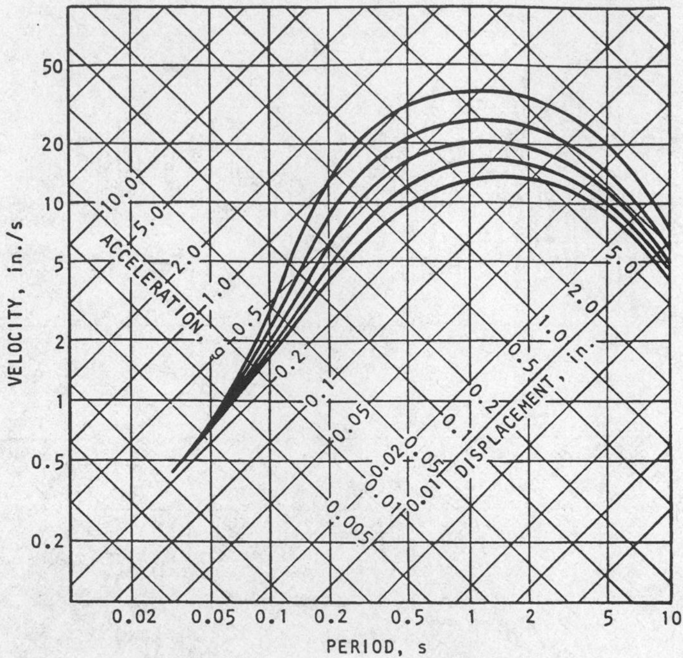

FIGURE 41.2 Example of smoothed design spectrum.

experience during its service life. Therefore, the design spectrum, which incorporates response spectra for several earthquakes and hence represents a kind of "average" response, is generally used in seismic design. These design spectra generally appear to be smooth or to consist of a series of straight lines. Detailed discussion of the construction and use of design spectra is beyond the scope of this chapter; further information can be found in References [5,6]. It suffices to note for the purpose of this chapter that design spectra may be used in seismic design to determine the response of a structure to a design earthquake with given intensity (maximum effective ground acceleration) from the natural period of the structure, its damping level, and other factors. Figure 41.2 shows a smoothed design spectrum curve based on the average shapes of response spectra of several strong earthquakes.

41.2.2 Structural Dynamic Response Modifications

By observing the response/design spectra in Figures 41.1a, it is seen that manipulating the natural period and/or the damping level of a structure can effectively modify its dynamic response. By inserting a relatively flexible isolation bearing in place of a conventional bridge bearing between a bridge superstructure and its supporting substructure, seismic isolation bearings are able to lengthen the natural period of the bridge from a typical value of less than 1 second to 3 to 5 s. This will usually result in a reduction of earthquake-induced response and force by factors of 3 to 8 from those of fixed-support bridges [7].

As for the effect of damping, most bridge structures have very little inherent material damping, usually in the range of 1 to 5% of critical. The introduction of nonstructural damping becomes necessary to reduce the response of a structure.

Some kind of a damping device or mechanism is also a necessary component of any successful seismic isolation system. As mentioned earlier, in an isolated structural system deformation mainly occurs across the isolator. Many factors limit the allowable deformation taking place across an

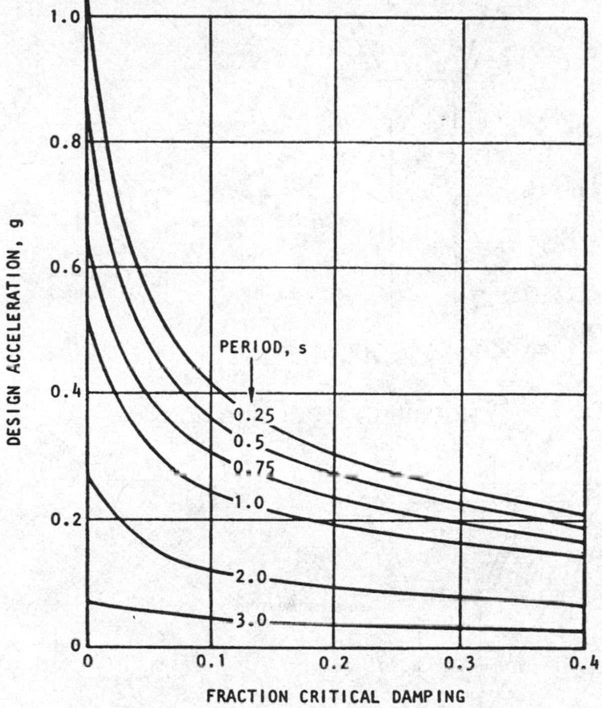

FIGURE 41.3 Effect of damping on response spectrum.

isolator, e.g., space limitation, stability requirement, etc. To control deformation of the isolators, supplemental damping is often introduced in one form or another into isolation systems.

It should be pointed out that the effectiveness of increased damping in reducing the response of a structure decreases beyond a certain damping level. Figure 41.3 illustrates this point graphically. It can be seen that, although acceleration always decreases with increased damping, its rate of reduction becomes lower as the damping ratio increases. Therefore, in designing supplemental damping for a structure, it needs to be kept in mind that there is a most-cost-effective range of added damping for a structure. Beyond this range, further response reduction will come at a higher cost.

41.2.3 Modeling of Seismically Isolated Structures

A simplified SDOF model of a bridge structure is shown in Figure 41.4. The mass of the superstructure is represented by *m*, pier stiffness by spring constant k_0, and structural damping by a viscous damping coefficient c_0. The equation of motion for this SDOF system, when subjected to an earthquake ground acceleration excitation, is expressed as:

$$m_0\ddot{x} + c_0\dot{x} + k_0 x = -m_0\ddot{x}_g \tag{41.1}$$

The natural period of motion T_0, time required to complete one cycle of vibration, is expressed as

$$T_0 = 2\pi\sqrt{\frac{m_0}{k_0}} \tag{41.2}$$

Addition of a seismic isolator to this system can be idealized as adding a spring with spring constant k_i and a viscous damper with damping coefficient c_i, as shown in Figure 41.5. The combined stiffness of the isolated system now becomes

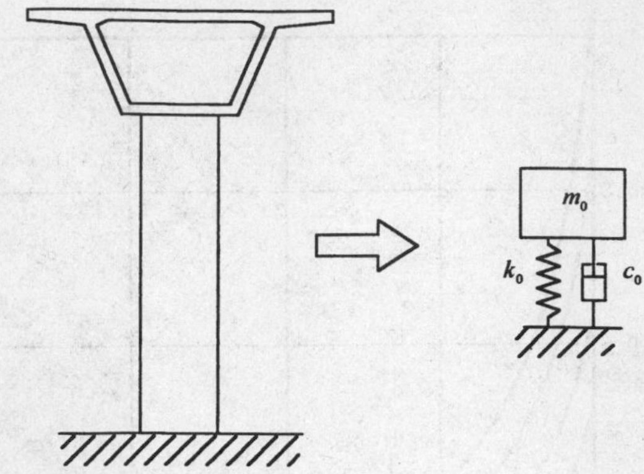

FIGURE 41.4 SDOF dynamic model.

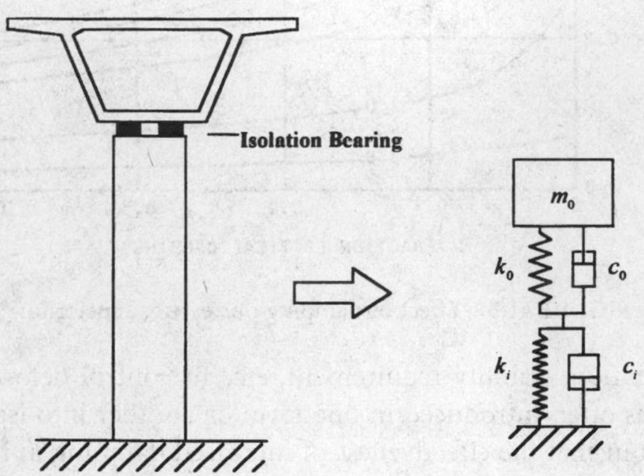

FIGURE 41.5 SDOF system with seismic isolator.

$$K = \frac{k_0 k_i}{k_0 + k_i} \tag{41.3}$$

Equation (41.1) is modified to

$$m_0 \ddot{x} + (c_0 + c_i)\dot{x} + Kx = -m_0 \ddot{x}_g \tag{41.4}$$

and the natural period of vibration of the isolated system becomes

$$T = 2\pi \sqrt{\frac{m_0}{K}} = 2\pi \sqrt{\frac{m_0 (k_0 + k_i)}{k_0 k_i}} \tag{41.5}$$

When the isolator stiffness is smaller than the structural stiffness, K is smaller than k_0; therefore, the natural period of the isolated system T is longer than that of the original system. It is of interest to note that, in order for the isolator to be effective in modifying the the natural period of the structure, k_i should be smaller than k_0 to a certain degree. For example, if k_i is 50% of k_0, then T will be about 70% larger than T_0. If k_i is only 10% of k_0, then T will be more than three times of T_0.

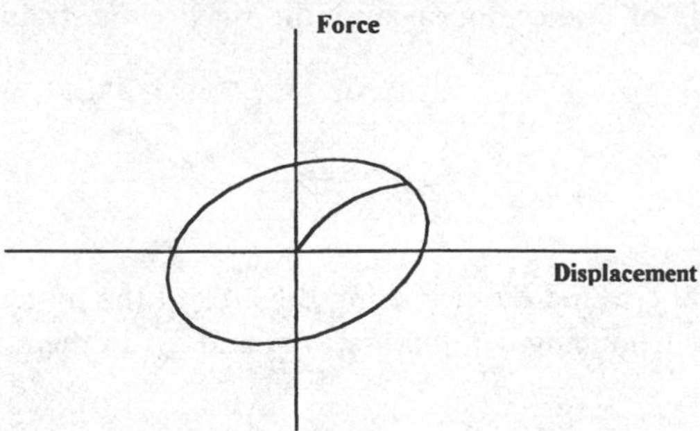

FIGURE 41.6 Generic damper hysteresis loops.

More complex structural systems will have to be treated as multiple-degree-of-freedom (MDOF) systems; however, the principle is the same. In these cases, spring elements will be added to appropriate locations to model the stiffness of the isolators.

41.2.4 Effect of Energy Dissipation on Structural Dynamic Response

In discussing energy dissipation, the terms *damping* and *energy dissipation* will be used interchangeably. Consider again the simple SDOF system used in the previous discussion. In the theory of structural dynamics [8], critical value of damping coefficient c_c is defined as the amount of damping that will prevent a dynamic system from free oscillation response. This critical damping value can be expressed in terms of the system mass and stiffness:

$$c_c = 2\sqrt{m_0 k_o} \tag{41.6}$$

With respect to this critical damping coefficient, any amount of damping can now be expressed in a relative term called damping ratio ξ, which is the ratio of actual system damping coefficient over the critical damping coefficient. Thus,

$$\xi = \frac{c_0}{c_c} = \frac{c_0}{2\sqrt{m_0 k_0}} \tag{41.7}$$

Damping ratio is usually expressed as a percentage of the critical. With the use of damping ratio, one can compare the amount of damping of different dynamic systems.

Now consider the addition of an energy dissipation device. This device generates a force $f(x,\dot{x})$ that may be a function of displacement or velocity of the system, depending on the energy dissipation mechanism. Figure 41.6 shows a hysteresis curve for a generic energy dissipation device. Equation (41.1) is rewritten as

$$x + \frac{c_0}{m_0}x + \frac{k_0}{m_0}x + \frac{f(x,x)}{m_0} = -x_g \tag{41.8}$$

There are different approaches to modeling the effects damping devices have on the dynamic response of a structure. The most accurate approach is linear or nonlinear time history analysis by modeling the true behavior of the damping device. For practical applications, however, it will often be accurate enough to represent the effectiveness of a damping mechanism by an equivalent viscous damping ratio. One way to define the equivalent damping ratio is in terms of energy E_d dissipated by the device in one cycle of cyclic motion over the maximum strain energy E_{ms} stored in the structure [8]:

$$\xi_{eq} = \frac{E_d}{4_\pi E_{ms}} \tag{41.9}$$

For a given device, E_d can be found by measuring the area of the hysteresis loop. Equation (41.9) can now be rewritten by introducing damping ratio ξ_0 and ξ_{eq}, in the form

$$\ddot{x} + 2\sqrt{\frac{k_0}{m}}\left(\xi + \xi_{eq}\right)\dot{x} + \frac{k_0}{m_0}x = -\ddot{x}_g \tag{41.10}$$

This concept of equivalent viscous damping ratio can also be generalized to use for MDOF systems by considering ξ_{eq} as modal damping ratio and E_d and E_{ms} as dissipated energy and maximum strain energy in each vibration mode [9]. Thus, for the *i*th vibration mode of a structure, we have

$$\xi_{eq}^i = \frac{E_d^i}{4\pi E_{ms}^i} \tag{41.11}$$

Now the dynamic response of a structure with supplemental damping can be solved using available linear analysis techniques, be it linear time history analysis or response spectrum analysis.

41.3 Seismic Isolation and Energy Dissipation Devices

Many different types of seismic isolation and supplemental energy dissipation devices have been developed and tested for seismic applications over the last three decades, and more are still being investigated. Their basic behaviors and applications for some of the more widely recognized and used devices will be presented in this section.

41.3.1 Elastomeric Isolators

Elastomeric isolators, in their simplest form, are elastomeric bearings made from rubber, typically in cylindrical or rectangular shapes. When installed on bridge piers or abutments, the elastomeric bearings serve both as vertical bearing devices for service loads and lateral isolation devices for seismic load. This requires that the bearings be stiff with respect to vertical loads but relatively flexible with respect to lateral seismic loads. In order to be flexible, the isolation bearings have to be made much thicker than the elastomeric bearing pads used in conventional bridge design. Insertion of horizontal steel plates, as in the case of steel reinforced elastomeric bearing pads, significantly increases vertical stiffness of the bearing and improves stability under horizontal loads. The total rubber thickness influences essentially the maximum allowable lateral displacement and the period of vibration.

For a rubber bearing with given bearing area A, shear modulus G, height h, allowable shear strain γ, shape factor S, and bulk modulus K, its horizontal stiffness and period of vibration can be expressed as

$$K = \frac{GA}{h} \tag{41.12}$$

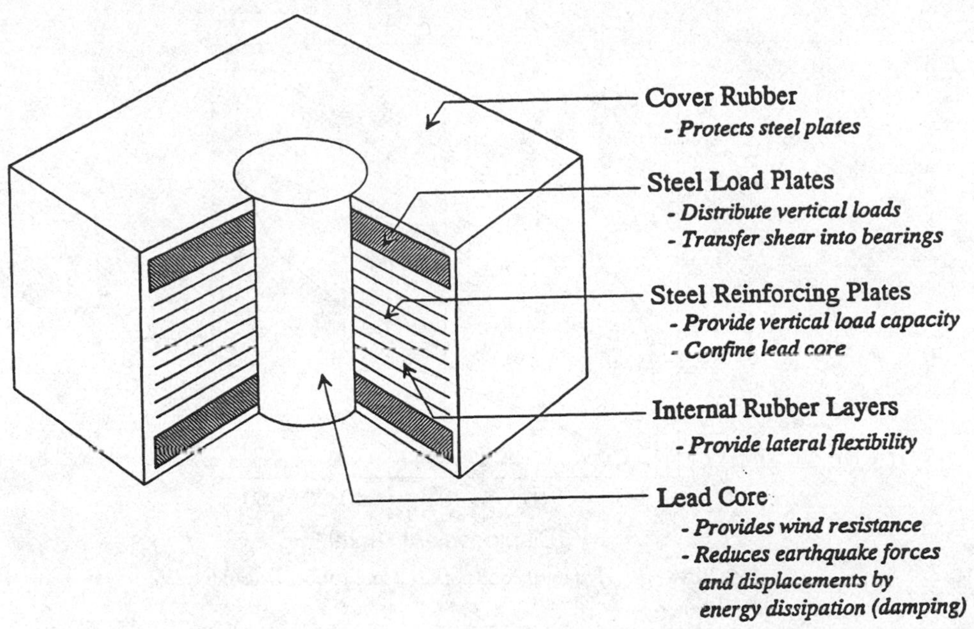

FIGURE 41.7 Typical construction of a lead core rubber bearing.

$$T_b = 2\pi\sqrt{\frac{M}{K}} = 2\pi\sqrt{\frac{Sh\gamma A'}{Ag}} \tag{41.13}$$

where A' is the overlap of top and bottom areas of a bearing at maximum displacement. Typical values for bridge elastomeric bearing properties are $G = 1$ MPa (145 psi), $K = 200$ MPa (290 psi), $\gamma = 0.9$ to 1.4, $S = 3$ to 40. The major variability lies in S, which is a function of plan dimension and rubber layer thickness.

One problem associated with using pure rubber bearings for seismic isolation is that the bearing could easily experience excessive deformation during a seismic event. This will, in many cases, jeopardize the stability of the bearing and the superstructure it supports. One solution is to add an energy dissipation device or mechanism to the isolation bearing. The most widely used energy dissipation mechanism in elastomeric isolation bearing is the insertion of a lead core at the center of the bearing. Lead has a high initial shear stiffness and relatively low shear yielding strength. It essentially has elastic–plastic behavior with good fatigue properties for plastic cycles. It provides a high horizontal stiffness for service load resistance and a high energy dissipation for strong seismic load, making it ideal for use with elastomeric bearings.

This type of lead core elastomeric isolation, also known as lead core rubber bearing (LRB), was developed and patented by the Dynamic Isolation System (DIS). The construction of a typical lead core elastomeric bearing is shown in Figure 41.7. An associated hysteresis curve is shown in Figure 41.8. Typical bearing sizes and their load bearing capacities are given in Table 41.1 [7].

Lead core elastomeric isolation bearings are the most widely used isolation devices in bridge seismic design applications. They have been used in the seismic retrofit and new design in hundreds of bridges worldwide.

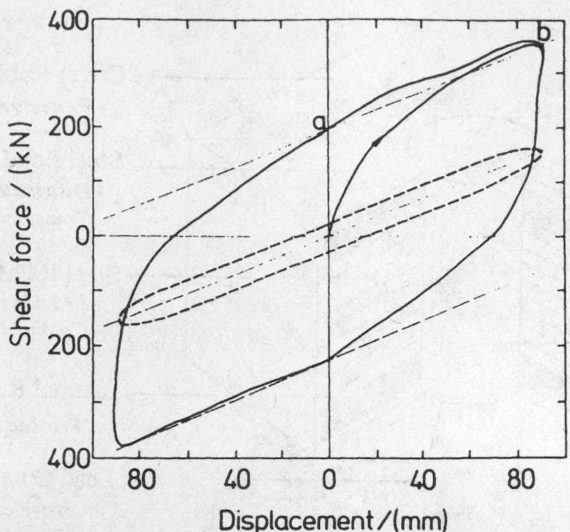

FIGURE 41.8 Hysteresis loops of lead core rubber bearing.

41.3.2 Sliding Isolators

Sliding-type isolation bearings reduce the force transferred from superstructure to the supporting substructure when subject to earthquake excitations by allowing the superstructure to slide on a low friction surface usually made from stainless steel-PTFE. The maximum friction between the sliding surfaces limits the maximum force that can be transferred by the bearing. The friction between the surfaces will also dissipate energy. A major concern with relying only on simple sliding bearings for seismic application is the lack of centering force to restore the structure to its undisplaced position together with poor predictability and reliability of the response. This can be addressed by combining the slider with spring elements or, as in the case of friction pendulum isolation (FPI) bearings, by making the sliding surface curved such that the self-weight of the structure will help recenter the superstructure. In the following, the FPI bearings by Earthquake Protection Systems (EPS) will be presented as a representative of sliding-type isolation bearings.

The FPI bearing utilizes the characteristics of a simple pendulum to lengthen the natural period of an isolated structure. Typical construction of an FPI bearing is shown in Figure 41.9. It basically consists of a slider with strength-bearing spherical surface and a treated spherical concave sliding surface housed in a cast steel bearing housing. The concave surface and the surface of the slider have the same radius to allow a good fit and a relatively uniform pressure under vertical loads. The operation of the isolator is the same regardless of the direction of the concave surface. The size of the bearing is mainly controlled by the maximum design displacement.

The concept is really a simple one, as illustrated in Figure 41.10. When the superstructure moves relative to the supporting pier, it behaves like a simple pendulum. The radius, R, of the concave surface controls the isolator period,

$$T = 2\pi\sqrt{\frac{R}{g}} \tag{41.14}$$

where g is the acceleration of gravity. The fact that the isolator period is independent of the mass of the supported structure is an advantage over the elastomeric isolators because fewer factors are involved in selecting an isolation bearing. For elastomeric bearings, in order to lengthen the period

TABLE 41.1 Total Dead Plus Live-Load Capacity of Square DIS Bearings (kN)

Plan Size		Bonded Area	Rubber Layer Thickness, mm			
W(mm)	*B*(mm)	(mm²)	6.5	9.5	12.5	19
229	229	52,258	236	160	125	85
254	254	64,516	338	227	173	120
279	279	78,064	463	311	236	165
305	305	92,903	614	414	311	214
330	330	109,032	796	534	405	276
356	356	126,451	1,010	676	512	351
381	381	145,161	1,263	845	641	436
406	406	165,161	1,552	1,041	783	529
432	432	186,451	1,882	1,259	952	641
457	457	209,032	2,255	1,508	1,139	770
483	483	232,903	2,678	1,793	1,348	912
508	508	258,064	3,149	2,104	1,583	1,068
533	533	284,516	3,674	2,455	1,846	1,241
559	559	312,257	4,252	2,842	2,135	1,437
584	584	341,290	4,888	3,265	2,455	1,650
610	610	371,612	5,582	3,727	2,802	1,882
635	635	403,225	6,343	4,234	3,185	2,135
660	660	436,128	7,170	4,786	3,598	2,411
686	686	470,322	8,064	5,382	4,043	2,713
711	711	505,805	9,029	6,027	4,528	3,034
737	737	542,580	10,070	6,721	5,048	3,380
762	762	580,644	11,187	7,464	5,609	3,754
787	787	619,999	12,383	8,264	6,205	4,154
813	813	660,644	13,660	9,118	6,845	4,581
838	838	702,579	15,025	10,026	7,530	5,040
864	864	745,805	16,480	10,995	8,255	5,524
889	889	790,321	18,023	12,023	9,029	6,040
914	914	836,127	19,660	13,117	9,848	6,587

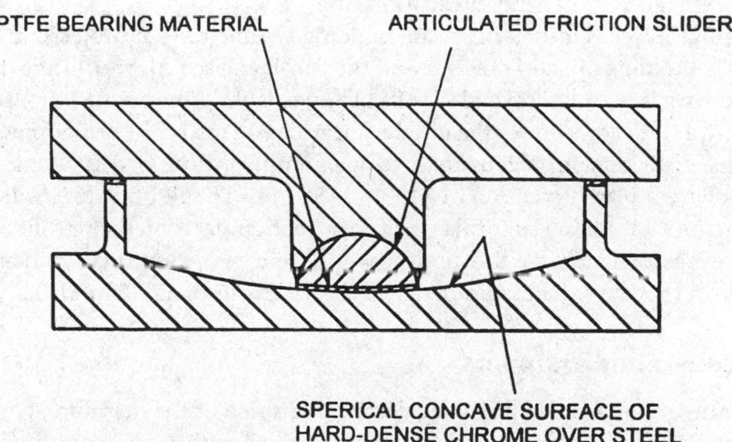

FIGURE 41.9 Typical construction of a FPI.

of an isolator without varying the plan dimensions, one has to increase the height of the bearing which is limited by stability requirement. For FPI bearings, one can vary the period simply by changing the radius of the concave surface. Another advantage the FPI bearing has is high vertical load-bearing capacity, up to 30 million lb (130,000 kN) [10].

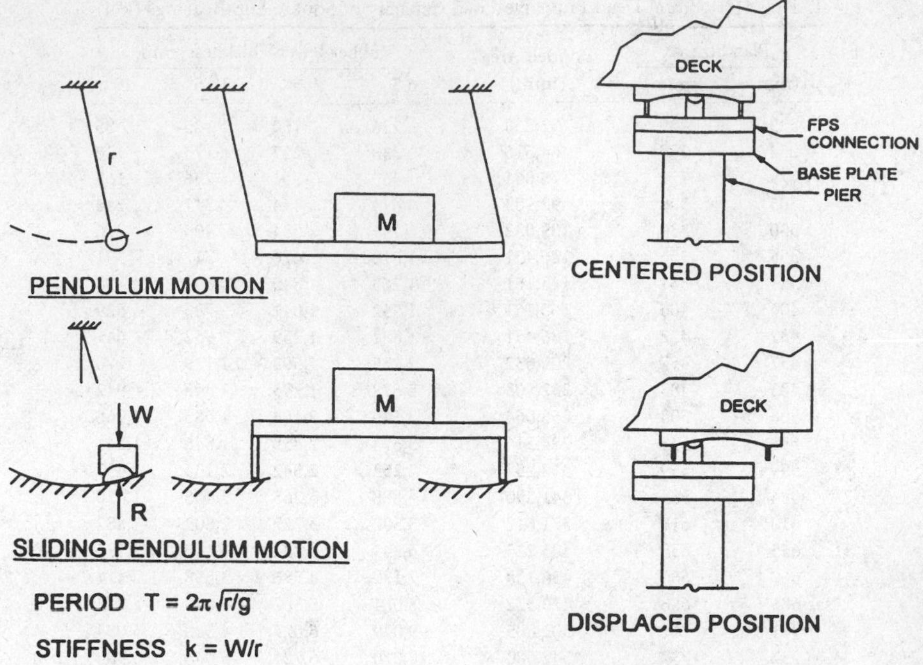

FIGURE 41.10 Basic operating principle of FPI.

The FPI system behaves rigidly when the lateral load on the structure is less than the friction force, which can be designed to be less than nonseismic lateral loads. Once the lateral force exceeds this friction force, as is the case under earthquake excitation, it will respond at its isolated period. The dynamic friction coefficient can be varied in the range of 0.04 to 0.20 to allow for different levels of lateral resistance and energy dissipation.

The FPI bearings have been used in several building seismic retrofit projects, including the U.S. Court of Appeals Building in San Francisco and the San Francisco Airport International Terminal. The first bridge structure to be isolated by FPI bearings is the American River Bridge in Folsom, California. Figure 41.11 shows one of the installed bearings on top of the bridge pier. The maximum designed bearing displacement is 250 mm, and maximum vertical load is about 16,900 kN. The largest bearings have a plan dimension of 1150 × 1150 mm. The FPI bearings will also be used in the Benicia–Martinez Bridge in California when construction starts on the retrofit of this mile-long bridge. The bearings designed for this project will have a maximum plan dimension of 4500 × 4500 mm to accommodate a maximum designed displacement of 1200 mm [11].

41.3.3 Viscous Fluid Dampers

Viscous fluid dampers, also called hydraulic dampers in some of the literature, typically consist of a piston moving inside the damper housing cylinder filled with a compound of silicone or oil. Figure 41.12 shows typical construction of a Taylor Device's viscous fluid damper and its corresponding hysteresis curve. As the piston moves inside the damper housing, it displaces the fluid which in turn generates a resisting force that is proportional to the exponent of the velocity of the moving piston, i.e.,

$$F = cV^k \tag{41.15}$$

FIGURE 41.11 A FPI bearing installed on a bridge pier.

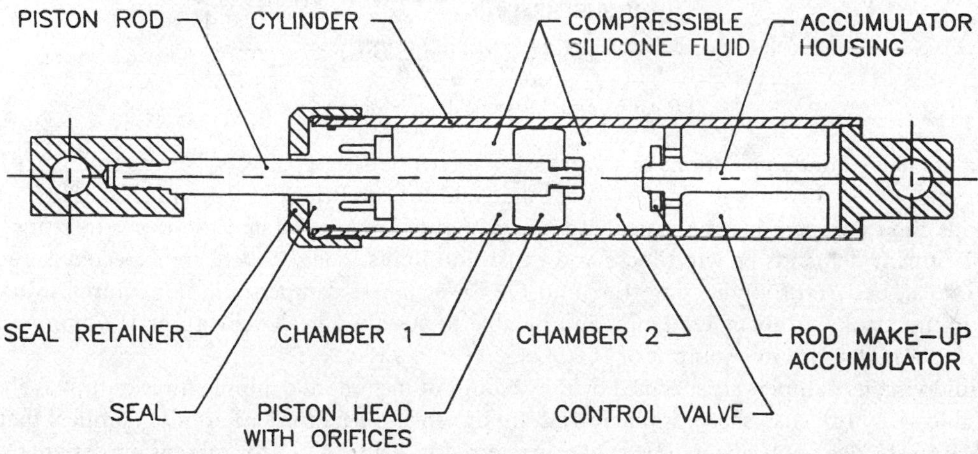

FIGURE 41.12 Typical construction of a taylor devices fluid viscous damper.

where c is the damping constant, V is the velocity of the piston, and k is a parameter that may be varied in the range of 0.1 to 1.2, as specified for a given application. If k equals 1, we have a familiar linear viscous damping force. Again, the effectiveness of the damper can be represented by the amount of energy dissipated in one complete cycle of deformation:

$$E_d = \int F dx \tag{41.16}$$

The earlier applications of viscous fluid dampers were in the vibration isolation of aerospace and defense systems. In recent years, theoretical and experimental studies have been performed in an effort to apply the viscous dampers to structure seismic resistant design [4,12]. As a result, viscous

TABLE 41.2 Fluid Viscous Damper Dimension Data (mm)

Model	A	B	C	D	E	F
100 kips (445 kN)	3327	191	64	81	121	56
200 kips (990 kN)	3353	229	70	99	127	61
300 kips (1335 kN)	3505	292	76	108	133	69
600 kips (2670 kN)	3937	406	152	191	254	122
1000 kips (4450 kN)	4216	584	152	229	362	122
2000 kips (9900 kN)	4572	660	203	279	432	152

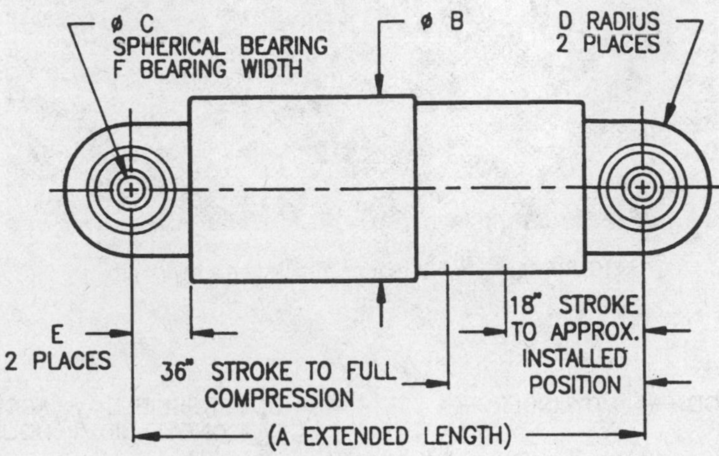

FIGURE 41.13 Viscous damper dimension.

dampers have found applications in several seismic retrofit design projects. For example, they have been considered for the seismic upgrade of the Golden Gate Bridge in San Francisco [13], where viscous fluid dampers may be installed between the stiffening truss and the tower to reduce the displacement demands on wind-locks and expansion joints. The dampers are expected to reduce the impact between the stiffening truss and the tower. These dampers will be required to have a maximum stroke of about 1250 mm, and be able to sustain a peak velocity of 1880 mm/s. This requires a maximum force output of 2890 kN.

Fluid viscous dampers are specified by the amount of maximum damping force output as shown in Table 41.2 [14]. Also shown in Table 41.2 are dimension data for various size dampers that are typical for bridge applications. The reader is referred to Figure 41.13 for dimension designations.

41.3.4 Viscoelastic Dampers

A typical viscoelastic damper, as shown in Figure 41.14, consists of viscoelastic material layers bonded with steel plates. Viscoelastic material is the general name for those rubberlike polymer materials having a combined feature of elastic solid and viscous liquid when undergoing deformation. Figure 41.14 also shows a typical hysteresis curve of viscoelastic dampers. When the center plate moves relative to the two outer plates, the viscoelastic material layers undergo shear deformation. Under a sinusoidal cyclic loading, the stress in the viscoelastic material can be expressed as

$$\sigma = \gamma_0 \left(G' \sin \omega t + G'' \cos \omega t \right) \tag{41.17}$$

where γ_0 represents the maximum strain, G' is shear storage modulus, and G'' is the shear loss modulus, which is the primary factor determining the energy dissipation capability of the viscoelastic material.

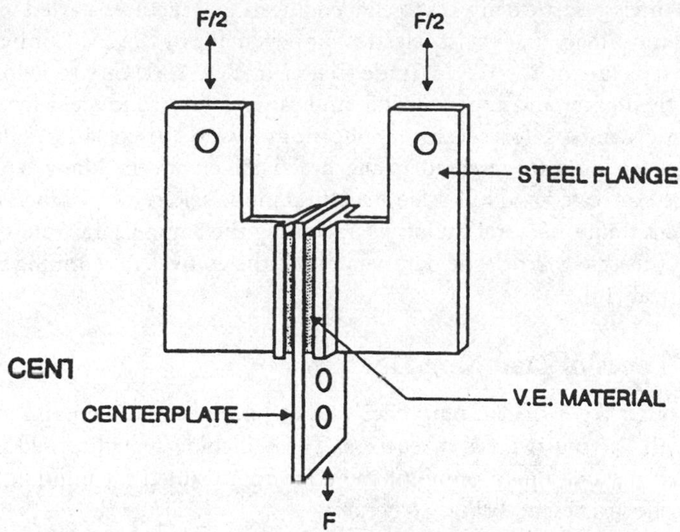

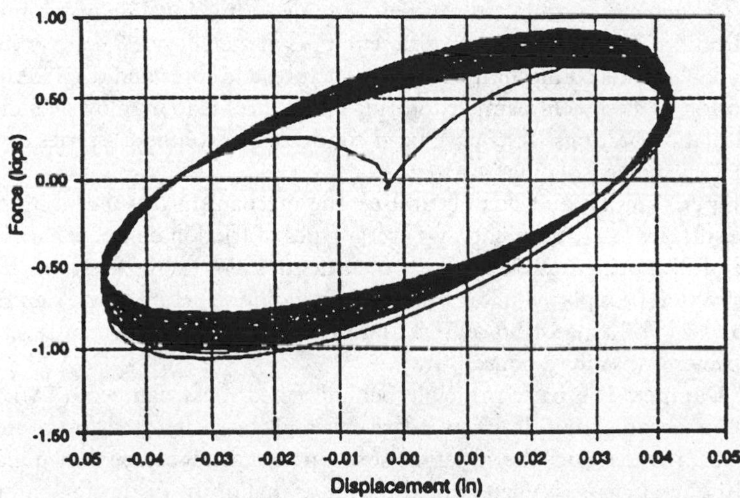

FIGURE 41.14 Typical viscoelastic damper and its hysteresis loops.

After one complete cycle of cyclic deformation, the plot of strain vs. stress will look like the hysteresis shown in Figure 41.14. The area enclosed by the hysteresis loop represents the amount of energy dissipated in one cycle per unit volume of viscoelastic material:

$$e_d = \pi\gamma_0^2 G' \tag{41.18}$$

The total energy dissipated by viscoelastic material of volume V can be expressed as

$$E_d = \pi\gamma_0^2 G'' V \tag{41.19}$$

The application of viscoelastic dampers to civil engineering structures started more than 20 years ago, in 1968, when more than 20,000 viscoelastic dampers made by the 3M Company were installed in the twin-frame structure of the World Trade Center in New York City to help resist wind load.

The application of viscoelastic dampers to civil engineering structures started more than 20 years ago, in 1968, when more than 20,000 viscoelastic dampers made by the 3M Company were installed in the twin-frame structure of the World Trade Center in New York City to help resist wind load. In the late 1980s, theoretical and experimental studies were first conducted for the possibility of applying viscoelastic dampers for seismic applications [9,15]. Viscoelastic dampers have since received increased attention from researchers and practicing engineers. Many experimental studies have been conducted on scaled and full-scale structural models. Recently, viscoelastic dampers were used in the seismic retrofit of several buildings, including the Santa Clara County Building in San Jose, California. In this case, viscoelastic dampers raised the equivalent damping ratio of the structure to 17% of critical [16].

41.3.5 Other Types of Damping Devices

There are several other types of damping devices that have been studied and applied to seismic resistant design with varying degrees of success. These include metallic yield dampers, friction dampers, and tuned mass dampers. Some of them are more suited for building applications and may be of limited effectiveness to bridge structures.

Metallic Yield Damper. Controlled use of sacrificial metallic energy dissipating devices is a relatively new concept [17]. A typical device consists of one or several metallic members, usually made of mild steel, which are subjected to axial, bending, or torsional deformation depending on the type of application. The choice between different types of metallic yield dampers usually depends on location, available space, connection with the structure, and force and displacement levels. One possible application of steel yield damper to bridge structures is to employ steel dampers in conjunction with isolation bearings. Tests have been conducted to combine a series of cantilever steel dampers with PTFE sliding isolation bearing.

Friction Damper. This type of damper utilizes the mechanism of solid friction that develops between sliding surfaces to dissipate energy. Several types of friction dampers have been developed for the purpose of improving seismic response of structures. For example, studies have shown that slip joints with friction pads placed in the braces of a building structure frame significantly reduced its seismic response. This type of braced friction dampers has been used in several buildings in Canada for improving seismic response [4,18].

Tuned Mass Damper. The basic principle behind tuned mass dampers (TMD) is the classic dynamic vibration absorber, which uses a relatively small mass attached to the main mass via a relatively small stiffness to reduce the vibration of the main mass. It can be shown that, if the period of vibration of the small mass is tuned to be the same as that of the disturbing harmonic force, the main mass can be kept stationary. In structural applications, a tuned mass damper may be installed on the top floor to reduce the response of a tall building to wind loads [4]. Seismic application of TMD is limited by the fact that it can only be effective in reducing vibration in one mode, usually the first mode.

41.4 Performance and Testing Requirements

Since seismic isolation and energy dissipation technologies are still relatively new and often the properties used in design can only be obtained from tests, the performance and test requirements are critical in effective applications of these devices. Testing and performance requirements, for the most part, are prescribed in project design criteria or construction specifications. Some nationally recognized design specifications, such as AASHO *Guide Specifications for Seismic Isolation Design* [19], also provide generic testing requirements.

Almost all of the testing specified for seismic isolators or energy dissipation devices require tests under static or simple cyclic loadings only. There are, however, concerns about how well will properties obtained from these simple loading tests correlate to behaviors under real earthquake loadings. Therefore, a major earthquake simulation testing program is under way. Sponsored by the Federal Highway Administration and the California Department of Transportation, manufacturers of isolation and energy dissipation devices were invited to provide their prototype products for testing under earthquake loadings. It is hoped that this testing program will lead to uniform guidelines for prototype and verification testing as well as design guidelines and contract specifications for each of the different systems. The following is a brief discussion of some of the important testing and performance requirements for various systems.

41.4.1 Seismic Isolation Devices

For seismic isolation bearings, performance requirements typically specify the maximum allowable lateral displacements under seismic and nonseismic loadings, such as thermal and wind loads; horizontal deflection characteristics such as effective and maximum stiffnesses; energy dissipation capacity, or equivalent damping ratio; vertical deflections; stability under vertical loads; etc. For example, the AASHTO *Guide Specifications for Seismic Isolation Design* requires that the design and analysis of isolation system prescribed be based on prototype tests and a series of verification tests as briefly described in the following:

Prototype Tests:

I. Prototype tests need to be performed on two full-size specimens. These tests are required for each type and size similar to that used in the design.
II. For each cycle of tests, the force–deflection and hysteresis behavior of the specimen need to be recorded.
III. Under a vertical load similar to the typical average design dead load, the specimen need to be tested for
   A. Twenty cycles of lateral loads corresponding to the maximum nonseismic loads;
   B. Three cycles of lateral loading at displacements equaling 25, 50, 75, 100, and 125% of the total design displacement;
   C. Not less than 10 full cycles of loading at the total design displacement and a vertical load similar to dead load.
IV. The stability of the vertical load-carrying element needs to be demonstrated by one full cycle of displacement equaling 1.5 times the total design displacement under dead load plus or minus vertical load due to seismic effect.

System Characteristics Tests:

I. The force–deflection characteristics need to be based on cyclic test results.
II. The effective stiffness of an isolator needs to be calculated for each cycle of loading as

$$k_{\text{eff}} = \frac{F_p - F_n}{\Delta_p - \Delta_n} \tag{41.20}$$

where F_p and F_n are the maximum positive and negative forces, respectively, and Δ_p and Δ_n are the maximum positive and negative displacements, respectively.
III. The equivalent viscous damping ratio ξ of the isolation system needs to be calculated as

$$\xi = \frac{\text{Total Area}}{4\pi \sum \frac{kd^2}{2}} \tag{41.21}$$

where Total Area shall be taken as the sum of areas of the hysteresis loops of all isolators; the summation in the denominator represents the total strain energy in the isolation system.

In order for a specimen to be considered acceptable, the results of the tests should show positive incremental force-carrying capability, less than a specified amount of variation in effective stiffness between specimens and between testing cycles for any given specimen. The effective damping ratio also needs to be within certain range [19].

41.4.2 Testing of Energy Dissipation Devices

As for energy dissipation devices, there have not been any codified testing requirements published. The Federal Emergency Management Agency 1994 NEHRP Recommended Provisions for Seismic Regulation for New Buildings contain an appendix that addresses the use of energy dissipation systems and testing requirements [20]. There are also project-specific testing requirements and proposed testing standards by various damper manufacturers.

Generally speaking, testing is needed to obtain appropriate device parameters for design use. These parameters include the maximum force output, stroke distance, stiffness, and energy dissipation capability. In the case of viscous dampers, these are tested in terms of damping constant C, exponential constant, maximum damping force, etc. Most of the existing testing requirements are project specific. For example, the technical requirements for viscous dampers to be used in the retrofit of the Golden Gate Bridge specify a series of tests to be carried out on model dampers [13,21]. Prototype tests were considered to be impractical because of the limitation of available testing facilities. These tests include cyclic testing of model dampers to verify their constitutive law and longevity of seals and a drop test of model and prototype dampers to help relate cyclic testing to the behavior of the actual dampers. Because the tests will be on model dampers, some calculations will be required to extrapolate the behavior of the prototype dampers.

41.5 Design Guidelines and Design Examples

In the United States, design of seismic isolation for bridges is governed by the *Guide Specifications for Seismic Isolation Design* (hereafter known as "Guide Specifications") published by AASHTO in 1992. Specifications for the design of energy dissipation devices have not been systematically developed, while recommended guidelines do exist for building-type applications.

In this section, design procedure for seismic isolation design and a design example will be presented mainly based on the AASHTO Guide Specifications. As for the design of supplemental energy dissipation, an attempt will be made to summarize some of the guidelines for building-type structures and their applicability to bridge applications.

41.5.1 Seismic Isolation Design Specifications and Examples

The AASHTO Guide Specifications were written as a supplement to the AASHTO *Standard Specifications for Highway Bridges* [22] (hereafter known as "Standard Specifications"). Therefore, the seismic performance categories and site coefficients are identical to those specified in the Standard Specifications. The response modification factors are the same as in the Standard Specifications except that a reduced R factor of 1.5 is permitted for essentially elastic design when the design intent of seismic isolation is to eliminate or significantly reduce damage to the substructure.

General Requirements

There are two interrelated parts in designing seismic isolation devices for bridge applications. First of all, isolation bearings must be designed for all nonseismic loads just like any other bearing devices. For example, for lead core rubber isolation bearings, both the minimum plan size and the thickness of individual rubber layers are determined by the vertical load requirement. The minimum isolator

TABLE 41.3 Damping Coefficient *B*

Damping Ratio (ξ)	≤ 2%	5%	10%	20%	30%
B	0.8	1.0	1.2	1.5	1.7

Source: AASHTO, *Guide Specification for Seismic Isolation Design*, Washington, D.C., 1991. With permission.

height is controlled by twice the displacement due to combined nonseismic loads. The minimum diameter of the lead core is determined by the requirement to maintain elastic response under combined wind, brake, and centrifugal forces. Similar requirements can also be applied to other types of isolators. In addition to the above requirements, the second part of seismic isolation design is to satisfy seismic safety requirements. The bearing must be able to support safely the vertical loads at seismic displacement. This second part is accomplished through the analysis and design procedures described below.

Methods of Analysis

The Guide Specifications allow treatment of energy dissipation in isolators as equivalent viscous damping and stiffness of isolated systems as effective linear stiffness. This permits both the single and multimodal methods of analysis to be used for seismic isolation design. Exceptions to this are isolated systems with damping ratios greater than 30% and sliding type of isolators without a self-centering mechanism. Nonlinear time history analysis is required for these cases.

Single-Mode Spectral Analysis

In this procedure, equivalent static force is given by the product of the elastic seismic force coefficient C_s and dead load W of the superstructure supported by isolation bearings, i.e.,

$$F = C_s W \tag{41.22}$$

$$C_s = \frac{\sum k_{\text{eff}} \times d_i}{W} \tag{41.23}$$

$$C_s = \frac{AS_i}{T_e B} \tag{41.24}$$

where

$\sum k_{\text{eff}}$ = the sum of the effective linear stiffness of all bearings supporting the superstructure

$d_i = \dfrac{10AS_iT_e}{B}$ = displacement across the isolation bearings

A = the acceleration coefficient

B = the damping coefficient given in Table 41.3

$T_e = \sqrt{\dfrac{W}{g\sum k_{\text{eff}}}}$ = the period of vibration

The equivalent static force must be applied independently to the two orthogonal axes and combined per the procedure of the standard specifications. The effective linear stiffness should be calculated at the design displacement.

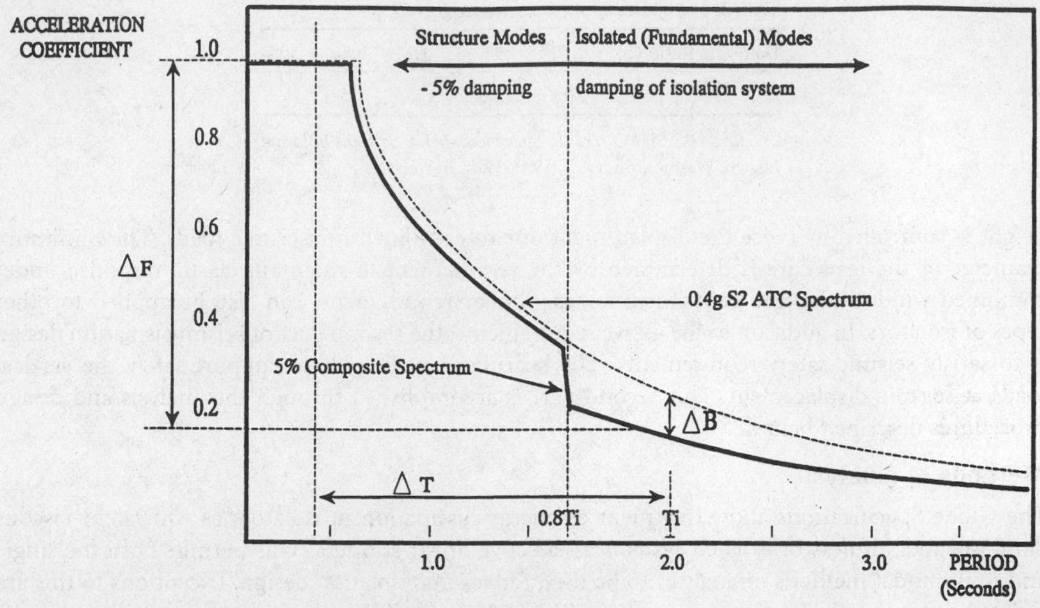

FIGURE 41.15 Modified input response spectrum.

Response Spectrum Analysis

This procedure is the same as specified in the Standard Specifications using the 5% damping ground motion response spectra with the following modifications:

1. The isolation bearings are represented by their effective stiffness values.
2. The response spectrum is modified to include the effect of higher damping of the isolated system. This results in a reduction of the response spectra values for the isolated modes. For all the other modes, the 5% damping response spectra should be used.

A typical modified response spectrum is shown in Figure 41.15.

Time History Analysis

As mentioned earlier, time history analysis is required for isolation systems with high damping ratio (>30%) or non-self-centering isolation systems. The isolation systems need to be modeled using nonlinear force–deflection characteristics of the isolator obtained from tests. Pairs of ground acceleration time history recorded from different events should be selected. These acceleration time histories should be frequency-scaled to match closely the appropriate response spectra for the site. Recommended methods for scaling are also given in the Guide Specifications. At least three pairs of time histories are required by the code. Each pair should be simultaneously applied to the model. The maximum response should be used for the design.

Design Displacement and Design Force

It is necessary to know and limit the maximum displacement of an isolation system resulting from seismic loads and nonseismic service loads for providing adequate clearance and design structural elements. The Guide Specifications require that the total design displacement be the greater of 50% of the elastomer shear strain in an elastomeric bearing system and the maximum displacement resulted from the combination of loads specified in the Standard Specifications.

Design forces for a seismically isolated bridge are obtained using the same load combinations as given for a conventionally designed bridge. Connection between superstructure and substructure shall be designed using force $F = k_{\text{eff}} d_i$. Columns and piers should be designed for the maximum force that may be developed in the isolators. The foundation design force needs not to exceed the elastic force nor the force resulted from plastic hinging of the column.

Other Requirements

It is important for an isolation system to provide adequate rigidity to resist frequently occurring wind, thermal, and braking loads. The appropriate allowed lateral displacement under nonseismic loads is left for the design engineer to decide. On the lateral restoring force, the Guide Specifications require a restoring force that is 0.25W greater than the lateral force at 50% of the design displacement. For systems not configured to provide a restoring force, more stringent vertical stability requirements have to be met.

The Guide Specifications recognize the importance of vertical stability of an isolated system by requiring a factor of safety not less than three for vertical loads in its undeformed state. A system should also be stable under the dead load plus or minus the vertical load due to seismic load at a horizontal displacement of 1.5 times the total design displacement. For systems without a lateral restoring force, this requirement is increased to three times the total design displacement.

Guidelines for Choosing Seismic Isolation

What the Guide Specifications do not cover are the conditions under which the application of seismic isolation becomes necessary or most effective. Still, some general guidelines can be drawn from various literatures and experiences as summarized below.

One factor that favors the use of seismic isolation is the level of acceptable damage to the bridge. Bridges at critical locations need to stay open to traffic following a seismic event with no damage or minor damages that can be quickly repaired. This means that the bridges are to be essentially designed elastically. The substructure pier and foundation cost could become prohibitive if using conventional design. The use of seismic isolation may be an economic solution for these bridges, if not the only solution. This may apply to both new bridge design and seismic upgrade of existing bridges.

Sometimes, it is desirable to reduce the force transferred to the superstructure, as in the case of seismic retrofit design of the Benicia–Martinez Bridge, in the San Francisco Bay, where isolation bearings were used to limit the forces in the superstructure truss members [11].

Another factor to consider is the site topography of the bridge. Irregular terrain may result in highly irregular structure configurations with significant pier height differences. This will result in uneven seismic force distributions among the piers and hence concentrated ductility demands. Use of seismic isolation bearings will make the effective stiffness and expected displacement of piers closer to each other resulting in a more even force distribution [23].

For seismic upgrading of existing bridges, isolation bearings can be an effective solution for understrength piers, insufficient girder support length, and inadequate bearings.

In some cases, there may not be an immediate saving from the use of seismic isolation over a conventional design. Considerations need to be given to a life-cycle cost comparison because the use of isolation bearings generally means much less damages, and hence lower repair costs in the long run.

Seismic Isolation Design Example

As an example, a three-span continuous concrete box-girder bridge structure, shown in Figure 41.16, will be used here to demonstrate the seismic isolation design procedure. Material and structure properties are also given in Figure 41.16. The bridge is assumed to be in a high seismic area with an acceleration coefficient A of 0.40, soil profile Type II, $S = 1.2$. For simplicity, let us use the single mode spectral analysis method for the analysis of this bridge. Assuming that the isolation bearings will be designed to provide an equivalent viscous damping of 20%, with a damping coefficient, B, of 1.5. The geometry and section properties of the bridge are taken from the worked example in the Standard Specifications with some modifications.

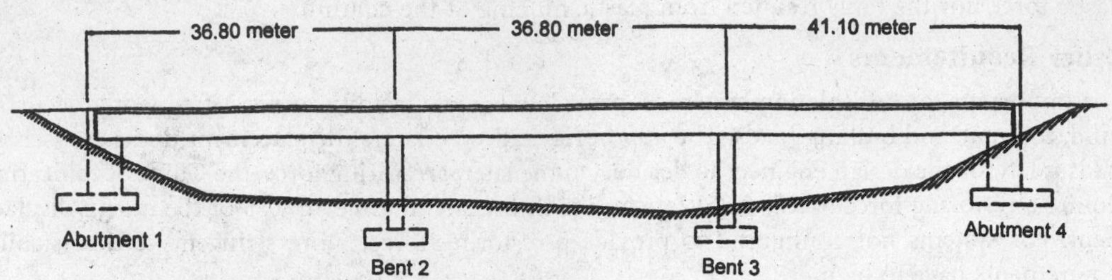

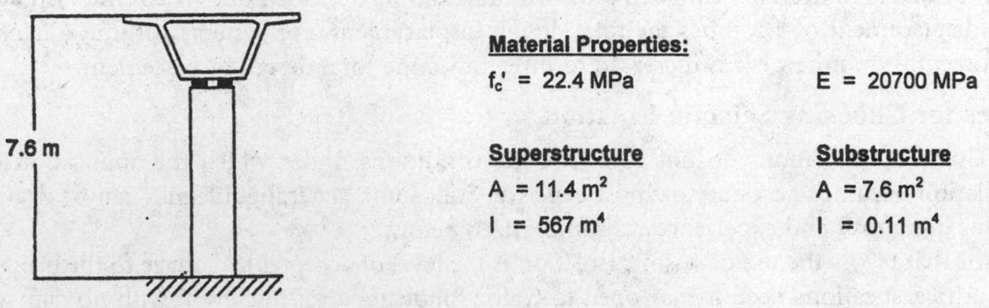

FIGURE 41.16 Example three-span bridge structure.

Force Analysis

Maximum tributary mass occurs at Bent 3, with a mass of 123 ft² × 150 lb/ft³ × 127.7 ft = 2356 kips (1,065,672 kg). Consider earthquake loading in the longitudinal direction. For fixed top of column support, the stiffness $k_0 = (12\ \mathrm{EI})/H^3 = 12 \times 432{,}000 \times 39/25^3 = 12{,}940$ kips/ft (189 kN/mm). This results in a fixed support period

$$T_0 = 2\pi\sqrt{\frac{W}{k_0 g}} = 2\pi\sqrt{\frac{2356}{12940\times 32.3}} = 0.47\mathrm{s}$$

The corresponding elastic seismic force

$$F_0 = C_s W = \frac{1.2AS}{T^{2/3}} = \frac{1.2\times0.4\times1.2}{0.47^{2/3}} W = 0.95W = 2238\ \text{kip}\ \ (9955\ \text{kN})$$

Now, let us assume that, with the introduction of seismic isolation bearings at the top of the columns, the natural period of the structure becomes 2.0 s, and damping $B = 1.5$. From Eqs. (41.22) and (41.24), the elastic seismic force for the isolated system,

$$F_i = C_s W = \frac{AS_i}{T_e B} W = \frac{0.4\times1.2}{2.0\times1.5} W = 0.16W = 377\ \text{kips}\ \ (1677\ \text{kN})$$

Displacement across the isolation bearing

$$d = \frac{10AS_i T_e}{B} = \frac{10\times0.4\times1.2\times2.0}{1.5} = 6.4\ \text{in.}\ (163\ \text{mm})$$

TABLE 41.4 Seismic Isolation Design Example Results

T_e (s)	0.5	1.0	1.5	2.0	2.5	3.0
k_{eff} (kips/in.)	306.48	76.62	34.05	19.16	12.26	8.51
(kN/mm)	(53.67)	(13.42)	(5.96)	(3.35)	(2.15)	(1.49)
d (in.)	1.60	3.20	4.80	6.40	8.00	9.60
(mm)	(40.64)	(81.28)	(121.92)	(162.56)	(203.20)	(243.84)
C_s	0.64	0.32	0.21	0.16	0.13	0.11
F_i (kip)	1507.84	753.92	502.61	376.96	301.57	251.31
(kN)	(6706.87)	(3353.44)	(2235.62)	(1676.72)	(1341.37)	(1117.81)

Table 41.4 examines the effect of isolation period on the elastic seismic force. For an isolated period of 0.5 s, which is approximately the same as the fixed support structure, the 30% reduction in elastic seismic force represents basically the effect of the added damping of the isolation system.

Isolation Bearing Design
Assume that four elastomeric (lead core rubber) bearings are used at each bent for this structure. Vertical local due to gravity load is $P = 2356/4 = 589$ kips (2620 kN). We will design the bearings such that the isolated system will have a period of 2.5 s.

$$T_e = \sqrt{\frac{W}{g\sum k_{eff}}}$$

and

$$k_{eff} = 4\left(\frac{GA}{T}\right)$$

where T is the total thickness of the elastomer. We have

$$\frac{GA}{T} = 3.06 \text{ kip/in. } (0.54 \text{ kN/mm})$$

Assuming a shear modulus $G = 145$ psi (1.0 MPa) and bearing thickness of $T = 18$ in. (457 mm) with thickness of each layer t_i equaling 0.5 in . This gives a bearing area $A = 380$ in^2 (245,070 mm^2). Hence, a plan dimension of 19.5 × 19.5 in. (495 × 495 mm).

Check shape factor:

$$S = \frac{ab}{2t_i(a+b)} = \frac{19.5\times 19.5}{2\times 0.5(19.5+19.5)} \qquad \text{OK}$$

Shear strain in the elastomer is the critical characteristic for the design of elastomeric bearings.

Three shear strain components make up the total shear strain; these are shear strains due to vertical compression, rotation, and horizontal shear deformation. In the Guide Specifications, the shear strain due to compression by vertical load is given by

$$\gamma_c = \frac{3SW}{2A_rG(1+2kS^2)}$$

where $A_r = 19.5 \times (19.5 \text{ in.} - 8.0 \text{ in.}) = 224.3 \text{ in.}^2$ is the reduced bearing area representing the effective bearing area when undergoing horizontal displacement. In this case horizontal displacement is 8.0 in. For the purpose of presenting a simple example, an approximation of the previous expression can be used:

$$\gamma_c = \frac{\sigma}{GS} = \frac{589 \times 1000}{224.3 \times 145 \times 9.75} = 1.85$$

Shear strain due to horizontal shear deformation

$$\gamma_s = \frac{d}{T} = \frac{8 \text{ in.}}{18 \text{ in.}} = 0.44$$

and shear strain due to rotation

$$\gamma_r = \frac{B^2\theta}{2t_i T} = \frac{19.5^2 \times 0.01}{2 \times 0.5 \times 18} = 0.21$$

The Guide Specifications require that the sum of all three shear strain components be less than 50% of the ultimate shear strain of the elatomer, or 5.0, whichever is smaller. In this example, the sum of all three shear strain components equals 2.50 < 5.0.

In summary, we have designed four elastomeric bearings at each bent with a plan dimension of 19.5 × 19.5 in. (495 × 495 mm) and 36 layers of 0.5 in. elastomer with $G = 145$ psi (1 MPa).

41.5.2 Guidelines for Energy Dissipation Devices Design

There are no published design guidelines or specifications for application of damping devices to bridge structures. Several recommended guidelines for application of dampers to building structures have been in development over the last few years [20,24,25]. It is hoped that a brief summary of these developments will be beneficial to bridge engineers.

General Requirements

The primary function of an energy dissipation device in a structure is to dissipate earthquake-induced energy. No special protection against structural or nonstructural damage is sought or implied by the use of energy dissipation systems.

Passive energy dissipation systems are classified as displacement-dependent, velocity-dependent, or other. The fluid damper and viscoelastic damper as discussed in Section 41.3 are examples of the velocity-dependent energy dissipation system. Friction dampers are displacement-dependent. Different models need to be used for different classes of energy dissipation systems. In addition to increasing the energy dissipation capacity of a structure, energy dissipation systems may also alter the structure stiffness. Both damping and stiffness effects need to be considered in designing energy dissipation systems.

Analysis Procedures

The use of linear analysis procedures is limited to viscous and viscoelastic energy dissipation systems. If nonlinear response is likely or hysteretic or other energy dissipaters are to be analyzed and designed, nonlinear analysis procedure must be followed. We will limit our discussion to linear analysis procedure.

Similar to the analysis of seismic isolation systems, linear analysis procedures include three methods: linear static, linear response spectrum, and linear time history analysis.

When using the linear static analysis method, one needs to make sure that the structure, exclusive of the dampers, remains elastic, that the combined structure damper system is regular, and that effective damping does not exceed 30%. The earthquake-induced displacements are reduced due to equivalent viscous damping provided by energy dissipation devices. This results in reduced base shears in the building structure.

The acceptability of the damped structure system should be demonstrated by calculations such that the sum of gravity and seismic loads at each section in each member is less than the member or component capacity.

The linear dynamic response spectrum procedure is used for more complex structure systems, where structures are modeled as MDOF systems. Modal response quantities are reduced based on the amount of equivalent modal damping provided by supplemental damping devices.

Detailed System Requirements

Other factors that need to be considered in designing supplemental damping devices for seismic applications are environmental conditions, nonseismic lateral loads, maintenance and inspection, and manufacturing quality control.

Energy dissipation devices need to be designed with consideration given to environmental conditions including aging effect, creep, and ambient temperature. Structures incorporated with energy-dissipating devices that are susceptible to failure due to low-cycle fatigue should resist the prescribed design wind forces in the elastic range to avoid premature failure. Unlike conventional construction materials that are inspected on an infrequent basis, some energy dissipation hardware will require regular inspections. It is, therefore, important to make these devices easily accessible for routine inspection and testing or even replacement.

41.6 Recent Developments and Applications

The last few years have seen significantly increased interest in the application of seismic isolation and supplemental damping devices. Many design and application experiences have been published. A shift from safety-only-based seismic design philosophy to a safety-and-performance-based philosophy has put more emphasis on limiting structural damage by controlling structural seismic response. Therefore, seismic isolation and energy dissipation have become more and more attractive alternatives to traditional design methods. Design standards are getting updated with the new development both in theory and technology. While the Guide Specifications referenced in this chapter addresses mainly elastomeric isolation bearing, new design specifications under development and review will include provisions for more types of isolation devices [26].

41.6.1 Practical Applications of Seismic Isolation

Table 41.5 lists bridges in North America that have isolation bearings installed. This list, as long as it looks, is still not complete. By some estimates, there have been several hundred isolated bridges worldwide and the number is growing. The Earthquake Engineering Research Center (EERC) at the University of California, Berkeley keeps a complete listing of the bridges with isolation and energy dissipation devices. Table 41.5 is based on information available from the EERC Internet Web site.

41.6.2 Applications of Energy Dissipation Devices to Bridges

Compared with seismic isolation devices, the application of energy dissipation devices as an independent performance improvement measure is lagging behind. This is due, in part, to the lack of code development and limited applicability of the energy dissipation devices to bridge-type structures as discussed earlier. Table 41.6 gives a list of bridge structures with supplemental damping devices against seismic and wind loads. This table is, again, based on information available from the EERC Internet Web site.

TABLE 41.5 Seismically-Isolated Bridges in North America

Bridge	Location		Owner	Engineer	Bridge Description	Bearing Type	Design Criteria
Dog River Bridge, New, 1992	AL	Mobile Co.	Alabama Hwy = 2E Dept.	Alabama Hwy. Dept.	Three-span cont. steel plate girders	LRB (DIS/Furon)	AASHTO Category A
Deas Slough Bridge, Retrofit, 1990	BC	Richmond (Hwy. 99 over Deas Slough),	British Columbia Ministry of Trans. & Hwys.	PBK Eng. Ltd.	Three-span cont. riveted haunched steel plate girders	LRB (DIS/Furon)	AASHTO $A = 0.2g$, Soil profile, Type III
Burrard Bridge Main Spans, Retrofit, 1993	BC	Vancouver (Burrard St. over False Cr.),	City of Vancouver	Buckland & Taylor Ltd.	Side spans are simple span deck trusses; center span is a Pratt through truss	LRB (DIS/Furon)	AASHTO $A = 0.21g$, Soil profile, Type I
Queensborough Bridge, Retrofit, 1994	BC	New Westminster (over N. arm of Fraser River),	British Columbia Ministry of Trans. & Hwys.	Sandwell Eng.	High-level bridge, three-span cont. haunched steel plate girders; two-girder system with floor beams	LRB (DIS/Furon)	AASHTO $A = 0.2g$, Soil profile, Type I
Roberts Park Overhead, New, 1996	BC	Vancouver (Deltaport Extension over BC Rail tracks)	Vancouver Port Corp.	Buckland & Taylor Ltd.	Five-span continuous curved steel plate girders, three girder lines	LRB	AASHTO $A = 0.26g$, Soil profile, Type II
Granville Bridge, Retrofit, 1996	BC	Vancouver, Canada	—	—	—	FIP	—
White River Bridge, 1997 (est.)	YU	Yukon, Canada	Yukon Trans. Services	—	—	FPS	—
Sierra Pt. Overhead, Retrofit, 1985	CA	S. San Francisco (U.S. 101 over S.P. Railroad)	Caltrans	Caltrans	Longitudinal steel plate girders, trans. steel plate bent cap girders	LRB (DIS/Furon)	Caltrans $A = 0.6g$, 0 to 10 ft alluvium
Santa Ana River Bridge, Retrofit, 1986	CA	Riverside	MWDSC	Lindvall, Richter & Assoc.	Three 180 ft simple span through trusses, 10 steel girder approach spans	LRB (DIS/Furon)	ATC $A = 0.4g$, Soil profile, Type II
Eel River Bridge, Retrofit, 1987	CA	Rio Dell (U.S. 101 over Eel River)	Caltrans	Caltrans	Two 300 ft steel through truss simple spans	LRB (DIS/Furon)	Caltrans $A = 0.5g$, < 150 ft alluvium
Main Yard Vehicle Access Bridge, Retrofit, 1987	CA	Long Beach (former RR bridge over Long Beach Freeway)	LACMTA	W. Koo & Assoc., Inc.	Two 128 ft simple span steel through plate girders, steel floor beams, conc. deck	LRB (DIS/Furon)	Caltrans $A = 0.5g$, 10 to 80 ft alluvium
All-American Canal Bridge, Retrofit, 1988	CA	Winterhaven, Imperial Co. (I-8 over All-American Canal)	Caltrans	Caltrans	Cont. steel plate girders (replacing former steel deck trusses)	LRB (DIS/Furon)	Caltrans $A = 0.6g$, >150 ft alluvium
Carlson Boulevard Bridge, New, 1992	CA	Richmond (part of 23rd St. Grade Separation Project)	City of Richmond	A-N West, Inc.	Simple span multicell conc. box girder	LRB (DIS/Furon)	Caltrans $A = 0.7g$, 80 to 150 ft alluvium
Olympic Boulevard Separation, New, 1993	CA	Walnut Creek (part of the 24/680 Reconstruction Project)	Caltrans	Caltrans	Four-span cont. steel plate girders	LRB (DIS/Furon)	Caltrans $A = 0.6g$, 10 to 80 ft alluvium

Alemany Interchange, Retrofit, 1994	CA	I-280/U.S. 101 Interchange, San Francisco	Caltrans	PBQD	Single and double deck viaduct, R.C. box girders and cols., 7-cont. units	LRB (DIS/Furon)	Caltrans $A = 0.5g$, 10 to 80 ft alluvium
Route 242/I-680 Separation, Retrofit, 1994	CA	Concord (Rte. 242 SB over I-680)	Caltrans	HDR Eng., Inc.	8 ft-deep cont. prestressed conc. box girder	LRB (DIS/Furon)	Caltrans $A = 0.53g$, 80 to 150 ft alluvium
Bayshore Boulevard Overcrossing, Retrofit, 1994	CA	San Francisco (Bayshore Blvd. over U.S. 101)	Caltrans	Winzler and Kelly	Continuous welded steel plate girders	LRB (DIS/Furon)	Caltrans $A = 0.53g$, 0 to 10 ft alluvium
1st Street over Figuero, Retrofit, 1995	CA	Los Angeles	City of Los Angeles	Kercheval Engineers	Continuous steel plate girders with tapered end spans	LRB	Caltrans $A = 0.6g$, 0 to 10 ft alluvium
Colfax Avenue over L.A. River, Retrofit, 1995	CA	Los Angeles	City of Los Angeles	Kercheval Engineers	Deck truss center span flanked by short steel beam spans	LRB (DIS)	Caltrans $A = 0.5g$, 10 to 80 ft alluvium
Colfax Avenue over L.A. River, Retrofit, 1995	CA	Los Angeles	City of Los Angeles	—	—	Eradiquake (RJ Watson)	—
3-Mile Slough, Retrofit, 1997 (est.)	CA	—	Caltrans	—	—	LRB (Skellerup)	—
Rio Vista, Retrofit, 1997 (est.)	CA	—	Caltrans	—	—	LRB (Skellerup)	—
Rio Mondo Bridge, Retrofit, 1997 (est.)	CA	—	Caltrans	—	—	FPS (EPS)	—
American River Bridge City of Folsom, New, 1997 (est.)	CA	Folsom	City of Folsom	-HDR	Ten-span, 2-frame continuous concrete box girder bridge	FPS (EPS)	Caltrans $A = 0.5g$, 10 to 80 ft alluvium
GGB North Viaduct, Retrofit, 1998 (est.)	CA	—	GGBHTD	—	—	LRB	—
Benicia–Martinez Bridge Retrofit, 1998 (est.)	CA	—	Caltrans	—	—	FPS (EPS)	—
Coronado Bridge, Retrofit, 1998 (est.)	CA	—	Caltrans	—	—	HDR (not selected)	—
Saugatuck River Bridge, Retrofit, 1994	CT	Westport (I-95 over Saugatuck R.)	ConnDOT	H.W. Lochner, Inc.	Three cont. steel plate girder units of 3, 4, and 3 spans	LRB (DIS/Furon)	AASHTO $A = 0.16g$, Soil profile, Type II
Lake Saltonstall Bridge, New, 1995	CT	E. Haven & Branford (I-95 over Lake Saltonstall)	ConnDOT	Steinman Boynton Gronquist & Birdsall	Seven-span cont. steel plate girders	LRB (DIS/Furon)	AASHTO $A = 0.15g$, Soil profile, Type III
RT 15 Viaduct, 1996	CT	Hamden	ConnDOT	Boswell Engineers	—	EradiQuake (RJ Watson)	—
Sexton Creek Bridge, New, 1990	IL	Alexander Co. (IL Rte. 3 over Sexton Creek)	ILDOT	ILDOT	Three-span cont. steel plate girders	LRB (DIS/Furon)	AASHTO $A = 0.2g$, Soil profile, Type III

TABLE 41.5 (continued) Seismically-Isolated Bridges in North America

Bridge	Location		Owner	Engineer	Bridge Description	Bearing Type	Design Criteria
Cache River Bridge, Retrofit, 1991	IL	Alexander Co. (IL Rte. 3 over Cache R. Diversion Channel)	ILDOT	ILDOT	Three-span cont. steel plate girders	LRB (DIS/Furon)	AASHTO $A = 0.2g$, Soil profile, Type III
Route 161 Bridge, New, 1991	IL	St. Clair Co.	ILDOT	Hurst-Rosche Engrs., Inc.	Four-span cont. steel plate girders	LRB (DIS/Furon)	AASHTO $A = 0.14g$, Soil profile, Type III
Poplar Street East Approach, Bridge #082-0005, Retrofit, 1992	IL	E. St. Louis (carrying I-55/70/64 across Mississippi R.)	ILDOT	Sverdrup Corp. & Hsiong Assoc.	Two dual steel plate girder units supported on multicol. or wall piers; piled foundations	LRB (DIS/Furon)	AASHTO $A = 0.12g$, Soil profile, Type III
Chain-of-Rocks Road over FAP 310, New, 1994	IL	Madison Co.	ILDOT	Oates Assoc.	Four-span cont. curved steel plate girders	LRB (DIS/Furon)	AASHTO $A = 0.13g$, Soil profile, Type III
Poplar Street East Approach, Roadway B, New, 1994	IL	E. St. Louis	ILDOT	Sverdrup Corp.	Three-, four- and five-span cont. curved steel plate girder units	LRB (DIS/Furon)	AASHTO $A = 0.12g$, Soil profile, Type III
Poplar Street East Approach, Roadway C, New, 1995	IL	E. St. Louis	ILDOT	Sverdrup Corp.	Three-, four- and five-span cont. curved steel plate girder units	LRB (DIS/Furon)	AASHTO $A = 0.12g$, Soil profile, Type III
Poplar Street Bridge, Retrofit, 1995	IL	—	ILDOT	—	—	—	—
RT 13 Bridge, 1996	IL	Near Freeburg	ILDOT	Casler, Houser & Hutchison	—	EradiQuake (RJ Watson)	—
Wabash River Bridge, New, 1991	IN	Terra Haute, Vigo Co. (U.S.-40 over Wabash R = 2E)	INDOT	Gannett Flemming	Seven-span cont. steel girders	LRB (DIS/Furon)	AASHTO $A = 0.1g$, Soil profile, Type II
US-51 over Minor Slough, New, 1992	KY	Ballard Co.	KTC	KTC	Three 121 ft simple span prestressed conc. I girders with cont. deck	LRB (DIS/Furon)	AASHTO $A = 0.25g$, Soil profile, Type II
Clays Ferry Bridge, Retrofit, 1995	KY	I-75 over Kentucky R.	KTC	KTC	Five-span cont. deck truss, haunched at center two piers	LRB (DIS/Furon)	AASHTO $A = 0 =$ 2E1g, Soil profile, Type I
Main Street Bridge, Retrofit, 1993	MA	Saugus (Main St. over U.S. Rte 1)	MHD	Vanasse Hangen Brustlin, Inc.	Two-span cont. steel beams with conc. deck	LRB (DIS/Furon)	AASHTO $A = 0.17g$, Soil profile, Type I
Neponset River Bridge, New, 1994	MA	New Old Colony RR over Neponset R. between Boston and Quincy	MBTA	Sverdrup Corp.	Simple span steel through girders; double-track ballasted deck	LRB (DIS/Furon)	AASHTO $A = 0.15g$, Soil profile, Type III

South Boston Bypass Viaduct, New, 1994	MA	S. Boston	MHDCATP	DRC Consult., Inc.	Conc. deck supported with three trapez. steel box girders; 10-span cont. unit with two curved trapez. steel box girders	LRB (DIS/Furon)	AASHTO $A = 0.17g$, Soil profile, Type III
South Station Connector, New, 1994	MA	Boston	MBTA	HNTB	Curved, trapezoidal steel ox girders.	LRB (DIS)	AASHTO $A = 0.18g$, Soil profile, Type III
North Street Bridge No. K-26, Retrofit, 1995	MA	Grafton (North Street over Turnpike)	MTA	The Maguire Group Inc.	Steel beams, two-span continuous center unit flanked by simple spans.	LRB (DIS)	AASHTO $A = 0.17g$, Soil profile, Type II
Old Westborough Road Bridge, Retrofit, 1995	MA	Grafton	MTA	The Maguire Group Inc.	Steel beams, two-span continuous center unit flanked by simple spans.	LRB (DIS)	AASHTO $A = 0.17g$, Soil profile, Type I
Summer Street Bridge, Retrofit, 1995	MA	Boston (over Fort Point Channel)	MHD	STV Group	Six-span continuous steel beams	LRB (DIS)	AASHTO $A = 0.17g$, Soil profile, Type III
West Street over I-93, Retrofit, 1995	MA	Wilmington	MHD	Vanesse Hangen Brustlin,	Four-span continuous steel beams with concrete deck.	LRB (DIS)	AASHTO $A = 0.17g$, Soil profile, Type I
Park Hill over Mass. Pike (I-90), 1995	MA	Millbury	Mass Turnpike	Purcell Assoc./HNTB	—	EradiQuake (RJ Watson)	—
RT 6 Swing Bridge, 1995	MA	New Bedford	MHD	Lichtenstein	—	EradiQuake (RJ Watson)	—
Mass Pike (I-90) over Fuller & North Sts., 1996	MA	Ludlow	Mass Turnpike	Maguire/HNTB	—	EradiQuake (RJ Watson)	—
Endicott Street over RT 128 (I-95), 1996	MA	Danvers	MHD	Anderson Nichols	—	EradiQuake (RJ Watson)	—
I-93 Mass Ave. Interchange, 1996	MA	S. Boston (Central Artery (I-93)/Tunnel (I-90))	MHD	Ammann & Whitney	—	HDR (SEP, formerly Furon)	—
Holyoke/South Hadley Bridge, 1996	MA	South Hadley, MA (Reconstruct over Conn. River & Canal St.)	MHD	Bayside Eng. Assoc., Inc.	—	LRB, NRB (SEP, formerly Furon)	—
NB I-170 Bridge, New, 1991	MO	St. Louis (Metrolink Light Rail over NB I-170)	BSDA	Booker Assoc., Inc. and Horner & Shifrin	Two-span cont. steel box girder flanked by short span steel box girders	LRB (DIS/Furon)	AASHTO $A = 0.1g$, Soil profile, Type I
Ramp 26 Bridge, New, 1991	MO	St. Louis (Metrolink Light Rail over Ramp 26)	BSDA	Booker Assoc., Inc. and Horner & Shifrin	Four-span cont. haunched conc. box girder	LRB (DIS/Furon)	AASHTO $A = 0.1g$, Soil profile, Type I
Springdale Bridge, New, 1991	MO	St. Louis (Metrolink Light Rail over Springdale Rd.)	BSDA	Booker Assoc., Inc. and Horner & Shifrin	Three-span cont. haunched conc. box girder	LRB (DIS/Furon)	AASHTO $A = 0.1g$, Soil profile, Type I
SB I-170/EB I-70 Bridge, New, 1991	MO	St. Louis (Metrolink Light Rail over SB I-170/EB I-70)	BSDA	Booker Assoc., Inc. and Horner & Shifrin	Simple span steel box girder, cont. haunched conc. box girder, cont. curved steel box girder	LRB (DIS/Furon)	AASHTO $A = 0.1g$, Soil profile, Type I

TABLE 41.5 (continued) Seismically-Isolated Bridges in North America

Bridge	Location		Owner	Engineer	Bridge Description	Bearing Type	Design Criteria
UMSL Garage Bridge, New, 1991	MO	St. Louis (Metrolink Light Rail over access to UMSL garage)	BSDA	Booker Assoc., Inc. and Horner & Shifrin	Three-span cont. haunched conc. box girder	LRB (DIS/Furon)	AASHTO $A = 0.1g$, Soil profile, Type I
East Campus Drive, Bridge New, 1991	MO	St. Louis (Metrolink Light Rail over E. Campus Dr.)	BSDA	Booker Assoc., Inc. and Horner & Shifrin	Four-span cont. haunched conc. box girder	LRB (DIS/Furon)	AASHTO $A = 0.1g$, Soil profile, Type I
Geiger Road Bridge, New, 1991	MO	St. Louis (Metrolink Light Rail over Geiger Rd.)	BSDA	Booker Assoc., Inc. and Horner & Shifrin	Equal cont. units: one tangent, one curved, four-span haunched conc. box girder	LRB (DIS/Furon)	AASHTO $A = 0.1g$, Soil profile, Type I
Hidalgo–San Rafael Distributor, New, 1995	MX	Mexico (north of Mexico City)	MTB	Dr. Melchor Rodriguez Caballero	Multispan continuous curved steel box girder.	LRB (DIS/Furon)	AASHTO $A = 0.48g$, Soil profile, Type II
Relocated NH Route 85 over NH Route 101, New, 1992	NH	Exeter-Stratham, Rockingham Co.	NHDOT	Webster-Martin, Inc.	Two-span cont. steel plate girders	LRB (DIS/Furon)	AASHTO $A = 0.15g$, Soil profile, Type I
Everett Turnpike over Nashua River & Canal, 1994	NH	Nashua	NHDOT	Fay Spofford & Thorndike	—	Eradiquake (RJ Watson)	—
Squamscott River Bridge, New, 1992	NH	Exeter (Relocated NH Rte. 101 over Squamscott R.)	NHDOT	Webster-Martin, Inc.	Six-span cont. steel plate girders	LRB (DIS/Furon)	AASHTO $A = 0.15g$, Soil profile, Type III
Pine Hill Road over Everett Turnpike, New, 1994	NH	Nashua	NHDOT	Costello Lomasney & de Napoli, Inc.	Two-span cont. steel plate girders	LRB (DIS/Furon)	AASHTO $A = 0.15g$, Soil profile, Type I
I-93 over Fordway Ext., 1997	NH	Derry	NHDOT	Clough Habour	—	Eradiquake (RJ Watson)	—
Pequannock River Bridge, New, 1991	NJ	Morris & Passaic Co. (I-287 over Pequannock R.,)	NJDOT	Goodkind & O'Dea, Inc.	Three cont. steel plate girder units of 2, 3, and 3 spans	LRB (DIS/Furon)	AASHTO $A = 0.12g$, Soil profile, Type II
Foundry Street Overpass 106.68, Retrofit, 1993	NJ	Newark (NJ Tpk. over Foundry St.)	NJTPA	Frederick R. Harris, Inc.	Simple span steel beams and conc. deck	LRB (DIS/Furon)	AASHTO $A = 0.18g$, Soil profile, Type II
Wilson Avenue Overpass W105.79SO, Retrofit, 1994	NJ	Newark (NJ Tpk. NSO-E over Wilson Ave.)	NJTPA	Frederick R. Harris, Inc.	Steel beams, three simple spans	LRB (DIS/Furon)	AASHTO $A = 0.18g$, Soil profile, Type I

Conrail Newark Branch Overpass E106.57, Retrofit, 1994	NJ	Newark (NJ Tpk. NB over Conrail-Newark Branch)	NJTPA	Gannett-Fleming, Inc.	Steel plate girders, four simple spans	LRB (DIS/Furon)	AASHTO $A = 0.18g$, Soil profile, Type II
Wilson Avenue Overpass E105.79SO, Retrofit, 1994	NJ	Newark (NJ Tpk. Relocated E-NSO & W-NSO over Wilson Ave.)	NJTPA	Frederick R. Harris, Inc.	Steel beams, three simple spans	LRB (DIS/Furon)	AASHTO $A = 0.18g$, Soil profile, Type I
Relocated E-NSO Overpass W106.26A, New, 1994	NJ	Newark (NJ Tpk. E-NSO ramp)	NJTPA	Frederick Harris, Inc.	Steel plate girders, cont. units of five and four spans	LRB (DIS/Furon)	AASHTO $A = 0.18g$, Soil profile, Type II
Berry's Creek Bridge, Retrofit, 1995	NJ	E. Rutherford (Rte. 3 over Berry's Cr. and NJ Transit)	NJDOT	Goodkind and O'Dea, Inc.	Cont. steel plate girders; units of three, four, three, and three spans	LRB (Furon)	AASHTO $A = 0.18g$, Soil profile, Type II
Conrail Newark Branch Overpass W106.57, Retrofit, 1995	NJ	Newark (NJ Tpk. Rd. NSW over Conrail-Newark Branch & access rd.)	NJTPA	Frederick R. Harris, Inc.	Steel beams, six simple spans	LRB (DIS)	AASHTO $A = 0.18g$, Soil profile, Type I
Norton House Bridge, Retrofit, 1996	NJ	Pompton Lakes Borough and Wayne Township, Passaic County	NJDOT	A.G. Lichtenstein & Assoc.	Three-span continuous steel beams	LRB (DIS)	AASHTO $A = 0.18g$, Soil profile, Type II
Tacony-Palmyra Approaches, 1996	NJ	Palmyra, NJ	Burlington County Bridge Comm.	Steinman/Parsons Engineers	—	LRB (SEP, formerly Furon)	—
Rt. 4 over Kinderkamack Rd., 1996	NJ	Hackensack, NJ (Widening & Bridge Rehabilitation)	NJDOT	A.G. Lichtenstein & Assoc.	—	LRB, NRB (SEP)	—
Baldwin Street/Highland Avenue, 1996	NJ	Glen Ridge, NJ Bridge over Conrail	NJDOT	A.G. Lichtenstein & Asso.	—	LRB NRB (SEP, formerly Furon)	—
I-80 Bridges B764E & W, Retrofit, 1992	NV	Verdi, Washoe Co. (I-80 over Truckee R. and a local roadway)	NDOT	NDOT	Simple span composite steel plate girders or rolled beams	LRB (DIS/Furon)	AASHTO $A = 0 =$ 2E37g, Soil profile, Type I
West Street Overpass, Retrofit, 1991	NY	Harrison, Westchester Co. (West St. over I-95 New England Thwy.)	NYSTA	N.H. Bettigole, P.C.	Four simple span steel beam structures	LRB (DIS/Furon)	AASHTO $A = 0.19g$, Soil profile, Type III
Aurora Expressway Bridge, Retrofit, 1993	NY	Erie Co. (SB lanes of Rte. 400 Aurora Expy. over Cazenovia Cr.)	NYSDOT	NYSDOT	Cont. steel beams with conc. deck	LRB (DIS/Furon)	AASHTO $A = 0.19g$, Soil profile, Type III
Mohawk River Bridge, New, 1994	NY	Herkimer	NYSTA	Steinman Boynton Gronquist & Birdsall	Three-span haunched riveted steel plate girders; simple span riveted steel plate girders or rolled beams	LRB (DIS/Furon)	AASHTO $A = 0.19g$, Soil profile, Type II

TABLE 41.5 (continued) Seismically-Isolated Bridges in North America

Bridge	Location		Owner	Engineer	Bridge Description	Bearing Type	Design Criteria
Moodna Creek Bridge, Retrofit, 1994	NY	Orange County (NYST over Moodna Cr. at MP52.83)	NYSTA	Ryan Biggs Assoc., Inc.	Three simple spans; steel plate girder center span; rolled beam side spans	LRB (DIS/Furon)	AASHTO $A = 0.15g$, Soil profile, Type II
Conrail Bridge, New, 1994	NY	Herkimer (EB and WB rdwys. of NYST over Conrail, Rte. 5, etc.)	NYSTA	Steinman Boynton Gronquist & Birdsall	Four-span cont. curved haunched welded steel plate girders.	LRB (DIS/Furon)	AASHTO $A = 0.19g$, Soil profile, Type II
Maxwell Ave. over I-95, 1995	NY	Rye	NYS Thruway Authority	Casler Houser & Hutchison	—	EradiQuake (RJ Watson)	—
JFK Terminal One Elevated Roadway, New, 1996	NY	JFK International Airport, New York City	Port Authority of New York & New Jersey	STV Group	Continuous and simple span steel plate girders	LRB	AASHTO $A = 0.19g$, Soil profile, Type III
Buffalo Airport Viaduct, 1996	NY	Buffalo	NFTA	Lu Engineers	—	EradiQuake (RJ Watson)	—
Yonkers Avenue Bridge, 1997	NY	Yonkers	NY DOT	Voilmer & Assoc.	—	EradiQuake (RJ Watson)	—
Clackamas Connector, New, 1992	OR	Milwaukie (part of Tacoma St. Interchange)	ODOT	ODOT	Eight-span cont=2E post-tensioned conc. trapez. box girder	LRB (DIS/Furon)	AASHTO $A = 0.29g$, Soil profile, Type III
Hood River Bridges, 1995	OR	Hood River, OR	ODOT	ODOT	—	NRB (Furon)	—
Marquam Bridge, Retrofit, 1995	OR	—	ODOT	—	—	FIP	—
Hood River Bridge, Retrofit, 1996	OR	Hood River, OR	ODOT	ODOT	—	FIP	—
Toll Plaza Road Bridge, New, 1990	PA	Montgomery Co. (Approach to toll plaza over Hwy. LR145)	PTC	CECO Assoc., Inc.	176 ft simple span composite steel plate girder	LRB (DIS/Furon)	AASHTO $A = 0.1g$, Soil profile, Type II
Montebella Bridge Relocation, 1996	PR	Puerto Rico	P.R. Highway Authority	Walter Ruiz & Assoc.	—	LRB, NRB (SEP, formerly Furon)	—
Blackstone River Bridge, New, 1992	RI	Woonsocket	RIDOT	R.A. Cataldo & Assoc.	Four-span cont. composite steel plate girders	LRB (DIS/Furon)	AASHTO $A = 0.1g$, Soil profile, Type II
Providence Viaduct, Retrofit, 1992	RI	Rte. I-95, Providence	RIDOT	Maguire Group	Five-span steel plate girders/hunched steel plate girder units	LRB (DIS/Furon)	AASHTO $A = 0.32g$, Soil profile, Type III
Seekonk River Bridge, Retrofit, 1995	RI	Pawtuckett (I-95 over Seekonk River)	RIDOT	A.G Lichenstein & Assoc.	Haunched steel, two-girder floor beam construction.	LRB (DIS)	AASHTO $A = 0.32g$, Soil profile, Type I

I-295 to Rt. 10, 1996	RI	Warwick/Cranston (Bridges 662 & 663)	RIDOT	Commowealth Engineers & Consultants	—	LRB (SEP, formerly Furon)	—
Chickahominy River Bridge, New, 1996	VA	Hanover-Hennico County Line (US1 over Chickahominy River)	VDOT	Alpha Corp.	Simple span prestress concrete I-girders with continuous deck.	LRB (DIS)	AASHTO $A = 0.13g$, Soil profile, Type I
Ompompanoosuc River Bridge, Retrofit, 1992	VT	Rte. 5, Norwich	VAT	VAT	Three-span cont. steel plate girders	LRB (DIS/Furon)	AASHTO $A = 0.25g$, Soil profile, Type III
Cedar River Bridge New, 1992	WA	Renton (I-405 over Cedar R. and BN RR)	WSDOT	WSDOT	Four-span cont. steel plate girders	LRB (DIS/Furon)	AASHTO $A = 0.25g$, Soil profile, Type II
Lacey V. Murrow Bridge, West Approach, Retrofit, 1992	WA	Seattle (Approach to orig. Lake Washington Floating Br.)	WSDOT	Arvid Grant & Assoc., Inc.	Cont. conc. box girders; cont. deck trusses; simple span tied arch	LRB (DIS/Furon)	AASHTO $A = 0.25g$, Soil profile, Type II
Coldwater Creek Bridge No. 11, New, 1994	WA	SR504 (Mt. St. Helens Hwy.) over Coldwater Lake Outlet	WSDOT	WSDOT	Three-span cont. steel plate girders	LRB (DIS/Furon)	AASHTO $A = 0.55g$, Soil profile, Type I
East Creek Bridge No. 14, New, 1994	WA	SR504 (Mt. St. Helens Hwy.) over East Cr.	WSDOT	WSDOT	Three-span cont. steel plate girders	LRB (DIS)	AASHTO $A = 0.55g$, Soil profile, Type I
Home Bridge, New, 1994	WA	Home (Key Penninsula Highway over Von Geldem Cove)	Pierce Co. Public Works/Road Dept.	Pierce Co. Public WorksDept.	Prestressed concrete girders; simple spans; continuous for live load.	LRB (DIS)	AASHTO $A = 0.25g$, Soil profile, Type II
Duwamish River Bridge, Retrofit, 1995	WA	Seattle (I-5 over Duwamish River)	WSDOT	Exceltech	Cont. curved steel plate girder unit flanked by curved concrete box girder end spans	LRB (DIS)	AASHTO $A = 0.27g$, Soil profile, Type II

TABLE 41.6 Bridges in North America with Supplemental Damping Devices

Bridge	Location	Type and Number of Dampers	Year	Notes
San Francisco–Oakland Bay Bridge	San Francisco, CA	Viscous dampers Total: 96	1998 (design)	Retrofit of West Suspension spans. 450~650 kips force output, 6~22 in. strokes
Gerald Desmond Bridge	Long Beach, CA	Viscous dampers (Enidine) Total: 258	1996	Retrofit, 258 × 50 kip shock absorbers, 6 in. stroke
Cape Girardeau Bridge	Cape Girardeau, MO	Viscous dampers (Taylor)	1997	New construction of a cable-stayed bridge; Dampers used to control longitudinal earthquake movement while allowing free thermal movement.
The Golden Gate Bridge	San Francisco, CA	Viscous dampers (to be det.) Total: 40	1999 (est.)	Retrofit, 40 × 650 kip nonlinear dampers, ± 24 in.
Santiago Creek Bridge	California	Viscous dampers (Enidine)	1997 (est.)	New construction; dampers at abutments for energy dissipation in longitudinal direction
Sacramento River Bridge at Rio Vista	Rio Vista, CA	Viscous dampers (Taylor)	1997 (est.)	Retrofit; eight dampers used to control uplift of lift-span towers
Vincent Thomas Bridge	Long Beach, CA	Viscous dampers (to be det.) Total: 16	—	Retrofit, 8 × 200 kip and 8 × 100 kip linear dampers, ± 12 in.
Montlake Bridge	Seattle, WA	Viscous dampers (Taylor)	1996	Protection of new bascule leafs from runaway
West Seattle Bridge	Seattle, WA	Viscous dampers (Taylor)	1990	Deck isolation for swing bridge.

41.7 Summary

An attempt has been made to introduce the basic concepts of seismic isolation and supplemental energy dissipation, their history, current developments, applications, and design-related issues. Although significant strides have been made in terms of implementing these concepts to structural design and performance upgrade, it should be mentioned that these are emerging technologies and advances are being made constantly. With more realistic prototype testing results being made available to the design community of seismic isolation and supplemental energy dissipation devices from the FHWA/Caltrans testing program, significant improvement in code development will continuously make design easier and more standardized.

Acknowledgments

The author would like to express his deepest gratitude to Professor T. T. Soong, State University of New York at Buffalo, for his careful, thorough review, and many valuable suggestions. The author is also indebted to Dr. Lian Duan, California State Department of Transportation, for his encouragement, patience, and valuable input.

References

1. Applied Technology Council, *Improved Seismic Design Criteria for California Bridges: Provisional Recommendations*, ATC-32, Applied Technology Council, Redwood City, 1996.
2. Housner, G. W., Bergman, L. A., Caughey, T. K., Chassiakos, A. G., Claus, R. O., Masri, S. F., Seklton, R. E., Soong, T. T., Spencer, B. F., and Yao, J. T. P., Structural control: past, present, and future, *J. Eng. Mech. ASCE*, 123(9), 897–971.
3. Soong, T. T., and Dargush, G. F., Passive energy dissipation and active control, in *Structure Engineering Handbook*, Chen, W. F., Ed., CRC Press, Boca Raton, FL, 1997.
4. Soong, T. T. and Dargush, G. F., *Passive Energy Dissipation Systems in Structural Engineering*, John Wiley & Sons, New York, 1997.
5. Housner, G. W. and Jennings, P. C., *Earthquake Design Criteria*, Earthquake Engineering Research Institute, 1982.
6. Newmark, N. M. and Hall, W. J., Procedures and Criteria for Earthquake Resistant Design, *Building Practice for Disaster Mitigation*, Department of Commerce, Feb. 1973.
7. Dynamic Isolation Systems, *Force Control Bearings for Bridges — Seismic Isolation Design*, Rev. 4, Lafayette, CA, Oct. 1994.
8. Clough, R. W. and Penzien, J., *Dynamics of Structures*, 2nd ed., McGraw-Hill, New York, 1993.
9. Zhang, R., Soong, T. T., and Mahmoodi, P., Seismic response of steel frame structures with added viscoelastic dampers, *Earthquake Eng. Struct. Dyn.*, 18, 389–396, 1989.
10. Earthquake Protection Systems, *Friction Pendulum Seismic Isolation Bearings*, Product Technical Information, Earthquake Protection Systems, Emeriville, CA, 1997.
11. Liu, D. W, Nobari, F. S., Schamber, R. A., and Imbsen, R. A., Performance based seismic retrofit design of Benicia–Martinez bridge, in *Proceedings, National Seismic Conference on Bridges and Highways,* Sacramento, CA, July 1997.
12. Constantinou, M. C. and Symans, M. D., Seismic response of structures with supplemental damping, *Struct. Design Tall Buildings*, 2, 77–92, 1993.
13. Ingham, T. J., Rodriguez, S., Nader, M. N., Taucer, F., and Seim, C., Seismic retrofit of the Golden Gate Bridge, in *Proceedings, National Seismic Conference on Bridges and Highways,* Sacramento, CA, July 1997.
14. Taylor Devices, Sample Technical Specifications for Viscous Damping Devices, Taylor Devices, Inc., North Tonawanda, NY, 1996.
15. Lin, R. C., Liang, Z., Soong, T. T., and Zhang, R., An Experimental Study of Seismic Structural Response with Added Viscoelastic Dampers, Report No. NCEER-88-0018, National Center For Earthquake Engineering Research, State University of New York at Buffalo, February, 1988.
16. Crosby, P., Kelly, J. M., and Singh, J., Utilizing viscoelastic dampers in the seismic retrofit of a thirteen story steel frame building, *Structure Congress,* XII, Atlanta, GA, 1994.
17. Skinner, R. I., Kelly, J. M., and Heine, A. J., Hysteretic dampers for earthquake resistant structures, *Earthquake Eng. Struct. Dyn.*, 3, 297–309, 1975.
18. Pall, A. S. and Pall, R., Friction-dampers used for seismic control of new and existing buildings in Canada, in *Proceedings ATC 17-1 Seminar on Isolation, Energy Dissipation and Active Control,* San Francisco, 1992.
19. AASHTO, *Guide Specifications for Seismic Isolation Design,* American Association of State Highway and Transportation Officials, Washington, D.C., June 1991.
20. FEMA, NEHRP Recommended Provisions for Seismic Regulations for New Buildings, 1994 ed., Federal Emergency Management Agency, Washington, D.C., May 1995.
21. EERC/Berkeley, Pre-qualification Testing of Viscous Dampers for the Golden Gate Bridge Seismic Rehabilitation Project, A Report to T. Y. Lin International, Inc., Report No. EERCL/95-03, Earthquake Engineering Research Center, University of California at Berkeley, December 1995.

22. AASHTO, *Standard Specifications for Highway Bridges*, 16th ed., American Association of State Highway and Transportaion Officials, Washington, D.C., 1996.
23. Priestley, M. J. N, Seible, F., and Calvi, G. M., *Seismic Design and Retrofit of Bridges*, John Wiley an&& Sons, New York, 1996.
24. Whittaker, A., Tentative General Requirements for the Design and Construction of Structures Incorporating Discrete Passive Energy Dissipation Devices, ATC-15-4, Redwood City, CA, 1994.
25. Applied Technology Council, BSSC Seismic Rehabilitation Projects, ATC-33.03 (Draft), Redwood City, CA, 1997. .
26. AASHTO-T3 Committee Task Group, *Guide Specifications for Seismic Isolation Design*, T3 Committee Task Group Draft Rewrite, America Association of State Highway and Transportation Officials, Washington, D.C., May, 1997.

42

Soil–Foundation–Structure Interaction

Wen-Shou Tseng
International Civil Engineering Consultants, Inc.

Joseph Penzien
International Civil Engineering Consultants, Inc.

42.1 Introduction

Prior to the 1971 San Fernando, California earthquake, nearly all damages to bridges during earthquakes were caused by ground failures, such as liquefaction, differential settlement, slides, and/or spreading; little damage was caused by seismically induced vibrations. Vibratory response considerations had been limited primarily to wind excitations of large bridges, the great importance of which was made apparent by failure of the Tacoma Narrows suspension bridge in the early 1940s, and to moving loads and impact excitations of smaller bridges.

The importance of designing bridges to withstand the vibratory response produced during earthquakes was revealed by the 1971 San Fernando earthquake during which many bridge structures collapsed. Similar bridge failures occurred during the 1989 Loma Prieta and 1994 Northridge, California earthquakes, and the 1995 Kobe, Japan earthquake. As a result of these experiences, much has been done recently to improve provisions in seismic design codes, advance modeling and analysis

0-8493-7434-0/00/$0.00+$.50

procedures, and develop more effective detail designs, all aimed at ensuring that newly designed and retrofitted bridges will perform satisfactorily during future earthquakes.

Unfortunately, many of the existing older bridges in the United States and other countries, which are located in regions of moderate to high seismic intensity, have serious deficiencies which threaten life safety during future earthquakes. Because of this threat, aggressive actions have been taken in California, and elsewhere, to retrofit such unsafe bridges bringing their expected performances during future earthquakes to an acceptable level. To meet this goal, retrofit measures have been applied to the superstructures, piers, abutments, and foundations.

It is because of this most recent experience that the importance of coupled soil–foundation–structure interaction (SFSI) on the dynamic response of bridge structures during earthquakes has been fully realized. In treating this problem, two different methods have been used (1) the "elastodynamic" method developed and practiced in the nuclear power industry for large foundations and (2) the so-called empirical *p–y* method developed and practiced in the offshore oil industry for pile foundations. Each method has its own strong and weak characteristics, which generally are opposite to those of the other, thus restricting their proper use to different types of bridge foundation. By combining the models of these two methods in series form, a hybrid method is reported herein which makes use of the strong features of both methods, while minimizing their weak features. While this hybrid method may need some further development and validation at this time, it is fundamentally sound; thus, it is expected to become a standard procedure in treating seismic SFSI of large bridges supported on different types of foundation.

The subsequent sections of this chapter discuss all aspects of treating seismic SFSI by the elastodynamic, empirical *p–y*, and hybrid methods, including generating seismic inputs, characterizing soil–foundation systems, conducting force–deformation demand analyses using the substructuring approach, performing force–deformation capacity evaluations, and judging overall bridge performance.

42.2 Description of SFSI Problems

The broad problem of assessing the response of an engineered structure interacting with its supporting soil or rock medium (hereafter called soil medium for simplicity) under static and/or dynamic loadings will be referred here as the soil–structure interaction (SSI) problem. For a building that generally has its superstructure above ground fully integrated with its substructure below, reference to the SSI problem is appropriate when describing the problem of interaction between the complete system and its supporting soil medium. However, for a long bridge structure, consisting of a superstructure supported on multiple piers and abutments having independent and often distinct foundation systems which in turn are supported on the soil medium, the broader problem of assessing interaction in this case is more appropriately and descriptively referred to as the soil–foundation–structure interaction (SFSI) problem. For convenience, the SFSI problem can be separated into two subproblems, namely, a soil–foundation interaction (SFI) problem and a foundation–structure interaction (FSI) problem. Within the context of SFSI, the SFI part of the total problem is the one to be emphasized, since, once it is solved, the FSI part of the total problem can be solved following conventional structural response analysis procedures. Because the interaction between soil and the foundations of a bridge makes up the core of an SFSI problem, it is useful to review the different types of bridge foundations that may be encountered in dealing with this problem.

42.2.1 Bridge Foundation Types

From the perspective of SFSI, the foundation types commonly used for supporting bridge piers can be classified in accordance with their soil-support configurations into four general types: (1) spread footings, (2) caissons, (3) large-diameter shafts, and (4) slender-pile groups. These types as described separately below are shown in Figure 42.1.

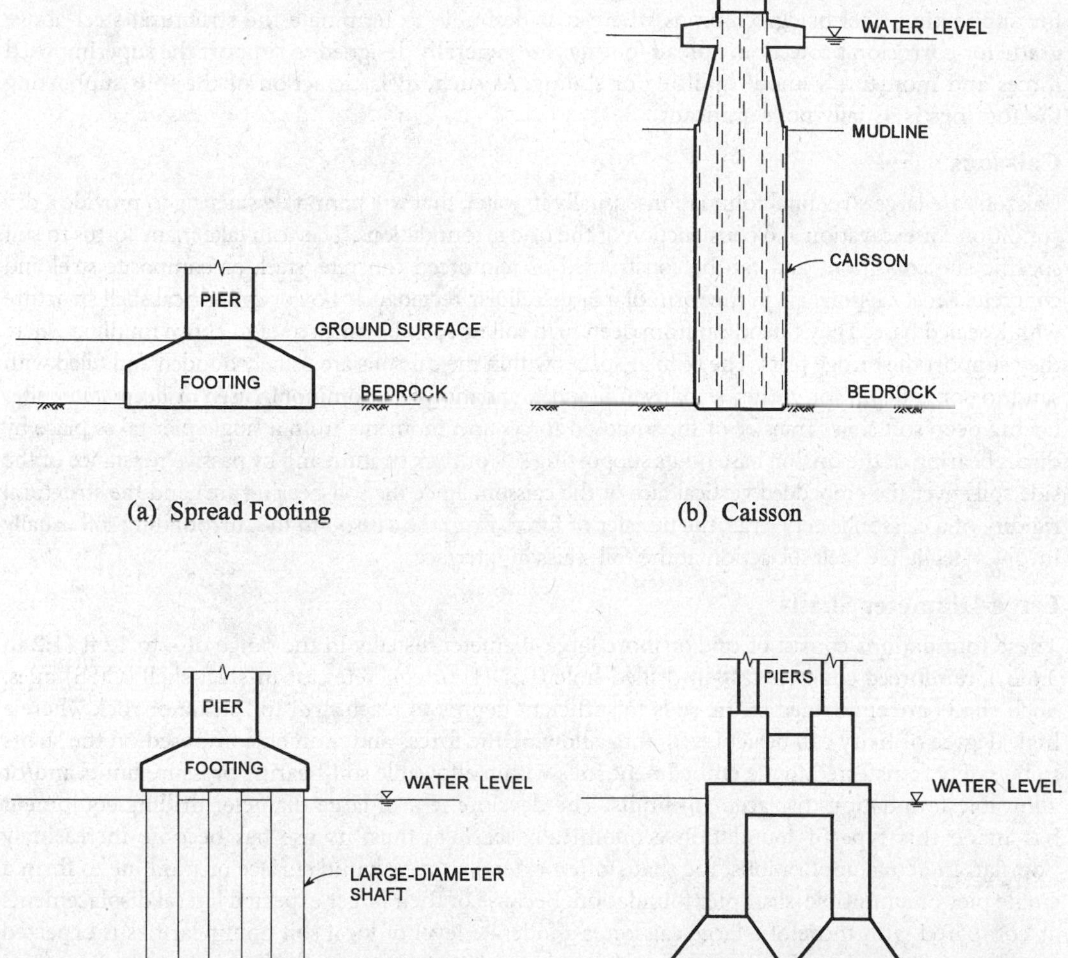

FIGURE 42.1 Bridge foundation types: (a) spread footing; (b) caisson; (c) large-diameter shafts; and (d) slender-pile group.

Spread Footings

Spread footings bearing directly on soil or rock are used to distribute the concentrated forces and moments in bridge piers and/or abutments over sufficient areas to allow the underlying soil strata to support such loads within allowable soil-bearing pressure limits. Of these loads, lateral forces are resisted by a combination of friction on the foundation bottom surface and passive soil pressure on its embedded vertical face. Spread footings are usually used on competent soils or rock which have high allowable bearing pressures. These foundations may be of several forms, such as (1) isolated footings, each supporting a single column or wall pier; (2) combined footings, each supporting two or more closely spaced bridge columns; and (3) pedestals which are commonly used for supporting steel bridge columns where it is desirable to terminate the structural steel above grade for corrosion protection. Spread footings are generally designed to support the superimposed forces and moments without uplifting or sliding. As such, inelastic action of the soils supporting the footings is usually not significant.

Caissons

Caissons are large structural foundations, usually in water, that will permit dewatering to provide a dry condition for excavation and construction of the bridge foundations. They can take many forms to suit specific site conditions and can be constructed of reinforced concrete, steel, or composite steel and concrete. Most caissons are in the form of a large cellular rectangular box or cylindrical shell structure with a sealed base. They extend up from deep firm soil or rock-bearing strata to above mudline where they support the bridge piers. The cellular spaces within the caissons are usually flooded and filled with sand to some depth for greater stability. Caisson foundations are commonly used at deep-water sites having deep soft soils. Transfer of the imposed forces and moments from a single pier takes place by direct bearing of the caisson base on its supporting soil or rock stratum and by passive resistance of the side soils over the embedded vertical face of the caisson. Since the soil-bearing area and the structural rigidity of a caisson is very large, the transfer of forces from the caisson to the surrounding soil usually involves negligible inelastic action at the soil–caisson interface.

Large-Diameter Shafts

These foundations consist of one or more large-diameter, usually in the range of 4 to 12 ft (1.2 to 3.6 m), reinforced concrete cast-in-drilled-hole (CIDH) or concrete cast-in-steel-shell (CISS) piles. Such shafts are embedded in the soils to sufficient depths to reach firm soil strata or rock where a high degree of fixity can be achieved, thus allowing the forces and moments imposed on the shafts to be safely transferred to the embedment soils within allowable soil-bearing pressure limits and/or allowable foundation displacement limits. The development of large-diameter drilling equipment has made this type of foundation economically feasible; thus, its use has become increasingly popular. In actual applications, the shafts often extend above ground surface or mudline to form a single pier or a multiple-shaft pier foundation. Because of their larger expected lateral displacements as compared with those of a large caisson, a moderate level of local soil nonlinearities is expected to occur at the soil–shaft interfaces, especially near the ground surface or mudline. Such nonlinearities may have to be considered in design.

Slender-Pile Groups

Slender piles refer to those piles having a diameter or cross-sectional dimensions less than 2 ft (0.6 m). These piles are usually installed in a group and provided with a rigid cap to form the foundation of a bridge pier. Piles are used to extend the supporting foundations (pile caps) of a bridge down through poor soils to more competent soil or rock. The resistance of a pile to a vertical load may be essentially by point bearing when it is placed through very poor soils to a firm soil stratum or rock, or by friction in case of piles that do not achieve point bearing. In real situations, the vertical resistance is usually achieved by a combination of point bearing and side friction. Resistance to lateral loads is achieved by a combination of soil passive pressure on the pile cap, soil resistance around the piles, and flexural resistance of the piles. The uplift capacity of a pile is generally governed by the soil friction or cohesion acting on the perimeter of the pile. Piles may be installed by driving or by casting in drilled holes. Driven piles may be timber piles, concrete piles with or without prestress, steel piles in the form of pipe sections, or steel piles in the form of structural shapes (e.g., H shape). Cast-in-drilled-hole piles are reinforced concrete piles installed with or without steel casings. Because of their relatively small cross-sectional dimensions, soil resistance to large pile loads usually develops large local soil nonlinearities that must be considered in design. Furthermore, since slender piles are normally installed in a group, mutual interactions among piles will reduce overall group stiffness and capacity. The amounts of these reductions depend on the pile-to-pile spacing and the degree ofsoil nonlinearity developed in resisting the loads.

42.2.2 Definition of SFSI Problem

For a bridge subjected to externally applied static and/or dynamic loadings on the aboveground portion of the structure, the SFSI problem involves evaluation of the structural performance (demand/capacity ratio) of the bridge under the applied loadings taking into account the effect of SFI. Since in this case the ground has no initial motion prior to loading, the effect of SFI is to provide the foundation–structure system with a flexible boundary condition at the soil–foundation interface location when static loading is applied and a compliant boundary condition when dynamic loading is applied. The SFI problem in this case therefore involves (1) evaluation of the soil–foundation interface boundary flexibility or compliance conditions for each bridge foundation, (2) determination of the effects of these boundary conditions on the overall structural response of the bridge (e.g., force, moment, or deformation) demands, and (3) evaluation of the resistance capacity of each soil–foundation system that can be compared with the corresponding response demand in assessing performance. That part of determining the soil–foundation interface boundary flexibilities or compliances will be referred to subsequently in a gross term as the "foundation stiffness or impedance problem"; that part of determining the structural response of the bridge as affected by the soil–foundation boundary flexibilities or compliances will be referred to as the "foundation–structure interaction problem"; and that part of determining the resistance capacity of the soil–foundation system will be referred to as the "foundation capacity problem."

For a bridge structure subjected to seismic conditions, dynamic loadings are imposed on the structure. These loadings, which originate with motions of the soil medium, are transmitted to the structure through its foundations; therefore, the overall SFSI problem in this case involves, in addition to the foundation impedance, FSI, and foundation capacity problems described above, the evaluation of (1) the soil forces acting on the foundations as induced by the seismic ground motions, referred to subsequently as the "seismic driving forces," and (2) the effects of the free-field ground-motion-induced soil deformations on the soil–foundation boundary compliances and on the capacity of the soil–foundation systems. In order to evaluate the seismic driving forces on the foundations and the effects of the free-field ground deformations on compliances and capacities of the soil–foundation systems, it is necessary to determine the variations of free-field motion within the ground regions which interact with the foundations. This problem of determining the free-field ground motion variations will be referred to herein as the "free-field site response problem." As will be shown later, the problem of evaluating the seismic driving forces on the foundations is equivalent to determining the "effective or scattered foundation input motions" induced by the free-field soil motions. This problem will be referred to here as the "foundation scattering problem."

Thus, the overall SFSI problem for a bridge subjected to externally applied static and/or dynamic loadings can be separated into the evaluation of (1) foundation stiffnesses or impedances, (2) foundation–structure interactions, and (3) foundation capacities. For a bridge subjected to seismic ground motion excitations, the SFSI problem involves two additional steps, namely, the evaluation of free-field site response and foundation scattering. When solving the total SFSI problem, the effects of the nonzero soil deformation state induced by the free-field seismic ground motions should be evaluated in all five steps mentioned above.

42.2.3 Demand vs. Capacity Evaluations

As described previously, assessing the seismic performance of a bridge system requires evaluation of SFSI involving two parts. One part is the evaluation of the effects of SFSI on the seismic-response demands within the system; the other part is the evaluation of the seismic force and/or deformation capacities within the system. Ideally, a well-developed methodology should be one that is capable of solving these two parts of the problem concurrently in one step using a unified suitable model for the system. Unfortunately, to date, such a unified method has not yet been developed. Because of the complexities of a real problem and the different emphases usually demanded of the solutions for the two parts, different solution strategies and methods of analysis are warranted for solving these two parts of the overall SFSI problem. To be more specific, evaluation on the demand side of the problem is concerned with the overall SFSI system behavior which is controlled by the mass, damping (energy dissipation), and stiffness properties, or, collectively, the impedance properties, of the entire system; and, the solution must satisfy the dynamic equilibrium and compatibility conditions of the global system. This system behavior is not sensitive, however, to approximations made on local element behavior; thus, its evaluation does not require sophisticated characterizations of the detailed constitutive relations of its local elements. For this reason, evaluation of demand has often been carried out using a linear or equivalent linear analysis procedure. On the contrary, evaluation of capacity must be concerned with the extreme behavior of local elements or subsystems; therefore, it must place emphasis on the detailed constitutive behaviors of the local elements or subsystems when deformed up to near-failure levels. Since only local behaviors are of concern, the evaluation does not have to satisfy the global equilibrium and compatibility conditions of the system fully. For this reason, evaluation of capacity is often obtained by conducting nonlinear analyses of detailed local models of elements or subsystems or by testing of local members, connections, or sub-assemblages, subjected to simple pseudo-static loading conditions.

Because of the distinct differences between effective demand and capacity analyses as described above, the analysis procedures presented subsequently differentiate between these two parts of the overall SFSI problem.

42.3 Current State-of-the-Practice

The evaluation of SFSI effects on bridges located in regions of high seismicity has not received as much attention as for other critical engineered structures, such as dams, nuclear facilities, and offshore structures. In the past, the evaluation of SFSI effects for bridges has, in most cases, been regarded as a part of the bridge foundation design problem. As such, emphasis has been placed on the evaluation of load-resisting capacities of various foundation systems with relatively little attention having been given to the evaluation of SFSI effects on seismic-response demands within the complete bridge system. Only recently has formal SSI analysis methodologies and procedures, developed and applied in other industries, been adopted and applied to seismic performance evaluations of bridges [1], especially large important bridges [2,3].

Even though the SFSI problems for bridges pose their own distinct features (e.g., multiple independent foundations of different types supported in highly variable soil conditions ranging from hard to very soft), the current practice is to adopt, with minor modifications, the same methodologies and procedures developed and practiced in other industries, most notably, the nuclear power and offshore oil industries. Depending upon the foundation type and its soil-support condition, the procedures currently being used in evaluating SFSI effects on bridges can broadly be classified into two main methods, namely, the so-called elastodynamic method that has been developed and practiced in the nuclear power industry for large foundations, and the so-called empirical *p–y* method that has been developed and practiced in the offshore oil industry for pile foundations. The bases and applicabilities of these two methods are described separately below.

42.3.1 Elastodynamic Method

This method is based on the well-established elastodynamic theory of wave propagation in a linear elastic, viscoelastic, or constant-hysteresis-damped elastic half-space soil medium. The fundamental element of this method is the constitutive relation between an applied harmonic point load and the corresponding dynamic response displacements within the medium called the dynamic Green's functions. Since these functions apply only to a linear elastic, visoelastic, or constant-hysteresis-damped elastic medium, they are valid only for linear SFSI problems. Since application of the elastodynamic method of analysis uses only mass, stiffness, and damping properties of an SFSI system, this method is suitable only for global system response analysis applications. However, by adopting the same equivalent linearization procedure as that used in the seismic analysis of free-field soil response, e.g., that used in the computer program SHAKE [4], the method has been extended to one that can accommodate global soil nonlinearities, i.e., those nonlinearities induced in the free-field soil medium by the free-field seismic waves [5].

Application of the elastodynamic theory to dynamic SFSI started with the need for solving machine–foundation vibration problems [6]. Along with other rapid advances in earthquake engineering in the 1970s, application of this theory was extended to solving seismic SSI problems for building structures, especially those of nuclear power plants [7–9]. Such applications were enhanced by concurrent advances in analysis techniques for treating soil dynamics, including development of the complex modulus representation of dynamic soil properties and use of the equivalent linearization technique for treating ground-motion-induced soil nonlinearities [10–12]. These developments were further enhanced by the extensive model calibration and methodology validation and refinement efforts carried out in a comprehensive large-scale SSI field experimental program undertaken by the Electric Power Research Institute (EPRI) in the 1980s [13]. All of these efforts contributed to advancing the elastodynamic method of SSI analysis currently being practiced in the nuclear power industry [5].

Because the elastodynamic method of analysis is capable of incorporating mass, stiffness, and damping characteristics of each soil, foundation, and structure subsystem of the overall SFSI system, it is capable of capturing the dynamic interactions between the soil and foundation subsystems and between the foundations and structure subsystem; thus, it is suitable for seismic demand analyses. However, since the method does not explicitly incorporate strength characteristics of the SFSI system, it is not suitable for capacity evaluations.

As previously mentioned in Section 42.2.1, there are four types of foundation commonly used for bridges: (1) spread footings, (2) caissons, (3) large-diameter shafts, and (4) slender-pile groups. Since only small local soil nonlinearities are induced at the soil–foundation interfaces of spread footings and caissons, application of the elastodynamic method of seismic demand analysis of the complete SFSI system is valid. However, the validity of applying this method to large-diameter shaft foundations depends on the diameter of the shafts and on the amplitude of the imposed loadings. When the shaft diameter is large so that the load amplitudes produce only small local soil nonlinearities, the method is reasonably valid. However, when the shaft diameter is relatively small, the larger-amplitude loadings will produce local soil nonlinearities sufficiently large to require that the method be modified as discussed subsequently. Application of the elastodynamic method to slender-pile groups is usually invalid because of the large local soil nonlinearities which develop near the pile boundaries. Only for very low amplitude loadings can the method be used for such foundations.

42.3.2 Empirical "*p-y*" Method

This method was originally developed for the evaluation of pile–foundation response due to lateral loads [14–16] applied externally to offshore structures. As used, it characterizes the lateral soil resistance per unit length of pile, p, as a function of the lateral displacement, y. The p–y relation is generally developed on the basis of an empirical curve which reflects the nonlinear resistance of the local soil surrounding the pile at a specified depth (Figure 42.2). Construction of the curve depends mainly on soil material strength parameters, e.g., the friction angle, ϕ, for sands and cohesion, c, for clays at the specified depth. For shallow soil depths where soil surface effects become important, construction of these curves also depends on the local soil failure mechanisms, such as failure by a passive soil resistance wedge. Typical p–y curves developed for a pile at different soil depths are shown in Figure 42.3. Once the set of p–y curves representing the soil resistances at discrete values

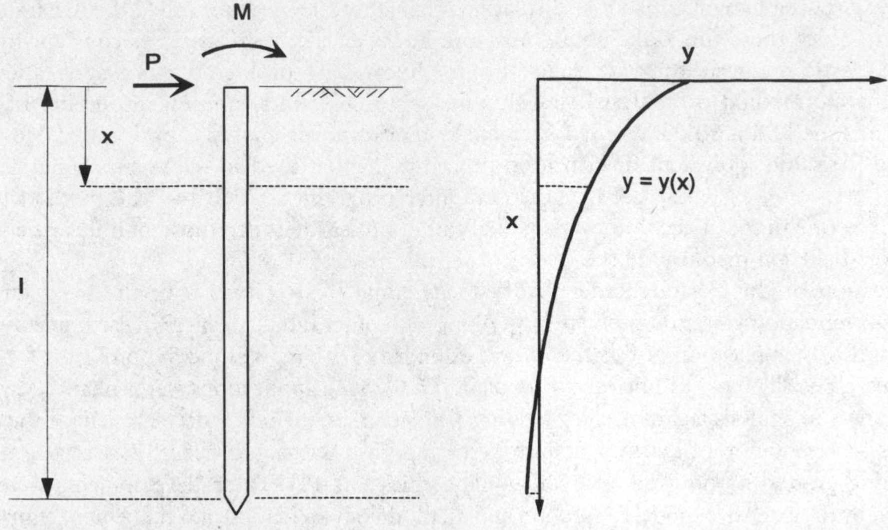

Pile Deflection Curve

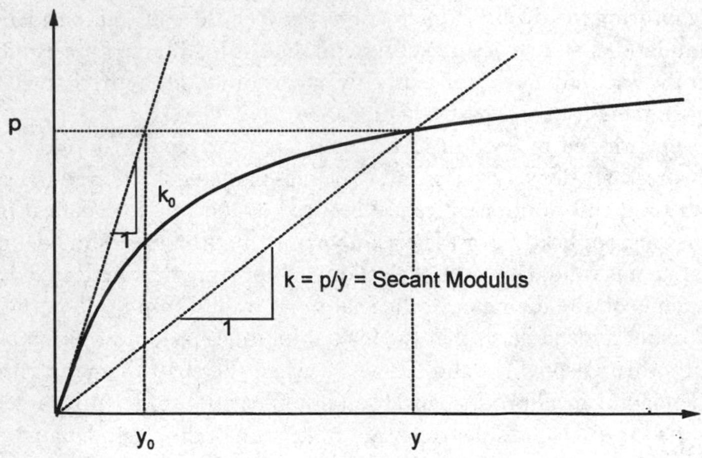

Lateral Soil Resistance "p-y" Curve at Depth x

FIGURE 42.2 Empirical *p–y* curves and secant modulus.

of depth along the length of the pile has been constructed, evaluation of pile response under a specified set of lateral loads is accomplished by solving the problem of a beam supported laterally on discrete nonlinear springs. The validity and applicability of this method are based on model calibrations and correlations with field experimental results [15,16].

Based on the same model considerations used in developing the *p–y* curves for lateral response analysis of piles, the method has been extended to treating the axial resistance of soils to piles per unit length of pile, *t*, as a nonlinear function of the corresponding axial displacement, z, resulting in the so-called axial *t–z* curve, and treating the axial resistance of the soils at the pile tip, Q, as a

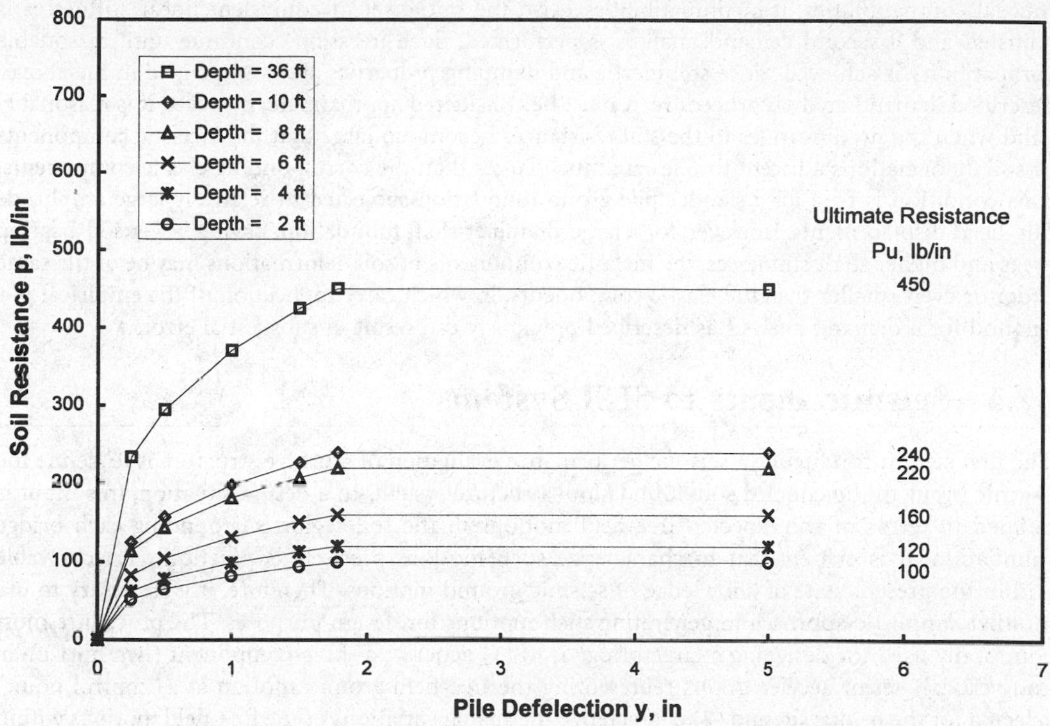

FIGURE 42.3 Typical *p–y* curves for a pile at different depths.

nonlinear function of the pile tip axial displacement, *d*, resulting in the so-called *Q–d* curve. Again, the construction of the *t–z* and *Q–d* curves for a soil-supported pile is based on empirical curvilinear forms and the soil strength parameters as functions of depth. By utilizing the set of *p–y*, *t–z*, and *Q–d* curves developed for a pile foundation, the response of the pile subjected to general three-dimensional (3-D) loadings applied at the pile head can be solved using the model of a 3-D beam supported on discrete sets of nonlinear lateral *p–y*, axial *t–z*, and axial *Q–d* springs. The method as described above for solving a soil-supported pile foundation subjected to applied loadings at the pile head is referred to here as the empirical *p–y* method, even though it involves not just the lateral *p–y* curves but also the axial *t–z* and *Q–d* curves for characterizing the soil resistances.

Since this method depends primarily on soil-resistance strength parameters and does not incorporate soil mass, stiffness, and damping characteristics, it is, strictly speaking, only applicable for capacity evaluations of slender-pile foundations and is not suitable for seismic demand evaluations because, as mentioned previously, a demand evaluation for an SFSI system requires the incorporation of the mass, stiffness, and damping properties of each of the constituent parts, namely, the soil, foundation, and structure subsystems.

Even though the *p–y* method is not strictly suited to demand analyses, it is current practice in performing seismic-demand evaluations for bridges supported on slender-pile group foundations to make use of the empirical nonlinear *p–y*, *t–z*, and *Q–d* curves in developing a set of equivalent linear lateral and axial soil springs attached to each pile at discrete elevations in the foundation. The soil–pile systems developed in this manner are then coupled with the remaining bridge structure to form the complete SFSI system for use in a seismic demand analysis. The initial stiffnesses of the equivalent linear *p–y*, *t–z*, and *Q–d* soil springs are based on secant moduli of the nonlinear *p–y*, *t–y*, and *Q–d* curves, respectively, at preselected levels of lateral and axial pile displacements, as shown schematically in Figure 42.2. After completing the initial demand analysis, the amplitudes of pile displacement are compared with the corresponding preselected amplitudes to check on their mutual compatibilities. If incompatibilities exist, the initial set of equivalent linear stiffnesses is adjusted and a second demand analysis is performed. Such iterations continue until reasonable compatibility is achieved. Since soil inertia and damping properties are not included in the above-described demand analysis procedure, it must be considered approximate; however, it is reasonably valid when the nonlinearities in the soil resistances become so large that the inelastic components of soil deformations adjacent to piles are much larger than the corresponding elastic components. This condition is true for a slender-pile group foundation subjected to relatively large amplitude pile-head displacements. However, for a large-diameter shaft foundation, having larger soil-bearing areas and higher shaft stiffnesses, the inelastic components of soil deformations may be of the same order or even smaller than the elastic components, in which case, application of the empirical *p*–*y* method for a demand analysis as described previously can result in substantial errors.

42.4 Seismic Inputs to SFSI System

The first step in conducting a seismic performance evaluation of a bridge structure is to define the seismic input to the coupled soil–foundation–structure system. In a design situation, this input is defined in terms of the expected free-field motions in the soil region surrounding each bridge foundation. It is evident that to characterize such motions precisely is practically unachievable within the present state of knowledge of seismic ground motions. Therefore, it is necessary to use a rather simplistic approach in generating such motions for design purposes. The procedure most commonly used for designing a large bridge is to (1) generate a three-component (two horizontal and vertical) set of accelerograms representing the free-field ground motion at a "control point" selected for the bridge site and (2) characterize the spatial variations of the free-field motions within each soil region of interest relative to the control motions.

The control point is usually selected at the surface of bedrock (or surface of a firm soil stratum in case of a deep soil site), referred to here as "rock outcrop," at the location of a selected reference pier; and the free-field seismic wave environment within the local soil region of each foundation is assumed to be composed of vertically propagating plane shear (*S*) waves for the horizontal motions and vertically propagating plane compression (*P*) waves for the vertical motions. For a bridge site consisting of relatively soft topsoil deposits overlying competent soil strata or rock, the assumption of vertically propagating plane waves over the depth of the foundations is reasonably valid as confirmed by actual field downhole array recordings [17].

The design ground motion for a bridge is normally specified in terms of a set of parameter values developed for the selected control point which include a set of target acceleration response spectra (ARS) and a set of associated ground motion parameters for the design earthquake, namely (1) magnitude, (2) source-to-site distance, (3) peak ground (rock-outcrop) acceleration (PGA), velocity (PGV), and displacement (PGD), and (4) duration of strong shaking. For large important bridges, these parameter values are usually established through regional seismic investigations coupled with site-specific seismic hazard and ground motion studies, whereas, for small bridges, it is customary to establish these values based on generic seismic study results such as contours of regional PGA values and standard ARS curves for different general classes of site soil conditions.

For a long bridge supported on multiple piers which are in turn supported on multiple foundations spaced relatively far apart, the spatial variations of ground motions among the local soil regions of the foundations need also be defined in the seismic input. Based on the results of analyses using actual earthquake ground motion recordings obtained from strong motion instrument arrays, such as the El Centro differential array in California and the SMART-1 array in Taiwan, the spatial variations of free-field seismic motions have been characterized using two parameters: (1) apparent horizontal wave propagation velocity (speed and direction) which controls the first-order spatial variations of ground motion due to the seismic wave passage effect and (2) a set of horizontal and vertical ground motion "coherency functions" which quantifies the second-order ground motion variations due to scattering and complex 3-D wave propagation [18]. Thus, in addition to the design ground motion parameter values specified for the control motion, characterizing the design seismic inputs to long bridges needs to include the two additional parameters mentioned above, namely, (1) apparent horizontal wave velocity and (2) ground motion coherency functions; therefore, the seismic input motions developed for the various pier foundation locations need to be compatible with the values specified for these two additional parameters.

Having specified the design seismic ground motion parameters, the steps required in establishing the pier foundation location-specific seismic input motions for a particular bridge are

1. Develop a three-component (two horizontal and vertical) set of free-field rock-outcrop motion time histories which are compatible with the design target ARS and associated design ground motion parameters applicable at a selected single control point location at the bridge site (these motions are referred to here simply as the "response spectrum compatible time histories" of control motion).
2. Generate response-spectrum-compatible time histories of free-field rock-outcrop motions at each bridge pier support location such that their coherencies relative to the corresponding components of the response spectrum compatible motions at the control point and at other pier support locations are compatible with the wave passage parameters and the coherency functions specified for the site (these motions are referred to here as "response spectrum and coherency compatible motions).
3. Carry out free-field site response analyses for each pier support location to obtain the time-histories of free-field soil motions at specified discrete elevations over the full depth of each foundation using the corresponding response spectrum and coherency compatible free-field rock-outcrop motions as inputs.

In the following sections, procedures will be presented for generating the set of response spectrum compatible rock-outcrop time histories of motion at the control point location and for generating the sets of response spectrum and coherency compatible rock-outcrop time histories of motion at all pier support locations, and guidelines will be given for performing free-field site response analyses.

42.4.1 Free-Field Rock-Outcrop Motions at Control-Point Location

Given a prescribed set of target ARS and a set of associated design ground motion parameters for a bridge site as described previously, the objective here is to develop a three-component set of time histories of control motion that (1) provides a reasonable match to the corresponding target ARS and (2) has time history characteristics reasonably compatible with the other specified associated ground motion parameter values. In the past, several different procedures have been used for developing rock-outcrop time histories of motion compatible with a prescribed set of target ARS. These procedures are summarized as follows:

1. *Response Spectrum Compatibility Time History Adjustment Method* [19–22] — This method as generally practiced starts by selecting a suitable three-component set of initial or "starting" accelerograms and proceeds to adjust each of them iteratively, using either a time-domain [21,23] or a frequency-domain [19,20,22] procedure, to achieve compatibility with the specified target ARS and other associated parameter values. The time-domain adjustment procedure usually produces only small local adjustments to the selected starting time histories, thereby producing response spectrum compatible time histories closely resembling the initial motions. The general "phasing" of the seismic waves in the starting time history is largely maintained while achieving close compatibility with the target ARS: minor changes do occur, however, in the phase relationships. The frequency-domain procedure as commonly used retains the phase relationships of an initial motion, but does not always provide as close a fit to the target spectrum as does the time-domain procedure. Also, the motion produced by the frequency-domain procedure shows greater visual differences from the initial motion.

2. *Source-to-Site Numerical Model Time History Simulation Method* [24–27] — This method generally starts by constructing a numerical model to represent the controlling earthquake source and source-to-site transmission and scattering functions, and then accelerograms are synthesized for the site using numerical simulations based on various plausible fault-rupture scenarios. Because of the large number of time history simulations required in order to achieve a "stable" average ARS for the ensemble, this method is generally not practical for developing a complete set of time histories to be used directly; rather it is generally used to supplement a set of actual recorded accelerograms, in developing site-specific target response spectra and associated ground motion parameter values.
3. *Multiple Actual Recorded Time History Scaling Method* [28,29] — This method starts by selecting multiple 3-component sets (generally ≥7) of actual recorded accelerograms which are subsequently scaled in such a way that the average of their response spectral ordinates over the specified frequency (or period) range of interest matches the target ARS. Experience in applying this method shows that its success depends very much on the selection of time histories. Because of the lack of suitable recorded time histories, individual accelerograms often have to be scaled up or down by large multiplication factors, thus raising questions about the appropriateness of such scaling. Experience also indicates that unless a large ensemble of time histories (typically >20) are selected, it is generally difficult to achieve matching of the target ARS over the entire spectral frequency (or period) range of interest.
4. *Connecting Accelerogram Segments Method* [55] — This method produces a synthetic time history by connecting together segments of a number of actual recorded accelerograms in such a way that the ARS of the resulting time history fits the target ARS reasonably well. It generally requires producing a number of synthetic time histories to achieve acceptable matching of the target spectrum over the entire frequency (or period) range of interest.

At the present time, Method 1 is considered most suitable and practical for bridge engineering applications. In particular, the time-domain time history adjustment procedure which produces only local time history disturbances has been applied widely in recent applications. This method as developed by Lilhanand and Tseng [21] in 1988, which is based on earlier work by Kaul [30] in 1978, is described below.

The time-domain procedure for time history adjustment is based on the inherent definition of a response spectrum and the recognition that the times of occurrence of the response spectral values for the specified discrete frequencies and damping values are not significantly altered by adjustments of the time history in the neighborhoods of these times. Thus, each adjustment, which is made by adding a small perturbation, $\delta a(t)$, to the selected initial or starting acceleration time history, $a(t)$, is carried out in an iterative manner such that, for each iteration, i, an adjusted acceleration time history, $a_i(t)$, is obtained from the previous acceleration time history, $a_{(i-1)}(t)$, using the relation

$$a_i(t) = a_{(i-1)}(t) + \delta a_i(t) \tag{42.1}$$

The small local adjustment, $\delta a_i(t)$, is determined by solving the integral equation

$$\delta Ri(\omega_{j,}\ \beta_k) = \int_0^{t_{jk}} \delta a_i(\tau) h_{jk}(t_{jk} - \tau) d\tau \tag{42.2}$$

which expresses the small change in the acceleration response value $\delta R_i(\omega_j, \beta_k)$ for frequency ω_j and damping β_k resulting from the local time history adjustment $\delta a_i(t)$. This equation makes use of the acceleration unit–impulse response function $h_{jk}(t)$ for a single-degree-of-freedom oscillator having a natural frequency ω_j and a damping ratio β_k. Quantity t_{jk} in the integral represents the time at which its corresponding spectral value occurs, and τ is a time lag.

By expressing $\delta a_i(t)$ as a linear combination of impulse response functions with unknown coefficients, the above integral equation can be transformed into a system of linear algebraic equations that can easily be solved for the unknown coefficients. Since the unit–impulse response functions decay rapidly due to damping, they produce only localized perturbations on the acceleration time history. By repeatedly applying the above adjustment, the desired degree of matching between the response spectra of the modified motions and the corresponding target spectra is achieved, while, in doing so, the general characteristics of the starting time history selected for adjustment are preserved.

Since this method of time history modification produces only local disturbances to the starting time history, the time history phasing characteristics (wave sequence or pattern) in the starting time history are largely maintained. It is therefore important that the starting time history be selected carefully. Each three-component set of starting accelerograms for a given bridge site should preferably be a set recorded during a past seismic event that has (1) a source mechanism similar to that of the controlling design earthquake, (2) a magnitude within about ±0.5 of the target controlling earthquake magnitude, and (3) a closest source-to-site distance within 10 km of the target source-to-site distance. The selected recorded accelerograms should have their PGA, PGV, and PGD values and their strong shaking durations within a range of ±25% of the target values specified for the bridge site and they should represent free-field surface recordings on rock, rocklike, or a stiff soil site; no recordings on a soft site should be used. For a close-in controlling seismic event, e.g., within about 10 km of the site, the selected accelerograms should contain a definite velocity pulse or the so-called fling. When such recordings are not available, Method 2 described previously can be used to generate a starting set of time histories having an appropriate fling or to modify the starting set of recorded motions to include the desired directional velocity pulse.

Having selected a three-component set of starting time histories, the horizontal components should be transformed into their principal components and the corresponding principal directions should be evaluated [31]. These principal components should then be made response spectrum compatible using the time-domain adjustment procedure described above or the standard frequency-domain adjustment procedure[20,22,32]. Using the latter procedure, only the Fourier amplitude spectrum, not the phase spectrum, is adjusted iteratively.

The target acceleration response spectra are in general identical for the two horizontal principal components of motion; however, a distinct target spectrum is specified for the vertical component. In such cases, the adjusted response spectrum compatible horizontal components can be oriented horizontally along any two orthogonal coordinate axes in the horizontal plane considered suitable for structural analysis applications. However, for bridge projects that have controlling seismic events with close-in seismic sources, the two horizontal target response spectra representing motions along a specified set of orthogonal axes are somewhat different, especially in the low-frequency (long-period) range; thus, the response spectrum compatible time histories must have the same definitive orientation. In this case, the generated three-component set of response spectrum compatible time histories should be used in conjunction with their orientation. The application of this three-component set of motions in a different coordinate orientation requires transforming the motions to the new coordinate system. It should be noted that such a transformation of the components will generally result in time histories that are not fully compatible with the original target response spectra. Thus, if response spectrum compatibility is desired in a specific coordinate orientation (such as in the longitudinal and transverse directions of the bridge), target response spectra in the specific orientation should be generated first and then a three-component set of fully response spectrum compatible time histories should be generated for this specific coordinate system.

As an example, a three-component set of response spectrum compatible time histories of control motion, generated using the time-domain time history adjustment procedure, is shown in Figure 42.4.

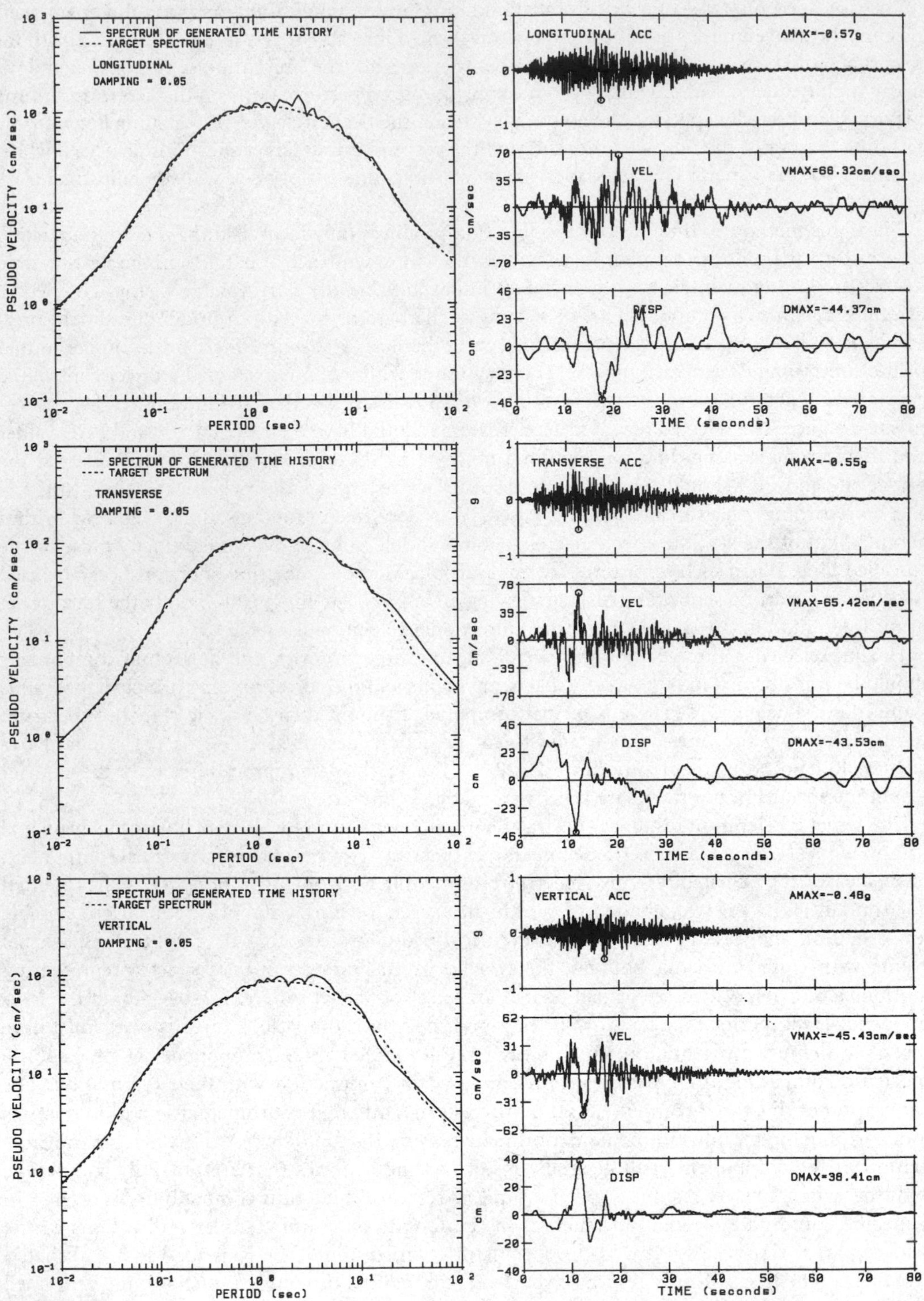

FIGURE 42.4 Examples of a three-component set of response spectrum compatible time histories of control motion.

42.4.2 Free-Field Rock-Outcrop Motions at Bridge Pier Support Locations

As mentioned previously, characterization of the spatial variations of ground motions for engineering purposes is based on a set of wave passage parameters and ground motion coherency functions. The wave passage parameters currently used are the apparent horizontal seismic wave speed, V, and its direction angle θ relative to an axis normal to the longitudinal axis of the bridge. Studies of strong- and weak-motion array data including those in California, Taiwan, and Japan show that the apparent horizontal speed of S-waves in the direction of propagation is typically in the 2 to 3 km/s range [18,33]. In applications, the apparent wave-velocity vector showing speed and direction must be projected along the bridge axis giving the apparent wave speed in that direction as expressed by

$$V_{\text{bridge}} = \frac{V}{\sin\theta} \tag{42.3}$$

To be realistic, when θ becomes small, a minimum angle for θ, say, 30°, should be used in order to account for waves arriving in directions different from the specified direction.

The spatial coherency of the free-field components of motion in a single direction at various locations on the ground surface has been parameterized by a complex coherency function defined by the relation

$$\Gamma_{ij}(i\omega) = \frac{S_{ij}(i\omega)}{\sqrt{S_{ii}(\omega)}\sqrt{S_{jj}(\omega)}} \qquad i, j = 1, 2, \ldots, n \text{ locations} \tag{42.4}$$

in which $S_{ij}(i\omega)$ is the smoothed complex cross-power spectral density function and $S_{ii}(\omega)$ and $S_{jj}(\omega)$ are the smoothed real power spectral density (PSD) functions of the components of motion at locations i and j. The notation $i\omega$ in the above equation is used to indicate that the coefficients $S_{ij}(i\omega)$ are complex valued (contain both real and imaginary parts) and are dependent upon excitation frequency ω. Based on analyses of strong-motion array data, a set of generic coherency functions for the horizontal and vertical ground motions has been developed [34]. These functions for discrete separation distances between locations i and j are plotted against frequency in Figure 42.5.

Given a three-component set of response spectrum compatible time histories of rock-outcrop motions developed for the selected control point location and a specified set of wave passage parameters and "target" coherency functions as described above, response spectrum compatible and coherency compatible multiple-support rock-outcrop motions applicable to each pier support location of the bridge can be generated using the procedure presented below. This procedure is based on the "marching method" developed by Hao et al. [32] in 1989 and extended by Tseng et al. [35] in 1993.

Neglecting, for the time being, ground motion attenuation along the bridge axis, the components of rock-outcrop motions at all pier support locations in a specific direction have PSD functions which are common with the PSD function $S_o(\omega)$ specified for the control motion, i.e.,

$$S_{ii}(\omega) = S_{jj}(\omega) = S_o(\omega) = |u_o(i\omega)|^2 \tag{42.5}$$

where $u_o(i\omega)$ is the Fourier transform of the corresponding component of control motion, $u_o(t)$. By substituting Eq. (42.5) into Eq. (42.4), one obtains

$$S_{ij}(i\omega) = \Gamma_{ij}(i\omega)\, S_o(\omega) \tag{42.6}$$

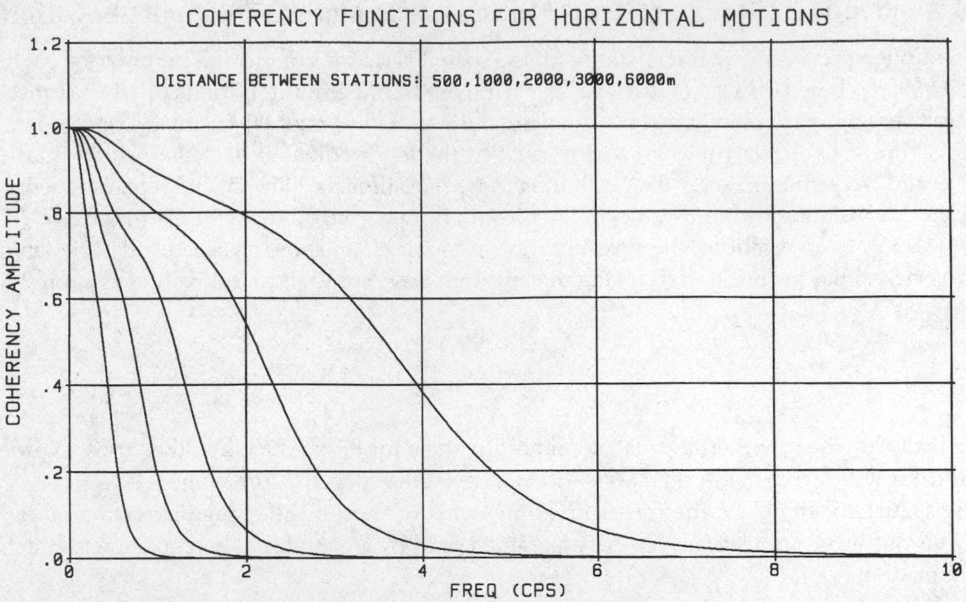

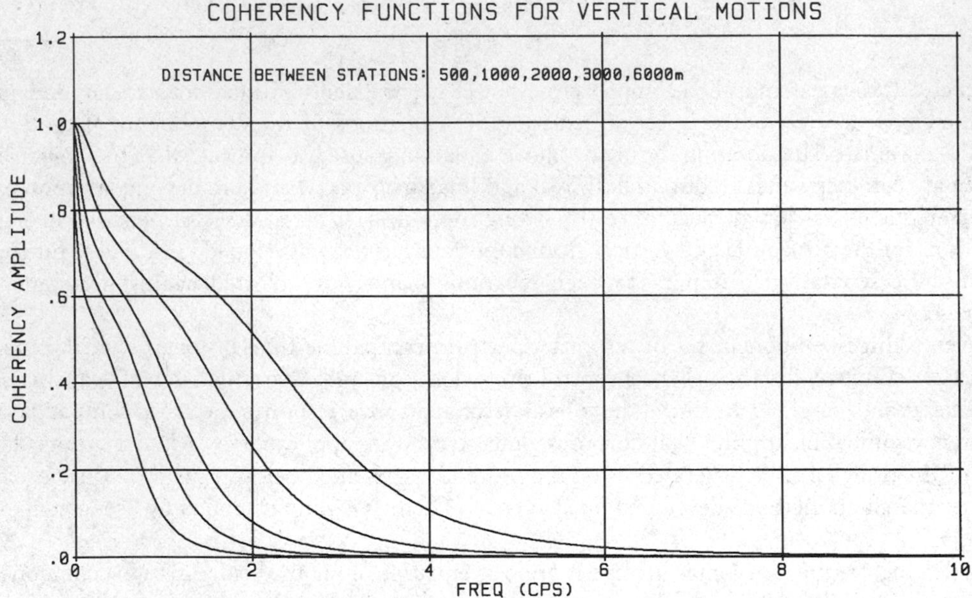

FIGURE 42.5 Example of coherency functions of frequency at discrete separation distances.

which can be rewritten in a matrix form for all pier support locations as follows:

$$\boldsymbol{S}(i\omega) = \boldsymbol{\Gamma}(i\omega)\ S_o(\omega) \tag{42.7}$$

Since, by definition, the coherency matrix $\boldsymbol{\Gamma}(i\omega)$ is an Hermitian matrix, it can be decomposed into a complex conjugate pair of lower and upper triangular matrices $\boldsymbol{L}(i\omega)$ and $\boldsymbol{L}^*(i\omega)^T$ as expressed by

$$\boldsymbol{\Gamma}(i\omega) = \boldsymbol{L}(i\omega)\ \ \boldsymbol{L}^*(i\omega)^T \tag{42.8}$$

in which the symbol * denotes complex conjugate. In proceeding, let

$$u(i\omega) = L(i\omega)\eta_{\phi_i}(i\omega)u_o(i\omega) \tag{42.9}$$

in which $u(\omega)$ is a vector containing components of motion $u_i(\omega)$ for locations, $i = 1, 2, \ldots, n$; and, $\eta_{\phi_i}(i\omega) = \{e^{i\phi_i(\omega)}\}$ is a vector containing unit amplitude components having random-phase angles $\phi_i(\omega)$. If $\phi_i(\omega)$ and $\phi_j(\omega)$ are uniformly distributed random-phase angles, the relations

$$E[\eta_{\phi_i}(i\omega)\eta^*_{\phi_j}(i\omega)] = 0 \quad \text{if } i \neq j$$

$$E[\eta_{\phi_i}(i\omega)\eta^*_{\phi_j}(i\omega)] = 1 \quad \text{if } i = j \tag{42.10}$$

will be satisfied, where the symbol $E[\ ]$ represents ensemble average. It can easily be shown that the ensemble of motions generated using Eq. (42.9) will satisfy Eq. (42.7). Thus, if the rock-outcrop motions at all pier support locations are generated from the corresponding motions at the control point location using Eq. (42.9), the resulting motions at all locations will satisfy, on an ensemble basis, the coherency functions specified for the site. Since the matrix $L(i\omega)$ in Eq. (42.9) is a lower triangular matrix having its diagonal elements equal to unity, the generation of coherency compatible motions at all pier locations can be achieved by marching from one pier location to the next in a sequential manner starting with the control pier location.

In generating the coherency compatible motions using Eq. (42.9), the phase angle shifts at various pier locations due to the single plane-wave passage at the constant speed V_{bridge} defined by Eq. (42.3) can be incorporated into the term $\eta_{\phi_i}(i\omega)$. Since the motions at the control point location are response spectrum compatible, the coherency compatible motions generated at all other pier locations using the above-described procedure will be approximately response spectrum compatible. However, an improvement on their response spectrum compatibility is generally required, which can be done by adjusting their Fourier amplitudes but keeping their Fourier phase angles unchanged. By keeping these angles unchanged, the coherencies among the adjusted motions are not affected. Consequently, the adjusted motions will not only be response spectrum compatible, but will also be coherency compatible.

In generating the response spectrum- and coherency-compatible motions at all pier locations by the procedure described above, the ground motion attenuation effect has been ignored. For a long bridge located close to the controlling seismic source, attenuation of motion with distance away from the control pier location should be considered. This can be achieved by scaling the generated motions at various pier locations by appropriate scaling factors determined from an appropriate ground motion attenuation relation. The acceleration time histories generated for all pier locations should be integrated to obtain their corresponding velocity and displacement time histories, which should be checked to ensure against having numerically generated baseline drifts. Relative displacement time histories between the control pier location and successive pier locations should also be checked to ensure that they are reasonable. The rock-outcrop motions finally obtained should then be used in appropriate site-response analyses to develop the corresponding free-field soil motions required in conducting the SFSI analyses for each pier location.

42.4.3 Free-Field Soil Motions

As previously mentioned, the seismic inputs to large bridges are defined in terms of the expected free-field soil motions at discrete elevations over the entire depth of each foundation. Such motions must be evaluated through location-specific site-response analyses using the corresponding previously described rock-outcrop free-field motions as inputs to appropriately defined soil–bedrock models. Usually, as mentioned previously, these models are based on the assumption that the horizontal and vertical free-field soil motions are produced by upward/downward propagation of one-dimensional shear and compression waves, respectively, as caused by the upward propagation of incident waves in the underlying rock or firm soil formation. Consistent with these types of motion, it is assumed that the local soil medium surrounding each foundation consists of uniform horizontal layers of infinite lateral extent. Wave reflections and refractions will occur at all interfaces of adjacent layers, including the soil–bedrock interface, and reflections of the waves will occur at the soil surface. Computer program SHAKE [4,44] is most commonly used to carry out the above-described one-dimensional type of site-response analysis. For a long bridge having a widely varying soil profile from end to end, such site-response analyses must be repeated for different soil columns representative of the changing profile.

The cyclic free-field soil deformations produced at a particular bridge site by a maximum expected earthquake are usually of the nonlinear hysteretic form. Since the SHAKE computer program treats a linear system, the soil column being analyzed must be modeled in an equivalent linearized manner. To obtain the equivalent linearized form, the soil parameters in the model are modified after each consecutive linear time history response analysis is complete, which continues until convergence to strain-compatible parameters are reached.

For generating horizontal free-field motions produced by vertically propagating shear waves, the needed equivalent linear soil parameters are the shear modulus G and the hysteretic damping ratio β. These parameters, as prepared by Vucetic and Dobry [36] in 1991 for clay and by Sun et al. [37] in 1988 and by the Electric Power Research Institute (EPRI) for sand, are plotted in Figures 42.6 and 42.7, respectively, as functions of shear strain γ. The shear modulus is plotted in its nondimensional form G/G_{max} where G_{max} is the *in situ* shear modulus at very low strains ($\gamma \leq 10^{-4}\%$). The shear modulus G must be obtained from cyclic shear tests, while G_{max} can be obtained using $G_{max} = \rho V_s^2$ in which ρ is mass density of the soil and V_s is the *in situ* shear wave velocity obtained by field measurement. If shear wave velocities are not available, G_{max} can be estimated using published empirical formulas which correlate shear wave velocity or shear modulus with blow counts and/or other soil parameters [38–43]. To obtain the equivalent linearized values of G/G_{max} and β following each consecutive time history response analysis, values are taken from the G/G_{max} vs. γ and β vs. γ relations at the effective shear strain level defined as $\gamma_{eff} = \alpha\gamma_{max}$ in which γ_{max} is the maximum shear strain reached in the last analysis and α is the effective strain factor. In the past, α has usually been assigned the value 0.65; however, other values have been proposed (e.g., Idriss and Sun [44]). The equivalent linear time history response analyses are performed in an iterative manner, with soil parameter adjustments being made after each analysis, until the effective shear strain converges to essentially the same value used in the previous iteration [45]. This normally takes four to eight iterations to reach 90 to 95% of full convergence when the effective shear strains do not exceed 1 to 2%. When the maximum strain exceeds 2%, a nonlinear site-response analysis is more appropriate. Computer programs available for this purpose are DESRA [46], DYNAFLOW [47], DYNAID [48], and SUMDES [49].

For generating vertical free-field motions produced by vertically propagating compression waves, the needed soil parameters are the low-strain constrained elastic modulus $E_p = \rho V_p^2$, where V_p is the compression wave velocity, and the corresponding damping ratio. The variations of these soil parameters with compressive strain have not as yet been well established. At the present time, vertical site-response analyses have generally been carried out using the low-strain constrained elastic moduli, E_p, directly and the strain-compatible damping ratios obtained from the horizontal response analyses, but limited to a maximum value of 10%, without any further strain-compatibility iterations. For soils submerged in water, the value of E_p should not be less than the compression wave velocity of water.

Having generated acceleration free-field time histories of motion using the SHAKE computer program, the corresponding velocity and displacement time histories should be obtained through

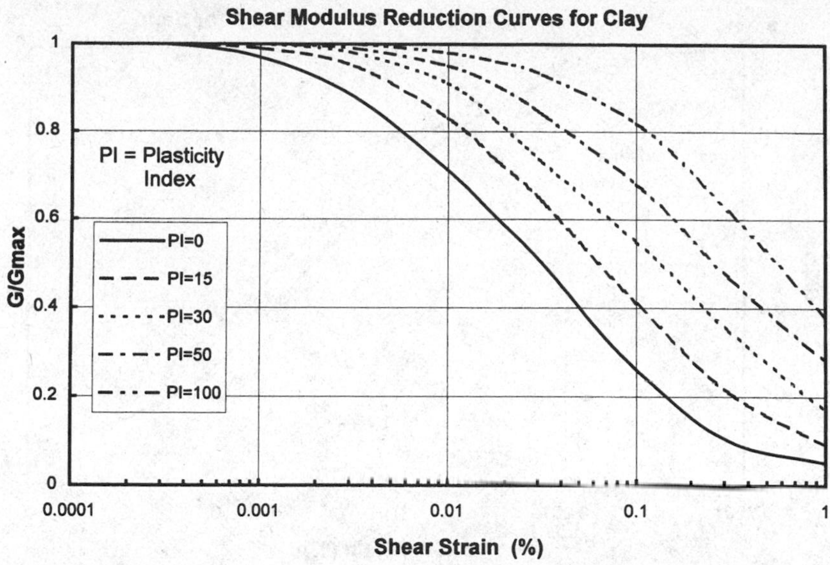

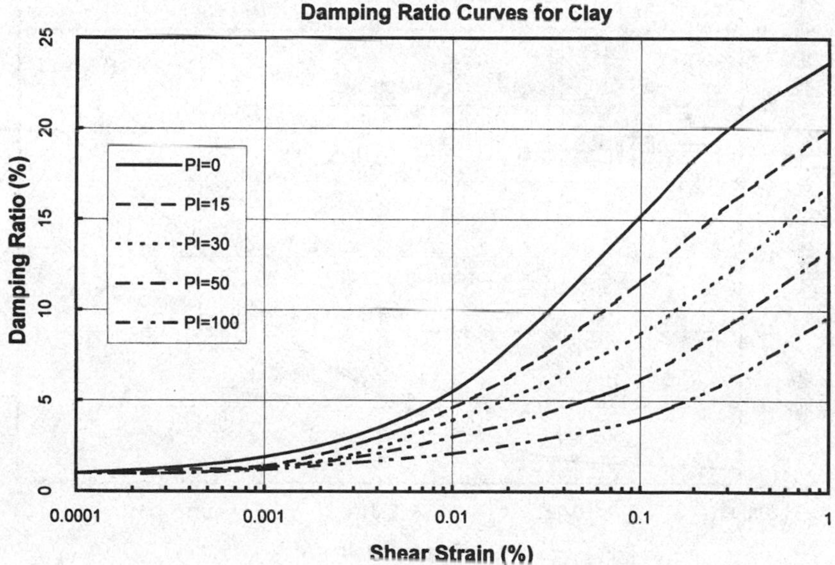

FIGURE 42.6 Equivalent linear shear modulus and hysteretic damping ratio as functions of shear strain for clay. (Source: Vucetic, M. and Dobry, R., *J. Geotech. Eng. ASCE*, 117(1), 89-107, 1991. With permission.)

single and double integrations of the acceleration time histories. Should unrealistic drifts appear in the displacement time histories, appropriate corrections should be applied. Should such drifts appear in a straight-line fashion, it usually indicates that the durations specified for Fourier transforming the recorded accelerograms are too short; thus, increasing these durations will usually correct the problem. If the baseline drifts depart significantly from a simple straight line, this tends to indicate that the analysis results may be unreliable; in which case, they should be carefully checked before being used. Time histories of free-field relative displacement between pairs of pier locations should also be generated and then be checked to judge the reasonableness of the results obtained.

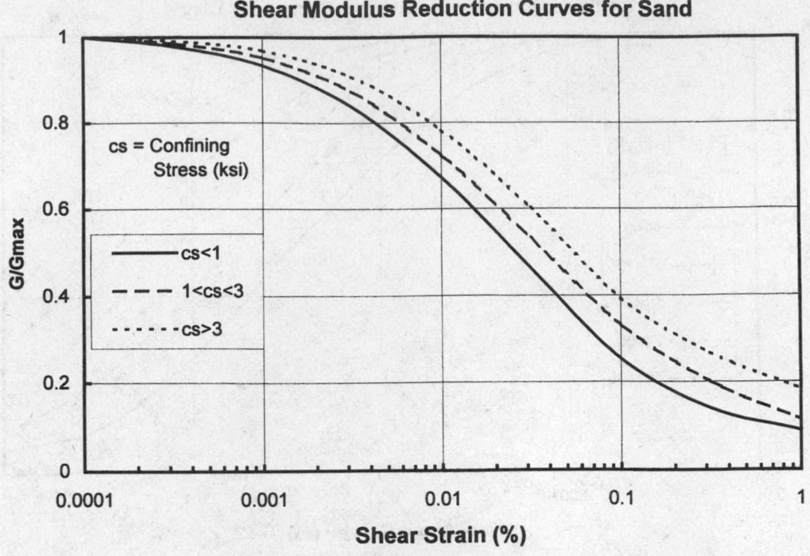

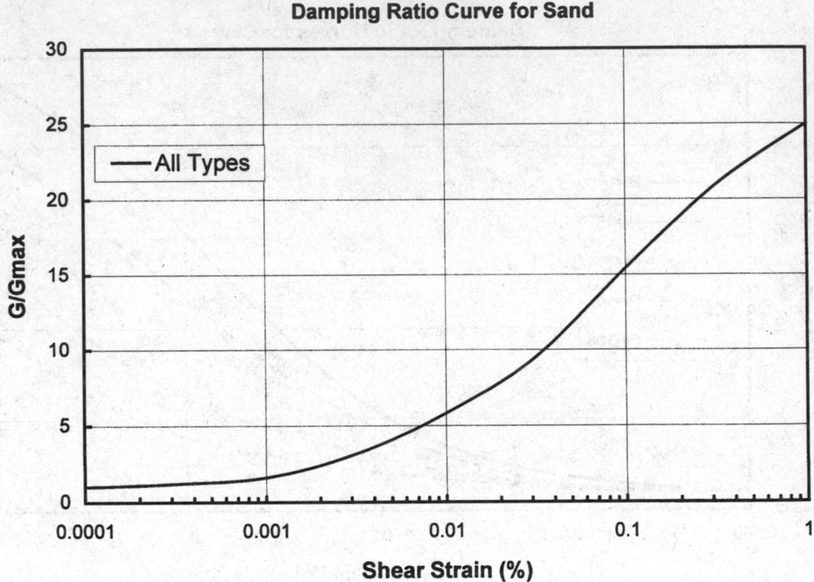

FIGURE 42.7 Equivalent linear shear modulus and hysteretic damping ratio as functions of shear strain for sand. (Source: Sun, J. I. et al., Reort No. UBC/EERC-88/15, Earthquake Engineer Research Center, University of California, Berkeley, 1988.)

42.5 Characterization of Soil–Foundation System

The core of the dynamic SFSI problem for a bridge is the interaction between its structure–foundation system and the supporting soil medium, which, for analysis purposes, can be considered to be a full half-space. The fundamental step in solving this problem is to characterize the constitutive relations between the dynamic forces acting on each foundation of the bridge at its interface boundary with the soil and the corresponding foundation motions, expressed in terms of the displacements, velocities, and accelerations. Such forces are here called the soil–foundation interaction forces. For a bridge subjected to externally applied loadings, such as dead, live, wind, and wave loadings, these SFI forces are functions of the foundation motions only; however, for a bridge subjected to seismic loadings, they are functions of the free-field soil motions as well.

Let h be the total number of degrees of freedom (DOF) of the bridge foundations as defined at their soil–foundation interface boundaries; $\boldsymbol{u}_h(t)$, $\dot{\boldsymbol{u}}_h(t)$, and $\ddot{\boldsymbol{u}}_h(t)$ be the corresponding foundation displacement, velocity, and acceleration vectors, respectively; and $\bar{\boldsymbol{u}}_h(t)$, $\dot{\bar{\boldsymbol{u}}}_h(t)$, and $\ddot{\bar{\boldsymbol{u}}}_h(t)$ be the free-field soil displacement, velocity, and acceleration vectors in the h DOF, respectively; and let $\boldsymbol{f}_h(t)$ be the corresponding SFI force vector. By using these notations, characterization of the SFI forces under seismic conditions can be expressed in the general vectorial functional form:

$$\boldsymbol{f}_h(t) = \Im_h\ (\boldsymbol{u}_h(t),\ \dot{\boldsymbol{u}}_h(t),\ \ddot{\boldsymbol{u}}_h(t),\ \bar{\boldsymbol{u}}_h(t)\ \ \dot{\bar{\boldsymbol{u}}}_h(t),\ \ddot{\bar{\boldsymbol{u}}}_h(t)) \tag{42.11}$$

Since the soils in the local region immediately surrounding each foundation may behave nonlinearly under imposed foundation loadings, the form of $\Im_h$ is, in general, a nonlinear function of displacements $\boldsymbol{u}_h(t)$ and $\bar{\boldsymbol{u}}_h(t)$ and their corresponding velocities and accelerations.

For a capacity evaluation, the nonlinear form of $\Im_h$ should be retained and used directly for determining the SFI forces as functions of the foundation and soil displacements. Evaluation of this form should be based on a suitable nonlinear model for the soil medium coupled with appropriate boundary conditions, subjected to imposed loadings which are usually much simplified compared with the actual induced loadings. This part of the evaluation will be discussed further in Section 42.8.

For a demand evaluation, the nonlinear form of $\Im_h$ is often linearized and then transformed to the frequency domain. Letting $\boldsymbol{u}_h(i\omega)$, $\dot{\boldsymbol{u}}_h(i\omega)$, $\ddot{\boldsymbol{u}}_h(i\omega)$, $\bar{\boldsymbol{u}}_h(i\omega)$, $\dot{\bar{\boldsymbol{u}}}_h(i\omega)$, $\ddot{\bar{\boldsymbol{u}}}_h(i\omega)$, and $\boldsymbol{f}_h(i\omega)$ be the Fourier transforms of $\boldsymbol{u}_h(t)$, $\dot{\boldsymbol{u}}_h(t)$, $\ddot{\boldsymbol{u}}_h(t)$, $\bar{\boldsymbol{u}}_h(t)$, $\dot{\bar{\boldsymbol{u}}}_h(t)$, $\ddot{\bar{\boldsymbol{u}}}_h(t)$, and $\boldsymbol{f}_h(t)$, respectively, and making use of the relations

$$\dot{\boldsymbol{u}}_h(i\omega) = i\omega\boldsymbol{u}_h(i\omega)\,; \quad \ddot{\boldsymbol{u}}_h(i\omega) = -\omega^2\boldsymbol{u}_h(i\omega)$$

and

$$\dot{\bar{\boldsymbol{u}}}_h(i\omega) = i\omega\bar{\boldsymbol{u}}_h(i\omega)\,; \quad \ddot{\bar{\boldsymbol{u}}}_h(i\omega) = -\omega^2\bar{\boldsymbol{u}}_h(i\omega)\,, \tag{42.12}$$

Equation (42.11) can be cast into the more convenient form:

$$\boldsymbol{f}_h(i\omega) = \Im_h\ (\,u_h(i\omega), \bar{\boldsymbol{u}}_h(i\omega)\,) \tag{42.13}$$

To characterize the linear functional form of $\Im_h$, it is necessary to solve the dynamic boundary-value problem for a half space soil medium subjected to force boundary conditions prescribed at the soil–foundation interfaces. This problem is referred to here as the "soil impedance" problem, which is a part of the foundation impedance problem referred to earlier in Section 42.2.2.

In linearized form, Eq. (42.13) can be expressed as

$$\boldsymbol{f}_h(i\omega) = \boldsymbol{G}_{hh}(i\omega)\ \ \{\boldsymbol{u}_h(i\omega) - \bar{\boldsymbol{u}}_h(i\omega)\} \tag{42.14}$$

in which $\boldsymbol{f}_h(i\omega)$ represents the force vector acting on the soil medium by the foundation and the matrix $\boldsymbol{G}_{hh}(i\omega)$ is a complex, frequency-dependent coefficient matrix called here the "soil impedance matrix."

Define a force vector $\bar{\boldsymbol{f}}_h(i\omega)$ by the relation

$$\bar{\boldsymbol{f}}_h(i\omega) = \boldsymbol{G}_{hh}(i\omega)\ \ \bar{\boldsymbol{u}}_h(i\omega) \tag{42.15}$$

This force vector represents the internal dynamic forces acting on the bridge foundations at their soil–foundation interface boundaries resulting from the free-field soil motions when the foundations are held fixed, i.e., $\boldsymbol{u}_h(i\omega) = \boldsymbol{0}$. The force vector $\bar{\boldsymbol{f}}_h(i\omega)$ as defined in Eq. (42.15) is the "seismic driving force" vector mentioned previously in Section 42.2.2. Depending upon the type of bridge foundation, the characterization of the soil impedance matrix $\boldsymbol{G}_{hh}(i\omega)$ and associated free-field soil input motion vector $\bar{\boldsymbol{u}}_h(i\omega)$ for demand analysis purposes may be established utilizing different soil models as described below.

42.5.1 Elastodynamic Model

As mentioned in Section 42.3.1, for a large bridge foundation such as a large spread footing, caisson, or single or multiple shafts having very large diameters, for which the nonlinearities occurring in the local soil region immediately adjacent to the foundation are small, the soil impedance matrix $\boldsymbol{G}_{hh}(i\omega)$ can be evaluated utilizing the dynamic Green's functions (dynamic displacements of the soil medium due to harmonic point-load excitations) obtained from the solution of a dynamic boundary-value problem of a linear damped-elastic half-space soil medium subjected to harmonic point loads applied at each of the h DOF on the soil–foundation interface boundaries. Such solutions have been obtained in analytical form for a linear damped-elastic continuum half-space soil medium by Apsel [50] in 1979. Because of complexities in the analytical solution, dynamic Green's functions have only been obtained for foundations having relatively simple soil–foundation interface geometries, e.g., rectangular, cylindrical, or spherical soil–foundation interface geometries, supported in simple soil media. In practical applications, the dynamic Green's functions are often obtained in numerical forms based on a finite-element discretization of the half-space soil medium and a corresponding discretization of the soil–foundation interface boundaries using a computer program such as SASSI [51], which has the capability of properly simulating the wave radiation boundary conditions at the far field of the half-space soil medium. The use of finite-element soil models to evaluate the dynamic Green's functions in numerical form has the advantage that foundations having arbitrary soil–foundation interface geometries can be easily handled; it, however, suffers from the disadvantage that the highest frequency, i.e., cutoff frequency, of motion for which a reliable solution can be obtained is limited by size of the finite element used for modeling the soil medium.

Having evaluated the dynamic Green's functions using the procedure described above, the desired soil impedance matrix can then be obtained by inverting, frequency-by-frequency, the "soil compliance matrix," which is the matrix of Green's function values evaluated for each specified frequency ω. Because the dynamic Green's functions are complex valued and frequency dependent, the coefficients of the resulting soil impedance matrix are also complex-valued and frequency dependent. The real parts of the soil impedance coefficients represent the dynamic stiffnesses of the soil medium which also incorporate the soil inertia effects; the imaginary parts of the coefficients represent the energy losses resulting from both soil material damping and radiation of stress waves into the far-field soil medium. Thus, the soil impedance matrix as developed reflects the overall dynamic characteristics of the soil medium as related to the motion of the foundation at the soil–foundation interfaces.

Because of the presence of the foundation excavation cavities in the soil medium, the vector of free-field soil motions $\bar{\boldsymbol{u}}_h(i\omega)$ prescribed at the soil–foundation interface boundaries has to be derived from the seismic input motions of the free-field soil medium without the foundation excavation cavities as described in Section 42.4. The derivation of the motion vector $\bar{\boldsymbol{u}}_h(i\omega)$ requires the solution of a dynamic boundary-value problem for the free-field half-space soil medium having foundation excavation cavities subjected to a specified seismic wave input such that the resulting solution satisfies the traction-free conditions at the surfaces of the foundation excavation cavities. Thus, the resulting seismic response motions, $\bar{\boldsymbol{u}}_h(i\omega)$, reflect the effects of seismic wave scattering due to the presence of the cavities. These motions are, therefore, referred to here as the "scattered free-field soil input motions."

The effects of seismic wave scattering depend on the relative relation between the characteristic dimension, ℓ_f, of the foundation and the specific seismic input wave length, λ, of interest, where

$\lambda = 2\pi V_s/\omega$ or $2\pi V_p/\omega$ for vertically propagating plane shear or compression waves, respectively; V_s and V_p are, as defined previously, the shear and compression wave velocities of the soil medium, respectively. If the input seismic wave length λ is much longer than the characteristic length ℓ_f, the effect of wave scattering will be negligible; on the other hand, when $\lambda \le \ell_f$, the effect of wave scattering will be significant. Since the wave length λ is a function of the frequency of input motion, the effect of wave scattering is also frequency dependent. Thus, it is evident that the effect of wave scattering is much more important for a large bridge foundation, such as a large caisson or a group of very large diameter shafts, than for a small foundation having a small characteristic dimension, such as a slender-pile group; it can also be readily deduced that the scattering effect is more significant for foundations supported in soft soil sites than for those in stiff soil sites.

The characterization of the soil impedance matrix utilizing an elastodynamic model of the soil medium as described above requires soil material characterization constants which include (1) mass density, ρ; (2) shear and constrained elastic moduli, G and E_p (or shear and compression wave velocities, V_s and V_p); and (3) constant-hysteresis damping ratio, β. As discussed previously in Section 42.4.3, the soil shear modulus decreases while the soil hysteresis damping ratio increases as functions of soil shear strains induced in the free-field soil medium due to the seismic input motions. The effects of these so-called global soil nonlinearities can be easily incorporated into the soil impedance matrix based on an elastodynamic model by using the free-field-motion-induced strain-compatible soil shear moduli and damping ratios as the soil material constants in the evaluation of the dynamic Green's functions. For convenience of later discussions, the soil impedance matrix, $\boldsymbol{G}_{hh}(i\omega)$, characterized using an elastodynamic model will be denoted by the symbol $\boldsymbol{G}^e_{hh}(i\omega)$.

42.5.2 Empirical *p–y* Model

As discussed in Section 42.3.2, for a slender-pile group foundation for which soil nonlinearities occurring in the local soil regions immediately adjacent to the piles dominate the behavior of the foundation under loadings, the characterization of the soil resistances to pile deflections has often relied on empirically derived *p–y* curves for lateral resistance and *t–z* and *Q–d* curves for axial resistance. For such a foundation, the characterization of the soil impedance matrix needed for demand analysis purposes can be made by using the secant moduli derived from the nonlinear *p–y*, *t–z*, and *Q–d* curves, as indicated schematically in Figure 42.2. Since the development of these empirical curves has been based upon static or pseudo-static test results, it does not incorporate the soil inertia and material damping effects. Thus, the resulting soil impedance matrix developed from the secant moduli of the *p–y*, *t–z*, and *Q–d* curves reflects only the static soil stiffnesses but not the soil inertia and soil material damping characteristics. Hence, the soil impedance matrix so obtained is a real-valued constant coefficient matrix applicable at the zero frequency ($\omega = 0$); it, however, is a function of the foundation displacement amplitude. This matrix is designated here as $\boldsymbol{G}^s_{hh}(0)$ to differentiate it from the soil impedance matrix $\boldsymbol{G}^e_{hh}(i\omega)$ defined previously. Thus, Eq. (42.14) in this case is given by

$$\boldsymbol{f}_h(i\omega) = \boldsymbol{G}^s_{hh}(0)\{\boldsymbol{u}_h(i\omega) - \bar{\boldsymbol{u}}_h(i\omega)\} \tag{42.16}$$

where $\boldsymbol{G}^s_{hh}(0)$ depends on the amplitudes of the relative displacement vector $\Delta\boldsymbol{u}_h(i\omega)$ defined by

$$\Delta\boldsymbol{u}_h(i\omega) = \boldsymbol{u}_h(i\omega) - \bar{\boldsymbol{u}}_h(i\omega) \tag{42.17}$$

As mentioned previously in Section 42.3.2, the construction of the *p–y*, *t–z*, and *Q–d* curves depends only on the strength parameters but not on the stiffness parameters of the soil medium; thus, the effects of global soil nonlinearities on the dynamic stiffnesses of the soil medium, as caused by soil shear modulus decrease and soil-damping increase as functions of free-field-motion-induced soil shear strains, cannot be incorporated into the soil impedance matrix developed from these curves. Furthermore, since these curves are developed on the basis of results from field tests in which there are no free-field ground-motion-induced soil deformations, the effects of such global soil nonlinearities on the soil strength characterization parameters and hence the *p–y*, *t–z*, and *Q–d* curves cannot be incorporated.

Because of the small cross-sectional dimensions of slender piles, the seismic wave-scattering effect due to the presence of pile cavities is usually negligible; thus, the scattered free-field soil input motions $\bar{u}_h(i\omega)$ in this case are often taken to be the same as the free-field soil motions when the cavities are not present.

42.5.3 Hybrid Model

From the discussions in the above two sections, it is clear that characterization of the SFI forces for demand analysis purposes can be achieved using either an elastodynamic model or an empirical *p–y* model for the soil medium, each of which has its own merits and deficiencies. The elastodynamic model is capable of incorporating soil inertia, damping (material and radiation), and stiffness characteristics, and it can incorporate the effects of global soil nonlinearities induced by the free-field soil motions in an equivalent linearized manner. However, it suffers from the deficiency that it does not allow for easy incorporation of the effects of local soil nonlinearities. On the contrary, the empirical *p–y* model can properly capture the effects of local soil nonlinearities in an equivalent linearized form; however, it suffers from the deficiencies of not being able to simulate soil inertia and damping effects properly, and it cannot treat the effects of global soil nonlinearities. Since the capabilities of the two models are mutually complementary, it is logical to combine the elastodynamic model with the empirical *p–y* model in a series form such that the combined model has the desired capabilities of both models. This combined model is referred to here as the "hybrid model."

To develop the hybrid model, let the relative displacement vector, $\Delta\boldsymbol{u}_h(i\omega)$, between the foundation displacement vector $\boldsymbol{u}_h(i\omega)$ and the scattered free-field soil input displacement vector $\bar{\boldsymbol{u}}_h(i\omega)$, as defined by Eq. (42.17), be decomposed into a component representing the relative displacements at the soil–foundation interface boundary resulting from the elastic deformation of the global soil medium outside of the soil–foundation interface, designated as $\Delta\boldsymbol{u}_h^e(i\omega)$, and a component representing the relative displacements at the same boundary resulting from the inelastic deformations of the local soil regions adjacent the foundation, designated as $\Delta\boldsymbol{u}_h^i(i\omega)$; thus,

$$\Delta\boldsymbol{u}_h(i\omega) = \Delta\boldsymbol{u}_h^i(i\omega) + \Delta\boldsymbol{u}_h^e(i\omega) \tag{42.18}$$

Let $\boldsymbol{f}_h^e(i\omega)$ represent the elastic force vector which can be characterized in terms of the elastic relative displacement vector $\boldsymbol{u}_h^e(i\omega)$ using the elastodynamic model, in which case

$$\boldsymbol{f}_h^e(i\omega) = \boldsymbol{G}_{hh}^e(i\omega)\Delta\boldsymbol{u}_h^e(i\omega) \tag{42.19}$$

where $\boldsymbol{G}_{hh}^e(i\omega)$ is the soil impedance matrix as defined previously in Section 42.5.1, which can be evaluated using an elastodynamic model. Let $\boldsymbol{f}_h^i(i\omega)$ represent the inelastic force vector which is assumed to be related to $\Delta\boldsymbol{u}_h^i(i\omega)$ by the relation

$$\boldsymbol{f}_h^i(i\omega) = \boldsymbol{G}_{hh}^i(i\omega)\Delta\boldsymbol{u}_h^i(i\omega) \tag{42.20}$$

The characterization of the matrix $\boldsymbol{G}_{hh}^i(i\omega)$ can be accomplished by utilizing the soil secant stiffness matrix $\boldsymbol{G}_{hh}^s(0)$ developed from the empirical *p–y* model by the procedure discussed below.

Solving Eqs. (42.19) and (42.20) for $\Delta\boldsymbol{u}_h^e(i\omega)$ and $\Delta\boldsymbol{u}_h^i(i\omega)$, respectively, substituting these relative displacement vectors into Eq. (42.18), and making use of the force continuity condition that $f_h^e(i\omega) = f_h^i(i\omega)$, since the elastodynamic model and the inelastic local model are in series, one obtains

$$f_h(i\omega) = \left\{[G_{hh}^i(i\omega)]^{-1} + [G_{hh}^e(i\omega)]^{-1}\right\}^{-1} \Delta u_h(i\omega) \tag{42.21}$$

Comparing Eq. (42.14) with Eq. (42.21), one finds that by using the hybrid model, the soil impedance matrix is given by

$$G_{hh}(i\omega) = \left\{[G_{hh}^i(i\omega)]^{-1} + [G_{hh}^e(i\omega)]^{-1}\right\}^{-1} \tag{42.22}$$

Since the soil impedance matrix $G_{hh}^s(i\omega)$ is formed by the static secant moduli of the nonlinear *p–y, t–z,* and *Q–d* curves when $\omega = 0$, Eq. (42.22) becomes

$$G_{hh}^s(0) = \left\{[G_{hh}^i(0)]^{-1} + [G_{hh}^e(0)]^{-1}\right\}^{-1} \tag{42.23}$$

where $G_{hh}^s(0)$ is the soil stiffness matrix derived from the secant moduli of the nonlinear *p–y, t–z,* and *Q–d* curves. Solving Eq. (42.23), for $G_{hh}^i(0)$ gives

$$G_{hh}{}^i(0) = \left\{[G_{hh}^s(0)]^{-1} - [G_{hh}^e(0)]^{-1}\right\}^{-1} \tag{42.24}$$

Thus, Eq. (42.22) can be expressed in the form

$$G_{hh}(i\omega) = \left\{[G_{hh}^i(0)]^{-1} + [G_{hh}^e(i\omega)]^{-1}\right\}^{-1} \tag{42.25}$$

From Eq. (42.25), it is evident that when $\Delta u_h{}^i(i\omega) << \Delta u_h{}^e(i\omega)$, $G_{hh}(i\omega) \rightarrow G_{hh}^e(i\omega)$; however, when $\Delta u_h{}^i(i\omega) >> \Delta u_h{}^e(i\omega)$, $G_{hh}(i\omega) \rightarrow G_{hh}^i(0) \rightarrow G_{hh}^s(0)$. Thus, the hybrid model represented by this equation converges to the elastodynamic model when the local inelastic soil deformations are relatively small, as for the case of a large footing, caisson, or very large diameter shaft foundation, whereas it converges to the empirical *p–y* model when the local inelastic soil deformations are relatively much larger, as for the case of a slender-pile group foundation. For a moderately large diameter shaft foundation, the local inelastic and global elastic soil deformations may approach a comparable magnitude; in which case, the use of a hybrid model to develop the soil impedance matrix as described above can properly represent both the global elastodynamic and local inelastic soil behaviors.

As local soil nonlinearities are induced by the relative displacements between the foundation and the scattered free-field soil input motions, they do not affect the scattering of free-field soil motions due to the traction-free conditions present at the surface of the foundation cavities. Therefore, in applying the hybrid model described above, the scattered free-field soil input motion vector $\bar{u}_h(i\omega)$ should still be derived using the elastodynamic model described in Section 42.5.1.

42.6 Demand Analysis Procedures

42.6.1 Equations of Motion

The seismic response of a complete bridge system involves interactions between the structure and its supporting foundations and between the foundations and their surrounding soil media. To develop the equations of motion governing the response of this system in discrete (finite-element) form, let s denote the number of DOF in the structure, excluding its f DOF at the structure/foundation interface locations, and let g denote the number of DOF in the foundations, also excluding the f DOF but including the h DOF at all soil–foundation interfaces as defined in Section 42.5. Corresponding with those DOF, let vectors $\boldsymbol{u}_s(t)$, $\boldsymbol{u}_f(t)$, and $\boldsymbol{u}_g(t)$ contain the total displacement time histories of motion at the DOF s, f, and g, respectively.

Linear Modeling

Since the soil medium surrounding all foundations is continuous and of infinite extent, a rigorous model of a complete bridge system must contain stiffness and damping coefficients which are dependent upon the excitation (or response) frequencies. Such being the case, the corresponding equations of motion of the complete system having n DOF ($n = s + f + g$) must rigorously be represented in the frequency domain.

Considering the coupled structure–foundation system as a free-free (no boundary constraints) system having externally applied forces $-\boldsymbol{f}_h(t)$ acting in the h DOF, its equations of motion can be expressed in the frequency-domain form:

$$\begin{bmatrix} \boldsymbol{D}_{ss}(i\omega) & \boldsymbol{D}_{sf}(i\omega) & \boldsymbol{0} \\ \boldsymbol{D}_{sf}^T(i\omega) & \boldsymbol{D}_{ff}(i\omega) & \boldsymbol{D}_{fg}(i\omega) \\ \boldsymbol{0} & \boldsymbol{D}_{sf}^T(i\omega) & \boldsymbol{D}_{gg}(i\omega) \end{bmatrix} \begin{Bmatrix} \boldsymbol{u}_s(i\omega) \\ \boldsymbol{u}_f(i\omega) \\ \boldsymbol{u}_g(i\omega) \end{Bmatrix} = \begin{Bmatrix} \boldsymbol{0} \\ \boldsymbol{0} \\ \boldsymbol{f}_g(i\omega) \end{Bmatrix} \tag{42.26}$$

in which $\boldsymbol{u}_s(i\omega)$, $\boldsymbol{u}_f(i\omega)$, $\boldsymbol{u}_g(i\omega)$, and $\boldsymbol{f}_g(i\omega)$ are the Fourier transforms of vectors $\boldsymbol{u}_s(t)$, $\boldsymbol{u}_f(t)$, $\boldsymbol{u}_g(t)$, and $\boldsymbol{f}_g(t)$, respectively; and matrices $\boldsymbol{D}_{ij}(i\omega)$, $i, j = s, f, g$, are the corresponding impedance (dynamic stiffness) matrices. The g components in vectors $\boldsymbol{u}_g(i\omega)$ and $\boldsymbol{f}_g(i\omega)$ are ordered such that their last h components make up vectors $\boldsymbol{u}_h(i\omega)$ and $-\boldsymbol{f}_h(i\omega)$, respectively, with all other components being equal to zero.

For a viscously damped linear structure–foundation system, the impedance matrices $\boldsymbol{D}_{ij}(i\omega)$ are of the form:

$$\boldsymbol{D}_{ij}(i\omega) = \boldsymbol{K}_{ij} + i\omega\boldsymbol{C}_{ij} - \omega^2\boldsymbol{M}_{ij} \qquad i,\ j = s,\ f,\ g \tag{42.27}$$

in which $\boldsymbol{K}_{ij}$, $\boldsymbol{C}_{ij}$, and $\boldsymbol{M}_{ij}$ are the standard stiffness, damping, and mass matrices, respectively, which would appear in the equations of motion of the system if expressed in the time domain. For a constant-hysteresis-damped linear system, the impedance matrices are given by

$$\boldsymbol{D}_{ij}(i\omega) = \boldsymbol{K}_{ij}^* - \omega^2\boldsymbol{M}_{ij} \qquad i,\ j = s,\ f,\ g \tag{42.28}$$

in which $\boldsymbol{K}_{ij}^*$ is a complex stiffness matrix obtained by assembling individual finite-element matrices $\boldsymbol{K}^{*(m)}$ of the form

$$\boldsymbol{K}^{*(m)} \equiv \left\{1 - 2(\beta^{(m)})^2 + 2i\beta^{(m)}\sqrt{1-(\beta^{(m)})^2}\right\}\boldsymbol{K}^{(m)} \doteq (1 + 2i\beta^{(m)})\boldsymbol{K}^{(m)} \tag{42.29}$$

where $\boldsymbol{K}^{(m)}$ denotes the standard elastic stiffness matrix for finite element m as used in the assembly process to obtain matrix $\boldsymbol{K}_{ij}$ and $\beta^{(m)}$ is a damping ratio specified appropriately for the material used in finite-element m [56].

The hysteretic form of damping represented in Eq. (42.28) is the more appropriate form to use for two reasons: (1) it is easy to accommodate different damping ratios for the different materials used in the system and (2) the resulting modal damping is independent of excitation (or response) frequency ω, consistent with test evidence showing that real damping is indeed essentially independent of this frequency. As noted by the form of Eq. (42.27), viscous damping is dependent upon frequency ω, contrary to test results; thus, preference should definitely be given to the use of hysteretic damping for linear systems which can be solved in the frequency domain. Hysteretic damping is unfortunately incompatible with solutions in the time domain.

Vector $-\boldsymbol{f}_h(i\omega)$, which makes up the last h components in force vector $\boldsymbol{f}_g(i\omega)$ appearing in Eq. (42.26), represents, as defined in Section 42.5, the internal SFI forces at the soil–foundation interfaces when the entire coupled soil–foundation–structure system is responding to the free-field soil input motions. Therefore, to solve the SFSI problem, this vector must be characterized in terms of the foundation displacement vector $\boldsymbol{u}_h(i\omega)$ and the scattered free-field soil displacement vector $\bar{\boldsymbol{u}}_h(i\omega)$. As discussed previously in Section 42.5, for demand analysis purposes, this vector can be linearized to the form

$$-\boldsymbol{f}_h(i\omega) = \boldsymbol{G}_{hh}(i\omega)\{\bar{\boldsymbol{u}}_h(i\omega) - \boldsymbol{u}_h(i\omega)\} \tag{42.30}$$

in which $-\boldsymbol{f}_h(i\omega)$ represents the force vector acting on the foundations from the soil medium and $\boldsymbol{G}_{hh}(i\omega)$ is the soil impedance matrix which is complex valued and frequency dependent.

Substituting Eq. (42.30) into Eq. (42.26), the equations of motion of the complete bridge system become

$$\begin{bmatrix} \boldsymbol{D}_{ss}(i\omega) & \boldsymbol{D}_{sf}(i\omega) & \boldsymbol{0} \\ \boldsymbol{D}_{sf}^T(i\omega) & \boldsymbol{D}_{ff}(i\omega) & \boldsymbol{D}_{fg}(i\omega) \\ \boldsymbol{0} & \boldsymbol{D}_{fg}^T(i\omega) & [\boldsymbol{D}_{gg}(i\omega)+\boldsymbol{G}_{gg}(i\omega)] \end{bmatrix} \begin{Bmatrix} \boldsymbol{u}_s(i\omega) \\ \boldsymbol{u}_f(i\omega) \\ \boldsymbol{u}_g(i\omega) \end{Bmatrix} = \begin{Bmatrix} \boldsymbol{0} \\ \boldsymbol{0} \\ \bar{\boldsymbol{f}}_g(i\omega) \end{Bmatrix} \tag{42.31}$$

in which

$$\boldsymbol{G}_{gg}(i\omega) = \begin{bmatrix} \boldsymbol{0} & \boldsymbol{0} \\ \boldsymbol{0} & \boldsymbol{G}_{hh}(i\omega) \end{bmatrix}; \quad \bar{\boldsymbol{f}}_g(i\omega) = \begin{Bmatrix} \boldsymbol{0} \\ \bar{\boldsymbol{f}}_h(i\omega) \end{Bmatrix} \tag{42.32}$$

Vector $\bar{\boldsymbol{f}}_h(i\omega)$ is the free-field soil "seismic driving force" vector defined by Eq. (42.15), in which the free-field soil displacements in vector $\bar{\boldsymbol{u}}_h(i\omega)$ result from scattering of incident seismic waves propagating to the bridge site as explained previously in Section 42.5.

Nonlinear Modeling

When large nonlinearities develop in the structure–foundation subsystem during a seismic event, evaluation of its performance requires nonlinear modeling and analysis in the time domain. In this case, the standard linear equations of motion of the complete system as expressed by

$$\begin{bmatrix} \boldsymbol{M}_{ss} & \boldsymbol{M}_{sf} & \boldsymbol{0} \\ \boldsymbol{M}_{sf}^T & \boldsymbol{M}_{ff} & \boldsymbol{M}_{fg} \\ \boldsymbol{0} & \boldsymbol{M}_{fg}^T & \boldsymbol{M}_{gg} \end{bmatrix} \begin{Bmatrix} \ddot{\boldsymbol{u}}_s(t) \\ \ddot{\boldsymbol{u}}_f(t) \\ \ddot{\boldsymbol{u}}_g(t) \end{Bmatrix} + \begin{bmatrix} \boldsymbol{C}_{ss} & \boldsymbol{C}_{sf} & \boldsymbol{0} \\ \boldsymbol{C}_{sf}^T & \boldsymbol{C}_{ff} & \boldsymbol{C}_{fg} \\ \boldsymbol{0} & \boldsymbol{C}_{fg}^T & \boldsymbol{C}_{gg} \end{bmatrix} \begin{Bmatrix} \dot{\boldsymbol{u}}_s(t) \\ \dot{\boldsymbol{u}}_f(t) \\ \dot{\boldsymbol{u}}_g(t) \end{Bmatrix} + \begin{bmatrix} \boldsymbol{K}_{ss} & \boldsymbol{K}_{sf} & \boldsymbol{0} \\ \boldsymbol{K}_{sf}^T & \boldsymbol{K}_{ff} & \boldsymbol{K}_{fg} \\ \boldsymbol{0} & \boldsymbol{K}_{fg}^T & \boldsymbol{K}_{gg} \end{bmatrix} \begin{Bmatrix} \boldsymbol{u}_s(t) \\ \boldsymbol{u}_f(t) \\ \boldsymbol{u}_g(t) \end{Bmatrix} = \begin{Bmatrix} \boldsymbol{0} \\ \boldsymbol{0} \\ \boldsymbol{f}_g(t) \end{Bmatrix} \tag{42.33}$$

must be modified appropriately to characterize the nonlinearities for use in a step-by-step numerical solution. Usually, it is the third term on the left-hand side of this equation that must be modified to represent the nonlinear hysteretic force–deformation behavior taking place in the individual finite elements of the system. The second term in this equation, representing viscous damping forces, is usually retained in its linear form with the full viscous damping matrix $\boldsymbol{C}$ being expressed in the Rayleigh form

$$\boldsymbol{C} = \alpha_R \boldsymbol{M} + \beta_R \boldsymbol{K} \tag{42.34}$$

in which $\boldsymbol{M}$ and $\boldsymbol{K}$ are the full mass and elastic-stiffness matrices shown in Eq. (42.33) and α_R and β_R are constants assigned numerical values which will limit the modal damping ratios to levels within acceptable bounds over a range of modal frequencies dominating the seismic response.

For a time-domain solution of Eq. (42.33) in its modified nonlinear form, all parameters in the equation must be real (no imaginary parts) and frequency independent. It remains therefore to modify the soil impedance matrix $\boldsymbol{G}_{hh}(i\omega)$ so that when introduced into Eq. (42.30), the inverse Fourier transform of $-\boldsymbol{f}_h(i\omega)$ to the time domain will yield a vector $-\boldsymbol{f}_h(t)$ having no frequency-dependent parameters. To accomplish this objective, separate $\boldsymbol{G}_{hh}(i\omega)$ into its real and imaginary parts in accordance with

$$\boldsymbol{G}_{hh}(i\omega) = \boldsymbol{G}_{hh}^R(\omega) + i\boldsymbol{G}_{hh}^I(\omega) \tag{42.35}$$

in which $\boldsymbol{G}_{hh}^R(\omega)$ and $\boldsymbol{G}_{hh}^I(\omega)$ are real functions of ω. Then approximate these functions using the relations

$$\boldsymbol{G}_{hh}^R(\omega) \doteq \overline{\boldsymbol{K}}_{hh} - \omega^2 \overline{\boldsymbol{M}}_{hh}\,; \quad \boldsymbol{G}_{hh}^I(\omega) \doteq \omega \overline{\boldsymbol{C}}_{hh} \tag{42.36}$$

where the real constants in matrices $\overline{\boldsymbol{K}}_{hh}$, $\overline{\boldsymbol{M}}_{hh}$, and $\overline{\boldsymbol{C}}_{hh}$ are assigned numerical values to provide best fits to the individual frequency-dependent functions in matrices $\boldsymbol{G}_{hh}^R(\omega)$ and $\boldsymbol{G}_{hh}^I(\omega)$ over the frequency range of major influence on seismic response. Typically, applying these best fits to the range $0 < \omega < 4\pi$ radians/second, corresponding to the range $0 < f < 2$ Hz, where $f = \omega/2\pi$, is adequate for most large bridges. In this fitting process, it is sufficient to treat $\overline{\boldsymbol{M}}_{hh}$ as a diagonal matrix, thus affecting only the diagonal functions in matrix $\boldsymbol{G}_{hh}^R(\omega)$. The reason for selecting the particular frequency-dependent forms of Eqs. (42.36) is that when they are substituted into Eq. (42.35), which in turn is substituted into Eq. (42.30), the resulting expression for $\boldsymbol{f}_h(i\omega)$ can be Fourier transformed to the time domain yielding

$$-\boldsymbol{f}_h(t) = \overline{\boldsymbol{K}}_{hh}\{\overline{\boldsymbol{u}}_h(t) - \boldsymbol{u}_h(t)\} + \overline{\boldsymbol{C}}_{hh}\{\dot{\overline{\boldsymbol{u}}}_h(t) - \dot{\boldsymbol{u}}_h(t)\} + \overline{\boldsymbol{M}}_{hh}\{\ddot{\overline{\boldsymbol{u}}}_h(t) - \ddot{\boldsymbol{u}}_h(t)\} \tag{42.37}$$

Substituting $-\boldsymbol{f}_h(t)$ given by this equation for the last h components in vector $\boldsymbol{f}_g(t)$, with all other components in $\boldsymbol{f}_g(t)$ being equal to zero, and then substituting the resulting vector $\boldsymbol{f}_g(t)$ into Eq. (42.33) gives

$$\begin{bmatrix} \boldsymbol{M}_{ss} & \boldsymbol{M}_{sf} & \boldsymbol{0} \\ \boldsymbol{M}_{sf}^T & \boldsymbol{M}_{ff} & \boldsymbol{M}_{fg} \\ \boldsymbol{0} & \boldsymbol{M}_{fg}^T & [\boldsymbol{M}_{gg} + \overline{\boldsymbol{M}}_{gg}] \end{bmatrix} \begin{Bmatrix} \ddot{\boldsymbol{u}}_s(t) \\ \ddot{\boldsymbol{u}}_f(t) \\ \ddot{\boldsymbol{u}}_g(t) \end{Bmatrix} + \begin{bmatrix} \boldsymbol{C}_{ss} & \boldsymbol{C}_{sf} & \boldsymbol{0} \\ \boldsymbol{C}_{sf}^T & \boldsymbol{C}_{ff} & \boldsymbol{C}_{fg} \\ \boldsymbol{0} & \boldsymbol{C}_{fg}^T & [\boldsymbol{C}_{gg} + \overline{\boldsymbol{C}}_{gg}] \end{bmatrix} \begin{Bmatrix} \dot{\boldsymbol{u}}_s(t) \\ \dot{\boldsymbol{u}}_f(t) \\ \dot{\boldsymbol{u}}_g(t) \end{Bmatrix} + \tag{42.38}$$

$$\begin{bmatrix} \boldsymbol{K}_{ss} & \boldsymbol{K}_{sf} & \boldsymbol{0} \\ \boldsymbol{K}_{sf}^T & \boldsymbol{K}_{ff} & \boldsymbol{K}_{fg} \\ \boldsymbol{0} & \boldsymbol{K}_{fg}^T & [\boldsymbol{K}_{gg} + \overline{\boldsymbol{K}}_{gg}] \end{bmatrix} \begin{Bmatrix} \boldsymbol{u}_s(t) \\ \boldsymbol{u}_f(t) \\ \boldsymbol{u}_g(t) \end{Bmatrix} = \begin{Bmatrix} \boldsymbol{0} \\ \boldsymbol{0} \\ \overline{\boldsymbol{K}}_{gg}\overline{\boldsymbol{u}}_g(t) + \overline{\boldsymbol{C}}_{gg}\dot{\overline{\boldsymbol{u}}}_g(t) + \overline{\boldsymbol{M}}_{gg}\ddot{\overline{\boldsymbol{u}}}_g(t) \end{Bmatrix}$$

in which

$$\overline{M}_{gg} = \begin{bmatrix} 0 & 0 \\ 0 & \overline{M}_{hh} \end{bmatrix}; \ \overline{K}_{gg} = \begin{bmatrix} 0 & 0 \\ 0 & \overline{K}_{hh} \end{bmatrix}; \ \overline{C}_{gg} = \begin{bmatrix} 0 & 0 \\ 0 & \overline{C}_{hh} \end{bmatrix} \tag{42.39}$$

showing that no frequency-dependent parameters remain in the equations of motion, thus allowing the standard time-domain solution procedure to be used for solving them. Usually, the terms $\overline{C}_{gg}\,\dot{\overline{u}}_g(t)$ and $\overline{M}_{gg}\,\ddot{\overline{u}}_g(t)$ on the right-hand side of Eq. (42.38) have small effects on the solution of this equation; however, the importance of their contributions should be checked. Having modified the third term on the left-hand side of Eq. (42.38) to its nonlinear hysteretic form, the complete set of coupled equations can be solved for displacements $u_s(t)$, $u_f(t)$, $u_g(t)$ using standard step-by-step numerical integration procedures.

42.6.2 Solution Procedures

One-Step Direct Approach

In this approach, the equations of motion are solved directly in their coupled form. If the system is treated as being fully linear (or equivalent linear), the solution can be carried out in the frequency domain using Eq. (42.31). In doing so, the complete set of complex algebraic equations is solved separately for discrete values of ω over the frequency range of interest yielding the corresponding sets of displacement vectors $u_s(i\omega)$, $u_f(i\omega)$, and $u_g(i\omega)$. Having obtained these vectors for the discrete values of ω, they are inverse Fourier transformed to the time domain giving vectors $u_s(t)$, $u_f(t)$, $u_g(t)$. The corresponding time histories of internal forces and/or deformations in the system can then be obtained directly using standard finite-element procedures.

If the structure–foundation subsystem is modeled as a nonlinear system, the solution can be carried out in the time domain using Eq. (42.38). In this case, the coupled nonlinear equations of motion are solved using standard step-by-step numerical integration procedures.

This one-step direct approach is simple and straightforward to implement for a structural system supported on a single foundation, such as a building. However, for a long, multiple-span bridge supported on many independent foundations, a very large system of equations and an associated very large number of seismic free-field inputs in vector $\overline{u}_g\ (i\omega)$ result, making the solution computationally impractical, especially when large nonlinearities are present in the equations of motion. In this case, it is desirable to simplify the problem by finding separate solutions to a set of smaller problems and then combine the solutions in steps so as to achieve the desired end result. The multiple-step substructuring approach described subsequently is ideally suited for this purpose.

Multiple-Step Substructuring Approach

For long bridges supported on multiple foundations, the support separation distances are sufficiently large so that each foundation subsystem can be treated as being independent of the others; therefore, the soil impedance matrix for each foundation will be uncoupled from those of the other foundations. In this case, to simplify the overall problem, each foundation subsystem can be analyzed separately to obtain a boundary impedance matrix called the foundation impedance matrix and a consistent boundary force vector called the foundation driving-force vector, both of which are associated with the DOF at its structure–foundation interface. Having obtained the foundation impedance matrix and associated driving force vector for each foundation subsystem, all such matrices and vectors can be combined into the equations of motion for the total structure as a free-free system, resulting in $(s + f)$ DOF present in the structure–foundation subsystem rather than the $(s + f + g)$ DOF present in the complete soil–structure–foundation system. This reduced set of equations having $(s + f)$ DOF can be solved much more efficiently than solving the equations for the complete system having $(s + f + g)$ DOF as required by the one-step direct approach.

Referring to Eq. (42.31), it is seen that the linear equations of motion for each independent foundation system j can be expressed in the frequency-domain form:

$$\begin{bmatrix} \boldsymbol{D}_{ff}^{j}(i\omega) & \boldsymbol{D}_{fg}^{j}(i\omega) \\ \boldsymbol{D}_{fg}^{j}(i\omega)^{T} & [\boldsymbol{D}_{gg}^{j}(i\omega)+\boldsymbol{G}_{gg}^{j}(i\omega)] \end{bmatrix} \begin{Bmatrix} \boldsymbol{u}_{f}^{j}(i\omega) \\ \boldsymbol{u}_{g}^{j}(i\omega) \end{Bmatrix} = \begin{Bmatrix} \boldsymbol{0} \\ \overline{\boldsymbol{f}}_{g}^{j}(i\omega) \end{Bmatrix} \tag{42.40}$$

in which

$$\overline{\boldsymbol{f}}_{g}^{j}(i\omega)=\boldsymbol{G}_{gg}^{j}(i\omega)\overline{\boldsymbol{u}}_{g}^{j}(i\omega) \tag{42.41}$$

Solving the second of Eqs. (42.40) for $\boldsymbol{u}_{g}^{j}(i\omega)$ gives

$$\boldsymbol{u}_{g}^{j}(i\omega)=\left[\boldsymbol{D}_{gg}^{j}(i\omega)+\boldsymbol{G}_{gg}^{j}(i\omega)\right]^{-1}\left[-\boldsymbol{D}_{fg}^{j}(i\omega)^{T}\boldsymbol{u}_{f}^{j}(i\omega)+\overline{\boldsymbol{f}}_{g}^{j}(i\omega)\right] \tag{42.42}$$

Substituting this equation into the first of Eqs. (42.40) yields

$$\left[\boldsymbol{D}_{ff}^{j}(i\omega)+\boldsymbol{F}_{ff}^{j}(i\omega)\right]\boldsymbol{u}_{f}^{j}(i\omega)=\overline{\boldsymbol{f}}_{f}^{j}(i\omega) \tag{42.43}$$

where

$$\boldsymbol{F}_{ff}^{j}(i\omega)\equiv-\boldsymbol{D}_{fg}^{j}(i\omega)[\boldsymbol{D}_{gg}^{j}(i\omega)+\boldsymbol{G}_{gg}^{j}(i\omega)]^{-1}\boldsymbol{D}_{fg}^{j}(i\omega)^{T} \tag{42.44}$$

$$\overline{\boldsymbol{f}}_{f}^{j}(i\omega)\equiv-\boldsymbol{D}_{fg}^{j}(i\omega)[\boldsymbol{D}_{gg}^{j}(i\omega)+\boldsymbol{G}_{gg}^{j}(i\omega)]^{-1}\overline{\boldsymbol{f}}_{g}^{j}(i\omega) \tag{42.45}$$

Matrix $\boldsymbol{F}_{ff}^{j}(i\omega)$ and vector $\overline{\boldsymbol{f}}_{f}^{j}(i\omega)$ will be referred to here as the foundation impedance matrix and its associated foundation driving-force vector, respectively, for the jth foundation. For convenience, a foundation motion vector $\overline{\boldsymbol{u}}_{f}^{j}(i\omega)$ will now be defined as given by

$$\overline{\boldsymbol{u}}_{f}^{j}(i\omega)\equiv\boldsymbol{F}_{ff}^{j}(i\omega)^{-1}\overline{\boldsymbol{f}}_{f}^{j}(i\omega) \tag{42.46}$$

so that the driving-force vector $\overline{\boldsymbol{f}}_{f}^{j}(i\omega)$ can be expressed in the form

$$\overline{\boldsymbol{f}}_{f}^{j}(i\omega)=\boldsymbol{F}_{ff}^{j}(i\omega)\overline{\boldsymbol{u}}_{f}^{j}(i\omega) \tag{42.47}$$

The motion vector $\overline{\boldsymbol{u}}_{f}^{j}(i\omega)$ given by Eq. (42.46) will be referred to subsequently as the "effective (scattered) foundation input motion" vector. Conceptually, this is the vector of foundation motions which, when multiplied by the foundation impedance matrix $\boldsymbol{F}_{ff}^{j}(i\omega)$, yields the foundation driving-force vector $\overline{\boldsymbol{f}}_{f}^{j}(i\omega)$ resulting from the prescribed scattered free-field soil input motions contained in vector $\overline{\boldsymbol{u}}_{g}^{j}(i\omega)$.

Combining Eqs. (42.43) for all foundation subsystems with the equations of motion for the complete free-free structure subsystem yields the desired reduced matrix equation of motion for the entire structure–foundation system in the linear form

$$\begin{bmatrix} \boldsymbol{D}_{ss}(i\omega) & \boldsymbol{D}_{sf}(i\omega) \\ \boldsymbol{D}_{sf}(i\omega)^{T} & [\boldsymbol{D}_{ff}^{s}(i\omega)+\boldsymbol{F}_{ff}(i\omega)] \end{bmatrix} \begin{Bmatrix} \boldsymbol{u}_{s}(i\omega) \\ \boldsymbol{u}_{f}(i\omega) \end{Bmatrix} = \begin{Bmatrix} \boldsymbol{0} \\ \overline{\boldsymbol{f}}_{f}(i\omega) \end{Bmatrix} \tag{42.48}$$

in which $D_{ss}(i\omega)$ and $D_{sf}(i\omega)$ are given by Eqs. (42.27) and (42.28) directly, $D^s_{ff}(i\omega)$ is that part of $D_{ff}(i\omega)$ given by these same equations as contributed by the structure only, and

$$\bar{f}_f(i\omega) = F_{ff}(i\omega)\bar{u}_f(i\omega) \tag{42.49}$$

The solution of Eq. (42.48) for discrete values of ω over the frequency range of interest gives the desired solutions for $u_s(i\omega)$ and $u_f(i\omega)$. To obtain the corresponding solution $u^j_g(i\omega)$ for each foundation subsystem *j*, a backsubstitution is required. This is done by substituting the solution $u^j_f(i\omega)$ for each foundation subsystem *j* into Eq. (42.42) and computing the corresponding response motions in vector $u^j_g(i\omega)$. This step will be called the "foundation feedback" analysis.

When large nonlinearities develop in the structure during a seismic event, the reduced equations of motion representing the coupled structure–foundation system must be expressed in the time domain. To do so, consider the structure alone as a free-free linear system having externally applied forces $f_f(t)$ acting in the *f* DOF. The equations of motion for this system can be expressed in the frequency domain form:

$$\begin{bmatrix} D_{ss}(i\omega) & D_{sf}(i\omega) \\ D_{sf}(i\omega)^T & D^s_{ff}(i\omega) \end{bmatrix} \begin{Bmatrix} u_s(i\omega) \\ u_f(i\omega) \end{Bmatrix} = \begin{Bmatrix} 0 \\ f_f(i\omega) \end{Bmatrix} \tag{42.50}$$

in which $f_f(i\omega)$ is the Fourier transform of vector f_f(t). If Eq. (42.50) is to represent the coupled structure–foundation system, then $f_f(i\omega)$ must satisfy the relation

$$f_f(i\omega) = F_{ff}(i\omega)\left\{\bar{u}_f(i\omega) - u_f(i\omega)\right\} \tag{42.51}$$

in which matrix $F_{ff}(i\omega)$ is an assembly of the individual foundation impedance matrices $F^j_{ff}(i\omega)$ given by Eq. (42.44) for all values of *j* and vector $\bar{u}_f(i\omega)$ is the corresponding complete foundation-motion vector containing all individual vectors $\bar{u}^j_f(i\omega)$ given by Eq. (42.46).

Equation (42.50) can be converted to the time-domain form:

$$\begin{bmatrix} M_{ss} & M_{sf} \\ M^T_{sf} & M^s_{ff} \end{bmatrix} \begin{Bmatrix} \ddot{u}_s(t) \\ \ddot{u}_f(t) \end{Bmatrix} + \begin{bmatrix} C_{ss} & C_{sf} \\ C^T_{sf} & C^s_{ff} \end{bmatrix} \begin{Bmatrix} \dot{u}_s(t) \\ \dot{u}_f(t) \end{Bmatrix} + \begin{bmatrix} K_{ss} & K_{sf} \\ K^T_{sf} & K^s_{ff} \end{bmatrix} \begin{Bmatrix} u_s(t) \\ u_f(t) \end{Bmatrix} = \begin{Bmatrix} 0 \\ f_f(t) \end{Bmatrix} \tag{42.52}$$

in which K^s_{ff}, C^s_{ff}, and M^s_{ff} are the standard stiffness, damping, and mass matrices contributed by the structure only (no contributions from the foundation) and $f_f(t)$ is the inverse Fourier transform of $f_f(i\omega)$ given by Eq. (42.51). In order for $f_f(t)$ to have no frequency-dependent parameters, as required by a time-domain solution, matrix $\mathbf{F}_{ff}(i\omega)$ should be separated into its real and imaginary parts in accordance with

$$F_{ff}(i\omega) = F^R_{ff}(\omega) + iF^I_{ff}(\omega) \tag{42.53}$$

in which $F^R_{ff}(\omega)$ and $F^I_{ff}(\omega)$ can be approximated using the relations

$$F^R_{ff}(\omega) \doteq (\bar{K}_{ff} - \omega^2\bar{M}_{ff}); \quad F^I_{ff}(\omega) \doteq \omega\bar{C}_{ff} \tag{42.54}$$

where the real constants in matrices $\bar{K}_{ff}$, $\bar{M}_{ff}$, and $\bar{C}_{ff}$ are assigned numerical values to provide best fits to the individual frequency-dependent functions in matrices $F^R_{ff}(\omega)$ and $F^I_{ff}(\omega)$ over the frequency range of major influence on seismic response; usually the range $0 < \omega < 4\pi$ rad/s. is adequate for large bridges. In this fitting process, it is sufficient to treat $\bar{M}_{ff}$ as a diagonal matrix, thus affecting only the diagonal functions in matrix $F^R_{ff}(\omega)$.

Substituting Eqs. (42.54) into Eq. (42.53) and the resulting Eq. (42.53) into Eq. (42.51), this latter equation can be inverse Fourier transformed, giving

$$f_t(t) = \overline{K}_{ff}\left\{\overline{u}_f(t) - u_f(t)\right\} + \overline{C}_{ff}\left\{\dot{\overline{u}}_f(t) - \dot{u}_f(t)\right\} + \overline{M}_{ff}\left\{\ddot{\overline{u}}_f(t) - \ddot{u}_f(t)\right\} \tag{42.55}$$

which when introduced into Eq. (42.50) yields the desired reduced linear equations of motion in the time-domain form

$$\begin{bmatrix} \boldsymbol{M}_{ss} & \boldsymbol{M}_{ff} \\ \boldsymbol{M}_{sf}^T & [\boldsymbol{M}_{ff}^s + \overline{\boldsymbol{M}}_{ff}] \end{bmatrix} \begin{Bmatrix} \ddot{\boldsymbol{u}}_s(t) \\ \ddot{\boldsymbol{u}}_f(t) \end{Bmatrix} + \begin{bmatrix} \boldsymbol{C}_{ss} & \boldsymbol{C}_{sf} \\ \boldsymbol{C}_{sf}^T & [\boldsymbol{C}_{ff}^s + \overline{\boldsymbol{C}}_{ff}] \end{bmatrix} \begin{Bmatrix} \dot{\boldsymbol{u}}_s(t) \\ \dot{\boldsymbol{u}}_f(t) \end{Bmatrix} +$$

$$\begin{bmatrix} \boldsymbol{K}_{ss} & \boldsymbol{K}_{sf} \\ \boldsymbol{K}_{sf}^T & [\boldsymbol{K}_{ff}^s + \overline{\boldsymbol{K}}_{ff}] \end{bmatrix} \begin{Bmatrix} \boldsymbol{u}_s(t) \\ \boldsymbol{u}_f(t) \end{Bmatrix} = \begin{bmatrix} \boldsymbol{0} \\ \overline{\boldsymbol{K}}_{ff}\overline{\boldsymbol{u}}_f(t) + \overline{\boldsymbol{C}}_{ff}\dot{\overline{\boldsymbol{u}}}_f(t) + \overline{\boldsymbol{M}}_{ff}\ddot{\overline{\boldsymbol{u}}}_f(t) \end{bmatrix} \tag{42.56}$$

showing that no frequency-dependent parameters remain in the equations of motion, thus satisfying the time-domain solution requirement. Again, as explained previously, the full viscous damping matrix in this equation is usually expressed in the Rayleigh form given by Eq. (42.34) in which constants α_R and β_R are assigned numerical values to limit the modal damping ratios to levels within acceptable bounds over the range of frequencies dominating seismic response. As explained previously for Eq. (42.38), the damping and mass terms on the right-hand side of Eq. (42.56) usually have small effects on the solution; however, their importance should be checked.

Having modified the third term on the left-hand side of Eq. (42.56) to its nonlinear hysteretic form, the complete set of coupled equations can be solved for displacements $\boldsymbol{u}_s(t)$ and $\boldsymbol{u}_f(t)$ using standard step-by-step numerical integration procedures.

To obtain the corresponding $\boldsymbol{u}_g^j(i\omega)$ for each foundation subsystem *j*, the previously defined foundation feedback analyses must be performed. To do so, each subvector $\boldsymbol{u}_f^j(t)$ contained in vector $\boldsymbol{u}_f(t)$, must be Fourier transformed to obtain $\boldsymbol{u}_f^j(i\omega)$. Having these subvectors for all values of *j*, each one can be substituted separately into Eq. (42.42) giving the corresponding subvector $\boldsymbol{u}_g^j(i\omega)$. Inverse Fourier transforming each of these subvectors yields the corresponding vectors $\boldsymbol{u}_f^j(t)$ for all values of *j*.

42.7 Demand Analysis Examples

This section presents the results of three example solutions to illustrate applications of the demand analysis procedures described in the previous section, in particular, the multiple-step substructuring approach. These examples have been chosen from actual situations to illustrate application of the three methods of soil–foundation modeling: (1) the elastodynamic method, (2) the empirical *p–y* method, and (3) the hybrid method.

42.7.1 Caisson Foundation

The first example is chosen to illustrate application of the elastodynamic method of modeling and analysis to a deeply embedded caisson foundation of a large San Francisco Bay crossing bridge. The foundation considered is a large reinforced concrete cellular caisson, 80 ft (24.4 m) long, 176 ft (53.6 m) wide, and 282 ft (86.0 m) tall, located at a deep soil site and filled with water. The configuration of the caisson and its supporting soil profile and properties are shown in Figure 42.8. The soil properties are the shear-strain-compatible equivalent linear properties obtained from free-field site-response analyses using SHAKE with the seismic input motions prescribed at the bedrock surface

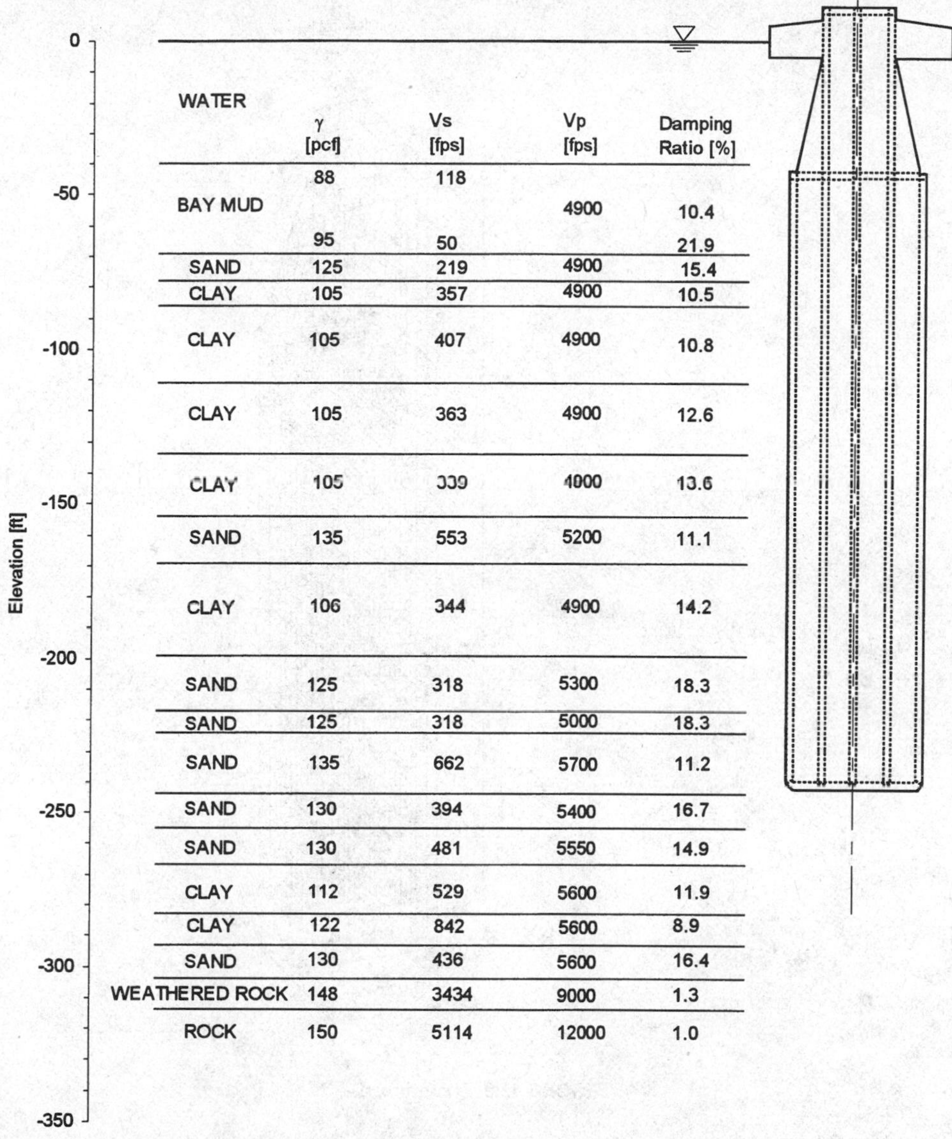

FIGURE 42.8 Configuration and soil profile and properties of the cassion foundation at its SASSI half-model.

in the form of rock-outcrop motion. Thus, these properties have incorporated stiffness degradation effects due to global soil nonlinearities induced in the free-field by the selected seismic input.

Since the caisson is deeply embedded and has large horizontal dimensions, the local soil nonlinearities that develop near the soil–caisson interface are relatively small; therefore, they were neglected in the demand analysis. The soil–caisson system was modeled using the elastodynamic method; i.e., the system was modeled by an elastic foundation structure embedded in a damped-elastic soil medium having the properties shown in Figure 42.8. This model, developed using the finite-element SASSI computer program for one quarter of the soil–caisson system, is shown in Figure 42.8. Using this model, the foundation impedance matrix, i.e., $\boldsymbol{F}^j_{ff}(i\omega)$ defined by Eq. (42.44), and its associated effective (scattered) foundation input motion vector, i.e., $\bar{\boldsymbol{u}}^j_f(i\omega)$ defined by Eq. (42.46), were evaluated consistent with the free-field seismic input using SASSI. The foundation impedance matrix associated with the six DOF of the node located at the top of the caisson El. 40 ft. (12.2 m) was

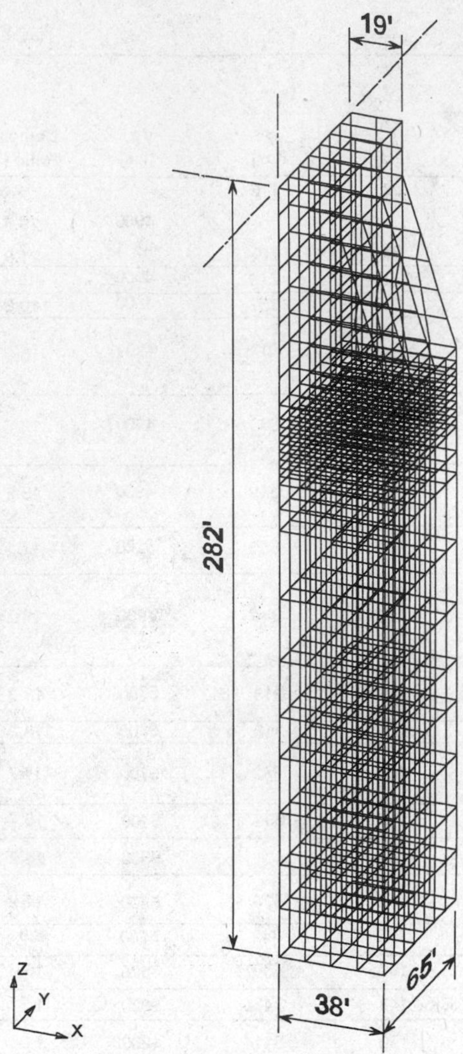

FIGURE 42.8 (continued)

computed following the procedure described in Section 42.6. The individual impedance functions in this matrix are shown in Figure 42.9. The amplitudes of the transfer functions for longitudinal response motions of the caisson relative to the corresponding seismic input motion, as computed for several elevations, are shown in Figure 42.10. The 5% damped acceleration response spectra computed for these motions are also shown in Figure 42.10 where they can be compared with the 5% damped response spectra for the corresponding seismic input motion prescribed at the bedrock level and the corresponding free-field soil motion at the mudline elevation.

As indicated in Figure 42.10, the soil–caisson interaction system alone, without pier tower and superstructure of the bridge being present, has characteristic translational and rocking mode frequencies of 0.7 and 1.4 Hz (periods 1.4 and 0.7 s), respectively. The longitudinal scattered foundation motion associated with the foundation impedance matrix mentioned above is the motion represented by the response spectrum for El. 40 ft (12.2 m) as shown in Figure 42.10.

The response spectra shown in Figure 42.10 indicate that, because of the 0.7-s translational period of the soil–caisson system, the scattered foundation motion at the top of the caisson where the bridge pier tower would be supported, exhibits substantial amplifications in the neighborhood of

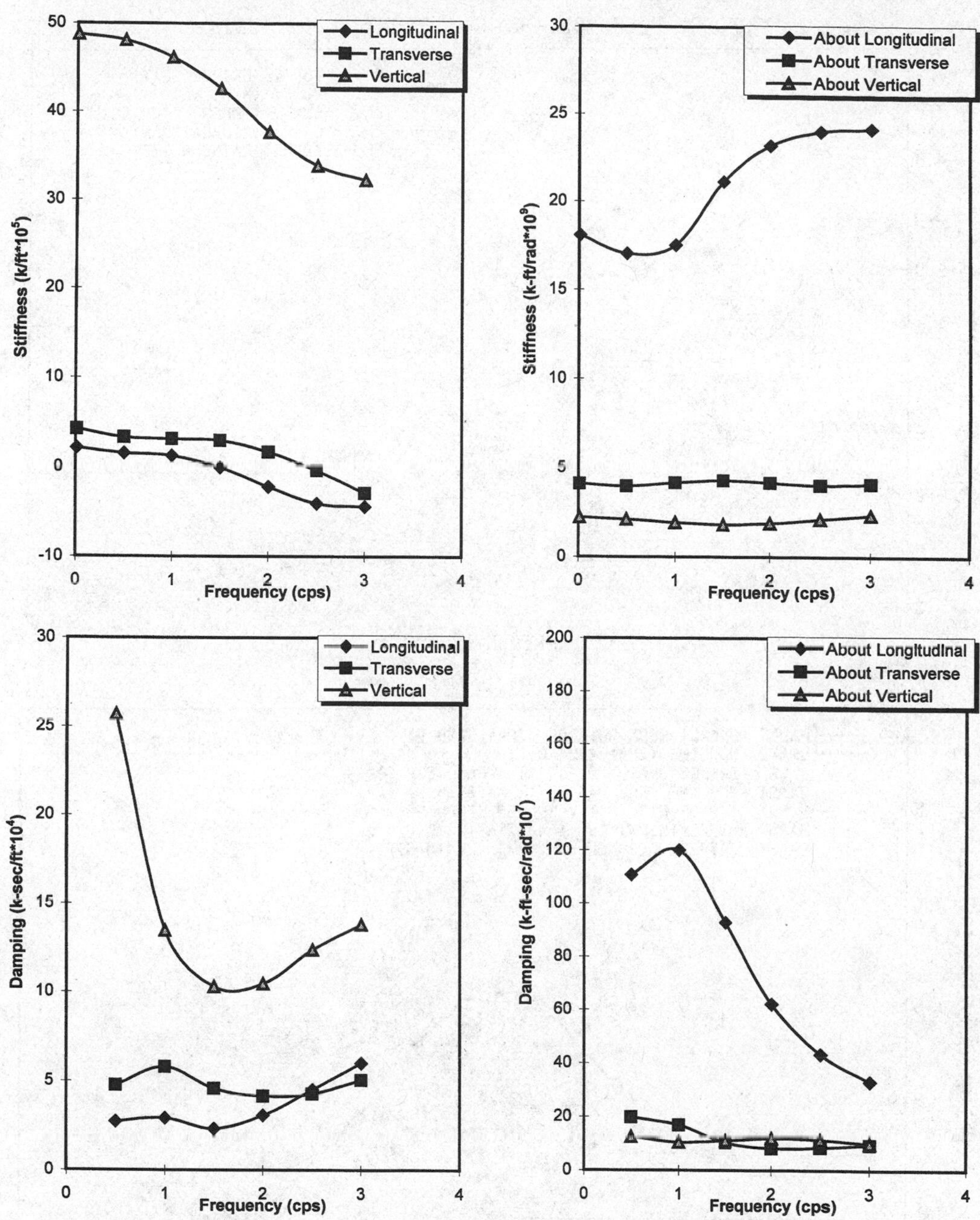

FIGURE 42.9 Foundation impedance functions at the top of the caisson considered.

this period. In the period range longer than 2.0 s, in which the major natural vibration frequencies of the bridge system are located, the spectral values for the scattered foundation input motion are seen to be smaller than the corresponding values for the free-field mudline motion. The above results point out the importance of properly modeling both the stiffness and the inertial properties of the soil–caisson system so that the resulting scattered foundation motions to be used as input to the foundation–structure system will appropriately represent the actual dynamic characteristics of the soil–caisson interaction system.

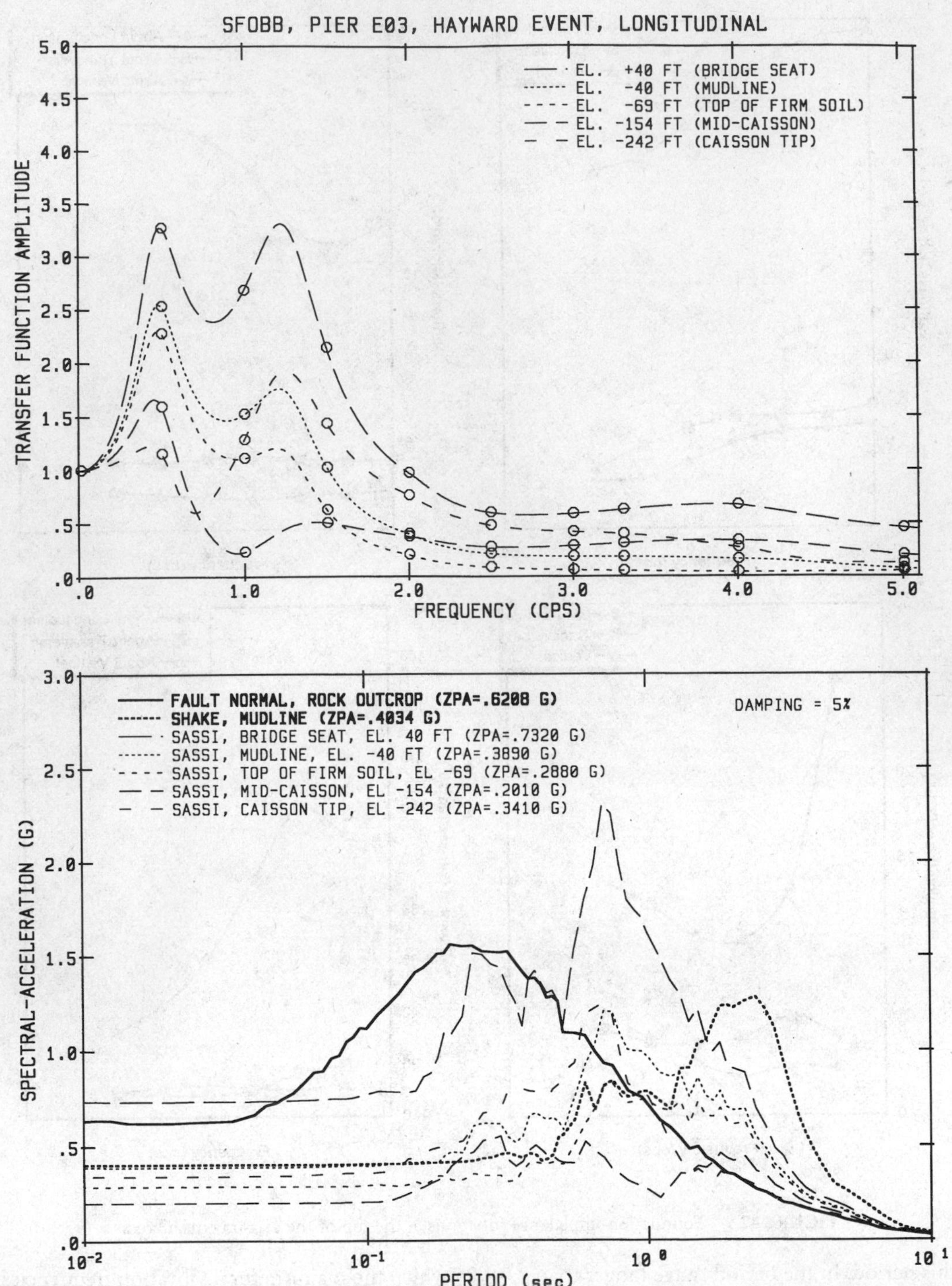

FIGURE 42.10 Transfer function amplitudes and 5% damped response spectra for scattered foundation motions of the caisson at several elevations.

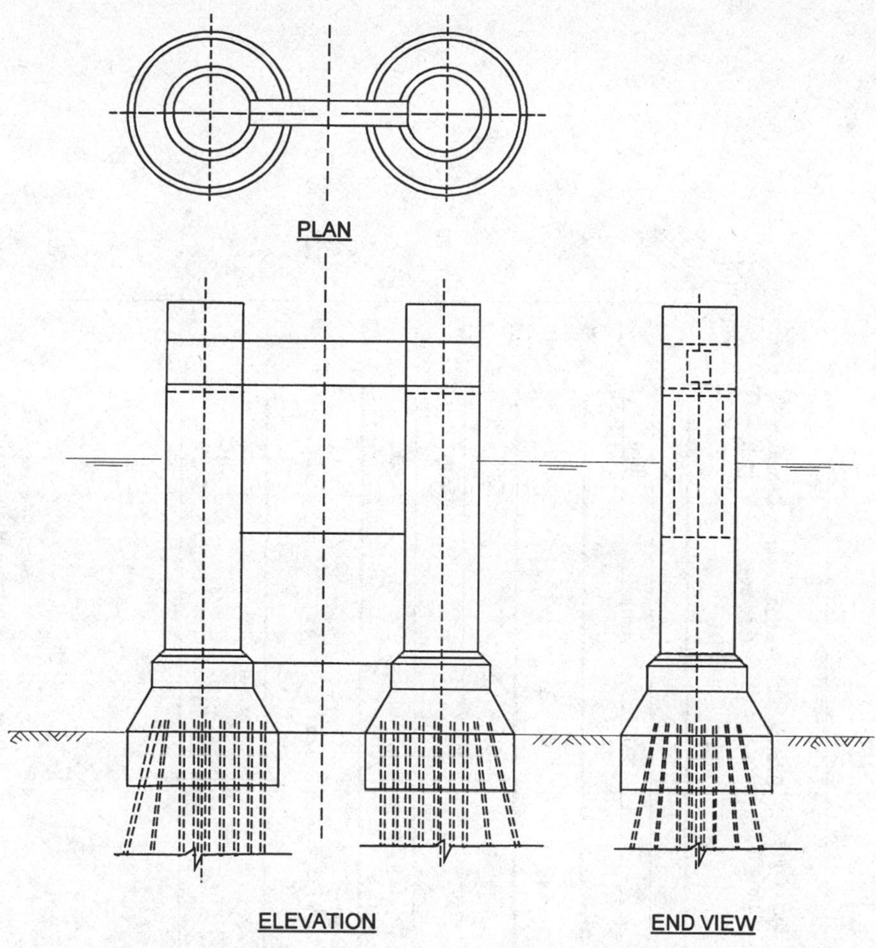

FIGURE 42.11 Configuration of the slender-pile group foundation considered.

42.7.2 Slender-Pile Group Foundation

The second example is to illustrate application of the empirical *p–y* method in a demand analysis of a slender-pile group foundation constructed at a deep soil site. The pile group foundation selected is one of 78 pier foundations of a long, water-crossing steel truss bridge. The foundation is constructed of two 24-ft (7.32-m)-diameter bell-shaped precast reinforced concrete pile caps, which are linked together by a deep cross-beam, as shown in Figure 42.11. Each bell-shaped pile cap is supported on a group of 28 steel 14BP89 H-pipes, giving a total of 56 piles supporting the combined two-bell pile cap. The piles in the outer ring and in the adjacent inner ring are battered at an angle of 4 to 1 and 6 to 1, respectively, leaving the remaining piles as vertical piles. The top ends of all piles are embedded with sufficient lengths into the concrete that fills the interior space of the bell-shaped pile caps such that these piles can be considered as fixed-head piles. The piles penetrate deep into the supporting soil medium to an average depth of 147 ft (44.8 m) below the mudline, where they encounter a thick dense sand layer. The soil profile and properties at this foundation location are shown in Figure 42.12. As indicated in this figure, the top 55 ft (16.8 m) of the site soil is composed of a 35-ft (10.7-m) layer of soft bay mud overlying a 20-ft (6.1-m) layer of loose silty sand.

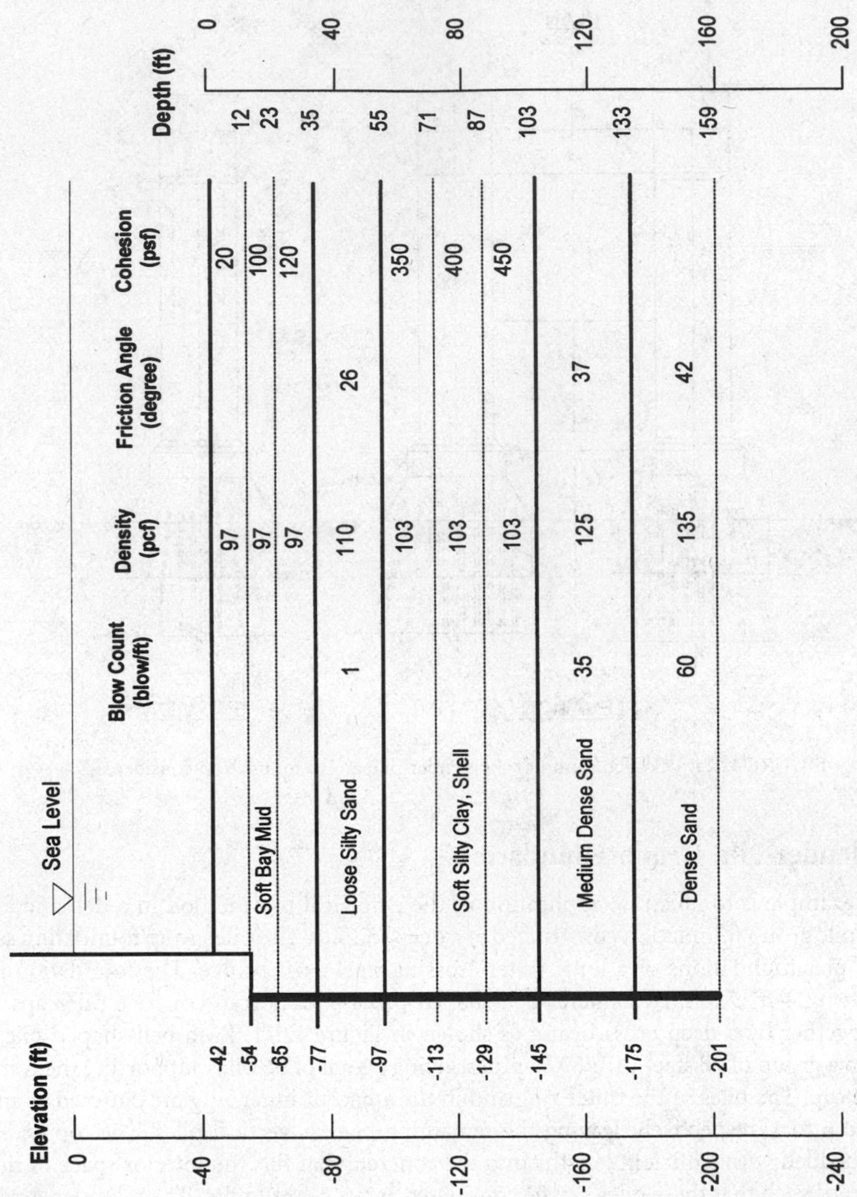

FIGURE 42.12 Soil profile and properties at the slender-pile group foundation considered.

Because of the soft topsoil layers and the slender piles used, the foundation under seismic excitations is expected to undergo relatively large foundation lateral displacements relative to the free-field soil. Thus, large local soil nonlinearities are expected to occur at the soil–pile interfaces. To model the nonlinear soil resistances to the lateral and axial deflections of the piles, the empirically derived lateral *p–y* and axial *t–z,* and the pile-tip *Q–d* curves for each pile were used. Typical *p–y* and *t–z* curves developed for the piles are shown in Figure 42.13. Using the nonlinear *p–y, t–z,* and *Q–d* curves developed, evaluation of the foundation impedance matrix, $\boldsymbol{F}^f_{ff}(i\omega)$, and the associated scattered foundation input motion vector, $\bar{\boldsymbol{u}}_f(i\omega)$, were obtained following the procedures described below:

1. Determine the pile group deflected shape using a nonlinear analysis program such as GROUP [52], LPIPE [53], and APILE2 [54], as appropriate, under an applied set of monotonically increasing axial and lateral forces and an overturning moment.
2. Select target levels of axial and lateral deflections at each selected soil depth corresponding to a selected target level of pile cap displacement and determine the corresponding secant moduli from the applicable nonlinear *p–y, t–z,* and *Q–d* curves.
3. Develop a model of a group of elastic beams supported on elastic axial and lateral soil springs for the pile group using the elastic properties of the piles and the secant moduli of the soil resistances obtained in Step 2 above.
4. Compute the foundation impedance matrix and associated scattered foundation input motion vector for the model developed in Step 3 using Eqs. (42.44) and (42.46).

Since the *p–y, t–z,* and *Q–d* curves represent pseudo-static force–deflection relations, the resulting foundation impedance matrix computed by the above procedure is a real (not complex) frequency-independent pseudo-static stiffness matrix, i.e., $\boldsymbol{F}^j_{ff}(i\omega) = \boldsymbol{F}^j_{ff}(0)$. For the pile group foundation considered in this example, the beam-on-elastic-spring model shown schematically in Figure 42.14 was used. The foundation stiffness matrix is associated with the six DOF of the nodal point located at the bottom center of the pile cap is shown in Figure 42.14. The scattered foundation motions in the longitudinal, transverse, and vertical directions associated with this foundation stiffness matrix are represented by their 5% damped acceleration response spectra shown in Figure 42.15. These spectra can be compared with the corresponding spectra for the seismic input motion prescribed at the pile tip elevation and the free-field mudline motions computed from free-field site-response analyses using SHAKE. As shown in Figure 42.15, the spectral values for the scattered pile cap motions, which would be used as input to the foundation–structure system, are lower than the spectral values for the free-field mudline motions. This result is to be expected for two reasons: (1) the soft topsoil layers present at the site are not capable of driving the pile group foundation and (2) the battered piles, acting with the vertical piles, resist lateral loads primarily through stiff axial truss action, in which case, the effective input motions at the pile cap are controlled more by the free field soil motions at depth, where more competent soil resistances are present, than by the soil motions near the surface.

42.7.3 Large-Diameter Shaft Foundation

The third example illustrates the application of the demand analysis procedure using the hybrid method of modeling. This method is preferred for a foundation constructed of a group of large-diameter CISS or CIDH shafts. Because of the large horizontal dimensions and substantial masses associated with the shafts in this type of foundation, the dynamic interaction of the shafts with the surrounding soil medium is more appropriately modeled and analyzed using the elastodynamic method; however, because the shafts resist loadings in a manner like piles, the local soil nonlinearities present in the soil–shaft interface regions near the ground surface where soft soils are usually present may be sufficiently large that they should be explicitly considered using a method such as the empirical *p–y* method.

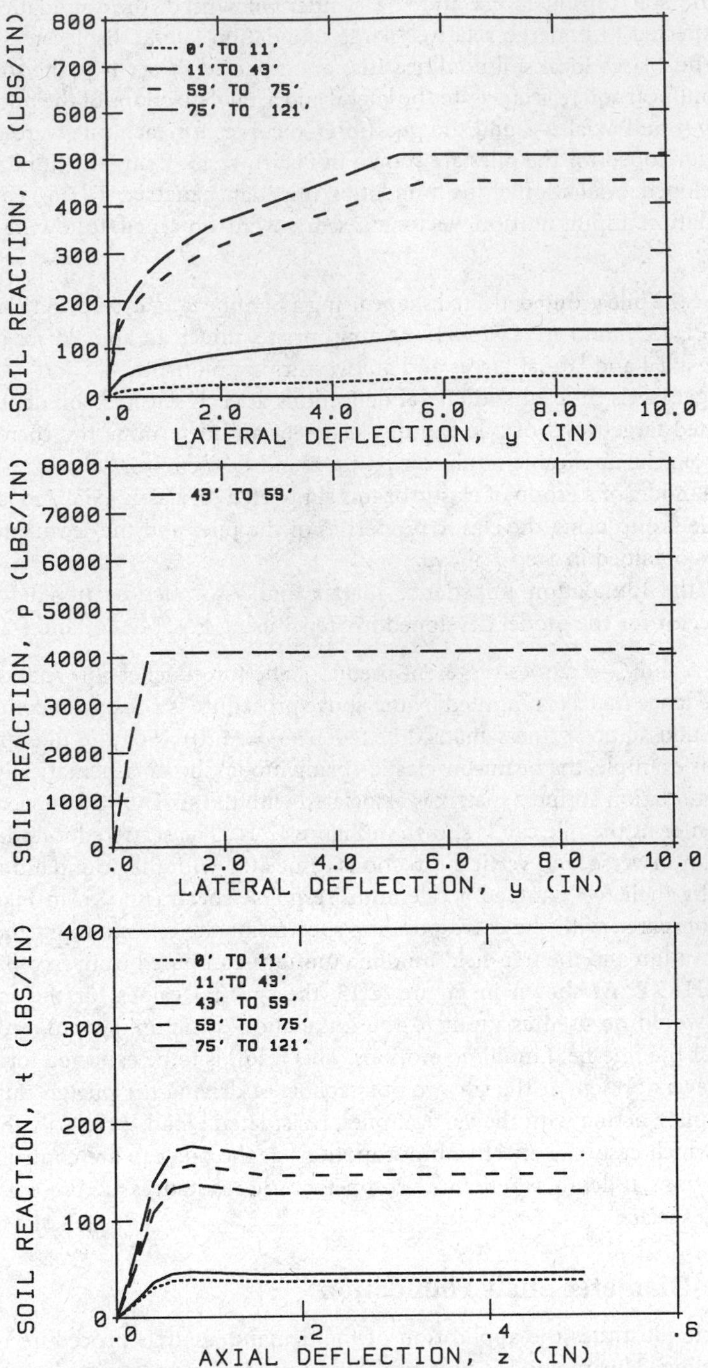

FIGURE 42.13 Typical *p–y* and *t–z* curves for the piles of the slender-pile group foundation considered.

The foundation selected for this example is composed of two 10.5-ft (3.2 m)-diameter shafts 150 ft (45.7 m) long, each consisting of a steel shell of wall thickness 1.375 in. (34.9 mm) filled with concrete. These two shafts are designed to be used as seismic retrofit shear piles for adding lateral stiffnesses and lateral load-resistance capacities to the H-pile group foundation considered in the second example discussed previously. The two shafts are to be linked to the existing pile group at

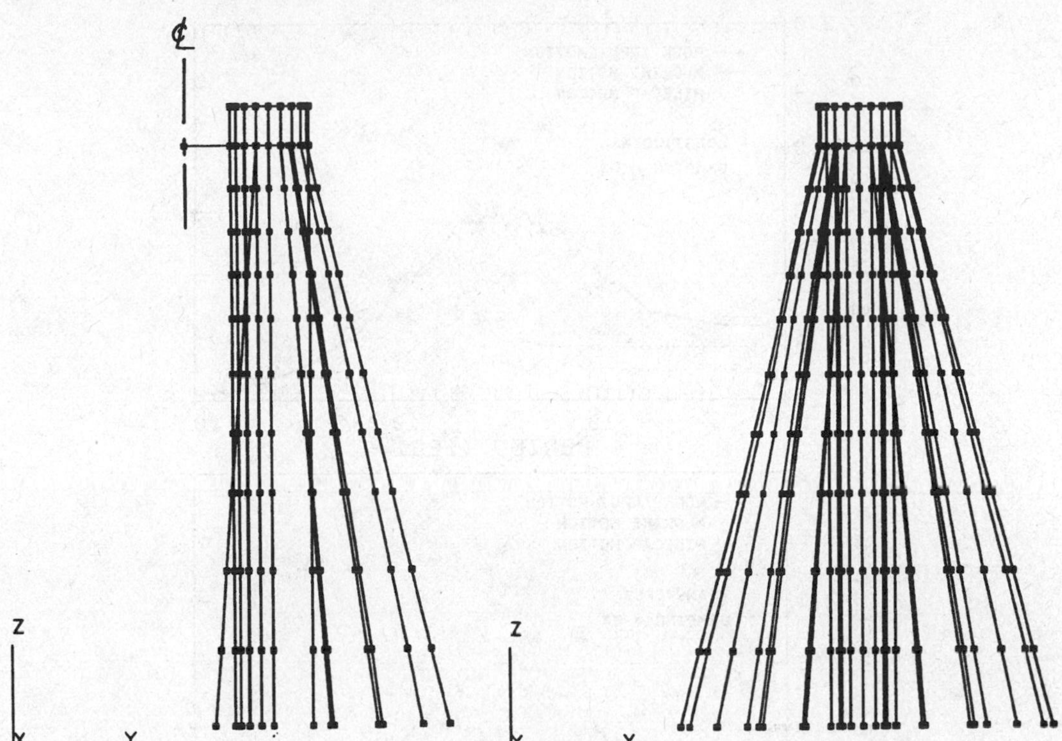

FIGURE 42.14 Beam-on-elastic-foundation half-model for the slender-pile group foundation considered.

the pile cap through a pile cap extension which permits the shafts to resist only horizontal shear loads acting on the pile cap, not axial loads and overturning moments. These shear piles have been designed to resist seismic horizontal shear loads acting on the pile head up to 3000 kips (13,344 kN) each.

To determine the foundation impedance matrix and the scattered pile cap motion vector associated with the horizontal displacements of the shafts at the pile cap, an SASSI model of one half of the soil–shaft system is developed, as shown in Figure 42.16. The soil properties used in this model are the strain-compatible properties shown in Table 42.1, which were obtained from the free-field site-response analyses using SHAKE; thus, the effects of global soil nonlinearities induced in the free-field soil by the design seismic input have been incorporated. To model the local soil nonlinearities occurring near the soil–shaft interface, three-directional (two lateral and one axial) soil springs are used to connect the beam elements representing the shafts to the soil nodes located at the boundary of the soil–shaft interfaces. The stiffnesses of these springs are derived in such a manner that they match the secant moduli of the empirical *p–y*, *t–z*, and *Q–d* curves developed for the shafts, as described previously in Section 42.5.3. Using the complete hybrid model shown in Figure 42.16, foundation compliances as functions of frequency were developed for harmonic pile-head shear loads varying from 500 (2224) to 3000 kips (13,334 kN). The results obtained are shown in Figure 42.17. It is seen that by incorporating local soil nonlinearities using the hybrid method, the resulting foundation compliance coefficients are not only frequency dependent due to the soil and shaft inertias and soil-layering effects as captured by the elastodynamic method, but they are also load–deflection amplitude dependent due to the local soil nonlinearities, as captured by the empirical *p–y* method. The shear load–deflection curves obtained at the pile head in the low-frequency range (≤1.0 Hz) are shown in Figure 42.18. The deflection curve for zero frequency, i.e., the static loading case, compares well with that obtained from a nonlinear analysis using LPILE [53], as indicated in Figure 42.18.

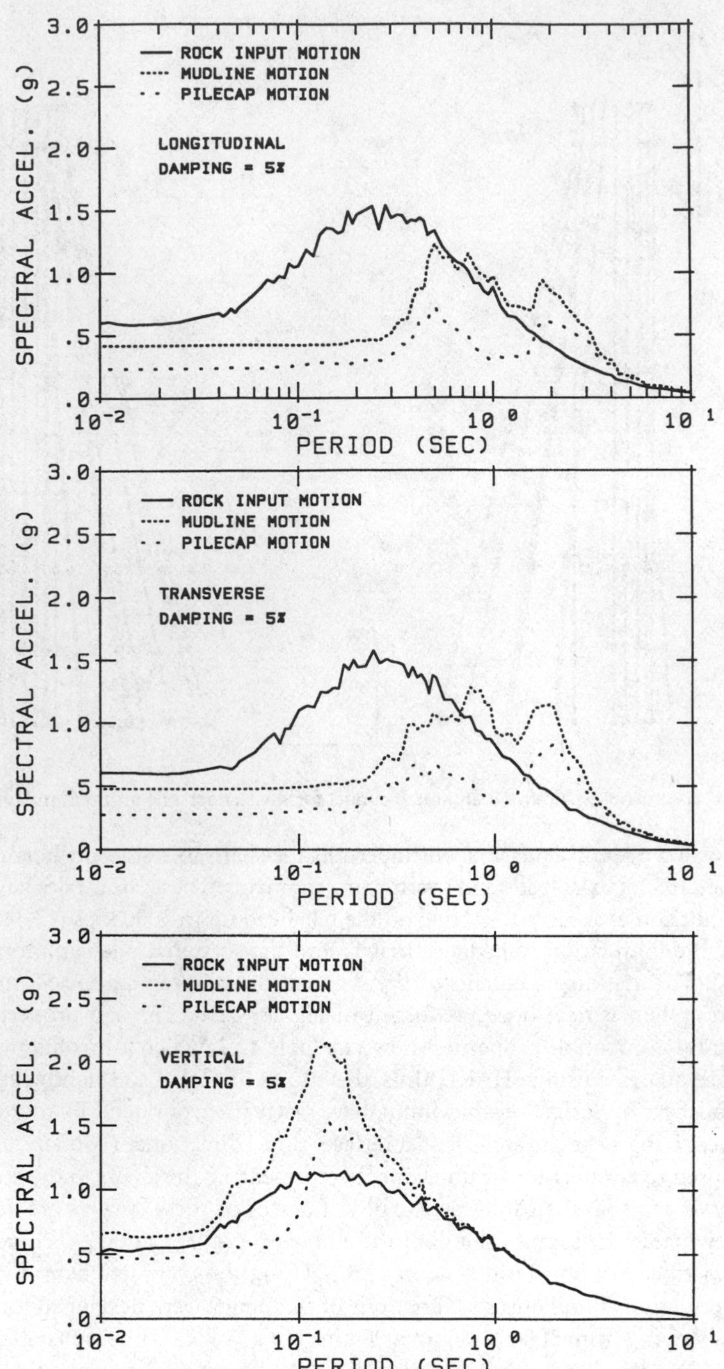

FIGURE 42.15 Comparisons of 5% damped response spectra for the rock input, mudline, and scattered pile cap motions in longitudinal, transverse, and vertical directions.

Subjecting the foundation to the design seismic input motions prescribed at the pile tip elevation and the corresponding free-field soil motions over its full depth, scattered foundation motions in the longitudinal and transverse directions of the bridge at the bottom center of the pile cap were

FIGURE 42.16 SASSI half-model of the large-diameter shaft foundation considered.

TABLE 42.1 Strain-Compatible Soil Properties for the Large-Diameter Shaft Foundation

El. ft (m)	Depth; ft (m)	Thickness; ft (m)	Unit Wt., $\frac{k}{ft^3}\left(\frac{kN}{m^3}\right)$	Shear Wave Velocity $\frac{ft}{s}\left(\frac{m}{s}\right)$	Shear Wave Damping Ratio	Compression Wave Velocity, $\frac{ft}{s}\left(\frac{m}{s}\right)$	Compression Wave Damping Ratio
–50 (–15.2)	0 (0.0)	10 (3.05)	0.096 (15.1)	202.1 (61.6)	0.10	4,800 (1,463)	0.09
–60 (–18.3)	10 (3.05)	10 (3.05)	0.096 (15.1)	207.5 (63.3)	0.15	5,000 (1,524)	0.10
–70 (–21.3)	20 (6.10)	10 (3.05)	0.096 (15.1)	217.7 (66.4)	0.17	5,000 (1,524)	0.10
–80 (–24.4)	30 (9.15)	20 (6.10)	0.110 (17.3)	137.5 (41.9)	0.25	4,300 (1,311)	0.10
–100 (–30.5)	50 (15.2)	10 (3.05)	0.096 (15.1)	215.7 (65.8)	0.20	4,800 (1,463)	0.10
–110 (–33.5)	60 (18.3)	20 (6.10)	0.096 (15.1)	218.4 (66.6)	0.20	4,300 (1,311)	0.10
–130 (–39.6)	80 (24.4)	20 (6.10)	0.096 (15.1)	233.0 (71.0)	0.20	4,900 (1,494)	0.10
–150 (–45.7)	100 (30.5)	20 (6.10)	0.120 (18.8)	420.4 (128.2)	0.20	5,500 (1,677)	0.10
–170 (–51.8)	120 (36.6)	10 (3.05)	0.120 (18.8)	501.0 (152.7)	0.19	6,000 (1,829)	0.10
–180 (–54.9)	130 (39.6)	10 (3.05)	0.120 (18.8)	532.7 (162.4)	0.19	5,800 (1,768)	0.10
–190 (–57.9)	140 (42.7)	20 (6.10)	0.125 (19.6)	607.2 (185.1)	0.18	5,800 (1,768)	0.10
–210 (–64.0)	160 (48.8)	20 (6.10)	0.128 (20.1)	806.9 (246.0)	0.16	5,800 (1,768)	0.10
–230 (–70.1)	180 (54.9)	10 (3.05)	0.133 (20.9)	1,374.4 (419.0)	0.11	6,400 (1,951)	0.10
–240 (–78.2)	190 (57.9)	5 (1.52)	0.140 (21.9)	2,844.9 (867.3)	0.02	12,000 (3,658)	0.02
–245 (–74.7)	195 (59.5)	halfspace	0.145 (22.8)	6,387.2 (1,947.3)	0.01	12,000 (3,658)	0.01

obtained as shown in terms of their 5% damped acceleration response spectra in Figure 42.19, where they can be compared with the corresponding response spectra for the seismic input motions and the free-field mudline motions. It is seen that, because of the substantial masses of the shafts, the spectral amplitudes of the scattered motions are higher than those of the free-field mudline motions

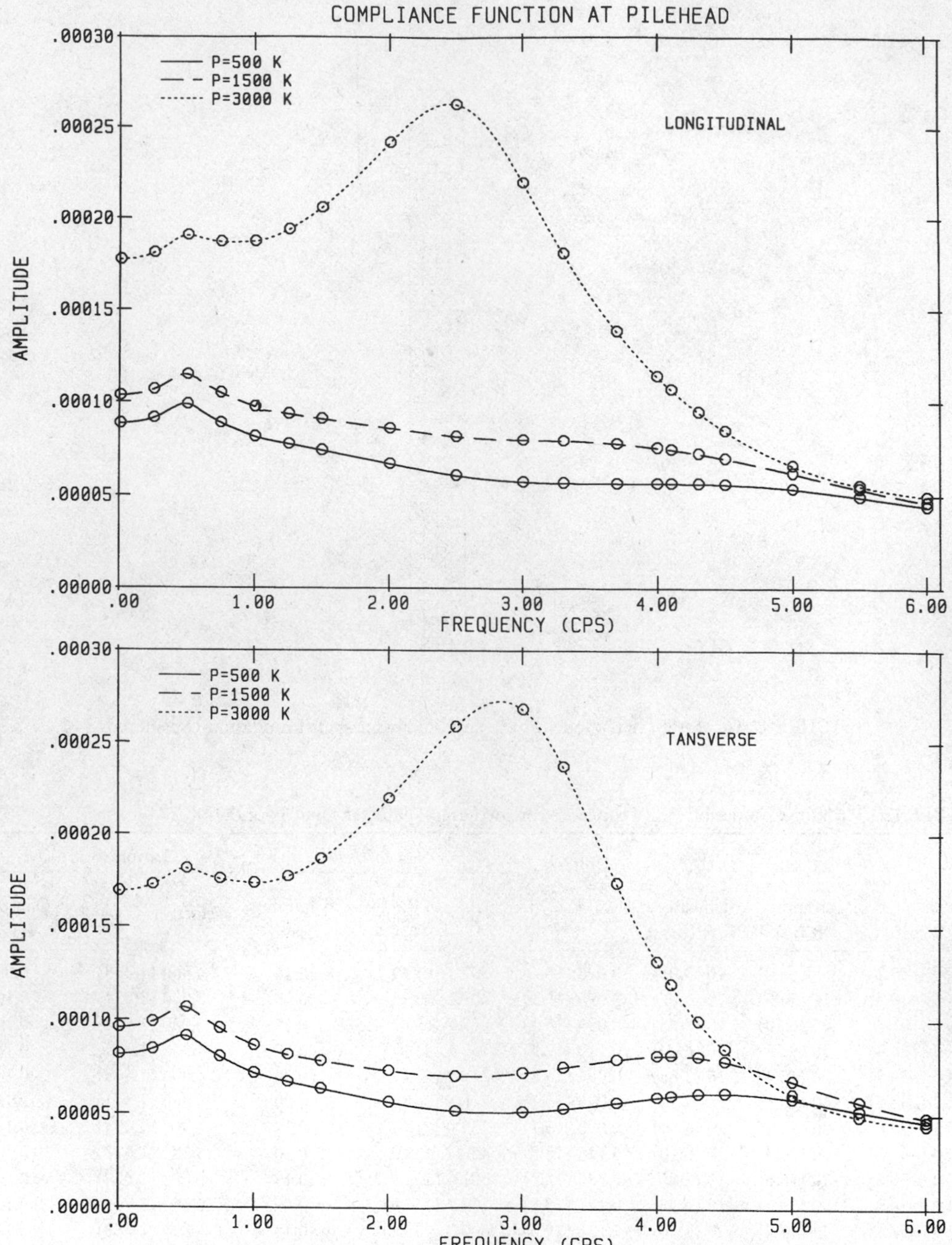

FIGURE 42.17 Foundation compliance functions at discrete values of shear load applied at the top of the shaft foundation.

for frequencies in the neighborhood of the soil–shaft system frequencies. Thus, for large-diameter shaft foundations constructed in deep, soft soil sites, it is important that the soil and shaft inertias be properly included in the SFI. Neglecting the shaft masses will result in underestimating the scattered pile cap motions in the longitudinal and transverse directions of the bridge, as represented in Figure 42.20.

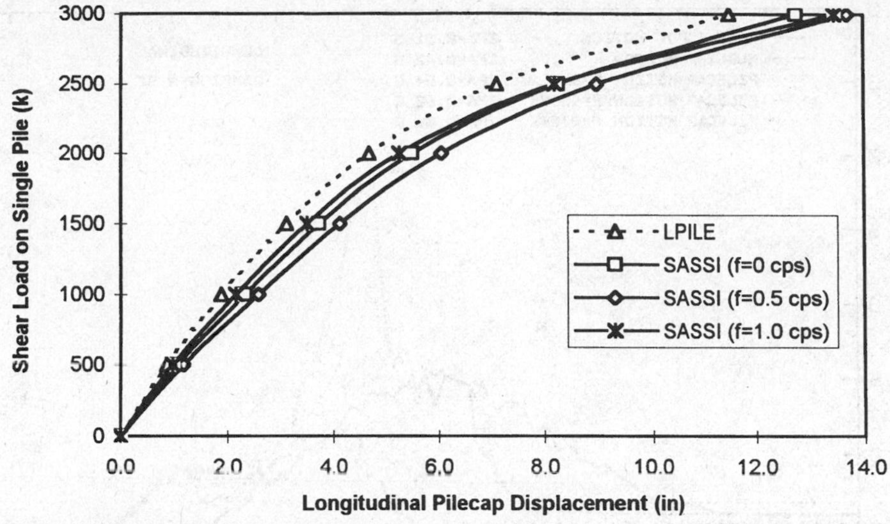

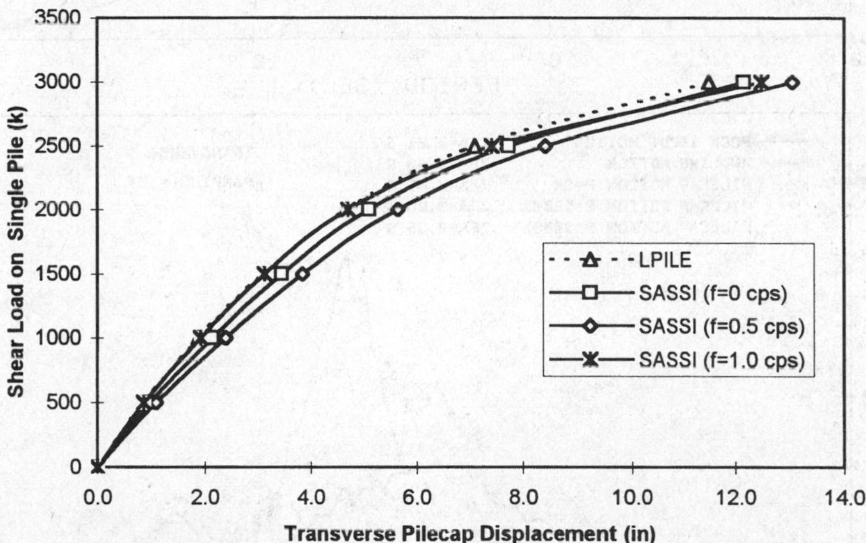

FIGURE 42.18 Typical sher load–deflection curves at several forcing frequencies.

42.8 Capacity Evaluations

The objective of the capacity evaluation is to determine the most probable levels of seismic resistance of the various elements, components, and subsystems of the bridge. The resistance capacities provided by this evaluation, along with the corresponding demands, provide the basis for judging seismic performance of the complete bridge system during future earthquakes. In the domain of SFSI as discussed here, the capacity evaluation focuses on soil–foundation systems.

For a bridge subjected to static loadings, the soil–foundation capacities of interest are the load resistances and the associated foundation deflections and settlements. Their evaluation constitutes the bulk of the traditional foundation design problem. When the bridge is subjected to oscillatory dynamic loadings, including seismic, the static capacities mentioned above are, alone, insufficient in the process of judging soil–foundation performance. In this case, it is necessary to assess the

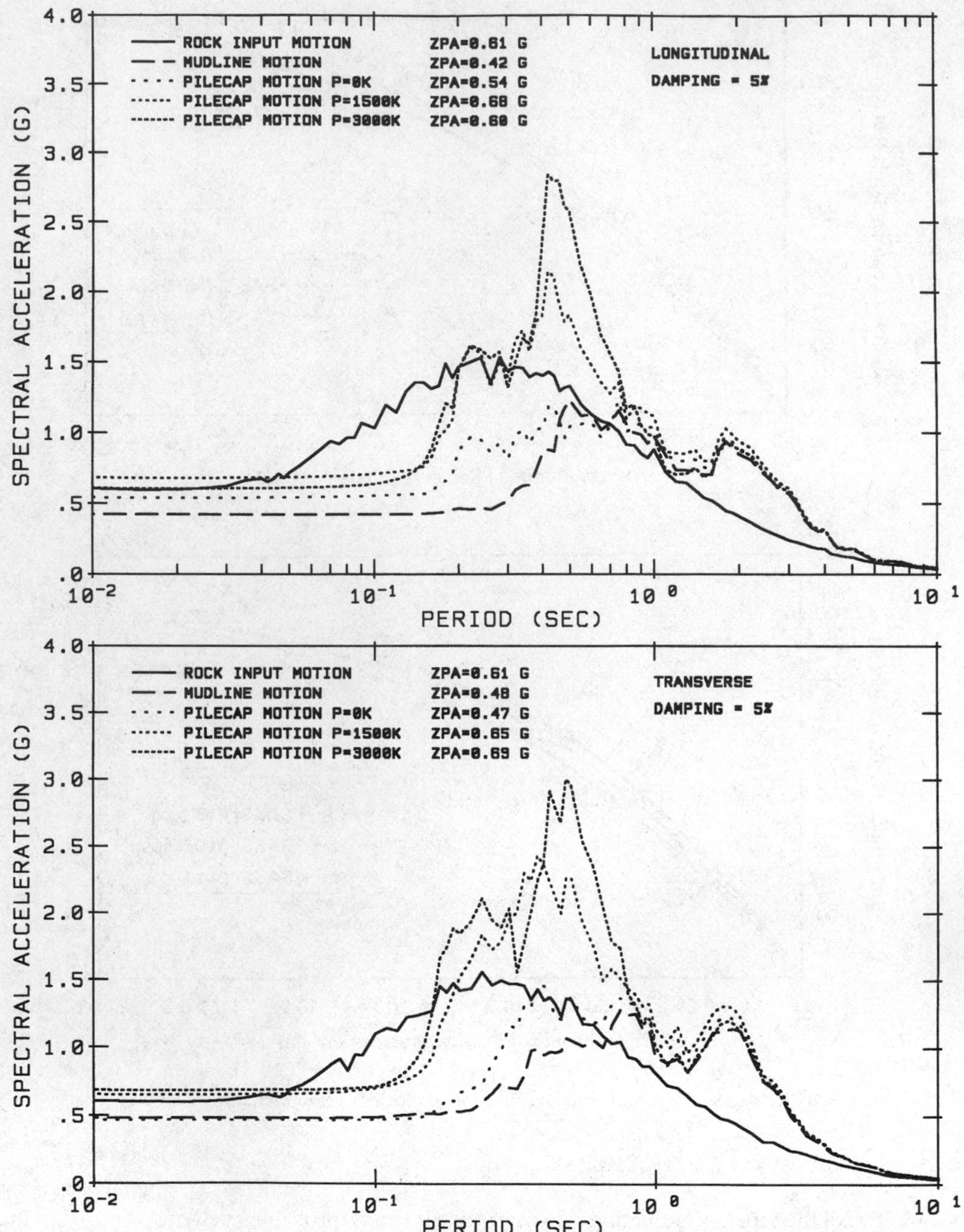

FIGURE 42.19 Comparisons of 5% damped response spectra for the longitudinal and transverse rock input, mudline, and scattered pile cap motions for the shaft foundations at several shear load levels.

entire load–deflection relationships, including their cyclic energy dissipation characteristics, up to load and/or deformation limits approaching failure conditions in the soil–foundation system. Because of the complexity of this assessment, the capacity evaluation must be simplified in order to make it practical. This is usually done by treating each soil–foundation system independently and by subjecting it to simplified pseudo-static monotonic and/or cyclic deformation-controlled step-by-step patterns of loading, referred to here as "push-over" analysis.

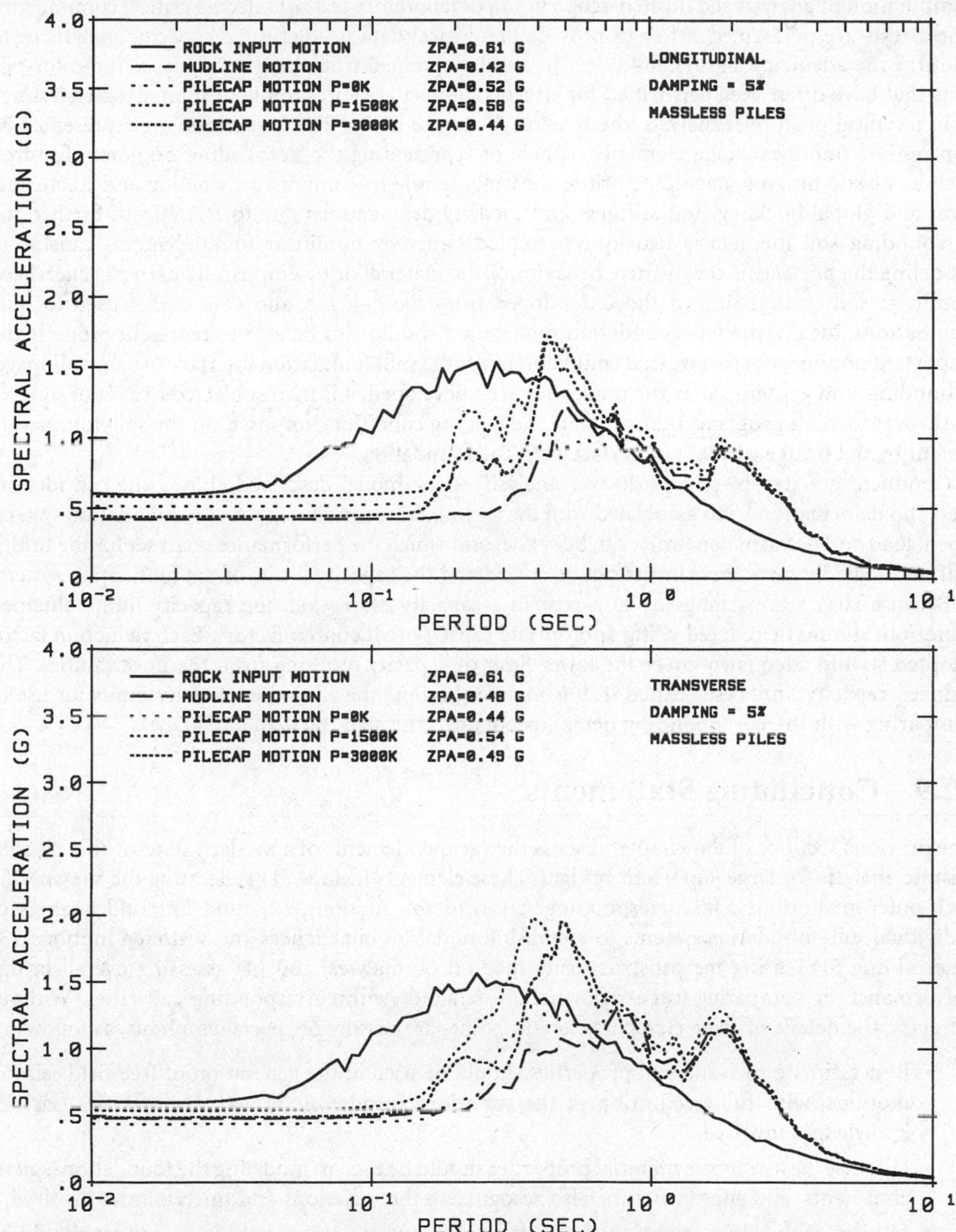

FIGURE 42.20 Comparisons of 5% damped response spectra for the longitudinal and transverse rock input, mudline, and scattered pile cap motions for the shaft foundation without masses in the shafts.

Because near-failure behavior of a soil–foundation system is involved in the capacity evaluation, it necessarily involves postelastic nonlinear behavior in the constituent components of the system, including the structural elements and connections of the foundation and its surrounding soil medium. Thus, ideally, a realistic evaluation of the capacities should be based on *in situ* tests conducted on prototypical foundation systems. Practical limitations, however, generally do not allow the conduct of such comprehensive tests. It is usually necessary, therefore, to rely solely on a combination of analysis and limited-scope *in situ* or laboratory tests of selected critical components. These tests are performed either to provide the critical data needed for a capacity analysis or to confirm the adequacy and reliability of the results obtained from such an analysis. Indicator-pile tests that have often been performed for a bridge project are an example of limited-scope testing.

In a typical push-over analysis, the structural components of the foundation are represented by appropriate nonlinear finite elements capable of representing the near-failure nonlinear features, such as plastic hinging, ductile or brittle shearing, tensile or compressive yielding and fracturing, local and global buckling, and stiffness and capacity degradations due to P-Δ effects; further, the surrounding soil medium is usually represented either by nonlinear finite elements capable of modeling the postelastic constitutive behavior of the material or by empirically derived generalized nonlinear soil springs such as those developed from the p–y, t–z, and Q–d curves used for pile foundations. Ideally, the soil–foundation model used should also be able to represent properly the important nonlinear behaviors that could develop at the soil–foundation interfaces, such as slippage, debonding, and gapping. After the model has been developed, it is then subjected to a set of suitable push-over loading programs that simulate the loading conditions imposed on the soil–foundation system by the bridge pier at its interface with the foundation.

Conducting a step-by-step push-over analysis of the model described above, one can identify load and deformation levels associated with the various failure modes in the soil–foundation system. Then, load and deformation limits can be set beyond which the performance goals set for the bridge will no longer be met. These limits can be considered the capacity limits of the foundation system.

Because large uncertainties usually exist in a capacity evaluation, the capacity limits obtained therefrom should be reduced using appropriate capacity reduction ϕ factors. Each reduction factor adopted should adequately cover the lower limit of capacity resulting from the uncertainties. The reduced capacity limits established in this manner become the allowable capacity limits for use in comparing with the corresponding demands obtained through the demand analysis.

42.9 Concluding Statements

The previous sections of this chapter discuss the various elements of a modern state-of-the-art SFSI seismic analysis for large important bridges. These elements include (1) generating the site-specific rock-outcrop motions and corresponding free-field soil motions, (2) modeling and analysis of individual soil–foundation systems to establish foundation impedances and scattered motions, (3) determining SFSI using the substructuring method of analyses, and (4) assessing overall bridge performance by comparing force–deformation demands with corresponding capacities. Without retracing the details of these elements, certain points are worthy of special emphasis, as follows:

- Best-estimate rock and soil properties should be used in the generation of free-field seismic motions, with full recognition of the variations (randomness) and uncertainties (lack of knowledge) involved.
- Likewise, best-estimate material properties should be used in modeling the foundations, piers, abutments, and superstructure, also recognizing the variations and uncertainties involved.
- In view of the above-mentioned variations and uncertainties, sensitivity analyses should be conducted to provide a sound basis for judging overall seismic performance.
- Considering the current state of development, one should clearly differentiate between the requirements of a seismic force–deformation demand analysis and the corresponding capacity evaluation. The former is concerned with global system behavior; thus, it must satisfy only global dynamic equilibrium and compatibility. The latter, however, places emphasis on the behavior of local elements, components, and subsystems, requiring that equilibrium and compatibility be satisfied only at the local level within both the elastic and postelastic ranges of deformation.

- In conducting a demand analysis, equivalent linear modeling, coupled with the substructuring method of analysis, has the advantages that (1) the results are more controllable and predictable, (2) the uncertainties in system parameters can easily be evaluated separately, and (3) the SFSI responses can be assessed at stages. These advantages lead to a high level of confidence in the results when the nonlinearities are relatively weak. However, when strong nonlinearities are present, nonlinear time history analyses should be carried out in an iterative manner so that system response is consistent with the nonlinearities.
- When strong nonlinearities are present in the overall structural system, usually in the piers and superstructure, multiple sets of seismic inputs should be used separately in conducting the demand analyses; since, such nonlinearities cause relatively large dispersions of the maximum values of critical response.
- The clastodynamic method of treating SFSI is valid for foundations having large horizontal dimensions, such as large spread footings and caissons; while the empirical *p–y* method is valid only for slender-pile foundations subjected to large-amplitude deflections. For foundations intermediate between these two classes, e.g., those using large-diameter shafts, both of these methods are deficient in predicting SFSI behavior. In this case, the hybrid method of modeling has definitive advantages, including its ability to treat all classes of foundations with reasonable validity.
- The *p–y* method of treating SFSI in both demand analyses and capacity evaluations needs further development, refinement, and validation through test results, particularly with regard to establishing realistic *p–y*, t–z, and *Q–d* curves. For seismic applications, changes in the characteristics of these curves, due to global soil nonlinearities induced by the free-field ground motions, should be assessed.
- The hybrid method of treating SFSI, while being fundamentally sound, also needs further development, refinement, and test validation to make it fully acceptable for bridge applications.
- Systematic research and development efforts, involving laboratory and field tests and analytical correlation studies, are required to advance the SFSI analysis methodologies for treating bridge foundations.

The state of the art of SFSI analysis of large bridge structures has been rapidly changing in recent years, a trend that undoubtedly will continue on into the future. The reader is therefore encouraged to take note of new developments as they appear in future publications.

Acknowledgment

The authors wish to express their sincere thanks and appreciation to Joseph P. Nicoletti and Abbas Abghari for their contributions to Sections 42.2 and 42.4, respectively.

References

1. Mylonakis, G., Nikolaou, A., and Gazetas, G., Soil-pile-bridge seismic interaction: kinematic and inertia effects. Part I: Soft soil, *Earthquake Eng. Struct. Dyn.*, 26, 337–359, 1997.
2. Tseng, W. S., Soil-foundation-structure interaction analysis by the elasto-dynamic method, in *Proc. 4th Caltrans Seismic Research Workshop*, Sacramento, July 9–11, 1996.
3. Lam, I. P. and Law, H., Soil-foundation-structure interaction — analytical considerations by empirical *p–y* method, in *Proc. 4th Caltrans Seismic Research Workshop*, Sacramento, July 9–11, 1996.
4. Schnabel, P. B., Lysmer, J., and Seed, H. B., SHAKE — A Computer Program for Earthquake Response Analysis of Horizontally Layered Sites, Report No. EERC 72-12, Earthquake Engineering Research Center, University of California, Berkeley, 1972.

5. Tseng, W. S. and Hadjian, A. H., Guidelines for Soil-Structure Interaction Analysis, EPRI NP-7395, Electric Power Research Institute, Palo Alto, CA, October 1991.
6. Richart, F. E., Hall, J. R., Jr., and Woods, R. D., *Vibrations of Soils and Foundations*, Prentice-Hall, Englewood Cliffs, NJ, 1970.
7. Veletsos, A. S. and Wei, Y. T., Lateral and rocking vibration of footings, *J. Soil Mech. Found. Div. ASCE*, 97(SM9), 1227–1249, 1971.
8. Kausel, E. and Roësset, J. M., Soil-structure interaction problems for nuclear containment structures, in *Proc. ASCE Power Division Specialty Conference*, Denver, CO, August 1974.
9. Wong, H. L. and Luco, J. E., Dynamic response of rigid foundations of arbitrary shape, *Earthquake Eng. Struct. Dyn.*, 4, 587–597, 1976.
10. Seed, H. B. and Idriss, I. M., Soil Moduli and Damping Factors for Dynamic Response Analysis, Report No. EERC 70-10, Earthquake Engineering Research Center, University of California, Berkeley, 1970.
11. Waas, G., Analysis Method for Footing Vibrations through Layered Media, Ph.D. dissertation, University of California, Berkeley, 1972.
12. Lysmer, J., Udaka, T., Tsai, C. F., and Seed, H. B., FLUSH — A Computer Program for Approximate 3-D Analysis of Soil-Structure Interaction, Report No. EERC75-30, Earthquake Engineering Research Center, University of California, Berkeley, 1975.
13. Tang, Y. K., Proceedings: EPRI/NRC/TPC Workshop on Seismic Soil-Structure Interaction Analysis Techniques Using Data from Lotung Taiwan, EPRI NP-6154, Electric Power Research Institute, Palo Alto, CA, March 1989.
14. Matlock, H. and Reese, L. C., Foundation analysis of offshore pile supported structures, in *Proc. 5th Int. Conf. on Soil Mech. and Found. Eng.*, Paris, France, July 17–22, 1961.
15. Matlock, H., Correlations for design of laterally loaded piles in soft clay, in *Proc. Offshore Technology Conference*, Paper No. OTC1204, Dallas, TX, April 22–24, 1970.
16. Reese, L. C., Cox, W. R., and Koop, F. D., Analysis of laterally loaded piles in sand, in *Proc. Offshore Technology Conference*, Paper No. OTC2080, Dallas, TX, May 6–8, 1974.
17. Chang, C. Y., Tseng, W. S., Tang, Y. K., and Power, M. S., Variations of earthquake ground motions with depth and its effects on soil–structure interaction, *Proc. Second Department of Energy Natural Phenomena Hazards Mitigation Conference*, Knoxville, Tennessee, October 3–5, 1989.
18. Abrahamson, N. A., Spatial Variation of Earthquake Ground Motion for Application to Soil-Structure Interaction, Report No. TR-100463, Electric Power Research Institute, Palo Alto, CA, March 1992.
19. Gasparini, D. and Vanmarcke, E. H., SIMQKE: A program for Artificial Motion Generation, Department of Civil Engineering, Massachusetts Institute of Technology, Cambridge, 1976.
20. Silva, W. J. and Lee, K., WES RASCAL Code for Synthesizing Earthquake Ground Motions, State-of-the-Art for Assessing Earthquake Hazards in United States, Report 24, Army Engineers Waterway Experimental Station, Miscellaneous Paper 5-73-1, 1987.
21. Lilhanand, K. and Tseng, W. S., Development and application of realistic earthquake time histories compatible with multiple-damping design response spectra, in *Proc. 9th World Conference of Earthquake Engineers*, Tokyo-Kyoto, Japan, August 2–9, 1988.
22. Bolt, B. A. and Gregor, N. J., Synthesized Strong Ground Motions for the Seismic Condition Assessment of Eastern Portion of the San Francisco Bay Bridge, Report No. EERC 93-12, Earthquake Engineering Research Center, University of California, Berkeley, 1993.
23. Abrahamson, N. A., Nonstationary spectral matching, *Seismol. Res. Lett.*, 63(1), 1992.
24. Boore D. and Akinson, G., Stochastic prediction of ground motion and spectral response parameters at hard-rock sites in eastern North America, *Bull. Seismol. Soc. Am.*, 77, 440–467, 1987.
25. Sommerville, P. G. and Helmberger, D. V., Modeling earthquake ground motion at close distance, in *Proc. EPRI/Stanford/USGS Workshop on Modeling Earthquake Ground Motion at Close Distances*, August 23, 1990.

26. Papageorgiou, A. S. and Aki, K., A specific barrier model to the quantitative description of inhomogeneous faulting and the prediction of strong motion, *Bull. Seismol. Soc. Am.*, Vol. 73, 693–722, 953–978, 1983.
27. Bolt, B. A., Ed., *Seismic Strong Motion Synthetics*, Academic Press, New York, 1987.
28. U.S. Nuclear Regulatory Commission, *Standard Review Plan*, Section 3.71., Revision 2, Washington, D.C., August, 1989.
29. International Conference of Building Officials, *Uniform Building Code*, Whittier, CA, 1994.
30. Kaul, M. K., Spectrum-consistent time-history generation, *J. Eng. Mech. Div. ASCE*, 104(EM4), 781, 1978.
31. Penzien, J. and Watabe, M., Simulation of 3-dimensional earthquake ground motions, *J. Earthquake Eng. Struct. Dyn.*, 3(4), 1975.
32. Hao, H., Oliviera, C. S., and Penzien, J., Multiple-station ground motion processing and simulation based on SMART-1 array data, *Nuc. Eng. Des.*, III(6), 2229–2244, 1989.
33. Chang, C. Y., Power, M. S., Idriss, I. M., Sommerville, P. G., Silva, W., and Chen, P. C., Engineering Characterization of Ground Motion, Task II: Observation Data on Spatial Variations of Earthquake Ground Motion, NUREG/CR-3805, Vol. 3, U.S. Nuclear Regulatory Commission, Washington, D.C., 1986.
34. Abrahamson, N. A., Schneider, J. F., and Stepp, J. C., Empirical spatial coherency functions for application to soil-structure interaction analyses, *Earthquake Spectra*, 7, 1, 1991.
35. Tseng, W. S., Lilhanand, K., and Yang, M. S., Generation of multiple-station response-spectra-and-coherency-compatible earthquake ground motions for engineering applications, in *Proc. 12th Int. Conference on Struct. Mech. in Reactor Technology*, Paper No. K01/3, Stuttgart, Germany, August 15–20, 1993.
36. Vucetic, M. and Dobry, R., Effects of soil plasticity on cyclic response, *J. Geotech. Eng. ASCE*, 117(1), 89–107, 1991.
37. Sun, J. I., Golesorkhi, R., and Seed, H. B., Dynamic Moduli and Damping Ratios for Cohesive Soils, Report No. UBC/EERC-88/15, Earthquake Engineer Research Center, University of California, Berkeley, 1988.
38. Hardin, B. O. and Drnevich, V., Shear modulus and damping in soils: design equations and curves, *J. Soil Mech. Found. Div. ASCE*, 98(7), 667–691, 1972.
39. Seed, H. B., and Idriss, I. M., Soil Moduli and Damping Factors for Dynamic Response Analyses, Report No. EERC 70-10, Earthquake Engineering Research Center, University of California, Berkeley, 1970.
40. Seed, H. B., Wong, R. T., Idriss, I. M., and Tokimatsu, K., Moduli and Damping Factors for Dynamic Analyses of Cohesionless Soils, Report No. UCB/EERC-84/14, Earthquake Engineering Research Center, University of California, Berkeley, 1984.
41. Hardin, B. O. and Black, W. L., Vibration modulus of normally consolidated clay, *J. Soil Mech. Found. Div. ASCE*, 94(2), 353–369, 1968.
42. Hardin, B. O., The nature of stress-strain behavior for soils, *Proc. ASEC Geotech. Eng. Div. Specialty Conference on Earthquake Eng. and Soil Dyn.*, Vol. 1, 3-90, 1978.
43. Dickenson, S. E., Dynamic Response of Soft and Deep Cohesive Soils During the Loma Prieta earthquake of October 17, 1989, Ph.D. dissertation, University of California, Berkeley, 1994.
44. Idriss, I. M. and Sun, J. I., User's Manual for SHAKE 91, Center for Geotechnical Modeling, University of California, Davis, 1992.
45. Seed, H. B. and Idriss, I. M., Influence of soil conditions on ground motions during earthquakes, *J. Soil Mech. Found. Div.*, 95(SM1), 99–138, 1969.
46. Lee, M. K. W. and Finn, W. D. L., DESRA-2: Dynamic Effective Stress Response Analysis of Soil Deposits with Energy Transmitting Boundary Including Assessment of Liquification Potential, Report No. 38, Soil Mechanics Series, Department of Civil Engineering, University of British Columbia, Vancouver, 1978.

47. Prevost, J. H. DYNAFLOW: A Nonlinear Transient Finite Element Analysis Program, Department of Civil Engineering and Operations Research, Princeton University, Princeton, NJ, 1981; last update, 1993.
48. Prevost, J. H., DYNAID: A Computer Program for Nonlinear Seismic Site Response Analysis, Report No. NCEER-89-0025, National Center for Earthquake Engineering Research, Buffalo, New York, 1989.
49. Li, X. S., Wang, Z. L., and Shen, C. K., SUMDES, A Nonlinear Procedure for Response Analysis of Horizontally-Layered Sites, Subjected to Multi-directional Earthquake Loading, Department of Civil Engineering, University of California, Davis, 1992.
50. Apsel, R. J., Dynamic Green's Function for Layered Media and Applications to Boundary Value Problems, Ph.D. thesis, University of California, San Diego, 1979.
51. Lysmer, J., Tabatabaie-Raissai, M., Tajirian, F., Vahdani, S., and Ostadan, F., SASSI — A System for Analysis of Soil-Structure Interaction, Report No. UCB/GT/81-02, Department of Civil Engineering, University of California, Berkeley, 1981.
52. Reese, L. C., Awoshirka, K., Lam, P. H. F., and Wang, S. T., Documentation of Computer Program GROUP — Analysis of a Group of Piles Subjected to Axial and Lateral Loading, Ensoft, Inc., Austin, TX, 1990.
53. Reese, L. C. and Wang, S. T., Documentation of Computer Program LPILE — A Program for Analysis of Piles and Drilled Shafts under Lateral Loads," Ensoft, Inc., Austin, TX, 1989; latest update, Version 3.0, May 1997.
54. Reese, L. C. and Wang, S. T., Documentation of Computer Program APILE2 — Analysis of Load Vs. Settlement for an Axially Loaded Deep Foundation, Ensoft, Inc., Austin, TX, 1990.
55. Seed, H. B. and Idriss, I. M., Rock Motion Accelerograms for High-Magnitude Earthquakes, Report No. EERC 69-7, Earthquake Engineering Research Center, University of California, Berkeley, 1969.
56. Clough, R. W. and Penzien, J., *Dynamics of Structures,* 2nd ed., McGraw-Hill, New York, 1993.

43

Seismic Retrofit Technology

Kevin I. Keady
California Department of Transportation

Fadel Alameddine
California Department of Transportation

Thomas E. Sardo
California Department of Transportation

43.1 Introduction

Upon completion of the seismic analysis and the vulnerability study for existing bridges, the engineer must develop a retrofit strategy to achieve the required design criteria. Depending on the importance of structures, there are two levels of retrofit. For ordinary structures, a lower level of retrofit may be implemented. The purpose of this level of retrofit is to prevent collapse. With this level of retrofit, repairable damage is generally expected after a moderate earthquake. Following a major earthquake, extensive damage is expected and replacement of structures may be necessary. For important structures, a higher level of retrofit could be required at a considerably higher cost. The purpose of this level of retrofit is not only to prevent collapse, but also to provide serviceability after a major earthquake.

There are two basic retrofit philosophies for a concrete bridge. The first is to force plastic hinging into the columns and keep the superstructure elastic. This is desirable because columns can be more easily inspected, retrofitted, and repaired than superstructures. The second is to allow plastic hinging in the superstructure provided that ductility levels are relatively low and the vertical shear load-carrying capacity is maintained across the hinge. This is desirable when preventing hinging in the superstructure is either prohibitively expensive or not possible. In other words, this strategy is permissible provided that the hinge in the superstructure does not lead to collapse. To be conservative, the contribution of concrete should be ignored and the steel stirrups need to be sufficient to carry 1.5 times the dead-load shear reaction if hinging is allowed in the superstructure.

There are two basic retrofit philosophies for steel girder bridges. The first is to let the bearings fail and take retrofit measures to ensure that the spans do not drop off their seats and collapse. In this scenario, the bearings act as a "fuse" by failing at a relatively small seismic force and thus protecting the substructure from being subjected to any potential larger seismic force. This may be

0-8493-7434-0/00/$0.00+$.50

TABLE 43.1 Seismic Performance Criteria

Ground Motion at Site	Minimum performance level	Important bridge performance level
Functional Evaluation	Immediate service level repairable damage	Immediate service level minimal damage
Safety Evaluation	Limited service level significant damage	Limited service level repairable damage

the preferred strategy if the fusing force is low enough such that the substructure can survive with little or no retrofit. The second philosophy is to make sure that the bearings do not fail. It implies that the bearings transfer the full seismic force to the substructure and retrofitting the substructure may be required. The substructure retrofit includes the bent caps, columns or pier walls, and foundations. In both philosophies, a superstructure retrofit is generally required, although the extent is typically greater with the fixed bearing scheme.

The purpose of this chapter is to identify potential vulnerabilities to bridge components and suggest practical retrofit solutions. For each bridge component, the potential vulnerabilities will be introduced and retrofit concepts will be presented along with specific design considerations.

43.2 Analysis Techniques for Bridge Retrofit

For ordinary bridges, a dynamic modal response spectrum analysis is usually performed under the input earthquake loading. The modal responses are combined using the complete quadratic combination (CQC) method. The resulting orthogonal responses are then combined using the 30% rule. Two cases are considered when combining orthogonal seismic forces. Case 1 is the sum of forces due to transverse loading plus 30% of forces due to longitudinal loading. Case 2 is the sum of forces due to longitudinal loading plus 30% of forces due to transverse loading.

A proper analysis should consider abutment springs and trusslike restrainer elements. The soil foundation structure interaction should be considered when deemed important. Effective properties of all members should be used. Typically, two dynamic models are utilized to bound the assumed nonlinear response of the bridge: a "tension model" and a "compression model." As the bridge opens up at its joints, it pulls on the restrainers. In contrast, as the bridge closes up at its joints, its superstructure elements go into compression.

For more important bridges, a nonlinear time history analysis is often required. This analysis can be of uniform support excitation or of multiple supports excitation depending on the length of the bridge and the variability of the subsurface condition.

The input earthquake loading depends on the type of evaluation that is considered for the subject bridge. Table 43.1 shows the seismic performance criteria for the design and evaluation of bridges developed by the California Department of Transportation [1]. The safety evaluation response spectrum is obtained using:

1. Deterministic ground motion assessment using maximum credible earthquake, or
2. Probabilistically assessed ground motion with a long return period.

The functional evaluation response spectrum is derived using probabilistically assessed ground motions which have a 60% probability of not being exceeded during the useful life of the bridge. A separate functional evaluation is usually required only for important bridges.

With the above-prescribed input earthquake loading and using an elastic dynamic multimodal response spectrum analysis, the displacement demand can be computed. The displacement capacity of various bents may be then calculated using two-dimensional or three-dimensional nonlinear static push-over analysis with strain limits associated with expected damage at plastic hinge locations. When performing a push-over analysis, a concrete stress–strain model that considers effects of transverse confinement, such as Mander's model, and a steel stress–strain curve are used for considering material nonlinearity [2]. Limiting the concrete compressive strain to a magnitude smaller than the confined concrete ultimate compressive strain and the steel strain to a magnitude

TABLE 43.2 Strain Limits

	Significant Damage	Repairable Damage	Minimal Damage
Concrete strain limit ε_c	ε_{cu}	2/3 * ε_{cu}	The greater of 1/3 * ε_{cu} or 0.004
Grade 430 bar #29 to #57 steel strain limit ε_s	0.09	0.06	0.03
Grade 280 bar #29 to #57 steel strain limit ε_s	0.12	0.08	0.03
Grade 430 bar #10 to #25 steel strain limit ε_s	0.12	0.08	0.03
Grade 280 bar #10 to #25 steel strain limit ε_s	0.16	0.10	0.03

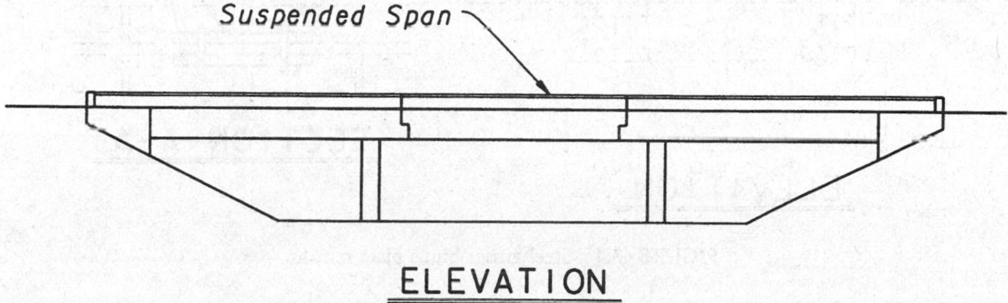

FIGURE 43.1 Suspended span.

smaller than steel rupture strain results in lesser curvature of the cross section under consideration. Smaller curvatures are usually associated with smaller cracks in the plastic hinge region.

Table 43.2 shows the general guidelines on strain limits that can be considered for a target level of damage in a plastic hinge zone. These limits are applied for the ultimate concrete strain ε_{cu} and the ultimate strain in the reinforcing steel ε_{su}. The ultimate concrete strain can be computed using a concrete confinement model such as Mander's model. It can be seen that for a poorly confined section, the difference between minimal damage and significant damage becomes insignificant. With displacement demands and displacement capacities established, the demand-to-capacity ratios can be computed showing adequacy or inadequacy of the subject bridge.

43.3 Superstructure Retrofits

Superstructures can be categorized into two different categories: concrete and steel. After the 1971 San Fernando, California earthquake, the primary failure leading to collapse was identified as unseating of superstructures at the expansion joints and abutments, a problem shared by both types of superstructures. Other potential problems that may exist with steel superstructures are weak cross-bracing and/or diaphragms. Concrete bridge superstructures have the potential to form plastic hinges during a longitudinal seismic response which is largely dependent upon the amount of reinforcement used and the way it is detailed.

43.3.1 Expansion Joints and Hinges

During an earthquake, adjacent bridge frames will often vibrate out of phase, causing two types of displacement problems. The first type is a localized damage caused by the frames pounding together at the hinges. Generally, this localized damage will not cause bridge collapse and is therefore not a major concern. The second type occurs when the hinge joint separates, possibly allowing the adjacent spans to become unseated if the movement is too large. Suspended spans (i.e., two hinges within one span) are especially vulnerable to becoming unseated (Figure 43.1).

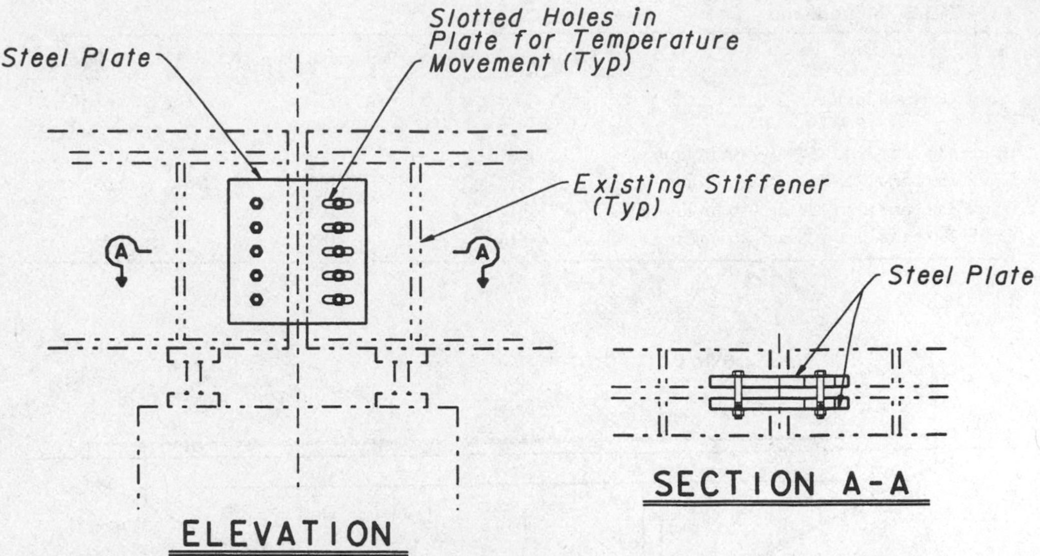

FIGURE 43.2 Steel girder hinge plate retrofit.

43.3.1.1 Simply Supported Girders

The most common problem for simply supported structures is girders falling off their seats due to a longitudinal response. If the seismic force on the structure is large enough to fail the bearings, then the superstructure becomes vulnerable to unseating at the supports.

There are several ways of retrofitting simply supported steel girders and/or precast concrete girders. The most common and traditional way is to use cable restrainers, since the theory is fundamentally the same for both types of girders. For more about cable restrainers, refer to Section 43.3.1.3. Care should be taken when designing the cables to intrude as little as possible on the vertical clearance between the girders and the roadway. Note that the cable retrofit solution for simply supported girders can be combined with a cap seat extension if expected longitudinal displacements are larger than the available seat width.

Another possible solution for steel girders is to make the girders continuous over the bents by tying the webs together with splice plates (Figure 43.2). The splice plate should be designed to support factored dead-load shears assuming the girder becomes unseated. The splice plate is bolted to the girder webs and has slotted or oversized holes to allow for temperature movement. This retrofit solution will usually work for most regular and straight structures, but not for most irregular structures. Any situation where the opposing girders do not line up will not work. For example, bridges that vary in width or are bifurcated may have different numbers of girders on opposite sides of the hinge. Bridges that are curved may have the girders at the same location but are kinked with respect to each other. In addition, many structures may have physical restrictions such as utilities, bracing, diaphragms, stiffeners, etc. which need to be relocated in order for this strategy to work.

43.3.1.2 Continuous Girders with In-Span Hinges

For continuous steel girders, the hinges are typically placed near the point of zero moment which is roughly at 20% of the span length. These hinges can be either seat type as shown in Figure 43.3 or hanger type as shown in Figure 43.4. The hanger-type hinges are designed for vertical dead and live loads. These loads are typically larger than forces that can be imparted onto the hanger bar from a longitudinal earthquake event, and thus retrofitting the hanger bar is generally unnecessary. Hanger-type hinges typically have more seismic resistance than seat-type hinges but may still be subjected to seismic damage. Hanger bars are tension members that are vulnerable to differential

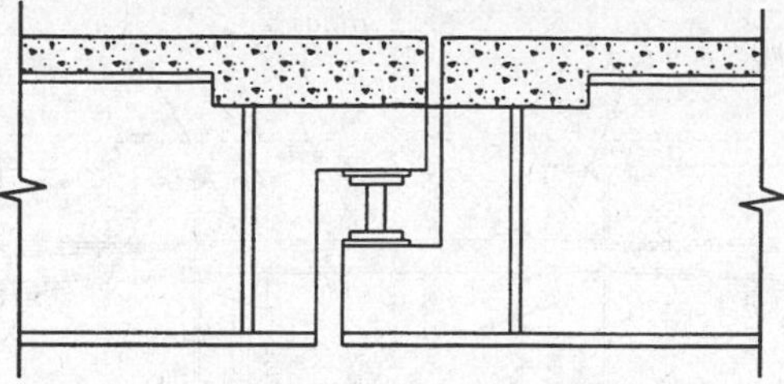

FIGURE 43.3 Seat-type hinge.

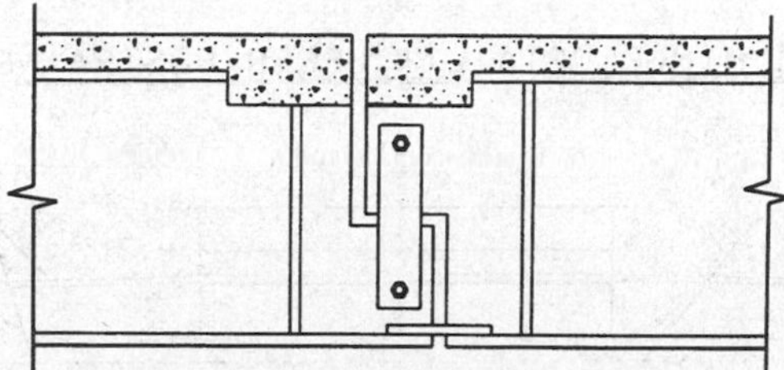

FIGURE 43.4 Hanger-type hinge.

transverse displacement on either side of the hinge. The differential displacement between the girders causes the hanger bars to go into bending plus tension. These hinges often have steel bars or angles that bear against the opposite web, or lugs attached to the flanges, which were designed to keep the girders aligned transversely for wind forces. These devices are usually structurally inadequate and are too short to be effective with even moderate seismic shaking. Consideration should be given to replacing them or adding supplemental transverse restrainers [3]. Cross-bracing or diaphragms on both sides of the hinge may have to be improved in conjunction with the transverse restrainer.

It can generally be assumed that any seat-type hinge used with steel girders will need additional transverse, longitudinal, and vertical restraint in even moderately severe seismic areas [3].

Continuous concrete box girders typically have in-span-type hinges. These hinge seats were typically 150 to 200 mm, while some were even less on many of the older bridges. Because of the localized damage that occurs at hinges (i.e., spalling of the concrete, etc.), the actual length of hinge seat available is much less than the original design. Therefore, a means of providing a larger hinge seat and/or tying the frames together is necessary.

43.3.1.3 Restrainers

Restrainers are used to tie the frames together, limiting the relative displacements from frame to frame and providing a load path across the joint. The main purpose is to prevent the frames from falling off their supports. There are two basic types of restrainers, cables and rods. The choice between cables and rods is rather arbitrary, but some factors to consider may be structure period, flexibility, strength of hinge/bent diaphragm, tensile capacity of the superstructure, and, to some degree, the geometry of the superstructure.

There are various types of longitudinal cable restraining devices as shown in Figures 43.5 to 43.7. Cable restraining units, such as the ones shown in Figures 43.5 and 43.6 generally have an advantage over high-strength rods because of the flexibility with its usage for varying types of superstructures. For simply supported girders, cables may be anchored to a bracket mounted to the underside of the girder flange and wrapped around the bent cap and again anchored as shown in Figure 43.8. This is the preferred method. Another possibility is to have the cables anchored to a bracket mounted to the bottom flange and simply attached to an opposing bracket on the other side of the hinge, as

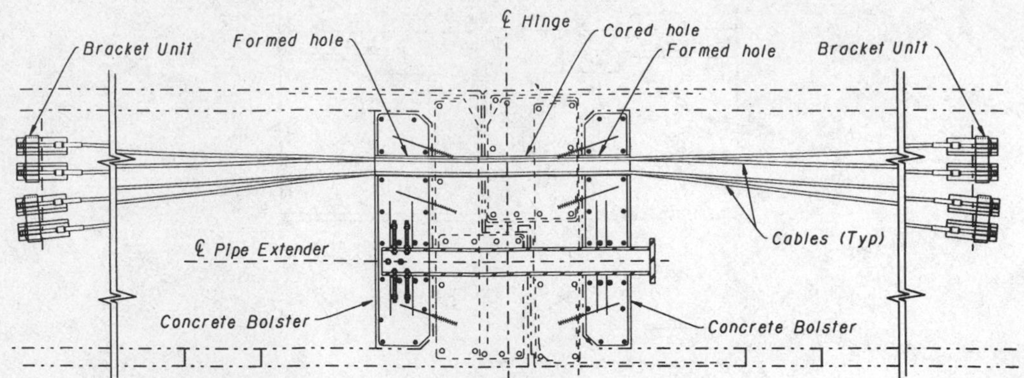

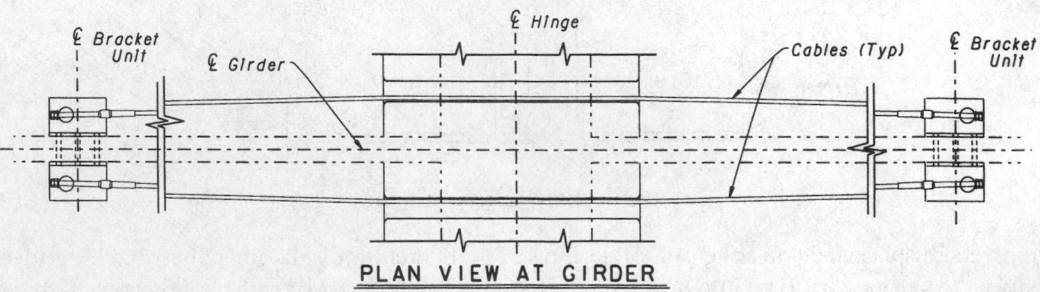

FIGURE 43.5 Type 1 hinge restrainer.

shown in Figure 43.9. The latter example is generally used for shorter bridges with a larger seat area at the bent cap or in situations where vertical clearances may be limited. Moreover, cables have the advantage of using a variety of lengths, since the anchorage devices can be mounted anywhere along the girders, whether it be steel or concrete, in addition to being anchored to the nearest bent cap or opposite side of the hinge diaphragm. For example, if the restrainer is relatively short, this may shorten the period, possibly increasing the demand to the adjacent bridge frames. Therefore, for this example, it may be desirable to lengthen the restrainer keeping the force levels to within the capacities of the adjacent segments [4]. On the other hand, if the restrainer is too long, unseating can thus occur, and an additional means of extending the seat length becomes necessary.

High-strength rods are another option for restricting the longitudinal displacements and can be used with short seats without the need for seat extenders. Unlike cables, when high-strength rods are used, shear keys or pipes are generally used in conjunction since rods can be sheared with transverse movements at hinge joints. Geometry may be a limiting factor when using high-strength rods. For example, if a box-girder bridge is shallow, it may not be possible to install a long rod through a narrow access opening. For both cables and rods, the designer needs to consider symmetry when locating restraining devices.

43.3.1.4 Pipe Seat Extenders

When a longer restraining device is preferred, increased longitudinal displacements will result and may cause unseating. It is therefore necessary to incorporate pipe seat extenders to be used in conjunction with longer restrainers when unseating will result. A 200 mm (8 in.), XX strong pipe is used for the pipe seat extender which is placed in 250 mm (10 in.) cored (and formed) hole (Figure 43.10). A 250-mm cored hole allows vertical jacking if elastomeric pads are present and replacement is required after the pad fails. Longitudinal restraining devices (namely, cables and rods) must be strain compatible with the seismic deflections imposed upon the hinge joint. In other

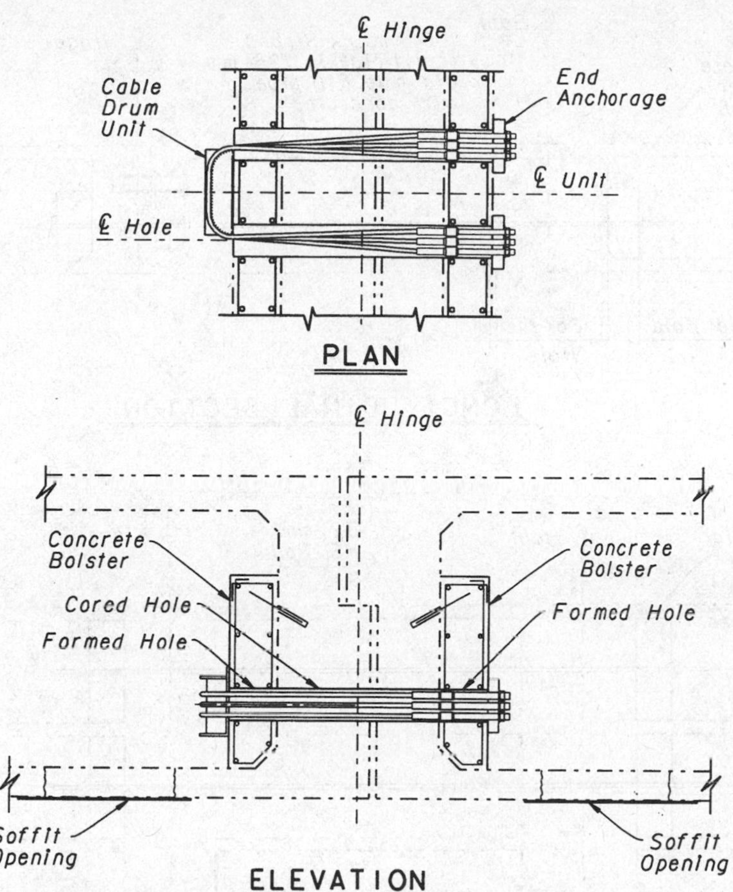

FIGURE 43.6 C-1 hinge restrainer.

words, if the longitudinal restrainers were too short, the device would have yielded long before the pipe seat extender was mobilized, deeming the longitudinal restrainers useless. To limit the number of cored holes in the diaphragm, a detail has been developed to place the restrainer cables through the pipes as shown in Figure 43.11. The pipes are not only used for vertical load-carrying capacity, but can also be used successfully as transverse shear keys.

43.3.2 Steel Bracing and Diaphragms

Lateral stiffening between steel girders typically consists of some type of cross-bracing system or channel diaphragm. These lateral bracing systems are usually designed to resist wind loads, construction loads, centrifugal force from live loads and seismic loads. The seismic loads prescribed by older codes were a fraction of current code seismic loads and, in some cases, may not have controlled bracing design. In fact, in many cases, the lateral bracing system is not able to withstand the "fusing" forces of the bearing capacity and/or shear key capacity. As a result, bracing systems may tend to buckle and, if channel diaphragms are not full depth of the girder, the webs could cripple. In general, the ideal solution is to add additional sets of bracing, stiffeners, and/or full-depth channel diaphragms as close to the bearings as physically possible.

Retrofit solutions chosen will depend on space restrictions. New bracing or diaphragms must be placed to not interfere with existing bracing, utilities, stiffeners, cable restrainers and to leave enough access for maintenance engineers to inspect the bearings. Skewed bents further complicate space

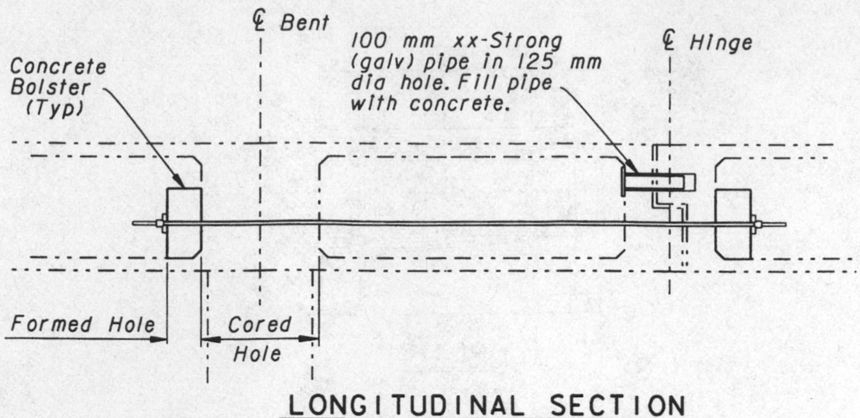

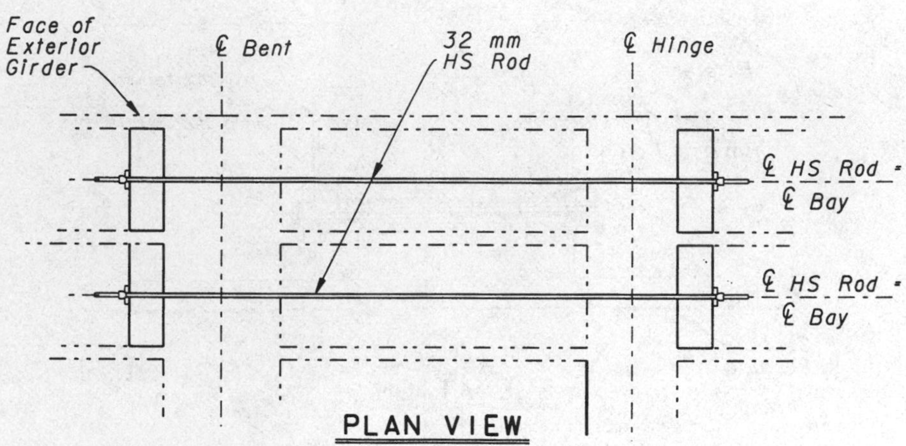

FIGURE 43.7 HS rod restrainer.

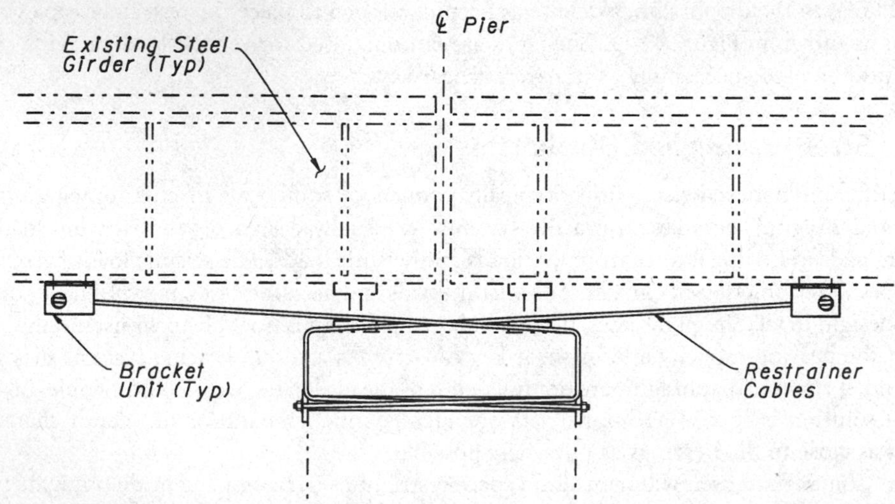

FIGURE 43.8 Cable restrainer through bent cap.

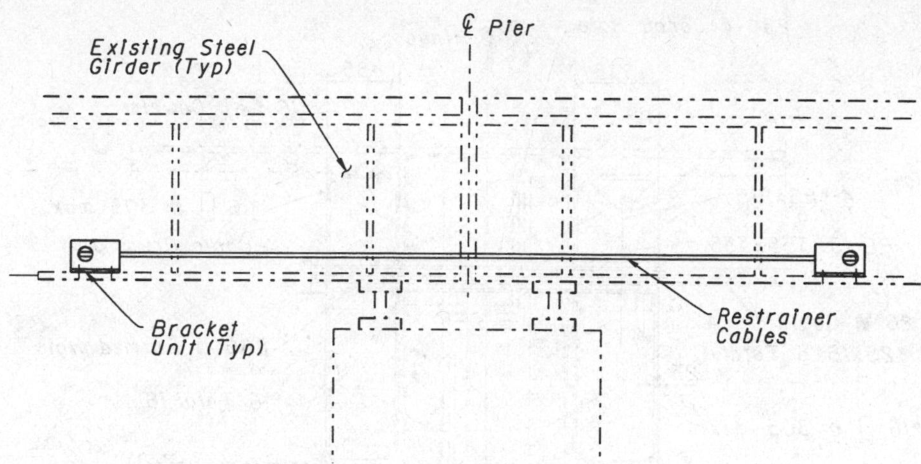

FIGURE 43.9 Girder-to-girder cable restrainer.

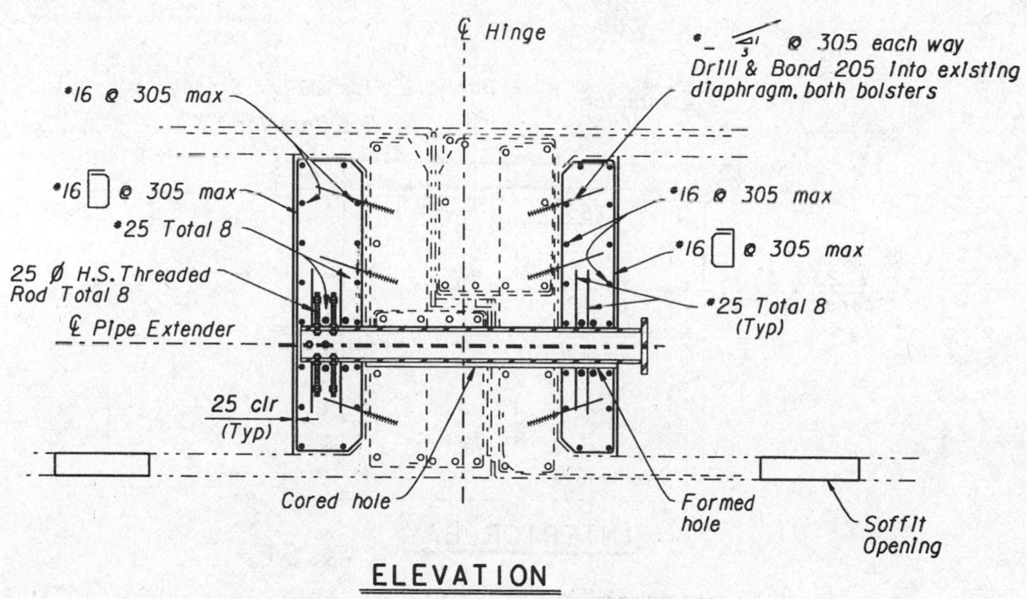

FIGURE 43.10 Pipe seat extender.

restrictions. When choosing a retrofit solution, the engineer must keep in mind that the retrofit will be constructed while the structure is carrying traffic. Stresses in a bracing member tend to cycle under live load which makes it difficult for the engineer to assess actual member stresses. As a result, any retrofit solution that requires removing and replacing existing members is not recommended. In addition, careful consideration should be given to bolted vs. welded connections. As previously mentioned, structures are under live loads during the retrofit operation and thus members are subjected to cyclic stresses. It may be difficult to achieve a good-quality weld when connecting to a constantly moving member. If bolted connections are used, preference should be given to end-bearing connections over friction connections. Friction connections have more stringent surface preparation requirements which are difficult and expensive to achieve in the field, as is the case with lead-based paint.

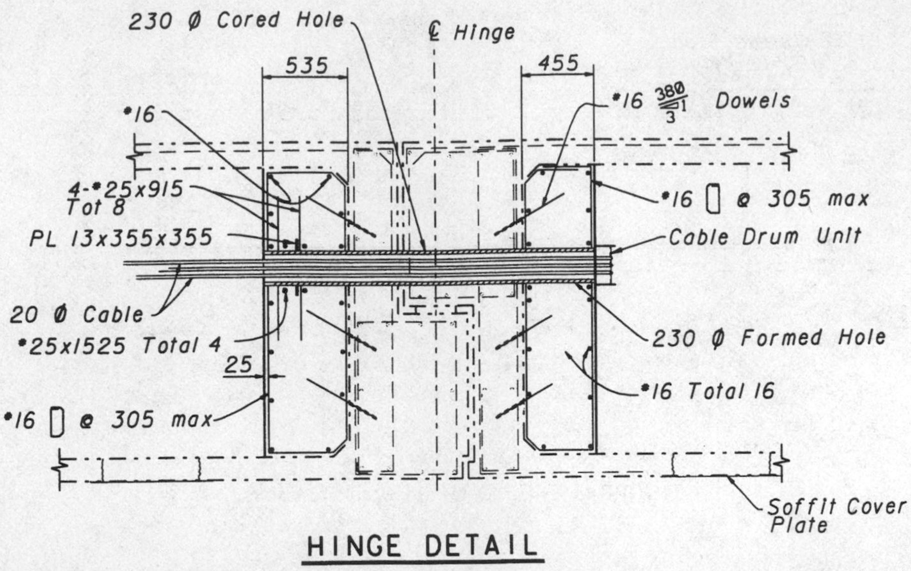

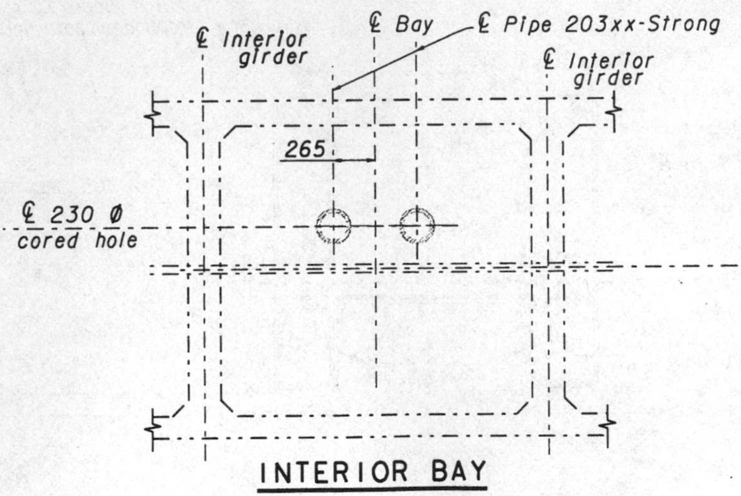

FIGURE 43.11 Restrainers/pipe seat extender.

43.3.3 Concrete Edge Beams

Edge beams are used to enhance the longitudinal capacity of a concrete bridge. These beams link consecutive bents together outside the existing box structure. In the United States, edge beams have been used to retrofit double-deck structures with long outriggers.

In a single-level bridge structure, outriggers are vulnerable in torsion under longitudinal excitation. Two retrofit alternatives are possible. The first alternative is to strengthen the outrigger cap while maintaining torsional and flexural fixity to the top of the column. The second alternative is to pin the top of the column; thus reducing the torsional demand on the vulnerable outrigger cap. Using the second alternative, the column bottom fixity needs to be ensured by means of a full footing retrofit.

In a double-deck structure, pinning the connection between the lower level outrigger cap and the column is not possible since fixity at that location is needed to provide lateral support to the upper deck. In situations where the lower deck is supported on a long outrigger cap, the torsional softening of that outrigger may lead to loss of lateral restraint for the upper-deck column. This weakness can be remedied by using edge beams to provide longitudinal lateral restraint. The edge beams need to be stiff and strong enough to ensure plastic hinging in the column and reduce torsional demand on the lower-deck outrigger cap.

43.4 Substructure Retrofits

Most earthquake damage to bridge structures occurs at the substructure. There are many types of retrofit schemes to increase the seismic capacity of existing bridges and no one scheme is necessarily more correct than another. One type of retrofit scheme may be to encase the columns and add an overlay to footings. Another might be to attract forces into the abutments and out of the columns and footings. Listed below are some different concepts to increase the capacities of individual members of the substructure.

43.4.1 Concrete Columns

Bridge columns constructed in the United States prior to 1971 are generally deficient in shear, flexure, and/or lateral confinement. Stirrups used were typically #13 bars spaced at 300 mm on center (#4 bars at 12 in.) for the entire column length including the regions of potential plastic hinging. Typically, the footings were constructed with footing dowels, or starter bars, with the longitudinal column reinforcement lapped onto the dowels. As the force levels in the column approach yield, the lap splice begins to slip. At the onset of yielding, the lap splice degrades into a pin-type condition and within the first few cycles of inelastic bending, the load-carrying capacity degrades significantly. This condition can be used to allow a "pin" to form and avoid costly footing repairs. Various methods have been successfully used to both enhance the shear capacity and ductility by increasing the lateral confinement of the plastic hinge zone for bridge columns with poor transverse reinforcement details. Following is a list of these different types and their advantages and/or disadvantages.

43.4.1.1 Column Casings

The theory behind any of the column casing types listed below is to enhance the ductility, shear, and/or flexural capacity of an existing reinforced concrete column and, in some cases, to limit the radial dilating strain within the plastic hinge zone. Because of the lap splice detail employed in older columns, one of the issues facing column retrofitting is to maintain fixity at the column base. The lateral confining pressure developed by the casing is capable of limiting the radial dilating strain of the column, enough to "clamp" the splice together, preventing any slippage from occurring. Tests have shown that limiting the radial dilating strains to less than 0.001, the lap splice will remain fixed and is capable of developing the full plastic moment capacity of the section [5]. Contrary to limiting the radial strains is to permit these strains to take place (i.e., radial dilating strains greater than 0.001) allowing a "pin" to form while providing adequate confinement throughout the plastic hinge region.

Steel Casings

There are three types of retrofit schemes that are currently employed to correct the problems of existing columns through the use of steel jacketing. The first type is typically known as a class F type of column casing retrofit, as shown in Figure 43.12. This type of casing is fully grouted and is placed the full height of the column. It is primarily used for a column that is deficient in shear and flexure. It will limit the radial dilating strain to less than 0.001, effectively fixing the lap splice from

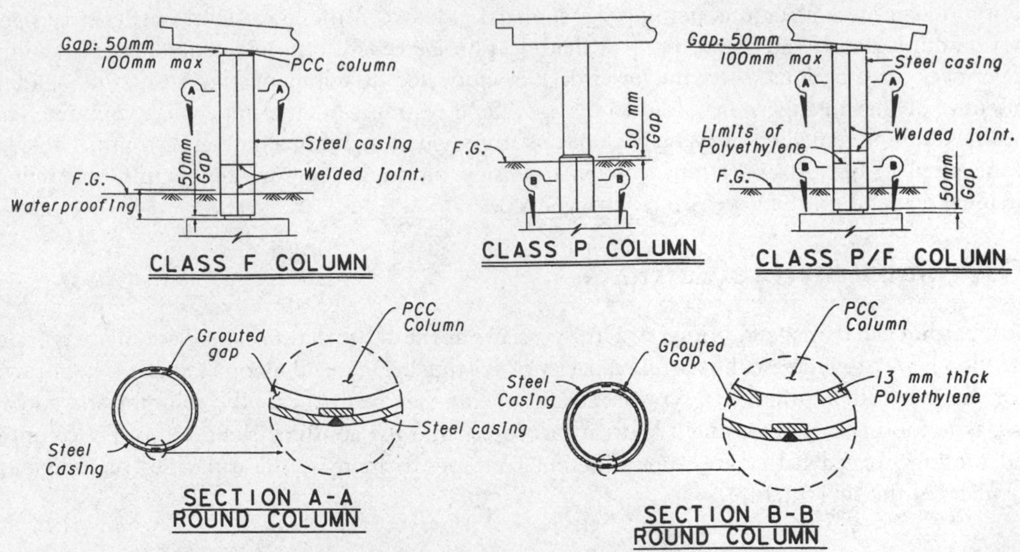

FIGURE 43.12 Steel column casings.

slipping. The lateral confining pressure for design is taken as 2.068 MPa (300 psi) and, when calculated for a 13-mm-thick steel casing with A36 steel, is equivalent to a #25 bar (#8 bar) at a spacing of about 38 mm. It can therefore be seen that the confinement, as well as the shear, is greatly enhanced. The allowable displacement ductility ratio for a class F column casing is typically 6, allowing up to 8 in isolated locations. It has been tested well beyond this range; however, the allowable ductility ratio is reduced to prohibit fracturing of the longitudinal bars and also to limit the level of load to the footing.

The second type of casing is a class P type and is only a partial height casing; therefore, it does not help a column that is deficient in shear. As can be seen from Figure 43.12, the main difference between a class F and a class P type column casing is the layer of polyethylene between the grout and the column. This will permit the column to dilate outward, allowing the strain to exceed 0.001, forming a pin at the base of the column. It should be noted that the casing is still required in this condition to aid in confining the column. The limit of this type of retrofit is typically taken as 1.5 times the columns diameter or to where the maximum moment has decreased to 75%.

The third type of steel column casing is a combination of the first two and, hence, is known as a class P/F, also shown in Figure 43.12. It is used like a class P casing, but for a column with a shear deficiency. All of these casings can be circular (for circular or square columns) or oblong (for rectangular columns). If a rectangular column is deficient in shear only, it is sometimes permitted to use flat steel plates if the horizontal clearances are limited and it is not possible to fit an oblong column in place. For aesthetic purposes, the class P casing may be extended to full height in highly visible areas. It can be unsightly if an oblong casing is only partial height. It is important to mention that the purpose of the 50-mm gap at the ends of the column is to prevent the casing from bearing against the supporting member acting as compression reinforcement, increasing the flexural capacity of the column. This would potentially increase the moments and shears into the footing and/or bent cap under large seismic loads.

Concrete Casings

When retrofitting an unusually shaped column without changing the aesthetic features of the geometry of the column, a concrete casing may be considered as an alternative. Existing columns are retrofitted by placing hoops around the outer portion of the column and then drilling and bonding bars into the column to enclose the hoops. The reinforcement is then encased with concrete,

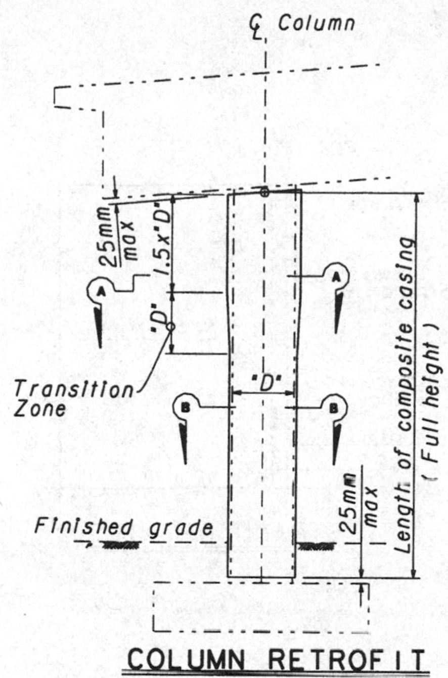

FIBER VOLUME = 35% min

E-GLASS		
ROUND COLUMN		
COLUMN DIA	t1(min)	t2(min)
305mm	5mm	5mm
610mm	9mm	5mm
915mm	13mm	7mm
1220mm	17mm	9mm
1525mm	22mm	11mm
1830mm	25mm	13mm

Note A:
Epoxy Resin-Glass Composite

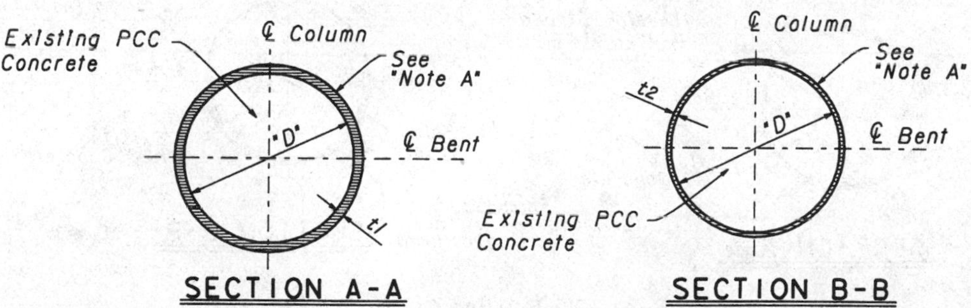

FIGURE 43.13 Advanced composite column casing.

thus maintaining the original shape of the column. The design of a concrete jacket follows the requirements of a new column. Although this method increases the shear and flexural capacities of the column and provides additional confinement without sacrificing aesthetics, it is labor intensive and therefore can be costly.

Advanced Composite Casings

Recently, there has been significant research and development using advanced composites in bridge design and retrofit. Similar to steel casings, advanced composite casings increase the confinement and shear capacity of existing concrete bridge columns. This type of column retrofit has proved to be competitive with steel casings when enhancing column shear capacity and may also provide an economic means of strengthening bridge columns (Figure 43.13). However, currently, composites are not economic when limiting lap splice slippage inside expected plastic hinge zones. The advantage of using some types of composite casings is that the material can be wrapped to the column without changing its geometric shape. This is important when aesthetics are important or lateral clearances at roadways are limited.

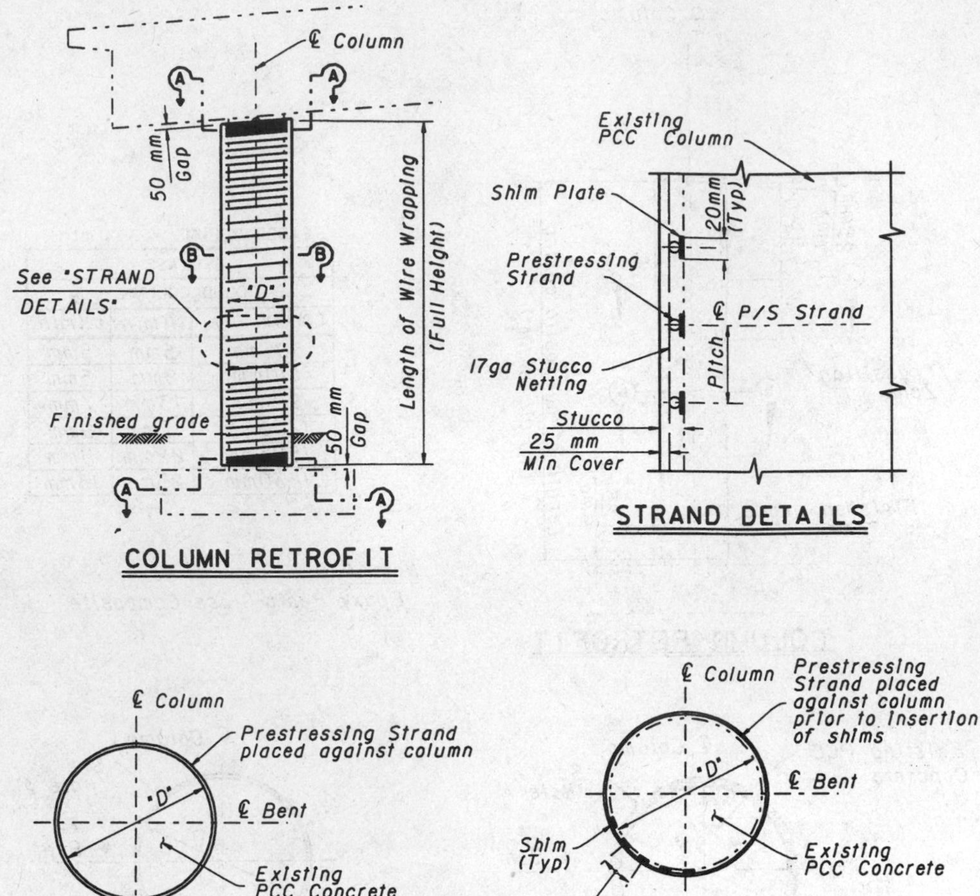

FIGURE 43.14 Column wire wrap.

Wire Wrap Casings

Another type of system that was recently approved by the California Department of Transportation is a "wire wrap" system. It consists of a prestressing strand hand wrapped onto a column; then wedges are placed between the strand and the column, effectively prestressing the strand and actively confining the column, as shown in Figure 43.14. The advantage of this type of system is that, like the advanced composites, it can be wrapped to any column without changing the geometric shape. Its basic disadvantage is that it is labor intensive and currently can only be applied to circular columns.

43.4.1.2 In-Fill Walls

Reinforced concrete in-fill walls may also be used as an alternative for multicolumn bridge bents, as shown in Figure 43.15. This has two distinct advantages: it increases the capacity of the columns in the transverse direction and limits the transverse displacements. By limiting the displacements

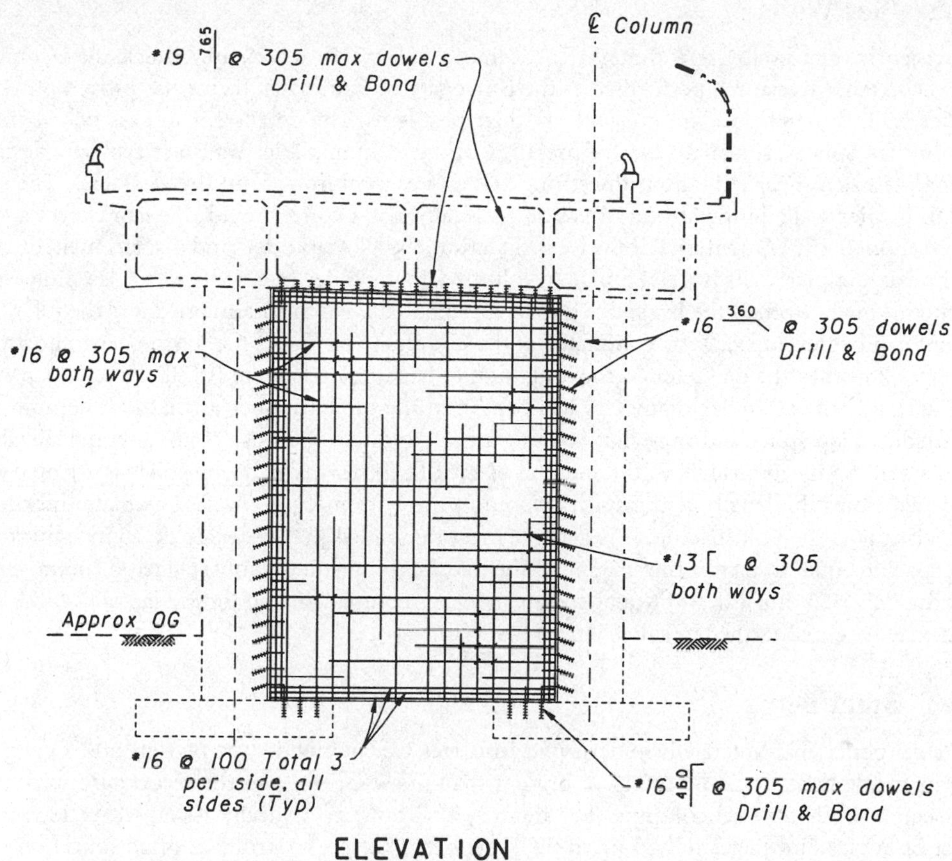

FIGURE 43.15 In-fill wall.

transversely, the potential for plastic hinge formation in the bent cap is eliminated. Therefore, cost may prove to be less than some other retrofit alternatives mentioned earlier. It is important to note that the in-fill wall is not effective in the longitudinal capacity of bridge bents with little or no skew.

43.4.1.3 Link Beams

Link beams are used to enhance the transverse capacity of a concrete bent. The placement of a link beam over a certain height above ground level determines its function.

A link beam can be placed just below the soffit level and acts as a substitute to a deficient existing bent cap. The main function of this kind of a link beam is to protect the existing superstructure and force hinging in the column.

In other cases, a link beam is placed somewhere between the ground level and the soffit level in order to tune the transverse stiffness of a particular bent. This type of retrofit can be encountered in situations where a box superstructure is supported on bents with drastically unequal stiffnesses. In this case, the center of mass and the center of rigidity of the structure are farther apart. This eccentricity causes additional displacement on the outer bents which can lead to severe concentrated ductility demands on just a few bents of the subject bridge. This behavior is not commonly preferred in seismically resistant structures and the use of link beams in this case can reduce the eccentricity between the center of mass and the center of rigidity. This structural tuning is important in equally distributing and reducing the net ductility demands on all the columns of the retrofitted bridge.

43.4.2 Pier Walls

Until recently, pier walls were thought to be more vulnerable to seismic attack than columns. However, extensive research performed at the University of California, Irvine has proved otherwise [6]. Pier walls, nevertheless, are not without their problems. The details encompass poor confinement and lap splices, similar to that of pre-1971 bridge columns. Pier walls are typically designed and analyzed as a shear wall about the strong axis and as a column about the weak axis. The shear strength of pier walls in the strong direction is usually not a concern and one can expect a shear stress of about 0.25 $\sqrt{f_c'}$ (MPa). For the weak direction, the allowable demand displacement ductility ratio in existing pier walls is 4.0. Similar to columns, many older pier walls were also built with a lap splice at the bottom. If the lap splice is long enough, fixity will be maintained and the full plastic moment can be developed. Tests conducted at the University of California, Irvine, have shown that lap splices 28 times the diameter of the longitudinal bar to be adequate [6]. However, a lap splice detail with as little as 16 bar diameters will behave in the same manner as that of a column with an inadequate lap splice and may slip, forming a pin condition. Because of the inherent flexibility of a pier wall about its weak axis, the method of retrofit for this type of lap splice is a plate with a height two times the length of the splice, placed at the bottom of the wall. The plate thickness is not as critical as the bolt spacing. It is generally recommended that the plate be 25 mm thick with a bolt spacing equal to that of the spacing of the main reinforcement, only staggered (not to exceed approximately 355 mm). If additional confinement is required for the longer lap splice, the plate height may be equal to the lap splice length.

43.4.3 Steel Bents

Most steel bents encountered in older typical bridges can be divided in two groups. One group contains trestle bents typically found in bridges spanning canyons, and the second group contains open-section built-up steel columns. The built-up columns are typically I-shaped sections which consist of angles and plates bolted or riveted together. The second group is often found on small bridges or elevated viaducts.

Trestle steel bents are commonly supported on pedestals resting on rock or relatively dense foundation. In general, the truss members in these bents have very large slenderness ratios which lead to very early elastic buckling under low-magnitude earthquake loading. Retrofitting of this type of bent consists basically of balancing between member strengthening and enhancing the tensile capacity of the foundation and keeping connection capacities larger than member capacities. In many situations, foundation retrofit is not needed where bent height is not large and a stable rocking behavior of the bent can be achieved. Strengthening of the members can be obtained by increasing the cross-sectional area of the truss members or reducing the unsupported length of the members.

Figure 43.16 shows the retrofit of Castro Canyon bridge in Monterey County, California. The bent retrofit consists of member strengthening and the addition of a reinforced concrete block around the bent-to-pedestal connection. In this bridge, the pedestals were deeply embedded in the soil which added to the uplift capacity of the foundation.

For very tall trestle bents (i.e., 30 m high), foundation tie-downs, in addition to member strengthening, might be needed in order to sustain large overturning moments. Anchor bolts for base plates supported on top of pedestals are usually deficient. Replacement of these older bolts with high-strength bolts or the addition of new bolts can be done to ensure an adequate connection capable of developing tension and shear strength. The addition of new bolts can be achieved by coring through the existing base plate and pedestal or by enlarging the pedestal with a concrete jacket surrounding the perimeter bolts. The use of sleeved anchor bolts is desirable to induce some flexibility into the base connection.

The second group of steel bents contains open section built-up columns. These members may fail because of yielding, local buckling, or lateral torsional buckling. For members containing a single I-shaped section, lateral torsional buckling typically governs. Retrofit of this type of column

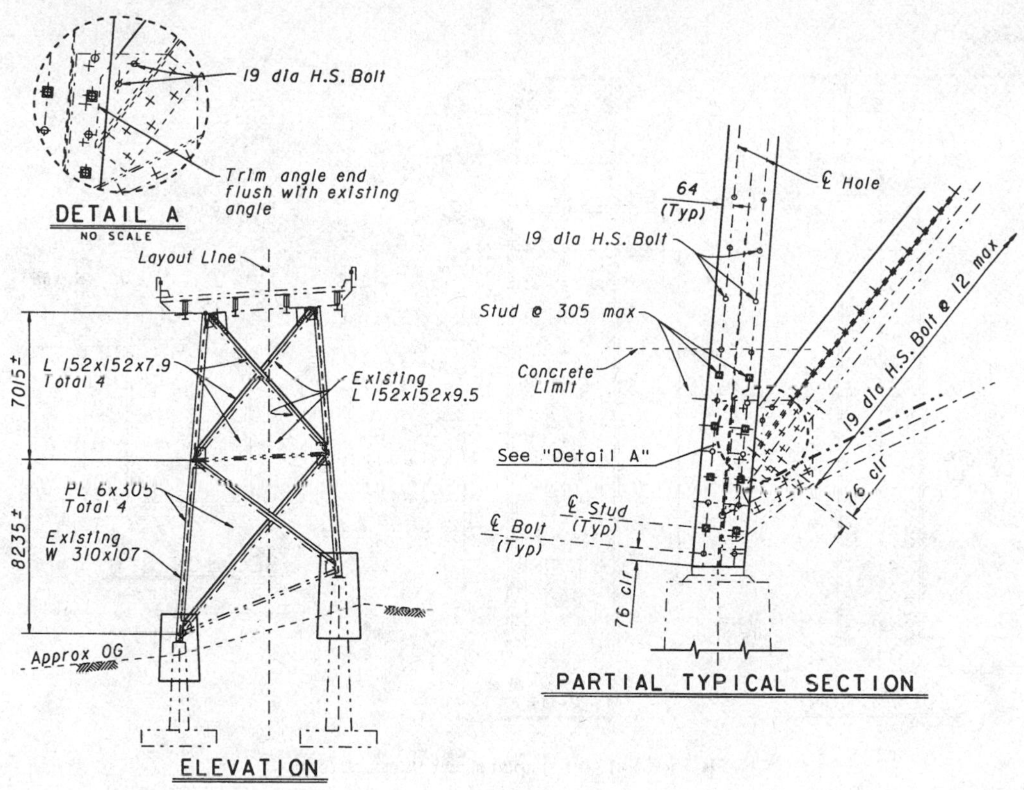

FIGURE 43.16 Trestle bent retrofit.

consists of enclosing the section by bolting channel sections to the flanges. Figure 43.17 shows this type of retrofit. Installation of these channels is made possible by providing access through slotted holes. These holes are later covered by tack welding plates or left open. For larger members with an open section as seen in Figure 43.18, retrofit consists of altering the existing cross section to a multicell box section.

The seismic behavior of the multicell box is quite superior to an open section of a single box. These advantages include better torsional resistance and a more ductile postelastic behavior. In a multicell box, the outside plates sustain the largest deformation. This permits the inside plates to remain elastic in order to carry the gravity load and prevent collapse during an earthquake. To maintain an adequate load path, the column base connection and the supporting foundation should be retrofitted to ensure the development of the plastic hinge just above the base connection. This requires providing a grillage to the column base as seen in Figure 43.18 and a footing retrofit to ensure complete load path.

43.4.4 Bearing Retrofit

Bridge bearings have historically been one of the most vulnerable components in resisting earthquakes. Steel rocker bearings in particular have performed poorly and have been damaged by relatively minor seismic shaking. Replacement of any type of bearing should be considered if failure will result in collapse of the superstructure. Bearing retrofits generally consist of replacing steel rocker-type bearings with elastomeric bearings. In some cases, where a higher level of serviceability is required, base isolation bearings may be used as a replacement for steel bearings. For more information on base isolation, see the detailed discussion in Chapter 41. Elastomeric bearings are preferred over steel rockers because the bridge deck will only settle a small amount when the bearings

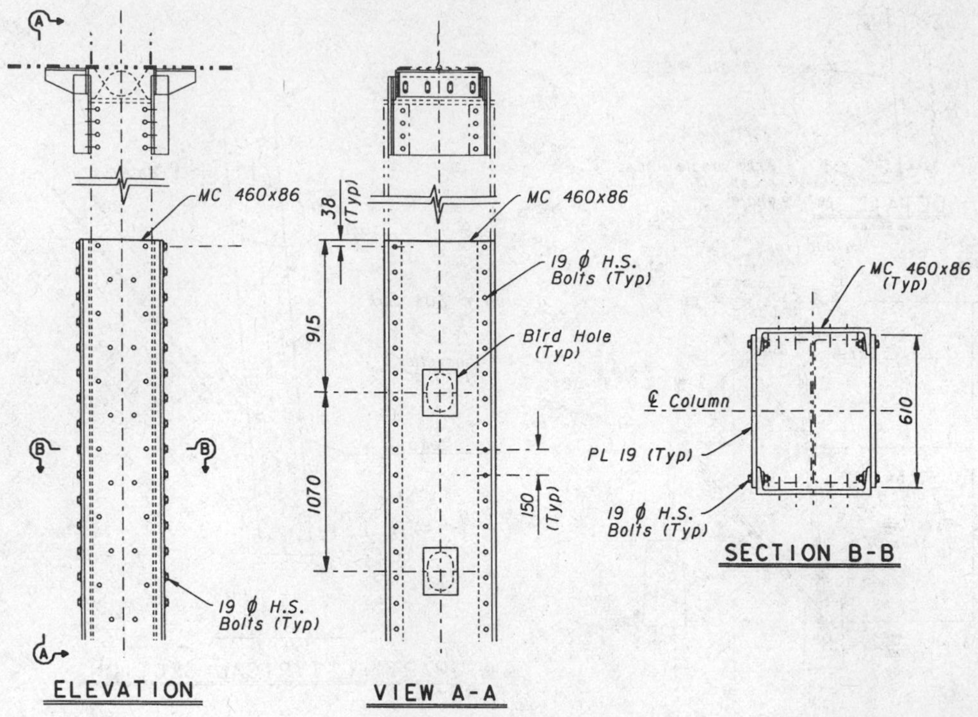

FIGURE 43.17 I-shaped steel column retrofit.

fail, whereas the deck will settle several inches when rockers fail. Elastomeric bearings also have more of a base isolation effect than steel rockers. Both types of bearings may need catchers, seat extenders, or some other means of providing additional support to prevent the loss of a span. Although elastomeric bearings perform better than steel rockers, it is usually acceptable to leave the existing rockers in place since bearings replacement is more expensive than installing catchers to prevent collapse during an earthquake.

43.4.5 Shear Key Retrofit

The engineer needs to consider the ramifications of a shear key retrofit. The as-built shear keys may have been designed to "fuse" at a certain force level. This fusing will limit the amount of force transmitted to the substructure. Thus, if the shear keys are retrofitted and designed to be strong enough to develop the plastic capacity of the substructure, this may require a more expensive substructure retrofit. In many cases, it is rational to let the keys fail to limit forces to the substructure and effectively isolate the superstructure. Also note that superstructure lateral bracing system retrofits will also have to be increased to handle increased forces from a shear key retrofit. In many cases, the fusing force of the existing shear keys is large enough to require a substructure retrofit. In these situations, new or modified shear keys need to be constructed to be compatible with the plastic capacity of the retrofitted substructure. There are other situations that may require a shear key retrofit. Transverse movements may be large enough so that the external girder displaces beyond the edge of the bent cap and loses vertical support. For a multiple-girder bridge, it is likely that the side of the bridge may be severely damaged and the use of a shoulder or lane will be lost, but traffic can be routed over a portion of the bridge with few or no emergency repairs. This is considered an acceptable risk. On the other hand, if the superstructure of a two- or three-girder bridge is displaced transversely so that one line of girders loses its support, the entire bridge may collapse. Adequate transverse restraint, commonly in the form of shear keys, should be provided.

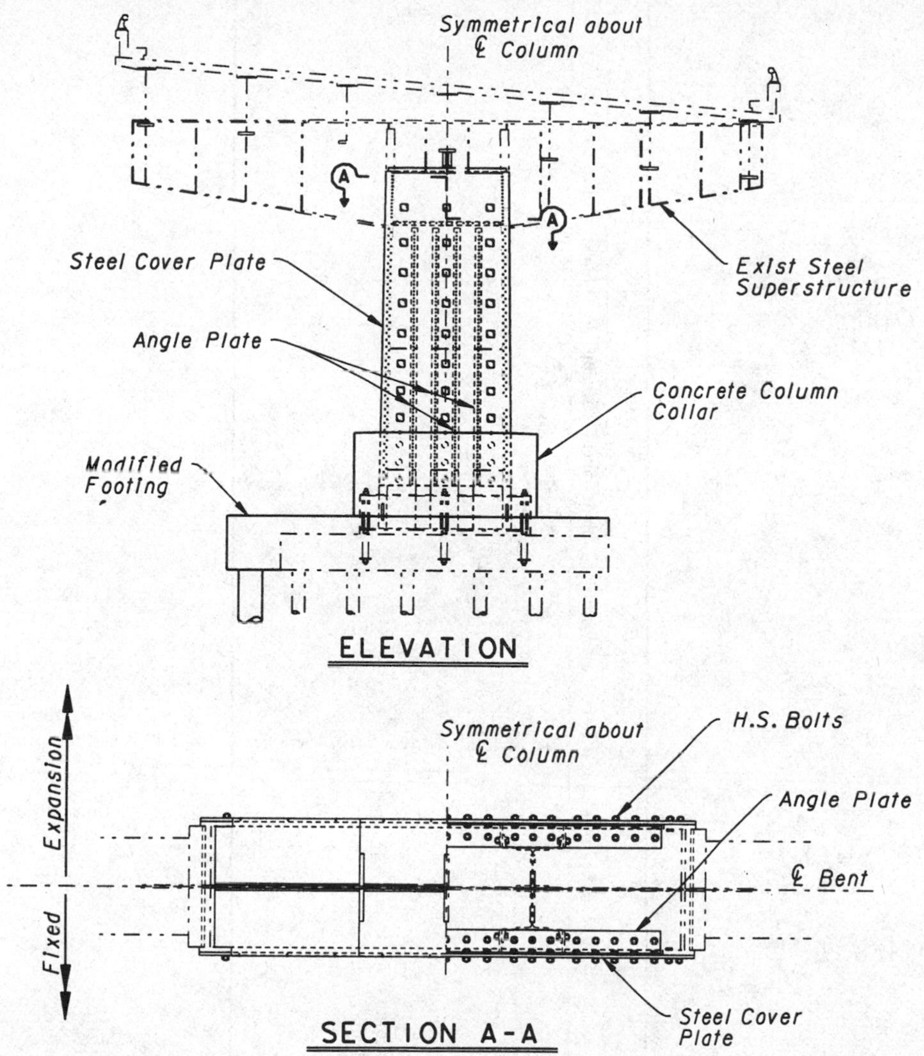

FIGURE 43.18 Open-section steel column retrofit.

43.4.6 Cap Beam Retrofit

There are several potential modes of failure associated with bent caps. Depending on the type of bent cap, these vulnerabilities could include bearing failures, shear key failures, inadequate seat widths, and cap beam failures. Table 43.3 lists several types of bent caps and their associated potential vulnerabilities.

Cap beam modes of failure may include flexure, shear, torsion, and joint shear. Prior to the 1989 Loma Prieta earthquake, California, there was very little emphasis placed on reinforcement detailing of bent cap beams for lateral seismic loads in the vicinity of columns. As a result, cap beams supported on multiple columns were not designed and detailed to handle the increased moment and shear demands that result from lateral transverse framing action. In addition, the beam–column joint is typically not capable of developing the plastic capacity of the column and thus fails in joint shear. For cap beams supported by single columns, although they do not have framing action in the transverse direction and are not subjected to moment and shear demands that are in addition

TABLE 43.3 Potential Bent Cap Vulnerabilities

Cap Type	Bearings	Shear Keys	Seat Width	Cap Beam				
				Moment	Shear	Torsion	Joint Shear	Bolted Cap/Col Connection
Concrete drop cap — single column bent	x	x	x				x	
Concrete drop cap — multicolumn bents	x	x	x	x	x		x	
Integral concrete cap — single column bent							x	
Integral concrete cap — multicolumn bent				x	x	x	x	
Inverted T — simple support for dead load, continuous for live load — single col. bent		x					x	
Inverted T — simple support for dead load, continuous for live load — multi col. bent		x					x	
Inverted T — simple support for both dead load, live load — single column bent	x	x	x				x	
Inverted T — simple support for both dead load and live load — multiple column bent	x	x	x	x	x	x	x	
Steel bent cap — single column bent	x	x	x					x
Steel bent cap — ulticolumn bent	x	x	x					x
Integral outrigger bent						x	x	
Integral C bent						x	x	

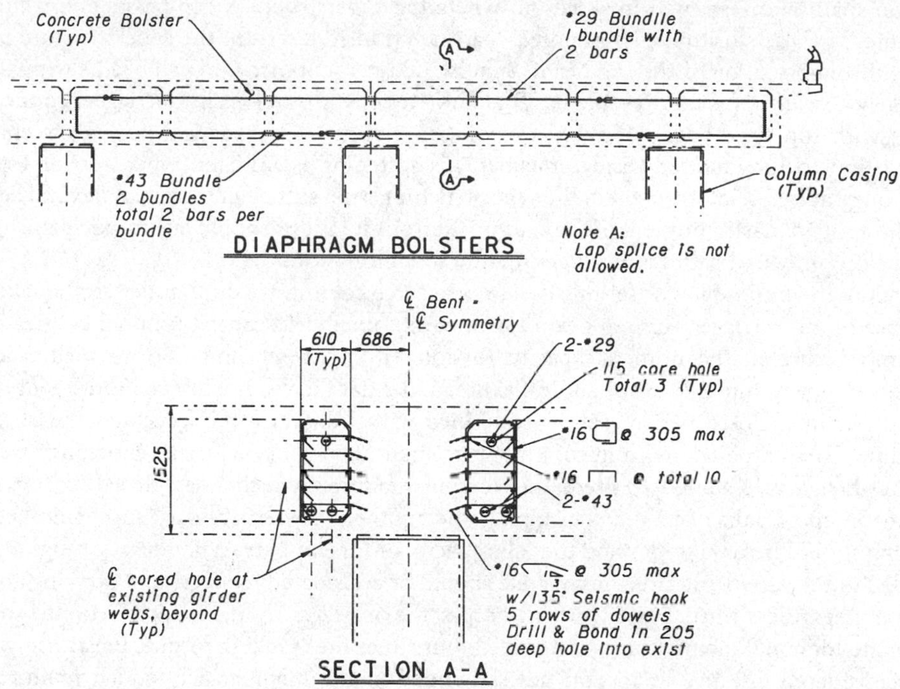

FIGURE 43.19 Bent cap retrofit.

to factored vertical loads, joint shear must still be considered as a result of longitudinal seismic response. In these situations, retrofit of the superstructure is not common since single-column bents are typically fixed to the footing and fixity at top of the column is not necessary.

Retrofit solutions that address moment and shear deficiencies typically include adding a bolster to the existing cap (Figure 43.19). Additional negative and positive moment steel can be placed on the top and bottom faces of the bolsters as required to force plastic hinging into the columns. These bolsters will also contain additional shear stirrups and steel dowels to ensure a good bond with the existing cap for composite action. Prestressing can also be included in the bolsters. In fact, prestressing has proved to be an effective method to enhance an existing cap moment, shear, and joint shear capacity. This is particularly true for bent caps that are integral with the superstructure.

Special consideration should be given to the detailing of bolsters. The engineer needs to consider bar hook lengths and bending radii to make sure that the stirrups will fit into the bolsters. Although, the philosophy is to keep the superstructure elastic and take all inelastic action in the columns, realistically, the cap beams will have some yield penetration. Thus, the new bolster should be detailed to provide adequate confinement for the cap to guarantee ductile behavior. This suggests that bolsters should not be just doweled onto the existing bent cap. There should be a continuous or positive connection between the bolsters through the existing cap. This can be achieved by coring through the existing cap. The hole pattern should be laid out to miss the existing top and bottom steel of the cap. It is generally difficult to miss existing shear stirrups so the engineer should be conservative when designing the new stirrups by not depending on the existing shear steel. The steel running through the existing cap that connects the new bolsters can be continuous stirrups or high-strength rods which may or may not be prestressed.

The cap retrofit is much easier with an exposed cap but can be done with integral caps. In order to add prestressing or new positive and negative steel and to add dowels to make sure that the bolsters in adjacent bays are continuous or monolithic, the existing girders have to be cored. Care must be taken to avoid the main girder steel and/or prestressing steel.

Torsion shall be investigated in situations where the superstructure, cap beam, and column are monolithic. In these situations, longitudinal loads are transferred from the superstructure into the columns through torsion of the cap beam. Superstructures supported on cap beams with bearings are unlikely to cause torsional problems. Torsion is mainly a problem in outriggers connected to columns with top fixed ends. However, torsion can also exist in bent cap beams susceptible to softening due to longitudinal displacements. This softening is initiated when top or especially bottom longitudinal reinforcement in the superstructure is not sufficient to sustain flexural demands due to the applied plastic moment of the column. Retrofit solutions should ensure adequate member strength along the load path from superstructure to column foundation.

In general, the philosophy of seismic design is to force column yielding under earthquake loads. In the case of an outrigger, the cap beam torsional nominal yield capacity should be greater than the column flexural plastic moment capacity. Torsion reinforcement should be provided in addition to reinforcement required to resist shear, flexure, and axial forces. Torsion reinforcement consists of closed stirrups, closed ties, or spirals combined with transverse reinforcement for shear, and longitudinal bars combined with flexural reinforcement. Lap-spliced stirrups are considered ineffective in outriggers, leading to a premature torsional failure. In such cases, closed stirrups should not be made up of pairs of U-stirrups lapping one another. Where necessary, mechanical couplers or welding should be used to develop the full capacity of torsion bars. When plastic hinging cannot be avoided in the superstructure, the concrete should be considered ineffective in carrying any shear or torsion. Regardless where plastic hinging occurs, the outrigger should be proportioned such that the ultimate torsional moment does not exceed four times the cracking torque. Prestressing should not be considered effective in torsion unless bonded in the member. Unbonded reinforcement, however, can be used to supply axial load to satisfy shear friction demands to connect outrigger caps to columns and superstructure. Bonded tendons should not be specified in caps where torsional yielding will occur. Designers must consider effects of the axial load in caps due to transverse column plastic hinging when satisfying shear and torsion demands.

43.4.7 Abutments

Abutments are generally classified into two types: seat type and monolithic. The monolithic type of abutment is commonly used for shorter bridges, whereas longer bridges typically use a seat type. Contrary to the seat-type abutment, the monolithic abutment has the potential for heavy damage. This is largely due to the fact that the designer has more control through the backwall design. The backwall behaves as a fuse to limit any damage to the piles. However, since this damage is not a collapse mechanism, it is therefore considered to be acceptable damage. Additionally, the monolithic abutment has proven itself to perform very well in moderate earthquakes, sustaining little or no damage. Some typical problems encountered in older bridges are

- Insufficient seat length for seat-type abutments;
- Large gallery, or gap, between the backwall and superstructure end diaphragm;
- Insufficient longitudinal and/or transverse shear capacity;
- Weak end diaphragms at monolithic abutments.

Following are some of the more common types of retrofits used to remedy the abutment problems mentioned above.

43.4.7.1 Seat Extenders

Seat extenders at abutments and drop caps generally consist of additional concrete scabbed onto the existing face (Figures 43.20 and 43.21). The design of seat extenders that are attached to existing abutment or bent cap faces should be designed like a corbel. When designing the connecting steel between the new seat extender and the existing concrete, shear friction for vertical loads should be considered. Tensile forces caused by friction should also be considered when the girder moves in

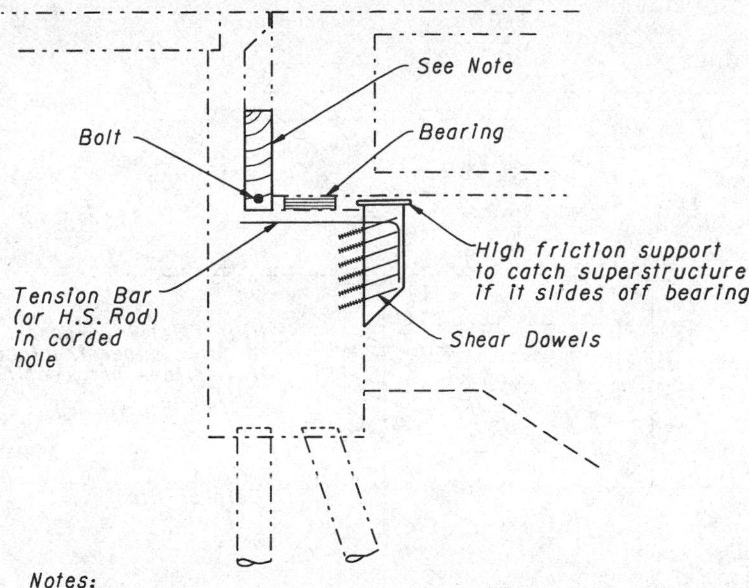

Notes:

Gap filler to mobilize backwall and embankment soil. Filler can be:

A) Steel or hardwood strips inserted by slipping them horizontally in space above seat and rotating to vertical. Bolt together.

B) Fill space with concrete through top use polystyrene between new concrete and bridge superstructure requires traffic control.

Hardwood filler (Option A) shown in sketch above.

FIGURE 43.20 Seat extender at abutment.

the longitudinal direction and pulls the new concrete away from the existing bent cap or abutment. Compression strut and bearing loads under the girder also need to be considered. Note that the face of the existing concrete should be intentionally roughened before the new concrete is placed to ensure a good bond.

If bearing failure results in the superstructure dropping 150 mm or more, catchers could be added to minimize the drop. Catchers generally are designed to limit the superstructure drop to 50 mm and can provide additional seat width. In other words, catchers are basically seat extenders that are detailed to reduce the amount the superstructure is allowed to drop. The design procedure is similar for both seat extenders and catchers. In some cases, an elastomeric bearing pad is placed on top of the catcher to provide a landing spot for the girder after bearing failure. The friction factor for concrete on an elastomeric pad is less than for concrete on concrete so the tension force in the corbel could possibly be reduced. One special consideration for catchers is to make sure to leave enough room to access the bearing for inspection and replacement.

43.4.7.2 Fill galleries with timber, concrete or steel

Some seat-type abutments typically have a gallery, or a large gap, between the superstructures end diaphragm and the backwall. It is important to realize that the columns must undergo large deformations before the soil can be mobilized behind the abutment if this gap is not filled. Therefore, as a means of retrofit, the gallery is filled with concrete, steel, or timber to engage the backwall and, hence, the soil (Figure 43.22). However, timber is a potential fire hazard and in some parts of the

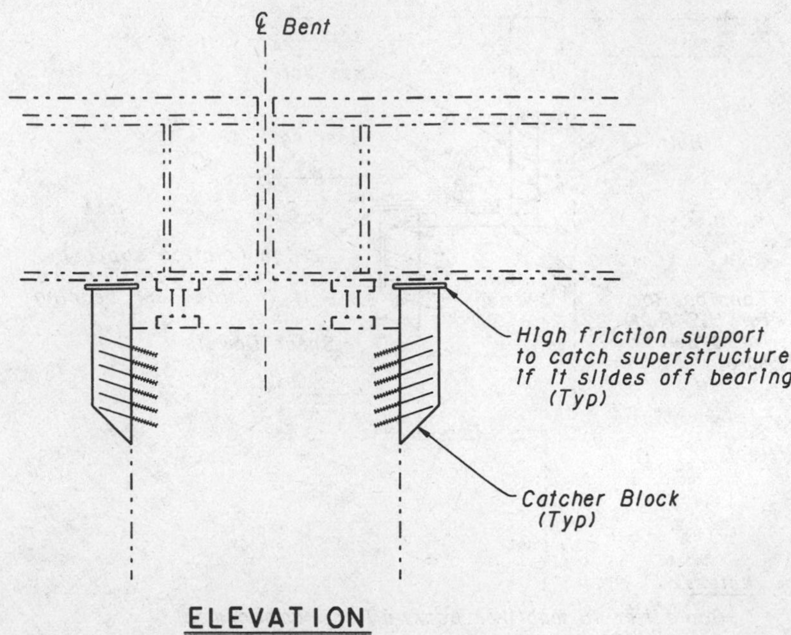

FIGURE 43.21 Seat extender at bent cap.

United States may be susceptible to termite attack. When filling this gap, the designer should specify the expected thermal movements, rather than the required thickness. This prevents any problems that may surface if the backwall is not poured straight.

43.4.7.3 L Brackets on Superstructure Soffit

Similar in theory to filling the gallery behind the backwall is adding steel angles (or brackets) to the flanges of steel I-girders, as seen in Figure 43.23. These brackets act as "bumpers" that transfer the longitudinal reaction from the superstructure into the abutment, and then into the soil.

43.4.7.4 Shear Keys, Large CIDH Piles, Anchor Slabs, and Vertical Pipes

For shorter bridges, an effective retrofit scheme is to attract the forces away from the columns and footings and into the abutments. This usually means modifying and/or strengthening the abutment, thereby "locking up" the abutment, limiting the displacements and, hence, attracting most of the loads. Although this type of retrofit is more effective in taking the load out of columns for shorter structures, the abutments still may require strengthening, in addition to retrofitting the columns, for longer bridges.

Methods that are intended to mobilize the abutment and the soil behind the abutment may consist of vertical pipes, anchor piles, seismic anchor slabs, or shear keys as shown in Figures 43.24 and 43.25. For heavily skewed or curved bridges, anchor piles or vertical pipes are generally the preferred method due to the added complication from geometry. For instance, as the bridge rotates away from the abutment, there is nothing to resist this movement. By adding an anchor pile at the acute corners of the abutment, the rotation is prohibited and the anchor pile then picks up the load.

43.4.7.5 Catchers

When the superstructure is founded on tall bearings at the abutments, the bearings, as mentioned earlier, are susceptible to damage. If tall enough, the amount of drop the superstructure would undergo can significantly increase the demands at the bents. To remedy this, catcher blocks are constructed next to the bearings to "catch" the superstructure and limit the amount of vertical displacement.

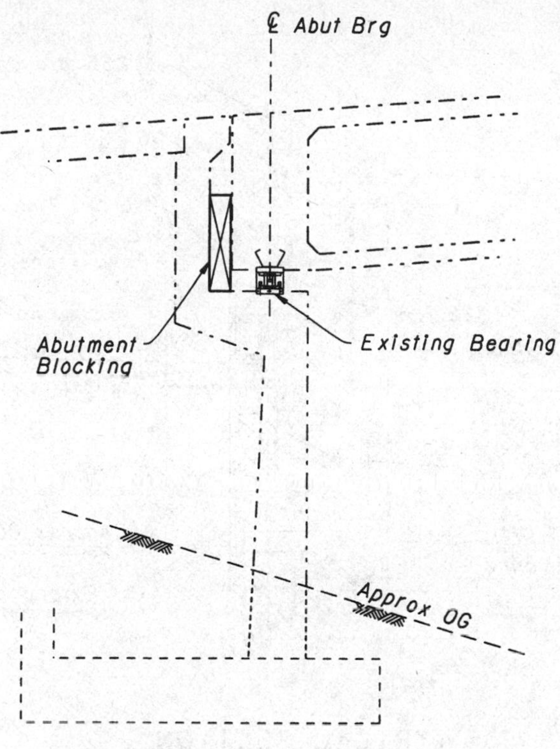

FIGURE 43.22 Abutment blocking.

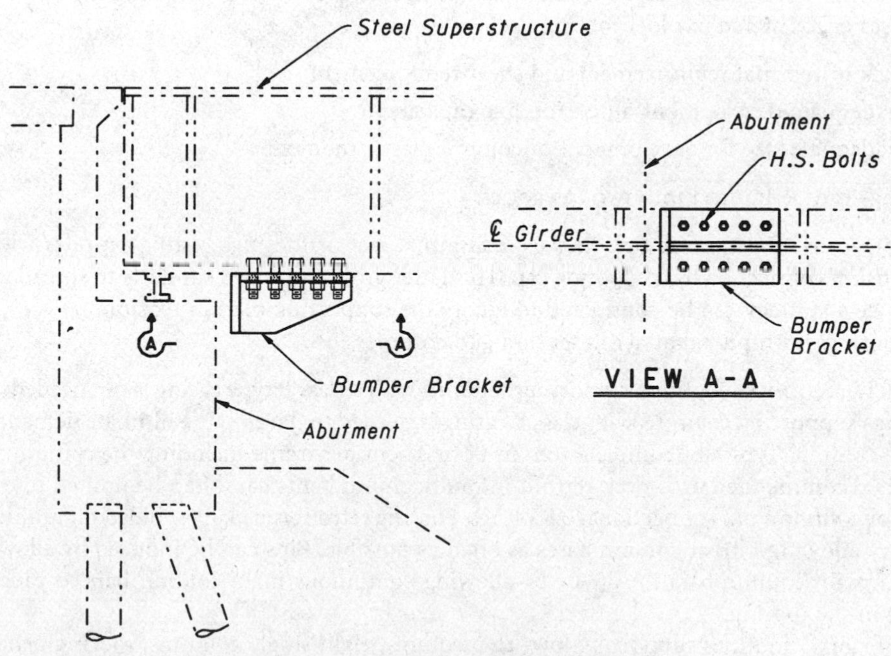

FIGURE 43.23 Bumper bracket at abutment.

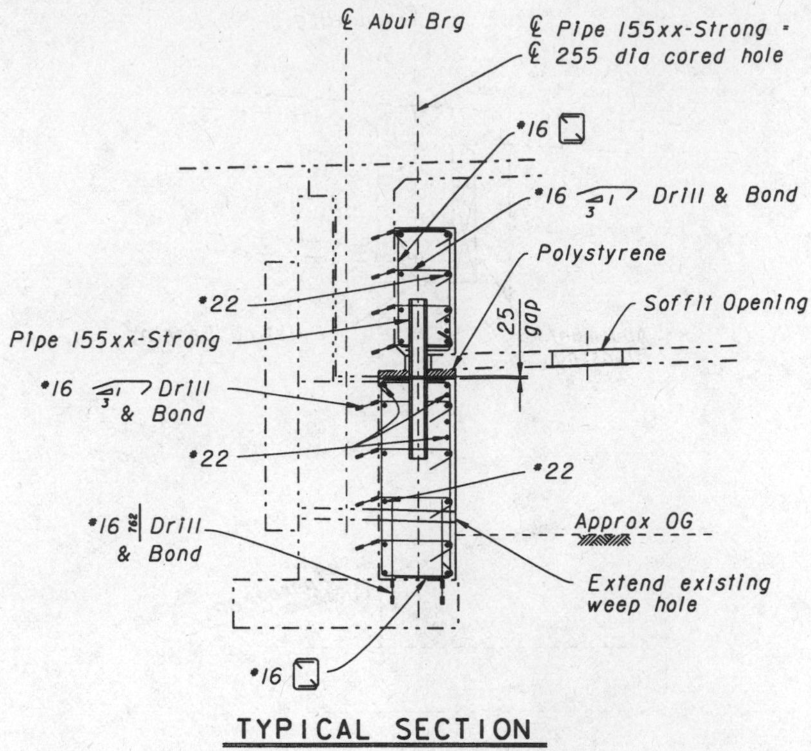

FIGURE 43.24 Vertical pipe at abutment.

43.4.8 Foundations

Older footings have many vulnerabilities that can lead to failure. The following is a list of major weaknesses encountered in older footings:

- Lack of top mat reinforcement and shear reinforcement;
- Inadequate development of tension pile capacity;
- Inadequate size for development of column plastic moment.

Footings can be lumped into two categories:

I. Spread footings resting on relatively dense material or footings resting on piles with weak tension connection to the footing cap. This latter group is treated similarly to spread footings since a strategy can be considered to ignore the supporting piles in tension.
II. Footings with piles that act in tension and compression.

In general, retrofit of footings supporting columns with a class P type casing is not needed; retrofit of footings supporting columns with class F casing is needed to develop the ultimate demand forces from the column. Typically, complete retrofit of one bent per frame including the column and the footing is recommended. However, retrofit in multicolumn bents can often be limited to columns because of common pin connections to footings. Footing retrofit is usually avoided on multicolumn bridges by allowing pins at column bases as often as possible. Pins can be induced by allowing lap splices in main column bars to slip, or by allowing continuous main column bars to cause shear cracking in the footing.

For category I footings supporting low- to medium-height single columns, rocking behavior of the bent should be investigated for stability and the footing capacity can be compared against the

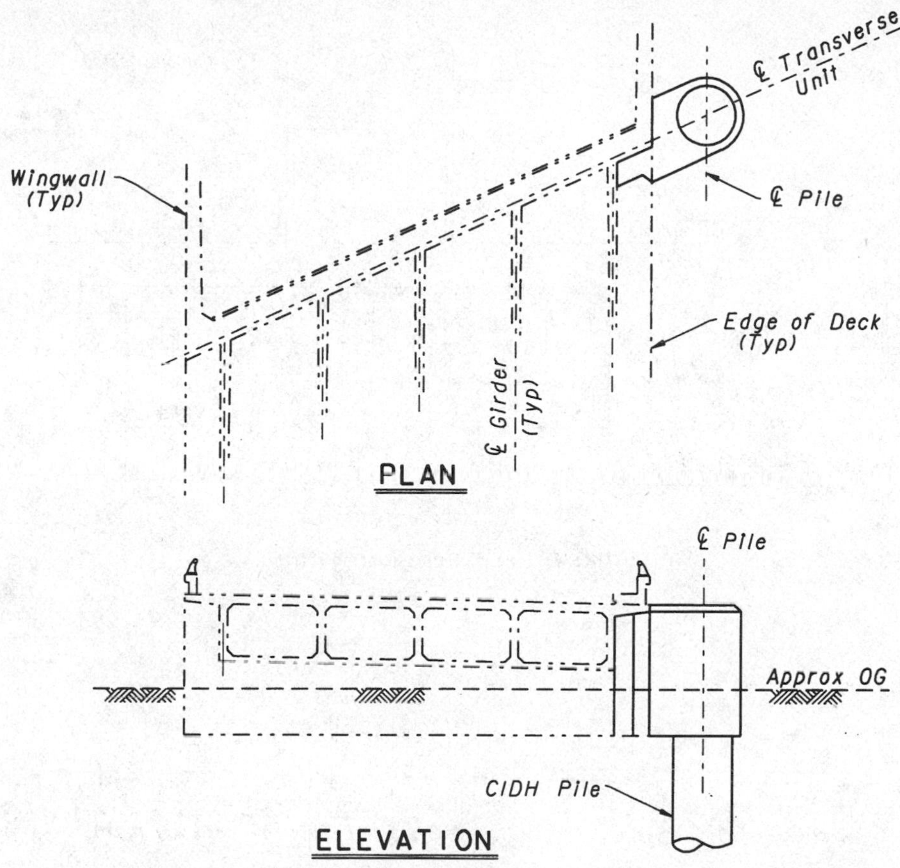

FIGURE 43.25 Anchor pile at abutment.

resulting forces from the rocking analysis. These forces can be of lesser magnitude than forces induced by column plastic hinging. Typically, retrofits for this case consist of adding an overlay to enhance the footing shear capacity or even widening of the footing to gain a larger footprint for stability and increase of the flexural moment capacity. The new concrete is securely attached to the old footing. This is done by chipping away at the concrete around the existing reinforcement and welding or mechanically coupling the new reinforcement to the old one. Holes are then drilled and the dowels are bonded between the faces of the old and new concrete as shown in Figure 43.26.

For category I footings supporting tall, single columns, rocking behavior of the bent can lead to instability, and some additional piles might be needed to provide stability to the tall bent. This type of modification leads to increased shear demand and increased tension demand on the top fiber of the existing footing that requires addition of a top mat reinforcement (Figure 43.26). The top mat is tied to the existing footing with dowels, and concrete is placed over the new piles and reinforcement. Where high compressive capacity piles are added, reinforcement with an extension hook is welded or mechanically coupled to existing bottom reinforcement. The hook acts to confine the concrete in the compression block where the perimeter piles are under compression demand.

When tension capacity is needed, the use of standard tension/compression piles is preferred to the use of tie-downs. In strong seismic events, large movements in footings are associated with tie-downs. Generally, tie-downs cannot be prestressed to reduce movements without overloading existing piles in compression. The tie-down movements are probably not a serious problem with short columns where P–Δ effects are minimal. Also, tie-downs should be avoided where groundwater could affect the quality of installation.

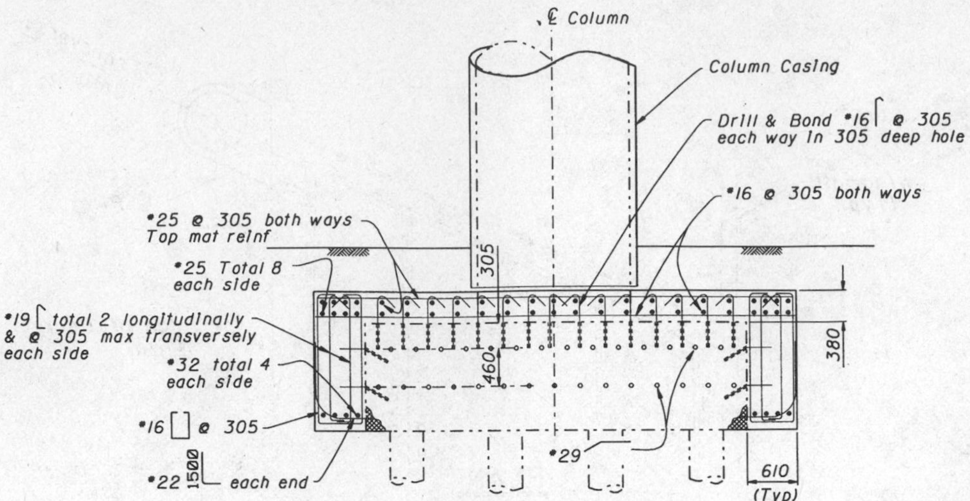

FIGURE 43.26 Widening footing retrofit.

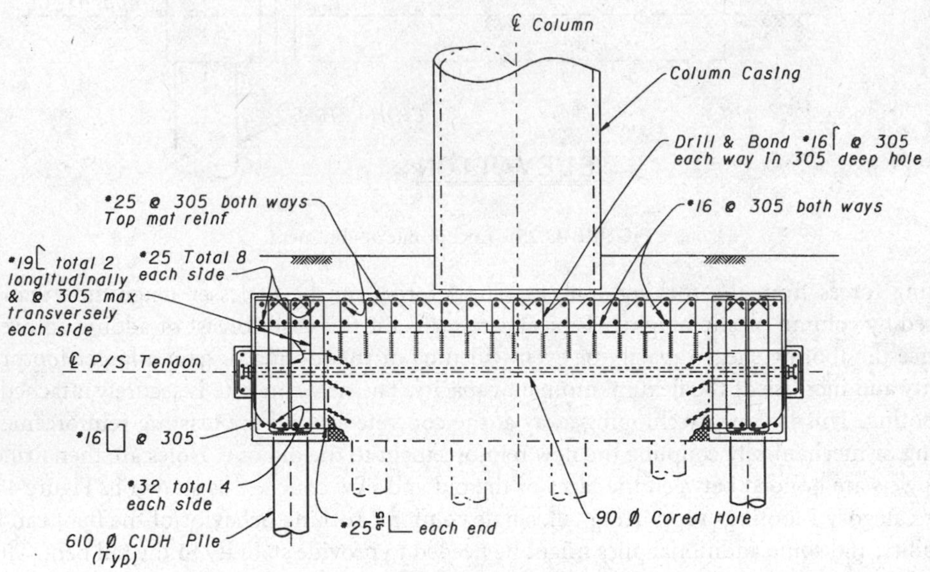

FIGURE 43.27 Footing retrofit using prestressing.

For category II footings, the ability of existing piles to cause tension on the top fiber of the footing where no reinforcement is present can lead to footing failure. Therefore, adding a top overlay in conjunction with footing widening might be necessary.

In sites where soft soil exists, the use of larger piles (600 mm and above) may be deemed necessary. These larger piles may induce high flexural demands requiring additional capacity from the bottom reinforcement. In this situation, prestressing of the footing becomes an alternative solution since it enhances the footing flexural capacity in addition to confining the concrete where perimeter piles act in compression (Figure 43.27). This retrofit is seldom used and is considered a last recourse.

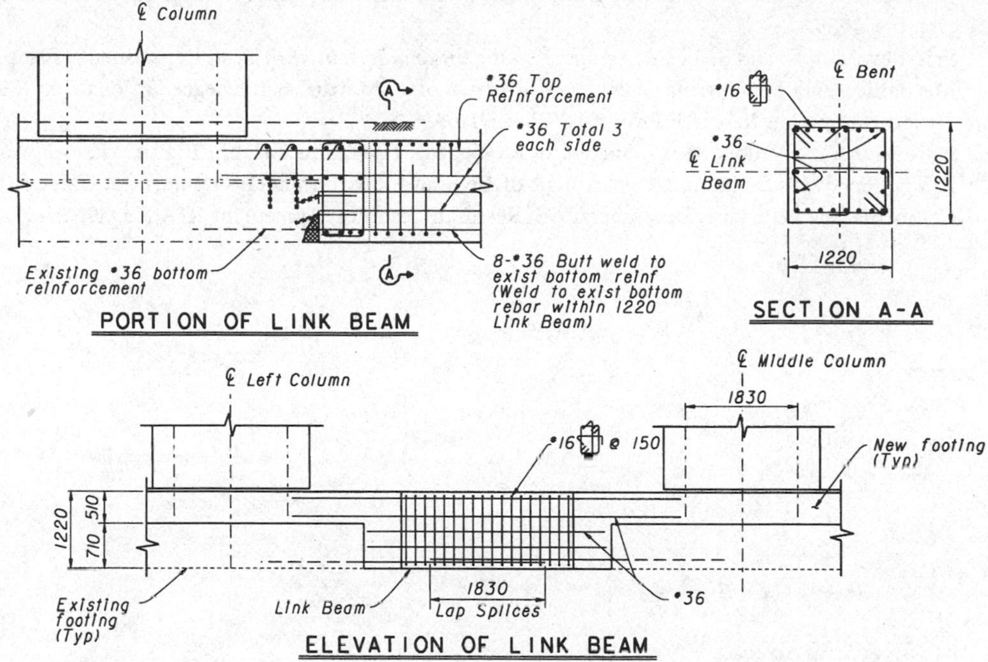

FIGURE 43.28 Link beam footing retrofit.

A rare but interesting situation occurs when a tall multicolumn bent has pinned connections to the superstructure instead of the usual monolithic connections and is resting on relatively small spread footings. This type of bent is quite vulnerable under large overturning moments and the use of link beams, as shown in Figure 43.28, is considered economical and sufficient to provide adequate stability and load transfer mechanism in a seismic event.

43.5 Summary

The seismic-resistant retrofit design of bridges has been evolving dramatically in the last decade. Many of the retrofit concepts and details discussed in this chapter have emerged as a result of research efforts and evaluation of bridge behavior in past earthquakes. This practice has been successfully tested in relatively moderate earthquakes but has not yet seen the severe test of a large-magnitude earthquake. The basic philosophy of current seismic retrofit technology in the U.S. is to prevent collapse by providing sufficient seat for displacement to take place or by allowing ductility in the supporting members. The greatest challenge to this basic philosophy will be the next big earthquake. This will serve as the utmost test to current predictions of earthquake demands on bridge structures.

References

1. Department of Transportation, State of California, Bridge Design Specifications, unpublished revision, Sacramento, 1997.
2. Mander, J. B., Priestley, M. J. N., and Park, R., *Theoretical stress-strain model for confined concrete, J. Struct. Eng. ASCE,* 114(8), 1804–1826,
3. Department of Transportation, State of California, Caltrans Memo to Designers, Sacramento, May 1994.

4. Department of Transportation, State of California, Caltrans Bridge Design Aids, Sacramento, October 1989.
5. Priestley, M. J. N. and Seible, F., *Seismic Assessment and Retrofit of Bridges*, Department of Applied Mechanics and Engineering Sciences, University of California, San Diego, La Jolla, July 1991, Structural Systems Research Project SSRP — 91/03.
6. Haroun, M. A., Pardoen, G. C., Shepherd, R., Haggag, H. A., and Kazanjy, R. P., *Cyclic Behavior of Bridge Pier Walls for Retrofit*, Department of Civil and Environmental Engineering, University of California, Irvine, Irvine, December 1993, Research Technical Agreement RTA No. 59N974.

44

Seismic Design Practice in Japan

Shigeki Unjoh
Public Works Research Institute

Nomenclature

The following symbols are used in this chapter. The section number in parentheses after definition of a symbol refers to the section where the symbol first appears or is defined.

a space of tie reinforcement (Section 44.4.4)
A_{CF} sectional area of carbon fiber (Figure 44.19)
A_h area of tie reinforcements (Section 44.4.4)
A_w sectional area of tie reinforcement (Section 44.4.4)
b width of section (Section 44.4.4)
c_B coefficient to evaluate effective displacement (Section 44.4.7)
c_B modification coefficient for clearance (Section 44.4.11)
c_{df} modification coefficient (Section 44.4.2)
c_c modification factor for cyclic loading (Section 44.4.4)
c_D modification coefficient for damping ratio (Section 44.4.6)
c_e modification factor for scale effect of effective width (Section 44.4.4)
c_E modification coefficient for energy-dissipating capability (Section 44.4.7)
c_P coefficient depending on the type of failure mode (Section 44.4.2)
c_{pt} modification factor for longitudinal reinforcement ratio (Section 44.4.4)

0-8493-7434-0/00/$0.00+$.50

c_R	factor depending on the bilinear factor r (Section 44.4.2)
c_W	corrective coefficient for ground motion characteristics (Section 44.4.9)
c_Z	modification coefficient for zone (Section 44.4.3)
d	effective width of tie reinforcements (Section 44.4.4)
d	height of section (Section 44.4.4)
D	a width or a diameter of a pier (Section 44.4.4)
D_E	coefficient to reduce soil constants according to F_L value (Section 44.4.11)
E_c	elastic modules of concrete (Section 44.4.4)
E_{CF}	elastic modulus of carbon fiber (Figure 44.19)
E_{des}	gradient at descending branch (Section 44.4.4)
F_L	liquefaction resistant ratio (Section 44.4.9)
$F(u)$	restoring force of a device at a displacement u (Section 44.4.7)
h	height of a pier (Section 44.4.4)
h_B	height of the center of gravity of girder from the top of bearing (Figure 44.13)
h_B	equivalent damping of a Menshin device (Section 44.4.7)
h_i	damping ratio of *i*th mode (Section 44.4.6)
h_{ij}	damping ratio of *j*th substructure in *i*th mode (Section 44.4.6)
h_{Bi}	damping ratio of *i*th damper (Section 44.4.7)
h_{Pi}	damping ratio of *i*th pier or abutment (Section 44.4.7)
h_{Fui}	damping ratio of *i*th foundation associated with translational displacement (Section 44.4.7)
$h_{F\theta i}$	damping ratio of *i*th foundation associated with rotational displacement(Section 44.4.7)
H	distance from a bottom of pier to a gravity center of a deck (Section 44.4.7)
H_0	shear force at the bottom of footing (Figure 44.12)
I	importance factor (Section 44.5.2)
k_{hc}	lateral force coefficient (Section 44.4.2)
k_{hc}	design seismic coefficient for the evaluation of liquefaction potential (Section 44.4.9)
k_{hc0}	standard modification coefficient (Section 44.4.3)
k_{hcm}	lateral force coefficient in Menshin design (Section 44.4.7)
k_{he}	equivalent lateral force coefficient (Section 44.4.2)
k_{hem}	equivalent lateral force coefficient in Menshin design (Section 44.4.7)
k_{hp}	lateral force coefficient for a foundation (Section 44.4.2)
k_j	stiffness matrix of *j*th substructure (Section 44.4.6)
K	stiffness matrix of a bridge (Section 44.4.6)
K_B	equivalent stiffness of a Menshin device (Section 44.4.7)
K_{Pi}	equivalent stiffness of *i*th pier or abutment (Section 44.4.7)
K_{Fui}	translational stiffness of *i*th foundation (Section 44.4.7)
$K_{F\theta i}$	rotational stiffness of *i*th foundation (Section 44.4.7)
L	shear stress ratio during an earthquake (Section 44.4.9)
L_A	redundancy of a clearance (Section 44.4.11)
L_E	clearance at an expansion joint (Section 44.4.11)
L_P	plastic hinge length of a pier (Section 44.4.4)
M_0	moment at the bottom of footing (Figure 44.12)
P_a	lateral capacity of a pier (Section 44.4.2)
P_s	shear capacity in consideration of the effect of cyclic loading (Section 44.4.4)
P_{s0}	shear capacity without consideration of the effect of cyclic loading (Section 44.4.4)
P_u	bending capacity (Section 44.4.2)
r	bilinear factor defined as a ratio between the first stiffness (yield stiffness) and the second stiffness (postyield stiffness) of a pier (Section 44.4.2)
r_d	modification factor of shear stress ratio with depth (Section 44.4.9)
R	dynamic shear strength ratio (Section 44.4.9)
R	priority (Section 44.5.2)
R_D	dead load of superstructure (Section 44.4.11)
R_{heq} and R_{veq}	vertical reactions caused by the horizontal seismic force and vertical force (Section 44.4.11)
R_L	cyclic triaxial strength ratio (Section 44.4.9)
R_U	design uplift force applied to the bearing support (Section 44.4.11)
s	space of tie reinforcements (Section 44.4.4)
S	earthquake force (Section 44.5.2)
S_c	shear capacity shared by concrete (Section 44.4.4)
S_{I} and S_{II}	acceleration response spectrum for Type-I and Type-II ground motions (Section 44.4.6)

S_{I0} and S_{II0} standard acceleration response spectrum for Type-I and Type-II ground motions (Section 44.4.6)
S_E seat length (Section 44.4.11)
S_{EM} minimum seat length (cm) (Section 44.4.11)
S_s shear capacity shared by tie reinforcements (Section 44.4.4)
T natural period of fundamental mode (Table 44.3)
ΔT difference of natural periods (Section 44.4.11)
T_1 and T_2 natural periods of the two adjacent bridge systems (Section 44.4.11)
u_B design displacement of isolators (Section 44.4.7)
u_{Be} effective design displacement (Section 44.4.7)
u_{Bi} design displacement of *i*th Menshin device (Section 44.4.7)
u_G relative displacement of ground along the bridge axis (Section 44.4.11)
u_R relative displacement (cm) developed between a superstructure and a substructure (Section 44.4.11)
V_0 vertical force at the bottom of footing (Figure 44.12)
V_T structural factor (Section 44.5.2)
V_{RP1} design specification (Section 44.5.2)
V_{RP2} pier structural factor (Section 44.5.2)
V_{RP3} aspect ratio (Section 44.5.2)
V_{MP} steel pier factor (Section 44.5.2)
V_{FS} unseating device factor (Section 44.5.2)
V_F foundation factor (Section 44.5.2)
w_v weighting factor on structural members (Section 44.5.2)
W equivalent weight (Section 44.4.2)
W elastic strain energy (Section 44.4.7)
W_P weight of a pier (Section 44.4.2)
W_U weight of a part of superstructure supported by the pier (Section 44.4.2)
ΔW energy dissipated per cycle (Section 44.4.7)
α safety factor (Section 44.4.4)
α, β coefficients depending on shape of pier (Section 44.4.4)
α_m safety factor used in Menshin design (Section 44.4.7)
δ_y yield displacement of a pier (Section 44.4.2)
δ_R residual displacement of a pier after an earthquake (Section 44.4.2)
δ_{Ra} allowable residual displacement of a pier (Section 44.4.2)
δ_u ultimate displacement of a pier (Section 44.4.4)
ε_c strain of concrete (Section 44.4.4)
ε_{cc} strain at maximum strength (Section 44.4.4)
ε_G ground strain induced during an earthquake along the bridge axis (Section 44.4.11)
ε_s strain of reinforcements (Section 44.4.4)
ε_{sy} yield strain of reinforcements (Section 44.4.4)
θ angle between vertical axis and tie reinforcement (Section 44.4.4)
θ_{pu} ultimate plastic angle (Section 44.4.4)
μ_a allowable displacement ductility factor of a pier (Section 44.4.2)
μ_m allowable ductility factor of a pier in Menshin design (Section 44.4.7)
μ_R response ductility factor of a pier (Section 44.4.2)
ρ_s tie reinforcement ratio (Section 44.4.4)
σ_c stress of concrete (Section 44.4.4)
σ_{cc} strength of confined concrete (Section 44.4.4)
σ_{CF} stress of carbon fiber (Figure 44.19)
σ_{ck} design strength of concrete (Section 44.4.4)
σ_s stress of reinforcements (Section 44.4.4)
σ_{sy} yield strength of reinforcements (Section 44.4.4)
σ_v total loading pressure (Section 44.4.9)
σ'_v effective loading pressure (Section 44.4.9)
τ_c shear stress capacity shared by concrete (Section 44.4.4)
ϕ_{ij} mode vector of *j*th substructure in *i*th mode (Section 44.4.6)
ϕ_i mode vector of a bridge in *i*th mode (Section 44.4.6)
ϕ_y yield curvature of a pier at bottom (Section 44.4.4)
ϕ_u ultimate curvature of a pier at bottom (Section 44.4.4)

44.1 Introduction

Japan is one of the most seismically disastrous countries in the world and has often suffered significant damage from large earthquakes. More than 3000 highway bridges have suffered damage since the 1923 Kanto earthquake. The earthquake disaster prevention technology for highway bridges has been developed based on such bitter damage experiences. Various provisions for designing bridges have been developed to prevent damage due to the instability of soils such as soil liquefaction. Furthermore, design detailings including unseating prevention devices are implemented. With progress in improving seismic design provisions, damage to highway bridges caused by the earthquakes has been decreasing in recent years.

However, the Hyogo-ken Nanbu earthquake of January 17, 1995 caused destructive damage to highway bridges. Collapse and near collapse of superstructures occurred at nine sites, and other destructive damage occurred at 16 sites [1]. The earthquake revealed that there are a number of critical issues to be revised in the seismic design and seismic retrofit of bridges [2,3].

This chapter presents technical developments for seismic design and seismic retrofit of highway bridges in Japan. The history of the earthquake damage and development of the seismic design methods is first described. The damage caused by the 1995 Hyogo-ken Nanbu earthquake, the lessons learned from the earthquake, and the seismic design methods introduced in the 1996 *Seismic Design Specifications for Highway Bridges* are then described. Seismic performance levels and design methods as well as ductility design methods for reinforced concrete piers, steel piers, foundations, and bearings are described. Then the history of the past seismic retrofit practices is described. The seismic retrofit program after the Hyogo-ken-Nanbu earthquake is described with emphasis on the seismic retrofit of reinforced concrete piers as well as research and development on the seismic retrofit of existing highway bridges.

44.2 History of Earthquake Damage and Development of Seismic Design Methods

A year after the 1923 Great Kanto earthquake, consideration of the seismic effect in the design of highway bridges was initiated. The Civil Engineering Bureau of the Ministry of Interior promulgated "The Method of Seismic Design of Abutments and Piers" in 1924. The seismic design method has been developed and improved through bitter experience in a number of past earthquakes and with progress of technical developments in earthquake engineering. Table 44.1 summarizes the history of provisions in seismic design for highway bridges.

In particular, the seismic design method was integrated and upgraded by compiling the "Specifications for Seismic Design of Highway Bridges" in 1971. The design method for soil liquefaction and unseating prevention devices was introduced in the Specifications. It was revised in 1980 and integrated as "Part V: Seismic Design" in Design Specifications of Highway Bridges. The primitive check method for ductility of reinforced concrete piers was included in the reference of the Specifications. It was further revised in 1990 and ductility check of reinforced concrete piers, soil liquefaction, dynamic response analysis, and design detailings were prescribed. It should be noted here that the detailed ductility check method for reinforced concrete piers was first introduced in the 1990 Specifications.

However, the Hyogo-ken Nanbu earthquake of January 17, 1995, exactly 1 year after the Northridge earthquake, California, caused destructive damage to highway bridges as described earlier. After the earthquake the Committee for Investigation on the Damage of Highway Bridges Caused by the Hyogo-ken Nanbu Earthquake (chairman, Toshio Iwasaki, Executive Director, Civil Engineering Research Laboratory) was established in the Ministry of Construction to investigate the damage and to identify the factors that caused the damage.

TABLE 44.1 History of Seismic Design Methods

		1926 Details of Road Structure (draft) Road Law, MIA	1939 Design Specifications of Steel Highway Bridges (draft) MIA	1956 Design Specifications of Steel Highway Bridges, MOC	1964 Design Specifications of Substructures (Pile Foundations), MOC	1964 Design Specifications of Steel Highway Bridges, MOC	1966 Design Specifications of Substructures (Survey and Design), MOC	1968 Design Specifications of Substructures (Piers and Direct Foundations), MOC	1970 Design Specifications of Substructures (Caisson Foundations), MOC	1971 Specifications for Seismic Design of Highway Bridges, MOC	1972 Design Specifications of Substructures (Cast-in-Piles), MOC	1975 Design Specifications of Substructures (Pile Foundations), MOC	1980 Design Specifications of Highway Bridges, MOC	1990 Design Specifications of Highway Bridges, MOC
Seismic loads	Seismic coefficient	Largest seismic loads Varied dependent on the site	$k_h = 0.2$	$k_h = 0.1–0.35$ Varied dependent on the site and ground condition					Standardization of seismic coefficient provision of modified seismic coefficient method	$k_h = 0.1–0.3$			Revision of application range of modified seismic coefficient method	$k_h = 0.1–0.3$ Integration of seismic coefficient method and modified one.
	Dynamic earth pressure	Equations proposed by Mononobe and Okabe were supposed to be used					Provision of dynamic earth pressure							
	Dynamic hydraulic pressure	Less effect on piers except high piers in deep water					Provision of hydraulic pressure			Provision of dynamic hydraulic pressure				
Reinforced concrete column	Bending at bottom	Supposed to be designed in a similar way provided in current design Specifications						Provisions of Definite Design Method						
	Shear	Less effect on RC piers except those with smaller section area such as RC frame and hollow section						Check of shear strength					Provision of definite design method, decreasing of allowable shear stress	
	Termination of Main Reinforcement at Midheight											Elongation of anchorage length of terminated reinforcement at midheight		
	Bearing capacity for lateral force							Less effect on RC piers with larger section area				Ductility check		Check for bearing capacity for lateral force
Footing							Provisions of definite design method (designed as a cantilever plate)						Provisions of effective width and check of shear strength	
Pile foundation		Bearing capacity in vertical direction was supposed to be checked				Provisions of Definite Design Method (bearing capacity in vertical and horizontal directions)					Special Condition (Foundation on Slope, Consolidation Settlement, Lateral Movement)		Provisions of Design Details for Pile Head	
Direct foundation		Stability (overturning and slip) was supposed to be checked					Provisions of Definite Design Method (bearing capacity, stability analysis)							
Caisson foundation				Supposed to be designed in a similar way provided in Design Specification of Caisson Foundation of 1959					Provisions of Definite Design Method					
Soil Liquefaction									Provisions of soil layers of which bearing capacity shall be ignored in seismic design			Provisions of evaluation method of soil liquefaction and the treatment in seismic design		Consideration of effect of fine sand content
Bearing support	Bearing support		Provisions of Design Methods for steel bearing supports (bearing, roller, anchor bolt)						Provision of transmitting method of seismic load at bearing					
	Devices preventing falling-off of superstructure							Provision of bearing seat length S	Provisions of stopper at movable bearings, devices for preventing superstructure from falling (seat length S, connection of adjacent decks)			Provisions of stopper at movable bearings, devices for preventing superstructure from falling (seat length S_e devices)		

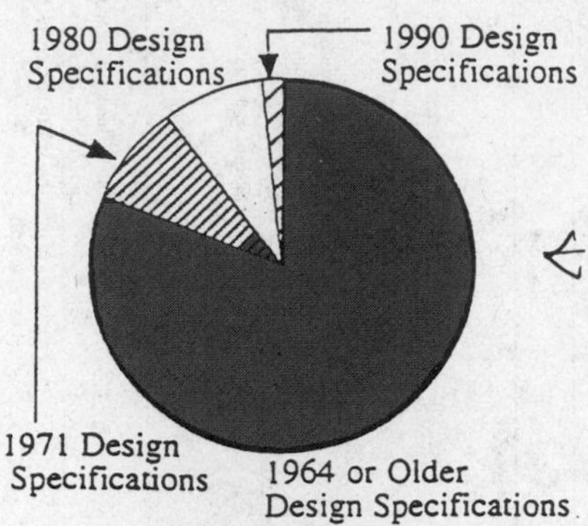

FIGURE 44.1 Design specifications referred to in design of Hanshin Expressway [2].

On February 27, 1995, the Committee approved the "Guide Specifications for Reconstruction and Repair of Highway Bridges Which Suffered Damage Due to the Hyogo-ken Nanbe Earthquake," [4], and the Ministry of Construction announced on the same day that the reconstruction and repair of the highway bridges which suffered damage in the Hyogo-ken Nanbu earthquake should be made by the Guide Specifications. It was decided by the Ministry of Construction on May 25, 1995 that the Guide Specifications should be tentatively used in all sections of Japan as emergency measures for seismic design of new highway bridges and seismic strengthening of existing highway bridges until the Design Specifications of Highway Bridges is revised.

In May, 1995, the Special Sub-Committee for Seismic Countermeasures for Highway Bridges (chairman, Kazuhiko Kawashima, Professor of the Tokyo Institute of Technology) was established in the Bridge Committee (chairman, Nobuyuki Narita, Professor of the Tokyo Metropolitan University), Japan Road Association, to draft the revision of the Design Specifications of Highway Bridges. The new Design Specifications of Highway Bridges [5,6] was approved by the Bridge Committee, and issued by the Ministry of Construction on November 1, 1996.

44.3 Damage of Highway Bridges Caused by the Hyogo-ken Nanbu Earthquake

The Hyogo-ken Nanbu earthquake was the first earthquake to hit an urban area in Japan since the 1948 Fukui earthquake. Although the magnitude of the earthquake was moderate (M7.2), the ground motion was much larger than anticipated in the codes. It occurred very close to Kobe City with shallow focal depth.

Damage was developed at highway bridges on Routes 2, 43, 171, and 176 of the National Highway, Route 3 (Kobe Line) and Route 5 (Bay Shore Line) of the Hanshin Expressway, and the Meishin and Chugoku Expressways. Damage was investigated for all bridges on national highways, the Hanshin Expressway, and expressways in the area where destructive damage occurred. The total number of piers surveyed reached 3396 [1]. Figure 44.1 shows Design Specifications referred to in the design of the 3396 highway bridges. Most of the bridges that suffered damage were designed according to the 1964 Design Specifications or the older Design Specifications. Although the seismic design methods have been improved and amended several times since 1926, only a requirement for lateral force coefficient was provided in the 1964 Design Specifications or the older Specifications.

Figure 44.2 compares damage of piers (bridges) on the Route 3 (Kobe Line) and Route 5 (Bay Shore Line) of the Hanshin Expressway. Damage degree was classified as A_s (collapse), A (nearly collapse), B (moderate damage), C (damage of secondary members), and D (minor or no damage). Substructures on Route 3 and Route 5 were designed with the 1964 Design Specifications and the 1980 Design Specifications, respectively. It should be noted in this comparison that the intensity of

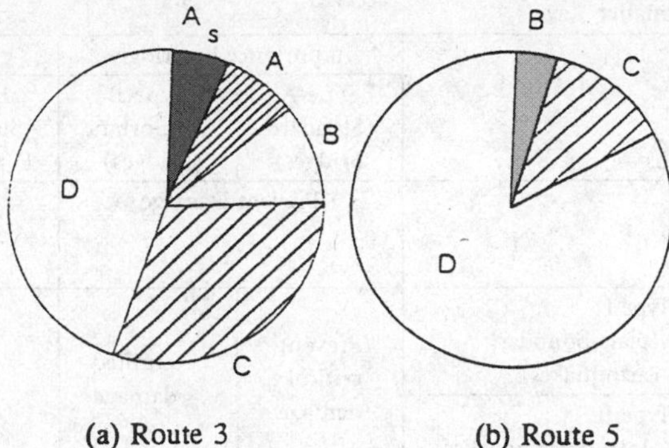

FIGURE 44.2 Comparison of damage degree between Route 3 (a) and Route 5 (b) (As: collapse, A: near collapse, B: moderate damage, C: damage of secondary members, D: minor or no damage) [2].

ground shaking in terms of response spectra was smaller at the Bay Area than the narrow rectangular area where JMA seismic intensity was VII (equivalent to modified Mercalli intensity of X-XI). Route 3 was located in the narrow rectangular area, while Route 5 was located in the Bay Area. Keeping in mind such differences in ground motion, it is apparent in Figure 44.2 that about 14% of the piers on Route 3 suffered As or A damage while no such damage was developed in the piers on Route 5.

Although damage concentrated on the bridges designed with the older Design Specifications, it was thought that essential revision was required even in the recent Design Specifications to prevent damage against destructive earthquakes such as the Hyogo-ken Nanbu earthquake. The main modifications were as follows:

1. To increase lateral capacity and ductility of all structural components in which seismic force is predominant so that ductility of a total bridge system is enhanced. For such purpose, it was required to upgrade the "Check of Ductility of Reinforced Concrete Piers," which has been used since 1990, to a "ductility design method" and to apply the ductility design method to all structural components. It should be noted here that "check" and "design" are different; the check is only to verify the safety of a structural member designed by another design method, and is effective only to increase the size or reinforcements if required, while the design is an essential procedure to determine the size and reinforcements.
2. To include the ground motion developed at Kobe in the earthquake as a design force in the ductility design method.
3. To specify input ground motions in terms of acceleration response spectra for dynamic response analysis more actively.
4. To increase tie reinforcements and to introduce intermediate ties for increasing ductility of piers. It was decided not to terminate longitudinal reinforcements at midheight to prevent premature shear failure, in principle.
5. To adopt multispan continuous bridges for increasing number of indeterminate of a total bridge system.
6. To adopt rubber bearings for absorbing lateral displacement between a superstructure and substructures and to consider correct mechanism of force transfer from a superstructure to substructures.
7. To include the Menshin design (seismic isolation).
8. To increase strength, ductility, and energy dissipation capacity of unseating prevention devices.
9. To consider the effect of lateral spreading associated with soil liquefaction in design of foundations at sites vulnerable to lateral spreading.

TABLE 44.2 Seismic Performance Levels

Type of Design Ground Motions		Importance of Bridges		Design Methods	
		Type-A (Standard Bridges)	Type-B (Important Bridges)	Equivalent Static Lateral Force Methods	Dynamic Analysis
Ground motions with high probability to occur		Prevent Damage		Seismic coefficient method	Step by Step analysis or Response spectrum analysis
Ground motions with low probability to occur	Type I (plate boundary earthquakes)	Prevent critical damage	Limited damage	Ductility design method	
	Type II (Inland earthquakes)				

44.4 1996 Seismic Design Specifications of Highway Bridges

44.4.1 Basic Principles of Seismic Design

The 1995 Hyogo-ken Nanbu earthquake, the first earthquake to be considered that such destructive damage could be prevented due to the progress of construction technology in recent years, provided a large impact on the earthquake disaster prevention measures in various fields. Part V: Seismic Design of the Design Specifications of Highway Bridges (Japan Road Association) was totally revised in 1996, and the design procedure moved from the traditional seismic coefficient method to the ductility design method. The revision was so comprehensive that the past revisions of the last 30 years look minor.

A major revision of the 1996 Specifications is the introduction of explicit two-level seismic design consisting of the seismic coefficient method and the ductility design method. Because Type I and Type II ground motions are considered in the ductility design method, three design seismic forces are used in design. Seismic performance for each design force is clearly defined in the Specifications.

Table 44.2 shows the seismic performance level provided in the 1996 Design Specifications. The bridges are categorized into two groups depending on their importance: standard bridges (Type A bridges) and important bridges (Type B bridges). The seismic performance level depends on the importance of the bridge. For moderate ground motions induced in earthquakes with a high probability of occurrence, both A and B bridges should behave in an elastic manner without essential structural damage. For extreme ground motions induced in earthquakes with a low probability of occurrence, Type A bridges should prevent critical failure, whereas Type B bridges should perform with limited damage.

In the ductility design method, two types of ground motions must be considered. The first is the ground motions that could be induced in plate boundary-type earthquakes with a magnitude of about 8. The ground motion at Tokyo in the 1923 Kanto earthquake is a typical target of this type of ground motion. The second is the ground motion developed in earthquakes with magnitude of about 7 to 7.2 at very short distance. Obviously, the ground motions at Kobe in the Hyogo-ken Nanbu earthquake is a typical target of this type of ground motion. The first and the second ground motions are called Type I and Type II ground motions, respectively. The recurrence time of Type II ground motion may be longer than that of Type I ground motion, although the estimation is very difficult.

The fact that lack of near-field strong motion records prevented serious evaluation of the validity of recent seismic design codes is important. The Hyogo-ken Nanbu earthquake revealed that the history of strong motion recording is very short, and that no near-field records have yet been measured by an earthquake with a magnitude on the order of 8. It is therefore essential to have sufficient redundancy and ductility in a total bridge system.

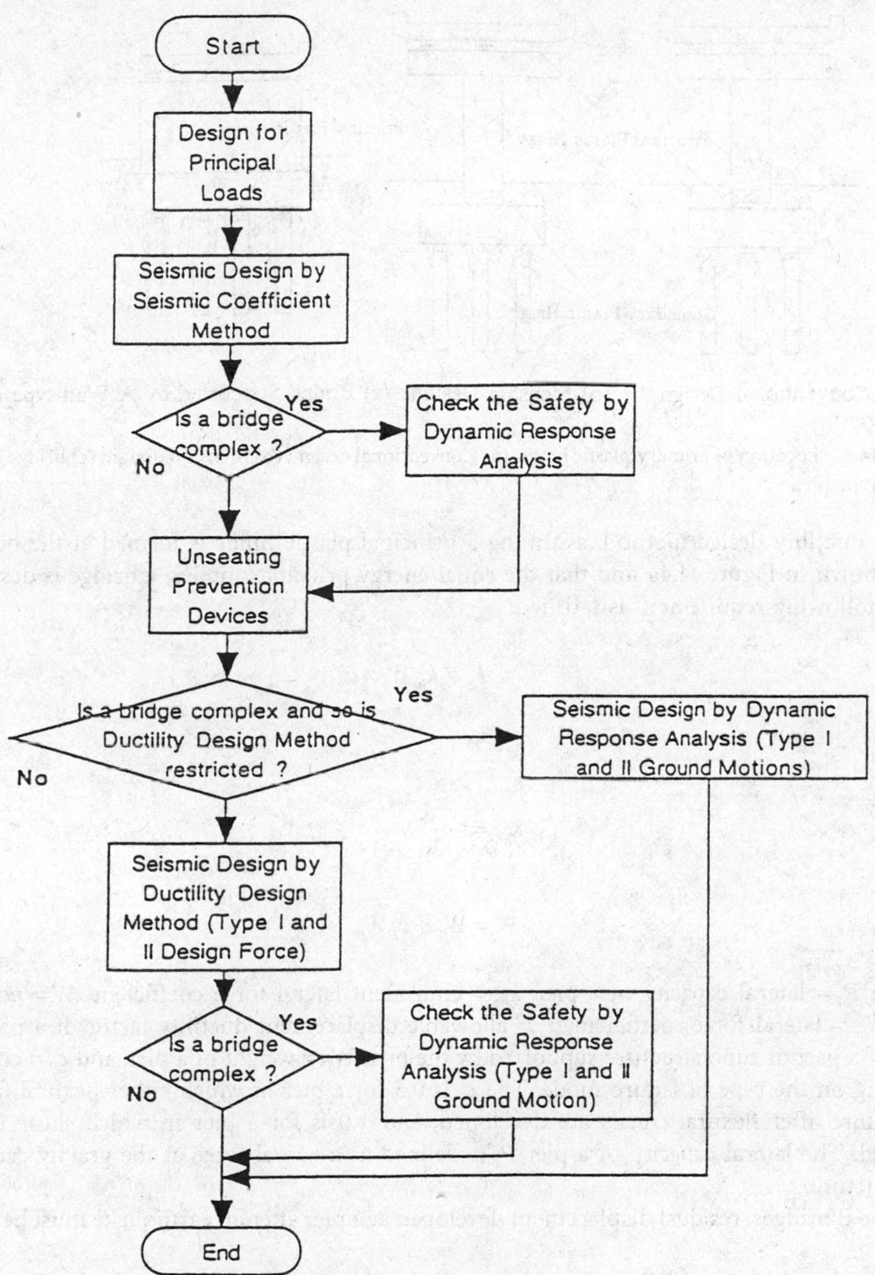

FIGURE 44.3 Flowchart of seismic design.

44.4.2 Design Methods

Bridges are designed by both the seismic coefficient method and the ductility design method as shown in Figure 44.3. In the seismic coefficient method, a lateral force coefficient ranging from 0.2 to 0.3 has been used based on the allowable stress design approach. No change has been made since the 1990 Specifications in the seismic coefficient method.

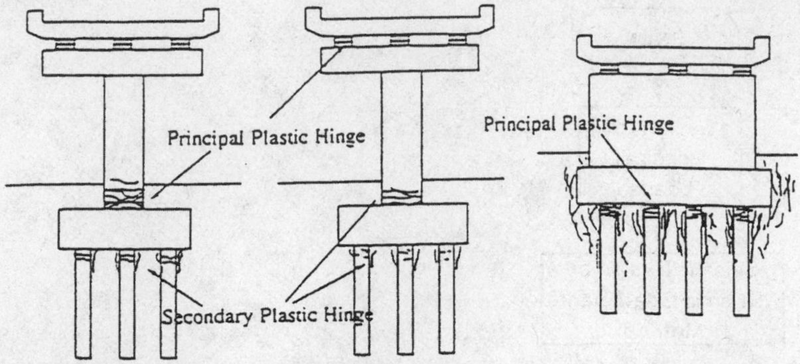

FIGURE 44.4 Location of primary plastic hinge. (a) Conventional design; (b) Menshin design; (c) bridge supported by a wall-type pier.

In the ductility design method, assuming a principal plastic hinge is formed at the bottom of pier as shown in Figure 44.4a and that the equal energy principle applies, a bridge is designed so that the following requirement is satisfied:

$$P_a > k_{he}W \tag{44.1}$$

where

$$k_{he} = \frac{k_{hc}}{\sqrt{2\mu_a - 1}} \tag{44.2}$$

$$W = W_U + c_P W_P \tag{44.3}$$

in which P_a = lateral capacity of a pier, k_{he} = equivalent lateral force coefficient, W = equivalent weight, k_{hc} = lateral force coefficient, μ_a = allowable displacement ductility factor of a pier, W_U = weight of a part of superstructure supported by the pier, W_P = weight of a pier, and c_P = coefficient depending on the type of failure mode. The c_P is 0.5 for a pier in which either flexural failure or shear failure after flexural cracks are developed, and 1.0 is for a pier in which shear failure is developed. The lateral capacity of a pier P_a is defined as a lateral force at the gravity center of a superstructure.

In Type B bridges, residual displacement developed at a pier after an earthquake must be checked as

$$\delta_R < \delta_{Ra} \tag{44.4}$$

where

$$\delta_R = c_R\left(\mu_R - 1\right)(1-r)\delta_y \tag{44.5}$$

$$\mu_R = 1/2\left\{\left(k_{hc}\cdot W/P_a\right)^2 + 1\right\} \tag{44.6}$$

in which δ_R = residual displacement of a pier after an earthquake, δ_{Ra} = allowable residual displacement of a pier, r = bilinear factor defined as a ratio between the first stiffness (yield stiffness) and the second stiffness (postyield stiffness) of a pier, c_R = factor depending on the bilinear factor r, μ_R = response ductility factor of a pier, and δ_y = yield displacement of a pier. The δ_{Ra} should be 1/100 of the distance between the bottom of a pier and the gravity center of a superstructure.

In a bridge with complex dynamic response, the dynamic response analysis is required to check the safety of a bridge after it is designed by the seismic coefficient method and the ductility design method. Because this is only for a check of the design, the size and reinforcements of structural members once determined by the seismic coefficient method and the ductility design methods may be increased if necessary. It should be noted, however, that under the following conditions in which the ductility design method is not directly applied, the size and reinforcements can be determined based on the results of a dynamic response analysis as shown in Figure 44.3. Situations when the ductility design method should not be directly used include:

1. When principal mode shapes that contribute to bridge response are different from the ones assumed in the ductility design methods
2. When more than two modes significantly contribute to bridge response
3. When principal plastic hinges form at more than two locations, or principal plastic hinges are not known where to be formed
4. When there are response modes for which the equal energy principle is not applied

In the seismic design of a foundation, a lateral force equivalent to the ultimate lateral capacity of a pier P_u is assumed to be a design force as

$$k_{hp} = c_{df}\,P_u/W \tag{44.7}$$

in which k_{hp} = lateral force coefficient for a foundation, c_{df} = modification coefficient (= 1.1), and W = equivalent weight by Eq. (44.3). Because the lateral capacity of a wall-type pier is very large in the transverse direction, the lateral seismic force evaluated by Eq. (44.7) in most cases becomes excessive. Therefore, if a foundation has sufficiently large lateral capacity compared with the lateral seismic force, the foundation is designed assuming a plastic hinge at the foundation and surrounding soils as shown in Figure 44.4c.

44.4.3 Design Seismic Force

Lateral force coefficient k_{hc} in Eq. (44.2) is given as

$$k_{hc} = c_z \cdot k_{hc0} \tag{44.8}$$

in which c_Z = modification coefficient for zone, and is 0.7, 0.85, and 1.0 depending on the zone, and k_{hc0} = standard modification coefficient. Table 44.3 and Figure 44.5 show the standard lateral force coefficients k_{hc0} for Type I and Type II ground motions. Type I ground motions have been used since 1990 (1990 Specifications), while Type II ground motions were newly introduced in the 1996 Specifications. It should be noted here that the k_{hc0} at stiff site (Group I) has been assumed smaller than the k_{hc0} at moderate (Group II) and soft soil (Group III) sites in Type I ground motions as well as the seismic coefficients used for the seismic coefficient method. Type I ground motions were essentially estimated from an attenuation equation for response spectra that is derived from a statistical analysis of 394 components of strong motion records. Although the response spectral accelerations at short natural period are larger at stiff sites than at soft soil sites, the tendency has not been explicitly included in the past. This was because damage has been more developed at soft sites than at stiff sites. To consider such a fact, the design force at stiff sites is assumed smaller than

TABLE 44.3 Lateral Force Coefficient k_{hc0} in the Ductility Design Method

Soil Condition	Lateral Force Coefficient k_{hc0}		
	Type I Ground Motion		
Group I (stiff)		$k_{hc0} = 0.7$ for $T \leqq 1.4$	$k_{hc0} = 0.876T^{2/3}$ for $T > 1.4$
Group II (moderate)	$k_{hc0} = 1.51T^{1/3}$ ($k_{hc0} \geqq 0.7$) for $T < 0.18$	$k_{hc0} = 0.85$ for $0.18 \leqq T \leqq 1.6$	$k_{hc0} = 1.16T^{2/3}$ for $T > 1.6$
Group III (soft)	$k_{hc0} = 1.51T^{1/3}$ ($k_{hc0} \geqq 0.7$) for $T < 0.29$	$k_{hc0} = 1.0$ for $0.29 \leqq T \leqq 2.0$	$k_{hc0} = 1.59T^{2/3}$ for $T > 2.0$
	Type II Ground Motion		
Group I (stiff)	$k_{hc0} = 4.46T^{2/3}$ for $T \leqq 0.3$	$k_{hc0} = 2.00$ for $0.3 \leqq T \leqq 0.7$	$k_{hc0} = 1.24T^{4/3}$ for $T > 0.7$
Group II (moderate)	$k_{hc0} = 3.22T^{2/3}$ for $T < 0.4$	$k_{hc0} = 1.75$ for $0.4 \leqq T \leqq 1.2$	$k_{hc0} = 2.23T^{4/3}$ for $T > 1.2$
Group III (soft)	$k_{hc0} = 2.38T^{2/3}$ for $T < 0.5$	$k_{hc0} = 1.50$ for $0.5 \leqq T \leqq 1.5$	$k_{hc0} = 2.57T^{4/3}$ for $T > 1.5$

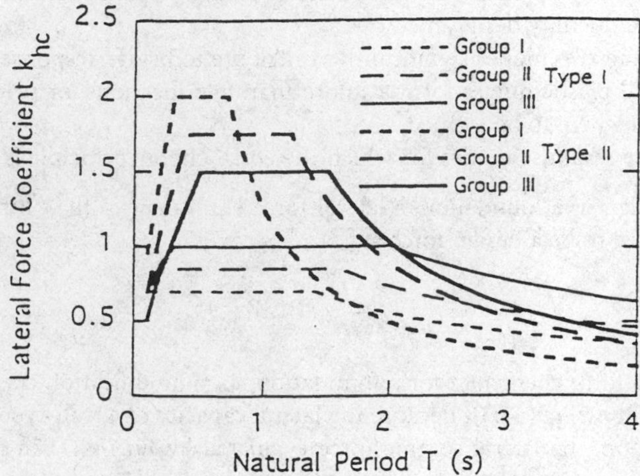

FIGURE 44.5 Type I and Type II ground motions in the ductility design method.

that at soft sites even at short natural period. However, being different from such a traditional consideration, Type II ground motions were determined by simply taking envelopes of response accelerations of major strong motions recorded at Kobe in the Hyogo-ken Nanbu earthquake.

Although the acceleration response spectral intensity at short natural period is higher in Type II ground motions than in Type I ground motions, the duration of extreme accelerations excursion is longer in Type I ground motions than Type II ground motions. As will be described later, such a difference of the duration has been taken into account to evaluate the allowable displacement ductility factor of a pier.

44.4.4 Ductility Design of Reinforced Concrete Piers

44.4.4.1 Evaluation of Failure Mode

In the ductility design of reinforced concrete piers, the failure mode of the pier is evaluated as the first step. Failure modes are categorized into three types based on the flexural and shear capacities of the pier as

1. $P_u \leqq P_s$ bending failure
2. $P_s \leq P_u \leqq P_{s0}$ bending to shear failure
3. $P_{s0} \leq P_u$ shear failure

in which P_u = bending capacity, P_s = shear capacity in consideration of the effect of cyclic loading, and P_{s0} = shear capacity without consideration of the effect of cyclic loading.

The ductility factor and capacity of the reinforced concrete piers are determined according to the failure mode as described later.

44.4.4.2 Displacement Ductility Factor

Th allowable displacement ductility factor of a pier μ_a in Eq. (44.2) is evaluated as

$$\mu_a = 1 + \frac{\delta_u - \delta_y}{\alpha \delta_y} \tag{44.9}$$

in which α = safety factor, δ_y = yield displacement of a pier, and δ_u = ultimate displacement of a pier. As well as the lateral capacity of a pier P_a in Eq. (44.1), the δ_y and δ_u are defined at the gravity center of a superstructure. In a reinforced concrete single pier as shown in Figure 44.4a, the ultimate displacement δ_u is evaluated as

$$\delta_u = \delta_y + (\phi_u - \phi_y) L_P (h - L_P/2) \tag{44.10}$$

in which ϕ_y = yield curvature of a pier at bottom, ϕ_u = ultimate curvature of a pier at bottom, h = height of a pier, and L_P = plastic hinge length of a pier. The plastic hinge length is given as

$$L_P = 0.2\,h - 0.1\,D \left(0.1\,D \leqq L_P \leqq 0.5D\right) \tag{44.11}$$

in which D is a width or a diameter of a pier.

The yield curvature ϕ_y and ultimate curvature ϕ_u in Eq. (44.10) are evaluated assuming a stress–strain relation of reinforcements and concrete as shown in Figure 44.6. The stress σ_c – strain ε_c relation of concrete with lateral confinement is assumed as

$$\sigma_c = \begin{cases} E_c \varepsilon_c \left(1 - \dfrac{1}{n}\left(\dfrac{\varepsilon_c}{\varepsilon_{cc}}\right)^{n-1}\right) & (0 \leqq \varepsilon_c \leqq \varepsilon_{cc}) \\ \sigma_{cc} - E_{des}(\varepsilon_c - \varepsilon_{cc}) & (\varepsilon_{cc} < \varepsilon_c \leqq \varepsilon_{cu}) \end{cases} \tag{44.12}$$

$$n = \frac{E_c \varepsilon_{cc}}{E_c \varepsilon_{oo} - \sigma_{cc}} \tag{44.13}$$

in which σ_{cc} = strength of confined concrete, E_c = elastic modules of concrete, ε_{cc} = strain at maximum strength, and E_{des} = gradient at descending branch. In Eq. (44.12), σ_{cc}, ε_{cc}, and E_{des} are determined as

$$\sigma_{cc} = \sigma_{ck} + 3.8\alpha\rho_s\sigma_{sy} \tag{44.14}$$

$$\varepsilon_{cc} = 0.002 + 0.033\,\beta\frac{\rho_s\sigma_{sy}}{\sigma_{ck}} \tag{44.15}$$

$$E_{des} = 11.2\frac{\sigma_{ck}^2}{\rho_s\sigma_{sy}} \tag{44.16}$$

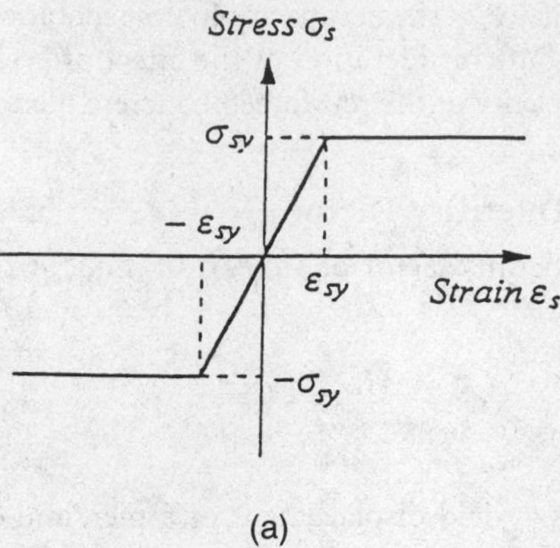

(a)

(b) Concrete

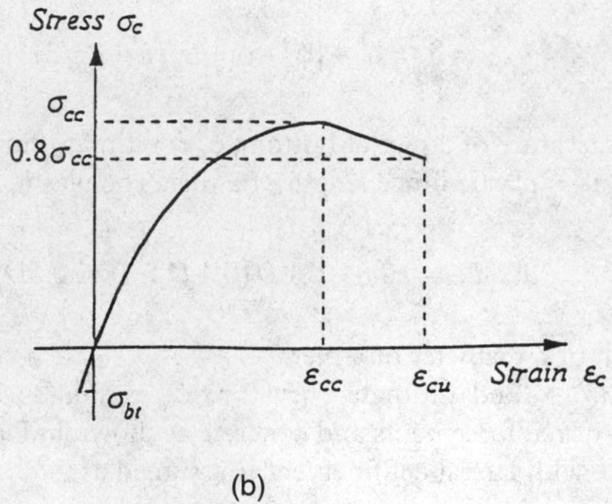

(b)

FIGURE 44.6 Stress and strain relation of confined concrete and reinforcing bars. (a) Steel (b) concrete.

in which σ_{ck} = design strength of concrete, σ_{sy} = yield strength of reinforcements, α and β = coefficients depending on shape of pier ($\alpha = 1.0$ and $\beta = 1.0$ for a circular pier, and $\alpha = 0.2$ and $\beta = 0.4$ for a rectangular pier), and ρ_s = tie reinforcement ratio defined as

$$\rho_s = \frac{4A_h}{sd} \leqq 0.018 \tag{44.17}$$

in which A_h = area of tie reinforcements, s = space of tie reinforcements, and d = effective width of tie reinforcements.

The ultimate curvature ϕ_u is defined as a curvature when concrete strain at longitudinal reinforcing bars in compression reaches an ultimate strain ε_{cu} defined as

$$\varepsilon_{cu} = \begin{cases} \varepsilon_{cc} & \text{for Type I ground motions} \\ \varepsilon_{cc} + \dfrac{0.2\sigma_{cc}}{E_{des}} & \text{for Type II ground motions} \end{cases} \tag{44.18}$$

It is important to note that the ultimate strain ε_{cu} depends on the types of ground motions; the ε_{cu} for Type II ground motions is larger than that for Type I ground motions. Based on a loading test,

TABLE 44.4 Safety Factor α in Eq. 44.9

Type of Bridges	Type I Ground Motion	Type II Ground Motion
Type B	3.0	1.5
Type A	2.4	1.2

TABLE 44.5 Modification Factor on Scale Effect for Shear Capacity Shared by Concrete

Effective Width of Section d (m)	Coefficient c_c
$d \leqq 1$	1.0
$d = 3$	0.7
$d = 5$	0.6
$d \geqq 10$	0.5

it is known that a certain level of failure in a pier such as a sudden decrease of lateral capacity occurs at smaller lateral displacement in a pier subjected to a loading hysteresis with a greater number of load reversals. To reflect such a fact, it was decided that the ultimate strain ε_{cu} should be evaluated by Eq. (44.18), depending on the type of ground motions. Therefore, the allowable ductility factor μ_a depends on the type of ground motions; the μ_a is larger in a pier subjected to Type II ground motions than a pier subjected to Type I ground motions.

It should be noted that the safety factor α in Eq. (44.9) depends on the type of bridges as well as the type of ground motions as shown in Table 44.4. This is to preserve higher seismic safety in the important bridges, and to take account of the difference of recurrent time between Type I and Type II ground motions.

44.4.4.3 Shear Capacity

Shear capacity of reinforced concrete piers is evaluated by a conventional method as

$$P_s = S_c + S_s \tag{44.19}$$

$$S_c = c_c c_e c_{pt} \tau_c bd \tag{44.20}$$

$$S_s = \frac{A_w \sigma_{sy} d\,(\sin\theta + \cos\theta)}{1.15a} \tag{44.21}$$

in which P_s = shear capacity; S_c = shear capacity shared by concrete; S_s = shear capacity shared by tie reinforcements, τ_c = shear stress capacity shared by concrete; c_c = modification factor for cyclic loading (0.6 for Type I ground motions; 0.8 for Type II ground motions); c_e = modification factor for scale effect of effective width; c_{pt} = modification factor for longitudinal reinforcement ratio; b and d = width and height of section, A_w = sectional area of tie reinforcement; σ_{sy} = yield strength of tie reinforcement, θ = angle between vertical axis and tie reinforcement, and a = space of tie reinforcement.

The modification factor on the scale effect of effective width, c_e, was based on experimental study of loading tests of beams with various effective heights and was newly introduced in the 1996 Specifications. Table 44.5 shows the modification factor on scale effect.

44.4.4.4 Arrangement of Reinforcement

Figure 44.7 shows a suggested arrangement of tie reinforcement. Tie reinforcement should be deformed bars with a diameter equal or larger than 13 mm, and it should be placed in most bridges

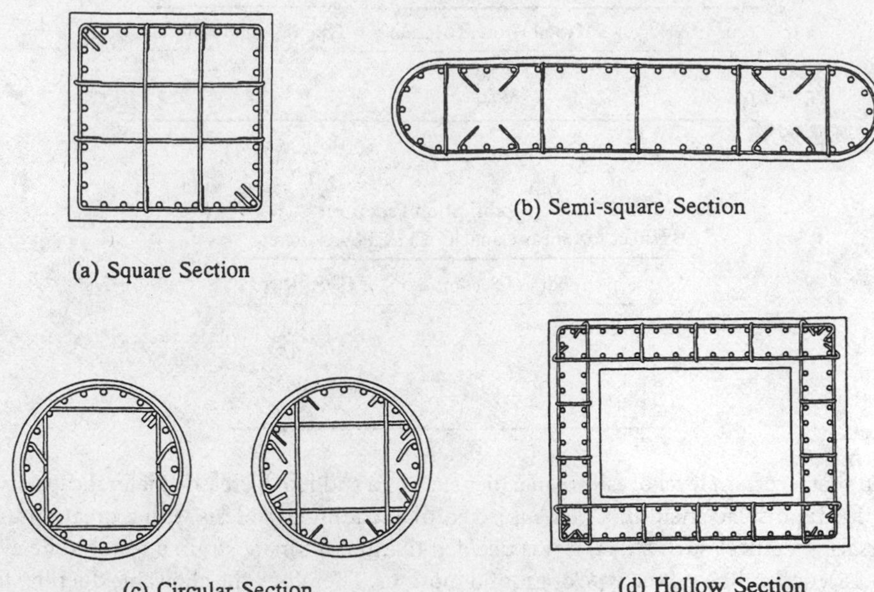

FIGURE 44.7 Confinement of core concrete by tie reinforcement. (a) Square section; (b) semisquare section; (c) circular section; (d) hollow section.

at a distance of no longer than 150 mm. In special cases, such as bridges with pier height taller than 30 m, the distance of tie reinforcement may be increased at height so that pier strength should not be sharply decreased at the section. Intermediate ties should be also provided with the same distance with the ties to confine the concrete. Space of the intermediate ties should be less than 1 m.

44.4.4.5 Two-Column Bent

To determine the ultimate strength and ductility factor for two-column bents, it is modeled as a frame model with plastic hinges at both ends of a lateral cap beam and columns as shown in Figure 44.8. Each elastic frame member has the yield stiffness which is obtained based on the axial load by the dead load of the superstructure and the column. The plastic hinge is assumed to be placed at the end part of a cap beam and the top and bottom part of each column. The plastic hinges are modeled as spring elements with a bilinear moment–curvature relation. The location of plastic hinges is half the distance of the plastic hinge length off from the end edge of each member, where the plastic hinge length L_P is assumed to be Eq. (44.11).

When the two-column bent is subjected to lateral force in the transverse direction, axial force developed in the beam and columns is affected by the applied lateral force. Therefore, the horizontal force–displacement relation is obtained through the static push-over analysis considering axial force N/moment M interaction relation. The ultimate state of each plastic hinge is obtained by the ultimate plastic angle θ_{pu} as

$$\theta_{pu} = \left(\phi_u/\phi_y - 1\right)L_P\phi_y \tag{44.22}$$

in which ϕ_u = ultimate curvature and ϕ_y = yield curvature.

The ultimate state of the whole two-bent column is determined so that all four plastic hinges developed reach the ultimate plastic angle.

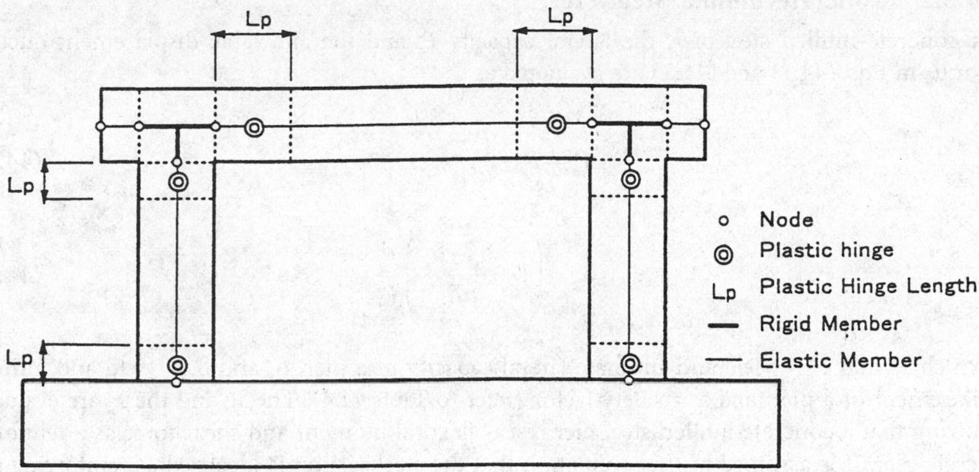

FIGURE 44.8 Analytical idealization of a two-column bent.

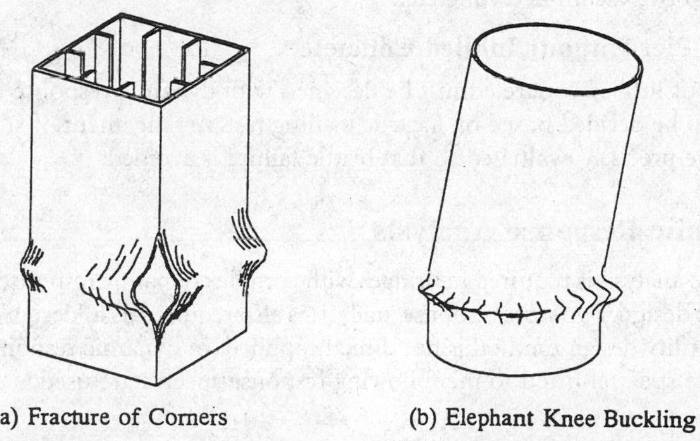

FIGURE 44.9 Typical brittle failure modes of steel piers. (a) Fracture of corners; (b) elephant knee buckling.

44.4.5 Ductility Design of Steel Piers

44.4.5.1 Basic Concept

To improve seismic performance of a steel pier, it is important to avoid specific brittle failure modes. Figure 44.9 shows the typical brittle failure mode for rectangular and circular steel piers. The following are the countermeasures to avoid such brittle failure modes and to improve seismic performance of steel piers:

1. Fill the steel column with concrete.
2. Improve structural parameters related to buckling strength.
   - Decrease the width–thickness ratio of stiffened plates of rectangular piers or the diameter–thickness ratio of steel pipes;
   - Increase the stiffness of stiffeners;
   - Reduce the diaphragm spacing;
   - Strengthen corners using the corner plates;
3. Improve welding section at the corners of rectangular section
4. Eliminate welding section at the corners by using round corners.

44.4.5.2 Concrete-Infilled Steel Pier

In a concrete-infilled steel pier, the lateral capacity P_a and the allowable displacement ductility factor μ_a in Eqs. (44.1) and (44.2) are evaluated as

$$P_a = P_y + \frac{P_u - P_y}{\alpha} \tag{44.23}$$

$$\mu_a = \left(1 + \frac{\delta_u - \delta_y}{\alpha \delta_y}\right) \frac{P_u}{P_a} \tag{44.24}$$

in which P_y and P_u = yield and ultimate lateral capacity of a pier; δ_y and δ_u = yield and ultimate displacement of a pier; and α = safety factor (refer to Table 44.4). The P_a and the μ_a are evaluated idealizing that a concrete-infilled steel pier resists flexural moment and shear force as a reinforced concrete pier. It is assumed in this evaluation that the steel section is idealized as reinforcing bars and that only the steel section resists axial force. A stress vs. strain relation of steel and concrete as shown in Figure 44.10 is assumed. The height of infilled concrete has to be decided so that bucking is not developed above the infilled concrete.

44.4.5.3 Steel Pier without Infilled Concrete

A steel pier without infilled concrete must be designed with dynamic response analysis. Properties of the pier need to be decided based on a cyclic loading test. Arrangement of stiffness and welding at corners must be precisely evaluated so that brittle failure is avoided.

44.4.6 Dynamic Response Analysis

Dynamic response analysis is required in bridges with complex dynamic response to check the safety factor of the static design. Dynamic response analysis is also required as a "design" tool in the bridges for which the ductility design method is not directly applied. In dynamic response analysis, ground motions which are spectral-fitted to the following response spectra are used;

$$S_I = c_Z \cdot c_D \cdot S_{I0} \tag{44.25}$$

$$S_{II} = c_Z \cdot c_D \cdot S_{II0} \tag{44.26}$$

in which S_{I} and S_{II} = acceleration response spectrum for Type I and Type II ground motions; $S_{\text{I}\,0}$ and $S_{\text{II}\,0}$ = standard acceleration response spectrum for Type I and Type II ground motions, respectively; c_Z = modification coefficient for zone, refer to Eq. (44.8); and c_D = modification coefficient for damping ratio given as

$$c_D = \frac{1.5}{40h_i + 1} + 0.5 \tag{44.27}$$

Table 44.6 and Figure 44.11 show the standard acceleration response spectra (damping ratio h = 0.05) for Type I and Type II ground motions.

It is recommended that at least three ground motions be used per analysis and that an average be taken to evaluate the response.

In dynamic analysis, modal damping ratios should be carefully evaluated. To determine the modal damping ratios, a bridge may be divided into several substructures in which the energy-dissipating mechanism is essentially the same. If one can specify a damping ratio of each substructure for a given mode shape, the modal damping ratio for the *i*th mode, h_i, may be evaluated as

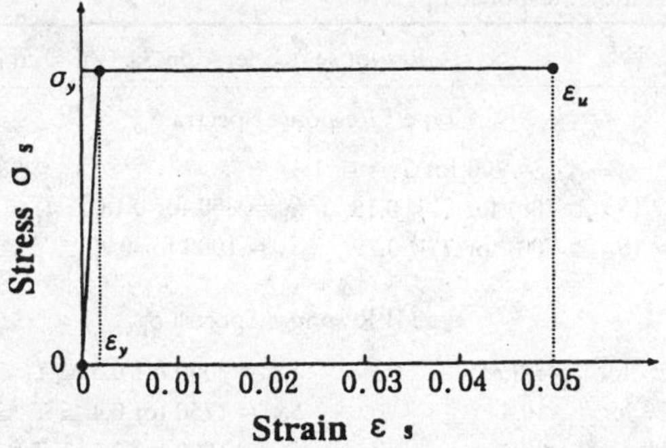

(a)

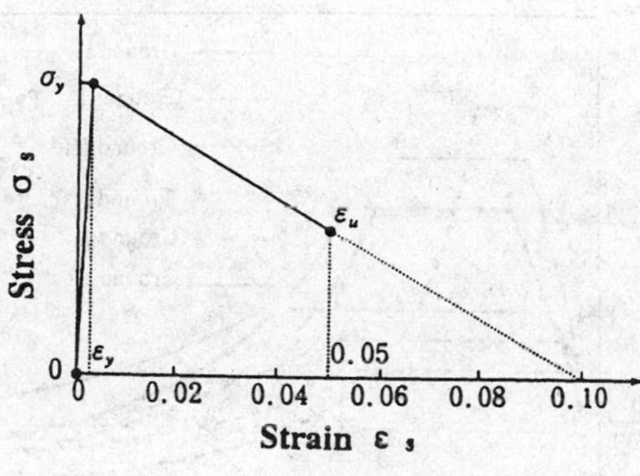

(b)

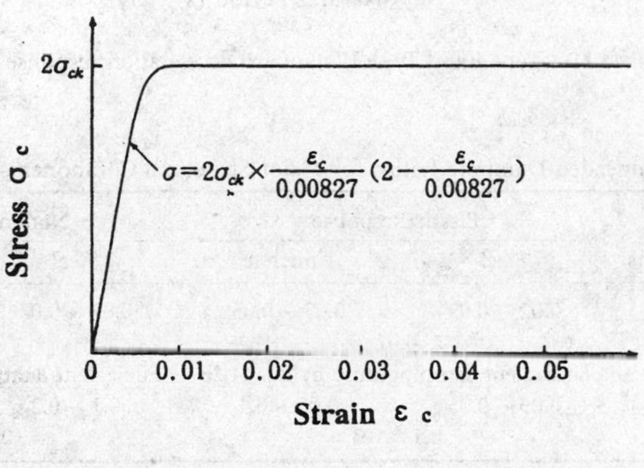

(c)

FIGURE 44.10 Stress–strain relation of steel and concrete. (a) Steel (tension); (b) steel (compression); (c) concrete.

$$h_i = \frac{\sum_{j=1}^{n} \phi_{ij}^T \cdot h_{ij} \cdot K_j \cdot \phi_{ij}}{\Phi_i^T \cdot K \cdot \Phi_i} \tag{44.28}$$

TABLE 44.6 Standard Acceleration Response Spectra

Soil Condition	Response Acceleration S_{10} (gal = cm/s²)		
	Type I Response Spectra S_{10}		
Group I	$S_{10} = 700$ for $T_i = \leqq 1.4$		$S_{10} = 980/T_i$ for $T_i > 1.4$
Group II	$S_{10} = 1505 T_i^{1/3}$ ($S_{10} \geqq 700$) for $T_i < 0.18$	$S_{10} = 850$ for $0.18 \leqq T_i \leqq 1.6$	$S_{10} = 1360/T_i$ for $T_i > 1.6$
Group III	$S_{10} = 1511 T_i^{1/3}$ ($S_{10} \geqq 700$) for $T_i < 0.29$	$S_{10} = 1000$ for $0.29 \leqq T_i \leqq 2.0$	$S_{10} = 2000/T_i$ for $T_i > 2.0$
	Type II Response Spectra S_{110}		
Group I	$S_{110} = 4463 T_i^{2/3}$ for $T_i \leqq 0.3$	$S_{110} = 2000$ for $0.3 \leqq T_i \leqq 0.7$	$S_{110} = 1104/T_i^{5/3}$ for $T_i > 0.7$
Group II	$S_{110} = 3224 T_i^{2/3}$ for $T_i < 0.4$	$S_{110} = 1750$ for $0.4 \leqq T_i \leqq 1.2$	$S_{110} = 2371/T_i^{5/3}$ for $T_i > 1.2$
Group III	$S_{110} = 2381 T_i^{2/3}$ for $T_i < 0.5$	$S_{110} = 1500$ for $0.5 \leqq T_i \leqq 1.5$	$S_{110} = 2948 T_i^{5/3}$ for $T_i > 1.5$

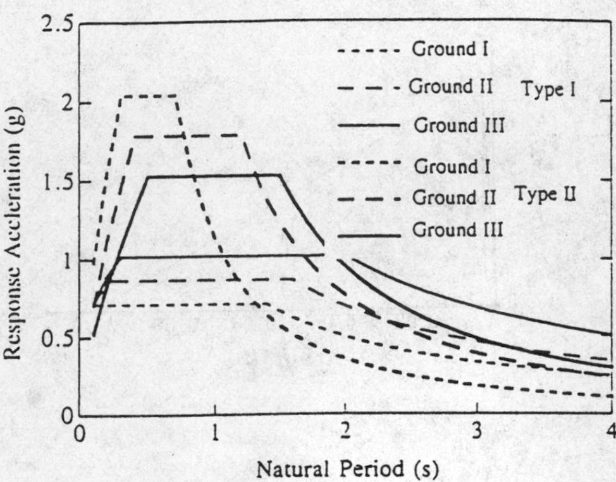

FIGURE 44.11 Type I and Type II standard acceleration response spectra.

TABLE 44.7 Recommended Damping Ratios for Major Structural Components

	Elastic Response		Nonlinear Response	
Structural Components	Steel	Concrete	Steel	Concrete
Superstructure	0.02 ~ 0.03	0.03 ~ 0.05	0.02 ~ 0.03	0.03 ~ 0.05
Rubber bearings	0.02		0.02	
Menshin bearings	Equivalent damping ratio by Eq. 44.26		Equivalent damping ratio by Eq. 44.46	
Substructures	0.03 ~ 0.05	0.05 ~ 0.1	0.1 ~ 0.2	0.12 ~ 0.2
Foundations	0.1 ~ 0.3		0.2 ~ 0.4	

in which h_{ij} = damping ratio of the *j*th substructure in the *i*th mode, ϕ_{ij} = mode vector of the *j*th substructure in the *i*th mode, k_j = stiffness matrix of the *j*th substructure, K = stiffness matrix of a bridge, and Φ_i = mode vector of a bridge in the *i*th mode, which is given as

$$\Phi_i^T = \left\{\phi_{i1}^T, \phi_{i2}^T, \ldots, \phi_{in}^T\right\} \tag{44.29}$$

Table 44.7 shows recommended damping ratios for major structural components.

TABLE 44.8 Modification Coefficient for Energy Dissipation Capability

Damping Ratio for First Mode h	Coefficient c_E
$h < 0.1$	1.0
$0.1 \leqq h < 0.12$	0.9
$0.12 \leqq h < 0.15$	0.8
$h \geqq 0.15$	0.7

44.4.7 Menshin (Seismic Isolation) Design

44.4.7.1 Basic Principle

Implementation of Menshin bridges should be carefully chosen from the point of view not only of seismic performance but also of function for traffic and maintenance, based on the advantage and disadvantage of increasing natural period. The Menshin design should not be adopted in the following situations:

1. Sites vulnerable to loss of bearing capacity due to soil liquefaction and lateral spreading;
2. Bridges supported by flexible columns;
3. Soft soil sites where potential resonance with surrounding soils could be developed by increasing the fundamental natural period; and
4. Bridges with uplift force at bearings.

It is suggested that the design be made with an emphasis on an increase of energy-dissipating capability and a distribution of lateral force to as many substructures as possible. To concentrate the hysteretic deformation not at piers, but at bearings, the fundamental natural period of a Menshin bridge should be about two times or more longer than the fundamental natural period of the same bridge supported by conventional bearings. It should be noted that an elongation of natural period aiming to decrease the lateral force should not be attempted.

44.4.7.2 Design Procedure

Menshin bridges are designed by both the seismic coefficient method and the ductility design method. In the seismic coefficient method, no reduction of lateral force from the conventional design is made.

In the ductility design method, the equivalent lateral force coefficient k_{hem} in the Menshin design is evaluated as

$$k_{hem} = \frac{k_{hcm}}{\sqrt{2\mu_m - 1}} \tag{44.30}$$

$$k_{hcm} = c_E \cdot k_{hc} \tag{44.31}$$

in which k_{hcm} = lateral force coefficient in Menshin design, μ_m = allowable ductility factor of a pier, c_E = modification coefficient for energy-dissipating capability (refer to Table 44.8), and k_{hc} = lateral force coefficient by Eq. (44.8). Because the k_{hc} is the lateral force coefficient for a bridge supported by conventional bearings, Eq. (44.31) means that the lateral force in the Menshin design can be reduced, as much as 30%, by the modification coefficient c_E depending on the modal damping ratio of a bridge.

Modal damping ratio of a menshin bridge h for the fundamental mode is computed as Eq. (44.32). In Eq. (44.32), h_{Bi} = damping ratio of the ith damper, h_{Pi} = damping ratio of the ith pier or abutment, h_{Fui} = damping ratio of the ith foundation associated with translational displacement, $h_{F\phi i}$ = damping ratio of the ith foundation associated with rotational displacement, K_{Pi} = equivalent stiffness of the ith pier or abutment, K_{Fui} = translational stiffness of the ith foundation, $K_{F\phi i}$ = rotational stiffness of the ith foundation, u_{Bi} = design displacement of the ith Menshin device, and H = distance from the bottom of a pier to a gravity center of a deck.

In the Menshin design, the allowable displacement ductility factor of a pier μ_m in Eq. (44.30) is evaluated by

$$h = \frac{\sum K_{Bi} \cdot u_{Bi}^2 \left(h_{Bi} + \frac{h_{Pi} \cdot K_{Bi}}{K_{Pi}} + \frac{h_{Fui} \cdot K_{Bi}}{K_{Fui}} + \frac{h_{F\theta i} \cdot K_{Bi} \cdot H^2}{K_{Fei}} \right)}{\sum K_{Bi} \cdot u_{Bi}^2 \left(1 + \frac{K_{Bi}}{K_{Pi}} + \frac{K_{Bi}}{K_{Fui}} + \frac{K_{Bi} \cdot H^2}{K_{F\theta i}} \right)} \tag{44.32}$$

$$\mu_m = 1 + \frac{\delta_u - \delta_y}{\alpha_m \delta_y} \tag{44.33}$$

in which α_m is a safety factor used in Menshin design and is given as

$$\alpha_m = 2\alpha \tag{44.34}$$

where α is the safety factor in the conventional design (refer to Table 44.4). Equation (44.34) means that the allowable displacement ductility factor in the Menshin design μ_m should be smaller than the allowable displacement ductility factor μ_a by Eq. (44.2) in the conventional design. The reason for the smaller allowable ductility factor in the Menshin design is to limit the hysteretic displacement of a pier at the plastic hinge zone so that the principal hysteretic behavior occurs at the Menshin devices, as shown in Figure 44.4b.

44.4.7.2 Design of Menshin Devices

Simple devices that can resist extreme earthquakes must be used. The bearings have to be anchored to a deck and substructures with bolts, and should be replaceable. Clearance has to be provided between a deck and an abutment or between adjacent decks.

Isolators and dampers must be designed for a desired design displacement *uB*. The design displacement u_B is evaluated as

$$u_B = \frac{k_{hem} W_U}{K_B} \tag{44.35}$$

in which k_{hem} = equivalent lateral force coefficient by Eq. (44.31), K_B = equivalent stiffness, and W_U = dead weight of a superstructure. It should be noted that, because the equivalent lateral force coefficient k_{hem} depends on the type of ground motions, the design displacement u_B also depends on the same.

The equivalent stiffness K_B and the equivalent damping ratio h_B of a Menshin device are evaluated as

$$K_B = \frac{F(u_{Be}) - F(-u_{Be})}{2u_{Be}} \tag{44.36}$$

$$h_B = \frac{\Delta W}{2\pi W} \tag{44.37}$$

$$u_{Be} = c_B \cdot u_B \tag{44.38}$$

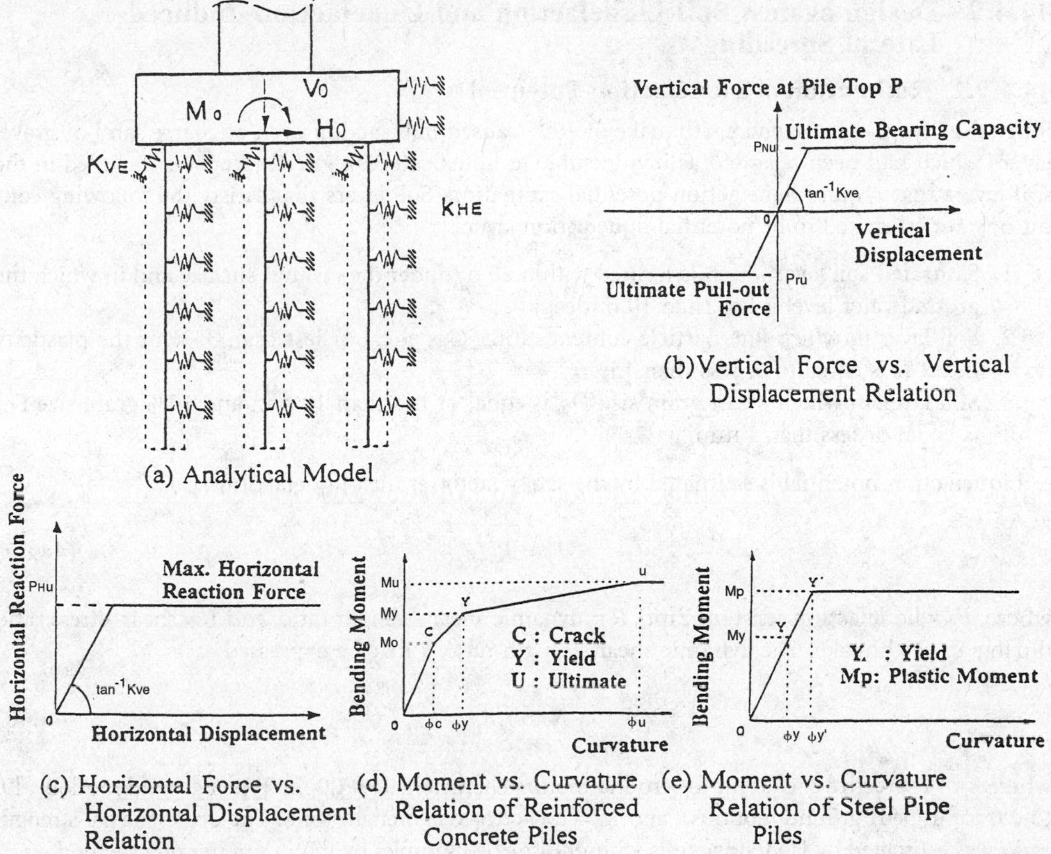

FIGURE 44.12 Idealized nonlinear model of a pile foundation. (a) Analybical model; (b) vertical force vs. vertical displacement relation; (c) horizontal force vs. horizontal displacement relation; (d) moment vs. curvature relation of reinforced concrete piles; (e) moment vs. curvature relation of steel pipe piles.

in which $F(u)$ = restoring force of a device at a displacement u, u_{Be} = effective design displacement, ΔW = energy dissipated per cycle, W = elastic strain energy, and c_B = coefficient to evaluate effective displacement (= 0.7).

44.4.8 Design of Foundations

The evaluation methods of ductility and strength of foundations such as pile foundations and caisson foundations were newly introduced in the 1996 Specifications.

For a pile foundation, a foundation should be so idealized that a rigid footing is supported by piles which are supported by soils. The flexural strength of a pier defined by Eq. (44.7) is to be applied as a seismic force to foundations at the bottom of the footing together with the dead-weight superstructure, pier, and soils on the footing. Figure 44.12 shows the idealized nonlinear model of a pile foundation. The nonlinearity of soils and piles is considered in the analysis.

The safety of the foundation is to be checked so that (1) the foundation does not reach its yield point; (2) if the primary nonlinearity is developed in the foundations, the response displacement is less than the displacement ductility limit; and (3) the displacement developed in the foundation is less than the allowable limit. The allowable ductility and the allowable limit of displacement were noted as 4 in displacement ductility, 40 cm in horizontal displacement, and 0.025 rad in rotation angle.

For a caisson-type foundation, the foundation should be modeled as a reinforced concrete column that is supported by soil spring model; the safety is checked in the same way as the pile foundations.

44.4.9 Design against Soil Liquefaction and Liquefaction-Induced Lateral Spreading

44.4.9.1 Estimation of Liquefaction Potential

Since the Hyogo-ken Nanbu earthquake of 1995 caused liquefaction even at coarse sand or gravel layers which had been regarded as invulnerable to liquefication, a gravel layer was included in the soil layers that require liquefaction potential estimation. Soil layers that satisfy the following conditions are estimated to be potential liquefaction layers:

1. Saturated soil layer which is located within 20 m under the ground surface and in which the groundwater level is less than 10 m deep;
2. Soil layer in which fine particle content ratio *FC* is equal or less than 35% or the plasticity index I_P is equal to or less than 15;
3. Soil layer in which mean grain size D_{50} is equal or less than 10 mm and 10% grain size D_{10} is equal or less than 1 mm.

Liquefaction potential is estimated by the safety factor against liquefaction F_L as

$$F_L = R/L \tag{44.39}$$

where, F_L = liquefaction resistant ratio, R = dynamic shear strength ratio, and L = shear–stress ratio during an earthquake. The dynamic shear strength ratio R may be expressed as

$$R = c_W R_L \tag{44.40}$$

where c_W = corrective coefficient for ground motion characteristics (1.0 for Type I ground motions, 1.0 to 2.0 for Type II ground motions), and R_L = cyclic triaxial strength ratio. The cyclic triaxial strength ratio was estimated by laboratory tests with undisturbed samples by the *in situ* freezing method.

The shear–stress ratio during an earthquake may be expressed as

$$L = r_d k_{hc} \sigma_v / \sigma'_v \tag{44.41}$$

where r_d = modification factor shear–stress ratio with depth, k_{hc} = design seismic coefficient for the evaluation of liquefaction potential, σ_v = total loading pressure, σ'_v = effective loading pressure.

It should be noted here that the design seismic coefficient for the evaluation of liquefaction potential k_{hc} ranges from 0.3 to 0.4 for Type I ground motions, and from 0.6 to 0.8 for Type II ground motions.

44.4.9.2 Design Treatment of Liquefaction for Bridge Foundations

When liquefaction occurs, the strength and the bearing capacity of a soil decreases. In the seismic design of highway bridges, soil constants of a sandy soil layer which is judged liable to liquefy are reduced according to the F_L value. The reduced soil constants are calculated by multiplying the coefficient D_E in Table 44.9 to the soil constants estimated on an assumption that the soil layer does not liquefy.

44.4.9.3 Design Treatment of Liquefaction-Induced Ground Flow for Bridge Foundations

The influence of liquefaction-induced ground flow was included in the revised Design Specifications in 1996. The case in which ground flow that may affect bridge seismicity is likely to occur is generally that the ground is judged to be liquefiable and is exposed to biased Earth pressure, for example, the ground behind a seaside protection wall. The effect of liquefaction-induced ground flow is

TABLE 44.9 Reduction Coefficient for Soil Constants Due to Soil Liquefaction

Range of F_L	Depth from the Present Ground Surface x (m)	Dynamic Shear Strength Ratio R	
		$R \leqq 0.3$	$0.3 < R$
$F_L \leqq 1/3$	$0 \leqq x \leqq 10$	0	1/6
	$10 < x \leqq 20$	1/3	1/3
$1/3 < F_L \leqq 2/3$	$0 \leqq x \leqq 10$	1/3	2/3
	$10 < x \leqq 20$	2/3	2/3
$2/3 < F_L \leqq 1$	$0 \leqq x \leqq 10$	2/3	1
	$10 < x \leqq 20$	1	1

considered as the static force acting on a structure. This method premises that the surface soil is of the nonliquefiable and liquefiable layers, and the forces equivalent to the passive Earth pressure and 30% of the overburden pressure are applied to the structure in the nonliquefiable layer and liquefiable layer, respectively.

The seismic safety of a foundation is checked by confirming that the displacement at the top of foundation caused by ground flow does not exceed an allowable value, in which a foundation and the ground are idealized as shown in Figure 44.12. The allowable displacement of a foundation may be taken as two times the yield displacement of a foundation. In this process, the inertia force of structure is not necessary to be considered simultaneously, because the liquefaction-induced ground flow may take place after the principal ground motion.

44.4.10 Bearing Supports

The bearings are classified into two groups: Type A bearings resisting the seismic force considered in the seismic coefficient method, and Type B bearings resisting the seismic force of Eq. (44.2). Seismic performance of Type B bearings is, of course, much higher than that of Type A bearings. In Type A bearings, a displacement-limiting device, which will be described later, has to be coinstalled in both longitudinal and transverse directions, while it is not required in Type B bearings. Because of the importance of bearings as one of the main structural components, Type B bearings should be used in Menshin bridges.

The uplift force applied to the bearing supports is specified as

$$R_U = R_D - \sqrt{R_{heq}^2 + R_{veq}^2} \tag{44.42}$$

in which R_U = design uplift force applied to the bearing support, R_D = dead load of superstructure, R_{heq} and R_{veq} are vertical reactions caused by the horizontal seismic force and vertical force, respectively. Figure 44.13 shows the design forces for the bearing supports.

44.4.11 Unseating Prevention Systems

Unseating prevention measures are required for highway bridges. Unseating prevention systems consist of enough seat length, a falling-down prevention device, a displacement-limiting device, and a settlement prevention device. The basic requirements are as follows:

1. The unseating prevention systems have to be so designed that unseating of a superstructure from its supports can be prevented even if unpredictable failures of structural members occur;
2. Enough seat length must be provided and a falling-down prevention device must be installed at the ends of a superstructure against longitudinal response. If Type A bearings are used, a displacement-limiting device has to be further installed at not only the ends of a superstructure but at each intermediate support in a continuous bridge; and

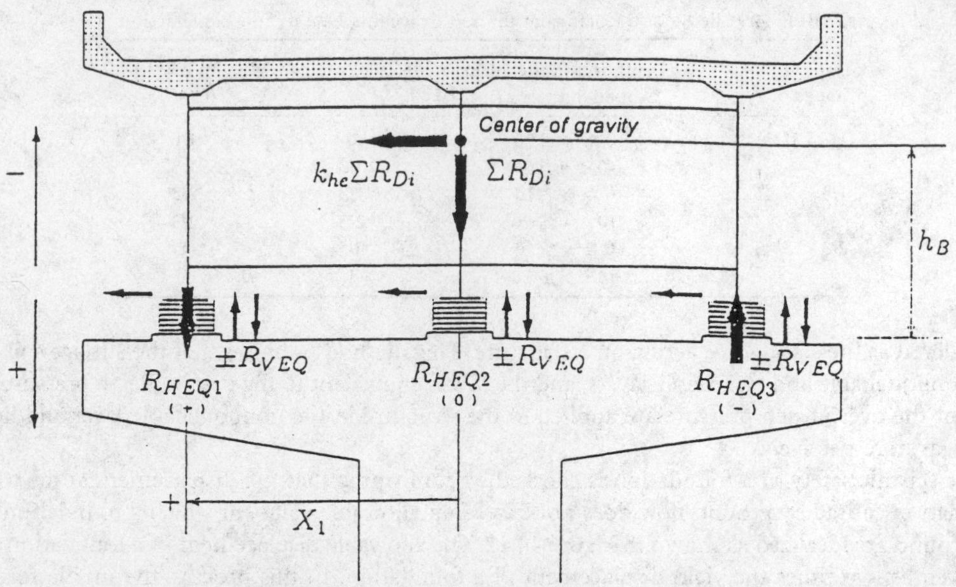

FIGURE 44.13 Design forces for bearing supports.

3. If Type A bearings are used, a displacement-limiting device is required at each support against transverse response. The displacement-limiting device is not generally required if Type B bearings are used. But, even if Type B bearings are adopted, it is required in skewed bridges, curved bridges, bridges supported by columns with narrow crests, bridges supported by few bearings per pier, and bridges constructed at sites vulnerable to lateral spreading associated with soil liquefaction.

The seat length S_E is evaluated as

$$S_E = u_R + u_G \geqq S_{EM} \tag{44.43}$$

$$S_{EM} = 70 + 0.5l \tag{44.44}$$

$$u_G = 100 \cdot \varepsilon_G \cdot L \tag{44.45}$$

in which u_R = relative displacement (cm) developed between a superstructure and a substructure subjected to a seismic force equivalent to the equivalent lateral force coefficient k_{hc} by Eq. (44.2); u_G = relative displacement of ground along the bridge axis; S_{EM} = minimum seat length (cm); ε_G = ground strain induced during an earthquake along the bridge axis, which is 0.0025, 0.00375, and 0.005 for Group I, II, and III sites, respectively; L = distance that contributes to the relative displacement of ground (m); and 1 = span length (m). If two adjacent decks are supported by a pier, the larger span length should be l in evaluating the seat length.

In the Menshin design, in addition to the above requirements, the following considerations have to be made.

1. To prevent collisions between a deck and an abutment or between two adjacent decks, enough clearance must be provided. The clearance between those structural components S_B should be evaluated as

$$S_B = \begin{cases} u_B + L_A & \text{between a deck and an abutment} \\ c_B \cdot u_B + L_A & \text{between two adjacent decks} \end{cases} \tag{44.46}$$

TABLE 44.10 Modification Coefficient for Clearance c_B

$\Delta\ T/T_1$	c_B
$0 \leqq \Delta\ T/T_1 < 0.1$	1
$0.1 \leqq \Delta\ T/T_1 < 0.8$	$\sqrt{2}$
$0.8 \leqq \Delta\ T/T_1 \leqq 1.0$	1

in which u_B = design displacement of Menshin devices (cm) by Eq. (44.39), L_A = redundancy of a clearance (generally ± 1.5 cm), and c_B = modification coefficient for clearance (refer to Table 44.10). The modification coefficient c_B was determined based on an analysis of the relative displacement response spectra. It depends on a difference of natural periods $\Delta T = T_1 - T_2$ ($T_1 > T_2$), in which T_1 and T_2 represent the natural period of the two adjacent bridge systems.

2. The clearance at an expansion joint L_E is evaluated as

$$L_E = u_B + L_A \tag{44.47}$$

in which u_B = design displacement of Menshin devices (cm) by Eq. (44.39), and L_A = redundancy of a clearance (generally ± 1.5 cm).

44.5 Seismic Retrofit Practices for Highway Bridges

44.5.1 Past Seismic Retrofit Practices

The Ministry of Construction has conducted seismic evaluations of highway bridges throughout the country five times since 1971 as a part of the comprehensive earthquake disaster prevention measures for highway facilities. Seismic retrofit for vulnerable highway bridges had been successively made based on the seismic evaluations. Table 44.11 shows the history of past seismic evaluations [7,8].

The first seismic evaluation was made in 1971 to promote earthquake disaster prevention measures for highway facilities. The significant damage of highway bridges caused by the 1971 San Fernando earthquake in the United States triggered the seismic evaluation. Highway bridges with span lengths longer than or equal to 5 m on all systems of national expressways and highways were evaluated. Attention was paid to detect deterioration such as cracks of reinforced concrete structures, tilting, sliding, settlement, and scouring of foundations. Approximately 18,000 highway bridges in total were evaluated and approximately 3200 bridges were found to require retrofit.

Following the first, seismic evaluations had been subsequently made in 1976, 1979, 1986, and 1991 with gradually expanding highways and evaluation items. The seismic evaluation in 1986 was made with the increase of social needs to ensure seismic safety of highway traffic after the damage caused by the Urakawa-oki earthquake in 1982 and the Nihon-kai-chubu earthquake in 1983. The highway bridges with span lengths longer than or equal to 15 m on all systems of national expressways, national highways and principal local highways, and overpasses were evaluated. The evaluation items included deterioration, unseating prevention devices, strength of substructures, and stability of foundations. Approximately 40,000 bridges in total were evaluated and approximately 11,800 bridges were found to require retrofit. The latest seismic evaluation was made in 1991. The number of highways to be evaluated has increased from the number evaluated in 1986. Approximately 60,000 bridges in total were evaluated and approximately 18,000 bridges were found to require retrofit. Through a series of seismic retrofit works, approximately 32,000 bridges were retrofitted by the end of 1994.

TABLE 44.11 Past Seismic Evaluations of Highway Bridges

Year	Highways Inspected	Inspection Items	Number of Bridges		
			Inspected	Require Strengthening	Strenghtened
1971	All sections of national expressways and national highways, and sections of others (bridge length ≧ 5m)	1. Deterioration 2. Bearing seat length *S* for bridges supported by bent piles	18,000	3,200	1,500
1976	All sections of national expressways and national highways, and sections of others (Bridge Length ≧ 15m or Overpass Bridges)	1. Deterioration of substructures, bearing supports, and girders/slabs 2. Bearing seat length *S* and devices for preventing falling-off of superstructure	25,000	7,000	2,500
1979	All sections of national expressways, national highways, and principallocal highways, and sections of others (bridge length ≧ 15 m or overpass bridges)	1. Deterioration of substructures and bearing supports 2. Devices for preventing falling-off of superstructure 3. Effect of soil liquefaction 4. Bearing capacity of soils and piles 5. Strength of RC piers 6. Vulnerable foundations (bent pile and RC frame on two independent caisson founcations)	35,000	16,000	13,000
1986	All sections of national expressways, national highways and principal local highways, and sections of others (bridge length ≧ 15 m or overpass bridges)	1. Deterioration of substructures, bearing supports, and concrete girders 2. Devices for preventing falling-off of superstructure 3. Effect of soil liquefaction 4. Strength of RC piers (bottom of piers and termination zone of main reinforcement) 5. Bearing capacity of piles 6. Vulnerable foundations (bent piles and RC frame on two independent caisson foundations)	40,000	11,800	8,000
1991	All sections of national expressways, national highways and principal local highways, and sections of others (bridge length ≧ 15 m or overpass bridges)	1. Deterioration of substructures, bearing supports, and concrete girders 2. Devices for preventing falling-off of superstructure 3. Effect of soil liquifaction 4. Strength of RC piers (piers and termination zone of main reinforcement) 5. Vulnerable foundations (bent piles and RC frame on two independent caisson foundations)	60,000	18,000	7,000 (as of the end of 1994)

Note: Number of bridges inspected, number of bridges that required strengthening, and number of bridges strengthened are approximate numbers.

TABLE 44.12 Application of the Guide Specifications

Types of Roads and Bridges	Double Deckers, Overcrossings on Roads and Railways, Extremely Important Bridges from Disaster Prevention and Road Network	Others
Expressways, urban expressways, designated urban expressway, Honshu–Shikoku Bridges, designated national highways	Apply all items, in principle	Apply all items, in principle
Nondesignated national highways, prefectural roads, city, town, and village roads	Apply all items, in principle	Apply partially, in principle

The seismic evaluations in 1986 and 1991 were made based on a statistical analysis of bridges damaged and undamaged in the past earthquakes [9]. Because the collapse of bridges tends to develop because of excessive relative movement between the superstructure and the substructures and the failure of substructures associated with inadequate strength, the evaluation was made based on both the relative movement and the strength of the substructure.

Emphasis had been placed on installing unseating prevention devices in the past seismic retrofit. Because the installation of the unseating prevention devices was being completed, it had become important to promote strengthening of those substructures with inadequate strength and lateral stiffness.

44.5.2 Seismic Retrofit after the Hyogo-ken Nanbu Earthquake

44.5.2.1 Reference for Applying Guide Specifications to New Highway Bridges and Seismic Retrofit of Existing Highway Bridges

After the 1995 Hyogo-ken Nanbu earthquake, the "Part V: Seismic Design" of the "Design Specifications of Highway Bridges" (Japan Road Association) was completely revised in 1996 as discussed in the previous sections.

Because most of the substructures designed and constructed before 1971 do not meet the current seismic requirements, it is urgently needed to study the level of seismic vulnerability requiring retrofit. Upgrading the reliability of predictions of possible failure modes in future earthquakes is also very important. Since the seismic retrofit of substructures requires more cost, it is necessary to develop and implement effective and inexpensive retrofit measures and to design methods to provide for the next event.

For increasing seismic safety of the highway bridges that suffered damage by the Hyogo-ken Nanbu earthquake, various new drastic changes were tentatively introduced in the "Guide Specifications for Reconstruction and Repair of Highway Bridges Which Suffered Damage Due to the Hyogo-ken Nanbu Earthquake." Although intensified review of design could be made when it was applied to the bridges only in the Hanshin area, it may not be so easy for field design engineers to follow up the new Guide Specifications when the Guide Specifications is used for seismic design of all new highway bridges and seismic strengthening of existing highway bridges. Based on such demand, the "Reference for Applying the Guide Specifications to New Bridges and Seismic Strengthening of Existing Bridges" [10] was issued on June 30, 1995 by the Sub-Committee for Seismic Countermeasures for Highway Bridges, Japan Road Association.

The Reference classified the application of the Guide Specifications as shown in Table 44.12 based on the importance of the roads. All items of the Guide Specifications are applied for bridges on extremely important roads, while some items which prevent brittle failure of structural components are applied for bridges on important roads. For example, for bridges on important roads, the items for Menshin design, tie reinforcements, termination of longitudinal reinforcements, type of bearings, unseating prevention devices and countermeasures for soil liquefaction are applied, while the remaining items such as the design force, concrete-infilled steel bridges, and ductility check for foundations, are not applied.

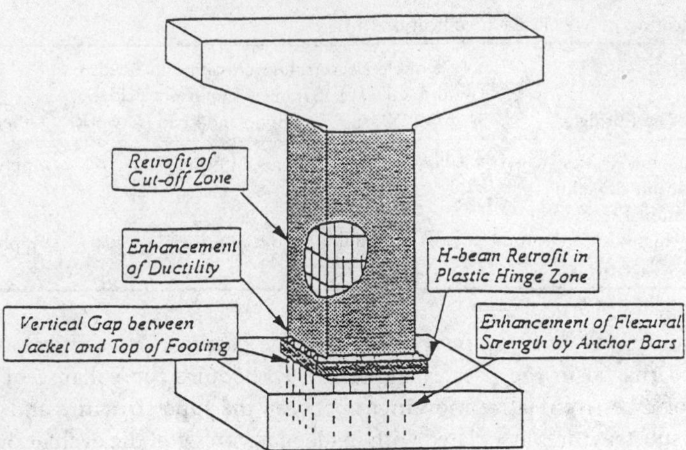

FIGURE 44.14 Seismic retrofit of reinforced concrete piers by steel jacket with controlled increase of flexural strength.

Because damage was concentrated in single reinforced concrete piers/columns with small concrete sections, a seismic retrofit program has been initiated for those columns that were designed according to the pre-1980 Design Specifications, at extremely important bridges such as bridges on expressways, urban expressways, and designated highway bridges, and also double-deckers and overcrossings, etc. which significantly affect highway functions once damaged. In the 3-year program, approximately 30,000 piers will be evaluated and retrofitted. Unseating devices also should be installed for these extremely important bridges.

The main purpose of the seismic retrofit of reinforced concrete columns is to increase their shear strength, in particular in piers with termination of longitudinal reinforcements without enough anchoring length. This increases the ductility of columns, because premature shear failure can be avoided.

However, if only ductility of piers is increased, residual displacement developed at piers after an earthquake may increase. Therefore, the flexural strength should also be increased. However, the increase of flexural strength of piers tends to increase the seismic force transferred from the piers to the foundations. It was found from an analysis of various types of foundations that failure of the foundations by increasing the seismic force may not be significant if the increasing rate of the flexural strength of piers is less than two. It is therefore suggested to increase the flexural strength of piers within this limit so that it does not cause serious damage to foundations.

For such requirements, seismic strengthening by steel jackets with controlled increase of flexural strength was suggested [10, 11]. This uses a steel jacket surrounding the existing columns as shown in Figure 44.14. Epoxy resin or nonshrinkage concrete mortar are injected between the concrete surface and the steel jacket. A small gap is provided at the bottom of piers between the steel jacket and the top of the footing. This prevents excessive increase in the flexural strength.

To increase the flexural strength of columns in a controlled manner, anchor bolts are provided at the bottom of the steel jacket. They are drilled into the footing. By selecting an appropriate number and size of the anchor bolts, the degree of increase of the flexural strength of piers may be controlled. The gap is required to trigger the flexural failure at the bottom of columns. A series of loading tests are being conducted at the Public Works Research Institute to check the appropriate gap and number of anchor bolts. Table 44.13 shows a tentatively suggested thickness of steel jackets and size and number of anchor bolts. They are for reinforced concrete columns with a/b less than 3, in which a and b represent the width of a column in transverse and longitudinal direction, respectively. The size and number of anchor bolts were evaluated so that the increasing rate of flexural strength of columns is less than about 2.

TABLE 44.13 Tentative Retrofit Method by Steel Jacketing

Column/Piers	Steel Jackets	Anchor Bolts
$a/b \leqq 2$	SM400, $t = 9$ mm	SD295, D35 ctc 250 mm
$2 < a/b \leqq 3$ Column supporting lateral force of a continuous girder through fixed bearing and with $a/b \leqq 3$	SM400, $t = 12$ mm	

Conventional reinforced concrete jacketing methods are also applied for the retrofit of reinforced concrete piers, especially for piers that require an increase of strength. It should be noted here that the increase of the strength of the pier should be carefully designed in consideration with the strength of foundations and footings.

44.5.2.2 Research and Development on Seismic Evaluation and Retrofit of Highway Bridges

Prioritization Concept for Seismic Evaluation

The 3-year retrofit program was completed in the 1997 fiscal year. In the program, the single reinforced concrete piers/columns with small concrete section which were designed by the pre-1980 Design Specifications on important highways have been evaluated and retrofitted and other bridges with wall-type piers, steel piers, and frame piers, and so on, as well as the bridges on the other highways, should be evaluated and retrofitted if required in the next retrofit program. Since there are approximately 200,000 piers, it is required to develop prioritization methods and methods to evaluate vulnerability for the intentional retrofit program.

Figure 44.15 shows the simple flowchart to prioritize the retrofit work to bridges. The importance of the highway, structural factors, member vulnerability (reinforced concrete piers, steel piers, unseating prevention devices, foundations) are the factors to be considered for prioritization.

Priority R of each bridge may be evaluated by Eq. (44.48).

$$R = I \cdot S \cdot V_T \cdot w_v \cdot \left[f\left(V_{RP1}, V_{RP2}, V_{RP3}\right), V_{MP}, V_{FS}, V_F \right] \times 100 \tag{44.48}$$

$$f\left(V_{RP1}, V_{RP2}, V_{RP3}\right) = V_{RP1} \cdot V_{RP2} \cdot V_{RP3} \tag{44.49}$$

in which R = priority, I = importance factor, S = earthquake force, V_T = structural factor, w_v = weighting factor on structural members, V_{RP1} = design specification, V_{RP2} pier structural factor, V_{RP3} = aspect ratio, V_{MP} = steel pier factor, V_{FS} = unseating device factor, and V_F = foundation factor. Each item and category with a weighting number is tentatively shown in Table 44.14. If this prioritization method is to applied the bridges damaged during the Hyogo-ken Nanbu earthquake, the categorization number is given as shown in Table 44.14.

Seismic Retrofit of Wall-Type Piers

The steel-jacketing method as described in the above was applied for reinforced concrete with circular section or rectangular section of $a/b < 3$. It is required to develop the seismic retrofit method for a wall-type pier. The confinement of concrete was provided by a confinement beam such as the H-shaped steel beam for rectangular piers. However, since the size of the confinement beam becomes very large, the confinement may be provided by other measures, such as intermediate anchors for a wall-type pier.

The seismic retrofit concept for a wall-type pier is the same as that for rectangular piers. It is important to increase the flexural strength and ductility capacity with the appropriate balance. Generally, the longitudinal reinforcement ratio is smaller than that for rectangular piers; therefore, the flexural strength is smaller. Thus, it is essential to increase the flexural strength appropriately. Since the longitudinal

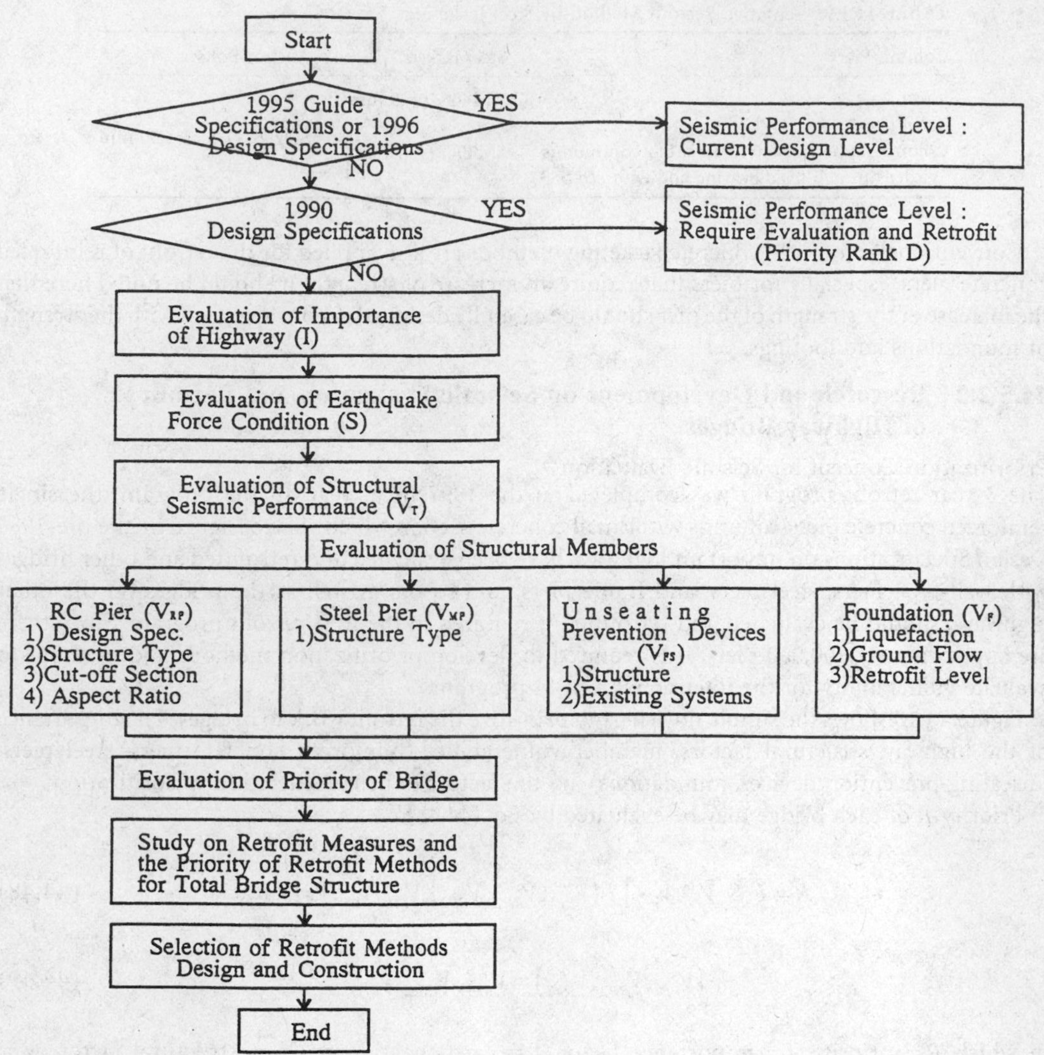

FIGURE 44.15 Prioritization concept of seismic retrofit of highway bridges.

reinforcement was generally terminated at midheight without appropriate anchorage length, it is also important to strengthen both the flexural and shear strength midheight section.

Figure 44.16 shows the possible seismic retrofit method for wall-type piers. To increase the flexural strength, the additional reinforcement by rebars or anchor bars are fixed to the footing. The number of reinforcements is designed to give the necessary flexural strength. It should be noted here that anchoring of additional longitudinal reinforcement is controlled to develop plastic hinge to the bottom of pier rather than the midheight section with termination of longitudinal reinforcement. And the increase of strength should be carefully designed considering the effect on the foundations and footings. The confinement in the plastic hinge zone is provided by steel bars for prestressed concrete or rebars which were installed inside of the column section.

Seismic Retrofit of Two-Column Bents

During the Hyogo-ken Nanbu earthquake, some two-column bents were damaged in the longitudinal and transverse directions. The strength and ductility characteristics of the two-column bents have been studied and the analysis and design method was introduced in the 1996 Design Specifications [12].

TABLE 44.14 Example of Prioritization Factors for Seismic Retrofit of Highway Bridges

Item	Category	Evaluation Point
Importance of highway (I)	1. Emergency routes	1.0
	2. Overcrossing with emergency routes	0.9
	3. Others	0.6
Earthquake force (S)	1. Ground condition Type I	1.0
	2. Ground condition Type II	0.9
	3. Ground condition Type III	0.8
Structural factor (V_τ)	1. Viaducts	1.0
	2. Supported by abutments at both ends	0.5
Weighting factor on structural members (V_τ)	1. Reinforced concrete pier	1.0
	2. Steel pier	0.95
	3. Unseating prevention devices	0.9
	4. Foundation	0.8
Reinforced concrete pier 1. Design specification (V_{RP1})	1. Pre-1980 Design Specifications	1.0
	2. Post-1980 Design Specifications	0.7
2. Pier structure (V_{RP2})	1. Single column	1.0
	2. Wall-type column	0.8
	3. Two-column bent	0.7
3. Aspect ratio (V_{RP3})	1. h/D ≦ 3	1.0
	2. 3 < h/D < 4 with cutoff section	0.9
	3. H/D ≧ 4 with cutoff section	0.9
	4. $3 < h/D < 4$ without cutoff section	0.7
	5. $H/D \geqq 4$ without cutoff section	0.7
Steel pier (V_{MP})	1. Single column	1.0
	2. Frame structure	0.8
Unseating prevention devices (V_{FS})	1. Without unseating devices	1.0
	2. With one device	0.9
	3. With two devices	0.8
Foundations (V_F)	1. Vulnerable to Ground Flow (without unseating devices)	1.0
	2. Vulnerable to Ground Flow	0.9
	3. Vulnerable to Liquefaction (without unseating devices)	0.7
	4. Vulnerable to Liquefaction	0.6
Evaluation of the priority R	1. $R \geqq 0.8$	Priority Rank A
	2. $0.7 \leqq R < 0.8$	Priority Rank B
	3. $R < 0.7$	Priority Rank C

The strength and ductility of existing two-column bents were studied both in the longitudinal and transverse directions. In the longitudinal direction, the same as a single column, it is required to increase the flexural strength and ductility with appropriate balance. In the transverse direction, the shear strength of the columns or the cap beam is generally not enough in comparison with the flexural strength.

Figure 44.17 shows the possible seismic retrofit methods for two-column bents. The concept of the retrofit is to increase flexural strength and ductility as well as shear capacity for columns and cap beams. Since axial force in the cap beam is much smaller than that in the columns, increasing the shear capacity is essential for the retrofit of the cap beam. It should be noted that since the jacketing of cap beam is difficult because of the existing bearing supports and construction space, it is required to develop more effective retrofit measures for cap beams such as application of jacketing by new materials with high modulus of elasticity and high strength and out-cable pre-stressing, etc.

Seismic Retrofit Using New Materials

Retrofit work is often restricted because construction space is limited to open the structure for public traffic, particularly for the seismic retrofit of highway bridges in urban areas [13]. Therefore, there are sites where conventional steel jacketing and reinforced concrete jacketing methods are

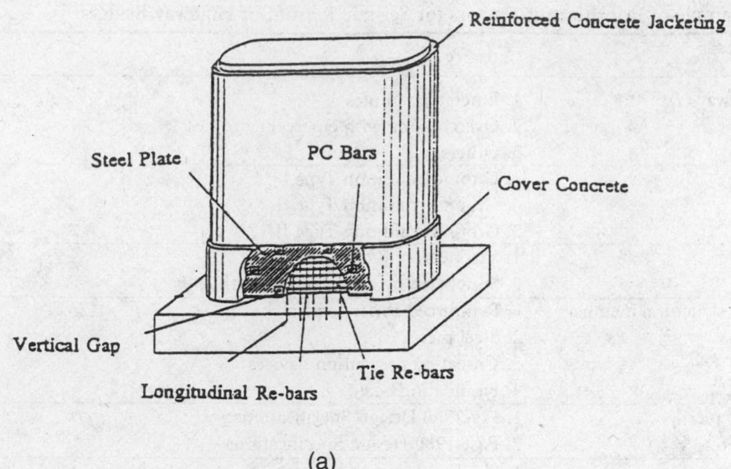

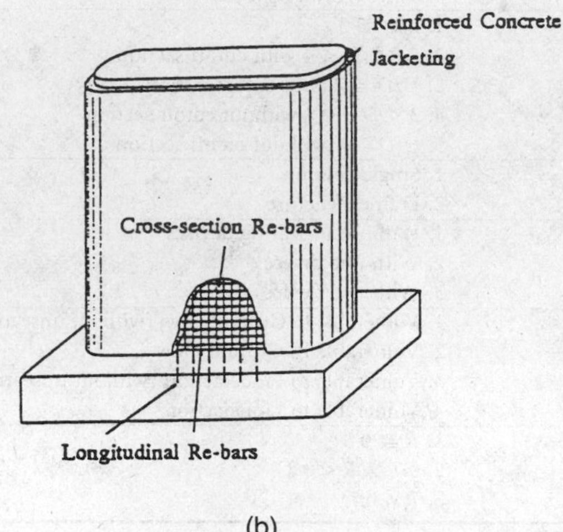

FIGURE 44.16 Seismic retrofit of wall-type piers. (a) Integrated seismic retrofit method with reinforced concrete and steel jacketing; (b) reinforced concrete jacketing.

difficult to apply. New materials such as carbon fiber sheets and aramid fiber sheets are attractive for application in the seismic retrofit of such bridges with construction restrictions as shown in Figure 44.18. The new materials such as fiber sheets are very light, do not need machines for use, and are easy to construct using glue bond as epoxy resin.

There are various studies on seismic retrofit methods using fiber sheets. Figure 44.19 shows the cooperative effect between fiber sheets and reinforcement for shear strengthening of a single reinforced concrete column. When carbon fiber sheets, which have almost the same elasticity and 10 times the failure strength as those of a reinforcing bar, are assumed to be applied, it is important to design the effects of carbon fiber sheets to achieve the required performance of seismic retrofit. In particular, strengthening of flexural, shear capacities, and ductility for reinforced concrete columns should be carefully evaluated. Based on experimental studies, it is essential to evaluate appropriately the effect of materials on the strengthening, carefully considering the material properties such as the modulus of elasticity and strength.

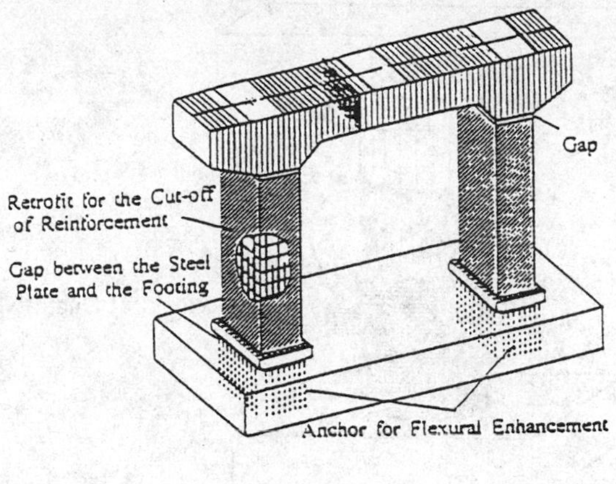

(a)

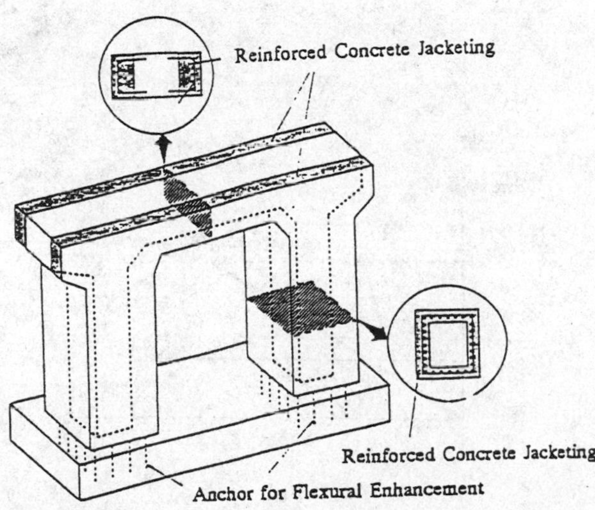

(b)

FIGURE 44.17 Seismic retrofit of two-column bents. (a) Steel jacketing; (b) reinforced concrete jacketing.

Acknowledgments

Drafting of the revised version of the "Part V: Seismic Design" of the "Design Specifications of Highway Bridges" was conducted by the Special Sub-Committee for Seismic Countermeasures for Highway Bridges" and was approved by the Bridge Committee, Japan Road Association. Dr. Kazuhiko Kawashima, Professor of the Tokyo Institute of Technology, chairman of the Special Sub-Committee and all other members of the Special Sub-Committee and the Bridge Committee are gratefully acknowledged.

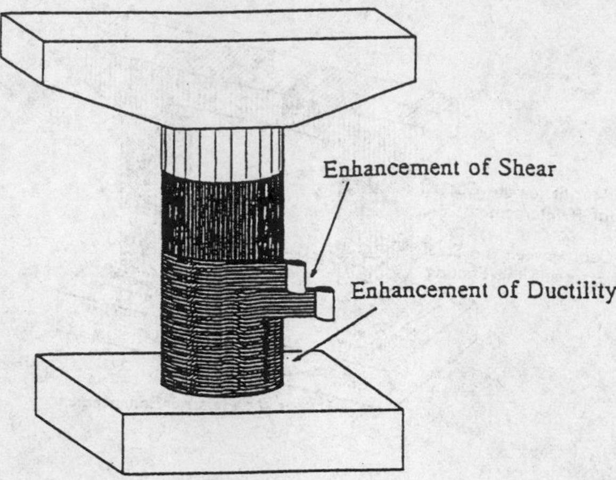

FIGURE 44.18 Application to new materials for seismic retrofit of reinforced column.

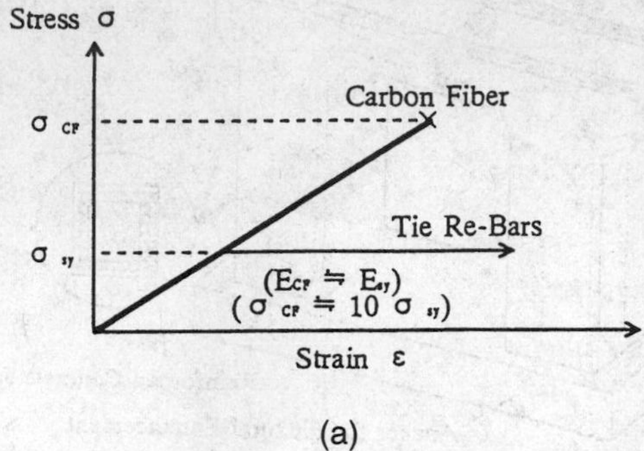

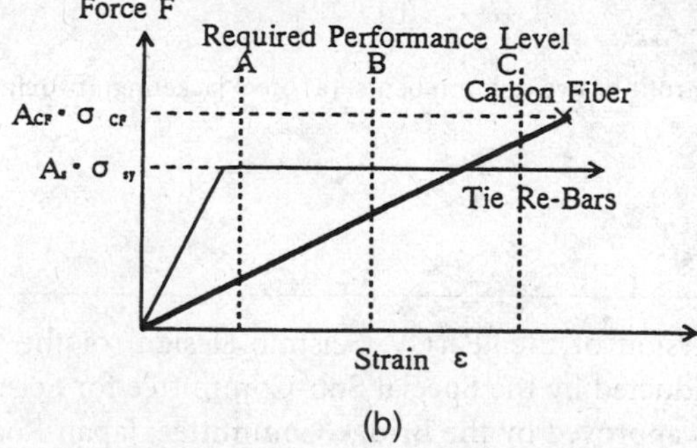

FIGURE 44.19 Cooperative effect between tie reinforcement and carbon fiber sheets. (a) Stress–strain relation; (b) force–strain relation.

References

1. Ministry of Construction, Report on the Damage of Highway Bridges by the Hyogo-ken Nanbu Earthquake, Committee for Investigation on the Damage of Highway Bridges Caused by the Hyogo-ken Nanbu Earthquake, 1995 [in Japanese].
2. Kawashima, K., Impact of Hanshin/Awaji Earthquake on Seismic Design and Seismic Strengthening of Highway Bridges, Report No. TIT/EERG 95-2, Tokyo Institute of Technology, 1995.
3. Kawashima, K. and Unjoh, S., The damage of highway bridges in the 1995 Hyogo-ken Nanbu earthquake and its impact on Japanese seismic design, *J. Earthquake Eng.*, 1(3), 1997.
4. Ministry of Construction, Guide Specifications for Reconstruction and Repair of Highway Bridges Which Suffered Damage Due to the Hyogo-ken Nanbu Earthquake, 1995 [in Japanese].
5. Japan Road Association, Design Specifications of Highway Bridges, Part I: Common Part, Part II: Steel Bridges, Part III: Concrete Bridges. Part IV: Foundations. Part V: Seismic Design, 1996 [in Japanese].
6. Kawashima, K., Nakano, M., Nishikawa, K., Fukui, J., Tamura, K., and Unjoh, S., The 1996 seismic design specifications of highway bridges, *Proceedings of the 29th Joint Meeting of U.S.–Japan Panel on Wind and Seismic Effects,* UJNR, Technical Memorandum of PWRI, No. 3524, 1997.
7. Kawashima, K., Unjoh, S., and Mukai, H., Seismic strengthening of highway bridges, in *Proceedings of the 2nd U.S.–Japan Workshop on Seismic Retrofit of Bridges,* Berkeley, January 1994, Technical Memorandum of PWRI, No. 3276.
8. Unjoh, S., Terayama, T., Adachi, Y., and Hoshikuma, J., Seismic retrofit of existing highway bridges in Japan, in *Proceedings of the 29th Joint Meeting of U.S.–Japan Panel on Wind and Seismic Effects,* UJNR, Technical Memorandum of PWRI, No. 3524, 1997.
9. Kawashima, K. and Unjoh, S., An inspection method of seismically vulnerable existing highway bridges, *Struct. Eng. Earthquake Eng.*, 7(7), *Proc. JSCE,* April 1990.
10. Japan Road Association, Reference for Applying Guided Specifications to New Highway Bridge and Seismic Strengthening of Existing Highway Bridges, June, 1995.
11. Hoshikuma. J., Otsuka, H., and Nagaya, K., Seismic retrofit of square RC Column by steel jacketing, *Proceedings of the 3rd U.S. Japan Workshop on Seismic Retrofit of Bridges,* Osaka, Japan, December 1996, Technical Memorandum of PWRI, No. 3481.
12. Terayama, T. and Otsuka, H., Seismic evaluation and retrofit of existing multi-column bents, in *Proceedings of the 3rd U.S.–Japan Workshop on Seismic Retrofit of Bridges,* Osaka, Japan, December 1996, Technical Memorandum of PWRI, No. 3481.
13. Public Works Research Center, Research Report by the Committee on Seismic Retrofit Methods Using Carbon Fiber Sheet, September 1996 [in Japanese].

Section V

Construction and Maintenance

45

Steel Bridge Construction

Jackson Durkee

Consulting Structural Engineer, Bethlehem, Pa.

0-8493-7434-0/00/$0.00+$.50

45.1 Introduction

This chapter addresses some of the principles and practices applicable to the construction of medium- and long-span steel bridges — structures of such size and complexity that construction engineering becomes an important or even the governing factor in the successful fabrication and erection of the superstructure steelwork.

We begin with an explanation of the fundamental nature of construction engineering, then go on to explain some of the challenges and obstacles involved. The basic considerations of cambering are explained. Two general approaches to the fabrication and erection of bridge steelwork are described, with examples from experience with arch bridges, suspension bridges, and cable-stayed bridges.

The problem of erection-strength adequacy of trusswork under erection is considered, and a method of appraisal offered that is believed to be superior to the standard working-stress procedure.

Typical problems with respect to construction procedure drawings, specifications, and practices are reviewed, and methods for improvement suggested. The need for comprehensive bridge erection-engineering specifications, and for standard conditions for contracting, is set forth, and the design-and-construct contracting procedure is described.

Finally, we take a view ahead, to the future prospects for effective construction engineering in the U.S.

The chapter also contains a large number of illustrations showing a variety of erection methods for several types of major steel bridges.

45.2 Construction Engineering in Relation to Design Engineering

With respect to bridge steelwork the differences between construction engineering and design engineering should be kept firmly in mind. Design engineering is of course a concept and process well known to structural engineers; it involves preparing a set of plans and specifications — known as the contract documents — that define the structure in its completed configuration, referred to as the geometric outline. Thus, the design drawings describe to the contractor the steel bridge superstructure that the owner wants to see in place when the project is completed. A considerable design engineering effort is required to prepare a good set of contract documents.

Construction engineering, however, is not so well known. It involves governing and guiding the fabrication and erection operations needed to produce the structural steel members to the proper cambered or "no-load" shape, and get them safely and efficiently "up in the air" in place in the structure, so that the completed structure under the deadload conditions and at normal temperature will meet the geometric and stress requirements stipulated on the design drawings.

Four key considerations may be noted: (1) design engineering is widely practiced and reasonably well understood, and is the subject of a steady stream of technical papers; (2) construction engineering is practiced on only a limited basis, is not as well understood, and is hardly ever discussed; (3) for medium- and long-span bridges, the construction engineering aspects are likely to be no less important than design engineering aspects; and (4) adequately staffed and experienced construction-engineering offices are a rarity.

45.3 Construction Engineering Can Be Critical

The construction phase of the total life of a major steel bridge will probably be much more hazardous than the service-use phase. Experience shows that a large bridge is more likely to suffer failure during erection than after completion. Many decades ago, steel bridge design engineering had progressed to the stage where the chance of structural failure under service loadings became altogether remote. However, the erection phase for a large bridge is inherently less secure, primarily because of the prospect of inadequacies in construction engineering and its implementation at the job site. The hazards associated with the erection of large steel bridges will be readily apparent from a review of the illustrations in this chapter.

Figure 45.1 Failure of a steel girder bridge during erection, 1995. Steel bridge failures such as this one invite suspicion that the construction engineering aspects were not properly attended to. (Newspaper archive photo.)

For significant steel bridges the key to construction integrity lies in the proper planning and engineering of steelwork fabrication and erection. Conversely, failure to attend properly to construction engineering constitutes an invitation to disaster. In fact, this thesis is so compelling that whenever a steel bridge failure occurs during construction (see for example Figure 45.1), it is reasonable to assume that the construction engineering investigation was either inadequate, not properly implemented, or both.

45.4 Premises and Objectives of Construction Engineering

During the erection sequences the various components of steel bridges may be subjected to stresses that are quite different from those which will occur under the service loadings and which have been provided for by the designer. For example, during construction there may be a derrick moving and working on the partially erected structure, and the structure may be cantilevered out some distance causing tension-designed members to be in compression and vice versa. Thus, the steelwork contractor needs to engineer the bridge members through their various construction loadings, and strengthen and stabilize them as may be necessary. Further, the contractor may need to provide temporary members to support and stabilize the structure as it passes through its successive erection configurations.

In addition to strength problems there are also geometric considerations. The steelwork contractor must engineer the construction sequences step by step to ensure that the structure will fit properly together as erection progresses, and that the final or closing members can be moved into position and connected. Finally, of course, the steelwork contractor must carry out the engineering studies needed to ensure that the geometry and stressing of the completed structure under normal temperature will be in accordance with the requirements of the design plans and specifications.

45.5 Fabrication and Erection Information Shown on Design Plans

Regrettably, the level of engineering effort required to accomplish safe and efficient fabrication and erection of steelwork superstructures is not widely understood or appreciated in bridge design offices, nor indeed by many steelwork contractors. It is only infrequently that we find a proper level of capability and effort in the engineering of construction.

The design drawings for an important bridge will sometimes display an erection scheme, even though most designers are not experienced in the practice of erection engineering and usually expend only a minimum or even superficial effort on erection studies. The scheme portrayed may not be practical, or may not be suitable in respect to the bidder or contractor's equipment and experience. Accordingly, the bidder or contractor may be making a serious mistake if he relies on an erection scheme portrayed on the design plans.

As an example of misplaced erection effort on the part of the designer, there have been cases where the design plans show cantilever erection by deck travelers, with the permanent members strengthened correspondingly to accommodate the erection loadings; but the successful bidder elected to use water-borne erection derricks with long booms, thereby obviating the necessity for most or all of the erection strengthening provided on the design plans. Further, even in those cases where the contractor would decide to erect by cantilevering as anticipated on the plans, there is hardly any way for the design engineer to know what will be the weight and dimensions of the contractor's erection travelers.

45.6 Erection Feasibility

Of course, the bridge designer does have a certain responsibility to his client and to the public in respect to the erection of the bridge steelwork. This responsibility includes: (1) making certain, during the design stage, that there is a feasible and economical method to erect the steelwork; (2) setting forth in the contract documents any necessary erection guidelines and restrictions; and (3) reviewing the contractor's erection scheme, including any strengthening that may be needed, to verify its suitability. It may be noted that this latter review does not relieve the contractor from responsibility for the adequacy and safety of the field operations.

Bridge annals include a number of cases where the design engineer failed to consider erection feasibility. In one notable instance the design plans showed the 1200 ft (366 m) main span for a long crossing over a wide river as an esthetically pleasing steel tied-arch. However, erection of such a span in the middle of the river was impractical; one bidder found that the tonnage of falsework required was about the same as the weight of the permanent arch-span steelwork. Following opening of the bids, the owner found the prices quoted to be well beyond the resources available, and the tied-arch main span was discarded in favor of a through-cantilever structure, for which erection falsework needs were minimal and practical.

It may be noted that design engineers can stand clear of serious mistakes such as this one, by the simple expedient of conferring with prospective bidders during the preliminary design stage of a major bridge.

45.7 Illustrations of Challenges in Construction Engineering

Space does not permit comprehensive coverage of the numerous and difficult technical challenges that can confront the construction engineer in the course of the erection of various types of major steel bridges. However, some conception of the kinds of steelwork erection problems, the methods available to resolve them, and the hazards involved can be conveyed by views of bridges in various stages of erection; refer to the illustrations in the text.

45.8 Obstacles to Effective Construction Engineering

There is an unfortunate tendency among design engineers to view construction engineering as relatively unimportant. This view may be augmented by the fact that few designers have had any significant experience in the engineering of construction.

Further, managers in the construction industry must look critically at costs, and they can readily develop the attitude that their engineers are doing unnecessary theoretical studies and calculations, detached from the practical world. (And indeed, this may sometimes be the case.) Such management apprehension can constitute a serious obstacle to staff engineers who see the need to have enough money in the bridge tender to cover a proper construction engineering effort for the project. There is the tendency for steelwork construction company management to cut back the construction engineering allowance, partly because of this apprehension and partly because of the concern that other tenderers will not be allotting adequate money for construction engineering. This effort is often thought of by company management as "a necessary evil" at best — something they would prefer not to be bothered with or burdened with.

Accordingly, construction engineering tends to be a difficult area of endeavor. The way for staff engineers to gain the confidence of management is obvious — they need to conduct their investigations to a level of technical proficiency that will command management respect and support, and they must keep management informed as to what they are doing and why it is necessary. As for management's concern that other bridge tenderers will not be putting into their packages much money for construction engineering, this concern is no doubt often justified, and it is difficult to see how responsible steelwork contractors can cope with this problem.

45.9 Examples of Inadequate Construction Engineering Allowances and Effort

Even with the best of intentions, the bidder's allocation of money to construction engineering can be inadequate. A case in point involved a very heavy, long-span cantilever truss bridge crossing a major river. The bridge superstructure carried a contract price of some \$30 million, including an allowance of \$150,000, or about one-half of 1%, for construction engineering of the permanent steelwork (i.e., not including such matters as design of erection equipment). As fabrication and erection progressed, many unanticipated technical problems came forward, including brittle-fracture aspects of certain grades of the high-strength structural steel, and aerodynamic instability of H-shaped vertical and diagonal truss members. In the end the contractor's construction engineering effort mounted to about \$1.3 million, almost nine times the estimated cost.

Another significant example — this one in the domain of buildings — involved a design-and-construct project for airplane maintenance hangars at a prominent international airport. There were two large and complicated buildings, each 100 × 150 m (328 × 492 ft) in plan and 37 m (121 ft) high with a 10 m (33 ft) deep space-frame roof. Each building contained about 2450 tons of structural steelwork. The design-and-construct steelwork contractor had submitted a bid of about \$30 million, and included therein was the magnificent sum of \$5,000 for construction engineering, under the expectation that this work could be done on an incidental basis by the project engineer in his "spare time."

As the steelwork contract went forward it quickly became obvious that the construction engineering effort had been grossly underestimated. The contractor proceeded to staff-up appropriately and carried out in-depth studies, leading to a detailed erection procedure manual of some 270 pages showing such matters as erection equipment and its positioning and clearances; falsework requirements; lifting tackle and jacking facilities; stress, stability, and geometric studies for gravity and wind loads; step-by-step instructions for raising, entering, and connecting the steelwork components; closing and swinging the roof structure and portal frame; and welding guidelines and procedures. This erection procedure manual turned out to be a key factor in the success of the fieldwork. The cost of this construction engineering effort amounted to about ten times the estimate, but still came to a mere one-fifth of 1% of the total contract cost.

In yet another example a major steelwork general contractor was induced to sublet the erection of a long-span cantiliever truss bridge to a reputable erection contractor, whose quoted *price* for the work was less than the general contractor's estimated *cost*. During the erection cycle the general contractor's engineers made some visits to the job site to observe progress, and were surprised and disconcerted to observe how little erection engineering and planning had been accomplished. For example, the erector had made no provision for installing jacks in the bottom-chord jacking points for closure of the main span; it was left up to the field forces to provide the jack bearing components inside the bottom-chord joints and to find the required jacks in the local market. When the job-built installations were tested it was discovered that they would not lift the cantilevered weight, and the job had to be shut down while the field engineer scouted around to find larger-capacity jacks. Further, certain compression members did not appear to be properly braced to carry the erection loadings; the erector had not engineered those members, but just assumed they were adequate. It became obvious that the erector had not appraised the bridge members for erection adequacy and had done little or no planning and engineering of the critical evolutions to be carried out in the field.

Many further examples of inadequate attention to construction engineering could be presented. Experience shows that the amounts of money and time allocated by steelwork contractors for the engineering of construction are frequently far less than desirable or necessary. Clearly, effort spent on construction engineering is worthwhile; it is obviously more efficient and cheaper, and certainly much safer, to plan and engineer steelwork construction in the office in advance of the work, rather than to leave these important matters for the field forces to work out. Just a few bad moves on site, with the corresponding waste of labor and equipment hours, will quickly use up sums of money much greater than those required for a proper construction engineering effort — not to mention the costs of any job accidents that might occur.

The obvious question is "Why is construction engineering not properly attended to?" Do not contractors learn, after a bad experience or two, that it is both necessary and cost effective to do a thorough job of planning and engineering the construction of important bridge projects? Experience and observation would seem to indicate that some steelwork contractors learn this lesson, while many do not. There is always pressure to reduce bid prices to the absolute minimum, and to add even a modest sum for construction engineering must inevitably reduce the prospect of being the low bidder.

45.10 Considerations Governing Construction Engineering Practices

There are no textbooks or manuals that define how to accomplish a proper job of construction engineering. In bridge construction (and no doubt in building construction as well) the engineering of construction tends to be a matter of each firm's experience, expertise, policies, and practices. Usually there is more than one way to build the structure, depending on the contractor's ingenuity and engineering skill, his risk appraisal and inclination to assume risk, the experience of his fabrication and erection work forces, his available equipment, and his personal preferences. Experience shows that each project is different; and although there will be similarities from one bridge of a given type to another, the construction engineering must be accomplished on an individual project basis. Many aspects of the project at hand will turn out to be different from those of previous similar jobs, and also there may be new engineering considerations and requirements for a given project that did not come forward on previous similar work.

During the estimating and bidding phase of the project the prudent, experienced bridge steelwork contractor will "start from scratch" and perform his own fabrication and erection studies, irrespective of any erection schemes and information that may be shown on the design plans. These studies can involve a considerable expenditure of both time and money, and thereby place that contractor at a disadvantage in respect to those bidders who are willing to rely on hasty, superficial studies, or — where the design engineer has shown an erection scheme — to simply assume that it has been engineered correctly and proceed to use it. The responsible contractor, on the other hand, will appraise the feasible construction methods and evaluate their costs and risks, and then make his selection.

After the contract has been executed the contractor will set forth how he intends to fabricate and erect, in detailed plans that could involve a large number of calculation sheets and drawings along with construction procedure documents. It is appropriate for the design engineer on behalf of his client to review the contractor's plans carefully, perform a check of construction considerations, and raise appropriate questions. Where the contractor does not agree with the designer's comments the two parties get together for review and discussion, and in the end they concur on essential factors such as fabrication and erection procedures and sequences, the weight and positioning of erection equipment, the design of falsework and other temporary components, erection stressing and strengthening of the permanent steelwork, erection stability and bracing of critical components, any erection check measurements that may be needed, and span closing and swinging operations.

The design engineer's approval is needed for certain fabrication plans, such as the cambering of individual members; however, in most cases the designer should stand clear of actual *approval* of the contractor's construction plans since he is not in a position to accept construction responsibility, and too many things can happen during the field evolutions over which the designer has no control.

It should be emphasized that even though the design engineer has usually has no significant experience in steelwork construction, the contractor should welcome his comments and evaluate them carefully and respectfully. In major bridge projects many construction matters can be improved on or get out of control or can be improved upon, and the contractor should take advantage of every opportunity to improve his prospects and performance. The experienced contractor will make sure that he works constructively with the design engineer, standing well clear of antagonistic or confrontational posturing.

45.11 Camber Considerations

One of the first construction engineering problems to be resolved by the steel bridge contractor is the cambering of individual bridge components. The design plans will show the "geometric outline" of the bridge, which is its shape under the designated load condition — commonly full dead load — at normal temperature. The contractor, however, fabricates the bridge members under the no-load condition, and at the "shop temperature" — the temperature at which the shop measuring tapes have been standardized and will have the correct length. The difference between the shape of a member under full dead load and normal temperature, and its shape at the no-load condition and shop temperature, is defined as member camber.

While camber is inherently a simple concept, it is frequently misunderstood; indeed, it is often not correctly defined in design specifications and contract documents. For example, beam and girder camber has been defined in specifications as "the convexity induced into a member to provide for vertical curvature of grade and to offset the anticipated deflections indicated on the plans when the member is in its erected position in the structure. Cambers shall be measured in this erected position..." This definition is not correct, and reflects a common misunderstanding of a key structural engineering term. Camber of bending members is not convexity, nor does it have anything to do with grade vertical curvature, nor is it measured with the member in the erected position. Camber — of a bending member, or any other member — is the *difference in shape* of the member under its no-load fabrication outline as compared with its geometric outline; and it is "measured" — i.e., the cambered dimensions are applied to the member — not when it is in the *erected* position (whatever that might be), but rather, when it is in the *no-load* condition.

In summary, camber is a *difference* in shape and not the shape itself. Beams and girders are commonly cambered to compensate for deadload bending, and truss members to compensate for deadload axial force. However, further refinements can be introduced as may be needed; for example, the arch-rib box members of the Lewiston-Queenston bridge (Fig. 45.4) were cambered to compensate for deadload axial force, bending, and shear.

A further common misunderstanding regarding cambering of bridge members involves the effect of the erection scheme on cambers. The erection scheme may require certain members to be strengthened, and this in turn will affect the cambers of those members (and possibly of others as well, in the case of statically indeterminate structures). However, the fabricator should address the matter of cambering only after the final sizes of all bridge members have been determined. Camber is a function of member properties, and there is no merit to calculating camber for members whose cross-sectional areas may subsequently be increased because of erection forces.

Thus, the erection scheme may affect the required member properties, and these in turn will affect member cambering; but the erection scheme does not *of itself* have any effect on camber. Obviously, the temporary stress-and-strain maneuvers to which a member will be subjected, between its no-load condition in the shop and its full-deadload condition in the completed structure, can have no bearing on the camber calculations for the member.

To illustrate the general principles that govern the cambering procedure, consider the main trusses of a truss bridge. The first step is to determine the erection procedure to be used, and to augment the strength of the truss members as may be necessary to sustain the erection forces. Next, the bridge deadload weights are determined, and the member deadload forces and effective cross-sectional areas are calculated.

Consider now a truss chord member having a geometric length of 49.1921 ft panel-point-to-panel-point and an effective cross-sectional area of 344.5 in.², carrying a deadload compressive force of 4230 kips. The bridge normal temperature is 45F and the shop temperature is 68F. We proceed as follows:

1. Assume that the chord member is in place in the bridge, at the full dead load of -4230 kips and the normal temperature of 45F.
2. Remove the member from the bridge, allowing its compressive force to fall to zero. The member will increase in length by an amount ΔL_s:

$$\Delta L_s = \frac{SL}{AE} = \frac{4230 \ kips \times 49.1921 \ ft}{344.5 \ in^2 \times 29000 \ kips/in^2}$$

$$= 0.0208 \, ft$$

3. Now raise the member temperature from 45F to 68F. The member will increase in length by an additional amount ΔL_t:

$$\Delta L_t = L\omega t = (49.1921 = 0.0208) \ ft \times$$

$$0.0000065/\text{deg} \times (68-45)\text{deg}$$

$$= 0.0074 \ ft$$

4. The total increase in member length will be:

$$\Delta L = \Delta L_s + \Delta L_t = 0.0208 + 0.0074$$

$$= 0.0282 \ ft$$

5. The theoretical cambered member length — the no-load length at 68F — will be:

$$L_{tc} = 49.1921 + 0.0282 = 49.2203 \ ft$$

6. Rounding L_{tc} to the nearest 1/32 in., we obtain the cambered member length for fabrication as:

$$L_{fc} = 49 \ ft \ 2\frac{21}{32} in$$

Accordingly, the general procedure for cambering a bridge member of any type can be summarized as follows:

1. Strengthen the structure to accommodate erection forces, as may be needed.

2. Determine the bridge deadload weights, and the corresponding member deadload forces and effective cross-sectional areas.
3. Starting with the structure in its geometric outline, remove the member to be cambered.
4. Allow the deadload force in the member to fall to zero, thereby changing its shape to that corresponding to the no-load condition.
5. Further change the shape of the member to correspond to that at the shop temperature.
6. Accomplish any rounding of member dimensions that may be needed for practical purposes.
7. The total change of shape of the member — from geometric (at normal temperature) to no-load at shop temperature — constitutes the member camber.

It should be noted that the gusset plates for bridge-truss joints are always fabricated with the connecting-member axes coming in at their *geometric* angles. As the members are erected and the joints fitted-up, secondary bending moments will be induced at the truss joints under the steel-load-only condition; but these secondary moments will disappear when the bridge reaches its full-deadload condition.

45.12 Two General Approaches to Fabrication and Erection of Bridge Steelwork

As has been stated previously, the objective in steel bridge construction is to fabricate and erect the structure so that it will have the geometry and stressing designated on the design plans, under full deadload at normal temperature. This geometry is known as the geometric outline. In the case of steel bridges there have been, over the decades, two general procedures for achieving this objective:

1. The "field adjustment" procedure — Carry out a continuing program of steelwork surveys and measurements in the field as erection progresses, in an attempt to discover fabrication and erection deficiencies; and perform continuing steelwork adjustments in an effort to compensate for such deficiencies and for errors in span baselines and pier elevations.
2. The "shop control" procedure — Place total reliance on first-order surveying of span baselines and pier elevations, and on accurate steelwork fabrication and erection augmented by meticulous construction engineering; and proceed with erection without any field adjustments, on the basis that the resulting bridge deadload geometry and stressing will be as good as can possibly be achieved.

Bridge designers have a strong tendency to overestimate the capability of field forces to accomplish accurate measurements and effective adjustments of the partially erected structure, and at the same time they tend to underestimate the positive effects of precise steel bridgework fabrication and erection. As a result, we continue to find contract drawings for major steel bridges that call for field evolutions such as the following:

1. **Continuous trusses and girders** — At the designated stages, measure or "weigh" the reactions on each pier, compare them with calculated theoretical values, and add or remove bearing-shoe shims to bring measured values into agreement with calculated values.
2. **Arch bridges** — With the arch ribs erected to midspan and only the short, closing "crown sections" not yet in place, measure thrust and moment at the crown, compare them with calculated theoretical values, and then adjust the shape of the closing sections to correct for errors in span-length measurements and in bearing-surface angles at skewback supports, along with accumulated fabrication and erection errors.
3. **Suspension bridges** — Following erection of the first cable wire or strand across the spans from anchorage to anchorage, survey its sag in each span and adjust these sags to agree with calculated theoretical values.
4. **Arch bridges and suspension bridges** — Carry out a deck-profile survey along each side of the bridge under the steel-load-only condition, compare survey results with the theoretical profile,

Figure 45.2 Erection of arch ribs, Rainbow Bridge, Niagara Falls, New York, 1941. Bridge span is 950 ft (290 m), with rise of 150 ft 46 m); box ribs are 3 × 12 ft (0.91 × 3.66 m). Tiebacks were attached starting at the end of the third tier and jumped forward as erection progressed (see Figure 45.3). Much permanent steelwork was used in tieback bents. Derricks on approaches load steelwork onto material cars that travel up arch ribs. Travelers are shown erecting last full-length arch-rib sections, leaving only the short, closing crown sections to be erected. Canada is at right, the U.S. at left. (Courtesy of Bethlehem Steel Corporation.)

and shim the suspender sockets so as to render the bridge floorbeams level in the completed structure.

5. **Cable-stayed bridges** — At each deck-steelwork erection stage, adjust tensions in the newly erected cable stays so as to bring the surveyed deck profile and measured stay tensions into agreement with calculated theoretical data.

There are two prime obstacles to the success of "field adjustment" procedures of whatever type: (1) field determination of the actual geometric and stress conditions of the partially erected structure and its components will not necessarily be definitive, and (2) calculation of the corresponding "proper" or "target" theoretical geometric and stress conditions will most likely prove to be less than authoritative.

45.13 Example of Arch Bridge Construction

In the case of the arch bridge closing sections referred to heretofore, experience on the construction of two major fixed-arch bridges crossing the Niagara River gorge from the U.S. to Canada — the Rainbow and the Lewiston-Queenston arch bridges (see Figures 45.2 through 45.5) — has demonstrated the difficulty, and indeed the futility, of attempts to make field-measured geometric and stress conditions agree with calculated theoretical values. The broad intent for both structures was to make such adjustments in the shape of the arch-rib closing sections at the crown (which were nominally about 1ft [0.3m] long) as would bring the arch-rib actual crown moments and thrusts into agreement with the calculated theoretical values, thereby correcting for errors in span-length measurements, errors in bearing-surface angles at the skewback supports, and errors in fabrication and erection of the arch-rib sections.

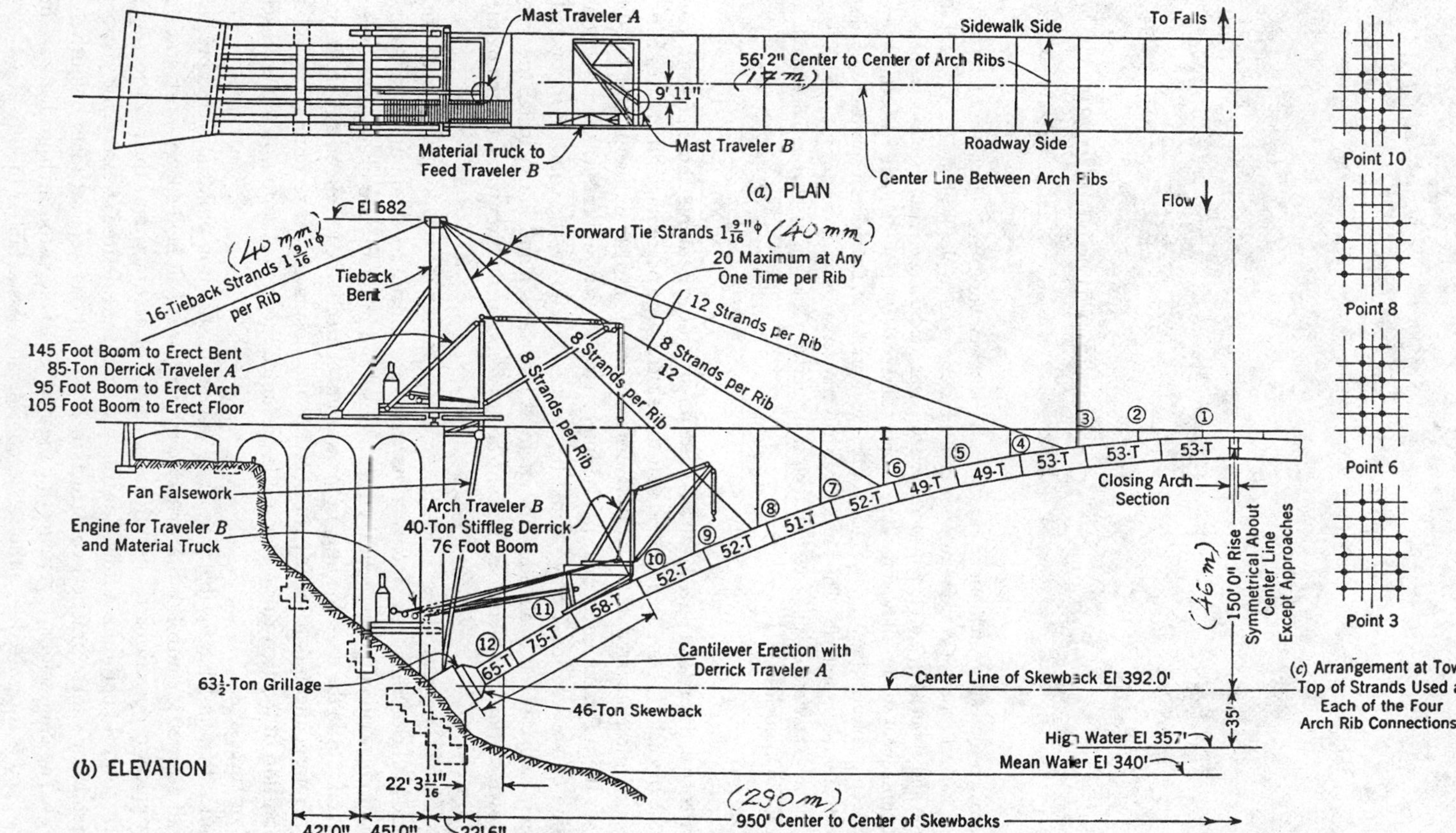

Figure 45.3 Rainbow Bridge, Niagara Falls, New York, showing successive arch tieback positions. Arch-rib erection geometry and stressing were controlled by means of measured tieback tensions in combination with surveyed arch-rib elevations.

Figure 45.4 Lewiston-Queenston arch bridge, near Niagara Falls, New York, 1962. The longest fixed-arch span in the U.S. at 1000 ft (305 m); rise is 159 ft (48 m). Box arch-rib sections are typically about 3 × 13-1/2 ft (0.9 × 4.1 m) in cross-section and about 44-1/2 ft (13.6 m) long. Job was estimated using erection tiebacks (same as shown in Figure 45.3), but subsequent studies showed the long, sloping falsework bents to be more economical (even if less secure looking). Much permanent steelwork was used in the falsework bents. Derricks on approaches load steelwork onto material cars that travel up arch ribs. The 115-ton-capacity travelers are shown erecting the last full-length arch-rib sections, leaving only the short, closing crown sections to be erected. Canada is at left, the U.S. at right. (Courtesy of Bethlehem Steel Corporation.)

Following extensive theoretical investigations and on-site measurements the steelwork contractor found, in the case of each Niagara arch bridge, that there were large percentage differences between the field-measured and the calculated theoretical values of arch-rib thrust, moment, and line-of-thrust position, and that the measurements could not be interpreted so as to indicate what corrections to the theoretical closing crown sections, if any, should be made. Accordingly, the contractor concluded that the best solution in each case was to abandon any attempts at correction and simply install the theoretical-shape closing crown sections. In each case, the contractor's recommendation was accepted by the design engineer.

Points to be noted in respect to these field-closure evolutions for the two long-span arch bridges are that accurate jack-load closure measurements at the crown are difficult to obtain under field conditions; and calculation of corresponding theoretical crown thrusts and moments are likely to be questionable because of uncertainties in the dead loading, in the weights of erection equipment, and in the steelwork temperature. Therefore, attempts to adjust the shape of the closing crown sections so as to bring the actual stress condition of the arch ribs closer to the presumed theoretical condition are not likely to be either practical or successful.

It was concluded that for long, flexible arch ribs, the best construction philosophy and practice is (1) to achieve overall geometric control of the structure by performing all field survey work and steelwork fabrication and erection operations to a meticulous degree of accuracy, and then (2) to rely on that overall geometric control to produce a finished structure having the desired stressing and geometry. For the Rainbow arch bridge, these practical construction considerations were set forth definitively by the contractor in [2]. The contractor's experience for the Lewiston-Queenston arch bridge was similar to that on Rainbow, and was reported — although in considerably less detail — in [10].

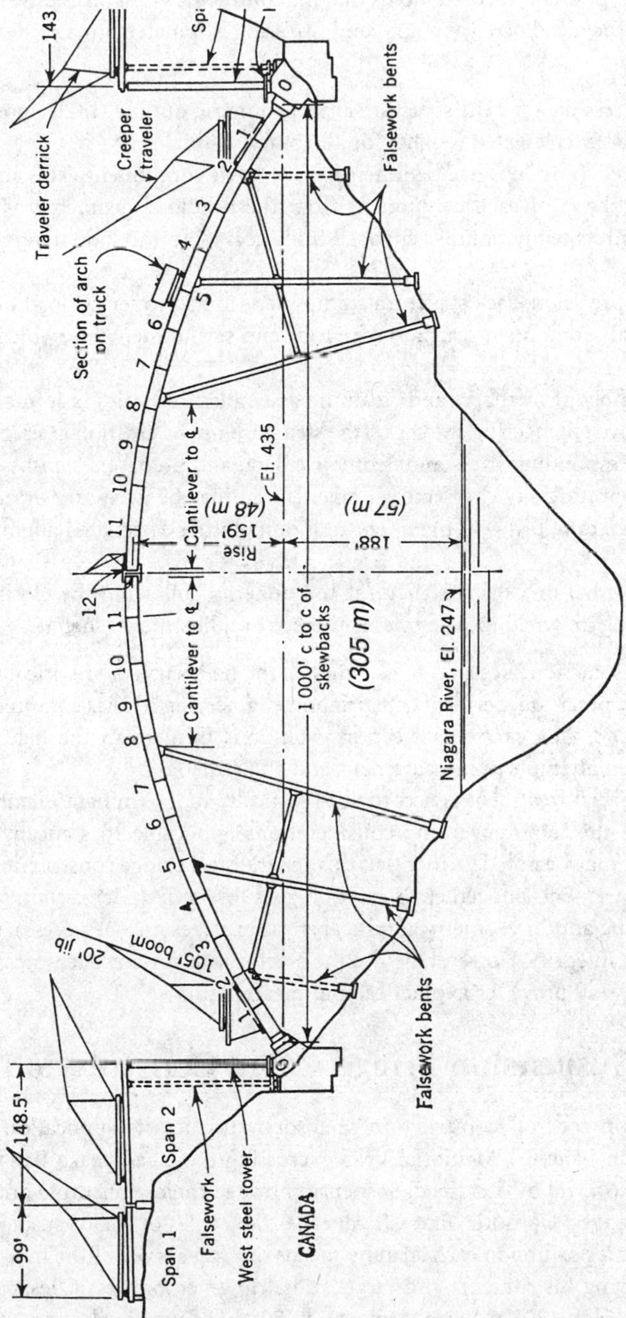

Figure 45.5 Lewiston-Queenston arch bridge near Niagara Falls, New York. Crawler cranes erect steelwork for spans 1 and 6 and erect material derricks theron. These derricks erect traveler derricks, which move forward and erect supporting falsework and spans 2, 5, and 4. Traveler derricks erect arch-rib sections 1 and 2 and supporting falsework at each skewback, then set up creeper derricks, which erect arches to midspan.

45.14 Which Construction Procedure Is To Be Preferred?

The contractor's experience on the construction of the two long-span fixed-arch bridges is set forth at length since it illustrates a key construction theorem that is broadly applicable to the fabrication and erection of steel bridges of all types. This theorem holds that the contractor's best procedure for achieving, in the completed structure, the deadload geometry and stressing stipulated on the design plans, is generally as follows:

1. Determine deadload stress data for the structure at its geometric outline (under normal temperature), based on accurately calculated weights for all components.
2. Determine the cambered (i.e., "no-load") dimensions of each component. This involves determining the change of shape of each component from the deadload geometry, as its deadload stressing is removed and its temperature is changed from normal to the shop temperature. (Refer to Section 45.11)
3. Fabricate, with all due precision, each structural component to its proper no-load dimensions — except for certain flexible components such as wire rope and strand members, which may require special treatment.
4. Accomplish shop assembly of members and "reaming assembled" of holes in joints, as needed.
5. Carry out comprehensive engineering studies of the structure under erection at each key erection stage, determining corresponding stress and geometric data, and prepare a step-by-step erection procedure plan, incorporating any check measurements that may be necessary or desirable.
6. During the erection program, bring all members and joints to the designated alignment prior to bolting or welding.
7. Enter and connect the final or closing structural components, following the closing procedure plan, without attempting any field measurements thereof or adjustments thereto.

In summary, the key to construction success is to accomplish the field surveys of critical baselines and support elevations with all due precision, perform construction engineering studies comprehensively and shop fabrication accurately, and then carry the erection evolutions through in the field without any second guessing and ill-advised attempts at measurement and adjustment.

It may be noted that no special treatment is accorded to statically indeterminate members; they are fabricated and erected under the same governing considerations applicable to statically determinate members, as set forth above. It may be noted further that this general steel bridge construction philosophy does not rule out check measurements altogether, as erection goes forward; under certain special conditions, measurements of stressing and/or geometry at critical erection stages may be necessary or desirable in order to confirm structural integrity. However, before the erector calls for any such measurements he should make certain that they will prove to be practical and meaningful.

45.15 Example of Suspension Bridge Cable Construction

In order to illustrate the "shop control" construction philosophy further, its application to the main cables of the first Wm. Preston Lane, Jr., Memorial Bridge, crossing the Chesapeake Bay in Maryland, completed in 1952 (Figure 45.6), will be described. Suspension bridge cables constitute one of the most difficult bridge erection challenges. Up until "first Chesapeake" the cables of major suspension bridges had been adjusted to the correct position in each span by means of a sag survey of the first-erected cable wires or strands, using surveying instruments and target rods. However, on first Chesapeake, with its 1600 ft (488 m) main span, 661 ft (201 m) side spans, and 450 ft (137 m) back spans, the steelwork contractor recommended abandoning the standard cable-sag survey and adopting the "setting-to-mark" procedure for positioning the guide strands — a significant new concept in suspension bridge cable construction.

The steelwork contractor's rationale for "setting to marks" was spelled out in a letter to the design engineer (see Figure 45.7). (The complete letter is reproduced because it spells out significant construction

Figure 45.6 Suspension spans of first Chesapeake Bay Bridge, Maryland, 1952. Deck steelwork is under erection and is about 50% complete. A typical four-panel through-truss deck section, weighing about 100 tons, is being picked in west side span, and also in east side span in distance. Main span is 1600 ft (488 m) and side spans are 661 ft (201 m); towers are 324 ft (99 m) high. Cables are 14 in. (356 mm) in diameter and are made up of 61 helical bridge strands each (see Figure 45.8).

philosophies.) This innovation was accepted by the design engineer. It should be noted that the contractor's major argument was that setting to marks would lead to more accurate cable placement than would a sag survey. The minor arguments, alluded to in the letter, were the resulting savings in preparatory office engineering work and in the field engineering effort, and most likely in construction time as well.

Each cable consisted of 61 standard helical-type bridge strands, as shown in Figure 45.8. To implement the setting-to-mark procedure each of three bottom-layer "guide strands" of each cable (i.e., strands 1, 2, and 3) was accurately measured in the manufacturing shop under the simulated full-deadload tension, and circumferential marks were placed at the four center-of-saddle positions of each strand. Then, in the field, the guide strands (each about 3955 ft [1205 m] long) were erected and positioned according to the following procedure:

1. Place the three guide strands for each cable "on the mark" at each of the four saddles and set normal shims at each of the two anchorages.
2. Under conditions of uniform temperature and no wind, measure the sag differences among the three guide strands of each cable, at the center of each of the five spans.
3. Calculate the "center-of-gravity" position for each guide-strand group in each span.
4. Adjust the sag of each strand to bring it to the center-of gravity position in each span. This position was considered to represent the correct theoretical guide-strand sag in each span.

The maximum "spread" from the highest to the lowest strand at the span center, prior to adjustment, was found to be 1-3/4 in (44 mm) in the main span, 3-1/2 in. (89 mm) in the side spans, and 3-3/4 in (95 mm) in the back spans. Further, the maximum change of perpendicular sag needed to bring the guide strands to the center-of-gravity position in each span was found to be 15/16 in (24 mm) for the main span, 2-1/16 in (52 mm) for the side spans, and 2-1/16 in (52 mm) for the back spans. These small adjustments testify to the accuracy of strand fabrication and to the validity of the setting-to-mark strand adjustment procedure, which was declared to be a success by all parties concerned. It seems doubtful that such accuracy in cable positioning could have been achieved using the standard sag-survey procedure.

With the first-layer strands in proper position in each cable, the strands in the second and subsequent layers were positioned to hang correctly in relation to the first layer, as is customary and proper for suspension bridge cable construction.

This example provides good illustration that the construction engineering philosophy referred to as the shop-control procedure can be applied advantageously not only to typical rigid-type steel structures, such as continuous trusses and arches, but also to flexible-type structures, such as suspension bridges.

July 6th, 1951
JJ:MM
C-1756

[To the design engineer]

Gentlemen: Attention of Mr. ____________________
Re: Chesapeake Bay Bridge — Suspension Span Cables

In our studies of the method of cable erection, we have arrived at the conclusion that setting of the guide strands to measured marks, instead of to surveyed sag, is a more satisfactory and more accurate method. Since such a procedure is not in accordance with the specifications, we wish to present for your consideration the reasoning which has led us to this conclusion, and to describe in outline form our proposed method of setting to marks.

On previous major suspension bridges, most of which have been built with parallel-wire instead of helical-strand cables, the thought has evidently been that setting the guide wire or guide strand to a computed sag, varying with the temperature, would be the most accurate method. This is associated with the fact that guide wires were never measured and marked to length. These established methods were carried over when strand-type cables came into use. An added reason may have been the knowledge that a small error in length results in a relatively large error in sag; and on the present structure the length-error to sag-error ratios are 1:2.4 and 1:1.5 for the main span and side spans, respectively.

However, the reading of the sag in the field is a very difficult operation because of the distances involved, the slopes of the side spans and backstays, the fact that even slight wind causes considerable motion to the guide strand, and for other practical reasons. We also believe that even though readings are made on cloudy days or at night, the actual temperature of all portions of the structure which will affect the sag cannot be accurately known. We are convinced that setting the guide strands according to the length marks thereon, which are place under what amount to laboratory or ideal conditions at the manufacturing plant, will produce more accurate results than would field measurement of the sag.

To be specific, consider the case of field determination of sag in the main span, where it is necessary to establish accessible platforms, and an H.I. and a foresight somewhat below the desired sag elevation; and then to sight on the foresight and bring a target, hung from the guide strand, down to the line-of-sight. In the present case it is 1600 ft (488 m) to the foresight and 800 ft (244 m) to the target. Even if the line-of-sight were established just right, it would be only under perfect conditions of temperature and air — if indeed then — that such a survey would be precise. The difficulties are still greater in the side spans and back spans, where inclined lines-of-sight must be established by a series of offset measurements from distant bench marks. There is always the danger, particularly in the present location and at the time now scheduled, that days may be lost in waiting for the right conditions of weather to make an instrument survey feasible.

There is a second factor of doubt involved. The strand is measured under a known stress and at a known modulus, with "mechanical stretch" taken out. It is then reeled to a relatively small diameter and unreeled at the bridge site. Under its own weight, and until the full dead load has been applied, there is an indeterminable loss in mechanical set, or loss of modulus. A strand set to proper sag for the final modulus will accordingly be set too low, and the final cable will be below plan elevation. This possible error can only be on the side that is less desirable. Evidently, also, it could be on the order of 1-1/2 in (40 mm) of sag increase for 1% of temporary reduction in modulus. If the strand were set to sag based on the assumed smaller modulus than will exist for the fully loaded condition, we doubt whether this smaller modulus could be chosen closely enough to ensure that the final sag would be correct. We are assured, however, by our manufacturing plant, that even though the modulus under bare-cable weight may be subject to unknown variation, the modulus which existed at the pre-stressing bed under the measuring tension will be duplicated when this same tension is

Figure 45.7 Setting cable guide strands to marks.

reached under dead load. Therefore, if the guide stand is set to measured marks, the doubt as to modulus is eliminated.

A third source of error is temperature. In past practice the sag has been adjusted, by reference to a chart, in accordance with the existing temperature. Granted that the adjustment is made in the early morning (the fog having risen but the sun not), it is hard to conceive that the actual average temperature in 3955 ft (1205 m) of strand will be that recorded by any thermometer. The mainspan sag error is about 0.7 in. (18 mm) per deg C of temperature.

These conditions are all greatly improved at the strand pre-stressing bed. There seems to be no reason to doubt that the guide strands can be measured and marked to an insignificant degree of error, at a stipulated stress and under a well-soaked and determinable temperature. Any errors in sag level must result from something other than the measured length of the guide strand.

There is one indispensable condition which, however, holds for either method of setting. That is, that the total distance from anchorage to anchorage, and the total calculated length of strand under its own-weight stress, must agree within the limits of shimming provided in the anchorages. Therefore, this distance in the field must be checked to close agreement. While the measured length of strand will be calculated with precision, it is interesting to note that in this calculation, it is not essential that the modulus be known with exactness. The important factor is that the strand length under the final deadload stress will be calculated exactly; and since that length is measured under the corresponding average strand stress, knowledge of the modulus is not a consideration. If the modulus at deadload stress is not as assumed, the only effect will be a change of deflection under live load, and this is minor. We emphasize again that the stand length under dead load, and the length as measured in the prestressing bed, will be identical regardless of the modulus.

The calculations for the bare-cable position result in pulled-back positions for the tops of the towers and cable bents, in order to control the unbalanced forces tending to slip the strands in the saddles. These pullback distances may be slightly in error without the slipping forces overcoming friction and thereby becoming apparent. Such errors would affect the final sags of strands set to sag. However, they would have no effect on the final sags of strands set-to-mark at the saddles; these errors change the temporary strand sags only, and under final stress the sags and the shaft leans will be as called for by the design plans.

It sometimes has happened that a tower which at its base is square to the bridge axis, acquires a slight skew as it rises. The amount of this skew has never, so far as we know, been important. If it is disregarded and the guide strands are attached without any compensating change, then the final loading will, with virtual certainty, pull the tower square. All sources of possible maladjustment have now been discussed except one — the errors in the several span lengths at the base of the towers and bents. The intention is to recognize and accept these, by performing the appropriate check measurements; and to correct for them by slipping the guide strands designated amounts through the saddles such that the center-of-saddle mark on the strand will be offset by that same amount from the centerline of the saddle.

If we have left unexplained herein any factor that seems to you to render our procedure questionable, we are anxious to know of it and discuss it with you in the near future; and we will be glad to come to your offices for this purpose. The detailed preparations for observing strand sags would require considerable time, and we are not now doing any work along those lines.

Yours very truly,
Chief Engineer

Figure 45.7 *(Continued)* Setting cable guide strands to marks.

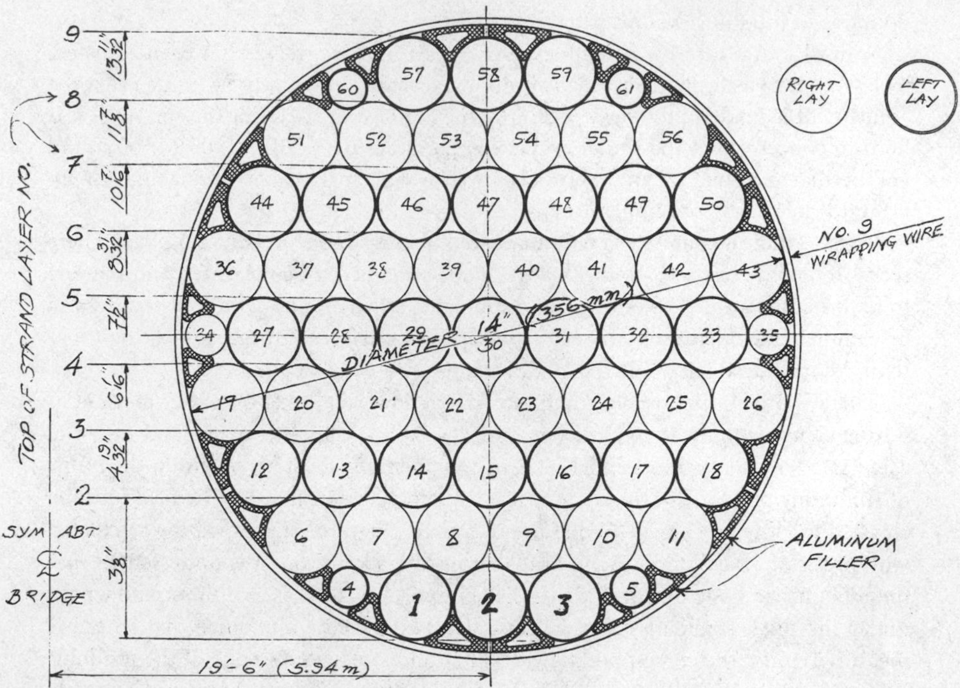

Figure 45.8 Main cable of first Chesapeake Bay suspension bridge, Maryland. Each cable consists of 61 helical-type bridge strands, 55 of 1-11/16 in (43 mm) and 6 of 29/32 in. (23 mm) diameter. Strands 1, 2, and 3 were designated "guide strands' and were set to mark at each saddle and to normal shims at anchorages.

There is, however, an important caveat: the steelwork contractor must be a firm of suitable caliber and experience.

45.16 Example of Cable-Stayed Bridge Construction

In the case cable-stayed bridges, the first of which were built in the 1950s, it appears that the governing construction engineering philosophy calls for field measurement and adjustment as the means for control of stay-cable and deck-structure geometry and stressing. For example, we have seen specifications calling for the completed bridge to meet the following geometric and stress requirements:

1. The deck elevation at midspan shall be within 12 in (305 mm) of theoretical.
2. The deck profile at each cable attachment point shall be within 2 in (50mm) of a parabola passing through the actual (i.e., field-measured) midspan point.
3. Cable-stay tensions shall be within 5% of the "corrected theoretical" values.

Such specification requirements introduce a number of problems of interpretation, field measurement, calculation, and field correction procedure, such as the following:

1. Interpretation:
   - The specifications are silent with respect to transverse elevation differentials. Therefore, two deck-profile control parabolas are presumably needed, one for each side of the bridge.
2. Field measurement of actual deck profile:
   - The temperature will be neither constant nor uniform throughout the structure during the survey work.

- The survey procedure itself will introduce some inherent error.

3. Field measurement of cable-stay tensions:
   - Hydraulic jacks, if used, are not likely to be accurate within 2%, perhaps even 5%; further, the exact point of "lift off" will be uncertain.
   - Other procedures for measuring cable tension, such as vibration or strain gaging, do not appear to define tensions within about 5%.
   - All cable tensions cannot be measured simultaneously; an extended period will be needed, during which conditions will vary and introduce additional errors.
4. Calculation of "actual" bridge profile and cable tensions:
   - Field-measured data must be transformed by calculation into "corrected actual" bridge profiles and cable tensions, at normal temperature and without erection loads.
   - Actual dead weights of structural components can differ by perhaps 2% from nominal weights, while temporary erection loads probably cannot be known within about 5%.
   - The actual temperature of structural components will be uncertain and not uniform.
   - The mathematical model itself will introduce additional error.
5. "Target condition" of bridge:
   - The "target condition" to be achieved by field adjustment will differ from the geometric condition, because of the absence of the deck wearing surface and other such components; it must therefore be calculated, introducing additional error.
6. Determining field corrections to be carried out by erector, to transform "corrected actual" bridge into "target condition" bridge:
   - The bridge structure is highly redundant, and changing any one cable tension will send geometric and cable-tension changes throughout the structure. Thus, an iterative correction procedure will be needed.

It seems likely that the total effect of all these practical factors could easily be sufficient to render ineffective the contractor's attempts to fine tune the geometry and stressing of the as-erected structure in order to bring it into agreement with the calculated bridge target condition. Further, there can be no assurance that the specifications requirements for the deck-profile geometry and cable-stay tensions are even compatible; it seems likely that *either* the deck geometry *or* the cable tensions may be achieved, but not *both.*

Specifications clauses of the type cited seem clearly to constitute unwarranted and unnecessary field-adjustment requirements. Such clauses are typically set forth by bridge designers who have great confidence in computer-generated calculation, but do not have a sufficient background in and understanding of the practical factors associated with steel bridge construction. Experience has shown that field procedures for major bridges developed unilaterally by design engineers should be reviewed carefully to determine whether they are practical and desirable and will in fact achieve the desired objectives.

In view of all these considerations, the question comes forward as to what design and construction principles should be followed to ensure that the deadload geometry and stressing of steel cable-stayed bridges will fall within acceptable limits. Consistent with the general construction-engineering procedures recommended for other types of bridges, we should abandon reliance on field measurements followed by adjustments of geometry and stressing, and instead place prime reliance on proper geometric control of bridge components during fabrication, followed by accurate erection evolutions as the work goes forward in the field.

Accordingly, the proper construction procedure for cable-stayed steel bridges can be summarized as follows:

1. Determine the actual bridge baseline lengths and pier-top elevations to a high degree of accuracy.
2. Fabricate the bridge towers, cables, and girders to a high degree of geometric precision.
3. Determine, in the fabricating shop, the final residual errors in critical fabricated dimensions, including cable-stay lengths after socketing, and positions of socket bearing surfaces or pinholes.
4. Determine "corrected theoretical" positioning for each individual cable stay.
5. During erection, bring all tower and girder structural joints into shop-fabricated alignment, with fair holes, etc.
6. At the appropriate erection stages, install "corrected theoretical" positional for each cable stay.
7. With the structure in the all-steel-erected condition (or other appropriate designated condition), check it over carefully to determine whether any significant geometric or other discrepancies are in evidence. If there are none, declare conditions acceptable and continue with erection.

This construction engineering philosophy can be summarized by stating that if the steelwork fabrication and erection are properly engineered and carried out, the geometry and stressing of the completed structure will fall within acceptable limits; whereas, if the fabrication and erection are not properly done, corrective measurements and adjustments attempted in the field are not likely to improve the structure, or even to prove satisfactory. Accordingly, in constructing steel cable-stayed bridges we should place full reliance on accurate shop fabrication and on controlled field erection, just as is done on other types of steel bridges, rather than attempting to make measurements and adjustments in the field to compensate for inadequate fabrication and erection.

45.17 Field Checking at Critical Erection Stages

As has been stated previously, the best governing procedure for steel bridge construction is generally the shop control procedure, wherein full reliance is placed on accurate fabrication of the bridge components as the basis for the integrity of the completed structure. However, this philosophy does not rule out the desirability of certain checks in the field as erection goes forward, with the objective of providing assurance that the work is on target and no significant errors have been introduced.

It would be impossible to catalog those cases during steel bridge construction where a field check might be desirable; such cases will generally suggest themselves as the construction engineering studies progress. We will only comment that these field-check cases, and the procedures to be used, should be looked at carefully, and even skeptically, to make certain that the measurements will be both desirable and practical, producing meaningful information that can be used to augment job integrity.

45.18 Determination of Erection Strength Adequacy

Quite commonly, bridge member forces during the erection stages will be altogether different from those that will prevail in the completed structure. At each critical erection stage the bridge members must be reviewed for strength and stability, to ensure structural integrity as the work goes forward. Such a construction engineering review is typically the responsibility of the steelwork erector, who carries out thorough erection studies of the structure and calls for strengthening or stabilizing of members as needed. The erector submits the studies and recommendations to the design engineer for review and comment, but normally the full responsibility for steelwork structural integrity during erection rests with the erector.

In the U.S., bridgework design specifications commonly require that stresses in steel structures under erection shall not exceed certain multiples of design allowable stresses. Although this type of erection stress limitation is probably safe for most steel structures under ordinary conditions, it is not necessarily adequate for the control of the erection stressing of large monumental-type bridges. The key point to be understood here is that fundamentally, there is no logical fixed relationship between design allowable stresses, which are based upon somewhat uncertain long-term service loading requirements along with some degree of assumed structural deterioration, and stresses that are safe and economical during the

Figure 45.9 Cable-stayed orthotropic-steel-deck bridge over Mississippi River at Luling, La., 1982; view looking northeast. The main span is 1222 ft (372 m); the A-frame towers are 350 ft (107 m) high. A barge-mounted ringer derrick erected the main steelwork, using a 340 ft (104 m) boom with a 120 ft (37 m) jib to erect tower components weighing up to 183 tons, and using a shorter boom for deck components. Cable stays at the ends of projecting cross girders are permanent; others are temporary erection stays. Girder section 16-west of north portion of bridge, erected a few days previously, is projecting at left; companion girder section 16-east is on barge ready for erection (see Figure 45.10).

bridge erection stages, where loads and their locations are normally well defined and the structural material is in new condition. Clearly, the basic premises of the two situations are significantly different, and "factored design stresses" must therefore be considered unreliable as a basis for evaluating erection safety.

There is yet a further problem with factored design stresses. Large truss-type bridges in various erection stages may undergo deflections and distortions that are substantial compared with those occurring under service conditions, thereby introducing apprehension regarding the effect of the secondary bending stresses that result from joint rigidity.

Recognizing these basic considerations, the engineering department of a major U.S. steelwork contractor went forward in the early 1970s to develop a logical philosophy for erection strength appraisal of large structural steel frameworks, with particular reference to long-span bridges, and implemented this philosophy with a stress analysis procedure. The effort was successful and the results were reported in a paper published by the American Society of Civil Engineers in 1977[6]. This stress analysis procedure, designated the erection rating factor (ERF) procedure, is founded directly upon basic structural principles, rather than on bridge-member design specifications, which are essentially irrelevant to the problem of erection stressing.

It may be noted that a significant inducement toward development of the ERF procedure was the failure of the first Quebec cantilever bridge in 1907 (see Figures 45.11 and 45.12). It was quite obvious that evaluation of the structural safety of the Quebec bridge at advanced cantilever erection stages such

Figure 45.10 Luling Bridge deck steelwork erection, 1982; view looking northeast (refer to Figure 45.9). The twin box girders are 14 ft (4.3 m) deep; the deck plate is 7/16 in. (11 mm) thick. Girder section 16-east is being raised into position (lower right) and will be secured by large-pin hinge bars prior to fairing-up of joint holes and permanent bolting. Temporary erection stays are jumped forward as girder erection progresses.

as that portrayed in Figure 45.11, by means of the factored-design-stress procedure, would inspire no confidence and would not be justifiable.

The erection rating factor (ERF) procedure for a truss bridge can be summarized as follows:

1. Assume either (a) pin-ended members (no secondary bending), (b) plane-frame action (rigid truss joints, secondary bending in one plane), or (c) space-frame action (bracing-member joints also rigid, secondary bending in two planes), as engineering judgement dictates.
2. Determine, for each designated erection stage, the member primary forces (axial) and secondary forces (bending) attributable to gravity loads and wind loads.
3. Compute the member stresses induced by the combined erection axial forces and bending moments.
4. Compute the ERF for each member at three or five locations: at the middle of the member; at each joint, inside the gusset plates (usually at the first row of bolts); and, where upset member plates or gusset plates are used, at the stepped-down cross-section outside each joint.
5. Determine the minimum computed ERF for each member and compare it with the stipulated minimum value.
6. Where the computed minimum ERF equals or exceeds the stipulated minimum value, the member is considered satisfactory. Where it is less, the member may be inadequate; reevaluate the critical part of it in greater detail and recalculate the ERF for further comparison with the stipulated minimum. (Initially calculated values can often be increased significantly.)
7. Where the computed minimum ERF remains less than the stipulated minimum value, strengthen the member as required.

Note that member forces attributable to wind are treated the same as those attributable to gravity loads. The old concept of "increased allowable stresses" for wind is not considered to be valid for erection

Figure 45.11 First Quebec railway cantilever bridge, 23 August 1907. Cantilever erection of south main span, six days before collapse. The tower traveler erected the anchor span (on falsework) and then the cantilever arm; then erected the top-chord traveler, which is shown erecting suspended span at end of cantilever arm. The main span of 1800 ft (549 m) was the world's longest of any type. The sidespan bottom chords second from pier (arrow) failed in compression because latticing connecting chord corner angles was deficient under secondary bending conditions.

conditions and is not used in the ERF procedure. Maximum acceptable ℓ/r and b/t values are included in the criteria. ERFs for members subjected to secondary bending moments are calculated using interaction equations.

45.19 Philosophy of the Erection Rating Factor

In order that the structural integrity and reliability of a steel framework can be maintained throughout the erection program, the minimum probable (or "minimum characteristic") strength value of each member must necessarily be no less than the maximum probable (or "maximum characteristic") force value, under the most adverse erection condition. In other words, the following relationship is required:

$$S - \Delta S \geq F + \Delta F \tag{45.1}$$

where

S = computed or nominal strength value for the member

ΔS = maximum probable member strength underrun from the computed or nominal value

F = computed or nominal force value for the member

ΔF = maximum probable member force overrun from the computed or nominal value

Equation 45.1 states that in the event the actual strength of the structural member is less than the nominal strength, S, by an amount ΔS, while at same time the actual force in the member is greater than the nominal force, F, by an amount ΔF, the member strength will still be no less than the member force,

Figure 45.12 Wreckage of south anchor span of first Quebec railway cantilever bridge, 1907. View looking north from south shore a few days after collapse of 29 August 1907, the worst disaster in the history of bridge construction. About 20,000 tons of steelwork fell into the St. Lawrence River, and 75 workmen lost their lives.

and so the member will not fail during erection. This equation provides a direct appraisal of erection realities, in contrast to the allowable-stress approach based on factored design stresses.

Proceeding now to rearrange the terms in Equation 45.1, we find that

$$S\left(1-\frac{\Delta S}{S}\right) \geq F\left(1+\frac{\Delta F}{F}\right); \quad \frac{S}{F} \geq \frac{1+\dfrac{\Delta F}{F}}{1-\dfrac{\Delta S}{S}} \tag{45.2}$$

The ERF is now defined as

$$ERF \equiv \frac{S}{F} \tag{45.3}$$

that is, the nominal strength value, S, of the member divided by its nominal force value, F. Thus, for erection structural integrity and reliability to be maintained, it is necessary that

$$ERF \geq \frac{1+\dfrac{\Delta F}{F}}{1-\dfrac{\Delta S}{S}} \tag{45.4}$$

45.20 Minimum Erection Rating Factors

In view of possible errors in (1) the assumed weight of permanent structural components, (2) the assumed weight and positioning of erection equipment, and (3) the mathematical models assumed for purposes of erection structural analysis, it is reasonable to assume that the actual member force for a given erection condition may exceed the computed force value by as much as 10%; that is, it is reasonable to take $\Delta F/F$ as equal to 0.10.

For tension members, uncertainties in (1) the area of the cross-section, (2) the strength of the material, and (3) the member workmanship, indicate that the actual member strength may be up to 15% less than the computed value; that is, $\Delta S/S$ can reasonably be taken as equal to 0.15. The additional uncertainties associated with compression member strength suggest that $\Delta S/S$ be taken as 0.25 for those members. Placing these values into Equation 45.4, we obtain the following minimum ERFs:

$$\text{Tension members:} \quad ERF_{t,\min} = (1+0.10)/(1-0.15)$$

$$= 1.294, \textit{say}\ 1.30$$

$$\text{Compressionmember:} \quad ERF_{c,\min} = (1+0.10)/(1-0.25)$$

$$= 1.467, \textit{say}\ 1.45$$

The proper interpretation of these expressions is that if, for a given tension (compression) member, the ERF is calculated as 1.30 (1.45) or more, the member can be declared safe for the particular erection condition. Note that higher, or lower, values of erection rating factors may be selected if conditions warrant.

The minimum ERFs determined as indicated are based on experience and judgment, guided by analysis and test results. They do not reflect any specific probabilities of failure and thus are not based on the concept of an acceptable risk of failure, which might be considered the key to a totally rational approach to structural safety. This possible shortcoming in the ERF procedure might be at least partially overcome by evaluating the parameters $\Delta F/F$ and $\Delta S/S$ on a statistical basis; however, this would involve a considerable effort, and it might not even produce significant results.

It is important to recognize that the ERF procedure for determining erection strength adequacy is based directly on fundamental strength and stability criteria, rather than being only indirectly related to such criteria through the medium of a design specification. Thus, the procedure gives uniform results for the erection rating of framed structural members irrespective of the specification that was used to design the members. Obviously, the end use of the completed structure is irrelevant to its strength adequacy during the erection configurations, and therefore the design specification should not be brought into the picture as the basis for erection appraisal.

Experience with application of the ERF procedure to long-span truss bridges has shown that it places the erection engineer in much better contact with the physical significance of the analysis than can be obtained by using the factored-design-stress procedure. Further, the ERF procedure takes account of secondary stresses, which have generally been neglected in erection stress analysis.

Although the ERF procedure was prepared for application to truss bridge members, the simple governing structural principle set forth by Equation 45.1 could readily be applied to bridge members and components of any type.

45.21 Deficiencies of Typical Construction Procedure Drawings and Instructions

At this stage of the review it is appropriate to bring forward a key problem in the realm of bridge construction engineering: the strong tendency for construction procedure drawings to be insufficiently

Figure 45.13 Visiting the work site. It is of first-order importance for bridge construction engineers to visit the site regularly and confer with the job superintendent and his foremen regarding practical considerations. Construction engineers have much to learn from the work forces in shop and field, and vice versa. (Courtesy of Bethlehem Steel Corporation.)

clear, and for step-by-step instructions to be either lacking or less than definitive. As a result of these deficiencies it is not uncommon to find the contractor's shop and field evolutions to be going along under something less than suitable control.

Shop and field operations personnel who are in a position to speak frankly to construction engineers will sometimes let them know that procedure drawings and instructions often need to be clarified and upgraded. This is a pervasive problem, and it results from two prime causes: (1) the fabrication and erection engineers responsible for drawings and instructions do not have adequate on-the-job experience, and (2) they are not sufficiently skilled in the art of setting forth on the documents, clearly and concisely, exactly what is to be done by the operations forces — and, sometimes of equal importance, what *is not* to be done.

This matter of clear and concise construction procedure drawings and instructions may appear to be a pedestrian matter, but it is decidedly not. *It is a key issue of utmost importance to the success of steel bridge construction.*

45.22 Shop and Field Liaison by Construction Engineers

In addition to the need for well-prepared construction procedure drawings and instructions, it is essential for the staff engineers carrying out construction engineering to set up good working relations with the shop and field production forces, and to visit the work sites and establish effective communication with the personnel responsible for accomplishing what is shown on the documents.

Construction engineers should review each projected operation in detail with the work forces, and upgrade the procedure drawings and instructions as necessary, as the work goes forward. Further, engineers should be present at the work sites during critical stages of fabrication and erection. As a component of these site visits, the engineers should organize special meetings of key production personnel to go over critical operations in detail — complete with slides and blackboard as needed — thereby providing the work forces with opportunities to ask questions and discuss procedures and potential problems, and providing engineers the opportunity to determine how well the work forces understand the operations to be carried out.

This matter of liaison between the office and the work sites — like the preceding issue of clear construction procedure documents — may appear to be somewhat prosaic; again, however, i*t is a matter of paramount importance.* Failure to attend to these two key issues constitutes a serious problem in steel bridge construction, and opens the door to high costs and delays, and even to erection accidents.

45.23 Comprehensive Bridge Erection-Engineering Specifications

The erection rating factor (ERF) procedure for determination of erection strength adequacy, as set forth heretofore for bridge trusswork, could readily be extended to cover bridge members and components of any type under erection loading conditions. Bridge construction engineers should work toward this objective, in order to release erection strength appraisal from the limitations of the commonly used factored-design-stress procedure.

Looking still further ahead, it is apparent that there is need in the bridge engineering profession for comprehensive erection engineering specifications for steel bridge construction. Such specifications should include guidelines for such matters as devising and evaluating erection schemes, determining erection loads, evaluating erection strength adequacy of all types of bridge members and components, designing erection equipment, and designing temporary erection members such as falsework, tiedowns, tiebacks, and jacking struts. The specifications might also cover contractual considerations associated with construction engineering.

The key point to be recognized here is that the use of bridge *design* specifications as the basis for erection engineering studies, as is currently the custom, is not appropriate. Erection engineering is a related but different discipline, and should have its own specifications. However, given the current fragmented state of construction engineering in the U.S. (refer to Section 45.26), it is difficult to envision how such erection engineering specifications could be prepared. Proprietary considerations associated with each erection firm's experience and procedures could constitute an additional obstacle.

45.24 Standard Conditions for Contracting

A further basic problem in respect to the future of steel bridge construction in the U.S. lies in the absence of standard conditions for contracting.

On through the 19th century both the design and the construction of a major bridge in the U.S. were frequently the responsibility of a single prominent engineer, who could readily direct and coordinate the work and resolve problems equitably. Then, over, the first thirty years or so of the 20th century this system was progressively displaced by the practice of competitive bidding on plans and specifications prepared by a design engineer retained by the owner. As a result the responsibility for the structure previously carried by the designer-builder became divided, with the designer taking responsibility for service integrity of the completed structure while prime responsibility for structural adequacy and safety during construction was assumed by the contractor. Full control over the preparation of the plans and specifications — the contract documents — was retained by the design engineer.

This divided responsibility has resulted in contract documents that may not be altogether equitable, since the designer is inevitably under pressure to look after the immediate financial interests of his client, the owner. Documents prepared by only one party to a contract can hardly be expected to reflect the appropriate interest of the other party. However, until about mid-20th-century design and construction responsibilities for major bridgework, although divided between the design engineer and the construction engineer, were nonetheless usually under the control of leading members of the bridge engineering profession who were able to command the level of communication and cooperation needed for resolution of inevitable differences of opinion within a framework of equity and good will.

Since the 1970s there has been a trend away from this traditional system of control. The business and management aspects of design firms have become increasingly important, while at the same time steelwork construction firms have become more oriented toward commercial and legal considerations. Professional design and construction engineers have lost stature correspondingly. As a result of these adverse trends, bridgework specifications are being ever more stringently drawn, bidding practices are becoming increasingly aggressive, claims for extra reimbursement are proliferating, insurance costs for all concerned are rising, and control of bridge engineering and construction is being influenced to an increasing extent by administrators and attorneys. These developments have not benefited the bridge owners, the design engineering profession, the steelwork construction industry, or the public — which must ultimately pay all of the costs of bridge construction.

It seems clear that in order to move forward out of this unsatisfactory state of affairs, a comprehensive set of standard conditions for contracting should be developed to serve as a core document for civil engineering construction — a document that would require only the addition of special provisions in order to constitute the basic specifications for any major bridge construction project. Such standard conditions would have to be prepared "off line" by a group of high-level engineering delegates having well established engineering credentials.

A core contract document such as the one proposed has been in general use in Great Britain since 1945, when the first edition of the *Conditions of Contract and Forms of Tender, Agreement and Bond for Use in Connection with Works of Civil Engineering Construction* was published by The Institution of Civil Engineers (ICE). This document, known informally as *The ICE Conditions of Contract*, is now in its 6th edition [1]. It is kept under review and revised as necessary by a permanent Joint Contracts Committee consisting of delegates from The Institution of Civil Engineers, The Association of Consulting Engineers, and The Federation of Civil Engineering Contractors. This document is used as the basis for the majority of works of civil engineering construction that are contracted in Great Britain, including steel bridges.

Further comments on the perceived need for U.S. standard conditions for contracting can be found in [7].

45.25 Design-and-Construct

As has been mentioned, design-and-construct was common practice in the U.S. during the 19th century. Probably the most notable example was the Brooklyn Bridge, where the designer-builders were John A. Roebling and his son Washington A. Roebling. Construction of the Brooklyn Bridge was begun in 1869 and completed in 1883. Design-and-construct continued in use through the early years of the 20th century; the most prominent example from that era may be the Ambassador suspension bridge between Detroit, Michigan and Windsor, Ontario, Canada, completed in 1929. The Ambassador Bridge was designed and built by the McClintic Marshall Construction Co., Jonathan Jones, chief engineer; it has an 1850 ft main span, at that time the world's record single span.

Design-and-construct has not been used for a major steel bridge in the U.S. since the Ambassador Bridge. However, the procedure has seen significant use throughout the 20th century for bridges in other countries and particularly in Europe; and most recently design-construct-operate-maintain has come into the picture. Whether these procedures will find significant application in the U.S. remains to be seen.

The advantages of design-and-construct are readily apparent:

1. More prospective designs are likely to come forward, than when designs are obtained from only a single organization.
2. Competitive designs are submitted at a preliminary level, making it possible for the owner to provide some input to the selected design between the preliminary stage and design completion.
3. The owner knows the price of the project at the time the preliminary design is selected, as compared with design-bid where the price is not known until the design is completed and bids are received.
4. As the project goes forward the owner deals with only a single entity, thereby reducing and simplifying his administrative effort.

5. The design-and-construct team members must work effectively together, eliminating the antagonisms and confrontations that can occur on a design-bid project.

A key requirement in the design-and-construct system for a project is the meticulous preparation of the request-for-proposals (RFP). The following essentials should be covered in suitable detail and clarity:

1. Description of project to be constructed.
2. Scope of work.
3. Structural component types and characteristics: which are required, which are acceptable, and which are not acceptable.
4. Minimum percentages of design and construction work that must be performed by the team's own forces.
5. Work schedules; time incentives and disincentives.
6. Procedure to be followed when actual conditions are found to differ from those assumed.
7. Quality control and quality assurance factors.
8. Owner's approval prerogatives during final-design stage and construction stage.
9. Applicable local, state and federal regulations.
10. Performance and payment bonding requirements.
11. Warranty requirements.
12. Owner's procedure for final approval of completed project.

In preparing the RFP, the owner should muster all necessary resources from both inside and outside his organization. Political considerations should be given due attention. Document drafts should receive the appropriate reviews, and an RFP brought forward that is in near-final condition. Then, at the start of the contracting process, the owner will typically proceed as follows:

1. Announce the project and invite prospective teams to submit qualifications.
2. Prequalify a small number of teams, perhaps three to five, and send the draft RFP to each.
3. Hold a meeting with the prequalified teams for informal exchange of information and to discuss questions.
4. Prepare the final RFP and issue it to each prequalified team, and announce the date on which proposals will be due.

The owner will customarily call for the proposals to be submitted in two separate components: the design component, showing the preliminary design carried to about the 25% level; and the monetary component, stating the lump-sum bid. Before the bids are opened the owner will typically carry out a scoring process for the preliminary designs, not identifying the teams with their designs, using a 10 point or 100 point grading scale and giving consideration to the following factors:

1. Quality of the design,
2. Bridge aesthetics.
3. Fabrication and erection feasibility and reliability.
4. Construction safety aspects.
5. Warranty and long-term maintenance considerations.
6. User costs.

Using these and other such scoring factors (which can be assigned weights if desired), a final overall design score is assigned to each preliminary design. Then the lump-sum bids are opened. A typical procedure is to divide each team's bid price by its design score, yielding and overall price rating, and to award the contract to the design-and-construct team having the lowest price rating.

Following the contract award the successful team will proceed to bring its preliminary design up to the final-design level, with no site work permitted during this interval. It is customary for the owner to award each unsuccessful submitting team a stipend to partially offset the costs of proposal preparation.

45.26 Construction Engineering Procedures and Practices — The Future

The many existing differences of opinion and procedures in respect to proper governance of steelwork fabrication and erection for major steel bridges raises the question: How do proper bridge construction guidelines come into existence and find their way into practice and into bridge specifications? Looking back over the period roughly from 1900 to 1975, we find that the major steelwork construction companies in the U.S. developed and maintained competent engineering departments that planned and engineered large bridges (and smaller ones as well) through the fabrication and erection processes with a high degree of proficiency. Traditionally, the steelwork contractor's engineers worked in cooperation with design-office engineers to develop the full range of bridgework technical factors, including construction procedure and practices.

However, times have changed; since the 1970's major steel bridge contractors have all but disappeared in the U.S., and further, very few bridge design offices have on their staffs engineers experienced in fabrication and erection engineering. As a result, construction engineering often receives less attention and effort than it needs and deserves, and this is not a good omen for the future of the design and construction of large bridges in the U.S.

Bridge construction engineering is not a subject that is or can be taught in the classroom; it must be learned on the job with major steelwork contractors. The best route for an aspiring young construction engineer is to spend significant amounts of time in the fabricating shop and at the bridge site, interspersed with time doing construction-engineering technical work in the office. It has been pointed out previously that although construction engineering and design engineering are related, they constitute different practices and require diverse backgrounds and experience. Design engineering can essentially be learned in the design office; construction engineering, however, cannot — it requires a background of experience at work sites. Such experience, it may be noted, is valuable also for design engineers; however, it is not as necessary for them as it is for construction engineers.

The training of future steelwork construction engineers in the U.S. will be handicapped by the demise of the "Big Two" steelwork contractors in the 1970s. Regrettably, it appears that surviving steelwork contractors in the U.S. generally do not have the resources for supporting strong engineering departments, and so there is some question as to where the next generation of steel bridge construction engineers in the U.S. will be coming from.

45.27 Concluding Comments

In closing this review of steel bridge construction it is appropriate to quote from the work of an illustrious British engineer, teacher, and author, the late Sir Alfred Pugsley [15]:

> A further crop of [bridge] accidents arose last century from overloading by traffic of various kinds, but as we have seen, engineers today concentrate much of their effort to ensure that a margin of strength is provided against this eventuality. But there is one type of collapse that occurs almost as frequently today as it has over the centuries: collapse at a late stage of erection.
>
> The erection of a bridge has always presented its special perils and, in spite of ever-increasing care over the centuries, few great bridges have been built without loss of life. Quite apart from the vagaries of human error, with nearly all bridges there comes a critical time near completion when the success of the bridge hinges on some special operation. Among such are … the fitting of a last section … in a steel arch, the insertion of the closing central [members] in a cantilever bridge, and the lifting of the roadway deck [structure] into

position on a suspension bridge. And there have been major accidents in many such cases. It may be wondered why, if such critical circumstances are well known to arise, adequate care is not taken to prevent an accident. Special care is of course taken, but there are often reasons why there may still be "a slip bewixt cup and lip". Such operations commonly involve unusually close cooperation between constructors and designers, and between every grade of staff, from the laborers to the designers and directors concerned; and this may put a strain on the design skill, on detailed inspection, and on practical leadership that is enough to exhaust even a Brunel.

In such circumstances it does well to … recall [the] dictum … that "it is essential not to have faith in human nature. Such faith is a recent heresy and a very disastrous one." One must rely heavily on the lessons of past experience in the profession. Some of this experience is embodied in professional papers describing erection processes, often (and particularly to young engineers) superficially uninteresting. Some is crystallized in organizational habits, such as the appointment of resident engineers from both the contracting and [design] sides. And some in precautions I have myself endeavored to list …

It is an easy matter to list such precautions and warnings, but quite another for the senior engineers responsible for the completion of a bridge to stand their ground in real life. This is an area of our subject that depends in a very real sense on the personal qualities of bridge engineers … At bottom, the safety of our bridges depends heavily upon the integrity of our engineers, particularly the leading ones.

45.28 Further Illustrations of Bridges Under Construction, Showing Erection Methods

Figure 45.14 Royal Albert Bridge across River Tamar, Saltash, England, 1857. The two 455 ft (139 m) main spans, each weighing 1060 tons, were constructed on shore, floated out on pairs of barges, and hoisted about 100 ft (30 m) to their final position using hydraulic jacks. Pier masonry was built up after each 3 ft (1 m) lift.

Figure 45.15 Eads Bridge across the Mississippi River, St. Louis, Mo., 1873. The first important metal arch bridge in the U.S., supported by four planes of hingeless trussed arches having chrome-steel tubular chords. Spans are 502-520-502 ft (153-158-153 m). During erection, arch trusses were tied back by cables passing over temporary towers built on the piers. Arch ribs were packed in ice to effect closure.

Figure 45.16 Glasgow (Missouri) railway truss bridge, 1879. Erection on full supporting falsework was commonplace in the 19th century. The world's first all-steel bridge, with five 315 ft (96 m) through-truss simple spans, crossing the Missouri River.

Figure 45.17 Niagara River railway cantilever truss bridge, near Niagara Falls, New York, 1883. Massive wood erection traveler constructed side span on falsework, then cantilevered half of main span to midspan. Erection of other half of bridge was similar. First modern-type cantilever bridge, with 470 ft (143 m) clear main span having a 120 ft (37 m) center suspended span.

a. The massive cantilevers of the Forth bridge, shown under erection, were conceived in the shadow of the Tay bridge disaster.

Figure 45.18 Construction of monumental Forth Bridge, Scotland, 1888. Numerous small movable booms were used, along with erection travelers for cantilevering the two 1710 ft (521 m) main spans. The main compression members are tubes 12 ft (3.65 m) in diameter; many other members are also tubular. Total steelwork weight is 51,000 tons. Records are not clear regarding such essentials as cambering and field fitting of individual members in this heavily redundant railway bridge. The Forth is arguably the world's greatest steel structure.

Figure 45.19 Pecos River railway viaduct, Texas, 1892. Erection by massive steam-powered wood traveler having many sets of falls and very long reach. Cantilever-truss main span has 185 ft (56 m) clear opening.

Figure 45.20 Raising of suspended span, Carquinez Strait Bridge, California, 1927. The 433 ft (132 m) suspended span, weighting 650 tons, was raised into position in 35 min., driven by four counterweight boxes having a total weight of 740 tons.

Figure 45.21 First Cooper River cantilever bridge, Charleston, S.C., 1929. Erection travelers constructed 450 ft (137 m) side spans on falsework, then went on to erect 1050 ft (320 m) main span (including 437.5 ft [133 m] suspended span) by cantilevering to midspan.

Figure 45.22 Erecting south tower of Golden Gate Bridge, San Francisco, 1935. A creeper traveler with two 90 ft (27 m) booms erects a tier of tower cells for each leg, then is jumped to the top of that tier and proceeds to erect the next tier. The tower legs are 90 ft (27 m) center-to-center and 690 ft (210 m) high. When the traveler completed the north tower (in background) it erected a Chicago boom on the west tower leg, which dismantled the creeper, erected tower-top bracing, and erected two small derricks (one shown) to service cable erection. Each tower contains 22,200 tons of steelwork.

Figure 45.23 Balanced-cantilever erection, Governor O.K. Allen railway/highway cantilever bridge, Baton Rouge, La., 1939. First use of long balanced-cantilever erection procedure in the U.S. On each pier 650 ft (198 m) of steelwork, about 4000 tons, was balanced on the 40 ft (12 m) base formed by a sloping falsework bent. The compression load at the top of the falsework bent was measured at frequent intervals and adjusted by positioning a counterweight car running at bottom-chord level. The main spans are 848-650-848 ft (258-198-258 m); 650 ft span shown. (Courtesy of Bethlehem Steel Corporation.)

Figure 45.24 Tower erection, second Tacoma Narrows Bridge, Washington, 1949. This bridge replaced first Tacoma Narrows bridge, which blew down in a 40 mph (18 m/sec) wind in 1940. The tower legs are 60 ft (18 m) on centers and 462 ft (141 m) high. The creeper traveler is shown erecting the west tower, in background. On the east tower, the creeper erected a Chicago boom at the top of the south leg: this boom dismantled the creeper, then erected the tower-top bracing and a stiffleg derrick, which proceeded to dismantle the Chicago boom. The tower manhoist can be seen at the second-from-topmost landing platform. Riveting cages are approaching the top of the tower. Note tower-base erection kneebraces, required to ensure tower stability in free-standing condition (see Figure 45.27).

Figure 45.25 Aerial spinning of parallel-wire main cables, second Tacoma Narrows suspension bridge, Washington, 1949. Each main cable consists of 8702 parallel galvanized high-strength wires of 0.196 in (4.98 mm) diameter, laid up as 19 strands of mostly 460 wires each. Following compaction the cable became a solid round mass of wires with a diameter of 20-1/4 in (514 mm).

Figure 45.25a Tramway starts across from east anchorage carrying two wire loops. Three 460-wire strands have been spun, with two more under construction. Tramway spinning wheels pull wire loops across the three spans from east anchorage to west anchorage. Suspended footbridges provide access to cables. Spinning goes on 24 hours per day.

Figure 45.25b Tramway arrives at west anchorage. Wire loops shown in Figure 45.25a are removed from spinning wheels and placed around strand shoes at west anchorage. This tramway then returns empty to east anchorage, while tramway for other "leg" of endless hauling rope brings two wire loops across for second strand that is under construction for this cable.

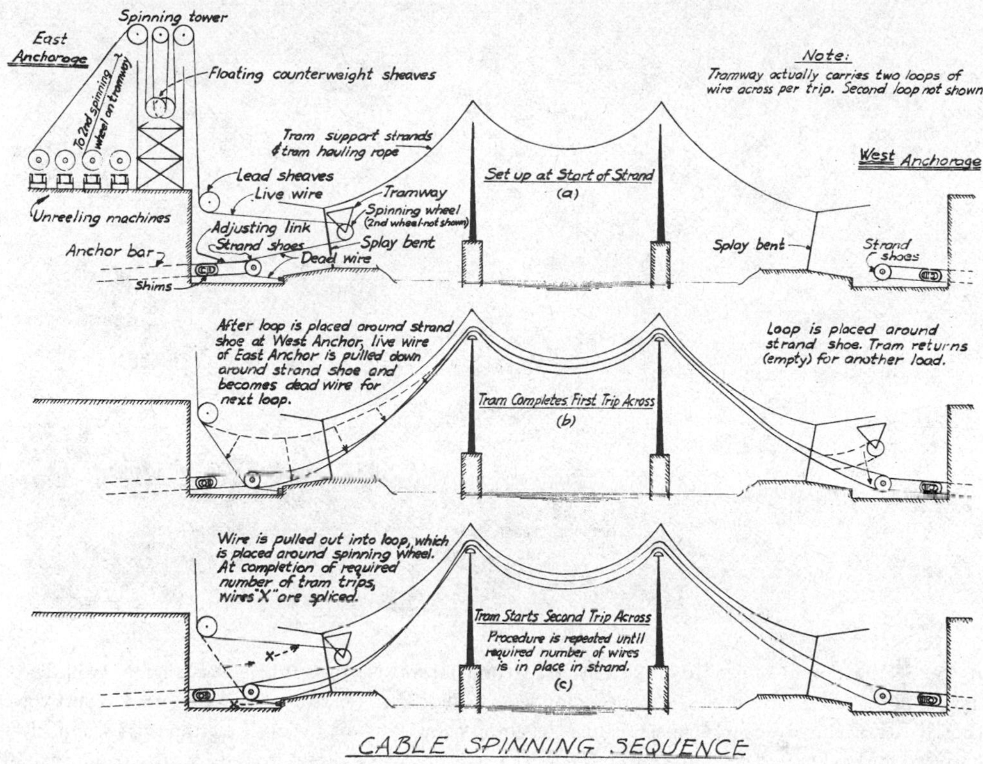

Figure 45.26a Erection of individual wire loops.

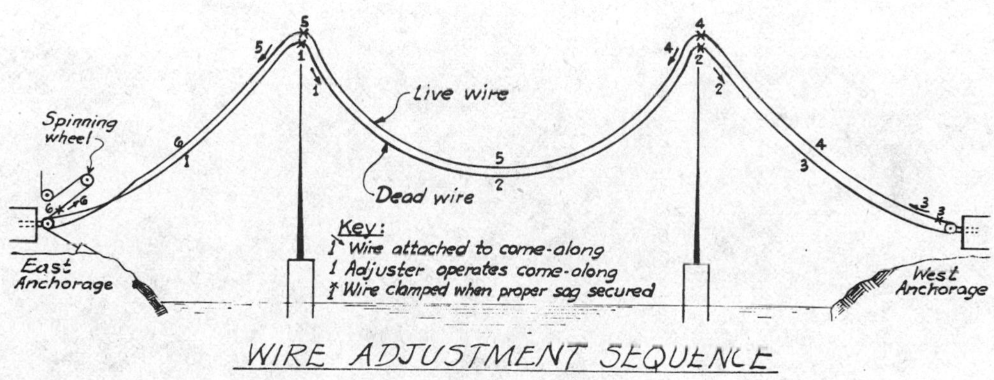

Figure 45.26b Adjustment of individual wire loops.

Figure 45.26 Cable-spinning procedure for constructing suspension bridge-parallel-wire main cables, showing details of aerial spinning method for forming individual 5 mm wires into strands containing 400 to 500 wires. Each wire loop is erected as shown in Figure 45.26a (refer to Figure 45.25), then adjusted to the correct sag as shown in Figure 45.26b. Each completed strand is banded with tape, then adjusted to the correct sag in each span. With all strands in place, they are compacted to form a solid round homogeneous mass of wires. The aerial spinning method was developed by John Roebling in the mid-19th century.

Figure 45.27 Erection of suspended deck steelwork, second Tacoma Narrows Bridge, Washington, 1950. The Chicago boom on the tower raises deck steelwork components to deck level, where they are transported to deck travelers by material cars. Each truss double panel is connected at top-chord level to previously erected trusses, and left open at bottom-chord level to permit temporary upward deck curvature, which results from the partial loading condition of the main suspension cables. The main span (at right) is 2800 ft (853 m), and side spans are 100 ft (335 m). The stiffening trusses are 33 ft (10 m) deep and 60 ft (18 m) on centers. Tower-base kneebraces (see Figure 45.24) show clearly here.

Figure 45.28 Moving deck traveler forward, second Tacoma Narrows Bridge, Washington, 1950. The traveler pulling-falls leadline passes around the sheave beams at the forward end of the stringers, and is attached to the front of the material car (at left). The material car is pulled back toward the tower, advancing the traveler two panels to its new position at the end of the deck steelwork. Arrows show successive positions of material car. (a) Traveler at star of move, (b) traveler advanced one panel, and (c) traveler at end of move.

Figure 45.29 Erecting closing girder sections of Passaic River Bridge, New Jersey Turnpike, 1951. Huge double-boom travelers, each weighing 270 tons, erect closing plate girders of the 375 ft (114 m) main span. The closing girders are 14 ft (4.3 m) deep and 115 ft (35 m) long and weigh 146 tons each. Sidewise entry was required (as shown) because of long projecting splice material. Longitudinal motion was provided at one pier, where girders were jacked to effect closure. Closing girders were laterally stable without floor steel fill-in, such that derrick falls could be released immediately. (Courtesy of Bethlehem Steel Corporation.)

Figure 45.30 Floating-in erection of a truss span, first Chesapeake Bay Bridge, Maryland, 1951. Erected 300 ft (91 m) deck-truss spans form erection dock, providing a work platform for two derrick travelers. A permanent deck-truss span serves as a falsework truss supported on barges and is shown carrying the 470 ft (143 m) anchor arm of the through-cantilever truss. This span is floated to its permanent position, then landed onto its piers by ballasting the barges. (a) Float leaves erection dock, and (b) float arrives at permanent position. (Courtesy of Bethlehem Steel Corporation.)

Figure 45.31 Floating-in erection of a truss span, first Chesapeake Bay Bridge, Maryland, 1952. A 480 ft (146 m) truss span, weighting 850 tons, supported on falsework consisting of a permanent deck-truss span along with temporary members, is being floated-in for landing onto its piers. Suspension bridge cables are under construction in background. (Courtesy of Bethlehem Steel Corporation.)

Figure 45.32 Erection of a truss span by hoisting, first Chesapeake Bay Bridge, Maryland, 1952. A 360 ft (110 m) truss span is floated into position on barges and picked clear using four sets of lifting falls. Suspension bridge deck is under construction at right. (Courtesy of Bethlehem Steel Corporation.)

Figure 45.33 Erection of suspension bridge deck structure, first Chesapeake Bay Bridge, Maryland, 1952. A typical four-panel through-truss deck section, weighting 99 tons, has been picked from the barge and is being raised into position using four sets of lifting falls attached to main suspension cables. The closing deck section is on the barge, ready to go up next. (Courtesy of Bethlehem Steel Corporation.)

Figure 45.34 Greater New Orleans cantilever bridge, Louisiana, 1957. Tall double-boom deck travelers started at ends of main bridge and erected anchor spans on falsework, then the 1575 ft (480 m) main span by cantilevering to midspan. (Courtesy of Bethlehem Steel Corporation.)

Figure 45.35 Tower erection, second Delaware Memorial Bridge, Wilmington, Del., 1966. The tower erection traveler has reached the topmost erecting position and swings into place the 23 ton closing top-strut section. The tower legs were jacked apart about 2 in (50 mm) to provide entering clearance. The traveler jumping beams are in the topmost working position, above the cable saddles. The tower steelwork is about 418 ft (127 m) high. Cable anchorage pier is under construction at right. First Delaware Memorial Bridge (1951) is at left. The main span of both bridges is 2150 ft (655 m). (Courtesy of Bethlehem Steel Corporation.)

Figure 45.36 Erecting orthotropic-plate decking panel, Poplar Street Bridge, St. Louis, Mo., 1967. A five-span, 2165 ft (660 m) continuous box-girder bridge, main span 600 ft (183 m). Projecting box ribs are 5 1/2 × 17 ft (1.7 × 5.2 m) in cross-section, and decking section is 27 × 50 ft (8.2 × 15.2 m). Decking sections were field welded, while all other connections were field bolted. Box girders are cantilevered to falsework bents using overhead "positioning travelers" (triangular structure just visible above deck at left) for intermediate support. (Courtesy of Bethlehem Steel Corporation.)

Figure 45.37 Erection of parallel-wire-stand (PWS) cables, the Newport Bridge suspension spans, Narragansett Bay, R.I., 1968. Bridge engineering history was made at Newport with the development and application of shop-fabricated parallel-wire socketed strands for suspension bridge cables. Each Newport cable was formed of seventy-six 61-wire PWS, each 4512 ft (1375 m) long and weighing 15 tons. Individual wires are 0.202 in. (5.13 mm) in diameter and are zinc coated. Parallel-wire cables can be constructed of PWS faster and at lower cost than by traditional air spinning of individual wires (see Figures 45.25 and 45.26). (Courtesy of Bethlehem Steel Corporation.)

Figure 45.37a Aerial tramway tows PWS from west anchorage up side span, then on across other spans to east anchorage. Strands are about 1 3/4 in (44 mm) in diameter.

Figure 45.37b Cable formers maintain strand alignment in cables prior to cable compaction. Each finished cable is about 15-1/4 in .(387 mm) in diameter. (Courtesy of Bethlehem Steel Corporation.)

Figure 45.38 Pipe-type anchorage for parallel-wire-strand (PWS) cables, the Newport Bridge suspension spans, Narragansett Bay, R.I., 1967. Pipe anchorages shown will be embedded in anchorage concrete. The socketed end of each PWS is pulled down its pipe from the upper end, then seated and shim-adjusted against the heavy bearing plate at the lower end. The pipe-type anchorage is much simpler and less costly than the standard anchor-bar type used with aerial-spun parallel-wire cables (see Figure 45.25b). (Courtesy of Bethlehem Steel Corporation.)

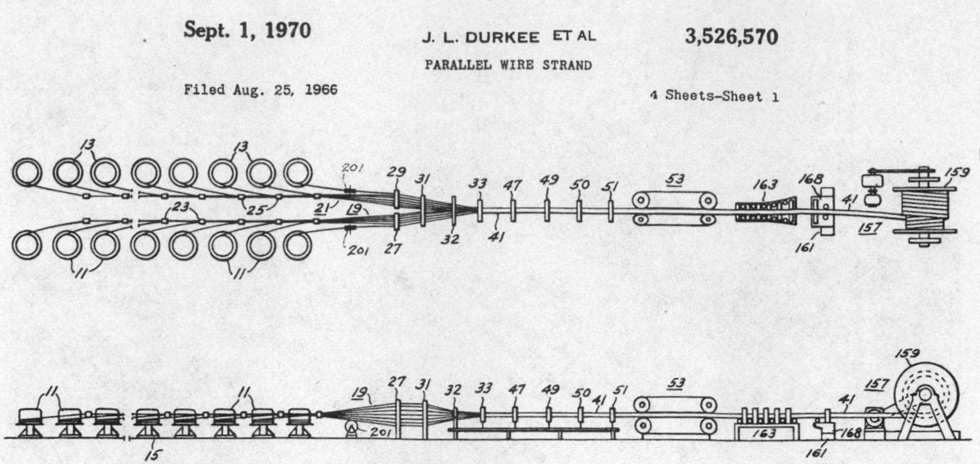

Figure 45.39 Manufacturing facility for production of shop-fabricated parallel-wire strands (PWS). Prior to 1966, parallel-wire suspension bridge cables had to be constructed wire-by-wire in the field using the aerial spinning procedure developed by John Roebling in the mid-19th century (refer to Figures 45.25 and 45.26). In the early 1960s a major U.S. steelwork contractor originated and developed a procedure for manufacturing and reeling parallel-wire strands, as shown in these patent drawings. A PWS can contain up to 127 wires (see Figures 45.45 and 45.46). (a) Plan view of PWS facility. Turntables 11 contain "left-hand" coils of wire and turntables 13 contain "right-hand" coils, such that wire cast is balanced in the formed strand. Fairleads 23 and 25 guide the wires into half-layplates 27 and 29, followed by full layplates 31 and 32 whose guide holes delineate the hexagonal shape of final strand 41. (b) Elevation view of PWS facility. Hexagonal die 33 contains six spring-actuated rollers that form the wires into regular-hexagon shape; and similar roller dies 47, 49, 50, and 51 maintain the wires in this shape as PWS 41 is pulled along by hexagonal dynamic clamp 53. The PWS is bound manually with plastic tape at about 3 ft (1 m) intervals as it passes along between roller dies. The PWS passes across roller table 163, then across traverse carriage 168, which is operated by traverse mechanism 161 to direct the PWS properly onto reel 159. Finally, the reeled PWS is moved off-line for socketing. Note that wire measuring wheels 201 can be installed and used for control of strand length.

Figure 45.40 Suspended deck steelwork erection, the Newport Bridge suspension spans, Narragansett Bay, R.I., 1968. The closing mainspan deck section is being raised into position by two cable travelers, each made up of a pair of 36 in (0.91 m) wide-flange rolled beams that ride the cables on wooden wheels. The closing section is 40 1//2 ft (12 m) long at top-chord level, 66 ft (20 m) wide and 16 ft (5 m) deep, and weighs about 140 tons. (Courtesy of Bethlehem Steel Corporation.)

Figure 45.41 Erection of Kansas City Southern Railway box-girder bridge, near Redland, Okla., by "launching," 1970. This nine-span continuous box-girder bridge is 2110 ft (643 m) long, with a main span of 330 ft (101 m). Box cross-section is 11 × 14.9 ft (3.35 × 4.54 m). The girders were launched in two "trains," one from the north end and one from the south end. A "launching nose" was used to carry the leading end of each girder train up onto the skidway supports as the train was pushed out onto successive piers. Closure was accomplished at center of main span. (Courtesy of Bethlehem Steel Corporation.)

Figure 45.41a Leading end of north girder train moves across 250 ft (76 m) span 4, approaching pier 5. Main span 330 ft (101 mm) is to right of pier 5.

Figure 45.41b Launching nose rides up onto pier 5 skidway units, removing girder-train leading-end sag.

Figure 45.41c Leading end of north girder train is now supported on pier 5.

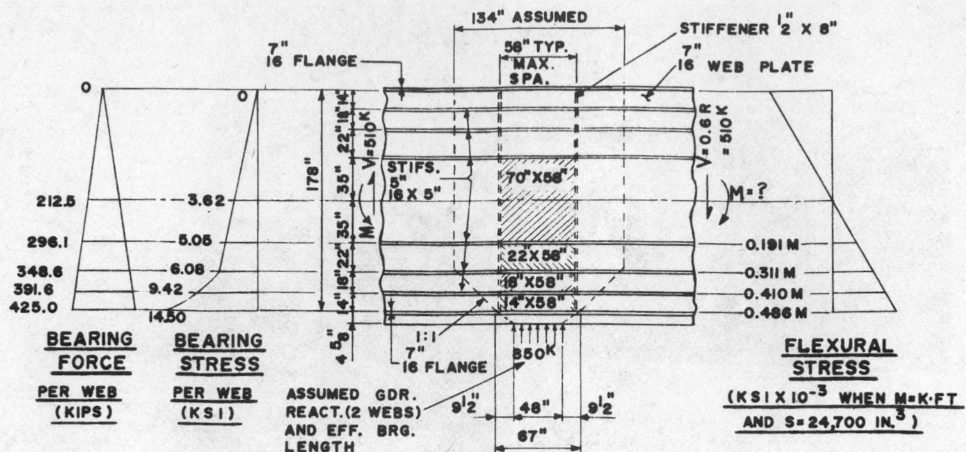

Figure 45.42a Typical assumed erection loading of box-girder web panels in combined moment, shear, and transverse compression.

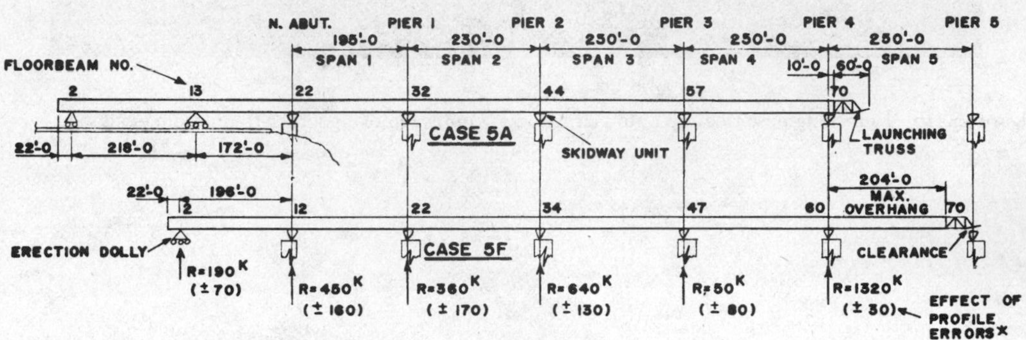

Figure 45.42b Launch of north girder train from pier 4 to pier 5.

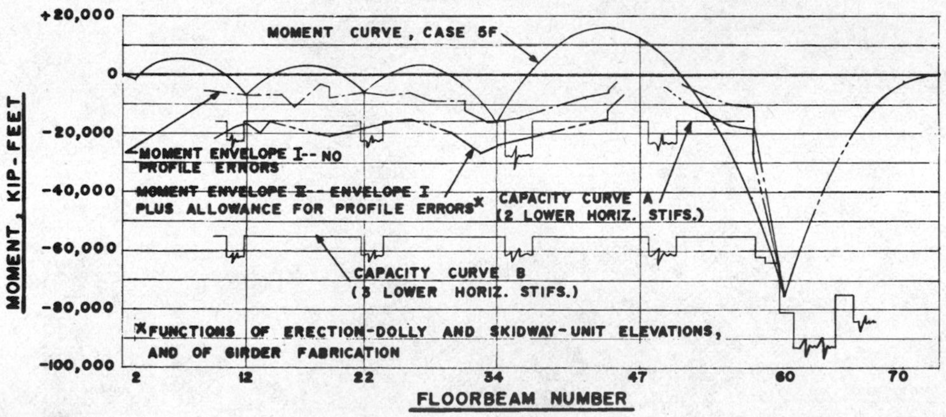

Figure 45.42c Negative-moment envelopes occurring simultaneously with reaction, for launch of north girder train to pier 5.

Figure 45.42 Erection strengthening to withstand launching, Kansas City Southern Railway box-girder bridge, near Redland, Okla. (see Figure 45.41).

Figure 45.43 Erection of west arch span of twin-arch Hernando de Soto Bridge, Memphis, Tenn., 1972. The two 900 ft (274 m) continuous-truss tied arch spans were erected by a high-tower derrick boat incorporating a pair of barges. West-arch steelwork (shown) was cantilevered to midspan over two pile-supported falsework bents. Projecting east-arch steelwork (at right) was then cantilevered to midspan (without falsework) and closed with falsework-supported other half-arch. (Courtesy of Bethlehem Steel Corporation.)

Figure 45.44 Closure of east side span, Commodore John Barry cantilever truss bridge, Chester, Pa., 1973. A high-tower derrick boat (in background) started erection of trusses at both main piers, supported on falsework; then erected top-chord travelers for main and side spans. The sidespan traveler carried steelwork erection to closure, as shown, and the falsework bent was then removed. The east-mainspan traveler then cantilevered the steelwork (without falsework) to midspan, concurrently with cantilever erection by the west-half mainspan traveler, and the trusses were closed at midspan. Commodore Barry has a 1644 ft (501 m) main span, the longest cantilever span in the U.S., and 822 ft (251 m) side spans. (Courtesy of Bethlehem Steel Corporation.)

Figure 45.45 Reel of parallel-wire strand (PWS), Akashi Kaikyo suspension bridge, Kobe, Japan, 1994. Each socketed PWS is made up of 127 0.206 in. (5.23 mm) wires, is 13,360 ft (4073 m) long, and weighs 96 tons. Plastic-tape bindings secure the strand wires at 1 m intervals. Sockets can be seen on right side of reel. These PWS are the longest and heaviest ever manufactured. (Courtesy of Nippon Steel — Kobe Steel.)

Figure 45.46 Parallel-wire strand main cable, Akashi Kaikyo suspension bridge, Kobe, Japan, 1994. The main span is 6532 ft (1991 m), by far the world's longest. The PWS at right is being towed across the spans, supported on rollers. The completed cable is made up of 290 PWS, making a total of 36,830 wires, and has a diameter of 44.2 in (1122 mm) following compaction — the largest bridge cables built to date. Each 127-wire PWS is about 2-3/8 in (60 mm) in diameter. (Courtesy of Nippon Steel — Kobe Steel.)

Figure 45.47 Artist's rendering of proposed Messina Strait suspension bridge connecting Sicily with mainland Italy. The Messina Strait crossing has been under discussion since about 1870, under investigation since about 1955, and under active design since about 1975. The first realistic proposals for a crossing were made in 1969 in response to an international competition sponsored by the Italian government. There were 158 submissions — eight American, three British, three French, one German, one Swedish, and the remaining Italian. Forty of the submissions showed a single-span or multi-span suspension bridge. The enormous bridge shown has a single span of 10827 ft (3300 m) and towers 1250 ft (380 m) high. The bridge construction problems for such a span would be tremendously challenging. (Courtesy of Stretto di Messina, S.p.A.)

References

1. Conditions of Contract and Forms of Tender, Agreement and Bond for Use in Connection with Works of Civil Engineering Construction, 6th ed. (commonly known as "ICE Conditions of Contract"), Inst. Civil Engrs. (U.K.), 1991.
2. Copp, J.I., de Vries, K., Jameson, W.H., and Jones, J. 1945. Fabrication and Erection Controls, Rainbow Arch Bridge Over Niagara Gorge — a Symposium, *Transactions ASCE,* vol. 110.
3. Durkee, E.L., 1945. Erection of Steel Superstructure, Rainbow Arch Bridge Over Niagara Gorge — A Symposium, *Transactions ASCE,* vol. 110.
4. Durkee, J.L., 1966. Advancements in Suspension Bridge Cable Construction, *Proceedings, International Symposium on Suspension Bridges,* Laboratorio Nacional de Engenharia Civil, Lisbon.
5. Durkee, J.L., 1972. Railway Box-Girder Bridge Erected by Launching, J. *Struct. Div., ASCE,* July.
6. Durkee, J.L., and Thomaides, S.S., 1977. Erection Strength Adequacy of Long Truss Cantilevers, *J. Struct. Div., ASCE,* January.
7. Durkee, J.L., 1977. Needed: U.S. Standard Conditions for Contracting, *J. Struct. Div., ASCE,* June.
8. Durkee, J.L., 1982. Bridge Structural Innovation: A Firsthand Report, *J. Prof. Act., ASCE,* July.
9. Enquiry into the Basis of Design and Methods of Erection of Steel Box Girder Bridges. Final Report of Committee, 4 vols. (commonly known as "The Merrison Report"), HMSO (London), 1973/4.
10. Feidler, L.L., Jr., 1962. Erection of the Lewiston-Queenston Bridge, *Civil Engrg., ASCE,* November.
11. Freudenthal, A.M., Ed., 1972. The Engineering Climatology of Structural Accidents, *Proceedings of the International Conference on Structural Safety and Reliability,* Pergamon Press, Elmsford, N.Y.
12. Holgate, H., Kerry, J.G.G., and Galbraith, J., 1908. Royal Commission Quebec Bridge Inquiry Report, Sessional Paper No. 154, vols. I and II, S.E. Dawson, Ottawa, Canada.
13. Leto, I.V., 1994. Preliminary design of the Messina Strait Bridge, *Proc. Inst. Civil Engrs.* (U.K.), vol. 102(3), August.
14. Petroski, H., 1993. Predicting Disaster, *American Scientist,* vol. 81, March.
15. Pugsley, A., 1968. The Safety of Bridges, *The Structural Engineer,* U.K., July.
16. Ratay, R.T., Ed., 1996. *Handbook of Temporary Structures in Construction,* 2nd ed., McGraw-Hill, New York.
17. Schneider, C.C., 1905. The Evolution of the Practice of American Bridge Building, *Transactions ASCE,* vol. 54.
18. Sibly, P.G. and Walker, A.C., 1977. Structural Accidents and Their Causes, *Proc. Inst. Civil Engrs.* (U.K.), vol. 62(1), May.
19. Smith, D.W., 1976. Bridge Failures, *Proc. Inst. Civil Engrs.* (U.K.), vol. 60(1), August.

46

Concrete Bridge Construction

Simon A. Blank
California Department of Transportation

Michael M. Blank
U.S. Army Corps of Engineers

Luis R. Luberas
U.S. Army Corps of Engineers

46.1 Introduction

This chapter will focus on the principles and practices related to construction of concrete bridges in which construction engineering contributes greatly to the successful completion of the projects. We will first present the fundamentals of construction engineering and analyze the challenges and obstacles involved in such processes and then introduce the problems in relation to design, construction practices, project planning, scheduling and control, which are the ground of future factorial improvements in effective construction engineering in the United States. Finally, we will discuss prestressed concrete, high-performance concrete, and falsework in some detail.

46.2 Effective Construction Engineering

The construction industry is a very competitive business and many companies who engage in this marketplace develop proprietary technology in their field. In reality, most practical day-to-day issues are very common to the whole industry. Construction engineering is a combination of art and

0-8493-7434-0/00/$0.00+$.50

science and has a tendency to become more the art of applying science (engineering principles) and approaches to the construction operations. Construction engineering includes design, construction operation, and project management. The final product of the design team effort is to produce drawings, specifications, and special provisions for various types of bridges. A fundamental part of construction engineering is construction project management (project design, planning, scheduling, controlling, etc.).

Planning starts with analysis of the type and scope of the work to be accomplished and selection of techniques, equipment, and labor force. Scheduling includes the sequence of operations and the interrelation of operations both at a job site and with external aspects, as well as allocation of manpower and equipment. Controlling consists of supervision, engineering inspection, detailed procedural instructions, record maintenance, and cost control. Good construction engineering analysis will produce more valuable, effective, and applicable instructions, charts, schedules, etc.

The objective is to plan, schedule, and control the construction process such that every construction worker and every activity contributes to accomplishing tasks with minimum waste of time and money and without interference. All construction engineering documents (charts, instructions, and drawings) must be clear, concise, definitive, and understandable by those who actually perform the work. As mentioned before, the bridge is the final product of design team efforts. When all phases of construction engineering are completed, this product — the bridge — is ready for to take service loading. In all aspects of construction engineering, especially in prestressed concrete, design must be integrated for the most effective results. The historical artificial separation of the disciplines — design and construction engineering — was set forth to take advantage of the concentration of different skills in the workplace. In today's world, the design team and construction team must be members of one team, partners with one common goal. That is the reason partnering represents a new and powerful team-building process, designed to ensure that projects become positive, ethical, and win–win experiences for all parties involved.

The highly technical nature of a prestressing operation makes it essential to perform preconstruction planning in considerable detail. Most problems associated with prestressed concrete could have been prevented by properly planning before the actual construction begins. Preconstruction planning at the beginning of projects will ensure that the structure is constructed in accordance with the plans, specifications, special provisions, and will also help detect problems that might arise during construction. It includes (1) discussions and conferences with the contractor, (2) review of the responsibilities of other parties, and (3) familiarization with the plans, specifications, and special provisions that relate to the planned work, especially if there are any unusual conditions. The preconstruction conference might include such items as scheduling, value of engineering, grade control, safety and environmental issues, access and operational considerations, falsework requirements, sequence of concrete placement, and concrete quality control and strength requirements. Pre-construction planning has been very profitable and in many has cases resulted in substantial reduction of labor costs. More often in prestressed concrete construction, the details of tendon layout, selection of prestressing system, mild-steel details, etc. are left up to general contractors or their specialized subcontractors, with the designer showing only the final prestress and its profile and setting forth criteria. And contractors must understand the design consideration fully to select the most efficient and economical system. Such knowledge may in many cases provide a competitive edge, and construction engineering can play a very important role in it.

46.3 Construction Project Management

46.3.1 General Principles

Construction project management is a fundamental part of construction engineering. It is a feat that few, if any, individuals can accomplish alone. It may involve a highly specialized technical field or science, but it always includes human interactions, attitudes and aspects of leadership, common sense, and resourcefulness. Although no one element in construction project management will create success, failure in one of the foregoing elements will certainly be enough to promote failure and to escalate costs. Today's construction environment requires serious consultation and management of the following life-cycle elements: design (including specifications, contract clauses, and drawings), estimating, budgeting, scheduling, procurement, biddability–constructibility–operability (BCO) review, permits and licenses, site survey, assessment and layout, preconstruction and mutual understanding conference, safety, regulatory requirements, quality control (QC), construction acceptance, coordination of technical and special support, construction changes and modifications, maintenance of progress drawings (redlines), creating as-built drawings, project records, among other elements.

Many construction corporations are becoming more involved in environmental restoration either under the Resource Conservation and Recovery Act (RCRA) or under the Comprehensive Environmental Response, Compensation and Liability Act (CERCLA, otherwise commonly known as the Superfund) This new involvement requires additional methodology and considerations by managers. Some elements that would otherwise be briefly covered or completely ignored under normal considerations may be addressed and required in a site Specific Health and Environmental Response Plan (SHERP). Some elements of the SHERP may include site health and safety staff, site hazard analysis, chemical and analytical protocol, personal protective equipment requirements and activities, instrumentation for hazard detection, medical surveillance of personnel, evacuation plans, special layout of zones (exclusion, reduction and support), and emergency procedures.

Federal government contracting places additional demands on construction project management in terms of added requirements in the area of submittals and transmittals, contracted labor and labor standards, small disadvantaged subcontracting plans, and many other contractual certification issues, among others. Many of these government demands are recurring elements throughout the life cycle of the project which may require adequate resource allocation (manpower) not necessary under the previous scenarios.

The intricacies of construction project management require the leadership and management skills of a unique individual who is not necessarily a specialist in any one of the aforementioned elements but who has the capacity to converse and interface with specialists in the various fields (i.e., chemists, geologists, surveyors, mechanics, etc.). An individual with a combination of an engineering undergraduate degree and a graduate business management degree is most likely to succeed in this environment. Field management experience can substitute for an advanced management degree.

It is the purpose of this section to discuss and elaborate elements of construction project management and to relate some field experiences and considerations. The information presented here will only promote further discussion and is not intended to be all-inclusive.

46.3.2 Contract Administration

Contract administration focuses on the relationships between the involved parties during the contract performance or project duration. Due to the nature of business, contract administration embraces numerous postaward and preaward functions. The basic goals of contract administration are to assure that the owner is satisfied and all involved parties are compensated on time for their efforts. The degree and intensity of contract administration will vary from contact to contract depending upon the size and complexity of the effort to be performed. Since money is of the essence, too many resources can add costs and expenditures to the project, while insufficient resources may also cost in loss of time, in inefficiencies, and in delays. A successful construction project management program is one that has the vision and flexibility to allocate contract administrative personnel and resources wisely and that maintains a delicate balance in resources necessary to sustain required efficiencies throughout the project life cycle.

46.3.3 Project Design

Project design is the cornerstone of construction project management. In this phase, concepts are drawn, formulated, and created to satisfy a need or request. The design is normally supported by sound engineering calculations, estimates, and assumptions. Extensive reviews are performed to minimize unforeseen circumstances, avoiding construction changes or modifications to the maximum extent possible in addition to verifying facts, refining or clarifying concepts, and dismissing assumptions. This phase may be the ideal time for identification and selection of the management team.

Normally, 33, 65, 95, and 100% design reviews are standard practice. The final design review follows the 95% design review which is intended for the purpose of assuring that review comments have been either incorporated into the design or dismissed from consideration. Reviews include design analysis reviews and BCO reviews. It can be clearly understood from the nomenclature that a BCO encompasses all facets of a project. Biddability relates to how the contact requirements are worded to assure clarity of purpose or intent and understanding by potential construction contractors. Constructibility concentrates on how components of the work or features of the work are assembled and how they relate to the intended final product. The main purpose of the constructibility review is to answer questions, such as whether it can be built in the manner represented in the contact drawings and specifications. Interaction between mechanical, civil, electrical, and other related fields is also considered here. Operability includes aspects of maintenance and operation, warranties, services, manpower, and resource allocation during the life of the finished work.

The finished product of the design phase should include construction drawings illustrating dimensions, locations, and details of components; contract clauses and special clauses outlining specific needs of the construction contractor; specifications for mechanical, civil, and electrical or special equipment; a bidding and payment schedule with details on how parties will be compensated for work performed or equipment produced and delivered; responsibilities; and operation and maintenance (O&M) requirements. In many instances, the designer is involved throughout the construction phase for design clarification or interpretation, incorporation of construction changes or modifications to the project, and possible O&M reviews and actions. It is not uncommon to have the designer perform contract management services for the owner.

There are a number of computer software packages readily available to assist members of the management team in writing, recording, transmitting, tracking, safekeeping, and incorporating BCO comments. Accuracy of records and safekeeping of documentation regarding this process has proved to be valuable when a dispute, claim, design deficiency, or liability issue are encountered later during the project life cycle.

46.3.4 Planning and Scheduling

Planning and schedulings are ongoing tasks throughout the project until completion and occupancy by a certain date occur. Once the design is completed and the contractor selected to perform the work, the next logical step may be to schedule and conduct a preconstruction conference. Personnel representing the owner, designer, construction contractor, regulatory agencies, and any management/oversight agency should attend this conference. Among several key topics to discuss and understand, construction planning and scheduling is most likely to be the main subject of discussion. It is during this conference that the construction contractor may present how the work will be executed. The document here is considered the "baseline schedule." Thereafter, the baseline schedule becomes a living document by which progress is recorded and measured. Consequently, the baseline schedule can be updated and reviewed in a timely manner and becomes the construction progress schedule. As stated previously, the construction progress schedule is the means by which the construction contractor records progress of work, anticipates or forecasts requirements so proper procurement and allocation of resources can be achieved, and reports the construction status of work upwardly to the owner or other interested parties. In addition, the construction contractor may use progress schedule information to assist in increasing efficiencies or to formulate the basis of payment for services provided or rendered and to anticipate cash flow requirements. The construction progress schedule can be updated as needed, or mutually agreed to by the parties, but for prolonged projects it is normally produced monthly.

A dedicated scheduler, proper staffing, and adequate computer and software packages are important to accomplish this task properly. On complex projects, planning and scheduling is a full-time requirement.

46.3.5 Safety and Environmental Considerations

Construction of any bridge is a hazardous activity by nature. No person may be required to work in surroundings or under conditions that are unsafe or dangerous to his or her health. The construction project management team must initiate and maintain a safety and health program and perform a job hazard analysis with the purpose of eliminating or reducing risks of accidents, incidents, and injuries during the performance of the work. All features of work must be evaluated and assessed in order to identify potential hazards and implement necessary precautions or engineer controls to prevent accidents, incidents, and injuries.

Frequent safety inspections and continued assessment are instrumental in maintaining the safety aspects and preventive measures and considerations relating to the proposed features of work. In the safety area, it is important for the manager to be able to distinguish between accidents/incidents and injuries. Lack of recorded work-related injuries is not necessarily a measure of how safe the work environment is on the project site. The goal of every manager is to complete the job in an accident/incident- and injury-free manner, as every occurrence costs time and money.

Today's construction operational speed, government involvement, and community awareness are placing more emphasis, responsibilities, and demands on the designer and construction contractor to protect the environment and human health. Environmental impact statements, storm water management, soil erosion control plans, dust control plan, odor control measures, analytical and disposal requirements, Department of Transportation (DOT) requirements for overland shipment, activity hazard analysis, and recycling are some of the many aspects that the construction project management team can no longer ignore or set aside. As with project scheduling and planning, environmental and safety aspects of construction may require significant attention from a member of the construction management team. When not properly coordinated and executed, environmental considerations and safety requirements can delay the execution of the project and cost significant amounts of money.

46.3.6 Implementation and Operations

Construction implementation and operations is the process by which the construction project manager balances all construction and contract activities and requirements in order to accomplish the tasks. The bulk of construction implementation and operations occurs during the construction phase of the project. The construction project management team must operate in synchronization and maintain good communication channels in order to succeed in this intense and demanding phase. Many individuals in this field may contend that the implementation and operation phase of the construction starts with the site mobilization. Although it may be an indicator of actual physical activity taking place on site, construction implementation and operations may include actions and activities prior to the mobilization to the project site.

Here, a delicate balance is attempted to be maintained between all activities taking place and those activities being projected. Current activities are performed and accomplished by field personnel with close monitoring by the construction management staff. Near (approximately 1 week ahead), intermediate (approximately 2 to 4 weeks), and distant future (over 4 weeks) requirements are identified, planned, and scheduled in order to procure equipment and supplies, schedule work crews, and maintain efficiencies and progress. Coordinating progress and other meetings and conferences may take place during the implementation and operation phase.

46.3.7 Value Engineering

Some contracts include an opportunity for contractors to submit a value engineering (VE) recommendation. This recommendation is provided to either the owner or designer. The purpose of the VE is to promote or increase the value of the finished product while reducing the dollars spent or invested; in other words, to provide the desired function for minimum cost(s). VE is not intended to reduce performance, reliability, maintainability, or life expectancy below the level required to perform the basic function. Important VE evaluation criteria performed are in terms of "collateral savings" — the measurable net reductions in the owner's/agency's overall costs of construction, operations, maintenance, and/or logistics support. In most cases, collateral savings are shared between the owner/agency and the proponent of the VE by reducing the contract price or estimated cost in the amount of the instant contract savings and by providing the proponents of the VE a share of the savings by adding the amount calculated to the contract price or fee.

46.3.8 Quality Management

During the construction of a bridge, construction quality management (CQM) play a major role in quality control and assurance. CQM refers to all control measures and assurance activities instituted by the parties to achieve the quality established by the contract requirements and specifications. It encompasses all phases of the work, such as approval of submittals, procurements, storage of materials and equipment, coordination of subcontractor activities, and the inspections and the tests required to ensure that the specified materials are used and that installations are acceptable to produce the required product. The key elements of the CQM are the contractor quality control (CQC) and quality assurance (QA). To be effective, there must be a planned program of actions, and lines of authority and responsibilities must be established. CQC is primarily the construction contractor's responsibily while QA is primarily performed by an independent agency (or other than the construction contractor) on behalf of the designer or owner. In some instances, QA may be performed by the designer. In this manner, a system of checks and balances is achieved minimizing the conflicts between quality and efficiency normally developed during construction. Consequently, CQM is a combined responsibility.

In the CQC, the construction contractor is primarily responsible for (1) producing the quality product on time and in compliance with the terms of the contract; (2) verifying and checking the adequacy of the construction contractor's quality control program of the scope and character necessary to achieve the quality of construction outlined in the contract; and (3) producing and maintaining acceptable records of its QC activities. In the QA, the designated agency is primarily responsible for (1) establishing standards and QC requirements; (2) verifying and checking adequacy of the construction contractor's QC (QA for acceptance), performing special tests and inspections as required in the contract, and determining that reported deficiencies have been corrected; and (3) assuring timely completion.

46.3.9 Partnership and Teamwork

A great deal of construction contract success, as discussed before, is attributable to partnering. Partnering should be undertaken and initiated at the earliest stage during the construction project management cycle. Some contracts may have a special clause which is intended to encourage the construction contractor to establish clear channels of communication and effective working relationships. The best approach to partnering is for the parties to volunteer to participate.

Partnering differs from the team-building concept. Team building may encourage establishing open communications and relationships when all parties share liabilities, risk, and money exposure, but not necessarily share costs of risks. The immediate goal of partnering is to establish mutual agreement(s) at the initial phases of the project on the following areas: identification of common goals; identification of common interests; establishment of lines of communication; establishment of lines of authority and decision making; commitment to cooperative problem solving, among others.

Partnering takes the elements of luck, hope, and personality out of determining project success. It facilitates workshops in which stakeholders in a specific project or program come together as a team that results in breakthrough success for all parties involved. For example, the Office of Structure Construction (OSC) of the California Department of Transportation (Caltrans) has a vision of delivery of structure construction products of the highest possible quality in partnership with their clients. And this work is not only of high quality, but is delivered in the safest, most cost-effective, and fastest manner possible. In partnership with the districts or other clients, the Office of Structure Construction (OSC) does the following to fulfill its purpose:

- Administers and inspects the construction of the Caltrans transportation structures and related facilities in a safe and efficient manner;
- Provides specialized equipment and training, standards, guidelines, and procedural manuals to ensure consistency of inspection and administration by statewide OSC staff;
- Provides consultations on safety for OSC staff and district staff performing structure construction inspection work;
- Conducts reviews and provides technical consultation and assistance for trenching and shoring temporary support and falsework construction reviews;
- Provides technical recommendations on the preparations of structure claims and the contract change orders (CCOs);
- Provides construction engineering oversight on structure work on non-state-administrated projects;
- Conducts BCO review.

46.3.10 Project Completion and Turnover of Facility

Success in construction project management may be greatly impacted during project completion and turnover of the facilities to the user or owner. The beginning of the project completion and turnover phase may be identified by one of the following: punch list developed, prefinal inspections scheduled, support areas demobilized, site restoration initiated, just to mention a few. Many of the problems encountered during this last phase may be avoided or prevented with proper user or owner participation and involvement during the previous phases, particularly during the construction where changes and modifications may have altered the original design. A good practice in preventing conflicts during the completion and turnover of the facilities is to invite the owner or user to all construction progress meetings and acceptance inspections. In that manner, the user or owner is completely integrated during the construction with ample opportunity to provide feedback and be part of the decision-making process. In addition, by active participation, the owner or user is being informed and made aware of changes, modifications, and/or problems associated with the project.

46.4 Major Construction Considerations

Concrete bridge construction involves site investigation; structure design; selection of materials — steel, concrete, aggregates, and mix design; workmanship of placment and curing of concrete; handling and maintenance of the structure throughout its life. Actually, site investigations are made of any structure, regardless of how insignificant it may be. The site investigation is very important for intelligent design of the bridge structures and has a significant influence on selection of the material and mix. A milestone is to investigate the fitness of the location to satisfy the requirements of the bridge structure. Thus, investigation of the competence of the foundation to carry the service load safely and an investigation of the existence of forces or substances that may attack the concrete structure can proceed. Of course, the distress or failure may have several contributing causal factors: unsuitable materials, construction methods, loading conditions; faulty mix design; design mistakes; conditions of exposure; curing condition, or environmental factors.

46.5 Structural Materials

46.5.1 Normal Concrete

Important Properties

Concrete is the only material that can be made on site, and is practically the most dependable and versatile construction material used in bridge construction. Good durable concrete is quality concrete that meets all structural and aesthetic requirements for a period of structure life at minimum cost. We are looking for such properties as workability in the fresh condition; strength in accordance with design, specifications, and special provisions; durability; volume stability; freedom from blemishes (scaling, rock pockets, etc.); impermeability; economy; and aesthetic appearance. Concrete when properly designed and fabricated can actually be crack-free not only under normal service loads, but also under moderate overload, which is very attractive for bridges that are exposed to an especially corrosive atmosphere.

The codes and specifications usually specify the minimum required strength for various parts of a bridge structure. The required concrete strength is determined by design engineers. For cast-in-place concrete bridges, a compressive strength of 3250 to 5000 psi (22 to 33 MPa) is usual. For precast structure compressive strength of 4000 to 6000 psi (27 to 40 MPa) is often used. For special precast, prestressed structures compressive strength of 6000 to 8000 psi (40 to 56 MPa) is used. Other properties of concrete are related to the strength, although not necessarily dependent on the strength.

Workability is the most important property of fresh concrete and depends on the properties and proportioning of the materials: fine and coarse aggregates, cement, water, and admixtures. Consistency, cohesiveness, and plasticity are elements of workability. Consistency is related to the fluidity of mix. Just adding water to a batch of concrete will make the concrete more fluid or "wetter," but the quality of the concrete will diminish. Consistency increases when water is added and an average of 3% in total water per batch will change the slump about 1 in. (2.54 cm). The research and practice show that workability is a maximum in concrete of medium consistency, between 3 in. (7.62 cm) and 6 in. (15.24 cm) slump. Very dry or wet mixes produce less-workable concrete. Use of relatively harsh and dry mixes is allowed in structures with large cross sections, but congested areas containing much reinforcement steel and embedded items require mixes with a high degree of workability.

A good and plastic mixture is neither harsh nor sticky and will not segregate easily. Cohesiveness is not a function of slump, as very wet (high-slump) concrete lacks plasticity. On the other hand, a low-slump mix can have a high degree of plasticity. A harsh concrete lacks plasticity and cohesiveness and segregates easily.

Workability has a great effect on the cost of placing concrete. Unworkable concrete, not only requires more labor and effort in placing, but also produces rock pockets and sand streaks, especially in small congested forms. It is a misconception that compaction or consolidation of concrete in the form can be done with minimum effort if concrete is fluid or liquid to flow into place. It is obvious that such concrete will flow in place but segregate badly, so that large aggregate will settle out of the mortar and excess water will rise to the top surface. And unfortunately, this error in workmanship will become apparent after days, even months later, showing up as cracks, low strength, and general inferiority of concrete. The use of high-range water-reducing admixtures (superplasticizers) allows placing of high-slump, self-leveling concrete. They increase strength of concrete and provide great workability without adding an excessive amount of water. As an example of such products used in the Caltrans is PolyHeed 997 which meets the requirements for a Type A, water-reducing admixture specified in ASTM C 494-92, Corps of Engineers CRD-C 87-93, and AASHTO M 194-87, the Standard Specifications for chemical admixtures for concrete.

Special Consideration for Cold-Weather Construction

Cold weather can damage a concrete structure by freezing of fresh concrete before the cement has achieved final set and by repeated cycles of freezing of consequent expansion of water in pores and openings in hardened concrete. Causes of poor frost resistance include poor design of construction joints, segregation of concrete during placement; leaky formwork; poor workmanship, resulting in honeycomb and sand streaks; insufficient or absent drainage, permitting water to accumulate against concrete. In order to provide resistance against frost adequate drainage should be designed. If horizontal construction joints are necessary, they should be located below the low-water or above the high-water line about 2 to 3 ft (0.6 to 1 m). Previously placed concrete must be cleaned up completely. Concrete mix should have a 7% (max) air for ½ in. (12.7 mm) or ¾ in. (19 mm) (max) aggregate, ranging down to 3 to 4% for cobble mixes. It is essential to use structurally sound aggregates with low porosity. The objective of frost-resistant concrete mix is to produce good concrete with smooth, dense, and impermeable surface. This can be implemented by good construction techniques used in careful placement of concrete as near as possible to its final resting place, avoiding segregation, sand streaks, and honeycomb under proper supervision, quality control, and assurance.

Sudden changes in temperature can stress concrete and cause cracking or crazing. A similar condition exists when cold water is applied to freshly stripped warm concrete, particularly during hot weather. For the best results, the temperature difference should not exceed 25°F between concrete and curing water. In cases when anchor bolt holes were left exposed to weather, filled with water, freezing of water exerted sufficient force to crack concrete. This may happen on the bridge pier cap under construction.

Concrete Reinforcement and Placement

The optimum conditions for structural use is a medium slump of concrete and compaction by vibrators. A good concrete with low slump for the placing conditions can be ruined by insufficient or improper consolidation. Even workable concrete may not satisfy the needs of the bridge structure if it is not properly consolidated, preferably by vibration. An abrupt change in size, and congestion of reinforcement not only makes proper placing of concrete difficult but also causes cracks to develop. Misplacement of reinforcement within concrete will greatly contribute to development of structural cracks. The distress and failure of concrete are mostly caused by ignorance, carelessness, wrong assumptions, etc.

Concrete Mix and Trial Batches

The objective of concrete mix designs and trial batches is to produce cost-effective concrete with sufficient workability, strength, durability, and impermeability to meet the conditions of placing, finishing characteristics, exposure, loading, and other requirements of bridge structures. A complete discussion of concrete mixes and materials can be found in many texts such as *Concrete Manual* by Waddel [1]. The purpose of trial batches is to determine strength, water–cement ratio, combined grading of aggregates, slump, type and proportioning of cement, aggregates, entrained air, and admixtures as well as scheduling of trial batches and uniformity. Trial batches should always be made for bridge structures, especially for large and important ones. They should also be made in cases where there is no adequate information available for existing materials used in concrete mixes, and they are subjected to revision in the field as conditions require.

Consideration to Exposure Condition

Protection of waterfront structures should be considered when they are being designed. Designers often carefully consider structural and aesthetic aspects without consideration of exposure conditions. Chemical attack is aggravated in the presence of water, especially in transporting the chemiclas into the concrete through cracks, honeycombs, or pores in surfaces. Use of chamfers and fillers is good construction practice. Chamfering helps prevent spalling and chipping from moving objects. Fillets in reentrant corners eliminate possible scours or cracking. Reinforcement should be well covered with sound concrete and in most cases the 3 in. (7.62 cm) coverage is specified. First-class nonreactive and well-graded aggregates in accordance with the UBC standard should be used. Cement Type II or Type Y with a low of C_3 should be used. Careful consideration should be given to the use of an approved pozzolan with a record of successfully usage in a similar exposure. Mix design should contain an adequate amount of entrained air and other parameters in accordance with specifications or a special provision for a particular project. The concrete should be workable with slump and water–cement ratio as low as possible and containing at least 560 pcy (332 kg/m^3). To reduce mixing water for the same workability and, by the same token, to enhance strength and durability, a water-reducing admixture is preferred. The use of calcium chloride and Type III cement for acceleration of hardening and strength development is precluded. Concrete should be handled and placed with special care to avoid segregation and prevent honeycomb and sand streaks. The proper cure should be taken for at least seven days before exposure.

46.5.2 High-Performance Concrete

High-performance concrete (HPC) is composed of the same materials used in normal concrete, but proportioned and mixed to yield a stronger, more durable product. HPC structures last much longer and suffer less damage from heavy traffic and climatic condition than those made with conventional concrete. To promote the use of HPC in highway structures in the United States, a group of concrete experts representing the state DOTs, academia, the highway industry, and the Federal Highway Administration (FHWA) has developed a working definition of HPC, which includes performance criteria and the standard tests to evaluate performance when specifying an HPC mixture. The designer determines what level of strength, creep, shrinkage, elasticity, freeze/thaw durability, abrasion resistance, scaling resistance, and chloride permeability are needed. The definition specifies what tests grade of HPC satisfies those requirements and what tests to perform to confirm that the concrete meets that grade.

An example of the mix design for the 12,000-psi high-strength concrete used in the Orange County courthouse in Florida follows:

Gradient	Weight (pounds)
Cement, Type 1	900
Fly ash, Class F	72
Silica fume	62
Natural sand	980
No. 8 granite aggregate	1,780
Water	250
Water reducer	2 oz per cubic hundredweight
Superplasticizer	35 oz per cubic hundredweight

The Virginia and Texas DOTs have already started using HPC that is ultra-high-strength concrete 12,000 to 15,000 psi (80 to 100 MPa) in bridge construction and rehabilitation of the existing bridges [2].

46.5.3 Steel

All reinforcing steel for bridges is required to conform to specifications of ASTM Designation A615, Grade 60 or low-alloy steel deformed bars conforming to ASTM Designation A706. Prestressing steel: high-tensile wire conforming to ASTM Designations: A421, including Supplement I, High-tensile wire strand A416, Uncoated high-strength steel bars: A722, are usually used. All prestressing steel needs to be protected against physical damage and rust or other results of corrosion at all times from manufacture to grouting or encasing in concrete. Prestressing steel that has physical damage at any time needs to be rejected. Prestressing steel for post-tensioning that is installed in members prior to placing and curing of the concrete needs to be continuously protected against rust or other corrosion until grouted, by means of a corrosion inhibitor placed in the ducts or applied to the steel in the duct.

The corrosion inhibitor should conform to the specified requirements. When steam curing is used, prestressing steel for post-tensioning should not be installed until the stem curing is completed. All water used for flushing ducts should contain either quick lime (calcium oxide) or slaked lime (calcium hydroxide) in the amount of 0.01 kg/l. All compressed air used to blow out ducts should be oil free.

46.6 Construction Operations

46.6.1 Prestressing Methods

If steel reinforcement in reinforced concrete structures is tensioned against the concrete, the structure becomes a prestressed concrete structure. This can be accomplished by using pretensioning and post-tensioning methods.

Pretensioning

Pretensioning is accomplished by stressing tendons, steel wires, or strands to a predetermined amount. While stress is maintained in the tendons, concrete is placed in the structure. After the concrete in the structure has hardened, the tendons are released and the concrete bonded to the tendons becomes prestressed.

Widely used in pretensioning techniques are hydraulic jacks and strands composed of several wires twisted around a straight center wire. Pretensioning is a major method used in the manufacture of prestressed concrete in the United States. The basic principles and some of the methods currently used in the United States were imported from Europe, but much has been done in the United States to develop and adapt manufacturing procedures. One such adaptation employs pretensioned tendons which do not pass straight through the concrete member, but are deflected or draped into a trajectory that approximates a curve. This method is very widely practiced in the fabrication of precast bridge girders in the United States.

Post-Tensioning

A member is called as posttensioned when the tendons are tensioned after the concrete has hardened and attained sufficient strength (usually 70% final strength) to withstand the prestressing force, and each end of the tendons are anchored. Figure 46.1 shows a typical post-tensioning system. A common method used in the United States to prevent tendons from bonding to the concrete during placing and curing of the concrete is to encase the tendon in a mortar-tight metal tube or flexible metal hose before placing it in the forms. The metal hose or tube is referred to as a sheath or duct and remains in the structure. After the tendons have been stressed, the void between the tendons and the duct is filled with grout. The tendons become bonded to the structural concrete and protected from corrosion [3].Construction engineers can utilize prestressing very effectively to overcome excessive temporary stresses or deflections during construction, for example, using cantilevering techniques in lieu of falsework.

Prestressing is not a fixed state of stress and deformation, but is time dependent. Both concrete and steel may be deformed inelastically under continued stress. After being precompressed, concrete continues to shorten with time (creep). Loss of moisture with time also contribute to a shortening (shrinkage). In order to reduce prestress losses due to creep and shrinkage and to increase the level of precompression, use of higher-strength not only steel but also higher-strength concrete, that has low creep, shrinkage, and thermal response is recommended. New chemical admixtures such as high-range water-reducing admixtures (superplasticizers) and slag used for producing high-performance concrete and for ultra-high-strength concrete. The new developments are targeted to producing high-strength steel that is "stabilized" against stress relaxation which leads to a reduction of stress in tendons, thus reducing the prestress in concrete.

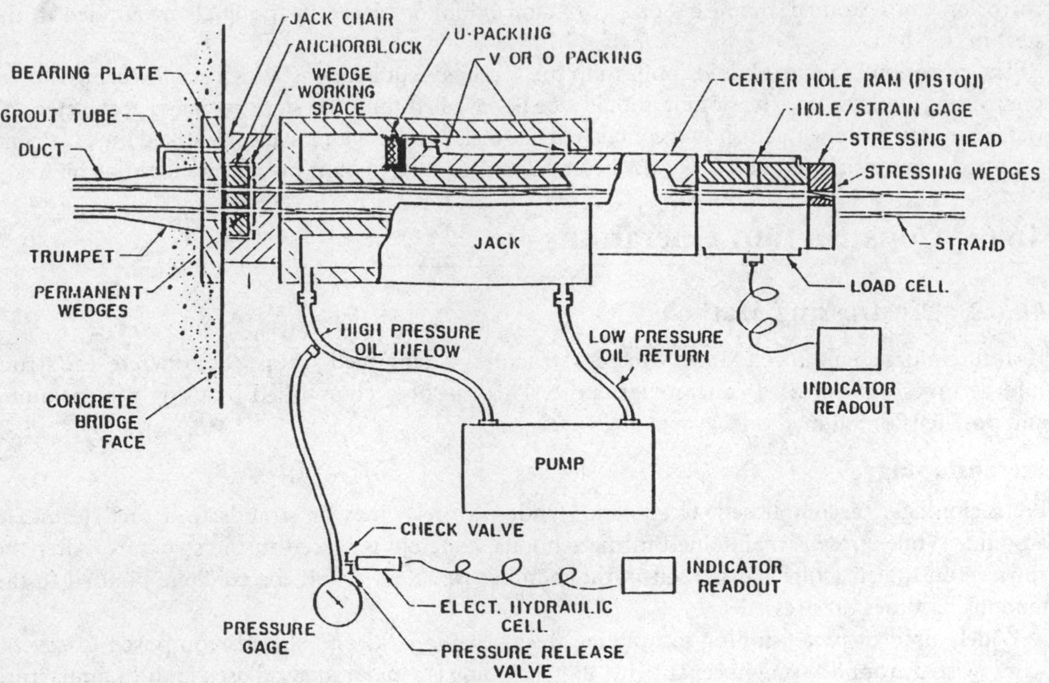

FIGURE 46.1 Typical post-tensioning system.

46.6.2 Fabrication and Erection Stages

During construction, not all elements of a bridge have the same stresses they were designed for. That is the reason it is a very important part of construction engineering to be aware of this and to make sure that appropriate steps have been taken. For example, additional reinforcement will be added to the members in the fabrication stage and delivered to the job site for erection.

In the case of cast-in-place box-girder bridge construction the sequences of prestressing tendons have to be engineered step-by-step to ensure that the structure will have all parameters for future service load after completion of this stage.

The sequence of the erection itself may produce additional stresses that structures or portions of the structures were not designed for. These stresses and the stability of structures during erection are a big concern that is often overlooked by designers and contractors — construction sequences play a very important role in the erection of a segmental type of bridge. It seems that we have to give more attention to analysis of the role of the construction engineering implementation of such erections. And, yes, sometimes the importance of construction engineering to accomplish safe and efficient fabrication and erection of bridge structures (precast, prestressed girders, cast-in-pile) is not sufficiently emphasized by design engineers and/or fabrication, erection contractors.

Unfortunately, we have to admit that the design set of drawings even for an important bridge does not include the erection scheme. And, of course, we can show many examples of misplaced erection efforts on the part of the designer, but our goal is to show why it happened and to make efforts to pay more attention to the fabrication and erection stages. Even if such an erection scheme is included in the design drawings, contractors are not supposed to rely solely on what is provided by the designer's erection plan.

Sometimes a design can be impractical, or it may not be suitable in terms of the erection contractor's equipment and experience. Because the erection plans usually are very generalized and because not enough emphasis is given to the importance of this stage, it is important that the

FIGURE 46.2 Pine Valley Creek Bridge — construction at Pier 4.

designer understand the contractor's proposed method so that the designer can determine if these methods are compatible with the plans, specifications, and requirements of the contract. This is the time that any differences should be resolved. The designer should also discuss any contingency plan in case the contractor has problems. In many instances, the designer is involved throughout the construction phase for design and specification clarification or interpretation, incorporation of construction changes or modifications to the project, and possible O&M reviews/action.

46.6.3 Construction of Segmental Bridges

The first precast segmental box-girder bridge was built by Jean Muller, the Choisy-le-Roi Bridge crossing the Seine River in 1962. In North America (Canada), a cast-in-place segmental bridge on the Laurentian Autoroute, near Ste. Adele, Quebec, in 1964 and a precast segmental bridge crossing the Lievre River near Notre Dame du Laus also in Quebec in 1967 were constructed. In the United States, the first precast segmental bridge was completed in 1973 in Corpus Christi (Texas). The Pine Valley Creek Bridge with five spans (270 + 340 + 450 + 380 + 270 ft) supported by 340-ft-high pier as shown in Figures 46.2 to 46.5 is the first cast-in-place segmental bridge constructed in the United States in 1974 using the cantilever method. The ends of the bridge are skewed to fit the bridge into the canyon. The superstructure consists of two parallel box structures each providing a roadway width of 40 ft between railings. The superstructures are separated by a 38-ft median.

Segmental cantilever construction is a fairly recent development, and the concept has been improved and used successfully to build bridges throughout the world. Its unique characteristic of needing no ground-supported falsework makes the method attractive for use over congested streets, waterways, deep gorges, or ocean inlets. It has been used for spans of less than 100 ft, all the way to the current record span of 755 ft over the Urato River in Japan. Another advantage of the method lies in its economy and efficiency of material use. Construction of segmental bridges can be classified by three methods: balanced cantilever, span-by-span, and progressive placement or incremental launching. For detailed discussion see Chapter 11.

FIGURE 46.3 Pine Valley Creek Bridge — pier construction

FIGURE 46.4 Pine Valley Creek Bridge — girder construction

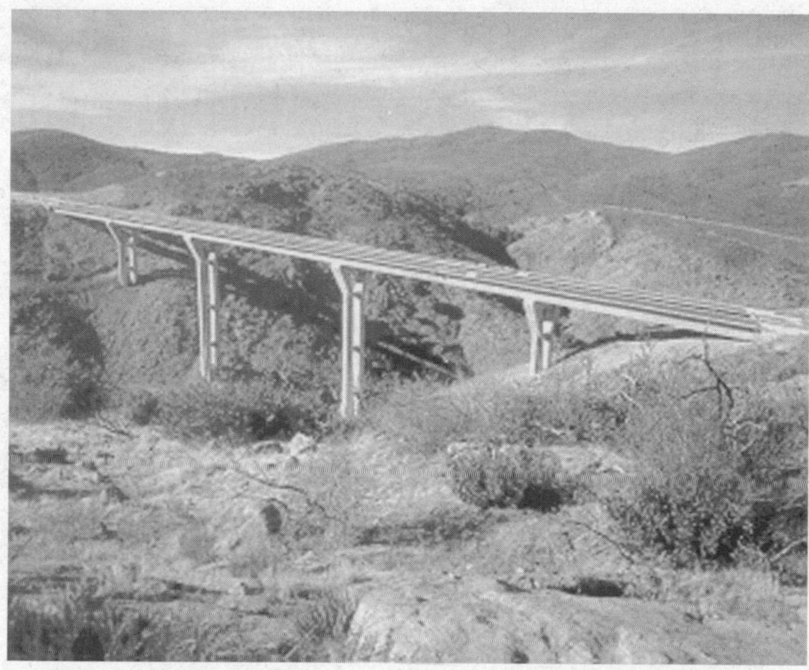

FIGURE 46.5 Pine Valley Creek Bridge — construction completion.

46.6.4 Construction of HPC Bridges

The first U.S. bridge was built with HPC under the Strategic Highway Research Program (SHRP) in Texas in 1996. The FHWA and the Texas DOT in cooperation with the Center for Transportation Research (CTR) at the University of Texas at Austin sponsored a workshop to showcase HPC for bridges in Houston in 1996. The purpose of the event was to introduce the new guidelines to construction professionals and design engineers, and to show how HPC was being used to build more durable structures. It was also focused on the pros and cons of using HPC, mix proportioning, structural design, HPC in precast prestressed and cast-in-place members, long-term performance, and HPC projects in Nebraska, New Hampshire, and Virginia. The showcase had a distinctly regional emphasis because local differences in cements, aggregates, and prestressing fabricators have a considerable impact on the design and construction of concrete structures. In Texas, concrete can be produced with compressive strength of 13,000 to 15,000 psi (900 to 1000 MPa).

The Louetta Road Overpass using HPC is expected to have a useful life of 75 to 100 years, roughly double the average life of a standard bridge. A longer life span means not only lower user cost, but motorists will encounter fewer lane closures and other delays caused by maintenance work. At the present time 15 HPC bridge have been built in the United States.

The first bridge to utilize HPC fully in all aspects of design and construction is the Louetta Road Overpass on State Highway 249 in Houston. The project consists of two U-beam bridges carrying two adjacent lanes of traffic. The spans range from 121.5 to 135.5 ft (37 to 41.3 m) long. The HPC is about twice as strong as conventional concrete. It costs an average of $260/m^2 ($24/ft^2) of deck area, a price compatible with the 12 conventional concrete bridges on the same project. The second Texas HPC bridge located in San Angelo carries the eastbound lanes of U.S. Route 67 over the North Concho River, U.S. 87, and the South Orient railroad. The 954-ft (291-m) HPC I-beam bridge runs parallel to a conventional concrete bridge. The HPC was chosen for the east-bound lanes because the span crossing the North Concho River was 157 ft (48 m) long. This distance exceeds the capacity of Texas conventional prestressed concrete U-beam simple-span construction. The San Angelo Bridge presents an ideal opportunity for comparing HPC and conventional concrete. The first spans

FIGURE 46.6 Falsework at I-80 HOV construction, Richmond, CA.

of two bridges are the same length and width making it easy to compare the cost and performance between HPC and conventional concrete. The comparison indicated that conventional concrete lanes of the first span required seven beams with 5.6 ft (1.7 m) spacing, while the HPC span required only four beams with 11 ft (3.4 m) spacing.

46.7 Falsework

Falsework may be defined as a temporary framework on which the permanent structure is supported during its construction. The term *falsework* is universally associated with the construction of cast-in-place concrete structures, particularly bridge superstructures. The falsework provides a stable platform upon which the forms may be built and furnish support for the bridge superstructure.

Falsework is used in both building and bridge construction. The temporary supports used in building work are commonly referred to as "shoring." It is also important to note the difference between "formwork" and "falsework." Formwork is used to retain plastic concrete in its desired shape until it has hardened. It is designed to resist the fluid pressure of plastic concrete and additional pressure generated by vibrators. Because formwork does not carry dead load of concrete, it can be removed as soon as the concrete hardens. Falsework does carry the dead load of concrete, and therefore it has to remain in place until the concrete becomes self-supporting. Plywood panels on the underside of a concrete slab serve both as a formwork and as a falsework member. For design, however, such panels are considered to be forms in order to meet all design and specification requirements applied to them.

Bridge falsework can be classified in two types: (1) conventional systems (Figure 46.6) in which the various components (beams, posts, caps, bracings, etc.)
are erected individually to form the completed system and (2) proprietary systems in which metal components are assembled into modular units that can be stacked, one above the other, to form a series of towers that compose the vertical load-carrying members of the system.

The contractor is responsible for designing and constructing safe and adequate falsework that provides all necessary rigidity, supports all load composed, and produces the final product (structure) according to the design plans, specifications, and special provisions. It is very important also to keep in mind that approval by the owner of falsework working drawings or falsework inspection will in no way relieve the contractor of full responsibility for the falsework. In the state of California, any falsework height that exceeds 13 ft (4 m) or any individual falsework clear span that exceeds 17 ft (5 m) or where provision for vehicular, pedestrian, or railroad traffic through the falsework is made, the drawings have to be signed by the registered civil engineer in the state of California. The design drawings should include details of the falsework removal operations, methods and sequences of removal, and equipment to be used. The drawings must show the size of all load-supporting members, connections and joints, and bracing systems. For box-girder structures, the drawings must show members supporting sloping exterior girders, deck overhangs, and any attached construction walkway. All design-controlling dimensions, including beam length and spacing, post locations and spacing, overall height of falsework bents, vertical distance between connectors in diagonal bracing must be shown.

It is important that falsework construction substantially conform to the falsework drawings. As a policy consideration, minor deviations to suit field conditions or the substitution of materials will be permitted if it is evident by inspection that the change does not increase the stresses or deflections of any falsework members beyond the allowable values, nor reduce the load-carrying capacity of the overall falsework system. If revision is required, the approval of revised drawings by the state engineer is also required. Any change in the approved falsework design, however minor it may appear to be, has the potential to affect adversely the structural integrity of the falsework system. Therefore, before approving any changes, the engineer has to be sure that such changes will not affect the falsework system as a whole.

References

1. Waddel, J. J., *Concrete Manual*, 1994.
2. *Focus Mag.*, May 1996.
3. Libby, J., *Modern Prestressed Concrete*, 1994.
4. Gerwick, B.C., Jr., *Construction of Prestressed Concrete Structures*, 1994.
5. Fisk, E. R., *Construction Project Administration*, John Wiley & Sons, New York, 1994.
6. Blank, M. M., *Selected Published Articles from 1986 to 1995*, Naval Air Warfare Center, Warmminster, PA, 1996.
7. Blank, M. M. and Blank, S. A., *Effective Construction Management Tools*, Management Division, Naval Air Warfare Center, Warmminster, PA, 1995.
8. Godfrey, K. A., *Partnering in Design and Construction*, McGraw-Hill, New York, 1995.
9. Blank, M. M. et al., Partnering: the key to success, *Found. Drilling*, March/April, 1997.
10. Kubai, M. T. *Engineering Quality in Construction Partnering and TQM*, McGraw-Hill, New York, 1994.
11. Rubin, D.K., A burning sensation in Texas, *Eng. News Rec.*, July 12, 1993.
12. Partnering Guide for Environmental Missions of the Air Force, Army and Navy, Prepared by a Tri-Service Committee, Air Force, Army, and Navy, July 1996.
13. Post, R. G., Effective partnering, *Construction*, June 1996.
14. Schriener, J., Partnering, TQM, ADF, low insurance cost, *Eng. News Rec.*, January 15, 1995.
15. Caltrans, *Standard Specifications*, California Department of Transportation, Sacramento, 1997.
16. Caltrans, *A Guide for Field Inspection of Cast-in-Place Post-Tensioned Structures*, California Department of Transportation, Sacramento, 1992.

47

Substructures of Major Overwater Bridges

Ben C. Gerwick, Jr.
Ben C. Gerwick Inc. and University of California, Berkeley

47.1 Introduction

The design and construction of the piers for overwater bridges present a series of demanding criteria. In service, the pier must be able to support the dead and live loads successfully, while resisting environmental forces such as current, wind, wave, sea ice, and unbalanced soil loads, sometimes even including downslope rock fall. Earthquake loadings present a major challenge to design, with cyclic reversing motions propagated up through the soil and the pier to excite the superstructure. Accidental forces must also be resisted. Collision by barges and ships is becoming an increasingly serious hazard for bridge piers in waterways, both those piers flanking the channel and those of approaches wherever the water depth is sufficient.

0-8493-7434-0/00/$0.00+$.50

Soil–structure foundation interaction controls the design for dynamic and impact forces. The interaction with the superstructure is determined by the flexibility of the entire structural system and its surrounding soil.

Rigid systems attract very high forces: under earthquake, the design forces may reach 1.0 *g*, whereas flexible structures, developing much less force at longer periods, are subject to greater deflection drift. The design must endeavor to obtain an optimal balance between these two responses. The potential for scour due to currents, amplified by vortices, must be considered and preventive measures instituted.

Constructibility is of great importance, in many cases determining the feasibility. During construction, the temporary and permanent structures are subject to the same environmental and accidental loadings as the permanent pier, although for a shorter period of exposure and, in most cases, limited to a favorable time of the year, the so-called weather window. The construction processes employed must therefore be practicable of attainment and completion. Tolerances must be a suitable compromise between practicability and future performance. Methods adopted must not diminish the future interactive behavior of the soil–structure system.

The design loadings for overwater piers are generally divided into two limit states, one being the limit state for those loadings of high probability of occurrence, for which the response should be essentially elastic. Durability needs to be considered in this limit state, primarily with respect to corrosion of exposed and embedded steel. Fatigue is not normally a factor for the pier concepts usually considered, although it does enter into the considerations for supplementary elements such as fender systems and temporary structures such as dolphins if they will be utilized under conditions of cyclic loading such as waves. In seismic areas, moderate-level earthquakes, e.g., those with a return period of 300 to 500 years, also need to be considered.

The second limit state is that of low-probability events, often termed the "safety" or "extreme" limit state. This should include the earthquake of long return period (1000 to 3000 years) and ship collision by a major vessel. For these, a ductile response is generally acceptable, extending the behavior of the structural elements into the plastic range. Deformability is essential to absorb these high-energy loads, so some damage may be suffered, with the provision that collapse and loss of life are prevented and, usually, that the bridge can be restored to service within a reasonable time.

Plastic hinging has been adopted as a principle for this limit state on many modern structures, designed so that the plastic hinging will occur at a known location where it can be most easily inspected and repaired. Redundant load paths are desirable: these are usually only practicable by the use of multiple piles.

Bridge piers for overwater bridges typically represent 30 to 40% of the overall cost of the bridge. In cases of deep water, they may even reach above 50%. Therefore, they deserve a thorough design effort to attain the optimum concept and details.

Construction of overwater bridge piers has an unfortunate history of delays, accidents, and even catastrophes. Many construction claims and overruns in cost and time relate to the construction of the piers. Constructibility is thus a primary consideration.

The most common types of piers and their construction are described in the following sections.

47.2 Large-Diameter Tubular Piles

47.2.1 Description

Construction of steel platforms for offshore petroleum production as well as deep-water terminals for very large vessels carrying crude oil, iron, and coal, required the development of piling with high axial and lateral capacities, which could be installed in a wide variety of soils, from soft sediments to rock. Lateral forces from waves, currents, floating ice, and earthquake as well as from berthing dominated the design. Only large-diameter steel tubular piles have proved able to meet these criteria (Figures 47.1 and 47.2).

FIGURE 47.1 Large-diameter steel tubular pile, Jamuna River Bridge, Bangladesh

FIGURE 47.2 Driving large-diameter steel tubular pile.

FIGURE 47.3 Steel tubular pile being installed from jack-up barge. Socket will be drilled into rock and entire pile filled with tremie concrete.

Such large piling, ranging from 1 to 3 m in diameter and up to over 100 m in length required the concurrent development of very high energy pile-driving hammers, an order of magnitude higher than those previously available. Drilling equipment, powerful enough to drill large-diameter sockets in bedrock, was also developed (Figure 47.3).

Thus when bridge piers were required in deeper water, with deep sediments of varying degrees or, alternatively, bare rock, and where ductile response to the lateral forces associated with earthquake, ice, and ship impact became of equal or greater importance than support of axial loads, it was only natural that technology from the offshore platform industry moved to the bridge field.

The results of this "lateral" transfer exceeded expectations in that it made it practicable and economical to build piers in deep waters and deep sediments, where previously only highly expensive and time-consuming solutions were available.

47.2.2 Offshore Structure Practice

The design and construction practices generally follow the Recommended Practice for Planning, Designing and Constructing Fixed Offshore Platforms published by the American Petroleum Institute, API-RP2A [1]. This recommended practice is revised frequently, so the latest edition should always be used. Reference [2] presents the design and construction from the construction contractor's point of view.

There are many variables that affect the designs of steel tubular piles: diameter, wall thickness (which may vary over the length), penetration, tip details, pile head details, spacing, number of piles, geometry, and steel properties. There must be consideration of the installation method and its effect on the soil–pile interaction. In special cases, the tubular piles may be inclined, i.e., "raked" on an angle from vertical.

In offshore practice, the piles are almost never filled with concrete, whereas for bridge piers, the designer's unwillingness to rely solely on skin friction for support over a 100-year life as well as concern for corrosion has led to the practice of cleaning out and filling with reinforced concrete. A recent advance has been to utilize the steel shell along with the concrete infill in composite action to increase strength and stiffness. The concrete infill is also utilized to resist local buckling under overload and extreme conditions. Recent practice is to fill concrete in zones of high moment.

Tubular piles are used to transfer the superimposed axial and lateral loads and moments to the soil. Under earthquake, the soil imparts dynamic motions to the pile and hence to the structure. These interactions are highly nonlinear. To make matters even more complex, the soils are typically nonuniform throughout their depth and have different values of strength and modulus.

In design, axial loads control the penetration while lateral load transfer to the soil determines the pile diameter. Combined pile stresses and installation stresses determine the wall thickness. The interaction of the pile with the soil is determined by the pile stiffness and diameter. These latter lead to the development of a P–y curve, P being the lateral shear at the head of the pile and y being the deflection along the pile. Although the actual behavior is very complex and can only be adequately solved by a computerized final design, an initial approximation of three diameters can give an assumed "point of fixity" about which the top of the pile bends.

Experience and laboratory tests show that the deflection profile of a typical pile in soft sediments has a first point of zero deflection about three diameters below the mudline, followed by deflection in reverse bending and finally a second point of zero displacement. Piles driven to a tip elevation at or below this second point have been generally found to develop a stable behavior in lateral displacement even under multiple cycles of high loading.

If the deflection under extreme load is significant, P–Δ effects must also be considered. Bridge piers must not only have adequate ultimate strength to resist extreme lateral loads but must limit the displacement to acceptable values. If the displacement is too great, the P–Δ effect will cause large additional bending moments in the pile and consequently additional deflection.

The axial compressive behavior of piles in bridge piers is of dominant importance. Settlement of the pile under service and extreme loads must be limited. The compressive axial load is resisted by skin friction along the periphery of the pile, by end bearing under the steel pile tip, and by the end bearing of the soil plug in the pile tip. This latter must not exceed the skin friction of the soil on the inside of the pile, since otherwise the plug will slide upward. The actual characteristics of the soil plug are greatly affected by the installation procedures, and will be discussed in detail later.

Axial tension due to uplift under extreme loads such as earthquakes is resisted by skin friction on the periphery and the deadweight of the pile and footing block.

Pile group action usually differs from the summation of individual piles and is influenced by the stiffness of the footing block as well as by the applied bending moments and shears. This group action and its interaction with the soil are important in the final design, especially for dynamic loading such as earthquakes.

API-RP2A Section G gives a design procedure for driven steel tubular piles as well as for drilled and grouted piles.

Corrosion and abrasion must be considered in determining the pile wall thickness. Corrosion typically is most severe from just below the waterline to just above the wave splash level at high tide, although another vulnerable location is at the mudline due to the oxygen gradient. Abrasion typically is most severe at the mudline because of moving sands, although suspended silt may cause abrasion throughout the water column.

Considering a design lifetime of a major bridge of 100 years or more, coatings are appropriate in the splash zone and above, while sacrificial anodes may be used in the water column and at the mudline. Additional pile wall thickness may serve as sacrificial steel: for seawater environment, 10 to 12 mm is often added.

47.2.3 Steel Pile Design and Fabrication

Tubular steel piles are typically fabricated from steel plate, rolled into "cans" with the longitudinal seam being automatically welded. These cans are then joined by circumferential welds. Obviously, these welds are critical to the successful performance of the piles. During installation by pile hammer, the welds are often stressed very highly under repeated blows: defective welds may crack in the weld or the heat-affected zone (HAZ). Welds should achieve as full joint penetration as practicable, and the external weld profile should merge smoothly with the base metal on either side.

API-RP2A, section L, gives guidance on fabrication and welding. The fabricated piles should meet the specified tolerances for both pile straightness and for cross section dimensions at the ends. These latter control average diameter and out-of-roundness. Out-of-roundness is of especial concern as it affects the ability to match adjacent sections for welding.

Inspection recommendations are given in API-RP2A, section N. Table N.4-1, with reference to structural tubulars, calls for 10% of the longitudinal seams to be verified by either ultrasonic (UT) or radiography (RT). For the circumferential weld seams and the critical intersection of the longitudinal and circumferential seams, 100% UT or RT is required.

Because of the typically high stresses to which piles supporting bridge piers are subjected, both under extreme loads and during installation, as well as the need for weldability of relatively thick plates, it is common to use a fine-grained steel of 290 to 350 MPa yield strength for the tubular piles.

Pile wall thickness is determined by a number of factors. The thickness may be varied along the length, being controlled at any specific location by the loading conditions during service and during installation.

The typical pile used for a bridge pier is fixed at the head. Hence, the maxima combined bending and axial loads will occur within the 1½ diameters immediately below the bottom of the footing. Local buckling may occur. Repeated reversals of bending under earthquake may even lead to fracture. This area is therefore generally made of thicker steel plate. Filling with concrete will prevent local buckling. General column buckling also needs to be checked and will usually be a maximum at a short distance below the mudline.

Installation may control the minimum wall thickness. The hammer blows develop high compressive waves which travel down the pile, reflecting from the tip in amplified compression when high tip resistance is encountered. When sustained hard driving with large hammers is anticipated, the minimum pile wall thickness should be $t = 6.35 + D/100$ where t and D are in millimeters. The drivability of a tubular pile is enhanced by increasing the wall thickness. This reduces the time of driving and enables greater penetration to be achieved.

During installation, the weight of the hammer and appurtenances may cause excessive bending if the pile is being installed on a batter. Hydraulic hammers usually are fully supported on the pile, whereas steam hammers and diesel hammers are partially supported by the crane.

If the pile is cleaned out during driving in order to enable the desired penetration to be achieved, external soil pressures may develop high circumferential compression stresses. These interact with the axial driving stresses and may lead to local buckling.

The tip of the pile is subject to very high stresses, especially if the pile encounters boulders or must be seated in rock. This may lead to distortion of the tip, which is then amplified during successive blows. In extreme cases, the tip may "tear" or may "accordion" in a series of short local axial buckles. Cast steel driving shoes may be employed in such cases; they are usually made of steels of high toughness as well as high yield strength. The pile head also must be thick enough to withstand both the local buckling and the bursting stresses due to Poisson's effect.

The transition between sections of different pile wall thickness must be carefully detailed. In general, the change in thickness should not be more than 12 mm at a splice and the thicker section should be beveled on a 1:4 slope.

FIGURE 47.4 Large-diameter tubular steel pile being positioned.

47.2.4 Transportation and Upending of Piles

Tubular piles may be transported by barge. For loading, they are often simply rolled onto the barge, then blocked and chained down. They may also be transported by self-flotation. The ends are bulkheaded during deployment. The removal of the bulkheads can impose serious risks if not carefully planned. One end should be lifted above water for removal of that bulkhead, then the other. If one bulkhead is to be removed underwater by a diver, the water inside must first be equalized with the outside water; otherwise the rush of water will suck the diver into the pipe. Upending will produce high bending moments which limit the length of the sections of a long pile (Figure 47.4). Otherwise the pile may be buckled.

47.2.5 Driving of Piles

The driving of large-diameter tubular piles [2] is usually done by a very large pile hammer. The required size can be determined by both experience and the use of a drivability analysis, which incorporates the soil parameters.

Frequently, the tubular pile for a bridge pier is too long or too heavy to install as a single section. Hence, piles must be spliced during driving. To assist in splicing, stabbing guides may be preattached to the tip of the upper segment, along with a backup plate. The tip of the upper segment should be prebeveled for welding.

FIGURE 47.5 Arrangement of internal jet piping and "spider" struts in large-diameter tubular pile.

Splicing is time-consuming. Fortunately, on a large-diameter pile of 2 to 4 m diameter, there is usually space to work two to three crews concurrently. Weld times of 4 to 8 h may be required. Then the pile must cool down (typically 2 h) and NDT performed. Following this, the hammer must be repositioned on top of the pile. Thus a total elapsed time may be 9 to 12 h, during which the skin friction on the pile sides "sets up," increasing the driving resistance and typically requiring a number of blows to break the pile loose and resume penetration.

When very high resistance is encountered, various methods may be employed to reduce the resistance so that the design pile tip may be reached. Care must be taken that these aids do not lessen the capacity of the pile to resist its design loads.

High resistance of the tubular pile is primarily due to plugging of the tip; the soil in the tip becomes compacted and the pile behaves as a displacement pile instead of cutting through the soil. The following steps may be employed.

1. *Jetting internally to break up the plug, but not below the tip.* The water level inside must be controlled, i.e., not allowed to build up much above the outside water level, in order to prevent piping underneath. Although a free jet or arrangement of jets may be employed, a very effective method is to manifold a series of jets around the circumference and weld the down-going pipes to the shell (Figure 47.5). Note that these pipes will pick up parasitic stresses under the pile hammer blows.
2. *Clean out by airlift.* This is common practice when using large-diameter tubular piles for bridge piers but has serious risks associated with it. The danger arises from the fact that an airlift can remove water very rapidly from the pile, creating an unbalanced head at the tip, and allowing run-in of soil. Such a run-in can result in major loss of resistance, not only under the tip in end bearing but also along the sides in skin friction.

   Unfortunately, this problem has occurred on a number of projects! The prevention is to have a pump operating to refill the pile at the same rate as the airlift empties it — a very difficult matter to control. If structural considerations allow, a hole can be cut in the pile wall so that the water always automatically balances. This, of course, will only be effective when the hole is below water. The stress concentrations around such a hole need to be carefully evaluated. Because of the risks and the service consequences of errors in field control, the use of an airlift is often prohibited. The alternative method, one that is much safer, is the use of a grab bucket (orange peel bucket) to remove the soil mechanically. Then, the water level can be controlled with relative ease.
3. *Drilling ahead a pilot hole, using slurry.* If the pile is kept full of slurry to the same level as the external water surface, then a pilot hole, not to exceed 75% of the diameter, may be drilled ahead from one to two diameters. Centralizers should be used to keep the drilled hole properly

aligned. Either bentonite or a polymer synthetic slurry may be used. In soils such as stiff clay or where a binder prevents sloughing, seawater may be used. Reverse circulation is important to prevent erosion of the soils due to high-velocity flow. Drilling ahead is typically alternated with driving. The final seating should be by driving beyond the tip of the drilled hole to remobilize the plug resistance.

4. *External jetting.* External jetting relieves the skin friction during driving but sometimes permanently reduces both the lateral and axial capacity. Further, it is of only secondary benefit as compared with internal jetting to break up the plug. In special cases, it may still be employed. The only practicable method to use with long and large tubular piles is to weld the piping on the outside or inside with holes through the pile wall. Thus, the external jetting resembles that used on the much larger open caissons. As with them, low-pressure, high-volume water flow is most effective in reducing the skin friction. After penetration to the tip, grout may be injected to partially restore the lateral and axial capacity.

47.2.6 Utilization of Piles in Bridge Piers

There are several possible arrangement for tubular piles when used for bridge piers. These differ in some cases from those used in offshore platforms.

1. The pile may be driven to the required penetration and left with the natural soil inside. The upper portion may then be left with water fill or, in some cases, be purposely left empty in order to reduce mass and weight; in this case it must be sealed by a tremie concrete plug. To ensure full bond with the inside wall, that zone must be thoroughly cleaned by wire brush on a drill stem or by jet.

For piles fixed at their head, at least 2 diameters below the footing are filled with concrete to resist local buckling. Studs are installed in this zone to ensure shear transfer.

2. The pile, after driving to final penetration, is cleaned out to within one diameter of the tip. The inside walls are cleaned by wire brush or jet. A cage of reinforcing steel may be placed to augment the bending strength of the tubular shell. Centralizers should be used to ensure accurate positioning. The pile is then filled with tremie concrete. Alternatively, an insert steel tubular with plugged tip may be installed with centralizers, and the annular space filled with tremie grout. The insert tubular may need to be temporarily weighted and/or held down to prevent flotation in the grout.

Complete filling of a tubular pile with concrete is not always warranted. The heat of hydration is a potential problem, requiring special concrete mix design and perhaps precooling.

The reasons for carrying out this practice, so often adopted for bridge piers although seldom used in offshore structures, are

a. Concern over corrosion loss of the steel shell over the 100-year lifetime;
b. A need to ensure positively the ability of the permanent plug to sustain end bearing;
c. Prevention of local buckling near the mudline and at the pile head;
d. To obtain the benefits of composite behavior in stiffness and bending capacity.

If no internal supplemental reinforcement is required, then the benefits of (b), (c), and (d) may be achieved by simple filling with tremie concrete. To offset the heat of hydration, the core may be placed as precast concrete blocks, subsequently grouted into monolithic behavior. Alternatively, an insert pile may be full length. In this case, only the annulus is completely filled. The insert pile is left empty except at the head and tip.

The act of cleaning out the pile close to the tip inevitably causes stress relaxation in the soil plug below the clean-out. This will mean that under extreme axial compression, the pile will undergo a small settlement before it restores its full resistance. To prevent this, after the concrete plug has hardened, grout may be injected just beneath the plug, at a pressure that will restore the compactness of the soil but not so great as to pipe under the tip or fracture the foundation, or the pile may be re-seated by driving.

3. The tubular pile, after being installed to design penetration, may be filled with sand up to two diameters below the head, then with tremie concrete to the head. Reinforcing steel may be placed in the concrete to transfer part of the moment and tension into the footing block. Studs may be pre-installed on that zone of the pile to ensure full shear transfer. The soil and sand plug will act to limit local buckling at the mudline under extreme loads.

4. A socket may be drilled into rock or hard material beyond the tip of the driven pile, and then filled with concrete. Slurry is used to prevent degradation of the surface of the hole and sloughing. Seawater may be used in some rocks but may cause slaking in others such as shales and siltstone. Bentonite slurry coats the surface of the hole; the hole should be flushed with seawater just before concreting. Synthetic slurries are best, since they react in the presence of the calcium ion from the concrete to improve the bond. Synthetic polymer slurries biodegrade and thus may be environmentally acceptable for discharge into the water.

When a tubular pile is seated on rock and the socket is then drilled below the tip of the pile, it often is difficult to prevent run-in of sands from around the tip and to maintain proper circulation. Therefore, after landing, a hole may be drilled a short distance, for example, with a churn drill or down-the-hole drill and then the pile reseated by the pile hammer.

Either insert tubulars or reinforcing steel cages are placed in the socket, extending well up into the pile. Tremie concrete is then placed to transfer the load in shear. In the case where a tubular insert pile is used, its tip may be plugged. Then grout may be injected into the annular space to transfer the shear.

Grout should not be used to fill sockets of large-diameter tubulars. The heat of hydration will damage the grout, reducing its strength. Tremie concrete should be used instead, employing small-size coarse aggregate, e.g., 15 mm, to ensure workability and flowability.

Although most sockets for offshore bridge piers have been cylindrical extensions of the tubular pile, in some offshore oil platforms belled footings have been constructed to transfer the load in end bearing. Hydraulically operated belling tools are attached to the drill string. Whenever transfer in end bearing is the primary mechanism, the bottom of the hole must be cleaned of silt just prior to the placement of concrete.

47.2.7 Prestressed Concrete Cylinder Piles

As an alternative to steel tubular piling, prestressed concrete cylinder piles have been used for a number of major overwater bridges, from the San Diego–Coronado and Dunbarton Bridges in California to bridges across Chesapeake Bay and the Yokohama cable-stayed bridge (Figures 47.6 and 47.7). Diameters have ranged from 1.5 to 6 m and more. They offer the advantage of durability and high axial compressive capacity. To counter several factors producing circumferential strains, especially thermal strains, spiral reinforcement of adequate cross-sectional area is required. This spiral reinforcement should be closely spaced in the 2-m zone just below the pile cap, where sharp reverse bending occurs under lateral loading.

Pile installation methods vary from driving and jetting of the smaller-diameter piles to drilling in the large-diameter piling (Figure 47.8).

47.2.8 Footing Blocks

The footing block constructed at the top of large-diameter tubular piles serves the purpose of transmitting the forces from the pier shaft to the piles. Hence, it is subjected to large shears and significant moments. The shears require extensive vertical reinforcement, for both global shear (from the pier shaft) and local shear (punching shear from the piles). Large concentrations of reinforcement are required to distribute the moments. Post-tensioned tendons may be effectively utilized.

FIGURE 47.6 Large-diameter prestressed concrete pile, Napa River Bridge, California.

FIGURE 47.7 Prestressed concrete cylinder pile for Oosterschelde Bridge, the Netherlands.

Although the primary forces typically produce compression in the upper surface of the footing block, secondary forces and particularly high temporary stresses caused by the heat of hydration produce tension in the top surface. Thus, adequate horizontal steel must be provided in the top and bottom in both directions.

FIGURE 47.8 Installing concrete cylinder pile by internal excavation, jetting, and pull-down force from barge.

The heat of hydration of the cemetitious materials in a large footing block develops over a period of several days. Due to the mass of the block, the heat in the core may not dissipate and return to ambient for several weeks.

The outside surface meantime has cooled and contracted, producing tension which often leads to cracking. Where inadequate reinforcement is provided, the steel may stretch beyond yield, so that the cracks become permanent. If proper amounts of reinforcement are provided, then the cracking that develops will be well distributed, individual cracks will remain small, and the elastic stress in the reinforcement will tend to close the cracks as the core cools.

Internal laminar cracking may also occur, so vertical reinforcement and middepth reinforcement should also be considered.

Footing blocks may be constructed in place, just above water, with precast concrete skirts extending down below low water in order to prevent small boats and debris from being trapped below. In this case, the top of the piles may be exposed at low water, requiring special attention to the prevention of corrosion.

Footing blocks may be constructed below water. Although cofferdams may be employed, the most efficient and economical way is usually to prefabricate the shell of the footing block. This is then floated into place. Corner piles are then inserted through the structure and driven to grade. The prefabricated box is then lowered down by ballasting, supported and guided by the corner piles. Then the remaining piles are threaded through holes in the box and driven. Final connections are made by tremie concrete.

Obviously, there are variations of the above procedure. In some cases, portions of the box have been kept permanently empty, utilizing their buoyancy to offset part of the deadweight.

Transfer of forces into the footing block requires careful detailing. It is usually quite difficult to transfer full moment by means of reinforcing inside the pile shell. If the pile head can be dewatered,

FIGURE 47.9 Large steel sheet pile cofferdam for Second Delaware Memorial Bridge, showing bracing frames.

reinforcing steel bars can be welded to the inside of the shell. Cages set in the concrete plug at the head may employ bundled bars with mechanical heads at their top. Alternatively the pile may be extended up through the footing block. Shear keys can be used to transfer shear. Post-tensioning tendons may run through and around the pile head.

47.3 Cofferdams for Bridge Piers

47.3.1 Description

The word *cofferdam* is a very broad term to describe a construction that enables an underwater site to be dewatered. As such, cofferdams can be large or small. Medium-sized cofferdams of horizontal dimensions from 10 to 50 m have been widely used to construct the foundations of bridge piers in water and soft sediments up to 20 m in depth; a few have been larger and deeper (Figure 47.9). Typical bridge pier cofferdams are constructed of steel sheet piles supported against the external pressures by internal bracing.

A few very large bridge piers, such as anchorages for suspension bridges, have utilized a ring of self-supporting sheet pile cells. The interior is then dewatered and excavated to the required depth. A recent such development has been the building of a circular ring wall of concrete constructed by the slurry trench method (Figures 47.10 and 47.11). Concrete cofferdams have also used a ring wall of precast concrete sheet piles or even cribs.

47.3.2 Design Requirements

Cofferdams must be designed to resist the external pressures of water and soil [3]. If, as is usual, a portion of the external pressures is designed to be resisted by the internal passive pressure of the soil, the depth of penetration must be selected conservatively, taking into account a potential sudden reduction in passive pressure due to water flow beneath the tip as a result of unbalanced water pressures or jetting of piles. The cofferdam structure itself must have adequate vertical support for self-load and equipment under all conditions.

In addition to the primary design loads, other loading conditions and scenarios include current and waves, debris and ice, overtopping by high tides, flood, or storm surge. While earthquake-induced loads, acting on the hydrodynamic mass, have generally been neglected in the past, they

FIGURE 47.10 Slurry wall cofferdam for Kawasaki Island ventilation shaft, Trans-Tokyo Bay Tunnels and Bridge.

FIGURE 47.11 Concrete ring wall cofferdam constructed by slurry trench methods.

are now often being considered on major cofferdams, taking into account the lower input accelerations appropriate for the reduced time of exposure and, where appropriate, the reduced consequences.

Operating loads due to the mooring of barges and other floating equipment alongside need to be considered. The potential for scour must be evaluated, along with appropriate measures to reduce the scour. When the cofferdam is located on a sloping bank, the unbalanced soil loads need to be properly resisted. Accidental loads include impact from boats and barges, especially those working around the site.

FIGURE 47.12 Dewatering the cofferdam for the main tower pier, Second Delaware Memorial Bridge.

The cofferdam as a whole must be adequately supported against the lateral forces of current waves, ice, and moored equipment, as well as unbalanced soil loads. While a large deep-water cofferdam appears to be a rugged structure, when fully excavated and prior to placement of the tremie concrete seal, it may be too weak to resist global lateral forces. Large tubular piles, acting as spuds in conjunction with the space-frame or batter piles may be needed to provide stability.

The cofferdam design must be such as to integrate the piling and footing block properly. For example, sheet piles may prevent the installation of batter piles around the periphery. To achieve adequate penetration of the sheet piles and to accommodate the batter piles, the cofferdam may need to be enlarged. The arrangement of the bracing should facilitate any subsequent pile installation.

To enable dewatering of the cofferdam (Figure 47.12), a concrete seal is constructed, usually by the tremie method. This seal is designed to resist the hydrostatic pressure by its own buoyant weight and by uplift resistance provided by the piling, this latter being transferred to the concrete seal course by shear (Figure 47.13).

In shallow cofferdams, a filter layer of coarse sand and rock may permit pumping without a seal. However, in most cases, a concrete seal is required. In some recent construction, a reinforced concrete footing block is designed to be constructed underwater, to eliminate the need for a separate concrete seal. In a few cases, a drainage course of stone is placed below the concrete seal; it is then kept dewatered to reduce the uplift pressure. Emergency relief pipes through the seal course will prevent structural failure of the seal in case the dewatering system fails.

The underwater lateral pressure of the fresh concrete in the seal course and footing block must be resisted by external backfill against the sheet piles or by internal ties.

47.3.3 Internally Braced Cofferdams

These are the predominant type of cofferdams. They are usually rectangular in shape, to accommodate a regular pattern of cross-lot bracing.

FIGURE 47.13 Pumped-out cofferdam showing tremie concrete seal and predriven steel H-piles.

The external wall is composed of steel sheet piles of appropriate section modulus to develop bending resistance. The loading is then distributed by horizontal wales to cross-lot struts. These struts should be laid out on a plan which will permit excavation between them, to facilitate the driving of piling and to eliminate, as far as practicable, penetration of bracing through the permanent structure.

Wales are continuous beams, loaded by the uniform bearing of sheet piles against them. They are also loaded axially in compression when they serve as a strut to resist the lateral loads acting on them end-wise. Wales in turn deliver their normal loads to the struts, developing concentrated local bearing loads superimposed upon the high bending moments, tending to produce local buckling. Stiffeners are generally required.

While stiffeners are readily installed on the upperside, they are difficult to install on the underside and difficult to inspect. Hence, these stiffeners should be pre-installed during fabrication of the members.

The wales are restrained from global buckling in the horizontal plane by the struts. In the vertical plane they are restrained by the friction of the sheet piles, which may need to be supplemented by direct fixation. Blocking of timber or steel shims is installed between the wales and sheet piles to fit the irregularities in sheet pile installation and to fill in the needed physical clearances.

Struts are horizontal columns, subject to high axial loading, as well as vertical loads from self-weight and any equipment that is supported by them. Their critical concern is stability against buckling. This is countered in the horizontal plane by intersecting struts but usually needs additional support in the vertical plane, either by piling or by trussing two or more levels of bracing.

The orthogonal horizontal bracing may be all at one elevation, in which case the intersections of the struts have to be accommodated, or they may be vertically offset, one level resting on top of the other. This last is normally easier since, otherwise, the intersections must be detailed to transmit the full loads across the joint. This is particularly difficult if struts are made of tubular pipe sections. If struts are made of wide-flanged or H-section members, then it will usually be found preferable to construct them with the weak axis in the vertical plane, facilitating the detailing of strut-to-strut intersections as well as strut-to-wale intersections. In any event, stiffeners are required to prevent buckling of the flanges.

For deep-water piers, the cofferdam bracing is best constructed as a space-frame, with two or more levels joined together by posts and diagonals in the vertical plane. This space-frame may be completely prefabricated and set as a unit, supported by vertical piles. These supporting piles are typically of large-diameter tubular members, driven through sleeves in the bracing frame and connected to it by blocking and welding.

The setting of such a space-frame requires a very large crane barge or equivalent, with both adequate hoisting capacity and reach. Sometimes, therefore, the bracing frame is made buoyant, to be partially or wholly self-floating. Tubular struts can be kept empty and supplemental buoyancy can be provided by pontoons.

Another way to construct the bracing frame is to erect one level at a time, supported by large tubular piles in sleeves. The lower level is first erected, then the posts and diagonal bracing in the vertical plane. The lower level is then lowered by hoists or jacks so that the second level can be constructed just above water and connections made in the dry.

A third way is to float in the prefabricated bracing frame on a barge, drive spud piles through sleeves at the four corners, and hang the bracing frame from the piles. Then the barge is floated out at low tide and the bracing frame lowered to position.

47.3.4 Circular Cofferdams

Circular cofferdams are also employed, with ring wales to resist the lateral forces in compression. The dimensions are large, and the ring compression is high. Unequal loading is frequently due to differential soil pressures. Bending moments are very critical, since they add to the compression on one side. Thus the ring bracing must have substantial strength against buckling in the horizontal plane.

47.3.5 Excavation

Excavation should be carried out in advance of setting the bracing frame or sheet piles, whenever practicable. Although due to side slopes the total volume of excavation will be substantially increased, the work can be carried out more efficiently and rapidly than excavation within a bracing system.

When open-cut excavation is not practicable, then it must be carried out by working through the bracing with a clamshell bucket. Struts should be spaced as widely as possible so as to permit use of a large bucket. Care must be taken to prevent impact with the bracing while the bucket is being lowered and from snagging the bracing from underneath while the bucket is being hoisted. These accidental loads may be largely prevented by temporarily standing up sheet piles against the bracing in the well being excavated, to act as guides for the bucket.

Except when the footing course will be constructed directly on a hard stratum or rock, overexcavation by 1 m or so will usually be found beneficial. Then the overexcavation can be backfilled to grade by crushed rock.

47.3.6 Driving of Piles in Cofferdams

Pilings can be driven before the bracing frame and sheet piles are set. They can be driven by underwater hammers or followers. To ensure proper location, the pile driver be equipped with telescopic leads, or a template be set on the excavated river bottom or seafloor.

Piling may alternatively be driven after the cofferdam has been installed, using the bracing frame as a template. In this case, an underwater hammer presents problems of clearance due to its large size, especially for batter piles. Followers may be used, or, often, more efficiently, the piles may be lengthened by splicing to temporarily extend all the way to above water. They are then cut off to grade after the cofferdam has been dewatered. This procedure obviates the problems occasioned if a pile fails to develop proper bearing since underwater splices are not needed. It also eliminates cutoff waste. The long sections of piling cutoff after dewatering can be taken back to the fabrication yard and re-spliced for use on a subsequent pier.

All the above assumes driven steel piling, which is the prevalent type. However, on several recent projects, drilled shafts have been constructed after the cofferdam has been excavated. In the latter case, a casing must be provided, seated sufficiently deep into the bottom soil to prevent run-in or blowout [see Section 47.2.6, item (4)].

Driven timber or concrete piles may also be employed, typically using a follower to drive them below water.

47.3.7 Tremie Concrete Seal

The tremie concrete seal course functions to resist the hydrostatic uplift forces to permit dewatering. As described earlier, it usually is locked to the foundation piling to anchor the slab. It may be reinforced in order to enable it to distribute the pile loads and to resist cracking due to heat of hydration.

Tremie concrete is a term derived from the French to designate concrete placed through a pipe. The term has subsequently evolved to incorporate both a concrete mix and a placement procedure. Underwater concreting has had both significant successes and significant failures. Yet the system is inherently reliable and concrete equal or better than concrete placed in the dry has been produced at depths up to 250 m. The failures have led to large cost overruns due to required corrective action. They have largely been due to inadvertently allowing the concrete to flow through or be mixed with the water, which has caused washout of the cement and segregation of the aggregate.

Partial washout of cement leads to the formation of a surface layer of laitance which is a weak paste. This may harden after a period of time into a brittle chalklike substance.

The tremie concrete mix must have an adequate quantity of cementitious materials. These can be a mixture of portland cement with either fly ash or blast furnace slag (BFS). These are typically proportioned so as to reduce the heat of hydration and to promote cohesiveness. A total content of cementitious materials of 400 kg/m^3 (~700 lb/cy) is appropriate for most cases.

Aggregates are preferably rounded gravel so they flow more readily. However, crushed coarse aggregates may be used if an adequate content of sand is provided. The gradation of the combined aggregates should be heavy toward the sand portion — a 45% sand content appears optimum for proper flow. The maximum size of coarse aggregate should be kept small enough to flow smoothly through the tremie pipe and any restrictions such as those caused by reinforcement. Use of 20 mm maximum size of coarse aggregate appears optimum for most bridge piers.

A conventional water-reducing agent should be employed to keep the water/cementitious material ratio below 0.45. Superplasticizers should not normally be employed for the typical cofferdam, since the workability and flowability may be lost prematurely due to the heat generated in the mass concrete. Retarders are essential to prolong the workable life of the fresh mix if superplasticizers are used.

Other admixtures are often employed. Air entrainment improves flowability at shallow water depths but the beneficial effects are reduced at greater depths due to the increased external pressure. Weight to reduce uplift is also lost.

Microsilica may be included in amounts up to 6% of the cement to increase the cohesiveness of the mix, thus minimizing segregation. It also reduces bleed. Antiwashout admixtures (AWA) are also employed to minimize washout of cementitious materials and segregation. They tend to promote self-leveling and flowability. Both microsilica and AWA may require the use of superplasticizers in which case retarders are essential. However, a combination of silica fume and AWA should be avoided as it typically is too sticky and does not flow well.

Heat of hydration is a significant problem with the concrete seal course, as well as with the footing block, due to the mass of concrete. Therefore, the concrete mix is often precooled, e.g., by chilling of the water or the use of ice. Liquid nitrogen is sometimes employed to reduce the temperature of the concrete mix to as low as 5°C. Heat of hydration may be reduced by incorporating substantial amounts of fly ash to replace an equal portion of cement. BFS–cement can also be used to reduce

FIGURE 47.14 Placing underwater concrete through hopper and tremie pipe, Verrazano Narrows Bridge, New York.

heat, provided the BFS is not ground too fine, i.e., not finer than 2500 cm^2/g and the proportion of slag is at least 70% of the total.

The tremie concrete mix may be delivered to the placement pipe by any of several means. Pumping and conveyor belts are best because of their relatively continuous flow. The pipe for pumping should be precooled and insulated or shielded from the sun; conveyor belts should be shielded. Another means of delivery is by bucket. This should be air-operated to feed the concrete gradually to the hopper at the upper end of the tremie pipe. *Placement down the tremie pipe should be by gravity feed only* (Figure 47.14).

Although many placements of tremie concrete have been carried out by pumping, there have been serious problems in large placements such as cofferdam seals. The reasons include:

1. Segregation in the long down-leading pipe, partly due to formation of a partial vacuum and partly due to the high velocity;
2. The high pressures at discharge;
3. The surges of pumping.

Since the discharge is into fresh concrete, these phenomena lead to turbulence and promote intermixing with water at the surface, forming excessive laitance.

These discharge effects can be contrasted with the smooth flow from a gravity-fed pipe in which the height of the concrete inside the tremie pipe automatically adjusts to match the external pressure of water vs. the previously placed concrete. For piers at considerable depths, this balance point will be about half-way down. The pipe should have an adequate diameter in relation to the maximum size of coarse aggregate to permit remixing: a ratio of 8 to 1 is the minimum. A slight inclination of the tremie pipe from the vertical will slow the feed of new concrete and facilitate the escape of entrapped air.

For starting the tremie concrete placement, the pipe must first be filled slightly above middepth. This is most easily done by plugging the end and placing the empty pipe on the bottom. The empty pipe must be negatively buoyant. It also must be able to withstand the external hydrostatic pressure as well as the internal pressure of the underwater concrete. Joints in the tremie pipe should be gasketed and bolted to prevent water being sucked into the mix by venturi action. To commence placement, with the tremie pipe slightly more than half full, it is raised 150 mm off the bottom. The temporary plug then comes off and the concrete flows out. The above procedure can be used both for starting and for resuming a placement, as, for example, when the tremie is relocated, or after a seal has been inadvertently lost.

The tremie pipe should be kept embedded in the fresh concrete mix a sufficient distance to provide backpressure on the flow (typically 1 m minimum), but not so deep as to become stuck in the concrete due to its initial set. This requires adjustment of the retarding admixture to match the rate of concrete placement and the area of the cofferdam against the time of set, keeping in mind the acceleration of set due to heat as the concrete hydrates.

Another means for initial start of a tremie concrete placement is to use a pig which is forced down the pipe by the weight of the concrete, expelling the water below. This pig should be round or cylindrical, preferably the latter, equipped with wipers to prevent leakage of grout and jamming by a piece of aggregate. An inflated ball, such as an athletic ball (volleyball or basketball) must never be used; these collapse at about 8 m water depth! A pig should not be used to restart a placement, since it would force a column of water into the fresh concrete previously placed.

Mixes of the tremie concrete described will flow outward on a slope of about 1 on 8 to 1 on 10. With AWAs, an even flatter surface can be obtained.

A trial batch with underwater placement in a shallow pit or tank should always be done before the actual placement of the concrete seal. This is to verify the cohesiveness and flowability of the mix. Laboratory tests are often inadequate and misleading, so a large-scale test is important. A trial batch of 2 to 3 m^3 has often been used.

The tremie concrete placement will exert outward pressure on the sheet piles, causing them to deflect. This may in turn allow new grout to run down past the already set concrete, increasing the external pressure. To offset this, the cofferdam can be partially backfilled before starting the tremie concreting and tied across the top. Alternatively, dowels can be welded on the sheets to tie into the concrete as it sets; the sheet piles then have to be left in place.

Due to the heat of hydration, the concrete seal will expand. Maximum temperature may not be achieved for several days. Cooling of the mass is gradual, starting from the outside, and ambient temperature may not be achieved for several weeks. Thus an external shell is cooling and shrinking while the interior is still hot. This can produce severe cracking, which, if not constrained, will create permanent fractures in the seal or footing. Therefore, in the best practice, reinforcing steel is placed in the seal to both provide a restraint against cracking and to help pull the cracks closed as the mass cools.

After a relatively few days, the concrete seal will usually have developed sufficient strength to permit dewatering. Once exposed to the air, especially in winter, the surface concrete will cool too fast and may crack. Placing insulation blankets will keep the temperature more uniform. They will, of course, have to be temporarily moved to permit the subsequent work to be performed.

47.3.8 Pier Footing Block

The pier footing block is next constructed. Reinforcement on all faces is required, not only for structural response but also to counteract thermal strains. Reference is made to Section 31.2.8 in which reinforcement for the footing block is discussed in more detail.

The concrete expands as it is placed in the footing block due to the heat of hydration. At this stage it is either still fresh or, if set, has a very low modulus. Then it hardens and bonds to the tremie concrete below. The lock between the two concrete masses is made even more rigid if piling protrudes through the top of the tremie seal, which is common practice. Now the footing block cools and tries to shrink but is restrained by the previously placed concrete seal. Vertical cracks typically form. Only if there is sufficient bottom reinforcement in both directions can this shrinkage and cracking be adequately controlled. Note that these tensile stresses are permanently locked into the bottom of the footing block and the cracks will not close with time, although creep will be advantageous in reducing the residual stresses.

After the footing block has hardened, blocking may be placed between it and the sheet piles. This, in turn, may permit removal of the lower level of bracing. As an alternative to bracing, the footing block may be extended all the way to the sheet piles, using a sheet of plywood to prevent adhesion.

47.3.9 Pier Shaft

The pier shaft is then constructed. Block-outs may be required to allow the bracing to pass through. The internal bracing is removed in stages, taking care to ensure that this does not result in overloading a brace above. Each stage of removal should be evaluated.

Backfill is then placed outside the cofferdam to bring it up to the original seabed. The sheet piles can then be removed. The first sheets are typically difficult to break loose and may require driving or jacking in addition to vibration. Keeping in mind the advantage of steel sheet piles in preventing undermining of the pier due to scour, as well as the fact that removal of the sheets always loosens the surrounding soil, hence reducing the passive lateral resistance, it is often desirable to leave the sheet piles in place below the top of the footing. They may be cut off underwater by divers; then the tops are pulled by vibratory hammers.

Antiscour stone protection is now placed, with an adequate filter course or fabric sheet in the case of fine sediments.

47.4 Open Caissons

47.4.1 Description

Open caissons have been employed for some of the largest and deepest bridge piers [4]. These are an extension of the "wells" which have been used for some 2000 years in India. The caisson may be constructed above its final site, supported on a temporary sand island, and then sunk by dredging out within the open wells of the caisson, the deadweight acting to force the caisson down through the overlying soils (Figure 47.15). Alternatively, especially in sites overlain by deep water, the caisson may be prefabricated in a construction basin, floated to the site by self-buoyancy, augmented as necessary by temporary floats or lifts, and then progressively lowered into the soils while building up the top.

Open caissons are effective but costly, due to the large quantity of material required and the labor for working at the overwater site. Historically, they have been the only means of penetrating thorough deep overlying soils onto a hard stratum or bedrock. However, their greatest problem is maintaining stability during the early phases of sinking, when they are neither afloat nor firmly embedded and supported. Long and narrow rectangular caissons are especially susceptible to tipping, whereas square and circular caissons of substantial dimensions relative to the water depth are inherently more stable. Once the caisson tips, it tends to drift off position. It is very difficult to bring it back to the vertical without overcorrecting.

When the caisson finally reaches its founding elevation, the surface of rock or hard stratum is cleaned and a thick tremie concrete base is placed. Then the top of the caisson is completed by casting a large capping block on which to build the pier shaft.

FIGURE 47.15 Open-caisson positioned within steel jackets on "pens."

47.4.2 Installation

The sinking of the cofferdam through the soil is resisted by skin friction along the outside and by bearing on the cutting edges. Approximate values of resistance may be obtained by multiplying the friction factor of sand on concrete or steel by the at-rest lateral force at that particular stage, $f = \O K_0 wh^2$ where f is the unit frictional resistance, Ø the coefficient of friction, w the underwater unit weight of sand, K_0 the at-rest coefficient of lateral pressure, and h the depth of sand at that level. f is then summed up over the embedded depth. In clay, the cohesive shear controls the "skin friction." The bearing value of the cutting edges is generally the "shallow bearing value," i.e., five times the shear strength at that elevation.

These resistances must be overcome by deadweight of the caisson structure, reduced by the buoyancy acting on the submerged portions. This deadweight may be augmented by jacking forces on ground anchors.

The skin friction is usually reduced by lubricating jets causing upward flow of water along the sides. Compressed air may be alternated with water through the jets; bentonite slurry may be used to provide additional lubrication. The bearing on the cutting edges may be reduced by cutting jets built into the walls of the caisson or by free jets operating through holes formed in the walls. Finally, vibration of the soils near and around the caisson may help to reduce the frictional resistance.

When a prefabricated caisson is floated to the site, it must be moored and held in position while it is sunk to and into the seafloor. The moorings must resist current and wave forces and must assist in maintaining the caisson stable and in a vertical attitude. This latter is complicated by the need to build up the caisson walls progressively to give adequate freeboard, which, of course, raises the center of gravity.

Current force can be approximately determined by the formula

$$F = CA\rho\frac{V^2}{2g}$$

where C varies from 0.8 for smooth circular caissons to 1.3 for rectangular caissons, A is the area, ρ is the density of water, and V is the average current over the depth of flotation. Steel sheet piles develop high drag, raising the value of C by 20–30%.

As with all prismatic floating structures, stability requires that a positive metacentric height be maintained. The formula for metacentric height, $\overline{GM}$ is

$$\overline{GM} = \overline{KB} - \overline{KG} + \overline{BM}$$

where $\overline{KB}$ is the distance from the base to the center of buoyancy, $\overline{KG}$ the distance to the center of gravity, and $\overline{BM} = I/V$.

I is the moment of inertia on the narrowest (most sensitive) axis, while V is the displaced volume of water. For typical caissons, a $\overline{GM}$ of +1 m or more should be maintained.

The forces from mooring lines and the friction forces from any dolphins affect the actual attitude that the structure assumes, often tending to tip it from vertical. When using mooring lines, the lines should be led through fairleads attached near the center of rotation of the structure. However, this location is constantly changing, so the fairlead attachment points may have to be shifted upward from time to time.

Dolphins and "pens" are used on many river caissons, since navigation considerations often preclude mooring lines. These are clusters of piles or small jackets with pin piles and are fitted with vertical rubbing strips on which the caisson slides.

Once the caisson has been properly moored on location, it is ballasted down. As it nears the existing river or harbor bottom, the current flow underneath increases dramatically. When the bottom consists of soft sediments, these may rapidly scour away in the current. To prevent this, a mattress should be first installed.

Fascine mattresses of willow, bamboo, or wood with filter fabric attached are ballasted down with rock. Alternatively, a layer of graded sand and gravel, similar to the combined mix for concrete aggregate, can be placed. The sand on top will scour away, but the final result will be a reverse filter.

In order to float a prefabricated caisson to the site initially, false bottoms are fitted over the bottom of the dredging wells. These false bottoms are today made of steel, although timber was used on many of the famous open caissons from the 19th and the early part of the 20th centuries. They are designed to resist the hydrostatic pressure plus the additional force of the soils during the early phases of penetration. Once the caisson is embedded sufficiently to ensure stability, the false bottoms are progressively removed so that excavation can be carried out through the open wells. This removal is a very critical and dangerous stage, hazardous both to the caisson and to personnel. The water level inside at this stage should be slightly higher than that outside. Even then, when the false bottom under a particular well is loosened, the soil may suddenly surge up, trapping a diver. The caisson, experiencing a sudden release of bearing under one well, may plunge or tip.

Despite many innovative schemes for remote removal of false bottoms, accidents have occurred. Today's caissons employ a method for gradually reducing the pressure underneath and excavating some of the soil before the false bottom is released and removed. For such constructions, the false bottom is of heavily braced steel, with a tube through it, typically extending to the water surface. The tube is kept full of water and capped, with a relief valve in the cap. After the caisson has penetrated under its own weight and come to a stop, the relief valve is opened, reducing the pressure to the hydrostatic head only. Then the cap is removed. This is done for several (typically, four) wells in a balanced pattern. Then jets and airlifts may be operated through the tube to remove the soil under those wells. When the caisson has penetrated sufficiently far for safety against tipping, the wells are filled with water; the false bottoms are removed and dredging can be commenced.

47.4.3 Penetration of Soils

The penetration is primarily accomplished by the net deadweight, that is, the total weight of concrete steel and ballast less the buoyancy. Excavation within the wells is carried down in a balanced pattern until the bearing stratum is reached. Then tremie concrete is placed, of sufficient depth to transfer the design bearing pressures to the walls.

The term *cutting edge* is applied to the tips of the caisson walls. The external cutting edges are shaped as a wedge while the interior ones may be either double-wedge or square. In the past, concern over concentrated local bearing forces led to the practice of making the cutting edges of heavy and expensive fabricated steel. Today, high-strength reinforced concrete is employed, although if obstructions such as boulders, cobbles, or buried logs are anticipated or if the caisson must penetrate rock, steel armor should be attached to prevent local spalling.

The upper part of the caisson may be replaced by a temporary cofferdam, allowing the pier shaft dimensions to be reduced through the water column. This reduces the effective driving force on the caisson but maintains and increases its inherent stability.

The penetration requires the progressive failure of the soil in bearing under the cutting edges and in shear along the sides. Frictional shear on the inside walls is reduced by dredging while that on the outside walls is reduced by lubrication, using jets as previously described.

Controlling the penetration is an essentially delicate balancing of these forces, attempting to obtain a slight preponderance of sinking force. Too great an excess may result in plunging of the caisson and tipping or sliding sidewise out of position. That is why pumping down the water within the caisson, thus reducing buoyancy, is dangerous; it often leads to sudden inflow of water and soil under one edge, with potentially catastrophic consequences.

Lubricating jets may be operated in groups to limit the total volume of water required at any one time to a practicable pump capacity. In addition to water, bentonite may be injected through the lubricating jets, reducing the skin friction. Compressed air may be alternated with water jetting.

Other methods of aiding sinking are employed. Vibration may be useful in sinking the caisson through sands, especially when it is accompanied by jetting. This vibration may be imparted by intense vibration of a steel pile located inside the caisson or even by driving on it with an impact hammer to liquefy the sands locally.

Ground anchors inserted through preformed holes in the caisson walls may be jacked against the caisson to increase the downward force. They have the advantage that the actual penetration may be readily controlled, both regarding force exerted and displacement.

Since all the parameters of resistance and of driving force vary as the caisson penetrates the soil, and because the imbalance is very sensitive to relatively minor changes in these parameters, it is essential to plan the sinking process in closely spaced stages, typically each 2 to 3 m. Values can be precalculated for each such stage, using the values of the soil parameters, the changes in contact areas between soil and structure, the weights of concrete and steel and the displaced volume. These need not be exact calculations; the soil parameters are estimates only since they are being constantly modified by the jetting. However, they are valuable guides to engineering control of the operations.

There are many warnings from the writings of engineers in the past, often based on near-failures or actual catastrophes.

1. Verify structural strength during the stages of floating and initial penetration, with consideration for potentially high resistance under one corner or edge.
2. In removing false bottoms, be sure the excess pressure underneath has first been relieved.
3. Do not excavate below cutting edges.
4. Check outside soundings continually for evidence of scour and take corrective steps promptly.
5. Blasting underneath the cutting edges may blow out the caisson walls. Blasting may also cause liquefaction of the soils leading to loss of frictional resistance and sudden plunging. If blasting is needed, do it before starting penetration or, at least, well before the cutting edge reaches the hard strata so that a deep cushion of soil remains over the charges.
6. If the caisson tips, avoid drastic corrections. Instead, plan the correction to ensure a gradual return to vertical and to prevent the possibility of tipping over more seriously on the other side. Thus steps such as digging deeper on the high side and overballasting on the high side should be last resorts and, then, some means mobilized to arrest the rotation as the caisson nears vertical. Jacking against an external dolphin is a safer and more efficient method for correcting the tipping (Figure 47.16).

FIGURE 47.16 Open caisson for Sunshine Bridge across the Mississippi River. This caisson tipped during removal of false bottoms and is shown being righted by jacking against dolphins.

Sinking the caisson should be a continuous process, since once it stops the soil friction and/or shear increase significantly and it may be difficult to restart the caisson's descent.

47.4.4 Founding on Rock

Caissons founded on bare rock present special difficulties. The rock may be leveled by drilling, blasting, and excavation, although the blasting introduces the probability of fractures in the underlying rock. Mechanical excavation may therefore be specified for the last meter or two. Rotary drills and underwater road-headers can be used but the process is long and costly. In some cases, hydraulic rock breakers can be employed; in other cases a hammer grab or star chisel may be used. For the Prince Edward Island Bridge, the soft rock was excavated and leveled by a very heavy clamshell bucket. Hydraulic backhoes and dipper dredges have been used elsewhere. A powerful cutter-head dredge has been planned for use at the Strait of Gibraltar.

47.4.5 Contingencies

The planning should include methods for dealing with contingencies. The resistance of the soil and especially of hard strata may be greater than anticipated. Obstructions include sunken logs and even sunken buried barges and small vessels, as well as cobbles and boulders. The founding rock or stratum may be irregular, requiring special means of excavating underneath the cutting edge at high spots or filling in with concrete in the low spots. One contingency that should always be addressed is what steps to take if the caisson unexpectedly tips.

Several innovative solutions have been used to construct caissons at sites with especially soft sediments. One is a double-walled self-floating concept, without the need for false bottoms. Double-walled caissons of steel were used for the Mackinac Strait Bridge in Michigan. Ballast is progressively filled into the double-wall space while dredging is carried out in the open wells.

In the case of extremely soft bottom sediments, the bottom may be initially stabilized by ground improvement, for example, with surcharge by dumped sand, or by stone columns, so that the caisson may initially land on and penetrate stable soil. Great care must, of course, be exercised to maintain control when the cutting edge breaks through to the native soils below, preventing erratic plunging.

FIGURE 47.17 Excavating within pressurized working chamber of pneumatic caisson for Rainbow Suspension Bridge, Tokyo. (Photo courtesy of Shiraishi Corporation.)

This same principle holds true for construction on sand islands, where the cutting edge and initial lifts of the caisson may be constructed on a stratum of gravel or other stable material, then the caisson sunk through to softer strata below. Guides or ground anchors will be of benefit in controlling the sinking operation.

47.5 Pneumatic Caissons

47.5.1 Description

These caissons differ from the open caisson in that excavation is carried out beneath the base in a chamber under air pressure. The air pressure is sufficient to offset some portion of the ambient hydrostatic head at that depth, thus restricting the inflow of water and soil.

Access through the deck for workers and equipment and for the removal of the excavated soil is through an airlock. Personnel working under air pressure have to follow rigid regimes regarding duration and must undergo decompression upon exit. The maximum pressures and time of exposure under which personnel can work is limited by regulations. Many of the piers for the historic bridges in the United States, e.g., the Brooklyn Bridge, were constructed by this method.

47.5.2 Robotic Excavation

To overcome the problems associated with working under air pressure, the health hazards of "caisson disease," and the high costs involved, robotic cutters [5] have been developed to excavate and remove the soil within the chamber without human intervention. These were recently implemented on the piers of the Rainbow Suspension Bridge in Tokyo (Figures 47.17 and 47.18).

The advantage of the pneumatic caisson is that it makes it possible to excavate beneath the cutting edges, which is of special value if obstructions are encountered. The great risk is of a "blowout" in which the air escapes under one edge, causing a rapid reduction of pressure, followed by an inflow of water and soil, endangering personnel and leading to sudden tilting of the caisson. Thus, the use of pneumatic caissons is limited to very special circumstances.

FIGURE 47.18 Excavating beneath cutting edge of pneumatic caisson. (Photo courtesy of Shiraishi Corporation.)

47.6 Box Caissons

47.6.1 Description

One of the most important developments of recent years has been the use of box caissons, either floated in or set in place by heavy-lift crane barges [4]. These box caissons, ranging in size from a few hundred tons to many thousands of tons, enable prefabrication at a shore site, followed by transport and installation during favorable weather "windows" and with minimum requirements for overwater labor. The development of these has been largely responsible for the rapid completion of many long overwater bridges, cutting the overall time by a factor of as much as three and thus making many of these large projects economically viable.

The box caisson is essentially a structural shell that is placed on a prepared underwater foundation. It is then filled with concrete, placed by the tremie method previously described for cofferdam seals. Alternatively, sand fill or just ballast water may be used.

Although many box caissons are prismatic in shape, i.e., a large rectangular base supporting a smaller rectangular column, others are complex shells such as cones and bells. When the box caisson is seated on a firm foundation, it may be underlain by a meter or two of stone bed, consisting of densified crushed rock or gravel that has been leveled by screeding. After the box has been set, underbase grout is often injected to ensure uniform bearing.

47.6.2 Construction

The box caisson shell is usually the principal structural element although it may be supplemented by reinforcing steel cages embedded in tremie concrete. This latter system is often employed when joining a prefabricated pier shaft on top of a previously set box caisson.

Box caissons may be prefabricated of steel; these were extensively used on the Honshu–Shikoku Bridges in Japan (Figures 47.19 and 47.20). After setting, they were filled with underwater concrete, in earlier cases using grout-intruded aggregate, but in more recent cases tremie concrete.

For reasons of economy and durability, most box caissons are made of reinforced concrete. Although they are therefore heavier, the concurrent development of very large capacity crane barges and equipment has made their use fully practicable. The weight is advantageous in providing stability in high currents and waves.

FIGURE 47.19 Steel box caisson being positioned for Akashi Strait Bridge, Japan, despite strong currents.

FIGURE 47.20 Akashi Strait Suspension Bridge is founded on steel box caissons filled with tremie concrete.

47.6.3 Prefabrication Concepts

Prefabrication of the box caissons may be carried out in a number of interesting ways. The caissons(s) may be constructed on the deck of a large submersible barge. In the case of the two concrete caissons for the Tsing Ma Bridge in Hong Kong, the barge then moved to a site where it could submerge to launch the caissons. They were floated to the site and ballasted down onto the predredged rock base. After sealing the perimeter of the cutting edge, they were filled with tremie concrete.

In the case of the 66 piers for the Great Belt Western Bridge in Denmark, the box caissons were prefabricated on shore in an assembly-line process. (Figures 47.21 and 47.22). They were progressively moved out onto a pier from which they could be lifted off and carried by a very large crane barge to their site. They were then set onto the prepared base. Finally, they were filled with sand and antiscour stone was placed around their base.

FIGURE 47.21 Prefabrication of box caisson piers for Great Belt Western Bridge, Denmark.

FIGURE 47.22 Prefabricated concrete box caissons are moved by jacks onto pier for load-out.

A similar procedure has been followed for the approach piers on the Oresund Bridge between Sweden and Denmark and on the Second Severn Bridge in Southwest England. For the Great Belt Eastern Bridge, many of the concrete box caissons were prefabricated in a construction basin (Figure 47.23). Others were fabricated on a quay wall.

For the Prince Edward Island Bridge, bell-shaped piers, with open bottom, weighing up to 8000 tons, were similarly prefabricated on land and transported to the load-out pier and onto a barge, using transporters running on pile-supported concrete beams (Figures 47.24 through 47.26). Meanwhile, a shallow trench had been excavated in the rock seafloor, in order to receive the lower end of the bell. The bell-shaped shell was then lowered into place by the large crane barge. Tremie concrete was placed to fill the peripheral gap between bell and rock.

47.6.4 Installation of Box Caissons by Flotation

Large concrete box caissons have been floated into location, moored, and ballasted down onto the prepared base (Figure 47.27). During this submergence, they are, of course, subject to current, wave, and wind forces. The moorings must be sufficient to control the location; "taut moorings" are therefore used for close positioning.

FIGURE 47.23 Large concrete box caissons fabricated in construction basin for subsequent deployment to site by self-flotation, Great Belt Eastern Bridge, Denmark.

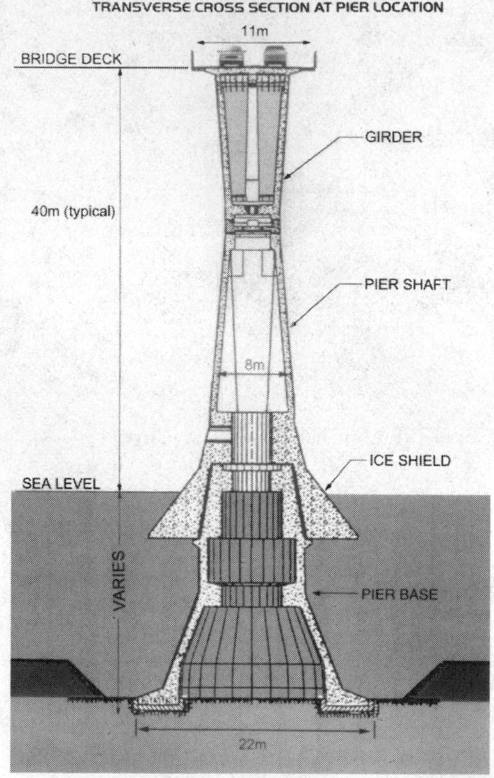

FIGURE 47.24 Schematic representation of substructure for Prince Edward Island Bridge, Canada. Note ice shield, designed to reduce forces from floating ice in Northumberland Strait.

The taut moorings should be led through fairleads on the sides of the caisson, in order to permit lateral adjustment of position without causing tilt. In some cases where navigation requirements prevent the use of taut moorings, dolphins may be used instead. These can be faced with a vertical rubbing strip or master pile. Tolerances must be provided in order to prevent binding.

FIGURE 47.25 Prefabrication of pier bases, Prince Edward Island Bridge, Canada.

FIGURE 47.26 Prefabricated pier shaft and icebreaker, Prince Edward Island Bridge, Canada.

Stability is of critical importance for box caissons which are configured such that the water plane diminishes as they are submerged. It is necessary to calculate the metacentric height, $\overline{GM}$, at every change in horizontal cross section as it crosses the water plane, just as previously described for open caissons.

During landing, as during the similar operation with open caissons, the current under the caisson increases, and scour must be considered. Fortunately, in the case of box caissons, they are being

FIGURE 47.27 Prefabricated box caisson is floated into position, Great Belt Eastern Bridge, Denmark. Note temporary cofferdam above concrete caisson.

landed either directly on a leveled hard stratum or on a prepared bed of densified stone, for which scour is less likely.

As the base of the caisson approaches contact, the prism of water trapped underneath has to escape. This will typically occur in a random direction. The reaction thrust of the massive water jet will push the caisson to one side. This phenomenon can be minimized by lowering the last meter slowly.

Corrections for the two phenomena of current scour and water-jet thrust are in opposition to one another, since lowering slowly increases the duration of exposure to scour. Thus it is essential to size and compact the stone of the stone bed properly and also to pick a time of low current, e.g., slack tide for installation.

47.6.5 Installing Box Caissons by Direct Lift

In recent years, very heavy lift equipment has become available. Jack-up, floating crane barges, and catamaran barges, have all been utilized (Figures 48.28 through 48.31). Lifts up to 8000 tons have been made by crane barge on the Great Belt and Prince Edward Island Bridges.

The box caissons are then set on the prepared bed. Where it is impracticable to screed a stone bed accurately, landing seats may be preset to exact grade under water and the caisson landed on them, and tremie concrete filled in underneath.

Heavy segments, such as box caissons, are little affected by current — hence can be accurately set to near-exact location in plan. Tolerances of the order of 20 to 30 mm are attainable.

47.6.6 Positioning

Electronic distance finders (EDF), theodolites, lasers, and GPS are among the devices utilized to control the location and grade. Seabed and stone bed surveys may be by narrow-beam high-frequency sonar and side-scan sonar. At greater depths, the sonar devices may be incorporated in an ROV to get the best definition.

47.6.7 Grouting

Grouting or concreting underneath is commonly employed to ensure full bearing. It is desirable to use low-strength, low-modulus grout to avoid hard spots. The edges of the caisson have to be sealed by penetrating skirts or by flexible curtains which can be lowered after the caisson is set in place,

FIGURE 47.28 Installing precast concrete box caissons for Second Severn Crossing, Bristol, England. Extreme tidal range of 10 m and high tidal current imposed severe demands on installation procedures and equipment.

FIGURE 47.29 Lifting box caisson from quay wall on which it was prefabricated and transporting it to site while suspended from crane barge.

since otherwise the tremie concrete will escape, especially if there is a current. Heavy canvas or submerged nylon, weighted with anchor chain and tucked into folds, can be secured to the caisson during prefabrication. When the caisson is finally seated, the curtains can be cut loose; they will restrain concrete or grout at low flow pressures. Backfill of stone around the edges can also be used to retain the concrete or grout.

Heat of hydration is also of concern, so the mix should not contain excessive cement. The offshore industry has developed a number of low-heat, low modulus, thixotropic mixes suitable for this use. Some of them employ seawater, along with cement, fly ash, and foaming agents. BFS cement has also been employed.

Box caissons may be constituted of two or more large segments, set one on top of the other and joined by overlapping reinforcement encased in tremie concrete. The segments often are match-cast to ensure perfect fit.

FIGURE 47.30 Setting prefabricated box caisson on which is mounted a temporary cofferdam, Great Belt Eastern Bridge, Denmark.

FIGURE 47.31 Setting 7000-ton prefabricated box caisson, Great Belt Western Bridge, Denmark.

47.7 Present and Future Trends

47.7.1 Present Practice

There is a strong incentive today to use large prefabricated units, either steel or concrete, that can be rapidly installed with large equipment, involving minimal on-site labor to complete. On-site operations, where required, should be simple and suitable for continuous operation. Filling prefabricated shells with tremie concrete is one such example.

Two of the concepts previously described satisfy these current needs. The first, a box caisson — a large prefabricated concrete or steel section — can be floated in or lifted into position on a hard seafloor. The second, large-diameter steel tubular piles, can be driven through soft and variable soils to be founded in a competent stratum, either rock or dense soils. These tubular piles are especially suitable for areas of high seismicity, where their flexibility and ductility can be exploited to reduce the acceleration transmitted to the superstructure (Figure 47.32). In very deep water, steel-framed

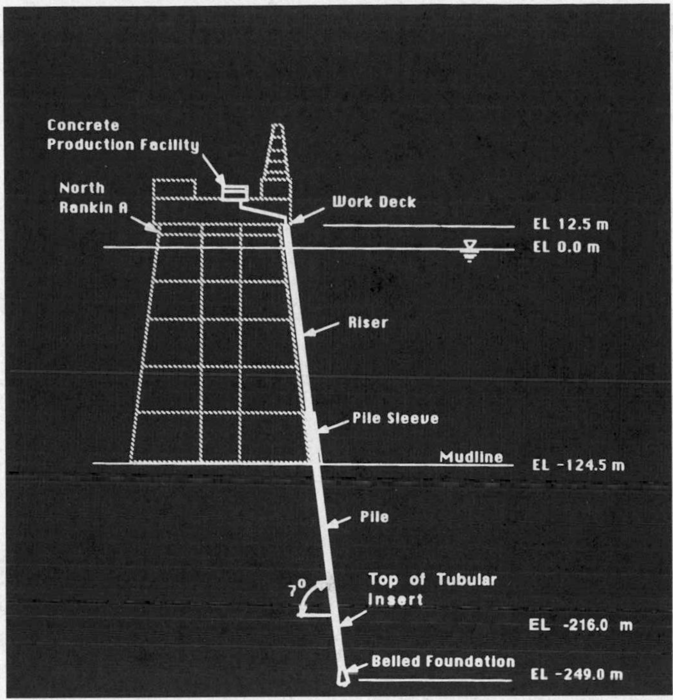

FIGURE 47.32 Conceptual design for deep-water bridge pier, utilizing prefabricated steel jacket and steel tubular pin piles.

jackets may be employed to support the piles through the water column (Figure 47.33). The box caisson, conversely, is most suitable to resist the impact forces from ship collision. The expanding use of these two concepts is leading to further incremental improvements and adaptations which will increase their efficiency and economy.

Meanwhile cofferdams and open caissons will continue to play an important but diminishing role. Conventional steel sheet pile cofferdams are well suited to shallow water with weak sediments, but involve substantial overwater construction operations.

47.7.2 Deep Water Concepts

The Japanese have had a study group investigating concepts for bridge piers in very deep water, and soft soils. One initial concept that has been pursued is that of the circular cofferdam constructed of concrete by the slurry wall process [3]. This was employed on the Kobe anchorage for the Akashi Strait Bridge and on the Kawasaki Island Ventilation Structure for the Trans-Tokyo Bay tunnels, the latter with a pumped-out head of 80 m, in extremely soft soils in a zone of high seismicity (see Section 47.3.1).

Floating piers have been proposed for very deep water, some employing semisubmersible and tension "leg-platform" concepts from the offshore industry. While technically feasible, the entire range of potential adverse loadings, including accidental flooding, ship impact, and long-period swells, need to be thoroughly considered. Tethered pontoons of prestressed concrete have been successfully used to support a low-level bridge across a fjord in Norway.

Most spectacular of all proposed bridge piers are those designed in preliminary feasibility studies for the crossing of the Strait of Gibraltar. Water depths range from 305 m for a western crossing to 470 m for a shorter eastern crossing. Seafloor soils are highly irregular and consist of relatively weak sandstone locally known as flysch. Currents are strong and variable. Wave and swell exposure is significant. For these depths, only offshore platform technology seems appropriate.

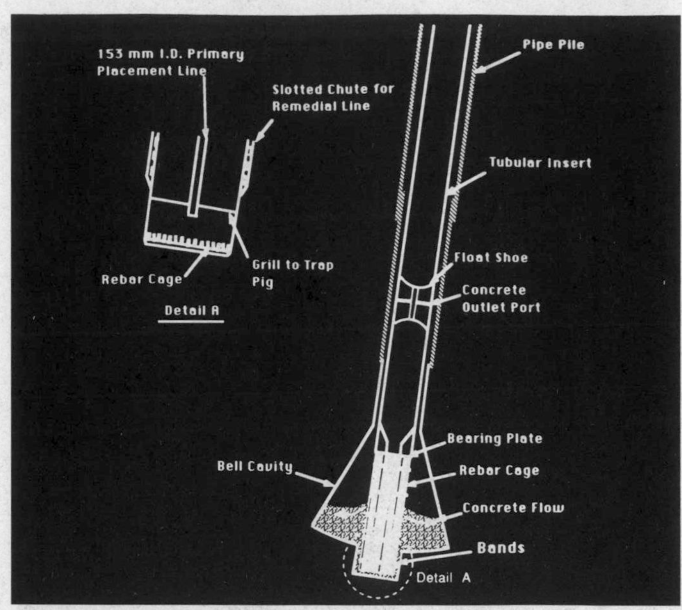

FIGURE 47.33 Belled footing provides greater bearing area for driven-and-drilled steel tubular pile.

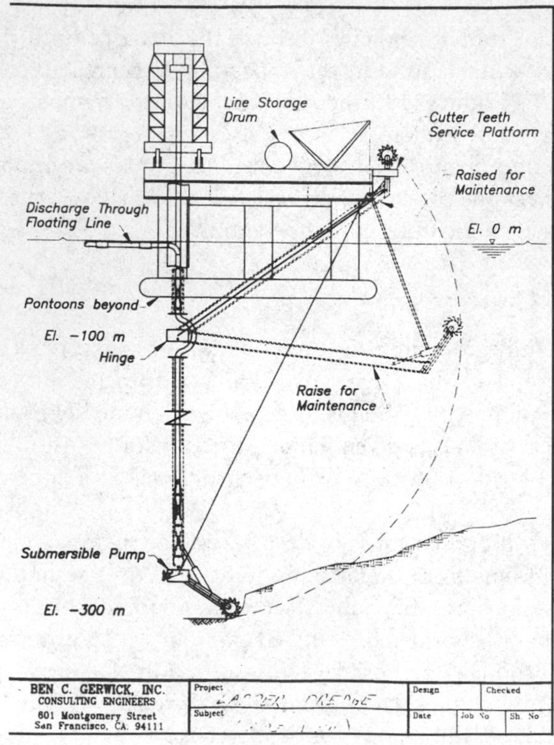

FIGURE 47.34 Concept for preparation of seabed for seating of prefabricated box piers in 300 m water depth, Strait of Gibraltar.

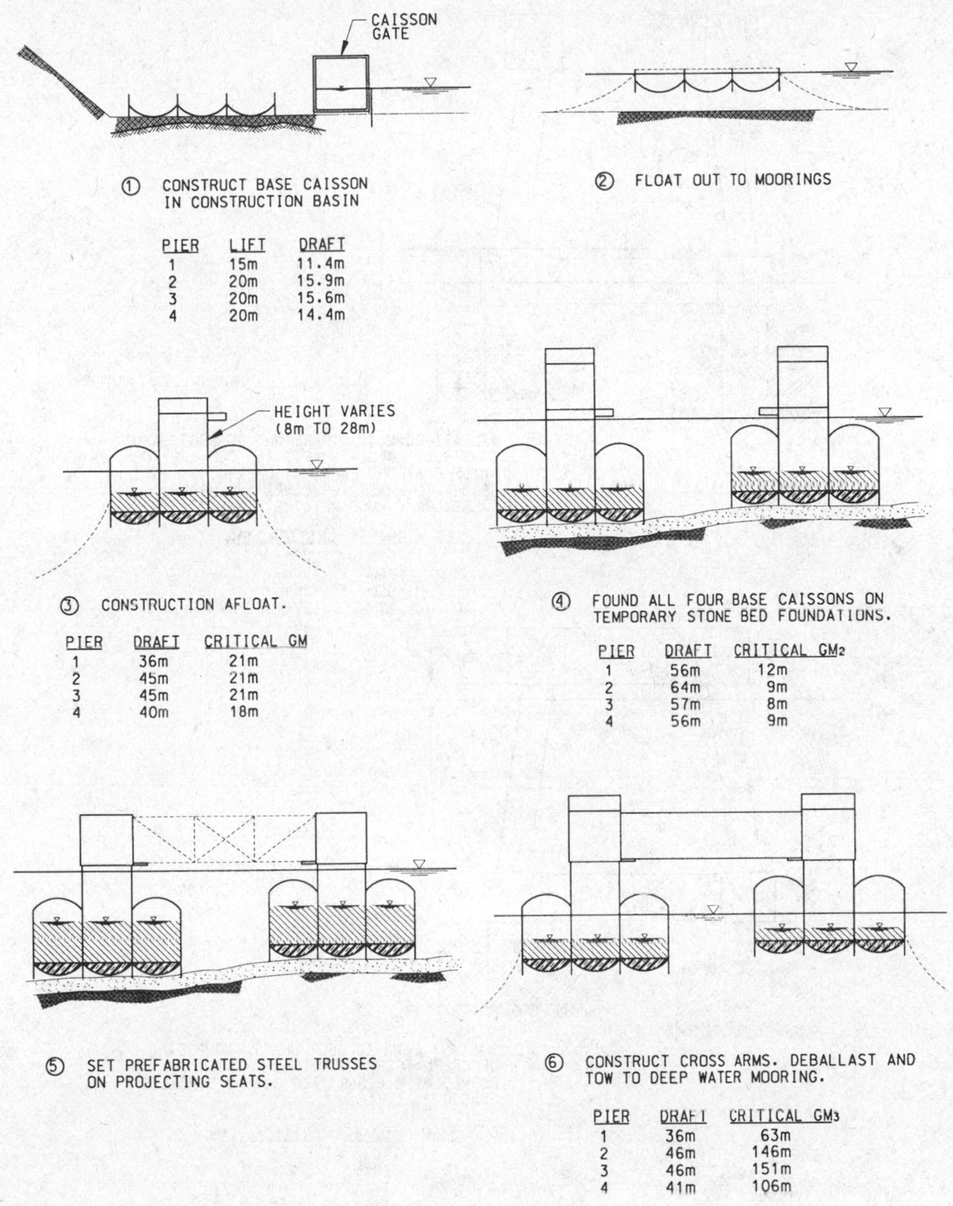

FIGURE 47.35 (a,b,c) Fabrication and installation concept for piers in 300 m water depth for crossing of Strait of Gibraltar.

Both steel jackets with pin piles and concrete offshore structures were investigated. Among the other criteria that proved extremely demanding were potential collision by large crude oil tankers and, below water, by nuclear submarines.

These studies concluded that the concrete offshore platform concept was a reasonable and practicable extension of current offshore platform technology. Leveling and preparing a suitable foundation is the greatest challenge and requires the integration and extension of present systems of dredging well beyond the current state of the art (Figure 47.34). Conceptual systems for these structures have been developed which indicate that the planned piers are feasible by employing an extension of the concepts successfully employed for the offshore concrete platforms in the North Sea (Figure 47.35).

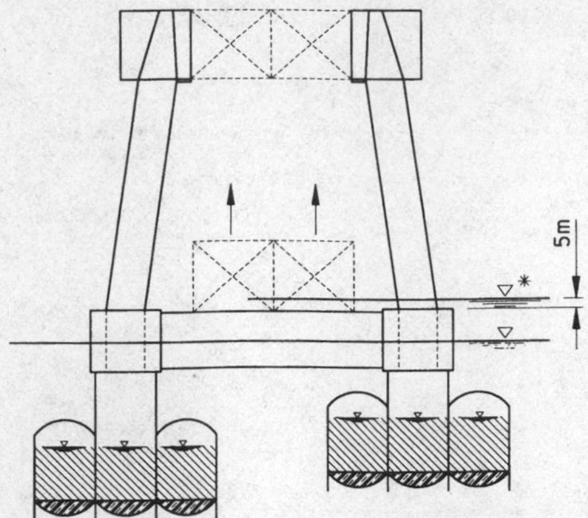

⑦ BALLAST DOWN TO MID-HEIGHT OF CROSS-ARMS. CONSTRUCT ALL 4 SHAFTS AND CONICAL TOPS. RAISE PLATFORM FABRICATION TRUSSES AND SECURE TO SHAFT TOP CONES AND UPPER FALSEWORK (TRUSS WEIGTHS ~ 2000T)

PIER	DRAFT*	CRITICAL GM[4]
1	96m	22m
2	106m	34m
3	95m	36m
4	97m	27m

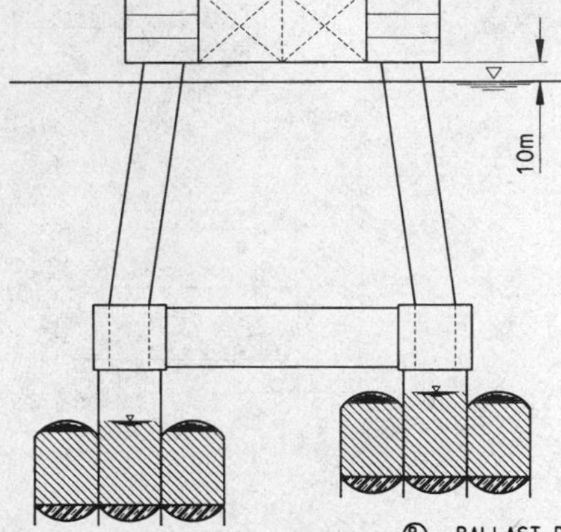

⑧ BALLAST DOWN TO DEEP DRAFT. COMPLETE UPPER PLATFORM, VERTICAL CROSS WALLS FORMED WITH REBAR TIED BUT NO CONCRETE PLACED.

PIER	DRAFT*	CRITICAL GM[5]
1	103m	5m
2	281m	21m
3	246m	16m
4	196m	6m

FIGURE 47.35 (continued)

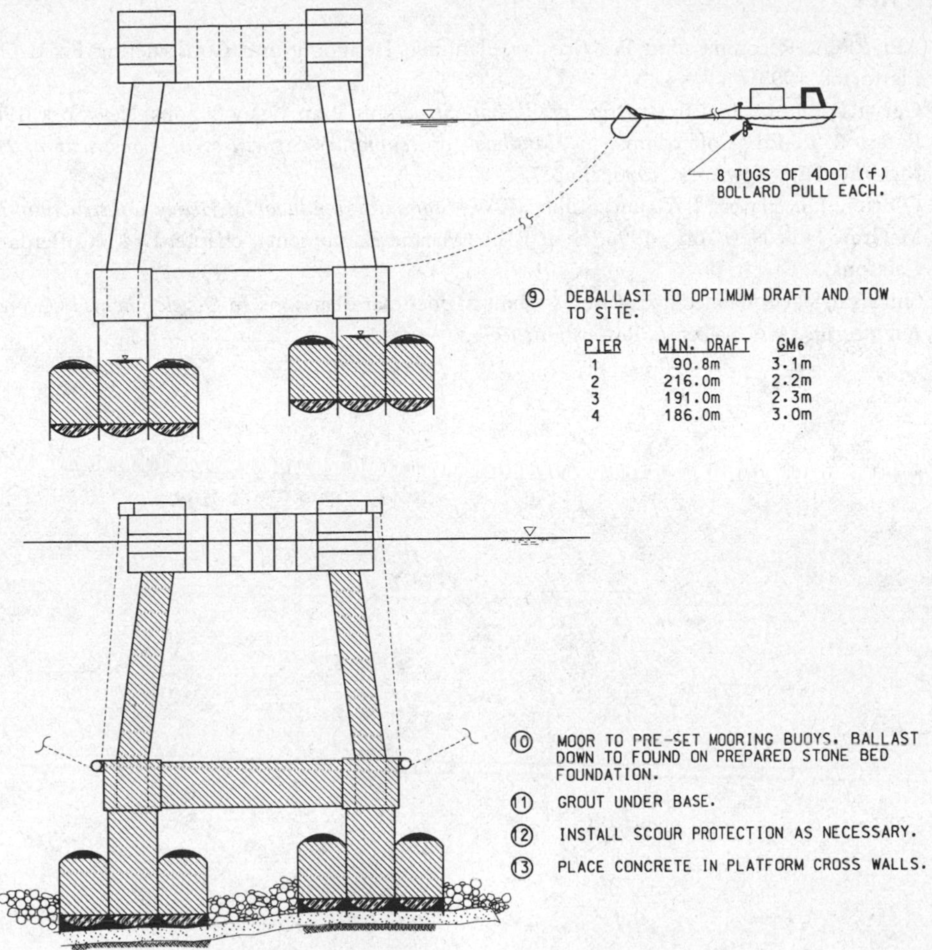

8 TUGS OF 400T (f) BOLLARD PULL EACH.
⑨ DEBALLAST TO OPTIMUM DRAFT AND TOW TO SITE.
PIER
MIN. DRAFT
GM6
1
90.8m
3.1m
2
216.0m
2.2m
3
191.0m
2.3m
4
186.0m
3.0m
⑩ MOOR TO PRE-SET MOORING BUOYS. BALLAST DOWN TO FOUND ON PREPARED STONE BED FOUNDATION.
⑪ GROUT UNDER BASE.
⑫ INSTALL SCOUR PROTECTION AS NECESSARY.
⑬ PLACE CONCRETE IN PLATFORM CROSS WALLS.

FIGURE 47.35 (continued)

Reference

1. API-RP2A, Recommended Practice for Planning, Designing and Constructing Fixed Offshore Platforms, 1993.
2. Gerwick, Ben C., Jr., *Construction of Offshore Structures*, John Wiley & Sons, New York, 1986.
3. Ratay, R. T., Ed., Cofferdams, in *Handbook of Temporary Structures in Construction,* 2nd ed., McGraw-Hill, New York, 1996, chap. 7.
4. O'Brien, J. J., Havers, J. A., and Stubbs, F. W., *Standard Handbook of Heavy Construction,* 3rd ed., McGraw-Hill, New York, 1996, chap. B-11 (Marine Equipment), chapter D-4, (Cofferdams and Caissons).
5. Shiraishi, S., Unmanned excavation systems in pneumatic caissons, in *Developments in Geotechnical Engineering,* A.A. Balkema, Rotterdam, 1994.

48

Bridge Construction Inspection

Mahmoud Fustok
California Department of Transportation

Masoud Alemi
California Department of Transportation

48.1 Introduction

Bridge construction inspection provides quality assurance for building bridges and plays a very important role in the bridge industry. Bridge construction involves two types of structures: permanent and temporary structures. Permanent structures, including foundations, abutments, piers, columns, wingwalls, superstructures, and approach slabs, are those that perform the structural functions of a bridge during its service life. Temporary structures, such as shoring systems, guying systems, forms and falsework, are those that support the permanent structure during its erection and construction.

This chapter discusses inspection principles, followed by the guidelines for inspecting materials, construction operations, component construction, and temporary structures. It also touches on safety considerations and documentation.

0-8493-7434-0/00/$0.00+$.50

48.2 Inspection Objectives and Responsibilities

48.2.1 Objectives

The objective of construction inspection is to ensure that the work is being performed according to the project plans, specifications [1,2], and the appropriate codes including AASHTO, AWS [3,4] as necessitated by the project. The specifications describe the expected quality of materials; standard methods of work; methods and frequency of testing; and the variation or tolerance allowed.. Design and construction of temporary structures should also meet the requirements of construction manuals [5–8].

48.2.2 Responsibilities of the Inspector

The inspector's primary responsibility is to make sure that permanent structures are constructed in accordance with project plans and specifications and to ensure that the operations and/or products meet the quality standard. The inspector is also responsible to determine the design adequacy of temporary structures proposed for use by the contractor. The qualified inspector should have a thorough knowledge of specifications and should exercise good judgment. The inspector should keep a detailed diary of daily observations, noting particularly all warnings and instructions given to the contractor. The inspector should maintain continual communication with the contractor and resolve issues before they become problems.

48.3 Material Inspection

48.3.1 Concrete

At the beginning of the project, a set of concrete mix designs should be proposed by the contractor for use in the project. These mix designs are based on the specification requirements, the desired workability of the mix, and availability of local resources. The proposed concrete mix designs should be reviewed and approved by the inspector.

Method and frequency of sampling and testing of concrete are covered in the specifications. Concrete cylinders are sampled, cured, and tested to determine their compressive strength. The following tests are conducted to check some other concrete properties including:

- Cement content;
- Cleanness value of the coarse aggregate;
- Sand equivalent of the fine aggregate;
- Fine, coarse, and combined aggregate grading;
- Uniformity of concrete.

48.3.2 Reinforcement

Reinforcing steel properties and fabrication should conform to the specifications. A Certificate of Compliance and a copy of the mill test report for each heat and size of reinforcing steel should be furnished to the inspector. These reports should show the physical and chemical analysis of the reinforcing bars.

Bars should not be bent in a manner that damages the material. Check for cracking on Grade 60 reinforcing steel where radii of hooks have been bent too tight. Bars with kinks or improper bends must not be used in the project. Hooks and bends must conform to the specifications.

For epoxy-coated reinforcing bars, a Certificate of Compliance conforming to the specifications must be furnished for each shipment. Epoxy coating material must be tested for specification compliance. Bars are tested for coating thickness and for adhesion requirements. Any damage to the coating caused by shipment and/or installation must be repaired with patching material compatible with the coating material.

48.3.3 Structural Steel

The contractor furnishes to the inspector a copy of all mill orders, certified mill test reports, and a Certificate of Compliance for all fabricated structural steel to be used in the work [9]. In addition, the test reports for steels with specified impact values include the results of Charpy V-notch impact tests. Structural steel which used to fabricate fracture critical members should meet the more stringent Charpy V-notch requirements [4].

48.4 Operation Inspection

48.4.1 Layout and Grades

Bridge inspectors spend a major part of their time making sure that the structure is being built at the correct locations and elevations as shown on the project plans. Lines and grades are provided at reference points by surveyors. Based on these reference points the contractor establishes the lines and grades for building the structure. Horizontal alignment usually consists of a series of circular curves connected with tangents. Vertical alignment usually consists of a series of parabolic curves that also are connected with tangents. When a bridge is on a horizontal curve, its cross section is sloped to counteract the centrifugal forces. This slope is called the superelevation. Therefore, the inspector uses the reference points and basic geometric principles for horizontal and vertical curves to check the bridge geometry. The following layout and grades are to be checked in the field:

1. Pile locations and cutoff elevations;
2. Footings location and grades;
3. Column pour grades;
4. Abutments and wing walls pour grades;
5. Falsework grade points;
6. Lost deck grade points;
7. Overhang jacks and edge of deck.

The deck contour sheet (DCS) is a scaled topographic plan that shows the top of the bridge deck elevations [10]. With some manipulation, the DCS provides elevations for various components and construction stages of a bridge, including abutments, columns, falsework, lost deck, and edge of deck. Keeping tight control on alignment and grades with the DCS will produce smooth vehicle ridability and an aesthetically pleasing bridge.

Falsework (FW) grades are adjusted to meet the soffit elevations. FW points should be placed at locations where the FW will be adjusted, and may be shot at each stage. For example, FW points should not be set on girder centerlines of a cast-in-place prestressed box-girder bridge; otherwise they will be covered by the prestress ducts.

Lost deck dowels (LDD) are the control points to construct the top deck of the bridge. Accurate field layout and DCS layout of the LDD points are essential. Before the soffit and stem pour, the contractor places LDD at predetermined points. After the girder pour, the inspector measures the elevations of the LDD and compares them to those picked from the DCS; then the amount of adjustment needed to the deck grades is calculated.

Inspection of the edge of deck (EOD) grades is very critical for controlling and smooth operation of deck-finishing tools. Additionally, EOD controls the thickness of the concrete slab. It is recommended to locate EOD grade points at points of form adjustment.

FW bents are erected at elevations higher than the theoretical grades to offset the anticipated FW settlements. For FW used to construct cast-in-place concrete bridges, camber strips are usually placed on top of FW stringers to offset (1) deflection of the FW beam under its own weight and the actual load imposed, (2) difference between beam profile and bridge profile grade, and (3) difference between beam profile and any permanent camber.

Steel plate girders are fabricated with built-in upward camber to offset vertical deflections due to the girder weight and other dead loads such as deck and barrier, deflection caused by concrete creep, and to provide any vertical curve required by the profile. The total values are usually called the web camber. Header camber is the web camber minus the girder deflection due to its own weight. If an FW bent is needed to erect a field splice for a steel girder, the grades of the FW bent should be set to no-load elevations. No-load elevation is equal to the plan elevation plus the sum of deflections minus the depth of the superstructure.

48.4.2 Concrete Pour

Ready-mixed concrete is usually delivered to the field by concrete trucks. Each load of ready-mixed concrete should be accompanied by a ticket showing volume of concrete, mix number, time of batch, and reading of the revolution counter. The inspector should ensure the concrete meets the specification requirements regarding:

- Elapse time of batch;
- Number of drum revolutions;
- Concrete temperature;
- Concrete slump (penetration).

Addition of water to the delivered concrete, at the job site, should be approved by the inspector and it should follow the specifications.

The inspector should ensure that concrete is consolidated using vibrators by methods that will not cause segregation of aggregates and will result in a dense homogeneous concrete free of voids and rock pockets. The vibrator should not be dragged horizontally over the top of the concrete surface. Special care must be taken in vibrating areas where there is a high concentration of reinforcing steel. To prevent concrete from segregation caused by excessive free fall, a double-belting hopper or an elephant trunk should be used to guide concrete placing for piles, foundations, and walls.

While placing concrete into a footing, pouring concrete at one location and using vibrators to spread it should not be allowed since it causes aggregate segregation. Care should be taken when topping off column pour to the proper grade. For a cast-in-place box-girder bridge, columns are usually poured 30 mm higher than the theoretical grade to help butt soffit plywood and hide the joint.

While placing concrete, soffit thickness and height of concrete in stems should be checked. Concrete in bent caps should be placed at the proper grade to allow room for minimum concrete clearance over the main cap top reinforcement as shown on the project plans. The appropriate rebar concrete cover is essential in preventing steel rusting.

During bridge deck placement, the pour front should not exceed 3 to 4 m ahead of the finishing machine. Application of curing compound should be performed by power-operated equipment, and it should follow the finishing machine closely.

48.4.3 Reinforcement Placing

Reinforcement should be properly placed as shown on the plans in terms of grade, size, quantity, and location of steel rebar. Reinforcement should be firmly and securely held in position by wiring at intersections and splices with wire ties. The "pigtails" on wire ties should be flattened so that they maintain the minimum concrete coverage from formed surfaces. Rust bleeding can occur from pigtails which extend to the concrete surface.

Positioning of reinforcing steel in the forms is usually accomplished by the use of precast mortar blocks. The use of metal, plastic, and wooden support chairs is normally not permitted. For soffit and deck, blocks should be sufficient to keep mats off plywood a distance equal to the specified rebar clearance. For short- to medium-height columns, reinforcement clearance can be checked from the top of the form with a mirror or flashlight. For foundations, adequate blocking should be provided to hold the bottom mat in proper position. If column steel rests on the bottom mat, extra blocking may be required. Reinforcing steel should be protected from bond breaker substances.

Splicing of reinforcing bars can be done by lapping, butt welding, mechanical butt splice, or mechanical lap splicing.

For lap splicing A 615 steel rebar, the following splicing length is recommended [1]:

For Grade 300:

- Bar No. 25 and smaller is 30*D*
- Bar Nos. 30 and 35 is 45*D*

For Grade 400:

- Bar No. 25 or smaller is 45*D*
- Bar Nos. 30 and 35 is 60*D*

where *D* is diameter of the smaller bar joined.

Reinforcing bars larger than No. 35 should not be spliced by lapping.

For welded rebar, welds should be the right size and be free of cracks, lack of fusion, undercutting, and porosity. Butt welds should be made with multiple passes while flare welds may be made in one pass. For quality assurance, radiographic examinations might be performed on full-penetration butt-welded splices. Radiographs can be made by either X ray or gamma ray.

For mechanical butt splices, splicing should be performed in accordance with the manufacturer's recommendations. Tests of sample splices should be done for quality assurance.

Reinforcement for abutment and wingwalls should be checked before forms are buttoned up.

Proper reinforcing steel placement in deck, especially truss bars, is very important, since the moment-carrying capacity of a bridge deck is greatly sensitive to the effective depth of the section. String lines are used between grade points to check steel clearance and deck thickness.

48.4.4 Welding of Structural Steel

The inspector is responsible for the quality assurance inspection (QAI) of all welding. The inspector should ascertain that equipment, procedures, and techniques are in accordance with specifications. Contract specifications usually refer to welding codes such as AWS D1.5 [4]. All welding should be performed in accordance with an approved welding procedure and by a certified welder. All welding materials such as electrodes, fluxes, and shielding gases must be properly packaged, stored, and dried. Quality of the welds is largely dependent on the welding equipment. The equipment should be checked to ensure that it is in good working condition. Travel speed and rate of flow of shielding gases should be monitored. The actual heat input should not exceed the maximum heat input that was tested and approved. Preheat and interpass temperature specifications should be adhered to as they affect cooling rate and heat input. When postheat is specified, the temperature and duration must be monitored.

Base metal at the welding area (root face, groove face) must be within allowable roughness tolerances. All mill scale should be removed from surfaces where girder web to flange welds are made.

Inspectors must verify that all nondestructive testing (NDT) has been performed and passed the specified requirements [11,12]. The inspector should maintain a record of all locations of inspected areas with the NDT report and findings, together with the method of repairs and NDT test results of weld repairs.

Fillet weld profile should be within final dimensional requirements for leg and throat size and surface contour. Bumps and craters due to starts and stops, weld rollover, and insufficient leg and throat must be ground and repaired to acceptable finish.

48.4.5 High-Strength Bolts

The contact surfaces of all high-strength bolted connections should be thoroughly cleaned of rust, mill scale, dirt, grease, paint, lacquer, or other material. Before the installation of fasteners, the inspector should check the marking, surface condition of bolts, nuts, and washers for compliance with the specifications. Nuts and bolts that are not galvanized should be clean and dry or lightly lubricated. Nuts for high-strength galvanized bolts should be overtapped after galvanizing, and then treated with a lubricant [13].

High-strength bolts may be tensioned by use of calibrated power wrench, a manual torque wrench, the turn-of-nut method, or by tightening and using a direct tension indicator. The inspector should observe calibration and testing procedures required to confirm that the selected procedure is properly applied and the required tensions are provided [14]. The inspector should monitor the installation in the work to ensure that the selected procedure, as demonstrated in the testing, is routinely properly applied.

To inspect completed joints, the following procedure is recommended [15]. A representative sample of bolts from the diameter, length, and grade of the bolts used in the work should be tightened in the tension-measuring device by any convenient means to an initial condition equal to approximately 15% of the required fastener tension and then to the minimum tension specified. Tightening beyond the initial condition must not produce greater nut rotation than 1.5 times that permitted in the specifications. The inspection wrench should be applied to the tightened bolts and the torque necessary to turn the nut or head 5° should be determined.

The Coronet load indicator (CLI) is a simple and accurate aid for tightening and inspecting high-strength bolts. The CLI is a hardened round washer with bumps on one face. As the bolt and nut are tightened, the clamping force flattens the bumps, placed against the underside of the bolt head. The nut should be tightened until the gap is reduced to 0.38 mm. This requirement applies to both A325 and A490 bolts. Once the gap is reduced to the required dimension, the bolt and nut are properly tightened. Visual gap inspection is usually adequate by comparing them against gaps which were checked with the feeler gauge. All CLI for A325 are round. CLI for A490 has three ears or the letter V stamped at three places.

Reuse of ASTM A490 and galvanized ASTM A325 bolts is not allowed. Reuse of ASTM A325 bolts may be allowed if it is approved by the inspector. Reuse does not include retightening bolts which may have been loosened by the tightening of adjacent bolts.

48.5 Component Inspection

48.5.1 Foundation

Conventional bridge foundations can be classified in three types: (1) spread footing foundation, (2) pile foundation, (3) special-case foundations such as pile shafts, tiebacks, soil nails, and tie-downs. The first two types of foundations are more common, and therefore, their inspections will be discussed. Problems that may be encountered during foundation construction should be discussed with the designer and the engineering geologist who performed the foundation studies.

FIGURE 48.1 Foundation forms.

48.5.1.1 Spread Footing Foundations

Spread footing foundations support the load by bearing directly on the foundation stratum. The conformity of the foundation material with the log of the test boring should be checked. Additionally, the bearing surface should be free of disturbed material and be compacted if it is necessary.

Foundations are poured using forms (Figure 48.1) or using the soil as a form in neat-line excavation method. Inspection of the forms is covered in Section 48.6.4. The "neat-line" excavation method is usually done for column and retaining wall footings. The toe of retaining wall should be placed against undisturbed material. Depth and dimensions of footing need to be checked. Standing water and all sloughed material in the excavation must be removed prior to placement of concrete into the foundation. The foundation material should be wet down but not saturated. Footings more than 760 mm vertical dimension, and with a top layer of reinforcement, should be reconsolidated by a vibrator for a depth of 300 mm. Reconsolidation should be done not less than 15 min after the initial leveling of the top of the foundation has been completed. A curing compound will usually be used on top of the footing. If the construction joint is sandblasted before the completion of the curing period, the exposed area should be cured using an alternative method for the remaining time.

48.5.1.2 Pile Foundations

Pile foundations transmit design loads into adjacent soil through pile friction, end bearing, or both. Tops of piles are never exact. Determine the pile with highest elevation and block up bottom-mat steel reinforcement to horizontal accordingly. The highest pile may have to be cut off if the grade is unreasonable.

There are two major types of piles: cast-in-drilled-hole piles and driven piles.

Cast-in-Drilled-Hole Concrete Piles

For cast-in-drilled-hole piles, the inspector should check the following:

- Diameter, depth, and straightness of drilled holes;
- Cleanness of the bottom of holes from water and loose materials.

Material encountered during drilling should be compared with that shown on the log of test borings. If there is a significant difference, the designer should be informed.

For 400-mm-diameter piles, the top 4500 mm of concrete should be vibrated; for larger-diameter piles the full length should be vibrated [16].

Using steel casing is one method to prevent soil cave-ins and intrusion of groundwater. Casing is pulled when placing concrete keeping its bottom below the concrete surface. Waiting too long to pull the casing may cause the concrete to set up and may lead to the following problems:

- The concrete comes up with the casing.
- The casing cannot be removed.
- The concrete may not fill the voids left by the casing.

Use of a concrete mix with fluidity at the high end of the allowable range will help to mitigate these problems.

A slurry displacement method can be used to prevent cave-in of unstable soil and intrusion of groundwater into the drilled hole. The drilling slurry remains in the drilled hole until it is displaced by concrete. Concrete is placed using a delivery system with rigid tremie tube or a rigid pump tube, starting at the bottom of the drilled hole. Sampling and testing of drilling slurry is an important quality control requirement. The following properties of drilling slurries should be monitored: density, sand content, pH value, and viscosity [16].

Inspection tubes are installed inside the spiral or hoop reinforcement in a straight alignment in order to facilitate pile testing. Inspection tubes permit the insertion of a testing probe that measures the density of the pile concrete. A radiographic technique, commonly called gamma ray scattering, is used to measure the density. If the pile is accepted, the inspection tubes are cleaned and filled with grout.

Driven Piles

Prior to start-up of a pile-driving operation, the inspector should check the hammer type and the pile size. Piles should be marked for logging. During the pile-driving operation, the inspector should monitor plumbness or batter of the pile, and log the pile penetration.

Charts of calibration curves are developed for different pile capacity and hammer. By using the energy theory, a penetration-per-blow chart which corresponds to the specified capacity of the piles can be developed. One of the commonly used formulas to determine the bearing capacity of a pile is the ENR formula:

$$p = \frac{E}{6(s + 2.54)} \quad q \tag{48.1}$$

where

p = safe load in kilonewtons

E = manufacturer's rating for energy developed by the hammer in joules

s = average penetration per blow in millimeters

On a large structure which necessitates long piles to be driven in order to satisfy the design load-bearing requirements, monitoring tests may be performed to determine the pile-driving chart.

For pipe steel piles, piles should have the specified diameter, length, and wall thickness as shown on the project plans. If the piles are to be spliced, welding should be performed by a certified welder and the quality of the welded joints should be checked. The method of pipe splicing should be in accordance with project plans and specification requirements. Steel shells may be driven open or closed ended. For shells that are driven open ended, soil should be augured out, and the pile should be cleaned before it is filled with concrete.

For precast concrete piles, the following should be checked:

FIGURE 48.2 Crane used to hold the column cage; column is being cured with plastic wrap.

- Pile is free of damage or cracks;
- Size and length of piles;
- Age of pile, minimum 14 days before it is allowed to be driven [1].

If piles are driven through new embankment greater than 1500 mm, predrilling is required [1]. Problems with the driving operation can be categorized into three types:

1. Hard driving occurs if the soil is too dense or the hammer does not have enough energy to drive the pile. This problem can be solved by predrilling, jetting, or using a larger low-velocity hammer. If a pile is undergoing hard driving and suddenly experiences a large movement, this could indicate a fracture of the pile belowground. In this case the pile should be extracted and replaced or a replacement pile should be driven next to it.
2. Soft pile occurs when the pile is driven to the specified tip elevation but has not attained the specified bearing capacity. This pile is set for a minimum of 12 h, then retapped. If the retapped pile will not attain the bearing capacity, then the contractor has to furnish longer piles.
3. Alignment of piles: if the pile begins to move out of plumbness, correction should be made. The pile may have to be pulled and redriven.

48.5.2 Concrete Columns and Pile Shaft

For medium to long columns, column cages should be held at the top with a crane until footing concrete is set as insurance for the guying system (Figure 48.2). For extremely tall columns, a crane holds the cage until the column is poured.

For fixed columns, make sure ties to the bottom mat are placed in accordance with the project plans to prevent the cage moving during placement of concrete into the footing. For pinned columns, check key details and verify that adequate blocking is provided to support the steel cage at the proper height above a key.

Check the sequence of attaching and removing the guying system to ensure it is done in accordance with the approved plans. Location of utilities should be checked prior to forming a column.

FIGURE 48.3 Wingwall and abutment construction.

Column forms are usually removed a few days after the concrete pour. Columns should be covered with plastic wrap until they are cured (Figure 48.2).

Pile shafts are encountered in bedrock material usually close to canyons or hillside areas with limited room for footing foundations. They are primarily a cast-in-drilled-hole footing with neat-line excavation. The column has the same size extension as the pile shaft or a slightly smaller section. Shoring is required in areas that are not solid rock and any excavation out of the neat area should be filled with concrete. Since blasting is the most common method of excavation, extreme caution is necessary to protect workers and the public.

48.5.3 Abutment and Wingwalls

Abutment and wingwalls are normally formed and poured at the same time for seat-type abutments (Figure 48.3). For the diaphragm abutments of cast-in-place, prestressed-concrete box girders, wingwalls should be placed after stressing. Utility openings in wingwalls and/or abutments should be checked in accordance with the project plans. Bearing pad and internal key layout are usually checked after pour strips are in place.

48.5.4 Superstructure

The following is a summary of the construction inspection for concrete box and steel plate girders which are commonly used for building short- to medium-size bridges.

48.5.4.1 Cast-in-Place Concrete Box Girder

Concrete box girders are constructed in two stages. First, FW is erected and stem and soffit are formed and poured (Figure 48.4). Second, lost deck is built and then the concrete for the bridge deck is poured (Figure 48.5). For prestressed concrete, once the bridge deck is cured, the frame is prestressed, and then the FW is removed. Erection of stem and soffit will be discussed here, and bridge prestressing and deck erection will be discussed in following sections.

The following items need to be checked during construction of soffit and stems:

- Size of camber strips placed on top of FW stringers;
- Location of utility and/or soffit access openings and the corresponding reinforcing details;

FIGURE 48.4 Bridge stem and soffit forms for concrete box girder.

FIGURE 48.5 Bridge deck reinforcement with bidwell finishing machine.

- Location and elevation of block-outs for deck drain pipes;
- Size and profile of the prestressing ducts;
- Smoothness of duct-to-flare connection at bearing plate for alignment of line profile (Figure 48.6);
- Tattletales readings; readings exceeding those anticipated should be investigated to determine if they are due to FW failure or due to excessive soil settlement; corrective measure should be taken accordingly.

FIGURE 48.6 Bridge soffit and girder reinforcements with prestressing duct profile.

Inspectors should make sure that

- Trumpets are properly secured to the bearing plates;
- Ducts and intermediate vents are secured in place;
- Proper gap is maintained between snap ties and ducts to prevent any damage to prestressing ducts;
- Tendon openings are sealed to prevent water or debris from entering;
- Size and location of prestress bearing plates at abutment or hinge diaphragms are in accordance with the approved prestressing working drawings;
- Elastomeric bearing pads are installed properly at abutments and the remainder of the abutment seat area is covered with expanded polystyrene of the same thickness as the pads and that joints are sealed with tape to prevent grout leakage;
- Curing compound is applied appropriately and on time.

Due to the tremendous amount of force involved in prestressing a concrete bridge, careful consideration should be given to this operation. Caution should be exercised around the prestressing jack during the stressing operation and around ducts while they are not grouted (Figure 48.7).

Prior to placement of tendons in prestressing ducts, the following should be verified:

- Ducts are unobstructed and free of water and debris;
- Strands are free of rust.

Prior to stressing, the contractor should submit the required calibration curves for specific jack/gauge combinations. It is the inspector's responsibilitu to determine

- That tendons are installed in accordance with plans
- Whether stressing should be done from one end or two ends
- That stressing sequence is performed in accordance with plans

During stressing, strands are painted on both ends to check slippage. Prestressing calibration curves are plotted, and measured elongation is compared to calculated elongation using the actual area and modulus of elasticity [17].

FIGURE 48.7 Prestressing strands and anchor set.

During the grouting of tendons, the following needs to be checked:

- Water/cement ratio;
- Efflux time;
- That grout is continuously agitated.

48.5.4.2 Steel Girder

Structural steel is usually inspected at the fabrication site. In the field, girders should be checked for damage that may have occurred during transportation to the job. All bearing assemblies should be set level and to the elevations shown on the plans [18]. Full bearing on concrete should be obtained under bearing assemblies (Figure 48.8).

Fracture-critical members (FCM) are tension members or tension components of bending members, the failure of which may result in collapse of the bridge. FMCs are usually identified on the plans or described in the contract documents. FCMs are subject to the additional provisions of Section 12 of AASHTO/AWS D1.5. All welds to FCMs are considered fracture critical and should conform to the requirements of the fracture control plan.

All surfaces of structural steel that are to be painted should be blast-cleaned to produce a dense, uniform surface with an angular anchor pattern [19]. On the same day that blast cleaning is done, structures should be painted with undercoats prior to their erection. Steel surfaces that are inaccessible for painting after erection should be fully painted before they are erected. Steel areas where paint has been damaged due to erection should be cleaned and painted with undercoats before the application of any subsequent paint. Subsequent painting should not be performed until the cleaned surfaces are dry. Succeeding applications of paint should be of such shade to provide contrast with the paint being covered. Paint thickness is measured with a thickness gauge that is calibrated by a magnetic film.

48.5.4.3 Precast Prestressed Girders

Like steel beams, precast prestressed girders are inspected at the fabricator site. Precast concrete members are usually cured by the water or steam method. Upon arrival to the project site, girders should be inspected for damage that may have occurred while being transported to the project site.

FIGURE 48.8 Bearing assembly of steel plate girder.

Precast girders should be handled, transported, and erected using extreme care to avoid twisting, racking, or other distortion that would result in cracking or damage. Girders are placed on elastomeric pads at certain locations shown in the plans. Girders are braced and held together by temporary wooden blocking.

The top of the girder elevation is determined by profiling each girder. The profile grade will determine the location of the finished grade of the top slab and the location of the slab forms. After girders have been profiled and finished grades have been determined, placing forms for the top slab can start. Prestressed concrete panels are a type of slab form that is left in place and becomes the bottom part of the concrete slab. They are normally 100 mm thick [20], rectangular shaped, and vary in width and length.

48.5.4.4 Concrete Deck

The bridge top deck is probably the most critical part in terms of smooth vehicle ridability and aesthetically pleasing bridge. Smoothness of the bridge surface and the approach slabs should be tested by a bridge profilograph. Two profiles will be obtained in each lane. Surfaces that fail to conform to the smoothness tolerances should be ground until these tolerances are achieved. Grinding should not reduce the concrete cover on reinforcing steel to less than 40 mm [1]. The following items need to be checked during the construction of bridge deck:

1. Adequacy of sandblasting on top of girders of cast-in-place concrete bridges;
2. Tops of girders are free of dust;
3. Block-outs for joint seal assemblies;
4. Overhang chamfer and screed pipes have smooth line;
5. Clearance of finishing machine roller to the steel mat (see Figure 48.5), and height of deck drain inlets in relation to finishing machine roller;
6. Tattletales for additional FW settlement;
7. The curing rugs periodically for dampness after completion of deck pour.

Inspectors should ensure that:

1. Soffit vents, access openings, drains, and their support systems are clear from steel rebar and located in accordance with the project plans [21];

FIGURE 48.9 Bridge deck is being cured with moist rugs.

2. Application of rugs or mats is begun within 4 h after completion of deck finishing, and no later than the following morning (Figure 48.9);
3. Prior to prestressing or release of FW, deck surface is inspected for crack intensity to ensure that it meets the allowable tolerance set forth in the specification.

48.6 Temporary Structures

48.6.1 Falsework

FW is used to provide temporary support to the superstructure during construction. To construct a cast-in-place concrete bridge, a complete FW system is needed. Such a system consists of FW foundation, bents, stringers, joists, and forms (Figure 48.10). To erect a steel girder, a simple FW system might be needed. Such a system consists of FW bent and foundation.

FW is designed by the contractor and approved by the inspector. FW is designed to resist vertical loads, as well as longitudinal and transverse horizontal loads. The inspector needs to have a basic understanding of design of timber members, steel members, cables, and to be familiar with application of the FW manual to check the submitted calculations and plans accordingly.

Temporary bracing or other means are needed to hold FW in stable condition during erection and removal.

Typical FW includes timber pads, corbels, steel cells, timber or steel posts, steel or timber FW bent caps, steel stringers, and timber joists. Cables are usually used to resist lateral load in the longitudinal direction. Timber, cable, or steel bar bracing is used to resist lateral loads in the transverse direction. Forms are placed on top of the joist.

When FW pad foundations are used, soil-bearing capacity may be approximated based on observed soil classification or by performing soil load tests. FW pads must be placed on a level and firm material. Pad foundations should be protected from flooding and from undermining by surface runoff. Continuous pads should be inspected to ensure the pad joints are located according to the approved FW plans.

FIGURE 48.10 FW system for cast-in-place concrete bridge.

Timber FW materials should be inspected for defects. Bolts and/or nails can be used to connect timber framing. Inspectors should check edge distance, end distance, and minimum spacing as required by the FW manual [5]. For nailed or spiked connections, ensure that adequate penetrations are provided. If the grades of FW bent are adjusted at the bottom of the post, check diagonal bracing and its connection for any distortion caused by differential movement. Wedging might be needed to ensure full bearing at contact surfaces.

The use of worn or kinked cable should not be permitted. Cables should be looped around a thimble with a minimum diameter corresponding to the cable diameter. Proper clip installation is critical and should be inspected carefully. Check preloading of cables used for internal bracing. Preloading of cables in a frame should be done simultaneously to prevent distortion.

FW with traffic openings should be in accordance with approved FW plans and specifications. In addition, adequacy of vertical and horizontal clearance as shown on the FW plans and as specified in the project specifications should be checked. Inspect the FW lighting system to ensure that the system is detailed according to the approved FW lighting plans.

During concrete pour, inspect FW for the following: excessive settlement, crushing of wedges, and deflection of bracing or distortion of its connection.

48.6.2 Shoring, Sloping, and Benching

Shoring, sloping, and benching are methods to prevent excavation cave-ins. Application of these methods depends on soil type, depth, and size of excavation.

Shoring systems are used to support the sides of excavations from cave-in. Steel members, timber members, or a combination of both are used to construct shoring systems. Trenching is similar to shoring, but the excavation is narrow relative to its length; the width at bottom is less than 5 m. During excavation, verify that the soil properties are the same as was anticipated in the design. Make sure that shoring members have the size and spacing as shown in the approved plans. If a tieback system is used, cables should be preloaded. Inspections should be made after a rain storm or other hazardous conditions. The inspector should ensure that ladders, ramps, or other safe means are provided in excavated areas for providing safe access.

FIGURE 48.11 Temporary deck form for steel bridge.

48.6.3 Guying Systems

Guying systems are temporary structures used to stabilize column cages during construction. Guying systems usually consist of a set of cables connecting the cage or the form to heavy loads such as deadmen or K-rail. This system is designed to resist an assumed wind load applied to the cage. The inspector should check the guying system calculations submitted by the contractor for the adequacy of the system and also to ensure that the erection sequence and the removing of guys to place forms is performed in accordance with the approved plans.

48.6.4 Concrete Forms

Concrete forms should be mortar-tight and with sufficient strength to prevent excessive deflection during placement of concrete. Forms are used to hold the concrete in its plastic state until it is hardened. Forms should be cleaned of all dirt, mortar, debris, nails, and wires. Forms that will later be removed should be thoroughly coated with form oil prior to use (Figure 48.11). Forms should be wet down before placing concrete.

For foundations, attention should be given to bracing of forms to prevent any movement during concrete placement. The bottom of forms should be checked for gaps that may cause excessive leakage of concrete.

48.7 Safety

The primary responsibility of the inspector is to ensure that a safe working environment and practices are maintained at the project site. They should set an example by following the code of safe practice and also by using personal safety equipment including hard hats, gloves, and protective clothing. In addition, they must enforce the safety issues as specified in the contract specifications. This may involve monitoring the operation of equipment and other construction equipment including barricades, warning lights, and reflectors to ensure that they are installed in accordance with the plans and specifications [22].

Prior to entering elevated or excavated areas, the inspector should ascertain that safe access is provided and proper worker protection is in place.

48.8 Record Keeping and As-Built Plans

In addition to construction inspection, the inspector is also responsible for maintaining an accurate and complete record of work that is being performed by contractors. Project records and reports are necessary to determine that contract requirements have been met so that payments can be made to the contractor. Project records should be kept current, complete, accurate, and should be submitted on time.

It is critical that the inspector keep a written diary of the activities that take place in the field. The diary should contain information concerning the work being inspected, including unusual incidents and important conversations. This information may become very critical in case of legal action for litigation involving construction claims or job failure.

As-built plans should reflect any deviation that may exist between the project plans and what was built in the field. Accurate and complete as-built plans are very important and useful for maintaining the bridge, and any future work on the bridge. The as-built plans also provide input and information for future seismic retrofit of the bridge.

48.9 Summary

This chapter discusses the construction inspection of new bridges. It provided guidelines for inspecting the main materials commonly used in building bridges and the major construction operations. Although more emphasis is placed on typical short- to medium-length bridges, the same principles are applied to other type of bridges.

References

1. Caltrans, *Standard Specifications*, California Department of Transportation, Sacramento, 1995.
2. Caltrans, *Standard Plans*, California Department of Transportation, Sacramento, 1995.
3. AASHTO, *LRFD Bridge Design Specifications*, American Association of State Highway and Transportation Officials, Washington, D.C., 1994.
4. AWS, *Bridge Welding Code*, ANSI/AASHTO/AWS D1.5-95, American Welding Society, Miami, FL, 1995.
5. Caltrans, *Falsework Manual*, Office of Structure Construction, California Department of Transportation, Sacramento, 1988.
6. Caltrans, *Trenching & Shoring Manual*, Office of Structure Construction, California Department of Transportation, Sacramento, 1990.
7. FHWA, Lateral Support Systems and Underpinning, Vol. 1: Design and Construction, Report No. FHWA-RD-75-128, U.S. Federal Highway Administration, Washington, D.C., 1976.
8. Mehtlan, J., *Outline of Field Construction Procedure*, Office of Structure Construction, California Department of Transportation, Sacramento, 1988.
9. Barsom, J. M., Properties of bridge steels, Vol. I, Chap. 3, in *Highway Structures Design Handbook*, American Institute of Steel Construction, 1994.
10. Caltrans, *Bridge Construction Survey Guide*, Office of Structure Construction, California Department of Transportation, Sacramento, 1991.
11. AWS, *Welding Inspection*, 2nd ed., American Welding Society, Miami, FL, 1980.
12. AWS, *Guide for the Visual Inspection of Welds*, ANSI/AWS B1.11-88, American Welding Society, Miami, FL, 1997.
13. FHWA, High-Strength Bolts for Bridges, FHWA-SA-91-031, U.S. Department of Transportation, Washington, D.C., 1991.

14. AISC, Mechanical fasteners for steel bridges, Vol. I, Chap. 4A, in *Highway Structures Design Handbook,* American Institute of Steel Construction, Chicago, 1996.
15. Caltrans, *Bridge Construction Record & Procedures,* California Department of Transportation, Sacramento, 1994.
16. Caltrans, *Foundation Manual,* Office of Structure Construction California Department of Transportation, Sacramento, 1996.
17. Caltrans, *Prestress Manual,* California Department of Transportation, Sacramento, 1992.
18. AISC, Steel erection for highway, railroad and other bridge structures, Vol. I, Chap. 14, in *Highway Structures Design Handbook,* American Institute of Steel Construction, Chicago, 1994.
19. Caltrans, *Source Inspection Manual,* Office of Materials Engineering and Testing Services, California Department of Transportation, Sacramento, 1995.
20. TxDOT, *Bridge Construction Inspection,* Texas Department of Transportation, Austin, TX, 1997.
21. Caltrans, *Bridge Deck Construction Manual,* Office of Structure Construction, California Department of Transportation, Sacramento, 1991.
22. FHWA, Bridge Inspector's Training Manual/90, FHWA-PD-91-015, U.S. Department of Transportation, Washington, D.C., 1991.

49

Maintenance Inspection and Rating

Murugesu Vinayagamoorthy
California Department of Transportation

49.1 Introduction

Before the 1960s, little emphasis was given to inspection and maintenance of bridges in the United States. After the 1967 tragic collapse of the Silver Bridge at Point Pleasant in West Virginia, national interest in the inspection and maintenance rose considerably. The U.S. Congress passed the Federal Highway Act of 1968 which resulted in the establishment of the National Bridge Inspection Standard (NBIS). The NBIS sets the national policy regarding bridge inspection procedure, inspection frequency, inspector qualifications, reporting format, and rating procedures. In addition to the establishment of NBIS, three manuals — FHWA Bridge Inspector's Training Manual 70 [1], AASHO Manual for Maintenance Inspection of Bridges [2], and FHWA Recording and Coding Guide for the Structure Inventory and Appraisal of the Nation's Bridges [3] — have been developed and

0-8493-7434-0/00/$0.00+$.50

updated [4–10] since the 1970s. These manuals along with the NBIS provide definitive guidelines for bridge inspection. Over the past three decades, the bridge inspection program evolved into one of the most-sophisticated bridge management systems. This chapter will focus only on the basic, fundamental requirements for maintenance inspection and rating.

49.2 Maintenance Documentation

Each bridge document needs to have items such as structure information, structural data and history, description on and below the structure, traffic information, load rating, condition and appraisal ratings, and inspection findings. The inspection findings should have the signature of the inspection team leader.

All states in the United States are encouraged, but not mandated, to use the codes and instructions given in the Recording and Coding Guide [8,9] while documenting the bridge inventory. In order to maintain the nation's bridge inventory, FHWA requests all state agencies to submit data on the Structure Inventory and Appraisal (SI&A) Sheet. The SI&A sheet is a tabulation of pertinent information about an individual bridge. The information on SI&A sheet is a valuable aid to establish maintenance and replacement priorities and to determine the maintenance cost of the nation's bridges.

49.3 Fundamentals of Bridge Inspection

49.3.1 Qualifications and Responsibilities of Bridge Inspectors

The primary purpose of bridge inspection is to maintain the public safety, confidence, and investment in bridges. Ensuring public safety and investment decision requires a comprehensive bridge inspection. To this end, a bridge inspector should be knowledgeable in material and structural behavior, bridge design, and typical construction practices. In addition, inspectors should be physically strong because the inspection sometimes requires climbing on rough, steep, and slippery terrain, working at heights, or working for days.

Some of the major responsibilities of a bridge inspector are as follows:

- Identifying minor problems that can be corrected before they develop into major repairs;
- Identifying bridge components that require repairs in order to avoid total replacement;
- Identifying unsafe conditions;
- Preparing accurate inspection records, documents, and recommendation of corrective actions; and
- Providing bridge inspection program support.

In the United States, NBIS requires a field leader for highway bridge inspection teams. The field team leader should be either a professional engineer or a state certified bridge inspector, or a Level III bridge inspector certified through the National Institute for Certification of Engineering Technologies. It is the responsibility of the inspection team leader to decide the capability of individual team members and delegate their responsibilities accordingly. In addition, the team leader is responsible for the safety of the inspection team and establishing the frequency of bridge inspections.

49.3.2 Frequency of Inspection

NBIS requires that each bridge that is opened to public be inspected at regular intervals not exceeding 2 years. The underwater components that cannot be visually evaluated during periods of low flow or examined by feel for their physical conditions should be inspected at an interval not exceeding 5 years.

The frequency, scope, and depth of the inspection of bridges generally depend on several parameters such as age, traffic characteristics, state of maintenance, fatigue-prone details, weight limit posting level, and known deficiencies. Bridge owners may establish the specific frequency of inspection based on the above factors.

49.3.3 Tools for Inspection

In order to perform an accurate and comprehensive inspection, proper tools must be available. As a minimum, an inspector needs to have a 2-m (6-ft) pocket tape, a 30-m (100-ft) tape, a chipping hammer, scrapers, flat-bladed screwdriver, pocketknife, wire brush, field marking crayon, flashlight, plumb bob, binoculars, thermometer, tool belt with tool pouch, and a carrying bag. Other useful tools are a shovel, vernier or jaw-type calipers, lighted magnifying glass, inspection mirrors, dye penetrant, 1-m (4-ft) carpenter's level, optical crack gauge, paint film gauge, and first-aid kits. Additional special inspection tools are survey, nondestructive testing, and underwater inspection equipment.

Inspection of a bridge prompts several unique challenges to bridge inspectors. One of the challenges to inspectors is the accessibility of bridge components. Most smaller bridges can be accessed from below without great effort, but larger bridges need the assistance of accessing equipment and vehicles. Common access equipment are ladders, rigging, boats or barges, floats, and scaffolds. Common access vehicles are manlifts, snoopers, aerial buckets, and traffic protection devices. Whenever possible, it is recommended to access the bridge from below since this eliminates the need for traffic control on the bridge. Setting up traffic control may create several problems, such as inconvenience to the public, inspection cost, and safety of the public and inspectors.

49.3.4 Safety during Inspection

During the bridge inspection, the safety of inspectors and of the public using the bridge or passing beneath the bridge should be given utmost importance. Any accident can cause pain, suffering, permanent disability, family hardship, and even death. Thus, during the inspection, inspectors are encouraged to follow the standard safety guidelines strictly.

The inspection team leader is responsible for creating a safe environment for inspectors and the public. Inspectors are always encouraged to work in pairs. As a minimum, inspectors must wear safety vests, hard hats, work gloves, steel-toed boots, long-sleeved shirts, and long pants to ensure their personal safety. Other safety equipment are safety goggles, life jackets, respirator, gloves, and safety belt. A few other miscellaneous safety items include walkie-talkies, carbon monoxide detectors, and handheld radios.

Field clothes should be appropriate for the climate and the surroundings of the inspection location. When working in a wooded area, appropriate clothing should be worn to protect against poisonous plants, snakes, and disease-carrying ticks. Inspectors should also keep a watchful eye for potential hazardous environments around the inspection location. When entering a closed bridge box cells, air needs to be checked for the presence of oxygen and toxic or explosive gases. In addition, care should be taken when using existing access ladders and walkways since the ladder rungs may be rusted or broken. When access vehicles such as snoopers, booms, or rigging are used, the safe use of this equipment should be reviewed before the start of work.

49.3.5 Reports of Inspection

Inspection reports are required to establish and maintain a bridge history file. These reports are useful in identifying and assessing the repair requirements and maintenance needs of bridges. NBIS requires that the findings and results of a bridge inspection be recorded on standard inspection forms. Actual field notes and numerical conditions and appraisal ratings should be included in the standard inspection form. It is also important to recognize that these inspection reports are legal documents and could be used in future litigation.

Descriptions in the inspection reports should be specific, detailed, quantitative, and complete. Narrative descriptions of all signs of distress, failure, or defects with sufficient accuracy should be noted so that another inspector can make a comparison of condition or rate of disintegration in the future. One example of a poor description is, "Deck is in poor condition." A better description would be, "Deck is in poor condition with several medium to large cracks and numerous spalls." The seriousness and the amount of all deficiencies must be clearly stated in an inspection report.

In addition to inspection findings about the various bridge components, other important items to be included in the report are any load, speed, or traffic restrictions on the bridge; unusual loadings; high water marks; clearance diagram; channel profile; and work or repairs done to the bridge since the last inspection.

When some improvement or maintenance work alters the dimensions of the structure, new dimensions should be obtained and reported. When the structure plans are not in the history file, it may be necessary to prepare plans using field measurements. These measurements will later be used to perform the rating analysis of the structure.

Photographs and sketches are the most effective ways of describing a defect or the condition of structural elements. It is therefore recommended to include sketches and/or photographs to describe or illustrate a defect in a structural element. At least two photographs for each bridge for the record are recommended.

Other tips on photographs are

- Place some recognizable items that will allow the reviewer to visualize the scale of the detail;
- Include plumb bob to show the vertical line; and
- Include surrounding details so one could relate other details with the specific detail.

After inspecting a bridge, the inspector should come to a reasonable conclusion. When the inspector cannot interpret the inspection findings and determine the cause of a specific finding (defect), the advice of more-experienced personnel should be sought. Based on the conclusion, the inspector may need to make a practical recommendation to correct or preclude bridge defects or deficiencies. All instructions for maintenance work, stress analysis, posting, further inspection, and repairs should be included in the recommendation. Whenever recommendations call for bridge repairs, the inspector must carefully describe the type of repairs, the scope of the work, and an estimate of the quantity of materials.

49.4 Inspection Guidelines

49.4.1 Timber Members

Common damage in timber members is caused by fungi, parasites, and chemical attack. Deterioration of timber can also be caused by fire, impact or collisions, abrasion or mechanical wear, overstress, and weathering or warping.

Timber members can be inspected by both visual and physical examination. Visual examination can detect the following: fungus decay, damage by parasites, excessive deflection, checks, splits, shakes, and loose connections. Once the damages are detected visually, the inspector should investigate the extent of each damage and properly document them in the inspection report. Deterioration of timber can also be detected using sounding methods — a nondestructive testing method. Tapping on the outside surface of the member with a hammer detects hollow areas, indicating internal decay. There are a few advanced nondestructive and destructive techniques available. Two of the commonly used destructive tests are boring or drilling and probing. And, two of the nondestructive tests are Pol-Tek and ultrasonic testing. The Pol-Tek method is used to detect low-density regions and ultrasonic testing is used to measure crack and flaw size.

49.4.2 Concrete Members

Common concrete member defects include cracking, scaling, delamination, spalling, efflorescence, popouts, wear or abrasion, collision damage, scour, and overload. Brief descriptions of common damages are given in this section.

Cracking in concrete is usually large enough to be seen with the naked eye, but it is recommended to use a crack gauge to measure and classify the cracks. Cracks are classified as hairline, medium, or wide cracks. Hairline cracks cannot be measured by simple means such as pocket ruler, but simple means can be used for the medium and wide cracks. Hairline cracks are usually insignificant to the capacity of the structure, but it is advisable to document them. Medium and wide cracks are significant to the structural capacity and should be recorded and monitored in the inspection reports. Cracks can also be grouped into two types: structural cracks and nonstructural cracks. Structural cracks are caused by the dead- and live-load stresses. Structural cracks need immediate attention, since they affect the safety of the bridge. Nonstructural cracks are usually caused by thermal expansion and shrinkage of the concrete. These cracks are insignificant to the capacity, but these cracks may lead to serious maintenance problems. For example, thermal cracks in a deck surface may allow water to enter the deck concrete and corrode the reinforcing steel.

Scaling is the gradual and continuing loss of surface mortar and aggregate over an area. Scaling is classified into four categories: light, medium, heavy, and severe.

Delamination occurs when layers of concrete separate at or near the level of the top or outermost layer of reinforcing steel. The major cause of delamination is the expansion or the corrosion of reinforcing steel due to the intrusion of chlorides or salts. Delaminated areas give off a hollow sound when tapped with a hammer. When a delaminated area completely separates from the member, a roughly circular or oval depression, which is termed as spall, will be formed in the concrete.

The inspection of concrete should include both visual and physical examination. Two of the primary deteriorations noted by visual inspections are cracks and rust stains. An inspector should recognize the fact that not all cracks are of equal importance. For example, a crack in a prestressed concrete girder beam, which allows water to enter the beam, is much more serious than a vertical crack in the backwall. A rust stain on the concrete members is one of the signs of corroding reinforcing steel in the concrete member. Corroded reinforcing steel produces loss of strength within concrete due to reduced reinforced steel section, and loss of bond between concrete and reinforcing steel. The length, direction, location, and extent of the cracks and rust stains should be measured and reported in the inspection notes.

Some common types of physical examination are hammer sounding and chain drag. Hammer sounding is used to detect areas of unsound concrete and usually used to detect delaminations. Tapping the surfaces of a concrete member with a hammer produces a resonant sound that can be used to indicate concrete integrity. Areas of delamination can be determined by listening for hollow sounds. The hammer sounding method is impractical for the evaluation of larger surface areas. For larger surface areas, chain drag can be used to evaluate the integrity of the concrete with reasonable accuracy. Chain drag surveys of decks are not totally accurate, but they are quick and inexpensive.

There are other advanced techniques — destructive and nondestructive — available for concrete inspection. Core sampling is one of the destructive techniques of concrete inspection. Some of the nondestructive inspection techniques are

- Delamination detection machinery to identify the delaminated deck surface;
- Copper sulfate electrode, nuclear methods to determine corrosion activity;
- Ground-penetrating radar, infrared thermography to detect deck deterioration;
- Pachometer to determine the position of reinforcement; and
- Rebound and penetration method to predict concrete strength.

49.4.3 Steel and Iron Members

Common steel and iron member defects include corrosion, cracks, collision damage, and overstress. Cracks usually initiate at the connection detail, at the termination end of a weld, or at a corroded location of a member and then propagate across the section until the member fractures. Since all of the cracks may lead to failure, bridge inspectors need to look at each and every one of these potential crack locations carefully. Dirt and debris usually form on the steel surface and shield the defects on the steel surface from the naked eye. Thus, the inspector should remove all dirt and debris from the metal surface, especially from the surface of fracture-critical details, during the inspection of defects.

The most recognizable type of steel deterioration is corrosion. The cause, location, and extent of the corrosion need to be recorded. This information can be used for rating analysis of the member and for taking preventive measures to minimize further deterioration. Section loss due to corrosion can be reported as a percentage of the original cross section of a component. The corrosion section loss is calculated by multiplying the width of the member and the depth of the defect. The depth of the defect can be measured using a straightedge ruler or caliper.

One of the important types of damage in steel members is fatigue cracking. Fatigue cracks develop in bridge structures due to repeated loadings. Since this type of cracking can lead to sudden and catastrophic failure, the bridge inspector should identify fatigue-prone details and should perform a thorough inspection of these details. For painted structures, breaks in the paint accompanied by rust staining indicate the possible existence of a fatigue crack. If a crack is suspected, the area should be cleaned and given a close-up visual inspection. Additionally, further testing such as dye penetrant can be done to identify the crack and to determine its extent. If fatigue cracks are discovered, inspection of all similar fatigue details is recommended.

Other types of damage may occur due to overstress, vehicular collision, and fire. Symptoms of damage due to overstress are inelastic elongation (yielding) or decrease in cross section (necking) in tension members, and buckling in compression members. The causes of the overstress should be investigated. The overstress of a member could be the result of several factors such as loss of composite action, loss of bracing, loss of proper load-carrying path, and failure or settlement of bearing details.

Damage due to vehicular collision includes section loss, cracking, and shape distortion. These types of damage should be carefully documented and repair work process should be initiated. Until the repair work is completed, restriction of vehicular traffic based on the rating analysis results is recommended.

Similar to timber and concrete members, there are advanced destructive and nondestructive techniques available for steel inspection. Some of the nondestructive techniques used in steel bridges are

- Acoustic emissions testing to identify growing cracks;
- Computer tomography to render the interior defects;
- Dye penetrant to define the size of the surface flaws; and
- Ultrasonic testing to detect cracks in flat and smooth members.

49.4.4 Fracture-Critical Members

Fracture-critical members (FCM) or member components are defined as tension components of members whose failure would be expected to result in collapse of a portion of a bridge or an entire bridge [7,8]. A redundant steel bridge that has multiple load-carrying mechanisms is seldom categorized as a fracture-critical bridge.

Since the failure to locate defects on FCMs in a timely manner may lead to catastrophic failure of a bridge, it is important to ensure that FCMs are inspected thoroughly. Hands-on involvement of the team leader is necessary to maintain the proper level of inspection and to make independent checks of condition appraisals. In addition, adequate time to conduct a thorough inspection should be allocated by the team leader. Serious problems in FCMs must be addressed immediately by restricting traffic on the bridge and repairing the defects under an emergency contract. Less serious problems requiring repairs or retrofit should be placed on the programmed repair work so that they will be incorporated into the maintenance schedule.

Bridge inspectors need to identify the FCMs using the guidelines provided in the Inspection of Fracture Critical Bridge Members [7,8]. There are several vulnerable fracture-critical locations in a bridge. Some of the obvious locations are field welds, nonuniform welds, welds with unusual profile, and intermittent welds along the girder. Other possible locations are insert plate termination points, floor beam to girder connections, diaphragm connection plates, web stiffeners, areas that are vulnerable to corrosion, intersecting weld location, sudden change in cross section, and coped sections. Detailed descriptions of each of these fracture-critical details are listed in the Inspection of Fracture Critical Bridge Members [7,8]. Once the FCM is identified in a bridge structure, information such as location, member components, likelihood to have fatigue- or corrosion related damage, needs to be gathered. The information gathered on the member should become a permanent record and the condition of the member should be updated on every subsequent inspection.

FCMs can be inspected by both visual and physical examination. During the visual inspection, the inspector performs a close-up, hands-on inspection using standard, readily available tools. During the physical examination, the inspector uses the most-sophisticated nondestructive testing methods. Some of the FCMs may have details that are susceptible to fatigue cracking and others may be in poor condition due to corrosion. The inspection procedures of corrosion- and fatigue-prone members are described in Section 49.4.3.

49.4.5 Scour-Critical Bridges

Bridges spanning over waterways, especially rivers and streams, sometimes provide major maintenance challenges. These bridges are susceptible to scour of the riverbed. When the scoured riverbed elevation falls below the top of the footing, the bridge is referred to as scour critical.

The rivers, whether small or large, could significantly change their size over the period of the lifetime of a bridge. A riverbed could be altered in several ways and thereby jeopardize the stability of the bridges. A few of the possible types of riverbed alterations are scour, hydraulic opening, channel misalignment, and bank erosion. Scour around the bridge substructures poses potential structural stability concerns. Scour at bridges depends on the hydraulic features upstream and downstream, riverbed sediments, substructure section profile, shoreline vegetation, flow velocities, and potential debris. The estimation of the overall scour depth will be used to identify scour-prone and scour-critical bridges. Guidance for the scour evaluation process is provided in Evaluating Scour at Bridges [11].

A typical scour evaluation process falls into two phases: inventory phase and evaluation phase. The main goal of the inventory phase is to identify those bridges that are vulnerable to scour (scour-prone bridges). Evaluation during this phase is made using the available bridge records, inspection records, history of the bridge, original stream location, evidence of scour, deposition of debris, geology, and general stability of the streambed. Once the scour-prone bridges are identified, the evaluation phase needs to be performed. The scour evaluation phase requires in-depth field review to generate data for estimation of the hydraulics and scour depth. The procedure of scour estimation is outlined in Evaluating Scour at Bridges [11]. The scour depths are then compared with the existing foundation condition. When the scour depth is above the top of the footing, the bridge would require no action. However, when the scour depth is within the limits of the footing or piles, a structural stability analysis is needed. If the scour depth is below the pile tips or spread footing base, monitoring of the bridge is required. These results obtained from the scour evaluation process are entered into the bridge inventory.

49.4.6 Underwater Components

Underwater components are mostly substructure members. Since the accessibility of these members is difficult, special equipment is necessary to inspect these underwater components. Also, visibility during the underwater inspection is generally poor, and therefore a thorough inspection of the members will not be possible. Underwater inspection is classified as visual (Level 1), detailed (Level 2), and comprehensive (Level 3) to specify the level of effort of inspection. Details of these various levels of inspection are discussed in the Manual for Maintenance Inspection of Bridges [2] and Evaluating Scour at Bridges [11].

Underwater steel structure components are susceptible to corrosion, especially in the low to high water zone. Some of the defects observed in underwater timber piles are splitting, decay or rot, marine borers, decay of timber at connections, and corrosion of connectors. It is important to recognize that the timber piles may appear sound on the outside shell but be severely damaged inside. Some of the most common defects in underwater concrete piles are cracking, spalls, exposed reinforcing, sulfate attack, honeycombing, and scaling. When cracking, spalls, and exposed reinforcing are detected, structural analysis may be required to ensure the safety of the bridge.

49.4.7 Decks

The materials typically used in the bridge structures are concrete, timber, and steel. Sections 49.4.1 to 49.4.3 discuss some of the defects associated with each of these materials. In this section, the damage most likely to occur in bridge decks is discussed.

Common defects in steel decks are cracked welds, broken fasteners, corrosion, and broken connections. In a corrugated steel flooring system, section loss due to corrosion may affect the load-carrying capacity of the deck and thus the actual amount of remaining materials needs to be evaluated and documented.

Common defects in timber decks are crushing of the timber deck at the supporting floor system, flexure damages such as splitting, sagging, and cracks in tension areas, and decay of the deck due to biological organisms, especially in the areas exposed to drainage.

Common defects in concrete decks are wear, scaling, delamination, spalls, longitudinal flexure cracks, transverse flexure cracks in the negative moment regions, corrosion of the deck rebars, cracks due to reactive aggregates, and damage due to chemical contamination. The importance of a crack varies with the type of concrete deck. A large to medium crack in a noncomposite deck may not affect the load-carrying capacity of the main load-carrying member. On the other hand, several cracks in a composite deck will affect the structural capacity. Thus, an inspector must be able to identify the functions of the deck while inspecting it.

Sometimes a layer of asphalt concrete (AC) overlay will be placed to provide a smooth driving and wearing surface. Extra care is needed during the inspection, because AC overlay prevents the inspector's ability to inspect the top surface of the deck visually for damage.

49.4.8 Joint Seals

Damage to the joint seals is caused by vehicle impact, extreme temperature, and accumulation of dirt and debris. Damage from debris and vehicles such as snowplows could cause the joint seals to be torn, pulled out of anchorage, or removed altogether. Damage from extreme temperature could break the bond between the joint seal and deck and consequently result in pulling out the joint seal altogether.

The primary function of deck joints is to accommodate the expansion and contraction of the bridge superstructure. These deck joints also provide a smooth transition from the approach roadway to the bridge deck. Deck joints are placed at hinges between two decks of adjacent structures, and between the deck sections and abutment backwall. The joint seals used in the bridge industry can be divided into two groups: open joints and closed joints. Open joints allow water and debris to pass through the joints. Dripping water through open joints usually damages the bearing details. Closed joints do not allow water and debris to pass through them. A few of the closed joints are compression seal, poured joint seal, sliding plate joint, plank seal, sheet seal, and strip seal.

In the case of closed joints, damage to the joint seal material will cause the water to drip on the bearing seats and consequently damage the bearing. Accumulation of dirt and debris may prevent normal thermal expansion and contraction, which may in turn cause cracking in the deck, backwall, or both. Cracking in the deck may affect the ride quality of the bridge, may produce larger impact load from vehicles, and may reduce the live-load-carrying capacity of the bridge.

49.4.9 Bearings

Bearings used in bridge structures could be categorized into two groups: metal and elastomeric. Metal bearings sometimes become inoperable (sometimes referred as "frozen") due to corrosion, mechanical bindings, buildup of debris, or other interference. Frozen bearings may result in bending, buckling, and improper alignment of members. Other types of damage are missing fasteners, cracked welds, corrosion on the sliding surface, sole plate rests only on a portion of the masonry plate, and binding of lateral shear keys.

Damage in elastomeric bearing pads is excessive bulging, splitting or tearing, shearing, and failure of bond between sole and masonry plate. Excessive bulging indicates that the bearing might be too tall. When the pad is under excessive strain for a long period, the pad will experience shearing failure.

Inspectors need to assess the exact condition of the bearing details and to recommend corrective measures that allow the bearing details to function properly. Since the damage to the bearings will affect the other structural members as time passes, repair of bearing damage needs to be considered as a preventive measure.

49.5 Fundamentals of Bridge Rating

49.5.1 Introduction

Once a bridge is constructed, it becomes the property of the owner or agency. The evaluation or rating of existing bridges is a continuous activity of the agency to ensure the safety of the public. The evaluation provides necessary information to repair, rehabilitate, post, close, or replace the existing bridge.

In the United States, since highway bridges are designed for the AASHTO design vehicles, most U.S. engineers tend to believe that the bridge will have adequate capacity to handle the actual present traffic. This belief is generally true if the bridge was constructed and maintained as shown in the design plan. However, changes in a few details during the construction phase, failure to attain the recommended concrete strength, unexpected settlements of the foundation after construction, and unforeseen damage to a member could influence the capacity of the bridge. In addition, old bridges might have been designed for a lighter vehicle than is used at present, or a different design code. Also, the live-load-carrying capacity of the bridge structure may have altered as a result of deterioration, damage to its members, aging, added dead loads, settlement of bents, or modification to the structural member.

Sometimes, an industry would like to transport their heavy machinery from one location to another location. These vehicles would weigh much more than the design vehicles and thus the bridge owner may need to determine the current live-load-carrying capacity of the bridge. In the following sections, establishing the live load-carrying capacity and the bridge rating will be discussed.

49.5.2 Rating Principles

In general, the resistance of a structural member (R) should be greater than the demand (Q) as follows:

$$R \geq Q_d + Q_l + \sum_i Q_i \tag{49.1}$$

where Q_d is the effect of dead load, Q_l is the effect of live load, and Q_i is the effect of load i.

Eq. (49.1) applies to design as well as evaluation. In the bridge evaluation process, maximum allowable live load needs to be determined. After rearranging the above equation, the maximum allowable live load will become

$$Q_l \leq R - \left(Q_d + \sum_i Q_i\right) \tag{49.2}$$

Maintenance engineers always question whether a fully loaded vehicle (rating vehicle) can be allowed on the bridge and, if not, what portion of the rating vehicle could be allowed on a bridge. The portion of the rating vehicle will be given by the ratio between the available capacity for live-load effect and the effect of the rating vehicle. This ratio is called the rating factor (RF).

$$\text{RF} = \frac{\text{Available capacity for the live-load effect}}{\text{Rating vehicle load demand}} = \frac{R - \left(Q_d + \sum_i Q_i\right)}{Q_l} \tag{49.3}$$

When the rating factor equals or exceeds unity, the bridge is capable of carrying the rating vehicle. On the other hand, when the rating factor is less than unity the bridge may be overstressed while carrying the rating vehicle.

The capacity of a member is usually independent of the live-load demand. Thus, Eq. (49.3) is generally a linear expression. However, there are cases where the capacity of a member dependent on the live-load forces. For example, available moment capacity depends on the total axial load in biaxial bending members. In a biaxially loaded member, the Eq. (49.3) will be a second-order expression.

Thermal, wind, and hydraulic loads may be neglected in the evaluation process because the likelihood of occurrence of extreme values during the relatively short live-load loading is small. Thus, the effects of the dead and live loads are the only two loads considered in the evaluation process.

49.5.3 Rating Philosophies

During the structural evaluation process, the location and type of critical failure modes are first identified; Eq. (49.3) is then solved for each of these potential failures. Although the concept of evaluation is the same, the mathematical relationship of this basic equation for allowable stress design (ASD), load factor design (LFD), and Load and resistance factor design (LRFD) differs. Since the resistance and load effect can never be established with certainty, engineers use safety factors to give adequate assurance against failure. ASD includes safety factors in the form of allowable stresses of the material. LFD considers the safety factors in the form of load factors to account for the uncertainty of the loadings and resistance factors to account for the uncertainty of structural response. LRFD treats safety factors in the form of load and resistance factors that are based on the probability of the loadings and resistances.

For ASD, the rating factor expression Eq. (49.3) can be written as

$$\mathrm{RF} = \frac{R - \left(\sum D + \sum_i L_i (1+I) \right)}{L(1+I)} \tag{49.4}$$

For LFD, the rating factor expression Eq. (49.3) can be written as

$$\mathrm{RF} = \frac{\phi R_n - \sum \gamma_D D - \sum_{i=1}^{n} \gamma_{Li} L_i (1+I)}{\gamma_L L(1+I)} \tag{49.5}$$

For LFRD, the rating factor expression Eq. (49.3) can be written as

$$\mathrm{RF} = \frac{\phi R_n - \sum \gamma_D D - \sum_{i=1}^{n} \gamma_{Li} L_i (1+I)}{\gamma_L L(1+I)} \tag{49.6}$$

where R is the allowable stress of the member; ϕR_n is nominal resistance; D is the effect of dead loads; L_i is the live-load effect for load i other than the rating vehicle; L the nominal live-load effect of the rating vehicle; I is the impact factor for the live-load effect; γ_D, γ_{Li}, and γ_L are dead- and live-load factors, respectively.

Researchers are now addressing the LRFD method, and thus the LRFD approach may be revised in the near future. Since the LRFD method is being developed at this time, the LRFD method is not discussed further in this chapter.

In order to use the above equations (Eqs. 49.4 to 49.6) in determining the rating factors, one needs to estimate the effects of individual live-load vehicles. The effect of individual live-load vehicles on structural member could only be obtained by analyzing the bridge using a three-dimensional analysis. Thus, obtaining the rating factor using the above expressions is very difficult and time-consuming.

To simplify the above equations, it is assumed that similar rating vehicles will occupy all the possible lanes to produce the maximum effect on the structure. This assumption allows us to use the AASHTO live-load distribution factor approach to estimate the live-load demand and eliminate the need for the three-dimensional analysis.

And the simplified rating factor equations become as follows:

$$\text{For ASD:} \quad \mathrm{RF} = \frac{R - D}{L(1+I)} \tag{49.7}$$

$$\text{For LFD:} \quad \mathrm{RF} = \frac{\phi R_n - \gamma_D D}{\gamma_L L(1+I)} \tag{49.8}$$

$$\text{For LRFD:} \quad \mathrm{RF} = \frac{\phi R_n - \gamma_D D}{\gamma_L L(1+I)} \tag{49.9}$$

In the derivation of the above equations (Eqs. 49.7 to 49.9), it is assumed that the resistance of the member is independent of the loads. A few exceptions to this assumption are beam–column members and beams with high moment and shear. In a beam–column member, axial capacity or moment capacity depends on the applied moment or applied axial load on the member. Thus, as the live-load forces in the member increase, the capacity of the member would decrease. In other words, the numerator of the above equations (available live-load capacity) will drop as the live load increases. Thus, the rating factor will no longer be a constant value, and will be a function of live load.

49.5.4 Level of Ratings

There are two levels of rating for bridges: inventory and operating. The rating that reflects the absolute maximum permissible load that can be safely carried by the bridge is called an operating rating. The load that can be safely carried by a bridge for indefinite period is called an inventory rating.

The life of a bridge depends on the fatigue life or serviceability limits of bridge materials. Higher frequent loading and unloading may affect the fatigue life or serviceability of a bridge component and thereby the life of the bridge. Thus, in order to maintain a bridge for an indefinite period, live-load-carrying capacity available for frequently passing vehicles needs to be estimated at service. This process is referred to as inventory rating.

Less frequent vehicles may not affect the fatigue life or serviceability of a bridge, and thus live-load-carrying capacity available for less frequent vehicles need not be estimated using serviceability criteria. In addition, since less frequent vehicles do not damage the bridge structure, bridge structures could be allowed to carry higher loads. This process is referred to as operating rating.

49.5.5 Structural Failure Modes

In the ASD approach, when a portion of a structural member is stressed beyond the allowable stress, the structure is considered failed. In addition, since any portion of the structural member material never reaches its yield, the deflections or vibrations will always be satisfied. Thus, the serviceability of a bridge is assured when the allowable stress method is used to check a bridge member. In other words, in the ASD approach, serviceability and strength criteria are satisfied automatically. The inventory and operating allowable stresses for various types of failure modes are given in the AASHTO Manual for Condition Evaluation of Bridges 1994 [12] (Rating Manual).

In the LFD approach, failure could occur at two different limit states: serviceability and strength. When the load on a member reaches the ultimate capacity of the member, the structure is considered failed at its ultimate strength limit state. When the structure reaches its maximum allowable serviceability limits, the structure is considered failed at its serviceability limit state. In LFD approach, satisfying one of the limit states will not automatically guarantee the satisfaction of the other limit state. Thus, both serviceability and strength criteria need to be checked in the LFD method. However, when the operating rating is estimated, the serviceability limits need not be checked.

49.6 Superstructure Rating Examples

In this section, several problems are illustrated to show the bridge rating procedures. In the following examples, AASHTO *Standard Specification for Highway Bridges,* 16th ed.[13] is referred to as Design Specifications and AASHTO *Manual for Condition Evaluation of Bridges* 1994 [12] is referred to as Rating Manual. All the notations used in these examples are defined in either the Design Specifications or the Rating Manual.

49.6.1 Simply Supported Timber Bridge

Given

Typical cross section of a 16-ft (4.88-m) long simple-span timber bridge is shown in Figure 49.1. 13.4 × 16 in. (101.6 × 406.4 mm) timber stringers are placed at 18 in. (457 mm) spacing. 4 × 12 in.

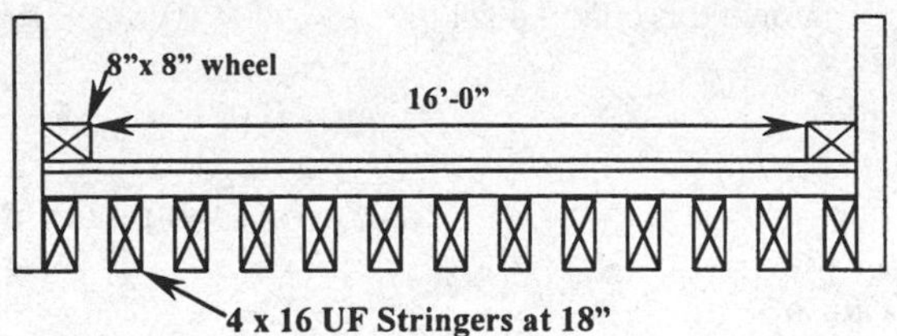

FIGURE 49.1 Typical cross section detail of simply supported timber bridge example.

(101.6 × 305 mm) timber planks are used as decking. 8 × 8 in. (203 × 203 mm) timber is used as wheel guard. Barrier rails (10 lb/ft or 0.1 N/mm) are placed at either side of the bridge. The traffic lane width of the bridge is 16 ft (4.88 m). Assume that the allowable stresses at operating level are as follows: F_b for stringer as 1600 psi (11 MPa) and F_v of stringer level as 115 psi (0.79 MPa).

Requirement
Determine the critical rating factors for interior stinger for HS20 vehicle using the ASD approach.

Solution
For this simply supported bridge, the critical locations for ratings will be the locations where shear and moments are higher.

According to Design Specifications Section 13.6.5.2, shear needs to be checked at a distance (s) $3d$ or $0.25L$ from the bearing location for vehicle live loads; thus,

$$s = 3d = 3 \times 16 \text{ in.}/12 = 4.0 \text{ ft or}$$

$$= 0.25L = 0.25 \times 16 \text{ ft} = 4.0 \text{ ft. Thus, } s \text{ is taken as } 4.0 \text{ ft } (1.22 \text{ m}).$$

Maximum dead- and live-load shear will occur at this point and thus in the following calculations, shear is estimated at this critical location.

1. Dead Load Calculations

Self-weight of the stringer $= 0.05 \times 4 \times 16 \times \dfrac{1}{144} = 0.022 \text{ kips/ft}$

Weight of deck (using tributary area) $= 1.5 \times 4 \times 12 \times \dfrac{1}{144} \times 0.05 = 0.025 \text{ kips/ft}$

Weight of 1.5 in. AC on the deck $= 1.5 \times \dfrac{1.5}{12} \times 0.144 = 0.027 \text{ kips/ft}$

Barrier rail and curb $= \left(10 + 50 \times \dfrac{8 \times 8}{144}\right) \times \dfrac{2}{13 \times 1000} = 0.004 \text{ kips/ft}$

Total uniform dead load on the stringer $= 0.022 + 0.025 + 0.027 + 0.004$

$= \underline{0.078 \text{ kips/ft}}$

Maximum dead load moment at midspan $= \dfrac{wl^2}{8} = \dfrac{0.078 \times 16^2}{8} = 2.5 \text{ kip-ft (3390 N-m)}$

Maximum dead load shear at this critical point $= w \times (0.5L - s)$

$= 0.078 \times (0.5 \times 16 - 4)$

$= 0.31$ kips (1.38 kN)

2. Live-Load Calculations

The travel width is less than 18 ft. Thus, according to Section 6.7.2.2 of the Rating Manual, this bridge needs to be rated for one traffic lane. From Designs Specifications Table 3.23.1,

$$\text{Number of wheels on the stringer} = \frac{S}{4} = \frac{1.5}{4} = 0.38$$

Maximum moment due to HS20 loading (Appendix A3, Rating Manual)

= (64) (0.38)
= 24.32 kip-ft (33,000 N-m)

In order to estimate the live-load shear, we need to estimate the shear due to undistributed and distributed HS20 loadings. (See Design Specifications 13.6.5.2 for definition of V_{LU} and V_{LD})

Shear due to undistributed HS20 loadings $= V_{LU}$ $= 16 \times 12/16 = 12$ kips (53.4 kN)

Shear due to distributed HS20 loading $= V_{LD}$ $= 16 \times 12/16 \times (0.38)$

$= 4.56$ kips (20.3 kN)

Thus, shear due to HS20 live load $= 0.5(0.6V_{LU} + V_{LD})$ $= 5.88$ kips (26.1 kN)

3. Capacity Calculations

a. *Moment capacity at midspan*:
Moment capacity of the timber stringer at *Operating level* =

$$F_b S_x = 1600 \times \frac{1}{6} \times 4 \times 16^2 \times \frac{1}{12{,}000} = 22.8 \text{ kip-ft (30,900 N-m)}$$

According to Section 6.6.2.7 of Rating Manual, the operating level stress of a timber stringer can be taken as 1.33 times the inventory level stress.
Thus, moment capacity of the timber stringer at *Inventory level*

= 22.8/1.33
= 17.1 kip-ft (23,200 N-m)

b. *Shear capacity at support*:
Shear capacity of the timber section (controlled by horizontal shear) = $(2/3) bdf_v$:

V_c at *operating level* = (2/3) × 4 × 16 × 115 psi × 1/1000 = 4.91 kips (21.8 kN)

V_c at *inventory level* = 14.91/1.33 = 3.69 kips (16.4 kN)

4. Rating Calculations

$$\text{Rating factor based on ASD method} = \text{RF} = \frac{R - D}{L(1+I)}$$

By substituting appropriate values, the rating factor can be determined.

a. *Based on moment at midspan*:

$$\text{Inventory rating factor } RF_{INV\text{-}MOM} = \frac{17.1 - 2.5}{24.32} = 0.600$$

$$\text{Operating rating factor } RF_{OPR\text{-}MOM} = \frac{22.8 - 2.5}{24.32} = 0.835$$

b. *Based on shear at the support*:

$$\text{Inventory rating factor } RF_{INV\text{-}SHE} = \frac{3.69 - 0.31}{5.88} = 0.575$$

$$\text{Operating rating factor } RF_{OPR\text{-}SHE} = \frac{4.91 - 0.31}{5.88} = 0.782$$

5. **Summary**
   It is found that the critical rating factor is controlled by shear in the stringers. The critical inventory and operating rating of the bridge will be 0.575 and 0.782, respectively.

49.6.2 Simply Supported T-Beam Concrete Bridge

Given
A bridge, which was built in 1929, consists of three simple-span reinforced concrete T-beams on concrete bents and abutments. The span lengths are 16 ft (4.88 m), 50 ft (15.24 m), and 10 ft (3.05 m). Typical cross section and girder details are shown in Figure 49.2. General notes given in the plan indicate that $f_c = 1000$ psi (6.9 MPa) and $f_s = 18{,}000$ psi (124.1 MPa). Assume the weight of each barrier rail as 250 lb/ft (3.6 N/mm).

Requirement
Determine the critical rating factor of the interior girder of the second span (50 ft. or 15.24 m) for HS20 vehicles assuming no deterioration of materials occurred.

Solution

1. **Dead-Load Calculations**

Self-weight of the girder = (3.5) (1.333) (0.15) = 0.700 kips/ft

(4 × 4 in.) Fillets between girder and slab = 2(1/2) (4/12) (4/12) (0.15) = 0.017 kips/ft

Slab weight (based on tributary area) = (6.667)(8/12) (0.15) = 0.667 kips/ft

Contribution from barrier rail (equally distributed among girders) = 2 (0.25/3) = 0.167 kips/ft

$$\text{Thus, total uniform load on the interior girder} = \underline{1.551 \text{ kips/ft}} \,(22.6 \text{ N/mm})$$

$$\text{Dead-load moment at midspan} = \underline{484.6 \text{ kips/ft} \,(0.657 \text{ MN/m})}$$

$$\text{Dead load shear at a distance } d \text{ from support} = \underline{32.31 \text{ kips} \,(143.7 \text{ kN})}$$

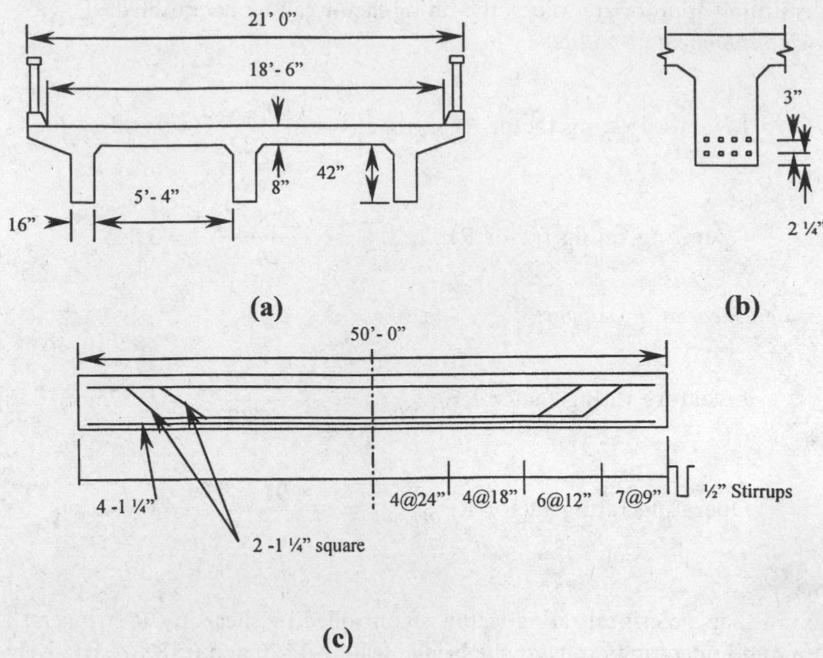

FIGURE 49.2 Details of simply supported T-beam concrete bridge example. (a) Typical cross section; (b) reinforcement locations; (c) T-beam girder details.

2. **Live-Load Calculations**
   The traffic lane width of this bridge is 18.5 ft. According to Design Specifications, any bridge with a minimum traffic lane width of 18 ft needs to carry two lanes. Hence, the number of live-load wheels will be based on two traffic lanes. From Table 3.23.1A of Design Specifications for two traffic lanes for T-beams is given by *S*/6.0

   Number of live-load wheel line = 6.667/6.0 = $\underline{1.111}$

   AASHTO standard impact factor for moment = 50/(125 + 50) = $\underline{0.286}$

   AASHTO standard impact factor for shear at support = 50/(125 + 50) = $\underline{0.286}$

   The live-load moments and shear tables listed in the Rating Manual are used to determine the live-load demand.

Maximum HS20 moment for 50 ft span without impact/wheel line = 298.0 kips-ft

Thus, HS20 moment with impact at midspan = (1.286) (1.111) (298.0)

= $\underline{\text{425.7 kips-ft (0.58 MN-m)}}$

Maximum HS20 shear at a distance *d* from the support/wheel line = 28.32 kips

Thus, maximum HS20 shear = (1.286) (1.111) (28.32) = $\underline{\text{40.46 kips (180.0 kN)}}$

3. Capacity Calculations

Strengths of concrete and rebars are first determined (see Rating Manual Section 6.6.2.3):

$$f_c' = \frac{f_c}{0.4} \quad \text{thus } f_c' = 2500 \text{ psi}$$

and $f_s = 18{,}000$ psi and thus $f_y = 33{,}000$ psi.

a. *Moment capacity at midspan:*

Total area of the steel (note these bars are 1¼ square bars) = (8) (1.25) (1.25) = 12.5 in.^{2}

Centroid of the rebars from top deck = 42 + 8 – 3.75 = 46.250 in.

Effective width of the deck b_{eff} = minimum of $12t_s + b_w$ = 112 in.

Span/4 = 150 in.

Spacing = 80 in. (Controls)

Uniform stress block depth = $a = \dfrac{A_s f_y}{0.85 f_c' b_{\text{eff}}} = 2.426$ in. $< t_s = 8$ in.

$$M_u = \phi A_s f_y \left(d - \frac{a}{2}\right) = 0.9 \times 12.5 \times 33 \left(46.25 - \frac{2.426}{2}\right) \times \left(\frac{1}{12}\right) = 1393.3 \text{ kips-ft (1.88 MN-m)}$$

b. *Shear capacity at support:*

According to AASHTO specification, shear at a distance d (50 in.) from the support needs to be designed. Thus, the girder is rated at a distance d from the support. From the girder details, it is estimated that ½ in. ϕ stirrups were placed at a spacing of 12 in. and two 1¼ square bars were bent up. The effects of these bent-up bars are ignored in the shear capacity calculations.

Shear capacity due to concrete section:

$$V_c = 2\sqrt{f_c'}\, b_w d = 2\sqrt{2500} \times 16 \times 46.25 \times \left(\frac{1}{1000}\right) = 74 \text{ kips (329 kN)}$$

Shear capacity due to shear reinforcement:

$$V_s = 2A_v \frac{F_y d_s}{S} = 2 \times 0.20 \times \frac{33 \times 46.25}{12} = 50.88 \text{ kips (226 kN)}$$

Total shear capacity:

$$V_u = \phi (V_s + V_c) = 0.85\,(74.0 + 50.88) = 106.2 \text{ kips (472 kN)}$$

4. Rating Calculations

$$\text{Rating factor} = \frac{\phi R_n - \gamma_D D}{\gamma_L \beta_L L(1+I)}$$

TABLE 49.1 Rating Calculations of Simply Supported T-Beam Concrete Bridge Example

Location	Description	Inventory Rating	Operating Rating
Midspan	Moment	$\frac{1393.3-1.3\times 484.6}{1.3\times 1.67\times 425.7}=0.825$	$\frac{1393.3-1.3\times 484.6}{1.3\times 425.7}=1.38$
At support	Shear	$\frac{106.2-1.3\times 32.31}{1.3\times 1.67\times 40.46}=0.731$	$\frac{106.2-1.3\times 32.31}{1.3\times 40.46}=1.22$

According to Rating Manual, γ_D is 1.3, γ_L is 1.3, and β_L is 1.67 and 1.0 for inventory and operating factors, respectively. By substituting these values and appropriate load effect values, the moment and shear rating could be estimated. The calculations and results are given in Table 49.1.

5. **Summary**
Critical rating of the interior girder will then be 0.731 at inventory level and 1.22 at operating rating level for HS20 vehicle.

49.6.3 Two-Span Continuous Steel Girder Bridge

Given
Typical section of a two-span continuous steel girder bridge, which was built in 1967, is shown in Figure 49.3a. Steel girder profile is given in Figure 49.3b. The general plan states that f_s = 20,000 psi (137.9 MPa) and f_c = 1200 psi (8.28 MPa). Assume that (a) each barrier rail weighs 250 lb/ft (3.6 N/mm); (b) girders were not temporarily supported during the concrete pour; (c) girder is composite for live loads; (d) girder is braced every 15 ft and the weight of bracing per girder is 330 lb.

Requirement
Determine the rating factors of interior girders using ASD method.

Solution

1. **Dead Load Calculations**

Deck weight (tributary area approach) = (6.625/12) (6.625) (0.15) = 0.549 kips/ft

Average uniform self-weight for the analysis = 1431 kips/90 ft = 0.159 kips/ft

Average diaphragm load (uniformly distributed) = (0.33) (4/90) = 0.015 kips/ft

$$\text{Thus, total uniform dead load on the girder} = \underline{0.723\ \text{kips/ft}\ (10.5\ \text{N/mm})}$$

$$\text{Barrier rail load (equally distributed among all girders)} = (2)(250)/14 = 0.0358\ \text{kips/ft}$$

$$\text{Thus, total additional dead load on the girder} = \underline{0.0358\ \text{kips/ft}}(0.56\ \text{N/mm})$$

2. **Live Load Calculations**

$$\text{Number of wheels per girder (for two or more lanes)} = S/5.5 = 6.625/5.5 = \underline{1.206}$$

Analysis Results: Analysis is done using two-dimensional program and the moments and shears at critical locations are listed in the Table 49.2.

Section properties at 0.4th and 1.0th points are estimated and the results are given in Table 49.3.

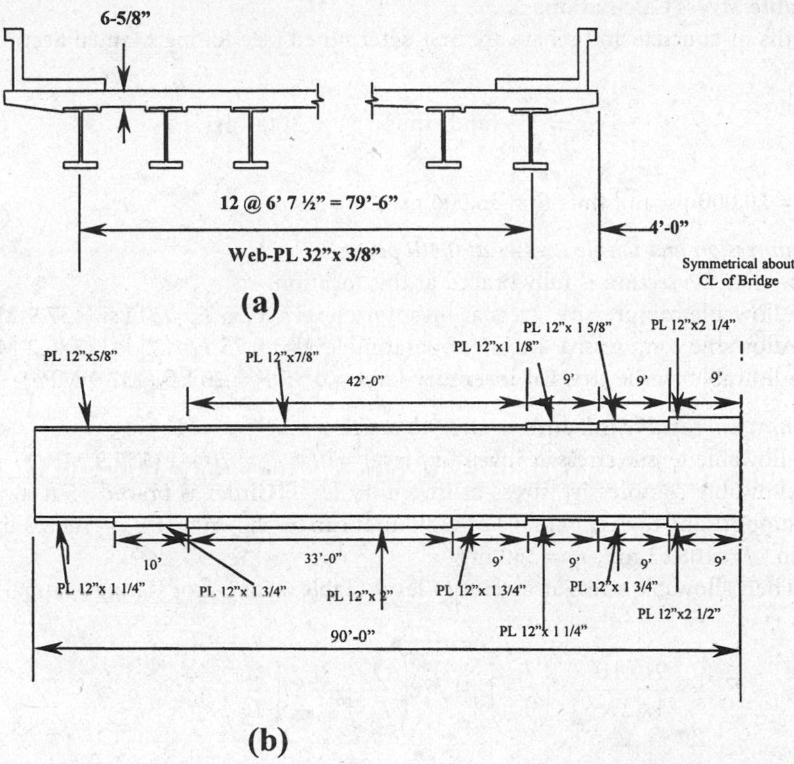

FIGURE 49.3 Details of two-span continuous steel girder bridge example. (a) Typical section; (b) girder elevation.

TABLE 49.2 Load Demands at 0.4th and 1.0th Point for Steel Girder Bridge Example

Description	0.4th Point (36.1 ft)	At Support (90 ft)
Dead load moments in kip-ft	410.0	–732.0
Dead load shear in kips	–1.7	–40.7
Additional dead load moment in kip-ft	22.0	–39.0
Additional dead load shear in kips	–0.1	–2.2
HS20 maximum positive moment in kip-ft	807.0	0.0
HS20 maximum negative moment in kip-ft	–177.0	–714.0
HS20 max. positive shear force in kips	21.8	0.0
HS20 max. negative shear force in kips	–19.3	–49.9

TABLE 49.3 Section Properties of Girder Sections for Steel Girder Bridge Example

	I_{gg} (in.^{4})	Y_b (in.)	Y_t (in.)	S_{xb} (in.^{3})	S_{xt} (in.^{3})
Section at 0.4th Point					
For dead loads	9,613.9	12.94	21.94	743.11	438.24
For additonal dead loads	17,406.5	19.69	15.19	884.18	1,146.10
For live loads	24,782.7	26.02	8.86	952.59	2,797.50
Section at 1.0th Point					
For dead loads	17,852.1	17.70	19.05	1,008.50	937.20
For additional dead loads	17,852.1	17.70	19.05	1,008.50	937.20
For live loads	17,852.1	17.70	19.05	1,008.50	937.20

3. Allowable Stress Calculations

Strengths of concrete and rebars are first determined (see Rating Manual Section 6.6.2.3):

$$f_c' = \frac{f_c}{0.4} \text{ and, thus } f_c' = 3000 \text{ psi}$$

and $f_s = 20{,}000$ psi and thus $F_y = 36{,}000$ psi.

a. *Compression and tensile stresses at 0.4th point*:
   Note that the section is fully braced at this location.
   i. Allowable compressive stress at inventory level = 0.55 F_y = 20 ksi (137.9 MPa)
   ii. Allowable compressive stress at operating level = 0.75 F_y = 27 ksi (186.2 MPa)
   iii. Allowable tensile stress at inventory level = 0.55 F_y = 20 ksi (137.9 MPa)

b. *Compression and tensile stresses at 1.0th point*:
   i. allowable tensile stress at inventory level = 0.55 F_y = 20 ksi (137.9 MPa)
   ii. allowable compressive stress at inventory level: Girder is braced 15 ft away from the support and thus $L_b = 15 \times 12 = 180$ in. It can be shown that $S_{xc} = 1008.3$ in^3; $d = 36.75$ in.; $J = 108.63$ in.^{4}; $I_{yc} = 360$ in.^{4}
   Then allowable stress at inventory level (Table 6.6.2.1-1 of Rating Manual):

$$F_b = \frac{91\times10^6\, C_b}{1.82\times S_{xc}}\left(\frac{I_{yc}}{L_b}\right)\sqrt{0.772\left(\frac{J}{I_{yc}}\right)+9.87\left(\frac{d}{L_b}\right)^2}$$

$$= \frac{91\times10^6\,(1.00)}{1.82\times1008.3}\left(\frac{360}{180}\right)\sqrt{0.772\left(\frac{108.63}{360}\right)+9.87\left(\frac{36.75}{180}\right)^2}\left(\frac{1}{1000}\right)$$

$= 79.5 > 0.55F_y = 20$ ksi (Note that C_b is conservatively assumed as 1.0.)

   Thus, $F_b = 20$ ksi (137.9 MPa)

   iii. Allowable compressive stress at operating level: The allowable stress at operating level is given:

$$F_b = \frac{91\times10^6\, C_b}{1.34\times S_{xc}}\left(\frac{I_{yc}}{L_b}\right)\sqrt{0.772\left(\frac{J}{I_{yc}}\right)+9.87\left(\frac{d}{L_b}\right)^2}$$

(Table 6.6.2.1-2, Rating Manual)

$= 108.0 > 0.75\ F_y = 27$ ksi

   Thus, $F_b = 27$ ksi (186.2 MPa)

c. *Allowable inventory shear stresses at 0.4th and 1.0th point*:

$$D/t_w = 32/0.375 = 85.33$$

   Girder is unstiffened and thus $k = 5$;

$$\frac{6000\sqrt{k}}{\sqrt{F_y}} = 70.71 < D/t_w < \frac{7500\sqrt{k}}{\sqrt{F_y}} = 88.3$$

TABLE 49.4 Estimated Stress Demands for Steel Girder Bridge Example

Load Description	At 0.4th Point	At 1.0th Point	Fiber Location
DL moment	–6.62	8.71	At bottom fiber
ADL moment	–0.30	0.463	At bottom fiber
LL + *I* moment	–10.16	8.49	At bottom fiber
DL moment	11.23	–9.37	At top fiber
ADL moment	0.23	–0.49	At top fiber
LL + *I* moment	3.46	–9.14	At top fiber
DL shear	0.129	2.95	Shear stress
ADL shear	0.007	0.15	Shear stress
LL + *I* shear	1.667	3.62	Shear stress

Thus,

$$C = \frac{6000\sqrt{k}}{\left(\frac{D}{t_w}\right)\sqrt{F_y}} = 0.828$$

$$F_v = \frac{F_y}{3}\left(C + \frac{0.87(1-C)}{\sqrt{1+\left(\frac{d_o}{D}\right)^2}}\right) = 11.76 \text{ ksi } (81.1 \text{ MPa})$$

d. *Allowable operating shear stresses at 0.4th and 1.0th point*:

$$F_v = 0.45F_y\left(C + \frac{0.87(1-C)}{\sqrt{1+\left(\frac{d_o}{D}\right)^2}}\right) = 15.88 \text{ ksi } (109.5 \text{ MPa})$$

4. **Load Stress Calculations**
Bending stress calculations are made using appropriate section modulus and moments. Results are reported in Table 49.4. The sign convention used in Table 49.4 is as follows: compressive stress is positive and tensile stress is negative. Also, estimated shear stresses are given in Table 49.4.

5. **Rating Calculations**
The rating factor in ASD approach is given by

$$\frac{R-D}{L(1+I)}$$

and the rating calculations are made and given in Table 49.5.

6. **Summary**
The critical rating factor of the girder is controlled by tensile stress on the top fiber at the 1.0th point. The critical inventory and operating rating factors are 1.11 and 1.87, respectively.

TABLE 49.5 Rating Calculations Using ASD Method for Steel Girder Bridge Example

Location	Description	Inventory Rating	Operating Rating
0.4th point	Shear	$\frac{11.76-(0.129+0.007)}{1.667}=6.97$	$\frac{15.88-(0.129+0.007)}{1.667}=9.44$
	Stress at top fiber	$\frac{20-(11.23+0.23)}{3.46}=2.46$	$\frac{27-(11.23+0.23)}{3.46}=4.49$
	Stress at bottom fiber	$\frac{20-(6.62+0.30)}{10.16}=1.28$	$\frac{27-(6.62+0.30)}{10.16}=1.97$
1.0th point	Shear	$\frac{11.76-(2.95+0.15)}{3.62}=2.39$	$\frac{15.88-(2.95+0.15)}{3.62}=3.53$
	Stress at top fiber	$\frac{20-(9.37+0.50)}{9.14}=1.11$	$\frac{27-(9.37+0.50)}{9.14}=1.87$
	Stress at bottom fiber	$\frac{20-(8.71+0.463)}{8.49}=1.28$	$\frac{27-(8.71+0.463)}{8.49}=2.10$

49.6.4 Two-Span Continuous Prestressed, Precast Concrete Box Beam Bridge

Given

Typical section and elevation of three continuous-span precast, prestressed box-girder bridge is shown in Figure 49.4. The span length of each span is 120 ft (36.5 m), 133 ft (40.6 m), and 121 ft (36.9 m). Total width of the bridge is 82 ft (25 m) and a number of precast, prestressed box girders are placed at a spacing of 10 ft (3.1 m). The cross section of the box beam and the tendon profile of the girder are shown in Figure 49.4. Each barrier rail weighs 1268 lb/ft (18.5 N/mm). Information gathered from the plans is (a) f_c' of the girder and slab is 5500 and 3500 psi, respectively; (b) working force (total force remaining after losses including creep) = 2020 kips; (c) x at midspan = 9 in. Assume that (1) the bridge was made continuous for live loading; (2) no temporary supports were used during the erection of the precast box beams; (3) properties of the precast box are area = 1375 in^2; moment of inertia = 30.84 ft^4; Y_t = 28.58 in.; Y_b = 34.4 in.; (4) F_y of reinforcing steel is 40 ksi.

Requirement

Rate the interior girder of Span 2 for HS20 vehicle.

Solution

1. **Dead Load Calculations**

Self-weight of the box beam = (1375/144) (0.15) = 1.43 kips/ft

Weight of Slab (tributary area approach) = (6.75/12) (10) (0.15) = 0.85 kips/ft

Total dead weight on the box beam $= \underline{2.28 \text{ kips/ft } (33.2 \text{ N/mm})}$

Contribution of barrier rail on box beam = 2(1.268/8) = 0.318 kips/ft

Thus, total additional dead load on the box beam $= \underline{0.318 \text{ kips/ft } (4.6 \text{ N/mm})}$

Girder is simply supported for dead loads; thus maximum dead load moment = (2.28) (133^2/8)

$= \underline{4926 \text{ kips/ft} (6.68 \text{ MN/m})}$

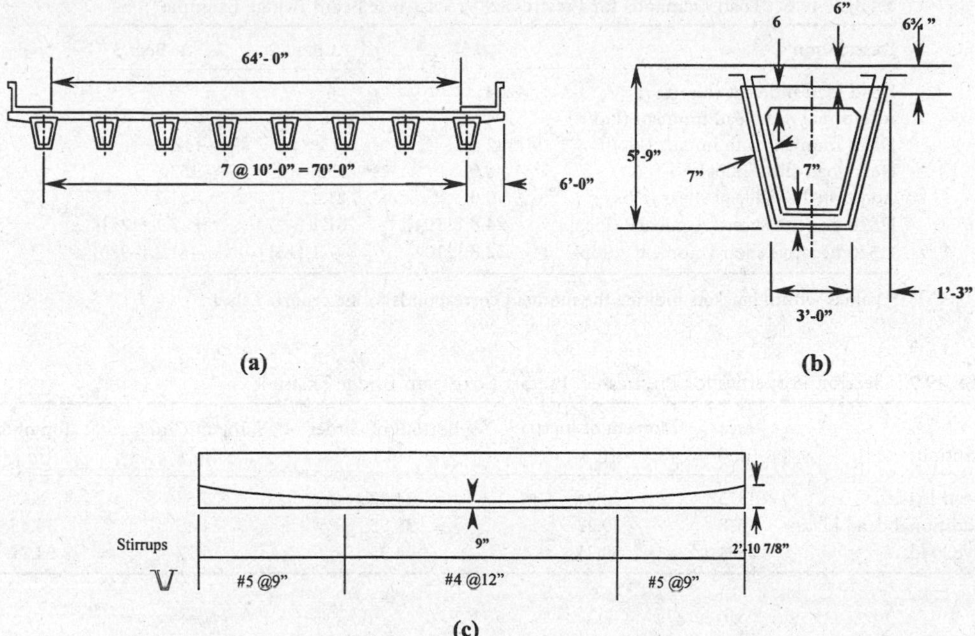

FIGURE 49.4 Details of two-span continuous prestressed box beam bridge example. (a) Typical section; (b) beam section details; (c) prestressing tendon profile.

2. **Live Load Calculations**
   According to Article 3.28 of Design Specifications, distribution factor (DF) for interior spread box beam is given by

$$DF = \left(\frac{2\,N_L}{N_B}\right) + k\left(\frac{S}{L}\right)$$

where N_L = number of traffic lanes = 64/12 = 5 (no fractions); N_B = Number of beams = 8; S = girder spacing = 8 ft; L = span length = 133 ft; W = roadway width = 64 ft

$$k = 0.07\ W - N_L\ (0.10\ N_L - 0.26) - 0.2\ N_B - 0.12 = 1.56$$

$$\text{Thus, DF} = \left(\frac{2\times 5}{8}\right) + 1.56\left(\frac{10}{133}\right) = 1.37 \text{ wheels}$$

3. **Demands on the Girder**
   Load demands are estimated using a two-dimensional analysis, and a summary is given in Table 49.6.

4. **Section Property Calculations**
   In order to estimate the stresses on the prestress box beam, the section properties for composite girder need to be estimated. Calculations of the composite girder properties are done separately and the final results are listed here in Table 49.7.

TABLE 49.6 Load Demands for Prestressed Precast Box Beam Bridge Example

Description	0.5L	At Bent 2	At Bent 3
Dead load moment (kip/ft)	4224	0	0
Additional dead load moment (kip/ft)	194	–506	–513
HS20 moment with impact (kip/ft)	1142	–1313	–1322
Dead load shear (kips)	0.0	153.6	–153.6
Additional dead load shear (kips)	0.0	21.1	–21.2
HS20 positive shear (moment)[a] (kips)	24.8 [1104]	61.1 [–974]	7.1 [127]
HS20 negative shear (moment)[a] (kips)	–24.8 [1104]	–7.1 [131]	–61.2 [–980]

[a] Values within brackets indicate the moment corresponds to the reported shear.

TABLE 49.7 Section Properties for Prestressed, Precast Box Beam Bridge Example

Description	Area (in.^{2})	Moment of Inertia (ft^4)	Y Bottom of Girder (in.)	Y Top of Girder (in.)	Y Top of Slab (in.)
For dead loads	1375	30.84	34.42	28.58	NA
For additional dead loads	1578	39.22	38.55	24.45	30.45
For live loads	1984	50.75	44.23	18.77	24.77

TABLE 49.8 Stresses at Midspan for Prestressed, Precast Box Beam Bridge Example

Location = Midspan	Stresses in the Box Beam (psi)			
Load Description	At Top Concrete Fiber	At Bottom Concrete Fiber	At Centroid of Composite Box Beam Concrete Fiber	At Prestress Tendon
Dead load (self + slab)	2265	–2728	777	20.15
Prestress P_{eff} = 2020 kips e = 25.42 in.	–1615	3443	–108	147.1
Additional dead (barrier)	70	–110	16	0.845
Live load	244	–575	0	4.59
Live load moment for shear	236	–556	0	4.43

TABLE 49.9 Stresses at Bent 2 for Prestressed, Precast Box Beam Bridge Example

Location = Bent 2	Stresses in the Box Beam (psi)				
Load Description	At Top Concrete Fiber	At Bottom Concrete Fiber	At Centroid of Composite Box Beam Concrete Fiber	At Top of Slab Fiber	At Prestress Tendon
Dead load (self + slab)	0	0	0	0	0
Prestress P_{eff} = 2020 kips e = 12 in.	680	680	680	0	167.5
Additional dead (barrier)	–183	288	–4	–228	–0.3
Live load	–281	662	0	–371	–1.47
Live load moment for positive shear	–208	491	0	–274	–1.08

5. **Stress Calculations**
   Stresses at different fiber locations are calculated using

$$\left(\frac{P}{A}\right)+\left(\frac{M\,c}{I}\right)$$

   expression. The summary of the results at midspan and at Bent 2 locations is given in Tables 49.8 and Table 49.9, respectively.

6. **Capacity Calculations**

   a. *Moment capacity at midspan*:
   The actual area of steel could only be obtained from the shop plans. Since the shop plans are not readily available, the following approach is used. Assume the total loss including the creep loss = 35 ksi (241.3 MPa).

$$\text{Thus, the area of prestressing steel} = \frac{\text{Working force}}{0.75 \times 270 - 35} = \frac{2020}{167.5} = 12.06 \text{ in.}^2 \ \ (7781 \text{ mm}^2)$$

$$b_{\text{eff}} = 120 \text{ in.}; \quad t_s = 6.75 \text{ in.}; \quad d_p = (5.75)(12) - 9 \text{ in.} = 60 \text{ in.}; \quad b_w = 14 \text{ in.}$$

$$\rho^* = \frac{A_s^*}{bd} = \frac{12.06}{120 \times 60} = 0.001675$$

$$f_{su}^* = f_s\left(1 - \frac{0.5\rho^* f_s'}{f_c'}\right) = 270\left(1 - \frac{0.5 \times 0.001675 \times 270}{5.5}\right) = 258.9 \text{ ksi} \ \ (1785 \text{ MPa})$$

$$\text{Neutral axis location} = 1.4\, d\rho^* \frac{f_{su}^*}{f_c'} = 1.4 \times 60 \times 0.001675 \times \frac{258.9}{5.5} = 6.62 \text{ in.} < t_s = 6.75 \text{ in.}$$

   Since the neutral axis falls within the slab, this girder can be treated as a rectangular section for moment capacity calculations.

$$R = \phi\, M_n = \phi A_s^* f_{su}^* d\left(1 - 0.6\,\rho \frac{f_{su}^*}{f_c'}\right) \quad \text{and} \quad \phi = 1.00$$

$$= \underline{14873.1 \text{ kips/ft } (20.17 \text{ MN/m})}$$

   b. *Moment capacity at the face of the support*:
   15 #11 bars are used at top of the bent; thus, the total area of steel = (15)(1.56) = 23.4 in.^{2} Depth of the reinforcing steel from the top of compression fiber = 69 – 1.5 – 1.41/2 = 66.795 in. (1696.6 mm). F_y = 60 ksi. Resistance reduction factor ϕ = 0.90. Then, the moment capacity

$$\phi\, \text{M}_\text{n} = \underline{6547.2 \text{ kip/ft } (8.88 \text{ MN/m})} \ \ \text{(based on T section)}$$

   c. *Shear capacity at midspan*:
   Design Specification's Section 9.20 addresses the shear capacity of a section. Shear capacity depends on the cracking moment of the section. When the live load causes tension at bottom fiber, cracking moment is to be calculated based on the bottom fiber stress. On the other hand, when the live load causes tension at the top fiber of the beam, cracking moment is to be calculated based on the top fiber stress.

   At midspan location, the moment reported with the maximum live-load shear is positive. Positive moments will induce tension at the bottom fiber and thus cracking moment is to be based on the stress at bottom fiber.

TABLE 49.10 Rating Calculations Prestressed, Precast Box Beam Bridge Example

Location	Description	Inventory Rating	Operating Rating
Midspan	Maximum moment	$\frac{14873.1-1.3\times(4224+194)}{1.3\times1.67\times1142}=3.69$	$\frac{14873.1-1.3\times(4224+194)}{1.3\times1142}=6.16$
	Maximum shear	$\frac{179-1.3\times(0+0)}{1.3\times1.67\times24.8}=3.33$	$\frac{179-1.3\times(0+0)}{1.3\times24.8}=5.56$
Bent 2	Maximum moment	$\frac{6544.2-1.3\times(0+506)}{1.3\times1.67\times1313}=2.06$	$\frac{6544.2-1.3\times(0+506)}{1.3\times1313}=3.45$
	Maximum shear	$\frac{766-1.3\times(153.6+21.1)}{1.3\times1.67\times61.1}=4.07$	$\frac{766-1.3\times(153.6+21.1)}{1.3\times61.1}=6.80$

f'_c = 5500 psi and from Table 49.10; f_{pe} at midspan bottom fiber = 3443 psi

f_d at bottom fiber = –2728 – 110 = –2838 psi; f_{pc} at centroid = 777 – 108 + 16 = 685 psi

$$M_{cr}=\frac{I}{Y_t}\left(6\sqrt{f'_c}+f_{pe}-f_d\right)=\frac{50.75\times12^4}{44.23}\left(6\sqrt{5500}+3443-2838\right)\left(\frac{1}{12{,}000}\right)=2081\text{ kips/ft}$$

Factored total moment:

$$M_{\max}=1.3\,M_D+(1.3)(1.67)\,M_{LL+I}$$

$$=1.3\,(4224+194)+2.167\,(1104)=8136\text{ kips-ft}$$

Factored total shear:

$$V_i=1.3\,(0+0)+2.167\,(24.8)=53.7\text{ kips}$$

V_d = 0 kips; b_w = 14 in.; d = 60 in.; f_{pc} = 685 psi

$$V_{ci}=0.6\sqrt{f'_c}\,b_w d+V_d+\frac{V_i M_{cr}}{M_{\max}}=0.6\sqrt{5500}\times14\times60\times\left(\frac{1}{1000}\right)+0+\frac{53.7\times2081}{8136}$$

= 51.2 kips (227.7 kN) (Controls — since smaller than V_{cw})

$$V_{cw}=\left(3.5\sqrt{f'_c}+0.3f_{pc}\right)b_w d+V_p=\left(3.5\sqrt{5500}+0.3\times685\right)14\times60\times\left(\frac{1}{1000}\right)+0$$

= 390 kips (1734 kN)

V_c = 51.2 kips (227.7 kN) (smaller of V_{ci} and V_{cw})

$$V_s=2A_v\frac{F_y d_s}{S}=4\times0.20\times\frac{40\times60}{12}=160\text{ kips}\ \ (711.7\text{ kN})$$

Shear capacity at midspan:

$$V_u=\phi\,(V_c+V_s)=0.85\,(51.2+160)=179\text{ kips (796.1 kN)}$$

d *Shear capacity at the face of support at Bent 2:*
Negative shear reported at this location is so small and thus rating will not be controlled by the negative shear at Bent 2. Moment reported with the positive shear is negative, and thus, the following calculations are based on the stress at top fiber. From the Table 49.11, f_d at top of slab fiber = –228 psi, and f_{pe} at support top of slab fiber (slab poured after prestressing) = 0 psi.

$$M_{cr} = \frac{I}{Y_t}\left(6\sqrt{f_c'} + f_{pe} - f_d\right) = \frac{50.75\times 12^4}{44.23}\left(6\sqrt{3500} + 0 - 228\right) = 252 \text{ kips/ft}$$

$V_d = 153.6 + 21.1 = 174.7$ kips; $b_w = 14$ in.; $d = 69 - 1.5 - 1.41/2 = 66.795$ in.; $f_{pc} = 676$ psi

Factored total moment:

$$M_{max} = 1.3\ M_D + (1.3)(1.67)\ M_{LL+I}$$
$$= 1.3(0 + -506) + 2.167(-974) = -2769 \text{ kips-ft}$$

Factored total shear:

$$V_i = 1.3 \times (153.6 + 21.1) + 2.167 \times (61.1) = 360 \text{ kips}$$

$$V_{ci} = 0.6\sqrt{5500}\times 14\times 66.795\left(\frac{1}{1000}\right) + 0 + \frac{360\times 251.7}{2769} = 74.3 \text{ kips}$$

$$V_{cw} = \left(3.5\sqrt{f_c'} + 0.3f_{pc}\right)b_w d + V_p = \left(3.5\sqrt{5500} + 0.3\times 676\right)14\times 66.795\left(\frac{1}{1000}\right) + 0 = 432 \text{ kips}$$

$V_c = 74.3$ kips (330.4 kN) (smaller of the V_{cw} and V_{ci})

$$V_s = 2A_v\frac{F_y d_s}{S} = 4\times 0.31\times\frac{60\times 66.695}{6} = 827 \text{ kips (3678.5 kN)}$$

Shear capacity at Bent 2:

$$V_u = \phi\ (V_c + V_s) = 0.85\ (74.3 + 827) = 766 \text{ kips (3408 kN)}$$

7. **Rating Calculations**
As discussed in Section 49.5.4, the rating calculations for load factor method need to be done using strength and serviceability limit states. Serviceability level rating needs not be done at the operating level.
   a. *Rating calculations based on serviceability limit state:*
   Serviceability conditions are listed in AASHTO Design Specification Sections 9.15.1 and 9.15.2.2. These conditions are duplicated in the Rating Manual.
   i. Using the compressive stress under all load combination:

The general expression will be $RF_{INV\text{-}COMALL} = \dfrac{0.6f_c' - f_d - f_p + f_s}{f_l}$

At midspan $RF_{INV\text{-}COMALL} = \dfrac{0.6\times 5500 - (2265 + 70) - (-1615) + 0}{244} = 10.57$

At Bent 2 support $\text{RF}_{\text{INV-COMALL}} = \dfrac{0.6 \times 5500 - (0 + 288) - 680 + 0}{662} = 3.52$

ii. Using the compressive stress of live load, half the prestressing and permanent dead load:

The general expression will be $RF_{INV\text{-}COMLIVE} = \dfrac{0.4 f_c' - f_d - 0.5 f_p + 0.5 f_s}{f_l}$

At midspan $RF_{INV\text{-}COMLIVE} = \dfrac{0.4 \times 5500 - (2265 + 70) - 0.5(-1615) + 0.5(0)}{244} = 2.76$

At Bent 2 support $RF_{INV\text{-}COMLIVE} = \dfrac{0.4 \times 5500 - (0 + 288) - 0.5(680) + 0.5(0)}{662} = 2.37$

iii. Using the allowable tension in concrete:

The general expression will be $RF_{INV\text{-}CONTEN} = \dfrac{6\sqrt{f_c'} - f_d - f_p - f_s}{f_l}$

At midspan $RF_{INV\text{-}CONTEN} = \dfrac{6\sqrt{5500} - (2728 + 110) - (-3443) - 0}{575} = 1.826$

At Bent 2 support $RF_{INV\text{-}CONTEN} = \dfrac{6\sqrt{5500} - (0 + 183) - (-680) - 0}{281} = 3.352$

iv. Using the allowable prestressing steel tension at service level:

The general expression will be $RF_{INV\text{-}PRETEN} = \dfrac{0.8 f_y^* - f_d - f_p - f_s}{f_l}$

At midspan $RF_{INV\text{-}PRETEN} = \dfrac{0.8 \times 270 - 20.99 - (147.1) - 0}{4.59} = 10.43$

At Bent 2 support $RF_{INV\text{-}PRETEN} = \dfrac{0.8 \times 270 - (-3.08) - 167.5 - 0}{1.468} = 30.94$

b. Rating calculations based on strength limit state:

$$\text{The general expression for Rating factor} = \frac{\phi R_n - \gamma_D D}{\gamma_L \beta_L L(1 + I)}$$

According to AASHTO Rating Manual, γ_D is 1.3, γ_L is 1.3, and β_L is 1.67 and 1.0 for inventory and operating factor, respectively. Rating calculations are made and given in Table 49.10.

8. **Summary**
The critical inventory rating of the interior girder is controlled by the tensile stress on concrete at midspan location. The critical operating rating of the girder is controlled by moment at Bent 2 location.

49.6.5 Bridges without Plans

There are some old bridges in service without plans. Establishing safe live-load-carrying capacity is essential to have a complete bridge document. When an inspector comes across a bridge without plans, sufficient field physical dimensions of each member and overall bridge geometry should be taken and recorded. In addition, information such as design year, design vehicle, designer, live-load history, and field condition of the bridge needs to be collected and recorded. This information will be very helpful to determine the safe live-load-carrying capacity. Also, bridge inspectors need to establish the material strength either using the design year or coupon testing.

Design vehicle information could be established based on the designer (state or local agency) and the design year. For example, all state bridges have been designed using the HS20 vehicle since 1944 and all local agency bridges have been designed using the H15 vehicle since 1950.

In steel girder bridges, section properties of the members could be determined based on the field dimensions. During the estimation of the moment capacity, it is recommended to assume that the steel girders are noncomposite with the slab unless substantial evidence is gathered to prove otherwise.

In concrete girder bridges, field dimensions help to estimate the dead loads on the girders. Since the area of reinforcing steels is not known or is difficult to establish, determining the safe live load poses challenges to bridge owners. The live-load history and field condition of a bridge could be used to establish the safe load capacity of the bridge. For example, if a particular bridge has been carrying several heavy vehicles for years without damaging the bridge, this bridge could be left open for all legal vehicles.

49.7 Posting of Bridges

Bridge inspection and the strength evaluation process are two integral parts of bridge posting. The purpose of bridge inspection is to obtain the information that is necessary to evaluate the bridge capacity and the adequacy of the bridge properly. When a bridge is found to have inadequate capacity for legal vehicles, engineers need to look at several alternatives prior to closing the bridge to the public. Some of the possible alternatives are imposing speed limits, reducing vehicular traffic, limiting or posting for vehicle weight, restricting the vehicles to certain lanes, recommending possible small repairs to alleviate the problem. In addition, when the evaluations show that the structure is marginally inadequate, frequent inspections to monitor the physical condition of the bridge and traffic flow may be recommended.

Standard evaluation methods described in the Section 49.5 may be overly conservative. When a more accurate answer is required, a more-detailed analysis, such as three-dimensional analysis or physical load testing can be performed.

The weight and axle configuration of vehicles allowed to use highways without special permits is governed by the statutory law. Thus, the traffic live loads used for posting purposes should be representative of the actual vehicles using the bridge. The representative vehicles vary with each state in the United States. Several states use the three hypothetical legal vehicle configurations given in the Rating Manual [12]. Whereas a few states use their own specially developed legal truck configurations, AASHTO H or HS design trucks, or some combination of truck types. NBIS requires that posting of a bridge must be done when the operating rating for three hypothetical legal vehicles listed in the Rating Manual [12] is less than unity. Furthermore, the NBIS requirement allows the bridge owner to post a bridge for weight limits between inventory and operating level. Because of this flexible NBIS requirement, there is a considerable variation in posting practices among various state and local jurisdictions.

Although engineers may recommend one or a combination of the alternatives described above, it is the owner, not the engineer, who ultimately makes the decision. Many times, bridges are posted for reasons other than structural evaluation, such as posting at a lower weight level to limit vehicular or truck traffic, posting at a higher weight level when the owner believes a lower posting would not be prudent and is willing to accept a higher level of risk. Weight limit posting may cause inconvenience and hardship to the public. In order to reduce inconvenience to the public, the owner needs to look at the weight limit posting as a last resort. In addition, it is sometimes in the public interest to allow certain overweight vehicles such as firefighting equipment and snow removal equipment on a posted bridge. This is usually done through the use of special permits.

References

1. FHWA, Bridge Inspector's Training Manual 70, U.S. Department of Transportation, Washington, D.C., 1970.
2. AASHO, *Manual for Maintenance Inspection of Bridges*, American Association of State Highway Officials, Washington, D.C., 1970.
3. FHWA, Recording and Coding Guide for the Structure Inventory and Appraisal of the Nation's Bridges, U.S. Department of Transportation, Washington, D.C., 1972.
4. FHWA, Bridge Inspector's Manual for Movable Bridges, (Supplement to Manual 70), U.S. Department of Transportation, Washington, D.C., 1970.
5. FHWA, Culvert Inspection Manual, (Supplement to Manual 70), U.S. Department of Transportation, Washington, D.C., 1970.
6. FHWA, Inspection of Fracture Critical Bridge Members, U.S. Department of Transportation, Washington, D.C., 1970.
7. FHWA, Inspection of Fracture Critical Bridge Members, U.S. Department of Transportation, Washington, D.C., 1986.
8. FHWA, Recording and Coding Guide for the Structure Inventory and Appraisal of the Nation's Bridges, U.S. Department of Transportation, Washington, D.C., 1979.
9. FHWA, Recording and Coding Guide for the Structure Inventory and Appraisal of the Nation's Bridges, U.S. Department of Transportation, Washington, D.C., 1988.
10. FHWA, Bridge Inspector's Training Manual 90, U.S. Department of Transportation, Washington, D.C., 1991.
11. FHWA, Hydraulic Engineering Circular (HEC) No.18, Evaluating Scour at Bridges, U.S. Department of Transportation, Washington, D.C., 1990.
12. AASHTO, *Manual for Condition Evaluation of Bridges 1994*, American Association of State Highway and Transportation Officials, Washington, D.C., 1994.
13. AASHTO, *Standard Specification for Highway Bridges*, 16th ed., American Association of State Highway and Transportation Officials, Washington, D.C., 1996.

50

Strengthening and Rehabilitation

F. Wayne Klaiber
Iowa State University

Terry. J. Wipf
Iowa State University

50.1 Introduction

About one half of the approximately 600,000 highway bridges in the United States were built before 1940, and many have not been adequately maintained. Most of these bridges were designed for lower traffic volumes, smaller vehicles, slower speeds, and lighter loads than are common today. In addition, deterioration caused by environmental factors is a growing problem. According to the Federal Highway Administration (FHWA), almost 40% of the nation's bridges are classified as deficient and in need of rehabilitation or replacement. Many of these bridges are deficient because their load-carrying capacity is inadequate for today's traffic. Strengthening can often be used as a cost-effective alternative to replacement or posting.

0-8493-7434-0/00/$0.00+$.50

The live-load capacity of various types of bridges can be increased by using different methods, such as (1) adding members, (2) adding supports, (3) reducing dead load, (4) providing continuity, (5) providing composite action, (6) applying external post-tensioning, (7) increasing member cross section, (8) modifying load paths, and (9) adding lateral supports or stiffeners. Some methods have been widely used, but others are new and have not been fully developed.

All strengthening procedures presented in this chapter apply to the superstructure of bridges. Although bridge span length is not a limiting factor in the various strengthening procedures presented, the majority of the techniques apply to short-span and medium-span bridges. Several of the strengthening techniques, however, are equally effective for long-span bridges. No information is included on the strengthening of existing foundations because such information is dependent on soil type and conditions, type of foundation, and forces involved.

The techniques used for strengthening, stiffening, and repairing bridges tend to be interrelated so that, for example, the stiffening of a structural member of a bridge will normally result in its being strengthened also. To minimize misinterpretation of the meaning of strengthening, stiffening, and repairing, the authors' definitions of these terms are given below. In addition to these terms, definitions of maintenance and rehabilitation, which are sometimes misused, are also given.

Maintenance: The technical aspect of the upkeep of the bridges; it is preventative in nature. Maintenance is the work required to keep a bridge in its present condition and to control potential future deterioration.

Rehabilitation: The process of restoring the bridge to its original service level.

Repair: The technical aspect of rehabilitation; action taken to correct damage or deterioration on a structure or element to restore it to its original condition.

Stiffening: Any technique that improves the in-service performance of an existing structure and thereby eliminates inadequacies in serviceability (such as excessive deflections, excessive cracking, or unacceptable vibrations).

Strengthening: The increase of the load-carrying capacity of an existing structure by providing the structure with a service level higher than the structure originally had (sometimes referred to as upgrading).

In recent years the FHWA and National Cooperative Highway Research Program (NCHRP) have sponsored several studies on bridge repair, rehabilitation, and retrofitting. Inasmuch as some of these procedures also increase the strength of a given bridge, the final reports on these investigations are excellent references. These references, plus the strengthening guidelines presented in this chapter, will provide information an engineer can use to resolve the majority of bridge strengthening problems. The FHWA and NCHRP final reports related to this investigation are References [1–13].

Four of these references, [1,2,11,12] are of specific interest in strengthening work. Although not discussed in this chapter, the live-load capacity of a given bridge can often be evaluated more accurately by using more-refined analysis procedures. If normal analytical methods indicate strengthening is required, frequently more-sophisticated analytical methods (such as finite-element analysis) may result in increased live-load capacities and thus eliminate the need to strengthen or significantly decrease the amount of strengthening required.

By load testing bridges, one frequently determines live-load capacities considerably larger than what one would determine using analytical procedures. Load testing of bridges makes it possible to take into account several contributions (such as end restraint in simple spans, structural contributions of guardrails, etc.) that cannot be included analytically. In the past few years, several states have started using load testing to establish live-load capacities of their bridges. An excellent reference on this procedure is the final report for NCHRP Project 12-28(13)A [14]. Most U.S. states have some type of bridge management system (BMS). To the authors' knowledge, very few states are using their BMS to make bridge strengthening decisions. At the present time, there are not sufficient base line data (first cost, life cycle costs, cost of various strengthening procedures, etc.) to make strengthening/replacement decisions.

Examination of National Bridge Inventory (NBI) bridge records indicates that the bridge types with greatest potential for strengthening are steel stringer, timber stringer, and steel through-truss. If rehabilitation and strengthening cannot be used to extend their useful lives, many of these bridges will require replacement in the near future. Other bridge types for which there also is potential for strengthening are concrete slab, concrete T, concrete stringer, steel girder floor beam, and concrete deck arch. In this chapter, information is provided on the more commonly used strengthening procedures as well as a few of the new procedures that are currently being researched.

50.2 Lightweight Decks

50.2.1 Introduction

One of the more fundamental approaches to increase the live-load capacity of a bridge is to reduce its dead load. Significant reductions in dead load can be obtained by removing an existing heavier concrete deck and replacing it with a lighter-weight deck. In some cases, further reduction in dead load can be obtained by replacing the existing guardrail system with a lighter-weight guardrail. The concept of strengthening by dead-load reduction has been used primarily on steel structures, including the following types of bridges: steel stringer and multibeam, steel girder and floor beam, steel truss, steel arch, and steel suspension bridges; however, this technique could also be used on bridges constructed of other materials.

Lightweight deck replacement is a feasible strengthening technique for bridges with structurally inadequate, but sound, steel stringers or floor beams. If, however, the existing deck is not in need of replacement or extensive repair, lightweight deck replacement would not be economically feasible.

Lightweight deck replacement can be used conveniently in conjunction with other strengthening techniques. After an existing deck has been removed, structural members can readily be strengthened, added, or replaced. Composite action, which is possible with some lightweight deck types, can further increase the live-load carrying capacity of a deficient bridge.

50.2.2 Types

Steel grid deck is a lightweight flooring system manufactured by several firms. It consists of fabricated, steel grid panels that are field-welded or bolted to the bridge superstructure. The steel grids may be filled with concrete, partially filled with concrete, or left open (Figure 50.1).

Open-Grid Steel Decks

Open-grid steel decks are lightweight, typically weighing 15 to 25 psf (720 to 1200 Pa) for spans up to 5 ft (1.52 m). Heavier decks, capable of spanning up to 9 ft (2.74 m), are also available; the percent increase in live-load capacity is maximized with the use of an open-grid steel deck. Rapid installation is possible with the prefabricated panels of steel grid deck. Open-grid steel decks also have the advantage of allowing snow, water, and dirt to wash through the bridge deck, thus eliminating the need for special drainage systems.

A disadvantage of the open grids is that they leave the superstructure exposed to weather and corrosive chemicals. The deck must be designed so water and debris do not become trapped in the grids that rest on the stringers. Other problems associated with open-steel grid decks include weld failure and poor skid resistance. Weld failures between the primary bearing bars of the deck and the supporting structure have caused maintenance problems with some open-grid decks. The number of weld failures can be minimized if the deck is properly erected.

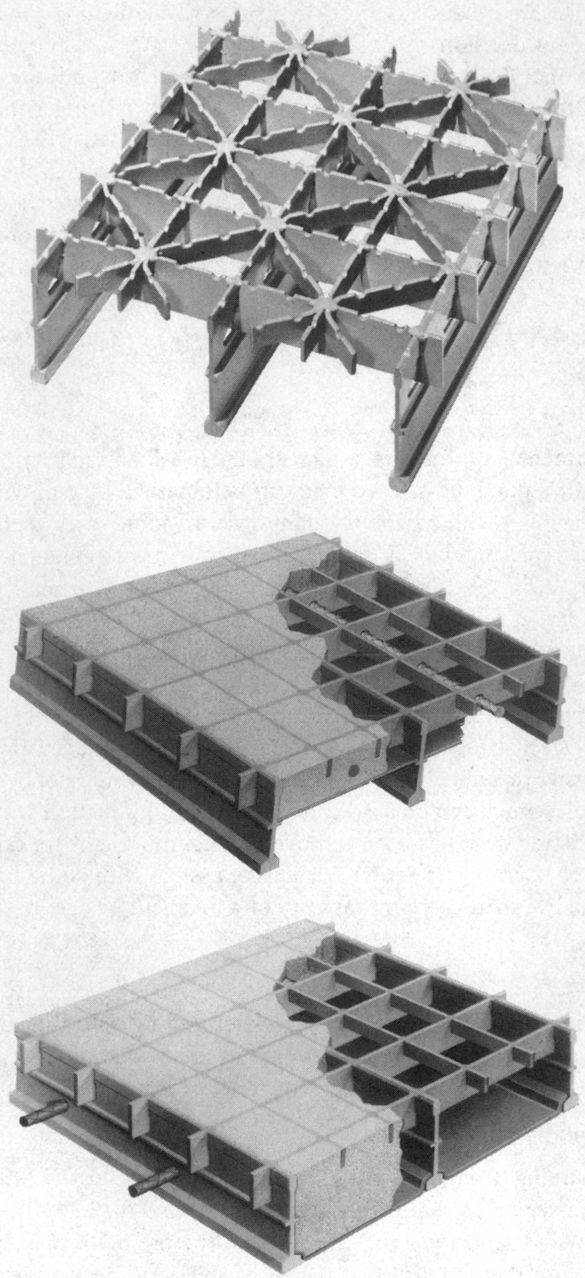

FIGURE 50.1 Steel-grid bridge deck. Top photo shows open steel grid deck; center photo shows half-filled steel grid deck; bottom photo shows filled steel grid deck. (*Source:* Klaiber, F.W. et al., NCHRP 293, Transportation Research Board, 1987. With permission.)

In an effort to improve skid resistance, most open-grid decks currently on the market have serrated or notched bars at the traffic surface. Small studs welded to the surface of the steel grids have also been used to improve skid resistance. While these features have improved skid resistance, they have not eliminated the problem entirely [12]. Open-grid decks are often not perceived favorably by the general public because of the poor riding quality and increased tire noise.

Concrete-Filled Steel Grid Decks

Concrete-filled steel grid decks weigh substantially more, but have several advantages over the open-grid steel decks, including increased strength, improved skid resistance, and better riding quality. The steel grids can be either half or completely filled with concrete. A 5-in. (130-mm)-thick, half-filled steel grid weighs 46 to 51 psf (2.20 to 2.44 kPa), less than half the weight of a reinforced concrete deck of comparable strength. Typical weights for 5-in. (130-mm) thick steel grid decks, filled to full depth with concrete, range from 76 to 81 psf (3.64 to 3.88 kPa). Reduction in the deadweight resulting from concrete-filled steel grid deck replacement alone only slightly improves the live-load capacity; however, the capacity can be further improved by providing composite action between the deck and stringers.

Steel grid panels that are filled or half-filled with concrete may either be precast prior to erection or filled with concrete after placement. With the precast system, only the grids that have been left open to allow field welding of the panels must be filled with concrete after installation. The precast system is generally used when erection time must be minimized.

A problem that has been associated with concrete-filled steel grid decks, addressed in a study by Timmer [15], is the phenomenon referred to as deck growth — the increase in length of the filled grid deck caused by the rusting of the steel I-bar webs. The increase in thickness of the webs due to rusting results in comprehensive stresses in the concrete fill. Timmer noted that in the early stages of deck growth, a point is reached when the compression of the concrete fill closes voids and capillaries in the concrete. Because of this action, the amount of moisture that reaches the resting surfaces is reduced and deck growth is often slowed down or even halted. If, however, the deck growth continues beyond this stage, it can lead to breakup of the concrete fill, damage to the steel grid deck, and possibly even damage to the bridge superstructure and substructure. Timmer's findings indicate that the condition of decks that had been covered with some type of wearing surface was superior to those that had been left unsurfaced. A wearing surface is also recommended to prevent wearing and eventual cupping of the concrete between the grids.

Exodermic Deck

Exodermic deck is a recently developed, prefabricated modular deck system that has been marketed by major steel grid deck manufacturers. The first application of Exodermic deck was in 1984 on the Driscoll Bridge located in New Jersey [16]. As shown in Figure 50.2, the bridge deck system consists of a thin upper layer, 3 in. (76 mm) minimum, of prefabricated concrete joined to a lower layer of steel grating. The deck weighs from 40 to 60 psf (1.92 to 2.87 kPa) and is capable of spanning up to 16 ft (4.88 m).

Exodermic decks have not exhibited the fatigue problems associated with open-grid decks or the growth problems associated with concrete-filled grid decks. As can be seen in Figure 50.2, there is no concrete fill and thus no grid corrosion forces. This fact, coupled with the location of the neutral axis, minimizes the stress at the top surface of the grid.

Exodermic deck and half-filled steel grid deck have the highest percent increase in live-load capacity among the lightweight deck types with a concrete surface. As a prefabricated modular deck system, Exodermic deck can be quickly installed. Because the panels are fabricated in a controlled environment, quality control is easier to maintain and panel fabrication is independent of the weather or season.

Laminated Timber Deck

Laminated timber decks consist of vertically laminated 2-in. (51-mm) (nominal) dimension lumber. The laminates are bonded together with a structural adhesive to form panels that are approximately 48 in. (1.22 m) wide. The panels are typically oriented transverse to the supporting structure of the bridge (Figure 50.3). In the field, adjacent panels are secured to each other with steel dowels or stiffener beams to allow for load transfer and to provide continuity between the panels.

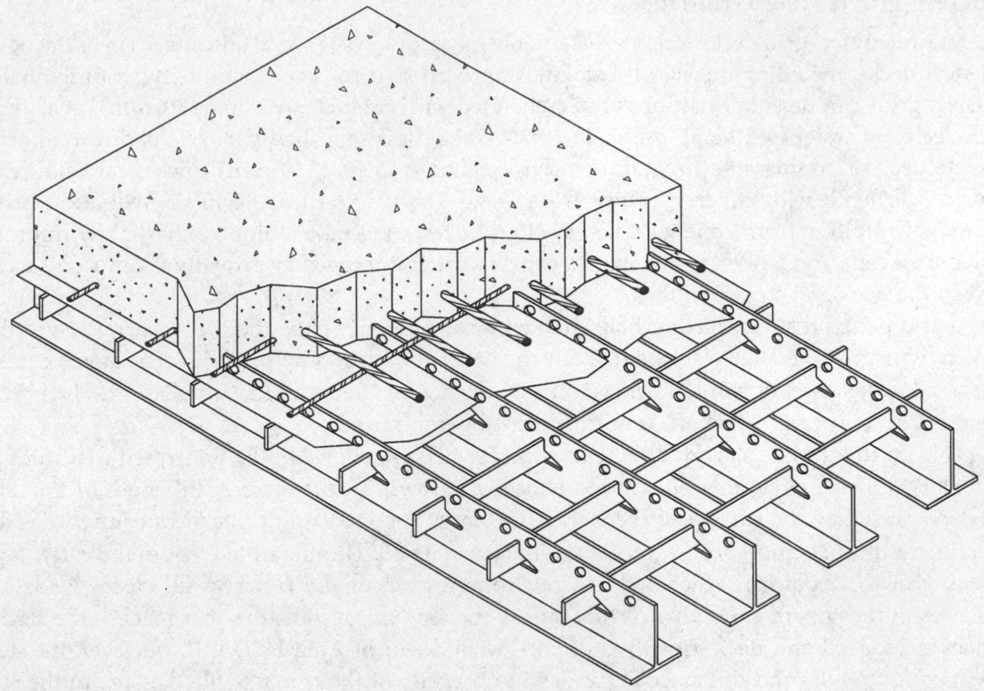

FIGURE 50.2 Exodermic deck system. (*Source:* Exodermic Bridge Deck Inc., Lakeville, CT, 1999. With permission.)

A steel–wood composite deck for longitudinally oriented laminates has been developed by Bakht and Tharmabala [17]. Individual laminates are transversely post-tensioned in the manner developed by Csagoly and Taylor [18]. The use of shear connectors provides partial composite action between the deck and stringers. Because the deck is placed longitudinally, diaphragms mounted flush with the stringers may be required for support. Design of this type of timber deck is presented in References [19–21].

The laminated timber decks used for lightweight deck replacement typically range in depth from 3⅛ to 6¾ in. (79 to 171 mm) and from 10.4 to 22.5 psf (500 to 1075 Pa) in weight. A bituminous wearing surface is recommended.

Wood is a replenishable resource that offers several advantages: ease of fabrication and erection, high strength-to-weight ratio, and immunity to deicing chemicals. With the proper treatment, heavy timber members also have excellent thermal insulation and fire resistance [22]. The most common problem associated with wood as a structural material is its susceptibility to decay caused by living fungi, wood-boring insects, and marine organisms. With the use of modern preservative pressure treatments, however, the expected service life of timber decks can be extended to 50 years or more.

Lightweight Concrete Deck

Structural lightweight concrete, concrete with a unit weight of 115 pcf (1840 kg/m^3) or less, can be used to strengthen steel bridges that have normal-weight, noncomposite concrete decks. Special design considerations are necessary for lightweight concrete. Its modulus of elasticity and shear strength are less than that of normal-weight concrete, whereas its creep effects are greater [23]. The durability of lightweight concrete has been a problem in some applications.

Lightweight concrete for deck replacement can be either cast in place or installed in the form of precast panels. A cast-in-place lightweight concrete deck can easily be made to act compositely with the stringers. The main disadvantage of a cast-in-place concrete deck is the length of time required for concrete placement and curing.

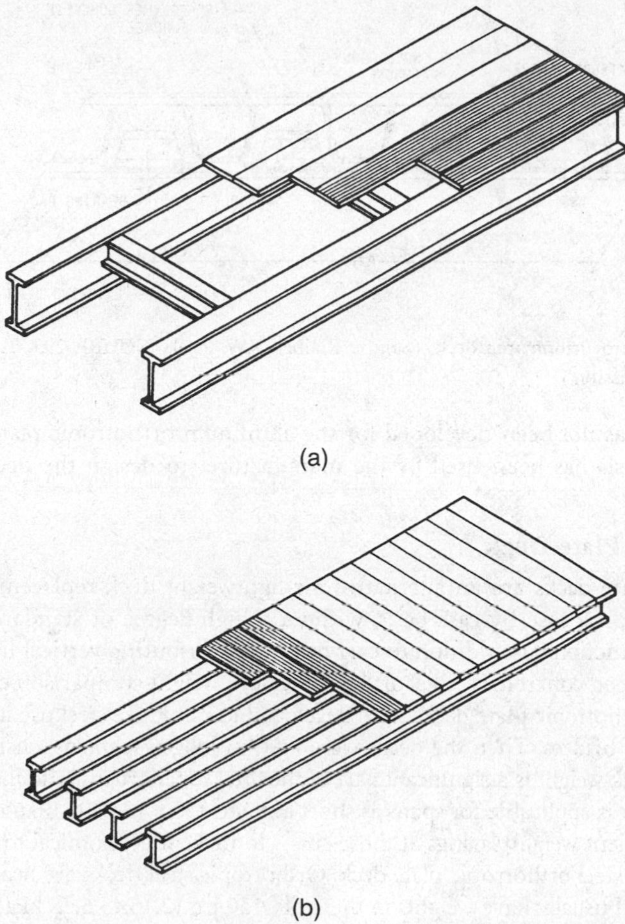

FIGURE 50.3 Laminated timber deck. (a) Longitudenal orietation; (b) transverse orientation. (*Source:* Klaiber, F.W. et al., NCHRP 293, Transportation Research Board, 1987. With permission.)

Lightweight precast panels, fabricated with either mild steel reinforcement or transverse prestressing, have been used in deck replacement projects to help minimize erection time and resulting interruptions to traffic. Precast panels require careful installation to prevent water leakage and cracking at the panel joints. Composite action can be attained between the deck and the superstructure; however, some designers have chosen not to rely on composite action when designing a precast deck system.

Aluminum Orthotropic Plate Deck

Aluminum orthotropic deck is a structurally strong, lightweight deck weighing from 20 to 25 psf (958 to 1197 Pa). A proprietary aluminum orthotropic deck system that is currently being marketed is shown in Figure 50.4. The deck is fabricated from highly corrosion-resistant aluminum alloy plates and extrusions that are shop-coated with a durable, skid-resistant, polymer wearing surface. Panel attachments between the deck and stringer must not only resist the upward forces on the panels, but also allow for the differing thermal movements of the aluminum and steel superstructure. For design purposes, the manufacturer's recommended connection should not be considered to provide composite action.

The aluminum orthotropic plate is comparable in weight to the open-grid steel deck. The aluminum system, however, eliminates some of the disadvantages associated with open grids: poor ridability and acoustics, weld failures, and corrosion caused by through drainage. A wheel-load

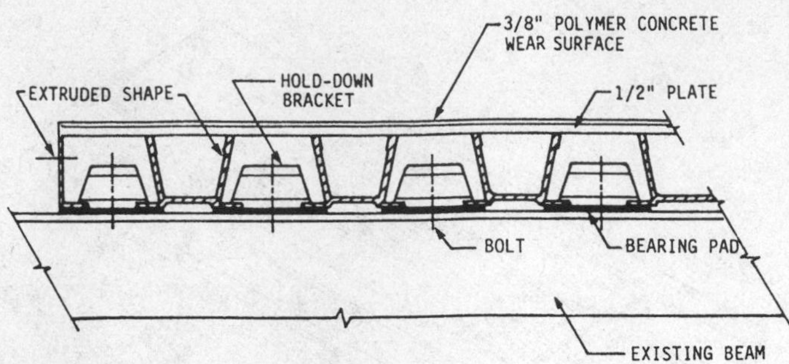

FIGURE 50.4 Aluminum orthotropic deck. (*Source:* Klaiber, F.W. et al., NCHRP 293, Transportation Research Board, 1987. With permission.)

distribution factor has not been developed for the aluminum orthotropic plate deck at this time. Finite-element analysis has been used by the manufacturer to design the deck on a project-by-project basis.

Steel Orthotropic Plate Deck

Steel orthotropic plate decks are an alternative for lightweight deck replacements, that generally have been designed on a case-by-case basis, without a high degree of standardization. The decks often serve several functions in addition to carrying and distributing vertical live loads and, therefore, a simple reinforced concrete vs. steel orthotropic deck weight comparison could be misleading.

Originally, steel orthotropic plate decks were developed to minimize steel use in 200- to 300-ft (61- to 91-m) span girder bridges. Then the decks were used in longer-span suspension and cable-stayed bridges where the deck weight is a significant part of the total superstructure design load. Although the steel orthotropic deck is applicable for spans as short as 80 to 120 ft (24.4 to 36.6 m), it is unlikely that there would be sufficient weight savings at those spans to make it economical to replace a reinforced concrete deck with a steel orthotropic plate deck. Orthotropic steel decks are heavier than aluminum orthotropic decks and usually have weights in the 45 to 130 psf (2.15 to 6.22 kPa) range.

50.2.3 Case Studies

Steel Grid Deck

The West Virginia Department of Highways was one of the first to develop a statewide bridge rehabilitation plan using open-grid steel deck [24]. By 1974, 25 bridges had been renovated to meet or exceed AASHTO requirements. Deteriorated concrete decks were replaced with lightweight, honeycombed steel grid decks fabricated from ASTM A588 steel. The new bridge floors are expected to have a 50-year life and to require minimal maintenance.

In 1981, the West Virginia Department of Highways increased the live-load limit on a 1794-ft (546.8-m)-long bridge over the Ohio River from 3 tons (26.69 kN) to 13 tons (115.65 kN) by replacing the existing reinforced concrete deck with an open steel grid deck [25, 26]. The existing deck was removed and the new deck installed in sections allowing half of the bridge to be left open for use by workers, construction vehicles, and equipment, and, if needed, emergency vehicles.

The strengthening of the 250-ft (76.2-m)-long Old York Road Bridge in New Jersey in the early 1980s combined deck replacement with the replacement of all of the main framing members and the modernization of the piers and abutments [27]. The existing deck was replaced with an ASTM A588 open-grid steel deck. The posted 10-ton (89-kN) load limit was increased to 36 tons (320 kN) and the bridge was widened from 18 ft (5.49 m) to 26 ft (7.92 m).

Exodermic Deck

The first installation of Exodermic deck was in 1984 on the 4400-ft (1340-m)-long Driscoll Bridge located in New Jersey [16]. The deck, weighing 53 psf (2.54 kPa), consisted of a 3-in. (76-mm) upper layer of prefabricated reinforced concrete joined to a lower layer of steel grating. Approximately 30,000 ft^2 (2790 m^2) of deck was replaced at this site.

Exodermic deck was also specified for the deck replacement on a four-span bridge which overpasses the New York State Thruway [28]. The bridge was closed to traffic during deck removal and replacement. Once the existing deck has been removed, it is estimated that approximately 7500 ft^2 (697 m^2) of Exodermic deck will be installed in 3 working days.

Lightweight Concrete Deck

Lightweight concrete was used as early as 1922 for new bridge construction in the United States. Over the years, concrete made with good lightweight aggregate has generally performed satisfactorily; however, some problems related to the durability of the concrete have been experienced. The Louisiana Department of Transportation has experienced several deck failures on bridges built with lightweight concrete in the late 1950s and early 1960s. The deck failures have typically occurred on bridges with high traffic counts and have been characterized by sudden and unexpected collapse of sections of the deck.

Lightweight concrete decks can either be cast in place or factory precast. Examples of the use of lightweight concrete for deck replacement follow.

Cast-in-Place Concrete

New York state authorities used lightweight concrete to replace the deck on the north span of the Newburgh–Beacon Bridge [8, 29]. The existing deck was replaced with 6½ in. (165 mm) of cast-in-place lightweight concrete that was surfaced with a 1½ in. (38 mm) layer of latex modified concrete. Use of the lightweight concrete allowed the bridge to be widened from two to three lanes with minimal modifications to the substructure. A significant reduction in the cost of widening the northbound bridge was attributed to the reduction in dead load.

Precast Concrete Panels

Precast modular-deck construction has been used successfully since 1967 when a joint study, conducted by Purdue University and Indiana State Highway Commission, found precast, prestressed deck elements to be economically and structurally feasible for bridge deck replacement [30, 31].

Precast panels, made of lightweight concrete, 115 pcf (1840 kg/m^3), were used to replace and widen the existing concrete deck on the Woodrow Wilson Bridge, located on Interstate 95 south of Washington, D.C. [32,33]. The precast panels were transversely prestressed and longitudinally post-tensioned. Special sliding steel-bearing plates were used between the panels and the structural steel to prevent the introduction of unwanted stresses in the superstructure. The Maryland State Highway Commission required that all six lanes of traffic be maintained during the peak traffic hours of the morning and evening. Two-way traffic was maintained at night when the removal and replacement of the deck was accomplished.

Aluminum Orthotropic Plate Deck

The 104-year-old Smithfield Street Bridge in Pittsburgh, Pennsylvania has undergone two lightweight deck replacements, both involving aluminum deck [34]. The first deck replacement occurred in 1933 when the original heavyweight deck was replaced with an aluminum deck and floor framing system. The aluminum deck was coated with a 1½-in. (38-mm) asphaltic cement wearing surface. The new deck, weighing 30 psf (1.44 kPa), eliminated 751 tons (6680 kN) of deadweight and increased the live-load capacity from 5 tons (44.5 kN) to 20 tons (178 kN).

Excessive corrosion of some of the deck panels and framing members necessitated the replacement of the aluminum deck on the Smithfield Street Bridge in 1967. At that time, a new aluminum orthotropic plate deck with a ⅜-in. (9.5-mm)-thick polymer concrete wearing surface was installed. This new deck weighed 15 psf (718 Pa) and resulted in an additional 108-ton (960-kN) reduction in deadweight. The panels were originally attached to the structure with anodized aluminum bolts, but the bolts were later replaced with galvanized steel bolts after loosening and fracturing of the aluminum bolts became a problem. The aluminum components of the deck have shown no significant corrosion; however, because of excessive wear, the wearing surface had to be replaced in the mid-1970s. The new wearing surface consisted of aluminum-expanded mesh filled with epoxy resin concrete. This wearing surface has also experienced excessive wear, and thus early replacement is anticipated.

Steel Orthotropic Plate Deck

Steel orthotropic plate decks were first conceived in the 1930s for movable bridges and were termed battledecks. Steel orthotropic decks were rapidly developed in the late 1940s in West Germany for replacement of bridges destroyed in World War II during a time when steel was in short supply, and replacement of bridge decks with steel orthotropic plate decks became a means for increasing the live-load capacity of medium- to long-span bridges in West Germany in the 1950s.

In 1956 Woelting and Bock [35] reported the rebuilding of a wrought iron, 536-ft (163-m) span bridge near Kiel. The two-hinged, deck arch bridge, which carried both rail and highway traffic, was widened and strengthened through rebuilding essentially all of the bridge except the arches and abutments. The replacement steel orthotropic deck removed approximately 190 tons of dead load from the bridge, improved the deck live-load capacity, and was constructed in such a way as to replace the original lateral wind bracing truss.

The live-load class of a bridge near Darmstadt was raised by means of a replacement steel orthotropic deck also in the mid-1970s [36]. The three-span, steel-through-truss bridge had been repaired and altered twice since World War II, but the deck had finally deteriorated to the point where it required replacement. The existing reinforced concrete deck was then replaced with a steel orthotropic plate deck, and the reduction in weight permitted the bridge to be reclassified for heavier truck loads.

50.3 Composite Action

50.3.1 Introduction

Modification of an existing stringer and deck system to a composite system is a common method of increasing the flexural strength of a bridge. The composite action of the stringer and deck not only reduces the live-load stresses but also reduces undesirable deflections and vibrations as a result of the increase in the flexural stiffness from the stringer and deck acting together. This procedure can also be used on bridges that only have partial composite action, because the shear connectors originally provided are inadequate to support today's live loads.

The composite action is provided through suitable shear connection between the stringers and the roadway deck. Although numerous devices have been used to provide the required horizontal shear resistance, the most common connection used today is the welded stud.

50.3.2 Applicability and Advantages

Inasmuch as the modifications required for providing composite action for continuous spans and simple spans are essentially the same, this section is written for simple spans. Composite action can effectively be developed between steel stringers and various deck materials, such as normal-weight reinforced concrete (precast or cast in place), lightweight reinforced concrete (precast or cast in place), laminated timber, and concrete-filled steel grids. These are the most common materials used in composite decks; however, there are some instances in which steel deck plates have been made

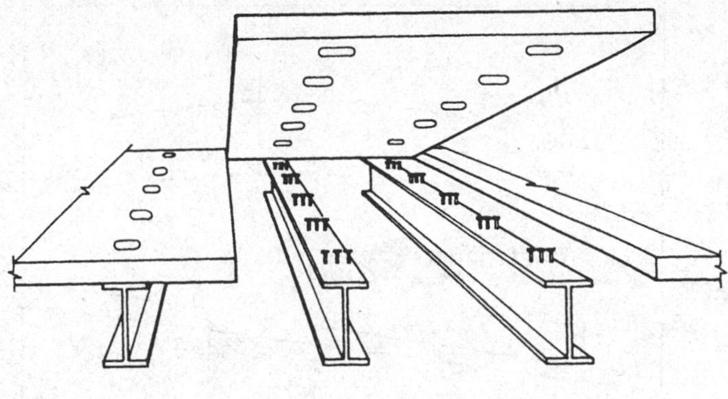

NOTE: SHEAR STUDS SHOWN ARE
ACTUALLY ADDED AFTER
PRECAST DECK IS POSITIONED.

FIGURE 50.5 Precast deck with holes. (*Source:* Klaiber, F.W. et al., NCHRP 293, Transportation Research Board, 1987. With permission.)

composite with steel stringers. In the following paragraphs these four common deck materials will be discussed individually.

Because steel stringers are normally used for support of all the mentioned decks, they are the only type of superstructure reviewed. The condition of the deck determines how one can obtain composite action between the stringers and an existing concrete deck. If the deck is badly deteriorated, composite action is obtained by removing the existing deck, adding appropriate shear connectors to the stringers, and recasting the deck. This was done in Blue Island, Illinois, on the 1500-ft (457-m)-long steel plate girder Burr Oak Avenue Viaduct [37].

If it is desired to reduce interruption of traffic, precast concrete panels are one of the better solutions. The panels are made composite by positioning holes formed in the precast concrete directly over the structural steel. Welded studs are then attached through the preformed holes. This procedure was used on an I-80 freeway overpass near Oakland, California [38]. As shown in Figure 50.5, panels 30 ft (9.1 m) to 40 ft (12.2 m) long, with oblong holes 12 in. (305 mm) × 4 in. (100 mm) were used to replace the existing deck. Four studs were welded to the girders through each hole. Composite action was obtained by filling the holes, as well as the gaps between the panels and steel stringers, with fast-curing concrete.

If the concrete deck does not need replacing, composite action can be obtained by coring through the existing concrete deck to the steel superstructure. Appropriate shear connectors are placed in the holes; the desired composite action is then obtained by filling the holes with nonshrink grout. This procedure was used in the reconstruction of the Pulaski Skyway near the Holland Tunnel linking New Jersey and New York [38]. After removing an asphalt overlay and some of the old concrete, the previously described procedure with welded studs placed in the holes was used. The holes were then grouted and the bridge resurfaced with latex-modified concrete.

Structural lightweight concrete has been used in both precast panels and in cast-in-place bridge decks. Comments made on normal-weight concrete in the preceding paragraphs essentially apply to lightweight concrete also. However, since the shear strength, fatigue strength, and modulus of elasticity of lightweight concrete are less than that of normal-weight concrete, these lesser values must be taken into account in design.

The advantages of composite action can be seen in Figure 50.6. Shown in this graph is the decrease in the top flange stress as a result of providing composite action on a simply supported single-span bridge with steel stringers and an 8-in. concrete deck. As may be seen in this figure, two stringer spacings, 6 ft (1.8 m) and 8 ft (2.4 m) are held constant, while the span length was varied from 20 ft (6.1 m) to 70 ft (21.3 m). These stresses are based on the maximum moment that results from either

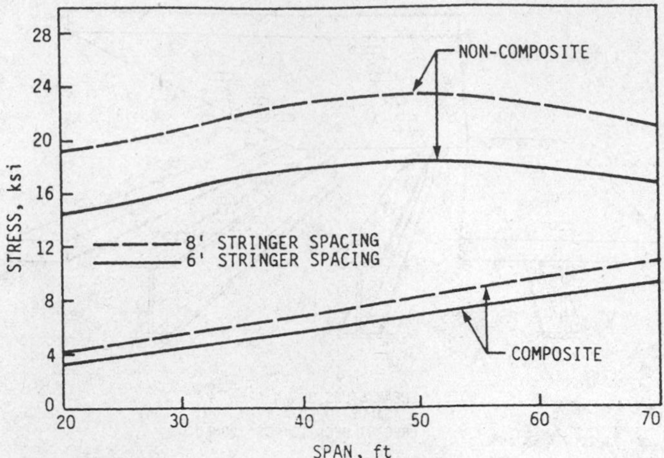

FIGURE 50.6 Stress in top flange of stringer, composite action vs. noncomposite action. (*Source:* Klaiber, F.W. et al., NCHRP 293, Transportation Research Board, 1987. With permission.)

the standard truck loading (HS20-44) or the standard lane loading, whichever governs. Concrete stresses were considerably below the allowable stress limit; composite action reduced the stress in the bottom flange 15 to 30% for long and short spans, respectively. As may be seen in Figure 50.6 for a 40-ft (12.2-m) span with 8-ft (2.4-m) stringer spacing, composite action will reduce the stress in the top flange 68%, 22 ksi (152 Mpa) to 7 ksi (48 MPa). Composite action is slightly more beneficial in short spans than in long spans, and the larger the stringer spacings, the more stress reduction when composite action is added. Results for other types of deck are similar but will depend on the type and size of deck, amount of composite action obtained, type of support system, and the like.

50.3.3 Types of Shear Connectors

As previously mentioned, in order to create composite action between the steel stringers and the bridge deck some type of shear connector is required. In the past, several different types of shear connectors were used in the field; these connectors can be seen in Figure 50.7. Of these, because of the advancements and ease in application, welded studs have become the most commonly used shear connector today. In the strengthening of an existing bridge, frequently one of the older types of shear connectors will be encountered. A strength evaluation must be undertaken to ensure that the shear connectors present are adequate. The following references can be used to obtain the ultimate strength of various types of shear connectors. A method for calculating the strength of a flat bar can be found in Cook [39]; also, work done by Klaiber et al. [40] can be used in evaluating the strength of stiffened angles. Older AASHTO standard specifications can be used to obtain ultimate strength of shear connectors; for example, values for spirals can be found in the AASHTO standard specifications from 1957 to 1968. The current AASHTO specifications only give ultimate-strength equations for welded studs and channels; thus, if shear connectors other than these two are encountered, the previously mentioned references should be consulted.

The procedure employed for using high-strength bolts as shear connectors (Figure 50.8) is very similar to that used for utilizing welded studs in existing concrete, except for the required holes in the steel stringer. To minimize slip, the hole in the steel stringer is made the same size as the diameter of the bolt. Dedic and Klaiber [41] and Dallam [42,43] have shown that the strength and stiffness of high-strength bolts are essentially the same as those of welded shear studs. Thus, existing AASHTO ultimate-strength formulas for welded stud connectors can be used to estimate the ultimate capacity of high-strength bolts.

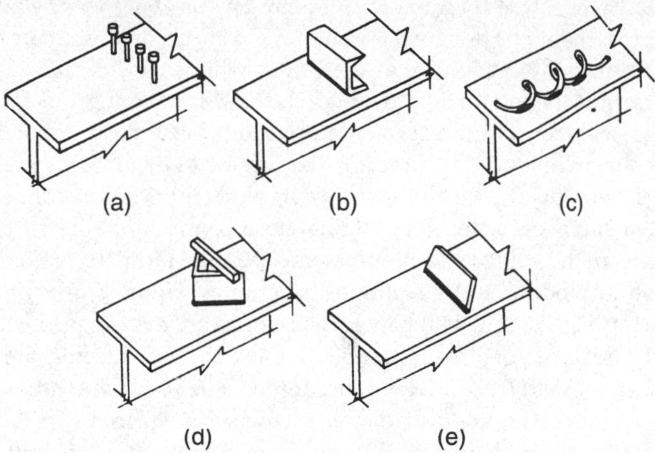

FIGURE 50.7 Common shear connectors. (a) Welded studs; (b) channel; (c) spiral; (d) stiffened angle; (e) inclined flat bar. (*Source:* Klaiber, F.W. et al., NCHRP 293, Transportation Research Board, 1987. With permission.)

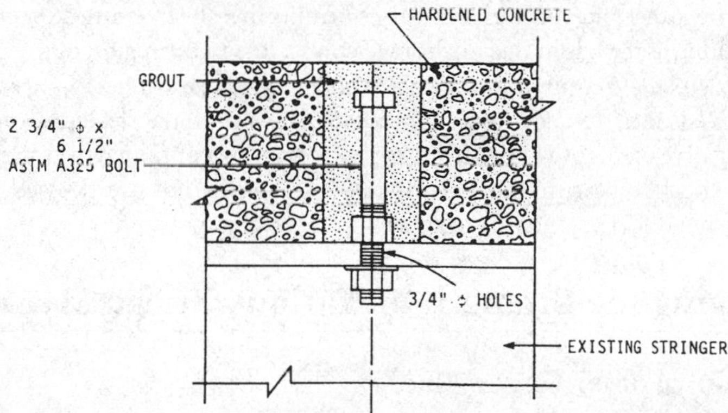

FIGURE 50.8 Details of double-nutted high-strength bolt shear connector. (*Source:* Klaiber, F.W. et al., NCHRP 293, Transportation Research Board, 1987. With permission.)

50.3.4 Design Considerations

The means of obtaining composite action will depend on the individual bridge deck. If the deck is in poor condition and needs to be replaced, the following variables should be considered: (1) weldability of steel stringers, (2) type of shear connector, and (3) precast vs. cast in place.

To determine the weldability of the shear connector, the type of steel in the stringers must be known. If the type of steel is unknown, coupons may be taken from the stringers to determine their weldability. If it is found that welding is not possible, essentially the only alternative for shear connection is high-strength bolts. Although the procedure is rarely done, bolts could be used to attach channels to the stringers for shear connection. When welding is feasible, either welded studs or channels can be used. Because of the ease of application of the welded studs, channels are rarely used today. In older constructions where steel cover plates were riveted to the beam flanges, an option that may be available is to remove the rivets connecting the top cover plate to the top flange of the beam and replace the rivets with high-strength bolts in a manner similar to that which is shown in Figure 50.8.

According to the current AASHTO manual, *Standard Specifications for Highway Bridges,* in new bridges shear connectors should be designed for fatigue and checked for ultimate strength. However, in older bridges, the remaining fatigue life of the bridge will be considerably less than that of the new shear connectors; thus, one only needs to design the new shear connectors for ultimate strength. If an existing bridge with composite action requires additional shear connectors, the ultimate strength capacity of the original shear connector (connector #1) and new shear connectors (connectors #2) can be simply added even though they are different types of connectors. Variation in the stiffness of the new shear connectors and original shear connectors will have essentially no effect on the elastic behavior of the bridge and nominal effect on the ultimate strength [44].

The most common method of creating composite action when one works with precast concrete decks is to preform slots in the individual panels. These slots are then aligned with the stringers for later placement of shear connectors (see Figure 50.5). Once shear connectors are in place, the holes are filled with nonshrink concrete. A similar procedure can be used with laminated timber except the holes for the shear connectors are drilled after the panels are placed.

When it is necessary to strengthen a continuous span, composite action can still be employed. One common approach is for the positive moment region to be designed using the same procedure as that for simple span bridges. When designing the negative moment region, the engineer has two alternatives. The engineer can continue the shear connectors over the negative moment region, in which case the longitudinal steel can be used in computing section properties in the negative moment region. The other alternative is to discontinue the shear connectors over the negative moment region. As long as the additional anchorage connectors in the region of the point of dead-load contraflexure are provided, as required by the code, continuous shear connectors are not needed. When this second alternative is used, the engineer cannot use the longitudinal steel in computing the section properties in the negative moment region. If shear connectors are continued over the negative moment region, one should check to be sure that the longitudinal steel is not overstressed. Designers should consult the pertinent AASHTO standards to meet current design guidelines.

50.4 Improving the Strength of Various Bridge Members

50.4.1 Addition of Steel Cover Plates

Steel Stringer Bridges

Description

One of the most common procedures used to strengthen existing bridges is the addition of steel cover plates to existing members. Steel cover plates, angles, or other sections may be attached to the beams by means of bolts or welds. The additional steel is normally attached to the flanges of existing sections as a means of increasing the section modulus, thereby increasing the flexural capacity of the member. In most cases the member is jacked up during the strengthening process, relieving dead-load stresses on the existing member. The new cover plate section is then able to accept both live-load and dead-load stresses when the jacks are removed, which ensures that less steel will be required in the cover plates. If the bridge is not jacked up, the cover plate will carry only live-load stresses, and more steel will be required.

Applicability, Advantages, and Disadvantages

The techniques described in this section are widely applicable to steel members whose flexural capacity is inadequate. Members in this category include steel stringers (both composite and noncomposite), floor beams, and girders on simply supported or continuous bridges. Note, however, that cover plating is most effective on composite members.

There are a number of advantages to using steel cover plates as a method of strengthening existing bridges. This method can be quickly installed and requires little special equipment and minimal labor and materials. If bottom flange stresses control the design, cover plating is effective even if the deck is not replaced. In this case, it is more effective when applied to noncomposite construction. In addition, design procedures are straightforward and thus require minimal time to complete.

In certain instances these advantages may be offset by the costly problems of traffic control and jacking of the bridge. As a minimum, the bridge may have to be closed or separate traffic lanes established to relieve any stresses on the bridge during strengthening. In addition, significant problems may develop if part of the slab must be removed in order to add cover plates to the top of the beams. When cover plates are attached to the bottom flange, the plates should be checked for underclearance if the situation requires it. Still another potential problem if welding is used is that the existing members may not be compatible with current welding materials.

The most commonly reported problem encountered with the addition of steel cover plates is fatigue cracking at the top of the welds at the ends of the cover plates. In a study by Wattar et al. [45], it was suggested that bolting be used at the cover plate ends. Tests showed that bolting the ends raises the fatigue category of the member from stress Category E to B and also results in material savings by allowing the plates to be cut off at the theoretical cutoff points.

Another method for strengthening this detail is to grind the transverse weld to a 1:3 taper [46]. This is a practice of the Maryland State Highway Department. Using an air hammer to peen the toe of the weld and introduce compressive residual stresses is also effective in strengthening the connection [46]. The fatigue strength can be improved from stress category E to D by using this technique. Either solution has been shown to reduce significantly the problem of fatigue cracking at the cover plate ends.

Materials other than flange cover plates may be added to stringer flanges for strengthening. For example, the Iowa Department of Transportation prefers to attach angles to the webs of steel I-beam bridges (either simply supported or continuous spans) with high-strength bolts as a means to reduce flexural live load stresses in the beams. Figure 50.9 shows a project completed by the Iowa Department of Transportation involving the addition of angles to steel I-beams using high-strength bolts. In some instances the angles are attached only near the bottom flange. Normally, the bridge is not jacked up during strengthening, and only the live loads are removed from the particular I-beam being strengthened. Because the angles are bolted on, problems of fatigue cracking that could occur with welding are eliminated. This method does have one potential problem, however: the possibility of having to remove part of a web stiffener should one be crossed by an angle.

Another method of adding material to existing members for strengthening is shown in Figure 50.10 where structural Ts were bolted to the bottom flanges of the existing stringers using structural angles. This idea represented a design alternative recommended by Howard, Needles, Tammen and Bergendoff as one method of strengthening a bridge comprising three 50-ft (15.2-m) simple spans. Each of the four stringers per span was strengthened in a similar manner.

Design Procedure

The basic design steps required in the design of steel cover plates follow:

1. Determine moment and shear envelopes for desired live-load capacity of each beam.
2. Determine the section modulus required for each beam.
3. Determine the optimal amount of steel to achieve desired section modulus–strength requirement, fatigue requirement.
4. Design connection of cover plates to beam strength requirement, fatigue requirement.
5. Determine safe cutoff point for cover plates.

In addition to the foregoing design steps, the following construction considerations may prove helpful:

1. Grinding the transverse weld to a 1:3 taper or bolting the ends of plates rather than welding reduces fatigue cracking at the cover plate ends [46,47].
2. In most cases a substantial savings in steel can be made if the bridge is jacked to relieve dead-load stresses prior to adding cover plates.

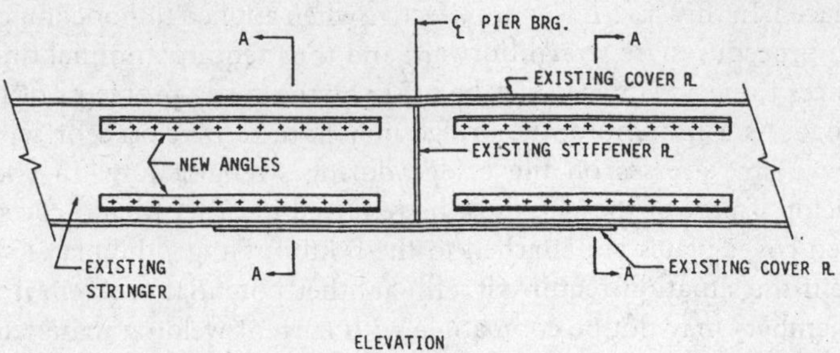

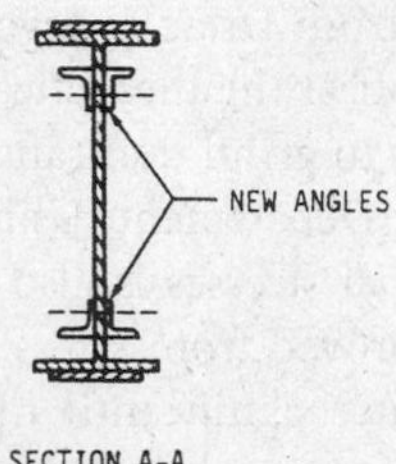

FIGURE 50.9 Iowa DOT method of adding angles to steel I-beams. (*Source:* Klaiber, F.W. et al., NCHRP 293, Transportation Research Board, 1987. With permission.)

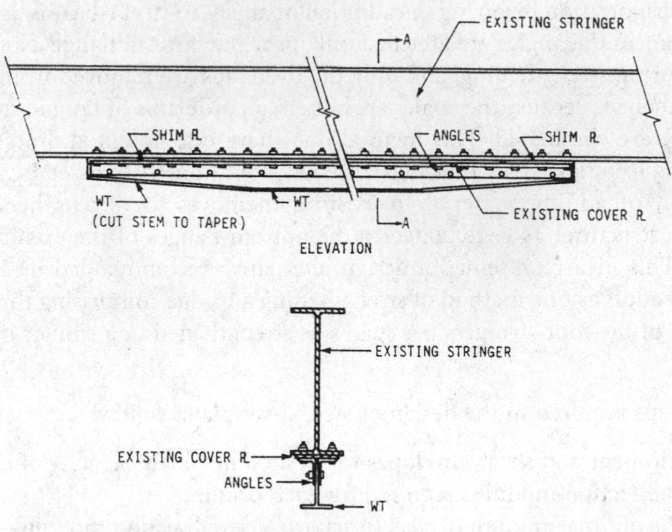

FIGURE 50.10 Strengthening of existing steel stringer by addition of structural T section. (*Source:* Klaiber, F.W. et al., NCHRP 293, Transportation Research Board, 1987. With permission.)

3. The welding of a cover plate should be completed within a working day. This minimizes the possibility of placing a continuous weld at different temperatures and inducing stress concentrations.
4. Shot blasting of existing beams to clean welding surface may be necessary.

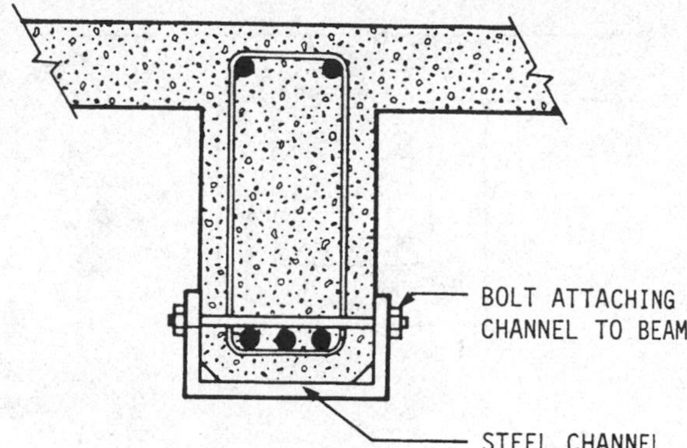

FIGURE 50.11 Addition of a steel channel to an existing reinforced concrete beam. (*Source:* Klaiber, F.W. et al., NCHRP 293, Transportation Research Board, 1987. With permission.)

Reinforced Concrete Bridges

Description

One method of increasing flexural capacity of a reinforced concrete beam is to attach steel cover plates or other steel shapes to the tension face of the beam. The plates or shapes are normally attached by bolting, keying, or doweling to develop continuity between the old beam and the new material. If the beam is also inadequate in shear, combinations of straps and cover plates may be added to improve both shear and flexural capacity. Because a large percentage of the load in most concrete structures is dead load, for cover plating to be most effective, the structure should be jacked prior to cover plating to reduce the dead-load stresses of the member. The addition of steel cover plates may also require the addition of concrete to the compression face of the member.

Applicability

A successful method of strengthening reinforced concrete beams has involved the attachment of a steel channel to the stem of a beam. This technique is shown in Figure 50.11. Taylor [48] performed tests on a section using steel channels and found it to be an effective method of strengthening. An advantage to this method is that rolled channels are available in a variety of sizes, require little additional preparation prior to attachment, and provide a ready formwork for the addition of grouting. The channels can also be easily reinforced with welded cover plates if additional strength is required. Prefabricated channels are an effective substitute when rolled sections of the required size are not available. It should be noted that the bolts are placed above the longitudinal steel so that the stirrups can carry shear forces transmitted by the channels. If additional sheer capacity is required, external stirrups should also be installed. It is also recommended that an epoxy resin grout be used between the bolts and concrete. The epoxy resin grout provides greater penetration in the bolt holes, thereby reducing slippage and improving the strength of the composite action.

Bolting steel plates to the bottom and sides of beam sections has also been performed successfully, as documented by Warner [49]. Bolting may be an expensive and time-consuming method, because holes usually have to be drilled through the old concrete. Bolting is effective, however, in providing composite action between the old and new material.

The placement of longitudinal reinforcement in combination with a concrete sleeve or concrete cover is another method for increasing the flexural capacity of the member. This method is shown in Figures 50.12a and b as outlined in an article on strengthening by Westerberg [50]. Warner [49] presents a similar method that is shown in Figure 50.12c.

Developing a bond between the old and new material is critical to developing full continuity. Careful cleaning and preparation of the old concrete and the application of a suitable epoxy-resin

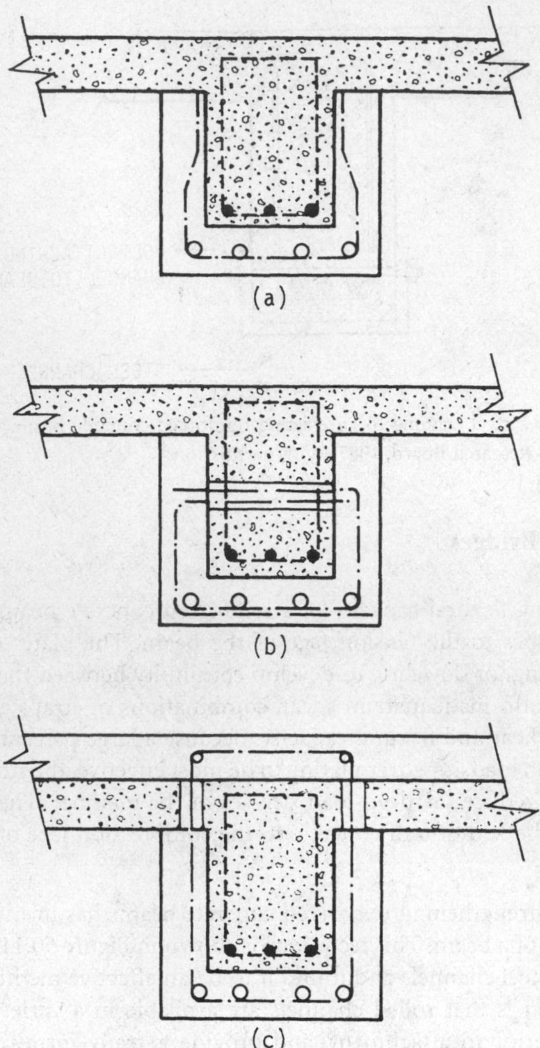

FIGURE 50.12 Techniques for increasing the flexural capacity of reinforced concrete beams with reinforced concrete sleeves. (*Source:* Klaiber, F.W. et al., NCHRP 293, Transportation Research Board, 1987. With permission.)

primer prior to adding new concrete should provide adequate bonding. Stirrups should also be added to provide additional shear reinforcement and to support the added longitudinal bars.

Design and Analysis Procedure

The design of steel cover plates for concrete members is dependent on the amount of continuity assumed to exist between the old and new material. If one assumes that full continuity can be achieved and that strains vary linearly throughout the depth of the beam, calculations are basically straightforward. As stated earlier, much of the load in concrete structures is dead load, and jacking of the deck during cover plating will greatly reduce the amount of new steel required. It should also be pointed out that additional steel could lead to an overreinforced section. This could be compensated for by additional concrete or reinforcing steel in the compression zone.

Case Studies

Steel cover plates can be used in a variety of situations. They can be used to increase the section modulus of steel, reinforced concrete, and timber beams. Steel cover plates are also an effective method of strengthening compression members in trusses by providing additional cross-sectional area and by reducing the slenderness ratio of the member.

Mancarti [51] reported the use of steel cover plates to strengthen floor beams on the Pit River Bridge and Overhead in California. The truss structure required strengthening of various other components to accommodate increased dead load. Stringers in this bridge were strengthened by applying prestressing tendons near the top flange to reduce tensile stress in the negative moment region. This prestressing caused increased compressive stresses in the bottom flanges, which in turn required the addition of steel bars to the tops of the stringer bottom flange.

In a report by Rodriguez et al. [52], a number of cases of cover-plating existing members of old railway trusses were cited. These case studies included the inspection of 109 bridges and a determination of their safety. Some strengthening techniques included steel-cover-plating beam members as well as truss members. Cover plates used to reinforce existing floor beams on a deficient through-truss were designed to carry all live-load bending moment. Deficient truss members were strengthened with box sections made up of welded plates. The box was placed around the existing member and connected to it by welding.

50.4.2 Shear Reinforcement

External Shear Reinforcement for Concrete, Steel, and Timber Beams

The shear strength of reinforced concrete beams or prestressed concrete beams can be improved with the addition of external steel straps, plates, or stirrups. Steel straps are normally wrapped around the member and can be post-tensioned. Post-tensioning allows the new material to share both dead and live loads equally with the old material, resulting in more efficient use of the material added. A disadvantage of adding steel straps is that cutting the deck to apply the straps leaves them exposed on the deck surface and thus difficult to protect. By contrast, adding steel plates does not require cutting through the deck. The steel plates are normally attached to the beam with bolts or dowels.

External stirrups may also be applied with different configurations. Figure 50.13a shows a method of attaching vertical stirrups using channels at the top and bottom of the beam. The deck (not shown in either figure) provides protection for the upper steel channel [53]. Adding steel sections at the top of the beam web and attaching stirrups is shown in Figure 50.13b. In this manner, cutting holes through the deck is eliminated. External stirrups can also be post-tensioned in most situations if desired.

Another method of increasing shear strength is shown in Figure 50.14. This method is a combination of post-tensioning and the addition of steel in the form of prestressing tendons. As recommended in a strengthening manual by the OECD [54], tendons may be added in a vertical or inclined orientation and may be placed either within the beam web or inside the box as shown in the figure. Care should be taken to avoid overstressing parts of the structure when prestressing. If any cracks exist in the member, it is a good practice to inject them with an epoxy before applying the prestressing forces. Documentation of this type of reinforcement technique is made also by Audrey and Suter [55] and Dilger and Ghali [56]. Figure 50.15 illustrates the technique used by Dilger and Ghali [56] where web thickening was added to the inside of the box web before adding external reinforcement consisting of stressed steel bars. The thickening was required to reduce calculated tensile stresses at the outside of the web due to prestressing the reinforcement.

West [57] makes reference to a number of methods of attaching steel plates to deficient steel I-beam girder webs as a means of increasing their shear strength. The steel plates are normally of panel size and are attached between stiffeners by bolting or welding. Where shear stresses are high, the plates should fit tightly between the stiffeners and girder flanges. West indicates that one advantage of this method is that it can be applied under traffic conditions.

Timber stringers with inadequate shear capacity can be strengthened by adding steel cover plates. NCHRP Report 222 [11] demonstrates a method of repairing damaged timber stringers with

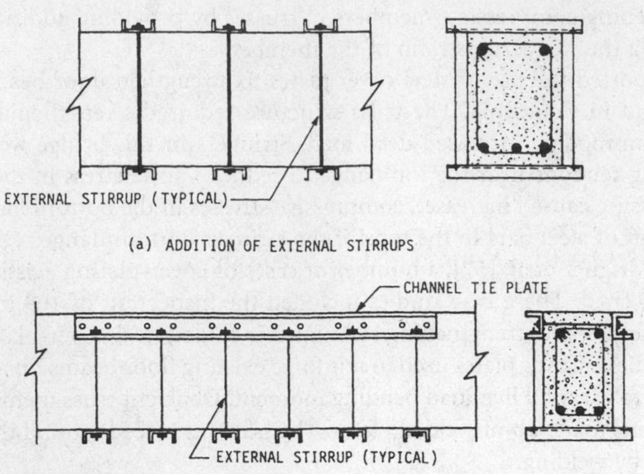

FIGURE 50.13 Methods of adding external shear reinforcement to reinforced concrete beams. (*Source:* Klaiber, F.W. et al., NCHRP 293, Transportation Research Board, 1987. With permission.)

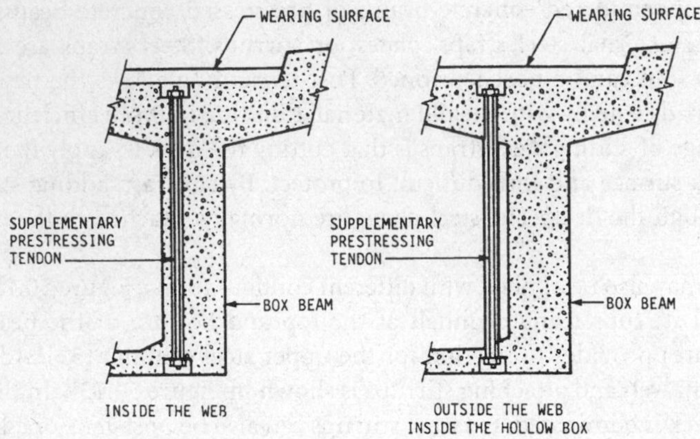

FIGURE 50.14 External shear reinforcement of box beam girders. (*Source:* Klaiber, F.W. et al., NCHRP 293, Transportation Research Board, 1987. With permission.)

inadequate shear capacity. The procedure involves attaching steel plates to the bottom of the beam in the deficient region and attaching it with draw-up bolts placed on both sides of the beam. Holes are drilled through the top of the deck, and a steel strap is placed at the deck surface and at the connection to the bolts.

Epoxy Injection and Rebar Insertion

The Kansas Department of Transportation has developed and successfully used a method for repairing reinforced concrete girder bridges. The bridges had developed shear cracks in the main longitudinal girders [58]. The procedure used by the Kansas Department of Transportation not only prevented further shear cracking but also significantly increased the shear strength of the repaired girders.

The method involves locating and sealing all of the girder cracks with silicone rubber, marking the girder centerline on the deck, locating the transverse deck reinforcement, vacuum drilling 45° holes that avoid the deck reinforcement, pumping the holes and cracks full of epoxy, and inserting reinforcing bars into the epoxy-filled holes. A typical detail is shown in Figure 50.16.

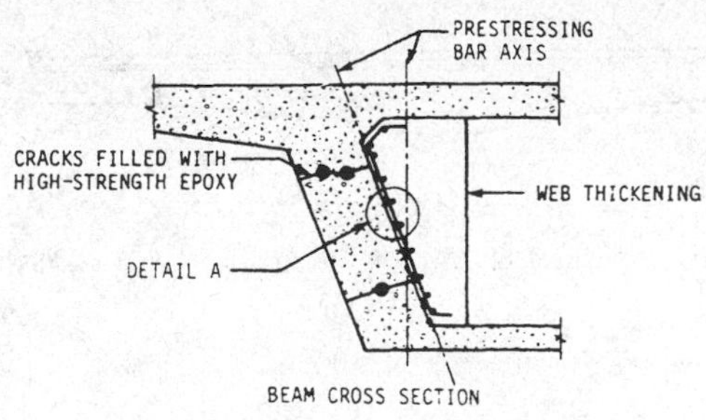

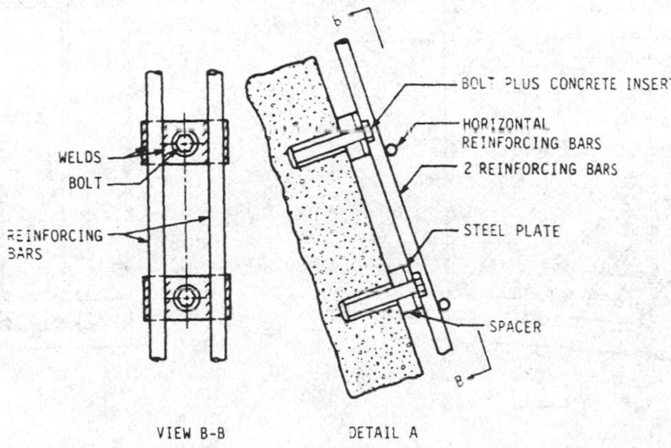

FIGURE 50.15 Details of web reinforcement to strengthen box beam in shear. (*Source:* Klaiber, F.W. et al., NCHRP 293, Transportation Research Board, 1987. With permission.)

An advantage of using the epoxy repair and rebar insertion method is its wide application to a variety of bridges. Although the Kansas Department of Transportation reported using this strengthening method on two-girder, continuous, reinforced concrete bridges, this method can be a practical solution on most types of prestressed concrete beam and reinforced concrete girder bridges that require additional shear strength. The essential equipment requirements needed for this strengthening method may limit its usefulness, however. Prior to drilling, the transverse deck steel must be located. The drilling unit and vacuum pump required must be able to drill quickly straight holes to a controlled depth and keep the holes clean and free of dust.

Addition of External Shear Reinforcement

Strengthening a concrete bridge member that has a deficient shear capacity can be performed by adding external shear reinforcement. The shear reinforcement may consist of steel side plates or steel stirrup reinforcement. This method has been applied on numerous concrete bridge systems.

A method proposed by Warner [49] involves adding external stirrups. The stirrups consist of steel rods placed on both sides of the beam section and attached to plates at the top and bottom of the section. In some applications, channels are mounted on both sides at the top of the section to attach the stirrups. This eliminates drilling through the deck to make the connection to a plate.

In a study by Dilger and Ghali [56], external shear reinforcement was used to repair webs of prestressed concrete bridges. Although the measures used were intended to bring the deficient members to their original flexural capacity, the techniques applied could be used for increasing the shear strength of existing members. Continuous box girders in the 827-ft (252-m)-long bridges had become severely cracked when prestressed. The interior box beam webs were strengthened by the addition of 1-in. (25-mm)-diameter steel rods placed on both sides of the web. Holes were drilled

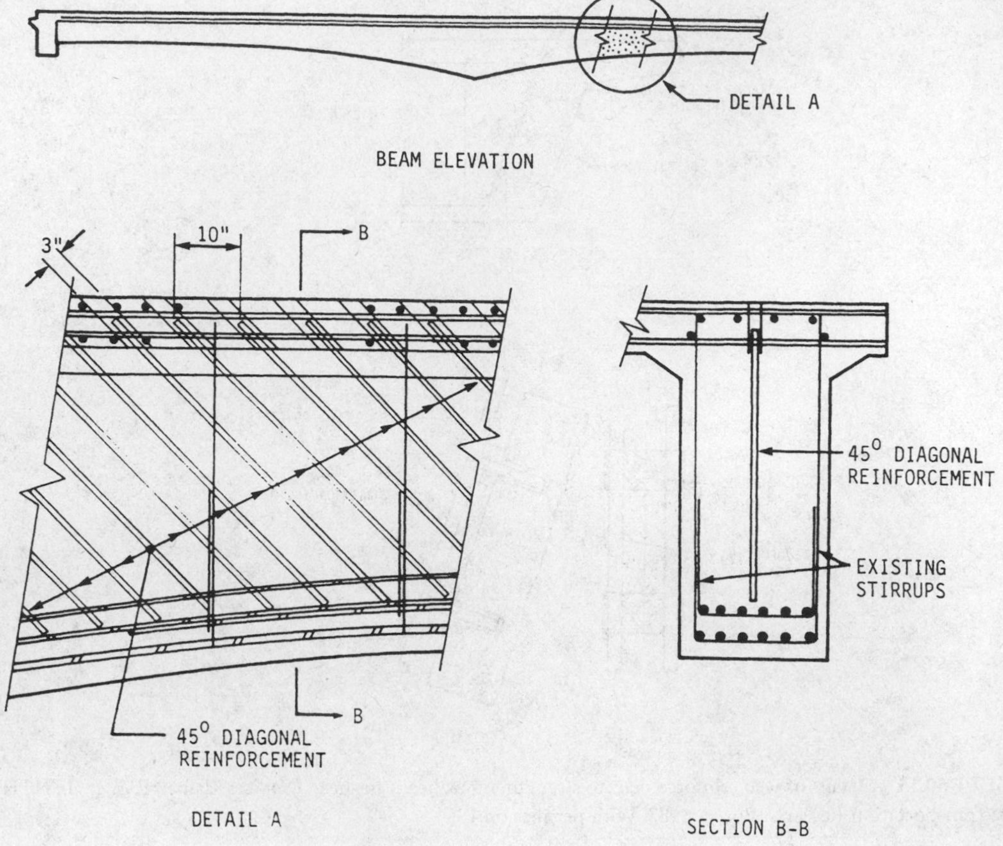

FIGURE 50.16 Kansas DOT shear strengthening procedure. (*Source:* Klaiber, F.W. et al., NCHRP 293, Transportation Research Board, 1987. With permission.)

in the upper and lower slabs as close as possible to the web to minimize local bending stresses in the slabs. Post-tensioning tendons were placed through the holes, stressed, and then anchored.

The slanted outside webs were strengthened with reinforcing steel. Before the bars were added, the inside of the web was "thickened" and the reinforcement was attached with anchor bolts placed through steel plates that were welded to the reinforcement. The web thickening was necessary because the prestressing would have produced substantial tensile stresses at the outside face of the web.

50.4.3 Jacketing of Timber or Concrete Piles and Pier Columns

Improving the strength of timber or concrete piles and pier columns can be achieved by encasing the column in concrete or steel jackets. The jacketing may be applied to the full length of the column or only to severely deteriorated sections. The jacketing increases the cross-sectional area of the column and reduces the slenderness ratio of the column. Partial encasement of a column can also be particularly effective when an unbalanced moment acts on the column. Figure 50.17 illustrates two such concepts for member addition that were noted from work on strengthening reinforced concrete structures in Europe [50].

Completely encasing the existing column in a concrete jacket has been a frequently used method of strengthening concrete pier columns. Normally, the reinforcement is placed around the existing column perimeter inside the jacket and "ramset" to the existing member [50]. The difficulty most often observed with this technique is developing continuity between the old and new material. This

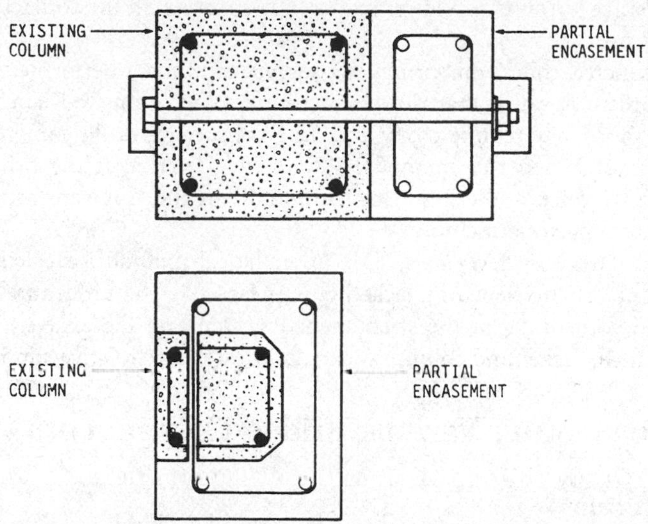

FIGURE 50.17 Partial jacketing of an existing column. (*Source:* Klaiber, F.W. et al., NCHRP 293, Transportation Research Board, 1987. With permission.)

is critical if part of the load is to be transferred to the new material. Work by Soliman [59] on repair of reinforced concrete columns by jacketing has included an experimental investigation of the bond stresses between the column and jacket. The first step is normally surface preparation of the existing concrete column. Consideration should also be given at this time to jacking of the superstructure and placing temporary supports on either side of the column. Solimann [59] concludes that this is an important step, since the shrinkage phenomenon causes compressive stresses on the column that will be reduced if the existing column is unloaded. In addition, supports will be necessary if the column shows significant signs of deterioration. This procedure will also allow the new material to share equally both dead and live loads after the supports are removed. Additional longitudinal reinforcing bars and stirrups are then placed around the column. Spiral stirrup reinforcement should be used because it will provide greater strength and ductility than normal stirrups [59]. An epoxy resin is then applied to the old concrete to increase the bonding action between the old concrete and the concrete to be added. Formwork is then erected to form the jacket, and concrete is placed and compacted.

Jacketing techniques have been used extensively for seismic retrofitting of existing pier columns and this topic is discussed in Chapter 43. A recent report by Wipf et al. [60] provides an extensive list and discussion of various retrofit methods for reinforced concrete bridge columns, including the use of steel jackets and fiber-reinforced polymer wraps.

Modification Jacketing

Increasing the load-carrying capacity of bridge pier columns or timber piles supporting bent caps is normally achieved through the addition of material to the existing cross section. Jacketing or adding a sleeve around the column perimeter can be performed a number of ways.

In a paper by Karamchandani [61], various concepts for jacketing existing members are illustrated. These include addition of reinforcement and concrete around three sides of rectangular beams as well as placement only at the bottom of the beam web. Additional schemes are also illustrated for column members. The effectiveness of this method depends on the degree of adhesion between new and existing concrete, which can vary between 30 and 80% of the total strength of the *in situ* concrete. The author suggests welding new reinforcing to the existing reinforcement and using concrete with a slump of 3 to 4 in. (75 to 100 mm). The use of rapid hardening cements is not recommended, since it results in a lower strength of concrete on the contact surface because of high contraction stresses.

The addition of concrete collars on reinforced concrete columns is performed most efficiently by using circular reinforcement rather than dowels or shear keys according to Klein and Gouwens [62]. While the other methods may require costly and time-consuming drilling and/or cutting, circular reinforcement does not. When this method is used, shear-friction is the primary load-transfer mechanism between the collar and the existing column. Klein and Gouwens have outlined a design procedure for this strengthening method.

In a paper by Syrmakezis and Voyatzis [63], an analytical method for calculating the stiffness coefficients of columns strengthened by jacketing is presented. The procedure uses compatibility conditions for the deformations of the strengthened system and the analysis can consider rigid connections between the jacket and column on a condition where relative slip is allowed.

50.5 Post-Tensioning Various Bridge Components

50.5.1 Introduction

Since the 19th century, timber structures have been strengthened by means of king post and queen post-tendon arrangements; these forms of strengthening by post-tensioning are still used today. Since the 1950s, post-tensioning has been applied as a strengthening method in many more configurations to almost all common bridge types. The impetus for the recent surge in post-tensioning strengthening is undoubtedly a result of its successful history of more than 40 years and the current need for strengthening of bridges in many countries.

Post-tensioning can be applied to an existing bridge to meet a variety of objectives. It can be used to relieve tension overstresses with respect to service load and fatigue-allowable stresses. These overstresses may be axial tension in truss members or tension associated with flexure, shear, or torsion in bridge stringers, beams, or girders.

Post-tensioning also can reduce or reverse undesirable displacements. These displacements may be local, as in the case of cracking, or global, as in the case of excessive bridge deflections. Although post-tensioning is generally not as effective with respect to ultimate strength as with respect to service-load-allowable stresses, it can be used to add ultimate strength to an existing bridge. It is possible to use post-tensioning to change the basic behavior of a bridge from a series of simple spans to continuous spans. All of these objectives have been fulfilled by post-tensioning existing bridges, as documented in the engineering literature.

Most often, post-tensioning has been applied with the objective of controlling longitudinal tension stresses in bridge members under service-loading conditions. Figure 50.18 illustrates the axial forces, shear forces, and bending moments that can be achieved with several simple tendon configurations. The concentric tendon in Figure 50.18a will induce an axial compression force that, depending on magnitude, can eliminate part or all of an existing tension force in a member or even place a residual compression force sufficient to counteract a tension force under other loading conditions. The amount of post-tensioning force that can safely be applied, of course, is limited by the residual-tension dead-load force in the member.

The tendon configuration in Figure 50.18a is generally used only for tension members in trusses, whereas the remaining tendon configurations in Figure 50.18 would be used for stringers, beams, and girders. The eccentric tendon in Figure 50.18b induces both axial compression and negative bending. The eccentricity of the tendon may be varied to control the proportions of axial compression vs. bending applied to the member. Length of the tendon also may be varied to apply post-tensioning only to the most highly stressed portion of the member. The polygonal tendon profile in Figure 50.18c also induces axial compression and negative bending, but the negative bending is nonuniform within the post-tensioned region. Locations of bends on the tendon and eccentricities of the attachments at the bends can be set to control the moments caused by the post-tensioning.

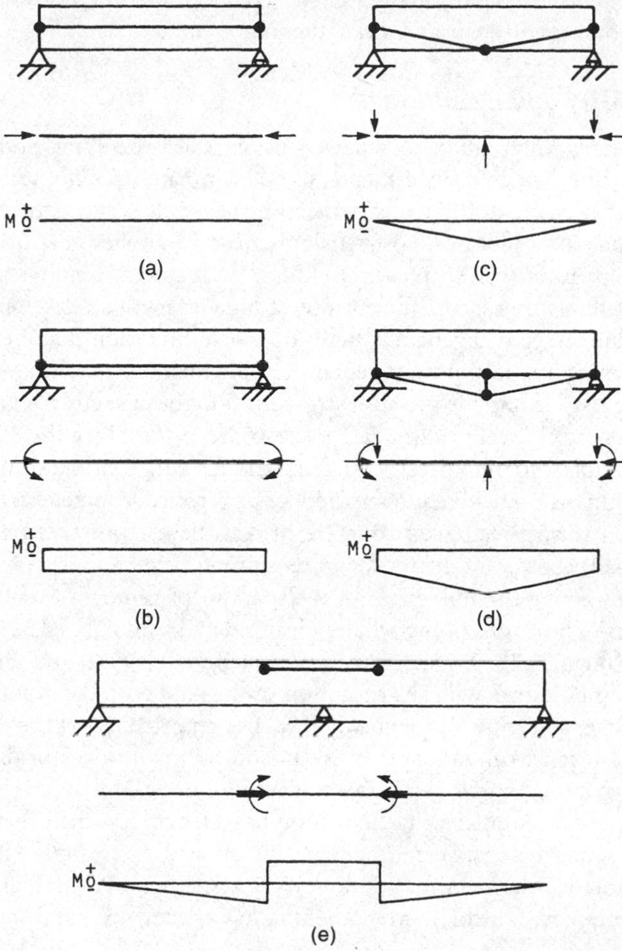

FIGURE 50.18 Forces and moment induced by longitudinal post-tensioning. (a) Concentric tendon; (b) eccentric tendon; (c) polygonal tendon; (d) king post; (e) eccentric tendon, two-span member. (*Source:* Klaiber, F.W. et al., NCHRP 293, Transportation Research Board, 1987. With permission.)

The polygonal tendon also induces shear forces that are opposite to those applied by live and dead loads.

The king post tendon configuration in Figure 50.18d is a combination of the eccentric and polygonal tendon configurations. Because the post is beyond the profile of the original member, the proportion of moment to axial force induced in the member to be strengthened will be large.

The tendon configuration in Figure 50.18e is an eccentric tendon attached over the central support of a two-span member. In this configuration, the amount of positive moment applied in the central support region depends not only on the force in the tendon and its eccentricity, but also on the locations of the anchorages on the two spans. If the anchorages are moved toward the central support, the amount of positive moment applied will be greater than if the anchorages are moved away from the central support. This fact and the fact that there is some distribution of moment and force among parallel post-tensioned members have not always been correctly recognized, and there are published errors in the literature.

The axial force, shear force, and bending moment effects of post-tensioning described above have enough versatility in application so as to meet a wide variety of strengthening requirements. Probably this is the only strengthening method that can actually reverse undesirable behavior in an existing bridge rather than provide a simple patching effect. For both these reasons, post-tensioning has become a very commonly used repair and strengthening method.

50.5.2 Applicability and Advantages

Post-tensioning has many capabilities: to relieve tension, shear, bending, and torsion-overstress conditions; to reverse undesirable displacements; to add ultimate strength; to change simple span to continuous span behavior. In addition, post-tensioning has some very practical advantages. Traffic interruption is minimal; in some cases, post-tensioning can be applied to a bridge with no traffic interruption. Few site preparations, such as scaffolding, are required. Tendons and anchorages can be prefabricated. Post-tensioning is an efficient use of high strength steel. If tendons are removed at some future date, the bridge will generally be in no worse condition than before strengthening.

To date, post-tensioning has been used to repair or strengthen most common bridge types. Most often, post-tensioning has been applied to steel stringers, floor beams, girders, and trusses, and case histories for strengthening of steel bridges date back to the 1950s. Since the 1960s, external post-tensioning has been applied to reinforced concrete stringer and T bridges. In the past 20 years, external post-tensioning has been added to a variety of prestressed, concrete stringer and box beam bridges. Many West German prestressed concrete bridges have required strengthening by post-tensioning due to construction joint distress. Post-tensioning even has been applied to a reinforced concrete slab bridge by coring the full length of the span for placement of tendons [63].

Known applications of post-tensioning will be idealized and summarized as Schemes A through L in Figures 50.19 through 50.22. Typical schemes for stringers, beams, and girders are contained in Figure 50.19. The simplest and, with the exception of the king post, the oldest scheme is Scheme A: a straight, eccentric tendon shown in Figure 50.19a. Lee reported use of the eccentric tendon for strengthening of British cast iron and steel highway and railway bridges in the early 1950s [64]. Since then, Scheme A has been applied to many bridges in Europe, North America, and other parts of the world. Scheme A is most efficient if the tendon has a length less than that of the member, so that the full post-tensioning negative moment is not applied to regions with small dead-load moments. The variation on Scheme A for continuous spans, Scheme AA in Figure 50.19e, has been reported in use for deflection control or strengthening in Germany [65] and the United States [66] since the late 1970s.

The polygonal tendon, Scheme B in Figure 50.19b and its extension to continuous spans, Scheme BB in Figure 50.19f, has been in use since at least the late 1960s. Vernigora et al. [67] reported the use of Scheme BB for a five-span, reinforced-concrete T-beam bridge in 1969 [67]. The bridge over the Welland Canal in Ontario, Canada, was converted from simple-span to continuous-span behavior by means of external post-tensioning cables.

Scheme C in Figure 50.19c provided the necessary strengthening for a steel plate, girder railway bridge in Czechoslovakia in 1964 [68]. The tendons and compression struts for the bridge were fabricated from steel T sections, and the tendons were stressed by deflection at bends rather than by elongation as is the usual case. The tendons for the plate girder bridge were given a three-segment profile to apply upward forces at approximately the third points of the span, so that the existing dead-load moments could be counteracted efficiently. In the late 1970s in the United States, Kandall [69] recommended use of Scheme C for strengthening because it does not place additional axial compression in the existing structure. For other schemes, the additional axial compression induced by post-tensioning will add compressive stress to regions that may be already overstressed in compression.

Scheme D in Figure 50.19d, was used in Minnesota in 1975 to strengthen temporarily a steel stringer bridge [70]. It was possible to strengthen that bridge economically with scrap timber and cable for the last few years of its life before it was replaced.

The tendon schemes in Figure 50.19, in general, appear to be very similar to reinforcing bar patterns for concrete beams. Thus, it is not surprising that post-tensioning also has been used for

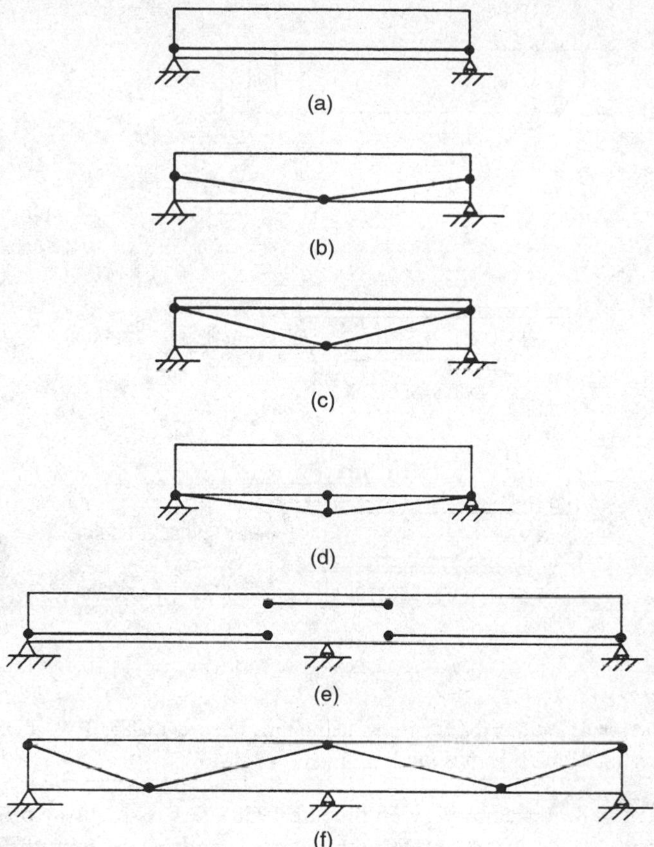

FIGURE 50.19 Tendon configurations for flexural post-tensioning of beams. (a) Scheme A, eccentric tendon; (b) Scheme B, polygonal tenton; (c) Scheme C, polygonal tendon with compression strut; (d) Scheme D, king post; (e) Scheme AA, eccentric tendons; (f) Scheme BB, polygonal tendons. (*Source:* Klaiber, F.W. et al., NCHRP 293, Transportation Research Board, 1987. With permission.)

shear strengthening, in patterns very much like those for stirrups in reinforced concrete beams. Scheme E in Figure 50.20a illustrates a pattern of external stirrups for a beam in need of shear strengthening. Types of post-tensioned external stirrups have been used or proposed for timber beams [11], reinforced concrete beams and, as illustrated in Figure 50.20b, for prestressed concrete box-girder bridges [71].

Post-tensioning was first applied to steel trusses for purposes of strengthening in the early 1950s [64], at about the same time that it was first applied to steel stringer and steel girder, floor beam bridges. Typical strengthening schemes for trusses are presented in Figure 50.21. Scheme F, concentric tendons on individual members, shown in Figure 50.21a, was first reported for the proposed strengthening of a cambered truss bridge in Czechoslovakia in 1964 [68]. For that bridge it was proposed to strengthen the most highly stressed tension diagonals by post-tensioning. Scheme F tends to be uneconomical because it requires a large number of anchorages, and very few truss members benefit from the post-tensioning.

Scheme G in Figure 50.21b, a concentric tendon on a series of members, has been the most widely used form of post-tensioning for trusses. Lee [64] describes the use of this scheme for British railway bridges in the early 1950s, and there have been a considerable number of bridges strengthened with this scheme in Europe.

The polygonal tendon in Scheme H, Figure 50.21c, has not been reported for strengthening purposes, but it has been used in the continuous-span version of Scheme I in Figure 50.21d for a

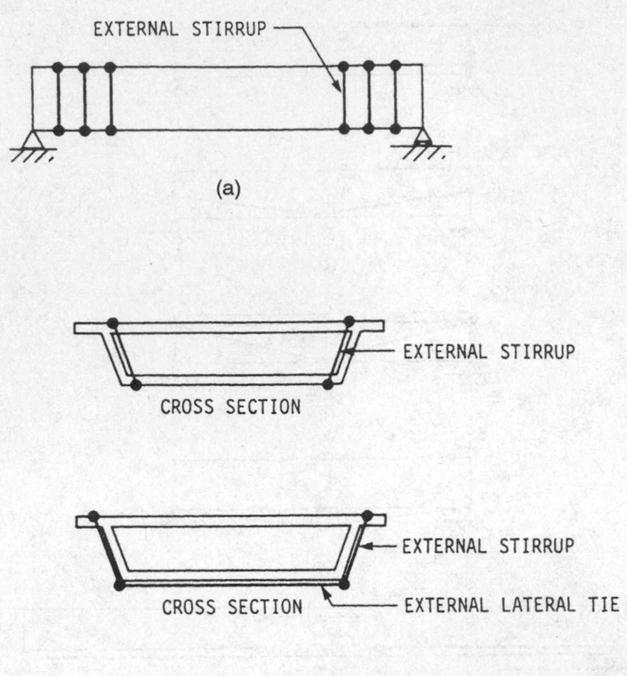

FIGURE 50.20 Tendon configurations for shear post-tensioning. (*Source:* Klaiber, F.W. et al., NCHRP 293, Transportation Research Board, 1987. With permission.)

two-span truss bridge in Switzerland [72]. In the late 1960s, a truss highway bridge in Aarwangen, Switzerland, was strengthened by means of four-segment tendons on each of the two spans. The upper chord of each truss was unable to carry the additional compression force induced by the post-tensioning, and, therefore, a free-sliding compression strut was added to each top chord to take the axial post-tensioning force.

Scheme J, the king post in Figure 50.21e, has been suggested for new as well as existing trusses [7]; however, cases of its actual use for strengthening have not been reported in the literature. Because most trusses are placed on spans greater than 100 ft (30.5 m), the posts below the bridge could extend down quite far and severely reduce clearance under the bridge. The king post or queen post would thus be in a very vulnerable position and would not be appropriate in many situations.

Most uses of post-tensioning for strengthening have been on the longitudinal members in bridges; however, post-tensioning has also been used for strengthening in the transverse direction. After the deterioration of the lateral load distribution characteristics of laminated timber decks was noted in Canada in the mid-1970s [73], Scheme K in Figure 50.22a was used to strengthen the deck. A continuous-steel channel waler at each edge of the deck spreads the post-tensioning forces from threadbar tendons above and below the deck, thereby preventing local overstress in the timber. A similar tendon arrangement, shown in Figure 50.22b, was used in an Illinois bridge [74] to tie together spreading, prestressed concrete box beams.

The overview of uses of post-tensioning for bridge strengthening given above identifies the most important concepts that have been used in the past and indicates the versatility of post-tensioning as a strengthening method.

50.5.3 Limitations and Disadvantages

When post-tensioning is used as a strengthening method, it increases the allowable stress range by the magnitude of the applied post-tensioning stress. If maximum advantage is taken of the increased

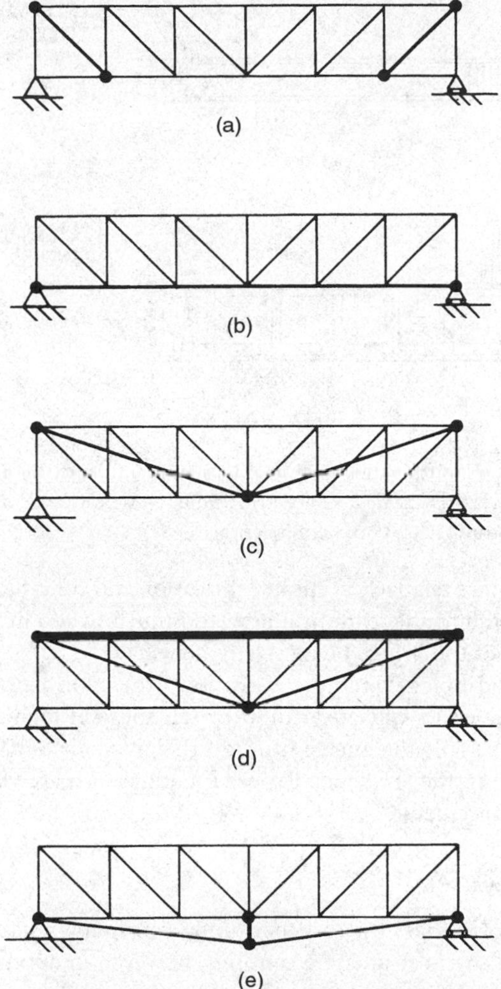

FIGURE 50.21 Tendon configurations for post-tensioning trusses. (a) Scheme F, concentric tendons on individual members; (b) Scheme G, concentric tendon on a series of members; (c) Scheme H, polygonal tendon; (d) Scheme I, polygonal tendon with compression strut; (e) Scheme J, king post. (*Source:* Klaiber, F.W. et al., NCHRP 293, Transportation Research Board, 1987. With permission.)

allowable stress range, the factor of safety against ultimate load will be reduced. The ultimate-load capacity thus will not increase at the same rate as the allowable-stress capacity. For short-term strengthening applications, the reduced factor of safety should not be a limitation, especially in view of the recent trend toward smaller factors of safety in design standards. For long-term strengthening applications, however, the reduced factor of safety may be a limitation.

At anchorages and brackets where tendons are attached to the bridge structure, there are high local stresses that require consideration. Any cracks initiated by holes or expansion anchors in the structure will spread with live-load dynamic cycling.

Because post-tensioning of an existing bridge affects the entire bridge (beyond the members that are post-tensioned), consideration must be given to the distribution of the induced forces and moments within the structure. If all parallel members are not post-tensioned, if all parallel members are not post-tensioned equally, or if all parallel members do not have the same stiffness, induced forces and moments will be distributed in some manner different from what is assumed in a simple analysis.

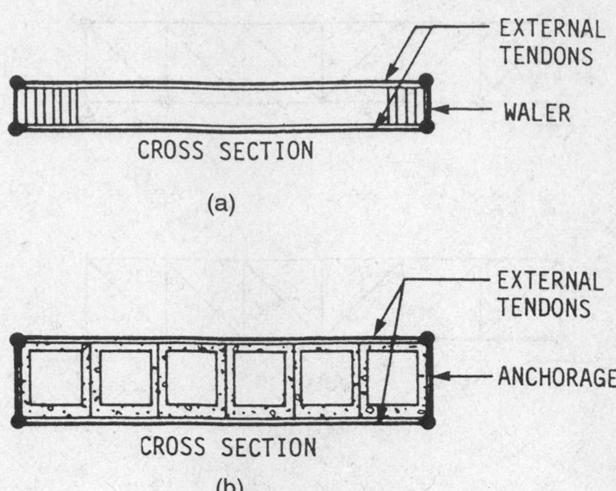

FIGURE 50.22 Tendon configuration for transverse post-tensioning of decks. (a) Scheme K, concentric tendons and walers, laminated timber deck; (b) Scheme K, concentric tendons, box beams. (*Source:* Klaiber, F.W. et al., NCHRP 293, Transportation Research Board, 1987. With permission.)

Post-tensioning does require relatively accurate fabrication and construction and relatively careful monitoring of forces locked into the tendons. Either too much or too little tendon force can cause overstress in the members of the bridge being strengthened.

Tendons, anchorages, and brackets require corrosion protection because they are generally in locations that can be subjected to saltwater runoff or salt spray. If tendons are placed beyond the bridge profile, they are vulnerable to damage from overheight vehicles passing under the bridge or vulnerable to damage from traffic accidents. Exposed tendons also are vulnerable to damage from fires associated with traffic accidents.

50.5.4 Design Procedures

In general, strengthening of bridges by post-tensioning can follow established structural analysis and design principles. The engineer must be cautious, however, in applying empirical design procedures as they are established only for the conditions of a particular strengthening problem.

Every strengthening problem requires careful examination of the existing structure. Materials in an existing bridge were produced to some previous set of standards and may have deteriorated due to exposure over many years. The existing steel in steel members may not be weldable with ordinary procedures, and steel shapes are not likely to be dimensioned to current standards. Shear connectors and other parts may have unknown capacities due to unusual configurations.

Strengthening an existing bridge involves more than strengthening individual members. Even a simple-span bridge is indeterminate, and post-tensioning and other strengthening will affect the behavior of the entire bridge. If the indeterminate nature of the bridge is not recognized during analysis, the post-tensioning applied for strengthening purposes may not have the desired stress-relieving effects and may actually cause overstress.

Post-tensioning involves application of relatively large forces to regions of a structure that were not designed for such large forces. There is more likelihood of local overstress at tendon anchorages and brackets than at conventional member connections. Brackets need to be designed to distribute the concentrated post-tensioning forces over sufficiently large portions of the existing structure.

Members and bridges subjected to longitudinal post-tensioning will shorten axially and, depending on the tendon configuration, also will shorten and elongate with flexural stresses. These shortening and elongation effects must be considered, so that the post-tensioning has its desired effect. Frozen bridge bearings require repair and lubrication, and support details should be checked for restraints.

External tendons, whether cable or threadbar, are relatively vulnerable to corrosion, damage from overheight vehicles, traffic accidents, or fires associated with accidents. Corrosion protection and placement of the tendons are thus very important with respect to the life of the post-tensioning. Safety is also a consideration because a tendon that ruptures suddenly can pose a hazard.

For the past few years, the authors and other Iowa State University colleagues have been investigating the use of external post-tensioning (Scheme A and AA in Figure 50.19) for strengthening existing single-span and continuous-span steel stringer bridges. The research, which has been recently completed, involved laboratory testing, field implementation, and the development of design procedures. The strengthening procedures that were developed are briefly described in the following sections.

50.5.5 Longitudinal Post-Tensioning of Stringers

Simple Spans

Essentially all single-span composite steel stringer bridges constructed in Iowa between 1940 and 1960 have smaller exterior stringers. These stringers are significantly overstressed for today's legal loads; interior stringers are also overstressed to a lesser degree. Thus, the post-tension system developed is only applied to the exterior stringers; through lateral load distribution a stress reduction is also obtained in the interior stringers.

By analyzing an undercapacity bridge, an engineer can determine the overstress in the interior and exterior stringers. This overstress is based on the procedure of isolating each bridge stringer from the total structure. The amount of post-tensioning required to reduce the stress in the stringers can then be determined if the amount of post-tensioning force remaining on the exterior stringers is known; this force can be quantified with force and moment fractions. A force fraction, FF, is the ratio of the axial force that remains on a post-tensioned stringer at midspan to the sum of the axial forces for all bridge stringers at midspan, while a moment fraction, MF, is the moment remaining on the post-tensioned stringer divided by the sum of midspan moments for all bridge stringers. Knowing these fractions, the required post-tensioning force may be determined by utilizing the following relationship:

$$f = FF\left[\frac{P}{A}\right] + MF\left[\frac{Pec}{I}\right] \tag{50.1}$$

where

f = desired stress reduction in stringer lower flange
P = post-tensioning force required on each exterior stringer
A = cross-sectional area of exterior stringers
e = eccentricity of post-tensioning force measured from the neutral axis of the bridge
c = distance from neutral axis of stringer to lower flange
I = moment of inertia of exterior stringer at section being analyzed

Force fractions and moment fractions as well as other details on the procedure may be found in Reference [75].

Span length and relative beam stiffness were determined to be the most significant variables in the moment fractions. As span length increases, exterior beams retain less moment; exterior beams that are smaller than the interior beams retain less post-tensioning moment than if the beams were all the same size.

The strengthening procedure and design methodology just described have been used on several bridges in the states of Iowa, Florida, and South Dakota. In all instances, the procedure was employed by local contractors without any significant difficulties. Application of this strengthening procedure to a 72-ft (34.0-m)-long 45° skewed bridge in Iowa is shown in Figure 50.23.

FIGURE 50.23 Single-span bridge strengthened by post-tensioning.

Continuous Spans

Similar to the single-span bridges, Iowa has a large number of continuous-span composite steel stringer bridges that also have excessive flexural stresses. Through laboratory tests, it was determined that the desired stress reduction could be obtained by post-tensioning the positive moment regions of the various stringers in most situations. In the cases in which there are excessive overstresses in the negative moment regions, it may be necessary to use superimposed trusses (see Figure 50.24) on the exterior stringers in addition to post-tensioning the positive moment regions. Similar to single-span bridges, it was decided to use force fractions and moment fractions to determine the distribution of strengthening forces in a given bridge. As one would expect, the design procedure is considerably more involved for continuous-span bridges as one has to consider transverse and longitudinal distribution of forces.

The required strengthening forces and final stringer envelopes should be calculated. The various strengthening schemes that can be used are shown in Figure 50.25. A designer selects the schemes required for obtaining the desired stress reduction. For additional details on the strengthening procedure the reader is referred to Reference [76]. Shown in Figure 50.26 is a three-span continuous bridge near Mason City, IA, that has been strengthened using the schemes shown in Figure 50.25.

50.6 Developing Additional Bridge Continuity

50.6.1 Addition of Supplemental Supports

Description

Supplemental supports can be added to reduce span length and thereby reduce the maximum positive moment in a given bridge. By changing a single-span bridge to a continuous, multiple-span bridge,

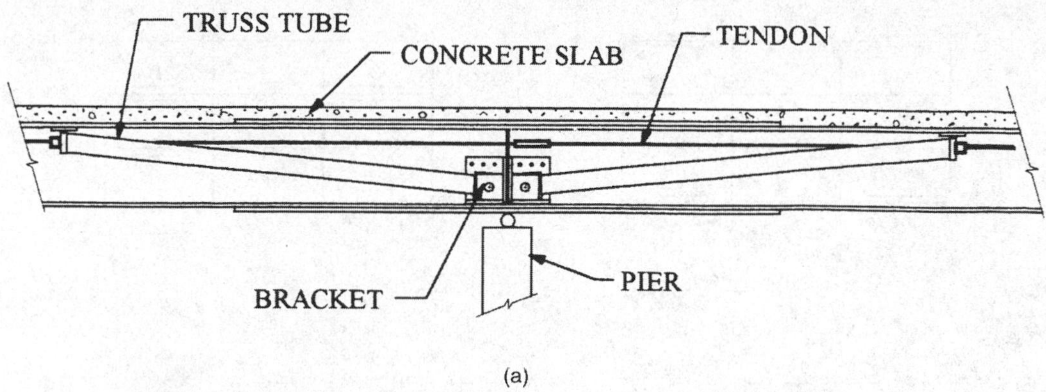

(a)

(b)

FIGURE 50.24 Superimposed truss system. (a) Superimposed truss; (b) photograph of superimposed truss.

stresses in the bridge can be altered dramatically, thereby improving the maximum live-load capacity of the bridge. Even though this method may be quite expensive because of the cost of adding an additional pier(s), it may still be desirable in certain situations.

Applicability and Advantages

This method is applicable to most types of stringer bridges, such as steel, concrete, and timber, and has also been used on truss bridges [7]. Each of these types of bridges has distinct differences.

If a supplemental center support is added to the center of an 80-ft (24.4-m)-long steel stringer bridge that has been designed for HS20-44 loading, the maximum positive live-load moment is reduced from 1164.9 ft-kips (1579.4 kN·m) to 358.2 ft-kips (485.7 kN·m), which is a reduction of over 69%. At the same time, however, a negative moment of 266.6 ft-kips (361.5 kN·m) is created which must be taken into account. In situations where the added support cannot be placed at the center, reductions in positive moments are slightly less.

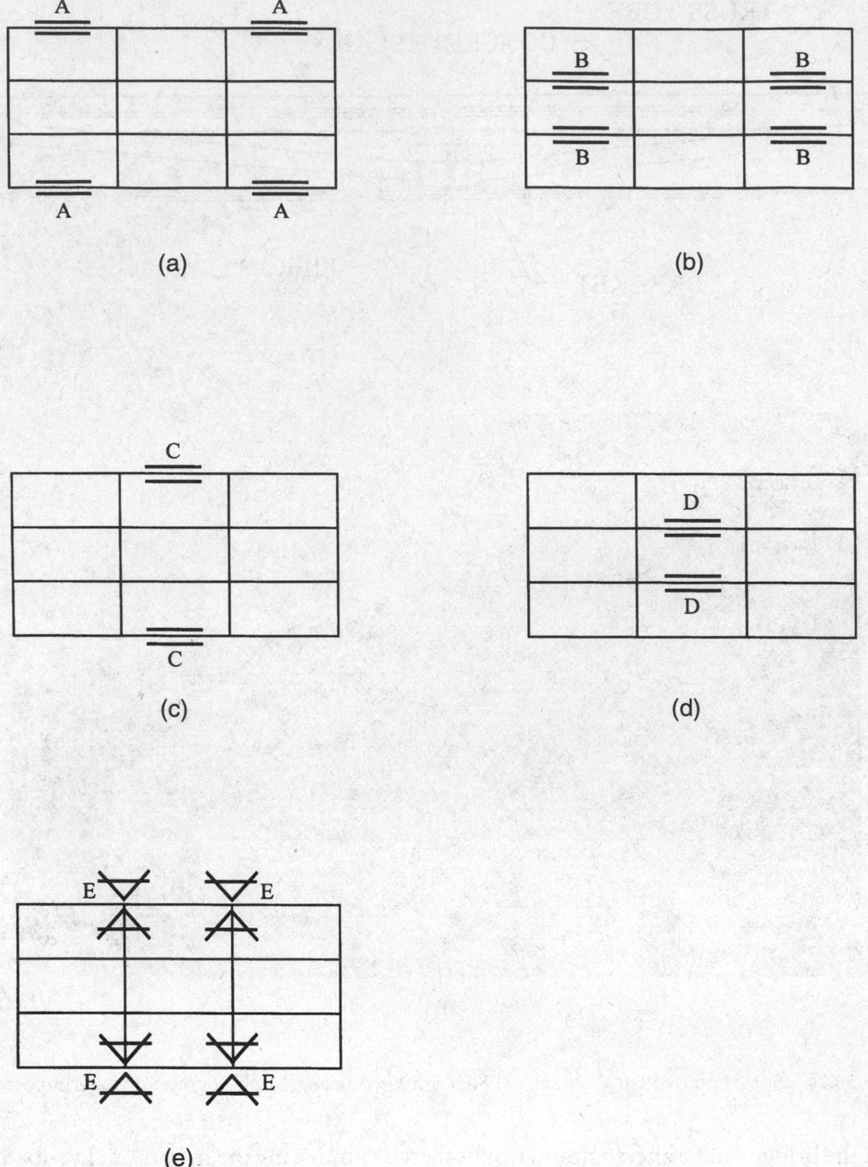

FIGURE 50.25 Strengthening schemes for continuous-span bridge. (a) Strengthening Scheme A; post-tensioning end spans of the exterior stringers; (b) strengthening Scheme B: post-tensioning end spans of the interior stringers; (c) strengthening Scheme C: post-tensioning center spans of the exterior stringers; (d) strengthening Scheme D: post-tensioning center spans of the interior strangers; (e) strengthening Scheme E: superimposed trusses at the piers of the exterior stringers.

Limitations and Disadvantages

Depending on the type of bridge, there are various limitations in this method of strengthening. First, because of conditions directly below the existing bridge, there may not be a suitable location for the pier, as, for example, when the bridge requiring strengthening passes over a roadway or railroad tracks. Other constraints, such as soil conditions, the presence of a deep gorge, or stream velocity, could greatly increase the length of the required piles, making the cost prohibitive.

FIGURE 50.26 Photograph of three-span continuous bridge strengthened with post-tensioning and superimposed trusses.

This method is most cost-effective with medium- to long-span bridges. This eliminates most timber stringer bridges because of their short lengths. In truss bridges, the trusses must be analyzed to determine the effect of adding an additional support. All members would have to be examined to determine if they could carry the change in force caused by the new support. Of particular concern would be members originally designed to carry tension, but which because of the added support must now carry compressive stresses. Because of these problems, the emphasis in this section will be on steel and concrete stringer bridges.

Design Considerations

Because the design of each intermediate pier system is highly dependent on many variables such as the load on pier, width and height of bridge, and soil conditions, it is not feasible to include a generalized design procedure for piers. The engineer should use standard pier design procedures. A brief discussion of several of the more important considerations (condition of the bridge, location of pier along bridge, soil condition, type of pier, and negative moment reinforcement) is given in the following paragraphs.

Providing supplemental support is quite expensive; therefore, the condition of the bridge is very important. If the bridge is in good to excellent condition and the only major problem is that the bridge lacks sufficient capacity for present-day loading, this method of strengthening should be considered. On the other hand, if the bridge has other deficiencies, such as a badly deteriorated deck or insufficient roadway width, a less expensive strengthening method with a shorter life should be considered.

The type of pier system employed greatly depends on the loading and also the soil conditions. The most common type of pier system used in this method is either steel H piles or timber piles with a steel or timber beam used as a pier cap. A method employed by the Florida Department of Transportation [77] can be used to install the piles under the bridge with limited modification to the existing bridge. This method consists of cutting holes through the deck above the point of application of the piles. Piles are then driven into position through the deck. The piles are then cut off so that a pier cap and rollers can be placed under the stringers. Other types of piers, such as concrete pile bents, solid piers, or hammerhead piers, can also be used; however, cost may restrict their use.

Another major concern with this method is how to provide reinforcement in the deck when the region in the vicinity of the support becomes a negative moment region. With steel stringers the bridge may either be composite or noncomposite. If noncomposite, the concrete deck is not required to carry any of the negative moment and therefore needs no alteration. On the other hand, if composite action exists, the deck in the negative moment region should be removed and replaced with a properly reinforced deck. For concrete stringer bridges the deck in the negative moment region should be removed. Reinforcement to ensure shear connection between the stringers and deck must be installed and the deck replaced with a properly reinforced deck. This method, although expensive and highly dependent on the surroundings, may be quite effective in the right situation.

50.6.2 Modification of Simple Spans

Description

In this method of strengthening, simply supported adjacent spans are connected together with a moment and shear-type connection. Once this connection is in place, the simple spans become one continuous span, which alters the stress distribution. The desired decrease in the maximum positive moment, however, is accompanied by the development of a negative moment over the interior supports.

Applicability and Advantages

This method can be used primarily with steel and timber bridges. Although it could also be used on concrete stringer bridges, the difficulties in structural connecting to adjacent reinforced concrete beams result in the method being impractical. The stringer material and the type of deck used will obviously dictate construction details. Thus, the main advantage of this procedure is that it is possible to reduce positive moments (obviously the only moments present in simple spans) by working over the piers and not near the midspan of the stringers. This method also reduces future maintenance requirements because it eliminates a roadway joint and one set of bearings at each pier where continuity is provided [12].

Limitations and Disadvantages

The main disadvantage of modifying simple spans is the negative moment developed over the piers. To provide continuity, regardless of the type of stringers or deck material, one must design for and provide reinforcement for the new negative moments and shears. Providing continuity also increases the vertical reactions at the interior piers; thus, one must check the adequacy of the piers to support the increase in axial load.

Design Considerations

The main design consideration for both types of stringers (steel and timber) concerns how to ensure full connection (shear and moment) over the piers. The following sections will give some insight into how this may be accomplished.

Steel Stringers

Berger [12] has provided information, some of which is summarized here, on how to provide continuity in a steel stringer concrete deck system. If the concrete deck is in sound condition, a portion of it must be removed over the piers. A splice, which is capable of resisting moment as well as shear, is then installed between adjacent stringers. Existing bearings are removed and a new bearing assembly is installed. In most instances, it will be necessary to add new stiffener plates and diaphragms at each interior pier. After the splice plates and bearing are in place, the reinforcement required in the deck over the piers is added and a deck replaced. Such a splice is shown in Figure 50.27.

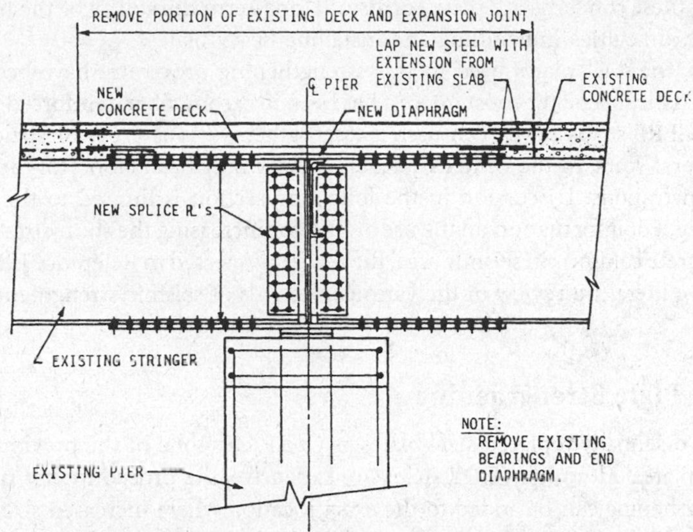

FIGURE 50.27 Conceptual details of a moment- and shear-type connection. (*Source:* Klaiber, F.W. et al., NCHRP 293, Transportation Research Board, 1987. With permission.)

Recently, the Robert Moses Parkway Bridge in Buffalo, New York [78] which originally consisted of 25 simply supported spans ranging from 63 ft (19.2 m) to 77 ft (23.5 m) in length was seismically retrofitted. Moment and shear splices were added to convert the bridge to continuous spans: one two-span element, one three-span element, and five four-span elements. This modification not only strengthened the bridge, but also provided redundancy and improved its earthquake resistance.

Timber Stringers
When providing continuity in timber stringers, steel plates can be placed on both sides and on the top and bottom of the connection and then secured in place with either bolts or lag screws. When adequate plates are used, this provides the necessary moment and shear transfer required. Additional strength can be obtained at the joint by injecting epoxy into the timber cracks as is suggested by Avent et al. [79]. Although adding steel plates requires the design and construction of a detailed connection, significant stress reduction can be obtained through its use.

50.7 Recent Developments

50.7.1 Epoxy Bonded Steel Plates

Epoxy-bonded steel plates have been used to strengthen or repair buildings and bridges in many countries around the world including Australia, South Africa, Switzerland, the United Kingdom, Japan, to mention a few.

The principle of this strengthening technique is rather simple: an epoxy adhesive is used to bond steel plates to overstressed regions of reinforced concrete members. The steel plates are typically located in the tension zone of a beam; however, plates located in the compression and shear zones have also been utilized. The adhesive provides a shear connection between the reinforced concrete beam and the steel plate, resulting in a composite structural member. The addition of plates in the tension zone not only increases the area of tension steel, but also lowers the neutral axis, resulting in a reduction of live-load stresses in the existing reinforcement. The tension plates effectively increase the flexural stiffness, thereby reducing cracking and deflection of the member.

Although this procedure has been used on dozens of bridges in other countries, to the authors' knowledge, it has not been used on any bridges in the United States due to concerns with the method. Some of these concerns are plate corrosion, long-term durability of the bond connection, plate peeling, and difficulties in handling and installing heavy plates.

In recent years, the steel plates used in this strengthening procedure have been replaced with fiber-reinforced plastic sheets; the most interest has been in carbon fiber-reinforced polymer (CFRP) strips. Although CFRP strips have been used to strengthen various types of structures in Europe and Japan for several years, in the United States there have only been laboratory investigations and some field demonstrations. Discussion in the following section is limited to the use of CFRP in plate strengthening. For information on the use of FRP for increasing the shear strength and ductility of reinforced concrete columns in seismic area, the reader is referred to Reference [91]. This reference is a comprehensive literature review of the various methods of seismic strengthening of reinforced columns.

50.7.2 CFRP Plate Strengthening

CFRP strips have essentially replaced steel plates as CFRP has none of the previously noted disadvantages of steel plates. Although CFRP strips are expensive, the procedure has many advantages: less weight, strengthening can be added to the exact location where increased strength is required, strengthening system takes minimal space, material has high tensile strength, no corrosion problems, easy to handle and install, and excellent fatigue properties. As research is still in progress in Europe, Japan, Canada, and the United States on this strengthening procedure, and since the application of CFRP strips obviously varies from structure to structure, rather than providing details on this procedure, several examples of its application will be described in the following paragraphs.

In 1994, legal truck loads in Japan were increased by 25% to 25 tons. After a review of several concrete slab bridges, it was determined that they were inadequate for this increased load. Approximately 50 of these bridges were strengthened using CFRP sheets bonded to the tension face. The additional material not only reduced the stress in the reinforcing bars, it also reduced the deflections in the slabs due to the high modulus of elasticity of the CFRP sheets.

Recently, a prestressed concrete (P/C) beam in West Palm Beach, Florida, which had been damaged by being struck by an overheight vehicle, was repaired using CFRP. This repair was accomplished in 15 hours by working three consecutive nights with minimal disruption of traffic. The alternative to this repair technique was to replace the damaged P/C with a new P/C beam. This procedure would have taken close to 1 month, and would have required some road closures.

The Oberriet–Meiningen three-span continuous bridge was completed in 1963. This bridge over the Rhine River connects Switzerland and Austria. Due to increased traffic loading, it was determined that the bridge needed strengthening. Strengthening was accomplished in 1996 by increasing the deck thickness 3.1 in. (8 cm) and adding 160 CFRP strips 13.1 ft (4 m) long on 29.5-in. (75-cm) intervals to the underside of the deck. The combination of these two remedies increased the capacity of the bridge so that it is in full compliance with today's safety and load requirements.

Three severely deteriorated 70-year-old reinforced concrete frame bridges near Dreselou, Germany, have recently been strengthened (increased flexure and shear capacity) using CFRP plates. Prior to strengthening, the bridges were restricted to 2-ton vehicles. With strengthening, 16-ton vehicles are now permitted to use the bridges. Prior to implementing the CFRP strengthening procedure, laboratory tests were completed on this strengthening technique at the Technical University in Brauwschweigs, Germany.

50.8 Summary

The purpose of this chapter is to identify and evaluate the various methods of strengthening existing highway bridges and to a lesser extent railroad bridges. Although very few references have been made to railroad bridges, the majority of the strengthening procedures presented could in most situations be applied to railroad bridges.

In this chapter, information on five strengthening procedures (lightweight deck replacement, composite action, strengthening of various bridge members, post-tensioning, and development of bridge continuity) have been presented. A brief introduction to using CFRP strips in strengthening has also been included.

In numerous situations, strengthening a given bridge, rather than replacing it or posting it, is a viable economical alternative which should be given serious consideration.

For additional information on bridge strengthening/rehabilitation, the reader is referred to References [1,2,60] which have 208, 379, and 199 references, respectively, on the subject.

Acknowledgments

This chapter was primarily based on NCHRP 12-28(4) "Methods of Strengthening Existing Highway Bridges" (NCHRP Report 293 [2]) which the authors and their colleagues, Profs. K. F. Dunker and W. W. Sanders, completed several years ago. Information from this investigation was supplemented with information from NCHRP Synthesis 249 "Methods for Increasing Live Load Capacity of Existing Highway Bridges" [1], literature reviews, and the results of research projects the authors have completed since submitting the final report to NCHRP 12-28(4).

We wish to gratefully acknowledge the Transportation Research Board, National Research Council, Washington, D.C., who gave us permission to use material from NCHRP 293 and NCHRP Synthesis 249. Information, opinions, and recommendations are from the authors. The Transportation Research Board, the American Association of State Highway and Transportation Officials, and the Federal Highway Administration do not necessarily endorse any particular products, methods, or procedures presented in this chapter.

References

1. Dorton, R. A. and Reel, R., Methods of Increasing Live Load Capacity of Existing Highway Bridges, Synthesis of Highway Practice 249, National Academy Press, Washington, D.C., 1997, 66 pp.
2. Klaiber, F. W., Dunker, K. F., Wipf, T. J., and Sanders, W. W., Jr., Methods of Strengthening Existing Highway Bridges, NCHRP 293, Transportation Research Board, 1987, 114 pp.
3. Sprinkle, M. M., Prefabricated Bridge Elements and Systems. NCHRP Synthesis of Highway Practice 119, Aug. 1985, 75 pp.
4. Shanafelt, G. O. and Horn, W. B., Guidelines for Evaluation and Repair of Prestressed Concrete Bridge Members, NCHRP Report 280, Dec. 1985, 84 pp.
5. Shanafelt, G. O. and Horn, W. B., Guidelines for Evaluation and Repair of Damaged Steel Bridge Members, NCHRP Report 271, June 1984, 64 pp.
6. Applied Technology Council (ATC), Seismic Retrofitting Guidelines for Highway Bridges. Federal Highway Administration, Final Report, Report No. FHWA-RD-83-007, Dec. 1983, 219 pp.
7. Sabnis, G. M., Innovative Methods of Upgrading Structurally and Geometrically Deficient through Truss Bridges, FHWA, Report No. FHWA-RD-82-041, Apr. 1983, 130 pp.
8. University of Virginia Civil Engineering Department, Virginia Highway and Transportation Research Council, and Virginia Department of Highways and Transportation, Bridges on Secondary Highways and Local Roads — Rehabilitation and Replacement, NCHRP Report 222, May 1980, 73 pp.
9. Mishler, H. W. and Leis, B. N., Evaluation of Repair Techniques for Damaged Steel Bridge Members: Phase I, Final Report, NCHRP Project 12-17, May 1981, 131 pp.
10. Shanafelt, G. O. and Horn, W. B., Damage Evaluation and Repair Methods for Prestressed Concrete Bridge Members, NCHRP Report 226, Nov. 1980, 66 pp.
11. University of Virginia Civil Engineering Department, Virginia Highway and Transportation Research Council, and Virginia Department of Highways and Transportation, Bridges on Secondary Highways and Local Roads — Rehabilitation and Replacement. NCHRP Report 222, May 1980, 73 pp.

12. Berger, R. H., Extending the Service Life of Existing Bridges by Increasing Their Load Carrying Capacity, FHWA Report No. FHWA-RD-78-133, June 1978, 75 pp.
13. Fisher, J. W., Hausamann, H., Sullivan, M. D., and Pense, A. W., Detection and Repair of Fatigue Damage in Welded Highway Bridges, NCHRP Report 206, June 1979, 85 pp.
14. Lichenstein, A. G., Bridge Rating through Nondestructive Load Testing, NCHRP Project 12-28(13)A, 1993, 117 pp.
15. Timmer, D. H., A study of the concrete filled steel grid bridge decks in Ohio, in *Bridge Maintenance and Rehabilitation Conference,* Morgantown, WV, Aug. 13–16, 1980, 422–475.
16. DePhillips, F. C., Bridge deck installed in record time, *Public Works,* 116(1), 76–77, 1985.
17. Bakht, B. and Tharmabala, T., Steel-wood composite bridges and their static load performance, in *Can. Soc. Civ. Eng. Annual Conference,* Saskatoon, May 27–31, 1985, 99–118.
18. Csagoly, P. F. and Taylor, R. J., A structural wood system for highway bridges, in *International Association for Bridge and Structural Engineering 11th Congress,* Final Report, Italy, Aug. 31–Sept. 5, 1980, 219–225.
19. Taylor, R. J., Batchelor, B. D., and Vandalen, K., Prestressed wood bridges, *Proc. International Conference on Short and Medium Span Bridges,* Toronto, Aug. 8–12, 1982, 203–218.
20. Ministry of Transportation and Communications, Design of Wood Bridge Using the Ontario Highway Bridge Design Code, Ontario, Canada, 1983, 72 pp.
21. Ministry of Transportation and Communications, Design of Prestressed Wood Bridges Using the Ontario Highway Bridge Design Code, Ontario, Canada, 1983, 30 pp.
22. Muchmore, F. W., Techniques to bring new life to timber bridges, *ASCE J. Struct. Eng.,* 110(8), 1832–1846, 1984.
23. Mackie, G. K., Recent uses of structural lightweight concrete, *Concrete Constr.,* 30(6), 497–502, 1985.
24. Steel grids rejuvenate old bridges, in *The Construction Advisor,* Associated General Contractors of Missouri, Jefferson City, MO, May 1974, 18–19.
25. "Lightweight decking rehabs downrated Ohio River span, *Rural Urban Roads,* 20(4), 25–26, 1982.
26. CAWV Members Join Forces to Reinforce Bridge, Reprint from *West Virginia Construction News,* 1982, 3 pp.
27. The rehabilitation of the Old York Road Bridge, *Rural Urban Roads,* 21(4), 22–23, 1983.
28. Campisi, V. N., Exodermic deck systems: a recent development in replacement bridge decks, *Modern Steel Constr.,* 26(3), 28–30, 1986.
29. Holm, T. A., Structural lightweight concrete for bridge redecking, *Concrete Constr.,* 30(8), 667–672, 1985.
30. Ford, J. H., Use of Precast, Prestressed Concrete for Bridge Decks, Progress Report, Purdue University and Indiana State Highway Commission, Joint Highway Research Project No. C-36-56N, No. 20, July 1969, 161 pp.
31. Kropp, P. K., Milinski, E. L., Gutzwiller, M. J., and Lee, R. H., Use of Precast-Prestressed Concrete for Bridge Decks, Final Report, Purdue University and Indiana State Highway Commission, Joint Highway Research Project No. C-36-56N, July 1975, 85 pp.
32. Greiner Engineering Sciences, Inc., Widening and replacement of concrete deck of Woodrow Wilson Memorial Bridge, Paper presented at Session 187 (Modular Bridge Decks), 62nd Annual Transportation Research Board Meeting, Jan. 20, 1983, 16 pp.
33. Nickerson, R. L., Bridge rehabilitation–construction view expediting bridge redecking, in *Proc. 2nd Annual International Bridge Conference,* Pittsburgh, PA, June 17–19, 1985, 5–9.
34. Stemler, J. R., Aluminum orthotropic bridge deck system, Paper presented at 3rd Annual International Bridge Conference, Pittsburgh, PA, June 2–4, 1986, 62–65.
35. Woeltinger, O. and Bock, F., The alteration of the High Bridge Levensau over the North-East Canal [Der Umbau der Hochbruecke Levensau über den Nord-Ostsee-Kanal.] *Stahlbau,* West Germany, 23(12), 295–303, 1956 [in German].
36. Freudenberg, G., Raising the load capacity of an old bridge, [Erhoehung der Tragfaehigkeit einer alten Bruecke], *Stahlau,* West Germany, 48(3), 76–78, 1979 [in German].

37. Bridge rebuilt with composite design, *Eng. News Rec.*, 165(20), 91, 1960.
38. Collabella, D., Rehabilitation of the Williamsburg and Queensboro Bridges — New York City, *Municipal Eng. J.*, 70 (Summer), 1984, 30 pp.
39. Cook, J. P., The shear connector, *Composite Construction Methods,* John Wiley & Sons, New York, 1977, 168–172.
40. Klaiber, F. W., Dedic, D. J., Dunker, K. F., and Sanders, W. W., Jr., Strengthening of Existing Single Span Steel Beam and Concrete Deck Bridges, Final Report — Part I, Engineering Research Institute Project 1536, ISU-ERI-Ames-83185, Iowa State University, Feb. 1983, 185 pp.
41. Dedic, D. J. and Klaiber, F. W., High strength bolts as shear connectors in rehabilitation work, *Concrete Int. Des. Constr.* 6(7), 41–46, 1984.
42. Dallam, L. N., Pushout Tests with High-Strength Bolt Shear Connectors, Missouri Cooperative Highway Research Program Report 68-7, Engineering Experiment Station, University of Missouri-Columbia, 1968, 66 pp.
43. Dallam, L. N., Static and Fatigue Properties of High-Strength Bolt Shear Connectors, Missouri Cooperative Highway Research Program Report 70-2, Engineering Experiment Station, University of Missouri-Columbia, 1970, 49 pp.
44. Dunker, K. F., Klaiber, F. W., Beck, B. L., and Sanders, W. W., Jr., Strengthening of Existing Single-Span Steel-Beam and Concrete Deck Bridges, Final Report — Part II, Engineering Research Institute Project 1536, ISU-ERI-Ames-85231, Iowa State University, March 1985, 146 pp.
45. Watter, F., Albrecht, P., and Sahli, A. H., End bolted cover plates, *ASCE J. Struct. Eng.*, 111(6), 1235–1249, 1985.
46. Park, S. H., *Bridge Rehabilitation and Replacement* (*Bridge Repair Practice*), S. H. Park, Trenton, NJ, 1984, 818 pp.
47. Albrecht, P., Watter, F., and Sahli, A., Toward fatigue-proofing cover plate ends, in *Proc. W. H. Munse Symposium on Behavior of Metal Structures, Research to Practice,* ASCE National Convention, Philadelphia, PA, May 17, 1983, 24–44.
48. Taylor, R., Strengthening of reinforced and prestressed beams, *Concrete,* 10(12), 28–29, 1976.
49. Warner, R. F., Strengthening, stiffening and repair of concrete structures, *Int. Assoc. Bridge Struct. Eng. Surv.*, 17 May, 25–41, 1981.
50. Westerberg, B., Strengthening and repair of concrete structures, [Forstarkning och reparation av betongkonstruktioner], *Nord. Betong,* Sweden, 7–13, 1980 [in Swedish].
51. Mancarti, G. D., Resurfacing, restoring and rehabilitating bridges in California, in *Proc. International Conference on Short and Medium Span Bridges,* Toronto, Aug. 8–12, 1982, 344–355.
52. Rodriguez, M., Giron H., and Zundelevich, S., Inspection and design for the rehabilitation of bridges for Mexican railroads, *Proc. 2nd Annual International Bridge Conference,* Pittsburgh, PA, June 17–19, 1985, 12 pp.
53. Warner, R. F., Strengthening, stiffening and repair of concrete structures, in *Proc. International Symposium on Rehabilitation of Structures,* Maharastra Chapter of the American Concrete Institute, Bombay, Dec. 1981, 187–197.
54. OECD, Organization for Economic Co-Operation and Development Scientific Expert Group, *Bridge Rehabilitation and Strengthening,* Paris, 1983, 103 pp.
55. Suter, R. and Andrey, D., Rehabilitation of bridges, [Assainissement de ponts] *Inst. Statique et Structures Beton Arme et Precontraint,* Switzerland, 106 (Mar.) 105–115, 1985, [in French].
56. Dilger, W. H. and Ghali, A., Remedial measures for cracked webs of prestressed concrete bridges, *J. PCI,* 19(4), 76–85, 1984.
57. West, J. D., Some methods of extending the life of bridges by major repair or strengthening, *Proc. ICE,* 6, (Session 1956–57) 183–215, 1957.
58. Stratton, F. W., Alexander, R., and Nolting, W., Development and Implementation of Concrete Girder Repair by Post-Reinforcement, Kansas Department of Transportation, May 1982, 31 pp.
59. Soliman, M. I., Repair of distressed reinforced concrete columns, in *Canadian Society for Civil Engineering Annual Conference,* Saskatoon, Canada, May 27–31, 1985, 59–78.

60. Wipf, T. J., Klaiber, F. W., and Russo, F. M., Evaluation of Seismic Retrofit Methods for Reinforced Concrete Bridge Columns, Technical Report NCEER-97-0016 National Center for Earthquake Engineering Research, Buffalo, NY, Dec. 1997, 168 pp.
61. Karamchandani, K. C., Strengthening of Reinforced Concrete Members, in *Proc. International Symposium on Rehabilitation of Structures,* Maharastra Chapter of the American Concrete Institute, Bombay, Dec. 1981, 157–159.
62. Klein, G. J. and Gouwens, A. J., Repair of columns using collars with circular reinforcement, *Concrete Int. Des. Constr.,* 6(7) 23–31, 1984.
63. Rheinisches Strassenbauamt Moenchengladbach, Rehabilitation of Structure 41 in Autobahnkreuz Holz, [Erfahrungsbericht–Sanierung des Bauwerks Nr. 41 in Autobahnkreuz Holz] Rheinisches Strassenbauamt Moenchengladbach, Germany, 1983, 11 pp. [in German].
64. Lee, D. H., Prestressed concrete bridges and other structures, *Struct. Eng.,* 30(12), 302–313, 1952.
65. Jungwirth, D. and Kern, G., Long-term maintenance of prestressed concrete structures — prevention, detection and elimination of defects, [Langzeitverhalten von Spannbeton — Konstruktionen Verhueten, Erkennen und Beheben von Schaden], *Beton- Stahlbetonbau,* West Germany, 75(11),262–269, 1980, [in German].
66. Mancarti, G. D., Strengthening California's Steel Bridges by Prestressing, TRB Record 950, Transportation Research Board (1984) 183–187.
67. Vernigora, E., Marcil, J. R. M., Slater, W. M., and Aiken, R. V., Bridge rehabilitation and strengthening by continuous post-tensioning, *J. PCI,* 14(2) 88–104, 1969.
68. Ferjencik, P. and Tochacek, M., *Prestressing in Steel Structures,* [Die Vorspannung in Stahlbau], Wilhelm Ernst & Sohn, West Germany, 1975, 406 [in German].
69. Kandall, C., Increasing the load-carrying capacity of existing steel structures, *Civ. Eng.,* 38(10), 48–51, 1968.
70. Benthin, K., Strengthening of Bridge No. 3699, Chaska, Minnesota, Minnesota Department of Transportation, 1975, 11 pp.
71. Andrey, D. and Suter, R., *Maintenance and Repair of Construction Works,* [Maintenance et reparation de ouvrages d'art], Ecole Polytechnique Federale de Lausanne, Lausanne, Switzerland, 1986, [in French].
72. Mueller, T., Alteration of the highway bridge over the Aare River in Aarwangen, [Umbau der Strassenbruecke über die Aare in Aarwangen] *Schweiz. Bauz.,* 87(11), 199–203, 1969, [in German].
73. Taylor, R. J. and Walsh, H., Prototype Prestressed Wood Bridge, *TRB Report 950,* Transportation Research Board, 1984, 110–122.
74. Lamberson, E. A., Post-Tensioning Concepts for Strengthening and Rehabilitation of Bridges and Special Structures: Three Case Histories of Contractor Initiated Bridge Redesigns, Dywidag Systems International, Lincoln Park, NJ, 1983, 48 pp.
75. Dunker, K. F., Klaiber, F. W., and Sanders, W. W., Jr., Post-tensioning distribution in composite bridges, *J. Struct. Eng. ASCE,* 112 (ST11), 2540–2553, 1986.
76. El-Arabaty, H. A., Klaiber, F. W., Fanous, F. S., and Wipf, T. J., Design methodology for strengthening of continuous-span composite bridges, *J. Bridge Eng., ASCE,* 1(3), 104–111, 1996.
77. Roberts, J., Manual for Bridge Maintenance Planning and Repair Methods, Florida Department of Transportation, 1978, 282 pp.
78. Malik, A. H., Seismic Retrofit of the Robert Moses Parkway Bridge, in *Proc. 12th U.S.–Japan Bridge Engineering Workshop,* Buffalo, NY, Oct. 1996, 215–228.
79. Avent, R. R., Emkin, L. Z., Howard, R. H., and Chapman, C. L., Epoxy-repaired bolted timber connections, *J. Struct. Div. Proc. ASCE,* 102(ST4), 821–838, 1976.

Section VI

Special Topics

51

Applications of Composites in Highway Bridges

Joseph M. Plecnik
California State University, Long Beach

Oscar Henriquez
California State University, Long Beach

51.1 Introduction

Building a functional transportation infrastructure is a high priority for any nation. Equally important is maintaining and upgrading its integrity to keep pace with increasing usage, higher traffic loads, and new technologies. At present, in the United States a great number of bridges are considered structurally deficient, and many are restricted to lighter traffic loads and lower speeds. Such bridges need to be repaired or replaced. This task may be achieved by using the same or similar technologies and materials used originally for their initial construction many years ago. However, new materials and technologies may provide beneficial alternatives to traditional materials in upgrading existing bridges, and in the construction of new bridges. Composite materials offer unique properties that may justify their gradual introduction into bridge repair and construction.

The difference between industrial or commercial composites and advanced composites is vague but based primarily on the quality of materials. Advanced composites utilize fibers, such as graphite and Kevlar®, and matrix materials of higher strength and modulus of elasticity than industrial composites, which usually are fabricated with E-glass or S-glass fibers and with polyester or vinyl ester matrices. Advanced composites use polymer matrix materials, such as modified epoxies and polyimides, or ceramic and metal matrices. More-sophisticated manufacturing techniques are generally required to produce advanced composites. Industrial composites require little or no special curing processes, such as the use of autoclaves or vacuum techniques for advanced composites.

Composites are herein limited to materials fabricated with thin fibers or filaments and bonded together in layers or lamina with a polymer matrix. The polymers discussed are primarily polyesters and epoxies. The fibers considered include glass (E and S type), graphite, and Kevlar. The filaments may be of short fiber length (such as chopped fibers which may be less than 25 mm long) or continuous filaments. The ability to orient the fibers in any desired direction is one of the truly great advantages of composite materials as opposed to isotropic materials such as steel. The anisotropic nature of composite materials enables an engineer literally to custom design each element within a structure to achieve the optimum use of material properties.

Composite materials have been successfully utilized in many other industries in the past 50 years. The leisure industry, primarily boating, was probably the first industry to successfully and overwhelmingly adopt composite materials in the construction of pleasure craft and small ships. In the industrial application fields, pipes, tanks, pressure vessels, and a variety of other components manufactured primarily with fiberglass, composite materials have been used for over 50 years. In the defense and aerospace industry, more-advanced composites have been increasingly used since the early 1960s. In all of these industries, one or several unique properties of composites were successfully exploited to replace conventional materials.

The initial study [1] on the use of composites in bridges was performed for the U.S. Federal Highway Administration in the early 1980s and had as its main objective the determination of the feasibility of adapting composites to highway bridges. This study considered the adaptability of composites to major bridge components, given the unique characteristics of these types of materials. The study concluded that bridge decks and cables are the most suitable bridge components for use of composite materials.

The purpose of this chapter is to introduce the current and future technologies and the most feasible applications of composites in highway bridge infrastructure. Basic composite material properties are presented and their advantages and disadvantages discussed. The applications of composites in bridges presented in this chapter include beams and girders, cables, reinforcing bars, decks and wearing surface, and techniques to repair or retrofit existing bridge structures. The methods and significance of nondestructive evaluation techniques are also discussed relative to the feasibility of incorporating composite components into bridge systems.

51.2 Material Properties

51.2.1 Reinforcing Fibers

Fibers provide the reinforcement for the matrix of composite materials. Fiber reinforcement can be found in many forms, from short fibers to very long strands, and from individual fibers to cloth and braided material. The fibers provide most of the strength of the composites since most matrix materials have relatively low strength properties. Thus, fibers in composites function as steel in reinforced concrete.

The most typical fiber materials used in civil engineering composite structures are glass, aramid (Kevlar), and graphite (carbon). A variation in mechanical properties can be achieved with different types of fiber configurations. A comparison of typical values of mechanical properties for common reinforcing fibers is provided in Table 51.1.

TABLE 51.1 Typical Fiber Design Properties

Property	E-Glass (Strand)	S-Glass (Strand)	Kevlar-49 (Yarn)	High-Modulus Graphite (Tow)	High-Strength Graphite (Tow)
Tensile strength (MPa)	3100	3800	3400	2200	3600
Tensile modulus (GPa)	72	86	124	345	235
Specific gravity	2.60	2.50	1.44	1.90	1.80
Tensile elongation (%)	4.9	5.7	2.8	0.6	1.4

TABLE 51.2 Typical Properties of Polymer Resins

Property	Polyester	Epoxy	Phenolic
Tensile strength (MPa)	55	27–90	35–50
Tensile modulus (GPa)	2.0	0.70–3.4	7.0–9.7
Specific gravity	1.25–1.45	1.1–1.4	1.4–1.9
Elongation (%)	5–300	3–50	—
Coefficient of thermal expansion (10^{-6} m/m/K)	70–145	18–35	27–40
Water absorption (% in 24 h)	0.08–0.09	0.08–0.15	0.30–0.50

Glass fiber has been the most common type of reinforcement for polymer matrix. Glass fibers, which are silica based, were the first synthetic fibers commercially available with relatively high modulus. Two common types of glass fibers are designated as E-glass and S-glass. E-glass fibers are good electrical insulators. S-glass fibers, which have a higher silica content, possess slightly better mechanical properties than E-glass. Some applications require fibers with better strength or elastic modulus than glass. Graphite (carbon) and aramid fibers can provide these desired properties. The use of these fibers is generally selective in civil engineering applications, given their higher cost compared with glass fibers.

51.2.2 Matrix Materials

Thermosetting polymer resins are the type of matrix material commonly used for civil engineering applications. Polymers are chainlike molecules built up from a series of monomers. The molecular size of the polymer helps to determine its mechanical properties. Thermosetting polymers, unlike thermoplastic polymers, do not soften or melt on heating, but they decompose. Other matrix materials, such as ceramics and metals, are used for more-specialized applications.

The most common thermosetting resins used in civil engineering applications are polyesters, epoxies, and to a lesser degree, phenolics. A summary of typical properties for resins is provided in Table 51.2. Polyester resins are relatively inexpensive, and provide adequate resistance to a variety of environmental factors and chemicals. Epoxies are more expensive but also have better properties than polyesters. Some of the advantages of epoxies over polyesters are higher strength, slightly higher modulus, low shrinkage, good resistance to chemicals, and good adhesion to most fibers. Phenolic resin is generally used for high-temperature (150 to 200°C) applications and relatively mild corrosive environments.

51.3 Advantages and Disadvantages of Composites in Bridge Applications

The rapid rise in the use of composites in many industries, such as aerospace, leisure, construction, and transportation, is due primarily to significant advantages of composites over conventional materials, such as metals, concrete, or unreinforced plastics. The following presents a brief discussion on the probable advantages and disadvantages of composites in highway bridge type applications.

The first and primary advantage of composites in bridge structures will probably be a significant reduction in weight, due to the higher specific strength (strength/density) of composites over conventional materials, such as steel and concrete. The lightweight advantage of composites in bridge decks is clearly illustrated in Table 51.3. In most short bridge applications, the lighter structural system, if adequate from the structural point of view, will probably not affect the dynamic performance of the bridge. In longer bridges, it is conceivable that a lighter-weight system may require additional design considerations to avoid dynamic behavioral problems.

The second and equally important advantage of composites is their superior corrosion resistance in all environments typically experienced by bridges throughout the world. Corrosion resistance of composites can be further enhanced by the use of premium resin systems, such as vinyl esters or epoxies in comparison with conventional resins, such as polyesters. The excellent corrosion resistance characteristics of composites, and the lower maintenance costs, may result in lower life-cycle costs than those of bridge components manufactured with steel or concrete materials. The lower life-cycle costs may be the third significant advantage of composite bridge components. However, it is anticipated that the initial cost of such composite bridge components will be considerably greater than that of conventional materials.

The fourth significant advantage of composites in bridge applications is their modular construction. It is envisioned that composite bridge deck components will be fabricated in large modules, either in the shop or at the bridge site, then assembled at the bridge site to form a desired structural system. Such modular construction will not only reduce construction costs, but also reduce the time of construction. Fifth, it is envisioned that the initial usage of composites in bridges will involve rehabilitation or retrofitting of existing bridges in large urban areas. The modular construction described above will greatly reduce the time required for retrofitting, thus reducing traffic congestion, accidents, and time delays for commuters in heavily traveled urban areas. The layered structure of composites is also an advantage that may be highly beneficial for fatigue-type loads in bridges. By placing fibers in appropriate directions, both the strength and fatigue resistance of the composite laminate is greatly enhanced. The fatigue behavior of composites when properly designed is superior to that of ductile materials, such as the conventional A36 steel.

The disadvantages of utilizing composites in infrastructure applications such as bridges are considerable, but not overwhelming. The first, but not necessarily the most significant, disadvantage of composites is their relatively high initial costs. This topic was discussed in the previous section relative to initial vs. life-cycle costs. Although graphite and other advanced fibers will probably reduce in cost with increased volume of consumption, it is very doubtful that the cost of glass fibers can be significantly reduced with increasing volume of consumption. The cost of matrices, such as polymer-based resins, will also not be reduced significantly with increased consumption.

The second disadvantage of composite structural systems is the lack of highly efficient mechanical connections. The mechanical bolted connections in composite applications are not as efficient or as easily designed as in the case of steel-type welded and bolted connections used in steel structures. To reduce mechanical-type connections, adhesive-type joints are required. However, adhesion of one part to another requires detailed knowledge of the adhesive and the bond surfaces, as well as quality control. All of these factors generally result in relatively low allowable adhesive stresses. Furthermore, many engineers tend to dislike adhesive-type connections in the presence of fatigue and vibration-type loads.

TABLE 51.3 Comparison of Dead Load (D.L.) of Deck Systems and Superstructure

Bridge Type	Bascule with 1.22 m Stringer Spacing (span = 76.2 m, width = 18.9 m)				Deck on Steel I-Girder 2.13 m Girder Spacing (span = 16.3 m, width = 8.5 m)		Deck on AASHTO Type III Prestressed Girders Spaced at 2.13 m (span = 16.3 m, width = 8.5 m)	
Deck Type	127 mm Open Steel Grid	152 mm Deep X-Shaped FRP with Sand Layer Wearing Surface	127 mm Concrete-Filled Steel Grid	152 mm Deep X-Shaped FRP with Sand Layer Wearing Surface	165 mm Thick Concrete with 5 mm Wearing Surface	229 mm Deep X-Shaped FRP with Sand Layer Wearing Surface	178 mm Thick Concrete with 5 mm Wearing Surface	229 mm Deep X-Shaped FRP with Sand Layer Wearing Surface
Deck weight only (KN)	1379	1155	5654	1155	541.4	157	583.2	157
Deck D.L. % reduction	16		80		71		73	
Girder weight (KN)	3610		3610		111.8		693.5	
Curbs and railing (KN)	300.3		300.3		166.6		166.6	
Future wearing surface (KN)	0	17.8	0	17.8	166.6	2.2	166.6	2.2
Details, stiffeners, etc. (KN)	290.9		290.9		35.7			23.8
Inspection walkway (KN)	66.7		66.7		None		None	
Total D.L. (KN)	5647	5441	9922	5441	1022	473.3	1634	1043
Total D.L. % reduction	3.6		45		54		36	

The third disadvantage is the relatively low modulus of glass fiber composites. Unless all fibers are oriented in a single direction, the modulus of elasticity of glass-type composites (E- or S-glass) will be somewhat similar to that of concrete. Since design of bridges is often governed by deflection or stiffness criteria, as opposed to strength, the cross-sectional properties of the fiberglass component would have to be nearly identical to that of concrete. The use of high-modulus fibers, such as graphite, enhances the modulus or stiffness characteristics of composites. However, even if all the graphite fibers are placed in the same direction (unidirectional laminate), the modulus of elasticity of the composite may not approach that of steel. Only with the use of very high modulus fibers (above 350 GPa), will the tensile modulus of the composite approach that of steel. To alleviate this very significant stiffness disadvantage, composite structural systems must generally be designed differently when stiffness criteria govern the design.

The fourth significant disadvantage is the relatively low fire resistance of structural composites where polymer-based matrices are used, which represent the bulk of the composites utilized outside of the aerospace industry. This disadvantage has effectively disallowed the use of polymer-based composites in fire-critical applications such as buildings. In bridge applications, fire is a relatively infrequent phenomenon. Elevated temperatures, such as in the southwestern part of the United States, may, however, affect the structural properties of composites on bridge applications.

Several additional disadvantages of composites include relatively complex material properties and current lack of codes and specifications, which tend to dissuade engineers from understanding and utilizing such materials. The presence of local defects, which are difficult and perhaps impossible to detect on a large structural system, are also viewed as a significant quality control problem.

51.4 Pultruded Composite Shapes and Composite Cables

51.4.1 Pultruded Composite Shapes

Composites are commercially available in a variety of pultruded shapes [2–4]. Some of the most common shapes available for construction purposes are I-beams, W-sections, angles, channels, square and rectangular tubes, round tubes, and solid bars. However, almost any shape of constant cross section can be pultruded.

Composite pultruded beam shapes have a potential use in bridges. However, the relatively low modulus of glass and graphite composite shapes limits their use. The effect of modulus of elasticity can be seen with the following comparison between A36 steel beams with modulus of elasticity $E =$ 200 GPa, fiber-reinforced polymer (FRP) beams with $E = 17.2$ GPa, graphite beams with $E = 103.5$ GPa, and glass fiber–reinforced polymer (GFRP)/graphite hybrid beams for a two-lane, 16.76-m-span bridge. Assuming full lateral support, a total of five beams spaced at 2.29 m, and a 178-mm-thick concrete slab, the following results are obtained.

For the case of noncomposite action between beams and concrete slab, a steel beam, W36 × 194, with cross-sectional area, $A = 36{,}770\ \text{mm}^2$, satisfies all AASHTO requirements for HS20-44 loading [5,6]. Using GFRP beams, the deflection requirement of $L/800$, which controls the design, cannot possibly be satisfied using a depth of 914 mm and a flange width of 457 mm. A GFRP I48 × 24 × 3.25 with $A = 187{,}700\ \text{mm}^2$ or I60 × 30 × 1.5 with $A = 113{,}200\ \text{mm}^2$ beam is necessary. If an all-graphite beam is used, an I36 × 18 × 1.25 with $A = 56{,}060\ \text{mm}^2$ will satisfy all requirements. A hybrid beam with a 1320-mm total depth, 660-mm flange width, and 45.7-mm web and flange thickness also satisfies stiffness requirements. For this example, the hybrid beam should have a 7.6-mm-thick layer of graphite in the center of both flanges, and for the total width of the flange.

If composite action is achieved between the concrete slab and the beams, a W30 × 99 with $A =$ 18,770 mm^2 steel beam is adequate for this bridge. An all-FRP I48 × 24 × 1.0 beam with $A =$ 60,650 mm^2 or I42 × 21 × 1.5 with $A = 78{,}390\ \text{mm}^2$ will also meet stiffness and stress requirements when composite action is included. A comparison of sizes for all the beam cross sections used in this example is presented in Figure 51.1.

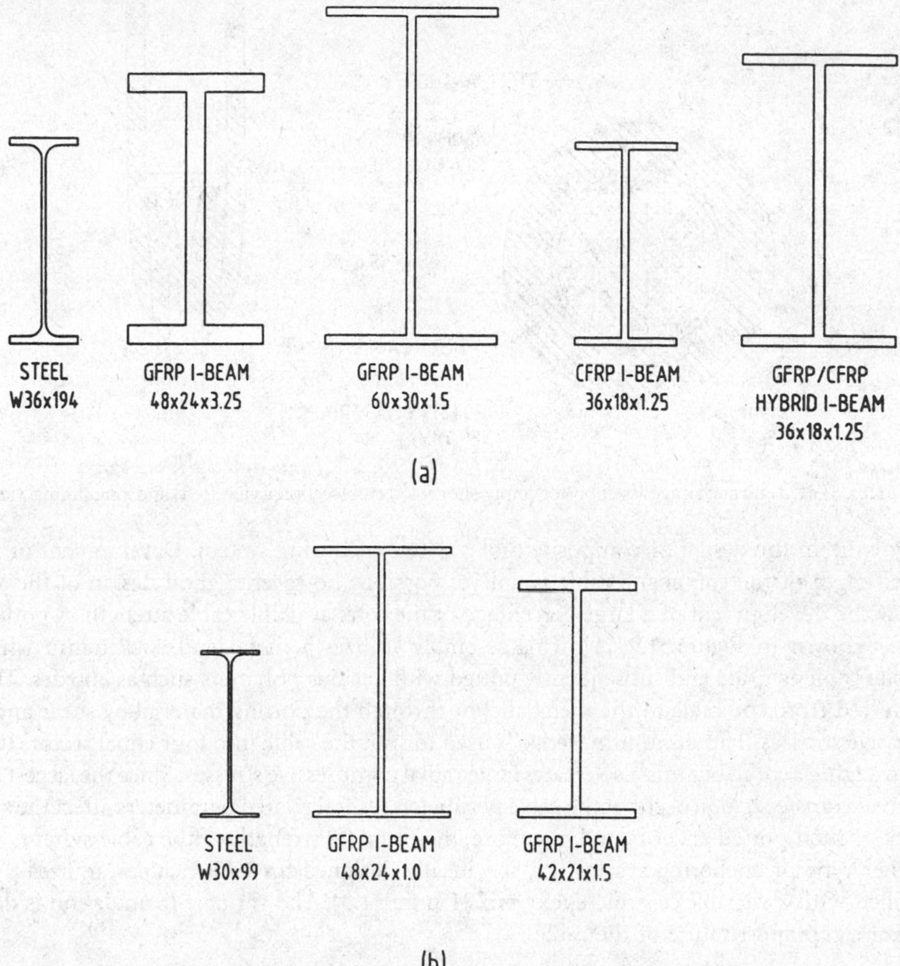

FIGURE 51.1 Comparison of different beams for a two-lane 16.80-m-span bridge, with a total of five beams spaced at 2.30 m, and a 180-mm-thick concrete slab. (a) Noncomposite action between beams and slab; (b) composite action between beams and slabs.

51.4.2 Composite Cables

Composites in the form of cables, strands, and rods have potential applications in bridges. Among these applications are suspension and stay cables and prestressing tendons. High tensile strength, corrosion resistance, and light weight are the most important characteristics that make composites strong candidates to replace steel for these types of applications. Corrosion of traditional steel cables and tendons may impose a significant maintenance cost for bridges. Composite cables, with proper selection of materials and design, may exceed the useful life of traditional bridge cables.

Carbon fiber–reinforced polymer (CFRP) composite cables have been used for cable-stay bridges [7]. Compared with steel, carbon composites can provide the equivalent tensile strength with only a fraction of the weight. GFRP tendons have been used to prestress concrete bridge girders. The computation of section strength using GFRP tendons is very similar to the methods used for steel tendons. In the case of post-tensioned structures, an adequate anchorage system must be used to minimize prestressing losses. One of the advantages of GFRP compared with steel tendons is that a lower modulus of elasticity is translated into lower prestressing losses, due to creep and shrinkage of the concrete.

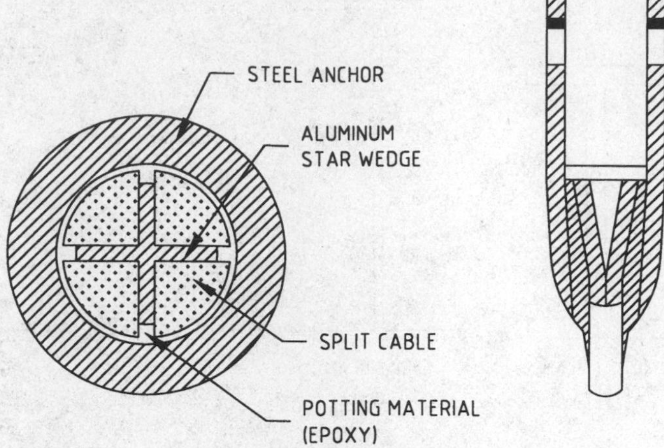

FIGURE 51.2 Potted end anchorage assembly for composite cables; cross sectional view (left) and longitudinal view (right)

A key issue in the design of composite cables is the anchoring system. Development of the full tensile strength of the composite cable is not yet possible; however, a good design of the anchors can allow the development of a large percentage of the total available cable strength. A potted-type anchor is shown in Figure 51.2 [1]. This assembly utilizes a metal end socket into which the composite cable is fitted and subsequently potted with various polymers such as epoxies. The load is transferred from the cable to the metal anchor through the potting material by shear and radial compressive stresses. The aluminum wedge is used to split the cable into four equal sectors to create greater wedging action, but this also creates large radial compressive stresses. Since the largest stresses at the stress transfer region occur at the cable perimeter, several related parameters affect the strength property of such potted anchors and, therefore, the ultimate strength of the cable system.

Another type of anchoring system [7], specifically designed for CFRP cables, utilizes a conical cavity filled with a variable ceramic/epoxy mix (Figure 51.3). The variable formulation is designed to control creep and rupture of the cable.

51.5 FRP Reinforcing Bars for Concrete

Fibers such as glass, aramid, and carbon can be used as reinforcing bars (rebars) for concrete beams. The use of these fibers can increase the longevity of this type of structural element, given the corrosive deterioration of steel reinforcement in reinforced concrete members. Tests have shown that a higher ultimate strength can be achieved with FRP rebars than with mild steel rebars. This strength can be achieved due to the high tensile strength of most fibers. The lower stiffness of FRP fibers, such as glass, will result in larger deflections compared with steel-reinforced concrete.

An important factor in the use of FRP bars is the bond between the bar and the concrete [8]. The use of smooth FRP bars results in a significant reduction of flexural capacity. Thus, smooth FRP bars must be surface-treated to improve bonding by methods such as sand coating. Test results have also shown that smaller-diameter FRP rebars are more effective for flexural capacity than larger-diameter bars. However, in general, bond characteristics are variable due to the variations in FRP reinforcing bar products. Other factors that affect the bond characteristics are concrete strength, concrete confinement, type of loading, time-dependent effects, amount of concrete cover, and type and volume of fiber and matrix. In the State-of-the-Art Report 440R-96 on FRP Reinforcement for Concrete Structures [9], the American Concrete Institute (ACI) recognizes the need for additional testing data to develop expressions that will be valid for different conditions, and can be included in a design code. Some expressions for FRP bar development lengths have been proposed recently.

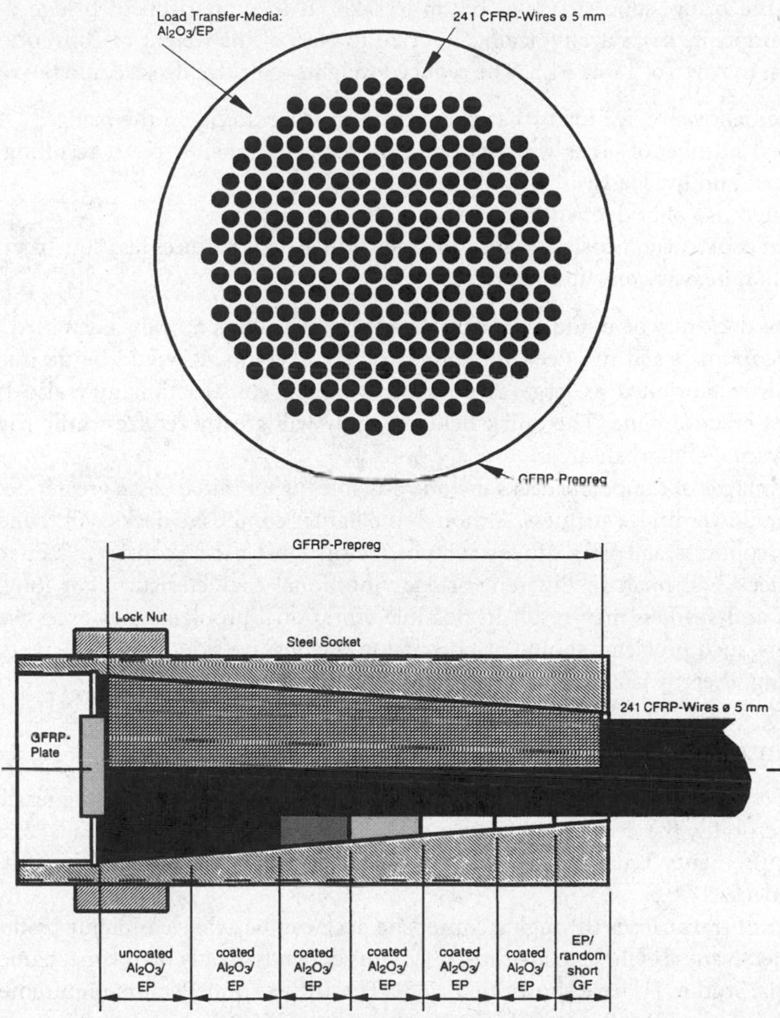

FIGURE 51.3 Specifically designed conical-shaped anchor enhances load transfer to the carbon cable.

51.6 Composite Bridge Decks

51.6.1 Advantages and Disadvantages

The bridge deck appears to be one of the most suitable bridge components for use of structural composites in highway applications [1]. The primary advantages of composite bridge decks are their relative lighter weight, corrosion resistance, and fabrication in modular units which may be rapidly installed without the need for shoring and formwork.

Reference [1] provides the results of a study of bridge dead loads with various types of conventional and composite bridge decks. Table 51.3 provides a summary of this comparison for a 76.2-m-long bascule bridge and two 16.3-m-span conventional bridges. The deck used for the comparison is the X-shaped cross section yielding truss-type deck behavior. The last row of Table 51.3 indicates that for bascule-type bridges with an open steel grid deck, the composite deck would not appreciably reduce the total dead loads of the bridge superstructure (deck, stringers, and girders). However, if the steel grid is filled with concrete, the total dead-load reduction with a composite deck is 45%. Similarly, for conventional bridges, the composite deck reduces the total dead load of the bridge superstructure by up to 54%. If a comparison of bridge decks alone is considered, composite decks are typically about 20 to 30% of the weight of conventional concrete decks as shown in row 3 of Table 51.3. The reduced weights of bridge decks could be translated into:

1. Increased allowable live loads that result from moving traffic on the bridge;
2. Increased number of lanes with the same girders, columns, or piers, resulting in the same total dead and live loads;
3. Continued use of bridge without reducing its load capacity;
4. Reduced construction costs, because a lighter bridge deck requires less construction time and effort than heavier conventional decks.

A composite deck may be made of prefabricated modular units quickly assembled at the bridge site. Due to economics and the need for minimization of joints, it would be desirable that deck sections could be fabricated as large as possible. Modular construction may also translate into relatively short erection time. The quick field assembly will greatly reduce traffic routing costs, a significant advantage in urban areas.

The disadvantages of composite decks include possible higher initial costs, greater deck and girder deflections, and lower bridge stiffness. Although the lighter composite decks will reduce dead loads on the girders, columns, and piers, other structural factors must also be considered. First, the reduced mass of the deck will result in different bridge vibrational characteristics. For long bridges, the reduced mass and stiffness may result in possible vibrational problems and excessive deflections. For short spans, such problems should not occur. On the positive side, composite materials provide higher damping, thereby reducing these vibrational tendencies.

51.6.2 Composite Deck Systems

The choice of a deck configuration should be made on structural and economic feasibility considerations. Structurally, the deck should carry dead loads and specified live loads, and also satisfy deflection requirements. Economically, a composite deck should be cost-effective if it is to replace conventional decks.

The transfer of traffic loads through a composite deck can be achieved mainly by flexure or truss action. The effectiveness of these load-transfer systems depends greatly on the mechanical properties of the materials. Studies [1] have shown that AASHTO stiffness or deflection requirements for decks are difficult to satisfy with low-modulus materials such as GFRP. It was also shown that truss-type load-transfer elements (Figure 51.4a through e) are preferable to sandwich or flexural-type structural elements (Figure 51.4f).

51.6.3 Truss-Type Deck System

A composite deck system that transfers the traffic loads to the stringers and girders mainly by truss action in the transverse direction of the bridge has been developed [1]. Several shapes were studied and evaluated based on AASHTO requirements in order to determine an economical and structurally efficient deck cross section. A deck with a total depth of 229 mm satisfied the stress and stiffness requirements for all the cross sections considered and shown in Figure 51.4. The X-shaped cross section (Figure 51.4b) is the optimum design from the viewpoint of stiffness and dead load. The X-shaped deck transfers live loads primarily by truss action, which provides less deflection than flexure-type members.

In addition to analytical studies, an extensive experimental program has proved the feasibility of the X-shaped cross section for a composite deck [1,10,11]. Specimens were fabricated and subjected to static and fatigue testing under AASHTO loads and the heavier "alternate military load." These specimens sustained over 30 million fatigue cycles without failure or degradation, and suffered only minor overall stiffness loss.

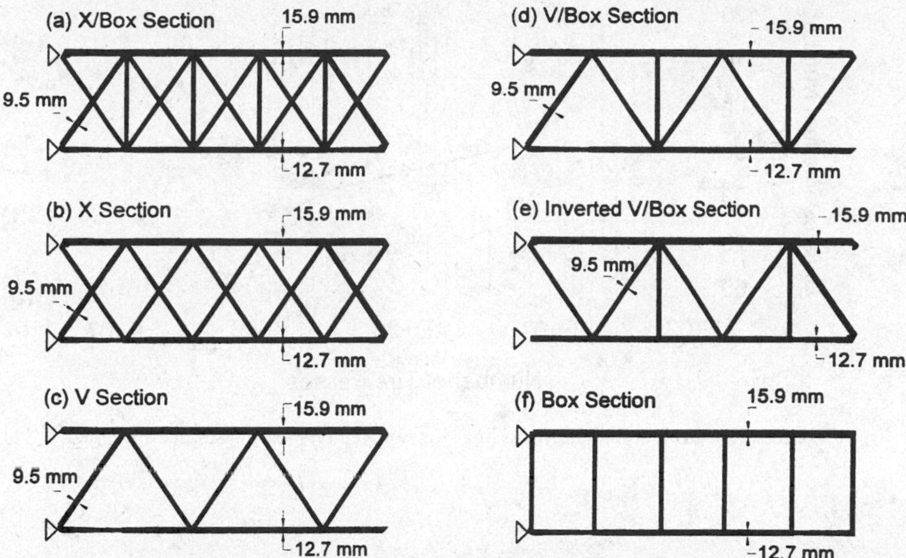

FIGURE 51.4 Shapes considered for design of a truss-type composite bridge deck. (a) X/box section; (b) X section; (c) V section; (d) V/box section; (e) inverted V/box section; (f) box section.

51.7 Wearing Surface for a Composite Deck

The wearing surface of a composite bridge deck should provide laminate protection, adequate skid resistance, and safety against hydroplaning. A thin layer of sand–epoxy mix has been developed for this purpose. The mix consists of sand retained between No. 8 and No. 30 sieves and a matrix-type epoxy. The mix is applied directly to the top surface of the composite FRP deck to a thickness of 1.5 to 3 mm after surface preparation.

The performance of this wearing surface has been evaluated with a series of tests. Freeze/thaw cycling and high-temperature tests were performed to determine the response of the system to weather conditions expected in most parts of the United States.

Simulated truck traffic was applied to evaluate the performance in terms of particle loss and abrasion. Specimens were tested using the accelerated loading facility (ALF), which consists of a frame with a set of truck tires that run along a stretch of pavement. Specimens with the sand–epoxy wearing surface were embedded in the pavement, and their integrity observed during the test. When there is a loss of sand particles, the skid resistance and texture depth of the wearing surface decreases, and the risk of skidding and hydroplaning increases. The surface deterioration was monitored using the British Pendulum method and the sand patch test [12]. Variation of the average British Pendulum Number (BPN) with number of tire passes is shown in Figure 51.5. A stabilized BPN above the minimum acceptable BPN of 60, after 1 million cycles, may indicate that the wearing surface will maintain its serviceability for an extended amount of time. The sand patch readings, shown in Figure 51.6, also indicate a reduction in the rate of mean average texture height loss at an acceptable level.

51.8 Composite Bridge Structural Systems

The use of composite materials to build an entire bridge superstructure is a possibility that is being explored by engineers. An all-composite bridge, designed to meet AASHTO HS25 loading, was built and installed near Russell, Kansas in 1996 [13]. The net span is 7.08 m and the width is 8.45 m. Three side-by-side panels connected by interlocking longitudinal joints were used to construct the

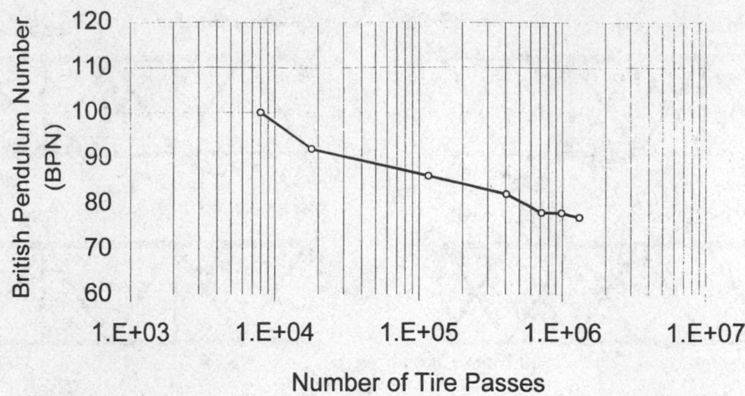

FIGURE 51.5 British pendulum readings for sand–epoxy wearing surface specimen.

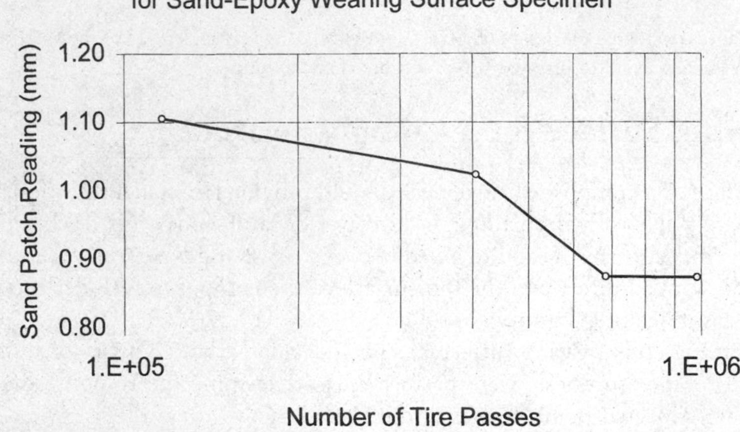

FIGURE 51.6 Sand patch test readings for sand–epoxy wearing surface specimen.

bridge. A 56-cm-deep sandwich construction was used for the panels, whose facing thicknesses were 13 and 19 mm for the top and bottom, respectively. A combination of chopped strand mat and uniaxial fibers was used with polyester resin to build the facings, which are attached by a honeycomb core. The wearing surface was a 19-mm-thick gravel–polyester resin mix.

Another all-composite 10-m-wide by 30-m-long bridge was installed in Hamilton, Ohio in 1997 [14]. This bridge, designed to meet the AASHTO HS20 specification, was fabricated with polyester resin matrix and E-glass fiber reinforcement. The bridge was delivered to the site in three sections, which consisted of two main components that formed the final box beam type of structure. The first component was a tapered U-shaped beam of approximately 0.6 m depth. The webs and the lower flange of the beam were reinforced with stitched triaxial and biaxial fabrics. The flange was also reinforced with additional unidirectional fibers. Beams, the second main component of the structure, were integral with the composite deck. A sandwich panel construction (approximately 15 cm deep) was used for the deck, with flat composite facing plates and a core of pultruded rectangular tubes oriented in the transverse direction of the bridge. The total weight of the composite bridge was approximately 100 kN, including the guardrails, but excluding the asphalt wearing surface.

51.9 Column Wrapping Using Composites

A unique application of composite materials in bridge infrastructure is bridge column wrapping or jacketing. This procedure involves the application of multiple layers of a composite around the perimeter of columns. Since the late 1980s, column wrapping with composites was seen as an alternative to the conventional steel jackets used to retrofit reinforced concrete columns of bridges in California. Column wrapping may also be used to repair columns that suffered a limited amount of damage. Column wrapping with composites may have some advantages over steel jacketing, such as reduced maintenance, improved durability, speed of installation, and reduced interference with ongoing operations, including traffic.

A reinforced concrete column can be retrofitted using the wrapping technique to increase its flexural ductility and shear strength. A proper confinement of the concrete core and longitudinal reinforcement is highly desirable for a ductile design. Confinement has been used to prevent the longitudinal bars from buckling, even after a plastic hinge has formed in the confined region, and to improve the performance of lapped longitudinal reinforcement in regions of plastic hinge formation. The shear capacity of the column can also be increased by wrapping a column using composites.

Several materials have been used to retrofit bridges with the wrapping method. The most common types of fibers used are glass and carbon. Glass fibers have been used with a polyester, vinyl ester, or epoxy matrix, and carbon fibers are used with epoxy resins. Fibers may be applied in various forms, such as individual rovings, mats, and woven fabrics. The California Department of Transportation (Caltrans) uses its Composite Specification to establish standard procedures for selection of the system to be used, material properties, and application.

The most common techniques for application of composite column jackets are wet wrap, prepreg wrap, and precured shells. In the wet wrap technique, a fiberglass fabric is wrapped around the column as many times as required to achieve the design thickness. The fabric is saturated with resin just before or during the application process, and then allowed to cure at ambient temperature. This system is manually applied and does not require special equipment.

The second technique, prepreg wrap, involves the use of continuous prepreg carbon fiber/epoxy tow that is mechanically wound onto the column. External heating equipment may be used to cure the composite.

Another column-wrapping system uses a precured shell. These shells, usually with glass fiber reinforcement, are fabricated with the same curvature as the column. A longitudinal cut is made on one side of the shell, or it may be cut in two longitudinal sections. The cut shell is then fitted and bonded onto the column.

Experimental studies [15] have been performed to determine the effectiveness of the column-wrapping systems using composite materials as compared to steel jacketing. However, even though favorable results have been published, acceptance by bridge owners and design engineers is not yet universal.

51.10 Strengthening of Bridge Girders Using CFRP Laminates

Strengthening and repair of bridge girders have been recently achieved with the use of CFRP. This technique was initially developed in the late 1980s [16] and applied to bridge structures in the early 1990s.

The use of CFRP for girder strengthening is similar to the attachment of steel plates onto concrete girders. However, CFRP presents the advantages of easier handling, higher corrosion resistance, elimination of welded connections, and excellent fatigue behavior. An important factor to be considered when using this technique is the adhesion between the beam and the CFRP strip. The contact surfaces must be adequately prepared, and an effective bond must be developed [17].

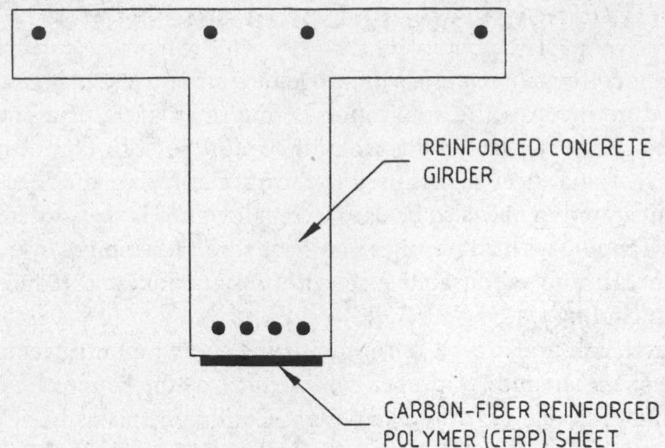

FIGURE 51.7 Strengthening of reinforced concrete girder using CFRP laminate.

In principle, reinforcing with CFRP laminates consists of a CFRP strip bonded onto the tension surface of the girder to increase or restore its original flexural capacity. The CFRP strip can be applied either nontensioned or tensioned. The high strength and stiffness of the CFRP allow the use of very thin layers to achieve the desired capacity, which can be calculated using procedures similar to those used to design traditional concrete beams. Figure 51.7 illustrates this technique.

51.11 Composite Highway Light Poles

Composite light poles were originally developed for nonhighway applications such as parking lots. Circular uniform taper along its length is the most common design. Typical pole lengths vary from 4.30 to 13.7 m. The outside pole diameter typically varies from 73 to 133 mm at the top, and from 122 to 311 mm at the base.

The most common manufacturing process for light poles is filament winding. In this process, the fiberglass filaments are wrapped around a tapered steel mandrel at specified angles to obtain the design shape with a polyester matrix. The filament winding angles vary according to the design requirements.

The primary advantages of FRP light poles are reduced weight and higher corrosion and weathering resistance compared with steel or aluminum poles. The reduced weight allows for lower shipping costs and lower installation costs. The higher corrosion resistance results in reduced maintenance costs and longer life expectancy as opposed to metal poles.

The primary disadvantages of the FRP poles are the complexity of the design and manufacturing process. For light poles, the maximum stresses are in the axial direction. Hence, the filament winding process requires more material than a process which places most or all of the fibers in the axial direction. Nevertheless, fiberglass is the most cost-effective material for use in highway light poles for heights of 7.60 to 10.7 m.

51.12 Nondestructive Evaluation of Composite Bridge Systems

The success or failure of composites in adaptation to various components in bridges greatly depends on the ability to evaluate both the short- and the long-term behavior of such composite structural elements using nondestructive evaluation (NDE) techniques. NDE and visual inspection are routinely performed on existing bridges, and new NDE technology is being developed for conventional bridge materials such as concrete and steel. Due to the greater complexity of composite materials in comparison with conventional steel and concrete, it is anticipated that existing or new NDE technology will have to be adapted to composite bridge components in order to justify their usage in critical structural elements.

NDE techniques in composites have been developed primarily in the defense and advanced technology industries. These techniques have evolved over the past 30 to 40 years, and their effectiveness in evaluating the performance of composite materials is quite impressive. However, most of these techniques require relatively sophisticated equipment, and are generally localized. Such localized NDE techniques will only describe the current and possibly predict the future behavior of the composite in a very small or localized region. In large-scale structures or structural components such as bridges, such localized techniques have limited significance in terms of the overall behavior of the structure. Therefore, NDE techniques which can evaluate the behavior and performance on a large or global scale are preferable. However, it is unfortunate that at the present time, such global techniques are not sufficiently accurate, too expensive, or not well developed technologically.

The strain gauge method of determining localized stresses and strains is well understood in civil engineering, and is widely used in the analysis of structures such as bridges, both in the laboratory and the field. The strain gauge technique is a localized type of an NDE technique and, therefore, may yield only the stress and strain levels in the lamina to which the strain gauge is attached. Strain gauge data may reveal very little about the possible delamination of inner lamina or the presence of defects within the laminate. The strain gauge method is currently used for evaluation of stresses and strains in composite tanks, pressure vessels, buildings, and various composite bridge applications which have been described here. As in other materials, the strain gauge technique is extremely beneficial in determining stress concentrations at critical locations such as the radii regions of the sections shown in Figure 51.4.

The second NDE technique is acoustic emissions (AE), which was developed more than 40 years ago, but has been successfully adapted to the evaluation of composite materials only within the last 20 years. Although this technique is described as nondestructive, the sounds or the energy emitted by the composite occur when some form of degradation of the laminate is occurring at the time of the applied loads. In simplistic terms, if no AEs are recorded, no degradation of the composite laminate is occurring. This method has been successfully utilized in the evaluation of many aerospace composite components, as well as civil engineering types of composite elements or structures, such as stacks, tanks, pressure vessels, building components, and tanker trucks. In the last application, composite tanker trucks have been evaluated on a regular basis for a period of 15 years using the AE technique. The results from these AE studies have shown the feasibility of predicting the future behavior of composite systems under fatigue loading. The bridge deck in Figure 51.4b has also been evaluated with the AE technique and the results have indicated that it is possible to predict the future behavior of such a composite bridge element utilizing AE data collected at different fatigue cycles as discussed in Section 51.6. Since AE sensors are attached at localized points on a composite structure, the data that are gathered only define the behavior of the bridge deck in a relatively localized region. However, the significant advantage of AE over strain gauges is that the behavior of the entire thickness of the laminate can be evaluated.

The continuous graphite filaments technique is a relatively simple and a global NDE method. This method essentially utilizes graphite filaments, which are electrically conductive, embedded in a glass type of composite. Glass composites are nonelectrically conductive. Since the graphite fibers can be chosen with a modulus of more than 10 times that of glass, and with a strain at failure of much less than the corresponding glass filaments, the graphite fiber will fail first within a glass fiber composite. This method is relatively inexpensive and global. The graphite filaments may be embedded anywhere within the glass fiber laminate during the fabrication of the composite structure. Additional graphite filaments may be bonded onto the surfaces of the glass fiber composite before, during, and/or after installation of the structure. This technique has been successfully utilized in determining the critical stress locations on a global scale for the extremely complex bridge deck system shown in Figure 51.4b.

When the graphite filament is broken, the electrical circuit is also broken, thus indicating high stress levels. However, the location of these high stress levels on any single graphite filament circuit is nearly impossible to predict at this time. The open electrical circuit indicates that the strain level within the fiberglass structure is excessive, but failure of the overall structure will not occur. Thus, the primary intent of such a graphite filament technique is to signal existing degradation, and possible future failure of the bridge, many truck cycles before it actually occurs. This technique may be utilized to provide a warning to the public, and cause a bridge to be shut down prior to impending failure.

The visual inspection method for composites has also been codified into an ASTM specification. Such inspections would be very similar to current periodic visual bridge inspections of steel and concrete bridges. Visual inspections are global in nature but cannot detect any possible degradation of the interior lamina within a laminate. This is a distinctive drawback to any method that involves visual inspection of external surfaces only.

Ultrasonic NDE has been widely used as an NDE technique in advanced and aerospace composites. In industrial composites, such as tanks, pipes, etc., the ultrasonic technique has been limited to determining thicknesses, detecting localized defects or voids, crack formations, and delaminations. The ultrasonic technique is a localized NDE technique and relatively time-consuming and labor-intensive. Extensive computer imaging is possible with this and other techniques discussed below which can greatly enhance the accuracy of this method.

The fiber-optics technique is analogous to the continuous graphite filament concept. The fiber-optics technology utilizes continuous fiber-optic cables which can be embedded in the laminate or on exterior surfaces. The presence of localized stress concentrations results in reduced transmission of light through the fiber-optic cable which can be related to the level of localized stresses. This technique may also predict the location of high stress levels, which continuous graphite filaments cannot do.

A variety of advanced NDE techniques are utilized in evaluation of advanced composites but are seldom used in industrial composites. Thermal NDE methods essentially use the theory of heat flow in laminates where the presence of defects, voids, or delaminations will alter the heat transfer properties. Radiographic NDE techniques utilize the transmission of electromagnetic waves through materials, and the knowledge that the presence of defects, voids, or delaminations will result in alteration of such wave transmissions. Both of these two methods may be used in local or global applications. Computer imaging may greatly enhance the effectiveness of both these techniques. However, due to the current cost, both of these techniques are economically prohibitive for periodic evaluation of large composite components as envisioned for bridges.

Other advanced NDE techniques are also available in the advanced composites industry. However, most of these techniques are currently cost-prohibitive or impractical for field evaluation of composite bridge components under less than laboratory type conditions.

51.13 Summary

The inclusion of composites into highway bridges will probably occur gradually over the next decade. Due to the strict stiffness and safety requirements, the use of composites in all structural elements of highway bridges may not be feasible in the near future. Therefore, the initial use of composites in bridges will probably be limited to those bridge elements where the unique properties of composites will result in more favorable design than with the use of conventional materials.

References

1. Plecnik, J. M. and Ahmad, S. H., Transfer of Composites Technology to Design and Construction of Bridges, Final Report Prepared for the U.S. Department of Transportation, Federal Highway Administration, Sept. 1989.
2. Creative Pultrusions, Inc., Design Guide — Standard and Custom Fiberglass-Reinforced Structural Shapes, Alum Bank, PA.
3. Morrison Molded Fiberglass Company (MMFG), Engineering Manual — EXTREN® Fiberglass Structural Shapes, Bristol, VA, 1989.
4. NUPLA Corporation, Pultruded Shapes Catalog, Sun Valley, CA, 1995.
5. American Association of State Highway and Transportation Officials, *Standard Specifications for Highway Bridges,* 16th ed., AASHTO, Washington, D.C., 1996.
6. American Institute of Steel Construction, *Manual of Steel Construction — Allowable Stress Design,* 8th ed., AISC, Chicago, IL, 1989.
7. Meier, U. and Meier, H., CFRP finds use in cable support for bridge, *Mod. Plast.*, 73(4), 87, 1996.
8. Ehsani, M. R., Saadatmanesh, H., and Tao, S., Design recommendations for bond of GFRP rebars to concrete, *J. Struct. Eng.*, 122(3), 247, 1996.
9. American Concrete Institute, State-of-the-Art Report on Fiber Reinforced Plastic (FRP) Reinforcement for Concrete Structures, Reported by ACI Committee 440, 1996.
10. Plecnik, J. and Henriquez, O., Composite bridges and NDE applications, in *Proceedings: Conference on Nondestructive Evaluation of Bridges,* Arlington, VA, August 22–27, 1992.
11. Plecnik, J. M., Henriquez, O. E., Cooper, J., and Munley, E., Development of an FRP system for bridge deck replacement, paper presented at U.S.–Canada–Europe Workshop on Bridge Engineering, Zurich, Switzerland, July 1997.
12. Lemus, J., Wearing Surface Studies on Accelerated Loading Facility (ALF), California State University, Long Beach, Report No. 97-10-16, Long Beach, CA, 1997.
13. Plunkett, J. D., Fiber-Reinforced Polymer Honeycomb Short Span Bridge for Rapid Installation, IDEA Project Final Report, Transportation Research Board, National Research Council, June 1997.
14. Dumlao, C., Lauraitis, K., Abrahamson, E., Hurlbut, B., Jacoby, M., Miller, A., and Thomas, A., Demonstration low-cost modular composite highway bridge, paper presented at 1st International Conference on Composites in Infrastructure, Tucson, AZ, 1996.
15. Seible, F., Priestley, M. J. N., Hegemier, G. A., and Innamorato, D., Seismic retrofit of RC columns with continuous carbon fiber jackets, *J. Composites Constr.*, 1(2), 52, 1997.
16. Meier, U., Deuring, M., Meier, H., and Schwegler, G., Strengthening of structures with CFRP laminates: research and applications in Switzerland, in *Advanced Composite Materials in Bridges and Structures,* Canadian Society for Civil Engineers, 1992, 243.
17. Arduini, M. and Nanni, A., Behavior of precracked RC beams strengthened with carbon FRP sheets, *J. Composites Constr.*, 1(2), 63, 1997.

52

Effective Length of Compression Members

Lian Duan
California Department of Transportation

Wai-Fah Chen
Purdue University

52.1 Introduction *

The concept of *effective length factor* or *K factor* plays an important role in compression member design. Although great efforts have been made in the past years to eliminate the *K* factor in column design, *K* factors are still popularly used in practice for routine design [1]

Mathematically, the effective length factor or the *elastic K* factor is defined as

$$K = \sqrt{\frac{P_e}{P_{cr}}} = \sqrt{\frac{\pi^2 EI}{L^2 P_{cr}}} \tag{52.1}$$

where P_e is Euler load, elastic buckling load of a pin-ended column, P_{cr} is elastic buckling load of an end-restrained framed column, E is modulus of elasticity, I is moment of inertia in the flexural buckling plane, and L is unsupported length of column.

* Much of the material of this chapter was taken from Duan, L. and Chen, W. F., Chapter 17: Effective length factors of compression members, in *Handbook of Structural Engineering,* Chen, W. F., Ed., CRC Press, Boca Raton, FL, 1997.

0-8493-7434-0/00/$0.00+$.50

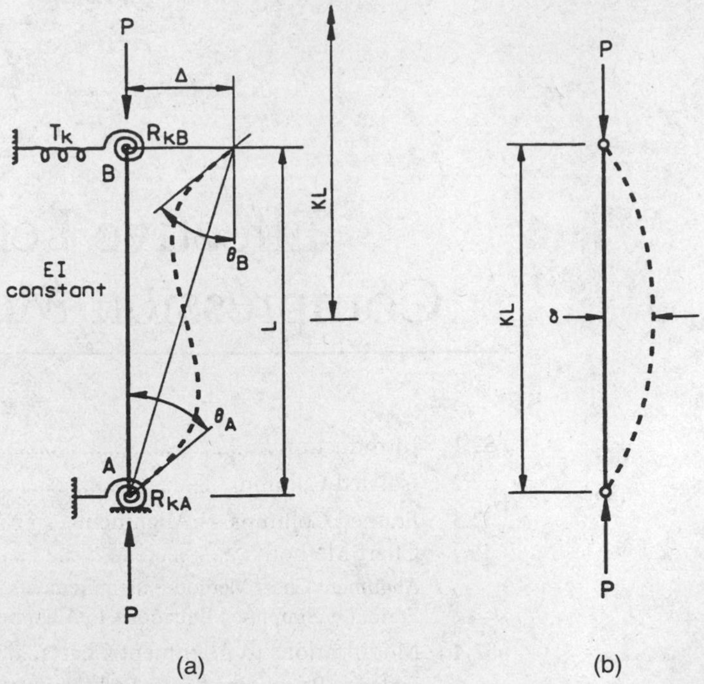

FIGURE 52.1 Isolated columns. (a) End-restrained columns; (b) pin-ended columns.

Physically, the K factor is a factor that, when multiplied by actual length of the end-restrained column (Figure 52.1a), gives the length of an equivalent pin-ended column (Figure 52.1b) whose buckling load is the same as that of the end-restrained column. It follows that the *effective length KL* of an end-restrained column is the length between adjacent inflection points of its pure flexural buckling shape.

Practically, design specifications provide the resistance equations for pin-ended columns, while the resistance of framed columns can be estimated through the K factor to the pin-ended column strength equations. Theoretical K factor is determined from an elastic eigenvalue analysis of the entire structural system, while practical methods for the K factor are based on an elastic eigenvalue analysis of selected subassemblages. This chapter presents the state-of-the-art engineering practice of the effective length factor for the design of columns in bridge structures.

52.2 Isolated Columns

From an eigenvalue analysis, the general K factor equation of an end-restrained column as shown in Figure 52.1 is obtained as

$$\textbf{det}\begin{vmatrix} C+\dfrac{R_{kA}L}{EI} & S & -(C+S) \\ S & C+\dfrac{R_{kB}L}{EI} & -(C+S) \\ -(C+S) & -(C+S) & 2(C+S)-\left(\dfrac{\pi}{K}\right)^2+\dfrac{T_k L^3}{EI} \end{vmatrix} = 0 \tag{52.2}$$

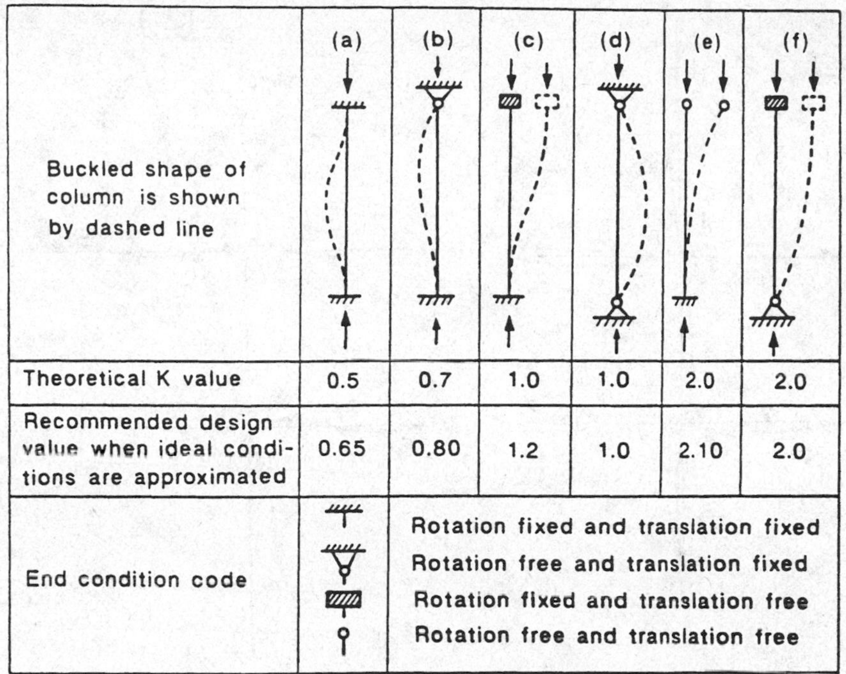

(a) (b) (c) (d) (e) (f)
Buckled shape of column is shown by dashed line
Theoretical K value 0.5 0.7 1.0 1.0 2.0 2.0
Recommended design value when ideal conditions are approximated 0.65 0.80 1.2 1.0 2.10 2.0
End condition code
Rotation fixed and translation fixed
Rotation free and translation fixed
Rotation fixed and translation free
Rotation free and translation free

FIGURE 52.2 Theoretical and recommended K factors for isolated columns with idealized end conditions. (*Source:* American Institute of Steel Construction. *Load and Resistance Factor Design Specification for Structural Steel Buildings*, 2nd ed., Chicago, IL, 1993. With permission. Also from Johnston, B. G., Ed., Structural Stability Research Council, *Guide to Stability Design Criteria for Metal Structures*, 3rd ed., John Wiley & Sons, New York, 1976. With permission.)

where the stability function C and S are defined as

$$C = \frac{(\pi / K)\sin(\pi / K) - (\pi / K)^2 \cos(\pi / K)}{2 - 2\cos(\pi / K) - (\pi / K)\sin(\pi / K)} \tag{52.3}$$

$$S = \frac{(\pi / K)^2 - (\pi / K)\sin(\pi / K)}{2 - 2\cos(\pi / K) - (\pi / K)\sin(\pi / K)} \tag{52.4}$$

The largest value of K satisfying Eq. (52.2) gives the elastic buckling load of an end-retrained column.

Figure 52.2 summarizes the theoretical K factors for columns with some idealized end conditions [2,3]. The recommended K factors are also shown in Figure 52.2 for practical design applications. Since actual column conditions seldom comply fully with idealized conditions used in buckling analysis, the recommended K factors are always equal or greater than their theoretical counterparts.

52.3 Framed Columns — Alignment Chart Method

In theory, the effective length factor K for any columns in a framed structure can be determined from a stability analysis of the entire structural analysis — eigenvalue analysis. Methods available for stability analysis include slope–deflection method [4], three-moment equation method [5], and energy methods [6]. In practice, however, such analysis is not practical, and simple models are often used to determine the effective length factors for farmed columns [7~10]. One such practical procedure that provides an approximate value of the elastic K factor is the alignment chart method [11]. This procedure has been adopted by the AASHTO [2] and AISC [3]. Specifications and the

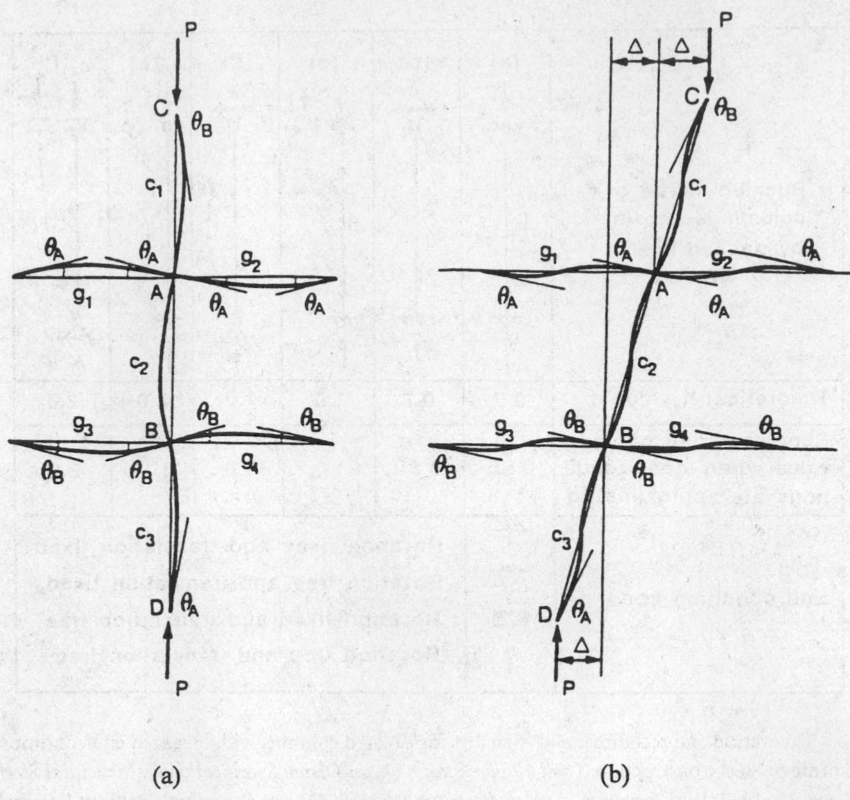

FIGURE 52.3 Subassemblage models for *K* factors of framed columns. (a) Braced frames; (b) unbraced frames.

ACI-318-95 Code [12], among others. At present, most engineers use the alignment chart method in lieu of an actual stability analysis.

52.3.1 Alignment Chart Method

The structural models employed for determination of *K* factors for framed columns in the alignment chart method are shown in Figure 52.3 The assumptions [2,4] used in these models are

1. All members have constant cross section and behave elastically.
2. Axial forces in the girders are negligible.
3. All joints are rigid.
4. For braced frames, the rotations at near and far ends of the girders are equal in magnitude and opposite in direction (i.e., girders are bent in single curvature).
5. For unbraced frames, the rotations at near and far ends of the girders are equal in magnitude and direction (i.e., girders are bent in double curvature).
6. The stiffness parameters $L\sqrt{P/EI}$, of all columns are equal.
7. All columns buckle simultaneously.

By using the slope–deflection equation method and stability functions, the effective length factor equations of framed columns are obtained as follows:

For columns in braced frames:

$$\frac{G_A G_B}{4}(\pi/K)^2 + \left(\frac{G_A + G_B}{2}\right)\left(1 - \frac{\pi/K}{\tan(\pi/K)}\right) + \frac{2\tan(\pi/2K)}{\pi/K} - 1 = 0 \qquad (52.5)$$

For columns in unbraced frames:

$$\frac{G_A G_B (\pi / K)^2 - 36^2}{6(G_A + G_B)} - \frac{\pi / K}{\tan(\pi / K)} = 0 \tag{52.6}$$

where G is stiffness ratios of columns and girders, subscripts A and B refer to joints at the two ends of the column section being considered, and G is defined as

$$G = \frac{\sum (E_c I_c / L_c)}{\sum (E_g I_g / L_g)} \tag{52.7}$$

where Σ indicates a summation of all members rigidly connected to the joint and lying in the plane in which buckling of the column is being considered; subscripts c and g represent columns and girders, respectively.

Eqs. (52.5) and (52.6) can be expressed in form of alignment charts as shown in Figure 52.4. It is noted that for columns in braced frames, the range of K is $0.5 \le K \le 1.0$; for columns in unbraced frames, the range is $1.0 \le K \le \infty$. For column ends supported by but not rigidly connected to a footing or foundations, G is theoretically infinity, but, unless actually designed as a true friction-free pin, may be taken as 10 for practical design. If the column end is rigidly attached to a properly designed footing, G may be taken as 1.0.

Example 52.1

Given

A four-span reinforced concrete bridge is shown in Figure 52.5. Using the alignment chart, determine the K factor for Column DC. $E = 25{,}000$ MPa.

Section Properties are

Superstructure:	$I = 3.14\ (10^{12})\ \text{mm}^4$	$A = 5.86\ (10^6)\ \text{mm}^2$
Columns:	$I = 3.22\ (10^{11})\ \text{mm}^4$	$A = 2.01\ (10^6)\ \text{mm}^2$

Solution

1. **Calculate G factor for Column DC.**

$$G_D = \frac{\sum_D (E_c I_c / L_c)}{\sum_D (E_g I_g / L_g)} = \frac{3.22(10^{12}) / 12{,}000}{2(3.14)(10^{12}) / 55{,}000} = 0.235$$

$G_D = 1.0$ (Ref. [3])

2. **From the alignment chart in Figure 52.4b, $K = 1.21$ is obtained.**

52.3.2 Requirements for Braced Frames

In stability design, one of the major decisions engineers have to make is the determination of whether a frame is braced or unbraced. The AISC-LRFD [3] states that a frame is braced when "lateral stability is provided by diagonal bracing, shear walls or equivalent means." However, there is no specific provision for the "amount of stiffness required to prevent sidesway buckling" in the AISC, AASHTO, and other specifications. In actual structures, a completely braced frame seldom exists.

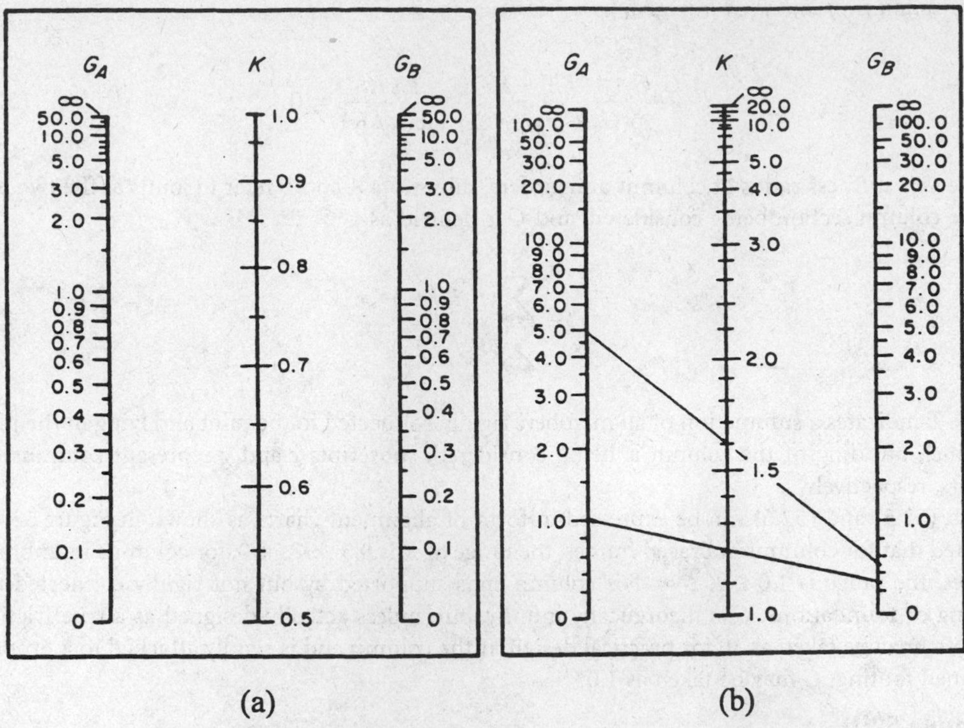

FIGURE 52.4 Alignment charts for effective length factors of framed columns. (a) Braced frames; (b) unbraced frames. (*Source:* American Institute of Steel Construction, *Load and Resistance Factor Design Specifications for Structural Steel Buildings,* 2nd ed., Chicago, IL, 1993. With permission. Also from Johnston, B. G., Ed., Structural Stability Research Council, *Guide to Stability Design Criteria for Metal Structures,* 3rd ed., John Wiley & Sons, New York, 1976. With permission.)

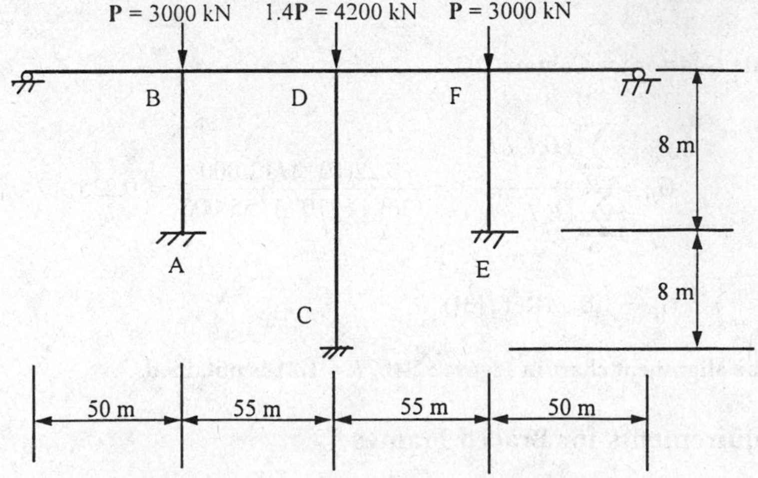

FIGURE 52.5 A four-span reinforced concrete bridge.

But in practice, some structures can be analyzed as braced frames as long as the lateral stiffness provided by bracing system is large enough. The following brief discussion may provide engineers with the tools to make engineering decisions regarding the basic requirements for a braced frame.

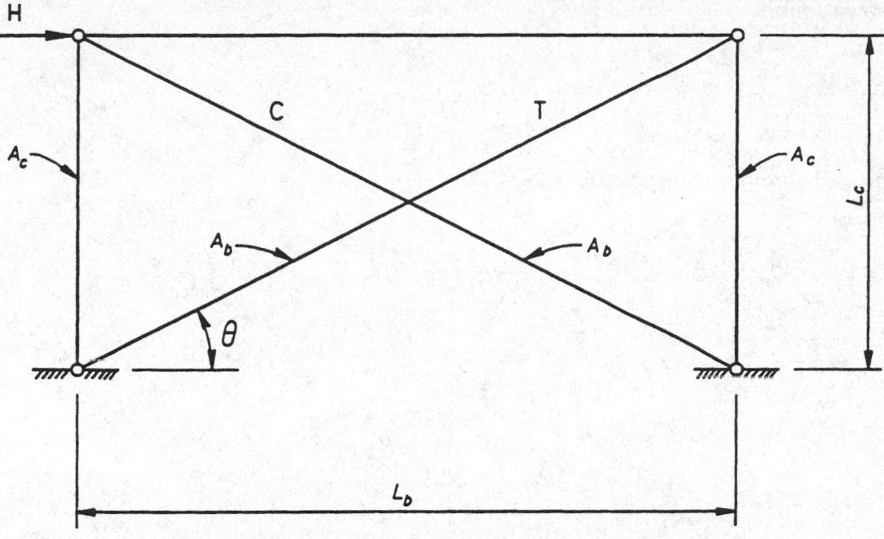

FIGURE 52.6 Diagonal cross-bracing system.

52.3.2.1 Lateral Stiffness Requirement

Galambos [13] presented a simple conservative procedure to estimate the minimum lateral stiffness provided by a bracing system so that the frame is considered braced.

$$\text{Required Lateral Stiffness } T_k = \frac{\sum P_n}{L_c} \tag{52.8}$$

where $\sum$ represents summation of all columns in one story, P_n is nominal axial compression strength of column using the effective length factor $K = 1$, and L_c is unsupported length of the column.

52.3.2.2 Bracing Size Requirement

Galambos [13] employed Eq. (52.8) to a diagonal bracing (Figure 52.6) and obtained minimum requirements of diagonal bracing for a braced frame as

$$A_b = \frac{\left[1 + (L_b / L_c)^2\right]^{3/2} \sum P_n}{(L_b / L_c)^2 E} \tag{52.9}$$

where A_b is cross-sectional area of diagonal bracing and L_b is span length of beam.

A recent study by Aristizabal-Ochoa [14] indicates that the size of diagonal bracing required for a totally braced frame is about 4.9 and 5.1% of the column cross section for "rigid frame" and "simple farming," respectively, and increases with the moment inertia of the column, the beam span and with beam to column span ratio L_b/L_c.

52.3.3 Simplified Equations to Alignment Charts

52.3.3.1. Duan–King–Chen Equations

A graphical alignment chart determination of the K factor is easy to perform, while solving the chart Eqs. (52.5) and (52.6) always involves iteration. To achieve both accuracy and simplicity for design purpose, the following alternative K factor equations were proposed by Duan, King, and Chen [15].

For braced frames:

$$K = 1 - \frac{1}{5+9G_A} - \frac{1}{5+9G_B} - \frac{1}{10+G_AG_B} \tag{52.10}$$

For unbraced frames:

$$\text{For } K < 2 \quad K = 4 - \frac{1}{1+0.2G_A} - \frac{1}{1+0.2G_B} - \frac{1}{1+0.01G_AG_B} \tag{52.11}$$

$$\text{For } K \geq 2 \quad K = \frac{2\pi a}{0.9+\sqrt{0.81+4ab}} \tag{52.12}$$

where

$$a = \frac{G_AG_B}{G_A+G_B} + 3 \tag{52.13}$$

$$b = \frac{36}{G_A+G_B} + 6 \tag{52.14}$$

52.3.3.2 French Equations

For braced frames:

$$K = \frac{3G_AG_B + 1.4(G_A + G_B) + 0.64}{3G_AG_B + 2.0(G_A + G_B) + 1.28} \tag{52.15}$$

For unbraced frames:

$$K = \sqrt{\frac{1.6G_AG_B + 4.0(G_A + G_B) + 7.5}{G_A + G_B + 7.5}} \tag{52.16}$$

Eqs. (52.15) and (52.16) first appeared in the French Design Rules for Steel Structure [16] in 1966, and were later incorporated into the *European Recommendations for Steel Construction*[17]. They provide a good approximation to the alignment charts [18].

52.4 Modifications to Alignment Charts

In using the alignment charts in Figure 52.4 and Eqs. (52.5) and (52.6), engineers must always be aware of the assumptions used in the development of these charts. When actual structural conditions differ from these assumptions, unrealistic design may result [3,19,20]. SSRC Guide [19] provides methods enabling engineers to make simple modifications of the charts for some special conditions, such as, for example, unsymmetrical frames, column base conditions, girder far-end conditions, and flexible conditions. A procedure that can be used to account for far ends of restraining columns being hinged or fixed was proposed by Duan and Chen [21~23], and Essa [24]. Consideration of effects of material inelasticity on the *K* factor for steel members was developed originally by Yura

[25] and expanded by Disque [26]. LeMessurier [27] presented an overview of unbraced frames with or without leaning columns. An approximate procedure is also suggested by AISC-LRFD [3]. Several commonly used modifications for bridge columns are summarized in this section.

52.4.1 Different Restraining Girder End Conditions

When the end conditions of restraining girders are not rigidly jointed to columns, the girder stiffness (I_g/L_g) used in the calculation of G factor in Eq. (52.7) should be multiplied by a modification factor α_k given below:

For a braced frame:

$$\alpha_k = \begin{cases} 1.0 & \text{rigid far end} \\ 2.0 & \text{fixed far end} \\ 1.5 & \text{hinged far end} \end{cases} \tag{52.17}$$

For a unbraced frame:

$$\alpha_k = \begin{cases} 1.0 & \text{rigid far end} \\ 2/3 & \text{fixed far end} \\ 0.5 & \text{hinged far end} \end{cases} \tag{52.18}$$

52.4.2 Consideration of Partial Column Base Fixity

In computing the K factor for monolithic connections, it is important to evaluate properly the degree of fixity in foundation. The following two approaches can be used to account for foundation fixity.

52.4.2.1. Fictitious Restraining Beam Approach

Galambos [28] proposed that the effect of partial base fixity can be modeled as a fictitious beam. The approximate expression for the stiffness of the fictitious beam accounting for rotation of foundation in the soil has the form:

$$\frac{I_s}{L_B} = \frac{q\,BH^3}{72\,E_{\text{steel}}} \tag{52.19}$$

where q is modulus of subgrade reaction (varies from 50 to 400 lb/in.^{3}, 0.014 to 0.109 N/mm^3); B and H are width and length (in bending plane) of foundation, and E_{steel} is modulus of elasticity of steel.

Based on Salmon et al. [29] studies, the approximate expression for the stiffness of the fictitious beam accounting for the rotations between column ends and footing due to deformation of base plate, anchor bolts, and concrete can be written as

$$\frac{I_s}{L_B} = \frac{b\,d^2}{72\,E_{\text{steel}}\,/\,E_{\text{concrete}}} \tag{52.20}$$

where b and d are width and length of the base plate, subscripts concrete and steel represent concrete and steel, respectively. Galambos [28] suggested that the smaller of the stiffness calculated by Eqs. (52.25) and (52.26) be used in determining K factors.

52.4.2.2 AASHTO-LRFD Approach

The following values are suggested by AASHTO-LRFD [2]:

$G = 1.5$ footing anchored on rock
$G = 3.0$ footing not anchored on rock
$G = 5.0$ footing on soil
$G = 1.0$ footing on multiple rows of end bearing piles

Example 52.2

Given
Determine K factor for the Column AB as shown in Figure 52.5 by using the alignment chart with the necessary modifications. Section and material properties are given in Example 52.1 and spread footings are on soil.

Solution

1. **Calculate G factor with Modification for Column AB.**
   Since the far end of restraining girders are hinged, girder stiffness should be multiplied by 0.5. Using section properties in Example 52.1, we obtain:

$$G_B = \frac{\sum_B (E_c I_c / L_c)}{\sum_B \alpha_k (E_g I_g / L_g)}$$

$$= \frac{3.22(10^{12})/8{,}000}{(3.14)(10^{12})/55{,}000 + 0.5(3.14)(10^{12})/50{,}000} = 0.454$$

$$G_A = 5.0 \quad \text{(Ref. [2])}$$

2. **From the alignment chart in Figure 52.4b, $K = 1.60$ is obtained.**

52.4.3 Column Restrained by Tapered Rectangular Girders

A modification factor α_T was developed by King et al. [30] for those framed columns restrained by tapered rectangular girders with different far-end conditions. The following modified G factor is introduced in connection with the use of alignment charts:

$$G = \frac{\sum (E_c I_c / L_c)}{\sum \alpha_T (E_g I_g / L_g)} \tag{52.21}$$

where I_g is moment of inertia of the girder at the near end. Both closed-form and approximate solutions for modification factor α_T were derived. It is found that the following two-parameter power-function can describe the closed-form solutions very well:

$$\alpha_T = \alpha_k (1 - r)^{\beta} \tag{52.22}$$

in which the parameter α_k is a constant (Eqs. 52.17 and 52.18) depending on the far-end conditions, and β is a function of far-end conditions and tapering factor a and r as defined in Figure 52.7.

1. For a linearly tapered rectangular girder (Figure 52.7a):

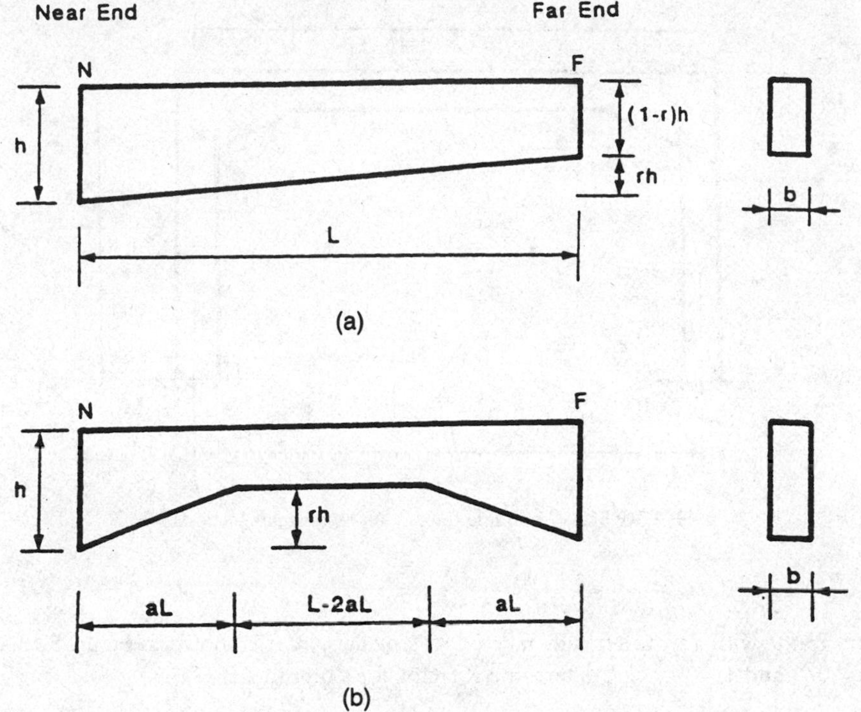

FIGURE 52.7 Tapered rectangular girders. (a) Linearly tapered girder. (b) symmetrically tapered girder.

For a braced frame:

$$\beta = \begin{cases} 0.02 + 0.4r & \text{rigid far end} \\ 0.75 - 0.1r & \text{fixed far end} \\ 0.75 - 0.1r & \text{hinged far end} \end{cases} \tag{52.23}$$

For an unbraced frame:

$$\beta = \begin{cases} 0.95 & \text{rigid far end} \\ 0.70 & \text{fixed far end} \\ 0.70 & \text{hinged far end} \end{cases} \tag{52.24}$$

2. For a symmetrically tapered rectangular girder (Figure 52.7b)
For a braced frame:

$$\beta = \begin{cases} 3 - 1.7a^2 - 2a & \text{rigid far end} \\ 3 + 2.5a^2 - 5.55a & \text{fixed far end} \\ 3 - a^2 - 2.7a & \text{hinged far end} \end{cases} \tag{52.25}$$

For an unbraced frame:

$$\beta = \begin{cases} 3 + 3.8a^2 - 6.5a & \text{rigid far end} \\ 3 + 2.3a^2 - 5.45a & \text{fixed far end} \\ 3 - 0.3a & \text{hinged far end} \end{cases} \tag{52.26}$$

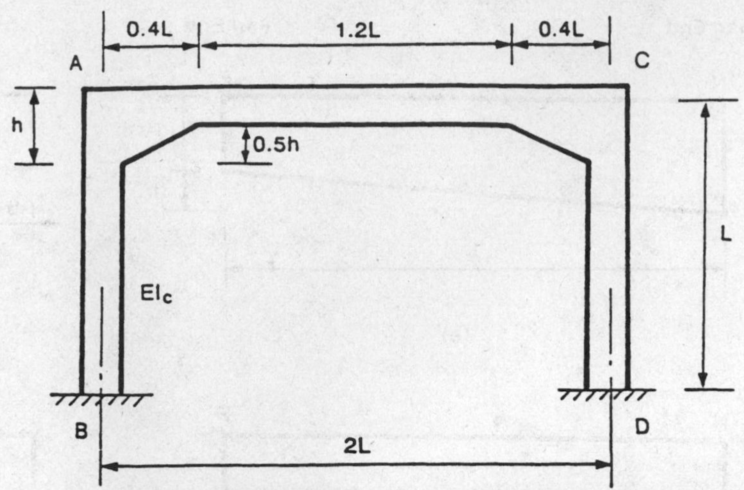

FIGURE 52.8 A simple frame with rectangular sections.

Example 52.3

Given

A one-story frame with a symmetrically tapered rectangular girder is shown in Figure 52.8. Assuming $r = 0.5$, $a = 0.2$, and $I_g = 2I_c = 2I$, determine K factor for Column AB.

Solution

1. **Use the Alignment Chart with Modification**

   For joint A, since the far end of girder is rigid, use Eqs. (52.26) and (52.22)

$$\beta = 3 + 3.8(0.2)^2 - 6.5(0.2) = 1.852$$

$$\alpha_T = (1 - 0.5)^{1.852} = 0.277$$

$$G_A = \frac{\sum E_c I_c / L_c}{\sum \alpha_T E_g I_g / L_g} = \frac{EI/L}{0.277\ E(2I)/2L} = 3.61$$

$$G_B = 1.0 \quad \text{(Ref. [3])}$$

   From the alignment chart in Figure 52.4b, $K = 1.59$ is obtained

2. **Use the Alignment Chart without Modification**

   A direct use of Eq. (52.7) with an average section (0.75h) results in

$$I_g = 0.75^3\ (2I) = 0.844\,I$$

$$G_A = \frac{EI / L}{0.844\,EI / 2L} = 2.37$$

$$G_B = 1.0$$

   From the alignment chart in Figure 52.4b, $K = 1.50$, *or* (1.50 – 1.59)/1.59 = –6% in error on the less conservative side.

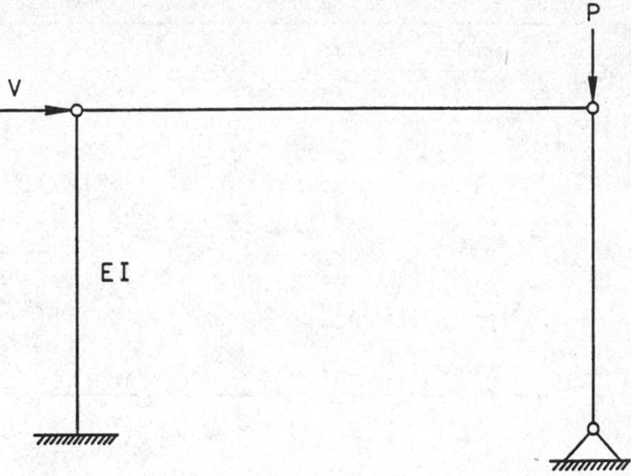

FIGURE 52.9 Subassemblage of LeMessurier method.

52.5 Framed Columns — Alternative Methods

52.5.1 LeMessurier Method

Considering that all columns in a story buckle simultaneously and strong columns will brace weak columns (Figure 52.9), a more accurate approach to calculate K factors for columns in a side-sway frame was developed by LeMessurier [27]. The K_i value for the ith column in a story can be obtained by the following expression:

$$K_i = \sqrt{\frac{\pi^2 EI_i}{L_i^2 P_i}\left(\frac{\sum P + \sum C_L P}{\sum P_L}\right)} \tag{52.27}$$

where P_i is axial compressive force for member i, and subscript i represents the ith column and ΣP is the sum of axial force of all columns in a story.

$$P_L = \frac{\beta EI}{L^2} \tag{52.28}$$

$$\beta = \frac{6(G_A + G_B) + 36}{2(G_A + G_B) + G_A G_B + 3} \tag{52.29}$$

$$C_L = \left(\beta \frac{K_o^2}{\pi^2} - 1\right) \tag{52.30}$$

in which K_o is the effective length factor obtained by the alignment chart for unbraced frames and P_L is only for those columns that provide side-sway stiffness.

Example 52.4

Given

Determine K factors for bridge columns shown in Figure 52.5 by using the LeMessurier method. Section and material properties are given in Example 52.1.

TABLE 52.1 Example 52.4 — Detailed Calculations by LeMessurier Method

Members	AB and EF	CD	Sum	Notes
I (mm^4 × 10^{11})	3.217	3.217	—	
L (mm)	8,000	12,000	—	
G_{top}	0.454	0.235	—	Eq. (52.7)
G_{bottom}	0.0	0.0	—	Eq. (52.7)
β	9.91	10.78	—	Eq. (52.29)
K_{io}	1.082	1.045	—	Alignment chart
C_L	0.176	0.193	—	Eq. (52.30)
P_L	50,813E	24,083E	123,709E	Eq. (52.28)
P	P	1.4P	3.4P	P = 3,000 kN
C_LP	0.176P	0.270P	0.622P	P = 3,000 kN

Solutions

The detailed calculations are listed in Table 52.1 By using Eq. (52.32), we obtain:

$$K_{AB} = \sqrt{\frac{\pi^2 EI_{AB}}{L_{AB}^2 P_{AB}} \left(\frac{\sum P + \sum C_L P}{\sum P_L} \right)}$$

$$= \sqrt{\frac{\pi^2 E(3.217)(10^{11})}{(8.000)^2 (P)} \left(\frac{3.4P + 0.622P}{123,709E} \right)} = 1.270$$

$$K_{CD} = \sqrt{\frac{\pi^2 EI_{CD}}{L_{CD}^2 P_{CD}} \left(\frac{\sum P + \sum C_L P}{\sum P_L} \right)}$$

$$= \sqrt{\frac{\pi^2 E(3.217)(10^{11})}{(12,000)^2 (1.4P)} \left(\frac{3.4P + 0.622P}{123,709E} \right)} = 0.715$$

52.5.2 Lui Method

A simple and straightforward approach for determining the effective length factors for framed columns without the use of alignment charts and other charts was proposed by Lui [31]. The formulas take into account both the member instability and frame instability effects explicitly. The K factor for the ith column in a story was obtained in a simple form:

$$K_i = \sqrt{\left(\frac{\pi^2 EI_i}{P_i L_i^2} \right) \left[\left(\sum \frac{P}{L} \right) \left(\frac{1}{5\sum \eta} + \frac{\Delta_1}{\sum H} \right) \right]} \tag{52.31}$$

where $\Sigma(P/L)$ represents the sum of axial-force-to-length ratio of all members in a story; ΣH is the story lateral load producing Δ_1, Δ_1 is the first-order interstory deflection; η is member stiffness index and can be calculated by

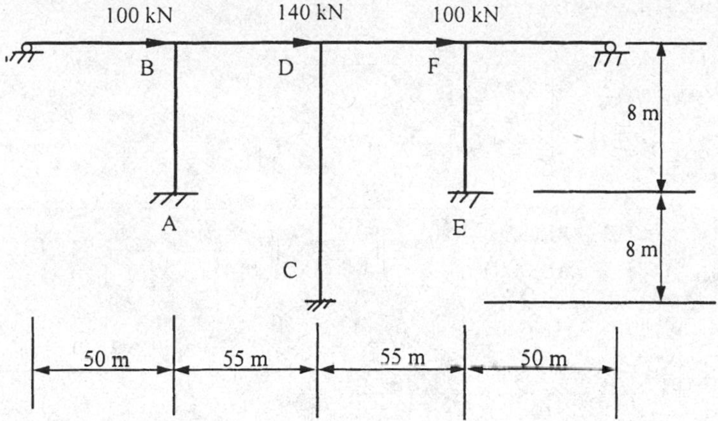

FIGURE 52.10 A bridge structure subjected to fictitious lateral loads.

TABLE 52.2 Example 52.5 — Detailed Calculations by Lui Method

Members	AB and EF	CD	Sum	Notes
I (mm^4 × 10^{11})	3.217	3.217	—	
L (mm)	8,000	12,000	—	
H (kN)	150	210	510	
Δ_I (mm)	0.00144	0.00146	—	
$\Delta_I/\Sigma H$ (mm/kN)	—	—	2.843 (10^{-6})	Average
M_{top} (kN-m)	–476.9	–785.5	—	
M_{bottom} (kN-m)	–483.3	–934.4	—	
m	0.986	0.841	—	
η (kN/mm)	185,606	46,577	417,789	Eq. (52.32)
P/L (kN/mm)	P/8,000	1.4 P/12,000	1.1P/3,000	P = 3,000 kN

$$\eta = \frac{(3 + 4.8m + 4.2m^2)EI}{L^3} \tag{52.32}$$

in which m is the ratio of the smaller to larger end moments of the member; it is taken as positive if the member bends in reverse curvature, and negative for single curvature.

It is important to note that the term ΣH used in Eq. (52.36) is not the actual applied lateral load. Rather, it is a small disturbing or fictitious force (taken as a fraction of the story gravity loads) to be applied to each story of the frame. This fictitious force is applied in a direction such that the deformed configuration of the frame will resemble its buckled shape.

Example 52.5

Given
Determine the K factors for bridge columns shown in Figure 52.5 by using the Lui method. Section and material properties are given in Example 52.1.

Solutions
Apply fictitious lateral forces at B, D, and F (Figure 52.10) and perform a first-order analysis. Detailed calculation is shown in Table 52.2.

By using Eq. (52.31), we obtain

$$K_{AB} = \sqrt{\left(\frac{\pi^2 E I_{AB}}{P_{AB}\, L_{AB}^2}\right)\left[\left(\sum \frac{P}{L}\right)\left(\frac{1}{5\sum \eta} + \frac{\Delta_1}{\sum H}\right)\right]}$$

$$= \sqrt{\left(\frac{\pi^2 (25{,}000)(3.217)(10^{11})}{P(8{,}000)^2}\right)\left[\left(\frac{1.1P}{3{,}000}\right)\left(\frac{1}{5(417{,}789)} + 2.843(10^{-6})\right)\right]}$$

$$= 1.229$$

$$K_{CD} = \sqrt{\left(\frac{\pi^2 E I_{CD}}{P_{CD}\, L_{CD}^2}\right)\left[\left(\sum \frac{P}{L}\right)\left(\frac{1}{5\sum \eta} + \frac{\Delta_1}{\sum H}\right)\right]}$$

$$= \sqrt{\left(\frac{\pi^2 (25{,}000)(3.217)(10^{11})}{1.4P(12{,}000)^2}\right)\left[\left(\frac{1.1P}{3{,}000}\right)\left(\frac{1}{5(417{,}789)} + 2.843(10^{-6})\right)\right]}$$

$$= 0.693$$

52.5.3 Remarks

For a comparison, Table 52.3 summarizes the K factors for the bridge columns shown in Figure 52.5 obtained from the alignment chart, LeMessurier and Lui methods, as well as an eigenvalue analysis. It is seen that errors of alignment chart results are rather significant in this case. Although the K factors predicted by Lui's formulas and LeMessurier's formulas are almost the same in most cases, the simplicity and independence of any chart in the case of Lui's formula make it more desirable for design office use [32].

TABLE 52.3 Comparison of K Factors for Frame in Figure 52.5

Columns	Theoretical	Alignment Chart	Lui Eq. (52.31)	LeMessurier Eq. (52.27)
AB	1.232	1.082	1.229	1.270
CD	0.694	1.045	0.693	0.715

52.6 Crossing Bracing Systems

Picard and Beaulieu [33,34] reported theoretical and experimental studies on double diagonal cross-bracings (Figure 52.6) and found that

1. A general effective length factor equation is given as

$$K = \sqrt{0.523 - \frac{0.428}{C/T}} \geq 0.50 \qquad (52.33)$$

where C and T represent compression and tension forces obtained from an elastic analysis, respectively.

2. When the double diagonals are continuous and attached at an intersection point, the *effective length* of the compression diagonal is 0.5 times the diagonal length, i.e., $K = 0.5$, because the C/T ratio is usually smaller than 1.6.

El-Tayem and Goel [35] reported a theoretical and experimental study about the X-bracing system made from single equal-leg angles. They concluded that

1. Design of X-bracing system should be based on an exclusive consideration of one half diagonal only.
2. For X-bracing systems made from single equal-leg angles, an effective length of 0.85 times the half-diagonal length is reasonable, i.e., $K = 0.425$.

52.7 Latticed and Built-Up Members

It is a common practice that when a buckling model involves relative deformation produced by shear forces in the connectors, such as lacing bars and batten plates, between individual components, a modified effective length factor K_m or effective slenderness ratio $(KL/r)_m$ is used in determining the compressive strength. K_m is defined as

$$K_m = \alpha_v K \tag{52.34}$$

in which K is the usual effective length factor of a latticed member acting as a unit obtained from a structural analysis; and α_v is the shear factor to account for the effect of shear deformation on the buckling strength, Details of the development of the shear factor α_v can be found in textbooks by Bleich [5] and Timoshenko and Gere [36]. The following section briefly summarizes α_v formulas for various latticed members.

52.7.1 Latticed Members

By considering the effect of shear deformation in the latticed panel on buckling load, shear factor α_v of the following form has been introduced:

Laced Compression Members (Figures 52.11a and b)

$$\alpha_v = \sqrt{1 + \frac{\pi^2 EI}{(KL)^2} \frac{d^3}{A_d E_d a b^2}} \tag{52.35}$$

Compression Members with Battens (Figure 52.11c)

$$\alpha_v = \sqrt{1 + \frac{\pi^2 EI}{(KL)^2} \left(\frac{ab}{12 E_b I_b} + \frac{a^2}{24 E I_f} \right)} \tag{52.36}$$

Laced-Battened Compression Members (Figure 52.11d)

$$\alpha_v = \sqrt{1 + \frac{\pi^2 EI}{(KL)^2} \left(\frac{d^3}{A_d E_d a b^2} + \frac{b}{a A_b E_b} \right)} \tag{52.37}$$

Compression Members with Perforated Cover Plates (Figure 52.11e)

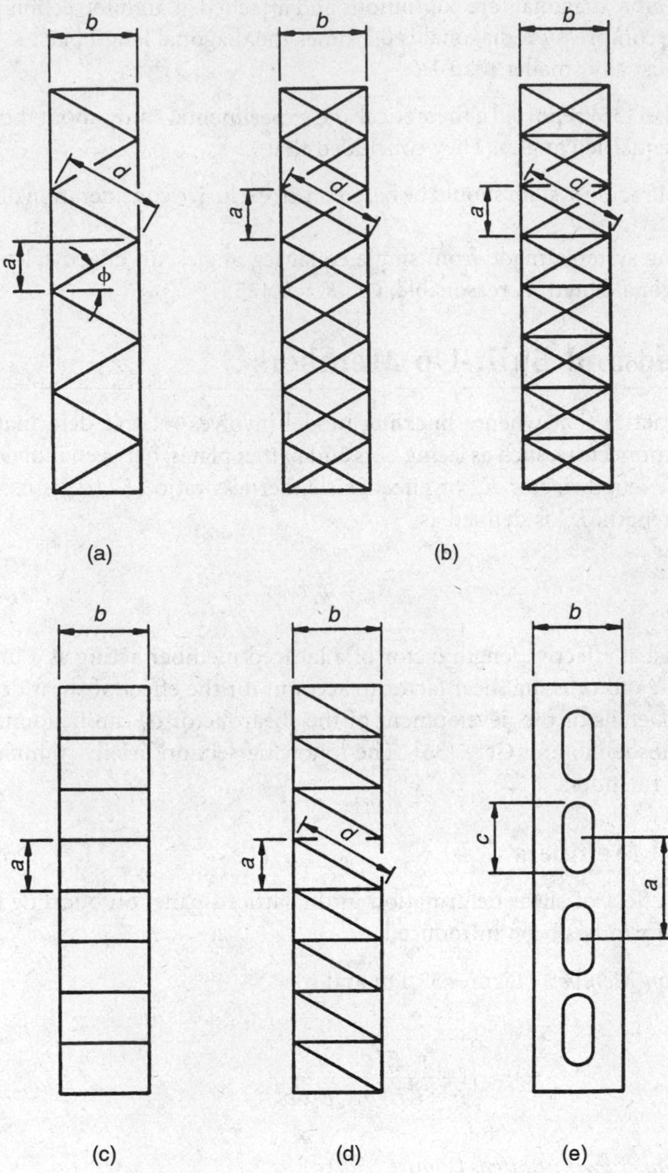

FIGURE 52.11 Typical configurations of latticed members. (a) Single lacing; (b) double lacing; (c) battens; (d) lacing-battens; (e) perforated cover plates.

$$\alpha_v = \sqrt{1 + \frac{\pi^2 EI}{(KL)^2}\left(\frac{9c^3}{64aEI_f}\right)} \tag{52.38}$$

where E_d is modulus of elasticity of materials for lacing bars; E_b is modulus of elasticity of materials for batten plates; A_d is cross-sectional area of all diagonals in one panel; I_b is moment inertia of all battens in one panel in the buckling plane, and I_f is moment inertia of one side of main components taken about the centroid axis of the flange in the buckling plane; *a, b, d* are height of panel, depth of member, and length of diagonal, respectively; and *c* is the length of a perforation.

The Structural Stability Research Council [37] suggested that a conservative estimating of the influence of 60° or 45° lacing, as generally specified in bridge design practice, can be made by modifying the overall effective length factor K by multiplying a factor α_v, originally developed by Bleich [5] as follows:

$$\text{For } \frac{KL}{r} > 40, \quad \alpha_v = \sqrt{1 + 300/(KL/r)^2} \tag{52.39}$$

$$\text{For } \frac{KL}{r} \leq 40, \quad \alpha_v = 1.1 \tag{52.40}$$

It should be pointed out that the usual K factor based on a solid member analysis is included in Eqs. (52.35) through (52.38). However, since the latticed members studied previously have pin-ended conditions, the K factor of the member in the frame was not included in the second terms of the square root of the above equations in their original derivations [5,36].

52.7.5 Built-Up Members

AISC-LRFD [3] specifies that if the buckling of a built-up member produces shear forces in the connectors between individual component members, the usual slenderness ratio KL/r for compression members must be replaced by the modified slenderness ratio $(KL/r)_m$ in determining the compressive strength.

1. *For snug-tight bolted connectors:*

$$\left(\frac{KL}{r}\right)_m = \sqrt{\left(\frac{KL}{r}\right)_o^2 + \left(\frac{a}{r_i}\right)^2} \tag{52.41}$$

2. *For welded connectors and for fully tightened bolted connectors:*

$$\left(\frac{KL}{r}\right)_m = \sqrt{\left(\frac{KL}{r}\right)_o^2 + 0.82\frac{\alpha^2}{(1+\alpha^2)}\left(\frac{a}{r_{ib}}\right)^2} \tag{52.42}$$

where $(KL/r)_o$ is the slenderness ratio of built-up member acting as a unit, $(KL/r)_m$ is modified slenderness ratio of built-up member, a/r_i is the largest slenderness ratio of the individual components, a/r_{ib} is the slenderness ratio of the individual components relative to its centroidal axis parallel to axis of buckling, a is the distance between connectors, r_i is the minimum radius of gyration of individual components, r_{ib} is the radius of gyration of individual components relative to its centroidal axis parallel to member axis of buckling, α is the separation ratio $= h/2r_{ib}$, and h is the distance between centroids of individual components perpendicular to the member axis of buckling.

Eq. (52.41) is the same as that used in the current Italian code, as well as in other European specifications, based on test results [38]. In this equation, the bending effect is considered in the first term in square root, and shear force effect is taken into account in the second term. Eq. (52.42) was derived from elastic stability theory and verified by test data [39]. In both cases, the end connectors must be welded or slip-critical-bolted.

52.8 Tapered Columns

The state-of-the-art design for tapered structural members was provided in the SSRC guide [37]. The charts as shown in Figure 52.12 can be used to evaluate the effective length factors for tapered

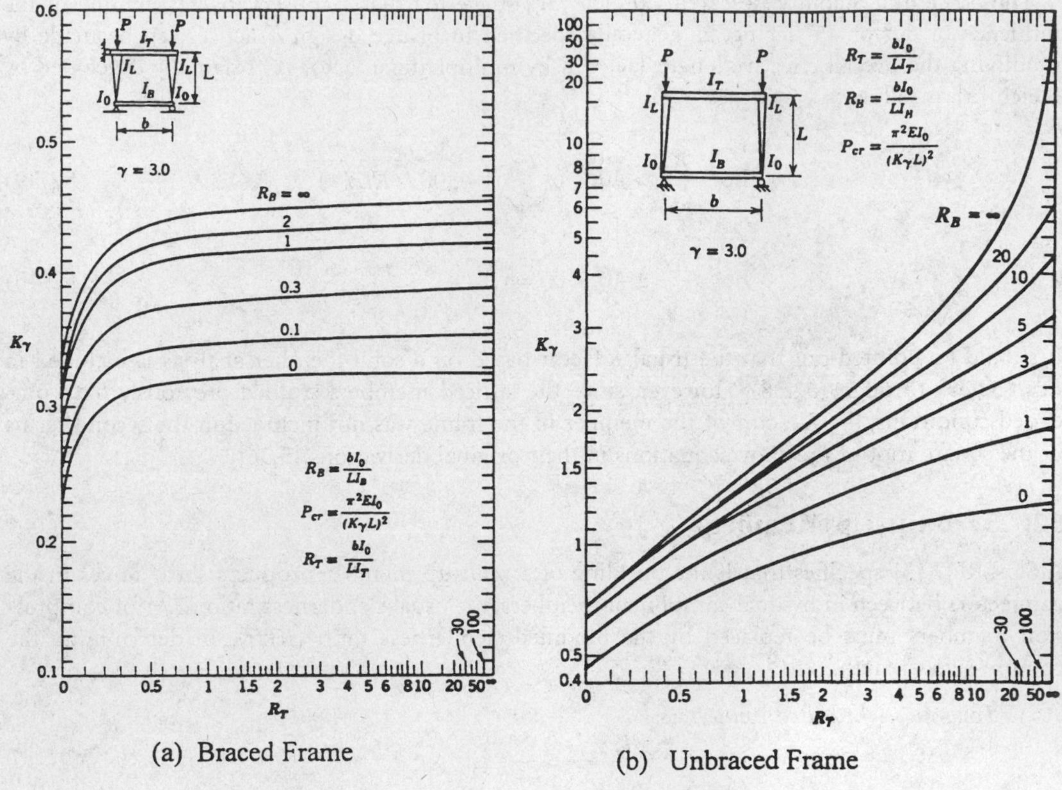

FIGURE 52.12 Effective length factor for tapered columns. (a) Braced frame; (b) unbraced frame. (*Source:* Galambos, T. V., Ed., Structural Stability Research Council Guide to Stability Design Criteria for Metal Structures, 4th ed., John Wiley & Sons, New York, 1988. With permission.)

column restrained by prismatic beams [37]. In these figures, I_T and I_B are the moment of inertia of top and bottom beam, respectively; b and L are length of beam and column, respectively; and γ is tapering factor as defined by

$$\gamma = \frac{d_1 - d_o}{d_o} \tag{52.43}$$

where d_o and d_1 are the section depth of column at the smaller and larger end, respectively.

52.9 Summary

This chapter summarizes the state-of-the-art practice of the effective length factors for isolated columns, framed columns, diagonal bracing systems, latticed and built-up members, and tapered columns. Design implementation with formulas, charts, tables, and various modification factors adopted in current codes and specifications, as well as those used in bridge structures, are described. Several examples are given to illustrate the steps of practical applications of these methods.

References

1. McGuire, W., Computers and steel design, *Modern Steel Constr.*, 32(7), 39, 1992.
2. AASHTO, *LRFD Bridge Design Specifications*, American Association of State Highway and Transportation Officials, Washington, D.C., 1994.
3. AISC, *Load and Resistance Factor Design Specification for Structural Steel Buildings*, 2nd ed., American Institute of Steel Construction, Chicago, IL, 1993.
4. Chen, W. F. and Lui, E. M., *Stability Design of Steel Frames*, CRC Press, Boca Raton, FL, 1991.
5. Bleich, F., *Buckling Strength of Metal Structures*, McGraw-Hill, New York, 1952.
6. Johnson, D. E., Lateral stability of frames by energy method, *J. Eng. Mech. ASCE*, 95(4), 23, 1960.
7. Lu, L. W., A survey of literature on the stability of frames, *Weld. Res. Counc. Bull.*, New York, 1962.
8. Kavanagh, T. C., Effective length of framed column, *Trans. ASCE*, 127(II) 81, 1962.
9. Gurfinkel, G. and Robinson, A. R., Buckling of elasticity restrained column, *J. Struct. Div. ASCE*, 91(ST6), 159, 1965.
10. Wood, R. H., Effective lengths of columns in multi-storey buildings, *Struct. Eng.*, 50(7–9), 234, 295, 341, 1974.
11. Julian, O. G. and Lawrence, L. S., Notes on J and L Nomograms for Determination of Effective Lengths, unpublished report, 1959.
12. ACI, *Building Code Requirements for Structural Concrete* (ACI 318-95) and Commentary (ACI 318R-95), American Concrete Institute, Farmington Hills, MI, 1995.
13. Galambos, T. V., Lateral support for tier building frames, *AISC Eng. J.*, 1(1), 16, 1964.
14. Aristizabal-Ochoa, J. D., K-factors for columns in any type of construction: nonparadoxical approach, *J. Struct. Eng. ASCE*, 120(4), 1272, 1994.
15. Duan, L., King, W. S., and Chen, W. F., *K* factor equation to alignment charts for column design, *ACI Struct. J.*, 90(3), 242, 1993.
16. *Regles de Cacul des Constructions en acier*, CM66, Eyrolles, Paris, 1975.
17. ECCS, *European Recommendations for Steel Construction*, European Convention for Construction Steelworks, 1978.
18. Dumonteil, P., Simple equations for effective length factors, *AISC Eng. J.*, 29(3), 111, 1992.
19. Johnston, B. G., Ed., Structural Stability Research Council, *Guide to Stability Design Criteria for Metal Structures*, 3rd ed., John Wiley & Sons, New York, 1976.
20. Liew, J. Y. R., White, D. W., and Chen, W. F., Beam-column design in steel frameworks — insight on current methods and trends, *J. Constr. Steel. Res.*, 18, 269, 1991.
21. Duan, L. and Chen, W. F., Effective length factor for columns in braced frames, *J. Struct. Eng. ASCE*, 114(10), 2357, 1988.
22. Duan, L. and Chen, W. F., Effective length factor for columns in unbraced frames, *J. Struct. Eng. ASCE*, 115(1), 150, 1989.
23. Duan, L. and Chen, W. F., 1996. Errata of paper: effective length factor for columns in unbraced frames, *J. Struct. Eng. ASCE*, 122(1), 224, 1996.
24. Essa, H. S., Stability of columns in unbraced frames, *J. Struct. Eng., ASCE*, 123(7), 952, 1997.
25. Yura, J. A., The effective length of columns in unbraced frames, *AISC Eng. J.*, 8(2), 37, 1971.
26. Disque, R. O., Inelastic *K* factor in design, *AISC Eng. J.*, 10(2), 33, 1973.
27. LeMessurier, W. J., A practical method of second order analysis, part 2 — rigid frames, *AISC Eng. J.*, 14(2), 50, 1977.
28. Galambos, T. V., Influence of partial base fixity on frame instability, *J. Struct. Div.* ASCE, 86(ST5), 85, 1960.
29. Salmon, C. G., Schenker, L., and Johnston, B. G., Moment-rotation characteristics of column anchorage, *Trans. ASCE*, 122, 132, 1957.
30. King, W. S., Duan, L., Zhou, R. G., Hu, Y. X., and Chen, W. F., *K* factors of framed columns restrained by tapered girders in U.S. codes, *Eng. Struct.*, 15(5), 369, 1993.

31. Lui, E. M., A novel approach for K-factor determination. *AISC Eng. J.*, 29(4), 150, 1992.
32. Shanmugam, N. E. and Chen, W. F., An assessment of K factor formulas, *AISC Eng. J.*, 32(3), 3, 1995.
33. Picard, A. and Beaulieu, D., Design of diagonal cross bracings, part 1: theoretical study, *AISC Eng. J.*, 24(3), 122, 1987.
34. Picard, A. and Beaulieu, D., Design of diagonal cross bracings, part 2: experimental study, *AISC Eng. J.*, 25(4), 156, 1988.
35. El-Tayem, A. A. and Goel, S. C., Effective length factor for the design of X-bracing systems, *AISC Eng. J.*, 23(4), 41, 1986.
36. Timoshenko, S. P. and Gere, J. M., *Theory of Elastic Stability*, 2nd ed., McGraw-Hill, New York, 1961.
37. Galambos, T. V., Ed., *Structural Stability Research Council, Guide to Stability Design Criteria for Metal Structures*, 4th ed., John Wiley & Sons, New York, 1988.
38. Zandonini, R., Stability of compact built-up struts: experimental investigation and numerical simulation, *Constr. Met.*, 4, 1985 [in Italian].
39. Aslani, F. and Goel, S. C., An analytical criteria for buckling strength of built-up compression members, *AISC Eng. J.*, 28(4), 159, 1991.

53

Fatigue and Fracture

Robert J. Dexter
University of Minnesota

John W. Fisher
Lehigh University

53.1 Introduction

Bridges do not usually fail due to inadequate load capacity, except when an overweight truck is illegally driven onto an old bridge with very low load rating. When bridge superstructures "fail," it is usually because of excessive deterioration by corrosion and/or fatigue cracking rather than inadequate load capacity. Although most deterioration can be attributed to lack of proper maintenance, there are choices made in design that also can have an impact on service life. Yet the design process for bridges is focused primarily on load capacity rather than durability.

This chapter of the handbook will inform the reader about a particular aspect of durability, i.e., the fatigue and fracture failure mode, and about detailing for improved resistance to fatigue and fracture. Only aspects of fatigue and fracture that are relevant to design or assessment of bridge deck and superstructure components are discussed. Concrete and aluminum structural components are discussed briefly, but the emphasis of this section is on steel structural components.

The fatigue and fracture design and assessment procedures outlined in this chapter are included in the American Association of State Highway and Transportation Officials (AASHTO) specifications for bridges [1]. Some of the bridges built before the mid-1970s (when the present fatigue-design specifications were adopted) may be susceptible to fatigue cracking. There are valuable lessons that can be learned from the problems that these bridges experienced, and several examples will be used in this chapter to illustrate various points. These lessons have been incorporated into the present AASHTO specifications [2,3]. As a result, steel bridges that have been built in the last few decades have not and will not have any significant problems with fatigue and fracture [2].

These case histories of fatigue cracking should not create the false impression that there is an inherent fatigue problem with steel bridges. The problems that occur are confined to older bridges. These problems are, for the most part, relatively minor and can be corrected with inexpensive retrofits. The problems are even easier to avoid in new designs. Therefore, because there are some

0-8493-7434-0/00/$0.00+$.50

FIGURE 53.1 View of cracked girder of Lafayette Street Bridge in St. Paul, MN showing fatigue crack originating from backing bars and lack of fusion on the weld attaching the lateral bracing attachment plate to the web and to the transverse stiffener.

fatigue problems with older bridges, one should not get the impression that there are ongoing fatigue problems with modern bridges designed by the present fatigue-design specifications.

Detailing rules are perhaps the most important part of the fatigue and fracture design and assessment procedures. The detailing rules are intended to avoid notches and other stress concentrations. These detailing rules are useful for the avoidance of brittle fracture as well as fatigue. Because of the detailing rules, modern steel bridges are detailed in a way that appears much cleaner than those built before the 1970s. There are fewer connections and attachments in modern bridges, and the connections use more fatigue-resistant details such as high-strength bolted joints.

For example, AWS D1.5 (Bridge Welding Code) does not permit backing bars to be left in place on welds. This rule is a result of experience such as that shown in Figure 53.1. Figure 53.1 shows lateral gusset plates on the Lafayette St. Bridge in St. Paul, MN that cracked and led to a fracture of a primary girder in 1976 [3,4]. In this detail, backing bars were left in place under the groove welds joining the lateral gusset plate to the transverse stiffener and to the girder web. The backing bars create a cracklike notch, often accompanied by a lack-of-fusion defect. Fatigue cracks initiate from this cracklike notch and the lack of fusion in the weld to the transverse stiffener because, in this case, the plane of the notch is perpendicular to the primary fluctuating stress.

The Bridge Welding Code AWS D1.5 at present requires that backing bars be removed from all bridge welds to avoid these notches. Prior to 1994, this detailing rule was not considered applicable to seismic moment-resisting building frames. Consequently, many of these frames fractured when the 1994 Northridge earthquake loaded them. Backing bars left on the beam flange-to-column welds of these frames created a built-in cracklike notch. This notch contributed to the Northridge fractures, along with lack-of-fusion defects and low-toughness welds [5–7].

Figure 53.2 shows a detail where a primary girder flange penetrates and is continuous through the web of a cross girder of the Dan Ryan Elevated structures in Chicago [8]. In this case, the short vertical welds at the sides of the flange were defective. Fatigue cracks initiated at these welds,

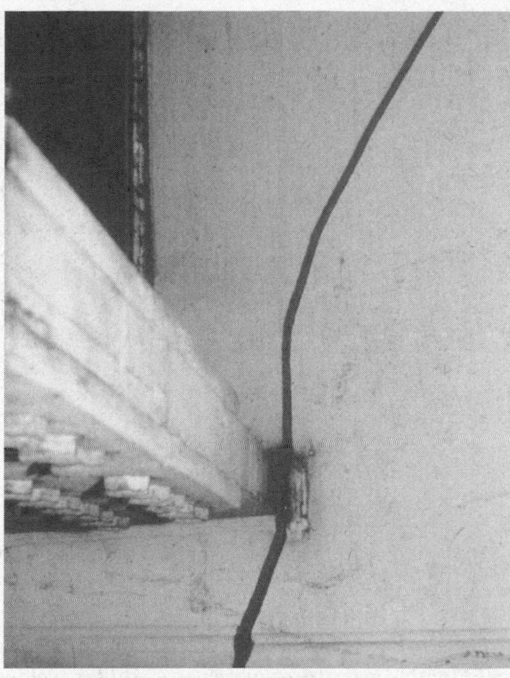

FIGURE 53.2 View of cracked cross girder of Dan Ryan elevated structure in Chicago showing cracking originating from short vertical welds which are impossible to make without lack of fusion defects.

which led to fracture of the cross girder. It is unlikely that good welds could have been made for this detail. A better alternative would have been to have cope holes at the ends of the flange. Note that in Figures 53.1 and 53.2, the fractures did not lead to structural collapse. The reason for this reserve tolerance to large cracks will be discussed in Section 53.2.

In bridges, there are usually a large number of cycles of significant live load, and fatigue will almost always precede fracture. Therefore, controlling fatigue is practically more important than controlling fracture. The civil engineering approach for fatigue is explained in Section 53.3. The fatigue life (N) of particular details is determined by the nominal stress range (S) from S–N curves. The nominal stress S–N curves are the lower-bound curves to a large number of full-scale fatigue test data. The full-scale tests empirically take into account a number of variables with great uncertainty, e.g., residual stress, weld profile, environment, and discontinuities in the material from manufacturing. Consequently, the variability of fatigue life data at a particular stress range is typically about a factor of 10.

Usually, the only measures taken in design that are primarily intended to assure fracture resistance are to specify materials with minimum specified toughness values, such as a Charpy V-Notch (CVN) test requirement. As explained in Section 53.4, toughness is specified so that the structure is resistant to brittle fracture despite manufacturing defects, fatigue cracks, and/or unanticipated loading. These material specifications are less important for bridges than the S–N curves and detailing rules, however.

Steel structures have exhibited unmatched ductility and integrity when subjected to seismic loading. Modern steel bridges in the United States which are designed to resist fatigue and fracture from truck loading have not exhibited fractures in earthquakes. It would appear that the modern bridge design procedures which consider fatigue and fracture from truck loading are also adequate to assure resistance to brittle fracture under seismic loading. Although rare, fractures of bridge structural elements have occurred during earthquakes outside the United States. For example, brittle fractures occurred on several types of steel bridge piers during the 1995 Hyogo-ken Nanbu earthquake in Japan [9].

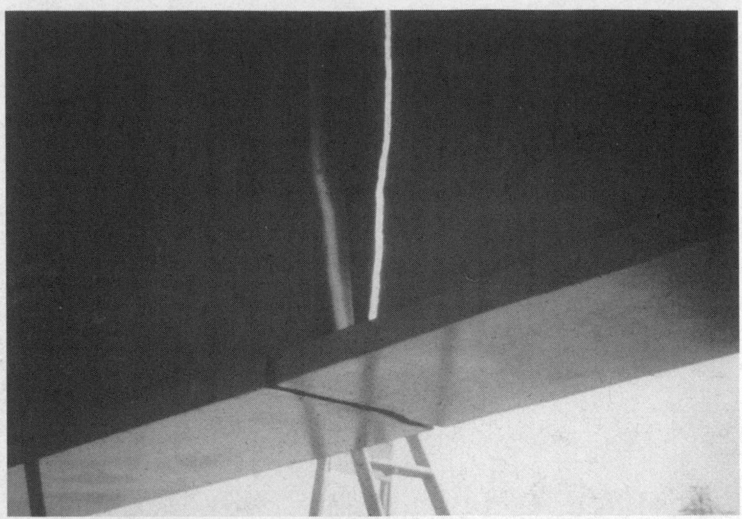

FIGURE 53.3 View of cracked girder of I-79 Bridge at Neville Island in Pittsburgh as an example of a bridge that is sufficiently redundant to avoid collapse despite a fracture of the tension flange and the web.

These fatigue and fracture design and assessment procedures for bridges are also applicable to many other types of cyclically loaded structures which use similar welded and bolted details, e.g., cranes, buildings, chimneys, transmission towers, sign, signal, and luminaire support structures, etc. In fact, these procedures are similar to those in the American Welding Society AWS D1.1, "Structural Welding Code — Steel" [10], which is applicable to a broad range of welded structures.

This "civil engineering" approach to fatigue and fracture could also be applied to large welded and bolted details in structures outside the traditional domain of civil engineering, including ships, offshore structures, mobile cranes, and heavy vehicle frames. However, the civil engineering approach to fatigue presented here is different from traditional mechanical engineering approaches. The mechanical engineering approaches are well suited to smooth machined parts and other applications where a major portion of the fatigue life of a part is consumed in forming an initial crack. In the mechanical engineering approaches, the fatigue strength is proportional to the ultimate tensile strength of the steel. The experimental data show this is not true for welded details, as discussed below.

53.2 Redundancy, Ductility, and Structural Collapse

Fatigue is considered a serviceability limit state for bridges because the fatigue cracks and fractures that have occurred have mostly not been significant from the standpoint of structural integrity. Redundancy and ductility of steel bridges have prevented catastrophic collapse. Only in certain truly nonredundant structural systems can fatigue cracking lead to structural collapse.

The I-79 Bridge at Neville Island in Pittsburgh is an example of the robustness of even so-called fracture critical or nonredundant two-girder bridges. In 1977, one of the girders developed a fatigue crack in the tension flange at the location of a fabrication repair of an electroslag weld splice [3]. As shown in Figure 53.3, the crack completely fractured the bottom flange and propagated up the web of this critical girder. A tugboat captain happened to look up and notice the crack extending as he passed under the bridge.

Although two-girder bridges are considered nonredundant, other elements of the bridge, particularly the deck, are usually able to carry the loads and prevent collapse as in the case of the I-79 bridge. Today, because of the penalties in design and fabrication for nonredundant or fracture-critical members, simple and low-cost two-girder bridges are seldom built. Note that the large cracks shown in the bridges in Figures 53.1 and 53.2 also did not lead to structural collapse. Unfortunately, this built-in redundancy shown by these structures is difficult to predict and is not explicitly recognized in design.

The beneficial effects of redundancy on fatigue and fracture are best explained in terms of the boundary conditions on the structural members. The truck loads and wind loads on bridges are essentially "fixed-load" or "load-control" boundary conditions. On a local scale, however, most individual members and connections in redundant structures are essentially under "displacement-control" boundary conditions. In other words, because of the stiffness of the surrounding structure, the ends of the member have to deform in a way that is compatible with nearby members. A cracked member in parallel with other similar but uncracked members will experience a decreasing load range and nominal stress range as the stiffness of the cracked member decreases. This behavior under displacement control is referred to as load shedding and it can slow down the rate of fatigue crack propagation.

If a fatigue crack forms in one element of a bolted or riveted built-up structural member, the crack cannot propagate directly into neighboring elements. Usually, a riveted member will not fail until additional cracks form in one or more additional elements. Therefore, riveted built-up structural members are inherently redundant. Once a fatigue crack forms, it can propagate directly into all elements of a continuous welded member and cause failure at service loads. Welded structures are not inferior to bolted or riveted structures; they require more attention to design, detailing, and quality.

Ductility is required in order for redundancy to be completely effective. As the net section of a cracking bridge member decreases, the plastic moment capacity of the member decreases. If a member is sufficiently ductile, it can tolerate a crack so large that the applied moment exceeds the plastic moment for the net section and a mechanism will form in the member [11–13]. If the member can then deform to several times the yield rotation, the load will be shed to the deck and other members.

Minimum levels of fracture toughness are necessary to achieve ductility, but are not sufficient. The fracture toughness assures that brittle fracture does not occur before general yielding of the net cross section. However, net section yielding is not very ductile unless the yielding can spread to the gross section, which requires strain hardening in the stress–strain relationship of the steel, or a reasonably low yield-to-tensile ratio [12–14].

53.3 Fatigue Resistance

Low-cycle fatigue is a possible failure mode for structural members or connections which are cycled into the inelastic region for a small number of cycles (less that 1000) [15,16]. For example, bridge pier structures may be subjected to low-cycle fatigue in an earthquake [9]. Brittle fractures occurred in Japan in steel piers that underwent large plastic strain cycles during the 1995 Hyogo-ken Nanbu earthquake in Japan [9]. However, in order to focus on the more common phenomenon of high-cycle fatigue, low-cycle fatigue is not discussed further in this section.

Truck traffic is the primary cause of high-cycle fatigue of bridges. Wind loads may also be a fatigue design consideration in bridges. Wind-induced vibration has caused numerous fatigue problems in sign, signal, and luminaire support structures [17].

Although cracks can form in structures cycled in compression, they arrest and are not structurally significant. Therefore, only members or connections for which the stress cycle is at least partially in tension need to be assessed.

In most bridges, the ratio of the fatigue design truck load to the strength design load is large enough that fatigue may control the design of much of the structure. In long-span bridges, the load on much of the superstructure is dominated by the dead load, with the fluctuating live-load part relatively small. These members will not be sensitive to fatigue. However, the deck, stringers, and floor beams of bridges are subjected to primarily live load and therefore will be controlled by

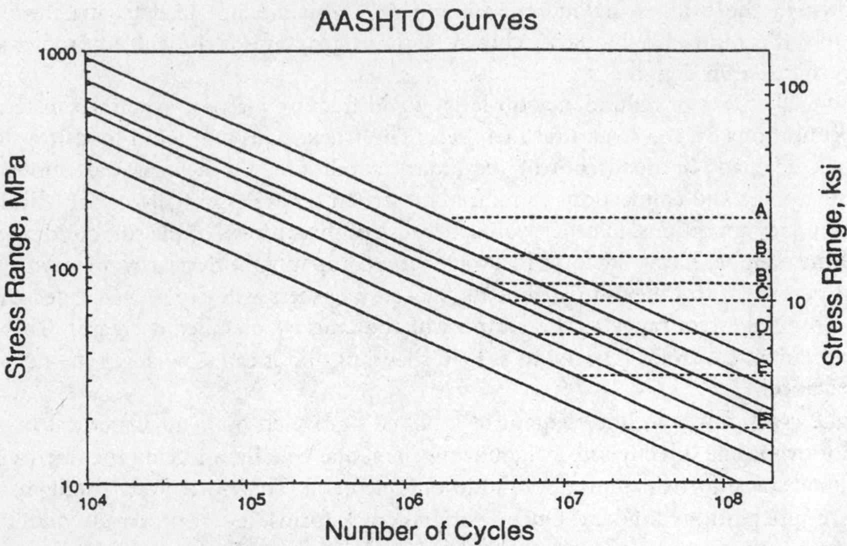

FIGURE 53.4 The lower-bound *S–N* curves for the seven primary fatigue categories from the AASHTO, AREA, AWS, and AISC specifications. The dotted lines are the CAFL and indicate the detail category.

fatigue. Fortunately, the deck, stringers, and floor beams are secondary members which, if they failed, would not lead to structural collapse.

When information about a specific crack is available, a fracture mechanics crack growth rate analysis should be used to calculate remaining life [23–25]. However, in the design stage, without specific initial crack size data, the fracture mechanics approach is not any more accurate than the *S–N* curve approach [25]. Therefore, the fracture mechanics crack growth analysis will not be discussed further.

Welded and bolted details for bridges and buildings are designed based on the nominal stress range rather than the local "concentrated" stress at the weld detail. The nominal stress is usually obtained from standard design equations for bending and axial stress and does not include the effect of stress concentrations of welds and attachments. Since fatigue is typically only a serviceability problem, fatigue design is carried out using service loads as discussed in Section 53.3.5. Usually, the nominal stress in the members can be easily calculated without excessive error. However, the proper definition of the nominal stresses may become a problem in regions of high stress gradients [26,27].

It is standard practice in fatigue design of welded structures to separate the weld details into categories having similar fatigue resistance in terms of the nominal stress. Each category of weld details has an associated *S–N* curve. The *S–N* curves for steel in the AASHTO [1], AISC [28], ANSI/AWS [10], and American Railway Engineers Association (AREA) provisions are shown in Figure 53.4. *S–N* curves are presented for seven categories of weld details — A through E′, in order of decreasing fatigue strength. These *S–N* curves are based on a lower bound to a large number of full-scale fatigue test data with a 97.5% survival limit.

The slope of the regression line fit to the test data for welded details is typically in the range 2.9 to 3.1 [20,21]. Therefore, in the AISC and AASHTO codes as well as in Eurocode 3 [29], the slopes have been standardized at 3.0. The effect of the welds and other stress concentrations are reflected in the ordinate of the *S–N* curves for the various detail categories.

Figure 53.4 shows the fatigue threshold or constant amplitude fatigue limits (CAFL) for each category as horizontal dashed lines. When constant-amplitude tests are performed at stress ranges below the CAFL, noticeable cracking does not occur. The number of cycles associated with the CAFL is whatever number of cycles corresponds to that stress range on the *S–N* curve for that

TABLE 53.1 Constant-Amplitude Fatigue Limits for AASHTO and Aluminum Association *S–N* Curves

Detail Category	CAFL for Steel (MPa)	CAFL for Aluminum (MPa)
A	165	70
B	110	41
B′	83	32
C	69	28
D	48	17
E	31	13
E′	18	7

category or class of detail. The CAFL occurs at an increasing number of cycles for lower fatigue categories or classes. Sometimes, different details, which share a common *S–N* curve (or category) in the finite-life regime, have different CAFL.

Typically, small-scale specimen tests will result in longer apparent fatigue lives. Therefore, the *S–N* curve must be based on tests of full-size structural components such as girders. Testing on full-scale welded members has indicated that the primary effect of constant-amplitude loading can be accounted for in the live-load stress range; i.e., the mean stress is not significant [18–21]. The reason that the dead load has little effect on the lower bound of the results is that, locally, there are very high residual stresses from welding. Mean stress may be important for some details that are not welded, such as anchor bolts [17,22]. In order to be conservative for nonwelded details, in which there may be a significant effect of the mean stress, the fatigue test data should be generated under loading with a high tensile mean stress.

The strength and type of steel have only a negligible effect on the fatigue resistance expected for a particular detail [18–21]. The welding process also does not typically have an effect on the fatigue resistance [18–21]. The independence of the fatigue resistance from the type of steel greatly simplifies the development of design rules for fatigue since it eliminates the need to generate data for every type of steel.

The full-scale fatigue experiments have been carried out in moist air and therefore reflect some degree of environmental effect or corrosion fatigue. Full-scale fatigue experiments in seawater do not show significantly lower fatigue lives [30], provided that corrosion is not so severe that it causes pitting. The fatigue lives seem to be more significantly influenced by the stress concentration at the toe of welds and the initial discontinuities. Therefore, these lower-bound *S–N* curves can be used for design of bridges in any natural environmental exposure, even near salt spray. However, pitting from severe corrosion may become a fatigue-critical condition and should not be allowed [31,32].

Similar *S–N* curves have been proposed by the Aluminum Association [56] for welded aluminum structures. Table 53.1 summarizes the CAFL for steel and aluminum for categories A through E′. The design procedures are based on associating weld details with specific categories. For both steel and aluminum, the separation of details into categories is approximately the same.

The categories in Figure 53.4 range from A to E′ in order of decreasing fatigue strength. There is an eighth category, F, in the specifications (not shown in Figure 53.4) which applies to fillet welds loaded in shear. However, there have been very few if any failures related to shear, and the stress ranges are typically very low such that fatigue rarely would control the design. Therefore, the shear stress Category F will not be discussed further. In fact, there have been very few if any failures which have been attributed to details which have a fatigue strength greater than Category C.

53.3.1 Classification of Details in Metal Structural Components

Details must be associated with one of the drawings in the specification [1] to determine the fatigue category. The following is a brief, simplified overview of the categorization of fatigue details. In some cases, this overview has left out some details so the specification should always be checked for the appropriate detail categorization. The AISC specification [28] has a somewhat better presentation of the sketches and explanation of the detail categorization than the AASHTO specifications. Also, several reports have been published which show a large number of illustrations of details and their categories [33,34]. In addition, the Eurocode 3 [29] and the British Standard 7608 [35] have more detailed illustrations for their categorization than does the AISC or AASHTO specifications. A book by Maddox [36] discusses categorization of many details in accordance with BS 7608, from which roughly equivalent AISC categories can be inferred.

Small holes are considered Category D details. Therefore, riveted and mechanically fastened joints (other than high-strength bolted joints) loaded in shear are evaluated as Category D in terms of the net section nominal stress. Properly tensioned high-strength bolted joints loaded in shear may be classified as Category B. Pin plates and eyebars are designed as Category E details in terms of the stress on the net section.

Welded joints are considered longitudinal if the axis of the weld is parallel to the primary stress range. Continuous longitudinal welds are Category B or B′ details. However, the terminations of longitudinal fillet welds are more severe (Category E). (The termination of full-penetration groove longitudinal welds requires a ground transition radius but gives greater fatigue strength, depending on the radius.) If longitudinal welds must be terminated, it is better to terminate at a location where the stress ranges are less severe.

Attachments normal to flanges or plates that do not carry significant load are rated Category C if less than 51 mm long in the direction of the primary stress range, D if between 51 and 101 mm long, and E if greater than 101 mm long. (The 101 mm limit may be smaller for plate thinner than 9 mm). If there is not at least 10-mm edge distance, then Category E applies for an attachment of any length. The Category E′, slightly worse than Category E, applies if the attachment plates or the flanges exceed 25 mm in thickness.

Transverse stiffeners are treated as short attachments (Category C). Transverse stiffeners that are used for cross-bracing or diaphragms are also treated as Category C details with respect to the stress in the main member. In most cases, the stress range in the stiffener from the diaphragm loads is not considered, because these loads are typically unpredictable. However, the detailing of attachment plates is critical to avoid distortion-induced fatigue, as discussed in Section 53.3.2.

In most other types of load-carrying attachments, there is interaction between the stress range in the transverse load-carrying attachment and the stress range in the main member. In practice, each of these stress ranges is checked separately. The attachment is evaluated with respect to the stress range in the main member, and then it is separately evaluated with respect to the transverse stress range. The combined multiaxial effect of the two stress ranges is taken into account by a decrease in the fatigue strength; i.e., most load-carrying attachments are considered Category E details.

The fatigue strength of longitudinal attachments can be increased if the ends are given a radius and the fillet or groove weld ends are ground smooth. For example, a longitudinal attachment (load-bearing or not) with a transition radius greater than 50 mm can be considered Category D. If the transition radius of a groove-welded longitudinal attachment is increased to greater than 152 mm (with the groove-weld ends ground smooth), the detail (load bearing or not) can be considered Category C.

53.3.2 Detailing to Avoid Distortion-Induced Fatigue

It is clear from the type of cracks that occur in bridges that a significant proportion of the cracking is due to distortion that results from such secondary loading [37]. The solution to the problem of fatigue cracking due to secondary loading usually relies on the qualitative art of good detailing [38]. Often, the best solution to distortion cracking problems may be to stiffen the structure. Typically, the better connections are more rigid.

FIGURE 53.5 View of floor beam of Throgg's Neck Bridge in New York showing crack in the cope that has been repaired by drilling a stop hole. The crack is caused by incompatibility between the curvature of the superstructure and the orthotropic steel deck that is bolted onto the floor beams.

One of the most overlooked secondary loading problems occurs at the interface of structures with different flexural rigidities and curvatures [39,40]. Figure 53.5 shows a typical crack at the floor beam flange cope in the Throg's Neck Bridge in New York. One of the closed trapezoidal ribs of an orthotropic steel deck is visible in Figure 53.5. The orthotropic deck was added to the structure to replace a deteriorating deck by bolting onto the floor beams. However, the superstructure has curvature that is incompatible with the stiff deck. The difference in curvature manifests as out-of-plane rotation of the flange of the floor beam. The crack is caused by out-of-plane bending of the floor beam web at the location of the cope, which has many built-in discontinuities due to the flame cutting.

Another example of secondary loading from out-of-plane distortion may occur at attachment plates for transverse bracing or for a floor beam. These attachment plates, which may have distortion-induced out-of-plane loads, should be welded directly to both flanges as well as the web. In older bridges, it was common practice not to weld transverse stiffeners and attachment plates to the tension flange of welded I-girders and box girders. The practice of not allowing transverse fillet welds on the tension flange is not necessary and is due to unwarranted concern about brittle fracture of the tension flange [37,38]. Unfortunately, this practice is not harmless, because numerous fatigue cracks have occurred due to distortion in the "web gap," i.e., the narrow gap between the termination of the attachment plate fillet welds and the flange [37,38]. Figure 53.6 shows an example of a crack that formed along the fillet weld that attaches a diaphragm connection plate to the web of a box girder.

In most cases, these web-gap-cracking problems can be solved by rigidly attaching the attachment plate to the tension flange. To retrofit existing bridges, a very thick T or angle may be high-strength-bolted in to join the attachment plate to the tension flange [38]. The cracked detail shown in Figure 53.6 was retrofit this way. In other cases a better solution is to make the detail more flexible. This flexibility can be accomplished by increasing the size of the gap, allowing the distortion to take place over a greater length so that lower stresses are created.

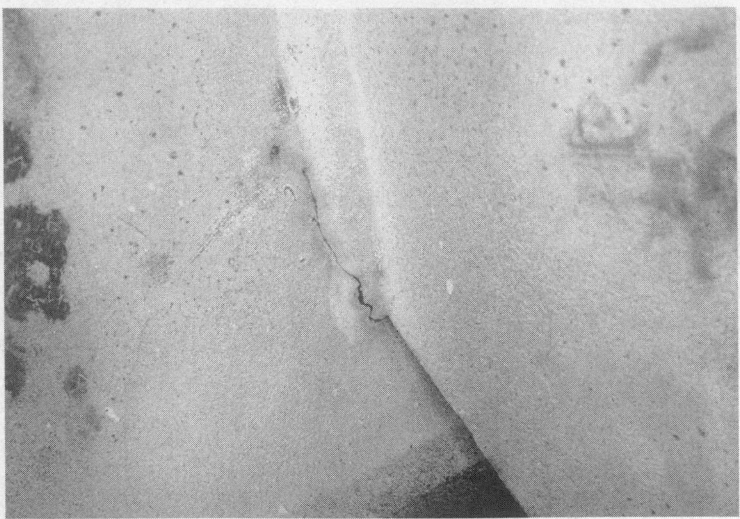

FIGURE 53.6 View of crack in the fillet weld joining the diaphragm connection plate to the web of a box section in the Washington Metro elevated structures. The crack is caused by distortion in the small gap between the bottom of the attachment plate and the box girder flange.

FIGURE 53.7 Connection angle from stringer to floor beam connection of a bridge over the St. Croix River on I-94 that cracked because it was too stiff.

The flexibility approach is used to prevent cracking at the terminations of transverse stiffeners that are not welded to the bottom flange. If there is a narrow web gap between the end of a transverse stiffener and the bottom flange, cracking can occur due to distortion of the web gap from inertial loading during handling and shipping. To prevent this type of cracking, the gap between the flange and the end of the stiffener should be between four and six times the thickness of the web [38].

Another example where the best details are more flexible is connection angles for "simply supported" beams. Despite our assumptions, such simple connections transmit up to 40% of the theoretical fixed-end moment, even though they are designed to transmit only shear forces. This unintentional end moment may crack the connection angles. The cracked connection angle shown in Figure 53.7 was from the stringer to floor beam connection on a bridge that formerly was on I-94 over the St. Croix River (it was recently replaced).

FIGURE 53.8 Close-up view of a crack originating at the termination of a flange on a floor beam with a "simple" connection to an attachment plate in the Dresbach Bridge in Minnesota. The crack is an indication of an end moment in the simple connection that is not predicted in design.

For a given load, the moment in the connection decreases significantly as the rotational stiffness of the connection decreases. The increased flexibility of connection angles allows the limited amount of end rotation to take place with reduced bending stresses. A criterion has been developed for the design of these angles to provide sufficient flexibility [41]. The criterion states that the angle thickness (t) must be

$$t < 12\,(g^2/L) \tag{53.1}$$

where (g) is the gauge and L is the span length. For example, using S.I. units, for connection angles with a gauge of 76 mm and a beam span of 7000 mm, the angle thickness should be just less than 10 mm. To solve a connection-angle-cracking problem in service, the topmost rivet or bolt may be removed and replaced with a loose bolt to ensure the shear capacity. For loose bolts, steps are required to ensure that the nuts do not back off.

Another result of unintended end moment on a floor beam is shown in Figure 53.8 from the Dresbach Bridge 9320 in Minnesota. In this bridge, the web of the welded built-up floor beam extends beyond the flanges in order to bolt to the floor beam connection plate on the girder. The unintended moment at the end of the floor beam is enough to cause cracking in the web plate at the termination of the flange fillet welds. There is a large stress concentration at this location caused by the abrupt change in section at the end of the flange; i.e., this is a Category E′ detail.

Significant stresses from secondary loading are often in a different direction than the primary stresses. Fortunately, experience with multiaxial loading experiments on large-scale welded structural details indicates the loading perpendicular to the local notch or the weld toe dominates the fatigue life. The cyclic stress in the other direction has no effect if the stress range is below 83 MPa and only a small influence above 83 MPa [26,37]. The recommended approach for multiaxial loads [26] is

1. To decide which loading (primary or secondary) dominates the fatigue cracking problem (typically the loading perpendicular to the weld axis or perpendicular to where cracks have previously occurred in similar details); and,
2. To perform the fatigue analysis using the stress range in this direction (i.e., ignore the stresses in the orthogonal directions).

53.3.3 Classification of Details in Concrete Structural Components

Concrete structures are typically less sensitive to fatigue than welded steel and aluminum structures. However, fatigue may govern the design when impact loading is involved, such as pavement, bridge decks, and rail ties. Also, as the age of concrete girders in service increases, and as the applied stress ranges increase with increasing strength of concrete, the concern for fatigue in concrete structural members has also increased.

According to ACI Committee Report 215R-74 in the Manual of Standard Practice [42], the fatigue strength of plain concrete at 10 million cycles is approximately 55% of the ultimate strength. However, even if failure does not occur, repeated loading may contribute to premature cracking of the concrete, such as inclined cracking in prestressed beams. This cracking can then lead to localized corrosion and fatigue of the reinforcement [43].

The fatigue strength of straight, unwelded reinforcing bars and prestressing strand can be described (in terms of the categories for steel details described above) with the Category B *S–N* curve. ACI Committee 215 suggests that members be designed to limit the stress range in the reinforcing bar to 138 MPa for high levels of minimum stress (possibly increasing to 161 MPa for less minimum stress). Fatigue tests show that previously bent bars had only about half the fatigue strength of straight bars, and failures have occurred down to 113 MPa [44]. Committee 215 recommends that half of the stress range for straight bars be used, i.e., 69 MPa for the worst-case minimum stress. Equating this recommendation to the *S–N* curves for steel details, bent reinforcement may be treated as a Category D detail.

Provided the quality is good, butt welds in straight reinforcing bars do not significantly lower the fatigue strength. However, tack welds reduce the fatigue strength of straight bars about 33%, with failures occurring as low as 138 MPa. Fatigue failures have been reported in welded wire fabric and bar mats [45].

If prestressed members are designed with sufficient precompression that the section remains uncracked, there is not likely to be any problem with fatigue. This is because the entire section is resisting the load ranges and the stress range in the prestessing strand is minimal. Similarly, for unbonded prestressed members, the stress ranges will be very small. However, there is reason to be concerned for bonded prestressing at cracked sections because the stress range increases locally. The concern for cracked sections is even greater if corrosion is involved. The pitting from corrosive attack can dramatically lower the fatigue strength of reinforcement [43].

Although the fatigue strength of prestressing strand in air is about equal to Category B, when the anchorages are tested as well, the fatigue strength of the system is as low as half the fatigue strength of the wire alone (i.e., about Category E). When actual beams are tested, the situation is very complex, but it is clear that much lower fatigue strength can be obtained [46,47]. Committee 215 has recommended the following for prestressed beams:

1. The stress range in prestressed reinforcement, determined from an analysis considering the section to be cracked, is not to exceed 6% of the tensile strength of the reinforcement (this is approximately equivalent to Category C).

2. Without specific experimental data, the fatigue strength of unbonded reinforcement and their anchorages is to be taken as half of the fatigue strength of the prestressing steel. (This is approximately equivalent to Category E.) Lesser values are to be used at anchorages with multiple elements.

53.3.4 Classification of Stay Cables

The Post-Tensioning Institute has issued "Recommendations for Stay Cable Design and Testing" [48]. The PTI recommends that uncoupled bar stay cables are Category B details, while coupled (glued) bar stay cables are Category D. The fatigue strengths of stay cables are verified through fatigue testing. Two types of tests are performed: (1) fatigue testing of the strand, and (2) testing of relatively short lengths of the assembled cable with anchorages. The recommended test of the system is 2 million cycles at a stress range (158 MPa) which is 35 MPa greater than the fatigue allowable for Category B at 2 million cycles. This test should pass with less than 2% wire breaks. A subsequent proof test must achieve 95% of the actual ultimate tensile strength of the tendons.

53.3.5 Characterization of Truck Loading for Fatigue

An actual service load history is likely to consist of cycles with a variety of different load ranges, i.e., variable-amplitude loading. However, the *S–N* curves that are the basis of the fatigue design provisions are based on constant-amplitude loading. A procedure is shown below to convert variable stress ranges to an equivalent constant-amplitude stress range with the same number of cycles. This procedure is based on the damage summation rule jointly credited to Palmgren and Miner (referred to as Miner's rule) [49]. If the slope of the *S–N* curve is equal to 3, then the relative damage of stress ranges is proportional to the cube of the stress range. Therefore, the effective stress range (S_{Re}) is equal to the cube root of the mean cube (rmc) of the stress ranges, i.e.,

$$S_{Re} = [(n_i / N_{\text{total}})\, S_i^3]^{1/3} \tag{53.2}$$

where n_i is the number of stress ranges of magnitude S_i and N_{total} is the total number of stress ranges.

The fatigue design truck in the LRFD version of the AASHTO bridge design specification and in the present 16th edition of the Standard Specifications is a three-axle HS20 truck. The front axle has a weight of 36 kN and the rear two axles each have a weight of 142 kN. It is very important to note that the single rear axles of the HS20 fatigue truck are actually intended to represent pairs of tandem axles [50,51]. This simplification eases design of main members by decreasing the number of axles (loads) which must be considered. Representation as a single axle is reasonable for design of bridge main members, since the close spacing of tandem axles (about 1.2 m) effectively generates only one stress cycle.

This simplified three-axle truck is not appropriate for the design of deck elements and even some floor beams. Each axle of the tandem axle groups creates a unique stress cycle in a deck element. When the entire distribution of trucks is considered, this results in approximately 4.5 axles or cycles of loading on average for every truck. However, the LRFD code does not clearly indicate that this should be taken into account in the fatigue design of these elements. If the HS20 tandem axle load is split, the effective axle load is 71 kN for each axle.

The HS20 truck is used in strength calculations with a load factor greater than 1.0. However, there is a load factor of 0.75 for fatigue in the LRFD specification, which implies that the actual fatigue design truck is an HS15, i.e., an axle load of 107 kN (really a pair of tandem 53 kN axles). Thus, the intent of the AASHTO LRFD Specifications is to use an HS15 loading for fatigue. The load factor was invented so that an additional and possibly confusing design truck would not have to be defined.

The use of an HS15 truck as representative of the rmc of the variable series of trucks is based on extensive weigh-in-motion (WIM) data and was recommended in NCHRP Report 299 "Fatigue Evaluation Procedures for Steel Bridges" [50]. A constant axle spacing of 30 ft was found to best approximate the axle spacing of typical four and five axle trucks responsible for most fatigue damage to bridges. This HS15 truck is supposed to represent the effective or rmc gross vehicle weight for a distribution of future trucks.

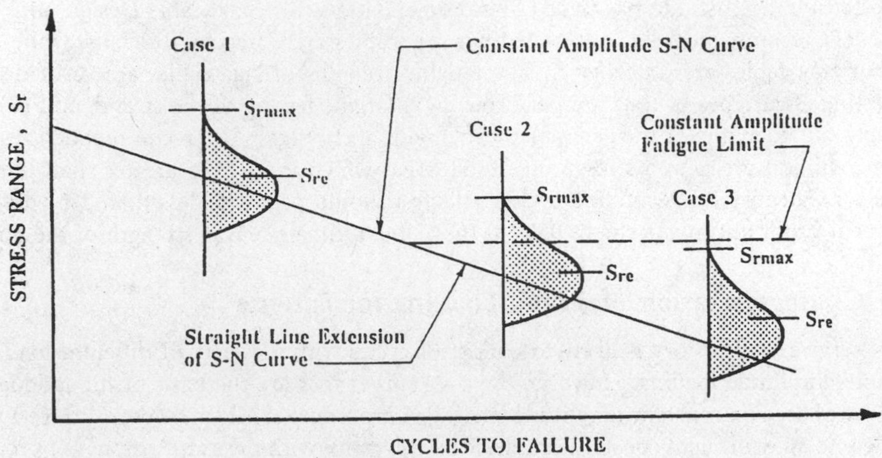

FIGURE 53.9 Schematic of three different cases of the relationship between the spectrum of applied stress ranges and the *S–N* curve. S_{re} is the effective or rmc stress range and S_{rmax} is the fatigue limit-state stress range with an exceedance of 1:10,000. Case 1 and 2 are in the finite-life regime, whereas case 3 illustrates the infinite-life regime.

In the AASHTO LRFD specification, the effective stress range is computed from this effective load, and the effective stress range is compared with an allowable stress range. The allowable stress range is obtained from the constant-amplitude *S–N* curve, using the same number of cycles as the number of variable cycles used to compute the rmc stress range.

This concept is illustrated schematically in Figure 53.9. Figure 53.9 shows the lower part of an *S–N* curve with three different variable stress-range distributions superposed. The effective stress range, shown as S_{re} in Figure 53.9, is the rmc of the stress ranges for a particular distribution, as defined in Eq. 53.2. The effective variable-amplitude stress range is used with *S–N* curves the same way as a constant-amplitude stress range is used, as shown for Case 1 in Figure 53.9.

Full-scale variable-amplitude fatigue tests show that if more than 0.01% of the stress ranges in a distribution are above the CAFL (Case 2 in Figure 53.9), fatigue cracking will still occur [52]. The few cycles that are above the CAFL seem to keep the process going, such that even the stress ranges that are below the CAFL apparently contribute to the fatigue crack growth. The effective stress range and number of cycles to failure data from these tests fell along a straight-line extrapolation of the constant-amplitude curve. Therefore, the approach in the AASHTO LRFD specification is to use the straight-line extrapolation of the *S–N* curve below the CAFL to compute the allowable stress range. There is a lower cutoff of this straight-line extrapolation, which will be discussed later.

The British fatigue design standard BS 7608 [35] uses this same approach for variable amplitude loading; i.e., an rmc stress range is used with an extrapolation of the *S–N* curve below the CAFL. Eurocode 3 [29] also uses the effective stress range concept; however, it uses a different slope for the extrapolation of the *S–N* curve below the CAFL.

Case 1 and Case 2 in Figure 53.9 are in what is called the finite-life regime because the calculations involve a specific number of cycles. Case 3 in Figure 53.9 represents what is referred to as the infinite-life regime. In the infinite-life regime, essentially all of the stress ranges are below the CAFL. The full-scale variable-amplitude fatigue tests referred to earlier [52] show that as many as 0.01% of the stress ranges can exceed the CAFL without resulting in cracking. Consequently, the stress range associated with an exceedance probability of 0.01% is often referred to as the "fatigue-limit-state" stress range.

The lower cutoff of the straight-line extrapolation of the *S–N* curve in the AASHTO LRFD specification is related to this infinite-life phenomenon. Specifically, the AASHTO LRFD specification cuts off the extrapolation of the *S–N* curve at a stress range equal to half the CAFL. The intent of this limit is actually to assure that the fatigue-limit-state stress range is just below the CAFL. The AASHTO LRFD procedure assumes that there is a fixed relationship between the rmc or effective stress range and the limit-state stress range, specifically that the effective stress range is equal to half the limit-state stress range.

Depending on the type of details, the infinite-life regime begins at about 35 million cycles. When designing for a number of cycles larger than this limit, it is no longer important to quantify the precise number of cycles. Any design where the rmc stress range is less than half the CAFL, or where the fatigue-limit-state stress range is less than the CAFL, should theoretically result in essentially infinite life.

The ratio of the fatigue-limit-state stress range to the effective stress range is assumed to be the same as the ratio of the fatigue-limit-state load range to the effective load range, although this is only an approximation at best. Therefore, the fatigue-limit-state load range is defined as having a probability of exceedance over the lifetime of the structure of 0.01%. A structure with millions of cycles is likely to see load ranges exceeding this magnitude hundreds of times, therefore, the fatigue-limit-state load range is not as large as the extreme loads used to check ultimate strength. The fatigue-limit-state load range is assumed to be about twice the effective rmc load range in the LRFD specification.

Recall that the HS20 loading with a load factor of 0.75, i.e., an HS15 loading, was defined as the effective load in the AASHTO LRFD Specification. Therefore, the LRFD Specifications implies that the fatigue-limit-state truck would be HS30. The designer computes a stress range using the HS20 truck with a load factor of 0.75. By assuring that this computed stress range is less than half the CAFL, the AASHTO LRFD specification is supposed to assure that the stress ranges from random variable traffic will not exceed the CAFL more than 0.01% of the time.

The ratio of the effective gross vehicle weight (GVW) to the GVW of the fatigue-limit-state truck in the measured spectrum is referred to in the literature as the "alpha factor" [25]. (This unfortunate choice of nomenclature should not be confused with another alpha factor related to the ratio of the actual stress ranges in the bridge to computed stress ranges.) The LRFD Specifications imply that alpha equals 0.5.

The alpha factor of 0.5 implied by the LRFD Specifications is not consistent with the findings of NCHRP Report 299 [50] and the Guide Specification for Fatigue Design of Steel Bridges [53], which imply the alpha factor should be closer to 0.33. With alpha of 0.33, the fatigue limit state truck is about three times heavier than the effective fatigue truck, or about HS45. This finding was based on a reliability analysis, comparison with the original AASHTO fatigue limit check, review of nationwide WIM data, and a study of alpha factors from measured stress histograms.

According to the statistics of the GVW histograms [50], this HS45 fatigue-limit-state truck has only a 0.023% probability (about 1 in 5000) of exceedance, which is almost consistent with the recommendation from NCHRP 354 that the fatigue-limit-state stress range has an exceedance of less than 1:10,000. If the alpha of 0.33 is correct, the stress range produced by the fatigue truck should be compared to ⅓ the CAFL rather than ½. The HS30 fatigue-limit-state truck implied by the AASHTO LRFD provisions clearly has a much higher probability of exceedance.

The apparent inconsistency in the exceedance level of the fatigue-limit-state truck weight was intentional and was a result of "calibrating" the LRFD bridge specifications to give fatigue design requirements which are similar to those of preceding specifications. The theoretically low exceedance level of the fatigue-limit-state truck weight implied by the LRFD code was used because it was felt that other aspects of the design process are overconservative, such as the assumptions in the structural analysis [54]. The Guide Specification for Fatigue Design of Steel Bridges [53] apparently resulted in overly conservative estimates of fatigue life when compared with observed field behavior. As a result, the LRFD Specification was "calibrated" to match existing field experience.

Measured axle load data [51,55] show axle loads which substantially exceed the 107 kN fatigue-limit-state axle load (half of the tandem axle load from the HS30 truck) implied by the AASHTO LRFD Specifications. The shape of the axle load spectra is essentially the same as the GVW spectra; i.e., the alpha factor is about 0.33. This alpha factor would suggest the fatigue-limit-state axle load with an exceedance level of about 1:10,000 would be approximately 160 kN. In fact, measured axle load data [55] show axle loads exceeding 160 kN in some cases. The appropriate axle load to use in general design specifications is very uncertain, however, because the measured axle load spectra are very site specific in this extreme tail of the distributions and the data are very sparse.

Aside from the fatigue-limit-state axle load or GVW, the rest of the loading spectrum does not matter when using the infinite-life approach. Also, the precise number of cycles does not have to be forecast. Rather, it is only necessary to establish that the total number of cycles exceeds the number of cycles associated with the CAFL. Therefore, it is not necessary to know precisely the expected life of the deck and future traffic volumes. Thus, despite the uncertainty in the appropriate value for the fatigue-limit-state axle load, the infinite-life approach is considerably simpler than trying to account for the cumulative damage of the whole distribution of future axle loads, which is even more uncertain.

The infinite-life approach relies upon the CAFL as the parameter determining the fatigue resistance. The emphasis in fatigue testing of details should therefore be on defining the CAFL. Unfortunately, there is a need for additional testing to define these CAFL better. Many of the CAFL values in Table 53.1 were based on judgment rather than specific test data at stress ranges down near the CAFL. Additional research should be performed to investigate the validity of many of the CAFL.

Clearly, most structures carry enough truck traffic to justify an infinite-life fatigue design approach, especially for the deck elements. For example, assuming 25-year life and a Category C detail it can be shown that the maximum permissible ADTT is about 850 trucks/day if the number of cycles associated with half the CAFL is just slightly exceeded in a deck element. In most cases, designing for infinite fatigue life rather than designing for a finite number of cycles adds little cost. This infinite-life approach has also recently been applied in developing AASHTO fatigue design specifications for wind-loaded sign, signal, and luminaire support structures [17] and modular bridge expansion joints [57–59].

In the fatigue design of bridge deck elements according to the AASHTO LRFD specifications, the fatigue design stress range is obtained from a static analysis where the wheel loads (half the axle loads) are applied in patches. The load patches in the LRFD Specifications are calculated in a manner which differs from the calculation in the Standard Specifications. However, in both methods, the patches increase in size as the load increases, which results in an applied pressure on continuous deck surfaces which is approximately the same as typical truck tire pressures (700 kPa). In the AASHTO LRFD Specifications, the patch has a fixed width of 508 mm and a tire pressure of 860 kPa. The patch increases in length as the load increases. At present, the LRFD specifications are not clear as to how load factors are to be used in calculating the patch size. It is not conservative to use the strength load factors in calculating the patch size, since this tends to spread the load over a greater area. Ideally, the patches for the fatigue loading should be calculated on the basis of the limit-state HS30 loading, although it would be conservative to use either the HS20 or factored HS15 loading.

53.4 Fracture Resistance

As explained previously, it is considered more important to focus on the prevention of fatigue cracks in bridges than to focus on the resistance to fracture. However, for structural components which are not subjected to significant cyclic loading, fracture could still possibly occur without prior fatigue crack growth. The primary tension chords of long-span truss would be one example. Usually, this would occur as the loads are applied for the first time during construction.

Unlike fatigue, fracture behavior depends strongly on the type and strength level of the steel or filler metal. In general, fracture toughness has been found to decrease with increasing yield strength of a material, suggesting an inverse relationship between the two properties. In practice, however, fracture toughness is more complex than implied by this simple relationship since steels with similar strength levels can have widely varying levels of fracture toughness.

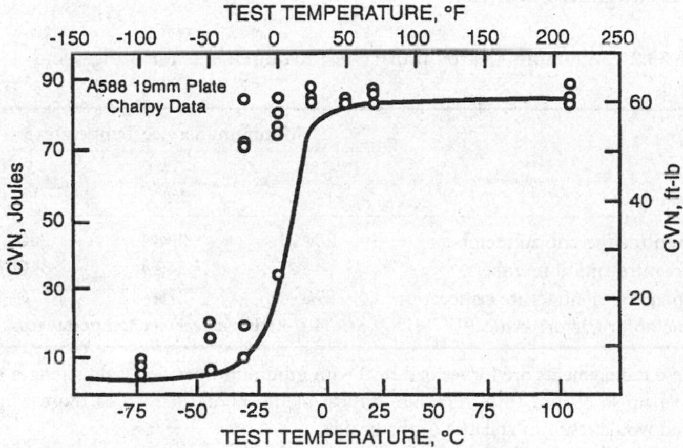

FIGURE 53.10 Charpy Energy Transition Curve for A588 Grade 50 (350 MPa yield strength) Structural Steel.

Steel exhibits a transition from brittle to ductile fracture behavior as the temperature increases. For example, Figure 53.10 shows a plot of the energy required to fracture CVN impact test specimens of A588 structural steel at various temperatures. These results are typical for ordinary hot-rolled structural steel. The transition phenomenon shown in Figure 53.10 is a result of changes in the underlying microstructural fracture mode.

There are three types of fracture with different behavior:

1. Brittle fracture is associated with cleavage of individual grains on select crystallographic planes. This type of fracture occurs at the lower end of the temperature range, although the brittle behavior can persist up to the boiling point of water in some low-toughness materials. This part of the temperature range is called the lower shelf because the minimum toughness is fairly constant up to the transition temperature. Brittle fracture may be analyzed with linear elastic fracture mechanics [2,23,24] because the plasticity that occurs is negligible.
2. Transition-range fracture occurs at temperatures between the lower shelf and the upper shelf and is associated with a mixture of cleavage and fibrous fracture on a microstructural scale. Because of the mixture of micromechanisms, transition-range fracture is characterized by extremely large variability.
3. Ductile fracture is associated with a process of void initiation, growth, and coalescence on a microstructural scale, a process requiring substantial energy. This higher end of the temperature range is referred to as the upper shelf because the toughness levels off and is essentially constant for higher temperatures. Ductile fracture is also called fibrous fracture

due to the fibrous appearance of the fracture surface, or shear fracture due the usually large slanted shear lips on the fracture surface.

Ordinary structural steel such as A36 or A572 is typically only hot-rolled, while to achieve very high toughness steels must be controlled-rolled, i.e., rolled at lower temperatures, or must receive some auxiliary heat treatment such as normalization. In contrast to the weld metal, the cost of the steel is a major part of total costs. The expense of the high-toughness steels has not been found to be warranted for most bridges, whereas the cost of high-toughness filler metal is easily justifiable. Hot-rolled steels, which fracture in the transition region at the lowest service temperatures, have sufficient toughness for the required performance of most welded buildings and bridges.

ASTM specifications for bridge steel (A709) provide for minimum CVN impact test energy levels. Structural steel specified by A36, A572, or A588, without supplemental specifications, does not require the Charpy test to be performed. The results of the CVN test, impact energies, are often referred to as "notch-toughness" values.

TABLE 53.2 Minimum Charpy Impact Test Requirements for Bridges and Buildings

	Minimum Service Temperature		
Material	–18°C, J@°C	–34°C, J@°C	–51°C, J@°C
Steel: nonfracture-critical members[a,b]	20@21	20@4	20@–12
Steel: fracture-critical members[a,b]	34@21	34@4	34@–12
Weld metal for nonfracture critical[a]	27@–18	27@–18	27@–29
Weld metal for fracture critical[a,b]	34@–29 for all service temperatures		

[a] These requirements are for welded steel with minimum specified yield strength up to 350 MPa up to 38 mm thick. Fracture-critical members are defined as those which if fractured would result in collapse of the bridge.

[b] The requirements pertain only to members subjected to tension or tension due to bending.

The CVN specification works by assuring that the transition from brittle to ductile fracture occurs at some temperature less than service temperature. This requirement ensures that brittle fracture will not occur as long as large cracks do not develop. Because the Charpy test is relatively easy to perform, it will likely continue to be the measure of toughness used in steel specifications. Often 34 J (25 ft-lbs), 27 J (20 ft-lbs), or 20 J (15 ft-lbs) are specified at a particular temperature. The intent of specifying any of these numbers is the same, i.e., to make sure that the transition starts below this temperature.

Some Charpy toughness requirements for steel and weld metal for bridges are compared in Table 53.2. This table is simplified and does not include all the requirements.

Note that the bridge steel specifications require a CVN at a temperature that is 38°C greater than the minimum service temperature. This "temperature shift" accounts for the effect of strain rates, which are lower in the service loading of bridges (on the order of 10^{-3}) than in the Charpy test (greater than 10^{1}) [23]. It is possible to measure the toughness using a Charpy specimen loaded at a strain rate characteristic of bridges, called an intermediate strain rate, although the test is more difficult and the results are more variable. When the CVN energies from an intermediate strain rate are plotted as a function of temperature, the transition occurs at a temperature about 38°C lower for materials with yield strength up to 450 MPa.

As shown in Table 53.2, the AWS D1.5 Bridge Welding Code specifications for weld metal toughness are more demanding than the specifications for base metal. The extra margin of fracture toughness in the weld metal is reasonable because the weld metal is always the location of discontinuities and high tensile residual stresses. Because the cost of filler metal is relatively small in comparison with the overall cost of materials, it is usually worth the cost to get high-toughness filler metal.

The minimum CVN requirements are usually sufficient to assure damage tolerance, i.e., to allow cracks to grow quite long before fracture occurs. Fatigue cracks grow at an exponentially increasing rate; therefore, most of the fatigue life transpires while the crack is very small. Additional fracture toughness, greater than the minimum specified values, will allow the crack to grow to a larger size before sudden fracture occurs. However, the crack is growing so rapidly at the end of life that the additional toughness may increase the life only insignificantly. Therefore, specification of toughness levels exceeding these minimum levels is usually not worth the increased cost of the materials.

The fractures of steel connections that occurred in the Northridge earthquake of 1994 is an example of what can happen if there are no filler metal toughness specifications. At that time, there were no requirements for weld metal toughness in AWS D1.1 for buildings, even for seismic welded steel moment frames (WSMF). This lack of requirements was rationalized because typically the weld deposits are higher toughness than the base metal. However, this is not always the case; e.g., the self-shielded flux-cored arc welds (FCAW-S) used in many of the WSMF that fractured in the Northridge earthquake were reported to be very low toughness [5–7].

ASTM A673 has specifications for the frequency of Charpy testing. The *H* frequency requires a set of three CVN specimens to be tested from one location for each heat or about 50 tons. These tests can be taken from a plate with thickness up to 9 mm different from the product thickness if it is rolled from the same heat. The *P* frequency requires a set of three specimens to be tested from one end of every plate, or from one shape in every 15 tons of that shape. For bridge steel, the AASHTO specifications require CVN tests at the *H* frequency as a minimum. For fracture-critical members, the guide specifications require CVN testing at the *P* frequency. In the AISC specifications, CVN tests are required at the *P* frequency for thick plates and jumbo sections. A special test location in the core of the jumbo section is specified, as well as the requirement that the section tested be produced from the top of the ingot.

Even the *P* testing frequency may be insufficient for as-rolled structural steel. In a recent report for NCHRP [60], CVN data were obtained from various locations on bridge steel plates. The data show that because of extreme variability in CVN across as-rolled plates, it would be possible to miss potentially brittle areas of plates if only one location per plate is sampled. For plates that were given a normalizing heat treatment, the excessive variability was eliminated.

Quantitative means for predicting brittle fracture are available, i.e., fracture mechanics [2,6,11–13,23,24,61]. These quantitative fracture calculations are typically not performed in design, but are often used in service to assess a particular defect. There is at best only about plus or minus 30% accuracy in these fracture predictions, however. Several factors contributing to this lack of accuracy include (1) variability of material properties; (2) changes in apparent toughness values with changes in test specimen size and geometry; (3) differences in toughness and strength of the weld zone; (4) complex residual stresses; (5) high gradients of stress in the vicinity of the crack due to stress concentrations; and (6) the behavior of cracks in complex structures of welded intersecting plates.

50.5 Summary

1. Structural elements where the live load is a large percentage of the total load are potentially susceptible to fatigue. Many factors in fabrication can increase the potential for fatigue including notches, misalignment, and other geometric discontinuities, thermal cutting, weld joint design (particularly backing bars), residual stress, nondestructive evaluation and weld defects, intersecting welds, and inadequate weld access holes. The fatigue design procedures in the AASHTO Specifications are based on control of the stress range and knowledge of

the fatigue strength of the various details. Using these specifications, it is possible to identify and avoid details which are expected to have low fatigue strength.

2. The simplified fatigue design method for infinite life is justified because of the uncertainty in predicting the future loading on a structure. The infinite-life fatigue design philosophy requires that essentially all the stress ranges are less than the CAFL. One advantage of this approach for structures with complex stress histories is that it is not necessary to predict accurately the entire future stress range distribution. The fatigue design procedure simply requires a knowledge of the stress range with an exceedance level of 1:10,000. The infinite-life approach relies upon the CAFL as the parameter determining the fatigue resistance. The emphasis in fatigue testing of details should therefore be on defining the CAFL. Additional research should be performed to investigate the validity of many of the CAFL in the present specification.
3. Welded connections and thermal-cut holes, copes, blocks, or cuts are potentially susceptible to brittle fracture. Many interrelated design variables can increase the potential for brittle fracture including lack of redundancy, large forces and moments with dynamic loading rates, thick members, geometric discontinuities, and high constraint of the connections. Low temperature can be a factor for exposed structures. The factors mentioned above that influence the potential for fatigue have a similar effect on the potential for fracture. In addition, cold work, flame straightening, weld heat input, and weld sequence can also affect the potential for fracture. The AASHTO Specifications require a minimum CVN "notch toughness" at a specified temperature for the base metal and the weld metal of members loaded in tension or tension due to bending. Almost two decades of experience with these bridge specifications have proved that they are successful in significantly reducing the number of brittle fractures.

References

1. "AASHTO, *LRFD Bridge Design Specifications*, American Association of State Highway Transportation Officials, Washington, D.C., 1994.
2. Fisher, J. W., The evolution of fatigue resistant steel bridges, in *1997 Distinguished Lectureship*, Transportation Research Board, 76th Annual Meeting, Washington, D.C., Jan. 12–16, 1997, Paper No. 971520: 1–22.
3. Fisher, J. W., *Fatigue and Fracture in Steel Bridges*, John Wiley & Sons, New York, 1984.
4. Fisher, J. W., Pense, A. W., and Roberts, R., Evaluation of fracture of Lafayette Street Bridge, *J. Struct. Div. ASCE*, 103(ST7), 1977.
5. Kaufmann, E. J., Fisher, J. W., Di Julio, R. M., Jr., and Gross, J. L., Failure Analysis of Welded Steel Moment Frames Damaged in the Northridge Earthquake, NISTIR 5944, National Institute of Standards and Technology, Gaithersburg, MD, January 1997.
6. Fisher, J. W., Dexter, R. J. and Kaufmann, E. J., Fracture mechanics of welded structural steel connections, in Background Reports: Metallurgy, Fracture Mechanics, Welding, Moment Connections, and Frame Systems Behavior, Report No. SAC 95-09, FEMA-288, March 1997.
7. Kaufmann, E. J., Xue, M., Lu, L.-W., and Fisher, J. W., Achieving ductile behavior of moment connections, *Mod. Steel Constr.*, 36(1), 30–39, January 1996; see also, Xue, M., Kaufmann, E. J., Lu, L.-W., and Fisher, J. W., Achieving ductile behavior of moment connections — Part II, *Mod. Steel Constr.*, 36(6), 38–42, June 1996.
8. Engineers Investigate Cracked El, *Eng. News Rec.*, 200(3), January 19, 1979.
9. Miki, C., Fractures in seismically loaded bridges, *Prog. Struct. Eng. Mat.*, 1(1), 115–121, September 1997.
10. ANSI/AWS D1.1, Structural Welding Code — Steel, American Welding Society, Miami, 1996.

11. Dexter, R. J., Load-deformation behavior of cracked high-toughness steel members, in *Proceedings of the 14th International Conference on Offshore Mechanics and Arctic Engineering Conference (OMAE)*, 18–22 June 1995, Copenhagen Denmark, Salama et al., Eds., Vol. III, *Materials Engineering*, AMSE, 1995, 87–91.
12. Dexter, R. J. and Gentilcore, M. L., Evaluation of Ductile Fracture Models for Ship Structural Details, Report SSC-393, Ship Structure Committee, Washington, D.C., 1997.
13. Dexter, R. J. and Gentilcore, M. L., Predicting extensive stable tearing in structural components, in *Fatigue and Fracture Mechanics: 29th Volume, ASTM STP 1321*, T. L. Panontin and S. D. Sheppard, Eds., American Society for Testing and Materials, Philadelphia, 1998.
14. Dexter, R. J., Significance of strength undermatching of welds in structural behaviour, in *Mis-Matching of Interfaces and Welds,* K.-H. Schwalbe and M. Kocak, Eds., GKSS Research Center, Geesthacht, Germany, 1997, 55–73.
15. Castiglioni, C. A., Cumulative damage assessment in structural steel details, in *IABSE Symposium, Extending the Lifespan of Structures,* San Francisco, IABSE, 1995, 1061–1066.
16. Krawinkler, H. and Zohrei, M., Cumulative damage in steel structures subjected to earthquake ground motion, *Comput. Struct.*, 16 (1–4), 531–541, 1983.
17. Kaczinski, M. R., Dexter, R. J., and Van Dien, J. P., Fatigue-Resistant Design of Cantilevered Signal, Sign, and Light Supports, NCHRP Report 412, National Cooperative Highway Research Program,Transportation Research Board, Washington, D.C., 1998
18. Fisher, J. W., Frank, K. H., Hirt, M. A., and McNamee, B. M., Effect of Weldments on the Fatigue Strength of Steel Beams, National Cooperative Highway Research Program (NCHRP) Report 102, Highway Research Board, Washington, D.C., 1970.
19. Fisher, J. W., Albrecht, P. A., Yen, B. T., Klingerman, D. J., and McNamee, B. M., Fatigue Strength of Steel Beams with Welded Stiffeners and Attachments, National Cooperative Highway Research Program (NCHRP) Report 147, Transportation Research Board, Washington, D.C., 1974.
20. Dexter, R. J., Fisher, J. W. and Beach, J. E., Fatigue behavior of welded HSLA-80 members, in *Proceedings, 12th International Conference on Offshore Mechanics and Arctic Engineering,* Vol. III, Part A, Materials Engineering, ASME, New York, 1993, 493–502.
21. Keating, P. B. and Fisher, J. W., Evaluation of Fatigue Tests and Design Criteria on Welded Details, NCHRP Report 286, National Cooperative Highway Research Program, September 1986.
22. VanDien, J. P., Kaczinski, M. R., and Dexter, R. J., Fatigue testing of anchor bolts, in *Building an International Community of Structural Engineers,* Vol. 1, *Proc. of Structures Congress XIV,* Chicago, 1996, 337–344.
23. Barsom, J. M. and Rolfe, S. T., *Fracture and Fatigue Control in Structures,* 2nd ed., Prentice-Hall, Englewood Cliffs, NJ, 1987.
24. Broek, D. *Elementary Fracture Mechanics,* 4th ed., Martinis Nijhoff Publishers, Dordrecht, Netherlands, 1987.
25. Kober, Dexter, R. J., Kaufmann, E. J., Yen, B. T., and Fisher, J. W., The effect of welding discontinuities on the variability of fatigue life, in *Fracture Mechanics, Twenty-Fifth Volume, ASTM STP 1220,* F. Erdogan and Ronald J. Hartranft, Eds., American Society for Testing and Materials, Philadelphia, 1994.
26. Dexter, R. J., Tarquinio, J. E., and Fisher, J. W., An application of hot-spot stress fatigue analysis to attachments on flexible plate, in *Proceedings of the 13th International Conference on Offshore Mechanics and Arctic Engineering,* American Society of Mechanical Engineers (ASME), New York, 1994.
27. Yagi, J., Machida, S., Tomita, Y., Matoba, M., and Kawasaki, T., Definition of Hot-Spot Stress in Welded Plate Type Structure for Fatigue Assessment, International Institute of Welding, IIW-XIII-1414-91, 1991.
28. AISC, *Load and Resistance Factor Design Specification for Structural Steel Buildings,* 2nd ed., American Institute of Steel Construction (AISC), Chicago, 1993.
29. ENV 1993-1-1, Eurocode 3: Design of steel structures — Part 1.1: General rules and rules for buildings, European Committee for Standardization (CEN), Brussels, April 1992.

30. Roberts, R. et al., Corrosion Fatigue of Bridge Steels, Vol. 1–3, Reports FHWA/RD-86/165, 166, and 167, Federal Highway Administration, Washington, D.C., May 1986.
31. Outt, J. M. M., Fisher, J. W., and Yen, B. T., Fatigue Strength of Weathered and Deteriorated Riveted Members, Report DOT/OST/P-34/85/016, Department of Transportation, Federal Highway Administration, Washington, D.C., October 1984.
32. Albrecht, P. and Shabshab, C., Fatigue strength of weathered rolled beam made of A588 steel, *J. Mater. Civil Eng.* (*ASCE*), 6(3), 407–428, 1994.
33. Demers, C. and Fisher, J. W., Fatigue Cracking of Steel Bridge Structures, Vol. I: A Survey of Localized Cracking in Steel Bridges — 1981 to 1988, Report No. FHWA-RD-89-166; also, Volume II, A Commentary and Guide for Design, Evaluation, and Investigating Cracking, Report No. FHWA-RD-89-167, FHWA, McLean, VA, March 1990.
34. Yen, B. T., Huang, T., Lai, L.-Y., and Fisher, J. W., Manual for Inspecting Bridges for Fatigue Damage Conditions, Report No. FHWA-PA-89-022 + 85-02, Fritz Engineering Laboratory Report No. 511.1, Pennsylvania Department of Transportation, Harrisburg, PA., January 1990.
35. BS 7608, Code of Practice for Fatigue Design and Assessment of Steel Structures, British Standards Institute, London, 1994.
36. Maddox, S. J., *Fatigue Strength of Welded Structures,* 2nd ed., Abington Publishing, Cambridge, U.K., 1991.
37. Fisher, J. W., Jian, J., Wagner, D. C., and Yen, B. T., Distortion-Induced Fatigue Cracking in Steel Bridges, National Cooperative Highway Research Program (NCHRP) Report 336, Transportation Research Board, Washington, D.C., 1990.
38. Fisher, J. W. and Keating, P. B., Distortion-induced fatigue cracking of bridge details with web gaps, *J. Constr. Steel Res.*, 12, 215–228, 1989.
39. Fisher, J. W., Kaufmann, E. J., Koob, M. J., and White, G., Cracking, fracture assessment, and repairs of Green River Bridge, I-26, in *Proc. of Fourth International Bridge Engineering Conference*, San Francisco, Aug. 28–30, 1995, Vol 2, National Academy Press, Washington, D.C., 1995, 3–14.
40. Fisher, J. W., Yen, B. T., Kaufmann, E. J., Ma, Z. Z. and Fisher, T. A., Cracking evaluation and repair of cantilever bracket tie plates of Edison Bridge, in *Proc. of Fourth International Bridge Engineering Conference*, San Francisco, Aug. 28–30, 1995, Vol 2, National Academy Press, Washington, D.C. 1995, 15–25.
41. Yen, B. T. et al., Fatigue behavior of stringer-floorbeam connections, in *Proc. of the Eighth International Bridge Conference*, Paper IBC-91-19, Engineers' Society of Western Pennsylvania, 1991, 149–155.
42. ACI Committee 215, Considerations for Design of Concrete Structures Subjected to Fatigue Loading, ACI 215R-74 (Revised 1992), ACI Manual of Standard Practice, Vol. 1, 1996.
43. Hahin, C., Effects of Corrosion and Fatigue on the Load-Carrying Capacity of Structural Steel and Reinforcing Steel, Illinois Physical Research Report No. 108, Illinois Department of Transportation, Springfield, March 1994.
44. Pfister, J. F. and Hognestad, E., High strength bars as concrete reinforcement. Part 6, fatigue tests, *J. PCA Res. Dev. Lab.*, 6(1), 65–84, 1964.
45. Sternberg, F., Performance of Continuously Reinforced Concrete Pavement, I-84 Southington, Connecticut State Highway Department, June 1969.
46. Rabbat, B. G. et al., Fatigue tests of pretensioned girders with blanketed and draped strands, *J. Prestressed Concrete Inst.*, 24(4), 88–115, 1979.
47. Overnman, T. R., Breen, J. E., and Frank, K. H., Fatigue Behavior of Pretensioned Concrete Girders, Research Report 300-2F, Center for Transportation Research, The University of Texas at Austin, November 1984.
48. Ad hoc Committee on Cable-Stayed Bridges, Recommendations for Stay Cable Design and Testing, Post-Tensioning Institute, Phoenix, January 1986.
49. Miner, M. A., Cumulative damage in fatigue, *J. Appl. Mech.*, 12, A-159, 1945.

50. Moses, F., Schilling, C. G., and Raju, K. S., Fatigue Evaluation Procedures for Steel Bridges, NCHRP Report 299, National Cooperative Highway Research Program, 1987.
51. Schilling, C. G., Variable Amplitude Load Fatigue, Task A — *Literature Review*, Vol. I: — Traffic Loading and Bridge Response, Publ. No. FHWA-RD-87-059, Federal Highway Administration, July 1990.
52. Fisher, J. W., et al., Resistance of Welded Details under Variable Amplitude Long-Life Fatigue Loading, National Cooperative Highway Research Program Report 354, Transportation Research Board, Washington, D.C., 1993.
53. American Association of State Highway and Transportation Officials, Guide Specifications for Fracture Critical Non-Redundant Steel Bridge Members, American Association of State Highway and Transportation Officials, Washington, D.C., 1989 (with interims).
54. Dexter, R. J. and Fisher, J. W., The effect of unanticipated structural behavior on the fatigue reliability of existing bridge structures, in *Structural Reliability in Bridge Engineering*, D. M. Frangopol and G. Hearn, Eds., Proceedings of a Workshop, University of Colorado at Boulder, 2–4 Oct. 1996, McGraw-Hill, New York, 1996, 90–100.
55. Nowak, A. S. and Laman, J. A., Monitoring Bridge Load Spectra, in *IABSE Symposium, Extending the Lifespan of Structures*, San Francisco, 1995.
56. Menzemer, C. C. and Fisher, J. W., Revisions to the Aluminum Association fatigue design specifications, 6th Int. Conf. on Aluminum Weldments, April 3–5, 1995, Cleveland. Available from AWS, Miami, FL, 1995, 11–23.
57. Dexter, R. J., Connor, R. J., and Kaczinski, M. R., Fatigue Design of Modular Bridge Expansion Joints, NCHRP Report 402, National Cooperative Highway Research Program, Transportation Research Board, Washington, D.C., 1997.
58. Dexter, R. J., Kaczinski, M. R., and Fisher, J. W., Fatigue testing of modular expansion joints for bridges, in *IABSE Symposium*, 1995, *Extending the Lifespan of Structures*, Vol. 73/2, San Francisco, 1995, 1091–1096.
59. Kaczinski, M. R., Dexter, R. J., and Connor, R. J., Fatigue design and testing of modular bridge expansion joints, in *Proceedings of the Fourth World Congress on Joint Sealing and Bearing Systems for Concrete Structures*, Sacramento, September 1996.
60. Frank, K. H., et al., Notch Toughness Variability in Bridge Steel Plates, NCHRP Report 355, National Cooperative Highway Research Program, 1993.
61. PD 6493, Guidance on the Methods for Assessing the Acceptability of Flaws in Fusion Welded Structures, PD 6493: 1991, British Standards Institution (BSI), London, 1991.

54

Statistics of Steel Weight of Highway Bridges

Shouji Toma

Hokkai-Gakuen University, Japan

54.1 Introduction

In this chapter, a database of steel highway bridges is formed to assess designs by analyzing them statistically. No two bridges are exact replicas of each other because of the infinite variety of site conditions. Each bridge meets specific soil, traffic, economic, and aesthetics conditions. The structural form, the support conditions, the length, width, and girder spacing, pedestrian lanes, and the materials, all depend on a unique combination of design criteria. Even if the stipulated criteria are identical, the final bridges are not, as they naturally reflect the individual intentions of different designers. Therefore, steel weight is a major interest to engineers.

Steel weight of highway bridges is one of the most important of the many factors that influence bridge construction projects. The weight gives a good indication of structural, economic, and safety

0-8493-7434-0/00/$0.00+$.50

features of the bridge. Generally, the weight is expressed by as a force per square unit of road surface area (tonf/m^2 or kN/m^2). Stochastic distribution of the weight includes many influential factors to designs that cause scatter. The analysis of this scatter may suggest the characteristics of the bridges. As a general rule, simple bridges are lighter than more complex ones, bridges with high safety margins are heavier, and composite construction results in a lighter bridge overall. A designer thereby gets insight into the characteristics of a bridge. As bridge design also requires the estimate of steel weight in advance, the data collected here are useful.

In Japan, many steel bridges have been constructed in the past few decades. The weight of steel used in these bridges has been collected into a single database. The bridges are all Japanese, but engineers from other countries use similar structural and economic considerations and can usefully employ these in their designs. In this chapter, Japanese design criteria are presented first. The live loads and material properties are described in special detail to clarify differences that other countries may note. Then, the computer database is explained and used to make comparisons between plate and box girders, truss and frame bridges, simply supported and continuously supported bridges, reinforced concrete slab deck and steel deck, and more.

54.2 Design Criteria

54.2.1 Live Loads

The strength required for a bridge to sustain largely depends on the live load, and the live load generally differs from country to country. Since the weight information used here follows Japanese specifications, those will be the ones explained. The last version of the bridge design specification was published in 1996 [1], and is based on a truck weight of 25 tonf (245 kN). However, the bridges studied here were designed using an old version of the code [2], and thus used a truck load of 20 tonf (196 kN).

The 20 t live load (TL-20) takes the two forms shown in Figure 54.1a. The T-load is used to design local components such as the slab or the floor system and the L-load is used for global ones such as the main girders. The T-load is the concentrated wheel loads and the L-load is further subdivided. A partially distributed load (caused by the truck) and a load distributed along the length of the bridge (corresponding to the average traffic load) comprises the L-load. Most of the bridges were designed for TL-20, but on routes, such as those near harbor ports, heavy truck loads are expected and these were designed for TT-43 (Figure 54.1b). In this database the difference is not considered.

When a bridge has side lanes for pedestrian traffic, and the live load (the crowd load) is small compared to vehicular traffic loads, usually less steel is required. However, the difference of the weight for pedestrian and vehicular lanes is not considered in this database. The surface area of the sidewalk is considered equally as heavy as the area in the vehicle lanes.

54.2.2 Materials

The strength of steel varies widely. A mild steel may have a yield strength of about 235 N/mm^2 and is commonly used in bridge design but higher strengths of 340 or 450 N/mm^2 are also used, often in large bridges. Various strength of steel are considered in this study. Clearly, when higher-strength steels are used, the weight of steel required goes down. However, the difference in strength level of steel is not distinguished in the database. Aa a selection of strength level is made considering rationality of design, it will generally result in similar decisions for many bridges. In other words, similar bridge designs specify similar material strengths. The effect of strength is thus included implicitly in the database.

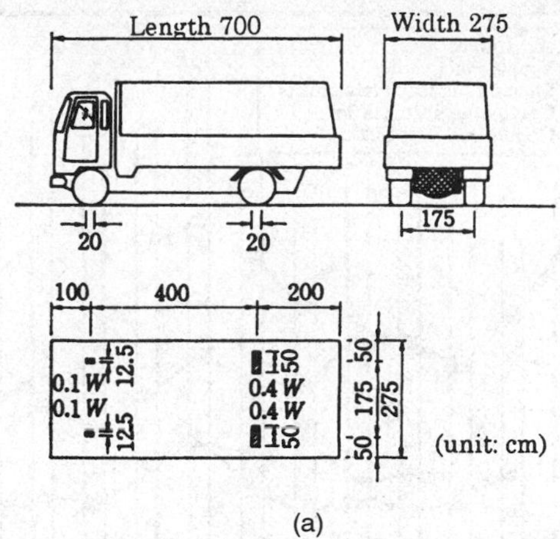

(a)

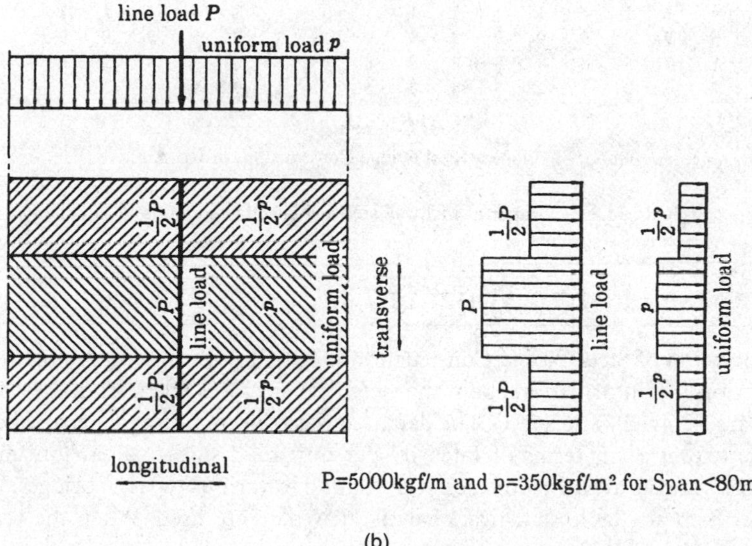

(b)

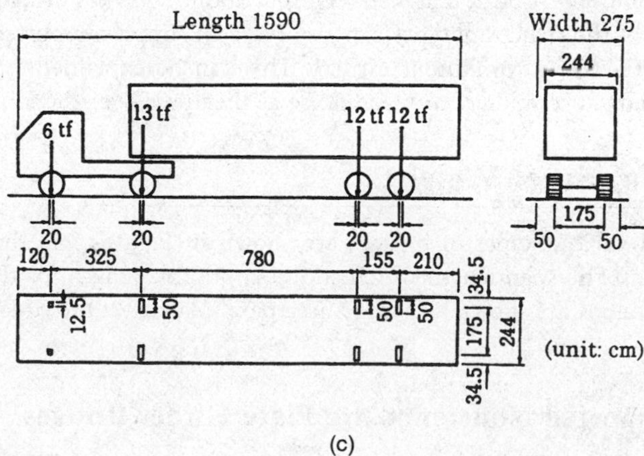

(c)

FIGURE 54.1 Live load (TL-20). (a) T-Load (W = 20 tf); (b) L-Load; (c) TT-43 (W = 43 tf).

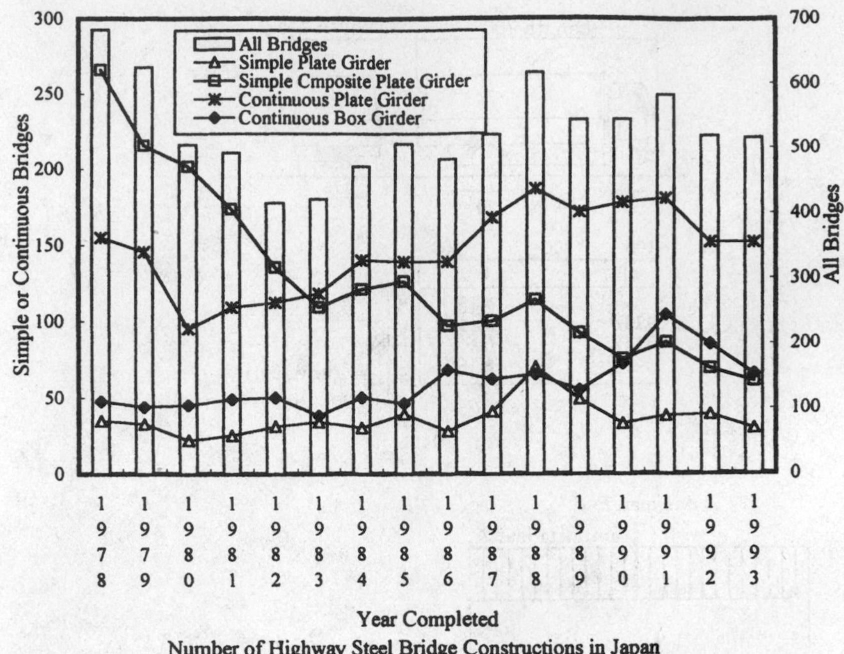

FIGURE 54.2 Number of highway steel bridge constructions in Japan.

54.3 Database of Steel Weights

The Japan Association of Steel Bridge Construction (JASBC) publishes an annual report on steel bridge construction [3]. Information about the weight of steel was taken from these reports over a period of 15 years (from 1978 to 1993). The database was collected using a personal computer [4]. The weight was expressed in terms of intensity per unit road surface area (tonf/m^2). Table 54.1 shows the quantity of data available for each year relating to various types of bridges. When enough data exist to perform a reliable statistical analysis, new data are used. When the year's sample is small, all the data are included.

The data in Table 54.1 are plotted in Figure 54.2, which also shows the number of steel bridges constructed in Japan. From Figure 54.2, it can be seen that about 500 steel bridges are constructed each year. The tendency of the structural types can also be seen: simply supported composite plate girders are gradually replaced by continuous girders. This can be explained as expansion joints damage the pavement and cause vehicles to make noise as they pass over the joints.

54.4 Statistics of Steel Weights

Weight distributions for various types of bridges are shown in Figures 54.3 through 54.13. The weights are plotted against the span length which shows applicable length for the type of bridge. In the figures the mean values are shown by a line and a parabola curve; the equations are given in Table 54.2.

54.4.1 Simply Supported Noncomposite Plate Girder Bridges

In Figure 54.3 the distributions for simply supported plate girder bridges with reinforced concrete (RC) slab and steel decks are shown. The steel weight varies considerably, from which one can investigate the peculiarity of the bridge.

TABLE 54.1 Number of Input Data

Type of Bridge	Year Completed																Total
	1978	1979	1980	1981	1982	1983	1984	1985	1986	1987	1988	1989	1990	1991	1992	1993	
Simple plate girder	35	33	22	25	31	34	30	39	28	41	70	49	33	38	39	30	577
Simple plate girder (steel deck)	6	2	5	4	6	9	6	8	9	11	12	9	14	5	4	8	118
Simple composite plate girder	266	216	202	174	135	109	121	126	97	100	114	92	75	86	69	61	2043
Simple box girder	30	29	34	24	24	12	29	33	24	36	41	36	35	40	32	44	503
Simple box girder (steel deck)	15	12	6	6	4	7	6	10	16	5	16	14	14	21	20	28	200
Simple composite box girder	42	36	18	23	9	13	10	13	12	17	21	18	11	8	6	10	267
Continuous plate girder	155	146	95	109	112	118	140	139	139	168	187	172	178	180	147	150	2335
Continuous plate girder (steel deck)	0	4	4	0	0	5	6	6	6	1	4	5	6	5	0	2	54
Continuous box girder	48	44	45	49	50	38	50	46	68	62	65	55	72	104	85	66	947
Continuous box girder (steel deck)	9	18	19	16	11	16	19	24	23	17	25	20	23	27	28	42	337
Simple truss	16	26	15	7	11	16	11	14	9	15	15	10	17	8	10	11	211
Continuous truss	10	13	9	10	0	6	12	8	6	12	7	6	12	5	6	2	124
Langer	19	12	8	12	7	10	7	12	4	5	3	7	5	4	8	11	134
Trussed Langer	2	9	4	5	2	2	0	3	2	1	5	2	4	1	4	1	47
Lohse	11	12	12	10	9	11	11	8	19	7	8	11	13	17	17	7	183
Nielsen Lohse	2	0	0	0	1	4	4	2	4	3	4	5	5	7	5	7	53
Rigid frame (Rahmen)	16	12	5	15	3	9	8	10	10	12	17	15	10	8	14	18	182
Rigid frame (π type)	3	6	4	6	4	4	4	5	2	5	6	6	7	6	8	7	83
Arch bridge	—	—	—	—	—	—	—	—	—	—	—	2	4	3	4	2	15
Cable-stayed bridge (steel deck)	0	0	2	2	0	2	1	5	4	5	2	4	5	5	5	6	48
Total	685	630	509	497	419	425	475	511	482	523	622	538	543	578	511	513	8461

TABLE 54.2 Coefficients of Regression Equations

Type of Bridge	a ($\times 10^{-2}$)	b	Standard Deviation (1)	α ($\times 10^{-4}$)	β ($\times 10^{-2}$)	γ	Standard Deviation (2)	Year	No. of Data	Correlation Coefficient	Fig. No.
Simple plate girder	0.5866	0.0124	0.0325	0.4621	0.2075	0.0881	0.0324	1989–1993	189	0.758	54.3a
Simple plate girder (steel deck)	0.3504	0.2499	0.0420	0.1228	–0.5853	0.4252	0.0419	1978–1993	118	0.353	54.3b
Simple composite plate girder	0.6084	–0.0306	0.0249	0.3824	0.2985	0.0307	0.0249	1989–1993	383	0.830	54.4
Simple box girder	0.5917	0.0778	0.0410	0.4350	0.1488	0.1866	0.0409	1989–1993	187	0.803	54.5a
Simple box girder (steel deck)	0.3019	0.2738	0.0709	0.0616	0.2303	0.2930	0.0709	1978–1993	200	0.556	54.5b
Simple composite box girder	0.4765	0.1007	0.0412	0.3329	0.1290	0.1887	0.0411	1981–1993	171	0.714	54.6
Continuous plate girder	0.3729	0.0533	0.0331	–0.3092	0.6425	–0.0035	0.0330	1991–1993	477	0.653	54.7a
Continuous plate girder (steel deck)	0.2329	0.2464	0.0484	–0.2413	0.4482	0.2022	0.0481	1978–1993	54	0.508	54.7b
Continuous box girder	0.3029	0.1510	0.0499	0.0099	0.2906	0.1546	0.0499	1989–1993	382	0.665	54.8a
Continuous box girder (steel deck)	0.1516	0.3110	0.0634	0.0213	0.1080	0.3307	0.0633	1978–1993	337	0.593	54.8b
Simple truss	0.2993	0.1421	0.0504	0.3711	–0.2355	0.3284	0.0493	1978–1993	211	0.592	54.9a
Continuous truss	0.2221	0.1633	0.0602	0.0959	0.4830	0.0257	0.0567	1978–1993	124	0.799	54.9b
Langer	0.2907	0.1433	0.0632	–0.0135	0.3140	0.1338	0.0632	1978–1993	134	0.675	54.10a
Trussed Langer	0.2696	0.1700	0.0609	0.1693	–0.1173	0.3794	0.0592	1978–1993	47	0,741	54.10b
Lohse	0.2372	0.1956	0.0942	0.0110	0.2128	0.2076	0.0941	1978–1993	183	0.676	54.11a
Nielsen Lohse	0.2372	0.1956	0.1019	0.0110	0.2128	0.2076	0.1018	1978–1993	53	0.735	54.11b
Rigid frame (Rahmen)	0.4326	0.0542	0.0737	0.4399	–0.1004	0.2024	0.0711	1978–1993	182	0.659	54.14a
Rigid frame (π type)	0.4982	0.0050	0.0555	0.2477	0.1528	0.1160	0.0544	1978–1993	83	0.813	54.14b
Cable-stayed bridge (steel deck)	0.2102	0.2944	0.2056	0.0407	–0.0014	0.4736	0.1937	1978–1993	48	0.784	54.15
Equations (tf/m^2)	$aL + b$ … (1)			$\alpha L^2 + \beta L + \gamma$ … (2)					L = span (m)		

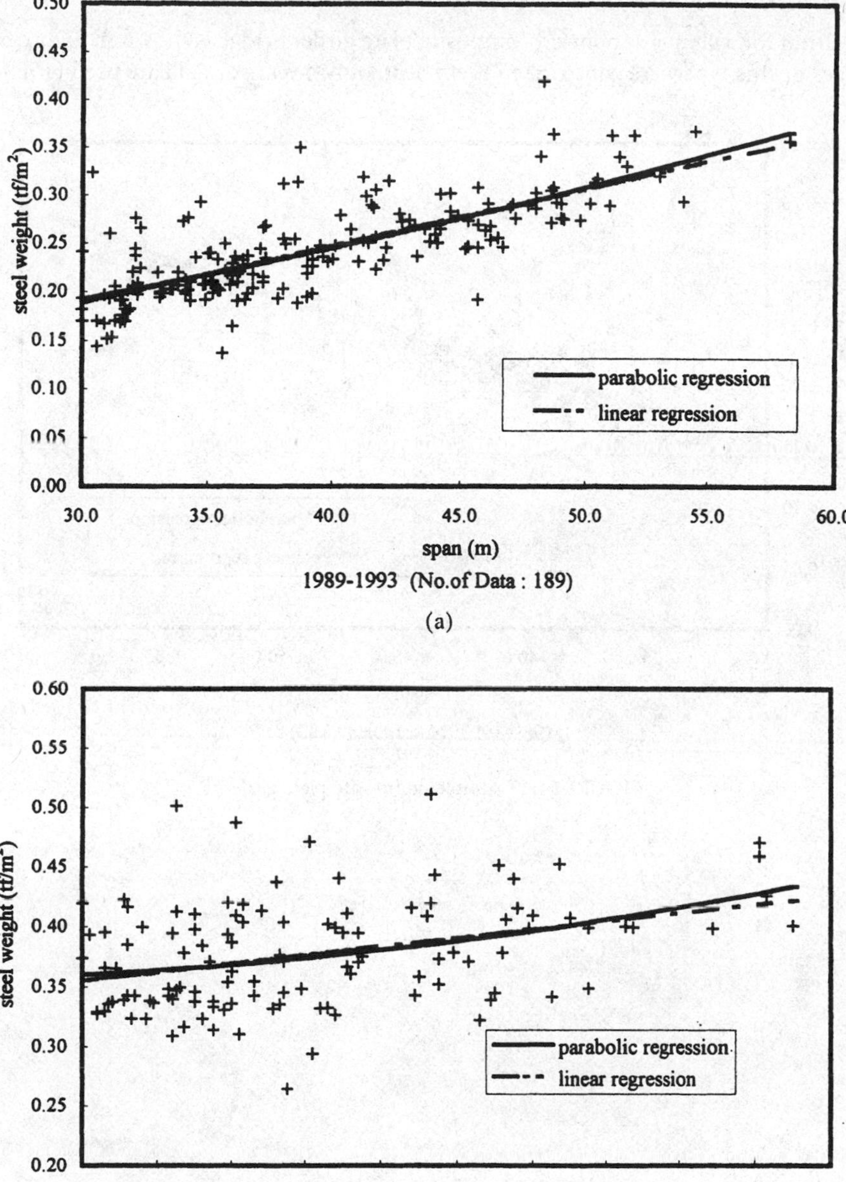

FIGURE 54.3 Simple noncomposite plate girders. (a) RC slab deck; (b) steel deck.

54.4.2 Simply Supported Composite Plate Girder Bridges

The distribution for a simply supported composite plate girder bridge is shown in Figure 54.4. Since many bridges of this type were constructed every year, only 4 years of data are used (1989 to 1993).

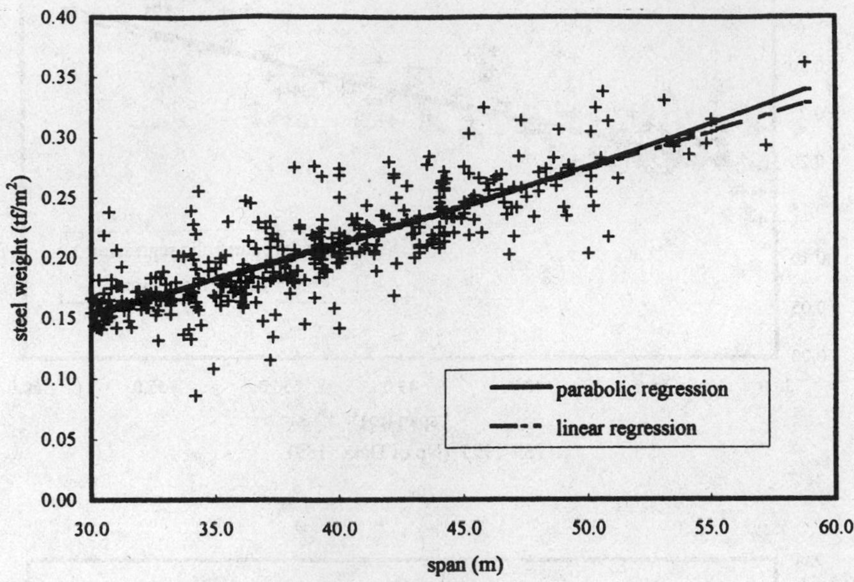

FIGURE 54.4 Simple composite plate girders.

54.4.3 Simply Supported Box-Girder Bridges

The distribution for a simply supported box-girder bridge (noncomposite) for RC slab and steel decks is plotted in Figure 54.5. Steel deck bridges show more variation than RC deck bridges. A simply supported composite box-girder bridge is plotted in Figure 54.6.

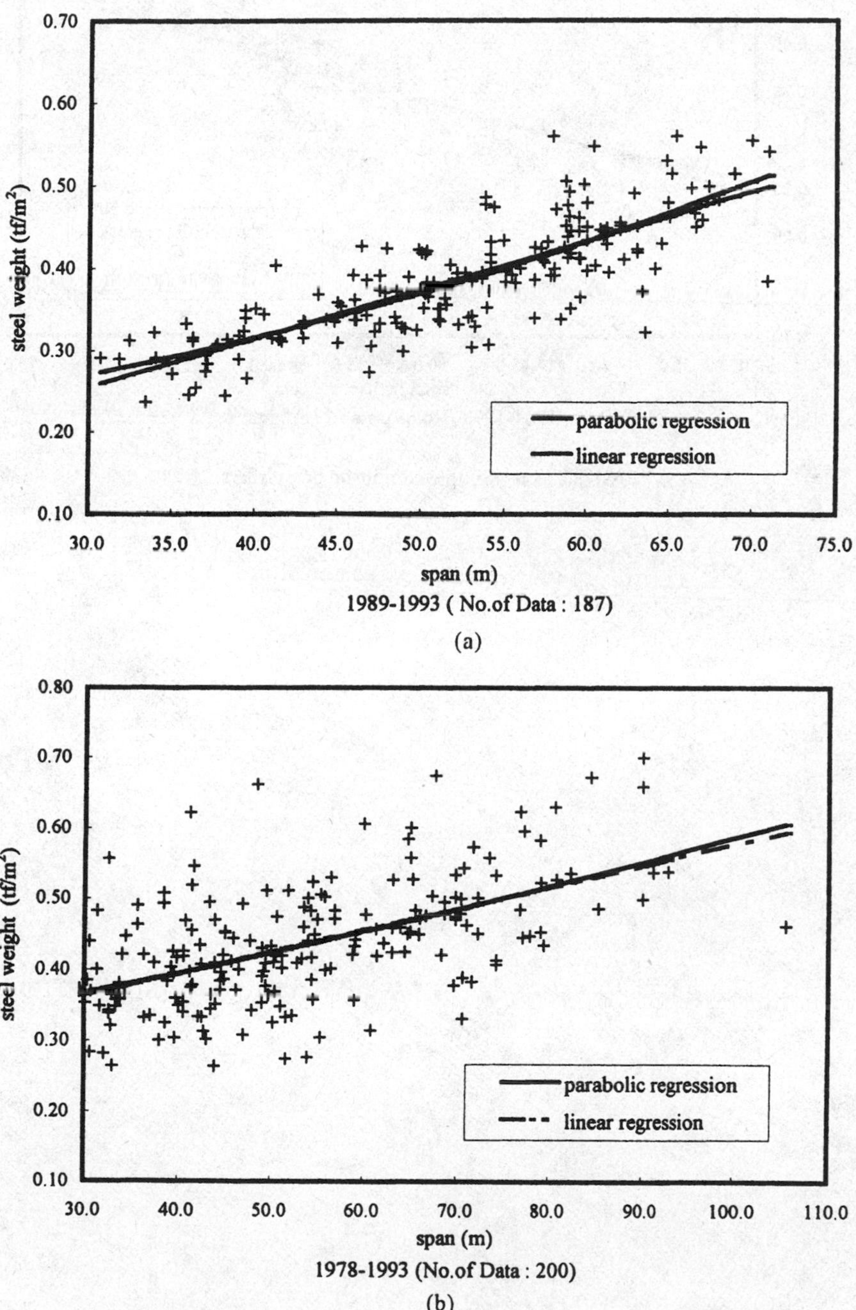

FIGURE 54.5 Simple noncomposite box girders. (a) RC slab deck; (b) steel deck.

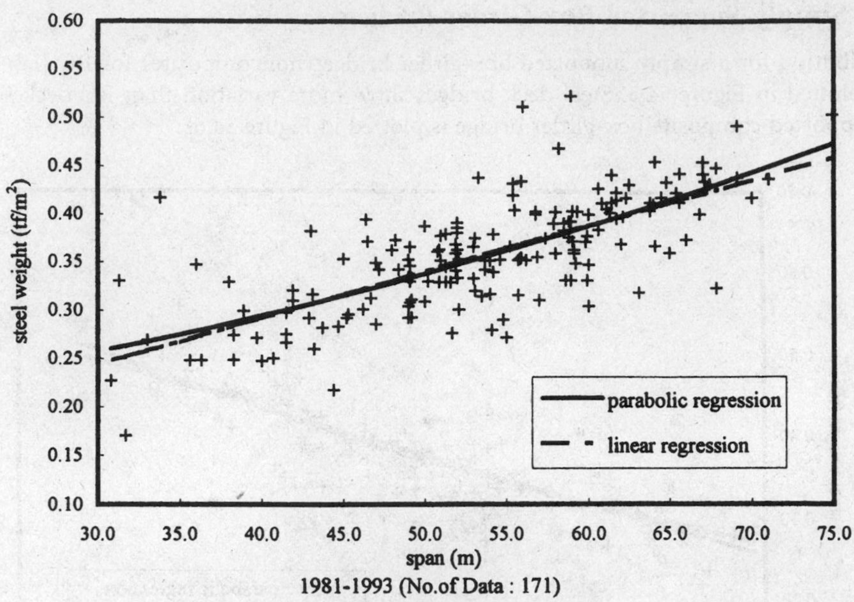

FIGURE 54.6 Simple composite box girders.

54.4.4 Continuously Supported Plate Girder Bridges

Recently, continuous bridges are gaining popularity as defects caused by expansion joints are avoided. Steel weights for continuous bridges with RC slab deck (noncomposite) constructed in the 3 years 1991 to 1993 and with steel deck constructed in the 15 years 1978 to 1993 are plotted in Figure 54.7. The steel deck has only few data and shows wide scatter.

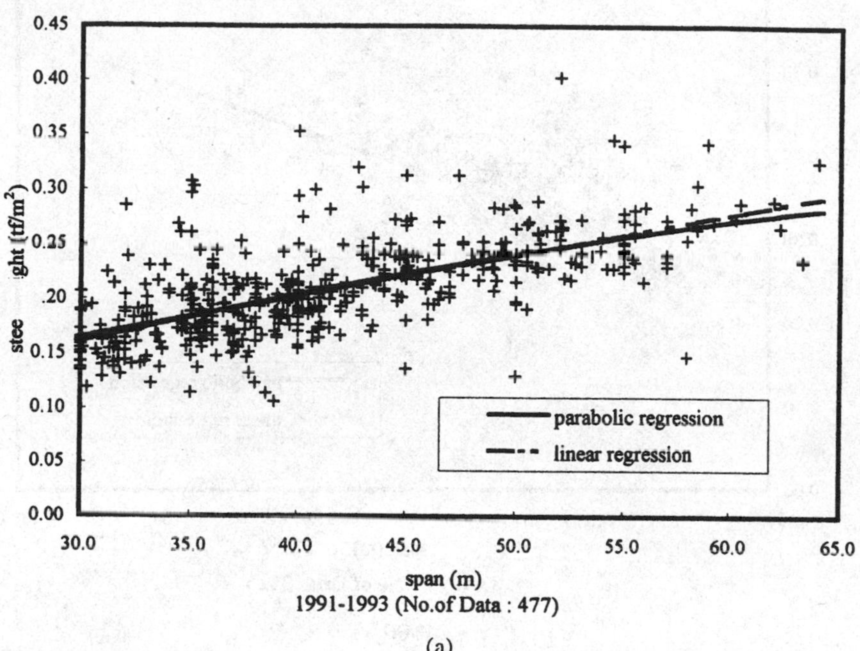

(a)

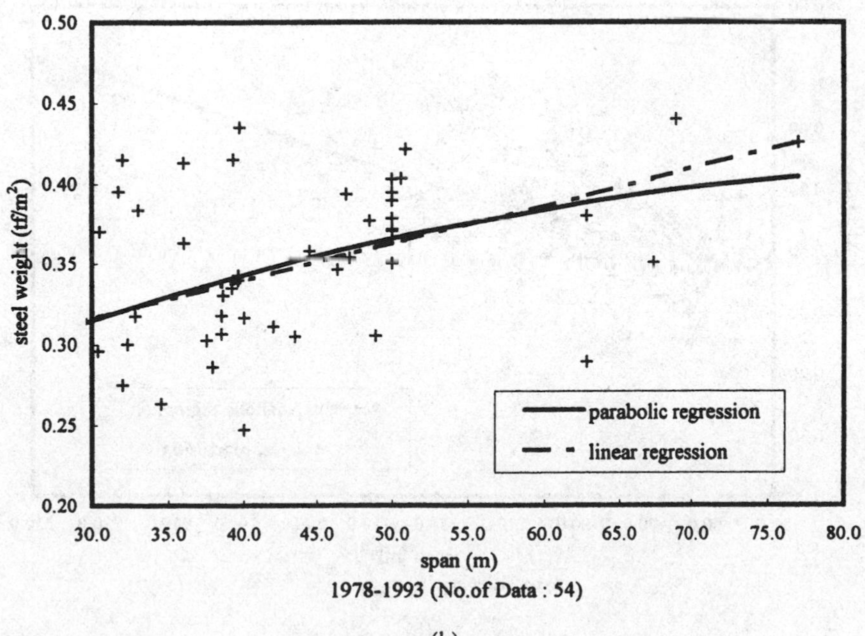

(b)

FIGURE 54.7 Continuous plate girders. (a) RC slab deck; (b) steel deck.

54.4.5 Continuously Supported Box-Girder Bridges

Figure 54.8 shows the distribution for a continuous box-girder bridge with RC slab deck and steel deck. This type has a relatively wide scatter. It can be seen that the applicable span length of steel deck bridges (Figure 54.8b) is much longer than RC slab deck bridges (Figure 54.8a).

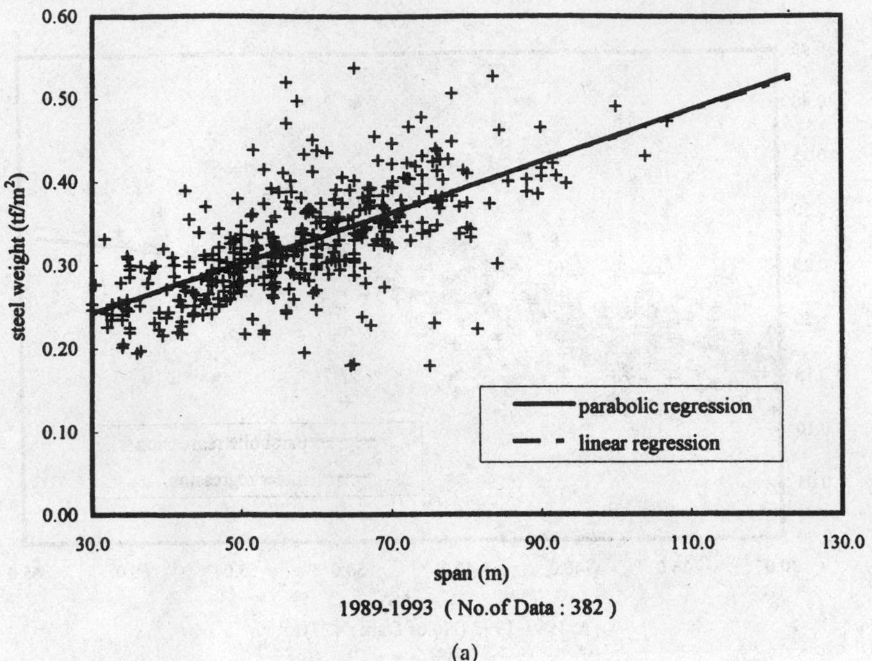

(a)

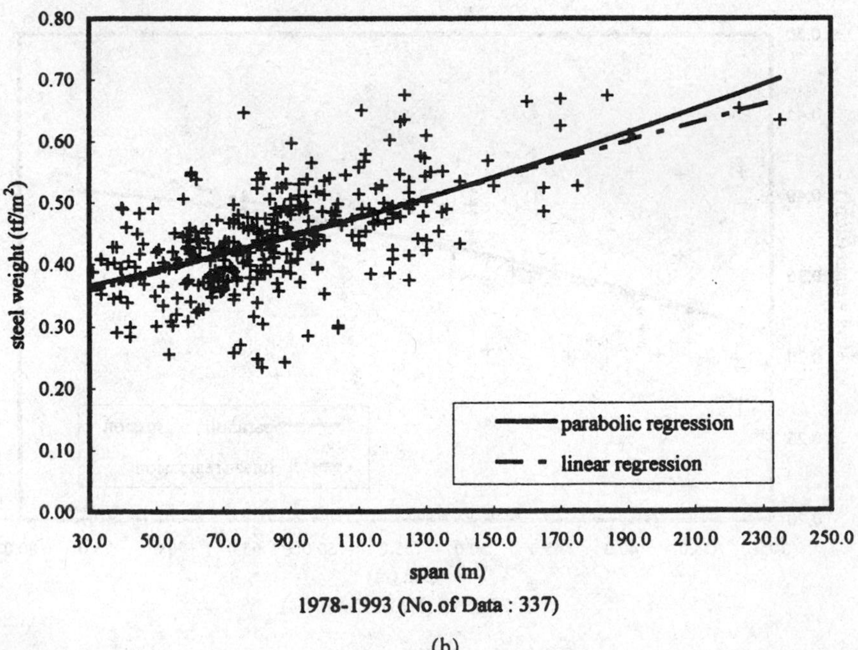

(b)

FIGURE 54.8 Continuous box girders. (a) RC slab deck; (b) steel deck.

54.4.6 Truss Bridges

Figure 54.9 is for simply and continuously supported truss bridges. The data cluster at moderate span length making prediction for the weight of truss bridges for short or long spans not accurate.

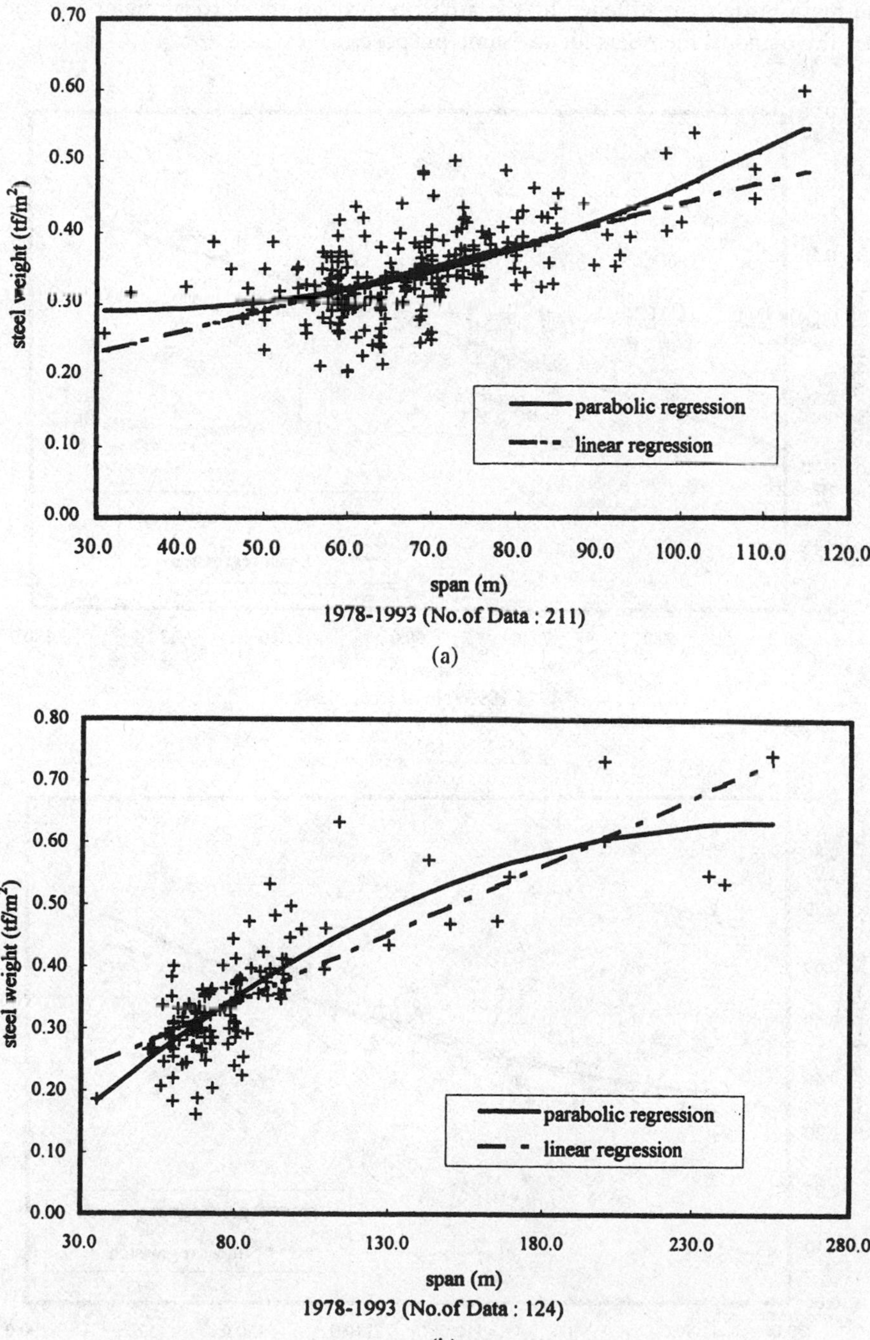

FIGURE 54.9 Truss bridges. (a) Simple truss; (b) continuous truss.

54.4.7 Arch Bridges

Figures 54.10 and 54.11 are the distributions for two arch types; Langer bridges and Lohse bridges. It is assumed in the structural analysis that the arch rib of Lohse bridge carries bending moment, shear force, and axial compression while Langer bridge only carries axial compression. In the Langer bridge, the main girders are stiffened by the arch rib through the vertical members. The trussed Langer uses the diagonal members for the same purpose.

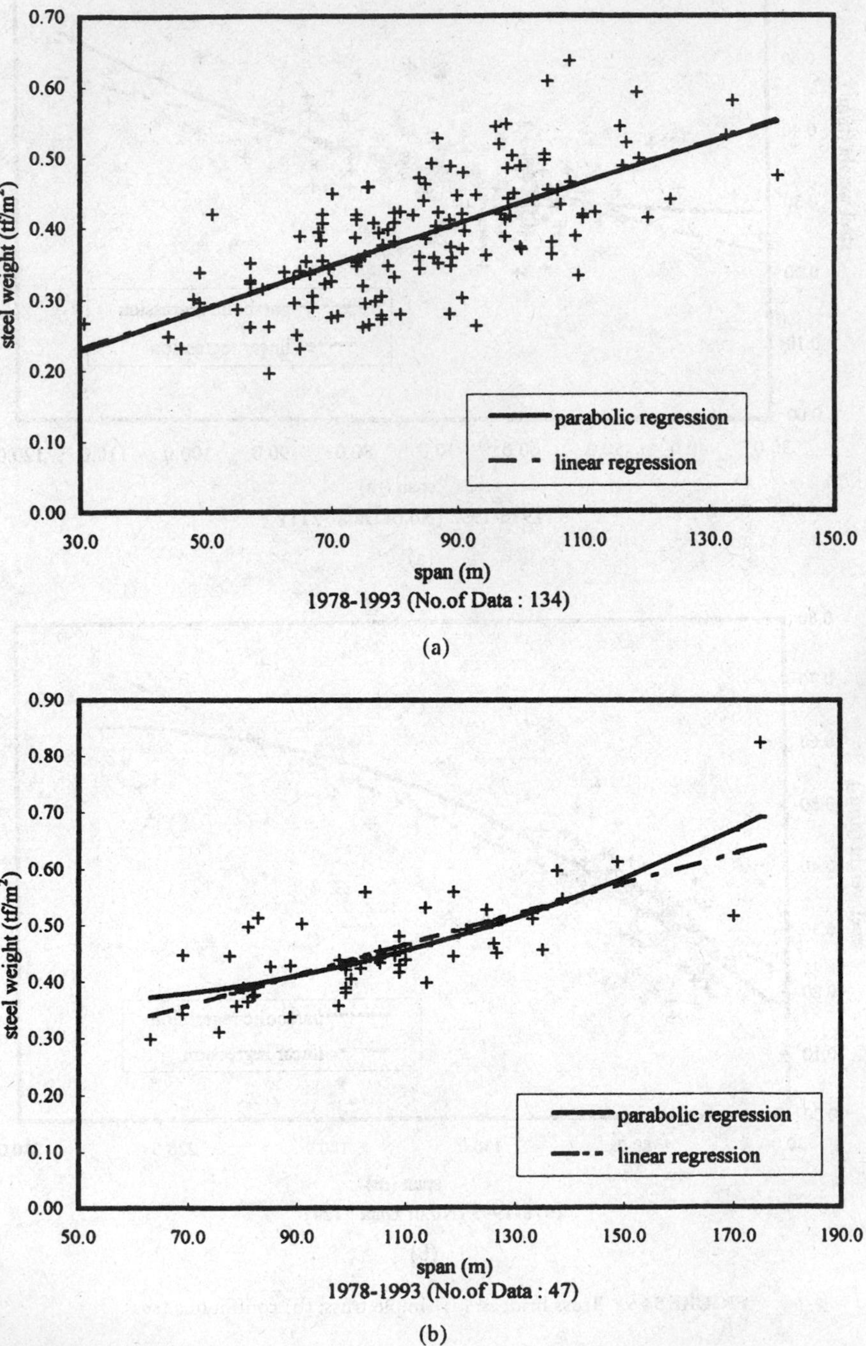

FIGURE 54.10 Langer bridges. (a) Langer; (b) trussed langer.

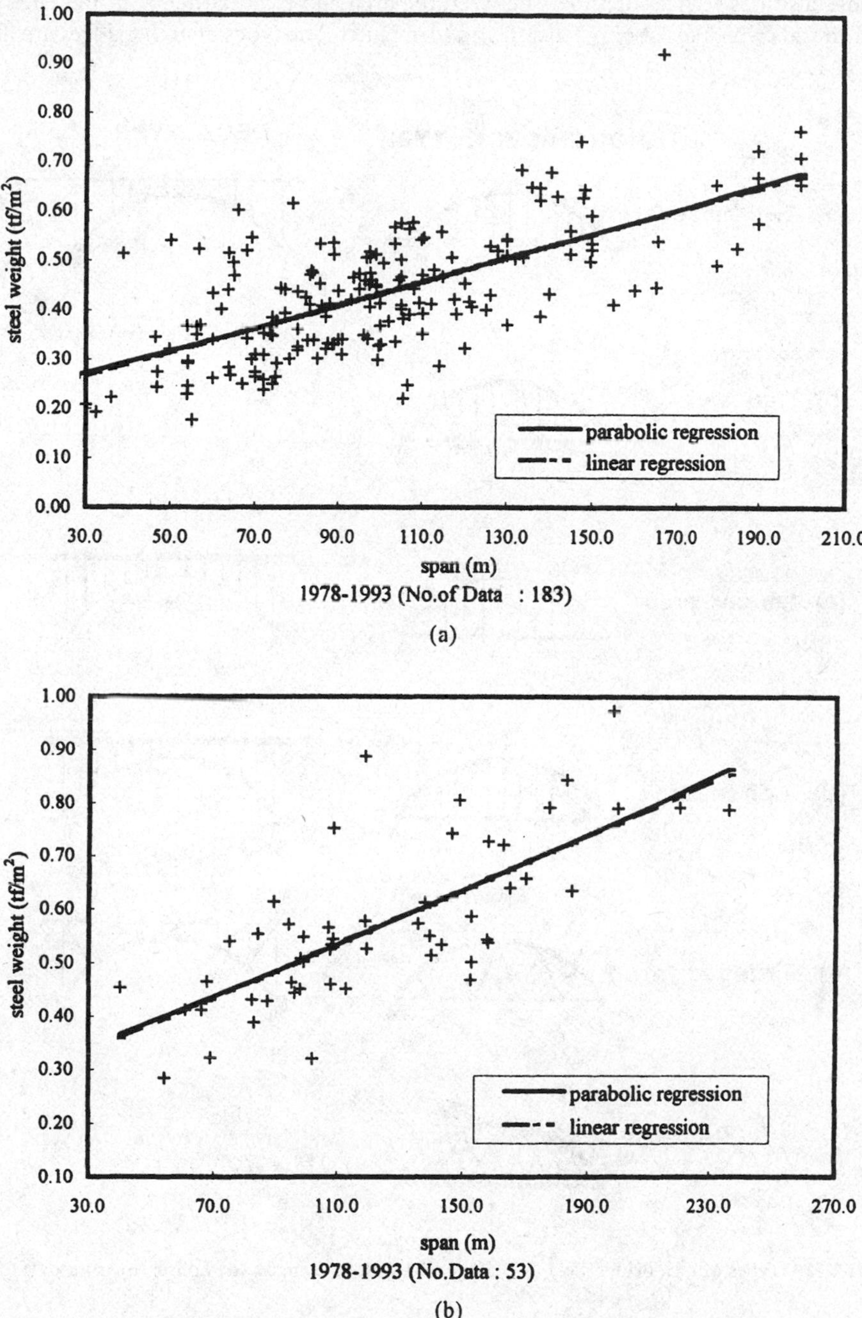

FIGURE 54.11 Lohse bridges. (a) Lohse; (b) Nielsen Lohse.

The Lohse also has vertical members between the arch and main girders, but the Nielsen Lohse has only thin rods which resist only tension and form a net. The types of arch bridges are illustrated in Figure 54.12.

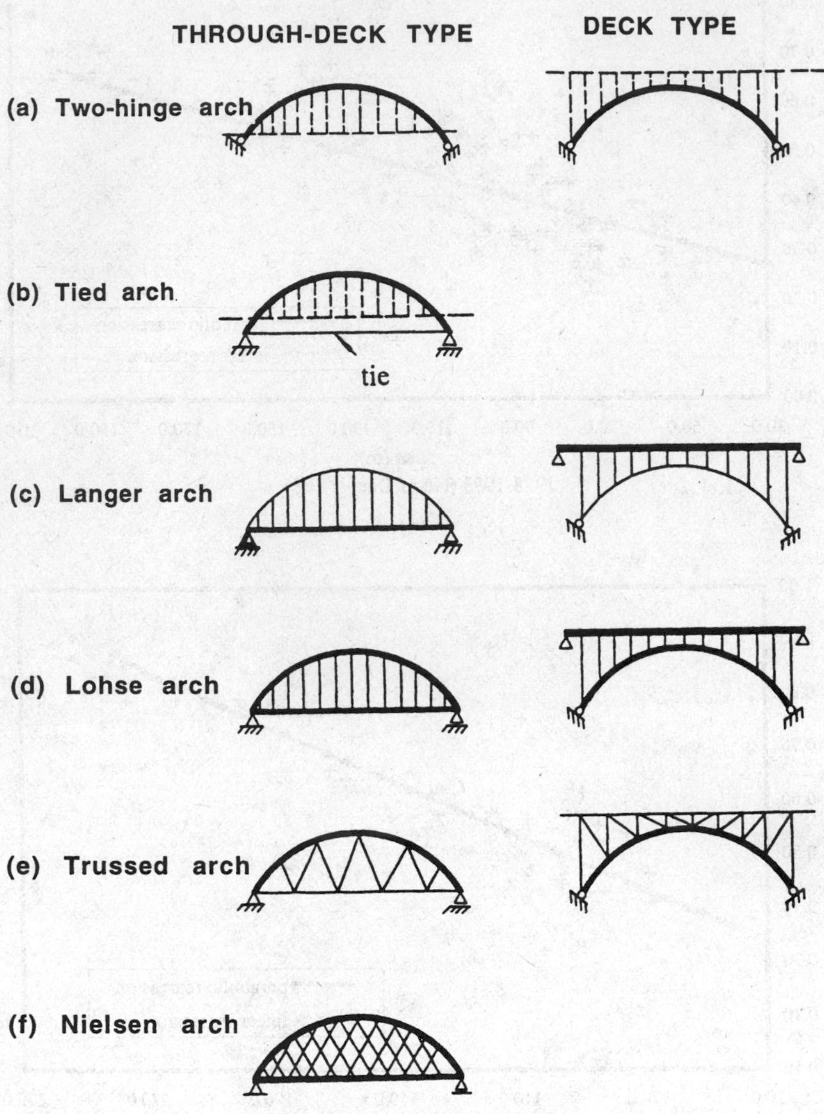

FIGURE 54.12 Types of arch bridges. (a) Two hinge; (b) tied; (c) Langer; (d) Lohse; (e) trussed; (f) Nielson.

54.4.8 Rahmen Bridges (Rigid Frames)

The Rahmen bridge is a frame structure in which all members carry bending moment and axial and shear forces. There are many variations of structural form for this type of construction as shown in Figure 54.13. Figure 54.14 shows the weight distribution for typical π-Rahmen and other types.

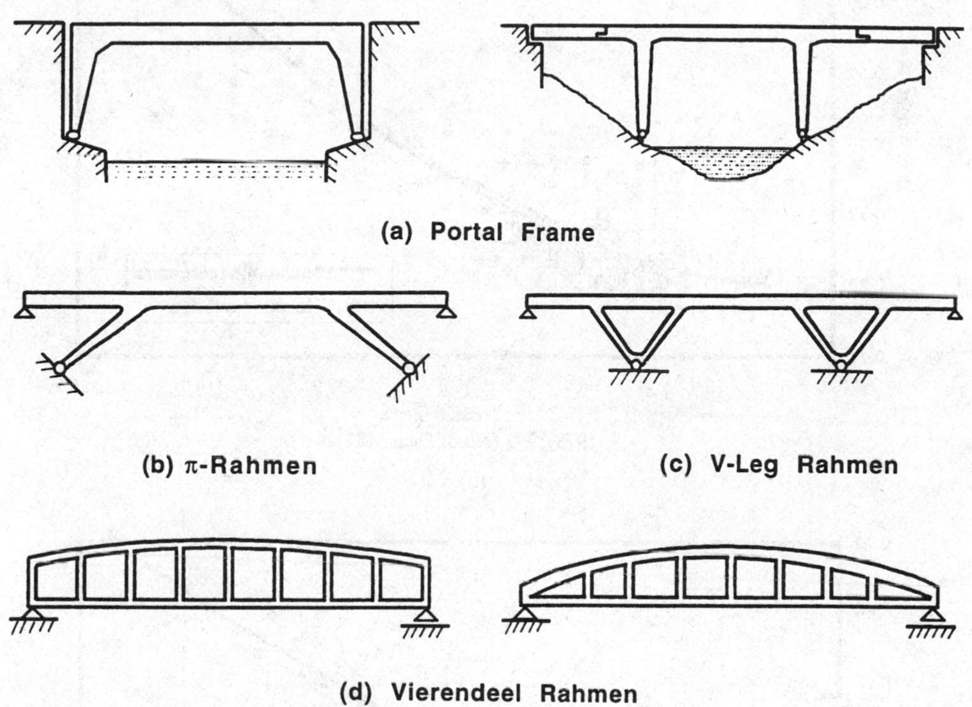

FIGURE 54.13 Types of Rahmen bridges. (a) Portal frame; (b) π-Rahmen; (c) V-leg Rahmen; (d) Vierendeel Rahmen.

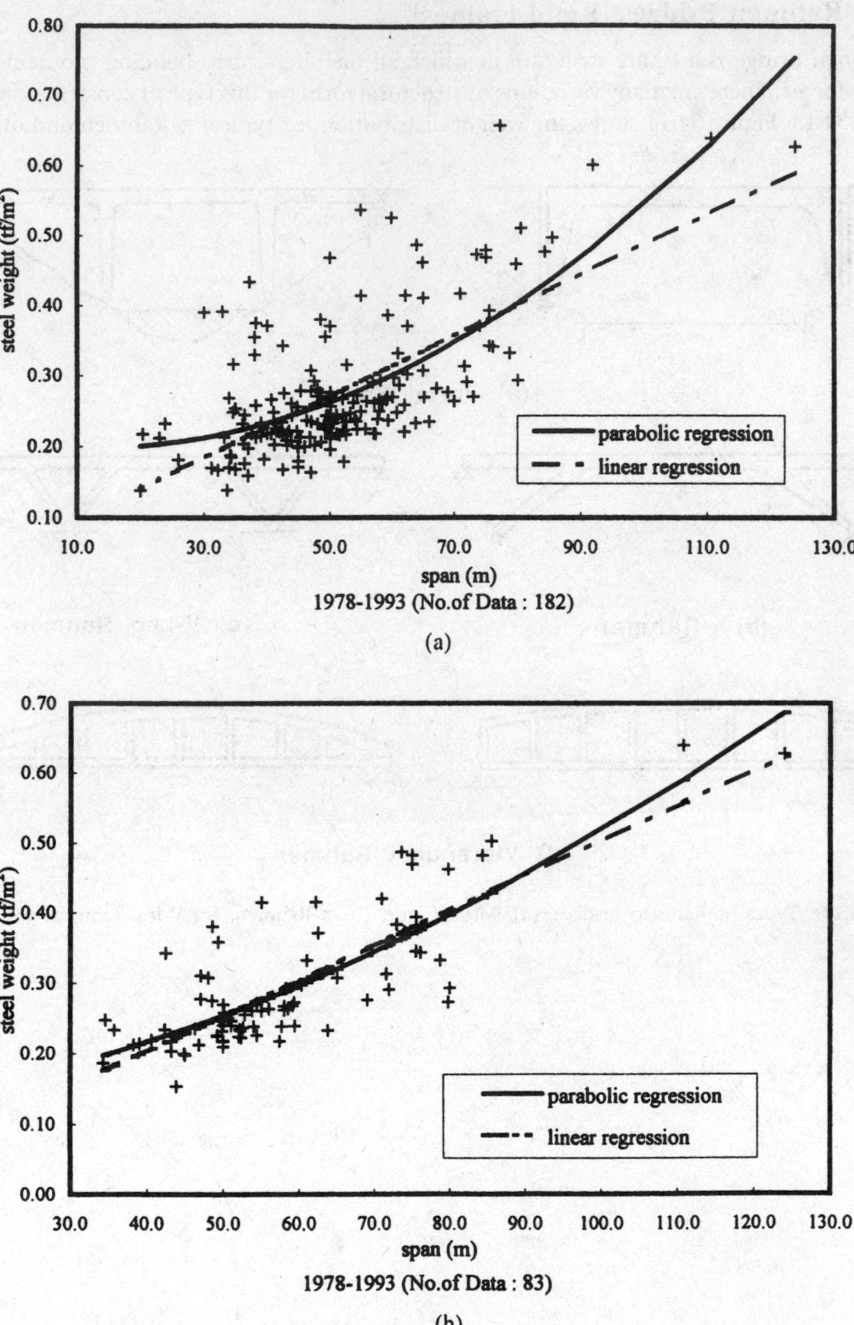

FIGURE 54.14 Rigid frames (Rahmen). (a) Rigid frame (general type); (b) π-Rahmen.

54.4.9 Cable-Stayed Bridges

Figure 54.15 shows the weight of cable-stayed bridges. The data may not be sufficient for statistical analysis. The scatter is more significant at long spans.

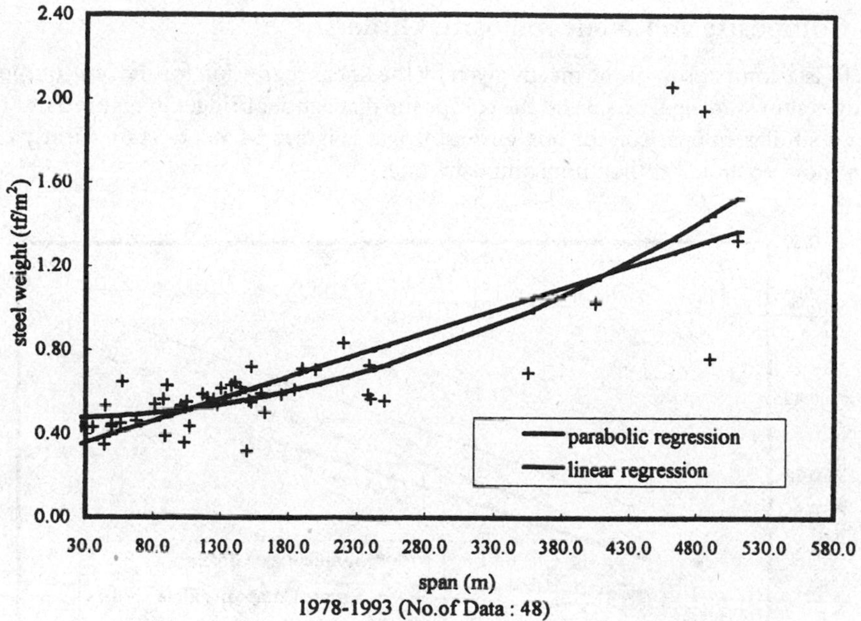

FIGURE 54.15 Cable-stayed bridges (steel deck).

54.5 Regression Equations

The two lines in the distribution figures shown previously in Figures 54.3 through 54.13 are the mean values obtained by linear regression using the least-squares method. They are linear and parabolic. It seems that the parabolic curve does not always give a better prediction. Table 54.2 gives the coefficients of the regression equations to give designers the information necessary for estimating steel weight and assessing designs.

54.6 Comparisons

The weight distributions in Figures 54.3 through 54.13 are compared from various points of view in the following.

54.6.1 Composite and Noncomposite Girders

Figure 54.16 is a comparison of the means given by the linear regression for the noncomposite plate girder bridges shown in Figure 54.3 and the composite plate girder bridges in Figure 54.4. The figure also shows a similar comparison for box-girder bridges (Figures 54.5 and 54.6). Clearly composite girders are more economical than noncomposite ones.

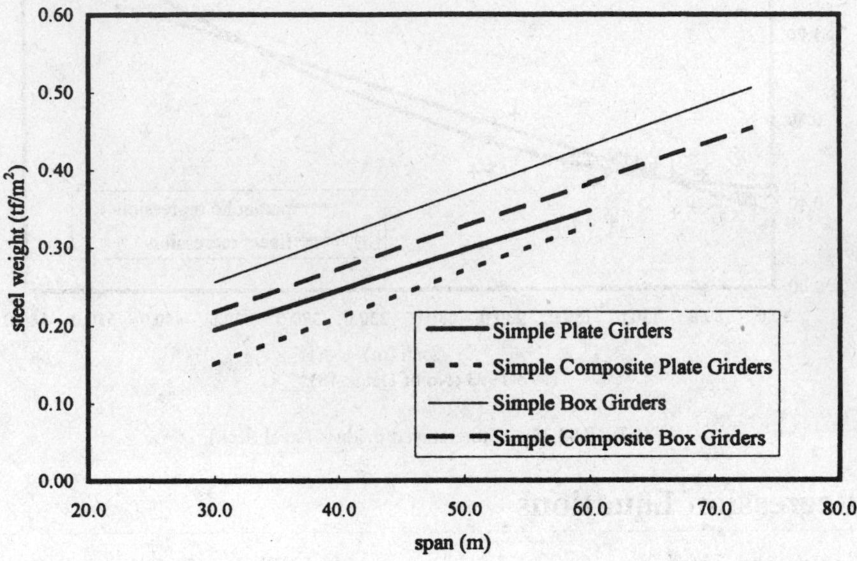

FIGURE 54.16 Comparison between composite and noncomposite plate girders.

54.6.2 Simply and Continuously Supported Girders

The difference caused by variation in support conditions is shown in Figure 54.17 for plate and box girders. The figures shown are for bridges with RC slab and steel decks. It is judged that continuous girders are more advantageous when the spans are long. There is no significant difference between simple plate and box girders for steel deck bridges. Continuous box girders can be used in long-span bridges.

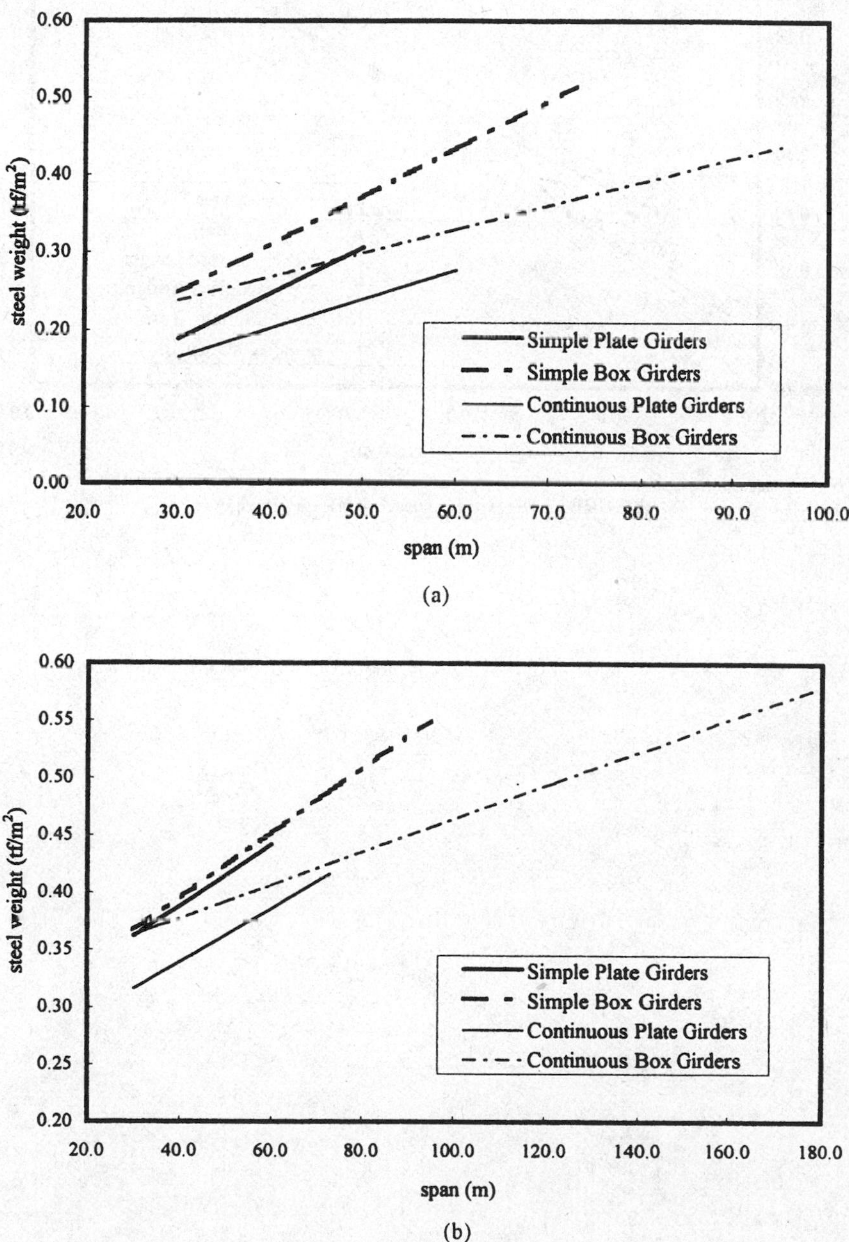

FIGURE 54.17 Comparison of girder bridges. (a) RC slab deck; (b) steel deck.

54.6.3 Framed Bridges

Six types of framed bridges are compared in Figure 54.18. The Nielsen bridge is the heaviest. The Nielsen and Lohse bridges, as well as the trussed Langer, are best suited to long spans.

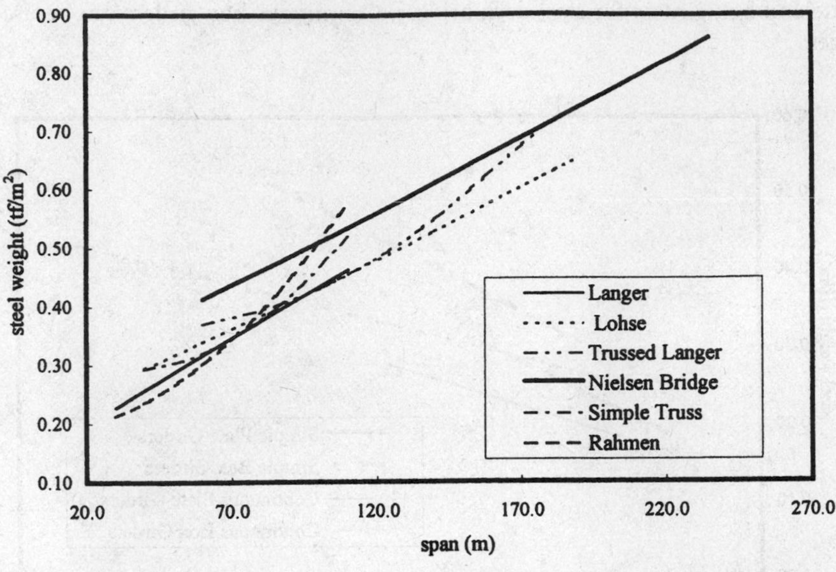

FIGURE 54.18 Comparison of framed bridges.

54.6.4 RC Slab Deck and Steel Deck

Figure 54.19a shows a comparison between the mean values of plate girder bridges with RC slab and steel decks. Bridges with steel decks are naturally much heavier than those with RC slab decks because the weight of the decks is included.

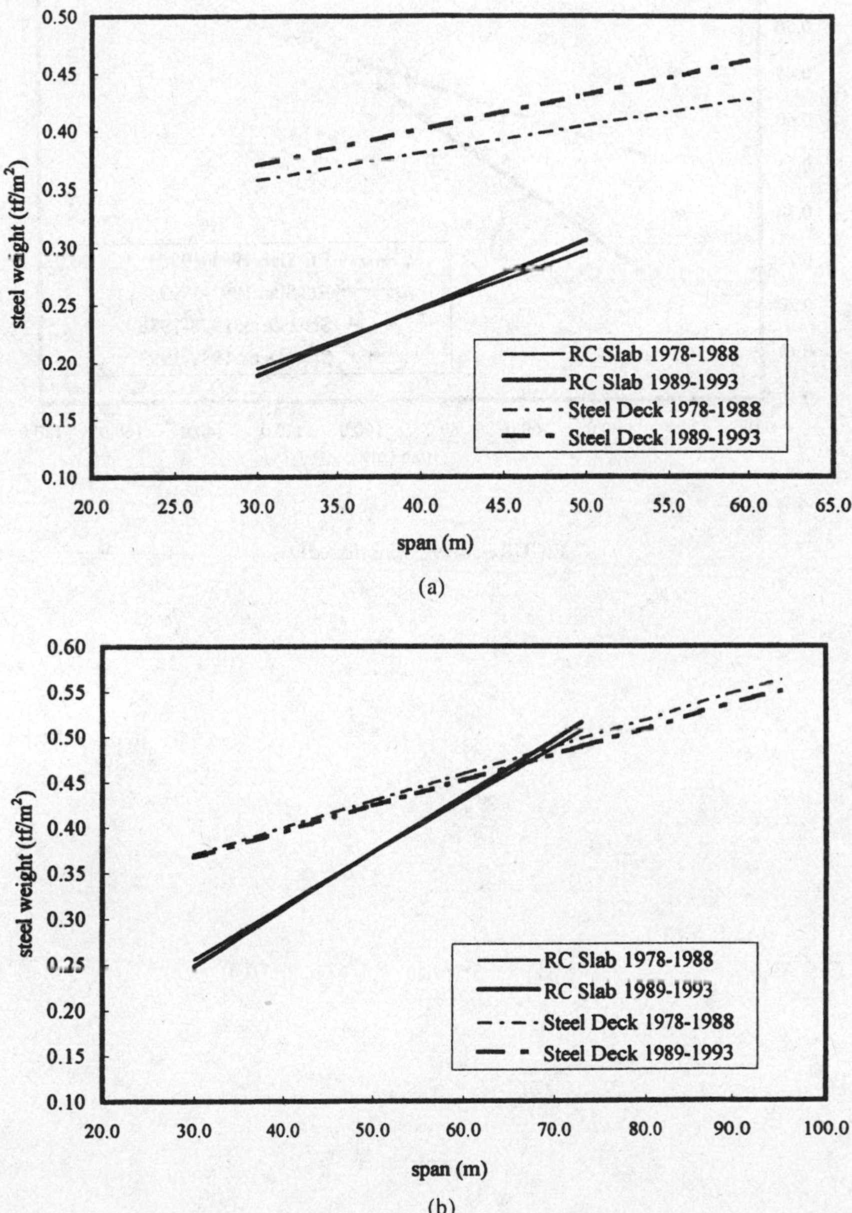

FIGURE 54.19 Comparison between RC slab and steel deck bridges. (a) Simple plate girders; (b) simple box girders; (c) continuous box girders.

A similar comparison for the box girder is shown in Figure 54.19(b). The difference gets smaller as the span length increases implying that steel deck bridges are economical when spans are long.

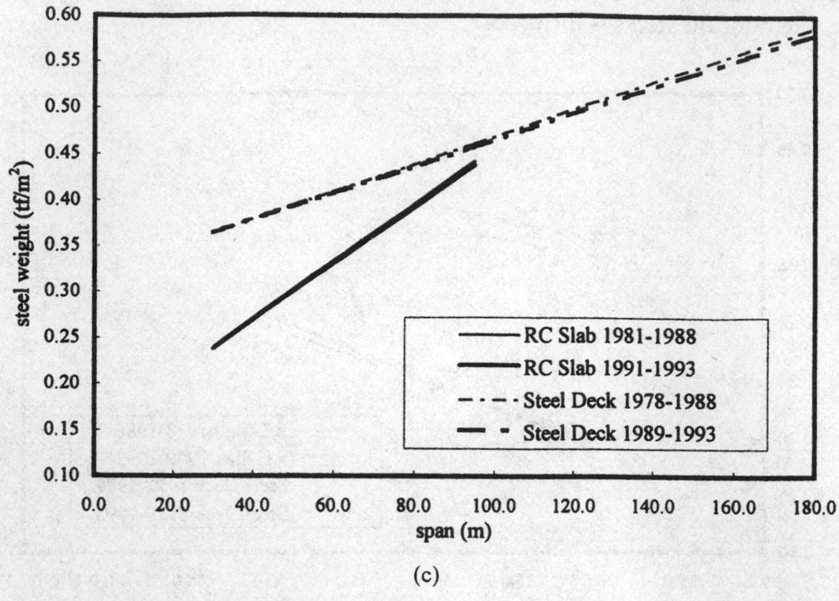

(c)

FIGURE 54.19 (continued)

54.7 Assessment of Bridge Design

54.7.1 Deviation

The distribution of the weights can be expressed by standard Gaussian techniques giving a mean value of 50 and a standard deviation of 10 as shown in Figure 54.20. The mean value $X(L)$ is calculated by the regression equations in Table 54.2 and converted to 50. The standard deviation σ can also be obtained from the regression equations table (Table 54.2), and converted to 10 using standard Gaussian procedures.

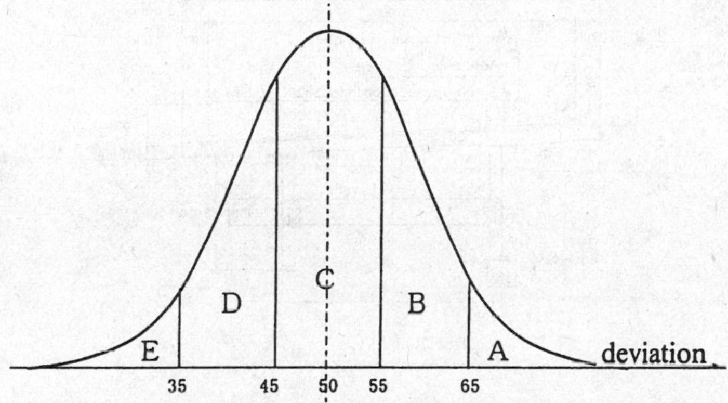

FIGURE 54.20 Classification of distribution.

The deviation (H) of the designed steel weight (X) is obtained using the equation

$$H = \frac{X - X(L)}{\sigma} \times 10 + 50 \tag{54.1}$$

H can be used as an index to compare the designs statistically and perform simple assessments of designs.

54.7.2 Assessment of Design

An example assessment of a typical design is discussed in the following. The labor and maintenance cost of bridges have become a major consideration in all countries. To solve this, a new design concept is proposed using only two girders with wide girder spacing. Figure 54.21 is one of the two-girder bridges that were constructed in Japan. It is a two-span continuous bridge with each span length 53 m. The road width is 10 m and the girder spacing 6 m. In this bridge, the section of the girder is not changed in an erection block to reduce welding length, thus reducing the labor cost.

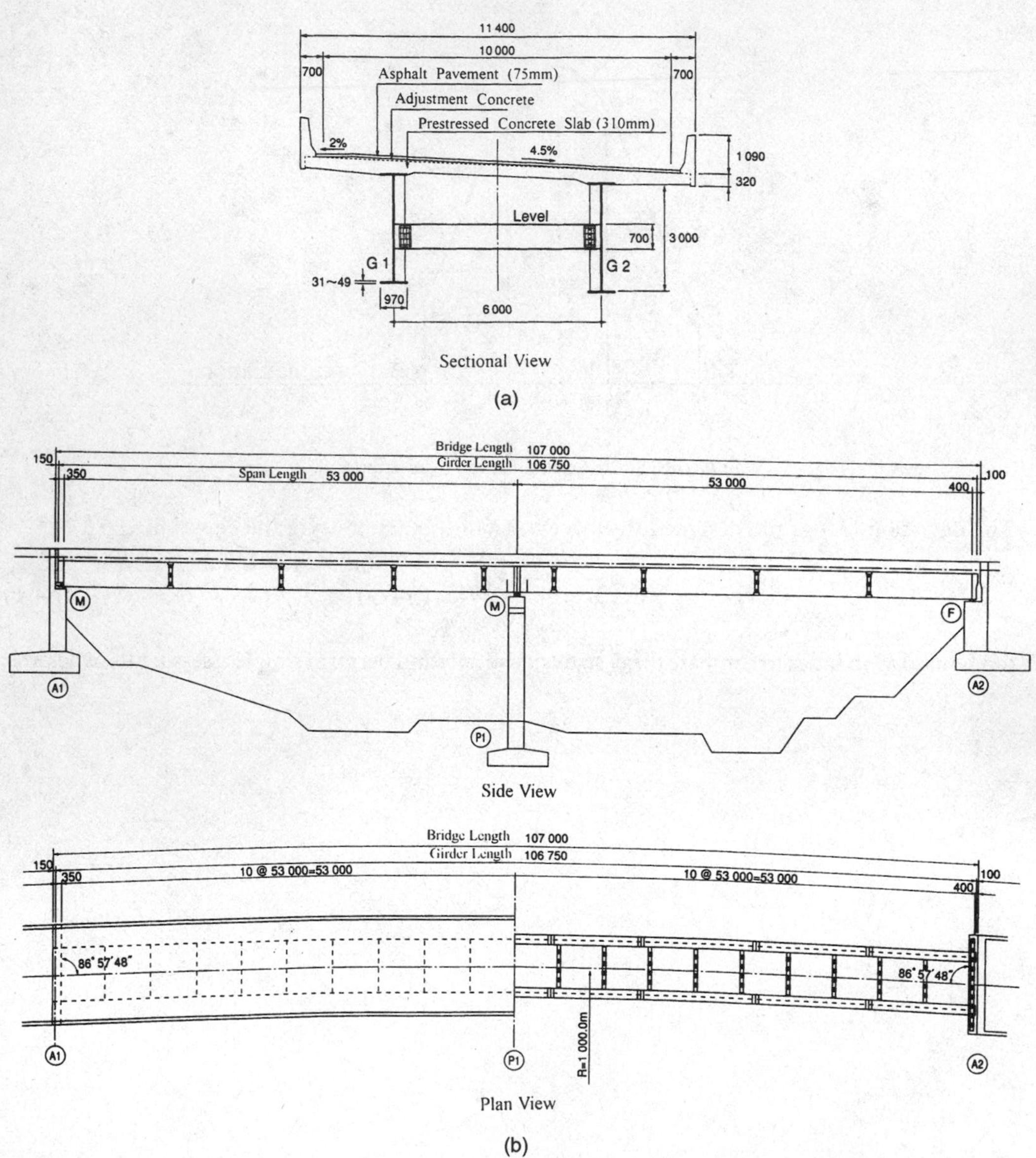

FIGURE 54.21 General plan of two-girder bridge. (a) Sectional view; (b) plan view. (*Bridges in Japan 1995-96*, JSCE)

The steel weight of this bridge is plotted in Figure 54.22. The deviation in this case is H = 62.8 (Rank B) using Eq. (54.1). In the calculation, the mean and the standard deviations are shown in Table 54.2. Note that most of the continuous bridges in Figure 54.22 are three-span continuous bridges. In addition, the design of this bridge follows the new code [1]. Those make the deviation for this case tend to be higher. From these deviation values the steel weight of a similar bridge can be estimated.

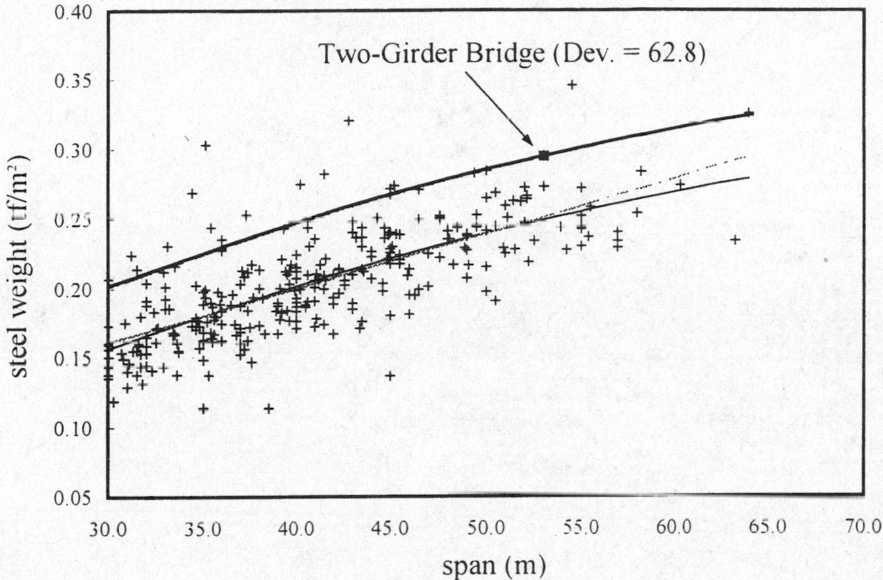

FIGURE 54.22 Two-girder bridge in continuous bridges.

54.8 Summary

The steel weight of bridges is a general indication of the design which tells an overall result. It reflects every influential design factor. A database has been put together to allow assessment of designs and prediction for the steel weight of various types of highway bridges. The distributions are plotted and shown for each type of bridge. From the figures, comparisons are made from various points of view to see the differences in each type of bridge. The regression equations for mean weight are derived, from which designers can estimate the steel weight for their own design or see economical or safety features of the bridge as compared with others.

References

1. Japan Road Association (JRA), Specifications for Highway Bridges, Vol. 1 Common Part, December 1996 [in Japanese].
2. Japan Road Association (JRA), Specifications for Highway Bridges, Vol. 1 Common Part, February 1990 [in Japanese].
3. Japan Association of Steel Bridge Construction (JASBC), Annual Report of Steel Bridge Construction, 1978 to 1993 [in Japanese].
4. Toma, S. and Honda, Y., Database of steel weight for highway bridges, *Bridge Eng.*, 29(8), 1993 [in Japanese].

55

Weigh-in-Motion Measurement of Trucks on Bridges

Andrzej S. Nowak
University of Michigan

Sangjin Kim
Kyungpook National University, Korea

55.1 Introduction

Knowledge of the past and current load spectra, together with predicted future loads, is essential in the evaluation and fatigue analysis of existing bridges. Many trucks carry loads in excess of design limits. This may lead to fatigue failure. The information concerning actual load is very important for the rating of bridges. Therefore, there is a need for accurate and inexpensive methods to determine the actual loads, the strength of the bridge, and its remaining life. There is also a need for verification of live load used for the development of a new generation of bridge design codes [1,2]. It has been confirmed that truck loads are strongly site specific [3,4,5]. Some bridges carry heavy truck traffic (volume and magnitude); others carry only lighter traffic. Furthermore, load effects such as bending moment, shear, and/or stress are component specific [6,7]. This observation is important in evaluation of the fatigue damage and prediction of remaining life. This chapter presents some of the practical procedures used for field measurement of truck weights and resulting strains.

55.2 Weigh-in-Motion Truck Weight Measurement

55.2.1 Weigh-in-Motion Equipment

The bridge live load is the load caused by truck traffic. In the past, truck data were collected by truck surveys, which had limitations. The most common survey method consisted of weighing trucks using static scales installed in weigh stations at fixed locations along major highways. The usefulness of the data obtained, however, is limited because many drivers of overloaded trucks intentionally avoid the scales, and therefore the results are biased to lighter trucks.

0-8493-7434-0/00/$0.00+$.50

Therefore, research effort was focused on developing weigh-in-motion (WIM) methodology which can provide unbiased truck data, including axle weight, axle spacing, vehicle speed, multiple presence of trucks, and average daily truck traffic (ADTT). Very good results were obtained by using a WIM system with a bridge as a scale. Sensors measure strains in girders, and this is used to calculate the truck parameters at the highway speed.

The WIM system provides instrumentation invisible to truck drivers, and, therefore, the drivers do not try to avoid the scale. The system is portable and can be easily installed on a bridge to obtain site-specific truck data.

The bridge WIM system consists of three basic components: strain transducers, axle detectors (tape switches or infrared sensors), and data acquisition and processing system (Figure 55.1). The analog front end (AFE) acts as a signal conditioner and amplifier with a capacity of eight input channels. Each channel can condition and amplify signals from strain transducers. During data acquisition, the AFE maintains the strain signals at zero. The autobalancing of the strain transducers is activated when the first axle of the vehicle crosses the first axle detector. As the truck crosses the axle detectors the speed and axle spacing are determined. When the vehicle enters the bridge, the strain sampling is activated. As the last axle of the vehicle has exited the instrumented bridge span, the strain sampling is turned off. Data received from strain transducers are digitized and sent to the computer where axle weights are determined by an influence line algorithm. These data do not include dynamic loads. This process takes from 1.5 to 3.0 s, depending on the instrumented span length, vehicle length, number of axles, and speed. The data are then saved to memory.

The WIM equipment is calibrated using calibration trucks. The readings are verified and calibration constants are determined by running a truck with known axle loads over the bridge several times in each lane. The comparison of the results indicates that the accuracy of measurements is within 13% for 11-axle trucks. Gross vehicle weight (GVW) accuracy for five-axle trucks is within 5%, however, the accuracy is within 20% for axle loads [5].

55.2.2 Testing Procedure

The WIM system provides truck axle weights, gross vehicle weights, and axle spacings. Strains are measured in lower flanges of the girders and the strain time history is decomposed using influence lines to determine vehicle axle weights, as shown in Figure 55.1. The strain transducer can be clamped to the upper or lower surface of the bottom flange of the steel girder as shown in Figure 55.2. All transducers are placed on the girders at the same distance from the abutment, in the middle third of a simple span. The vehicle speed, time of arrival, and lane of travel are obtained using lane sensors on the roadway placed before the instrumented span of the bridge (Figure 55.3).

Two types of lane sensors can be used depending on the site conditions: tape switches and infrared sensors. Tape switches consist of two metallic strips that are held out of contact in the normal condition. As a vehicle wheel passes over the tape it forces the metallic strips into contact and grounds a switch. If a voltage is impressed across the switch, a signal is obtained at the instant the vehicle crosses the tape. This signal is fed to a computer whereby the speed, axle spacing, and number of axles are determined. The tape switches are placed perpendicular to the traffic flow and used to trigger the strain data collection. All cables used to connect tape switches and strain transducer to the AFE are five-pin wire cables.

The major problem with tape switches is their vulnerability to damage by moving traffic, particularly if the pavement is wet. Various alternative devices can be considered. Infrared sensors can be used to replace the tape switches. The infrared system consists of a source of infrared light beam and a reflector. The light source is installed on the side of the road. The reflector is installed in the center of the traffic lane. However, the problem of their vulnerability to damage by moving traffic has not been resolved. The infrared system is more difficult to install and trucks can easily move the reflector and interrupt the operation (the light beam must be aligned).

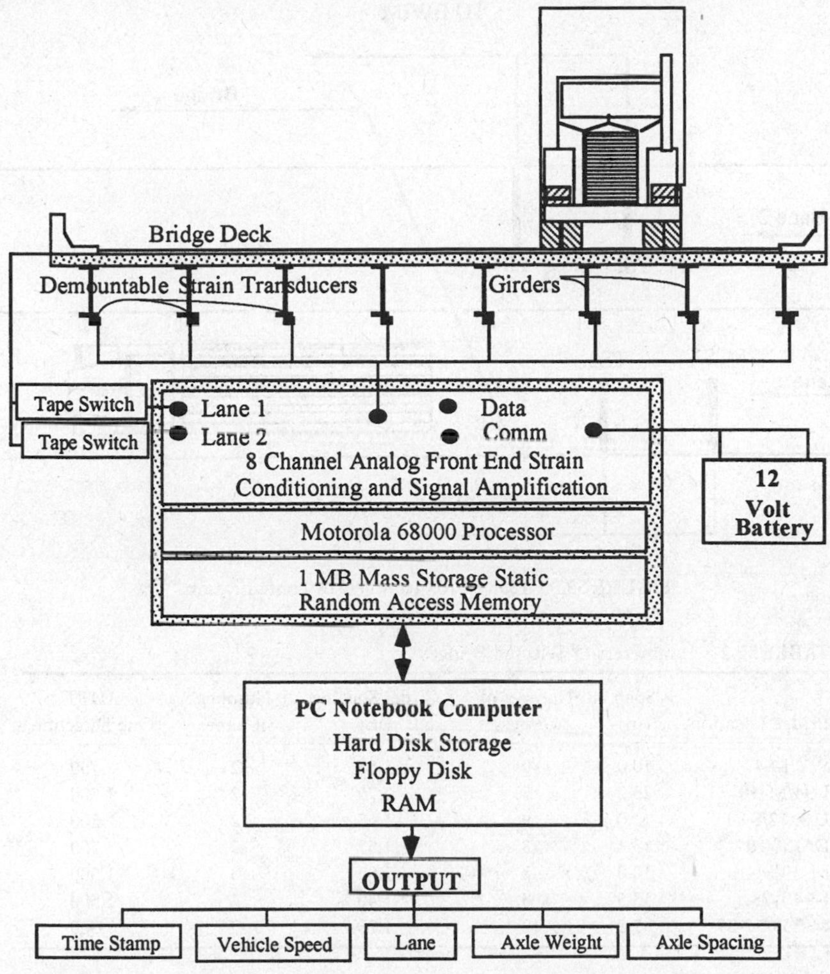

FIGURE 55.1 WIM truck measurement system.

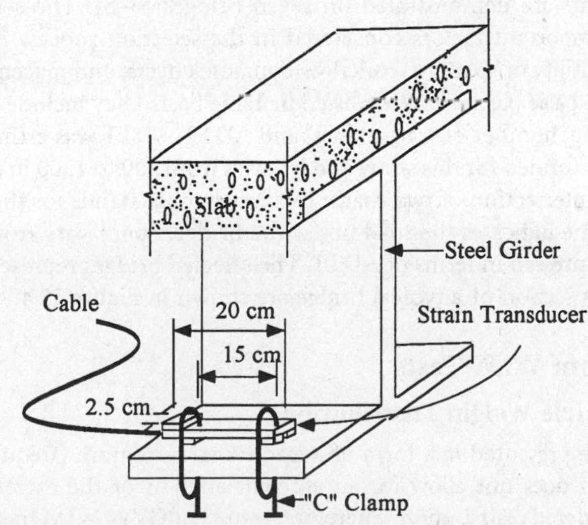

FIGURE 55.2 Demountable strain transducer mounted to the lower flange.

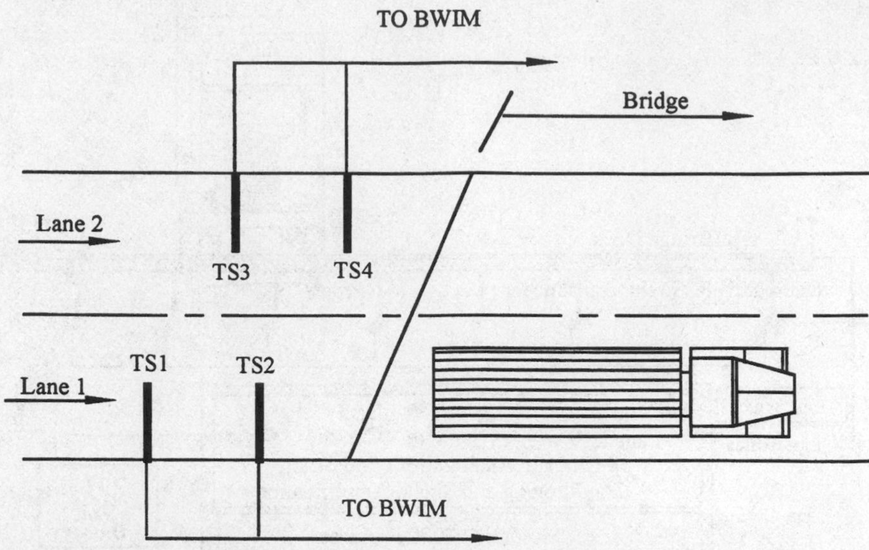

FIGURE 55.3 Plan of roadway sensor configuration.

TABLE 55.1 Parameters of Selected Bridges

Bridge Location	Span (m)	Number of Girders	Girder Spacing (m)	Number of Lanes	ADTT (One Direction)
WY/I-94	10.0	9	1.55	2	750
I-94/M-10	23.2	5	2.70	2	1,500
U.S. 12/I-94	12.0	9	1.68	2	500
DA/M-10	13.1	8	1.57	2	750
M-39/M-10	10.0	8	1.85	3	1,500
I-94/I-75	13.5	8	1.40	2	1,500
M-153/M-39	9.5	12	1.75	3	500

55.2.3 Selection of Bridges for Testing

The WIM measurements are demonstrated on seven bridges [3–5]. The selected structures are located in Michigan. Important factors considered in the selection process included accessibility from the ground, availability of space to work, low dynamic effects, and placement of tape switches or infrared sensors. The basic parameters are listed in Table 55.1. They include span length, number of girders, girder spacing, number of traffic lanes, and ADTT. ADTT was estimated on the basis of truck measurements performed for this study and it varies from 500 to 1500 in one direction. Bridge location is denoted by intersection of two roads; the first symbol stands for the road carried by the bridge, and the other one indicates the road under the bridge. Spans vary from about 10 to 25 m. The traffic volume is expressed in terms of ADTT. The selected bridges represent typical structures. The elevation and cross section of a typical bridge are shown in Figure 55.4.

55.2.4 Results from WIM Tests

55.2.4.1 Gross Vehicle Weight Distributions

The WIM results can be presented in a form of a traditional histogram (frequency or cumulative). However, this approach does not allow for an efficient analysis of the extreme values (upper or lower tails) of the considered distribution. Therefore, results of GVW WIM measurements for seven bridges are shown in Figure 55.5 in the form of cumulative distribution functions (CDFs) on the

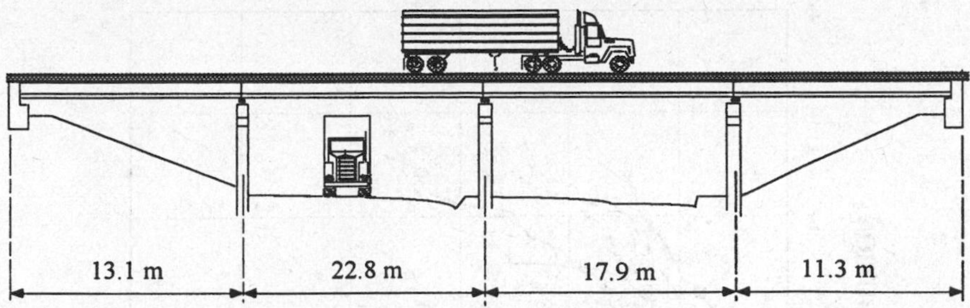

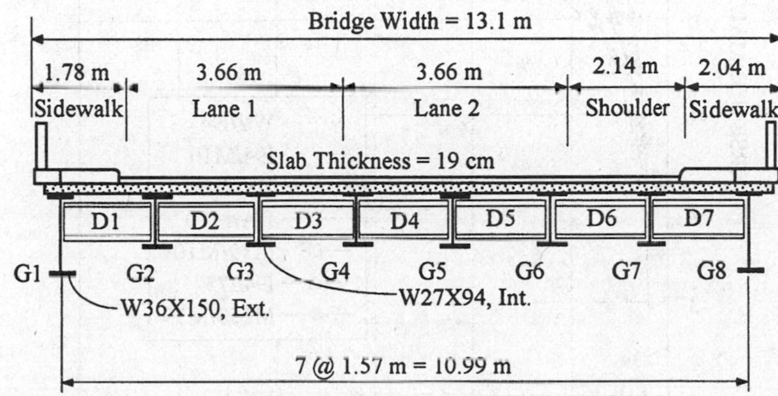

FIGURE 55.4 Bridge DA/M-10.

normal probability paper. CDFs are used to present and compare the critical extreme values of the data. They are plotted on normal probability paper [8]. The horizontal scale is in terms of the considered truck parameter (e.g., GVW, axle weight, lane moment, or shear force). The vertical scale represents the probability of being exceeded, p. Then, the probability of being exceeded (vertical scale) is replaced with the inverse standard normal distribution function, $\Phi^{-1}(p)$. For example, $\Phi^{-1}(p) = 0$, corresponds to the probability of being exceeded, $p = 0.5$; $\Phi^{-1}(p) = 1$, corresponds to $p = 0.159$; and $\Phi^{-1}(p) = -1$ corresponds to $p = 0.841$; and so on.

The distribution of truck type by number of axles will typically bear a direct relationship to the GVW distribution; the larger the population of multiple-axle vehicles (greater than five axles) the greater the GVW load spectra. Past research has indicated that 92 to 98% of trucks are four- and five-axle vehicles. The data obtained in this study indicate that between 40 and 80% of the truck population are five-axle vehicles, depending considerably on the location of the bridge. Three- and four-axle vehicles are often configured similarly to five-axle vehicles, and when included with five-axle vehicles account for between 55 and 95% of the truck population. Between 0 and 7.4% of the trucks are 11-axle vehicles in Michigan.

Most states in the United States allow a maximum GVW of 355 kN where up to five axles per vehicle are permitted. The State of Michigan legal limit allows for an 11-axle truck of up to 730 kN, depending on axle configuration. There were a number of illegally loaded trucks measured during data collection at several of the sites. Maximum WIM truck weights (1192 kN) exceeded legal limits by as much as 63%.

55.2.4.2 Axle Weight Distributions

Potentially more important for bridge fatigue and pavement design are the axle weights and spacing for trucks passing over the bridge. Figure 55.6 presents the distributions of the axle weights of the measured vehicles. All distributions include axles with weights greater than 22 kN. The maximum axle weights vary from 90 to 225 kN, and average values from 30 kN to 60 kN.

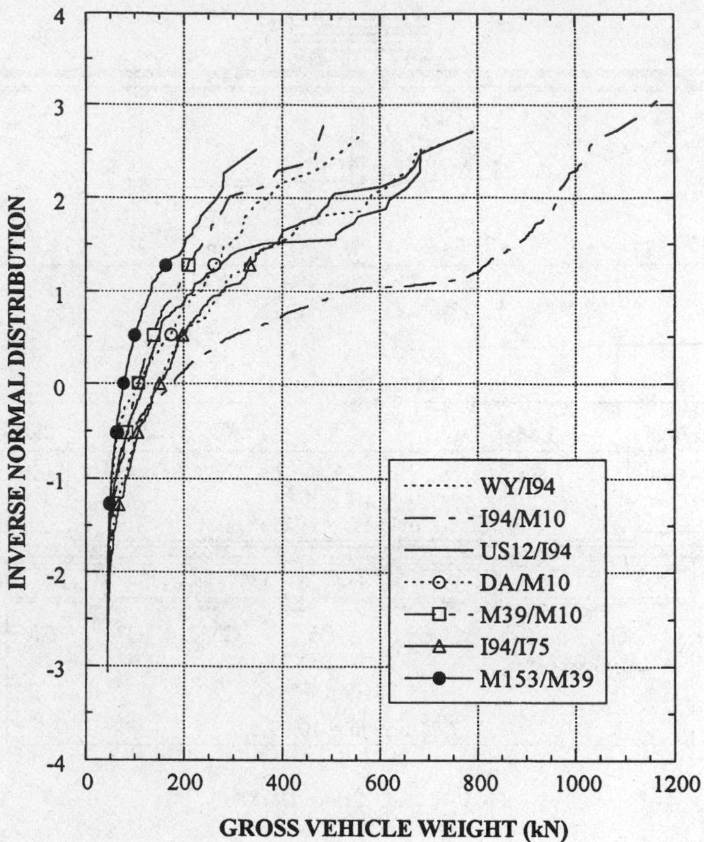

FIGURE 55.5 CDFs of GVW for the considered bridges.

55.2.4.3 Lane Moment and Shear Distributions

The structure is affected by load effects. Therefore, for the measured trucks, lane moments and shears were calculated for various spans. The resulting CDFs for 27-m span are shown in Figure 55.7 for lane moment and Figure 55.8 for lane shear. Each truck in the database is analytically driven across the bridge to determine the maximum static bending moment (shear) per lane. The CDFs of the lane moments (shears) for a span of 27 m are then determined. As a point of reference, the calculated moments (shears) are divided by design moment (shear) specified by the new AASHTO LRFD Specification [1]. The design live load according to the AASHTO LRFD is a superposition of a truck weighing 320 kN (three axles: 35, 145, and 145 kN spaced 4.25 m) and a uniformly distributed load of 9.3 kN/m. The lane moments in Figure 55.7 show a wide variation among bridges. Maximum values of the ratio of lane moment to LRFD moment vary from 0.6 at M-153/M-39 to 2.0 at I-94/M-10. All sites have a median lane moment between 0.16 and 0.34 times LRFD moment, which corresponds to an inverse normal value of 0. The variation of lane shears in Figure 55.8 is similar to that of lane moments. For I-94/M-10, the extreme value exceeds 2.0. For other bridges, the maximum shears vary from 0.65 at M-153/M-39 to 1.5 at I-94/I-75.

55.3 Fatigue Load Measurement

55.3.1 Testing Equipment

The Stress Measuring System (SMS) with the main unit manufactured by the SoMat Corporation is shown in Figure 55.9. The SMS compiles stress histograms for the girders and other components.

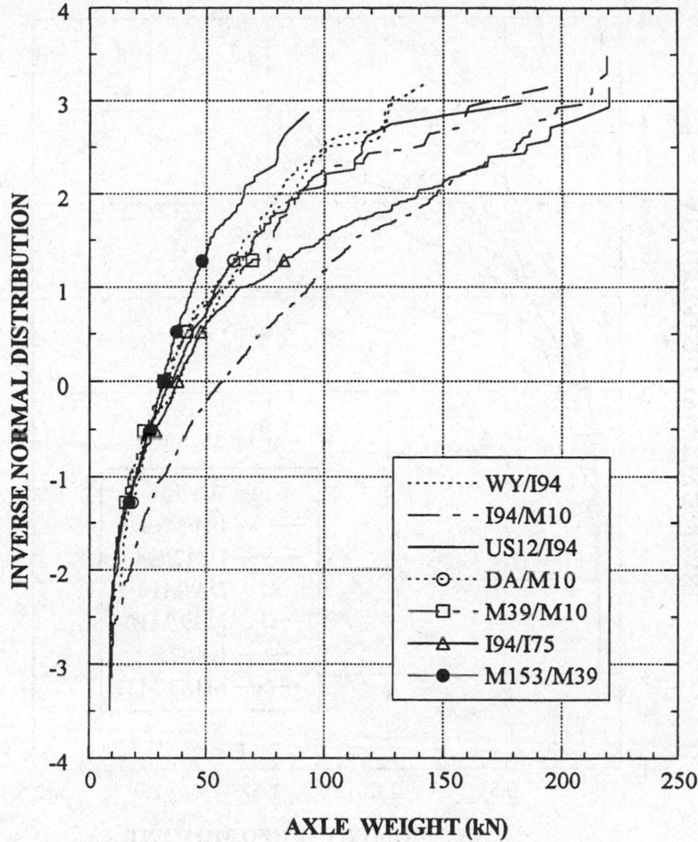

FIGURE 55.6 CDFs of axle weight for the considered bridges.

The SMS collects the strain history under normal traffic and assembles the stress cycle histogram by the rainflow method of cycle counting, and other counting methods. The data are then stored to memory and downloaded at the conclusion of the test period. The rainflow method counts the number, *n*, of cycles in each predetermined stress range, S_i, for a given stress history. The SMS is capable of recording up to 4 billion cycles per channel for extended periods in an unattended mode. Strain transducers were attached to all girders at the lower, midspan flanges of a bridge. Dynamic strain cycles were measured under normal traffic using the rainflow algorithm.

The SoMat Corporation system for its Strain Gauge Module is shown in Figure 55.9. It includes a power/processor/communication module, 1 MB CMOS extended memory unit, and eight strain gauge signal conditioning modules. The system is designed to collect strains through eight channels in both attended and unattended modes with a range of 2.1 to 12.5 mV. A second notebook computer is used to communicate with the SoMat system for commands regarding data acquisition mode, calibration, initialization, data display, and downloading of data. The SoMat system has been configured specifically for the purpose of collecting stress–strain histories and statistical analysis for highway bridges. This is possible due to the modular component arrangement of the system.

The data-acquisition system consists of five major components totaling 12 modules — eight strain transducer signal conditioning modules and four for Battery Pack, Power/Communications, 1-MB CMOS Extended Memory, and Model 2100 NSC 80180 Processor (see Figure 55.9). Regulated power is supplied by a rechargeable 11.3 to 13.4 V electrically isolated DC–DC converter. This unit powers all modules as well as provides excitation for strain transducers. Serial communications via RS 232C connector and battery backup for memory protection are provided by the Power/Communications

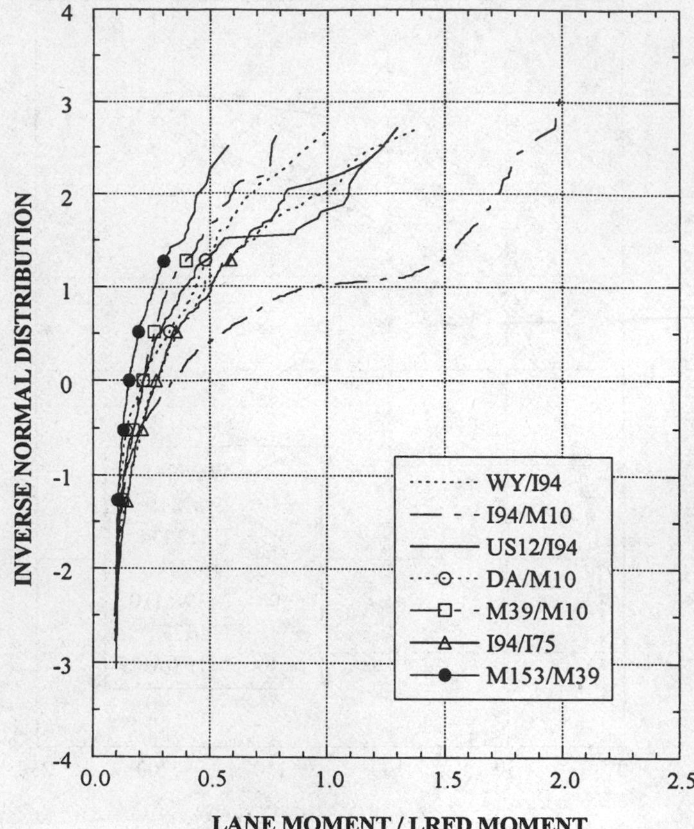

FIGURE 55.7 CDFs of lane moment for the span length of 27 m.

module. An Extended Memory Module of 1 Mbyte, high-speed, low-power CMOS RAM with backup battery for data protection is included for data storage. Eight strain gauge conditioning modules each provide 5-v strain transducer excitation, internal shunt calibration resistors, and an 8-bit, analog-to-digital converter.

Strain measurement range is ±2.1 mV minimum and ±12.5 mV maximum. The processor module consists of 32 kbytes of programmable memory and an NSC 80180 high-speed processor capable of sampling data in simultaneous mode resulting in a maximum sampling rate of 3000 Hz. Communication to the PC is via RS 232C at 57,600 baud. Data acquisition modes include time history, burst time history, sequential peak valley, time at level matrix, rainflow matrix, and peak valley matrix. Following collection, data are reviewed and downloaded to the PC hard drive for storage, processing, analysis, and plotting.

55.3.2 Rainflow Method of Cycle Counting

Development of a probabilistic fatigue load model requires collection of actual dynamic stress time histories of various members and components. Following the collection of time histories, data must be processed into a usable form. This section presents the characteristics of the dynamic stress time history commonly found in steel girder highway bridges as a random process, and rainflow method of counting fatigue damage.

Commonly occurring load histories in fatigue analysis often are categorized as either narrowband or wideband processes. Narrowband processes are characterized by an approximately constant period, such as that shown in Figure 55.10a. Wideband processes are characterized by higher

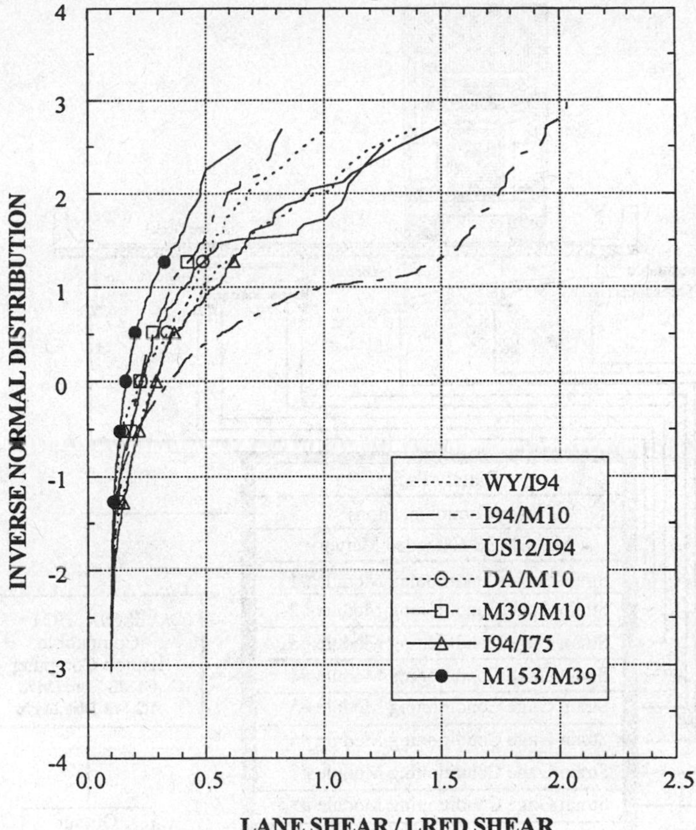

FIGURE 55.8 CDFs of lane shear for the span length of 27 m.

frequency small excursions superimposed on a lower, variable frequency process, such as that shown in Figure 55.10b. For steel girder highway bridges, where the loading is both random and dynamic, the stress histories are wideband in nature.

Stress histories that are wideband in nature do not allow for simple cycle counting. The cycles are irregular with variable frequencies and amplitudes. Several cycle-counting methods are available for the case of wideband and nonstationary processes, each successful to a degree in predicting the fatigue life of a structure. The rainflow method is preferred due to the identification of stress ranges within the variable amplitude and frequency stress histogram, which are associated with closed hysteresis loops. This is important when comparing the counted cycles with established fatigue test data obtained from constant-amplitude stress histories.

The rainflow method counts the number, n, of cycles in each predetermined stress range, S_i, for a given stress history. Rules of counting are applied to the stress history after orienting the trace vertically, positive time axis pointing downward. This convention facilitates the flow of "rain" due to gravity along the trace and is merely a device to aid in understanding of the method. Following are rules for the rainflow method (see Figure 55.11):

1. All positive peaks are evenly numbered.
2. A rainflow path is initiated at the inside of each stress peak and trough.
3. A rainflow progresses along a slope and "drips" down to the next slope.
4. A rainflow is permitted to continue unless the flow was initiated at a minimum more negative than the minimum opposite the flow and similarly for a rainflow initiated at a maximum. For example, path 1–8, 9–10, 2–3, 4–5, and 6–7.

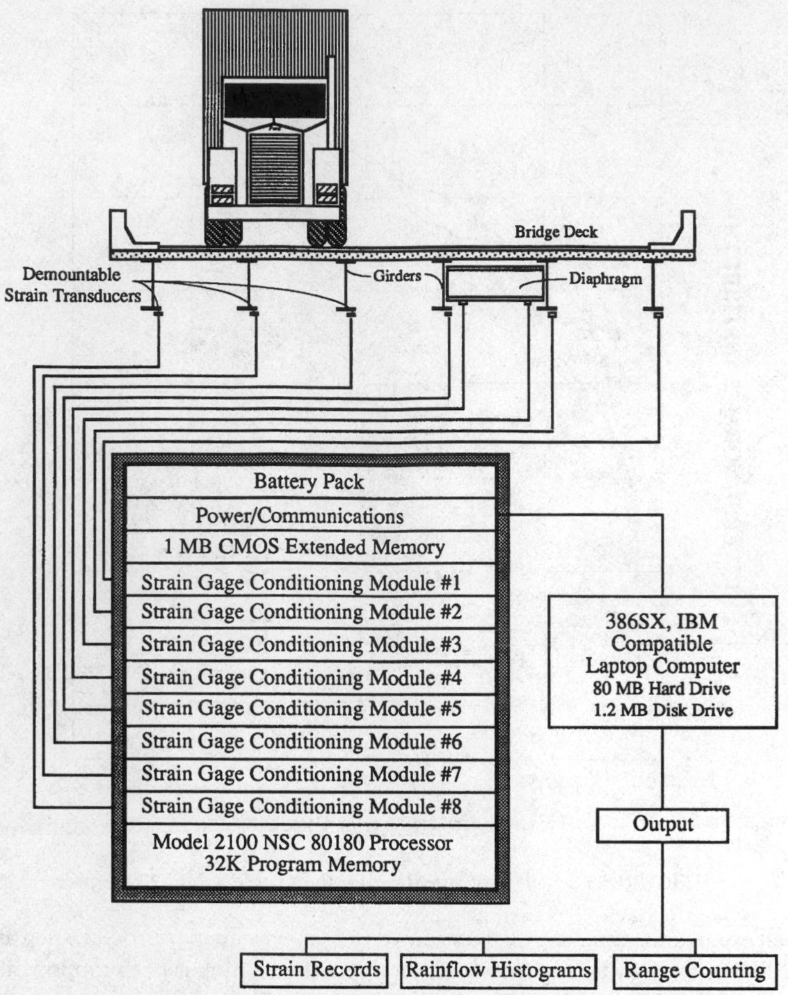

FIGURE 55.9 Somat strain data acquisition system.

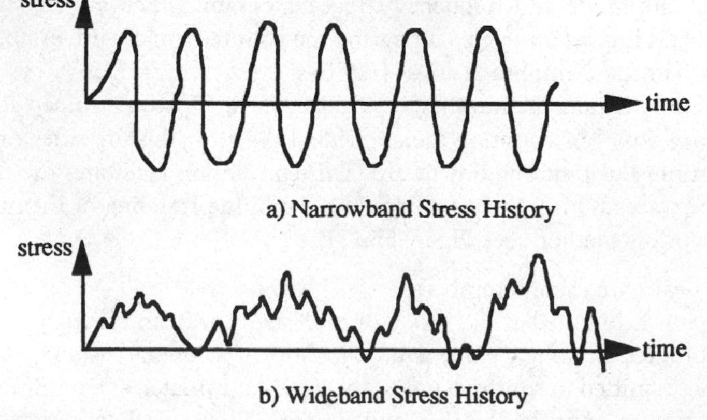

FIGURE 55.10 Example of narrowband (a) and wideband (b) stress histories.

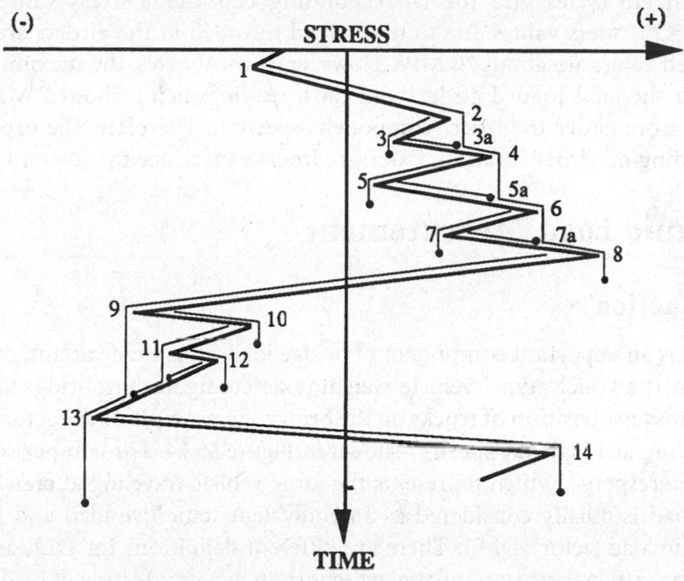

FIGURE 55.11 Rainflow counting diagram.

5. A rainflow must stop if it meets another flow that flows from above. For example, path 3–3a, 5–5a, and 7–7a.
6. A rainflow is not initiated until the preceding flow has stopped.

Following the above procedure each segment of the history is counted only once. Half cycles are counted between the most negative minimum and positive maximum, as well as the half cycles or interruptions between the maximum and minimum. As shown in Figure 55.11, all negative trough-initiated half cycles will eventually be paired with a peak initiated cycle of equal magnitude. For a more-detailed explanation and discussion of the rainflow method and others see an introductory text on fatigue analysis [9].

55.3.3 Results of Strain Spectra Testing

Strain histories were collected continuously for 1-week periods and reduced using the rainflow algorithm [5,6]. Data were collected for each girder in the bridge. The data are presented here in the form of CDFs and represent strain cycles due to 7 days of normal traffic. For an easier interpretation of results, the CDFs are plotted on normal probability paper (Figure 55.12).

For each bridge, the CDFs are shown for strains in girders numbered from 1 (exterior, on the right-hand side looking in the direction of the traffic). The number of girders for measured bridges varies from 6 to 10. The average strain is less than 50×10^{-6} for all girders and all bridges; however, the largest strains were observed in girders supporting the right traffic lane (girder numbers G3, G4, and G5) and nearest the left wheel of traffic in the right lane. As expected, the exterior girders of each bridge experience the lowest strain extremes in the spectrum.

As a means of comparison of fatigue live load, the equivalent stress, s_{eq}, is calculated for each girder using the following root mean cube (RMC) formula:

$$s_{eq} = \sqrt[3]{p_i S_i^3} \tag{55.1}$$

where S_i = midpoint of the stress interval i and p_i = the relative frequency of cycle counts for interval i. The stress, S_i, is calculated as a product of strain and modulus of elasticity of steel.

The CDFs of strain cycles and the corresponding equivalent stress values are shown in Figures 55.12 to 55.23. Stress values due to traffic load recorded in the girders are rather low. The maximum observed values are about 70 MPa. However, in most cases, the maximum values do not exceed 35 MPa for the most loaded girder, with most readings being about 7 MPa. Stress spectra considerably vary from girder to girder (component-specific). Therefore, the expected fatigue life is different depending on girder location. Exterior girders experience the lowest load spectra.

55.4 Dynamic Load Measurement

55.4.1 Introduction

The dynamic load is an important component of bridge loads. It is time variant, random in nature, and it depends on the vehicle type, vehicle weight, axle configuration, bridge span length, road roughness, and transverse position of trucks on the bridge. An example of the actual bridge response for a vehicle traveling at a highway speed is shown in Figure 55.24. For comparison, also shown is an equivalent static response, which represents the same vehicle traveling at crawling speed.

The dynamic load is usually considered as an equivalent static live load and it is expressed in terms of a dynamic load factor (DLF). There are different definitions for DLF, as summarized by Bakht and Pinjarkar [10] in their state-of-the-art report on dynamic testing of bridges. In this study, DLF is taken as the ratio of dynamic and static responses [7,11]:

$$\text{DLF} = D_{\text{dyn}}/D_{\text{stat}} \tag{55.2}$$

where D_{dyn} = the maximum dynamic response (e.g., stress, strain, or deflection) measured from the test data, $D_{\text{dyn}} = D_{\text{total}} - D_{\text{stat}}$; D_{total} = total response; and D_{stat} = the maximum static response obtained from the filtered dynamic response.

55.4.2 Measured Dynamic Load

Field measurements are performed to determine the actual truck load effects and to verify the available analytical models [7,11,12]. The tests are carried out on steel girder bridges. Measurements are taken using a WIM system with strain transducers. For each truck passage, the dynamic response is monitored by recording strain data. The truck weight, speed, axle configuration, and lane occupancy are also determined and recorded. A numerical procedure is developed to filter and process collected data. The DLF is determined under normal truck traffic for various load ranges and axle configurations.

An example of the actual static and dynamic stresses is shown in Figure 55.25. CDF of the static stress is plotted on the normal probability paper. For each value of static stress, the corresponding dynamic stress is also shown. The stress due to dynamic load is nearly constant and is not dependent on truck weight. Figure 55.26 shows DLFs as a function of static strains. Also shown in the figure is the power curve fit, which approximately represents mean values of DLFs. In general, the DLF decreases as the static strain increases. Therefore, the DLF is reduced for heavier trucks.

55.5 Summary

On the basis of WIM measurements, site-specific load spectra are presented for several bridge sites. Component stress spectra are presented for girders in the form of CDFs and the equivalent stresses are calculated for comparison. Site-specific statistics required for fatigue analysis based on current models are presented. Strains are measured to determine component-specific load spectra.

The truck load spectra for bridges are strongly site specific. Bridges located on major routes between large industrial metropolitan areas will experience the highest extreme loads. Routes where

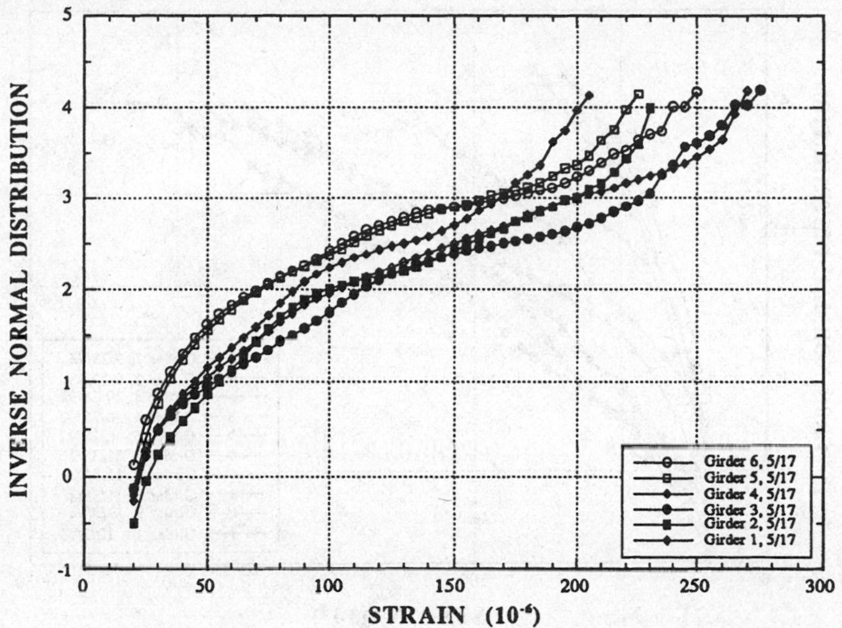

FIGURE 55.12 Strains for girders in bridge U.S. 23/HR.

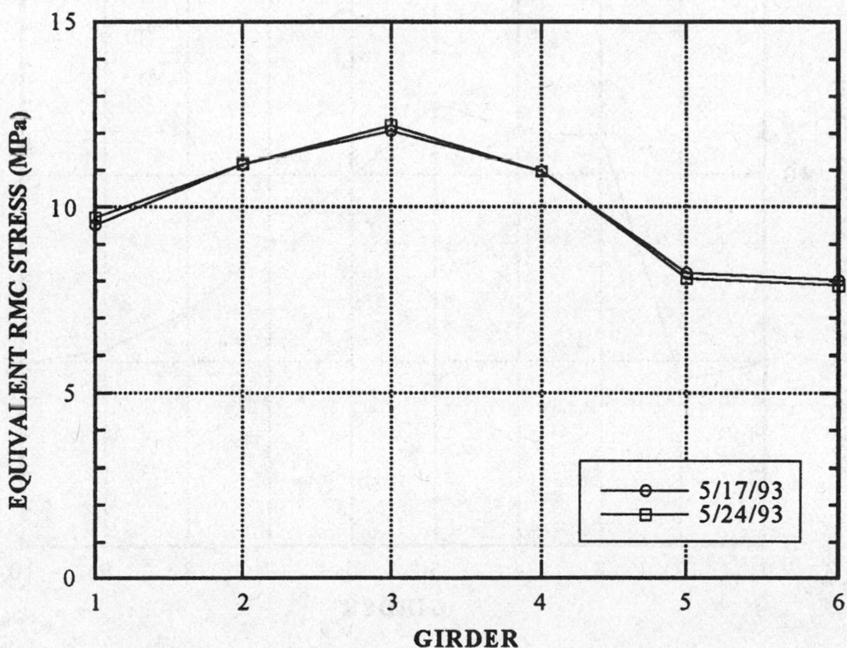

FIGURE 55.13 Equivalent stresses for girders in bridge U.S. 23/HR.

vehicles are able to circumvent stationary weigh stations will have very high extreme loads. Bridges not on a major route and those that are very near a weigh station experience lower extreme loads.

Live-load stress spectra are strongly component specific. Each component may have a different distribution of strain cycle range. The girder that is nearest the left wheel track of vehicles traveling

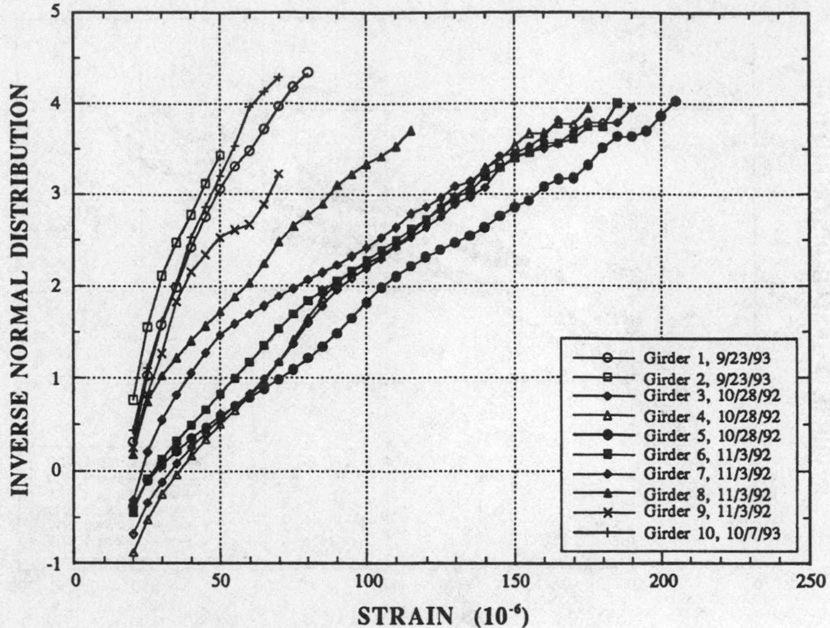

FIGURE 55.14 Strains for girders in bridge U.S. 23/SR.

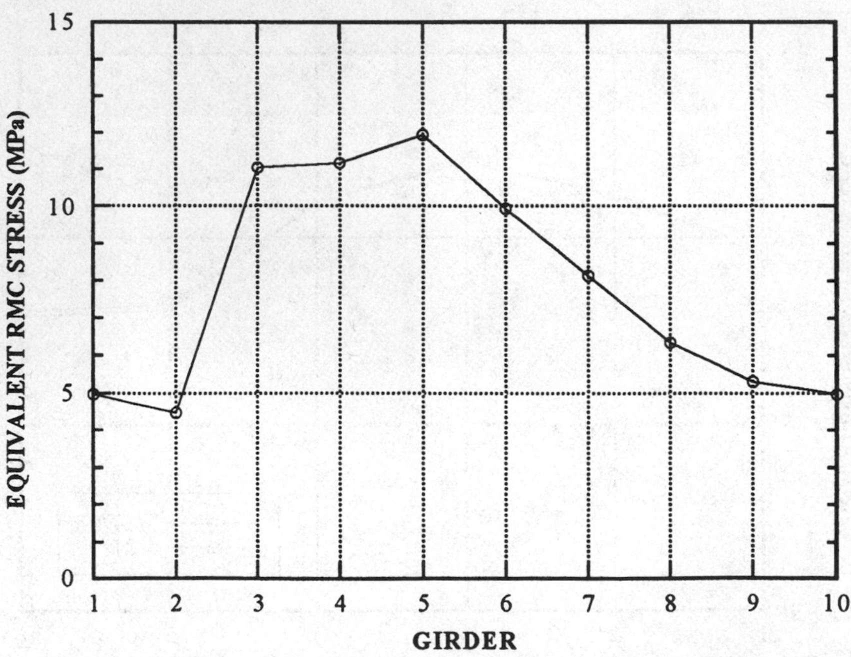

FIGURE 55.15 Equivalent stresses for girders in bridge U.S. 23/SR.

in the right lane experiences the highest stresses in the stress spectra. The stresses decrease as a function of the distance from this location. This information can be useful to target bridge inspection efforts to the critical members.

The stress due to dynamic load is nearly constant and it is not dependent on truck weight. In general, DLF decreases as the static strain increases. Therefore, the DLF is reduced for heavier trucks.

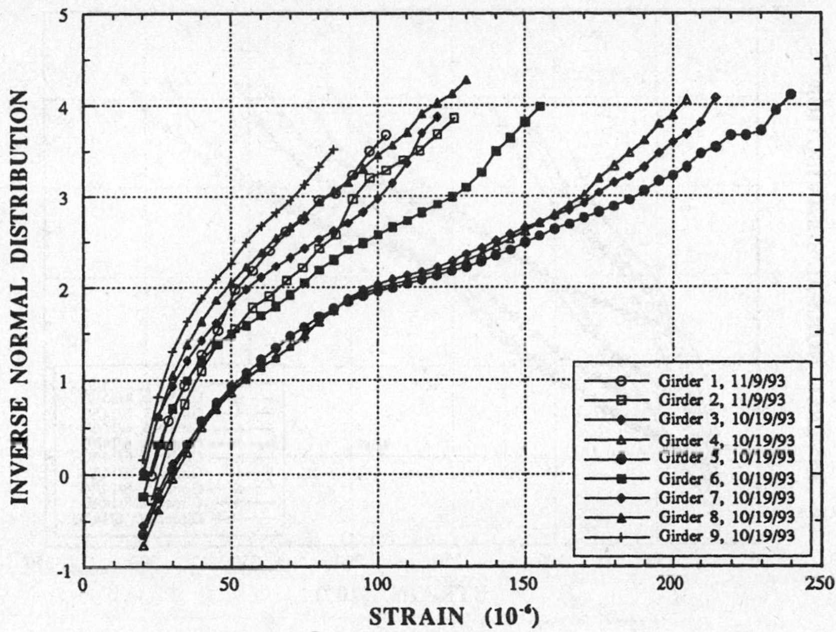

FIGURE 55.16 Strains for girders in bridge I-94/JR.

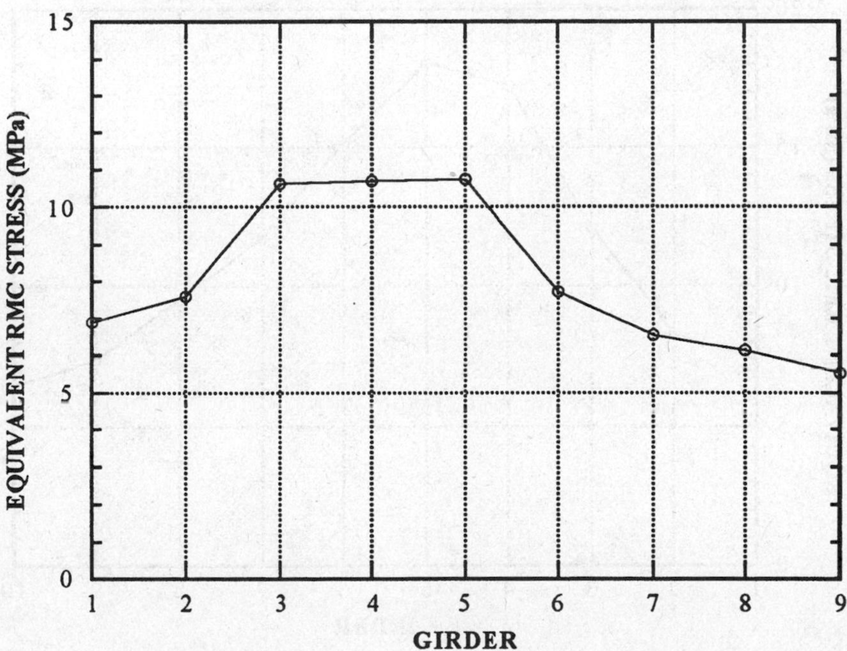

FIGURE 55.17 Equivalent stresses for girders in bridge I-94/JR.

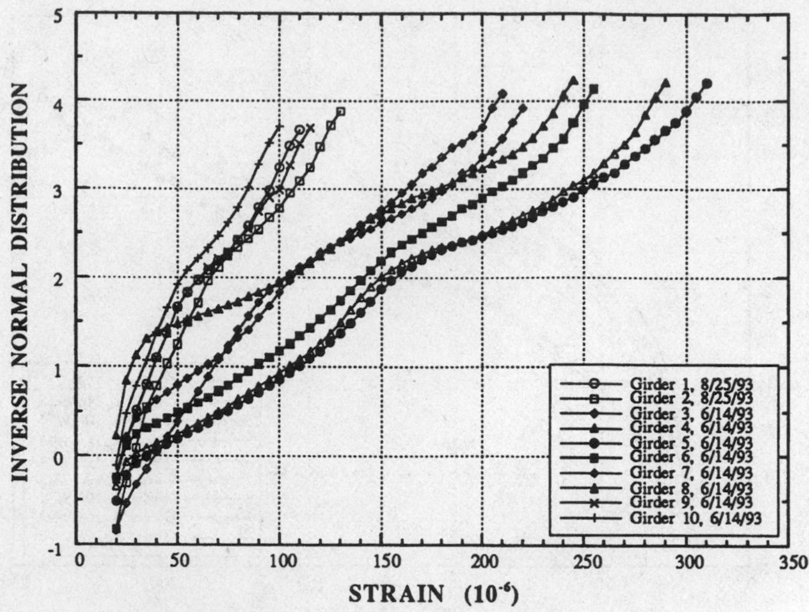

FIGURE 55.18 Strains for girders in bridge 1-94/PR.

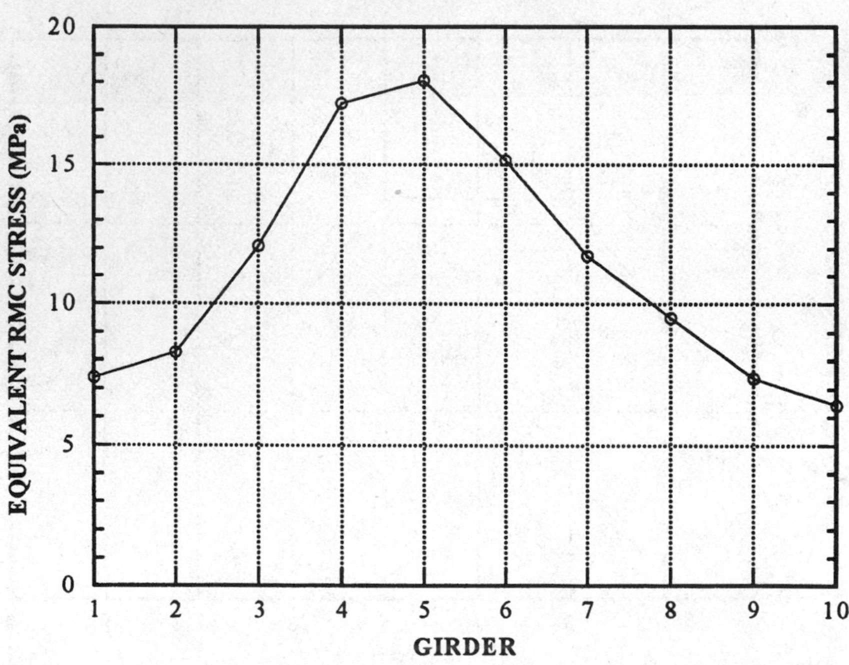

FIGURE 55.19 Equivalent stresses for girders in bridge I-94/PR.

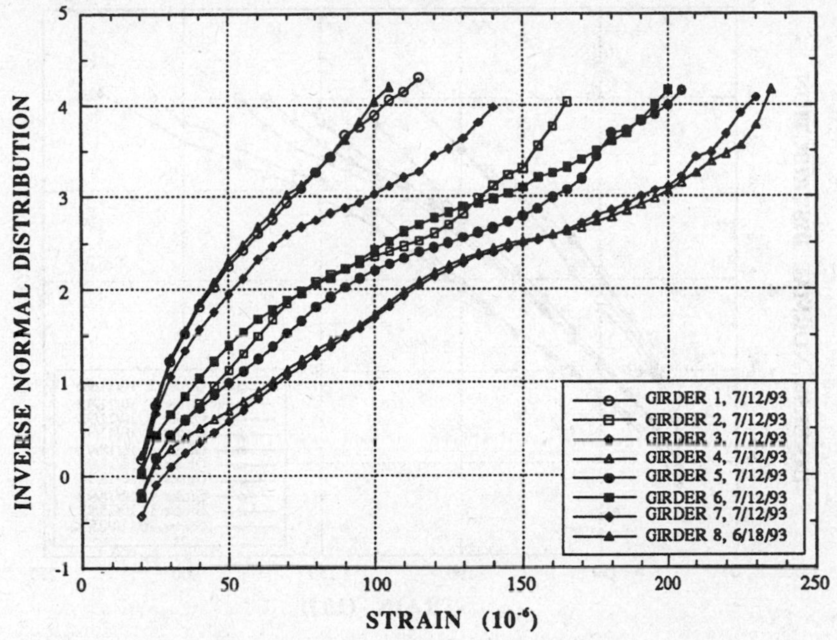

FIGURE 55.20 Strains for girders in bridge M-14/NY.

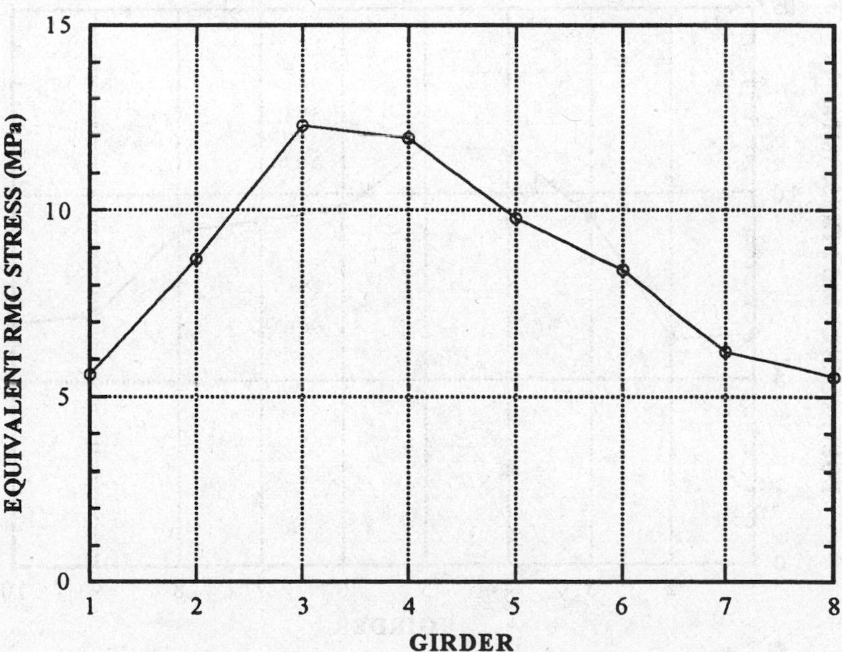

FIGURE 55.21 Equivalent stresses for girders in bridge M-14/NY.

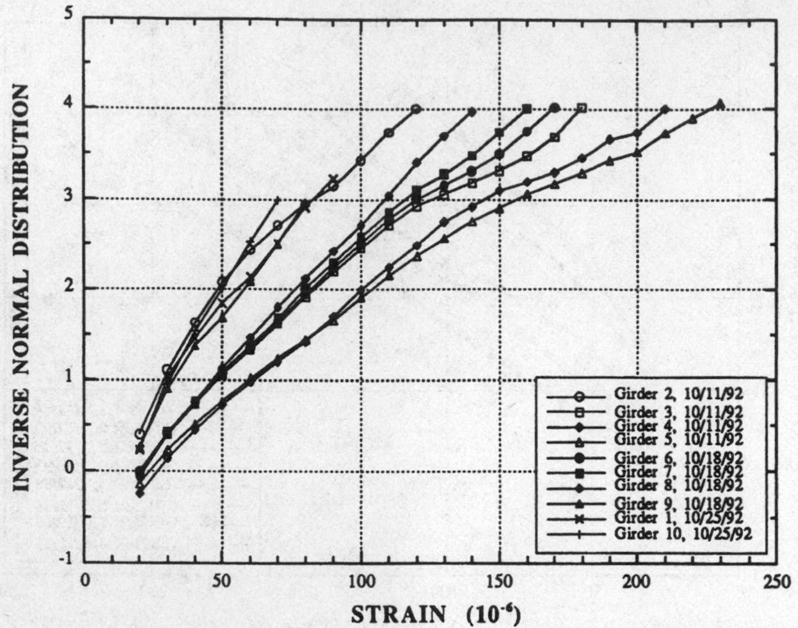

FIGURE 55.22 Strains for girders in bridge I-75/BC.

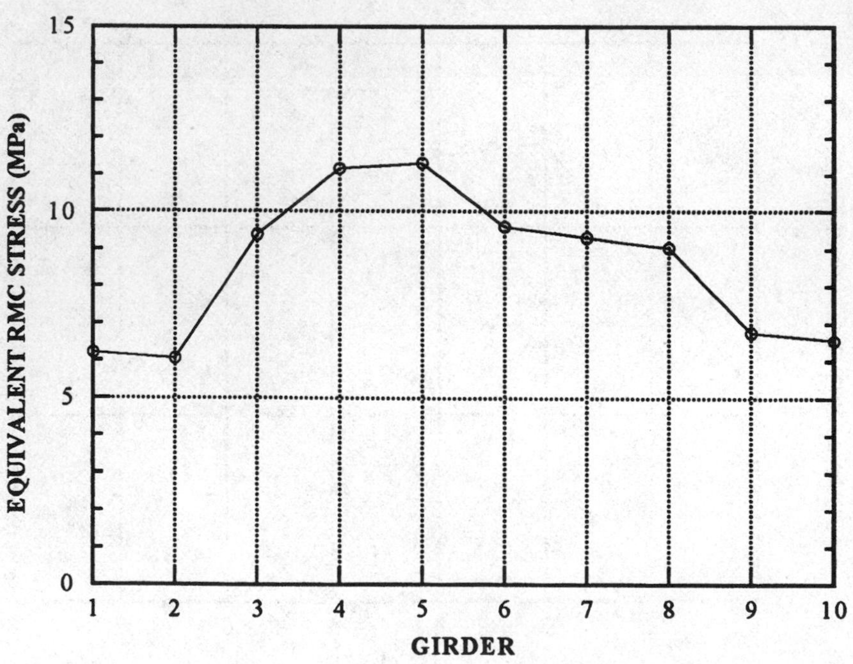

FIGURE 55.23 Equivalent stresses for girders in bridge I-75/BC.

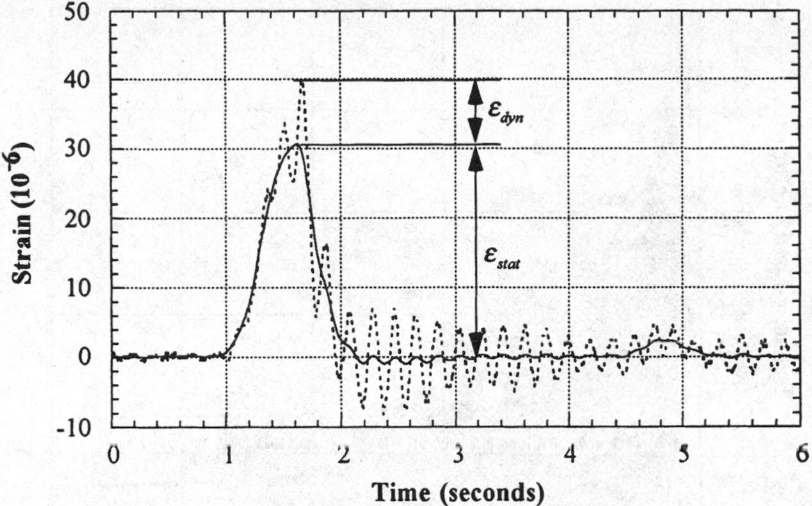

FIGURE 55.24 Dynamic and static strain under a truck traveling at highway speed.

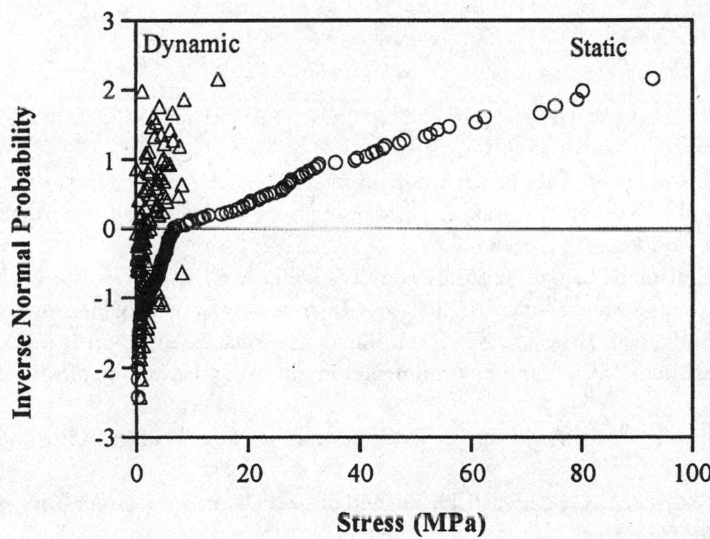

FIGURE 55.25 Typical CDF of static stress and corresponding dynamic stress.

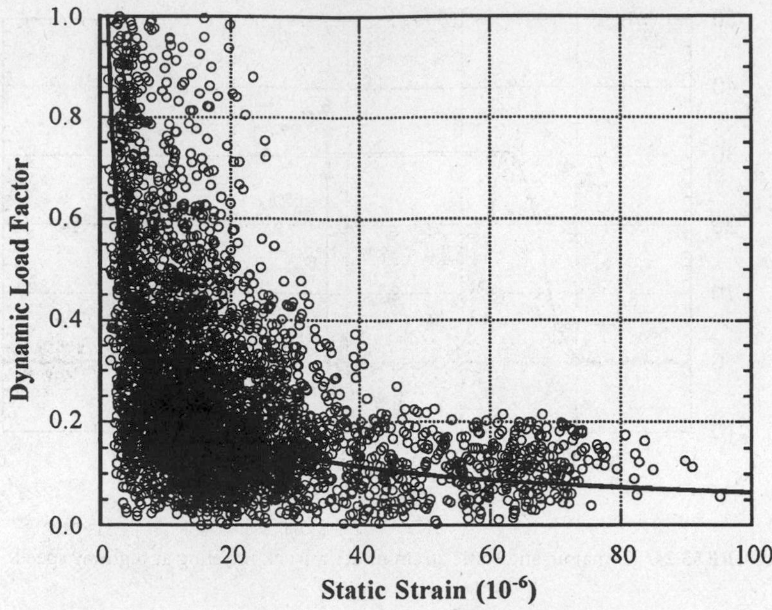

FIGURE 55.26 DLF vs. Static strain.

References

1. AASHTO, *LRFD Bridge Design Specifications*, 2nd ed., American Association of State and Transportation Officials, Washington, D.C, 1998.
2. Nowak, A. S. and Hong, Y-K., Bridge live-load models, *ASCE J. Struct. Eng.*, 117, 2757, 1991.
3. Kim, S., Sokolik, A. F., and Nowak, A. S., Measurement of truck load on bridges in the Detroit area, *Transp. Res. Rec.*, 1541, 58, 1996.
4. Nowak, A. S., Kim, S., Laman, J., Saraf, V., and Sokolik, A. F., Truck Loads on Selected Bridges in the Detroit Area, Research Report UMCE 94-34, University of Michigan, Ann Arbor, 1994.
5. Nowak, A. S., Laman, J. A., and Nassif, H., Effect of Truck Loading on Bridges, Report UMCE 94-22, Department of Civil and Environmental Engineering, University of Michigan, Ann Arbor, 1994.
6. Laman, J. A. and Nowak, A. S., Fatigue-load models for girder bridges, *ASCE J. Struct. Eng.*, 122, 726, 1996.
7. Kim, S. and Nowak, A. S., Load distribution and impact factors for I-girder bridges, *ASCE J. Bridge Eng.*, 2, 97, 1997.
8. Benjamin, J. R. and Cornell, C. A., *Probability, Statistics and Decision for Civil Engineers*, McGraw-Hill, New York, 1970.
9. Bannantine, J. A., Comer, J. J., and Handrock, J. L., *Fundamentals of Metal Fatigue Analysis*, Prentice-Hall, Englewood Cliffs, NJ, 1992.
10. Bakht, B. and Pinjarkar, S. G., Dynamic testing of highway bridges — a review, *Transp. Res. Rec.*, 1223, 93, 1989.
11. Nassif, H. and Nowak, A. S., Dynamic load spectra for girder bridges, *Transp. Res. Rec.*, 1476, 69, 1995.
12. Hwang, E. S. and Nowak, A. S., Simulation of dynamic load for bridges, *ASCE J. Struct. Eng.*, 117, 1413, 1991.

56

Impact Effect of Moving Vehicles

Mingzhu Duan
Quincy Engineering, Inc.

Philip C. Perdikaris
Case Western Reserve University

Wai-Fah Chen
Purdue University

56.1 Introduction

Vehicles such as trucks and trains passing bridges at certain speeds will cause dynamic effects, among them global vibration and local hammer effects. The dynamic loads for moving vehicles are considered "impact" in bridge engineering because of the relatively short duration. The magnitude of the dynamic response depends on the bridge span, stiffness and surface roughness, and vehicle dynamic characteristics such as moving speed and isolation system. Unlike earthquake loads which can cause vibration in bridge longitudinal, transverse, and vertical directions, moving vehicles mainly excite vertical vibration of the bridge. Impact effect has influence primarily on the superstructure and some of substructure members above the ground because the energy will be dissipated effectively in members underground by the bearing soils.

Although the interaction between moving vehicles and bridges is rather complex, the dynamic effects of moving vehicles on bridges are accounted for by a dynamic load allowance, *IM*, in addition to static live load (*LL*) in the current bridge design specifications [1–3]. According to the American Association of State Highway and Transportation Officials (AASHTO) and the American Railway Engineering Association (AREA) specifications,

$$IM = \frac{D_{\text{dyn}}}{D_{\text{st}}} - 1 \tag{56.1}$$

where D_{dyn} is the maximum dynamic response for deflection, moment, or shear of the structural members and D_{st} is the corresponding maximum static response. The total live-load effect, *LL*, can then be expressed as

$$LL = AF \times D_{\text{st}} \tag{56.2}$$

0-8493-7434-0/00/$0.00+$.50

and

$$AF = 1 + IM \tag{56.3}$$

where AF is the amplification factor representing the dynamic amplification of the static load effect and IM is the impact factor determined by an empirical formula in design code. No dynamic analysis is thus required in the design practice.

Most early research work on the dynamic bridge behavior of bridges under moving vehicles focused on an analytical approach modeling a bridge as a simply supported beam [5] or a simply supported plate [8] under constant or pulsating moving loads (moving load model). The dynamic effects under different speeds of the moving loads and different damping ratios were studied. It was found that the speed of vehicles and the fundamental period of the bridge dominate the dynamic behavior of the bridge. Since most bridges consist of both beams and plates such as girder deck bridges, the above simplified model has limited validity. Along with analytical study, numerical methods such as finite-element analysis and the finite-difference method have been used recently in studying the dynamic response of a vehicle–bridge system [10,21]. Two sets of equations of motion were developed for the bridge and vehicle, respectively. These equations are coupled at the contacting points between bridge and vehicle and the contact points are time and space dependent due to vehicles moving along a rough surface. An iteration procedure should be used to solve the coupled equations. Field measurements are another alternative to investigate the dynamic effect [13] which disclosed the range of live-load effect for steel I-girder bridges under truck load.

Based on analytical analysis and field measurement studies, major characteristics of the bridge dynamic response under moving vehicles can be summarized as follows:

1. Measured impact factors [13], IM, on highway bridges vary significantly, e.g., with the mean of about 0.12 and the standard deviation of about 0.05 for steel I-girder bridges. The measured impact factors are well below those of the AASHTO specifications.
2. Impact factor increases as vehicle speed increases in most cases.
3. Impact factor decreases as bridge span increases.
4. Under the conditions of "very good" road surface roughness (amplitude of highway profile curve is less than 1 cm), the impact factor is well below that in design specifications. But the impact factor increases tremendously with increasing road surface roughness from "good" to "poor" (the amplitude of highway profile curve is more than 4 cm) and can be well beyond the impact factor in design specifications.
5. Impact factor decreases as vehicles travel in more than one lane. The chance of maximum dynamic response occurring at the same time for all vehicles is small.
6. Impact factor for exterior girders is much larger than for interior girders because the excited torsion mode shapes contribute to the dynamic response of exterior girders.
7. The first mode shape of the bridge is dominant in most cases, especially for the dynamic effect in the interior girder in single-span bridges.

The impact factor, IM, is a well-accepted measurement for the dynamic effect of bridges under moving vehicles and is used in design specifications worldwide. Consideration of impact effect for highway and railway bridges in design practice will be introduced through examples in Sections 56.2 and 56.3, respectively. Free vibration and forced vibration by a moving vehicle will be introduced in Sections 56.4 and 56.5 to disclose the dynamic behavior of the bridge vibration.

56.2 Consideration of Impact Effect in Highway Bridge Design

Since the impact effect on bridges by moving vehicles is influenced by factors such as bridge span, stiffness, surface roughness, and speed and suspension system of moving vehicles, the impact factor varies within a large range. While the actual modeling of this effect is complex, the calculation of impact effect is greatly simplified in the bridge design practice by avoiding any analysis of vehicle-induced vibration. In general, the dynamic effect is contributed by two sources: (1) local hammer effect by the vehicle wheel assembly riding surface discontinuities such as deck joints, cracks, delaminations, and potholes; and (2) global vibration caused by vehicles moving on long undulations in the roadway pavement, such as those caused by settlement of fill, or by resonant excitation of the bridge. The first source has a local impact effect on bridge joints and expansions. On the other hand, the second source will have influence on most of the superstructure members and some of the substructure members. A variety of considerations and design formulas are proposed worldwide for the second source, which means that the bridge community has not reached a consensus on this issue [4]. The differences among various code specifications worldwide for the dynamic amplification factor vs. the fundamental frequency are large. A large dynamic effect is considered in some countries for the bridge frequency ranging from 1.0 to 5.0 Hz, which is the frequency range of the fundamental frequencies of most truck suspension systems. It is largely an attempt to penalize the bridge designed within this frequency range. But the accurate evaluation of the first frequency of a bridge can be hardly performed in the design stage.

In the AASHTO *Standard Specifications of Highway Bridges* (1996) in the United States, the impact factor due to bridge vibration for members in Group A including superstructure, piers, and those portions of concrete and steel piles above the ground that support the superstructure, is simply expressed as a function of bridge span:

$$IM = \frac{50}{L+125} \leq 0.30 \tag{56.4}$$

where L (in ft) is the length of span loaded to create maximum stress.

In the AASHTO *LRFD Bridge Design Specifications* (1994), the static effects of the design truck or tandem shall be increased by the impact effect in the percentage specified in Table 56.1.

In Table 56.1, 75% of the impact effect is considered for deck joints for all limit states due to local hammer effect, 15% for fatigue and fracture limit states for members vulnerable to cyclic loading such as shear connectors and welding members, and 33% for all other members influenced by global vibration. Field tests indicate that in the majority of highway bridges, the dynamic component of the response does not exceed 25% of the static response to vehicles. Since the specified live-load combination of the design truck and lane load represents a group of exclusion vehicles which are at least 4/3 of those caused by the design truck alone on short- and medium-span bridges, the specified value of 33% in Table 56.1 is the product of 4/3 and the basic 25%. The impact effect is not considered for retaining walls not subjected to vertical reactions from superstructure, and for underground foundation components due to the damping effect of soil.

The dynamic effect caused by vehicles is accounted for in bridge design practice in the live-load calculation, as shown in the following example.

Example 56.1 — Live-Load Calculation in Highway Bridge Design

Given

The bridge shown in Figure 56.2 is a steel girder-concrete deck bridge with a span of 50 ft (15.24 m). The bridge carrying two lanes of traffic loads consists of four girders with girder-to-girder spacing of 7 ft (2.13 m). The steel girders are W36 × 150 and the average concrete deck thickness is 9 in. (3.54 m). The bridge is designed to carry a specified truck of HS20-44.

Solution 1

Using AASHTO *Standard Specifications of Highway Bridges* (1996) to calculate the design live-load moment for the interior steel girder under design truck HS20-44.

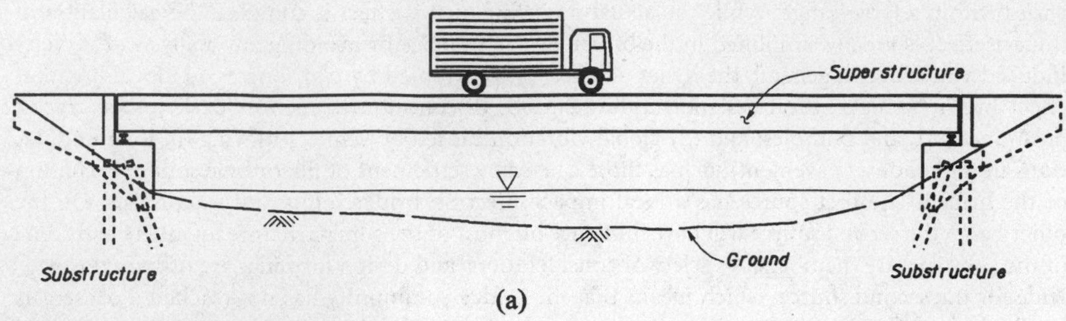

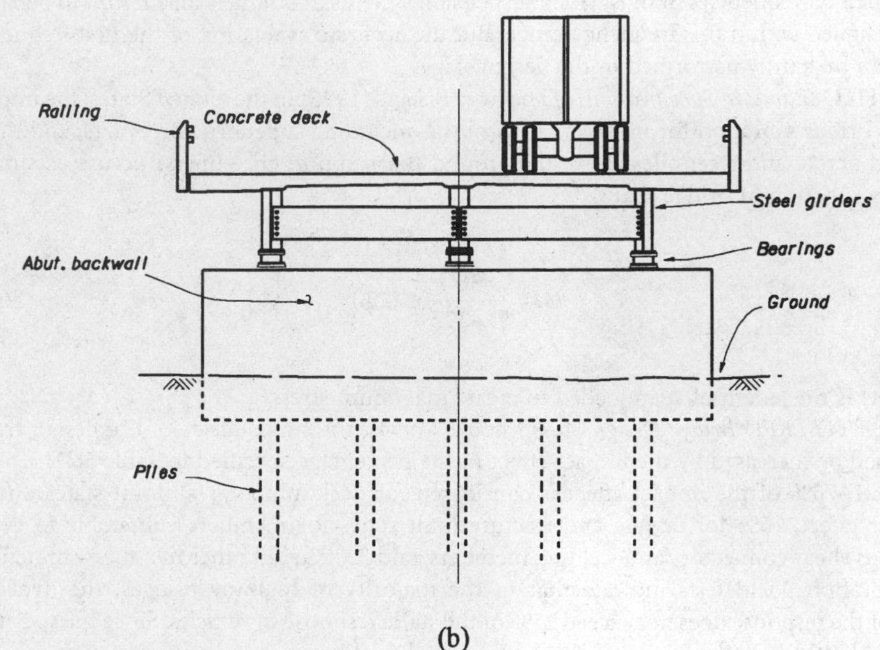

FIGURE 56.1 (a) Moving vehicle on a bridge. (b) Section view of the bridge and the vehicle.

TABLE 56.1 Dynamic Load Allowance, *IM*

Component	*IM*
Deck joint — All limit states	75%
All other components:	
• Fatigue and fracture limit state	15%
• All other limit states	33%

1. **Calculate the Static Design Moment in the Interior Stringer under Truck HS20-44**
   a. *Compute the static maximum truck load moment in a lane at the critical section of the superstructure induced by truck loading.*
   There are two types of vehicle live loading: truck and lane loading. For bending moments caused by HS20-44 loading for a span length of up to 140 ft, truck loading will govern.

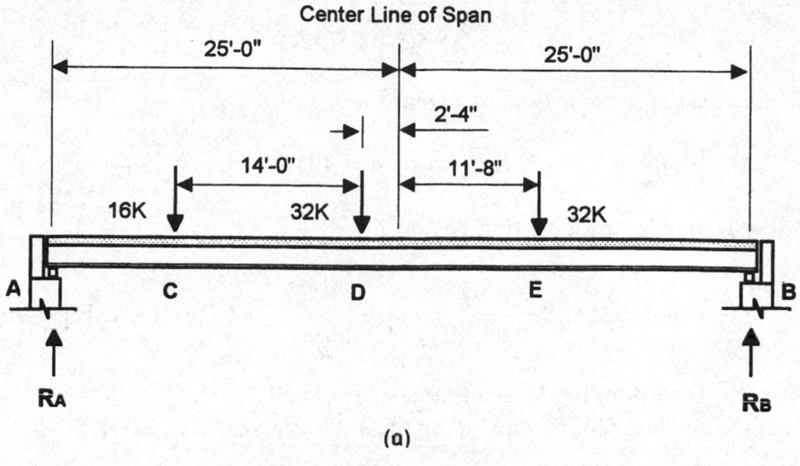

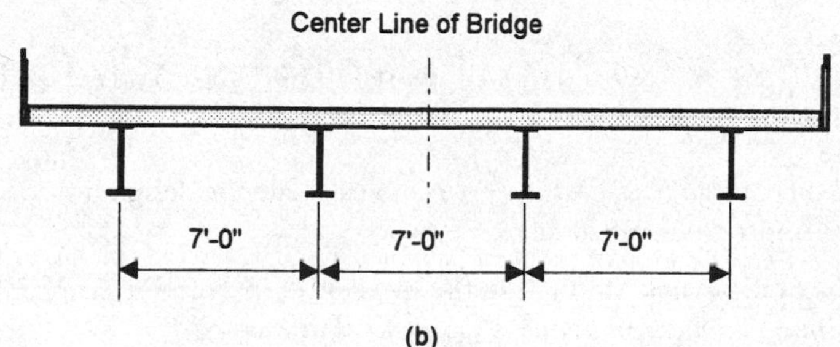

FIGURE 56.2 (a) Elevation of the steel girder bridge under truck load HS20-44. (b) Section of the steel girder bridge.

The maximum live-load moment is induced at the most adverse truck position. In fact, the maximum moment occurs at the position of the second concentrated load when the centerline of the span is midway between the center of gravity of loads and the second concentrated load as shown in Figure 56.2a. The reaction R_A at the end A is determined by using the equilibrium equation of the sum of the moments about point B:

$$\sum M_B = 0$$

Thus,

$$50\ R_A - 8\ \times\ 41.33 - 32\ \times\ 27.33 - 32\ \times\ 13.33 = 0,\ \text{and}$$

$$R_A = \frac{1632}{50} = 32.64\ \text{kip}\ (145.25\ \text{kN})$$

The maximum live-load moment at Point D (see Figure 56.2) is

$$M_{\max} = 22.67\ R_A\ - 8\ \times\ 4 = 627.95\ \text{kips-ft}\ (851.50\ \text{kN-m})$$

b. *Compute the axle load distribution factor, DF, for the stringers*:
For the two lane concrete floor with steel I-beam girders spacing at 7 ft (2.13 m), the distribution factor, *DF*, according to the AASHTO-96 is

$$DF = \frac{S}{2\times 5.5} = \frac{7}{2\times 5.5} = 0.64$$

c. *Calculate the design static bending moment in a stringer:*

$$M_{LL} = DF(M_{\max}) = 627.95 \times 0.64 = 401.89 \text{ kips-ft } (544.96 \text{ kN-m})$$

2. Determine Dynamic Amplification Factor, *AF*

$$AF = 1.0 + IM = 1.0 + \frac{50}{L+125} = 1.0 + \frac{50}{50+125} = 1.29 \quad \text{(AASHTO-96)}$$

3. Calculate the Total Bending Moment under Live Load

The total live bending moment including the amplification factor is

$$M_{LL+IM} = M_{LL}\,AF$$

$$= 401.89 \times 1.29 = 518.44 \text{ kips-ft } (703.00 \text{ kN-m}) \quad \text{(AASHTO-96)}$$

Solution 2

Using AASHTO-LRFD *Design Specifications* (1994) to calculate the design live-load moment for an interior steel girder under truck HS20-44.

1. Calculate the Static Truck Moment in the Interior Girder under Truck HS20-44

a. *The maximum moment in a lane is the same as in Method 1:*

$$M_{\max} = 627.95 \text{ kips-ft} \quad (851.50 \text{ kN-m})$$

b. *Compute the moment load distribution factor, DF, for the steel girders:*

For the two-lane concrete floor with steel I-girders spacing at 7 ft (2.13 m), the distribution factor, *DF*, is

$$DF = 0.075 + \left(\frac{S}{9.5}\right)^{0.6}\left(\frac{S}{L}\right)^{0.2}\left(\frac{K_g}{12.0\,Lt_s^3}\right)^{0.1} = 0.075 + \left(\frac{7.0}{9.5}\right)^{0.6}\left(\frac{7.0}{L}\right)^{0.2} \times 1.0 = 0.63$$

(for preliminary design use)

c. *Calculate the design static bending moment in a stringer:*

$$M_{LL} = DF(M_{\max}) = 627.95 \times 0.63 = 395.64 \text{ kips-ft } (538.07 \text{ kN-m})$$

2. Determine Dynamic Amplification Factor, *AF*

$$AF = 1.0 + IM = 1.0 + 0.33 = 1.33 \quad \text{(AASHTO-94)}$$

3. Calculate the Total Bending Moment in an Interior Stringer under Truck Load

The total bending moment under truck load including the amplification factor is

$$M_{LL+IM}\text{ (truck)} = M_{LL}\,AF = 395.64 \times 1.33 = 526.20 \text{ kips-ft } (715.64 \text{ kN-m}) \quad \text{(AASHTO-94)}$$

4. The vehicular live-load moment is the combination of the design truck load with impact allowance and design lane load without impact allowance (0.64 kip/ft over 10.0 ft per lane, 0.88 kN/m over 3.0 m):

$$M_{LL+IM}\text{ (total)} = 526.20 + 0.63 \times (0.64 \times 50/2 \times 22.67 - 1/2 \times 0.64 \times 22.67^2)$$
$$= 651.10 \text{ kips-ft (885.50 kN-m)}$$

56.3 Consideration of Impact Effect in Railway Bridge Design

For railway bridges the ratio of live load caused by moving vehicles such as locomotive and trains to the dead load is mostly higher than that in highway bridges. Similarly, as in highway bridge design, static live-load effect by vehicles should be increased by the impact factor to account for the dynamic amplification effect. The most important sources of bridge impact are

1. Initial vehicle bounce and roll,
2. Vehicle speed, and
3. Bridge dynamic properties and track-surface roughness.

In the AREA specifications [3], the impact loads specified are based on investigations and tests of railroad bridges in service under passage of locomotive and train loads. The vibration for the bridge is the most dominant dynamic effect in railway bridge design. In the vibration, the vertical vibration effect will be coupled with the rocking effect (*RE*) caused by vehicle pitch movement (transverse vehicle rotation). Thus, a couple with 10% of axle load acting down on one rail and up on the other rail should be added into the vertical impact effect. The rocking effect, *RE*, should be expressed as a percentage; either 10% of the axle load or 20% of the wheel load. The total impact effect can be calculated as

1. Percentage of live load for rolling equipment without hammer blow, such as diesels and electric locomotives, etc.,

$$\begin{aligned} IM &= RE + 40 - \frac{3L^2}{1600} \quad \text{if } L < 50 \text{ ft (15.24 m)} \\ IM &= RE + 16 - \frac{600}{L-30} \quad \text{if } L \geq 80 \text{ ft (24.39 m)} \end{aligned} \tag{56.5}$$

2. Percentage of live load for steam locomotives with hammer blow:
   a. For beam spans, stringers, girders, floor beams, parts of deck truss span carrying load from floor beam only:

$$\begin{aligned} IM &= RE + 40 - \frac{L^2}{500} \quad \text{if } L < 100 \text{ ft (30.48 m)} \\ IM &= RE + 10 - \frac{1600}{L-40} \quad \text{if } L \geq 100 \text{ ft (30.48 m)} \end{aligned} \tag{56.6}$$

   b. For truss spans:

$$IM = RE + 10 - \frac{4000}{L+25} \tag{56.7}$$

where L is the effective span length (ft).

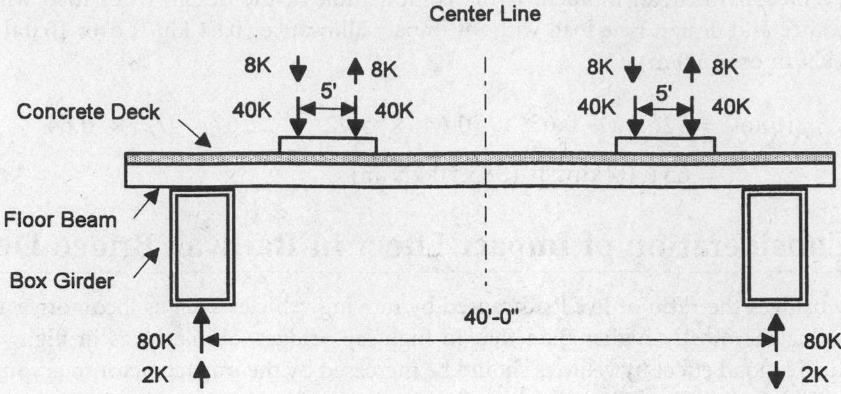

FIGURE 56.3 Rocking effect model for railway bridge.

Tests have shown that the impact load on ballasted deck bridges can be reduced to 90% of that specified for open-deck bridges because of the damping effect which results from a ballasted deck bridge. It was found from a parametric study [9] that the impact ranges from 24.9 to 26.0%, with an average of 25.6%, except for hangers. The following example shows how to account for impact effect in live-load calculation in railway bridge design.

Example 56.2 — Live Load Calculation in Railway Bridge Design

Given

A ballasted 105-ft-long (32.01-m-long) deck bridge is supported by two box girders. The box girder spacing is 40 ft (12.20 m) to carry two tracks, as shown in Figure 56.3. Live load is Cooper E-80 without hammer blow. The maximum bending moment M_{LL} at midspan under static live load is 15,585 kips-ft (21,195.60 kN-m) per track based on AREA values. The bending moment M_{LL+IM} at midspan under live load can be calculated as follows.

1. **Determine Rocking Effect (*RE*)**
   The maximum axle load is 80 kips (356.00 kN) (40 kips per rail, 178.00 kN per rail).
   The couple force generated by the rocking effect:

$$20\% \times 40 = 8 \text{ kip } (35.60 \text{ kN})$$

   The total couple force at the box girders is:

$$\text{total couple} = 8 \times 5 \text{ ft (rail spacing)} \times 2 \text{ tracks} = 40 \text{ kips-ft } (54.24 \text{ kN-m})$$

$$\text{total couple force} = \frac{\text{total couple}}{\text{stringer spacing}} = \frac{80}{40} = 2.0 \text{ kip } (8.9 \text{ kN})$$

   The rocking effect on the stringers is the ratio of rocking reaction to static reaction:

$$RE = \frac{2.0}{80} = 0.025 = 2.5\%$$

2. **Determine the Percentage of Impact Effect from AREA Specifications**
   For $L > 80$ ft (24.39 m) and no hammer effect case:

$$IM = RE + \left(16 + \frac{600}{L-30}\right)\% = 2.5 + \left(16 + \frac{600}{108-30}\right)\% = 26.21\%$$

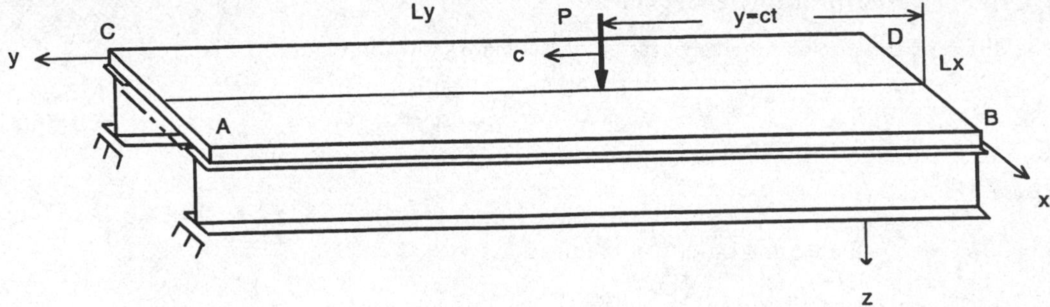

FIGURE 56.4 Girder deck bridge under moving constant load.

For ballasted deck bridges, 90% reduction should be considered, that is:

$$IM = 90\% \times 26.21\% = 0.24$$

3. Determine Live-Load Effect of the Maximum Bending Moment at Midspan

$$AF = 1.0 + IM = 1.0 + 0.24 = 1.24$$

$$M_{\text{LL+IM}} = AF\ (M_{LL}) = 1.24 \times 15{,}585 = 19{,}264 \text{ kips-ft } (26{,}199.04 \text{ kN-m})$$

56.4 Free Vibration Analysis

56.4.1 Structural Models

In the following two sections, the bridge vibration under moving load will be discussed to investigate the dynamic response of bridges. There have been basically two types of analysis methods: numerical analysis (sprung mass model) and analytical analysis (moving load model). The numerical analysis models the interaction between vehicle and bridge and expresses the dynamic behavior numerically. On the other hand, analytical analysis greatly simplifies vehicle interaction with bridge and models a bridge as a plate or beam but expresses the dynamic behavior explicitly. Good accuracy can be obtained using the analytical model when the ratio of live-load to self-weight of the superstructure is less than 0.3. The analytical analysis method will be presented for a bridge with beam-plate system.

The structure shown in Figure 56.4 represents a plate with two opposite edges AC and BD simply supported by rigid ground and the other two edges AB and CD simply supported on two beams. The assumptions made are

a. The stress–strain relationship for the beam and plate material is linear elastic;
b. There exists a neutral plane surface in the plate and the existence of the beams does not have any influence on the position of this neutral plane, i.e., the beam can only provide a vertical force reaction to the edges AB and CD for the plate. The structure is a noncomposite girder deck bridge; and
c. Thin plate and simple beam theories are applicable.

56.4.2 Free Vibration Analysis

The differential equation for free vibration of the plate shown in Figure 56.4 is

$$D\ \nabla^{(4)} W + m\ W_{tt}^{(2)} = 0 \tag{56.8}$$

where

$D = \dfrac{Eh^3}{12(1-\mu)}$ = the flexural rigidity of the plate
m = is the mass per unit area of the plate
W = is the vertical deflection of the plate

The boundary conditions for the plate are as follows:

At $y = 0$ and $y = L_y$:

$$W\ (x,\ y,\ t) = 0 \tag{56.9a}$$

$$M_y(x,\ y,\ t) = 0 \tag{56.9b}$$

At $x = 0$:

$$M_x\ (x,\ y,\ t) = 0 \tag{56.9c}$$

$$Q_x + \frac{\partial M_{xy}}{\partial y} = E_b I_b \frac{\partial^4 W}{\partial y^4} + m_b \frac{\partial^2 W}{\partial t^2} \tag{56.9d}$$

At $x = L_x$:

$$M_x\ (x,\ y,\ t) = 0 \tag{56.9e}$$

$$Q_x + \frac{\partial M_{xy}}{\partial y} = -E_b I_b \frac{\partial^4 W}{\partial y^4} - m_b \frac{\partial^2 W}{\partial t^2} \tag{56.9f}$$

where $E_b I_b$ and m_b is the flexural rigidity and mass per unit beam length, respectively, and M_x, M_y are the transverse and longitudinal bending moments, respectively. M_{xy} is the torque and Q_y is the shear force at the edges of the plate, respectively. Their signs are shown in Figure 56.5. The terms at the right-hand side in Eqs. (56.9d) and (56.9f) represent the interaction forces at the edges AB and CD between the plate and the beams. It is seen that the participation of the beams is taken into account in the partial differential equations through the boundary conditions for the plate.

Assuming that the structure vibrates in a mode so that the deflected shape is described by

$$W = \sin\frac{i\pi y}{L_y} X_j(x)\ \sin(\omega_{ij} t + q) \tag{56.10}$$

the following equations can be derived by substituting Eq. (56.10) into Eq. (56.8):

$$X_j^{(4)} - 2\frac{i^2\pi^2}{L_y^2} X_j^{(2)} + \frac{i^4\pi^4}{L_y^4} X_j - \frac{m}{D}\omega_{ij} X_j = 0 \tag{56.11}$$

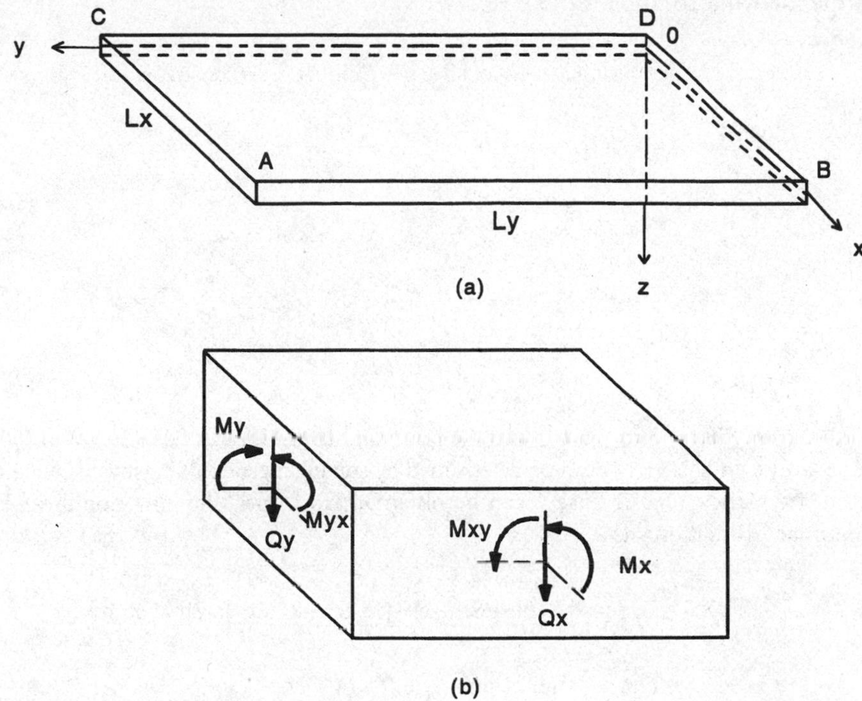

FIGURE 56.5 (a) A three-dimensional plate with length L_y and width L_x. (b) A volume element with inter forces acting on its sides.

The solution to the homogeneous differential equation above is

$$X_j = e^{\lambda\, x/Lx}$$

where λ must satisfy the characteristic equation:

$$\frac{l^4}{L_x^4} - 2\frac{i^2\pi^2}{L_y^2}\frac{l^2}{L_x^2} + \left(\frac{i^4\pi^4}{L_y^4} - \frac{m\omega_{ij}^2}{D}\right) = 0 \tag{56.12}$$

There are four roots for this equation. According to the signs of the roots, X_j (x) will take two different forms, which will be discussed separately.

Case 1

$$i^2 > pr_{ij}^2$$

where

$$pr_{ij}^2 = \frac{\omega_{ij}^2 mL_y^2}{D\pi^4} \tag{56.13}$$

We have the following solution for Eq. (56.11)

$$X_j = \sinh \lambda_1 u + A\cosh \lambda_1 u + B\sinh \lambda_2 u + C\cosh \lambda_2 u$$
$$\lambda_{1,2} = \frac{\pi}{\gamma}\left[i^2 \mp pr_{ij}^2\right] \tag{56.14}$$

where

$$u = x/L_x \tag{56.15.a}$$

$$\gamma = L_y / L_x \tag{56.15.b}$$

By substituting Eq. (56.14) and (56.10) into the boundary conditions in Eqs. (56.9a) to (56.9f), the constants *A, B,* and *C* in Eq. (56.14) can be determined by solving a group of eigenvalue equations [7].

The natural frequency ω_{ij} in Case 1 can be obtained from the following nonlinear equation from the boundary conditions as

$$2\lambda_1\lambda_2 D_1^2 D_2^2 (\cosh \lambda_1 \cosh \lambda_2 - 1) - \left(\lambda_1^2 D_1^4 + \lambda_2^2 D_2^4\right)\sinh \lambda_1 \sinh \lambda_2$$
$$-\left(D_1 + D_2\right)^2 Q^2 \sinh \lambda_1 \sinh \lambda_2 + 2Q(D_1 + D_2) \tag{56.16}$$
$$\left(\lambda_2 D_2^2 \sinh \lambda_1 \cosh \lambda_2 - \lambda_1 D_1^2 \cosh \lambda_1 \sinh \lambda_2\right) = 0$$

where

$$D_{1,2} = \frac{\pi^2}{\gamma^2}\left((1-\mu)i^2 + pr_{ij}^2\right)$$

$$Q = (E_b D_b \frac{i^4\pi^4}{L_y^4} - m_b\omega_{ij}^2)/DL_y^3$$

and *Q* in Eq. (56.16) represents the interaction between the plate and the beams.

Case 2

$$i^2 < pr_{ij}^2$$

In this case, the solution for Eq. (56.11) would be

$$X_j = \sin \lambda_1 u + A\cos \lambda_1 u + B\sinh \lambda_2 u + C\cosh \lambda_2 u \tag{56.17}$$

where

$$l_{1,2} = \frac{\pi}{\gamma}\left[pr_{ij}^2 \mp i^2\right]$$

By substituting Eqs. (56.17) and (56.11) into the boundary conditions in Eqs. (56.9a) to (56.9f), we can obtain the constants in Eq. (56.17) as in Case 1 by solving a group of eigenvalue equations [7].

The natural frequency ω_{ij} in Case 2 can be obtained from the following equation:

$$\begin{aligned}&2\lambda_1\lambda_2 D_1^2 D_2^2(\cos\lambda_1\cosh\lambda_2-1)-\left(\lambda_1^2 D_1^4+\lambda_2^2 D_2^4\right)\sin\lambda_1\sinh\lambda_2\\&-\left(D_1+D_2\right)^2 Q^2\sin\lambda_1\sinh\lambda_2+2Q(D_1+D_2)\\&\left(\lambda_2 D_2^2\sin\lambda_1\cosh\lambda_2-\lambda_1 D_1^2\cos\lambda_1\sinh\lambda_2\right)=0\end{aligned} \tag{56.18}$$

Example 56.3 Free Vibration Analysis for a Beam-Plate Bridge

A one-lane bridge deck structure is shown in Figure 56.4. The plate is made of an isotropic material representing reinforced concrete, and the beams are made of steel with a W36 × 150 section. The bridge span length is 80 ft (24.39 m) and the bridge width is 10 ft (3.05 m). The thickness of the deck plate is 8 in. (3.15 cm). The elasticity modulus of the steel girder $E_b = 29.0 \times 10^3$ ksi (200.0×10^3 MPa) and the elasticity modulus of the concrete plate is $E = 4.38 \times 10^3$ ksi (30.18 MPa). The mass density of the plate $m = 0.8681$ lb/in.^{2} g (0.61 kN/cm^2 g) and the mass density of the beam is $m_b = 12.50$ lb/in.$^2 g$ (22.0 N/cm g) where g is the gravitational acceleration. The moment of inertia of the beam is $I_b = 9030$ in.^{4} (375,800 cm^4). All other properties are calculated as:

Bending rigidity of the plate in y direction:

$$L_x D = \frac{L_x E t^3}{12(1.0-\mu)} = 4.56\times10^{10}\ \text{lb.in.}^2\ (1.28\times10^{12}\,\text{kN.cm}^2)$$

Bending rigidity ratio between beam and plate:

$$R_{ej} = \frac{2.0 E_b I_b}{L_x D} = 11.5$$

Mass ratio between beam and plate:

$$R_m = \frac{m_b}{L_x\, m} = 0.12$$

First natural frequency of the plate as a beam:

$$\omega_{11}^p = \frac{p^2}{L_y^2}\sqrt{\frac{D}{m}} = 4.40\ \text{rad/s}$$

First natural frequency of the beam:

$$\omega_1^b = \frac{p^2}{L_y^2}\sqrt{\frac{E_b I_b}{m_b}} = 30.45\ \text{rad/s}$$

Subscripts i and j represent the ith and jth mode shape in the y- and x-direction, respectively. The natural frequencies, ω_{ij}, can be determined numerically using Eqs. (56.16) and (56.18). Some of the first several natural frequencies are shown in Table 56.2.

The mode shapes of the bridge are shown in Figures 56.6 and 56.7 using normalized dimensions in all three directions. It is seen that W_{12}, W_{14}, W_{22}, and W_{24} are all asymmetric mode shapes in the

TABLE 56.2 Natural Frequencies of One-Lane Bridge Deck (rad/s; plate aspect ratio = 8.0)

	$j = 1$	$j = 2$	$j = 3$
$i = 1$	14.53	35.43	482.65
$i = 2$	60.09	110.62	506.27
$i = 3$	125.67	217.44	556.53
$i = 4$	206.05	354.98	647.51

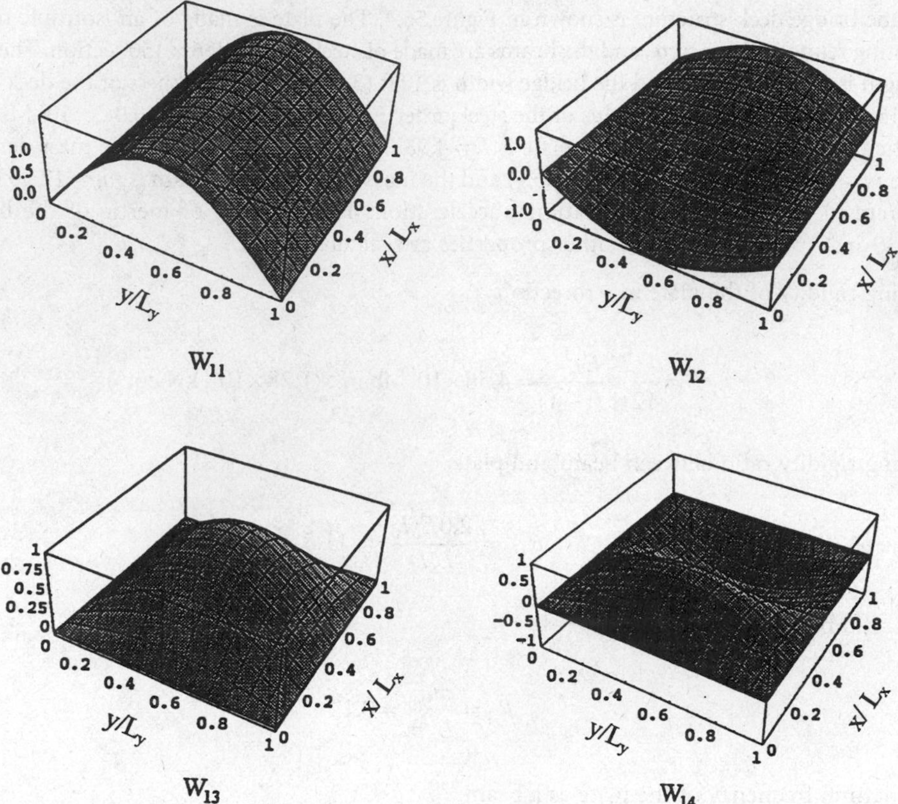

FIGURE 56.6 Normalized three-dimensional mode shapes W_{ij}, $i = 1$, $j = 1,2,3,4$, for a one-lane bridge deck. (subscripts i-j is the mode shape number in y- and x-directions, respectively).

x-direction about the center line $x = L_x/2$. When a moving load traverses the plate along this line, these mode shapes would not be excited. Thus, there are no contributions from these mode shapes.

The first mode shape of the beam–plate system is nearly a constant in the x-direction, as shown in this example of a rather high plate aspect ratio. This means that the beams have the same first mode shape as the plate. In this case, the first natural frequency of the beam–plate system can be approximately evaluated as

$$\omega_{11}^2 = \frac{2R_m \omega_1^{b\,2} + \omega_{11}^{p\,2}}{1 + 2R_m} \tag{56.19}$$

where ω_1^b and ω_1^p are fundamental frequencies of the beam and plate as a beam, respectively.

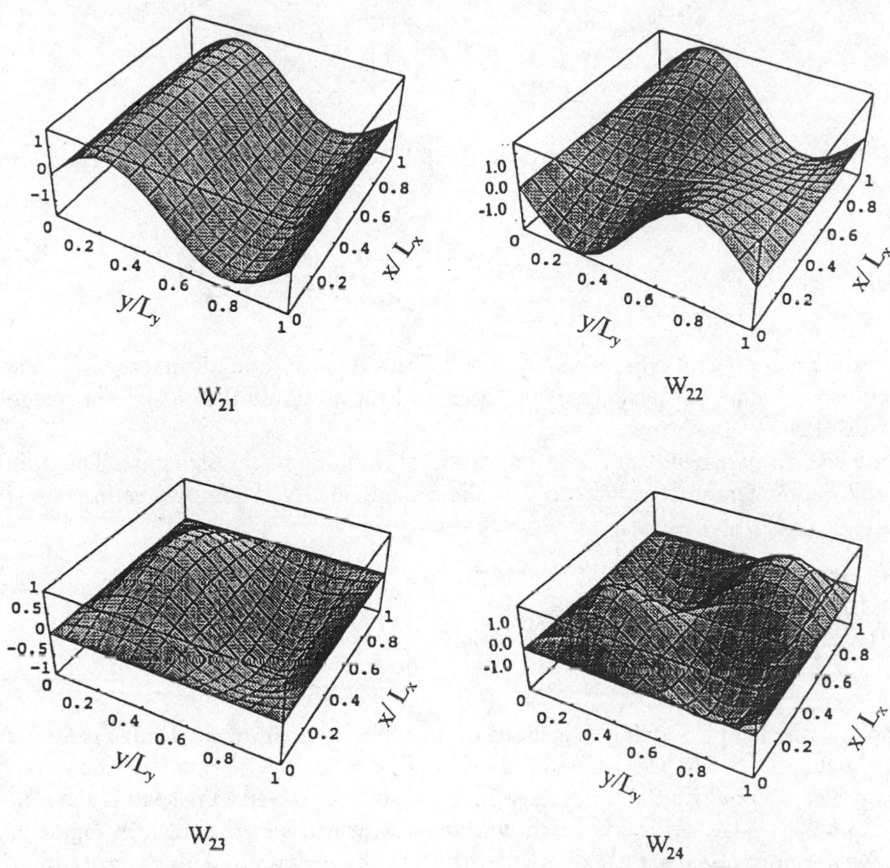

FIGURE 56.7 Normalized three-dimensional mode shapes W_{ij}, $i = 2$, $j = 1,2,3,4$, for a one-lane bridge deck (subscripts i-j is the mode shape number in y- and x-directions, respectively).

56.5 Forced Vibration Analysis under Moving Load

56.5.1 Dynamic Response Analysis

The governing equation for the plate with smooth surface supported by two beams under a moving constant load shown in Figure 56.4 is

$$\nabla^4 W + \frac{m}{D}\frac{\partial^2 W}{\partial t^2} = \frac{1}{D}p(x,y,t) \quad (0 < t < L_y/c) \tag{56.20}$$

where

$$p(x,y,t) = P\,\delta(y - ct)\,\delta(x - L_x/2)$$

and c is the speed of the moving load, and P the magnitude of the moving load. A Dirac delta function represents a unit concentrated force acting on the deck.

The method of modal superposition is used to get the dynamic response by assuming

$$W = \sum_{i=1}^{+\infty} \sum_{j=1}^{+\infty} W_{ij}(t) \sin \frac{i\pi y}{L_y} X_j(x) \tag{56.21}$$

By substituting Eq. (56.21) into Eq. (56.20), the following ordinary differential equation can be derived:

$$W_{ij,t}^{(2)}(t) + 2\xi_{ij}\omega_{ij} W_{ij,t}^{(1)} + \omega_{ij}^2 W_{ij} = \frac{2P}{m L_x L_y} X_{ij}(L_x/2) \sin \frac{i\pi c}{L_y} t \quad (0 < t < L_y/c) \tag{56.22}$$

where ξ_{ij} is the damping ratio for mode shape *i–j*. The initial condition is that the bridge structure is in a static state before the load enters the span and the structure is in a state of free vibration after the load traverses the bridge.

There are several parameters affecting the dynamic response of the structure. The influence of some typical parameters such as the speed of the moving load and damping ratio is presented in the following.

The vehicle speed is normalized as

$$\alpha = \frac{\pi c / L_y}{\omega_{11}} \tag{56.23}$$

For example, if the load is moving at a speed of 60 mph, the span of the bridge is 80 ft and first natural frequency is 14 Hz, the normalized speed is $\alpha = 0.25$. A normalized speed of $\alpha = 0.5$ represents the case for which the time needed for a vehicle to traverse the span is the same as the first period of the bridge. Typical normalized dynamic deflections are shown in Figure 56.8. It is seen that for the normalized speed ranging $0 < \alpha < 1.0$, the maximum deflection occurs when the load is on the bridge while if $\alpha > 1.0$, the maximum deflection occurs after the load traverses the bridge. When the load is on the plate ($ct/L_y < 1.0$), the deflections at the center are usually positive. But the deflections are negative after the load has traversed the plate ($ct/L_y > 1.0$), especially for the case of $0 < \alpha < 0.5$. The change of sign for deflection results in a curvature change and, hence, in a stress sign change.

The normalized deflection at the center of the plate is defined as the dynamic deflection caused by moving constant load divided by the corresponding static deflection caused by the concentrated constant force. Figure 56.9 shows its spectra for an aspect ratio of 8.0. The normalized speed of the moving load affects the maximum dynamic response significantly. For small α values, e.g., less than 0.20, which refers to long-span bridges or slow-speed vehicle, there is little dynamic amplification. Very low speed values or very long bridges do not result in much dynamic response. When the normalized speed is greater than 1.3, corresponding to very short span bridges, the dynamic effect is also not significant because the duration of excitation is extremely short. The maximum dynamic response happens for normalized speeds ranging from 0.45 to 0.65, when load frequency is near the bridge fundamental frequency. It is also found that the maximum bending moment amplification factors are 1.2 for M_y and 1.1 for M_x at a normalized speed of 0.4, while the maximum deflection amplification is about 1.5 at a normalized speed of 0.55. The dynamic amplification for bending moments is less than the dynamic amplification for deflections.

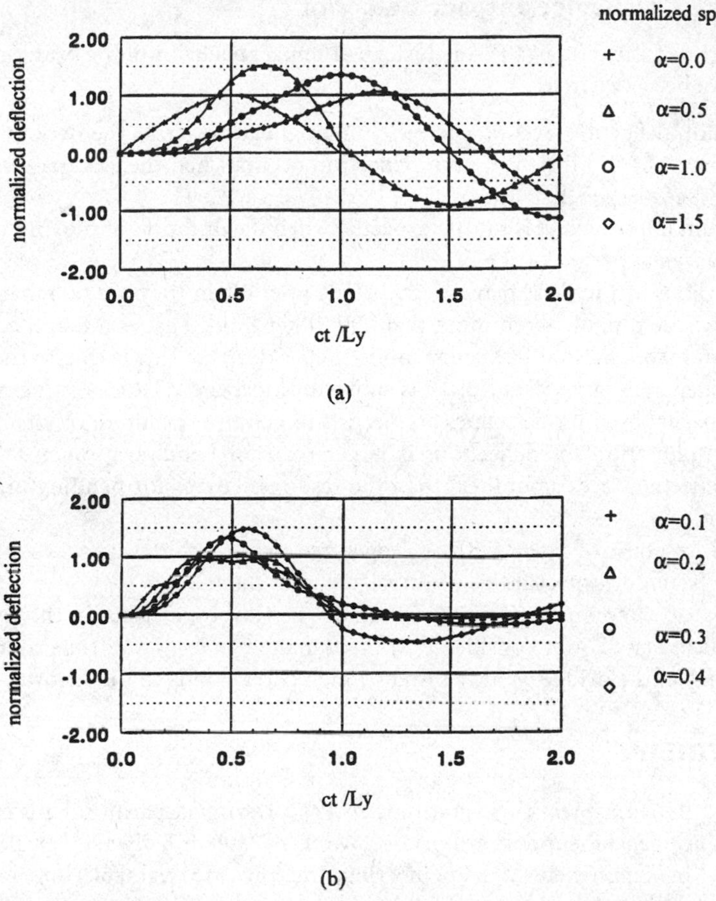

FIGURE 56.8 (a) Normalized deflections at the center of the deck with normalized speed, α = 0.0, 0.5, 1.0, and 1.5. (b) Normalized deflections at the center of the deck with normalized speed, α = 0.1, 0.2, 0.3, and 0.4.

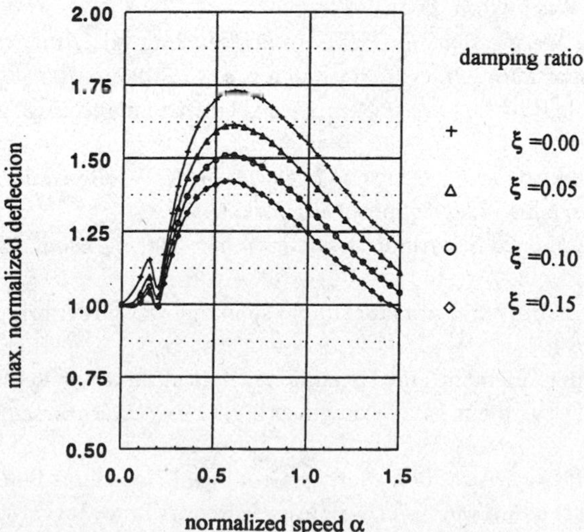

FIGURE 56.9 Normalized deflection spectra at the center of the deck for different damping ratios and normalized speeds.

56.5.2 Summary of Bridge Impact Behavior

For one-lane bridge deck structures, the conclusions of impact behavior of the bridge under moving constant loads have been drawn as:

1. The maximum deflection occurs when the moving load traverses the deck at a normalized speed less than 1.0, and the maximum deflection occurs when the load passes the deck at a normalized speed larger than 1.0.
2. The maximum impact effect is mostly expected when the duration of moving vehicle is close to the fundamental period of the bridge.
3. The aspect ratios of the deck play an important role. When they are less than 4.0, the first mode shape is dominant, when more than 8.0, other mode shapes are excited. The contributions from higher natural frequency mode shapes decrease slowly due to the fact that the natural frequency, ω_{ij}, increases slowly as subscript i increases. Thus, a sufficient number of terms of superimposed mode shapes are needed to get more accurate results.
4. Dynamic amplification for deflections is larger than for bending moments. The response curves of deflections are "smoother" than the response curves for bending moments in the time domain.
5. The dynamic response of a plate with two edges free and the other two edges simply supported is close to a beam for aspect ratios of the plate larger than 2.0.
6. The analysis of a moving constant load model usually overestimates the dynamic effect because vehicle mass is not considered in the dynamic analysis and thus overestimates the first frequency of the bridge–vehicle system which corresponds to the "shorter-span" bridge.

Acknowledgments

The authors would like to express their gratitude to Prof. Dario Gasparini for his endeavors and opinions and to the financial support from Case Western Reserve University in performing the dynamic analysis. Mr. Kang Chen, MG Engineering, Inc. provided valuable information for the railway bridge part, which is greatly appreciated.

References

1. AASHTO, *LRFD Bridge Design Specifications*, American Association of State Highway and Transportation Officials, Washington, D.C., 1994.
2. AASHTO, *Standard Specifications for Highway Bridges*, 16th ed., American Association of State Highway and Transportation Officials, Washington, D.C., 1996.
3. AREA, *Manual for Railway Engineering*, American Railway Engineering Association, Washington, D.C., 1996.
4. Barker, R. M. and Puckett, J. A., *Design of Highway Bridges — Based on AASHTO LRFD, Bridge Design Specifications*, John Wiley & Sons, New York, 1996.
5. Biggs, J. M., *Introduction to Structural Dynamics under Moving Loads*, McGraw-Hill, New York, Inc., 1964.
6. Chang, D. and Lee, H., Impact factors for simple-span highway girder bridges, *J. Struct. Eng. ASCE*, 120(3), 880–889, 1994.
7. Duan, M., Static Finite Element and Dynamic Analytical Study of Reinforced Concrete Bridge Decks, M.S. thesis, Department of Civil Engineering, Case Western Reserve University, Cleveland, OH, 1994.
8. Fryba, L., *Introduction to Structural Dynamics*, Groningen Noordhoff Futern, 1972.
9. Garg, V. K., *Dynamics of Railway Vehicle Systems*, Harcourt Brace Jovanovich, New York, 1984.
10. Huang, D., Wang, T.-L., and Shahawy, M., Impact analysis of continuous multi-girder bridges due to moving vehicles, *J. Struct. Eng. ASCE*, 118(12), 3427–3443, 1992.

11. Hino, J., Yoshimura, T., and Konishi, K., A finite element method prediction of the vibration of a bridge subjected to a moving vehicle load, *J. Sound Vibration,* 96(6), 45–53, 1984.
12. Huang, D., Wang, T.-L., and Shahawy, M., Vibration of thin-walled box-girder bridges exited by vehicles, *J. Struct. Eng. ASCE,* 121(9), 1330–1337, 1995.
13. Kim, S. and Nowak, A., Load distribution and impact factors for I-girder bridges, *J. Bridge Eng. ASCE,* 2(3), 1997.
14. Lin, Y. H. and Trethewey, M. W., Finite element analysis of elastic beams subjected to moving loads, *J. Sound Vibration,* 136(2), 323–342, 1990.
15. Petrou, M. F., Perdikaris, P. C., and Duan, M., Static behavior of noncomposite concrete bridge decks under concentrated loads, *J. Bridge Eng. ASCE,* 1(4), 143–154, 1996.
16. Scheling, D. R., Galdos, N. H., and Sahin, M. A., Evaluation of impact factors for horizontally curved steel box bridges, *J. Struct. Eng. ASCE,* 118(11), 3203–3221, 1992.
17. Timoshenko, S. P. and Woinowsky-Krieger, S., *Theory of Plates and Shells,* 2nd ed., McGraw-Hill, New York, 1959.
18. Wang, T.-L., Huang, D., and Shahawy, M., Dynamic response of multi-girder bridges, *J. Struct. Eng. ASCE,* 118(8), 2222–2238, 1992.
19. Xanthakos, P., *Theory and Design of Bridges,* John Wiley & Sons, New York, 1994.
20. Yang, Y.-B. and Lin, B.-H., Vehicle-bridge interaction analysis by dynamic condensation method, *J. Struct. Eng. ASCE,* 121(2), 1636–1643, 1995.
21. Yang, Y.-B. and Yau, J.-D., Vehicle-bridge interaction element for dynamic analysis, *J. Struct. Eng. ASCE,* 118(11), 1512–1518, 1997.

57

Wind Effects on Long-Span Bridges

Chun S. Cai
Florida Department of Transportation

Serge Montens
Jean Muller International, France

57.1 Introduction

The development of modern materials and construction techniques has resulted in a new generation of lightweight flexible structures. Such structures are susceptible to the action of winds. Suspension bridges and cable-stayed bridges shown in Figure 57.1 are typical structures susceptible to wind-induced problems.

The most renowned bridge collapse due to winds is the Tacoma Narrows suspension bridge linking the Olympic Peninsula with the rest of the state of Washington. It was completed and opened to traffic on July 1, 1940. Its 853-m main suspension span was the third longest in the world. This bridge became famous for its serious wind-induced problems that began to occur soon after it opened. "Even in winds of only 3 to 4 miles per hour, the center span would rise and fall as much as four feet…, and drivers would go out of their way either to avoid it or cross it for the roller coaster thrill of the trip. People said you saw the lights of cars ahead disappearing and reappearing as they bounced up and down. Engineers monitored the bridge closely but concluded that the motions were predictable and tolerable" [1].

On November 7, 1940, 4 months and 6 days after the bridge was opened, the deck oscillated through large displacements in the vertical vibration modes at a wind velocity of about 68 km/h. The motion changed to a torsional mode about 45 min later. Finally, some key structural members became overstressed and the main span collapsed.

0-8493-7434-0/00/$0.00+$.50

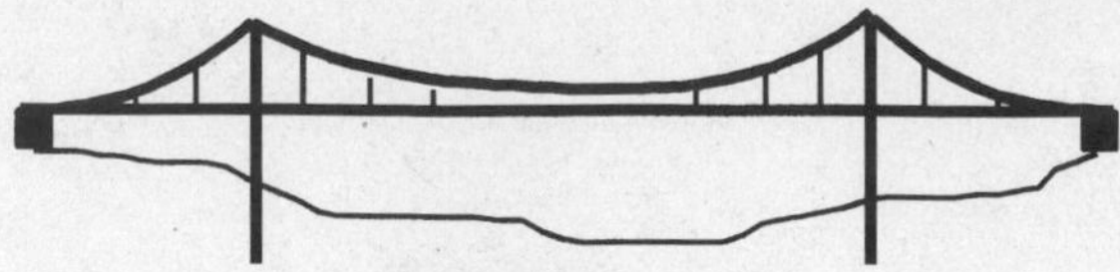

Suspension Bridge

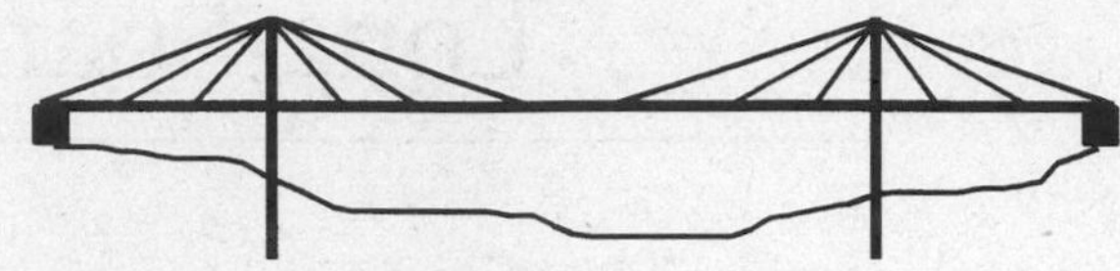

Cable-Stayed Bridge

FIGURE 57.1 Typical wind-sensitive bridges.

Some bridges were destroyed by wind action prior to the failure of the Tacoma Narrows bridge. However, it was this failure that shocked and intrigued bridge engineers to conduct scientific investigations of bridge aerodynamics. Some existing bridges, such as the Golden Gate suspension bridge in California with a main span of 1280 m, have also experienced large wind-induced oscillations, although not to the point of collapse. In 1953, the Golden Gate bridge was stiffened against aerodynamic action [2].

Wind-induced vibration is one of the main concerns in a long-span bridge design. This chapter will give a brief description of wind-induced bridge vibrations, experimental and theoretical solutions, and state-of-the-art applications.

57.2 Winds and Long-Span Bridges

57.2.1 Description of Wind at Bridge Site

The atmospheric wind is caused by temperature differentials resulting from solar radiations. When the wind blows near the ground, it is retarded by obstructions making the mean velocity at the ground surface zero. This zero-velocity layer retards the layer above and this process continues until the wind velocity becomes constant. The distance between the ground surface and the height of constant wind velocity varies between 300 m and 1 km. This 1-km layer is referred to as the boundary layer in which the wind is turbulent due to its interaction with surface friction. The variation of the mean wind velocity with height above ground usually follows a logarithmic or exponential law.

The velocity of boundary wind is defined by three components: the along-wind component consisting of the mean wind velocity, $\overline{U}$, plus the turbulent component $u(t)$, the cross-wind turbulent component $v(t)$, and the vertical turbulent component $w(t)$. The turbulence is described in terms of turbulence intensity, integral length, and spectrum [3].

The turbulence intensity I is defined as

$$I = \frac{\sigma}{\overline{U}} \tag{57.1}$$

where σ = the standard deviation of wind component $u(t)$, $v(t)$, or $w(t)$; $\overline{U}$ = the mean wind velocity.

Integral length of turbulence is a measurement of the average size of turbulent eddies in the flow. There are a total of nine integral lengths (three for each turbulent component). For example, the integral length of $u(t)$ in the x-direction is defined as

$$L_u^x = \frac{1}{\sigma_u^2} \int_0^\infty R_{u1u2}(x)dx \qquad (57.2)$$

where $R_{u1u2}(x)$ = cross-covariance function of $u(t)$ for a spatial distance x.

The wind spectrum is a description of wind energy vs. wind frequencies. The von Karman spectrum is given in dimensionless form as

$$\frac{nS(n)}{\sigma^2} = \frac{4\frac{nL}{\bar{U}}}{\left[1 + 70.8\left(\frac{nL}{\bar{U}}\right)^2\right]^{5/6}} \qquad (57.3)$$

where n = frequency (Hz); S = autospectrum; and L = integral length of turbulence. The integral length of turbulence is not easily obtained. It is usually estimated by curve fitting the spectrum model with the measured field data.

57.2.2 Long-Span Bridge Responses to Wind

Wind may induce instability and excessive vibration in long-span bridges. Instability is the onset of an infinite displacement granted by a linear solution technique. Actually, displacement is limited by structural nonlinearities. Vibration is a cyclic movement induced by dynamic effects. Since both instability and vibration failures in reality occur at finite displacement, it is often hard to judge whether a structure failed due to instability or excessive vibration-induced damage to some key elements.

Instability caused by the interaction between moving air and a structure is termed aeroelastic or aerodynamic instability. The term *aeroelastic* emphasizes the behavior of deformed bodies, and *aerodynamic* emphasizes the vibration of rigid bodies. Since many problems involve both deformation and vibration, these two terms are used interchangeably hereafter. Aerodynamic instabilities of bridges include divergence, galloping, and flutter. Typical wind-induced vibrations consist of vortex shedding and buffeting. These types of instability and vibration may occur alone or in combination. For example, a structure must experience vibration to some extent before flutter instability starts.

The interaction between the bridge vibration and wind results in two kinds of forces: motion-dependent and motion-independent. The former vanishes if the structures are rigidly fixed. The latter, being purely dependent on the wind characteristics and section geometry, exists whether or not the bridge is moving. The aerodynamic equation of motion is expressed in the following general form:

$$[M]\{\ddot{Y}\} + [C]\{\dot{Y}\} + [K]\{Y\} = \{F(Y)\}_{md} + \{F\}_{mi} \qquad (57.4)$$

where $[M]$ = mass matrix; $[C]$ = damping matrix; $[K]$ = stiffness matrix; $\{Y\}$ = displacement vector; $\{F(Y)\}_{md}$ = motion-dependent aerodynamic force vector; and $\{F\}_{mi}$ = motion-independent wind force vector.

The motion-dependent force causes aerodynamic instability and the motion-independent part together with the motion-dependent part causes deformation. The difference between short-span and long-span bridge lies in the motion-dependent part. For the short-span bridges, the motion-dependent part is insignificant and there is no concern about aerodynamic instability. For flexible structures like long-span bridges, however, both instability and vibration need to be carefully investigated.

57.3 Experimental Investigation

Wind tunnel testing is commonly used for "wind-sensitive" bridges such as cable-stayed bridges, suspension bridges, and other bridges with span lengths or structure types significantly outside of the common ranges. The objective of a wind tunnel test is to determine the susceptibility of the bridges to various aerodynamic phenomena.

The bridge aerodynamic behavior is controlled by two types of parameters, i.e., structural and aerodynamic. The structural parameters are the bridge layout, boundary condition, member stiffness, natural modes, and frequencies. The aerodynamic parameters are wind climate, bridge section shape, and details. The design engineers need to provide all the information to the wind specialist to conduct the testing and analysis.

57.3.1 Scaling Principle

In a typical structural test, a prototype structure is scaled down to a scale model according to mass, stiffness, damping, and other parameters. In testing, the wind blows in different vertical angles (attack angles) or horizontal angles (skew angles) to cover the worst case at the bridge site. To obtain reliable information from a test, similarity must be maintained between the specimen and the prototype structure. The geometric scale λ_L, a basic parameter which is controlled by the size of an available wind tunnel, is denoted as the ratio of the dimensions of model (B_m) to the dimensions of prototype bridge (B_p) as [4]

$$\lambda_L = \frac{B_m}{B_p} \tag{57.5}$$

where subscripts m and p indicate model and prototype, respectively.

To maintain the same Froude number for both scale model and prototype bridge requires,

$$\left(\frac{U^2}{Bg}\right)_m = \left(\frac{U^2}{Bg}\right)_p \tag{57.6}$$

where g is the air gravity, which is the same for the model and prototype bridge. From Eqs. (57.5) and (57.6) we have the wind velocity scale λ_v as

$$\lambda_V = \frac{U_m}{U_p} = \sqrt{\lambda_L} \tag{57.7}$$

Reynolds number equivalence requires

$$\left(\frac{\rho UB}{\mu}\right)_m = \left(\frac{\rho UB}{\mu}\right)_p \tag{57.8}$$

where μ = viscosity and ρ = wind mass density. Equations (57.5) and (57.8) give the wind velocity scale as

$$\lambda_V = \frac{1}{\lambda_L} \tag{57.9}$$

which contradicts Eq. (57.7). It is therefore impossible in model scaling to satisfy both the Froude number equivalence and Reynolds number equivalence simultaneously. For bluff bodies such as bridge decks, flow separation is caused by sharp edges and, therefore, the Reynolds number is not important except it is too small. The too-small Reynolds number can be avoided by careful selection of λ_L. Therefore, the Reynolds number equivalence is usually sacrificed and Froude number equivalence is maintained.

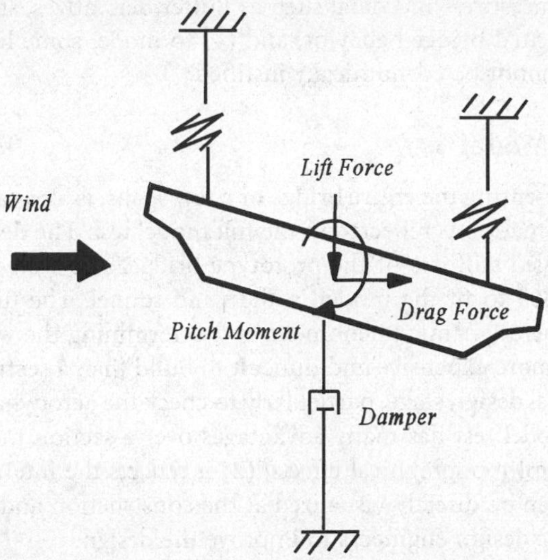

FIGURE 57.2 End view of section model.

To apply the flutter derivative information to the prototype analysis, nondimensional reduced velocity must be the same, i.e.,

$$\left(\frac{U}{NB}\right)_m = \left(\frac{U}{NB}\right)_p \tag{57.10}$$

Solving Eqs. (57.5), (57.7) and (57.10) gives the natural frequency scale as

$$\lambda_N = \frac{N_m}{N_p} = \frac{1}{\sqrt{\lambda_L}} \tag{57.11}$$

The above equivalence of reduced velocity between the section model and prototype bridge is the basis to use the section model information to prototype bridge analysis. Therefore, it should be strictly satisfied.

57.3.2 Section model

A typical section model represents a unit length of a prototype deck with a scale from 1:25 to 1:100. It is usually constructed from materials such as steel, wood, or aluminum to simulate the scaled mass and moment of inertia about the center of gravity. The section model represents only the outside shape (aerodynamic shape) of the deck. The stiffness and the vibration characteristics are represented by the spring supports.

By rigidly mounding the section in the wind tunnel, the static wind forces, such as lift, drag, and pitch moment, can be measured. To measure the aerodynamic parameters such as the flutter derivatives, the section model is supported by a spring system and connected to a damping source as shown in Figure 57.2. The spring system can be adjusted to simulate the deck stiffness in vertical and torsional directions, and therefore simulate the natural frequencies of the bridges. The damping characteristics are also adjustable to simulate different damping.

A section model is less expensive and easier to conduct than a full model. It is thus widely used in (1) the preliminary study to find the best shape of a bridge deck; (2) to identify the potential wind-induced problems such as vortex-shedding, flutter, and galloping and to guide a more-sophisticated full model study; (3) to measure wind data, such as flutter derivatives, static force coefficients for analytical prediction of actual bridge behavior; and (4) to model some less important bridges for which a full model test cannot be economically justified.

57.3.3 Full Bridge Model

A full bridge model, representing the entire bridge or a few spans, is also called an aeroelastic model since the aeroelastic deformation is reflected in the full model test. The deck, towers, and cables are built according to the scaled stiffness of the prototype bridge. The scale of a full bridge model is usually from 1:100 to 1:300 to fit the model in the wind tunnel. The full model test is used for checking many kinds of aerodynamic phenomena and determining the wind loading on bridges.

A full bridge model is more expensive and difficult to build than a section model. It is used only for large bridges at the final design stage, particularly to check the aerodynamics of the construction phase. However, a full model test has many advantages over a section model: (1) it simulates the three-dimensional and local topographical effects; (2) it reflects the interaction between vibration modes; (3) wind effects can be directly visualized at the construction and service stages; and (4) it is more educational to the design engineers to improve the design.

57.3.4 Taut Strip Model

For this model, taut strings or tubes are used to simulate the stiffness and dynamic characteristics of the bridge such as the natural frequencies and mode shapes for vertical and torsional vibrations. A rigid model of the deck is mounted on the taut strings. This model allows, for example, to represent the main span of a deck. The taut strip model falls between section model and full model with respect to cost and reliability. For less important bridges, the taut strip model is a sufficient and economical choice. The taut strip model is used to determine critical wind velocity for vortex shedding, flutter, and galloping and displacement and acceleration under smooth or turbulent winds.

57.4 Analytical Solutions

57.4.1 Vortex Shedding

Vortex shedding is a wake-induced effect occurring on bluff bodies such as bridge decks and pylons. Wind flowing against a bluff body forms a stream of alternating vortices called a von Karman vortex street shown in Figure 57.3a. Alternating shedding of vortices creates an alternative force in a direction normal to the wind flow. This alternative force induces vibration. The shedding frequency of vortices from one surface, in either torsion or lift, can be described in terms of a nondimensional Strouhal number, *S*, as

$$S = \frac{ND}{\overline{U}} \tag{57.12}$$

where N = shedding frequency and D = characteristic dimension such as the diameter of a circular section or depth of a deck.

The Strouhal number (ranging from 0.05 to 0.2 for bridge decks) is a constant for a given section geometry and details. Therefore, the shedding frequency (N) increases with the wind velocity to maintain a constant Strouhal value (S). The bridge vibrates strongly but self-limited when the frequency of vortex shedding is close to one of the natural frequencies of a bridge, say, N_1 as shown in Figure 57.3. This phenomenon is called lock-in and the corresponding wind velocity is called critical velocity of vortex shedding.

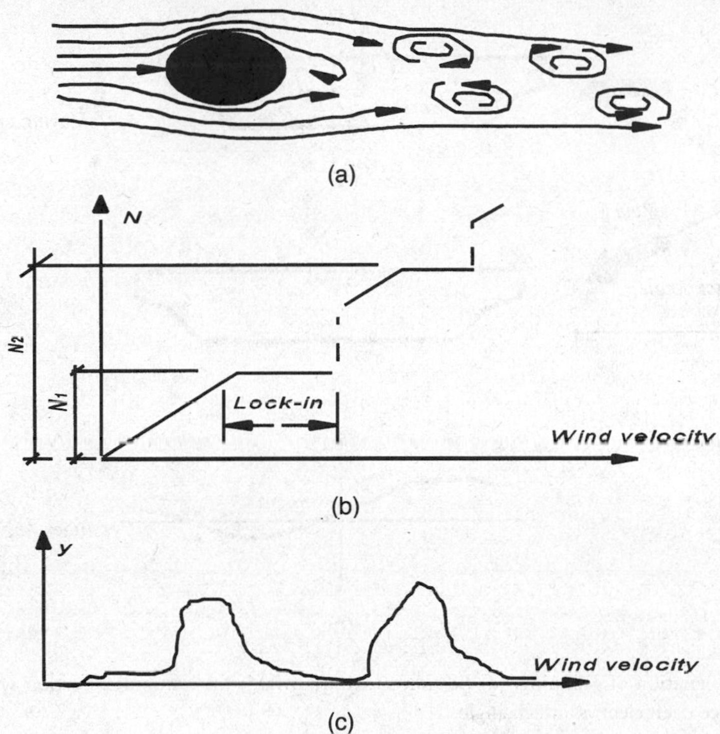

FIGURE 57.3 Explanation of vortex shedding. (a) Von Karman Street; (b) lock-in phenomenon; (c) bridge vibration

The lock-in occurs over a small range of wind velocity within which the Strouhal relation is violated since the increasing wind velocity and a fixed shedding frequency results in a decreasing Strouhal number. The bridge natural frequency, not the wind velocity, controls the shedding frequency. As wind velocity increases, the lock-in phenomenon disappears and the vibration reduces to a small amplitude. The shedding frequency may lock in another higher natural frequency (N_2) at higher wind velocity. Therefore, many wind velocities cause vortex shedding.

To describe the above experimental observation, much effort has been made to find an expression for forces resulting from vortex shedding. Since the interaction between the wind and the structure is very complex, no completely successful model has yet been developed for bridge sections. Most models deal with the interaction of wind with circular sections. A semiempirical model for the lock-in is given as [3]

$$m\ddot{y}+c\dot{y}+ky=\frac{1}{2}\rho U^2(2D)\left[Y_1(K)\left(1-\varepsilon\frac{y^2}{D^2}\right)\frac{\dot{y}}{D}+Y_2(K)\frac{y}{D}+\frac{1}{2}C_L(K)\sin(\omega t+\phi)\right] \tag{57.13}$$

where $k = B\omega/\overline{U}$ = reduced frequency; Y_1, Y_2, ε, and C_L = parameters to be determined from experimental observations. The first two terms of the right side account for the motion-dependent force. More particularly, the $\dot{y}$ term accounts for aerodynamic damping and y term for aerodynamic stiffness. The ε accounts for the nonlinear aerodynamic damping to ensure the self-limiting nature of vortex shedding. The last term represents the instantaneous force from vortex shedding alone which is sinusoidal with the natural frequency of bridge. Solving the above equation gives the vibration y.

Vortex shedding occurs in both laminar and turbulent flow. According to some experimental observations, turbulence helps to break up vortices and therefore helps to suppress the vortex shedding response. A more complete analytical model must consider the interaction between modes, the spanwise correlation of aerodynamic forces and the effect of turbulence.

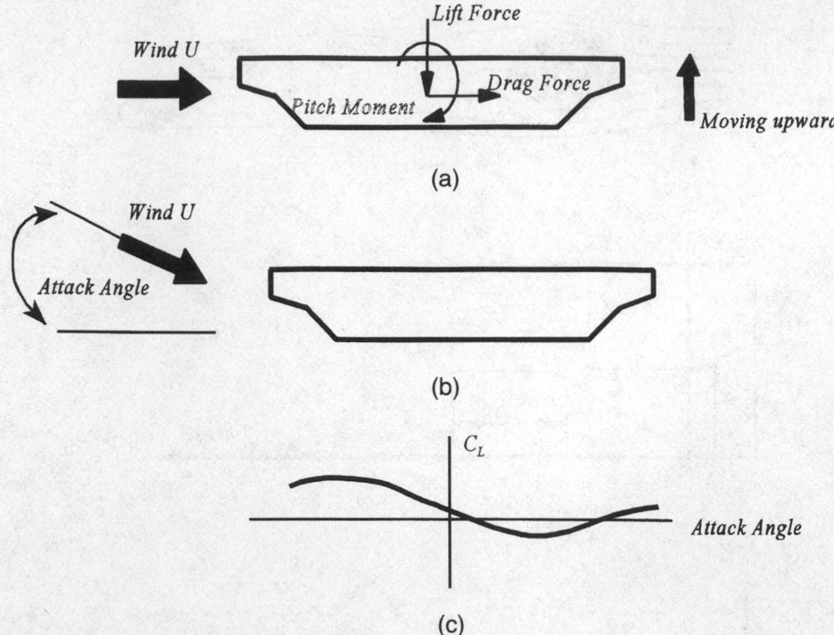

FIGURE 57.4 Explanation of galloping. (a) Section moving upward; (b) motionless section with a wind attack angle; (c) static force coefficient vs. attack angle.

For a given section shape with a known Strouhal number and natural frequencies, the lock-in wind velocities can be calculated with Eq. (57.12). The calculated lock-in wind velocities are usually lower than the maximum wind velocity at bridge sites. Therefore, vortex shedding is an inevitable aerodynamic phenomenon. However, vibration excited by vortex shedding is self-limited because of its nonlinear nature. A relatively small damping is often sufficient to eliminate, or at least reduce, the vibrations to acceptable limits.

Although there are no acceptance criteria for vortex shedding in the design specifications and codes in the United States, there is a common agreement that limiting acceleration is more appropriate than limiting deformation. It is usually suggested that the acceleration of vortex shedding is limited to 5% of gravity acceleration when wind speed is less than 50 km/h and 10% of gravity acceleration when wind speed is higher. The acceleration limitation is then transformed into the displacement limitation for a particular bridge.

57.4.2 Galloping

Consider that in Figure 57.4 (a) a bridge deck is moving upward with a velocity $\dot{y}$ under a horizontal wind *U*. This is equivalent to the case of Figure 57.4b that the deck is motionless and the wind blows downward with an attack angle α ($\tan(\alpha) = \dot{y}/U$). If the measured static force coefficient of this case is negative (upward), then the deck section will be pushed upward further resulting in a divergent vibration or galloping. Otherwise, the vibration is stable. Galloping is caused by a change in the effective attack angle due to vertical or torsional motion of the structure. A negative slope in the plot of either static lift or pitch moment coefficient vs. the angle of attack, shown in Figure 57.4c, usually implies a tendency for galloping. Galloping depends mainly on the quasi-steady behavior of the structure.

The equation of motion describing this phenomenon is

$$m\ddot{y} + c\dot{y} + ky = -\frac{1}{2}\rho U^2 B\left(\frac{dC_L}{d\alpha} + C_D\right)_{\alpha=0}\frac{\dot{y}}{U} \tag{57.14}$$

The right side represents the aerodynamic damping and C_L and C_D are static force coefficients in the lift and drag directions, respectively. If the total damping is less than zero, i.e.,

$$c+\frac{1}{2}\rho UB\left(\frac{dC_L}{d\alpha}+C_D\right)_{\alpha=0}\leq 0 \tag{57.15}$$

then the system tends toward instability. Solving the above equation gives the critical wind velocity for galloping. Since the mechanical damping c is positive, the above situation is possible only if the following Den Hartog criterion [5] is satisfied

$$\left(\frac{dC_L}{d\alpha}+C_D\right)_{\alpha=0}\leq 0 \tag{57.16}$$

Therefore, a wind tunnel test is usually conducted to check against Eq. (57.16) and to make necessary improvement of the section to eliminate the negative tendency for the possible wind velocity at a bridge site.

Galloping rarely occurs in highway bridges, but noted examples are pedestrian bridges, pipe bridges, and ice-coated cables in power lines. There are two kinds of cable galloping: cross-wind galloping, which creates large-amplitude oscillations in a direction normal to the flow, and wake galloping caused by the wake shedding of the upwind structure.

57.4.3 Flutter

Flutter is one of the earliest recognized and most dangerous aeroelastic phenomena in airfoils. It is created by self-excited forces that depend on motion. If a system immersed in wind flow is given a small disturbance, its motion will either decay or diverge depending on whether the energy extracted from the flow is smaller or larger than the energy dissipated by mechanical damping. The theoretical line dividing decaying and diverging motions is called the critical condition. The corresponding wind velocity is called the critical wind velocity for flutter or simply the flutter velocity at which the motion of the bridge deck tends to grow exponentially as shown in Figure 57.5a.

When flutter occurs, the oscillatory motions of all degrees of freedom in the structure couple to create a single frequency called the flutter frequency. Flutter is an instability phenomenon; once it takes place, the displacement is infinite by linear theory. Flutter may occur in both laminar and turbulent flows.

The self-excited forces acting on a unit deck length are usually expressed as a function of the flutter derivatives. The general format of the self-excited forces written in matrix form [2,6] for finite element analysis is

$$\begin{Bmatrix} L_{se} \\ D_{se} \\ M_{se} \end{Bmatrix} = \frac{1}{2}\rho U^2(2B)\left(\begin{bmatrix} \frac{k^2H_4^*}{B} & \frac{k^2H_6^*}{B} & k^2H_3^* \\ \frac{k^2P_4^*}{B} & \frac{k^2P_6^*}{B} & k^2P_3^* \\ k^2A_4^* & k^2A_6^* & k^2A_3^*B \end{bmatrix}\begin{Bmatrix} h \\ p \\ \alpha \end{Bmatrix} + \begin{bmatrix} \frac{kH_1^*}{U} & \frac{kH_5^*}{U} & \frac{kH_2^*B}{U} \\ \frac{kP_1^*}{U} & \frac{kP_5^*}{U} & \frac{kP_2^*B}{U} \\ \frac{kA_1^*B}{U} & \frac{kA_5^*B}{U} & \frac{kA_2^*B^2}{U} \end{bmatrix}\begin{Bmatrix} \dot{h} \\ \dot{p} \\ \dot{\alpha} \end{Bmatrix}\right) \tag{57.17}$$

$$= U^2\left[F_d\right]\{q\} + U^2\left[F_v\right]\{\dot{q}\}$$

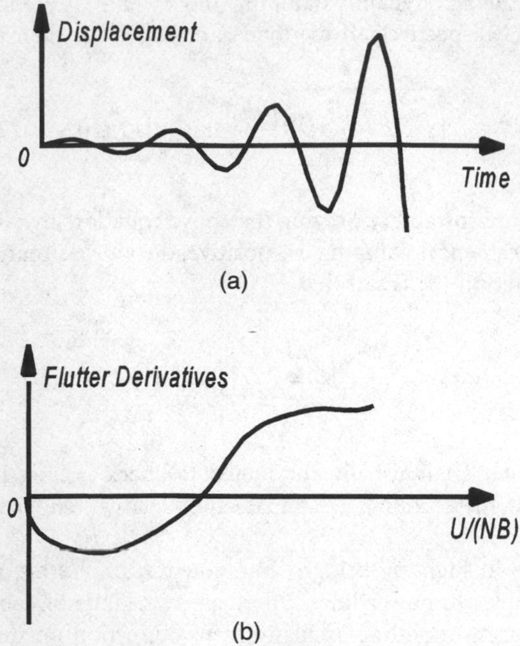

FIGURE 57.5 Explanation of flutter. (a) Bridge flutter vibration; (b) typical flutter derivations.

where L_{se}, D_{se}, and M_{se} = self-excited lift force, drag force, and pitch moment, respectively; h, p, and α = displacements at the center of a deck in the directions corresponding to L_{se}, D_{se}, and M_{se}, respectively; ρ = mass density of air; B = deck width; H_i^*, P_i^*, and A_i^* (i = 1 to 6) = generalized flutter derivatives; $k = B\omega/\bar{U}$ = reduced frequency; ω = oscillation circular frequency; $\bar{U}$ = mean wind velocity; and $[F_d]$ and $[F_v]$ = flutter derivative matrices corresponding to displacement and velocity, respectively.

While the flutter derivatives H_i^* and A_i^* have been experimentally determined for i = 1 to 4, the term P_i^* is theoretically derived in state-of-the-art applications. The other flutter derivatives (for i = 5 and 6) have been neglected in state-of-the-art analysis.

In linear analyses, the general aerodynamic motion equations of bridge systems are expressed in terms of the generalized mode shape coordinate $\{\xi\}$

$$[M*]\{\ddot{\xi}\}+([D*]-U^2[AD*])\{\dot{\xi}\}+([K*]-U^2[AS*])\{\xi\}=0 \tag{57.18}$$

where $[M^*]$, $[D^*]$, and $[K^*]$ = generalized mass, damping, and stiffness matrices, respectively; and $[AS^*]$ and $[AD^*]$ = generalized aerodynamic stiffness and aerodynamic damping matrices, respectively. Matrices $[M^*]$, $[D^*]$, and $[K^*]$ are derived the same way as in the general dynamic analysis. Matrices $[AS^*]$ and $[AD^*]$, corresponding to $[F_d]$ and $[F_v]$ in Eq. (57.17), respectively, are assembled from aerodynamic element forces. It is noted that even the structural and dynamic matrices $[K^*]$, $[M^*]$, and $[D^*]$ are uncoupled between modes, the motion equation is always coupled due to the coupling of aerodynamic matrices $[AS^*]$ and $[AD^*]$.

Flutter velocity, U, and flutter frequency, ω, are obtained from the nontrivial solution of Eq. (57.18) as

$$\left|\left(-\omega^2[M*]+[K*]-\bar{U}^2[AS*]+\omega([D*]-\bar{U}^2[AD*])\,i\right)\right|=0 \tag{57.19}$$

For a simplified uncoupled single degree of freedom, the above equation reduces to

$$\omega^2 = \frac{[K*] - \overline{U}^2[AS*]}{[M*]} \tag{57.20}$$

and

$$U_{cr}^2 = \frac{[D*]}{[AD*]} \tag{57.21}$$

Since the aerodynamic force $[AS^*]$ is relatively small, it can be seen that the flutter frequency in Eq. (57.20) is close to the natural frequency $[K^*]/[M^*]$. Equation (57.21) can be also derived from Eq. (57.18) as the zero-damping condition. Zero-damping cannot occur unless $[AD^*]$ is positive. The value of $[AD^*]$ depends on the flutter derivatives. An examination of the flutter derivatives gives a preliminary judgment of the flutter behavior of the section. Necessary section modifications should be made to eliminate the positive flutter derivatives as shown in Figure 57.5b, especially the A_2^* and H_1^*. The A_2^* controls the torsional flutter and the H_1^* controls the vertical flutter. It can be seen from Eq. (57.21) that an increase in the mechanical damping $[D^*]$ increases the flutter velocity. It should be noted that for a coupled flutter, zero-damping is a sufficient but not a necessary condition.

A coupled flutter is also called stiffness-driven or classical flutter. An uncoupled flutter is called damping driven flutter since it is caused by zero-damping. Since flutter of a suspension bridge is usually controlled by its first torsional mode, the terminology *flutter* was historically used for a torsional aerodynamic instability. Vertical aerodynamic instability is traditionally treated in a quasi-static approach, i.e., as is galloping. In recent literature, flutter is any kind of aerodynamic instability due to self-excited forces, whether vertical, torsional, or coupled vibrations.

Turbulence is assumed beneficial for flutter stability and is usually ignored. Some studies include turbulence effect by treating along-wind velocity U as mean velocity, $\overline{U}$, plus a turbulent component, $u(t)$. The random nature of $u(t)$ results in an equation of random damping and stiffness. Complicated mathematics, such as stochastic differentiation, need to be involved to solve the equation [7].

Time history and nonlinear analyses can be conducted on Eq. (57.18) to investigate postflutter behavior and to include the effects of both geometric and material nonlinearities. However, this is not necessary for most practical applications.

57.4.4 Buffeting

Buffeting is defined as the forced response of a structure to random wind and can only take place in turbulent flows. Turbulence resulting from topographical or structural obstructions is called oncoming turbulence. Turbulence induced by bridge itself is called signature turbulence. Since the frequencies of signature turbulence are generally several times higher than the important natural frequencies of the bridge, its effect on buffeting response is usually small.

Buffeting is a random vibration problem of limited displacement. The effects of buffeting and vortex shedding are similar, except that vibration is random in the former and periodic in the latter. Both buffeting and vortex shedding influence bridge service behavior and may result in fatigue damage that could lead to a eventual collapse of a bridge. Buffeting also influences ultimate strength behavior.

Similar to Eq. (57.17), the buffeting forces are expressed in the matrix form [2] for finite element analysis as

$$\begin{Bmatrix} L_b \\ D_b \\ M_b \end{Bmatrix} = \frac{1}{2}\rho \bar{U}^2 B \begin{bmatrix} 2C_L & \left(\dfrac{dC_L}{d\alpha}+C_D\right) \\ 2C_D & \dfrac{dC_D}{d\alpha} \\ 2C_M B & \dfrac{dC_M}{d\alpha}B \end{bmatrix} \begin{Bmatrix} \dfrac{u(t)}{\bar{U}} \\ \dfrac{w(t)}{\bar{U}} \end{Bmatrix} = \bar{U}^2[C_b]\{\eta\} \tag{57.22}$$

where C_D, C_L, and C_M = static aerodynamic coefficients for drag, lift, and pitch moment, respectively; α = angle of wind attack; $[C_b]$ = static coefficient matrix; and $\{\eta\}$ = vector of turbulent wind components normalized by mean wind velocity.

The equation of motion for buffeting is similar to Eq. (57.18), but with one more random buffeting force as

$$[M^*]\{\ddot{\xi}\}+([D^*]-U^2[AD^*])\{\dot{\xi}\}+([K^*]-U^2[AS^*])\{\xi\}=\bar{U}^2[f^*]_b\{\eta\} \tag{57.23}$$

Fourier transform of Eq. (57.23) yields

$$\mathcal{F}(\{\xi\})=\bar{U}^2[G_1]\{f^*\}_b * \mathcal{F}(\{\eta\}) \tag{57.24}$$

where

$$[G_1]=\frac{1}{\left(-\omega^2[M^*]+[K^*]-\bar{U}^2[AS^*]+\omega\left([D^*]-\bar{U}^2[AD^*]\right)\right)} \tag{57.25}$$

Similarly, taking the conjugate transform of Eq. (57.23) yields

$$\overline{\mathcal{F}(\{\xi\})^T}=\bar{U}^2\overline{\mathcal{F}(\{\eta\})^T}\{f^*\}_b^T[G_2]^T \tag{57.26}$$

where

$$[G_2]=\frac{1}{\left(-\omega^2[M^*]+[K]^*-\bar{U}^2[AS^*]-\omega\left([D^*]-\bar{U}^2[AD^*]\right)\right)} \tag{57.27}$$

The superscript T represents for the matrix transpose, and the overbar stands for the Fourier conjugate transform for the formula above. Multiplying Eqs. (57.24) and (57.26) gives the following spectral density of generalized coordinates

$$\begin{bmatrix} S_{\xi_1\xi_1} & \dots & S_{\xi_1\xi_m} \\ S_{\xi_i\xi_1} & \dots & S_{\xi_i\xi_m} \\ S_{\xi_m\xi_1} & \dots & S_{\xi_m\xi_m} \end{bmatrix} = \bar{U}^4[G_1]\{f^*\}_b \begin{bmatrix} S_{\eta_1\eta_1} & S_{\eta_1\eta_2} \\ S_{\eta_2\eta_1} & S_{\eta_2\eta_2} \end{bmatrix} \{f^*\}_b^T[G_2]^T \tag{57.28}$$

where $S_{\eta i \eta j}$ = spectral density of normalized wind components. The mean square of the modal and physical displacements can be derived from their spectral densities. Once the displacement is known,

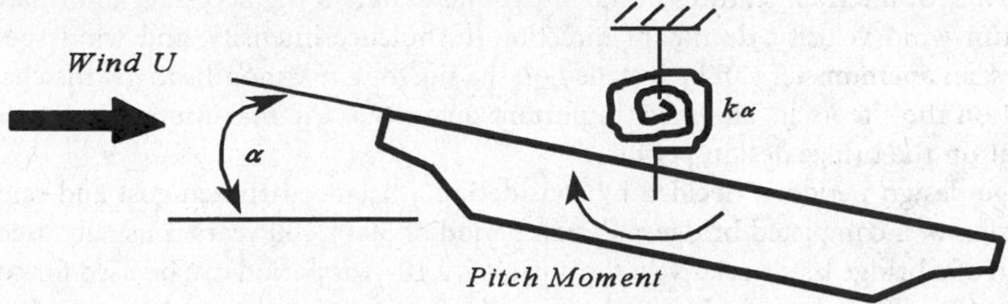

FIGURE 57.6 Explanation of torsional divergence.

the corresponding forces can be derived. The aerodynamic study should ascertain that no structural member is overstressed or overdeformed such that the strength and service limits are exceeded. For very long span bridges, a comfort criterion must be fulfilled under buffeting vibration.

57.4.5 Quasi-Static Divergence

Wind flowing against a structure exerts a pressure proportional to the square of the wind velocity. Wind pressure generally induces both forces and moments in a structure. At a critical wind velocity, the edge-loaded bridge may buckle "out-of-plane" under the action of a drag force or torsionally diverge under a wind-induced moment that increases with a geometric twist angle. In reality, divergence involves an inseparable combination of lateral buckling and torsional divergence.

Consider a small rotation angle as shown in Figure 57.6, the pitch moment resulting from wind is [3]

$$
\begin{aligned}
M_\alpha &= \frac{1}{2}\rho U^2 B^2 C_M(\alpha) \\
&= \frac{1}{2}\rho U^2 B^2 \left[C_{M0} + \frac{dC_m}{d\alpha}\Big|_{\alpha=0}\alpha \right]
\end{aligned}
\tag{57.29}
$$

When the pitch moment caused by wind exceeds the resisting torsional capacity, the displacement of the bridge diverges. Equating the aerodynamic force to the internal structural capacity gives

$$
k_\alpha \alpha - \frac{1}{2}\rho U^2 B^2 \left[C_{M0} + \frac{dC_m}{d\alpha}\Big|_{\alpha=0}\alpha \right] = 0 \tag{57.30}
$$

where k_α = spring constant of torsion. For an infinite α, we have the critical wind velocity for torsional divergence as

$$
U_{cr} = \sqrt{\frac{2k_\alpha}{\rho B^2 \dfrac{dC_M}{d\alpha}\Big|_{\alpha=0}}} \tag{57.31}
$$

57.5 Practical Applications

57.5.1 Wind Climate at Bridge Site

The wind climate at a particular bridge site is usually not available, but it is commonly decided according to the historical wind data of the nearest airport. The wind data are then analyzed considering the local terrain features of the bridge site to obtain the necessary information such as the maximum wind velocity, dominant direction, turbulence intensity, and wind spectrum. For large bridges, an anemometer can be installed on the site for a few months to get the characteristics of the wind on the site itself. The most important quantity is the maximum wind velocity, which is dependent on the bridge design period.

The bridge design period is decided by considering a balance between cost and safety. For the strength design of a completed bridge, a design period of 50 or 100 years is usually used. Since the construction of a bridge lasts a relatively short period, a 10-year period can be used for construction strength checking. This is equivalent to keeping the same design period but reducing the safety factor during construction.

Flutter is an instability phenomenon. Once it occurs, its probability of failure is assumed to be 100%. A failure probability of 10^{-5} per year for completed bridges represents an acceptable risk, which is equivalent to a design period of 100,000 years. Similarly, the design period of flutter during construction can be reduced to, say, 10,000 years. It should be noted that the design period does not represent the bridge service life, but a level of failure risk.

Once the design period has been decided, the maximum wind velocity is determined. Increasing the design period by one order of magnitude usually raises the wind velocity only by a few percent, depending on the wind characteristics. Wind velocity for flutter (stability) design is usually about 20% larger than that for buffeting (strength) design, although the design period for the former is several orders higher. Once wind characteristics and design velocity are available, a wind tunnel/analytical investigation is conducted.

57.5.2 Design Consideration

The aerodynamic behavior of bridges depends mainly on four parameters: structural form, stiffness, cross section shape and its details, and damping. Any significant changes that may affect these parameters need to be evaluated by a wind specialist.

1. *Structural form*: Suspension bridges, cable-stayed bridges, arch bridges, and truss bridges, due to the increase of rigidity in this order, generally have aerodynamic behaviors from worst to best. A truss-stiffened section, because it blocks less wind, is more favorable than a girder-stiffened one. But a truss-stiffened bridge is generally less stiff in torsion.
2. *Stiffness*: For long-span bridges, it is not economical to add more material to increase the stiffness. However, changing the boundary conditions, such as deck and tower connections in cable-stayed bridges, may significantly improve the stiffness. Cable-stayed bridges with A-shaped or inverted Y-shaped towers have higher torsional frequency than the bridges of H-shaped towers.
3. *Cross section shape and its details*: A streamlined section blocks less wind, thus has better aerodynamic behavior than a bluff section. Small changes in section details may significantly affect the aerodynamic behavior.
4. *Damping*: Concrete bridges have higher damping ratios than steel bridges. Consequently, steel bridges have more wind-induced problems than concrete bridges. An increase of damping can reduce aerodynamic vibration significantly.

Major design parameters are usually determined in the preliminary design stage, and then the aerodynamic behavior is evaluated by a wind specialist. Even if the bridge responds poorly under aerodynamic excitation, it is undesirable to change the major design parameters for reasons of scheduling and funding. The common way to improve its behavior is to change the section details. For example, changing the solid parapet to a half-opened parapet or making some venting slots [8] on the bridge deck may significantly improve the aerodynamic behavior. To have more choices on how to improve the aerodynamic behavior of long-span bridges, to avoid causing delays in the schedule, and to achieve an economical design, aerodynamics should be considered from the beginning.

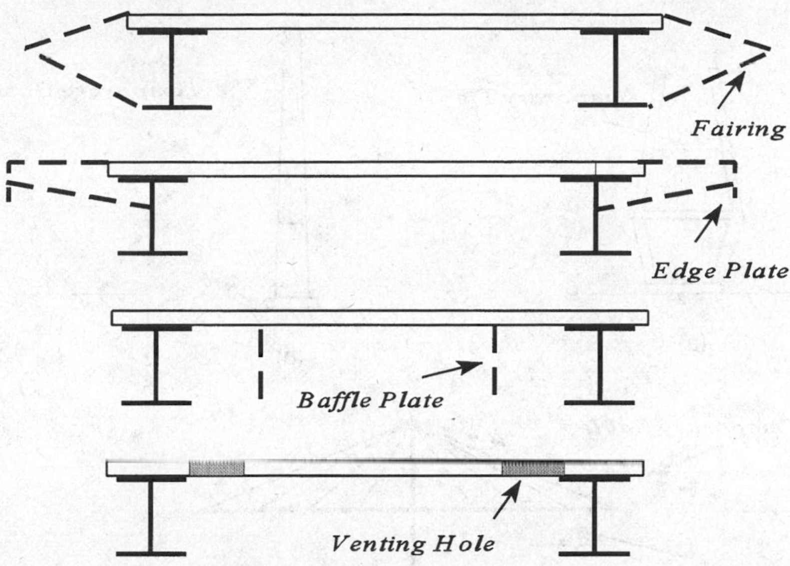

FIGURE 57.7 Typical aerodynamic modifications.

Although a streamlined section is always favorable for aerodynamic behavior, there have recently been many composite designs, due to their construction advantages, for long-span bridges. The composite section shapes, with the concrete deck on steel girders, are bluff and thus not good for aerodynamics, but can be improved by changing section details as shown in Figure 57.7 [9].

57.5.3 Construction Safety

The most common construction method for long-span bridges is segmental (staged) construction, such as balanced cantilever construction of cable-stayed bridges, tie-back construction of arch bridges, and float-in construction of suspended spans. These staged constructions result in different structural configurations during the construction time. Since some construction stages have lower stiffness and natural frequency than the completed bridges, construction stages are often more critical in terms of either strength of structural members or aerodynamic instability.

In the balanced cantilever construction of cable-stayed bridges, three stages are usually identified as critical, as shown in Figure 57.8. They are tower before completion, completed tower, and the stage with a longest cantilever arm. Reliable analytical solutions are not available yet, and wind tunnel testing is usually conducted to ensure safety. Temporary cables, tie-down, and bent are common countermeasures during the construction stages.

57.5.4 Rehabilitation

Aerodynamic design is a relatively new consideration in structural design. Some existing bridges have experienced wind problems because aerodynamic design was not considered in the original design. There are many measures to improve their aerodynamic behavior, such as structural stiffening, section streamlining, and installation of a damper. In the early days, structural stiffening was the major measure for this purpose. For example, the girders of the Golden Gate Bridge were stiffened in the 1950s, and Deer Isle Bridge in Maine has been stiffened since the 1940s by adding stays, cross-bracings, and strengthening the girders [10,11].

Although structural stiffening may have helped existing bridges survive many years of service, section streamlining has been commonly used recently. Streamlining the section is more efficient and less expensive than structural stiffening. Figure 57.9 shows the streamlined section of Deer Isle Bridge which has been proven very efficient [2,10].

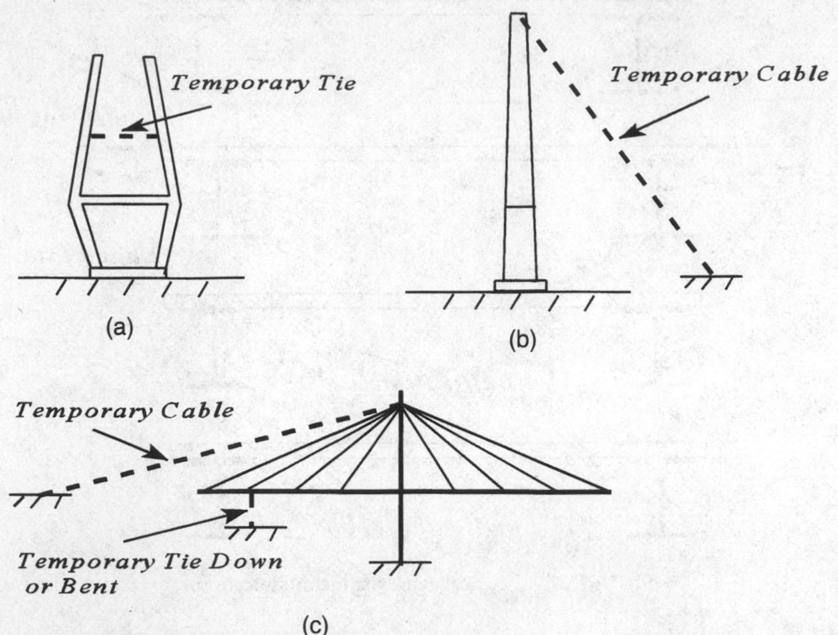

FIGURE 57.8 Typical construction stages. (a) Tower before completion; (b) completed tower; (c) stage with longest cantilever

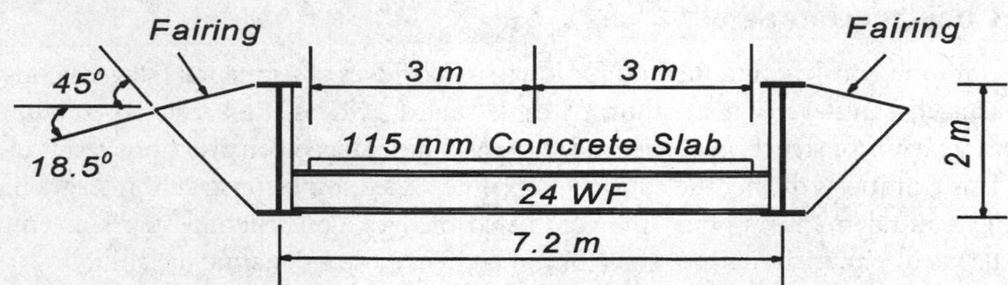

FIGURE 57.9 Deck section and fairings of Deer Isle Bridge.

57.5.5 Cable Vibration

A common wind-induced problem in long-span bridges is cable vibration. There are a number of wind-induced vibrations in cables, individually or as a group, such as vortex excitation, wake galloping, excitation of a cable by imposed movement of its extremities, rain/wind- and ice/wind-induced vibrations and buffeting of cables in strong turbulent winds.

While the causes of the cable vibrations are different from each other and the theoretical solutions complicated, some mitigating measures for these cable vibrations are shared:

1. *Raise damping*: This is an effective way for all kinds of cable vibrations. The cables are usually flexible and inherently low in damping; an addition of relatively small damping (usually at the cable ends) to the cable can dramatically reduce the vibration.
2. *Raise natural frequency*: The natural frequency depends on the cable length, the tension force, and the mass. Since the cable force and the mass are determined from the structural design, commonly the cable length is reduced by using spacers or cross cables.

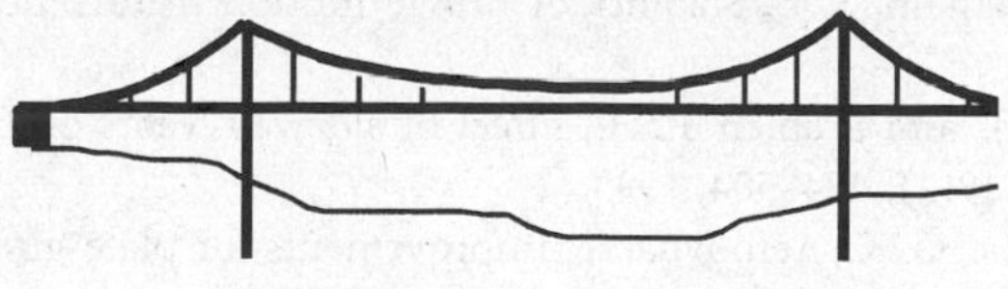

Suspension Bridge

FIGURE 57.10 Explanation of tuned mass damper.

3. *Change cable shape*: A change in the cable shape characteristics by increasing the surface roughness or adding protrusions to the cable surface reduces the rain/wind- and ice/wind-induced vibrations.
4. *Use other techniques*: Rearranging the cables or raising the cable mass density can also be used, but these are usually limited by other design constraints. Raising the mass may reduce the natural frequency, but it increases the damping and Scruton number ($m\zeta/\rho D^2$), and is overall beneficial.

57.5.6 Structural Control

Another way to improve aerodynamic behavior is to install either a passive or active control system on the bridges. A common practice in long-span bridges is the tuned mass damper (TMD). An example is the Normandy cable-stayed bridge in France. This bridge has a main span of 856 m. To reduce the horizontal vibration during construction due to buffeting, a TMD was installed. Wind tunnel testing showed that the TMD reduced the vibration by 30% [12,13].

The basic principles of a passive TMD are explained with an example shown in Figure 57.10. A TMD with spring stiffness k_2 and mass m_2 is attached to a structural mass m_1 which is excited by an external sinusoidal force $F\sin(\omega t)$. The vibration amplitude of this two-mass system is

$$X_1 = \frac{F\omega_m^2(\omega_d^2 - \omega^2)}{(\omega_d^2 - \omega^2)\left[(k_1 + k_2)\,\omega_m^2 - k_1\omega^2\right] - k_2\omega_d^2\omega_m^2}, \quad X_2 = \frac{F\omega_m^2\omega_d^2}{(\omega_d^2 - \omega^2)\left[(k_1 + k_2)\,\omega_m^2 - k_1\omega^2\right] - k_2\omega_d^2\omega_m^2} \tag{57.32}$$

where $\omega_m^2 = k_1/m_1$ and $\omega_d^2 = k_2/m_2$. It can be seen from Eq. (57.32) that by selection of the stiffness k_2 and mass m_2 such that ω_d equals ω, then the structure vibration X_1 is reduced to zero. Since the wind is not a single-frequency excitation, the TMD can reduce the vibration of bridges, but not to zero.

The performance of the passive TMD system can be enhanced by the addition of an active TMD, which can be done by replacing the passive damper device with a servo actuator system. The basic principle of active TMD is the feedback concept as used in modern control theory.

References

1. Berreby, D., The great bridge controversy, *Discover*, Feb., 26–33, 1992.
2. Cai, C. S., Prediction of Long-Span Bridge Response to Turbulent Wind, Ph.D. dissertation, University of Maryland, College Park, 1993.
3. Simiu, E. and Scanlan, R. H., *Wind Effects on Structures*, John Wiley & Sons, 2nd ed., New York, 1986.
4. Scanlan, R. H., State-of-the-Art Methods for Calculating Flutter, Vortex-Induced, and Buffeting Response of Bridge Structures, Report No. FHWA/RD-80/050, Washington, D.C., 1981.
5. Scruton, C., *An Introduction to Wind Effects on Structures*, Oxford University Press, New York, 1981.
6. Namini, A., Albrecht, P., and Bosch, H., Finite element-based flutter analysis of cable-suspended bridges, *J. Struct. Eng. ASCE*, 118(6), 1509–1526, 1992.

7. Lin, Y. K. and Ariaratnam, S. T., Stability of bridge motion in turbulent winds, *J. Struct. Mech.*, 8(1), 1–15, 1980.
8. Ehsan, F., Jones, N. J., and Scanlan, R. H., Effect of sidewalk vents on bridge response to wind, *J. Struct. Eng. ASCE,* 119(2), 484–504, 1993.
9. Irwin, P. A. and Stone, G. K., Aerodynamic improvements for plate-girder bridges, in *Proceedings, Structures Congress,* ASCE, San Francisco, CA, 1989.
10. Bosch, H. R., A Wind Tunnel Investigation of the Deer Isle-Sedgwick Bridge, Report No. FHWA-RD-87-027, Federal Highway Administration, McLean, VA, 1987.
11. Kumarasena, T., Scanlan, R. H., and Ehsan, F., Wind-induced motions of Deer Isle Bridge, *J. Struct. Eng. ASCE,* 117(11), 3356–3375, 1991.
12. Sorensen, L., The Normandy Bridge, the steel main span, in *Proc. 12th Annual International Bridge Conference,* Pittsburgh, PA, 1995.
13. Montens, S., Gusty wind action on balanced cantilever bridges, in *New Technologies in Structural Engineering,* Lisbon, Portugal, 1997.

58

Cable Force Adjustment and Construction Control

Danjian Han
South China University of Technology

Quansheng Yan
South China University of Technology

58.1 Introduction

Due to their aesthetic appeal and economic advantages, many cable-stayed bridges have been built over the world in the last half century. With the advent of high-strength materials for use in the cables and the development of digital computers for the structural analysis and the cantilever construction method, great progress has been made in cable-stayed bridges[1,2]. The Yangpu Bridge in China with a main span of 602 m completed in 1993, is the longest cable-stayed bridge with a composite deck. The Normandy Bridge in France, completed in 1994, with main span of 856 m is now the second-longest-span cable-stayed bridge. The Tatara Bridge in Japan, with a main span of 890 m, was opened to traffic in 1999. More cable-stayed bridges with larger spans are now in the planning.

Cable-stayed bridges are featured for their ability to have their behavior adjusted by cable stay forces [3–5]. Through the adjustment of the cable forces, the internal force distribution can be optimized to a state where the girder and the towers are compressed with little bending. Thus, the performance of material used for deck and pylons can be efficiently utilized.

0-8493-7434-0/00/$0.00+$.50

During the construction of a cable-stayed bridge there are two kinds of errors encountered frequently,[6,13]: one is the tension force error in the jacking cables, and the other is the geometric error in controlling the elevation of the deck. During construction the structure must be monitored and adjusted; otherwise errors may accumulate, the structural performance may be substantially influenced, or safety concerns may arise. With the widespread use of innovative construction methods, construction control systems play a more and more important role in construction of cable-stayed bridges [18,19].

There are two ways of adjustment: adjustment of the cable forces and adjustment of the girder elevations [7]. The cable-force adjustment may change both the internal forces and the configuration of the structure, while the elevation adjustment only changes the length of the cable and does not induce any change in the internal forces of the structure.

This chapter deals with two topics: cable force adjustment and construction control. The methods for determing the cable forces are discussed in Section 58.2, then a presentation of the cable force adjustment is given in Section 58.3. A simulation method for a construction process of prestressed concrete (PC) cable-stayed bridge is illustrated in Section 58.4, and a construction control system is introduced in Section 58.5.

58.2 Determination of Designed Cable Forces

For a cable-stayed bridge the permanent state of stress in a structure subjected to dead load is determined by the tension forces in the cable stays. The cable tension can be chosen so that bending moments in the girders and pylons are eliminated or at least reduced as much as possible. Thus the deck and pylon would be mainly under compression under the dead loads [3,10].

In the construction period the segment of deck is corbeled by cable stays and each cable placed supports approximately the weight of one segment, with the length corresponding to the longitudinal distance between the two stays. In the final state the effects of other dead loads such as wearing surface, curbs, fence, etc., as well as the traffic loads, must also be taken into account. For a PC cable-stayed bridge, the long-term effects of concrete creep and shrinkage must also be considered [4].

There are different methods of determining the cable forces and these are introduced and discussed in the following.

58.2.1 Simply Supported Beam Method

Assuming that each stayed cable supports approximately the weight of one segment, corresponding to the longitudinal distance between two stays, the cable forces can be estimated conveniently [3,4]. It is necessary to take into account the application of other loads (wearing surface, curbs, fences, etc.). Also, the cable is placed in such a way that the new girder element is positioned correctly, with a view to having the required profile when construction is finished.

Due to its simplicity and easy hand calculation, the method of the simply supported beam is usually used by designers in the tender and preliminary design stage to estimate the cable forces and the area of the stays. For a cable-stayed bridge with an asymmetric arrangement of the main span and side span or for the case that there are anchorage parts at its end, the cable forces calculated by this method may not be evenly distributed. Large bending moments may occur somewhere along the deck and/or the pylons which may be unfavorable.

58.2.2 Method of Continuous Beam on Rigid Supports

By assuming that under the dead load the main girder behaves like a continuous beam and the inclined stay cables provide rigid supports for the girder, the vertical component of the forces in stay cables are equal to the support reactions calculated on this basis [4,10]. The tension in the anchorage cables make it possible to design the pylons in such a way that they are not subjected to large bending moments when the dead loads are applied.

This method is widely used in the design of cable-stayed bridges. Under the cable forces calculated by this method, the moments in the deck are small and evenly distributed. This is especially favorable for PC cable-stayed bridges because the redistribution of internal force due to the effects of concrete creep could be reduced.

58.2.3 Optimization Method

In the optimization method of determining the stresses of the stay cables under permanent loads, the criteria (objective functions) are chosen so the material used in girders and pylons is minimized [8,11]. When the internal forces, mainly the bending moments, are evenly distributed and small, the quantity of material reaches a minimum value. Also the stresses in the structure and the deflections of the deck are limited to prescribed tolerances.

In a cable-stayed bridge, the shear deformations in the girder and pylons are neglected, the strain energy can be represented by

$$U = \frac{1}{2}\int_0^L \frac{M^2}{2\,EI}dx + \frac{1}{2}\int_0^L \frac{N^2}{2\,EA}dx \tag{58.1}$$

where EI is the bending stiffness of girder and pylons and EA is the axial stiffness.

It can be given in a discrete form when the structure is simulated by a finite-element model as

$$U = \sum_{i=1}^{N} \frac{L_i}{4\,E_i}\left(\frac{M_{il}^2 + M_{ir}^2}{I_i} + \frac{N_{il}^2 + N_{ir}^2}{A_i}\right) \tag{58.2}$$

where N is the total number of the girder and pylon elements, L_i is the length of the ith element, E is the modulus of elasticity, I_i and A_i are the moment of inertia and the sections area, respectively. M_{ir}, M_{il}, N_{ir}, N_{il} are the moments and the normal forces in the left and right end section of the ith element, respectively.

Under the application of dead loads and cable forces the bending moments and normal forces of the deck and pylon are given by

$$\{M\} = \{M_D\} + \{M_P\} = \{M_D\} + [S_M]*\{P_0\} \tag{58.3a}$$

$$\{N\} = \{N_D\} + \{N_P\} = \{N_D\} + [S_N]*\{P_0\} \tag{58.3b}$$

where $\{M_D\}$ and $\{M_P\}$ are the bending moment vectors induced by dead loads and the cable forces, respectively; $[S_M]$ is the moment influence matrix; $[S_N]$ is the normal force influence matrix, the component S_{ij} of influence matrix represents changes of the moment or the normal force in the ith element induced by the jth unit cable force. And $\{N_D\},\{N_P\}$ are the normal force vectors induced by dead loads and cable forces, respectively. $\{P_0\}$ is the vector of cable forces.

The corresponding displacements in deck and pylon are given as

$$\{F\} = \{F_D\} + \{F_P\} = \{F_D\} + [S_F]*\{P_0\} \tag{58.4}$$

where $\{F\}$ is the displacement vector, $\{S_F\}$ is the displacement influence matrix, $\{F_D\}$ and $\{F_P\}$ are the displacement vectors induced by dead loads and by cable forces respectively.

Substitute Eqs. (58.3a) and (58.3b) into Eq. (58.2), and replace the variables by

$$\{\overline{M}\}=[A]\{M\},\{\overline{N}\}=[B]\{N\} \tag{58.5}$$

in which $[A]$ and $[B]$ are diagonal matrices:

$$[A]=\text{Diag}\left[\sqrt{L_1/4E_1I_1},\sqrt{L_2/4E_2I_2},\ldots,\sqrt{L_n/4E_nI_n}\right]$$

$$[B]=Diag\left[\sqrt{L_1/4E_1A_1},\sqrt{L_2/4E_2A_2},\ldots,\sqrt{L_n/4E_nA_n}\right]$$

Then the strain energy of the cable-stayed bridge can be represented in matrix form as

$$U=\{P_0\}^T[\overline{S}]^T[\overline{S}]\{P_0\}+2\{\overline{P}_D\}^T[\overline{S}]\{P_0\}+\{\overline{P}_D\}^T\{\overline{P}_D\} \tag{58.6}$$

in which $[\overline{S}]=(\overline{S}_M,\overline{S}_N)^T=[A,B](S_M,S_N)^T$, $\{\overline{P}_D\}=\{M_D,N_D\}^T$.

Now, we want to minimize the strain energy of structure, i.e., to let

$$\partial U/\partial P_0=0 \tag{58.7}$$

under the following constraint conditions:

1. The stress range in girders and pylons must satisfy

$$\{\sigma\}_L\leq\{\sigma\}\leq\{\sigma\}_U \tag{58.8}$$

   in which $\{\sigma\}$ is the maximum stress value vector. And $\{\sigma\}_L,\{\sigma\}_U$ are vectors of the lower and upper bounds.
2. The stresses in stay cables are limited so that the stays can work normally.

$$\{\sigma\}_{LC}\leq\left\{\tfrac{P_{0C}}{A_C}\right\}\leq\{\sigma\}_{UC} \tag{58.9}$$

   in which A_C is the area of a stay, P_{0C} is the cable force and $\{\sigma\}_{LC},\{\sigma\}_{UC}$ represent the lower and upper bounds, respectively.
3. The displacements in the deck and pylon satisfy

$$\{|D_i|\}\leq\{\Delta\} \tag{58.10}$$

   in which the left hand side of Eqs. (58.10) is the absolute value of maximum displacement vector and the right-hand side is the allowable displacement vector.

Eqs. (58.6) and (58.7) in conjunction with the conditions (58.8) through (58.10) is a standard quadric programming problem with constraint conditions. It can be solved by standard mathematical methods.

Since the cable forces under dead loads determined by the optimization method are equivalent to the cable force under which the redistribution effect in the structure due to concrete creep is minimized [8], the optimization method is used more widely in the design of PC cable-stayed bridges.

58.2.4 Example

For a PC cable-stayed bridge as shown in Figures 58.1 and 58.2, the forces of cable stays under permanent loads (not taking into account the creep and shrinkage) can be determined by the above methods. The results obtained are shown in Figures 58.3 and 58.4. In these figures SB represents the "Simply Supported Beam Method," CBthe "Continuous Beam on Rigid Support Method," OPT the "Optimization Method," and M represents middle span, S side span numbering from the pylon location.

As can be seen, because the two ends of the cable-stayed bridge have anchored parts the cable forces located in these two regions obtained by the method of simply supported beam (SB) and by the method of continuous beam on rigid supports (CB) are not evenly distributed. The cable forces in the region near the pylon are very different with the three methods. In the other regions there is no prominent difference among the cable forces obtained by SB, CB, and OPT .

Generally speaking, the differences of cable forces under dead loads obtained by the above methods are not so significant. The method of simply supported beam is the most convenient and the easiest to use. The method of continuous beam on rigid supports is suitable to use in the design of PC cable-stayed bridges. The optimization method is based on a rigorous mathematical model. In practical engineering applications the choice of the above methods is very much dependent on the design stage and designer preference.

58.3 Adjustment of the Cable Forces

58.3.1 General

During the construction or service stage, many factors may induce errors in the cable forces and elevation of the girder, such as the operational errors in tensioning stays or the errors of elevation in laying forms [14,16,17]. Further, the discrepancies of parameter values between design and reality such as the modules of elasticity, the mass density of concrete, the weight of girder segments may give rise to disagreements between the real structural response and the theoretical prediction [13].If the structure is not adjusted to reduce the errors during construction, they may accumulate and the structure may deviate away from the intended design aim. Moreover, if the errors are greater than the allowable limits, they may give rise to the unfavorable effects to the structure. Through cable force adjustment, the construction errors can be eliminated or reduced to an allowable tolerance. In the service stage, because of concrete creep effects, cable force may need to be adjusted; thus, an optimal structural state can be reached or recovered.

58.3.2 Influence Matrix of the Cable Forces

Assuming that a unit amount of cable force is adjusted in one cable stay, the deformations and internal forces of the structure can be calculated by finite-element model. The vectors of change in deformations and internal forces are defined as influence vectors. In this way, the influence matrices can be formed for all the stay cables.

58.3.3 Linear Programming Method [7]

Assume that there are n cable stays whose cable forces are going to be adjusted, the adjustments are T_i $(i = 1,2,\ldots,n)$, these values form a vector of cable force adjustment $\{T\}$ as

$$\{T\} = \{T_1, T_2, \ldots, T_n\}^T \tag{58.11}$$

Denote internal force influence vector $\{P_l\}$ as

$$\{P_l\} = (P_{l1}, P_{l2}, \ldots, P_{ln})^T \quad (l = 1,2,\ldots,m) \tag{58.12}$$

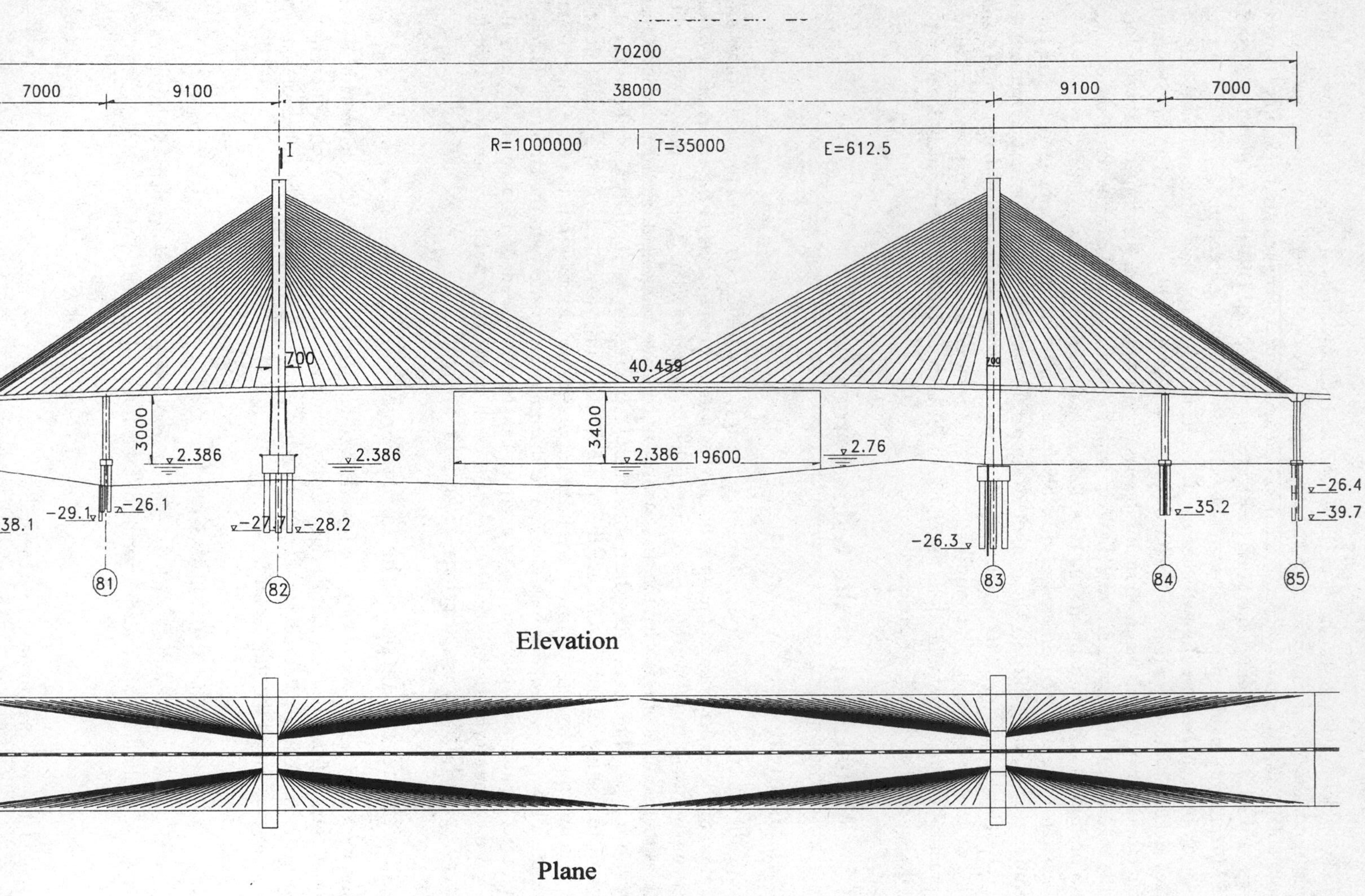

FIGURE 58.1 General view of a PC cable-stayed bridge.

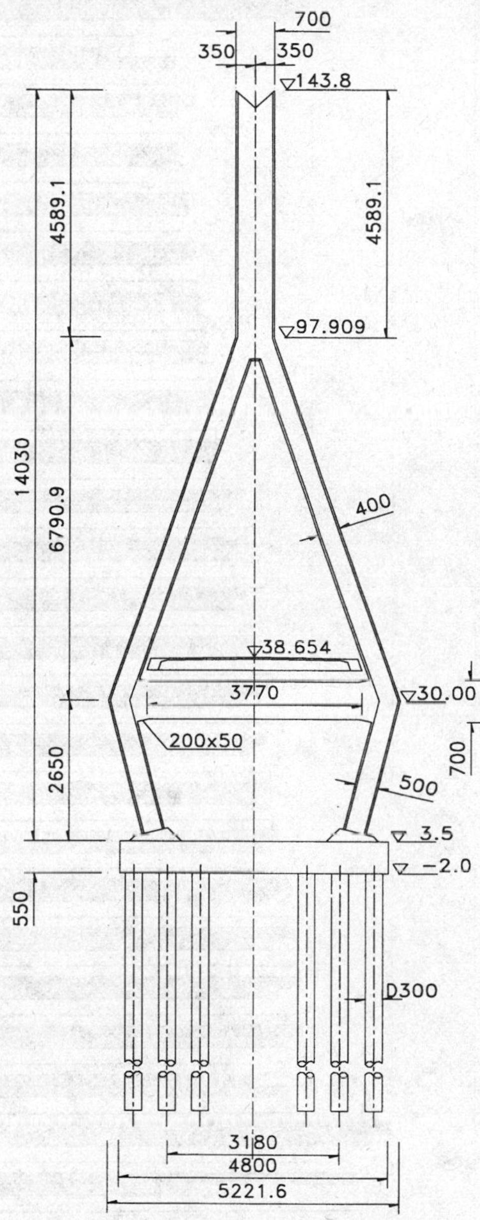

FIGURE 58.2 Side view of tower.

in which m is the number of sections of interest, P_{ij} is the internal force increment due to a unit tension of the jth cable. Denote displacement influence vector $\{D_i\}$ as

$$\{D_i\} = (D_{i1}, D_{i2}, \ldots, D_{in})^T \qquad (i = 1, 2, \ldots, k) \tag{58.13}$$

in which k is the number of sections of interest, D_{ij} is the displacement increment at section i due to a unit tension of the jth cable. Thus, the influence matrices of internal forces and displacements are given by

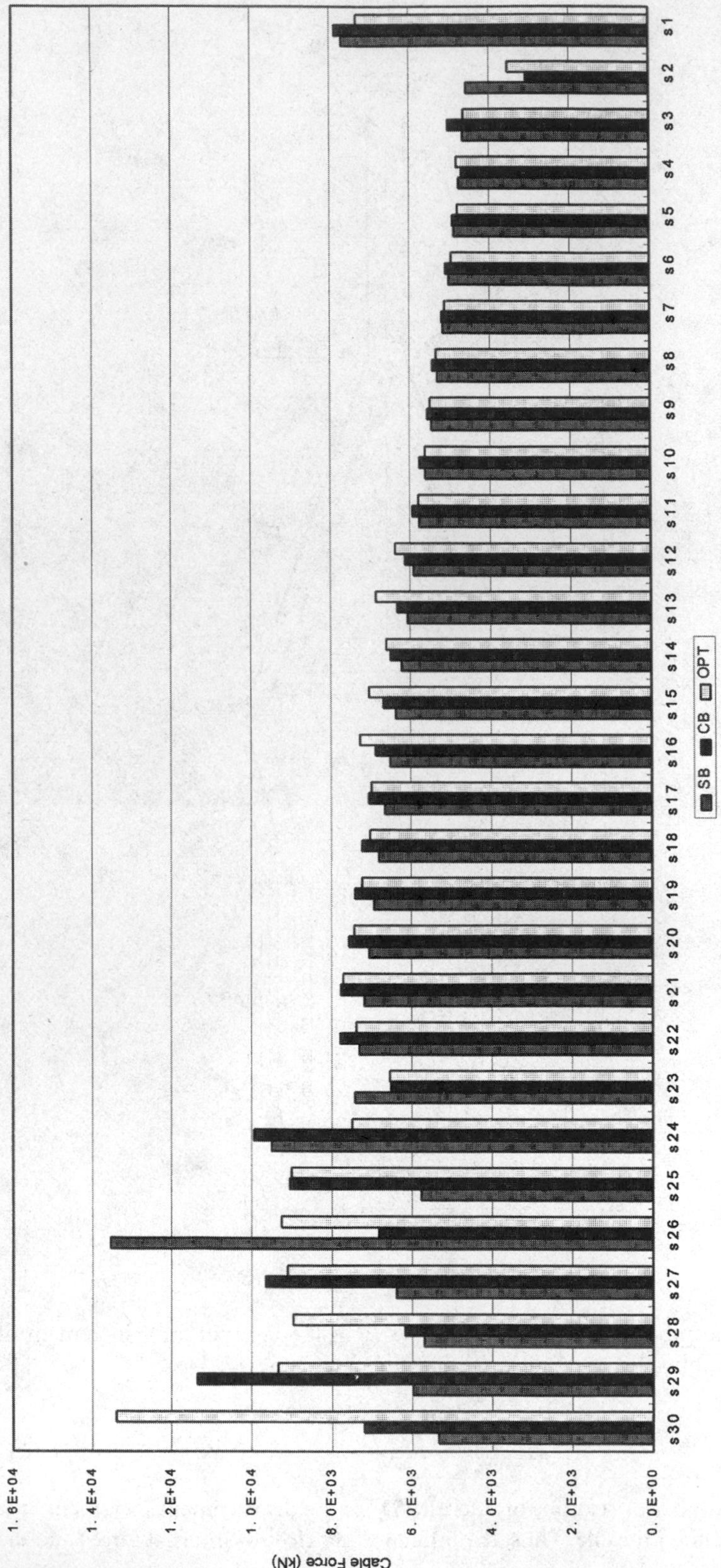

FIGURE 58.3 Comparison of the cable forces (kN) (side span).

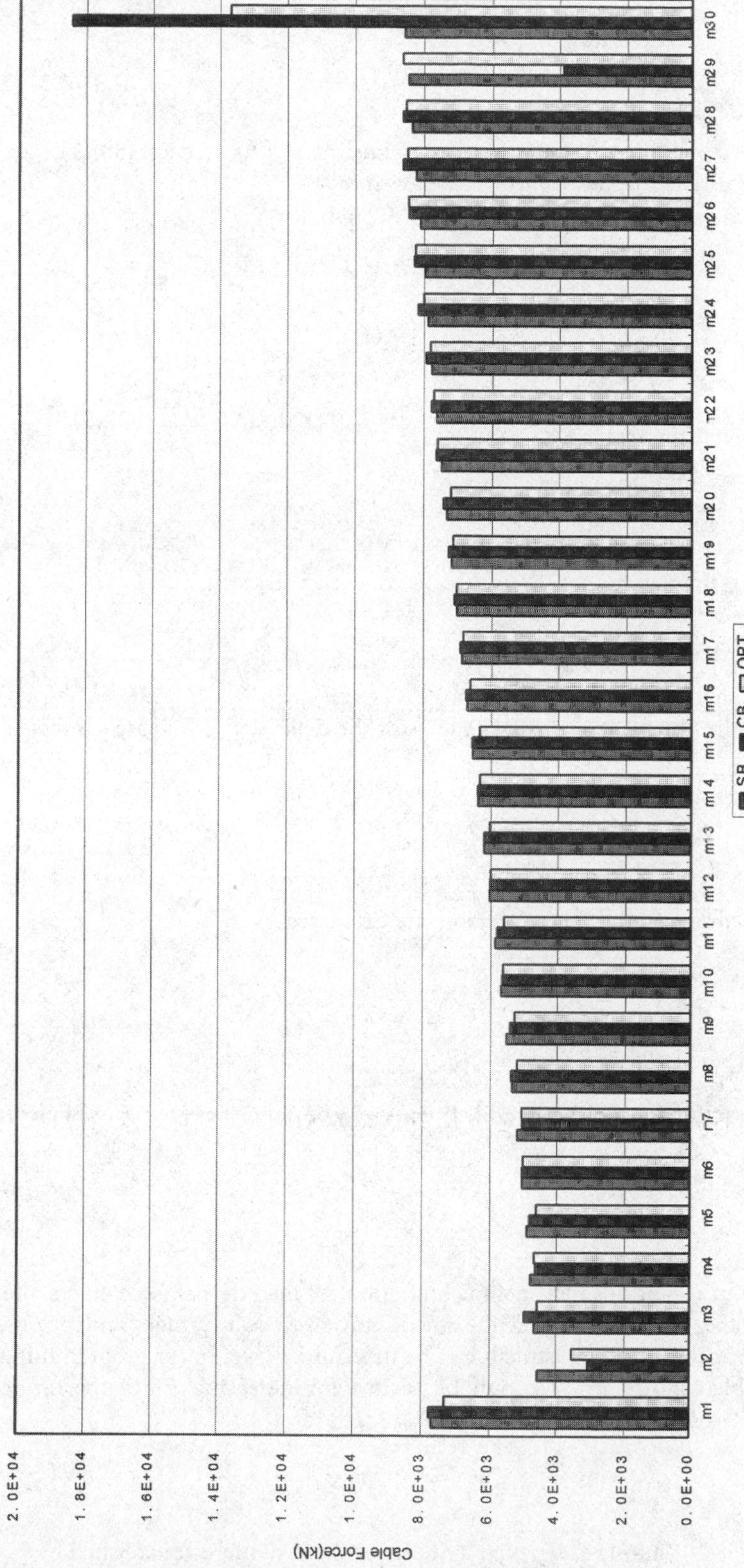

FIGURE 58.4 Comparison of the cable force (middle span).

$$\{P\} = (P_1, P_2, ..., P_m)^T \tag{58.14a}$$

$$\{D\} = (D_1, D_2, ..., D_k)^T \tag{58.14b}$$

respectively. Under application of the cable force adjustment $\{T\}$ (see Eq. (58.11)), the increment of the internal forces and displacements can be given by

$$\{\Delta P\} = \{P\}^T \{T\} \tag{58.15a}$$

$$\{\Delta D\} = \{D\}^T \{T\} \tag{58.15b}$$

respectively.

Denote deflection error vector $\{H\}$ as

$$\{H\} = (h_1, h_2, ..., h_m)^T \tag{58.16}$$

Denote vector of internal force $\{N\}$ as

$$\{N\} = (N_1, N_2, ..., N_m)^T \tag{58.17}$$

After cable force adjustments, the absolute values of the deflection errors are expressed by

$$|\lambda_k| = \left| \sum_{i=1}^{n} D_{ik} T_i - h_k \right| \tag{58.18}$$

and the absolute values of internal force errors are expressed by

$$|q_l| = \left| \sum_{i=1}^{n} P_{il} T_i - N_l \right| \tag{58.19}$$

The objective function for cable force adjustments may be defined as the errors of girder elevation, i.e.,

$$\min |\lambda_k| \tag{58.20}$$

and the constraint conditions may include limitations of the internal force errors, the upper and lower bounds of the cable forces, and the maximum stresses in girders and pylons. Then the optimum values of cable force adjustment can be determined by a linear programming model.

The value of cable adjustment $\{T\}$ could be positive for increasing or negative for decreasing of the cable forces. Introduce two auxiliary variables T_{1i}, T_{2i} as

$$T_i = T_{1i} - T_{2i} \qquad T_{1i} \geq 0,\ T_{2i} \geq 0 \tag{58.21}$$

Substitute Eq. (58.18) into (58.20), then a linear program model is established by

$$\text{min:} \quad \lambda_k \tag{58.22}$$

$$\text{subject to:} \quad \sum_{i=1}^{n} p_{li}(T_{1i} - T_{2i}) \geq N_l - \xi\overline{p}_l \qquad (l = 1,2,\ldots m) \tag{58.23a}$$

$$\sum_{i=1}^{n} p_{li}(T_{1i} - T_{2i}) \leq N_l + \xi\overline{p}_l \qquad (l = 1,2,\ldots m) \tag{58.23b}$$

$$\sum_{i=1}^{n} D_{ji}(T_{1i} - T_{2i}) \geq h_j - d_j \qquad (j = 1,2,\ldots k) \tag{58.23c}$$

$$\sum_{i=1}^{n} D_{ji}(T_{1i} - T_{2i}) \geq h_j - d_j \qquad (j = 1,2,\ldots k) \tag{58.23d}$$

$$T_{1i} - T_{2i} \leq \eta\overline{T}_i, \quad T_{1i} - T_{2i} \geq -\eta\overline{T}_i \qquad (I = 1,2,\ldots,n) \tag{58.23e}$$

in which $\overline{p}_l$ is the design value of internal force at section l, ξ is the allowable tolerance in percentage of the internal force. $\overline{T}_i$ is the design value of the cable force, η is the allowable tolerance in percentage of the cable forces.

Equations (58.22) and (58.23) form a standard linear programming problem which can be solved by mathematical software.

58.3.4 Order of Cable Adjustment

The adjustment values can be determined by the above method; however, the adjustments must be applied at the same time to all cables, and a great number of jacks and workers are needed [7]. In performing the adjustment, it is preferred that the cable stays are tensioned one by one.

When adjusting the cable force individually, the influence of the other cable forces must be considered. And since any cable must be adjusted only one time, the adjustment values can be calculated through the influence matrix of cable force.

$$\{\overline{T}\} = [S]\{T\} \tag{58.24}$$

where $\{\overline{T}\} = \{\overline{T}_1, \overline{T}_2, \ldots \overline{T}_n\}$ is the vector of actual adjustment value of cable tension. $[S]$ is the influence matrix of cable tension, whose component S_{ij} represents tension change of the *j*th cable when the *i*th cable changes a unit amount of force.

58.4 Simulation of Construction Process

58.4.1 Introduction

Segmental construction techniques have been widely used in construction of cable-stayed bridges. In this technique, the pylon(s) is built first; then the girder segments are erected one by one and supported by the inclined cables hung from the pylon(s). It is evident that the profile of the main girder and the final tension forces in the cables are strongly related to the erection method and the construction scheme. It is therefore important that the designer should be aware of the construction process and the necessity to look into the structural performance at every stage of construction [9,12].

In any case, structural safety is the most important issue. Since the stresses in the girder and pylon(s) are related to the cable tensions. Thus the cable forces are of great concern. Further, during construction, the geometric profile of the girder is also very important. It is clear that if the profile of the girder were not smooth or, finally, the cantilever ends could not meet together, then the construction might experience some trouble. The profile of the girder or the elevation of the bridge segments is mainly controlled by the cable lengths. Therefore, the cable length must be appropriately set at the erection of each segment. It also should be noted that in the construction process, the internal forces of the structure and the elevation of the girder could vary because usually the bridge segments are built by a few components at a time and the erection equipment is placed at different positions during construction and because some errors such as the weight of the segment and the tension force of the cable, etc. may occur. Thus, monitoring and adjustment are absolutely needed.

To reach the design aim, an effective and efficient simulation of the construction process step by step is very necessary. The objectives of the simulation analysis are [4,12]:

1. To determine the forces required in cable stays at each construction stage;
2. To set the elevation of the girder segment;
3. To find the consequent deformation of the structure at each construction stage;
4. To check the stresses in the girder and pylon sections.

The simulation methods are introduced and discussed in detail in the following sections. In Section 58.4.2, the technique of forward analysis is presented to simulate the assemblage process. Creep effects can be considered; however, the design aim may not be successfully achieved by such simulation because it is not so easy to determine the appropriate lengths of the cable stays which make the final elevation to achieve the design profile automatically. Another technique presented in Section 58.4.3 is the backward disassemblage analysis, which starts with the final aim of the structural state and disassembles segment by segment in a reverse way. The disadvantage of this method is that the creep effects may not be able to be defined. However, values obtained from the assemblage process may be used in this analysis. These two methods may be alternatively applied until convergence is reached.

It is noted that the simulation is only limited to that of the erection of the superstructure.

58.4.2 Forward Assemblage Analysis

Following the known erection procedure, a simulation analysis can be carried out by the finite-element method. This is the so-called forward assemblage analysis. It has been used to simulate the erection process for PC bridges built by the free cantilever method.

Concerning finite-element modeling, the structure may be treated as a plane frame or a space frame [4]. A plane frame model may be good enough for construction simulation because transverse loads, such as wind, can generally be ignored. In a plane frame model, the pylon(s) and the girder are modeled by some beam elements, while the stays are modeled as two-node bar elements with Ernst modules [3,4] by which the effects of cable sag can be taken into account. The structural configuration is changed stage by stage. Typically, in one assemblage stage, a girder segment treated as one or several beam elements is connected to the existing structure, while its weight is treated as a load to apply to the element. Also, the cable force is applied. Then an analysis is performed and the structure is changed to a new configuration.

In finite-element modeling, several factors such as the construction loads (weight of equipment and traveling carriage) and effects of concrete creep and shrinkage, must be considered in detail.

Traveling carriages are specially designed for construction of a particular bridge project. Generally there are two kinds of carriages. One is cantilever type (Figure 58.5a). The traveling carriages are mounted near the ends of girders, like a cantilever to support the next girder segment. In this case, the weight of the carriage can be treated as an external load applied to the end of the girder.

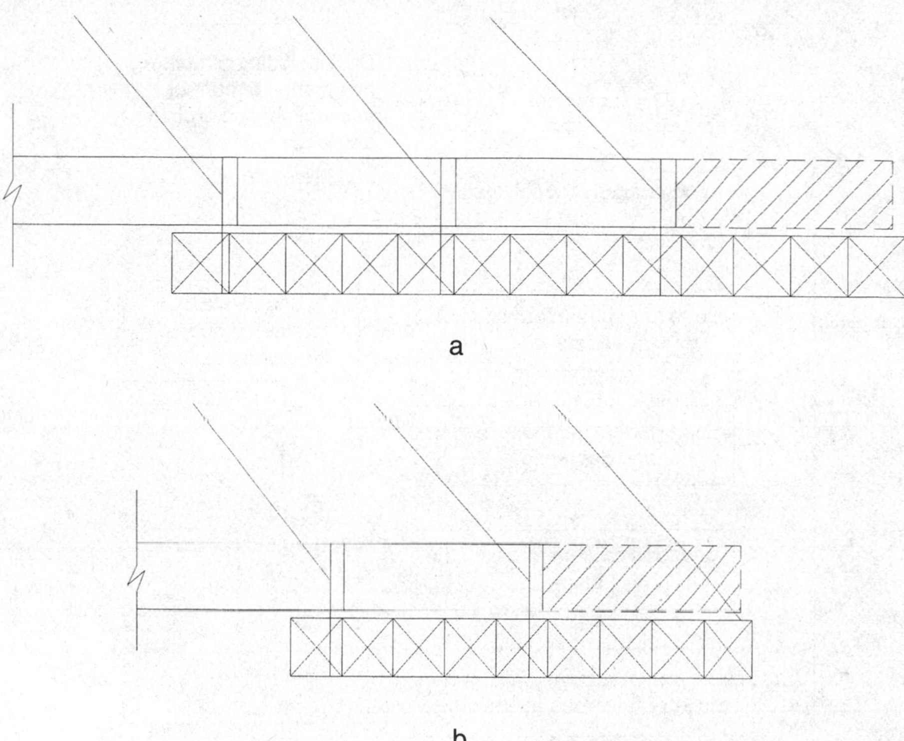

FIGURE 58.5 (a) Cantilever carriage; (b) cable-supported cantilever carriage.

With the development of multiple cable systems, the girder with lower height becomes more flexible. The girder itself is not able to carry the cantilever weights of the carriage and the segment. Then an innovative erection technique was proposed[1]. And another type of carriage is developed. This new idea is to use permanent stays to support the form traveler (Figure 58.5b) so that the concrete can be poured *in situ* [1,9]. This method enjoys considerable success at present because of the undeniable economic advantages. Its effectiveness has been demonstrated by many bridge practice. For the erecting method using the later type of carriage, the carriage works as a part of the whole structure when the segmental girder is poured *in situ.* Thus, the form traveler must be included in the finite-element model to simulate construction. A typical flowchart of forward assemblage analysis is shown in Figure 58.6.

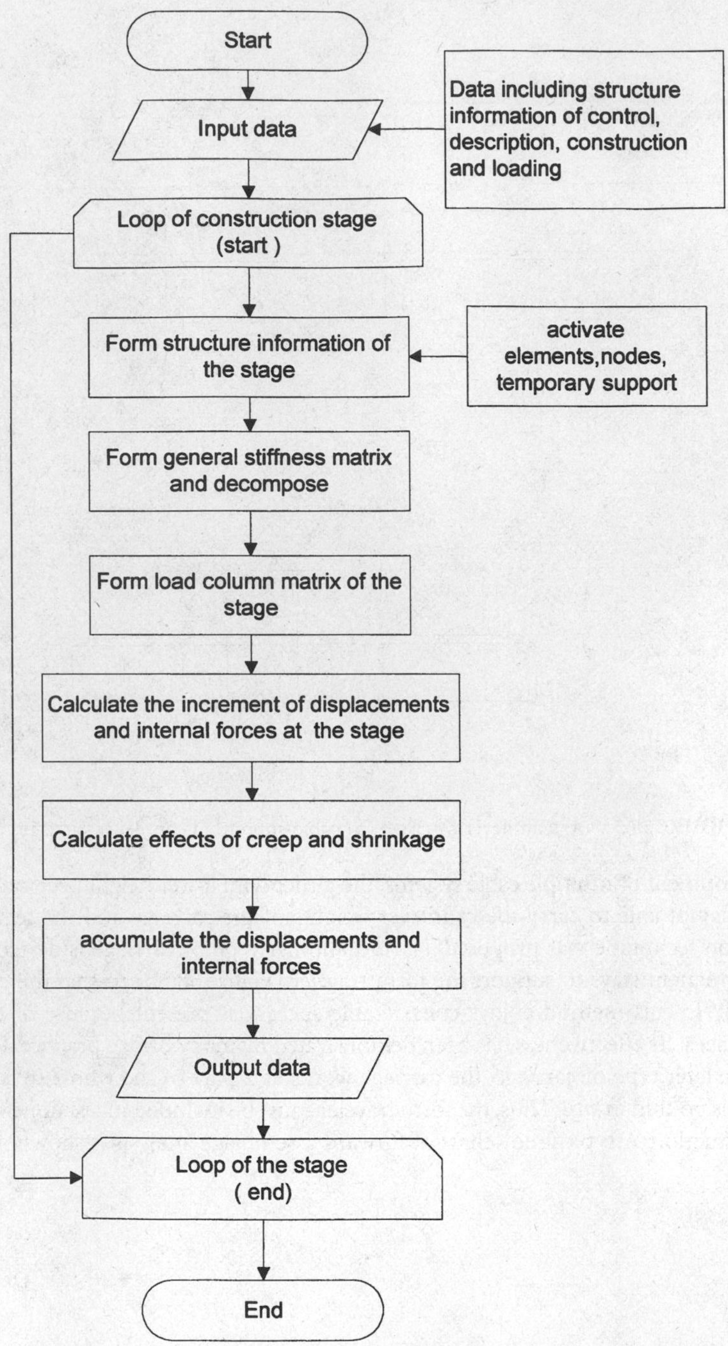

FIGURE 58.6 Flowchart of forward assemblage analysis.

With the forward assemblage analysis, the construction data can be worked out. And the actual permanent state of cable-stayed bridges can be reached. Further, if the erection scheme were modified during the construction period or in the case that significant construction error occurred, then the structural parameters or the temporary erection loads would be different from the values used in the design. It is possible to predict the cable forces and the sequential deformations at each stage by utilizing the forward assemblage analysis.

58.4.3 Backward Disassemblage Analysis

Following a reverse way to simulate the disassemblage process stage by stage, a backward analysis can be carried out also by finite-element method [11,12]. Not only the elevations of deck but also the length of cable stays and the initial tension of stays can be worked out by this method. And the completed state of structure at each stage can be evaluated.

The backward disassemblage analysis starts with a very ideal structural state in which it is assumed that all the creep and shrinkage deformation of concrete be completed, i.e., a state 5 years or 1500 days after the completion of the bridge construction. The structural deformations and internal forces at each stage are considered ideal reference states for construction of the bridge [11]. The backward analysis procedure for a PC cable-stayed bridge may be illustrated as follows:

Step 1. Compute the permanent state of the structure.
Step 2. Remove the effects of the creep and shrinkage of concrete of 1500 days or 5 years.
Step 3. Remove the second part of the dead loads, i.e., the weights of wearing surfacing, curbs and fence, etc.
Step 4. Apply the traveler and other temporary loads and supports.
Step 5. Remove the center segment, to analyze the semistructure separately.
Step 6. Move the form traveler backward.
Step 7. Remove the weight of the concrete of a pair of segments.
Step 8. Remove the cable stay.
Step 9. Remove the corresponding elements.

Repeat the Steps 6 to 9 until all the girder segments are disassembled. A flowchart for backward disassemblage analysis is shown in Figure 58.7.

As mentioned above, for the erecting method using conventional form traveler cast-in-place or precast concrete segments, the crane or the form traveler may be modeled as external loads. Thus the carriage moving is equivalent to a change of the loading position. However, for an erection method utilizing cable-supported traveling carriage the cable stays first work as supports of the carriage and later, after curing is finished, the cable stays are connected with the girder permanently. In backward disassemblage analysis, the form traveler moving must be related to a change of the structural system.

The backward analysis procedure can establish the necessary data for the erection at each stage such as the elevations of deck, the cable forces, the deformations of structure, and the stresses at critical sections of deck and pylon.

One of the disadvantages of backward analysis is that creep effects are not able to be estimated; therefore, forward and backward simulations should be used alternately to determine the initial tension and the length of stay cables.

58.5 Construction Control

58.5.1 Objectives and Control Means

Obviously, the objective of construction control is to build a bridge that achieves the design aim with acceptable error. During the construction of a cable-stayed bridge, some discrepancies may occur between the actual state and the state of design expectation [14,15]. The discrepancies may arise from elevation error in laying forms, errors in stressing cable stays by jacks, errors of the first part of the dead load, i.e., the self-weights of the girder segments, and the second part of the dead load, i.e., the self-weights of the surfacing, curbs and fencing, etc. On the other hand, a system error may occur in measuring the deflection of the girder and the pylons. It is impossible to eliminate all the errors. Actually, there are two basic requirements for the completed structure [12]: (1) the geometric profile matches the designed shape well and (2) the internal forces are within the designed

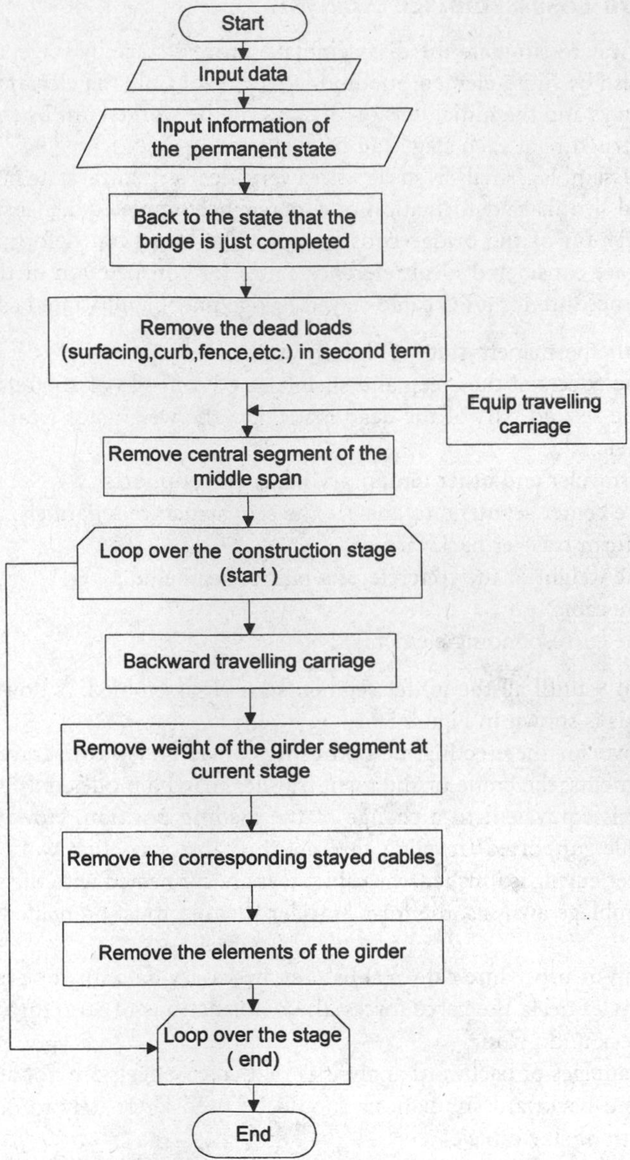

FIGURE 58.7 Flowchart of backward disassemblage analysis.

envelope values; specifically, the bending moments of the girder and the pylons are small and evenly distributed.

Since the internal forces of the girder and the pylons are closely related to the cable forces, the basic method of construction control is to adjust the girder elevation and the cable forces. If the error of the girder elevation deviated from the design value is small, such error can be reduced or eliminated by adjusting the elevation of the segment without inducing an angle between two adjacent deck segments. In this way we only change the geometric position of the girder without changing the internal force state of the structure. When the errors are not small, it is necessary to adjust the cable forces. In this case, both the geometric position change and the changes of internal forces occur in the structure.

Nevertheless, cable force adjustment are not preferable because they may take a lot of time and money. The general exercise at each stage is to find out the correct length of the cables and set the elevation of the segment appropriately. Cable tensioning is performed for the new stays only. Generally, a comprehensive adjustment of all the cables is only applied before connecting the two cantilever ends [21]. In case a group of cables needs to be adjusted, careful planning for the adjustment based on a detailed analysis is absolutely necessary.

58.5.2 Construction Control System

To guarantee structural safety and to reach the design aim, a monitoring and controlling system is important [13,15,19]. A typical construction controlling system consists of four subsystems: measuring subsystem, error and sensitivity analysis subsystem, control/prediction subsystem, and new design value calculation subsystem. An example of a construction control system for a PC cable-stayed bridge [18,20] is shown in Figure 58.8.

1. Measuring subsystem — The measuring items mainly include the elevation/deflection of the girder, the cable forces, the horizontal displacement of the pylon(s), the stresses of sections in the girder and the pylon(s), the modulus of elasticity and mass density of concrete, the creep and shrinkage of concrete, the temperature/temperature gradient in the structure.
2. Error and sensitivity analysis subsystem — In this subsystem, the temperature effects are first determined and removed. Then the sensitivity of structural parameter such as the elasticity modulus of concrete, self-weight, stiffness of the girder segment or pylon, etc., are analyzed. Through the analysis, the causes of errors can be found so that the corresponding adjustment steps are utilized.
3. Control/prediction subsystem — Compare the measuring values with those of the design expectation; if the differences are lower than the prescribed limits, then go to the next stage. And the elevation is determined appropriately. Otherwise, it is necessary to find out the reasons, then to eliminate or reduce those errors through proper measures. The magnitude of cable tension adjustment can be determined by a linear programming model.
4. New design value calculation subsystem — Since the structural parameters for the completed part have deviated from the designed values, the design expectation must be updated with the changed state of the structure. And the sequential construction follows new design values so that the final state of structure can be achieved optimally.

58.6 An Engineering Example

The general view of a PC cable-stayed bridge is shown in Figure 58.1. The cable-stayed portion of the bridge has a total length of 702 m. The main span between towers is 380 m and the side anchor span is 161 m. The side anchor spans consist of two spans of 90 and 71 m with an auxiliary pier. The main girder is composed with two edge girders and a deck plate. The edge girders are laterally stiffened by a T-shaped PC girder with 6-m spacing. The edge girder is a solid section whose height is 2.2 m and width varies from 2.6 m at intersection of girder and pylon to 2.0 m at middle span. The deck plate is 28 cm thick. The width of the deck is 37.80 m out-to-out with eight traffic lanes. Spatial 244 stay cables are arranged in a semifan configuration. The pylon is shaped like a diamond with an extension mast (see Figure 58.1). All the cable stays are anchored in the mast part of the pylon. The stay cables are attached to the edge girders at 6 m spacing.

At the side anchor span an auxiliary pier is arranged to increase the stiffness of the bridge. And an anchorage segment of deck is set up to balance the lifting forces from anchorage cables.

58.6.1 Construction Process

The bridge deck structure is erected by the balance cantilevering method utilizing cable-supported form travelers. The construction process is briefly described as follows.

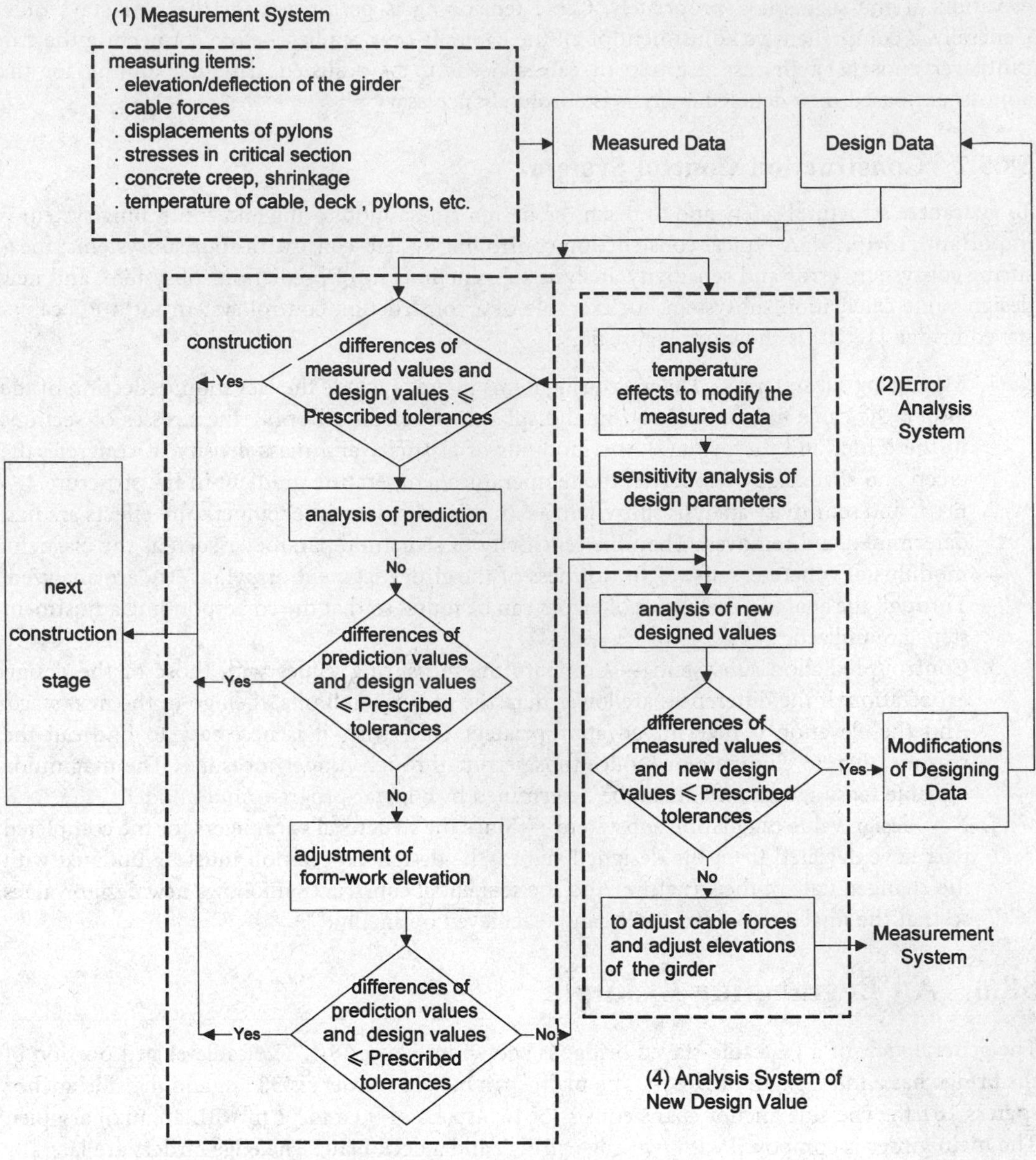

FIGURE 58.8 A typical construction control system for a PC cable-stayed bridge.

- Build the towers.
- Cast in place the first segment on timbering support.
- Erect the No. 1 cable and stress to its final length.
- Hoist the traveling carriages and positioning.
- Erect the girder segments one by one on the two sides of the pylons.
- Connect the cantilever ends of the side span with the anchorage parts.
- Continue to erect the girder segments in the center span.

- Connect the cantilever ends of the center span.
- Remove traveling carriages and temporary supports.
- Connect the girder with the auxiliary piers.
- Cast pavement and set up fence, etc.

A typical erection stage of one segment is described as follows:

- Move the traveling carriage forward and set up the form at proper levels.
- Erect and partially stress the stay cables attached to the traveler.
- Place reinforcement, post-tensioning bars and couple the stressed bars with those of the previously completed deck segment.
- Cast in place the deck concrete.
- Stress the stay cables to adjust the girder segments to proper levels.
- Cure deck panel and stress the longitudinal and lateral bars and strands.
- Loosen the connection between the stay cable and traveling carriage.
- Stress cable stays to the required value.

The above erection steps are repeated until the bridge is closed at the middle span.

58.6.2 Construction Simulation

The above construction procedure can be simulated stage by stage as illustrated in Section 58.4.2. Since creep and shrinkage occur and the second part of the dead weight will be loaded on the bridge girder after completion of the structure, a downward displacement is induced. Therefore, as the erection is just finished the elevation of the girder profile should be set higher than that of the design profile and the pylons should be leaning toward the side spans. In this example, the maximum value which is set higher than the designed profile in the middle of the bridge is about 35.0 cm, while the displacement of pylon top leaning to anchorage span is about 9.0 cm. The initial cable forces are listed in Table 58.1 to show the effects of creep. As can be seen, considering the long-term effects of concrete creep, the initial cable forces are a little greater than those without including the time-dependent effects.

58.6.3 Construction Control System

In the construction practice of this PC cable-stayed bridge, a construction control system is employed to control the cable forces and the elevation of the girder. Before starting concrete casting the reactions of the cable-supported form traveler are measured by strain gauge equipment. Thus the weights of the four travelers used in this bridge are known.

At each stage the mass density of concrete and the elasticity modulus of concrete are tested in the laboratory *in situ*. The calculation of construction is carried out with the measured parameters. In several sections of the deck and pylon, strain gauges are embedded to measure the strains of the structure during the whole construction period, thus the stress of the structure can be monitored.

In this example the main flowchart of the construction control system for a typical erecting segment is shown in Figure 58.9.

TABLE 58.1 Predicted Initial Cable Forces (kN)

No.	S1	S2	S3	S4	S5	S6	S7	S8	S9	S10
NCS	7380	3733	4928	5130	5157	5373	5560	5774	5969	6125
CS	7745	3854	5101	5331	5348	5549	5694	5873	6050	6185
No.	S11	S12	S13	S14	S15	S16	S17	S18	S19	S20
NCS	6320	6870	7193	7283	7326	7519	7317	7478	7766	8035
CS	6367	6908	7209	7331	7467	7639	7429	7580	7856	8114
No.	S21	S22	S23	S24	S25	S26	S27	S28	S29	S30
NCS	8366	8169	7693	8207	9579	9649	9278	9197	9401	12820
CS	8442	8246	7796	8354	9778	9791	9509	9296	9602	12930
No.	M1	M2	M3	M4	M5	M6	M7	M8	M9	M10
NCS	7334	3662	4782	4970	5035	5426	5479	5628	5778	6108
CS	7097	3433	4591	4877	5006	5477	5567	5736	5904	6231
No.	M11	M12	M13	M14	M15	M16	M17	M18	M19	M20
NCS	6139	6572	6627	6918	6973	7353	7540	7742	7882	7978
CS	6263	6706	6756	7031	7088	7422	7624	7813	7942	8060
No.	M21	M22	M23	M24	M25	M26	M27	M28	M29	M30
NCS	8384	8479	8617	8833	9175	9359	9394	9480	9641	13440
CS	8452	8561	8694	8931	9212	9413	9459	9570	9716	13570

No.: Cable number; M: middle span; S: side span; CS: with the effects of creep and shrinkage of concrete; NCS: without the effects of creep and shrinkage of concrete.

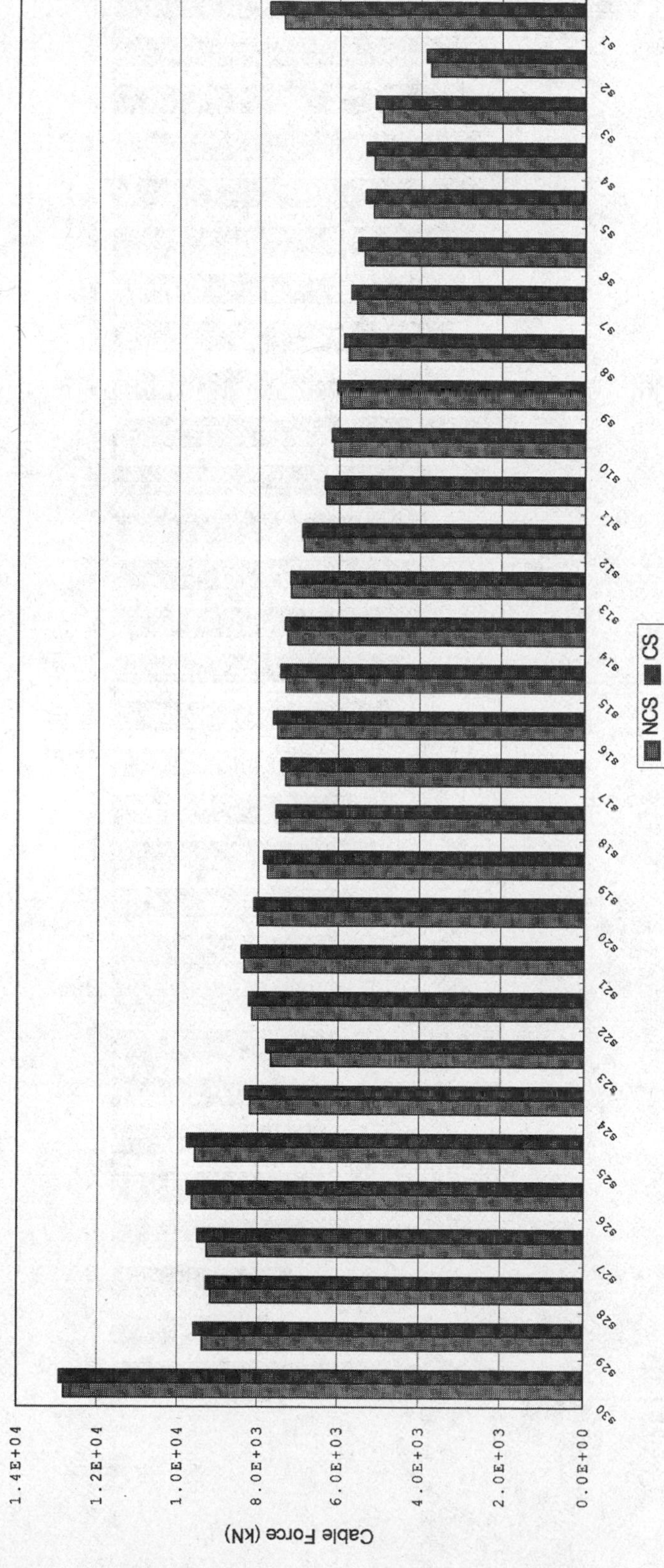

FIGURE 58.9 (a) Initial cable forces in side span determined by simulation analysis. (b) Initial cable forces in middle span determined by simulation analysis.

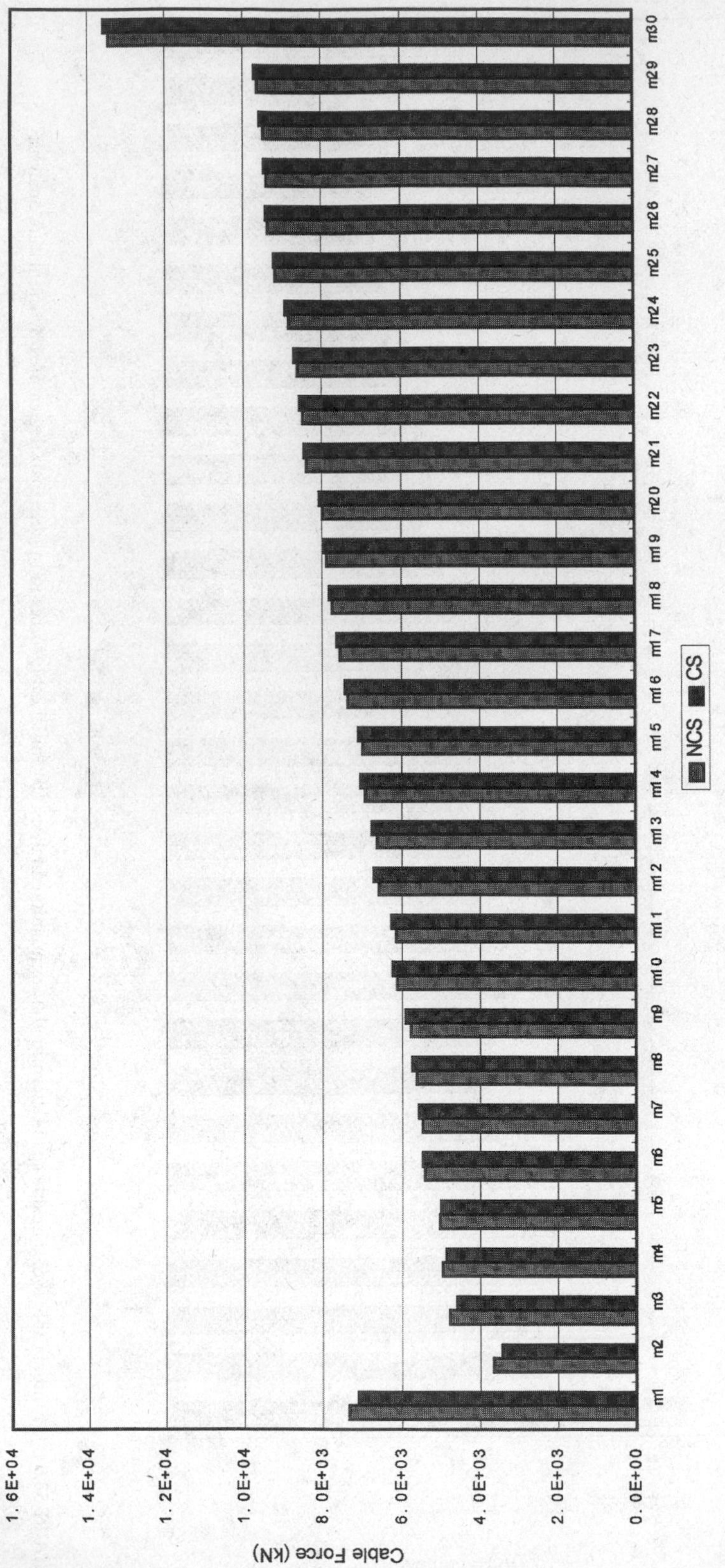

FIGURE 58.9 (continued)

References

1. Tang, M.C. The 40-year evolution of cable-stayed bridges, in *International Symposium on Cable-Stayed Bridges,* Lin Yuanpei et al., Eds., Shanghai, 1994, 30–11.
2. Leonhardt, F. and Zellner, W., Past, present and future of cable-stayed bridges, in *Cable-Stayed Bridges, Recent Developments and Their Future,* M. Ito et al., Eds., Elsevier Science Publishers, New York, 1991.
3. Podolny, W. and Scalmi, J., *Construction and Design of Cable-Stayed Bridges,* John Wiley & Sons, New York, 1983.
4. Walther, R., Houriet, B., Lsler, W., and Moia, P., *Cable-Stayed Bridges,* Thomas Telford, London, 1988.
5. Gimsing, N. J., *Cable Supported Bridges, Concept and Design,* John Wiley & Sons, New York, 1983.
6. Kasuga, A., Arai, H.,Breen, J. E., and Furukawa, K., Optimum cable-force adjustment in concrete cable-stayed bridges, *J. Struct. Eng., ASCE,* 121(4), 685–694, 1995.
7. Ma, W. T., Cable Force Adjustment and Construction Control of PC Cable Stayed Bridges, Ph.D. dissertation of Department of Civil Engineering, South China University of Technology, 1997 [in Chinese].
8. Wang, X. W. et al., A study of determination of cable tension under dead loads, *Bridge Constr.*, 4, 1–5, 1996 [in Chinese].
9. Yan, D. H. et al., Simulation analysis of Tongling Cable-Stayed Bridge for construction control, in *National Symposium on Highway Bridge,* Dai Jing, Ed., Beijing Renming Jiaotong Press, Guangzhou, 347–355, 1995. [in Chinese].
10. Zhou, L. X. et al. Prestressed Concrete Cable-Stayed Bridges, Beijing Renming Jiaotong Press, 1989 [in Chinese].
11. Xiao, R. C., Ling, P. Application of computational structural mechanics in construction design and control of bridge structures, *Comput. Struct. Mech. Appl.,* 10(1) 92–98, 1993 [in Chinese].
12. Fang, Z. and Liu, G. D., A Study of Construction Control System of Cable-Stayed Bridges, Research Report of Department of Civil Engineering, Hunan University, 1995 [in Chinese].
13. Chen, D. W., Xiang, H. F., and Zheng, X. G., Construction control of PC cable-stayed bridge, J. Civil Eng., 26(1) 1–11, 1993 [in Chinese].
14. Yoshimura, M., Ueki, Y., and Imai, Y., Design and construction of a prestressed concrete cable-stayed bridge: the Tsukuhara Ohashi Bridge, *J. Jpn. Prestressed Concrete Eng. Assoc.*, Tokyo, Japan, 29(1) 1987 [in Japanese].
15. Fujisawa, N. and Tomo, H., Computer-aided cable adjustment of cable-stayed bridges, *IABSE Proc.*, P-92/85, 1985.
16. Furukawa, K., Inoue, K., Nakkkayama, H., and Ishido, K., Studies on the management system of cable-stayed bridges under construction using multi-objective programming method. *Proc. JSCE,* Tokyo, Japan, 374(6), 1986 [in Japanese].
17. Furuta, H. et al., Application on fuzzy mathematical programming to cable tension adjustment of cable-stayed bridges, in *International Symposium on Cable-Stayed Bridges,* Lin Yuanpei et al. Eds., Shanghai, 1994, 584–595.
18. Takuwa, I. et al., Prestressed concrete cable-stayed bridge constructed on an expressway — the Tomei Ashigra Bridge, in *Cable-Stayed Bridges, Recent Developments and Their Future.* M. Ito et al., Eds., Elsevier Science Publishers, New York, 1991.
19. Yasuhiro, K. et al., Construction of Tokachi Ohashi Bridge Superstructure (PC cable-stayed bridge), *Bridge Found.,* 1, 7–15, 1995 [in Japanese].
20. Hidemi, O. et al., Construction of Ikara Bridge superstructure (PC cable-stayed bridge), *Bridge Found.,* No. 11, 7–14, 1995 [in Japanese].
21. Fushimi, T., et al., Erection of the Tsurumi Fairway Bridge superstructure, *Bridge Found.*, 10, 2–10, 1994 [in Japanese].

59

Active Control in Bridge Engineering

Zaiguang Wu
California Department of Transportation

59.1 Introduction

In bridge engineering, one of the constant challenges is to find new and better means to design new bridges or to strengthen existing ones against destructive natural effects. One avenue, as a traditional way, is to design bridges based on strength theory. This approach, however, can sometimes be untenable both economically and technologically. Other alternatives, as shown in Chapter 41, include installing isolators to isolate seismic ground motions or adding passive energy dissipation devices to dissipate vibration energy and reduce dynamic responses. The successful application of these new design strategies in bridge structures has offered great promise [11]. In comparison with passive energy dissipation, research, development, and implementation of active control technology has a more recent origin. Since an active control system can provide more control authority and adaptivity than a passive system, the possibility of using active control systems in bridge engineering has received considerable attention in recent years.

Structural control systems can be classified as the following four categories [6]:

- **Passive Control** — A control system that does not require an external power source. Passive control devices impart forces in response to the motion of the structure. The energy in a passively controlled structural system cannot be increased by the passive control devices.

0-8493-7434-0/00/$0.00+$.50

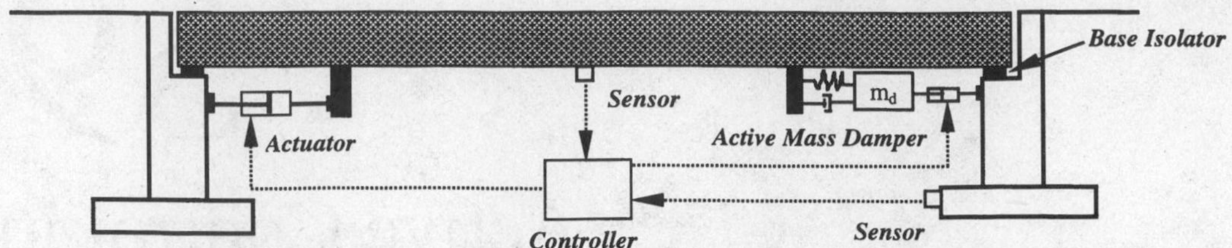

FIGURE 59.1 Base-isolated bridge with added active control system.

- **Active Control** — A control system that does require an external power source for control actuator(s) to apply forces to the structure in a prescribed manner. These controlled forces can be used both to add and to dissipate energy in the structure. In an active feedback control system, the signals sent to the control actuators are a function of the response of the system measured with physical sensors (optical, mechanical, electrical, chemical, etc.).
- **Hybrid Control** — A control system that uses a combination of active and passive control systems. For example, a structure equipped with distributed viscoelastic damping supplemented with an active mass damper on or near the top of the structure, or a base-isolated structure with actuators actively controlled to enhance performance.
- **Semiactive Control** — A control system for which the external energy requirements are an order of magnitude smaller than typical active control systems. Typically, semiactive control devices do not add mechanical energy to the structural system (including the structure and the control actuators); therefore, bounded-input and bounded-output stability is guaranteed. Semiactive control devices are often viewed as controlled passive devices.

Figure 59.1 shows an active bracing control system and an active mass damper installed on each of the abutments of a seismically isolated concrete box-girder bridge [8]. As we know, base isolation systems can increase the chances of the bridge surviving a seismic event by reducing the effects of seismic vibrations on the bridge. These systems have the advantages of simplicity, proven reliability, and no need for external power for operation. The isolation systems, however, may have difficulties in limiting lateral displacement and they impose severe constraints on the construction of expansion joints. Instead of using base isolation, passive energy dissipation devices, such as viscous fluid dampers, viscoelastic dampers, or friction dampers, can also be employed to reduce the dynamic responses and improve the seismic performance of the bridge. The disadvantage of passive control devices, on the other hand, is that they only respond passively to structural systems based on their designed behaviors.

The new developed active systems, a typical example as shown in Figure 59.1, have unique advantages. Based on the changes of structural responses and external excitations, these intelligent systems can actively adapt their properties and controlling forces to maximize the effectiveness of the isolation system, increase the life span of the bridge, and allow it to withstand extreme loading effects. Unfortunately, in an active control system, the large forces required from the force generator and the necessary power to generate these forces pose implementation difficulties. Furthermore, a purely active control system may not have proven reliability. It is natural, therefore, to combine the active control systems (Figure 59.1) with abutment base isolators, which results in the so-called hybrid control. A hybrid control system is more reliable than a purely active system, since the passive devices can still protect the bridge from serious damage if the active portion fails during the extreme earthquake events. But the installation and maintenance of the two different systems are the major shortcoming in a hybrid system. Finally, if the sliding bearings are installed at the bridge abutments and if the pressure or friction coefficient between two sliding surfaces can be adjusted actively based on the measured bridge responses, this kind of controlled bearing will then be known as semiactive control devices. The required power supply essential for signal processing and mechanical operation

TABLE 59.1 Bridge Control Systems

Systems	Typical Devices	Advantages	Disadvantages
Passive	Elastomeric bearings Lead rubber bearings Metallic dampers Friction dampers Viscoelastic dampers Tuned mass dampers Tuned liquid dampers	Simple Cheap Easy to install Easy to maintain No external energy Inherently stable	Large displacement Unchanged properties
Active	Active tendon Active bracing Active mass damper	Smart system	Need external energy May destabilize system Complicated system
Hybrid	Active mass damper + bearing Active bracing + bearing Active mass damper + VE damper	Smart and reliable	Two sets of systems
Semiactive	Controllable sliding bearings Controllable friction dampers Controllable fluid dampers	Inherently stable Small energy required Easy to install	Two sets of systems

is very small in a semiactive control system. A portable battery may have sufficient capacity to store the necessary energy before an earthquake event. This feature thus enables the control system to remain effective regardless of a major power supply failure. Therefore, the semiactive control systems seem quite feasible and reliable.

The various control systems with their advantages and disadvantages are summarized in Table 59.1.

Passive control technologies, including base isolation and energy dissipation, are discussed in Chapter 41. The focus of this chapter is on active, hybrid, and semiactive control systems. The relationships among different stages during the development of various intelligent control technologies are organized in Figure 59.2. Typical control configurations and control mechanisms are described first in Section 59.2. Then, the general control strategies and typical control algorithms are presented in Section 59.3, along with discussions of practical concerns in actual bridge applications of active control strategies. The analytical development and numerical simulation of various control systems applied on different types of bridge structures are shown as case studies in Section 59.4. Remarks and conclusions are given in Section 59.5.

59.2 Typical Control Configurations and Systems

As mentioned above, various control systems have been developed for bridge vibration control. In this section, more details of these systems are presented. The emphasis is placed on the motivations behind the development of special control systems to control bridge vibrations.

59.2.1 Active Bracing Control

Figure 59.3 shows a steel truss bridge with several actively braced members [1]. Correspondingly, the block diagram of the above control system is illustrated in Figure 59.4. An active control system generally consists of three parts. First, *sensors*, like human eyes, nose, hands, etc., are attached to the bridge components to measure either external excitations or bridge response variables. Second, *controllers*, like the human brain, process the measured information and compute necessary actions needed based on a given control algorithm. Third, *actuators*, usually powered by external sources, produce the required control forces to keep bridge vibrations under the designed safety range.

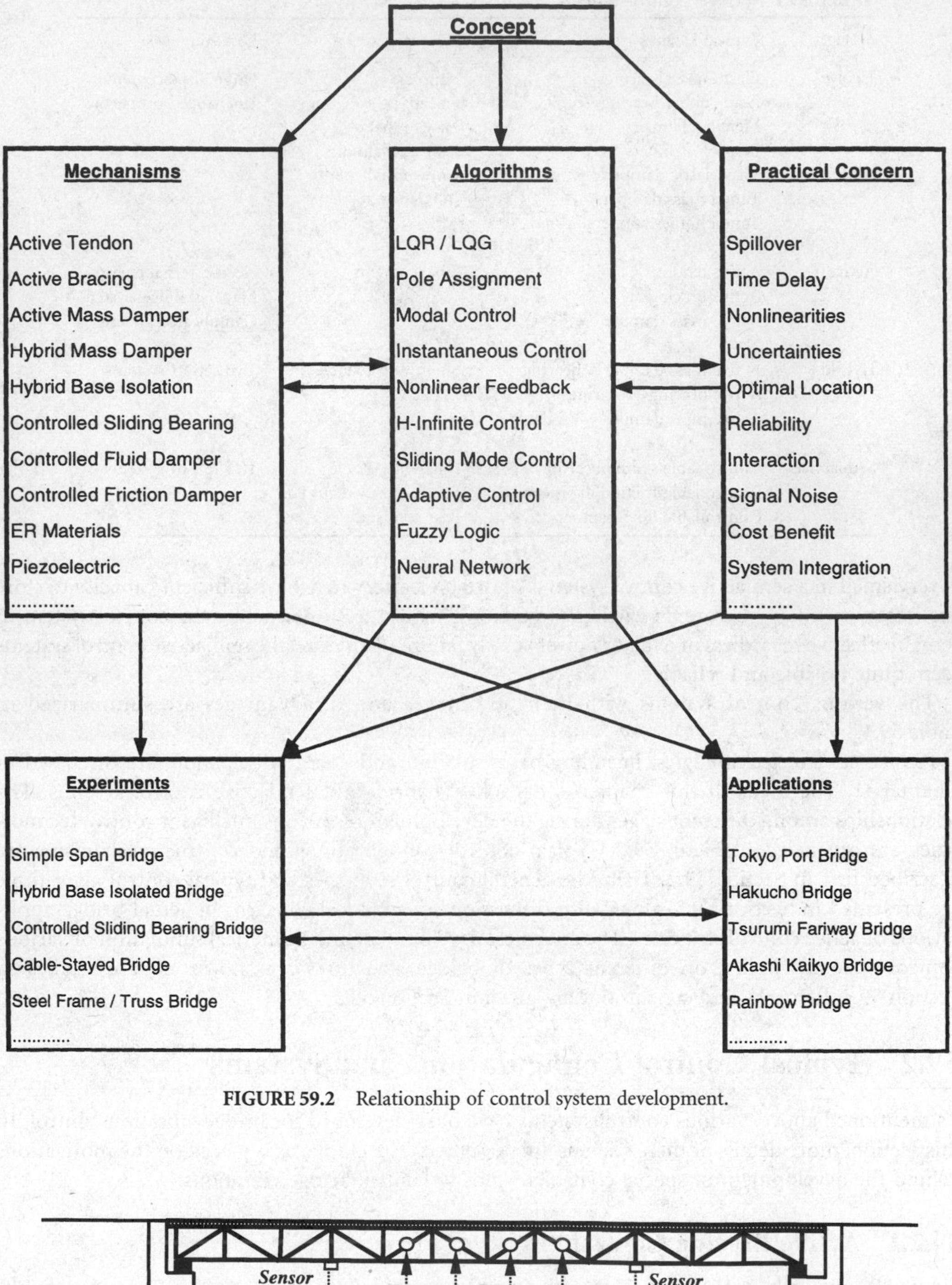

Concept

Mechanisms

Active Tendon
Active Bracing
Active Mass Damper
Hybrid Mass Damper
Hybrid Base Isolation
Controlled Sliding Bearing
Controlled Fluid Damper
Controlled Friction Damper
ER Materials
Piezoelectric
....

Algorithms

LQR / LQG
Pole Assignment
Modal Control
Instantaneous Control
Nonlinear Feedback
H-Infinite Control
Sliding Mode Control
Adaptive Control
Fuzzy Logic
Neural Network
....

Practical Concern

Spillover
Time Delay
Nonlinearities
Uncertainties
Optimal Location
Reliability
Interaction
Signal Noise
Cost Benefit
System Integration
....

Experiments

Simple Span Bridge
Hybrid Base Isolated Bridge
Controlled Sliding Bearing Bridge
Cable-Stayed Bridge
Steel Frame / Truss Bridge
....

Applications

Tokyo Port Bridge
Hakucho Bridge
Tsurumi Fariway Bridge
Akashi Kaikyo Bridge
Rainbow Bridge
....

FIGURE 59.2 Relationship of control system development.

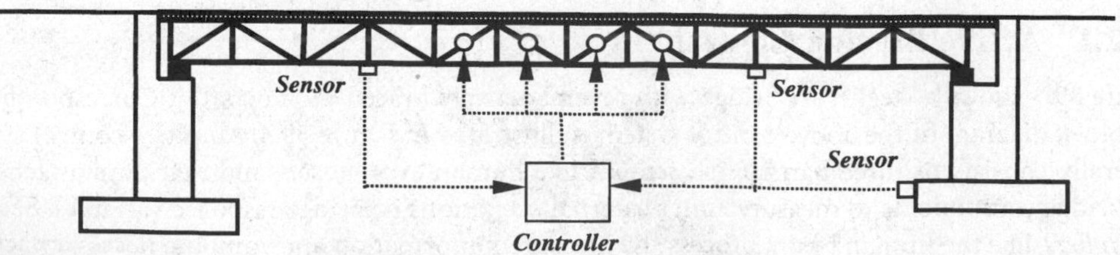

FIGURE 59.3 Active bracing control for steel truss bridge.

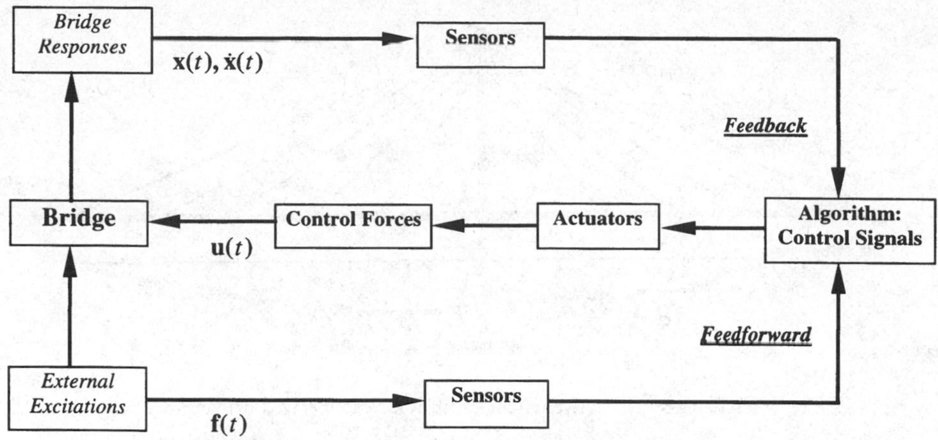

FIGURE 59.4 Block diagram of active control system.

Based on the information measured, in general, an active control system may be classified as three different control configurations. When only the bridge response variables are measured, the control configuration is referred to as *feedback control* since the bridge response is continually monitored and this information is used to make continuous corrections to the applied control forces. On the other hand, when only external excitations, such as earthquake accelerations, are measured and used to regulate the control actions, the control system is called *feedforward control.* Of course, if the information on both the response quantities and excitation are utilized for control design, combining the previous two terms, we get a new term, *feedback/feedforward control.* A bridge equipped with an active control system can adapt its properties based on different external excitations and self-responses. This kind of self-adaptive ability makes the bridge more effective in resisting extraordinary loading and relatively insensitive to site conditions and ground motions. Furthermore, an active control system can be used in multihazard mitigation situations, for example, to control the vibrations induced by wind as well as earthquakes.

59.2.2 Active Tendon Control

The second active control configuration, as shown in Figure 59.5, is an active tendon control system controlling the vibrations of a cable-stayed bridge [17,18]. Cable-stayed bridges, as typical flexible bridge structures, are particularly vulnerable to strong wind gusts. When the mean wind velocity reaches a critical level, referred to as the flutter speed, a cable-stayed bridge may exhibit vibrations with large amplitude, and it may become unstable due to bridge flutter. The mechanism of flutter is attributed to "vortex-type" excitations, which, coupled with the bridge motion, generate motion-dependent aerodynamic forces. If the resulting aerodynamic forces enlarge the motion associated with them, a self-excited oscillation (flutter) may develop. Cable-stayed bridges may also fail as a result of excessively large responses such as displacement or member stresses induced by strong earthquakes or heavy traffic loading. The traditional methods to strength the capacities of cable-stayed bridges usually yield a conservative and expensive design. Active control devices, as an alternative solution, may be feasible to be employed to control vibrations of cable-stayed bridges. Actuators can be installed at the anchorage of several cables. The control loop also includes sensors, controller, and actuators. The vibrations of the bridge girder induced by strong wind, traffic, or earthquakes are monitored by various sensors placed at optimal locations on the bridge. Based on the measured amplitudes of bridge vibrations, the controller will make decisions and, if necessary, require the actuators to increase or decrease the cable tension forces through hydraulic servomechanisms. Active tendon control seems ideal for the suppression of vibrations in a cable-stayed bridge since the existing stay cables can serve as active tendons.

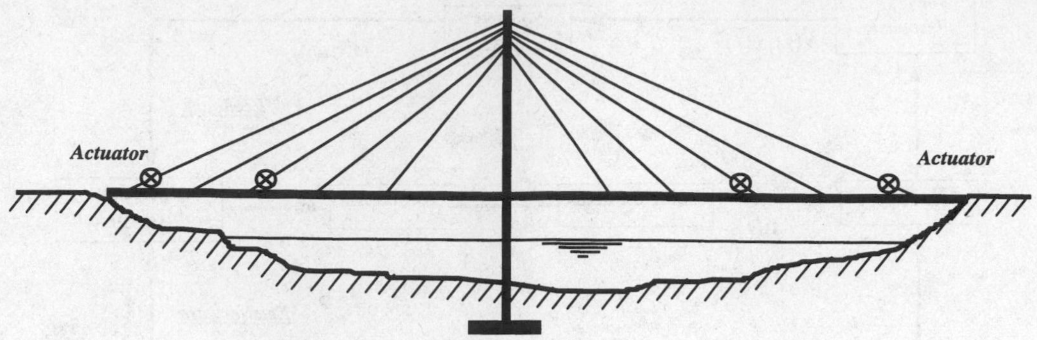

FIGURE 59.5 Active tendon control for cable-stayed bridge.

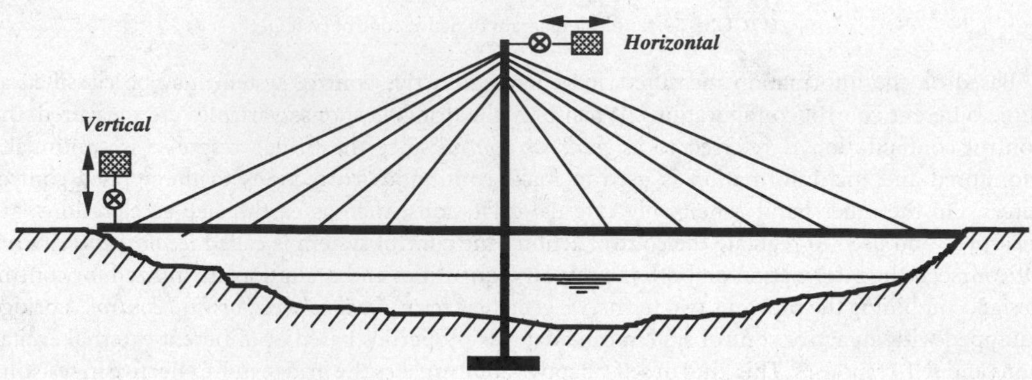

FIGURE 59.6 Active mass damper on cable-stayed bridge.

59.2.3 Active Mass Damper

Active mass damper, which is a popular control mechanism in the structural control of buildings, can be the third active control configuration for bridge structures. Figure 59.6 shows the application of this system in a cable-stayed bridge [12]. Active mass dampers are very useful to control the wind-induced vibrations of the bridge tower or deck during the construction of a cable-stayed bridge. Since cable-stayed bridges are usually constructed using the cantilever erection method, the bridge under construction is a relatively unstable structure supported only by a single tower. There are certain instances, therefore, where special attention is required to safeguard against the external dynamic forces such as strong wind or earthquake loads. Active mass dampers can be especially useful for controlling this kind of high tower structure. The active mass damper is the extension of the passive tuned mass damper by installing the actuators into the system. Tuned mass dampers (Chapter 41) are in general tuned to the first fundamental period of the bridge structure, and thus are only effective for bridge control when the first mode is the dominant vibration mode. For bridges under seismic excitations, however, this may not be always the case since the vibrational energy of an earthquake is spread over a wider frequency band. By providing the active control forces through the actuators, multimodal control can be achieved, and the control efficiency and robustness will be increased in an active mass damper system.

59.2.4 Seismic Isolated Bridge with Control Actuator

An active control system may be added to a passive control system to supplement and improve the performance and effectiveness of the passive control. Alternatively, passive devices may be installed in an active control scheme to decrease its energy requirements. As combinations of active and passive systems, hybrid control systems can sometimes alleviate some of the limitations and restrictions that exist in either an active or a passive control system acting alone. Base isolators are finding more and more applications in bridge engineering. However, their shortcomings are also becoming clearer. These include (1) the relative displacement of the base isolator may be too large to satisfy the design requirements, (2) the fundamental frequency of the base-isolated bridge cannot vary to respond favorably to different types of earthquakes with different intensities and frequency contents, and (3) when bridges are on a relatively soft ground, the effectiveness of the base isolator is limited. The active control systems, on the other hand, are capable of varying both the fundamental frequency and the damping coefficient of the bridge instantly in order to respond favorably to different types of earthquakes. Furthermore, the active control systems are independent of the ground or foundation conditions and are adaptive to external ground excitations. Therefore, it is natural to add the active control systems to the existing base-isolated bridges to overcome the above shortcomings of base isolators. A typical setup of seismic isolators with a control actuator is illustrated at the left abutment of the bridge in Figure 59.1 [8,19].

59.2.5 Seismic Isolated Bridge with Active Mass Damper

Another hybrid control system that combines isolators with active mass dampers is installed on the right abutment of the bridge in Figure 59.1 [8,19]. In general, either base isolators or tuned mass dampers are only effective when the responses of the bridge are dominated by its fundamental mode. Adding an actuator to this system will give the freedom to adjust the controllable frequencies based on different types of earthquakes. This hybrid system utilizes the advantages of both the passive and active systems to extend the range of applicability of both control systems to ensure integrity of the bridge structure.

59.2.6 Friction-Controllable Sliding Bearing

Currently, two classes of seismic base isolation systems have been implemented in bridge engineering: elastomeric bearing system and sliding bearing system. The elastomeric bearing, with its horizontal flexibility, can protect a bridge against strong earthquakes by shifting the fundamental frequency of the bridge to a much lower value and away from the frequency range where the most energy of the earthquake ground motion exists. For the bridge supported by sliding bearings, the maximum forces transferred through the bearings to the bridge are always limited by the friction force at the sliding surface, regardless of the intensity and frequency contents of the earthquake excitation. The vibrational energy of the bridge will be dissipated by the interface friction. Since the friction force is just the product of the friction coefficient and the normal pressure between two sliding surfaces, these two parameters are the critical design parameters of a sliding bearing. The smaller the friction coefficient or normal pressure, the better the isolation performance, due to the correspondingly small rate of transmission of earthquake acceleration to the bridge. In some cases, however, the bridge may suffer from an unacceptably large displacement, especially the residual displacement, between its base and ground. On the other hand, if the friction coefficient or normal pressure is too large, the bridge will be isolated only under correspondingly large earthquakes and the sliding system will not be activated under small to moderate earthquakes that occur more often. In order to substantially alleviate these shortcomings, therefore, the ideal design of a sliding system should vary its friction coefficient or normal pressure based on measured earthquake intensities and bridge responses. To this purpose, a friction-controllable sliding bearing has been developed, and Figure 59.7 illustrates one of its applications in bridge engineering [4,5]. It can be seen from Figure 59.7 that the friction forces in the sliding bearings are actively controlled by adjusting the fluid pressure in the fluid chamber located inside the bearings.

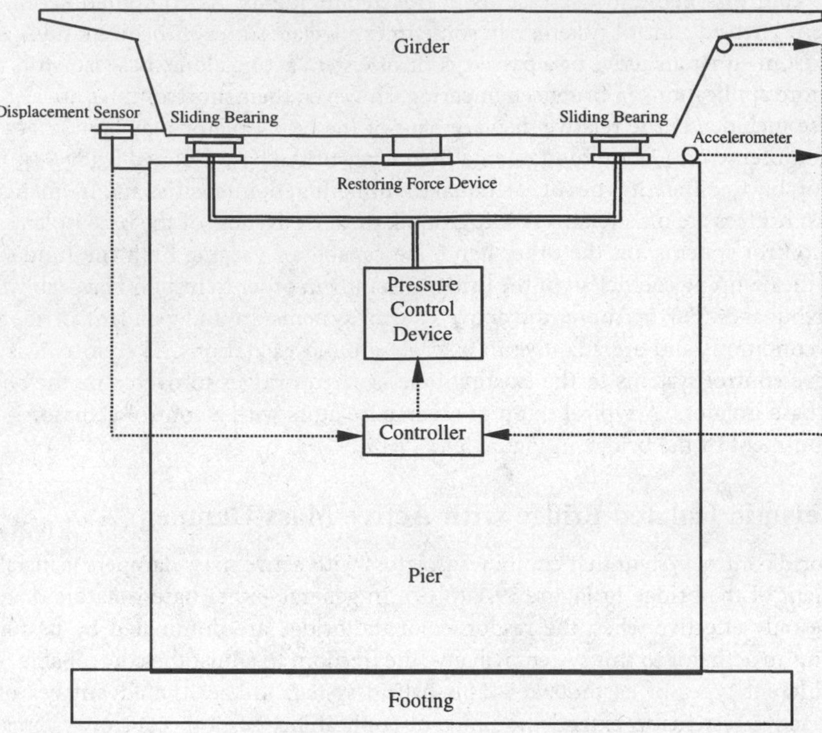

FIGURE 59.7 Controllable sliding bearing.

59.2.7 Controllable Fluid Damper

Dampers are very effective in reducing the seismic responses of bridges. Various dampers, as discussed in Chapter 41, have been developed for bridge vibration control. One of them is fluid damper, which dissipates vibrational energy by moving the piston in the cylinder filled with viscous material (oil). Depending on the different function provided by the dampers, different damping coefficients may be required. For example, one may set up a large damping coefficient to prevent small deck vibrations due to braking loads of vehicles or wind effects. However, when bridge deck responses under strong earthquake excitations exceed a certain threshold value, the damping coefficient may need to be reduced in order to maximize energy dissipation. Further, if excessive deck responses are reached, the damping coefficient needs to be set back to a large value, and the damper will function as a stopper. As we know, it is hard to change the damping coefficient after a passive damper is designed and installed on a bridge. The multifunction requirements for a damper have motivated the development of semiactive strategy. Figure 59.8 shows an example of a semiactive controlled fluid damper. The damping coefficient of this damper can be controlled by varying the amount of viscous flow through the bypass based on the bridge responses. The new damper will function as a damper stopper at small deck displacement, a passive energy dissipator at intermediate deck displacement, and a stopper with shock absorber for excessive deck displacement.

59.2.8 Controllable Friction Damper

Friction dampers, utilizing the interface friction to dissipate vibrational energy of a dynamic system, have been widely employed in building structures. A few feasibility studies have also been performed to exploit their capacity in controlling bridge vibrations. One example is shown in Figure 59.9, which has been utilized to control the vibration of a cable-stayed bridge [20]. The interface pressure

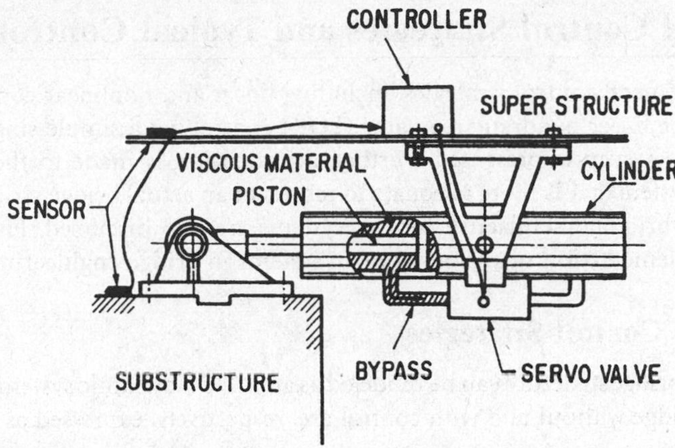

FIGURE 59.8 Controllable fluid damper. (*Source: Proceedings of the Second US–Japan Workshop on Earthquake Protective Systems for Bridges*. p. 481, 1992. With permission.)

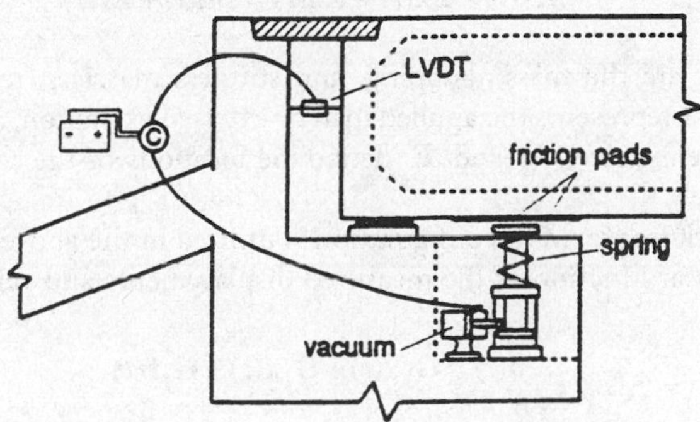

FIGURE 59.9 Controllable friction damper.

of this damper can be actively adjusted through a prestressed spring, a vacuum cylinder, and a battery-operated valve. Since a cable-stayed bridge is a typical flexible structure with relatively low vibration frequencies, its acceleration responses are small due to the isolation effect of flexibility, and short-duration earthquakes do not have enough time to generate large structural displacement responses. In order to take full advantage of the isolation effect of flexibility, it is better not to impose damping force in this case since the increase of large damping force will also increase bridge effective stiffness. On the other hand, if the earthquake excitation is sufficiently long and strong, the displacement of this flexible structure may be quite large. Under this condition, it is necessary to impose large friction forces to dissipate vibrational energy and reduce the moment demand at the bottom of the towers. Therefore, a desirable control system design will be a multistage control system having friction forces imposed at different levels to meet different needs of response control.

The most attractive advantage of the above semiactive control devices is their lower power requirement. In fact, many can be operated on battery power, which is most suitable during seismic events when the main power source to the bridge may fail. Another significant characteristic of semiactive control, in contrast to pure active control, is that it does not destabilize (in the bounded input/bounded output sense) the bridge structural system since no mechanical energy is injected into the controlled bridge system (i.e., , including the bridge and control devices) by the semiactive control devices. Semiactive control devices appear to combine the best features of both passive and active control systems. That is the reason this type of control system offers the greatest likelihood of acceptance in the near future of control technology as a viable means of protecting civil engineering structural systems against natural forces.

59.3 General Control Strategies and Typical Control Algorithms

In this section, the general control strategies, including linear and nonlinear controllers, are introduced first. Then, the linear quadratic regulator (LQR) controlling a simple single-degree-of-freedom (SDOF) bridge system is presented. Further, an extension is made to the multi-degree-of-freedom (MDOF) system that is more adequate to represent an actual bridge structure. The specific characteristics of hybrid and semiactive control systems are also discussed. Finally, the practical concerns about implementation of various control systems in bridge engineering are addressed.

59.3.1 General Control Strategies

Theoretically, a real bridge structure can be modeled as an MDOF dynamic system and the equations of motion of the bridge without and with control are, respectively, expressed as

$$\mathbf{M}\ddot{\mathbf{x}}(t)+\mathbf{C}\dot{\mathbf{x}}(t)+\mathbf{K}\mathbf{x}(t)=\mathbf{E}\mathbf{f}(t) \tag{59.1}$$

$$\mathbf{M}\ddot{\mathbf{x}}(t)+\mathbf{C}\dot{\mathbf{x}}(t)+\mathbf{K}\mathbf{x}(t)=\mathbf{D}\mathbf{u}(t)+\mathbf{E}\mathbf{f}(t) \tag{59.2}$$

where $\mathbf{M}$, $\mathbf{C}$, and $\mathbf{K}$ are the mass, damping, and stiffness matrices, respectively, $\mathbf{x}(t)$ is the displacement vector, $\mathbf{f}(t)$ represents the applied load or external excitation, and $\mathbf{u}(t)$ is the applied control force vector. The matrices $\mathbf{D}$ and $\mathbf{E}$ define the locations of the control force vector and the excitation, respectively.

Assuming the feedback/feedforward configuration is utilized in the above controlled system and the control force is a linear function of the measured displacements and velocities, i.e.,

$$\mathbf{u}(t)=\mathbf{G}_x\mathbf{x}(t)+\mathbf{G}_{\dot{x}}\dot{\mathbf{x}}(t)+\mathbf{G}_f\mathbf{f}(t) \tag{59.3}$$

where $\mathbf{G}_x$, $G_{\dot{x}}$, and $\mathbf{G}_f$ are known as control gain matrices.

Substituting Eq. (59.3) into Eq. (59.2), we obtain

$$\mathbf{M}\ddot{\mathbf{x}}(t)+(\mathbf{C}-\mathbf{D}\mathbf{G}_{\dot{x}})\dot{\mathbf{x}}(t)+(\mathbf{K}-\mathbf{D}\mathbf{G}_x)\mathbf{x}(t)=(\mathbf{E}+\mathbf{D}\mathbf{G}_f)\mathbf{f}(t) \tag{59.4}$$

Alternatively, it can be written as

$$\mathbf{M}\ddot{\mathbf{x}}(t)+\mathbf{C}_c(t)\dot{\mathbf{x}}(t)+\mathbf{K}_c(t)\mathbf{x}(t)=\mathbf{E}_c(t)\mathbf{f}(t) \tag{59.5}$$

Comparing Eq. (59.5) with Eq. (59.1), it is clear that the result of applying a control action to a bridge is to modify the bridge properties and to reduce the external input forces. Also this modification, unlike passive control, is real-time adaptive, which makes the bridge respond more favorably to the external excitation.

It should be mentioned that the above control effect is just an ideal situation: linear bridge structure with linear controller. Actually, physical structure/control systems, such as a hybrid base-isolated bridge, are inherently nonlinear. Thus, all control systems are nonlinear to a certain extent. However, if the operating range of a control system is small and the involved nonlinearities are smooth, then the control system may be reasonably approximated by a linearized system, whose dynamics is described by a set of linear differential equations, for instance, Eq. (59.5).

In general, nonlinearities can be classified as *inherent* (natural) and *intentional* (artificial). Inherent nonlinearities are those that naturally come with the bridge structure system itself. Examples

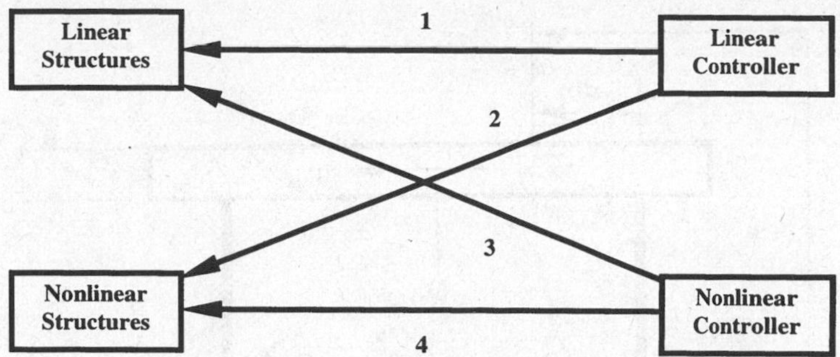

FIGURE 59.10 General control strategies.

of inherent nonlinearities include inelastic deformation of bridge components, seismic isolators, friction dampers, etc. Intentional nonlinearities, on the other hand, are artificially introduced into bridge structural systems by the designer [14,16]. Nonlinear control laws, such as optimal bang–bang control, sliding mode control, and adaptive control, are typical examples of intentional nonlinearities.

According to the properties of the bridge itself and properties of the controller selected, general control strategies may be classified into the following four categories, shown in Figure 59.10 [13].

- **Inherent linear control strategy**: A linear controller controlling a linear bridge structure. This is a simple and popular control strategy, such as LQR/LQG control, pole assignment/mode space control, etc. The implication of this kind of control law is based on the assumption that a controlled bridge will remain in the linear range. Thus, designing a linear controller is the simplest yet reasonable solution. The advantages of linear control laws are well understood and easy to design and implement in actual bridge control applications.
- **Intentional linearization strategy**: A linear controller controlling a nonlinear structure. This belongs to the second category of control strategy, as shown in Figure 59.10. Typical examples of this kind of control laws include instantaneous optimal control, feedback linearization, and gain scheduling, etc. This control strategy retains the advantages of the linear controller, such as simplicity in design and implementation. However, since linear control laws rely on the key assumption of small-range operation, when the required operational range becomes large, a linear controller is likely to perform poorly or sometimes become unstable, because nonlinearities in the system cannot be properly compensated.
- **Intentional nonlinearization strategy**: A nonlinear controller controlling a linear structure. Basically, if undesirable performance of a linear system can be improved by introducing a nonlinear controller intentionally, instead of using a linear controller, the nonlinear one may be preferable. This is the basic motivation for developing intentional nonlinearization strategy, such as optimal bang–bang control, sliding mode control, and adaptive control.
- **Inherent nonlinear control strategy**: A nonlinear controller controlling a nonlinear structure. It is reasonable to control a nonlinear structure by using a nonlinear controller, which can handle nonlinearities in large-range operations directly. Sometimes a good nonlinear control design may be simple and more intuitive than its linear counterparts since nonlinear control designs are often deeply rooted in the physics of the structural nonlinearities. However, since nonlinear systems can have much richer and more complex behaviors than linear systems, there are no systematic tools for predicting the behaviors of nonlinear systems, nor are there systematic procedures for designing nonlinear control systems. Therefore, how to identify and describe structural nonlinearities accurately and then design a suitable nonlinear controller based on those specified nonlinearities is a difficult and challenging task in current nonlinear bridge control applications.

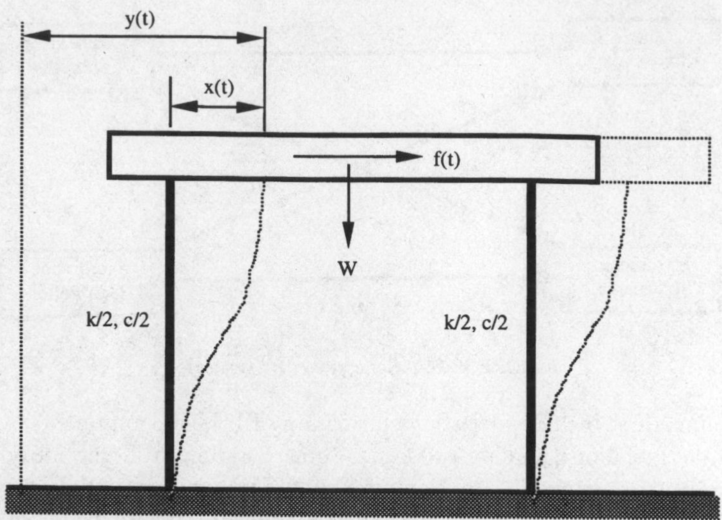

FIGURE 59.11 Simplified bridge model — SDOF system.

59.3.2 Single-Degree-of-Freedom Bridge System

Figure 59.11 shows a simplified bridge model represented by an SDOF system. The equation of motion for this SDOF system can be expressed as

$$m\ddot{x}(t) + c\dot{x}(t) + kx(t) = f(t) \tag{59.6}$$

where m represents the total mass of the bridge, k and c are the linear elastic stiffness and viscous damping provided by the bridge columns and abutments, $f(t)$ is an external disturbance, and $x(t)$ denotes the lateral movement of the bridge. For a specified disturbance, $f(t)$, and with known structural parameters, the responses of this SDOF system can be readily obtained by any step-by-step integration method.

In the above, $f(t)$ represents an arbitrary environmental disturbance such as earthquake, traffic, or wind. In the case of an earthquake load,

$$f(t) = -m\ddot{x}_g(t) \tag{59.7}$$

where $\ddot{x}_g(t)$ is earthquake ground acceleration. Then Eq. (59.6) can be alternatively written as

$$\ddot{x}(t) + 2\xi\ \omega\ \dot{x}(t) + \omega^2 x(t) = -\ddot{x}_0(t) \tag{59.8}$$

in which ω and ξ are the natural frequency and damping ratio of the bridge, respectively.

If an active control system is now added to the SDOF system, as indicated in Figure 59.12, the equation of motion of the extended SDOF system becomes

$$\ddot{x}(t) + 2\xi\ \omega\ \dot{x}(t) + \omega^2 x(t) = u(t) - \ddot{x}_0(t) \tag{59.9}$$

where $u(t)$ is the normalized control force per unit mass. The central topic of control system design is to find an optimal control force $u(t)$ to minimize the bridge responses. Various control strategies,

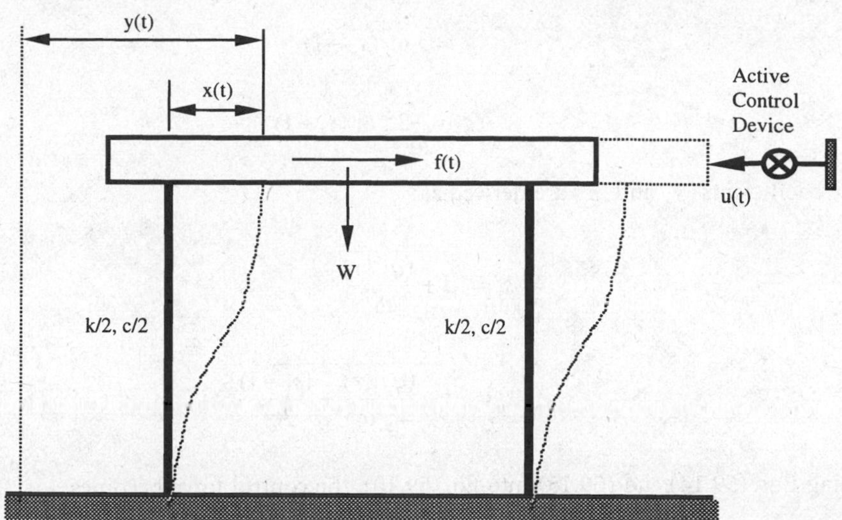

FIGURE 59.12 Simplified bridge with active control system.

as discussed before, have been proposed and implemented to control different structures under different disturbances. Among them, the LQR is the simplest and most widely used control algorithm [10,13].

In LQR, the control force $u(t)$ is designed to be a linear function of measured bridge displacement, $x(t)$, and measured bridge velocity, $\dot{x}(t)$,

$$u(t) = g_x x(t) + g_{\dot{x}} \dot{x}(t) \tag{59.10}$$

where g_x and $g_{\dot{x}}$ are two constant feedback gains which can be found by minimizing a performance index:

$$J = \frac{1}{2} \int_0^\infty [q_x x^2(t) + q_{\dot{x}} \dot{x}^2(t) + r u^2(t)] dt \tag{59.11}$$

where q_x, $q_{\dot{x}}$, and r are called weighting factors. In Eq. (59.11), the first term represents bridge vibration strain energy, the second term is the kinetic energy of the bridge, and the third term is the control energy input by external source powers. Minimizing Eq. (59.11) means that the total bridge vibration energy will be minimized by using minimum input control energy, which is an ideal optimal solution.

The role of weighting factors in Eq. (59.11) is to apply different penalties on the controlled responses and control forces. The assignment of large values to the weight factors q_x and $q_{\dot{x}}$ implies that a priority is given to response reductions. On the other hand, the assignment of a large value to weighting factor r means that the control force requirement is the designer's major concern. By varying the relative magnitudes of q_x, $q_{\dot{x}}$, and r, one can synthesize the controllers to achieve a proper trade-off between control effectiveness and control energy consumption. The effects of these weighting factors on the control responses of bridge structures will be investigated in the next section of case studies.

It has been found [15] that analytical solutions of the feedback constant gains, g_x and $g_{\dot{x}}$, are

$$g_x = -\omega^2 (s_x - 1) \tag{59.12}$$

$$g_{\dot{x}} = -2\xi\ \omega\ (s_{\dot{x}} - 1) \tag{59.13}$$

where the coefficients s_x and $s_{\dot{x}}$ are derived as

$$s_x = \sqrt{1 + \frac{(q_x / r)}{\omega^4}} \tag{59.14}$$

$$s_{\dot{x}} = \sqrt{1 + \frac{(q_{\dot{x}} / r)}{4\xi^2\omega^2} + \frac{(s_x - 1)}{2\xi^2}} \tag{59.15}$$

Substituting Eqs. (59.14) and (59.15) into Eq. (59.10), the control force becomes

$$u(t) = -\omega^2 (s_x - 1) x(t) - 2\xi\ \omega\ (s_{\dot{x}} - 1) \dot{x}(t) \tag{59.16}$$

Inserting the above control force into Eq. (59.9), one obtains the equation of motion of the controlled system as

$$\ddot{x}(t) + 2\xi\ \omega\ s_{\dot{x}} \dot{x}(t) + \omega^2 s_x x(t) = -\ddot{x}_0(t) \tag{59.17}$$

It is interesting to compare Eq. (59.8), which is an uncontrolled system equation, with Eq. (59.17), which is a controlled system equation. It can be seen that the coefficient s_x reflects a shift of the natural frequency caused by applying the control force, and the coefficient $s_{\dot{x}}$ indicates a change in the damping ratio due to control force action.

The concept of active control is clearly exhibited by Eq. (59.17). On the one hand, an active control system is capable of modifying properties of a bridge in such a way as to react to external excitations in the most favorable manner. On the other hand, direct reduction of the level of excitation transmitted to the bridge is also possible through an active control if a feedforward strategy is utilized in the control algorithm.

Major steps to design an SDOF control system based on LQR are

- Calculate the responses of the uncontrolled system from Eq. (59.8) by response spectrum method or the step-by-step integration, and decide whether a control action is necessary or not.
- If a control system is needed, then assign the values to the weighting factors q_x, $q_{\dot{x}}$, and r, and evaluate the adjusting coefficients s_x and $s_{\dot{x}}$ from Eqs. (59.14) and (59.15) directly.
- Find the responses of the controlled system and control force requirement from Eq. (59.17) and Eq. (59.16), respectively.
- Make the final trade-off decision based on concern about the response reduction or control energy consumption and, if necessary, start the next iterative process.

59.3.3 Multi-Degree-of-Freedom Bridge System

An actual bridge structure is much more complicated than the simplified model shown in Figure 59.11, and it is hard to model as an SDOF system. Therefore, a MDOF system will be introduced next to handle multispan or multimember bridges. The equation of motion for an MDOF system without and with control has been given in Eqs. (59.1) and (59.2), respectively. In the control system design, Eq. (59.2) is generally transformed into the following state equation for convenience of derivation and expression:

$$\dot{\mathbf{z}}(t) = \mathbf{A}\mathbf{z}(t) + \mathbf{B}\mathbf{u}(t) + \mathbf{W}\mathbf{f}(t) \tag{59.18}$$

where

$$\mathbf{z}(t) = \begin{bmatrix} \mathbf{x}(t) \\ \dot{\mathbf{x}}(t) \end{bmatrix}; \quad \mathbf{A} = \begin{bmatrix} \mathbf{0} & \mathbf{1} \\ -\mathbf{M}^{-1}\mathbf{K} & -\mathbf{M}^{-1}\mathbf{C} \end{bmatrix}; \quad \mathbf{B} = \begin{bmatrix} \mathbf{0} \\ \mathbf{M}^{-1}\mathbf{D} \end{bmatrix}; \quad \mathbf{W} = \begin{bmatrix} \mathbf{0} \\ \mathbf{M}^{-1}\mathbf{E} \end{bmatrix} \tag{59.19}$$

Similar to SDOF system design, the control force vector $\mathbf{u}(t)$ is related to the measured state vector $\mathbf{z}(t)$ as the following linear function:

$$\mathbf{u}(t) = \mathbf{G}\mathbf{z}(t) \tag{59.20}$$

in which $\mathbf{G}$ is a control gain matrix which can be found by minimizing the performance index [10]:

$$J = \frac{1}{2}\int_{0}^{\infty} [\mathbf{z}^T(t)\mathbf{Q}\mathbf{z}(t) + \mathbf{u}^T(t)\mathbf{R}\mathbf{u}(t)]dt \tag{59.21}$$

where $\mathbf{Q}$ and $\mathbf{R}$ are the weighting matrices and have to be assigned by the designer. Unlike an SDOF system, an analytical solution of control gain matrix $\mathbf{G}$ in Eq. (59.21) is currently not available. However, the matrix numerical solution is easy to find in general control program packages. Theoretically, designing a linear controller to control an MDOF system based on LQR principle is easy to accomplish. But the implementation of a real bridge control is not so straightforward and many challenging issues still remain and need to be addressed. This will be the last topic of this section.

59.3.4 Hybrid and Semiactive Control System

It should be noted from the previous section that most of the hybrid or semiactive control systems are intrinsically nonlinear systems. Development of control strategies that are practically implementable and can fully utilize the capacities of these systems is an important and challenging task. Various nonlinear control strategies have been developed to take advantage of the particular characteristics of these systems, such as optimal instantaneous control, bang–bang control, sliding mode control, etc. Since different hybrid or semiactive control systems have different unique features, it is impossible to develop a universal control law, like LQR, to handle all these nonlinear systems. The particular control strategy for a particular nonlinear control system will be discussed as a case study in the next section.

59.3.5 Practical Considerations

Although extensive theoretical developments of various control strategies have shown encouraging results, it should be noted that these developments are largely based on idealized system descriptions. From theoretical development to practical application, engineers will face a number of important issues; some of these issues are listed in Figure 59.2 and are discussed in this section.

59.3.5.1 Control Single Time Delay

As shown in Figure 59.1, from the measurement of vibration signal by the sensor to the application of a control action by the actuator, time has to be consumed in processing measured information, in performing online computation, and in executing the control forces as required. However, most of the current control algorithms do not incorporate this time delay into the programs and assume that all operations can be performed instantaneously. It is well understood that missing time delay may render the control ineffective and, most seriously, may cause instability of the system. One example is discussed here. Suppose:

1. The time periods consumed in processing measurement, computation, and force action are 0.01, 0.2, and 0.3 s, respectively;
2. The bridge vibration follows a harmonic motion with a period of 1.02 s; and
3. The sensor picks up a positive peak response of the bridge vibration at 5.0 s.

After the control system finishes all processes and applies a large control force onto the bridge, the time is 5.51 s. At this time, the bridge vibration has already changed its phase and reached the negative peak response. It is evident that the control force actually is not controlling the bridge but exciting the bridge. This kind of excitation action is very dangerous and may lead to an unstable situation. Therefore, the time delay must be compensated for in the control system implementation. Various techniques have been developed to compensate for control system time delay. The details can be found in Reference [10].

59.3.5.2 Control and Observation Spillover

Although actual-bridge structures are distributed parameter systems, in general, they are modeled as a large number of degrees of freedom discretized system, referred as the full-order system, during the analytical and simulation process. Further, it is difficult to design a control system based on the full-order bridge model due to online computation process and full state measurement. Hence, the full-order model is further reduced to a small number of degrees of freedom system, referred as a reduced-order system. Then, the control design is performed based on the reduced-order bridge model. After finishing the design, however, the implementation of the designed control system is applied on the actual distributed parameter bridge. Two problems may result. First, the designed control action can only control the reduced-order modes and may not be effective with the residual (uncontrolled) modes, and sometimes even worse to excite the residual modes. This kind of action is called *control spillover*, i.e., the control actions spill over to the uncontrolled modes and enhance the bridge vibration. Second, the control design is based on information observed from the reduced-order model. But, in reality, it is impossible to isolate the vibration signals from residual modes and the measured information must be contaminated by the residual modes. After the contaminated information is fed back into the control system, the control action, originally based on the "pure" measurements, may change, and the control performance may be degraded seriously. This is the so-called *observation spillover*. Again, all spillover effects must be compensated for in the control system implementation [10].

59.3.5.3 Optimal Actuator and Sensor Locations

Because a large number of degrees of freedom are usually involved in the bridge structure, it is impractical to install sensors on each degree-of-freedom location and measure all state variables. Also, in general, only fewer (often just one) control actuators are installed at the critical control locations. Two problems are (1) How many sensors and actuators are required for a bridge to be completely observable and controllable? (2) Where are the optimal locations to install these sensors and actuators in order to measure vibration signals and exert control forces most effectively? Actually, the vibrational control, property identification, health monitoring, and damage detection are closely related in the development of optimal locations. Various techniques and schemes have been successfully developed to find optimal sensor and actuator locations. Reference [10] provides more details about this topic.

59.3.5.4 Control–Structure Interaction

Like bridge structures, control actuators themselves are dynamic systems with inherent dynamic properties. When an actuator applies control forces to the bridge structure, the structure is in turn applying the reaction forces on the actuator, exciting the dynamics of the actuator. This is the so-called *control–structure interaction.* Analytical simulations and experimental verifications have indicated that disregarding the control–structure interaction may significantly reduce both the achievable control performance and the robustness of the control system. It is important to model the dynamics of the actuator properly and to account for the interaction between the structure and the actuator [3,9].

59.3.5.5 Parameter Uncertainty

Parameter identification is a very important part in the loop of structural control design. However, due to limitations in modeling and system identification theory, the exact identification of structural parameters is virtually impossible, and the parameter values used in control system design may deviate significantly from their actual values. This type of *parameter uncertainty* may also degrade the control performance. The sensitivity analysis and robust control design are effective means to deal with the parameter uncertainty and other modeling errors [9].

The above discussions only deal with a few topics of practical considerations in real bridge control implementation. Some other issues that must be investigated in the design of control system include the stability of the control, the noise in the digitized instrumentation signals, the dynamics of filters required to attenuate the signal noise, the potential for actuator saturation, any system nonlinearities, control system reliability, and cost-effectiveness of the control system. More-detailed discussions of these topics are beyond the scope of this chapter. A recent state-of-the-art paper is a very useful resource that deals with all the above topics [6].

59.4 Case Studies

59.4.1 Concrete Box-Girder Bridge

59.4.1.1 Active Control for a Three-Span Bridge

The first example of case studies is a three-span concrete box-girder bridge located in a seismically active zone. Figure 59.13a shows the elevation view of this bridge. The bridge has the span lengths of 38, 38, and 45 m, respectively. The width of the bridge is 32 m and the depth is 2.1 m. The column heights are 15 and 16 m at Bents 2 and 3, respectively. Each span has four oblong-shaped columns with 1.67 × 2.51 m cross section. The columns are monolithically connected with bent cap at the top and pinned with footing at the bottom. The bridge has a total weight of 81,442 kN or a total mass of 8,302,000 kg. The longitudinal stiffness, including abutments and columns, is 82.66 kN/mm. Two servo-hydraulic actuators are installed on the bridge abutments and controlled by the same controller to keep both actuators in the same phase during the control operation. The objective of using the active control system is to reduce the bridge vibrations induced by strong earthquake excitations. Only longitudinal movement will be controlled.

The analysis model of this bridge is illustrated in Figure 59.13b, and a simplified SDOF model is shown in Figure 59.13c. The natural frequency of the SDOF system $\omega = 19.83$ rad/s, and damping ratio $\xi = 5\%$. Without loss of the generality, the earthquake ground motion, $\ddot{x}_0(t)$, is described as a stationary random process. The well-known Kanai–Tajimi spectrum is utilized to represent the power spectrum density of the input earthquake, i.e.,

$$G_{\ddot{x}_0}(\omega) = \frac{G_0[1 + 4\xi_g{}^2(\omega/\omega_g)^2]}{[1-(\omega/\omega_g)^2]^2 + 4\xi_g{}^2(\omega/\omega_g)^2} \tag{59.22}$$

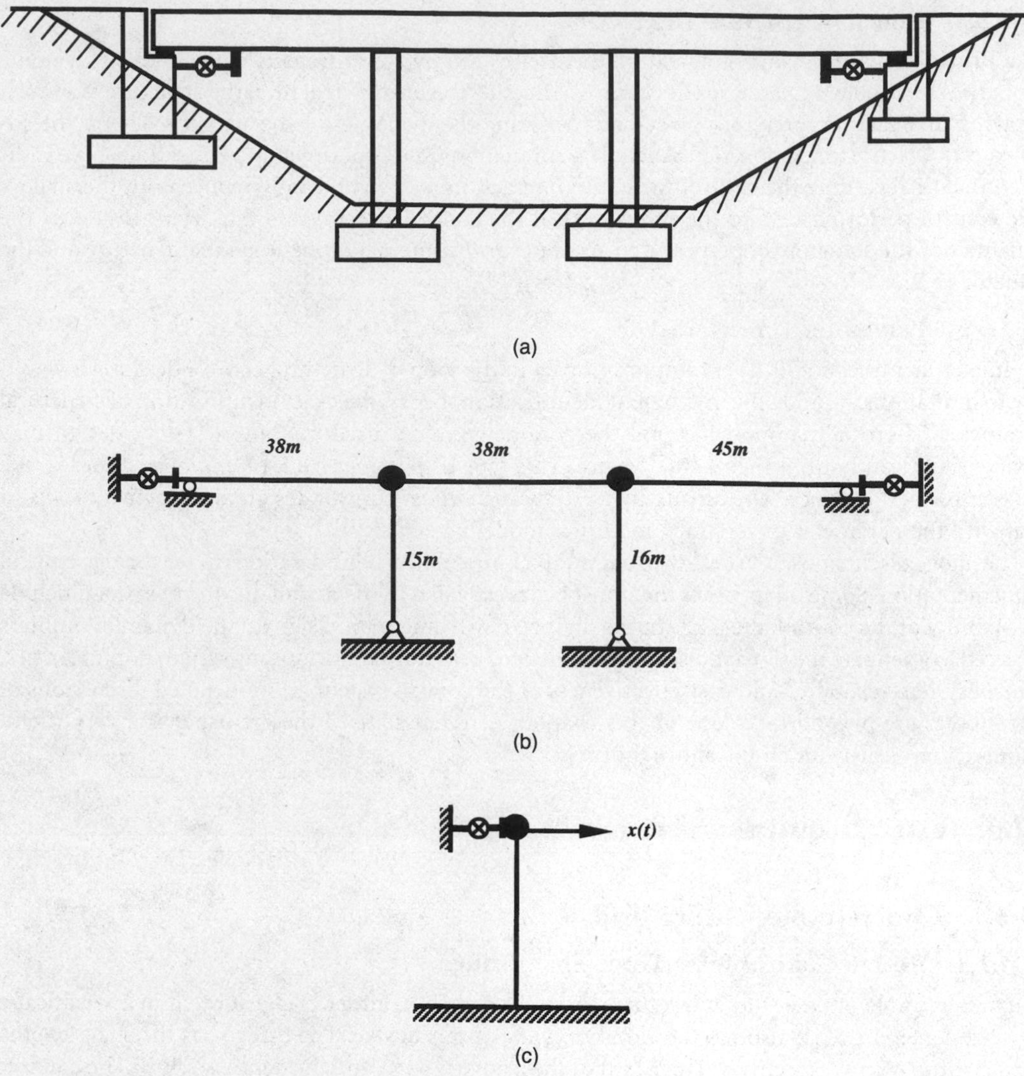

FIGURE 59.13 Three-span bridge with active control system. (a) Actual bridge; (b) bridge model for analysis; (c) SDOF system controlled by actuator.

where ω_g and ξ_g are, respectively, the frequency and damping ratio of the soil, whose values are taken as $\omega_g = 22.9$ rad/s and $\xi_g = 0.34$ for average soil condition. The parameter G_0 is the spectral density related to the maximum earthquake acceleration a_{max} [15]. At this bridge site, the maximum ground acceleration $a_{max} = 0.4\ g$.

The maximum response of an SDOF system with natural frequency ω and damping ratio ξ under $\ddot{x}_0(t)$ excitation can be estimated as

$$x_{max}(\omega,\xi) = \gamma_p \sigma_x \tag{59.23}$$

in which γ_p is a peak factor and σ_x is the root-mean-square response which can be determined by random vibration theory [2].

From Eq. (59.17), it is known that the frequency and damping ratio of a controlled system are

$$\omega_c = \sqrt{s_x}\,\omega; \quad \xi_c = (s_{\dot{x}} / \sqrt{s_x})\xi \tag{59.24}$$

TABLE 59.2 Summary of Three-Span Bridge Control

r	s_x	$s_{\dot{x}}$	ω_c *(rad/s)*	ξ (%)	d_{max} (cm)	d_{max} Redu (%)	a_{max} (g)	a_{max} Redu (%)	u_{max} (kN)	u_{max} Weight (%)
1E+07	1.000	1.001	19.83	0.05	3.15	0	1.23	0	10	0
100,000	1.000	1.103	19.83	0.06	2.86	9	1.12	9	1046	1
10,000	1.000	1.777	19.83	0.09	2.30	27	0.90	27	7894	10
5,000	1.001	2.305	19.83	0.12	1.97	37	0.77	37	13258	16
1,000	1.003	4.750	19.85	0.24	1.30	59	0.51	59	38097	47
500	1.005	6.643	19.88	0.33	0.72	77	0.28	77	57329	70

where s_x and $s_{\dot{x}}$ can be found from Eq. (59.14) and Eq. (59.15), respectively, once the weighting factors q_x, $q_{\dot{x}}$, and r are assigned by the designer. The maximum response of the controlled system is obtained from Eq. (59.23).

In this case study, the weighting factors are assigned as $q_{\dot{x}} = 100m$ and $q_x = k$. Through varying the weight factor r, one can obtain different control efficiencies by applying different control forces. Table 59.2 lists the control coefficients, controlled frequencies, damping ratios, maximum bridge responses, and maximum control force requirements based on various assignments of the weight factor r.

It can be seen from Table 59.2 that no matter how small the weighting factor r is, the coefficient s_x is always close to 1, which means that the structural natural frequency is hard to shift by LQR algorithm. However, the coefficient $s_{\dot{x}}$ increases significantly with decrease of the weighting factor r, which means that the major effect of LQR algorithm is to modify structural damping. This is just what we wanted. In fact, extensive simulation results have shown the same trend as indicated in Table 59.2 [13]. The maximum acceleration of the bridge deck is 1.23 *g* without control. If the control force is applied on the bridge with maximum value of 13,258 kN (16% bridge weight), the maximum acceleration response reduces to 0.77 *g*, the reduction factor is 37%. The larger the applied control force, the larger the response reduction. But, in reality, current servo-hydraulic actuators may not generate such a large control force.

59.4.1.2 Hybrid Control for a Simple-Span Bridge

The second example of the case studies, as shown in Figure 59.14, is a simple-span bridge equipped with rubber bearings and active control actuators between the bridge girder and columns [19]. The bridge has a span length of 30 m and column height of 22 m. The bridge is modeled as a nine-degree-of-freedom system, as shown in Figure 59.14b. Duo to symmetry, it is further reduced to a four-degree-of-freedom system, as shown in Figure 59.14c. The mass, stiffness, and damping properties of this bridge can be found in Reference [19].

The bridge structure is considered to be linear elastic except the rubber bearings. The inelastic stiffness restoring force of the rubber bearing is expressed as

$$F_s = \alpha k x(t) + (1-\alpha) k D_y \nu \tag{59.25}$$

in which $x(t)$ is the deformation of the rubber bearing, k is the elastic stiffness, α is the ratio of the postyielding to preyielding stiffness, D_y is the yield deformation, and ν is the hysteretic variable with $|\nu| \le 1$, where

$$\dot{\nu} = D_y^{-1}\{A\dot{x} - \beta|\dot{x}||\nu|^{n-1}\nu - \gamma\dot{x}|\nu|^n\} \tag{59.26}$$

In Eq. (59.26), the parameters A, β, γ, and n govern the scale, general shape, and smoothness of the hysteretic loop. It can be seen from Eq. (59.25) that if $\alpha = 1.0$, then the rubber bearing has a linear stiffness, i.e., $F_s = kx(t)$.

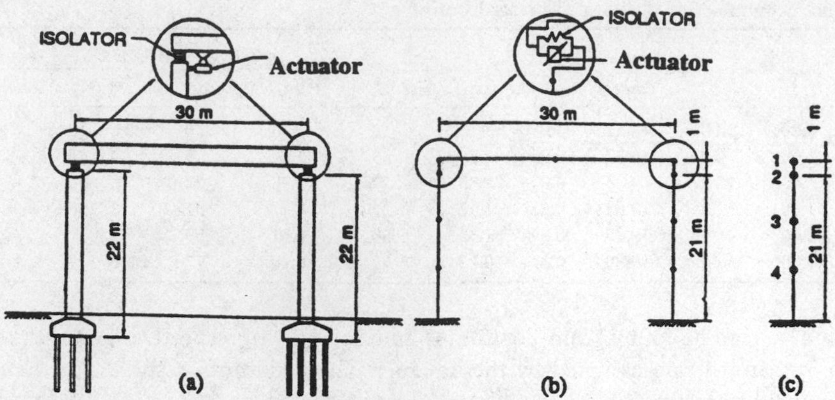

FIGURE 59.14 Simple-span bridge with hybrid control system. (a) Actual bridge; (b) lumped mass system; (c) four-degree-of-freedom system. (*Source: Proceedings of the Second U.S.–Japan Workshop on Earthquake Protective Systems for Bridges*, p. 482, 1992. With permission.)

TABLE 59.3 Summary of Simple-Span Bridge Control

Control System	d_{1max} (cm)	d_{2max} (cm)	d_{3max} (cm)	d_{4max} (cm)	a_{1max} (g)	V_{bmax} (kN)	u_{max} (% W_1)
Passive	24.70	3.96	3.07	1.25	1.31	1648	0
Hybrid	5.53	1.46	1.14	0.46	0.48	628	41

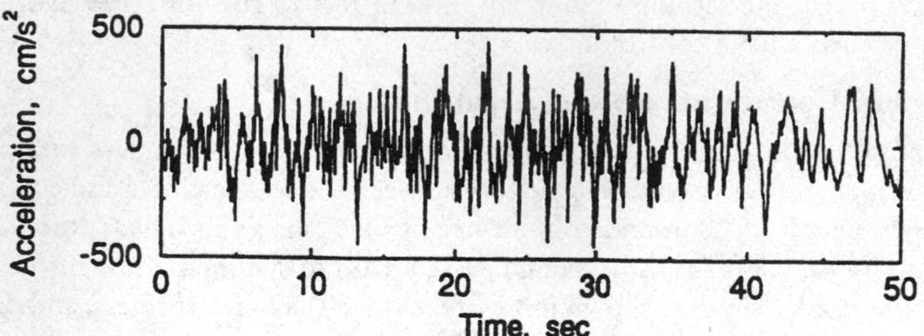

FIGURE 59.15 Simulated earthquake ground acceleration.

The LQR algorithm is incapable of handling the nonlinear structure control problem, as indicated in Eq. (59.25). Therefore, the sliding mode control (SMC) is employed to develop a suitable control law in this example. The details of SMC can be found from Reference [19].

The input earthquake excitation is shown in Figure 59.15, which is simulated such that the response spectra match the target spectra specified in the Japanese design specification for highway bridges. The maximum deformations (d_{1max}, d_{2max}, d_{3max}, and d_{4max}), maximum acceleration (a_{1max}), maximum base shear of the column (V_{bmax}), and maximum actuator control force (u_{max}) are listed in Table 59.3. It is clear that adding an active control system can significantly improve the performance and effectiveness of the passive control. Comparing with passive control alone, the reductions of displacement and acceleration at the bridge deck can reach 78 and 63%, respectively. The base shear of the column can be reduced to 38%. The cost is that each actuator has to provide the maximum control force up to 20% of the deck weight.

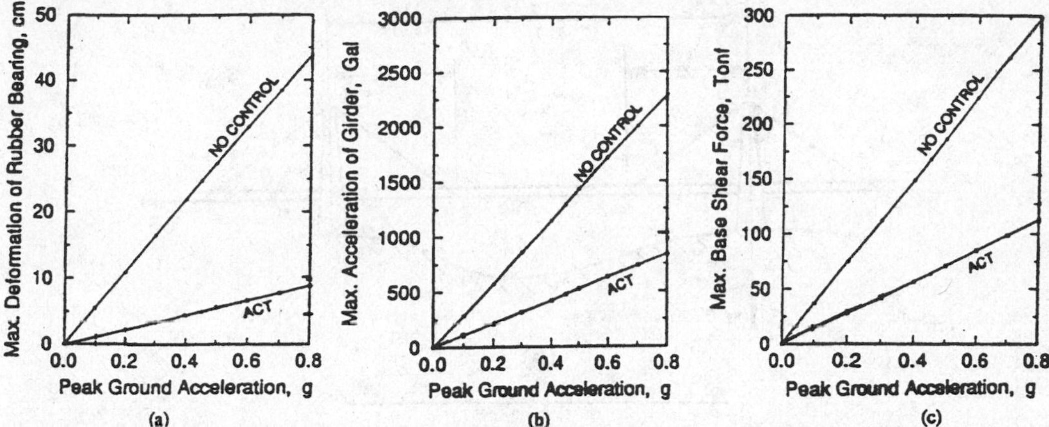

FIGURE 59.16 Bridge maximum responses. (a) deformation of rubber bearing; (b) acceleration of girder; (c) base shear force of pier.

In order to evaluate and compare the effectiveness of a hybrid control system over a wide range of earthquake intensities, the design earthquake shown in Figure 59.15 is scaled uniformly to different peak ground acceleration to be used as the input excitations. The peak response quantities for the deformation of rubber bearing, the acceleration of the bridge deck, and the base shear of the column are presented as functions of the peak ground acceleration in Figure 59.16. In this figure, "no control" means passive control alone, and "act" denotes hybrid control. Obviously, the hybrid control is much more effective over passive control alone within a wide range of earthquake intensities.

59.4.2 Cable-Stayed Bridge

59.4.2.1 Active Control for a Cable-Stayed Bridge

Cable-supported bridges, as typical flexible bridge structures, are particularly vulnerable to strong wind gusts. Extensive analytical and experimental investigations have been performed to increase the "critical wind speed" since wind speeds higher than the critical will cause aerodynamic instability in the bridge. One of these studies is to install an active control system to enhance the performance of the bridge under strong wind gusts [17,18].

Figure 59.17 shows the analytical model of the Sitka Harbor Bridge, Sitka, Alaska. The midspan length of the bridge is 137.16 m. Only two cables are supported by each tower and connected to the bridge deck at distance $a = l/3 = 45.72m$. The two-degree-of-freedom system is used to describe the vibrations of the bridge deck. The fundamental frequency in flexure $\omega_g = 5.083$ rad/s, and the fundamental frequency in torsion $\omega_f = 8.589$ rad/s. In this case study, the four existing cables, which are designed to carry the dead load, are also used as active tendons to which the active feedback control systems (hydraulic servomechanisms) are attached. The vibrational signals of the bridge are measured by the sensors installed at the anchorage of each cable, and then transmitted into the feedback control system. The sensed motion, in the form of electric voltage, is used to regulate the motion of hydraulic rams in the servomechanisms, thus generating the required control force in each cable.

Suppose that the accelerometer is used to measure the bridge vibration. Then the feedback voltage $v(t)$ is proportional to the bridge acceleration $\ddot{w}(t)$:

$$v(t) = p\ddot{w}(t) \tag{59.27}$$

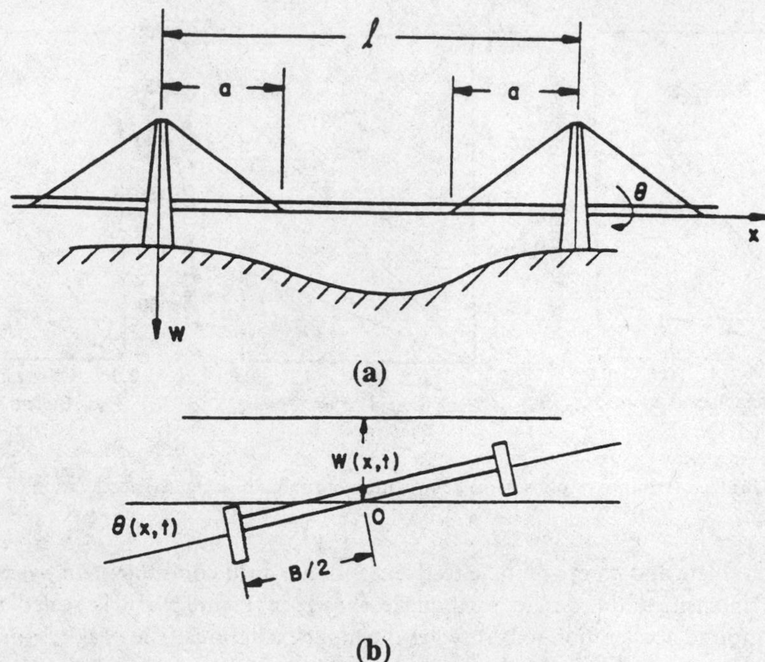

FIGURE 59.17 Cable-stayed bridge with active tendon control. (a) Side view with coordinate system; (b) two-degree-of-freedom model. (*Source*: Yang, J.N. and Giannopolous, F., *J. Eng. Mech. ASCE*, 105(5), 798–810, 1979. With permission.)

where p is the proportionality constant associated with each sensor. For active tendon configuration, the displacement $s(t)$ of hydraulic ram, which is equal to the additional elongation of the tendon (cable) due to active control action, is related to the feedback voltage $v(t)$ through the first-order differential equation:

$$\dot{s}(t) + R_1 s(t) = \frac{R_1}{R} v(t) \tag{59.28}$$

in which R_1 is the loop gain and R is the feedback gain of the servomechanism. The cable control force generated by moving the hydraulic ram is

$$u(t) = ks(t) \tag{59.29}$$

where k is the cable stiffness.

Combining Eq. (59.27) and Eq. (59.29), we have

$$u(t) = g(R_1, R)\ddot{w}(t) \tag{59.30}$$

It is obvious that Eq. (59.30) represents an acceleration feedback control and the control gain $g(R_1, R)$ depends on the control parameters R_1 and R which will be assigned by the designer. Further, two nondimensional parameters ε and τ are introduced to replace R_1 and R

$$\varepsilon = \frac{R_1}{\omega_f} \quad \text{and} \quad \tau = \frac{p\omega_f^{\ 2}}{R} \tag{59.31}$$

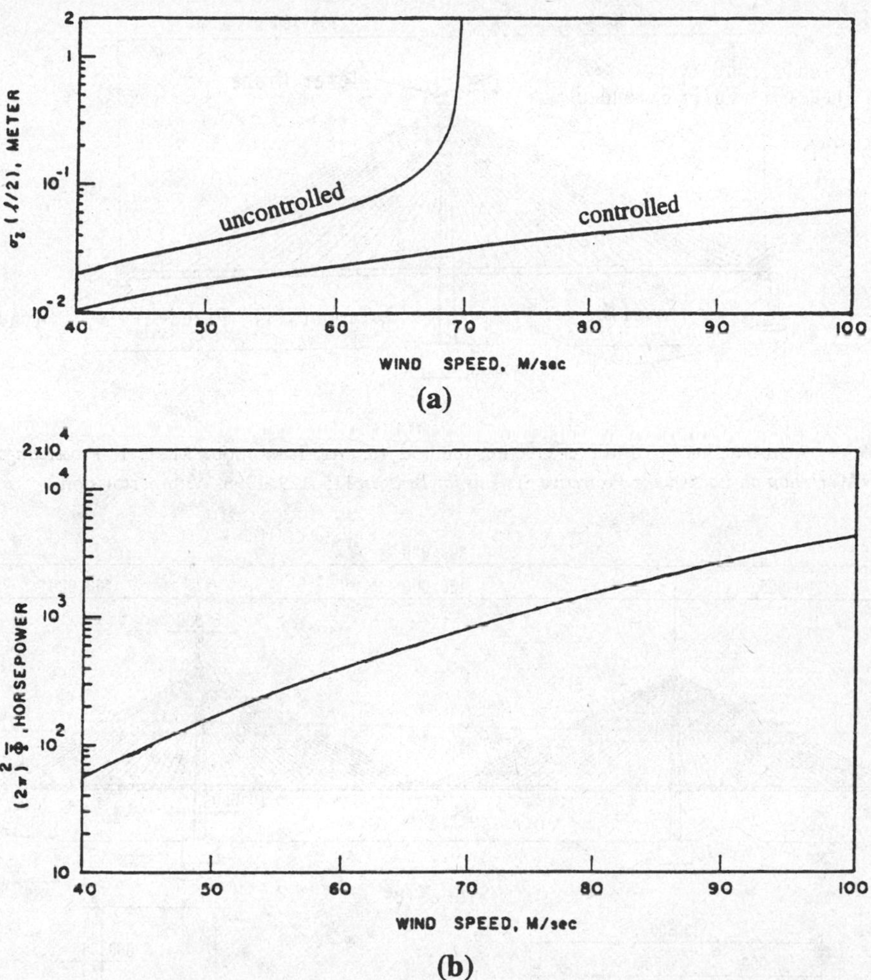

FIGURE 59.18 Root-mean-square displacement and average power requirement. (a) Root-mean-square displacement of bridge deck; (b) average power requirement. (*Source*: Yang, J.N., and Giannnopolous, F., *J. Eng. Mech., ASCE*, 105(5) 798-810, 1979. With permission.)

Finally, the critical wind speed and the control power requirement are all related to the control parameters ε and τ.

Figure 59.18a shows the root-mean-square displacement response of the bridge deck without and with control. In the control case the parameter $\varepsilon = 0.1, \quad \tau = 10$. Correspondingly, the average power requirement to accomplish active control is illustrated in Figure 59.18b. It can be seen that the bridge response is reduced significantly (up to 80% of the uncontrolled case) with a small power requirement by the active devices. In terms of critical wind speed, the value without control is 69.52 m/s, while with control it can be raised to any desirable level provided that the required control forces are realizable. Based on the studies, it appears that the active feedback control is feasible for applications to cable-stayed bridges.

59.4.2.2 Active Mass Damper for a Cable-Stayed Bridge under Construction

Figure 59.19 shows a cable-stayed bridge during construction using the cantilever erection method. It can be seen that not only the bridge weight but also the heavy equipment weights are all supported by a single tower. Under this condition, the bridge is a relatively unstable structure, and special

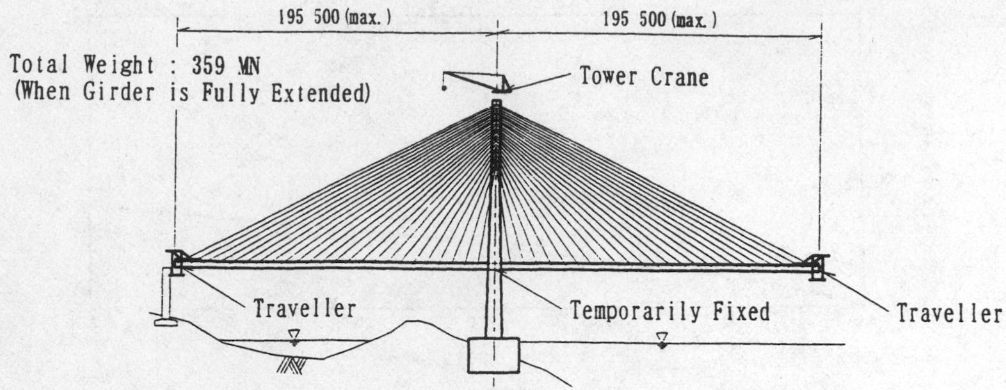

FIGURE 59.19 Construction by cantilever erection method. (*Source*: Tsunomoto, M., et al. *Proceedings of Fourth U.S.–Japan Workshop on Earthquake Protective Systems for Bridges*, 115–129, 1996. With permission.)

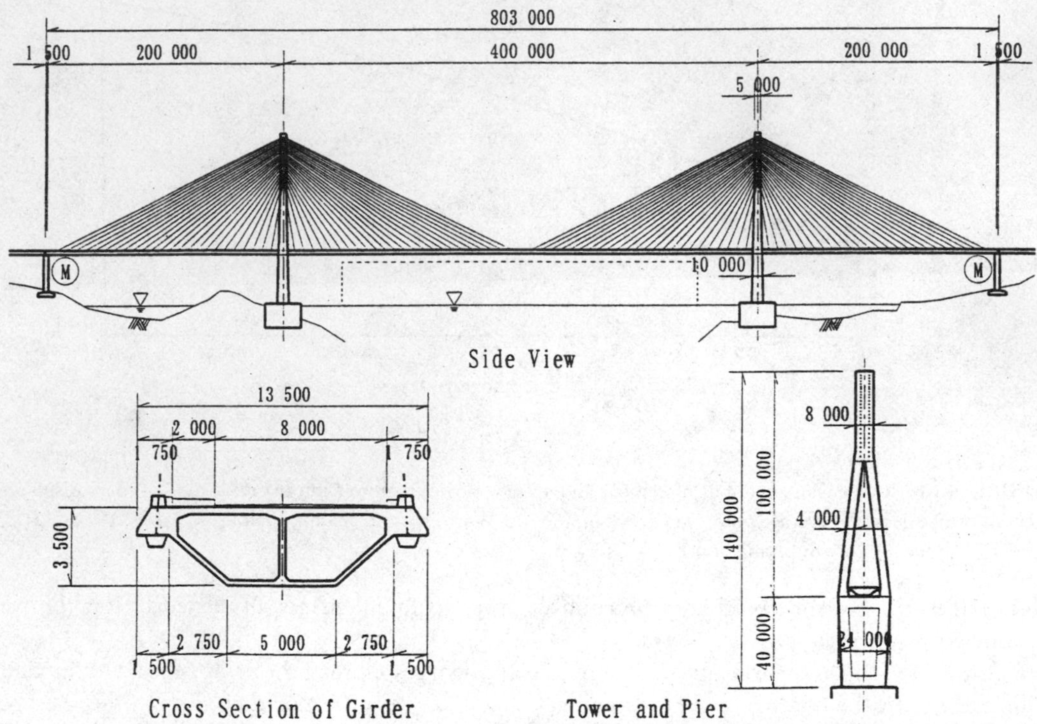

FIGURE 59.20 General view of cable-stayed bridge studied. (*Source*: Tsunomoto, M., et al. Proceedings of Fourth U.S.–Japan Workshop on Earthquake Protective Systems for Bridges, 115–129, 1996. With permission.)

attention is required to safeguard against dynamic external forces such as earthquake and wind loads. Since movable sections are temporarily fixed during the construction, the seismic isolation systems that will be adopted after the completion of the construction are usually ineffective for the bridge under construction. Active tendon control by using the bridge cable is also difficult to install on the bridge at this period. However, active mass dampers, as shown in Figure 59.6, have proved to be effective control devices in reducing the dynamic responses of the bridge under construction [12].

The bridge in this case study is a three-span continuous prestressed concrete cable-stayed bridge with a central span length of 400 m, as shown in Figure 59.20. When the girder is fully extended,

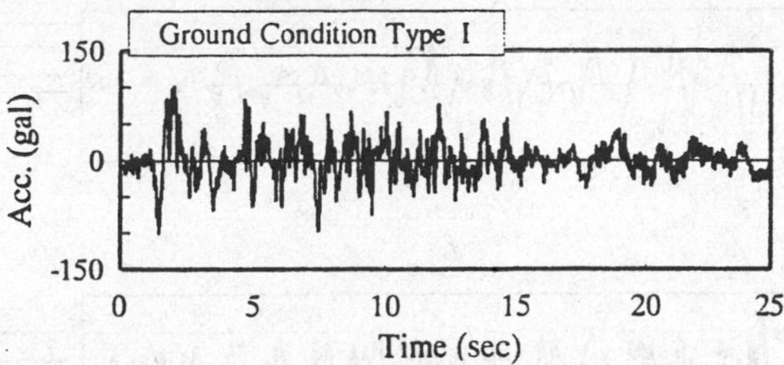

FIGURE 59.21 Input earthquake ground motion. (*Source*: Tsunomoto, M., et al. *Proceedings of Fourth U.S.–Japan Workshop on Earthquake Protective Systems for Bridges*, 115–129, 1996. With permission.)

the total weight is 359 MN, including bridge self-weight, traveler weight (1.37 MN at each end of the girder), and crane weight (0.78 MN at the top of the tower). The damping ratio for dynamic analysis is 1%. The ground input acceleration is shown in Figure 59.21. Since the connections between pier and footing and between girder and pier are fixed during construction, the moments at pier bottom and at tower bottom are the critical response parameters to evaluate the safety of the bridge at this period. Two control cases are investigated. In the first case, the active mass damper (AMD) is installed at the tower top and operates in the longitudinal direction. In the second case, the AMD is installed at the cantilever girder end and operates in the vertical direction. The AMD is controlled by the direct velocity feedback algorithm, in which the control force is only related to the measured velocity response at the location of the AMD. By changing the control gain, the maximum control force is adjusted to around 3.5 MN, which is about 1% of the total weight of the bridge.

Figures 59.22 and 59.23 show the time histories of the bending moments and control forces in Case 1 and Case 2, respectively. In Case 1, the maximum bending moment at the pier bottom is reduced by about 15%, but the maximum bending moment at the tower bottom is reduced only about 5%. In Case 2, the reduction of the bending moment at the tower bottom is the same as that in Case 1, but the reduction of the bending moment at the pier bottom is about 35%, i.e., 20% higher than the reduction in Case 1. The results indicate that the AMD is an effective control device to reduce the dynamic responses of the bridge under construction. Installing an AMD at the girder end of the bridge is more effective than installing it at the tower top. The response control in reducing the bending moment at the tower bottom is less effective than that at the pier bottom.

59.5 Remarks and Conclusions

Various structural protective systems have been developed and implemented for vibration control of buildings and bridges in recent years. These modern technologies have generated a strong impact in the traditional structural design and construction fields. The entire structural engineering discipline is undergoing a major change. It now seems desirable to encourage structural engineers and architects to seriously consider exploiting the capabilities of structural control systems for retrofitting existing structures and also enhancing the performance of prospective new structures.

The basic concepts of various control systems are introduced in this chapter. The emphasis is put on active control, hybrid control, and semiactive control for bridge structures. The different bridge control configurations are presented. The general control strategies and typical control algorithms are discussed. Through several case studies, it is shown that the active, hybrid, and semiactive control systems are quite effective in reducing bridge vibrations induced by earthquake, wind, or traffic.

It is important to recognize that although significant progress has been made in the field of active response control to bridge structures, we are now still in the study-and-development stage and await coming applications. There are many topics related to the active control of bridge structures that need research and resolution before the promise of smart bridge structures is fully realized. These topics are

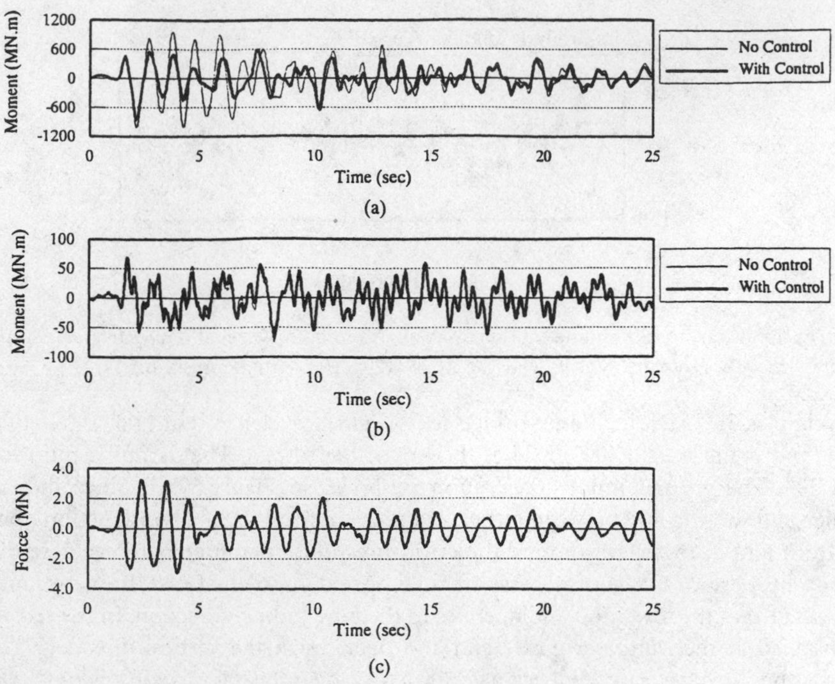

FIGURE 59.22 Bridge responses and control force with AMD at tower top. (a) moment at pier bottom; (b) moment at tower bottom; (c) control force of AMD. (*Source*: Tsunomoto, M., et al. *Proceedings of Fourth U.S.–Japan Workshop on Earthquake Protective Systems for Bridges*, 115–129, 1996. With permission.)

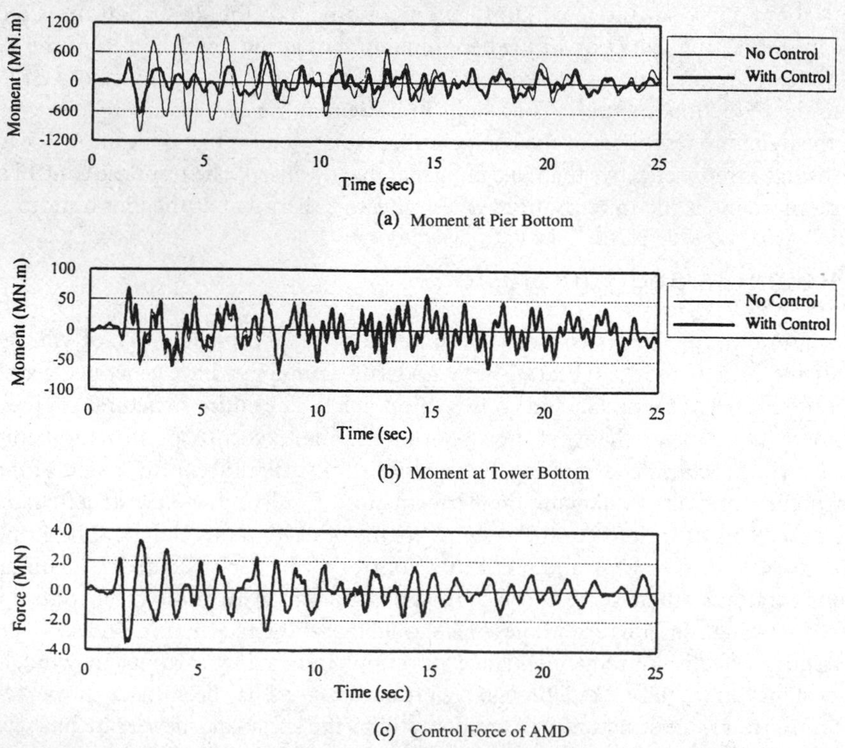

FIGURE 59.23 Bridge responses and control force with AMD at girder end. (a) moment at pier bottom; (b) moment at tower bottom; (c) control force of AMD. (*Source*: Tsunomoto, M., et al. *Proceedings of Fourth U.S.–Japan Workshop on Earthquake Protective Systems for Bridges*, 115–129, 1996. With permission.)

- Algorithms for active, hybrid, and semiactive control of nonlinear bridge structures;
- Devices with energy-efficient features able to handle strong inputs;
- Integration of control devices into complex bridge structures;
- Identification and modeling of nonlinear properties of bridge structures;
- Standardized performance evaluation and experimental verification;
- Development of design guidelines and specifications;
- Implementation on actual bridge structures.

References

1. Adeli, H. and Saleh, A., Optimal control of adaptive/smart structures, *J. Struct. Eng. ASCE,* 123(2), 218–226, 1997.
2. Ben-Haim, Y., Chen, G., and Soong, T. T., Maximum structural response using convex models, *J. Eng. Mech. ASCE,* 122(4), 325–333, 1996.
3. Dyke, S. J., Spencer, B. F., Quast, P., and Sain, M. K., The role of control-structure interaction in protective system design, *J. Eng. Mech. ASCE,* 121(2), 322–338, 1995.
4. Feng, M. Q., Shinozuka, M., and Fujii, S., Friction-controllable sliding isolation system, *J. Eng. Mech. ASCE,* 119(9), 1845–1864, 1993.
5. Feng, M. Q., Seismic response variability of hybrid-controlled bridges, *Probabilistic Eng. Mech.*, 9, 195–201, 1994.
6. Housner, G. W., Bergman, L. A., Caughey, T. K., Chassiakos, A. G., Claus, R. O., Masri, S. F., Skeleton, R. E., Soong, T. T., Spencer, B. F. and Yao, J. T. P., Structural control: past, present, and future, *J. Eng. Mech. ASCE,* 123(9), 897–971, 1997.
7. Kawashima, K. and Unjoh, S., Seismic response control of bridges by variable dampers, *J. Struct. Eng. ASCE,* 120(9), 2583–2601, 1994.
8. Reinhorn, A. M. and Riley, M., Control of bridge vibrations with hybrid devices. Proceedings of First World Conference on Structural Control, II, TA2, 1994, 50–59.
9. Riley, M., Reinhorn, A. M., and Nagarajaiah, S., Implementation issues and testing of a hybrid sliding isolation system, *Eng. Struct.*, 20(3), 144–154, 1998.
10. Soong, T. T., *Active Control: Theory and Practice,* Longman Scientific and Technical, Essex, England, and Wiley, New York, 1990.
11. Soong, T. T. and Dargush, G. F., *Passive Energy Dissipation Systems in Structural Engineering,* John Wiley & Sons, London, 1997.
12. Tsunomoto, M., Otsuka, H., Unjoh, S., and Nagaya, K., Seismic response control of PC cable-stayed bridge under construction by active mass damper, in *Proceedings of Fourth U.S. — Japan Workshop on Earthquake Protective Systems for Bridges,* 115–129, 1996.
13. Wu, Z., Nonlinear Feedback Strategies in Active Structural Control, Ph.D. dissertation, State University of New York at Buffalo, Buffalo, 1995.
14. Wu, Z., Lin. R. C., and Soong, T. T., Nonlinear feedback control for improved response reduction, *Smart Mat. Struct.*, 4(1), A140–A148, 1995.
15. Wu, Z. and Soong, T. T., Design spectra for actively controlled structures based on convex models, *Eng. Struct.*, 18(5), 341–350, 1996.
16. Wu, Z. and Soong, T. T., Modified bang-bang control law for structural control implementation, *J. Struct. Eng. ASCE,* 122(8), 771–777, 1996.
17. Yang, J. N. and Giannopolous, F., Active control and stability of cable-stayed bridge, *J. Eng. Mech. ASCE,* 105(4), 677–694, 1979.
18. Yang, J. N. and Giannopolous, F., Active control of two-cable-stayed bridge, *J. Eng. Mech. ASCE,* 105(5), 795–810, 1979.
19. Yang, J. N., Wu, J. C., Kawashima, K., and Unjoh, S., Hybrid control of seismic-excited bridge structures, *Earthquake Eng. Struct. Dyn.*, 24, 1437–1451, 1995.
20. Yang, C. and Lu, L. W., Seismic response control of cable-stayed bridges by semiactive friction damping, in *Proceedings of Fifth U.S. National Conference on Earthquake Engineering,* Vol. I, 1994, 911–920.

60

Vessel Collision Design of Bridges

Michael Knott
Moffatt & Nichol Engineers

Zolan Prucz
Modjeski and Masters, Inc.

Notations

The following symbols are used in this chapter. The section number in parentheses after definition of a symbol refers to the section or figure number where the symbol first appears or is identified.

AF annual frequency of bridge element collapse (Section 60.5.2)
B_M beam (width) of vessel (Figure 60.2)
B_P width of bridge pier (Figure 60.2)
DWT size of vessel based on deadweight tonnage (one tonne = 2205 lbs = 9.80 kN) (Section 60.4.1)
H ultimate bridge element strength (Section 60.5.2)
N number of one-way vessel passages through the bridge (Section 60.5.2)
P vessel collision impact force (Section 60.5.2)
P_{BH} ship collision impact force for head-on collision between ship bow and a rigid object (Section 60.6.1)
P_{DH} ship collision impact force between ship deckhouse and a rigid superstructure (Section 60.6.1)

0-8493-7434-0/00/$0.00+$.50

P_{MT} ship collision impact force between ship mast and a rigid superstructure (Section 60.6.1)
P_S ship collision impact force for head-on collision between ship bow and a rigid object (Section 60.6.1)
PA probability of vessel aberrancy (Section 60.5.2)
PC probability of bridge collapse (Section 60.5.2)
PG geometric probability of vessel collision with bridge element (Section 60.5.2)
R_{BH} ratio of exposed superstructure depth to the total ship bow depth (Section 60.6.1)
R_{DH} reduction factor for ship deckhouse collision force (Section 60.6.1)
V design impact speed of vessel (Section 60.6.1)
x distance to bridge element from the centerline of vessel transit path (Figure 60.2)
ϕ angle between channel and bridge centerlines (Figure 60.2)

60.1 Introduction

60.1.1 Background

It was only after a marked increase in the frequency and severity of vessel collisions with bridges that studies of the vessel collision problem have been initiated in recent years. In the period from 1960 to 1998, there have been 30 major bridge collapses worldwide due to ship or barge collision, with a total loss of life of 321 people. The greatest loss of life occurred in 1983 when a passenger ship collided with a railroad bridge on the Volga River, Russia; 176 were killed when the aberrant vessel attempted to transit through a side span of the massive bridge. Most of the deaths occurred when a packed movie theater on the top deck of the passenger ship was sheared off by the low vertical clearance of the bridge superstructure.

Of the bridge catastrophes mentioned above, 15 have occurred in the United States, including the 1980 collapse of the Sunshine Skyway Bridge crossing Tampa Bay, Florida, in which 396 m of the main span collapsed and 35 lives were lost as a result of the collision by an empty 35,000 DWT (deadweight tonnage) bulk carrier (Figure 60.1).

One of the more publicized tragedies in the United States involved the 1993 collapse of a CSX Railroad Bridge across Bayou Canot near Mobile, Alabama. During dense fog, a barge tow became lost and entered a side channel of the Mobile River where it struck a railroad bridge causing a large displacement of the structure. The bridge collapsed a few minutes later when a fully loaded Amtrak passenger train attempted to cross the damaged structure; 47 fatalities occurred as a result of the collapse and the train derailment.

It should be noted that there are numerous vessel collision accidents with bridges which cause significant damage, but do not necessarily result in collapse of the structure. A study of river towboat collisions with bridges located on the U.S. inland waterway system during the short period from 1970 to 1974 revealed that there were 811 accidents with bridges costing $23 million in damages and 14 fatalities. On the average, some 35 vessel collision incidents are reported every day to U.S. Coast Guard Headquarters in Washington, D.C.

A recent accident on a major waterway bridge occurred in Portland, Maine in September 1996 when a loaded tanker ship (171 m in length and 25.9 m wide) rammed the guide pile fender system of the existing Million Dollar Bridge over the Fore River. A large portion of the fender was destroyed; the flair of the ship's bow caused significant damage to one of the bascule leafs of the movable structure (causing closure of the bridge until repairs were made); and 170,000 gallons of fuel oil were spilled in the river due to a 9-m hole ripped in the vessel hull by an underwater protrusion of the concrete support pier (a small step in the footing). Although the main cause of the accident was attributed to pilot error, a contributing factor was certainly the limited horizontal clearance of the navigation opening through the bridge (only 29 m).

The 1980 collapse of the Sunshine Skyway Bridge was a major turning point in awareness and increased concern for the safety of bridges crossing navigable waterways. Important steps in the development of modern ship collision design principles and specifications include:

FIGURE 60.1 Sunshine Skyway Bridge, May 9, 1980 after being struck by the M/V *Summit Venture.*

- In 1983, a "Committee on Ship/Barge Collision," appointed by the Marine Board of the National Research Council in Washington, D.C., completed a study on the risk and consequences of ship collisions with bridges crossing navigable coastal waters in the United States [1].
- In June 1983, a colloquium on "Ship Collision with Bridges and Offshore Structures" was held in Copenhagen, Denmark under the auspices of the International Association for Bridge and Structural Engineering (IABSE), to bring together and disseminate the latest developments on the subject [2].
- In 1984, the Louisiana Department of Transportation and Development incorporated criteria for the design of bridge piers with respect to vessel collision for structures crossing waterways in the state of Louisiana [3,4].
- In 1988, a pooled-fund research project was sponsored by 11 states and the Federal Highway Administration to develop vessel collision design provisions applicable to all of the United States. The final report of this project [5] was adopted by AASHTO as a Vessel Collision Design Guide Specification in February, 1991 [6].
- In 1993, the International Association for Bridge and Structural Engineering (IABSE) published a comprehensive document that included a review of past and recent developments in the study of ship collisions and the interaction between vessel traffic and bridges [7].
- In 1994, AASHTO adopted the recently developed LRFD bridge design specifications [8], which incorporate the vessel collision provisions developed in Reference [6] as an integral part of the bridge design criteria.
- In December 1996, the Federal Highway Administration sponsored a conference on "The Design of Bridges for Extreme Events" in Atlanta, Georgia to discuss developments in design

loads (vessel collision, earthquake, and scour) and issues related to the load combinations of extreme events [9].

- In May 1998, an international symposium on "Advances in Bridge Aerodynamics, Ship Collision Analysis, and Operation & Maintenance" was held in Copenhagen, Denmark in conjunction with the opening of the record-setting Great Belt Bridge to disseminate the latest developments on the vessel collision subject [10].

Current highway bridge design practices in the United States follow the AASHTO specifications [6,8]. The design of railroad bridge protection systems against vessel collision is addressed in the American Railway Engineering and Maintenance-of-Way Association (AREMA) Manual for Railway Engineering [11]. Research and development work in the area of vessel collision with bridges continues. Several aspects, such as the magnitude of the collision loads to be used in design, and the appropriate combination of extreme events (such as collision plus scour) are not yet well established and understood. As further research results become available, appropriate code changes and updates can be expected.

60.1.2 Basic Concepts

The vulnerability of a bridge to vessel collision is affected by a variety of factors, including:

- Waterway geometry, water stage fluctuations, current speeds, and weather conditions;
- Vessel characteristics and navigation conditions, including vessel types and size distributions, speed and loading conditions, navigation procedures, and hazards to navigation;
- Bridge size, location, horizontal and vertical geometry, resistance to vessel impact, structural redundancy, and effectiveness of existing bridge protection systems;
- Serious vessel collisions with bridges are extreme events associated with a great amount of uncertainty, especially with respect to the impact loads involved. Since designing for the worst-case scenario could be overly conservative and economically undesirable, a certain amount of risk must be considered as acceptable. The commonly accepted design objective is to minimize (in a cost-effective manner) the risk of catastrophic failure of a bridge component, and at the same time reduce the risk of vessel damage and environmental pollution.

The intent of vessel collision provisions is to provide bridge components with a "reasonable" resistance capacity against ship and barge collisions. In navigable waterway areas where collision by merchant vessels may be anticipated, bridge structures should be designed to prevent collapse of the superstructure by considering the size and type of vessel, available water depth, vessel speed, structure response, the risk of collision, and the importance classification of the bridge. It should be noted that damage to the bridge (even failure of secondary structural members) is usually permitted as long as the bridge deck carrying motorist traffic does not collapse (i.e., sufficient redundancy and alternate load paths exist in the remaining structure to prevent collapse of the superstructure).

60.1.3 Application

The vessel collision design recommendations provided in this chapter are consistent with the AASHTO specifications [6,8] and they apply to all bridge components in navigable waterways with water depths over 2.0 ft (0.6 m). The vessels considered include merchant ships larger than 1000 DWT and typical inland barges.

60.2 Initial Planning

It is very important to consider vessel collision aspects as early as possible in the planning process for a new bridge, since they can have a significant effect on the total cost of the bridge. Decisions related to the bridge type, location, and layout should take into account the waterway geometry, the navigation channel layout, and the vessel traffic characteristics.

60.2.1 Selection of Bridge Site

The location of a bridge structure over a waterway is usually predetermined based on a variety of other considerations, such as environmental impacts, right-of-way, costs, roadway geometry, and political considerations. However, to the extent possible, the following vessel collision guidelines should be followed:

- Bridges should be located away from turns in the channel. The distance to the bridge should be such that vessels can line up before passing the bridge, usually at least eight times the length of the vessel. An even larger distance is preferable when high currents and winds are likely to occur at the site.
- Bridges should be designed to cross the navigation channel at right angles and should be symmetrical with respect to the channel.
- An adequate distance should exist between bridge locations and areas with congested navigation, port facilities, vessel berthing maneuvers, or other navigation problems.
- Locations where the waterway is shallow or narrow so that bridge piers could be located out of vessel reach are preferable.

60.2.2 Selection of Bridge Type, Configuration, and Layout

The selection of the type and configuration of a bridge crossing should consider the characteristics of the waterway and the vessel traffic, so that the bridge would not be an unnecessary hazard to navigation. The layout of the bridge should maximize the horizontal and vertical clearances for navigation, and the bridge piers should be placed away from the reach of vessels. Finding the optimum bridge configuration and layout for different bridge types and degrees of protection is an iterative process which weighs the costs involved in risk reduction, including political and social aspects.

60.2.3 Horizontal and Vertical Clearance

The horizontal clearance of the navigation span can have a significant impact on the risk of vessel collision with the main piers. Analysis of past collision accidents has shown that bridges with a main span less than two to three times the design vessel length or less than two times the channel width are particularly vulnerable to vessel collision.

The vertical clearance provided in the navigation span is usually based on the highest vessel that uses the waterway in a ballasted condition and during periods of high water level. The vertical clearance requirements need to consider site-specific data on actual and projected vessels, and must be coordinated with the Coast Guard in the United States. General data on vessel height characteristics are included in References [6,7].

60.2.4 Approach Spans

The initial planning of the bridge layout should also consider the vulnerability of the approach spans to vessel collision. Historical vessel collisions have shown that bridge approach spans were damaged in over 60% of the total number of accidents. Therefore, the number of approach piers exposed to vessel collision should be minimized, and horizontal and vertical clearance considerations should also be applied to the approach spans.

60.2.5 Protection Systems

Bridge protection alternatives should be considered during the initial planning phase, since the cost of bridge protection systems can be a significant portion of the total bridge cost. Bridge protection systems include fender systems, dolphins, protective islands, or other structures designed to redirect, withstand, or absorb the impact force and energy, as described in Section 60.8.

60.3 Waterway Characteristics

The characteristics of the waterway in the vicinity of the bridge site such as the width and depth of the navigation channel, the current speed and direction, the channel alignment and cross section, the water elevation, and the hydraulic conditions, have a great influence on the risk of vessel collision and must be taken into account.

60.3.1 Channel Layout and Geometry

The channel layout and geometry can affect the navigation conditions, the largest vessel size that can use the waterway, and the loading condition and speed of vessels.

The presence of bends and intersections with other waterways near the bridge increases the probability of vessels losing control and become aberrant. The navigation of downstream barge tows through bends is especially difficult.

The vessel transit paths in the waterway in relation to the navigation channel and the bridge piers can affect the risk of aberrant vessels hitting the substructure.

60.3.2 Water Depth and Fluctuations

The design water depth for the channel limits the size and draft of vessels using the waterway. In addition, the water depth plays a critical role in the accessibility of vessels to piers outside the navigation channel. The vessel collision analysis must include the possibility of ships and barges transiting ballasted or empty in the waterway. For example, a loaded barge with a 6 m draft would run aground before it could strike a pier in 4 m of water, but the same barge empty with a 1 m draft could potentially strike the pier.

The water level along with the loading condition of vessels influences the location on the pier where vessel impact loads are applied, and the susceptibility of the superstructure to vessel hits. The annual mean high water elevation is usually the minimum water level used in design. In waterways with large water stage fluctuations, the water level used can have a significant effect on the structural requirements for the pier and/or pier protection design. In these cases, a closer review of the water stage statistics at the bridge site is necessary in order to select an appropriate design water level.

60.3.3 Current Speed and Direction

Water currents at the location of the bridge can have a significant effect on navigation and on the probability of vessel aberrancy. The design water currents commonly used represent annual average values rather than the occasional extreme values that occur only a few times per year, and during which vessel traffic restrictions may also apply.

60.4 Vessel Traffic Characteristics

60.4.1 Physical and Operating Characteristics

General knowledge on the operation of vessels and their characteristics is essential for safe bridge design. The types of commercial vessels encountered in navigable waterways may be divided into ships and barge tows.

60.4.1.1 Ships

Ships are self-propelled vessels using deep-draft waterways. Their size may be determined based on the DWT. The DWT is the weight in metric tonnes (1 tonne = 2205 lbs = 9.80 kN) of cargo, stores, fuel, passenger, and crew carried by the ship when fully loaded. There are three main classes of merchant ships: bulk carriers, product carriers/tankers, and freighter/containers. General information on ship profiles, dimensions, and sizes as a function of the class of ship and its DWT is provided in References [6,7]. The dimensions given in References [6,7] are typical values, and due to the large variety of existing vessels, they should be regarded as general approximations.

The steering of ships in coastal waterways is a difficult process. It involves constant communications between the shipmaster, the helmsman, and the engine room. There is a time delay before a ship starts responding to an order to change speed or course, and the response of the ship itself is relatively slow. Therefore, the shipmaster has to be familiar with the waterway and be aware of obstructions and navigation and weather conditions in advance. Very often local pilots are used to navigate the ships through a given portion of a coastal waterway. When the navigation conditions are difficult, tugboats are used to assist ships in making turns. Ships need speed to be able to steer and maintain rudder control. A minimum vessel speed of about 5 knots (8 km/h) is usually needed to maintain steering. Fully loaded ships are more maneuverable, and in deep water they are directionally stable and can make turns with a radius equal to one to two times the length of the ship. However, as the underkeel clearance decreases to less than half the draft of the ship, many ships tend to become directionally unstable, which means that they require constant steering to keep them traveling in a straight line. In the coastal waterways of the United States, the underkeel clearance of many laden ships may be far less than this limit, in some cases as small as 5% of the draft of the ship. Ships riding in ballast with shallow draft are less maneuverable than loaded ships, and, in addition, they can be greatly affected by winds and currents. Historical accident data indicate that most bridge accidents involve empty or ballasted vessels.

60.4.1.2 Barge Tows

Barge tows use both deep-draft and shallow-draft waterways. The majority of the existing bridges cross shallow draft waterways where the vessel fleet comprises barge tows only. The size of barges in the United States is usually defined in terms of the cargo-carrying capacity in short tons (1 ton = 2000 lbs = 8.90 kN). The types of inland barges include open and covered hoppers, tank barges, and deck barges. They are rectangular in shape and their dimensions are quite standard so they can travel in tows. The number of barges per tow can vary from one to over 20, and their configuration, is affected by the conditions of the waterway. In most cases barges are pushed by a towboat. Information on barge dimensions and capacity, as well as on barge tow configurations is included in References [6,7]. A statistical analysis of barge tow types, configurations, and dimensions, which utilizes barge traffic data from the Ohio River, is reported in Reference [12].

It is very difficult to control and steer barge tows, especially in waterways with high stream velocities and cross currents. Taking a turn in a fast waterway with high current is a serious undertaking. In maneuvering a bend, tows experience a sliding effect in a direction opposite to the direction of the turn, due to inertial forces, which are often coupled with the current flow. Sometimes, bridge piers and fenders are used to line up the tow before the turn. Bridges located in a high-velocity waterway near a bend in the channel will probably be hit by barges numerous times during their lifetime. In general, there is a high likelihood that any bridge element that can be reached by a barge will be hit during the life of the bridge.

60.4.2 Vessel Fleet Characteristics

The vessel data required for bridge design include types of vessels and size distributions, transit frequencies, typical vessel speeds, and loading conditions. In order to determine the vessel size distribution at the bridge site, detailed information on both present and projected future vessel traffic is needed. Collecting data on the vessel fleet characteristics for the waterway is an important and often time-consuming process.

Some of the sources in the United States for collecting vessel traffic data are listed below:

- U.S. Army Corps of Engineers, District Offices
- Port authorities and industries along the waterway

- Local pilot associations and merchant marine organizations
- U.S. Coast Guard, Marine Safety & Bridge Administration Offices
- U.S. Army Corps of Engineers, "Products and Services Available to the Public," Water Resources Support Center, Navigation Data Center, Fort Belvoir, Virginia, NDC Report 89-N-1, August 1989
- U.S. Army Corps of Engineers, "Waterborne Commerce of the United States (WCUS), Parts 1 thru 5," Water Resources Support Center (WRSC), Fort Belvoir, Virginia
- U.S. Army Corps of Engineers, "Lock Performance Monitoring (LPM) Reports," Water Resources Support Center (WRSC), Fort Belvoir, Virginia
- Shipping registers (American Bureau of Shipping Register, New York; and Lloyd's Register of Shipping, London)
- Bridge tender reports for movable bridges

Projections for anticipated vessel traffic during the service life of the bridge should address both changes in the volume of traffic and in the size of vessels. Factors that need to be considered include:

- Changes in regional economics;
- Plans for deepening or widening the navigation channel;
- Planned changes in alternate waterway routes and in navigation patterns;
- Plans for increasing the size and capacity of locks leading to the bridge;
- Port development plans.

Vessel traffic projections that are made by the Maritime Administration of the U.S. Department of Transportation, Port Authorities, and U.S. Army Corps of Engineers in conjunction with planned channel-deepening projects or lock replacements are also good sources of information for bridge design. Since a very large number of factors can affect the vessel traffic in the future, it is important to review and update the projected traffic during the life of the bridge.

60.5 Collision Risk Analysis

60.5.1 Risk Acceptance Criteria

Bridge components exposed to vessel collision could be subjected to a very wide range of impact loads. Due to economic and structural constraints bridge design for vessel collision is not based on the worst-case scenario, and a certain amount of risk is considered acceptable.

The risk acceptance criteria consider both the probability of occurrence of a vessel collision and the consequences of the collision. The probability of occurrence of a vessel collision is affected by factors related to the waterway, vessel traffic, and bridge characteristics. The consequences of a collision depend on the magnitude of the collision loads and the bridge strength, ductility, and redundancy characteristics. In addition to the potential for loss of life, the consequences of a collision can include damage to the bridge, disruption of motorist and marine traffic, damage to the vessel and cargo, regional economic losses, and environmental pollution.

Acceptable risk levels have been established by various codes and for individual bridge projects [2–10]. The acceptable annual frequencies of bridge collapse values used generally range from 0.001 to 0.0001. These values were usually determined in conjunction with the risk analysis procedure recommended, and should be used accordingly.

The AASHTO provisions [6,8] specify an annual frequency of bridge collapse of 0.0001 for critical bridges and an annual frequency of bridge collapse of 0.001 for regular bridges. These annual frequencies correspond to return periods of bridge collapse equal to 1 in 10,000 years, and 1 in 1000 years, respectively. Critical bridges are defined as those bridges that are expected to continue to function after a major impact, because of social/survival or security/defense requirements.

60.5.2 Collision Risk Models

60.5.2.1 General Approach

Various collision risk models have been developed to achieve design acceptance criteria [2–10]. In general, the occurrence of a collision is separated into three events: (1) a vessel approaching the bridge becomes aberrant, (2) the aberrant vessel hits a bridge element, and (3) the bridge element that is hit fails. Collision risk models consider the effects of the vessel traffic, the navigation conditions, the bridge geometry with respect to the waterway, and the bridge element strength with respect to the impact loads. They are commonly expressed in the following form [6,8]:

$$AF = (N)\,(PA)\,(PG)\,(PC) \tag{60.1}$$

where *AF* is the annual frequency of collapse of a bridge element; *N* is the annual number of vessel transits (classified by type, size, and loading condition) which can strike a bridge element; *PA* is the probability of vessel aberrancy; *PG* is the geometric probability of a collision between an aberrant vessel and a bridge pier or span; *PC* is the probability of bridge collapse due to a collision with an aberrant vessel.

60.5.2.2 Vessel Traffic Distribution, *N*

The number of vessels, *N*, passing the bridge based on size, type, and loading condition and available water depth has to be developed for each pier and span component to be evaluated. All vessels of a given type and loading condition have to be divided into discrete groupings of vessel size by DWT to determine the contribution of each group to the annual frequency of bridge element collapse. Once the vessels are grouped and their frequency distribution is established, information on typical vessel characteristics may be obtained from site-specific data, or from published general data such as References [6,7].

60.5.2.3 Probability of Aberrancy, *PA*

The probability of vessel aberrancy reflects the likelihood that a vessel is out of control in the vicinity of a bridge. Loss of control may occur as a result of pilot error, mechanical failure, or adverse environmental conditions. The probability of aberrancy is mainly related to the navigation conditions at the bridge site. Vessel traffic regulations, vessel traffic management systems, and aids to navigation can improve the navigation conditions and reduce the probability of aberrancy.

The probability of vessel aberrancy may be evaluated based on site-specific information that includes historical data on vessel collisions, rammings, and groundings in the waterway, vessel traffic, navigation conditions, and bridge/waterway geometry. This has been done for various bridge design provisions and specific bridge projects worldwide [2,3,7,9,12]. The probability of aberrancy values determined range from 0.5×10^{-4} to over 7.0×10^{-4}.

As an alternative, the AASHTO provisions [6,8] recommend base rates for the probability of vessel aberrancy that are multiplied by correction factors for bridge location relative to bends in the waterway, currents acting parallel to vessel transit path, crosscurrents acting perpendicular to vessel transit path, and the traffic density of vessels using the waterway. The recommended base rates are 0.6×10^{-4} for ships, and 1.2×10^{-4} for barges.

60.5.2.4 Geometric Probability, *PG*

The geometric probability is the probability that a vessel will hit a particular bridge pier given that it has lost control (i.e., is aberrant) in the vicinity of the bridge. It is mainly a function of the geometry of the bridge in relation to the waterway. Other factors that can affect the likelihood that an aberrant vessel will strike a bridge element include the original vessel transit path, course, rudder position, velocity at the time of failure, vessel type, size, draft and maneuvering characteristics, and the hydraulic and environmental conditions at the bridge site. Various geometric probability models, some based on simulation studies, have been recommended and used on different bridge projects

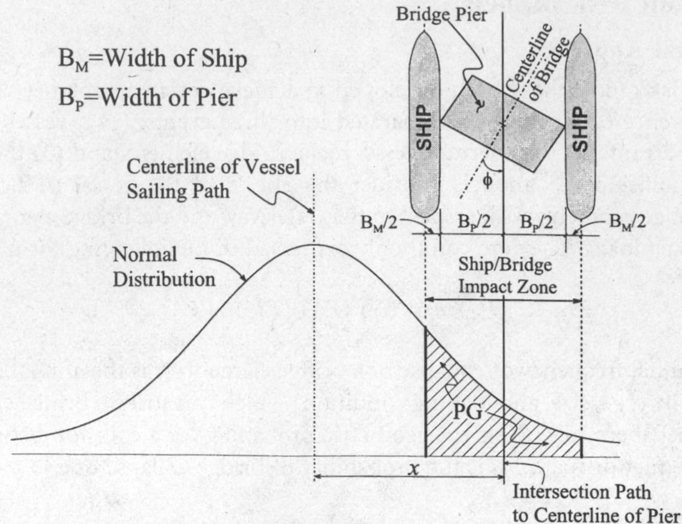

FIGURE 60.2 Geometric probability of pier collision.

[2,3,7]. The AASHTO provisions [6,8] use a normal probability density function about the centerline of the vessel transit path for estimating the likelihood of an aberrant vessel being within a certain impact zone along the bridge axis. Using a normal distribution accounts for the fact that aberrant vessels are more likely to pass under the bridge closer to the navigation channel than farther away from it. The standard deviation of the distribution equals the length of the design vessel considered. The probability that an aberrant vessel is located within a certain zone is the area under the normal probability density function within that zone (Figure 60.2).

Bridge elements beyond three times the standard deviation from the centerline of vessel transit path are designed for specified minimum impact load requirements, which are usually associated with an empty vessel drifting with the current.

60.5.2.5 Probability of Collapse, *PC*

The probability of collapse, *PC*, is a function of many variables, including vessel size, type, forepeak ballast and shape, speed, direction of impact, and mass. It is also dependent on the ultimate lateral load strength of the bridge pier (particularly the local portion of the pier impacted by the bow of the vessel). Based on collision damages observed from numerous ship–ship collision accidents which have been correlated to the bridge–ship collision situation [2], an empirical relationship has been developed based on the ratio of the ultimate pier strength, *H*, to the vessel impact force, *P*. As shown in Figure 60.3, for H/P ratios less than 0.1, *PC* varies linearly from 0.1 at $H/P = 0.1$ to 1.0 at $H/P = 0.0$. For H/P ratios greater than 0.1, *PC* varies linearly from 0.1 at $H/P = 0.1$ to 0.0 at $H/P = 1.0$.

60.6 Vessel Impact Loads

60.6.1 Ship Impact

The estimation of the load on a bridge pier during a ship collision is a very complex problem. The actual force is time dependent, and varies depending on the type, size, and construction of the vessel; its velocity; the degree of water ballast in the forepeak of the bow; the geometry of the collision; and the geometry and strength characteristics of the bridge. There is a very large scatter among the collision force values recommended in various vessel collision guidelines or used in various bridge projects [2–10].

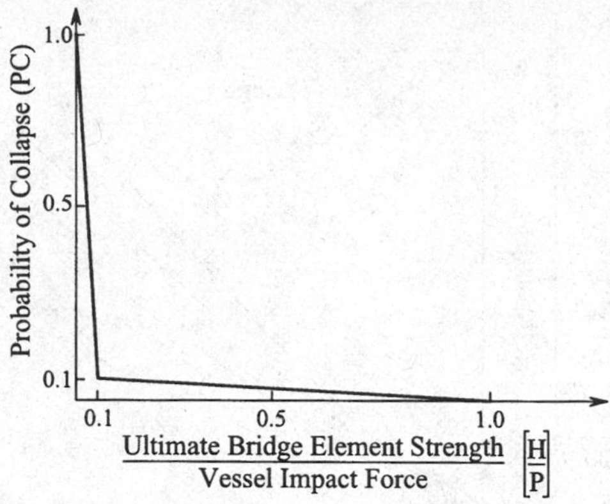

FIGURE 60.3 Probability of collapse distribution.

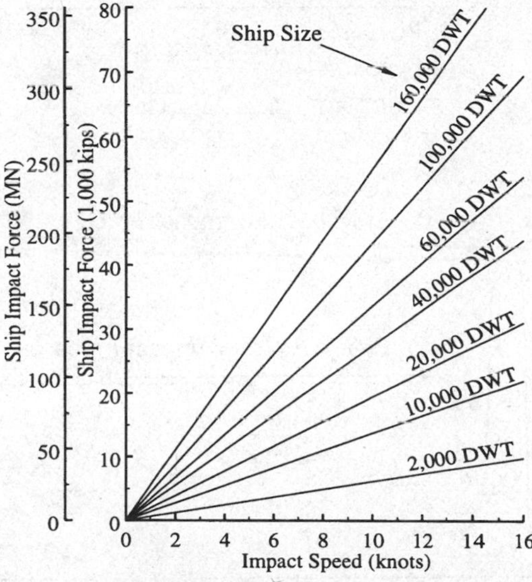

FIGURE 60.4 Ship impact force.

Ship collision forces are commonly applied as equivalent static loads. Procedures for evaluating dynamic effects when the vessel force indentation behavior is known are included in References [3,4,10,13,14]. The AASHTO provisions [6,8] use the following formula for estimating the static head-on ship collision force, P_S, on a rigid pier:

$$P_s = 0.98(DWT)^{\frac{1}{2}}(V/16) \tag{60.2}$$

where P_S is the equivalent static vessel impact force (MN); DWT is the ship deadweight tonnage in tonnes; and V is the vessel impact velocity in knots (Figure 60.4). This formulation was primarily developed from research conducted by Woisin in West Germany during 1967 to 1976 on physical ship models to generate data for protecting the reactors of nuclear power ships from collisions with other ships. A schematic representation of a typical impact force time history is shown in Figure 60.6 based on Woisin's test data. The scatter in the results of these tests is of the order of ±50%. The formula recommended (Eq. 60.2) uses a 70% fractile of an assumed triangular distribution with zero values at 0% and 100% and a maximum value at the 50% level (Figure 60.7).

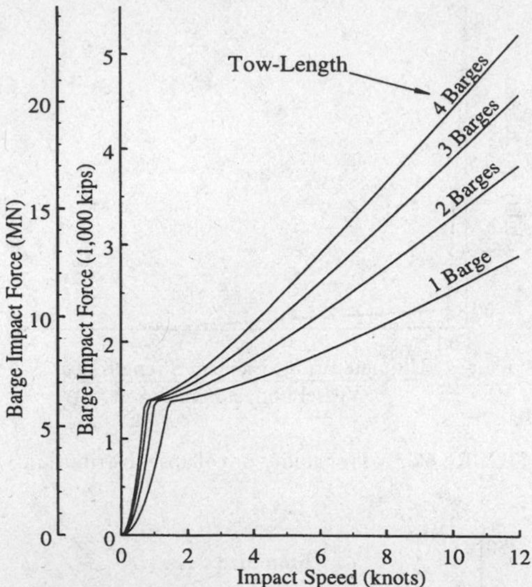

FIGURE 60.5 Barge impact force.

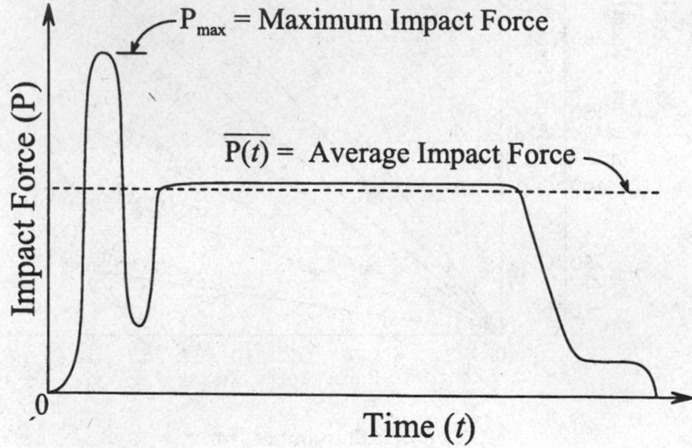

FIGURE 60.6 Typical ship impact force time history by Woisin.

Formulas for computing design ship collision loads on a bridge superstructure are given in the AASHTO provisions [6,8] as a function of the design ship impact force, P_S, as follows:

- Ship Bow Impact Force, P_{BH}:

$$P_{BH} = (R_{BH})\ (P_S) \tag{60.3}$$

where R_{BH} is a reduction coefficient equal to the ratio of exposed superstructure depth to the total bow depth.

- Ship Deckhouse Impact Force, P_{DH}:

$$P_{DH} = (R_{DH})\ (P_S) \tag{60.4}$$

where R_{DH} is a reduction coefficient equal to 0.10 for ships larger than 100,000 DWT, and

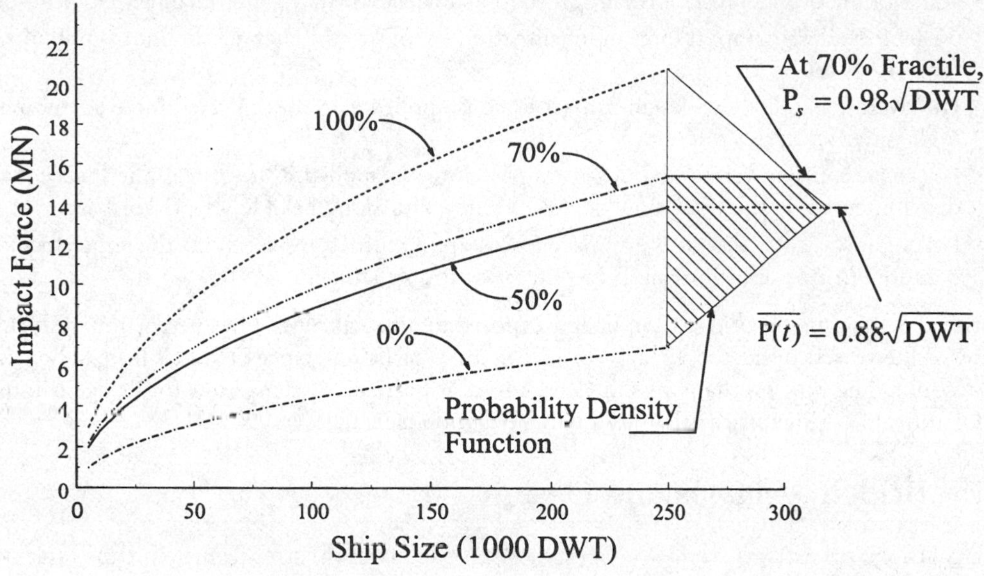

FIGURE 60.7 Probability density function of ship impact force.

$$0.2 - \frac{DWT}{100{,}000}(0.10)$$

for ships under 100,000 DWT.

- Ship Mast Impact Force, P_{MT}:

$$P_{MT} = 0.10\ P_{DH} \tag{60.5}$$

where P_{DH} is the ship deckhouse impact force.

The magnitude of the impact loads computed for ship bow and deckhouse collisions are quite high relative to the strength of most bridge superstructure designs. Also, there is great uncertainty associated with predicting ship collision loads on superstructures because of the limited data available and the ship–superstructure load interaction effects. It is therefore suggested that superstructures, and also weak or slender parts of the substructure, be located out of the reach of a ship's hull or bow.

60.6.2 Barge Impact

The barge collision loads recommended by AASHTO for the design of piers are shown in Figure 60.5 as a function of the tow length and the impact speed. Numerical formulations for deriving these relationships may be found in References [6,8].

The loads in Figure 60.5 were computed using a standard 59.5 × 10.7 m hopper barge. The impact force recommended for barges larger than the standard hopper barge is determined by increasing the standard barge impact force by the ratio of the width of the wider barge to the width of the standard hopper barge.

60.6.3 Application of Impact Forces

Collision forces on bridge substructures are commonly applied as follows:

- 100% of the design impact force in a direction parallel to the navigation channel (i.e., head-on);
- 50% of the design impact force in the direction normal to the channel (but not simultaneous with the head-on force);
- For overall stability, the design impact force is applied as a concentrated force at the mean high water level;
- For local collision forces, the design impact force is applied as a vertical line load equally distributed along the ship's bow depth for ships, and along head log depth for barges;
- For superstructure design the impact forces are applied transversely to the superstructure component in a direction parallel to the navigation channel.

When determining the bridge components exposed to physical contact by any portion of the hull or bow of the vessel considered, the bow overhang, rake, or flair distance of vessels have to be taken into account. The bow overhang of ships and barges is particularly dangerous for bridge columns and for movable bridges with relatively small navigation clearances.

60.7 Bridge Analysis and Design

Vessel collisions are extreme events with a very low probability of occurrence; therefore the limit state considered is usually structural survival. Depending on the importance of the bridge, various degrees of damage are allowed— provided that the structure maintains its integrity, hazards to traffic are minimized, and repairs can be made in a relatively short period of time. When the design is based on more frequent but less severe collisions, structural damage and traffic interruptions are not allowed.

Designing for vessel collision is commonly based on equivalent static loads that include global forces for checking overall capacity and local forces for checking local strength of bridge components. A clear load path from the location of the vessel impact to the bridge foundation needs to be established and the components and connections within the load path must be adequately designed and detailed. The design of individual bridge components is based on strength and stability criteria. Overall stability, redundancy, and ductility are important criteria for structural survival.

The contribution of the superstructure to the transfer of loads to adjacent substructure units depends on the capacity of the connection of the superstructure to substructure and the relative stiffness of the substructure at the location of the impact. Analysis guidelines for determining the distribution of collision loads to adjacent piers are included in Reference [15]. To find out how much of the transverse impact force is taken by the pier and how much is transferred to the superstructure, two analytical models are typically used. One is a two-dimensional or a three-dimensional model of the complete pier, and the other is a two-dimensional model of the superstructure projected on a horizontal plane. The projected superstructure may be modeled as a beam with the moment of inertia referred to a vertical axis through the center of the roadway, and with hinges at expansion joint locations. The beam is supported at pier locations by elastic horizontal springs representing the flexibility of each pier. The flexibility of the piers is obtained from pier models using virtual forces. The superstructure model is loaded with a transverse virtual force acting at the place where the pier under consideration is located. The spring in the model at that place is omitted to obtain a flexibility coefficient of the superstructure at the location of the top of the pier under consideration. Thus, the horizontal displacement of the top of the pier due to the impact force on the pier (usually applied at mean high water level) is equal to the true displacement of the superstructure due to the transmitted part of the impact force. The magnitude of the force transmitted to the superstructure is obtained by equating the total true displacement of the top of the pier from the pier model to the displacement of the superstructure. However, in order to consider partial transfer of lateral forces to the superstructure, positive steel or concrete connections of superstructure to substructure, such as shear keys must be provided. Similarly, for partial transfer to the superstructure of the longitudinal component of the impact force the shear capacity of the bearings must be adequate. When elastomeric bearings are used their longitudinal flexibility may be added to the longitudinal flexibility of the piers. If the ultimate capacity of the bearings is exceeded, then the pier must take the total longitudinal force and be treated as a cantilever.

The modeling of pile foundations could vary from the simple assumption of a point of fixity to nonlinear soil–structure interaction models, depending on the limit state considered and the sensitivity of the response to the soil conditions. Lateral load capacity analysis methods for pile groups that include nonlinear behavior are recommended in References [15,16] and the features of a finite-element analysis computer program developed for bridge piers composed of pier columns and cap supported on a pile cap and nonlinear piles and soil are presented in Reference [17]. Transient foundation uplift or rocking involving separation from the subsoil of an end bearing foundation pile group or the contact area of a foundation footing could be allowed under impact loading provided sufficient consideration is given to the structural stability of the substructure.

60.8 Bridge Protection Measures

The cost associated with protecting a bridge from catastrophic vessel collision can be a significant portion of the total bridge cost, and must be included as one of the key planning elements in establishing a bridge's type, location, and geometry. The alternatives listed below are usually evaluated in order to develop a cost-effective solution for a new bridge project:

- Design the bridge piers, foundations, and superstructure to withstand directly the vessel collision forces and impact energies;
- Design a pier fender system to reduce the impact loads to a level below the capacity of the pier and foundation;
- Increase span lengths and locate piers in shallow water out of reach of large vessels in order to reduce the impact design loads; and
- Protect piers from vessel collision by means of physical protection systems.

60.8.1 Physical Protection Systems

Piers exposed to vessel collision can be protected by special structures designed to absorb the impact loads (forces or energies), or redirect the aberrant vessel away from the pier. Because of the large forces and energies involved in a vessel collision, protection structures are usually designed for plastic deformation under impact (i.e., they are essentially destroyed during the head-on design collision and must be replaced). General types of physical protection systems include:

Fender Systems. These usually consist of timber, rubber, steel, or concrete elements attached to a pier to fully, or partially, absorb vessel impact loads. The load and energy absorbing characteristics of such fenders is relatively low compared with typical vessel impact design loads.

Pile-Supported Systems. These usually consist of pile groups connected by either flexible or rigid caps to absorb vessel impact forces. The piles may be vertical (plumb) or battered depending on the design approach followed, and may incorporate relatively large-diameter steel pipe or concrete pile sizes. The pile-supported protection structure may be either freestanding away from the pier, or attached to the pier itself. Fender systems may be attached to the pile structure to help resist a portion of the impact loads.

Dolphin Protection Systems. These usually consist of large-diameter circular cells constructed of driven steel sheet piles, filled with rock or sand, and topped by a thick concrete cap. Vessel collision loads are absorbed by rotation and lateral deformation of the cell during impact.

Island Protection Systems. These usually consist of protective islands built of a sand or quarry-run rock core and protected by outer layers of heavy rock riprap for wave, current, and ice protection. The island geometry is developed to stop an aberrant vessel from hitting a pier by forcing it to run aground. Although extremely effective as protection systems, islands are often difficult to use due to adverse environmental impacts on river bottoms (dredge and fill permits) and river currents (increase due to blockage), as well as impacts due to settlement and downdrag forces on the bridge piers.

Floating Protection Systems. These usually consist of cable net systems suspended across the waterway to engage and capture the bow of an aberrant vessel, or floating pontoons anchored in front of the piers. Floating protection systems have a number of serious drawbacks (environmental, effectiveness, maintenance, cost, etc.) and are usually only considered for extremely deep water situations where other protection options are not practicable.

The AASHTO Guide Specification [6] provides examples and contains a relatively extensive discussion of various types of physical protection systems, such as fenders, pile-supported structures, dolphins, protective islands, and floating structures. However, the code does not include specific procedures and recommendations on the actual design of such protection structures. Further research is needed to establish consistent analysis and design methodologies for protection structures, particularly since these structures undergo large plastic deformations during the collision.

60.8.2 Aids to Navigation Alternatives

Since 60 to 85% of all vessel collisions are caused by pilot error, it is important that all aspects of the bridge design, siting, and aids to navigation with respect to the navigation channel be carefully evaluated with the purpose of improving or maintaining safe navigation in the waterway near the bridge. Traditional aids include buoys, range markers, navigation lighting, and radar reflectors, as well as standard operating procedures and regulations specifically developed for the waterway by government agencies and pilot associations. Modern aids include advanced vessel traffic control systems (VTS) using shore-based radar surveillance and radio-telephone communication systems; special electronic transmitters known as Raycon devices mounted to bridge spans for improved radar images indicating the centerline of the channel; and advanced navigation positioning systems based on shipboard global positioning satellite (GPS) receivers using differential signal techniques to improve location accuracy.

Studies have indicated that improvements in the aids to navigation near a bridge can provide extremely cost-effective solutions to reducing the risk of collisions to acceptable levels. The cost of such aid to navigation improvements and shipboard electronic navigation systems is usually a fraction of the cost associated with expensive physical protection alternatives. However, few electronic navigation systems have ever been implemented (worldwide) due to legal complications arising from liability concerns; impacts on international laws governing trade on the high seas; and resistance by maritime users.

It should be noted that the traditional isolation of the maritime community must come to an end. In addition to the bridge costs, motorist inconvenience, and loss of life associated with a catastrophic vessel collision, significant environmental damage can also occur due to spilled hazardous or noxious cargoes in the waterway. The days when the primary losses associated with an accident rested with the vessel and her crew are over. The $13 million value of the *M/V Summit Venture* was far below the $250 million replacement cost of the Sunshine Skyway Bridge which the vessel destroyed. The losses associated with the 11 million gallons of crude oil spilled from the *M/V Exxon Valdez* accident off the coast of Alaska in 1989 are over $3.5 billion. Both of these accidents could have been prevented using shipboard advanced electronic navigation systems.

60.9 Conclusions

Experience to date has shown that the use of the vessel impact and bridge protection requirements (such as the AASHTO specifications [6,8]) for planning and design of new bridges has resulted in a significant change in proposed structure types over navigable waterways. Incorporation of the risk of vessel collision and cost of protection in the total bridge cost has almost always resulted in longer-span bridges being more economical than traditional shorter span structures, since the design goal for developing the bridge pier and span layout is the least cost of the total structure (including the protection costs). Typical costs for incorporating vessel collision and protection issues in the planning stages of a new bridge have ranged from 5% to 50% of the basic structure cost without protection.

Experience has also shown that it is less expensive to include the cost of protection in the planning stages of a proposed bridge, than to add it after the basic span configuration has been established without considering vessel collision concerns. Typical costs for adding protection, or for retrofitting an existing bridge for vessel collision, have ranged from 25% to over 100% of the existing bridge costs.

It is recognized that vessel collision is but one of a multitude of factors involved in the planning process for a new bridge. The designer must balance a variety of needs including political, social, and economic in arriving at an optimal bridge solution for a proposed highway crossing. Because of the relatively high bridge costs associated with vessel collision design for most waterway crossings, it is important that additional research be conducted to improve our understanding of vessel impact mechanics, the response of the structure, and the development of cost-effective protection systems.

References

1. National Research Council, *Ship Collisions with Bridges — The Nature of the Accidents, Their Prevention and Mitigation,* National Academy Press, Washington, D.C., 1983.
2. IABSE, *Ship Collision with Bridges and Offshore Structures,* International Association for Bridge and Structural Engineering, Colloquium Proceedings, Copenhagen, Denmark, 3 vols. (Introductory, Preliminary, and Final Reports), 1983.
3. Modjeski and Masters, Criteria for the Design of Bridge Piers with Respect to Vessel Collision in Louisiana Waterways, Report prepared for Louisiana Department of Transportation and Development and the Federal Highway Administration, 1984.
4. Prucz, Z. and Conway, W. B., Design of bridge piers against ship collision, in *Bridges and Transmission Line Structures,* L. Tall, Ed., ASCE, New York, 1987, 209–223.
5. Knott, M. A. and Larsen, O. D. 1990. *Guide Specification and Commentary for Vessel Collision Design of Highway Bridges,* U.S. Department of Transportation, Federal Highway Administration, Report No. FHWA-RD-91-006.
6. AASHTO, *Guide Specification and Commentary for Vessel Collision Design of Highway Bridges,* American Association of State Highway and Transportation Officials, Washington, D.C., 1991.
7. Larsen, O. D., Ship Collision with Bridges: The Interaction between Vessel Traffic and Bridge Structures, IABSE Structural Engineering Document 4, IABSE-AIPC-IVBH, Zurich, Switzerland, 1993.
8. AASHTO, *LRFD Bridge Design Specifications and Commentary,* American Association of State Highway and Transportation Officials, Washington, D.C., 1994.
9. FHWA, *The Design of Bridges for Extreme Events,* Proceedings of Conference in Atlanta, Georgia, December 3–6, 1996.
10. *International Symposium on Advances in Bridge Aerodynamics, Ship Collision Analysis, and Operation & Maintenance,* Copenhagen, Denmark, May 10–13, Balkema Publishers, Rotterdam, Netherlands, 1998.
11. AREMA, *Manual for Railway Engineering,* Chapter 8, Part 23, American Railway Engineering Association, Washington, D.C., 1999.
12. Whitney, M. W., Harik, I. E., Griffin, J. J., and Allen, D. L. Barge collision design of highway bridges, *J. Bridge Eng. ASCE,* 1(2), 47–58, 1996.
13. Prucz, Z. and Conway, W. B., Ship Collision with Bridge Piers — Dynamic Effects, Transportation Research Board Paper 890712, Transportation Research Board, Washington, D.C., 1989.

14. Grob, B. and Hajdin, N., Ship impact on inland waterways, *Struct. Eng. Int.*, IABSE, Zürich, Switzerland, 4, 230–235, 1996.
15. Kuzmanovic, B. O. and Sanchez, M. R., Design of bridge pier pile foundations for ship impact, *J. Struct. Eng. ASCE*, 118(8), 2151–2167, 1992.
16. Brown, D. A. and Bollmann, H. T. Pile supported bridge foundations designed for impact loading, Transportation Research Record 1331, TRB, National Research Council, Washington, D.C., 87–91, 1992.
17. Hoit, M., McVay, M., and Hays, C., Florida Pier Computer Program for Bridge Substructure Analysis: Models and Methods, Conference Proceedings, Design of Bridges for Extreme Events, FHWA, Washington, D.C., 1996.

61

Bridge Hydraulics

Jim Springer
California Department of Transportation

Ke Zhou
California Department of Transportation

61.1 Introduction

This chapter presents bridge engineers basic concepts, methods, and procedures used in bridge hydraulic analysis and design. It involves hydrology study, hydraulic analysis, on-site drainage design, and bridge scour evaluation.

Hydrology study for bridge design mainly deals with the properties, distribution, and circulation of water on and above the land surface. The primary objective is to determine either the peak discharge or the flood hydrograph, in some cases both, at the highway stream crossings. Hydraulic analysis provides essential methods to determine runoff discharges, water profiles, and velocity distribution. The on-site drainage design part of this chapter is presented with the basic procedures and references for bridge engineers to design bridge drainage.

Bridge scour is a big part of this chapter. Bridge engineers are systematically introduced to concepts of various scour types, presented with procedures and methodology to calculate and evaluate bridge scour depths, provided with guidelines to conduct bridge scour investigation and to design scour preventive measures.

61.2 Bridge Hydrology and Hydraulics

61.2.1 Hydrology

61.2.1.1 Collection of Data

Hydraulic data for the hydrology study may be obtained from the following sources: as-built plans, site investigations and field surveys, bridge maintenance books, hydraulic files from experienced report writers, files of government agencies such as the U.S. Corps of Engineers studies, U.S. Geological Survey (USGS), Soil Conservation Service, and FEMA studies, rainfall data from local water agencies, stream gauge data, USGS and state water agency reservoir regulation, aerial photographs, and floodways, etc.

Site investigations should always be conducted except in the simplest cases. Field surveys are very important because they can reveal conditions that are not readily apparent from maps, aerial

0-8493-7434-0/00/$0.00+$.50

photographs and previous studies. The typical data collected during a field survey include high water marks, scour potential, stream stability, nearby drainage structures, changes in land use not indicated on maps, debris potential, and nearby physical features. See HEC-19, Attachment D [16] for a typical Survey Data Report Form.

61.2.1.2 Drainage Basin

The area of the drainage basin above a given point on a stream is a major contributing factor to the amount of flow past that point. For given conditions, the peak flow at the proposed site is approximately proportional to the drainage area.

The shape of a basin affects the peak discharge. Long, narrow basins generally give lower peak discharges than pear-shaped basins. The slope of the basin is a major factor in the calculation of the time of concentration of a basin. Steep slopes tend to result in shorter times of concentration and flatter slopes tend to increase the time of concentration. The mean elevation of a drainage basin is an important characteristic affecting runoff. Higher elevation basins can receive a significant amount of precipitation as snow. A basin orientation with respect to the direction of storm movement can affect peak discharge. Storms moving upstream tend to produce lower peaks than those moving downstream.

61.2.1.3 Discharge

There are several hydrologic methods to determine discharge. Most of the methods for estimating flood flows are based on statistical analyses of rainfall and runoff records and involve preliminary or trial selections of alternative designs that are judged to meet the site conditions and to accommodate the flood flows selected for analysis.

Flood flow frequencies are usually calculated for discharges of 2.33 years through the overtopping flood. The frequency flow of 2.33 years is considered to be the mean annual discharge. The base flood is the 100-year discharge (1% frequency). The design discharge is the 50-year discharge (2% frequency) or the greatest of record, if practical. Many times, the historical flood is so large that a structure to handle the flow becomes uneconomical and is not warranted. It is the engineer's responsibility to determine the design discharge. The overtopping discharge is calculated at the site, but may overtop the roadway some distance away from the site.

Changes in land use can increase the surface water runoff. Future land-use changes that can be reasonably anticipated to occur in the design life should be used in the hydrology study. The type of surface soil is a major factor in the peak discharge calculation. Rock formations underlying the surface and other geophysical characteristics such as volcanic, glacial, and river deposits can have a significant effect on runoff. In the United States, the major source of soil information is the Soil Conservation Service (SCS). Detention storage can have a significant effect on reducing the peak discharge from a basin, depending upon its size and location in the basin.

The most commonly used methods to determine discharges are

1. Rational method
2. Statistical Gauge Analysis Methods
3. Discharge comparison of adjacent basins from gauge analysis
4. Regional flood-frequency equations
5. Design hydrograph

The results from various methods of determining discharge should be compared, not averaged.

61.2.1.3.1 Rational Method

The rational method is one of the oldest flood calculation methods and was first employed in Ireland in urban engineering in 1847. This method is based on the following assumptions:

TABLE 61.1 Runoff Coefficients for Developed Areas

Type of Drainage Area	Runoff Coefficient
Business	
Downtown areas	0.70–0.95
Neighborhood areas	0.50–0.70
Residential areas	
Single-family areas	0.30–0.50
Multiunits, detached	0.40–0.60
Multiunits, attached	0.60–0.75
Suburban	0.25–0.40
Apartment dwelling areas	0.50–0.70
Industrial	
Light areas	0.50–0.80
Heavy areas	0.60–0.90
Parks, cemeteries	0.10–0.25
Playgrounds	0.20–0.40
Railroad yard areas	0.20–0.40
Unimproved areas	0.10–0.30
Lawns	
Sandy soil, flat, 2%	0.05–0.10
Sandy soil, average, 2–7%	0.10–0.15
Sandy soil, steep, 7%	0.15–0.20
Heavy soil, flat, 2%	0.13–0.17
Heavy soil, average, 2–7%	0.18–0.25
Heavy soil, steep, 7%	0.25–0.35
Streets	
Asphaltic	0.70–0.95
Concrete	0.80–0.95
Brick	0.70–0.85
Drives and walks	0.75–0.85
Roofs	0.75–0.95

1. Drainage area is smaller than 300 acres.
2. Peak flow occurs when all of the watershed is contributing.
3. The rainfall intensity is uniform over a duration equal to or greater than the time of concentration, T_c.
4. The frequency of the peak flow is equal to the frequency of the rainfall intensity.

$$Q = CiA \tag{61.1}$$

where

Q = discharge, in cubic foot per second

C = runoff coefficient (in %) can be determined in the field and from Tables 61.1 and 61.2 [5,16] or a weighted C value is used when the basin has varying amounts of different cover. The weighted C value is determined as follows:

$$C = \frac{\sum C_j A_j}{\sum A_j} \tag{61.2}$$

i = rainfall intensity (in inches per hour) can be determined from either regional IDF maps or individual IDF curves

A = drainage basin area (in acres) is determined from topographic map

(*Note*: 1 sq. mile = 640 acres = 0.386 sq. kilometer)

TABLE 61.2 Runoff Coefficients for Undeveloped Area Watershed Types

Soil	0.12–0.16 No effective soil cover, either rock or thin soil mantle of negligible infiltration capacity	0.08–0.12 Slow to take up water, clay or shallow loam soils of low infiltration capacity, imperfectly or poorly drained	0.06–0.08 Normal, well-drained light or medium-textured soils, sandy loams, silt and silt loams	0.04–0.06 High, deep sand or other soil that takes up water readily, very light well-drained soils
Vegetal Cover	0.12–0.16 No effective plant cover, bare or very sparse cover	0.08–0.12 Poor to fair; clean cultivation crops, or poor natural cover, less than 20% of drainage area over good cover	0.06–0.08 Fair to good; about 50% of area in good grassland or woodland, not more than 50% of area in cultivated crops	0.04–0.06 Good to excellent; about 90% of drainage area in good grassland, woodland or equivalent cover
Surface Storage	0.10–0.12 Negligible surface depression few and shallow, drainageways steep and small, no marshes	0.08–0.10 Low, well-defined system of small drainageways; no ponds or marshes	0.06–0.08 Normal; considerable surface depression storage; lakes and pond marshes	0.04–0.06 High; surface storage, high; drainage system not sharply defined; large floodplain storage or large number of ponds or marshes

The time of concentration for a pear-shaped drainage basin can be determined using a combined overland and channel flow equation, the Kirpich equation:

$$T_c = 0.0195(L/S^{0.5})^{0.77} \tag{61.3}$$

where

T_c = Time of concentration in minutes
L = Horizontally projected length of watershed in meters
S = H/L (H = difference in elevation between the most remote point in the basin and the outlet in meters)

61.2.1.3.2 Statistical Gauge Analysis Methods

The following two methods are the major statistical analysis methods which are used with stream gauge records in the hydrological analysis.

1. Log Pearson Type III method
2. Gumbel extreme value method

The use of stream gauge records is a preferred method of estimating discharge/frequencies since they reflect actual climatology and runoff. Discharge records, if available, may be obtained from a state department of water resources in the United States. A good record set should contain at least 25 years of continuous records.

It is important, however, to review each individual stream gauge record carefully to ensure that the database is consistent with good statistical analysis practice. For example, a drainage basin with a large storage facility will result in a skewed or inconsistent database since smaller basin discharges will be influenced to a much greater extent than large discharges.

The most current published stream gauge description page should be reviewed to obtain a complete idea of the background for that record. A note should be given to changes in basin area over time, diversions, revisions, etc. All reliable historical data outside of the recorded period should be included. The adjacent gauge records for supplemental information should be checked and utilized to extend the record if it is possible. Natural runoff data should be separated from later controlled data. It is known that high-altitude basin snowmelt discharges are not compatible with rain flood discharges. The zero years must also be accounted for by adjusting the final plot positions, not by inclusion as minor flows. The generalized skew number can be obtained from the chart in Bulletin No.17 B [8].

Quite often the database requires modification for use in a Log Pearson III analysis. Occasionally, a high outlier, but more often low outliers, will need to be removed from the database to avoid skewing results. This need is determined for high outliers by using $Q_H = \bar{Q}_H + K\,S_H$, and low outliers by using $Q_L = \bar{Q}_L + K\,S_L$, where K is a factor determined by the sample size, $\bar{Q}_H$ and $\bar{Q}_L$ are the high and low mean logarithm of systematic peaks, Q_H and Q_L are the high and low outlier thresholds in log units, S_H and S_L are the high and low standard deviations of the logarithmic distribution. Refer to FHWA HEC-19, Hydrology [16] or USGS Bulletin 17B [8] for this method and to find the values of K.

The data to be plotted are "PEAK DISCHARGE, Q (CFS)" vs. "PROBABILITY, Pr" as shown in the example in Figure 61.1. This plot usually results in a very flat curve with a reasonably straight center portion. An extension of this center portion gives a line for interpolation of the various needed discharges and frequencies.

The engineer should use an adjusted skew, which is calculated from the generalized and station skews. Generalized skews should be developed from at least 40 stations with each station having at least 25 years of record.

The equation for the adjusted skew is

$$G_w = \frac{MSE_{G_S}(G_L) + MSE_{G_L}(G_S)}{MSE_{G_S} + MSE_{G_L}} \tag{61.4}$$

where

G_w = weighted skew coefficient
G_S = station skew
G_L = generalized skew
MSE_{G_S} = mean square error of station skew
MSE_{G_L} = mean square error of generalized skew

The entire Log Pearson type III procedure is covered by Bulletin No. 17B, "Guidelines for Determining Flood Flow Frequency" [8].

The Gumbel extreme value method, sometimes called the double-exponential distribution of extreme values, has also been used to describe the distribution of hydrological variables, especially the peak discharges. It is based on the assumption that the cumulative frequency distribution of the largest values of samples drawn from a large population can be described by the following equation:

$$f(Q) = e^{-e^{a(Q-b)}} \tag{61.5}$$

where

$a = \dfrac{1.281}{S}$

$b = \overline{Q} - 0.450\,S$

S = standard deviation
$\overline{Q}$ = mean annual flow

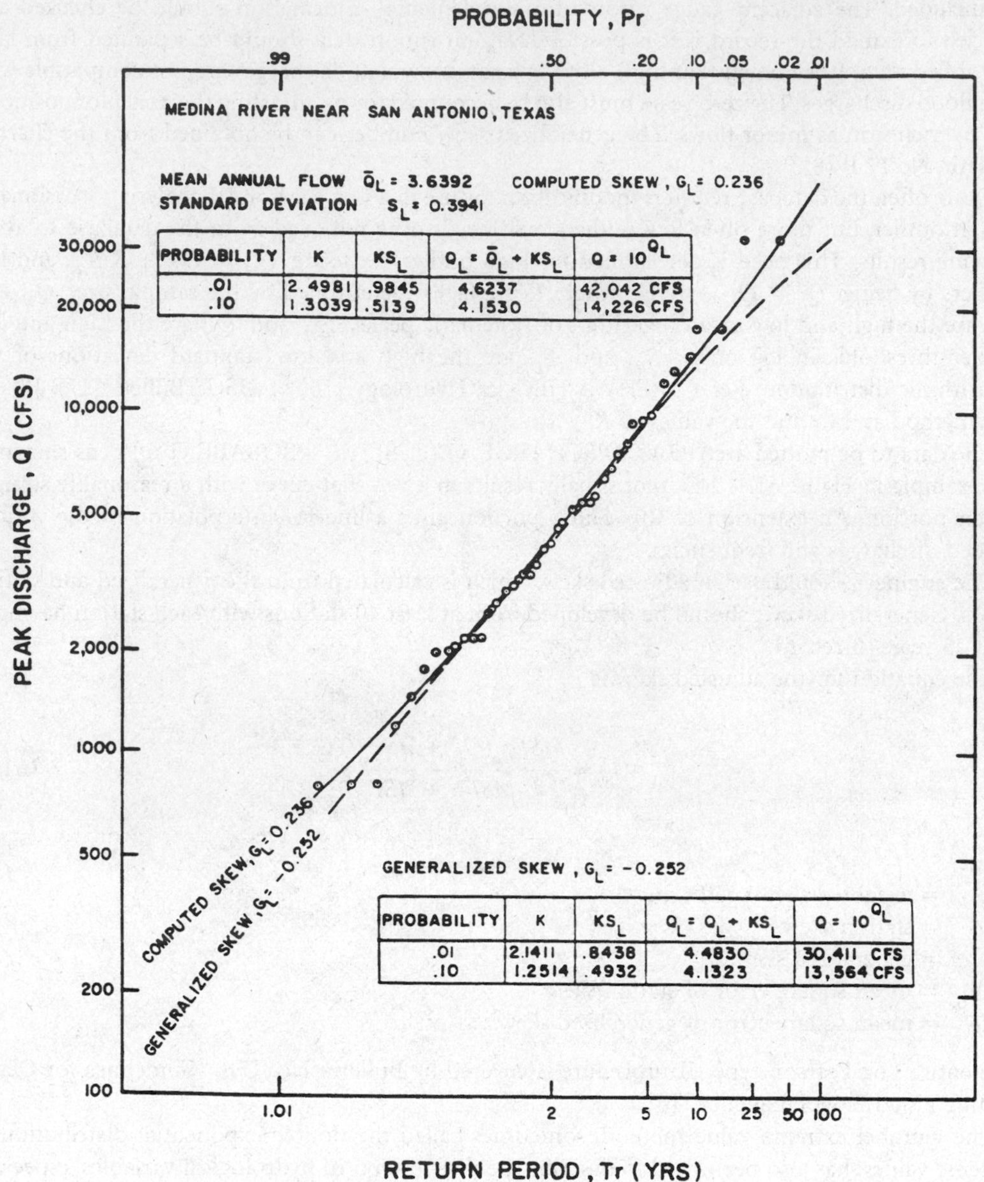

FIGURE 61.1 Log Pearson type III distribution analysis, Medina River, TX.

Values of this distribution function can be computed from Eq. (61.5). Characteristics of the Gumbel extreme value distribution are that the mean flow, $\overline{Q}$, occurs at the return period of T_r = 2.33 years and that it is skewed toward the high flows or extreme values as shown in the example of Figure 61.2. Even though it does not account directly for the computed skew of the data, it does predict the high flows reasonably well. For this method and additional techniques, please refer to USGS Water Supply Paper 1543-A, Flood-Frequency Analysis, and Manual of Hydrology Part 3.

The Gumbel extreme value distribution is given in "Statistics of Extremes" by E.J. Gumbel and is also found in HEC-19, p.73. Results from this method should be plotted on special Gumbel paper as shown in Figure 61.2.

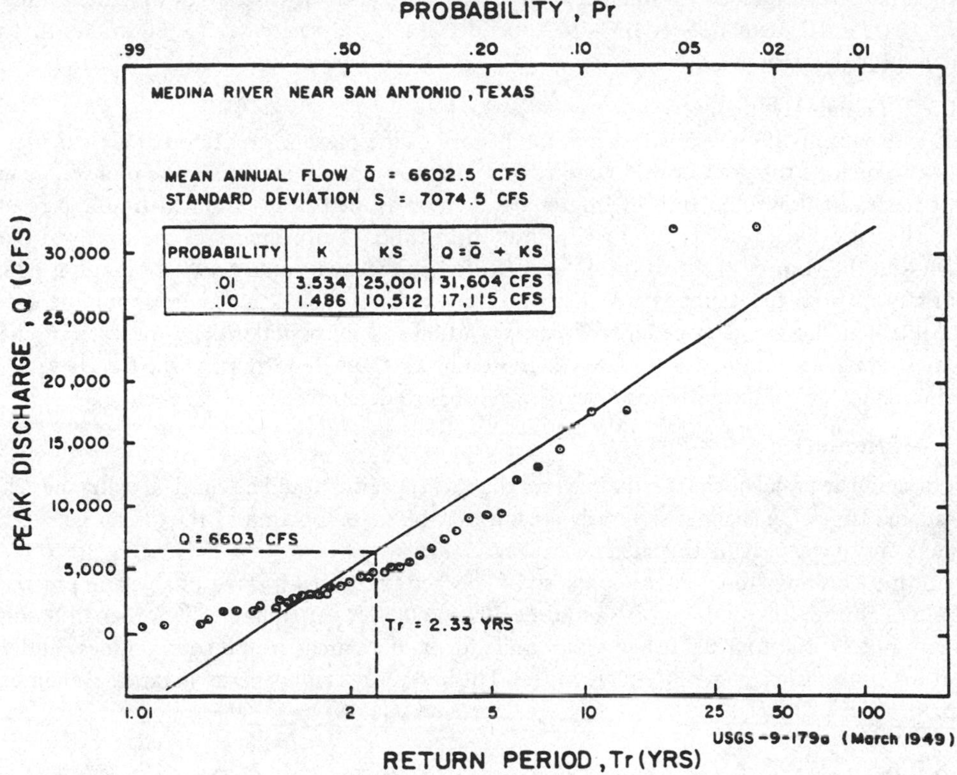

FIGURE 61.2 Gumbel extreme value frequency distribution analysis, Medina River, TX.

61.2.1.3.3 Discharge Comparison of Adjacent Basins

HEC 19, Appendix D [16] contains a list of reports for various states in the United States that have discharges at gauges that have been determined for frequencies from 2-year through 100-year frequencies. The discharges were determined by the Log Pearson III method. The discharge frequency at the gauges should be updated by the engineer using Log Pearson III and the Gumbel extreme value method.

The gauge data can be used directly as equivalent if the drainage areas are about the same (within less than 5%). Otherwise, the discharge determination can be obtained by the formula:

$$Q_u = Q_g (A_u / A_g)^b \tag{61.6}$$

where

Q_u = discharge at ungauged site

Q_g = discharge at gauged site

A_u = area of ungauged site

A_g = area of gauged site

b = exponent of drainage area

61.2.1.3.4 Regional Flood-Frequency Equations

If no gauged site is reasonably nearby, or if the record for the gauge is too short, then the discharge can be computed using the applicable regional flood-frequency equations. Statewide regional regression equations have been established in the United States. These equations permit peak flows to be estimated for return periods varying between 2 and 100 years. The discharges were determined by the Log Pearson III method. See HEC-19, Appendix D [16] for references to the studies that were conducted for the various states.

61.2.1.3.5 Design Hydrographs

Design hydrographs [9] give a complete time history of the passage of a flood at a particular site. This would include the peak flow. A runoff hydrograph is a plot of the response of a watershed to a particular rainfall event. A unit hydrograph is defined as the direct runoff hydrograph resulting from a rainfall event that lasts for a unit duration of time. The ordinates of the unit hydrograph are such that the volume of direct runoff represented by the area under the hydrograph is equal to 1 in. of runoff from the drainage area. Data on low water discharges and dates should be given as it will control methods and procedures of pier excavation and construction. The low water discharges and dates can be found in the USGS Water Resources Data Reports published each year. One procedure is to review the past 5 or 6 years of records to determine this.

61.2.1.4 Remarks

Before arriving at a final discharge, the existing channel capacity should be checked using the velocity as calculated times the channel waterway area. It may be that a portion of the discharge overflows the banks and never reaches the site.

The proposed design discharge should also be checked to see that it is reasonable and practicable. As a rule of thumb, the unit runoff should be 300 to 600 s-ft per square mile for small basins (to 20 square miles), 100 to 300 s-ft per square mile for median areas (to 50 square miles) and 25 to 150 s-ft for large basins (above 50 square miles). The best results will depend on rational engineering judgment.

61.2.2 Bridge Deck Drainage Design (On-Site Drainage Design)

61.2.2.1 Runoff and Capacity Analysis

The preferred on-site hydrology method is the rational method. The rational method, as discussed in Section 61.2.1.3.1, for on-site hydrology has a minimum time of concentration of 10 min. Many times, the time of concentration for the contributing on-site pavement runoff is less than 10 min. The initial time of concentration can be determined using an *overland flow* method until the runoff is concentrated in a curbed section. Channel flow using the roadway-curb cross section should be used to determine velocity and subsequently the time of flow to the first inlet. The channel flow velocity and flooded width is calculated using Manning's formula:

$$V = \frac{1.486}{n} A R^{2/3} S_f^{1/2} \tag{61.7}$$

where

V = velocity
A = cross-sectional area of flow
R = hydraulic radius
S_f = slope of channel
n = Manning's roughness value [11]

The intercepted flow is subtracted from the initial flow and the bypass is combined with runoff from the subsequent drainage area to determine the placement of the next inlet. The placement of inlets is determined by the allowable flooded width on the roadway.

Oftentimes, bridges are in sump areas, or the lowest spot on the roadway profile. This necessitates the interception of most of the flow before reaching the bridge deck. Two overland flow equations are as follows.

1. **Kinematic Wave Equation:**

$$t_o = \frac{6.92L^{0.6}n^{0.6}}{i^{0.4}S^{0.3}} \tag{61.8}$$

2. **Overland Equation:**

$$t_o = \frac{3.3(1.1-C)(L)^{1/2}}{(100\,S)^{1/3}} \tag{61.9}$$

where
t_o = overland flow travel time in minutes
L = length of overland flow path in meters
S = slope of overland flow in meters
n = manning's roughness coefficient [12]
i = design storm rainfall intensity in mm/h
C = runoff coefficient (Tables 61.1 and 61.2)

61.2.2.2 Select and Size Drainage Facilities

The selection of inlets is based upon the allowable flooded width. The allowable flooded width is usually outside the traveled way. The type of inlet leading up to the bridge deck can vary depending upon the flooded width and the velocity. Grate inlets are very common and, in areas with curbs, curb opening inlets are another alternative. There are various monographs associated with the type of grate and curb opening inlet. These monographs are used to determine interception and therefore the bypass [5].

61.2.3 Stage Hydraulics

High water (HW) stage is a very important item in the control of the bridge design. All available information should be obtained from the field and the Bridge Hydrology Report regarding HW marks, HW on upstream and downstream sides of the existing bridges, high drift profiles, and possible backwater due to existing or proposed construction.

Remember, observed high drift and HW marks are not always what they seem. Drift in trees and brush that could have been bent down by the flow of the water will be extremely higher than the actual conditions. In addition, drift may be pushed up on objects or slopes above actual HW elevation by the velocity of the water or wave action. Painted HW marks on the bridge should be searched carefully. Some flood insurance rate maps and flood insurance study reports may show stages for various discharges. Backwater stages caused by other structures should be included or streams should be noted.

Duration of high stages should be given, along with the base flood stage and HW for the design discharge. It should be calculated for existing and proposed conditions that may restrict the channel producing a higher stage. Elevation and season of low water should be given, as this may control design of tremie seals for foundations and other possible methods of construction. Elevation of overtopping flow and its location should be given. Normally, overtopping occurs at the bridge site, but overtopping may occur at a low sag in the roadway away from the bridge site.

61.2.3.1 Waterway Analysis

When determining the required waterway at the proposed bridge, the engineers must consider all adjacent bridges if these bridges are reasonably close. The waterway section of these bridges should be tied into the stream profile of the proposed structure. Structures that are upstream or downstream of the proposed bridge may have an impact on the water surface profile. When calculating the effective waterway area, adjustments must be made for the skew and piers and bents. The required waterway should be below the 50-year design HW stage.

If stream velocities, scour, and erosive forces are high, then abutments with wingwall construction may be necessary. Drift will affect the horizontal clearance and the minimum vertical clearance line of the proposed structure. Field surveys should note the size and type of drift found in the channel. Designs based on the 50-year design discharge will require drift clearance. On major streams and rivers, drift clearance of 2 to 5 m above the 50-year discharge is needed. On smaller streams 0.3 to 1 m may be adequate. A formula for calculating freeboard is

$$\text{Freeboard} = 0.1Q^{0.3} + 0.008V^2 \tag{61.10}$$

where
Q = discharge
V = velocity

61.2.3.2 Water Surface Profile Calculation

There are three prominent water surface profile calculation programs available [1,2]. The first one is HEC-2 which takes stream cross sections perpendicular to the flow. WSPRO is similar to HEC-2 with some improvements. SMS is a new program that uses finite-element analysis for its calculations. SMS can utilize digital elevation models to represent the streambeds.

61.2.2.3 Flow Velocity and Distribution

Mean channel, overflow velocities at peak stage, and localized velocity at obstructions such as piers should be calculated or estimated for anticipated high stages. Mean velocities may be calculated from known stream discharges at known channel section areas or known waterway areas of bridge, using the correct high water stage.

Surface water velocities should be measured roughly, by use of floats, during field surveys for sites where the stream is flowing. Stream velocities may be calculated along a uniform section of the channel using Manning's formula Eq. (61.7) if the slope, channel section (area and wetted perimeter), and roughness coefficient (n) are known.

At least three profiles should be obtained, when surveying for the channel slope, if possible. These three slopes are bottom of the channel, the existing water surface, and the HW surface based on drift or HW marks. The top of low bank, if overflow is allowed, should also be obtained. In addition, note some tops of high banks to prove flows fall within the channel. These profiles should be plotted showing existing and proposed bridges or other obstruction in the channel, the change of HW slope due to these obstructions and possible backwater slopes.

The channel section used in calculating stream velocities should be typical for a relatively long section of uniform channel. Since this theoretical condition is not always available, however, the nearest to uniform conditions should be used with any necessary adjustments made for irregularities.

Velocities may be calculated from PC programs, or calculator programs, if the hydraulic radius, roughness factor, and slope of the channel are known for a section of channel, either natural or artificial, where uniform stream flow conditions exists. The hydraulic radius is the waterway area divided by the wetted perimeter of an average section of the uniform channel. A section under a bridge whose piers, abutments, or approach fills obstruct the uniformity of the channel cannot be used as there will not be uniform flow under the structure. If no part of the bridge structure seriously obstructs or restricts the channel, however, the section at the bridge could be used in the above uniform flow calculations.

The roughness coefficient n for the channel will vary along the length of the channel for various locations and conditions. Various values for n can be found in the References [1,5,12,17].

At the time of a field survey the party chief should estimate the value of n to be used for the channel section under consideration. Experience is required for field determination of a relatively close to actual *n* value. In general, values for natural streams will vary between 0.030 and 0.070. Consider both low and HW *n* value. The water surface slope should be used in this plot and the slope should be adjusted for obstructions such as bridges, check dams, falls, turbulence, etc.

The results as obtained from this plot may be inaccurate unless considerable thought is given to the various values of slope, hydraulic radius, and *n*. High velocities between 15 and 20 ft/s (4.57 and 6.10 m/s through a bridge opening may be undesirable and may require special design considerations. Velocities over 20/ 6.10 m/s should not be used unless special design features are incorporated or if the stream is mostly confined in rock or an artificial channel.

61.3 Bridge Scour

61.3.1 Bridge Scour Analysis

61.3.1.1 Basic Scour Concepts

Scour is the result of the erosive action of flowing water, excavating and carrying away material from the bed and banks of streams. Determining the magnitude of scour is complicated by the cyclic nature of the scour process. Designers and inspectors need to study site-specific subsurface information carefully in evaluating scour potential at bridges. In this section, we present bridge engineers with the basic procedures and methods to analyze scour at bridges.

Scour should be investigated closely in the field when designing a bridge. The designer usually places the top of footings at or below the total potential scour depth; therefore, determining the depth of scour is very important. The total potential scour at a highway crossing usually comprises the following components [11]: aggradation and degradation, stream contraction scour, local scour, and sometimes with lateral stream migration.

61.3.1.1.1 Long-Term Aggradation and Degradation

When natural or human activities cause streambed elevation changes over a long period of time, aggradation or degradation occurs. Aggradation involves the deposition of material eroded from the channel or watershed upstream of the bridge, whereas degradation involves the lowering or scouring of the streambed due to a deficit in sediment supply from upstream.

Long-term streambed elevation changes may be caused by the changing natural trend of the stream or may be the result of some anthropogenic modification to the stream or watershed. Factors that affect long-term bed elevation changes are dams and reservoirs up- or downstream of the bridge, changes in watershed land use, channelization, cutoffs of meandering river bends, changes in the downstream channel base level, gravel mining from the streambed, diversion of water into or out of the stream, natural lowering of the fluvial system, movement of a bend, bridge location with respect to stream planform, and stream movement in relation to the crossing. Tidal ebb and flood may degrade a coastal stream, whereas littoral drift may cause aggradation. The problem for the bridge engineer is to estimate the long-term bed elevation changes that will occur during the lifetime of the bridge.

61.3.1.1.2 Stream Contraction Scour

Contraction scour usually occurs when the flow area of a stream at flood stage is reduced, either by a natural contraction or an anthropogenic contraction (like a bridge). It can also be caused by the overbank flow which is forced back by structural embankments at the approaches to a bridge. There are some other causes that can lead to a contraction scour at a bridge crossing [11]. The decreased flow area causes an increase in average velocity in the stream and bed shear stress through the contraction reach. This in turn triggers an increase in erosive forces in the contraction. Hence, more bed material is removed from the contracted reach than is transported into the reach. The natural streambed elevation is lowered by this contraction phenomenon until relative equilibrium is reached in the contracted stream reach.

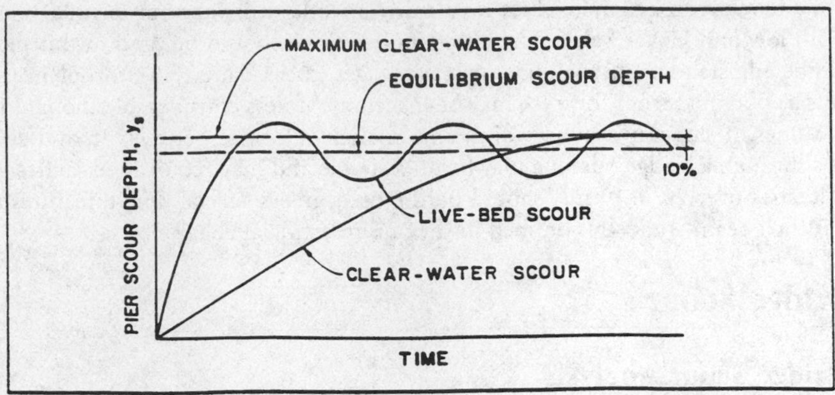

FIGURE 61.3 Illustrative pier scour depth in a sand-bed stream as a function of time.

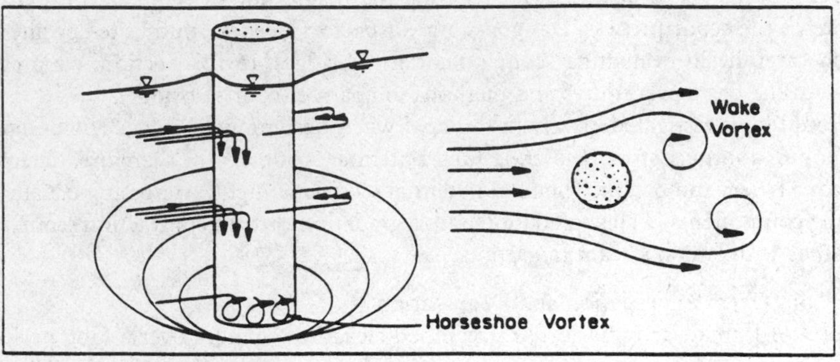

FIGURE 61.4 Schematic representation of local scour at a cylindrical pier.

There are two forms of contraction scour: live-bed and clear-water scours. Live-bed scour occurs when there is sediment being transported into the contracted reach from upstream. In this case, the equilibrium state is reached when the transported bed material out of the scour hole is equal to that transported into the scour hole from upstream. Clear-water scour occurs when the bed sediment transport in the uncontracted approach flow is negligible or the material being transported in the upstream reach is transported through the downstream at less than the capacity of the flow. The equilibrium state of the scour is reached when the average bed shear stress is less than that required for incipient motion of the bed material in this case (Figure 61.3).

61.3.1.1.3 Local Scour

When upstream flow is obstructed by obstruction such as piers, abutments, spurs, and embankments, flow vortices are formed at their base as shown in Figure 61.4 (known as horseshoe vortex). This vortex action removes bed material from around the base of the obstruction. A scour hole eventually develops around the base. Local scour can also be either clear-water or live-bed scour. In considering local scour, a bridge engineer needs to look into the following factors: flow velocity, flow depth, flow attack angle to the obstruction, obstruction width and shape, projected length of the obstruction, bed material characteristics, bed configuration of the stream channel, and also potential ice and debris effects [11, 13].

61.3.1.1.4 Lateral Stream Migration

Streams are dynamic. The lateral migration of the main channel within a floodplain may increase pier scour, embankment or approach road erosion, or change the total scour depth by altering the flow angle of attack at piers. Lateral stream movements are affected mainly by the geomorphology of the stream, location of the crossing on the stream, flood characteristics, and the characteristics of the bed and bank materials [11,13].

61.3.1.2 Designing Bridges to Resist Scour

It is obvious that all scour problems cannot be covered in this special topic section of bridge scour. A more-detailed study can be found in HEC-18, "Evaluating Scour at Bridges" and HEC-20, "Stream Stability at Highway Structures" [11,18]. As described above, the three most important components of bridge scour are long-term aggradation or degradation, contraction scour, and local scour. The total potential scour is a combination of the three components. To design a bridge to resist scour, a bridge engineer needs to follow the following observation and investigation steps in the design process.

1. **Field Observation** — Main purposes of field observation are as follows:
   - Observe conditions around piers, columns, and abutments (Is the hydraulic skew correct?),
   - Observe scour holes at bends in the stream,
   - Determine streambed material,
   - Estimate depth of scour, and
   - Complete geomorphic factor analysis.

   There is usually no fail-safe method to protect bridges from scour except possibly keeping piers and abutments out of the HW area; however, proper hydraulic bridge design can minimize bridge scour and its potential negative impacts.
2. **Historic Scour Investigation** — Structures that have experienced scour in the past are likely to continue displaying scour problems in the future. The bridges that we are most concerned with include those currently experiencing scour problems and exhibiting a history of local scour problems.
3. **Problem Location Investigation** — Problem locations include "unsteady stream" locations, such as near the confluence of two streams, at the crossing of stream bends, and at alluvial fan deposits.
4. **Problem Stream Investigation** — Problem streams are those that have the following characteristics of aggressive tendencies: indication of active degradation or aggradation; migration of the stream or lateral channel movement; streams with a steep lateral slope and/or high velocity; current, past, or potential in-stream aggregate mining operations; and loss of bank protection in the areas adjacent to the structure.
5. **Design Feature Considerations** — The following features, which increase the susceptibility to local scour, should be considered:
   - Inadequate waterway opening leads to inadequate clearance to pass large drift during heavy runoff.
   - Debris/drift problem: Light drift or debris may cause significant scour problems, moderate drift or debris may cause significant scour but will not create severe lateral forces on the structure, and heavy drift can cause strong lateral forces or impact damage as well as severe scour.
   - Lack of overtopping relief: Water may rise above deck level. This may not cause scour problems but does increase vulnerability to severe damage from impact by heavy drift.
   - Incorrect pier skew: When the bridge pier does not match the channel alignment, it may cause scour at bridge piers and abutments.
6. **Traffic Considerations** — The amount of traffic such as average daily traffic (ADT), type of traffic, the length of detour, the importance of crossings, and availability of other crossings should be taken into consideration.

7. **Potential for Unacceptable Damage** — Potential for collapse during flood, safety of traveling public and neighbors, effect on regional transportation system, and safety of other facilities (other bridges, properties) need to be evaluated.
8. **Susceptibility of Combined Hazard of Scour and Seismic** — The earthquake prioritization list and the scour-critical list are usually combined for bridge design use.

61.3.1.3 Scour Rating

In the engineering practice of the California Department of Transportation, the rating of each structure is based upon the following:

1. **Letter grading** — The letter grade is related to the potential for scour-related problems at this location.
2. **Numerical grading** — The numerical rating associated with each structure is a determination of the severity for the potential scour:

   A-1 No problem anticipated
   A-2 No problem anticipated/new bridge — no history
   A-3 Very remote possibility of problems
   B-1 Slight possibility of problems
   B-2 Moderate possibility of problems
   B-3 Strong possibility of problems
   C-1 Some probability of problems
   C-2 Moderate probability of problems
   C-3 Very strong probability of problems

Scour effect of storms is usually greater than design frequency, say, 500-year frequency. FHWA specifies 500-year frequency as 1.7 times 100-year frequency. Most calculations indicate 500-year frequency is 1.25 to 1.33 times greater than the 100-year frequency [3,8]; the 1.7 multiplier should be a maximum. Consider the amount of scour that would occur at overtopping stages and also pressure flows. Be aware that storms of lesser frequency may cause larger scour stress on the bridge.

61.3.2 Bridge Scour Calculation

All the equations for estimating contraction and local scour are based on laboratory experiments with limited field verification [11]. However, the equations recommended in this section are considered to be the most applicable for estimating scour depths. Designers also need to give different considerations to clear-water scour and live-bed scour at highway crossings and encroachments.

Prior to applying the bridge scour estimating methods, it is necessary to (1) obtain the fixed-bed channel hydraulics, (2) determine the long-term impact of degradation or aggradation on the bed profile, (3) adjust the fixed-bed hydraulics to reflect either degradation or aggradation impact, and (4) compute the bridge hydraulics accordingly.

61.3.2.1 Specific Design Approach

Following are the recommended steps for determining scour depth at bridges:

Step 1: Analyze long-term bed elevation change.
Step 2: Compute the magnitude of contraction scour.
Step 3: Compute the magnitude of local scour at abutments.
Step 4: Compute the magnitude of local scour at piers.
Step 5: Estimate and evaluate the total potential scour depths.

The bridge engineers should evaluate if the individual estimates of contraction and local scour depths from Step 2 to 4 are reasonable and evaluate the total scour derived from Step 5.

61.3.2.2 Detailed Procedures

1. **Analyze Long-Term Bed Elevation Change**— The face of bridge sections showing bed elevation are available in the maintenance bridge books, old preliminary reports, and sometimes in FEMA studies and U.S. Corps of Engineers studies. Use this information to estimate aggradation or degradation.
2. **Compute the Magnitude of Contraction Scour**— It is best to keep the bridge out of the normal channel width. However, if any of the following conditions are present, calculate contraction scour.
   a. Structure over channel in floodplain where the flows are forced through the structure due to bridge approaches
   b. Structure over channel where river width becomes narrow
   c. Relief structure in overbank area with little or no bed material transport
   d. Relief structure in overbank area with bed material transport

   The general equation for determining contraction scour is

   $$y_s = y_2 - y_1 \tag{61.11}$$

   where
   y_s = depth of scour
   y_1 = average water depth in the main channel
   y_2 = average water depth in the contracted section

   Other contraction scour formulas are given in the November 1995 HEC-18 publication— also refer to the workbook or HEC-18 for the various conditions listed above [11]. The detailed scour calculation procedures can be referenced from this circular for either live-bed or clear-water contraction scour.
3. **Compute the Magnitude of Local Scour at Abutments**— Again, it is best to keep the abutments out of the main channel flow. Refer to publication HEC-18 from FHWA [13]. The scour formulas in the publication tend to give excessive scour depths.
4. **Compute the Magnitude of Local Scour at Piers**— The pier alignment is the most critical factor in determining scour depth. Piers should align with stream flow. When flow direction changes with stages, cylindrical piers or some variation may be the best alternative. Be cautious, since large-diameter cylindrical piers can cause considerable scour. Pier width and pier nose are also critical elements in causing excessive scour depth.

Assuming a sand bed channel, an acceptable method to determine the maximum possible scour depth for both live-bed and clear-water channel proposed by the Colorado State University [11] is as follows:

$$\frac{y_s}{y_1} = 2.0 K_1 K_2 K_3 \left(\frac{a}{y_1}\right)^{0.65} F_r^{0.43} l \tag{61.12}$$

where
y_s = scour depth
y_1 = flow depth just upstream of the pier
K_1 = correction for pier shape from Figure 61.5 and Table 61.3
K_2 = correction for angle of attack of flow from Table 61.4
K_3 = correction for bed condition from Table 61.5
a = pier width
l = pier length
F_r = Froude number = $\frac{V}{(gy)}$ (just upstream from bridge)

Drift retention should be considered when calculating pier width/type.

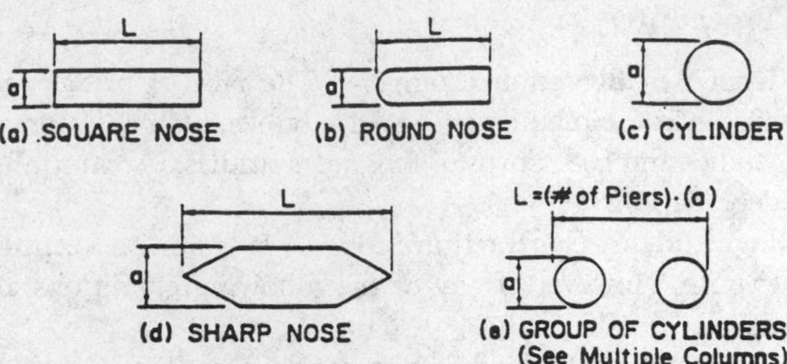

FIGURE 61.5 Common pier shapes.

TABLE 61.3 Correction Factor, K_1, for Pier Nose Shape

Shape of Pier Nose	K_1
Square nose	1.1
Round nose	1.0
Circular cylinder	1.0
Sharp nose	0.9
Group of cylinders	1.0

TABLE 61.4 Correction Factor, K_2, for Flow Angle of Attack

Angle	$L/a = 4$	$L/a = 8$	$L/a = 12$
0	1.0	1.0	1.0
15	1.5	2.0	2.5
30	2.0	2.75	3.5
45	2.3	3.3	4.3
90	2.5	3.9	5

TABLE 61.5 Increase in Equilibrium Pier Scour Depths K_3 for Bed Conditions

Bed Conditions	Dune Height H, ft	K_3
Clear-water scour	N/A	1.1
Plane bed and antidune flow	N/A	1.1
Small dunes	$10 > H > 2$	1.1
Medium dunes	$30 > H > 10$	1.1–1.2
Large dunes	$H > 30$	1.3

61.3.2.3 Estimate and Evaluate Total Potential Scour Depths

Total potential scour depths is usually the sum of long-term bed elevation change (only degradation is usually considered in scour computation), contraction scour, and local scour. Historical scour depths and depths of scourable material are determined by geology. When estimated depths from the above methods are in conflict with geology, the conflict should be resolved by the hydraulic engineer and the geotechnical engineer; based on economics and experience, it is best to provide for maximum anticipated problems.

61.3.3 Bridge Scour Investigation and Prevention

61.3.3.1 Steps to Evaluate Bridge Scour

It is recommended that an interdisciplinary team of hydraulic, geotechnical, and bridge engineers should conduct the evaluation of bridge scour. The following approach is recommended for evaluating the vulnerability of existing bridges to scour [11]:

Step 1. Screen all bridges over waterways into five categories: (1) low risk, (2) scour-susceptible, (3) scour-critical, (4) unknown foundations, or (5) tidal. Bridges that are particularly vulnerable to scour failure should be identified immediately and the associated scour problem addressed. These particularly vulnerable bridges are

1. Bridges currently experiencing scour or that have a history of scour problems during past floods as identified from maintenance records, experience, and bridge inspection records
2. Bridges over erodible streambeds with design features that make them vulnerable to scour
3. Bridges on aggressive streams and waterways
4. Bridges located on stream reaches with adverse flow characteristics

Step 2. Prioritize the scour-susceptible bridges and bridges with unknown foundations by conducting a preliminary office and field examination of the list of structures compiled in Step 1 using the following factors as a guide:

1. The potential for bridge collapse or for damage to the bridge in the event of a major flood
2. The functional classification of the highway on which the bridge is located.
3. The effect of a bridge collapse on the safety of the traveling public and on the operation of the overall transportation system for the area or region

Step 3. Conduct office and field scour evaluations of the bridges on the prioritized list in Step 2 using an interdisciplinary team of hydraulic, geotechnical, and bridge engineers:

1. In the United States, FHWA recommends using 500-year flood or a flow 1.7 times the 100-year flood where the 500-year flood is unknown to estimate scour [3,6]. Then analyze the foundations for vertical and lateral stability for this condition of scour. The maximum scour depths that the existing foundation can withstand are compared with the total scour depth estimated. An engineering assessment must be then made whether the bridge should be classified as a scour-critical bridge.
2. Enter the results of the evaluation study in the inventory in accordance with the instructions in the FHWA "Bridge Recording and Coding Guide" [7].

Step 4. For bridges identified as scour critical from the office and field review in Steps 2 and 3, determine a plan of action for correcting the scour problem (see Section 61.3.3.3).

61.3.3.2 Introduction to Bridge Scour Inspection

The bridge scour inspection is one of the most important parts of preventing bridge scour from endangering bridges. Two main objectives to be accomplished in inspecting bridges for scour are

1. To record the present condition of the bridge and the stream accurately; and
2. To identify conditions that are indicative of potential problems with scour and stream stability for further review and evaluation by other experts.

In this section, the bridge inspection practice recommended by U.S. FHWA [6,10] is presented for engineers to follow as guidance.

61.3.3.2.1 Office Review

It is highly recommended that an office review of bridge plans and previous inspection reports be conducted prior to making the bridge inspection. Information obtained from the office review provides a better foundation for inspecting the bridge and the stream. The following questions should be answered in the office review:

- Has an engineering scour evaluation been conducted? If so, is the bridge scour critical?
- If the bridge is scour-critical, has a plan of action been made for monitoring the bridge and/or installing scour prevention measures?
- What do comparisons of streambed cross sections taken during successive inspections reveal about the stream bed? Is it stable? Degrading? Aggrading? Moving laterally? Are there scour holes around piers and abutments?
- What equipment is needed to obtain stream-bed cross sections?
- Are there sketches and aerial photographs to indicate the planform locations of the stream and whether the main channel is changing direction at the bridge?
- What type of bridge foundation was constructed? Do the foundations appear to be vulnerable to scour?
- Do special conditions exist requiring particular methods and equipment for underwater inspections?
- Are there special items that should be looked at including damaged riprap, stream channel at adverse angle of flow, problems with debris, etc.?

61.3.3.2.2 Bridge Scour Inspection Guidance

The condition of the bridge waterway opening, substructure, channel protection, and scour prevention measures should be evaluated along with the condition of the stream during the bridge inspection. The following approaches are presented for inspecting and evaluating the present condition of the bridge foundation for scour and the overall scour potential at the bridge.

Substructure is the key item for rating the bridge foundations for vulnerability to scour damage. Both existing and potential problems with scour should be reported so that an interdisciplinary team can make a scour evaluation when a bridge inspection finds that a scour problem has already occurred. If the bridge is determined to be scour critical, the rating of the substructures should be evaluated to ensure that existing scour problems have been considered. The following items should be considered in inspecting the present condition of bridge foundations:

- Evidence of movement of piers and abutments such as rotational movement and settlement;
- Damage to scour countermeasures protecting the foundations such as riprap, guide banks, sheet piling, sills, etc.;
- Changes in streambed elevation at foundations, such as undermining of footings, exposure of piles; and
- Changes in streambed cross section at the bridge, including location and depth of scour holes.

In order to evaluate the conditions of the foundations, the inspectors should take cross sections of the stream and measure scour holes at piers and abutments. If equipment or conditions do not permit measurement of the stream bottom, it should be noted for further investigation.

To take and plot measurement of stream bottom elevations in relation to the bridge foundations is considered the single most important aspect of inspecting the bridge for actual or potential damage from scour. When the stream bottom cannot be accurately measured by conventional means, there are other special measures that need to be taken to determine the condition of the substructures or foundations such as using divers and using electronic scour detection equipment. For the purposes of evaluating resistance to scour of the substructures, the questions remain essentially the same for foundations in deep water as for foundations in shallow water [7] as follows:

- How does the stream cross section look at the bridge?
- Have there been any changes as compared with previous cross section measurements? If so, does this indicate that (1) the stream is aggrading or degrading or (2) is local or contraction scour occurring around piers and abutments?
- What are the shapes and depths of scour holes?
- Is the foundation footing, pile cap, or the piling exposed to the stream flow, and, if so, what is the extent and probable consequences of this condition?
- Has riprap around a pier been moved or removed?

Any condition that a bridge inspector considers to be an emergency or of a potentially hazardous nature should be reported immediately. This information as well as other conditions, which do not pose an immediate hazard but still warrant further investigation, should be conveyed to the interdisciplinary team for further review.

61.3.3.3 Introduction to Bridge Scour Prevention

Scour prevention measures are generally incorporated after the initial construction of a bridge to make it less vulnerable to damage or failure from scour. A plan of preventive action usually has three major components [11]:

1. Timely installation of temporary scour prevention measures;
2. Development and implementation of a monitoring program;
3. A schedule for timely design and construction of permanent scour prevention measures.

For new bridges [11], the following is a summary of the best solutions for minimizing scour damage:

1. Locating the bridge to avoid adverse flood flow patterns;
2. Streamlining bridge elements to minimize obstructions to the flow;
3. Designing foundations safe from scour;
4. Founding bridge pier foundations sufficiently deep to not require riprap or other prevention measures; and
5. Founding abutment foundations above the estimated local scour depth when the abutment is protected by well-designed riprap or other suitable measures.

For existing bridges, the available scour prevention alternatives are summarized as follows:

1. Monitoring scour depths and closing the bridge if excessive bridge scour exists;
2. Providing riprap at piers and/or abutments and monitoring the scour conditions;
3. Constructing guide banks or spur dikes;
4. Constructing channel improvements;
5. Strengthening the bridge foundations;
6. Constructing sills or drop structures; and
7. Constructing relief bridges or lengthening existing bridges.

These scour prevention measures should be evaluated using sound hydraulic engineering practice. For detailed bridge scour prevention measures and types of prevention measures, refer to Chapter 7 of "Evaluating Scour at Bridges" from U.S. FHWA. [10,11,18,19].

References

1. AASHTO, *Model Drainage Manual*, American Association of State Highway and Transportation Officials, Washington, D.C., 1991.
2. AASHTO, *Highway Drainage Guidelines*, American Association of State Highway and Transportation Officials, Washington, D.C., 1992.
3. California State Department of Transportation, *Bridge Hydraulics Guidelines*, Caltrans, Sacramento
4. California State Department of Transportation, *Highway Design Manual*, Caltrans, Sacramento,
5. Kings, *Handbook of Hydraulics*, Chapter 7 (*n* factors).
6. U.S. Department of the Interior, Geological Survey (USGS), Magnitude and Frequency of Floods in California, Water-Resources Investigation 77–21.
7. U.S. Department of Transportation, Recording and Coding Guide for the Structure Inventory and Appraisal of the Nation's Bridges, FHWA, Washington D.C., 1988.
8. U.S. Geological Survey, Bulletin No. 17B, Guidelines for Determining Flood Flow Frequency.
9. U.S. Federal Highway Administration, Debris-Control Structures, Hydraulic Engineering Circular No. 9, 1971.
10. U.S. Federal Highway Administration, Design of Riprap Revetments, Hydraulic Engineering Circular No. 11, 1989.
11. U.S. Federal Highway Administration, Evaluating Scour at Bridges, Hydraulic Engineering Circular No. 18, Nov. 1995.
12. U.S. Federal Highway Administration, Guide for Selecting Manning's Roughness Coefficient (*n* factors) for Natural Channels and Flood Plains, Implementation Report, 1984.
13. U.S. Federal Highway Administration, Highways in the River Environment, Hydraulic and Environmental Design Considerations, Training & Design Manual, May 1975.
14. U.S. Federal Highway Administration, Hydraulics in the River Environment, Spur Dikes, Sect. VI-13, May 1975.
15. U.S. Federal Highway Administration, Hydraulics of Bridge Waterways, Highway Design Series No. 1, 1978.
16. U.S. Federal Highway Administration, Hydrology, Hydraulic Engineering Circular No. 19, 1984.
17. U.S. Federal Highway Administration, Local Design Storm, Vol. I–IV (*n* factor) by Yen and Chow.
18. U.S. Federal Highway Administration, Stream Stability at Highway Structures, Hydraulic Engineering Circular No. 20, Nov. 1990.
19. U.S. Federal Highway Administration, Use of Riprap for Bank Protection, Implementation Report, 1986.

62

Sound Walls and Railings

Farzin Lackpour
Parsons Brinckerhoff-FG, Inc.

Fuat S. Guzaltan
Parsons Brinckerhoff-FG, Inc.

62.1 Sound Walls

62.1.1 Introduction

62.1.1.1 Need for Sound Walls

Population growth experienced during past decades in metropolitan areas has prompted the expansion and improvement of highway systems. As a direct result of these improvements, currently 90 million people in the United States live close to high-volume, high-speed highways. Rush-hour traffic on a typical high-volume, high-speed urban highway generates noise levels in the 80 to 90 dBA range. Within 50 to 100 yd (45 to 90 m) from the highway, due to absorption by the ground cover, the noise level dissipates to about 70 to 80 dBA. This ambient noise level, in comparison with a 50 to 55 dBA noise level in an average quiet house, is very intrusive to the majority of people, and should be further reduced to at least 60 to 70 dBA level by implementing noise abatement measures.

62.1.1.2 Design Noise Levels

In 1982, the Federal Highway Administration (FHWA) published the "Procedures for Abatement of Highway Traffic Noise and Construction Noise" in the Federal Aid Highway Program Manual, and therein established the acceptable noise levels at the location of the receivers (houses, schools, etc.) after the installation of the sound walls. This publication regulates the average allowable noise levels, $L_{eq}(h)$, and the peak allowable noise levels, $L_{10}(h)$ (the noise level that is exceeded more than 10% of the given period of time used to measure the allowable noise level) (Table 62.1)[1].

0-8493-7434-0/00/$0.00+$.50

TABLE 62.1 Noise Abatement Design Criteria

Activity Category	L_{eq}(h), dBA	L_{10}(h), dBA	Land Use Category
A	57	60 (exterior)	Tracts of lands in which serenity and quiet are of extraordinary significance and serve an important public need and where the preservation of those quantities is essential if the area is to continue to serve its intended purpose; such areas could include amphitheaters, particular parks or portions of parks, or open spaces which are dedicated or recognized by appropriate local officials for activities requiring special quantities of serenity and quiet
B	67	70 (exterior)	Residences, motels, hotels, public meeting rooms, schools, churches, libraries, hospitals, picnic areas, playgrounds, active sports areas, and parks
C	72	75 (exterior)	Developed lands, properties, or activities not included in categories A and B above
D			Undeveloped lands; for requirements see paragraphs 5.a (5) and (6) of Publication PPM 90-2
E		55 (interior)	Residences, motels, hotels, public meeting rooms, schools, churches, libraries, hospitals, and auditoriums

62.1.2 Selection of Sound Walls

62.1.2.1 Sound Wall Materials

When a sound barrier is inserted in the line of sight between a noise source and a receiver, the intensity of the noise diminishes on the receiver side of the wall. This reduction in the noise intensity is referred to as insertion loss. The main factors that contribute to the insertion loss are the diffraction and reflection of the noise by the sound wall, and transmission loss as noise travels through the wall material. The amount of diffraction and reflection can be controlled by varying the height and inclination of the wall, installing specially shaped closure pieces at the top of the wall, or coating the wall surface with a sound absorbent material. The transmission loss can be controlled by varying the thickness and density of the wall material. The transmission loss levels for several common construction materials are given in Table 62.2 [2].

Earth berms, concrete, timber, and to a certain extent steel have been the traditional choices of material for sound walls. Other materials such as composite plaster panels, concrete blocks, bricks, and plywood panels have also been successfully utilized in smaller quantities in comparison to the traditional materials. In recent years, the awareness and need have risen to recycle materials rather than bury them in landfills. This trend has led to the use of recycled tires, glass, and plastics as sound wall material.

Given the variety of materials available for use in sound wall construction, selection of an appropriate type of sound wall becomes a difficult task. An intelligent decision can only be made after investigating the major factors contributing to successful implementation of a sound wall project such as cost, aesthetics, durability/life cycle, constructibility, etc.

62.1.2.2 Decision Matrix

A decision matrix is a convenient way of comparing the performance of different sound wall alternatives. The first step in building a decision matrix is to determine the parameters that will be the basis of the evaluation and selection process. The most important parameters are cost, aesthetics, durability/life cycle, and constructibility. The cost of the sound wall should include the cost of the surface finish or treatment, landscaping, utility relocation, drainage system, right-of-way, environmental mitigation, maintenance, and future replacement. Better durability and longer life cycle almost always translate into higher initial construction cost and lower maintenance cost. Aesthetic treatment of the sound wall should provide visual compatibility with the surrounding environment. Restrictions such as right-of-way limitations and presence of nearby residential areas may affect the constructibility of certain sound wall types. Other parameters such as construction access, and

TABLE 62.2 Sound Wall Materials

Materials	Thickness, in. (mm)	Transmission Loss (TL)[a], dBA
	Woods[b]	
Fir	½ (13)	17
	1 (25)	20
	2 (50)	24
Pine	½ (13)	16
	1 (25)	19
	2 (50)	23
Redwood	½ (13)	16
	1 (25)	19
	2 (50)	23
Cedar	½ (13)	15
	1 (25)	18
	2 (50)	22
Plywood	½ (13)	20
	1 (25)	23
Particle Board[c]	½ (13)	20
	Metals[d]	
Aluminum	1/16 (1.6)	23
	⅛ (3)	25
	¼ (6)	27
Steel	24 ga (0.6)	18
	20 ga (0.9)	22
	16 ga (15)	25
Lead	1/16 (1.6)	28
	Concrete, Masonry, etc.	
Light concrete	4 (100)	36
	6 (150)	39
Dense concrete	4 (100)	40
Concrete block	4 (100)	32
	Composites	
Aluminum-faced plywood[e]	¾ (20)	21–23
Aluminum-faced particle board[e]	¾ (20)	21–23
Plastic lamina on plywood	¾ (20)	21–23
Plastic lamina on particle board	¾ (20)	21–23
	Miscellaneous	
Glass (safety glass)	¼ (6)	22
Plexiglas (shatterproof)	—	22–25
Masonite	½ (13)	20
Fiber glass/resin	¼ (6)	20
Stucco on metal lath	1 (25)	32
Polyester with aggregate surface[f]	3 (75)	20–30

[a] A weighted TL based on generalized truck spectrum.
[b] Tongue-and-groove boards recommended to avoid leaks (for fir, pine, redwood, and cedar).
[c] Should be treated for water resistance.
[d] May require treatment to reduce glare (for aluminum and steel).
[e] Aluminum is 0.01 in. thick. Special care is necessary to avoid delamination (for all composites).
[f] TL depends on surface density of the aggregate.

impacts on residences, parks, utilities, drainage systems, traffic, and environment should also be considered. In this step, the crucial issue is the identification of the relevant parameters in collaboration with the project owner. If the project owner is willing, receiving input from the local governments, residents, and the traveling public can be an invaluable asset in the success and acceptance of the project.

The second step in the process is assigning a percent weight to each parameter that is considered to be relevant in the first step (Table 62.3).

The third step involves assigning a rating ranging from 1 to 10 to each parameter. A rating of 10 represents the most desirable case, and a rating of 1 represents the least desirable case. For parameters that are associated with costs (Items 1, 3, 5, 7, 8, and 9 in Table 62.3), the rating can be based on the following formula once these costs are determined for each sound wall alternative:

$$\frac{\text{Cost of Least Expensive Alternative} \times 10}{\text{Cost of Alternative Considered}}$$

For the rating of less quantitative items such as aesthetics, constructibility, and construction access, the best approach is to define the factors which will give satisfactory results for the parameter in question. For instance, if a sound wall is considered atop an existing retaining wall, we may select balance (a pleasing proportion between the heights of the proposed sound wall and existing retaining wall), integration (presence of a fully integrated appearance between the proposed sound wall and existing retaining wall), and tonal value (uniformity of color and a pleasing contrast in textures between the proposed sound wall and existing retaining wall) as desirable parameters. We can assign a 10 rating to a sound wall alternative that displays all three factors, a 9 rating to the alternative that displays any two of the three factors, and an 8 rating to the alternative that satisfies only one of the factors.

The next step is to sum up the scores for each sound wall alternative, and rank them from the highest score to the lowest score (Table 62.3). The alternative with the highest score should be selected and recommended for design and construction.

62.1.3 Design Considerations

AASHTO *Guide Specifications for Structural Design of Sound Barriers* [3] is currently the main reference for the design loads, load combinations, and design criteria for concrete, steel, and masonry sound walls.

62.1.3.1 Design Loads

The loads that should be considered in the design of the sound barriers are dead load, wind load, seismic load, earth pressure, traffic impact, and ice and snow loads.

Dead Loads

The weight of all the components making up the sound wall are to be applied at the center of gravity of each component.

Wind Loads

Wind loads are to be applied perpendicular to the wall surface and at the centroid of the exposed surface. Minimum wind pressure is to be computed by the following formula [3]:

$$P = 0.00256\ (1.3V)^2\ C_d\ C_c \quad (0.0000473\ (1.3V)^2\ C_d\ C_c) \tag{62.1}$$

where

P = wind pressure in pounds per square foot (kilopascals)

V = wind speed in miles per hour (km/h) based on 50-year mean recurrence interval

TABLE 62.3 Decision Matrix for Sound Wall Alternatives

Rating Parameters	Relative weight, %	Generic Post and Panel Concrete Sound Wall				Proprietary Concrete Panel Sound Wall			
		Alternative I (Sound wall at the top of a retaining wall)		Alternative II (Sound wall in front of a retaining wall)		Alternative I (Sound wall at the top of a retaining wall)		Alternative II (Sound wall in front of a retaining wall)	
		Rating	Score	Rating	Score	Rating	Score	Rating	Score
1. Initial construction cost	40	7.8	3.12	10.00	4.00	6.20	2.48	7.30	2.92
2. Aesthetics	15	10.0	1.50	9.00	1.35	8.00	1.20	7.00	1.05
3. Right-of-way impact	10	10.0	1.00	10.0	1.00	10.0	1.00	10.0	1.00
4. Constructibility	10	7.00	0.70	10.0	1.00	5.00	0.50	8.00	0.80
5. Drainage impact	5	9.00	0.45	9.00	0.45	9.00	0.45	9.00	0.45
6. Construction access	5	9.00	0.45	8.00	0.40	9.00	0.45	8.00	0.40
7. Utility impact	5	10.00	0.50	9.00	0.45	10.00	0.50	9.00	0.45
8. Maintenance cost	5	10.00	0.50	8.00	0.40	10.00	0.50	8.00	0.40
9. Maintenance and protection of traffic	5	10.00	0.50	8.00	0.40	10.00	0.50	8.00	0.40
Total Score			8.72		9.45		7.58		7.87
Ranking			2		1		4		3

TABLE 62.4 Coefficient C_c

Exposure Category	Height Zone[a]		
	0 < H Û 14 (4)	14 (4) < H Û 29 (9)	Over 29 (9)
Exposure Bl — Urban and suburban areas with numerous closely spaced obstructions having the size of single-family dwellings or larger that prevail in the upwind direction from the sound wall for a distance of at least 1500 ft. (450 m); for sound walls not located on structures	0.37	0.50	0.59
Exposure B2 — Urban and suburban areas with more open terrain not meeting the requirements of Exposure Bl; for sound walls not located on structures.	0.59	0.75	0.85
Exposure C — Open terrain with scattered obstructions; this category includes flat, open country and grasslands; this exposure is to be used for sound walls located on bridge structures, retaining walls, or traffic barriers	0.80	1.00	1.10

[a] Given as the distance from average level of adjacent ground surface to centroid of loaded area in ft (m).

$(1.3V)$ = gust speed, 30% increase in design wind velocity
C_d = drag coefficient (1.2 for sound barriers)
C_c = combined height, exposure, and location coefficient

The three exposure categories and related C_c values shown in Table 62.4 are to be considered for determining the wind pressure.

Seismic Loads
The following load applies to sound walls if the structures in the same area are designed for seismic loads:

$$\text{Seismic Load} = \text{EQD} = \text{A} \times \text{f} \times \text{D} \tag{62.2}$$

where
EQD = seismic dead load
D = dead load of sound wall
A = acceleration coefficient
f = dead-load coefficient (use 0.75 for dead load, except on bridges; 2.50 for dead load on bridges; 8.0 for dead load for connections of non-cast-in-place walls to bridges; 5.0 for dead loads for connections of non-cast-in-place walls to retaining walls)

The product of A and f is not to be taken as less than 0.10.

Earth Loads
Earth loads that are applied to any portion of the sound wall and its foundations should conform to AASHTO *Standard Specifications for Highway Bridges,* Section 3.20 — Earth Pressure, except that live-load surcharge is not to be combined with seismic loads.

Traffic Loads
It will not be necessary to apply traffic impact loads to sound walls unless they are combined with concrete traffic barriers. The foundation systems for those sound wall and traffic barrier combinations that are located adjacent to roadway side slopes are not to be less than that required for the traffic impact load alone.

When a sound wall and traffic barrier combination is supported on a bridge superstructure, the design of the traffic barrier attachment details are based on the group loads that apply or the traffic load as given in AASHTO *Standard Specifications for Highway Bridges,* whichever controls.

TABLE 62.5 Load Combinations

Working Stress Design (WSD) Load Groups		Allowable Over-stress as % of Unit Stress	Load Factor Design (LFD) Load Groups	
Group I:	D + E + SC	100%	Group I:	β × D + 1.7E + 1.7SC
Group II:	D + W + E + SC	133%	Group II:	β × D + 1.7E + 1.3W + 1.3I
Group III:	D + EQD + E	133%	Group III:	β × D + 1.3 E + 1.3 EQE
Group IV:	D + W + E +I	133%	Group IV:	β × D + 1.3 E + 1.3 EQD
			Group V:	β × D + 1.1 E + 1.1 (EQE + EQD)

β = 1.0 or 1.3, whichever controls the design; D = dead load; E = lateral earth pressure; SC = live-load surcharge; W = wind load; EQD = seismic dead load; EQE = seismic earth load; I = ice and snow loads

Ice and Snow Loads
Where snow drifts are encountered, their effects need to be considered.

Bridge Loads
When a sound wall is supported by a bridge superstructure, the wind or seismic load to be transferred to the superstructure and substructure of the bridge is to be as specified above under Wind Loads and Seismic Loads. Additional reinforcement may be required in traffic barriers and deck overhangs to resist the loads transferred by the sound wall.

62.1.3.2 Load Combinations

The groups in Table 62.5 represent various combinations of loads to which the sound wall structure may be subjected. Each part of the wall and its foundation is to be designed for these load groups.

62.1.3.3 Functional Requirements

The basic functional requirements for sound walls are as follows:

- To prevent vehicular impacts the sound walls should be located as far away as possible from the roadway clear zone. At locations where right-of-way is limited, a guide rail or concrete barrier curb should be utilized in front of the sound wall.
- A sound wall, especially along curved alignments, should not block the line of sight of the driver, and therefore reduce the driver's sight distance to less than the distance required for safe stopping.
- To avoid undesirable visual impacts on the aesthetic features of the surrounding area, the minimum sound wall height should not be less than the height of the right-of-way fence, and walls higher than 15 ft (4.5 m) should be avoided.
- To prevent icing on the roadway, the sound walls should not be located within a distance of less than one and a half times the height to the traveled roadway.
- To prevent saturation of the sloped embankments and avoid unstable soil conditions, transverse and longitudinal drainage facilities should be provided along the sound wall.
- To control fire or chemical spills on the highway, fire hose connections should be provided through the sound wall to the fire hydrants on the opposite side.

62.1.3.4 Maintenance Considerations

Sound walls should be placed as close as possible to the right-of-way line to avoid creating a strip of land behind the sound wall and adjacent to the right-of-way line. If this is not practical, then consideration should be given to accommodating independent maintenance and landscaping functions behind the wall. In cases where the access to the right-of-way side of the sound wall is not possible via local streets, then access through the sound wall should be provided at set intervals along the wall by using a solid door or overlapping two parallel sound walls. Parallel sound walls

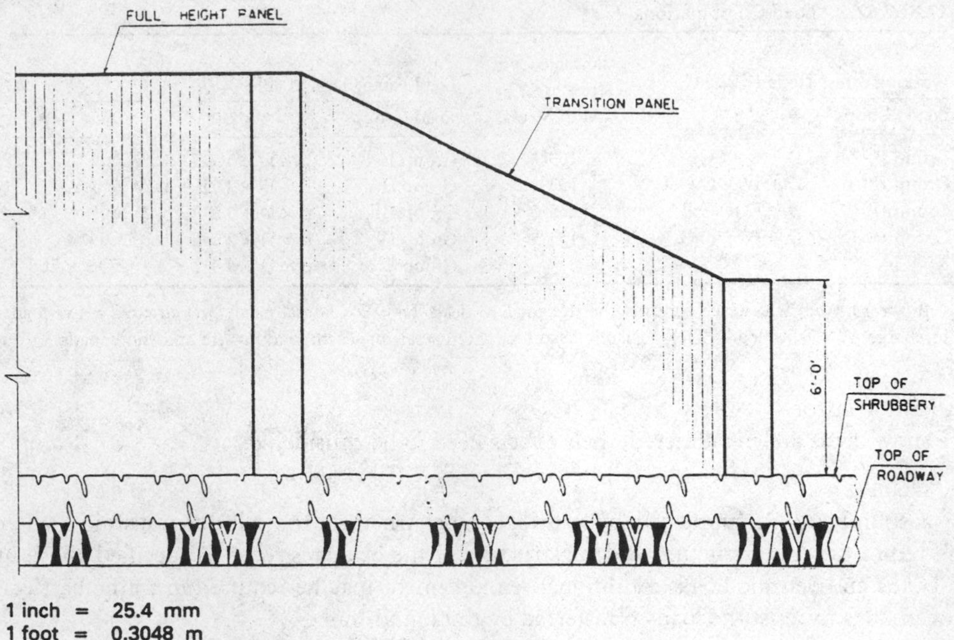

FIGURE 62.1 Typical transition at wall ends.

concealing an access opening should be overlapped a minimum of four times the offset distance in order to maintain the integrity of the noise attenuation of the main sound wall.

In urban settings, sound walls may be targets for graffiti. As a deterrence, the surface texture on the residential side of the wall should be selected rough and uneven so as to make the placement of graffiti difficult, or very smooth to facilitate easy removal of the graffiti. Sound walls with rough textures and dark colors are known to discourage graffiti.

62.1.3.5 Aesthetic Considerations

The selected sound wall alternative should address two aesthetic requirements: visual quality of the sound wall as a dynamic whole viewed from a vehicle in motion, and as a stationary form and texture as seen by the residents [4,5]. The appearance of the sound wall should avoid being monotonous to drivers; neither should it be too distractive. There are several ways of achieving a pleasing dynamic balance:

- Using discrete but balanced drops — 1 to 2 ft (0.3 to 0.6 m) — at the top of the walls to break the linear monotony.
- Implementing a gradual transition from the ground to the top of the sound wall by utilizing low-level slow-growing shrubbery in front of the wall and tapering wall panels at the ends of the wall (see Figure 62.1).
- Creating landmarks to give a sense of distance and location to the drivers. This can be achieved by utilizing distinct landscaping features with trees and plantings, or creating gateways with distinct architectural features such as planter boxes, wall niches, and terraces (Figure 62.2), or special surface finishes or textures using form liners (Figure 62.3). Gateways can also be used to delineate the limits of the individual communities along the sound wall by using a unique gateway design for each community.

As for the stationary view of the noise barrier, as seen by the residents, surface texturing and coloring are the most commonly used tools to gain acceptance by the public. By using a textured

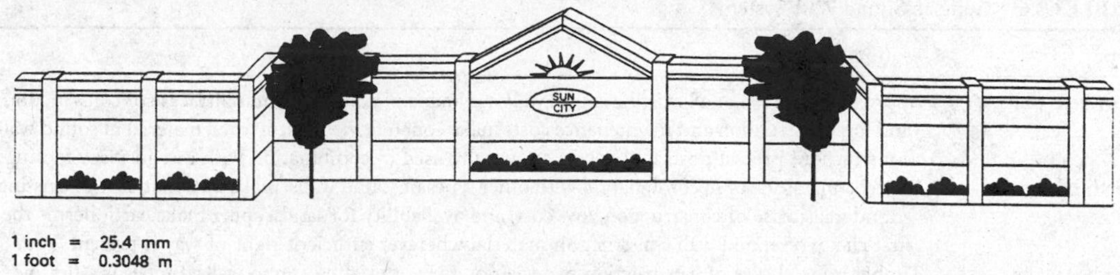

FIGURE 62.2 Gateway.

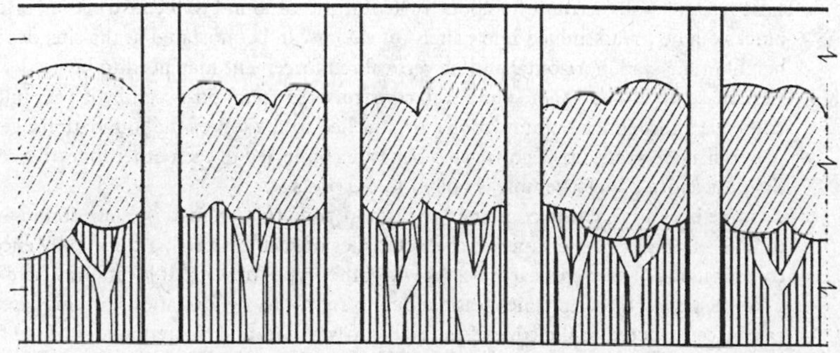

FIGURE 62.3 Architectural finish using form liners.

finish, it is possible to obtain different levels of light reflection on the wall surfaces and to evoke a sense of a third dimension. The most commonly used texturing method is raking the exposed face of the panel after concrete is placed in the formwork. Stamping a pattern in the fresh concrete surface is another texturing method. Coloring of concrete can be achieved by adding pigments to the concrete mix (internal coloring) or coating the surface of the panel with a water-based stain (external coloring). Although internal coloring would require less maintenance during the life cycle of the wall, achieving color consistency among panels is extremely difficult due to variations in the cement color and pigment dispersion rates. Staining offers uniformity in color. However, restaining of panels may be required after 10 to 15 years of service.

62.1.4 Ground-Mounted Sound Walls

62.1.4.1 Generic Sound Wall Systems

Generic sound wall systems (Table 62.6) make use of common construction materials such as earth, concrete, brick, masonry blocks, metal, and wood. With the exception of earth, all other materials are usually fabricated into post and panel systems in the shop, and installed on precast or cast-in-place concrete foundations at the site.

62.1.4.2 Proprietary Sound Wall Systems

In the late 1970s and early 1980s the regulatory actions by the Congress, Environmental Protection Agency (EPA), and Federal Highway Administration (FHWA) effectively launched a new industry. Ever since, a number of proprietary sound wall systems (Table 62.7) have been introduced and have become successful.

62.1.4.3 Foundation Types

The following foundation types are commonly used for sound walls:

TABLE 62.6 Generic Sound Wall Systems

Type	Features
Concrete walls	Approximately 45% of all existing sound walls are made of concrete; durability, ease of construction, and low construction and maintenance costs make concrete the most favored material in sound wall construction; precast posts and panels are usually used in combination with cast-in-place footings
Earth berms	Earth berms, alone or in combination with other types of sound walls, make up about 25% of existing sound walls; ease of construction, low cost, and availability for landscaping make earth berms the first choice of sound wall construction material wherever sufficient right-of-way is present
Timber walls	Timber is the choice of construction material for 15% of existing sound walls; timber is a flexible construction material, and can be used in a variety of ways in sound wall construction; timber posts can be solid sawn, glue laminated, or round pole type; the posts can be driven or embedded in concrete footings, and timber planks or plywood panels can be nailed or bolted to the posts at the site
Brick and concrete masonry block walls	These types of walls account for about 10% of the total sound wall construction; brick and masonry blocks can be preassembled into panels off site or can be mortared at the site; depending on the height of the wall, horizontal and/or vertical reinforcement may need to be used
Metal walls	These make up about 5% of sound wall construction; metal posts can be driven into the ground, embedded into concrete foundations, or attached to the top of the foundations; generally, metal panels that are made up of corrugated pans are connected to each other to form a solid surface, and then the entire panel assembly is bolted to the posts
Combination walls	It is sometimes advisable to combine two or more construction materials in a sound wall construction to take advantage of the superior characteristics of each material; a commonly encountered combination is the use of earth berms with other types of walls at locations where construction of a full-height earth berm is not feasible due to right-of-way limitations; in this case, an earth berm can be constructed to the edge of the right-of-way line, and the remainder of the height required for a full-level sound abatement can be provided by using a concrete, timber, or metal wall; another most commonly practiced combination in the field is the use of steel posts with concrete, timber, or composite wall panels or planks; the inherent advantage in this combination is the ability to make quick-bolted or welded connections between the steel posts and foundations.

Pile Foundations

Timber, steel, and concrete piles can be driven into the ground to act as a foundation, and also as a post for sound walls. One shortcoming of this foundation system is the problem of controlling the plumbness and location of driven posts. Also, damage to the pile end may often require trimming and/or repairs. On the other hand, the advantages of pile foundations are the ease and economy of installation into almost any kind of soil except rock, and completion of foundation and post installation in a single-step process.

Caisson (Bored) Foundations

Caissons are the most frequently used type of foundation in sound wall construction. It involves excavating a round hole using augering equipment (use of a metal casing may be required to prevent the collapse of the hole walls), installing a reinforcement cage, inserting the post or, alternatively, installing anchor bolt assemblies for the post connection, and finally placement of concrete. The advantages of caisson foundations are the ease of installation into any kind of soil (except soils containing large boulders), the convenience of performing the construction in tight spaces with a minor amount of disturbance to the surrounding environment, and ability to locate posts accurately in the horizontal plane and plumb in the vertical plane. The disadvantages are the high possibility of water intrusion into the excavated holes in areas with a high water table, interference of boulders with the augering process, and the presence of concurrent construction activities, such as augering, placement of reinforcement, insertion of post, and placement of concrete. In a caisson construction, all these tasks should be carefully orchestrated to achieve a cost-effective and expedited operation.

Spread Footings

Spread footings are the ideal choice where the construction site is suitable for a continuous trench-type excavation using heavy excavation equipment. Once the trench excavation is completed to the bottom of the proposed foundation, cast-in-place footings can be constructed on the ground, or

TABLE 62.7 Proprietary Sound Wall Systems

Type	Features
Siera Wall	This sound wall system consists of precast wall panels and cast-in-place foundations; each panel is precast integrally with a pilaster post along one edge, and attached to the adjacent panels with a tongue-and-groove connection; the connection between the wall panels and foundation is secured by welding the steel plates embedded at the base of the posts to the steel plates mounted on top of the foundations
Port-O-Wall	This is another sound wall system which utilizes precast panels and cast-in-place foundations; each precast concrete wall panel is secured to the adjacent panel with a tongue-and-groove connection; the panels are also mechanically connected by horizontal tie bars at the top and bottom; a rectangular hole at the bottom of each panel allows the installation of the transverse reinforcement for the construction of a continuous cast-in-place concrete footing
Fan Wall	This precast concrete wall system is a castellated freestanding wall that does not require a concrete footing; a rotatable and interlocking joint system allows the joining of panels at any angle; the joint along the sides of each panel features mating concave and convex edges and stainless steel aircraft-type cable connector assemblies
Sound Zero	This lightweight (8 lb/sf) (380 Pa) panel system is fabricated to the full wall height, and installed on top of cast-in-place footings or bridge parapets using concrete or steel posts
Carsonite© Sound Barrier	This panel system features tongue-and-groove modular sections made from a fiberglass-reinforced polymer composite shell that is filled with ground, recycled tires; the lightweight (7.5 lb/sf) (360 Pa) panels are preassembled off site and installed between posts anchored into cast-in-place footings, traffic barriers, or bridge parapets
Contech Noise Walls	This wall system consists of hot-rolled steel posts and cold-formed interlocking steel panels; all wall components are galvanized, and the panels are additionally protected by a choice of colored coating systems
Evergreen Noise Abatement Walls	This wall system consists of precast concrete units, and is supported on individual or continuous cast-in-place concrete foundations; the precast units are stacked up to form a freestanding wall; a select granular material is placed in each wall unit and compacted prior to the installation of the next higher unit; the trays of the wall units are filled with topsoil to support the planting and growth of evergreen and deciduous plants
Maccaferri Gabion Sound Walls	This sound wall system utilizes stacked-up gabion baskets to form a freestanding sound wall; zinc or PVC-coated wire baskets are filled with rock to blend with the natural environment; the cores of the baskets can also be filled with topsoil to allow planting of vegetation

precast footings can be erected. If precast footings are used, roughening of the bottom of footing in combination with placement of a crushed stone layer with a thin cement grout topping under the footing may be necessary to achieve a desired safety factor against sliding.

Tie-Down Foundations

At locations where rock is at or close to the surface, augering for caissons or excavating for spread footings may be too costly and time-consuming. A more practical solution in this case is the use of concrete pedestals anchored into rock with post-tensioned tie-downs. A pedestal detail similar to the stem of the spread footing can be constructed to allow insertion of the post into a recess in the stem, or installation of anchorage assemblies for the connection of the post to the top of the stem.

62.1.4.4 Typical Details

The sketches shown in this chapter depict a concrete sound wall system consisting of precast posts and panels supported on spread footings [6].

Precast panels (Figure 62.4) are 5 in. (130 mm) thick and reinforced with wire mesh fabric to sustain the wind loading. Additional mild reinforcement is placed along the periphery of the panels to carry the vertical loads. Two cast-in-place inserts are provided at the top of the panels for handling and erection. The panels are 4 ft (1.2 m) high and span between the posts spaced at 15 ft (4.6 m) on centers. Overlapping joints between the panels provide a leakproof surface. The roadway side of the panels is finished smooth and the residential side of the panel displays a raked finish to deter graffiti.

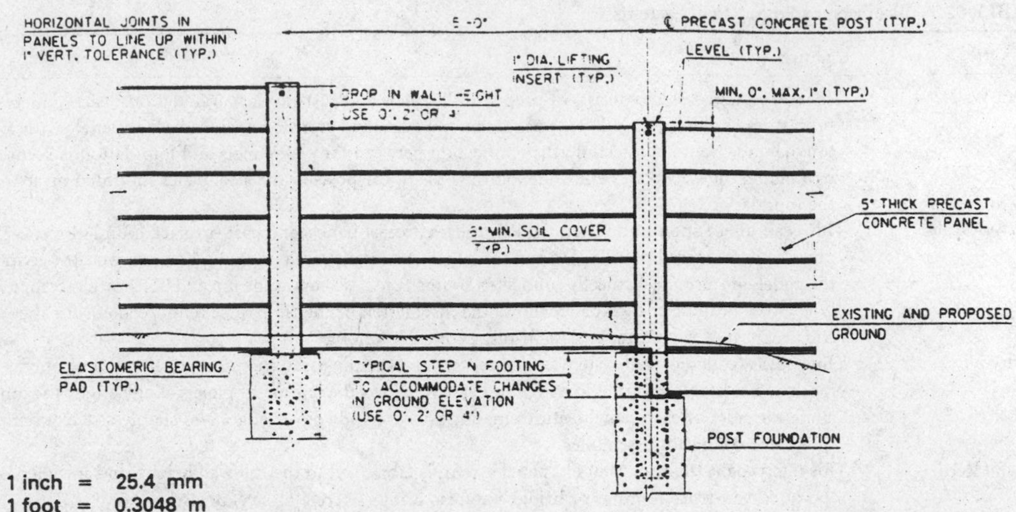

FIGURE 62.4 Precast concrete post and panel sound walls. (*Source*: Guzaltan, F., *PCI J.*, 27(4), 60, 1992. With permission.)

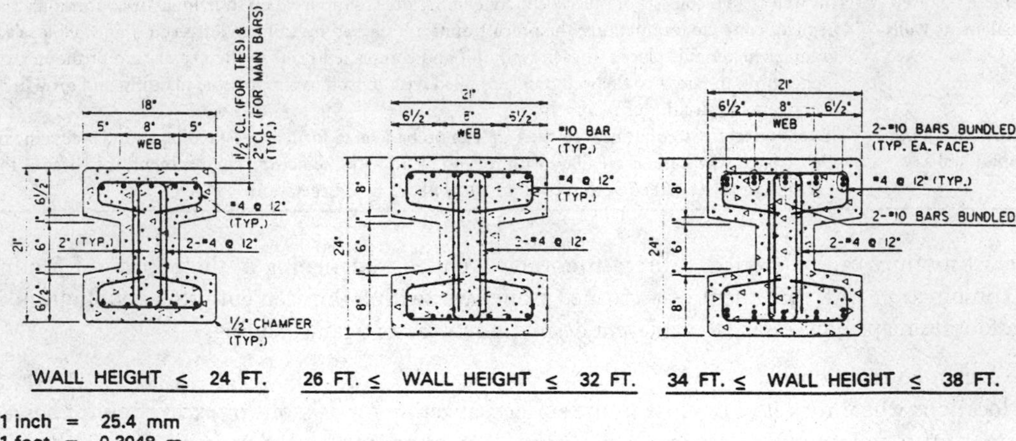

FIGURE 62.5 Concrete precast post details. (*Source*: Guzaltan, F., *PCI J.*, 27(4), 61, 1992. With permission.)

The precast concrete posts (Figure 62.5) are H-shaped to accept the insertion of panels and allow approximately a 10° angle change in the orientation of panels. Both flanges are reinforced with mild steel reinforcement to carry the wind loads that are transmitted from the panels. For posts longer than 34 ft (10 m), the use of prestressing strands may need to be considered to prevent reinforcement congestion in the flanges. Shear reinforcement wrapping the outline of the web and flanges is also provided.

Four types of foundations for use in different soil and site conditions are featured: caisson foundations (Figure 62.6) for sites without boulders; cast-in-place and precast spread footings (Figure 62.6) at sites where large boulders are frequent; and tie-down foundations (Figure 62.7) at locations where rock is close to the surface.

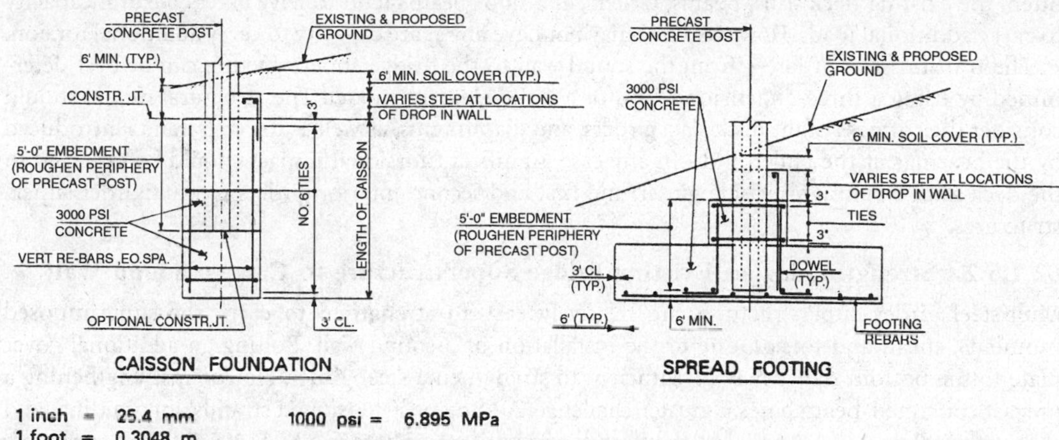

FIGURE 62.6 Caisson foundation and spread footing. (*Source*: Guzaltan, F., *PCI J.*, 27(4), 61, 1992. With permission.)

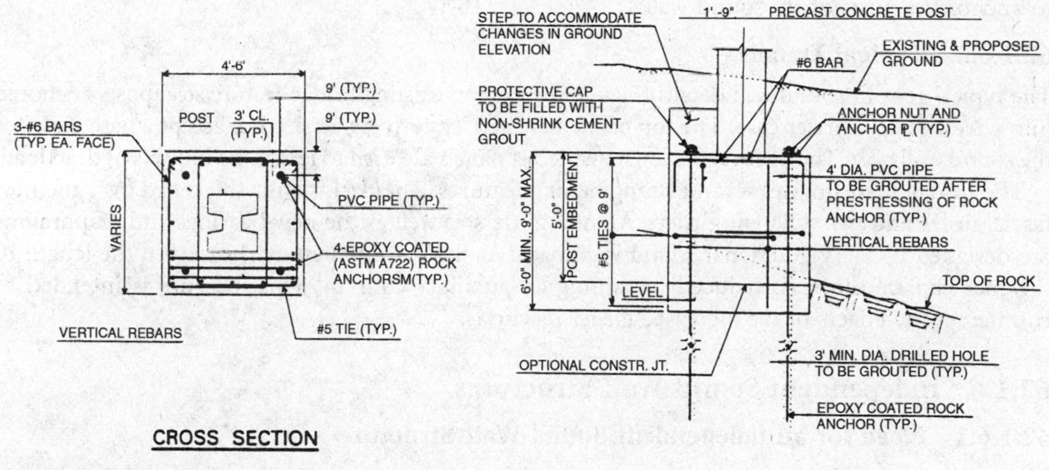

FIGURE 62.7 Tie-down foundation. (*Source*: Guzaltan, F., *PCI J.*, 27(4), 62, 1992. With permission.)

62.1.5 Bridge-Mounted Sound Walls

62.1.5.1 Assessment of an Existing Bridge to Carry a Sound Wall

The capacity of the existing superstructure components (barrier curb, deck slab, girders, and diaphragms) should be checked prior to deciding to attach sound barriers to an existing bridge. Furthermore, the capacity of the existing bearings and piers may also be investigated due to increases in the girder reactions. The main forces to be considered are the dead load of the sound wall, wind and ice loads on the panels and posts, and torsion created by the eccentricity of these forces. Quite often, the existing deck slabs, beams, girders, and floor beams tend to have excess flexural capacity to carry additional loads. However, they may not have any spare capacity to carry additional torsion.

The transmission of forces from the sound wall to the bridge superstructure can be best determined by using a three-dimensional grid or a finite-element model. The structural model should consider the stiffness of the deck slab, girders, and diaphragms, as well as the constraints introduced by the bearings at the girder ends. In any case, significant torsional impact should be expected in the deck slab, diaphragms, fascia girder, and first and second interior girders in multigirder superstructures.

62.1.5.2 Strengthening an Existing Bridge Superstructure to Carry a Sound Wall

Multisteel girder superstructures are relatively easy to strengthen to carry the superimposed moments, shear, and torsion due to the installation of a sound wall. Bolting an additional cover plate to the bottom flange may be sufficient to strengthen a steel girder. However, strengthening a precast concrete I-beam poses a greater challenge. Adding post-tensioned strands, longitudinal and shear reinforcement, and encasing the strands and reinforcing bars in a high-strength concrete mass may be a feasible way of strengthening an existing I-beam. Nevertheless, the time and cost involved in strengthening concrete I-beams should always be compared with the cost of replacing the superstructure before a strengthening alternative is seriously pursued.

Even if the deck slab and girders of an existing bridge are found to be adequate or amenable to strengthening, there may be still a need to replace at least the bridge parapets in order to be able to anchor the posts of the sound wall.

62.1.5.3 Typical Details

The typical retrofit sound wall details (Figure 62.8) on an existing bridge feature steel posts anchored into a New Jersey barrier curb. The top of the new barrier curb is oversized to accept a base plate for the sound wall post. The lightweight concrete precast panels are used to lessen the impacts of dead load.

The bridge incorporates several strengthening features. The deck slab is thickened over the new fascia girder and two adjacent girders. All new girders as well as the new bearings and diaphragms are designed to carry additional sound wall loads. At the existing hammerhead pier, the length of the pier cap cantilever is reduced by adding an auxiliary column. This measure is intended to counteract the effects of the increased girder reactions.

62.1.6 Independent Sound Wall Structures

62.1.6.1 Need for an Independent Sound Wall Structure

When it is not feasible to strengthen or replace the components of an existing bridge to carry a proposed sound wall, the alternative is to provide a freestanding structure to support the sound wall along an existing bridge.

In the design of an independent sound wall structure the following guidelines may be followed:

- There should be no noise leakage between the bridge and the independent structure. A closure device should be provided between two structures to block the noise leakage and, at the same time, permit thermal movements and differential settlement of structures without adversely affecting each other.
- In order to maintain consistency in the appearance of the sound wall, the panels on the independent structure should match the panels of the adjacent ground-mounted sound wall at least in texture and appearance.
- The independent structure should mimic the appearance of the existing bridge as much as possible and follow the span arrangement of the existing bridge. The substructure components of the independent structure should also utilize the same material and architectural treatment used on the adjacent bridge.

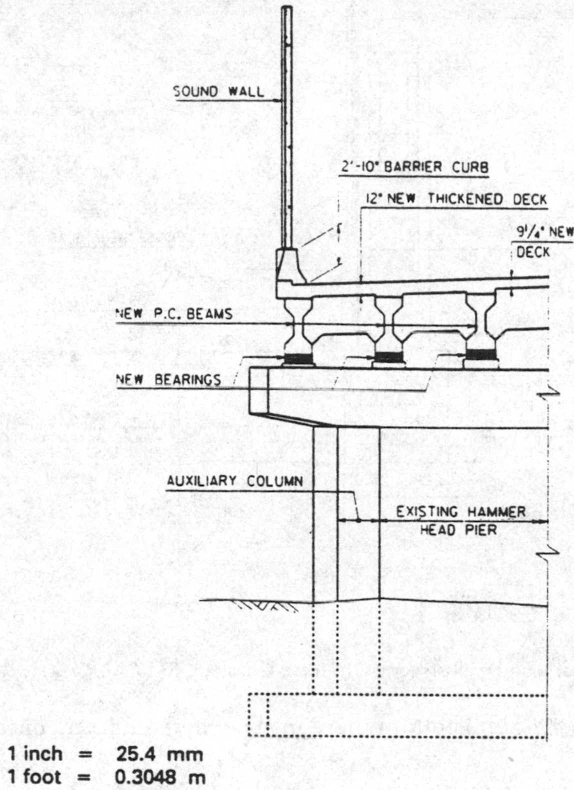

FIGURE 62.8 Bridge-mounted sound wall.

- The superstructure of the independent structure that carries the sound panels should match the horizontal lines of the bridge and be capable of carrying the dead load of the panels and the wind load acting on the panels.
- Effects of fatigue, due to the reversal of wind direction, should be considered in the design of the structural steel components of the independent structure as well as their connections.
- The independent sound barrier structure should not infringe on the lateral and vertical clearance envelope of the existing bridge.
- Since an independent sound wall structure usually carries small vertical loads (dead and ice loads) and highly eccentric transverse loads (wind loads), it should either be founded on piles with high tensile capacity or be connected with dowels to the abutments and piers of the existing bridge, providing that these bridge components display adequate capacity.

62.1.6.2 Typical Details

The following are the typical features of an independent sound wall structure (Figure 62.9) [6]:

- Cast-in-place reinforced concrete footings and pedestals that are cast against and doweled into the existing bridge abutments and piers in order to prevent the rocking of the independent sound wall structure in the transverse direction under large wind loads.
- A steel tower made of a pair of structural steel tubes is supported on the concrete pedestal and carries a twin Vierendeel truss.
- A Vierendeel truss is made of structural steel tubing, and consists of continuous top and bottom chords and groove welded struts spanning between the chords. A twin truss is formed

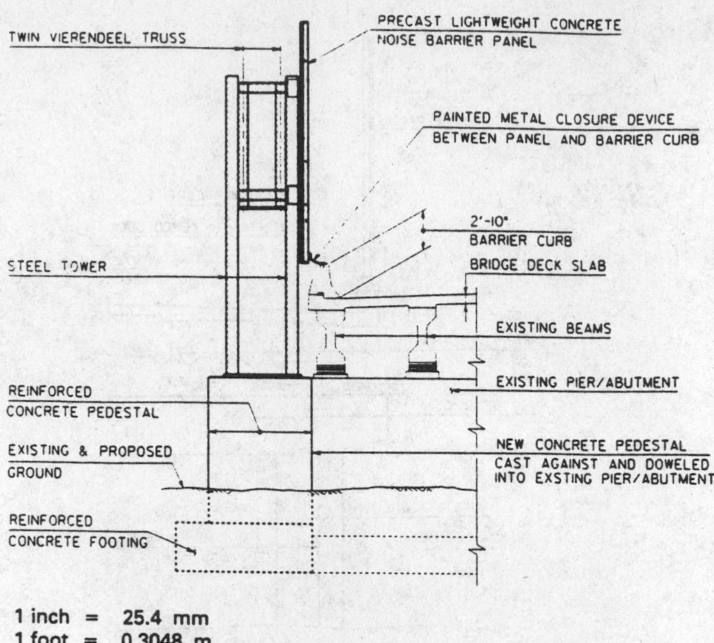

FIGURE 62.9 Independent sound wall structure. (*Source*: Guzaltan, F., *PCI J.*, 27(4), 62, 1992. With permission.)

by connecting each top and bottom chord member to the adjacent ones using groove-welded horizontal struts.

- The sound wall consists of 5-in. (130-mm)-thick lightweight reinforced concrete panels bolted to vertical posts (W shaped) spaced at 5-ft (1.5-m) intervals. The vertical posts are bolted to two spacers (W shaped) at the level of the top and bottom chords. These spacers are in turn shop-welded to the vertical struts of the truss.
- The gap between the wall panels and bridge parapet is closed using an L-shaped bent steel plate. The horizontal leg of the steel plate is bolted to the top of the parapet. The vertical leg is positioned upward and overlaps the wall panel approximately 10 in. (250 mm). A 2-in. (50-mm) gap is provided between the wall panels and upright leg of the bent plate to allow the independent movement of the bridge and sound wall.

62.2 Bridge Railings

62.2.1 Introduction

Railings are provided along the edges of structures in order to protect pedestrians, bicyclists, and vehicular traffic. Depending on the function they are designed to serve, bridge railings are classified as: pedestrian railing, bicycle railing, traffic railing, and combination railing. For each category, the AASHTO *Standard Specifications for Highway Bridges*, 16th ed. [7], has defined specific geometric requirements and the loads to be applied at various elements of the railings. An alternative approach for determining the geometries and the loads of unique or new railings is presented in the AASHTO *Guide Specifications for Bridge Railings* [8].

In the United States most states have their own standards for bridge railing geometry and design criteria. These standards generally follow the AASHTO Standard or Guide Specifications. In instances where a special design must be provided for a new railing or an existing railing must be upgraded, the following discussions are presented to help in understanding the AASHTO Standard and Guide Specifications and their differences.

By necessity, throughout the following section there are statements that paraphrase the contents of these references. Similarly, the figures and tables are those of AASHTO presented in a slightly different format to better serve the text. This treatise is not intended to substitute for the AASHTO Standard and Guide Specifications; rather it is meant to facilitate the AASHTO codes and their intention, as the author sees them.

As related to traffic railings and combination traffic and pedestrian railings, AASHTO Standard Specifications specifies only one set of loads applicable to all classes of vehicles and all types of roadways and traffic; it does not recognize variations in the vehicle type, percent of truck usage, design speeds, average daily traffic (ADT), etc. These variables are addressed in the AASHTO Guide Specifications which establishes guidelines for crash testing and evaluation of bridge railings based on three performance levels. Each performance level is based on the ADT of the roadway for which the railing is being considered. But other variables, such as design speed, percent truck use, rail offset to travel lane, and the type of highway (divided, undivided, and one-way), are also considered in the selection of the performance levels.

For each performance level the Guide Specifications establishes a certain crash test procedure and evaluation criteria. The testing procedure includes the type, weight, size, and geometry of the test vehicle as well as its speed and impact angle. Based on the crash test results, the railing is evaluated in accordance with the crash test evaluation criteria, which establishes the pertinent performance level and the type of traffic railing to suit the functional needs of a site.

Recognizing that crash tests are expensive and time-consuming, the Guide Specifications provide a table (Table 62.8), based on the results of actual crash tests conducted by the National Cooperative Highway Research Program (NCHRP). In this table design loads and traffic rail geometries are given for the three standard performance levels and two optional levels. These optional levels relate to heavier trucks and higher railings. The loads and dimensions provided in this table are applicable to four conceptual traffic rails that are representative of traffic rails or combination rails in the United States.

Table 62.8 is generally used to design the prototype railings that are to be crash tested. It is also used for the design of one-of-a-kind railings where the cost of crash testing cannot be justified. The highway agencies are encouraged to conduct crash test programs when specific traffic/combination railing is being considered for the first time. Railings that meet the crash test criteria for a desired performance level are exempt from the requirements of AASHTO Standard Specifications.

For bicycle and pedestrian railings the AASHTO Standard Specifications and Guide Specifications provide similar design requirements with minor differences in geometry. It is important to note that where pedestrian or bicycle traffic is expected, this traffic must be separated from travel lanes by a traffic railing or barrier. The height of this railing above the sidewalk or bikeway surface should be no less than 24 in. (610 mm). Also the face of the railing should have a smooth surface to prevent any snag potential. Where it is desirable to raise the height of the traffic barrier to prevent the bicycles from falling over the railing onto the roadway, or to improve the level of comfort, a traffic railing or a modified combination railing may be used.

Where a raised sidewalk curb with a width greater than 3.5 ft (1.067 m) is provided, the Guide Specifications would consider it acceptable to use only a crash-tested combination railing along the edge of the bridge. The curb must be included in the crash test that is used for determining the combination railing design. The Standard Specifications, however, makes no specific reference to raised sidewalks. Where the roadway curb projects more than 9 in. (229 mm) from traffic face of railing (construed to be a safetywalk or a sidewalk) the Standard Specifications allow the use of a combination traffic and pedestrian railing along the edge of the structure. On urban expressways where the curb projects less than 9 in. (229 mm), the Standard Specifications call for a combination railing to separate pedestrian walkways from the adjacent roadway. On rural expressways it calls for a traffic railing or barrier to separate pedestrians from vehicular traffic. A pedestrian railing must be provided along the edge of the structure where the pedestrian walkway is separated from the roadway.

TABLE 62.8 Bridge Railing Design Information — Bridge Railing Loads, Load Distribution, and Location

Quantity Designations	Railing Performance Level				
	PL-1	PL-2	PL-3	Optional PL-4	Optional PL-4T
			Group I[a] Loads (Body and Wheels)		
F_{BWH}	30 (133) kips	80 (356) kips	140 (623) kips	200 (890) kips	200 (890) kips
F_{BWL}	±9 (±40) kips	±24 (±107) kips	±42 (±187) kips	±60 (267) kips	±60 (±267) kips
F_{BWV}	±12 (±53) kips (down) –4 (–18) kips (up)	15 (67) kips (down) –5 (–22) kips (up)	+18 (80) kips (down) –6 (–27) kips (up)	+18 (80) kips (down) –6 (–27) kips (up)	+18 (80) kips (down) –6 (–27) kips (up)
			Group II[a] Loads (Trailer Floor)		
F_{FH}	—	—	—	240 (890) kips	200 (890) kips
F_{FL}	—	—	—	±60 (267) kips	±50 (±222) kips
F_{FV}	—	—	—	–18 (80) kips (down) – 6 (–27) kips (up)	+18 (80) kips (down) –6 (–27) kips (up)
			Group III[a] Loads (Tank Trailer)		
F_{TH}	—	—	—	—	200 (890) kips
F_{TL}	—	—	—	—	±50 (±222) kips
F_{TV}	—	—	—	—	+18 (80) kips (down) –6 (–27) kips (up)

Load Distribution Pattern Dimensions					
a	24 in. (610)	28 in. (711)	32 in. (813)	36 in. (914)	36 in. (914)
b	12 in. (305)	14 in. (356)	16 in. (406)	18 in. (457)	18 in. (457)
c	—	—	—	12 in. (305)	12 in. (305)
d	—	—	—	6 in. (152)	6 in. (152)
e	—	—	—	—	36 in. (914)
f	—	—	—	—	8 in. (203)
Load Locations					
h_{BW}	16 in. thru (H-6 in.) [(406) thru (H-152)]	17 in. thru (H-7 in.) [(432) thru (H-178)]	18 in. thru (H-8 in.) [(457) thru (H-203)]	19 in. thru (H_{BW}-9 in.) [(483) thru (H_{BW}-229)]	19 in. thru (H_{BW}-9 in.) [(483) thru (H_{BW}-229)]
h_F	—	—	—	51 in. (1295)	51 in. (1295)
h_T	—	—	—	—	74 in. (min) 84 in. (max) (1.880 m) (2.134 m)
Railing Geometry Dimensions					
H	27 in. (686) (min)	32 in. (813) (min)	42 in. (1067) (min)	54 in. (1372) (min)	78 in. (1981) (min)
H_A	10 in. (254) (max)	10 in. (254) (max)	10 in. (1254) (max)	10 in. (254) (max)	10 in. (254) (max)
H_{BW}	27 in. (686) (min)	32 in. (813) (min)	42 in. (1067) (min)	32 in. to 42 in. (813 to 1067)	32 in. to 42 in. (813 to 1067)
H_{BWR}	12 in. (305) (min)	12 in. (305) (min)	12 in. (305) (min)	12 in. (305) (min)	12 in. (305) (min)
H_F	—	—	—	54 in. (1372) (min)	54 in. (1372) (min)
H_{FR}	—	—	—	6 in. (152) (min)	6 in. (152) (min)
H_T	—	—	—	—	78 in. (1981) (min)
H_{TR}	—	—	—	—	8 in. (152) (min)

Note: PL = Performance Level.Where kips are indicated values in () indicate metric equivalent in kilonewtons; where inches are indicated values in () indicate metric equivalent in millimeters.

[a] Each set of Group Loads to be applied separately.

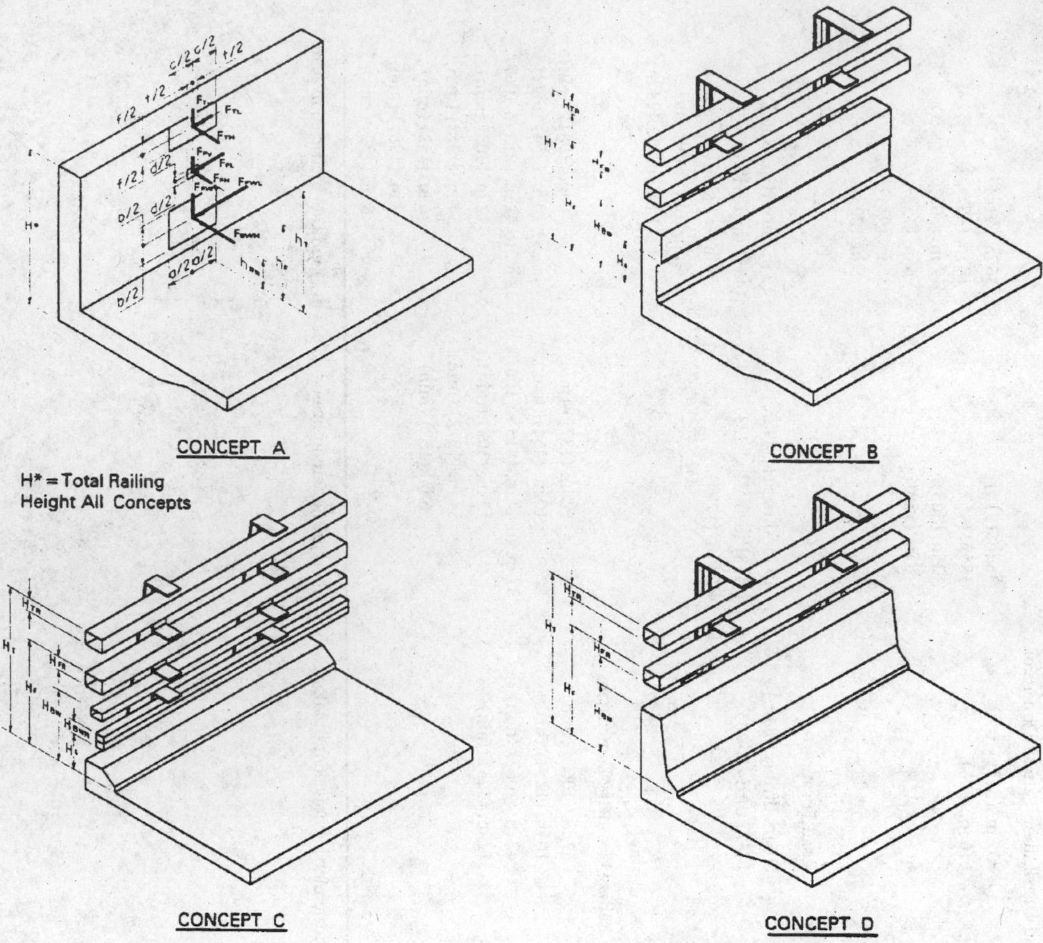

FIGURE 62.10 Bridge railing concepts — configuration and loading patterns. (*Source*: AASHTO, *Guide Specifications for Bridge Railings,* American Association of State Highway and Transportation Officials, Washington, D.C., 1989. With permission.)

62.2.2 Vehicular Railing

62.2.2.1 Geometry

Bridge railings are primarily provided to contain the vehicles using the bridge, but they are also required to (1) protect the occupants of an errant vehicle in a collision with the railing; (2) protect other vehicles near the collision; (3) protect the people and properties on the roadways or other areas underneath the structure; and (4) have the appearance and freedom of view from passing vehicles.

Figure 62.10 shows four conceptual railing configurations identified in the Guide Specifications. The railing dimensions and the magnitude of design loads to be applied at the points of load application are given in Table 62.8. All three standard performance levels and two optional performance levels are represented in Figure 62.10. Railings for each performance level can be constructed from this figure by assuming that railing elements for which no dimension is given in Table 62.8 do not exist.

The dimensions and configurations shown for these railings are designed to provide a smooth and continuous surface for the traffic side of the railings. The geometry of the rail system should

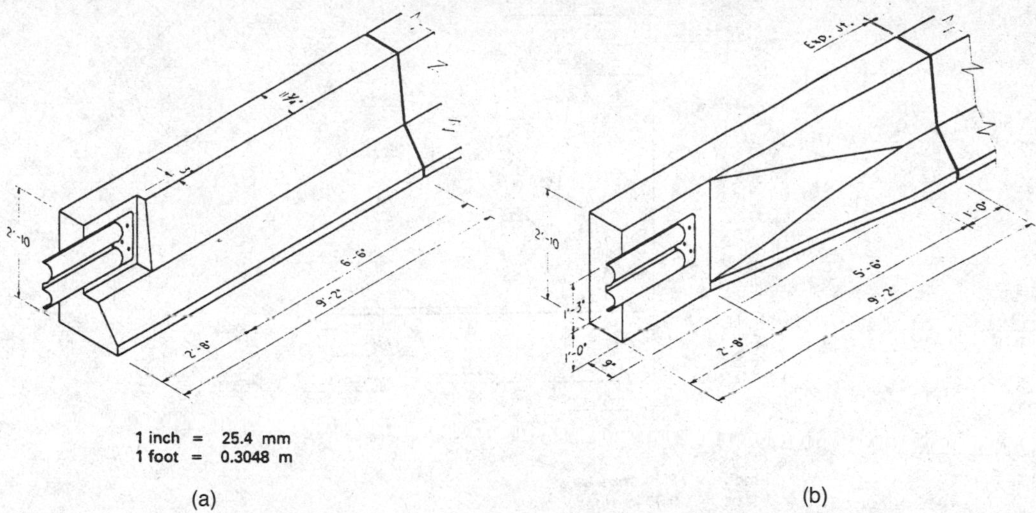

FIGURE 62.11 Guide rail attachment at end of bridge.

be such as to preclude any potential contact with the posts by major vehicle parts, should there be a penetration or opening through the railing. The Guide Specifications requires a minimum of 10 in. (254 mm) between the face of the railing and the face of the posts, where snagging is an obvious possibility.

At bridge ends where an open-face railing (guide rail) meets the bridge parapet or barrier, a transition is normally required to provide a smooth flow of traffic while eliminating any snag potential. The Guide Specifications requires that the close face railing (a parapet or a barrier) be flared a maximum of 3.5 longitudinal to one lateral, starting a minimum of 10 in. (254 mm) back of the open-face railing (a guide rail).

Figure 62.11 represents the standard details used in New Jersey for the attachment of guide rails to concrete barriers. These details can be applied at the ends of a parallel wingwall, the ends of a pylon created to provide the transition from guide rail to barrier, or at the ends of a bridge parapet where the thermal movements of the bridge can be accommodated by slotted holes in the guide rail, i.e., small movements.

Where posts are used in a bridge traffic railing, post spacing should not exceed 10 ft 0 in. (3.05 m).

The height of traffic railing is measured from the top of the highest rail or parapet to the top of the roadway, the top of the future overlay, or the top of a raised sidewalk. A raised sidewalk can be defined as a raised roadway curb located along the edge of the bridge, sufficiently wide to accommodate passage of two pedestrians shoulder to shoulder. While 4 ft (1.219 m) is the nominal minimum width, the Guide Specifications allows an absolute minimum width of 3 ft 6 in. (1.067 m).

In the past, safetywalks were used as a means of providing access for infrequent pedestrians or maintenance personnel along a long bridge where a sidewalk could not be justified. These safetywalks, varying in width from 1 ft 6 in. to 3 ft (457 mm to 914 mm), are no longer used since they do not provide sufficient protection for the pedestrians and traffic. Raised curbs supporting a railing, however, are still in use. These curbs generally have a horizontal projection of 9 in. (229 mm) or less from the railing face.

The minimum height of traffic railings or the traffic portion of combination railings is 2 ft 3 in. (686 mm). Where a parapet has a sloping face, intended to allow redirecting of vehicles back to the roadway, the minimum height allowed by the Standard Specifications is 2 ft 8 in. (813 mm). The Guide Specifications, however, lists a 2 ft 3-in. (686-mm) minimum dimension for the same height for Performance Level 1 (see Table 62.8).

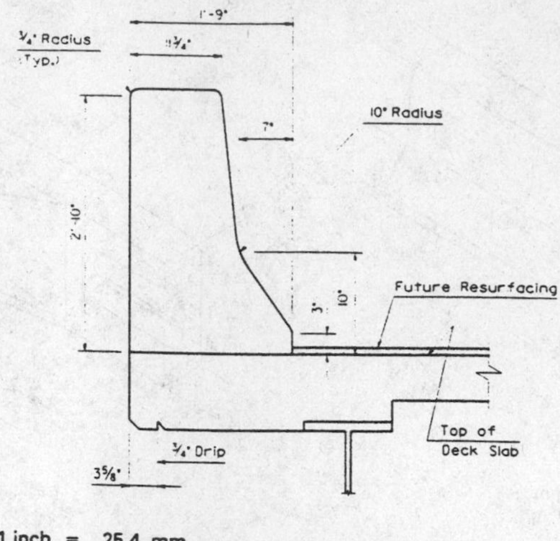

FIGURE 62.12 Standard New Jersey barrier bridge parapet to be used where there is no walkway on the bridge.

PL = Performance Level.Where kips are indicated values in () indicate metric equivalent in kilonewtons; where inches are indicated values in () indicate metric equivalent in millimeters.

The minimum height of a traffic barrier in some states exceeds the AASHTO requirements. For instance, in New Jersey, where the barrier is along the edge of the bridge, the minimum height is 2 ft 10 in. (864 mm), as shown in Figure 62.12. Where the barrier is used to separate pedestrians from highway traffic, the minimum height is 2 ft 8 in. (813 mm).

In a combination railing where the lower element is a parapet, or in a traffic railing, the height of the lower element should be no less than 1 ft 6 in. (457 mm). If a rail is the lower element, its height, measured from its center to reference surface, should be between 15 and 20 in. (381 and 508 mm).

The Guide Specifications requires that the clear distance between the bottom rail and the reference surface be no less than 10 in. (254 mm). The maximum clearance for the same dimension is given as 17 in. (432 mm) by the Standard Specifications. The maximum distance between adjacent rails should not exceed 15 in. (381 mm). Additionally, the traffic face of all rails should be within 1 in. (25 mm) of a vertical plane through the traffic face of the rail closest to traffic.

Thermal movements of the rails are normally provided for by the use of joints in sleeves in the rails. For short-length bridges these joints can be located anywhere along the bridge length. But for long-span bridges, where thermal movements are expected to exceed 2 to 3 in. (51 to 76 mm), it is prudent to place the splices at the expansion joint locations. To eliminate snag potential at these joints, sleeves for pipes (or steel hoods for concrete parapets) must be provided. The projection or depression of the rails at rail joints or steel hoods should not have a depth (thickness) greater than the rail wall thickness or ⅜ in. (10 mm), whichever is less.

62.2.2.3 Loads

Method 1 — Guide Specifications

The Guide Specifications provides criteria for the selection of performance levels of various sites based on the ADT, design speed, percent truck use, bridge rail offset from traffic, and the type of highway under consideration. Upon the selection of the performance level, Table 62.8 can be used to determine the magnitude of loads to be applied to the railing at locations shown in Figure 62.11. While the performance levels PL-1, PL-2, and PL-3 represent small automobiles, pickup trucks, medium-size single unit trucks, and van-type tractor-trailers, the optional performance levels PL-4 and PL-4T are used where heavier and larger trucks at higher volumes are likely to use the roadway. The optional level PL-4 is given for 54 in. (1.372 m) high and higher railings, and where truck volumes and truck type, size, and weight would be greater than PL-3. The optimal level, PL-4T, represents railings that have a minimum height of 6 ft 6 in. (1.981 m), and where truck volumes and highway alignment and use would justify such high railings, e.g., closed face barrier curbs over electrified tracks in high-speed, high-volume curved alignments.

The loads to be applied at the lowest level of the railing, F_{BWH}, F_{BWL}, and F_{BWW}, represent the impact loads of the body and wheels of an errant vehicle; they are given under Group 1 Loads for all five performance levels. The loads to be applied at the midrail of the railing, F_{FH}, F_{FL}, and F_{FV}, represent the impact loads of a trailer floor; they are given under Group II for optional performance levels PL-4 and PL-4T. The loads to be applied at the top level of performance level PL-T4, representing the potential impact of a tank trailer, are given under Group III Loading. While all three loads in each group should be applied simultaneously, and distributed over the designated area, only one group of loading should be applied at a time. All three loads in each group should be distributed evenly over the loaded areas. Dimensions of the load areas over which the forces for each loading category must be evenly distributed are given in Table 62.8.

Where the railing width is less than the related load area, the entire load should be distributed over the available width. Where a load area bears on more than one rail, the load applied to each rail should be prorated based on the distance from each rail to the reference surface. It is to be noted that Load F_{BWH} must be applied over a range of heights as shown in Table 62.8.

The loads on posts are transmitted through the longitudinal rail elements. These loads are to be distributed to no more than three posts.

As stated previously, the loading criteria outlined above are to be used for the design of prototype railings that are to be crash tested and for the design of one-of-a-kind railings where the cost of crash testing cannot be justified. Otherwise, the best way to ensure suitability of a particular railing for a given site is to subject it to applicable crash tests and evaluate the results in accordance with the performance criteria for a desired performance level. The crash test procedures are described in the NCHRP Report 350, Recommended Procedures for the Safety Performance Evaluation of Highway Features [9]. The criteria for evaluation of crash test results and the procedure for the selection of performance levels are provided in the *Guide Specifications for Bridge Railings.*

Since performance level selection is based on the assumption that a railing will be near its ultimate strength when subjected to its specific maximum containment load, it is recommended that ultimate strength approach be used in the analysis and design of the railings, posts, and supporting deck slab.

Where a railing is selected and successfully crash tested in accordance with the provisions of the Guide Specifications and NCHRP 350, it does not need to meet the requirements of AASHTO Standard Specifications, as described below.

Method 2 — Standard Specifications

The nominal transverse load P = 10 kips (44.5 kN) is to be applied and distributed as shown in Figure 62.13.

- Where the height of the top traffic rail exceeds 2 ft 9 in. (838 mm), the rail and the post is to be designed for a transverse load of CP, where

$$C = 1 + \frac{h-33}{18} > 1 \left(C = \frac{h-381}{457.2} \right)$$

  and h = height of top rail from reference surface in inches (millimeters). However, the maximum load applied to any element is not to exceed P.
- Where the rail face is more than 1 in. (25.4 mm) behind a vertical plane through the face of traffic rail closest to traffic, or where the rail center is less than 1 ft 3 in. (38 mm) from the reference surface, the rail should be designed for a $P/2$ or what is applied to an adjacent traffic rail, whichever is less.

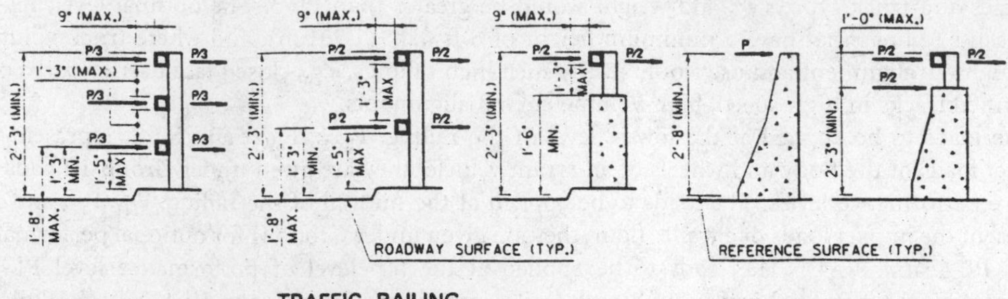

FIGURE 62.13 Traffic railing to be used where there is no curb or curb projects 9 ft or less from the traffic face of the railing. (*Source*: AASHTO, *Standard Specifications for Highway Bridges*, 16th ed., American Association of State Highway and Transportation Officials, Washington, D.C., 1996. With permission.)

- The posts are to be designed for *P* or *CP*, as shown in Figure 62.13. Simultaneous with the transverse loads a longitudinal load equal to $P/2$ or $CP/2$ is to be applied, divided among a maximum of four posts, assuming that railing is continuous. Also, posts are to be designed for an inward load, equal to ¼ of the outward loads.
- The rail attachment to the post is to be designed for a vertical load of ¼ of transverse loading, to be applied either upward or downward. This attachment should be also designed for an inward load of ¼ of transverse load.
- The rail members are to be designed for a moment of $P'L/6$, at the center of the panel and at the posts, where *L* is the post spacing and *P′* is *P*, *P*/2, or *P*/3, as modified by the factor *C*, where required. The handrail members of combination railings are to be designed for a moment of 0.1 wL^2 at the center and at the posts, where *w* is the pedestrian loading per unit length of rail; $w = 50$ lb/ft (729 N/m).

Where a concrete parapet or barrier curb is used, the transverse load of *P* or *CP* should be spread over a 5-ft (1.5-m) length of the parapet. Since AASHTO does not specify the location of load application it can be construed that even where the parapet/barrier curb ends the 5-ft (1.5-m) distribution length applies.

Where possible, bridge railings should preferably provide continuity for moment and shear throughout their length. To meet this goal, continuity transfer splices and expansion devices (capable of handling moments) and end anchorages (for transferring shear) must be provided for beam-and-post railings. Providing continuity transfer sleeves for concrete parapets may be more difficult because of the frequency of transverse joints in the parapets. These joints are normally provided at 10 to 20 ft (3 to 6 m) intervals to arrest potential temperature and shrinkage cracks that may otherwise develop, and to prevent parapet participation in the composite behavior of the fascia girders.

In designing the deck slab and distributing the loads from the posts to the deck slab, it is highly desirable to design the deck slab such that it would not sustain any damage due to a potential destruction of the post. This may be accomplished by providing additional reinforcement in the deck slab in order to distribute the concentrated loads from the posts over a larger area of the deck slab.

62.2.3 Bicycle Railing

62.2.3.1 Geometry

Bicycle railings are provided along the edges of the structure to contain the bicyclist where bicycle use is anticipated. Such railings are to be designed to provide safety while meeting the aesthetic

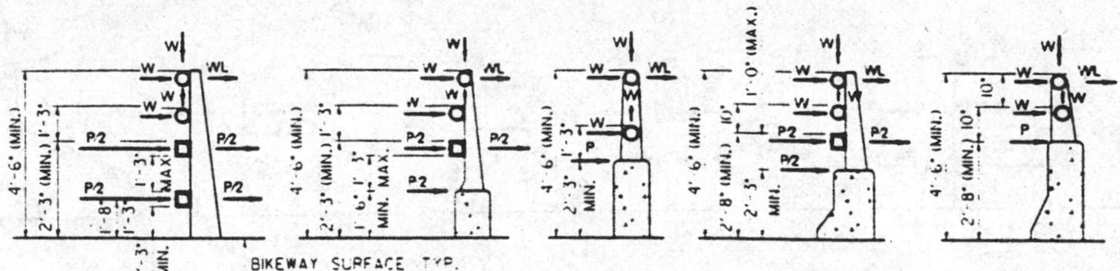

COMBINATION TRAFFIC AND BICYCLE RAILING

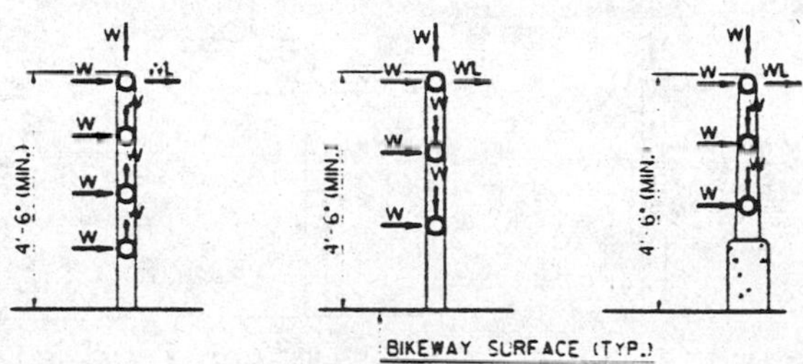

1 inch = 25.4 mm
1 foot = 0.3048 m

BICYCLE RAILING

NOTES:
Loads on left are applied to rails
Loads on right are applied to posts
W = Pedestrian loading per unit length of rail
L = Post Spacing

FIGURE 62.14 Bicycle railings. (*Source*: AASHTO, *Standard Specifications for Highway Bridges*, 16th ed., American Association of State Highway and Transportation Officials, Washington, D.C., 1996. With permission.)

requirements of the bridge owner. Where it is used in conjunction with vehicular traffic, freedom of view from passing vehicles is to be maintained.

As shown in Figure 62.14, the minimum height of a bicycle railing is 4 ft 6 in. (1372 mm), measured from the top of the riding surface to the top of the top rail.

The Standard Specifications requires that from the bikeway surface to 2 ft 3 in. (686 mm) above it, all railing elements be spaced such that a 6-in. (152-mm) sphere cannot pass through the railing. From 2 ft 3 in. to 4 ft 6 in. (686 to 1372 mm) from the bikeway surface an 8-in. (203 mm) sphere should not be able to pass through the railing. Where the railing is made up of horizontal and vertical elements, the spacing requirements apply to one or the other, but not to both.

The Guide Specifications requires that from the bikeway surface to 4 ft 6 in. (1372 mm) above it, horizontal elements of the railing have a clear spacing of 1 ft 3 in. (381 mm) and vertical elements a clear spacing of 8 in. (203 mm). Where the railing has both horizontal and vertical elements, the spacing requirements will apply to one or the other, but not to both.

62.2.3.2 Loads

As shown on Figure 62.14, the horizontal elements (rails) in a bicycle railing are to be designed for a minimum design loading of $w = 50$ lb/ft (729 N/m) acting laterally and vertically at the same time. The vertical elements (posts) should be designed for wL (where L is the post spacing) acting at the center of the upper rail, but at a height not greater than 4 ft 6 in. (1372 mm). Where a rail is located more than this limit, the design load will be determined by the designer.

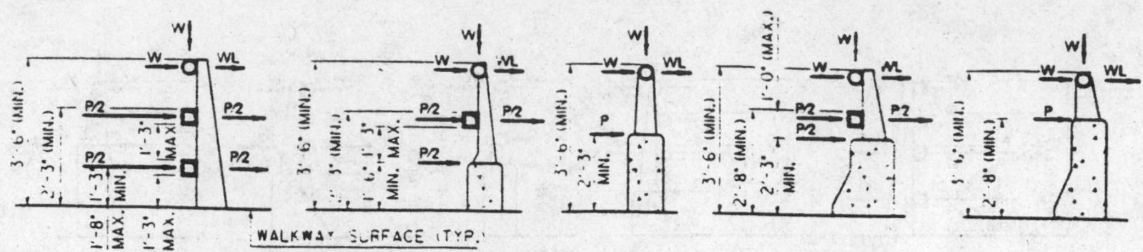
WALKWAY SURFACE (TYP.)
COMBINATION TRAFFIC AND PEDESTRIAN RAILING
To be used when curb projects more then
9" from the traffic face of railing

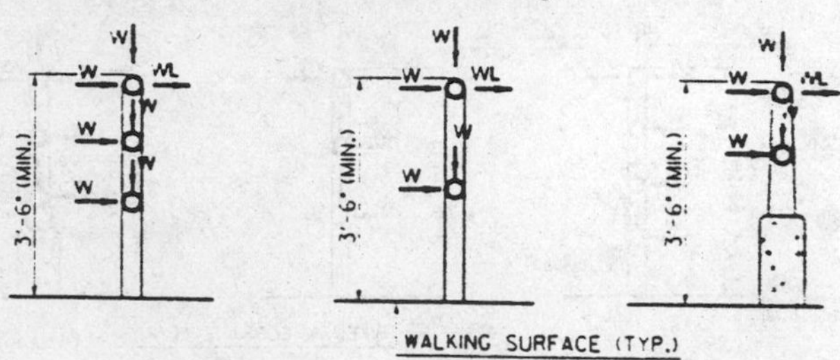
WALKING SURFACE (TYP.)
PEDESTRIAN RAILING
To be used on the outer edge of a sidewalk
when highway traffic is separeted from
pedestrian traffic by a traffic railing.
1 inch = 25.4 mm
1 foot = 0.3048 m
NOTES:
Loads on left are applied to rails
Loads on right are applied to posts
W = Pedestrian loading per unit length of rail
L = Post Spacing

FIGURE 62.15 Pedestrian railings. (*Source*: AASHTO, *Standard Specification for Highway Bridges*, 16th ed., American Association of State Highway and Transportation Officials, Washington, D.C., 1996. With permission).

Where vehicular traffic and bicycles are contained by a single combination railing, the Standard Specifications provides the geometry and loading requirements for five railing options, as reproduced in Figure 62.14.

62.2.4 Pedestrian Railing

62.2.4.1 Geometry

Pedestrian railings are provided along the outer edge of a sidewalk to contain pedestrians. Where the sidewalk is not raised, a traffic railing (a concrete barrier or a guide rail) must separate the pedestrians from highway traffic. In an urban setting, where there is a raised sidewalk, a combination traffic and pedestrian railing, along the outer edge of the sidewalk will be required. Such railings will be designed to provide safety while meeting the aesthetic requirements of the bridge owner. Consideration should also be given to freedom of view from passing vehicles.

As shown in Figure 62.15, the minimum height of a pedestrian or a combination railing is to be 3 ft 6 in. (1067 mm), measured from the top of the walkway to the top of the upper rail.

The Standard Specifications requires that, from the walkway surface to 27 in. (686 mm) above it, all railing elements be spaced such that a 6-in. (152 mm) sphere cannot pass through any opening in the railing. From 2 ft 3 in. to3 ft 6 in. (686 to 1067 mm) from the walkway surface, an 8-in. (203-mm) sphere should not be able to pass through the railing.

The Guide Specifications requires that, within the 3 ft 6-in. (1067 mm) height of the railing, the horizontal elements have a maximum clear spacing of 1 ft 3 in. (381 mm) while the vertical elements have a maximum clear spacing of 8 in. (203 mm). Where the railing has both horizontal and vertical elements, the spacing requirements will apply to one or the other, but not to both.

62.2.4.2 Loads

As shown on Figure 62.15, the horizontal elements (rails) in a pedestrian railing are to be designed for a minimum design loading of $w = 50$ lb/ft (729 N/m), acting laterally and vertically at the same time. Rail members located more than 5 ft (1.524 m) above the walkway are excluded from this requirement. The vertical elements (posts) are to be designed for wL (where L is the post spacing) acting at the center of the upper rail.

Where combination traffic and pedestrian railing is to be provided, the geometry and the loads may be obtained from one of the five options shown in the Standard Specifications and reproduced in Figure 62.15.

62.2.5 Structural Specifications and Guidelines for Bicycle and Pedestrian Railings

Bicycle and pedestrian railings are to be designed by the elastic method to the allowable stresses for the appropriate material. The following requirements are those specified in the AASHTO Standard Specifications, used by permission.

For aluminum alloys the design stresses given in the *Specifications for Aluminum Structures* 5th ed., December 1986, published by the Aluminum Association, Inc., for "Bridge and Similar Type Structures" apply. For alloys 6061-T6 (Table A.6), 6351-T5 (Table A.6), and 6063-T6 (Table A.8) apply, and for cast aluminum alloys the design stresses given for alloys A444.0-T4 (Table A.9), A356.0-T61 (Table A.9), and A356.0-T6 (Table A.9) apply.

For fabrication and welding of aluminum railing see Article 11.5 of the AASHTO *Standard Specifications for Highway Bridges.*

The allowable unit stresses for steel are as given in Article 10.32 of the AASHTO Standard Specifications, except as modified below.

For steels not generally covered by the Standard Specifications but having a guaranteed yield strength, F_y, the allowable unit stress, is derived by applying the general formulas as given in the Standard Specifications under "Unit Stresses" except as indicated below.

The allowable unit stress for shear is $F_v = 0.33\ F_y$

Round or oval steel tubes may be proportioned using an allowable bending stress, $F_b = 0.66\ F_y$, provided the R/t ratio (radius/thickness) is less than or equal to 40.

Square and rectangular steel tubes and steel W and I sections in bending with tension and compression on extreme fibers of laterally supported compact sections having an axis of symmetry in the plane of loading may be designed for an allowable stress $F_b = 0.60\ F_y$.

The requirements for a compact section are as follows:

In the above formulas b, t, and ℓ are in inches (millimeters) and f_a, F_a, and F_y are in psi (Mpa).

	English Units	S.I Units	
1. The width-to-thickness ratio of projecting elements of the compression flange of W and I sections not to exceed:	$\frac{b}{t} \leq \frac{1600}{\sqrt{F_y}}$	$\left(\frac{b}{t} \leq \frac{133}{\sqrt{Fy}}\right)$	(62.3)
2. The width-to-thickness ratio of the compression flange of square or rectangular tubes is not to exceed:	$\frac{b}{t} \leq \frac{6000}{\sqrt{F_y}}$	$\left(\frac{b}{t} \leq \frac{499}{\sqrt{Fy}}\right)$	(62.4)
3. The D/t ratio of webs is not to exceed:	$\frac{D}{t} \leq \frac{13{,}300}{\sqrt{F_y}}$	$\left(\frac{D}{t} \leq \frac{1106}{\sqrt{Fy}}\right)$	(62.5)
4. If subject to combined axial force and bending, the D/t ratio of webs is not to exceed:	$\frac{D}{t} < \frac{13{,}300\left[1-1.43\left(\frac{f_a}{F_a}\right)\right]}{\sqrt{F_y}}$	$\left(\frac{1106\left[1-1.43\,\frac{fa}{Fa}\right]}{\sqrt{Fy}}\right)$	(62.6)
but need not be less than:	$\frac{D}{t} < \frac{7000}{\sqrt{F_y}}$	$\left(\frac{D}{t} < \frac{581}{\sqrt{F_y}}\right)$	(62.7)
5. The distance between lateral supports in inches of W or I sections is not to exceed:	$\ell \leq \frac{2400b}{\sqrt{F_y}}$	$\left(\leq \frac{199.26}{\sqrt{F_y}}\right)$	(62.8)
or:	$\ell \leq \frac{20{,}000{,}000\,A_f}{dF_y}$	$\left(\ell \leq \frac{137{,}640\,A_f}{dF_y}\right)$	(62.9)

References

1. Procedures for Abatement of Highway Traffic Noise and Construction Noise, Federal Aid Highway Program Manual, Vol. 7, Chap. 7, Section 3, 1982.
2. Simpson, M. A., Noise Barrier Design Handbook, Publication No. FHWA-RD-76-58, U.S. Department of Transportation, 1976.
3. AASHTO, *Guide Specifications for Structural Design of Sound Barriers*, American Association of State Highway and Transportation Officials, Washington, D.C., 1989.
4. Blum, R. F. A Guide to Visual Quality in Noise Barrier Design, Publication No. FHWA-HI-94-039, U.S. Department of Transportation, 1976.
5. Highway Noise Barriers, National Cooperative Highway Research Program, Synthesis of Highway Practice 87, 1981.
6. Guzaltan, F., Precast concrete noise barrier walls for New Jersey Interstate Route 80, *PCI J.*, 27(4), 1992.
7. AASHTO, *Standard Specifications for Highway Bridges*, 16th ed., American Association of State Highway and Transportation Officials, Washington, D.C., 1996.
8. AASHTO, *Guide Specifications for Bridge Railings*, American Association of State Highway and Transportation Officials, Washington, D.C., 1989.
9. Recommended Procedures for Safety Evaluation of Highway Features, National Cooperative Highway Research Program Report 350.

Section VII
Worldwide Practice

63

Bridge Design Practice in China

Guohao Li
Tongji University
Rucheng Xiao
Tongji University

63.1 Introduction

63.1.1 Historical Evolution

With a recorded history of about 5000 years, China has a vast territory, topographically higher in the northwest and lower in the southeast. Networked with rivers, China has the well-known valleys of the Yangtze River, the Yellow River, and the Pearl River, which are the cradle of the Chinese nation and culture. Throughout history, the Chinese nation erected thousands of bridges, which form an important part of Chinese culture.

0-8493-7434-0/00/$0.00+$.50

FIGURE 63.1 Anji Bridge.

Ancient Chinese bridges are universally acknowledged and have enjoyed high prestige in world bridge history. They can be classified into four categories: beam, arch, cable suspension, and pontoon bridges.

The earliest reference to the beam bridge in Chinese history is the Ju Bridge dating from the Shang Dynasty (16th to 11th century B.C.). During the Song Dynasty (A.D. 960 to 1279), a large number of stone pier and stone-beam bridges were constructed. In Quanzhou alone, as recorded in ancient books, 110 bridges were erected during the two centuries, including 10 well-known ones. For example, the 362-span Anping Bridge was known for its length of 2223 m, a national record for over 700 years. To elongate the span, either the timber beams or the stone ones were placed horizontally on top of each other, the upper layer cantilevering over the lower one, thus supporting the simple beam in the middle. The extant single-span timber cantilever bridge, the Yinping Bridge built in Qing Dynasty (A.D. 1644 to 1911) has a span of more than 60 m with a covered housing on it.

The oldest arch bridge in China, which still survives and is well preserved, is the Anji Bridge, also known as the Zhaozhou Bridge, at Zhouxian, Hebei Province, built in the Sui Dynasty (Figure 63.1). It is a single segmental stone arch, composed of 28 individual arches bonded transversely, 37.02 m in span and rising 7.23 m above the chord line. Narrower in the upper part and wider in the lower, the bridge averages 9 m in width. The main arch ring is 1.03 m thick with protective arch stones on it. Each of its spandrels is perforated by two small arches, 3.8 and 2.85 m in clear span, respectively, so that flood can be drained and the bridge weight is lightened as well. The Anji Bridge has a segmental deck and the parapets are engraved with dragons and other animals. Its construction started in the 15th year of the reign of Kaithuang (A.D. 595) and was completed in the first year of Day's reign (A.D. 605) of the Sui Dynasty. To date, it has survived for 1393 years. The bridge, exquisite in workmanship, unique in structure, well proportioned and graceful in shape, with its meticulous yet lively engraving, has been regarded as one of the greatest achievements in China. Great attention has been paid to its preservation through successive dynasties. In 1991, the Anji Bridge was named among the world cultural relics.

Stone arches in China vary in accordance with different land transport and different natures between the north and south waterways. In the north, what prevails is the flat-deck bridge with solid spandrels, thick piers, and arch rings, whereas in the south crisscrossed with rivers, the hump-shaped bridge with thin piers and shell arches prevails.

In the southeastern part of China, Jiangsu and Zhejiang Provinces, networked with navigable rivers, boats were the main means of transportation. As bridges were to be built over tidal waters and their foundations laid in soft soil, even the stone arch bridge had to be built with thin piers and shell arches in order that its weight could be reduced as much as possible. The thinnest arch

FIGURE 63.2 Suzhou Baodai Bridge.

ring is merely 1/67 of the span, whereas for an average the depth of the arch ring is 1/20 of the span. The longest surviving combined multispan bridge with shell arches and thin piers is the Baodai Bridge (Figure 63.2) in Suzhou, Jiangsu Province. Built in the Tang Dynasty (A.D. 618 to 907) and having undergone a series of renovations in successive dynasties, the bridge is now 316.8 m long, 4.1 m wide, with 53 spans in all, the three central arches being higher than the rest for boats to pass through. Both ends of the bridge are ornamented with lions or pavilions and towers, all of stone.

Cable suspension bridges vary in kind according to the material of which the cables are made: rattan, bamboo, leather, and iron chain. According to historical records, 285 B.C. saw the Zha Bridge (bamboo cable bridge). Li Bin of the Qin State, who guarded Shu (256 to 251 B.C.), superintended the establishment of seven bridges in Gaizhou (now Chengdu, Sichuan Province), one of which was built of bamboo cables. The Jihong Bridge at Yongping County, Yunnan Province, is the oldest and broadest bridge with the mostly iron chains in China today. Spanning the Lanchang River, it is 113.4 m long, 4.1 m wide, and 57.3 m in clear span. There are 16 bottom chains and a handrail chain on each side. The bridge is situated on the ancient road leading to India and Burma.

The Luding Iron-Chain Bridge (Figure 63.3) in Sichuan Province, the most exquisite of the extant bridges of the same type, spans the Dadu River and has served as an important link between Sichuan Province and Tibet. It is 104 m in clear span, 2.8 m in width, with boards laid on the bottom chains. There are nine bottom chains, each about 128 m long, and 2 handrail chains on each side. On each bank, there is a stone abutment, whose deadweight balances the pulling force of the iron chains. Its erection began in 1705 and was completed in the following year.

According to historical records, a great number of pontoon bridges were built at nine and five different places over the Yangtze and the Yellow Rivers, respectively, in ancient times. In 1989 unearthed in Yongji, Shanxi Province, were four iron oxen, weighing over 10 tons each, and four life-size iron men, all with lively charm, exquisitely cast. They were intended to anchor the iron chains on the east bank of the Pujing Floating Bridge in the Tang Dynasty.

Ancient Chinese bridges, with various structures, exquisite workmanship, and reasonable details are the fruit of practical experience. Calculations and analyses by modern means prove that the great majority is in conformity with scientific principles. Ancient Chinese bridges are of great artistic and scientific value and have made remarkable achievements, from which we can assimilate rich nourishment to give birth to new and future bridges.

Comparatively speaking, the construction of modern bridges in China started late. Before the 1950s, many bridges were invested, designed, and constructed by foreigners. Most highway bridges were made up of wood. After the 1950s, China's bridge construction entered a new era. In 1956, the first prestressed concrete highway bridge was constructed. After 1 year, Wuhan Yangtze River Bridge was erected, which ended the history of the Yangtze River having no bridges. Nanjing Yangtze

FIGURE 63.3 Luding Iron-Chain Bridge.

River bridge was completed in 1969. In the 1960s, China began to adopt cantilever construction technology to construct T-type rigid frame bridges. During the 1970s, more prestressed concrete continuous bridges were constructed. China also began to practice new construction technology such as the lift-push launching method, the traveling formwork method, the span-by-span erecting method, etc. Two reinforced concrete cable-stayed bridges were constructed in 1975, which signified the start of cable-stayed bridge construction in China. Since 1980, China began to develop long-span bridges. One after another, many long-span bridges such as Humen Bridge (prestressed concrete continous rigid frame) in Guangdong Province with a main span of 270 m, Wanxian Yangtze River Bridge (arch reinforced concrete) in Shichuan Province with a main span of 420 m, Yangpu Bridge (cable-stayed) in Shanghai City with a main span of 602 m, etc. have been completed. The Jiangying Yangtze River (suspension) Bridge with a main span of 1385 m is under construction. The first two bridges mentioned above have the longest spans of their respective types in the world. Today, five large-scale and across-sea projects for high-class road arteries along the coast are under planning by the Ministry of Communications of China. From north to south, the road arteries cut across Bohai Strait, Yangtze Seaport, Hangzhou Bay, Pearl Seaport, Lingdingyang Ocean, and Qiongzhou strait. A large number of long-span bridges have to be constructed in these projects. The Lingdingyang long-span bridge project across Pearl Seaport has started.

63.1.2 Bridge Design Techniques

63.1.2.1 Design Specifications and Codes

There are two series of bridge design specifications and codes in China. One is for highway bridges [3] and the other for railway bridges [4]. In addition, there are design guides such as the wind-resistant guide for bridges [6]. Design Specifications for Highway Bridges are mainly for concrete bridges, which are widely constructed in China. Here only these specifications are presented because of space limitations.

The current Design Specifications for Highway Bridges [3], which were issued by the Ministry of Communications of the People's Republic of China in 1989, include six parts. They are the General Design Specification for Bridges, the Design Specification for Masonry Bridges, the Design Specification for Reinforced and Prestressed Concrete Bridges, the Design Specification for Footing and Foundations of Bridges, the Design Specification for Steel and Timber Members of Bridges, and the Seismic Design Specification for Bridges. The design philosophies and loads are provided in the General Design Specification.

In the specifications, two design philosophies are adopted: load and resistance factor design (RFD) theory for reinforced prestressed concrete members and allowable stress design (ASD) theory for steel and timber members.

Three basic requirements for strength, rigidity, and durability need to be checked for all bridge members. For a bridge member that may be subjected to bending, axial tension, or compression, combined bending and axial forces etc. should be checked in accordance with its loading states. To ensure its strength requirement, the rigidity of a bridge is evaluated according to the displacement range at the midspan or cantilever end. By checking the widths of cracks and taking some measurements, the durability of structures may be ensured.

63.1.2.2 Analysis Theories and Methods

The analysis of a bridge structure in terms of service is based on the assumption of linear elastic theory and general mechanics of materials. According to design requirements, the enveloping curves of internal forces and displacements of members of a bridge are calculated. Then, checking for strength, rigidity, and durability is done carefully in accordance with the design specifications. For simple structures, they are usually simplified as plane structures but they can also be analyzed more accurately by 3D-FEM.

For example, simply supported girder bridges are usually simplified in the following way. According to the cross section shape and the construction method, the bridge may be divided into several longitudinal basic members such as T-girders or hollow plate girders or box girders. The internal forces of the basic members caused by dead loads are calculated under an assumption of every basic member carrying the same loads. In order to consider the effect of space structure under live loads, the influence surfaces of internal forces and displacements are approximately simplified as two univariant curves; one is the influence line of internal forces or displacements of a basic member and another is the influence line of the transverse load distribution.

To prove the feasibility and reliability of the approximate method, extensive tests and theoretical studies have been conducted. Several methods to determine the influence lines of transverse load distribution for different structures and construction methods have been developed [5]. In the current practice, the transversely hinge-connected slab (or beam) method, rigid-connected beam method, rigid cross beam method, and lever principle method are used according to structures and construction methods. They may satisfy the design requirement for a lot of bridges. With computer programs, these simplified analysis methods have become very easy.

However, some bridges, such as irregular skewed bridges, curved bridges, and composite bridges, cannot be divided into several longitudinal girders that mainly have behaviors of vertical plane structures. They are not suited to the simplified analysis methods mentioned above. For those complex space structures, the influence surfaces of internal forces and displacements due to dead load are obtained by the static finite-element method and the maximal impact responses of internal forces and displacements caused by live loads can be obtained using dynamic analysis proceedures.

63.1.2.3 Theories and Methods for Long-Span Bridges

Long-span bridges are usually expensive to construct and are flexible in structural nature. In view of the economic and functional requirements, the problems of structural optimization, nonlinear analysis, stability analysis, and construction control become especially important to long-span bridges. Chinese bridge experts who participate in the study and design of China's long-span bridges have put forward many theories and methods to solve the problems mentioned above. In respect to the nonlinear analysis of long-span bridges, they developed an influence area method for geometric nonlinear analysis of live loads, nonlinear adjustment calculation method, and nonlinear construction simulation calculation method, for construction control [8]. Using finite displacement theory, a three-dimensional nonlinear analysis system considering dead load, live load, and construction stage and methods was developed [9]. Stability problems of truss, frame, and arch bridge have been studyed extensively [1]. A stability analysis approach was developed for the wind effect on long-span bridges. Optimization theory and techniques have been applied to all kinds of bridges successfully. The accuracy and efficiency of those methods developed have been verified by practical application.

63.1.2.4 Bridge CAD Techniques

Since the late 1970s, computer technologies have been widely employed for structural analysis in bridge design practice in China. Many special-purpose structural analysis programs for bridge design were developed. With full concern for the special feature of bridge design, for example, the Synthetical Bridge Program [9], provided the capability of construction stage transferring, concrete creep and shrinkage analysis, prestress calculation, etc. To a certain extent, widespread adoption of this program reflected the application status of computational technology in the field of highway bridge design in China during the years from the late 1970s to the early 1980s.

Since the 1980s, the popularization of computer graphics devices, such as the rolled drafting plotter and digitizer, have brought computational application from merely structural analyzing to aided design including both structural analysis and detail drafting. With the development of the highway system, standardized simply supported bridges have spread over China. Based on the microcomputer platform, many researchers and engineers began to develop automated CAD systems integrating structural analysis and detail drafting. The "Automated Medium and Short-Span Bridge CAD System on Micro-computer" cooperatively developed by the membership of China Highway Computer Application Association, for example, has the capabilities to accomplish all processes of simply supported T-beam and plate bridge design. With the aid of this system, only a few primary pieces of information are required to be input, and the computer will automatically produce a set of design documents including both specifications and drawings in a short time. The design efficiency is excellent compared with the traditional manner. Many design institutes and firms employed this system to design medium- and short-span bridges.

During the7th Five Year Plan of China (1985 to 1990), to develop a new highway bridge system, a special task group consisting of more than 40 practical bridge engineers and scholars was formed and organized by the Ministry of Communications. As a national key scientific research project, the allied group invested $2 million of RMB to research and develop the CAD techniques applied in the construction of highway bridges. In 1991, the "Highway Bridge CAD System (JT-HBCADS)" was successfully developed. More than 10 large highway bridge design institutes have installed this system and fulfilled the design of about 10 large bridges such as Nanpu Bridge, Yangpu Bridge, etc.

During the years from 1991 to 1995, the increase in personal computer (PC) hardware performance and software technology has issued a critical challenge to the development of research and application of bridge CAD techniques. Many advanced software development techniques, such as kernel database accessing, object-oriented programming, application visualizing, and rapid application developing, were entirely developed and made available for the personal computer, which brought forth lots of chances that had never appeared before in developing the new generation of integrated and intelligent bridge CAD systems.

With full regard to, and on the basis of, experience and acquaintance with the development of JT-HBCADS and many newly available support software technologies, the developing ideas of integrated bridge CAD system (BICADS) has been brought up, and the new generation BICADS was successfully developed thoroughly under the guidance of this thought. Taking the Windows NT operating system as the platform, the system architectural design of BICADS entirely adopted the kernel database accessing techniques to avoid the difficulties of system maintenance and upgrading the innate and unavoidable weakness caused by the traditional file system. The first version of BICADS consists of five subsystems including the Design Documentation, Pre-Processing of Bridge FEM, Bridge FEM Kernel, Post-Processing of Bridge FEM, and the Preliminary Design of Box Girder Bridges. Several detailed design subsystems of other commonly used bridges can be included by employing a good integrating and expanding mechanism in the main system. Additionally, the research of some fundamental problems in the field of bridge intelligent CAD techniques and the development of bridge experts system tools with graphics processing abilities have already yielded considerable promise. It is predicted that, motivated by the rapid development of computer technologies by the end of this century, a new generation in China's bridge CAD techniques application and research is being opened.

63.1.3 Experimental Research of Dynamic and Seismic Loads

Model Tests for Bridges
To establish the dynamic behavior base line for health monitoring bridge structures, the model tests are usually done just after construction of bridges. Experimental procedures that have been used in the past include (1) impact tests and (2) ambient vibrations. For large bridges, such as Shanghai Yangpu Bridge (cable-stayed bridge) and Shanghai Fengpu Bridge (continuous box-girder bridge), the method of using test vehicles (controlled traffic) for exciting bridges was successfully verified.

Shaking Table Test of Bridge Models
The tests of a simply supported beam and a continuous girder bridge model were performed on the shaking table (made by the MTS Co.). These tests were to evaluate the effect of ductility and seismic isolation on bridges, in which the viaduct of Shanghai Inner Ring Road was regarded as the background of the continuous girder bridge model; meanwhile, the analytical models of bridges and elements were verified.

Ductility Performance and Seismic Retrofitting Techniques for Bridge Piers
Recently, high-strength concrete with cylindrical compressive strength up to 100 MPa or higher can be made with locally obtainable materials, such as ordinary cement, sand, crushed stone, a water-reducing superplasticizer, standard mixing methods, and careful quality control in production. There are many characteristics for high-strength concrete that are beneficial in civil engineering, but, on the other hand, there are some shortcomings to the increasing use of high-strength concrete. For instance, brittle features and less postpeak deformability may cause brittle failure during earthquakes or under other conditions. Much work, theoretical and experimental, has been done by Chinese researchers for ductility design and improving design code of bridges. Through the tests and analyses, some important conclusions may be summarized briefly as follows:

1. Test results indicate that for high-strength concrete columns, very large ductility could be achieved by using lateral confining reinforcement.
2. All retrofitted piers using steel jackets, steel fiber concrete, expoxy concrete, and fiberglass-expoxy performed extremely satisfactorily. Good ductility, energy-dissipation capacity, and stable-deformation behavior were achieved.

Dynamic Behavior Test of Isolation Devices

To meet the requirements of earthquake resistance design of bridge, seismic design of isolated bridge and optimization have been widely used in China. The dynamic properties of elastomeric pad bearings (EP bearings) has been evaluated, including the shear modulus, hysteretic behavior, and sliding friction coefficient of EP bearings and Teflon plate-coated sliding bearings (TPCS bearings). The tests were done on an electro-hydraulic fatigue machine (made by INSTRON Co.) with an auxiliary clamping apparatus. These results may be summarized as follows:

1. At constant shear strain amplitude, the shear modulus of EP bearings increases with the increase in frequency. At constant frequency, the shear modulus obviously decreases with the increase in shear strain amplitude. Sizes and compression have no obvious effect on dynamic shear modulus.
2. At constant compression and sliding displacement amplitude, the hysteretic energy of TPCS bearings increases with the increase in frequency. At constant sliding displacement amplitude and frequency, the increased compression results in an increase in the hysteretic energy of TPCS bearings.
3. The friction coefficient of TPCS bearing decreases with the increase in compression.

Based on experimental research of rubber bearings and steel damping, a system of seismic isolation and energy absorption, composed of curved steel-strip energy absorbers and TPCS bearings, was developed, and then a seismic rubber bearing with curved mild-steel strip, was invented.

Recently, some kinds of improved seismic bearings have come out. A great number of dynamic experiments show that these types of bearings have better hysteretic characteristics than elastomeric laminated bearings. To avoid span failures of bridges upon impact, restricting blocks are usually placed at the end of beams. To compare the behavior of the blocks, three kinds of blocks [4] have been manufactured and an experiment has been conducted on these blocks: (1) " T-type" rubber blocks, (b) "bowl-type" rubber blocks, and (3) cubic reinforced concrete blocks. During the tests, the impact hammer freely fell from a given height and contact forces between the block and high-strength concrete hammer were recorded. The test results show it is very obvious that T-type rubber blocks have the best energy absorption capacity and the impact force of T-type rubber blocks is much lower than that of concrete blocks.

63.1.4 Wind Tunnel Test Techniques

Since the 1980s, with the building of long-span cable-stayed and suspension bridges, China has made great progress in wind engineering. For example, there are three boundary-layer wind tunnels in the National Key Laboratory for Disaster Reduction in Civil Engineering at Tongji University. TJ-1, TJ-2, and TJ-3 BLWTs, which have been put into service only for several years, have working sections of 1.2 m (width), 1.8 m (height); 3 m (width), 2.5 m (height); and 15 m (width), 2 m (height), respectively. The maximum wind speeds of these are 32, 17, and 65 m/s, respectively. Until now, about 30 model tests have been carried out in these wind tunnels. Wind-resistant researches on about 40 cable-stayed bridges and suspension bridges have been carried out mainly at Tongji University, Shanghai, China. More than 10 full-scale aeroelastic bridge model tests have been performed. To meet the requirements of the wind-resistant design of highway bridges with increasing spans, a Chinese Wind Resistant Design Guideline of Highway Bridge was compiled. Some achievements of flutter analysis, buffeting analysis, and wind-induced vibration control have been made and are introduced in the following.

Flutter Analysis

As is well known, the critical flutter velocity is the first factor that controls the design for a long-span bridge, especially located in typhoon areas. Precision of torsional frequency in the calculation is very important. The traditional single-beam model test of bridge deck usually gives estimates of torsional frequencies lower than the actual ones and may make a lower critical flutter velocity estimation. A three-beam model of a bridge deck which was developed by Xiang et al., [6] has been proved to be efficient in improving the precision of torsional frequency to a great extent.

The state-space method for flutter analysis overcomes the shortcomings of Scanlan's method for flutter analysis in which only one vertical mode and one torsional mode can be considered. A multimode flutter phenomenon was found. Participation of more than two modes in flutter make the critical velocity higher than that from Scanlan's method.

Buffeting Analysis

With the increase in span length, bridge structures tend to become more flexible. Excessive buffeting in near-ground turbulent wind, although not destructive, may cause fatigue problems due to high frequency of occurrence and traffic discomfort. Davenport and Scanlan et al., proposed buffeting analysis methods in the 1960s and 1970s, respectively. Since then, refinement studies on these methods have been made. It is possible to establish practical methods for buffeting response spectrum and buffeting-based selection.

Aerodynamic selection of deck cross section shape is important in the preliminary design stage of a long-span bridge. In the past year, this selection aimed mainly at flutter-based selection. The concept of "buffeting-based selection" and the corresponding method were used in the wind-resistant design of the Jiangying Yangtze River Bridge and the Humen Bridge, a suspension bridge with a main span of 888 m.

To investigate the nonlinear response characteristics of long-span bridges, a nonlinear buffeting analysis method in the time domain has been used to analyze the Jiangying Yangtze River Bridge and the Shantou Bay Bridge, etc. Analysis results show that for long-span suspension bridges the aerodynamic and structural nonlinear effects on the buffeting response should be considered.

Wind-Induced Vibration Control

In practice today, the increment of critical flutter velocity of a long-span bridge is usually achieved using aerodynamic measures. The theoretical analysis and experiments indicate that passive TMD may also be an effective device for flutter control. A couple of TMDs with proper parameters can increase the critical flutter velocity of the Humen-Gate Bridge with wind screens on the deck (for improving vehicle moving condition) by 50%, although the efficiency, duration, and reliability of the device for long-time-period use still have some problems to be solved.

The buffeting response increases with wind speed, and may become very strong at high wind speed. Two new methods were proposed for determination of optimal parameters of the TMD system for controlling buffeting response with only the vertical mode and with coupling the vertical and torsional modes, respectively.

63.1.5 Bridge Construction Techniques

63.1.5.1 Constructional Materials

According to the design specifications for bridges in China, the maximum strength of concrete is 60 MPa; the prestressing tendons include hard-drawn steel bars, high-strength steel wires, and high-strength strands, the strengths of which are from 750 to 1860 MPa; the general reinforcement bars are made of A3, 16Mn, etc.; the steel plate is made of A3 or 16Mn or 15MnVN, etc. In normal designs, the concrete used in prestressed bridges should have a strength higher than 40 MPa; the prestressing tendons used in pretensioned slab girders are hard-drawn 45 SiMnV bars with the strength of 750 MPa or steel strands with strength of 1860 MPa; the high tensile strength and low relaxation strands are widely used in post-tensioned concrete bridges. Now a viaduct usually has a lower depth of girders so high-strength concrete over 50 MPa is often adopted. Concrete having a strength of over 60 MPa and tensile wires and strands will be used in bridges in the future.

63.1.5.2 Prestressing Techniques

Prestressing techniques including internal and external prestressing have been used for about 40 years in China. Not only were the full and partial prestressed bridges constructed speedily, but also the preflex prestressed girders and double-prestressed girders have been used in viaducts and separation structures. The high tensile strength and low relaxation strands, the reliable anchorages, such as the OVM system, and the high-tonnage jacks have been widely used in many bridges including continuous girder bridges, T-frame bridges, cable-stayed bridges, and suspension bridges. The design and construction of prestressed concrete structures is a normal process in China. The external prestressing tendons, including unbonded tendons, have been used in new bridges and in the strengthening of many old bridges. Now, several external prestressed long-span composite bridges are being built in China.

63.1.5.3 Precast Techniques of Concrete and Steel Girders

Most simply supported girder bridges are made with fabricated methods in China, and factory production is usually adopted. When the span is shorter than about 22 m, the pretensioned, prestressed voided slab girder is often the best choice, and the high-strength and low-relaxation strands are used as the prestressed reinforcement. When the span is over about 25 m, the post-tensioned T-girder may be used, in which the strands are arranged with curved profiles. In the construction of some bridges and urban viaducts and in precasting yards, steam curing is often used to increase the strength of concrete early and to raise the working efficiency. Usually the weight and length of a precast girder are limited to below about 1200 kN and 50 m to ease transport and erection.

Segmental bridges are usually built using the cantilever casting method, or other casting methods; nevertheless, only a few segmental bridges are constructed with the cantilever erection method. We usually cast in place because it is noticed that the rusting of prestressing strands at the segment joints may cut down the service life of bridges. The high anticorrosive external prestressing tendon or strand cable is not widely adopted yet in post-tensioned segmental bridges.

In China, complete riveting techniques have been replaced by welding and high-strength bolting techniques. Complete welded box and composite girders have been used in urban viaducts, separation structures, and cable-stayed bridges; techniques adopted in shipbuilding, such as computer layout and precision cutting, are being introduced.

63.1.5.4 Cable Fabrication Techniques

About 10 to 20 years ago, the stay cable in China was fabricated mainly on the construction site and consisted of 5-mm-diameter or 7-mm-diameter parallel galvanized steel wires. It was protected with PE casing pipe grouted with cement, or with corrosion paint and three layers of glass fibers coated by epoxy resin. A lot of cable-stayed bridges have been built in the last decade and the cable fabrication techniques have developed rapidly. With the construction of Shanghai Nanpu Bridge in 1988, the first factory, which mechanically produced long-lay spiral parallel wire cables with a hot-extruded PE or PE and PU sheath, was established. Since then, the quality of stay cables has greatly improved, especially in resistance to corrosion. Now the maximum working tension of stay cables is over 10,000 kN and high-quality anchorage has been developed. In recent years, the parallel and spiral strand cables of factory production with maximum working tensions at over 10,000 kN have been frequently used in cable-stayed bridges.

At the same time, the main cables of Santo and Humen (suspension) Bridges were successfully fabricated in China; the parallel wire strand consisted of 127ϕ 5.2-mm zinc-coated steel wires and had a length of over 1600 m; the mean square root error in the length of wires was lower than 1/36,000. Now, Jiangyin Yangtze River Bridge, having the longest span, close to 1400 m, in China, is under construction; its main cables will also be prefabricated.

63.1.5.5 Construction Techniques of Large-Diameter Piles

In China, bored piles are usually adopted for large bridges. When the ground is poor or the rock formation is near the Earth's surface or riverbed, piles have to be built in the rock and they become the bearing piles. Normally, the diameter of bearing piles is about 0.8 to 2.5 m. A large-diameter pile can be adopted to replace the pile group in order to reduce material construction time. Usually this large-diameter pile has a diameter of 2.5 to 7 m, is hollow, and consists of two or three segments. The first segment of the pile is a double-wall steel and concrete composite drive pipe which is driven into a weathered layer as a cofferdam; the second segment is a hollow concrete bearing pile which has a smaller diameter than the first segment, and the pier shaft is connected on the top of this segment; the last segment has a minimum diameter or, similar to the second, it is built in the rock. As a result, construction is easy, and no platform or hollow pile uses up a lot of concrete and steel.

63.1.5.6 Advanced Construction Techniques

With the development of transportation in China, more and more large bridges have been built and new construction techniques have been developed. Continuous curved bridges have been built with the incremental launching method, and the speed of the cantilever casting construction method is about 5 or 6 days per segment. The cable-stayed composite bridges, whose composite girders are composed of prefabricated, wholly welded steel girders and precast reinforced concrete deck slabs, were constructed with the cantilever erection method — for example, the 602-m Shanghai Yangpu Bridge, built in 1993. For prestressed concrete cable-stayed bridges, the tensions of stay cables and alignment of girder can easily achieve their best states by using computer-automated control techniques. The construction method of modern long-span suspension bridges was a new technique in China several years ago, most using PWS (prefabricated parallel wire strand) methods.

The improvement of construction techniques is not only in continuous girder bridges, rigid frame bridges, cable-stayed bridges, and suspension bridges. In a deep valley or flood river, the stiff reinforcement skeleton consisting of steel pipes is used as the reinforcement of a long-span concrete arch ring; after the stiff reinforcement skeleton is erected and closed up at midspan, the concrete is pumped into the steel pipes; then, by using the traveling form, which is supported on the stiff reinforcement skeleton, the concrete is cast and the reinforced concrete box arch ring is formed. Another construction method used in long-span composite arch bridges is the swing method. The two halves of the arch are separately erected on each side of river embankments or hillsides; then, by using jacks, they are rotated around their supports under arch seats and closed at midspan; finally, the concrete is pumped into the pipe arch. In order to keep the balance of a half arch, water containers are usually used as the ballast weights.

The progress of construction techniques has not only been made for superstructures but also for substructures. The height of reinforced, prestressed hollow piers and precast piers used in deep valleys has reached over 80 m. Large-diameter hollow piles and large concrete and steel caissons and double-wall steel and concrete composite cofferdams are adopted in river or sea depths over 50 m.

63.2 Beam Bridges

63.2.1 General Description

Simple in structure, convenient to fabricate and erect, easy to maintain, and with less construction time and low cost, beam structures have found wide application in short- to medium-span bridges. In 1937, over the Qiantang River, in the city of Hangzhou, a railway-highway bipurpose bridge was erected , with a total length of 1453 m, the longest span being 67 m. When completed, it was a remarkable milestone of the beam bridge designed and built by Chinese engineers themselves.

Reinforced concrete beam structures are most commonly used for short- to medium-span bridges. A representative masterpiece is the Rong River Bridge completed in 1964 in the city of Nanning, the capital of Guangxi Zhuangzu Autonomous Region. The bridge, with a main span of 55 m and a cross section of a thin-walled box with continuous cells, designed in accordance with closed thin-walled member theory, is the first of its kind in China.

Prestressed concrete beam bridges are a new type of structure. China began to research and develop their construction in the 1950s. In early 1956, a simply supported prestressed concrete beam railway bridge with a main span of 23.9 m was erected over the Xinyi River along the Longhai Railway. Completed at the same time, the first prestressed concrete highway bridge was the Jingzhou Highway Bridge. The longest simply supported prestressed concrete beam which reaches 62 m in span is the Feiyun River Bridge in Ruan'an, Zhejiang Province, built in 1988. Another example is the 4475.09-m Yellow River Bridge, built in the city of Kaifeng, Henan Province in 1989. Its 77 spans are 50-m simply supported prestressed concrete beams and its continuous deck extends to 450 m. It is also noticeable that the Kaifeng Yellow River Bridge is designed on the basis of partially prestressed concrete theory. Representative of prestressed concrete continuous girder railway bridges, the second Qiantang River Bridge (completed in 1991) boasts its large span and its great length, its main span being 80 m long and continuous over 18 spans. Its erection was an arduous task as the piers were subjected to a wave height of 1.96 m and a tidal pressure of 32 kPa when under construction. The extensive construction of continuous beam bridges has led to the application of the incremental launching method especially to straight and plane curved bridges. In addition, large capacity (500-t) floating crane installation and movable slip forms as well as span erection schemes have also attained remarkable advancement.

Beam bridges are also used widely in overcrossings. In the 1980s, with the growth of urban construction and the development of highway transportation, numerous elevated freeways were built, which provide great traffic capacity and allow high vehicle speed, for instance, Beijing's Second and Third Freeway and East City Freeway, the Intermediate and Outer Freeway in Tianjin, and Guangzhou's Inner and Outer Freeway and viaduct. In Shanghai, the elevated inner beltway was completed in 1996. Subsequently, there has appeared an upsurge of erecting different-sized grade separation structures on urban main streets and express highways. Uutil now, in Beijing alone, 80-odd large overcrossings have been erected, which makes the city rank the first in the whole country in number and scale.

To optimize the bridge configuration, to reduce the peak moment value at supports, and to minimize the constructional depth of girders, V-shaped or Y-shaped piers are developed for prestressed concrete continuous beam, cantilever, or rigid frame bridges. The prominent examples are the Zhongxiao Bridge (1981) in Taiwan Province and the Lijiang Bridge (1987) at Zhishan in the city of Guilin.

63.2.2 Examples of Beam Bridges

Kaifeng Yellow River Bridge

Kaifeng Yellow River Bridge (Figure 63.4) is an extra large highway bridge, located at the northwest part of Kaifeng City, Henan Province. It consists of 108 spans ($77 \times 50 + 31 \times 20$) m, its total length reaching 4475.09 m.

Simply supported prestressed concrete T-girders are adopted for its superstructure. The deck is 18.5 m wide, including 12.3 m for motor vehicle traffic and two sidewalks 3.1 m wide each on both sides. Substructure applies single-row double-column piers, which rest on 2200-mm large-diameter bored pile foundations.

The bridge is of the same type as those built earlier over the Yellow River in Luoyang and Zhengzhou. Kaifeng Bridge has obtained an optimized design scheme, with its construction cost reduced and schedule shortened. The main characteristics of the bridge are as follows:

FIGURE 63.4 Kaifeng Yellow River Bridge.

1. Adoption of partial prestress concrete in the design of T-girder;
2. Modification of the beams over central piers as prestressed concrete structure;
3. Increase in the continuous length of the deck reaching 450 m.

The bridge was designed by Highway Planning, Survey and Design Institute of Henan Province, and constructed by Highway Engineering Bureau of Henan Province. It was opened in 1989.

Xuzhuangzi Overcrossing

Xuzhuangzi Overcrossing (Figure 63.5), a long bell-mouth interchange grade crossing on the freeway connecting Beijing-Tianjin and Tangshan, is a main entrance to the city of Tianjin.

The overcrossing has a total length of 4264 m. The superstructure consists of simply supported prestressed concrete T-griders and multispan continuous box girders. The 1.5-m-diameter bored piles and invested trapezoidal piers are adopted for the substructure.

The bridge was designed by the first Highway Survey and Design Institute, Tianjin Municipal Engineering Co. and constructed by the first Highway Co., Ministry of Communications, Kumagai Co., Ltd, Japan. It was opened to traffic in 1992.

Liuku Nu River Bridge

Liuku Nu River Bridge (Figure 63.6), the longest prestressed concrete continuous bridge in China at present, is located in the Nu River Lisu Autonomous Prefecture, Yunnan Province. It has three spans of length (85 + 154 + 85) m. The superstructure is a single-box single-cell girder with two 2.5-m-wide overhangs on both sides. The beam depth at the support is 8.5 m, i.e., $\frac{1}{18}$ of the span, while at the midspan it is only 2.8 m, i.e., $\frac{1}{55}$ of the span. The whole bridge has only two diaphragms at the hammer-headed block.

FIGURE 63.5 Xuzhuangzi Overcrossing.

FIGURE 63.6 Liuku Nu River Bridge.

Three-way prestress is employed. A large tonnage strand group anchorage system is applied. With tendons installed only in the top and bottom slabs, no bent-up or bent-down tendon is needed and the widening of the web is avoided, which makes the construction very convenient. Vertical prestress is provided by Grade 4 high-strength rolled screwed rebars with diameter of 32 mm, which also served as the rear anchorage devices of the form traveler during cantilever casting. For the substructure hollow piers supported by bored piles foundation on rock stratum were adopted.

The bridge was completed in 1993, designed by Highway Survey and Design Institute of Yunnan Province and constructed by Chongqing Bridge Engineering Co.

FIGURE 63.7 Second Qiantang River Bridge.

The Second Qiantang River Bridge

The second Qiantang River Bridge, located on Sibao in Hangzhou, Zhejiang Province, is a parallel and separate highway–railway bipurpose bridge (Figure 63.7). The 11.4-m-wide railway bridge carries two tracks, with a total length of 2861.4 m. The highway bridge, which was designed according to freeway standard, is 20 m wide and 1792.8 m long, carrying four-lane traffic. Both main bridges are of prestressed concrete continuous box girders, and the continuous beams reach a total length of 1340 m, i.e., 45 + 65 + 14 × 80 + 65 + 45 m, the longest in China at present.

To obtain the 506 mm expansion magnitude of the main bridge, composite expansion joints were applied in the highway bridge, whereas transition beams and expansion rails were used for the railway bridge. Pot neoprene bearings were specially designed to accommodate the large displacement and to offer sufficient vertical resistance.

Three-way prestress was introduced to the box girder. Strands and group anchorage system were adopted longitudinally, with the maximum stretching force in excess of 2000 kN. The cantilever casting method was used for the main construction of the bridge, while the bored piles foundation was constructed at river sections of rare strong tidal surge with a height of 1.96 m and a pressure reaching 32 kPa. The bridge was designed and constructed by Major Bridge Engineering Bureau, Ministry of Railway. It was completed in November 1991.

63.3 Arch Bridges

63.3.1 General Description

Of all types of bridges in China, the arch bridge takes the leading role in variety and magnitude. Statistics from all the sources available show that close to 60% of highway bridges are arch bridges. China is renowned for its mountains with an abundant supply of stone. Stone has been used as the main construction material for arch bridges. The Wuchao River Bridge in Hunan Province, for instance, with a span of 120 m is the longest stone arch bridge in the world. However, reinforced concrete arch bridges are also widely used in various forms and styles.

Most of the arches used in China fall into the following categories: box arch, two-way curved arch, ribbed arch, trussed arch, and rigid framed arch. The majority of these structures are deck bridges with wide clearance, and it costs less to build such bridges. The box arch is especially suitable for long-span bridges. The longest stone arch ever built in China is the Wu River Bridge in Beiling, Sichuan Province, whose span is as long as 120 m. The Wanxian Yangtze River Bridge in Wanxian, Sichuan Province with a spectacular span of 420 m set a world record in the concrete arch literature. A unique and successful improvement of the reinforced concrete arch, the two-way curved arch structure, which originated in Wuxi, Jiangsu Province, has found wide application all over the country, because of its advantages of saving labor and falsework. The largest span of this type goes to the 150-m-span Qianhe River Bridge in Henan Province, built in 1969. This trussed arch with light deadweight performs effectively on soft subsoil foundations. It has been adopted to improve the composite action between the rib and the spandrel. On the basis of the truss theory, a light and congruous reinforced concrete arch bridge has been gradually developed for short and medium spans. Through prestressing and with the application of cantilevering erection process, a special type of bridge known as a "cantilever composite trussed arch bridge" has come into use. An example of this type is the 330 m-span Jiangjie River Bridge in Guizhou Province. The Yong River Bridge, located in Yunnan Province, is a half-through ribbed arch bridge with a span of 312 m, the longest of its kind. With a simplified spandrel construction, the rigid framed arch bridge has a much better stress distribution on the main rib by means of inclined struts, which transfer to the springing point the force induced by the live load on the critical position. In the city of Wuxi, Jiangsu Province, three such bridges with a span of 100 m each were erected in succession across the Great Canal. Many bridges, quite a number of which are ribbed arch bridges, have been built either with tied-arches or with Langer's girders. The recently completed Wangcang Bridge in Sichuan Province and the Gaoming Bridge in Guangdong Province are both steel pipe arch bridges. The former has a 115-m prestressed tied-arch, while the latter has a 110-m half-through fixed rib arch. A few steel arch bridges and slant-legged rigid frame bridges have also been constructed.

In building arch bridges of short and medium spans, precast ribs are used to serve as temporary falsework. And sometimes a cantilever paving process is used. Large-span arch bridges are segmented transversely and longitudinally. With precast ribs, a bridge can be erected without scaffolding, its components being assembled complemented by cast-in-place concrete. Also, successful experience has been accumulated on arch bridge erection, particularly erection by the method of overall rotation without any auxiliary falsework or support.

Along with the construction of reinforced concrete arch bridges, research on the following topics has been carried out: optimum arch axis locus, redistribution of internal forces between concrete and reinforcement caused by concrete creep, analytical approach to continuous arch, and lateral distribution of load between arch ribs.

63.3.2 Examples of Masonry Arch Bridge

Longmen Bridge

Longmen Bridge (Figure 63.8), 12 km south of Luoyang City, Henan Province, is an entrance of the Longmen Grottoes over Yihe River. It is a 60 + 90 + 60 m three-span stone arch bridge, with a width of 12.6 m. A catenary of 1:8 rise-to-span ratio was chosen as the arch axis. The main arch ring has a constant cross section, with a depth of 1.1 m. Two stone arches of 6 m long each were arranged on either bank providing under crossing traffic. The bridge was constructed on steel truss falsework supported by temporary piers. It was designed and constructed by Highway Engineering Bureau, Communications Department of Henan Province and completed in 1961.

FIGURE 63.8 Longmen Bridge.

Wuchao River Bridge

Wuchao River Bridge (Figure 63.9), a structure on Fenghuang County Highway Route, Hunan Province, spans the valley of the Wuchao River with a total length of 241 m. To use local materials, a masonry arch bridge scheme was adopted. On the basis of the experience accumulated in the last 20 years of construction of masonry arch bridges in China, the bridge has a main span of 120 m, which is a world record for this type of bridge.

The bridge is 8 m wide. There are nine spandrel spans of 13 m each over the main spans; three spans of 13 m each for the south approach; a single span of 15 m for the north approach. The main arch ring is a structure of twin separated arch ribs, connected by eight reinforced concrete floor beams. A catenary of $m = 1.543$ was chosen as the arch axis, with a rise-to-span ratio of 1:5. The arch rib has a variable width and a uniform depth of 1.6 m. It is made up of block stone with a strength of 100 kPa and ballast concrete of 20 Mpa.

The lateral stability of the bridge was checked. Because the masonry volume of its superstructure is only 1.36 m^3/m^2, the structure achieves a slim and graceful aesthetic effect. The bridge was designed and constructed by Communication Bureau of Fenghuang County, Hunan Province. It was completed in 1990.

Heyuan DongRiver Bridge

Heyuan DongRiver Bridge (Figure 63.10) is on the Provincial Route near Heyuan County. It is a 6×50 m multispan masonry arch bridge with a width of $7 + 2 \times 1$ m and a total length of 420.06 m. The rise-to-span ratio of the arch ring is 1:6.

A transversely cantilevered setting method was applied for its arch ring construction. The arch ring was divided into several arch ribs, and each rib was longitudinally divided into several precast

FIGURE 63.9 Wuchao River Bridge.

FIGURE 63.10 Heyuan DongRiver Bridge.

concrete hollow blocks. Side ribs were erected by transversely setting with the support of the erected central rib. The bridge was designed by Highway Survey and Design Institute of Guangdong Province and constructed by Highway Engineering Department of Guangdong Province. It was completed in 1972.

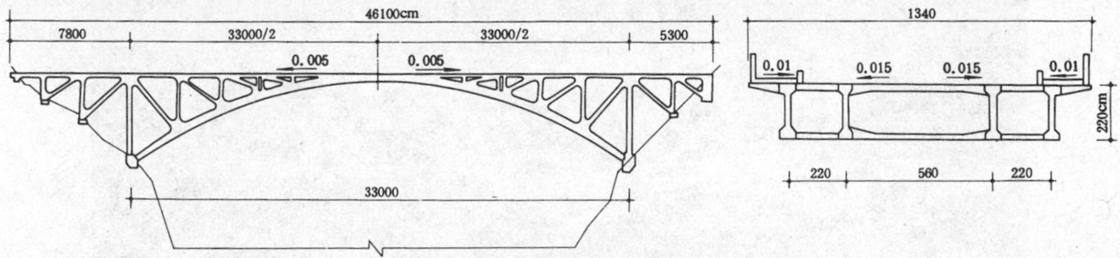

FIGURE 63.11 Jiangjie River Bridge.

FIGURE 63.12 Jinkui Grand Canal Bridge.

63.3.3 Examples of Prestressed Concrete, Reinforced Concrete, and Arch Bridges

Jiangjie River Bridge

Jiangjie River Bridge (Figure 63.11), located in Weng'an County, Guizhou Province, is a prestressed concrete truss arch bridge crossing Wujiang Valley at a height of 270 m above normal water level. It has a record-breaking main span of 330 m in China. Its side truss spans, 30 + 20 m on one side and 30 + 25 + 20 m for the other, are arranged along the mountain slopes. The total length of the bridge is 461 m.

The most obvious characteristics of the bridge are the use of batholite as the lower chords of the side spans and the anchoring of the prestress bars in tensile diagonals on the batholite. The deck is 13.4 m wide with 9 m for lanes and two pedestrian walkways of 1.5 m each. The arch depth is 2.7 m, L/122, and its width is 10.56 m, L/31.3, with a rise-to-span ratio being 1:6. The bridge was constructed by cantilever assembling. A derrick mast with a hoisting duty of 1200 kN was used. The bridge was designed by Communication Department of Guizhou Province and constructed by Bridge Engineering Co. of Guizhou Province.

Jinkui Grand Canal Bridge

Jinkui Grand Canal Bridge (Figure 63.12), with a main span reaching 100 m, is one of the longest rigid-framed prestressed concrete arch bridges on soft-soil foundation. It crosses the Grand Canal in Wuxi County, Jiangsu Province.

The bridge has a rise-to-span ratio of 1:10. The arch rib is of the I type with a constant cross section, while the solid spandrel segment has a variable cross section. Only two inclined braces are

FIGURE 63.13 Taibai Bridge.

arranged on either side to get a aesthetic effect. In order to reduce the deadweight, ribbed slabs are employed for the deck. The substructure includes combined-type thin-wall abutments, which are designed to resist the horizontal thrusts from superstructure by boring piles and slide-resistant slabs working jointly. The bridge was designed by Shanghai Urban Construction College and constructed by Bridge Engineering Co. of Wuxi County. It was completed in 1980.

Taibai Bridge

Taibai Bridge (Figure 63.13), a rigid-framed reinforced concrete arch highway bridge with a span of 130 m, is located in Dexi copper mining area, Jiangxi Province. The bridge was constructed by the swing method. After assembling steel bar skeletons and casting 100 mm bottom slab on simple scaffoldings, 42 25-mm tensile bars were stretched to get the structure separate from the scaffoldings. The whole swing system, with a total weight of 18,100 kN, was supported by a reinforced concrete spherical hinge on abutment foundation. The bridge was designed by Nanchang Non-ferrous Metallurgical Design Institute and constructed by Huachang Engineering Co. It was completed in March 1993.

Wanxian Yangtze River Bridge

The bridge located in Huangniu Kong, 7 km upstream from Wanxian, is an important structure on the No. 318 national highway (Figure 63.14). It is 864.12 m long. A reinforced concrete box arch with a rise-to-span ratio of 1:5 offers a single span of 420 m. Steel pipes are used to form stiffening arch skeletons before the erection of the main arch; there are 14 spans of 30 m prestressed concrete. Simply supported T-girders make up the spandrel structure, while 13 spans of the same girders are for the approaches. The continuous deck is 24 m wide, providing 2 × 7.75 m lanes for motor vehicle traffic and two sidewalks of 3.0 m each. A longitudinal slope of 1% is arranged from the midspan to either side with a radius of vertical curve being 5000 m, while the cross slope is 2%. The bridge was designed by Highway Survey and Design Institute of Sichuan Province and constructed by Highway Engineering Company of Sichuan Province. It was completed in 1997.

63.4 T-Type and Continuous Rigid Frame Bridges

63.4.1 General Description

The prestressed concrete rigid T-frame bridge was primarily developed and built in China in the 1960s. This kind of structure is most suitable to be erected by balanced cantilever construction

FIGURE 63.14 Wanxian Yangtze River Bridge.

process, either by cantilever segmental concreting with suspended formwork or by cantilever erection with segments of precast concrete. The first example of cantilever erection is the Wei River Bridge (completed in 1964) in Wuling, Henan Province, while the Liu River Bridge (completed in 1967) in Liuzhou in Guangxi Zhuangzu Autonomous Region is the first by cantilever casting. The Yangtze River Highway Bridge at Chongqing (completed in 1980), having a main span of 174 m, is regarded as the largest of this kind at present.

From prestressed concrete rigid T-frame bridges were developed multiple prestressed concrete continuous beam and continuous rigid frame bridges, which can have longer spans and offer better traffic conditions. Among others, the Luoxi Bridge in Guangzhou, Guangdong Province (completed in 1988) features a 180-m main span. The Huangshi Yangtze River Bridge in Hubei Province has a main span of 245 m. And the Humen Continuous Rigid Frame Bridge in Guangdong Province (completed in 1997), which has a 270-m main span, is regarded as the largest of this kind in the world.

63.4.2 Examples of T-Type Rigid Frame Bridges

Qingtongxia Yellow River Highway Bridge

Qingtongxia Highway Bridge (Figure 63.15) is 80 km south of Yinchuan, Ningxia. It is 743 m long and 14 m wide. The spans arrangement is 4 × 30 + 60 + 3 × 90 + 60 + 6 × 30 + 20 m. Prestressed concrete T-girders were adopted for the three main spans, while prestressed concrete simply supported beams were used for approaches. The T-frame is a two-cell single-box thin-wall structure, which was built by cantilever casting. The substructure consists of thin-wall hollow box piers, resting on elevated bored pile foundations, the piles having a diameter of 1.5 m. The bridge was designed by Highway Survey and Design Institute of Ningxia Province and constructed by Highway Engineering Bureau of Ningxia Province. It was completed in October 1991.

Huanglingji Bridge

Huanglingji Bridge (Figure 63.16), located in Hanyang County, Hubei Province, is a prestressed concrete truss T-frame highway bridge. It has spans of 7 × 20 + 53 + 90 +53 + 2 × 20 m, with a

FIGURE 63.15 Qingtongxia Yellow River Highway Bridge.

FIGURE 63.16 Huanglingji Bridge.

total length of 380.19 m. The 90-m-long main span is composed of two cantilever arms of 37 m each and a 16-m-long suspended span.

Caisson foundations and box piers were adopted for the substructure. Its superstructure consists of two trusses, with prestressed concrete simply supported slab on the top, which serves as upper bracing after transverse prestressing has been introduced. Prestressing tendons are used for tensile members, while common rebars for compressive members. Longitudinal prestress tendons are arranged in open channels which makes stretching convenient. The cantilever assembling method was employed. The bridge type featuresa slim configuration and saves construction materials. The bridge was designed at Tongji University in cooperation with Highway Engineering Bureau of Hubei Province. The construction unit was Road and Bridge Co. of Hubei Province. It was completed in 1979.

Hongtang Bridge

Hongtang Bridge (Figure 63.17), the longest highway bridge over Min River, is west of Fuzhou City, Fujian Province. It is 1843 m long and 12 m wide. The main span is a three-hinge connected lower chord supported prestressed concrete truss T-frame, which synthesizes the virtues of cable-stayed bridges, truss bridges, and T-frame bridges.

The bridge was erected by cantilever assembling with cable cranes. On the side shoal 31 spans are of prestressed concrete continuous girders erected by adopting nonglued segmental assembling span by span, a new technology first applied in Chinese bridge construction. Spans on the banks are of simply supported prestressed concrete beams.

The substructure of the bridge is prestressed concrete V-type hollow piers on bored piles foundation for the main span and dual-column bored piles foundation for the side spans. The bridge

FIGURE 63.17 Hongtang Bridge.

was designed by Communication Planning and Design Institute of Fujian Province and constructed by the Second Highway Engineering Co. of Fujian Province. It was completed in December of 1990.

63.4.3 Examples of Continuous Rigid Frame Bridges

Luoxi Bridge

Luoxi bridge (Figure 63.18), the longest prestressed concrete continuous rigid frame bridge in China, spans Pearl River in Guangzhou, Guangdong Province. It is 1916.04 m long and 15.5 m wide. The main bridge has spans of 65 + 125 + 180 + 110 m, providing a navigation clearance of 34 × 120 m.

The single-cell box beam has a variable depth, 10 m (i.e., 1/8 of the main span) at root and 3 m (i.e., 1/6 of the main span) at midspan. Three-way prestresses were introduced. A great tonnage group anchorage system with a post-tension force of 4275 kN for each group, which set a record in China, was employed longitudinally, with the tendons reaching 190 m long.

The superstructure was erected by cantilever casting. As the thickness is only 500 mm, the dual-wall hollow box piers of the main span have rather small thrust-resistant rigidity. Artificial islands were constructed around the piers to safeguard against the collision of passing vessels. The top diameter of each island is 23 and 28 m at bottom, with a height of 20 m. Two types of spans, 16 and 32 m, were chosen for the 1376.24-m-long approach mainly based on economical consideration, thus achieving a rather low construction cost. The bridge was designed by Highway Survey and Design Institute of Guangdong Province. It was constructed by Highway Engineering Department of Guangdong Province and completed in August 1988.

Huangshi Yangtze River Bridge

Huangshi Yangtze River Bridge (Figure 63.19) is located in Huangshi, Hubei Province, with its total length reaching 2580.08 m. A 162.5 + 3 × 245 + 162.5 m prestressed concrete continuous box girder rigid frame bridge was designed for the main bridge. The deck is 20 m wide, providing 15 m for motor vehicle traffic and 2.5 m on both sides for non-motor-vehicle traffic.

The approach along the Huangshi bank is 840.7 m long, consisting of continuous bridges and simply supported T-girder bridges with continuous decks, while the approach along the Xishui bank is 679.21 m, being single supported T-girder bridges with continuous decks.

FIGURE 63.18 Luoxi Bridge.

FIGURE 63.19 Huangshi Yangtze River Bridge.

FIGURE 63.20 Humen Bridge (continuous rigid frame).

A 28-m-diameter double-wall steel cofferdam with 16 3-m-diameter bored piles foundation was employed for piers of the main span, which provided enough capacity to resist impact force of ships. The navigation clearance of the bridge is 200 × 24 m, which allows the navigation of a vessel of 5000 tons. The bridge was designed by Highway Planning and Design Institute affiliated with the Ministry of Communications. It was constructed by China Road and Bridge Corporation and completed in 1996.

Humen Bridge

The Humen Bridge (Figure 63.20), an extra major highway bridge over Pearl River, is on the freeway connecting Guangzhou, Zhuhai and Shenzhen. It is composed of bridges of different types.

A rigid frame bridge (150 + 270 + 150 m) is arranged over the auxiliary navigation channel, with its main span reaching 270 m, a world record of the same type. The superstructure of the bridge consists of two separate bridges, each a single-box, single-cell prestressed concrete continuous rigid frame. The 24-m-wide deck provides 214.25 m for motor vehicle traffic. Adoption of 15.24 mm VSL prestress system makes thinner top slabs and no bottom slabs of the box girder possible, the single-box single-cell thin-wall section offers a greater moment of inertia per unit area, and the depth of main girder at the supports is 14.8 m (1/18 of the main span) and 5 m (1/54 of the main span) in midspan. The substructure consists of double thin-wall piers resting upon group piles foundations. The symmetrical cantilever casting method is employed for the erection of the superstructure. The bridge was designed by GuangDong Highway Planning and Design Institute and constructed by Highway Engineering Construction Ltd, Guangdong Province. It was opened to traffic in July 1997.

63.5 Steel Bridges

63.5.1 General Description

Steel structures are employed primarily for railway and railway–highway bipurpose bridges. In 1957, in the city of Wuhan, a railway–highway bipurpose bridge was erected over the Yangtze River, another milestone in China's bridge construction history. The bridge has continuous steel trusses with 128 m main spans. The rivet-connected truss is made of grade No. 3 steel. A newly developed cylinder shaft 1.55 m in diameter was initially used in the deep foundation. (Later in 1962, a 5.8-m cylinder shaft foundation was laid in the Ganjiang South Bridge in Nanchang, Jiangxi Province.) In 1968, another such bridge over the Yangtze River — the Nanjing Yangtze River Bridge — came into being. The whole project, including its material, design, and installation, was completed through the

Chinese own efforts. It is a rivet-connected continuous truss bridge with 160-m main spans. The material used is high-quality steel of 16 Mn. In construction, a deep-water foundation was developed. Open caissons were sunk to a depth of 54.87 m, and pretensioned concrete cylinder shafts 3.6 m in diameter were laid, thus forming a new type of compound foundation. And underwater cleaning was performed in a depth of 65 m. China's longest steel highway bridge is the Beige Yellow River Bridge in Shandong Province (1972), its main span being 113 m long. It has a continuous truss of bolt-connected welded members. The foundation is composed of 1.5-m-diameter concrete bored piles, whose penetration depth into subsoil reaches 107 m, the deepest pile ever drilled in China. A new structure of field-bolting welded box girder paved with orthotropic steel deck was first introduced in the North River Highway Bridge at Mafang, Guangdong Province, which was completed in 1980.

Another attractive and gigantic structure standing over the Yangtze River is the Jiujiang railway–highway Bridge completed in 1992. Chinese-made 15 MnVN steel was used and shop-welded steel plates 56 mm thick were bolted on site. The main span reaches 216 m. The continuous steel truss is reinforced by flexible stiffening arch ribs. In laying the foundation, a double-walled sheet piling cofferdam was built, in which a concrete bored pile was cast in place. When erecting the steel beams, double suspended cable frame took the place of a single one, which is another innovation.

Since the 1980s, stell girder or composite girder bridges have been adopted in the construction of long-span or complex-structure city bridges in China. For example, they were applied in Guangzhong Road Flyover and East Yanan Road Viaduct in Shanghai.

63.5.2 Examples of Steel Bridges

Nanjing Yangtze River Bridge

Nanjing Yangtze River Bridge (Figure 63.21) is a highway and railway double-deck continuous steel truss bridge in Nanjing, Jiangsu Province. On the upper deck there are four lanes of highway traffic, which are 15 m wide, plus two sidewalks of 2.25 m wide each, and on the lower deck two tracks for railway. The main bridge is 1576 m long. If approaches are taken into account, the length of the railway bridge reaches 6772 m and the highway bridge is 4588 m long.

Ten spans were arranged for the main bridge, including a side span of 128 m being simply supported steel truss and nine spans of 160 m each being continuous steel trusses, continued every three spans. The main truss is a parallel chord rhombic truss with reinforcing bottom chord. It was erected by cantilever assembling.

Considering the complex geologic conditions at the bridge site, different types of foundations were used: heavy concrete caissons with a depth of penetration reaching 54.87 m for areas with shallow water and deep coverings; a floating-type steel caisson combined with pipe column foundations was used for the first time at sites of deep water. The bridge was designed and constructed by the Major Bridge Engineering Bureau, Ministry of Railways. It was completed in December 1968.

Jiujiang Yangtze River Bridge

Jiujiang Bridge (Figure 63.22), on the border of Hubei Province and Jiangxi Province, is a double-deck highway and railway bipurpose bridge, with the longest truss span, 216 m, in China at present. The four-lane highway is on the upper deck, with a width of 14 m for motor vehicles and two sidewalks of 2 m each on both sides, while the double-track railway is carried on the lower deck.

The main bridge, divided into 11 spans, is all of steel. The three main spans (180 + 216 + 180 m) are combined truss-arch system, which consists of continuous steel truss beams and flexible steel stiffening arches. Two continuous steel truss beams of 3 × 162 m each are for the side span on the northern bank, while a 2 × 126 m continuous steel beam is over the south bank.

The main truss is of parallel chord triangular type with reinforcing bottom chord members, and its depth is 16 m and doubled at the supports. The stiffening arch over main span has a rise of 32 m and those over the side spans have a rise of 24 m. All steel structures are bolted and welded. The

FIGURE 63.21 Nanjing Yangtze River Bridge.

FIGURE 63.22 Jiujiang Yangtze River Bridge.

first-used 15 MnVN steel has a yield strength of 420 MPa. Double layers of suspenders were applied for assembling the truss beams.

There are different types of foundations employed in the bridge: circular reinforced concrete caisson for No. 1 pier in shoal, application of clay slurry lubricating jacket makes the depth of penetration reach 50 m; double-wall steel cofferdam and bored piles foundation in deep water with favorable rock conditions. The bridge was designed and constructed by the Major Bridge Engineering Bureau, Ministry of Railways. It was completed in May 1992.

63.6 Cable-Stayed Bridges

63.6.1 General Description

Cable-stayed bridges were first introduced into China in the early 1960s. Two trial bridges, the Xinwu Bridge with a main span of 54 m in Shanghai and the Tangxi Bridge with a span of 75.8 m in Yuyang, Sichuan Province — are both reinforced concrete cable-stayed bridges and were completed in 1975.

In 1977 the construction of long-span cable-stayed bridges began. The Jinan Bridge across the Yellow River with a main span of 220 m was completed in 1982. In the 1980s, the construction of cable-stayed bridges developed rapidly over a wide area in China. More than 30 bridges of various types were built in different provinces and municipalities. Among them, the Yong River Bridge in Tianjin has a main span of 260 m, and the Dongying Bridge in Shandong Province has span of 288 m, China's first steel cable-stayed bridge. In addition, the Haiying Bridge in Guangzhou has a 35-m-wide deck, single cable plane and double thin-walled pylon piers; the Jiujiang Bridge in Nanhai of Guangdong Province was erected by a floating crane with a capacity of 5000 kN; the Shimen Bridge in Chongqing has an asymmetrical single cable plane arrangement and a 230 m cantilever cast in place; and the attractive-looking Xiang River North Bridge in Changsha, Hunan Province, was completed in 1990 with light traveling formwork. All are representative of this period with their respective features.

At the beginning of the 1990s, with the completion of the Nanpu Bridge in Shanghai in 1991, a new high tide of construction of cable-stayed bridges began to surge in China. Now more then 20 cable-stayed bridges with a span of over 400 m have been completed, and a large number of long-span cable-stayed bridges are under design and construction. The most outstanding is the Yangpu Bridge with a main span of 602 m, a composite deck cable-stayed bridge in Shanghai.

63.6.2 Examples of Cable-Stayed Bridges

Laibin Hongshui River Railway Bridge

Laibin Hongshui River Bridge (Figure 63.23), a structure 398 m long over Hongshui River, is on the second Xianggui Railway Route. The main bridge, having three spans of 48 + 96 + 48 m, is the first prestressed concrete cable-stayed railway bridge built in China, with two pylons and an H-type cable configuration. The main girder, with a box section of two cells dimensioning 4.8 m (width) 3.2 m (depth), is prestressed longitudinally by 245 mm tendons whose $\sigma_p = 1600$ MPa.

The pylons are 29 m high, rigidly connected with main girder by strong box cross beams. Pot neoprene bearings were employed in the bridge. Three groups of parallel cables were installed on either side of each pylon, and each group consists of six bunches of 705-mm steel strands. To guarantee sufficient fatigue strength, key-grooved composite anchorage was specially designed, which also made adjustment and replacement of a cable possible.

The girders over side spans were cast on scaffoldings, while the middle span was constructed by cantilever casting. During the design and construction, special studies and tests were carried out to obtain the characteristics of the structure under railway loads. The bridge was designed by China Academy of Railway Sciences. It was constructed by Liuzhou Railway Bureau and completed in 1981.

Dongying Yellow River Bridge

Dongying Yellow River Bridge (Figure 63.24), the first steel cable-stayed bridge built in China, is on the highway route along the northern coast in Shandong Province. The total length of the bridge is 2817.46 m. The main bridge, a continuous steel cable-stayed bridge, has five spans of 60.5 + 136.5 + 288 + 136.5 + 60.5 m, while the 2135.46-m approaches are 71 spans of pretensioning prestressed concrete box girders, each being 30 m long.

FIGURE 63.23 Laibin Bridge.

FIGURE 63.24 Dongying Yellow River Bridge.

FIGURE 63.25 Shanghai Yangpu Bridge.

The deck is 19.5 m wide, among which 16 m are for vehicle traffic. Steel box girders with orthotropic plate deck form the main cable-stayed spans. It was erected by cantilever assembling. Each segment for assembling is about 12 m on average and consists of two side boxes, four plate decks, and cross beam. The H-type pylons are 69.7 m high. A fan-type cable configuration was adopted. Ten pairs of cables were installed on either side of each pylon, with an anchorage distance of 12 m on the deck. Each cable consists of 73/127ϕ7 galvanized steel wires with hot-squeezed sheath for protection. Bored piles foundation was employed in the substructure. The pylon rests on separate elevated pile caps which are supported by 22 piles with a diameter of 1.5 m and a length of 96.5 m.

The bridge was designed by Communication Planning and Design Institute of Shandong Province. It was constructed by Communication Engineering Co. of Shandong Province and completed in September 1987.

Yangpu Bridge

Shanghai Yangpu Bridge (Figure 63.25), located in Shanghai Yangpu District, is an important bridge in an urban district, which spans Huangpu River and connects Puxi old district with Pudong Development Zone. It is an essential component of the Inner Ring Elevated Viaduct. The bridge site is 11 km from the Nanpu Bridge, a cable-stayed bridge with 423 m main span completed in 1991.

The overall length of the bridge is 8354 m, including main spans, approach spans, and guide passage spans. The width of the bridge is 30.35 m. The main span is a dual-pylon, space dual-cable

FIGURE 63.26 Wuhan Yangtze River Highway Bridge.

plane, steel–concrete composite structure. As pylons rigidly connected with piers and separate from the girder, the superstructure is a suspended system longitudinally, with displacement-resistant and anti-seismic devices installed transversely.

The 200-m-high diamond-shaped pylons are of reinforced concrete, resting on steel pipe piles foundations. Columnar piers supported by precast concrete piles have found wide application in auxiliary piers, anchor piers, and side piers as well. Steel side box girders and I-type steel cross beams composite with precast reinforced concrete slabs make up the girder over the main span and two side spans. The center-to-center distance between the two main steel side box girders is 25 m, while that between steel cross beams is 4.5 m. For transitional spans, simply supported prestressed concrete T-girders are used. The stayed cables are 256 in number and there are 32 pairs of them on either side of each pylon. The maximum length of stays is 330 m, and 312ϕ7 high-strength parallel wires form the maximum cross section of the stays.

The bridge was designed by Shanghai Municipal Engineering Design Institute in cooperation with Tongji University, Shanghai Urban Construction Design Institute, and Shanghai Urban Construction College. The construction of the bridge was presided over by the Headquarters of Shanghai Huangpujiang Bridge Engineering Construction. It was completed and opened to traffic in 1993.

Wuhan Yangtze River Highway Bridge

The bridge (Figure 63.26) is a 4687.73-m-long structure in Wuhan, Hubei Province. Its cable-stayed bridge consists of prestressed concrete girder with t spans of 180 + 400 + 180 m. The cable-stayed bridge is a suspended system, with its longitudinal displacement restrained by the devices installed at the intersection parts of the girder and the pylons. The deck is 29.4 m wide, carrying six lanes 23 m wide and two pedestrian walks of 1.75 m each.

An open section with two side boxes is adopted, which is 3 m in depth and stiffened by cross beams every 4 m. The H-type reinforced concrete pylons are 94 m high, and a fan-type multicable configuration is adopted. There are a total of 392 cables, which are made up of 7-mm-diameter galvanized parallel steel wires and protected by hot-squeezed PE sheath. The maximum cable force reaches 5000 kN. Double-wall steel cofferdam bored piles foundations are employed. The bridge was designed and constructed by Major Bridge Engineering Bureau, Ministry of Railways. It was completed in 1995.

Huangshan Taiping Lake Bridge

The bridge (Figure 63.27) is a prestressed concrete cable-stayed bridge with a single pylon and a single cable plane. It has two spans of 190 m each. The pylon is 86.6 m high. A fan-type cable configuration with a cable distance of 6 m on the girder was adopted. Four lanes are arranged on

FIGURE 63.27 Huangshan Taiping Lake Bridge.

the deck, which is 18.2 m wide. The main girder is a three-cell prestressed concrete box girder with skew webs. It has a uniform depth of 3.5 m. The thickness of the top slab and the bottom slab are 220 and 200 mm, respectively. Three-way prestress was introduced. The bridge was designed by Lin & Li Consultants Shanghai Ltd., constructed by Major Bridge Engineering Bureau, Ministry of Railways, and opened to traffic in 1996.

63.7 Suspension Bridges

63.7.1 General Description

The construction of modern suspension bridges in China started in the 1960s. Some flexible suspension bridges with spans less than 200 m were built in the mountain areas of southwestern China, the Chaoyang Bridge in Chongqing, Sichuan Province, being the most famous one. However, the Dazi Bridge in Tibet completed in 1984 has a span of 500 m.

The upsurge of transportation engineering construction in the 1990s has led to a new stage of modern suspension bridges. The Shantou Bay Bridge in Shantou, Guangdong Province, was completed in 1995, having a 452 m concrete stiffening girder. The Humen Pearl River Bridge, a steel box girder suspension bridge with a main span of 888 m, was completed in 1997. The Jiangying Yangtze River Bridge with a main span of 1385 m, is now under construction.

63.7.2 Examples of Suspension Bridges

Chaoyang Bridge

Chaoyang Bridge (Figure 63.28), a highway bridge crossing Jialin River, is located in Beipei District, Chongqing. The bridge has three spans with a total length of 233.2 m. A double-chain reinforced girder suspension bridge, over 186 m long, is the main span, and two reinforced concrete slim curved beams are for the two side spans of 21.6 m each. The deck is 8.5 m wide, providing 7 m for motor vehicle traffic.

FIGURE 63.28 Chaoyang Bridge.

There are four cables in total and every two make a chain that is installed on either side. Each cable is made up of 19ϕ42 steel ropes; ϕ 42 steel pipe and ϕ42 steel rope are used as upper hanger and lower hanger, respectively. The stiffening girder, with depth of 2.0 m, is a single open-steel box composited with a reinforced concrete deck slab. The 63.8-m-high pylons are reinforced concrete portal frames. A tunnel-type anchorage system was adopted, and the tunnel length reached 15 m. The anchorage slabs are 1.8 m in depth and anchored in rock stratum.

The bridge was designed by Chongqing Communication Research Institute and Chongqing Communication Institute, and was built by Chongqing Bridge Engineering Corporation. The bridge was completed in 1969.

Dazi Bridge

Dazi Bridge (Figure 63.29), a 500-m suspension bridge crossing Lasa River, is located in Dazi, 25 km east of Lasa, Tibet. It is 4.5 m wide, providing only one lane for highway traffic. The main cables are made up of 5-mm-diameter parallel wires and its sag-to-span ratio is 1:5. The main girder is an open welded steel truss with a depth of 1.5 m. A simple orthotropic deck was formed by the composition of the longitudinal girder and steel plates. Four trucks of 200 kN each were used as design loads.

The bridge was designed by Highway Planning, Survey and Design Institute of Tibet Autonomous Region and constructed by Highway Bureau of Tibet Autonomous Region. It was completed in December 1984.

Shantou Bay Bridge

Shantou Bay Bridge (Figure 63.30), a 2420-m-long structure, crosses Shantou Bay in Shantou, Grangdong Province. The main bridge is a prestressed concrete suspension bridge, with a central span of 452 m and two side spans of 154 m each. Four spans of 25 m prestressed concrete T-girders

FIGURE 63.29 Dazi Bridge.

FIGURE 63.30 Shantou Bay Bridge.

connect the main bridge with the bank on either side. The width of the bridge varies from approaches of 27.3 m to the main bridge of 23.8 m.

The stiffening girder is a prestressed concrete flat box with three cells. It is prestressed longitudinally by tendons inside the top slab and those outside the bottom slab. Two prestressed concrete beams, which are specially designed to connect the central span and two side spans, provide a smooth transition of the deck. A pair of flexible steel piers was installed between each beam and the lower cross beam of a pylon, to get the stiffening girder restrained elastically and maintain the structural flexible characteristics.

The pylons are three-layer reinforced concrete rigid frame structures, each resting on two separate caisson foundations connected by a strong bracing beam on the top. The center-to-center distance

FIGURE 63.31 Hong Kong Tsing Ma Bridge.

between the two cable saddles on one pylon is 25.2 m. Roller bearings were employed during the erection of cables, which were fixed to the saddles later.

The main cables have a sag-to-span ratio of 1:10. Each cable consists of 110 bunches of 9ϕ.5-mm parallel wires, which makes its diameter reach approximately 550 to 630 mm. Every 6 m a hanger, which is composed of a pair of ϕ42 steel ropes, is installed. The segment assembling scheme was adopted for the girder erection. A 5.7-m long precast segment, weighing 16,000 kN, was connected with the completed girder by wet joint.

The bridge was designed and constructed by Major Bridge Engineering Bureau, Ministry of Railways. It was open to traffic in 1995.

Hong Kong Tsing Ma Bridge

Tsing Ma Bridge (Figure 63.31), a highway–railway bipurpose suspension bridge on the freeway between the new airport and the urban district, connecting the Tsing Yi Island and Mawan Island in Hong Kong, is the world's longest of its kind. The main span is 1377 m. The design of its main girder was mainly based on consideration of its aerodynamic stability and a truss type was finally adopted: The cross beams are of Vierendeel truss. The whole longitudinal girder can be treated as a composite structure.

There are six lanes of highway traffic on the upper deck and three passages on the lower deck, the central one on lower deck for railway while the other opened to highway traffic under severe weather condition. The main cable is of 80 × 368 + 11 × 360 steel wires, each wire having a diameter of 5.35 mm. The construction of the bridge was started in 1992 and it was opened to traffic in 1997.

Jiangyin Yangtze River Bridge

Jiangyin Yangtze River Bridge (Figure 63.32) is located on the planned North–South Principal Highway System in the coastal area between Jiangyin and Jinjiang in Jiangsu Province. It is a large suspension bridge with a central span of 1385 m, and will be the first bridge with a span in excess of 1000 m designed and built by Chinese engineers. Its total length reaches nearly 3 km. The deck is designed to carry six-lane highway traffic, while median and emergency parking strips are also considered, with two 1.5-m-wide pedestrian walks on the central span.

Flat steel box girder with wind fairing is adopted, whose depth and width are 3 and 37.7 m, respectively. The two main cables, with sag-to-span ratio being 1:10.5, are composed of five galvanized high-strength wires, and will be erected by the PWS method. The bridge provides a navigation clearance 50 m high. The 190-m-high pylons are reinforced concrete structures. The northern pylon, located in the shallow water area outside the north bank, rests on piles foundation constructed by the sand island method, whereas the southern one is on rock stratum of the bank.

FIGURE 63.32 Jiangyin Yangtze River Bridge.

The south anchorage is of gravity type embedded on rock bed in comparison with the north anchorage gravity-friction type on soft-soil ground. The north anchorage is a massive concrete caisson, measuring 69 by 51 m in plan and 58 m in depth. It is the largest concrete caisson in the world. All the approaches are prestressed concrete beam bridges. A multispan prestressed concrete continuous rigid frame structure was chosen for the northern side span.

The project was designed by Highway Planning and Design Institute, Ministry of Communications in cooperation with Communication Design Institute of Jiangsu Province and Tongji University. It was scheduled to be completed in 1999.

References

1. Guohao Li, *Stability and Vibration of Bridge Structures,* 2nd ed., China Railway Publishing House, 1996.
2. Haifang Xiang, *Bridges in China,* Tongji University Press, A&U Publication Ltd., Hong Kong, 1993.
3. Chinese Design Code for Highway Bridges, People's Communication Press House of China, 1991.
4. Chinese Design Code for Railway Bridges, China Railway Publishing House, 1988.
5. Guohao Li, Computation of Transverse Load Distribution for Highway Bridges, People's Communication Press House of China, 1977.
6. Haifang Xiang, Wind Resistance Design Guideline for Highway Bridges, People's Communication Press House of China, 1996.

7. Yuan Wancheng and Fan Lichu, Study on energy absorption of a new aseismic rubber bearing for bridge, *J. Vibration Shock*, 14(3), 1995 [in Chinese].
8. Rucheng Xiao, Influence matrix method for structural adjustment calculation of internal forces and displacements of concern sections, *Comput. Struct. Mech. Appl.* 9(1), 1992 [in Chinese].
9. Rucheng Xiao, Synthetical bridge nonlinear analysis program system, *China Comp. Appl. Highway Eng.*, 4(1), 1994.

64

Design Practice in Europe

Jean M. Muller

Jean Muller International, France

64.1 Introduction

Europe is one of the birthplaces of bridge design and technology, beginning with masonry bridges and aqueducts built under the Roman Empire throughout Europe. The Middle Ages also produced many innovative bridges. The modern role of the engineer in bridge design appeared in France in the 18th century. The first bridge made of cast iron was built in England at the end of the same century. Prestressed concrete was born in France before extending throughout the world. Cantilever construction and incremental launching of concrete decks were devised in Germany, as well as modern cable-stayed bridges. The streamlined box-girder deck for long-span suspension bridges was born in England. The variety of bridges in Europe is enormous, from the point of view of both their age and their type.

Outstanding works of bridge history in Europe can be presented as follows.

0-8493-7434-0/00/$0.00+$.50

Bridge	Year	Country	Designer	Comments
Unknown	600 B.C.	I	Etruscans	Probable use of vaults for bridge construction
Gardon River Bridge *	13 B.C.	F	Romans	Aqueduct 49 m high, with three rows of superposed arches
Céret Bridge over the River Tech	1339	F	Unknown	Masonry bridge spanning 42 m
Wettingen Bridge	1764	CH	Johann Ulrich Grubenmann	Biggest wooden bridge in Europe with a 61 m span
Coalbrookdale Bridge	1779	GB	Abraham Darby III	First metallic bridge: cast iron structure
Sunderland Bridge	1796	GB	Rowland Burdon	Six cast iron arches, each made up of 105 segments
Saint-Antoine Bridge	1823	CH	Guillaume Henri Dufour	First permanent suspension bridge with metallic cables in the world
Britannia Bridge	1850	GB	Robert Stephenson	First tubular straight girder, spanning 140 m, consisting of wrought iron sheets
Crumlin Viaduct	1857	GB	Charles Liddell	First metallic truss girder viaduct
Bridge over the River Isar	1857	D	Von Pauli, Gerber, Werder	Welded and bolted iron truss girder
Royal Albert Bridge	1859	GB	Isambard Kingdom Brunel	Metal truss girder, first of a whole modern generation of railway bridges
Maria Pia Bridge over the River Douro	1877	P	Gustave Eiffel	Arch spanning 160 m, made up of metal structure
Antoinette Bridge	1884	F	Paul Séjourné	Culmination of masonry bridges
Firth of Forth Bridge *	1890	GB	Sir John Fowler and Sir Benjamin Baker	First large steel bridge in the world — two main spans 520 m long
Alexandre III Bridge *	1900	F	Jean Résal	15 very slender arches composed of molded steel segments
Salginatobel Bridge	1930	CH	Robert Maillard	Arch marking the concrete box-girder birth
Albert Louppe Bridge *	1930	F	Eugène Freyssinet	Three reinforced concrete vaults, each spanning 188 m — wooden formwork spanning 170 m
Linz Bridge over the River Danube	1938	AUT	A. Sarlay and R. Riedl	First welded girder 250 m long — three spans
Luzancy Bridge	1946	F	Eugène Freyssinet	Concrete bridge prestressed in three directions, made up of precast segments
Cologne Deutz Bridge	1948	D	Fritz Leonhardt	Composite steel plate-concrete box-girder bridge spanning 184 m
Percha Bridge	1949	D	Dyckerhoff and Widmann	First reinforced concrete large span cantilever construction
Donzère Mondragon Bridge	1952	F	Albert Caquot	First cable-stayed bridge — 81 m long main span
Düsseldorf Northern Bridge	1957	D	Fritz Leonhardt	First modern cable-stayed metallic bridge
Bendorf Bridge *	1964	D	Ulrich Finsterwalder	Cast-in-place balanced cantilever girder bridge — 208 m long main span
Choisy Bridge	1965	F	Jean Muller	First prestressed concrete bridge consisting of precast segments with match-cast epoxy joints
First Severn Bridge *	1966	GB	William Brown	Decisive stage: deck aerodynamic study in a low- and high-speed wind tunnel
Weitingen Viaduct	1975	D	Fritz Leonhardt	Steel span world record: 263-m-long span
Saint-Nazaire Bridge	1975	F	Jean-Claude Foucriat	Steel cable-stayed bridge world record — 400-m-long main span
Brotonne Bridge	1977	F	Jean Muller	Prestressed concrete cable-stayed bridge world record — 320-m-long main span
Kirk Bridge	1980	Croatie	Ilija Stojadinovic	World record — prestressed concrete arch spanning 390 m
Ganter Bridge	1980	CH	Christian Menn	174-m-long cable stayed span — stay planes protected by concrete walls
Normandie Bridge *	1995	F	Michel Virlogeux	World record — cable-stayed bridge with a 856-m-long main span
Storebaelt Bridge *	1998	DK	Cowi Consult	6.6- and 6.8-km-long bridges including a suspension bridge with a 1624-m long central span
Tagus Bridge	1998	P	Campenon Bernard	13-km-long bridge including a cable-stayed bridge with a 420-m-long main span
Gibraltar Straight Bridge	Project	E	Not yet known	Suspension bridge: 3.5- to 5-km long main spans
Messina Straight Bridge	Project	I	Not yet known	Suspension bridge: 3.3-km-long main span

* A brief description of these bridges are given later with a photograph.

FIGURE 64.1 Gard Bridge over the River Gardon. (*Source*: Leonhardt, F., Ponts/Puentes — 1986 Presses Polytechniques Romandes. With permission.)

If we could choose only eight outstanding bridges, they would be as follows.

1. *Gardon River Bridge* (*13* B.C.) — The Gardon River Bridge, also named Gard bridge, located in France, is an aqueduct consisting of three rows of superposed arches, composed of big blocks of stone assembled without mortar. Its total length is 360 m, and its main arches are 23 m long between pillar axes. It fully symbolizes Roman engineering expertise from 50 B.C. to 50 A.D. (Figure 64.1). Built with large rectangular stones, the bridge surprises by its architectural simplicity. Repetitivity, symmetry, proportions, solidity reach perfection, although the overall impression is that this work is lacking spirit.
2. *Firth of Forth Bridge* (*1890*) — The Forth Railway Bridge, located in Scotland, Great Britain, was the first large steel bridge built in the world. Its gigantic girder span of 521 m, longer than the main span length of the greatest suspension bridges of the time, made this bridge a technical achievement (Figure 64.2). In all, 55,000 tons of steel and 6,500,000 rivets were necessary to build this structure costing more than 3 million sterling pounds. The very strong stiff structure, made of riveted tubes connected at nodes, consists of three balanced slanting elements and two suspended spans, with two approach spans formed of truss girders. The total bridge length is 2.5 km.
3. *Alexandre III Bridge* (*1900*) — This roadway bridge over the River Seine in Paris, France, designed by Jean Résal, bears on 15 parallel arches made up of molded steel segments assembled by bolts. These arches are rather shallow, the ratio is 1/17, and so, massive abutments are necessary. The River Seine is crossed by a single span, 107 m long; the bridge deck is 40 m wide (Figure 64.3).
4. *Albert Louppe Bridge* (*1930*) — This bridge, located in France, is the most beautiful expression of Eugène Freyssinet's reinforced concrete works. The three arches, each spanning 186.40 m

FIGURE 64.2 Firth of Forth Bridge. (Courtesy of J. Arthur Dixon.)

FIGURE 64.3 Alexandre III Bridge. (Courtesy of SETRA.)

(Figure 64.4) crossed the River Elorn for half the cost of a conventional metal bridge. The arches are three cell box girders, 9.50 m wide and 5.00 m deep on average. The deck is a girder with reinforced concrete truss webs. The formwork used for casting the three vaults, moved on two 35 by 8 m reinforced concrete barges, was the greatest and the most daring wooden structure in construction history with its 10-m-wide huge vault spanning 170 m.

FIGURE 64.4 Albert Louppe Bridge. (Courtesy of Jean Muller International.)

5. *Bendorf Bridge (1964)* — Built in 1964 near Koblenz, Germany, this structure has a total length of 1029.7 m with a navigation span 208 m long over the River Rhine. Designed by Ulrich Finsterwalder, it is an early and outstanding example of the cast-in-place balanced cantilever bridge (Figure 64.5). The continuous seven-span main river structure consists of twin independent single-cell box girders. Total width of the bridge cross section is 30.86 m. Girder depth is 10.45 m at the pier and 4.4 m at midspan. The main navigation span has a hinge at midspan, and the superstructure is cast monolithically with the main piers. The structure is three-dimensionally prestressed.
6. *First Severn Bridge (1966)* — The suspension bridge over the River Severn, Wales, Great Britain, designed and constructed in 1966, marks a distinct change in suspension bridge shape during the second half of the 20th century (Figure 64.6). William Brown, the main design engineer, created a 988-m-long central span. The deck is a stiff and streamlined box girder. Its aerodynamic stability was improved in a wind tunnel, with high-speed wind tests under compressed airflow. Since the opening of the bridge, many designers have been drawn from afar to its shape, new at the time, but now looked upon as classical.
7. *Normandie Bridge (1995)* — The cable-stayed bridge, crossing the River Seine near its mouth, in northern France, is 2140 m long. Its 856-m-long main span constitutes a world record for this kind of structure, although the bridge in principle does not bring much innovation in comparison with the Brotonne bridge from which it is derived (Figure 64.7). The central 624 m of the main span is made of steel, whereas the rest of the deck is made of prestressed concrete. The deck is designed specially to reduce the impact of wind blowing at 180 km/h. Reversed Y-shaped pylons are 200 m high. The stays, whose lengths vary from 100 to 440 m, have been the subject of an advanced aerodynamic study because they represent 60% of the bridge area on which the wind is applied.

FIGURE 64.5 Bendorf Bridge. (*Source*: Leonhardt, F., Ponts/Puentes — 1986 Presses Polytechniques Romandes. With permission.)

FIGURE 64.6 First Severn Bridge. (*Source*: Leonhardt, F., Ponts/Puentes — 1986 Presses Polytechniques Romandes. With permission.)

FIGURE 64.7 Normandie Bridge. (Courtesy of Campenon Bernard.)

8. *Great Belt Strait Crossing (1998)* — The Storebælt suspension bridge, located in Denmark, has a central span of 1624 m. It is the main piece of a complex comprising a combined highway and railway bridge 6.6 km long, a twin tube tunnel 8 km long, and a 6.8-km-long highway bridge (Figure 64.8). This link is part of one of the most ambitious projects in Europe, to join Sweden and the Danish archipelago to the European Continent by a series of bridges, viaducts, and tunnels, which can accommodate highway and railway traffic.

64.2. Design

64.2.1 Philosophy

To allow for the single internal market setup, the European legislation includes two directive types:

1. Directives "products," whose purpose is to unify the national rules in order to remove the obstacles in the way of the free product movement.
2. Directives "public markets," aiming to avoid national or even local behaviors from owners or public buyers.

By experience, the only means of ensuring that a bid based on a calculation method practiced in another state is not dismissed is to have a common set of calculation rules. These rules do not necessarily require the same numerical values.

Consequently, the European Community Commission has undertaken to set up a complex of harmonized technical rules with regard to building and civil engineering design, to propose an alternative to different codes and standards used by the individual member states, and finally to replace them. These technical rules are commonly referred to as "Structural Eurocodes."

The Eurocodes, common rules for structural design and justification, are the result of technical opinion and competence harmonization. These norms have a great commercial significance. The

FIGURE 64.8 Storebælt Bridge. (Courtesy of Cowi Consult.)

Eurocodes preparation began in 1976, and drafts of the four first Eurocodes were proposed during the 1980s. In 1990 the European Economical Community put the European Normalization Committee in charge of developing, publishing, and maintaining the Eurocodes.

In general, the Eurocode refers to an Interpretative Document. This is a very general text which makes a technical statement. In the European Community countries the mechanical resistance and stability verifications are generally based on consideration of limit states and on format of partial safety factors, without excluding the possibility of defining safety levels using other methods, for example, probability theory of reliability.

From this document which heads them up, the Eurocodes deal with projects and work execution modes. Numerical data included are given for well-defined application fields. Therefore, the Eurocodes are not only frameworks that define a philosophy allowing the various countries the possibility to tailor the contents individually, they are something completely unique in the normalization field.

A norm defines tolerances, materials, products, performances. The Eurocodes are entirely different because they attempt to be design norms, i.e., norms that define what is right and what is wrong. That is a unique venture of its kind.

The transformation of the Eurocodes into European norms was begun in 1996 and will be reality in 2001 for the first ones. For about 5 years before their final adoption, both the Eurocodes and the national norms will stay applicable.

Of course, there exists a need for connection between Eurocodes and various national rules. Variable numerical values and the possibility of defining certain specifications differently allow this adaptation. From 2007 to 2008 national norms will be progressively withdrawn. Concerning bridges, from 2008 to 2009 only the Eurocodes will be applicable.

These texts are completely coherent, thus it is possible to go from one to the other with coherent combinations. This coherence expands to the building field where its importance is more significant. Moreover, these texts are merely a part of vast normative whole which refers to construction norms, product norms, and test norms.

The Eurocodes are written by teams constituted of experts from the main European Union countries, who work unselfishly for the benefit of future generations. For this reason they are the fruit of a synthesis of different technical cultures. They constitute an open whole. Texts have been written with a clear distinction between principles of inviolable nature and applications rules. The latter can be modulated within certain limits, so that they do not act as a brake upon innovation, and appear as a decisive progress factor. They allow, by constituting an efficient rule of the game, the establishment of competition on intelligent and indisputable grounds.

The Eurocodes applicable to bridge design are as follows.

Eurocode 1: Basis of design and actions on structures [1]
  Part 2 Loads: dead loads, water, snow, temperature, wind, fire, etc
  Part 3 Traffic loads on bridges
Eurocode 2: Concrete structure design [2]
  Part 2: Concrete bridges
Eurocode 3: Steel structure design [3]
Eurocode 4: Steel–concrete composite structure design and dimensioning [4]
Eurocode 5: Wooden work design [5]
Eurocode 6: Masonry structure design [6]
Eurocode 7: Geotechnical design [7]
Eurocode 8: Earthquake-resistant structure design [8]
Eurocode 9: Aluminum alloy structure design [9]

64.2.2 Loads

The philosophy of Eurocode 1 is to realize a partial unification of concepts used to determine the representative values of the actions. In this way, most of the natural actions are based on a return period of 50 years. These actions are generally multiplied by a ULS (ultimate limit state) factor taken as 1.5. The return period depends on the reference duration of the action and the probability of exceeding it. This return period is generally 50 years for buildings and 100 years for bridges. This definition is rather conventional. At the moment, the Eurocode is a temporary norm. Consequently, the Eurocode 1 annex make it possible to use a formula which allows one to change the return period. With regard to traffic loads, Eurocodes constitute a completely new code, not inspired by another code. That means the elaboration was done as scientifically as possible.

The database of traffic loads consists of real traffic recordings. The highway section chosen is representative of European traffic in terms of vehicle distribution. On these real data, a certain number of mathematical processes are realized. But not all data were processed by mathematics and probability. Some situations allow definition of the characteristic load. These are obstruction situations, hold-up situations on one lane with a heavy but freely flowing traffic on the other lane, and so forth, i.e., realistic situations.

All these elements were mathematically extrapolated so that they correspond to a 1000-year return period, that is to say, a 10% probability of exceeding a certain level in 100 years. The axle distribution curve leads one to take into account a 1.35 ULS factor instead of 1.5 for a heavy axle. Concerning abnormal vehicles, the Eurocode gives a catalog from which the client chooses. The Eurocode defines as well, how an abnormal vehicle can use the bridge while traffic is kept on other lanes, which is rather realistic.

With regard to loads on railway bridges, the UIC models were revised in the Eurocode. Loads corresponding to a high-speed passenger train were also introduced in the Eurocode.

There are no military loads in Eurocodes. This type of loads is the client responsibility.

Concerning the wind, the speed measured at 10 m above the ground averaged over 10 min, with a 50-year return period, is taken into account. This return period seems to be somewhat conventional, because this speed is transformed into pressure by models and factors themselves including safety margin.

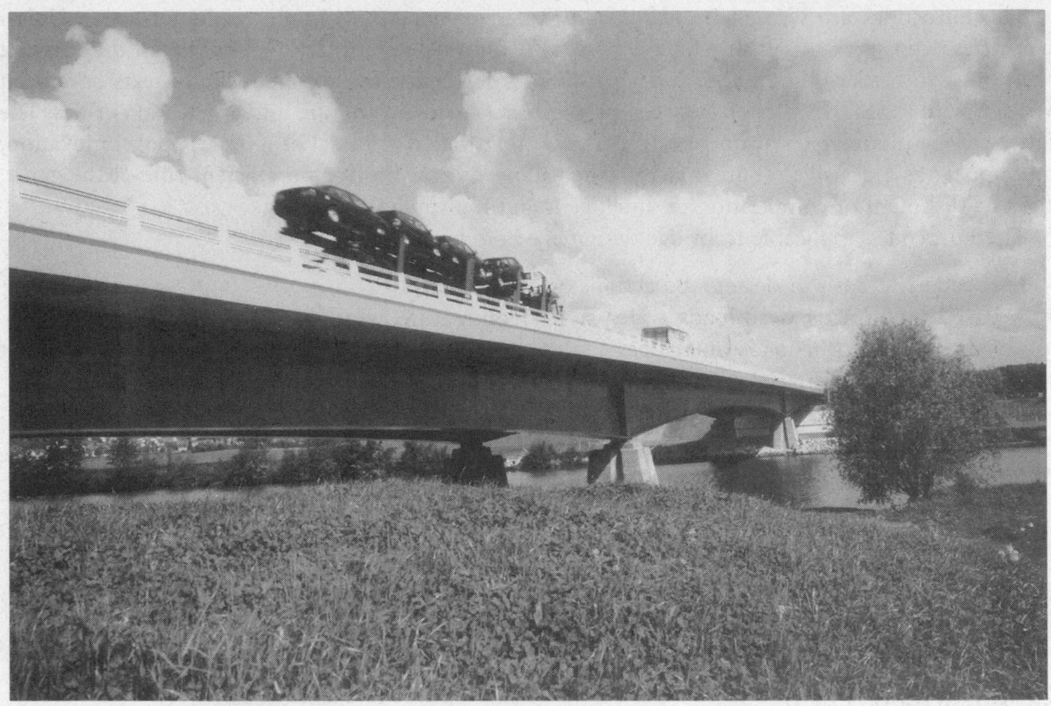

FIGURE 64.9 Oise Bridge. (Courtesy of Fred Boucher, SANEF.)

The most detailed studies show that the return period of the characteristic wind pressure value is rather contained by the interval between 100 and 200 years. After multiplication by the 1.5 ULS factor, this characteristic value has a return period indeed contained by the interval between 1000 and 10,000 years. The code also defines a dynamic amplification coefficient, which depends on the geometric characteristics of the element, its vibration period, and its structural and aerodynamic damping.

With regard to snow loads, the Eurocodes give maps for each European country. These maps show the characteristic depth of snow on the ground corresponding to a 50-year return period. Then this snow depth is transformed into snow weight taking into account additional details.

It is the same case for temperature. The characteristic value is the temperature corresponding to a 50-year return period. The characteristic value for earthquake loads, in Eurocode 8, corresponds as well to a 10% probability of exceeding the load in 50 years.

Therefore, the philosophy is rather clear with regard to loads. Some people wish to go toward greater unification, but it seems to be difficult to realize. Nevertheless, the load definition constitutes a comprehensible and homogeneous whole which is finally satisfactory.

64.3 Short- and Medium-Span Bridges

64.3.1. Steel and Composite Bridges

64.3.1.1 Oise River Bridge

In France, the Paris Boulogne highway link crosses the River Oise on a single steel concrete composite bridge (Figure 64.9). The bridge is 219 m long with a 105-m-long main span over the river and two symmetric side spans. The foundation of the bridge consists of 14 2.80-m-long, about 30-m-deep, diaphragm walls with variable thickness. Pier and abutment design is standard.

FIGURE 64.10 Roize Bridge. (Courtesy of Jean Muller International.)

The bridge deck is a composite structure, 2.50 m deep at midspan and on abutments, and 4.50 m deep on the piers. The steel main girders are spaced 11.40 m. The main girder bottom and top flange widths are constant, but their thicknesses vary continuously from 40 to 140 mm. The concrete slab has an effective width of 18 m. It is transversely prestressed with 4T15 cables, six units every 2.50 m.

The deck steel structure was assembled in halves, one behind each abutment on the embankment. Each half was launched over the river and welded together at midspan. The concrete deck slab was poured using two traveling formworks. The midspan area was poured first, followed by the pier areas.

Since 1994, the link has carried two traffic lanes, which will continue until the foreseen construction of a second parallel bridge.

64.3.1.2 Roize River Bridge

The Roize Bridge carries one of the French highway A49 link roads. Its deck was designed by Jean Muller (Figure 64.10). The choice made was a result of 10 years of studies on reducing the weight of medium-span bridge decks. Here the weight saving was obtained by replacing prestressed concrete cores by steel trusses constituting two triangulation planes (Warren-type) inclined and intersecting at the centerline of the bottom flange, by using a bottom flange formed of a welded-up hexagonal steel tube, and by reducing the thickness of the top slab by the use of high-strength concrete prestressed by bonded strands. The bridge was completed in 1990.

Indeed, innovation of this structure lies in its modular design. The steel structure is composed of tetrahedrons built in the factory, brought to site, and then assembled. The concrete slab also consists of prefabricated elements assembled *in situ*.

The deck is prestressed longitudinally by external tendons to keep a normal compression force in the upper slab on the piers, and to reduce the steel area of the bottom. It is also prestressed transversely.

FIGURE 64.11 Saint Pierre Bridge. (Courtesy of Albert Berenguier, Egis Group.)

The Roize Bridge structure has several advantages: light weight, low consumption of structural steel, industrialized fabrication, ease and speed of assembly, adaptability to complex geometric profile, durability. The basic characteristics are length = 112 m; width = 12.20 m; equivalent thickness of B80 concrete = 0.18 m; structural steel = 112 kg/m^2 of deck; pretensioned prestress = 17 kg/m^3; transverse prestress = 15 kg/m^3; longitudinal prestress = 32 kg/m^3.

64.3.1.3 Saint Pierre Bridge

This bridge is located in the historical center of Toulouse in the southwest of France. Its architecture is inspired by 19th century metal truss bridges with variable depth, while using modern technologies for the execution (Figure 64.11). The bridge is a 240 m long steel–concrete composite structure, partially prestressed. The span lengths are the following: 36.88 m, 3 × 55.00 m, 36.88 m.

It is founded on 1.80-m-diameter molded piles. Each pair of piles is linked by a reinforced concrete box girder. This structure supports a pier consisting of two elements. The deck rests on inclined elastomeric bearings so that the bridge works as a frame in longitudinal direction.

The longitudinal composite structure is made up of two lateral metal truss girders. These girders of variable depth are spaced 11.4 m apart with a cross-beam joining them every 14 m. Both main girders and cross-beams are connected to the concrete slab. The concrete slab is 25 cm thick on the central part bearing the traffic lanes. Toward the edges the slab is 27 cm thick and is placed 75 cm higher than the central part, accommodating the sidewalks.

The structure is prestressed longitudinally by 4K15 cables constituted by greased strands located toward the edges of the slab. Transversely, it is prestressed by greased monostrands located in the slab central part. The steel deck structure is erected from the piers supporting on temporary piling. The concrete slab is poured *in situ* with formwork supported by the now self-supporting steel structure.

This bridge is perfectly integrated into its environment of historic monuments, and opened to traffic in 1987.

FIGURE 64.12 A1 highway overpasses. (Courtesy of J. P. Houdry, Egis Group.)

64.3.2. Concrete Bridges

64.3.2.1 Channel Bridges: Overpasses over Highway A1

A new segmental design for overpasses was developed in France in 1992 to 1993, taking into account the necessity of standardization. The bridges have decks comprising a single transverse slab supported by two longitudinal lateral ribs (Figure 64.12).

This concept, suitable for a wide variety of bridge types with span lengths of between 15 and 35 m, is encompassed in the following ideas:

- The deck is built using precast segments, match-cast, and longitudinally prestressed.
- The segments are transversely prestressed using greased monostrands.
- The lateral ribs are used as barriers.

The main advantages of this type of concept are the possibility of building the overpass without disruption of traffic very quickly, with longer spans, thus fewer spans (two instead of four spans), than for the usual precast conventional overpasses.

64.3.2.2 Progressively Placed Segmental Bridges

Fontenoy Bridge

Fontenoy Bridge is 621 m long and open to traffic in 1979. It allows the crossing of the River Moselle in the north east of France with the following spans: 43.12 m, 10 × 52.70 m, 50.80 m. The foundations are either coarse aggregate concrete footings or bored piles, depending on the resisting substratum. On typical piers the bearings are of the elastomeric type, and on the abutments they are of the sliding type. The deck is a simply supported concrete box girder, 10.50 m wide, with two inclined webs and a constant depth of 2.75 m.

FIGURE 64.13 Fontenoy Bridge. (Courtesy of Campenon Bernard.)

The progressive placement method is used to build the deck, starting at one end of the structure, proceeding continuously to the other end (Figure 64.13). A movable temporary stay arrangement is used to limit the cantilever stresses during construction. The temporary tower is located over the preceding pier. All stays are continuous through the tower and anchored in the previously completed deck structure.

Precast segments are transported over the completed portion of the deck to the tip of the cantilever span under construction, where they are positioned by a swivel crane that proceeds from one segment to the next. The box girder is longitudinally prestressed by internal 12T13 units.

Les Neyrolles Bridge

Nantua and Neyrolles Viaducts allow the A40 highway to link Geneva, Switzerland, to Macon, France. The Neyrolles Viaducts have a total length of 985.5 m divided into three independent structures. It is composed of 20 spans of 51 m approximately, except for one span of 62 m which crosses the "Bief du Mont" stream (Figure 64.14). The deck is a concrete box girder approximately 11 m wide. The box girder was erected of precast match-cast segments.

The assembly was performed by asymmetric cantilevering by means of temporary stays and a deck-mounted swivel crane. The mast ensured the stability through the back stays carried by the previous span. The mast allowed erection of spans up to 60 m. The side spans at the abutments could not be assembled likewise because of the absence of a balancing span. Consequently, these span segments were placed on falsework and finally each span was prestressed and put on its definitive supports by means of jacks. The largest span (62 m) was assembled by both methods of construction mentioned.

The first phase consisted of assembly by stay-supported asymmetric cantilevering until the last stay available. The second phase consisted in erecting the last precast segments on falsework. The bridge was completed in 1995.

FIGURE 64.14 The second Neyrolles viaduct. (Courtesy of Campenon Bernard.)

64.3.2.3 Rotationally Constructed Bridges

Gilly Bridge

The Gilly Bridge, close to Albertville in France, consisting of two perpendicular decks was opened to traffic in 1991. The main bridge crosses the river Isère and the access road to the Olympic site resorts (Figure 64.15).

It is a prestressed concrete cable-stayed bridge, with two spans, 102 m long above the river and 60 m long above the road. The A-shaped pylon is tilted backward 20°. The other bridge supports are a standard abutment on the left bank and a massive abutment acting as counterweight on the right bank. Transversely, the 12-m concrete deck consists of two 1.90-m-deep and 1.10-m-wide lateral ribs with cross-beams spaced 3.0 m supporting the top slab.

The A-shaped pylon was built vertically. It was tilted to its definite position by pivoting around two temporary hinges located at its basis, the pylon being held back by two 19T15 cables. After tilting, hinges were frozen by prestressing and concreting.

FIGURE 64.15 The Gilly Bridge. (Courtesy of Razel.)

The 162-m-long main bridge deck was concreted on a general formwork located on the right bank, parallel to the river. After concreting and cable-stay tensioning, the deck was placed in its definite position by a 90° rotation around a vertical axis. During the deck rotation the whole structure weighing 6000 t is supported on three points. Vertical reactions are measured continuously by electronic equipment to check dynamic effects.

Resorting to original construction methods has allowed realization of a bridge of high quality both structurally and aesthetically.

Ben Ahin Bridge

The Ben Ahin Bridge crossing the river Meuse in Belgium is a cable-stayed asymmetric bridge, 341 m in overall length (Figure 64.16), constructed in 1988. The reinforced concrete bridge deck, partially prestressed, is suspended by 40 cables anchored to a single tower structure. The central span is 168 m long. The deck girder has a box section, 21.80 m wide at the top fiber and 8.70 m at the bottom fiber. The depth, constant along the whole bridge, is 2.90 m.

The entire structure consisting of the tower structure, the stay cables, and the deck girder was constructed on the left bank of the river. After completion it was rotated by 70° relative to the tower axis, in order to swing the bridge around to its final definite position (Figure 64.17). Two pairs of jacks, each 500 ton force, located underneath the pylon sliding on Teflon, and four jacks each 300 ton force, located 45 m from the pylon underneath a stability metal frame, allowed the rotation of the 16,000 ton structure.

This method, already used in France for lighter bridges, was in this case designed to set a world record.

64.3.3. Truss Bridges

64.3.3.1 Sylans Bridge

The Sylans Viaduct runs through the French Jura Mountain complex. In this location, along the shores of a lake, difficulty lies in the uncertainty of the foundation soil since the route runs along a very steep slope whose 30-m-thick surface stratum comprises an eroded and fractured material of very doubtful stability.

FIGURE 64.16 Ben Ahin Bridge. (Courtesy of Daylight for Greisch.)

FIGURE 64.17 Ben Ahin Bridge during rotation. (Courtesy of Photo Studio 9 for Greisch.)

FIGURE 64.18 Sylans Bridge — two parallel decks. (Courtesy of Bouygues.)

The 1266-m-long viaduct comprises 21 60-m-long spans, each composed of two identical parallel decks 15 m apart and staggered 10 m in height (Figure 64.18), and was constructed in 1988. The deck is a prestressed concrete space truss structure 10.75 m wide and 4.17 m deep all along the bridge. It consists of 586 precast segments, i.e., 14 segments for each viaduct span.

Each typical concrete segment consists of two slabs linked by four inclined planes of diagonal prestressed concrete braces of 20 cm^2 cross section, assembled in pairs in the form of Xs. For every segment the diagonal braces are precast separately with a concrete of 65 MPa cylinder strength, and assembled with the segment-reinforcing cage. Then, the top and bottom slabs are poured with 50-MPa concrete. Finally, the diagonals are prestressed.

The deck segments are put in place by the cantilever method using a 135 m long launching girder. The deck prestressing consists of four families:

- Cantilever cables located below the top slab: 4T15 units;
- Strongly inclined cables from pier to withstand the shear force: 12T15 units;
- Horizontal continuity cables on and inside the bottom slab: 12T15 units;
- Horizontal cables in the top and the bottom slabs: respectively, 4T15 and 7T15 units.

The deck bears on its piers through reinforced elastomeric bearings.

Piers are supported by 6- to 35-m-tall, 4-m-diameter caissons. A circular concrete cap is cast on the caissons and anchored to the hard bedrock. In all, 3.5 years were necessary to build this bridge designed with the intent of achieving the maximum lightness possible.

64.3.3.2 Boulonnais Bridges

The three Boulonnais Viaducts are located on A16 highway which links Great Britain to the urban area of Paris, France, via the Channel Tunnel, and was completed in 1998. Their characteristics are as follows:

Name	Length, m	Span Distribution	Height above the Valley Floor, m
Quéhen	474	44.50 + 5 × 77.00 + 44.50	30
Herquelingue	259	52.50 + 2 × 77.00 + 52.50	25
Echinghen	1300	44.50 + 3 × 77.00 + 93.50	75
		5 × 110.00 + 93.50 + 3 × 77.00 + 44.50	

FIGURE 64.19 Boulonnais Bridges — pier transparency. (Courtesy of Jean Muller International.)

The foundations consist of diaphragm walls to a depth of 42 m. The typical pier is based on four diaphragm walls, whereas tallest piers are founded on eight diaphragm walls. These diaphragm walls were realized using drilling mud. Quantities are 3800 m of diaphragm walls, a third of which was excavated with a cutting bit; 10,000 m^3 of concrete; 870 tons of reinforcing steel.

Each pier consists of two slender shafts, of diamond shape. These are linked on top by an aesthetically pleasing pier cap, on which the deck is supported (Figure 64.19).

The gap between the two pier shafts increases the bridge transparency created by the truss at deck level. The four tallest pier shafts are linked on their lower part by a transverse wall to increase the buckling stability.

The deck is a composite structure made of match-cast segments, assembled by cantilever method. The three bridges are formed by 524 segments. The deck structure consists of two prestressed concrete slabs, joined by four inclined V-shaped steel planes. Six inclined planes improve the transverse behavior of the deck near bridge supports.

The 23-cm-thick top slab is stiffened by four 70-cm-deep longitudinal ribs located in the diagonal planes. The top slab is prestressed transversely. The 27-cm-thick bottom slab is stiffened by longitudinal ribs and by two transverse beams per segment.

The deck is built by the cantilever method using a 132-m-long launching gantry weighing 500 tons. Segments, weighing 125 tons at the minimum, are put in place symmetrically in pairs. Imbalance between both cantilevers during erection never exceeds 20 tons.

FIGURE 64.20 Dole Bridge. (Courtesy of Campenon Bernard.)

The Echinghen Viaduct is located on a very windy site, a few kilometers from the Channel shore. Gusts of wind exceed 57 km/h 103 days a year, and 100 km/h 3 days a year. A project-specific calculation taking into account the turbulent wind was developed to study the bridge construction phases. This calculation led to imposition of very rigorous cantilever construction kinematics.

Moreover, a wind screen was designed for the windward side of the deck in prevailing wind to avoid very strict traffic limitations.

64.4. Long-Span Bridges

64.4.1 Girder Bridges

64.4.1.1 Dole Bridge

The Dole Bridge, completed in 1995, crossing the River Doubs in France, is 496 m long. It is a continuous seven-span box girder with variable depth. The typical span is 80 m long (Figure 64.20). The deck is erected by the balanced cantilever method using a traveling formwork.

The deck is a composite structure, 14.5 m wide, with two concrete slabs and two corrugated steel webs. The webs are welded to connection plates fixed to the top and bottom slabs by connection angles. Pier and abutment segments are strictly concrete segments.

The deck is longitudinally prestressed by three tendon families:

- Cantilever tendons, anchored on the top slab fillets: 12T15 tendons;
- Continuity tendons, located in the bottom slab in the central area of each span: 12T15 tendons;
- External prestressing, tensioned after completion of the deck, with a trapezoidal layout. The technology used allows removal and replacement of any tendon.

The Dole Bridge is the fourth bridge with corrugated steel webs erected in France.

FIGURE 64.21 Nantua Viaduct. (Courtesy of Campenon Bernard.)

64.4.1.2 Nantua Bridge

Nantua and Neyrolles Viaducts allow the A40 highway to link Geneva, Switzerland, to Macon, France. The Nantua viaduct is 1003 m long, divided in 10 spans. It was constructed in 1986. Its height above the ground varies from 10 to 86 m (Figure 64.21).

The western viaduct extremity is a 124-m-long span supported in a tunnel bored through the cliff. To balance this span, a concrete counterweight had to be constructed inside the cliff in a tunnel extension. The counterweight translates on sliding bearings of unusual size. The relatively large spans (approximately 100 m long) necessitated a variable-depth concrete box girder.

The construction principle for the deck is segments cast *in situ* symmetrically on mobile equipment. The 11.65-m-wide deck, for the first two-way roadway section of the highway, is longitudinally prestressed by cables located inside the concrete.

Various foundation methods were used, necessitated by differences in the soil bearing capacity.

64.4.2 Arch Bridges

64.4.2.1 Kirk Bridges

These concrete arch bridges were designed to provide a link between the Continent and the Isle of Kirk (former Yugoslavia). The two arches have spans of 244 and 390 m, respectively (Figure 64.22). The largest span represents a world record in its category. The box-girder arches are 8 m (width) × 4 m (height) and 13 m (width) × 6.50 (height), respectively.

The construction was carried out in two phases: In the first phase a box-girder arch, constituting the central part of the bridge, was made by using onshore precast segments. The assembly was performed by cantilevering from both banks by means of a mobile gantry (which was carried by the part of the arch already constructed) and of temporary stays. The use of precasting provided a better quality of concrete, a more precise tolerance of fabrication and reduced construction time. The keystone of the arch was likewise placed by means of a mobile gantry. The closure of the two

FIGURE 64.22 Kirk Bridges. (*Source*: Leonhardt, F., Ponts/Puentes — 1986 Presses Polytechniques Romandes. With permission.)

FIGURE 64.23 La Roche Bernard Bridge. (Courtesy of Campenon Bernard.)

semiarches was controlled by means of hydraulic jacks. The second phase of construction consisted of placing the lateral parts of the bridge, composed of large beams connected to the central arches. An *in situ* concreting of the joints between the precast segments and vertical and transversal prestressing ensure the monolithic integrity of the structure.

64.4.2.2 La Roche Bernard Bridge

La Roche Bernard Bridge, completed in 1996, is 376 m long and 20.80 m wide. It crosses the River Vilaine in Brittany, France, by an arch spanning 201 m and small approach spans (Figure 64.23).

The deck is a composite structure consisting of a steel box girder, 1.67 m deep with a trapezoidal shape, covered by a thin 23-cm-thick prestressed concrete slab. It is supported on four piers founded on the ground and six small piers fixed on the arch. The piers are spaced between 32 and 36 m. Like many other composite decks, the box girder is launched using a launching nose (20 m long); the slab is cast afterward. The concrete arch is 8 m wide with a height varying from 3.50 m at the springing to 2.90 m at the crown.

For the erection, the balanced cantilever process was applied using traveling formwork. Moreover, three temporary bents with 500 t jacks and two temporary pylons were successively used. The temporary bents were located below the segments S3 (the third), S5, or S15, and the temporary pylons were located on the riverbank or on the top of segment S15.

Except for segments S0 (springing segment) to S6 using the temporary, all other segments were erected by use of temporary pylons and temporary stays (11T15 and 13T15 units). The segments S7 to S13 were erected by means of stays fixed to the pylon on the riverbank and the temporary bent below S5.

The other segments S17 to 27 were erected by the use of stays fixed on the main pylon and by the use of bents below segments S5 and S15. The main pylon was placed on segment S15 and anchored in the previously erected segments

While the number of stays fixed on the main pylon increased during erection, the number of stays on the other pylon decreased. Consequently, when the segment S20 was supported by the temporary stays, fixed to the main pylon, all stays on the other pylon had been removed.

64.4.2.3 Millau Bridge

To allow the highway A75, in France, to link two plateaus separated by the Tarn Valley five different crossings were designed. One of the proposals for traversing the 300-m-deep and 2500-m-wide valley was developed by JMI and consisted on a large arch and two approach viaducts. Two types of structures were designed for the deck: the basic scheme was based on a concrete box girder, while the alternative project was based on a steel–concrete composite box girder. Many features are common for the two designs, which is the reason only the basic project is described below:

The crossing is divided into three viaducts:

- The north approach viaduct: 486.50 m long, with four spans of between 66.50 and 168 m;
- The main viaduct: one arch spanning the 602 m over the river (Figure 64.24);
- The south approach viaduct: 1445.5 m long, with eight spans of 168 m and one shorter span of 101.50 m.

The 24 m wide roadway is carried by a 8-m-wide concrete box girder whose depth varies from 4 m at midspan to 10 m on pier, except at the central part of the arch where the depth is constant and equal to 4 m. Transversely, both 8-m-wide cantilevers are supported by struts, spaced 3.50 m. The box-girder webs are vertical and 500 mm thick. The bottom slab thickness decreases from 600 mm on pier to 300 mm at midspan.

For the approach viaducts and the first spans on the arch, the balanced cantilever method using traveling formwork is applied. Two families of PT are used: internal PT split in cantilever or continuity units and external PT for general continuity units.

Due to the great length of this bridge, an expansion joint is placed at midspan between P12 and P13, about 1500 m from the north abutment. This joint is equipped with two longitudinal steel girders simply supported on either side of the joint, which allow partial transfer of the bending moment and transfer of the shear force while reducing the deflections.

64.4.3. Truss Bridges

Bras de la Plaine Bridge

The future bridge, located on Isle of La Réunion, in the Indian Ocean (France), will span over the Bras de la Plaine valley which has highly inclined slopes (80°) and reaches a depth of 110 m.

The single-span prestressed composite truss deck, 270 m long, has an innovative static scheme: two cantilevers are restrained in counterweight abutments and linked at midspan by a hinge

FIGURE 64.24 Millau Bridge. (J. P. Houdry, Courtesy of Alain Spielmann.)

FIGURE 64.25 Bras de la Plaine Bridge. (Courtesy of Jean Muller International.)

(Figure 64.25). The deck structure, 17 m deep near the abutments and 4 m deep at midspan, comprises two concrete slabs linked by two inclined truss planes.

The upper 60 MPa (cylinder) concrete slab is 12 m wide. The lower 60 MPa (cylinder) concrete slab has a parabolic profile with variable thickness and width. Each truss panel consists of circular steel diagonals connected directly to the concrete slabs.

FIGURE 64.26 Theodor Heuss Bridge. (*Source*: Beyer, E., Bruckenbau, Beton Verlag, 1971. With permission.)

At midspan, four girders allow transmission of vertical and horizontal shear force and horizontal bending moment. The prestressing system is composed of internal tendons only located in the upper slab. Deck erection will begin by the end of 1999 using the standard cantilever erection method.

64.4.4. Cable-Stayed Bridges

64.4.4.1 Theodor-Heuss bridge

This bridge, also called "Northbridge", belongs to a family of three steel structures on the Rhine River in Düsseldorf, Germany. Northbridge is the first of the two, built in 1957, and belongs to the first generation of cable-stayed bridges.

This type of bridge was conceived to allow the crossing of large spans without intermediate ground support using cables to support the deck elastically in construction (Figure 64.26). The steel deck is 26.60 m wide and 476 m long divided into two approach spans of 108 m and the main span of 260 m. On the flooded riverbank, a five-span approach bridge extends the cable-stayed bridge. The four pylons are 41 m high, slender (1.90 long vs. 1.55 m wide) and spaced 17.60 m.

The main span is supported by four pairs of three cables fixed to the pylons. The three cables are parallel, set out like a harp in a single vertical plane and anchored in each edge of the deck with a spacing 36 m. Due to this spacing, the deck must be stiff, hence a depth of 3.14 m. This depth is extended further on to the approach bridge.

Regarding its erection, it was one of the first times that the balanced cantilever method was applied. The first cantilever segment of 36 m long was erected with the deck elastically supported with one pair of stays. The second segment and the others were erected at midspan.

64.4.4.2 Saint Nazaire Bridge

The bridge of St. Nazaire near the mouth of the Loire River in France, is approximately 3350 m long (Figure 64.27). It is composed of a central part, a 720-m cable-stayed steel bridge, and of two approach viaducts consisting, respectively, of 22 and 30 spans made up of precast concrete girders, each span being 50 m long.

The cable-stayed bridge has a central span 404 m long and two 158 m lateral spans. It is composed of steel box girders, 15 m wide. The construction of the cable stayed bridge, completed in 1975, was carried out in three phases.

FIGURE 64.27 Saint Nazaire Bridge. (Courtesy of Jean Muller International.)

1. The first phase consisted of the construction of the side spans. The steel box girders were assembled in the factory to pieces of 96 m. Then two segments each 96 m were assembled on site by welded joints and transported by two barges to be ultimately hoisted up to their final position.
2. In the second phase, the segments constituting the pylons were assembled on the bridge deck. Then the pylon was lifted by rotation to reach its definitive position.
3. The third phase consisted in erecting the central span as two cantilevers of 197.20 m of length with closure joint at midspan. The segments were lifted from barges with beam-and-winch system.

64.4.4.3 Brotonne Bridge

The Brotonne Bridge was designed to cross the River Seine downstream from Rouen, France (Figure 64.28). It was opened to traffic in 1977. It is composed of two approach viaducts and a cable-stayed structure with a 320-m-long central span. The deck consists of a prestressed concrete box girder 3.97 m deep and 19.20 m wide (Figure 64.29). The stays and the pylon (Figure 64.30) are placed in a single plane along the longitudinal axis of the bridge. The approaches and the main bridge were erected in the same way. In both cases a cantilever construction was used with success. The length of the segments was 3 m.

The segments were cast in place except for the webs which were precast and prestressed. The erection of the deck-girder consisted of extending the bottom slab form of the traveling formwork carried by the previous completed segment, then placing the precast webs that formed the basic shape and acted as a guide for the remaining traveling formwork. The webs were transported and lifted by a tower crane. Concerning the main bridge, the stays were tensioned in every two segments and were anchored in the top slab axis. For the segments, two inclined internal stiffeners were provided to transfer vertical loading generated by the stays. These stiffeners were prestressed.

FIGURE 64.28 Brotonne Bridge. (Courtesy of Campenon Bernard.)

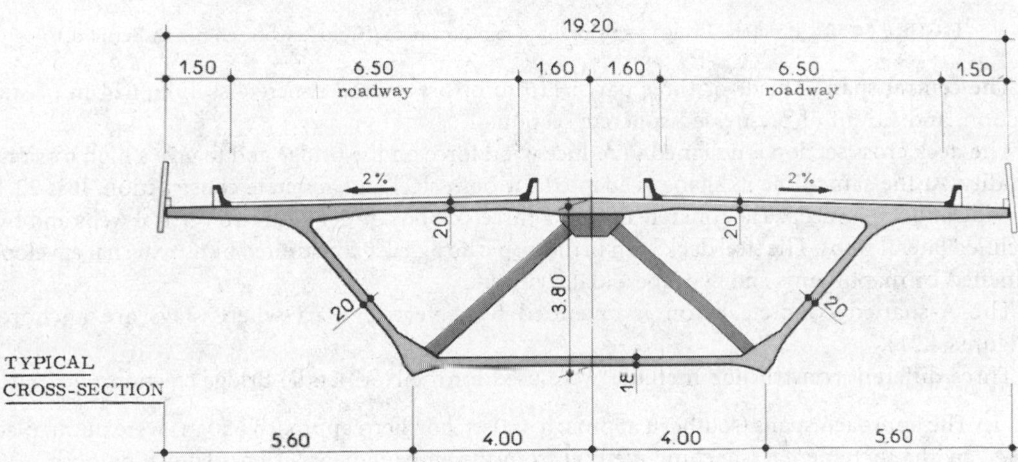

FIGURE 64.29 Brotonne Bridge — typical cross section. (Courtesy of Campenon Bernard.)

64.4.4.4 Normandie Bridge

Since 1994, the Normandie Bridge has allowed the A29 highway to pass over the River Seine near its mouth in northern France (see Figure 64.7). It is a cable-stayed bridge, 2141 m long with the following spans:

$$27.75 \text{ m} + 32.50 \text{ m} + 9 \times 43.50 \text{ m} + 96.0 \text{ m} + 856 \text{ m}$$
$$\text{(longest cable stayed span in the world)} + 96.00 \text{ m} + 14 \times 43.50 \text{ m} + 32.50 \text{ m}$$

FIGURE 64.30 Brotonne Bridge — pylon base reinforcement. (Courtesy of Campenon Bernard.)

The central span is made of three parts: 116 m of prestressed concrete section, 624 m of steel section, and 116 m of prestressed concrete section.

The deck cross section is designed to reduce wind force on the bridge and to give a high torsional rigidity. At the same time its shape is adapted for both steel and concrete construction. It is 22.30 m wide and 3.0 m deep. The concrete deck is a three-cell box girder with two vertical webs and two inclined lateral webs. The steel deck is an orthotropic box girder constituted by an external envelope, stiffened by diaphragms and by trapezoidal stringers.

The A-shaped concrete pylon is extended by a vertical part where stays are anchored (Figure 64.31).

Three different construction methods were used for the Normandie Bridge erection:

1. The approach spans (southern approach 460 m, northern approach 650 m) were put in place by the incremental launching method from the embankment, using a launching nose.
2. On both sides of the 200-m-tall pylons, the superstructure was built by the cable-stayed balanced cantilever method with segments cast *in situ* in a traveling formwork. From the 90-m-long cantilevers, the 96-m side span was joined to the incrementally launched spans. Then the construction of the concrete deck was finalized with an additional 20 m of cast *in situ* cable-stayed cantilever on the river side.
3. The central part of the main span was erected by 19.65-m-long steel segments supplied by barge, lifted up by crane, and finally welded to the previous segment. A pair of cables was tensioned before moving the crane to lift the following segment.

The bridge foundations are the following:

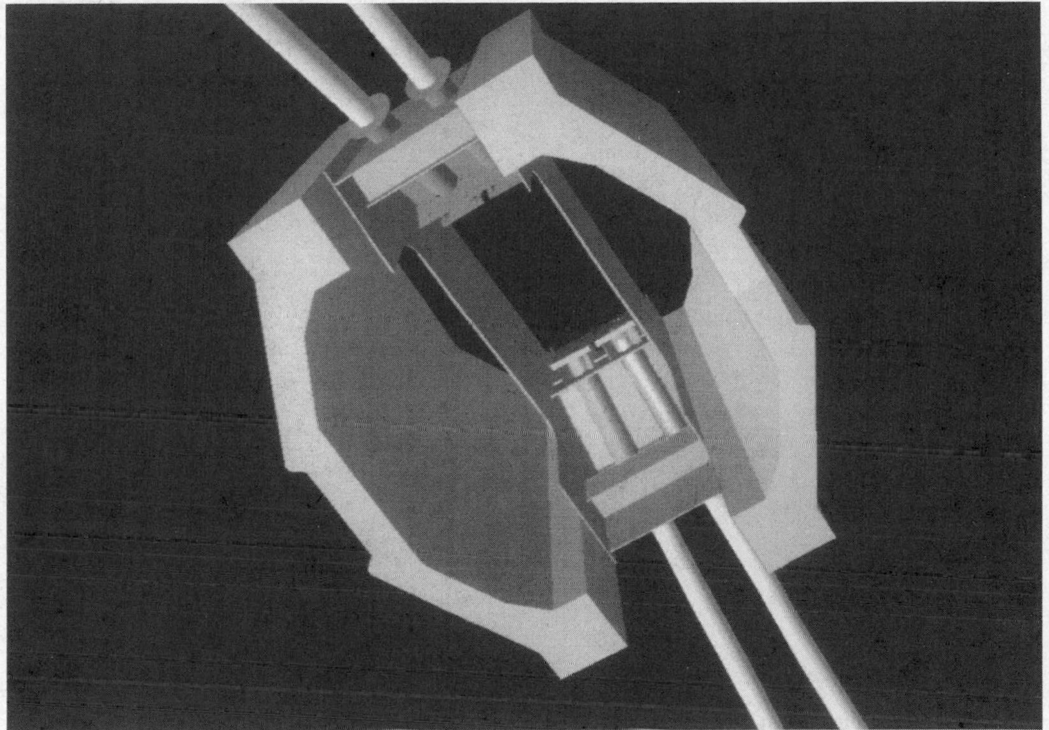

FIGURE 64.31 Normandie Bridge — cable stay anchorages. (Courtesy of Campenon BErnard.)

- Piers and abutments are founded on 1.50-m-diameter, 40-m-deep bored piles — four or five piles per pier.
- The towers are founded on 2.10-m-diameter, 50-m-deep bored piles — 14 piles for each pylon leg.

64.4.4.5 Bi-Stayed Bridge

The clear span of a conventional cable-stayed bridge is limited by the capacity of the deck to resist the axial compressive loads near the pylons created by the horizontal component of the stay forces. For the current materials (70 MPa high-strength concrete, for example), the limit span is between 1200 and 1500 m, depending upon the imagination and the boldness of the designer. Beyond this limit, only suspension bridges allow spanning very large crossings. This situation has now changed, thanks to the new so-called bi-stayed concept.

Deck construction still proceeds in the same fashion as for conventional cable-stayed bridges; starting from the pylons outwardly in a symmetrical sequence, the deck is suspended by successive stays. At a certain stage of construction [for a deck length equal to "a_1," Figure 64.32: (13a)] on either side of each pylon, for example), the deck axial load will have absorbed the full capacity of the materials (with provision for the future effect of live loads). No additional deck length may be added, without exceeding the allowable stresses.

At this stage, a second family of stays is installed [(Figure 64.32 (13b)], assigned to suspend the center portion of the main span. These additional stays are symmetrical with one another with regard to the main span centerline and no more with regard to the pylon. Furthermore, they are no longer anchored in the deck itself, but rather in outside earth abutments at both ends of the bridge, much in the same way as the main cables of a suspension bridge. The vertical load assigned to each stay is now balanced along a continuous tension chain, made up of the center portion of

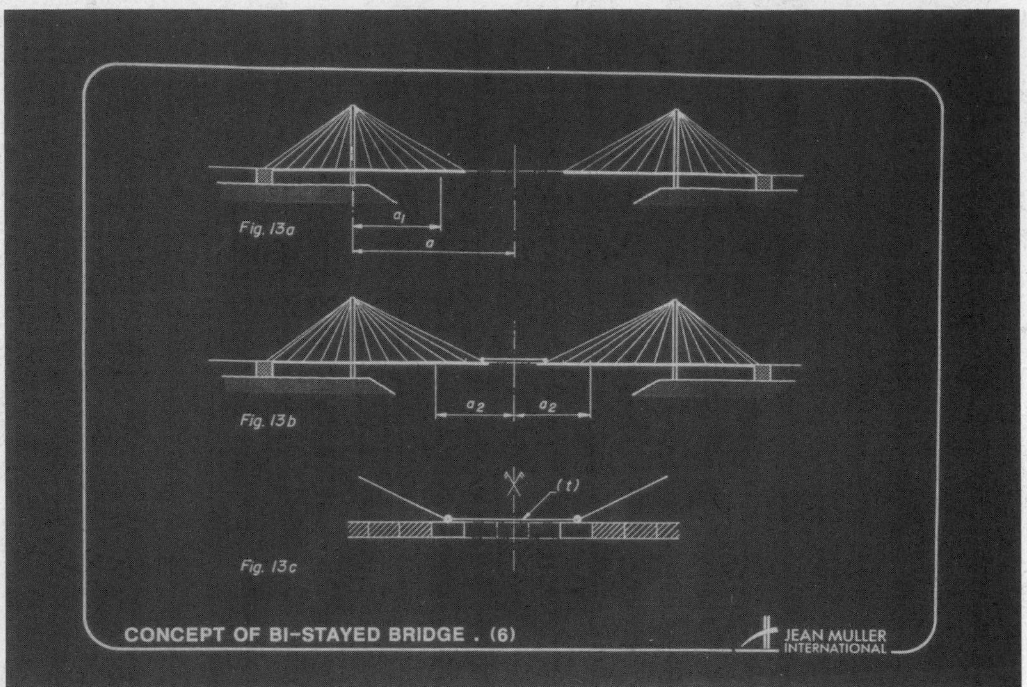

FIGURE 64.32 Bi-stayed bridge. (Courtesy of Jean Muller International.)

the deck (subjected to tension loads), associated with two symmetrical stays, deviated above the pylon heads, to be finally anchored outside the bridge deck.

Along the deck, an axial compression load appears in the vicinity of the pylons (created by the first family of stays), changed into a tension axial load at the centerline of the main span (created by the second family of stays).

In this first application of the new concept, one may increase the maximum clear span in the ratio (a_1 + a_2/a_1), i.e., about 1.5.

In fact, it is possible to go much beyond that stage, while improving the quality of the structure, by using prestressing [Figure 64.32 (13c)]. On the deck length suspended by the second family of stays, prestressing tendons are installed to offset at least all axial tension forces due to dead and live loads. When no live load is applied, the deck is subjected to a compression load, which vanishes when the bridge is fully loaded.

With the usual proportions of dead to live loads, it is easily demonstrated that the maximum span length can be multiplied by 2.5. One can now consider with confidence the construction of a clear span of 3000 m.

A practical example of the new concept was prepared for an exceptional crossing in Southeast Asia with a 1 200 m clear main span. The deck carried six lanes of highway traffic, two train tracks, and two special lanes for emergency vehicles. The bridge was also subject to typhoons.

64.5. Large Projects

64.5.1. Second Severn Bridge

The second Severn Bridge provides a faster link between England and Wales. The structure, 5126 m long (Figure 64.33), consists of three parts: the eastern viaduct, 2103 m long; the main cable-stayed bridge, 946.6 m long; and the western viaduct, 2077 m long. From east to west the bridge span lengths are:

FIGURE 64.33 Second Severn Bridge at twilight time. (Courtesy of G.T.M.)

32 m, 58 m, 23 × 98.12 m, 456 m, 23 × 98.12 m, 65 m

1. The approach viaducts are founded on multicellular reinforced concrete caissons, one per pier, precast on land, with a weight varying from 1100 to 2000 according to the piers. These caissons are transported by barge from the precasting yard to the relevant site. The barge is equipped with a pair of crawler tractors of 1500-ton loading capacity.

Of the 47 concrete piers, 38 are precast, representing 338 concrete elements. Three to seven elements joined by wax-grouted vertical prestressing are necessary to build one pier. Two rectangular pier shafts are erected on each foundation caisson.

The approach viaduct deck consists of two parallel monocellular prestressed concrete box girders connected by the upper slab to provide a 33.20-m-wide platform on most of the bridge length. The deck depth varies from 7.0 m on pier to 3.95 m in the span central part.

The approach viaduct deck is divided into about 500-m-long sections. Expansion joints are located at midspan. The typical deck section consists of four spans and two cantilevers supported by five piers.

All spans are made of 3.643-m-long precast segments; these match-cast segments are put in place by the balanced cantilever method with epoxy joints. For this construction a 230-m-long launching gantry weighing 850 tons is used. All prestressing cables are external with all tendons individually protected in wax-grouted HDPE (high density polyethylene) sheaths. Four prestressing cable families can be distinguished:

- Cantilever tendons: 11 to 12 pairs per cantilever
- Continuity bottom tendons: 3 to 4 pairs per span
- Continuity top tendons: 1 to 2 pairs for span
- General continuity tendons: 5 to 6 pairs per span, spread over two spans

2. The bridge environment is particularly constraining: the Severn estuary is subject to the second strongest tides in the world which represents a differential capable of exceeding 14 m, with strong currents of 8 to 10 knots in certain places and occasionally strong winds. Furthermore, 80% of the foundations are exposed at low tide.

This means that the key to this challenge of the tides is a maximum use of prefabricated components. That explains choices made for the approach viaducts: precasting of foundation caissons, piers at sea, deck segments. That explains as well the main bridge pylon cross-beam and the anchorage block precasting, and precasting of the cable-stayed bridge deck elements.

3. The cable-stayed bridge is 946.60 m long. It is a symmetric work with the following span lengths: 49.06 m, 2 × 98.12 m, 456 m, 2 × 98.12 m, 49.08 m. The bridge towers are founded on precast multicellular caissons. Each 137-m-tall tower consists of two rectangular hollow concrete shafts: reinforced concrete for the typical section and prestressed concrete for the stay anchorage area.

Each pylon caisson is equipped with a 45 m^3/h capacity ready-mix plant, and two metallic platforms to store reinforcement and formwork. This equipment allows one to give maximum autonomy to pylon teams. Pylon shafts are concreted *in situ* with a climbing formwork in 3.80-m-long sections. The cross-beams are precast on land and weigh 1300 and 900 tons, respectively. The lower cross-beam is lifted in place by a crane barge and then linked by concreting to the pylon legs. The upper tie beam is lifted and put down on the lower one, and then lifted to its definite position by jacking.

The first cross-beam is located at a 40-m height above the highest tide, the other forming a frame on the level of the stay-cable anchorage area. The main bridge deck is simply supported on the lower pylon cross-beams with transverse stops. It is supported on four secondary piers on both side spans with antiuplift bearings, and last simply supported on the access viaduct extremities.

The deck is a composite structure consisting of

- Two 2.50-m-deep I girders linked every 3.65 m by a truss beam. The distance between the two main girders is 25.2 m.
- A reinforced concrete slab about 35 m wide and 20 cm or 22.5 cm thick for a typical section.

The deck is assembled of 128 precast elements, 34.60 m wide and 7 m long. The steel structure is assembled by bolts at the precasting yard; then the concrete slab is poured except at the connection joint between two consecutive segments. Each standard segment weighing about 170 t is positioned by trailer and transported by barge to the site. The deck segments are lifted and positioned by a pair of mobile cranes located at the end of each cantilever, and bolted to the previous segment. Then two stays are tensioned and the joint with the previous segment is concreted.

The bridge deck is supported by four stay planes, each made up of 60 stays from 19 to 75 T15 strands with a length varying from 35 to 243 m.

4. The second Severn River crossing bridge provides three traffic lanes in each direction, emergency lanes, safety barriers, and lateral wind screens. The construction of this new bridge is financed by private sector. The existing and the new toll bridges are managed by a concessionary group which takes responsibility for design, construction, financing, operating, and maintenance of both bridges.

Over 2 years of study and 4 years of work on site, challenged by the extreme tides, were necessary to build this bridge, located 5 km downstream from the suspension bridge erected in 1966, 30 years earlier.

64.5.2. Great Belt Bridges

The construction of the fixed link across the Great Belt Strait is a bridge and a tunnel project of exceptional dimensions. The Great Belt fixed link consists of three major projects:

FIGURE 64.34 Great Belt Bridges — the railway tunnel. (Courtesy of Jean Muller International.)

1. The railway tunnel under the eastern channel between Zealand and the island of Sprogø, in the middle of the Belt. It is a bored tunnel comprising two single-track tubes each with an internal diameter of 7.7 m and an external diameter of 8.50 m (Figure 64.34). The total tunnel length is 8 km.

Four 220-m-long boring machines have worked down to 75 m below sea level. The twin tunnel tubes are lined with interlocking concrete rings made of precast concrete segments each of a width of 1.65 m in the direction of boring. Each ring consists of six circle segments and a smaller key segment. A total of 62,000 tunnel segments were manufactured.

The twin tunnel tubes are connected at 250-m intervals by cross-passages with an internal diameter of 4.5 m, lined with cast iron segments assembled as rings. Each cross-passage consists of 22 rings of each 18 elements.

The railway tunnel is the second longest underwater bored tunnel, the tunnel beneath the English Channel being the longest.

2. The highway bridge across the eastern channel is 6790 m long. It consists of a suspension bridge with a 1624-m-long main span (see Figure 64.8) and two 535-m-long side spans, and of 23 approach spans totaling 4096 m (14 spans for the eastern approach and 9 spans for the western approach).

The bridge towers are founded on concrete caissons weighing 32,000 tons, placed on the seabed. The two legs of the pylons are cast in climbing formwork from the base to the pylon top 254 m above sea level. Cross-beams interconnect the pylon legs at heights of 125 and 240 m.

The anchor blocks for the suspension cables are also founded on concrete caissons weighing 55,000 tons. The rest of the two anchor blocks, including the special distribution chambers in which the main cables are anchored, are cast *in situ* by a conventional method to the top height of 63.4 m above sea level.

Among the bridge piers, the most part, i.e., 18, are prefabricated. Each pier consists of three elements: a caisson, a lower pier shaft, and finally a top pier shaft. The bridge piers weigh 6000 t

FIGURE 64.35 Great Belt Bridges — 193-m-long approach span. (Courtesy of Cowi Consult.)

on average. Conventional floating cranes are used for assembly of both the caissons and the pier shafts.

The steel superstructure of the main span comprises a fully welded box girder 4 m deep and 31 m wide. After floating the 48-m-long segments to a position under the main cables, they are hoisted into place by winches, and then welded to the previous section.

The two main cables each have diameter of 85 cm and a length of approximately 3 km. Each main cable includes 148 cables, and each cable includes 126 wires with a diameter of 5.13 mm; 20,000 tons of the steel representing a length of 112,000 km constitute the suspension of the bridge.

The steel superstructure of the approach spans comprises a fully welded box girder with a constant girder depth of 6.7 m, a width of 26 m, and a typical span of 193 m (Figure 64.35). The cross section has the same wing shape as the main span girder. The steel girders, each weighing about 2300 tons, are hoisted from a barge by a large floating crane.

The steel panels for the road girders are manufactured in Italy and then shipped from Livorno to Sines, Portugal. Here they are processed into bridge sections, which are floated to Aalborg (Northern Denmark) and welded together into complete bridge spans.

3. The combined road and railway bridge crosses the western channel between Funen and Sprogø. This west bridge is a 6.6-km-long all-concrete bridge with separate decks for rail and highway traffic. The bridge consists of six continuous bridge sections of a length of about 1100 m; the individual bridge sections are linked by expansion joints and hydraulic dampers that transmit only instantaneous forces.

The box girder underneath the rail track is only 12.3 m wide, compared to the roadway girder width of 24.1 m. However, the railway girder is 1.36 m deeper than the roadway girder.

The piers of the west bridge are founded on precast caissons. Each caisson receives two pier shafts, one for roadway girder, the other for railway girder. Each of the 110.4-m-long girder elements is cast in fixed steel shuttering in five sections. These sections are progressively linked by prestressing at the precasting yard. A special vessel, *Svanen,* a self-propelled floating crane with a lift capacity of 7123 tons, was used to transport the foundation caissons to the relevant position, to lift the pier shafts into place and to place the 110 m long girders on the top on the pier shafts.

In addition to the bridge and tunnel sections, the Great Belt fixed link, opened to traffic in 1998, also includes new road and railway sections on land, connecting the existing highways and railways with the fixed link.

64.5.3. Tagus River Bridges

The Tagus Bridge, also named the Vasco da Gama Bridge, is a 17.2-km-long structure connecting the northern and southern banks of the Tagus estuary. This project will solve a great part of the traffic problems in Lisbon by creating a link between new highway systems in the north and in the south of the city, and makes the traffic flow more easily between the northern and the southern parts of Portugal (Figure 64.36).

The Vasco da Gama project is divided into seven distinct sections, five of which are bridges and viaducts, representing 12.3 km.

1. The northern viaduct, with a total length 488 m of 11 spans, crosses the northern railway line of the Portuguese Railway Company (C.P.) and several local junctions. The deck width is variable to accommodate connection to local roads by slip roads. Span lengths vary from 42 to 47 m. The deck, 3.50 m deep, is cast *in situ* span by span. The typical span is 29.3 m wide and made up of four T-shaped concrete beams.

2. The Exhibition viaduct, with a total length 672 m of 12 spans, is also situated on the northern bank of the Tagus. It crosses the area where the 1998 World Exhibition took place. The bridge span lengths are the following: 2 × 46.2 m, 3 × 52.3 m, 55.3 m, 6 × 61.3 m. The deck, 29.3 m wide, is made up of twin prestressed concrete box girders, connected by the upper slab. Each box girder consists of precast segments put in place by mobile cranes using the balanced cantilever method. After pouring the cantilever closure joints, external prestressing is tensioned inside the box girders to ensure deck continuity.

The deck is supported by concrete piers founded on 1700-mm-diameter piles through a 4.5-m-thick concrete pile cap.

3. The main bridge is a cable-stayed bridge, 829 m long with a 420-m-long main span. The H-shaped pylons are founded on 2.2-m-diameter piles; these 44 bored piles are 53 m long. A very stiff and robust pile cap allows the foundation to withstand impact from a 30,000 ton vessel traveling at 8 knots.

The pylons comprise two legs and reach a height of 150 m. These legs, with a cross section varying from 12 × 7.7 m to 5.5 × 4.7 m, are slip formed. They are linked by a 10-m-deep prestressed box girder at base level, and by a transverse cross beam 87 m above the base, poured *in situ* in four stages. The upper part of the pylon, above the cross-beam, consists of a composite steel–concrete structure in which stays are anchored.

The deck, 31.28 m wide, consists of two longitudinal 2.50-m-deep and 1.30-m-wide concrete girders, connected every 4.4 m by 2.0-m-deep steel cross-beams. This composite structure is completed by a 25-cm concrete top slab (Figure 64.37). The 8.83-m-long deck segments are cast *in situ* using traveling formworks by the balanced cantilever method. Two points should be noticed: during segment concreting the final stays are used as temporary stays and the traveling formwork is designed to pass beyond the rear piers.

Like all the other structures, the main bridge is designed to withstand violent earthquake effects without damage. Consequently, there is no fixed link between the deck and its supports. Dampers are installed, steel dampers transversely, and longitudinally steel dampers outfitted with hydraulic couplers. On top of that, damping guide deviators are placed at each end of the cable stays.

4. The central viaduct is 6531 m long of 80 spans and cross the Tagus estuary above sandbanks and two shipping channels. The deck for the most part of its length is less than 14 m above sea

FIGURE 64.36 Tagus Bridge — artist's view. (Courtesy of Campenon Bernard.)

level. The typical span is 78.62 m long, but over the two shipping channels the span length rises to 130 m and the height above sea level rises to over 30 m. The viaduct span lengths are the following: 79.62 m, 3 × 78.62 m/93.53 m, 130.00 m, 93.53 m/60 × 78.62 m/93.53 m, 130.00 m, 93.53 m/11 × 78.62 m.

The deck, 29.3 m wide, consists of two parallel prestressed concrete box girders with two webs of constant height of 3.95 m, connected by the upper slab. Over the shipping channel, the girder depth is variable from 3.95 m at midspan to 7.95 m on piers.

Every span is precast in eighths; these segments, with a unit weight about 240 tons, are assembled on a bench by prestressing after adjustment. Each 1800 ton to 2000 ton girder is lifted and stored by a gantry crane with 2200 tons capacity load.

To transport and place the girders at up to 50 m above sea level, a special catamaran is used, equipped with two cranes with a capacity of 1400 tons at a radius of 25 m, 82 m tall. The rhythm of transport and placing is at a standard rate of one beam every 2 days. Prestressing cables ensure

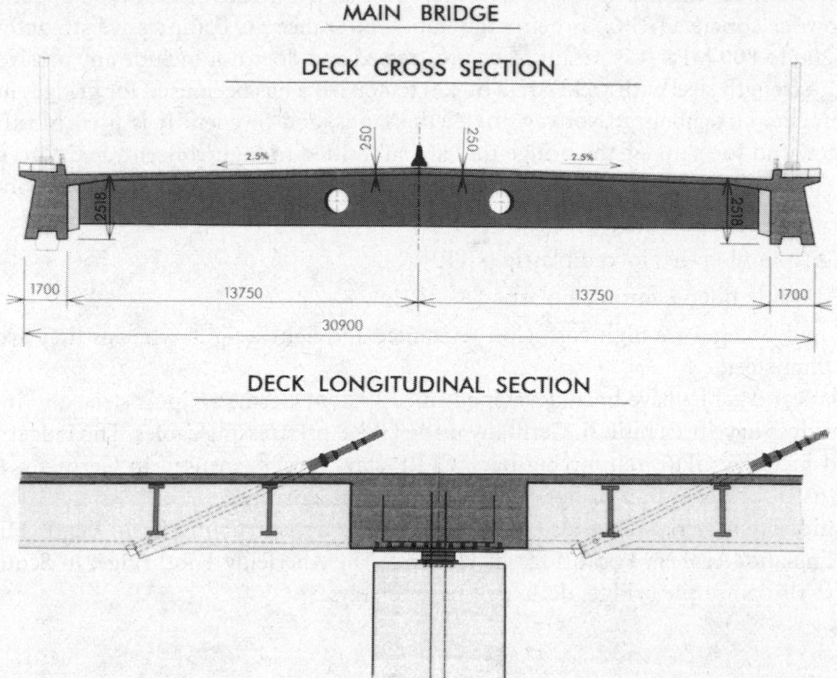

FIGURE 64.37 Tagus Bridge — main bridge deck cross section. (Courtesy of Campenon Bernard.)

longitudinal continuity of the deck. After concreting of continuity slab between the two parallel box girders, the transverse prestressing cables are tensioned. The deck is supported by concrete piers founded on 1700-mm-diameter deep piles.

5. The southern viaduct comprises 85 spans, each 45 m long, totaling 3825 m. As in the northern viaduct, the deck is composed by four T-shaped 3.50-m-deep concrete girders. The deck is cast *in situ* span by span with four mobile casting gantries working above the deck on two casting fronts. The deck is supported by concrete piers founded on bored piles for the land piers and on driven piles for the river piers.

The Vasco da Gama Bridge construction began in February 1995 and was finished in March 1998. It was privately financed and represents a cost of approximately $ 1 billion.

64.6 Future European Bridges

Future trends in bridge design can be classified in four categories:

1. Development of existing materials
2. Development of new materials
3. New structural association of materials
4. Structural control

1. The main materials used for bridges — concrete and steel — are still under development; their strength is always increasing. High-performance concrete (HPC) has been used for bridges for the first time in France and in the Scandinavian countries. Concrete with a compressive strength of 60 MPa (cylinder) at 28 days is becoming common for large bridges, especially for long spans and high piers. However, the advantage of HPC is not only strength, but durability, because this concrete is much more compact and much less porous than ordinary concrete. A new type of concrete called reactive powder concrete (RPC) is being developed in France; its compressive strength at 28 days can reach 200 to 800 MPa. It is meant to be prestressed and does not include any passive reinforcement. High-strength steel with yield stress of 420 to 460 MPa has been used for bridges in Germany, Finland, France, Luxembourg, Norway, the Netherlands, and Sweden. It is used mostly for long-span bridges, and for parts of the bridge that are submitted to high concentrated forces.

2. New composite materials are being developed for bridges in Europe. The main ones are:

- Glass fiber–reinforced plastic (GFRP),
- Carbon fiber–reinforced plastic (CFRP),
- Aramide fiber–reinforced plastic (AFRP).

Their main advantages are high corrosion resistance and light weight, whereas they are still more expensive than steel.

GFRP bars and cables have been developed since 1980 in Germany, in Austria, and in France. At least five bridges have been built in Germany using GFRP prestressing cables. The Fidget Footbridge in England includes GFRP reinforcing bars. CFRP stays have been used in Germany. AFRP stays have been used for pedestrian bridges in Holland and in Norway.

New composite materials have also been used for the deck structure itself: Bends Mill movable bridge in England, Arnhem Footbridge in Holland. The Aberfeldy Footbridge, in Scotland, is the world's first all-composite bridge: deck, pylons, and stays.

FIGURE 64.38 Chavanon Bridge. (Courtesy of Jean Muller International.)

3. The association of steel I-girders with a concrete slab has become very common for medium-span bridges. We think that this association of steel and concrete will be developed for a large variety of composite structures in the future: truss decks, arches, pylons, piers. A number of such innovative projects have been built in France and Switzerland, for example. The use of each material to its best capability will lead to more efficient and economical structures. A significant example of such a structure is implemented on the highway A89 which will link Clermond Ferrand to Bordeaux in southern France. To cross the deep valley of the River Chavanon respecting the natural environment, a suspension bridge is being built (Figure 64.38). The bridge deck is a steel concrete composite structure 22.4 m wide and 3.0 m deep. It is suspended by a single plane of suspension cables located at the cross section axis. The inverted V-shaped pylon straddles over the deck and leaves it free of any support. Its top is 52 m above the deck. This bridge, with a 300-m-long main span is an innovative, efficient, and very aesthetic projec**4.** With the development of high-strength materials, and possibly lightweight materials, bridges will become more and more slender and light, hence more sensitive to fatigue and dynamic problems, especially for long-span bridges. Consequently, it will become necessary to control vibrations due to traffic loads, wind, and earthquakes. This control can be achieved through passive devices, such as dampers, and active devices such as active pre-stressing tendons, active stays, active aerodynamic appendages. This will be the road toward "intelligent" bridges of the future...

An existing bridge could easily be equipped with an active device. Such a device was implemented in the Rogerville Viaduct, opened to traffic in 1996, located in northern France, on highway A29 not far from the Normandie Bridge (Figure 64.39). It is a continuous steel box girder, placed across the expansion joint, between two adjacent cantilever arms (Figure 64.40). It rests on two diaphragms on either side and may be adjusted before the bridge is opened to traffic to transfer shear force and bending moment, and consequently to compensate subsequent effects of steel relaxation and concrete creep. At the moment, the continuity girder is a passive device

This connection could be equipped to transfer (long term under dead load, and short term under live load), shear force and moment in an active fashion at all times (Figure 64.41). In other words, the magnitude of shear load and moment across the joint may be monitored and adjusted at the designer's request to restore all the geometric and mechanical properties of a continuous deck across the expansion joint.

FIGURE 64.39 Rogerville Viaduct. (Courtesy of Jean Muller International.)

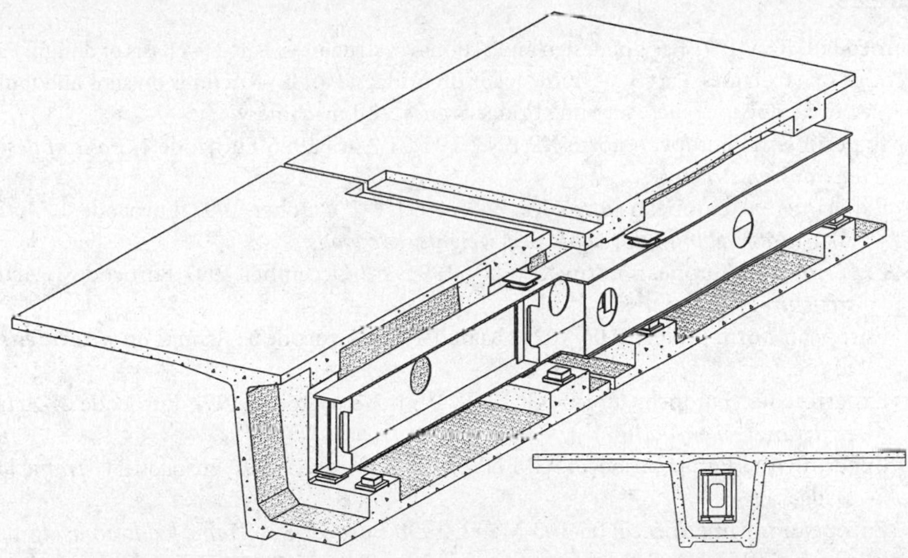

FIGURE 64.40 Rogerville Viaduct — expansion joint device. (Courtesy of Jean Muller International.)

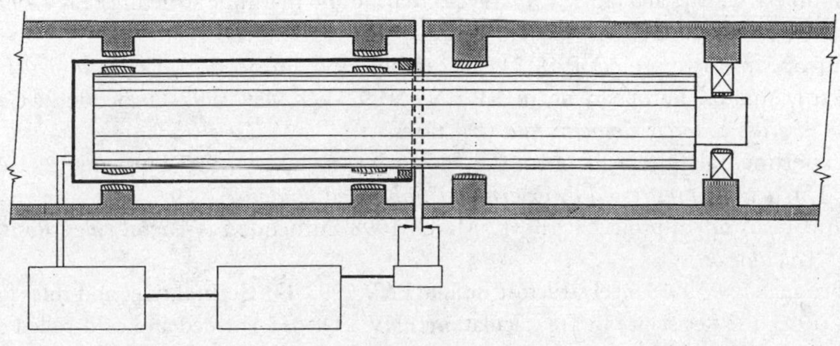

FIGURE 64.41 Active connection. (Courtesy of Jean Muller International.)

References

1. Eurocode 1. ENV 1991 Basis of design and actions on structures: Part 1 — Basis of design; Part 2 — Actions on structures; Part 3 — Traffic loads on bridges; Part 4 — Actions on silos and tanks; Part 5 — Actions due to cranes, traveling bridge cranes, and machinery.

   Experimental European norm XP ENV 1991-1 April 1996 Eurocode 1, *Basis of design and actions on structures.*

   Experimental European norm XP ENV 1991-2-1 October 1997 Eurocode 1, *Actions on structures: voluminal weights, self-weights, live loads.*

   Experimental European norm XP ENV 1991-2-2 December 1997 Eurocode 1, *Actions on structures exposed to fire.*

   European norm project P 06-102-2 March 1998 Eurocode 1. *Actions on structures exposed to fire.*

   Experimental European norm XP ENV 1991-2-3 October 1997 Eurocode 1. *Actions on structures: snow loads.*

   Experimental European norm XP ENV 1991-3 October 1997 Eurocode 1, *Traffic loads on bridges.*

   European norm project P 06-103 March 1998 Eurocode 1, *Traffic loads on bridges.*

2. Eurocode 2, ENV 1992 Concrete structure design; ENV 1992-1-1 General rules and rules for buildings; ENV 1992-1-2 Resistance to fire calculation; ENV 1992-1-3 Precast concrete elements and structures; ENV 1992-1-4 Lightweight concrete; ENV 1992-1-5 Structures prestressed by external or unbonded tendons; ENV 1992-1-6 Nonreinforced concrete structures; ENV 1992-2 Reinforced and prestressed concrete bridges; ENV 1992-3 Concrete foundations; ENV 1992-4 Retaining structures and tanks; ENV 1992-5 Marine and maritime structures; ENV 1992-6 Massive structures; ENV 1992-X Post-tension systems

   European norm project P 18-711 December 1992 Eurocode 2, *Concrete structure design.*

   Experimental European norm XP ENV 1992-1-3 May 1997 Eurocode 2, *General rules. Precast concrete elements and structures.*

   Experimental European norm XP ENV 1992-1-5 May 1997 Eurocode 2, *General rules. Structures prestressed by external or unbonded tendons.*

   European norm project P 18-712 March 1998 Eurocode 2, *General rules. Resistance to fire calculation.*

3. Eurocode 3, ENV 1993 Steel structure design; ENV 1993-1-1 General rules and rules for buildings; ENV 1993-1-2 Resistance to fire calculation; ENV 1993-1-3 Formed and cold-rolled element use; ENV 1993-1-4 Stainless steel use; ENV 1993-2 Plate-shaped bridges and structures; ENV 1993-3 Towers, masts, and chimneys; ENV 1993-4 Tanks, silos, and pipelines; ENV 1993-5 Piles and sheet piles; ENV 1993-6 Crane structures; ENV 1993-7 Marine and maritime structures; ENV 1993-8 Agricultural structures.

   Experimental European norm project P 22-311 December 1992 Eurocode 3, *Steel structure design.*

   Experimental European norm XP ENV 1993-1-2 December 1997 Eurocode 3, *General rules. Behavior under the action of fire.*

4. Eurocode 4, ENV 1994 — Steel–concrete composite structure design; ENV 1994-1-1 General rules for buildings; ENV 1994-1-2 Resistance to fire calculation; ENV 1994-2 Bridges.

   Experimental European norm project P 22-391 September 1994 Eurocode 4, *Steel–concrete composite structure design and dimensioning.*

   Experimental European norm XP ENV 1994-1-2 December 1997 Eurocode 4, *General rules. Behavior under the action of fire.*

   European norm project PROJECT P 22-392 March 1998 Eurocode 4, *General rules. Behavior under the action of fire.*

5. Eurocode 5, ENV 1995 Wooden work design; ENV 1995-1-1 General rules and rules for buildings; ENV 1995-1-2 Resistance to fire calculation; ENV 1995-2 Wooden bridges.

   Experimental European norm XP ENV 1995-1-1 August 1995 and AMDT.1 February 1998 Eurocode 5, *Wooden work design. General rules.*

   European norm project PROJECT P 21-712 March 1998 Eurocode 5, *General rules. Behavior under the action of fire.*

6. Eurocode 6, ENV 1996 Masonry structure design; ENV 1996-1-1 General rules and rules for reinforced or nonreinforced masonry; ENV 1996-1-2 Resistance to fire calculation; ENV 1996-1-X Cracking and deformation checking; ENV 1996-1-X Detailed rules for lateral loads; ENV 1996-1-X Complex-shaped section in masonry structures; ENV 1995-2 Guide for design, material choice, and construction of masonry structures; ENV 1995-3 Simple and simplified rules for masonry structures; ENV 1995-4 Masonry structures with low requirements.

   Experimental European norm XP ENV 1996-1-2 December 1997 Eurocode 6, *General rules. Behavior under the action of fire.*

   European norm project P 10-611B March 1998 Eurocode 6, *General rules. Rules for reinforced or nonreinforced masonry.*

   European norm project P 10-612 March 1998 Eurocode 6, *General rules. Behavior under the action of fire.*

7. Eurocode 7, ENV 1997 Geotechnical design; ENV 1997-1 General rules; ENV 1997-2 Laboratory test norms; ENV 1997-3 Sampling and test *in situ* norms; ENV 1997-4 Additional rules for special elements and structures.

   Experimental European norm XP ENV 1997-1 December 1996 Eurocode 7, *Geotechnical design. General rules.*

   European norm project PROJECT P 94-250-1 May 1997 Eurocode 7, *Geotechnical design. General rules.*

8. Eurocode 8, ENV 1998 Earthquake-resistant structure design and dimensioning; ENV 1998-1-1 General rules: seismic actions and general requirements for structures; ENV 1998-1-2 General rules: general rules for buildings; ENV 1998-1-3 General rules: special rules for various elements and materials; ENV 1998-1-4 General rules: Building strengthening and repairing; ENV 1998-2 Bridges; ENV 1998-3 Towers, masts, and chimneys; ENV 1998-4 Silos, tanks, and pipes; ENV 1998-5 Foundations, retaining structures, and geotechnical aspects.

   European norm project PROJECT P 06-031-1 March 1998 Eurocode 8, *Earthquake-resistant structure design and dimensioning.*

9. Eurocode 9, ENV 1999 Aluminum alloy structure design; ENV 1999-1-1 General rules and rules for buildings; ENV 1999-1-2 Additional rules for aluminum alloy structure design under the action of fire; ENV 1999-2 Rules concerning fatigue.

65

Design Practice in Japan

Masatsugu Nagai
Nagaoka University of Technology

Tetsuya Yabuki
University of Ryukyu

Shuichi Suzuki
Honshu-Shikoku Bridge Authority

65.1 Design

Tetsuya Yabuki

65.1.1 Design Philosophy

In the current Japanese bridge design practice [1], there are two design philosophies: ultimate strength design and working stress design.

1. Ultimate strength design considering structural nonlinearities compares the ultimate load-carrying capacity of a structure with the estimated load demands and maintains a suitable ratio between them. Generally, this kind of design philosophy is applied to the long-span bridge structures with spans of more than 200 m, i.e., arches, cable-stayed girder bridges, stiffened suspension bridges, etc.
2. Working stress design relies on an elastic linear analysis of the structures at normal working loads. The strength of the structural member is assessed by imposing a factor of safety between the maximum stress at working loads and the critical stress, such as the tension yield stress

0-8493-7434-0/00/$0.00+$.50

TABLE 65.1 Loading Combinations and Their Multiplier Coefficients for Allowable Stresses

No.	Loading Combination	Multiplier Coefficient for Allowable Stresses
1	P + PP + T	1.15
2	P + PP + W	1.25
3	P + PP + T + W	1.35
4	P + PP + BK	1.25
5	P + PP + CO	1.70 for steel members 1.50 for reinforced concrete members
6	W	1.2
7	BK	1.2
8	P except L and I + EQ	1.5
9	ER	1.25

of material and the shear yield stress of or the compression buckling stress of material (see Section 65.1.4).

65.1.2 Load

The Japanese Association of Highways, the Standard Specification of Highway Bridges [1] (JAH-SSHB) defines all load systems in terms four load systems as follows:

1. Primary loads (*P*) — dead load (*D*), live load (*L*), impact load (*I*), prestressed forces (*PS*), creep (*CR*), shrinkage (*SH*), earth pressure (*E*), hydraulic pressure (*HP*), uplift force by buoyancy (*U*).
2. Secondary loads (*S*) — wind load (*W*), thermal force (*T*), effect of earthquakes (*EQ*).
3. Particular loads corresponding to the primary load (*PP*) — snow load (*SW*), effect of displacement of ground (*GD*), effect of displacement of support (*SD*), wave pressure (*WP*), centrifugal force (*CF*).
4. Particular loads (*PA*) — raking force (*BK*), tentative forces at the erection (*ER*), collision force(CO), etc.

The combinations of loads and forces to which a structure may be subjected and their multiplier coefficients for allowable stresses are specified as shown in Table 65.1. The most severe combination of loads and forces for a structure within combinations given in Table 65.1 is to be taken as the design load system. Details on the loads have not been given here and the reader should refer to the specification [1].

Limiting values of deflection are expressed as a ratio of spans for the individual superstructure types and span lengths.

65.1.3 Theory

In most cases, design calculations for both concrete and steel bridges are based on the assumptions of linear behavior (i.e., elastic stress–strain) and small deflection theory. It may be unreasonable, however, to apply linear analysis to a long-span structure causing the large displacements. The JAH-SSHB specifies that the ideal design procedure including nonlinear analyses at the ultimate loads should be used for the large deformed structure.

Bridges with flat stiffening decks raise some anxieties for the wind resistance. The designer needs to test to ensure the resistances for wind forces and/or the aerodynamic instabilities. In Japan, wind tunnel model testing including the full model and sectional model test is often applied for these verifications . The methods of model testing include full-model tests and sectional-model tests. The vibrations induced by vehicles, rain winds, and earthquakes are usually controlled by oil dampers, high damping rubbers, and/or vane dampers.

65.1.4 Stability Check

The JAH-SSHB specifies the strength criteria on stabilities for fundamental compression, shear plate, and arch/frame elements. The strength criteria for the stability of those elements are presented as follows:

1. Compressive strength for plate element — Fundamental plate material strength under uniform compression is mentioned here but details have not been given in all cases.

$$\left.\begin{aligned} \frac{\sigma_{cl}}{\sigma_Y} &= 1.0 && \text{for } R \le 0.7 \\ &= \frac{0.5}{R^2} && \text{for } 0.7 \le R \end{aligned}\right\} \tag{65.1}$$

where σ_{cl} = plate strength under uniform compression, R = equivalent slenderness parameter defined as

$$R = \frac{b}{t}\sqrt{\frac{\sigma_Y}{E}} \cdot \sqrt{\frac{12\left(1-\mu^2\right)}{\pi k^2}} \tag{65.2}$$

and b = width of plate, t = thickness of plate, μ = Poisson's ratio, k = coefficient applied in elastic plate buckling.

2. Compressive strength for axially loaded member — The column strength for overall instability is specified as

$$\left.\begin{aligned} \frac{\sigma_{cg}}{\sigma_Y} &= 1.0 && \text{for } \lambda \le 0.2 \\ &= 1.109 - 0.545\bar{\lambda} && \text{for } 0.2 \le \bar{\lambda} \le 1.0 \\ &= 1.0/\left(0.773 + \bar{\lambda}^2\right) && \text{for } 1.0 \le \bar{\lambda} \end{aligned}\right\} \tag{65.3}$$

where σ_{cg} = column strength, σ_Y = yield-stress level of material, $\bar{\lambda}$ = slenderness ratio parameter defined as follows

$$\bar{\lambda} = \frac{1}{\pi}\sqrt{\frac{\sigma_Y}{E}} \cdot \frac{\ell}{r} \tag{65.4}$$

and r = radius of gyration of column member and ℓ = effective column length. Thus, the ultimate stress of axially effective material, σ_c, is specified as

$$\sigma_c = \frac{\sigma_{cg} \cdot \sigma_{cl}}{\sigma_Y} \tag{65.5}$$

3. Bending compressive strength — The ultimate strength for bending compression is specified, based on the lateral-torsional stability strength of beam under uniform bending moment as follows:

$$\left.\begin{aligned}\frac{\sigma_{bg}}{\sigma_Y} &= 1.0 && \text{for } \alpha \le 0.2\\ &= 1.0 - 0.412(\alpha - 0.2) && \text{for } 0.2 \le \alpha\end{aligned}\right\} \tag{65.6}$$

where σ_{bg} = lateral-torsional stability strength of beam under uniform moment, α = equivalent slenderness parameter defined as

$$\left.\begin{aligned}\alpha &= \frac{2}{\pi} K \sqrt{\frac{\sigma_Y}{E}} \cdot \frac{\ell}{b}\\ K &= 2 && \text{for } A_w / A_c \le 2\\ &= \sqrt{3 + 0.5 A_w / A_c} && \text{for } 2 \le A_w / A_c\end{aligned}\right\} \tag{65.7}$$

and A_w = gross area of web plate, A_c = gross area of compression flange, ℓ = laterally unbraced length, b = width of compression flange. The effect of nonuniform bending is estimated by the multiplier coefficient, *m,* as follows

$$m = \frac{M}{M_{eq}} \tag{65.8}$$

in which M = bending moment at a reference cross section, M_{eq} = equivalent conversion moment given as

$$M_{eq} = 0.6M_1 + 0.4M_2 \quad \text{or} \quad M_{eq} = 0.4M_1 \quad \text{where} \quad M_1 \ge M_2 \tag{65.9}$$

65.1.5 Fabrication and Erection

Fabrication and erection procedures depend on the structural system of the bridge, the site conditions, dimensions of the shop-fabricated bridge units, equipment, and other factors characteristic of a particular project. This includes methods of shop cutting and welding, the selection of lifting equipment and tackle, method of transporting materials and components, the control of field operation such as concrete placement, and alignment and completion of field joints in steel, and also the detailed design of special erection details such as those required at the junctions of an arch, a cantilever erection, and a cable-stayed erection. Therefore, for each structure, it is specified that the contractor should check

1. Whether each product has its specified quality or not.
2. Whether the appointed erection methods are used or not.

As a matter of course, the field connections of main members of the steel structure should be assembled in the shop.

Details on the inspections have not been given here and the reader should refer to the specifications [1].

FIGURE 65.1 Tennyo-bashi.

65.2 Stone Bridges

Tetsuya Yabuti

It is possible that stone bridges were built in very ancient times but that through lack of careful maintenance and/or lack of utility they were destroyed so that no trace remains. Since stone masonry is generally suited to compressive stresses, it is usually used for arch spans. Therefore, most stone bridges that have survived to the present are arch bridges. Generally, stone arch bridges are classified into two types: the European voussoirs are built of bricks and the Chinese type where each voussoir in arch is curved and behaves as a rib element.

Figure 65.1 shows Tennyo-bashi (span length, 9.5 m) located in Okinawa prefecture. This is the only area in Japan that has the Chinese type. This bridge is the oldest Chinese-type stone arch bridge in Japan that has survived to the present time; it was originally constructed in 1502. Figure 65.2 shows Tsujyun Bridge located in Kumamoto prefecture (length of span, 75.6 m; raise of arch, 20.2 m; width of bridge, 6.3 m). This bridge is typical of aqueduct stone arch bridges that have survived to the present in Japan and was originally constructed in 1852. Figure 65.3 shows Torii-bashi Bridge located in Oita prefecture which is one of the multispanned stone arch bridges constructed early in the 20th century. This bridge is a five-span arch bridge (length of bridge, 55.15 m; width of bridge; 4.35 m; height of bridge, 14.05 m) constructed in 1916.

Separate stones sometimes have enough tensile strength to permit their being used for beams and slabs as seen in Hojyo-bashi which is the clapper bridge shown in Figure 65.4. This bridge located in Okinawa prefecture has a span of 5.5 m and was originally constructed in 1498.

65.3 Timber Bridges [2,3]

Masatsugu Nagai

Since 1990, the number of timber bridges constructed has increased. Most of them use glue-laminated members and many are pedestrian bridges. To date, about 10 bridges have been constructed to carry 14 or 20 tf trucks. All were constructed on a forest road. We have no design code for timber bridges. However, there is a manual for designing and constructing timber bridges.

The following is an introduction to timber arch, cable-stayed, and suspension bridges in Japan.

FIGURE 65.2 Tsujyun Bridge.

FIGURE 65.3 Torii-bashi.

1. Arch Bridges — Table 65.2 shows nine arch bridges. Figure 65.5 shows Hiraoka bridge, which, of timber arch bridges, has the longest span in Japan. Figure 65.6 shows Kaminomori bridge [4]. It is the first arch bridge, which carries a 20 tf truck load.
2. Cable-Stayed Bridges — Table 65.3 shows three cable-stayed bridges. Figure 65.7 shows Yokura bridge. It has a world record span length of 77.0 m, and has a concrete tower.
3. Suspension Bridges — Table 65.4 shows two suspension bridges. Figure 65.8 shows Momosuke bridge. Momosuke bridge is an oldest timber suspension bridge in Japan. It was constructed in 1922, and reconstructed in 1993.

FIGURE 65.4 Hojyo-bashi.

TABLE 65.2 Arch Bridges

Name	Span and Width (m)	Construction Year	Remarks
Yunomata	13.0 6.0	1990	Tied arch bridge for 14 tf truck loading
Kisoohashi	33.0 6.0	1991	Fixed arch pedestrian bridge
Deai	39.0 2.0	1992	Tied arch pedestrian bridge
Hiroaka	45.0 3.0	1993	Three hinged arch pedestrian bridge
Yasuraka	30.0 1.5	1993	Nielsen Lohse type pedestrian bridge
Chuo	21 5.0	1993	Lohse type pedestrian bridge
Kaminomori	23 5.0	1994	Two hinged arch bridge for 20 tf truck loading
Awaiido	24.0 8.0	1994	Lohse type bridge for 20 tf truck loading
Meoto	20 1.5	1994	Two hinged arch pedestrian bridge

65.4 Steel Bridges

Tetsuya Yabuti

In Japan, metal as a structural material began with cast and/or wrought iron used on bridges after the 1870s. Through lack of these utilities in urban areas, however, almost all of those bridges have broken down. Since 1895, steel has replaced wrought iron as the principal metallic bridge material.

After the great Kanto earthquake disaster in 1923, high tensile steels have been positively adopted for bridge structural uses and Kiyosu Bridge (length of bridge, 183 m; width of bridge, 22 m) shown in Figure 65.9 is a typical example. This eyebar-chain-bridge over the Sumida river in Tokyo is a self-anchored suspension bridge and a masterpiece among riveted bridges. It was completed in 1928.

FIGURE 65.5 Hiraoka Bridge.

FIGURE 65.6 Kaminomori Bridge.

Figure 65.10 shows one of the curved tubular girder bridges located in the metropolitan expressway. Curved girder bridges have become an essential feature of highway interchanges and urban expressways now common in Japan.

TABLE 65.3 Cable-Stayed Bridges

Name	Span and Width (m)	Construction Year	Remarks
Midori Kakehashi	27.5 2.0	1991	Two-span continuous pedestrian bridge
Yokura	77.0 5.0	1992	Three-span continuous pedestrian bridge
Himehana	21.5 1.5	1995	Three-span continuous pedestrian bridge

FIGURE 65.7 Yokura Bridge.

TABLE 65.4 Suspension Bridges

Name	Span and Width (m)	Construction Year	Remarks
Momosuke	104.5 2.3	1993	Four-span continuous pedestrian bridge
Fujikura	32.0 1.8	1994	Single-span pedestrian bridge

Figure 65.11 shows Katashinagawa Bridge (length of span; 1033.8 m = 116.9 + 168.9 + 116.9; width of bridge, 18 m) located in Gunma prefecture. This bridge is the longest curved-continuous truss bridge in Japan and was completed in 1985. Figure 65.12 shows Tatsumi Bridge (length of bridge, 544 m; width of bridge, 8 m) located in Tokyo. This viaduct bridge in the metropolitan expressway is a typical example of rigid frame bridges in an urban area and was completed in 1977. Figure 65.13 shows a typical π-shaped rigid frame bridge. This structural type is used as a viaduct over a highway or a highway bridge in mountain areas and is common in Japan.

FIGURE 65.8 Momosuke Bridge.

FIGURE 65.9 Kiyosu Bridge.

FIGURE 65.10 A curved tubular girder bridge in a metropolitan express highway.

FIGURE 65.11 Katashinagawa Bridge.

FIGURE 65.12 Tatsumi Bridge.

FIGURE 65.13 A typical π-shaped rigid frame bridge.

FIGURE 65.14 Saikai Bridge.

Saikai Bridge located in Nagasaki prefecture (length of span, 216 m; width of bridge, 7.5 m) shown in Figure 65.14 was completed in 1955. Construction of bridges in Japan after the World War II began in earnest with the Saikai Bridge. This bridge is a fixed arch bridge and the stress condition was improved by prestressing when the main arch was finally completed.

Figure 65.15 shows Ooyano Bridge located in Kumamoto prefecture. This bridge is a typical through-type arch bridge (length of bridge, 156 m; width of bridge, 6.5 m) and was constructed in 1966.

Figure 65.16 shows Ikuura Bridge located in Mie prefecture and completed in 1973. The main span of this bridge (length of span, 197 m; width of bridge, 8.3 m) is a tied arch with inclined hangers of the Nielsen system that is one of the most favored bridge types in Japan, along with the cable-stayed bridge.

Figure 65.17 shows Tsurumi-Tsubasa Bridge (length of bridge, 1021 m = center span of 510 m + two side spans of 255 m; height of towers, 136.7 m) located in Yokohama Bay, Kanagawa prefecture. This bridge is a single-plane, cable-stayed bridge with continuous three spans. It is the longest bridge of this type including those under design all over the world and was completed in 1994.

Figure 65.18 shows Iwagurojima Bridge (length of bridge, 790 m = center span of 420 m + two side spans of 185 m; height of towers, 148.1 m; width of bridge, 22.5 m) located Kagawa prefecture. This bridge completed in 1988 is a double-plane cable-stayed bridge with continuous three spans and is a combined bridge with highway and railway traffic. It has four express railways. The cable-stayed bridge with four express railways is unprecedented, including those under design worldwide. Figure 65.19 shows Kanmon Bridge (length of bridge, 1068 m = center span of 420 m + two side spans of 185 m) completed in 1973. This bridge spans over the Kanmon channel and links Moji in Fukuoka prefecture, Kyushu Island, and Shimonoseki in Yamaguchi prefecture, main island. It is the first bridge in Japan spanning a channel.

The Japan Association of Steel Bridge Construction has contributed to the preparation of some photographs in this section.

FIGURE 65.15 Ooyano Tsubasa Bridge.

FIGURE 65.16 Ikuura Bridge.

FIGURE 65.17 Tsurumu-Tsubasa Bridge.

FIGURE 65.18 Iwakuro Bridge.

FIGURE 65.19 Kanmon Bridge.

65.5 Concrete Bridges

Tetsuya Yabuki

Construction of reinforced concrete bridges began in the 1900s in Japan but has gradually become useless because of change of the utility conditions in urban areas. Since the 1950s, the use of prestressing spread to nearly every type of simple structural element and spans of concrete bridges became much longer. Probably the most significant observable feature of prestressed concrete is its crack-free surface under service loads. Especially, when the structure is exposed to weather conditions, elimination of cracks prevents corrosion. Many reinforced concrete bridges constructed previously are being replaced by prestressed concrete ones in Japan.

Figure 65.20 shows a typical reinforced concrete bridge damaged by corrosion. Most of these kinds of bridges have been replaced by prestressed concrete structures. Figure 65.21 shows Chousei Bridge (length of bridge, 10.8 m = three continuous spans of 3.6 m) located in Ishikawa prefecture. This bridge is a pretensioned simple composite slab bridge. It was completed in 1952 and is the first prestressed concrete bridge in Japan. Figure 65.22 shows Ranzan Bridge (length of bridge, 75 m = main span of 51.2 + two side spans of 11.9 m) located in Kanagawa prefecture. This bridge is a rigid frame bridge composed by three spans with a hinge. It was completed in 1959 and is the first bridge in Japan that was constructed by the cantilever erection.

Figure 65.23 shows International Expo No. 9 Bridge (length of bridge, 27 m; width of bridge, 5.5 m; thickness of slab, 0.1 m) located in Osaka. This bridge is a pedestrian bridge. It was completed in 1969 and is the first suspended slab bridge in Japan. Figure 65.24 shows Takashimadaira Bridge (length of bridge, 230 m = 75 + 75 + 80; width of bridge, 18 m) located in Tokyo. This viaduct in the metropolitan expressway is composed by linking three bridges and completed in 1973. Each one is a continuous three span-bridge (length of spans, 25 m + 25 + 25 m).

FIGURE 65.20 A typical reinforced concrete bridge damaged by corrosion.

FIGURE 65.21 Chosei-bashi.

FIGURE 65.22 Ranzan Bridge.

FIGURE 65.23 International Expo No. 9 Bridge.

FIGURE 65.24 Takashimadaira Bridge.

Figure 65.25 shows Akayagawa Bridge (length of bridge, 298 m; arch span, 116 m) located in Gunma prefecture. This rib arch bridge is the longest concrete arch railway bridge in Japan and was completed in 1979. Its arch rib is composed of a plate with thickness of 0.8 m and mainly receives compressive stress. Figure 65.26 shows Omotogawa Bridge located in Iwate prefecture. This bridge was completed in 1979 and is the first prestressed concrete stayed railway bridge in the world.

The Japan Prestressed Concrete Association has contributed to preparation of some of the photographs in this section.

65.6 Hybrid Bridges

Masatsugu Nagai

Hybrid bridges consist of composite and compound bridges. Composite bridges have a cross section of steel and concrete connected by shear connectors. Compound bridges consist of different materials, such as steel and concrete. In Japan, many composite girder bridges have been constructed. However, since 1980, the number of composite girder bridges has decreased. One of main reasons is the damage of concrete decks due to overloading by heavy trucks. In recent years, for economic reasons, the choice of composite girder construction has been reconsidered. In bridge systems, prestressed precast concrete slabs are used to attain higher durability. The following is an introduction of the practices and plans of the hybrid bridges of Japan Highway Public Corporation.

Figure 65.27 shows Hontani Bridge (total span length, 197 m = 44 + 97 + 56 m; width, 11.4 m) constructed in Gifu prefecture in 1998. It has a box section with corrugated steel plate used as a web between upper and lower concrete slabs. To reduce the total weight of the concrete box girder, instead of concrete webs, a steel web was used. This kind of structural system was first employed in the Cognac Bridge in France. However, for the connection between concrete and steel plate, a simple system with reinforcing bars attached to the corrugated plate and without steel flange was used.

FIGURE 65.25 Akayagawa Bridge.

FIGURE 65.26 Omotogawa Bridge.

FIGURE 65.27 Hontani Bridge. (Courtesy of Japan Highway Public Corporation.)

FIGURE 65.28 Tomoegawa Bridge. (Courtesy of Japan Highway Public Corporation.)

Figure 65.28 shows Tomoegawa Bridge (total span length, 475 m = 59 + 3 × 119 + 59 m; width, 16.5 m) which will be constructed on the route of the New Tomei Expressway discussed in Section 65.8.2. It has a box section, and a steel truss is used as a web between the upper and lower concrete slabs, whose main purpose is to reduce the total weight of the superstructures. When the span length becomes long, if a steel plate is used as the web, a horizontal connection in the bridge

FIGURE 65.29 Ibigawa Bridge. (Courtesy of Japan Highway Public Corporation.)

longitudinal direction is necessary for transportation from the shop to the site. To avoid the problem, a truss member is planned to be used. This kind of bridge system was first used in Arbois Bridge in France. However, in this bridge, a new connecting system between truss member and concrete slab will be used.

Figure 65.29 shows Ibigawa Bridge (total span length, 1397 m = 154 + 4 × 271.5 + 157 m; width 33 m) which will be constructed on the route of the New Meishin Expressway. It crosses the Ibi River near Nagoya City. Kisogawa Bridge, which crosses the Kiso River running parallel to the Ibi River, has a structural form similar to that of Ibigawa Bridge. The bridge consists of concrete and steel girders. The steel box girders are used for the middle part with a length of 100 m in the central four spans, and are connected to the concrete girder. The concrete girder is suspended by diagonal stays from concrete towers. The height of the tower from the deck level is lower than that used in conventional cable-stayed bridges. Since the span length is 271.5 m, if the concrete bridges are used, considerably greater depth is inevitable. By suspending the concrete girder and using a steel girder with lighter weight in the central portion of the bridge, reduction of depth is attained.

Figure 65.30 shows Shinkawa Bridge (total span length, 278 m = 2 × 40 + 113 + 2 × 40 m; width 21.4 m) which will be constructed on the highway in Shikoku Island. It consists of the concrete and steel box girders. Since the side span length is planned to be short, a concrete box girder is used and connected to the steel girder in the side span adjacent to the main span. The purpose of this countermeasure is to balance the weight.

Figure 65.31 shows Kitachikumagawa Bridge (total span length, 346.7 m = 84.35 + 2 × 89 + 84.35 m; width, 10.4 m) which is located near Nagano City and was constructed in 1997. In this bridge, reinforced concrete piers were connected to the steel box girders. By employing a rigid frame structure, the bearings are avoided and good performance against earthquake is attained.

Shigehara Bridge (total span length; 166.8 m = 47.4 + 72 + 47.4 m; width, 10.4 m) was constructed on the highway in Kyushu Island in 1995. It has composite piers, in which steel pipes are encased instead of reinforcing bars as shown in Figure 65.32. This system is developed to reduce the volume of the reinforcing bars, and resulted in the reduced construction work.

FIGURE 65.30 Shinkawa Bridge. (Courtesy of Japan Highway Public Corporation.)

FIGURE 65.31 Kitachikuma Bridge. (Courtesy of Japan Highway Public Corporation.)

FIGURE 65.32 Piers of Shigehara Bridge. (Courtesy of Japan Highway Public Corporation.)

65.7 Long-Span Bridges (Honshu–Shikoku Bridge Project)

Masatsugu Nagai and Shuichi Suzuki

The Honshu–Shikoku Bridge Project is a national project to link the Honshu and Shikoku Islands. Construction of the long-span bridge started in 1975 and was completed in 1999. Figure 65.33 shows three routes, in which many long-span cable-supported bridges are constructed. The following is an introduction of the super- and substructures of these cable-supported bridges.

65.7.1 Kobe–Naruto Route

This route is 89 km long, and two suspension bridges are arranged. Figure 65.34 shows the Akashi Kaikyo Bridge [5] (total span length, 3911 m = 960 + 1991 + 960 m) which is a three-span, two-hinged suspension bridge, and has a world-record span length of 1991 m. The distance between two cables is 35.5 m. This bridge was opened to traffic in 1998. The original plan was to carry both rail and road traffic. In 1985, this plan was changed so that the bridge carries highway traffic only. It is known, in the design of long-span suspension bridges, that ensuring safety against static and dynamic instabilities under wind load is an important issue. Aerodynamic stability was investigated through boundary layer

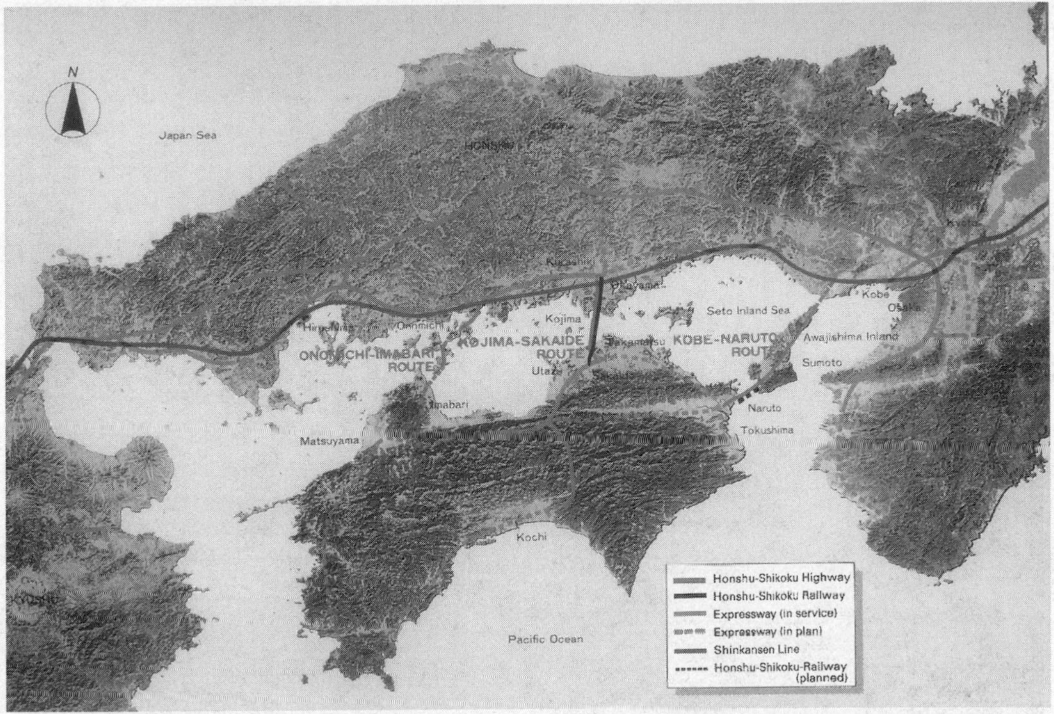

FIGURE 65.33 Honshu–Shikoku Bridge connecting route.

wind tunnel test with a 1/100 full model, and the test was conducted as a cooperative study between the Honshu–Shikoku Bridge Authority and the Ministry of Construction. Through the test, it was confirmed that flutter occurred at a wind velocity exceeding 78 m/s, a value was well above the required wind velocity. To avoid two main cables on each side, wires with a tensile strength of 1760 MPa (higher than that used in other suspension bridges) was used. Further, since the ratio of live load to dead load is small, the factor of safety of 2.2 against tensile strength was utilized.

The laying-down caisson method was adopted for the main tower foundations. The method is to install a prefabricated steel caisson shown in Figure 65.35 on a dredged seabed and subsequently complete a rigid foundation structure by filling concrete inside the caisson compartments. Desegregation concrete was developed for this purpose. The anchorage foundation was constructed with an underground slurry wall method on the Kobe side and the spread foundation method on the Awajishima side. Highly workable concrete was developed for the body of anchorages shown in Figure 65.36.

Figure 65.37 shows the Ohnaruto Bridge completed in 1985. The center and side spans are 876 and 330 m, respectively. The distance between two cables are 34 m. To achieve aerodynamic stability, vertical stabilizing plates were installed under the median strip of the deck to change the wind flow patterns. This kind of stabilizing countermeasure was also adopted in the stiffening truss of the Akashi Kaikyo Bridge.

The multicolumn method, aiming to avoid disrupting the famous Naruto Whirlpools in the Naruto Straits, was used for the main and side tower foundations.

65.7.2 Kojima–Sakaide Route

This route is 37 km long, and three suspension bridges and two cable-stayed bridges were constructed. Since the bridges carry both roadway and railway traffic, a truss girder was selected; its upper deck is used for the roadway and the lower deck for the railway.

FIGURE 65.34 Akashi Kaikyo Bridge.

The laying-down caisson method was utilized for 11 underwater foundations in this route. Prepacked concrete, in which coarse aggregate was at first packed inside the steel caisson and then mortar was injected into voids of the aggregate, was developed for the underwater concrete of the foundations.

Figure 65.38 shows the Shimotsui-Seto Bridge completed in 1988, which is a single-span bridge with 940 m in span length. The distance between two cables is 35 m, a value that is common to all suspension bridges on this route. The main cables of most long-span suspension bridges in Japan have been erected by the prefabricated strand (PS) method. However, this bridge employed the air spinning (AS) method. Since the cable is anchored to the rock directly (a tunnel anchor), the AS method enabled making the anchoring system small.

Figure 65.39 shows the Kita Bisan-Seto Bridge and the Minamai Bisan-Seto Bridge. These bridges have a three-span continuous truss girder. The center spans of theses two bridges are 990 and 1100 m, respectively, and their side spans are each 274 m. These bridges were constructed in 1988. The side view is similar to that of the San Francisco–Oakland Bay Bridge in the United States. The cables of two bridges are anchored to opposite sides of one common anchorage. Hence, the anchorage is subjected only to the difference between the horizontal components of tension in the cables of the two bridges. At the end the bridge, an expansion joint allowing 1.5 m movement was installed, and, also, a transition girder system was used to absorb large amounts of changes in inclination in the track for ensuring running stability of the train. The laying-down caisson method was used for the six underwater foundations. The Sikoku side anchorage foundation is the largest one in this route, reaching 50 m below sea level.

Figure 65.40 shows the Iwakurojima Bridge and the Hitsuishijima Bridge. Both bridges have center spans of 420 m and side spans of 185 m. These bridges were constructed in 1988. First, these bridges were designed as a Gerber-type truss girder. After carrying out a study on the

FIGURE 65.35 Steel caisson.

possibility of cable-stayed bridges, the bridges were changed to that type. Because the weight of the girder required to carry road and rail traffic is substantial, two parallel cables were anchored to an upper chord on one side of the truss girder. Due to the narrow distance between the two cables, wake galloping in the leeward cables was observed. To suppress oscillations, a damping device connecting two cables was used. At each end of the girders, elastic springs in the bridge longitudinal direction were installed. This elastic support adjusts the natural period of the bridge, resulting in a reduction of inertia force due to earthquake. The laying-down caisson method was used for five underwater foundations.

FIGURE 65.36 Body of anchorage.

65.7.3 Onomichi–Imabara Route

This route is 60 km long. In this route, five suspension bridges and four cable-stayed bridges were opened to traffic.

Figure 65.41 shows the Ohsima Bridge completed in 1988, which is a single-span bridge with a 560-m span length. The distance between two cables is 22.5 m. For the stiffening girder, a trapezoidal box section with a depth of 2.2 m was used because of its economical efficiency and maintenance. This is the first application of the box girder to the stiffening girder of a suspension bridge in Japan. A spread foundation was adopted for all the anchorages and the main towers which were constructed on land.

Figure 65.42 shows the Innoshima Bridge whose center and side spans are 770 and 250 m, respectively. The distance between two cables is 26 m. This bridge was constructed at an early stage of the grander project. One strand consisting of 127 wires, which are larger than the more commonly used size of 61 or 91, were developed. A reduced construction period was attained by using this strand. The same type of foundations as the Ohsima Bridge was adopted for this bridge.

Figure 65.43 shows the Kurishima Kaikyo Bridges, which are three consecutive suspension bridges. These bridges were opened to traffic in 1999. The center spans of the three bridges are 600, 1020, and 1030 m. The distance between two cables is 27 m. The girder has a streamlined box section with a depth of 4.3 m. The block of the girder is transported by a self-sailing barge, lifted up with the lifting machine, and finally fixed to hanger ropes. For connection of the block of some parts of the tower, tension bolts instead of friction bolts were employed. Two common anchorages that have a similar cable strand anchoring system as the Kita and Minami Bisan-Seto

FIGURE 65.37 Ohnaruto Bridge.

Bridges were constructed in these three bridges. The laying-down caisson method was adopted for six underwater foundations. A concrete caisson was used for the two relatively small caissons, while the other caissons were made of steel.

Figure 65.44 shows the Tarata Bridge [6,7], which has a world-record span of 890 m and side spans of 320 and 270 m. This bridge was opened to traffic in 1999. Since the side span is short, prestressed concrete girders with a span of 110 and 62.5 m from both abutments are used, and connected to steel girder. These prestressed concrete girders work to counter the large uplifting forces and contribute to an increased in-plane flexural rigidity of the bridge. The depth of the girder is 2.7 m and the ratio of the center span length to girder depth is around 330. Buckling instability of the girder was investigated through analytical and experimental studies. In the experimental study, a 1/50 full model was used. Aerodynamic stability was investigated by wind tunnel tests with a 1/70 full model and a 1/200 full model. In the latter model, the influence of the surrounding topography on aerodynamic behavior was also investigated. The laying-down caisson method was used for two of the tower foundations.

Figure 65.45 shows the Ikuchi Bridge completed in 1989. The center span is 490 m and the side span is 150 m. Again, since the side span length is short, prestressed concrete girders were used in both side spans and were connected to the steel girder. Various connecting methods were investigated, through which a combination of bearing plate and shear studs was used. Each cable consists of galvanized steel wires with a diameter of 7 mm, and is coated with polyethylene (PE) tube. A pile foundation was adopted for all piers since the ground was composed of weathered rock.

FIGURE 65.38 Shimotsui-Seto Bridge.

65.8 New Bridge Technology Relating to Special Bridge Projects

Masatsugu Nagai

65.8.1 New Material in the Tokyo Wan Aqua-Line Bridge [8]

The Tokyo Wan Aqua-Line with a 15.1 km length crosses the Tokyo Bay, and links Kanagawa and Chiba prefectures. It was opened to traffic in December 1997. It has a marine section of 14.3 km, and consists of an approximately 9.5-km-long tunnel and 4.4-km-long bridge structures. Here, an outline of the bridge and new material for corrosion protection of steel piers are introduced.

Figure 65.46 shows a general view of the bridges, which consists of 3-, 10-, 11-, 10-, and 9-span continuous steel box girder bridges. Since the maximum span length of 240 m is needed for a navigation channel and the strength of soil foundation is weak, steel box girders were designed and constructed. For erection of most of the bridges, floating cranes and deck barges were used. After completion of the superstructure, vortex-induced oscillation with an amplitude over 0.5 m was observed. To suppress it, 16 tuned mass dampers, as shown in Figure 65.47, were installed. Part of the steel deck was cut out and stiffened. After installing the tuned mass dampers, the removed plates were welded to the original position.

To attain higher durability of steel piers, titanium-clad steel is attached to the steel piers. Figure 65.48 shows the pier with the titanium-clad steel. It was used in the region of the steel piers affected by tidal movement and saltwater spray. By employing this system, a maintenance-free system more than 100 years is expected.

FIGURE 65.39 Kita-Bisan Seto and Minami-Bisan Seto Bridges.

65.8.2 New Bridge System in the New Tomei Meishin Expressway

Japan Highway Public Corporation started constructing the New Tomei (between Tokyo and Nagoya City) Meishin (between Nagoya and Kobe Cities) Expressway which is around 600 km long and links the big cities of Tokyo, Nagoya, Osaka, and Kobe.

Figure 65.49 shows a plate I-girder bridge with a small number of main girders, which is a new bridge system employed for Ohbu Viaduct near Nagoya City. This viaduct was constructed in 1998. Conventionally, the distance of the I-girder, which corresponds to the span of concrete slabs, has been designed to be less than 3 m. However, in this project, using prestressed, precast concrete slabs,

FIGURE 65.40 Iwakurojima and Hitsuishijima Bridges.

the span of the slab extended to 6 m. Hence, for a three-lane bridge with a width of around 15 m, the number of the I-girders is reduced from 6 to 3. Further, the I-girders are connected by simple beams arranged at a distance of around 10 m only. Conventionally, I-girders have been stiffened by cross-beams or bracing, which are installed at a distance less than 6 m, and a lateral bracing member, which is installed at a lower level of the girder. This simple bridge system can reduce the construction cost and also the painting area. Further, this system leads to easy inspection.

65.8.3 Superconducting Magnetic Leviation Vehicle System [9]

Japan Railway Corporation has a plan of constructing a new line, the Chou–Shinkansen line. Using a high-speed train, it will run through the central part of Japan from Tokyo to Osaka. Now, in a test line with a 18.4 km length constructed in Yamanashi prefecture, the running stability, etc. of the high-speed train is being tested.

Figure 65.50 shows the Maglev car running through Ogatayama Bridge which will be explained later. It levitates 100 mm from the ground and runs at a speed of 500 km/h. A repulsive force and an attractive force induced between the superconducting magnets on the vehicle and the coils on the side walls (propulsion coils and levitation coils) are used for propelling and levitating the car. The following are the major technical issues involved in designing the structures.

1. The deflection of the structures should be small to ensure running stability and riding comfort of the train.
2. High accuracy should be attained when positioning the coil on the side walls.
3. Magnetic drag force due to magnetic phenomenon produced between the magnet and steel should be small.

FIGURE 65.41 Ohshima Bridge.

The countermeasure of the third issue is to use a low-magnetic metal such as austenitic high-manganese steel when the metal is positioned within 1.5 m of the superconducting magnet.

Ogatayama Bridge completed in 1995, a Nielsen Lohse bridge, is shown in Figure 65.50. The span and arch rise are 136.5 and 23 m, respectively. The width between the center of the arch chords is 15 m at springing and 9.6 m at arch crown. In this bridge, the steel structures such as the arch chord and floor system are designed to be positioned 1.5 m apart from the magnet. However, for the reinforcing bars encased in the guide way, high-manganese low-magnetic steel is used.

65.8.4 Menshin Bridge on the Hanshin Expressway

Early in the morning on 17 January 1995, a huge earthquake shook the densely populated southern part of Hyogo prefecture. Many steel and concrete bridges fell due to the collapse of reinforced concrete piers. Many steel piers suffered buckling damage and two collapsed. Immediately after the earthquake, investigation began to identify the causes of damage and repair work, such as encasing the concrete piers and increasing longitudinal ribs in the steel piers. The concrete deck changed to the steel deck, and the collapsed prestressed concrete girders changed to steel bridges. This is to reduce the weight of the superstructures. In addition, metal bearings were also replaced by rubber supports for greater damping.

Figure 65.51 shows the typical rubber support. Almost all metal bearings were changed to this type. Figure 65.52 shows the rigid-frame-type piers. At the foot of a column, rubber bearings were installed to isolate earthquake acceleration. Bridges with these isolating system are called Menshin bridges.

FIGURE 65.42 Imnoshima Bridge.

FIGURE 65.43 Kurushima Kaikyo Bridges.

FIGURE 65.44 Tatara Bridge.

FIGURE 65.45 Ikuchi Bridge.

FIGURE 65.46 Steel bridges on Tokyo Wan Aqua-Line. (Courtesy of Trans-Tokyo Bay Highway Corporation.)

65.8.5 Movable Floating Bridge in Osaka City [10]

The world's first swing and floating bridge is under construction in the Port of Osaka. It will be completed in 2000. The main role of the bridge is to connect two reclaimed islands (the names are Maishima and Yumenoshima). The width of the waterway between the two islands is around 400 m. In case of the occurrence of unforeseen accidents in the main waterway of the Port of Osaka nearby, this warterway (subwaterway) will provide an alternative entrance. On such occasions, large-sized vessels will pass the subwaterway. In addition, the soil foundation is not strong enough to resist the loads of a conventional bridge. Hence, a movable floating bridge with two pontoon foundations and a swing type has been conceived. The bridge has a total length of 940 m and a width of 38.4 m. The floating part has a length of 410 m and a main span length of 280 m with a double-arch rib rigidly connected to two steel pontoons as shown in Figure 65.53. Safety against dynamic responses subjected to waves, winds, earthquake, and heavy track loading have been investigated through numerical and experimental studies.

65.9 Summary

Tetsuya Yabuki

A chronological table of the major revisions of the standard specification of highway bridges and the concrete standard specification in Japan during the latter half of the 20th century is shown in

FIGURE 65.47 Tuned mass damper. (Courtesy of Trans-Tokyo Bay Highway Corporation.)

TABLE 65.5 Chronological Table on the Revisions of the Standard Specifications and the Major Earthquake Disasters in Japan

Epoch	Occurrences
1948	Fukui earthquake disaster
1956	Revision of the Standard Specification of Highway Bridges
1964	Niigata earthquake disaster
1968	Tokachi offshore-earthquake disaster
1971	Establishment of self-editing on the seismic design specification in Standard Specification of Highway Bridges
1973	Revision of the Standard Specification of Highway Bridges
1978	Miyagi Prefecture offshore-earthquake disaster
1980	Revision of the Standard Specification of Highway Bridges
1986	Revision of the Concrete Standard Specification
1995	Great Hanshin-Awaji earthquake disaster
1996	Revision of the Standard Specification of Highway Bridges
	Revision of the Concrete Standard Specification

Table 65.5. The major earthquake disasters in Japan are also shown in Table 65.5 for reference purposes. The design specifications of bridges have been revised mainly whenever strong earthquakes have occurred as shown in Table 65.5. We may be able to make the statement that the most important influence on the evolution of bridges in Japan has been earthquakes. This shows that disasters have controlled bridge engineering. The evaluation of conditions to produce an efficient bridge for seismic motion is distinctly an engineering problem. Therefore, bridge engineers have to escalate their efforts from now on so that bridge engineering can control bridge disasters. The evolution of bridges will be achieved by bridge engineers' efforts to develop optimum structural performance with materials that will be in shorter supply, be longer in durability, give higher strength, and bring less dynamic inertia force.

FIGURE 65.48 Steel pier with titanium-clad steel. (Courtesy of Nippon Steel.)

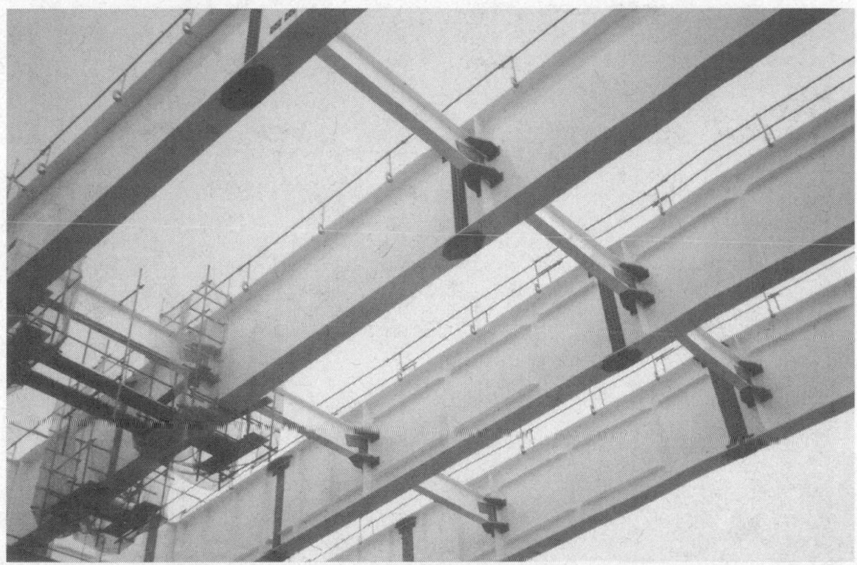

FIGURE 65.49 Plate I-girder with simple cross-beams.

FIGURE 65.50 Maglev car and Ogatayama Bridge. (Courtesy of Japan Railway Corporation.)

FIGURE 65.51 Rubber bearings.

FIGURE 65.52 Rigid frame pier of Benten section. (Courtesy of Hanshin Expressway Public Corporation.)

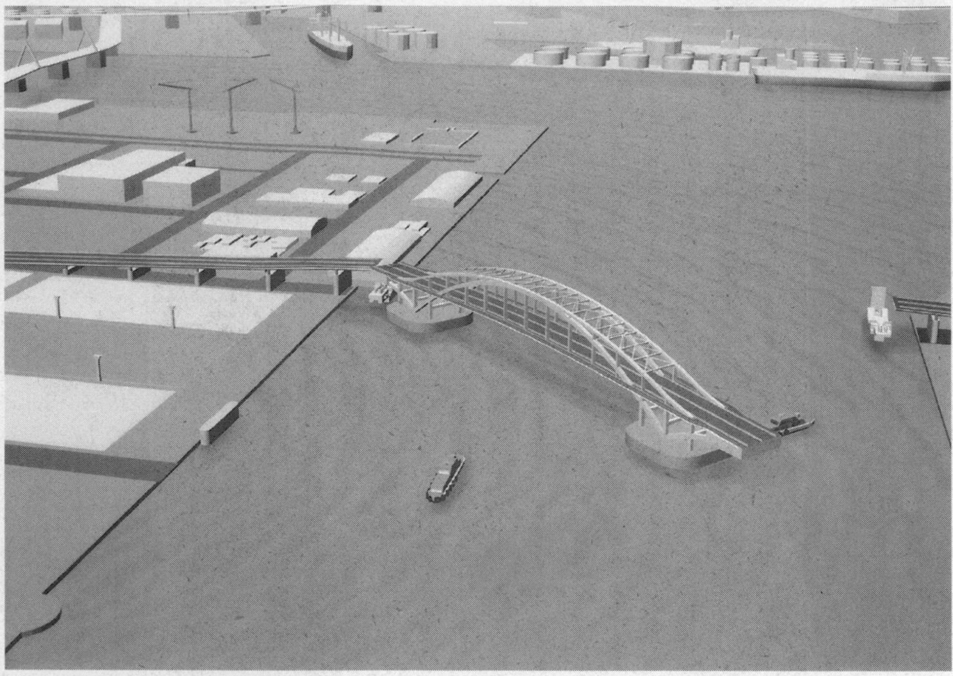

FIGURE 65.53 Yumeshima-Maishima Bridge. (Courtesy of Osaka Municipal Government.)

References

1. The Japanese Association of Highway, Standard Specification of Highway Bridges, rev. 1996.
2. S. Usuki and K. Komatsu, Two timber road bridges, *J. Struct. Eng. Int.*, IABSE, 23–24, 1998.
3. S. Usuki, Y. Horie, and K. Hasebe, Timber arch bridge on city roadway, in *Proc. of the 1990 Int. Timber Engineering Conference*, Vol. 2, Japan, 1990, 431–438.
4. S. Usuki, K. Komatsu, I. Kagiwada, and H. Abe, Drift pin connections with II-shaped internal steel plates for an arch bridge, in *Proc. of Pacific Timber Engineering Conference*, Australia, 1994, 51–58.
5. S. Saeki, Y. Fujino, K. Tada, M. Kitagawa, and T. Kanazaki, Technological aspect of the Akashi Kikyo Bridge, ASCE Structures Congress, Boston, 1995, 237–252.
6. T. Endo, T. Iijima, A. Okukawa, and M. Ito, The technical challenge of a long cable-stayed bridge — Tatara bridge, in *Cable-Stayed Bridges — Recent Development and Their Future*, Elsevier, New York, 417–436, 1991.
7. K. Tada, H. Sato, T. Kanazaki, and H. Katsuchi, Full model wind tunnel test of the Tatara bridge, in *Proc. of Bridges into the 21st Century*, Hong Kong, 1995, 671–678.
8. H. Ikeda, The Trans-Tokyo Bay Highway project, in *Proc. of the 6th East Asia–Pacific Conference on Structural Engineering & Construction*, Taiwan, 1998, 41–48.
9. H. Wakui and N. Matsumoto, Dynamic study on Twin-Beam-Guidway for JR Maglev, in *Proc. of Int. Conference on Speedup Technology for Railway and Maglev Vehicles*, Vol. 1, 1993, 126–131.
10. T. Maruyama, E. Watanabe, T. Utsunomiya, and H. Tanaka, A new movable floating bridge in Osaka Harbor, in *Proc. of the 6th East Asia–Pacific Conference on Structural Engineering & Construction*, Taiwan, 1998, 429–434.

66

Design Practice in Russia

Simon A. Blank

California Department of Transporation

Oleg A. Popov

Joint Stock Company Giprotransmost (Tramos), Russia

Vadim A. Seliverstov

Joint Stock Company Giprotransmost (Tramos), Russia

66.1 Introduction

Bridge design and construction practice in former USSR, especially Russia, is not much known by foreign engineers. Many advanced structural theories and construction practices have been established. In view of the global economy, the opportunities to apply such advanced theories to practice became available with the collapse of the iron curtain.

In 1931, Franklin D. Roosevelt said, "There can be little doubt that in many ways the story of bridge building is the story of civilization. By it, we can readily measure a progress in each particular country." The development of bridge engineering is based on previous experiences and historical aspects. Certainly, the Russian experience in bridge engineering has it own specifics.

0-8493-7434-0/00/$0.00+$.50

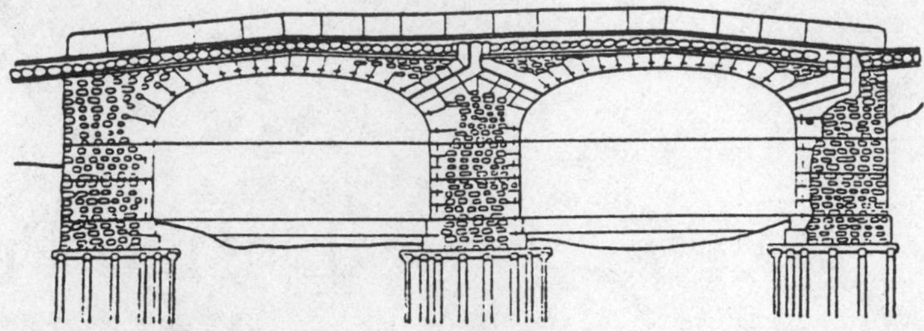

FIGURE 66.1 Typical masonry bridge (1786).

66.2 Historical Evolution

66.2.1 Masonry and Timber Bridges

The most widespread types of bridges in the old time were timber and masonry bridges. Because there were plenty of natural wood resources in ancient Russia, timber bridges were solely built up to the end of the 15th century.

For centuries masonry bridges (Figure 66.1) have been built on territories of such former republics of the USSR as Georgia and Armenia. From different sources it is known that the oldest masonry bridges in Armenia and Georgia were built in about the 4th to 6th centuries. One of the remaining old masonry bridges in Armenia is the Sanainsky Bridge over the River Debeda-chai built in 1234. The Red Bridge over the River Chram in Georgia was constructed in the 11th century. Probably the first masonry bridges built in Russia were in Moscow. The oldest constructed in 1516 was the Troitsky arch masonry bridge near the Troitsky Gate of the Kremlin. The largest masonry bridge over the Moskva River, named Bolshoi Kamenny, was designed by Yacobson and Kristler. The construction started in 1643 but after 2 years the construction was halted because of Kristler's death. Only in 1672 was this construction continued by an unknown Russian master and the bridge was mostly completed in 1689. Finally, the Russian Czar Peter I completed the construction of this bridge. Bolshoi Kamenny Bridge is a seven-span structure with total length of 140 m and width of 22 m [1]. This bridge was rebuilt twice (in 1857 and in 1939). In general, a masonry bridge cannot compete with the bridges of other materials, due to cost and duration of construction.

Ivan Kulibin (Russian mechanics engineer, 1735 to 1818) designed a timber arch bridge over the Neva River having a span of about 300 m, illustrating one of the attempts in searching for efficient structural form [2]. He tested a 1/10 scale model bridge to investigate the adequacy of members and found a large strength in the new structural system. However, for unknown reasons, the bridge was not built.

66.2.2 Iron and Steel Bridges

The extensive progress in bridge construction in the beginning of the 19th century was influenced by overall industrial development. A number of cast-iron arch bridges for roadway and railway traffic were built. At about the same time construction of steel suspension bridges was started in Petersburg. In 1824 the Panteleimonovsky Bridge over the Fontanka River having a span of about 40 m was built. In 1825 the pedestrian Potchtamsky Bridge over the Moika Bankovsky was built, and the Lion's and Egyptian Bridges over the Fontanka River having a span of 38.4 m were constructed in 1827.

The largest suspension bridges built in the 19th century were the chain suspension bridge over the Dnepr River in Kiev (Ukraine), with a total length 710 m including six spans (66.3 and 134.1 m) and

FIGURE 66.2 Bridge over the Luga River (1857).

approaches, constructed in 1853. In 1851 to 1853 two similar suspension bridges were constructed over the Velikaya River in Ostrov City with spans of 93.2 m. The development in suspension bridge systems was based on an invention of wire cables. In Russia, one of the first suspension bridges using wire cables was built in 1836 near the Brest-Litovsk fortress. This bridge crossed the River West Bug, having a span of about 89 m. However, suspension bridges of the first half of the 19th century, due to the lack of structural performance understanding, had inadequate stiffness in both vertical and horizontal directions. This appeared to be the main cause of a series of catastrophes with suspension bridges in different countries. Any occurrence of catastrophes in Russia was not noted. To reduce the flexibility of suspension bridges, at first timber and then steel stiffening trusses were applied. However, this innovation improved the performance only partly, and the development of beam bridges became inevitable in the second half of the 19th century.

The first I-beam bridge in Russia was the Semenovsky Bridge over the Fontanka River in Petersburg constructed in 1857 and the bridge over the Neman River in Kovno on the railway of Petersburg–Warsaw constructed in 1861. However, I-beams for large spans proved to be very heavy, and this dictated a wider application of truss systems.

The construction of the railway line in 1847 to 1851 between Petersburg and Moscow required a large number of bridges. Zhuravsky modified the structural system of timber truss implemented by Howe in 1840 to include continuous systems. These bridges over the Rivers Volga, Volhov, and some others had relatively large spans; e.g., the bridge over the Msta River had a 61.2 m span. This structural system (known in Russia as the Howe–Zhuravsky system) was further widely used for bridge construction up to the mid-20th century.

The steel truss bridges at first structurally repeated the types of timber bridges such as plank trusses or lattice trusses. A distinguished double-track railway bridge with deck truss system, designed by Kerbedz, was constructed in 1857 over the Luga River on the Petersburg–Warsaw railway line (Figure 66.2). The bridge consists of two continuous spans (each 55.3 m.). The P-shape cross sections were used for chord members, and angles for diagonals. Each track was carried by a separate superstructure which includes two planes of trusses spaced at 2.25 m. The other remarkable truss bridge built in 1861 was the Borodinsky Bridge over the Moskva River in Moscow, with span length of 42.7 m. The cornerstone of the 1860s was an introduction of a caisson foundation for bridge substructures.

Up to the 1880s, steel superstructures were fabricated of wrought steel. Cast-steel bridge superstructures appeared in Russia in 1883. And after the 1890s, wrought steel was no longer used for superstructures. In 1884, Belelubsky established the first standard designs of steel superstructures covering a span range from 54.87 to 109.25 m. For spans exceeding 87.78 m, polygonal trusses were designed. A typical superstructure having a 87.78-m span is shown in Figure 66.3. Developments in structural theory and technology advances in the steel industry expanded the capabilities of shop fabrication of steel structures and formed a basis for further simplification of truss systems and an increase of panel sizes. This improvement resulted in application of a triangular type of trusses. By the end of the 19th century, a tendency to transition from lattice truss to triangular truss was outlined, and Proskuryakov initiated using a riveted triangular truss system in bridge superstructures. The first riveted triangular truss bridge in Russia was constructed in 1887 on the Romny–Krmenchug railway line.

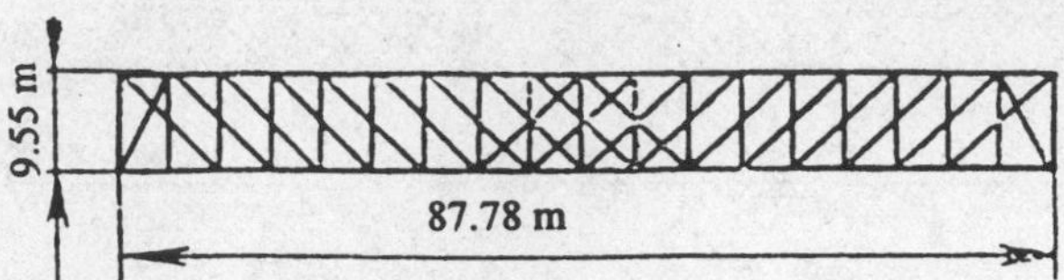

FIGURE 66.3 Standard truss superstructure.

Many beam bridges built in middle of the 19th century were of a continuous span type. Continuous bridge systems have economic advantages, but they are sensitive to pier settlement and have bigger movement due to temperature change. To take these aspects into account was a complicated task at that time. In order to transfer a continuous system to a statically determined cantilever system, hinges were arranged within spans. This was a new direction in bridge construction. The first cantilever steel railway bridge over the Sula River designed by Proskuryakov was built in 1888. The first steel cantilever highway bridge over the Dnepr River in Smolensk was constructed. The steel bridge of cantilever system over the Dnestr River, a combined railway and highway bridge, having a span of 102 m, was designed by Boguslavsky and built in 1894. In 1908, the steel bridge over the River Dnepr near Kichkas, carrying railway and highway traffic and with a record span of 190 m, was constructed. Development of new techniques for construction of deep foundations and an increase of live loads on bridges made the use of continuous systems more feasible compared with the other systems.

In the middle of the 19th century arch bridges were normally constructed of cast iron, but from the 1880s, steel arch bridges started to dominate the cast iron bridges. The first steel arch bridges were designed as fixed arches. Hinged arch bridges appeared later and became more widely used. The need to apply arches in plain areas led to the creation of depressed through-arch bridges.

66.3 Modern Development

The 20th century has been remarkable in the rapid spread of new materials (reinforced concrete, prestressed concrete, and high-strength steel), new structural forms (cable-stayed bridges), and new construction techniques (segmental construction) in bridge engineering. The steel depressed through arch truss railway bridge over the Moskva River built in 1904 is shown in Figure 66.4. To reduce a thrust, arches of cantilever system were used. For example, the Kirovsky Bridge over the Neva River having spans of 97 m was built in Petersburg in 1902. Building new railway lines required construction of many long, multispan bridges. In 1932, two distinguished arch steel bridges designed by Streletsky were constructed over the Old and New Dnepr River (Ukraine) having main spans of 224 m and 140 m, respectively (Figure 66.5).

The first reinforced concrete structures in Russian bridge construction practice were culverts at the Moscow–Kazan railway line (1892). In the early 20th century, the use of reinforced concrete was limited to small bridges having spans of up to 6 m. In 1903, the road ribbed arch bridge over the Kaslagach River was built (Figure 66.6). This bridge had a total length of 30.73 m and a length of arch span of 17 m. In 1904, the road bridge over the Kazarmen River was constructed. The bridge had a total length of 298.2 m and comprised 13 reinforced concrete arch spans of 21.3 m, having ribs of box section.

The existing transportation infrastructure of Russia is less developed compared with other European countries. The average density of railway and highway mileage is about five times less than that of the United States. For the past two decades, railway and highway construction activities have slowed down, but bridge design and construction have dramatically increased.

FIGURE 66.4 Steel depressed arch railway bridge over the Moskva River.

FIGURE 66.5 Steel arch bridge with a span of 224 m.

66.3.1 Standardization of Superstructures

An overview of the number and scale of bridges constructed in Russia shows that about 70% of railway bridges have a span length less than 33.6 m, and 80% of highway bridges have less than 42 m. Medium and small bridges are, therefore, predominant in construction practice and standard structural solutions have been developed and used efficiently, using standardized design features for modern bridges.

The current existing standardization covers the design of superstructures for certain bridge types. For railway bridges, standard designs are applicable to spans from 69 to 132.0 m. These are reinforced concrete superstructures of slab, stringer types, box girder; steel superstructures of slab, stringer types; composite superstructures; steel superstructures of through plate girder, deck truss, through-truss types. For highway bridges, standard designs cover the span range from 12 to 147 m. These are reinforced concrete superstructures of voided slab, stringer, channel (P-shaped) girder, solid web girder, box girder types, composite superstructures of steel web girder types, and steel superstructures of web and box girder types.

Modern highway bridges having spans up to 33 m and railway bridges up to 27.6 m are normally constructed with precast concrete simple beams. For highway bridges, continuous superstructures of solid web girder types of precast concrete segments are normally used for spans of 42 and 63 m, and box girder type of precast box segments are used for spans of 63 and 84 m. The weight of

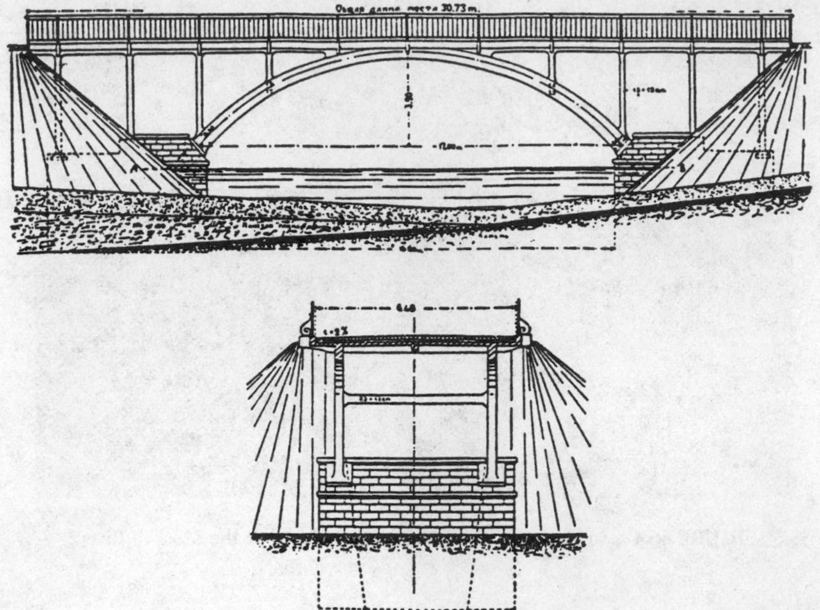

FIGURE 66.6 Reinforced concrete ribbed arch bridge over the Kashlagash River (1903).

precast segments does not exceed 60 tons and meets the requirements of railway and highway transportation clearances. For railway bridges, steel box superstructures of full span (33.6 m) shop-fabricated segments have become the most widespread in current practice.

The present situation in Russia is characterized by a relative increase in a scale of application of steel superstructures for bridges. After the 1990s, a large number of highway steel bridges were constructed. Their construction was primarily based on modularization of superstructure elements: shop-fabricated segments having a length of up to 21 m. Many of them were built in the city of Moscow, and on and over the Moscow Ring Road, as well as the bridges over the Oka River in Nizhni Novgorod, over the Belaya River in the Ufa city, and many others. A number of steel and composite highway bridges having spans ranging from 60 to 150 m are currently under design or construction. In construction of bridges over the Moskva, Dnepr, Oka, Volga, Irtish, Ob Rivers, on the peripheral highway around Ankara (Turkey), on the Moscow Ring Road, and some others, continuous steel superstructures of web and box girder types permanent structure depth are typically used. The superstructures are assembled with modularized, shop-fabricated elements, which are welded at the shop or construction site to form a complete cross section configuration. Erection is normally accomplished by incremental launching or cantilever segmental construction methods.

The extensive use of steel bridges in Russia is based on optimum structural solutions, which account for interaction of fabrication technology and erection techniques. The efficiency is proved by high-quality welded connections, which allow erection of large prefabricated segments, and a reduction in quantity of works and in the construction period. A low maintenance cost that can be predicted with sufficient accuracy is also an advantage, while superstructures of prestressed reinforced concrete in some cases require essential and frequently unpredictable expenses to ensure their capacity and durability.

66.3.2 Features of Substructure

Construction of bridge piers in Russia was mainly oriented on the use of precast concrete segments in combination with cast-in-place concrete. The usual practice is to construct piers with columns of uniform rectangular or circular sections fixed at the bottom of the foundations. For highway bridges with spans up to 33 m full-height precast rectangular columns of 50 × 80 cm are of standard design. For longer-span bridges, the use of precast contour segments forming an outer shape of piers and cast-in-place methods has become more widespread.

A typical practice is use of driven precast concrete piles for the pile foundation. Standard types of precast concrete piles are of square section 0.35 × 0.35, 0.40 × 0.40, 0.45 × 0.45 m and of circular hollow section of 0.60 m diameter. Also, in the last two decades, CIDH (cast-in-place drilled holes) piles of 0.80 to 1.7 m in diameter have been widely used for foundations.

An increasing tendency is to use pile shafts especially in urban areas, when superstructures are borne directly by piles extended above the ground level (as columns). The efficiency was reached by implementing piles of square sections 35 × 35, 40 × 40 cm, and of circular Section 80 cm in diameter. Bored and cast-in-place piles of large diameter ranging between 1.6 to 3.0 m drilled to a depth of up to 50 m and steel casings are widely employed in bridge foundations. In foundations for bridges over the rivers and reservoirs, bored piles using a nonwithdrawable steel casing within the zone of change in water level and scour depth are normally used. These foundation types were applied for construction of bridges over the Oka River on the peripheral road around Nizhni Novgorod, over the Volga River in Kineshma City, over the Ob River in Barnaul City, over the Volga River in Ulyanovsk City, and some others.

Open abutments are most commonly used for highway bridges. Typical shapes are bank seats, bank seats on piles, and buried skeleton (spill-through). Wall abutments and bank seat on piles are the types of substructure mainly used for railroad bridges. Wingwalls are typically constructed back from the abutment structure and parallel to the road.

66.4 Design Theory and Methods

66.4.1 Design Codes and Specifications

The Russian Bridge Code SNIP 2.05.03-84 [5] was first published in 1984, amended in 1991, and reissued in 1996. In Russia the new system of construction codes was adopted in 1995. In accordance with this system the bridge design must satisfy the requirements of the bridge code, local codes, and industry standards. The Standards introduce new requirements resolving inconsistencies found in the bridge code. The bridge code covers design of new and the rehabilitation of existing bridges and culverts for highways, railways, tramways, metro lines, and combined highway–railway bridges. The requirements specified are for the location of the structures in all climatic conditions in the former USSR, and for seismic regions of magnitude up to 9 on the Richter scale. The bridge code has seven main sections: (1) general provisions, (2) loads, (3) concrete and reinforced concrete structures, (4) steel structures, (5) composite structures, (6) timber structures, and (7) foundations.

In 1995, the Moscow City Department of Transportation developed and adapted "Additional Requirements for Design and Construction of Bridges and Overpasses on the Moscow Ring Highway" to supplement the bridge code for design of the highway widening and rehabilitation of the 50 bridge structures on the Moscow Ring Highway. The live load is increased by 27% in the "Additional Requirements." In 1998 the Moscow City Department issued the draft standard TSN 32 "Regional Building Norms for Design of Town Bridge Structures in Moscow" [6] on the basis of the "Additional Requirements." The new standard specifies an increased live load and abnormal loading and reflects the necessity to improve the reliability and durability of bridge structures. The final TSN 32 was issued in 1998.

66.4.2 Design Concepts and Philosophy

In the former Soviet Union, the ultimate strength design method (strength method) was adopted for design of bridges and culverts in 1962. Three limit states — (1) the strength at ultimate load,

(2) deformation at service load, and (3) cracks width at service load — were specified in the bridge design standard, a predecessor of the current bridge code. Later, the limit states 2 and 3 were combined in one group. The State Standard: GOST 27751-88 "Reliability of Constructions and Foundations" [4] specifies two limit states: strength and serviceability. The first limit state is related to the structural failure such as loss of stability of the structure or its parts, structural collapse of any character (ductile, brittle, fatigue) and development of the mechanism in a structural system due to material yielding or shear at connections. The second limit is related to the cracking (crack width), deflections of the structure and foundations, and vibration of the structure.

The main principles for design of bridges are specified in the Building Codes and Regulations — "Bridges and Culverts" SNIP 2.05.03-84 [5]. The ultimate strength is obtained from specified material strengths (e.g., the concrete at maximum strength and usually the steel yielding). In general, bridge structures should satisfy ultimate strength limit in the following format:

$$\gamma_d S_d + \gamma_l (1+\mu) S_l \eta \leq F\left(m_1, m_2, \gamma_n, \gamma_m, R_n, A\right) \tag{66.1}$$

where S_d and S_l are force effects due to dead load and live load, respectively; γ_d and γ_l are overload coefficients; μ is dynamic factor; η is load combination factor; F is function determining limit state of structure; m_1 is general working condition factor accounting for possible deviations of constructed structure from design dimensions and geometrical form; m_2 is coefficient characterizing uncertainties of structure behavior under load and inaccuracy of calculations; γ_n is coefficient of material homogeneity; γ_m *is* working condition factor of material; R_n is nominal resistance of material; A is geometric characteristic of structure element.

The serviceability limit state requirement is

$$f \leq \Delta \tag{66.2}$$

where f is design deformation or displacement and Δ is ultimate allowable deformation or displacement.

The analysis of bridge superstructure is normally implemented using three-dimensional analysis models. Simplified two-dimensional models considering interaction between the elements are also used.

66.4.3 Concrete Structure Design

Concrete structures are designed for both limit states. Load effects of statically indeterminate concrete bridge structures are usually obtained with consideration for inelastic deformation and cracking in concrete. A proper consideration is given to redistribution of effects due to creep and shrinkage of concrete, forces adjustment (if any), cracking, and prestressing which are applied using coefficients of reliability for loads equal to 1.1 or 0.9.

The analyses to the strength limit state include calculations for strength and stability at the conditions of operation, prestressing, transportation, storing, and erection. The fatigue analysis of bridge structures is made for operation conditions.

The analyses to the serviceability limit state comprise calculations for the same conditions as indicated above for the strength limit state. The bridge code stipulates five categories of requirements to crack resistance: no cracks; allowing a small probability of crack formation (width opening up to 0.015 mm) due to live-load action on condition that closing cracks perpendicular to longitudinal axis of element under the dead load is assured; allowing the opening of cracks after passing of live load over the bridge within the limitations of crack width opening 0.15, 0.20, 0.30 mm, respectively.

The bridge code also specifies that the ultimate elastic deflections of superstructures are not to exceed, for railroad bridges, $L/600$ and, for highway bridges, $L/400$. The new standard TSN 32 provides a more strict limit of $L/600$ to the deflections of highway bridges in Moscow City.

66.4.4 Steel Structure Design

Steel members are analyzed for both groups of limit states. Load effects in elements of steel bridge structures are determined usually using elastic small deformation theory. Geometric nonlinearity is required to be accounted for in the calculation of systems in which such an account causes a change in effects and displacements more than 5%. The strength limit states for steel members are limited to member strength, fatigue, general stability, and local buckling. The calculations for fatigue are obligatory for railroad and highway bridges.

For steel superstructures in calculations for strength and stability to the strength limit state the code requires consideration of physical nonlinearity in the elastoplastic stage. Maximum residual tensile strain is assumed as 0.0006, and shear strain is equal to 0.00105.

The net sections are used for strength design of high-strength bolt (friction) connections and the gross sections for fatigue, stability, and stiffness design. A development of limited plastic deformations of steel is allowed for flexural members in the strength limit state. The principles for design of steel bridges with consideration for plastic deformations are reviewed in a monograph [9]. Stability design checks include the global flexure and torsion buckling as well as flange and web local buckling.

A composite bridge superstructure is normally based on a hypothesis of plane sections. Elastic deformations are considered in calculations of effects that occur in elements of statically indeterminate systems as well as in calculations of strength and stability, fatigue, crack control, and ordinates of camber.

66.4.5 Stability Design

Piers and superstructures of bridges are required to be checked for stability with respect to overturning and sliding under the action of load combinations. The sliding stability is checked with reference to a horizontal plane. The working condition factors of more than unity should also be applied for overturning and driving and less than unity for resisting forces.

66.4.6 Temporary Structures

The current design criteria for temporary structures used for bridge construction are set forth in the Guideline BSN (Department Building Norms) 136-78 [10]. This departmental standard was developed mainly as an addition to the bridge code and also some other codes related to bridge construction. The BSN 136-78 is a single volume first published in 1978 and amended in 1984. These guidelines cover the design of various types of temporary structures (sheet piling, cofferdams, temporary piers, falsework, etc.) and devices required for construction of permanent bridge structures. It specifies loads and overload coefficients, working condition factors to be used in the design, and requirements for design of concrete and reinforced concrete, steel, and timber temporary structures. Also, it provides special requirements for devices and units of general purpose, construction of foundations, forms of cast-in-place structures, and erection of steel and composite superstructures.

The temporary structures are designed to the two limit states similar to the principles established for permanent bridge structures. Meanwhile, the overload coefficients and working condition factors have a lower value compared to that of permanent structures. The recent study has shown that the Guideline BSN 136-78 requires revision, and therefore initial recommendations to improve the specified requirements and some other aspects have been reviewed in Reference [11].

The BSN 136-78 specifies a 10-year frequency flood. Also on the basis of technical-economical justification up to 2-year return period may be taken in the design, but in this case special measures for high-water discharge and passing of ice are required. The methods of providing technical hydrologic justification for reliable functioning of temporary structures are reviewed in the Guideline [12]. To widen the existing structures, the range of the design flood return period for temporary structures in the direction of lower and higher probabilities of exceedance is also recommended [12].

66.5 Inspection and Test Techniques

The techniques discussed herein reflect current requirements set forth in SNIP 3.06.07-86, the rules of inspection and testing for bridges and culverts [13]. This standard covers inspection and test procedures of constructed (new) or rehabilitated (old) bridges. Also, this standard is applicable for inspection and tests of structures currently under operation or for bridges designed for special loads such as pipelines, canals, and others. Inspection and testing of bridges are implemented to determine conditions and to investigate the behavior of structures. These works are implemented by special test organizations and contractor or operation agencies.

All newly constructed bridges are to be inspected before opening to traffic. The main intention of inspection is to verify that a bridge meets the design requirements and the requirements on quality of works specified by SNIP 3.06.04-91 [14]. Inspection is carried out by means of technical check up, control measurements, and instrumentation of bridge structural parts. If so required, inspection may additionally comprise nondestructive tests, laboratory tests, and setting up long-term instrumentation, etc. Results obtained during inspection are then compared to allowable tolerances for fabrication and erection specified in SNIP 3.06.04-91. If tolerances or other standard requirements are breached, the influence of these noted deviations on bridge load-carrying capacity and the service state are estimated.

Prior to structural testing, various details should be precisely defined on the basis of inspection results. Load tests require elaborate safety procedures to protect both the structure and human life. Maximum load, taking into account the design criteria and existing structural deviations, needs to be established. The position of structure (before testing is started) for future identification of changes resulting from load tests needs to be recorded (marked up). For the purpose of dynamic load tests, conditions of load passing over the bridge are evaluated.

66.5.1 Static Load Tests

During static load testing, displacements and deformations of the structure and its parts, stresses in typical sections of elements, local deformations (crack opening, displacements at connections, etc.) are measured. Moreover, depending on the type of structure, field conditions, and the testing purpose, measurements of angle strain and load effects in stays or struts may be executed.

For static load tests a bridge is loaded by locomotives, rolling stock of railways, metro or tramway trains, trucks as live loads. In cases where separate bridge elements are tested or the stiffness of the structure is determined, jacks, winches, or other individual loads may be needed. Load effects in members obtained from tests should not exceed the effects of live loads considered in the design accounting for an overload factor of unit and the value of dynamic factor taken in the design. At the same time, load effects in members obtained from the test are not to be less than 70% of that due to design live loading. Weight characteristics of transportation units used for tests should be measured with an accuracy of at least 5%. When testing railroad, metro, tramway bridges, or bridges for heavy trucks, load effects in a member normally should not be less than those due to the heaviest live loads passing over a given bridge.

Quantities of static load tests depend on bridge length and complexity of structures. Superstructures of longer span are usually tested in detail. In multispan bridges having similar equal spans, only one superstructure is tested in detail; other superstructures are tested on the basis of a reduced program, and thus only deflections are measured.

During testing, live loads are positioned on deck in such a manner that maximum load effects (within the limits outlined above) occur in a member tested. Time for test load carrying at each position is to be determined by a stabilization of readings at measuring devices. Observed deformation increments within a period of 5 min should not exceed 5%. In order to improve the accuracy of measurements, time of loading, unloading, and taking readings is to be minimized as much as possible. Residual deformations in the structure are to be determined on the basis of the first test loading results. Loading of structures by test load should normally be repeated. The number of repeated loading is established considering the results that are obtained from the first loading.

66.5.2 Dynamic Load Tests

Dynamic load tests are performed in order to evaluate the dynamic influence (impact factor) of actual moving vehicles and to determine the main dynamic characteristics of the structure (free oscillation frequency and oscillation form, dynamic stiffness, and characteristics of damping oscillation). During dynamic load testing, overall structure displacements and deformations (e.g., midspan deflections, displacements of superstructure end installed over movable bearings), as well as, in special cases, displacements and stresses in individual members of the structure are measured.

The heavy vehicles that may really pass over the bridge are to be used in dynamic load tests to determine dynamic characteristics of structures and moving impact. Vibrating, wind, and other loads may be used for dynamic tests. To investigate oscillations excited by moving vehicles, the trucks are required to pass over the bridge at different speeds, starting from a speed of 10 to 15 km/h. This allows us to determine the behavior of the structure within a range of typical speeds. It is recommended to conduct at least 10 heats of trucks at various speeds and repeat those heats, when increased dynamic impact is noted. In some cases when motorway bridges are tested, to increase the influence of moving vehicles (e.g., to ascertain dynamic characteristics of the structure), a special measure may be applied. This measure is to imitate the deck surface roughness e.g., by laying planks (e.g., 4 cm thick) perpendicular to the roadway spaced at the same distance as the distance between the truck wheels.

When testing pedestrian bridges, excitation of structure free oscillations is made by throwing down a load or by a single pedestrian or a group of pedestrians walking or running over bridges. Throwing down a load (e.g., castings having a weight of 0.3 to 2 t) creates the impact load on a roadway surface typically from 0.5 to 2.0 m height. The location of test load application should coincide with the section where maximum deflections have occurred (midspan, cantilever end). To protect the roadway surface a sand layer or protective decking is placed. The load is dropped down several times and each time the height is adjusted. The results of these tests give diagrams of free oscillations of superstructures. Load effects in structural members during the test execution should not exceed those calculated in the design as stipulated in the section above.

66.5.3 Running in of Bridge under Load

To reveal an adequate behavior of structure under the heaviest operational loads, running in of bridges is conducted. Running in of railroad and metro bridges is implemented under heavy trains, the bridges designed for AB highway loading run-in by heavy trailers. Visual observations of structure behavior under load are performed. Also, midspan deflection may be measured by simple means such as leveling. A number of at least 12 load passes (shuttle-type) with different speeds are recommended in the running procedures of railroad and metro bridges. The first two or three passes are performed at a low speed of 5 to 10 km/h; if deflections measurements are required, the trains are stopped. Positioning trailers over marginal lane having 10 m spacing between the back and front wheels of adjacent units is recommended in running in bridges designed for AB highway loading of two or more lanes. It is recommended that single trailers pass over the free lane at a speed of 10 to 40 km/h, and a number of passes are normally taken, at least five. When visual observations are completed, trailers are moved to another marginal lane and single trailers pass over the lane, which is set free. For running in of single-lane bridges, the passes of single trailers only are used.

FIGURE 66.7 Typical through-truss bridge of 44 m span on the Baikal-Amour Railway Bridge Line.

66.6 Steel and Composite Bridges

In recent decades, there has been further development in design and construction of steel bridges in Russia. New systems provide a higher level of standardization and reduce construction cost.

66.6.1 Superstructures for Railway Bridges

Most railroad bridges are steel composite bridges with single-track superstructure. Standard superstructures [15] are applicable to spans from 18.2 to 154 m (steel girder spans 18.2 to 33.6 m; composite girder spans 18.2 to 55 m; deck truss spans 44 to 66 m; and through-truss 33 to 154 m). Figures 66.7 and 66.8 show typical railroad bridges. Steel box girders as shown in Figure 66.9 have a span of 33.6 m. Similar box girders can be assembled by connecting two prefabricated units. These units are shop-welded and field-bolted with high-strength bolts of friction type. More details are given in References [3] and [15]. For truss systems, the height of trusses is from 8.5 to 24 m with panels of 5.5 and 11 m. Two types of bridge deck — ballasted deck with roadbed of reinforced concrete or corrosion-resistant steel and ballastless deck with track over wooden ties or reinforced concrete slabs — are usually used. For longer than 154 m or double-track bridge, special design is required.

66.6.2 Superstructures for Highway Bridges

Standard steel superstructures covering spans from 42 to 147 m are based on modularization of elements: 10.5 m length box segment, 21 m double-T segment, and 10.5 m orthotropic deck segment. Typical standard cross sections are shown in Figure 66.10. Main technological features of shop production of steel bridge structures have been maintained and refined. Automatic double arc-welding machines are used in the 90% shop fabrication.

66.6.3 Construction Techniques

Steel and composite superstructures for railway bridges are usually erected by cantilever cranes (Figure 66.11) having a capacity up to 130 tons and boom cranes with a larger capacity. The superstructures of 55-m-span bridges may be erected by incremental launching method using a

FIGURE 66.8 Deck girder composite bridge of 55 m span over the Mulmuta River (Siberia).

FIGURE 66.9 Steel box superstructure of 33.6 m span for railway bridges.

nose. When the cantilever crane is applied to erection of the superstructure of 55 m span, a temporary pier is required. For truss superstructures, cantilever and semicantilever erection methods are widely implemented. Figure 66.12 shows the cantilever erection of bridge over the Lena River (span of 110 + 132 + 110 m) using derrick cranes.

For highway bridges, the incremental launching method has been the main erection method for plate girder and box girder bridges since 1970, although this method has been known in Russia for a long time. Cantilever and semicantilever methods are also used for girder bridges. Floating-in is an effective method to erect superstructures over waterways when a large quantity of assembled superstructure segments are required. The application of standard pontoons simplified the assembly of the erection floating system. Equipping of floating temporary piers by an air leveling system and other special equipment made this erection method more reliable and technological.

66.6.4 Typical Girder Bridges

Pavelesky Railroad Overhead

Design of overpasses in Moscow was a really challenge to engineers. It requires low construction depth and minimum interruption of traffic flow. Figure 66.13 shows the Pavelesky Overhead built

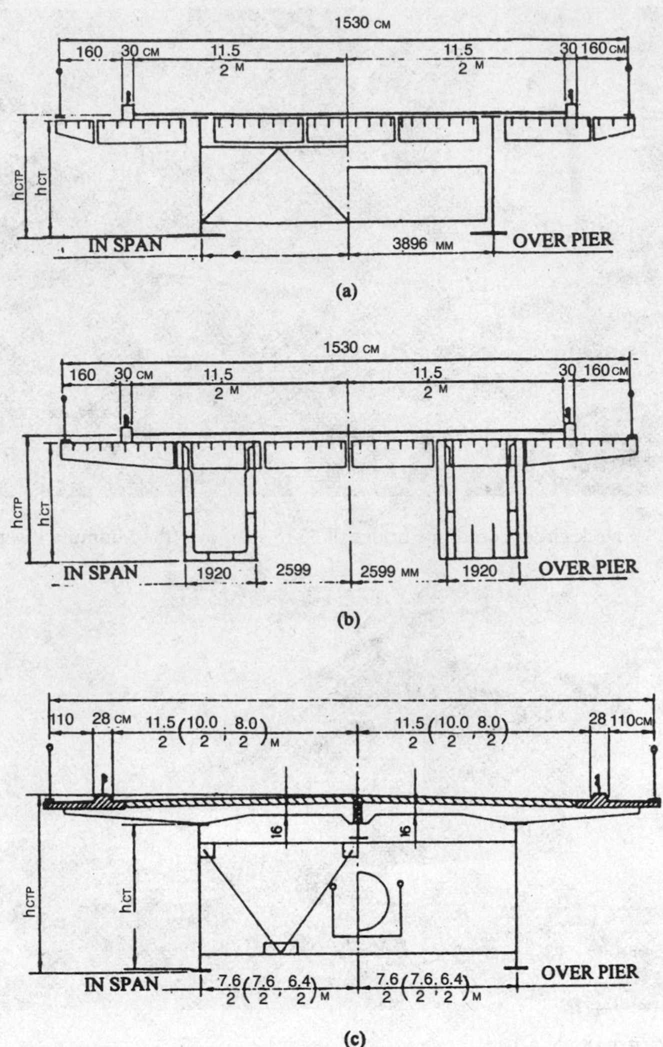

FIGURE 66.10 Typical highway superstructure cross sections. (a) Steel plate sirder; (b) box girder; (c) composite girder.

in 1996 over the widened Moscow Ring Road. The bridge has a horizontal curve of $R = 800$ m and carries a triple-track line. The steel superstructure (Figure 66.14) was designed of low construction depth to meet the specified 5.5 m highway clearance and to maintain the existing track level. This new four-spans (11 + 30 + 30 + 11 m) skewed structure was designed as simple steel double-T girders with a depth of 1.75 m. The orthotropic plate deck of a thickness of 20 mm with inverted T-ribs was used. An innovation of this bridge is the combination of a cross beam (on which longitudinal ribs of orthotropic deck are borne) and a vertical stiffener of the main girder connected to the bottom flange, thus forming a diaphragm spaced at 3 m. A deck cover sheet in contact with ballast was protected by metallized varnish coating.

Moskva River Bridge

Figure 66.15 shows the Moskva River Bridge in the Moscow region, built in 1983. The superstructure is a continuous three-span (51.2 + 96 + 51.2 m) twin box girders (depth of 2.53 m) with orthotropic deck. The bridge carries two lanes of traffic in each direction and sidewalks of 3 m. In transverse

FIGURE 66.11 Erection of steel box superstructure of 33.6 m by boom cranes.

FIGURE 66.12 Cantilever erection of typical truss bridge using derrick cranes.

FIGURE 66.13 Paveletsky Railroad Overhead.

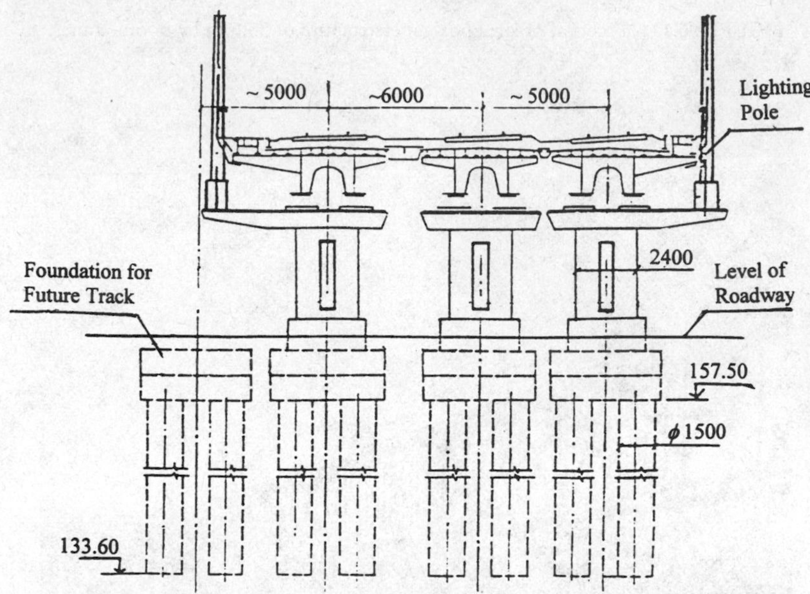

FIGURE 66.14 Typical cross-section of Paveletsky Railroad Overhead (all dimensions in mm).

direction, boxes are braced, connected by cross frames at 9 m. Piers are of reinforced concrete Y-shaped frames. Foundation piles are 40 × 40 cm section and driven to a depth of 16 m. The box girders were launched using temporary piers (Figure 66.16) from the right bank of the Moskva River. Four sliding devices were installed of each pier. The speed of launching reached 2.7 m/h.

Oka River Bridge

Oka River Bridge, near Nizhny Novgorod City, consisting of twin box section (Figure 66.17) formed with two L-shaped elements with orthotropic deck was open to traffic in 1993. A steel bridge of 988 m length is over the Oka River with a span arrangement of 2 × 85 + 5 × 126 + 2 × 84 m.

FIGURE 66.15 Moskva River Bridge in Krilatskoe.

FIGURE 66.16 Superstructure launching for Moskva River Bridge in Krilatskoe.

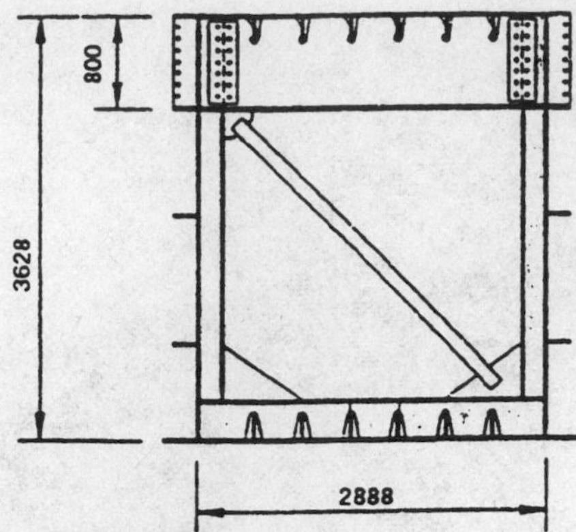

FIGURE 66.17 Cross section of main box girder formed of two L-shaped elements (all dimensions in mm).

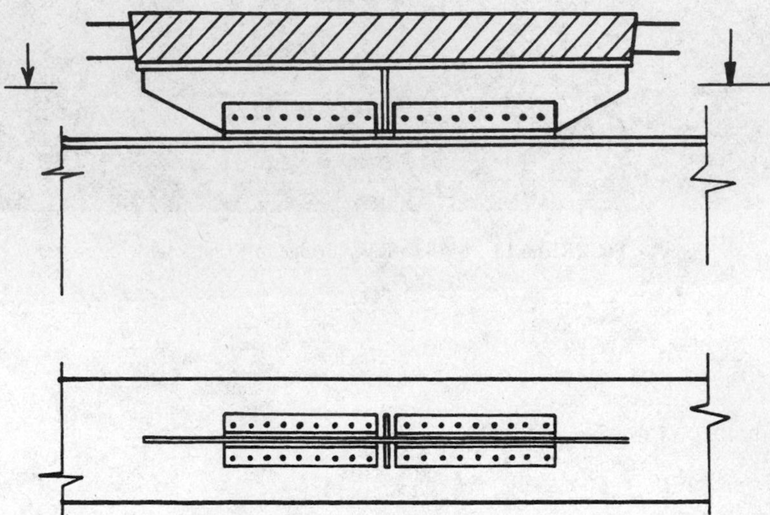

FIGURE 66.18 Connection details of precast reinforced concrete deck and steel girder.

Ural River Bridge

The Ural River Bridge near Uralsk City was completed in 1998. This five-span (84 + 3 × 105 + 84 m) continuous composite girder bridge (depth of 3.6 m) on a gradient of 2.6 m carries a single lane of highway with overall width of 14.8 m. Precast concrete deck segments are connected with steel girders by high-strength friction bolts (Figure 66.18). Connection details are presented in Reference [3].

Chusovaya River (Perm-Beresniki) Highway Bridge

The Chusovaya River Bridges, with a length of 1504 m and on convex vertical curves in radius of 8000 and 25,000 m was completed in 1997. The superstructure comprises two continuous steel composite girders (4 × 84 + 84 + 126 + 5 × 147 + 126 + 84 m). One main problem of bridge construction is the construction of pier foundations within the river under complex geologic and hydrologic conditions. All river piers were constructed under protection of sheet pilling with the use of scows. Piers were constructed using floating cranes. The superstructure segments having lengths of 94.8, 84, and 99.7 m were assembled on the right bank, and then slipped to the river and placed over the floating pier.

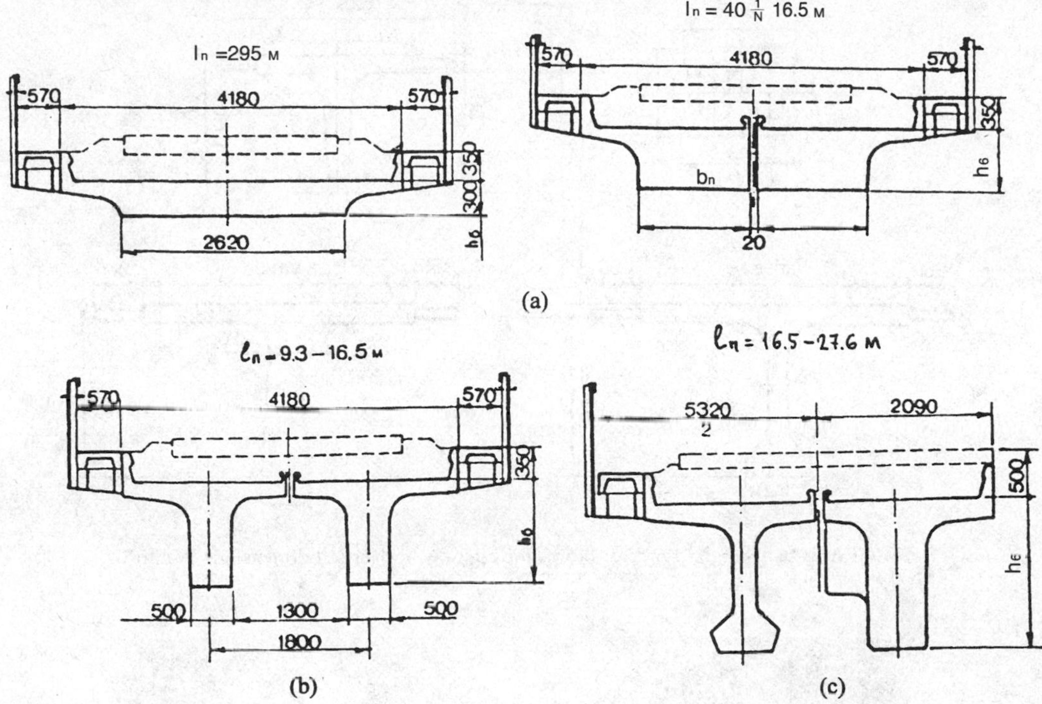

FIGURE 66.19 Typical cross sections of concrete superstructures for railway bridges (all other dimensions in mm). (a) Nonprestressed reinforced concrete slab; (b) nonprestressed reinforced concrete I-beam; (c) prestressed concrete T-beam.

66.7 Concrete Bridges

About 90% of modern bridges are concrete bridges. The recent decades have been characterized by intensive development of standardized precast concrete bridge elements. Main operational requirements and various conditions of construction are taken into account by the existing standardization. This allows a flexible approach in solving architectural and planning tasks and construction of bridges of various span lengths and clearances. Precast concrete bridges, both railway and highway, have been built in many environments, ranging from urban to rural areas. General characteristics of the common types of concrete bridge structures are provided in the following subsections.

66.7.1 Superstructures for Railway Bridges

For railway bridges, standardized shapes for girders and slabs have been widely used for span lengths of 2.95 to 27.0 m. The simplest type of bridge superstructure is the deck slab, which may be solid or voided. The standard structures are designed to carry live load S14 (single-track) and may be located on curved sections of alignment having a radius of 300 m and more. Typical nonprestressed concrete slabs with structural depth from 0.65 to 1.35 m are applicable for spans ranging 2.95 to 16.5 m. Nonprestressed precast T-beams with depth from 1.25 to 1.75 m are used for spans ranging from 9.3 to 16.5 m. Prestressed concrete T-beams with depths from 1.75 to 2.6 m are applicable for spans from 16.5 to 27.6 m. Single-track bridges consist of two precast full segments connected at diaphragms by welding of steel joint straps and then pouring concrete.

All three types of precast superstructure segments are normally fabricated at the shop and transported to a construction site. The waterproofing system is shop-applied. Longitudinal gaps between the segments are covered by steel plates. All superstructure segments starting from 1 to 5 m

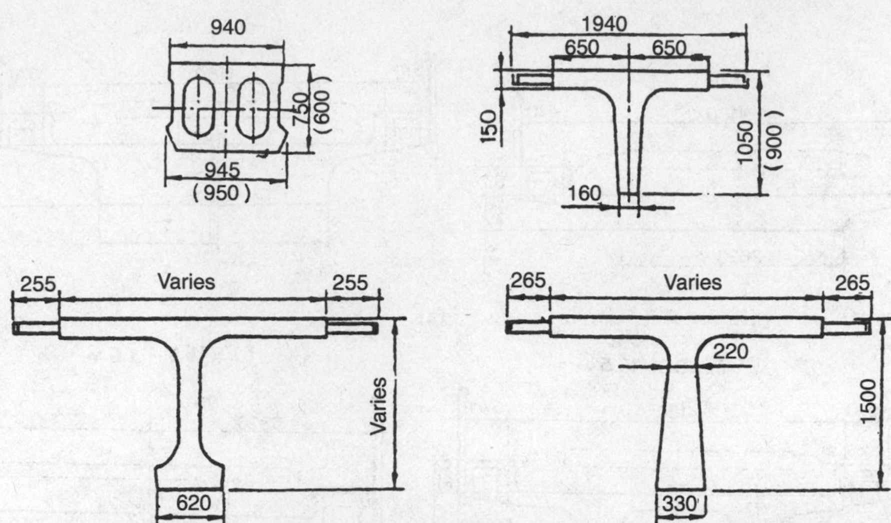

FIGURE 66.20 Standard precast beams for highway bridges (all dimensions in mm).

FIGURE 66.21 Typical section of overhead on the Moscow Ring Road.

length are placed on steel bearings. Superstructures may be connected into a partially continuous system, thus allowing adjustment of horizontal forces transferred to piers.

66.7.2 Superstructures for Highway Bridges

More than 80% of bridges on federal highway networks have span lengths not exceeding 33 m. Figure 66.20 shows typical standard cross sections for highway bridges. Nonprestressed concrete void slabs with structural depth from 0.6 to 0.75 m are applicable for spans ranging 12 to 18 m. Nonprestressed precast T-beams with depth from 0.9 to 1.05 m are used for spans ranging from 11.1 to 17.8 m. Prestressed concrete T-beams with depths from 1.2 to 1.7 m are applicable for spans from 32.3 to 41.5 m. Figure 66.21 shows a construction site of precast concrete beams on the Moscow Ring Road.

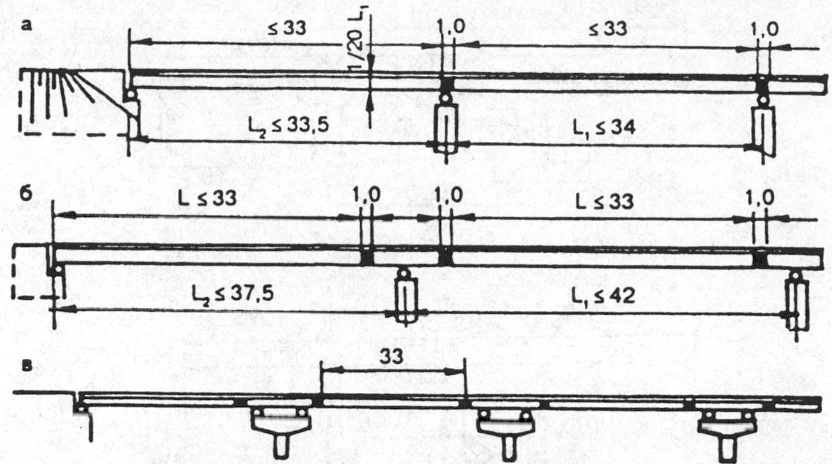

FIGURE 66.22 Typical continuous bridge schemes composed of standard beams (all dimensions in mm).

Fabrication of pretensioned standard beams are conducted at a number of specialized shops. However, by experience some technological difficulties have been noted. These are operations related to installation of reinforcement (space units) into molds of a complex configuration, e.g., bulbous bottom of the girder section, placing of concrete into and taking off the beam from these molds. To improve the fabrication procedure, a special shape of the beam web in the form of a drop has been developed. The standard design of a beam having a 27 m length has been elaborated. Precast prestressed T-beams which are the most widespread in the current construction practice have structural depth from 1.2 to 1.7 m with typical top width of 1.8 m and weight of one beam from 32.3 to 59 tons.

Where special transport facilities are not available or transportation limitations exist or manufacturing facilities such as stressing strands are expensive, precast-in-segment post-tensioned beams are considered to be beneficial. The designs of standard T-beams have been worked out considering both prestressing systems: post-tensioning and pretensioning. The standard T-beams have been designed precast in segments with subsequent post-tensioning for the span lengths of 24, 33, and 42 m.

Due to transportation limitations or restrictions or other reasons, a rational alternative for the post-tensioned beams is beams with transverse joints. In this case the beam consists of segments of limited length and weight (up to 11.8 tons) which may be transported by the usual means. Such segments may be precast either on site or in short lengths at the factory with subsequent post-tensioning at the site. The joints between segments may be implemented by filling the joint gap of 20 to 30 mm thickness with concrete (thin joint); by placing concrete of a minimum thickness of 70 mm (thick joint); or epoxy glued having a 5 mm thickness.

To reduce expansion joints and improve road conditions, the partially continuous system — simple beams at the erection stage and continuous beams at service stage — has been widely used for superstructure; spans up to 33 m. The girders are connected by casting the deck slab over support locations slabs are connected by welding of steel straps on the top of deck. Figure 66.22 shows a typical continuous bridge scheme composed of standard precast slabs or T-beams.

Figure 66.23 shows a standard design for continuous superstructures with double-T beams. The overall width of the precast segment may reach 20 m, but its length is limited to 3 m due to transportation constraints. Ducts for prestressing cables are placed in the web only.

In 1990, another standard design for box-girder continuous systems was initiated. A typical box section is shown in Figure 66.24. Each segment has 1.4 m at bent and 2.2 m within spans with weight up to 62 tons. Using standard precast segments, superstructures can have spans of 63, 71.8, 84, and 92.8 m. Due to financial difficulties the design work on standard box segmental superstructures has not been completed. However, the general idea has been implemented at a number of bridges.

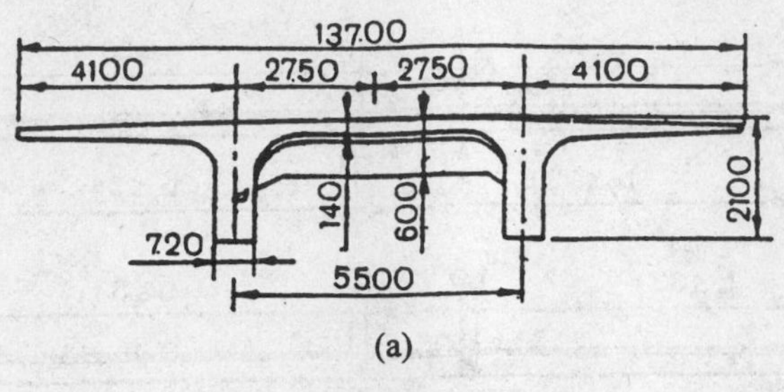

(a)

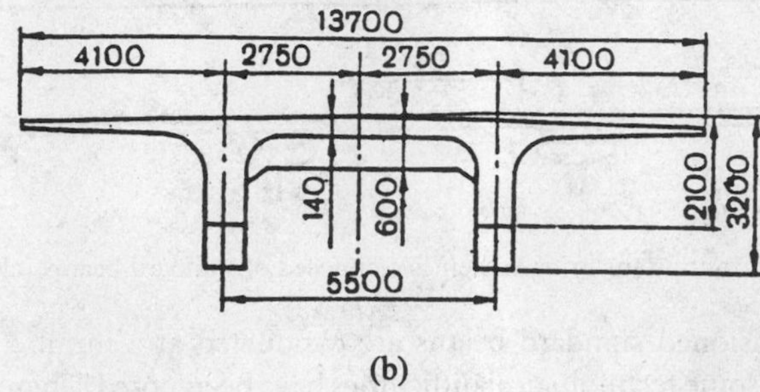

(b)

FIGURE 66.23 Typical cross sections of standard solid web girders. (a) For span arrangement of 33 + n(42) + 33 m; (b) for span arrangement of 42 + n(63) + 42 m (all dimensions in mm).

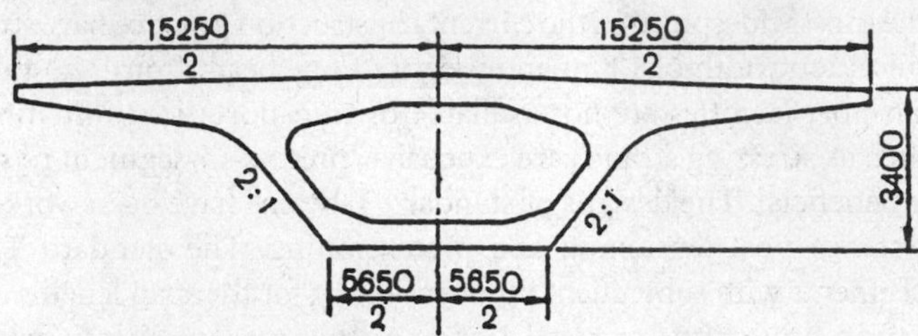

FIGURE 66.24 Typical cross section of precast box segment (all dimensions in mm).

66.7.3 Construction Techniques

The precast full-span beam segments are fabricated at specialized shops on the basis of standard designs. The erections of solid girder segments up to 33 m are conducted by mobile boom cranes of various capacities, gantry cranes, and launching gantries. Special scaffoldings have been designed to erect the solid girder of 24 to 63 m (Figure 66.25).

When large-span bridges are constructed, different cranes may be used simultaneously. At a low-water area of rivers, gantry cranes may be applied, but at deep water, cranes SPK-65 or others may be used. A typical erection scheme of precast cantilever bridge over the Volga River built in 1970 is shown in Figure 66.26.

Although concrete bridge superstructures are constructed mainly by precast segments, in recent years cast-in-place superstructure construction has been reviewed on a new technological level. By this method, construction of superstructure is organized in the area behind the abutment on the approaches to a bridge. The successive portion of the superstructure is cast against the preceding segment and prestressed to it before proceeding to erection by the method of incremental launching.

Cast-in-place concrete main girders and modified precast concrete decks are usually used. Traditionally, when precast deck slabs were used for bridge construction, these precast elements were fabricated with provision of holes for shear connectors that were later filled with concrete. This

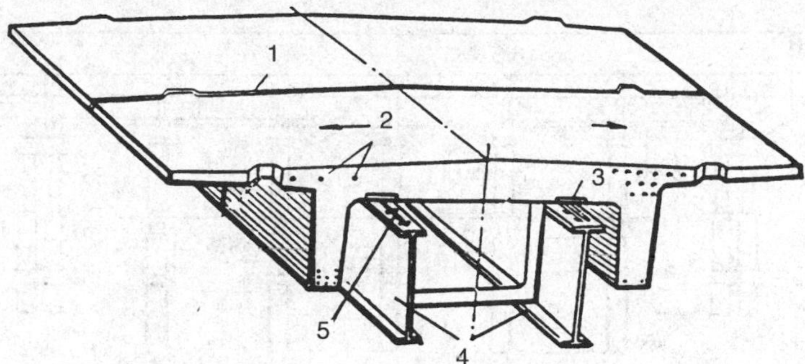

FIGURE 66.25 Position of precast segments of solid web girders over special scaffoldings. 1–glued joints; 2–ducts; 3–embeds; 4–movable special scaffoldings; 5–rails for segment moving.

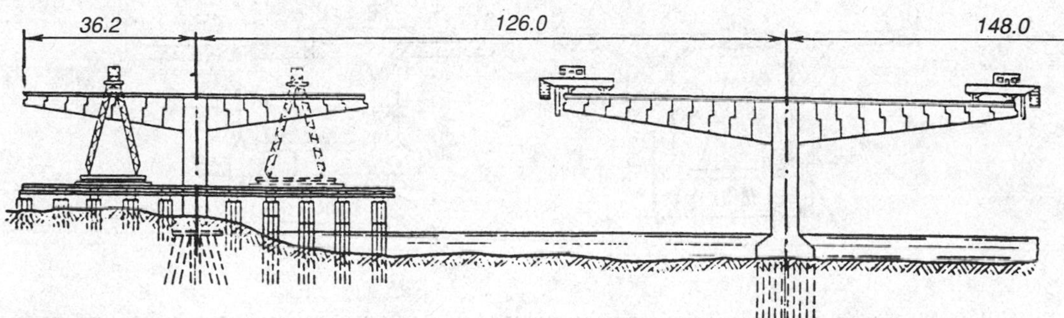

FIGURE 66.26 Typical erection scheme of the bridge over the Volga River (all dimensions in mm).

solution has several disadvantages. To improve the practice, new types of joint between structural members were introduced. The use of steel embeds in precast slabs allows connection of main girders by means of angles and high-strength bolts [3].

66.7.4 Typical Bridges

Komarovka Bridge

A railway bridge over the Komarovka River has recently been built in Ussuryisk (far east of Russia). The bridge of 106.85 m length has a span arrangement of 6 × 16.5 m (Figure 66.27). The superstructure is of reinforced concrete beams of standard design. The intermediate piers are of cast-in-place reinforced concrete. Abutments are spill-through type of precast elements and cast-in-place concrete. Foundations are on bored 1.5-m-diameter piles. Steel casings of 1.35 m diameter were placed in the top portion of the piles. Separate connections of the superstructure into the two systems reduced the temperature forces in the girt. Structurally, the girt (Figure 66.28) consists of two angles of 125 × 125 × 10 mm, which are jointed to the bearing nodes by means of gussets, installed between the bottom flange of superstructure and the sole plate of the fixed bearings. The girt is attached to each gusset plate by high-strength bolts. To reduce the temperature stresses, the girts were fixed to the superstructure and embeds over the abutments at 0° C.

Kashira Oka River Bridge

In 1995, a bridge of total length of 1.96 km crossing over the Oka River near Kashira was constructed. The main spans over the navigation channel are 44.1 + 5 × 85.5 + 42.12 m. The bridge carries three traffic lanes in each direction. Seven-span continuous superstructures are precast concrete box

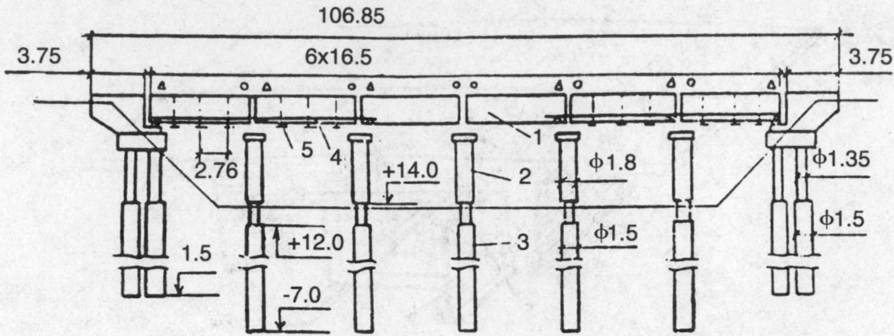

FIGURE 66.27 General arrangement of bridge over the Komarovka River. 1–RC super structure; 2–cast-in-place pier; 3–bored pile; 4–special girt; 5–key to limit sideward movement (all dimensions in mm).

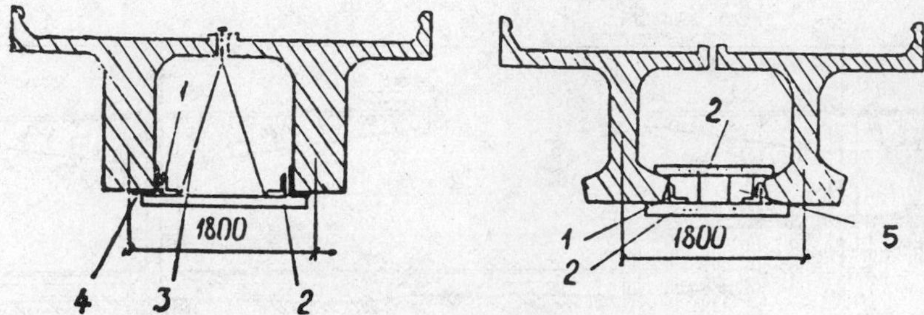

FIGURE 66.28 Special girt details. (Left) Nonprestressed and (right) prestressed girder. 1–girt; 2–Crosstie; 3–anchor element; 4–noftlen pad; 5–jointing plank (all dimensions in mm).

segments with depth of 3.4 m and width of 16 m constructed using a cantilever method with further locking in the middle of the spans. A typical cross section is shown in Figure 66.29. The precast segments vary from 1.5 to 1.98 m and are governed by a capacity of erection equipment limited to 60 tons. Piers are cast-in-place, slip-formed. Foundations are on bored piles of 1.5 and 1.7 m diameter.

Frame Bridge with Slender Legs

Development of structural forms and erection techniques for prestressed concrete led to a construction of fixed rigid frames for bridges. Compared to continuous-span frame bridge systems it is less commonly used. To form a frame system, precast superstructure and pier elements are concreted at overpier section (1 m along the bridge) and at the deck section (0.36 m wide in a transverse direction). Figure 66.30 shows a typical frame bridge with slender legs. The design foundations need an individual approach depending on the site geologic conditions. Simple forms of precast elements, low mass, clear erection scheme, and aesthetic appearance are the main advantages of such bridge systems.

Buisky Perevoz Vyatka River Bridge

This highway bridge (a nonconventional structure) as shown in Figure 66.31 was open to traffic in 1985. The bridge is a cantilever frame system with a suspended span of 32.3 m. The superstructure is a single box rectangular box girder with depth of 3.75 m and width of 8.66 m. Overall deck width is 10 m. The river piers are of cast-in-place concrete and the foundations are on bored 1.5 m piles penetrated to a depth of 16 m. One of the piers is founded on a caisson placed at a depth of 8 m. The 32-mm-diameter bars of pile reinforcement were stressed with 50 kN force per bar for better crack resistance.

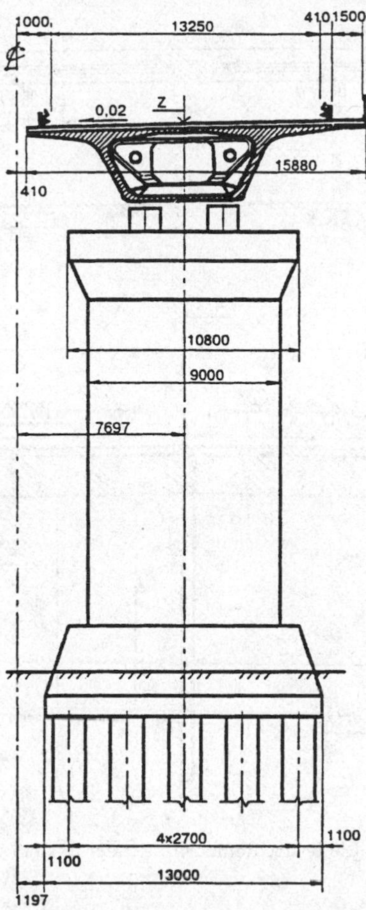

FIGURE 66.29 Typical section of Kashira Oka River Bridge (all dimensions in mm).

Penza Sura River Bridge

This highway bridge comprises a precast prestressed concrete frame of a two-hinge system (Figure 66.32) built in 1975. The superstructure is three boxes with variable sections (Figure 66.33) with a total of 66 prestressing strands. The inclined legs of frames have a box section of 25 × 1.5 m at the top and a solid section 1.45 × 0.7 m at the bottom. The legs of each frame are reinforced by 12 prestressing strands. Each strand diameter of 5 mm consists of 48 wires. Piers are cast-in-place concrete. Foundations are on driven hollow precast concrete piles of 0.6 m. Each frame structure was erected of 60% precast segments. These segments are 5.6 m wide and 2.7 to 3.3 m long. Each frame system superstructure was erected with 12 segments in a strict sequence, starting from the center of the span to the piers, The span segments are placed into the design position, glued, and prestressed. Leg segments are connected with span segments by cast-in-place concrete. A special sequence, as shown in Figure 66.34, for tensioning the strands was established for this bridge.

Moscow Moskva River Arch Bridge

The cantilever arch bridge (Figure 66.35) over the Moskva River on the Moscow Ring Road was built in 1962. The bridge is a three-span structure with 48.65 + 98 + 48.64 m. The overall road width is 21 m and sidewalks are 1.5 m on each side. Half arches are connected by a tie of 10 prestressing strands at the level of the roadway. The erection of the superstructure was implemented on steel scaffoldings. Arches are erected of precast elements weighing from 10 to 20 tons.

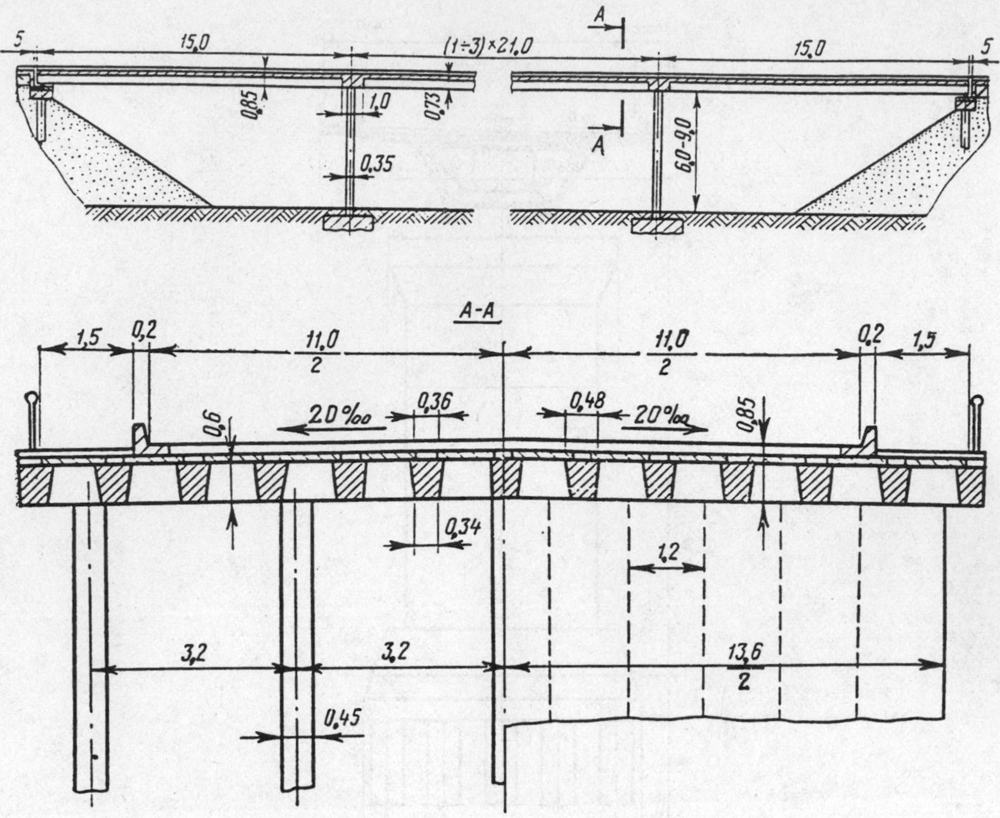

FIGURE 66.30 Typical bridge frame with slender legs (all dimensions in mm).

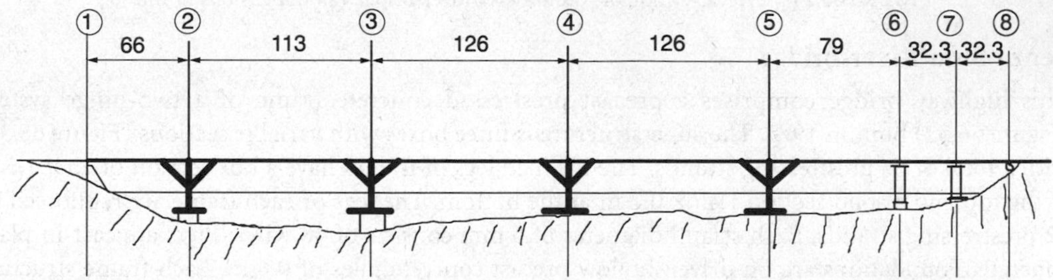

FIGURE 66.31 General scheme of Buisky Perevoz Vyatka River Bridge (all dimensions in mm).

66.8 Cable-Stayed Bridges

The first cable-stayed bridges were constructed in the former USSR during the period 1932–1936. The cable-stayed highway bridge with a span of 80 m designed by Kriltsov over the Magna River (former Georgian SSR) was constructed in 1932. The bridges over the Surhob River, having a span of 120.2 m, and over the Narin River, having a span of 132 m, were constructed in 1934 and 1936, respectively. Figure 66.36 shows a general view of the Narin River Bridge. A stiffening girder of steel truss system was adopted in these bridges.

FIGURE 66.32 Penza Sura River Bridge.

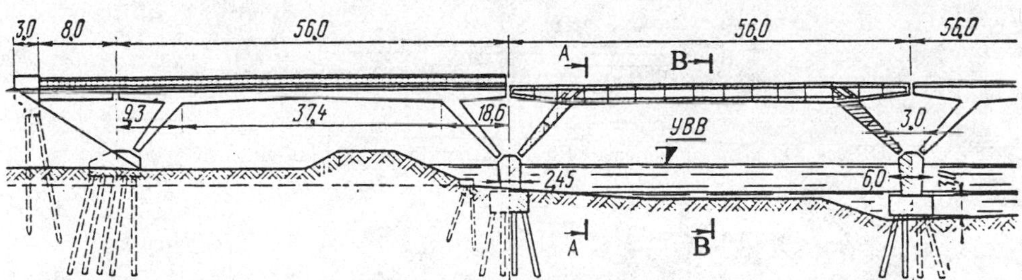

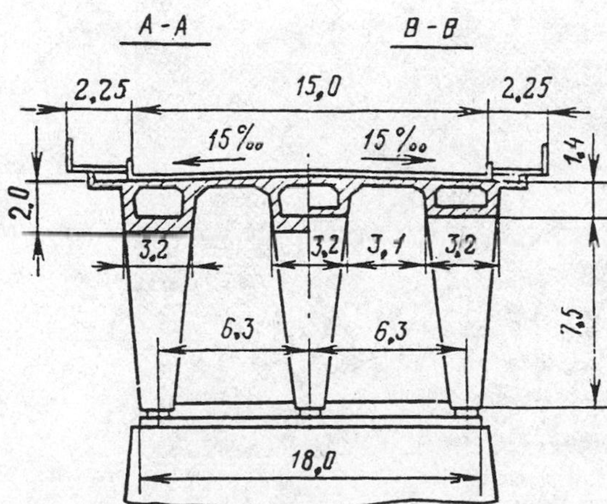

FIGURE 66.33 Typical cross section of Penza Sura River Bridge (all dimensions in mm).

The modern period of cable-stayed bridge construction may be characterized by the following projects: the Dnepr River Bridge in Kiev (1962), the Moscow Dnepr River Bridge in Kiev (1976), Cherepovets Scheksna River Bridge (1980), Riga Daugava River Bridge (1981), the Dnepr River South Bridge in Kiev (1991). Two cable-stayed bridges over the Volga River, Uiyanovsk, and over the Ob River near Surgut are currently under construction.

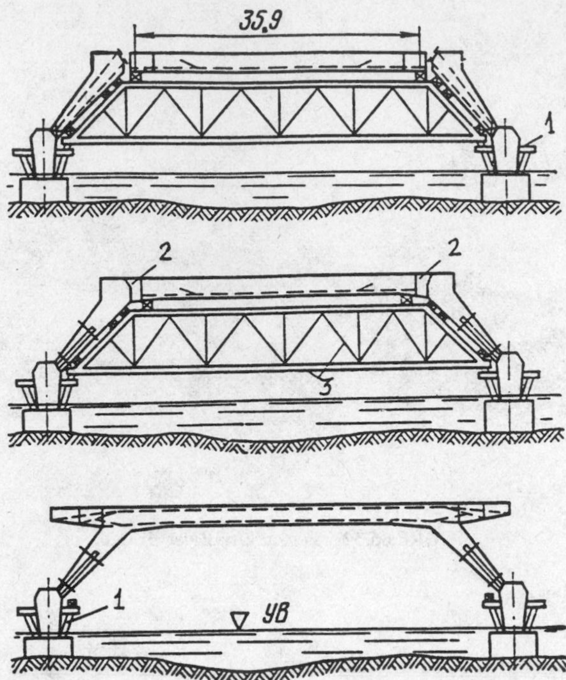

FIGURE 66.34 General sequence of frame structure erection (all dimensions in mm).

FIGURE 66.35 Moscow Moskva River arch bridge.

Kiev Dnepr River Bridge

The first concrete cable-stayed bridge (Figure 66.37) crossing over the harbor of the Dnepr River in Kiev was constructed in 1962. The three-span cable-stayed system has spans of 65.85 + 144 + 65.85 m. The bridge carries highway traffic having a width of roadway of 7 m and sidewalks of 1.5 m on each side. The superstructure comprises two main II-shaped prestressed concrete beams of 1.5 m deep, 1.4 m wide, and spaced at 9.6 m. Cable arrangement is of radiating shape. The stays are composed of strands of 73 and 55 mm in diameter. Towers are cast-in-place reinforced concrete structure.

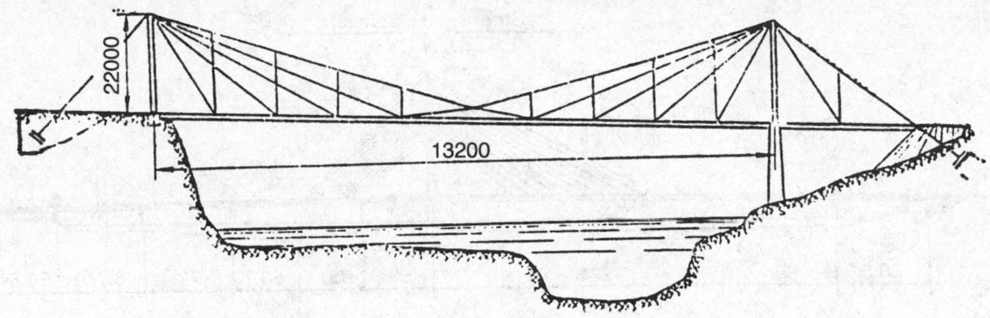

FIGURE 66.36 Narin River Bridge (all dimensions in mm).

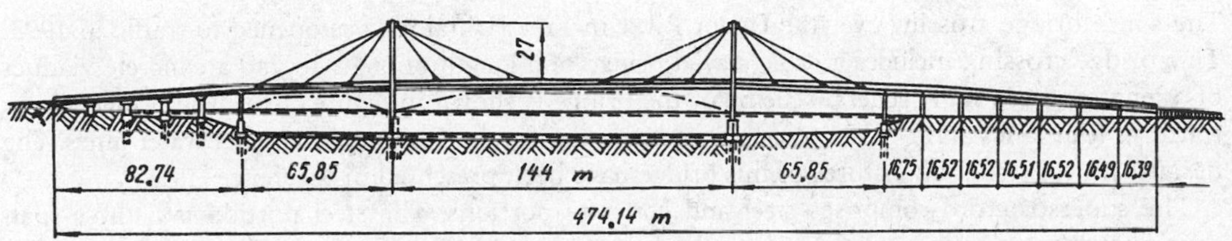

FIGURE 66.37 Kiev Dnepr River Bridge.

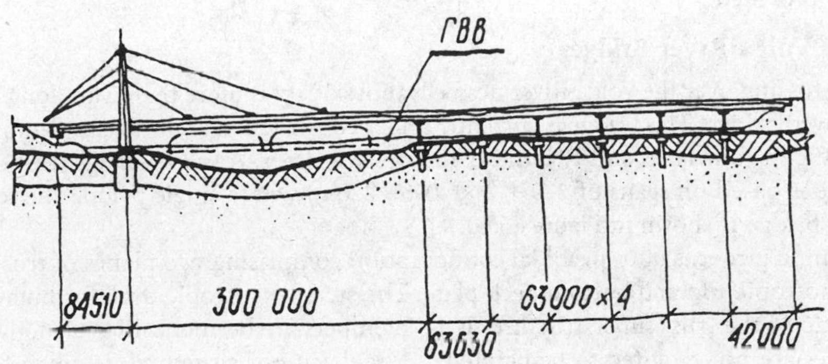

FIGURE 66.38 Moscow Dnepr River Bridge (all dimensions in mm).

Moscow Dnepr River Bridge

In 1976 the Moscow bridge (Figure 66.38) with a cable-stayed system was constructed. The bridge carries six lanes of traffic and five large-diameter pipes below the deck, and has an overall width of 31 m. The three-span continuous structure has a span arrangement of 84.5 + 300 + 63 m. The stiffening girder comprises twin steel box beams with orthotropic deck fabricated of 10 XCND low-alloyed steel grade. To meet the transportation clearances, the depth of the girder was limited to 3.6 m. In a cross section the main beams are 5.5 m wide with a distance between inner webs of adjacent girders equal to 20.2 m and diaphragms spaced at 12.5 m. The stiffening girder has a fixed connection to the abutment and movable connections at the pylon and intermediate pier. A shaped single reinforced concrete pylon 125 m high has a box section of its legs. Each stay is formed from 91 parallel galvanized wires (diameter 5 mm). The stays have a hexagonal section of 55 × 48 cm and are installed in two inclined planes.

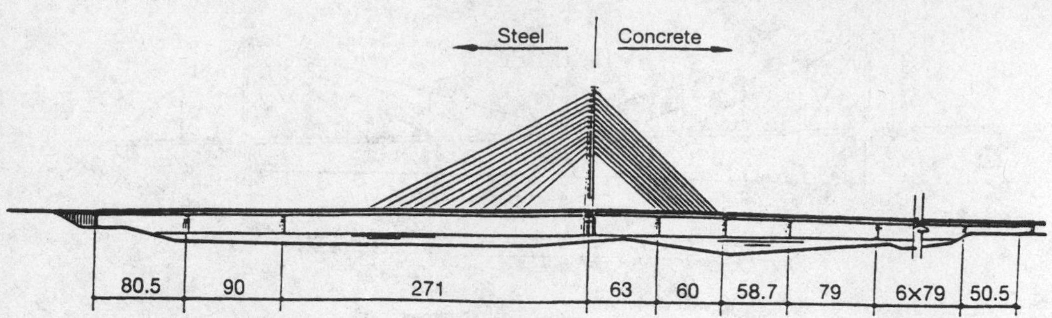

FIGURE 66.39 Dnepr River South Bridge (all dimensions in mm).

Dnepr River South Bridge

The south bridge crossing over the Dnepr River in Kiev (Ukraine) was opened to traffic in 1993. This bridge crossing includes a cable-stayed bridge of a length of 564.5 m and a concrete viaduct of a length of 662 m. A general scheme of the bridge is shown in Figure 66.39. The bridge allows traffic of four lanes and two rail tracks. The bridge also carries four large-diameter water pipes. The design and construction features of this bridge have been presented in Reference [19].

The superstructure comprises steel and concrete portions. The steel portion is a three-span continuous box girder formed of vertical I-beams with orthotropic deck. The concrete portion provides the required counterweight and was constructed of segmental prestressed concrete box sections. The pylon is a two-column cast-in-place concrete frame structure having cross struts between the columns. Cables stays are positioned in two planes, thus torsional rigidity of the bridge is effectively provided.

Ulyanovsk Volga River Bridge

The bridge crossing over the Volga River near Ulyanovsk City is more than 5 km long and includes the cable-stayed bridge. This bridge is currently under construction and will carry combined traffic: highway traffic of four lanes and two tracks for streetcar lines. A span arrangement of this cable-stayed bridge is based on spans of 220 + 2 × 407 + 220 m with a single pylon. The scheme of the cable-stayed bridge is shown in Figure 66.40.

The stiffening girder is of trapezoidal configuration comprising two planes of truss interacting with an orthotropic top and bottom deck plate. The steel orthotropic deck for highway traffic is located on top of the truss superstructure. Truss members are hermetically sealed; therefore their inner surfaces are not required to be painted. A peculiarity of structural detailing for the joints is that the node in a form of hermetically sealed welded box was fabricated at the shop. Connections of flanges and diagonals are moved from the node center. Longitudinal ribs of the orthortropic deck system are of box section; transverse beams are spaced at 5.5 m. The stiffening girder is supported by two planes of stays. Each stay includes cables of parallel wires with diameter of 7 mm. The number of wires in the cables varies from 127 to 271. The weight of stiffening girder of cable-stayed bridge is 20,115 t, and stays, 1900 t. The superstructure is erected by the cantilever method.

The approach steel superstructure is also of truss system with continuous spans of 2 × 220 m. This steel superstructure differs from that of the main span by a transverse rectangular shape only. The structural details are similar to the main-span superstructure. The weight of 2 × 20 m superstructure is 7640 t. The superstructure is assembled on the bank and then erected by 220 m spans using the floating-in method.

The cast-in-place pylon of 204 m height has an inverted Y-shape in the direction along the bridge and a frame of 2 h (H configuration) with cross struts in the transverse direction. For a general

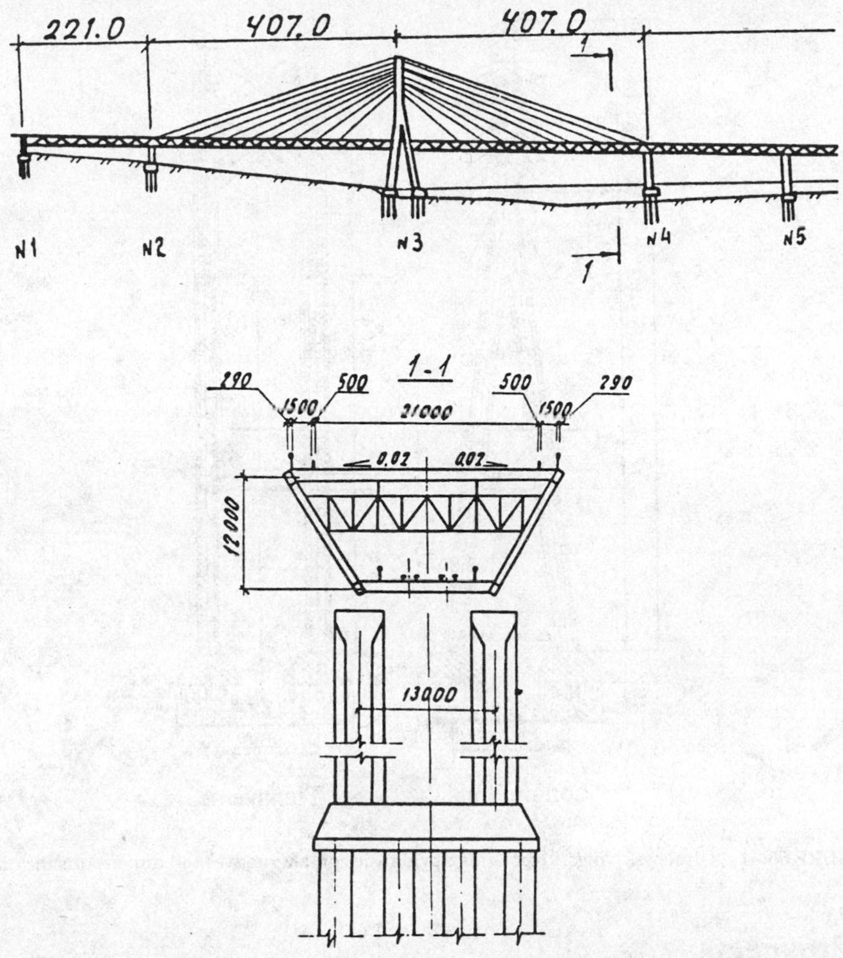

FIGURE 66.40 Ulyanovsk Volga River Bridge (all dimensions in mm).

view of the design alternative for the pylon see Figure 66.41. Towers have a hollow box section with thickness of walls equal to 0.8 to 1.0 m. This type of the pylon structure was influenced by symmetrical side spans of 407 m and absence of permanent guys. The pylon interacts with the superstructure under the unsymmetrical loading by means of its bending stiffness. Foundations are on bored piles of 1.7 m diameter having a bell shape of 3.5 m in diameter at the end.

Surgut Ob River Bridge

Recently, the construction of new bridge crossing the River Ob near Surgut City has started. The overall length of this bridge crossing is a little more than 2 km. A general scheme of the bridge is shown in Figure 66.42. The bridge has an overall width of 15.2 m and will allow traffic of two lanes. It is located in profile on a convex curve having a radius of 120,000 m. The superstructure is a single steel box girder with orthotropic deck. A single pylon of 146 m high is to be constructed in the bottom portion of precast segments forming an outer shape and cast-in-place concrete core, and in the upper portion of two parallel steel towers (transverse section) with struts creating a frame. Intermediate piers are constructed of precast segments with cast-in-place concrete core and with foundation on bored piles with steel casing of 1420 mm in diameter. Abutment are of cast-in-place concrete with foundation on reinforced concrete piles of hollow section 0.6 m in diameter and filled by concrete.

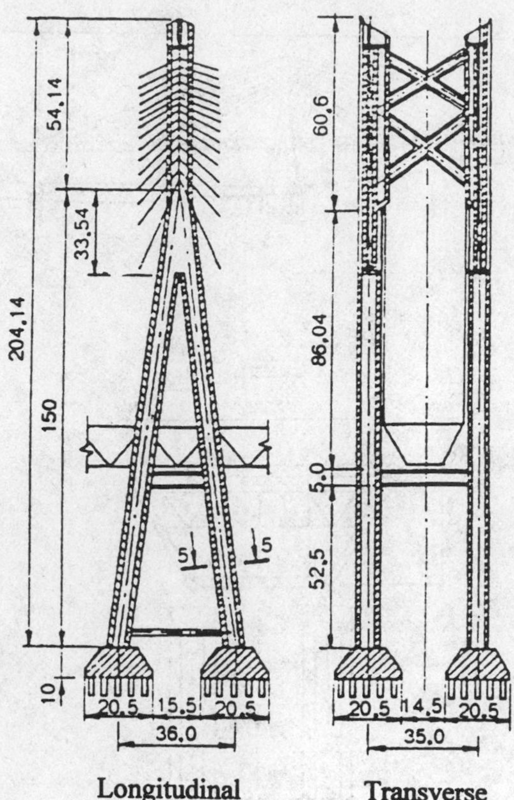

FIGURE 66.41 Ulyanovsk Volga River Bridge pylon (design alternative) (all dimensions in mm).

66.9 Prospects

In recent years, fewer new bridges have been designed and constructed in Russia. The demands for rehabilitation and strengthening of bridge structures are increasing every year. The future directions of bridge design practice are

- Revision and modification of national standards considering Eurocode and standards of other leading countries;
- Considerations for interactions of structural solutions with technological processes considering aesthetic, ecological, and operational requirements;
- Development of new structure forms such as precast or cast-in-place reinforced concrete and prestressing concrete, steel and composite structures for piers and superstructures to improve reliability and durability;
- Redesign of standard structures considering practical experience in engineering, fabrication, and erection practices;
- Unification of shop-fabricated steel elements ready for erection to a maximum dimension fitting the transportation requirements, improvement of precast concrete decks, improvement of corrosion protection systems to a life span up to 12 years; and
- Development of relevant mobile equipment and practical considerations of cast-in-place concrete bridges.

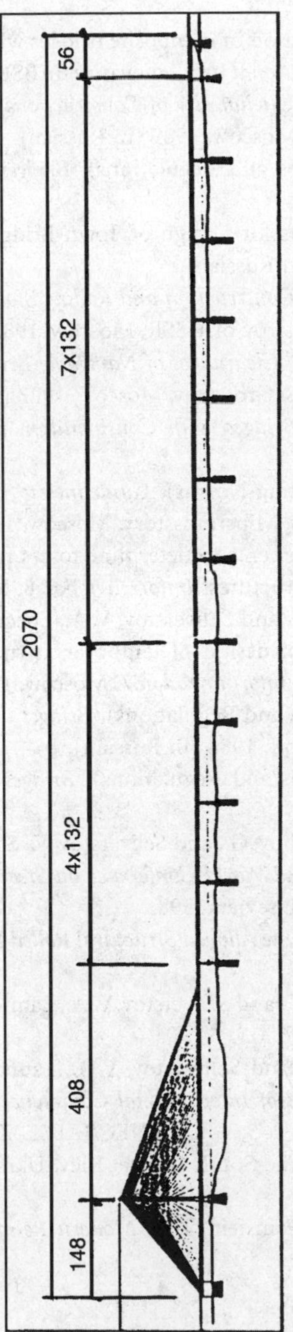

FIGURE 66.42 Surgut Ob River Bridge (all dimensions in mm).

References

1. Belyaev, A.V., *Moskvoretsky's Bridges*, Academy of Science of the USSR, Moscow, Lenningrad, 1945, chap. 1 [in Russian].
2. Evgrafov, G. K. and Bogdanov, N. N., *Design of Bridges*, Transport, Moscow, 1996, chap. 1, [in Russian].
3. Monov, B. and Seliverstov, V., Erection of composite bridges with precast deck slabs, in *Proceedings of Composite Construction Conventional and Innovative*, IABSE, Zurich, 1997, 531.
4. GOST, 27751-88 (State Standard), *Reliability of Constructions and Foundations, Principal Rules of Calculations*, Grosstroy of USSR, Moscow, 1989 [in Russian].
5. SNIP 2.05.03-84 (Building Norms and Regulations), *Bridges and Culverts*, Ministry of Russia, Moscow, 1996 [in Russian].
6. TSN 32, Regional Building Norms for Design of Town Bridge Structures in Moscow, 1st Draft, Giprotransmost, Moscow, 1997 [in Russian].
7. GOST, 9238-83 (State Standard), *Construction and Rolling Stock Clearance Diagrams for the USSR Railways of 1520 mm Gauge*, Grosstroy of USSR, Moscow, 1988 [in Russian].
8. GOST, 26775-97 (State Standard), *Clearances of Navigable Bridge Spans in the Inland Waterways, Norms and Technical Requirements*, Grosstroy, Moscow, 1997 [in Russian].
9. Potapkin, A. A., *Design of Steel Bridges with Consideration for Plastic Deformation*, Transport, Moscow, 1984 [in Russian].
10. BSN 136-78 (Departmental Building Norms), *Guidelines to Design of Temporary Structures and Devices for Construction of Bridges*, Minstransstory, Moscow, 1978, amended 1984 [in Russian].
11. Seliverstov, V. A., Specified requirements to determine forces from hydrologic and meteorological factors for design of temporary structures, *Imformavtodor*, 8, Moscow, 1997 [in Russian].
12. Perevoznikov, B. F., Ivanova, E. N., and Seliverstov, V. A., Specified requirements and recommendations to improve the process of design of temporary structures and methods of hydrologic justification of their functioning, *Imformavtodor*, 7, Moscow, 1997 [in Russian].
13. SNIP 3.06.07-86 (Building Norms and Regulations), *Bridges and Culverts, Rules of Inspection and Testing*, Grosstroy of USSR, Moscow, 1988 [in Russian].
14. SNIP 3.06.04-91 (Building Norms and Regulations), *Bridges and Culverts*, Grosstroy of USSR, Moscow, 1992 [in Russian].
15. Popov, O. A., Monov, B., Kornoukhov, G., and Seliverstov, V., Standard structural solutions in steel bridge design, in *Proceedings of 2nd World Conference on Steel in Construction*, May 11–13, 1998, The Steel Construction Institute, Elsevier, 1998.
16. GOST 6713-91 (State Standard), *Low Alloyed Structural Rolled Stock for Bridge Building*, Grosstroy, Moscow, 1992 [in Russian].
17. Zhuravov, L. N., Chemerinsky, O. I., and Seliverstov, V. A., Launching steel bridges in Russia, *Struct. Eng. Int.*, 6(3), IABSE, Zurich, 1996.
18. Popov, O. A., Chemerinsky, O. I., and Seliverstov, V. A., Launching construction of bridges: the Russian experience, in *Proceedings of International Conference on New Technologies in Structural Engineering*, Lisbon, Portugal, July 2–5, Vol. 2, LNEC/IABSE, Zurich, 1997.
19. Korniyiv M. H. and Fuks, G. B., The South Bridge: Kiev, Ukraine, *Struct. Eng. Int.*, 4(4), IABSE, Zurich, 1994.
20. Kriltsov, E. I., Popov, O. A., and Fainstein, I. S., *Modern Reinforced Concrete Bridges*, Transport, Moscow, 1974 [in Russian].

67

The Evolution of Bridges in the United States

Norman F. Root
California Department of Transportation

0-8493-7434-0/00/$0.00+$.50

67.1 Introduction

American civilization with its bridges is relatively recent compared with the ancient civilizations of Asia, Europe, and even South America. The Americas are the last continents to have become heavily populated and industrialized.

The evolution of bridges in the United States is probably not much different from anywhere else in the world. Civilizations have borrowed their bridging ideas from each other for centuries. Fallen logs across streams served as primitive bridges that led to the concept of girder spans in use today. Suspension spans across deep chasms is a primitive idea used throughout the world. The stone arch introduced by the ancient Romans is a naturally occurring, efficient, and pleasing structural shape that has been used with various evolving materials.

FIGURE 67.1 The aqueduct bridge at La Purisima Mission, Santa Barbara County, California, is an example of a primitive bridge, a short-span stone slab. Built in 1813, it is the oldest bridge in California. (Courtesy of California Department of Transportation.)

Bridge practice evolves as user needs, traffic, and vehicles change, technology progresses, and new materials are developed. But span length is still the primary determining factor for bridge type selection.

67.2 Early U.S. Bridges

The first recorded bridge in the United States was built at James Towne Island, Virginia in 1611. This is the site of one of the earliest European colonies. It was a timber structure, actually a wharf accessing ships anchored in deeper water (Figure 67.1).

67.3 The Canal Era

By water was an early method of heavy transport as the United States began to expand inland from the Eastern Seaboard. Canal builders in the late 1700s and early 1800s were the first to construct U.S. bridges of any consequence. The concept of stone arches, borrowed from Roman aqueducts, was common during this era. Besides, the stone arch readily adapts to the loads imposed (Figure 67.2).

FIGURE 67.2 Scholarie Creek Aqueduct is the Erie Canal over Scholarie Creek at Fort Hunter, New York. It was built by John Jervis in 1841. Canals were the first major users of bridges in the United States. (Courtesy of American Society of Civil Engineers.)

Turnpikes

Private toll roads during the colonial period, 1600s and 1700s, often built timber structures. Logs are natural beams and their ready availability made them natural materials for early bridges.

Timber Bridges

Timber is easy to work and build with. But timber bridges require constant maintenance; joints loosen as the wood shrinks and vibrates from traffic, and wood must be protected from the elements (Figure 67.3).

FIGURE 67.3 Dolan Creek Bridge on the Monterey Coast in California was built in 1932. This is one of only two three-pin timber arch bridges ever built on the California State Highway system. It lasted only a few years, and has since been replaced with a concrete bridge in 1961. (Courtesy of California Department of Transportation.)

Covered Timber Bridges

Many timber bridges of the 19th century were covered to protect the wood from the elements and in northern climates to keep snow off the decks (Figures 67.4 and 67.5).

FIGURE 67.4 The Bridgeport Covered Bridge in California may be the longest single-span, 70.1 m, covered bridge in the world. The superstructure is a Burr arch superimposed on a Howe truss. It was a toll bridge built by David Wood in 1862, and was later purchased by the Virginia Turnpike Company. (Courtesy of California Department of Transportation.)

FIGURE 67.5 The Cornish–Windsor Covered Bridge is a two-span town-lattice truss crossing the Connecticut River between Cornish, New Hampshire and Windsor, Vermont. Built in 1866 it is the longest covered bridge, 140.2 m, in the United States. It has been designated a National Civil Engineering Landmark by the American Society of Civil Engineers. (Courtesy of American Society of Civil Envineers.)

Iron Bridges

Cast-iron bridge members were first considered due to the proximity of several foundries near the National Road. The material turned out to be quite strong and very durable. Cast iron is resistant to normal corrosion associated with ferrous metals (Figures 67.6 and 67.7).

FIGURE 67.6 Dunlap's Creek Bridge, built in 1839, is the first iron bridge in the United States. It was built for the National Cumberland Road, at Brownsville, Pennsylvania, by Captain Richard Delafield of the Army Corps of Engineers. The bridge is still in service today. (Courtesy of Federal Highway Administration.)

FIGURE 67.7 Bow Bridge in Central Park, New York, is the oldest surviving wrought-iron bridge in the United States, built in 1862. It has the longest span, 26.5 m, of five ornately decorated bridges in the park, all designed by Calvert Vaux and Jacob Wrey Mould. (Courtesy of American Society of Civil Engineers.)

67.4 The Railroad Era

The age of steam ushered in an era where bridge building in the United States came of age. Railroads became the dominant mode of transportation for both passengers and freight. Easy grades required for railroads, in turn, required lots of bridges. Canals were all but forgotten and wagon roads went into a 50-year period of neglect (Figure 67.8).

FIGURE 67.8 Starrucca Viaduct, built in the form of the ancient Roman aqueducts, was designed by James Kirkwood for the New York and Erie Rail Road in 1848. It is located over the Starrucca Creek plain at Lanesboro, Pennsylvania. This was the first bridge to use a concrete foundation. This bridge is still in service. (Courtesy of American Society of Civil Engineers.)

Trusses

Squire Whipple and Herman Haupt, two American railroad bridge engineers, are credited with being the first to calculate methods for determining stresses in truss members and were thereby able to determine their appropriate sizes. Each worked independently of the other, in the mid-19th century, using ancient knowledge of mathematics, physics, and strength of materials.

The knowledge to engineer trusses made their construction popular. They provided strength with considerable savings in materials and weight. The concepts of rational principles are equally applicable to both timber and metal trusses. Many other engineers quickly embraced the concepts and patented various truss diagonal configurations for their own use. Many of their names are familiar today: Pratt, Parker, Howe, Burr, Fink, and Warren, to name a few.

Railroad Trestles

See Figures 67.9 through 67.11.

FIGURE 67.9 Theodore Judah took advantage of timber to build trestles quickly and move on, while racing to build the Central Pacific Railroad, the California end of the Transcontinental Railroad. He solved the long-term maintenance problem by later filling in the trestle with cut and tunnel spoil, forming an embankment which would remain long after the timber had rotted away. This is the Secrettown Trestle in the California Sierras, built in 1865, being buried in earth fill. (Courtesy of California State Library.)

FIGURE 67.10 The Devil's Gate High Bridge at Georgetown, Colorado, appears too spindly to support a railroad. But clever use of tension counters distributes the reversing loads throughout the towers. The bridge was prefabricated by Clark Reeves and Company of Phoenixville, Pennsylvania, for the Colorado Central Railroad in 1884. The trestle was in continuous use until torn down in 1939. A replica rebuilt in 1984 is now in use by the Georgetown Loop Mining and Railroad Park. (Courtesy of Missouri Historical Society.)

FIGURE 67.11 Keddie Wye is a unique steel tower trestle built by the Union Pacific Railroad in California's rugged Feather River Canyon in 1912. The wye trestle emerges from a tunnel in the south wall of the canyon splitting rail traffic over the river; one leg heads north to meet with the Burlington Northern Railroad and the other is the main line heading east toward Chicago. (Courtesy of the Feather River Rail Society.)

Steel Arch Bridges

See Figures 67.12 through 67.14.

FIGURE 67.12 Eads' Bridge over the Mississippi River at Washington Street in Saint Louis shattered engineering precedents of the time. It was the first extensive use of steel for bridge construction. The three 175+ m arch spans are each four 464-mm steel truss-stiffened wrought iron tubes. The spandrels are extensive steel truss and lattice work. Built by James B. Eads in 1874. Eads' Bridge is pictured on the two dollar denomination United States postage stamp series commemorating the Trans-Mississippi Exposition of 1896. (Courtesy of U.S. Bureau of Engraving and Printing.)

FIGURE 67.13 Navajo Bridge at Marble Canyon, near Lee's Ferry, Arizona, is the classic example of an arch sprung between canyon walls. This is also an example of a deck truss, an evolution for automobiles, beyond the through truss. When built, in 1929, it was the highest bridge in the world, 162.5 m, from deck to water. It was designed by Ralph Hoffman of the Arizona Highway Department. A parallel twin designed by Cannon Associates has since been constructed, in 1996. (Courtesy of American Society of Civil Engineers.)

FIGURE 67.14 The Cold Springs Canyon steel plate girder arch, in Santa Barbara County, California, is the longest arch span at 213.4 m, and a rise of 121.9 m. The bridge has won a Lincoln Foundation welding award, American Institute of Steel Construction beauty award, and the Governor's Design Award. Built in 1963, it was designed by the California Division of Highways, Marv Shulman, design engineer. (Courtesy of California Department of Transportation.)

Kit Bridges

During the late 19th and early 20th centuries, several bridge companies sold "American Standard," prefabricated wrought iron bridge pieces (bridge in a box), of given span lengths that could be erected on site. All one had to do was order a bridge from a catalog, build abutments for the appropriate span length, and assemble the pieces erector-set-style. Kit bridges are readily adaptable to disassembly, transport, and reuse elsewhere, as has been the case for many of these bridges still in use (Figures 67.15 and 67.16).

FIGURE 67.15 Laughery Creek Bridge, near Aurora, Indiana, was built by the Wrought Iron Bridge Company of Canton, Ohio, in 1878. Its 92-m span was unprecedented. This bridge appeared on the cover of the company's catalog in 1893. (Courtesy of American Society of Civil Engineers.)

FIGURE 67.16 This detail at Haupt Creek, in Sonoma County, California, shows a typical pin connection of a kit bridge and a "Phoenix Column," a patented cast-iron member built exclusively by the Phoenix Iron Works of Pennsylvania. This bridge was built in 1880. (Courtesy of California Department of Transportation.)

67.5 The Motor Car Era

Almost instantaneously, at the turn of the 20th century, the nation was swept up into the automotive age. Long-neglected wagon roads became important once again. State Highway Departments sprang up and road and bridge building, under the "Good Roads Movement," took on a new fervor. Railroad engineering became almost stagnant. Most new highway bridge engineers were former railroad bridge engineers, so many of the early highway bridges looked just like railroad through-truss bridges.

Steel Truss Bridges

See Figures 67.17 through 67.19.

FIGURE 67.17 The Carquinez Straits Bridge in California, built by the American Toll Bridge Company as a private toll bridge in 1927, is an example of a cantilevered truss with eye bar tension members. A parallel twin using welded hybrid high-strength steels was designed and built in 1954 by the California Division of Highways, Roger Sunbury, engineer. Steel truss bridges are considered by many to be ugly. Carquinez is not one of the worst examples, but when a candidate bridge architect interviewing for the California Department was shown a picture of the twin spans and asked for comments, he answered, "Why make the same mistake twice?" He got the job as Chief Bridge Architect. (Courtesy of California Department of Transportation.)

FIGURE 67.18 Coos Bay Bridge on the Oregon Coast Highway is one of several landmark bridges designed by Conde B. McCullough of the Oregon Highway Department. The 225.2 m main span is a classic example of a cantilever truss. Built in 1936, it is the largest of McCullough's coastal gems. The concrete arch end spans and spires are a McCullough trademark. The bridge is now named the McCullough Memorial Bridge in honor of the engineer. (Courtesy of American Society of Civil Engineers.)

FIGURE 67.19 The San Francisco–Oakland Bay Bridge east, is part of the longer 13.3 km crossing composed of the west suspension span, a tunnel through Yerba Buena Island, and this cantilever truss east span. The seismic retrofitting solution at this site is to replace the bridge. There is local controversy over the type of span to be used. There are cost concerns, fear by San Francisco that an east side signature span could overshadow their west suspension span, and aspirations by Oakland that their city is also deserving of a signature span on their side of the Bay. (Courtesy of California Department of Transportation.)

Reinforced Concrete

About the same time as the motor car era began, the turn of the 20th century, the concept of reinforced concrete was introduced. It was generally unaccepted until the San Francisco earthquake of 1906. The few reinforced concrete buildings were the only structures to survive. From that time on, reinforced concrete has been widely used (Figure 67.20).

FIGURE 67.20 Alvord Lake Bridge is the first reinforced concrete bridge, built by Ernest Ransome, the developer of reinforced concrete, in 1888. This bridge is still in service carrying State Route 1 over Golden Gate Park in San Francisco. The facia is hammered to resemble familiar stone arch work. The bridge is a National Historic Civil Engineering Landmark. (Courtesy of California Department of Transportation.)

Concrete Arches

Reinforced concrete arches were popular during the early part of the 20th century. Reinforced concrete was the modern material, and arches were a comfortable, tried, and true shape. Thousands of reinforced concrete arches were built until the 1950s (Figures 67.21 through 67.27).

FIGURE 67.21 The Colorado Street Bridge over the Arroyo Seco in Pasadena, California, is the highest scoring bridge for historical significance in the state. The main span is 46.6 m with a height of 45.7 m. The structure is highly adorned with Beaux Art ornamentation. It was designed in 1912 by John Waddell, the "Dean" of American bridge engineering. The bridge served the famed Route 66 for many years. Seismic retrofitting was a challenge in trying to maintain the bridge's historic aesthetic features. (Courtesy of California Department of Transportation.)

FIGURE 67.22 Fern Bridge near Ferndale, California, is a remarkable structure that has withstood the test of time. Six major floods since it was built have washed out other bridges on the lower Eel River, but Fernbridge still stands. It is composed of seven 61-m rubble-filled closed spandrel concrete arches, each on 250 timber piles. it was designed by John B. Leonard in 1911 for Humboldt County. It is now part of the California State Highway system. (Courtesy of California Department of Transportation.)

FIGURE 67.23 Harlan D. Miller (Dog Creek) Bridge is an example of state-of-the-art bridge development by the State of California under Bridge Engineer Harlan D. Miller in 1926. The State Legislature named the bridge in his honor for the great strides he accomplished with state bridges. Miller died only a week after receiving the honor, so the bridge became the Harlan D. Miller Memorial Bridge. (Courtesy of California Department of Transportation.)

FIGURE 67.24 Bixby Creek Bridge on the Monterey Coast in California is one of the most picturesque and photographed bridges in California. This Monterey Coast Highway was the first designated Scenic Highway in California, in 1961. The route is also the first to be designated an All American Road. Built in 1932, it has a main span of 109.7 m and is 79.2 m above the streambed. Construction required 26 stories of falsework. It was designed by Harvey Stover of the California Division of Highways. Seismic retrofitting is complicated due to aesthetic restrictions established by historical preservation codes. (Courtesy of California Department of Transportation.)

FIGURE 67.25 Conde McCullough, of the Oregon State Highway Department, designer of the two bridges shown in Figures 67.25 and 67.26, gained fame as the designer of several landmark bridges on the Oregon Coast Highway. The Rogue River Bridge at Gold Beach, Oregon, is a typical open spandrel concrete arch. The monumental spires at the abutment piers are a McCullough trademark. Both of these bridges were built in 1932. (Courtesy of Oregon Department of Transportation.)

FIGURE 67.26 The double-tiered concrete arch end spans at Cape Creek, on the Oregon Coast, are reminiscent of Roman aqueducts. The north-bound highway at this point emerges from a tunnel providing a picturesque view of the Heceta Head Lighthouse, as the traveler glides out over Cape Creek. (Courtesy of American Society of Civil Engineers.)

FIGURE 67.27 The Lilac Road arch gracefully frames the southern entrance to the fertile San Luis Rey Valley, of Southern California. It was built over Interstate Route 15 in San Diego County in 1978. The designer was Fred Michaels of the California Department of Transportation. (Courtesy of California Department of Transportation.)

Concrete Girders

See Figures 67.28 and 67.29.

FIGURE 67.28 Rockcut Bridge, owned by Stevens and Ferry Counties, earned a Portland Cement Association design award in 1997. The designer was Nicholls Engineering. (Courtesy of Portland Cement Association.)

FIGURE 67.29 The North Santiam (Gates) Bridge, in Marion County, Oregon, earned a Portland Cement Association design award in 1997. It was designed by the Oregon Department of Transportation. (Courtesy of Portland Cement Association.)

Canticrete

See Figure 67.30.

FIGURE 67.30 The landmark Alsea Bay Bridge at Waldport, Oregon, is the only Conde McCullough bridge that has required replacement due to deterioration. The new structure is reminiscent of McCullough's style and utilizes the best of the two most popular building materials, concrete and steel. The concrete-covered steel members are called canticrete. This new span, built in 1992, was designed by Howard Needles Tamman Bergendoff. It won awards from both the American Institute of Steel Construction and the Portland Cement Association. (Courtesy of Howard Needles Tamman Bergendoff.)

Suspension Bridges

Suspension bridges are one of the oldest concepts in the world. The first recorded suspension bridge in the United States was a chain-link catenary over Jacobs Creek in 1801 at Uniontown, Pennsylvania. Suspension bridges have continued to be a favored type into modern times. They are graceful and especially practical for long spans (Figures 67.31 through 67.35).

FIGURE 67.31 The Brooklyn Bridge is probably the best known of the classic U.S. bridges. It is one of the early uses of wire rope, being a combination suspension and cable-stayed span. Designed and built in 1883 by John and Washington Roebling for the City of New York. (Courtesy of American Society of Civil Engineers.)

FIGURE 67.32 The west span of the San Francisco–Oakland Bay Bridge is really two suspension bridges end to end with a central anchorage between the two. It is the only double-suspension bridge in the world. Opened in 1936, it is owned by the California Department of Transportation and designed by its predecessor, the California Division of Highways, Charles Andrew, Chief Bridge Engineer. (Courtesy of California Department of Transportation.)

FIGURE 67.33 The Golden Gate Bridge is one of the best-known landmarks in the United States. It spans the entrance to San Francisco Bay. It held the longest span, 1280 m, record for 27 years. Designed and built in 1937 by Charles B. Strauss. It is owned by the Golden Gate Bridge, Highway and Transportation District. (Courtesy of California Department of Transportation.)

FIGURE 67.34 The Mackinac Straits Bridge was the winner of the 1958 American Institute of Steel Construction's Artistic Bridge Award and several gold medals. Its design provided a level of aerodynamic stability never before attained in a suspension bridge. It has a main span of 1158 m. It was designed by David Steinman and is owned by the Mackinac Bridge Authority in northern Michigan. (Courtesy of David Steinman.)

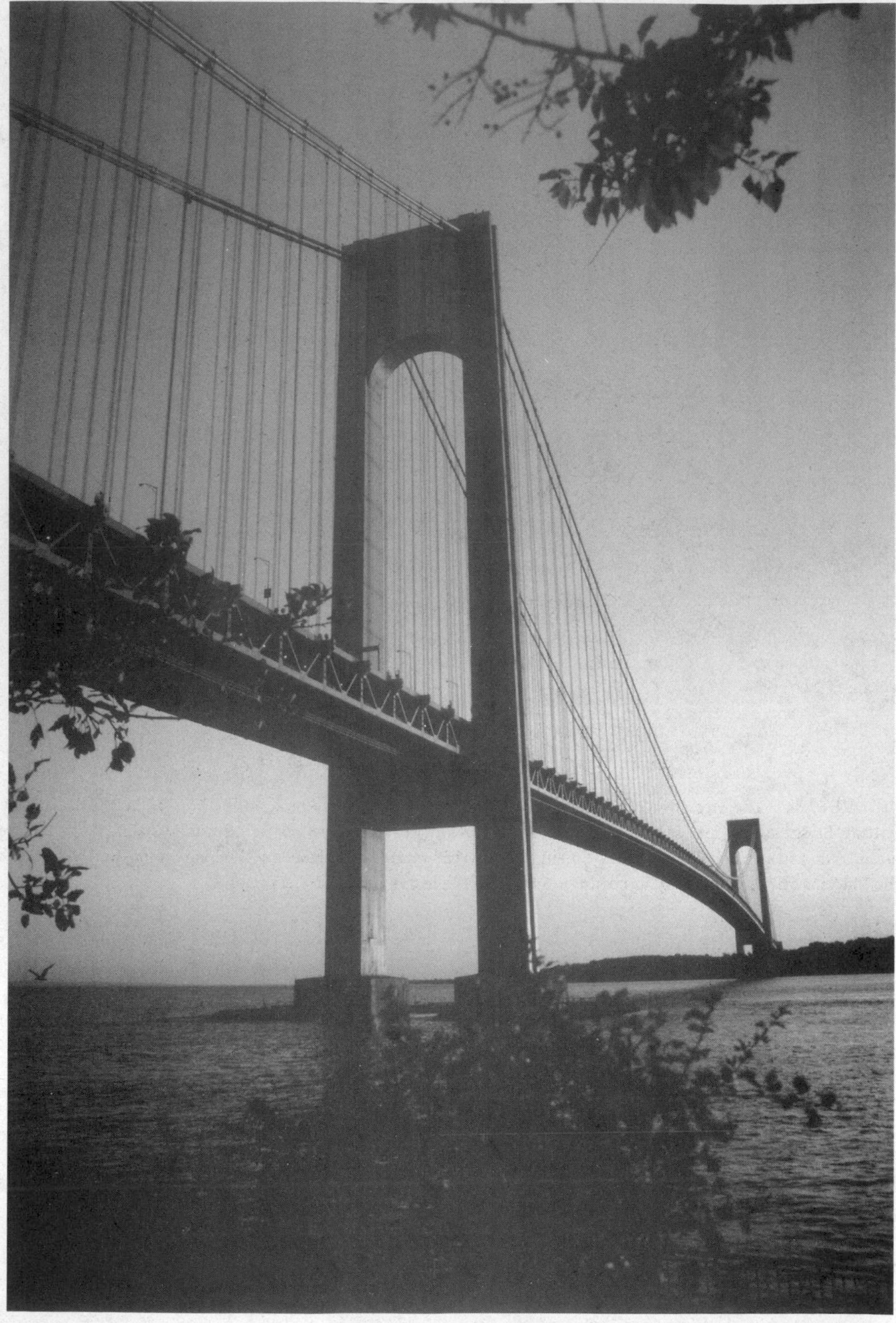

FIGURE 67.35 The Verrazano Narrows Bridge, in New York City, has the longest span, 1298 m, of any bridge in America. Designed by Amman and Whitney, it was opened in 1964. (Courtesy of Metropolitan Transportation Authority, New York.)

Movable Bridges

See Figures 67.36 and 67.37.

FIGURE 67.36 The Tower Bridge in Sacramento earned an AISC design award in 1936, the year in which it was built. This unique lift span is clad in steel plate to cover the moving parts. It was designed by Leonard Hollister of the California Division of Highways. (Courtesy of California Department of Transportation.)

FIGURE 67.37 The double-swing span George P. Coleman Bridge over the York River in Yorktown, Virginia, was recently widened. To minimize impacts on heavy traffic flows, new truss sections were built in dry dock and floated into place as the old were floated out. The replacement designed by Parsons, Brinckerhoff, Quade and Douglas won the 1997 George P. Richardson Medal for outstanding achievement. (Courtesy of Parsons Brinckerhoff.)

Floating Bridge

See Figure 67.38.

FIGURE 67.38 The 2377-m-long Lacey V. Murrow Floating Bridge across Lake Washington near Seattle is composed of hollow concrete pontoons. The depth of water, 45.7 m, precludes piers, but there are some bridge spans over shallow water near the shore that can pass small vessels. It was designed by Charles Andrew and Clark Elkridge in 1940. The bridge is listed on the National Register of Historic places. (Courtesy of American Society of Civil Engineers.)

67.6 The Interstate Era

The Federal System of Interstate and Defense Highways following World War II gave another boost to highway and bridge building. The system designed to be nonstop, separated, and controlled access requires many bridges in order to function as planned. Old-time bridge engineers had a difficult time trying to adapt. Their experience up until then had been to bridge the low spot in valleys crossing over waterways. Now, bridge engineers found themselves building bridges over dry land, at ridges, and over the highways themselves. Several new innovations were spawned during this prolific period. Composite steel, concrete box girders, and prestressed concrete became routine.

Concrete Box Girders

This superstructure type, developed by Jim Jurkovich of the California Division of Highways, has good torsional stability and provides exceptional wheel load distribution across the girders. Concrete structures evolved into the preferred types, starting in California. California has an abundant source of aggregates and cement. Contractors learned to build them at costs competitive with steel (Figures 67.39 and 67.40).

FIGURE 67.39 The Four Level Interchange in downtown Los Angeles, built in 1950, is the first multilevel interchange of two freeways. It is a reinforced concrete box girder, a type developed by, and to become the hallmark of, the California Division of Highways. (Courtesy of California Department of Transportation.)

FIGURE 67.39 Mission Valley Viaduct sweeps Interstate Route 805 over the San Diego River floodplain in southern California. (Courtesy of California Department of Transportation.)

Prestressed Concrete

Prestressed concrete is a natural evolution of concrete girders. It makes the best use of the compressive qualities of concrete and the tensile properties of steel. Prestressing allows shallower structure depth, and a tremendous savings in approach roadway earthwork for interstate separations. Prestressed concrete can be either pretensioned or post-tensioned, precast or cast in place. All of these options have their place under different situations. The California Division of Highways pioneered this system in the 1940s and has since made extensive use of cast-in-place post-tensioned concrete box girders. The type has become so prevalent that construction contractors are able to build them for the same or less cost than normal reinforced concrete structures (Figures 67.41 and 67.42).

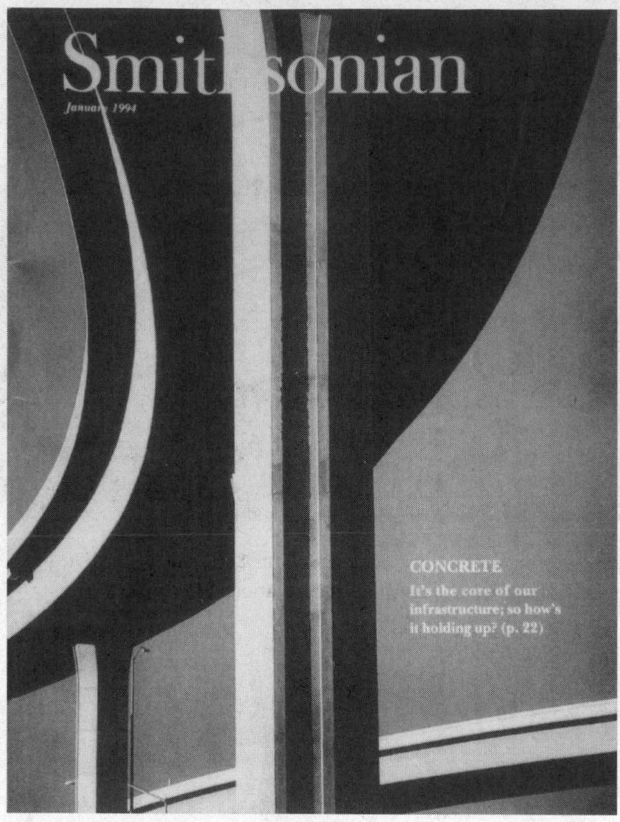

FIGURE 67.41 The Interstate Routes 105/110 Interchange in Los Angeles, California, is a massive forest of concrete columns supporting intertwined roadways. The *Smithsonian Magazine* highlighted the edifice as an artistic concrete creation in their January 1994 cover story. It was designed by Elweed Pomeroy of the California Department of Transportation. (Courtesy of Smithsonian Institution.)

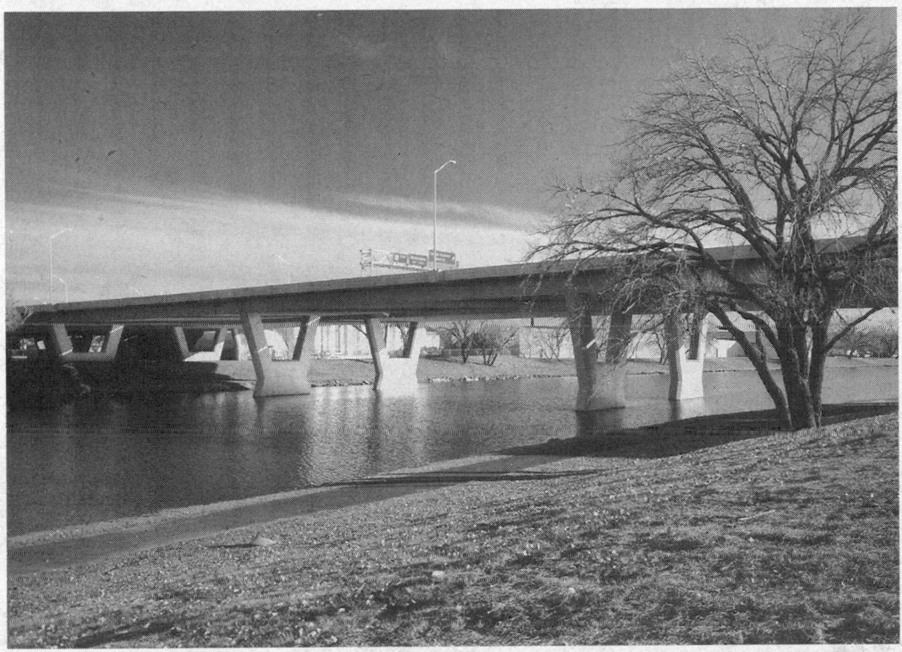

FIGURE 67.42 Kellogg Central Business District Viaduct in Wichita, Kansas, designed by Howard Needles Tamman Bergendoff, won a Portland Cement Association award in 1996. (Courtesy of Howard Needles Tamman Bergendoff.)

Composite Steel

Composite steel girders, where a concrete deck is attached to the top flange of a steel girder through mechanical connectors, utilizes the best advantages of the compressive properties of concrete and the tensile properties of steel. While concrete was dominating the California and western bridge scene, steel remained the primary building material in the eastern and midwestern states (Figures 67.43 and 67.44).

FIGURE 67.43 The South Fork of the Eel River Bridge in Northern California exemplifies the virtues of composite steel structures. This 1958 bridge won an AISC award that year. (Courtesy of California Department of Transportation.)

FIGURE 67.44 The Cuyahoga River Valley Bridge built in 1980 for the Ohio Turnpike Authority earned an AISC award that year. It was designed by Howard Needles Tamman Bergendoff. (Courtesy of American Institute of Steel Construction.)

A Resurgence of Steel

As the Interstate Highway program began to utilize more and more concrete structures, during the 1960s, the steel and welding industry struggled to maintain its share of the bridge market. Many innovations were introduced for the use of steel through this campaign, by the development of exotic steels, distribution of design aids and examples, and conducting of design contests.

Steel Girders

See Figures 67.45 and 67.46.

FIGURE 67.45 The Eugene A. Doran Memorial (San Mateo Creek–Crystal Springs Reservoir) Bridge is a prize-winning bridge in a park setting. Sloping exterior facia plates provide web stiffening and aesthetic treatment. This welded plate steel girder bridge, built in 1970, was designed by Bob Cassano of the California Division of Highways. (Courtesy of California Department of Transportation.)

FIGURE 67.46 The Sacramento River Bridge at Elkhorn is a steel girder utilizing high-strength steel. Built in 1970 for Interstate Route 5, the bridge earned an AISC award that year. It was designed by Bert Bezzone of the California Division of Highways. (Courtesy of California Department of Transportation.)

Steel Box Girders and Orthotropic Steel Decks

See Figures 67.47 through 67.49.

FIGURE 67.47 The Klamath River crossing at Orleans in Northern California is a picturesque setting on a back road. There have been seven structures at this site, one burned and five have been washed away during major floods. The current steel box girder suspension span has lasted longer than any of its predecessors. Built in 1967, it was designed by Bert Bezzone of the California Division of Highways. (Courtesy of California Department of Transportation.)

FIGURE 67.48 San Mateo–Hayward Bridge over the San Francisco Bay, not only has composite steel approach spans, but the main span has an orthotropic steel deck. Listed among distinctive bridges, it won an American Institute of Steel Construction prize in 1968. It was designed by the California Division of Bay Toll Crossings. (Courtesy of California Department of Transportation.)

FIGURE 67.49 The Coronado Island Bridge over San Diego Bay in Southern California is a steel box girder. It earned an AISC award in 1970. It was designed by the California Division of Bay Toll Crossings. (Courtesy of California Department of Transportation.)

67.7 Era of the Signature Bridge

With the energetic vision of the great Interstate Era virtually complete, and the ensuing rush of the Seismic Retrofit Age winding down, bridge engineers turned their imaginative minds toward the building of great monuments.

Segmental Prestressed Bridges

Advanced technology of high-strength concrete and prestressing allows the cantilevering of structures out over deep valleys and bodies of water (Figures 67.50 and 67.51).

FIGURE 67.50 The California Department of Transportation experimented with and built its only segmental bridge, on Interstate Route 8 over Pine Valley in San Diego County in 1974. Bert Bezzone was the design engineer. The bridge received an American Society of Civil Engineers award in 1974. (Courtesy of California Department of Transportation.)

FIGURE 67.51 This graceful arch by Figg Engineering, carries the historic Natchez Trace over the park in Tennessee. It is the first and longest, 317 m, precast segmental arch. It received a design award of excellence in 1996. (Courtesy of Figg Engineering.)

Cable-Stayed Bridges

See Figures 67.52 and 67.53.

FIGURE 67.52 The new cable-stayed Sunshine Skyway, built in 1987, clearly a signature bridge, makes a bright statement over the entrance to Tampa Bay, Florida. This structure by Figg Engineering replaced the former Sunshine Skyway truss brought down by an errant barge that weighed more than that bridge. The new piers are protected by caissons as big and heavy as that barge. (Courtesy of Figg Engineering.)

FIGURE 67.53 The Cheasapeake and Delaware Canal Bridge owned by Delaware Department of Transportation was designed by Figg Engineering. It received an Excellence in Design Award in 1996. (Courtesy of Figg Engineering.)

Composites

The new definition of composite bridges has nothing to do with steel or concrete. Composites in modern usage refer to groups of organic chemical polymers commonly known as plastics. These are still experimental materials as far as bridges are concerned, but have been used successfully in other industries for some time now. Composites are now being used with fiber-wrap bridge columns as a seismic retrofit technique. The California Department of Transportation is currently designing an experimental span which will be concrete-filled composite tube girders with a composite deck.

FIGURE 67.54 Laurel Lick Bridge is the second all-composite bridge to be completed It is owned by the West Virginia Department of Highways and was built experimentally in conjunction with West Virginia University, in 1997. (Courtesy of West Virginia Department of Highways.)

67.8 Epilogue

All superstructure types are seen in combination, and with many variations. Even though seemingly prevalent during evolving eras, type periods greatly overlap, with type selections being more dependent upon crossing length and foundation conditions. In fact, every superstructure type is still being built today in response to various needs.

Let us all admire and learn from those Americans who have contributed, pioneered, and those who have consistently created award-winning structures of which all in the bridge-building profession can be proud. These include Squire Whipple, James Eads, Theodore Cooper, Gustav Lindenthal, Othmar Amman, David Steinman, Ralph Modjeski, Leon Moisseiff, John and Washington Roebling, Joseph Strauss, John Waddell, Conde McCullough, T.Y. Lin, Eugene Figg, the California Department of Transportation, and Howard Needles Tamman Bergendoff.

Index

A

C

D

E

F

G

H

I

J

K

L

M

N

P

Q

R

S

T

U

V

W

X

Y

Z